D1675057

CHEMISTRY

THE CENTRAL SCIENCE 15TH EDITION

Photosynthesis is arguably the most important chemical reaction on Earth. Stimulated by sunlight, molecules of chlorophyll in plant cells convert CO_2 and water to oxygen gas and glucose. Without photosynthesis, living organisms would not be able to convert solar energy into the chemical energy that sustains life and the Earth's atmosphere would contain very little oxygen.

List of Elements with Their Symbols and Atomic Weights

Element	Symbol	Atomic Number	Atomic Weight
Actinium	Ac	89	227.03[a]
Aluminum	Al	13	26.981538
Americium	Am	95	243.06[a]
Antimony	Sb	51	121.760
Argon	Ar	18	39.948
Arsenic	As	33	74.92160
Astatine	At	85	209.99[a]
Barium	Ba	56	137.327
Berkelium	Bk	97	247.07[a]
Beryllium	Be	4	9.012183
Bismuth	Bi	83	208.98038
Bohrium	Bh	107	270.1[a]
Boron	B	5	10.81
Bromine	Br	35	79.904
Cadmium	Cd	48	112.414
Calcium	Ca	20	40.078
Californium	Cf	98	251.08[a]
Carbon	C	6	12.0107
Cerium	Ce	58	140.116
Cesium	Cs	55	132.905452
Chlorine	Cl	17	35.453
Chromium	Cr	24	51.9961
Cobalt	Co	27	58.933194
Copernicium	Cn	112	285.2[a]
Copper	Cu	29	63.546
Curium	Cm	96	247.07[a]
Darmstadtium	Ds	110	281.2[a]
Dubnium	Db	105	268.1[a]
Dysprosium	Dy	66	162.50
Einsteinium	Es	99	252.08[a]
Erbium	Er	68	167.259
Europium	Eu	63	151.964
Fermium	Fm	100	257.10[a]
Flerovium	Fl	114	289.2[a]
Fluorine	F	9	18.9984016
Francium	Fr	87	223.02[a]
Gadolinium	Gd	64	157.25
Gallium	Ga	31	69.723
Germanium	Ge	32	72.64
Gold	Au	79	196.966569
Hafnium	Hf	72	178.49
Hassium	Hs	108	269.1[a]
Helium	He	2	4.002602
Holmium	Ho	67	164.93033
Hydrogen	H	1	1.00794
Indium	In	49	114.818
Iodine	I	53	126.90447
Iridium	Ir	77	192.217
Iron	Fe	26	55.845
Krypton	Kr	36	83.80
Lanthanum	La	57	138.9055
Lawrencium	Lr	103	262.11[a]
Lead	Pb	82	207.2
Lithium	Li	3	6.941
Livermorium	Lv	116	293[a]
Lutetium	Lu	71	174.967
Magnesium	Mg	12	24.3050
Manganese	Mn	25	54.938044
Meitnerium	Mt	109	278.2[a]
Mendelevium	Md	101	258.10[a]
Mercury	Hg	80	200.59
Molybdenum	Mo	42	95.95
Moscovium	Mc	115	289.2[a]
Neodymium	Nd	60	144.24
Neon	Ne	10	20.1797
Neptunium	Np	93	237.05[a]
Nickel	Ni	28	58.6934
Nihonium	Nh	113	286.2[a]
Niobium	Nb	41	92.90637
Nitrogen	N	7	14.0067
Nobelium	No	102	259.10[a]
Oganesson	Og	118	294.2[a]
Osmium	Os	76	190.23
Oxygen	O	8	15.9994
Palladium	Pd	46	106.42
Phosphorus	P	15	30.973762
Platinum	Pt	78	195.078
Plutonium	Pu	94	244.06[a]
Polonium	Po	84	208.98[a]
Potassium	K	19	39.0983
Praseodymium	Pr	59	140.90766
Promethium	Pm	61	145[a]
Protactinium	Pa	91	231.03588
Radium	Ra	88	226.03[a]
Radon	Rn	86	222.02[a]
Rhenium	Re	75	186.207[a]
Rhodium	Rh	45	102.90550
Roentgenium	Rg	111	282.2[a]
Rubidium	Rb	37	85.4678
Ruthenium	Ru	44	101.07
Rutherfordium	Rf	104	267.1[a]
Samarium	Sm	62	150.36
Scandium	Sc	21	44.955908
Seaborgium	Sg	106	269.1[a]
Selenium	Se	34	78.97
Silicon	Si	14	28.0855
Silver	Ag	47	107.8682
Sodium	Na	11	22.989770
Strontium	Sr	38	87.62
Sulfur	S	16	32.065
Tantalum	Ta	73	180.9479
Technetium	Tc	43	98[a]
Tellurium	Te	52	127.60
Tennessine	Ts	117	293.2[a]
Terbium	Tb	65	158.92534
Thallium	Tl	81	204.3833
Thorium	Th	90	232.0377
Thulium	Tm	69	168.93422
Tin	Sn	50	118.710
Titanium	Ti	22	47.867
Tungsten	W	74	183.84
Uranium	U	92	238.02891
Vanadium	V	23	50.9415
Xenon	Xe	54	131.293
Ytterbium	Yb	70	173.04
Yttrium	Y	39	88.90584
Zinc	Zn	30	65.39
Zirconium	Zr	40	91.224

[a]Mass of longest-lived or most important isotope.

CHEMISTRY

THE CENTRAL SCIENCE 15TH EDITION

Theodore L. Brown
University of Illinois at Urbana-Champaign

H. Eugene LeMay, Jr.
University of Nevada, Reno

Bruce E. Bursten
Worcester Polytechnic Institute

Catherine J. Murphy
University of Illinois at Urbana-Champaign

Patrick M. Woodward
The Ohio State University

Matthew W. Stoltzfus
The Ohio State University

 Pearson

Content Development: Matt Walker, John Murdzek
Content Management: Jeanne Zalesky, Deborah Harden, Prudence Wei-Lin Huang
Content Production: Kristen Flathman, Beth Sweeten, Mary Tindle, Molly Montaro, Katie Foley, Jayne Sportelli, Lizette Faraji, Tod Regan, Chloe Veylit, Maria Guglielmo Walsh, Jerilyn Bockorick
Product Management: Chris Hess, Ian Desrosiers
Product Marketing: Candice Madden
Rights and Permissions: Ben Ferrini, Matthew Perry, Zosef Sotheo Herrero

Please contact https://support.pearson.com/getsupport/s/ with any queries on this content

Cover Image: Shutterstock

Copyright © 2023, 2018, 2015 by Pearson Education, Inc. or its affiliates, 221 River Street, Hoboken, NJ 07030. All Rights Reserved. Manufactured in the United States of America. This publication is protected by copyright, and permission should be obtained from the publisher prior to any prohibited reproduction, storage in a retrieval system, or transmission in any form or by any means, electronic, mechanical, photocopying, recording, or otherwise. For information regarding permissions, request forms, and the appropriate contacts within the Pearson Education Global Rights and Permissions department, please visit www.pearsoned.com/permissions/.

Acknowledgments of third-party content appear on page P-1, which constitutes an extension of this copyright page.

PEARSON, ALWAYS LEARNING, and Mastering® Chemistry are exclusive trademarks owned by Pearson Education, Inc. or its affiliates in the U.S. and/or other countries.

Unless otherwise indicated herein, any third-party trademarks, logos, or icons that may appear in this work are the property of their respective owners, and any references to third-party trademarks, logos, icons, or other trade dress are for demonstrative or descriptive purposes only. Such references are not intended to imply any sponsorship, endorsement, authorization, or promotion of Pearson's products by the owners of such marks, or any relationship between the owner and Pearson Education, Inc., or its affiliates, authors, licensees, or distributors.

Library of Congress Cataloging-in-Publication Data
Cataloging-in-Publication Data is available on file at the Library of Congress.

5 2022

Access Code Card
ISBN-10: 0137542747
ISBN-13: 9780137542741

Rental
ISBN-10: 0137493606
ISBN-13: 9780137493609

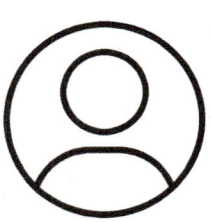

Pearson's Commitment to Diversity, Equity, and Inclusion

Pearson is dedicated to creating bias-free content that reflects the diversity, depth, and breadth of all learners' lived experiences.

We embrace the many dimensions of diversity, including but not limited to race, ethnicity, gender, sex, sexual orientation, socioeconomic status, ability, age, and religious or political beliefs.

Education is a powerful force for equity and change in our world. It has the potential to deliver opportunities that improve lives and enable economic mobility. As we work with authors to create content for every product and service, we acknowledge our responsibility to demonstrate inclusivity and incorporate diverse scholarship so that everyone can achieve their potential through learning. As the world's leading learning company, we have a duty to help drive change and live up to our purpose to help more people create a better life for themselves and to create a better world.

Our ambition is to purposefully contribute to a world where:

- Everyone has an equitable and lifelong opportunity to succeed through learning.
- Our educational content accurately reflects the histories and lived experiences of the learners we serve.

- Our educational products and services are inclusive and represent the rich diversity of learners.
- Our educational content prompts deeper discussions with students and motivates them to expand their own learning (and worldview).

Accessibility

We are also committed to providing products that are fully accessible to all learners. As per Pearson's guidelines for accessible educational Web media, we test and retest the capabilities of our products against the highest standards for every release, following the WCAG guidelines in developing new products for copyright year 2022 and beyond.

 You can learn more about Pearson's commitment to accessibility at
https://www.pearson.com/us/accessibility.html

Contact Us

While we work hard to present unbiased, fully accessible content, we want to hear from you about any concerns or needs with this Pearson product so that we can investigate and address them.

 Please contact us with concerns about any potential bias at
https://www.pearson.com/report-bias.html.

 For accessibility-related issues, such as using assistive technology with Pearson products, alternative text requests, or accessibility documentation, email the Pearson Disability Support team at **disability.support@pearson.com**

To our students,
whose enthusiasm and curiosity
have often inspired us,
and whose questions and suggestions
have sometimes taught us.

BRIEF CONTENTS

CONTENTS

3 Chemical Reactions and Reaction Stoichiometry 85

4 Reactions in Aqueous Solution 119

5 Thermochemistry 161

6 Electronic Structure of Atoms 211

7 Periodic Properties of the Elements 255

8 Basic Concepts of Chemical Bonding 297

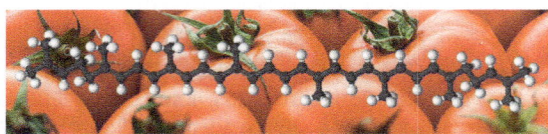

9 Molecular Geometry and Bonding Theories 335

10 Gases 391

11 Liquids and Intermolecular Forces 431

12 Solids and Modern Materials 469

13 Properties of Solutions 521

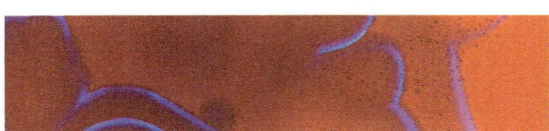

14 Chemical Kinetics 565

15 Chemical Equilibrium 621

16 Acid–Base Equilibria 661

17 Aqueous Equilibria: Buffers, Titrations, and Solubility 713

21 Nuclear Chemistry 891

22 Chemistry Nonmetals

Contents

23 Transition Metals and Coordination Chemistry 977

24 The Chemistry of Life: Organic and Biological Chemistry 1021

APPENDICES

CHEMICAL APPLICATIONS AND ESSAYS

A Closer Look

Chemistry and Sustainability

Chemistry and Life

Strategies for Success

PREFACE

To the Instructor

Philosophy

We authors of *Chemistry: The Central Science* are delighted and honored that you have chosen us as your instructional partners for your general chemistry class. Collectively we have taught general chemistry to multiple generations of students. Therefore, we understand the challenges and opportunities of teaching a class that so many students take. We have also been active researchers who appreciate both the learning and the discovery aspects of the chemical sciences. Our varied, wide-ranging experiences have formed the basis of the close collaborations we have enjoyed as coauthors. In writing our book, our focus is on the students: We try to ensure that the text is not only accurate and up-to-date but also clear and readable. We strive to convey the breadth of chemistry and the excitement that scientists experience in making new discoveries that contribute to our understanding of the physical world. We want the student to appreciate that chemistry is not a body of specialized knowledge that is separate from most aspects of modern life, but central to any attempt to address a host of societal concerns, including renewable energy, environmental sustainability, and improved human health. Most of all, we want to provide you and your students the most effective tools for teaching and learning.

Publishing the fifteenth edition of this text bespeaks an exceptionally long record of successful textbook writing. We are appreciative of the loyalty and support the book has received over the years, and we are mindful of our obligation to justify each new edition. We begin our approach to each new edition with an intensive author retreat in which we ask ourselves the deep questions that we must answer before we can move forward. What justifies yet another edition? What is changing in the world not only of chemistry, but with respect to science education and the qualities of the students we serve? How can we help your students not only learn the principles of chemistry, but also become critical thinkers who can think more like chemists?

Answers to these questions lie only partly in the changing face of chemistry itself. The introduction of many new technologies has changed the landscape in the teaching of sciences at all levels. The use of online resources in accessing information and presenting learning materials has markedly changed the role of the textbook as one element among many tools for student learning. Our challenge as authors is to maintain the text as the primary source of chemical knowledge and practice, while at the same time integrating it with the new avenues for learning made possible by technology. This book incorporates a number of technologies to improve pedagogy, including use of computer-based classroom tools, such as Mastering Chemistry, which is continually evolving to provide more effective means of testing and evaluating student performance, while giving the student immediate and helpful feedback. Video feedback for a select number of *Exam Prep* questions is also available in Mastering Chemistry, which is new to this edition

As authors, we want this text to be a central, indispensable learning tool for students. Whether as a physical book or in electronic form, it can be carried everywhere and used at any time. It is the best place students can go to obtain the information outside of the classroom needed for learning, skill development, reference, and test preparation. The text, more effectively than any other instrument, provides the depth of coverage and coherent background in modern chemistry that students need to serve their professional interests and, as appropriate, to prepare for more advanced chemistry courses.

If the text is to be effective in supporting your role as instructor, it must be addressed to the students. We strive to keep our writing clear and interesting, complemented by figures and illustrations wherever possible. The book has numerous in-text study aids for students, including student-friendly learning objectives and follow-up self-assessment exercises for each section of the book, carefully placed descriptions of problem-solving strategies, and a wealth of problems in varying formats at the end of each chapter. We hope our cumulative experience as teachers is evident in our pacing, choice of examples, and the types of study aids and motivational tools we have employed. We believe students are more enthusiastic about learning chemistry when they see its importance relative to their own goals and interests; therefore, we have highlighted many important applications of chemistry in everyday life. We hope you are able to make use of this material.

It is our philosophy, as authors, that the text and all the supplementary materials provided to support its use must work in concert with you, the instructor. A textbook is only as useful to students as the instructor permits it to be. This book is replete with features that help students learn and that can guide them as they acquire both conceptual understanding and problem-solving skills. There is a great deal here for the students to use, too much for all of it to be absorbed by any student in a one-year course. You will be the guide to the best use of the book. Only with your active help will the students be able to utilize most effectively all that the text and its supplements offer. Students care about grades, of course, and with encouragement they will also become interested in the subject matter and care about learning. Please consider emphasizing features of the book that can enhance student appreciation of chemistry, such as the *Chemistry and Sustainability* and *Chemistry and Life* boxes that show how chemistry impacts modern life and its relationship to health and life processes. Also consider emphasizing conceptual understanding (placing less emphasis on simple manipulative, algorithmic problem solving) and urging students to use the rich on-line resources available.

Organization and Contents

The first five chapters give a largely macroscopic, phenomenological view of chemistry. The basic concepts introduced—such as nomenclature, stoichiometry, and thermochemistry—provide necessary background for many of the laboratory experiments performed in general chemistry. We believe an early introduction to thermochemistry is desirable because so much of our understanding of chemical processes is based on considerations of energy changes. By incorporating bond enthalpies in the Thermochemistry chapter we aim to emphasize the connection between the macroscopic properties of substances and the submicroscopic world of atoms and bonds. We believe we have produced an effective, balanced approach to teaching thermodynamics in general chemistry, as well as providing students with an introduction to some of the global issues involving energy production and consumption. It is no easy matter to walk the narrow pathway between trying to teach too much at too high a level versus resorting to oversimplifications. For the book as a whole, the emphasis has been on imparting *conceptual* understanding, as opposed to presenting equations into which students are supposed to plug numbers.

The next four chapters (Chapters 6–9) deal with electronic structure and bonding. For more advanced students, *A Closer Look* boxes in Chapters 6 and 9 highlight radial probability functions and the phases of orbitals. Placing this latter discussion in *A Closer Look* box in Chapter 9 enables those who wish to cover this topic to do so, while others may decide to bypass it. For those instructors who wish to take an "atoms first" approach to teaching chemistry, starting with Chapters 1, 2, and 6 (while filling in concepts and problem-solving skills from Chapters 3–5 as they come up, especially in the laboratory) may work well.

In Chapters 10–13, the focus of the text changes to the next level of the organization of matter: examining the states of matter. Chapters 10 and 11 deal with gases, liquids, and intermolecular forces, while Chapter 12 is devoted to solids, presenting a contemporary view of the solid state as well as of modern materials accessible to general chemistry students. Chapter 12 provides an opportunity to show how abstract chemical bonding concepts impact real-world applications. The modular organization of the chapter allows you to tailor your coverage to focus on the materials (semiconductors, polymers, nanomaterials, and so forth) that are most relevant to your students and your own interests. This section of the book concludes with Chapter 13, which covers the formation and properties of solutions.

The next several chapters examine the factors that determine the speed and extent of chemical reactions: kinetics (Chapter 14), equilibria (Chapters 15–17), thermodynamics (Chapter 19), and electrochemistry (Chapter 20). Also in this section is a chapter on environmental chemistry (Chapter 18), in which the concepts developed in preceding chapters are applied to a discussion of the atmosphere and hydrosphere. This chapter has increasingly become focused on green chemistry and the impacts of human activities on Earth's water and atmosphere, and expands on many of the concepts first introduced in the *Chemistry and Sustainability* boxes.

After a discussion of nuclear chemistry (Chapter 21), the book ends with three survey chapters. Chapter 22 deals with nonmetals, Chapter 23 with the chemistry of transition metals, including coordination compounds, and Chapter 24 with the chemistry of organic compounds and elementary biochemical themes. These final four chapters are developed in an independent, modular fashion and can be covered in any order.

Our chapter sequence provides a fairly standard organization, however we recognize that not everyone teaches all the topics in the order we have chosen. We have, therefore, made sure instructors can make common changes in teaching sequence with no loss in student comprehension. In particular, many instructors prefer to introduce gases (Chapter 10) after stoichiometry (Chapter 3) rather than with states of matter. The chapter on gases has been written to permit this change with *no* disruption in the flow of material. It is also possible to treat balancing redox equations (Sections 20.1 and 20.2) earlier, after the introduction of redox reactions in Section 4.4. Finally, some instructors like to cover organic chemistry (Chapter 24) right after bonding (Chapters 8 and 9). This, too, is a largely seamless move.

We have brought students into greater contact with descriptive organic and inorganic chemistry by integrating examples throughout the text. You will find pertinent and relevant examples of "real" chemistry woven into all the chapters to illustrate principles and applications. Some chapters, of course, more directly address the "descriptive" properties of elements and their compounds, especially Chapters 4, 7, 11, 18, and 22–24. We also incorporate descriptive organic and inorganic chemistry in the end-of-chapter exercises.

Major Global Changes in This Edition

As with every new edition of *Chemistry: The Central Science,* the book has undergone a great many changes as we strive to keep the content current, and to improve the clarity and effectiveness of the text, the art, and the exercises. Among the myriad changes there are certain points of emphasis that we use to organize and guide the revision process. In creating the fifteenth edition, our revision was organized around the following points:

- We continue to use our in-classroom experiences with today's students to develop new tools that make the learning of chemistry more effective for students. In particular, we continue to develop new ways in which we can make our text a better, more indispensable learning tool for students. First, we have added *Learning Objectives* to each section of the text. These targeted goals are written in a student-friendly, easily understood manner which emphasizes the important concepts that provide students with achievable goals in their studies. Up to six learning objectives are placed in a margin box at the beginning of each section for ease of use. Then, at the end of each section, we have added a set of *Self-Assessment Exercises* for the students to complete as the equivalent of a low-stakes quiz as they proceed through the text. The author team was careful to ensure that each learning objective is covered by a self-assessment exercise. The self-assessment exercises, assignable in

Mastering, are structured as multiple-choice questions with wrong answers (distractors) chosen to probe common misconceptions and errors that students tend to make. Responses, both correct and incorrect, contain feedback written by the author team to help students recognize their mistakes and provides hints to get on the right track if they select the wrong answer. This organizational structural change has meant that some sections were combined with others, compared to previous editions of the book.

- At the end of each chapter are a series of *Exam Prep* questions. These multiple-choice questions, assignable in Mastering, are tied to each chapter's *Learning Objectives* and constitute what amounts to a practice exam for students to take on their own. As with the self-assessment exercises each answer contains clear, concise feedback written by the authors.

- Extensive effort has gone into creating enhanced content for the eText version of this book to make it so much more than just an electronic copy of the physical textbook. A select number of the Exam Prep questions include worked out solution videos, which allow students to check their work and solidify their understanding as they prepare for their midterm exams. These questions can also be assigned in Mastering Chemistry.

- Chemistry as the central science is intimately tied to larger issues such as climate change, humanity's use of energy, abundance of clean water, food insecurity, and more. Therefore, many of the former *Chemistry Put to Work* boxes have been reimagined and rewritten as *Chemistry and Sustainability* boxes, to showcase the means by which chemists contribute to the understanding of, and movement toward, a sustainable society. We maintain our focus on the positive aspects of chemistry without neglecting problems that can arise in an increasingly technological world. Our goal is to help students appreciate the real-world perspective of chemistry and the ways in which chemistry affects their lives. To address some of the most pressing societal issues of our time many new boxes have been added, including an introduction to these boxes in Chapter 1, which highlights the *UN Sustainability Development Goals*.

- The *Additional Exercises* at the end of each chapter are no longer separated into "additional exercises," and "integrative exercises," nor are brackets used to indicate problems that are abnormally difficult. For the student, this simulates the more real-world environment of an examination where such distinctions are not made.

- Throughout the text, updates to the periodic table and numerical constants have been implemented; for instance, in 2019, the values of Avogadro's number and a number of other physical constants were redefined, and we have included the most current values.

- Astute readers may notice the *Give It Some Thought* feature has been removed. The removal of this feature makes the reading experience less fragmented. The best *Give It Some Thought* questions have been reimagined as multiple-choice questions that are now part of the Self-Assessment Exercises and Exam Prep questions.

Chapter-Level Changes in This Edition

Chapter 1 continues the trend from the fourteenth edition to emphasize, early on, the importance of energy. The inclusion of energy in the opening chapter provides much greater flexibility for the order in which subsequent chapters can be covered. The *Chemistry and Sustainability* boxes are introduced in this chapter to frame the understanding that, historically, chemistry has had both positive and negative impacts on sustainability. More depth has been added to the discussion of significant figures compared to previous editions.

In Chapter 2, the treatment of organic chemistry nomenclature has been expanded to better match the long-standing section on inorganic nomenclature. This change involved moving some material, including end-of-chapter exercises, from Chapter 24. Inorganic acids such as HCl and organic acids such as acetic acid are now clearly distinguished.

In Chapter 5, the discussion of energy has been updated and expanded, especially regarding the section on foods and fuels, to connect directly to the focus on sustainability in the text.

In Chapter 6, a new (nearly) full-page figure has been added that shows the relationship between the wave function, the probability density, and the radial probability function for the $1s$, $2s$ and $3s$ orbitals. By explicitly showing the wave functions, instructors who want to cover phases of orbitals have a better fundamental grounding with which to do so.

In Chapter 10, the average speed of gas molecules is now quantitatively described, in addition to the root-mean-square speed and the most probable speed. The discussion of diffusion and mean free path has been expanded to introduce the students to the concept of the random walk.

In Chapter 13, the explanation of why the solubility of gaseous and solid solutes typically change in the opposite direction with temperature is significantly expanded. The effects of entropy are emphasized, to the extent they can be in this chapter where entropy is first introduced but not covered in full detail. In acknowledgment of the complex interplay of enthalpy and entropy, some details are left to Chapter 19. New material that explores the use of reverse osmosis to desalinate ocean water has been added (some of this material was previously in Chapter 18).

In Chapter 14, sections have been renamed to better reflect their content. A new *Closer Look* box on diffusion-controlled reactions and activation-controlled reactions, highly relevant to reaction mechanisms, has been added.

In Chapter 15, the discussion of the Haber process has been expanded to show both the positive and negative implications of the process with respect to sustainability: the importance of the Haber process in addressing food insecurity vis-à-vis the enormous energy consumption and carbon footprint of the process.

In Chapter 16, the discussion of Lewis acids and bases has been moved from the end of the chapter to the first section where the Arrhenius and Brønsted-Lowry definitions of acids and bases are introduced. This rearrangement makes for a more natural discussion of the acidic properties of small, highly charged cations in the section on the acid–base properties of salt solutions.

Chapter 17 has been renamed to make its content (buffer, titrations, solubility equilibria) more clear.

Many aspects of Chapter 18 (ozone hole, atmospheric CO_2 levels, acid rain, ocean acidification, etc.) are constantly changing. This material has been revised to reflect the most up-to-date data and scientific consensus on future trends.

In Chapter 19, we have substantially rewritten the early sections to help students better understand the concepts of spontaneous, nonspontaneous, reversible, and irreversible processes and their relationships. These improvements have led to a clearer definition of entropy. The topical box on "Entropy and Human Society" has been revised as a *Chemistry and Sustainability* feature, with a greater emphasis on the sustainability aspects of the second law of thermodynamics.

In Chapter 21, the discussion of the various means of generating electricity has been revised, and a section on the health hazards of environmental radon has been added.

Finally, in Chapter 24 a new Chemistry and Life box on COVID-19 and mRNA vaccines has been added.

To the Student

Chemistry: The Central Science, Fifteenth Edition, has been written to introduce you to modern chemistry. As authors, we have, in effect, been engaged by your instructor to help you learn chemistry. Based on the comments of students and instructors who have used this book in its previous editions, we believe that we have done that job well. Of course, we expect the text to continue to evolve through future editions. We invite you to write to tell us what you like about the book so we will know where we have helped you most. We would also like to learn of any shortcomings in an effort improve the book in subsequent editions. Our addresses are given at the end of the Preface.

Advice for Learning and Studying Chemistry

Learning chemistry requires both the assimilation of many concepts and the development of analytical skills. In this text, we have provided you with numerous tools to help you succeed in both tasks. If you are going to succeed in your chemistry course, you will have to develop good study habits. Science courses, and chemistry in particular, make different demands on your learning skills than do other types of courses. We offer the following tips for success in your study of chemistry:

Don't fall behind! As the course moves along, new topics will build on material already presented. If you don't keep up with your reading and problem solving, you will find it much harder to follow the lectures and discussions on current topics. Experienced teachers know that students who read the relevant sections of the text *before* coming to a class learn more from the class and retain greater recall. "Cramming" just before an exam has been shown to be an ineffective way to study any subject, chemistry included.

Focus your study. The amount of information you will be expected to learn may seem overwhelming. We have tried to help you by incorporating *Learning Objectives* into the beginning of each section within each chapter; with accompanying *Self-Assessment Exercises* at the end each section so you can test your knowledge. At the end of each chapter, we provide you with *Exam Prep* questions, which you can think of as a multiple-choice practice exam. During your time with your instructor in the classroom, it is essential to recognize those concepts and skills that are particularly important. Pay attention to what your instructor is emphasizing. As you work through the *Sample Exercises* and homework assignments, try to see what general principles and skills they employ. Use the *What's Ahead* feature at the beginning of each chapter to help orient yourself to what is important in each chapter. A single reading of a chapter will generally not be enough for successful learning of chapter concepts and problem-solving skills. You will often need to go over assigned materials more than once. Don't skip the *Go Figure* features, *Sample Exercises,* and *Practice Exercises.* These are your guides to whether you are learning the material. They are also good preparation for test-taking. The *Key Equations* at the end of the chapter will also help you focus your study.

Keep good lecture notes. Your lecture notes will provide you with a clear and concise record of what your instructor regards as the most important material to learn. Using your lecture notes in conjunction with this text is the best way to determine which material to study.

Skim topics in the text before they are covered in lecture. Reviewing a topic before lecture will make it easier for you to take good notes. First read the *What's Ahead* points and the end-of-chapter *Summary;* then quickly read through the chapter, skipping Sample Exercises and supplemental sections. Paying attention to the titles of sections and subsections gives you a feeling for the scope of topics. Try to avoid thinking that you must learn and understand everything right away.

Do a certain amount of preparation before lecture. More than ever, instructors are using the lecture period not simply as a one-way channel of communication from teacher to student. Rather, they expect students to come to class ready to work on problem solving and critical thinking. Coming to class unprepared is not a good idea for any lecture environment, but it certainly is not an option for an active learning classroom if you aim to do well in the course.

After lecture, carefully read the topics covered in class. As you read, pay attention to the concepts presented and to the application of these concepts in the *Sample Exercises.* Once you think you understand a *Sample Exercise,* test your understanding by working the accompanying *Practice Exercise.*

Learn the language of chemistry. As you study chemistry, you will encounter many new words. It is important to pay attention to these words and to know their meanings or the entities to which they refer. Knowing how to identify chemical substances from their names is an important skill; it can help you avoid painful mistakes on examinations. For example, "chlorine" and "chloride" refer to very different things.

Attempt the assigned end-of-chapter exercises. Working the exercises selected by your instructor provides necessary practice in recalling and using the essential ideas of the chapter. You cannot learn merely by observing; you must be a participant. In particular, try to resist checking the *Solutions Manual* (if you have one) until you have made a sincere effort to solve the exercise yourself. If you get stuck on an exercise, however, get help from your instructor, your teaching assistant, or another student. Spending more than 20 minutes on a single exercise is rarely effective unless you know that it is particularly challenging.

Learn to think like a scientist. This book is written by scientists who love chemistry. We encourage you to develop your critical thinking skills by taking advantage of features in this new edition, such as exercises that focus on conceptual learning and the *Design an Experiment* exercises.

Use online resources. Some things are more easily learned by discovery, and others are best shown in three dimensions. If your instructor has included Mastering Chemistry with your book, take advantage of the unique tools it provides to get the most out of your time in chemistry.

The bottom line is to work hard, study effectively, and use the tools available to you, including this textbook. We want to help you learn more about the world of chemistry and why chemistry is the central science. If you really learn chemistry, you can be the life of the party, impress your friends and parents, and . . . well, also pass the course with a good grade.

Answers to Go Figures, Practice Exercises and Self Assessment Exercises are available as PDF files within Mastering Chemistry. We invite instructors to share these resources as needed.

Acknowledgments

The production of a textbook is a team effort requiring the involvement of many people besides the authors who contributed hard work and talent to bring this edition to life. Although their names don't appear on the cover of the book, their creativity, time, and support have been instrumental in all stages of its development and production.

Each of us has benefited greatly from discussions with colleagues and from correspondence with instructors and students both here and abroad. Colleagues have also helped immensely by reviewing our materials, sharing their insights, and providing suggestions for improvements. For this edition, we were particularly blessed with an exceptional group of accuracy checkers who read through our materials looking for both technical inaccuracies and typographical errors.

Reviewers of Chemistry: *The Central Science*

S. K. Airee, *University of Tennessee*

John J. Alexander, *University of Cincinnati*

Robert Allendoerfer, *SUNY Buffalo*

Patricia Amateis, *Virginia Polytechnic Institute and State University*

Sandra Anderson, *University of Wisconsin*

John Arnold, *University of California*

Socorro Arteaga, *El Paso Community College*

Margaret Asirvatham, *University of Colorado*

Todd L. Austell, *University of North Carolina, Chapel Hill*

Yiyan Bai, *Houston Community College*

Melita Balch, *University of Illinois at Chicago*

Rebecca Barlag, *Ohio University*

Christine Barnes, *University of Tennessee, Knoxville*

Rosemary Bartoszek-Loza, *The Ohio State University*

Hafed Bascal, *University of Findlay*

Boyd Beck, *Snow College*

Kelly Beefus, *Anoka-Ramsey Community College*

Amy Beilstein, *Centre College*

Donald Bellew, *University of New Mexico*

Phil Bennett, *Santa Fe Community College*

Victor Berner, *New Mexico Junior College*

Narayan Bhat, *University of Texas, Pan American*

Jo Blackburn, *Richland College*

Merrill Blackman, *United States Military Academy*

Salah M. Blaih, *Kent State University*

Carribeth Bliem, *University of North Carolina, Chapel Hill*

Stephen Block, *University of Wisconsin, Madison*

James A. Boiani, *SUNY Geneseo*

John Bookstaver, *St. Charles Community College*

Leon Borowski, *Diablo Valley College*

Simon Bott, *University of Houston*

Kevin L. Bray, *Washington State University*

Daeg Scott Brenner, *Clark University*

Gregory Alan Brewer, *Catholic University of America*

Karen Brewer, *Virginia Polytechnic Institute and State University*

Ron Briggs, *Arizona State University*

Edward Brown, *Lee University*

Gary Buckley, *Cameron University*

Scott Bunge, *Kent State University*

William Butler, *Rochester Institute of Technology*

Carmela Byrnes, *Texas A&M University*

B. Edward Cain, *Rochester Institute of Technology*

Kim Calvo, *University of Akron*

Donald L. Campbell, *University of Wisconsin*

Rachel Campbell, *Florida Gulf Coast University*

Gene O. Carlisle, *Texas A&M University*

David Carter, *Angelo State University*

Elaine Carter, *Los Angeles City College*

Robert Carter, *University of Massachusetts at Boston Harbor*

Ann Cartwright, *San Jacinto Central College*

David L. Cedeño, *Illinois State University*

Dana Chatellier, *University of Delaware*

Stanton Ching, *Connecticut College*

Paul Chirik, *Cornell University*

Ted Clark, *The Ohio State University*

Tom Clayton, *Knox College*

William Cleaver, *University of Vermont*

Beverly Clement, *Blinn College*

Robert D. Cloney, *Fordham University*

Doug Cody, *Nassau Community College*

John Collins, *Broward Community College*

Edward Werner Cook, *Tunxis Community Technical College*

Elzbieta Cook, *Louisiana State University*

Enriqueta Cortez, *South Texas College*

Jason Coym, *University of South Alabama*

Thomas Edgar Crumm, *Indiana University of Pennsylvania*

Dwaine Davis, *Forsyth Tech Community College*

Michelle Dean, *Kennesaw State University*

Nancy De Luca, *University of Massachusetts, Lowell North Campus*

Angel de Dios, *Georgetown University*

John M. DeKorte, *Glendale Community College*

Michael Denniston, *Georgia Perimeter College*

Marian DeWane, *University California Irvine*

Daniel Domin, *Tennessee State University*

James Donaldson, *University of Toronto*

Patrick Donoghue, *Appalachian State University*

Bill Donovan, *University of Akron*

Tom Dowd, *Harper College*

Stephen Drucker, *University of Wisconsin-Eau Claire*

Ronald Duchovic, *Indiana University–Purdue University at Fort Wayne*

Robert Dunn, *University of Kansas*

David Easter, *Southwest Texas State University*

Joseph Ellison, *United States Military Academy*

George O. Evans II, *East Carolina University*

Emmanue Ewane, *Houston Community College*

Jordan Fantini, *Denison University*

James M. Farrar, *University of Rochester*

Debra Feakes, *Texas State University at San Marcos*

Gregory M. Ferrence, *Illinois State University*

Clark L. Fields, *University of Northern Colorado*

Jennifer Firestine, *Lindenwood University*

Jan M. Fleischner, *College of New Jersey*

Paul A. Flowers, *University of North Carolina at Pembroke*

Michelle Fossum, *Laney College*

Roger Frampton, *Tidewater Community College*

Joe Franek, *University of Minnesota*

David Frank, *California State University*

Cheryl B. Frech, *University of Central Oklahoma*

Ewa Fredette, *Moraine Valley College*

Kenneth A. French, *Blinn College*

Karen Frindell, *Santa Rosa Junior College*

John I. Gelder, *Oklahoma State University*

Robert Gellert, *Glendale Community College*

Luther Giddings, *Salt Lake Community College*

Paul Gilletti, *Mesa Community College*

Peter Gold, *Pennsylvania State University*

Eric Goll, *Brookdale Community College*

John Gorden, *Auburn University*

James Gordon, *Central Methodist College*

John Gorden, *Auburn University*

Palmer Graves, *Florida International University*

Thomas J. Greenbowe, *University of Oregon*

Michael Greenlief, *University of Missouri*

Eric P. Grimsrud, *Montana State University*

Nathan Grove, *University of North Carolina, Wilmington*

Brian Gute, *University of Minnesota, Duluth*

Margie Haak, *Oregon State University*

John Hagadorn, *University of Colorado*

Randy Hall, *Louisiana State University*

John M. Halpin, *New York University*

Marie Hankins, *University of Southern Indiana*

Robert M. Hanson, *St. Olaf College*

Daniel Haworth, *Marquette University*

Michael Hay, *Pennsylvania State University*

Inna Hefley, *Blinn College*

David Henderson, *Trinity College*

Brad Herrick, *Colorado School of Mines*

Paul Higgs, *Barry University*

Carl A. Hoeger, *University of California, San Diego*

Gary G. Hoffman, *Florida International University*

Deborah Hokien, *Marywood University*

Robin Horner, *Fayetteville Tech Community College*

Roger K. House, *Moraine Valley College*

Amanda Howell, *Appalachian State University*

Michael O. Hurst, *Georgia Southern University*

William Jensen, *South Dakota State University*

Jeff Jenson, *University of Findlay*

Janet Johannessen, *County College of Morris*

Milton D. Johnston, Jr., *University of South Florida*

Andrew Jones, *Southern Alberta Institute of Technology*

Booker Juma, *Fayetteville State University*

Ismail Kady, *East Tennessee State University*

Siam Kahmis, *University of Pittsburgh*

Steven Keller, *University of Missouri*

John W. Kenney, *Eastern New Mexico University*

Neil Kestner, *Louisiana State University*

Angela King, *Wake Forest University*

Jesudoss Kingston, *Iowa State University*

Leslie Kinsland, *University of Louisiana*

Louis J. Kirschenbaum, *University of Rhode Island*

Donald Kleinfelter, *University of Tennessee, Knoxville*

Daniela Kohen, *Carleton University*

David Kort, *George Mason University*

Jeffrey Kovac, *University of Tennessee*

George P. Kreishman, *University of Cincinnati*

Paul Kreiss, *Anne Arundel Community College*

Manickham Krishnamurthy, *Howard University*

Sergiy Kryatov, *Tufts University*

Brian D. Kybett, *University of Regina*

William R. Lammela, *Nazareth College*

John T. Landrum, *Florida International University*

Richard Langley, *Stephen F. Austin State University*

Russ Larsen, *University of Iowa*

Joe Lazafame, *Rochester Institute of Technology*

N. Dale Ledford, *University of South Alabama*

Ernestine Lee, *Utah State University*

David Lehmpuhl, *University of Southern Colorado*

Robley J. Light, *Florida State University*

Donald E. Linn, Jr., *Indiana University–Purdue University Indianapolis*

David Lippmann, *Southwest Texas State*

Patrick Lloyd, *Kingsborough Community College*

Encarnacion Lopez, *Miami Dade College, Wolfson*

Ramón López de la Vega, *Florida International University*

Charity Lovett, *Seattle University*

Arthur Low, *Tarleton State University*

Rosemary Loza, *The Ohio State University*

Michael Lufaso, *University of North Florida*

Gary L. Lyon, *Louisiana State University*

Preston J. MacDougall, *Middle Tennessee State University*

Jeffrey Madura, *Duquesne University*

Larry Manno, *Triton College*

Asoka Marasinghe, *Moorhead State University*

Earl L. Mark, *ITT Technical Institute*

Pamela Marks, *Arizona State University*

Albert H. Martin, *Moravian College*

Przemyslaw Maslak, *Pennsylvania State University*

Hilary L. Maybaum, *ThinkQuest, Inc.*

Armin Mayr, *El Paso Community College*

Marcus T. McEllistrem, *University of Wisconsin*

Craig McLauchlan, *Illinois State University*

Jeff McVey, *Texas State University at San Marcos*

William A. Meena, *Valley College*

Joseph Merola, *Virginia Polytechnic Institute and State University*

Stephen Mezyk, *California State University*

Gary Michels, *Creighton University*

Diane Miller, *Marquette University*

Eric Miller, *San Juan College*

Gordon Miller, *Iowa State University*

Shelley Minteer, *Saint Louis University*

Massoud (Matt) Miri, *Rochester Institute of Technology*

Mohammad Moharerrzadeh, *Bowie State University*

Tracy Morkin, *Emory University*

Barbara Mowery, *York College*

Kathleen E. Murphy, *Daemen College*

Kathy Nabona, *Austin Community College*

Robert Nelson, *Georgia Southern University*

Al Nichols, *Jacksonville State University*

Ross Nord, *Eastern Michigan University*

Jessica Orvis, *Georgia Southern University*

Mark Ott, *Jackson Community College*

Jason Overby, *College of Charleston*

Robert H. Paine, *Rochester Institute of Technology*

Robert T. Paine, *University of New Mexico*

Sandra Patrick, *Malaspina University College*

Mary Jane Patterson, *Brazosport College*

Tammi Pavelec, *Lindenwood University*

Albert Payton, *Broward Community College*

Lee Pedersen, *University of North Carolina*

Christopher J. Peeples, *University of Tulsa*

Kim Percell, *Cape Fear Community College*

Gita Perkins, *Estrella Mountain Community College*

Richard Perkins, *University of Louisiana*

Nancy Peterson, *North Central College*

Robert C. Pfaff, *Saint Joseph's College*

John Pfeffer, *Highline Community College*

Lou Pignolet, *University of Minnesota*

Bernard Powell, *University of Texas*

Bob Pribush, *Butler University*

Jeffrey A. Rahn, *Eastern Washington University*

Steve Rathbone, *Blinn College*

Bhavna Rawal, *Houston Community College*

Scott Reeve, *Arkansas State University*

John Reissner Helen Richter Thomas Ridgway, *University of North Carolina, University of Akron, University of Cincinnati*

Al Rives, *Wake Forest University*

Gregory Robinson, *University of Georgia*

Mark G. Rockley, *Oklahoma State University*

Lenore Rodicio, *Miami Dade College*

Amy L. Rogers, *College of Charleston*

Jimmy R. Rogers, *University of Texas at Arlington*

Kathryn Rowberg, *Purdue University at Calumet*

Steven Rowley, *Middlesex Community College*

Kresimir Rupnik, *Louisiana State University*

Joel Russell, *Oakland University*

James E. Russo, *Whitman College*

Theodore Sakano, *Rockland Community College*

Michael J. Sanger, *University of Northern Iowa*

Jerry L. Sarquis, *Miami University*

James P. Schneider, *Portland Community College*

Mark Schraf, *West Virginia University*

Melissa Schultz, *The College of Wooster*

Gray Scrimgeour, *University of Toronto*

Paula Secondo, *Western Connecticut State University*

Stacy Sendler, *Arizona State University*

Michael Seymour, *Hope College*

Kathy Thrush Shaginaw, *Villanova University*

Susan M. Shih, *College of DuPage*

David Shinn, *University of Hawaii at Hilo*

Lewis Silverman, *University of Missouri at Columbia*

Vince Sollimo, *Burlington Community College*

David Soriano, *University of Pittsburgh-Bradford*

Richard Spinney, *The Ohio State University*

Eugene Stevens, *Binghamton University*

Jerry Suits, *University Northern Colorado*

James Symes, *Cosumnes River College*

Greg Szulczewski, *University of Alabama, Tuscaloosa*

Matt Tarr, *University of New Orleans*

Dennis Taylor, *Clemson University*

Iwao Teraoka, *Polytechnic University*

Domenic J. Tiani, *University of North Carolina, Chapel Hill*

Edmund Tisko, *University of Nebraska at Omaha*

Richard S. Treptow, *Chicago State University*

Harold Trimm, *Broome Community College*

Michael Tubergen, *Kent State University*

Claudia Turro, *The Ohio State University*

James Tyrell, *Southern Illinois University*

Michael J. Van Stipdonk, *Wichita State University*

Philip Verhalen, *Panola College*

Ann Verner, *University of Toronto at Scarborough*

Edward Vickner, *Gloucester County Community College*

John Vincent, *University of Alabama*

Maria Vogt, *Bloomfield College*

Emanuel Waddell, *University of Alabama, Huntsville*

Tony Wallner, *Barry University*

Lichang Wang, *Southern Illinois University*

Thomas R. Webb, *Auburn University*

Clyde Webster, *University of California at Riverside*

Karen Weichelman, *University of Louisiana-Lafayette*

Paul G. Wenthold, *Purdue University*

Laurence Werbelow, *New Mexico Institute of Mining and Technology*

Wayne Wesolowski, *University of Arizona*

Sarah West, *University of Notre Dame*

Linda M. Wilkes, *University at Southern Colorado*

Charles A. Wilkie, *Marquette University*

Darren L. Williams, *West Texas A&M University*

Kurt Winklemann, *Florida Institute of Technology*

Klaus Woelk, *University of Missouri, Rolla*

Steve Wood, *Brigham Young University*

Troy Wood, *SUNY Buffalo*

Kimberly Woznack, *California University of Pennsylvania*

Thao Yang, *University of Wisconsin*

David Zax, *Cornell University*

Bob Zelmer, *The Ohio State University*

Dr. Susan M. Zirpoli, *Slippery Rock University*

Edward Zovinka, *Saint Francis University*

We would also like to express our gratitude to our partners at Pearson whose hard work, imagination, and commitment have contributed so greatly to the final form of this edition: Deb Harden and Ian Desrosiers, our Content Analyst and Product Manager, respectively, for their unflagging enthusiasm, continuous encouragement, and support; John Murdzek, our Development Editor, whose depth of experience, deep knowledge of chemistry, and careful attention to detail were invaluable to this revision; Matt Walker, Content Development Manager, who has brought his experience and insight to oversight of the entire project; and Beth Sweeten, Senior Content Producer, for keeping us on task and helping us to prioritize the many tasks that go into a revision of this scope. This revision would not have been possible without their hard work and guidance.

There are many others who also deserve special recognition, including the following: Mary Tindle, our production editor, whose expertise, advice, flexibility, and commitment to accuracy have allowed us to produce a textbook that is both beautiful and of high quality; and Roxy Wilson (University of Illinois), who so ably coordinated the difficult job of working out solutions to the end-of-chapter exercises. Finally, we wish to thank our families and friends for their love, support, encouragement, and patience as we brought this fifteenth edition to completion.

Theodore L. Brown
Department of Chemistry
University of Illinois at Urbana-Champaign
Urbana, IL 61801
tlbrown@illinois. edu or tlbrown1@ earthlink.net

H. Eugene LeMay, Jr.
Department of Chemistry
University of Nevada Reno, NV 89557
lemay@unr.edu

Bruce E. Bursten
Department of Chemistry and Biochemistry
Worcester Polytechnic Institute
Worcester, MA 01609
bbursten@wpi.edu

Catherine J. Murphy
Department of Chemistry
University of Illinois at Urbana-Champaign
Urbana, IL 61801
murphycj@illinois. edu

Patrick M. Woodward
Department of Chemistry and Biochemistry
The Ohio State University
Columbus, OH 43210
woodward.55@ osu.edu

Matthew W. Stoltzfus
Department of Chemistry and Biochemistry
The Ohio State University
Columbus, OH 43210
stoltzfus.5@osu. edu

ABOUT THE AUTHORS

The Brown/Lemay/Bursten/Murphy/Woodward/Stoltzfus Author Team values collaboration as an integral component to overall success. While each author brings unique talent, research interests, and teaching experiences, the team works together to review and develop the entire text. It is this collaboration that keeps the content ahead of educational trends and contributes to continuous innovations in teaching and learning throughout the text and technology. Some of the new key features in the fifteenth edition and accompanying Mastering Chemistry course are highlighted on the upcoming pages.

Theodore L. Brown received his Ph.D. from Michigan State University in 1956. Since then, he has been a member of the faculty of the University of Illinois, Urbana-Champaign, where he is now Professor of Chemistry, Emeritus. He served as Vice Chancellor for Research, and Dean of The Graduate College, from 1980 to 1986, and as Founding Director of the Arnold and Mabel Beckman Institute for Advanced Science and Technology from 1987 to 1993. Professor Brown has been an Alfred P. Sloan Foundation Research Fellow and has been awarded a Guggenheim Fellowship. In 1972 he was awarded the American Chemical Society Award for Research in Inorganic Chemistry and received the American Chemical Society Award for Distinguished Service in the Advancement of Inorganic Chemistry in 1993. He has been elected a Fellow of the American Association for the Advancement of Science, the American Academy of Arts and Sciences, and the American Chemical Society.

H. Eugene Lemay, Jr., received his B.S. degree in Chemistry from Pacific Lutheran University (Washington) and his Ph.D. in Chemistry in 1966 from the University of Illinois, Urbana-Champaign. He then joined the faculty of the University of Nevada, Reno, where he is currently Professor of Chemistry, Emeritus. He has enjoyed Visiting Professorships at the University of North Carolina at Chapel Hill, at the University College of Wales in Great Britain, and at the University of California, Los Angeles. Professor LeMay is a popular and effective teacher, who has taught thousands of students during more than 40 years of university teaching. Known for the clarity of his lectures and his sense of humor, he has received several teaching awards, including the University Distinguished Teacher of the Year Award (1991) and the first Regents' Teaching Award given by the State of Nevada Board of Regents (1997).

Bruce E. Bursten received his Ph.D. in Chemistry from the University of Wisconsin in 1978. After two years as a National Science Foundation Postdoctoral Fellow at Texas A&M University, he joined the faculty of The Ohio State University, where he rose to the rank of Distinguished University Professor. In 2005, he moved to the University of Tennessee, Knoxville, as Distinguished Professor of Chemistry and Dean of the College of Arts and Sciences. In 2015, he moved to Worcester Polytechnic Institute as Provost and Professor of Chemistry and Biochemistry. Professor Bursten has been a Camille and Henry Dreyfus Foundation Teacher-Scholar and an Alfred P. Sloan Foundation Research Fellow, and he is a Fellow of both the American Association for the Advancement of Science and the American Chemical Society. At Ohio State he received the University Distinguished Teaching Award in 1982 and 1996, the Arts and Sciences Student Council Outstanding Teaching Award in 1984, and the University Distinguished Scholar Award in 1990. He received the Spiers Memorial Prize and Medal of the Royal Society of Chemistry in 2003, the Morley Medal of the Cleveland Section of the American Chemical Society in 2005, and the American Chemical Society Award for Distinguished Service in the Advancement of Inorganic Chemistry in 2020. He was President of the American Chemical Society for 2008 and Chair of the Section on Chemistry of the American Association for the Advancement of Science in 2015. Professor Bursten's research program focuses on theoretical studies of compounds of the transition-metal and actinide elements.

Catherine J. Murphy received two B.S. degrees, one in Chemistry and one in Biochemistry, from the University of Illinois, Urbana-Champaign, in 1986. She received her Ph.D. in Chemistry from the University of Wisconsin in 1990. She was a National Science Foundation and National Institutes of Health Postdoctoral Fellow at the California Institute of Technology from 1990 to 1993. In 1993, she joined the faculty of the University of South Carolina, Columbia, becoming the Guy F. Lipscomb Professor of Chemistry in 2003. In 2009 she moved to the University of Illinois, Urbana-Champaign, as the Peter C. and Gretchen Miller Markunas Professor of Chemistry. Professor Murphy has been honored for both research and teaching as a Camille Dreyfus Teacher-Scholar, an Alfred P. Sloan Foundation Research Fellow, a Cottrell Scholar of the Research Corporation, a National Science Foundation CAREER Award winner, and a subsequent NSF Award for Special Creativity. She has also received a USC Mortar Board Excellence in Teaching Award, the USC Golden Key Faculty Award for Creative Integration of Research and Undergraduate Teaching, the USC Michael J. Mungo Undergraduate Teaching Award, and the USC Outstanding Undergraduate Research Mentor Award. From 2006–2011, Professor Murphy served as a Senior Editor for the *Journal of Physical Chemistry*; in 2011, she became the Deputy Editor for the *Journal of Physical Chemistry C*. She is an elected Fellow of the American Association for the Advancement of Science (2008), the American Chemical Society (2011), the Royal Society of Chemistry (2014), and the U.S. National Academy of Sciences (2015). Professor Murphy's research program focuses on the synthesis, optical properties, surface chemistry, biological applications, and environmental implications of colloidal inorganic nanomaterials.

Patrick M. Woodward received B.S. degrees in both Chemistry and Engineering from Idaho State University in 1991. He received a M.S. degree in Materials Science and a Ph.D. in Chemistry from Oregon State University in 1996. He spent two years as a postdoctoral researcher in the Department of Physics at Brookhaven National Laboratory. In 1998, he joined the faculty of the Chemistry Department at The Ohio State University where he currently holds the rank of Professor. He has been a visiting professor at Durham University in the UK, the University of Sydney in Australia, and the University of Bordeaux in France. Professor Woodward is a Fellow of the American Chemical Society and has been recognized with an Alfred P. Sloan Foundation Research Fellowship and a National Science Foundation CAREER Award. He has served as Vice Chair for Undergraduate Studies in the Department of Chemistry and Biochemistry at Ohio State University, and director of the Ohio REEL program. Professor Woodward's research program focuses on understanding the links between bonding, structure, and properties of solid-state inorganic materials.

Matthew W. Stoltzfus received his B.S. degree in Chemistry from Millersville University in 2002 and his Ph.D. in Chemistry in 2007 from The Ohio State University. He spent two years as a teaching postdoctoral assistant for the Ohio REEL program, an NSF-funded center that works to bring authentic research experiments into the general chemistry lab curriculum in 15 colleges and universities across the state of Ohio. In 2009, he joined the faculty of Ohio State where he currently holds the position of Senior Lecturer. In addition to lecturing on general chemistry, he served as a Faculty Fellow for the Digital First Initiative, inspiring instructors to offer engaging digital learning content to students through emerging technology. Through this initiative, he developed an iTunes U general chemistry course, which has attracted over 220,000 students from all over the world. The iTunes U course, along with the videos at www.drfus.com, are designed to supplement the text and can be used by any general chemistry student. Stoltzfus has received several teaching awards, including the inaugural Ohio State University 2013 Provost's Award for Distinguished Teaching by a Lecturer, and he is recognized as an Apple Distinguished Educator.

INTRODUCTION: MATTER, ENERGY, AND MEASUREMENT

▲ THE MANUFACTURE OF SYNTHETIC PIGMENTS is one of the oldest examples of industrial chemistry. The impressionist artists made extensive use of the bold colors of the newly available pigments, as exemplified in van Gogh's painting *Road with Cyprus and Star*.

The title of this book is *Chemistry: The Central Science* because much of what goes on in the world around us involves chemistry. **Chemistry** is the study of matter, its properties, and the changes that matter undergoes. As you progress in your study, you will come to see how chemical principles operate in all aspects of our daily lives, including the ways our bodies process the food that we eat, the production of energy to power our vehicles and portable electronic devices, the brilliant color changes in fall leaves, and critical issues in our environment. We will also see that the properties of substances can be tailored for specific applications by controlling their composition and structure.

This first chapter provides an overview of what chemistry is about and what chemists do. The "What's Ahead" list in this and all of the chapters gives an overview of the chapter organization and of some of the ideas we will consider.

1.1 | The Study of Chemistry

Chemistry is at the heart of many changes we see in the world around us, and it accounts for the different properties we see in matter. To understand how these changes and properties arise, we need to look far beneath the surfaces of our everyday observations.

The Atomic and Molecular Perspective of Chemistry

Chemistry is the study of the properties and behavior of matter. **Matter** is the physical material of the universe; it is anything that has mass and occupies space. A **property** is any characteristic that allows us to recognize a particular type of matter and to distinguish it from other types. This book, your body, the air you are breathing, and the clothes you are wearing are all samples of matter. We observe a tremendous variety of matter in our world, but countless experiments have shown that all matter is composed of combinations of a little over 100 substances called **elements**. One of our major goals will be to relate the properties of matter to its composition—that is, to the specific elements it contains.

Chemistry also provides a background for understanding the properties of matter in terms of **atoms**, the exceedingly small building blocks of matter. Each element is composed of a unique kind of atom. We will see that the properties of matter relate to both the kinds of atoms the matter contains (*composition*) and the arrangements of these atoms (*structure*).

In a **molecule**, two or more atoms are joined in specific shapes. Throughout this text we represent molecules using colored spheres to show how the atoms are connected (**Figure 1.1**). The color provides a convenient way to distinguish between atoms of different elements. For example, notice that the molecules of ethanol and ethylene

Go Figure Which molecule has the most carbon atoms? How many carbon atoms does it contain?

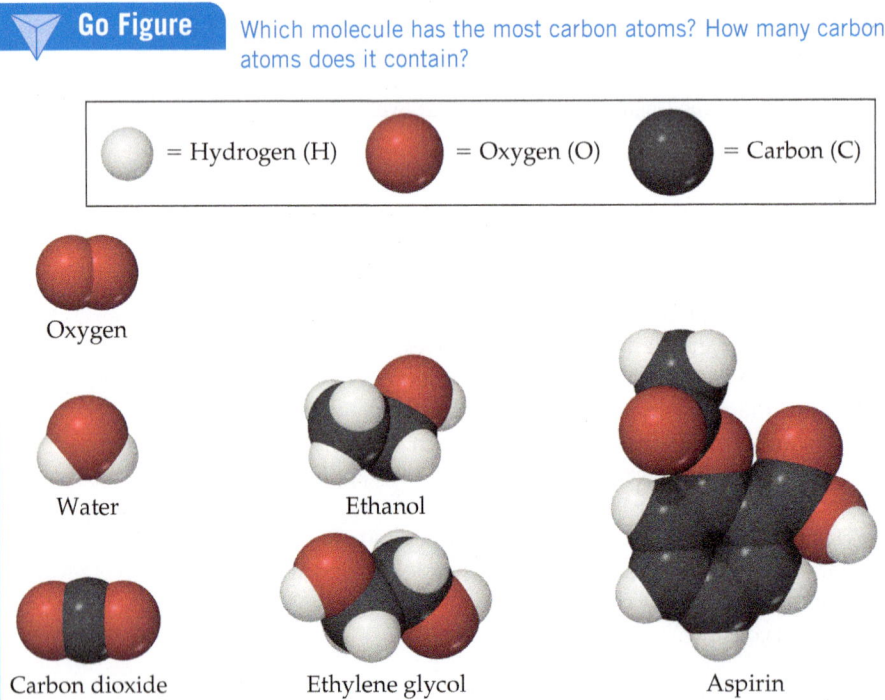

▲ **Figure 1.1 Molecular models.** The white, black, and red spheres represent atoms of hydrogen, carbon, and oxygen, respectively.

 Learning Objectives

When you finish **Section 1.1**, you should be able to:
- Explain the concepts of matter, atoms, and molecules.
- Demonstrate how molecules and the atoms that compose them are represented by molecular models.

glycol in Figure 1.1 have different compositions and structures. Ethanol contains one oxygen atom, depicted by one red sphere, whereas ethylene glycol contains two oxygen atoms.

Even apparently minor differences in the composition or structure of molecules can cause profound differences in properties. For example, ethanol and ethylene glycol (Figure 1.1) appear to be quite similar. Ethanol is the alcohol in beverages such as beer and wine, whereas ethylene glycol is a viscous liquid used as automobile antifreeze. The properties of these two substances differ in many ways, as do their biological activities. Ethanol is consumed throughout the world, but ethylene glycol is highly toxic. One of the challenges chemists undertake is to alter the composition or structure of molecules in a controlled way, creating new substances with different properties. For example, the common drug aspirin (Figure 1.1) was first synthesized in 1897 in a successful attempt to improve on a natural product extracted from willow bark that had long been used to alleviate pain.

Every change in the observable world—from boiling water to the changes that occur as our bodies combat invading viruses—has its basis in the world of atoms and molecules. Thus, as we proceed with our study of chemistry, we will find ourselves thinking in two realms: the *macroscopic* realm of ordinary-sized objects (*macro* means large) and the *submicroscopic* realm of atoms and molecules. We make our observations in the macroscopic world, but to understand that world, we must visualize how atoms and molecules behave at the submicroscopic level. Chemistry is the science that seeks to understand the properties and behavior of matter by studying the properties and behavior of atoms and molecules.

Why Study Chemistry?

Chemistry is all around us. Examples include the household chemicals used for cleaning and disinfecting that became so important during the COVID-19 pandemic (**Figure 1.2**). The chemical industry in the United States is nearly a $600 billion enterprise that employs over 500,000 people and accounts for nearly 10% of all U.S. exports.

▲ **Figure 1.2 Household chemicals.** The cleansing and disinfecting properties of these household products used so extensively during the COVID 19 pandemic are due to the chemicals they contain.

Chemistry lies near the heart of many matters of public concern, such as improvement of health care, conservation of natural resources, protection of the environment, and the supply of energy needed to keep society running. Using chemistry, we have discovered and continually improved upon pharmaceuticals, fertilizers and pesticides, plastics, solar panels, light-emitting diodes (LEDs), and building materials. We have also discovered that some chemicals are harmful to our health or the environment. This means that we must be sure that the materials with which we come into contact are safe. As a citizen and consumer, it is in your best interest to understand the effects, both positive and negative, that chemicals can have—we want you to have a balanced outlook regarding their uses.

You may be studying chemistry because it is an essential part of your curriculum. Your major might be chemistry, or it could be biology, engineering, pharmacy, agriculture, geology, or some other field. Chemistry is central to a fundamental understanding of governing principles in many science-related fields. For example, our interactions with the material world raise basic questions about the materials around us. We will see that chemistry is central to most realms of modern life.

Self-Assessment Exercises

SAE 1.1   Which of the following statements is *false*? (**a**) All matter is composed of atoms of the elements. (**b**) The atoms of different elements must be different. (**c**) A molecule must contain atoms from two or more elements. (**d**) Different molecules can be made from the same elements. (**e**) Matter has mass and occupies space.

Propylene glycol

SAE 1.2   The molecule shown here is *propylene glycol*, a substance that is used extensively in the chemical industry. The color key is: White = hydrogen, red = oxygen, black = carbon. How many carbon atoms are in a molecule of propylene glycol? (**a**) 2 (**b**) 3 (**c**) 5 (**d**) 8 (**e**) 13

SAE 1.3   The molecule shown here is called *acetamide*. How many different elements and how many atoms are in a molecule of acetamide? (**a**) 3 elements, 4 atoms (**b**) 3 elements, 9 atoms (**c**) 4 elements, 4 atoms (**d**) 4 elements, 9 atoms (**e**) There is not enough information to answer the question.

Acetamide

1.2 | Classifications of Matter

As we progress through this text, we will continue to learn more about the properties of matter and how the atoms and elements that comprise matter affect those properties. Let's begin our study of chemistry by examining two fundamental ways in which matter is classified. Matter is typically characterized by its physical state (gas, liquid, or solid) and its composition (whether it is an element, a *compound*, or a *mixture*).

States of Matter

Think about what happens when liquid water freezes into ice. Both liquid water and ice consist of molecules of water, and yet we know there is a difference between them—one is liquid and one is solid. A sample of matter can be a gas, a liquid, or a solid. These three forms, called the **states of matter**, differ in some of their observable properties.

Learning Objectives

When you finish Section 1.2, you should be able to:

▶ Compare and contrast the different states of matter: solid, liquid, and gas.

▶ Distinguish among elements, compounds, and mixtures.

▶ Identify the atomic symbols of common elements.

- A **gas** (also known as vapor) has no fixed volume or shape; rather, it uniformly fills its container. A gas can be compressed to occupy a smaller volume, or it can expand to occupy a larger one.
- A **liquid** has a distinct volume independent of its container, assumes the shape of the portion of the container it occupies, and is not easily compressed.
- A **solid** has both a definite shape and a definite volume and is not easily compressed.

The properties of the states of matter can be understood on the molecular level (**Figure 1.3**). In a gas, the molecules are far apart and moving at high speeds, colliding repeatedly with one another and with the walls of the container. Compressing a gas decreases the amount of space between molecules and increases the frequency of collisions between molecules but does not alter the size or shape of the molecules. In a liquid, the molecules are packed closely together but still move rapidly. The rapid movement allows the molecules to slide over one another; thus, a liquid pours easily. In a solid, the molecules are held tightly together, usually in definite arrangements in which the molecules can wiggle only slightly in their otherwise fixed positions. Thus, the distances between molecules are similar in the liquid and solid states, but while the molecules are for the most part locked in place in a solid, they retain considerable freedom of motion in a liquid. Changes in temperature and/or pressure can convert one state of matter to another, illustrated by such familiar processes as ice melting or water vapor condensing. We discuss these conversions from one state to another in greater detail in Chapter 11.

Pure Substances

Most forms of matter we encounter—the air we breathe (a gas), the gasoline we burn in our cars (a liquid), and the sidewalk we walk on (a solid)—are not chemically pure. We can, however, separate these forms of matter into pure substances. A **pure substance** (usually referred to simply as a *substance*) is matter that has distinct properties and a composition that does not vary from sample to sample. Water and table salt (sodium chloride) are examples of pure substances.

All substances are either elements or compounds.

- *Elements* are substances that cannot be decomposed into simpler substances. On the molecular level, each element is composed of only one kind of atom [Figure 1.4(**a**) and (**b**)].
- **Compounds** are substances composed of two or more elements; they contain two or more kinds of atoms [Figure 1.4(**c**)]. Water, for example, is a compound composed of two elements: hydrogen and oxygen.

Figure 1.4(**d**) shows a mixture of substances. **Mixtures** are combinations of two or more substances in which each substance retains its chemical identity.

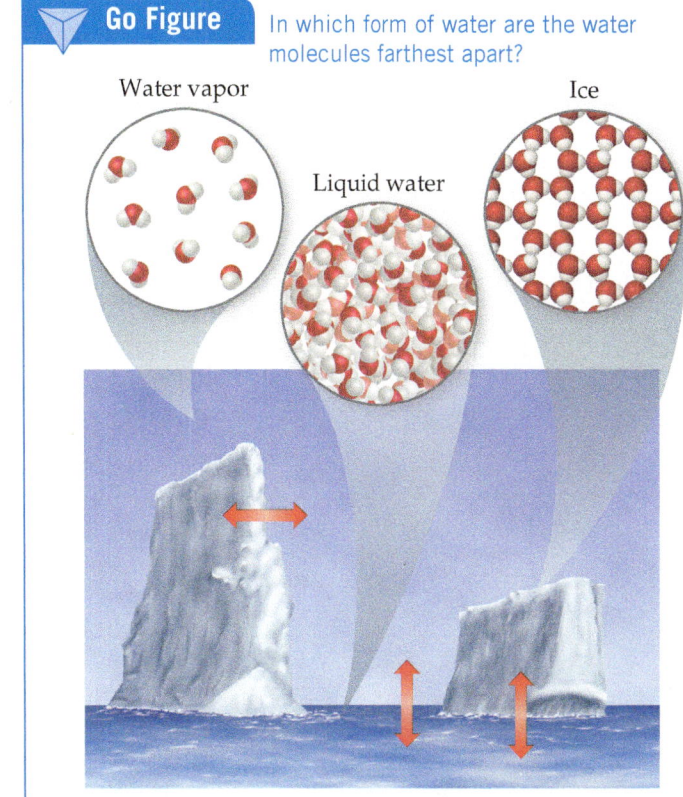

Go Figure In which form of water are the water molecules farthest apart?

Water vapor   Liquid water   Ice

▲ **Figure 1.3 The three physical states of water—water vapor, liquid water, and ice.** We can see the liquid and solid states but not the gas (vapor) state. The red arrows show that the three states of matter can interconvert.

Go Figure How do the molecules of a compound differ from the molecules of an element?

| (a) Atoms of an element | (b) Molecules of an element | (c) Molecules of a compound | (d) Mixture of elements and a compound |

Elements are composed of only one kind of atom.

Compounds must have at least two kinds of atoms.

▲ **Figure 1.4 Molecular comparison of elements, compounds, and mixtures.**

Go Figure

If the lower pie chart were drawn as the percentage in terms of number of atoms rather than the percentage in terms of mass, would the hydrogen slice of the pie get larger or smaller?

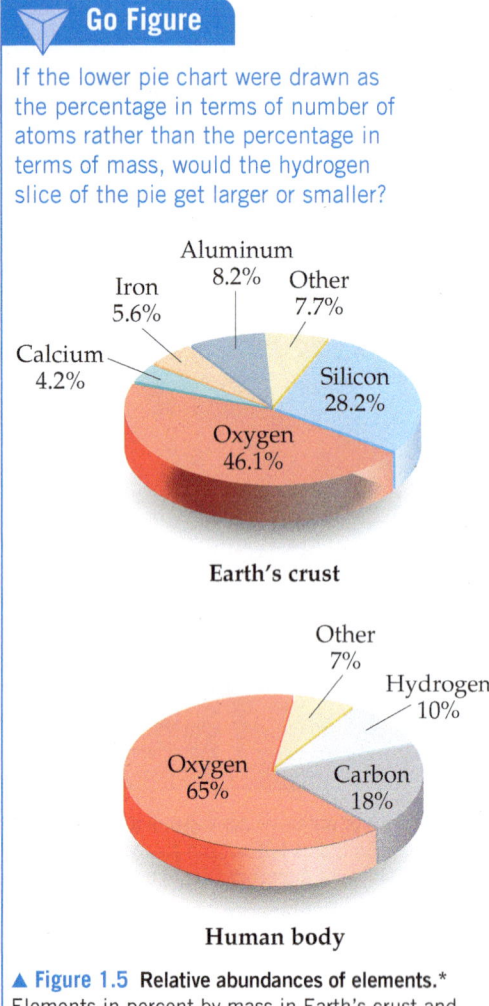

Earth's crust

Human body

▲ **Figure 1.5 Relative abundances of elements.*** Elements in percent by mass in Earth's crust and in the human body.

TABLE 1.1		**Some Common Elements and Their Symbols**			
Carbon	C	Aluminum	Al	Copper	Cu (from cuprum)
Fluorine	F	Bromine	Br	Iron	Fe (from ferrum)
Hydrogen	H	Calcium	Ca	Lead	Pb (from plumbum)
Iodine	I	Chlorine	Cl	Mercury	Hg (from hydrargyrum)
Nitrogen	N	Helium	He	Potassium	K (from kalium)
Oxygen	O	Lithium	Li	Silver	Ag (from argentum)
Phosphorus	P	Magnesium	Mg	Sodium	Na (from natrium)
Sulfur	S	Silicon	Si	Tin	Sn (from stannum)

Elements

Currently, 118 elements are known, though they vary widely in abundance. Hydrogen constitutes about 74% of the mass in the Milky Way galaxy, and helium constitutes 24%. Closer to home, only five elements—oxygen, silicon, aluminum, iron, and calcium—account for over 90% of the mass of Earth's crust, and only three—oxygen, carbon, and hydrogen—account for over 90% of the mass of the human body (**Figure 1.5**).

Table 1.1 lists some common elements, along with the chemical *symbols* used to denote them. The symbol for each element consists of one or two letters, with the first letter capitalized. These symbols are derived mostly from the English names of the elements, but sometimes they are derived from a foreign name instead (see the last column in Table 1.1). You will need to know these symbols and learn others as we encounter them in the text.

All of the known elements and their symbols are listed on the front inside cover of this text in a table known as the *periodic table*. In the periodic table, the elements are arranged in columns so that closely related elements are grouped together. We describe the periodic table in more detail in Section 2.5 and consider the periodically repeating properties of the elements in Chapter 7.

Compounds

Most elements can interact with other elements to form compounds. For example, when hydrogen gas burns in oxygen gas, the elements hydrogen and oxygen combine to form the compound water. Because each molecule of water contains two hydrogen atoms and one oxygen atom, we denote the molecule as H_2O. The subscript 2 indicates that there are two H atoms in the molecule. When there is only one atom of an element in a molecule, as is the case for O in water, we do not explicitly use the subscript 1.

Water can be decomposed back into its elements by passing an electrical current through it (**Figure 1.6**).

Decomposing pure water into its constituent elements shows that it contains 11% hydrogen and 89% oxygen by mass, regardless of its source. This ratio is constant because every water molecule is composed of two hydrogen atoms and one oxygen atom:

 Hydrogen atom (written H)

 Oxygen atom (written O)

 Water molecule (written H_2O)

The amount of oxygen by mass is greater than the amount of hydrogen by mass because oxygen atoms are heavier than hydrogen atoms.

*_____

**CRC Handbook of Chemistry and Physics*, 97th ed. (2016–2017), pp. 14–17.

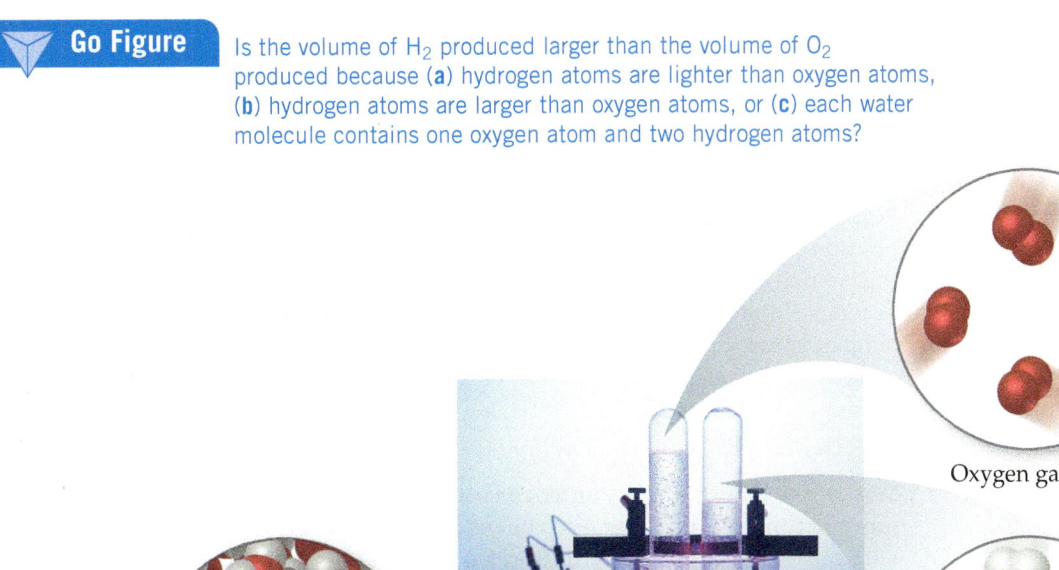

Go Figure Is the volume of H_2 produced larger than the volume of O_2 produced because (**a**) hydrogen atoms are lighter than oxygen atoms, (**b**) hydrogen atoms are larger than oxygen atoms, or (**c**) each water molecule contains one oxygen atom and two hydrogen atoms?

Oxygen gas, O_2

Hydrogen gas, H_2

Water, H_2O

▲ **Figure 1.6 Electrolysis of water.** Water decomposes into its component elements, hydrogen and oxygen, when an electrical current is passed through it. The volume of hydrogen, collected in the right test tube, is twice the volume of oxygen.

The elements hydrogen and oxygen themselves exist naturally as *diatomic* (two-atom) molecules:

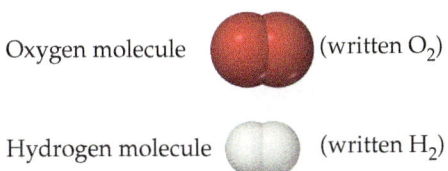

Oxygen molecule (written O_2)

Hydrogen molecule (written H_2)

As seen in Table 1.2, the properties of water bear no resemblance to the properties of its component elements. Hydrogen, oxygen, and water are each a unique substance, a consequence of the uniqueness of their respective molecules.

TABLE 1.2 Comparison of Water, Hydrogen, and Oxygen

	Water	Hydrogen	Oxygen
State[a]	Liquid	Gas	Gas
Normal boiling point	100 °C	−253 °C	−183 °C
Density[a]	1000 g/L	0.084 g/L	1.33 g/L
Flammable	No	Yes	No

[a]At room temperature and atmospheric pressure.

The observation that the elemental composition of a compound is always the same is known as the **law of constant composition** (or the **law of definite proportions**). French chemist Joseph Louis Proust (1754–1826) first stated the law in about 1800. Although this law has been known for 200 years, the belief persists among some people that a fundamental difference exists between compounds prepared in the laboratory and the corresponding compounds found in nature. This simply is not true. Regardless of its source—nature or a laboratory—a pure compound has the same composition and properties under the same conditions. Both chemists and nature must use the same elements and operate under the same natural laws. When two materials differ in composition or properties, either they are composed of different compounds or they differ in purity.

Mixtures

Most of the matter we encounter consists of mixtures of different substances. Each substance in a mixture retains its chemical identity and properties. In contrast to a pure substance, which by definition has a fixed composition, the composition of a mixture can vary. A cup of sweetened coffee, for example, can contain either a little sugar or a lot. The substances making up a mixture are called *components* of the mixture.

Some mixtures do not have the same composition, properties, and appearance throughout. Rocks and wood, for example, vary in texture and appearance in any typical sample. Such mixtures are *heterogeneous* [Figure 1.7(**a**)]. Mixtures that are uniform throughout are *homogeneous*. Air is a homogeneous mixture of nitrogen, oxygen, and smaller amounts of other gases. The nitrogen in air has all the properties of pure nitrogen because both the pure substance and the mixture contain the same nitrogen molecules. Salt, sugar, and many other substances dissolve in water to form homogeneous mixtures [Figure 1.7(**b**)]. Homogeneous mixtures are also called **solutions**. Although the term *solution* conjures up an image of a liquid, solutions can be solids, liquids, or gases.

In Figure 1.8 we summarize the classification of matter into elements, compounds, and mixtures. You should be comfortable in classifying substances among these three categories.

(a)

(b)

▲ **Figure 1.7 Mixtures.** (a) Many common materials, including rocks, are heterogeneous mixtures. This photograph of granite shows a heterogeneous mixture of silicon dioxide and other metal oxides. (b) Homogeneous mixtures are called solutions. Many substances, including the blue solid shown here [copper(II) sulfate pentahydrate], dissolve in water to form solutions.

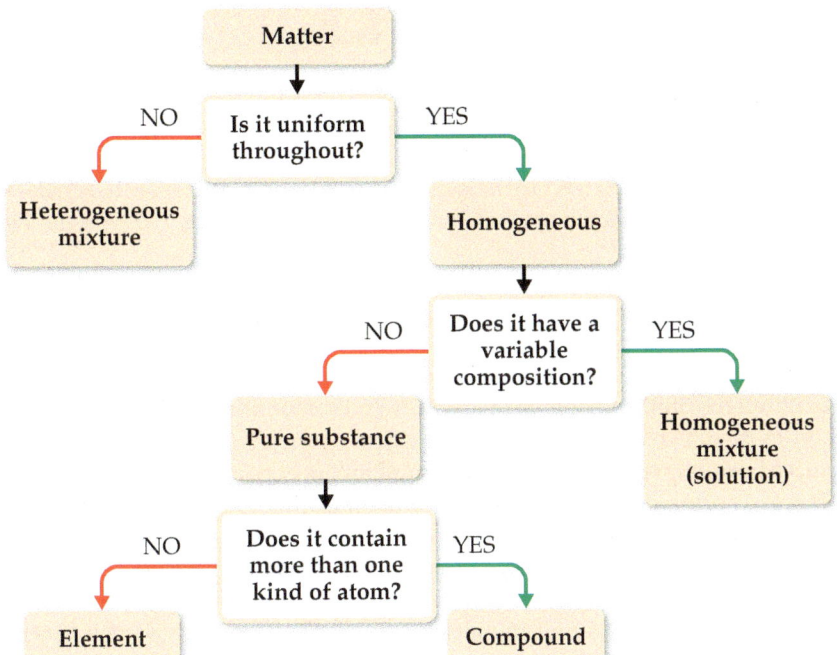

◀ **Figure 1.8 Classification of matter.** All pure matter is classified ultimately as either an element or a compound.

 ## Sample Exercise 1.1

Distinguishing among Elements, Compounds, and Mixtures

Classify each of the following as an element, a compound, a homogeneous mixture, or a heterogeneous mixture: (**a**) molten iron; (**b**) a chocolate chip cookie; (**c**) a container of pure ethylene glycol (see Figure 1.1); (**d**) a cup of water with a teaspoon of sugar dissolved in it.

SOLUTION

We can follow the flowchart in Figure 1.8 to classify each substance: (**a**) Molten iron is simply iron heated to the point at which it melts into a liquid. It still contains only iron atoms and is therefore a pure substance that is an element. (**b**) A cookie contains several different substances in different amounts. It is not uniform throughout and is therefore a heterogeneous mixture. (**c**) As seen in Figure 1.1, an ethylene glycol molecule contains C, H, and O atoms. The sample of pure ethylene glycol is therefore a pure

substance that is a compound. (**d**) The sugar water that results from dissolving sugar in water has the same composition throughout. It is a homogeneous mixture, also known as a solution.

▶ **Practice Exercise**

"White gold" contains gold and a "white" metal, such as palladium. Two samples of white gold differ in the relative amounts of gold and palladium they contain. Both samples are uniform in composition throughout. Use Figure 1.8 to classify white gold.

 ## Self-Assessment Exercises

SAE 1.4 Which of the following processes is best described as a liquid turning into a gas? (**a**) snow melting (**b**) molten metal hardening upon cooling (**c**) rubbing alcohol evaporating from your skin (**d**) steam condensing on a cold surface (**e**) dry ice, which is frozen CO_2, evaporating

SAE 1.5 Which of the following characterizations is *incorrect*? (**a**) Chicken noodle soup is a heterogeneous mixture. (**b**) Aspirin is composed of 60.0% carbon, 4.5% hydrogen, and 35.5% oxygen by mass, regardless of its source. Aspirin is a compound. (**c**) The tanks that scuba divers use contain nitrogen and oxygen gas. The tanks contain a homogeneous mixture. (**d**) Yellow sulfur consists of molecules that contain eight-membered rings of sulfur atoms. Yellow sulfur is a compound. (**e**) The graphite in lead pencils consists entirely of sheets of carbon atoms. Graphite is a form of an element.

SAE 1.6 The compound called *heme* is present in red blood cells. Heme contains carbon, hydrogen, iron, nitrogen, and oxygen. Which of the following is the correct list of the symbols of the elements in a heme molecule? (**a**) Ca, Hy, Ir, Ni, Ox (**b**) C, H, Fe, N, O (**c**) C, H, I, N, O (**d**) C, H₂, Fe, N₂, O₂

SAE 1.7 A certain material has a fixed composition of S and Cl atoms regardless of its source. Which of the following statements about the material is *true*? (**a**) The material is a homogeneous mixture. (**b**) The material contains atoms of silicon and chlorine. (**c**) The material must contain equal numbers of S and Cl atoms. (**d**) The material is a heterogeneous mixture. (**e**) The material is a compound that contains two elements.

1.3 | Properties of Matter

⚠ **Learning Objectives**

When you finish Section 1.3, you should be able to:

▶ Distinguish between chemical and physical properties, and between intensive and extensive properties.

▶ Differentiate between chemical and physical changes.

▶ Describe how filtration, distillation, and chromatography can be used to separate mixtures of substances.

Every substance has unique properties. For example, the properties listed in Table 1.2 allow us to distinguish hydrogen, oxygen, and water from one another. The properties of a substance include everything that describes the substance, such as its color, its physical state, its mass, its ability to react with other substances, and numerous other characteristics. In this section, we look more closely at the types of properties that we typically use in describing substances

Types of Properties of Substances

The properties of matter can be categorized as physical or chemical. **Physical properties** can be observed without changing the identity and composition of the substance. These properties include color, odor, density, melting point, boiling point, and hardness. **Chemical properties** describe the way a substance may change, or *react*, to form other substances. A common chemical property is flammability, the ability of a substance to burn in the presence of oxygen.

The properties of substances are further separated into two categories depending on whether the property depends on how much of the substance is being considered. Some properties, such as temperature and melting point, are *intensive properties*. **Intensive properties** do *not* depend on the amount of sample being examined and are particularly useful in chemistry because many intensive properties can be used to *identify* substances. **Extensive properties**, such as mass and volume, depend on the amount of sample. Extensive properties relate to the *amount* of substance present.

Physical and Chemical Changes

The changes substances undergo are either physical or chemical. During a **physical change**, a substance changes its physical appearance but not its composition: It is the same substance before and after the change. The evaporation of water is a physical change. When water evaporates, it changes from the liquid state to the gas state, but it is still composed of water molecules, as depicted in Figure 1.3. All **changes of state** (for example, from liquid to gas or from liquid to solid) are physical changes.

In a **chemical change** (also called a **chemical reaction**), a substance is transformed into a chemically different substance. When hydrogen burns in air, for example, it undergoes a chemical change because it combines with oxygen to form water (**Figure 1.9**).

Chemical changes can be dramatic. In the account given in **Figure 1.10**, Ira Remsen (1846–1927), author of a popular chemistry text, describes his first experiences with chemical reactions.

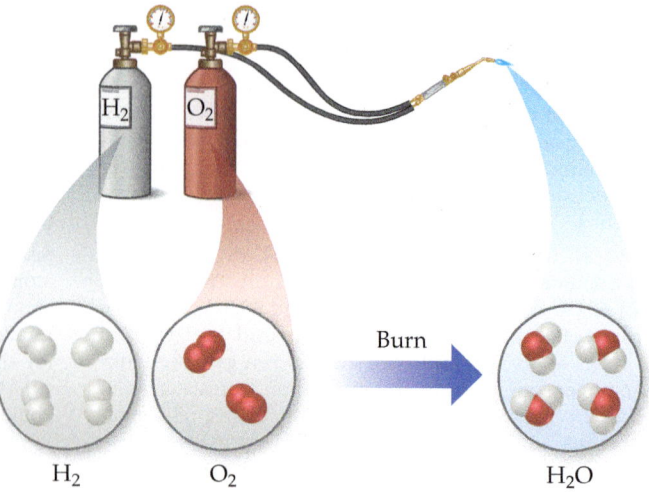

H₂   O₂ $\xrightarrow{\text{Burn}}$ H₂O

▲ Figure 1.9 **A chemical reaction.**

While reading a textbook of chemistry, I came upon the statement "nitric acid acts upon copper," and I determined to see what this meant. Having located some nitric acid, I had only to learn what the words "act upon" meant. In the interest of knowledge I was even willing to sacrifice one of the few copper cents then in my possession. I put one of them on the table, opened a bottle labeled "nitric acid," poured some of the liquid on the copper, and prepared to make an observation. But what was this wonderful thing which I beheld? The cent was already changed, and it was no small change either. A greenish-blue liquid foamed and fumed over the cent and over the table. The air became colored dark red. How could I stop this? I tried by picking the cent up and throwing it out the window. I learned another fact: nitric acid acts upon fingers. The pain led to another unpremeditated experiment. I drew my fingers across my trousers and discovered nitric acid acts upon trousers. That was the most impressive experiment I have ever performed. I tell of it even now with interest. It was a revelation to me. Plainly the only way to learn about such remarkable kinds of action is to see the results, to experiment, to work in the laboratory.

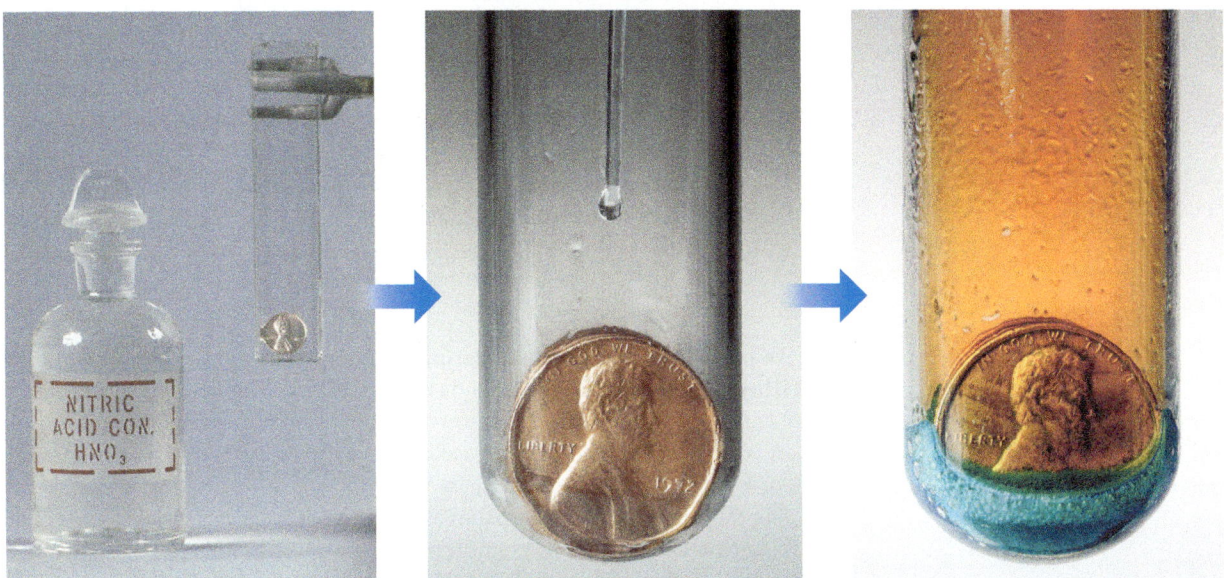

▲ **Figure 1.10 The chemical reaction between a copper penny and nitric acid.** The dissolved copper produces the blue-green solution; the reddish-brown gas produced is nitrogen dioxide.

Separation of Mixtures

We can separate a mixture into its components by taking advantage of differences in their properties. For example, a heterogeneous mixture of iron filings and gold filings could be sorted by color into iron and gold. A less tedious approach would be to use a magnet to attract the iron filings, leaving the gold ones behind. We can also take advantage of an important chemical difference between these two metals: Many acids dissolve iron but not gold. Thus, if we put our mixture into an appropriate acid, the acid would dissolve the iron and the solid gold would be left behind. The two could then be separated by **filtration** (Figure 1.11). We would have to use other chemical reactions, which we describe later, to transform the dissolved iron back into metal.

◀ Figure 1.11 **Separation by filtration.** A mixture of a solid and a liquid is poured through filter paper. The liquid passes through the paper while the solid remains on the paper.

▶ **Figure 1.12 Distillation.** Apparatus for separating a sodium chloride solution (salt water) into its components.

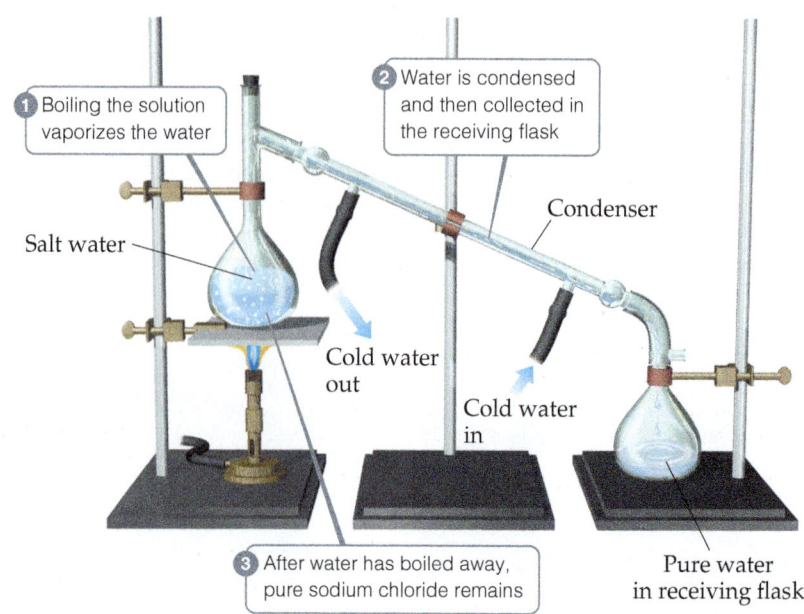

① Boiling the solution vaporizes the water

② Water is condensed and then collected in the receiving flask

Condenser

Salt water

Cold water out

Cold water in

③ After water has boiled away, pure sodium chloride remains

Pure water in receiving flask

An important method of separating the components of a homogeneous mixture is **distillation**, a process that depends on the different abilities of substances to form gases. For example, if we boil a solution of salt and water, the water evaporates, forming a gas, and the salt is left behind. The gaseous water can be converted back to a liquid on the walls of a condenser, as shown in **Figure 1.12**.

The differing abilities of substances to adhere to the surfaces of solids can also be used to separate mixtures. For example, a mixture of substances can be separated by placing a sample of the mixture at the top of a column filled with a porous solid, as shown in **Figure 1.13**. When a suitable solvent is added to the top of the column, the mixture separates into its different components. This separation technique is called **chromatography** (literally "the writing of colors").

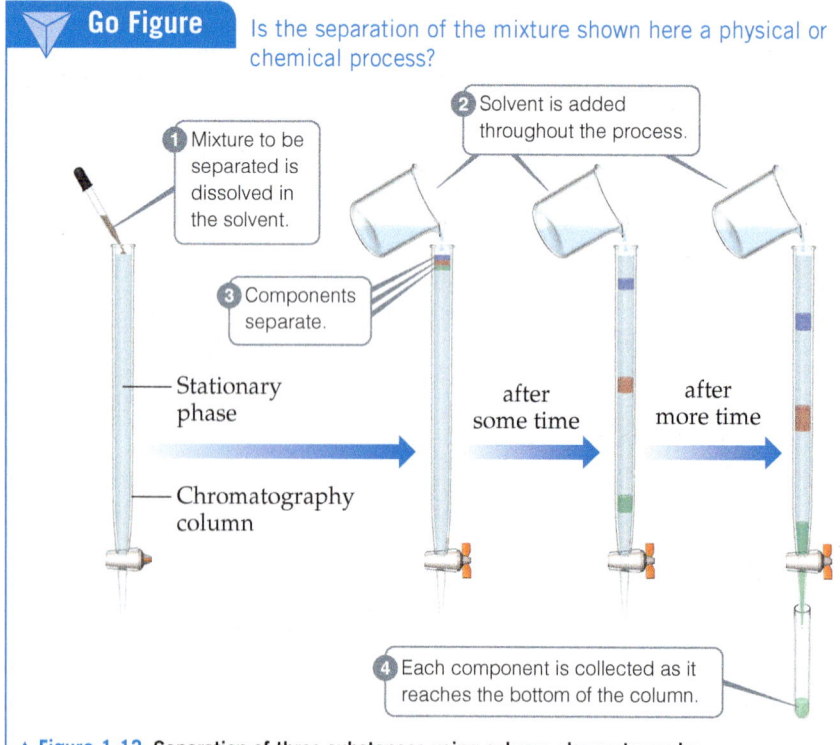

▼ **Go Figure**   Is the separation of the mixture shown here a physical or chemical process?

① Mixture to be separated is dissolved in the solvent.

② Solvent is added throughout the process.

③ Components separate.

Stationary phase

Chromatography column

after some time

after more time

④ Each component is collected as it reaches the bottom of the column.

▲ **Figure 1.13 Separation of three substances using column chromatography.**

Self-Assessment Exercises

SAE 1.8 Imagine you have a gallon of the substance called *methanol*. Under usual conditions, methanol is a flammable colorless liquid. Which of the following statements about the properties of methanol is *correct*? (**a**) The fact that methanol is colorless is a chemical property. (**b**) The mass of the gallon of methanol is an intensive property. (**c**) The fact that methanol is flammable is a physical property. (**d**) The temperature at which the methanol freezes is an intensive property.

SAE 1.9 Each of the following describes either a physical change or a chemical change of a substance:

   (**i**) An iron nail rusts when exposed to damp air.
  (**ii**) Water vapor condenses on a cold window to form frost.
 (**iii**) Plants make sugar from carbon dioxide and water.

Which of these processes are *chemical* changes? (**a**) Only one of the three is a chemical change (**b**) i and ii (**c**) i and iii (**d**) ii and iii (**e**) All three are chemical changes.

SAE 1.10 Which of the following statements about separation techniques is *incorrect*? (**a**) Filtration can be used to separate a solid substance from a liquid. (**b**) Distillation is a technique that depends on the differences in the evaporation of different substances. (**c**) Solid A dissolves in water and solid B does not dissolve in water. After adding a mixture of the two solids to water, filtration would be an effective way of separating a mixture of solids A and B. (**d**) Chromatography is a technique that depends on the differences in the rate of travel of different substances through a solid medium. (**e**) Chromatography can be used to separate a compound into its elements.

1.4 | The Nature of Energy

All objects in the universe are made of matter, but matter alone is not enough to describe the behavior of the world around us. The water in an alpine lake and a pot of boiling water are both made from the same substance, but your body will experience a very different sensation if you put your hand in each. The difference between the two is their energy content; boiling water has more energy than chilled water. To understand chemistry, we must also understand energy and the changes in energy that accompany chemical processes.

 **Learning Objectives**

When you finish Section 1.4, you should be able to:

▶ Describe the concepts of *energy*, *work*, and *heat*.

▶ Distinguish between *kinetic* and *potential* energy.

Work and Heat

Unlike matter, energy does not have mass and cannot be held in our hands, but its effects can be observed and measured. **Energy** is defined as *the capacity to do work or transfer heat*. Both work and heat are means by which energy is transferred from one object to another. **Work** is *the energy transferred when a force exerted on an object causes a displacement of that object*. **Heat** is *the energy transferred to cause the temperature of an object to increase*. The concepts of work and heat are illustrated in the examples shown in Figure 1.14.

Although the temperature of an object is intuitive to most people, the definition of work is less apparent. We define work, w, as the product of the force exerted on the object, F, and the distance, d, that it moves:

$$w = F \times d \qquad [1.1]$$

The **force** F is defined as any push or pull exerted on the object.* Familiar examples include gravity and the attraction between opposite poles of a bar magnet. It takes work to lift an object off of the floor or to pull apart two magnets that have come together at the opposite poles. When you type on a keyboard, you are doing work on the keys as they are displaced by the force of your fingers.

Kinetic Energy and Potential Energy

To understand energy, we need to grasp its two fundamental forms, kinetic energy and potential energy. Objects, whether they are automobiles, baseballs, or molecules, can

---

*In using this equation, only the component of the force that is acting in the same direction as the distance traveled is used. That will generally be the case for problems we will encounter in this chapter.

Work done by player on ball to make ball move

(a)

Heat added by burner to water makes water temperature rise

(b)

▲ **Figure 1.14 Work and heat, two forms of energy.** (**a**) *Work* is energy used to cause an object to move against an opposing force. (**b**) *Heat* is energy used to increase the temperature of an object.

possess **kinetic energy**, the energy of *motion*. The magnitude of the kinetic energy, E_k, of an object depends on its mass, m, and velocity, v:

$$E_k = \frac{1}{2} mv^2 \qquad [1.2]$$

Thus, the kinetic energy of an object increases as its velocity or speed* increases. For example, a car has greater kinetic energy moving at 65 miles per hour (mi/h) than it does at 25 mi/h. For a given velocity, the kinetic energy increases with increasing mass. Thus, a large truck traveling at 65 mi/h has greater kinetic energy than a motorcycle traveling at the same velocity because the truck has the greater mass.

In chemistry, we are interested in the kinetic energy of atoms and molecules. Although these particles are too small to be seen, they have mass and are in motion, so they possess kinetic energy. When a substance is heated, be it a pot of water on the stove or a metal bench sitting in the Sun, the atoms and molecules in that substance gain kinetic energy and their average speed increases. Hence, we see that the transfer of heat is simply the transfer of kinetic energy at the molecular level. We return to this concept in later chapters of the book.

All other forms of energy—the energy stored in a stretched spring, in a weight held above your head, or in a chemical bond—are classified as potential energy. An object has **potential energy** by virtue of its position relative to other objects. Potential energy is, in essence, the "stored" energy that arises from the attractions and repulsions an object experiences in relation to other objects.

We are familiar with many instances in which potential energy is converted into kinetic energy. For example, think of a cyclist poised at the top of a hill (**Figure 1.15**). Because of the attractive force of gravity, the gravitational potential energy of the bicycle is greater at the top of the hill than at the bottom. As a result, the bicycle easily rolls down the hill with increasing speed. As it does so, gravitational potential energy is converted into kinetic energy. The gravitational potential energy decreases as the bicycle rolls down the hill, while at the same time its kinetic energy increases as it picks up speed (Equation 1.2). This example illustrates that kinetic and potential energy are interconvertible.

Gravitational forces play a negligible role in the ways that atoms and molecules interact with one another. Forces that arise from electrical charges are more important when dealing with atoms and molecules. One of the most important forms of potential

High potential energy, zero kinetic energy

Decreasing potential energy, increasing kinetic energy

▲ **Figure 1.15 Potential energy and kinetic energy.** The potential energy initially stored in the motionless bicycle and rider at the top of the hill is converted to kinetic energy as the bicycle moves down the hill and loses potential energy.

---

*Strictly speaking, velocity is a *vector* quantity that has a direction—that is, it tells you how fast an object is moving and in what direction. Speed is a *scalar* quantity that tells you how fast an object is moving but not the direction of the motion. Unless otherwise stated, we will not be concerned with the direction of motion, so velocity and speed are used interchangeably in this book.

energy in chemistry is *electrostatic potential energy*, which arises from the interactions between charged particles. Opposite charges attract each other, whereas like charges repel. The strength of this interaction increases as the magnitudes of the charges increase, and it decreases as the distance between charges increases. We return to electrostatic energy several times throughout the book.

One of our goals in chemistry is to relate the energy changes seen in the macroscopic world to the kinetic or potential energy of substances at the molecular level. Many substances, fuels for example, release energy when they react. The *chemical energy* of a fuel is due to the potential energy stored in the arrangements of its atoms. As we explain in later chapters, *chemical energy is released when bonds between atoms are formed, and it is consumed when bonds between atoms are broken.* When a fuel burns, some bonds are broken and others are formed, but the net effect is to convert chemical potential energy to thermal energy, the energy associated with temperature.

 Self-Assessment Exercises

SAE 1.11   Which of the following statements about energy, work, and heat is *false*? (**a**) Energy can be classified into two fundamental forms: kinetic energy and potential energy. (**b**) Heat is energy transferred to increase the temperature of an object. (**c**) Energy is the capacity to do work or transfer heat. (**d**) Work causes the displacement of an object. (**e**) Energy has mass and occupies space.

SAE 1.12   A rocket takes off from the surface of Earth and accelerates into the sky. Which of the following correctly describes the changes in the kinetic and gravitational potential energy of the rocket as it takes off? (**a**) Its kinetic energy decreases and its gravitational potential energy decreases. (**b**) Its kinetic energy increases and its gravitational potential energy decreases. (**c**) Its kinetic energy decreases and its gravitational potential energy increases. (**d**) Its kinetic energy increases and its gravitational potential energy

increases. (**e**) The kinetic and gravitational potential energies of the rocket are unchanged.

SAE 1.13   Consider the following three vehicles in motion:

(**i**)   A compact car that weighs 2000 pounds and is traveling at 40 miles per hour.
(**ii**) A medium-sized car that weighs 3000 pounds and is traveling at 40 miles per hour.
(**iii**) A truck that weighs 4000 pounds and is traveling at 20 miles per hour.

Which of the following is the correct ordering of the vehicles from smallest to largest kinetic energy? (**a**) iii < i < ii (**b**) i < ii < iii (**c**) i = iii < ii (**d**) i < iii < ii (**e**) ii < iii = i

CHEMISTRY AND SUSTAINABILITY   An Introduction

As the global population grows, we continue to consume Earth's resources and put strains on the capacity of the human race to provide essential needs to sustain its existence. The notion of *sustainability* is used to describe the intersection of economic, social, and environmental factors to determine the best ways in which we can exist on Earth while being responsible stewards of it.

In 1991, the World Conservation Union, the United Nations Environment Programme, and the World Wildlife Fund for Nature published a report entitled "Caring for the Earth: A Strategy for Sustainable Living."* In this report, they proposed a set of principles for a sustainable society, which began with the following sentences: "Living sustainably depends on accepting a duty to seek harmony with other people and with nature. The guiding rules are that people must share with each other and care for the Earth. Humanity must take no more from nature than nature can replenish." These guiding principles have become increasingly relevant in a world that is subject to greater climate change, food shortages, increasing scarcity of clean water, and global pandemics.

Undeniably, some applications of chemistry have had negative impacts on sustainability, such as industrial chemical spills, the use of chemical substances that were later found to be toxic, and the generation of chemical waste as technology has advanced.

However, chemistry has also had some of the most positive effects in increasing the sustainability of our world. It promises even more as we head to the future.

Sustainability entails more than scientific and technological advances. In 2016, the United Nations introduced its 17 Sustainable Development Goals (**Figure 1.16**). These far-reaching goals address important issues of social justice and global economic issues. Chemistry, as the central science, has the capacity to improve global conditions in many of these important areas and will therefore have a major role in achieving these ambitious goals.

In this edition of our textbook, we are introducing a new set of feature boxes entitled "Chemistry and Sustainability." In these boxes, we highlight some of the ways in which applications of chemistry and related fields (such as chemical engineering and materials science) are making ours a more sustainable planet. We also present some of the new developments that promise to enhance sustainability in the near future. Many of you who are students of chemistry now will likely end up as practitioners of the chemical sciences and may very well end up working on these exciting new discoveries that make ours a better, and more sustainable, world. We hope you find these examples both instructive and inspiring.

Related Exercises: 1.41, 1.91

Continued

*David Munro and Martin Holdgate, eds. 1991. "Caring for the Earth: A Strategy for Sustainable Living," Gland, CH: International Union for the Conservation of Nature, United Nations Environmental Program, and World Wide Fund for Nature. Available at https://portals.iucn.org/library/node/6439.

▲ **Figure 1.16 The 17 UN Sustainable Development Goals.** In 2016, the United Nations proposed 17 goals for global sustainable development that span social, economic, and environmental issues. The UN has challenged the world to achieve these goals by 2030: https://www.un.org/sustainabledevelopment/sustainable-development-goals.

1.5 | Units of Measurement

Many properties of matter are *quantitative;* that is, they are associated with numbers. When a number represents a measured quantity, the units of that quantity must be specified. To say that the length of a pencil is 17.5 is meaningless. Expressing the number with its units, 17.5 centimeters (cm), properly specifies the length. In this section, we look more closely at the units we use for measurement in science.

The Metric System and SI Units

The units used for scientific measurements are those of the **metric system**, developed in France during the late eighteenth century. Most countries use the metric system as their system of measurement. The United States has traditionally used the English system, although use of the metric system has become more common (**Figure 1.17**).

In 1960, an international agreement was reached specifying a particular choice of metric units for use in scientific measurements. These preferred units are called **SI units**, after the French *Système International d'Unités*. This system has seven *base units* from which all other units are derived (**Table 1.3**). In this chapter , we discuss the SI base units for length, mass, and temperature. The SI units for other measures, such as volume, can be derived from these fundamental base units.

With SI units, prefixes are used to indicate decimal fractions or multiples of various units. For example, the prefix *milli-* represents a 10^{-3} fraction, one-thousandth, of a unit: A milligram (mg) is 10^{-3} gram (g), a millimeter (mm) is 10^{-3} meter (m), and so forth. **Table 1.4** lists the prefixes used in SI units; the prefixes most commonly encountered in chemistry are kilo-, centi-, milli-, micro-, and nano. In using SI units and in working

Learning Objectives

When you finish Section 1.5, you should be able to:

▶ Identify the seven base units and the common prefixes used in the *metric system.*

▶ Convert temperatures between the Fahrenheit, Celsius, and Kelvin scales.

▶ Distinguish metric base units from derived units, such as volume.

▶ Interconvert among mass, volume, and density, given two of the three quantities.

▶ Calculate energy quantities in joules.

▲ Figure 1.17 **Metric units.** Metric measurements are increasingly common in the United States, as exemplified by the volume printed on this soda can in both metric units (liters, L) and English units (quarts, qt, and fluid ounces, fl oz).

TABLE 1.3 SI Base Units

Physical Quantity	Name of Unit	Abbreviation
Length	Meter	m
Mass	Kilogram	kg
Temperature	Kelvin	K
Time	Second	s or sec
Amount of substance	Mole	mol
Electric current	Ampere	A or amp
Luminous intensity	Candela	cd

TABLE 1.4 Prefixes Used in the Metric System and with SI Units

Prefix	Abbreviation	Meaning	Example	
Peta	P	10^{15}	1 petawatt (PW)	$= 1 \times 10^{15}$ watts[a]
Tera	T	10^{12}	1 terawatt (TW)	$= 1 \times 10^{12}$ watts
Giga	G	10^{9}	1 gigawatt (GW)	$= 1 \times 10^{9}$ watts
Mega	M	10^{6}	1 megawatt (MW)	$= 1 \times 10^{6}$ watts
Kilo	k	10^{3}	1 kilowatt (kW)	$= 1 \times 10^{3}$ watts
Deci	d	10^{-1}	1 deciwatt (dW)	$= 1 \times 10^{-1}$ watt
Centi	c	10^{-2}	1 centiwatt (cW)	$= 1 \times 10^{-2}$ watt
Milli	m	10^{-3}	1 milliwatt (mW)	$= 1 \times 10^{-3}$ watt
Micro	μ[b]	10^{-6}	1 microwatt (μW)	$= 1 \times 10^{-6}$ watt
Nano	n	10^{-9}	1 nanowatt (nW)	$= 1 \times 10^{-9}$ watt
Pico	p	10^{-12}	1 picowatt (pW)	$= 1 \times 10^{-12}$ watt
Femto	f	10^{-15}	1 femtowatt (fW)	$= 1 \times 10^{-15}$ watt
Atto	a	10^{-18}	1 attowatt (aW)	$= 1 \times 10^{-18}$ watt
Zepto	z	10^{-21}	1 zeptowatt (zW)	$= 1 \times 10^{-21}$ watt

[a]The watt (W) is the SI unit of power, which is the rate at which energy is either generated or consumed. The SI unit of energy is the joule (J); $1\,J = 1\,kg \cdot m^2/s^2$ and $1\,W = 1\,J/s$.

[b]Greek letter mu, pronounced "mew."

problems throughout this text, you must be comfortable using exponential notation. If you are unfamiliar with exponential notation or want to review it, refer to Appendix A.1.

Although non-SI units are being phased out, some are still commonly used by scientists. Whenever we first encounter a non-SI unit in the text, the SI unit will also be given. The relationships between the non-SI and SI units we use most frequently in this text appear on the back inside cover. We discuss how to convert from one to the other in Section 1.7.

Length and Mass

The SI base unit of *length* is the meter, which is a distance slightly longer than a yard. **Mass*** is a measure of the amount of material in an object. The SI base unit of mass is the kilogram (kg), which is equal to about 2.2 pounds (lb). This base unit is unusual because it uses a prefix, *kilo-*, instead of the word *gram* alone. We obtain other units for mass by adding prefixes to the word *gram*.

Sample Exercise 1.2
Using SI Prefixes

What is the name of the unit that equals (**a**) 10^{-9} gram, (**b**) 10^{-6} second, (**c**) 10^{-3} meter?

SOLUTION

We can find the prefix related to each power of ten in Table 1.4: (**a**) nanogram, ng; (**b**) microsecond, μs; (**c**) millimeter, mm.

▶ **Practice Exercise**

(**a**) How many picometers are there in 1 m? (**b**) Express 6.0×10^3 m using a prefix to replace the power of ten. (**c**) Use exponential notation to express 4.22 mg in grams. (**d**) Which of the following lengths is greatest: 12.0 m, 2.0×10^3 mm, 1.5×10^{-3} km, or 1.8×10^{10} nm?

Temperature

Temperature, a measure of the hotness or coldness of an object, is a physical property that determines the direction of heat flow. Heat always flows spontaneously from a substance at higher temperature to one at lower temperature. Thus, the influx of heat we feel when we touch a hot object tells us that the object is at a higher temperature than our hand.

The temperature scales commonly employed in science are the Celsius and Kelvin scales. The **Celsius scale** was originally based on the assignment of 0 °C to the freezing point of water and 100 °C to its boiling point at sea level (**Figure 1.18**).

The **Kelvin scale** is the SI temperature scale, and the SI unit of temperature is the *kelvin* (K). Zero on the Kelvin scale is the temperature at which all thermal motion ceases, a temperature referred to as **absolute zero**. On the Celsius scale, absolute zero has the value −273.15 °C. The Celsius and Kelvin scales have equal-sized units; that is, a kelvin is the same size as a degree Celsius. Thus, the Kelvin and Celsius scales are related according to

$$K = °C + 273.15 \qquad [1.3]$$

The freezing point of water, 0 °C, is 273.15 K (Figure 1.18). Notice that we do not use a degree sign (°) with temperatures on the Kelvin scale.

The common temperature scale in the United States is the *Fahrenheit scale*, which is not generally used in science. Water freezes at 32 °F and boils at 212 °F. The Fahrenheit and Celsius scales are related according to

$$°C = \frac{5}{9}(°F - 32) \quad \text{or} \quad °F = \frac{9}{5}(°C) + 32 \qquad [1.4]$$

---

*Mass and weight are not the same thing. Mass is a measure of the amount of matter, whereas weight is the force exerted on this mass by gravity. For example, an astronaut weighs less on the Moon than on Earth because the Moon's gravitational force is less than Earth's. The astronaut's mass on the Moon, however, is the same as it is on Earth.

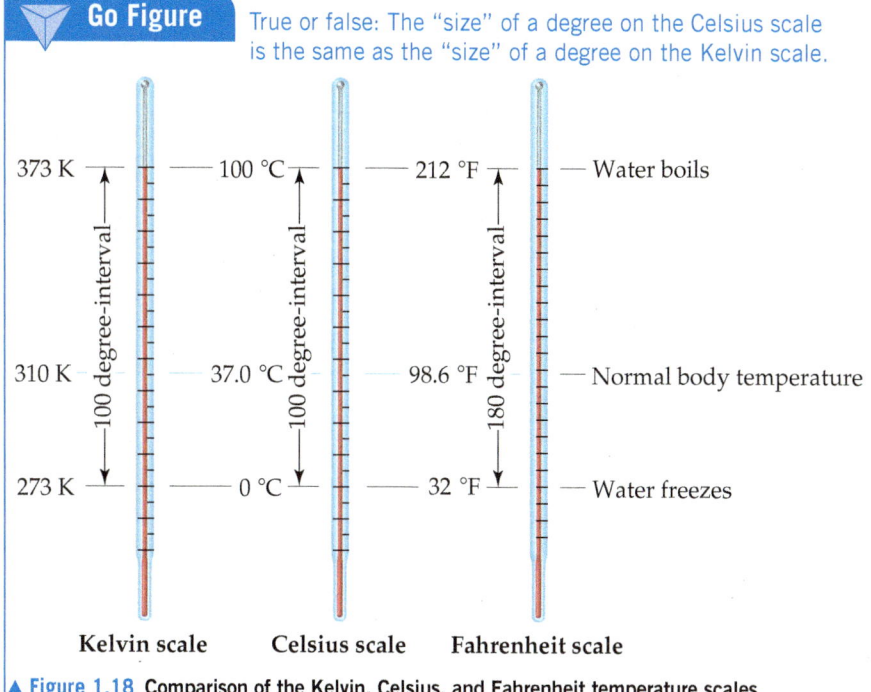

Go Figure   True or false: The "size" of a degree on the Celsius scale is the same as the "size" of a degree on the Kelvin scale.

Kelvin scale   Celsius scale   Fahrenheit scale

▲ **Figure 1.18 Comparison of the Kelvin, Celsius, and Fahrenheit temperature scales.**

Derived SI Units

The SI base units are used to formulate *derived units*. A **derived unit** is obtained by multiplication or division of one or more of the base units. We begin with the defining equation for a quantity and, then substitute the appropriate base units. For example, *speed* is defined as the ratio of distance traveled to elapsed time. Thus, the derived SI unit for speed is the SI unit for distance (length), m, divided by the SI unit for time, s, which gives m/s, read "meters per second." Two common derived units in chemistry are those for volume and density.

The *volume* of a cube is its length cubed, length^3. Thus, the derived SI unit of volume is the SI unit of length, m, raised to the third power. The cubic meter, m^3, is the volume of a cube that is 1 m on each edge (**Figure 1.19**). Smaller units, such as cubic centimeters, cm^3 (sometimes written cc), are more frequently used in chemistry. The volume unit most commonly used in chemistry is the *liter* (L), which equals a cubic decimeter, dm^3, and is slightly larger than a quart. (The liter is the first metric unit we have encountered that is *not* an SI unit.) There are 1000 milliliters (mL) in a liter, and 1 mL is the same volume as 1 cm^3: 1 mL = 1 cm^3.

In the lab, you will likely use the devices in **Figure 1.20** to measure and deliver volumes of liquids. Syringes, burettes, and pipettes deliver amounts of liquids with more precision than graduated cylinders. Volumetric flasks are used to contain specific volumes of liquid.

Go Figure

How many 1-L bottles are required to contain 1 m^3 of liquid?

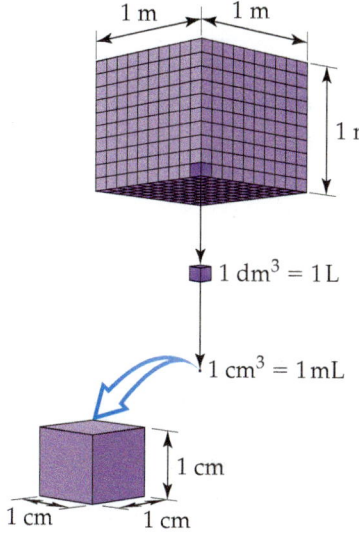

$1 \text{ dm}^3 = 1 \text{ L}$

$1 \text{ cm}^3 = 1 \text{ mL}$

▲ **Figure 1.19 Volume relationships.** The volume occupied by a cube 1 m on each edge is one cubic meter, 1 m^3. Each cubic meter contains 1000 dm^3, 1 m^3 = 1000 dm^3. One liter is the same volume as one cubic decimeter, 1 L = 1 dm^3. Each cubic decimeter contains 1000 cubic centimeters, 1 dm^3 = 1000 cm^3. One cubic centimeter equals one milliliter, 1 cm^3 = 1 mL.

Sample Exercise 1.3
Converting Units of Temperature

A weather forecaster predicts the temperature will reach 30 °C. What is this temperature **(a)** in K and **(b)** °F?

SOLUTION

(a) Using Equation 1.3, we have K = 30 + 273 = 303 K.

(b) Using Equation 1.4, we have

$$°F = \frac{9}{5}(30) + 32 = 54 + 32 = 86 °F.$$

▶ **Practice Exercise**
Ethylene glycol, the major ingredient in antifreeze, freezes at 260 K. What is the freezing point in **(a)** °C and **(b)** °F?

▶ **Figure 1.20 Common volumetric glassware.**

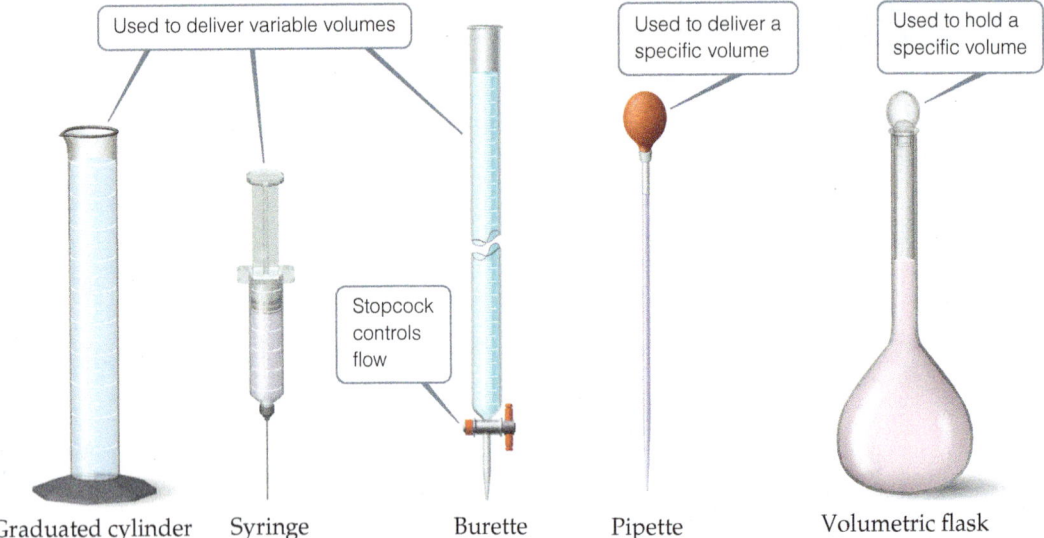

Used to deliver variable volumes

Used to deliver a specific volume

Used to hold a specific volume

Stopcock controls flow

Graduated cylinder   Syringe   Burette   Pipette   Volumetric flask

TABLE 1.5 Densities of Selected Substances at 25 °C

Substance	Density (g/cm³)
Air	0.001
Balsa wood	0.16
Ethanol	0.79
Olive oil	0.92
Water	1.00
Ethylene glycol	1.09
Table sugar	1.59
Table salt	2.16
Iron	7.9
Gold	19.32

Density is defined as the amount of mass in a unit volume of a substance:

$$\text{density} = \frac{\text{mass}}{\text{volume}} \qquad [1.5]$$

The densities of solids and liquids are commonly expressed in either grams per cubic centimeter (g/cm^3) or grams per milliliter (g/mL). The densities of some common substances are listed in **Table 1.5**. It is no coincidence that the density of water is $1.00\ g/mL$; the gram was originally defined as the mass of 1 mL of water at a specific temperature. Because most substances change volume when they are heated or cooled, densities are temperature dependent, and so temperature should be specified when reporting densities. If no temperature is reported, we assume 25 °C, close to normal room temperature.

The terms *density* and *weight* are sometimes confused. A person who says that iron weighs more than air generally means that iron has a higher density than air—1 kg of air has the same mass as 1 kg of iron, but the iron occupies a smaller volume, thereby giving it a higher density. If we combine two liquids that do not mix, such

Sample Exercise 1.4

Determining Density and Using Density to Determine Volume or Mass

(a) Calculate the density of mercury if $1.00 \times 10^2\ g$ occupies a volume of $7.36\ cm^3$.

(b) Calculate the volume of 65.0 g of liquid methanol (wood alcohol) if its density is 0.791 g/mL.

(c) What is the mass in grams of a cube of gold (density = $19.32\ g/cm^3$) if the length of the cube is 2.00 cm?

SOLUTION

(a) We are given mass and volume, so Equation 1.3 yields

$$\text{Density} = \frac{\text{mass}}{\text{volume}} = \frac{1.00 \times 10^2\ g}{7.36\ cm^3} = 13.6\ g/cm^3$$

(b) Solving Equation 1.3 for volume and then using the given mass and density gives

$$\text{Volume} = \frac{\text{mass}}{\text{density}} = \frac{65.0\ g}{0.791\ g/mL} = 82.2\ mL$$

(c) We can calculate the mass from the volume of the cube and its density. The volume of a cube is given by its length cubed:

$$\text{Volume} = (2.00\ cm)^3 = (2.00)^3\ cm^3 = 8.00\ cm^3$$

Solving Equation 1.3 for mass and substituting the volume and density of the cube, we have

$$\text{Mass} = \text{volume} \times \text{density} = (8.00\ cm^3)(19.32\ g/cm^3) = 155\ g$$

▶ **Practice Exercise**

(a) Calculate the density of a 374.5-g sample of copper if it has a volume of $41.8\ cm^3$. (b) A student needs 15.0 g of ethanol for an experiment. If the density of ethanol is 0.789 g/mL, how many milliliters of ethanol are needed?

as olive oil and water, the less dense liquid (in this case the oil) will float on the denser liquid (the water).

Units of Energy

The SI unit for energy is the **joule** (pronounced "jool"), J, in honor of James Joule (1818–1889), a British scientist who investigated work and heat. If we return to Equation 1.2, where kinetic energy was defined, we see that the units of energy are (mass) × (velocity)^2. Thus, it follows that joules are a derived unit, $1\,J = 1\,(kg) \times (m/s)^2 = 1\,kg\text{-}m^2/s^2$. Numerically, a 2-kg mass moving at a velocity of 1 m/s possesses a kinetic energy of 1 J:

$$E_k = \frac{1}{2}mv^2 = \frac{1}{2}(2\,kg)(1\,m/s)^2 = 1\,kg\text{-}m^2/s^2 = 1\,J$$

Because a joule is not a very large amount of energy, we often use *kilojoules* (kJ) in discussing the energies associated with chemical reactions. For example, the amount of heat released when hydrogen and oxygen react to form 1 g of water is 16 kJ.

It is still quite common in chemistry, biology, and biochemistry to find energy changes associated with chemical reactions expressed in the non-SI unit of calories. A **calorie** (cal) was originally defined as the amount of energy required to raise the temperature of 1 g of water from 14.5 to 15.5 °C. It has since been defined in terms of a joule:

$$1\,cal = 4.184\,J\ (exactly)$$

A related energy unit that is familiar to anyone who has read a food label is the nutritional *Calorie* (note the capital C), which is 1000 times larger than *calorie* with a lowercase c: $1\,Cal = 1000\,cal = 1\,kcal$.

Sample Exercise 1.5
Identifying and Calculating Energy Changes

A standard propane (C_3H_8) tank used in an outdoor grill holds approximately 9.0 kg of propane. When the grill is operating, propane reacts with oxygen to form carbon dioxide and water. For every gram of propane that reacts with oxygen in this way, 46 kJ of energy is released as heat. **(a)** How much energy is released if the entire contents of the propane tank react with oxygen? **(b)** As the propane reacts, does the potential energy stored in chemical bonds increase or decrease? **(c)** If you were to store an equivalent amount of potential energy by pumping water to an elevation of 75 m above the ground, what mass of water would be needed? (Note: The force due to gravity acting on the water, which is the water's weight, is $F = m \times g$, where m is the mass of the object and g is the gravitational constant, $g = 9.8\,m/s^2$.)

SOLUTION

(a) We can calculate the amount of energy released from the propane as heat by converting the mass of propane from kg to g and then using the fact that 46 kJ of heat are released per gram:

$$E = 9.0\,kg \times \frac{1000\,g}{1\,kg} \times \frac{46\,kJ}{1\,g} = 4.1 \times 10^5\,kJ = 4.1 \times 10^8\,J$$

(b) When propane reacts with oxygen, the potential energy stored in the chemical bonds is converted to an alternate form of energy, heat. Therefore, the potential energy stored as chemical energy must decrease.

(c) The amount of work done to pump the water to a height of 75 m can be calculated using Equation 1.1:

$$w = F \times d = (m \times g) \times d$$

Equation 1.1 can then be rearranged to solve for the mass of water:

$$m = \frac{w}{g \times d} = \frac{4.1 \times 10^8\,J}{(9.8\,m/s^2)(75\,m)} = \frac{4.1 \times 10^8\,kg\text{-}m^2/s^2}{(9.8\,m/s^2)(75\,m)}$$

$$= 5.6 \times 10^5\,kg$$

At 25 °C, this mass of water would have a volume of 560,000 L, or roughly 150,000 gallons. Thus, we see that large amounts of potential energy can be stored as chemical energy.

▶ **Practice Exercise**

A 12-oz vanilla milkshake at a fast-food restaurant contains 547 Calories. What quantity of energy is this in joules?

A CLOSER LOOK | The Scientific Method

Where does scientific knowledge come from? How is it acquired? How do we know it is reliable? How do scientists add to it or modify it?

There is nothing mysterious about how scientists work. The first idea to keep in mind is that scientific knowledge is gained through observations of the natural world. A principal aim of the scientist is to organize these observations by identifying patterns and regularity, making measurements, and associating one set of observations with another.

After an observation is made, the next step is to ask *why* nature behaves in the manner we observe. To answer this question, the scientist constructs a model, known as a *hypothesis*, to explain the observations. Initially, the hypothesis is likely to be pretty tentative. There could be more than one reasonable hypothesis. If a hypothesis is correct, then certain results and observations should follow from it. In this way, hypotheses can stimulate the design of experiments to learn more about the system being studied. Scientific creativity comes into play in thinking of hypotheses that are fruitful in suggesting good experiments to do, ones that will shed new light on the nature of the system.

As more information is gathered, the initial hypotheses get winnowed down. Eventually, just one may stand out as most consistent with a body of accumulated evidence. We then begin to call this hypothesis a *theory*, a model that has predictive powers and that accounts for all the available observations. A theory also generally is consistent with other, perhaps larger and more general theories. For example, a theory of what goes on inside a volcano has to be consistent with more general theories regarding heat transfer, chemistry at high temperature, and so forth.

We will encounter many theories as we proceed through this book. Some of them have been found over and over again to be consistent with observations. However, no theory can be proven to be absolutely true. We can treat it as though it is, but there always remains a possibility that there is some respect in which a theory is wrong. A famous example is Isaac Newton's theory of mechanics, which yielded such precise results for the mechanical behavior of matter that no exceptions to it were found before the twentieth century. But Albert Einstein showed that Newton's theory of the nature of space and time is incorrect. Einstein's theory of relativity represented a fundamental shift in how we think of space and time. He predicted where the exceptions to predictions based on Newton's theory might be found. Although only small departures from Newton's theory were predicted, they *were* observed. Einstein's theory of relativity became accepted as the correct model. For most uses, however, Newton's laws of motion are quite accurate enough.

The overall process we have just considered, illustrated in Figure 1.21, is often referred to as *the scientific method*. But there is no single scientific method. Many factors play a role in advancing scientific knowledge. The one unvarying requirement is that our explanations must be consistent with observations and that they must depend solely on natural phenomena.

What can we do when we observe nature behaving a certain way and test our theories with additional observations? When nature behaves in a certain way over and over again, under all sorts of different conditions, we can summarize that behavior in a *scientific law*. For example, it has been repeatedly observed that in a chemical reaction there is no change in the total mass of the materials reacting as compared with the materials that are formed; we call this observation the *law of conservation of mass*. It is important to make a distinction between a theory and a scientific law. On the one hand, a scientific law is a statement of what always happens, to the best of our knowledge. A theory, on the other hand, is an *explanation* for what happens. If we discover some law fails to hold true, then we must assume the theory underlying that law is wrong in some way.

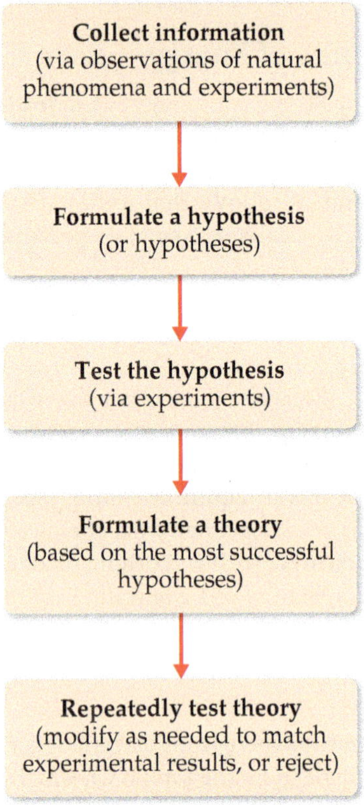

Related Exercises: 1.70, 1.93

▲ Figure 1.21 **The scientific method.**

Self-Assessment Exercises

SAE 1.14 Which of the following might you expect for the playing time of a typical pop song?

(**a**) 1.8×10^2 ms

(**b**) 9.5×10^{-4} km

(**c**) 2×10^3 Ms

(**d**) 2.1×10^{14} ps

(**e**) 1.6×10^{-5} μs

SAE 1.15 The elements cadmium (Cd), gallium (Ga), and sodium (Na) melt at the following temperatures:

Cd: 321 °C; Ga: 303 K; Na: 98 °C

Which of these elements will melt on a very hot day when the air temperature is 100 °F? (**a**) Na (**b**) Ga (**c**) Cd and Ga (**d**) Ga and Na (**e**) All three elements will melt.

SAE 1.16 Which of the following represents a measurement of density?

(**a**) 3.5×10^2 cm^3

(**b**) 6.7×10^3 g/L

(**c**) 4.0×10^{-1} kg-m^2/s^2

(**d**) 1.7 m^2

(**e**) 8.9×10^{-2} g/s

SAE 1.17 The density of gold is 19.3 g/cm^3. What is the mass of a cube of gold that is 2.50 cm long on each edge? (**a**) 302 g (**b**) 15.6 g (**c**) 121 g (**d**) 48.3 g (**e**) 1.24 g

SAE 1.18 What is the kinetic energy in J of a 360-kg motorcycle moving at 88 km/h?

(**a**) 1.1×10^{-1} J

(**b**) 4.4×10^3 J

(**c**) 1.1×10^5 J

(**d**) 2.2×10^5 J

(**e**) 1.4×10^6 J

1.6 | Uncertainty in Measurement and Significant Figures

Much of what we cover throughout this text involves various quantitative aspects of chemistry. Exploring these quantitative aspects requires that we learn how to properly deal with numbers, whether they are measured quantities or calculated ones. In this section, we begin to examine the rules and vocabulary associated with such quantities.

Exact and Inexact Numbers

Two kinds of numbers are encountered in scientific work: *exact numbers* (those whose values are known exactly) and *inexact numbers* (those whose values have some uncertainty). Most of the exact numbers we will encounter in this book have defined values. For example, there are exactly 12 eggs in a dozen, exactly 1000 g in a kilogram, and exactly 2.54 cm in an inch. The number 1 in any conversion factor, such as 1 m = 100 cm or 1 kg = 2.2046 lb, is an exact number. Exact numbers can also result from counting objects. For example, we can count the exact number of marbles in a jar or the exact number of people in a classroom.

Numbers obtained by measurement are always *inexact*. The equipment used to measure quantities always has inherent limitations (equipment errors), and there are differences in how different people make the same measurement (human errors). Suppose 10 students with 10 balances are to determine the mass of the same dime. The 10 measurements will probably vary slightly for various reasons. The balances might be calibrated slightly differently, and there might be differences in how each student reads the mass from the balance. Remember: *Uncertainties always exist in measured quantities.*

Precision and Accuracy

The terms *precision* and *accuracy* are often used in discussing the uncertainties of measured values. **Precision** is a measure of how closely individual measurements agree with one another. **Accuracy** refers to how closely individual measurements agree with the correct, or "true," value. The dart analogy in **Figure 1.22** illustrates the difference between these two concepts.

In the laboratory, we often perform several "trials" of an experiment and average the results. The precision of the measurements is often expressed in terms of the *standard deviation* (Appendix A.5), which reflects how much the individual measurements differ from the average. We gain confidence in our measurements if we obtain nearly the same value each time—that is, when the standard deviation is small. Figure 1.22 reminds us, however, that precise measurements can be inaccurate. For example, in the laboratory portion of your course, you will likely use a high-precision balance, such as the one shown in **Figure 1.23**, to measure quantities of substances. Such balances can typically measure masses to 0.1 mg accuracy; the glass enclosure that surrounds the weighing pan keeps the measurement from being affected by any motion of the air in the room. However, if a very sensitive balance is poorly calibrated, the masses we measure will be consistently either high or low. They will be inaccurate even if they are precise.

Significant Figures

Suppose you determine the mass of a dime on a balance capable of measuring to the nearest 0.0001 g. You could report the mass as 2.2405 ± 0.0001 g. The ± notation (read "plus or minus") expresses the magnitude of the uncertainty of your measurement. In much scientific work, we drop the ± notation with the understanding that *there is always some uncertainty in the last digit reported for any measured quantity.*

Figure 1.24 shows a thermometer with its liquid column between two scale marks. We can read the certain digits from the scale and estimate the uncertain one. Seeing that the liquid is between the 25 °C and 30 °C marks, we estimate the temperature to be 27 °C, being uncertain of the second digit of our measurement. By *uncertain* we mean that the temperature is reliably 27 °C and not 28 °C or 26 °C, but we can't say that it is *exactly* 27 °C.

Learning Objectives

When you finish Section 1.6, you should be able to:

▶ Distinguish between *exact* and *inexact* numbers in scientific calculations.

▶ Explain the difference between the *precision* and *accuracy* of a measurement.

▶ Demonstrate the use of significant figures, exponential notation, and SI units in calculations.

Go Figure

How would the darts be positioned on the target for the case of "good accuracy, poor precision"?

Good accuracy
Good precision

Poor accuracy
Good precision

Poor accuracy
Poor precision

▲ Figure 1.22 **Precision and accuracy.**

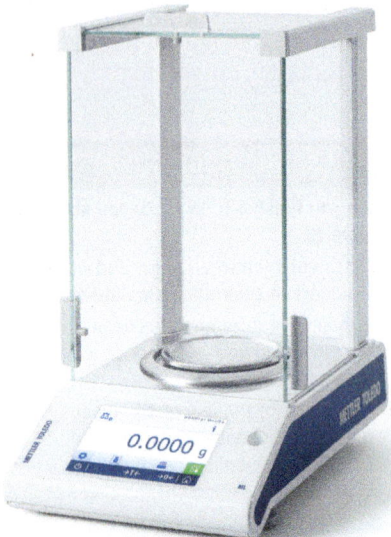

▲ **Figure 1.23 High-precision digital balance used in a chemistry laboratory.** Digital balances provide more precision than those that involve pans and weights. This one shows masses to 0.0001 g. The glass housing around the balance pan keeps air currents from affecting the measurement.

All digits of a measured quantity, including the uncertain one, are called **significant figures**. A measured mass reported as 2.2 g has two significant figures, whereas one reported as 2.2405 g has five significant figures. The greater the number of significant figures, the greater the precision implied for the measurement.

To determine the number of significant figures in a reported measurement, read the number from left to right, counting the digits starting with the first digit that is not zero. *In any measurement that is properly reported, all nonzero digits are significant.* Because zeros can be used either as part of the measured value or merely to locate the decimal point, they may or may not be significant:

- Zeros *between* nonzero digits are always significant—1005 kg (four significant figures); 7.03 cm (three significant figures).

- Zeros *at the beginning* of a number are never significant; they merely indicate the position of the decimal point—0.02 g (one significant figure); 0.0026 cm (two significant figures).

- Zeros *at the end* of a number are significant if the number contains a decimal point—0.0200 g (three significant figures); 3.0 cm (two significant figures).

A question arises when a number ends with zeros but contains no decimal point. In such cases, it is normally assumed that the zeros are *not* significant. Exponential notation (Appendix A.1) can be used to indicate whether end zeros are significant. For example, a mass of 10,300 g can be written to show three, four, or five significant figures, depending on how the measurement is obtained:

1.03×10^4 g	(three significant figures)
1.030×10^4 g	(four significant figures)
1.0300×10^4 g	(five significant figures)

In these numbers, all the zeros to the right of the decimal point are significant (rules 1 and 3). (The exponential term 10^4 does not add to the number of significant figures.) Notice the advantage of using exponential notation: Reporting a mass as 10,300 g suggests that only three digits are significant even if the measurement has greater precision than that.

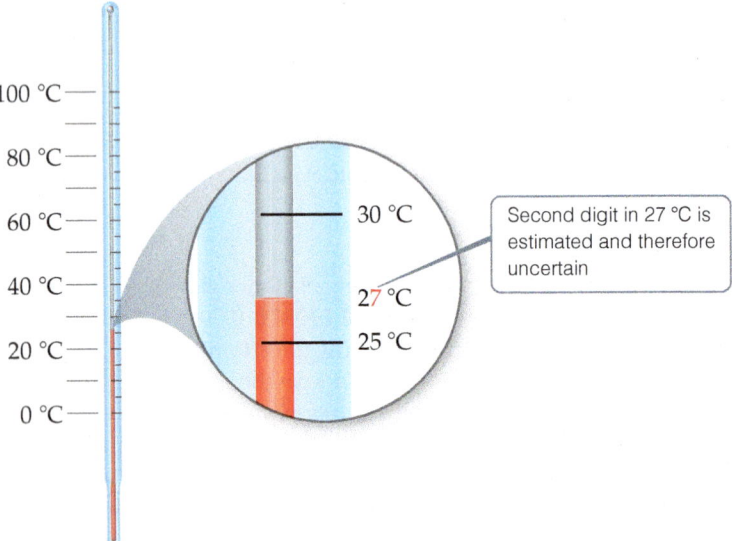

▲ **Figure 1.24 Uncertainty and significant figures in a measurement.**

Sample Exercise 1.6
Assigning Appropriate Significant Figures

The state of Colorado is listed in a road atlas as having a population of 5,546,574 and an area of 104,091 square miles. Do the numbers of significant figures in these two quantities seem reasonable? If not, what seems to be wrong with them?

SOLUTION

The population of Colorado must vary from day to day as people move in or out, are born, or die. Thus, the reported number suggests a much higher degree of *accuracy* than is possible. Moreover, it would not be feasible to actually count every individual resident in the state at any given time. Thus, the reported number suggests far greater *precision* than is possible. A reported number of 5,500,000 would better reflect the actual state of knowledge.

The area of Colorado does not normally vary from time to time, so the question here is whether the accuracy of the measurements is

good to six significant figures. It would be possible to achieve such accuracy using satellite technology, provided the legal boundaries are known with sufficient accuracy.

▶ **Practice Exercise**
A 3-lb bag of onions at the grocery store is found to have 14 onions in it. Are the two numbers in this statement—3 and 14—exact or inexact numbers?

Sample Exercise 1.7
Determining the Number of Significant Figures in a Measurement

How many significant figures are in each of the following numbers (assume that each number is a measured quantity)? **(a)** 5000, **(b)** 6.023×10^{23}, **(c)** 4.003.

SOLUTION

(a) One; we assume that the zeros are not significant when there is no decimal point shown. If the number has more significant figures, a decimal point should be employed or the number should be written in exponential notation. Thus, 5000. has four significant figures, and 5.00×10^3 has three, whereas 5000, with no decimal point, has

only one. **(b)** Four; the exponential term does not add to the number of significant figures. **(c)** Four; the zeros are significant figures.

▶ **Practice Exercise**
How many significant figures are in each of the following measurements? **(a)** 3.549 g, **(b)** 2.3×10^4 cm, **(c)** 0.00134 m³.

Significant Figures in Calculations

Apply the following rule when carrying measured quantities through calculations:

> *The least certain measurement limits the certainty of the calculated quantity and thereby determines the number of significant figures in the final answer.*

The final answer should be reported with only one uncertain digit. To keep track of significant figures in calculations, we make frequent use of two rules—one for addition and subtraction, and another for multiplication and division.

1. *For addition and subtraction,* the result has the same number of decimal places as the measurement with the *fewest decimal places.* When the result contains more than the correct number of significant figures, it must be rounded to the correct number. Consider the following example in which the uncertain digits appear in color:

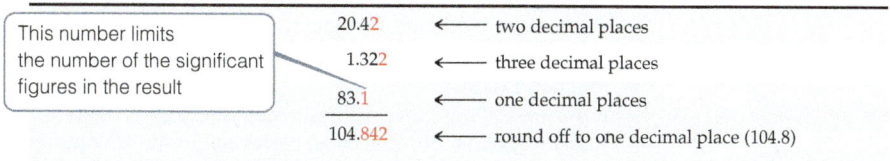

We report the result as 104.8 because 83.1 has only one digit past the decimal point.

Note that whole numbers without decimal points have *zero* decimal places. We look at the first nonzero digit going from right to left to determine what is significant. Thus:

$54 + 137 = 191$ (the "ones digit" is significant in both numbers that are being added)

$120 + 18 = 138$, which must be rounded to 140 (the last significant figure in 120 is in the "tens digit")

2. *For multiplication and division*, the result contains the same number of significant figures as the measurement with the *fewest significant figures*. When the result contains more than the correct number of significant figures, it must be rounded. For example, the area of a rectangle whose measured edge lengths are 6.221 and 5.2 cm, is the product of the lengths of the two sides. The area should be reported with two significant figures, 32 cm², even though a calculator shows the product to have more digits:

$$Area = (6.221 \text{ cm})(5.2 \text{ cm}) = 32.3492 \text{ cm}^2 \Rightarrow \text{round off to } 32 \text{ cm}^2$$

because 5.2 has two significant figures.

In determining the final answer for a calculated quantity, *exact numbers* are assumed to have an infinite number of significant figures. Thus, when we say, "There are 12 inches in 1 foot," the number 12 is exact, and we need not worry about the number of significant figures in it.

When *rounding numbers*, look at the leftmost digit to be removed:

- If the leftmost digit removed is less than 5, the preceding number is left unchanged. Thus, rounding 7.248 to two significant figures gives 7.2.

- If the leftmost digit removed is 5 or greater, the preceding number is increased by 1. Rounding 4.735 to three significant figures gives 4.74, and rounding 2.376 to two significant figures gives 2.4.*

Sample Exercise 1.8
Determining the Number of Significant Figures in a Calculated Quantity

A vessel containing a gas at 25 °C is weighed, emptied, and then reweighed as depicted in **Figure 1.25**. From the data provided, calculate the density of the gas at 25 °C.

SOLUTION

To calculate the density, we must know both the mass and the volume of the gas. The mass of the gas is just the difference in the masses of the full and empty container:

$$\begin{array}{r} 837.63 \text{ g} \\ -836.25 \text{ g} \\ \hline 1.38 \text{ g} \end{array}$$

Pump out gas

Volume: 1.05×10^3 cm³
Mass: 837.63 g

Mass: 836.25 g

▲ **Figure 1.25** Uncertainty and significant figures in a measurement.

In subtracting numbers, we determine the number of significant figures in our result by counting decimal places in each quantity. In this case, each quantity has two decimal places. Thus, the mass of the gas, 1.38 g, has two decimal places. Notice that although each of the individual weights has five significant figures, the difference in the weights has only three significant figures.

Using the volume given in the question, 1.05×10^3 cm³, and the definition of density, we have

$$Density = \frac{mass}{volume} = \frac{1.38 \text{ g}}{1.05 \times 10^3 \text{ cm}^3}$$

$$= 1.31 \times 10^{-3} \text{ g/cm}^3 = 0.00131 \text{ g/cm}^3$$

In dividing numbers, we determine the number of significant figures our result should contain by counting the number of significant figures in each quantity. There are three significant figures in our answer, corresponding to the number of significant figures in the two numbers that form the ratio. In this example, then, following the rules for determining significant figures gives an answer containing only three significant figures, even though the measured masses contain five significant figures.

▶ **Practice Exercise**
If the mass of the container in the sample exercise (Figure 1.25) were measured to three decimal places before and after pumping out the gas, could the density of the gas then be calculated to four significant figures?

---

*Your instructor may want you to use a slight variation on the rule when the leftmost digit to be removed is exactly 5, with no following digits or only zeros following. One common practice is to round up to the next higher number if that number will be even and down to the next lower number otherwise. Thus, 4.7350 would be rounded to 4.74, and 4.7450 would also be rounded to 4.74.

To complete our discussion of significant figures, let's look a little more closely at how we handle them when numbers are expressed in exponential notation. When addition and subtraction are involved, we must make sure that each number has the same exponent. For example, consider adding 1.50×10^3 and 2.38×10^2. Before we do so, we will need to express the second number with the same exponent as the first number:

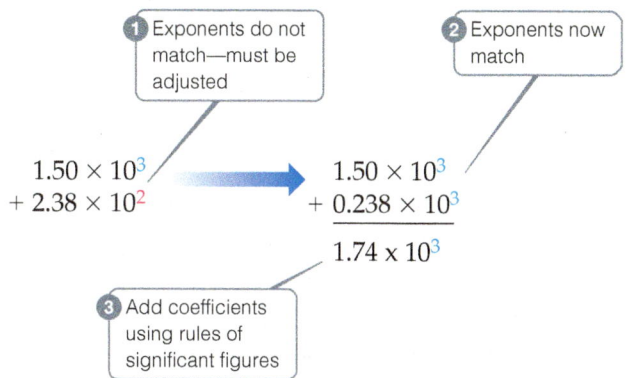

Notice that when we add the numbers, the 8 in the second number is no longer significant because the first number has only two digits past the decimal. We round the final result to the correct number of significant figures past the decimal. You should demonstrate to yourself that we would have gotten the same answer if we had adjusted the exponent of the first number rather than the second.

Multiplying and dividing numbers in exponential notation follow the same rules for multiplication and division that we described before. The number of significant figures is equal to that in the number with the smaller number of significant figures:

$$(6.743 \times 10^{-5}) \times (5.26 \times 10^2) = (6.743 \times 5.26) \times (10^{-5} \times 10^2) = 35.5 \times 10^{-3} = 3.55 \times 10^{-2}$$

Notice that the final answer has three significant figures, the same as 5.26×10^2. Moreover, we have adjusted the exponent in the final answer to get a number between 1 and 10, as is usual for exponential notation.

Finally, you will likely face situations in which you have to do more than one calculation to get the final answer. In general, *when a calculation involves two or more steps and you write answers for intermediate steps, retain at least one nonsignificant digit for the intermediate answers.* This procedure ensures that small errors from rounding at each step do not combine to affect the final result. When using a calculator, you may enter the numbers one after another, rounding only the final answer. Accumulated rounding-off errors may account for small differences among results you obtain and answers given in the text for numerical problems.

Sample Exercise 1.9
Significant Figures and Exponential Notation

Calculate the result of the following to the correct number of significant figures:

$$(9.60 \times 10^4 + 8.3 \times 10^3)/(5.2687 \times 10^2)$$

SOLUTION

To get the answer, we need to carry out two successive calculations. First, we sum the numbers in the numerator, then divide by the number in the denominator. To add the numbers in the numerator, we adjust the second number so that it has the same exponent as the first:

$$(9.60 \times 10^4) + (8.3 \times 10^3) =$$
$$(9.60 \times 10^4) + (0.83 \times 10^4) = 10.43 \times 10^4 = 1.043 \times 10^5$$

      match exponents   add   adjust exponent for exponential notation

Notice that we first rewrite 8.3×10^3 as 0.83×10^4 in order to have the same exponent as the 9.60×10^4. We then add the numbers. Because both have two digits after the decimal, the result has

Continued

two digits past the decimal as well. We then rewrite the result in standard exponential notation—namely, a number between 1 and 10 multiplied by an exponent.

We now divide the sum in the numerator by the number in the denominator:

$$(1.043 \times 10^5)/(5.2687 \times 10^2) = 0.1980 \times 10^3 = 1.980 \times 10^2$$

The numerator of this quotient has four significant figures, and the denominator has five. We therefore express the answer to four significant figures, the smaller number.

▶ **Practice Exercise**

Determine the value of the following expression to the correct number of significant figures:

$$146.3/(4.3 \times 10^4 + 6.75 \times 10^5)$$

 Self-Assessment Exercises

SAE 1.19 Consider the following three quantities:

(**i**) the dimensions of your lecture hall
(**ii**) the mass of a piece of paper
(**iii**) the number of nanometers in a meter

Which of these quantities is or are *inexact*? (**a**) Only one of them is inexact (**b**) i and ii (**c**) i and iii (**d**) ii and iii (**e**) All three are inexact.

SAE 1.20 A sensitive balance is miscalibrated by using an incorrect standard weight. You make four trial weights of a sample using this balance and get the following values: 1.5201 g, 1.5203 g, 1.5199 g, and 1.5202 g. Which of the following describes the precision and accuracy of your measurements? (**a**) good precision, good accuracy (**b**) poor precision, poor accuracy (**c**) poor precision, good accuracy (**d**) good precision, poor accuracy.

SAE 1.21 A volumetric flask has a reported volume of 0.00250 L. How many significant figures are there in this measurement? (**a**) 2 (**b**) 3 (**c**) 4 (**d**) 5 (**e**) 6

SAE 1.22 A hiker has a timer to check her speed during a hike. At a time of 115 min into the hike, she passes a distance post marked 9.4 km. Later, at 275 min into her hike, she passes a post marked 22.9 km. What is her average speed in km/hr between these two posts given to the correct number of significant figures? (**a**) 5 km/h (**b**) 5.1 km/h (**c**) 5.06 km/h (**d**) 5.063 km/h (**e**) 5.0625 km/h

---

STRATEGIES FOR SUCCESS | **Estimating Answers**

Calculators are wonderful devices; they enable you to get to the wrong answer very quickly! Fortunately, you can take certain steps to avoid putting that wrong answer into your homework set or your exam paper. First, keep track of the units in a calculation and use the correct conversion factors. Second, do a quick mental check to be sure that your answer is reasonable by making a "ballpark" estimate or a "back-of-the-envelope" calculation.

A ballpark estimate involves making a rough calculation using numbers that are rounded off in such a way that the arithmetic can be done without a calculator. Even though this approach does not give an exact answer, it gives one that is roughly the correct size. By using dimensional analysis and by estimating answers, you can readily check the reasonableness of your calculations.

You can get better at making estimates by practicing in everyday life. How far is it from your dorm room to the chemistry lecture hall? How much do your parents pay for gasoline per year? How many bikes are there on campus? If you respond "I have no idea" to these questions, you're giving up too easily. Try estimating familiar quantities and you'll get better at making estimates in science and in other aspects of your life where a misjudgment can be costly.

1.7 | Dimensional Analysis

 Learning Objectives

When you finish Section 1.7, you should be able to:

▶ Use and, as needed, combine conversion factors in order to arrive at the desired units for an answer.

▶ Perform calculations in which we raise a known conversion factor by a power.

Because measured quantities have units associated with them, it is important to keep track of units as well as numerical values when using the quantities in calculations. In this section, we show you how to use the units as a guide to solving problems correctly by applying a technique called *dimensional analysis*.

In **dimensional analysis**, units are multiplied together or divided into each other along with the numerical values that accompany them. By canceling equivalent units, we can keep track of both the nature and quantity of what has been calculated. Using dimensional analysis helps ensure that solutions to problems yield the proper units. Moreover, it provides a systematic way of solving many numerical problems and of checking solutions for possible errors.

Conversion Factors

The key to successful dimensional analysis is the correct use of *conversion factors* to change one unit into another. A **conversion factor** is a fraction whose numerator and denominator are the same quantity expressed in different units. For example, 2.54 cm and 1 in. are, by definition, the same length: 2.54 cm = 1 in. This relationship allows us to write two conversion factors:

$$\frac{2.54 \text{ cm}}{1 \text{ in.}} \quad \text{and} \quad \frac{1 \text{ in.}}{2.54 \text{ cm}}$$

We use the first factor to convert inches to centimeters. For example, the length in centimeters of an object that is 8.50 in. long is

$$\text{Number of centimeters} = (8.50 \text{ in.}) \underbrace{\frac{2.54 \text{ cm}}{1 \text{ in.}}}_{\text{Given unit}}^{\text{Desired unit}} = 21.6 \text{ cm}$$

The unit of inches in the denominator of the conversion factor (1 *inch*) cancels the unit of inches in the given data (8.50 *inches*), so that the centimeters unit in the numerator of the conversion factor becomes the unit of the final answer. Because the numerator and denominator of a conversion factor are equal, multiplying any quantity by a conversion factor is equivalent to multiplying by the number 1 and so does not change the intrinsic value of the quantity. The length 8.50 in. is the same as the length 21.6 cm.

In general, we begin any conversion by examining the units of the given data and the units we desire. We then ask ourselves what conversion factors we have available to take us from the units of the given quantity to those of the desired one. When we multiply a quantity by a conversion factor, the units multiply and divide as follows:

$$\text{Given unit} \times \frac{\text{desired unit}}{\text{given unit}} = \text{desired unit}$$

If the desired units are not obtained in a calculation, an error must have been made somewhere. Careful inspection of units often reveals the source of the error.

Sample Exercise 1.10
Converting Units

If a weightlifting barbell has a mass of 125 lb, what is its mass in grams? (Note: 1 lb = 453.59 g)

SOLUTION

Because we are given the mass in pounds and want to express the mass in grams, we use a relationship between these units of mass. The conversion factor is given: 1 lb = 453.59 g. To cancel pounds and leave grams, we write the conversion factor with grams in the numerator and pounds in the denominator:

$$\text{Mass in grams} = (125 \text{ lb})\left(\frac{453.59 \text{ g}}{1 \text{ lb}}\right) = 5.67 \times 10^4 \text{ g}$$

The answer is given to three significant figures, which is the number of significant figures in 125 lb. The process we have used is diagrammed in the drawing shown.

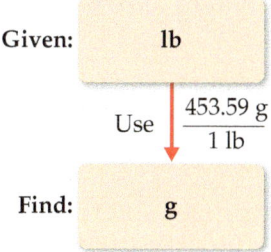

▶ **Practice Exercise**
By using a conversion factor from the back inside cover, determine the length in kilometers of a 500.0-mi automobile race.

Using Two or More Conversion Factors

It is often necessary to use several conversion factors in solving a problem. As an example, let's convert the length of an 8.00-m rod to inches. The table on the back inside cover does not give the relationship between meters and inches. It *does*, however, give the relationship between centimeters and inches (1 in. = 2.54 cm). From our knowledge of SI prefixes, we know that $1 \text{ cm} = 10^{-2} \text{ m}$. Thus, we can convert step by step, first from meters to centimeters and then from centimeters to inches:

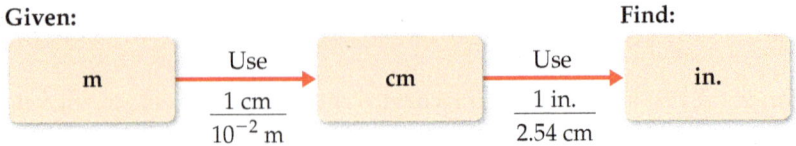

Given: **Find:**

Combining the given quantity (8.00 m) and the two conversion factors, we have

$$\text{Number of inches} = (8.00 \text{ m})\left(\frac{1 \text{ cm}}{10^{-2} \text{ m}}\right)\left(\frac{1 \text{ in.}}{2.54 \text{ cm}}\right) = 315 \text{ in.}$$

The first conversion factor is used to cancel meters and convert the length to centimeters. Thus, meters are written in the denominator and centimeters in the numerator. The second conversion factor is used to cancel centimeters and convert the length to inches, so it has centimeters in the denominator and inches, the desired unit, in the numerator. The final answer is given to three significant figures, which is the number of significant figures in 8.00 m. Both of the conversion factors are exact numbers, so we do not need to consider those in determining the number of significant figures in the answer.

Note that you could also have used $100 \text{ cm} = 1 \text{ m}$ as a conversion factor in the second parentheses. As long as you keep track of your given units and cancel them properly to obtain the desired units, you will be successful in your calculations.

 Sample Exercise 1.11

Converting Units Using Two or More Conversion Factors

The average speed of a nitrogen molecule in air at 25 °C is 515 m/s. Convert this speed to miles per hour.

SOLUTION

To go from the given units, m/s, to the desired units, mi/h, we must convert meters to miles and seconds to hours. From our knowledge of SI prefixes we know that $1 \text{ km} = 10^3 \text{ m}$. From the relationships given on the back inside cover of the book, we find

that $1 \text{ mi} = 1.6093 \text{ km}$. Thus, we can convert m to km and then convert km to mi. From our knowledge of time, we know that $60 \text{ s} = 1 \text{ min}$ and $60 \text{ min} = 1 \text{ h}$. Thus, we can convert s to min and then convert min to h. The overall process is

Given: **Find:**

Applying first the conversions for distance and then those for time, we can set up one long equation in which unwanted units are canceled:

$$\text{Speed in mi/h} = \left(515 \frac{\text{m}}{\text{s}}\right)\left(\frac{1 \text{ km}}{10^3 \text{ m}}\right)\left(\frac{1 \text{ mi}}{1.6093 \text{ km}}\right)\left(\frac{60 \text{ s}}{1 \text{ min}}\right)\left(\frac{60 \text{ min}}{1 \text{ h}}\right)$$

$$= 1.15 \times 10^3 \text{ mi/h}$$

Our answer has the desired units. We can check our calculation using the estimating procedure described in the "Strategies in Success" box. The given speed is about 500 m/s. Dividing by 1000 converts m to km, giving 0.5 km/s. Because 1 mi is about 1.6 km, this speed corresponds to $0.5/1.6 = 0.3 \text{ mi/s}$. Multiplying by 60 gives about $0.3 \times 60 = 20 \text{ mi/min}$. Multiplying again by 60 gives $20 \times 60 = 1200 \text{ mi/h}$. The approximate solution (about 1200 mi/h) and the detailed solution (1150 mi/h) are satisfyingly close. The answer

to the detailed solution has three significant figures, corresponding to the number of significant figures in the given speed in m/s.

▶ **Practice Exercise**
A car travels 32 mi per gallon of gasoline. What is the mileage in kilometers per liter?

Calculations in Which a Conversion Factor Is Raised to a Power

The conversion factors previously noted convert from one unit of a given measure to another unit of the same measure, such as from length to length. We also have conversion factors that convert from one measure to a different one. The density of a substance, for example, can be treated as a conversion factor between mass and volume. Suppose we want to know the mass in grams of 2 cubic inches ($2.00\ \text{in.}^3$) of gold, which has a density of $19.3\ \text{g/cm}^3$. The density gives us the conversion factors:

$$\frac{19.3\ \text{g}}{1\ \text{cm}^3} \quad \text{and} \quad \frac{1\ \text{cm}^3}{19.3\ \text{g}}$$

Because we want a mass in grams, we use the first factor, which has mass in grams in the numerator. To use this factor, however, we must first convert cubic inches to cubic centimeters. The relationship between in.^{3} and cm^3 is not given on the back inside cover, but the relationship between inches and centimeters is given: 1 in. = 2.54 cm (exactly). By cubing this conversion factor, we obtain the desired conversion:

$$\left(\frac{2.54\ \text{cm}}{1\ \text{in.}}\right)^3 = \frac{(2.54)^3\ \text{cm}^3}{(1)^3\ \text{in.}^3} = \frac{16.39\ \text{cm}^3}{1\ \text{in.}^3}$$

Notice that both the numbers and the units are cubed. Also, because 2.54 is an exact number, we can retain as many digits of $(2.54)^3$ as we need. We have used four, one more than the number of digits in the density ($19.3\ \text{g/cm}^3$). Applying our conversion factors, we can now solve the problem:

$$\text{Mass in grams} = (2.00\ \text{in.}^3)\left(\frac{16.39\ \text{cm}^3}{1\ \text{in.}^3}\right)\left(\frac{19.3\ \text{g}}{1\ \text{cm}^3}\right) = 633\ \text{g}$$

The procedure is diagrammed here. The final answer is reported to three significant figures, the same number of significant figures as in $2.00\ \text{in.}^3$ and $19.3\ \text{g/cm}^3$.

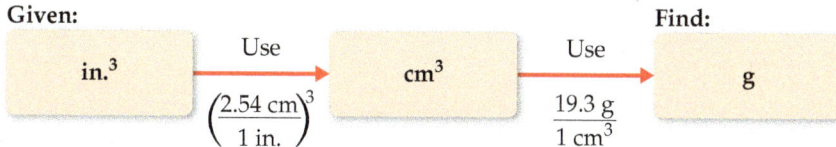

Sample Exercise 1.12

Converting Volume Units

Earth's oceans contain approximately $1.36 \times 10^9\ \text{km}^3$ of water. Calculate the volume in liters.

SOLUTION

From the back inside cover, we find $1\ \text{L} = 10^{-3}\ \text{m}^3$, but there is no relationship listed involving km^3. From our knowledge of SI prefixes, however, we know that $1\ \text{km} = 10^3\ \text{m}$, so we can use this relationship between lengths to write the desired conversion factor between volumes:

$$\left(\frac{10^3\ \text{m}}{1\ \text{km}}\right)^3 = \frac{10^9\ \text{m}^3}{1\ \text{km}^3}$$

Thus, converting from km^3 to m^3 to L, we have

$$\text{Volume in liters} = (1.36 \times 10^9\ \text{km}^3)\left(\frac{10^9\ \text{m}^3}{1\ \text{km}^3}\right)\left(\frac{1\ \text{L}}{10^{-3}\ \text{m}^3}\right)$$
$$= 1.36 \times 10^{21}\ \text{L}$$

▶ **Practice Exercise**

The surface area of Earth is $510 \times 10^6\ \text{km}^2$, and 71% of this is ocean. Using the data from the sample exercise, calculate the average depth of the world's oceans in feet.

Sample Exercise 1.13

Conversions Involving Density

What is the mass in grams of 2.00 gal of water? The density of water is 1.00 g/mL.

SOLUTION

Before we begin solving this exercise, we note the following:

(1) We are given 2.00 gal of water (the known, or given, quantity) and asked to calculate its mass in grams (the unknown).

(2) We have the following conversion factors either given, commonly known, or available on the back inside cover of the text:

$$\frac{1.00\ \text{g water}}{1\ \text{mL water}} \quad \frac{1\ \text{L}}{1000\ \text{mL}} \quad \frac{1\ \text{L}}{1.057\ \text{qt}} \quad \frac{1\ \text{gal}}{4\ \text{qt}}$$

The first of these conversion factors must be used as written (with grams in the numerator) to give the desired result, whereas the last conversion factor must be inverted in order to cancel gallons:

$$\text{Mass in grams} = (2.00 \text{ gal})\left(\frac{4 \text{ qt}}{1 \text{ gal}}\right)\left(\frac{1 \text{ L}}{1.057 \text{ qt}}\right)\left(\frac{1000 \text{ mL}}{1 \text{ L}}\right)\left(\frac{1.00 \text{ g}}{1 \text{ mL}}\right)$$

$$= 7.57 \times 10^3 \text{ g water}$$

The unit of our final answer is appropriate, and we have taken care of our significant figures. We can further check our calculation by estimating. We can round 1.057 off to 1. Then focusing on the numbers that do not equal 1 gives $8 \times 1000 = 8000$ g, in agreement with the detailed calculation.

You should also use common sense to assess the reasonableness of your answer. In this case we know that most people can lift two gallons of milk, although it would be tiring to carry them around all day. Milk is mostly water and will have a density not too different from that of water. Therefore, we might estimate that a gallon of water has mass that is more than 5 lb but less than 50 lb. The mass we have calculated, $7.57 \text{ kg} \times 2.2 \text{ lb/kg} = 16.7$ lb, is thus reasonable.

▶ **Practice Exercise**
The density of the organic compound *benzene* is 0.879 g/mL. Calculate the mass in grams of 1.00 qt of benzene.

 Self-Assessment Exercises

SAE 1.23 The mass of Earth is estimated to be 5.97×10^{27} g. What is the mass of Earth in lb to three significant figures? (Note: 1 kg = 2.2046 lb.)

(a) 1.32×10^{25} lb

(b) 2.71×10^{24} lb

(c) 1.32×10^{28} lb

(d) 2.71×10^{27} lb

(e) 5.97×10^{24} lb

SAE 1.24 A new hybrid vehicle achieves highway gas distance that is listed as 25 km/L. What is the highway gas distance in miles per gallon? (Note: 1 mi = 1.609 km and 1 gal = 3.785 L.) (a) 4.1 mi/gal (b) 40 mi/gal (c) 59 mi/gal (d) 95 mi/gal (e) 150 mi/gal

SAE 1.25 In 2018, the population density in Singapore was 21,400 residents per square mile. What is the population density of Singapore in residents per square kilometer? (Note: 1 mi = 1.609 km.)

(a) 8.27×10^3 residents/km^2

(b) 1.33×10^4 residents/km^2

(c) 2.34×10^4 residents/km^2

(d) 3.44×10^4 residents/km^2

(e) 5.54×10^4 residents/km^2

STRATEGIES FOR SUCCESS | The Importance of Practice—and How to Use the Book

If you have ever played a musical instrument or participated in athletics, you know that the keys to success are practice and discipline. You cannot learn to play a piano merely by listening to music, and you cannot learn how to play basketball merely by watching games on television. Likewise, you cannot learn chemistry by merely watching your instructor give lectures. Simply reading this book, listening to lectures, or reviewing notes will not usually be sufficient when exam time comes around. Your task is to master chemical concepts to a degree that you can put them to use in solving problems and answering questions. Solving problems correctly takes practice—actually, a fair amount of it. You will do well in your chemistry course if you embrace the idea that you need to master the materials presented and then learn how to apply them in solving problems. Even if you're a brilliant student, this will take time; it's what being a student is all about. Almost no one fully absorbs new material on a first reading, especially when new concepts are being presented. You are sure to master the content of the chapters more fully by reading them through at least twice, even more for passages that present you with difficulties in understanding.

Throughout the book, we have provided sample exercises in which the solutions are shown in detail. For practice exercises, we supply only the answer, at the back of the book. It is important that you use these exercises to test yourself.

Each chapter of the book is broken into a number of sections, each of which can be treated as a module for learning. At the beginning of each section, we present a bulleted list of learning goals for you, introduced with the suggestion "When you finish this section, you should be able to." These will serve as a checklist to keep you on track in your learning of the material in the section. At the end of each section we have provided a brief set of multiple-choice Self-Assessment Exercises (SAEs). The SAEs are multiple-choice questions that link directly to the learning goals for the section. Research

has shown that interleaving passive learning (for example, reading) with active problem solving is an effective approach to mastering concepts and skills, so you will find it helpful to complete the SAEs for a given section before moving on to the next section.

At the end of each chapter, we also provide you with a set of Exam Prep questions, which will help you test yourself on your understanding of the chapter as a whole. You will find it helpful to attempt these Exam Prep questions before moving forward to the next chapter.

The practice exercises in this text and the homework assignments given by your instructor provide the minimal practice that you will need to succeed in your chemistry course. Only by working all the assigned problems will you face the full range of difficulty and coverage that your instructor expects you to master for exams. There is no substitute for a determined and perhaps lengthy effort to work problems on your own. If you are stuck on a problem, however, ask for help from your instructor, a teaching assistant, a tutor, or a fellow student. Spending an inordinate amount of time on a single exercise is rarely effective unless you know that it is particularly challenging and is expected to require extensive thought and effort.

Finally, if you haven't done so as yet, we suggest you read the TO THE STUDENT material in the Preface of the book. There we provide you with guidance on how text features such as "What's Ahead," Key Terms, Learning Outcomes, and Key Equations will help you remember what you have learned. You will also find out more about other key features in the book, such as the Visualizing Concepts exercises at the end of each chapter as well as the various other categories within the end-of-chapter exercises.

We want your journey in chemistry to be a productive and even an enjoyable one! We hope you find the tools we provide to assist you in your learning of chemistry to be helpful and enriching. It is a wonderful, if at times daunting, journey, and we want you to be successful in it.

Chapter Summary and Key Terms

THE STUDY OF CHEMISTRY (SECTION 1.1) **Chemistry** is the study of the composition, structure, properties, and changes of **matter**. The composition of matter relates to the kinds of **elements** it contains. The structure of matter relates to the ways the **atoms** of these elements are arranged. A **property** is any characteristic that gives a sample of matter its unique identity. A **molecule** is an entity composed of two or more atoms, with the atoms attached to one another in a specific way.

CLASSIFICATIONS OF MATTER (SECTION 1.2) Matter exists in three physical states—**gas**, **liquid**, and **solid**—which are known as the **states of matter**. There are two kinds of **pure substances**: **elements** and **compounds**. Each element has a single kind of atom and is represented by a chemical symbol consisting of one or two letters, with the first letter capitalized. Compounds are composed of two or more elements joined chemically. The **law of constant composition**, also called the **law of definite proportions**, states that the elemental composition of a pure compound is always the same. Most matter consists of a mixture of substances. **Mixtures** have variable compositions and can be either homogeneous or heterogeneous; homogeneous mixtures are called **solutions**.

PROPERTIES OF MATTER (SECTION 1.3) Each substance has a unique set of **physical properties** and **chemical properties** that can be used to identify it. During a **physical change**, matter does not change its composition. **Changes of state** are physical changes. In a **chemical change (chemical reaction)**, a substance is transformed into a chemically different substance. **Intensive properties** are independent of the amount of matter examined and are used to identify substances. **Extensive properties** relate to the amount of substance present. Differences in physical and chemical properties are used to separate substances. Three common separation techniques are **filtration**, **distillation**, and **chromatography**.

THE NATURE OF ENERGY (SECTION 1.4) **Energy** is defined as the capacity to do work or transfer heat. **Work** is the energy transferred when a force exerted on an object causes a displacement of that object. **Heat** is the energy used to cause the temperature of an object to increase. An object can possess energy in two forms: **kinetic energy**, which is the energy associated with the motion of an object, and **potential energy**, which is the energy that an object possesses by virtue of its position relative to other objects. Important forms of potential energy include gravitational energy and electrostatic energy.

UNITS OF MEASUREMENT (SECTION 1.5) Measurements in chemistry are made using the **metric system**. Special emphasis is placed on **SI units**, which are based on the meter (m), the kilogram (kg), and the second (s) as the basic units of length, **mass**, and time, respectively. SI units use prefixes to indicate fractions or multiples of base units. The SI **temperature** scale is the **Kelvin scale**, although the **Celsius scale** is frequently used as well. **Absolute zero** is the lowest temperature attainable. It has the value 0 K. A **derived unit** is obtained by multiplication or division of SI base units. Derived units are needed for defined quantities such as speed or volume. **Density** is an important derived unit that equals mass divided by volume. The SI unit of energy is the **joule** (J): $1 J = 1 kg\text{-}m^2/s^2$. The **calorie** (cal) is a commonly used non-SI unit of energy: $1 cal = 4.18 J$.

UNCERTAINTY IN MEASUREMENT AND SIGNIFICANT FIGURES (SECTION 1.6) All measured quantities are inexact to some extent. The **precision** of a measurement indicates how closely different measurements of a quantity agree with one another. The **accuracy** of a measurement indicates how well a measurement agrees with the accepted or "true" value. The **significant figures** in a measured quantity include one estimated digit, the last digit of the measurement. The significant figures indicate the extent of the uncertainty of the measurement. Certain rules must be followed so that a calculation involving measured quantities is reported with the appropriate number of significant figures.

DIMENSIONAL ANALYSIS (SECTION 1.7) In the **dimensional analysis** approach to problem solving, we keep track of units as we carry measurements through calculations. The units are multiplied together, divided into each other, or canceled like algebraic quantities. Obtaining the proper units for the final result is an important means of checking the method of calculation. When converting units and when carrying out several other types of problems, **conversion factors** are used. These factors are ratios constructed from valid relations between equivalent quantities.

Key Equations

- $w = F \times d$ [1.1]   Work done by a force in the direction of displacement

- $E_k = \dfrac{1}{2}mv^2$ [1.2]   Kinetic energy

- $K = °C + 273.15$ [1.3]   Converting between Celsius (°C) and Kelvin (K) temperature scales

- $°C = \dfrac{5}{9}(°F - 32)$ or $°F = \dfrac{9}{5}(°C) + 32$ [1.4]   Converting between Celsius (°C) and Fahrenheit (°F) temperature scales

- $Density = \dfrac{mass}{volume}$ [1.5]   Definition of density

Exam Prep

Here are some exam-style problems designed to test your understanding of the material in this chapter.

EP 1.1 Which of the following best describes the situation depicted in the figure? (**a**) molecules of a single pure element, (**b**) molecules of a mixture of two different elements, (**c**) molecules of a pure compound, or (**d**) a mixture of molecules of a compound and those of an element.

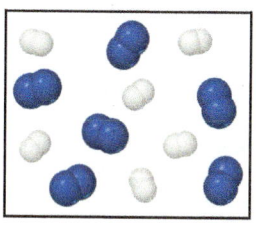

EP 1.2 Which of the following statements about the states of matter is *incorrect*? (**a**) The molecules in the gas phase are farther apart than those in a liquid phase of the same substance. (**b**) The molecules in a liquid have greater freedom to move than the molecules in a solid. (**c**) A solid can be readily compressed. (**d**) A gas will completely fill its container.

EP 1.3 Which of the following is a homogeneous mixture? (**a**) the air that we breathe (**b**) a bottle of pure water (**c**) molten iron (**d**) a bowl of oatmeal

EP 1.4 A certain mineral is described as containing the elements Ca, Fe, Mg, Si, and O. Which of the following elements is *not* in the mineral? (**a**) iron (**b**) silicon (**c**) calcium (**d**) carbon (**e**) magnesium

EP 1.5 A piece of metal is described as silvery in color, which is a _____ property, with a mass of 10 lb, which is a an _____ property. Which of the following best fills in the blanks in the sentence? (**a**) physical, intensive (**b**) physical, extensive (**c**) chemical, intensive (**d**) chemical, extensive

EP 1.6 The following describe either physical or chemical changes of a substance:

   (**i**)   When heated, a mixture of hydrogen and oxygen gases reacts to form water.

   (**ii**)   When heated, liquid water turns into steam.

   (**iii**)   When heated, a piece of paper ignites and burns.

Which of these processes are *physical* changes? (**a**) Only one of the three is a physical change. (**b**) i and ii (**c**) i and iii (**d**) ii and iii (**e**) All three are physical changes.

EP 1.7 Which of the following statements about separation techniques is *true*? (**a**) Filtration can be used to separate a homogeneous mixture of two liquids. (**b**) Chromatography is a good technique for separating a mixture of two solids. (**c**) Separation techniques involve separating molecules into atoms. (**d**) Distillation is a technique that depends on the differences in the tendencies of substances to form gases.

EP 1.8 In a steam engine, water is converted to steam in a boiler and the steam is used to drive the wheels. Which of the following statements about this engine is or are correct?

   (**i**)   Both work and heat are forms of energy.

   (**ii**)   The energy used to convert the water to steam is heat.

   (**iii**)   When the steam drives the wheels, work is being done.

(**a**) Only one statement is correct. (**b**) Statements i and ii are correct. (**c**) Statements i and iii are correct. (**d**) Statements ii and iii are correct. (**e**) All three statements are correct.

EP 1.9 An apple falls from a tree to the ground. Which of the following correctly describes the changes in the apple's kinetic energy and gravitational potential energy as it first begins to fall from the tree? (**a**) Its kinetic energy increases and its gravitational potential energy increases. (**b**) Its kinetic energy decreases and its gravitational potential energy increases. (**c**) Its kinetic energy increases and its gravitational potential energy decreases. (**d**) Its kinetic energy decreases and its gravitational potential energy decreases.

EP 1.10 Which of the following weights would you expect to be suitable for weighing on an ordinary bathroom scale?

(**a**) 2.0×10^9 mg   (**c**) 5.0×10^{-4} kg   (**e**) 5.5×10^8 dg

(**b**) 2500 μg   (**d**) 4.0×10^6 cg

EP 1.11 Liquids A, B, and C boil at 80.1 °C, 181.4 °F, and 341.8 K, respectively. Which of the following is the correct ordering of the boiling points, from lowest to highest?

(**a**) A < B < C   (**c**) B < C < A

(**b**) C < A < B   (**d**) A < C < B

EP 1.12 Platinum, Pt, is one of the rarest of the metals. Worldwide annual production is only about 130 tons. Platinum has a density of 21.4 g/cm^3. If thieves were to steal platinum from a bank using a small truck with a maximum payload capacity of 900 lb, how many 1 L bars of the metal could they take? (**a**) 19 bars (**b**) 2 bars (**c**) 42 bars (**d**) 1 bar (**e**) 47 bars

EP 1.13 The density of copper is 8.96 g/cm^3. What is the mass in kg of a cube of copper metal that is 12.0 cm on each side? (**a**) 0.108 kg (**b**) 1.29 kg (**c**) 15.5 kg (**d**) 1.55×10^7 kg

EP 1.14 What is the kinetic energy of a 5.0-kg object moving at a speed of 20 m/s? (**a**) 50 J (**b**) 100 J (**c**) 250 J (**d**) 1000 J (**e**) 2000 J

EP 1.15 How many of the following quantities are *inexact* numbers: the number of μg in a gram; the number of square yards in a square mile; the height of a building; the density of aluminum; the number of seconds in a day; the number of cal in one J? (**a**) 1 (**b**) 2 (**c**) 3 (**d**) 4 (**e**) 5

EP 1.16 Which of the following statements about precision and accuracy of a scientific instrument is or are correct?

   (**i**)   An instrument can produce measurements that have good precision but poor accuracy.

   (**ii**)   An instrument that produces measurements with good accuracy must have good precision.

   (**iii**)   An instrument with good accuracy will lead to measurements close to the correct value.

(**a**) Only one statement is correct. (**b**) Statements i and ii are correct. (**c**) Statements i and iii are correct. (**d**) Statements ii and iii are correct. (**e**) All three statements are correct.

EP 1.17 An object is determined to have a mass of 0.01080 g. How many significant figures are there in this measurement? (**a**) 2 (**b**) 3 (**c**) 4 (**d**) 5 (**e**) 6

EP 1.18 A typical 400 m race in the Olympics is completed in a time between 40 and 50 s. A digital stopwatch is listed as having precision ± 0.01 s. How many significant figures should be used in reporting the times for the race in s? (**a**) 1 (**b**) 2 (**c**) 3 (**d**) 4 (**e**) 5

EP 1.19 You are asked to determine the mass of a piece of copper using its reported density, 8.96 g/mL, and a 150-mL graduated cylinder. First, you add 105 mL of water to the graduated cylinder; then you place the piece of copper in the cylinder and record a volume of 137 mL. What is the mass of the copper reported with the correct number of significant figures?

(**a**) 287 g   (**c**) 286.72 g/mL   (**e**) 2.9×10^2 g

(**b**) 3.5×10^{-3} g   (**d**) 3.48×10^{-3} g

EP 1.20 To how many significant figures should the following expression be reported?

$$[(2.67 \times 10^3) - (4.7 \times 10^2)] \times (3.154 \times 10^4)$$

(**a**) 1 (**b**) 2 (**c**) 3 (**d**) 4 (**e**) 5

EP 1.21 A commercial Boeing 757 airliner has a typical cruising speed of 528 miles per hour. What is this speed in m/s? (Note: 1 mi = 1.609 km.) (**a**) 91.2 m/s (**b**) 236 m/s (**c**) 850 m/s (**d**) 3060 m/s

EP 1.22 The density of a particular flooring material is reported as 60.0 lb/ft^3. What is the density in kg/L? (**a**) 138 kg/L (**b**) 0.961 kg/L (**c**) 259 kg/L (**d**) 15.8 kg/L (**e**) 11.5 kg/L

EP 1.23 A ream of 8.5 in × 11 in "24 pound" printer paper contains exactly 500 sheets and has a mass of 6.00 lb. What is the mass in mg of a 1.00 cm × 1.00 cm square of the paper? (Note: 1 kg = 2.2046 lb and 1 in. = 2.54 cm.) (**a**) 1.50 mg (**b**) 9.02 mg (**c**) 22.9 mg (**d**) 58.2 mg

Exercises

Visualizing Concepts

1.1 The molecule shown here is *leucine*, one of the amino acids that are the building blocks of proteins. The color key for the atoms is carbon (black), hydrogen (light gray), oxygen (red), and nitrogen (blue). (**a**) How many different elements are in a molecule of leucine? (**b**) How many carbon atoms are in a molecule of leucine? (**c**) Which element has the largest number of atoms in a leucine molecule? (**d**) What is the total number of atoms in a molecule of leucine? [Section 1.1]

1.2 Which of the following figures represents (**a**) a pure element, (**b**) a mixture of two elements, (**c**) a pure compound, (**d**) a mixture of an element and a compound? (More than one picture might fit each description.) [Section 1.2]

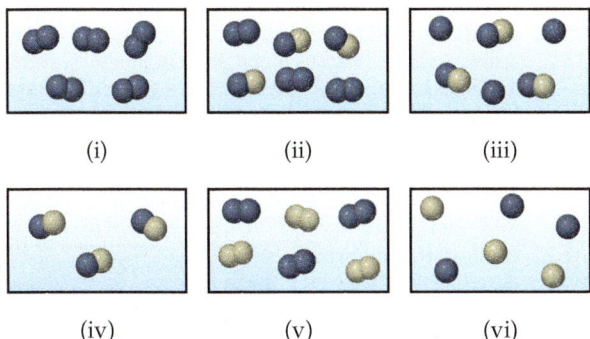

(i)   (ii)   (iii)

(iv)   (v)   (vi)

1.3 Which of the following diagrams represents a chemical change? [Section 1.3]

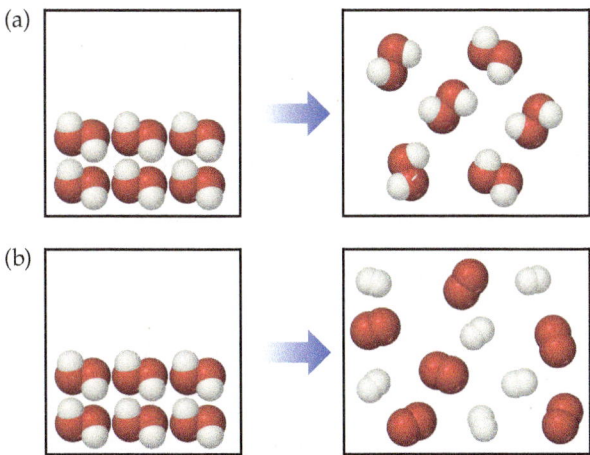

(a)

(b)

1.4 Musical instruments such as trumpets and trombones are made from an alloy called *brass*. Brass is composed of copper and zinc atoms and appears homogeneous under an optical microscope. The approximate composition of most brass objects is a 2:1 ratio of copper to zinc atoms, but the exact ratio varies somewhat from one piece of brass to another. (**a**) Would you classify brass as an element, a compound, a homogeneous mixture, or a heterogeneous mixture? (**b**) Would it be correct to say that brass is a solution? [Section 1.2]

1.5 Which of the following is the best description for the formation of frost on windows in the winter: (**i**) a solid is turning into a gas, (**ii**) a liquid is turning into a solid, (**iii**) a gas is turning into a liquid, (**iv**) a gas is turning into a solid, or (**v**) a solid is turning into a liquid? [Section 1.2]

1.6 Consider the two spheres shown here, one made of silver and the other of aluminum. (**a**) What is the mass of each sphere in kg? (**b**) The force of gravity acting on an object is $F = mg$, where m is the mass of an object and g is the acceleration of gravity (9.8 m/s^2). How much work do you do on each sphere if you raise it from the floor to a height of 2.2 m? (**c**) Does the act of lifting the sphere off the ground increase the potential energy of the aluminum sphere by a larger, a smaller, or the same amount as the silver sphere? (**d**) If you release the spheres simultaneously, they will have the same velocity when they hit the ground. Will they have the same kinetic energy? If not, which sphere will have more kinetic energy? [Section 1.4]

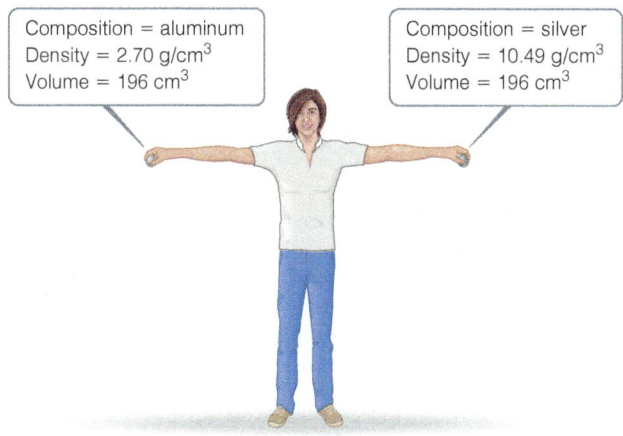

Composition = aluminum
Density = 2.70 g/cm³
Volume = 196 cm³

Composition = silver
Density = 10.49 g/cm³
Volume = 196 cm³

1.7 Is the separation method used in brewing a cup of coffee best described as distillation, filtration, or chromatography? [Section 1.3]

1.8 (**a**) Three spheres of equal size are composed of aluminum (density = 2.70 g/cm³), silver (density = 10.49 g/cm³), and nickel (density = 8.90 g/cm³). List the spheres from lightest to heaviest. (**b**) Three cubes of equal mass are composed of gold (density = 19.32 g/cm³), platinum (density = 21.45 g/cm³), and lead (density = 11.35 g/cm³). List the cubes from smallest to largest. [Section 1.5]

1.9 The three targets from a rifle range shown below were produced by: (**A**) the instructor firing a newly acquired target rifle; (**B**) the instructor firing his personal target rifle; and (**C**) a student who has fired his target rifle only a few times. (**a**) Comment on the accuracy and precision for each of these three sets of results. (**b**) For the A and C results in the future to look like those in B, what needs to happen? [Section 1.6]

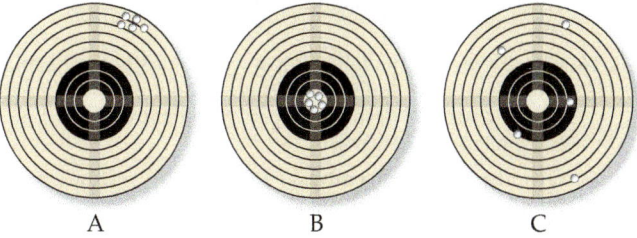

A   B   C

1.10 (**a**) What is the length of the pencil in the following figure if the ruler reads in centimeters? How many significant figures are there in this measurement? (**b**) An automobile speedometer with circular scales reading both miles per hour and kilometers per hour is shown. What speed is indicated, in both units? How many significant figures are in the measurements? [Section 1.6]

1.11 **(a)** How many significant figures should be reported for the volume of the metal bar shown here? **(b)** If the mass of the bar is 104.72 g, how many significant figures should be reported when its density is determined using the calculated volume? [Section 1.6]

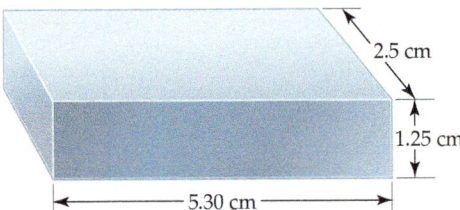

2.5 cm

1.25 cm

5.30 cm

1.12 Consider the jar of jelly beans in the photo. To get an estimate of the number of beans in the jar you weigh six beans and obtain masses of 3.15, 3.12, 2.98, 3.14, 3.02, and 3.09 g. Then you weigh the jar with all the beans in it, and obtain a mass of 2082 g. The empty jar has a mass of 653 g. Based on these data, estimate the number of beans in the jar. Justify the number of significant figures you use in your estimate. [Section 1.6]

Classification and Properties of Matter (Sections 1.1, 1.2, and 1.3)

1.13 Which of the following statements about atoms and molecules is *false*? **(a)** Atoms and molecules are forms of matter. **(b)** A molecule contains two or more atoms. **(c)** The atoms of two different elements can be the same. **(d)** Atoms and molecules have mass and occupy space. **(e)** A molecule can contain atoms from a single element.

1.14 The molecular model of a molecule has four black spheres, ten light gray spheres, and one red sphere. Which of the following statements is or are true? **(i)** There are 15 atoms in the molecule. **(ii)** The black spheres are atoms of the same element. **(iii)** There are atoms of three different elements in the molecule.

1.15 Classify each of the following as a pure substance or a mixture. If a mixture, indicate whether it is homogeneous or heterogeneous: **(a)** rice pudding, **(b)** seawater, **(c)** magnesium, **(d)** crushed ice.

1.16 Classify each of the following as a pure substance or a mixture. If a mixture, indicate whether it is homogeneous or heterogeneous: **(a)** gasoline vapor, **(b)** sugar crystals, **(c)** a hardboiled egg, **(d)** sawdust.

1.17 Give the chemical symbol or name for the following elements, as appropriate: **(a)** sulfur **(b)** gold **(c)** potassium **(d)** chlorine **(e)** copper **(f)** U **(g)** Ni **(h)** Na **(i)** Al **(j)** Si.

1.18 Give the chemical symbol or name for each of the following elements, as appropriate: **(a)** carbon **(b)** nitrogen **(c)** titanium **(d)** zinc **(e)** iron **(f)** P **(g)** Ca **(h)** He **(i)** Pb **(j)** Ag.

1.19 A solid white substance A is heated strongly in the absence of air. It decomposes to form a new white substance B and a gas C. The gas has exactly the same properties as the product obtained when carbon is burned in an excess of oxygen. Based on these observations, can we determine whether solids A and B and gas C are elements or compounds?

1.20 You are hiking in the mountains and find a shiny gold nugget. It might be the element gold, or it might be "fool's gold,"

which is a nickname for iron pyrite, FeS_2. Which of the following physical properties do you think would help determine if the shiny nugget is really gold—appearance, melting point, density, or physical state?

1.21 In the process of attempting to characterize a substance, a chemist makes the following observations: The substance is a silvery white, lustrous metal. It melts at 649 °C and boils at 1105 °C. Its density at 20 °C is 1.738 g/cm³. The substance burns in air, producing an intense white light. It reacts with chlorine to give a brittle white solid. The substance can be pounded into thin sheets or drawn into wires. It is a good conductor of electricity. Which of these characteristics are physical properties, and which are chemical properties?

1.22 **(a)** Read the following description of the element zinc and indicate which are physical properties and which are chemical properties. Zinc melts at 420 °C. When zinc granules are added to dilute sulfuric acid, hydrogen is given off and the metal dissolves. Zinc

has a hardness on the Mohs scale of 2.5 and a density of 7.13 g/cm³ at 25 °C. It reacts slowly with oxygen gas at elevated temperatures to form zinc oxide, ZnO. **(b)** Which properties of zinc can you describe from the photo? Are these physical or chemical properties?

1.23 Label each of the following as either a physical process or a chemical process: **(a)** rusting of a metal can, **(b)** boiling a cup of water, **(c)** pulverizing an aspirin, **(d)** digesting a candy bar, **(e)** exploding of nitroglycerin?

1.24 A match is lit and held under a cold piece of metal. The following observations are made: **(a)** The match burns. **(b)** The metal gets warmer. **(c)** Water condenses on the metal. **(d)** Soot (carbon) is deposited on the metal. Which of these occurrences are due to physical changes, and which are due to chemical changes?

1.25 For each of the following processes, would filtration, distillation, or chromatography be the most effective separation technique: **(a)** removing the pulp from freshly squeezed orange juice, **(b)** separating a food dye into its individual components, **(c)** desalinating seawater?

1.26 Two beakers contain clear, colorless liquids. When the contents of the beakers are mixed, a white solid is formed. **(a)** Is this an example of a chemical or a physical change? **(b)** What would be the most convenient way to separate the newly formed white solid from the liquid mixture—filtration, distillation, or chromatography?

The Nature of Energy (Section 1.4)

1.27 Which of the following statements about energy, work, and heat is or are true? **(i)** Work and heat are both forms of energy. **(ii)** In order for work to be performed on an object, it must not move. **(iii)** Heat is energy that causes a change in temperature.

1.28 A soccer ball is dropped out of a window on the second floor of a dormitory to the ground below. Which of the following statements is or are true? **(i)** The kinetic energy of the ball is greatest at the time it is dropped out the window. **(ii)** As the ball falls, potential energy is converted into kinetic energy. **(iii)** The potential energy of the ball is due to the force of gravity acting on it.

1.29 **(a)** Calculate the kinetic energy (in joules) of a 1200-kg automobile moving at 18 m/s. **(b)** Convert this energy to calories. **(c)** If a truck weighing 2100 kg has the same kinetic energy as the automobile in part (a), what is its speed in m/s?

1.30 The mass of a helium atom is roughly four times that of a hydrogen atom. The mass of an oxygen atom is roughly 16 times that of a hydrogen atom. **(a)** For each of the following pairs, choose the one that has the greater kinetic energy: (i) a H atom moving at 1000 m/s or a He atom moving at 400 m/s, (ii) a H atom moving at 1000 m/s or an O atom moving at 400 m/s, (iii) a He atom moving at 1000 m/s or an O atom moving at 400 m/s. **(b)** A He atom is moving at 800 m/s. What is the speed of an O atom that has the same kinetic energy as the He atom?

1.31 Two positively charged particles are first brought close together and then released. Once released, the repulsion between particles causes them to move away from each other. **(a)** This is an example of potential energy being converted into what form of energy? **(b)** Does the electrostatic potential energy of the two particles increase or decrease as the distance between them is increased? **(c)** If the positive charge on the two particles were doubled at a fixed distance, would the electrostatic potential energy of the two particles increase, decrease, or stay the same?

1.32 For each of the following processes, does the potential energy of the object(s) increase or decrease? **(a)** The distance between two oppositely charged particles is increased. **(b)** Water is pumped from ground level to the reservoir of a water tower 30 m above the ground. **(c)** The bond in a chlorine molecule, Cl_2, is broken to form two chlorine atoms.

Units of Measurement (Section 1.5)

1.33 What exponential notation do the following abbreviations represent? **(a)** d **(b)** c **(c)** f **(d)** μ **(e)** M **(f)** k **(g)** n **(h)** m **(i)** p

1.34 Use appropriate metric prefixes to write the following measurements without use of exponents:
(a) 8.6×10^{-9} m **(d)** 1.81×10^{-2} s **(f)** 2.75×10^{-5} g
(b) 3.55×10^{4} g **(e)** 4.95×10^{6} m **(g)** 5.1×10^{2} cm
(c) 6.48×10^{-7} L

1.35 Make the following conversions: **(a)** 72 °F to °C, **(b)** 216.7 °C to °F, **(c)** 233 °C to K, **(d)** 315 K to °F, **(e)** 2500 °F to K, **(f)** 0 K to °F.

1.36 **(a)** The temperature on a warm summer day is 87 °F. What is the temperature in °C? **(b)** Many scientific data are reported at 25 °C. What is this temperature in kelvins and in degrees Fahrenheit? **(c)** Suppose that a recipe calls for an oven temperature of 400 °F. Convert this temperature to degrees Celsius and to kelvins. **(d)** Liquid nitrogen boils at 77 K. Convert this temperature to degrees Fahrenheit and to degrees Celsius.

1.37 **(a)** A sample of tetrachloroethylene, a liquid used in dry cleaning that is being phased out because of its potential to cause cancer, has a mass of 40.55 g and a volume of 25.0 mL at 25 °C. What is its density at this temperature? Will tetrachloroethylene float on water? (Materials that are less dense than water will float.) **(b)** Carbon dioxide (CO_2) is a gas at room temperature and pressure. However, carbon dioxide can be put under pressure to become a "supercritical fluid" that is a much safer dry-cleaning agent than tetrachloroethylene. At a certain pressure, the density of supercritical CO_2 is 0.469 g/cm³. What is the mass of a 25.0-mL sample of supercritical CO_2 at this pressure?

1.38 **(a)** A cube of an unknown metal that is 1.200 cm on a side has a mass of 17.66 g at 25 °C. What is its density in g/cm³ at this temperature? **(b)** The density of magnesium metal is 1.74 g/cm³ at 25 °C. What mass of magnesium displaces 75.0 mL of water at 25 °C? **(c)** Ethylene glycol is a viscous liquid with a density of 1.114 g/mL at 20 °C. Calculate the mass of 0.3750 L of ethylene glycol at this temperature.

1.39 **(a)** To identify a liquid substance, a student determined its density. Using a graduated cylinder, she measured out a 45-mL sample of the substance. She then measured the mass of the sample, finding that it weighed 38.5 g. She knew that the substance had to be either isopropyl alcohol (density 0.785 g/mL) or toluene (density 0.866 g/mL). What are the calculated density and the probable identity of the substance? **(b)** An experiment requires 78.1 g of benzene, a liquid whose density is 0.876 g/mL. Rather than weigh the sample on a balance, a chemist chooses to dispense the liquid using a graduated cylinder. What volume of the liquid should he use? **(c)** A cubic piece of metal measures 5.00 cm on each edge. If the metal is nickel, whose density is 8.90 g/cm³, what is the mass of the cube?

1.40 **(a)** After the label fell off a bottle containing a clear liquid believed to be a solvent called tetrahydrofuran, a chemist measured the density of the liquid to verify its identity. A 25.0-mL portion of the liquid had a mass of 22.08 g. A chemistry handbook lists the density of tetrahydrofuran at 25 °C as 0.8833 g/mL. Is the calculated density in agreement with the tabulated value? **(b)** An experiment requires 50.0 g of a substance called *n*-hexane, whose density at 25 °C is 0.6606 g/mL. What volume of *n*-hexane should be used? **(c)** The density of titanium metal is 4.506 g/cm³ at 20 °C. What is the mass of a spherical ball of titanium with a diameter of 2.00 cm. [The volume of a sphere is $(4/3)\pi r^3$, where r is the radius.]

1.41 For January through April 2020, at the height of the global coronavirus pandemic, it is estimated that global emissions of carbon dioxide (CO_2) decreased by 940 million metric tons (1 metric ton = 1000 kg) compared to the same period in 2019. **(a)** Express this mass of CO_2 in grams without exponential notation, using an appropriate metric prefix. **(b)** Combustion of a gallon of gasoline generates about 8900 g of CO_2. How many gallons of gasoline would need to be burned to generate 940 million metric tons of CO_2? **(c)** Under normal room temperature conditions, the density of carbon dioxide gas is about 1.8 g/L. Express the volume of 940 million metric tons of CO_2 without exponential notation, using an appropriate metric prefix.

1.42 Silicon for computer chips is grown in large cylinders called "boules" that are 300 mm in diameter and 2 m in length, as shown. The density of silicon is 2.33 g/cm³. Silicon wafers for making integrated circuits are sliced from a 2.0-m boule and are typically 0.75 mm thick and 300 mm in diameter. **(a)** How many wafers can be cut from a single boule? **(b)** What is the mass of a silicon wafer? (The volume of a cylinder is given by $\pi r^2 h$, where r is the radius and h is its height.)

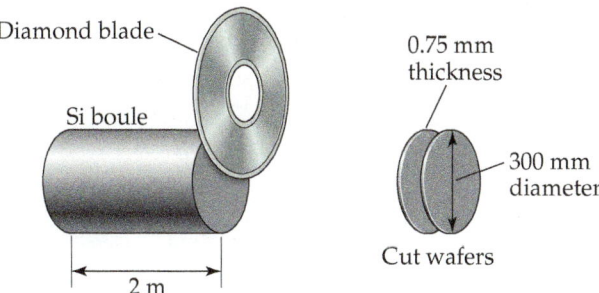

Uncertainty in Measurement (Section 1.6)

1.43 Indicate which of the following are exact numbers: **(a)** the mass of a 3- by 5-in. index card, **(b)** the number of ounces in a pound, **(c)** the volume of a cup of Starbucks coffee, **(d)** the number of inches in a mile, **(e)** the number of microseconds in a week, **(f)** the number of pages in this book.

1.44 Indicate which of the following are exact numbers: (**a**) the mass of a 32-oz bag of sugar, (**b**) the number of students in your chemistry class, (**c**) the temperature of the surface of the Sun, (**d**) the mass of a postage stamp, (**e**) the number of milliliters in a cubic meter of water, (**f**) the average height of NBA basketball players.

1.45 What is the number of significant figures in each of the following measured quantities? (**a**) 820 g (**b**) 32.40 s (**c**) 90.01 mm (**d**) 0.00404 L (**e**) 3.50×10^3 cm³ (**f**) 400 kg

1.46 Indicate the number of significant figures in each of the following measured quantities: (**a**) 9.09 kg, (**b**) 6.040×10^{-18} m³, (**c**) 0.0030 cm, (**d**) 290.00 K, (**e**) 12.8690 g, (**f**) 6.40×10^4 m/s.

1.47 Round each of the following numbers to four significant figures and express the result in standard exponential notation: (**a**) 102.53070, (**b**) 656.980, (**c**) 0.008543210, (**d**) 0.000257870, (**e**) −0.0357202.

1.48 (**a**) The diameter of Earth at the equator is 7926.381 mi. Round this number to three significant figures and express it in standard exponential notation. (**b**) The circumference of Earth through the poles is 40,008 km. Round this number to four significant figures and express it in standard exponential notation.

1.49 Carry out the following operations and express the answers with the appropriate number of significant figures.
(**a**) $14.3505 + 2.65$
(**b**) $952.7 - 140.7389$
(**c**) $(3.29 \times 10^4)(0.2501)$
(**d**) $0.0588/0.677$

1.50 Carry out the following operations and express the answer with the appropriate number of significant figures.
(**a**) $320.5 - (6104.5/2.3)$
(**b**) $[(2.853 \times 10^3) - (1.200 \times 10^3)] \times 2.8954$
(**c**) $(0.0045 \times 20,000.0) + (2813 \times 12)$
(**d**) $863 \times [1255 - (3.45 \times 108)]$

1.51 Carry out the following operations and express the answers in exponential notation with the appropriate number of significant figures.
(**a**) $6.754 \times 10^4 + 3.12 \times 10^5$
(**b**) $7.93 \times 10^6 - 1.6405 \times 10^4$
(**c**) $(8.67 \times 10^3 + 4.3 \times 10^1) \times 1.924 \times 10^{-2}$
(**d**) $(5.37 \times 10^{-3} - 8.44 \times 10^{-2})/(9.113 \times 10^{-3} - 7.46 \times 10^{-4})$

1.52 Carry out the following operations and express the answers in exponential notation with the appropriate number of significant figures.
(**a**) $2.791 \times 10^4 + 8.76 \times 10^3$
(**b**) $4.67 \times 10^2 - 5.4437 \times 10^4$
(**c**) $(2.481 \times 10^{-2} + 7.33 \times 10^{-4}) \times (1.924 \times 10^{-2} + 6.70)$
(**d**) $(1.3 \times 10^{-4} - 3.746 \times 10^{-2})/(1.3 \times 10^2 - 3.746 \times 10^4)$

1.53 You weigh an object on a balance and read the mass in grams according to the picture. How many significant figures are in this measurement?

1.54 You have a graduated cylinder that contains a liquid (see photograph). Write the volume of the liquid, in milliliters, using the proper number of significant figures.

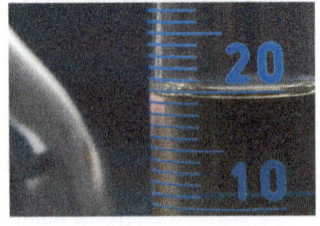

Dimensional Analysis (Section 1.7)

1.55 What are the conversion factors needed to convert the following? (**a**) mm to nm (**b**) mg to kg (**c**) km to ft (**d**) in.³ to cm³

1.56 Determine the appropriate conversion factors for the following: (**a**) μm to mm, (**b**) ms to ns, (**c**) mi to km, (**d**) ft³ to L.

1.57 (**a**) A bumblebee flies with a ground speed of 15.2 m/s. Calculate its speed in km/h. (**b**) The lung capacity of the blue whale is 5.0×10^3 L. Convert this volume into gallons. (**c**) The Statue of Liberty is 151 ft tall. Calculate its height in meters. (**d**) Bamboo can grow up to 60.0 cm/day. Convert this growth rate into inches per hour.

1.58 (**a**) The speed of light in a vacuum is 2.998×10^8 m/s. Calculate its speed in miles per hour. (**b**) The Sears Tower in Chicago is 1454 ft tall. Calculate its height in meters. (**c**) The Vehicle Assembly Building at the Kennedy Space Center in Florida has a volume of 3,666,500 m³. Convert this volume to liters and express the result in standard exponential notation. (**d**) An individual suffering from a high cholesterol level in her blood has 242 mg of cholesterol per 100 mL of blood. If the total blood volume of the individual is 5.2 L, how many grams of total blood cholesterol does the individual's body contain?

1.59 Perform the following conversions: (**a**) 5.00 days to s, (**b**) 0.0550 mi to m, (**c**) \$1.89/gal to dollars per liter, (**d**) 0.510 in./ms to km/h, (**e**) 22.50 gal/min to L/s, (**f**) 0.02500 ft³ to cm³.

1.60 Carry out the following conversions: (**a**) 0.105 in. to mm, (**b**) 0.650 qt to mL, (**c**) 8.75 μm/s to km/h, (**d**) 1.955 m³ to yd³, (**e**) \$3.99/lb to dollars per kg, (**f**) 8.75 lb/ft³ to g/mL.

1.61 (**a**) How many liters of wine can be held in a wine barrel whose capacity is 31 gal? (**b**) The recommended adult dose of Elixophyllin®, a drug used to treat asthma, is 6 mg/kg of body mass. Calculate the dose in milligrams for a 185-lb person. (**c**) If an automobile is able to travel 400 km on 47.3 L of gasoline, what is the gas mileage in miles per gallon? (**d**) When the coffee is brewed according to directions, a pound of coffee beans yields 50 cups of coffee (4 cups = 1 qt). How many kg of coffee are required to produce 200 cups of coffee?

1.62 (**a**) If an electric car is capable of going 225 km on a single charge, how many charges will it need to travel from Seattle, Washington, to San Diego, California, a distance of 1257 mi, assuming that the trip begins with a full charge? (**b**) If a migrating loon flies at an average speed of 14 m/s, what is its average speed in mi/h? (**c**) What is the engine piston displacement in liters of an engine whose displacement is listed as 450 in.³? (**d**) In March 1989, the *Exxon Valdez* ran aground and spilled 240,000 barrels of crude petroleum off the coast of Alaska. One barrel of petroleum is equal to 42 gal. How many liters of petroleum were spilled?

1.63 The concentration of carbon monoxide in an urban apartment is 2800 μg/m³. What mass of carbon monoxide in grams is present in a room measuring 13.5 ft × 17.1 ft × 10.0 ft?

1.64 The density of air at ordinary atmospheric pressure and 25 °C is 1.19 g/L. Each 1.00 g of air contains 0.21 g of oxygen gas. What is the mass, in kilograms, of the oxygen gas in a room that measures 15.1 ft × 19.6 ft × 9.0 ft?

1.65 The density of tungsten metal is 19.35 g/cm³. (**a**) What is the density of tungsten in pounds per cubic foot? (**b**) A rectangular block of tungsten has a length of 4.00 cm, a width of 2.00 cm, and a height of 1.50 cm. Using the correct number of significant figures, what is the mass of the block in kg?

1.66 Aluminum has a density of 2.70 g/cm³. (**a**) What is the density of aluminum in kg/m³? (**b**) A package of aluminum foil contains 50 ft² of foil, which weighs approximately 8.0 oz. What is the approximate thickness of the foil in millimeters?

1.67 Gold can be hammered into extremely thin sheets called gold leaf. An architect wants to cover a 100 ft × 82 ft ceiling with gold leaf that is five-millionths of an inch thick. The density of gold is 19.32 g/cm³. If gold costs $1768 per troy ounce (1 troy ounce = 31.10348 g), how much will it cost the architect to buy the necessary gold?

1.68 A copper refinery produces a copper ingot weighing 150 lb. If the copper is drawn into wire whose diameter is 7.50 mm, how many feet of copper can be obtained from the ingot? The density of copper is 8.94 g/cm³. (Assume that the wire is a cylinder whose volume $V = \pi r^2 h$, where r is its radius and h is its height or length.)

Additional Exercises

1.69 Classify each of the following as a pure substance, a solution, or a heterogeneous mixture: (**a**) a gold ingot, (**b**) a cup of coffee, (**c**) a wood plank, (**d**) a container of plain yogurt, (**e**) a block of dry ice, which is frozen carbon dioxide, (**f**) the gas in a scuba tank filled with 30% oxygen and 70% nitrogen.

1.70 (**a**) Which is more likely to eventually be shown to be incorrect: an hypothesis or a theory? (**b**) A(n) _____ reliably predicts the behavior of matter, while a(n) _____ provides an explanation for that behavior.

1.71 A sample of ascorbic acid (vitamin C) is synthesized in the laboratory. It contains 1.50 g of carbon and 2.00 g of oxygen. Another sample of ascorbic acid isolated from citrus fruits contains 6.35 g of carbon. According to the law of constant composition, how many grams of oxygen does it contain?

1.72 Ethyl chloride is sold as a liquid (see photo) under pressure for use as a local skin anesthetic. Ethyl chloride boils at 12 °C at atmospheric pressure. When the liquid is sprayed onto the skin, it boils off, cooling and numbing the skin as it vaporizes. (**a**) What changes of state are involved in this use of ethyl chloride?  (**b**) What is the boiling point of ethyl chloride in degrees Fahrenheit? (**c**) The bottle shown contains 103.5 mL of ethyl chloride. The density of ethyl chloride at 25 °C is 0.765 g/cm³. What is the mass of ethyl chloride in the bottle?

1.73 Two students determine the percentage of iron in a sample as a laboratory exercise. The true percentage is 34.43%. The students' results for three determinations are as follows:

(**1**) 34.44, 34.41, 34.46

(**2**) 34.51, 34.56, 34.48

(**a**) Calculate the average percentage for each set of data and state which set is the more accurate based on the average. (**b**) Precision can be judged by examining the average of the deviations from the average value for that data set. (Calculate the average value for each data set; then calculate the average value of the absolute deviations of each measurement from the average.) Which set is more precise?

1.74 What type of quantity (for example, length, volume, density) do the following units indicate? (**a**) mL (**b**) cm² (**c**) mm³ (**d**) mg/L (**e**) ps (**f**) nm (**g**) K

1.75 Give the derived SI units for each of the following quantities in base SI units:

(**a**) acceleration = distance/time²

(**b**) force = mass × acceleration

(**c**) work = force × distance

(**d**) pressure = force/area

(**e**) power = work/time

(**f**) velocity = distance/time

(**g**) energy = mass × (velocity)²

1.76 Substance A melts at 350 °C, substance B melts at 580 K, and substance C melts at 650 °F. What is the correct order of the melting point temperatures of A, B, and C from lowest to highest?

1.77 The distance from Earth to the Moon is approximately 240,000 mi. (**a**) What is this distance in meters? (**b**) The peregrine falcon has been measured as traveling up to 350 km/h in a dive. If this falcon could fly to the Moon at this speed, how many seconds would it take? (**c**) The speed of light is 3.00 × 10⁸ m/s. How long does it take for light to travel from Earth to the Moon and back again? (**d**) Earth travels around the Sun at an average speed of 29.783 km/s. Convert this speed to miles per hour.

1.78 Which of the following would you characterize as a pure or nearly pure substance? (**a**) baking powder (**b**) lemon juice (**c**) propane gas, used in outdoor gas grills (**d**) aluminum foil (**e**) ibuprofen (**f**) bourbon whiskey (**g**) helium gas (**h**) clear water pumped from a deep aquifer

1.79 (**a**) A baseball weighs 5.13 oz. What is the kinetic energy, in joules, of this baseball when it is thrown by a Major League pitcher at 95.0 mi/h? (**b**) By what factor will the kinetic energy change if the speed of the baseball is decreased to 55.0 mi/h? (**c**) What happens to the kinetic energy when the baseball is caught by the catcher? Is it converted mostly to heat or to some form of potential energy?

1.80 The U.S. quarter has a mass of 5.67 g and is approximately 1.55 mm thick. (**a**) How many quarters would have to be stacked to reach 575 ft, the height of the Washington Monument? (**b**) How much would this stack weigh? (**c**) How much money would this stack contain? (**d**) The U.S. National Debt Clock showed the outstanding national debt to be $2.625 × 10¹³ in June 2020. How many stacks like the one described would be necessary to pay off this debt?

1.81 An Olympic swimming pool is 164 ft long and 82 ft wide. Assume that it is filled to a depth of 3.0 m, which is the Olympic recommendation. (**a**) What volume of water in gal is needed to fill the pool? (**b**) What volume of water in L is needed to fill the pool? (**c**) The U.S. Geological Survey estimates that in 2015 U.S. domestic use of water was 82 gal/day per person. Using this value for daily usage, for what length of time could the water used to fill an Olympic swimming pool provide the domestic water usage for a community of 25,000 residents?

1.82 A watt is a measure of power (the rate of energy change) that is equal to 1 J/s. (**a**) Calculate the number of joules in a kilowatt-hour. (**b**) An adult person radiates heat to the surroundings at about the same rate as a 100-watt electric incandescent light bulb. What is the total amount of energy in kcal radiated to the surroundings by an adult over a 24-h period?

1.83 By using estimation techniques, determine which of the following is the heaviest and which is the lightest: a 5-lb bag of potatoes, a 5-kg bag of sugar, or 1 gal of water (density =1.0 g/mL).

1.84 The liquid substances mercury (density = 13.6 g/mL), water (1.00 g/mL), and cyclohexane (0.778 g/mL) do not form a solution when mixed but separate in distinct layers. Sketch how the liquids would position themselves in a test tube.

1.85 Two spheres of equal volume are placed on the scales as shown. (**a**) Which one is more dense? (**b**) When a blue sphere with a diameter that is 1.93 times the diameter of the red sphere is put on the balance, the two spheres have the same mass. What is the ratio of the density of the blue sphere to that of the red sphere? (The volume of a sphere is $(4/3)\pi r^3$, where r is the radius.)

1.86 Water has a density of 0.997 g/cm^3 at $25\,°C$; ice has a density of 0.917 g/cm^3 at $-10\,°C$. **(a)** If a soft-drink bottle whose volume is 1.50 L is completely filled with water and then frozen to $-10\,°C$, what volume does the ice occupy? **(b)** Can the ice be contained within the bottle?

1.87 The mass of an empty oil drum is listed as 2.0×10^4 g. After a quantity of biodiesel fuel is added to the drum, the total mass is 7.05×10^5 g. **(a)** To how many significant figures should the mass of the biodiesel in the drum be reported? **(b)** The density of the biodiesel fuel is 0.875 g/cm^3. What is the volume of biodiesel that has been added to the drum, in L?

1.88 A 32.65-g sample of a solid is placed in a flask. Toluene, in which the solid is insoluble, is added to the flask so that the total volume of solid and liquid together is 50.00 mL. The solid and toluene together weigh 58.58 g. The density of toluene at the temperature of the experiment is 0.864 g/mL. What is the density of the solid?

1.89 A thief plans to steal a platinum sphere with a radius of 28.9 cm from a museum. The density of platinum is 21.5 g/cm^3. **(a)** What is the mass of the sphere in pounds? [The volume of a sphere is $V = (4/3)\pi r^3$.] **(b)** Is the thief likely to be able to walk off with the platinum sphere unassisted?

1.90 Automobile batteries contain sulfuric acid, which is commonly referred to as "battery acid." Calculate the number of grams of sulfuric acid in 1.00 gal of battery acid if the solution has a density of 1.28 g/mL and is 38.1% sulfuric acid by mass.

1.91 The total rate at which power is used by humans worldwide is approximately 18.4 TW (terawatts). The solar flux averaged over the sunlit half of Earth is 680 W/m^2 (assuming no clouds). The area of Earth's disc as seen from the Sun is $1.28 \times 10^{14} \text{ m}^2$. The surface area of Earth is approximately 197,000,000 square miles. How much of Earth's surface would we need to cover with solar energy collectors to power the planet for use by all humans? Assume that the solar energy collectors can convert only 15% of the available sunlight into useful power.

1.92 The potential energy of an object subject to gravity is given by $E_p = mgh$, where m is the mass of the object, g is the acceleration due to gravity (9.8 m/s^2), and h is the distance of the object from the surface of Earth. What is the kinetic energy and velocity of the silver sphere in Problem 1.6 at the moment it hits the ground? Assume that all of the sphere's initial potential energy is converted to kinetic energy at impact.

1.93 In 2005, J. Robin Warren and Barry J. Marshall shared the Nobel Prize in Physiology or Medicine for discovering the bacterium *Helicobacter pylori* and for establishing experimental proof that it plays a major role in gastritis and peptic ulcer disease. The story began when Warren, a pathologist, noticed that bacilli were associated with the tissues taken from patients suffering from ulcers. Look up the history of this case and describe Warren's first hypothesis. What sorts of evidence did it take to create a credible theory based on it?

1.94 A 25.0-cm-long cylindrical glass tube, sealed at one end, is filled with ethanol. The mass of ethanol needed to fill the tube is found to be 45.23 g. The density of ethanol is 0.789 g/mL. Calculate the inner diameter of the tube in centimeters.

1.95 Gold is alloyed (mixed) with other metals to increase its hardness in making jewelry. **(a)** Consider a piece of gold jewelry that weighs 9.85 g and has a volume of 0.675 cm^3. The jewelry contains only gold and silver, which have densities

of 19.3 and 10.5 g/cm^3, respectively. If the total volume of the jewelry is the sum of the volumes of the gold and silver that it contains, calculate the percentage of gold (by mass) in the jewelry. **(b)** The relative amount of gold in an alloy is commonly expressed in units of carats. Pure gold is 24 carat, and the percentage of gold in an alloy is given as a percentage of this value. For example, an alloy that is 50% gold is 12 carat. State the purity of the gold jewelry in carats.

1.96 Paper chromatography is a simple but reliable method for separating a mixture into its constituent substances. You have a purple vegetable dye, and you want to determine whether the dye consists of multiple components. You try two different paper chromatography procedures and achieve the separations shown in the figure. Which of the following statements is or are true? **(i)** Trial 2 has achieved greater separation than has Trial 1. **(ii)** These experiments prove that the dye contains exactly two components. **(iii)** The separation of the two components in Trial 2 depends on how long the experiment is conducted.

 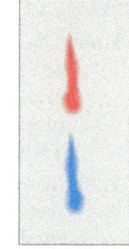

Trial 1   Trial 2

1.97 Judge the following statements as true or false. If you believe a statement to be false, provide a corrected version.

(a) Air and water are both elements.

(b) All mixtures contain at least one element and one compound.

(c) Compounds can be decomposed into two or more other substances; elements cannot.

(d) Elements can exist in any of the three states of matter.

(e) When yellow stains in a kitchen sink are treated with bleach water, the disappearance of the stains is due to a physical change.

(f) A hypothesis is more weakly supported by experimental evidence than a theory.

(g) The number 0.0033 has more significant figures than 0.033.

(h) Conversion factors used in converting units always have a numerical value of one.

(i) Compounds always contain at least two different elements.

1.98 You are assigned the task of separating a desired granular material with a density of 3.62 g/cm^3 from an undesired granular material that has a density of 2.04 g/cm^3. You want to do this by shaking the mixture in a liquid in which the heavier material will fall to the bottom and the lighter material will float. A solid will float on any liquid that is more dense. Using an Internet-based source or a handbook of chemistry, find the densities of the following substances: carbon tetrachloride, hexane, benzene, and diiodomethane. Which of these liquids will serve your purpose, assuming no chemical interaction takes place between the liquid and the solids?

1.99 In 2009, a team from Northwestern University and Western Washington University reported the preparation of a new "spongy" material composed of nickel, molybdenum, and sulfur that excels at removing mercury from water. The density of this new material is 0.20 g/cm^3, and its surface area is 1242 m^2 per gram of material. **(a)** Calculate the volume of a 10.0-mg sample of this material. **(b)** Calculate the surface area for a 10.0-mg sample of this material. **(c)** A 10.0-mL sample of contaminated water had 7.748 mg of mercury in it. After treatment with 10.0 mg of the new spongy material, 0.001 mg of mercury remained in the contaminated water. What percentage of the mercury was removed from the water? **(d)** What is the final mass of the spongy material after the exposure to mercury?

ATOMS, MOLECULES, AND IONS

▲ The wonderful variety of colors and forms in nature result from only a dozen or so elements.

WHAT'S AHEAD

2.1 ▶ **The Atomic Theory of Matter** Learn how scientists were able to postulate that atoms are the smallest building block of matter, long before they could be seen directly.

2.2 ▶ **The Discovery of Atomic Structure** Examine key experiments that led to the discovery of *electrons* and to the *nuclear model* of the atom.

2.3 ▶ **The Modern View of Atomic Structure** Learn how *atomic number* and *mass number* can be used to express the number of each subatomic particle— protons, neutrons, and electrons—in a given atom.

2.4 ▶ **Atomic Weights** Learn about the concept of *atomic weights* and how they are derived from the masses and abundances of individual atoms.

2.5 ▶ **The Periodic Table** Examine the organization of the *periodic table*, in which elements are put in order of increasing atomic number and grouped by chemical similarity.

2.6 ▶ **Molecules and Molecular Compounds** Explore the assemblies of atoms called *molecules* and learn how their compositions are represented by *empirical* and *molecular formulas*.

2.7 ▶ **Ions and Ionic Compounds** Realize that atoms can gain or lose electrons to form *ions* and learn how to use the periodic table to predict the charges on ions as well as the empirical formulas of *ionic* compounds.

2.8 ▶ **Naming Inorganic Compounds** Consider the systematic way in which substances are named, called *nomenclature*, and how this nomenclature is applied to inorganic compounds.

2.9 ▶ **Some Simple Organic Compounds** Become familiar with simple families of organic compounds, compounds containing carbon and hydrogen, and the nomenclature of these compounds.

Everything in our universe is formed from only about 100 different kinds of fundamental building blocks, called **atoms**. In a sense, these different atoms are like the 26 letters of the English alphabet that join in different combinations to form the immense number of words in our language. But what rules govern the ways in which atoms combine? How do the properties of a substance relate to the kinds of atoms it contains? Indeed, what is an atom like, and what makes the atoms of one element different from those of another?

In this chapter, we introduce the basic structure of atoms and discuss the formation of molecules and ions. This knowledge provides you with the foundation you need to understand the chapters that follow.

2.1 | The Atomic Theory of Matter

⚠ **Learning Objectives**

When you finish **Section 2.1**, you should
be able to:

▶ List the basic postulates of Dalton's
atomic theory.

▶ Use the laws derived from Dalton's
atomic theory to explain simple
chemical reactions.

Philosophers from the earliest times speculated about the nature of the fundamental "stuff" from which the world is made. Democritus (460–370 BCE) and other early Greek philosophers described the material world as being made up of tiny indivisible particles that they called *atomos*, meaning "indivisible" or "uncuttable." Later, however, Plato and Aristotle formulated the notion that there can be no ultimately indivisible particles, and the "atomic" view of matter faded for many centuries during which Aristotelean philosophy dominated Western culture.

The notion of atoms reemerged in Europe during the seventeenth century. As chemists learned to measure the amounts of elements that reacted with one another to form new substances, the ground was laid for an atomic theory that linked the idea of elements with the idea of atoms. That theory came from the work of John Dalton during the period 1803–1807. Dalton's atomic theory was based on four postulates (see **Figure 2.1**).

A good theory explains known facts; Dalton's theory explained several laws of chemical combination known at the time.

- The *law of constant composition* (Section 1.2), based on postulate 4:

    In a given compound, the relative numbers and kinds of atoms are constant.

- The **law of conservation of mass,** based on postulate 3:

    The total mass of materials present after a chemical reaction is the same as the total mass present before the reaction.

A good theory also predicts new facts; Dalton used his theory to deduce

- The **law of multiple proportions:**

    If two elements A and B combine to form more than one compound, the masses of B that can combine with a given mass of A are in the ratio of small whole numbers.

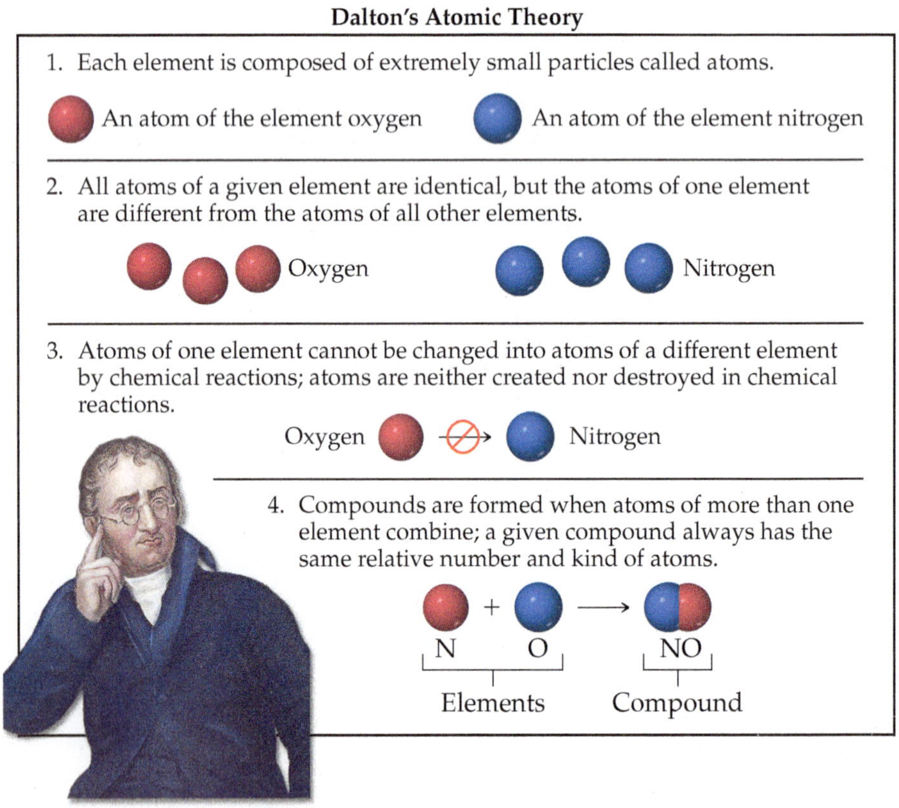

Dalton's Atomic Theory

1. Each element is composed of extremely small particles called atoms.

    ● An atom of the element oxygen   ● An atom of the element nitrogen

2. All atoms of a given element are identical, but the atoms of one element are different from the atoms of all other elements.

    ●●● Oxygen   ●●● Nitrogen

3. Atoms of one element cannot be changed into atoms of a different element by chemical reactions; atoms are neither created nor destroyed in chemical reactions.

    Oxygen ● –⊘→ ● Nitrogen

4. Compounds are formed when atoms of more than one element combine; a given compound always has the same relative number and kind of atoms.

    ● + ● ⟶ ●●
    N   O   NO
    ⎣___Elements___⎦ Compound

▶ **Figure 2.1 The four postulates of Dalton's atomic theory.*** John Dalton (1766–1844), the son of a poor English weaver, began teaching at age 12. He spent most of his years in Manchester, where he taught both grammar school and college. His lifelong interest in meteorology led him to study gases, then chemistry, and eventually atomic theory. Despite his humble beginnings, Dalton gained a strong scientific reputation during his lifetime.

*Dalton, "Atomic Theory" 1844.

We can illustrate this law by considering water and hydrogen peroxide, both of which consist of the elements hydrogen and oxygen. In forming water, 8.0 g of oxygen combines with 1.0 g of hydrogen. In forming hydrogen peroxide, 16.0 g of oxygen combines with 1.0 g of hydrogen. Thus, the ratio of the masses of oxygen per gram of hydrogen in the two compounds is 2:1. Using Dalton's atomic theory, we conclude that hydrogen peroxide contains twice as many atoms of oxygen per hydrogen atom than does water.

 Self-Assessment Exercises

SAE 2.1 Which of the following statements is or are true?

(**i**) Atoms are the fundamental building blocks of matter.
(**ii**) According to Dalton, atoms of the same element can be different from one another.
(**iii**) According to Dalton, if two atoms are different, they must be atoms of different elements.

(**a**) i only (**b**) ii only (**c**) iii only (**d**) i and ii (**e**) i and iii

SAE 2.2 You are working at an elemental analysis lab. You decompose a 5.62 g sample of ammonia, NH_3, into its elements, yielding 4.63 g of nitrogen and 0.99 g of hydrogen. The next sample that arrives for analysis weighs 86.3 g and is supposed to contain only nitrogen and hydrogen. You decompose this sample and obtain 75.5 g of nitrogen and 10.8 g of hydrogen. Which of the following choices describes the best conclusion from these data?

(**a**) The second sample is ammonia because it only contains nitrogen and hydrogen. (**b**) The second sample cannot be ammonia because it must contain elements besides nitrogen and hydrogen. (**c**) The second sample cannot be ammonia because the ratio of masses of nitrogen to hydrogen is different from that for ammonia. (**d**) The second sample is ammonia because the ratio of masses of nitrogen to hydrogen is the same as that for the first ammonia sample.

2.2 | The Discovery of Atomic Structure

Dalton based his conclusions about atoms on chemical observations made in the laboratory. By assuming the existence of atoms, he was able to account for the laws of constant composition and multiple proportions. But neither Dalton nor those who followed him for the next hundred years had any direct evidence for the existence of atoms. Today, however, we can measure the properties of individual atoms and even obtain images of them (**Figure 2.2**).

As scientists developed methods for probing the nature of matter, the supposedly indivisible atom began to show signs of a more complex structure, and today we know that the atom is composed of **subatomic particles**. Before we summarize the current model, we briefly consider a few of the landmark discoveries that led to that model. We will see that the atom is composed in part of electrically charged particles, some with a positive charge and some with a negative charge. As we discuss the development of our current model of the atom, keep in mind this fact: *Particles with the same charge repel one another, whereas particles with opposite charges attract one another.*

Cathode Rays and Electrons

During the mid-1800s, scientists began to study electrical discharge through a glass tube pumped almost empty of air (**Figure 2.3**). When a high voltage was applied to the electrodes in the tube, radiation was produced between the electrodes. This radiation, called **cathode rays**, originated at the negatively charged electrode (the *cathode*) and traveled to the positively charged electrode (the *anode*). Although the rays could not be seen, their presence was detected because they caused certain materials to *fluoresce*—that is, to give off light.

Experiments showed that cathode rays are deflected by electric or magnetic fields in a way consistent with there being a stream of negative electrical charge. The British scientist J. J. Thomson (1856–1940) observed that cathode rays are the same regardless of the identity of the cathode material. In a paper published in 1897, Thomson described cathode rays as streams of negatively charged particles that we now call **electrons**.

 Learning Objectives

When you finish **Section 2.2**, you should be able to:

▶ Describe the experiments that led to the discovery of the electron and its charge.
▶ Describe types of radioactivity and their properties.
▶ Describe the nuclear model of the atom and the experiments that led to its discovery.

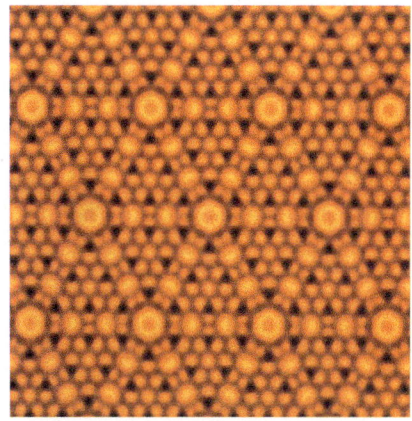

▲ **Figure 2.2 The surface of silicon**, magnified by a factor of more than 2,000,000. This image was obtained from a technique called transmission electron microscopy. The silicon atoms have rearranged themselves to minimize their energy, resulting in this beautifully symmetric pattern.

Go Figure Does this experiment show that electrons flow from the cathode to the anode, or is it possible for electrons to flow from the anode to the cathode?

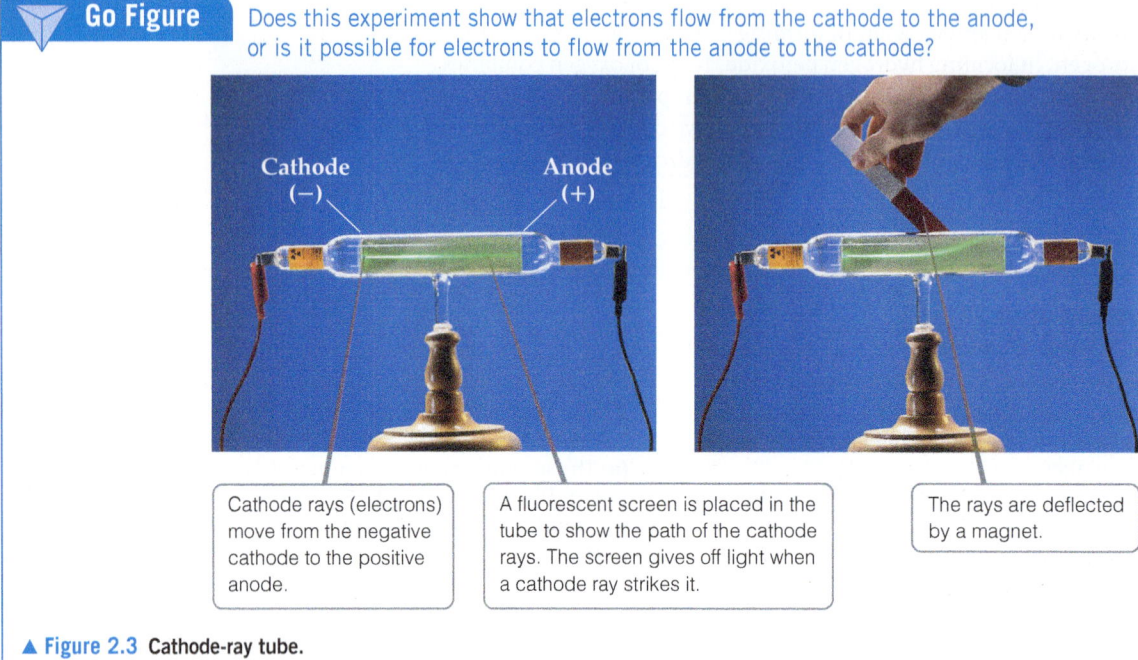

Cathode rays (electrons) move from the negative cathode to the positive anode.

A fluorescent screen is placed in the tube to show the path of the cathode rays. The screen gives off light when a cathode ray strikes it.

The rays are deflected by a magnet.

▲ Figure 2.3 **Cathode-ray tube.**

Thomson constructed a cathode-ray tube having a hole in the anode through which the cathode rays could pass. Electrically charged plates and a magnet were positioned perpendicular to the beam, and a fluorescent screen that would give off light when struck with a cathode ray was located at one end (**Figure 2.4**). Because the electron is a negatively charged particle, the electric field deflected the rays in one direction, whereas the magnetic field deflected them in the opposite direction. Thomson adjusted the strengths of the fields so that the effects balanced each other, allowing the electrons to travel in a straight path to the screen. Knowing the strengths that resulted in the straight path made

Go Figure If no magnetic field were applied, would you expect the electron beam to be deflected upward or downward by the electric field?

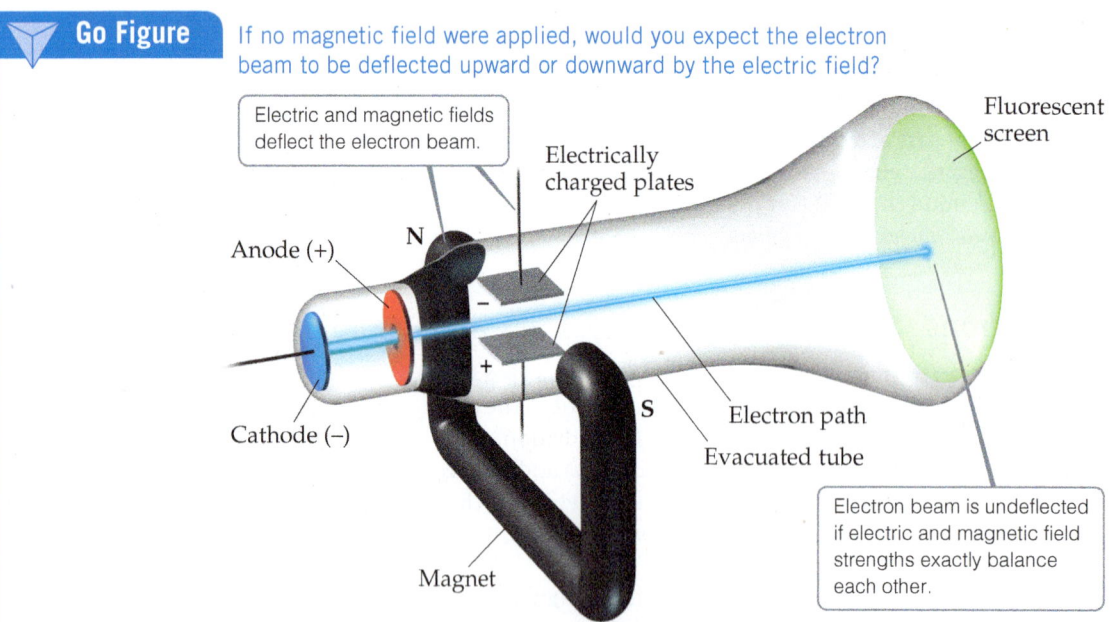

Electric and magnetic fields deflect the electron beam.

Electrically charged plates

Fluorescent screen

Anode (+)

N

Cathode (−)

+

S

Electron path

Evacuated tube

Magnet

Electron beam is undeflected if electric and magnetic field strengths exactly balance each other.

▲ Figure 2.4 **Cathode-ray tube with perpendicular magnetic and electric fields.** The cathode rays (electrons) originate at the cathode and are accelerated toward the anode, which has a hole in its center. A narrow beam of electrons passes through the hole and travels to the fluorescent screen that glows when struck by a cathode ray.

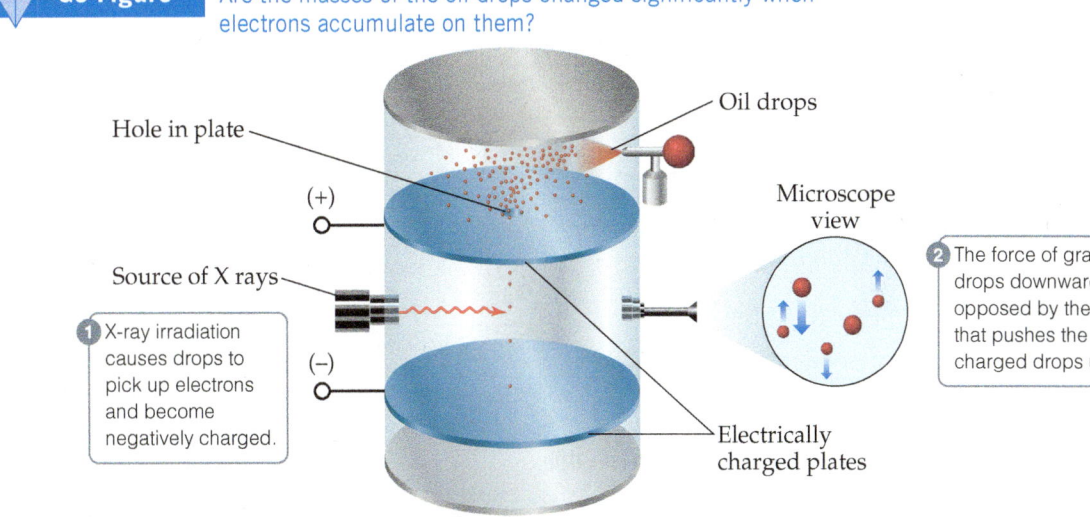

Go Figure Are the masses of the oil drops changed significantly when electrons accumulate on them?

Hole in plate

Oil drops

(+)

Microscope view

Source of X rays

1 X-ray irradiation causes drops to pick up electrons and become negatively charged.

(−)

2 The force of gravity pulls drops downward but is opposed by the electric field that pushes the negatively charged drops upward.

Electrically charged plates

▲ Figure 2.5 **Millikan's oil-drop experiment to measure the charge of the electron.** Small drops of oil are allowed to fall between electrically charged plates. Millikan measured how varying the voltage between the plates affected the rate of fall. From these data he calculated the negative charge on the drops. Because the charge on any drop was always some integral multiple of 1.602×10^{-19} C, Millikan deduced this value to be the charge of a single electron.

it possible to calculate a value of 1.76×10^8 coulombs* per gram for the ratio of the electron's electrical charge to its mass.

Once the charge-to-mass ratio of the electron was known, measuring either quantity allowed scientists to calculate the other. In 1909, Robert Millikan (1868–1953) of the University of Chicago succeeded in measuring the charge of an electron by performing the experiment described in **Figure 2.5**. He then calculated the mass of the electron by using his experimental value for the charge, 1.602×10^{-19} C, and Thomson's charge-to-mass ratio, 1.76×10^8 C/g:

$$\text{Electron mass} = \frac{1.602 \times 10^{-19} \, \cancel{C}}{1.76 \times 10^8 \, \cancel{C}/\text{g}} = 9.10 \times 10^{-28} \, \text{g}$$

This result agrees well with the currently accepted value for the electron mass, 9.10938×10^{-28} g. This mass is about 2000 times smaller than that of hydrogen, the lightest atom.

Radioactivity

In 1896, the French scientist Henri Becquerel (1852–1908) discovered that a compound of uranium spontaneously emits high-energy radiation. This spontaneous emission of radiation is called **radioactivity**. At Becquerel's suggestion, Marie Curie (**Figure 2.6**) and her husband, Pierre, began experiments to identify and isolate the source of radioactivity in the compound. They concluded that it was the uranium atoms.

Further study of radioactivity, principally by the British scientist Ernest Rutherford (1871–1937), revealed three types of radiation: alpha (α), beta (β), and gamma (γ). Rutherford was a very important figure in this period of atomic science. After working at Cambridge University with J. J. Thomson, he moved to McGill University in Montreal, where he did research on radioactivity that led to his 1908 Nobel Prize in Chemistry. In 1907, he returned to England as a faculty member at Manchester University, where he conducted the α-particle scattering experiments that we describe in the next subsection.

▲ Figure 2.6 **Marie Sklodowska Curie (1867–1934).** In 1903, Henri Becquerel, Marie Curie, and her husband, Pierre, were jointly awarded the Nobel Prize in Physics for their pioneering work on radioactivity (a term she introduced). In 1911, Marie Curie won a second Nobel Prize, this time in Chemistry, for her discovery of the elements polonium and radium.

*The coulomb (C) is the SI unit for electrical charge.

Go Figure   What is the significance of the observation that the alpha and beta rays
are bent in opposite directions?

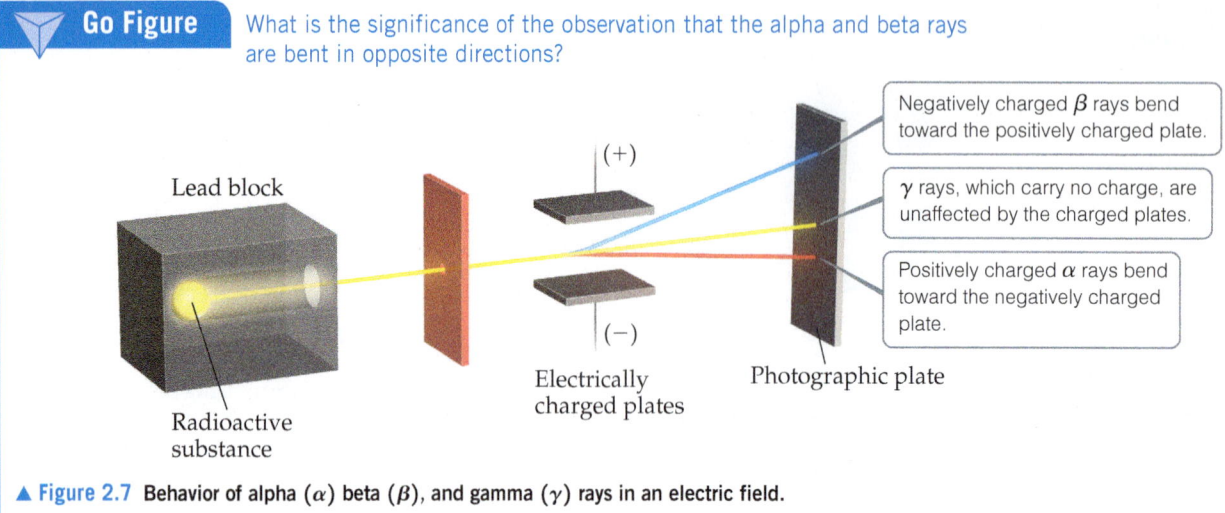

Negatively charged β rays bend
toward the positively charged plate.

γ rays, which carry no charge, are
unaffected by the charged plates.

Positively charged α rays bend
toward the negatively charged
plate.

Lead block

(+)

(−)

Electrically
charged plates

Photographic plate

Radioactive
substance

▲ **Figure 2.7** Behavior of alpha (α) beta (β), and gamma (γ) rays in an electric field.

Rutherford showed that the paths of α and β radiation are bent by an electric field, although in opposite directions, while γ radiation is unaffected by the field (**Figure 2.7**). From this finding he concluded that α and β rays consist of fast-moving electrically charged particles. In fact, β particles are nothing more than high-speed electrons that can be considered the radioactive equivalent of cathode rays. Because of their negative charge, they are attracted to a positively charged plate. The α particles have a positive charge and are attracted to a negative plate. In units of the charge of the electron, β particles have a charge of 1−, whereas α particles have a charge of 2+. Each α particle has a mass about 7400 times that of an electron. Gamma radiation is high-energy electromagnetic radiation similar to X rays; it does not consist of particles and it carries no charge.

The Nuclear Model of the Atom

With growing evidence that the atom was composed of smaller particles, scientists tried to explain how the particles fit together. During the early 1900s, Thomson reasoned that because electrons contribute only a very small fraction of an atom's mass, they probably are responsible for an equally small fraction of the atom's size. He proposed that the atom consists of a uniform positive sphere of matter in which the mass is evenly distributed and in which the electrons are embedded like raisins in a pudding or seeds in a watermelon (**Figure 2.8**). This *plum-pudding model*, named after a traditional English dessert, was very short-lived.

In 1910, Rutherford was studying the angles at which α particles were deflected, or *scattered*, as they passed through a thin sheet of gold foil (**Figure 2.9**). He discovered that almost all the particles passed directly through the foil without deflection, with a few particles deflected about 1°, consistent with Thomson's plum-pudding model. For the sake of completeness, Rutherford suggested that Ernest Marsden (1889–1970), an undergraduate student working in the laboratory, look for scattering at large angles. To everyone's surprise, a small amount of scattering was observed at large angles, with some particles scattered back in the direction from which they had come. The explanation for these results was not immediately obvious, but they were clearly inconsistent with Thomson's plum-pudding model.

Rutherford explained the results by postulating the **nuclear model** of the atom, in which most of the mass of each gold atom and all of its positive charge reside in a very small, extremely dense region that he called the **nucleus**. He postulated further that most of the volume of an atom is empty space in which electrons move around the nucleus. In the α-scattering experiment, most of the particles passed through the foil unscattered because they did not encounter the minute nucleus of any gold atom. Occasionally, however, an α particle came close to a gold nucleus. In such encounters,

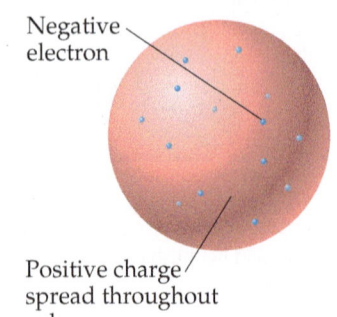

Negative
electron

Positive charge
spread throughout
sphere

▲ **Figure 2.8** J. J. Thomson's plum-pudding
model of the atom. Ernest Rutherford and
Ernest Marsden proved this model wrong.

Go Figure Predict where the spots on the circular fluorescent screen would be if the plum-pudding model of the atom was correct.

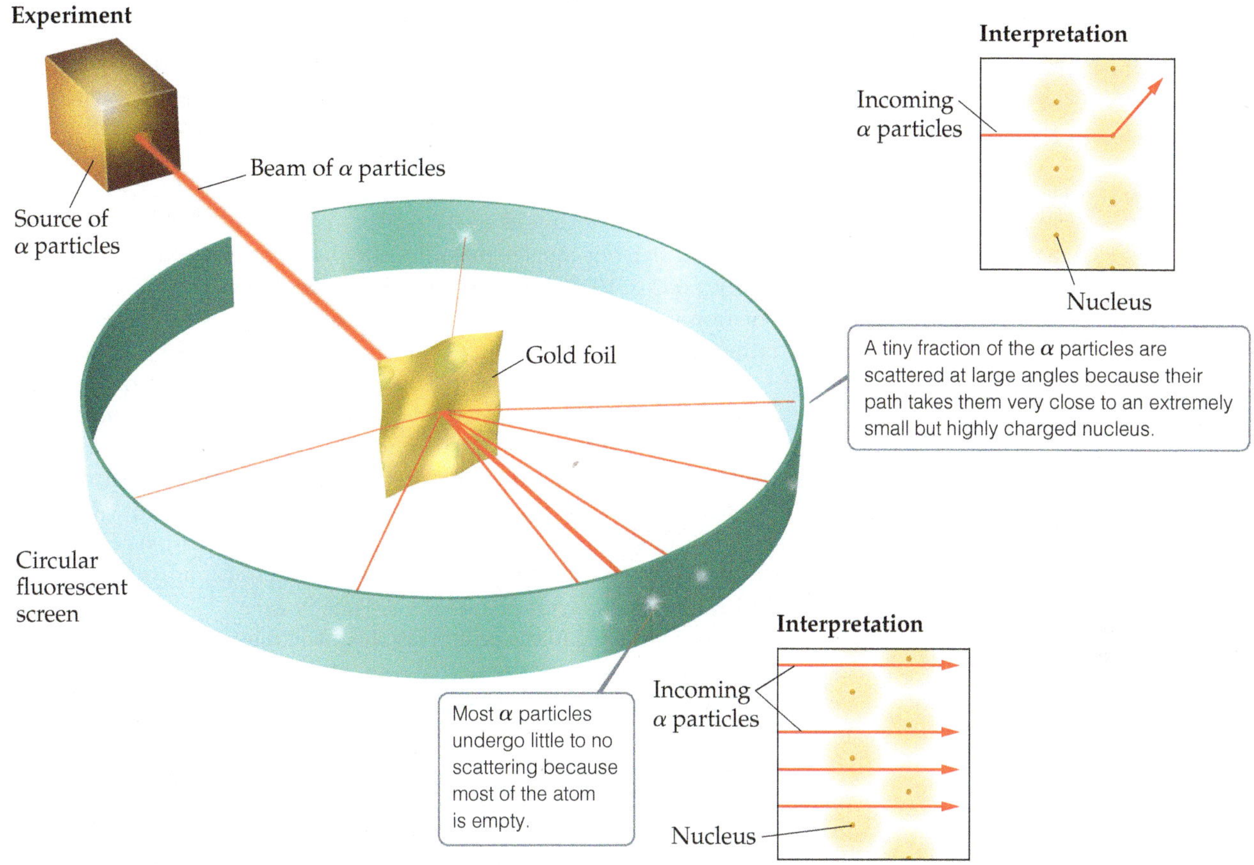

Experiment

Interpretation

Source of
α particles

Beam of α particles

Incoming
α particles

Nucleus

A tiny fraction of the α particles are
scattered at large angles because their
path takes them very close to an extremely
small but highly charged nucleus.

Gold foil

Circular
fluorescent
screen

Interpretation

Most α particles
undergo little to no
scattering because
most of the atom
is empty.

Incoming
α particles

Nucleus

▲ **Figure 2.9 Rutherford's α-scattering experiment.** When α particles pass through a gold foil, most
pass through undeflected, but some are scattered, a few at very large angles. The nuclear model of
the atom explains why a few α particles are deflected at large angles. Although the nuclear atom
has been depicted here as a yellow sphere, it is important to realize that most of the space around
the nucleus contains only the low-mass electrons.

the repulsion between the highly positive charge of the gold nucleus and the positive
charge of the α particle was strong enough to deflect the particle, as shown in
Figure 2.9.

Subsequent experiments led to the discovery of positive particles (**protons**) and
neutral particles (**neutrons**) in the nucleus. Protons were discovered in 1919 by
Rutherford and neutrons in 1932 by British scientist James Chadwick (1891–1972). Thus,
the atom is composed of electrons, protons, and neutrons.

 Self-Assessment Exercises

SAE 2.3 Which of the following statements is false? (**a**) Electrons
are subatomic particles that bear a negative charge. (**b**) Millikan's
oil-drop experiment showed that one electron has a charge of
1.602×10^{-19} coulombs. (**c**) Thomson's cathode-ray experiments
showed that electrons do have mass. (**d**) One electron has a mass of
about a picogram, 10^{-12} g.

SAE 2.4 What can you conclude from the observation that gamma
radiation is not deflected by an electric field? (**a**) Gamma rays have
no mass. (**b**) Gamma rays have no charge. (**c**) Gamma rays have neither mass nor charge.

SAE 2.5 Which of the following statements is false? (**a**) The
nucleus consists of protons and neutrons. (**b**) The nucleus is the
densest part of the atom. (**c**) Rutherford's gold foil experiment
proved that the plum-pudding model of the atom is correct. (**d**) To
remain electrically neutral, all atoms should contain equal numbers
of electrons and protons.

2.3 | The Modern View of Atomic Structure

Learning Objectives

When you finish Section 2.3, you should be able to:

▶ Compare the relative charges and masses of protons, neutrons, and electrons.

▶ Describe the interactions between protons, neutrons, and electrons that are responsible for holding atoms together.

▶ Correlate an isotope's chemical symbol with the number of protons, neutrons, and electrons it contains.

Go Figure

What is the approximate diameter of the nucleus in units of pm?

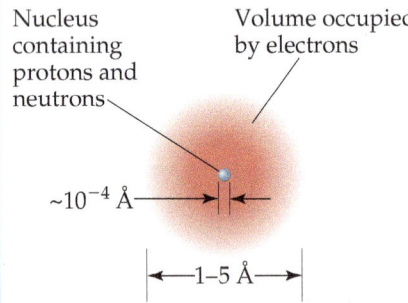

Nucleus containing protons and neutrons

Volume occupied by electrons

~10^{-4} Å

1–5 Å

▲ **Figure 2.10** **The structure of the atom.** A cloud of rapidly moving electrons occupies most of the volume of the atom. The nucleus occupies a tiny region at the center of the atom and is composed of the protons and neutrons. The nucleus contains virtually all the mass of the atom.

Since Rutherford's time, the list of particles that make up nuclei has grown and continues to increase as physicists have learned more and more about atomic nuclei. Examples of those subatomic particles include quarks, leptons, and bosons.* As chemists, however, we can take a simple view of the atom because only three subatomic particles—the proton, neutron, and electron—have a bearing on chemical behavior.

As noted earlier, the charge of an electron is -1.602×10^{-19} C. The charge of a proton is opposite in sign but equal in magnitude to that of an electron: $+1.602 \times 10^{-19}$ C. The quantity 1.602×10^{-19} C is called the **electronic charge**. For convenience, the charges of atomic and subatomic particles are usually expressed as multiples of this charge rather than as coulombs. Thus, the charge of an electron is 1− and that of a proton is 1+. Neutrons are electrically neutral (which is how they received their name). *Every atom has an equal number of electrons and protons, so atoms have no net electrical charge.*

Protons and neutrons reside in the tiny nucleus of the atom. The vast majority of an atom's volume is the space in which the electrons reside (**Figure 2.10**). Most atoms have diameters between 1×10^{-10} m (100 pm) and 5×10^{-10} m (500 pm). A convenient non-SI unit of length used for atomic dimensions is the **angstrom** (Å), where 1 Å $= 1 \times 10^{-10}$ m $= 100$ pm. Thus, atoms have diameters of approximately 1–5 Å. The diameter of a chlorine atom, for example, is 200 pm, or 2.0 Å.

Electrons are attracted to the protons in the nucleus by the electrostatic force that exists between particles of opposite electrical charge. In later chapters, we show that the strength of the attractive forces between electrons and nuclei can be used to explain many of the differences among different elements.

Atoms have extremely small masses. The mass of the heaviest known atom, for example, is approximately 4×10^{-22} g. Because it would be cumbersome to express such small masses in grams, we use the **atomic mass unit** (amu),* where 1 amu $= 1.66054 \times 10^{-24}$ g. A proton has a mass of 1.0073 amu, a neutron 1.0087 amu, and an electron 5.486×10^{-4} amu (Table 2.1). Because it takes 1836 electrons to equal the mass of one proton, and 1839 electrons to equal the mass of a single neutron, the nucleus accounts for nearly the entire mass of an atom.

TABLE 2.1   Comparison of the Proton, Neutron, and Electron

Particle	Charge	Mass (amu)
Proton	Positive (1+)	1.0073
Neutron	None (neutral)	1.0087
Electron	Negative (1−)	5.486×10^{-4}

The diameter of an atomic nucleus is approximately 10^{-4} Å, only a small fraction of the diameter of the atom as a whole. You can appreciate the relative sizes of the atom and its nucleus by imagining that if the hydrogen atom were as large as a football stadium, the nucleus would be the size of a marble. Because the tiny nucleus carries most of the mass of the atom in such a small volume, it has an incredibly high density—on the order of 10^{13}–10^{14} g/cm^3. A candy bar of that density would weigh over 2.5 billion tons!

Figure 2.10 incorporates the features we have just discussed. The significance of representing the region containing electrons as an indistinct cloud will become clear in later chapters when we consider the energies and spatial arrangements of the electrons. For now, however, we have all the information we need to discuss many topics that form the basis of everyday uses of chemistry.

---

*The electron is an elementary particle that cannot be divided into smaller particles, whereas protons and neutrons are made up of smaller particles called quarks.

*The SI abbreviation for the atomic mass unit is u. We use the more common abbreviation amu.

Atomic Numbers, Mass Numbers, and Isotopes

What makes an atom of one element different from an atom of another element? The atoms of each element have a *characteristic number of protons*. The number of protons in an atom of any particular element is called that element's **atomic number**. Because an atom has no net electrical charge, the number of electrons it contains must equal the number of protons. All atoms of carbon, for example, have six protons and six electrons, whereas all atoms of oxygen have eight protons and eight electrons. Thus, carbon has atomic number 6, whereas oxygen has atomic number 8. The atomic number of each element is listed with the name and symbol of the element on the front inside cover of the text.

Atoms of a given element can differ in the number of neutrons they contain and, consequently, in mass. For example, while most atoms of carbon have six neutrons, some have more and some have less. The symbol $^{12}_{6}C$ (read "carbon twelve," carbon-12) represents the carbon atom containing six protons and six neutrons, whereas carbon atoms that contain six protons and eight neutrons have mass number 14, are represented as $^{14}_{6}C$, and are referred to as carbon-14.

Sample Exercise 2.1
Atomic Size

The diameter of a U.S. dime is 17.9 mm, and the diameter of a silver atom is 2.88 Å. How many silver atoms could be arranged side by side across the diameter of a dime?

SOLUTION

The unknown is the number of silver (Ag) atoms. Using the relationship 1 Ag atom = 2.88 Å as a conversion factor relating number of atoms and distance, we start with the diameter of the dime, first converting this distance into angstroms and then using the diameter of the Ag atom to convert distance to number of Ag atoms:

$$\text{number of Ag atoms} = (17.9 \text{ mm})\left(\frac{10^{-3} \text{ m}}{1 \text{ mm}}\right)\left(\frac{1 \text{ Å}}{10^{-10} \text{ m}}\right)\left(\frac{1 \text{ Ag atom}}{2.88 \text{ Å}}\right)$$

$$= 6.22 \times 10^{7} \text{ Ag atoms}$$

That is, 62.2 million silver atoms could sit side by side across a dime!

▶ **Practice Exercise**

The diameter of a carbon atom is 1.54 Å. **(a)** Express this diameter in picometers. **(b)** How many carbon atoms could be aligned side by side across the width of a pencil line that is 0.20 mm wide?

A CLOSER LOOK   Basic Forces

Four basic forces are known in nature: (1) gravitational, (2) electromagnetic, (3) strong nuclear, and (4) weak nuclear. *Gravitational forces* are attractive forces that act between all objects in proportion to their masses. Gravitational forces between atoms or between subatomic particles are so small that they are of no chemical significance.

Electromagnetic forces are attractive or repulsive forces that act between either electrically charged or magnetic objects. The magnitude of the electric force between two charged particles is given by *Coulomb's law*: $F = kQ_1Q_2/d^2$, where Q_1 and Q_2 are the magnitudes of the charges on the two particles, d is the distance between their centers, and k is a constant determined by the units for Q and d. (Section 1.4) A negative value for the force indicates attraction, whereas a positive value indicates repulsion. Electric forces are of primary importance in determining the chemical properties of elements.

All nuclei except those of hydrogen atoms contain two or more protons. Because like charges repel, electrical repulsion would cause the protons to fly apart if the *strong nuclear force* did not keep them together. As the name implies, this force can be quite strong but only when particles are extremely close together, as are the protons and neutrons in a nucleus. At this distance, the attractive strong nuclear force is stronger than the positive–positive repulsive electric force and holds the nucleus together.

The *weak nuclear force* is weaker than the electric force and the strong nuclear force but stronger than the gravitational force. We are aware of its existence only because it shows itself in certain types of radioactivity.

Related Exercise: 2.114

The atomic number is indicated by the subscript; the superscript, called the **mass number**, is the number of protons plus neutrons in the atom:

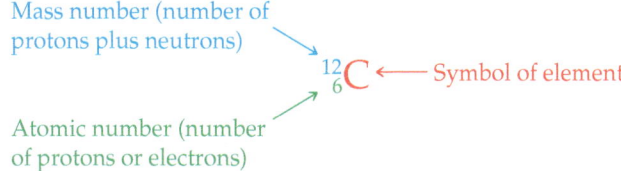

Because all atoms of a given element have the same atomic number, the subscript is redundant and is often omitted. Thus, the symbol for carbon-12 can be represented simply as ^{12}C.

Atoms with identical atomic numbers but different mass numbers (that is, the same number of protons but different numbers of neutrons) are called **isotopes** of one another. Several isotopes of carbon are listed in Table 2.2. We generally use the notation with superscripts only when referring to a particular isotope of an element. It is important to keep in mind that the isotopes of any given element are all alike chemically. A carbon dioxide molecule that contains a ^{13}C atom behaves for all practical purposes identically to one that contains a ^{12}C atom.

TABLE 2.2   Some Isotopes of Carbon[a]

Symbol	Number of Protons	Number of Electrons	Number of Neutrons
^{11}C	6	6	5
^{12}C	6	6	6
^{13}C	6	6	7
^{14}C	6	6	8

[a]Almost 99% of the carbon found in nature is ^{12}C.

Sample Exercise 2.2

Determining the Number of Subatomic Particles in Atoms

How many protons, neutrons, and electrons are in an atom of (a) ^{197}Au, (b) strontium-90?

SOLUTION

(a) The superscript 197 is the mass number (protons + neutrons). According to the list of elements given on the front inside cover, gold has atomic number 79. Consequently, an atom of ^{197}Au has 79 protons, 79 electrons, and $197 - 79 = 118$ neutrons.

(b) The atomic number of strontium is 38. Thus, all atoms of this element have 38 protons and 38 electrons. The strontium-90 isotope has $90 - 38 = 52$ neutrons.

▶ **Practice Exercise**

How many protons, neutrons, and electrons are in an atom of (a) ^{138}Ba, (b) phosphorus-31?

Sample Exercise 2.3

Writing Symbols for Atoms

Magnesium has three isotopes with mass numbers 24, 25, and 26. (a) Write the complete chemical symbol (superscript and subscript) for each. (b) How many neutrons are in an atom of each isotope?

SOLUTION

(a) Magnesium has atomic number 12, so all atoms of magnesium contain 12 protons and 12 electrons. The three isotopes are therefore represented by $^{24}_{12}Mg$, $^{25}_{12}Mg$, and $^{26}_{12}Mg$.

(b) The number of neutrons in each isotope is the mass number minus the number of protons. The numbers of neutrons in an atom of each isotope are therefore 12, 13, and 14, respectively.

▶ **Practice Exercise**

Give the complete chemical symbol for the atom that contains 82 protons, 82 electrons, and 126 neutrons.

Self-Assessment Exercises

SAE 2.6 Which pairing of subatomic particle with charge is correct? **(a)** proton/1− **(b)** neutron/1+ **(c)** electron/0 **(d)** electron/1+ **(e)** electron/0

SAE 2.7 The density of the nucleus of a particular atom is 1×10^{13} g/cm^3. Given that the diameter of a nucleus is 1×10^{-4} Å, estimate the mass of the nucleus, assuming it is a sphere. **(a)** 10^{13} g **(b)** 41889 g **(c)** 4.19×10^{-23} g **(d)** 5.23×10^{-24} g

SAE 2.8 If the nucleus is so dense and full of positively charged protons, why doesn't it fly apart due to electrostatic repulsions between protons? **(a)** The negatively charged electrons in the nucleus lead to electrostatic attractions that keep it together. **(b)** The neutrons act as electrostatic "glue" to reduce electrostatic repulsion between protons in the nucleus. **(c)** The strong nuclear force keeps the nucleus together in spite of electrostatic repulsions between protons. **(d)** Actually, nuclei do disintegrate around us all the time.

SAE 2.9 Which of the following is the correct chemical symbol for uranium-235, the radioactive isotope that is the basis for many nuclear weapons? **(a)** $^{235}_{92}$U **(b)** $^{92}_{235}$U **(c)** $^{143}_{92}$U **(d)** $^{235}_{92}$Ur

2.4 | Atomic Weights

Atoms are small pieces of matter, so they have mass. In this section, we discuss the mass scale used for atoms and introduce the concept of *atomic weights*.

The Atomic Mass Scale

Scientists of the nineteenth century were aware that atoms of different elements have different masses. They found, for example, that each 100.0 g of water contains 11.1 g of hydrogen and 88.9 g of oxygen. Thus, water contains 88.9/11.1 = 8 times as much oxygen, by mass, as hydrogen. Once scientists understood that water contains two hydrogen atoms for each oxygen atom, they concluded that an oxygen atom must have $2 \times 8 = 16$ times as much mass as a hydrogen atom. Hydrogen, the lightest atom, was arbitrarily assigned a relative mass of 1 (no units). Atomic masses of other elements were at first determined relative to this value. Thus, oxygen was assigned an atomic mass of 16.

Today we can determine the masses of individual atoms with a high degree of accuracy and precision. For example, we know that the ^{1}H atom has a mass of 1.6735×10^{-24} g and the ^{16}O atom has a mass of 2.6560×10^{-23} g. As we noted in Section 2.3, it is convenient to use the *atomic mass unit* when dealing with these extremely small masses:

$$1 \text{ amu} = 1.66054 \times 10^{-24} \text{ g} \quad \text{and} \quad 1 \text{ g} = 6.02214 \times 10^{23} \text{ amu}$$

The atomic mass unit is presently defined by assigning a mass of exactly 12 amu to an individual atom of the ^{12}C isotope of carbon. In these units, an ^{1}H atom has a mass of 1.0078 amu and an ^{16}O atom has a mass of 15.9949 amu.

Atomic Weight

Most elements occur in nature as mixtures of isotopes. We can determine the *average atomic mass* of an element, usually called the element's **atomic weight**, by summing (indicated by the uppercase Greek sigma, Σ) over the masses of its isotopes multiplied by their relative abundances:

$$\text{Atomic weight} = \sum_{\substack{\text{over all} \\ \text{isotopes of} \\ \text{the element}}} [(\text{fractional isotope abundance}) \times (\text{isotope mass})] \quad [2.1]$$

Naturally occurring carbon, for example, is composed of 98.93% ^{12}C and 1.07% ^{13}C. The masses of these isotopes are 12 amu (exactly) and 13.00335 amu, respectively, making the atomic weight of carbon

$$(0.9893)(12 \text{ amu}) + (0.0107)(13.00335 \text{ amu}) = 12.01 \text{ amu}$$

The atomic weights of the elements are listed in both the periodic table and the table of elements on the front inside cover of this text.

 Learning Objectives

When you finish Section 2.4 you should be able to:

▶ Describe the relationship between the atomic weight of an element and the weights and abundances of the naturally occurring isotopes of that element.

▶ Interconvert between atomic weights and isotopic abundances for an element.

▶ Describe the principles of mass spectrometry.

▶ Interpret/predict data from a mass spectrometer.

Sample Exercise 2.4
Calculating the Atomic Weight of an Element from Isotopic Abundances

Naturally occurring chlorine is 75.75% ^{35}Cl (atomic mass 34.969 amu) and 24.22% ^{37}Cl (atomic mass 36.966 amu). Calculate the atomic weight of chlorine.

SOLUTION

We can calculate the atomic weight by multiplying the abundance of each isotope by its mass and summing these products. Because 75.78% = 0.7578 and 24.22% = 0.2422, we have

Atomic wight = (0.7578)(34.969 amu) + (0.2422)(36.966 amu)

$$= 26.50 \text{ amu} + 8.953 \text{ amu}$$

$$= 35.45 \text{ amu}$$

This answer makes sense: The atomic weight, which is actually the average atomic mass, is between the masses of the two isotopes and is closer to the value of ^{35}Cl, the more abundant isotope.

▶ **Practice Exercise**

Three isotopes of silicon occur in nature: ^{28}Si (92.23%), atomic mass 27.97693 amu; ^{29}Si (4.68%), atomic mass 28.97649 amu; and ^{30}Si (3.09%), atomic mass 29.97377 amu. Calculate the atomic weight of silicon.

A CLOSER LOOK   The Mass Spectrometer

The most accurate means for determining atomic weights is provided by the **mass spectrometer** (**Figure 2.11**). There are various designs of mass spectrometers, but they all operate on similar principles. The first step is to get atoms or molecules into the gas phase. Sometimes the sample to be analyzed is already a gas, whereas in other cases heating, application of an electric field, or a pulse of laser light may be needed to create gas-phase atoms or molecules. Next, the gas-phase species must be converted to positively charged particles called *ions*. There are many approaches to creating ions, including bombardment with beams of high-energy electrons or chemical reactions with other gas-phase molecules. Once gas-phase ions have been produced, they are accelerated toward a negatively charged grid. After the ions pass through the grid, they encounter two slits that allow only a narrow beam of ions to pass. This beam then passes between the poles of a magnet, which deflects the ions into a curved path. For ions with the same charge, the extent of deflection depends on mass—the more massive the ion, the less the deflection. The ions are thereby separated according to their masses. By changing the strength of the magnetic field or the accelerating voltage on the grid, ions of various masses can be selected to enter the detector.

A graph of the intensity of the detector signal versus ion atomic mass is called a *mass spectrum* (**Figure 2.12**). Analysis of a mass spectrum gives both the masses of the ions reaching the detector and their relative abundances, which are obtained from the signal intensities. Knowing the atomic mass and abundance of each isotope allows us to calculate the atomic weight of an element, as shown in Sample Exercise 2.4.

Mass spectrometers are used extensively today to identify chemical compounds and analyze mixtures of substances. Any molecule that loses electrons can fall apart, forming an array of positively charged fragments. The mass spectrometer measures the masses of these fragments, producing a chemical "fingerprint" of the molecule and providing clues about how the atoms were connected in the original molecule. Thus, a chemist might use this technique to determine the molecular structure of a newly synthesized compound, to analyze proteins in the human genome, or to identify a pollutant in the environment.

Related Exercises: 2.37, 2.38, 2.40, 2.94, 2.100, 2.101

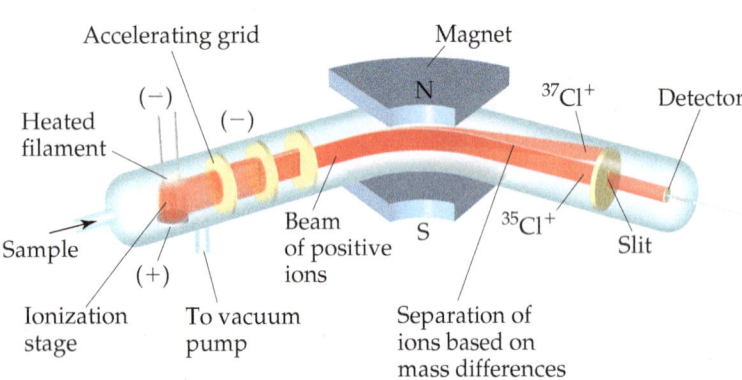

▲ **Figure 2.11 How a mass spectrometer works.** The example here is for chlorine. Cl atoms are first ionized to form Cl$^+$ ions, the ions are then accelerated with an electric field, and finally their path is directed by a magnetic field. The paths of the ions of the two Cl isotopes diverge as they pass through the magnetic field. Nearly all of the air in the instrument must be removed with the vacuum pump before the experiment starts, so the paths of the ions are uninterrupted.

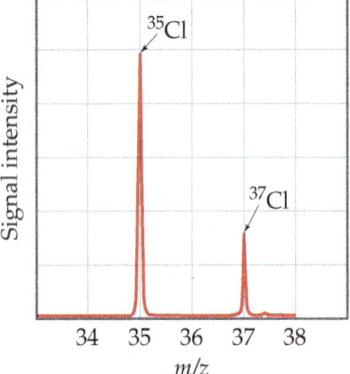

▲ **Figure 2.12 Mass spectrum of atomic chlorine.** The fractional abundances of the isotopes ^{35}Cl and ^{37}Cl are indicated by the relative signal intensities of the beams of Cl$^+$ reaching the detector of the mass spectrometer. Technically, the mass spectrometer measures the mass/charge ratio of ions; but most ions are only singly charged in the experiment, so you can read the horizontal axis as the atomic mass of the isotope in units of amu.

Self-Assessment Exercises

SAE 2.10 Boron has two naturally occurring isotopes: ^{10}B, with a mass of 10.01 amu, and ^{11}B, with a mass of 11.01 amu. Use the average atomic weight of boron from the periodic table to determine the abundance of the ^{11}B isotope. (**a**) 3.8% (**b**) 20.0% (**c**) 80.0% (**d**) 81.0%

SAE 2.11 A new element, Brownium, is discovered in a parallel universe. Brownium's average atomic weight is 28.00 amu; it has 14 protons in its nucleus; and its two isotopes have either 13 or 16 neutrons in the nucleus. What is the ratio of the 27-amu isotope to the 30-amu isotope? (**a**) 1:1 (**b**) 1:3 (**c**) 2:3 (**d**) 2:1 (**e**) Not enough information is given to answer this question.

SAE 2.12 Which statement is true about mass spectrometry? (**a**) The data from a mass spectrometer tells you the charges of ions. (**b**) Neutral atoms can be detected in the mass spectrometer. (**c**) The heavier a charged particle is, the more it will be deflected by a magnetic field. (**d**) Different isotopes of an element will be deflected by different amounts in a mass spectrometer. (**e**) More than one of statements (a), (b), (c), (d) are true.

SAE 2.13 Use the following data from a mass spectrometer of an unknown element to determine the probable identity of the element. (**a**) Th (**b**) Pa (**c**) Zr (**d**) Nb

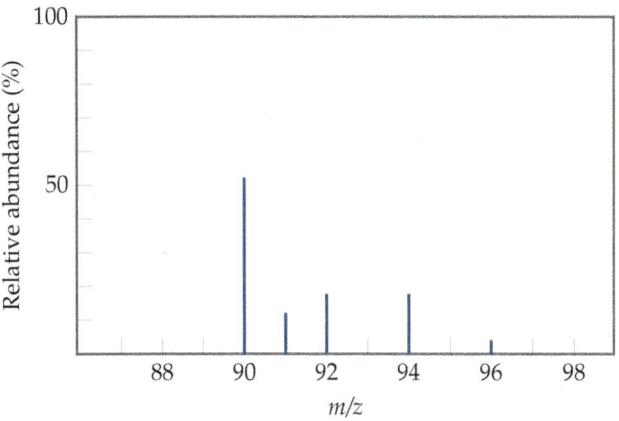

2.5 | The Periodic Table

As the list of known elements expanded during the early 1800s, attempts were made to find patterns in chemical behavior. These efforts culminated in the development of the periodic table in 1869. The periodic table plays a major role in later chapters of this text, but it is so important and useful that you should become acquainted with it now. In fact, *the periodic table is the most significant tool that chemists use for organizing and remembering chemical facts.*

Many elements show strong similarities to one another. The elements lithium (Li), sodium (Na), and potassium (K) are all soft, very reactive metals that produce heat and sparks upon mixing with water. The elements helium (He), neon (Ne), and argon (Ar) are all nonreactive gases. If the elements are arranged in order of increasing atomic number, their chemical and physical properties show a repeating, or *periodic*, pattern. For example, each of the soft, reactive metals—lithium, sodium, and potassium—comes immediately after one of the nonreactive gases—helium, neon, and argon, respectively—as shown in **Figure 2.13**.

 Learning Objectives

When you finish Section 2.5, you should be able to:

▶ Infer an element's general properties from its location in the periodic table.

▶ Describe the overall structure of the periodic table of the elements in terms of periods and groups.

 **Go Figure**   If F is a reactive nonmetal, which other element or elements shown here are likely to be reactive nonmetals?

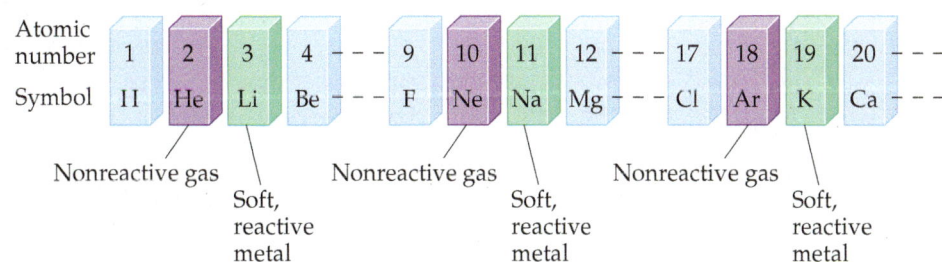

▲ **Figure 2.13** **Arranging elements by atomic number reveals a periodic pattern of properties.** This pattern is the basis of the periodic table.

The arrangement of elements in order of increasing atomic number, with elements having similar properties placed in vertical columns, is known as the **periodic table (Figure 2.14)**. The table shows the atomic number and atomic symbol for each element, and the atomic weight is often given as well, as in this typical entry for potassium:

19	← —— Atomic number
K	← —— Atomic symbol
39.0983	← —— Atomic weight

You might notice slight variations in periodic tables from one book to another or between those in the lecture hall and in the text. These are simply matters of style; there are no fundamental differences.

The horizontal rows of the periodic table are called **periods**. The first period consists of only two elements, hydrogen (H) and helium (He). The second and third periods consist of eight elements each. The fourth and fifth periods contain 18 elements. The sixth and seventh periods have 32 elements each, but in order to fit on a page, 14 of the elements from each period (atomic numbers 57–70 and 89–102) appear at the bottom of the table.

The vertical columns are called **groups**. The way in which the groups are labeled is somewhat arbitrary. Three labeling schemes are in common use, two of which are shown in Figure 2.14.

- The top set of labels, which have A and B designations, is widely used in North America. Roman numerals, rather than Arabic ones, are often employed in this scheme. Group 7A, for example, is often labeled VIIA.

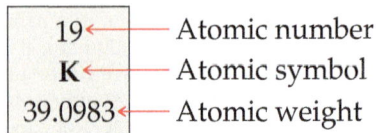

Go Figure Where are sodium and potassium in the periodic table? Do they belong to the same period? Do they belong to the same group?

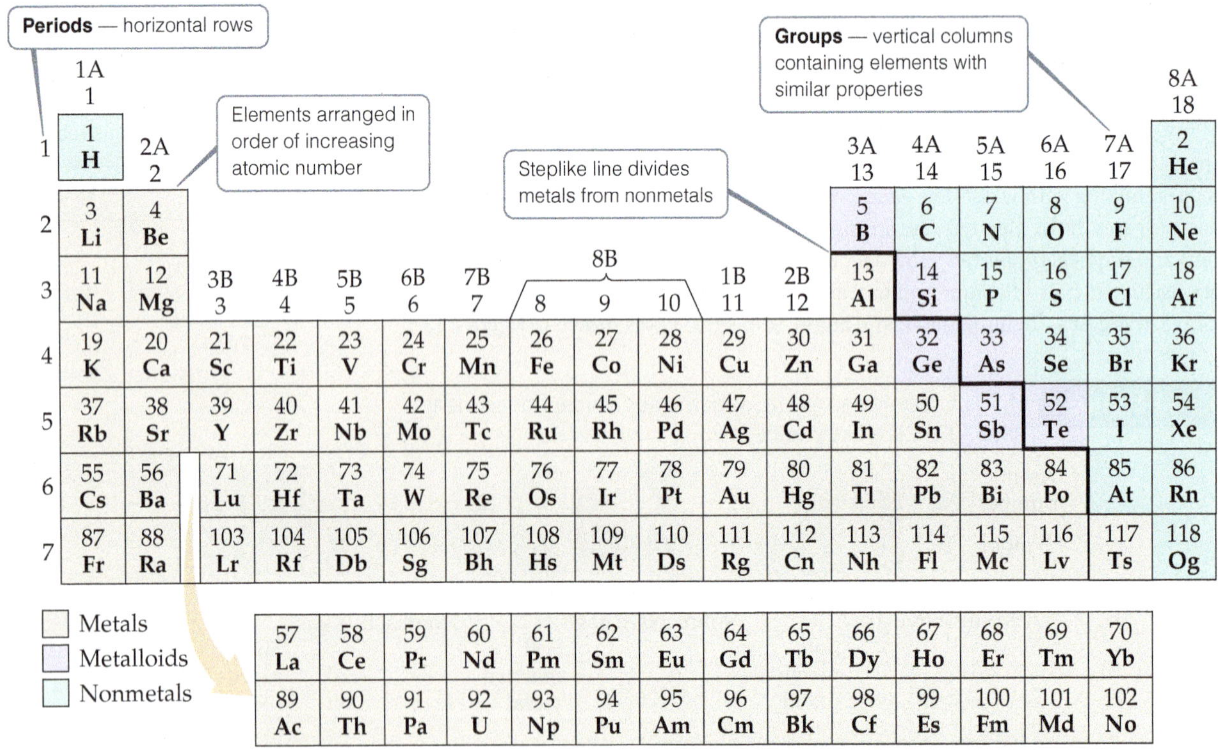

Periods — horizontal rows

Groups — vertical columns containing elements with similar properties

Elements arranged in order of increasing atomic number

Steplike line divides metals from nonmetals

	1A 1	2A 2	3B 3	4B 4	5B 5	6B 6	7B 7	8B 8	8B 9	8B 10	1B 11	2B 12	3A 13	4A 14	5A 15	6A 16	7A 17	8A 18
1	1 H																	2 He
2	3 Li	4 Be											5 B	6 C	7 N	8 O	9 F	10 Ne
3	11 Na	12 Mg											13 Al	14 Si	15 P	16 S	17 Cl	18 Ar
4	19 K	20 Ca	21 Sc	22 Ti	23 V	24 Cr	25 Mn	26 Fe	27 Co	28 Ni	29 Cu	30 Zn	31 Ga	32 Ge	33 As	34 Se	35 Br	36 Kr
5	37 Rb	38 Sr	39 Y	40 Zr	41 Nb	42 Mo	43 Tc	44 Ru	45 Rh	46 Pd	47 Ag	48 Cd	49 In	50 Sn	51 Sb	52 Te	53 I	54 Xe
6	55 Cs	56 Ba	71 Lu	72 Hf	73 Ta	74 W	75 Re	76 Os	77 Ir	78 Pt	79 Au	80 Hg	81 Tl	82 Pb	83 Bi	84 Po	85 At	86 Rn
7	87 Fr	88 Ra	103 Lr	104 Rf	105 Db	106 Sg	107 Bh	108 Hs	109 Mt	110 Ds	111 Rg	112 Cn	113 Nh	114 Fl	115 Mc	116 Lv	117 Ts	118 Og

☐ Metals
☐ Metalloids
☐ Nonmetals

57 La	58 Ce	59 Pr	60 Nd	61 Pm	62 Sm	63 Eu	64 Gd	65 Tb	66 Dy	67 Ho	68 Er	69 Tm	70 Yb
89 Ac	90 Th	91 Pa	92 U	93 Np	94 Pu	95 Am	96 Cm	97 Bk	98 Cf	99 Es	100 Fm	101 Md	102 No

▲ **Figure 2.14 The periodic table of the elements.**

- Europeans use a similar convention that assigns the A and B labels differently.
- In an effort to eliminate confusion, the International Union of Pure and Applied Chemistry (IUPAC) introduced a convention that numbers the groups from 1 through 18 with no A or B designations, as shown in Figure 2.14.

We use the traditional North American convention with Arabic numerals and the letters A and B.

Elements in a group often exhibit similarities in physical and chemical properties. For example, the "coinage metals"—copper (Cu), silver (Ag), and gold (Au)—belong to group 1B. These elements are less reactive than most metals, which is why they have been traditionally used throughout the world to make coins. Many other groups in the periodic table also have names, listed in **Table 2.3**.

TABLE 2.3   Names of Some Groups in the Periodic Table

Group	Name	Elements
1A	Alkali metals	Li, Na, K, Rb, Cs, Fr
2A	Alkaline earth metals	Be, Mg, Ca, Sr, Ba, Ra
6A	Chalcogens	O, S, Se, Te, Po
7A	Halogens	F, Cl, Br, I, At
8A	Noble gases	He, Ne, Ar, Kr, Xe, Rn

We explain in Chapters 6 and 7 that elements in a group have similar properties because they have the same arrangement of electrons at the periphery of their atoms. However, we need not wait until then to make good use of the periodic table; after all, the chemists who developed the table knew nothing about electrons! We can use the table, as they intended, to correlate behaviors of elements and to help us remember many facts.

The color code of Figure 2.14 shows that, except for hydrogen, all the elements on the left and in the middle of the table are **metallic elements**, or **metals**. All the metallic elements share characteristic properties, such as luster and high electrical and heat conductivity, and all of them except mercury (Hg) are solid at room temperature.* The metals are separated from the **nonmetallic elements**, or **nonmetals**, by a stepped line that runs from boron (B) to astatine (At). (Note that hydrogen, although on the left side of the table, is a nonmetal.) At room temperature and pressure, some of the nonmetals are gaseous, some are solid, and one is liquid. Nonmetals generally differ from metals in appearance (**Figure 2.15**) and in other physical properties. Many of

Go Figure Name two ways in which the metals shown here differ in general appearance from the nonmetals.

Metals

Iron (Fe)   Copper (Cu)   Aluminum (Al)

Silver (Ag)   Lead (Pb)   Gold (Au)

Nonmetals

Bromine (Br)   Carbon (C)

Sulfur (S)   Phosphorus (P)

▲ **Figure 2.15** Examples of metals and nonmetals.

*All metals become liquids if heated sufficiently. Hg simply has the lowest melting point of any metallic element. Although sodium (Na), potassium (K), rubidium (Rb), cesium (Cs), and gallium (Ga) are solids at room temperature, they all melt at temperatures below 100 °C.

the elements that lie along the line that separates metals from nonmetals have properties that fall between those of metals and nonmetals. These elements are often referred to as **metalloids**.

Sample Exercise 2.5

Using the Periodic Table

Which two of these elements would you expect to show the greatest similarity in chemical and physical properties: B, Ca, F, He, Mg, P?

SOLUTION

Elements in the same group of the periodic table are most likely to exhibit similar properties. We therefore expect Ca and Mg to be most alike because they are in the same group (2A, the alkaline earth metals).

▶ **Practice Exercise**

Locate Na (sodium) and Br (bromine) in the periodic table. Give the atomic number of each and classify each as metal, metalloid, or nonmetal.

Self-Assessment Exercises

SAE 2.14 Find manganese on the periodic table. Which of the following statements is true about this element? (**a**) Manganese has atomic number 12. (**b**) Manganese is in period 7. (**c**) Manganese is an alkaline earth metal. (**d**) None of statements (a), (b), (c) is true.

SAE 2.15 Which of the following statements is or are true?

(**i**) Elements in the same period of the periodic table are expected to have chemically similar properties.

(**ii**) Elements in the same group of the periodic table are expected to have chemically similar properties.
(**iii**) Elements in the same period of the periodic table are arranged in order of increasing atomic number.
(**a**) i only (**b**) ii only (**c**) iii only (**d**) i and iii (**e**) ii and iii

2.6 | Molecules and Molecular Compounds

Learning Objectives

When you finish Section 2.6, you should be able to:
▶ Write the molecular formula of a compound given a picture of its structure.
▶ Differentiate between the molecular and empirical formulas of a molecular compound.

Even though the atom is the smallest representative sample of an element, only the noble-gas elements are normally found in nature as isolated atoms. Most matter is composed of molecules or ions. We examine molecules here in Section 2.6 and ions in Section 2.7.

Molecules and Chemical Formulas

Several elements are found in nature in molecular form—two or more of the same type of atom bound together. For example, most of the oxygen in air consists of molecules that contain two oxygen atoms. As we saw in Section 1.2, we represent this molecular oxygen by the **chemical formula** O_2 (read "oh two"). The subscript tells us that two oxygen atoms are present in each molecule. A molecule made up of two atoms is called a **diatomic molecule**.

Oxygen also exists in another molecular form known as *ozone*. Molecules of ozone consist of three oxygen atoms, making the chemical formula O_3. Even though "normal" oxygen (O_2) and ozone (O_3) are both composed only of oxygen atoms, they exhibit very different chemical and physical properties. For example, O_2 is essential for life, but O_3 is toxic; O_2 is odorless, whereas O_3 has a sharp, pungent smell.

The elements that normally occur as diatomic molecules are hydrogen, oxygen, nitrogen, and the halogens ($H_2, O_2, N_2, F_2, Cl_2, Br_2,$ and I_2). Except for hydrogen, these diatomic elements are clustered on the right side of the periodic table.

Compounds composed of molecules that contain more than one type of atom are called **molecular compounds** or **molecular substances**. A molecule of the compound methane, for example, consists of one carbon atom and four hydrogen atoms and is therefore represented by the chemical formula CH_4. Lack of a subscript on the C indicates one atom of C per methane molecule. Several common molecules of both elements and compounds are shown in **Figure 2.16**. Notice how the composition of each substance is given by its chemical formula. Notice also that these substances are composed only of nonmetallic elements.

Most of the molecular substances we discuss contain only nonmetals.

Molecular and Empirical Formulas

Chemical formulas that indicate the actual numbers of atoms in a molecule, such as those in Figure 2.16, are called **molecular formulas**. Chemical formulas that give only the relative number of atoms of each type in a molecule are called **empirical formulas**. The subscripts in an empirical formula are always the smallest possible whole-number ratios. The molecular formula for hydrogen peroxide is H_2O_2, for example, whereas its empirical formula is HO. The molecular formula for ethylene is C_2H_4, so its empirical formula is CH_2. For many substances, the molecular formula and the empirical formula are identical, as in the case of water, H_2O.

Whenever we know the molecular formula of a compound, we can determine its empirical formula. The converse is not true, however. If we know the empirical formula of a substance, we cannot determine its molecular formula unless we have more information. Why, then, do chemists bother with empirical formulas? As we explain in Chapter 3, common experimental methods of analyzing substances lead to the empirical formula only. For example, if you decomposed hydrogen peroxide H_2O_2 into its elements and weighed them, you could determine that there were equal numbers of hydrogen and oxygen atoms, but you would not know if the molecular formula was HO, H_2O_2, H_3O_3, or the like. Once the empirical formula is known, additional experiments can give the information needed to convert the empirical formula to the molecular one. In addition, many substances do not exist as isolated molecules; major examples are the ionic compounds that are discussed later in this chapter. For these substances, we must rely on empirical formulas.

Hydrogen, H_2 Oxygen, O_2

Water, H_2O Hydrogen peroxide, H_2O_2

Carbon monoxide, CO   Carbon dioxide, CO_2

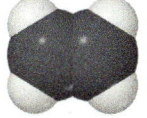

Methane, CH_4 Ethylene, C_2H_4

▲ Figure 2.16 **Molecular models.** Notice how the chemical formulas of these simple molecules correspond to their compositions.

 Sample Exercise 2.6

Relating Empirical and Molecular Formulas

Write the empirical formulas for (**a**) glucose, a substance also known as either blood sugar or dextrose—molecular formula $C_6H_{12}O_6$; and (**b**) nitrous oxide, a substance used as an anesthetic and commonly called laughing gas—molecular formula N_2O.

SOLUTION

(**a**) The subscripts of an empirical formula are the smallest whole-number ratios. The smallest ratios are obtained by dividing each subscript by the largest common factor, in this case 6. The resultant empirical formula for glucose is CH_2O.

(**b**) Because the subscripts in N_2O are already the lowest integral numbers, the empirical formula for nitrous oxide is the same as its molecular formula, N_2O.

▶ **Practice Exercise**
Give the empirical formula for *decaborane*, whose molecular formula is $B_{10}H_{14}$.

Picturing Molecules

The molecular formula of a substance does not show how its atoms are joined together. A **structural formula** is needed to convey that information, as in the following examples:

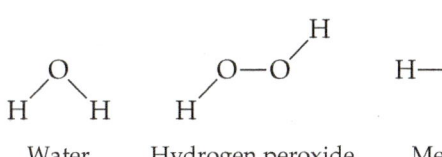

Water   Hydrogen peroxide   Methane

Go Figure Which molecular model, the ball-and-stick or the space-filling, more effectively shows the angles between bonds around a central atom?

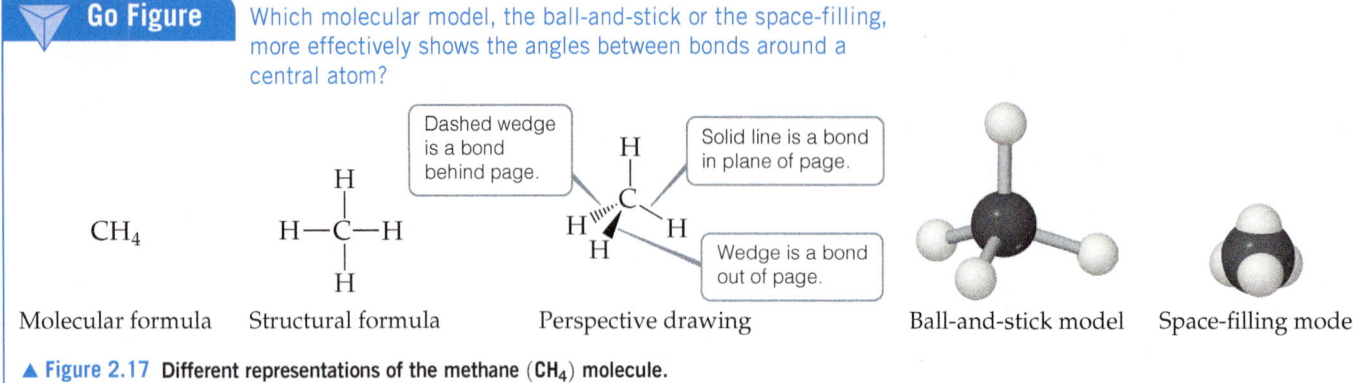

CH_4

Molecular formula   Structural formula   Perspective drawing   Ball-and-stick model   Space-filling model

Dashed wedge is a bond behind page.

Solid line is a bond in plane of page.

Wedge is a bond out of page.

▲ **Figure 2.17 Different representations of the methane (CH_4) molecule.**

The atoms are represented by their chemical symbols, and lines are used to represent the bonds that hold the atoms together. The observation that the angles between the bonds are not all the same indicates something about the detailed geometric structure of molecules, as we now discuss.

A structural formula does not typically depict the actual geometry of the molecule—that is, the actual angles at which atoms are joined; for that depiction, more sophisticated representations are needed (**Figure 2.17**).

- **Perspective drawings** use wedges and dashed wedges to depict bonds that are not in the plane of the paper. This gives a crude sense of the three-dimensional shape of a molecule. This is the most common drawing that chemists use. For a molecule in which all the atoms lie in one plane, such as H_2O, we do not need to use wedges and dashed wedges.

- **Ball-and-stick models** show atoms as spheres and bonds as sticks. This type of model has the advantage of accurately representing the angles at which the atoms are attached to one another in a molecule and the relative lengths of the bonds that hold the atoms together (Figure 2.17). Sometimes the chemical symbols of the elements are superimposed on the balls, but often the atoms are identified simply by color.

- **Space-filling models** depict what a molecule would look like if the atoms were scaled up in size (Figure 2.17). These models show the relative sizes of the atoms, but the angles between atoms, which help define their molecular geometry, are often more difficult to see than in ball-and-stick models. Because space-filling models give a good representation of the true size of a molecule, they are useful for picturing how two molecules might fit together or pack in the solid state. As with ball-and-stick models, the identities of the atoms are typically indicated by color.

Self-Assessment Exercises

SAE 2.16 What is the molecular formula of this compound?

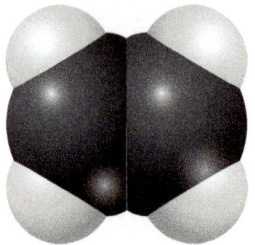

(**a**) CH_4 (**b**) C_2H_4 (**c**) C_2O_4 (**d**) O_2H_4

SAE 2.17 Oleic acid is a common fatty acid in nature. A shorthand way to write its chemical structure is $CH_3(CH_2)_7CHCH(CH_2)_7COOH$. What is oleic acid's empirical formula? (**a**) C_3H_5O (**b**) $C_6H_{10}O_2$ (**c**) $C_9H_{17}O$ (**d**) $C_{18}H_{34}O_2$

SAE 2.18 Acetylene's molecular formula is C_2H_2; benzene's is C_6H_6. Which statement about these two molecules is *false*? (**a**) Both acetylene and benzene have the same empirical formula. (**b**) The ratio of C/H masses in both benzene and acetylene is the same. (**c**) The number of carbon atoms in one benzene molecule is the same as the number of carbon atoms in one acetylene molecule. (**d**) Benzene is a bigger molecule than acetylene.

2.7 | Ions and Ionic Compounds

If electrons are removed from or added to an atom, a charged particle called an **ion** is formed. An ion with a positive charge is a **cation** (pronounced CAT-ion), whereas a negatively charged ion is an **anion** (AN-ion).

To see how ions form, consider the sodium atom, which has 11 protons and 11 electrons. This atom easily loses one electron. The resulting cation has 11 protons and 10 electrons, which means it has a net charge of 1+.

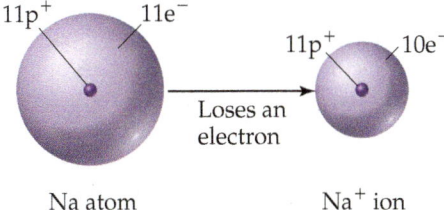

$11p^+$ $11e^-$

$11p^+$ $10e^-$

Loses an electron

Na atom Na^+ ion

The net charge on an ion is represented by a superscript. The superscripts +, 2+, and 3+, for instance, mean a net charge resulting from the *loss* of one, two, and three electrons, respectively. The superscripts −, 2−, and 3− represent net charges resulting from the *gain* of one, two, and three electrons, respectively. Chlorine, with 17 protons and 17 electrons, for example, can gain an electron in chemical reactions, producing the Cl^- ion:

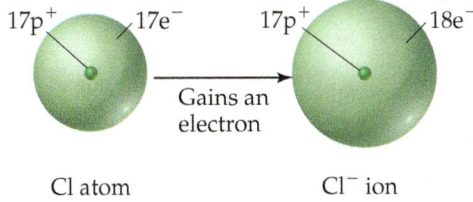

$17p^+$ $17e^-$

$17p^+$ $18e^-$

Gains an electron

Cl atom Cl^- ion

In general, metal atoms tend to lose electrons to form cations, whereas nonmetal atoms tend to gain electrons to form anions. Thus, ionic compounds tend to be composed of both metal cations and nonmetal anions, as in NaCl.

Learning Objectives

When you finish Section 2.7, you should be able to:

▶ Write chemical symbols for ions and predict the most common ionic charges for elements toward the left and right sides of the periodic table.

▶ Predict the empirical formula of an ionic compound given its constituent elements.

▶ Distinguish between molecular and ionic compounds based on the chemical formula of the compound.

Sample Exercise 2.7

Writing Chemical Symbols for Ions

Give the chemical symbol, including superscript indicating mass number, for (**a**) the ion with 22 protons, 26 neutrons, and 19 electrons; and (**b**) the ion of sulfur that has 16 neutrons and 18 electrons.

SOLUTION

(**a**) The number of protons is the atomic number of the element. According to the periodic table or a list of elements, the element with atomic number 22 is titanium (Ti). The mass number (protons plus neutrons) of this isotope of titanium is $22 + 26 = 48$. Because the ion has three more protons than electrons, it has a net charge of 3+ and is designated $^{48}Ti^{3+}$.

(**b**) The periodic table shows that sulfur (S) has an atomic number of 16. Thus, each atom or ion of sulfur contains 16 protons. We are told that the ion also has 16 neutrons, meaning the mass number is $16 + 16 = 32$. Because the ion has 16 protons and 18 electrons, its net charge is 2− and the ion symbol is $^{32}S^{2-}$.

In general, we focus on the net charges of ions and ignore their mass numbers unless the circumstances dictate that we specify a certain isotope.

▶ **Practice Exercise**

How many protons, neutrons, and electrons does the $^{79}Se^{2-}$ ion possess?

In addition to simple ions such as Na^+ and Cl^-, there are **polyatomic ions**, such as NH_4^+ (ammonium ion) and SO_4^{2-} (sulfate ion), which consist of atoms joined as in a molecule, but carrying a net positive or negative charge. Polyatomic ions are discussed in Section 2.8.

It is important to realize that the chemical properties of ions are very different from the chemical properties of the atoms from which the ions are derived. The addition or removal of one or more electrons produces a charged species with behavior very different from that of its associated atom or group of atoms. For example, sodium metal reacts violently with water, but an ionic compound containing sodium ions, such as NaCl, does not.

Predicting Ionic Charges

The noble gases (group 8A—see Table 2.3) are chemically nonreactive elements that form very few compounds. Many atoms gain or lose electrons to end up with the same number of electrons as the noble gas closest to them in the periodic table. We might deduce that atoms tend to acquire the electron arrangements of the noble gases because these electron arrangements are very stable. Nearby elements can obtain these same stable arrangements by losing or gaining electrons. For example, the loss of one electron from an atom of sodium leaves it with the same number of electrons as in a neon atom (10). Similarly, when chlorine gains an electron, it ends up with 18, the same number of electrons as in argon. This simple observation will be helpful for now to account for the formation of ions. A deeper explanation awaits us in Chapter 8, where we discuss chemical bonding.

Sample Exercise 2.8
Predicting Ionic Charge

Predict the charge expected for the most stable ion of barium and the most stable ion of oxygen.

SOLUTION

We will assume that barium and oxygen form ions that have the same number of electrons as the nearest noble-gas atom. From the periodic table, we see that barium has atomic number 56. The nearest noble gas is xenon, atomic number 54. Barium can attain a stable arrangement of 54 electrons by losing two electrons, forming the Ba^{2+} cation.

Oxygen has atomic number 8. The nearest noble gas is neon, atomic number 10. Oxygen can attain this stable electron arrangement by gaining two electrons, forming the O^{2-} anion.

▶ **Practice Exercise**
Predict the charge expected for the most stable ion of (**a**) aluminum and (**b**) fluorine.

The periodic table is useful for remembering ionic charges, especially those of elements on the left and right sides of the table. As **Figure 2.18** shows, the charges of these ions relate in a simple way to their positions in the table: The group 1A elements (alkali metals) form 1+ ions, the group 2A elements (alkaline earth metals) form 2+ ions, the group 7A elements (halogens) form 1− ions, and the group 6A elements form 2− ions. Many of the other groups do not lend themselves to such simple rules.

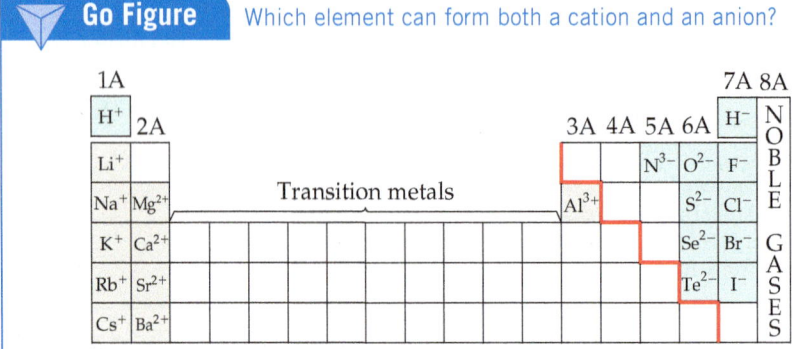

▲ **Figure 2.18 Predictable charges of some common ions.** Notice that the red stepped line that divides metals from nonmetals also separates cations from anions. Hydrogen forms both 1+ and 1− ions. The H anion is shown in group7A, but that is just for convenience to show the common charges and reactivity; an element only appears once in the full periodic table.

Ionic Compounds

A great deal of chemical activity involves the transfer of electrons from one substance to another. **Figure 2.19** shows that when elemental sodium is allowed to react with elemental chlorine, an electron transfers from a sodium atom to a chlorine atom, forming a Na^+ ion and a Cl^- ions. Because objects of opposite charges attract, the Na^+ and the Cl^- ions bind together to form the compound sodium chloride (NaCl). Sodium chloride, which we know better as common table salt, is an example of an **ionic compound**, a compound made up of cations and anions.

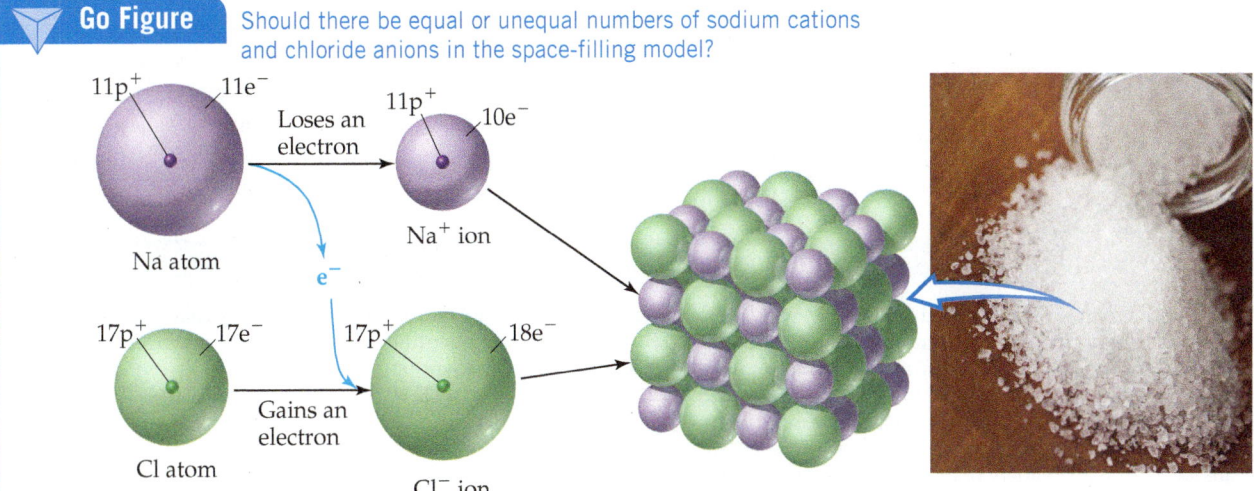

Go Figure Should there be equal or unequal numbers of sodium cations and chloride anions in the space-filling model?

$11p^+$ $11e^-$ Loses an electron $11p^+$ $10e^-$

Na atom e^- Na^+ ion

$17p^+$ $17e^-$ Gains an electron $17p^+$ $18e^-$

Cl atom Cl^- ion

▲ **Figure 2.19 Formation of an ionic compound.** The transfer of an electron from a sodium atom to a chlorine atom leads to the formation of a Na^+ ion and a Cl^- ion. These ions are arranged in a lattice in solid sodium chloride, NaCl. This representation of solid NaCl is a space-filling one that shows the relative sizes of the ions.

We can often tell whether a compound is ionic (consisting of ions) or molecular (consisting of molecules) from its composition. As a general rule, cations are metal ions and anions are nonmetal ions. Consequently,

Ionic compounds are generally combinations of metals and nonmetals, as in NaCl.

Sample Exercise 2.9

Identifying Ionic and Molecular Compounds

Which of these compounds would you expect to be ionic: N_2O, Na_2O, $CaCl_2$, SF_4?

SOLUTION

We predict that Na_2O and $CaCl_2$ are ionic compounds because they are composed of a metal combined with a nonmetal. We predict (correctly) that N_2O and SF_4 are molecular compounds because they are composed entirely of nonmetals.

▶ **Practice Exercise**
Explain why each of the following statements is probably true:
(a) Every compound of Rb with a nonmetal is ionic in character.
(b) Every compound of nitrogen with a halogen element is a molecular compound.
(c) The compound $MgKr_2$ does not exist.
(d) Na and K are very similar in the compounds they form with nonmetals.
(e) If contained in an ionic compound, calcium (Ca) will be in the form of the doubly charged ion, Ca^{2+}.

The ions in ionic compounds are arranged in three-dimensional structures, as Figure 2.19 shows for NaCl. Because there is no discrete "molecule" of NaCl, we are able to write only an empirical formula for this substance. This is true for most other ionic compounds.

We can write the empirical formula for an ionic compound if we know the charges of the ions. Because chemical compounds are always electrically neutral, the ions in an ionic compound always occur in such a ratio that the total positive charge equals the total negative charge. Thus, there is one Na^+ to one Cl^- in NaCl, one Ba^{2+} to two Cl^- in $BaCl_2$, and so forth.

As you consider these and other examples, you will see that if the charges on the cation and anion are equal, the subscript on each ion is 1. If the charges are unequal, the charge on one ion (without its sign) will become the subscript on the other ion. For example, the ionic compound formed from Mg (which forms Mg^{2+} ions) and N (which forms N^{3-} ions) is Mg_3N_2:

$$Mg^{2+} \quad N^{3-} \longrightarrow Mg_3N_2$$

There is one caveat when you are using this approach. Remember that the empirical formula should be the smallest possible whole-number ratio of the two elements. So the empirical formula for the ionic compound formed between Ti^{4+} and O^{2-} is TiO_2 rather than Ti_2O_4.

CHEMISTRY AND LIFE | **Elements Required by Living Organisms**

The elements essential to life are highlighted in color in **Figure 2.20**. More than 97% of the mass of most organisms is made up of just six of these elements—oxygen, carbon, hydrogen, nitrogen, phosphorus, and sulfur. Water is the most common compound in living organisms, accounting for at least 70% of the mass of most cells. In the solid components of cells, carbon is the most prevalent element by mass. Carbon atoms are found in a vast variety of organic molecules, bonded either to other carbon atoms or to atoms of other elements. Nearly all proteins, for example, contain the carbon-based group

which occurs repeatedly in the molecules.

In addition, 23 other elements have been found in various living organisms. Five are ions required by all organisms: Ca^{2+}, Cl^-, Mg^{2+}, K^+, and Na^+. Calcium ions, for example, are necessary for the formation of bone and the transmission of nervous system signals. Many other elements are needed in only very small quantities and consequently are called *trace* elements. For example, trace quantities of copper are required in the diet of humans to aid in the synthesis of hemoglobin.

Related Exercise: 2.104

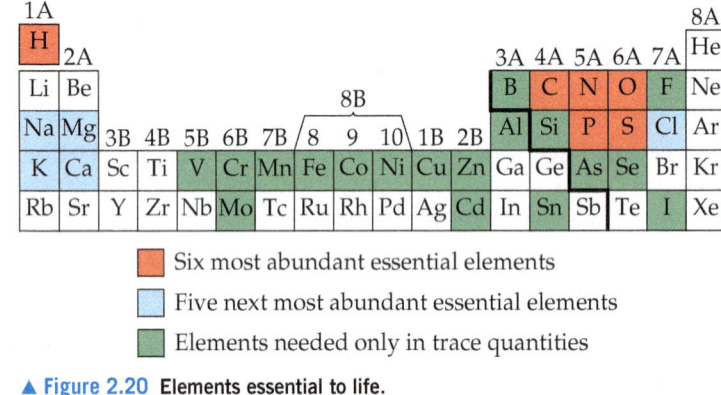

- ■ Six most abundant essential elements
- ■ Five next most abundant essential elements
- ■ Elements needed only in trace quantities

▲ **Figure 2.20 Elements essential to life.**

Sample Exercise 2.10

Using Ionic Charge to Write Empirical Formulas for Ionic Compounds

Write the empirical formula of the compound formed by **(a)** Al^{3+} and Cl^- ions, **(b)** Al^{3+} and O^{2-} ions, **(c)** Mg^{2+}, and NO_3^- ions.

SOLUTION

(a) Three Cl^- ions are required to balance the charge of one Al^{3+} ion, making the empirical formula $AlCl_3$.

(b) Two Al^{3+} ions are required to balance the charge of three O^{2-} ions. A 2:3 ratio is needed to balance the total positive charge of $6+$ and the total negative charge of $6-$. The empirical formula is Al_2O_3.

(c) Two NO_3^- ions are needed to balance the charge of one Mg^{2+}, yielding $Mg(NO_3)_2$. The formula for the polyatomic ion, NO_3^-, must be enclosed in parentheses so that it is clear that the subscript 2 applies to all the atoms of that ion.

▶ **Practice Exercise**

Write the empirical formula for the compound formed by **(a)** Na^+ and PO_4^{3-}, **(b)** Zn^{2+} and SO_4^{2-}, **(c)** Fe^{3+} and CO_3^{2-}.

Self-Assessment Exercises

SAE 2.19 Which pair of element and most common ionic charge is correct? (**a**) F/1+ (**b**) Na/1− (**c**) C/6+ (**d**) Ba/2+

SAE 2.20 Which of the following statements is *false*? (**a**) Elements tend to lose or gain electrons in order to have the same number of electrons as the nearest noble gas. (**b**) Nonmetals tend to form anions. (**c**) The group 7A elements tend for form 1+ ions. (**d**) Metals tend to form cations. (**e**) Ionic compounds generally contain both a metal and a nonmetal.

SAE 2.21 Predict the empirical formula of strontium chloride. (**a**) SrCl (**b**) $SrCl_2$ (**c**) $SrCl_3$ (**d**) SCl (**e**) SCl_2

SAE 2.22 Which pair of compounds/types is correct? (**a**) SO_2, ionic compound (**b**) MnO_2, molecular compound (**c**) TiO_2, molecular compound (**d**) CF_4, ionic compound (**e**) NaH, ionic compound

2.8 | Naming Inorganic Compounds

The names and chemical formulas of compounds are essential vocabulary in chemistry. The system used in naming substances is called **chemical nomenclature**, from the Latin words *nomen* (name) and *calare* (to call).

There are more than 163 million known chemical substances. Naming them all would be a hopelessly complicated task if each had a name independent of all others. Many important substances that have been known for a long time, such as water (H_2O) and ammonia (NH_3), do have traditional names (called *common names*). For most substances, however, we rely on a set of rules that leads to an informative and unique name that conveys the composition of the substance.

The rules for chemical nomenclature are based on the division of substances into categories. The major division is between organic and inorganic compounds. *Organic compounds* contain carbon and hydrogen, often in combination with oxygen, nitrogen, or other elements. All others are *inorganic compounds*. Early chemists associated organic compounds with plants and animals and inorganic compounds with the nonliving portion of our world. Although this distinction is no longer pertinent, the classification between organic and inorganic compounds continues to be useful. In this section, we consider the basic rules for naming three categories of inorganic compounds: ionic compounds, acids, and molecular compounds.

 Learning Objectives

When you finish Section 2.8, you should be able to:

▶ Interconvert between the name and empirical formula of an ionic compound.

▶ Interconvert between the name and chemical formula of an inorganic acid.

▶ Interconvert between the name and chemical formula of binary molecular compounds.

Names and Formulas of Ionic Compounds

Recall from Section 2.7 that ionic compounds usually consist of metal cations combined with nonmetal anions.

1. Cations

a. *Cations formed from metal atoms have the same name as the metal:*

Na^+ sodium ion Zn^{2+} zinc ion Al^{3+} aluminum ion

b. *If a metal can form cations with different charges, the positive charge is indicated by a Roman numeral in parentheses following the name of the metal:*

Fe^{2+}	iron(II) ion	Cu^+	copper(I) ion
Fe^{3+}	iron(III) ion	Cu^{2+}	copper(II) ion

Ions of the same element that have different charges have different chemical and physical properties, such as color (**Figure 2.21**).

Most metals that form cations with different charges are transition metals, elements that occur in the middle of the periodic table, from group 3B to group 2B (as indicated on the periodic table on the front inside cover of this book). The metals that form only one cation (only one possible charge) are those of group 1A and group 2A, as well as Al^{3+} (group 3A). The transition-metal ions Ag^+

▲ Figure 2.21 **Different ions of the same element have different properties.** Both substances shown are copper oxides. One is cuprous oxide, Cu_2O; the other is cupric oxide, CuO.

(group 1B), Zn^{2+} (group 2B), and Cd^{2+} (group 2B) also form only one cation in most situations. Charges are not normally expressed when naming these ions. However, if there is any doubt in your mind whether a metal forms more than one cation, use a Roman numeral to indicate the charge. It is never wrong to do so, even though it may be unnecessary.

An older method still widely used for distinguishing between differently charged ions of a metal uses the suffixes *-ous* and *-ic* added to the root of the element's Latin name:

Fe^{2+}	ferrous ion	Cu^+	cuprous ion
Fe^{3+}	ferric ion	Cu^{2+}	cupric ion

Although we only rarely use these older names in this text, you might encounter them elsewhere.

c. *Cations formed from molecules composed of nonmetal atoms have names that end in -ium:*

NH_4^+	ammonium ion	H_3O^+	hydronium ion

These two ions are the only ions of this kind that we encounter frequently in the text.

The names and formulas of some common cations are listed in **Table 2.4** and on the back inside cover of the text. Those on the left side in Table 2.4 are the monatomic ions that do not have more than one possible charge. Those on the right side are either polyatomic cations or cations with more than one possible charge. The Hg_2^{2+} ion is unusual because, even though it is a metal ion, it is not monatomic. It is called the mercury(I) ion because it can be thought of as two Hg^+ ions bound together. The cations that you will encounter most frequently in this text are shown in boldface. You should learn these cations first.

TABLE 2.4 Common Cations[a]

Charge	Formula	Name	Formula	Name
1+	**H^+**	**hydrogen ion**	**NH_4^+**	**ammonium ion**
	Li^+	lithium ion	Cu^+	copper(I) or cuprous ion
	Na^+	**sodium ion**		
	K^+	**potassium ion**		
	Cs^+	cesium ion		
	Ag^+	**silver ion**		
2+	**Mg^{2+}**	**magnesium ion**	Co^{2+}	cobalt(II) or cobaltous ion
	Ca^{2+}	**calcium ion**	**Cu^{2+}**	**copper(II)** or cupric ion
	Sr^{2+}	strontium ion	**Fe^{2+}**	**iron(II)** or ferrous ion
	Ba^{2+}	barium ion	Mn^{2+}	manganese(II) or manganous ion
	Zn^{2+}	**zinc ion**	Hg_2^{2+}	mercury(I) or mercurous ion
	Cd^{2+}	cadmium ion	Hg^{2+}	mercury(II) or mercuric ion
			Ni^{2+}	nickel(II) or nickelous ion
			Pb^{2+}	**lead(II)** or plumbous ion
			Sn^{2+}	tin(II) or stannous ion
3+	**Al^{3+}**	**aluminum ion**	Cr^{3+}	chromium(III) or chromic ion
			Fe^{3+}	**iron(III)** or ferric ion

[a]The ions we use most often in this book are in boldface. Learn them first.

2. Anions

a. *The names of monatomic anions are formed by replacing the ending of the name of the element with -ide:*

H^- hyd**ride** ion	O^{2-} ox**ide** ion	N^{3-} nit**ride** ion

A few polyatomic anions also have names ending in *-ide*:

OH^- hydrox**ide** ion	CN^- cyan**ide** ion	O_2^{2-} perox**ide** ion

b. *Polyatomic anions containing oxygen have names ending in either -ate or -ite and are called* **oxyanions**. The suffix *-ate* is used for the most common or representative oxyanion of an element, whereas *-ite* is used for an oxyanion that has the same charge but one O atom fewer:

NO_3^-	nit**rate** ion	SO_4^{2-}	sul**fate** ion
NO_2^-	nit**rite** ion	SO_3^{2-}	sul**fite** ion

Prefixes are used when the series of oxyanions of an element extends to four members, as with the halogens. The prefix *per-* indicates one more O atom than the oxyanion ending in *-ate*; *hypo-* indicates one O atom fewer than the oxyanion ending in *-ite*:

ClO_4^-	**per**chlor**ate** ion (one more O atom than chlorate)
ClO_3^-	chlor**ate** ion
ClO_2^-	chlor**ite** ion (one O atom fewer than chlorate)
ClO^-	**hypo**chlor**ite** ion (one O atom fewer than chlorite)

These rules are summarized in **Figure 2.22**.

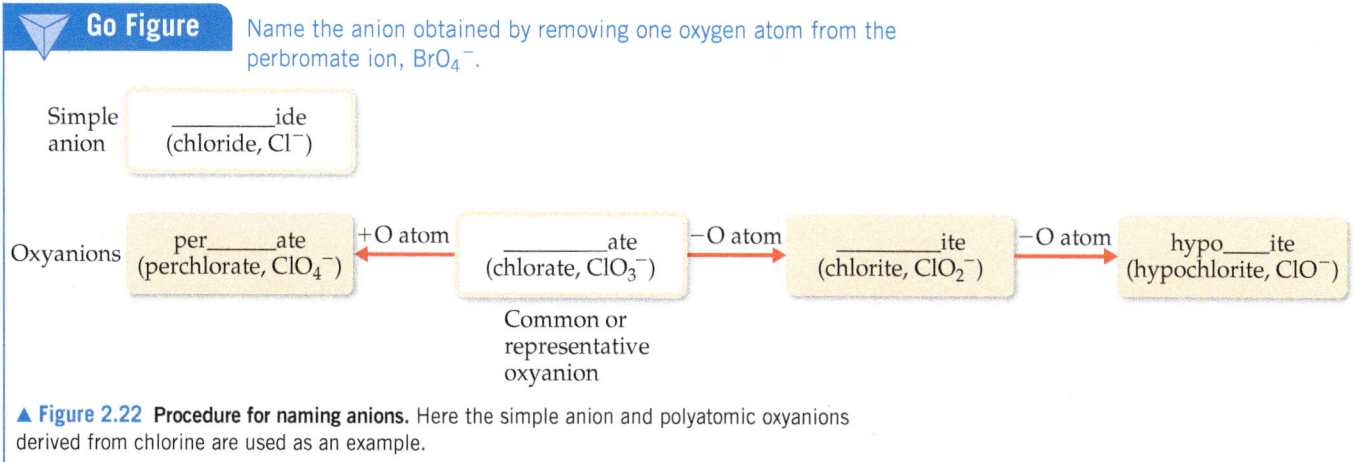

Go Figure   Name the anion obtained by removing one oxygen atom from the perbromate ion, BrO_4^-.

▲ **Figure 2.22 Procedure for naming anions.** Here the simple anion and polyatomic oxyanions derived from chlorine are used as an example.

Figure 2.23 can help you remember the charge and number of oxygen atoms in the various oxyanions. Notice that when the central atom is from the second period (C and N), the maximum number of oxygen atoms found in an oxyanion is three, whereas when the central atom is from period 3 (or greater), the maximum number of oxygen atoms increases to four.

Beginning at the lower right in Figure 2.23, note that ionic charge increases from right to left, from 1– for ClO_4^- to 3– for PO_4^{3-}. In the second period, the charges also increase from right to left, from 1– for NO_3^- to 2– for CO_3^{2-}. Notice also that although each of the anions in Figure 2.24 ends in *-ate*, the ClO_4^- ion also has a *per-* prefix.

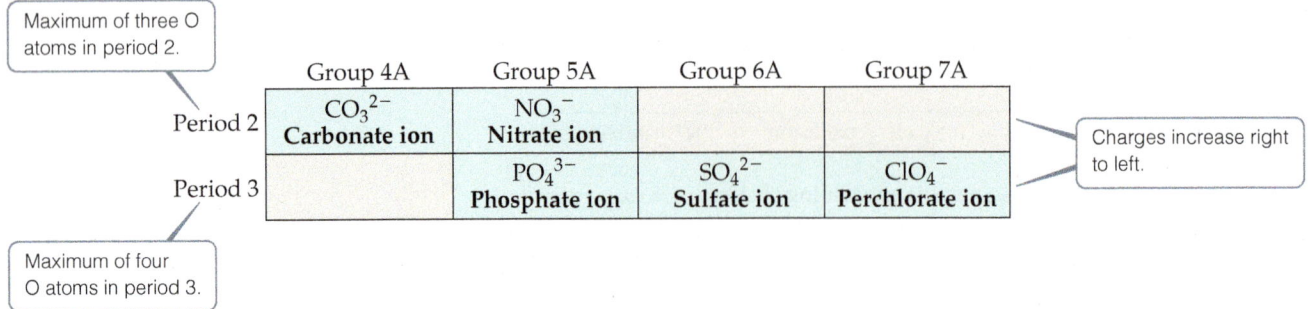

▲ **Figure 2.23 Common oxyanions.** The composition and charges of common oxyanions are related to their location in the periodic table.

Sample Exercise 2.11
Determining the Formula of an Oxyanion from Its Name

Based on the formula for the sulfate ion, predict the formula for (**a**) the selenate ion and (**b**) the selenite ion. (Sulfur and selenium are both in group 6A and form analogous oxyanions.)

SOLUTION

(**a**) The sulfate ion is SO_4^{2-}. The analogous selenate ion is therefore SeO_4^{2-}.

(**b**) The ending *-ite* indicates an oxyanion with the same charge but one O atom fewer than the corresponding oxyanion that ends in *-ate*. Thus, the formula for the selenite ion is SeO_3^{2-}.

▶ **Practice Exercise**

The formula for the bromate ion is analogous to that for the chlorate ion. Write the formula for the hypobromite and bromite ions.

c. *Anions derived by adding H^+ to an oxyanion are named by adding as a prefix the word* hydrogen *or* dihydrogen, *as appropriate:*

CO_3^{2-}	carbonate ion	PO_4^{3-}	phosphate ion
HCO_3^-	hydrogen carbonate ion	$H_2PO_4^-$	dihydrogen phosphate ion

Notice that each H^+ added reduces the negative charge of the parent anion by one. An older method for naming some of these ions uses the prefix *bi-*. Thus, the HCO_3^- ion is commonly called the bicarbonate ion, and HSO_4^- is sometimes called the bisulfate ion.

The names and formulas of the common anions are listed in **Table 2.5** and on the back inside cover of the text. Those anions whose names end in *-ide* are listed on the left portion of Table 2.5, whereas those whose names end in *-ate* are listed on the right. The most common of these ions are shown in boldface. You should learn the names and formulas of the boldfaced anions first. The formulas of the ions whose names end with *-ite* can be derived from those ending in *-ate* by removing an O atom. Notice the location of the monatomic ions in the periodic table. Those of group 7A always have a 1− charge (F^-, Cl^-, Br^-, and I^-), and those of group 6A have a 2− charge (O^{2-} and S^{2-}).

3. Ionic Compounds

Names of ionic compounds consist of the cation name followed by the anion name:

$CaCl_2$	calcium chloride
$Al(NO_3)_3$	aluminum nitrate
$Cu(ClO_4)_2$	copper(II) perchlorate (or cupric perchlorate)

In the chemical formulas for aluminum nitrate and copper(II) perchlorate, parentheses followed by the appropriate subscript are used because the compounds contain two or more polyatomic ions.

TABLE 2.5   Common Anions[a]

Charge	Formula	Name	Formula	Name
1−	H^-	hydride ion	**CH_3COO^-** (or $C_2H_3O_2^-$)	**acetate ion**
	F^-	**fluoride ion**	ClO_3^-	chlorate ion
	Cl^-	**chloride ion**	**ClO_4^-**	**perchlorate ion**
	Br^-	**bromide ion**	**NO_3^-**	**nitrate ion**
	I^-	**iodide ion**	MnO_4^-	permanganate ion
	CN^-	cyanide ion		
	OH^-	**hydroxide ion**		
2−	**O^{2-}**	**oxide ion**	**CO_3^{2-}**	**carbonate ion**
	O_2^{2-}	peroxide ion	CrO_4^{2-}	chromate ion
	S^{2-}	**sulfide ion**	$Cr_2O_7^{2-}$	dichromate ion
			SO_4^{2-}	**sulfate ion**
3−	N^{3-}	nitride ion	**PO_4^{3-}**	**phosphate ion**

[a]The ions we use most often are in boldface. Learn them first.

Sample Exercise 2.12

Determining the Names of Ionic Compounds from Their Formulas

Name the ionic compounds (**a**) K_2SO_4, (**b**) $Ba(OH)_2$, (**c**) $FeCl_3$.

SOLUTION

In naming ionic compounds, it is important to recognize polyatomic ions and to determine the charge of cations with variable charge.

(**a**) The cation is K^+, the potassium ion, and the anion is SO_4^{2-}, the sulfate ion, making the name *potassium sulfate*. (If you thought the compound contained S^{2-} and O^{2-} ions, you failed to recognize the polyatomic sulfate ion.)

(**b**) The cation is Ba^{2+}, the barium ion, and the anion is OH^-, the hydroxide ion: barium hydroxide.

(**c**) You must determine the charge of Fe in this compound because an iron atom can form more than one cation. Because the compound contains three chloride ions, Cl^-, the cation must be Fe^{3+}, the iron(III), or ferric, ion. Thus, the compound is iron(III) chloride or ferric chloride.

▶ **Practice Exercise**
Name the ionic compounds (**a**) NH_4Br, (**b**) Cr_2O_3, (**c**) $Co(NO_3)_2$.

Names and Formulas of Inorganic Acids

Acids are an important class of hydrogen-containing compounds, and they are named in a special way. For our present purposes, an *acid* is a substance whose molecules yield hydrogen ions (H^+) when dissolved in water. Acids may be inorganic, such as hydrochloric acid (HCl); or they may be organic, such as acetic acid (CH_3COOH, where the COOH is the fragment that produces hydrogen ions in water, also creating CH_3COO^- anions). Right now we focus only on inorganic acids. When we encounter the chemical formula for an inorganic acid at this stage of the course, it will be written with H as the first element, as in HCl and H_2SO_4.

An acid is composed of an anion connected to enough H^+ ions to neutralize, or balance, the anion's charge. Thus, the SO_4^{2-} ion requires two H^+ ions, forming H_2SO_4. The name of an acid is related to the name of its anion, as summarized in Figure 2.24.

How to Name Inorganic Acids

1. *Acids containing anions whose names end in* -ide *are named by changing the* -ide *ending to* -ic, *adding the prefix* hydro- *to this anion name, and then following with the word* acid:

Anion	Corresponding Acid
Cl^- (chloride)	HCl (hydrochloric acid)
S^{2-} (sulfide)	H_2S (hydrosulfuric acid)

2. *Acids containing anions whose names end in* -ate *or* -ite *are named by changing* -ate *to* -ic *and* -ite *to* -ous, *then adding the word* acid. *Prefixes in the anion name are retained in the name of the acid:*

Anion		Corresponding Acid	
ClO_4^-	(perchlorate)	$HClO_4$	(perchloric acid)
ClO_3^-	(chlorate)	$HClO_3$	(chloric acid)
ClO_2^-	(chlorite)	$HClO_2$	(chlorous acid)
ClO^-	(hypochlorite)	$HClO$	(hypochlorous acid)

▶ **Figure 2.24 Procedure for naming inorganic acids.** Here inorganic acids containing chlorine are used as an example. Prefixes used for oxyanions, such as, *per-* and *hypo-*, are retained in the acids derived from those anions.

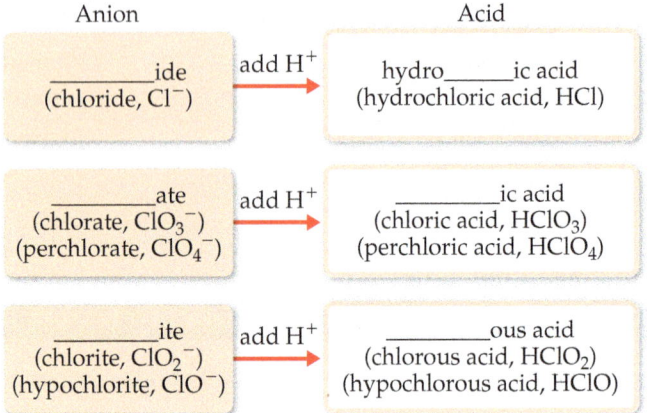

Sample Exercise 2.13

Relating the Names and Formulas of Inorganic Acids

Name the acids (a) HCN, (b) HNO_3, (c) H_2SO_4, (d) H_2SO_3.

SOLUTION

(a) The anion from which this acid is derived is CN^-, the cyanide ion. Because this ion has an *-ide* ending, the acid is given a *hydro-* prefix and an *-ic* ending: hydrocyanic acid. Only water solutions of HCN are referred to as hydrocyanic acid. The pure compound, which is a gas under normal conditions, is called hydrogen cyanide. Both hydrocyanic acid and hydrogen cyanide are *extremely* toxic.

(b) Because NO_3^- is the nitrate ion, HNO_3 is called nitric acid (the *-ate* ending of the anion is replaced with an *-ic* ending in naming the acid).

(c) Because SO_4^{2-} is the sulfate ion, H_2SO_4 is called sulfuric acid.

(d) Because SO_3^{2-} is the sulfite ion, H_2SO_3 is sulfurous acid (the *-ite* ending of the anion is replaced with an *-ous* ending).

▶ **Practice Exercise**
Give the chemical formulas for **(a)** hydrobromic acid, **(b)** carbonic acid.

Names and Formulas of Binary Molecular Compounds

The procedures used for naming *binary* (two-element) molecular compounds are similar to those used for naming ionic compounds:

How to Name Binary Molecular Compounds

1. *The name of the element farther to the left in the periodic table (closest to the metals) is usually written first. An exception occurs when the compound contains oxygen and chlorine, bromine, or iodine (any halogen except fluorine), in which case oxygen is written last.*

2. *If both elements are in the same group, the one closer to the bottom of the table is named first.*

3. *The name of the second element is given an* -ide *ending.*

4. *Greek prefixes* (Table 2.6) indicate the number of atoms of each element. (Exception: The prefix *mono-* is never used with the first element.) When the prefix ends in *a* or *o* and the name of the second element begins with a vowel, the *a* or *o* of the prefix is often dropped.

The following examples illustrate these rules:

Cl_2O	dichlorine monoxide	NF_3	nitrogen trifluoride
N_2O_4	dinitrogen tetroxide	P_4S_{10}	tetraphosphorus decasulfide

Rule 4 is necessary because we cannot predict formulas for most molecular substances the way we can for ionic compounds. Molecular compounds that contain hydrogen and one other element are an important exception, however. These compounds can be treated as if they were neutral substances containing H^+ ions and anions. Thus, you can predict that the substance named hydrogen chloride has the formula HCl, containing one H^+ to balance the charge of one Cl^-. (The name *hydrogen chloride* is used only for the pure compound; water solutions of HCl are called hydrochloric acid. The distinction, which is important, is explained in Section 4.1.) Similarly, the formula for hydrogen sulfide is H_2S because two H^+ ions are needed to balance the charge on S^{2-}.

TABLE 2.6   Prefixes Used in Naming Binary Compounds Formed between Nonmetals

Prefix	Meaning
mono-	1
di-	2
tri-	3
tetra-	4
penta-	5
hexa-	6
hepta-	7
octa-	8
nona-	9
deca-	10

Sample Exercise 2.14

Relating the Names and Formulas of Binary Molecular Compounds

Name the compounds (**a**) SO_2, (**b**) PCl_5, (**c**) Cl_2O_3.

SOLUTION

The compounds consist entirely of nonmetals, so they are molecular rather than ionic. Using the prefixes listed in Table 2.6, we have (**a**) sulfur dioxide, (**b**) phosphorus pentachloride, (**c**) dichlorine trioxide.

▶ **Practice Exercise**
Give the chemical formulas for (**a**) silicon tetrabromide, (**b**) disulfur dichloride, (**c**) diphosphorus hexaoxide.

Self-Assessment Exercises

SAE 2.23   What is the empirical formula for ammonium nitrite? (**a**) NH_3NO_3 (**b**) NH_4NO_3 (**c**) NH_3NO_2 (**d**) NH_4NO_2

SAE 2.24   Which acid has the most oxygen atoms in its chemical formula? (**a**) hypobromous acid (**b**) selenic acid (**c**) chlorous acid (**d**) nitric acid

SAE 2.25   Which molecular compound has the smallest number of atoms per molecule? (**a**) carbon disulfide (**b**) gallium trichloride (**c**) silicon tetrabromide (**d**) nitrogen monoxide (also called nitric oxide)

2.9 | Some Simple Organic Compounds

The study of compounds of carbon is called **organic chemistry**, and as noted earlier, compounds that contain carbon and hydrogen, often in combination with oxygen, nitrogen, or other elements, are called *organic compounds*. Organic compounds are a very important part of chemistry, far outnumbering all other types of chemical substances. We examine organic compounds in a systematic way in Chapter 24, but you will encounter many examples of them throughout the text. Here we present a brief introduction to some of the simplest organic compounds and the ways in which they are named.

Learning Objectives

When you finish Section 2.9, you should be able to:

▶ Interconvert between alkane name, condensed structural formula, structural formula, and molecular formula.

▶ Recognize and draw structural formulas for isomers of alkanes.

▶ Distinguish between alcohols and carboxylic acids.

▶ Recognize and draw structural formulas for isomers of alcohols and carboxylic acids.

Alkanes

Compounds that contain only carbon and hydrogen are called **hydrocarbons**. In the simplest class of hydrocarbons, **alkanes**, each carbon is bonded to four other atoms. The three smallest alkanes are methane (CH_4), ethane (C_2H_6), and propane (C_3H_8). The structural formulas of these three alkanes are as follows:

Methane   Ethane   Propane

Although hydrocarbons are binary molecular compounds, they are not named like the binary inorganic compounds discussed in Section 2.8. Instead, each alkane has a name that ends in *-ane*. The alkane with four carbons is called *butane*. For alkanes with five or more carbons, the names are derived from prefixes like those in Table 2.6. An alkane with eight carbon atoms, for example, is *octane* (C_8H_{18}), where the *octa-* prefix for eight is combined with the *-ane* ending for an alkane. The first 10 "straight-chain" alkanes, those with linear chains of carbons atoms, are listed in Table 2.7.

TABLE 2.7   **First 10 Members of the Straight-Chain Alkane Series**

Molecular Formula	Condensed Structural Formula	Name	Boiling Point (°C)
CH_4	CH_4	Methane	−161
C_2H_6	CH_3CH_3	Ethane	−89
C_3H_8	$CH_3CH_2CH_3$	Propane	−44
C_4H_{10}	$CH_3CH_2CH_2CH_3$	Butane	−0.5
C_5H_{12}	$CH_3CH_2CH_2CH_2CH_3$	Pentane	36
C_6H_{14}	$CH_3CH_2CH_2CH_2CH_2CH_3$	Hexane	68
C_7H_{16}	$CH_3CH_2CH_2CH_2CH_2CH_2CH_3$	Heptane	98
C_8H_{18}	$CH_3CH_2CH_2CH_2CH_2CH_2CH_2CH_3$	Octane	125
C_9H_{20}	$CH_3CH_2CH_2CH_2CH_2CH_2CH_2CH_2CH_3$	Nonane	151
$C_{10}H_{22}$	$CH_3CH_2CH_2CH_2CH_2CH_2CH_2CH_2CH_2CH_3$	Decane	174

The formulas for the alkanes listed in Table 2.7 are written in a notation called *condensed structural formulas*. This notation reveals the way in which atoms are bonded to one another but does not require drawing in all the bonds. For example, the structural formula and the condensed structural formulas for butane (C_4H_{10}) are

$$H_3C-CH_2-CH_2-CH_3$$

or

$$CH_3CH_2CH_2CH_3$$

Isomers

Compounds that have the same molecular formula but different arrangements of atoms are called **isomers**. There are many kinds of isomers in chemistry, as we explain throughout this book. Methane, ethane, and propane have only one isomer. But once we get to butane and larger alkanes, *branched-chain* hydrocarbons are possible and do exist (Table 2.8). These are also called **structural isomers**, meaning that the only difference between the compounds is their molecular structure.

TABLE 2.8   Isomers of C_4H_{10} and C_5H_{12}

Systematic Name (Common Name)	Structural Formula	Condensed Structural Formula	Space-Filling Model
Butane (*n*-butane)		$CH_3CH_2CH_2CH_3$	
2-Methylpropane (isobutane)		$CH_3{-}CH{-}CH_3$ $\quad\quad\quad\; CH_3$	
Pentane (*n*-pentane)		$CH_3CH_2CH_2CH_2CH_3$	
2-Methylbutane (isopentane)		$CH_3{-}CH{-}CH_2{-}CH_3$ $\quad\quad CH_3$	
2,2-Dimethylpropane (neopentane)		$CH_3{-}\overset{\displaystyle CH_3}{\underset{\displaystyle CH_3}{C}}{-}CH_3$	

The number of possible structural isomers increases rapidly with the number of carbon atoms in the alkane. There are 18 isomers with the molecular formula C_8H_{18}, for example, and 75 with the molecular formula $C_{10}H_{22}$.

Nomenclature of Alkanes

In the first column of Table 2.8, the names in parentheses are called the *common names*. The common name of the isomer with no branches begins with the letter *n* (indicating the "normal" structure). When one CH_3 group branches off the major chain, the common name of the isomer begins with *iso-*, and when two CH_3 groups branch off, the common name begins with *neo-*. As the number of isomers grows, however, it becomes impossible to find a suitable prefix to denote each isomer by a common name. The need for a systematic means of naming organic compounds was recognized as early as 1892, when an organization called the International Union of Chemistry met in Geneva to formulate rules for naming organic substances. Since that time, the task of updating the rules for naming compounds has fallen to the International Union of Pure and Applied

Chemistry (IUPAC). Chemists everywhere, regardless of their nationality, subscribe to a common system for naming compounds.

The IUPAC names for the isomers of butane and pentane are the ones given first in Table 2.8. These systematic names, as well as those of other organic compounds, have three parts:

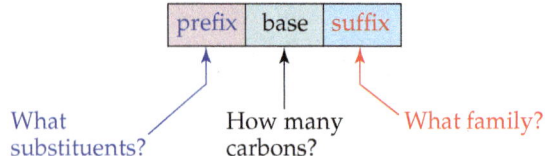

The following steps summarize the procedures used to name alkanes, all of which have names ending with *-ane*. We use a similar approach to write the names of other organic compounds.

How to Name Alkanes

1. *Find the longest continuous chain of carbon atoms, and use the name of this chain (given in Table 2.7) as the base name.* Be careful in this step because the longest chain may not be written in a straight line, as in the following structure:

$$CH_3 \overset{2}{-} CH \overset{1}{-} CH_3$$
$$\underset{3}{CH_2} \overset{}{-} \underset{4}{CH_2} \overset{}{-} \underset{5}{CH_2} \overset{}{-} \underset{6}{CH_3}$$

2-Methyl*hexane*

Because the longest continuous chain contains six C atoms, this isomer is named as a substituted hexane. Groups attached to the main chain are called *substituents* because they are substituted in place of an H atom on the main chain. In this molecule, the CH_3 group not enclosed by the blue outline is the only substituent in the molecule.

2. *Number the carbon atoms in the longest chain, beginning with the end nearest a substituent.* In our example, we number the C atoms beginning at the upper right because that places the CH_3 substituent on C2 of the chain. (If we had numbered from the lower right, the CH_3 would be on C5.) The chain is numbered from the end that gives the lower number to the substituent position.

3. *Name each substituent.* A substituent formed by removing an H atom from an alkane is called an **alkyl group**. Alkyl groups are named by replacing the *-ane* ending of the alkane name with *-yl*. The methyl group (CH_3), for example, is derived from methane (CH_4) and the ethyl group (C_2H_5) is derived from ethane (C_2H_6). Table 2.9 lists six common alkyl groups.

4. *Begin the name with the number or numbers of the carbon or carbons to which each substituent is bonded.* For our compound, the name *2-methylhexane* indicates the presence of a methyl group on C2 of a hexane (six-carbon) chain.

5. *When two or more substituents are present, list them in alphabetical order.* The presence of two or more of the same substituent is indicated by the prefixes *di-* (two), *tri-* (three), *tetra-* (four), *penta-* (five), and so forth. The prefixes are ignored in determining the alphabetical order of the substituents:

$$\overset{7}{CH_3}$$
$$CH_3 \overset{5}{-} CH \overset{6}{-} CH_2$$
$$\overset{4}{CH} \overset{3}{-} CH \overset{}{-} CH_2CH_3$$
$$CH_3 \overset{2}{-} CH \overset{}{-} CH_3$$
$$\overset{1}{CH_3}$$

3-Ethyl-2,4,5-trimethylheptane

TABLE 2.9 Condensed Structural Formulas and Common Names for Several Alkyl Groups

Group	Name
CH_3-	Methyl
CH_3CH_2-	Ethyl
$CH_3CH_2CH_2-$	Propyl
$CH_3CH_2CH_2CH_2-$	Butyl
$\begin{matrix} CH_3 \\ \vert \\ HC- \\ \vert \\ CH_3 \end{matrix}$	Isopropyl
$\begin{matrix} CH_3 \\ \vert \\ CH_3-C- \\ \vert \\ CH_3 \end{matrix}$	*tert*-Butyl

Sample Exercise 2.15
Naming Alkanes

What is the systematic name for the following alkane?

$$CH_3-CH_2-CH-CH_3$$
$$CH_3-CH-CH_2$$
$$CH_3-CH_2$$

SOLUTION

Analyze We are given the condensed structural formula of an alkane and asked to give its name.

Plan Because the hydrocarbon is an alkane, its name ends in *-ane*. The name of the parent hydrocarbon is based on the longest continuous chain of carbon atoms. Branches are alkyl groups, named after the number of C atoms in the branch and located by counting C atoms along the longest continuous chain.

Solve The longest continuous chain of C atoms extends from the upper left CH_3 group to the lower left CH_3 group and is seven C atoms long:

$$^1CH_3-^2CH_2-^3CH-CH_3$$
$$CH_3-^4CH-^5CH_2$$
$$^7CH_3-^6CH_2$$

The parent compound is thus heptane. There are two methyl groups branching off the main chain. Hence, this compound is a dimethylheptane. To specify the location of the two methyl groups, we must number the C atoms so that the methyl substituents have the smallest numbers. This means that we should start numbering at the upper left carbon. There is a methyl group on C3 and one on C4. The compound is thus 3,4-dimethylheptane.

▶ **Practice Exercise**
Name the following alkane:

$$CH_3-CH-CH_3$$
$$CH_3-CH-CH_2$$
$$CH_3$$

Sample Exercise 2.16
Writing Condensed Structural Formulas

Write the condensed structural formula for 3-ethyl-2-methylpentane.

SOLUTION

Analyze We are given the systematic name for a hydrocarbon and asked to write its condensed structural formula.

Plan Because the name ends in *-ane*, the compound is an alkane, meaning that all the carbon–carbon bonds are single bonds. The parent hydrocarbon is pentane, indicating five C atoms (Table 2.7). There are two alkyl groups specified, an ethyl group (two carbon atoms, C_2H_5) and a methyl group (one carbon atom, CH_3). Counting from left to right along the five-carbon chain, the name tells us that the ethyl group is attached to C3 and the methyl group is attached to C2.

Solve We begin by writing five C atoms attached by single bonds. These represent the backbone of the parent pentane chain:

$$C-C-C-C-C$$

We next place a methyl group on the second C and an ethyl group on the third C of the chain. We then add hydrogens to all the other C atoms to make four bonds to each carbon:

$$CH_3$$
$$CH_3-CH-CH-CH_2-CH_3$$
$$CH_2CH_3$$

The formula can be written more concisely as

$$CH_3CH(CH_3)CH(C_2H_5)CH_2CH_3$$

where the branching alkyl groups are indicated in parentheses.

▶ **Practice Exercise**
Write the condensed structural formula for 2,3-dimethylhexane.

Alcohols and Carboxylic Acids

Other classes of organic compounds are obtained when one or more hydrogen atoms in an alkane are replaced with *functional groups*, which are specific groups of atoms. An **alcohol**, for example, is obtained by replacing an H atom of an alkane with an

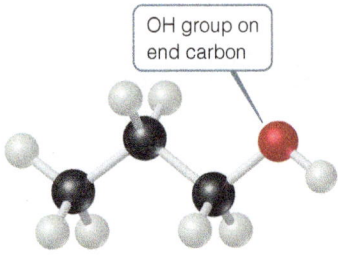

1-Propanol

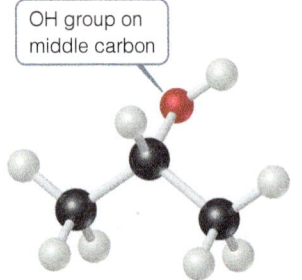

2-Propanol

▲ **Figure 2.25** **The two isomers of propanol.**

—OH group. The name of the alcohol is derived from that of the alkane by adding an *-ol* ending:

$$H-\underset{\underset{H}{|}}{\overset{\overset{H}{|}}{C}}-OH \qquad H-\underset{\underset{H}{|}}{\overset{\overset{H}{|}}{C}}-\underset{\underset{H}{|}}{\overset{\overset{H}{|}}{C}}-OH \qquad H-\underset{\underset{H}{|}}{\overset{\overset{H}{|}}{C}}-\underset{\underset{H}{|}}{\overset{\overset{H}{|}}{C}}-\underset{\underset{H}{|}}{\overset{\overset{H}{|}}{C}}-OH$$

Methan**ol**   Ethan**ol**   1-Propan**ol**

Alcohols have properties that are very different from those of the alkanes from which the alcohols are obtained. For example, methane, ethane, and propane are all colorless gases under normal conditions, whereas methanol, ethanol, and propanol are colorless liquids. We discuss the reasons for these differences in Chapter 11.

Isomers exist for alcohols and all other derivatives of alkanes. The prefix "1" in the name 1-propanol indicates that the replacement of H with OH has occurred at one of the "outer" carbon atoms rather than the "middle" carbon atom. A different compound, called either 2-propanol or isopropyl alcohol, is obtained when the OH functional group is attached to the middle carbon atom (**Figure 2.25**). If you put the —OH group on the "other" carbon, you would *not* have "3-propanol"; instead, you should just turn the molecule around and notice that it is 1-propanol, too.

Organic acids, most commonly known as **carboxylic acids**, contain the carboxylic acid group, —COOH, in place of an H in an alkane. The following are a few examples of carboxylic acids and their common names:

$$H-COOH$$
Formic acid

$$H-\underset{\underset{H}{|}}{\overset{\overset{H}{|}}{C}}-COOH$$
Acetic acid

$$H-\underset{\underset{H}{|}}{\overset{\overset{H}{|}}{C}}-\underset{\underset{H}{|}}{\overset{\overset{H}{|}}{C}}-\underset{\underset{H}{|}}{\overset{\overset{H}{|}}{C}}-COOH$$
Butyric acid

Proper names for acids are "*x*-anoic acid," where the "*x*" is the same prefix as in Table 2.9 for the relevant number of carbon atoms. Thus, the proper name of formic acid is methanoic acid; the proper name of acetic acid is ethanoic acid; and the proper name of butyric acid is butanoic acid.

Much of the richness of organic chemistry is possible because organic compounds can form long chains of carbon–carbon bonds. They also form rings of carbon atoms that produce compounds with very different properties. The series of alkanes that begins with methane, ethane, and propane and the series of alcohols that begins with methanol, ethanol, and propanol can both be extended for as long as we desire, in principle. The properties of alkanes, alcohols, and carboxylic acids change as the chains get longer. For example, octanes, which are alkanes with eight carbon atoms, are liquids under normal conditions. (Octanes are the principal components of gasoline.) If the alkane series is extended to tens of thousands of carbon atoms, we obtain *polyethylene*, a solid substance that is used to make thousands of plastic products, such as plastic bags, food containers, and laboratory equipment. We explore much more organic chemistry throughout the book, but especially in Chapter 24.

Sample Exercise 2.17
Writing Structural and Molecular Formulas for Hydrocarbons

Assuming the carbon atoms in *pentane* are in a linear chain, write (**a**) the structural formula and (**b**) the molecular formula for this alkane.

SOLUTION

(**a**) Alkanes contain only carbon and hydrogen, and each carbon is attached to four other atoms. The name *pentane* contains the prefix *penta-* for five (Table 2.6), and we are told that the carbons are in a linear chain. If we then add enough hydrogen atoms to make four bonds to each carbon, we obtain the structural formula

$$H-\overset{\overset{\displaystyle H}{|}}{\underset{\underset{\displaystyle H}{|}}{C}}-\overset{\overset{\displaystyle H}{|}}{\underset{\underset{\displaystyle H}{|}}{C}}-\overset{\overset{\displaystyle H}{|}}{\underset{\underset{\displaystyle H}{|}}{C}}-\overset{\overset{\displaystyle H}{|}}{\underset{\underset{\displaystyle H}{|}}{C}}-\overset{\overset{\displaystyle H}{|}}{\underset{\underset{\displaystyle H}{|}}{C}}-H$$

This form of pentane is often called *n*-pentane, where the *n*- stands for "normal" because all five carbon atoms are in one line in the structural formula.

(**b**) Once the structural formula is written, we determine the molecular formula by counting the atoms present. Thus, *n*-pentane has the molecular formula C_5H_{12}.

▶ **Practice Exercise**

These two compounds have "butane" in their name. Are they isomers?

$$H-\overset{\overset{\displaystyle H}{|}}{\underset{\underset{\displaystyle H}{|}}{C}}-\overset{\overset{\displaystyle H}{|}}{\underset{\underset{\displaystyle H}{|}}{C}}-\overset{\overset{\displaystyle H}{|}}{\underset{\underset{\displaystyle H}{|}}{C}}-\overset{\overset{\displaystyle H}{|}}{\underset{\underset{\displaystyle H}{|}}{C}}-H$$

Butane

Cyclobutane

Self-Assessment Exercises

SAE 2.26 What is the systematic name of the following compound?

$$CH_3-CH_2-\overset{\overset{\displaystyle CH_3}{|}}{\underset{\underset{\displaystyle CH_2}{\underset{|}{\underset{\displaystyle CH_3}{}}}}{C}}-CH_3$$

(**a**) 3-ethyl-3-methylbutane
(**b**) 2-ethyl-2-methylbutane
(**c**) 3,3-dimethylpentane
(**d**) isoheptane
(**e**) 1,2-dimethylneopentane

SAE 2.27 How many hydrogen atoms are in 2,2-dimethylhexane? (**a**) 6 (**b**) 8 (**c**) 16 (**d**) 18 (**e**) 20

SAE 2.28 How many isomers of C_6H_{14} are there? (**a**) 4 (**b**) 5 (**c**) 6 (**d**) 7 (**e**) 8

SAE 2.29 Which of these is an alcohol?

(**a**) CH_3OCH_3 (**c**) $HCOOH$
(**b**) $CH_3CH(OH)CH_3$ (**d**) CH_3COCH_3

SAE 2.30 How many isomers of butyric acid are there? (**a**) 1 (**b**) 2 (**c**) 3 (**d**) 4 (**e**) 5

STRATEGIES FOR SUCCESS | How to Take a Test

At about this time in your study of chemistry, you are likely to face your first examination. The best way to prepare is to study, do homework diligently, and get help from the instructor on any material that is unclear or confusing. (See the advice for learning and studying chemistry presented in the preface of the book.) Here are some general guidelines for taking tests.

Depending on the nature of your course, the exam could consist of a variety of different types of questions.

1. **Multiple-choice questions** In large-enrollment courses, the most common kind of test question is the multiple-choice question. Many of the Self-Assessment Exercises at the end of each section of this book are written in this format to give you practice at this style of question. When faced with this type of problem, the first thing to realize is that the instructor has written the question so that at first glance all the answers appear to be correct. Thus, you should not jump to the conclusion that because one of the choices looks correct, it must be correct.

If a multiple-choice question involves a calculation, do the calculation, check your work, and *only then* compare your answer with the choices. Keep in mind, though, that your instructor has anticipated the most common errors you might make in solving a given problem and has probably listed the incorrect answers resulting from those errors. Always double-check your reasoning and use dimensional analysis to arrive at the correct numeric answer and the correct units.

If you are not sure of the correct choice in a multiple-choice question that does *not* involve a calculation, then eliminate all the choices you know for sure to be incorrect. The reasoning you use in eliminating incorrect choices may offer insight into which of the remaining choices is correct.

2. **Calculations in which you must show your work** In questions of this kind, you may receive partial credit even if you do not arrive at the correct answer, depending on whether the instructor can follow your line of reasoning. It is
Continued

important, therefore, to be neat and organized in your calculations. Pay particular attention to what information is given and to what your unknown is. Think about how you can get from the given information to your unknown.

You may want to write a few words or a diagram on the test paper to indicate your approach. Then write out your calculations as neatly as you can. Show the units for every number you write down, and use dimensional analysis as much as you can, showing how units cancel.

3. **Questions requiring drawings** Questions of this kind will come later in the course, but it is useful to talk about

them here. (You should review this box before each exam to remind yourself of good exam-taking practices.) Be sure to label your drawing as completely as possible.

Finally, if you find that you simply do not understand how to arrive at a reasoned response to a question, do not linger over the question. Put a check next to it and go on to the next one. If time permits, you can come back to the unanswered questions, but lingering over a question when nothing is coming to mind is wasting time you may need to finish the exam.

Chapter Summary and Key Terms

THE ATOMIC THEORY OF MATTER; THE DISCOVERY OF ATOMIC STRUCTURE (SECTIONS 2.1 AND 2.2) **Atoms** are the basic building blocks of matter. They are the smallest units of an element that can combine with other elements. Atoms are composed of even smaller particles, called **subatomic particles**. Some of these subatomic particles are charged and follow the usual behavior of charged particles: Particles with the same charge repel one another, whereas particles with opposite charges are attracted to one another.

We considered some of the important experiments that led to the discovery and characterization of subatomic particles. Thomson's experiments on the behavior of **cathode rays** in magnetic and electric fields led to the discovery of the electron and allowed its charge-to-mass ratio to be measured. Millikan's oil-drop experiment determined the charge of the electron. Becquerel's discovery of **radioactivity**, the spontaneous emission of radiation by atoms, gave further evidence that the atom has a substructure. Rutherford's studies of how α particles scatter when passing through thin metal foils led to the **nuclear model** of the atom, showing that the atom has a dense, positively charged **nucleus**.

THE MODERN VIEW OF ATOMIC STRUCTURE (SECTION 2.3) Atoms have a nucleus that contains **protons** and **neutrons**; **electrons** move in the space around the nucleus. The magnitude of the charge of the electron, 1.602×10^{-19}C, is called the **electronic charge**. The charges of particles are usually represented as multiples of this charge—an electron has a 1− charge, and a proton has a 1+ charge. The masses of atoms are usually expressed in terms of **atomic mass units** ($1\,\text{amu} = 1.66054 \times 10^{-24}$g). The dimensions of atoms are often expressed in units of **angstroms** ($1\,\text{Å} = 10^{-10}$m).

Elements can be classified by **atomic number**, the number of protons in the nucleus of an atom. All atoms of a given element have the same atomic number. The **mass number** of an atom is the sum of the numbers of protons and neutrons. Atoms of the same element that differ in mass number are known as **isotopes**.

ATOMIC WEIGHTS (SECTION 2.4) The atomic mass scale is defined by assigning a mass of exactly 12 amu to a ^{12}C atom. The **atomic weight** (average atomic mass) of an element can be calculated from the relative abundances and masses of that element's isotopes. The **mass spectrometer** provides the most direct and accurate means of experimentally measuring atomic (and molecular) weights.

THE PERIODIC TABLE (SECTION 2.5) The **periodic table** is an arrangement of the elements in order of increasing atomic number. Elements with similar properties are placed in vertical columns. The elements in a column are known as a **group**. The elements in a horizontal row are known as a **period**. The **metallic elements (metals)**, which comprise the majority of the elements, dominate the left side and the middle of the table; the **nonmetallic elements (nonmetals)** are located on the upper right side. Many of the elements that lie along the line that separates metals from nonmetals are **metalloids**.

MOLECULES AND MOLECULAR COMPOUNDS (SECTION 2.6) Atoms can combine to form **molecules**. Compounds composed of

molecules (**molecular compounds**) usually contain only nonmetallic elements. A molecule that contains two atoms is called a **diatomic molecule**. The composition of a substance is given by its **chemical formula**. A molecular substance can be represented by its **empirical formula**, which gives the relative numbers of atoms of each kind, but is usually represented by its **molecular formula**, which gives the actual numbers of each type of atom in a molecule. **Structural formulas** show the order in which the atoms in a molecule are connected. **Ball-and-stick models** and **space-filling models** convey additional information about the shapes of molecules.

IONS AND IONIC COMPOUNDS (SECTION 2.7) Atoms can either gain or lose electrons, forming charged particles called **ions**. Metals tend to lose electrons, becoming positively charged ions (**cations**). Nonmetals tend to gain electrons, forming negatively charged ions (**anions**). Because **ionic compounds** are electrically neutral, containing both cations and anions, they usually contain both metallic and nonmetallic elements. Atoms that are joined together, as in a molecule, but carry a net charge are called **polyatomic ions**. The chemical formulas used for ionic compounds are empirical formulas, which can be written readily if the charges of the ions are known. The total positive charge of the cations in an ionic compound must equal the total negative charge of the anions.

NAMING INORGANIC COMPOUNDS (SECTION 2.8) The set of rules for naming chemical compounds is called **chemical nomenclature**. We studied the systematic rules used for naming three classes of inorganic substances: ionic compounds, inorganic acids, and binary molecular compounds. In naming an ionic compound, the cation is named first and then the anion. Cations formed from metal atoms have the same name as the metal. If the metal can form cations of differing charges, the charge is indicated using Roman numerals. Monatomic anions have names ending in *ide*. Polyatomic anions containing oxygen and another element (**oxyanions**) have names ending in *ate* or *ite*. In naming binary molecular compounds, Greek prefixes are used to denote the number of each element in the molecular formula, and the element farthest to the left in the periodic table (closest to metallic elements) is generally written first.

SOME SIMPLE ORGANIC COMPOUNDS (SECTION 2.9) **Organic chemistry** is the study of compounds that contain carbon. The simplest class of organic molecules is the **hydrocarbons**, which contain only carbon and hydrogen. Hydrocarbons in which each carbon atom is attached to four other atoms are called **alkanes**. Alkanes have names that end in *ane*, such as methane and ethane. Other organic compounds are formed when an H atom of a hydrocarbon is replaced with a functional group. An **alcohol**, for example, is a compound in which an H atom of a hydrocarbon is replaced by an OH functional group. Alcohols have names that end in *ol*, such as methanol and ethanol. **Organic acids** (also called **carboxylic acids**) contain the COOH functional group. Compounds with the same molecular formula but different bonding arrangements of their constituent atoms are called **isomers**.

Key Equations

- Atomic weight $= \sum\limits_{\substack{\text{over}\\\text{all isotopes}}}[(\text{fractional isotope abundance}) \times (\text{isotope mass})]$ [2.1] Calculating atomic weight as a fractionally weighted average of isotopic masses.

Exam Prep

EP 2.1 Which of the following factors determines the size of an atom? (**a**) the volume of the nucleus (**b**) the volume of space occupied by the electrons of the atom (**c**) the volume of a single electron, multiplied by the number of electrons in the atom (**d**) the total nuclear charge (**e**) the total mass of the electrons surrounding the nucleus.

EP 2.2 Which of these atoms has the largest number of neutrons? (**a**) ^{148}Eu (**b**) ^{157}Dy (**c**) ^{149}Nd (**d**) ^{162}Ho

EP 2.3 Which of the following is an incorrect representation for a neutral atom? (**a**) $^{6}_{3}$Li (**b**) $^{13}_{6}$C (**c**) $^{63}_{30}$Cu (**d**) $^{30}_{15}$P (**e**) $^{108}_{47}$Ag

EP 2.4 There are two stable isotopes of copper found in nature, ^{63}Cu and ^{65}Cu. If the atomic weight of copper Cu is 63.546 amu, which of the following statements are true? (**a**) ^{65}Cu contains two more protons than ^{63}Cu. (**b**) ^{63}Cu must be more abundant than ^{65}Cu. (**c**) All copper atoms have a mass of 63.546 amu.

EP 2.5 A biochemist who is studying the properties of certain sulfur (S)–containing compounds in the body wonders whether trace amounts of another nonmetallic element might have similar behavior. To which element should she turn her attention? (**a**) F (**b**) As (**c**) Se (**d**) Cr (**e**) P

EP 2.6 Tetracarbon dioxide is an unstable oxide of carbon with the following molecular structure:

What are the molecular and empirical formulas of this substance? (**a**) C_2O_2, CO_2 (**b**) C_4O, CO (**c**) C_4O_2, CO_2 (**d**) C_4O_2, C_2O (**e**) C_2O, CO_2

EP 2.7 In which of the following species is the difference between the number of protons and the number of electrons largest? (**a**) Ti^{2+} (**b**) P^{3-} (**c**) Mn (**d**) Se^{2-} (**e**) Ce^{4+}

EP 2.8 Although it is helpful to know that many ions have the electron arrangement of a noble gas, many elements, especially among the metals, form ions that do not have a noble-gas electron arrangement. Use the periodic table, Figure 2.14, to determine which of the following ions has a noble-gas electron arrangement: (**a**) Ti^{4+} (**b**) Mn^{2+} (**c**) Pb^{2+} (**d**) Zn^{2+}

EP 2.9 Which of these compounds is molecular? (**a**) CBr_4 (**b**) FeS (**c**) NaCl (**d**) PbF_2

EP 2.10 Which of the following nonmetals will form an ionic compound with Sc^{3+} that has a 1:1 ratio of cations to anions? (**a**) Ne (**b**) F (**c**) O (**d**) N

EP 2.11 Which of the following oxyanions is incorrectly named? (**a**) ClO_2^-, chlorate (**b**) IO_4^-, periodate (**c**) SO_3^{2-}, sulfite (**d**) IO_3^-, iodate (**e**) NO_2^-, nitrite

EP 2.12 Which of the following ionic compounds is incorrectly named? (**a**) $Zn(NO_3)_2$, zinc nitrate (**b**) $TeCl_4$, tellurium(IV) chloride (**c**) Fe_2O_3, diiron trioxide (**d**) BaO, barium oxide (**e**) $Mn_3(PO_4)_2$ manganese(II) phosphate

EP 2.13 Which of the following acids are incorrectly named? (**a**) hydrofluoric acid, HF (**b**) nitrous acid, HNO_3 (**c**) perbromic acid, $HBrO_4$ (**d**) hydroiodic acid, HI (**e**) selenic acid, H_2SeO_4

EP 2.14 What is the molecular formula for carbon disulfide? (**a**) CS_2 (**b**) CO (**c**) C_3O_2 (**d**) CBr_4 (**e**) CF

EP 2.15 What is the systematic name for this alkane:

$$CH_3-CH_2-\underset{\underset{\displaystyle CH_3}{|}}{CH}-CH_3$$

(**a**) 3,3-dimethylpropane (**b**) isobutane (**c**) 3-methylbutane (**d**) 2-methylbutane (**e**) pentane

EP 2.16 What is the empirical formula of hexane? (**a**) C_6H_6 (**b**) C_6H_{15} (**c**) C_3H_7 (**d**) C_2H_6

EP 2.17 How many isomers of n-butanol are there? (**a**) 1 (**b**) 2 (**c**) 3 (**d**) 4

Exercises

Visualizing Concepts

2.1 A charged particle moves between two electrically charged plates, as shown here.

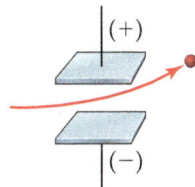

(**a**) What is the sign of the electrical charge on the particle? (**b**) As the charge on the plates is increased, would you expect the bending to increase, decrease, or stay the same? (**c**) As the mass of the particle is increased while the speed of the particles remains the same, would you expect the bending to increase, decrease, or stay the same? [Section 2.2]

2.2 The following diagram is a representation of 20 atoms of a fictitious element, which we will call nevadium (Nv). The red spheres are ^{293}Nv, and the blue spheres are ^{295}Nv. (**a**) Assuming that this sample is a statistically representative sample of the element, calculate the percent abundance of each element. (**b**) If the mass of ^{293}Nv is 293.15 amu and that of ^{295}Nv is 295.15 amu, what is the atomic weight of Nv? [Section 2.4]

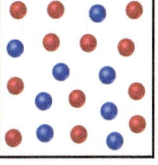

2.3 Four of the boxes in the following periodic table are colored. Which of these are metals and which are nonmetals? Which one is an alkaline earth metal? Which one is a noble gas? Which is a transition metal? [Section 2.5]

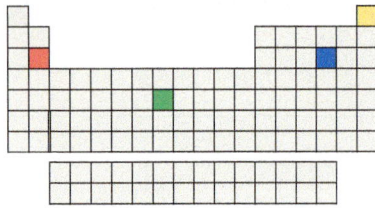

2.4 Does the following drawing represent a neutral atom or an ion? Write its complete chemical symbol, including mass number, atomic number, and net charge (if any). [Sections 2.3 and 2.7]

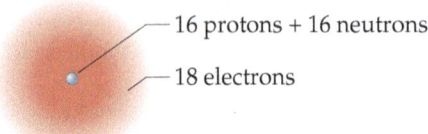

16 protons + 16 neutrons

18 electrons

2.5 (**a**) Which of the following diagrams most likely represents an ionic compound? (**b**) Which of the following diagrams most likely represents a molecular compound? [Sections 2.6 and 2.7]

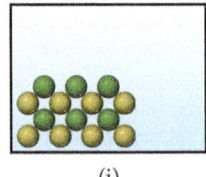 

(i)   (ii)

2.6 Write the chemical formula for the following compound. Is the compound ionic or molecular? Name the compound. [Sections 2.6 and 2.8]

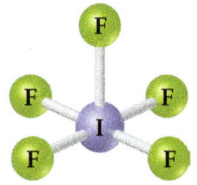

2.7 Five of the boxes in the following periodic table are colored. Predict the charge on the ion associated with each of these elements. [Section 2.7]

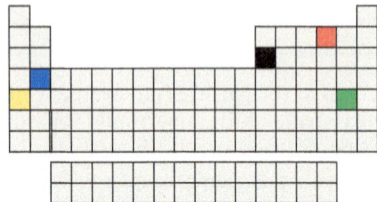

2.8 The following diagram represents an ionic compound in which the red spheres represent cations and the blue spheres represent anions. Which of the following compounds is consistent with the drawing? (**a**) potassium bromide (**b**) potassium sulfate (**c**) calcium nitrate (**d**) iron(III) sulfate [Sections 2.7 and 2.8]

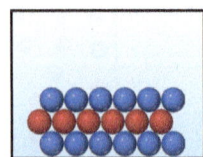

2.9 Are these two compounds isomers? [Section 2.9]

2.10 In the Millikan oil-drop experiment (see Figure 2.5), the tiny oil drops are observed through the viewing lens as rising, stationary, or falling, as shown here. The arrows indicate the rate of motion. (**a**) What causes their rate of fall to vary from their rate in the absence of an electric field? (**b**) Why do some drops move upward? [Section 2.2]

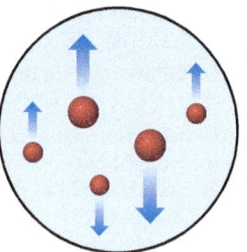

The Atomic Theory of Matter and the Discovery of Atomic Structure (Sections 2.1 and 2.2)

2.11 A 1.0-g sample of carbon dioxide (CO_2) is fully decomposed into its elements, yielding 0.273 g of carbon and 0.727 g of oxygen. (**a**) What is the ratio of the mass of O to C? (**b**) If a sample of a different compound decomposes into 0.429 g of carbon and 0.571 g of oxygen, what is its ratio of the mass of O to C? (**c**) According to Dalton's atomic theory, what is the empirical formula of the second compound?

2.12 Hydrogen sulfide is composed of two elements: hydrogen and sulfur. In an experiment, 6.500 g of hydrogen sulfide is fully decomposed into its elements. (**a**) If 0.384 g of hydrogen is obtained in this experiment, how many grams of sulfur must be obtained? (**b**) What fundamental law does this experiment demonstrate?

2.13 A chemist finds that 30.82 g of nitrogen will react with 17.60, 35.20, 70.40, or 88.00 g of oxygen to form four different compounds. (**a**) Calculate the mass of oxygen per gram of nitrogen in each compound. (**b**) How do the numbers in part (**a**) support Dalton's atomic theory?

2.14 In a series of experiments, a chemist prepared three different compounds that contain only iodine and fluorine and determined the mass of each element in each compound:

Compound	Mass of Iodine (g)	Mass of Fluorine (g)
1	4.75	3.56
2	7.64	3.43
3	9.41	9.86

(**a**) Calculate the mass of fluorine per gram of iodine in each compound. (**b**) How do the numbers in part (**a**) support the atomic theory?

2.15 Which of the subatomic particles in an atom does not have a charge (and was therefore the last to be discovered)?

2.16 An unknown particle is caused to move between two electrically charged plates, as illustrated in Figure 2.7. You hypothesize that the particle is a proton. (**a**) If your hypothesis is correct, would the particle be deflected in the same or opposite direction as the β rays? (**b**) Would it be deflected by a smaller or larger amount than the β rays?

2.17 Which set of statements is true about Rutherford's gold foil experiment?

(**i**) This is the main experiment that showed that atoms have a dense nucleus.

(**ii**) The data from the experiment showed that alpha particles scattered equally at all angles from the gold foil.

(**iii**) Electrons were emitted from the gold atoms in straight lines.

(**a**) i only (**b**) ii only (**c**) iii only (**d**) i and ii (**e**) i and iii

2.18 Millikan determined the charge on the electron by studying the static charges on oil drops falling in an electric field (Figure 2.5). A student carried out this experiment using several oil drops for her measurements and calculated the charges on the drops. She obtained the following data:

Droplet	Calculated Charge (C)
A	1.60×10^{-19}
B	3.15×10^{-19}
C	4.81×10^{-19}
D	6.31×10^{-19}

(**a**) What is the significance of the fact that the droplets carried different charges? (**b**) What conclusion can the student draw from these data regarding the charge of the electron? (**c**) What value (and to how many significant figures) should she report for the electronic charge?

The Modern View of Atomic Structure; Atomic Weights (Sections 2.3 and 2.4)

2.19 The radius of an atom of gold (Au) is about 1.35 Å. (**a**) Express this distance in nanometers (nm) and in picometers (pm). (**b**) How many gold atoms would have to be lined up to span 1.0 mm? (**c**) If the atom is assumed to be a sphere, what is the volume in cm^3 of a single Au atom?

2.20 An atom of rhodium (Rh) has a diameter of about 2.7×10^{-8} cm. (**a**) What is the radius of a rhodium atom in angstroms (Å) and in meters (m)? (**b**) How many Rh atoms would have to be placed side by side to span a distance of 6.0 μm? (**c**) If you assume that the Rh atom is a sphere, what is the volume in m^3 of a single atom?

2.21 Answer the following questions without referring to Table 2.1. (**a**) What are the main subatomic particles that make up the atom? (**b**) What is the relative charge (in multiples of the electronic charge) of each of the particles? (**c**) Which of the particles is the most massive? (**d**) Which is the least massive?

2.22 Determine whether each of the following statements is true or false. (**a**) The nucleus has most of the mass and comprises most of the volume of an atom. (**b**) Every atom of a given element has the same number of protons. (**c**) The number of electrons in an atom equals the number of neutrons in the atom. (**d**) The protons in the nucleus of the helium atom are held together by a force called the strong nuclear force.

2.23 Consider an atom of ^{10}B. (**a**) How many protons, neutrons, and electrons does this atom contain? (**b**) What is the symbol of the atom obtained by adding one proton to ^{10}B? (**c**) What is the symbol of the atom obtained by adding one neutron to ^{10}B? (**d**) Are either of the atoms obtained in parts (**b**) and (**c**) isotopes of ^{10}B? If so which one?

2.24 Consider an atom of ^{63}Cu. (**a**) How many protons, neutrons, and electrons does this atom contain? (**b**) What is the symbol of the ion obtained by removing two electrons from ^{63}Cu? (**c**) What is the symbol for the isotope of ^{63}Cu that possesses 36 neutrons?

2.25 (**a**) Define atomic number and mass number. (**b**) Which of these can vary without changing the identity of the element? (**c**) Which one of these is usually larger for a given element?

2.26 (**a**) Which two of the following are isotopes of the same element: $^{31}_{16}$X, $^{31}_{15}$X, $^{32}_{16}$X? (**b**) What is the identity of the element whose isotopes you have selected?

2.27 How many protons, neutrons, and electrons are in the following atoms? (**a**) ^{40}Ar (**b**) ^{65}Zn (**c**) ^{70}Ga (**d**) ^{80}Br (**e**) ^{184}W (**f**) ^{243}Am

2.28 Each of the following isotopes is used in medicine. Indicate the number of protons and neutrons in each isotope. (**a**) phosphorus-32 (**b**) chromium-51 (**c**) cobalt-60 (**d**) technetium-99 (**e**) iodine-131 (**f**) thallium-201

2.29 Fill in the gaps in the following table, assuming each column represents a neutral atom.

Symbol	^{79}Br			
Protons		25		82
Neutrons		30	64	
Electrons			48	86
Mass no.				222	207

2.30 Fill in the gaps in the following table, assuming each column represents a neutral atom.

Symbol	^{112}Cd			
Protons		38		92
Neutrons		58	49	
Electrons			38	36
Mass no.			81	235

2.31 Write the correct symbol, with both superscript and subscript, for each of the following. Use the list of elements in the front inside cover as needed. (**a**) the isotope of platinum that contains 118 neutrons (**b**) the isotope of krypton with mass number 84 (**c**) the isotope of arsenic with mass number 75 (**d**) the isotope of magnesium that has an equal number of protons and neutrons

2.32 Hydrogen is unusual in that its three isotopes have common chemical symbols: hydrogen (H), deuterium (D), and tritium (T). Based on these names, predict how many protons and neutrons are in the nuclei of H, D, and T.

2.33 The atomic weight of boron is reported as 10.81, yet no atom of boron has the mass of 10.81 amu. Which is the best explanation? (**a**) The measurement of atomic mass is only reliable to two significant figures. (**b**) The atomic weight is an average of many individual atoms. (**c**) The atomic weight is an average of many isotopes of the same nuclear composition.

2.34 (**a**) What is the mass in amu of a carbon-12 atom? (**b**) Why is the atomic weight of carbon reported as 12.011 in the table of elements and the periodic table in the front inside cover of this text?

2.35 Only two isotopes of copper occur naturally: ^{63}Cu (atomic mass = 62.9296 amu; abundance 69.17%) and ^{65}Cu (atomic mass = 64.9278 amu; abundance 30.83%). Calculate the atomic weight (average atomic mass) of copper.

2.36 Rubidium has two naturally occurring isotopes, rubidium-85 (atomic mass = 84.9118 amu; abundance = 72.15%) and rubidium-87 (atomic mass = 86.9092 amu; abundance = 27.85%). Calculate the atomic weight of rubidium.

2.37 (**a**) Thomson's cathode-ray tube (Figure 2.4) and the mass spectrometer (Figure 2.11) both involve the use of electric or magnetic fields to deflect charged particles. What are the charged particles involved in each of these experiments? (**b**) What are the labels on the axes of a mass spectrum? (**c**) To measure the mass spectrum of a sample of atoms, the atom must first lose one or more electrons. Which would you expect to be deflected more by the same setting of the electric and magnetic fields, a Cl^+ or a Cl^{2+} ion?

2.38 Consider the mass spectrometer shown in Figure 2.11. Determine whether each of the following statements is true or false. **(a)** The paths of neutral (uncharged) atoms are not affected by the magnet. **(b)** The height of each peak in the mass spectrum is inversely proportional to the mass of that isotope. **(c)** For a given element, the number of peaks in the spectrum is equal to the number of naturally occurring isotopes of that element.

2.39 Naturally occurring magnesium has the following isotopic abundances:

Isotope	Abundance (%)	Atomic mass (amu)
^{24}Mg	78.99	23.98504
^{25}Mg	10.00	24.98584
^{26}Mg	11.01	25.98259

(a) What is the average atomic mass of Mg? **(b)** Sketch the mass spectrum of Mg.

2.40 Mass spectrometry is more often applied to molecules than to atoms. We show in Chapter 3 that the *molecular weight* of a molecule is the sum of the atomic weights of the atoms in the molecule. The mass spectrum of H_2 is taken under conditions that prevent decomposition into H atoms. The two naturally occurring isotopes of hydrogen are ^{1}H (atomic mass = 1.00783 amu; abundance 99.9885%) and ^{2}H (atomic mass = 2.01410 amu; 0.0115%). **(a)** How many peaks will the mass spectrum have? **(b)** Give the relative atomic masses of each of these peaks. **(c)** Which peak will be the largest, and which the smallest?

The Periodic Table, Molecules, Molecular Compounds, Ions, and Ionic Compounds (Sections 2.5, 2.6, and 2.7)

2.41 For each of the following elements, write its chemical symbol, locate it in the periodic table, give its atomic number, and indicate whether it is a metal, metalloid, or nonmetal. **(a)** chromium **(b)** helium **(c)** phosphorus **(d)** zinc **(e)** magnesium **(f)** bromine **(g)** arsenic

2.42 Locate each of the following elements in the periodic table; give its name and atomic number, and indicate whether it is a metal, metalloid, or nonmetal. **(a)** Li **(b)** Sc **(c)** Ge **(d)** Yb **(e)** Mn **(f)** Sb **(g)** Xe

2.43 For each of the following elements, write its chemical symbol, determine the name of the group to which it belongs (Table 2.3), and indicate whether it is a metal, metalloid, or nonmetal. **(a)** potassium **(b)** iodine **(c)** magnesium **(d)** argon **(e)** sulfur

2.44 The elements of group 4A show an interesting change in properties moving down the group. Give the name and chemical symbol of each element in the group and label it as a nonmetal, metalloid, or metal.

2.45 The structural formulas of the compounds *n*-butane and isobutane are shown below. **(a)** Determine the molecular formula of each. **(b)** Determine the empirical formula of each. **(c)** Which formulas—empirical, molecular, or structural—allow you determine these are different compounds?

n-butane   isobutane

2.46 Ball-and-stick representations of benzene, a colorless liquid often used in organic chemistry reactions, and acetylene, a gas used as a fuel for high-temperature welding, are shown below. **(a)** Determine the molecular formula of each. **(b)** Determine the empirical formula of each.

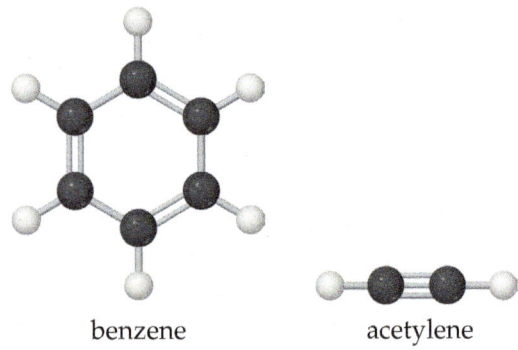

benzene   acetylene

2.47 What are the molecular and empirical formulas for each of the following compounds?

2.48 Two substances have the same molecular and empirical formulas. Does this mean that they must be the same compound?

2.49 Write the empirical formula corresponding to each of the following molecular formulas. **(a)** Al_2Br_6 **(b)** C_8H_{10} **(c)** $C_4H_8O_2$ **(d)** P_4O_{10} **(e)** $C_6H_4Cl_2$ **(f)** $B_3N_3H_6$

2.50 Determine the molecular and empirical formulas of the following. **(a)** the organic solvent *cyclohexane*, which has six carbon atoms and twelve hydrogen atoms **(b)** the compound *silicon tetrachloride*, which has a silicon atom and four chlorine atoms and is used in the manufacture of computer chips **(c)** the reactive substance *diborane*, which has two boron atoms and six hydrogen atoms **(d)** the sugar called *glucose*, which has six carbon atoms, twelve hydrogen atoms, and six oxygen atoms

2.51 How many hydrogen atoms are in each of the following? **(a)** C_2H_5OH **(b)** $Ca(C_2H_5COO)_2$ **(c)** $(NH_4)_3PO_4$

2.52 How many of the indicated atoms are represented by each chemical formula? **(a)** carbon atoms in $C_4H_9COOCH_3$ **(b)** oxygen atoms in $Ca(ClO_3)_2$ **(c)** hydrogen atoms in $(NH_4)_2HPO_4$

2.53 Write the molecular and structural formulas for the compounds represented by the following molecular models.

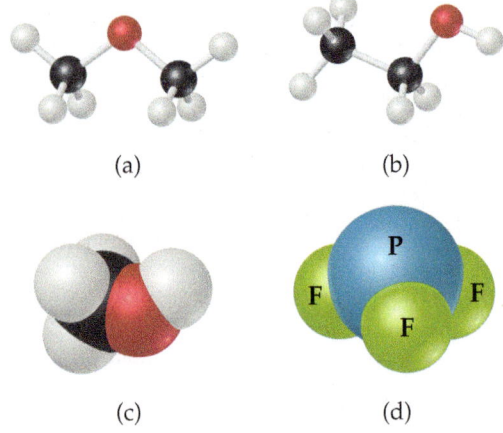

(a)   (b)

(c)   (d)

2.54 Write the molecular and structural formulas for the compounds represented by the following models.

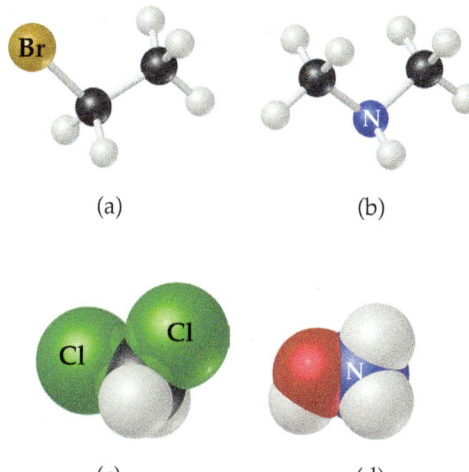

(a)   (b)

(c)   (d)

2.55 Fill in the gaps in the following table:

Symbol	$^{59}Co^{3+}$			
Protons		34	76	80
Neutrons		46	116	120
Electrons		36		78
Net charge			2+	

2.56 Fill in the gaps in the following table:

Symbol	$^{31}P^{3-}$			
Protons		34	50	
Neutrons		45	69	118
Electrons			46	76
Net charge		2−		3+

2.57 Each of the following elements is capable of forming an ion in chemical reactions. By referring to the periodic table, predict the charge of the most stable ion of each. (**a**) Mg (**b**) Al (**c**) K (**d**) S (**e**) F

2.58 Using the periodic table, predict the charges of the ions of the following elements. (**a**) Ga (**b**) Sr (**c**) As (**d**) Br (**e**) Se

2.59 Using the periodic table to guide you, predict the chemical formula and name of the compound formed by the following elements. (**a**) Ga and F (**b**) Li and H (**c**) Al and I (**d**) K and S

2.60 The most common charge associated with scandium in its compounds is 3+. Indicate the chemical formulas you would expect for compounds formed between scandium and (**a**) iodine, (**b**) sulfur, (**c**) nitrogen.

2.61 Predict the chemical formula for the ionic compound formed by (**a**) Ca^{2+} and Br^-, (**b**) K^+ and CO_3^{2-}, (**c**) Al^{3+} and CH_3COO^-, (**d**) NH_4^+ and SO_4^{2-}, (**e**) Mg^{2+} and PO_4^{3-}.

2.62 Predict the chemical formulas of the compounds formed by the following pairs of ions. (**a**) Cr^{3+} and Br^- (**b**) Fe^{3+} and O^{2-} (**c**) Hg_2^{2+} and CO_3^{2-} (**d**) Ca^{2+} and ClO_3^- (**e**) NH_4^+ and PO_4^{3-}

2.63 Complete the table by filling in the formula for the ionic compound formed by each pair of cations and anions, as shown for the first pair.

Ion	K^+	NH_4^+	Mg^{2+}	Fe^{3+}
Cl^-	KCl			
OH^-				
CO_3^{2-}				
PO_4^{3-}				

2.64 Complete the table by filling in the formula for the ionic compound formed by each pair of cations and anions, as shown for the first pair.

Ion	Na^+	Ca^{2+}	Fe^{2+}	Al^{3+}
O^{2-}	Na_2O			
NO_3^-				
SO_4^{2-}				
AsO_4^{3-}				

2.65 Predict whether each of the following compounds is molecular or ionic. (**a**) B_2H_6 (**b**) CH_3OH (**c**) $LiNO_3$ (**d**) Sc_2O_3 (**e**) CsBr (**f**) NOCl (**g**) NF_3 (**h**) Ag_2SO_4

2.66 Predict whether each of the following compounds is molecular or ionic. (**a**) PF_5 (**b**) NaI (**c**) SCl_2 (**d**) $Ca(NO_3)_2$ (**e**) $FeCl_3$ (**f**) LaP (**g**) $CoCO_3$ (**h**) N_2O_4

Naming Inorganic Compounds; Some Simple Organic Compounds (Sections 2.8 and 2.9)

2.67 Give the chemical formula for (**a**) chlorite ion, (**b**) chloride ion, (**c**) chlorate ion, (**d**) perchlorate ion, (**e**) hypochlorite ion.

2.68 Selenium, an element required nutritionally in trace quantities, forms compounds analogous to sulfur. Name the following ions. (**a**) SeO_4^{2-} (**b**) Se^{2-} (**c**) HSe^- (**d**) $HSeO_3^-$

2.69 Give the names and charges of the cation and anion in each of the following compounds. (**a**) CaO (**b**) Na_2SO_4 (**c**) $KClO_4$ (**d**) $Fe(NO_3)_2$ (**e**) $Cr(OH)_3$

2.70 Give the names and charges of the cation and anion in each of the following compounds. (**a**) CuS (**b**) Ag_2SO_4 (**c**) $Al(ClO_3)_3$ (**d**) $Co(OH)_2$ (**e**) $PbCO_3$

2.71 Name the following ionic compounds. (**a**) Li_2O (**b**) $FeCl_3$ (**c**) NaClO (**d**) $CaSO_3$ (**e**) $Cu(OH)_2$ (**f**) $Fe(NO_3)_2$ (**g**) $Ca(CH_3COO)_2$ (**h**) $Cr_2(CO_3)_3$ (**i**) K_2CrO_4 (**j**) $(NH_4)_2SO_4$

2.72 Name the following ionic compounds. (**a**) KCN (**b**) $NaBrO_2$ (**c**) $Sr(OH)_2$ (**d**) CoTe (**e**) $Fe_2(CO_3)_3$ (**f**) $Cr(NO_3)_3$ (**g**) $(NH_4)_2SO_3$ (**h**) NaH_2PO_4 (**i**) $KMnO_4$ (**j**) $Ag_2Cr_2O_7$

2.73 Write the chemical formulas for the following compounds. (**a**) aluminum hydroxide (**b**) potassium sulfate (**c**) copper(I) oxide (**d**) zinc nitrate (**e**) mercury(II) bromide (**f**) iron(III) carbonate (**g**) sodium hypobromite

2.74 Give the chemical formula for each of the following ionic compounds. (**a**) sodium phosphate (**b**) zinc nitrate (**c**) barium bromate (**d**) iron(II) perchlorate (**e**) cobalt(II) hydrogen carbonate (**f**) chromium(III) acetate (**g**) potassium dichromate

2.75 Give the name or chemical formula, as appropriate, for each of the following acids. (**a**) $HBrO_3$ (**b**) HBr (**c**) H_3PO_4 (**d**) hypochlorous acid (**e**) iodic acid (**f**) sulfurous acid

2.76 Provide the name or chemical formula, as appropriate, for each of the following acids. (**a**) hydroiodic acid (**b**) chloric acid (**c**) nitrous acid (**d**) H_2CO_3 (**e**) $HClO_4$ (**f**) CH_3COOH

2.77 Give the name or chemical formula, as appropriate, for each of the following binary molecular substances. (**a**) SF_6 (**b**) $SeCl_2$ (**c**) XeO_3 (**d**) dinitrogen tetroxide (**e**) hydrogen cyanide (**f**) tetraphosphorus hexasulfide

2.78 The oxides of nitrogen are very important components in urban air pollution. Name each of the following compounds. (**a**) N_2O (**b**) NO (**c**) NO_2 (**d**) N_2O_5 (**e**) N_2O_4

2.79 Write the chemical formula for each substance mentioned in the following word descriptions (use the front inside cover to find the symbols for the elements you do not know). (**a**) Zinc carbonate can be heated to form zinc oxide and carbon dioxide. (**b**) On treatment with hydrofluoric acid, silicon dioxide forms silicon tetrafluoride and water. (**c**) Sulfur dioxide reacts with water to form sulfurous acid. (**d**) The substance phosphorus trihydride, commonly called phosphine, is a toxic gas. (**e**) Perchloric acid reacts with cadmium to form cadmium(II) perchlorate. (**f**) Vanadium(III) bromide is a colored solid.

2.80 Assume that you encounter the following sentences in your reading. What is the chemical formula for each substance mentioned? (**a**) Sodium hydrogen carbonate is used as a deodorant. (**b**) Calcium hypochlorite is used in some bleaching solutions. (**c**) Hydrogen cyanide is a very poisonous gas. (**d**) Magnesium hydroxide is used as a cathartic. (**e**) Tin(II) fluoride has been used as a fluoride additive in toothpastes. (**f**) When cadmium sulfide is treated with sulfuric acid, fumes of hydrogen sulfide are given off.

2.81 (**a**) What elements are contained in hydrocarbons? (**b**) *n*-Octane is the alkane with a linear chain of eight carbon atoms. Write the structural formula for this compound. (**c**) What is the molecular formula of octane? (**d**) What is the empirical formula for octane?

2.82 (**a**) What is meant by the term *isomer*? (**b**) How many isomers does propane have? (**c**) How many isomers does butane have?

2.83 (**a**) What functional group characterizes an alcohol? (**b**) Write a structural formula for 1-pentanol.

2.84 Consider the following organic substances: ethanol, propane, hexane, and propanol. (**a**) Which of these molecules contains an OH group? (**b**) Which of these molecules contains three carbon atoms?

2.85 Your classmate writes out an incorrect condensed structural formula for methanol: CH_4OH. Why is this incorrect? (**a**) Alcohols should have the —COOH functional group. (**b**) There should be more than one carbon atom. (**c**) There are too many hydrogen atoms. (**d**) More than one of (a), (b), (c).

2.86 Your classmate says there are two isomers for ethanol. Are they right or not, and if so, why or why not? (**a**) The classmate is correct, because you can put the OH group on either carbon. (**b**) The classmate is correct, because you can put the COOH group on either carbon. (**c**) The classmate is not correct, there is only one isomer of ethanol. (**d**) The classmate is not correct, because there are more than two isomers of ethanol.

2.87 All the structures shown here have the molecular formula C_8H_{18}. Which structures are the same molecule? (*Hint:* One way to answer this question is to determine the chemical name for each structure.)

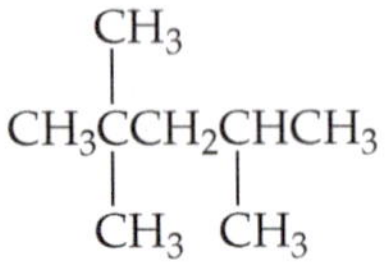

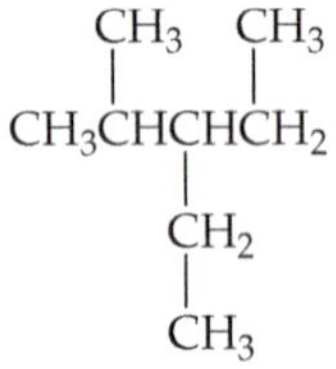

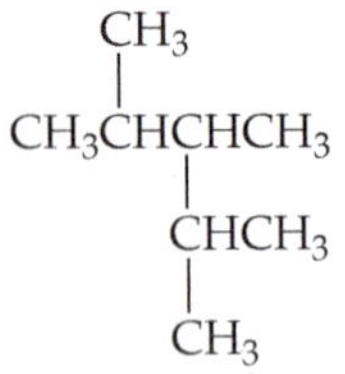

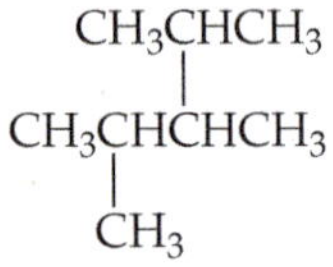

2.88 True or false: isopentane and hexane are isomers.

2.89 Give the name or condensed structural formula, as appropriate.

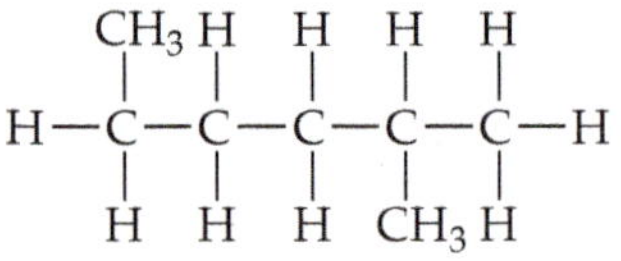

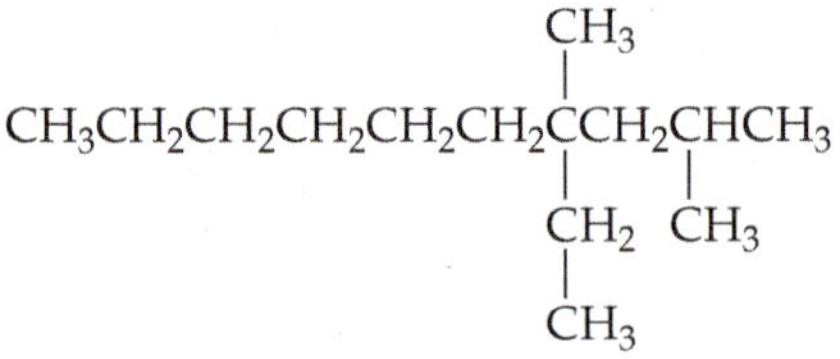

(**c**) 2-methylheptane

(**d**) 4-ethyl-2,3-dimethyloctane

2.90 Give the name or condensed structural formula, as appropriate.

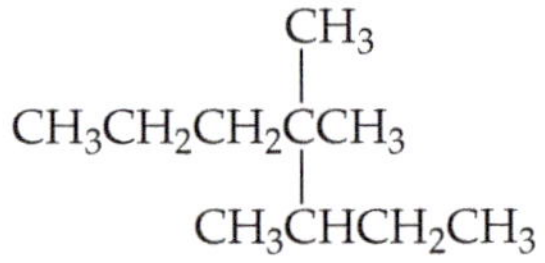

(**c**) 2,5,6-trimethylnonane

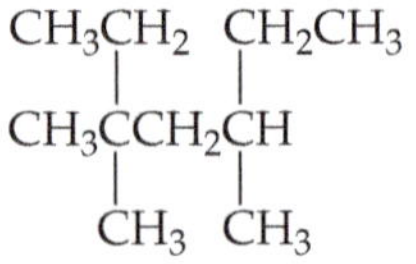

(**d**) 4-ethyl-5,6-dimethyldodecane

2.91 Two compounds are drawn below. Which statement is most correct? (**a**) These are two isomers of an alcohol. (**b**) These are two isomers of a carboxylic acid. (**c**) These are two isomers of a compound that is both an alcohol and a carboxylic acid. (**d**) These are two drawings of the same compound that is both an alcohol and a carboxylic acid.

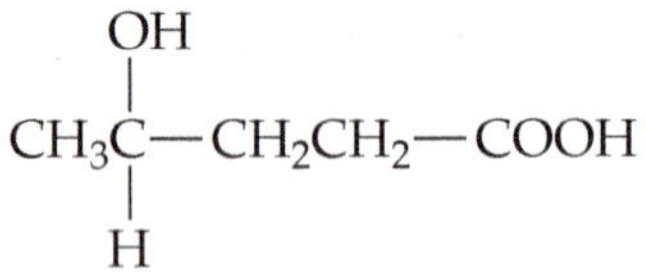

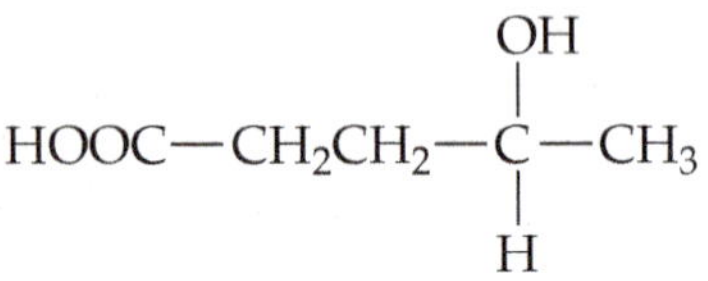

2.92 How many isomers are there of *n*-pentanol?

Additional Exercises

2.93 Suppose a scientist repeats the Millikan oil-drop experiment but reports the charges on the drops using an unusual (and imaginary) unit called the *warmomb* (wa). The scientist obtains the following data for four of the drops:

Droplet	Calculated Charge (wa)
A	3.84×10^{-8}
B	4.80×10^{-8}
C	2.88×10^{-8}
D	8.64×10^{-8}

(a) If all the droplets were the same size, which would fall most slowly through the apparatus? **(b)** From these data, what is the best choice for the charge of the electron in warmombs? **(c)** Based on your answer to part **(b)**, how many electrons are there on each of the droplets? **(d)** What is the conversion factor between warmombs and coulombs?

2.94 The natural abundance of ^{3}He is 0.000137%. **(a)** How many protons, neutrons, and electrons are in an atom of ^{3}He? **(b)** Based on the sum of the masses of their subatomic particles, which is expected to be more massive, an atom of ^{3}He or an atom of ^{3}H (which is also called *tritium*)? **(c)** Based on your answer to part **(b)**, what would need to be the precision of a mass spectrometer that is able to differentiate between peaks that are due to ^{3}He$^+$ and ^{3}H$^+$?

2.95 Identify the element represented by each of the following symbols and give the number of protons and neutrons in each. **(a)** $^{74}_{33}$X **(b)** $^{127}_{53}$X **(c)** $^{152}_{63}$X **(d)** $^{209}_{83}$X

2.96 The nucleus of ^{6}Li is a powerful absorber of neutrons. It exists in the naturally occurring metal to the extent of 7.5%. In the era of nuclear deterrence, large quantities of lithium were processed to remove ^{6}Li for use in hydrogen bomb production. The lithium metal remaining after removal of ^{6}Li was sold on the market. **(a)** What are the compositions of the nuclei of ^{6}Li and ^{7}Li? **(b)** The atomic masses of ^{6}Li and ^{7}Li are 6.015122 and 7.016004 amu, respectively. A sample of lithium depleted in the lighter isotope was found on analysis to contain 1.442% ^{6}Li. What is the average atomic weight of this sample of the metal?

2.97 The element oxygen has three naturally occurring isotopes, with 8, 9, and 10 neutrons in the nucleus, respectively. **(a)** Write the full chemical symbols for these three isotopes. **(b)** Which of these isotopes contains 10 electrons? **(c)** Oxygen-18 can be converted to fluorine-18 by a beam of high-energy protons. How many neutrons are in the nucleus of fluorine-18?

2.98 The element lead (Pb) consists of four naturally occurring isotopes with atomic masses 203.97302, 205.97444, 206.97587, and 207.97663 amu. The relative abundances of these four isotopes are 1.4, 24.1, 22.1, and 52.4%, respectively. From these data, calculate the atomic weight of lead.

2.99 Gallium (Ga) consists of two naturally occurring isotopes with masses of 68.926 and 70.925 amu. **(a)** How many protons and neutrons are in the nucleus of each isotope? Write the complete atomic symbol for each, showing the atomic number and mass number. **(b)** The average atomic mass of Ga is 69.72 amu. Calculate the abundance of each isotope.

2.100 There are two different isotopes of bromine atoms. Under normal conditions, elemental bromine consists of Br$_2$ molecules, and the mass of a Br$_2$ molecule is the sum of the masses of the two atoms in the molecule. The mass spectrum of Br$_2$ consists of three peaks:

m/z	Relative Peak Intensity
157.836	0.2569
159.834	0.4999
161.832	0.2431

(a) What is the origin of each peak (of what isotopes does each consist)? **(b)** What is the mass of each isotope? **(c)** Determine the average molecular mass of a Br$_2$ molecule. **(d)** Determine the average atomic mass of a bromine atom. **(e)** Calculate the abundances of the two isotopes.

2.101 It is common in mass spectrometry to assume that the mass of a cation is the same as that of its parent atom. **(a)** Using data in Table 2.1, determine the number of significant figures that must be reported before the difference in masses of ^{1}H and ^{1}H$^+$ is significant. **(b)** What percentage of the mass of an ^{1}H atom does the electron represent?

2.102 From the following list of elements—Ar, H, Ga, Al, Ca, Br, Ge, K, O—pick the one that best fits each description. Use each element only once. **(a)** an alkali metal **(b)** an alkaline earth metal **(c)** a noble gas **(d)** a halogen **(e)** a metalloid **(f)** a nonmetal listed in group 1A **(g)** a metal that forms a 3+ ion **(h)** a nonmetal that forms a 2− ion **(i)** an element that resembles aluminum

2.103 The first atoms of seaborgium (Sg) were identified in 1974. The longest-lived isotope of Sg has a mass number of 266. **(a)** How many protons, electrons, and neutrons are in an ^{266}Sg atom? **(b)** Atoms of Sg are very unstable, and it is therefore difficult to study this element's properties. Based on the position of Sg in the periodic table, what element should it most closely resemble in its chemical properties?

2.104 The explosion of an atomic bomb releases many radioactive isotopes, including strontium-90. Considering the location of strontium in the periodic table, suggest a reason for the fact that this isotope is particularly dangerous for human health.

2.105 From the molecular structures shown here, identify the one that corresponds to each of the following species. **(a)** chlorine gas **(b)** propane **(c)** nitrate ion **(d)** sulfur trioxide **(e)** methyl chloride, CH$_3$Cl

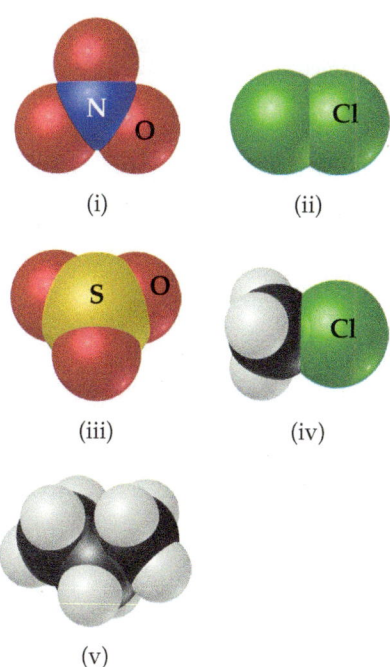

2.106 Name each of the following oxides. Assuming that the compounds are ionic, what charge is associated with the metallic element in each case? (**a**) NiO (**b**) MnO_2 (**c**) Cr_2O_3 (**d**) MoO_3

2.107 Fill in the blanks in the following table:

Cation	Anion	Formula	Name
			Lithium oxide
Fe^{2+}	PO_4^{3-}		
		$Al_2(SO_4)_3$	
			Copper(II) nitrate
Cr^{3+}	I^-		
		$MnClO_2$	
			Ammonium carbonate
			Zinc perchlorate

2.108 Instead of having three carbons in a row, the three carbons of the hydrocarbon cyclopropane form a ring, as shown in the following perspective drawing:

Cyclopropane was at one time used as an anesthetic, but its use was discontinued, in part because it is highly flammable. (**a**) What is the empirical formula of cyclopropane? (**b**) The three carbon atoms are necessarily in a plane. Is the entire molecule in a plane (flat)? (**c**) Can you use mass spectrometry to tell the difference between cyclopropane and propane, assuming the technique is reliable to 0.1 amu? (**d**) What change would you make to the structure shown to draw cyclopropanol? (**e**) How many isomers of cyclopropanol are there (ignore wedge positions for the purposes of this problem)?

2.109 Elements in the same group of the periodic table often form oxyanions with the same general formula. The anions are also named in a similar fashion. Based on these observations,

suggest a chemical formula or name, as appropriate, for each of the following ions. (**a**) BrO_4^- (**b**) SeO_3^{2-} (**c**) arsenate ion (**d**) hydrogen tellurate ion

2.110 Carbonic acid occurs in carbonated beverages. When allowed to react with lithium hydroxide, it produces lithium carbonate. Lithium carbonate is used to treat depression and bipolar disorder. Write chemical formulas for carbonic acid, lithium hydroxide, and lithium carbonate.

2.111 Give the chemical names of each of the following familiar compounds. (**a**) NaCl (table salt) (**b**) $NaHCO_3$ (baking soda) (**c**) NaOCl (in many bleaches) (**d**) NaOH (caustic soda) (**e**) $(NH_4)_2CO_3$ (smelling salts) (**f**) $CaSO_4$ (plaster of Paris)

2.112 Many familiar substances have common, unsystematic names. For each of the following, give the correct systematic name. (**a**) salt-peter, KNO_3 (**b**) soda ash, Na_2CO_3 (**c**) lime, CaO (**d**) muriatic acid, HCl (**e**) Epsom salts, $MgSO_4$ (**f**) milk of magnesia, $Mg(OH)_2$

2.113 Because many ions and compounds have very similar names, there is great potential for confusing them. Write the correct chemical formulas to distinguish between (**a**) calcium sulfide and calcium hydrogen sulfide, (**b**) hydrobromic acid and bromic acid, (**c**) aluminum nitride and aluminum nitrite, (**d**) iron(II) oxide and iron(III) oxide, (**e**) ammonia and ammonium ion, (**f**) potassium sulfite and potassium bisulfite, (**g**) mercurous chloride and mercuric chloride, and (**h**) chloric acid and perchloric acid.

2.114 In what part of the atom does the strong nuclear force operate?

2.115 Chemists work with so many hydrocarbons that they abbreviate the structures with lines and vertices. Hexane, for instance, can be drawn as follows:

In this drawing, each end or vertex is a carbon atom, and it is assumed that sufficient hydrogens are added to each carbon to give each carbon 4 connections: $CH_3CH_2CH_2CH_2CH_2CH_3$. Moreover, the zig-zag nature of the drawing provides information about the relative angles of the carbons to each other. What, then, is the systematic name of the following compound?

3

CHEMICAL REACTIONS AND REACTION STOICHIOMETRY

▲ THESE STAINLESS STEEL REACTORS are where yeast-driven reactions occur that convert a sugar-rich aqueous solution called wort into beer.

In this chapter, we begin exploring chemical reactions. We describe how to predict the products of a few simple classes of reactions—combination, decomposition, and combustion reactions. We show that the number of atoms or molecules of a substance can be determined from its identity and the masses of the atoms that make up the substance. Building on that knowledge, we develop tools to predict how much of each product and reactant will be present at the end of a reaction from the amounts present before the reaction starts.

WHAT'S AHEAD

3.1 ▶ Chemical Equations Use chemical formulas to write equations representing chemical reactions.

3.2 ▶ Simple Patterns of Chemical Reactivity Examine some simple chemical reactions: *combination reactions, decomposition reactions*, and *combustion reactions*.

3.3 ▶ Formula Weights Use *formula weights* to obtain quantitative information from chemical formulas.

3.4 ▶ Avogadro's Number and the Mole Use chemical formulas to relate the mass of a substance to the numbers of atoms, molecules, or ions contained in the substance, a relationship that leads to the crucially important concept of the *mole*, defined as 6.02×10^{23} objects (atoms, molecules, ions, and so on).

3.5 ▶ Empirical Formulas from Analyses Apply the mole concept to determine chemical formulas from the masses of each element in a given quantity of a compound.

3.6 ▶ Quantitative Information from Balanced Equations Use the quantitative information inherent in chemical formulas and equations together with the mole concept to predict the amounts of substances consumed or produced in chemical reactions.

3.7 ▶ Limiting Reactants Recognize that one reactant may be used up before others in a chemical reaction. This is the *limiting reactant*. Once the limiting reactant is used up, the reaction stops, leaving some excess of the other starting materials.

3.1 | Chemical Equations

⚠ **Learning Objectives**

When you finish Section 3.1, you should be able to:

▶ Explain the law of conservation of mass in terms of the reactants and products of a chemical equation.

▶ Balance chemical equations by writing out the chemical formulas (and appropriate coefficients) of the reactants and products.

Much of the energy that we use in our daily lives, including transportation, comes from chemical reactions. In the internal combustion engine, gasoline and air react to produce energy, along with gaseous by-products (mostly carbon dioxide and water). These substances exit the tailpipes of vehicles—an issue that is central to discussions about climate change. Suppose we wanted to know how many molecules of CO_2 are produced by burning a certain amount of gasoline. The tools of chemistry we explore in this chapter enable us to accurately determine how many molecules of hydrocarbon and oxygen are consumed and how many molecules of by-products are generated.

Stoichiometry (pronounced stoy-key-OM-uh-tree, roughly meaning "element measure" in Greek) is the area of study that examines the quantities of substances consumed and produced in chemical reactions. Chemists and chemical engineers use stoichiometry every day when running the reactions that drive the worldwide chemical industry. Stoichiometry is built on an understanding of atomic masses (Section 2.4), chemical formulas, and the **law of conservation of mass**. (Section 2.1) This important principle tells us that *atoms are neither created nor destroyed during a chemical reaction*. The changes that occur during any reaction merely rearrange the atoms. The same collection of atoms is present both before and after the reaction.

We represent chemical reactions by **chemical equations**. When the gas hydrogen (H_2) burns, for example, it reacts with oxygen (O_2) in the air to form water (H_2O). We write the chemical equation for this reaction as

$$2H_2 + O_2 \longrightarrow 2H_2O \qquad\qquad [3.1]$$

We read the + sign as "reacts with" and the arrow as "produces." The chemical formulas to the left of the arrow represent the starting substances, called **reactants**. The chemical formulas to the right of the arrow represent substances produced in the reaction, called **products**. The numbers in front of the formulas, called coefficients, indicate the relative numbers of molecules of each kind involved in the reaction. (As in algebraic equations, *the coefficient 1 is usually not written*.)

Because atoms are neither created nor destroyed in any reaction, a *balanced* chemical equation must have an equal number of atoms of each element on each side of the arrow. On the right side of Equation 3.1, for example, there are two molecules of H_2O, each composed of two atoms of hydrogen and one atom of oxygen (**Figure 3.1**). Thus, $2\,H_2O$ (read "two molecules of water") contains $2 \times 2 = 4$ H atoms and $2 \times 1 = 2$ O atoms. Thus, *the number of atoms is obtained by multiplying each subscript in a chemical formula by the coefficient for the formula*. Because there are four H atoms and two O atoms on each side of the equation, the equation is balanced.

Balancing Equations

Chemists write chemical equations to identify the reactants and products in a reaction. To determine the amount of product that can be made, or the amount of a reactant that is required, the equation needs to be *balanced* using the law of conservation of mass.

To construct a **balanced chemical equation**, we start by writing the formulas for the reactants on the left-hand side of the arrow and for the products on the right-hand side. We balance the equation by determining the coefficients that provide equal numbers of each type of atom on both sides of the equation. For most purposes, a balanced equation should contain the smallest possible whole-number coefficients.

In balancing an equation, you must distinguish between coefficients and subscripts. As **Figure 3.2** illustrates, changing a subscript in a formula—from H_2O to H_2O_2, for example—changes the identity of the substance. The substance H_2O_2, hydrogen peroxide, is quite different from the substance H_2O, water. *Never change subscripts when balancing an equation*. In contrast, placing a coefficient in front of a formula changes only the amount of the substance and not its identity. Thus, $2\,H_2O$ means two molecules of water, $3\,H_2O$ means three molecules of water, and so forth.

Reactants   Products

$2\,H_2 + O_2 \longrightarrow 2\,H_2O$

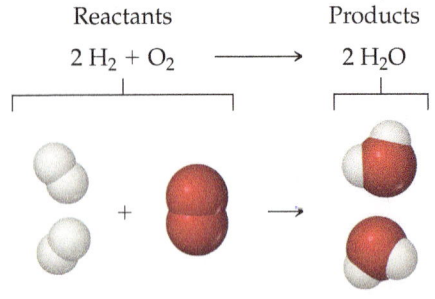

▲ **Figure 3.1** A balanced chemical equation.

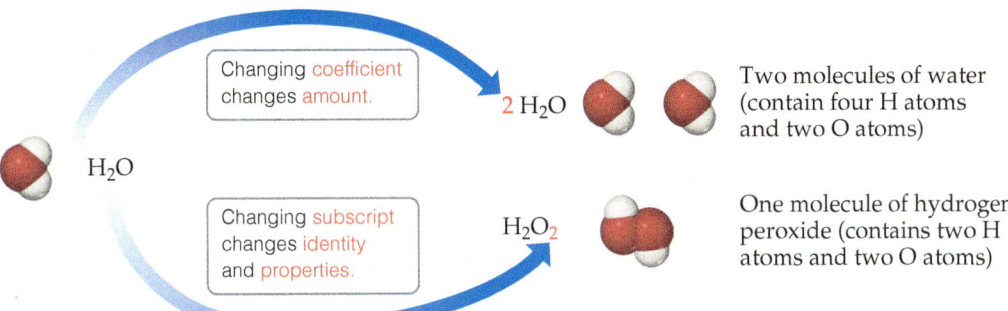

◄ **Figure 3.2** The difference between changing subscripts and changing coefficients in chemical equations.

Changing coefficient changes amount.

2 H₂O — Two molecules of water (contain four H atoms and two O atoms)

H₂O

Changing subscript changes identity and properties.

H₂O₂ — One molecule of hydrogen peroxide (contains two H atoms and two O atoms)

A Step-by-Step Example of Balancing a Chemical Equation

To illustrate the process of balancing an equation, consider the reaction that occurs when methane (CH_4), the principal component of natural gas, burns in air to produce carbon dioxide gas (CO_2) and water vapor (H_2O) (**Figure 3.3**). Both products contain oxygen atoms that come from O_2 in the air. Thus, O_2 is a reactant, and the unbalanced equation is

$$CH_4 + O_2 \longrightarrow CO_2 + H_2O \quad \text{(unbalanced)} \qquad [3.2]$$

It is usually best to balance first those elements that occur in the fewest chemical formulas in the equation. In our example, C appears in only one reactant (CH_4) and one product (CO_2). The same is true for H (CH_4 and H_2O). Notice, however, that O appears in one reactant (O_2) and two products (CO_2 and H_2O). So, let's begin with C. Because one molecule of CH_4 contains the same number of C atoms as one molecule of CO_2, the coefficients for these substances *must* be the same in the balanced equation. Therefore, we start by choosing the coefficient 1 (unwritten) for both CH_4 and CO_2.

Next we focus on H. On the left side of the equation we have CH_4, which has four H atoms, whereas on the right side of the equation we have H_2O, containing two H atoms. To balance the H atoms in the equation we place the coefficient 2 in front of H_2O. Now there are four H atoms on each side of the equation:

$$CH_4 + O_2 \longrightarrow CO_2 + 2 H_2O \quad \text{(unbalanced)} \qquad [3.3]$$

While the equation is now balanced with respect to hydrogen and carbon, it is not yet balanced for oxygen. Adding the coefficient 2 in front of O_2 balances the equation by giving four O atoms on each side [2×2 left, $2 + (2 \times 1)$ right]:

$$CH_4 + 2 O_2 \longrightarrow CO_2 + 2 H_2O \quad \text{(balanced)} \qquad [3.4]$$

The molecular view of the balanced equation is shown in **Figure 3.4**.

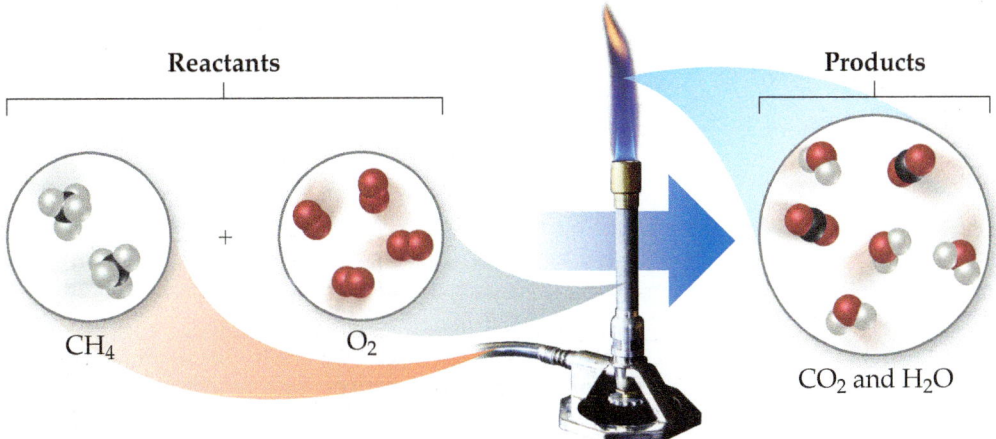

Reactants

+

CH_4 O_2

Products

CO_2 and H_2O

▲ **Figure 3.3** Methane reacts with oxygen in a Bunsen burner.

▶ **Figure 3.4** **Balanced chemical equation for the combustion of CH₄.**

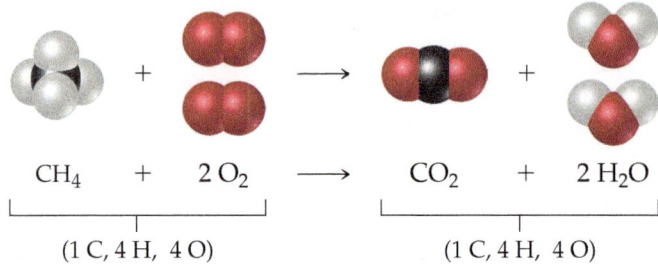

$$CH_4 \quad + \quad 2\,O_2 \quad \longrightarrow \quad CO_2 \quad + \quad 2\,H_2O$$

$$(1\,C, 4\,H, \ 4\,O) \qquad\qquad (1\,C, 4\,H, \ 4\,O)$$

 ## Sample Exercise 3.1

Interpreting and Balancing Chemical Equations

The following diagram represents a chemical reaction in which the red spheres are oxygen atoms and the blue spheres are nitrogen atoms. **(a)** Write the chemical formulas for the reactants and products. **(b)** Write a balanced equation for the reaction. **(c)** Is the diagram consistent with the law of conservation of mass?

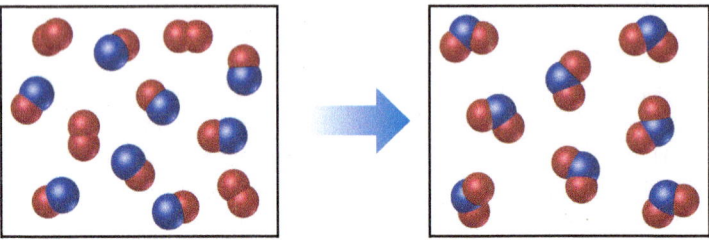

SOLUTION

(a) The left box, which represents reactants, contains two kinds of molecules, those composed of two oxygen atoms (O_2) and those composed of one nitrogen atom and one oxygen atom (NO). The right box, which represents products, contains only one kind of molecule, which is composed of one nitrogen atom and two oxygen atoms (NO_2).

(b) The unbalanced chemical equation is

$$O_2 + NO \longrightarrow NO_2 \quad \text{(unbalanced)}$$

An inventory of atoms on each side of the equation shows that there are one N and three O on the left side of the arrow and one N and two O on the right. To balance O, we must increase the number of O atoms on the right while keeping the coefficients for NO and NO_2 equal. Sometimes a trial-and-error approach is required; we need to go back and forth several times from one side of an equation to the other, changing coefficients first on one side of the equation and then the other until it is balanced. In our present case, let's start by increasing the number of O atoms on the right side of the equation by placing the coefficient 2 in front of NO_2:

$$O_2 + NO \longrightarrow 2\,NO_2 \quad \text{(unbalanced)}$$

Now the equation gives two N atoms and four O atoms on the right, so we go back to the left side. Placing the coefficient 2 in front of NO balances both N and O:

$$O_2 + 2\,NO \longrightarrow 2\,NO_2 \quad \text{(balanced)}$$

$$(2\,N, 4\,O) \quad (2\,N, 4\,O)$$

(c) The reactants box contains four O_2 and eight NO. Thus, the molecular ratio is one O_2 for each two NO, as required by the balanced equation. The products box contains eight NO_2, which means the number of NO_2 product molecules equals the number of NO reactant molecules, as the balanced equation requires.

There are eight N atoms in the eight NO molecules in the reactants box. There are also $4 \times 2 = 8$ O atoms in the O_2 molecules and 8 O atoms in the NO molecules, giving a total of 16 O atoms. In the products box, we find eight NO_2 molecules, which contain eight N atoms and $8 \times 2 = 16$ O atoms. Because there are equal numbers of N and O atoms in the two boxes, the drawing is consistent with the law of conservation of mass.

▶ **Practice Exercise**

In the following diagram, the white spheres represent hydrogen atoms, the black spheres carbon atoms, and the red spheres oxygen atoms.

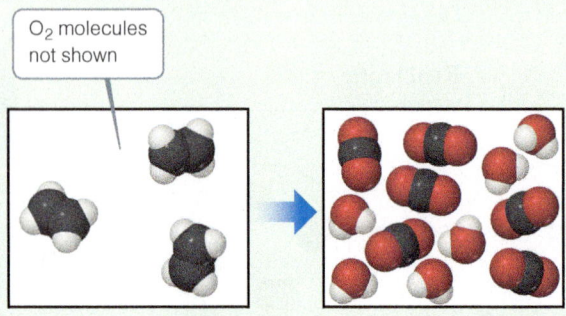

In this reaction, there are two reactants, ethylene, C_2H_4, which is shown, and oxygen, O_2, which is not shown, and two products, CO_2 and H_2O, both of which are shown. **(a)** Write a balanced chemical equation for the reaction. **(b)** How many O_2 molecules should be shown in the left (reactants) box?

Indicating the States of Reactants and Products

Symbols indicating the physical state of each reactant and product are often shown in chemical equations. We use the symbols (g), (l), (s), and (aq) for substances that are gases, liquids, solids, and dissolved in aqueous (water) solution, respectively. Thus, Equation 3.4 can be written

$$CH_4(g) + 2\,O_2(g) \longrightarrow CO_2(g) + 2\,H_2O(g) \qquad\qquad [3.5]$$

Sometimes symbols that represent the conditions under which the reaction proceeds appear above or below the reaction arrow. One example that we present later in this chapter involves the symbol Δ (Greek uppercase delta); a delta above the reaction arrow indicates the addition of heat.

Sample Exercise 3.2
Balancing Chemical Equations

Balance the equation

$$Na(s) + H_2O(l) \longrightarrow NaOH(aq) + H_2(g)$$

SOLUTION

Begin by counting each kind of atom on the two sides of the arrow. There are one Na, one O, and two H atoms on the left side, and one Na, one O, and three H atoms on the right. The Na and O atoms are currently balanced, but the number of H atoms is not. To increase the number of H atoms on the left, let's try placing the coefficient 2 in front of H_2O:

$$Na(s) + 2\,H_2O(l) \longrightarrow NaOH(aq) + H_2(g)$$

Although beginning this way does not balance H, it does increase the number of reactant H atoms, which we need to do. (Also, adding the coefficient 2 on H_2O unbalances O, but we will take care of that later.) Now that we have $2\,H_2O$ on the left, we balance H by putting the coefficient 2 in front of NaOH:

$$Na(s) + 2\,H_2O(l) \longrightarrow 2\,NaOH(aq) + H_2(g)$$

Balancing H in this way brings O into balance, but now Na is unbalanced, with one Na on the left and two on the right. To rebalance Na, we put the coefficient 2 in front of the reactant:

$$2\,Na(s) + 2\,H_2O(l) \longrightarrow 2\,NaOH(aq) + H_2(g)$$

We now have two Na atoms, four H atoms, and two O atoms on each side. The equation is balanced.

Comment Notice that we moved back and forth, placing a coefficient in front of H_2O, then NaOH, and finally Na. In balancing equations, we often find ourselves following this pattern of moving back and forth from one side of the arrow to the other, placing coefficients first in front of a formula on one side and then in front of a formula on the other side until the equation is balanced. You can always tell if you have balanced your equation correctly by checking that the number of atoms of each element is the same on the two sides of the arrow and that you've chosen the smallest set of coefficients that balances the equation.

▶ **Practice Exercise**
Balance these equations by providing the missing coefficients:
(a) __Fe(s) + __$O_2(g)$ $\longrightarrow$ __$Fe_2O_3(s)$
(b) __Al(s) + __HCl(aq) $\longrightarrow$ __$AlCl_3(aq)$ + __$H_2(g)$
(c) __$CaCO_3(s)$ + __HCl(aq) $\longrightarrow$ __$CaCl_2(aq)$ + __$CO_2(g)$ + __$H_2O(l)$

Self-Assessment Exercises

SAE 3.1 Carbon monoxide and oxygen react to form carbon dioxide according to the following balanced chemical equation:

$$2\,CO(g) + O_2(g) \longrightarrow 2\,CO_2(g)$$

If $CO(g)$ and $O_2(g)$ are combined in the ratio shown in the diagram below, where the black spheres represent carbon atoms and the red spheres represent oxygen atoms, which molecules will remain when the reaction is complete? **(a)** Only CO_2 molecules **(b)** CO_2 and CO molecules **(c)** CO_2 and O_2 molecules **(d)** CO and O_2 molecules

SAE 3.2 The unbalanced equation for the reaction between scandium metal and hydrochloric acid is:

$$\_\_Sc(s) + \_\_HCl(aq) \longrightarrow \_\_ScCl_3(aq) + \_\_H_2(g)$$

Once this equation is balanced with the smallest set of integer coefficients, what is the value of the coefficient in front of HCl(aq)? **(a)** 1 **(b)** 3 **(c)** 6 **(d)** 12

SAE 3.3 Sodium hydroxide and sodium metal react to form sodium oxide and hydrogen. Which of the following balanced equations represents this reaction?

(a) $NaOH(s) + Na(s) \longrightarrow Na_2O(s) + H_2(g)$
(b) $NaOH_2(s) + Na(s) \longrightarrow Na_2O(s) + H_2(g)$
(c) $2\,NaOH(s) + 2\,Na(s) \longrightarrow 2\,Na_2O(s) + H_2(g)$
(d) $NaOH(s) + Na(s) \longrightarrow Na_2O(s) + H(g)$

3.2 | Simple Patterns of Chemical Reactivity

⚠ Learning Objectives

When you finish Section 3.2, you should be able to:

▶ Identify combination reactions and predict their reaction products.
▶ Identify decomposition reactions and predict their reaction products.
▶ Identify combustion reactions and predict their reaction products.

One of the great triumphs of chemistry over the last hundred years is the development of fertilizers that enable us to feed the world. Ammonia, NH_3, is one of the principal chemicals farmers use to increase crop yield. The industrial process that is used to convert the elements nitrogen and hydrogen into ammonia, $N_2(g) + 3H_2(g) \rightarrow 2NH_3(g)$, may be a simple reaction, but it is one of the most important chemical reactions in the world. In this section, we introduce three broad classes of chemical reactions, including the combination reaction between nitrogen and hydrogen.

Combination and Decomposition Reactions

In **combination reactions**, two or more substances react to form one product (Table 3.1). For example, magnesium metal burns brilliantly in air to produce magnesium oxide (Figure 3.5):

$$2\,Mg(s) + O_2(g) \longrightarrow 2\,MgO(s) \qquad [3.6]$$

TABLE 3.1   Combination and Decomposition Reactions

Combination Reactions	
$A + B \longrightarrow C$	Two or more reactants combine to form a single product. Many elements react with one another in this fashion to form compounds.
$C(s) + O_2(g) \longrightarrow CO_2(g)$	
$N_2(g) + 3\,H_2(g) \longrightarrow 2\,NH_3(g)$	
$CaO(s) + H_2O(l) \longrightarrow Ca(OH)_2(aq)$	

Decomposition Reactions	
$C \longrightarrow A + B$	A single reactant breaks apart to form two or more substances. Many compounds react this way when heated.
$2\,KClO_3(s) \longrightarrow 2\,KCl(s) + 3\,O_2(g)$	
$PbCO_3(s) \longrightarrow PbO(s) + CO_2(g)$	
$Cu(OH)_2(s) \longrightarrow CuO(s) + H_2O(g)$	

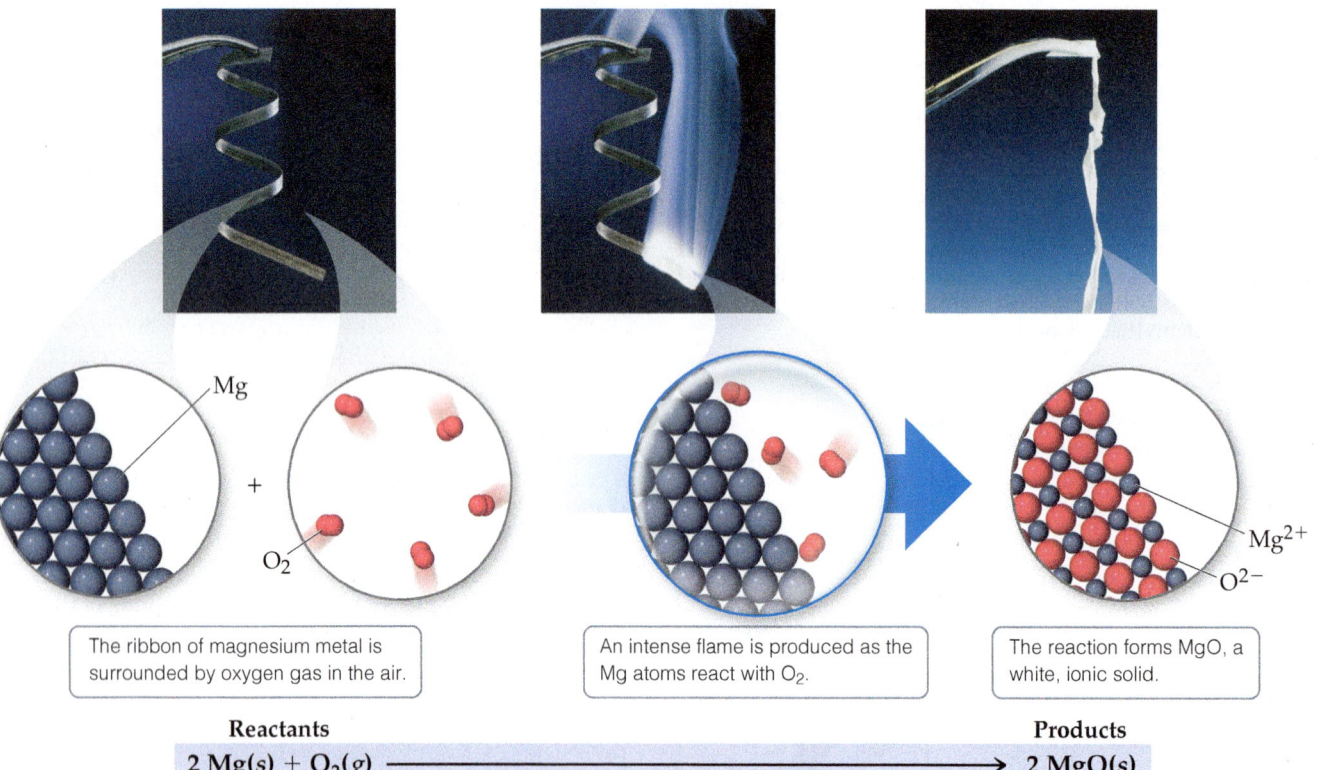

Mg

+

O_2

The ribbon of magnesium metal is surrounded by oxygen gas in the air.

An intense flame is produced as the Mg atoms react with O_2.

The reaction forms MgO, a white, ionic solid.

Mg^{2+}

O^{2-}

Reactants	Products
$2\,Mg(s) + O_2(g)$	$2\,MgO(s)$

▲ **Figure 3.5** The combustion of magnesium metal in air is a combination reaction.

This reaction is used to produce the bright flame generated by flares and some fireworks.

A combination reaction between a metal and a nonmetal, as in Equation 3.6, produces an ionic solid. Recall that the formula of an ionic compound can be determined from the charges of its ions. (Section 2.7) When magnesium reacts with oxygen, the magnesium loses electrons and forms the magnesium ion, Mg^{2+}. The oxygen gains electrons and forms the oxide ion, O^{2-}. Thus, the reaction product is MgO.

You should be able to recognize when a reaction is a combination reaction and to predict the products when the reactants are a metal and a nonmetal. Other examples of combination reactions include the formation of small gaseous molecules such as CO_2 and NH_3 from their elements, and the reaction of calcium oxide with water to produce calcium hydroxide:

$$C(s) + O_2(g) \longrightarrow CO_2(g) \qquad\qquad [3.7]$$

$$N_2(g) + 3H_2(g) \longrightarrow 2NH_3(g) \qquad\qquad [3.8]$$

$$CaO(s) + H_2O(l) \longrightarrow Ca(OH)_2(aq) \qquad\qquad [3.9]$$

In a **decomposition reaction**, one substance undergoes a reaction to produce two or more other substances (Table 3.1). For example, many metal carbonates decompose to form metal oxides and carbon dioxide when heated:

$$CaCO_3(s) \xrightarrow{\Delta} CaO(s) + CO_2(g) \qquad\qquad [3.10]$$

Decomposition of $CaCO_3$ is an important commercial process. Limestone or seashells, which are both primarily $CaCO_3$, are heated to prepare CaO, known as lime or quicklime. Tens of millions of tons of CaO are used in the United States each year in the production of glass and cement, in metallurgy where it is used to isolate the metals from their ores, and in steel manufacturing where it is used to remove impurities.

The decomposition of sodium azide (NaN_3) rapidly generates $N_2(g)$, and so this reaction is used to inflate safety air bags in automobiles (**Figure 3.6**):

$$2NaN_3(s) \longrightarrow 2Na(s) + 3N_2(g) \qquad\qquad [3.11]$$

The system is designed so that an impact ignites a detonator cap, which in turn causes NaN_3 to decompose explosively. A small quantity of NaN_3 (about 100 g) forms a large quantity of N_2 gas (about 50 L).

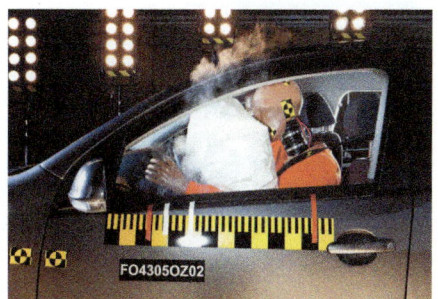

▲ **Figure 3.6** Decomposition of sodium azide, $NaN_3(s)$, produces $N_2(g)$ that inflates air bags in automobiles.

CHEMISTRY AND SUSTAINABILITY | Cement and CO_2 Emissions

Cement is produced on a scale that exceeds any other industrial material, approximately 4 billion metric tons per year. On a per mass basis, only water is consumed by society in larger quantities. Much of the cement is used as the binder in concrete, a composite of sand and gravel surrounded by a polymeric matrix containing hydrated oxides of calcium, silicon, aluminum, and iron. Cement has been used as a building material since Roman times, the Parthenon in Rome being one surviving example, but since 1950 the annual production of cement has increased more than 30-fold. This is a much faster rise than the approximate tripling of the world population that has occurred over the same period.

Five elements—oxygen, silicon, aluminum, iron, and calcium—account for more than 91% of Earth's crust. That cement is made primarily from these same five elements makes it possible to produce large volumes at a low cost, using local raw materials. In the most common type of cement, called Portland cement, calcium oxide accounts for 63% of the composition. The calcium oxide is obtained by heating limestone, which decomposes in what might be society's most important decomposition reaction, $CaCO_3(s) \longrightarrow CaO(s) + CO_2(g)$ (Equation 3.10). To drive the decomposition of $CaCO_3$ and trigger reactions between CaO and the oxides of Si, Al, and Fe, which mostly come from clays, the starting materials must be heated to temperatures near 1450 °C. Once this step is complete, the mixture is cooled and pulverized. The mixture at this point is called clinker, and it contains a number of oxides

such as Ca_3SiO_5, Cs_2SiO_4, $Ca_3Al_2O_6$, and Ca_2AlFeO_5. To complete the production process, a small amount of $CaSO_4 \cdot 2H_2O$ (gypsum) is added, and the solid is ground to a fine powder. When this powder is mixed with water, insoluble hydrate phases form around the larger aggregates that make up concrete, and the composite cures into a strong, durable structural material as the excess water evaporates.

There can be no argument that cement plays an outsized role in modern society; unfortunately, it also makes a surprisingly large contribution to global carbon dioxide emissions. It is estimated that approximately 8% of all human-made CO_2 emissions come from the manufacture of cement. Approximately 40% of these emissions come from the fuel consumed to heat the raw materials to such a high temperature, with the remainder coming from the liberation of CO_2 in the decomposition of limestone. Even if sources of renewable energy were tapped to provide all the energy needed to produce cement, there is no easy way to eliminate the CO_2 emissions that emanate from the use of limestone. Strategies that involve transitioning to cements that have lower calcium contents and/or formulations that improve the binding strength of cement, thereby reducing the percentage of cement in concrete, can help reduce emissions. However, to reduce CO_2 emissions to near zero will likely require capturing the CO_2 that is produced and either storing it or using it for other purposes.

Related Exercises: 3.105, 3.107

Sample Exercise 3.3

Writing Balanced Equations for Combination and Decomposition Reactions

Write a balanced equation for (**a**) the combination reaction between lithium metal and fluorine gas and (**b**) the decomposition reaction that occurs when solid barium carbonate is heated (two products form, a solid and a gas).

SOLUTION

(**a**) With the exception of mercury, all metals are solids at room temperature. Fluorine occurs as a diatomic molecule. Thus, the reactants are Li(*s*) and $F_2(g)$. The product will be composed of a metal and a nonmetal, so we expect it to be an ionic solid. Lithium ions have a 1+ charge, Li^+, whereas fluoride ions have a 1− charge, F^-. Thus, the chemical formula for the product is LiF. The balanced chemical equation is

$$2\,Li(s) + F_2(g) \longrightarrow 2\,LiF(s)$$

(**b**) The chemical formula for barium carbonate is $BaCO_3$. As mentioned, many metal carbonates decompose to metal

oxides and carbon dioxide when heated. In Equation 3.10, for example, $CaCO_3$ decomposes to form CaO and CO_2. Thus, we expect $BaCO_3$ to decompose to BaO and CO_2. Barium and calcium are both in group 2A in the periodic table, which further suggests they react in the same way:

$$BaCO_3(s) \longrightarrow BaO(s) + CO_2(g)$$

▶ **Practice Exercise**

Write a balanced equation for (**a**) solid mercury(II) sulfide decomposing into its component elements when heated and (**b**) aluminum metal combining with oxygen in the air.

Go Figure

Does this reaction produce or consume thermal energy (heat)?

▲ **Figure 3.7 Propane burning in air.** Liquid propane in the tank, C_3H_8, vaporizes and mixes with air as it escapes through the nozzle. The combustion reaction of C_3H_8 and O_2 produces a blue flame.

Combustion Reactions

Combustion reactions are rapid reactions that produce a flame. Most combustion reactions we observe involve O_2 from air as a reactant. The combustion of hydrocarbons (compounds that contain only carbon and hydrogen) in air is a major energy-producing process in our world. (Section 2.9)

Hydrocarbons combusted in air react with O_2 to form CO_2 and H_2O.* The number of molecules of O_2 required and the number of molecules of CO_2 and H_2O formed depend on the composition of the hydrocarbon, which acts as the fuel in the reaction. For example, the combustion of propane (C_3H_8, **Figure 3.7**), a gas used for cooking and home heating, is described by the equation

$$C_3H_8(g) + 5\,O_2(g) \longrightarrow 3\,CO_2(g) + 4\,H_2O(g) \qquad [3.12]$$

The state of the water in this reaction, $H_2O(g)$ or $H_2O(l)$ depends on the reaction conditions. In this case, $H_2O(g)$ is formed as the high-temperature flame burns in air.

Millions of compounds are made only of carbon, hydrogen, and oxygen. Notable classes of such molecules are sugars like sucrose ($C_{12}H_{22}O_{11}$) and alcohols like methanol (CH_3OH). Combustion of these oxygen-containing derivatives of hydrocarbons in air also produces CO_2, H_2O, and energy. Many of the substances that function as energy sources in metabolism, such as the sugar glucose ($C_6H_{12}O_6$), react with O_2 to ultimately form CO_2 and H_2O. In our bodies, however, the reactions take place in a series of intermediate steps that occur at body temperature. These reactions are called *oxidation reactions* rather than combustion reactions.

Sample Exercise 3.4

Writing Balanced Equations for Combustion Reactions

Write the balanced equation for the reaction that occurs when methanol, $CH_3OH(l)$, is burned in air.

SOLUTION

When any compound containing C, H, and O is combusted, it reacts with the $O_2(g)$ in air to produce $CO_2(g)$ and $H_2O(g)$. Thus, the unbalanced equation is

$$CH_3OH(l) + O_2(g) \longrightarrow CO_2(g) + H_2O(g)$$

The C atoms are balanced, one on each side of the arrow. Because CH_3OH has four H atoms, we place the coefficient 2 in front of H_2O to balance the H atoms:

$$CH_3OH(l) + O_2(g) \longrightarrow CO_2(g) + 2\,H_2O(g)$$

---

*When an insufficient quantity of O_2 is present, carbon monoxide (CO) is produced along with CO_2; this is called incomplete combustion. If the quantity of O_2 is severely restricted, the fine particles of carbon we call soot are produced. Complete combustion produces only CO_2 and H_2O. Unless stated to the contrary, we always take combustion to mean complete combustion.

Adding this coefficient balances H but gives four O atoms in the products. Because there are only three O atoms in the reactants, we are not finished. We can place the coefficient $\frac{3}{2}$ in front of O_2 to give four O atoms in the reactants ($\frac{3}{2} \times 2 = 3$ O atoms in $\frac{3}{2}O_2$):

$$CH_3OH(l) + \tfrac{3}{2}O_2(g) \longrightarrow CO_2(g) + 2H_2O(g)$$

Although this equation is balanced, it is not in its most conventional form because it contains a fractional coefficient.

Multiplying through by 2 removes the fraction and keeps the equation balanced:

$$2CH_3OH(l) + 3O_2(g) \longrightarrow 2CO_2(g) + 4H_2O(g)$$

▶ **Practice Exercise**

Write the balanced equation for the reaction that occurs when ethanol, $C_2H_5OH(l)$, burns in air.

 Self-Assessment Exercises

SAE 3.4 When Na and S undergo a combination reaction, what is the chemical formula of the product? **(a)** NaS **(b)** Na_2S **(c)** NaS_2 **(d)** Na_2S_3 **(e)** Na_3S_2

SAE 3.5 Using a process called electrolysis, we can decompose water into its constituent elements. Which of the following equations describes this reaction?

(a) $H_2O_2(l) \longrightarrow H_2(g) + O_2(g)$
(b) $H_2O(l) \longrightarrow H_2(g) + O_2(g)$
(c) $H_2O(l) \longrightarrow 2H(g) + O(g)$
(d) $2H_2O(l) \longrightarrow 2H_2(g) + O_2(g)$

SAE 3.6 What is the correct balanced equation for the reaction that occurs when diethyl ether, $CH_3CH_2OCH_2CH_3$, burns in air?

(a) $CH_3CH_2OCH_2CH_3(l) + 5O_2(g) \longrightarrow 4CO_2(g) + 5H_2O(g)$
(b) $CH_3CH_2OCH_2CH_3(l) + 6O_2(g) \longrightarrow 4CO_2(g) + 5H_2O(g)$
(c) $CH_3CH_2OCH_2CH_3(l) + 12O(g) \longrightarrow 4CO_2(g) + 5H_2O(g)$
(d) $2CH_3CH_2OCH_2CH_3(l) + 12O_2(g) \longrightarrow 8CO_2(g) + 10H_2O(g)$

3.3 | Formula Weights

Sulfuric acid, $H_2SO_4(l)$, is a common laboratory chemical that is used by the metric ton in the chemical industry. We can see from the molecular model in **Figure 3.8** that one molecule of H_2SO_4 contains one sulfur atom, four oxygen atoms, and two hydrogen atoms. In the lab, we most commonly dispense milliliters of its aqueous solution, $H_2SO_4(aq)$. While the atomic ratios of sulfur, oxygen, and hydrogen may be fixed by the chemical formula, how do we know how many molecules of H_2SO_4 are present in a sample we encounter in the lab? That is the topic we begin to explore in this section.

 **Learning Objectives**

When you finish **Section 3.3**, you should be able to:

▶ Calculate the formula weight of a substance from its empirical formula or its molecular weight from its molecular formula.

▶ Determine the elemental composition (by mass) of a compound from its empirical or molecular formula.

Formula and Molecular Weights

Although we cannot count individual atoms or molecules, we can indirectly determine their numbers if we know their masses. Therefore, we need to know more about the masses of substances. The **formula weight** (FW) of a substance is the sum of the atomic weights (AW) of the atoms in the chemical formula of the substance. Using atomic weights from the periodic table, we find, for example, that the formula weight of sulfuric acid, H_2SO_4, is 98.1 amu (atomic mass units):

$$\begin{aligned}
\text{FW of } H_2SO_4 &= 2(\text{AW of H}) + (\text{AW of S}) + 4(\text{AW of O}) \\
&= 2(1.0 \text{ amu}) + 32.1 \text{ amu} + 4(16.0 \text{ amu}) \\
&= 98.1 \text{ amu}
\end{aligned}$$

For convenience, we have rounded off the atomic weights to one decimal place, a practice we follow in most calculations in this book.

If the chemical formula is the chemical symbol of an element, such as Na, the formula weight equals the atomic weight of the element, in this case 23.0 amu. If the chemical formula is that of a molecule, the formula weight is also called the **molecular weight** (MW). The molecular weight of glucose ($C_6H_{12}O_6$), for example, is

$$\text{MW of } C_6H_{12}O_6 = 6(12.0 \text{ amu}) + 12(1.0 \text{ amu}) + 6(16.0 \text{ amu}) = 180.0 \text{ amu}$$

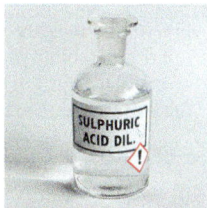

▲ **Figure 3.8 Sulfuric acid.** The sulfuric acid molecule, H_2SO_4, is represented here with yellow, red, and white spheres for sulfur, oxygen, and hydrogen, respectively. Also shown is a bottle of $H_2SO_4(aq)$ solution.

Because ionic substances exist as three-dimensional arrays of ions (see Figure 2.20), it is inappropriate to speak of *molecules* of these substances. Instead we use the empirical formula as the formula unit, and the formula weight of an ionic substance is determined by summing the atomic weights of the atoms in the empirical formula. For example, the formula unit of $CaCl_2$ consists of one Ca^{2+} ion and two Cl^- ions. Thus, the formula weight of $CaCl_2$ is

$$\text{FW of } CaCl_2 = 40.1 \text{ amu} + 2(35.5 \text{ amu}) = 111.1 \text{ amu}$$

Sample Exercise 3.5
Calculating Formula Weights

Calculate the formula weight of (**a**) sucrose, $C_{12}H_{22}O_{11}$ (table sugar); and (**b**) calcium nitrate, $Ca(NO_3)_2$.

SOLUTION

(**a**) By adding the atomic weights of the atoms in sucrose, we find the formula weight to be 342.0 amu:

$$12 \text{ C atoms} = 12(12.0 \text{ amu}) = 144.0 \text{ amu}$$
$$22 \text{ H atoms} = 22(1.0 \text{ amu}) = 22.0 \text{ amu}$$
$$11 \text{ O atoms} = 11(16.0 \text{ amu}) = \underline{176.0 \text{ amu}}$$
$$342.0 \text{ amu}$$

(**b**) If a chemical formula has parentheses, the subscript outside the parentheses is a multiplier for all atoms inside. Thus, for $Ca(NO_3)_2$ we have

$$1 \text{ Ca atom} = 1(40.1 \text{ amu}) = 40.1 \text{ amu}$$
$$2 \text{ N atoms} = 2(14.0 \text{ amu}) = 28.0 \text{ amu}$$
$$6 \text{ O atoms} = 6(16.0 \text{ amu}) = \underline{96.0 \text{ amu}}$$
$$164.1 \text{ amu}$$

▶ **Practice Exercise**
Calculate the formula weight of (**a**) $Al(OH)_3$, (**b**) CH_3OH, and (**c**) TaON.

Percentage Composition from Chemical Formulas

Chemists must sometimes calculate the *percentage composition* of a compound—that is, the percentage by mass contributed by each element in the substance. Let's say you are a forensic chemist, working in a crime lab, and your colleagues find a mysterious white powder at a crime scene. Is it salt, sugar, methamphetamine, cocaine, or something else?

One way to determine the identity of a substance is to measure its **elemental composition** and compare it to the calculated elemental compositions of possible candidate substances. The calculation depends on the formula weight of the substance, the atomic weight of the element of interest, and the number of atoms of that element in the chemical formula:

$$\begin{array}{c} \text{\% mass composition} \\ \text{of element} \end{array} = \frac{\left(\begin{array}{c}\text{number of atoms}\\\text{of element}\end{array}\right)\left(\begin{array}{c}\text{atomic weight}\\\text{of element}\end{array}\right)}{\text{formula weight of substance}} \times 100\% \qquad [3.13]$$

Sample Exercise 3.6
Calculating Percentage Composition

Calculate the percentage of carbon, hydrogen, and oxygen (by mass) in sucrose, $C_{12}H_{22}O_{11}$.

SOLUTION

Analyze We are given a chemical formula and asked to calculate the percentage by mass of each element.

Plan We use Equation 3.13, obtaining our atomic weights from a periodic table. We know the denominator in Equation 3.13, the formula weight of $C_{12}H_{22}O_{11}$, from Sample Exercise 3.5. We must use that value in three calculations, one for each element.

Solve

$$\text{\% carbon} = \frac{(12)(12.0 \text{ amu})}{342.0 \text{ amu}} \times 100\% = 42.1\%$$

$$\text{\% hydrogen} = \frac{(22)(1.0 \text{ amu})}{342.0 \text{ amu}} \times 100\% = 6.4\%$$

$$\text{\% oxygen} = \frac{(11)(16.0 \text{ amu})}{342.0 \text{ amu}} \times 100\% = 51.5\%$$

Check Our calculated percentages must add up to 100%, which they do. We could have used more significant figures for our atomic weights, giving more significant figures for our percentage composition, but we have adhered to our suggested guideline of rounding atomic weights to one digit beyond the decimal point.

▶ **Practice Exercise**
Calculate the percentage of potassium, by mass, in K_2PtCl_6.

The sum of all the mass percentages of each element in the compound must add up to 100%.

As an example, let's calculate the mass percentage of sulfur in sulfuric acid. Based on the molecular formula, we can see that for each H_2SO_4 molecule, one atom out of seven is sulfur. But that doesn't mean that 1/7 of the mass of the compound is sulfur, because the atoms that make up the molecule have different masses. Using Equation 3.13:

$$\% \text{ sulfur in } H_2SO_4 = \frac{(1)(32.1\,\text{amu})}{98.1\,\text{amu}} = 32.7\%$$

we find that almost a third of the mass of any sample of pure H_2SO_4 is due to sulfur.

STRATEGIES FOR SUCCESS | Problem Solving

Practice is the key to success in solving problems. As you practice, you can improve your skills by following these steps:

1. **Analyze the problem.** Read the problem carefully. What does it say? Draw a picture or diagram that will help you to visualize the problem. Write down both the data you are given and the quantity you need to obtain (the unknown).

2. **Develop a plan for solving the problem.** Consider a possible path between the given information and the unknown. What principles or equations relate the known data to the unknown? Recognize that some data may not be given explicitly in the problem; you may be expected to know certain quantities or to look them up in tables (such as atomic

weights). Recognize also that your plan may involve either a single step or a series of steps with intermediate answers.

3. **Solve the problem.** Use the known information and suitable equations or relationships to solve for the unknown. Dimensional analysis (Section 1.7) is a useful tool for solving a great number of problems. Be careful with significant figures, signs, and units.

4. **Check the solution.** Read the problem again to make sure you have found all the solutions asked for in the problem. Does your answer make sense? That is, is the answer outrageously large or small, or is it in the ballpark? Finally, are the units and significant figures correct?

 Self-Assessment Exercises

SAE 3.7 Which of the following is the correct formula weight for copper(II) chloride? (**a**) 99.0 amu (**b**) 134.5 amu (**c**) 162.6 amu (**d**) 233.5 amu

SAE 3.8 What is the formula weight of hypochlorous acid? (**a**) 36.5 amu (**b**) 52.5 amu (**c**) 67.4 amu (**d**) 100.5 amu

SAE 3.9 What is the percentage of phosphorus, by mass, in calcium phosphate? (**a**) 70.3% (**b**) 61.2% (**c**) 22.9% (**d**) 20.0% (**e**) 10.0%

SAE 3.10 A mysterious white powder found at a crime scene is analyzed and found to contain 66.9 ± 0.5% carbon by mass. One of the investigators hypothesizes that the substance is cocaine, $C_{17}H_{21}NO_4$. What is the percent carbon, by mass, in cocaine? (**a**) 39.5% (**b**) 64.3% (**c**) 67.3% (**d**) 70.6%

3.4 | Avogadro's Number and the Mole

Even the smallest samples we deal with in the laboratory contain enormous numbers of atoms, ions, or molecules. For example, a teaspoon of water (about 5 mL) contains 2×10^{23} water molecules, a number so large it almost defies comprehension. Chemists therefore have devised a counting unit for describing large numbers of atoms or molecules.

In everyday life we use such familiar counting units as dozen (12 objects) and gross (144 objects). In chemistry the counting unit for numbers of atoms, ions, or molecules in a laboratory-sized sample is the *mole*, often abbreviated mol. One **mole** is the amount of matter that contains as many objects (atoms, molecules, or whatever other objects we are considering) as the number of atoms in exactly 12 g of isotopically pure ^{12}C. From experiments, scientists have determined this number to be 6.0221415×10^{23}, which we usually round to 6.02×10^{23}. Scientists call this value **Avogadro's number**, N_A, in honor of the Italian scientist Amedeo Avogadro (1776–1856), and it is often cited with units of reciprocal moles, $6.02 \times 10^{23}\,\text{mol}^{-1}$.* The unit (read as either "inverse mole" or "per

 Learning Objectives

When you finish Section 3.4, you should be able to:

▶ Calculate the *molar mass* for a compound and relate this to its formula weight.

▶ Interconvert between grams, molecules, and moles of a substance.

*Avogadro's number is also referred to as the Avogadro constant. The latter term is the name adopted by agencies such as the National Institute of Standards and Technology (NIST), but Avogadro's number remains in widespread usage and is used in most places in this book.

mole") reminds us that there are 6.02×10^{23} objects per one mole. A mole of atoms, a mole of molecules, or a mole of anything else all contain Avogadro's number of objects:

$$1 \text{ mol } {}^{12}\text{C atoms} = 6.02 \times 10^{23} \, {}^{12}\text{C atoms}$$

$$1 \text{ mol } H_2O \text{ molecules} = 6.02 \times 10^{23} \, H_2O \text{ molecules}$$

$$1 \text{ mol } NO_3^- \text{ ions} = 6.02 \times 10^{23} \, NO_3^- \text{ ions}$$

Avogadro's number is so large that spreading 6.02×10^{23} marbles over Earth's surface would produce a layer about 3 miles thick. Avogadro's number of pennies placed side by side in a straight line would encircle Earth 300 trillion (3×10^{14}) times.

Sample Exercise 3.7
Estimating Numbers of Atoms

Without using a calculator, arrange these samples in order of increasing numbers of carbon atoms: 12 g ^{12}C, 1 mol C_2H_2, 9×10^{23} molecules of CO_2.

SOLUTION

Analyze We are given amounts of three substances expressed in grams, moles, and number of molecules, respectively, and are asked to arrange the samples in order of increasing numbers of C atoms.

Plan To determine the number of C atoms in each sample, we must convert 12 g ^{12}C, 1 mol C_2H_2 and 9×10^{23} molecules of CO_2 to numbers of C atoms. To make these conversions, we use the definition of mole and Avogadro's number.

Solve One mole is defined as the amount of matter that contains as many units of the matter as there are C atoms in exactly 12 g of ^{12}C. Thus, 12 g of ^{12}C contains 1 mol of C atoms = 6.02×10^{23} atoms.

One mol of C_2H_2 contains 6.02×10^{23} molecules. Because there are two C atoms in each molecule, this sample contains 12.04×10^{23} C atoms.

Because each CO_2 molecule contains one C atom, the CO_2 sample contains 9×10^{23} C atoms.

Hence, the order is 12 g ^{12}C (6×10^{23} atoms) $< 9 \times 10^{23} \, CO_2$ molecules (9×10^{23} C atoms) < 1 mol C_2H_2 (12×10^{23} C atoms).

Check We can check our results by comparing numbers of moles of C atoms in the samples because the number of moles is proportional to the number of atoms. Thus, 12 g of ^{12}C is 1 mol C, 1 mol of C_2H_2 contains 2 mol C, and 9×10^{23} molecules of CO_2 contain 1.5 mol C, giving the same order as stated previously.

▶ **Practice Exercise**
Without using a calculator, arrange these samples in order of increasing numbers of O atoms: 1 mol H_2O, 1 mol CO_2, 3×10^{23} molecules O_2.

Molar Mass

A dozen is the same number, 12, whether we have a dozen eggs or a dozen elephants. A dozen eggs has considerably less mass, however, than a dozen elephants. Similarly, a mole is always the *same number* (6.02×10^{23}) but 1-mol samples of different substances have *different masses*. Compare, for example, 1 mol of ^{12}C and 1 mol of ^{24}Mg. A single ^{12}C atom has a mass of 12 amu, whereas a single ^{24}Mg atom is twice as massive, 24 amu (to two significant figures). Because a mole of anything always contains the same number of particles, a mole of ^{24}Mg must be twice as massive as a mole of ^{12}C. Because a mole of ^{12}C has a mass of 12 g (by definition), a mole of ^{24}Mg must have a mass of 24 g. This example illustrates a general rule relating the mass of an atom to the mass of Avogadro's number (1 mol) of these atoms: *The atomic weight of an element in atomic mass units is numerically equal to the mass in grams of 1 mol of that element.* For example (the symbol $\Rightarrow$ means "therefore"),

Cl has an atomic weight of 35.5 amu $\Rightarrow$ 1 mol Cl has a mass of 35.5 g.

Au has an atomic weight of 197 amu $\Rightarrow$ 1 mol Au has a mass of 197 g.

For other kinds of substances, the same numerical relationship exists between formula weight and mass of 1 mol of a substance:

H_2O has a formula weight of 18.0 amu $\Rightarrow$ 1 mol H_2O has a mass of 18.0 g (**Figure 3.9**).

NaCl has a formula weight of 58.5 amu $\Rightarrow$ 1 mol NaCl has a mass of 58.5 g.

The mass in grams of one mole of a substance is called the **molar mass** of the substance. *The molar mass in grams per mole of any substance is numerically equal to its formula weight in atomic mass units.* For NaCl, for example, the formula weight is 58.5 amu and the molar mass is 58.5 g/mol. Mole relationships for several other substances are listed in **Table 3.2**, and **Figure 3.10** shows 1 mol quantities of three common substances.

The entries in Table 3.2 for N and N_2 point out the importance of stating the chemical form of a substance when using the mole concept. Suppose you read that 1 mol of nitrogen is produced in a particular reaction. You might interpret this statement to mean 1 mol of nitrogen atoms (14.0 g). Unless otherwise stated, however, what is probably meant is 1 mol of nitrogen molecules, N_2 (28.0 g), because N_2 is the most common chemical form of the element. To avoid ambiguity, it is important to state explicitly the chemical form being discussed. Using the chemical formula—N or N_2, for instance—avoids any confusion.

TABLE 3.2   Mole Relationships

Name of Substance	Formula	Formula Weight (amu)	Molar Mass (g/mol)	Number and Kind of Particles in One Mole
Atomic nitrogen	N	14.0	14.0	6.02×10^{23} N atoms
Molecular nitrogen or "dinitrogen"	N_2	28.0	28.0	$\begin{cases} 6.02 \times 10^{23}\ N_2\ \text{molecules} \\ 2(6.02 \times 10^{23})\ \text{N atoms} \end{cases}$
Silver	Ag	107.9	107.9	6.02×10^{23} Ag atoms
Silver ions	Ag^+	107.9[a]	107.9	$6.02 \times 10^{23}\ Ag^+$ ions
Barium chloride	$BaCl_2$	208.2	208.2	$\begin{cases} 6.02 \times 10^{23}\ BaCl_2\ \text{formula units} \\ 6.02 \times 10^{23}\ Ba^{2+}\ \text{ions} \\ 2(6.02 \times 10^{23})\ Cl^-\ \text{ions} \end{cases}$

[a]Recall that the mass of an electron is more than 1800 times smaller than the masses of the proton and the neutron; thus, ions and atoms have essentially the same mass.

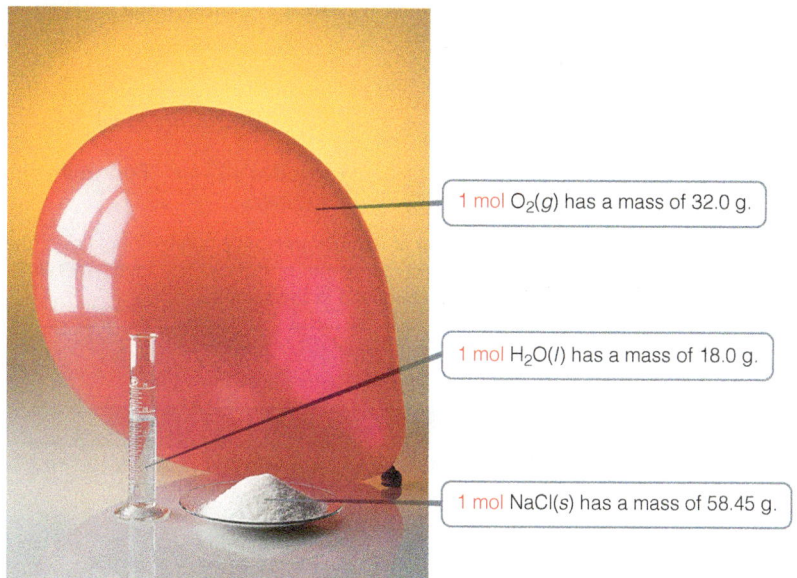

1 mol $O_2(g)$ has a mass of 32.0 g.

1 mol $H_2O(l)$ has a mass of 18.0 g.

1 mol NaCl(s) has a mass of 58.45 g.

▲ **Figure 3.10** **One mole each of a solid (NaCl), a liquid (H₂O), and a gas (O₂).** In each case, the mass in grams of 1 mol—that is, the molar mass—is numerically equal to the formula weight in atomic mass units. Each of these samples contains 6.02×10^{23} formula units.

Go Figure

What number do you get if you divide the mass of 1 mol of water by the mass of 1 molecule of water?

Single molecule

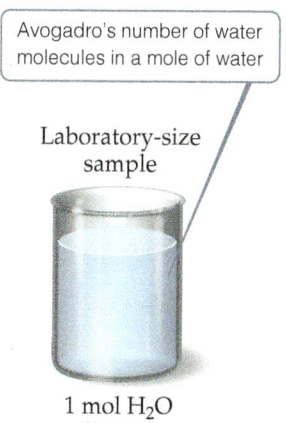

1 molecule H_2O
(18.0 amu)

Avogadro's number of water molecules in a mole of water

Laboratory-size sample

1 mol H_2O
(18.0 g)

▲ **Figure 3.9** **Comparing the mass of 1 molecule and 1 mol of H₂O.** Both masses have the same number but different units (atomic mass units and grams). Expressing both masses in grams indicates their huge difference: 1 molecule of H_2O has a mass of 2.99×10^{-23} g, whereas 1 mol of H_2O has a mass of 18.0 g.

Sample Exercise 3.8
Calculating Molar Mass

What is the molar mass of glucose, $C_6H_{12}O_6$?

SOLUTION

Analyze We are given a chemical formula and asked to determine its molar mass.

Plan Because the molar mass of any substance is numerically equal to its formula weight, we first determine the formula weight of glucose by adding the atomic weights of its component atoms. The formula weight will have units of amu, whereas the molar mass has units of grams per mole (g/mol).

Solve Our first step is to determine the formula weight of glucose:

$$
\begin{aligned}
6\ \text{C atoms} = 6(12.0\ \text{amu}) &= 72.0\ \text{amu} \\
12\ \text{H atoms} = 12(1.0\ \text{amu}) &= 12.0\ \text{amu} \\
6\ \text{O atoms} = 6(16.0\ \text{amu}) &= \underline{96.0\ \text{amu}} \\
&\ \ \ 180.0\ \text{amu}
\end{aligned}
$$

Because glucose has a formula weight of 180.0 amu, 1 mol of this substance (6.02×10^{23} molecules) has a mass of 180.0 g. In other words, $C_6H_{12}O_6$ has a molar mass of 180.0 g/mol.

Continued

Check A molar mass below 250 seems reasonable based on the earlier examples we have encountered, and grams per mole is the appropriate unit for the molar mass.

▶ **Practice Exercise**
Calculate the molar mass of $Ca(NO_3)_2$.

Interconverting Masses, Moles and Numbers of Particles

The mole concept provides the bridge between mass and number of particles (**Figure 3.11**). We can use this concept to relate the stoichiometries expressed in chemical equations to the amounts of chemicals used and produced in reactions. For example, let's calculate how many copper atoms are in an old copper penny. The penny has a mass of about 3 grams, and for simplicity we'll assume it is pure copper:

$$\text{Cu atoms} = (3 \text{ g Cu})\left(\frac{1 \text{ mol Cu}}{63.5 \text{ g Cu}}\right)\left(\frac{6.02 \times 10^{23} \text{ Cu atoms}}{1 \text{ mol Cu}}\right)$$
$$= 3 \times 10^{22} \text{ Cu atoms}$$

We have rounded our answer to one significant figure because we used only one significant figure for the mass of the penny. Notice how dimensional analysis provides a straightforward route from grams to numbers of atoms. The molar mass is used as a conversion factor to convert grams to moles; then Avogadro's number is used to convert moles to atoms (Figure 3.11). Notice also that our answer is a very large number. Any time you calculate the number of atoms, molecules, or ions in an ordinary sample of matter, you can expect the answer to be very large. In contrast, the number of moles in a sample will usually be small, often less than 1.

Go Figure What units would you assign to "molar mass" and "Avogadro's number" in this diagram?

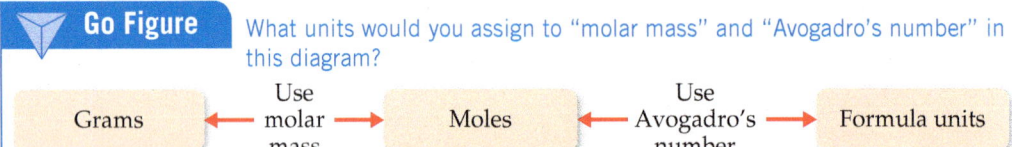

▲ **Figure 3.11 Procedure for interconverting mass and number of formula units.** The number of moles of the substance is central to the calculation. Thus, the mole concept can be thought of as the bridge between the mass of a sample in grams and the number of formula units contained in the sample.

 Sample Exercise 3.9
Converting Grams to Moles

Calculate the number of moles of glucose ($C_6H_{12}O_6$) in a 5.380-g sample.

SOLUTION

Analyze We are given the number of grams of a substance and its chemical formula and asked to calculate the number of moles.

Plan The molar mass of a substance provides the factor for converting grams to moles. The molar mass of $C_6H_{12}O_6$ is 180.0 g/mol (Sample Exercise 3.8).

Solve Using 1 mol $C_6H_{12}O_6$ = 180.0 g $C_6H_{12}O_6$ to write the appropriate conversion factor, we have

$$\text{Moles } C_6H_{12}O_6 = (5.380 \text{ g } C_6H_{12}O_6)\left(\frac{1 \text{ mol } C_6H_{12}O_6}{180.0 \text{ g } C_6H_{12}O_6}\right)$$
$$= 0.02989 \text{ mol } C_6H_{12}O_6$$

Check Because 5.380 g is less than the molar mass, an answer less than one mole is reasonable. The unit mol is appropriate. The original data had four significant figures, so our answer has four significant figures.

▶ **Practice Exercise**
A 508-g sample of sodium bicarbonate ($NaHCO_3$) contains how many moles of sodium bicarbonate?

 Self-Assessment Exercises

SAE 3.11 Two beakers contain equal masses of CCl_4 and NH_3, respectively. Which beaker contains more molecules? (**a**) The beaker with CCl_4 (**b**) The beaker with NH_3 (**c**) The two beakers contain the same number of molecules

SAE 3.12 An aqueous solution of caffeine contains 0.186 mol of caffeine. When the caffeine is separated from the rest of the solution, it is found to have a mass of 36.1 g. What is the molar mass of caffeine? (**a**) 0.00515 g/mol (**b**) 36.1 g/mol (**c**) 186 g/mol (**d**) 194 g/mol

3.5 | Empirical Formulas from Analyses

The empirical formula for a substance tells us the relative number of atoms of each element in the substance. (Section 2.6) The empirical formula H_2O shows that water contains two H atoms for each O atom. This ratio also applies on the molar level: 1 mol of H_2O contains 2 mol of H atoms and 1 mol of O atoms. Conversely, *the ratio of the numbers of moles of all elements in a compound gives the subscripts in the compound's empirical formula*. Thus, the mole concept provides a way of calculating empirical formulas from experimental data.

Mercury and chlorine combine to form a compound that is measured to be 74.0% mercury and 26.0% chlorine by mass. Thus, if we had a 100.0-g sample of the compound, it would contain 74.0 g of mercury and 26.0 g of chlorine. (Samples of any size can be used in problems of this type, but we will generally use 100.0 g to simplify the calculation of mass from percentage.) Using atomic weights to get molar masses, we calculate the number of moles of each element in the sample:

$$(74.0 \text{ g Hg})\left(\frac{1 \text{ mol Hg}}{200.6 \text{ g Hg}}\right) = 0.369 \text{ mol Hg}$$

$$(26.0 \text{ g Cl})\left(\frac{1 \text{ mol Cl}}{35.5 \text{ g Cl}}\right) = 0.732 \text{ mol Cl}$$

We then divide the larger number of moles by the smaller number to obtain the Cl:Hg mole ratio:

$$\frac{\text{moles of Cl}}{\text{moles of Hg}} = \frac{0.732 \text{ mol Cl}}{0.369 \text{ mol Hg}} = \frac{1.98 \text{ mol Cl}}{1 \text{ mol Hg}}$$

Because of experimental errors, calculated values for a mole ratio may not be exact whole numbers, as in the calculation here. The number 1.98 is very close to 2, however, and so we can confidently conclude that the empirical formula for the compound is $HgCl_2$. The empirical formula is correct because its subscripts are the smallest integers that express the *ratio* of atoms present in the compound.

The general procedure for determining empirical formulas is outlined in **Figure 3.12**.

Learning Objectives

When you finish Section 3.5, you should be able to:

▶ Calculate the empirical formula of a compound from the mass percentages of the elements that make up the compound.

▶ Determine the molecular formula of a molecular compound given its empirical formula and molecular weight.

▶ Determine the empirical formula of a compound containing only carbon, hydrogen, and oxygen from the results of combustion analysis.

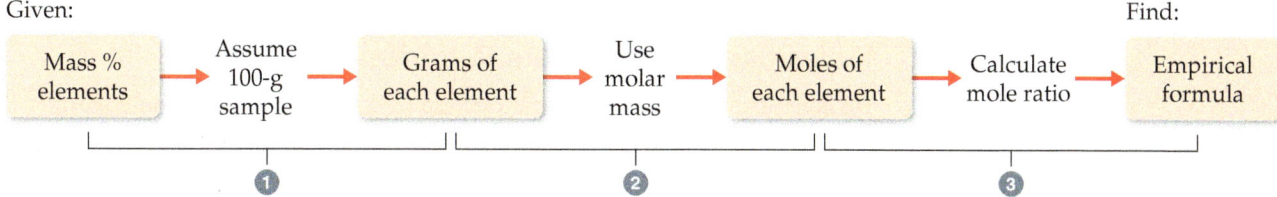

▲ Figure 3.12 Procedure for calculating an empirical formula from percentage composition.

Sample Exercise 3.10
Calculating an Empirical Formula

Ascorbic acid (vitamin C) contains 40.92% C, 4.58% H, and 54.50% O by mass. What is the empirical formula of ascorbic acid?

SOLUTION

Analyze We are to determine the empirical formula of a compound from the mass percentages of its elements.

Plan The strategy for determining the empirical formula involves the three steps given in Figure 3.12.

Solve
(1) For simplicity we assume we have exactly 100 g of material, although any other mass could also be used.

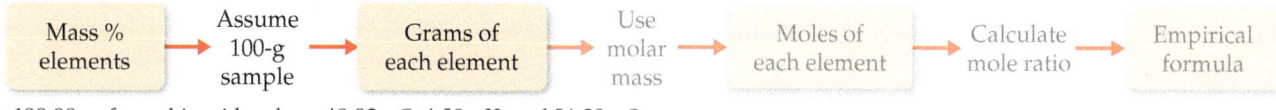

In 100.00 g of ascorbic acid we have 40.92 g C, 4.58 g H, and 54.50 g O.

Continued

(2) Next, we calculate the number of moles of each element. We use atomic masses with four significant figures to match the precision of our experimental masses.

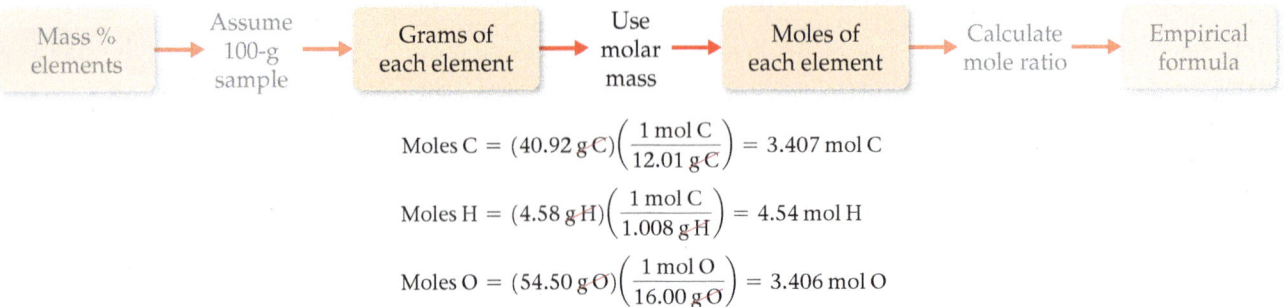

$$\text{Moles C} = (40.92 \text{ g C})\left(\frac{1 \text{ mol C}}{12.01 \text{ g C}}\right) = 3.407 \text{ mol C}$$

$$\text{Moles H} = (4.58 \text{ g H})\left(\frac{1 \text{ mol C}}{1.008 \text{ g H}}\right) = 4.54 \text{ mol H}$$

$$\text{Moles O} = (54.50 \text{ g O})\left(\frac{1 \text{ mol O}}{16.00 \text{ g O}}\right) = 3.406 \text{ mol O}$$

(3) We determine the simplest whole-number ratio of moles by dividing each number of moles by the smallest number of moles.

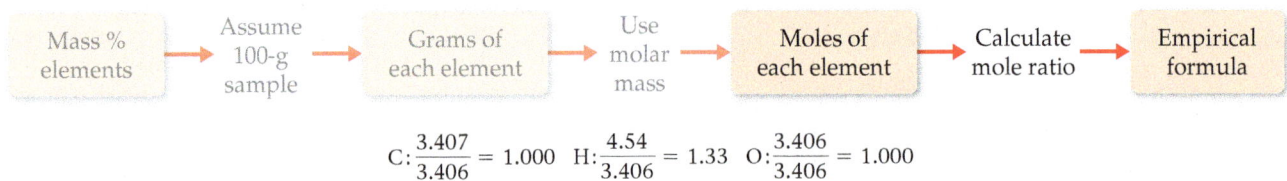

$$\text{C:}\frac{3.407}{3.406} = 1.000 \quad \text{H:}\frac{4.54}{3.406} = 1.33 \quad \text{O:}\frac{3.406}{3.406} = 1.000$$

The ratio for H is too far from 1 to attribute the difference to experimental error; in fact, it is quite close to $1\frac{1}{3}$. This suggests we should multiply the ratios by 3 to obtain whole numbers:

$$\text{C : H : O} = (3 \times 1 : 3 \times 1.33 : 3 \times 1) = (3 : 4 : 3)$$

Thus, the empirical formula is $C_3H_4O_3$.

Check It is reassuring that the subscripts are moderate-sized whole numbers. Also, calculating the percentage composition of $C_3H_4O_3$ gives values very close to the original percentages.

▶ **Practice Exercise**
A 5.325-g sample of methyl benzoate, a compound used in the manufacture of perfumes, contains 3.758 g of carbon, 0.316 g of hydrogen, and 1.251 g of oxygen. What is the empirical formula of this substance?

Molecular Formulas from Empirical Formulas

For molecular substances, the empirical formula and the molecular formula are often different. For example, benzene has a molecular formula of C_6H_6, but its empirical formula CH is the same as that of the gas acetylene, whose molecular formula is C_2H_2. Knowledge of the empirical formula is insufficient to differentiate these two very different compounds. Fortunately, we can obtain the molecular formula for any compound from its empirical formula if we know the molecular weight of the compound, which can be measured by a variety of methods, including mass spectrometry. (Section 2.4) *The subscripts in the molecular formula of a substance are always whole-number multiples of the subscripts in its empirical formula.* (Section 2.6) This multiple can be found by dividing the molecular weight by the empirical formula weight:

$$\text{Whole-number multiple} = \frac{\text{molecular weight}}{\text{empirical formula weight}} \quad \quad [3.14]$$

In Sample Exercise 3.10, for example, the empirical formula of ascorbic acid was determined to be $C_3H_4O_3$. This means the empirical formula weight is 3(12.0 amu) + 4(1.0 amu) + 3(16.0 amu) = 88.0 amu. The experimentally determined molecular weight is 176 amu. Thus, we find the whole-number multiple that converts the empirical formula to the molecular formula by dividing:

$$\text{Whole-number multiple} = \frac{\text{molecular weight}}{\text{empirical formula weight}} = \frac{176 \text{ amu}}{88.0 \text{ amu}} = 2$$

Consequently, we multiply the subscripts in the empirical formula by this multiple, giving the molecular formula: $C_6H_8O_6$.

Sample Exercise 3.11
Determining a Molecular Formula

Mesitylene, a hydrocarbon found in crude oil, has an empirical formula of C_3H_4 and an experimentally determined molecular weight of 121 amu. What is its molecular formula?

SOLUTION

Analyze We are given an empirical formula and a molecular weight of a compound and asked to determine its molecular formula.

Plan The subscripts in a compound's molecular formula are whole-number multiples of the subscripts in its empirical formula. We find the appropriate multiple by using Equation 3.14.

Solve The formula weight of the empirical formula C_3H_4 is

$$3(12.0\,amu) + 4(1.0\,amu) = 40.0\,amu$$

Next, we use this value in Equation 3.14:

$$\text{Whole-number multiple} = \frac{\text{molecular weight}}{\text{empirical formula weight}}$$

$$= \frac{121}{40.0} = 3.03$$

Only whole-number ratios make physical sense because molecules contain whole atoms. The 3.03 in this case could result from a small experimental error in the molecular weight. We therefore multiply each subscript in the empirical formula by 3 to give the molecular formula: C_9H_{12}.

Check We can have confidence in the result because dividing molecular weight by empirical formula weight yields nearly a whole number.

▶ **Practice Exercise**
Ethylene glycol, used in automobile antifreeze, is 38.7% C, 9.7% H, and 51.6% O by mass. Its molar mass is 62.1 g/mol. **(a)** What is the empirical formula of ethylene glycol? **(b)** What is its molecular formula?

Combustion Analysis

One technique for determining empirical formulas in the laboratory is *combustion analysis*, commonly used for compounds containing principally carbon and hydrogen.

When a compound containing carbon and hydrogen is completely combusted in an apparatus such as that shown in **Figure 3.13**, the carbon is converted to CO_2 and the hydrogen is converted to H_2O. (Section 3.2) From the masses of CO_2 and H_2O we can calculate the number of moles of C and H in the original sample and thereby the empirical formula. If a third element is present in the compound, its mass can be determined by subtracting the measured masses of C and H from the original sample mass.

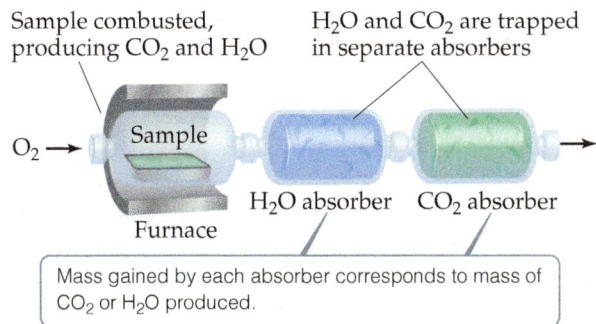

◀ **Figure 3.13 Apparatus for combustion analysis.**

Sample Exercise 3.12
Determining an Empirical Formula by Combustion Analysis

Isopropyl alcohol, sold as rubbing alcohol, is composed of C, H, and O. Combustion of 0.255 g of isopropyl alcohol produces 0.561 g of CO_2 and 0.306 g of H_2O. Determine the empirical formula of isopropyl alcohol.

SOLUTION

Analyze We are told that isopropyl alcohol contains C, H, and O atoms, and we are given the quantities of CO_2 and H_2O produced when a given quantity of the alcohol is combusted. We must determine the empirical formula for isopropyl alcohol, a task that requires us to calculate the number of moles of C, H, and O in the sample.

Plan We can use the mole concept to calculate grams of C in the CO_2 and grams of H in the H_2O—the masses of C and H in the alcohol before combustion. The mass of O in the compound equals the mass of the original sample minus the sum of the C and H masses. Once we have the C, H, and O masses, we can proceed as in Sample Exercise 3.10.

Continued

Solve Because all of the carbon in the sample is converted to CO_2, we can use dimensional analysis and the following steps to calculate the mass C in the sample.

$$\boxed{\begin{array}{c}\text{Mass } CO_2 \\ \text{produced}\end{array}} \xrightarrow{\begin{array}{c}\text{Molar} \\ \text{mass } CO_2 \\ \text{44.0 g/mol}\end{array}} \boxed{\begin{array}{c}\text{Moles } CO_2 \\ \text{produced}\end{array}} \xrightarrow{\begin{array}{c}\text{1 C atom} \\ \text{per } CO_2 \\ \text{molecule}\end{array}} \boxed{\begin{array}{c}\text{Moles of C} \\ \text{original sample}\end{array}} \xrightarrow{\begin{array}{c}\text{Molar} \\ \text{mass C} \\ \text{12.0 g/mol}\end{array}} \boxed{\begin{array}{c}\text{Mass C in} \\ \text{original sample}\end{array}}$$

Using the values given in this example, we see that the mass of C is

$$\text{Grams C} = (0.561 \text{ g } CO_2)\left(\frac{1 \text{ mol } CO_2}{44.0 \text{ g } CO_2}\right)\left(\frac{1 \text{ mol C}}{1 \text{ mol } CO_2}\right)\left(\frac{12.0 \text{ g C}}{1 \text{ mol C}}\right)$$

$$= 0.153 \text{ g C}$$

Because all of the hydrogen in the sample is converted to H_2O, we can use dimensional analysis and the following steps to calculate the mass H in the sample. We use three significant figures for the atomic mass of H to match the significant figures in the mass of H_2O produced.

$$\boxed{\begin{array}{c}\text{Mass } H_2O \\ \text{produced}\end{array}} \xrightarrow{\begin{array}{c}\text{Molar} \\ \text{mass } H_2O \\ \text{18.0 g/mol}\end{array}} \boxed{\begin{array}{c}\text{Moles } H_2O \\ \text{produced}\end{array}} \xrightarrow{\begin{array}{c}\text{2 H atoms} \\ \text{per } H_2O \\ \text{molecule}\end{array}} \boxed{\begin{array}{c}\text{Moles H in} \\ \text{original sample}\end{array}} \xrightarrow{\begin{array}{c}\text{Molar} \\ \text{mass H} \\ \text{1.01 g/mol}\end{array}} \boxed{\begin{array}{c}\text{Mass H in} \\ \text{original sample}\end{array}}$$

Using the values given in this example, we find that the mass of H is

$$\text{Grams H} = (0.306 \text{ g } H_2O)\left(\frac{1 \text{ mol } H_2O}{18.0 \text{ g } H_2O}\right)\left(\frac{2 \text{ mol H}}{1 \text{ mol } H_2O}\right)\left(\frac{1.01 \text{ g H}}{1 \text{ mol H}}\right) = 0.0343 \text{ g H}$$

The mass of the sample, 0.255 g, is the sum of the masses of C, H, and O. Thus, the O mass is

$$\text{Mass of O} = \text{mass of sample} - (\text{mass of C} + \text{mass of H}) = 0.255 \text{ g} - (0.153 \text{ g} + 0.0343 \text{ g}) = 0.068 \text{ g O}$$

The number of moles of C, H, and O in the sample is therefore

$$\text{Moles C} = (0.153 \text{ g C})\left(\frac{1 \text{ mol C}}{12.0 \text{ g C}}\right) = 0.0128 \text{ mol C}$$

$$\text{Moles H} = (0.0343 \text{ g H})\left(\frac{1 \text{ mol H}}{1.01 \text{ g H}}\right) = 0.0340 \text{ mol H}$$

$$\text{Moles O} = (0.068 \text{ g O})\left(\frac{1 \text{ mol O}}{16.0 \text{ g O}}\right) = 0.0043 \text{ mol O}$$

To find the empirical formula, we must compare the relative number of moles of each element in the sample, as illustrated in Sample Exercise 3.11.

$$\text{C}:\frac{0.0128}{0.0043} = 3.0 \quad \text{H}:\frac{0.0340}{0.0043} = 7.9 \quad \text{O}:\frac{0.0043}{0.0043} = 1.0$$

The first two numbers are close to the whole numbers 3 and 8, giving the empirical formula C_3H_8O.

▶ **Practice Exercise**
(a) Caproic acid, responsible for the odor of dirty socks, is composed of C, H, and O atoms. Combustion of a 0.225-g sample of this compound produces 0.512 g CO_2 and 0.209 g H_2O. What is the empirical formula of caproic acid?

(b) Caproic acid has a molar mass of 116 g/mol. What is its molecular formula?

Self-Assessment Exercises

SAE 3.13 In 2001, Japanese researchers found that a binary compound containing magnesium and boron conducts electricity with no resistance when cooled below 39 K. Elemental analysis showed that the compound is 52.9% magnesium and 47.1% boron by mass. What is the empirical formula of this compound? **(a)** MgB **(b)** Mg_2B **(c)** MgB_2 **(d)** Mg_2B_4

SAE 3.14 Styrene is the monomer used to make the polymer polystyrene. Elemental analysis shows that styrene is 92.3% carbon and 7.7% hydrogen by mass. Various techniques can be used to show that its molar mass is 104.2 g/mol. What is the molecular formula of styrene? **(a)** CH **(b)** C_8H_5 **(c)** C_8H_8 **(d)** C_7H_{20}

SAE 3.15 Isopentyl acetate is the molecule largely responsible for the aroma and taste of bananas. It is made up of carbon, hydrogen, and oxygen. Combustion of a 3.256-g sample of isopentyl acetate produces 7.706 g CO_2 and 3.156 g H_2O. Using these results, what is the correct empirical formula for isopentyl acetate? **(a)** C_3H_3O **(b)** CH_2 **(c)** $C_{3.5}H_7O$ **(d)** $C_7H_{14}O_2$

3.6 | Quantitative Information from Balanced Equations

When a chemical reaction is carried out, it's vital to understand how much of each product will be produced and how much of each reactant will be consumed. Carrying out chemical reactions without this knowledge can lead to unintended consequences. Maybe an expensive reactant will be wasted because much more of it was added than needed. A reaction might generate more gas than the reaction container can hold, leading to an explosion. In some reactions, particularly those involving solids, it can be difficult to separate the desired product from excess reactants. In this section, we show how to calculate the quantities of reactants consumed and products produced, given a balanced chemical equation representing the reaction.

The coefficients in a chemical equation represent the relative numbers of molecules in a reaction. The mole concept allows us to convert this information to the masses of the substances in the reaction. For instance, the coefficients in the balanced equation

$$2\,H_2(g) + O_2(g) \longrightarrow 2\,H_2O(l) \qquad [3.15]$$

indicate that two molecules of H_2 react with one molecule of O_2 to form two molecules of H_2O. It follows that the relative numbers of moles are identical to the relative numbers of molecules:

$2\,H_2(g)$	+	$O_2(g)$	$\longrightarrow$	$2\,H_2O(l)$
2 molecules		1 molecule		2 molecules
$2(6.02 \times 10^{23}$ molecules)		$1(6.02 \times 10^{23}$ molecules)		$2(6.02 \times 10^{23}$ molecules)
2 mol		1 mol		2 mol

We can generalize this observation to all balanced chemical equations: *The coefficients in a balanced chemical equation indicate both the relative numbers of molecules (or formula units) in the reaction and the relative numbers of moles.* **Figure 3.14** shows how this result corresponds to the law of conservation of mass.

 Learning Objective

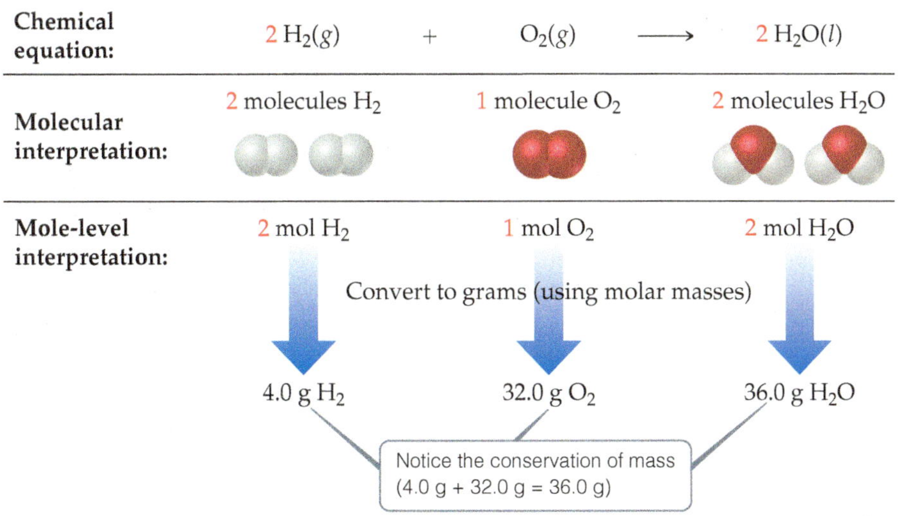

When you finish Section 3.6, you should be able to:

▶ Determine the quantities (grams, moles, or molecules) of products formed and/or reactants consumed in a chemical reaction where the quantity of one reactant that is identified limits the extent of the reaction.

Chemical equation:	$2\,H_2(g)$	+	$O_2(g)$	$\longrightarrow$	$2\,H_2O(l)$
Molecular interpretation:	2 molecules H_2		1 molecule O_2		2 molecules H_2O
Mole-level interpretation:	2 mol H_2		1 mol O_2		2 mol H_2O

Convert to grams (using molar masses)

| 4.0 g H_2 | 32.0 g O_2 | 36.0 g H_2O |

Notice the conservation of mass (4.0 g + 32.0 g = 36.0 g)

◀ **Figure 3.14 Interpreting a balanced chemical equation quantitatively.**

The quantities 2 mol H_2, 1 mol O_2, and 2 mol H_2O given by the coefficients in Equation 3.15 are called *stoichiometrically equivalent quantities*. The relationship between these quantities can be represented as

$$2\,\text{mol}\,H_2 \simeq 1\,\text{mol}\,O_2 \simeq 2\,\text{mol}\,H_2O$$

where the $\simeq$ symbol means "is stoichiometrically equivalent to." Stoichiometric relations such as these can be used to convert between quantities of reactants and products in a chemical reaction. For example, the number of moles of H_2O produced from 1.57 mol of O_2 is

$$\text{Moles } H_2O = (1.57 \text{ mol } O_2)\left(\frac{2 \text{ mol } H_2O}{1 \text{ mol } O_2}\right) = 3.14 \text{ mol } H_2O$$

As an additional example, consider the combustion of butane C_4H_{10}, the fuel in disposable lighters:

$$2 C_4H_{10}(l) + 13 O_2(g) \longrightarrow 8 CO_2(g) + 10 H_2O(g) \qquad [3.16]$$

Let's calculate the mass of CO_2 produced when 1.00 g of C_4H_{10} is burned. The coefficients in Equation 3.13 tell us how the amount of C_4H_{10} consumed is related to the amount of CO_2 produced: 2 mol $C_4H_{10} \simeq$ 8 mol CO_2. To use this stoichiometric relationship, we must convert grams of C_4H_{10} to moles using the molar mass of C_4H_{10}, 58.0 g/mol:

$$\text{Moles } C_4H_{10} = (1.00 \text{ g } C_4H_{10})\left(\frac{1 \text{ mol } C_4H_{10}}{58.0 \text{ g } C_4H_{10}}\right)$$

$$= 1.72 \times 10^{-2} \text{ mol } C_4H_{10}$$

We then use the stoichiometric factor from the balanced equation to calculate moles of CO_2:

$$\text{Moles } CO_2 = (1.72 \times 10^{-2} \text{ mol } C_4H_{10})\left(\frac{8 \text{ mol } CO_2}{2 \text{ mol } C_4H_{10}}\right)$$

$$= 6.88 \times 10^{-2} \text{ mol } CO_2$$

Finally, we use the molar mass of CO_2, 44.0 g/mol, to calculate the mass of CO_2 in grams:

$$\text{Grams } CO_2 = (6.88 \times 10^{-2} \text{ mol } CO_2)\left(\frac{44.0 \text{ g } CO_2}{1 \text{ mol } CO_2}\right)$$

$$= 3.03 \text{ g } CO_2$$

This conversion sequence involves three steps, as illustrated in **Figure 3.15**. These three conversions can be combined in a single equation:

$$\text{Grams } CO_2 = (1.00 \text{ g } C_4H_{10})\left(\frac{1 \text{ mol } C_4H_{10}}{58.0 \text{ g } C_4H_{10}}\right)\left(\frac{8 \text{ mol } CO_2}{2 \text{ mol } C_4H_{10}}\right)\left(\frac{44.0 \text{ g } CO_2}{1 \text{ mol } CO_2}\right)$$

$$= 3.03 \text{ g } CO_2$$

To calculate the amount of O_2 consumed in the reaction of Equation 3.16, we again rely on the coefficients in the balanced equation for our stoichiometric factor, 2 mol $C_4H_{10} \simeq$ 13 mol O_2:

$$\text{Grams } O_2 = (1.00 \text{ g } C_4H_{10})\left(\frac{1 \text{ mol } C_4H_{10}}{58.0 \text{ g } C_4H_{10}}\right)\left(\frac{13 \text{ mol } O_2}{2 \text{ mol } C_4H_{10}}\right)\left(\frac{32.0 \text{ g } O_2}{1 \text{ mol } O_2}\right)$$

$$= 3.59 \text{ g } O_2$$

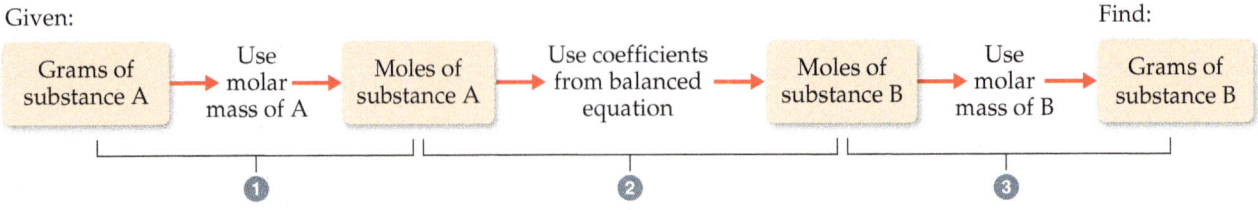

▲ **Figure 3.15 Procedure for calculating amounts of reactants consumed or products formed in a reaction.** The number of grams of a reactant consumed or product formed can be calculated in three steps, starting with the number of grams of any reactant or product.

Many chemical reactions either consume or produce heat (Figure 3.7). This heat is also a stoichiometric quantity. For instance, if a given reaction with a given number of reactant moles produces 100 J of energy in the form of heat, performing the reaction with twice the number of moles of reactants will produce 200 J of heat. We explore this concept further in Chapter 5.

Sample Exercise 3.13
Calculating Amounts of Reactants and Products

Determine how many grams of water are produced in the oxidation of 1.00 g of glucose, $C_6H_{12}O_6$:

$$C_6H_{12}O_6(s) + 6\,O_2(g) \longrightarrow 6\,CO_2(g) + 6\,H_2O(l)$$

SOLUTION

Analyze We are given the mass of a reactant and must determine the mass of a product in the given reaction.

Plan We follow the general strategy outlined in Figure 3.15:

(**1**) Convert grams of $C_6H_{12}O_6$ to moles using the molar mass of $C_6H_{12}O_6$.

(**2**) Convert moles of $C_6H_{12}O_6$ to moles of H_2O using the stoichiometric relationship 1 mol $C_6H_{12}O_6 \simeq 6$ mol H_2O.

(**3**) Convert moles of H_2O to grams using the molar mass of H_2O.

Solve

(**1**) First we convert grams of $C_6H_{12}O_6$ to moles using the molar mass of $C_6H_{12}O_6$.

$$\text{Moles } C_6H_{12}O_6 = (1.00 \text{ g } C_6H_{12}O_6)\left(\frac{1 \text{ mol } C_6H_{12}O_6}{180.0 \text{ g } C_6H_{12}O_6}\right)$$

(**2**) Next, we convert moles of $C_6H_{12}O_6$ to moles of H_2O using the stoichiometric relationship 1 mol $C_6H_{12}O_6 \simeq 6$ mol H_2O.

$$\text{Moles } H_2O = (1.00 \text{ g } C_6H_{12}O_6)\left(\frac{1 \text{ mol } C_6H_{12}O_6}{180.0 \text{ g } C_6H_{12}O_6}\right)\left(\frac{6 \text{ mol } H_2O}{1 \text{ mol } C_6H_{12}O_6}\right)$$

(**3**) Finally, we convert moles of H_2O to grams using the molar mass of H_2O.

$$\text{Grams } H_2O = (1.00 \text{ g } C_6H_{12}O_6)\left(\frac{1 \text{ mol } C_6H_{12}O_6}{180.0 \text{ g } C_6H_{12}O_6}\right)\left(\frac{6 \text{ mol } H_2O}{1 \text{ mol } C_6H_{12}O_6}\right)\left(\frac{18.0 \text{ g } H_2O}{1 \text{ mol } H_2O}\right)$$

$$= 0.600 \text{ g } H_2O$$

Check We can check the reasonableness of our result by doing a ballpark estimate of the mass of H_2O. Because the molar mass of glucose is 180 g/mol, 1 g of glucose equals 1/180 mol. Because 1 mol of glucose yields 6 mol H_2O, we would have 6/180 = 1/30 mol H_2O. The molar mass of water is 18 g/mol, so we have 1/30 × 18 = 6/10 = 0.6 g of H_2O, which agrees with the full calculation. The units, grams H_2O, are correct. The initial data had three significant figures, so three significant figures for the answer is correct.

▶ **Practice Exercise**
Decomposition of $KClO_3$ is sometimes used to prepare small amounts of O_2 in the laboratory: $2\,KClO_3(s) \longrightarrow 2\,KCl(s) + 3\,O_2(g)$. How many grams of O_2 can be prepared from 4.50 g of $KClO_3$?

Self-Assessment Exercises

SAE 3.16 When 5.00 g of methane, CH_4, reacts with an excess of oxygen in a combustion reaction, what quantity of H_2O is produced? (**a**) 2.81 g (**b**) 5.60 g (**c**) 10.0 g (**d**) 11.2 g

SAE 3.17 When heated sufficiently, lead(IV) oxide undergoes the following decomposition reaction: $2\,PbO_2(s) \longrightarrow 2\,PbO(s) + O_2(g)$. When a PbO_2 sample is heated, it produces 0.250 g of O_2. What is the mass of PbO_2 present prior to heating? (**a**) 0.0156 g (**b**) 0.500 g (**c**) 3.49 g (**d**) 3.74 g

3.7 | Limiting Reactants

⚠ **Learning Objectives**

When you finish **Section 3.7**, you should be able to:

▶ Determine the limiting reactant for a reaction from the quantities (grams, moles, or molecules) of each reactant present before the reaction starts.

▶ Calculate the theoretical yield for a reaction from the quantities (grams, moles, or molecules) of the limiting reactant present before the reaction starts.

▶ Calculate the percent yield of a chemical reaction from the actual yield and the theoretical yield of a given product.

Often the reactants used in a chemical reaction are not present in precise stoichiometric amounts. For example, a natural gas-fired power plant generates electricity by producing hot gases that drive turbines, predominantly through the following combustion reaction:

$$CH_4(g) + 2\,O_2(g) \longrightarrow CO_2(g) + 2\,H_2O(g) \qquad [3.17]$$

Power plants typically operate with an excess of $O_2(g)$ to extract the maximum energy from the hydrocarbon fuel and minimize production of harmful by-products, like carbon monoxide, that result from incomplete combustion. Consequently, the amount of CH_4 introduced determines how much CO_2 and H_2O are produced. In this section, we show how to do quantitative calculations for reactions where the reactants are not present in stoichiometrically equivalent quantities.

Suppose you wish to make several sandwiches using one slice of cheese and two slices of bread for each. Using Bd = bread, Ch = cheese, and Bd_2Ch = sandwich, we can represent the recipe for making a sandwich as a chemical equation:

$$2\,Bd + Ch \longrightarrow Bd_2Ch$$

If you have ten slices of bread and seven slices of cheese, you can make only five sandwiches and will have two slices of cheese left over. In this example, the amount of bread limits the number of sandwiches.

An analogous situation occurs in chemical reactions when one reactant is used up before the others. The reaction stops as soon as any reactant is totally consumed, leaving the excess reactants as leftovers. Suppose, for example, we have a mixture of 10 mol H_2 and 7 mol O_2, which react to form water:

$$2\,H_2(g) + O_2(g) \longrightarrow 2\,H_2O(g)$$

Because $2\,mol\,H_2 \,\hat{=}\, 1\,mol\,O_2$, the number of moles of O_2 needed to react with all the H_2 is

$$Moles\ O_2 = (10\ mol\ H_2)\left(\frac{1\ mol\ O_2}{2\ mol\ H_2}\right) = 5\ mol\ O_2$$

Because 7 mol O_2 is available at the start of the reaction, $7\,mol\,O_2 - 5\,mol\,O_2 = 2\,mol\,O_2$ is still present when the H_2 is fully consumed.

The reactant that is completely consumed in a reaction is called the **limiting reactant** because it determines, or limits, the amount of product formed. The other reactants are sometimes called *excess reactants*. In our example, shown in **Figure 3.16**, H_2 is the limiting reactant, which means that once all the H_2 has been consumed, the reaction stops. At that point some of the excess reactant O_2 is left over.

There are no restrictions on the starting amounts of reactants in any reaction. Indeed, many reactions are carried out using an excess of one or more reactants. The quantities of reactants consumed and products formed, however, are restricted by the quantity of the limiting reactant. For example, when a combustion reaction takes place in the open air, oxygen is plentiful and is therefore the excess reactant. If you run out of gasoline while driving, the car stops because the gasoline is the limiting reactant in the combustion reaction that moves the car.

Before we leave the example illustrated in Figure 3.16, let's summarize the data:

	2 H$_2$(g)	**+**	**O$_2$(g)**	**⟶**	**2 H$_2$O(g)**
Before reaction:	10 mol		7 mol		0 mol
Change (reaction):	−10 mol		−5 mol		+10 mol
After reaction:	0 mol		2 mol		10 mol

The second line in the table (Change) summarizes the amounts of reactants consumed (where this consumption is indicated by the minus signs) and the amount of the product formed (indicated by the plus sign). These quantities are restricted by the quantity of the limiting reactant and depend on the coefficients in the balanced equation. The mole ratio H_2:O_2:H_2O = 10:5:10 is a multiple of the ratio of the coefficients in the balanced equation, 2:1:2. The after-reaction quantities depend on the before-reaction

 **Go Figure** If the amount of H_2 is doubled, how many moles of H_2O will form?

Before reaction

After reaction

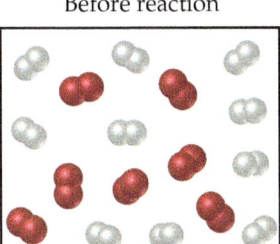

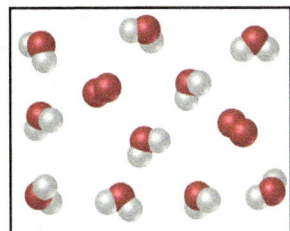

10 H_2 and 7 O_2

10 H_2O and 2 O_2 (no H_2 molecules)

▲ **Figure 3.16 Limiting reactant.** Because H_2 is completely consumed, it is the limiting reactant. Because some O_2 is left over after the reaction is complete, it is the excess reactant. The amount of H_2O formed depends on the amount of limiting reactant, H_2.

quantities and their changes. The after-reaction quantities are found by adding the before quantity and change quantity for each column. The amount of the limiting reactant (H_2) must be zero at the end of the reaction. What remains is 2 mol O_2 (excess reactant) and 10 mol H_2O (product).

 Sample Exercise 3.14

Calculating the Amount of Product Formed from a Limiting Reactant

The most important commercial process for converting N_2 from the air into nitrogen-containing compounds is based on the reaction of N_2 and H_2 to form ammonia (NH_3):

$$N_2(g) + 3H_2(g) \longrightarrow 2NH_3(g)$$

How many moles of NH_3 can be formed from 3.0 mol of N_2 and 6.0 mol of H_2?

SOLUTION

Analyze We are asked to calculate the number of moles of product, NH_3, given the quantities of each reactant, N_2 and H_2, available in a reaction. This is a limiting reactant problem.

Plan If we assume one reactant is completely consumed, we can calculate how much of the second reactant is needed. By comparing the calculated quantity of the second reactant with the amount available, we can determine which reactant is limiting. We then proceed with the calculation, using the quantity of the limiting reactant.

Solve The number of moles of H_2 needed for complete consumption of 3.0 mol of N_2 is

$$\text{Moles } H_2 = (3.0 \text{ mol } N_2)\left(\frac{3 \text{ mol } H_2}{1 \text{ mol } N_2}\right) = 9.0 \text{ mol } H_2$$

Because only 6.0 mol H_2 is available, we will run out of H_2 before the N_2 is gone, which tells us that H_2 is the limiting reactant. Therefore, we use the quantity of H_2 to calculate the quantity of NH_3 produced:

$$\text{Moles } NH_3 = (6.0 \text{ mol } H_2)\left(\frac{2 \text{ mol } NH_3}{3 \text{ mol } H_2}\right) = 4.0 \text{ mol } NH_3$$

Notice that we can calculate not only the number of moles of NH_3 formed but also the number of moles of each reactant remaining after the reaction. Notice also that although the initial number of moles of H_2 is greater than the initial number of moles of N_2, H_2 is nevertheless the limiting reactant because of its larger coefficient in the balanced equation.

Check Examine the change row of the summary table to see that the mole ratio of reactants consumed and product formed, 2:6:4, is a multiple of the coefficients in the balanced equation, 1:3:2. We confirm that H_2 is the limiting reactant because it is completely consumed in the reaction, leaving 0 mol at the end. Because 6.0 mol H_2 has two significant figures, our answer has two significant figures.

Comment It is useful to summarize the reaction data in a table:

	$N_2(g)$ +	$3H_2(g)$ $\longrightarrow$	$2NH_3(g)$
Before reaction:	3.0 mol	6.0 mol	0 mol
Change (reaction):	−2.0 mol	−6.0 mol	+4.0 mol
After reaction:	1.0 mol	0 mol	4.0 mol

▶ **Practice Exercise**
(a) When 1.50 mol of Al and 3.00 mol of Cl_2 combine in the reaction $2Al(s) + 3Cl_2(g) \longrightarrow 2AlCl_3(s)$, which is the limiting reactant? (b) How many moles of $AlCl_3$ are formed? (c) How many moles of the excess reactant remain at the end of the reaction?

Theoretical and Percent Yields

The quantity of product calculated to form when the limiting reactant is fully consumed is called the **theoretical yield**. The amount of product actually obtained, called the *actual yield*, is almost always less than (and can never be greater than) the theoretical yield. There are many reasons for this difference. Some of the reactants may not react, for example, or they may react in a way different from that desired, leading to side reactions. In addition, it is not always possible to recover all of the product from the reaction mixture. The **percent yield** is the actual yield divided by the theoretical yield, multiplied by 100% to convert to percent:

$$\text{Percent yield} = \frac{\text{actual yield}}{\text{theoretical yield}} \times 100\% \qquad [3.18]$$

Sample Exercise 3.15
Calculating Theoretical Yield and Percent Yield

Adipic acid, $H_2C_6H_8O_4$, used to produce nylon, is made commercially by a reaction between cyclohexane (C_6H_{12}) and oxygen (O_2):

$$2\,C_6H_{12}(l) + 5\,O_2(g) \longrightarrow 2\,H_2C_6H_8O_4(l) + 2\,H_2O(g)$$

(a) Assume that you carry out this reaction with 25.0 g of cyclohexane and that cyclohexane is the limiting reactant. What is the theoretical yield of adipic acid? **(b)** If you obtain 33.5 g of adipic acid, what is the percent yield for the reaction?

SOLUTION

Analyze We are given a chemical equation and the quantity of the limiting reactant (25.0 g of C_6H_{12}). We are asked to calculate the theoretical yield of a product $H_2C_6H_8O_4$ and the percent yield if only 33.5 g of product is obtained.

Plan

(a) The theoretical yield, which is the calculated quantity of adipic acid formed, can be calculated using the sequence of conversions shown in Figure 3.15.

(b) The percent yield is calculated by using Equation 3.18 to compare the given actual yield (33.5 g) with the theoretical yield.

Solve

(a) The theoretical yield is:

$$\text{Grams } H_2C_6H_8O_4 = (25.0\text{ g }C_6H_{12})\left(\frac{1\text{ mol }C_6H_{12}}{84.0\text{ g }C_6H_{12}}\right)\left(\frac{2\text{ mol }H_2C_6H_8O_4}{2\text{ mol }C_6H_{12}}\right)\left(\frac{146.0\text{ g }H_2C_6H_8O_4}{1\text{ mol }H_2C_6H_8O_4}\right) = 43.5\text{ g }H_2C_6H_8O_4$$

(b) The percent yield is:

$$\text{Percent yield} = \frac{\text{actual yield}}{\text{theoretical yield}} \times 100\% = \frac{33.5\text{ g}}{43.5\text{ g}} \times 100\% = 77.0\%$$

Check We can check our answer in **(a)** by doing a ballpark calculation. From the balanced equation we know that each mole of cyclohexane gives 1 mol adipic acid. We have $25/84 \approx 25/75 = 0.33$ mol hexane, so we expect 0.33 mol adipic acid, which equals about $0.33 \times 150 = 50$ g, roughly the same magnitude as the 43.5 g obtained in the more detailed calculation. In addition, our answer has the appropriate units and number of significant figures. In **(b)** the answer is less than 100%, as it must be from the definition of percent yield.

▶ **Practice Exercise**

Imagine you are working on ways to improve the process by which iron ore containing Fe_2O_3 is converted into iron:

$$Fe_2O_3(s) + 3\,CO(g) \longrightarrow 2\,Fe(s) + 3\,CO_2(g)$$

(a) If you start with 150 g of Fe_2O_3 as the limiting reactant, what is the theoretical yield of Fe? **(b)** If your actual yield is 87.9 g, what is the percent yield?

Self-Assessment Exercises

SAE 3.18 If eight molecules of CO and five molecules of O_2 react to form CO_2, what is the limiting reactant, and how many molecules of the excess reactant will remain when the reaction is complete?

(a) Limiting reactant = O_2, three molecules of CO remaining
(b) Limiting reactant = O_2, one molecule of CO remaining
(c) Limiting reactant = CO, one molecule of O_2 remaining
(d) Limiting reactant = CO, four molecules of O_2 remaining

SAE 3.19 If 1.75 g of titanium metal reacts with 1.25 g of oxygen gas to form titanium(IV) oxide, what is the theoretical yield of the product? **(a)** 1.25 g **(b)** 1.56 g **(c)** 2.92 g **(d)** 3.12 g

SAE 3.20 Ammonia reacts with hydrogen chloride in the gas phase to produce the ionic compound ammonium chloride via the reaction $NH_3(g) + HCl(g) \longrightarrow NH_4Cl(s)$. What is the percent yield of ammonium chloride if 0.81 g of NH_4Cl is formed when 0.75 g of NH_3 is reacted with 0.62 g HCl? **(a)** 34% **(b)** 59% **(c)** 89% **(d)** 91%

STRATEGIES FOR SUCCESS | Design an Experiment

One of the most important skills you can learn in school is how to think like a scientist. Questions such as: "What experiment might test this hypothesis?", "How do I interpret these data?", and "Do these data support the hypothesis?" are asked every day by chemists and other scientists as they go about their work.

We want you to become a good critical thinker as well as an active, logical, and curious learner. For this purpose, starting in this chapter, we include at the end of each chapter a special exercise called "Design an Experiment." Here is an example:

Is milk a pure liquid or a mixture of chemical components in water? Design an experiment to distinguish between these two possibilities.

You might already know the answer—milk is indeed a mixture of components in water—but the goal is to think of how to demonstrate this in practice. Upon thinking about it, you will likely realize that the key idea for this experiment is separation: You can prove that milk is a mixture of chemical components if you can figure out how to separate these components.

Testing a hypothesis is a creative endeavor. While some experiments may be more efficient than others, there is often more than one good way to test a hypothesis. Our question about milk, for example, might be explored by an experiment in which you boil a known quantity of milk until it is dry. Does a solid residue form in the bottom of the pan? If so, you could weigh it and calculate the percentage of solids in milk, which would offer good evidence that milk is a mixture. If there is no residue after boiling, then you still cannot distinguish between the two possibilities.

What other experiments might you do to demonstrate that milk is a mixture? You could put a sample of milk in a centrifuge, which you might have used in a biology lab, spin your sample, and observe if any solids collect at the bottom of the centrifuge tube; large molecules can be separated in this way from a mixture. Keep an open mind: Lacking a centrifuge, how else might you separate solids in the milk? You could consider using a filter with really tiny holes in it or perhaps even a fine strainer. You could propose that if milk were poured through this filter, some (large) solid components should stay on the top of the filter, while water (and really small molecules or ions) would pass through the filter. That result would be evidence that milk is a mixture. Does such a filter exist? Yes! But for our purposes, the existence of such a filter is not the point. The point is, can you use your imagination and your knowledge of chemistry to design a reasonable experiment? Don't worry too much about the exact apparatus you need for the Design an Experiment exercises. The goal is to imagine what you would need to do, or what kind of data you would need to collect, in order to answer the question. If your instructor allows it, you can collaborate with others in your class to develop ideas. Scientists discuss their ideas with other scientists all the time. We find that discussing ideas, and refining them, makes us better scientists and helps us collectively answer important questions.

The design and interpretation of scientific experiments is at the heart of the scientific method. Think of the Design an Experiment exercises as puzzles that can be solved in various ways, and enjoy your explorations!

Chapter Summary and Key Terms

CHEMICAL EQUATIONS (INTRODUCTION AND SECTION 3.1) The study of the quantitative relationships between chemical formulas and chemical equations is known as **stoichiometry**. One of the important concepts of stoichiometry is the **law of conservation of mass**, which states that the total mass of the products of a chemical reaction is the same as the total mass of the reactants. The same numbers of atoms of each type are present before and after a chemical reaction. A balanced **chemical equation** shows equal numbers of atoms of each element on each side of the equation. Equations are balanced by placing coefficients in front of the chemical formulas for the **reactants** and **products** of a reaction, not by changing the subscripts in chemical formulas.

SIMPLE PATTERNS OF CHEMICAL REACTIVITY (SECTION 3.2) Among the reaction types described in this chapter are (1) **combination reactions**, in which two reactants combine to form one product; (2) **decomposition reactions**, in which a single reactant forms two or more products; and (3) **combustion reactions** in oxygen, in which a substance, typically a hydrocarbon, reacts rapidly with O_2 to form CO_2 and H_2O.

FORMULA WEIGHTS (SECTION 3.3) Much quantitative information can be determined from chemical formulas and balanced chemical equations by using atomic weights. The **formula weight** of a compound equals the sum of the atomic weights of the atoms in its formula. If the formula is a molecular formula, the formula weight is also called the **molecular weight**. Atomic weights and formula weights can be used to determine the **elemental composition** of a compound.

AVOGADRO'S NUMBER AND THE MOLE (SECTION 3.4) A mole of any substance contains **Avogadro's number** (6.02×10^{23}) of formula units of that substance. The mass of a **mole** of atoms,

molecules, or ions (the **molar mass**) equals the formula weight of that material expressed in grams. The mass of 1 molecule of H_2O, for example, is 18.0 amu, so the mass of 1 mol of H_2O is 18.0 g. That is, the molar mass of H_2O is 18.0 g/mol.

EMPIRICAL FORMULAS FROM ANALYSIS (SECTION 3.5) The empirical formula of any substance can be determined from its percent composition by calculating the relative number of moles of each atom in an arbitrary-sized sample (typically 100 g) of the substance. If the substance is molecular in nature, its molecular formula can be determined from the empirical formula if the molecular weight is also known. Combustion analysis is a special technique for determining the empirical formulas of compounds containing only carbon, hydrogen, and/or oxygen.

QUANTITATIVE INFORMATION FROM BALANCED EQUATIONS AND LIMITING REACTANTS (SECTIONS 3.6 AND 3.7) The mole concept can be used to calculate the relative quantities of reactants and products in chemical reactions. The coefficients in a balanced equation give the relative numbers of moles of the reactants and products. To calculate the number of grams of a product from the number of grams of a reactant, first convert grams of reactant to moles of reactant. Then use the coefficients in the balanced equation to convert the number of moles of reactant to moles of product. Finally, convert moles of product to grams of product.

A **limiting reactant** is the reactant that is completely consumed in a reaction. When it is used up, the reaction stops, thus limiting the quantities of products formed. The **theoretical yield** of a reaction is the quantity of product calculated to form when the limiting reactant is fully consumed. The actual yield of a reaction is the amount of product experimentally obtained in a reaction; it is always less than the theoretical yield. The **percent yield** is the actual yield divided by the theoretical yield, converted to a percentage.

Key Equations

- $$\% \text{ mass composition of element} = \frac{\left(\begin{array}{c}\text{number of atoms} \\ \text{of element}\end{array}\right)\left(\begin{array}{c}\text{atomic weight} \\ \text{of element}\end{array}\right)}{\text{formula weight of compound}} \times 100\%$$ [3.13] This is the formula to calculate the mass percentage of each element in a compound. The sum of the percentages of all elements in a compound should add up to 100%.

- $$\text{Percent yield} = \frac{\text{actual yield}}{\text{theoretical yield}} \times 100\%$$ [3.18] This is the formula to calculate the percent yield of a reaction. The percent yield can never be more than 100%.

Exam Prep

EP 3.1 In the following diagram, the white spheres represent hydrogen atoms and the blue spheres represent nitrogen atoms.

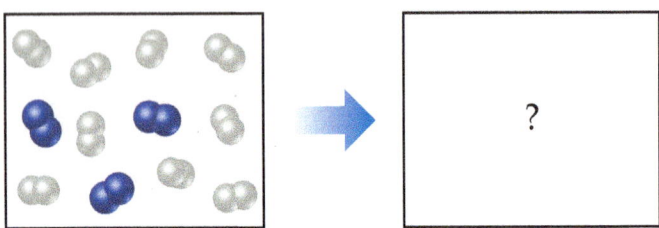

The two reactants combine to form a single product, ammonia, NH_3, which is not shown. Write a balanced chemical equation for the reaction. Based on the equation and the contents of the left (reactants) box, how many NH_3 molecules should be shown in the right (products) box? (**a**) 2 (**b**) 3 (**c**) 4 (**d**) 6 (**e**) 9

EP 3.2 The unbalanced equation for the reaction between methane and bromine is

$$\_CH_4(g) + \_Br_2(l) \longrightarrow \_CBr_4(s) + \_HBr(g)$$

Once this equation is balanced, what is the value of the coefficient in front of bromine, Br_2? (**a**) 1 (**b**) 2 (**c**) 3 (**d**) 4 (**e**) 6

EP 3.3 Which of the following reactions can be classified as a combination reaction?

(**a**) $2Mn_3O_4(s) \rightarrow 6MnO(s) + O_2(g)$

(**b**) $HCl(g) + NH_3(g) \rightarrow NH_4Cl(s)$

(**c**) $4NH_4ClO_4(s) \rightarrow 4HCl(g) + 2N_2(g) + 5O_2(g) + 6H_2O(g)$

(**d**) $2C_2H_2(g) + 3O_2(g) \rightarrow 4CO_2(g) + 2H_2O(g)$

EP 3.4 Which of the following reactions is the balanced equation that represents the decomposition reaction that occurs when silver(I) oxide is heated?

(**a**) $AgO(s) \longrightarrow Ag(s) + O(g)$

(**b**) $2AgO(s) \longrightarrow 2Ag(s) + O_2(g)$

(**c**) $Ag_2O(s) \longrightarrow 2Ag(s) + O(g)$

(**d**) $2Ag_2O(s) \longrightarrow 4Ag(s) + O_2(g)$

(**e**) $Ag_2O(s) \longrightarrow 2Ag(s) + O_2(g)$

EP 3.5 Write the balanced equation for the reaction that occurs when ethylene glycol, $C_2H_4(OH)_2$, burns in air.

(**a**) $C_2H_4(OH)_2(l) + 5O_2(g) \longrightarrow 2CO_2(g) + 3H_2O(g)$

(**b**) $2C_2H_4(OH)_2(l) + 5O_2(g) \longrightarrow 4CO_2(g) + 6H_2O(g)$

(**c**) $C_2H_4(OH)_2(l) + 3O_2(g) \longrightarrow 2CO_2(g) + 3H_2O(g)$

(**d**) $C_2H_4(OH)_2(l) + 5O(g) \longrightarrow 2CO_2(g) + 3H_2O(g)$

(**e**) $4C_2H_4(OH)_2(l) + 10O_2(g) \longrightarrow 8CO_2(g) + 12H_2O(g)$

EP 3.6 Which of the following is the correct formula weight for calcium phosphate? (**a**) 310.2 amu (**b**) 135.1 amu (**c**) 182.2 amu (**d**) 278.2 amu (**e**) 175.1 amu

EP 3.7 What is the percentage of nitrogen, by mass, in calcium nitrate? (**a**) 8.54% (**b**) 17.1% (**c**) 13.7% (**d**) 24.4% (**e**) 82.9%

EP 3.8 Which of the following samples contains the fewest sodium atoms? (**a**) 1 mol sodium oxide (**b**) 45 g sodium fluoride (**c**) 50 g sodium chloride (**d**) 1 mol sodium nitrate

EP 3.9 A sample of an ionic compound containing iron and chlorine is analyzed and found to have a molar mass of 126.8 g/mol. What is the charge of the iron in this compound? (**a**) 1+ (**b**) 2+ (**c**) 3+ (**d**) 4+

EP 3.10 How many chlorine atoms are in 12.2 g of CCl_4? (**a**) 4.77×10^{22} (**b**) 7.34×10^{24} (**c**) 1.91×10^{23} (**d**) 2.07×10^{23}

EP 3.11 Cyclohexane, a commonly used organic solvent, is 85.6% C and 14.4% H by mass with a molar mass of 84.2 g/mol. What is its molecular formula? (**a**) C_6H (**b**) CH_2 (**c**) C_5H_{24} (**d**) C_6H_{12} (**e**) C_4H_8

EP 3.12 The compound dioxane, which is used as a solvent in various industrial processes, is composed of C, H, and O atoms. Combustion of a 2.203-g sample of this compound produces 4.401 g CO_2 and 1.802 g H_2O. A separate experiment shows that it has a molar mass of 88.1 g/mol. Which of the following is the correct molecular formula for dioxane? (**a**) C_2H_4O (**b**) $C_4H_4O_2$ (**c**) CH_2 (**d**) $C_4H_8O_2$

EP 3.13 Sodium hydroxide reacts with carbon dioxide to form sodium carbonate and water:

$$2NaOH(s) + CO_2(g) \longrightarrow Na_2CO_3(s) + H_2O(l)$$

How many grams of Na_2CO_3 can be prepared from 2.40 g of NaOH? (**a**) 3.18 g (**b**) 6.36 g (**c**) 1.20 g (**d**) 0.0300 g

EP 3.14 Propane, C_3H_8, is a common fuel used for cooking and home heating. What mass of O_2 is consumed in the combustion of 1.00 g of propane? (**a**) 5.00 g (**b**) 0.726 g (**c**) 2.18 g (**d**) 3.63 g

EP 3.15 Suppose 24 mol of methanol and 15 mol of oxygen combine in the following combustion reaction:

$$2CH_3OH(l) + 3O_2(g) \longrightarrow 2CO_2(g) + 4H_2O(g)$$

What is the excess reactant, and how many moles of it remain at the end of the reaction? (**a**) 9 mol $CH_3OH(l)$ (**b**) 10 mol $CO_2(g)$ (**c**) 10 mol $CH_3OH(l)$ (**d**) 14 mol $CH_3OH(l)$ (**e**) 1 mol O_2

EP 3.16 Molten gallium reacts with arsenic to form the semiconductor gallium arsenide, GaAs, used in light emitting diodes and solar cells:

$$Ga(l) + As(s) \longrightarrow GaAs(s)$$

If 4.00 g of gallium is reacted with 5.50 g of arsenic, how many grams of the excess reactant are left at the end of the reaction? (**a**) 1.20 g As (**b**) 1.50 g As (**c**) 4.30 g As (**d**) 8.30 g Ga

EP 3.17 If 3.00 g of titanium metal is reacted with 6.00 g of chlorine gas, Cl_2, to form 7.7 g of titanium(IV) chloride in a combination reaction, what is the percent yield of the product? (**a**) 65% (**b**) 96% (**c**) 48% (**d**) 86%

EP 3.18 When a mixture of 52.0 g acetylene, C_2H_2, and 96.0 g oxygen, O_2, is ignited, the resultant combustion reaction produces CO_2 and H_2O. For this combustion reaction, the limiting reactant is _____ and the theoretical yield of CO_2 is _____. (**a**) O_2, 44.0 g (**b**) O_2, 106 g (**c**) C_2H_2, 44.0 g (**d**) C_2H_2, 88.0 g (**e**) O_2, 176 g

EP 3.19 Aluminum trichloride, $AlCl_3$, is formed from the reaction of $Al_2O_3(s)$ with HCl gas. The balanced reaction is:

$$Al_2O_3(s) + 6\,HCl(g) \rightarrow 2\,AlCl_3(s) + 3\,H_2O(g)$$

When 1.50 kg of Al_2O_3 is reacted with excess HCl until it is all consumed, 2.85 kg of $AlCl_3$ are recovered after the reaction is completed. What is the percent yield for this experiment? (**a**) 48.5% (**b**) 72.6% (**c**) 80.5% (**d**) 87.3% (**e**) 95.0%

Exercises

Visualizing Concepts

3.1 The reaction between reactant A (blue spheres) and reactant B (red spheres) is shown in the following diagram:

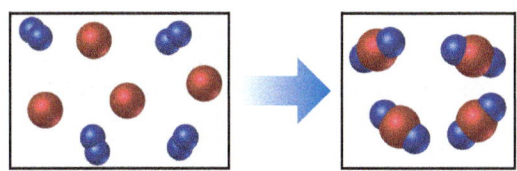

Based on this diagram, which equation best describes the reaction? [Section 3.1]

(**a**) $A_2 + B \longrightarrow A_2B$ (**c**) $2A + B_4 \longrightarrow 2AB_2$

(**b**) $A_2 + 4B \longrightarrow 2AB_2$ (**d**) $A + B_2 \longrightarrow AB_2$

3.2 The following diagram shows the combination reaction between hydrogen, H_2, and carbon monoxide, CO, to produce methanol, CH_3OH (white spheres are H, black spheres are C, red spheres are O). The correct number of CO molecules involved in this reaction is not shown. [Section 3.1] (**a**) Determine the number of CO molecules that should be shown in the left (reactants) box. (**b**) Write a balanced chemical equation for the reaction.

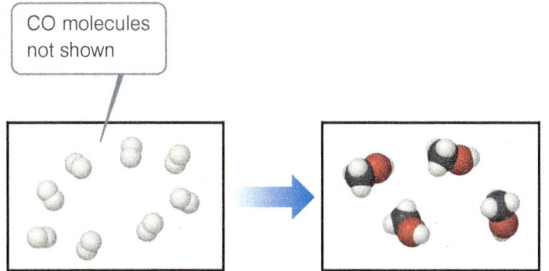

CO molecules not shown

3.3 The following diagram represents the collection of elements formed by a decomposition reaction. (**a**) If the blue spheres represent N atoms and the red ones represent O atoms, what was the empirical formula of the original compound? (**b**) Is it possible to determine the molecular formula of the reactant from the identities and quantities of the products? [Section 3.2]

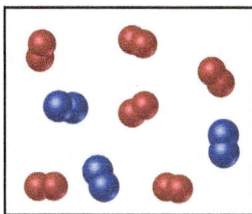

3.4 The following diagram represents the collection of CO_2 and H_2O molecules formed by complete combustion of a hydrocarbon. What is the empirical formula of the hydrocarbon? [Section 3.2]

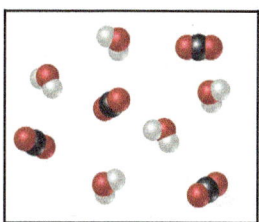

3.5 Glycine, an amino acid used by organisms to make proteins, is represented by the following molecular model. (**a**) Write its molecular formula. (**b**) Determine its molar mass. (**c**) Calculate how many moles of glycine are in a 100.0-g sample of glycine. (**d**) Calculate the percent nitrogen by mass in glycine. [Sections 3.3 and 3.5]

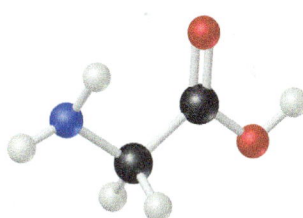

3.6 The following diagram represents a high-temperature reaction between CH_4 and H_2O. Based on this reaction, find how many moles of each product can be obtained starting with 4.0 mol CH_4. [Section 3.6]

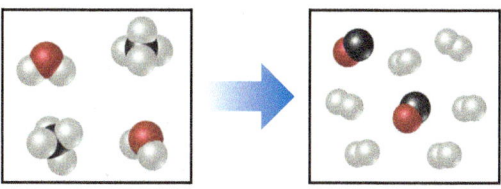

3.7 Nitrogen (N_2) and hydrogen (H_2) react to form ammonia (NH_3). Consider the mixture of N_2 and H_2 shown in the accompanying diagram. The blue spheres represent N, and the white ones represent H. (**a**) Write the balanced chemical equation for the reaction. (**b**) What is the limiting reactant? (**c**) How many molecules of ammonia can be made, assuming the reaction goes to completion, based on the diagram? (**d**) Are any reactant molecules left over, based on the diagram? If so, how many of which type are left over? [Section 3.7]

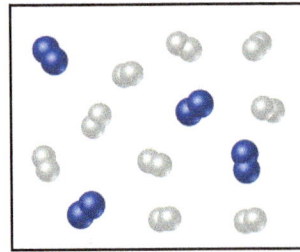

3.8 Nitrogen monoxide and oxygen react to form nitrogen dioxide. Consider the mixture of NO and O_2 shown in the accompanying diagram. The blue spheres represent N, and the red ones represent O. (**a**) How many molecules of NO_2 can be formed, assuming the reaction goes to completion? (**b**) What is the limiting reactant? (**c**) If the actual yield of the reaction was 75% instead of 100%, how many molecules of each kind would be present after the reaction was over?

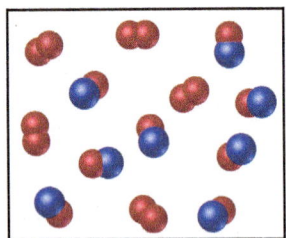

Chemical Equations (Section 3.1)

3.9 Write "true" or "false" for each statement. (**a**) We balance chemical equations as we do because energy must be conserved. (**b**) If the reaction $2\,O_3(g) \rightarrow 3\,O_2(g)$ goes to completion and all O_3 is converted to O_2, then the mass of O_3 at the beginning of the reaction must be the same as the mass of O_2 at the end of the reaction. (**c**) You can balance the "water-splitting" reaction $H_2O(l) \rightarrow H_2(g) + O_2(g)$ by writing it this way: $H_2O_2(l) \rightarrow H_2(g) + O_2(g)$.

3.10 A key step in balancing chemical equations is correctly identifying the formulas of the reactants and products. For example, consider the reaction between calcium oxide, $CaO(s)$, and $H_2O(l)$ to form aqueous calcium hydroxide. (**a**) Write a balanced chemical equation for this combination reaction, having correctly identified the product as $Ca(OH)_2(aq)$. (**b**) Is it possible to balance the equation if you incorrectly identify the product as $CaOH(aq)$, and if so, what is the equation?

3.11 Balance the following equations:
(**a**) $CO(g) + O_2(g) \longrightarrow CO_2(g)$
(**b**) $N_2O_5(g) + H_2O(l) \longrightarrow HNO_3(aq)$
(**c**) $CH_4(g) + Cl_2(g) \longrightarrow CCl_4(l) + HCl(g)$
(**d**) $Zn(OH)_2(s) + HNO_3(aq) \longrightarrow Zn(NO_3)_2(aq) + H_2O(l)$

3.12 Balance the following equations:
(**a**) $Li(s) + N_2(g) \longrightarrow Li_3N(s)$
(**b**) $TiCl_4(l) + H_2O(l) \longrightarrow TiO_2(s) + HCl(aq)$
(**c**) $NH_4NO_3(s) \longrightarrow N_2(g) + O_2(g) + H_2O(g)$
(**d**) $AlCl_3(s) + Ca_3N_2(s) \longrightarrow AlN(s) + CaCl_2(s)$

3.13 Balance the following equations:
(**a**) $Al_4C_3(s) + H_2O(l) \longrightarrow Al(OH)_3(s) + CH_4(g)$
(**b**) $C_5H_{10}O_2(l) + O_2(g) \longrightarrow CO_2(g) + H_2O(g)$

(**c**) $Fe(OH)_3(s) + H_2SO_4(aq) \longrightarrow Fe_2(SO_4)_3(aq) + H_2O(l)$
(**d**) $Mg_3N_2(s) + H_2SO_4(aq) \longrightarrow MgSO_4(aq) + (NH_4)_2SO_4(aq)$

3.14 Balance the following equations:
(**a**) $P_4O_{10}(s) + H_2O(l) \rightarrow H_3PO_4(aq)$
(**b**) $WCl_6(s) + Na_2S(s) \rightarrow WS_2(s) + NaCl(s) + S(s)$
(**c**) $NaHCO_3(s) + H_2SO_4(aq) \rightarrow CO_2(g) + H_2O(l) + Na_2SO_4(aq)$
(**d**) $NaN_3(s) + HNO_2(aq) \rightarrow N_2(g) + NO(g) + NaOH(aq)$

3.15 Write balanced chemical equations corresponding to each of the following descriptions: (**a**) Solid calcium carbide, CaC_2, reacts with water to form an aqueous solution of calcium hydroxide and acetylene gas, C_2H_2. (**b**) When solid potassium chlorate is heated, it decomposes to form solid potassium chloride and oxygen gas. (**c**) Solid zinc metal reacts with sulfuric acid to form hydrogen gas and an aqueous solution of zinc sulfate. (**d**) When liquid phosphorus trichloride is added to water, it reacts to form aqueous phosphorous acid, $H_3PO_3(aq)$, and aqueous hydrochloric acid. (**e**) When hydrogen sulfide gas is passed over solid hot iron(III) hydroxide, the resulting reaction produces solid iron(III) sulfide and gaseous water.

3.16 Write balanced chemical equations to correspond to each of the following descriptions: (**a**) When sulfur trioxide gas reacts with water, a solution of sulfuric acid forms. (**b**) Boron sulfide, $B_2S_3(s)$, reacts violently with water to form dissolved boric acid, H_3BO_3, and hydrogen sulfide gas. (**c**) Phosphine, $PH_3(g)$, combusts in oxygen gas to form water vapor and solid tetraphosphorus decaoxide. (**d**) When solid mercury(II) nitrate is heated, it decomposes to form solid mercury(II) oxide, gaseous nitrogen dioxide, and oxygen. (**e**) Copper metal reacts with hot concentrated sulfuric acid solution to form aqueous copper(II) sulfate, sulfur dioxide gas, and water.

Simple Patterns of Chemical Reactivity (Section 3.2)

3.17 (**a**) When the metallic element sodium combines with the nonmetallic element bromine, $Br_2(l)$, what is the chemical formula of the product? (**b**) Is the product a solid, liquid, or gas at room temperature? (**c**) In the balanced chemical equation for this reaction, what is the coefficient in front of the product?

3.18 (**a**) When a compound containing C, H, and O is completely combusted in air, what reactant besides the hydrocarbon is involved in the reaction? (**b**) What products form in this reaction? (**c**) What is the sum of the coefficients in the balanced chemical equation for the combustion of one mole of acetone, $C_3H_6O(l)$, in air?

3.19 Write a balanced chemical equation for the reaction that occurs when (**a**) $Mg(s)$ reacts with $Cl_2(g)$; (**b**) barium carbonate decomposes into barium oxide and carbon dioxide gas when heated; (**c**) the hydrocarbon styrene, $C_8H_8(l)$, is combusted in air; (**d**) dimethylether, $CH_3OCH_3(g)$, is combusted in air.

3.20 Write a balanced chemical equation for the reaction that occurs when (**a**) titanium metal reacts with $O_2(g)$; (**b**) silver(I) oxide decomposes into silver metal and oxygen gas when heated; (**c**) propanol, $C_3H_7OH(l)$ burns in air; (**d**) methyl *tert*-butyl ether, $C_5H_{12}O(l)$, burns in air.

3.21 Balance the following equations and indicate whether they are combination, decomposition, or combustion reactions:
(**a**) $C_3H_6(g) + O_2(g) \longrightarrow CO_2(g) + H_2O(g)$
(**b**) $NH_4NO_3(s) \longrightarrow N_2O(g) + H_2O(g)$
(**c**) $C_5H_6O(l) + O_2(g) \longrightarrow CO_2(g) + H_2O(g)$
(**d**) $N_2(g) + H_2(g) \longrightarrow NH_3(g)$
(**e**) $K_2O(s) + H_2O(l) \longrightarrow KOH(aq)$

3.22 Balance the following equations and indicate whether they are combination, decomposition, or combustion reactions:

(a) $PbCO_3(s) \longrightarrow PbO(s) + CO_2(g)$

(b) $C_2H_4(g) + O_2(g) \longrightarrow CO_2(g) + H_2O(g)$

(c) $Mg(s) + N_2(g) \longrightarrow Mg_3N_2(s)$

(d) $C_7H_8O_2(l) + O_2(g) \longrightarrow CO_2(g) + H_2O(g)$

(e) $Al(s) + Cl_2(g) \longrightarrow AlCl_3(s)$

Formula Weights (Section 3.3)

3.23 Determine the formula weights of each of the following compounds: (a) nitric acid, HNO_3 (b) $KMnO_4$ (c) $Ca_3(PO_4)_2$ (d) quartz, SiO_2 (e) gallium sulfide (f) chromium(III) sulfate (g) phosphorus trichloride.

3.24 Determine the formula weights of each of the following compounds: (a) phosgene, $COCl_2$, a colorless gas used in the manufacture of some plastics and as a chemical warfare agent during World War I (b) citric acid, $C_6H_8O_7$, a weak acid prevalent in citrus fruits (c) hydroxyapatite, $Ca_{10}(PO_4)_6$ $(OH)_2$, the principal component of human bones (d) myrcene, $C_{10}H_{16}$, an aromatic molecule found in herbs and spices such as thyme, hops, cannabis, and cardamom (e) benzaldehyde, C_6H_5CHO, the molecule largely responsible for the odor of almond extract.

3.25 Calculate the percentage by mass of oxygen in the following compounds: (a) morphine, $C_{17}H_{19}NO_3$ (b) codeine, $C_{18}H_{21}NO_3$ (c) cocaine, $C_{17}H_{21}NO_4$ (d) tetracycline, $C_{22}H_{24}N_2O_8$ (e) digitoxin, $C_{41}H_{64}O_{13}$ (f) vancomycin, $C_{66}H_{75}Cl_2N_9O_{24}$

3.26 Calculate the percentage by mass of the indicated element in the following compounds: (a) carbon in acetylene, C_2H_2, a gas used in welding (b) hydrogen in ascorbic acid, $HC_6H_7O_6$, also known as vitamin C (c) hydrogen in ammonium sulfate, $(NH_4)_2SO_4$, a substance used as a nitrogen fertilizer (d) platinum in $PtCl_2(NH_3)_2$, a chemotherapy agent called cisplatin (e) oxygen in the female sex hormone estradiol, $C_{18}H_{24}O_2$ (f) carbon in capsaicin, $C_{18}H_{27}NO_3$, the compound that gives the hot taste to chili peppers.

3.27 Based on the following structural formulas, calculate the percentage of carbon by mass present in each compound:

(a) Benzaldehyde (almond fragrance)

(b) Vanillin (vanilla flavor)

(c) Isopentyl acetate (banana flavor)

3.28 Calculate the percentage of carbon by mass in each of the compounds represented by the following models:

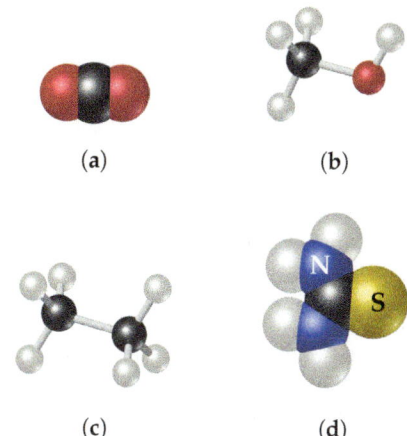

(a)   (b)

(c)   (d)

Avogadro's Number and the Mole (Section 3.4)

3.29 Characterize the following statements as either true or false. (a) A mole of horses contains a mole of horse legs. (b) A mole of water has a mass of 18.0 g. (c) The mass of one molecule of water is 18.0 g. (d) A mole of $NaCl(s)$ contains two moles of ions.

3.30 (a) What is the mass, in grams, of 1 mol of ^{12}C? (b) How many carbon atoms are present in 1 mol of ^{12}C?

3.31 Without doing any detailed calculations (but using a periodic table to give atomic weights), rank the following samples in order of increasing numbers of atoms: 0.50 mol H_2O, $23\,g$ Na, 6.0×10^{23} N_2 molecules.

3.32 Without doing any detailed calculations (but using a periodic table to give atomic weights), rank the following samples in order of increasing numbers of atoms: $42\,g$ of $NaHCO_3$, 1.5 mol CO_2, 6.0×10^{24} Ne atoms.

3.33 What is the mass, in kilograms, of an Avogadro's number of people, if the average mass of a person is 73 kg? How does this compare with the mass of Earth, $5.98 \times 10^{24}\,kg$?

3.34 If an Avogadro's number of pennies is divided equally among the 321 million men, women, and children in the United States, how many dollars would each receive? How does this compare with the gross domestic product (GDP) of the United States, which was $21.4 trillion in 2019? (GDP is the total market value of the nation's goods and services.)

3.35 Calculate the following quantities:

(a) mass, in grams, of 0.105 mol sucrose $(C_{12}H_{22}O_{11})$

(b) moles of $Zn(NO_3)_2$ in 143.50 g of this substance

(c) number of molecules in 1.0×10^{-6} mol CH_3CH_2OH

(d) number of N atoms in 0.410 mol NH_3

3.36 Calculate the following quantities:

(a) mass, in grams, of 1.50×10^{-2} mol CdS

(b) number of moles of NH_4Cl in 86.6 g of this substance

(c) number of molecules in 8.447×10^{-2} mol C_6H_6

(d) number of O atoms in 6.25×10^{-3} mol $Al(NO_3)_3$

3.37 (a) What is the mass, in grams, of 2.50×10^{-3} mol of ammonium phosphate?

(b) How many moles of chloride ions are in 0.2550 g of aluminum chloride?

(c) What is the mass, in grams, of 7.70×10^{20} molecules of caffeine, $C_8H_{10}N_4O_2$?

(d) What is the molar mass of cholesterol if 0.00105 mol has a mass of 0.406 g?

3.38 **(a)** What is the mass, in grams, of 1.223 mol of iron(III) sulfate?

(b) How many moles of ammonium ions are in 6.955 g of ammonium carbonate?

(c) What is the mass, in grams, of 1.50×10^{21} molecules of aspirin, $C_9H_8O_4$?

(d) What is the molar mass of diazepam (Valium®) if 0.05570 mol has a mass of 15.86 g?

3.39 The molecular formula of allicin, the compound responsible for the characteristic smell of garlic, is $C_6H_{10}OS_2$. **(a)** What is the molar mass of allicin? **(b)** How many moles of allicin are present in 5.00 mg of this substance? **(c)** How many molecules of allicin are in 5.00 mg of this substance? **(d)** How many S atoms are present in 5.00 mg of allicin?

3.40 The molecular formula of aspartame, the artificial sweetener marketed as NutraSweet®, is $C_{14}H_{18}N_2O_5$. **(a)** What is the molar mass of aspartame? **(b)** How many moles of aspartame are present in 1.00 mg of aspartame? **(c)** How many molecules of aspartame are present in 1.00 mg of aspartame? **(d)** How many hydrogen atoms are present in 1.00 mg of aspartame?

3.41 A sample of glucose, $C_6H_{12}O_6$, contains 1.250×10^{21} carbon atoms. **(a)** How many atoms of hydrogen does it contain? **(b)** How many molecules of glucose does it contain? **(c)** How many moles of glucose does it contain? **(d)** What is the mass of this sample in grams?

3.42 A sample of the male sex hormone testosterone, $C_{19}H_{28}O_2$, contains 3.88×10^{21} hydrogen atoms. **(a)** How many atoms of carbon does it contain? **(b)** How many molecules of testosterone does it contain? **(c)** How many moles of testosterone does it contain? **(d)** What is the mass of this sample in grams?

3.43 The allowable concentration level of vinyl chloride, C_2H_3Cl, in the atmosphere in a chemical plant is 2.0×10^{-6} g/L. How many moles of vinyl chloride in each liter does this represent? How many molecules per liter?

3.44 At least 25 μg of tetrahydrocannabinol (THC), the active ingredient in marijuana, is required to produce intoxication. The molecular formula of THC is $C_{21}H_{30}O_2$. How many moles of THC does this 25 μg represent? How many molecules?

Empirical Formulas from Analyses (Section 3.5)

3.45 Give the empirical formula of each of the following compounds if a sample contains **(a)** 0.0130 mol C, 0.0390 mol H, and 0.0065 mol O **(b)** 11.66 g iron and 5.01 g oxygen **(c)** 40.0% C, 6.7% H, and 53.3% O by mass.

3.46 Determine the empirical formula of each of the following compounds if a sample contains **(a)** 0.104 mol K, 0.052 mol C, and 0.156 mol O; **(b)** 5.28 g Sn and 3.37 g F; **(c)** 87.5% N and 12.5% H by mass.

3.47 Determine the empirical formulas of the compounds with the following compositions by mass: **(a)** 10.4% C, 27.8% S, and 61.7% Cl **(b)** 21.7% C, 9.6% O, and 68.7% F **(c)** 32.79% Na, 13.02% Al, and the remainder F

3.48 Determine the empirical formulas of the compounds with the following compositions by mass: **(a)** 55.3% K, 14.6% P, and 30.1% O **(b)** 24.5% Na, 14.9% Si, and 60.6% F **(c)** 62.1% C, 5.21% H, 12.1% N, and the remainder O

3.49 A compound whose empirical formula is XF_3 consists of 65% F by mass. What is the atomic mass of X?

3.50 The compound XCl_4 contains 75.0% Cl by mass. What is the element X?

3.51 What is the molecular formula of each of the following compounds?

(a) empirical formula CH_2, molar mass = 84.0 g/mol

(b) empirical formula NH_2Cl, molar mass = 51.5 g/mol

3.52 What is the molecular formula of each of the following compounds?

(a) empirical formula HCO_2, molar mass = 90.0 g/mol

(b) empirical formula C_2H_4O, molar mass = 88.0 g/mol

3.53 Determine the empirical and molecular formulas of each of the following substances: **(a)** Styrene, a compound used to make Styrofoam® cups and insulation, contains 92.3% C and 7.7% H by mass and has a molar mass of 104 g/mol. **(b)** Caffeine, a stimulant found in coffee, contains 49.5% C, 5.15% H, 28.9% N, and 16.5% O by mass and has a molar mass of 195 g/mol. **(c)** Monosodium glutamate (MSG), a flavor enhancer in certain foods, contains 35.51% C, 4.77% H, 37.85% O, 8.29% N, and 13.60% Na, and has a molar mass of 169 g/mol.

3.54 Determine the empirical and molecular formulas of each of the following substances: **(a)** Ibuprofen, a headache remedy, contains 75.69% C, 8.80% H, and 15.51% O by mass and has a molar mass of 206 g/mol. **(b)** Cadaverine, a foul-smelling substance produced by the action of bacteria on meat, contains 58.55% C, 13.81% H, and 27.40% N by mass; its molar mass is 102.2 g/mol. **(c)** Epinephrine (adrenaline), a hormone secreted into the bloodstream in times of danger or stress, contains 59.0% C, 7.1% H, 26.2% O, and 7.7% N by mass; its molar mass is about 180 amu.

3.55 **(a)** Combustion analysis of toluene, a common organic solvent, gives 5.86 mg of CO_2 and 1.37 mg of H_2O. If the compound contains only carbon and hydrogen, what is its empirical formula? **(b)** Menthol, the substance we can smell in mentholated cough drops, is composed of C, H, and O. A 0.1005-g sample of menthol is combusted, producing 0.2829 g of CO_2 and 0.1159 g of H_2O. What is the empirical formula for menthol? If menthol has a molar mass of 156 g/mol, what is its molecular formula?

3.56 **(a)** The characteristic odor of pineapple is due to ethyl butyrate, a compound containing carbon, hydrogen, and oxygen. Combustion of 2.78 mg of ethyl butyrate produces 6.32 mg of CO_2 and 2.58 mg of H_2O. What is the empirical formula of the compound? **(b)** Nicotine, a component of tobacco, is composed of C, H, and N. A 5.250-mg sample of nicotine was combusted, producing 14.242 mg of CO_2 and 4.083 mg of H_2O. What is the empirical formula for nicotine? If nicotine has a molar mass of 160 ± 5 g/mol, what is its molecular formula?

3.57 Valproic acid, used to treat seizures and bipolar disorder, is composed of C, H, and O. A 0.165-g sample is combusted to produce 0.166 g of water and 0.403 g of carbon dioxide. What is the empirical formula for valproic acid? If the molar mass is 144 g/mol, what is the molecular formula?

3.58 Propenoic acid, $C_3H_4O_2$, is a reactive organic liquid that is used in the manufacturing of plastics, coatings, and adhesives. An unlabeled container is thought to contain this liquid. A 0.275-g sample of the liquid is combusted to produce 0.102 g of water and 0.374 g carbon dioxide. Is the unknown liquid propenoic acid? Support your reasoning with calculations.

3.59 Washing soda, a compound used to prepare hard water for washing laundry, is a hydrate, which means that a certain number of water molecules are included in the solid structure. Its formula can be written as $Na_2CO_3 \cdot xH_2O$, where x is the number of moles of H_2O per mole of Na_2CO_3. When a 2.558-g sample of washing soda is heated at 125 °C, all the water of hydration is lost, leaving 0.948 g of Na_2CO_3. What is the value of x?

3.60 Epsom salts, a strong laxative used in veterinary medicine, is a hydrate, which means that a certain number of water molecules are included in the solid structure. The formula for Epsom salts can be written as $MgSO_4 \cdot xH_2O$, where x indicates the number of moles of H_2O per mole of $MgSO_4$. When 5.061 g of this hydrate is heated to 250 °C, all the water of hydration is lost, leaving 2.472 g of $MgSO_4$. What is the value of x?

Quantitative Information from Balanced Equations (Section 3.6)

3.61 Hydrofluoric acid, HF(aq), cannot be stored in glass bottles because compounds called silicates in the glass are attacked by the HF(aq). Sodium silicate (Na_2SiO_3), for example, reacts as follows:

$$Na_2SiO_3(s) + 8HF(aq) \longrightarrow H_2SiF_6(aq) + 2NaF(aq) + 3H_2O(l)$$

(**a**) How many moles of HF are needed to react with 0.300 mol of Na_2SiO_3? (**b**)How many grams of NaF form when 0.500 mol of HF reacts with excess Na_2SiO_3? (**c**) How many grams of Na_2SiO_3 can react with 0.800 g of HF?

3.62 The reaction between potassium superoxide, KO_2, and CO_2,

$$4KO_2 + 2CO_2 \longrightarrow 2K_2CO_3 + 3O_2$$

is used as a source of O_2 and an absorber of CO_2 in self-contained breathing equipment used by rescue workers.

(**a**) How many moles of O_2 are produced when 0.400 mol of KO_2 react in this fashion? (**b**) How many grams of KO_2 are needed to form 7.50 g of O_2? (**c**) How many grams of CO_2 are used when 7.50 g of O_2 are produced?

3.63 Several brands of antacids use $Al(OH)_3$ to react with stomach acid, which contains primarily HCl:

$$Al(OH)_3(s) + HCl(aq) \longrightarrow AlCl_3(aq) + H_2O(l)$$

(**a**) Balance this equation. (**b**) Calculate the number of grams of HCl that can react with 0.500 g of $Al(OH)_3$. (**c**) Calculate the number of grams of $AlCl_3$ and the number of grams of H_2O formed in part (a). (**d**) Show that your calculations in parts (b) and (c) are consistent with the law of conservation of mass.

3.64 An iron ore sample contains Fe_2O_3 together with other substances. Reaction of the ore with CO produces iron metal:

$$Fe_2O_3(s) + CO(g) \longrightarrow Fe(s) + CO_2(g)$$

(**a**) Balance this equation. (**b**) Calculate the number of grams of CO that can react with 0.350 kg of Fe_2O_3. (**c**) Calculate the number of grams of Fe and the number of grams of CO_2 formed when 0.350 kg of Fe_2O_3 reacts. (**d**) Show that your calculations in parts (b) and (c) are consistent with the law of conservation of mass.

3.65 Aluminum sulfide reacts with water to form aluminum hydroxide and hydrogen sulfide. (**a**) Write the balanced chemical equation for this reaction. (**b**) How many grams of aluminum hydroxide are obtained from 14.2 g of aluminum sulfide?

3.66 Calcium hydride reacts with water to form calcium hydroxide and hydrogen gas. (**a**) Write a balanced chemical equation for the reaction. (**b**) How many grams of calcium hydride are needed to form 4.500 g of hydrogen?

3.67 Automotive air bags inflate when sodium azide, NaN_3, rapidly decomposes to its component elements:

$$2NaN_3(s) \longrightarrow 2Na(s) + 3N_2(g)$$

(**a**) How many moles of N_2 are produced by the decomposition of 1.50 mol of NaN_3? (**b**) How many grams of NaN_3 are required to form 10.0 g of nitrogen gas? (**c**) How many grams of NaN_3 are required to produce $10.0 ft^3$ of nitrogen gas, about the size of an automotive air bag, if the gas has a density of 1.25 g/L?

3.68 The complete combustion of octane, C_8H_{18}, a component of gasoline, proceeds as follows:

$$2C_8H_{18}(l) + 25O_2(g) \longrightarrow 16CO_2(g) + 18H_2O(g)$$

(**a**) How many moles of O_2 are needed to burn 1.50 mol of C_8H_{18}? (**b**) How many grams of O_2 are needed to burn 10.0 g of C_8H_{18}? (**c**) Octane has a density of 0.692 g/mL at 20°C. How many grams of O_2 are required to burn 15.0 gal of C_8H_{18} (the capacity of an average fuel tank)? (**d**) How many grams of CO_2 are produced when 15.0 gal of C_8H_{18} are combusted?

3.69 A piece of aluminum foil $1.00 cm^2$ and 0.550-mm thick is allowed to react with bromine to form aluminum bromide.

(**a**) How many moles of aluminum were used? (The density of aluminum is $2.699 g/cm^3$.) (**b**) Write a balanced chemical equation for this reaction. (**c**) How many grams of aluminum bromide form, assuming the aluminum reacts completely?

3.70 Detonation of nitroglycerin proceeds as follows:

$$4C_3H_5N_3O_9(l) \longrightarrow$$
$$12CO_2(g) + 6N_2(g) + O_2(g) + 10H_2O(g)$$

(**a**) If a sample containing 2.00 mL of nitroglycerin (density = 1.592 g/mL) is detonated, how many moles of gas are produced? (**b**) If each mole of gas occupies 55 L under the conditions of the explosion, how many liters of gas are produced? (**c**) How many grams of N_2 are produced in the detonation?

3.71 The combustion of one mole of liquid ethanol, CH_3CH_2OH, produces 1367 kJ of heat. (**a**) Write a balanced chemical equation for the combustion of ethanol. (**b**) Calculate how much heat is produced when 235.0 g of ethanol are combusted.

3.72 The combustion of one mole of liquid octane, $CH_3(CH_2)_6CH_3$, produces 5470 kJ of heat. (**a**) Write a balanced chemical equation for the combustion of octane. (**b**) Calculate how much heat is produced if 1.000 gallon of octane is combusted. Octane has a density of 0.692 g/mL at 20°C.

Limiting Reactants (Section 3.7)

3.73 Imagine you were asked to assemble skateboards from decks, axles, and wheels, with each skateboard made from one deck, two axles, and four wheels. (**a**) What is the limiting reactant if you start from 25 decks, 48 axles, and 106 wheels? (**b**) What is the theoretical yield of skateboards? (**c**) If 22 boards are produced once the assembly is complete, what is the percent yield of skateboards?

3.74 Consider a recipe for making a club sandwich that consists of three slices toasted bread, 2 slices turkey breast, 2 pieces fried bacon, and one slice cheddar cheese. (**a**) If you start with 36 slices of bread, 20 slices of turkey breast, 24 pieces bacon, and 14 slices of cheese, what is the limiting reactant? (**b**) What is the theoretical yield of club sandwiches? (**c**) If you ask your roommate to prepare club sandwiches for a party and they finish with nine sandwiches, what is the percent yield of sandwiches?

3.75 Consider the mixture of ethanol, C_2H_5OH, and O_2 shown in the accompanying diagram. (**a**) Write a balanced equation for the combustion reaction that occurs between ethanol and oxygen. (**b**) Which reactant is the limiting reactant? (**c**) How many molecules of CO_2, H_2O, C_2H_5OH, and O_2 will be present if the reaction goes to completion?

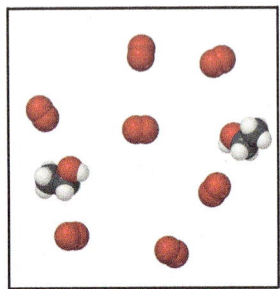

3.76 Consider the mixture of propane, C_3H_8, and O_2 shown here. (**a**) Write a balanced equation for the combustion reaction that occurs between propane and oxygen. (**b**) Which reactant is the limiting reactant? (**c**) How many molecules of CO_2, H_2O, C_3H_8, and O_2 will be present if the reaction goes to completion?

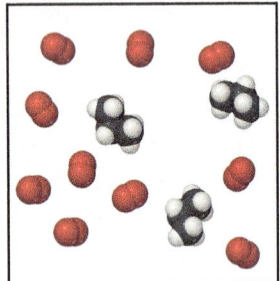

3.77 Sodium hydroxide reacts with carbon dioxide as follows:

$$2NaOH(s) + CO_2(g) \longrightarrow Na_2CO_3(s) + H_2O(l)$$

Which is the limiting reactant when 1.85 mol NaOH and 1.00 mol CO_2 are allowed to react? How many moles of Na_2CO_3 can be produced? How many moles of the excess reactant remain after the completion of the reaction?

3.78 Aluminum hydroxide reacts with sulfuric acid as follows:

$$2Al(OH)_3(s) + 3H_2SO_4(aq) \longrightarrow Al_2(SO_4)_3(aq) + 6H_2O(l)$$

Which is the limiting reactant when 0.500 mol $Al(OH)_3$ and 0.500 mol H_2SO_4 are allowed to react? How many moles of $Al_2(SO_4)_3$ can form under these conditions? How many moles of the excess reactant remain after completion of the reaction?

3.79 The fizz produced when an Alka-Seltzer tablet is dissolved in water is due to the reaction between sodium bicarbonate ($NaHCO_3$) and citric acid ($H_3C_6H_5O_7$):

$$3NaHCO_3(aq) + H_3C_6H_5O_7(aq) \longrightarrow$$
$$3CO_2(g) + 3H_2O(l) + Na_3C_6H_5O_7(aq)$$

In a certain experiment, 1.00 g of sodium bicarbonate and 1.00 g of citric acid are allowed to react. (**a**) Which is the limiting reactant? (**b**) How many grams of carbon dioxide form? (**c**) How many grams of the excess reactant remain after the limiting reactant is completely consumed?

3.80 One of the steps in the commercial process for converting ammonia to nitric acid is the conversion of NH_3 to NO:

$$4NH_3(g) + 5O_2(g) \longrightarrow 4NO(g) + 6H_2O(g)$$

In a certain experiment, 2.00 g of NH_3 reacts with 2.50 g of O_2. (**a**) Which is the limiting reactant? (**b**) How many grams of NO and H_2O form? (**c**) How many grams of the excess reactant remain after the limiting reactant is completely consumed? (**d**) Show that your calculations in parts (**b**) and (**c**) are consistent with the law of conservation of mass.

3.81 Solutions of sodium carbonate and silver nitrate react to form solid silver carbonate and a solution of sodium nitrate. A solution containing 3.50 g of sodium carbonate is mixed with a solution containing 5.00 g of silver nitrate. How many grams of sodium carbonate, silver nitrate, silver carbonate, and sodium nitrate are present after the reaction is complete?

3.82 Solutions of sulfuric acid and lead(II) acetate react to form solid lead(II) sulfate and a solution of acetic acid. If 5.00 g of sulfuric acid and 5.00 g of lead(II) acetate are mixed, calculate the number of grams of sulfuric acid, lead(II) acetate, lead(II) sulfate, and acetic acid present in the mixture after the reaction is complete.

3.83 When benzene (C_6H_6) reacts with bromine (Br_2), bromobenzene (C_6H_5Br) is obtained:

$$C_6H_6 + Br_2 \longrightarrow C_6H_5Br + HBr$$

(**a**) When 30.0 g of benzene reacts with 65.0 g of bromine, what is the theoretical yield of bromobenzene? (**b**) If the actual yield of bromobenzene is 42.3 g, what is the percentage yield?

3.84 When ethane (C_2H_6) reacts with chlorine (Cl_2), the main product is C_2H_5Cl, but other products containing Cl, such as $C_2H_4Cl_2$, are also obtained in small quantities. The formation of these other products reduces the yield of C_2H_5Cl. (**a**) Calculate the theoretical yield of C_2H_5Cl when 125 g of C_2H_6 reacts with 255 g of Cl_2, assuming that C_2H_6 and Cl_2 react only to form C_2H_2Cl and HCl. (**b**) Calculate the percent yield of C_2H_5Cl if the reaction produces 206 g of C_2H_5Cl.

3.85 Hydrogen sulfide is an impurity in natural gas that must be removed. One common removal method is called the Claus process, which relies on the reaction:

$$8\,H_2S(g) + 4\,O_2(g) \longrightarrow S_8(l) + 8\,H_2O(g)$$

Under optimal conditions, the Claus process gives 98% yield of S_8 from H_2S. If you started with 30.0 g of H_2S and 50.0 g of O_2, how many grams of S_8 would be produced, assuming 98% yield?

3.86 When hydrogen sulfide gas is bubbled into a solution of sodium hydroxide, the reaction forms sodium sulfide and water. How many grams of sodium sulfide are formed if 1.25 g of hydrogen sulfide is bubbled into a solution containing 2.00 g of sodium hydroxide, assuming that the sodium sulfide is made in 92.0% yield?

Additional Exercises

3.87 Write the balanced chemical equations for (**a**) the complete combustion of acetic acid (CH_3COOH), the main ingredient in vinegar; (**b**) the decomposition of solid calcium hydroxide into solid calcium oxide (lime) and water vapor; (**c**) the combination reaction between nickel metal and chlorine gas.

3.88 The effectiveness of nitrogen fertilizers depends on both their ability to deliver nitrogen to plants and the amount of nitrogen they can deliver. Four common nitrogen-containing fertilizers are ammonia, ammonium nitrate, ammonium sulfate, and urea $[(NH_2)_2CO]$. Rank these fertilizers in terms of the mass percentage nitrogen they contain.

3.89 (**a**) The molecular formula of acetylsalicylic acid (aspirin), one of the most common pain relievers, is $C_9H_8O_4$. How many moles of $C_9H_8O_4$ are in a 0.500-g tablet of aspirin? Assume the tablet is composed entirely of aspirin. (**b**) How many molecules of $C_9H_8O_4$ are in this tablet? (**c**) How many carbon atoms are in the tablet?

3.90 (**a**) One molecule of the antibiotic penicillin G has a mass of 5.342×10^{-21} g. What is the molar mass of penicillin G? (**b**) Hemoglobin, the oxygen-carrying protein in red blood cells, has four iron atoms per molecule and contains 0.340% iron by mass. Calculate the molar mass of hemoglobin.

3.91 Serotonin is a compound that conducts nerve impulses in the brain. It contains 68.2% C, 6.86% H, 15.9% N, and 9.08% O. Its molar mass is 176 g/mol. Determine its molecular formula.

3.92 The koala dines exclusively on eucalyptus leaves. Its digestive system detoxifies the eucalyptus oil, a poison to other animals. The chief constituent in eucalyptus oil is a substance called eucalyptol, which contains 77.87% C, 11.76% H, and the remainder O. (**a**) What is the empirical formula for this substance? (**b**) A mass spectrum of eucalyptol shows a peak at about 154 amu. What is the molecular formula of the substance?

3.93 Vanillin, the dominant flavoring in vanilla, contains C, H, and O. When 1.05 g of this substance is completely combusted, 2.43 g of CO_2 and 0.50 g of H_2O are produced. What is the empirical formula of vanillin?

3.94 An organic compound was found to contain only C, H, and Cl. When a 1.50-g sample of the compound was completely combusted in air, 3.52 g of CO_2 was formed. In a separate experiment, the chlorine in a 1.00-g sample of the compound was converted to 1.27 g of AgCl. Determine the empirical formula of the compound.

3.95 A compound, $KBrO_x$, where x is unknown, is analyzed and found to contain 52.92% Br. What is the value of x?

3.96 An element X forms an iodide (XI_3) and a chloride (XCl_3). The iodide is quantitatively converted to the chloride when it is heated in a stream of chlorine:

$$2\,XI_3 + 3\,Cl_2 \longrightarrow 2\,XCl_3 + 3\,I_2$$

If 0.5000 g of XI_3 is treated with chlorine, 0.2360 g of XCl_3 is obtained. (**a**) Calculate the atomic weight of the element X. (**b**) Identify the element X.

3.97 A method used by the U.S. Environmental Protection Agency (EPA) for determining the concentration of ozone in air is to pass the air sample through a "bubbler" containing sodium iodide, which removes the ozone according to the following equation:

$$O_3(g) + 2\,NaI(aq) + H_2O(l) \longrightarrow$$
$$O_2(g) + I_2(s) + 2\,NaOH(aq)$$

(**a**) How many moles of sodium iodide are needed to remove 5.95×10^{-6} mol of O_3? (**b**) How many grams of sodium iodide are needed to remove 1.3 mg of O_3?

3.98 A chemical plant uses electrical energy to decompose aqueous solutions of NaCl to give Cl_2, H_2, and NaOH:

$$2\,NaCl(aq) + 2\,H_2O(l) \longrightarrow 2\,NaOH(aq) + H_2(g) + Cl_2(g)$$

If the plant produces 1.5×10^6 kg (1500 metric tons) of Cl_2 daily, estimate the quantities of H_2 and NaOH produced.

3.99 The fat stored in a camel's hump is a source of both energy and water. Calculate the mass of H_2O produced by the metabolism of 1.0 kg of fat, assuming the fat consists entirely of tristearin ($C_{57}H_{110}O_6$), a typical animal fat, and assuming that during metabolism, tristearin reacts with O_2 to form only CO_2 and H_2O.

3.100 When hydrocarbons are burned in a limited amount of air, both CO and CO_2 form. When 0.450 g of a particular hydrocarbon was burned in air, 0.467 g of CO, 0.733 g of CO_2, and 0.450 g of H_2O were formed. (**a**) What is the empirical formula of the compound? (**b**) How many grams of O_2 were used in the reaction? (**c**) How many grams would have been required for complete combustion?

3.101 A mixture of $N_2(g)$ and $H_2(g)$ reacts in a closed container to form ammonia, $NH_3(g)$. The reaction ceases before either reactant has been totally consumed. At this stage 3.0 mol N_2, 3.0 mol H_2, and 3.0 mol NH_3 are present. How many moles of N_2 and H_2 were present originally?

3.102 A mixture containing $KClO_3$, K_2CO_3, $KHCO_3$, and KCl was heated, producing CO_2, O_2, and H_2O gases according to the following equations:

$$2\,KClO_3(s) \longrightarrow 2\,KCl(s) + 3\,O_2(g)$$
$$2\,KHCO_3(s) \longrightarrow K_2O(s) + H_2O(g) + 2\,CO_2(g)$$
$$K_2CO_3(s) \longrightarrow K_2O(s) + CO_2(g)$$

The KCl does not react under the conditions of the reaction. If 100.0 g of the mixture produces 1.80 g of H_2O, 13.20 g of CO_2, and 4.00 g of O_2, what was the composition of the original mixture? (Assume complete decomposition of the mixture.)

3.103 When a mixture of 10.0 g of acetylene (C_2H_2) and 10.0 g of oxygen (O_2) is ignited, the resulting combustion reaction produces CO_2 and H_2O. (**a**) Write the balanced chemical equation for this reaction. (**b**) Which is the limiting reactant? (**c**) How many grams of C_2H_2, O_2, CO_2, and H_2O are present after the reaction is complete?

3.104 (**a**) If an automobile travels 225 mi with a gas mileage of 20.5 mi/gal, how many kilograms of CO_2 are produced? Assume that the gasoline is composed of octane, $C_8H_{18}(l)$, whose density is 0.692 g/mL. (**b**) Repeat the calculation for a truck that has a gas mileage of 5 mi/gal.

3.105 **(a)** What is the theoretical yield of CO_2 if 1.00×10^3 kg (1 metric ton) of CaO is obtained by thermal decomposition of $CaCO_3$? **(b)** In ordinary Portland cement, the average CaO content of the clinker (the inorganic part of the dried cement prior to adding gypsum and any additives or fillers) is 63.5%. What quantity of CO_2 is emitted to produce 1.00×10^3 kg of clinker? **(c)** If the clinker makes up 95% of the mass of the cement, a good estimate for most cement produced in the twentieth century, what quantity of CO_2 is emitted in the production of 1.00×10^3 kg of cement? **(d)** In some parts of the world waste products, blast furnace slag from the manufacture of steel, or fly ash from coal-fired power plants are added to cement to reduce the amount of clinker in the cement. If the clinker makes up only 60% of the cement by mass, what quantity of CO_2 is emitted in the production of 1.00×10^3 kg of cement?

3.106 A particular coal contains 2.5% sulfur by mass. When this coal is burned at a power plant, the sulfur is converted into sulfur dioxide gas, which is a pollutant. To reduce sulfur dioxide emissions, calcium oxide (lime) is used. The sulfur dioxide reacts with calcium oxide to form solid calcium sulfite. **(a)** Write the balanced chemical equation for the reaction. **(b)** If the coal is burned in a power plant that uses 2000.0 tons of coal per day, what mass of calcium oxide is required daily to eliminate the sulfur dioxide? **(c)** How many grams of calcium sulfite does this power plant produce daily?

3.107 Two key components of cement clinker are Ca_3SiO_5, called tricalcium silicate or alite, and Ca_2SiO_4, called dicalcium silicate or belite. Both are formed when CaO produced from thermal decomposition of limestone ($CaCO_3$) reacts with SiO_2 present in the clays and other raw materials used to make cement. When water is added, both react to form hydrated phases that give cement much of its strength, though alite forms more rapidly and plays a more important role in the setting of the cement and its initial strength gain. **(a)** What is the mass percent Ca in alite? **(b)** What is the mass percent Ca in belite? **(c)** As the belite composition of cement increases, would it increase or decrease the quantity of CO_2 emitted in producing cement?

3.108 The source of oxygen that drives the internal combustion engine in an automobile is air. Air is a mixture of gases, principally $N_2(\sim79\%)$ and $O_2(\sim20\%)$. In the cylinder of an automobile engine, nitrogen can react with oxygen to produce nitric oxide gas, NO. As NO is emitted from the tailpipe of the car, it can react with more oxygen to produce nitrogen dioxide gas. **(a)** Write balanced chemical equations for both reactions. **(b)** Both nitric oxide and nitrogen dioxide are pollutants that can lead to acid rain and global warming; collectively, they are called NO_x gases. In 2009, the United States emitted an estimated 19 million tons of nitrogen dioxide into the atmosphere. How many grams of nitrogen dioxide is this? **(c)** The production of NO_x gases is an unwanted side reaction of the main engine combustion process that turns octane, C_8H_{18}, into CO_2 and water. If 85% of the oxygen in an engine is used to combust octane and the remainder is used to produce nitrogen dioxide, calculate how many grams of nitrogen dioxide would be produced during the combustion of 500 g of octane.

Design an Experiment

You will learn later in this book that sulfur is capable of forming two common oxides, SO_2 and SO_3. One question that we might ask is whether the direct reaction between sulfur and oxygen leads to the formation of SO_2, SO_3, or a mixture of the two. This question has practical significance because SO_3 can go on to react with water to form sulfuric acid, H_2SO_4, which is produced industrially on a very large scale. Consider also that the answer to this question may depend on the relative amount of each element that is present and the temperature at which the reaction is carried out. For example, on the one hand, carbon and oxygen normally react to form CO_2, but when not enough oxygen is present, CO can form. On the other hand, under normal reaction conditions, H_2 and O_2 react to form water, H_2O (rather than hydrogen peroxide H_2O_2), regardless of the starting ratio of hydrogen to oxygen.

Suppose you are given a bottle of sulfur, which is a yellow solid, a cylinder of O_2, a transparent reaction vessel that can be evacuated and sealed so that only sulfur, oxygen, and the product(s) of the reaction between the two are present, an analytical balance so that you can determine the masses of the reactants and/or products, and a furnace that can be used to heat the reaction vessel to 200°C where the two elements react. **(a)** If you start with 0.10 mol of sulfur in the reaction vessel, how many moles of oxygen would need to be added to form SO_2, assuming SO_2 forms exclusively? **(b)** How many moles of oxygen would be needed to form SO_3, assuming SO_3 forms exclusively? **(c)** Given the available equipment, how would you determine you had added the correct number of moles of each reactant to the reaction vessel? **(d)** What observation or experimental technique would you use to determine the identity of the reaction product(s)? Could differences in the physical properties of SO_2 and SO_3 be used to help identify the product(s)? Have any instruments been described in Chapters 1–3 that would allow you to identify the product(s)? **(e)** What experiments would you conduct to determine if the product(s) of this reaction (either SO_2 or SO_3 or a mixture of the two) can be controlled by varying the ratio of sulfur and oxygen that are added to the reaction vessel? What ratio(s) of S to O_2 would you test to answer this question?

4

REACTIONS IN AQUEOUS SOLUTION

▲ Seawater is a complex aqueous solution.

Water covers nearly two-thirds of our planet, and this simple substance has been the key to much of Earth's evolutionary history. Life almost certainly originated in water, and the need for water by all forms of life has helped determine diverse biological structures. Indeed, most of the mass of our bodies is water, and the chemical reactions responsible for life occur in water.

A solution in which water is the dissolving medium is called an **aqueous solution**. In this chapter we examine chemical reactions that take place in aqueous solutions. In addition, we extend the concepts of stoichiometry learned in Chapter 3 by considering how solution concentrations are expressed and used. Although the reactions we discuss in this chapter are relatively simple, they form the basis for understanding complex reaction cycles in biology, geology, and oceanography.

4.1 | General Properties of Aqueous Solutions

⚠ **Learning Objectives**

When you finish Section 4.1, you should be able to:

▶ Classify solutes as strong electrolytes, weak electrolytes, or nonelectrolytes in water.

▶ Describe how ionic compounds dissolve and are solvated in water.

▶ Relate strong and weak electrolyte behavior to chemical equilibria in solution.

A *solution* is a homogeneous mixture of two or more substances. (Section 1.2) The substance present in the greatest quantity is usually called the **solvent**, and the other substances are called **solutes**; they are said to be *dissolved in* the solvent. When a small amount of sodium chloride (NaCl) is dissolved in a large quantity of water, for example, water is the solvent and sodium chloride is the solute. The majority of the chemical reactions discussed in this book, as well as most of the reactions you will do in the laboratory, occur in aqueous solution.

Electrolytes and Nonelectrolytes

Most of us learn at a young age to keep electrical devices out of the bathtub so we don't shock ourselves. Although most of the water we encounter in daily life is electrically conducting, pure water is a very poor conductor of electricity. The conductivity of bathwater originates from the substances dissolved in the water, not from the water itself.

Not all substances that dissolve in water make the resulting solution conducting. Figure 4.1 shows a simple experiment to test the electrical conductivity of three solutions: pure water, a solution of table sugar (sucrose) in water, and a solution of table salt (NaCl) in water. A light bulb is connected to a battery-powered electrical circuit that contains two electrodes submerged in a beaker of each solution. In order for the light bulb to turn on, there must be an electrical current (that is, a flow of electrically charged particles) between the two electrodes immersed in the solution. Because the light bulb does not turn on in pure water, we conclude that there are not enough charged particles in pure water to create a circuit; water must then mostly exist as H_2O molecules. The solution containing sucrose ($C_{12}H_{22}O_{11}$) also does not turn on the light bulb; therefore, we conclude that the sucrose molecules in solution are uncharged. But the solution containing NaCl does provide enough charged particles to create an electrical circuit and turn on the light bulb. This is experimental evidence that Na^+ and Cl^- ions are formed in aqueous solution.

A substance (such as NaCl) whose aqueous solutions contain ions is called an **electrolyte**. A substance (such as $C_{12}H_{22}O_{11}$) that does not form ions in solution is called a **nonelectrolyte**. The different classifications of NaCl and $C_{12}H_{22}O_{11}$ arise largely because NaCl is an ionic compound, whereas $C_{12}H_{22}O_{11}$ is a molecular compound.

Pure water does not conduct electricity.

Pure water,
$H_2O(l)$

A **nonelectrolyte solution** does not conduct electricity.

Sucrose solution,
$C_{12}H_{22}O_{11}(aq)$

An **electrolyte solution** conducts electricity.

Sodium chloride solution,
NaCl(*aq*)

▲ **Figure 4.1** Completion of an electrical circuit with an electrolyte turns on the light.

How Compounds Dissolve in Water

Recall from Figure 2.19 that solid NaCl consists of an orderly arrangement of Na^+ and Cl^- ions. When NaCl dissolves in water, each ion separates from the solid structure and disperses throughout the solution [**Figure 4.2**(**a**)]. We say that the ionic solid *dissociates* into its component ions as it dissolves.

Water is a very effective solvent for ionic compounds. Although H_2O is an electrically neutral molecule, the O atom is rich in electrons and has a partial negative charge, while each H atom has a partial positive charge. The lowercase Greek letter delta (δ) is used to denote partial charge: A partial negative charge is denoted by δ^- ("delta minus"), and a partial positive charge is denoted by δ^+ ("delta plus"). Cations are attracted by the negative end of H_2O, whereas anions are attracted by the positive end.

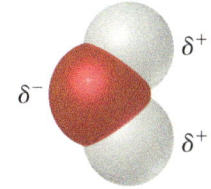

As an ionic compound dissolves, the ions become surrounded by H_2O molecules, as shown in Figure 4.2(a). The ions are said to be *solvated*. In chemical equations, we denote solvated ions by writing them as $Na^+(aq)$ and $Cl^-(aq)$, where *aq* is an abbreviation for "aqueous." (Section 3.1) **Solvation** helps stabilize the ions in solution and prevents cations and anions from recombining. Furthermore, because the ions and their shells of surrounding water molecules are free to move about, the ions become dispersed uniformly throughout the solution.

We can usually predict the nature of the ions in a solution of an ionic compound from the chemical name of the substance. Sodium sulfate (Na_2SO_4), for example, dissociates into sodium ions (Na^+) and sulfate ions (SO_4^{2-}). You should learn the formulas and charges of common ions (Tables 2.4 and 2.5) to understand the forms in which ionic compounds exist in aqueous solution.

▼ Go Figure Do both, just one, or neither of the solutions in this figure conduct electricity? If just one, which one?

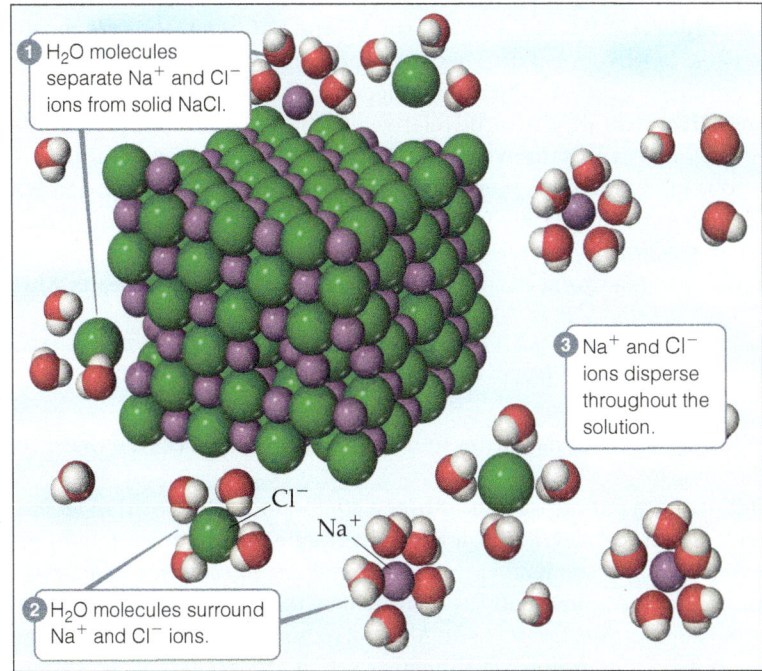

① H_2O molecules separate Na^+ and Cl^- ions from solid NaCl.

③ Na^+ and Cl^- ions disperse throughout the solution.

Cl^-

Na^+

② H_2O molecules surround Na^+ and Cl^- ions.

(a) Ionic compounds like sodium chloride, NaCl, form ions when they dissolve.

Methanol

(b) Molecular substances like methanol, CH_3OH, dissolve without forming ions.

▲ Figure 4.2 Dissolution in water. (a) When an ionic compound, such as sodium chloride, NaCl, dissolves in water, H_2O molecules separate, surround, and uniformly disperse the ions into the liquid. **(b)** Molecular substances that dissolve in water, such as methanol, CH_3OH, usually do so without forming ions. We can think of methanol in water as a simple mixing of two molecular species. In both (a) and (b), the water molecules have been moved apart so that the solute particles can be seen clearly.

When a molecular compound such as sucrose or methanol [Figure 4.2(**b**)] dissolves in water, the solution usually consists of intact molecules dispersed throughout the solution. Consequently, most molecular compounds are nonelectrolytes. A few molecular substances do have aqueous solutions that contain ions. Acids are the most important of these solutions. For example, when HCl(g) dissolves in water to form HCl(aq), the molecule *ionizes*; that is, it dissociates into H$^+$(aq) and Cl$^-$(aq) ions.

Strong and Weak Electrolytes

Electrolytes differ in the extent to which they conduct electricity. **Strong electrolytes** are those solutes that exist in solution completely or nearly completely as ions. Essentially all water-soluble ionic compounds (such as NaCl) and a few molecular compounds (such as HCl) are strong electrolytes. **Weak electrolytes** are those solutes that exist in solution mostly in the form of neutral molecules with only a small fraction in the form of ions. For example, in a solution of acetic acid (CH$_3$COOH), most of the solute is present as CH$_3$COOH(aq) molecules. Only a small fraction (about 1%) of the CH$_3$COOH has dissociated into H$^+$(aq) and CH$_3$COO$^-$(aq) ions.*

Do not confuse the extent to which an electrolyte dissolves (its solubility) with whether it is strong or weak. For example, CH$_3$COOH is extremely soluble in water but is a weak electrolyte. Ca(OH)$_2$, in contrast, is not very soluble in water, but the amount that does dissolve dissociates almost completely. Thus, Ca(OH)$_2$ is a strong electrolyte.

When a weak electrolyte, such as acetic acid, ionizes in solution, we write the reaction in the form

$$\text{CH}_3\text{COOH}(aq) \rightleftharpoons \text{CH}_3\text{COO}^-(aq) + \text{H}^+(aq) \qquad [4.1]$$

The half-arrows pointing in opposite directions mean that the reaction is significant in both directions. At any given moment, some CH$_3$COOH molecules are ionizing to form H$^+$ and CH$_3$COO$^-$ ions, and some of the H$^+$ and CH$_3$COO$^-$ ions are recombining to form CH$_3$COOH. The balance between these opposing processes determines the relative numbers of ions and neutral molecules. This balance produces a state of **chemical equilibrium** in which the relative numbers of each type of ion or molecule in the reaction are constant over time. Chemists use half-arrows pointing in opposite directions to represent reactions that go both forward and backward to achieve equilibrium, such as the ionization of weak electrolytes. In contrast, a single reaction arrow is used for reactions that largely go forward, such as the ionization of strong electrolytes. Because HCl is a strong electrolyte, we write the equation for the ionization of HCl as

$$\text{HCl}(aq) \longrightarrow \text{H}^+(aq) + \text{Cl}^-(aq) \qquad [4.2]$$

The absence of a left-pointing half-arrow indicates that the H$^+$ and Cl$^-$ ions have no tendency to recombine to form HCl molecules.

In the following sections, we look at how a compound's composition lets us predict whether it is a strong electrolyte, a weak electrolyte, or a nonelectrolyte. For the moment, you need only to remember that *water-soluble ionic compounds are strong electrolytes*. Ionic compounds can usually be identified by the presence of both metals and nonmetals [for example, NaCl, FeSO$_4$, and Al(NO$_3$)$_3$]. Ionic compounds containing the ammonium ion, NH$_4$$^+$ [for example, NH$_4$Br and (NH$_4$)$_2$CO$_3$], are exceptions to this rule of thumb.

---

*The chemical formula of acetic acid is sometimes written HC$_2$H$_3$O$_2$, so that the formula looks like that of other common acids such as HCl. The formula CH$_3$COOH conforms to the molecular structure of acetic acid, with the acidic H on the O atom at the end of the formula.

Sample Exercise 4.1
Relating Relative Numbers of Anions and Cations to Chemical Formulas

The accompanying diagram represents an aqueous solution of either $MgCl_2$, KCl, or K_2SO_4. Which solution does the drawing best represent?

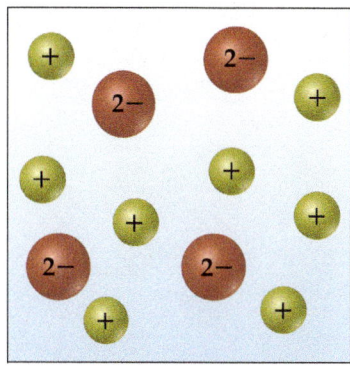

SOLUTION
Analyze We are asked to associate the charged spheres in the diagram with ions present in a solution of an ionic substance.

Plan We examine each ionic substance given to determine the relative numbers and charges of its ions. We then correlate these ionic species with the ones shown in the diagram.

Solve The diagram shows twice as many cations as anions, consistent with the formulation K_2SO_4.

Check Notice that the net charge in the diagram is zero, as it must be if it is to represent an ionic substance.

▶ **Practice Exercise**
If you were to draw diagrams representing aqueous solutions of (**a**) $NiSO_4$, (**b**) $Ca(NO_3)_2$, (**c**) Na_3PO_4, (**d**) $Al_2(SO_4)_3$, how many anions would you show if each diagram contained six cations?

 Self-Assessment Exercises

SAE 4.1 Which solution would contain 3 moles of ions for every mole of compound dissolved in water? (**a**) CH_3OH (**b**) NaCl (**c**) K_2SO_4 (**d**) $NaNO_3$ (**e**) $FeCl_3$

SAE 4.2 Consider the preparation of an aqueous solution of KBr from solid KBr and water. Which of the following statements is or are true?

(**i**) Because KBr is a strong electrolyte, the chemical equilibrium reaction would show a one-way arrow from solid to ions.
(**ii**) Once the solid is dissolved, the potassium ions are surrounded by water molecules, with the oxygen end of the H_2O molecule pointing at potassium.
(**iii**) This solution would not conduct electricity.

(**a**) i only (**b**) ii only (**c**) iii only (**d**) i and ii (**e**) i, ii, and iii

SAE 4.3 Are compounds that are very water-soluble always strong electrolytes? Choose the correct combination of statements:

(**i**) Yes, because a strong electrolyte is by definition very water-soluble.
(**ii**) Yes, because all compounds that dissolve completely in water are ionic compounds.
(**iii**) No, there are insoluble ionic compounds that are strong electrolytes.

(**a**) i only (**b**) ii only (**c**) iii only (**d**) both i and ii (**e**) None of i, ii, or iii is correct.

4.2 | Precipitation Reactions

Figure 4.3 shows two clear solutions being mixed. One solution contains potassium iodide, KI, dissolved in water, whereas the other contains lead nitrate, $Pb(NO_3)_2$, dissolved in water. The reaction between these two solutes produces a water-insoluble yellow solid. Reactions that result in the formation of an insoluble product are called **precipitation reactions**. A **precipitate** is an insoluble solid formed by a reaction in solution. In Figure 4.3 the precipitate is lead iodide (PbI_2), a compound that has a very low solubility in water (and so its phase is a solid, indicated by "s"):

$$Pb(NO_3)_2(aq) + 2\,KI(aq) \longrightarrow PbI_2(s) + 2\,KNO_3(aq) \qquad [4.3]$$

The other product of this reaction, potassium nitrate (KNO_3), remains in aqueous solution, so its phase is denoted as "aq."

Precipitation reactions occur when pairs of oppositely charged ions attract each other so strongly that they form an insoluble ionic solid. To predict whether certain combinations of ions form insoluble compounds, we must consider some guidelines concerning the solubilities of common ionic compounds.

 Learning Objectives

When you finish **Section 4.2**, you should be able to:
▶ Predict whether a compound is soluble or insoluble in water.
▶ Predict whether a precipitation reaction will occur when two aqueous solutions are mixed.
▶ Write balanced chemical equations including phases for metathesis reactions.
▶ Convert balanced molecular equations into full and net ionic equations.

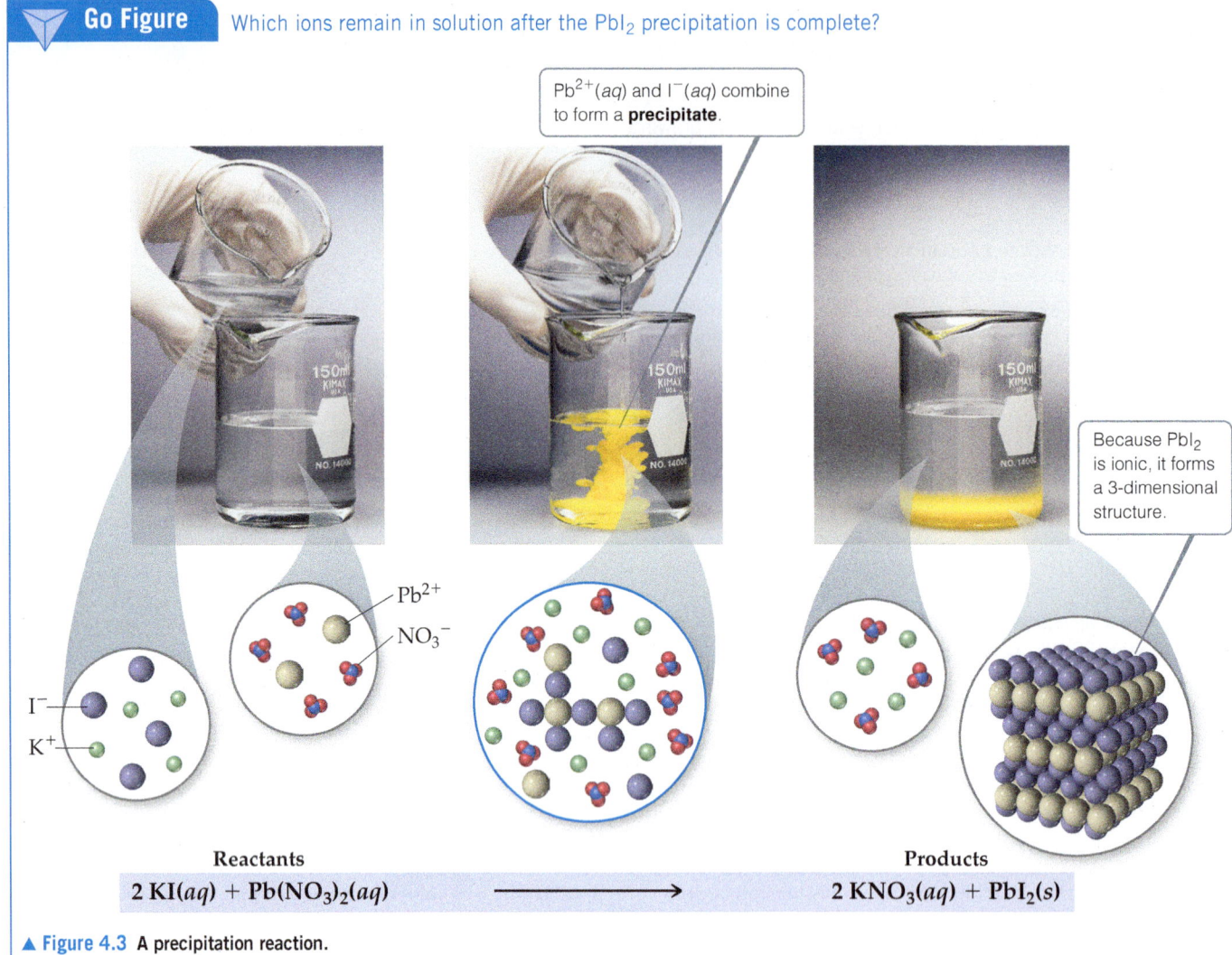

Go Figure   Which ions remain in solution after the PbI$_2$ precipitation is complete?

$Pb^{2+}(aq)$ and $I^-(aq)$ combine to form a **precipitate**.

Because PbI$_2$ is ionic, it forms a 3-dimensional structure.

Pb^{2+}

NO$_3^-$

I$^-$

K$^+$

Reactants		Products
$2 \text{ KI}(aq) + \text{Pb(NO}_3)_2(aq)$	$\longrightarrow$	$2 \text{ KNO}_3(aq) + \text{PbI}_2(s)$

▲ **Figure 4.3** A precipitation reaction.

Solubility Guidelines for Ionic Compounds

The **solubility** of a substance at a given temperature is the amount of the substance that can be dissolved in a given quantity of solvent at the given temperature. Any substance with a solubility less than 0.01 mol/L will be considered *insoluble*. In these cases, the attraction between the oppositely charged ions in the solid is too great for the water molecules to separate the ions to any significant extent; the substance remains largely undissolved.

Unfortunately, there are no rules based on simple physical properties such as ionic charge to guide us in predicting whether a particular ionic compound will be soluble. Experimental observations, however, have led to guidelines for predicting solubility for ionic compounds. For example, experiments show that all common ionic compounds that contain the nitrate anion, NO$_3^-$, are soluble in water. **Table 4.1** summarizes the solubility guidelines for common ionic compounds. The table is organized according to the anion in the compound, but it also reveals many important facts about cations. Note that *all common ionic compounds of the alkali metal ions (group 1A of the periodic table) and of the ammonium ions* (NH$_4^+$) *are soluble in water*.

How to Predict Whether a Precipitate Forms When Strong Electrolytes Mix

1. Note the ions present in the reactants.

2. Consider the possible cation–anion combinations.

3. Use Table 4.1 to determine if any of the combinations is insoluble.

TABLE 4.1   Solubility Guidelines for Common Ionic Compounds in Water

Soluble Ionic Compounds		Important Exceptions
Compounds containing	NO_3^-	None
	CH_3COO^-	None
	Cl^-	Compounds of Ag^+, Hg_2^{2+}, and Pb^{2+}
	Br^-	Compounds of Ag^+, Hg_2^{2+}, and Pb^{2+}
	I^-	Compounds of Ag^+, Hg_2^{2+}, and Pb^{2+}
	SO_4^{2-}	Compounds of Sr^{2+}, Ba^{2+}, Hg_2^{2+}, and Pb^{2+}

Insoluble Ionic Compounds		Important Exceptions
Compounds containing	S^{2-}	Compounds of NH_4^+, the alkali metal cations, Ca^{2+}, Sr^{2+}, and Ba^{2+}
	CO_3^{2-}	Compounds of NH_4^+, and the alkali metal cations
	PO_4^{3-}	Compounds of NH_4^+, and the alkali metal cations
	OH^-	Compounds of NH_4^+, the alkali metal cations, Ca^{2+}, Sr^{2+}, and Ba^{2+}

For example, will a precipitate form when solutions of $Mg(NO_3)_2$ and NaOH are mixed? Both substances are soluble ionic compounds and strong electrolytes. Mixing the solutions first produces a solution containing Mg^{2+}, NO_3^-, Na^+, and OH^- ions. Will either cation interact with either anion to form an insoluble compound? Knowing from Table 4.1 that $Mg(NO_3)_2$ and NaOH are both soluble in water, our only possibilities are Mg^{2+} with OH^- and Na^+ with NO_3^-. From Table 4.1 we see that hydroxides are generally insoluble. Because Mg^{2+} is not an exception, $Mg(OH)_2$ is insoluble and thus forms a precipitate. $NaNO_3$, however, is soluble, so Na^+ and NO_3^- remain in solution. The balanced equation for the precipitation reaction is

$$Mg(NO_3)_2(aq) + 2\,NaOH(aq) \longrightarrow Mg(OH)_2(s) + 2\,NaNO_3(aq) \qquad [4.4]$$

Exchange (Metathesis) Reactions

Notice in Equation 4.4 that the reactant cations exchange anions—Mg^{2+} ends up with OH^-, and Na^+ ends up with NO_3^-. The chemical formulas of the products are based on the charges of the ions—two OH^- ions are needed to give a neutral compound with Mg^{2+}, and one NO_3^- ion is needed to give a neutral compound with Na^+. (Section 2.7) *The equation can be balanced only after the chemical formulas of the products have been determined.*

Sample Exercise 4.2
Using Solubility Rules

Classify these ionic compounds as soluble or insoluble in water: (**a**) sodium carbonate, Na_2CO_3, (**b**) lead sulfate, $PbSO_4$.

SOLUTION

Analyze We are given the names and formulas of two ionic compounds and asked to predict whether they are soluble or insoluble in water.

Plan We can use Table 4.1 to answer the question. Thus, we need to focus on the anion in each compound because the table is organized by anions.

Solve

(**a**) According to Table 4.1, most carbonates are insoluble. But carbonates of the alkali metal cations (such as sodium ion) are an exception to this rule and are soluble. Thus, Na_2CO_3 is soluble in water.

(**b**) Table 4.1 indicates that although most sulfates are water soluble, the sulfate of Pb^{2+} is an exception. Thus, $PbSO_4$ is insoluble in water.

▶ **Practice Exercise**
Classify the following compounds as soluble or insoluble in water: (**a**) cobalt(II) hydroxide, (**b**) barium nitrate, (**c**) ammonium phosphate.

Reactions in which cations and anions appear to exchange partners conform to the general equation

$$AX + BY \longrightarrow AY + BX \qquad [4.5]$$

Example: $AgNO_3(aq) + KCl(aq) \longrightarrow AgCl(s) + KNO_3(aq)$

Such reactions are called either **exchange reactions** or **metathesis reactions** (meh-TATH-eh-sis, Greek for "to transpose"). Precipitation reactions conform to this pattern, as do many neutralization reactions between acids and bases, as we show in Section 4.3.

How to Balance a Metathesis Reaction

1. Use the chemical formulas of the reactants to determine which ions are present.

2. Write the chemical formulas of the products by combining the cation from one reactant with the anion of the other, using the ionic charges to determine the subscripts in the chemical formulas.

3. Check the water solubilities of the products. For a precipitation reaction to occur, at least one product must be insoluble in water.

4. Balance the equation.

Sample Exercise 4.3

Predicting a Metathesis Reaction

(a) Predict the identity of the precipitate that forms when aqueous solutions of $BaCl_2$ and K_2SO_4 are mixed. **(b)** Write the balanced chemical equation for the reaction.

SOLUTION

Analyze We are given two ionic reactants and asked to predict the insoluble product that they form.

Plan We need to write the ions present in the reactants and exchange the anions between the two cations. Once we have written the chemical formulas for these products, we can use Table 4.1 to determine which is insoluble in water. Knowing the products also allows us to write the equation for the reaction.

Solve

(a) The reactants contain Ba^{2+}, Cl^-, K^+, and SO_4^{2-} ions. Exchanging the anions gives us $BaSO_4$ and KCl. According to Table 4.1, most compounds of SO_4^{2-} are soluble, but those of Ba^{2+} are not. Thus, $BaSO_4$ is insoluble and will precipitate from solution. KCl is soluble.

(b) From part **(a)** we know the chemical formulas of the products, $BaSO_4$ and KCl. The balanced equation is

$$BaCl_2(aq) + K_2SO_4(aq) \longrightarrow BaSO_4(s) + 2\,KCl(aq)$$

▶ **Practice Exercise**
(a) What compound precipitates when aqueous solutions of $Fe_2(SO_4)_3$ and $LiOH$ are mixed? **(b)** Write a balanced equation for the reaction.

Ionic Equations and Spectator Ions

In writing equations for reactions in aqueous solution, it is often useful to indicate whether the dissolved substances are present predominantly as ions or as molecules. Let's reconsider the precipitation reaction between $Pb(NO_3)_2$ and 2 KI (Eq. 4.3):

$$Pb(NO_3)_2(aq) + 2\,KI(aq) \longrightarrow PbI_2(s) + 2\,KNO_3(aq)$$

An equation written in this fashion, showing the complete chemical formulas of reactants and products, is called a **molecular equation** because it shows chemical formulas without indicating ionic character. Because $Pb(NO_3)_2$, KI, and KNO_3 are all water-soluble ionic compounds and therefore strong electrolytes, we can write the equation in a form that indicates which species exist as ions in the solution:

$$Pb^{2+}(aq) + 2\,NO_3^-(aq) + 2\,K^+(aq) + 2\,I^-(aq) \longrightarrow$$
$$PbI_2(s) + 2\,K^+(aq) + 2\,NO_3^-(aq) \qquad [4.6]$$

An equation written in this form, with all soluble strong electrolytes shown as ions, is called a **complete ionic equation**.

Notice that $K^+(aq)$ and $NO_3^-(aq)$ appear on both sides of Equation 4.6. Ions that appear in identical forms on both sides of a complete ionic equation, called **spectator ions**, play no direct role in the reaction. Spectator ions, like algebraic quantities, can be canceled on either side of the reaction arrow because they are not reacting with anything. Once we cancel the spectator ions, we are left with the **net ionic equation**, which is one that includes only the ions and molecules directly involved in the reaction:

$$Pb^{2+}(aq) + 2\,I^-(aq) \longrightarrow PbI_2(s) \qquad [4.7]$$

Because charge is conserved in reactions, the sum of the ionic charges must be the same on both sides of a balanced net ionic equation. In this case, the 2+ charge of the cation and the two 1− charges of the anions add to zero, the charge of the electrically neutral product. *If every ion in a complete ionic equation is a spectator, no reaction occurs.*

Net ionic equations illustrate the similarities between various reactions involving electrolytes. For example, Equation 4.7 expresses the essential feature of the precipitation reaction between any strong electrolyte containing $Pb^{2+}(aq)$ and any strong electrolyte containing $I^-(aq)$: The ions combine to form a precipitate of PbI_2. Thus, a net ionic equation demonstrates that more than one set of reactants can lead to the same net reaction. For example, aqueous solutions of KI and MgI_2 share many chemical similarities because both contain I^- ions. Either solution when mixed with a $Pb(NO_3)_2$ solution produces $PbI_2(s)$. The complete ionic equation, however, identifies the actual reactants that participate in a reaction.

How to Write a Net Ionic Equation

1. Write a balanced molecular equation for the reaction.

2. Rewrite the equation to show the ions that form in solution when each soluble strong electrolyte dissociates into its ions. *Only strong electrolytes dissolved in aqueous solution are written in ionic form.*

3. Identify and cancel spectator ions.

Sample Exercise 4.4
Writing a Net Ionic Equation

Write the net ionic equation for the precipitation reaction that occurs when aqueous solutions of calcium chloride and sodium carbonate are mixed.

SOLUTION

Analyze Our task is to write a net ionic equation for a precipitation reaction, given the names of the reactants present in solution.

Plan We write the chemical formulas of the reactants and products and then determine which product is insoluble. We then write and balance the molecular equation. Next, we write each soluble strong electrolyte as separated ions to obtain the complete ionic equation. Finally, we eliminate the spectator ions to obtain the net ionic equation.

Solve Calcium chloride is composed of calcium ions, Ca^{2+}, and chloride ions, Cl^-; hence, an aqueous solution of the substance is $CaCl_2(aq)$. Sodium carbonate is composed of Na^+ ions and CO_3^{2-} ions; hence, an aqueous solution of the compound is $Na_2CO_3(aq)$. In the molecular equations for precipitation reactions, the anions and cations appear to exchange partners. Thus, we put Ca^{2+} and CO_3^{2-} together to give $CaCO_3$ and Na^+ and Cl^- together to give NaCl. According to the solubility guidelines in Table 4.1, $CaCO_3$ is insoluble and NaCl is soluble. The balanced molecular equation is

$$CaCl_2(aq) + Na_2CO_3(aq) \longrightarrow CaCO_3(s) + 2\,NaCl(aq)$$

In a complete ionic equation, *only* dissolved strong electrolytes (such as soluble ionic compounds) are written as separate ions. As the (aq) designations remind us, $CaCl_2$, Na_2CO_3, and NaCl

are all dissolved in the solution. Furthermore, they are all strong electrolytes. $CaCO_3$ is an ionic compound, but it is not soluble. We do not write the formula of any insoluble compound as its component ions. Thus, the complete ionic equation is

$$Ca^{2+}(aq) + 2\,Cl^-(aq) + 2\,Na^+(aq) + CO_3^{2-}(aq) \longrightarrow$$
$$CaCO_3(s) + 2\,Na^+(aq) + 2\,Cl^-(aq)$$

The spectator ions are Na^+ and Cl^-. Canceling them gives the following net ionic equation:

$$Ca^{2+}(aq) + CO_3^{2-}(aq) \longrightarrow CaCO_3(s)$$

Check We can check our result by confirming that both the elements and the electric charge are balanced. Each side has one Ca, one C, and three O, and the net charge on each side equals 0.

Comment If none of the ions in an ionic equation is removed from solution or is changed in some way, all ions are spectator ions and a reaction does not occur.

▶ **Practice Exercise**
Write the net ionic equation for the precipitation reaction that occurs when aqueous solutions of silver nitrate and potassium phosphate are mixed.

Self-Assessment Exercises

SAE 4.4   How many of the following compounds are soluble in water: CH_3COOBa, $MgCO_3$, FeS, $Mg_3(PO_4)_2$? **(a)** 0 **(b)** 1 **(c)** 2 **(d)** 3 **(e)** 4

SAE 4.5   What precipitate, if any, forms when a solution of ammonium sulfide is mixed with a solution of barium chloride? **(a)** NH_4Cl **(b)** BaS **(c)** $(NH_4)_2S$ **(d)** $BaCl_2$ **(e)** No precipitate forms.

SAE 4.6   An aqueous solution containing 3 millimoles of calcium chloride is mixed with an aqueous solution containing 2 millimoles of sodium phosphate. How many millimoles of calcium phosphate precipitate? **(a)** 0 **(b)** 1 **(c)** 2 **(d)** 3 **(e)** 6

SAE 4.7   An aqueous solution containing 2 millimoles of lead nitrate is mixed with an aqueous solution containing 4 millimoles of potassium iodide. How many millimoles of spectator ions, both cations and anions, are present in solution when the reaction is complete? **(a)** 0 **(b)** 4 **(c)** 8 **(d)** 12 **(e)** 14

4.3 | Acids, Bases, and Neutralization Reactions

Learning Objectives

When you finish Section 4.3, you should be able to:

▶ Describe the defining features of compounds that act as either acids or bases.

▶ Predict whether a compound is an acid or a base based on its chemical formula/structure.

▶ Differentiate between strong and weak acids/bases.

▶ Classify compounds as either strong or weak electrolytes based on their acid–base behavior.

▶ Write balanced chemical equations for neutralization reactions.

Chemists routinely examine the chemical formula and structure of a compound and predict its reactivity. Acids, bases, and acid–base reactions are among the most common type of compounds and reactivity that chemists use to develop new pharmaceuticals, new plastics, and many other useful substances.

Many acids and bases are industrial and household substances (**Figure 4.4**), and some are important components of biological fluids. Hydrochloric acid, for example, is both an important industrial chemical and the main constituent of gastric juice in your stomach. Acids and bases are also common electrolytes. In this section we examine the simplest description of acids and bases, which involve the presence of H^+ and OH^- ions in aqueous solution.

Acids

As noted in Section 2.8, **acids** are substances that ionize in aqueous solution to form hydrogen ions $H^+(aq)$. Because a hydrogen atom consists of a proton and an electron, H^+ is simply a proton. Thus, acids are often called *proton donors*. Molecular models of four common acids are shown in **Figure 4.5**.

Protons in aqueous solution are solvated by water molecules, just as other cations are [Figure 4.2(**a**)]. In writing chemical equations involving protons in water, therefore, we write $H^+(aq)$.

Molecules of different acids ionize to form different numbers of H^+ ions. Both HCl and HNO_3 are *monoprotic* acids, yielding one H^+ per molecule of acid. Sulfuric acid, H_2SO_4, is a *diprotic* acid, one that yields two H^+ per molecule of acid. The ionization of H_2SO_4 and other diprotic acids occurs in two steps:

$$H_2SO_4(aq) \longrightarrow H^+(aq) + HSO_4^-(aq) \qquad [4.8]$$

$$HSO_4^-(aq) \rightleftharpoons H^+(aq) + SO_4^{2-}(aq) \qquad [4.9]$$

▲ **Figure 4.4 Common household acids and bases.** Vinegar and lemon juice are common household acids. Ammonia and baking soda (sodium bicarbonate) are common household bases.

Hydrochloric acid, HCl

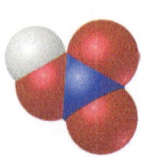
Nitric acid, HNO_3

Sulfuric acid, H_2SO_4

Acetic acid, CH_3COOH

H	O
Cl	C
N	S

▲ **Figure 4.5 Molecular models of four common acids.**

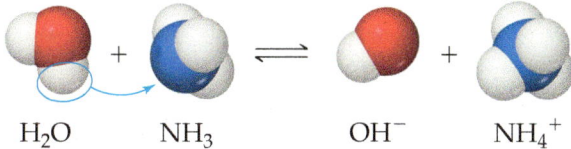

$$H_2O \qquad NH_3 \qquad\qquad OH^- \qquad NH_4^+$$

▲ **Figure 4.6 Proton transfer.** An H_2O molecule acts as a proton donor (acid), and NH_3 acts as a proton acceptor (base). In aqueous solutions, only a fraction of the NH_3 molecules react with H_2O. Consequently, NH_3 is a weak electrolyte.

Although H_2SO_4 is a strong electrolyte, only the first ionization (Equation 4.8) is complete. Thus, aqueous solutions of sulfuric acid contain a mixture of $H^+(aq)$, $HSO_4^-(aq)$, and $SO_4^{2-}(aq)$.

The molecule CH_3COOH (acetic acid) that we have mentioned frequently is the primary component in vinegar. Acetic acid has four hydrogens, as Figure 4.5 shows, but only one of them, the H that is bonded to an oxygen in the —COOH group, is ionized in water. Thus, the H in the COOH group breaks its O—H bond in water. The three other hydrogens in acetic acid are attached to carbon and do not break their C—H bonds in water. The reasons for this difference are very interesting and are discussed in Chapter 16.

Bases

Bases are substances that accept (react with) H^+ ions. Bases produce hydroxide ions (OH^-) when they dissolve in water. Ionic hydroxide compounds, such as NaOH, KOH, and $Ca(OH)_2$, are among the most common bases. When dissolved in water, they dissociate into ions, introducing OH^- ions into the solution.

Compounds that do not contain OH^- ions can also be bases. For example, ammonia (NH_3) is a common base. When added to water, it accepts an H^+ ion from a water molecule and thereby produces an OH^- ion (**Figure 4.6**):

$$H_2O(l) + NH_3(aq) \rightleftharpoons OH^-(aq) + NH_4^+(aq) \qquad\qquad [4.10]$$

Ammonia is a weak electrolyte because only about 1% of the NH_3 forms NH_4^+ and OH^- ions.

Strong and Weak Acids and Bases

Acids and bases that are strong electrolytes (completely ionized in solution) are **strong acids** and **strong bases**. Those that are weak electrolytes (partly ionized) are **weak acids** and **weak bases**. When reactivity depends only on $H^+(aq)$ concentration, strong acids are more reactive than weak acids. The reactivity of an acid, however, can depend on the anion as well as on $H^+(aq)$ concentration. For example, hydrofluoric acid (HF) is a weak acid (only partly ionized in aqueous solution), but it is very reactive and vigorously attacks many substances, including glass. This reactivity is due to the combined action of $H^+(aq)$ and $F^-(aq)$.

Table 4.2 lists the strong acids and bases we are most likely to encounter. You should commit this information to memory in order to correctly identify strong electrolytes and write net ionic equations. The list is brief because most acids are weak. (For H_2SO_4, as we noted earlier, only the first proton completely ionizes.) The only common strong bases

TABLE 4.2 Common Strong Acids and Bases

Strong Acids	Strong Bases
Hydrochloric acid, HCl	Group 1A metal hydroxides
Hydrobromic acid, HBr	(LiOH, NaOH, KOH, RbOH, CsOH)
Hydroiodic acid, HI	Heavy group 2A metal hydroxides
Chloric acid, $HClO_3$	$[Ca(OH)_2, Sr(OH)_2, Ba(OH)_2]$
Perchloric acid, $HClO_4$	
Nitric acid, HNO_3	
Sulfuric acid (first proton), H_2SO_4	

Sample Exercise 4.5
Comparing Acid Strengths

The following diagrams represent aqueous solutions of acids HX, HY, and HZ, with water molecules omitted for clarity. Rank the acids from strongest to weakest.

HX

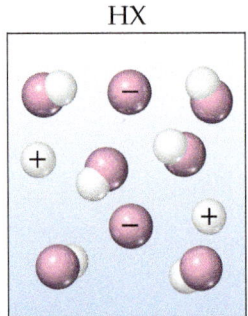

HY

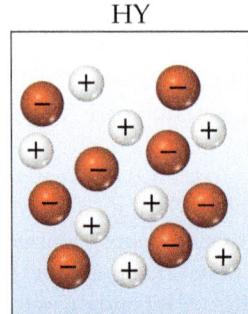

HZ

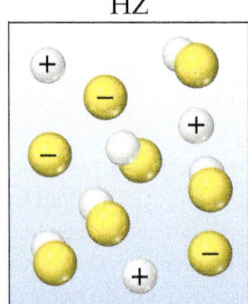

SOLUTION

Analyze We are asked to rank three acids from strongest to weakest, based on schematic drawings of their solutions.

Plan We can determine the relative numbers of uncharged molecular species in the diagrams. The strongest acid is the one with the most H$^+$ ions and the fewest undissociated molecules in solution. The weakest acid is the one with the largest number of undissociated molecules.

Solve The order is HY > HZ > HX. HY is a strong acid because it is totally ionized (no HY molecules in solution), whereas both HX and HZ are weak acids, whose solutions consist of a mixture of molecules and ions. Because HZ contains more H$^+$ ions and fewer molecules than HX, it is a stronger acid.

▶ **Practice Exercise**
When a substance HA is dissolved in water, the concentration of HA(*aq*) is much greater than the concentrations of H$^+$(*aq*) and A$^-$(*aq*). Is HA a strong acid, a strong base, a weak acid, or a weak base? What reasoning led to your decision?

are the common soluble metal hydroxides. The most common weak base is NH$_3$, which reacts with water to form OH$^-$ ions (Equation 4.10).

If our substance is a base, we can use Table 4.2 to determine whether it is a strong base. NH$_3$ is the only molecular base that we consider in this chapter, and it is a weak base; it is therefore a weak electrolyte (see **Table 4.3**). Finally, any molecular substance that we encounter in this chapter that is not an acid or NH$_3$ is probably a nonelectrolyte.

Identifying Strong and Weak Electrolytes

If we remember the common strong acids and bases (Table 4.2) and also remember that NH$_3$ is a weak base, we can make reasonable predictions about the electrolytic strength of a great number of *water-soluble* substances. Table 4.3 summarizes our observations about electrolytes. We first ask whether the substance is ionic or molecular. If it is ionic, it is a strong electrolyte. If the substance is molecular, we ask whether it is an acid or a base. (It is an acid if it either has H first in the chemical formula or contains a COOH group.) If it is an acid, we use Table 4.2 to determine whether it is a strong or weak electrolyte: All strong acids are strong electrolytes, and all weak acids are weak electrolytes. If an acid is not listed in Table 4.2, it is probably a weak acid and therefore a weak electrolyte.

TABLE 4.3   Summary of the Electrolytic Behavior of Common Soluble Ionic and Molecular Compounds

	Strong Electrolyte	Weak Electrolyte	Nonelectrolyte
Ionic	All	None	None
Molecular	Strong acids (see Table 4.2)	Weak acids, weak bases	All other compounds

Sample Exercise 4.6
Identifying Strong, Weak, and Nonelectrolytes

Classify these dissolved substances as a strong electrolyte, weak electrolyte, or nonelectrolyte: $CaCl_2$, HNO_3, C_2H_5OH (ethanol), HCOOH (formic acid), KOH.

SOLUTION

Analyze We are given several chemical formulas and asked to classify each substance as a strong electrolyte, weak electrolyte, or nonelectrolyte.

Plan The approach we take is outlined in Table 4.3. We can predict whether a substance is ionic or molecular based on its composition. As we saw in Section 2.7, most ionic compounds we encounter in this text are composed of a metal and a nonmetal, whereas most molecular compounds are composed only of nonmetals.

Solve Two compounds fit the criteria for ionic compounds: $CaCl_2$ and KOH. According to Table 4.3, all ionic compounds are strong electrolytes, so that is how we classify these two substances. The three remaining compounds are molecular. Two of these molecular substances, HNO_3 and HCOOH, are acids. Nitric acid, HNO_3, is a common strong acid (see Table 4.2), so it is a strong electrolyte. Because most acids are weak acids, our best guess would be that HCOOH is a weak acid (weak electrolyte), which is in fact the case.

The remaining molecular compound, C_2H_5OH, is neither an acid nor a base, so it is a nonelectrolyte.

Comment Although ethanol, C_2H_5OH, has an OH group, it is not a metal hydroxide and therefore not a base. Rather, ethanol is a member of a class of organic compounds that have C—OH bonds, which are known as alcohols. (Section 2.9) Organic compounds containing the COOH group are called carboxylic acids (Chapter 16). Molecules that have this group are weak acids.

▶ **Practice Exercise**
Consider solutions in which 0.1 mol of each of the following compounds is dissolved in 1 L of water: $Ca(NO_3)_2$ (calcium nitrate), $C_6H_{12}O_6$ (glucose), $NaCH_3COO$ (sodium acetate), and CH_3COOH (acetic acid). Rank the solutions in order of increasing electrical conductivity. (Note: The conductivity increases as the number of ions in solution increases.)

Neutralization Reactions and Salts

The properties of acidic solutions are quite different from those of basic solutions. Acids have a sour taste, whereas bases have a bitter taste.* Acids change the colors of certain dyes in a way that differs from the way bases affect the same dyes. This is the principle behind the indicator known as litmus paper (**Figure 4.7**). Acid–base chemistry is an important theme throughout all of chemistry that we begin to explore here.

When a solution of an acid and a solution of a base are mixed, a **neutralization reaction** occurs. The products of the reaction have none of the characteristic properties of either the acidic solution or the basic solution. For example, when hydrochloric acid is mixed with a solution of sodium hydroxide, the reaction is

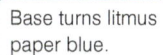
Base turns litmus paper blue.   Acid turns litmus paper red.

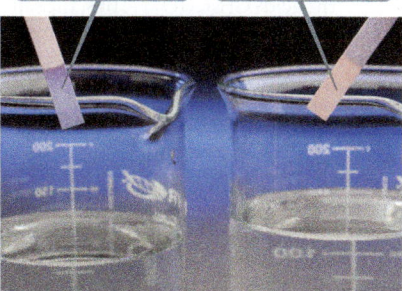

▲ **Figure 4.7 Litmus paper.** Litmus paper is coated with dyes that change color in response to exposure to either acids or bases.

$$HCl(aq) + NaOH(aq) \longrightarrow H_2O(l) + NaCl(aq) \qquad [4.11]$$
$$\text{(acid)} \qquad \text{(base)} \qquad \text{(water)} \qquad \text{(salt)}$$

Water and table salt, NaCl, are the products of the reaction. By analogy to this reaction, the term **salt** has come to mean any ionic compound whose cation comes from a base (for example, Na^+ from NaOH) and whose anion comes from an acid (for example, Cl^- from HCl). In general, *a neutralization reaction between an acid and a metal hydroxide produces water and a salt.*

Because HCl, NaOH, and NaCl are all water-soluble strong electrolytes, the complete ionic equation associated with Equation 4.11 is

$$H^+(aq) + Cl^-(aq) + Na^+(aq) + OH^-(aq) \longrightarrow$$
$$H_2O(l) + Na^+(aq) + Cl^-(aq) \qquad [4.12]$$

Therefore, the net ionic equation is

$$H^+(aq) + OH^-(aq) \longrightarrow H_2O(l) \qquad [4.13]$$

Equation 4.13 summarizes the main feature of the neutralization reaction between any strong acid and any strong base: $H^+(aq)$ and $OH^-(aq)$ ions combine to form $H_2O(l)$.

---

*Tasting chemical solutions is not a good practice. However, we have all had acids such as ascorbic acid (vitamin C), acetylsalicylic acid (aspirin), and citric acid (in citrus fruits) in our mouths, and we are familiar with their characteristic sour taste. Soaps, which are basic, have the characteristic bitter taste of bases.

Figure 4.8 shows the neutralization reaction between hydrochloric acid and the water-insoluble base $Mg(OH)_2$:

Molecular equation:

$$Mg(OH)_2(s) + 2\,HCl(aq) \longrightarrow MgCl_2(aq) + 2\,H_2O(l) \qquad [4.14]$$

Net ionic equation:

$$Mg(OH)_2(s) + 2\,H^+(aq) \longrightarrow Mg^{2+}(aq) + 2\,H_2O(l) \qquad [4.15]$$

Notice that the OH^- ions (this time in a solid reactant) and H^+ ions combine to form H_2O. Because the ions exchange partners, neutralization reactions between acids and metal hydroxides are metathesis reactions.

Sample Exercise 4.7

Writing Chemical Equations for a Neutralization Reaction

For the reaction between aqueous solutions of acetic acid (CH_3COOH) and barium hydroxide, $Ba(OH)_2$, write **(a)** the balanced molecular equation, **(b)** the complete ionic equation, and **(c)** the net ionic equation.

SOLUTION

Analyze We are given the chemical formulas for an acid and a base and are asked to write a balanced molecular equation, a complete ionic equation, and a net ionic equation for their neutralization reaction.

Plan As Equation 4.11 and the italicized statement that follows it indicate, neutralization reactions form two products, H_2O and a salt. We examine the cation of the base and the anion of the acid to determine the composition of the salt.

Solve

(a) The salt contains the cation of the base (Ba^{2+}) and the anion of the acid (CH_3COO^-). Thus, the salt formula is $Ba(CH_3COO)_2$. According to Table 4.1, this compound is soluble in water. The unbalanced molecular equation for the neutralization reaction is

$$CH_3COOH(aq) + Ba(OH)_2(aq)$$
$$\longrightarrow H_2O(l) + Ba(CH_3COO)_2(aq)$$

To balance this equation, we must provide two molecules of CH_3COOH to furnish the two CH_3COO^- ions and to supply the two H^+ ions needed to combine with the two OH^- ions of the base. The balanced molecular equation is

$$2\,CH_3COOH(aq) + Ba(OH)_2(aq)$$
$$\longrightarrow 2\,H_2O(l) + Ba(CH_3COO)_2(aq)$$

(b) To write the complete ionic equation, we identify the strong electrolytes and break them into ions. In this case, $Ba(OH)_2$ and $Ba(CH_3COO)_2$ are both water-soluble ionic compounds and hence strong electrolytes. Thus, the complete ionic equation is

$$2\,CH_3COOH(aq) + Ba^{2+}(aq) + 2\,OH^-(aq)$$
$$\longrightarrow 2\,H_2O(l) + Ba^{2+}(aq) + 2\,CH_3COO^-(aq)$$

(c) Eliminating the spectator ion, Ba^{2+}, and simplifying coefficients give the net ionic equation:

$$CH_3COOH(aq) + OH^-(aq) \longrightarrow H_2O(l) + CH_3COO^-(aq)$$

Check We can determine whether the molecular equation is balanced by counting the number of atoms of each kind on both sides of the arrow (10 H, 6 O, 4 C, and 1 Ba on each side). However, it is often easier to check equations by counting groups: There are 2 CH_3COO groups, as well as 1 Ba, and 4 additional H atoms and 2 additional O atoms on each side of the equation. The net ionic equation checks out because the numbers of each kind of element and the net charge are the same on both sides of the equation.

▶ **Practice Exercise**
For the reaction of aqueous potassium hydroxide with sulfuric acid, write **(a)** the balanced molecular equation and **(b)** the net ionic equation, assuming that both protons of sulfuric acid react.

Neutralization Reactions with Gas Formation

Many bases besides OH^- react with H^+ to form molecular compounds. Two of these that you might encounter in the laboratory are the sulfide ion and the carbonate ion. Both of these anions react with acids to form gases that have low solubilities in water.

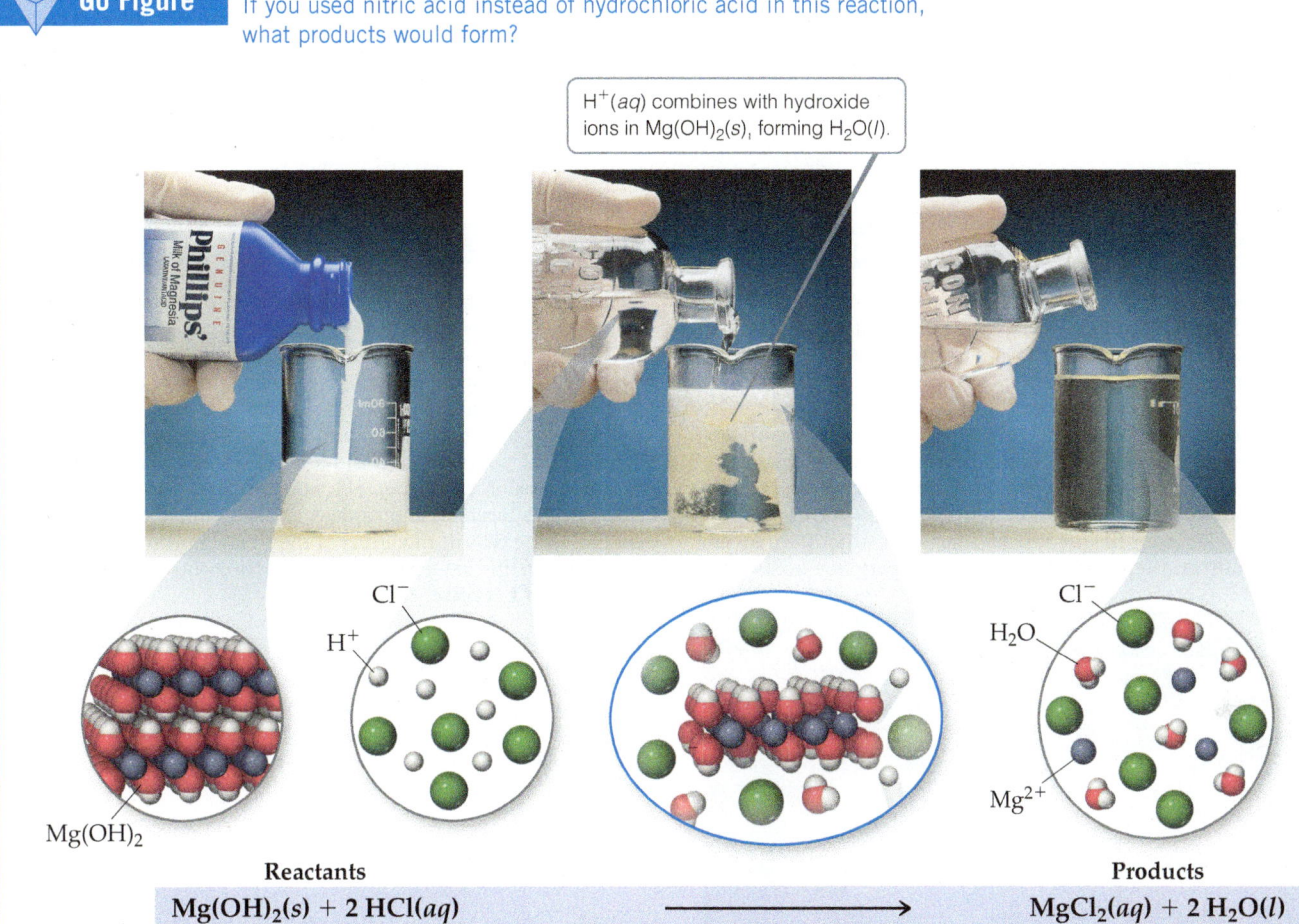

Go Figure If you used nitric acid instead of hydrochloric acid in this reaction, what products would form?

$H^+(aq)$ combines with hydroxide ions in $Mg(OH)_2(s)$, forming $H_2O(l)$.

Reactants

$$Mg(OH)_2(s) + 2\,HCl(aq) \longrightarrow MgCl_2(aq) + 2\,H_2O(l)$$

Products

▲ **Figure 4.8 Neutralization reaction between Mg(OH)$_2$(s) and hydrochloric acid.** Milk of magnesia is a suspension of water-insoluble magnesium hydroxide, $Mg(OH)_2(s)$, in water. When sufficient hydrochloric acid, $HCl(aq)$, is added, a reaction ensues that leads to an aqueous solution containing $Mg^{2+}(aq)$ and $Cl^-(aq)$ ions.

Hydrogen sulfide (H_2S), the substance that gives rotten eggs their foul odor, forms when an acid such as $HCl(aq)$ reacts with a metal sulfide such as Na_2S:

Molecular equation:

$$2\,HCl(aq) + Na_2S(aq) \longrightarrow H_2S(g) + 2\,NaCl(aq) \qquad [4.16]$$

Net ionic equation:

$$2\,H^+(aq) + S^{2-}(aq) \longrightarrow H_2S(g) \qquad [4.17]$$

Carbonates and bicarbonates react with acids to form $CO_2(g)$. Reaction of CO_3^{2-} or HCO_3^- with an acid first gives carbonic acid (H_2CO_3). For example, when hydrochloric acid is added to sodium bicarbonate, the reaction is

$$HCl(aq) + NaHCO_3(aq) \longrightarrow NaCl(aq) + H_2CO_3(aq) \qquad [4.18]$$

Carbonic acid is unstable. If present in solution in sufficient concentrations, it decomposes to H_2O and CO_2,. The CO_2 then escapes from the solution as a gas:

$$H_2CO_3(aq) \longrightarrow H_2O(l) + CO_2(g) \qquad [4.19]$$

The overall reaction is summarized by the following equations:

Molecular equation:

$$HCl(aq) + NaHCO_3(aq) \longrightarrow NaCl(aq) + H_2O(l) + CO_2(g) \qquad [4.20]$$

Net ionic equation:

$$H^+(aq) + HCO_3^-(aq) \longrightarrow H_2O(l) + CO_2(g) \qquad [4.21]$$

Both $NaHCO_3(s)$ and $Na_2CO_3(s)$ are used as neutralizers in acid spills; either salt is added until the fizzing caused by $CO_2(g)$ formation stops. Sometimes sodium bicarbonate is used as an antacid to soothe an upset stomach. In that case the HCO_3^- reacts with stomach acid to form $CO_2(g)$.

CHEMISTRY AND LIFE   Stomach Acids and Antacids

Your stomach secretes acids to help digest foods. These acids, which include hydrochloric acid, contain about 0.1 mol of H^+ per liter of solution. The stomach and digestive tract are normally protected from the corrosive effects of stomach acid by a mucosal lining. Holes can develop in this lining, however, allowing the acid to attack the underlying tissue, causing painful damage. These holes, known as ulcers, can be caused by the secretion of excess acids and/or by a weakness in the digestive lining. Many peptic ulcers are caused by infection by the bacterium *Helicobacter pylori*. Between 10 and 20% of Americans suffer from ulcers at some point in their lives. Many others experience occasional

indigestion, heartburn, or reflux due to digestive acids entering the esophagus.

The problem of excess stomach acid can be addressed by (1) removing the excess acid or (2) decreasing the production of acid. Substances that remove excess acid are called *antacids*, whereas those that decrease acid production are called *acid inhibitors*. Figure 4.9 shows several common over-the-counter antacids, which usually contain hydroxide, carbonate, or bicarbonate ions (Table 4.4). Antiulcer drugs, such as Tagamet® and Zantac®, are acid inhibitors. They act on acid-producing cells in the lining of the stomach. Formulations that control acid in this way are now available as over-the-counter drugs.

Related Exercise: 4.95

▲ **Figure 4.9 Antacids.** These products all serve as acid-neutralizing agents in the stomach.

TABLE 4.4   Some Common Antacids

Commercial Name	Acid-Neutralizing Agents
Alka-Seltzer®	$NaHCO_3$
Amphojel®	$Al(OH)_3$
Milk of Magnesia	$Mg(OH)_2$
Maalox®	$Mg(OH)_2$ and $Al(OH)_3$
Mylanta®	$Mg(OH)_2$ and $Al(OH)_3$
Rolaids®	$Mg(OH)_2$ and $CaCO_3$
Tums®	$CaCO_3$

Self-Assessment Exercises

SAE 4.8 How many of the following compounds are a weak acid: sulfuric acid, ammonia, nitric acid, acetic acid? **(a)** 0 **(b)** 1 **(c)** 2 **(d)** 3 **(e)** 4

SAE 4.9 Which statement about bases is *true*? **(a)** All bases produce hydroxide ions in water. **(b)** Ammonia is an example of a strong base. **(c)** Cesium hydroxide is not a base because it is not soluble in water. **(d)** Bases donate protons to water.

SAE 4.10 If you want to neutralize an aqueous solution of HCl with the smallest number of moles of reagent possible, what reagent is the best choice? **(a)** HNO_3 **(b)** NaOH **(c)** $Ca(OH)_2$ **(d)** Na_3PO_4 **(e)** NH_3

SAE 4.11 Which statement about neutralization reactions is *false*? **(a)** Neutralization reactions can be summarized as "acid plus base produces a salt plus water." **(b)** Neutralization reactions always require one mole of the acid for every mole of the base. **(c)** Neutral-

ization reactions can produce gases as products. **(d)** If your base is a solid, dripping a solution of acid on it will still produce a reaction.

SAE 4.12 What is the balanced net ionic equation for the neutralization reaction between acetic acid, CH_3COOH, and strontium hydroxide?

(a) $2\,CH_3COOH(aq) + Sr(OH)_2(aq)$
$$\longrightarrow (CH_3COO)_2Sr(aq) + 2\,H_2O(l)$$
(b) $2\,CH_3COOH(aq) + Sr^{2+}(aq) + 2\,OH^-(aq)$
$$\longrightarrow 2\,CH_3COO^-(aq) + Sr^{2+}(aq) + 2\,H_2O(l)$$
(c) $H^+(aq) + OH^-(aq) \longrightarrow H_2O(l)$
(d) $2\,H^+(aq) + 2\,OH^-(aq) \longrightarrow 2\,H_2O(l)$
(e) $CH_3COOH(aq) + OH^-(aq) \longrightarrow CH_3COO^-(aq) + H_2O(l)$

CHEMISTRY AND SUSTAINABILITY | **Acid Rain**

Raindrops are slightly acidic, due to dissolved carbon dioxide from the atmosphere:

$$CO_2(g) + H_2O(l) \longrightarrow H_2CO_3(aq)$$

This reaction is the reverse of Equation 4.19, and it occurs because a chemical equilibrium exists between carbonic acid and carbon dioxide in the raindrop.

The burning of coal and other fossil fuels produces CO_2 as well as various sulfur oxides (due to sulfur-containing impurities in the fossil fuels). One of these sulfur-containing compounds is SO_2, which can undergo further reaction to produce SO_3. SO_3 then reacts with water to produce the strong acid sulfuric acid:

$$SO_3(g) + H_2O(l) \longrightarrow H_2SO_4(aq)$$

Similarly, nitrogen oxides (NO, NO_2) that are by-products from fossil fuel combustion can react with water to produce nitric acid, HNO_3, another strong acid. Both sulfuric acid and nitric acid produced this way contribute to **acid rain**, rain that is 10 times more acidic than nature's rain. For many years, acid rain in some parts of the world caused buildings, plants, animals, and soil to suffer many adverse effects. But as scientists have figured out how to clean up the emissions as well as move to other sources of energy, acid rain is becoming less of a problem. We discuss acid rain in more detail in Chapter 18.

4.4 | Oxidation–Reduction Reactions

In precipitation reactions, cations and anions come together to form an insoluble ionic compound. In neutralization reactions, protons are transferred from one reactant to another. Now let's consider a third kind of reaction, one in which electrons are transferred from one reactant to another. Such reactions are called either **oxidation–reduction reactions** or **redox reactions**. In this section we concentrate on redox reactions where one of the reactants is a metal in its elemental form. Redox reactions are critical in understanding many biological and geological processes in the world around us; they also form the basis for energy-related technologies such as batteries and fuel cells (Chapter 20).

> ⚠ **Learning Objectives**
>
> **When you finish Section 4.4, you should be able to:**
> ► Determine the oxidation numbers (also called oxidation states) of elements in a compound.
> ► Identify which species are oxidized and which are reduced in an oxidation–reduction reaction.
> ► Write balanced chemical equations for oxidation–reduction reactions.
> ► Use the activity series to predict oxidation–reduction reactions between a metal and a metal cation or an acid (displacement reactions).

Oxidation and Reduction

One of the most familiar redox reactions is *corrosion* of a metal (**Figure 4.10**). In some instances, corrosion is limited to the surface of the metal, as is the case with the green coating that forms on copper roofs and statues. In other instances, the corrosion goes deeper, eventually compromising the structural integrity of the metal, as happens with the rusting of iron.

Corrosion is the conversion of a metal into a metal compound, by a reaction between the metal and some substance in its environment. When a metal corrodes, each metal atom loses one or more electrons to form a cation, which can combine with an anion to form an ionic compound. The green coating on the Statue of Liberty contains Cu^{2+} combined with carbonate and hydroxide anions; rust contains Fe^{3+} combined with oxide and hydroxide anions; and silver tarnish contains Ag^+ combined with sulfide anions.

(a) (b) (c)

▲ Figure 4.10 **Familiar corrosion products.** (**a**) A green coating forms when copper is oxidized. (**b**) Rust forms when iron corrodes. (**c**) A black tarnish forms as silver corrodes.

When an atom, ion, or molecule becomes more positively charged (that is, when it loses electrons), we say that it has been *oxidized*. Loss of electrons by a substance is called **oxidation**. The term *oxidation* is used because the first reactions of this sort to be studied were reactions with oxygen. Many metals react directly with O_2 in air to form metal oxides. In these reactions, the metal loses electrons to oxygen, forming an ionic compound of the metal ion and oxide ion. Rusting involves the reaction between iron metal and oxygen in the presence of water. In this process, Fe is *oxidized* (loses electrons) to form Fe^{3+}.

The reaction between iron and oxygen tends to be relatively slow, but other metals, such as the alkali and alkaline earth metals, react quickly upon exposure to air. **Figure 4.11** shows how the bright metallic surface of calcium tarnishes as CaO forms in the reaction

$$2\,Ca(s) + O_2(g) \longrightarrow 2\,CaO(s) \qquad [4.22]$$

In this reaction, Ca is oxidized to Ca^{2+} and neutral O_2 is transformed to O^{2-} ions. In Equation 4.22, the oxidation involves transfer of electrons from the calcium metal to the O_2, leading to formation of CaO. When an atom, ion, or molecule becomes more negatively charged (gains electrons), we say that it is *reduced*. The gain of electrons by a substance is called **reduction**. When one reactant loses electrons (that is, when it is oxidized), another reactant must gain them. In other words, oxidation of one substance must be accompanied by reduction of some other substance. In Equation 4.22, then, molecular oxygen is reduced to oxide (O^{2-}) ions.

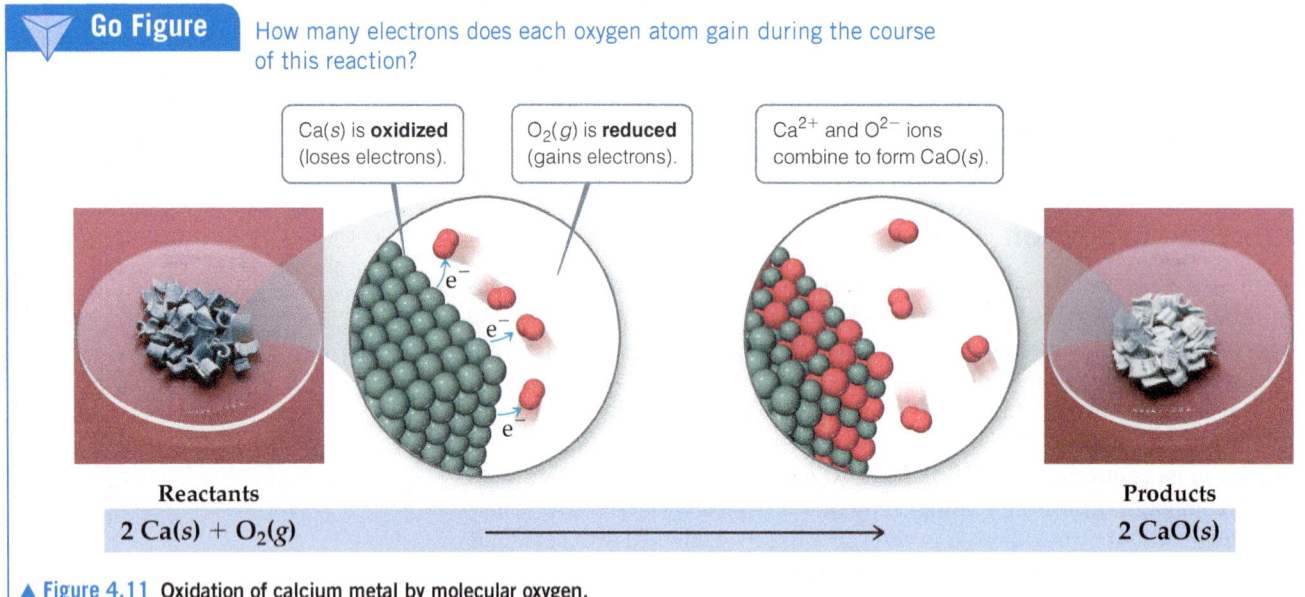

▼ **Go Figure** How many electrons does each oxygen atom gain during the course of this reaction?

Ca(*s*) is **oxidized** (loses electrons).

$O_2(g)$ is **reduced** (gains electrons).

Ca^{2+} and O^{2-} ions combine to form CaO(*s*).

Reactants
$2\,Ca(s) + O_2(g)$

Products
$2\,CaO(s)$

▲ **Figure 4.11** Oxidation of calcium metal by molecular oxygen.

Oxidation Numbers

Before we can identify an oxidation–reduction reaction, we must have a bookkeeping system—a way of keeping track of electrons gained by the substance being reduced and electrons lost by the substance being oxidized. Each atom in a neutral substance or ion is assigned an **oxidation number** (also known as an **oxidation state**). For monatomic ions, the oxidation number is the same as the charge. For neutral molecules and polyatomic ions, the oxidation number of a given atom is a hypothetical charge. This charge is assigned by artificially dividing up the electrons among the atoms in the molecule or ion. We use the following rules for assigning oxidation numbers:

1. *For an atom in its **elemental form**, the oxidation number is always zero.* Thus, each H atom in the H_2 molecule has an oxidation number of 0, and each P atom in the P_4 molecule has an oxidation number of 0.

2. *For any **monatomic ion**, the oxidation number equals the ionic charge.* Thus, K^+ has an oxidation number of $+1$, S^{2-} has an oxidation number of -2, and so forth.

In ionic compounds, the alkali metal ions (group 1A) always have a 1+ charge and therefore an oxidation number of $+1$. The alkaline earth metals (group 2A) are always $+2$, and aluminum (group 3A) is always $+3$ in ionic compounds. (In writing oxidation numbers, we write the sign before the number to distinguish them from the actual electronic charges, which we write with the number first.)

3. ***Nonmetals*** usually have negative oxidation numbers, although they can sometimes be positive:

   (a) *The oxidation number of **oxygen** is usually -2 in both ionic and molecular compounds.* The major exception is in compounds called peroxides, which contain the O_2^{2-} ion, giving each oxygen an oxidation number of -1.

   (b) *The oxidation number of **hydrogen** is usually $+1$ when bonded to nonmetals and -1 when bonded to metals* (for example, metal hydrides such as sodium hydride, NaH).

   (c) *The oxidation number of **fluorine** is -1 in all compounds.* The other **halogens** have an oxidation number of -1 in most binary compounds. When combined with oxygen, as in oxyanions, however, they have positive oxidation states.

4. ***The sum of the oxidation numbers*** *of all atoms in a neutral compound is zero. The sum of the oxidation numbers in a polyatomic ion equals the charge of the ion.* For example, in the hydronium ion H_3O^+, which is a more accurate description of $H^+(aq)$, the oxidation number of each hydrogen is $+1$ and that of oxygen is -2. Thus, the sum of the oxidation numbers is $3(+1) + (-2) = +1$, which equals the net charge of the ion. This rule is useful in obtaining the oxidation number of one atom in a compound or ion if you know the oxidation numbers of the other atoms, as illustrated in Sample Exercise 4.8.

Sample Exercise 4.8
Determining Oxidation Numbers

Determine the oxidation number of sulfur in (a) H_2S, (b) S_8, (c) SCl_2, (d) Na_2SO_3, (e) SO_4^{2-}.

SOLUTION

Analyze We are asked to determine the oxidation number of sulfur in two molecular species, in the elemental form, and in two substances containing ions.

Plan In each species, the sum of oxidation numbers of all the atoms must equal the charge on the species. We will use the rules outlined previously to assign oxidation numbers.

Solve

(a) When bonded to a nonmetal, hydrogen has an oxidation number of $+1$. Because the H_2S molecule is neutral, the sum of the oxidation numbers must equal zero. Letting x equal the oxidation number of S, we have $2(+1) + x = 0$. Thus, S has an oxidation number of -2.

(b) Because S_8 is an elemental form of sulfur, the oxidation number of S is 0.

(c) Because SCl_2 is a binary compound, we expect chlorine to have an oxidation number of -1. The sum of the oxidation numbers must equal zero. Letting x equal the oxidation number of S, we have $x + 2(-1) = 0$. Consequently, the oxidation number of S must be $+2$.

(d) Sodium, an alkali metal, always has an oxidation number of $+1$ in its compounds. Oxygen commonly has an oxidation state of -2. Letting x equal the oxidation number of S, we have $2(+1) + x + 3(-2) = 0$. Therefore, the oxidation number of S in this compound (Na_2SO_3) is $+4$.

(e) The oxidation state of O is -2. The sum of the oxidation numbers equals -2, the net charge of the SO_4^{2-} ion. Thus, we have $x + 4(-2) = -2$, in which case the oxidation number of S is $+6$.

Comment These examples illustrate that the oxidation number of a given element depends on the compound in which it occurs. The oxidation numbers of sulfur, as seen in these examples, range from -2 to $+6$.

▶ **Practice Exercise**
What is the oxidation state of the boldfaced element in (a) $\mathbf{P}_2O_5$, (b) $Na\mathbf{H}$, (c) $\mathbf{Cr}_2O_7^{2-}$, (d) $\mathbf{Sn}Br_4$, (e) $Ba\mathbf{O}_2$?

Oxidation of Metals by Acids and Salts

The reaction between a metal and either an acid or a metal salt conforms to the general pattern

$$A + BX \longrightarrow AX + B \qquad [4.23]$$

Examples: $\qquad Zn(s) + 2\,HBr(aq) \longrightarrow ZnBr_2(aq) + H_2(g)$

$$Mn(s) + Pb(NO_3)_2(aq) \longrightarrow Mn(NO_3)_2(aq) + Pb(s)$$

These reactions are also called **displacement reactions** because the ion in solution is *displaced* (replaced) through oxidation of an element.

Many metals undergo displacement reactions with acids, producing salts and hydrogen gas. For example, magnesium metal reacts with hydrochloric acid to form magnesium chloride and hydrogen gas (**Figure 4.12**):

$$Mg(s) + 2\,HCl(aq) \longrightarrow MgCl_2(aq) + H_2(g)$$

Oxidation number 0   +1 −1   +2 −1   0   [4.24]

The oxidation number of Mg changes from 0 to +2, an increase that indicates the atom has lost electrons and has therefore been oxidized. The oxidation number of H^+ in the acid decreases from +1 to 0, indicating that this ion has gained electrons and has therefore been reduced. Chlorine has an oxidation number of −1 both before and after the

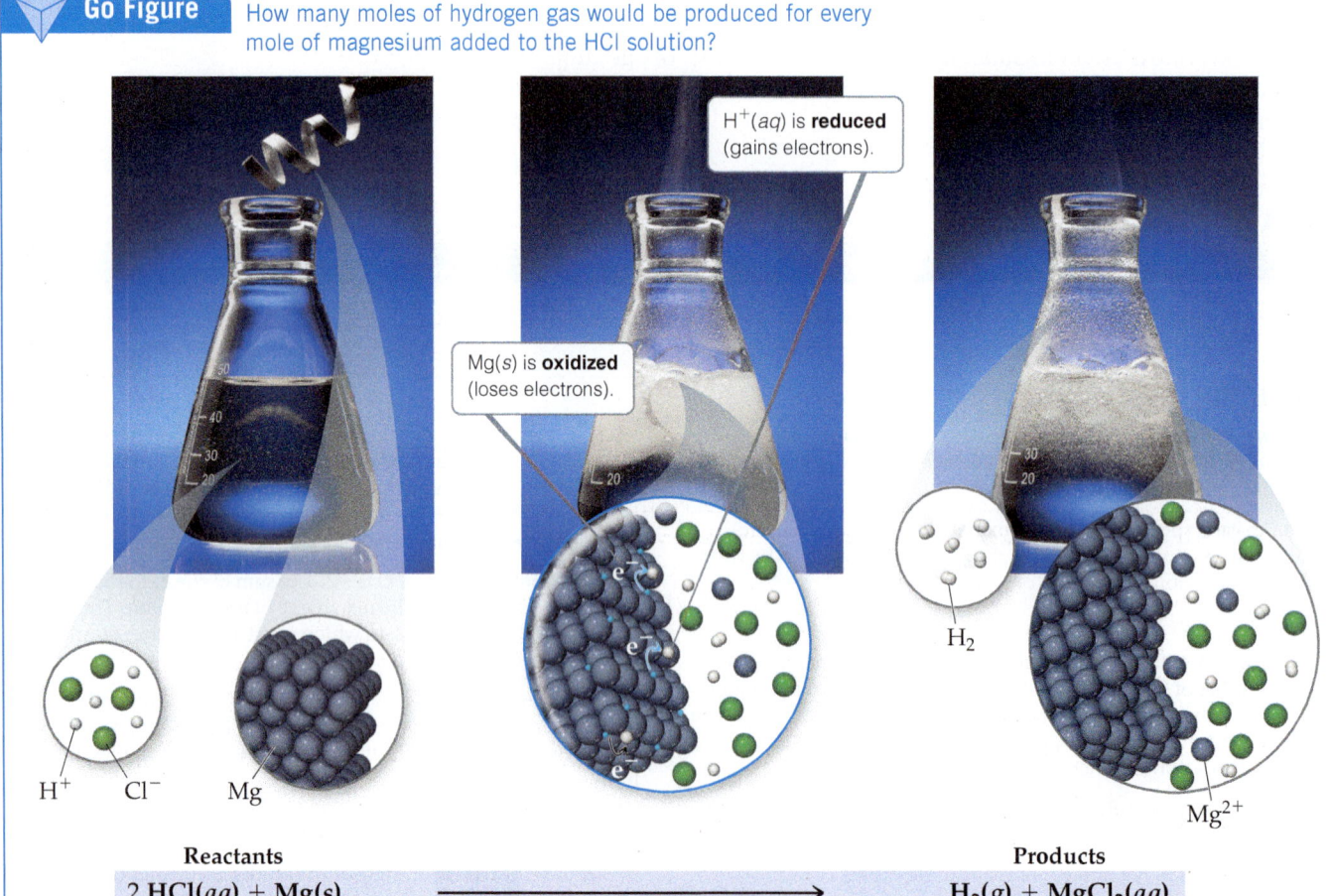

Go Figure   How many moles of hydrogen gas would be produced for every mole of magnesium added to the HCl solution?

$H^+(aq)$ is **reduced** (gains electrons).

Mg(s) is **oxidized** (loses electrons).

H_2

H^+ Cl^- Mg

Mg^{2+}

Reactants		Products
2 HCl(aq) + Mg(s)	$\longrightarrow$	**$H_2(g)$ + $MgCl_2(aq)$**

Oxidation   +1 −1   0   0   +2 −1
number

▲ Figure 4.12 **Reaction of magnesium metal with hydrochloric acid.** The metal is readily oxidized by the acid, producing hydrogen gas, $H_2(g)$, and $MgCl_2(aq)$.

reaction, indicating that it is neither oxidized nor reduced. In fact, the Cl^- ions are spectator ions, dropping out of the net ionic equation:

$$Mg(s) + 2H^+(aq) \longrightarrow Mg^{2+}(aq) + H_2(g) \qquad [4.25]$$

Metals can also be oxidized by aqueous solutions of various salts. Iron metal, for example, is oxidized to Fe^{2+} by aqueous solutions of Ni^{2+} such as $Ni(NO_3)_2(aq)$:

Molecular equation: $\quad Fe(s) + Ni(NO_3)_2(aq) \longrightarrow Fe(NO_3)_2(aq) + Ni(s) \qquad [4.26]$

Net ionic equation: $\qquad Fe(s) + Ni^{2+}(aq) \longrightarrow Fe^{2+}(aq) + Ni(s) \qquad [4.27]$

The oxidation of Fe to Fe^{2+} in this reaction is accompanied by the reduction of Ni^{2+} to Ni. Remember: *Whenever one substance is oxidized, another substance must be reduced.*

Sample Exercise 4.9
Writing Equations for Oxidation–Reduction Reactions

Write the balanced molecular and net ionic equations for the reaction of aluminum with hydrobromic acid.

SOLUTION

Analyze We must write two equations—molecular and net ionic— for the redox reaction between a metal and an acid.

Plan Metals react with acids to form salts and H_2 gas. To write the balanced equations, we must write the chemical formulas for the two reactants and then determine the formula of the salt, which is composed of the cation formed by the metal and the anion of the acid.

Solve The reactants are Al and HBr. The cation formed by Al is Al^{3+}, and the anion from hydrobromic acid is Br^-. Thus, the salt formed in the reaction is $AlBr_3$. Writing the reactants and products and then balancing the equation gives the molecular equation:

$$2 Al(s) + 6 HBr(aq) \longrightarrow 2 AlBr_3(aq) + 3 H_2(g)$$

Both HBr and $AlBr_3$ are soluble strong electrolytes. Thus, the complete ionic equation is

$$2 Al(s) + 6 H^+(aq) + 6 Br^-(aq) \longrightarrow$$
$$2 Al^{3+}(aq) + 6 Br^-(aq) + 3 H_2(g)$$

Because Br^- is a spectator ion, the net ionic equation is

$$2 Al(s) + 6 H^+(aq) \longrightarrow 2 Al^{3+}(aq) + 3 H_2(g)$$

Comment The substance oxidized is the aluminum metal because its oxidation state changes from 0 in the metal to +3 in the cation, thereby increasing in oxidation number. The H^+ is reduced because its oxidation state decreases from +1 in the acid to 0 in H_2.

▶ **Practice Exercise**
(a) Write the balanced molecular and net ionic equations for the reaction between magnesium and cobalt(II) sulfate.
(b) Which species is oxidized, and which is reduced in the reaction?

The Activity Series

Can we predict whether a certain metal will be oxidized either by an acid or by a particular salt? This question is of both practical importance and chemical interest. According to Equation 4.26, for example, it would be unwise to store a solution of nickel nitrate in an iron container because the solution would dissolve the container. When a metal is oxidized, it forms various compounds. Extensive oxidation can lead to the failure of metal machinery parts or the deterioration of metal structures.

Different metals vary in the ease with which they are oxidized. Zn is oxidized by aqueous solutions of Cu^{2+}, for example, but Ag is not. Zn, therefore, loses electrons more readily than Ag; that is, Zn is easier to oxidize than Ag.

A list of metals arranged in order of decreasing ease of oxidation, such as those in **Table 4.5**, is called an **activity series**. The metals at the top of the table, such as the alkali metals and the alkaline earth metals, are most easily oxidized; that is, they react most readily to form compounds. They are called the *active metals*. The metals at the bottom of the activity series, such as the transition elements from groups 8B and 1B, are very stable and form compounds less readily. These metals, which are used to make coins and jewelry, are called *noble metals* because of their low reactivity.

The activity series can be used to predict the outcome of reactions between metals and either metal salts or acids. *Any metal on the list can be oxidized by the ions of elements below it.*

TABLE 4.5   Activity Series of Metals in Aqueous Solution

Metal	Oxidation Reaction
Lithium	$Li(s) \longrightarrow Li^+(aq) + e^-$
Potassium	$K(s) \longrightarrow K^+(aq) + e^-$
Barium	$Ba(s) \longrightarrow Ba^{2+}(aq) + 2e^-$
Calcium	$Ca(s) \longrightarrow Ca^{2+}(aq) + 2e^-$
Sodium	$Na(s) \longrightarrow Na^+(aq) + e^-$
Magnesium	$Mg(s) \longrightarrow Mg^{2+}(aq) + 2e^-$
Aluminum	$Al(s) \longrightarrow Al^{3+}(aq) + 3e^-$
Manganese	$Mn(s) \longrightarrow Mn^{2+}(aq) + 2e^-$
Zinc	$Zn(s) \longrightarrow Zn^{2+}(aq) + 2e^-$
Chromium	$Cr(s) \longrightarrow Cr^{3+}(aq) + 3e^-$
Iron	$Fe(s) \longrightarrow Fe^{2+}(aq) + 2e^-$
Cobalt	$Co(s) \longrightarrow Co^{2+}(aq) + 2e^-$
Nickel	$Ni(s) \longrightarrow Ni^{2+}(aq) + 2e^-$
Tin	$Sn(s) \longrightarrow Sn^{2+}(aq) + 2e^-$
Lead	$Pb(s) \longrightarrow Pb^{2+}(aq) + 2e^-$
Hydrogen	$H_2(g) \longrightarrow 2H^+(aq) + 2e^-$
Copper	$Cu(s) \longrightarrow Cu^{2+}(aq) + 2e^-$
Silver	$Ag(s) \longrightarrow Ag^+(aq) + e^-$
Mercury	$Hg(l) \longrightarrow Hg^{2+}(aq) + 2e^-$
Platinum	$Pt(s) \longrightarrow Pt^{2+}(aq) + 2e^-$
Gold	$Au(s) \longrightarrow Au^{3+}(aq) + 3e^-$

Ease of oxidation increases

For example, copper is above silver in the series. Thus, copper metal is oxidized by silver ions:

$$Cu(s) + 2Ag^+(aq) \longrightarrow Cu^{2+}(aq) + 2Ag(s) \qquad [4.28]$$

The oxidation of copper to copper ions is accompanied by the reduction of silver ions to silver metal. The silver metal is evident on the surface of the copper wire in **Figure 4.13**. The copper(II) nitrate produces a blue color in the solution, as can be seen most clearly in the photograph on the right of Figure 4.13.

Only metals above hydrogen in the activity series are able to react with acids to form H_2. For example, Ni reacts with $HCl(aq)$ to form H_2:

$$Ni(s) + 2HCl(aq) \longrightarrow NiCl_2(aq) + H_2(g) \qquad [4.29]$$

Because elements below hydrogen in the activity series are not oxidized by H^+, Cu does not react with $HCl(aq)$. Interestingly, copper does react with nitric acid, as shown in Figure 1.11, but the reaction is not oxidation of Cu by H^+ ions. Instead, the metal is oxidized to Cu^{2+} by the nitrate ion, accompanied by the formation of brown nitrogen dioxide, $NO_2(g)$:

$$Cu(s) + 4HNO_3(aq) \longrightarrow Cu(NO_3)_2(aq) + 2H_2O(l) + 2NO_2(g) \qquad [4.30]$$

As the copper is oxidized in this reaction, NO_3^-, where the oxidation number of nitrogen is +5, is reduced to NO_2, where the oxidation number of nitrogen is +4. We examine reactions of this type in Chapter 20.

Go Figure   Why does this solution turn blue?

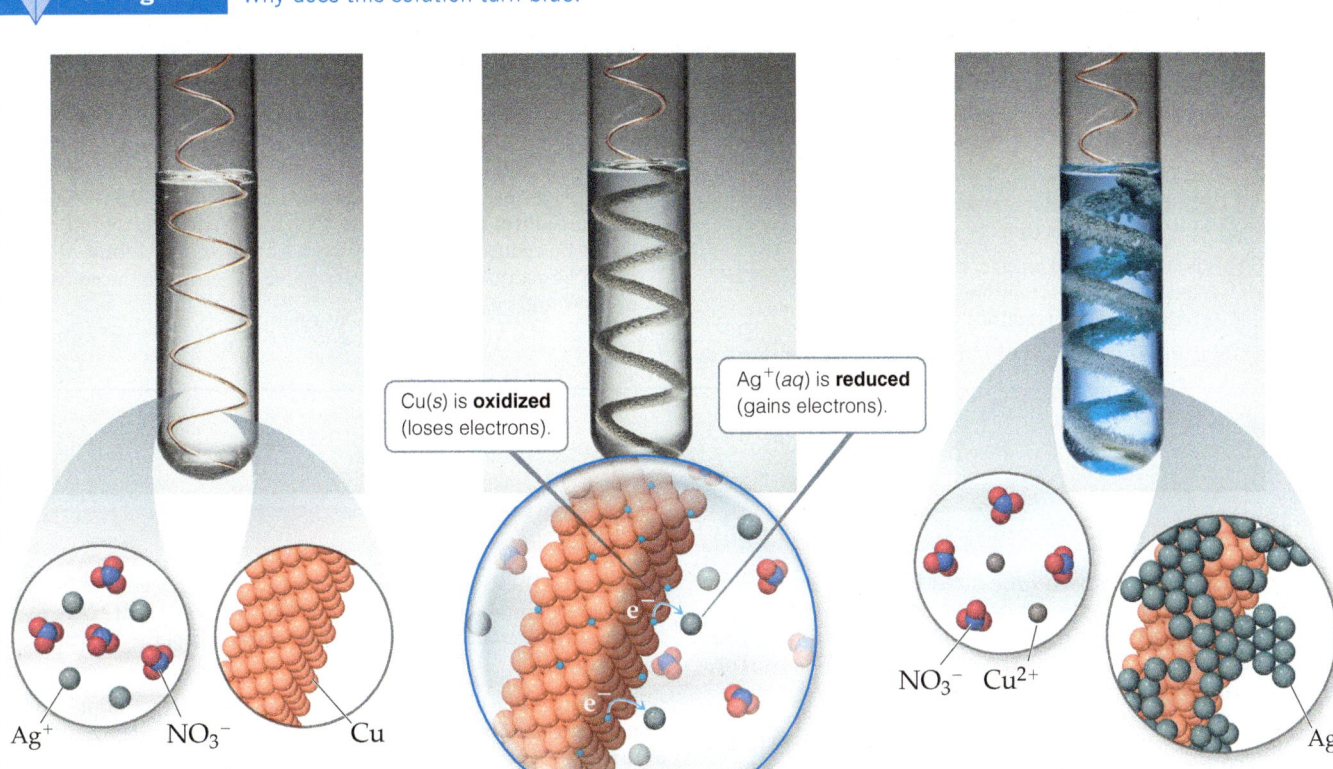

Cu(*s*) is **oxidized** (loses electrons).

Ag⁺(*aq*) is **reduced** (gains electrons).

e^-

e^-

Ag⁺   NO₃⁻   Cu

NO_3^- Cu^{2+}

Ag

Reactants   **Products**

$2\,AgNO_3(aq) + Cu(s)$ $\longrightarrow$ $Cu(NO_3)_2(aq) + 2\,Ag(s)$

▲ **Figure 4.13 Reaction of copper metal with silver ion.** When copper metal is placed in a solution of silver nitrate, a redox reaction forms silver metal and a blue solution of copper(II) nitrate.

Sample Exercise 4.10

Determining If an Oxidation–Reduction Reaction Will Occur

Will an aqueous solution of iron(II) chloride oxidize magnesium metal? If so, write the balanced molecular and net ionic equations for the reaction.

SOLUTION

Analyze We are given two substances—an aqueous salt, FeCl₂, and a metal, Mg—and are asked if they react with each other.

Plan A reaction occurs if the reactant that is a metal in its elemental form (Mg) is located *above* the reactant that is a metal in its oxidized form (Fe²⁺) in Table 4.5. If the reaction occurs, the Fe²⁺ ion in FeCl₂ is reduced to Fe, and the Mg is oxidized to Mg²⁺.

Solve Because Mg is *above* Fe in the table, the reaction occurs. To write the formula for the salt produced in the reaction, we must remember the charges on common ions. Magnesium is always present in compounds as Mg²⁺; the chloride ion is Cl⁻. The magnesium salt formed in the reaction is MgCl₂, meaning the balanced molecular equation is

$$Mg(s) + FeCl_2(aq) \longrightarrow MgCl_2(aq) + Fe(s)$$

Both FeCl₂ and MgCl₂ are soluble strong electrolytes and can be written in ionic form, which shows us that Cl⁻ is a spectator ion in the reaction. The net ionic equation is

$$Mg(s) + Fe^{2+}(aq) \longrightarrow Mg^{2+}(aq) + Fe(s)$$

The net ionic equation shows that Mg is oxidized and Fe²⁺ is reduced in this reaction.

Check The net ionic equation is balanced with respect to both charge and mass.

▶ **Practice Exercise**
Which of the following metals will be oxidized by an aqueous solution of Pb(NO₃)₂: Zn, Cu, Fe?

Self-Assessment Exercises

SAE 4.13   Order these compounds from the minimum to maximum oxidation number for iron: Fe_2O_3, Fe, $FeCl_2$.

(**a**) $FeCl_2 < Fe < Fe_2O_3$

(**b**) $Fe_2O_3 < FeCl_2 < Fe$

(**c**) $Fe < FeCl_2 < Fe_2O_3$

(**d**) $Fe < Fe_2O_3 < FeCl_2$

(**e**) $Fe_2O_3 < FeCl_2 < Fe$

SAE 4.14   Which of the following statements about the reaction $H_2(g) + O_2(g) \longrightarrow H_2O(l)$ is true?

(**i**) This is an example of an acid–base reaction.

(**ii**) O_2 is oxidized in this reaction.

(**iii**) H_2 is reduced in this reaction.

(**a**) i only (**b**) ii only (**c**) iii only (**d**) ii and iii (**e**) None of them are true.

SAE 4.15   If a compound in which Mn is initially in the +2 oxidation state in aqueous solution is reduced, what might the product contain? (**a**) Mn(s) (**b**) $Mn^{4+}(aq)$ (**c**) $Mn^{6+}(aq)$ (**d**) $Mn^{7+}(aq)$

SAE 4.16   Predict what would happen if you put an aqueous solution of $CoCl_2$ in contact with the metal aluminum. (**a**) There would be no reaction. (**b**) For every Co atom oxidized, one Al atom would be reduced. (**c**) For every three Co atoms oxidized, two Al atoms would be reduced. (**d**) For every Co atom reduced, one Al atom would be oxidized. (**e**) For every three Co atoms reduced, two Al atoms would be oxidized.

STRATEGIES FOR SUCCESS | Analyzing Chemical Reactions

In this chapter, you have been introduced to a great number of chemical reactions. It's not easy to get a "feel" for what happens when chemicals react. One goal of this textbook is to help you become more adept at predicting the outcomes of reactions. The key to gaining this "chemical intuition" is to learn how to categorize reactions.

Attempting to memorize individual reactions would be a futile task. It is far more fruitful to recognize patterns to determine the general category of a reaction, such as metathesis or oxidation–reduction. When faced with the challenge of predicting the outcome of a chemical reaction, ask yourself the following questions:

- What are the reactants?
- Are they electrolytes or nonelectrolytes?
- Are they acids or bases?
- If the reactants are electrolytes, will metathesis produce a precipitate? Water? A gas?

- If metathesis cannot occur, can the reactants engage in an oxidation–reduction reaction? This requires that there be both a reactant that can be oxidized and a reactant that can be reduced.

Being able to predict what happens during a reaction follows from asking basic questions like the ones above. Each question narrows the set of possible outcomes, steering you ever closer to a likely outcome. Your prediction might not always be entirely correct, but if you keep your wits about you, you will not be far off. As you gain experience, you will begin to look for reactants that might not be immediately obvious, such as water from the solution or oxygen from the atmosphere. Because proton transfer (acid–base) and electron transfer (oxidation–reduction) are involved in a huge number of chemical reactions, knowing the hallmarks of such reactions will mean you are well on your way to becoming an expert chemist.

4.5 | Concentrations of Solutions

Scientists use the term **concentration** to designate the amount of solute dissolved in a given quantity of solvent or quantity of solution. The greater the amount of solute dissolved in a certain amount of solvent, the more concentrated the resulting solution. In chemistry we often need to express the concentrations of solutions quantitatively.

Molarity

Molarity (symbol M) expresses the concentration of a solution as the number of moles of solute in a liter of solution (soln):

$$\text{Molarity} = \frac{\text{moles solute}}{\text{volume of solution in liters}} \qquad [4.31]$$

A 1.00 molar solution (written 1.00 M) contains 1.00 mol of solute in every liter of solution. **Figure 4.14** shows the preparation of 0.250 L of a 1.00 M solution of $CuSO_4$. The molarity of the solution is $(0.250 \text{ mol } CuSO_4)/(0.250 \text{ L soln}) = 1.00\ M$. Note that we use the abbreviation "soln" for "solution."

Learning Objectives

When you finish Section 4.5, you should be able to:

▶ Calculate the concentration of a solute in a solution in units of molarity.

▶ Interconvert between solute moles, solution volume, and solution molarity.

▶ Calculate the relative volumes of a concentrated stock solution and a solvent to mix in order to achieve a desired final volume and final concentration of a solution.

① Weigh out 39.9 g (0.250 mol) $CuSO_4$.

② Put $CuSO_4$ (solute) into 250-mL volumetric flask; add water and swirl to dissolve solute.

③ Add water until solution just reaches calibration mark on neck of flask, and swirl to mix.

▲ **Figure 4.14** Preparing 0.250 L of a 1.00 M solution of $CuSO_4$.

Sample Exercise 4.11
Calculating Molarity

Calculate the molarity of a solution made by dissolving 23.4 g of sodium sulfate (Na_2SO_4) in enough water to form 125 mL of solution.

SOLUTION

Analyze We are given the number of grams of solute (23.4 g), its chemical formula (Na_2SO_4), and the volume of the solution (125 mL), and we are asked to calculate the molarity of the solution.

Plan We can calculate molarity using Equation 4.31. To do so, we must convert the number of grams of solute to moles and the volume of the solution from milliliters to liters.

Solve

The number of moles of Na_2SO_4 is obtained by using its molar mass:

$$\text{Moles } Na_2SO_4 = (23.4 \text{ g } Na_2SO_4)\left(\frac{1 \text{ mol } Na_2SO_4}{142.1 \text{ g } Na_2SO_4}\right) = 0.165 \text{ mol } Na_2SO_4$$

Converting the volume of the solution to liters:

$$\text{Liters soln} = (125 \text{ mL})\left(\frac{1 \text{ L}}{1000 \text{ mL}}\right) = 0.125 \text{ L}$$

Thus, the molarity is

$$\text{Molarity} = \frac{0.165 \text{ mol } Na_2SO_4}{0.125 \text{ L soln}} = 1.32 \frac{\text{mol } Na_2SO_4}{\text{L soln}} = 1.32 \, M$$

Check Because the numerator is only slightly larger than the denominator, it is reasonable for the answer to be a little over 1 M. The units (mol/L) are appropriate for molarity, and three significant figures are appropriate for the answer because each of the initial pieces of data had three significant figures.

▶ **Practice Exercise**
Calculate the molarity of a solution made by dissolving 5.00 g of glucose ($C_6H_{12}O_6$) in sufficient water to form exactly 100 mL of solution.

Expressing the Concentration of an Electrolyte

In biology, the total concentration of ions in solution is very important in metabolic and cellular processes. When an ionic compound dissolves, the relative concentrations of the ions in the solution depend on the chemical formula of the compound. For example, a 1.0 M solution of NaCl is 1.0 M in Na^+ ions and 1.0 M in Cl^- ions, and a 1.0 M solution of Na_2SO_4 is 2.0 M in Na^+ ions and 1.0 M in SO_4^{2-} ions. Thus, the concentration of an electrolyte solution can be specified either in terms of the compound used to make the solution (1.0 M Na_2SO_4) or in terms of the ions in the solution (2.0 M Na^+ and 1.0 M SO_4^{2-}).

Sample Exercise 4.12
Calculating Molar Concentrations of Ions

What is the molar concentration of each ion present in a 0.025 M aqueous solution of calcium nitrate?

SOLUTION

Analyze We are given the concentration of the ionic compound used to make the solution and are asked to determine the concentrations of the ions in the solution.

Plan We can use the subscripts in the chemical formula of the compound to determine the relative ion concentrations.

Solve Calcium nitrate is composed of calcium ions (Ca^{2+}) and nitrate ions (NO_3^-), so its chemical formula is $Ca(NO_3)_2$. Because there are two NO_3^- ions for each Ca^{2+} ion, each mole of $Ca(NO_3)_2$ that dissolves dissociates into 1 mol of Ca^{2+} and 2 mol of NO_3^-. Thus, a solution that is 0.025 M in $Ca(NO_3)_2$ is 0.025 M in Ca^{2+} and 2 × 0.025 M = 0.050 M in NO_3^-:

$$\frac{\text{mol } NO_3^-}{L} = \left(\frac{0.025 \text{ mol } Ca(NO_3)_2}{L}\right)\left(\frac{2 \text{ mol } NO_3^-}{1 \text{ mol } Ca(NO_3)_2}\right)$$
$$= 0.050 \, M$$

Check The concentration of NO_3^- ions is twice that of Ca^{2+} ions, as the subscript 2 after the NO_3^- in the chemical formula $Ca(NO_3)_2$ suggests.

▶ **Practice Exercise**

What is the molar concentration of K^+ ions in a 0.015 M solution of potassium carbonate?

Interconverting Molarity, Moles, and Volume

If we know any two of the three quantities in the definition of molarity (Equation 4.31), we can calculate the third. For example, if we know the molarity of an HNO_3 solution to be 0.200 M, which means 0.200 mol of HNO_3 per liter of solution, we can calculate the number of moles of solute in a given volume, say 2.0 L. Molarity, therefore, is a conversion factor between volume of solution and moles of solute:

$$\text{Moles } HNO_3 = (2.0 \text{ L soln})\left(\frac{0.200 \text{ mol } HNO_3}{1 \text{ L soln}}\right) = 0.40 \text{ mol } HNO_3$$

To illustrate the conversion of moles to volume, let's calculate the volume of 0.30 M HNO_3 solution required to supply 2.0 mol of HNO_3:

$$\text{Liters soln} = (2.0 \text{ mol } HNO_3)\left(\frac{1 \text{ L soln}}{0.30 \text{ mol } HNO_3}\right) = 6.7 \text{ L soln}$$

Note that in this case we used the reciprocal of molarity in the conversion. If we examine the units in our conversion we see that:

$$\text{moles} \times (1/\text{molarity}) = \text{mol} \times (\text{L/mol}) = \text{L}$$

If one of the solutes is a liquid, we can use its density to convert its mass to volume and vice versa. For example, consider a beer that contains 5.0% ethanol (CH_3CH_2OH) by volume in water (along with other components). The density of ethanol is 0.789 g/mL. Therefore, if we wanted to calculate the molarity of ethanol (usually just called "alcohol" in everyday language) in this beer, we would first consider 1.00 L of beer. This 1.00 L of beer contains 0.950 L of water and 0.050 L of ethanol:

$$5.0\% = 5/100 = 0.050$$

Then we can calculate the moles of ethanol by proper cancellation of units, taking into account the density of ethanol and its molar mass (46.0 g/mol):

$$\text{Moles ethanol} = (0.050 \text{ L})\left(\frac{1000 \text{ mL}}{L}\right)\left(\frac{0.789 \text{ g}}{mL}\right)\left(\frac{1 \text{ mol}}{46.0 \text{ g}}\right) = 0.858 \text{ mol}$$

Because there are 0.858 moles of ethanol in 1.00 L of beer, the concentration of ethanol in beer is 0.86 M (taking into account significant figures).

Sample Exercise 4.13
Using Molarity to Calculate Grams of Solute

How many grams of Na_2SO_4 are required to make 0.350 L of 0.500 M Na_2SO_4?

SOLUTION

Analyze We are given the volume of the solution (0.350 L), its concentration (0.500 M), and the identity of the solute Na_2SO_4, and we are asked to calculate the number of grams of the solute in the solution.

Plan We can use the definition of molarity (Equation 4.31) to determine the number of moles of solute, and then convert moles to grams using the molar mass of the solute.

$$M_{Na_2SO_4} = \frac{\text{moles } Na_2SO_4}{\text{liters soln}}$$

Solve Calculating the moles of Na_2SO_4 using the molarity and volume of solution gives

$$M_{Na_2SO_4} = \frac{\text{moles } Na_2SO_4}{\text{liters soln}}$$

$$\text{Moles } Na_2SO_4 = \text{liters soln} \times M_{Na_2SO_4}$$

$$= (0.350 \text{ L soln})\left(\frac{0.500 \text{ mol } Na_2SO_4}{1 \text{ L soln}}\right)$$

$$= 0.175 \text{ mol } Na_2SO_4$$

Because each mole of Na_2SO_4 has a mass of 142.1 g, the required number of grams of Na_2SO_4 is

$$\text{Grams } Na_2SO_4 = (0.175 \text{ mol } Na_2SO_4)\left(\frac{142.1 \text{ g } Na_2SO_4}{1 \text{ mol } Na_2SO_4}\right)$$

$$= 24.9 \text{ g } Na_2SO_4$$

Check The magnitude of the answer, the units, and the number of significant figures are all appropriate.

▶ **Practice Exercise**
(a) How many grams of Na_2SO_4 are there in 15 mL of 0.50 M Na_2SO_4? **(b)** How many milliliters of 0.50 M Na_2SO_4 solution are needed to provide 0.038 mol of this salt?

Dilution

Solutions used routinely in the laboratory are often purchased or prepared in concentrated form (called *stock solutions*). Aqueous solutions of lower concentrations can then be obtained by adding water, a process called **dilution**.*

Let's see how we can prepare a dilute solution from a concentrated one. Suppose we want to prepare 250.0 mL (that is, 0.2500 L) of 0.100 M $CuSO_4$ solution by diluting a 1.00 M $CuSO_4$ stock solution. The main point to remember is that when solvent is added to a solution, the number of moles of solute remains unchanged:

$$\text{Moles solute before dilution} = \text{moles solute after dilution} \qquad [4.32]$$

Because we know both the volume (250.0 mL) and the concentration (0.100 mol/L) of the dilute solution, we can calculate the number of moles of $CuSO_4$ it contains:

$$\text{Moles } CuSO_4 \text{ in dilute soln} = (0.2500 \text{ L soln})\left(\frac{0.100 \text{ mol } CuSO_4}{\text{L soln}}\right)$$

$$= 0.0250 \text{ mol } CuSO_4$$

The volume of stock solution needed to provide 0.0250 mol $CuSO_4$ is therefore:

$$\text{Liters of conc soln} = (0.0250 \text{ mol } CuSO_4)\left(\frac{1 \text{ L soln}}{1.00 \text{ mol } CuSO_4}\right) = 0.0250 \text{ L}$$

Figure 4.15 shows the dilution carried out in the laboratory. Notice that the diluted solution is less intensely colored than the concentrated one.

In laboratory situations, calculations of this sort are often made with an equation derived by remembering that the number of moles of solute is the same in both the concentrated and dilute solutions and that moles = molarity × liters:

$$\text{Moles solute in conc soln} = \text{moles solute in dilute soln}$$

$$M_{conc} \times V_{conc} = M_{dil} \times V_{dil} \qquad [4.33]$$

---

*When diluting a concentrated acid or base, the acid or base should be added to water and then further diluted by adding more water. Adding water directly to concentrated acid or base can cause spattering because of the intense heat generated.

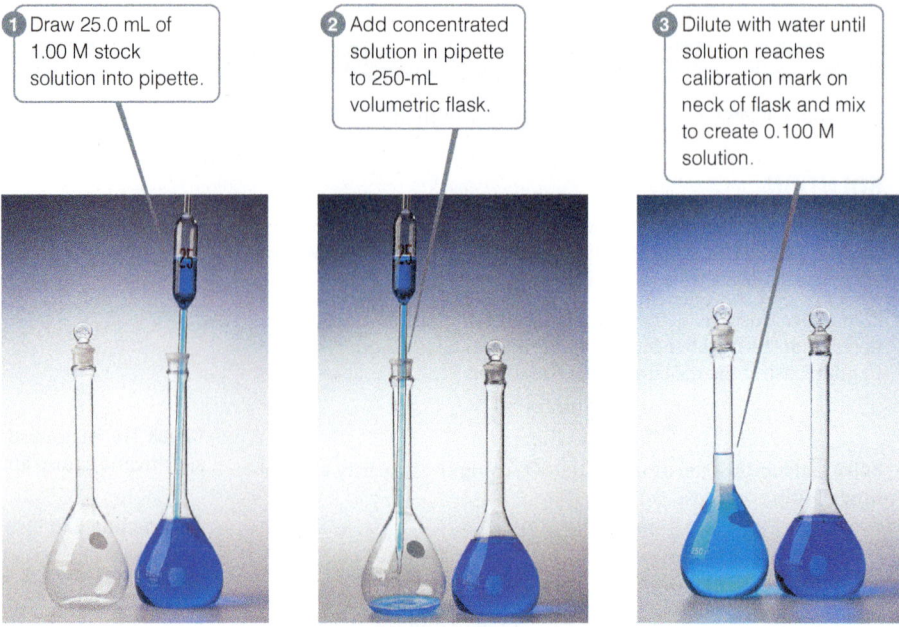

① Draw 25.0 mL of 1.00 M stock solution into pipette.

② Add concentrated solution in pipette to 250-mL volumetric flask.

③ Dilute with water until solution reaches calibration mark on neck of flask and mix to create 0.100 M solution.

▲ **Figure 4.15 Preparing 250.0 mL of 0.100 M CuSO$_4$ by dilution of 1.00 M CuSO$_4$.**

Although we derived Equation 4.33 in terms of liters, any volume unit can be used as long as it is used on both sides of the equation. For example, in the calculation we did for the CuSO$_4$ solution, we have

$$(1.00\,M)(V_{conc}) = (0.100\,M)(250.0\text{ mL})$$

Solving for V_{conc} gives $V_{conc} = 25.0$ mL as before.

Sample Exercise 4.14

Preparing a Solution by Dilution

How many milliliters of 3.0 M H$_2$SO$_4$ are needed to make 450 mL of 0.10 M H$_2$SO$_4$?

SOLUTION

Analyze We need to dilute a concentrated solution. We are given the molarity of a more concentrated solution (3.0 M) and the volume and molarity of a more dilute one containing the same solute (450 mL of 0.10 M solution). We must calculate the volume of the concentrated solution needed to prepare the dilute solution.

Plan We can calculate the number of moles of solute, H$_2$SO$_4$, in the dilute solution and then calculate the volume of the concentrated solution needed to supply this amount of solute. Alternatively, we can directly apply Equation 4.33. Let's compare the two methods.

Solve

Calculate the moles of H$_2$SO$_4$ in the dilute solution:

$$\text{Moles H}_2\text{SO}_4 \text{ in dilute solution} = (0.450\text{ L soln})\left(\frac{0.10\text{ mol H}_2\text{SO}_4}{1\text{ L soln}}\right)$$

$$= 0.045\text{ mol H}_2\text{SO}_4$$

Calculate the volume of the concentrated solution that contains 0.045 mol H$_2$SO$_4$:

Converting liters to milliliters gives 15 mL.

$$\text{Liters conc soln} = (0.045\text{ mol H}_2\text{SO}_4)\left(\frac{1\text{ L soln}}{3.0\text{ mol H}_2\text{SO}_4}\right) = 0.015\text{ L soln}$$

If we apply Equation 4.33, we get the same result:

$$(3.0\,M)(V_{conc}) = (0.10\,M)(450\text{ mL})$$

$$(V_{conc}) = \frac{(0.10\,M)(450\text{ mL})}{3.0\,M} = 15\text{ mL}$$

Either way, we see that if we start with 15 mL of 3.0 M H$_2$SO$_4$ and dilute it to a total volume of 450 mL, the desired 0.10 M solution will be obtained.

Check The calculated volume seems reasonable because a small volume of concentrated solution is used to prepare a large volume of dilute solution.

Comment The first approach can also be used to find the final concentration when two solutions of different concentrations are mixed, whereas the second approach, using Equation 4.33, can be used only for diluting a concentrated solution with pure solvent.

▶ **Practice Exercise**
(a) What volume of 2.50 M lead(II) nitrate solution contains 0.0500 mol of Pb^{2+}? (b) How many milliliters of 5.0 $M K_2Cr_2O_7$ solution must be diluted to prepare 250 mL of 0.10 M solution? (c) If 10.0 mL of a 10.0 M stock solution of NaOH is diluted to 250 mL, what is the concentration of the resulting stock solution?

Self-Assessment Exercises

SAE 4.17 What is the molarity of a solution that is made by dissolving 4.60 grams of methanol, CH_3OH, in water to a final volume of 1.85 L?

(a) 2.49 M (d) $7.77 \times 10^{-2} M$
(b) $1.44 \times 10^{-1} M$ (e) $1.26 \times 10^{-2} M$
(c) $1.38 \times 10^{-1} M$

SAE 4.18 Which solution will have the largest concentration of ions?
(a) 1.0 M NaCl (b) 1.4 M $NaNO_3$ (c) 1.0 M $CaCl_2$ (d) 0.6 M Na_3PO_4

SAE 4.19 A normal blood glucose ($C_6H_{12}O_6$) level is 85 mg glucose/dL. What is the concentration in units of molarity?

(a) 4.7 M (c) $4.7 \times 10^{-3} M$
(b) 2.1 M (d) $4.7 \times 10^{-4} M$

SAE 4.20 What volume of a $5.0 \times 10^{-1} M$ stock solution of Na_3PO_4 must be used to make 250 mL of a solution that is $3.8 \times 10^{-3} M$ in phosphate ions? (a) 0.0075 mL (b) 0.63 mL (c) 1.9 mL (d) 3.3 mL

4.6 | Solution Stoichiometry and Chemical Analysis

In Chapter 3 we learned that given the chemical equation for a reaction and the amount of one reactant consumed in the reaction, you can calculate the quantities of other reactants and products. In this section we extend this concept to reactions involving solutions.

Recall that the coefficients in a balanced equation give the relative number of moles of reactants and products. (Section 3.6) To use this information, we must convert the masses of substances involved in a reaction into moles. When dealing with pure substances, as we did in Chapter 3, we use molar mass to convert between grams and moles of the substances. This conversion is not valid when working with a solution because both solute and solvent contribute to its mass. However, if we know the solute concentration, we can use molarity and volume to determine the number of moles (moles solute = $M \times V$). **Figure 4.16** summarizes this approach to using stoichiometry for the reaction between a pure substance and a solution.

 Learning Objectives

When you finish Section 4.6, you should be able to:

▶ Calculate stoichiometric quantities for reactions in aqueous solutions.

▶ Use acid–base titrations to calculate quantities of solutes.

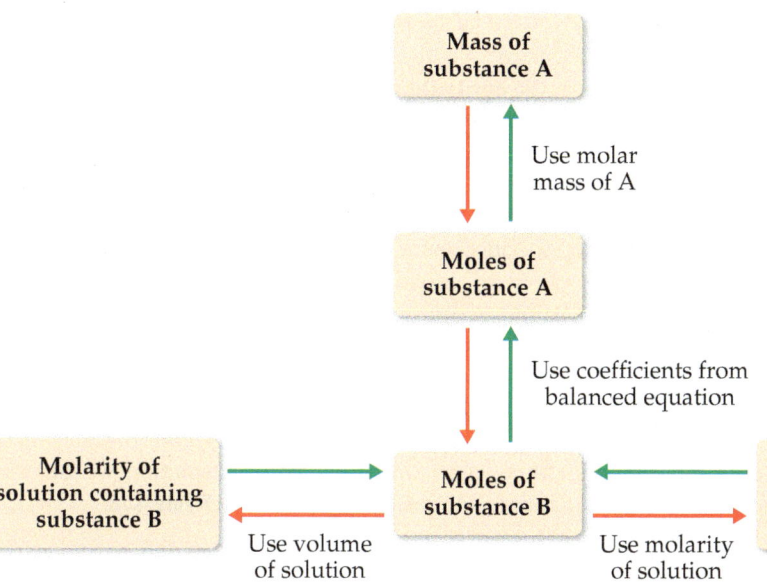

◀ **Figure 4.16 Procedure for solving stoichiometry problems involving reactions between a pure substance A and a solution containing a known concentration of substance B.** Starting from a known mass of substance A, we follow the red arrows to determine either the volume of the solution containing B (if the molarity of B is known) or the molarity of the solution containing B (if the volume of B is known). Starting from either a known volume or a known molarity of the solution containing B, we follow the green arrows to determine the mass of substance A.

Sample Exercise 4.15
Using Mass Relations in a Neutralization Reaction

How many grams of $Ca(OH)_2$ are needed to neutralize 25.0 mL of 0.100 M HNO_3?

SOLUTION

Analyze The reactants are an acid, HNO_3, and a base, $Ca(OH)_2$. The volume and molarity of HNO_3 are given, and we are asked how many grams of $Ca(OH)_2$ are needed to neutralize this quantity of HNO_3.

Plan Following the steps outlined by the green arrows in Figure 4.16, we use the molarity and volume of the HNO_3 solution (substance B in Figure 4.16) to calculate the number of moles of HNO_3. We then use the balanced equation to relate moles of HNO_3 to moles of $Ca(OH)_2$ (substance A). Finally, we use the molar mass to convert moles to grams of $Ca(OH)_2$:

$$V_{HNO_3} \times M_{HNO_3} \Rightarrow \text{mol } HNO_3 \Rightarrow \text{mol } Ca(OH)_2 \Rightarrow \text{g } Ca(OH)_2$$

Solve

The product of the molar concentration of a solution and its volume in liters gives the number of moles of solute:

$$\text{Moles } HNO_3 = V_{HNO_3} \times M_{HNO_3} = (0.0250 \text{ L})\left(\frac{0.100 \text{ mol } HNO_3}{\text{L}}\right)$$
$$= 2.50 \times 10^{-3} \text{mol } HNO_3$$

Because this is a neutralization reaction, HNO_3 and $Ca(OH)_2$ react to form H_2O and the salt containing Ca^{2+} and NO_3^-:

$$2\,HNO_3(aq) + Ca(OH)_2(s) \longrightarrow 2\,H_2O(l) + Ca(NO_3)_2(aq)$$

Thus, 2 mol $HNO_3 \simeq$ 1 mol $Ca(OH)_2$. Therefore, $\text{Grams } Ca(OH)_2 = (2.50 \times 10^{-3} \text{ mol } HNO_3) \times \left(\frac{1 \text{ mol } Ca(OH)_2}{2 \text{ mol } HNO_3}\right)\left(\frac{74.1 \text{ g } Ca(OH)_2}{1 \text{ mol } Ca(OH)_2}\right)$
$$= 0.0926 \text{ g } Ca(OH)_2$$

Check The answer is reasonable because a small volume of dilute acid requires only a small amount of base to neutralize it.

▶ **Practice Exercise**

(a) How many grams of NaOH are needed to neutralize 20.0 mL of 0.150 M H_2SO_4 solution? **(b)** How many liters of 0.500 M HCl(aq) are needed to react completely with 0.100 mol of $Pb(NO_3)_2(aq)$, forming a precipitate of $PbCl_2(s)$?

Titrations

To determine the concentration of a particular solute in a solution, chemists often carry out a **titration**, which involves combining a solution where the solute concentration is not known with a reagent solution of known concentration, called a **standard solution**. Just enough standard solution is added to completely react with the solute in the solution of unknown concentration. The point at which stoichiometrically equivalent quantities are brought together is known as the **equivalence point**. You will likely carry out at least one titration experiment in the laboratory portion of your course.

Titrations can be conducted using neutralization, precipitation, or oxidation–reduction reactions. **Figure 4.17** illustrates a typical neutralization titration, one between an HCl solution of unknown concentration and a standard NaOH solution. To determine the HCl concentration, we first add a specific volume of the HCl solution, 20.0 mL in this example, to a flask. Next we add a few drops of an acid–base **indicator**. The acid–base indicator is a dye that changes color on passing the equivalence point.* For example, the dye phenolphthalein is colorless in acidic solution but pink in basic solution. The standard solution is then slowly added until the solution turns pink, thereby indicating that the neutralization reaction between HCl and NaOH is complete. The standard solution is added from a *burette* so that we can accurately determine the added volume of NaOH solution. Knowing the volumes of both solutions and the concentration of the standard solution, we can calculate the concentration of the unknown solution as diagrammed in **Figure 4.18**.

---

*More precisely, the color change of an indicator signals the end point of the titration, which if the proper indicator is chosen lies very near the equivalence point. Acid–base titrations are discussed in more detail in Section 17.3.

Go Figure How would the volume of standard solution added change if that solution were Ba(OH)$_2$(aq) instead of NaOH(aq)?

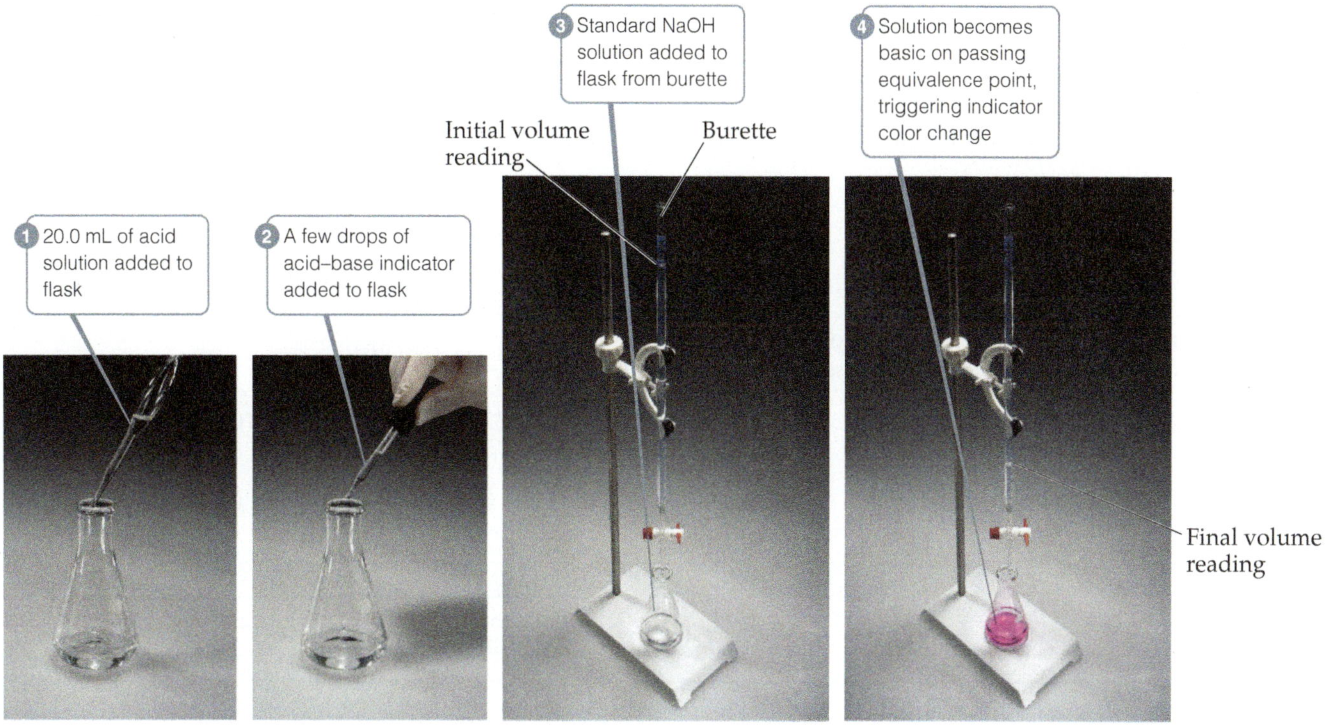

3 Standard NaOH solution added to flask from burette

4 Solution becomes basic on passing equivalence point, triggering indicator color change

Initial volume reading   Burette

1 20.0 mL of acid solution added to flask

2 A few drops of acid–base indicator added to flask

Final volume reading

▲ **Figure 4.17 Procedure for titrating an acid against a standard solution of NaOH.** The acid–base indicator, phenolphthalein, is colorless in acidic solution but takes on a pink color in basic solution.

◄ **Figure 4.18 Procedure for determining the concentration of a solution from titration with a standard solution.**

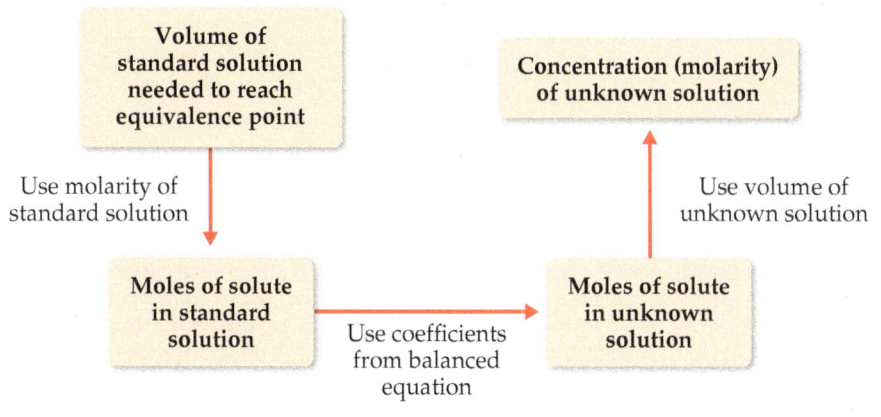

Volume of standard solution needed to reach equivalence point

Use molarity of standard solution

Moles of solute in standard solution

Use coefficients from balanced equation

Moles of solute in unknown solution

Use volume of unknown solution

Concentration (molarity) of unknown solution

Sample Exercise 4.16

Determining Solution Concentration by an Acid–Base Titration

One commercial method used to peel potatoes is to soak them in a NaOH solution for a short time and then remove the potatoes and spray off the peel. The NaOH concentration is normally 3 to 6 *M*, and the solution must be analyzed periodically. In one such analysis, 45.7 mL of 0.500 *M* H$_2$SO$_4$ is required to neutralize 20.0 mL of NaOH solution. What is the concentration of the NaOH solution?

SOLUTION

Analyze We are given the volume (45.7 mL) and molarity (0.500 *M*) of an H$_2$SO$_4$ solution (the standard solution) that reacts completely with 20.0 mL of NaOH solution. We are asked to calculate the molarity of the NaOH solution.

Plan Following the steps given in Figure 4.18, we use the H$_2$SO$_4$ volume and molarity to calculate the number of moles of H$_2$SO$_4$. Then we can use this quantity and the balanced equation for the reaction to calculate moles of NaOH. Finally, we can use moles of NaOH and the NaOH volume to calculate NaOH molarity.

Continued

Solve

The number of moles of H_2SO_4 is the product of the volume and molarity of this solution:

$$\text{Moles } H_2SO_4 = (45.7 \text{ mL soln})\left(\frac{1 \text{ L soln}}{1000 \text{ mL soln}}\right)\left(\frac{0.500 \text{ mol } H_2SO_4}{\text{L soln}}\right)$$

$$= 2.28 \times 10^{-2} \text{ mol } H_2SO_4$$

Acids react with metal hydroxides to form water and a salt. Thus, the balanced equation for the neutralization reaction is

$$H_2SO_4(aq) + 2\,NaOH(aq) \longrightarrow 2\,H_2O(l) + Na_2SO_4(aq)$$

According to the balanced equation, 1 mol $H_2SO_4 \simeq 2$ mol NaOH. Therefore,

$$\text{Moles NaOH} = (2.28 \times 10^{-2} \text{ mol } H_2SO_4)\left(\frac{2 \text{ mol NaOH}}{1 \text{ mol } H_2SO_4}\right)$$

$$= 4.56 \times 10^{-2} \text{ mol NaOH}$$

Knowing the number of moles of NaOH in 20.0 mL of solution allows us to calculate the molarity of this solution:

$$\text{Molarity NaOH} = \frac{\text{mol NaOH}}{\text{L soln}}$$

$$= \left(\frac{4.56 \times 10^{-2} \text{ mol NaOH}}{20.0 \text{ mL soln}}\right)\left(\frac{1000 \text{ mL soln}}{1 \text{ L soln}}\right)$$

$$= 2.28 \frac{\text{mol NaOH}}{\text{L soln}} = 2.28\,M$$

▶ **Practice Exercise**
What is the molarity of a NaOH solution if 48.0 mL neutralizes 35.0 mL of 0.144 $M\,H_2SO_4$?

Sample Exercise 4.17
Determining the Quantity of Solute by Titration

The quantity of Cl^- in a municipal water supply is determined by titrating the sample with Ag^+. The precipitation reaction taking place during the titration is

$$Ag^+(aq) + Cl^-(aq) \longrightarrow AgCl(s)$$

(a) How many grams of chloride ion are in a sample of the water if 20.2 mL of 0.100 $M\,Ag^+$ is needed to react with all the chloride in the sample? **(b)** If the sample has a mass of 10.0 g, what percentage of Cl^- does it contain?

SOLUTION

Analyze We are given the volume (20.2 mL) and molarity (0.100 M) of a solution of Ag^+ and the chemical equation for reaction of this ion with Cl^-. We are asked to calculate the number of grams of Cl^- in the sample and the mass percentage of Cl^- in the sample.

Plan (a) We can use the procedure outlined by the green arrows in Figure 4.16. We begin by using the volume and molarity of Ag^+ to calculate the number of moles of Ag^+ used in the titration. We then use the balanced equation to determine the moles of Cl^- in the sample and from that the grams of Cl^-. **(b)** To calculate the percentage of Cl^- in the sample, we compare the number of grams of Cl^- in the sample with the original mass of the sample, 10.0 g.

Solve

(a) Calculate the number of moles of Ag^+ used in the titration.

$$\text{Moles } Ag^+ = (20.2 \text{ mL soln})\left(\frac{1 \text{ L soln}}{1000 \text{ mL soln}}\right)\left(\frac{0.100 \text{ mol } Ag^+}{\text{L soln}}\right)$$

$$= 2.02 \times 10^{-3} \text{ mol } Ag^+$$

According to the balanced equation, 1 mol $Ag^+ \simeq 1$ mol Cl^-. Using this information and the molar mass of Cl, we have:

$$\text{Grams } Cl^- = (2.02 \times 10^{-3} \text{ mol } Ag^+)\left(\frac{1 \text{ mol } Cl^-}{1 \text{ mol } Ag^+}\right)\left(\frac{35.5 \text{ g } Cl^-}{\text{mol } Cl^-}\right)$$

$$= 7.17 \times 10^{-2} \text{ g } Cl^-$$

(b) Calculate the percentage of Cl^- used in the sample.

$$\text{Percent } Cl^- = \frac{7.17 \times 10^{-2} \text{ g}}{10.0 \text{ g}} \times 100\% = 0.717\% \; Cl^-$$

▶ **Practice Exercise**

A sample of an iron ore is dissolved in acid, and the iron is converted to Fe^{2+}. The sample is then titrated with 47.20 mL of 0.02240 M MnO_4^- solution. The oxidation–reduction reaction that occurs during titration is

$$MnO_4^-(aq) + 5\,Fe^{2+}(aq) + 8\,H^+(aq) \longrightarrow$$
$$Mn^{2+}(aq) + 5\,Fe^{3+}(aq) + 4\,H_2O(l)$$

(**a**) How many moles of MnO_4^- were added to the solution? (**b**) How many moles of Fe^{2+} were in the sample? (**c**) How many grams of iron were in the sample? (**d**) If the sample had a mass of 0.8890 g, what is the percentage of iron in the sample?

Self-Assessment Exercises

SAE 4.21 A 3.0-mL sample of a $6.42 \times 10^{-2}\,M\,NH_3$ solution is mixed with 2.0 mL of a $7.85 \times 10^{-2}\,M$ HCl solution. After purification, what is the theoretical yield of ammonium chloride? (**a**) 0 mg, because all components are soluble (**b**) 2.7 mg (**c**) 8.4 mg (**d**) 10.3 mg

SAE 4.22 How many kilograms of barium hydroxide are required to neutralize 50.0 L of a 0.125 M formic acid solution? (Formic acid, HCOOH, is a monoprotic acid.) (**a**) 0.535 kg (**b**) 1.07 kg (**c**) 34.3 kg (**d**) 68.5 kg

SAE 4.23 Biologically active weak bases include cocaine (molar mass 303.3 g/mol), the neurotransmitter dopamine (molar mass 153.2 g/mol), and the psychedelic drug 2,5-dimethoxy-4-methylamphetamine (molar mass 202.3 g/mol). A sample of 1.7 g of one of these substances is dissolved in 500.0 mL water, and this solution requires 56.0 mL of 0.100 M HCl to neutralize. Which weak base was in the solution? (**a**) cocaine (**b**) dopamine (**c**) 2,5-dimethoxy-4-methylamphetamine (**d**) It is not possible to figure this out with the given information.

Putting Concepts Together

A sample of 70.5 mg of potassium phosphate is added to 15.0 mL of 0.050 M silver nitrate, resulting in the formation of a precipitate. (**a**) Write the molecular equation for the reaction. (**b**) What is the limiting reactant in the reaction? (**c**) Calculate the theoretical yield, in grams, of the precipitate that forms.

SOLUTION

(**a**) Potassium phosphate and silver nitrate are both ionic compounds. Potassium phosphate contains K^+ and PO_4^{3-} ions, so its chemical formula is K_3PO_4. Silver nitrate contains Ag^+ and NO_3^- ions, so its chemical formula is $AgNO_3$. Because both reactants are strong electrolytes, the solution contains K^+, PO_4^{3-}, Ag^+, and NO_3^- ions before the reaction occurs. According to the solubility guidelines in Table 4.1, Ag^+ and PO_4^{3-} form an insoluble compound, so Ag_3PO_4 will precipitate from the solution. In contrast, K^+ and NO_3^- will remain in solution because KNO_3 is water soluble. Thus, the balanced molecular equation for the reaction is

$$K_3PO_4(aq) + 3\,AgNO_3(aq) \longrightarrow Ag_3PO_4(s) + 3\,KNO_3(aq)$$

(**b**) To determine the limiting reactant, we must examine the number of moles of each reactant. (Section 3.7) The number of moles of K_3PO_4 is calculated from the mass of the sample using the molar mass as a conversion factor. (Section 3.4) The molar mass of K_3PO_4 is $3(39.1) + 31.0 + 4(16.0) = 212.3\,g/mol$. Converting milligrams to grams and then to moles, we have:

$$(70.5\,\text{mg K}_3\text{PO}_4)\left(\frac{10^{-3}\,\text{g K}_3\text{PO}_4}{1\,\text{mg K}_3\text{PO}_4}\right)\left(\frac{1\,\text{mol K}_3\text{PO}_4}{212.3\,\text{g K}_3\text{PO}_4}\right)$$
$$= 3.32 \times 10^{-4}\,\text{mol K}_3\text{PO}_4$$

We determine the number of moles of $AgNO_3$ from the volume and molarity of the solution. (Section 4.5) Converting milliliters to liters and then to moles, we have:

$$(15.0\,\text{mL})\left(\frac{10^{-3}\,\text{L}}{1\,\text{mL}}\right)\left(\frac{0.050\,\text{mol AgNO}_3}{\text{L}}\right)$$
$$= 7.5 \times 10^{-4}\,\text{mol AgNO}_3$$

Comparing the amounts of the two reactants, we find that there are $(7.5 \times 10^{-4})/(3.32 \times 10^{-4}) = 2.3$ times as many moles of $AgNO_3$ as there are moles of K_3PO_4. According to the balanced equation, however, 1 mol K_3PO_4 requires 3 mol $AgNO_3$. Thus, there is insufficient $AgNO_3$ to consume the K_3PO_4, and $AgNO_3$ is the limiting reactant.

Continued

(c) The precipitate is Ag_3PO_4, whose molar mass is $3(107.9) + 31.0 + 4(16.0) = 418.7$ g/mol. To calculate the number of grams of Ag_3PO_4 that could be produced in this reaction (the theoretical yield), we use the number of moles of the limiting reactant, converting mol $AgNO_3 \Rightarrow$ mol $Ag_3PO_4 \Rightarrow$ g Ag_3PO_4. We use the coefficients in the balanced equation to convert moles of $AgNO_3$ to moles Ag_3PO_4, and we use the molar mass of Ag_3PO_4 to convert the number of moles of this substance to grams.

The answer has only two significant figures because the quantity of $AgNO_3$ is given to only two significant figures.

$$(7.5 \times 10^{-4} \text{ mol AgNO}_3)\left(\frac{1 \text{ mol Ag}_3\text{PO}_4}{3 \text{ mol AgNO}_3}\right)\left(\frac{418.7 \text{ g Ag}_3\text{PO}_4}{1 \text{ mol Ag}_3\text{PO}_4}\right)$$

$$= 0.10 \text{ g Ag}_3\text{PO}_4$$

Chapter Summary and Key Terms

GENERAL PROPERTIES OF AQUEOUS SOLUTIONS (INTRODUCTION AND SECTION 4.1) Solutions in which water is the dissolving medium are called **aqueous solutions**. The component of the solution that is present in the greatest quantity is the **solvent**. The other components are **solutes**.

Any substance whose aqueous solution contains ions is called an **electrolyte**. Any substance that forms a solution containing no ions is a **nonelectrolyte**. Electrolytes that are present in solution entirely as ions are **strong electrolytes**, whereas those that are present partly as ions and partly as molecules are **weak electrolytes**. Ionic compounds dissociate into ions when they dissolve, so they are **strong electrolytes**. The solubility of ionic substances is made possible by **solvation**, the interaction of ions with polar solvent molecules. Most molecular compounds are nonelectrolytes, although some are weak electrolytes and a few are strong electrolytes. When representing the ionization of a weak electrolyte in solution, half-arrows in both directions are used, indicating that the forward and reverse reactions can achieve a chemical balance called a **chemical equilibrium**.

PRECIPITATION REACTIONS (SECTION 4.2) **Precipitation reactions** are those in which an insoluble product, called a **precipitate**, forms. Solubility guidelines (see Table 4.1) help determine whether an ionic compound will be soluble in water. (The **solubility** of a substance is the amount that dissolves in a given quantity of solvent.) Reactions such as precipitation reactions, in which cations and anions appear to exchange partners, are called **exchange reactions**, or **metathesis reactions**.

Chemical equations can be written to show whether dissolved substances are present in solution predominantly as ions or molecules. When the complete chemical formulas of all reactants and products are used, the equation is called a **molecular equation**. A **complete ionic equation** shows all dissolved strong electrolytes as their component ions. In a **net ionic equation**, those ions that go through the reaction unchanged (**spectator ions**) are omitted.

ACIDS, BASES, AND NEUTRALIZATION REACTIONS (SECTION 4.3) Acids and bases are important electrolytes. **Acids** are proton donors; they increase the concentration of $H^+(aq)$ in aqueous solutions to which they are added. **Bases** are proton acceptors; they increase the concentration of $OH^-(aq)$ in aqueous solutions. Those acids and bases that are strong electrolytes are called **strong acids** and **strong bases**, respectively. Those that are weak electrolytes are **weak acids** and **weak bases**. When solutions of acids and bases are mixed, a neutralization reaction occurs. The **neutralization reaction** between an acid and a metal hydroxide produces water and a **salt**.

Gases can also be formed as a result of neutralization reactions. The reaction of a sulfide with an acid forms $H_2S(g)$; the reaction between a carbonate and an acid forms $CO_2(g)$.

OXIDATION–REDUCTION REACTIONS (SECTION 4.4) **Oxidation** is the loss of electrons by a substance, whereas **reduction** is the gain of electrons by a substance. **Oxidation numbers** (or **oxidation states**) keep track of electrons during chemical reactions and are assigned to atoms using specific rules. The oxidation of an element results in an increase in its oxidation number, whereas reduction is accompanied by a decrease in oxidation number. Oxidation is always accompanied by reduction, giving **oxidation–reduction**, or **redox, reactions**.

Many metals are oxidized by O_2, acids, and salts. The redox reactions between metals and acids as well as those between metals and salts are called **displacement reactions**. The products of these displacement reactions are always an element (H_2 or a metal) and a salt. Comparing such reactions allows us to rank metals according to their ease of oxidation. A list of metals arranged in order of decreasing ease of oxidation is called an **activity series**. Any metal on the list can be oxidized by ions of metals (or H^+) below it in the series.

CONCENTRATIONS OF SOLUTIONS (SECTION 4.5) The **concentration** of a solution expresses the amount of a solute dissolved in the solution. One of the common ways to express the concentration of a solute is in terms of molarity. The **molarity** of a solution is the number of moles of solute per liter of solution. Molarity makes it possible to interconvert solution volume and number of moles of solute. If the solute is a liquid, its density can be used in molarity calculations to convert between mass, volume, and moles. Solutions of known molarity can be formed either by weighing out the solute and diluting it to a known volume or by the **dilution** of a more concentrated solution of known concentration (a stock solution). Adding solvent to the solution (the process of dilution) decreases the concentration of the solute without changing the number of moles of solute in the solution ($M_{conc} \times V_{conc} = M_{dil} \times V_{dil}$).

SOLUTION STOICHIOMETRY AND CHEMICAL ANALYSIS (SECTION 4.6) In the common laboratory technique called **titration**, we combine a solution of known concentration (a **standard solution**) with a solution of unknown concentration to determine the unknown concentration or the quantity of solute in the unknown. The point in the titration at which stoichiometrically equivalent quantities of reactants are brought together is called the **equivalence point**. An indicator can be used to show the end point of the titration, which coincides closely with the equivalence point.

Key Equations

- Molarity = $\dfrac{\text{moles solute}}{\text{volume of solution in liters}}$ [4.31]   Molarity is the most commonly used unit of concentration in chemistry.

- $M_{conc} \times V_{conc} = M_{dil} \times V_{dil}$ [4.33]   When adding solvent to a concentrated solution to make a dilute solution, molarities and volumes of both concentrated and dilute solutions can be calculated if three of the quantities are known.

Exam Prep

EP 4.1 If you have an aqueous solution that contains 1.5 moles of HCl, how many moles of ions are in the solution? (**a**) 1.0 (**b**) 1.5 (**c**) 2.0 (**d**) 2.5 (**e**) 3.0

EP 4.2 Which of the following compounds is insoluble in water? (**a**) $(NH_4)_2S$ (**b**) $CaCO_3$ (**c**) NaOH (**d**) Ag_2SO_4 (**e**) $Pb(CH_3COO)_2$

EP 4.3 What precipitate forms when solutions of $Ba(NO_3)_2$ and KOH are mixed? (**a**) BaOH (**b**) KNO_3 (**c**) $K_2(NO_3)$ (**d**) $Ba(OH)_2$ (**e**) No precipitate will form.

EP 4.4 What happens when you mix an aqueous solution of sodium nitrate with an aqueous solution of barium chloride? (**a**) There is no reaction; all possible products are soluble. (**b**) Only barium nitrate precipitates. (**c**) Only sodium chloride precipitates. (**d**) Both barium nitrate and sodium chloride precipitate. (**e**) Nothing; barium chloride is not soluble and it remains as a precipitate.

EP 4.5 A set of aqueous solutions are prepared containing different acids at the same concentration: acetic acid, chloric acid, and hydrobromic acid. Which solution(s) are the most electrically conductive? (**a**) chloric acid (**b**) hydrobromic acid (**c**) acetic acid (**d**) both chloric acid and hydrobromic acid (**e**) all three solutions have the same electrical conductivity.

EP 4.6 Which of these substances, when dissolved in water, is a strong electrolyte? (**a**) ammonia (**b**) hydrofluoric acid (**c**) folic acid (**d**) sodium nitrate (**e**) sucrose.

EP 4.7 Which is the correct net ionic equation for the reaction of aqueous ammonia with nitric acid?

(**a**) $NH_4^+(aq) + H^+(aq) \longrightarrow NH_5^{2+}(aq)$
(**b**) $NH_3(aq) + NO_3^-(aq) \longrightarrow NH_2^-(aq) + HNO_3(aq)$
(**c**) $NH_2^-(aq) + H^+(aq) \longrightarrow NH_3(aq)$
(**d**) $NH_3(aq) + H^+(aq) \longrightarrow NH_4^+(aq)$
(**e**) $NH_4^+(aq) + NO_3^-(aq) \longrightarrow NH_4NO_3(aq)$

EP 4.8 In which compound is the oxidation state of oxygen -1? (**a**) O_2 (**b**) H_2O (**c**) H_2SO_4 (**d**) H_2O_2 (**e**) KCH_3COO

EP 4.9 Which of the following statements about the reaction between zinc and copper sulfate is true? (**a**) Zinc is oxidized, and copper ion is reduced. (**b**) Zinc is reduced, and copper ion is oxidized. (**c**) All reactants and products are soluble strong electrolytes. (**d**) The oxidation state of copper in copper sulfate is 0. (**e**) More than one of the previous choices are true.

EP 4.10 Which of these metals is the easiest to oxidize? (**a**) gold (**b**) lithium (**c**) iron (**d**) sodium (**e**) aluminum

EP 4.11 What is the molarity of a solution that is made by dissolving 3.68 g of sucrose ($C_{12}H_{22}O_{11}$) in sufficient water to form 275.0 mL of solution?

(**a**) 13.4 M (**d**) $7.43 \times 10^{-5}\,M$
(**b**) $7.43 \times 10^{-2}\,M$ (**e**) $3.91 \times 10^{-5}\,M$
(**c**) $3.91 \times 10^{-2}\,M$

EP 4.12 What is the ratio of the concentration of potassium ions to the concentration of carbonate ions in a 0.015 M solution of potassium carbonate? (**a**) 1:0.015 (**b**) 0.015:1 (**c**) 1:1 (**d**) 1:2 (**e**) 2:1

EP 4.13 What is the concentration of ammonia in a solution made by dissolving 3.75 g of ammonia in 120.0 L of water?

(**a**) $1.84 \times 10^{-3}\,M$ (**d**) 1.84 M
(**b**) $3.78 \times 10^{-2}\,M$ (**e**) 7.05 M
(**c**) 0.0313 M

EP 4.14 What volume of a 1.00 M stock solution of glucose must be used to make 500.0 mL of a $1.75 \times 10^{-2}\,M$ glucose solution in water? (**a**) 1.75 mL (**b**) 8.75 mL (**c**) 48.6 mL (**d**) 57.1 mL (**e**) 28,570 mL

EP 4.15 How many milligrams of sodium sulfide are needed to completely react with 25.00 mL of a 0.0100 M aqueous solution of cadmium nitrate, to form a precipitate of $CdS(s)$? (**a**) 13.8 mg (**b**) 19.5 mg (**c**) 23.5 mg (**d**) 32.1 mg (**e**) 39.0 mg

EP 4.16 What is the molarity of an HCl solution if 27.3 mL of it neutralizes 134.5 mL of 0.0165 M $Ba(OH)_2$? (**a**) 0.0444 M (**b**) 0.0813 M (**c**) 0.163 M (**d**) 0.325 M (**e**) 3.35 M

EP 4.17 A mysterious white powder is found at a crime scene. A simple chemical analysis concludes that the powder is a mixture of sugar and morphine ($C_{17}H_{19}NO_3$), a weak base similar to ammonia that has only one protonable nitrogen. The crime lab takes 10.00 mg of the mysterious white powder, dissolves it in 100.00 mL water, and titrates it to the equivalence point with 2.84 mL of a standard 0.0100 M HCl solution. What is the mass percentage of morphine in the white powder? (**a**) 8.10% (**b**) 17.3% (**c**) 32.6% (**d**) 49.7% (**e**) 81.0%.

Exercises

Visualizing Concepts

4.1 Which of the following schematic drawings best describes a solution of Li_2SO_4 in water (water molecules are not shown for simplicity)? [Section 4.1]

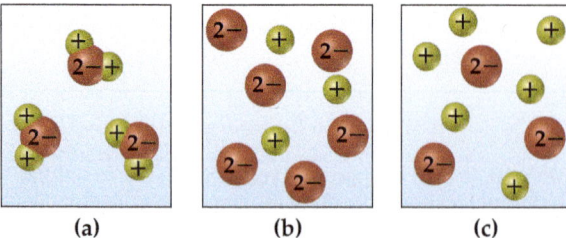

(a)   (b)   (c)

4.2 Aqueous solutions of three different substances, AX, AY, and AZ, are represented by the three accompanying diagrams. Identify each substance as a strong electrolyte, a weak electrolyte, or a nonelectrolyte. [Section 4.1]

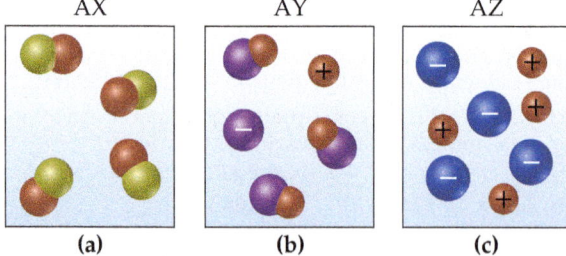

AX   AY   AZ

(a)   (b)   (c)

4.3 Use the molecular representations shown here to classify each compound as a nonelectrolyte, a weak electrolyte, or a strong electrolyte (see Figure 4.5 for the element color scheme). [Sections 4.1 and 4.3]

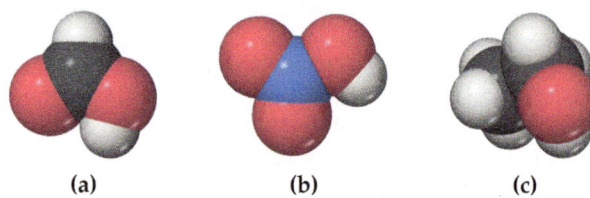

(a)   (b)   (c)

4.4 The concept of chemical equilibrium is very important. Which one of the following statements is the most correct way to think about equilibrium? (**a**) If a system is at equilibrium, nothing is happening. (**b**) If a system is at equilibrium, the rate of the forward reaction is equal to the rate of the back reaction. (**c**) If a system is at equilibrium, the product concentration is changing over time. [Section 4.1]

4.5 You are presented with a white solid and told that due to careless labeling it is not clear if the substance is barium chloride, lead chloride, or zinc chloride. When you transfer the solid to a beaker and add water, the solid dissolves to give a clear solution. Next a $Na_2SO_4(aq)$ solution is added and a white precipitate forms. What is the identity of the unknown white solid? [Section 4.2]

Add
H_2O

Add
$Na_2SO_4(aq)$

4.6 Which of the following ions will *always* be a spectator ion in a precipitation reaction? (**a**) Cl^- (**b**) NO_3^- (**c**) NH_4^+ (**d**) S^{2-} (**e**) SO^{2-} [Section 4.2]

4.7 When a metal strip is placed in $HCl(aq)$, a reaction occurs as indicated by the bubbles of gas formed on the surface of the metal. Which of the following could be the identities of the metal and the gas involved in this reaction? (**a**) Pt, H_2 (**b**) Pt, O_2 (**c**) Zn, H_2 (**d**) Zn, O_2 (**e**) Fe, Cl_2 [Section 4.4]

4.8 Which of these statements is true? (**a**) If a compound is oxidized, it is gaining electrons. (**b**) If a base is neutralized, it is gaining protons. (**c**) Elements that are metals cannot be oxidized. (**d**) If hydrogen gas is generated in a reaction, it must be an acid–base reaction. [Sections 4.3 and 4.4]

4.9 What kind of reaction is this? $N_2(g) + 3H_2(g) \longrightarrow 2NH_3(g)$ (**a**) an acid–base reaction (**b**) a metathesis reaction (**c**) a redox reaction (**d**) a precipitation reaction [Section 4.4]

4.10 An aqueous solution contains 1.2 mM of total ions. (**a**) If the solution is $NaCl(aq)$, what is the concentration of chloride ion? (**b**) If the solution is $FeCl_3(aq)$, what is the concentration of chloride ion? [Section 4.5]

4.11 Consider the following reagents shown below: zinc, copper, mercury (density 13.6 g/mL), silver nitrate solution, nitric acid solution. (**a**) Given a 500-mL Erlenmeyer flask and a balloon, how can you combine two or more of the reagents to initiate a chemical reaction that will inflate the balloon? Write a balanced chemical equation to represent this process. (**b**) What is the identity of the substance that inflates the balloon? (**c**) What is the theoretical yield of the substance that fills the balloon? [Section 4.5]

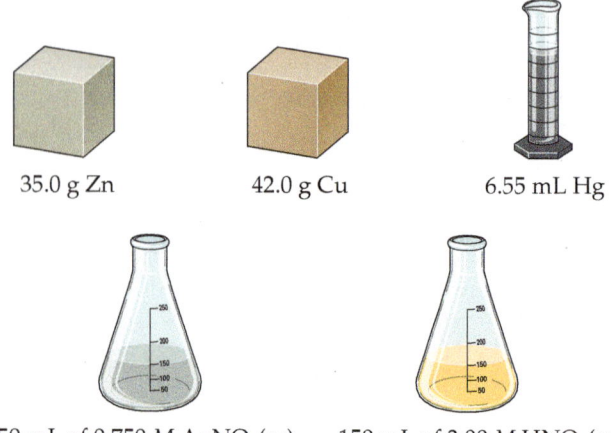

35.0 g Zn   42.0 g Cu   6.55 mL Hg

150 mL of 0.750 M $AgNO_3(aq)$ 150 mL of 3.00 M $HNO_3(aq)$

4.12 In a titration experiment, 50.0 mL of 0.075 *M* acetic acid, CH_3COOH, is titrated with the 0.250 *M* KOH(*aq*) that is in the burette. The drawing shows the level of the KOH in the burette before the titration begins. What will be the burette reading at the equivalence point? [Section 4.6]

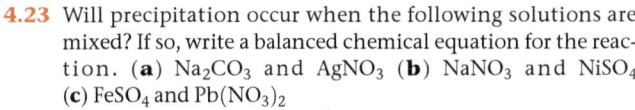

General Properties of Aqueous Solutions (Section 4.1)

4.13 State whether each of the following statements is true or false. (**a**) Electrolyte solutions conduct electricity because electrons are moving through the solution. (**b**) If you add a nonelectrolyte to an aqueous solution that already contains an electrolyte, the electrical conductivity will not change.

4.14 State whether each of the following statements is true or false. (**a**) When methanol, CH_3OH, is dissolved in water, a conducting solution results. (**b**) When acetic acid, CH_3COOH, dissolves in water, the solution is weakly conducting and acidic in nature.

4.15 We have learned in this chapter that many ionic solids dissolve in water as strong electrolytes, that is, as separated ions in solution. Which statement is most correct about this process? (**a**) Water is a strong acid and therefore is good at dissolving ionic solids. (**b**) Water is good at solvating ions because the hydrogen and oxygen atoms in water molecules bear partial charges. (**c**) The hydrogen and oxygen bonds of water are easily broken by ionic solids.

4.16 Would you expect that an anion would be physically closer to the oxygen or to the hydrogens of water molecules that surround it in solution?

4.17 Specify what ions are present in solution upon dissolving each of the following substances in water: (**a**) $FeCl_2$, (**b**) HNO_3, (**c**) $(NH_4)_2SO_4$, (**d**) $Ca(OH)_2$.

4.18 Specify what ions are present upon dissolving each of the following substances in water: (**a**) MgI_2, (**b**) K_2CO_3, (**c**) $HClO_4$, (**d**) $NaCH_3COO$.

4.19 Formic acid, $HCOOH$, is a weak electrolyte. What solutes are present in an aqueous solution of this compound? Write the chemical equation for the ionization of HCOOH.

4.20 Acetone, CH_3COCH_3, is a nonelectrolyte; hypochlorous acid, HClO, is a weak electrolyte; and ammonium chloride, NH_4Cl, is a strong electrolyte. (**a**) What are the solutes present in aqueous solutions of each compound? (**b**) If 0.1 mol of each compound is dissolved in solution, which one contains 0.2 mol of solute particles, which contains 0.1 mol of solute particles, and which contains somewhere between 0.1 and 0.2 mol of solute particles?

Precipitation Reactions (Section 4.2)

4.21 Using solubility guidelines, predict whether each of the following compounds is soluble or insoluble in water: (**a**) $MgBr_2$, (**b**) PbI_2, (**c**) $(NH_4)_2CO_3$, (**d**) $Sr(OH)_2$, (**e**) $ZnSO_4$.

4.22 Predict whether each of the following compounds is soluble in water: (**a**) AgI, (**b**) Na_2CO_3, (**c**) $BaCl_2$, (**d**) $Al(OH)_3$, (**e**) $Zn(CH_3COO)_2$.

4.23 Will precipitation occur when the following solutions are mixed? If so, write a balanced chemical equation for the reaction. (**a**) Na_2CO_3 and $AgNO_3$ (**b**) $NaNO_3$ and $NiSO_4$ (**c**) $FeSO_4$ and $Pb(NO_3)_2$

4.24 Identify the precipitate (if any) that forms when the following solutions are mixed, and write a balanced equation for each reaction. (**a**) $NaCH_3COO$ and HCl (**b**) KOH and $Cu(NO_3)_2$ (**c**) Na_2S and $CdSO_4$

4.25 Which ions remain in solution, unreacted, after each of the following pairs of solutions is mixed? (**a**) potassium carbonate and magnesium sulfate (**b**) lead nitrate and lithium sulfide (**c**) ammonium phosphate and calcium chloride

4.26 Write balanced net ionic equations for the reactions that occur in each of the following cases. Identify the spectator ion or ions in each reaction.

(**a**) $Cr_2(SO_4)_3(aq) + (NH_4)_2CO_3(aq) \longrightarrow$

(**b**) $Ba(NO_3)_2(aq) + K_2SO_4(aq) \longrightarrow$

(**c**) $Fe(NO_3)_2(aq) + KOH(aq) \longrightarrow$

4.27 Separate samples of a solution of an unknown salt are treated with dilute solutions of HBr, H_2SO_4, and NaOH. A precipitate forms in all three cases. Which of the following cations could be present in the unknown salt solution: K^+, Pb^{2+}, Ba^{2+}?

4.28 Separate samples of a solution of an unknown ionic compound are treated with dilute $AgNO_3$, $Pb(NO_3)_2$, and $BaCl_2$. Precipitates form in all three cases. Which of the following could be the anion of the unknown salt: Br^-, CO_3^{2-}, or NO_3^-?

4.29 You know that an unlabeled bottle contains an aqueous solution of one of the following: $AgNO_3$, $CaCl_2$, or $Al_2(SO_4)_3$. You take a portion of the solution and add an aqueous solution of $Ba(NO_3)_2$ to it, and observe that a white solid precipitates. Then you take another portion of the unlabeled solution and add an aqueous solution of NaCl to it; nothing appears to happen. What is the most likely identity of the solution in the unlabeled bottle: (**a**) silver nitrate, (**b**) calcium chloride, (**c**) aluminum sulfate?

4.30 Three solutions are mixed together to form a single solution; in the final solution, there are 0.2 mol $Pb(CH_3COO)_2$, 0.1 mol Na_2S, and 0.1 mol $CaCl_2$ present. What solid(s) will precipitate?

Acids, Bases, and Neutralization Reactions (Section 4.3)

4.31 Which of the following solutions is the most acidic? (**a**) 0.2 *M* LiOH (**b**) 0.2 *M* HI (**c**) 1.0 *M* methanol (CH_3OH)

4.32 Which of the following solutions is the most basic? (**a**) 0.6 *M* NH_3 (**b**) 0.150 *M* KOH (**c**) 0.100*M* $Ba(OH)_2$

4.33 State whether each of the following statements is true or false. Justify your answer in each case. (**a**) Sulfuric acid is a monoprotic acid. (**b**) HCl is a weak acid. (**c**) Methanol is a base.

4.34 State whether each of the following statements is true or false. Justify your answer in each case. (**a**) NH_3 contains no OH^- ions, and yet its aqueous solutions are basic. (**b**) HF is a strong acid. (**c**) Although sulfuric acid is a strong electrolyte, an aqueous solution of H_2SO_4 contains more HSO_4^- ions than SO_4^{2-} ions.

4.35 Label each of the following substances as an acid, base, salt, or none of the above. Indicate whether the substance exists in aqueous solution entirely in molecular form, entirely as ions, or as a mixture of molecules and ions. (**a**) HF (**b**) acetonitrile CH_3CN (**c**) $NaClO_4$ (**d**) $Ba(OH)_2$

4.36 An aqueous solution of an unknown solute is tested with litmus paper and found to be acidic. The solution is weakly conducting compared with a solution of NaCl of the same concentration. Which of the following substances could the unknown be: $KOH, NH_3, HNO_3, KClO_2, H_3PO_3, CH_3COCH_3$ (acetone)?

4.37 Classify each of the following substances as a nonelectrolyte, weak electrolyte, or strong electrolyte in water: (**a**) H_2SO_3, (**b**) CH_3CH_2OH (ethanol), (**c**) NH_3, (**d**) $KClO_3$, (**e**) $Cu(NO_3)_2$.

4.38 Classify each of the following aqueous solutions as a non-electrolyte, weak electrolyte, or strong electrolyte: (**a**) $LiClO_4$, (**b**) $HClO$, (**c**) $CH_3CH_2CH_2OH$ (propanol), (**d**) $HClO_3$, (**e**) $CuSO_4$, (**f**) $C_{12}H_{22}O_{11}$ (sucrose).

4.39 Complete and balance the following molecular equations, and then write the net ionic equation for each.

(**a**) $HBr(aq) + Ca(OH)_2(aq) \longrightarrow$

(**b**) $Cu(OH)_2(s) + HClO_4(aq) \longrightarrow$

(**c**) $Al(OH)_3(s) + HNO_3(aq) \longrightarrow$

4.40 Write the balanced molecular and net ionic equations for each of the following neutralization reactions: (**a**) Aqueous acetic acid is neutralized by aqueous barium hydroxide. (**b**) Solid chromium(III) hydroxide reacts with nitrous acid. (**c**) Aqueous nitric acid and aqueous ammonia react.

4.41 Write balanced molecular and net ionic equations for the following reactions, and identify the gas formed in each: (**a**) solid cadmium sulfide reacts with an aqueous solution of sulfuric acid; (**b**) solid magnesium carbonate reacts with an aqueous solution of perchloric acid.

4.42 Because the oxide ion is basic, metal oxides react readily with acids.

(**a**) Write the net ionic equation for the following reaction:

$$FeO(s) + 2 HClO_4(aq) \longrightarrow Fe(ClO_4)_2(aq) + H_2O(l)$$

(**b**) Based on the equation in part (**a**), write the net ionic equation for the reaction that occurs between $NiO(s)$ and an aqueous solution of nitric acid.

4.43 Magnesium carbonate and magnesium hydroxide are white solids that react with acidic solutions. Write the net ionic equations for the reaction that occurs when each substance reacts with a hydrochloric acid solution.

4.44 As K_2O dissolves in water, the oxide ion reacts with water molecules to form hydroxide ions. (**a**) Write the molecular and net ionic equations for this reaction. (**b**) Based on the definitions of acid and base, what ion is the base in this reaction? (**c**) What is the acid in the reaction? (**d**) What is the spectator ion in the reaction?

Oxidation–Reduction Reactions (Section 4.4)

4.45 True or false: (**a**) If a substance is oxidized, it is gaining electrons. (**b**) If an ion is oxidized, its oxidation number increases.

4.46 True or false: (**a**) Oxidation can occur without oxygen. (**b**) Oxidation can occur without reduction.

4.47 (**a**) Which region of the periodic table shown here contains elements that are easiest to oxidize? (**b**) Which region contains the least readily oxidized elements?

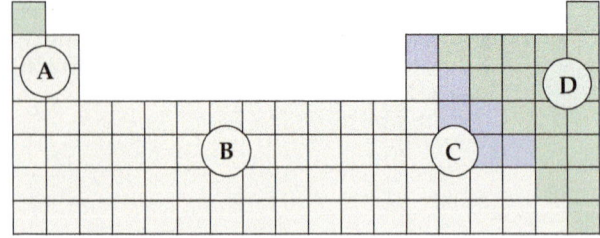

4.48 Determine the oxidation number of sulfur in each of the following substances: (**a**) barium sulfate, $BaSO_4$, (**b**) sulfurous acid, H_2SO_3, (**c**) strontium sulfide, SrS, (**d**) hydrogen sulfide, H_2S. (**e**) Locate sulfur in the periodic table in Exercise 4.47; what region is it in? (**f**) Which region(s) of the periodic table contains elements that can adopt both positive and negative oxidation numbers?

4.49 Determine the oxidation number for the indicated element in each of the following substances: (**a**) S in SO_2, (**b**) C in $COCl_2$, (**c**) Mn in $KMnO_4$, (**d**) Br in $HBrO$, (**e**) P in PF_3, (**f**) O in K_2O_2.

4.50 Determine the oxidation number for the indicated element in each of the following compounds: (**a**) Co in $LiCoO_2$, (**b**) Al in $NaAlH_4$, (**c**) C in CH_3OH (methanol), (**d**) N in GaN, (**e**) Cl in $HClO_2$, (**f**) Cr in $BaCrO_4$.

4.51 Which element is oxidized, and which is reduced in the following reactions?

(**a**) $N_2(g) + 3 H_2(g) \longrightarrow 2 NH_3(g)$

(**b**) $3 Fe(NO_3)_2(aq) + 2 Al(s) \longrightarrow$

$$3 Fe(s) + 2 Al(NO_3)_3(aq)$$

(**c**) $Cl_2(aq) + 2 NaI(aq) \longrightarrow I_2(aq) + 2 NaCl(aq)$

(**d**) $PbS(s) + 4 H_2O_2(aq) \longrightarrow PbSO_4(s) + 4 H_2O(l)$

4.52 Which of the following are redox reactions? For those that are, indicate which element is oxidized and which is reduced. For those that are not, indicate whether they are precipitation or neutralization reactions.

(**a**) $P_4(s) + 10 HClO(aq) + 6 H_2O(l) \longrightarrow$

$$4 H_3PO_4(aq) + 10 HCl(aq)$$

(**b**) $Br_2(l) + 2 K(s) \longrightarrow 2 KBr(s)$

(**c**) $CH_3CH_2OH(l) + 3 O_2(g) \longrightarrow 3 H_2O(l) + 2 CO_2(g)$

(**d**) $ZnCl_2(aq) + 2 NaOH(aq) \longrightarrow Zn(OH)_2(s) +$

$$2 NaCl(aq)$$

4.53 Write balanced net ionic equations for the reactions of (**a**) tin with hydrochloric acid, (**b**) aluminum with formic acid, $HCOOH$. *Hint*: These reactions produce a gas.

4.54 Write balanced net ionic equations for the reactions of (**a**) hydrochloric acid with nickel, (**b**) dilute sulfuric acid with iron. *Hint*: These reactions produce a gas.

4.55 Using the activity series (Table 4.5), write balanced chemical equations for the following reactions. If no reaction occurs, write NR. (**a**) iron metal is added to a solution of copper(II) nitrate (**b**) zinc metal is added to a solution of magnesium sulfate (**c**) hydrobromic acid is added to tin metal (**d**) hydrogen gas is bubbled through an aqueous solution of nickel(II) chloride (**e**) aluminum metal is added to a solution of cobalt(II) sulfate

4.56 Using the activity series (Table 4.5), write balanced chemical equations for the following reactions. If no reaction occurs, write NR. (**a**) nickel metal is added to a solution of copper(II) nitrate (**b**) a solution of zinc nitrate is added to a solution of magnesium sulfate (**c**) hydrochloric acid is added to gold metal (**d**) chromium metal is immersed in an aqueous solution of cobalt(II) chloride (**e**) hydrogen gas is bubbled through a solution of silver nitrate.

4.57 The metal cadmium tends to form Cd^{2+} ions. The following observations are made:

(**i**) When a strip of zinc metal is placed in $CdCl_2(aq)$, cadmium metal is deposited on the strip.

(**ii**) When a strip of cadmium metal is placed in $Ni(NO_3)_2(aq)$, nickel metal is deposited on the strip.

(a) Write net ionic equations to explain each of the preceding observations. (b) Is cadmium above or below zinc in the activity series? (c) Is cadmium above or below nickel in the activity series?

4.58 The following reactions (note that the arrows are pointing in only one direction) can be used to prepare an activity series for the halogens:

$$Br_2(aq) + 2\,NaI(aq) \longrightarrow 2\,NaBr(aq) + I_2(aq)$$
$$Cl_2(aq) + 2\,NaBr(aq) \longrightarrow 2\,NaCl(aq) + Br_2(aq)$$

(a) Which elemental halogen would you predict is the most stable, upon mixing with other halides? (b) Predict whether a reaction will occur when elemental chlorine and potassium iodide are mixed. (c) Predict whether a reaction will occur when elemental bromine and lithium chloride are mixed.

Concentrations of Solutions (Section 4.5)

4.59 (a) Is the concentration of a solution an intensive or an extensive property? (b) What is the difference between 0.50 mol HCl and 0.50 M HCl?

4.60 You make 1.000 L of an aqueous solution that contains 35.0 g of sucrose ($C_{12}H_{22}O_{11}$). (a) What is the molarity of sucrose in this solution? (b) How many liters of water would you have to add to this solution to reduce the molarity you calculated in part (a) by a factor of two?

4.61 (a) Calculate the molarity of a solution that contains 0.175 mol $ZnCl_2$ in exactly 150 mL of solution. (b) How many moles of protons are present in 35.0 mL of a 4.50 M solution of nitric acid? (c) How many milliliters of a 6.00 M NaOH solution are needed to provide 0.350 mol of NaOH?

4.62 (a) Calculate the molarity of a solution made by dissolving 12.5 grams of Na_2CrO_4 in enough water to form exactly 750 mL of solution. (b) How many moles of KBr are present in 150 mL of a 0.112 M solution? (c) How many milliliters of 6.1 M HCl solution are needed to obtain 0.150 mol of HCl?

4.63 The average adult human male has a total blood volume of 5.0 L. If the concentration of sodium ion in this average individual is 0.135 M, what is the mass of sodium ion circulating in the blood?

4.64 A person suffering from hyponatremia has a sodium ion concentration in the blood of 0.118 M and a total blood volume of 4.6 L. What mass of sodium chloride would need to be added to the blood to bring the sodium ion concentration up to 0.138 M, assuming no change in blood volume?

4.65 The concentration of alcohol (CH_3CH_2OH) in blood, called the "blood alcohol concentration" or BAC, is given in units of grams of alcohol per 100 mL of blood. The legal definition of intoxication, in many states of the United States, is that the BAC is 0.08 or higher. What is the molar concentration of alcohol in blood if the BAC is 0.08?

4.66 The average adult male has a total blood volume of 5.0 L. After drinking a few beers, he has a BAC of 0.10 (see Exercise 4.65). What mass of alcohol is circulating in his blood?

4.67 (a) How many grams of ethanol, CH_3CH_2OH, should you dissolve in water to make 1.00 L of vodka (which is an aqueous solution that is 6.86 M ethanol)? (b) Using the density of ethanol (0.789 g/mL), calculate the volume of ethanol you need to make 1.00 L of vodka.

4.68 One cup of fresh orange juice contains 124 mg of ascorbic acid (vitamin C, $C_6H_8O_6$). Given that one cup = 236.6 mL, calculate the molarity of vitamin C in orange juice.

4.69 (a) Which will have the highest concentration of potassium ion: 0.20 M KCl, 0.15 M K_2CrO_4, or 0.080 M K_3PO_4? (b) Which will contain the greater number of moles of potassium ion: 30.0 mL of 0.15 M K_2CrO_4 or 25.0 mL of 0.080 M K_3PO_4?

4.70 In each of the following pairs, indicate which has the higher concentration of I^- ion: (a) 0.10 M BaI_2 or 0.25 M KI solution, (b) 100 mL of 0.10 M KI solution or 200 mL of 0.040 M ZnI_2 solution, (c) 3.2 M HI solution or a solution made by dissolving 145 g of NaI in water to make 150 mL of solution.

4.71 Indicate the concentration of each ion or molecule present in the following solutions: (a) 0.25 M $NaNO_3$, (b) 1.3×10^{-2} M $MgSO_4$, (c) 0.0150 M $C_6H_{12}O_6$, (d) a mixture of 45.0 mL of 0.272 M NaCl and 65.0 mL of 0.0247 M $(NH_4)_2CO_3$. Assume that the volumes are additive.

4.72 Indicate the concentration of each ion present in the solution formed by mixing (a) 42.0 mL of 0.170 M NaOH with 37.6 mL of 0.400 M NaOH, (b) 44.0 mL of 0.100 M Na_2SO_4 with 25.0 mL of 0.150 M KCl, (c) 3.60 g KCl in 75.0 mL of 0.250 M $CaCl_2$ solution. Assume that the volumes are additive.

4.73 (a) You have a stock solution of 14.8 M NH_3. How many milliliters of this solution should you dilute to make 1000.0 mL of 0.250 M NH_3? (b) If you take a 10.0-mL portion of the stock solution and dilute it to a total volume of 0.500 L, what will be the concentration of the final solution?

4.74 (a) How many milliliters of a stock solution of 6.0 M HNO_3 would you have to use to prepare 110 mL of 0.500 M HNO_3? (b) If you dilute 10.0 mL of the stock solution to a final volume of 0.250 L, what will be the concentration of the diluted solution?

4.75 A medical lab is testing a new anticancer drug on cancer cells. The drug stock solution concentration is 1.5×10^{-9} M, and 1.00 mL of this solution will be delivered to a dish containing 2.0×10^5 cancer cells in 5.00 mL of aqueous fluid. What is the ratio of drug molecules to the number of cancer cells in the dish?

4.76 Calicheamicin gamma-1, $C_{55}H_{74}IN_3O_{21}S_4$, is one of the most potent antibiotics known: one molecule kills one bacterial cell. Describe how you would (carefully!) prepare 25.00 mL of an aqueous calicheamicin gamma-1 solution that could kill 1.0×10^8 bacteria, starting from a 5.00×10^{-9} M stock solution of the antibiotic.

4.77 Pure acetic acid, known as glacial acetic acid, is a liquid with a density of 1.049 g/mL at 25 °C. Calculate the molarity of a solution of acetic acid made by dissolving 20.00 mL of glacial acetic acid at 25 °C in enough water to make 250.0 mL of solution.

4.78 Glycerol, $C_3H_8O_3$, is a substance used extensively in the manufacture of cosmetics, foodstuffs, antifreeze, and plastics. Glycerol is a water-soluble liquid with a density of 1.2656 g/mL at 15 °C. Calculate the molarity of a solution of glycerol made by dissolving 50.000 mL glycerol at 15 °C in enough water to make 250.00 mL of solution.

Solution Stoichiometry and Chemical Analysis (Section 4.6)

4.79 You want to analyze a silver nitrate solution. (a) You could add HCl(aq) to the solution to precipitate out AgCl(s). What volume of a 0.150 M HCl(aq) solution is needed to precipitate the silver ions from 15.0 mL of a 0.200 M $AgNO_3$ solution? (b) You could add solid KCl to the solution to precipitate out AgCl(s). What mass of KCl is needed to precipitate the silver ions from 15.0 mL of 0.200 M $AgNO_3$ solution? (c) Given that a 0.150 M HCl(aq) solution costs $39.95 for 500 mL and that KCl costs $10/ton, which analysis procedure is more cost-effective?

4.80 You want to analyze a cadmium nitrate solution. What mass of NaOH is needed to precipitate the Cd^{2+} ions from 35.0 mL of 0.500 M $Cd(NO_3)_2$ solution?

4.81 (**a**) What volume of 0.115 M $HClO_4$ solution is needed to neutralize 50.00 mL of 0.0875 M NaOH? (**b**) What volume of 0.128 M HCl is needed to neutralize 2.87 g of $Mg(OH)_2$? (**c**) If 25.8 mL of an $AgNO_3$ solution is needed to precipitate all the Cl^- ions in a 785-mg sample of KCl (forming AgCl), what is the molarity of the $AgNO_3$ solution? (**d**) If 45.3 mL of a 0.108 M HCl solution is needed to neutralize a solution of KOH, how many grams of KOH must be present in the solution?

4.82 (**a**) How many milliliters of 0.120 M HCl are needed to completely neutralize 50.0 mL of 0.101 M $Ba(OH)_2$ solution? (**b**) How many milliliters of 0.125 M H_2SO_4 are needed to neutralize 0.200 g of NaOH? (**c**) If 55.8 mL of a $BaCl_2$ solution is needed to precipitate all the sulfate ion in a 752-mg sample of Na_2SO_4, what is the molarity of the $BaCl_2$ solution? (**d**) If 42.7 mL of 0.208 M HCl solution is needed to neutralize a solution of $Ca(OH)_2$, how many grams of $Ca(OH)_2$ must be in the solution?

4.83 Some sulfuric acid is spilled on a lab bench. You can neutralize the acid by sprinkling sodium bicarbonate on it and then mopping up the resulting solution. The sodium bicarbonate reacts with sulfuric acid according to:

$$2\,NaHCO_3(s) + H_2SO_4(aq) \longrightarrow Na_2SO_4(aq) + $$
$$2\,H_2O(l) + 2\,CO_2(g)$$

Sodium bicarbonate is added until the fizzing due to the formation of $CO_2(g)$ stops. If 27 mL of 6.0 M H_2SO_4 was spilled, what is the minimum mass of $NaHCO_3$ that must be added to the spill to neutralize the acid?

4.84 The distinctive odor of vinegar is due to acetic acid, CH_3COOH, which reacts with sodium hydroxide according to:

$$CH_3COOH(aq) + NaOH(aq) \longrightarrow$$
$$H_2O(l) + NaCH_3COO(aq)$$

If 3.45 mL of vinegar needs 42.5 mL of 0.115 M NaOH to reach the equivalence point in a titration, how many grams of acetic acid are in a 1.00-qt sample of this vinegar?

4.85 A 4.36-g sample of an unknown alkali metal hydroxide is dissolved in 100.0 mL of water. An acid–base indicator is added, and the resulting solution is titrated with 2.50 M HCl(aq) solution. The indicator changes color, signaling that the equivalence point has been reached, after 17.0 mL of the hydrochloric acid solution has been added. (**a**) What is the molar mass of the metal hydroxide? (**b**) What is the identity of the alkali metal cation: Li^+, Na^+, K^+, Rb^+, or Cs^+?

4.86 An 8.65-g sample of an unknown group 2A metal hydroxide is dissolved in 85.0 mL of water. An acid–base indicator is added and the resulting solution is titrated with 2.50 M HCl(aq) solution. The indicator changes color, signaling that the equivalence point has been reached, after 56.9 mL of the hydrochloric acid solution has been added. (**a**) What is the molar mass of the metal hydroxide? (**b**) What is the identity of the metal cation: Ca^{2+}, Sr^{2+}, or Ba^{2+}?

4.87 A solution of 100.0 mL of 0.200 M KOH is mixed with a solution of 200.0 mL of 0.150 M $NiSO_4$. (**a**) Write the balanced chemical equation for the reaction that occurs. (**b**) What precipitate forms? (**c**) What is the limiting reactant? (**d**) How many grams of this precipitate form? (**e**) What is the concentration of each ion that remains in solution?

4.88 A solution is made by mixing 15.0 g of $Sr(OH)_2$ and 55.0 mL of 0.200 M HNO_3. (**a**) Write a balanced equation for the reaction that occurs between the solutes. (**b**) Calculate the concentration of each ion remaining in solution. (**c**) Is the resulting solution acidic or basic?

4.89 A 0.5895-g sample of impure magnesium hydroxide is dissolved in 100.0 mL of 0.2050 M HCl solution. The excess acid then needs 19.85 mL of 0.1020 M NaOH for neutralization. Calculate the percentage by mass of magnesium hydroxide in the sample, assuming that it is the only substance reacting with the HCl solution.

4.90 A 1.248-g sample of limestone rock is pulverized and then treated with 30.00 mL of 1.035 M HCl solution. The excess acid then requires 11.56 mL of 1.010 M NaOH for neutralization. Calculate the percentage by mass of calcium carbonate in the rock, assuming that it is the only substance reacting with the HCl solution.

Additional Exercises

4.91 Uranium hexafluoride, UF_6, is processed to produce fuel for nuclear reactors and nuclear weapons. UF_6 is made from the reaction of elemental uranium with ClF_3, which also produces Cl_2 as a by-product. (**a**) Write the balanced molecular equation for the conversion of U and ClF_3 into UF_6 and Cl_2. (**b**) Is this a metathesis reaction? (**c**) Is this a redox reaction?

4.92 The accompanying photo shows the reaction between a solution of $Cd(NO_3)_2$ and one of Na_2S. (**a**) What is the identity of the precipitate? (**b**) What ions remain in solution? (**c**) Write the net ionic equation for the reaction. (**d**) Is this a redox reaction?

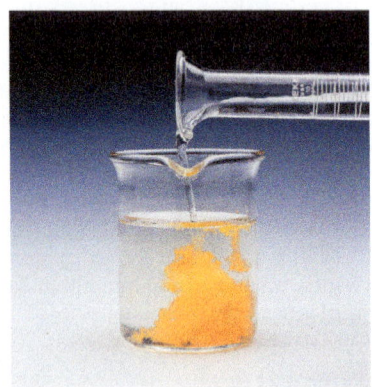

4.93 Suppose you have a solution that might contain any or all of the following cations: Ni^{2+}, Ag^+, Sr^{2+}, and Mn^{2+}. Addition of HCl solution causes a precipitate to form. After filtering off the precipitate, H_2SO_4 solution is added to the resulting solution and another precipitate forms. This is filtered off, and a solution of NaOH is added to the resulting solution. No precipitate is observed. Which ions are present in each of the precipitates? Which of the four ions listed above must be absent from the original solution?

4.94 You choose to investigate some of the solubility guidelines for two ions not listed in Table 4.1, the chromate ion (CrO_4^{2-}) and the oxalate ion ($C_2O_4^{2-}$). You are given 0.01 M solutions (A, B, C, D) of four water-soluble salts:

Solution	Solute	Color of Solution
A	Na_2CrO_4	Yellow
B	$(NH_4)_2C_2O_4$	Colorless
C	$AgNO_3$	Colorless
D	$CaCl_2$	Colorless

When these solutions are mixed, the following observations are made:

Experiment Number	Solutions Mixed	Result
1	A + B	No precipitate, yellow solution
2	A + C	Red precipitate forms
3	A + D	Yellow precipitate forms
4	B + C	White precipitate forms
5	B + D	White precipitate forms
6	C + D	White precipitate forms

(a) Write a net ionic equation for the reaction that occurs in each of the experiments. (b) Identify the precipitate formed, if any, in each of the experiments.

4.95 Antacids are often used to relieve pain and promote healing in the treatment of mild ulcers. Write balanced net ionic equations for the reactions between the aqueous HCl in the stomach and each of the following substances used in various antacids: (a) $Al(OH)_3(s)$, (b) $Mg(OH)_2(s)$, (c) $MgCO_3(s)$, (d) $NaAl(CO_3)(OH)_2(s)$, (e) $CaCO_3(s)$.

4.96 The commercial production of nitric acid involves the following chemical reactions:

$$4\,NH_3(g) + 5\,O_2(g) \longrightarrow 4\,NO(g) + 6\,H_2O(g)$$
$$2\,NO(g) + O_2(g) \longrightarrow 2\,NO_2(g)$$
$$3\,NO_2(g) + H_2O(l) \longrightarrow 2\,HNO_3(aq) + NO(g)$$

(a) Which of these reactions are redox reactions? (b) In each redox reaction identify the element undergoing oxidation and the element undergoing reduction. (c) How many grams of ammonia must you start with to make 1000.0 L of a 0.150 M aqueous solution of nitric acid? Assume all the reactions give 100% yield.

4.97 Neurotransmitters are molecules that are released by nerve cells to other cells in our bodies, and are needed for muscle motion, thinking, feeling, and memory. Dopamine is a common neurotransmitter in the human brain and is a weak base. Its molecular weight is 153.2 g/mol. (a) Patients with Parkinson's disease suffer from a shortage of dopamine and may need to take it to reduce symptoms. An IV (intravenous fluid) bag is filled with a solution that contains 400.0 mg dopamine per 250.0 mL of solution. What is the concentration of dopamine in the IV bag in units of molarity? (b) Experiments with rats show that if rats are dosed with 3.0 mg/kg of cocaine (that is, 3.0 mg cocaine per kg of animal mass), the concentration of dopamine in their brains increases by 0.75 μM after 60 seconds. Calculate how many molecules of dopamine would be produced in a rat (average brain volume 5.00 mm^3) after 60 seconds of a 3.0 mg/kg dose of cocaine.

4.98 Hard water contains Ca^{2+}, Mg^{2+}, and Fe^{2+}, which interfere with the action of soap and leave an insoluble coating on the insides of containers and pipes when heated. Water softeners replace these ions with Na^+. Keep in mind that charge balance must be maintained. (a) If 1500 L of hard water contains 0.020 M Ca^{2+} and 0.0040 M Mg^{2+}, how many moles of Na^+ are needed to replace these ions? (b) If the sodium is added to the water softener in the form of NaCl, how many grams of sodium chloride are needed?

4.99 Tartaric acid, $H_2C_4H_4O_6$, has two acidic hydrogens. The acid is often present in wines, and a salt derived from the acid precipitates from solution as the wine ages. A solution containing an unknown concentration of the acid is titrated with NaOH. It requires 24.65 mL of 0.2500 M NaOH solution to titrate both acidic protons in 50.00 mL of the tartaric acid solution. Write a balanced net ionic equation for the neutralization reaction, and calculate the molarity of the tartaric acid solution.

4.100 (a) A strontium hydroxide solution is prepared by dissolving 12.50 g of $Sr(OH)_2$ in water to make 50.00 mL of solution. What is the molarity of this solution? (b) Next the strontium hydroxide solution prepared in part (a) is used to titrate a nitric acid solution of unknown concentration. Write a balanced chemical equation to represent the reaction between strontium hydroxide and nitric acid solutions. (c) If 23.9 mL of the strontium hydroxide solution was needed to neutralize a 37.5 mL aliquot of the nitric acid solution, what is the concentration (molarity) of the acid?

4.101 A solid sample of $Zn(OH)_2$ is added to 0.350 L of 0.500 M aqueous HBr. The solution that remains is still acidic. It is then titrated with 0.500 M NaOH solution, and it takes 88.5 mL of the NaOH solution to reach the equivalence point. What mass of $Zn(OH)_2$ was added to the HBr solution?

4.102 Suppose you have 5.00 g of powdered magnesium metal, 1.00 L of 2.00 M potassium nitrate solution, and 1.00 L of 2.00 M silver nitrate solution. (a) Which one of the solutions will react with the magnesium powder? (b) What is the net ionic equation that describes this reaction? (c) What volume of solution is needed to completely react with the magnesium? (d) What is the molarity of the Mg^{2+} ions in the resulting solution?

4.103 (a) By titration, 15.0 mL of 0.1008 M sodium hydroxide is needed to neutralize a 0.2053-g sample of a weak acid. What is the molar mass of the acid if it is monoprotic? (b) An elemental analysis of the acid indicates that it is composed of 5.89% H, 70.6% C, and 23.5% O by mass. What is its molecular formula?

4.104 Gold is isolated from rocks by reaction with aqueous cyanide, CN$^-$: $4\,Au(s) + 8\,NaCN(aq) + O_2(g) + H_2O(l) \longrightarrow$ $4\,Na[Au(CN)_2](aq) + 4\,NaOH(aq)$ (a) Which atoms from which compounds are being oxidized, and which atoms from which compounds are being reduced? (b) The $[Au(CN)_2]^-$ ion can be converted back to Au(0) by reaction with Zn(s) powder. Write a balanced chemical equation for this reaction. (c) How many liters of a 0.200 M sodium cyanide solution would be needed to react with 40.0 kg of rocks that contain 2.00% by mass of gold?

4.105 A fertilizer railroad car carrying 34,300 gallons of commercial aqueous ammonia (30% ammonia by mass) tips over and spills. The density of the aqueous ammonia solution is 0.88 g/cm^3. What mass of citric acid, $C(OH)(COOH)(CH_2COOH)_2$ (which contains three acidic protons) is required to neutralize the spill? 1 gallon = 3.785 L.

4.106 A sample of 7.75 g of $Mg(OH)_2$ is added to 25.0 mL of 0.200 M HNO_3. (a) Write the chemical equation for the reaction that occurs. (b) Which is the limiting reactant in the reaction? (c) How many moles of $Mg(OH)_2$, HNO_3, and $Mg(NO_3)_2$ are present after the reaction is complete?

Design an Experiment

You are cleaning out a chemistry lab and find three unlabeled bottles, each containing white powder. Near these bottles are three loose labels: "Sodium sulfide," "Sodium bicarbonate," and "Sodium chloride." Let's design an experiment to figure out which label goes with which bottle.

(**a**) You could try to use the physical properties of the three solids to distinguish among them. Using an Internet resource or the *CRC Handbook of Chemistry and Physics*, look up the melting points, aqueous solubilities, or other properties of these salts. Are the differences among these properties for each salt large enough to distinguish among them? If so, design a set of experiments to distinguish each salt and therefore figure out which label goes on which bottle.

(**b**) You could use the chemical reactivity of each salt to distinguish it from the others. Which of these salts, if any, will act as an acid? A base? A strong electrolyte? Can any of these salts be easily oxidized or reduced? Can any of these salts react to produce a gas? Based on your answers to these questions, design a set of experiments to distinguish each salt and thus determine which label goes on which bottle.

5

THERMOCHEMISTRY

> ▲ **THERMITE REACTION.** The reaction between aluminum metal and iron oxide produces aluminum oxide and iron metal. It also generates enough heat to melt the iron metal formed in the reaction. The thermite reaction is a dramatic illustration of how potential energy stored in chemical bonds can be converted to heat.

With the exception of the energy from the Sun, most of the energy used in our daily lives comes from chemical reactions. The combustion of hydrocarbons, the use of batteries to power electronic devices, and the chemical reactions that provide the energy needed to sustain living organisms are examples of how chemical reactions produce energy.

In this chapter we begin to explore energy and the energy changes that accompany chemical reactions. The study of energy and its transformations is known as **thermodynamics** (Greek: *thérme-*, "heat"; *dy'namis*, "power"). This area of study began during the Industrial Revolution in order to develop the relationships among heat, work, and fuels in steam engines. Here we examine the relationships between chemical reactions and energy changes that involve heat. This portion of thermodynamics is called **thermochemistry**. We discuss additional aspects of thermodynamics in Chapter 19.

WHAT'S AHEAD

5.1 ▶ The Nature of Chemical Energy Recognize that chemical energy is a form of potential energy that arises largely from the electrostatic interactions of charged particles at the atomic level. Describe energy changes associated with chemical reactions by defining the reactants and products as the *system* and everything else in the universe as the *surroundings*.

5.2 ▶ The First Law of Thermodynamics Explore the *first law of thermodynamics*, which states that energy cannot be created or destroyed but can be transformed from one form to another or transferred between systems and surroundings. The energy possessed by a system is called its *internal energy*. Internal energy is a *state function*, a quantity whose value depends only on the current state of a system, not on how the system came to be in that state.

5.3 ▶ Enthalpy Define a state function called *enthalpy*, which is useful because the change in enthalpy measures the quantity of heat energy gained or lost by a system in a process occurring under constant pressure.

5.4 ▶ Enthalpies of Reaction Discover that the enthalpy change associated with a given chemical reaction is equal to the enthalpies of the products minus the enthalpies of the reactants. This quantity is directly proportional to the amount of reactant consumed in the reaction.

5.5 ▶ Calorimetry Describe *calorimetry*, an experimental technique used to measure heat changes in chemical processes.

5.6 ▶ Hess's Law Demonstrate that the enthalpy change for a given reaction can be calculated using appropriate enthalpy changes for related reactions. To do so, we apply *Hess's law*.

5.7 ▶ Enthalpies of Formation Establish standard values for enthalpy changes in chemical reactions, and use them to calculate enthalpy changes for reactions.

5.8 ▶ Bond Enthalpies Use average bond enthalpies to estimate gas-phase reaction enthalpies.

5.9 ▶ Foods and Fuels Describe how foods and fuels serve as sources of energy, and describe some related health and social issues.

5.1 | The Nature of Chemical Energy

Learning Objective

When you finish Section 5.1, you should be able to:

▶ Relate the energy changes associated with the making and breaking of chemical bonds to changes in the electrostatic potential energy that arises between charged particles.

What do we mean by chemical energy? As we learned in Section 1.4, energy is defined as the capacity to do work or transfer heat. Anyone who has sat by a fire or used a propane grill has witnessed chemical reactions that release heat [Figure 5.1(a)]. Some chemical reactions also absorb heat, such as those that occur when you cook. Chemical reactions can also do work in various ways.

To understand the use of a chemical reaction to do work, recall the definition of work from Chapter 1:

$$w = F \times d \qquad [5.1]$$

For example, the combustion reaction between gasoline and oxygen produces gases that expand and in the process do work that can be used to power an automobile. Champagne is made when yeasts use chemical reactions to ferment sugars into ethanol and carbon dioxide. By carrying out the last stages of fermentation in a corked bottle, pressure from the generated CO_2 builds up and can be used to do work when the cork is popped [Figure 5.1(**b**)]. In a battery, redox reactions produce electrical energy that can be used to do work.

All forms of energy can be classified as either kinetic or potential energy. (Section 1.4) The energy that originates from chemical reactions is associated mainly with changes in potential energy. This energy results from electrostatic interactions at the atomic level. Thus, if we are to understand the energy associated with chemical reactions, we must first understand electrostatic potential energy, which arises from the interactions between charged particles.

The electrostatic potential energy, E_{el}, associated with two charged particles is proportional to their electrical charges, Q_1 and Q_2, and is inversely proportional to the distance, d, separating them:

$$E_{el} = \frac{\kappa Q_1 Q_2}{d} \qquad [5.2]$$

where κ is a proportionality constant whose value is 8.99×10^9 J-m/C^2.* Recall from Chapter 1 that the units used to measure energy are joules, where $1\,J = 1\,kg\text{-}m^2/s^2$. At the atomic level, the charges Q_1 and Q_2 are typically on the order of magnitude of the charge of an electron (1.60×10^{-19} C), while distances range from tenths to tens of nanometers ($1\,nm = 1 \times 10^{-9}$ m). When we use Equation 5.2, we must also remember that Q_1 and Q_2 have signs that indicate whether the charges are positive (+) or negative (−).

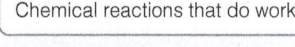

Chemical reactions that release heat   Chemical reactions that do work

(a)   (b)

▲ **Figure 5.1 Chemical reactions and energy.** Energy changes in chemical reactions can be used to transfer heat or do work.

---

*We read the combined units J-m/C^2 as joule-meters per coulomb squared. You may see combinations of units such as J-m/C^2 expressed with dots separating units instead of short dashes, J·m/C^2, or with dashes and dots totally absent, J m/C^2.

Go Figure   A positively charged particle and a negatively charged particle are initially far apart. What happens to their electrostatic potential energy as they are brought closer together?

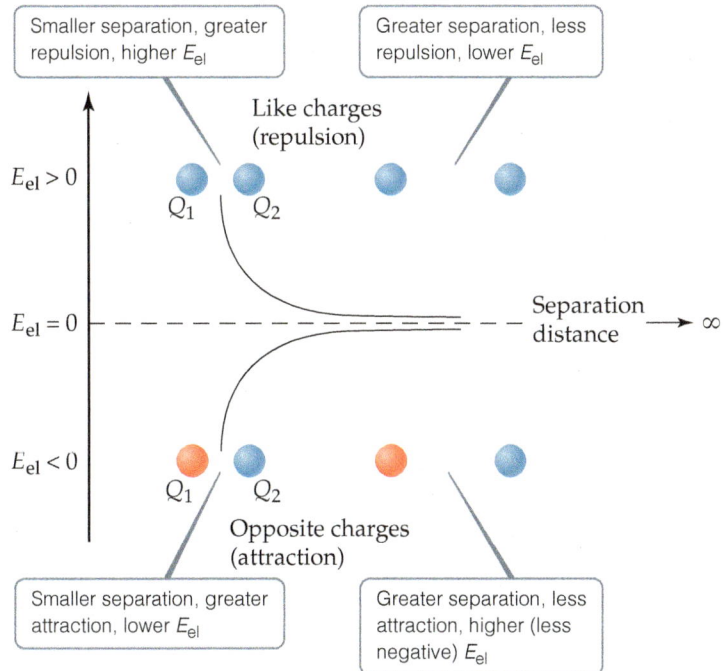

▲ **Figure 5.2 Electrostatic potential energy.** At finite separation distances, the electrostatic potential energy, E_{el}, is positive for objects with like charges and negative for objects that are oppositely charged. As the particles move farther apart, their electrostatic potential energy approaches zero.

According to Equation 5.2, electrostatic potential energy goes to zero as d becomes infinite. We define zero electrostatic potential energy as the potential energy at infinite separation of the charged particles. **Figure 5.2** illustrates how E_{el} behaves as the distance between two charges is varied. When Q_1 and Q_2 have the same sign (for example, both are positive), the two charged particles repel each other, and a repulsive force pushes them apart. To bring two positively charged objects close together, you have to do work to overcome the repulsive force that exists between the two. Upon releasing them, they move away from each other as potential energy is converted to kinetic energy. In this case, E_{el} is positive, and the potential energy decreases as the distance between particles increases. When Q_1 and Q_2 have opposite signs, the particles attract each other, and an attractive force pulls them toward each other. In this case E_{el} is negative, and the potential energy increases (becomes less negative) as the particles move apart.

To understand the relationship between electrostatic potential energy and the energy stored in chemical bonds, consider an ionic compound like NaCl. The Na^+ cations and Cl^- anions are held together by the electrostatic attraction between the oppositely charged ions. In chemistry we refer to this force as ionic bonding. We go into the details of ionic bonding in Chapter 8, but for now it is sufficient to understand that the ionic bonds that hold sodium and chloride ions together are based on the electrostatic attraction between cations and anions. To separate the ions, we must overcome, or break, the ionic bonds between Na^+ and Cl^-, increasing the potential energy, as illustrated in **Figure 5.3**. The energy to do this must come from some other source. The reverse process, where ions of opposite charge separated by a large distance are allowed to come together to form ionic bonds, lowers the potential energy and therefore releases energy. This illustrates a fundamental principle of thermochemistry:

Energy is released when chemical bonds are formed;
energy is consumed when chemical bonds are broken.

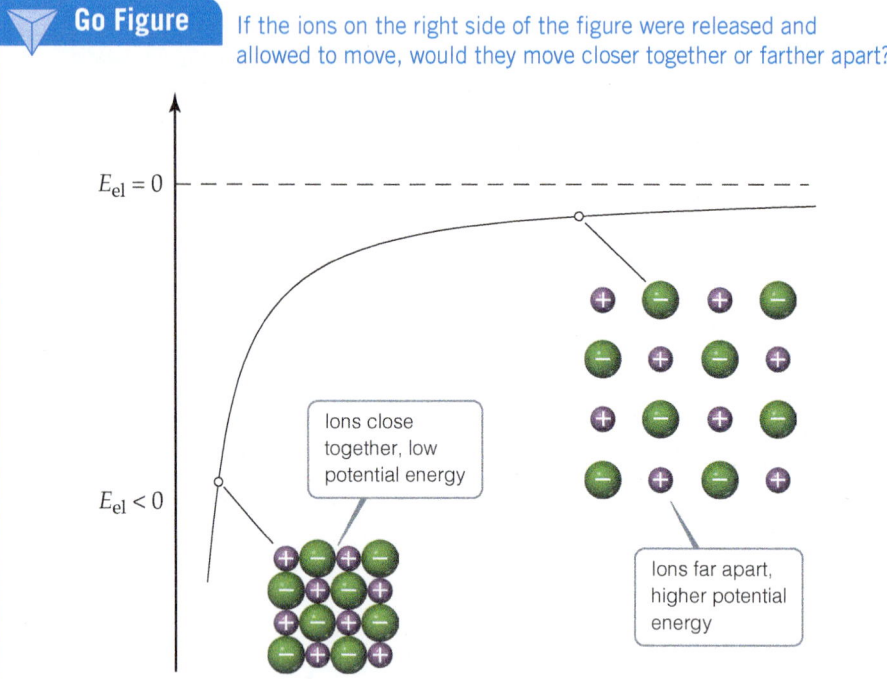

$E_{el} = 0$

Ions close together, low potential energy

$E_{el} < 0$

Ions far apart, higher potential energy

▲ **Figure 5.3 Electrostatic potential energy and ionic bonding.** As the separation between ions increases, the electrostatic potential energy increases (becomes less negative). As the distance separating the ions goes toward infinity, the electrostatic potential energy goes to zero. In real compounds, repulsions between core electrons place a lower limit on how closely the ions can approach each other.

These same principles apply to molecular substances such as water, H_2O, and methane, CH_4, that do not contain cations and anions. As we explain in Chapters 8 and 9, the atoms in a molecule are held together by covalent bonds. While the link to electrostatic potential energy is less obvious, the forces that underlie covalent bonds are also electrostatic in nature. Just like ionic bonds, energy must be provided to break covalent bonds and is released when they form.

 Self-Assessment Exercises

SAE 5.1 Two negative charges are separated by a distance d, as in the figure shown here. Which of the following statements is or are *true*?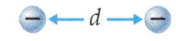

   (i) The two charges are attracted to one another.
   (ii) The electrostatic potential energy of this system is positive.
   (iii) As d increases, the electrostatic potential energy decreases.

(a) Only one of the statements is true. **(b)** i and ii are true. **(c)** i and iii are true. **(d)** ii and iii are true. **(e)** All three statements are true.

 Learning Objectives

When you finish Section 5.2, you should be able to:

▶ Relate the first law of thermodynamics to the energy changes that accompany chemical processes.

▶ Relate changes in internal energy to the transfer of heat (into or out of the system) and the work done (on or by the system).

▶ Distinguish quantities that are state functions and from those that are not.

5.2 | The First Law of Thermodynamics

We have seen that energy comes in many different forms, and we commonly observe processes that involve conversions from one form of energy to another. Dropping a stone down a deep well converts gravitational potential energy into kinetic energy. Heating your house by reacting natural gas with oxygen converts chemical energy into heat. In both cases—and every other one you can imagine—*energy can be converted from one form to another, but it is neither created nor destroyed.* This observation, one of the most important in all of science, is known as the **first law of thermodynamics**. In this section, we explore this fundamental concept in more detail.

System and Surroundings

To apply the first law of thermodynamics quantitatively, we need to divide the universe into a finite system of interest to us, and define the energy of that system more precisely. When analyzing energy changes, we focus on a limited and well-defined part of the universe, which is called the **system**; everything else is called the **surroundings**. When we study the energy change that accompanies a chemical reaction in a laboratory, the reactants and products constitute the system. The container and everything beyond it are considered the surroundings.

Systems may be open, closed, or isolated. An *open* system is one in which matter and energy can be exchanged with the surroundings. An uncovered pot of boiling water on a stove is an open system: Heat comes into the system from the stove, and water is released to the surroundings as steam.

Closed systems are the ones we can most readily study in thermochemistry. A *closed system* can exchange energy but not matter with its surroundings. For example, consider a mixture of hydrogen gas, H_2, and oxygen gas, O_2, in a cylinder fitted with a piston (**Figure 5.4**). The system is just the hydrogen and oxygen; the cylinder, piston, and everything beyond them (including us) are the surroundings. If the gases react to form water, energy is liberated:

$$2\,H_2(g) + O_2(g) \longrightarrow 2\,H_2O(g) + \text{energy}$$

Although the chemical form of the hydrogen and oxygen atoms in the system is changed by this reaction, the system has not lost or gained mass, which means it has not exchanged any matter with its surroundings. However, it can exchange energy with its surroundings in the form of *work* and *heat*.

An *isolated* system is one in which neither energy nor matter can be exchanged with the surroundings. An insulated Thermos containing hot coffee approximates an isolated system. The coffee eventually cools, however, so it is not perfectly isolated.

Internal Energy

The **internal energy**, E, of a system is the sum of *all* the kinetic and potential energies of the components of the system. For the system in Figure 5.4, for example, the internal energy includes not only the motions and interactions of the H_2 and O_2 molecules, but also the motions and interactions of their component nuclei and electrons. We generally do not know the numerical value of a system's internal energy. In thermodynamics, we are mainly concerned with the *change* in E (and, as we shall see, changes in other quantities as well) that accompanies a change in the system.

Imagine that we start with a system with an initial internal energy $E_{initial}$. The system then undergoes a change, which might involve work being done or heat being transferred. After the change, the final internal energy of the system is E_{final}. We define the *change* in internal energy, denoted ΔE (read "delta E"),* as follows:

$$\Delta E = E_{final} - E_{initial} \qquad [5.3]$$

We generally cannot determine the actual values of E_{final} and $E_{initial}$ for any system of practical interest. Nevertheless, we can determine the value of ΔE experimentally by applying the first law of thermodynamics.

Thermodynamic quantities such as ΔE have three parts:

1. a number

2. a unit

3. a sign

(1) and (2) together give the magnitude of the change, while (3) gives the direction. A *positive* value of ΔE results when $E_{final} > E_{initial}$, indicating that the system has gained energy from its surroundings. A *negative* value of ΔE results when $E_{final} < E_{initial}$, indicating that the system has lost energy to its surroundings. Notice that we are taking the

Go Figure

If the H_2 and O_2 molecules in the cylinder react to form H_2O, will the number of molecules in the cylinder change? Will the total mass in the cylinder change?

Energy can enter or leave system as heat or as work done on piston.

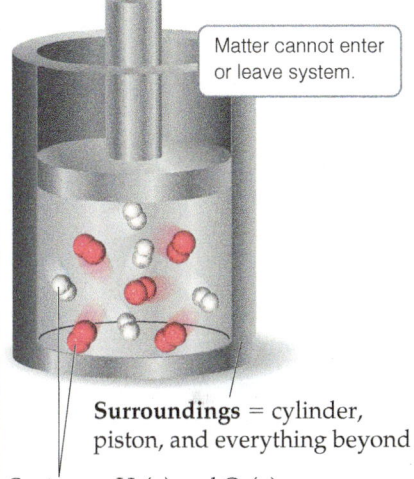

Matter cannot enter or leave system.

Surroundings = cylinder, piston, and everything beyond

System = $H_2(g)$ and $O_2(g)$

▲ **Figure 5.4** A closed system.

---

*The symbol Δ is commonly used to denote change. For example, a change in height, h, can be represented by Δh.

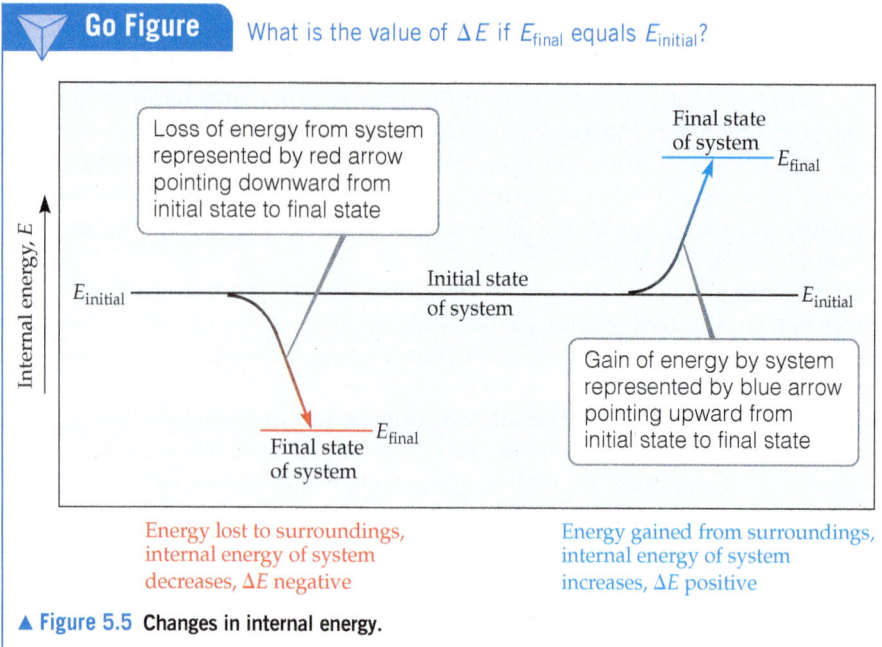

Go Figure   What is the value of ΔE if E_{final} equals $E_{initial}$?

▲ **Figure 5.5** Changes in internal energy.

Go Figure

Sketch the energy diagram representing the reaction $MgCl_2(s) \longrightarrow Mg(s) + Cl_2(g)$ knowing that the internal energy for a mixture of $Mg(s)$ and $Cl_2(g)$ is larger than that of $MgCl_2(s)$.

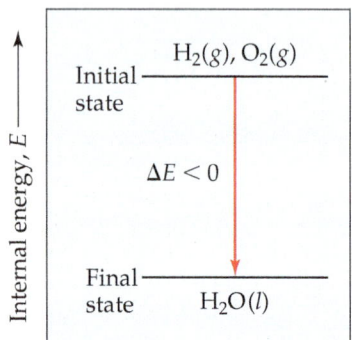

$E_{initial}$ greater than E_{final}; therefore, energy is released from system to surroundings during reaction and $\Delta E < 0$.

▲ **Figure 5.6** Energy diagram for the reaction $2\,H_2(g) + O_2(g) \longrightarrow 2\,H_2O(l)$.

point of view of the system rather than that of the surroundings in discussing the energy changes. Because energy is conserved, any increase in the energy of the system is accompanied by a decrease in the energy of the surroundings, and vice versa. These features of energy changes are summarized in **Figure 5.5**.

In a chemical reaction, the initial state of the system refers to the reactants and the final state refers to the products. In the reaction

$$2\,H_2(g) + O_2(g) \longrightarrow 2\,H_2O(l)$$

for instance, the initial state is $2\,H_2(g) + O_2(g)$ and the final state is $2\,H_2O(l)$. When hydrogen and oxygen form water at a given temperature, the system loses energy to the surroundings. Because energy is lost from the system, the internal energy of the products (final state) is less than that of the reactants (initial state), and ΔE for the process is negative. Thus, the *energy diagram* in **Figure 5.6** shows that the internal energy of the mixture of H_2 and O_2 is greater than that of the H_2O produced in the reaction.

Relating ΔE to Heat and Work

As we noted in **Section 5.1**, a system may exchange energy with its surroundings in two general ways: as heat or as work. The internal energy of a system changes in magnitude as heat is added to or removed from the system or as work is done on or by the system. If we think of internal energy as the system's bank account of energy, then "deposits" or "withdrawals" can be made in the form of either heat or work. Deposits increase the energy of the system (positive ΔE), whereas withdrawals decrease the energy of the system (negative ΔE).

We can use these ideas to write a useful algebraic expression of the first law of thermodynamics. When a system undergoes any chemical or physical change, the accompanying change in internal energy, ΔE, is the sum of the heat added to or liberated from the system, q, and the work done on or by the system, w:

$$\Delta E = q + w \qquad [5.4]$$

When heat is added to a system or work is done on a system, its internal energy increases. Therefore, when heat is transferred to the system from the surroundings, q has a positive value. Adding heat to the system is like making a deposit to the energy account—the energy of the system increases (**Figure 5.7**). Likewise, when work is done on the system by the surroundings, w has a positive value. Conversely, both the heat lost by the system to the surroundings and the work done by the system on the surroundings have negative values; that is, they lower the internal energy of the system. They are energy withdrawals and lower the amount of energy in the system's account.

Go Figure   Suppose a system receives a "deposit" of 50 J of work from the surroundings and loses a "withdrawal" of 85 J of heat to the surroundings. What are the magnitude and the sign of ΔE for this process?

System is interior of vault

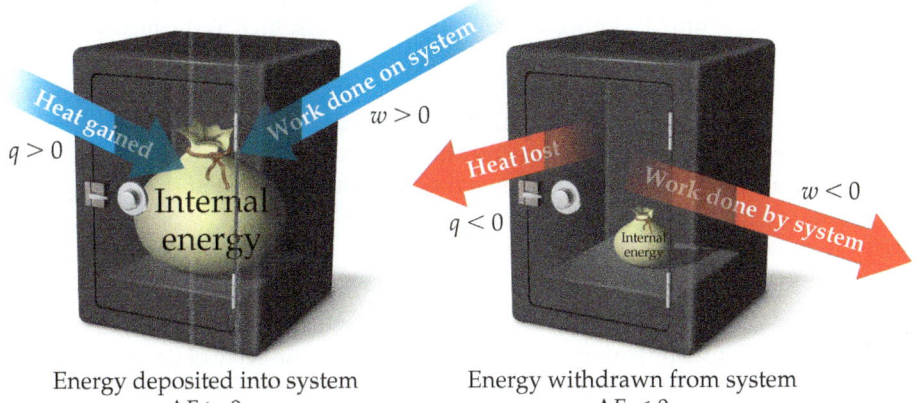

Energy deposited into system
$\Delta E > 0$

Energy withdrawn from system
$\Delta E < 0$

▲ **Figure 5.7 Sign conventions for heat and work.** Heat, q, transferred into a system and work, w, done on a system are both positive quantities, corresponding to "deposits" of internal energy into the system. Conversely, heat transferred out of a system to the surroundings and work done by the system on the surroundings are both "withdrawals" of internal energy from the system.

The sign conventions for q, w, and ΔE are summarized in **Table 5.1**. Notice that any energy entering the system as either heat or work carries a positive sign.

TABLE 5.1   Sign Conventions for q, w, and ΔE

For q	+ means system *gains* heat from surroundings	− means system *loses* heat to surroundings
For w	+ means work done *on* system by surroundings	− means work done *by* system on surroundings
For ΔE	+ means *net gain* of energy by system	− means *net loss* of energy by system

 ## Sample Exercise 5.1

Relating Heat and Work to Changes of Internal Energy

Gases A(g) and B(g) are confined in a cylinder-and-piston arrangement like the one shown in Figure 5.4 and react to form a solid product C(s): A(g) + B(g) $\longrightarrow$ C(s). As the reaction occurs, the system loses 1150 J of heat to the surroundings. The piston moves downward as the gases react to form a solid. As the volume of the gas decreases under the constant pressure of the atmosphere, the surroundings do 480 J of work on the system. What is the change in the internal energy of the system?

SOLUTION

Analyze The question asks us to determine ΔE, given information about q and w.

Plan Use Table 5.1 to determine the signs of q and w, and then use Equation 5.4, $\Delta E = q + w$, to calculate ΔE.

Continued

Solve Heat is transferred from the system to the surroundings, and work is done on the system by the surroundings, so q is negative and w is positive: $q = -1150$ J and $w = 480$ J. Thus,

$$\Delta E = q + w = (-1150\,\text{J}) + (480\,\text{J}) = -670\,\text{J}$$

The negative value of ΔE tells us that a net quantity of 670 J of energy has been transferred from the system to the surroundings.

Comment You can think of this change as a decrease of 670 J in the net value of the system's energy bank account (hence, the negative sign);

1150 J is withdrawn in the form of heat, while 480 J is deposited in the form of work. Notice that as the volume of the gases decreases, work is being done *on* the system *by* the surroundings, resulting in a deposit of energy.

▶ **Practice Exercise**
Calculate the change in the internal energy for a process in which a system absorbs 140 J of heat from the surroundings and does 85 J of work on the surroundings.

Endothermic and Exothermic Processes

Because transfer of heat to and from the system is central to our discussion in this chapter, we have some special terminology to indicate the direction of transfer. When a process occurs in which the system absorbs heat, the process is called **endothermic** (*endo-* means "into"). During an endothermic process, such as the melting of ice, heat flows *into* the system from its surroundings [**Figure 5.8**(**a**)]. If we, as part of the surroundings, touch a container in which ice is melting, the container feels cold to us because heat has passed from our hand to the container.

A process in which the system loses heat is called **exothermic** (*exo-* means "out of"). During an exothermic process, such as the combustion of gasoline, heat *exits* or flows *out* of the system into the surroundings [Figure 5.8(**b**)].

State Functions

Although we usually have no way of knowing the precise value of the internal energy of a system, E, it does have a fixed value for a given set of conditions. The conditions that influence internal energy include the temperature and pressure. Furthermore, the internal energy of a system is proportional to the total quantity of matter in the system because energy is an extensive property. (Section 1.3)

Suppose we define our system as 50 g of water at 25°C (**Figure 5.9**). The system could have reached this state by cooling 50 g of water from 100 to 25°C or by melting 50 g of ice and subsequently warming the water to 25°C. The internal energy of the water at 25°C is the same in either case. Internal energy is an example of a **state function**, that is, a property of a system that is determined by specifying the system's condition or state (in terms of temperature, pressure, and so forth). *The value of a state function depends only on the present state of the system, not on the path the system took to reach that state.* Because E is a state function, ΔE depends only on the initial and final states of the system, not on how the change occurs.

An analogy may help you understand the difference between quantities that are state functions and those that are not. Suppose you drive from Chicago, which is 596 ft above sea level, to Denver, which is 5280 ft above sea level. No matter which route you take, the altitude change is 4684 ft. The distance you travel, however, depends on your route. Altitude is analogous to a state function because the change in altitude is independent of the path taken. Distance traveled is not a state function.

Because E is a state function, the change in internal energy, ΔE, does not depend on the path taken between the initial and final states of the system. However, q and w are *not* state functions; their values depend on the path taken from the initial state to the final state. Because $\Delta E = q + w$, the sum of q and w does not depend on the path, even though q and w are not state functions. Thus, if changing the path by which a system goes from an initial state to a final state increases the value of q, that path change will also decrease the value of w by exactly the same amount. The result is that ΔE is the same for the two paths.

We can illustrate this principle using a flashlight battery as our system. As the battery is discharged, its internal energy decreases as the energy stored in the battery is

System = NH₄SCN + Ba(OH)₂·8H₂O

Heat flows from surroundings into system, temperature of beaker and surrounding air drops.

(a) $Ba(OH)_2 \cdot 8H_2O + 2\,NH_4SCN \longrightarrow Ba(SCN)_2 + 2\,NH_3 + 10\,H_2O$
An endothermic reaction

◀ **Figure 5.8 Endothermic and exothermic reactions.** In both instances, the system is defined as the reactants and products, while surroundings are the containers and everything else in the universe (including the temperature probe in the top reaction).

System = K + H₂O

Heat flows (violently) from system into surroundings, temperature of the water and surrounding air increases.

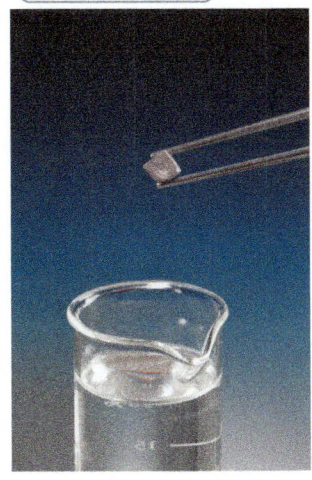 

(b) $2\,K + 2\,H_2O \longrightarrow 2\,KOH + H_2$
An exothermic reaction

50 g
H₂O(s)
0 °C

50 g
H₂O(l)
25 °C

50 g
H₂O(l)
100 °C

Ice initially at 0 °C warms up to liquid water at 25 °C.

Hot water initially at 100 °C cools to 25 °C.

Internal energy, E, of the water at 25 °C is the same regardless of the path taken to reach that state.

◀ **Figure 5.9 Internal energy, E, is a state function.** Any state function depends only on the present state of the system and not on the path by which the system arrived at that state.

▼ Go Figure

If the battery is defined as the system, what is the sign on *w* in part (**b**)?

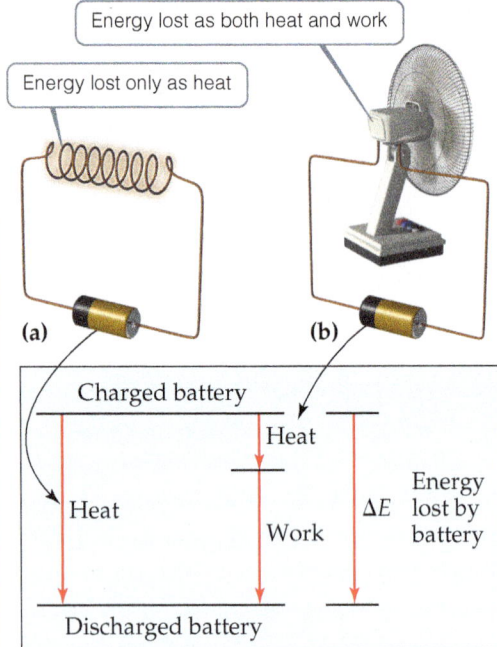

▲ **Figure 5.10 Internal energy is a state function, but heat and work are not.** (**a**) A battery shorted out by a wire loses energy to the surroundings only as heat; no work is performed. (**b**) A battery discharged through a motor loses energy as work (to make the fan turn) and also loses some energy as heat. The value of ΔE is the same for both processes, even though the values of *q* and *w* in (**a**) are different from those in (**b**).

released to the surroundings. In **Figure 5.10**, we consider two possible ways of discharging the battery. In Figure 5.10(a), a wire shorts out the battery, and no work is accomplished because nothing is moved against a force. All the energy lost from the battery is in the form of heat. (The wire gets warmer and releases heat to the surroundings.) In Figure 5.10(b), the battery is used to make a motor turn and the discharge produces work. Some heat is released, but not as much as when the battery is shorted out. The magnitudes of *q* and *w* must be different for these two cases. If the initial and final states of the battery are identical in the two cases, however, then $\Delta E = q + w$ must be the same in both cases because *E* is a state function. *Remember: ΔE depends only on the initial and final states of the system, not on the specific path taken from the initial state to the final state.*

▲ Self-Assessment Exercises

SAE 5.2 The internal energy for the initial and final states of a system are shown in the figure. Which of the following statements is *true*? (**a**) Heat must have been added to the system. (**b**) Work must have been done on the system. (**c**) The internal energy of the final state is greater than that of the initial state. (**d**) $\Delta E < 0$.

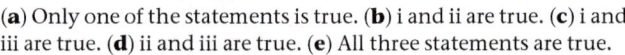

SAE 5.3 Consider a system consisting of a gas enclosed in a stoppered flask, as shown in the accompanying figure. The flask is initially in an ice bath at $T = 0\,°C$. It is then placed in a bath of boiling water at $T = 100\,°C$. Which of the following statements is or are *true*?

(**i**) The gas in the flask is a closed system.
(**ii**) $q > 0$ for this process.
(**iii**) $\Delta E > 0$ for this process.

(**a**) Only one of the statements is true. (**b**) i and ii are true. (**c**) i and iii are true. (**d**) ii and iii are true. (**e**) All three statements are true.

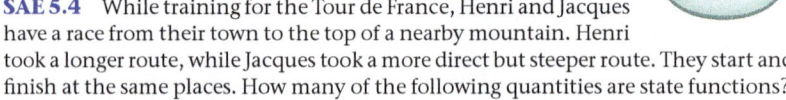

SAE 5.4 While training for the Tour de France, Henri and Jacques have a race from their town to the top of a nearby mountain. Henri took a longer route, while Jacques took a more direct but steeper route. They start and finish at the same places. How many of the following quantities are state functions?

(**i**) The elevation change that Henri experienced
(**ii**) The distance that Jacques traveled
(**iii**) The time it took for Henri to complete his route
(**iv**) The amount of work that Jacques expended during his route
(**v**) The change in temperature between the town and the top of the mountain

(**a**) 1 (**b**) 2 (**c**) 3 (**d**) 4 (**e**) 5

5.3 | Enthalpy

Many of the chemical and physical changes that occur around us, including most of the experiments you will do in the laboratory portion of your course, occur under the essentially constant pressure of Earth's atmosphere.* These changes can result in the release or absorption of heat and can be accompanied by work done by or on the system. In exploring these changes, it is useful to have a state function that relates mainly to heat flow. Under conditions of constant pressure, a thermodynamic quantity called *enthalpy* (from the Greek *enthalpein*, "to warm") provides such a function.

Enthalpy, which we denote by the symbol *H*, is defined as the internal energy plus the product of the *pressure, P*, and *volume, V*, of the system:

$$H = E + PV \qquad [5.5]$$

Like internal energy *E*, both *P* and *V* are state functions—they depend only on the current state of the system and not on the path taken to reach that state. Because energy, pressure, and volume are all state functions, enthalpy is also a state function.

▲ Learning Objectives

When you finish **Section 5.3**, you should be able to:

▶ Relate the thermodynamic quantity called enthalpy to the internal energy (*E*), pressure (*P*), and volume (*V*) of a system.

▶ Define pressure–volume ($P - V$) work and calculate the magnitude and sign of $P - V$ work for a process that occurs at constant pressure.

▶ Interpret the sign of ΔH with respect to whether a process is endothermic or exothermic.

―――――――――
*You may be familiar with the notion of atmospheric pressure from a previous course in chemistry. We discuss it in detail in Chapter 10. Here we need realize only that the atmosphere exerts a nearly constant pressure on the surface of Earth.

Pressure–Volume Work

To better understand the significance of enthalpy, recall from Equation 5.4 that ΔE involves not only the heat q added to or removed from the system but also the work w done by or on the system. Most commonly, the only kind of work produced by chemical or physical changes open to the atmosphere is the mechanical work associated with a change in volume. For example, consider the reaction of zinc metal with hydrochloric acid solution:

$$Zn(s) + 2\,H^+(aq) \rightarrow Zn^{2+}(aq) + H_2(g) \qquad [5.6]$$

When this reaction is run at constant pressure in the apparatus illustrated in **Figure 5.11**, the piston moves up or down to maintain a constant pressure in the vessel. If we assume for simplicity that the piston has no mass, the pressure in the apparatus is the same as atmospheric pressure. As the reaction proceeds, H_2 gas forms, and the piston rises. The gas within the flask is thus doing work on the surroundings by lifting the piston against the force of atmospheric pressure.

The work involved in the expansion or compression of gases is called **pressure–volume work** (P–V work). When the external pressure is constant in a process, as in our preceding example, the sign and magnitude of the pressure–volume work are given by

$$w = -P\Delta V \qquad [5.7]$$

where P is the external pressure and $\Delta V = V_{final} - V_{initial}$ is the change in volume of the system. The pressure P is always either a positive number or zero. If the volume of the system expands, then ΔV is positive as well. The negative sign in Equation 5.7 is necessary to conform to the sign convention for w (Table 5.1). When a gas expands, the system does work on the surroundings, as indicated by a negative value of w. On the other hand, if the gas is compressed, ΔV is negative (the volume decreases), and w is therefore positive according to Equation 5.7, meaning work is done on the system by the surroundings. The box "A Closer Look: Energy, Enthalpy, and P–V Work" discusses pressure–volume work in detail, but all you need to keep in mind for now is Equation 5.7, which applies to processes occurring at constant external pressure.

The units of work obtained by using Equation 5.7 will be those of pressure (usually atm) multiplied by those of volume (usually L). To express the work in the more familiar unit of joules, we use the conversion factor 1 L-atm = 101.3 J.

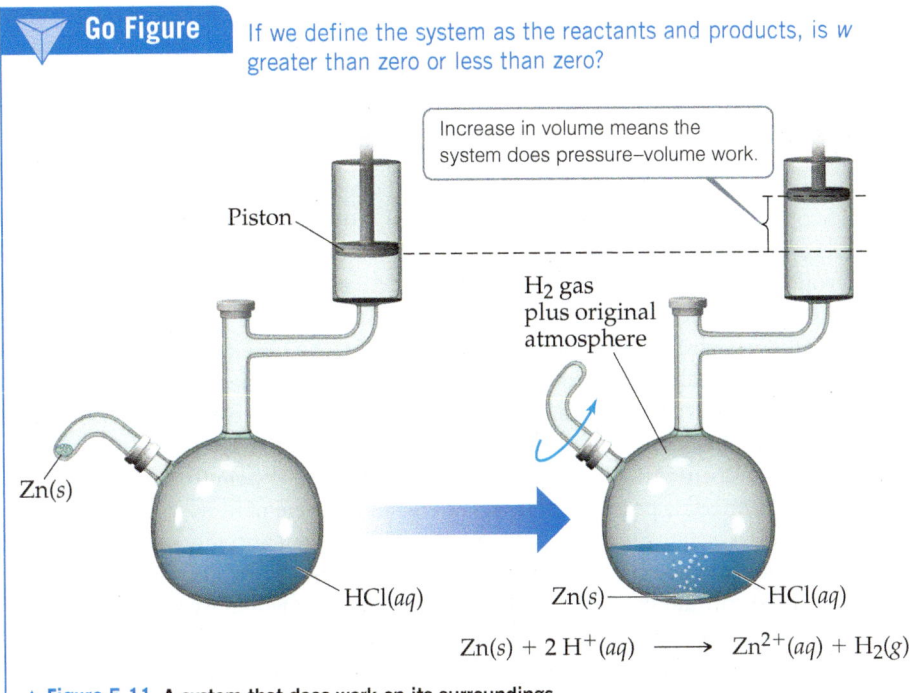

Go Figure If we define the system as the reactants and products, is w greater than zero or less than zero?

Increase in volume means the system does pressure–volume work.

Piston

H_2 gas plus original atmosphere

Zn(s)

HCl(aq)   Zn(s)   HCl(aq)

$$Zn(s) + 2\,H^+(aq) \longrightarrow Zn^{2+}(aq) + H_2(g)$$

▲ **Figure 5.11 A system that does work on its surroundings.**

Sample Exercise 5.2
Calculating Pressure–Volume Work

A fuel is burned in a cylinder equipped with a piston. The initial volume of the cylinder is 0.250 L, and the final volume is 0.980 L. If the piston expands against a constant pressure of 1.35 atm, how much work (in J) is done? (1 L-atm = 101.3 J)

SOLUTION

Analyze We are given an initial volume and a final volume from which we can calculate ΔV. We are also given the pressure, P. We are asked to calculate work, w.

Plan Equation 5.7, $w = -P\Delta V$, allows us to calculate the work done by the system from the given information.

Solve The volume change is

$$\Delta V = V_{final} - V_{initial} = 0.980\,L - 0.250\,L = 0.730\,L$$

Thus, the quantity of work is

$$w = -P\Delta V = -(1.35\,atm)(0.730\,L) = -0.986\,L\text{-atm}$$

Converting L-atm to J, we have

$$(-0.986\,\text{L-atm})\left(\frac{101.3\,J}{1\,\text{L-atm}}\right) = -99.9\,J$$

Check The significant figures are correct (3), and the units are the requested ones for energy (J). The negative sign is consistent with an expanding gas doing work on its surroundings.

▶ **Practice Exercise**
Calculate the work, in J, if the volume of a system contracts from 1.55 to 0.85 L at a constant pressure of 0.985 atm.

Enthalpy Change and Heat

Now let's find out what is so special about enthalpy. When a change occurs at constant pressure, the change in enthalpy, ΔH, is given by the relationship

$$\Delta H = \Delta(E + PV)$$
$$= \Delta E + P\Delta V \quad \text{(constant pressure)} \qquad [5.8]$$

That is, the change in enthalpy equals the change in internal energy plus the product of the constant pressure and the change in volume.

Recall that $\Delta E = q + w$ (Equation 5.4) and that the work involved in the expansion or compression of a gas is $w = -P\Delta V$ (at constant pressure). Substituting $-w$ for $P\Delta V$ and $q + w$ for ΔE into Equation 5.8, we have

$$\Delta H = \Delta E + P\Delta V = (q_P + w) - w = q_P \qquad [5.9]$$

The subscript P on q indicates that the process occurs at constant pressure. Thus, the equations show:

The change in enthalpy equals the heat q_P gained or lost by the system at constant pressure.

Because we can either measure or calculate q_P and because so many physical and chemical changes of interest to us occur at constant pressure, enthalpy is a more useful function for most reactions than is internal energy. In addition, for most common reactions, such as those you will do in lab, the difference in ΔH and ΔE is small because $P\Delta V$ is small compared to the heat involved.

When ΔH is positive (that is, when q_P is positive), the system has gained heat from the surroundings (Table 5.1), which means the process is endothermic. When ΔH is negative, the system has released heat to the surroundings, which means the process is exothermic. **Figure 5.12** continues the bank analogy of Figure 5.7. That is, under constant pressure, an endothermic process deposits energy in the system in the form of heat, whereas an exothermic process withdraws energy in the form of heat.

Because H is a state function, ΔH (which equals q_P) depends only on the initial and final states of the system, not on how the change occurs. At first glance this statement might seem to contradict our discussion in Section 5.2, in which we said that q is *not* a state function. There is no contradiction, however, because the relationship between ΔH and q_P requires that a specific path is taken between the initial and final states—namely, that only P–V work is involved and that the pressure is constant.

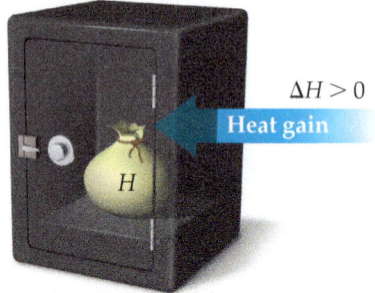

Constant pressure maintained in system

$\Delta H > 0$

Heat gain

H

(a) An endothermic reaction

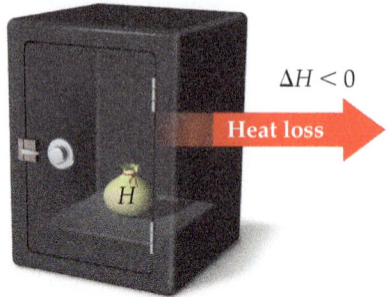

$\Delta H < 0$

Heat loss

H

(b) An exothermic reaction

ΔH is amount of heat that flows into or out of system under constant pressure.

▲ **Figure 5.12 Endothermic and exothermic processes. (a)** An endothermic process ($\Delta H > 0$) deposits heat into the system. **(b)** An exothermic process ($\Delta H < 0$) withdraws heat from the system.

Sample Exercise 5.3
Determining the Sign of ΔH

Indicate the sign of the enthalpy change, ΔH, in the following processes carried out under atmospheric pressure, and indicate whether each process is endothermic or exothermic: **(a)** An ice cube melts; **(b)** 1 g of butane (C_4H_{10}) is completely combusted to CO_2 and H_2O.

SOLUTION

Analyze Our goal is to determine whether ΔH is positive or negative for each process. Because each process occurs at constant pressure, the enthalpy change equals the quantity of heat absorbed or released, $\Delta H = q_P$.

Plan We must predict whether heat is absorbed or released by the system in each process. Processes in which heat is absorbed are endothermic and have a positive sign for ΔH; those in which heat is released are exothermic and have a negative sign for ΔH.

Solve In (a) the water that makes up the ice cube is the system. The ice cube absorbs heat from the surroundings as it melts, so ΔH is positive and the process is endothermic. In (b) the system is the 1 g of butane and the oxygen required to combust it. The combustion of butane in oxygen gives off heat, so ΔH is negative and the process is exothermic.

▶ **Practice Exercise**
Molten gold poured into a mold solidifies at atmospheric pressure. With the gold defined as the system, is the solidification an exothermic or endothermic process?

A CLOSER LOOK | Energy, Enthalpy, and *P–V* Work

In chemistry we are interested mainly in two types of work: electrical work and mechanical work done by expanding gases. We focus here on the latter, called pressure–volume, or *P–V*, work. Expanding gases in the cylinder of an automobile engine do *P–V* work on the piston; this work eventually turns the wheels. Expanding gases from an open reaction vessel do *P–V* work on the atmosphere. This work accomplishes nothing in a practical sense, but when monitoring energy changes in a system, we must keep track of all work, useful or not.

Let's consider a gas confined to a cylinder with a movable piston of cross-sectional area *A* (**Figure 5.13**). A downward force *F* acts on the piston. The pressure, *P*, on the gas is the force per area: $P = F/A$. We assume that the piston is massless and that the only pressure acting on it is the *atmospheric pressure* that is due to Earth's atmosphere, which we assume to be constant.

Suppose the gas expands and the piston moves a distance Δh. From Equation 5.1, the magnitude of the work done by the system is

$$\text{Magnitude of work} = \text{force} \times \text{distance} = F \times \Delta h \quad [5.10]$$

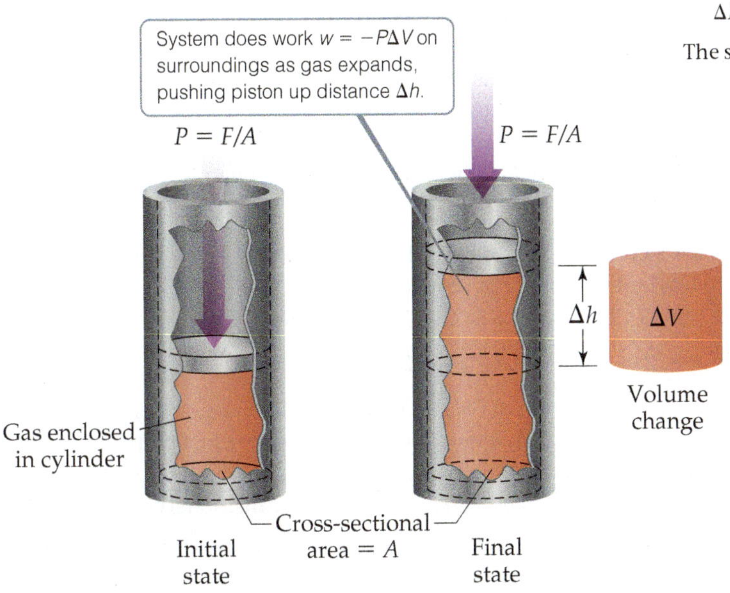

We can rearrange the definition of pressure, $P = F/A$, to $F = P \times A$. The volume change, ΔV, resulting from the movement of the piston is the product of the cross-sectional area of the piston and the distance it moves: $\Delta V = A \times \Delta h$. Substituting into Equation 5.10 gives

$$\text{Magnitude of work} = F \times \Delta h = P \times A \times \Delta h$$
$$= P \times \Delta V$$

Because the system (the confined gas) does work on the surroundings, the work is a negative quantity:

$$w = -P\Delta V \quad [5.11]$$

Now, if *P–V* work is the only work that can be done, we can substitute Equation 5.11 into Equation 5.4 to give

$$\Delta E = q + w = q - P\Delta V \quad [5.12]$$

When a reaction is carried out in a constant-volume container ($\Delta V = 0$), the heat transferred equals the change in internal energy:

$$\Delta E = q - P\Delta V = q - P(0) = q_V \quad \text{(constant volume)} \quad [5.13]$$

The subscript *V* indicates that the volume is constant.

Most reactions are run under constant pressure, so Equation 5.12 becomes

$$\Delta E = q_P - P\Delta V$$
$$q_P = \Delta E + P\Delta V \quad \text{(constant pressure)} \quad [5.14]$$

We see from Equation 5.8 that the right side of Equation 5.14 is the enthalpy change under constant-pressure conditions. Thus, $\Delta H = q_P$, as we saw in Equation 5.9.

In summary, the change in internal energy is equal to the heat gained or lost at constant volume, whereas the change in enthalpy is equal to the heat gained or lost at constant pressure. The difference between ΔE and ΔH is the amount of *P–V* work done by the system when the process occurs at constant pressure, $-P\Delta V$. The volume change accompanying many reactions is close to zero, which makes $P\Delta V$ small. That, in turn, makes the difference between ΔE and ΔH small. Under most circumstances, it is generally satisfactory to use ΔH as the measure of energy changes during most chemical processes.

Related Exercises: 5.35–5.38, 5.107

System does work $w = -P\Delta V$ on surroundings as gas expands, pushing piston up distance Δh.

$P = F/A$

$P = F/A$

Δh

ΔV

Volume change

Gas enclosed in cylinder

Cross-sectional area $= A$

Initial state

Final state

▲ **Figure 5.13 Pressure–volume work.** The amount of work done by the system on the surroundings is $w = -P\Delta V$.

Self-Assessment Exercises

SAE 5.5 Which of the following statements about enthalpy is correct? (**a**) Enthalpy is a short-hand term for internal energy. (**b**) A process for which $\Delta H > 0$ is called exothermic. (**c**) Because enthalpy relates to heat, the enthalpy change for a process depends on the path taken. (**d**) For a process run under constant pressure, the enthalpy change equals the amount of heat transferred into or out of the system.

SAE 5.6 A gas is confined to a piston equipped with a cylinder at constant temperature. Under a constant pressure on the cylinder of 1.60 atm, the gas is compressed from an initial volume of 2.00 L to a final volume of 0.80 L. What are the magnitude and sign of the work done during this process, in J? (Note: 1 L-atm = 101.3 J) (**a**) 0 J (**b**) −194 J (**c**) −130 J (**d**) +130 J (**e**) +194 J

SAE 5.7 When a system that consists of a block of dry ice, $CO_2(s)$, is placed on a wooden board at room temperature and under the constant pressure of the atmosphere, it sublimes into $CO_2(g)$. This process is _____ and has a _____ value of ΔH. (**a**) endothermic, positive (**b**) endothermic, negative (**c**) exothermic, positive (**d**) exothermic, negative

5.4 | Enthalpies of Reaction

In this section, we begin to explore one of the most common ways in which chemists use enthalpy—examining the enthalpy change that occurs during a chemical reaction.

Recall that $\Delta H = H_{final} - H_{initial}$. Hence, the enthalpy change for a chemical reaction is given by

$$\Delta H = H_{products} - H_{reactants} \quad\quad [5.15]$$

The enthalpy change that accompanies a reaction is called either the **enthalpy of reaction** or the *heat of reaction* and is sometimes written ΔH_{rxn}, where "rxn" is a commonly used abbreviation for "reaction."

When we give a numerical value for ΔH_{rxn}, we must specify the reaction involved. For example, when 2 mol $H_2(g)$ burn to form 2 mol $H_2O(g)$ at a constant pressure, the system releases 483.6 kJ of heat. We can summarize this information as

$$2\,H_2(g) + O_2(g) \longrightarrow 2\,H_2O(g) \quad \Delta H = -483.6\,kJ \quad [5.16]$$

The negative sign for ΔH tells us that this reaction is exothermic. Notice that ΔH is reported at the end of the balanced equation, without explicitly specifying the amounts of chemicals involved. In such cases, *the coefficients in the balanced equation represent the number of moles of reactants and products* producing the associated enthalpy change. Balanced chemical equations that show the associated enthalpy change in this way are called *thermochemical equations*.

The exothermic nature of this reaction is also shown in the *enthalpy diagram* in **Figure 5.14**. Notice that the enthalpy of the reactants is greater (more positive) than the enthalpy of the products. Thus, $\Delta H = H_{products} - H_{reactants}$ is negative.

The following guidelines are helpful when using thermochemical equations and enthalpy diagrams:

1. **Enthalpy is an extensive property.** The magnitude of ΔH is proportional to the amount of reactant consumed in the process. For example, 890 kJ of heat is produced when 1 mol of CH_4 is burned in a constant-pressure system:

$$CH_4(g) + 2\,O_2(g) \longrightarrow CO_2(g) + 2\,H_2O(l) \quad \Delta H = -890\,kJ \quad [5.17]$$

Because the combustion of 1 mol of CH_4 with 2 mol of O_2 releases 890 kJ of heat, the combustion of 2 mol of CH_4 with 4 mol of O_2 releases twice as much heat, 1780 kJ. Although chemical equations are usually written with whole-number coefficients, we will see examples of thermochemical equations that utilize fractional coefficients.

2. **The enthalpy change for a reaction is equal in magnitude, but opposite in sign, to ΔH for the reverse reaction.** For example, ΔH for the reverse of Equation 5.17 is +890 kJ:

$$CO_2(g) + 2\,H_2O(l) \longrightarrow CH_4(g) + 2\,O_2(g) \quad \Delta H = +890\,kJ \quad [5.18]$$

Learning Objectives

When you finish Section 5.4, you should be able to:

▶ Interpret the enthalpy of reaction, ΔH, using the enthalpies of the products and the reactants.

▶ Determine how ΔH changes when the equation for a balanced reaction is multiplied by the same number, or when the reaction is reversed.

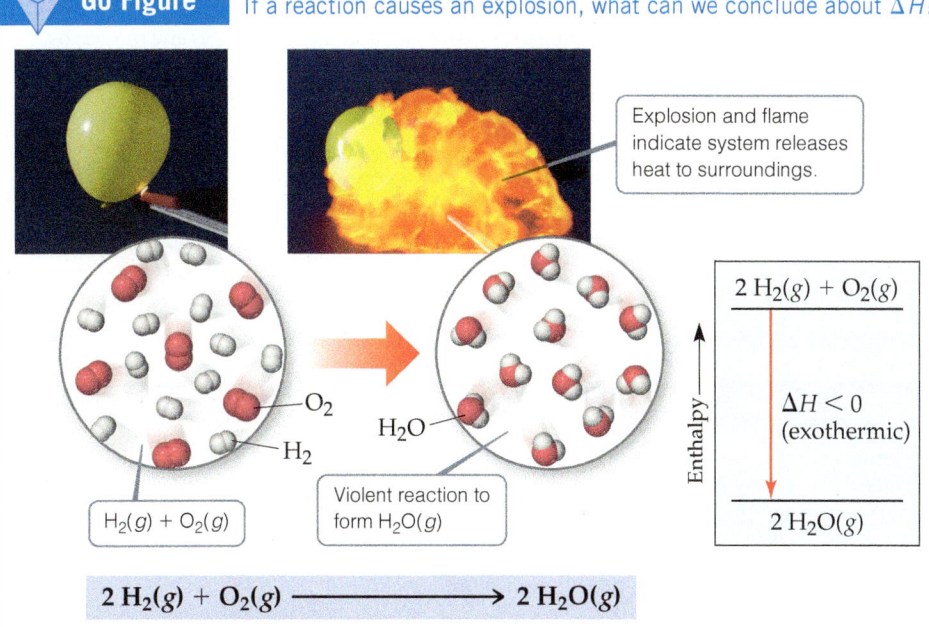

Go Figure   If a reaction causes an explosion, what can we conclude about ΔH?

Explosion and flame indicate system releases heat to surroundings.

$2 H_2(g) + O_2(g)$

$\Delta H < 0$ (exothermic)

$2 H_2O(g)$

O_2

H_2O

H_2

Violent reaction to form $H_2O(g)$

$H_2(g) + O_2(g)$

$$2 H_2(g) + O_2(g) \longrightarrow 2 H_2O(g)$$

▲ **Figure 5.14 Exothermic reaction of hydrogen with oxygen.** When a mixture of $H_2(g)$ and $O_2(g)$ is ignited to form $H_2O(g)$, the resulting explosion produces a ball of flame. Because the system releases heat to the surroundings, the reaction is exothermic, as indicated in the enthalpy diagram.

When we reverse a reaction, we reverse the roles of the products and the reactants. From Equation 5.15, reversing the products and reactants leads to the same magnitude of ΔH, but a change in sign (**Figure 5.15**).

3. **The enthalpy change for a reaction depends on the states of the reactants and products.** If the product in Equation 5.17 were $H_2O(g)$ instead of $H_2O(l)$, ΔH_{rxn} would be -802 kJ instead of -890 kJ. Less heat would be available for transfer to the surroundings because the enthalpy of $H_2O(g)$ is greater than that of $H_2O(l)$. One way to see this is to imagine that the product is initially liquid water. The liquid water must be converted to water vapor, and the conversion of 2 mol $H_2O(l)$ to 2 mol $H_2O(g)$ is an endothermic process that absorbs 88 kJ:

$$2 H_2O(l) \longrightarrow 2 H_2O(g) \qquad \Delta H = +88 \text{ kJ} \qquad [5.19]$$

Thus, it is important to specify the states of the reactants and products in thermochemical equations. In addition, we will generally assume that the reactants and products are both at the same temperature, 25°C, unless otherwise indicated.

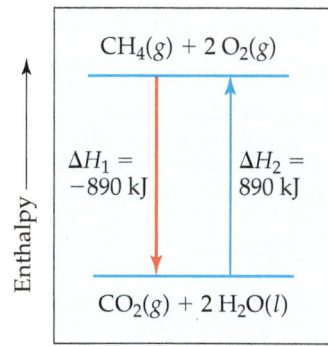

$CH_4(g) + 2 O_2(g)$

$\Delta H_1 = -890$ kJ

$\Delta H_2 = 890$ kJ

$CO_2(g) + 2 H_2O(l)$

▲ **Figure 5.15** ΔH **for the reverse of a reaction.** Reversing a reaction changes the sign but not the magnitude of the enthalpy change: $\Delta H_2 = -\Delta H_1$.

Sample Exercise 5.4

Relating ΔH to Quantities of Reactants and Products

How much heat is released when 4.50 g of methane gas is burned in a constant-pressure system? (Use the information given in Equation 5.17.)

SOLUTION

Analyze Our goal is to use a thermochemical equation to calculate the heat produced when a specific amount of methane gas is combusted. According to Equation 5.17, the system releases 890 kJ when 1 mol CH_4 is burned at constant pressure.

Plan Equation 5.17 provides us with a stoichiometric conversion factor: (1 mol $CH_4 \simeq -890$ kJ). Thus, we can convert moles of CH_4 to kJ of energy. First, however, we must convert grams of CH_4 to moles of CH_4.

Solve

The conversion sequence is:

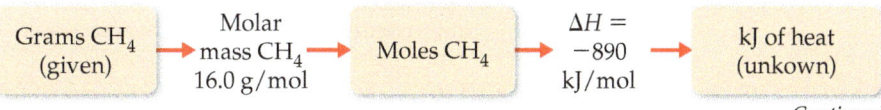

Grams CH_4 (given) → Molar mass CH_4 16.0 g/mol → Moles CH_4 → $\Delta H = -890$ kJ/mol → kJ of heat (unkown)

Continued

By adding the atomic weights of C and 4 H, we have 1 mol $CH_4 = 16.0$ g CH_4. We now use the appropriate conversion factors to convert grams of CH_4 to moles of CH_4 to kilojoules:

$$\text{Heat} = (4.50 \text{ g } \cancel{CH_4})\left(\frac{1 \text{ mol } \cancel{CH_4}}{16.0 \text{ g } \cancel{CH_4}}\right)\left(\frac{-890 \text{ kJ}}{1 \text{ mol } \cancel{CH_4}}\right) = -250 \text{ kJ}$$

The negative sign indicates that the system released 250 kJ into the surroundings.

▶ **Practice Exercise**

Hydrogen peroxide can decompose to water and oxygen by the reaction

$$2 \text{ H}_2\text{O}_2(l) \longrightarrow 2 \text{ H}_2\text{O}(l) + \text{O}_2(g) \qquad \Delta H = -196 \text{ kJ}$$

Calculate the quantity of heat released when 5.00 g of $H_2O_2(l)$ decomposes at constant pressure.

A CLOSER LOOK   Using Enthalpy as a Guide

If you hold a brick in the air and let it go, you know what happens: It falls as the force of gravity pulls it toward Earth. A process that is thermodynamically favored to happen, such as a brick falling to the ground, is called a *spontaneous* process. A spontaneous process can be either fast or slow; the rate at which processes occur is not governed by thermodynamics.

Chemical processes can be thermodynamically favored, or spontaneous, too. By spontaneous, however, we do not mean that the reaction will form products without any intervention. That can be the case, but often some energy must be imparted to get the process started. Just as it is spontaneous for a rock to roll down the side of a hill once started, so too can it be spontaneous for chemical reactions to proceed once enough energy is imparted to initiate the reaction (**Figure 5.16**). The enthalpy change in a reaction gives one indication as to whether the reaction is likely to be spontaneous. The combustion of $H_2(g)$ and $O_2(g)$, for example, is highly exothermic:

$$\text{H}_2(g) + \tfrac{1}{2}\text{O}_2(g) \longrightarrow \text{H}_2\text{O}(g) \qquad \Delta H = -242 \text{ kJ}$$

Hydrogen gas and oxygen gas can exist together in a volume indefinitely without noticeable reaction occurring. Once the reaction is initiated, however, energy is rapidly transferred from the system (the reactants) to the surroundings as heat, as shown previously in Figure 5.14. The system thus loses enthalpy by transferring the heat to the surroundings. (Recall that the first law of thermodynamics tells us that the total energy of the system plus the surroundings does not change; energy is conserved.)

Enthalpy change is not the only consideration in the spontaneity of reactions, nor is it a foolproof guide. For example, even though ice melting is an endothermic process,

$$\text{H}_2\text{O}(s) \longrightarrow \text{H}_2\text{O}(l) \qquad \Delta H = +6.01 \text{ kJ}$$

it is spontaneous at temperatures above the freezing point of water ($0\,°C$). The reverse process, water freezing, is spontaneous at temperatures below $0\,°C$. Thus, we know that ice at room temperature

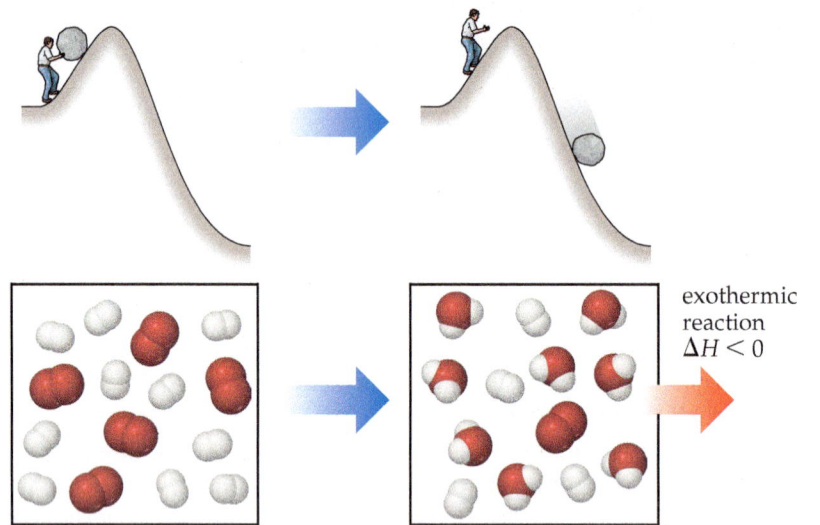

▲ **Figure 5.16 Exothermic reactions and spontaneity.** Chemical reactions that are highly exothermic (ΔH much less than 0) usually occur spontaneously. Such reactions often need some input of energy to get started, just like it can take a push to get a boulder rolling down a hill to a spot of lower potential energy. However, once they get started, such reactions spontaneously proceed to form the products, which have lower potential energy.

melts and that water put into a freezer at $-20\,°C$ turns into ice. Both processes are spontaneous under different conditions, even though they are the reverse of one another. In Chapter 19, we address the spontaneity of processes more fully. There we explain why a process can be spontaneous at one temperature but not at another, as is the case for the conversion of water to ice.

Despite these complicating factors, you should pay attention to the enthalpy changes in reactions. As a general observation, when the enthalpy change is large, it is the dominant factor in determining spontaneity. Thus, reactions for which ΔH is *large* (about 100 kJ or more) and *negative* tend to be spontaneous. Reactions for which ΔH is *large* and *positive* tend to be spontaneous only in the reverse direction.

Related Exercises: 5.47, 5.48

In many situations, we will find it valuable to know the sign and magnitude of the enthalpy change associated with a given chemical process. As we describe in the following sections, ΔH can be either determined directly by experiment or calculated from known enthalpy changes of other reactions.

Self-Assessment Exercises

SAE 5.8 For a hypothetical reaction $X(g) \rightarrow Y(g)$, you are told that the enthalpy of $Y(g)$ is greater than that of $X(g)$. Which of the following statements is *false*? (**a**) ΔH for the reaction is positive. (**b**) On an enthalpy diagram, the line for $Y(g)$ is below the line for $X(g)$. (**c**) The reaction is endothermic. (**d**) The enthalpies of $X(g)$ and $Y(g)$ could both have negative values.

SAE 5.9 The following is the balanced reaction and enthalpy change when one mole of carbon monoxide gas is formed from graphite and oxygen gas under constant pressure:

$$C(s) + {}^1\!/_2\, O_2(g) \longrightarrow CO(g) \quad \Delta H = -110.5 \text{ kJ}$$

What is the enthalpy change for the following reaction:

$$2\,CO(g) \longrightarrow 2\,C(s) + O_2(g) \quad \Delta H = ?$$

(**a**) -221.0 kJ (**b**) -55.25 kJ (**c**) $+55.25$ kJ (**d**) $+221.0$ kJ (**e**) More information is needed.

SAE 5.10 Solid sodium hydroxide dissolves in water to form aqueous sodium cations and hydroxide anions:

$$NaOH(s) \longrightarrow Na^+(aq) + OH^-(aq) \quad \Delta H = -44.5 \text{ kJ}$$

How much heat is released to the solution when 25.0 g of $NaOH(s)$ is dissolved in water under constant pressure? (**a**) 1.78 kJ (**b**) 27.8 kJ (**c**) 44.5 kJ (**d**) 71.2 kJ (**e**) 1110 kJ

5.5 | Calorimetry

The value of ΔH can be determined experimentally by measuring the heat flow accompanying a reaction at constant pressure. Typically, we can determine the magnitude of the heat flow by measuring the magnitude of the temperature change the heat flow produces. The measurement of heat flow is called **calorimetry**, and the device used to measure heat flow is a **calorimeter**.

Heat Capacity and Specific Heat

All substances get hotter as heat is added to them, but the magnitude of the temperature change produced by a given quantity of heat varies from substance to substance. The temperature change experienced by an object when it absorbs a certain amount of heat is determined by its **heat capacity**, denoted C. The heat capacity of an object is the amount of heat required to raise its temperature by 1 K (or 1 °C). The greater the heat capacity, the greater the heat required to produce a given increase in temperature.

For pure substances, the heat capacity is usually given for a specified amount of the substance. The heat capacity of one mole of a substance is called its **molar heat capacity**, C_m. The heat capacity of one gram of a substance is called its *specific heat capacity*, or merely its **specific heat**, C_s. The specific heat of a substance can be determined experimentally by measuring the temperature change, ΔT, that a known mass m of the substance undergoes when it gains or loses a specific quantity of heat q:

$$\text{Specific heat} = \frac{(\text{quantity of heat transferred})}{(\text{grams of substance}) \times (\text{temperature change})}$$

$$C_s = \frac{q}{m \times \Delta T} \qquad [5.20]$$

For example, 209 J is required to increase the temperature of 50.0 g of water by 1.00 K. Thus, the specific heat of water is

$$C_s = \frac{209 \text{ J}}{(50.0 \text{ g})(1.00 \text{ K})} = 4.18 \text{ J/g} - \text{K}$$

Notice how the units combine in the calculation. A temperature change in K is equal in magnitude to the temperature change in °C. (Section 1.5) Therefore, this specific heat for water can also be reported as 4.18 J/g-°C, where the unit is pronounced "Joules per gram-degree Celsius."

 Learning Objectives

When you finish Section 5.5, you should be able to:

▶ Interpret heat capacity and specific heat as measures of the heat needed to change the temperature of a substance.

▶ Use the relationships among C_s, q, m, and ΔT to calculate one of the values given the other three.

▶ Describe how a constant-pressure calorimeter works and analyze the results of constant-pressure calorimetry to determine heats of reactions.

▶ Describe how a constant-volume calorimeter works and analyze the results of constant-volume calorimetry to determine heats of reactions.

Go Figure

Is the process shown in the figure endothermic or exothermic?

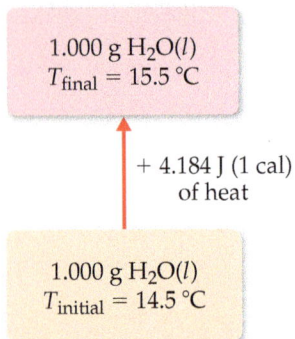

▲ **Figure 5.17** The specific heat of water is 4.184 J (1 cal) at 14.5 °C.

Because the specific heat values for a given substance can vary slightly with temperature, the temperature is often precisely specified. The 4.18 J/g-K value we use here for water, for instance, is for water initially at 14.5 °C (**Figure 5.17**). Water's specific heat at this temperature is used to define the calorie at the value given in Section 1.5: 1 cal = 4.184 J exactly.

When a sample absorbs heat (positive q), its temperature increases (positive ΔT). Rearranging Equation 5.20, we get

$$q = C_s \times m \times \Delta T \qquad [5.21]$$

Thus, we can calculate the quantity of heat a substance gains or loses by using its specific heat together with its measured mass and temperature change.

Table 5.2 lists the specific heats of several substances. Notice that the specific heat of liquid water is higher than the specific heats of the other substances listed. The high specific heat of water affects Earth's climate because it makes the temperatures of the oceans relatively resistant to change.

TABLE 5.2 Specific Heats of Some Substances at 298 K

Elements		Compounds	
Substance	Specific Heat (J/g-K)	Substance	Specific Heat (J/g-K)
$N_2(g)$	1.04	$H_2O(l)$	4.18
$Al(s)$	0.90	$CH_4(g)$	2.20
$Fe(s)$	0.45	$CO_2(g)$	0.84
$Hg(l)$	0.14	$CaCO_3(s)$	0.82

 Sample Exercise 5.5

Relating Heat, Temperature Change, and Heat Capacity

(**a**) How much heat is needed to warm 250 g of water (about 1 cup) from 22 °C (about room temperature) to 98 °C (near its boiling point)? (**b**) What is the molar heat capacity of water?

SOLUTION

Analyze In part (**a**) we must find the quantity of heat (q) needed to warm the water, given the mass of water (m), its temperature change (ΔT), and its specific heat C_s. In part (**b**) we must calculate the molar heat capacity (heat capacity per mole, C_m) of water from its specific heat (heat capacity per gram).

Plan (**a**) Given C_s, m, and ΔT, we can calculate the quantity of heat, q, using Equation 5.21. (**b**) We can use the molar mass of water and dimensional analysis to convert from heat capacity per gram to heat capacity per mole.

Solve

(**a**) The water undergoes a temperature change of:

$$\Delta T = 98°C - 22°C = 76°C = 76 \text{ K}$$

Using Equation 5.21, we have:

$$q = C_s \times m \times \Delta T$$

$$= (4.18 \text{ J/g-K})(250 \text{ g})(76 \text{ K}) = 7.9 \times 10^4 \text{ J}$$

(**b**) The molar heat capacity is the heat capacity of one mole of substance. Using the atomic weights of hydrogen and oxygen, we have:

$$1 \text{ mol } H_2O = 18.0 \text{ g } H_2O$$

From the specific heat used in part (**a**), we have:

$$C_m = \left(4.18 \frac{J}{g\text{-K}}\right)\left(\frac{18.0 \text{ g}}{1 \text{ mol}}\right) = 75.2 \text{ J/mol-K}$$

▶ **Practice Exercise**
(**a**) Large beds of rocks are used in some solar-heated homes to store heat. Assume that the specific heat of the rocks is 0.82 J/g-K. Calculate the quantity of heat absorbed by 50.0 kg of rocks if their temperature increases by 12.0 °C. (**b**) What temperature change would these rocks undergo if they emitted 450 kJ of heat?

Constant-Pressure Calorimetry

The techniques and equipment used in calorimetry depend on the nature of the process being studied. For many reactions, such as those occurring in solution, it is easy to control pressure so that ΔH is measured directly. A simple "coffee-cup" calorimeter (**Figure 5.18**) is often used in general chemistry laboratories to illustrate the principles of calorimetry.

Because the calorimeter is not sealed, the reaction occurs under the essentially constant pressure of the atmosphere.

Imagine adding two aqueous solutions, each containing a reactant, to a coffee-cup calorimeter. Once mixed, a reaction occurs. In this case, there is no physical boundary between the system and the surroundings. The reactants and products of the reaction are the system, and the water in which they are dissolved is part of the surroundings. (The calorimeter apparatus is also part of the surroundings.) If we assume that the calorimeter is perfectly insulated, then any heat released or absorbed by the reaction will raise or lower the temperature of the water in the solution. Thus, we measure the temperature change of the solution and assume that any changes are due to heat transferred from the reaction to the water (for an exothermic process) or transferred from the water to the reaction (for an endothermic process). In other words, by monitoring the temperature of the solution, we are seeing the flow of heat between the system (the reactants and products in the solution) and the surroundings (the water that forms the bulk of the solution).

For an exothermic reaction, heat is "lost" by the reaction and "gained" by the water in the solution, so the temperature of the solution rises. The opposite occurs for an endothermic reaction: Heat is gained by the reaction and lost by the water in the solution, so the temperature of the solution decreases. The heat gained or lost by the solution, q_{soln}, is therefore equal in magnitude but opposite in sign to the heat absorbed or released by the reaction, q_{rxn}: $q_{soln} = -q_{rxn}$. The value of q_{soln} is readily calculated from the mass of the solution, its specific heat, and the temperature change:

$$q_{soln} = (\text{specific heat of solution}) \times (\text{grams of solution}) \times \Delta T = -q_{rxn} \quad [5.22]$$

For dilute aqueous solutions, the specific heat of the solution is nearly the same as that of pure water, so it is normally safe to assume that the specific heat of the solution is 4.18 J/g-K.

Equation 5.22 makes it possible to calculate q_{rxn} from the temperature change of the solution in which the reaction occurs. A temperature increase ($\Delta T > 0$) means the reaction is exothermic ($q_{rxn} < 0$).

Go Figure

Why do you think two Styrofoam® cups are often used instead of just one?

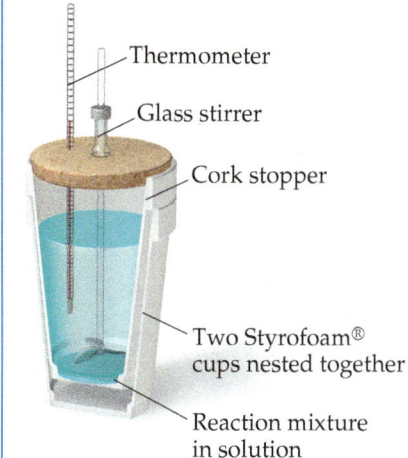

Thermometer

Glass stirrer

Cork stopper

Two Styrofoam® cups nested together

Reaction mixture in solution

▲ **Figure 5.18 Coffee-cup calorimeter.** This simple apparatus is used to measure temperature changes of reactions at constant pressure.

Sample Exercise 5.6

Measuring ΔH Using a Coffee-Cup Calorimeter

When a student mixes 50.0 mL of 1.0 M HCl and 50.0 mL of 1.0 M NaOH in a coffee-cup calorimeter, the temperature of the resultant solution increases from 21.0 to 27.5°C. Calculate the enthalpy change for the reaction in kJ per mole of HCl (kJ/mol HCl), assuming that the calorimeter loses only a negligible quantity of heat, that the total volume of the solution is 0.10 L, that its density is 1.0 g/mL, and that its specific heat is 4.18 J/g-K.

SOLUTION

Analyze Mixing solutions of HCl and NaOH results in an acid–base reaction:

$$HCl(aq) + NaOH(aq) \longrightarrow H_2O(l) + NaCl(aq)$$

We need to calculate the heat produced per mole of HCl, given the temperature increase of the solution, the number of moles of HCl and NaOH involved, and the density and specific heat of the solution.

Plan The total heat produced can be calculated using Equation 5.22. The number of moles of HCl consumed in the reaction must be calculated from the volume and molarity of this substance, and this amount is then used to determine the heat produced per mol HCl.

Solve

Because the total volume of the solution is 0.10 L (or 100 mL), its mass is:	$(100 \text{ mL})(1.0 \text{ g/mL}) = 100 \text{ g}$ (two significant figures)
The temperature change is:	$\Delta T = 27.5°C - 21.0°C = 6.5°C = 6.5 \text{ K}$
Using Equation 5.22, we have:	$q_{rxn} = -C_s \times m \times \Delta T$ $= -(4.18 \text{ J/g-K})(100 \text{ g})(6.5 \text{ K}) = -2.7 \times 10^3 \text{ J} = -2.7 \text{ kJ}$
Because the process occurs at constant pressure, we denote q_{rxn} as q_P:	$\Delta H = q_P = -2.7 \text{ kJ}$
To express the enthalpy change on a molar basis, we use the fact that the number of moles of HCl is given by the product of the volume (50.0 mL = 00500 L) and concentration (1.0 M = 1.0 mol/L) of the HCl solution:	$(0.0500 \text{ L})(1.0 \text{ mol/L}) = 0.050 \text{ mol}$
Thus, the enthalpy change per mole of HCl is:	$\Delta H = -2.7 \text{ kJ}/0.050 \text{ mol} = -54 \text{ kJ/mol}$

Continued

Check ΔH is negative (exothermic), consistent with the observed increase in the temperature. The magnitude of the molar enthalpy change seems reasonable.

▶ **Practice Exercise**

When 50.0 mL of 0.100 M AgNO$_3$ and 50.0 mL of 0.100 M HCl are mixed in a constant-pressure calorimeter, the temperature of the mixture increases from 22.30 to 23.11°C. The temperature increase is caused by the following reaction:

$$AgNO_3(aq) + HCl(aq) \longrightarrow AgCl(s) + HNO_3(aq)$$

Calculate ΔH for this reaction in kJ/mol AgNO$_3$, assuming that the combined solution has a mass of 100.0 g and a specific heat of 4.18 J/g-°C.

Bomb Calorimetry (Constant-Volume Calorimetry)

An important type of reaction studied using calorimetry is combustion, in which a compound reacts completely with excess oxygen. (Section 3.2) Combustion reactions are most accurately studied using a **bomb calorimeter** (**Figure 5.19**). The substance to be studied is placed in a small cup within an insulated sealed vessel called a *bomb*. The bomb, which is designed to withstand high pressures, has an inlet valve for adding oxygen and electrical leads for initiating the reaction. After the sample has been placed in the bomb, the bomb is sealed and pressurized with oxygen. It is then placed in the calorimeter and covered with an accurately measured quantity of water. The combustion reaction is initiated by passing an electrical current through a fine wire in contact with the sample. When the wire becomes sufficiently hot, the sample ignites.

The heat released when combustion occurs is absorbed by the water and the various components of the calorimeter (which all together make up the surroundings), causing the water temperature to rise. The change in water temperature caused by the reaction is measured very precisely.

To calculate the heat of combustion from the measured temperature increase, we must know the total heat capacity of the calorimeter, C_{cal}, which typically has the units kJ/°C. This quantity is determined by combusting a sample that releases a known quantity of heat and measuring the temperature change. For example, combustion of exactly 1 g of benzoic acid, C$_6$H$_5$COOH, in a bomb calorimeter produces 26.38 kJ of heat. Suppose 1.000 g of benzoic acid is combusted in a calorimeter, leading to a temperature increase of 4.857°C. The heat capacity of the calorimeter is then C_{cal} = 26.38 kJ/4.857°C = 5.431 kJ/°C. Once we know C_{cal}, we can measure temperature changes produced by other reactions, and from these we can calculate the heat evolved in the reaction, q_{rxn}:

$$q_{rxn} = -C_{cal} \times \Delta T \qquad [5.23]$$

Because reactions in a bomb calorimeter are carried out at constant volume, the heat transferred corresponds to the change in internal energy, ΔE, rather than the change in enthalpy, ΔH (Equation 5.13). For most reactions, however, the difference between ΔE and ΔH is very small. For the reaction discussed in Sample Exercise 5.7, for example, the difference between ΔE and ΔH is about 1 kJ/mol—a difference of less than 0.1%. It is possible to calculate ΔH from ΔE, but we need not concern ourselves with how these small corrections are made.

Go Figure

Is the water surrounding the reaction chamber part of the system or the surroundings?

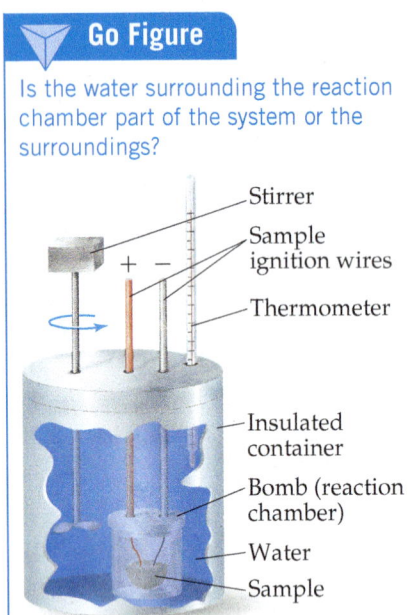

- Stirrer
- Sample ignition wires
- Thermometer
- Insulated container
- Bomb (reaction chamber)
- Water
- Sample

▲ **Figure 5.19** Bomb calorimeter.

Sample Exercise 5.7

Measuring q_{rxn} Using a Bomb Calorimeter

The combustion of methylhydrazine (CH$_6$N$_2$), a liquid rocket fuel, produces N$_2$(g), CO$_2$(g), and H$_2$O(l):

$$2\,CH_6N_2(l) + 5\,O_2(g) \longrightarrow 2\,N_2(g) + 2\,CO_2(g) + 6\,H_2O(l)$$

When 4.00 g of methylhydrazine is combusted in a bomb calorimeter, the temperature of the calorimeter increases from 25.00 to 39.50°C. In a separate experiment, the heat capacity of the calorimeter is measured to be 7.794 kJ/°C. Calculate the molar heat of reaction for the combustion of CH$_6$N$_2$.

SOLUTION

Analyze We are given ΔT, the heat capacity of the calorimeter (C_{cal}), and the amount of reactant combusted. Our goal is to calculate the enthalpy change per mole of methylhydrazine combusted.

Plan We will first calculate the heat evolved for the combustion of the 4.00-g sample. We will then convert this heat to a molar quantity.

Solve For combustion of the 4.00-g sample of methylhydrazine, the temperature change of the calorimeter is:

$$\Delta T = (39.50°C - 25.00°C) = 14.50°C$$

We can use ΔT and the value for C_{cal} to calculate the heat of reaction (Equation 5.23):

$$q_{rxn} = -C_{cal} \times \Delta T = -(7.794 \text{ kJ}/°C)(14.50°C) = -113.0 \text{ kJ}$$

We can readily convert this value to the heat of reaction for a mole of CH_6N_2:

$$\left(\frac{-113.0 \text{ kJ}}{4.00 \text{ g } CH_6N_2}\right) \times \left(\frac{46.1 \text{ g } CH_6N_2}{1 \text{ mol } CH_6N_2}\right) = -1.30 \times 10^3 \text{ kJ/mol } CH_6N_2$$

Check The units cancel properly, and the sign of the answer is negative, as it should be for an exothermic reaction. The magnitude of the answer seems reasonable.

▶ **Practice Exercise**

A 0.5865-g sample of lactic acid ($HC_3H_5O_3$) reacts with oxygen in a calorimeter whose heat capacity is 4.812 kJ/°C. The temperature increases from 23.10 to 24.95°C. Calculate the heat of combustion of lactic acid **(a)** per gram and **(b)** per mole.

CHEMISTRY AND LIFE   The Regulation of Body Temperature

Are you running a fever? For most of us, being asked that question was one of our first introductions to medical diagnosis. Indeed, a deviation in body temperature of only a few degrees indicates something is amiss. Maintaining a near-constant temperature is one of the primary physiological functions of the human body.

To understand how the body's heating and cooling mechanisms operate, we can view the body as a thermodynamic system. The body increases its internal energy content by ingesting foods from the surroundings. The foods, such as glucose ($C_6H_{12}O_6$), are metabolized—a process that is essentially controlled oxidation to CO_2 and H_2O:

$$C_6H_{12}O_6(s) + 6 O_2(g) \longrightarrow 6 CO_2(g) + 6 H_2O(l)$$
$$\Delta H = -2803 \text{ kJ}$$

Roughly 40% of the energy produced is ultimately used to do work in the form of muscle contractions and nerve cell activities. The remainder is released as heat, part of which is used to maintain body temperature. When the body produces too much heat, as in times of heavy physical exertion, it dissipates the excess to the surroundings.

Heat is transferred from the body to its surroundings primarily by *radiation*, *convection*, and *evaporation*. Radiation is the direct loss of heat from the body to cooler surroundings, much as a hot stovetop radiates heat to its surroundings. Convection is heat loss by virtue of heating air that is in contact with the body. The heated air rises and is replaced with cooler air, and the process continues. Warm clothing reduces convective heat loss in cold weather. Evaporative cooling occurs when perspiration is generated at the skin surface by the sweat glands (**Figure 5.20**). Heat is removed from the body as the perspiration evaporates. Perspiration is predominantly water, so the process is the endothermic conversion of liquid water into water vapor:

$$H_2O(l) \longrightarrow H_2O(g) \qquad \Delta H = +44.0 \text{ kJ}$$

The speed with which evaporative cooling occurs decreases as the atmospheric humidity increases, which is why we feel more sweaty and uncomfortable on hot, humid days.

When body temperature becomes too high, heat loss increases in two principal ways. First, blood flow near the skin surface increases, which allows for

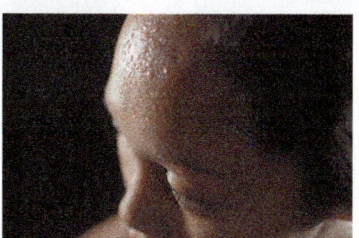
▲ **Figure 5.20** Perspiration, a mechanism to cool the human body.

increased radiational and convective cooling. The reddish, "flushed" appearance of a hot individual is due to this increased blood flow. Second, we sweat, which increases evaporative cooling. During extreme activity, the amount of perspiration can be as high as 2 to 4 liters per hour. As a result, the body's water supply must be replenished during these periods. If the body loses too much liquid through perspiration, it will no longer be able to cool itself and blood volume will decrease, which can lead to either *heat exhaustion* or the more serious *heat stroke*. However, replenishing water without replenishing the electrolytes lost during perspiration can also lead to serious problems. If the normal blood sodium level drops too low, dizziness and confusion set in, and the condition (called *hyponatremia*) can become critical. Drinking a sport drink that contains some electrolytes helps to prevent this problem.

When body temperature drops too low, blood flow to the skin surface decreases, thereby decreasing heat loss. The lower temperature also triggers small involuntary contractions of the muscles (shivering); the biochemical reactions that generate the energy to do this work also generate heat for the body. If the body is unable to maintain a normal temperature, the very dangerous condition called *hypothermia* can result.

▲ Self-Assessment Exercises

SAE 5.11 The specific heat of the metallic element rhodium is 0.240 J/g-K at 25 °C. If 586 J of heat is added to an 80.0 g block of rhodium metal that is initially at 25.0 °C, what is the expected final temperature of the block? **(a)** 26.8 °C **(b)** 30.5 °C **(c)** 32.3 °C **(d)** 55.5 °C **(e)** 2470 °C

SAE 5.12 When 20.0 g of aqueous solution B(*aq*) at 22.0 °C is added to 60.0 g of aqueous solution A(*aq*), also at 22.0 °C, in a coffee-cup calorimeter (Figure 5.18), the two substances react and the temperature of the mixed solution rises to 34.3 °C. What is the quantity of heat, in kJ, released during this reaction? Assume that the specific heat of the mixed solution is the same as that of pure water (4.18 J/g-K). **(a)** 3.08 kJ **(b)** 4.11 kJ **(c)** 7.36 kJ **(d)** 11.5 kJ

Continued

SAE 5.13 When a 4.25-g sample of solid sodium nitrate dissolves in 50.0 g of water in a coffee-cup calorimeter (Figure 5.18), the temperature rises from 24.00 °C to 28.78 °C. What is ΔH (in kJ/mol NaNO$_3$) for the dissolution of sodium nitrate:

$$NaNO_3(s) \longrightarrow Na^+(aq) + NO_3^-(aq)$$

Assume that the specific heat of the solution is the same as that of pure water (4.18 J/g-K). **(a)** −1.08 kJ/mol **(b)** −1.24 kJ/mol **(c)** −5.19 kJ/mol **(d)** −20.0 kJ/mol **(e)** −21.7 kJ/mol

SAE 5.14 A 3.50-g sample of naphthalene (C$_{10}$H$_8$) is burned in a bomb calorimeter whose total heat capacity is 9.46 kJ/°C. The temperature of the calorimeter increases from 24.1 to 39.0 °C. What is the heat of combustion per gram of naphthalene? **(a)** 40.3 kJ/g **(b)** 105 kJ/g **(c)** 141 kJ/g **(d)** 5160 kJ/g

5.6 | Hess's Law

In this section, we show that it is often possible to calculate the ΔH for a reaction from the tabulated ΔH values of other reactions. Thus, it is not necessary to make calorimetric measurements for all reactions.

Because enthalpy is a state function, the enthalpy change, ΔH, associated with any chemical process depends only on the amount of matter that undergoes change and on the nature of the initial state of the reactants and the final state of the products. This means that regardless of whether a particular reaction is carried out in one step or in a series of steps, the sum of the enthalpy changes associated with the individual steps must be the same as the enthalpy change associated with the one-step process. As an example, combustion of methane gas, CH$_4(g)$, to form CO$_2(g)$ and H$_2$O(l) can be thought of as occurring in one step, as represented on the left in **Figure 5.21**, or in two steps, as represented on the right in Figure 5.21: (1) combustion of CH$_4(g)$ to form CO$_2(g)$ and H$_2$O(g) and (2) condensation of H$_2$O(g) to form H$_2$O(l). The enthalpy change for the overall process is the sum of the enthalpy changes for these two steps:

$$CH_4(g) + 2\,O_2(g) \longrightarrow CO_2(g) + 2\,H_2O(g) \qquad \Delta H = -802\text{ kJ}$$

(Add) $\qquad 2\,H_2O(g) \longrightarrow 2\,H_2O(l) \qquad \Delta H = -88\text{ kJ}$

$$\overline{CH_4(g) + 2\,O_2(g) + 2\,H_2O(g) \longrightarrow CO_2(g) + 2\,H_2O(l) + 2\,H_2O(g) \quad \Delta H = -890\text{ kJ}}$$

The net equation is

$$CH_4(g) + 2\,O_2(g) \longrightarrow CO_2(g) + 2\,H_2O(l) \quad \Delta H = -890\text{ kJ}$$

Hess's law states that *if a reaction is carried out in a series of steps, ΔH for the overall reaction equals the sum of the enthalpy changes for the individual steps.* The overall enthalpy change for the process is independent of both the number of steps and the path by which the reaction is carried out. This law is a consequence of the fact that enthalpy is a state function. We can therefore calculate ΔH for any process as long as we find a route for which ΔH is known for each step. This means that a relatively small number of experimental measurements can be used to calculate ΔH for a vast number of reactions.

Hess's law provides a useful means of calculating energy changes that are difficult to measure directly. For instance, we can use the enthalpy changes for combustion reactions, which are fairly straightforward to measure, to determine the enthalpy change for a reaction that does not involve combustion. An example of this process is given in Sample Exercise 5.8, where we use combustion reactions to determine the enthalpy change when C(s) reacts with H$_2(g)$.

The key point of these examples is that H is a state function.

Because H is a state function, for a particular set of reactants and products, ΔH is the same whether the reaction takes place in one step or in a series of steps.

Learning Objectives

When you finish Section 5.6, you should be able to:

▶ Use Hess's law to determine ΔH for a reaction that is written as a series of steps.

Go Figure

What process corresponds to the −88 kJ enthalpy change?

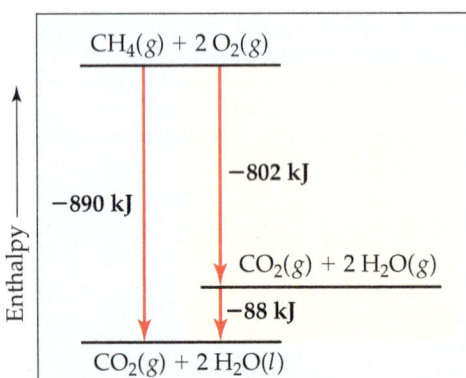

▲ **Figure 5.21 Enthalpy diagram for combustion of one mole of methane.** The enthalpy change of the one-step reaction equals the sum of the enthalpy changes of the reaction run in two steps: −890 kJ = −802 kJ + (−88 kJ).

Sample Exercise 5.8

Using Hess's Law with Three Equations to Calculate ΔH

Calculate ΔH for the reaction

$$2\,C(s) + H_2(g) \longrightarrow C_2H_2(g)$$

given the following chemical equations and their respective enthalpy changes:

$$C_2H_2(g) + \tfrac{5}{2}O_2(g) \longrightarrow 2\,CO_2(g) + H_2O(l) \quad \Delta H = -1299.6\,\text{kJ}$$
$$C(s) + O_2(g) \longrightarrow CO_2(g) \quad\quad\quad\quad\quad \Delta H = -393.5\,\text{kJ}$$
$$H_2(g) + \tfrac{1}{2}O_2(g) \longrightarrow H_2O(l) \quad\quad\quad\quad \Delta H = -285.8\,\text{kJ}$$

SOLUTION

Analyze We are given a chemical equation and asked to calculate its ΔH using three chemical equations and their associated enthalpy changes.

Plan We will use Hess's law, summing the three equations or their reverses and multiplying each by an appropriate coefficient so that they add to give the net equation for the reaction of interest. At the same time, we must keep track of the ΔH values, reversing their signs if the reactions are reversed and multiplying them by whatever coefficient is employed in the equation.

Solve Because the target equation has C_2H_2 as a product, we turn the first equation around; the sign of ΔH is therefore changed. The desired equation has $2\,C(s)$ as a reactant, so we multiply the second equation and its ΔH by 2. Because the target equation has H_2 as a reactant, we keep the third equation as it is. We then add the three equations and their enthalpy changes in accordance with Hess's law:

$$2\,CO_2(g) + H_2O(l) \longrightarrow C_2H_2(g) + \tfrac{5}{2}O_2(g) \quad \Delta H = +1299.6\,\text{kJ}$$

$$2\,C(s) + 2\,O_2(g) \longrightarrow 2\,CO_2(g) \quad\quad\quad \Delta H = -787.0\,\text{kJ}$$

$$H_2(g) + \tfrac{1}{2}O_2(g) \longrightarrow H_2O(l) \quad\quad\quad\quad \Delta H = -285.8\,\text{kJ}$$

$$2\,C(s) + H_2(g) \longrightarrow C_2H_2(g) \quad\quad\quad \Delta H = 226.8\,\text{kJ}$$

When the equations are added, the total amounts of CO_2, O_2, and H_2O are the same on both sides of the arrow, so these are canceled when writing the net equation.

Check The procedure must be correct because we obtained the correct net equation. In cases like this, you should go back over the numerical manipulations of the ΔH values to ensure that you did not make an inadvertent error with signs or coefficients.

▶ **Practice Exercise**

Calculate ΔH for the reaction

$$NO(g) + O(g) \longrightarrow NO_2(g)$$

given the following information:

$$NO(g) + O_3(g) \longrightarrow NO_2(g) + O_2(g) \quad \Delta H = -198.9\,\text{kJ}$$
$$O_3(g) \longrightarrow \tfrac{3}{2}O_2(g) \quad\quad\quad\quad\quad\quad \Delta H = -142.3\,\text{kJ}$$
$$O_2(g) \longrightarrow 2\,O(g) \quad\quad\quad\quad\quad\quad\quad \Delta H = 495.0\,\text{kJ}$$

We reinforce this point by giving one more example of an enthalpy diagram and Hess's law. Again we use combustion of methane to form CO_2 and H_2O, our reaction from Figure 5.21. This time we envision a different two-step path, with the initial formation of CO, which is then combusted to CO_2 (**Figure 5.22**). Even though the two-step path is different from that in Figure 5.21, the overall reaction again has $\Delta H_1 = -890$ kJ. Because H is a state function, both paths *must* produce the same value of ΔH. In Figure 5.22, that means $\Delta H_1 = \Delta H_2 + \Delta H_3$. We will soon see that breaking up reactions in this way allows us to derive the enthalpy changes for reactions that are hard to carry out in the laboratory.

 Go Figure Suppose the overall reaction were modified to produce $2\,H_2O(g)$ rather than $2\,H_2O(l)$. Would any of the values of ΔH in the diagram stay the same?

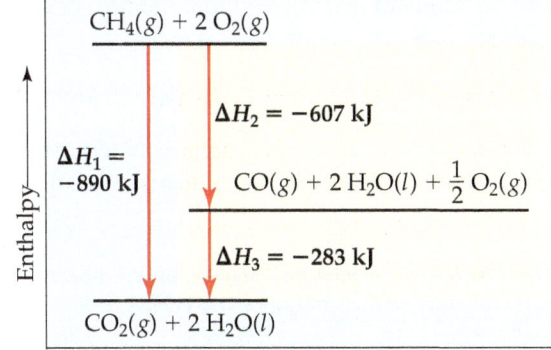

◀ **Figure 5.22 Enthalpy diagram illustrating Hess's law.** The net reaction is the same as in Figure 5.21, but here we imagine different reactions in our two-step version. As long as we can write a series of equations, each with a known value of ΔH that add up to the equation we need, we can calculate the overall ΔH.

Self-Assessment Exercises

SAE 5.15 Consider the following hypothetical reactions:

$$X \longrightarrow Y \quad \Delta H_1$$
$$Z \longrightarrow Y \quad \Delta H_2$$

Which of the following statements about these reactions is or are *true*?

    (i) The enthalpy change for the reaction $Y \longrightarrow X$ is $-\Delta H_1$.
    (ii) The enthalpy change for the reaction $2\,Z \longrightarrow 2\,Y$ is $2(\Delta H_2)$.
    (iii) The enthalpy change for the reaction $X \longrightarrow Z$ is $\Delta H_1 + \Delta H_2$.

(a) Only one of the statements is true. **(b)** i and ii are true. **(c)** i and iii are true. **(d)** ii and iii are true. **(e)** All three statements are true.

SAE 5.16 From the following enthalpies of reaction

$$2\,H_2(g) + O_2(g) \longrightarrow 2\,H_2O(l) \quad \Delta H = -572 \text{ kJ}$$

$$H_2(g) + O_2(g) \longrightarrow H_2O_2(l) \quad \Delta H = -188 \text{ kJ}$$

calculate ΔH for the reaction

$$2\,H_2O_2(l) \longrightarrow 2\,H_2O(l) + O_2(g).$$

(a) -196 kJ **(b)** -384 kJ **(c)** -760 kJ **(d)** -948 kJ

SAE 5.17 Given the following enthalpies of reaction

$$C_2H_4(g) + 3\,O_2(g) \longrightarrow 2\,CO_2(g) + 2\,H_2O(g) \quad \Delta H = -1323 \text{ kJ}$$

$$2\,CO(g) + O_2(g) \longrightarrow 2\,CO_2(g) \qquad\qquad \Delta H = -566 \text{ kJ}$$

$$H_2O(l) \longrightarrow H_2O(g) \qquad\qquad\qquad \Delta H = +44 \text{ kJ}$$

use Hess's law to calculate ΔH for the reaction

$$C_2H_4(g) + 2\,O_2(g) \longrightarrow 2\,CO(g) + 2\,H_2O(l).$$

(a) -757 kJ **(b)** -801 kJ **(c)** -845 kJ **(d)** -1977 kJ

5.7 | Enthalpies of Formation

We can use the methods just discussed to calculate enthalpy changes for a great many reactions from tabulated ΔH values. A particularly important process used for tabulating and using thermochemical data is the formation of a compound from its constituent elements. The enthalpy change associated with this process is called the **enthalpy of formation** (or *heat of formation*), ΔH_f, where the subscript *f* indicates that the substance has been *formed* from its constituent elements.

The magnitude of any enthalpy change depends on the temperature, pressure, and state (gas, liquid, or solid crystalline form) of the reactants and products. To compare enthalpies of different reactions, we must define a set of conditions, called a *standard state*, at which most enthalpies are tabulated. The standard state of a substance is its pure form at atmospheric pressure (1 atm) and the temperature of interest, which we usually choose to be 298 K (25°C).* The **standard enthalpy change** of a reaction is defined as the enthalpy change when all reactants and products are in their standard states. We denote a standard enthalpy change as $\Delta H°$, where the superscript ° indicates standard-state conditions.

The **standard enthalpy of formation** of a compound, $\Delta H_f°$, is the change in enthalpy for the reaction that forms one mole of the compound from its elements with all substances in their standard states:

$$\underset{\text{(in standard state)}}{\text{elements}} \longrightarrow \underset{\text{(1 mol in standard state)}}{\text{compound}} \qquad \Delta H_{rxn} = \Delta H_f° \qquad [5.24]$$

We usually report $\Delta H_f°$ values at 298 K. If an element exists in more than one form under standard conditions, the most stable form of the element is used for the formation reaction. For example, the standard enthalpy of formation for ethanol, C_2H_5OH, is the enthalpy change for the reaction

$$2\,C(graphite) + 3\,H_2(g) + \tfrac{1}{2}O_2(g) \longrightarrow C_2H_5OH(l) \qquad \Delta H_f° = -277.7 \text{ kJ} \quad [5.25]$$

The elemental source of oxygen is O_2, not O or O_3, because O_2 is the stable form of oxygen at 298 K and atmospheric pressure. Similarly, the elemental source of carbon is graphite

---

*The definition of the standard state for gases has been changed to 1 bar (1 atm = 1.013 bar), a slightly lower pressure than 1 atm. For most purposes, this change makes very little difference in the standard enthalpy changes.

Learning Objectives

When you finish Section 5.7, you should be able to:

▶ Describe what is meant by a standard enthalpy of formation, $\Delta H_f°$, of a substance.

▶ Use tabulated enthalpies of formation to calculate the enthalpies of reactions.

TABLE 5.3   Standard Enthalpies of Formation, $\Delta H_f°$, at 298 K

Substance	Formula	$\Delta H_f°$ (kJ/mol)	Substance	Formula	$\Delta H_f°$ (kJ/mol)
Acetylene	$C_2H_2(g)$	227.4	Hydrogen chloride	$HCl(g)$	−92.31
Ammonia	$NH_3(g)$	−45.94	Hydrogen fluoride	$HF(g)$	−273.30
Benzene	$C_6H_6(l)$	49.0	Hydrogen iodide	$HI(g)$	26.50
Calcium carbonate	$CaCO_3(s)$	−1207.1	Methane	$CH_4(g)$	−74.6
Calcium oxide	$CaO(s)$	−635.1	Methanol	$CH_3OH(l)$	−238.4
Carbon dioxide	$CO_2(g)$	−393.5	Propane	$C_3H_8(g)$	−103.85
Carbon monoxide	$CO(g)$	−110.5	Silver chloride	$AgCl(s)$	−127.0
Diamond	$C(s)$	1.88	Sodium bicarbonate	$NaHCO_3(s)$	−947.7
Ethane	$C_2H_6(g)$	−84.68	Sodium carbonate	$Na_2CO_3(s)$	−1130.8
Ethanol	$C_2H_5OH(l)$	−277.0	Sodium chloride	$NaCl(s)$	−411.1
Ethylene	$C_2H_4(g)$	52.4	Sucrose	$C_{12}H_{22}O_{11}(s)$	−2221
Glucose	$C_6H_{12}O_6(s)$	−1273	Water	$H_2O(l)$	−285.8
Hydrogen bromide	$HBr(g)$	−36.29	Water vapor	$H_2O(g)$	−241.8

and not diamond because graphite is the more stable (lower-energy) form at 298 K and atmospheric pressure. Likewise, the most stable form of hydrogen under standard conditions is $H_2(g)$, so this is used as the source of hydrogen in Equation 5.25.

The stoichiometry of formation reactions always indicates that one mole of the desired substance is produced, as in Equation 5.25. As a result, standard enthalpies of formation are reported in kJ/mol of the substance being formed. Some values are given in Table 5.3, and a more extensive table is provided in Appendix C.

By definition, *the standard enthalpy of formation of the most stable form of any element is zero.* Thus, the values of $\Delta H_f°$ for C(*graphite*), $H_2(g)$, $O_2(g)$, and the standard states of other elements are zero by definition.

Sample Exercise 5.9

Equations Associated with Enthalpies of Formation

For which of these reactions at 25°C does the enthalpy change represent a standard enthalpy of formation? For each that does not, what changes are needed to make it an equation whose ΔH is an enthalpy of formation?

(a) $2\,Na(s) + \frac{1}{2}O_2(g) \longrightarrow Na_2O(s)$

(b) $2\,K(l) + Cl_2(g) \longrightarrow 2\,KCl(s)$

(c) $C_6H_{12}O_6(s) \longrightarrow 6\,C(\textit{diamond}) + 6\,H_2(g) + 3\,O_2(g)$

SOLUTION

Analyze The standard enthalpy of formation is represented by a reaction in which each reactant is an element in its standard state and the product is one mole of the compound.

Plan We need to examine each equation to determine (1) whether the reaction is one in which one mole of substance is formed from the elements, and (2) whether the reactant elements are in their standard states.

Solve In (**a**) 1 mol Na_2O is formed from the elements sodium and oxygen in their proper states, solid Na and O_2 gas, respectively. Therefore, the enthalpy change for reaction (**a**) corresponds to a standard enthalpy of formation.

In (**b**) potassium is given as a liquid. It must be changed to the solid form, its standard state at room temperature. Furthermore, 2 mol KCl are formed, so the enthalpy change for the reaction as written is twice the standard enthalpy of formation

of $KCl(s)$. The equation for the formation reaction of 1 mol of $KCl(s)$ is

$$K(s) + \tfrac{1}{2}Cl_2(g) \longrightarrow KCl(s)$$

Reaction (**c**) does not form a substance from its elements. Instead, a substance decomposes to its elements, so this reaction must be reversed. Next, the element carbon is given as diamond, whereas graphite is the standard state of carbon at room temperature and 1 atm pressure. The equation that correctly represents the enthalpy of formation of glucose from its elements is

$$6\,C(\textit{graphite}) + 6\,H_2(g) + 3\,O_2(g) \longrightarrow C_6H_{12}O_6(s)$$

▶ **Practice Exercise**
Write the equation corresponding to the standard enthalpy of formation of liquid carbon tetrachloride (CCl_4) and look up $\Delta H_f°$ for this compound in Appendix C.

Using Enthalpies of Formation to Calculate Enthalpies of Reaction

We can use Hess's law and tabulations of ΔH_f° values, such as those in Table 5.3 and Appendix C, to calculate the standard enthalpy change for any reaction for which we know the ΔH_f° values for all reactants and products. For example, consider the combustion of propane under standard conditions:

$$C_3H_8(g) + 5\,O_2(g) \longrightarrow 3\,CO_2(g) + 4\,H_2O(l)$$

We can write this equation as the sum of three equations associated with standard enthalpies of formation:

$$C_3H_8(g) \longrightarrow 3\,C(s) + 4\,H_2(g) \qquad \Delta H_1 = -(1\,\text{mol}\,C_3H_8)[\Delta H_f^\circ \text{ for } C_3H_8(g)] \qquad [5.26]$$

$$3\,C(s) + 3\,O_2(g) \longrightarrow 3\,CO_2(g) \qquad \Delta H_2 = (3\,\text{mol}\,CO_2)[\Delta H_f^\circ \text{ for } CO_2(g)] \qquad [5.27]$$

$$4\,H_2(g) + 2\,O_2(g) \longrightarrow 4\,H_2O(l) \qquad \Delta H_3 = (4\,\text{mol}\,H_2O)[\Delta H_f^\circ \text{for } H_2O(l)] \qquad [5.28]$$

$$C_3H_8(g) + 5\,O_2(g) \longrightarrow 3\,CO_2(g) + 4\,H_2O(l) \quad \Delta H_{rxn}^\circ = \Delta H_1 + \Delta H_2 + \Delta H_3 \qquad [5.29]$$

It is sometimes useful to add subscripts to the enthalpy changes, as we have done here, to keep track of the associations between reactions and their ΔH values. Also, remember that we don't need to include ΔH_f° for $O_2(g)$ in our sum because it is zero by definition.

We can now use values from Table 5.3 to calculate ΔH_{rxn}°:

$$\Delta H_{rxn}^\circ = \Delta H_1 + \Delta H_2 + \Delta H_3$$

$$= -[1\,\text{mol}\,C_3H_8(g)](-103.85\,\text{kJ/mol}) + [3\,\text{mol}\,CO_2(g)](-393.5\,\text{kJ/mol})$$

$$+[4\,\text{mol}\,H_2O(l)](-285.8\,\text{kJ/mol})$$

$$= -2220\,\text{kJ} \qquad [5.30]$$

The enthalpy diagram in **Figure 5.23** shows the components of this calculation. In Step ❶ the reactants are decomposed into their constituent elements in their standard states. In Steps ❷ and ❸ the products are formed from the elements. Several aspects of how we use enthalpy changes in this process depend on the guidelines we discussed in Section 5.4.

❶ **Decomposition**. Equation 5.26 is the reverse of the formation reaction for $C_3H_8(g)$, so the enthalpy change for this decomposition reaction is the negative of the ΔH_f° value for the formation of 1 mol of propane: $-\Delta H_f^\circ[C_3H_8(g)]$.

❷ **Formation of CO_2**. Equation 5.27 is the formation reaction for 3 mol of $CO_2(g)$. Because enthalpy is an extensive property, the enthalpy change for this step is $3\Delta H_f^\circ[CO_2(g)]$.

❸ **Formation of H_2O**. The enthalpy change for Equation 5.28, the formation of 4 mol of H_2O, is $4\Delta H_f^\circ[H_2O(l)]$. The reaction specifies that $H_2O(l)$ is produced, so be careful to use the value of ΔH_f° for $H_2O(l)$ and not the value for $H_2O(g)$.

Note that in this analysis *the stoichiometric coefficients in the balanced equation represent the number of moles of each substance.* For Equation 5.29, therefore, $\Delta H_{rxn}^\circ = -2220\,\text{kJ}$ represents the enthalpy change for the reaction of 1 mol C_3H_8 and 5 mol O_2 to form 3 mol CO_2 and 4 mol H_2O.

We can break down any reaction into formation reactions as we have done here. When we do, we obtain the general result that the standard enthalpy change of a reaction is the sum of the standard enthalpies of formation of the products minus the standard enthalpies of formation of the reactants:

$$\Delta H_{rxn}^\circ = \Sigma n \Delta H_f^\circ(\text{products}) - \Sigma m \Delta H_f^\circ(\text{reactants}) \qquad [5.31]$$

The symbol Σ (sigma) means "the sum of," and n and m are the stoichiometric coefficients of the relevant chemical equation. The first term on the right in Equation 5.31 represents the formation reactions of the products, which are written in the "forward" direction in the chemical equation—that is, elements reacting to form products. This term is analogous to Equations 5.27 and 5.28. The second term on the right in Equation 5.31 represents the reverse of the formation reactions of the reactants, analogous to Equation 5.26, which is why this term is preceded by a minus sign. When applying

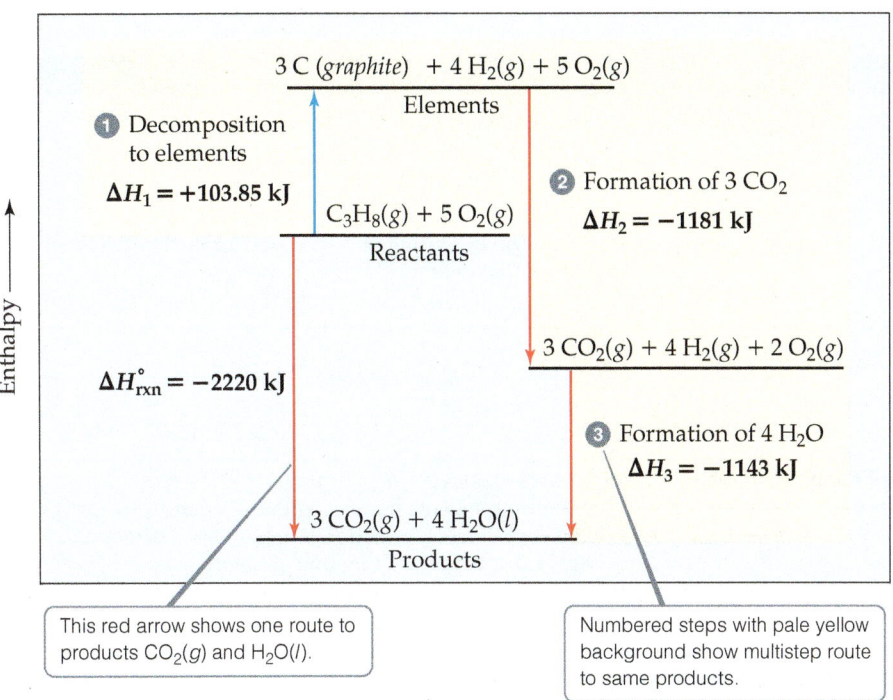

◀ **Figure 5.23 Enthalpy diagram for propane combustion.**

Equation 5.31, remember that ΔH_f° for any element in its most stable form [such as $O_2(g)$ in Equation 5.29] is zero and can therefore be essentially ignored in the summations.

 ## Sample Exercise 5.10
Calculating an Enthalpy of Reaction from Enthalpies of Formation

(a) Calculate the standard enthalpy change for the combustion of one mole of benzene, $C_6H_6(l)$, to $CO_2(g)$ and $H_2O(l)$.

(b) Compare the quantity of heat produced by the combustion of 1.00 g propane with that produced by 1.00 g benzene.

SOLUTION

Analyze (a) We are given a reaction [combustion of $C_6H_6(l)$ to form $CO_2(g)$ and $H_2O(l)$] and asked to calculate its standard enthalpy change, ΔH°. **(b)** We then need to compare the quantity of heat produced by combustion of 1.00 g C_6H_6 with that produced by 1.00 g C_3H_8, whose combustion was treated previously in the text. (See Equations 5.29 and 5.30.)

Plan (a) We first write the balanced equation for the combustion of C_6H_6. We then look up ΔH_f° values in Appendix C or in Table 5.3 and apply Equation 5.31 to calculate the enthalpy change for the reaction. **(b)** We use the molar mass of C_6H_6 to change the enthalpy change per mole to that per gram. We similarly use the molar mass of C_3H_8 and the enthalpy change per mole calculated in the text previously to calculate the enthalpy change per gram of that substance.

Solve

(a) We know that a combustion reaction involves $O_2(g)$ as a reactant. Thus, the balanced equation for the combustion reaction of 1 mol $C_6H_6(l)$ is:

We can calculate ΔH° for this reaction by using Equation 5.31 and data in Table 5.3. Remember to multiply the ΔH_f° value for each substance in the reaction by that substance's stoichiometric coefficient. The standard state of oxygen is $O_2(g)$, so $\Delta H_f^\circ[O_2(g)] = 0$.

$$C_6H_6(l) + \tfrac{15}{2} O_2(g) \longrightarrow 6\,CO_2(g) + 3\,H_2O(l)$$

$$\Delta H_{rxn}^\circ = [6\Delta H_f^\circ(CO_2) + 3\Delta H_f^\circ(H_2O)] - [\Delta H_f^\circ(C_6H_6) + \tfrac{15}{2}\Delta H_f^\circ(O_2)]$$

$$= [6(-393.5\text{ kJ}) + 3(-285.8\text{ kJ})] - [(49.0\text{ kJ}) + \tfrac{15}{2}(0\text{ kJ})]$$

$$= (-2361 - 857.4 - 49.0)\text{ kJ}$$

$$= -3267\text{ kJ}$$

(b) From the example worked in the text, $\Delta H^\circ = -2220$ kJ for the combustion of one mole of propane. In part (a) of this exercise, we determined that $\Delta H^\circ = -3267$ kJ for the combustion of one mole of benzene. To determine the heat of combustion per gram of each substance, we use the molar masses to convert moles to grams:

$C_3H_8(g)$: $(-2220\text{ kJ/mol})(1\text{mol}/44.1\text{ g}) = -50.3\text{ kJ/g}$
$C_6H_6(l)$: $(-3267\text{ kJ/mol})(1\text{mol}/78.1\text{ g}) = -41.8\text{ kJ/g}$

Comment Both propane and benzene are hydrocarbons. As a rule, the energy obtained from the combustion of a gram of hydrocarbon is between 40 and 50 kJ.

▶ **Practice Exercise**
Use Table 5.3 to calculate the enthalpy change for the combustion of 1 mol of ethanol:

$$C_2H_5OH(l) + 3\,O_2(g) \longrightarrow 2\,CO_2(g) + 3\,H_2O(l)$$

Sample Exercise 5.11
Calculating an Enthalpy of Formation Using an Enthalpy of Reaction

The standard enthalpy change for the reaction $CaCO_3(s) \longrightarrow CaO(s) + CO_2(g)$ is 178.1 kJ. Use Table 5.3 to calculate the standard enthalpy of formation of $CaCO_3(s)$.

SOLUTION

Analyze Our goal is to obtain $\Delta H_f°[CaCO_3]$.

Plan We begin by writing the expression for the standard enthalpy change for the reaction:

$$\Delta H°_{rxn} = \Delta H_f°[CaO] + \Delta H_f°[CO_2] - \Delta H_f°[CaCO_3]$$

Solve Inserting the given $\Delta H°_{rxn}$ and the known $\Delta H_f°$ values from Table 5.3 or Appendix C, we have:

$$178.1 \text{ kJ} = -635.5 \text{ kJ} - 393.5 \text{ kJ} - \Delta H_f°[CaCO_3]$$

Solving for $\Delta H_f°[CaCO_3]$ gives:

$$\Delta H_f°[CaCO_3] = -1207.1 \text{ kJ/mol}$$

Check We expect the enthalpy of formation of a stable solid such as calcium carbonate to be negative, as obtained.

▶ **Practice Exercise**

Given the following standard enthalpy change, use the standard enthalpies of formation in Table 5.3 to calculate the standard enthalpy of formation of $CuO(s)$:

$$CuO(s) + H_2(g) \longrightarrow Cu(s) + H_2O(l) \qquad \Delta H° = -129.7 \text{ kJ}$$

▲ Self-Assessment Exercises

SAE 5.18   Which of the following statements about standard enthalpies of formation is *false*? (**a**) Standard enthalpies of formation are always positive numbers or zero. (**b**) The standard enthalpy of formation for a compound is the enthalpy for the reaction in which one mole of the compound is made from its elements under standard conditions. (**c**) The standard enthalpy of formation of an element in its most stable form under standard conditions is zero. (**d**) The standard state of a pure substance is its form at atmospheric pressure and a specified temperature, which is usually 298 K. (**e**) The standard enthalpy of formation of a substance will be different for different phases of the substance.

SAE 5.19   Hypochlorous acid (HOCl) can decompose in the gas phase according to the following balanced equation:

$$2 \text{ HOCl}(g) \rightarrow 2 \text{ HCl}(g) + O_2(g) \quad \Delta H = -31.0 \text{ kJ} \ (T = 298 \text{ K}, P = 1 \text{ atm})$$

Using the enthalpy of this reaction and data in Table 5.3, calculate the standard enthalpy of formation for HOCl(g). (**a**) +76.8 kJ/mol (**b**) +61.3 kJ/mol (**c**) −76.8 kJ/mol (**d**) −123 kJ/mol (**e**) −154 kJ/mol

SAE 5.20   Carbon tetrachloride (CCl_4) can react with H_2 to form methane and hydrogen chloride:

$$CCl_4(g) + 4 \text{ H}_2(g) \rightarrow CH_4(g) + 4 \text{ HCl}(g)$$

$\Delta H_f°$ for $CCl_4(g)$ is −106.7 kJ/mol. Other values for $\Delta H_f°$ are given in Table 5.3. What is the standard enthalpy change for the above reaction? (**a**) −60.4 kJ (**b**) −337 kJ (**c**) −444 kJ (**d**) −551 kJ (**e**) More information is needed.

5.8 | Bond Enthalpies

The energy changes that accompany chemical reactions are closely related to the changes associated with forming and breaking chemical bonds—breaking bonds requires energy and forming bonds releases energy. In this section we show how enthalpy values can be assigned to individual bonds, which then can be used to estimate enthalpies of reaction.

By measuring the enthalpy of a reaction and keeping track of the bonds that are broken and formed, we can assign an enthalpy value to specific bonds. The **bond enthalpy** is the enthalpy change, ΔH, for the breaking of a particular bond in one mole of a gaseous substance. It is easiest to determine bond enthalpies from simple reactions where only one bond is broken, such as the dissociation of $Cl_2(g)$. A Cl_2 molecule is held together by a single covalent bond, which is represented as Cl—Cl. The dissociation of $Cl_2(g)$ into chlorine atoms results when the Cl—Cl bond is broken, as illustrated by the drawing in the margin:

$$Cl_2(g) \rightarrow 2 \text{ Cl}(g) \qquad \Delta H = 242 \text{ kJ}$$

Because the stoichiometric coefficients represent moles, in this reaction 1 mol of $Cl_2(g)$ produces 2 mol of Cl(g). In the course of the reaction, we are breaking 1 mol of Cl—Cl

▲ Learning Objectives

When you finish Section 5.8, you should be able to:

▶ Describe the concept of average bond enthalpies.

▶ Use average bond enthalpies to estimate the enthalpies of gas-phase reactions.

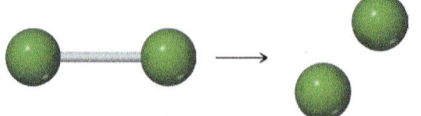

bonds. Hence, we state that the bond enthalpy for the Cl—Cl bond is 242 kJ/mol. The units kJ/mol for a bond enthalpy refer to the enthalpy *per mole of bonds broken*. The bond enthalpy is a positive number because energy must be supplied from the surroundings to break a bond. We use the letter D followed by the bond in question to represent bond enthalpies. For example, $D(\text{Cl}—\text{Cl})$ is the bond enthalpy for the Cl_2 bond, whereas $D(\text{H}—\text{Br})$ is the bond enthalpy for the HBr bond.

The previous example shows that it is straightforward to assign bond enthalpies for a reaction involving the breaking of a bond in a diatomic molecule. However, many important bonds, such as the C—H bond, exist only in polyatomic molecules. For these bonds, we usually use average bond enthalpies. For example, the enthalpy change for the process in which a methane molecule is decomposed into its five constituent atoms, as shown in the margin, can be used to define an average bond enthalpy for the C—H bond:

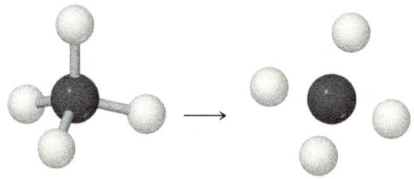

$$\text{CH}_4(g) \longrightarrow \text{C}(g) + 4\text{H}(g) \qquad \Delta H = 1660 \text{ kJ}$$

Because there are four equivalent C—H bonds in methane, the enthalpy of this reaction is four times the enthalpy needed to break a single C—H bond. Therefore, the average C—H bond enthalpy in CH_4 is $D(\text{C}—\text{H}) = (1660/4) \text{ kJ/mol} = 415 \text{ kJ/mol}$.

The exact bond enthalpy for a given pair of atoms, say C—H, depends on the rest of the molecule containing the atom pair. However, the variation from one molecule to another is generally small. If we consider C—H bond enthalpies in many different compounds, we find that the average bond enthalpy is 413 kJ/mol, close to the 415 kJ/mol value we just calculated from for the dissociation of CH_4.

As we explain in later chapters, atom pairs are sometimes held together by two bonds (a *double* bond) or three bonds (a *triple* bond). We can determine the bond enthalpies of multiple bonds in the same way we have just done for single bonds. For example, the oxygen atoms in $\text{O}_2(g)$ are held together by a double bond, O=O, rather than a single bond, O—O. If we break a molecule of $\text{O}_2(g)$ into two O(g) atoms, we are breaking one O=O double bond, as illustrated in the margin:

$$\text{O}=\text{O}(g) \rightarrow 2\text{ O}(g) \qquad \Delta H = 495 \text{ kJ}$$

Because we are breaking 1 mol of O=O double bonds in this process, the bond enthalpy for the O=O bond is 495 kJ/mol.

Table 5.4 lists average bond enthalpies for a number of atom pairs. Note that the bond enthalpy is always a positive quantity because *energy is required to break chemical bonds*. Conversely, *energy is released when a bond forms* between two atoms.

The magnitudes of the bond enthalpies give us important information about the relative strengths of chemical bonds: The greater the bond enthalpy, the stronger the bond. Thus, H—F bonds (567 kJ/mol) are much stronger than F—F bonds (155 kJ/mol). Notice also that double bonds between two atoms are stronger than single bonds between the same atoms, although they are not exactly twice as strong. We consider the variations in bond enthalpy for double and triple bonds in more detail in Chapter 8.

TABLE 5.4   Average Bond Enthalpies (kJ/mol)

C—H	413	N—H	391	O—H	463	F—F	155
C—C	348	N—N	163	O—O	146		
C=C	614	N—O	201	O=O	495	Cl—F	253
C—N	293	N—F	272	O—F	190	Cl—Cl	242
C—O	358	N—Cl	200	O—Cl	203		
C=O	799	N—Br	243	O—I	234	Br—F	237
C—F	485					Br—Cl	218
C—Cl	328	H—H	436			Br—Br	193
C—Br	276	H—F	567				
C—I	240	H—Cl	431			I—Cl	208
		H—Br	366			I—Br	175
		H—I	299			I—I	151

Bond Enthalpies and the Enthalpies of Reactions

Because enthalpy is a state function, we can use average bond enthalpies to estimate the enthalpies of reactions in which bonds are broken and new bonds are formed. This procedure allows us to estimate quickly whether a given reaction will be endothermic ($\Delta H > 0$) or exothermic ($\Delta H < 0$), even if we do not know ΔH_f° for all of the species involved.

Our strategy for estimating reaction enthalpies is a straightforward application of Hess's law. We use the fact that breaking bonds is always endothermic and that forming bonds is always exothermic. We therefore imagine that the reaction occurs in two steps:

1. We supply enough energy to break those bonds in the reactants that are not present in the products. The enthalpy of the system is increased by the sum of the bond enthalpies of the bonds that are broken.

2. We form the bonds in the products that were not present in the reactants. This step releases energy and therefore lowers the enthalpy of the system by the sum of the bond enthalpies of the bonds that are formed.

The enthalpy of the reaction, ΔH_{rxn}, is thus estimated as the sum of the bond enthalpies of the bonds broken minus the sum of the bond enthalpies of the bonds formed:

$$\Delta H_{rxn} = \Sigma \text{ (bond enthalpies} - \Sigma \text{ (bond enthalpies} \qquad \text{[5.32]}$$
$$\text{of bonds broken)} \qquad \text{of bonds formed)}$$

If the sum of the bond enthalpies of the bonds broken is greater than the sum of the bond enthalpies of the bonds formed, the reaction is endothermic ($\Delta H_{rxn} > 0$). Conversely, if the sum of the bond enthalpies of the bonds broken is less than the sum of the bond enthalpies of the bonds formed, the reaction is exothermic ($\Delta H_{rxn} < 0$).

Consider, for example, the gas-phase reaction between methane, CH_4, and chlorine to produce methyl chloride, CH_3Cl, and hydrogen chloride, HCl:

$$\text{H---CH}_3(g) + \text{Cl---Cl}(g) \longrightarrow \text{Cl---CH}_3(g) + \text{H---Cl}(g) \qquad \Delta H_{rxn} = ? \quad \text{[5.33]}$$

Our two-step procedure is outlined in **Figure 5.24**. The following bonds are broken and formed:

Bonds broken: 1 mol C—H, 1 mol Cl—Cl

Bonds formed: 1 mol C—Cl, 1 mol H—Cl

We first supply enough energy to break the C—H and Cl—Cl bonds, which raises the enthalpy of the system (indicated as $\Delta H_1 > 0$ in Figure 5.24). We then form the C—Cl and H—Cl bonds, which release energy and lower the enthalpy of the system ($\Delta H_2 < 0$ in Figure 5.24). Next, we use Equation 5.32 to estimate the enthalpy of the reaction:

$$\Delta H_{rxn} = [D(\text{C---H}) + D(\text{Cl---Cl})] - [D(\text{C---Cl}) + D(\text{H---Cl})]$$
$$\Delta H_{rxn} = (413 \text{ kJ} + 242 \text{ kJ}) - (328 \text{ kJ} + 431 \text{ kJ}) = -104 \text{ kJ}$$

The reaction is exothermic because the bonds in the products are stronger than the bond in the reactants.

Typically, bond enthalpies are used to estimate ΔH_{rxn} only if the needed ΔH_f° values are not readily available. For the preceding reaction, we cannot calculate ΔH_{rxn} from ΔH_f° values and Hess's law because ΔH_f° for $CH_3Cl(g)$ is not given in Appendix C. If we obtain the value of ΔH_f° for $CH_3Cl(g)$ from another source and use Equation 5.31, we find that $\Delta H_{rxn} = -99.8 \text{ kJ}$ for the reaction in Equation 5.33. The two values are slightly different because bond enthalpies are averaged over many compounds, but the use of average bond enthalpies provides a reasonably accurate estimate of the actual reaction enthalpy change.

It is important to remember that bond enthalpies are derived for *gaseous* molecules and that they are often *averaged* values. We will always get more precise enthalpies of reaction by using the ΔH_f° values if they are available. Nevertheless, average bond enthalpies are useful for obtaining quick estimates of reaction enthalpies for gas-phase reactions. In solids, liquids, and solutions, intermolecular forces between different molecules must also be taken into account, as we discuss in Chapter 11. For now it is sufficient to remember that bond enthalpies should not generally be used to estimate the enthalpies of reactions involving solids, liquids, or solutions.

Go Figure Is this reaction exothermic or endothermic?

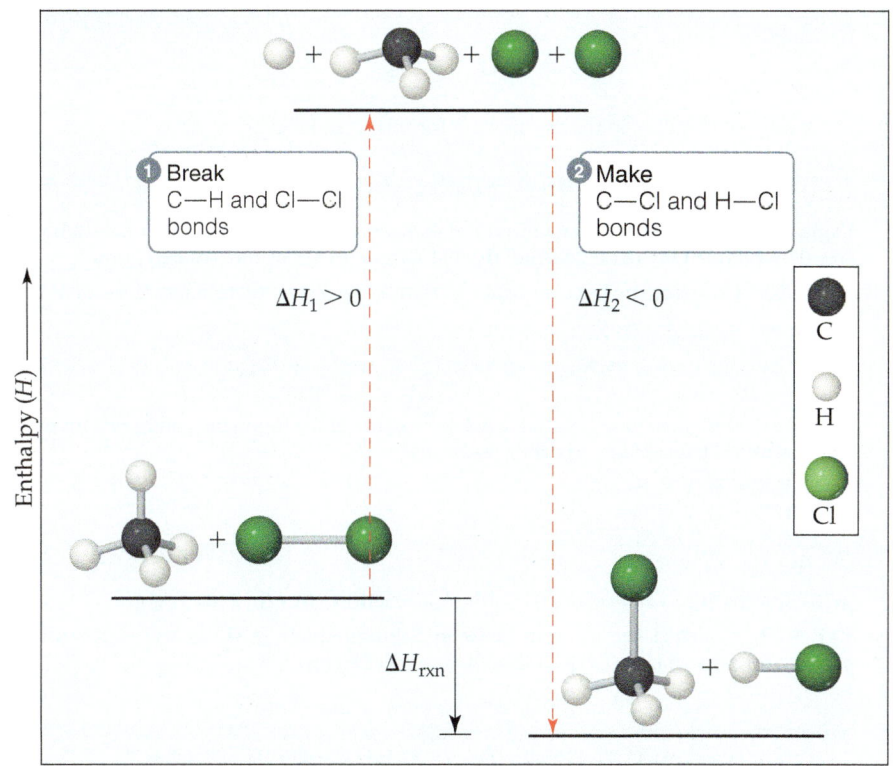

▲ **Figure 5.24 Using bond enthalpies to estimate ΔH_{rxn}.** Average bond enthalpies are used to estimate ΔH_{rxn} for the reaction of methane with chlorine to make methyl chloride and hydrogen chloride.

 Sample Exercise 5.12

Estimating Reaction Enthalpies from Bond Enthalpies

Use Table 5.4 to estimate ΔH for the following combustion reaction.

$$2\ \text{H—C—C—H}(g)\ +\ 7\ \text{O=O}(g)\ \longrightarrow\ 4\ \text{O=C=O}(g)\ +\ 6\ \text{H—O—H}(g)$$

SOLUTION

Analyze We are asked to use average bond enthalpies to estimate the enthalpy change for a chemical reaction.

Plan Energy is required to break twelve C—H bonds and two C—C bonds in the two molecules of C_2H_6, and seven O=O bonds in the seven O_2 molecules. Energy is released by forming eight C=O bonds (two for each of the four molecules of CO_2), and twelve O—H bonds (two per H_2O molecule).

Solve Using Equation 5.32 and data from Table 5.4, we have

$$\Delta H = [12D(\text{C—H}) + 2D(\text{C—C}) + 7D(\text{O=O})]$$
$$- [8D(\text{C=O}) + 12D(\text{O—H})]$$
$$= [12(413\ \text{kJ}) + 2(348\ \text{kJ}) + 7(495\ \text{kJ})]$$
$$- [8(799\ \text{kJ}) + 12(463\ \text{kJ})]$$
$$= 9117\ \text{kJ} - 11948\ \text{kJ}$$
$$= -2831\ \text{kJ}$$

Check This estimate can be compared with the value of -2856 kJ calculated from more accurate thermochemical data; the agreement is good.

▶ **Practice Exercise**
Use the average bond enthalpies in Table 5.4 to estimate ΔH for the combustion of ethanol.

$$\text{H—C—C—O—H}$$

▲ **Self-Assessment Exercises**

SAE 5.21   Molecules of gaseous hypobromous acid, HOBr(g), have one H—O bond and one O—Br bond:

$$H—O—Br$$

HOBr can react with H_2 in the gas phase to form H_2O and HBr:

$$HOBr(g) + H_2(g) \rightarrow H_2O(g) + HBr(g) \quad \Delta H = -216 \, kJ$$

Using this equation and the data in Table 5.4, determine a value for the bond enthalpy of the O—Br bond in HOBr. (**a**) 177 kJ/mol (**b**) 393 kJ/mol (**c**) 613 kJ/mol (**d**) 640 kJ/mol

SAE 5.22   Consider the hypothetical reaction between two diatomic molecules in the gas phase:

$$X—X \ (g) + Y—Y \ (g) \longrightarrow 2 \ X—Y \ (g)$$

Based on the bond enthalpies in Table 5.4, how many of the following combinations of X and Y are predicted to lead to exothermic reactions?

(**i**) X = H, Y = F
(**ii**) X = H, Y = Br
(**iii**) X = F, Y = Cl
(**iv**) X = Cl, Y = I

(**a**) 0 combinations will lead to an exothermic reaction. (**b**) 1 (**c**) 2 (**d**) 3 (**e**) 4

SAE 5.23   Consider the reaction between chloromethane (CH_3Cl) and water in the gas phase to form methanol (CH_3OH) and hydrogen chloride:

$$
\begin{array}{cccc}
\quad H & & H & \\
\quad | & & | & \\
H—C—Cl & + \ H—O—H \longrightarrow & H—C—O—H & + \ H—Cl \\
\quad | & & | & \\
\quad H & & H &
\end{array}
$$

Using the bond enthalpy data in Table 5.4, estimate the enthalpy change for this reaction. (**a**) −461 kJ (**b**) −439 kJ (**c**) +2 kJ (**d**) +465 kJ

5.9 | Foods and Fuels

▲ **Learning Objectives**

When you finish Section 5.9, you should be able to:

▶ Estimate the amount of energy released when a particular food is metabolized.

▶ Estimate the amount of energy released by the combustion of particular fuels.

We finish this chapter with a brief discussion of two practical ways chemical energy affects our daily lives: the energy we obtain from the food we eat, and the energy derived from the fuels used to generate electricity, heat, and transportation. We will see that most of the chemical reactions used for the production of energy are controlled or uncontrolled combustion reactions.

The energy released when one gram of any substance is combusted is the **fuel value** of the substance. The fuel value of any food or fuel can be measured by calorimetry. (Section 5.5)

Foods

Most of the energy our bodies need comes from carbohydrates and fats. The carbohydrates known as starches are decomposed in the intestines into glucose, $C_6H_{12}O_6$. Glucose is soluble in blood, and in the human body it is known as blood sugar. It is transported by the blood to cells where it reacts with O_2 in a series of steps, eventually producing $CO_2(g)$, $H_2O(l)$, and energy:

$$C_6H_{12}O_6(s) + 6 \, O_2(g) \longrightarrow 6 \, CO_2(g) + 6 \, H_2O(l) \quad \Delta H° = -2803 \, kJ$$

Notice that the products of this reaction (and others we see for foods) are associated with combustion reactions. (Section 3.2) Living organisms obtain energy by controlling these combustion reactions, via their metabolism, to release energy in small steps, which is more efficient and controllable.

When we discuss the fuel value of foods, we often use not only the unit of kJ but also kcal and the common unit of Calories (1 Cal = 1000 cal = 1 kcal). (Section 1.5) In the discussion that follows, we give fuel values in both kJ and kcal.

Because carbohydrates break down rapidly, their energy is quickly supplied to the body. However, the body stores only a very small amount of carbohydrates. The average fuel value of carbohydrates is 17 kJ/g (4 kcal/g). Although fuel values represent the heat released in a combustion reaction, by convention fuel values are reported as positive numbers.

Like carbohydrates, fats react with O_2 to produce CO_2 and H_2O when metabolized. The reaction of tristearin, $C_{57}H_{110}O_6$, a typical fat, is

$$2\,C_{57}H_{110}O_6(s) + 163\,O_2(g) \longrightarrow 114\,CO_2(g) + 110\,H_2O(l) \qquad \Delta H° = -71,609\ \text{kJ}$$

The body uses the chemical energy from foods to maintain body temperature (see the "Chemistry and Life" box in Section 5.5), to contract muscles, and to construct and repair tissues. Any excess energy is stored as fats. Fats are well suited to serve as the body's energy reserve for at least two reasons: (1) They are insoluble in water, which facilitates storage in the body, and (2) they produce more energy per gram than either proteins or carbohydrates, which makes them efficient energy sources on a mass basis. The average fuel value of fats is 38 kJ/g (9 kcal/g).

The combustion of carbohydrates and fats in a bomb calorimeter gives the same products as when they are metabolized in the body. The metabolism of proteins produces less energy than combustion in a calorimeter because the products are different. Proteins contain nitrogen, which is released in the bomb calorimeter as N_2. In the body this nitrogen ends up mainly as urea, $(NH_2)_2CO$. The body uses proteins mainly as building materials for organ walls, skin, hair, muscle, and so forth. On average, the metabolism of proteins produces 17 kJ/g (4 kcal/g), the same as for carbohydrates.

Fuel values for some common foods are listed in **Table 5.5**. Labels on packaged foods show the amounts of carbohydrate, fat, and protein contained in an average serving, as well as the amount of energy supplied by a serving (**Figure 5.25**).

The amount of energy our bodies require varies considerably, depending on such factors as weight, age, and muscular activity. About 100 kJ per kilogram of body mass per day is required to keep the body functioning at a minimal level. An average 70-kg (154-lb) person expends about 800 kJ/h when doing light work, and strenuous activity often requires 2000 kJ/h or more. When the fuel value, or caloric content, of the food we ingest exceeds the energy we expend, our body stores the surplus as fat.

Go Figure

Which value would change most if this label were for skim milk instead of whole milk: grams of fat, grams of total carbohydrate, or grams of protein?

Vitamin D Milk

Nutrition Facts
Serving Size 1 cup (236mL)

Amount Per Serving

Calories 150   Calories from Fat 70

	% Daily Value*
Total Fat 8g	**12%**
Saturated Fat 5g	**25%**
Trans Fat 0g	
Cholesterol 35mg	**11%**
Sodium 125mg	**5%**
Total Carbohydrate 12g	**4%**
Dietary Fiber 0g	**0%**
Sugars 12g	
Protein 8g	

Vitamin A 6%	•	Vitamin C 4%
Calcium 30%	•	Iron 0%
Vitamin D 25%		

* Percent Daily Values are based on a 2,000 calorie diet. Your daily values may be higher or lower depending on your calorie needs.

	Calories:	2,000	2,500
Total fat	Less than	65g	80g
Sat Fat	Less than	20g	25g
Cholesterol	Less than	300mg	300mg
Sodium	Less than	2,400mg	2,400mg
Total Carbohydrate		300g	375g
Dietary Fiber		25g	30g

Ingredients: Grade A Pasteurized Milk, Vitamin D3.

▲ **Figure 5.25** Nutrition label for whole milk.

TABLE 5.5   Compositions and Fuel Values of Some Common Foods

	Approximate Composition (% by Mass)			Fuel Value	
	Carbohydrate	Fat	Protein	kJ/g	kcal/g(Cal/g)
Carbohydrate	100	—	—	17	4
Fat	—	100	—	38	9
Protein	—	—	100	17	4
Apples	13	0.5	0.4	2.5	0.59
Beer[a]	1.2	—	0.3	1.8	0.42
Bread	52	3	9	12	2.8
Cheese	4	37	28	20	4.7
Eggs	0.7	10	13	6.0	1.4
Fudge	81	11	2	18	4.4
Green beans	7.0	—	1.9	1.5	0.38
Hamburger	—	30	22	15	3.6
Milk (whole)	5.0	4.0	3.3	3.0	0.74
Peanuts	22	39	26	23	5.5

[a]Beer typically contains 3.5% ethanol, which has fuel value.

Sample Exercise 5.13

Estimating the Fuel Value of a Food from Its Composition

(a) A 28-g (1-oz) serving of a popular breakfast cereal served with 120 mL of skim milk provides 8 g protein, 26 g carbohydrates, and 2 g fat. Using the average fuel values of these substances, estimate the fuel value (caloric content) of this serving.
(b) A person of average weight uses about 100 Cal/mi when running or jogging. How many servings of this cereal provide the fuel value requirements to run 3 mi?

SOLUTION

Analyze (a) The fuel value of the serving will be the sum of the fuel values of the protein, carbohydrates, and fat. **(b)** Here we are faced with the reverse problem, calculating the quantity of food that provides a specific fuel value.

Plan (a) We are given the masses of the protein, carbohydrates, and fat contained in a serving. We can use the data in Table 5.4 to convert these masses to their fuel values, which we can sum to get the total fuel value. **(b)** The problem statement provides a conversion factor between Calories and miles. The answer to part **(a)** provides us with a conversion factor between servings and Calories.

Solve

(a)

$$(8 \text{ g protein})\left(\frac{17 \text{ kJ}}{1 \text{ g protein}}\right) + (26 \text{ g carbohydrate})\left(\frac{17 \text{ kJ}}{1 \text{ g carbohydrate}}\right)$$

$$+ (2 \text{ g fat})\left(\frac{38 \text{ kJ}}{1 \text{ g fat}}\right) = 650 \text{ kJ (two significant figures)}$$

This corresponds to 160 kcal:

$$(650 \text{ kJ})\left(\frac{1 \text{ kcal}}{4.18 \text{ kJ}}\right) = 160 \text{ kcal}$$

The dietary Calorie is equivalent to 1 kcal, so the serving provides 160 Cal.

(b) We can use these factors in a straightforward dimensional analysis to determine the number of servings needed, rounded to the nearest whole number:

$$\text{Servings} = (3 \text{ mi})\left(\frac{100 \text{ Cal}}{1 \text{ mi}}\right)\left(\frac{1 \text{ serving}}{160 \text{ Cal}}\right) = 2 \text{ servings}$$

▶ **Practice Exercise**
(a) Dry red beans contain 62% carbohydrate, 22% protein, and 1.5% fat. Estimate the fuel value of these beans. **(b)** During a very light activity, such as reading or watching television, the average adult expends about 7 kJ/min. How many minutes of such activity can be sustained by the energy provided by a serving of chicken noodle soup containing 13 g protein, 15 g carbohydrate, and 5 g fat?

Fuels

During the complete combustion of fuels, carbon is converted to CO_2 and hydrogen is converted to H_2O, both of which have large negative enthalpies of formation. Consequently, the greater the percentage of carbon and hydrogen in a fuel, the higher its fuel value. Using Table 5.6, for example, compare the compositions and fuel values of bituminous coal and wood. The coal has a higher fuel value because of its greater carbon content.

In 2019 the United States consumed 1.06×10^{17} kJ of energy. This value corresponds to an average daily energy consumption per person of 8.8×10^5 kJ, roughly 100 times greater than the per capita food-energy needs. Figure 5.26 illustrates the sources of this energy.

Coal, petroleum, and natural gas, which are the world's major sources of energy, are known as **fossil fuels**. All have formed over millions of years from the decomposition of plants and animals and are being depleted far more rapidly than they are being formed.

Natural gas consists of gaseous hydrocarbons, which are compounds of hydrogen and carbon. It contains primarily methane (CH_4), with small amounts of ethane (C_2H_6), propane (C_3H_8), and butane (C_4H_{10}). We determined the fuel value of propane in Sample Exercise 5.10. Natural gas burns with far fewer by-products and produces less CO_2 than either petroleum or coal. **Petroleum** is a liquid composed of hundreds of compounds, most of which are hydrocarbons, with the remainder being chiefly organic compounds containing sulfur, nitrogen, or oxygen. **Coal**, which is solid, contains hydrocarbons of high molecular weight as well as compounds containing sulfur, oxygen, or nitrogen. Coal is the most abundant fossil fuel; current reserves are projected to last for well over 100 years at current consumption rates. However, the use of coal presents a number of problems.

Coal is a complex mixture of substances, and some of its components cause air pollution. When coal is combusted, the sulfur it contains is converted mainly to sulfur dioxide,

TABLE 5.6   Fuel Values and Compositions of Some Common Fuels

	Approximate Elemental Composition (Mass %)			Fuel Value (kJ/g)
	C	H	O	
Wood (pine)	50	6	44	18
Anthracite coal (Pennsylvania)	82	1	2	31
Bituminous coal (Pennsylvania)	77	5	7	32
Charcoal	100	0	0	34
Crude oil (Texas)	85	12	0	45
Gasoline	85	15	0	48
Natural gas	70	23	0	49
Hydrogen	0	100	0	142

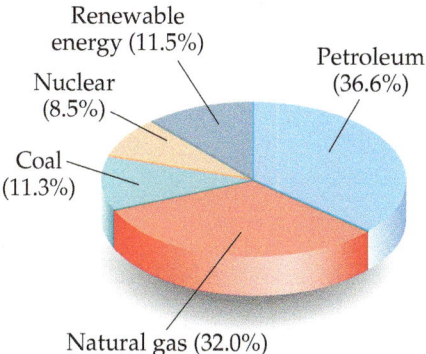

▲ **Figure 5.26 Energy consumption in the United States.*** In 2019 the United States consumed a total of 1.06×10^{17} kJ of energy.

SO_2, a troublesome air pollutant. Because coal is a solid, recovery from its underground deposits is expensive and often dangerous. Furthermore, coal deposits are not always close to locations of high-energy use, so there are often substantial shipping costs. U.S. dependence on coal as a primary energy source declined by roughly 5% from 2015 to 2019.

Fossil fuels release energy in combustion reactions, which ideally produce only CO_2 and H_2O. The production of CO_2 has become a major issue for science and public policy because of concerns that increasing concentrations of atmospheric CO_2 are causing global climate changes. We discuss the environmental aspects of atmospheric CO_2 in Chapter 18.

Other Energy Sources

Nuclear energy is the energy released in either the fission (splitting) or the fusion (combining) of atomic nuclei. Nuclear power based on nuclear fission currently produces about 19% of electric power in the United States and makes up about 8.5% of total U.S. energy production (Figure 5.26). On the one hand, nuclear energy is, in principle, free of the polluting emissions that are a major problem with fossil fuels. On the other hand, nuclear power plants produce radioactive waste products, and their use has therefore been controversial. We discuss issues related to the production of nuclear energy in Chapter 21.

Fossil fuels and nuclear energy are *nonrenewable* sources of energy: They are limited resources that we are consuming at a much greater rate than they can be regenerated. Eventually, these fuels will be expended, although estimates vary greatly as to when this will occur. Because nonrenewable energy sources will eventually be used up, a great deal of research is being conducted on **renewable energy sources**, that is, sources that are essentially inexhaustible. Renewable energy sources include *solar energy* from the Sun, *wind energy* harnessed by windmills, *geothermal energy* from the heat stored inside Earth, *hydroelectric energy* from flowing rivers, and *biomass energy* from crops and biological waste matter. Currently, renewable sources account for about 11.5% of the U.S. annual energy consumption. Electricity is typically generated by using a moving gas or liquid to turn a turbine that is connected to a generator, although there are some exceptions like solar cells. Of the 3.9×10^{16} kJ of electricity consumed in 2019 in the United States, approximately 62% came from burning fossil fuels. Other sources of electricity include nuclear energy (19%), hydroelectric power (6.8%), wind (7.1%), biomass (1.4%) and solar energy (2.6%).

Fulfilling our future energy needs will depend on developing the technology to harness solar energy with greater efficiency. Solar energy is the world's largest energy source. On a clear day, about 1 kJ of solar energy reaches each square meter of Earth's surface every second. The average solar energy falling on only 0.1% of U.S. land area is equivalent to all the energy this nation currently uses. Harnessing this energy is difficult because it is dilute (that is, distributed over a wide area) and varies with time of day and weather conditions. The effective use of solar energy will depend on the development of some

*"U.S. energy facts explained," U.S Energy Information Administration, U.S. Department of Energy, EIA.gov.

means of storing and distributing it. One practical way to do this is to use the Sun's energy to drive an endothermic chemical process that can be later reversed to release heat. One such reaction is

$$CH_4(g) + H_2O(g) + heat \longrightarrow CO(g) + 3 H_2(g)$$

This reaction proceeds in the forward direction at high temperatures, which can be obtained in a solar furnace. The CO and H_2 formed in the reaction could then be stored and allowed to react later, with the heat released being put to useful work.

CHEMISTRY AND SUSTAINABILITY | The Scientific and Political Challenges of Biofuels

One of the biggest challenges facing us in the twenty-first century is production of abundant sources of energy, both food and fuels. At the end of 2019, the global population was estimated to be 7.7 billion and growing at a rate of about 840 million per decade. A growing world population puts greater demands on the global food supply, especially in Asia and Africa, which together make up more than 76% of the world population.

A growing population also increases demands on the production of fuels for transportation, industry, electricity, heating, and cooling. As populous countries such as China and India have modernized, their per capita consumption of energy has increased significantly. In China, for instance, per capita energy consumption roughly doubled between 1990 and 2010, and in 2010 China passed the United States as the world's largest user of energy (although China's per capita consumption is still less than 35% of the U.S. figure).

Global energy consumption in 2019 was more than 4×10^{18} kJ, a staggeringly large number. More than 86% of current energy requirements comes from combustion of nonrenewable fossil fuels (petroleum, coal, and natural gas). The exploration of new fossil fuel sources often involves environmentally sensitive regions, making the search for new supplies of fossil fuels a major political and economic issue.

Petroleum has global importance largely because it provides liquid fuels, such as gasoline, that are critical to supplying transportation needs. One of the most promising—but controversial—alternatives to petroleum-based fuels is *biofuels*, liquid fuels derived from biological matter. The most common approach to producing biofuels is to transform plant sugars and other carbohydrates into combustible liquids.

The most commonly produced biofuel is *bioethanol*, which is ethanol (C_2H_5OH) made from fermentation of plant carbohydrates. The fuel value of ethanol is about two-thirds that of gasoline and is therefore comparable to that of coal (Table 5.6). The United States and Brazil dominate bioethanol production, together supplying 85% of the world's total.

In the United States, nearly all the bioethanol currently produced is made from yellow feed corn. Yeasts or other microorganisms are used to convert glucose ($C_6H_{12}O_6$) in the corn to ethanol and CO_2:

$$C_6H_{12}O_6(s) \longrightarrow 2 C_2H_5OH(l) + 2 CO_2(g) \quad \Delta H = 15.8 \text{ kJ}$$

This reaction is *anaerobic*—it does not involve $O_2(g)$—and the enthalpy change is positive and much smaller in magnitude than that for most combustion reactions. Other carbohydrates can be converted to ethanol in similar fashion. Because the carbon in growing plants comes from CO_2 in the air via photosynthesis, bioethanol contributes considerably less net CO_2 to the atmosphere and is therefore a more sustainable source of energy than are fossil fuels.

Producing bioethanol from corn is controversial for two main reasons. First, growing and transporting corn are both energy-intensive processes, and growing it requires the use of fertilizers. It is estimated that the *energy return* on corn-based bioethanol is only 34%—that is, for each 1.00 J of energy expended to produce the corn, 1.34 J of energy is produced in the form of bioethanol. Second, the use of corn as a starting material for making bioethanol competes with its use as an important component of the food chain (the so-called *food versus fuel* debate).

▲ **Figure 5.27** Sugarcane can be converted to a sustainable bioethanol product.

Much current research focuses on the formation of bioethanol from *cellulosic* plants, plants that contain the complex carbohydrate cellulose. Cellulose is not readily metabolized and so does not compete with the food supply. However, the chemistry for converting cellulose to ethanol is much more complex than that for converting corn. Cellulosic bioethanol could be produced from very fast-growing nonfood plants, such as prairie grasses and switchgrass, which readily renew themselves without the use of fertilizers.

The Brazilian bioethanol industry uses sugarcane as its feedstock (**Figure 5.27**). Sugarcane grows much faster than corn and needs neither fertilizers nor tending. Because of these differences, the energy return for sugarcane is much higher than that for corn. It is estimated that for each 1.0 J of energy expended in growing and processing sugarcane, 8.0 J of energy is produced as bioethanol. In 2019, Brazil produced more than 8600 million U.S. gallons of bioethanol.

While bioethanol is a sustainable fuel source, Brazil's expansion of the sugarcane industry has threatened environmentally sensitive regions, including the Amazon rainforest, via deforestation. These rainforests are among the most important ways in which the Earth reabsorbs the CO_2 that contributes to climate change. Thus, even as we develop ways to make certain aspects of our lives more sustainable, we must consider unintended consequences that adversely impact other areas of sustainability.

Other biofuels that are becoming a major part of the world economy include *biodiesel*, a substitute for petroleum-derived diesel fuel. Biodiesel is typically produced from crops that have a high oil content, such as soybeans and canola. It can also be produced from animal fats and waste vegetable oil from the food and restaurant industry.

Related Exercises: 5.99, 5.100, 5.120

Plants utilize solar energy in *photosynthesis*, the reaction in which the energy of sunlight is used to convert CO_2 and H_2O into carbohydrates and O_2:

$$6\,CO_2(g) + 6\,H_2O(l) + \text{sunlight} \longrightarrow C_6H_{12}O_6(s) + 6\,O_2(g) \qquad [5.34]$$

Photosynthesis is an important part of Earth's ecosystem because it replenishes atmospheric O_2, produces an energy-rich molecule that can be used as fuel, and consumes some atmospheric CO_2.

Perhaps the most direct way to use the Sun's energy is to convert it directly into electricity in photovoltaic devices, or *solar cells*, which we mentioned at the beginning of this chapter. The efficiencies of such devices have increased dramatically during the past few years. Technological advances have led to solar panels that last longer and produce electricity with greater efficiency at steadily decreasing unit cost. Indeed, the future of solar energy is, like the Sun itself, very bright.

Self-Assessment Exercises

Here are a few problems designed to test your understanding of the material.

SAE 5.24 A tablespoon of granulated table sugar contains 12.5 g of sucrose ($C_{12}H_{22}O_{11}$). The balanced reaction for the combustion of sucrose to $CO_2(g)$ and $H_2O(l)$ is:

$$C_{12}H_{22}O_{11}(s) + 12\,O_2(g) \rightarrow 12\,CO_2(g) + 11\,H_2O(l) \quad \Delta H = -5640\text{ kJ}$$

If we assume that the metabolism of sucrose leads to the same amount of heat as in the preceding equation, how many Calories are provided by complete metabolism of a tablespoon of sucrose? (**a**) 3.94 Cal (**b**) 49.3 Cal (**c**) 206 Cal (**d**) 1350 Cal

SAE 5.25 One component of biodiesel is a substance derived from fatty acids, called *ethyl stearate*. The formula of ethyl stearate is $C_{20}H_{40}O_2$, and ΔH for the combustion of 1.50 mol of the substance is $-18{,}900$ kJ. What is the fuel value in kJ/g of ethyl stearate? (**a**) 40.3 (**b**) 60.5 (**c**) 90.7 (**d**) 12,600

Putting Concepts Together

Trinitroglycerin, $C_3H_5N_3O_9$ (usually referred to simply as nitroglycerin), has been widely used as an explosive. Alfred Nobel used it to make dynamite in 1866. Rather surprisingly, it also is used as a medication, to relieve angina (chest pains resulting from partially blocked arteries to the heart) by dilating the blood vessels. At 1 atm pressure and 25°C, the enthalpy of decomposition of trinitroglycerin to form nitrogen gas, carbon dioxide gas, liquid water, and oxygen gas is -1541.4 kJ/mol.

(a) Write a balanced chemical equation for the decomposition of trinitroglycerin.

(b) Calculate the standard heat of formation of trinitroglycerin.

(c) A standard dose of trinitroglycerin for relief of angina is 0.60 mg. If the sample is eventually oxidized in the body (not explosively, though!) to nitrogen gas, carbon dioxide gas, and liquid water, what number of calories is released?

(d) One common form of trinitroglycerin melts at about 3°C. From this information and the formula for the substance, would you expect it to be a molecular or an ionic compound? Explain.

(e) Describe the various conversions of forms of energy when trinitroglycerin is used as an explosive to break rockfaces in highway construction.

SOLUTION

(a) The general form of the equation we must balance is

$$C_3H_5N_3O_9(l) \longrightarrow N_2(g) + CO_2(g) + H_2O(l) + O_2(g)$$

We go about balancing in the usual way. To obtain an even number of nitrogen atoms on the left, we multiply the formula for $C_3H_5N_3O_9$ by 2, which gives us 3 mol of N_2, 6 mol of CO_2 and 5 mol of H_2O. Everything is then balanced except for oxygen. We have an odd number of oxygen atoms on the right. We can balance the oxygen by using the coefficient $\frac{1}{2}$ for O_2 on the right:

$$2\,C_3H_5N_3O_9(l) \longrightarrow 3\,N_2(g) + 6\,CO_2(g) + 5\,H_2O(l) + \tfrac{1}{2}O_2(g)$$

We multiply through by 2 to convert all coefficients to whole numbers:

$$4\,C_3H_5N_3O_9(l) \longrightarrow 6\,N_2(g) + 12\,CO_2(g) + 10\,H_2O(l) + O_2(g)$$

(At the temperature of the explosion, water is a gas rather than a liquid, as shown in the preceding equation. The rapid expansion of the gaseous products creates the force of an explosion.)

Continued

(b) We can obtain the standard enthalpy of formation of nitroglycerin by using the heat of decomposition of trinitroglycerin, together with the standard enthalpies of formation of the other substances in the decomposition equation:

$$4\,C_3H_5N_3O_9(l) \longrightarrow 6\,N_2(g) + 12\,CO_2(g) + 10\,H_2O(l) + O_2(g)$$

The enthalpy change for this decomposition is $4(-1541.4\text{ kJ}) = -6165.6\text{ kJ}$. [We need to multiply by 4 because there are 4 mol of $C_3H_5N_3O_9(l)$ in the balanced equation.]

This enthalpy change equals the sum of the heats of formation of the products minus the heats of formation of the reactants, each multiplied by its coefficient in the balanced equation:

$$-6165.6\text{ kJ} = 6\Delta H_f^\circ[N_2(g)] + 12\,\Delta H_f^\circ[CO_2(g)] + 10\Delta H_f^\circ[H_2O(l)]$$
$$+ \Delta H_f^\circ[O_2(g)] - 4\Delta H_f^\circ[C_3H_5N_3O_9(l)]$$

The ΔH_f° values for $N_2(g)$ and $O_2(g)$ are zero, by definition. Using the values for $H_2O(l)$ and $CO_2(g)$ from Table 5.3 or Appendix C, we have

$$-6165.6\text{ kJ} = 12(-393.5\text{ kJ}) + 10(-285.8\text{ kJ}) - 4\Delta H_f^\circ[C_3H_5N_3O_9(l)]$$

$$\Delta H_f^\circ[C_3H_5N_3O_9(l)] = -353.6\text{ kJ/mol}$$

(c) Converting 0.60 mg $C_3H_5N_3O_9(l)$ to moles and using the fact that the decomposition of 1 mol of $C_3H_5N_3O_9(l)$ yields 1541.4 kJ,

we have:

$$(0.60 \times 10^{-3}\text{ g }C_3H_5N_3O_9)\left(\frac{1\text{ mol }C_3H_5N_3O_9}{227\text{ g }C_3H_5N_3O_9}\right)\left(\frac{1541.4\text{ kJ}}{1\text{ mol }C_3H_5N_3O_9}\right)$$

$$= 4.1 \times 10^{-3}\text{ kJ} = 4.1\text{ J}$$

(d) Because trinitroglycerin melts below room temperature, we expect that it is a molecular compound. With few exceptions, ionic substances are generally hard, crystalline materials that melt at high temperatures. (Sections 2.6 and 2.7) Also, the molecular formula suggests that it is a molecular substance because all of its constituent elements are nonmetals.

(e) The energy stored in trinitroglycerin is chemical potential energy. When the substance reacts explosively, it forms carbon dioxide, water, and nitrogen gas, which are of lower potential energy. In the course of the chemical transformation, energy is released in the form of heat; the gaseous reaction products are very hot. This high heat energy is transferred to the surroundings. Work is done as the gases expand against the surroundings, moving the solid materials and imparting kinetic energy to them. For example, a chunk of rock might be impelled upward. It has been given kinetic energy by transfer of energy from the hot, expanding gases. As the rock rises, its kinetic energy is transformed into potential energy. Eventually, it again acquires kinetic energy as it falls to Earth. When it strikes Earth, its kinetic energy is converted largely to thermal energy, though some work may be done on the surroundings as well.

Chapter Summary and Key Terms

CHEMICAL ENERGY (INTRODUCTION AND SECTION 5.1)
Thermodynamics is the study of energy and its transformations. In this chapter we have focused on **thermochemistry**, the transformation of energy—especially heat—during chemical reactions.

An object can possess energy in two forms: (1) **kinetic energy**, which is the energy due to the motion of the object, and (2) **potential energy**, which is the energy that an object possesses by virtue of its position relative to other objects. An electron in motion near a proton has kinetic energy because of its motion and potential energy owing to its electrostatic attraction to the proton.

Chemical energy originates largely from electrostatic interactions at the atomic level. Energy must be supplied to break chemical bonds leading to an increase in potential energy. Conversely, energy is released when chemical bonds form as the potential energy decreases.

THE FIRST LAW OF THERMODYNAMICS (SECTION 5.2)
When we study thermodynamic properties, we define a specific amount of matter as the **system**. Everything outside the system is the **surroundings**. When we study a chemical reaction, the system is generally the reactants and products. A closed system can exchange energy, but not matter, with the surroundings. The **internal energy** of a system is the sum of all the kinetic and potential energies of its component parts. The internal energy of a system can change because of energy transferred between the system and the surroundings.

According to the **first law of thermodynamics**, the change in the internal energy of a system, ΔE, is the sum of the heat, q, transferred into or out of the system and the work, w, done on or by the system: $\Delta E = q + w$. Both q and w have a sign that indicates the direction of energy transfer. When heat is transferred from the surroundings to the system, $q > 0$. Likewise, when the surroundings do work on the system, $w > 0$. In an **endothermic** process, the system absorbs heat from the surroundings; in an **exothermic** process, the system releases heat to the surroundings.

The internal energy, E, is a **state function**. The value of any state function depends only on the state or condition of the system and not on the details of how it came to be in that state. Heat, q, and work, w, are not state functions; their values depend on the particular way by which a system changes its state.

ENTHALPY (SECTIONS 5.3 AND 5.4)
When a gas is produced or consumed in a chemical reaction occurring at constant pressure, the system may perform **pressure–volume (P–V) work** against the prevailing pressure of the surroundings. For this reason, we define a new state function called **enthalpy**, H, which is related to energy: $H = E + PV$. In systems where only pressure–volume work is involved, the change in the enthalpy of a system, ΔH, equals the heat gained or lost by the system at constant pressure: $\Delta H = q_p$ (the subscript P denotes constant pressure). For an endothermic process, $\Delta H > 0$; for an exothermic process, $\Delta H < 0$.

In a chemical process, the **enthalpy of reaction** is the enthalpy of the products minus the enthalpy of the reactants: $\Delta H_{rxn} = H(\text{products}) - H(\text{reactants})$. Enthalpies of reaction follow some simple rules: (1) The enthalpy of reaction is proportional to the amount of reactant that reacts. (2) Reversing a reaction changes the sign of ΔH. (3) The enthalpy of reaction depends on the physical states of the reactants and products.

CALORIMETRY (SECTION 5.5)
The amount of heat transferred between the system and the surroundings is measured experimentally by **calorimetry**. A **calorimeter** measures the temperature change accompanying a process. The temperature change of a calorimeter depends on its **heat capacity**, the amount of heat required to raise its temperature by 1 K. The heat capacity for one mole of a pure substance is called its **molar heat capacity**; for one gram of the substance, we use the term **specific heat**. Water has a very high specific heat, 4.18 J/g-K. The amount of heat, q, absorbed by a substance is the product of its specific heat (C_s), its mass, and its temperature change: $q = C_s \times m \times \Delta T$.

If a calorimetry experiment is carried out under a constant pressure, the heat transferred provides a direct measure of the enthalpy change of the reaction. Constant-volume calorimetry is carried out in a vessel of fixed volume called a **bomb calorimeter**. The heat transferred under constant-volume conditions is equal to ΔE. Corrections can be applied to ΔE values to yield ΔH.

HESS'S LAW (SECTION 5.6) Because enthalpy is a state function, ΔH depends only on the initial and final states of the system. Thus, the enthalpy change of a process is the same whether the process is carried out in one step or in a series of steps. **Hess's law** states that if a reaction is carried out in a series of steps, ΔH for the reaction will be equal to the sum of the enthalpy changes for the steps. We can therefore calculate ΔH for any process, as long as we can write the process as a series of steps for which ΔH is known.

ENTHALPIES OF FORMATION (SECTION 5.7) The **enthalpy of formation**, ΔH_f, of a substance is the enthalpy change for the reaction in which the substance is formed from its constituent elements. Usually, enthalpies are tabulated for reactions where reactants and products are in their *standard states*. The standard state of a substance is its pure, most stable form at 1 atm and the temperature of interest (usually 298 K). Thus, the **standard enthalpy change** of a reaction, $\Delta H°$, is the enthalpy change when all reactants and products are in their standard states. The **standard enthalpy of formation**, $\Delta H_f°$, of a substance is the change in enthalpy for the reaction that forms one mole of the substance from its elements in their standard states. For any element in its standard state, $\Delta H_f° = 0$.

The standard enthalpy change for any reaction can be readily calculated from the standard enthalpies of formation of the reactants and products in the reaction:

$$\Delta H_{rxn}° = \sum n\Delta H_f°(\text{products}) - \sum m\Delta H_f°(\text{reactants})$$

BOND ENTHALPIES (SECTION 5.8) The strength of a covalent bond is measured by its bond enthalpy, which is the molar enthalpy change required to break a particular bond. Average bond enthalpies can be determined for a wide variety of covalent bonds. We can estimate the enthalpy changes during chemical reactions involving gaseous substances by adding the average bond enthalpies of the bonds that are broken and subtracting the average bond enthalpies of the bonds that are formed. When the energy needed to break bonds is larger than the energy released by forming bonds, the reaction enthalpy is positive; when the opposite situation holds, the reaction enthalpy is negative.

FOODS AND FUELS (SECTION 5.9) The **fuel value** of a substance is the heat released when one gram of the substance is combusted. Different types of foods have different fuel values and differing abilities to be stored in the body. The most common fuels are hydrocarbons found as **fossil fuels**, such as **natural gas**, **petroleum**, and **coal**. **Renewable energy sources** include solar energy, wind energy, biomass, and hydroelectric energy. Nuclear power does not utilize fossil fuels but does create controversial waste-disposal problems.

Key Equations

•	$w = F \times d$	[5.1]	Relation of work to force and distance
•	$E_{el} = \kappa Q_1 Q_2/d$	[5.2]	Electrostatic potential energy
•	$\Delta E = E_{final} - E_{initial}$	[5.3]	The change in internal energy
•	$\Delta E = q + w$	[5.4]	Relation of the change in internal energy to heat and work (the first law of thermodynamics)
•	$H = E + PV$	[5.5]	Enthalpy defined
•	$w = -P\Delta V$	[5.7]	The work done by an expanding gas at constant pressure
•	$\Delta H = \Delta E + P\Delta V = q_P$	[5.9]	Enthalpy change at constant pressure
•	$q = C_s \times m \times \Delta T$	[5.21]	Heat gained or lost based on specific heat, mass, and temperature change
•	$q_{rxn} = -C_{cal} \times \Delta T$	[5.23]	Heat exchanged between a reaction and calorimeter
•	$\Delta H_{rxn}° = \sum n\Delta H_f°(\text{products}) - \sum m\Delta H_f°(\text{reactants})$	[5.31]	Standard enthalpy change of a reaction
•	$\Delta H_{rxn} = \sum (\text{bond enthalpies of bonds broken}) - \sum (\text{bond enthalpies of bonds formed})$	[5.32]	The reaction enthalpy as a function of average bond enthalpies for reactions involving gas-phase molecules.

Exam Prep

EP 5.1 KBr consists of positively charged potassium ions and negatively charged bromide ions. Which one of the following statements is *false*? (**a**) The potassium and bromide ions are attracted to one another. (**b**) Energy must be added to separate the ions from one another. (**c**) The electrostatic potential energy between potassium and bromide ions is positive. (**d**) The strength of the interaction between the potassium and bromide ions depends on the distance between the ions.

EP 5.2 A mixture of gases A_2 and B_2 are introduced to a slender metal cylinder that has one end closed and the other fitted with a piston that makes a gas-tight seal so that the gases are a closed system. The cylinder is submerged in a large beaker of water whose temperature is 25°C, and a spark is used to trigger a reaction in the cylinder. At the completion of the reaction, the piston has moved downward, and the temperature of the water bath has increased to 28°C. If we define the system as the gases inside the cylinder, which of the following best describes the signs of q, w, and ΔE for this reaction? (**a**) $q < 0$, $w < 0$, $\Delta E < 0$ (**b**) $q < 0$, $w > 0$, $\Delta E < 0$ (**c**) $q < 0$, $w > 0$, the sign of ΔE cannot be determined from the information given (**d**) $q > 0$, $w > 0$, $\Delta E > 0$ (**e**) $q > 0$, $w < 0$, the sign of ΔE cannot be determined from the information given

EP 5.3   A system can change from State A to State B using two different paths, Path 1 and Path 2. Which of the following statements is or are *true*?

   (**i**) The value of q might be different for the two different paths.

   (**ii**) The value of ΔE might be different for the two different paths.

   (**iii**) The value of $q + w$ must be the same for the two different paths.

(**a**) Only one of the statements is true. (**b**) i and ii are true. (**c**) i and iii are true. (**d**) ii and iii are true. (**e**) All three statements are true.

EP 5.4   Which of the following statements about enthalpy is *false*? (**a**) Enthalpy is denoted H. (**b**) ΔH equals the heat gained or lost in a process that occurs at constant pressure. (**c**) Because E, P, and V are all state functions, H is also a state function. (**d**) Because ΔH relates to heat, the value of ΔH depends on the path taken between two states. (**e**) Under constant pressure, $\Delta H = \Delta E + P\Delta V$.

EP 5.5   If a balloon is expanded from 0.055 to 1.403 L against an external pressure of 1.02 atm, how many L-atm of work is done? (**a**) -0.056 L-atm (**b**) -1.37 L-atm (**c**) 1.43 L-atm (**d**) 1.49 L-atm (**e**) 139 L-atm

EP 5.6   A chemical reaction that gives off heat to its surroundings is said to be _____ and has a _____ value of ΔH. (**a**) endothermic, positive (**b**) endothermic, negative (**c**) exothermic, positive (**d**) exothermic, negative

EP 5.7   The figure shows an enthalpy diagram for the following hypothetical reaction run under constant pressure:

$$A(g) + B(g) \rightarrow C(g) + D(g)$$

Which of the following statements is or are *true*?

   (**i**) The reaction is exothermic.

   (**ii**) ΔH for the reaction is -200 kJ.

   (**iii**) The reaction is most likely a spontaneous process.

(**a**) Only one of the statements is true. (**b**) i and ii are true. (**c**) i and iii are true. (**d**) ii and iii are true. (**e**) All three statements are true.

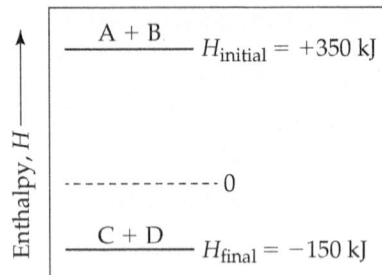

EP 5.8   The complete combustion of ethanol, C_2H_5OH (molar mass = 46.0 g/mol), proceeds as follows:

$$C_2H_5OH(l) + 3\,O_2(g) \rightarrow 2\,CO_2(g) + 3\,H_2O(l)\ \Delta H = -555\ kJ$$

What is the enthalpy change for combustion of 15.0 g of ethanol? (**a**) -12.1 kJ (**b**) -181 kJ (**c**) -422 kJ (**d**) -555 kJ (**e**) -1700 kJ

EP 5.9   The coinage metals (Group 1B) copper, silver, and gold have specific heats of 0.385, 0.233, and 0.129 J/g-K, respectively. Among this group, the specific heat _____ and the molar heat capacity _____ as the atomic weight increases. (**a**) increases, increases (**b**) increases, decreases (**c**) decreases, increases (**d**) decreases, decreases

EP 5.10   When 0.243 g of Mg metal is combined with enough HCl to make 100 mL of solution in a constant-pressure calorimeter, the following reaction occurs:

$$Mg(s) + 2\,HCl(aq) \longrightarrow MgCl_2(aq) + H_2(g)$$

If the temperature of the solution increases from 23.0 to 34.1°C as a result of this reaction, calculate ΔH in kJ/mol Mg. Assume that the solution has a specific heat of 4.18 J/g-°C and a density of 1.00 g/mL. (**a**) -19.1 kJ/mol (**b**) -111 kJ/mol (**c**) -191 kJ/mol (**d**) -464 kJ/mol (**e**) -961 kJ/mol

EP 5.11   The combustion of exactly 1.000 g of benzoic acid in a bomb calorimeter releases 26.38 kJ of heat. If the combustion of 0.550 g of benzoic acid causes the temperature of the calorimeter to increase from 22.01 to 24.27 °C, calculate the heat capacity of the calorimeter. (**a**) 0.660 kJ/°C (**b**) 6.42 kJ/°C (**c**) 14.5 kJ/°C (**d**) 21.2 kJ/°C (**e**) 32.7 kJ/°C

EP 5.12   Calculate ΔH for $2\,NO(g) + O_2(g) \longrightarrow N_2O_4(g)$, using the following information:

$$N_2O_4(g) \longrightarrow 2\,NO_2(g)\ \ \Delta H = +57.9\ kJ$$

$$2\,NO(g) + O_2(g) \longrightarrow 2\,NO_2(g)\ \ \Delta H = -113.1\ kJ$$

(**a**) 2.7 kJ (**b**) -55.2 kJ (**c**) -85.5 kJ (**d**) -171.0 kJ (**e**) $+55.2$ kJ

EP 5.13   Calculate ΔH for the reaction

$$C(s) + H_2O(g) \longrightarrow CO(g) + H_2(g)$$

given the following thermochemical equations:

$$C(s) + O_2(g) \longrightarrow CO_2(g)\ \ \ \ \Delta H_1 = -393.5\ kJ$$

$$2\,CO(g) + O_2(g) \longrightarrow 2\,CO_2(g)\ \ \ \Delta H_2 = -566.0\ kJ$$

$$2\,H_2(g) + O_2(g) \longrightarrow 2\,H_2O(g)\ \ \ \Delta H_3 = -483.6\ kJ$$

(**a**) 1443.1 kJ (**b**) 918.3 kJ (**c**) 131.3 kJ (**d**) 262.6 kJ (**e**) 656.1 kJ

EP 5.14   If the heat of formation of $H_2O(l)$ is -286 kJ/mol, which of the following thermochemical equations is correct?

(**a**) $2\,H(g) + O(g) \longrightarrow H_2O(l)$ $\Delta H = -286$ kJ

(**b**) $2\,H_2(g) + O_2(g) \longrightarrow 2\,H_2O(l)$ $\Delta H = -286$ kJ

(**c**) $H_2(g) + \frac{1}{2}O_2(g) \longrightarrow H_2O(l)$ $\Delta H = -286$ kJ

(**d**) $H_2(g) + O(g) \longrightarrow H_2O(g)$ $\Delta H = -286$ kJ

(**e**) $H_2O(l) \longrightarrow H_2(g) + \frac{1}{2}O_2(g)$ $\Delta H = -286$ kJ

EP 5.15   Calculate the enthalpy change for the reaction

$$2\,H_2O_2(l) \longrightarrow 2\,H_2O(l) + O_2(g)$$

using the following enthalpies of formation:

$$\Delta H_f^\circ[H_2O_2(l)] = -187.8\ kJ/mol\ \ \Delta H_f^\circ[H_2O(l)] = -285.8\ kJ/mol$$

(**a**) -98.0 kJ (**b**) -196.0 kJ (**c**) $+98.0$ kJ (**d**) $+196.0$ kJ (**e**) more information needed.

EP 5.16   If ΔH_{rxn}° is the enthalpy change for the reaction $2\,SO_2(g) + O_2(g) \longrightarrow 2\,SO_3(g)$, which of the following equations is correct?

(**a**) $\Delta H_f^\circ[SO_3] = \Delta H_{rxn}^\circ - \Delta H_f^\circ[SO_2]$

(**b**) $\Delta H_f^\circ[SO_3] = \Delta H_{rxn}^\circ + \Delta H_f^\circ[SO_2]$

(**c**) $2\Delta H_f^\circ[SO_3] = \Delta H_{rxn}^\circ + 2\Delta H_f^\circ[SO_2]$

(**d**) $2\,\Delta H_f^\circ[SO_3] = \Delta H_{rxn}^\circ - 2\,\Delta H_f^\circ[SO_2]$

(**e**) $2\Delta H_f^\circ[SO_3] = 2\Delta H_f^\circ[SO_2] - \Delta H_{rxn}^\circ$

EP 5.17 Which of the following statements about bond enthalpies is or are *true*?

(**i**) Bond enthalpies are always positive numbers.

(**ii**) The bond enthalpy for a double bond between two atoms is larger than that for a single bond between the same two atoms.

(**iii**) The larger the bond enthalpy, the stronger the bond.

(**a**) Only one of these statements is true. (**b**) i and ii are true. (**c**) i and iii are true. (**d**) ii and iii are true. (**e**) All three statements are true.

EP 5.18 Use the average bond enthalpies in Table 5.4 to estimate ΔH for the "water splitting reaction": $H_2O(g) \longrightarrow H_2(g) + \frac{1}{2}O_2(g)$. (**a**) 242 kJ (**b**) 417 kJ (**c**) 5 kJ (**d**) −5 kJ (**e**) −468 kJ

EP 5.19 A stalk of celery has a caloric content (fuel value) of 9.0 kcal. If 1.0 kcal is provided by fat and there is very little protein, estimate the number of grams of carbohydrate and fat in the celery.

(**a**) 2 g carbohydrate and 0.1 g fat (**b**) 2 g carbohydrate and 1 g fat (**c**) 1 g carbohydrate and 2 g fat (**d**) 32 g carbohydrate and 10 g fat

EP 5.20 Liquid methanol [$CH_3OH(l)$], liquid ethanol [$C_2H_5OH(l)$], and gaseous dimethyl ether [$CH_3OCH_3(g)$] are all fuels that contain only carbon, hydrogen, and oxygen atoms. The standard molar enthalpies for combustion of these three fuels to $CO_2(g)$ and $H_2O(g)$ are as follows:

Label	Compound	$\Delta H°_{combustion}$ (kJ/mol)
A	$CH_3OH(l)$	−638.5
B	$C_2H_5OH(l)$	−1235
C	$CH_3OCH_3(g)$	−1328

What is the correct order of the three compounds from lowest to highest fuel value in kJ/g? (**a**) A < B < C (**b**) C < B < A (**c**) B < C < A (**d**) A < B = C (**e**) A < C < B

Exercises

Visualizing Concepts

5.1 The accompanying photo shows a pipevine swallowtail caterpillar climbing up a twig. (**a**) As the caterpillar climbs, its potential energy is increasing. What source of energy has been used to effect this change in potential energy? (**b**) If the caterpillar is the system, can you predict the sign of q as the caterpillar climbs? (**c**) Does the caterpillar do work in climbing the twig? Explain. (**d**) Does the amount of work done in climbing a 12-inch section of the twig depend on the speed of the caterpillar's climb? (**e**) Does the change in potential energy depend on the caterpillar's speed of climb? [Section 5.1]

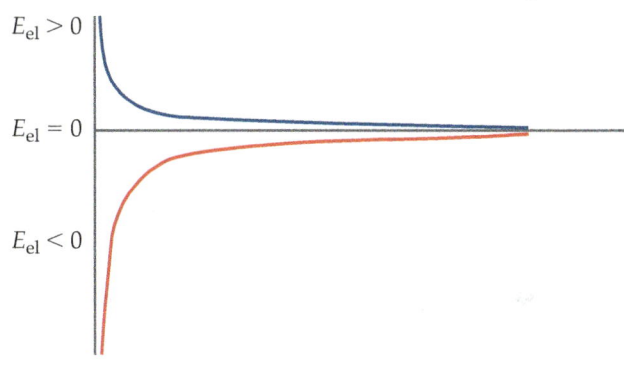

5.2 The blue curve in the plot here shows the potential energy of an ion A interacting with an ion B as a function of their separation distance. The red curve shows the interaction of ion A with a different ion (ion C) as a function of separation. (**a**) Do the charges on ions A and B have the same or different signs? (**b**) Do the charges on ions A and C have the same or different signs? (**c**) Is the magnitude of the charge on ion B greater than, equal to, or less than the magnitude of the charge on ion C? [Section 5.1]

5.3 (**a**) Does the accompanying energy diagram represent an increase or decrease in the internal energy of the system? (**b**) What sign is given for this process? (**c**) If there is no work associated with the process, is it exothermic or endothermic? [Section 5.2]

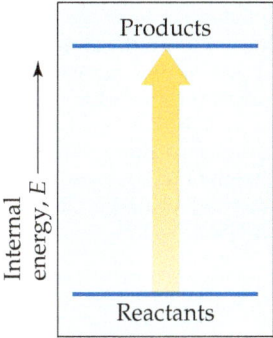

5.4 The contents of the closed box in each of the following illustrations represent a system, and the arrows show the changes to the system during some process. The lengths of the arrows represent the relative magnitudes of q and w. (**a**) Which of these processes is endothermic? (**b**) For which of these processes, if any, is $\Delta E < 0$? (**c**) For which process, if any, does the system experience a net gain in internal energy? [Section 5.2]

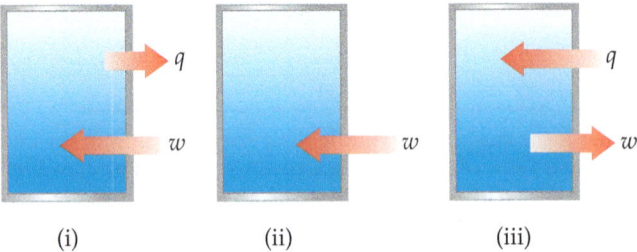

(i)   (ii)   (iii)

5.5 The diagram shows four states of a system, each with different internal energy, E. (**a**) Which of the states of the system has the greatest internal energy? (**b**) In terms of the ΔE values, write two expressions for the difference in internal energy between State A and State B. (**c**) Write an expression for the difference in energy between State C and State D. (**d**) Suppose there is another state of the system, State E, and its energy relative to State A is $\Delta E = \Delta E_1 + \Delta E_4$. Where would State E be on the diagram? [Section 5.2]

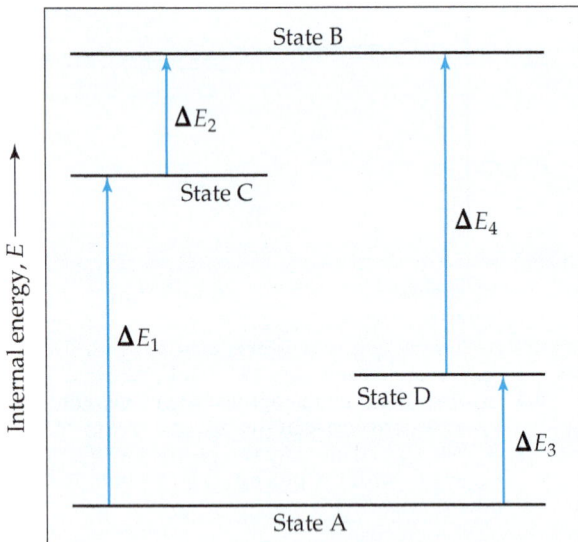

5.6 When you compress the air in a bicycle pump, the body of the pump gets warmer. (**a**) Assuming the pump and the air in it compose the system, what is the sign of w when you compress the air? (**b**) What is the sign of q for this process? (**c**) Based on your answers to parts (**a**) and (**b**), can you determine the sign of ΔE for compressing the air in the pump? If not, what would you expect for the sign of ΔE? What is your reasoning? [Section 5.2]

5.7 Imagine a container placed in a tub of water, as depicted in the accompanying diagram. (**a**) If the contents of the container are the system and heat is able to flow through the container walls, what qualitative changes will occur in the temperatures of the system and in its surroundings? From the system's perspective, is the process exothermic or endothermic? (**b**) If neither the volume nor the pressure of the system changes during the process, how is the change in internal energy related to the change in enthalpy? [Sections 5.2 and 5.3]

5.8 In the accompanying cylinder diagram, a chemical process occurs at constant temperature and pressure. (**a**) Is the sign of w indicated by this change positive or negative? (**b**) If the process is endothermic, does the internal energy of the system within the cylinder increase or decrease during the change, and is ΔE positive or negative? [Sections 5.2 and 5.3]

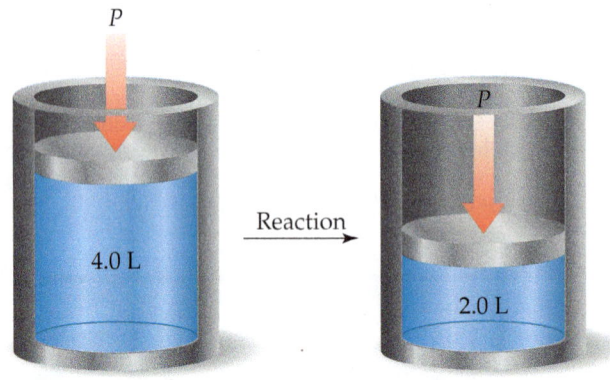

5.9 The gas-phase reaction shown, between N_2 and O_2, was run in an apparatus designed to maintain a constant pressure. (**a**) Write a balanced chemical equation for the reaction depicted and predict whether w is positive, negative, or zero. (**b**) Using data from Appendix C, determine ΔH for the formation of one mole of the product. [Sections 5.3 and 5.7]

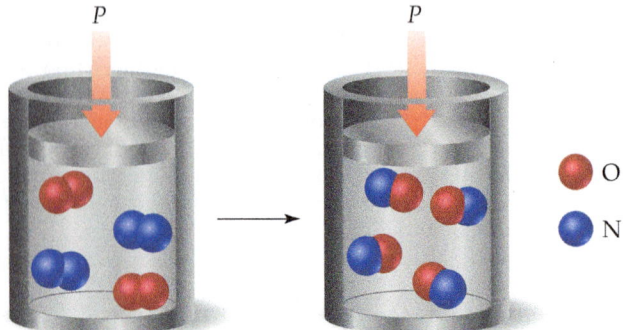

5.10 Consider the two diagrams that follow. (**a**) Based on (i), write an equation showing how ΔH_A is related to ΔH_B and ΔH_C. (**b**) Based on (ii), write an equation relating ΔH_Z to the other enthalpy changes in the diagram. (**c**) The equations you obtained in parts (**a**) and (**b**) are based on what law? (**d**) Would similar relationships hold for the work involved in each process? [Section 5.6]

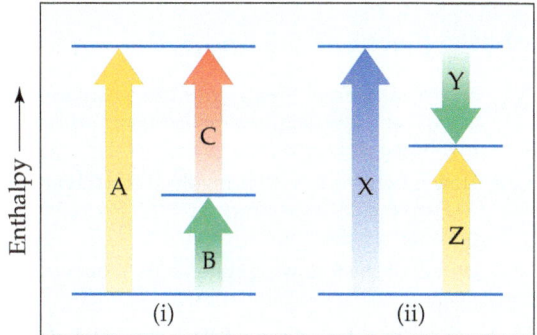

5.11 Consider the conversion of compound A into compound B: A $\longrightarrow$ B. For both compounds A and B, $\Delta H_f^\circ > 0$. (**a**) Sketch an enthalpy diagram for the reaction that is analogous to Figure 5.23. (**b**) Suppose the overall reaction is exothermic. What can you conclude? [Section 5.7]

5.12 Consider the reaction pictured below in which a molecule of X_2 reacts with a molecule of Y_2 to form two molecules of XY. The reaction is exothermic. Which of the following statements involving the bond enthalpies of X_2, Y_2, and XY is or are *true*?

   (**i**) $D(X\text{–}X)$, $D(Y\text{—}Y)$, and $D(X\text{—}Y)$ all have positive values.

   (**ii**) $2D(X\text{—}Y) < D(X\text{—}X) + D(Y\text{—}Y)$.

   (**iii**) $\Delta H = D(X\text{—}X) + D(Y\text{—}Y) - 2D(X\text{—}Y)$.

[Section 5.8]

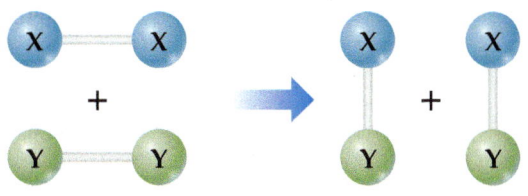

The Nature of Chemical Energy (Section 5.1)

5.13 (**a**) What is the electrostatic potential energy (in joules) between an electron and a proton that are separated by 53 pm? (**b**) What is the change in potential energy if the distance separating the electron and proton is increased to 1.0 nm? (**c**) Does the potential energy of the two particles increase or decrease when the distance is increased to 1.0 nm?

5.14 (**a**) What is the electrostatic potential energy (in joules) between two protons that are separated by 62 pm? (**b**) What is the change in potential energy if the distance separating the two is increased to 1.0 nm? (**c**) Does the potential energy of the two particles increase or decrease when the distance is increased to 1.0 nm?

5.15 (**a**) The electrostatic force (not energy) of attraction between two oppositely charged objects is given by the equation $F = \kappa(Q_1 Q_2/d^2)$ where $\kappa = 8.99 \times 10^9$ N-m^2/C^2, Q_1 and Q_2 are the charges of the two objects in coulombs, and d is the distance separating the two objects in meters. What is the electrostatic force of attraction (in newtons) between an electron and a proton that are separated by 1.00×10^2 pm? (**b**) The force of gravity acting between two objects is given by the equation $F = G(m_1 m_2/d^2)$, where $G = 6.674 \times 10^{-11}$ N-m^2/kg^2 is the gravitational constant, m_1 and m_2 are the masses of the two objects, and d is the distance separating them. What is the gravitational force of attraction (in newtons) between the electron and proton? (**c**) How many times larger is the electrostatic force of attraction?

5.16 Use the equations given in Problem 5.15 to calculate: (**a**) The electrostatic force of repulsion for two protons separated by 75 pm. (**b**) The gravitational force of attraction for two protons separated by 75 pm. (**c**) If allowed to move, will the protons be repelled or attracted to one another?

5.17 A sodium ion, Na$^+$, with a charge of 1.6×10^{-19} C and a chloride ion, Cl$^-$, with a charge of -1.6×10^{-19} C, are separated by a distance of 0.50 nm. How much work would be required to increase the separation of the two ions to an infinite distance?

5.18 A magnesium ion, Mg^{2+}, with a charge of 3.2×10^{-19} C and an oxide ion, O^{2-}, with a charge of -3.2×10^{-19} C, are separated by a distance of 0.35 nm. How much work would be required to increase the separation of the two ions to an infinite distance?

The First Law of Thermodynamics (Section 5.2)

5.19 (**a**) Which of the following cannot leave or enter a closed system: heat, work, or matter? (**b**) Which cannot leave or enter an isolated system? (**c**) What do we call the part of the universe that is not part of the system?

5.20 Classify each of the following as an open, closed, or isolated system: (**a**) The air in a balloon. (**b**) The water in a puddle that is evaporating outdoors. (**c**) A cold beverage in an insulated jug. (**d**) A chemical reaction in a stoppered flask that produces heat.

5.21 (**a**) According to the first law of thermodynamics, what quantity is conserved? (**b**) What is meant by the *internal energy* of a system? (**c**) By what means can the internal energy of a closed system increase?

5.22 (**a**) Write an equation that expresses the first law of thermodynamics in terms of heat and work. (**b**) Under what conditions will the quantities q and w be negative numbers?

5.23 Calculate ΔE and determine whether the process is endothermic or exothermic for the following cases: (**a**) $q = 0.763$ kJ and $w = -840$ J. (**b**) A system releases 66.1 kJ of heat to its surroundings while the surroundings do 44.0 kJ of work on the system.

5.24 For the following processes, calculate the change in internal energy of the system and determine whether the process is endothermic or exothermic: (**a**) A balloon is cooled by removing 0.655 kJ of heat. It shrinks on cooling, and the atmosphere does 382 J of work on the balloon. (**b**) A 100.0-g bar of gold is heated from 25°C to 50°C during which it absorbs 322 J of heat. Assume the volume of the gold bar remains constant.

5.25 A gas is confined to a cylinder fitted with a piston and an electrical heater, as shown here:

Suppose that current is supplied to the heater so that 100 J of energy is added. Consider two different situations. In case (1) the piston is allowed to move as the energy is added. In case (2) the piston is fixed so that it cannot move. (a) In which case does the gas have the higher temperature after addition of the electrical energy? (b) Identify the sign (positive, negative, or zero) of q and w in each case? (c) In which case is ΔE for the system (the gas in the cylinder) larger?

5.26 Consider a system consisting of two oppositely charged spheres hanging by strings and separated by a distance r_1, as shown in the accompanying illustration. Suppose they are separated to a larger distance r_2, by moving them apart. (a) What change, if any, has occurred in the potential energy of the system? (b) What effect, if any, does this process have on the value of ΔE? (c) What can you say about q and w for this process?

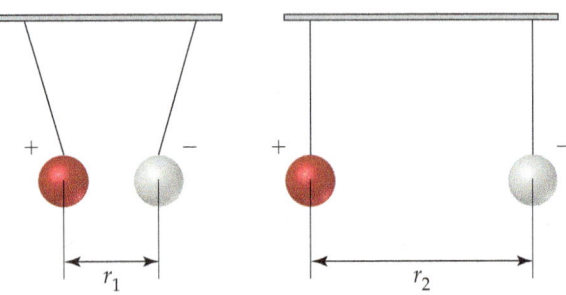

5.27 Imagine that you are climbing a mountain. Which of the following are state functions? (a) The distance you walk during your climb to the top (b) The change in elevation during the climb (c) The change in your gravitational potential energy during the climb (d) The number of calories you expend during the climb.

5.28 Indicate which of the following is independent of the path by which a change occurs. (a) the change in potential energy when a book is transferred from table to shelf (b) the heat evolved when a cube of sugar is oxidized to $CO_2(g)$ and $H_2O(g)$ (c) the work accomplished in burning a gallon of gasoline.

5.29 A system goes from one state to another by two different paths. For Path 1, the system does 140 J of work on the surroundings, and its internal energy increases by 330 J. For Path 2, the system absorbs 220 J of heat from the surroundings. (a) What is the value of q for Path 1? (b) What is the value of w for Path 2?

5.30 A system goes from State A to State B by two different paths: Path 1, for which $q_1 = 38$ kJ and $w_1 = -97$ kJ, and Path 2, for which $q_2 = -28$ kJ. (a) What is the value of ΔE for Path 1? (b) What is the value of w for Path 2?

Enthalpy (Sections 5.3 and 5.4)

5.31 During a normal breath, our lungs expand about 0.50 L against an external pressure of 1.0 atm. How much work is involved in this process (in J)?

5.32 How much work (in J) is involved in a chemical reaction if the volume decreases from 5.00 to 1.26 L against a constant pressure of 0.857 atm?

5.33 Which of the following statements about enthalpy is or are *true*?

  (i) Because E, P, and V are state functions, H is a state function.

  (ii) If a system changes its state under a constant external pressure, then $\Delta H = q$.

  (iii) Under constant-pressure conditions, if $\Delta H = 0$, then ΔE must also equal zero.

5.34 (a) Under what condition will the enthalpy change of a process equal the amount of heat transferred into or out of the system? (b) During a constant-pressure process, the system releases heat to the surroundings. Does the enthalpy of the system increase or decrease during the process? (c) In a constant-pressure process, $\Delta H = 0$. What can you conclude about ΔE, q, and w?

5.35 Assume that the following reaction occurs at constant pressure:

$$2\,Al(s) + 3\,Cl_2(g) \longrightarrow 2\,AlCl_3(s)$$

(a) If you are given ΔH for the reaction, what additional information do you need to determine ΔE for the process? (b) Which quantity is larger for this reaction? (c) Explain your answer to part (b).

5.36 Suppose that the gas-phase reaction $2\,NO(g) + O_2(g) \longrightarrow 2\,NO_2(g)$ were carried out in a constant-volume container at constant temperature. (a) Would the measured heat change represent ΔH or ΔE? (b) If there is a difference, which quantity is larger for this reaction? (c) Explain your answer to part (b).

5.37 A gas is confined to a cylinder under constant atmospheric pressure, as illustrated in Figure 5.4. When the gas undergoes a particular chemical reaction, it absorbs 824 J of heat from its surroundings and has 0.65 kJ of P–V work done on it by its surroundings. What are the values of ΔH and ΔE for this process?

5.38 A gas is confined to a cylinder under constant atmospheric pressure, as illustrated in Figure 5.4. When 0.49 kJ of heat is added to the gas, it expands and does 214 J of work on the surroundings. What are the values of ΔH and ΔE for this process?

5.39 The complete combustion of ethanol, $C_2H_5OH(l)$, to form $H_2O(g)$ and $CO_2(g)$ at constant pressure releases 1235 kJ of heat per mole of C_2H_5OH. (a) Write a balanced thermochemical equation for this reaction. (b) Draw an enthalpy diagram for the reaction.

5.40 The decomposition of $Ca(OH)_2(s)$ into $CaO(s)$ and $H_2O(g)$ at constant pressure requires the addition of 109 kJ of heat per mole of $Ca(OH)_2$. (**a**) Write a balanced thermochemical equation for the reaction. (**b**) Draw an enthalpy diagram for the reaction.

5.41 Ozone, $O_3(g)$, is a form of elemental oxygen that plays an important role in the absorption of ultraviolet radiation in the stratosphere. It decomposes to $O_2(g)$ at room temperature and pressure according to the following reaction:

$$2\,O_3(g) \longrightarrow 3\,O_2(g) \quad \Delta H = -284.6\,\text{kJ}$$

(**a**) What is the enthalpy change for this reaction per mole of $O_3(g)$? (**b**) Which has the higher enthalpy under these conditions, $2\,O_3(g)$ or $3\,O_2(g)$?

5.42 Without referring to tables, predict which of the following has the higher enthalpy in each case: (**a**) 1 mol $CO_2(s)$ or 1 mol $CO_2(g)$ at the same temperature (**b**) 2 mol of hydrogen atoms or 1 mol of H_2 (**c**) 1 mol $H_2(g)$ and 0.5 mol $O_2(g)$ at 25 °C or 1 mol $H_2O(g)$ at 25 °C (**d**) 1 mol $N_2(g)$ at 100 °C or 1 mol $N_2(g)$ at 300 °C

5.43 Consider the following reaction:

$$2\,Mg(s) + O_2(g) \longrightarrow 2\,MgO(s) \quad \Delta H = -1204\,\text{kJ}$$

(**a**) Is this reaction exothermic or endothermic? (**b**) Calculate the amount of heat transferred when 3.55 g of $Mg(s)$ reacts at constant pressure. (**c**) How many grams of MgO are produced during an enthalpy change of −234 kJ? (**d**) How many kilojoules of heat are absorbed when 40.3 g of $MgO(s)$ is decomposed into $Mg(s)$ and $O_2(g)$ at constant pressure?

5.44 Consider the following reaction:

$$2\,CH_3OH(g) \longrightarrow 2\,CH_4(g) + O_2(g) \quad \Delta H = +252.8\,\text{kJ}$$

(**a**) Is this reaction exothermic or endothermic? (**b**) Calculate the amount of heat transferred when 24.0 g of $CH_3OH(g)$ is decomposed by this reaction at constant pressure. (**c**) For a given sample of CH_3OH, the enthalpy change during the reaction is 82.1 kJ. How many grams of methane gas are produced? (**d**) How many kilojoules of heat are released when 38.5 g of $CH_4(g)$ react completely with $O_2(g)$ to form $CH_3OH(g)$ at constant pressure?

5.45 When solutions containing silver ions and chloride ions are mixed, silver chloride precipitates:

$$Ag^+(aq) + Cl^-(aq) \longrightarrow AgCl(s) \quad \Delta H = -65.5\,\text{kJ}$$

(**a**) Calculate ΔH for the production of 0.450 mol of AgCl by this reaction. (**b**) Calculate ΔH for the production of 9.00 g of AgCl. (**c**) Calculate ΔH when 9.25×10^{-4} mol of AgCl dissolves in water.

5.46 At one time, a common means of forming small quantities of oxygen gas in the laboratory was to heat $KClO_3$:

$$2\,KClO_3(s) \longrightarrow 2\,KCl(s) + 3\,O_2(g) \quad \Delta H = -89.4\,\text{kJ}$$

For this reaction, calculate ΔH for the formation of (**a**) 1.36 mol of O_2 and (**b**) 10.4 g of KCl. (**c**) Now consider the reverse reaction, in which $KClO_3$ is formed from KCl and O_2. What is ΔH for the formation of 19.1 g $KClO_3$ from KCl and O_2. (**d**) Would you expect the reaction in part (c) to occur spontaneously?

5.47 Consider the combustion of *isopropanol*, $C_3H_7OH(l)$, which is the primary component of rubbing alcohol:

$$C_3H_7OH(l) + 9/2\,O_2(g) \rightarrow 3\,CO_2(g) + 4\,H_2O(l) \quad \Delta H = -2248\,\text{kJ}$$

(**a**) What is the enthalpy change for the reverse reaction? (**b**) Balance the forward reaction with whole-number coefficients. What is ΔH for the reaction represented by this

equation? (**c**) Which is more likely to be thermodynamically favored, the forward reaction or the reverse reaction? (**d**) If the reaction were written to produce $H_2O(g)$ instead of $H_2O(l)$, would you expect the magnitude of ΔH to increase, decrease, or stay the same?

5.48 Consider the decomposition of liquid benzene, $C_6H_6(l)$, to gaseous acetylene, $C_2H_2(g)$:

$$C_6H_6(l) \longrightarrow 3\,C_2H_2(g) \quad \Delta H = +630\,\text{kJ}$$

(**a**) What is the enthalpy change for the reverse reaction? (**b**) What is ΔH for the formation of 1 mol of acetylene? (**c**) Which is more likely to be thermodynamically favored, the forward reaction or the reverse reaction? (**d**) If $C_6H_6(g)$ were consumed instead of $C_6H_6(l)$, would you expect the magnitude of ΔH to increase, decrease, or stay the same?

Calorimetry (Section 5.5)

5.49 (**a**) What are the units of molar heat capacity? (**b**) What are the units of specific heat? (**c**) If you know the specific heat of copper, what additional information do you need to calculate the heat capacity of a particular piece of copper pipe?

5.50 Two solid objects, A and B, are placed in boiling water and allowed to come to the temperature of the water. Each is then lifted out and placed in separate beakers containing 1000 g water at 10.0°C. Object A increases the water temperature by 3.50°C; B increases the water temperature by 2.60 °C. (**a**) Which object has the larger heat capacity? (**b**) What can you say about the specific heats of A and B?

5.51 (**a**) What is the specific heat of liquid water? (**b**) What is the molar heat capacity of liquid water? (**c**) What is the heat capacity of 185 g of liquid water? (**d**) How many kJ of heat are needed to raise the temperature of 10.00 kg of liquid water from 24.6 to 46.2°C?

5.52 (**a**) Which substance in Table 5.2 requires the smallest amount of energy to increase the temperature of 50.0 g of that substance by 10 K? (**b**) Calculate the energy needed for this temperature change.

5.53 The specific heat of ethanol, $C_2H_5OH(l)$, is 2.44 J/g-K. (**a**) How many J of heat are needed to raise the temperature of 80.0 g of ethanol from 10.0 to 25.0°C? (**b**) Which will require more heat, increasing the temperature of 1 mol of $C_2H_5OH(l)$ by a certain amount or increasing the temperature of 1 mol of $H_2O(l)$ by the same amount?

5.54 Consider the data about gold metal in Exercise 5.24(b). (**a**) Based on the data, calculate the specific heat of $Au(s)$. (**b**) Suppose that the same amount of heat is added to two 10.0-g blocks of metal, both initially at the same temperature. One block is gold metal, and the other is iron metal. Which block will have the greater rise in temperature after addition of the heat? (**c**) What is the molar heat capacity of $Au(s)$?

5.55 When a 5.10-g sample of solid sodium hydroxide dissolves in 100.0 g of water in a coffee-cup calorimeter (Figure 5.18), the temperature rises from 20.5 to 33.2 °C. (**a**) Calculate the quantity of heat (in kJ) released in the reaction. (**b**) Using your result from part (a), calculate ΔH (in kJ/mol NaOH) for the solution process. Assume that the specific heat of the solution is the same as that of pure water.

5.56 (**a**) When a 6.50-g sample of solid ammonium nitrate dissolves in 60.0 g of water in a coffee-cup calorimeter (Figure 5.18), the temperature drops from 21.8 to 14.0 °C. Calculate ΔH (in kJ/mol NH_4NO_3)) for the solution process:

$$NH_4NO_3(s) \longrightarrow NH_4^+(aq) + NO_3^-(aq)$$

Assume that the specific heat of the solution is the same as that of pure water. **(b)** Is this process endothermic or exothermic?

5.57 A 2.200-g sample of quinone ($C_6H_4O_2$) is burned in a bomb calorimeter whose total heat capacity is 7.854 kJ/°C. The temperature of the calorimeter increases from 23.44 to 30.57 °C. **(a)** What is the heat of combustion per gram of quinone? **(b)** What is the heat of combustion per mole of quinone?

5.58 A 1.800-g sample of phenol (C_6H_5OH) was burned in a bomb calorimeter whose total heat capacity is 11.66 kJ/°C. The temperature of the calorimeter plus contents increased from 21.36 to 26.37°C. **(a)** Write a balanced chemical equation for the bomb calorimeter reaction. **(b)** What is the heat of combustion per gram of phenol? **(c)** What is the heat of combustion per mole of phenol?

5.59 Under constant-volume conditions, the heat of combustion of benzoic acid (C_6H_5COOH) is 26.38 kJ/g. A 2.760-g sample of benzoic acid is burned in a bomb calorimeter. The temperature of the calorimeter increases from 21.60 to 29.93°C. **(a)** What is the total heat capacity of the calorimeter? **(b)** A 1.440-g sample of a new organic substance is combusted in the same calorimeter. The temperature of the calorimeter increases from 22.14 to 27.09°C. What is the heat of combustion per gram of the new substance? **(c)** Suppose that in changing samples, a portion of the water in the calorimeter was lost. In what way, if any, would this change the heat capacity of the calorimeter?

5.60 *Glycerol* (also called *glycerine*), $C_3H_8O_3(l)$, is a viscous, sweet-tasting liquid used in a variety of food and personal care products. Under constant-volume conditions, the combustion of 1.000 mol of glycerol releases 1653 kJ of heat to the surroundings. A 2.850-g sample of glycerol is burned in a bomb calorimeter. The temperature of the calorimeter increases from 22.11 to 27.42 °C. **(a)** Write a balanced equation for the combustion reaction, assuming the products are $CO_2(g)$ and $H_2O(l)$. **(b)** What is the total heat capacity of the calorimeter? **(c)** If the size of the glucose sample had been 4.000 g, what would the temperature change of the calorimeter have been? **(d)** Which of the following values about the combustion of glycerol under constant volume is or are correct: (i) $\Delta H = -1653$ kJ/mol; (ii) $\Delta E = -1653$ kJ/mol; (iii) $q_V = -1653$ kJ/mol?

Hess's Law (Section 5.6)

5.61 Consider the following hypothetical reactions:

$$A \longrightarrow B \quad \Delta H = +30 \text{ kJ}$$
$$B \longrightarrow C \quad \Delta H = +60 \text{ kJ}$$

(a) Use Hess's law to calculate the enthalpy change for the reaction $A \longrightarrow C$. **(b)** Construct an enthalpy diagram for substances A, B, and C, and show how Hess's law applies.

5.62 You are given the following hypothetical reactions:

$$2A \longrightarrow B + C \quad \Delta H = +50 \text{ kJ}$$
$$A \longrightarrow D \qquad \Delta H = -80 \text{ kJ}$$

Use Hess's law to calculate ΔH for the reaction $B + C \longrightarrow 2D$. **(b)** What is ΔH for the reaction $D \longrightarrow \frac{1}{2}B + \frac{1}{2}C$?

5.63 Calculate the enthalpy change for the reaction

$$P_4O_6(s) + 2O_2(g) \longrightarrow P_4O_{10}(s)$$

given the following enthalpies of reaction:

$$P_4(s) + 3O_2(g) \longrightarrow P_4O_6(s) \quad \Delta H = -1640.1 \text{ kJ}$$
$$P_4(s) + 5O_2(g) \longrightarrow P_4O_{10}(s) \quad \Delta H = -2940.1 \text{ kJ}$$

5.64 From the enthalpies of reaction

$$2C(s) + O_2(g) \longrightarrow 2CO(g) \qquad \Delta H = -221.0 \text{ kJ}$$
$$2C(s) + O_2(g) + 4H_2(g) \longrightarrow 2CH_3OH(g) \quad \Delta H = -402.4 \text{ kJ}$$

calculate ΔH for the reaction

$$CO(g) + 2H_2(g) \longrightarrow CH_3OH(g)$$

5.65 From the enthalpies of reaction

$$H_2(g) + F_2(g) \longrightarrow 2HF(g) \quad \Delta H = -537 \text{ kJ}$$
$$C(s) + 2F_2(g) \longrightarrow CF_4(g) \quad \Delta H = -680 \text{ kJ}$$
$$2C(s) + 2H_2(g) \longrightarrow C_2H_4(g) \quad \Delta H = +52.3 \text{ kJ}$$

calculate ΔH for the reaction of ethylene with F_2:

$$C_2H_4(g) + 6F_2(g) \longrightarrow 2CF_4(g) + 4HF(g)$$

5.66 Given the data

$$N_2(g) + O_2(g) \longrightarrow 2NO(g) \qquad \Delta H = +180.7 \text{ kJ}$$
$$2NO(g) + O_2(g) \longrightarrow 2NO_2(g) \qquad \Delta H = -113.1 \text{ kJ}$$
$$2N_2O(g) \longrightarrow 2N_2(g) + O_2(g) \quad \Delta H = -163.2 \text{ kJ}$$

use Hess's law to calculate ΔH for the reaction

$$N_2O(g) + NO_2(g) \longrightarrow 3NO(g)$$

5.67 We can use Hess's law to calculate enthalpy changes that cannot be measured. One such reaction is the conversion of methane to ethylene:

$$2CH_4(g) \longrightarrow C_2H_4(g) + 2H_2(g)$$

Calculate $\Delta H°$ for this reaction using the following thermochemical data:

$$CH_4(g) + 2O_2(g) \longrightarrow CO_2(g) + 2H_2O(l) \quad \Delta H° = -890.3 \text{ kJ}$$
$$C_2H_4(g) + H_2(g) \longrightarrow C_2H_6(g) \quad \Delta H° = -136.3 \text{ kJ}$$
$$2H_2(g) + O_2(g) \longrightarrow 2H_2O(l) \quad \Delta H° = -571.6 \text{ kJ}$$
$$2C_2H_6(g) + 7O_2(g) \longrightarrow 4CO_2(g) + 6H_2O(l) \quad \Delta H° = -3120.8 \text{ kJ}$$

5.68 Dimethyl ether, $CH_3OCH_3(g)$, is an industrially useful organic compound that is a gas at room temperature. Consider the following balanced thermochemical equations:

$$CH_3OCH_3(g) + 3O_2(g) \longrightarrow 2CO_2(g) + 3H_2O(g) \quad \Delta H° = -1328 \text{ kJ}$$
$$CH_4(g) + 2O_2(g) \longrightarrow CO_2(g) + 2H_2O(g) \quad \Delta H° = -802 \text{ kJ}$$
$$2C_2H_6(g) + 7O_2(g) \longrightarrow 4CO_2(g) + 6H_2O(g) \quad \Delta H° = -2855 \text{ kJ}$$

(a) All three of the above reactions produce $H_2O(g)$. If they were written to produce $H_2O(l)$ instead, would the values of $\Delta H°$ become more negative, become less negative, or stay the same? **(b)** By using Hess's law and the reactions provided, calculate $\Delta H°$ for the following balanced reaction between dimethyl ether and methane to produce ethane and water:

$$CH_3OCH_3(g) + 2CH_4(g) \longrightarrow 2C_2H_6(g) + H_2O(g)$$

Enthalpies of Formation (Section 5.7)

5.69 **(a)** What is meant by the term *standard conditions* with reference to enthalpy changes? **(b)** What is meant by the term *enthalpy of formation*? **(c)** What is meant by the term *standard enthalpy of formation*?

5.70 **(a)** What is the value of the standard enthalpy of formation of an element in its most stable form? **(b)** Write the chemical equation for the reaction whose enthalpy change is the standard enthalpy of formation of sucrose (table sugar), $C_{12}H_{22}O_{11}(s)$, $\Delta H_f°[C_{12}H_{22}O_{11}(s)]$.

5.71 For each of the following compounds, write a balanced thermochemical equation depicting the formation of one mole of the compound from its elements in their standard states and then look up $\Delta H_f°$ for each substance in Appendix C. **(a)** $N_2O(g)$ **(b)** $FeCl_3(s)$ **(c)** $P_4O_{10}(s)$ **(d)** $Ca(OH)_2(s)$

5.72 Write balanced equations that describe the formation of the following compounds from elements in their standard states, and then look up the standard enthalpy of formation for each substance in Appendix C. **(a)** $NH_4NO_3(s)$ **(b)** $K_2CO_3(s)$ **(c)** $SOCl_2(l)$ **(d)** $NaHCO_3(s)$

5.73 The following is known as the thermite reaction:

$$2\,Al(s) + Fe_2O_3(s) \longrightarrow Al_2O_3(s) + 2\,Fe(s)$$

This highly exothermic reaction is used for welding massive units, such as propellers for large ships. Using standard enthalpies of formation in Appendix C, calculate $\Delta H°$ for this reaction.

5.74 Many portable gas heaters and grills use propane, $C_3H_8(g)$, as a fuel. Using standard enthalpies of formation, calculate the quantity of heat produced when 10.0 g of propane is completely combusted in air under standard conditions.

5.75 Using values from Appendix C, calculate the standard enthalpy change for each of the following reactions:
(a) $2\,PCl_3(g) + O_2(g) \longrightarrow 2\,POCl_3(g)$
(b) $PbCO_3(s) \longrightarrow PbO(s) + CO_2(g)$
(c) $2\,FeCl_3(s) + 3\,H_2(g) \longrightarrow 2\,Fe(s) + 6\,HCl(g)$
(d) $2\,H_2O_2(l) \longrightarrow 2\,H_2O(l) + O_2(g)$

5.76 Using values from Appendix C, calculate the value of $\Delta H°$ for each of the following reactions:
(a) $NiO(s) + 2\,HCl(g) \longrightarrow NiCl_2(s) + H_2O(g)$
(b) $2\,NO_2(g) \longrightarrow N_2O_4(g)$
(c) $2\,HBr(g) + F_2(g) \longrightarrow 2\,HF(g) + Br_2(g)$
(d) $TiCl_4(l) + 2\,H_2O(l) \longrightarrow TiO_2(s) + 4\,HCl(aq)$

5.77 Complete combustion of 1 mol of acetone (C_3H_6O) liberates 1790 kJ:

$$C_3H_6O(l) + 4\,O_2(g) \longrightarrow 3\,CO_2(g) + 3\,H_2O(l)$$
$$\Delta H° = -1790\text{ kJ}$$

Using this information together with the standard enthalpies of formation of $O_2(g)$, $CO_2(g)$, and $H_2O(l)$ from Appendix C, calculate the standard enthalpy of formation of acetone.

5.78 Calcium carbide (CaC_2) reacts with water to form acetylene (C_2H_2) and $Ca(OH)_2$. From the following enthalpy of reaction data and data in Appendix C, calculate $\Delta H_f°$ for $CaC_2(s)$:

$$CaC_2(s) + 2\,H_2O(l) \longrightarrow Ca(OH)_2(s) + C_2H_2(g)$$
$$\Delta H° = -127.2\text{ kJ}$$

5.79 Gasoline is composed primarily of hydrocarbons, including many with eight carbon atoms, called *octanes*. One of the cleanest-burning octanes is a compound called 2,3,4-trimethylpentane, which has the following structural formula:

$$\begin{array}{ccc} CH_3 & CH_3 & CH_3 \\ | & | & | \\ H_3C-CH- & CH- & CH-CH_3 \end{array}$$

The complete combustion of one mole of this compound to $CO_2(g)$ and $H_2O(g)$ leads to $\Delta H° = -5064.9$ kJ.

(a) Write a balanced equation for the combustion of 1 mol of $C_8H_{18}(l)$. **(b)** By using the information in this problem and data in Table 5.3, calculate $\Delta H_f°$ for 2,3,4-trimethylpentane.

5.80 Diethyl ether, $C_4H_{10}O(l)$, a flammable compound that was once used as a surgical anesthetic, has the structure:

$$H_3C-CH_2-O-CH_2-CH_3$$

The complete combustion of 1 mol of $C_4H_{10}O(l)$ to $CO_2(g)$ and $H_2O(l)$ yields $\Delta H° = -2723.7$ kJ. **(a)** Write a balanced equation for the combustion of 1 mol of $C_4H_{10}O(l)$. **(b)** By using the information in this problem and the data in Table 5.3, calculate $\Delta H_f°$ for diethyl ether.

5.81 Ethanol (C_2H_5OH) is blended with gasoline as an automobile fuel. **(a)** Write a balanced equation for the combustion of liquid ethanol in air. **(b)** Calculate the standard enthalpy change for the reaction, assuming $H_2O(g)$ as a product. **(c)** Calculate the heat produced per liter of ethanol by combustion of ethanol under constant pressure. Ethanol has a density of 0.789 g/mL. **(d)** Calculate the mass of CO_2 produced per kJ of heat emitted.

5.82 Methanol (CH_3OH) is used as a fuel in race cars. **(a)** Write a balanced equation for the combustion of liquid methanol in air. **(b)** Calculate the standard enthalpy change for the reaction, assuming $H_2O(g)$ as a product. **(c)** Calculate the heat produced by combustion per liter of methanol. Methanol has a density of 0.791 g/mL. **(d)** Calculate the mass of CO_2 produced per kJ of heat emitted.

Bond Enthalpies (Section 5.8)

5.83 Which of the following statements about average bond enthalpies is or are *true*?
(i) It always requires energy to break a bond.
(ii) It requires more energy to break a single bond than to break a double bond.
(iii) Bond enthalpies can be used to estimate the reaction enthalpies of gas-phase reactions.

5.84 Which of the following statements about average bond enthalpies is or are *true*?
(i) Bond enthalpies are always positive values.
(ii) When a bond is made, energy is released that is equal in magnitude to the bond enthalpy.
(iii) The bond enthalpy of any element in its most stable state is zero.

5.85 **(a)** Use enthalpies of formation given in Appendix C to calculate ΔH for the reaction $CCl_4(g) \rightarrow C(g) + 4\,Cl(g)$, and use this value to estimate the bond enthalpy $D(C-Cl)$. **(b)** How large is the difference between the value calculated in part **(a)** and the value given in Table 5.4?

5.86 Ethane, C_2H_6, is an alkane with one $C-C$ bond and six $C-H$ bonds (Section 2.9). **(a)** Use enthalpies of formation given in Appendix C to calculate ΔH for the reaction $C_2H_6(g) \rightarrow 2\,C(g) + 6\,H(g)$. **(b)** Use the result from part **(a)** and the value of $D(C-H)$ from Table 5.4 to estimate the bond enthalpy $D(C-C)$. **(c)** How large is the difference between the value calculated for $D(C-C)$ in part **(b)** and the value given in Table 5.4?

5.87 Use bond enthalpies in Table 5.4 to estimate ΔH for each of the following reactions:
(a) $H-H(g) + Br-Br(g) \longrightarrow 2\,H-Br(g)$

(b)

$$2\,H-\overset{\displaystyle H}{\underset{\displaystyle H}{C}}-O-H \;+\; 3\,O{=}O \;\longrightarrow\; 2\,O{=}C{=}O \;+\; 4\,H-O-H$$

5.88 Use bond enthalpies in Table 5.4 to estimate ΔH for each of the following reactions:

(a)

$$Br-\overset{\displaystyle Br}{\underset{\displaystyle Br}{C}}-H \;+\; Cl-Cl \;\longrightarrow\; Br-\overset{\displaystyle Br}{\underset{\displaystyle Br}{C}}-Cl \;+\; Cl-H$$

(b)

$$H-\overset{\displaystyle H}{\underset{\displaystyle H}{C}}-H \;+\; 2\,O{=}O \;\longrightarrow\; O{=}C{=}O \;+\; 2\,H-O-H$$

5.89 Consider the reaction $2\,H_2(g) + O_2(g) \rightarrow 2\,H_2O(l)$. **(a)** Use the bond enthalpies in Table 5.4 to estimate ΔH for this reaction, ignoring the fact that water is in the liquid state. **(b)** Without doing a calculation, predict whether your estimate in part **(a)** is more negative or less negative than the true reaction enthalpy. **(c)** Use the enthalpies of formation in Appendix C to determine the true reaction enthalpy.

5.90 Consider the reaction $H_2(g) + I_2(s) \rightarrow 2HI(g)$. **(a)** Use the bond enthalpies in Table 5.4 to estimate ΔH for this reaction, ignoring the fact that iodine is in the solid state. **(b)** Without doing a calculation, predict whether your estimate in part **(a)** is more negative or less negative than the true reaction enthalpy. **(c)** Use the enthalpies of formation in Appendix C to determine the true reaction enthalpy.

Foods and Fuels (Section 5.9)

5.91 **(a)** What is meant by the term *fuel value*? **(b)** Which is a greater source of energy as food, 5 g of fat or 9 g of carbohydrate? **(c)** The metabolism of glucose produces $CO_2(g)$ and $H_2O(l)$. How does the human body expel these reaction products?

5.92 **(a)** Which releases the most energy when metabolized, 1 g of carbohydrates or 1 g of fat? **(b)** A particular chip snack food is composed of 12% protein, 14% fat, and the rest carbohydrate. What percentage of the calorie content of this food is fat? **(c)** How many grams of protein provide the same fuel value as 25 g of fat?

5.93 **(a)** A serving of a particular ready-to-serve chicken noodle soup contains 2.5 g fat, 14 g carbohydrate, and 7 g protein. Estimate the number of Calories in a serving. **(b)** According to its nutrition label, the same soup also contains 690 mg of sodium. Do you think the sodium contributes to the caloric content of the soup?

5.94 A pound of plain M&M® candies contains 96 g fat, 320 g carbohydrate, and 21 g protein. What is the fuel value in kJ in a 42-g (about 1.5 oz) serving? How many Calories does it provide?

5.95 The heat of combustion of fructose, $C_6H_{12}O_6$, is $-2812\,kJ/mol$. If a fresh golden delicious apple weighing 4.23 oz (120 g) contains 16.0 g of fructose, what caloric content does the fructose contribute to the apple?

5.96 The heat of combustion of ethanol, $C_2H_5OH(l)$, is $-1367\,kJ/mol$. A batch of Sauvignon Blanc wine contains 10.6% ethanol by mass. Assuming the density of the wine to be $1.0\,g/mL$, what is the caloric content due to the alcohol (ethanol) in a 6-oz glass of wine (177 mL)?

5.97 The standard enthalpies of formation of gaseous propyne (C_3H_4), propylene (C_3H_6), and propane (C_3H_8) are $+185.4$, $+20.4$, and $-103.8\,kJ/mol$, respectively. **(a)** Calculate the heat evolved per mole on combustion of each substance to yield $CO_2(g)$ and $H_2O(g)$. **(b)** Calculate the heat evolved on combustion of 1 kg of each substance. **(c)** Which is the most efficient fuel in terms of heat evolved per unit mass?

5.98 Methane, $CH_4(g)$, is the principal component of natural gas. Natural gas vehicles (NGVs) have become increasingly popular, in part because methane is a cleaner fuel than gasoline with regard to the production of atmospheric CO_2 as well as other pollutants. The principal component of gasoline is octane, $C_8H_{18}(l)$. The thermochemical equations for the combustion of methane and octane are:

$$CH_4(g) + 2\,O_2(g) \longrightarrow CO_2(g) + 2\,H_2O(l) \quad \Delta H = -890.4\,kJ$$
$$C_8H_{18}(l) + 25/2\,O_2(g) \longrightarrow 8\,CO_2(g) + 9\,H_2O(l) \quad \Delta H = -5471\,kJ$$

(a) What are the fuel values of methane and octane, in kJ/g? **(b)** For combustion of methane, what is the heat released per mol CO_2 generated? **(c)** For combustion of octane, what is the heat released per mol CO_2 generated? **(d)** A typical metropolitan bus requires about 23,200 kJ of energy from combustion per mile traveled. If an octane-powered bus travels 50.0 mi during a day, what mass of CO_2 in kg is produced? **(e)** If the bus in part **(d)** were powered by methane instead of octane, what mass of CO_2 in kg is produced?

5.99 At the end of 2020, global population was about 7.8 billion. What mass of glucose in kg would be needed to provide 1500 Cal/person/day of nourishment to the global population for one year? Assume that glucose is metabolized entirely to $CO_2(g)$ and $H_2O(l)$ according to the following thermochemical equation:

$$C_6H_{12}O_6(s) + 6\,O_2(g) \longrightarrow 6\,CO_2(g) + 6\,H_2O(l)$$
$$\Delta H° = -2803\,kJ$$

5.100 The automobile fuel called E85 consists of 85% ethanol and 15% gasoline. E85 can be used in the so-called flex-fuel vehicles (FFVs), which can use gasoline, ethanol, or a mix as fuel. Assume that gasoline consists of a mixture of octanes (different isomers of C_8H_{18}), that the average heat of combustion of $C_8H_{18}(l)$ is 5400 kJ/mol, and that gasoline has an average density of $0.70\,g/mL$. The density of ethanol is $0.79\,g/mL$. **(a)** By using the information given as well as data in Appendix C, compare the energy produced by combustion of 1.0 L of gasoline and of 1.0 L of ethanol. **(b)** Assume that the density and heat of combustion of E85 can be obtained by using 85% of the values for ethanol and 15% of the values for gasoline. How much energy could be released by the combustion of 1.0 L of E85? **(c)** How many gallons of E85 would be needed to provide the same energy as 10 gal of gasoline? **(d)** If gasoline costs $3.88 per gallon in the United States, what is the break-even price per gallon of E85 if the same amount of energy is to be delivered?

Additional Exercises

5.101 Two positively charged spheres, each with a charge of 2.0×10^{-5} C, a mass of 1.0 kg, and separated by a distance of 1.0 cm, are held in place on a frictionless track. **(a)** What is the electrostatic potential energy of this system? **(b)** If the spheres are released, will they move toward or away from each other? **(c)** What speed will each sphere attain as the distance between the spheres approaches infinity? [Section 5.1]

5.102 Identify whether work is being performed for each of the following. **(a)** a rock is thrown into the sky **(b)** two strong magnets are held at a fixed distance from one another **(c)** a bicyclist starts a ride by pushing on the pedals **(d)** a spring is compressed to half its original length **(e)** a gas expands into a vacuum

5.103 The air bags that provide protection in automobiles in the event of an accident expand because of a rapid chemical reaction. From the viewpoint of the chemical reactants as the system, what do you expect for the signs of q and w in this process?

5.104 An aluminum can of a soft drink is placed in a freezer. Later, you find that the can is split open and its contents have frozen. Work was done on the can in splitting it open. Where did the energy for this work come from?

5.105 Consider a system consisting of the following apparatus, in which gas is confined in one flask and there is a vacuum in the other flask. The flasks are separated by a valve. Assume that the flasks are perfectly insulated and will not allow the flow of heat into or out of the flasks to the surroundings. When the valve is opened, gas flows from the filled flask to the evacuated one. **(a)** Is work performed during the expansion of the gas? **(b)** Why or why not? **(c)** Can you determine the value of ΔE for the process?

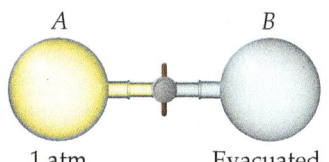

A   B

1 atm   Evacuated

5.106 A system consists of a sample of gas contained in a cylinder-and-piston arrangement. It undergoes the change in state shown in the drawing under two different situations: In Case 1, the cylinder and piston are perfect thermal insulators that do not allow heat to be transferred. In Case 2, the cylinder and piston are made up of a thermal conductor such as a metal, and during the state change, the cylinder gets warmer to the touch. Let q_1, w_1, and ΔE_1 be the values of q, w, and ΔE for Case 1, and let q_2, w_2, and ΔE_2 be the values for Case 2. **(a)** What is the value of q_1? **(b)** What is the sign of w_1? **(c)** What is the sign of ΔE_1? **(d)** What is the sign of q_2? **(e)** Is the value of ΔE_2 greater than, equal to, or less than the value of ΔE_1? **(f)** Is the final state of the gas the same for both cases?

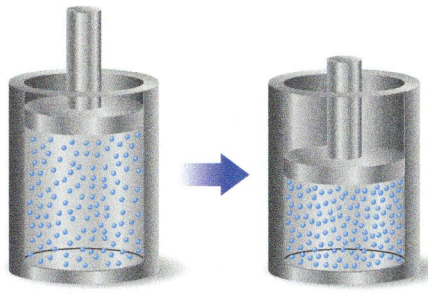

5.107 Limestone stalactites and stalagmites are formed in caves by the following reaction:

$$Ca^{2+}(aq) + 2\,HCO_3^-(aq) \longrightarrow CaCO_3(s) + CO_2(g) + H_2O(l)$$

If 1 mol of $CaCO_3$ forms at 298 K under 1 atm pressure, the reaction performs 2.47 kJ of P–V work, pushing back the atmosphere as the gaseous CO_2 forms. At the same time, 38.95 kJ of heat is absorbed from the environment. What are the values of ΔH and of ΔE for this reaction?

5.108 A house is designed to have passive solar energy features. Brickwork incorporated into the interior of the house acts as a heat absorber. Each brick weighs approximately 1.8 kg. The specific heat of the brick is 0.85 J/g-K. How many bricks must be incorporated into the interior of the house to provide the same total heat capacity as 1.7×10^3 gal of water?

5.109 A coffee-cup calorimeter of the type shown in Figure 5.18 contains 150.0 g of water at 25.1°C. A 121.0-g block of copper metal is heated to 100.4°C by putting it in a beaker of boiling water. The specific heat of $Cu(s)$ is 0.385 J/g-K. The Cu is added to the calorimeter, and after a time the contents of the cup reach a constant temperature of 30.1°C. **(a)** Determine the amount of heat, in J, lost by the copper block. **(b)** Determine the amount of heat gained by the water. The specific heat of water is 4.18 J/g-K. **(c)** The difference between your answers for **(a)** and **(b)** is due to heat loss through the Styrofoam® cups and the heat necessary to raise the temperature of the inner wall of the apparatus. The heat capacity of the calorimeter is the amount of heat necessary to raise the temperature of the apparatus (the cups and the stopper) by 1 K. Calculate the heat capacity of the calorimeter in J/K. **(d)** What would be the final temperature of the system if all the heat lost by the copper block were absorbed by the water in the calorimeter?

5.110 Propylene glycol, $C_3H_8O_2$, is a viscous liquid used in the production of polyester resins. A 1.500-g sample of propylene glycol is combusted to $CO_2(g)$ and $H_2O(l)$ in a bomb calorimeter for which the total heat capacity is 6.355 kJ/°C. The temperature of the calorimeter increases from 22.45 to 28.10°C. **(a)** Write a balanced equation for the combustion of propylene glycol to $CO_2(g)$ and $H_2O(l)$. **(b)** How much heat is evolved during the combustion of the 1.500-g sample of propylene glycol? **(c)** Calculate the enthalpy change for the combustion of one mole of propylene glycol to $CO_2(g)$ and $H_2O(l)$. **(d)** What is the value of ΔH_f° for propylene glycol?

5.111 **(a)** When a 0.235-g sample of benzoic acid is combusted in a bomb calorimeter (Figure 5.19), the temperature rises 1.642°C. When a 0.265-g sample of caffeine, $C_8H_{10}N_4O_2$, is burned, the temperature rises 1.525°C. Using the value 26.38 kJ/g for the heat of combustion of benzoic acid, calculate the heat of combustion per mole of caffeine at constant volume. **(b)** Assuming that there is an uncertainty of 0.002°C in each temperature reading and that the masses of samples are measured to 0.001 g, what is the estimated uncertainty in the value calculated for the heat of combustion per mole of caffeine?

5.112 Meals-ready-to-eat (MREs) are military meals that can be heated on a flameless heater. The heat is produced by the following reaction:

$$Mg(s) + 2\,H_2O(l) \longrightarrow Mg(OH)_2(s) + 2\,H_2(g)$$

(a) Calculate the standard enthalpy change for this reaction. **(b)** Calculate the number of grams of Mg needed for this reaction to release enough energy to increase the temperature of 75 mL of water from 21 to 79°C.

5.113 Burning methane in oxygen can produce three different carbon-containing products: soot (very fine particles of graphite), $CO(g)$, and $CO_2(g)$. **(a)** Write three balanced equations for the reaction of methane gas with oxygen to produce these three products. In each case, assume that $H_2O(l)$ is the only other product. **(b)** Determine the standard enthalpies for the reactions in part (a). **(c)** For which of the three potential products is the magnitude of ΔH the largest?

5.114 Gaseous iodine pentafluoride, IF_5, can be prepared by the reaction of gaseous iodine and gaseous fluorine according to the following balanced equation:

$$I_2(g) + 5\,F_2(g) \rightarrow 2\,IF_5(g)$$

(a) The value of ΔH_f° for $IF_5(g)$ is -843 kJ/mol. Using this value and data in Appendix C, calculate ΔH° for the preceding reaction. **(b)** The IF_5 molecule has five I—F single bonds. Use your answer to part (a) and data in Table 5.4 to estimate the bond enthalpy $D(\text{I—F})$.

5.115 Use average bond enthalpies from Table 5.4 to estimate ΔH for the following gas-phase reaction of ethylene, (C_2H_4), oxygen, and hydrogen to form ethylene glycol $(C_2H_6O_2)$, which is the principal component of automotive antifreeze:

5.116 Depending on their specific usage, fuels are judged in part on energy released per unit volume and energy released per unit mass. Three prospective fuels are listed in the following table, along with their densities and molar enthalpies of combustion. **(a)** Rank the three fuels according to their enthalpy produced per gram. **(b)** Rank them according to their enthalpy produced per cm^3:

Fuel	Density at 20°C (g/cm^3)	Molar Enthalpy of Combustion (kJ/mol)
Nitroethane, $C_2H_5NO_2(l)$	1.052	−1368
Ethanol, $C_2H_5OH(l)$	0.789	−1367
Methylhydrazine, $CH_6N_2(l)$	0.874	−1307

5.117 The hydrocarbons acetylene (C_2H_2) and benzene (C_6H_6) have the same empirical formula. Benzene is an "aromatic" hydrocarbon that is unusually stable because of its structure. **(a)** By using data in Appendix C, determine the standard enthalpy change for the reaction $3\,C_2H_2(g) \longrightarrow C_6H_6(l)$. **(b)** Which has greater enthalpy, 3 mol of acetylene gas or 1 mol of liquid benzene? **(c)** Determine the fuel value, in kJ/g, for acetylene and benzene.

5.118 **(a)** Ammonia (NH_3) boils at $-33°C$; at this temperature it has a density of 0.81 g/cm^3. The enthalpy of formation of $NH_3(g)$ is -46.2 kJ/mol, and the enthalpy of vaporization of $NH_3(l)$ is 23.2 kJ/mol. What is the enthalpy change when 1 L of liquid NH_3 is burned in air to give $N_2(g)$ and $H_2O(g)$? **(b)** Methanol, CH_3OH, is a liquid at room temperature, its density at 25 °C is 0.792 g/cm^3, and its $\Delta H_f^\circ = -239$ kJ/mol. What is the value of ΔH for the complete combustion of 1 L of liquid methanol to $CO_2(g)$ and $H_2O(g)$? **(c)** Based on the results of parts (a) and (b), which substance produces more heat per liter upon burning?

5.119 Three common hydrocarbons that contain four carbons are listed here, along with their standard enthalpies of formation:

Hydrocarbon	Formula	ΔH_f° (kJ/mol)
1,3-Butadiene	$C_4H_6(g)$	111.9
1-Butene	$C_4H_8(g)$	1.2
n-Butane	$C_4H_{10}(g)$	−124.7

(a) For each of these substances, calculate the molar enthalpy of combustion to $CO_2(g)$ and $H_2O(l)$. **(b)** Calculate the fuel value, in kJ/g, for each of these compounds. **(c)** For each hydrocarbon, determine the percentage of hydrogen by mass. **(d)** Based on your results, does the fuel value increase, stay the same, or decrease as the hydrogen content increases for these hydrocarbons?

5.120 The Sun supplies about 1.0 kilowatt of energy for each square meter of surface area (1.0 kW/m^2, where a watt $= 1\,J/s$). Plants produce the equivalent of about 0.20 g of sucrose $(C_{12}H_{22}O_{11})$ per hour per square meter. Assuming that the sucrose is produced as follows, calculate the percentage of sunlight used to produce sucrose.

$$12\,CO_2(g) + 11\,H_2O(l) \longrightarrow C_{12}H_{22}O_{11} + 12\,O_2(g)$$
$$\Delta H = 5645 \text{ kJ}$$

5.121 At 20°C (approximately room temperature), the average velocity of N_2 molecules in air is 1050 mph. **(a)** What is the average speed in m/s? **(b)** What is the kinetic energy (in J) of an N_2 molecule moving at this speed? **(c)** What is the total kinetic energy of 1 mol of N_2 molecules moving at this speed?

5.122 Consider two solutions: 50.0 mL of 1.00 M CuSO$_4$ and 50.0 mL of 2.00 M KOH. When the two solutions are mixed in a constant-pressure calorimeter, a precipitate forms and the temperature of the mixture rises from 21.5 to 27.7°C. **(a)** Before mixing, how many grams of Cu are present in the solution of CuSO$_4$? **(b)** Predict the identity of the precipitate in the reaction. **(c)** Write complete and net ionic equations for the reaction that occurs when the two solutions are mixed. **(d)** From the calorimetric data, calculate ΔH for the reaction that occurs on mixing. Assume that the calorimeter absorbs only a negligible quantity of heat, that the total volume of the solution is 100.0 mL, and that the specific heat and density of the solution after mixing are the same as those of pure water.

Design an Experiment

One of the important ideas of thermodynamics is that energy can be transferred in the form of heat or work. Imagine that you lived 180 years ago when the relationships between heat and work were not well understood. You have formulated a hypothesis that work can be converted to heat, with the same amount of work always generating the same amount of heat. To test this idea, you have designed an experiment using a device in which a falling weight is connected through pulleys to a shaft with an attached paddle wheel that is immersed in water. This is actually a classic experiment performed by James Joule in the 1840s. You can see various images of Joule's apparatus by searching the internet for "Joule experiment images."

(a) Using this device, what measurements would you need to make to test your hypothesis? **(b)** What equations would you use in analyzing your experiment? **(c)** Do you think you could obtain a reasonable result from a single experiment? Why or why not? **(d)** In what way could the precision of your instruments affect the conclusions that you make? **(e)** List ways that you could modify the equipment to improve the data you obtain if you were performing this experiment today instead of 180 years ago. **(f)** Give an example of how you could demonstrate the relationship between heat and a form of energy other than mechanical work.

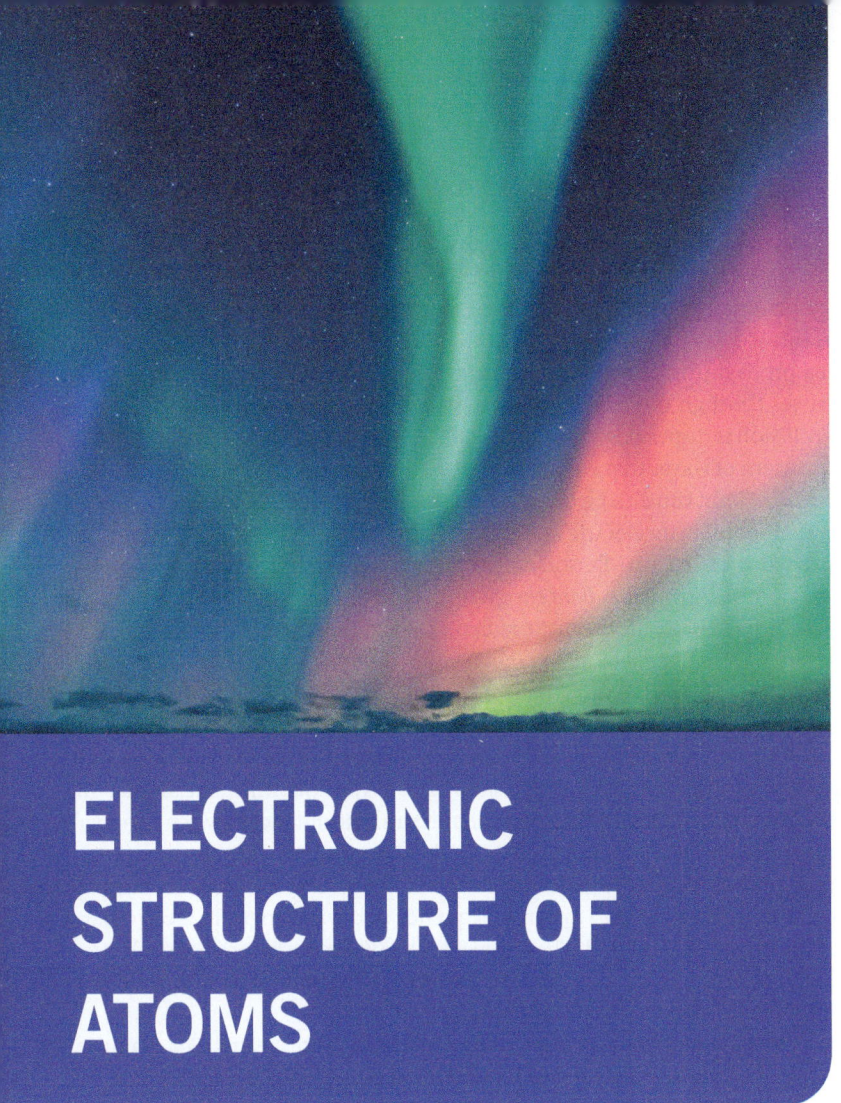

6

ELECTRONIC STRUCTURE OF ATOMS

▲ **THE AURORA BOREALIS OR NORTHERN LIGHTS** offer a brilliant display of color in the night sky. They result when ionized particles in the solar wind, mostly electrons and protons, excite atoms and molecules in the atmosphere. Light is produced when electrons in these excited atoms lose energy and fall back to their ground-state orbitals.

In this chapter, we explore *quantum theory*, which explains many aspects of the behavior of electrons in atoms, molecules, and all other forms of matter. We begin by looking at the nature of light and how our description of light was changed by quantum theory. We explore some of the tools used in *quantum mechanics*, a field developed in the first half of the twentieth century to describe the **electronic structure** of atoms—a term that refers to the number of electrons in the atom as well as their distribution around the nucleus and their energies. We will show that the quantum description of the electronic structure of atoms not only helps us to understand the arrangement of the elements in the periodic table, but also helps us understand why, for example, helium and neon are unreactive gases, whereas sodium and potassium are soft, reactive metals.

WHAT'S AHEAD

6.1 ▶ The Wave Nature of Light Learn that light (radiant energy, or *electromagnetic radiation*) has wave-like properties and is characterized by *wavelength*, *frequency*, and *speed*.

6.2 ▶ Quantized Energy and Photons Recognize that electromagnetic radiation also has particle-like properties and can be described as "particles" of light called *photons*.

6.3 ▶ Line Spectra and the Bohr Model Examine the light emitted by electrically excited atoms (*line spectra*), observations that led to the Bohr model of the atom.

6.4 ▶ The Wave Behavior of Matter Recognize that matter also has wave-like properties. As a result, it is impossible to determine simultaneously the exact position and the exact momentum of an electron in an atom (*Heisenberg's uncertainty principle*).

6.5 ▶ Quantum Mechanics and Atomic Orbitals Describe the behavior of the electron in a hydrogen atom by treating it as a wave. The *wave functions* that mathematically describe the electron's position and energy in an atom define *atomic orbitals*.

6.6 ▶ Representations of Orbitals Examine the sizes and shapes of orbitals and how they can be represented by graphs of electron density.

6.7 ▶ Many-Electron Atoms Learn that the energy levels of an atom having more than one electron are different from those of the hydrogen atom, and that each electron has an additional quantum-mechanical property called *spin*.

6.8 ▶ Electron Configurations Learn how the orbitals of the hydrogen atom can be used to describe the arrangements of electrons in many-electron atoms.

6.9 ▶ Electron Configurations and the Periodic Table Recognize that the electron configuration of an atom is related to the location of the element in the periodic table.

6.1 | The Wave Nature of Light

Learning Objectives

When you finish Section 6.1, you should be able to:

▶ Identify the different types of electromagnetic radiation and arrange them in order of their relative wavelength or frequency.

▶ Calculate the frequency of electromagnetic radiation from its wavelength and vice versa.

Much of our present understanding of the electronic structure of atoms and molecules has come from analysis of the light either emitted or absorbed by substances. To understand the way electrons behave in various substances, therefore, we must first learn more about light. In this section we learn about the wave-like properties of light.

A wave is a propagating disturbance that travels through a medium, transporting energy as it moves through the medium. When you drop a stone into a still pond, the energy of the collision is dispersed by vertical displacements of water molecules that lead to the familiar pattern of waves that emanate outward from the point of entry (**Figure 6.1**). Though less obvious, light can also be thought of as a wave of propagating energy. A light wave is made up of oscillating electric and magnetic fields, and thus it is referred to as **electromagnetic radiation**. Because electromagnetic radiation carries energy through space, it is also known as *radiant energy*.

There are many types of electromagnetic radiation in addition to visible light. These different types—radio waves, infrared radiation (heat), ultraviolet radiation, X rays—may seem very different from one another, but they all share certain fundamental characteristics.

All types of electromagnetic radiation move through a vacuum at 2.998×10^8 m/s, the *speed of light*. All have wave-like characteristics similar to those of waves that move through water. A cross section of a water wave (**Figure 6.2**) shows that it is *periodic*, which means that the pattern of peaks and troughs repeats itself at regular intervals. The distance between two adjacent peaks (or between two adjacent troughs) is called the **wavelength**. The number of complete wavelengths, or *cycles*, that pass a given point each second is the **frequency** of the wave.

Just as with water waves, we can assign a frequency and wavelength to electromagnetic waves, as illustrated in **Figure 6.3**. These and all other wave characteristics of electromagnetic radiation are due to the periodic oscillations in the intensities of the electric and magnetic fields associated with the radiation.

The speed of water waves can vary depending on how they are created. For example, the waves produced by a speed boat travel faster than those produced by a rowboat. In contrast, *all electromagnetic radiation moves at the same speed—namely, the speed of light*. As a result, the wavelength and frequency of electromagnetic radiation are always related in a straightforward way. If the wavelength is long, fewer cycles of the wave pass a given point per second, and so the frequency is low. Conversely, for a wave to have a high frequency, it must have a short wavelength. This inverse relationship between the frequency and wavelength of electromagnetic radiation is expressed by the equation

$$\lambda\nu = c \qquad [6.1]$$

where λ (lambda) is wavelength, ν (nu) is frequency, and c is the speed of light.

▲ **Figure 6.1 Water waves** are generated when an object is dropped into a pool of still water. The regular variation of peaks and troughs enables us to sense the motion of the waves as energy is dispersed from the point where the object impacted the water.

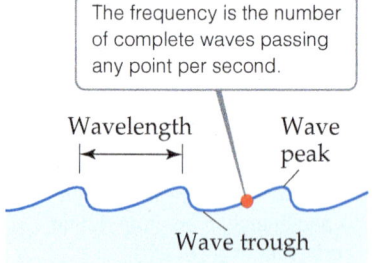

The frequency is the number of complete waves passing any point per second.

Wavelength ⊢——⊣   Wave peak

Wave trough

▲ **Figure 6.2 Water waves.** The *wavelength* is the distance between two adjacent peaks or two adjacent troughs.

Go Figure If wave (a) has a wavelength of 2.0 m and a frequency of 1.5×10^8 cycles/s, what are the wavelength and frequency of wave (b)?

Wavelength λ

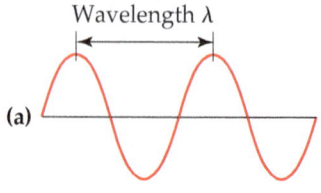

(a)

λ

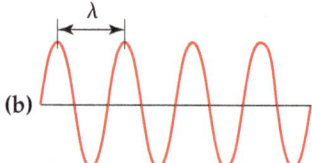

(b)

▲ **Figure 6.3 Electromagnetic waves.** Like water waves, electromagnetic radiation can be characterized by a wavelength. As the wavelength, λ, decreases, the frequency, ν, increases. The wavelength in (**b**) is *half* as long as that in (**a**), so the frequency of the wave in (**b**) is *twice* as great as that in (**a**).

Go Figure Is the wavelength of a microwave longer or shorter than the wavelength of visible light? By how many orders of magnitude do the two waves differ in wavelength?

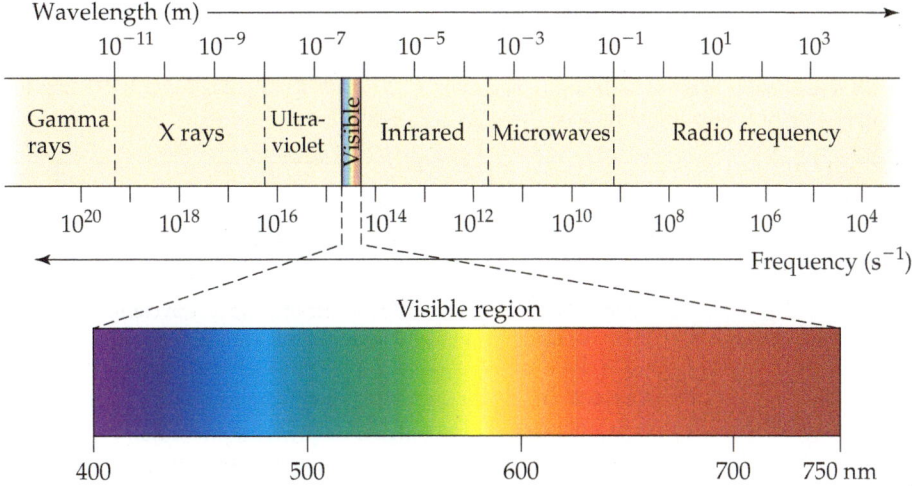

▲ **Figure 6.4 The electromagnetic spectrum.** Wavelengths in the spectrum range from very short gamma rays to very long radio waves.

Figure 6.4 shows the various types of electromagnetic radiation arranged in order of increasing wavelength, a display called the *electromagnetic spectrum*. The wavelengths span an enormous range. The wavelengths of gamma rays are comparable to the diameters of atomic nuclei, whereas the wavelengths of radio waves can be longer than a football field. Visible light (the light we can see with our eyes), which corresponds to wavelengths of about 400 to 750 nm (4.00×10^{-7} to 7.50×10^{-7} m), is an extremely small portion of the electromagnetic spectrum. The unit of length chosen to express wavelength depends on the type of radiation, as shown in **Table 6.1**.

Frequency is expressed in cycles per second, a unit also called a *hertz* (Hz). Because it is understood that cycles are involved, the units of frequency are normally given simply as "per second," which is denoted by s^{-1} or /s. For example, a frequency of 698 megahertz (MHz), a typical frequency for a cellular telephone, could be written as 698 MHz, 698,000,000 Hz, 698,000,000 s^{-1} or 698,000,000/s.

TABLE 6.1   Common Wavelength Units for Electromagnetic Radiation

Unit	Symbol	Length (m)	Type of Radiation
Angstrom	Å	10^{-10}	X ray
Nanometer	nm	10^{-9}	Ultraviolet, visible
Micrometer	μm	10^{-6}	Infrared
Millimeter	mm	10^{-3}	Microwave
Centimeter	cm	10^{-2}	Microwave
Meter	m	1	Television, radio
Kilometer	km	10^{3}	Radio

Sample Exercise 6.1

Calculating Frequency from Wavelength

The yellow light given off by a sodium vapor lamp used for public lighting has a wavelength of 589 nm. What is the frequency of this radiation?

SOLUTION

Analyze We are given the wavelength, λ, of the radiation and asked to calculate its frequency, ν.

Plan The relationship between the wavelength and the frequency is given by Equation 6.1. We can solve for ν and use the values

of λ and c to obtain a numerical answer. (The speed of light, c, is 3.00×10^{8} m/s to three significant figures.)

Solve Solving Equation 6.1 for frequency gives $\nu = c/\lambda$. When we insert the values for c and λ, we note that the units of length in these

Continued

two quantities are different. We use a conversion factor to convert the wavelength from nanometers to meters, so the units cancel:

$$\nu = \frac{c}{\lambda} = \left(\frac{3.00 \times 10^8 \text{ m/s}}{589 \text{ nm}}\right)\left(\frac{1 \text{ nm}}{10^{-9} \text{ m}}\right)$$

$$= 5.09 \times 10^{14} \text{ s}^{-1}$$

Check The high frequency is reasonable because of the short wavelength. The units are proper because frequency has units of "per second," or s^{-1}.

 Practice Exercise

(a) A laser used in orthopedic spine surgery produces radiation with a wavelength of 2.10 μm. Calculate the frequency of this radiation. **(b)** An FM radio station broadcasts electromagnetic radiation at a frequency of 103.4 MHz (megahertz; 1 MHz = 10^6 s^{-1}). Calculate the wavelength of this radiation. The speed of light is 2.998×10^8 m/s to four significant figures.

 Self-Assessment Exercises

SAE 6.1 If an electromagnetic wave has a frequency of 1.5×10^{10} s^{-1}, it falls in the _____ region of the electromagnetic spectrum, and its wavelength is _____. **(a)** radio frequency, 50 m **(b)** radio frequency, 2.0 cm **(c)** microwave, 50 m **(d)** microwave, 2.0 cm **(e)** gamma ray, 6.7×10^{-11} m

SAE 6.2 Which of the following statements is or are true?

  (i) Microwaves have a higher frequency than X rays.

 (ii) Microwaves have a longer wavelength than X rays.

(iii) Microwaves travel through a vacuum faster than X rays.

(a) only i **(b)** only ii **(c)** both i and iii **(d)** both ii and iii **(e)** All three statements are true.

 Learning Objectives

When you finish Section 6.2, you should be able to:

▶ Explain how blackbody radiation and the photoelectric effect provide evidence that electromagnetic radiation is quantized.

▶ Describe electromagnetic radiation as particles called photons and calculate the energy of a photon from either its frequency or its wavelength.

Go Figure

Which is at a higher temperature: the part of the nail glowing yellow or the part glowing red?

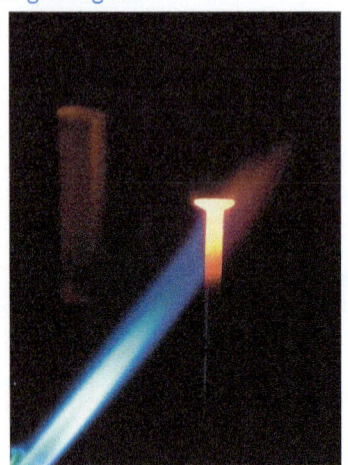

▲ **Figure 6.5 Color and temperature.** The color and intensity of the light emitted by a hot object, such as this nail, depend on the temperature of the object.

6.2 | Quantized Energy and Photons

Although the wave model of light explains many aspects of the behavior of light, several observations cannot be resolved by this model. Three of these observations are particularly pertinent to our understanding of how electromagnetic radiation and atoms interact: (1) the emission of light from hot objects (referred to as *blackbody radiation* because the objects studied appear black before heating), (2) the emission of electrons from metal surfaces on which light shines (the *photoelectric effect*), and (3) the emission of light from electronically excited gas atoms (*emission spectra*). These deviations from wave-like behavior led to the development of quantum theory in the early part of the twentieth century. This was a revolutionary advance that changed the way scientists understood both light and matter. In this section we examine the first two phenomena; emission spectra are examined in Section 6.3.

Hot Objects and the Quantization of Energy

When solids are heated, they emit radiation, as seen in the red glow of an electric stove burner or the bright white light of a tungsten light bulb. The wavelength distribution of the radiation depends on temperature; a red-hot object, for instance, is cooler than a yellowish or white-hot one (**Figure 6.5**). During the late 1800s, a number of physicists studied this phenomenon, trying to understand the relationship between the temperature and the intensity and wavelength of the emitted radiation. The prevailing laws of physics could not account for the observations.

In 1900, the German physicist Max Planck (1858–1947) solved the problem by making a daring assumption: He proposed that energy can be either released or absorbed by atoms only in discrete "chunks" of some minimum size. Planck gave the name **quantum** (meaning "fixed amount") to the smallest quantity of energy that can be emitted or absorbed as electromagnetic radiation. He proposed that the energy, E, of a single quantum is equal to a constant times the frequency of the radiation:

$$E = h\nu \qquad [6.2]$$

The proportionality constant h is called the **Planck constant** and has a value of 6.626×10^{-34} joule-second (J-s).

According to Planck's theory, matter can emit and absorb energy only in whole-number multiples of $h\nu$. If the quantity of energy emitted by an atom is $3h\nu$, for example, then we say that three quanta of energy have been emitted (*quanta* being the plural of *quantum*). Because the energy can be released only in specific amounts, we say that the allowed energies are *quantized*—their values are restricted to certain quantities. Planck's revolutionary proposal that energy is quantized proved to be correct, and he was awarded the 1918 Nobel Prize in Physics for his work on quantum theory.

If the notion of quantized energies seems strange, consider the difference between a ramp and a staircase (**Figure 6.6**). As you walk up a ramp, your potential energy increases in a uniform, continuous manner. When you climb a staircase, you can step only *on* individual stairs, not *between* them, so the increase in your potential energy is restricted to certain values and is therefore quantized.

If Planck's quantum theory is correct, why are its effects not obvious in our daily lives? Why do energy changes seem continuous rather than quantized, or "jagged"? The Planck constant is an extremely small number, so a quantum of energy, $h\nu$, is an extremely small quantity. Planck's rules regarding the gain or loss of energy are always the same, whether we are concerned with objects on the scale of our ordinary experience or with microscopic objects. With everyday objects, however, the gain or loss of a single quantum of energy is so small that it goes completely unnoticed. In contrast, when dealing with matter at the atomic level, the impact of quantized energies is far more significant.

Potential energy of person walking up ramp increases in uniform, continuous manner

Potential energy of person walking up steps increases in stepwise, quantized manner

▲ **Figure 6.6 Quantized versus continuous change in energy.**

The Photoelectric Effect and Photons

A few years after Planck presented his quantum theory, scientists began to see its applicability to many experimental observations. In 1905, Albert Einstein (1879–1955) used Planck's theory to explain the **photoelectric effect** (**Figure 6.7**). Light shining on a clean metal surface causes electrons to be emitted from the surface. A minimum frequency of light, different for different metals, is required for the emission of electrons. For example, light with a frequency of $4.60 \times 10^{14}\,\text{s}^{-1}$ or greater causes cesium metal to emit electrons, but if the light has frequency less than that, no electrons are emitted.

To explain the photoelectric effect, Einstein assumed that the radiant energy striking the metal surface behaves like a stream of tiny energy packets. Each packet, which is like a "particle" of energy, is called a **photon**. Extending Planck's quantum theory, Einstein deduced that each photon must have an energy equal to the Planck constant times the frequency of the light:

$$\text{Energy of photon}, E = h\nu \qquad [6.3]$$

Thus, radiant energy itself is quantized.

Under the right conditions, photons striking a metal surface can transfer their energy to electrons in the metal. A certain amount of energy—called the *work function*—is required for the electrons to overcome the attractive forces holding them in the metal. If the photons striking the metal have less energy than the work function, the electrons do not acquire sufficient energy to escape from the metal. Increasing the intensity of the light source doesn't lead to emission of electrons from the metal, but increasing the frequency of the incoming light can have that effect. We can understand this behavior if we think of light as being made up of photons with quantized energies. When the intensity (brightness) of the light is increased, the number of photons striking the surface per unit time increases, but the energy of each photon remains unchanged. When the frequency is increased, the energy of each photon increases, and when that energy is greater than the work function of a particular metal, electrons are emitted and the excess energy of the photon is converted into kinetic energy of the emitted electron. Einstein won the Nobel Prize in Physics in 1921 primarily for his explanation of the photoelectric effect.

To better understand what a photon is, imagine you have a light source that produces radiation of a single wavelength. Further suppose that you could switch the light on and off faster and faster to provide ever-smaller bursts of energy. Einstein's photon theory tells us that you would eventually come to the smallest energy burst, given by $E = h\nu$. This smallest burst consists of a single photon of light.

Go Figure

If the frequency of the incoming light is increased, will the energy of the ejected electrons increase, decrease, or stay the same?

Photon hits surface with energy $h\nu$

Electrons emitted from surface by energy of photon

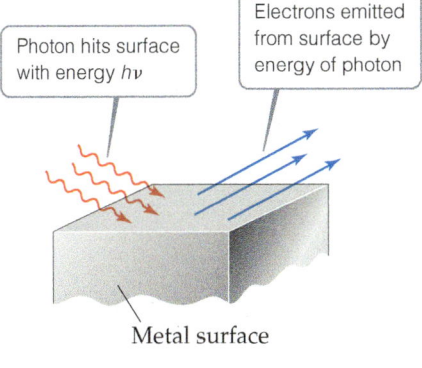

Metal surface

▲ **Figure 6.7 The photoelectric effect.**

Sample Exercise 6.2
Energy of a Photon

Calculate the energy of one photon of yellow light that has a wavelength of 589 nm.

SOLUTION

Analyze Our task is to calculate the energy, E, of a photon, given its wavelength, $\lambda = 589$ nm.

Plan We can use Equation 6.1 to convert the wavelength to frequency: $\nu = c/\lambda$. We can then use Equation 6.3 to calculate the energy of the photon: $E = h\nu$.

Solve The frequency, ν, is calculated from the given wavelength, as shown in Sample Exercise 6.1:

$$\nu = (3.00 \times 10^8 \, \text{m/s})/(589 \times 10^{-9} \, \text{m}) = 5.09 \times 10^{14} \, \text{s}^{-1}$$

The value of the Planck constant, h, is given both in the text and in the table of physical constants on the inside back cover of the text; thus, we can calculate E:

$$E = (6.626 \times 10^{-34} \, \text{J·s})(5.09 \times 10^{14} \, \text{s}^{-1}) = 3.37 \times 10^{-19} \, \text{J}$$

Comment If one photon of radiant energy supplies 3.37×10^{-19} J, we calculate that one mole of these photons will supply:

$$(6.022 \times 10^{23} \, \text{photons/mol})(3.37 \times 10^{-19} \, \text{J/photon})$$
$$= 2.03 \times 10^5 \, \text{J/mol}$$

▶ **Practice Exercise**

(a) A laser emits light that has a frequency of $4.69 \times 10^{14} \, \text{s}^{-1}$. What is the energy of one photon of this radiation? **(b)** If the laser emits a pulse containing 5.0×10^{17} photons of this radiation, what is the total energy of that pulse? **(c)** If the laser emits 1.3×10^{-2} J of energy during a pulse, how many photons are emitted?

The idea that the energy of light depends on its frequency helps us understand the diverse effects that different kinds of electromagnetic radiation have on matter. For example, because of their high frequency (short wavelength), photons of ultraviolet light possess more energy than photons of visible light (Figure 6.4). This explains why prolonged exposure to ultraviolet light causes tissue damage that can in some cases lead to skin cancer. X rays have much higher frequencies still, and the high energies of X-ray photons allow them to pass through the soft tissues of our bodies, an attribute that permits X rays to be used to image our bones and teeth. It also explains why medical and dental technicians take precautions to minimize their own and their patients' exposure to such high-energy photons.

Although Einstein's theory of light as a stream of photons rather than a wave explains the photoelectric effect and a great many other observations, it also poses a dilemma. Is light a wave, or does it consist of particles? The only way to resolve this dilemma is to adopt what might seem to be a bizarre position: We must consider that light possesses both wave-like and particle-like characteristics and, depending on the situation, will behave more like waves or more like particles. We will soon see that this dual wave-particle nature is also a characteristic trait of matter.

Self-Assessment Exercises

SAE 6.3   The work function of silver metal is 6.9×10^{-19} J. If a crystal of silver is exposed to a pulse of blue light with a wavelength of 450 nm, will electrons be ejected, and if so, what is the maximum kinetic energy they can have? **(a)** Electrons will not be ejected. **(b)** Electrons with a maximum kinetic energy of 4.4×10^{-19} J will be ejected. **(c)** Electrons with a maximum kinetic energy of 6.9×10^{-19} J will be ejected. **(d)** Electrons with a maximum kinetic energy of 2.5×10^{-19} J will be ejected.

SAE 6.4   Which form of electromagnetic radiation has the lowest energy photons? **(a)** ultraviolet light **(b)** infrared light **(c)** green light **(d)** red light

SAE 6.5   A laser gives off a pulse of infrared light with a wavelength of 1054 nm and a total energy of 2.5 J. How many photons are in this pulse of electromagnetic radiation?

(a) 1.3×10^{19} photons

(b) 1.9×10^{-19} photons

(c) 2.4×10^6 photons

(d) 5.3×10^{18} photons

6.3 | Line Spectra and the Bohr Model

The work of Planck and Einstein paved the way for understanding how electrons are arranged in atoms. In 1913, the Danish physicist Niels Bohr (**Figure 6.8**) offered a theoretical explanation of the *line spectrum* of the hydrogen atom, another phenomenon that had puzzled scientists during the nineteenth century.

Line Spectra

A particular source of radiant energy may emit a single wavelength, as in the light from a laser. Radiation composed of a single wavelength is *monochromatic*. However, most common radiation sources, including incandescent light bulbs and stars, produce radiation containing many different wavelengths—that is, *polychromatic* radiation. A **spectrum** is produced when radiation from a polychromatic source is separated into its component wavelengths, as shown in **Figure 6.9**. The resulting spectrum consists of a continuous range of colors—violet merges into indigo, indigo into blue, and so forth, with no (or very few) blank spots. This rainbow of colors, containing light of all wavelengths, is called a **continuous spectrum**. The most familiar example of a continuous spectrum is the rainbow produced when raindrops or mist acts as a prism for sunlight.

Not all radiation sources produce a continuous spectrum. When a high voltage is applied to tubes that contain different gases under reduced pressure, the gases emit different colors of light (**Figure 6.10**). The light emitted by neon gas is the familiar red-orange glow of many "neon" lights, whereas sodium vapor emits the yellow light characteristic of some modern streetlights. When light coming from such tubes is passed through a

Learning Objectives

When you finish **Section 6.3**, you should be able to:

▶ Explain the Bohr model of the hydrogen atom and use it to account for the line spectrum of hydrogen.

▶ Use the Bohr model to calculate the wavelengths (or frequencies) of light that can either be absorbed or emitted by the hydrogen atom.

▲ **Figure 6.8 Niels Bohr (1885–1962).** This Danish postage stamp, first issued in 1963, commemorates Bohr's atomic model.

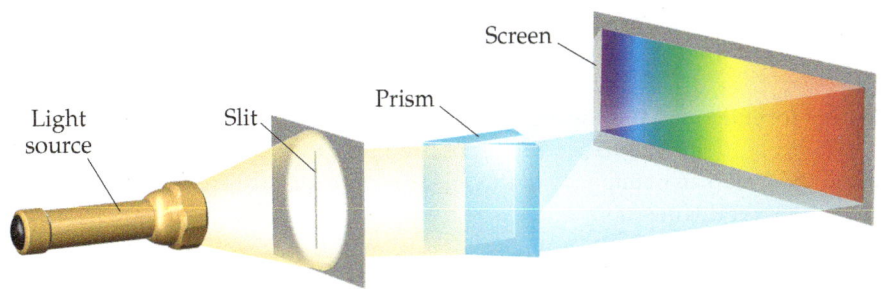

◀ **Figure 6.9 Creating a spectrum.** A continuous visible spectrum is produced when a narrow beam of white light is passed through a prism. The white light could be sunlight or light from an incandescent lamp.

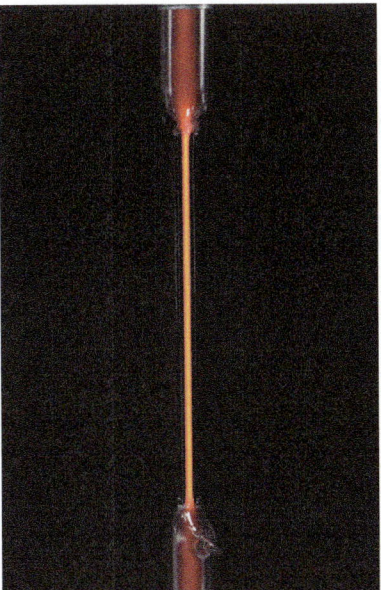
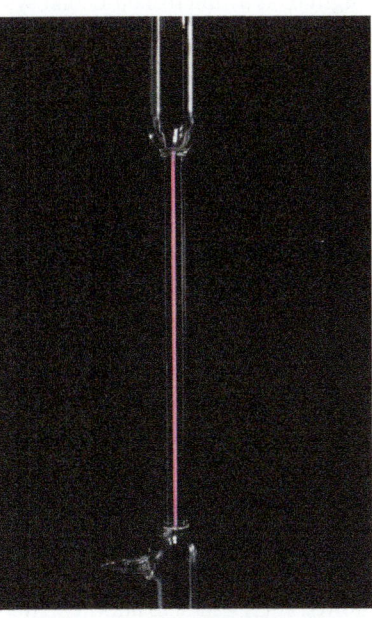

Neon (Ne)   Hydrogen (H)

◀ **Figure 6.10 Atomic emission of neon and hydrogen.** Different gases emit light of different characteristic colors when an electric current is passed through them.

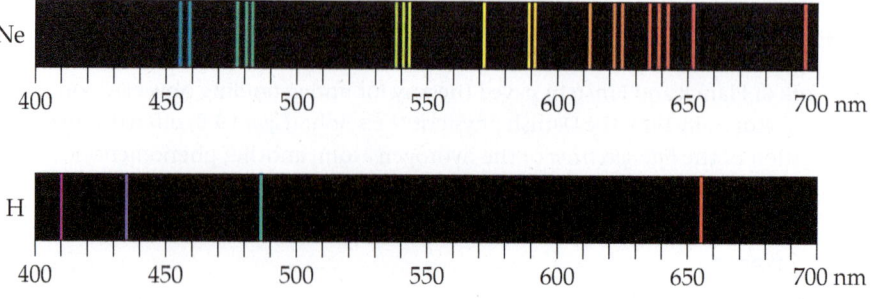

▲ **Figure 6.11 Line spectra of neon and hydrogen.** The colored lines occur at wavelengths present in the emission. The black regions are wavelengths for which no light is produced in the emission.

prism, only a few wavelengths are present in the resultant spectra (**Figure 6.11**). Each colored line in such spectra represents light of one wavelength. A spectrum containing radiation of only specific wavelengths is called a **line spectrum.**

When scientists first detected the line spectrum of hydrogen in the mid-1800s, they were fascinated by its simplicity. At that time, only four lines at wavelengths of 410 nm (violet), 434 nm (blue), 486 nm (blue-green), and 656 nm (red) were observed (Figure 6.11). In 1885, a Swiss schoolteacher named Johann Balmer showed that the wavelengths of these four lines fit an intriguingly simple formula that relates the wavelengths to integers. Later, additional lines were found in the ultraviolet and infrared regions of hydrogen's line spectrum. Soon Balmer's equation was extended to a more general one, called the *Rydberg equation*, which allows us to calculate the wavelengths of all the spectral lines of hydrogen:

$$\frac{1}{\lambda} = (R_H)\left(\frac{1}{n_1^2} - \frac{1}{n_2^2}\right) \qquad [6.4]$$

In this formula, λ is the wavelength of a spectral line, R_H is the *Rydberg constant* ($1.096776 \times 10^7 \text{ m}^{-1}$), and n_1 and n_2 are positive integers, with n_2 being larger than n_1. How could the remarkable simplicity of this equation be explained? It took nearly 30 more years to answer this question.

Bohr's Model

Rutherford's discovery of the nuclear atom (Section 2.2) suggested that an atom might be thought of as a "microscopic solar system" in which the electrons orbit the nucleus. To explain the line spectrum of hydrogen, Bohr assumed that electrons in hydrogen atoms move in circular orbits around the nucleus, but this assumption posed a problem. According to classical physics, a charged particle (such as an electron) moving in a circular path should continuously lose energy. As an electron loses energy, therefore, it should spiral into the positively charged nucleus. This behavior, however, does not happen—hydrogen atoms are stable. So how can we explain this apparent violation of the laws of physics? Bohr approached this problem in much the same way that Planck had approached the problem of the nature of the radiation emitted by hot objects: He assumed that the prevailing laws of physics were inadequate to describe all aspects of atoms. Furthermore, he adopted Planck's idea that energies are quantized.

Bohr based his model on three postulates:

1. Only orbits of certain radii, corresponding to certain specific energies, are permitted for the electron in a hydrogen atom.

2. An electron in a permitted orbit is in an "allowed" energy state. An electron in an allowed energy state does not radiate energy and, therefore, does not spiral into the nucleus.

3. Energy is emitted or absorbed by the electron only as the electron changes from one allowed energy state to another. This energy is emitted or absorbed as a photon that has energy $E = h\nu$.

The Energy States of the Hydrogen Atom

Starting with his three postulates and using classical equations for motion and for interacting electrical charges, Bohr calculated the energies corresponding to the allowed orbits for the electron in the hydrogen atom. Ultimately, the calculated energies fit the formula

$$E = (-hcR_H)\left(\frac{1}{n^2}\right) = (-2.18 \times 10^{-18}\,\text{J})\left(\frac{1}{n^2}\right) \qquad [6.5]$$

where h, c, and R_H are the Planck constant, the speed of light, and the Rydberg constant, respectively. The integer n, which can have whole-number values of 1, 2, 3, ..., ∞, is called the **principal quantum number**.

Each allowed orbit corresponds to a different value of n. The radius of the orbit gets larger as n increases. Thus, the first allowed orbit (the one closest to the nucleus) has $n = 1$, the next allowed orbit (the one second closest to the nucleus) has $n = 2$, and so forth. The electron in the hydrogen atom can be in any allowed orbit, and Equation 6.5 tells us the energy of the electron in that orbit.

The energies of the electron given by Equation 6.5 are negative for all values of n. The lower (more negative) the energy is, the more stable the atom. The energy is lowest (most negative) for $n = 1$. As n gets larger, the energy becomes less negative and therefore increases. We can liken the situation to a ladder in which the rungs are numbered from the bottom. The higher you climb (the greater the value of n), the higher the energy. The lowest-energy state ($n = 1$ analogous to the bottom rung) is called the **ground state** of the atom. When the electron is in a higher-energy state ($n = 2$ or higher), the atom is said to be in an **excited state**. **Figure 6.12** shows the allowed energy levels for the hydrogen atom for several values of n.

What happens as n becomes infinitely large? The radius increases and the energy of attraction between the electron and the nucleus approaches zero, so when $n = \infty$ the electron is completely separated from the nucleus and the energy of the electron is zero:

$$E = (-2.18 \times 10^{-18}\,\text{J})\left(\frac{1}{\infty^2}\right) = 0$$

The state in which the electron is completely separated from the nucleus is called the *reference*, or zero-energy, state of the hydrogen atom. Completely separating the electron from the nucleus corresponds to *ionizing* the hydrogen atom, a concept we return to in Chapter 7.

In his third postulate, Bohr assumed that the electron can "jump" from one allowed orbit to another by either absorbing or emitting photons whose radiant energy corresponds exactly to the energy difference between the two orbits. The electron must absorb energy in order to move to a higher-energy state (higher value of n). Conversely, radiant energy is emitted when the electron jumps to a lower-energy state (lower value of n).

Imagine a case in which the electron jumps from an initial state with principal quantum number n_i and energy E_i to a final state with principal quantum number n_f and energy E_f. Applying Equation 6.5, we see that the change in energy for this transition is

$$\Delta E = E_f - E_i = (-2.18 \times 10^{-18}\,\text{J})\left(\frac{1}{n_f^2} - \frac{1}{n_i^2}\right) \qquad [6.6]$$

What is the significance of the *sign* of ΔE? Notice that ΔE is positive when n_f is greater than n_i, which happens when the electron jumps to a higher-energy orbit. Conversely, ΔE is negative when n_f is less than n_i, which happens when the electron falls in energy to a lower-energy orbit.

As noted previously, transitions from one allowed state to another will involve a photon. *The energy of the photon (E_{photon}) must equal the difference in energy between the two states (ΔE).* When ΔE is positive, a photon must be *absorbed* as the electron jumps to a higher energy. When ΔE is negative, a photon is *emitted* as the electron falls to a lower-energy level. In both cases, the energy of the photon must match the energy difference between the states. Because the frequency ν is always a positive number, the energy

Go Figure

If the transition of an electron from the $n = 3$ state to the $n = 2$ state results in emission of visible light, is the transition from the $n = 2$ state to the $n = 1$ state more likely to result in the emission of infrared or ultraviolet radiation?

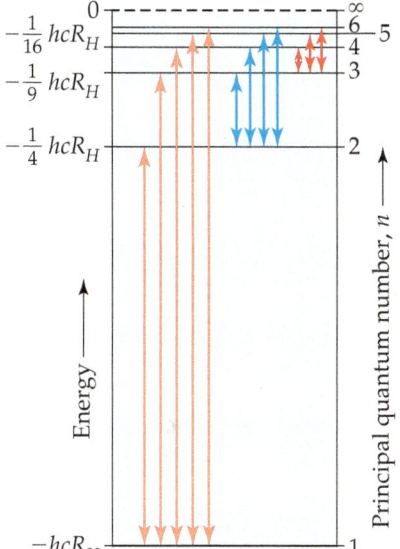

▲ **Figure 6.12 Energy levels in the hydrogen atom from the Bohr model.** The arrows refer to the transitions of the electron from one allowed energy state to another. The states shown are those for which $n = 1$ through $n = 6$ and the state for $n = \infty$ for which the energy, E, equals zero.

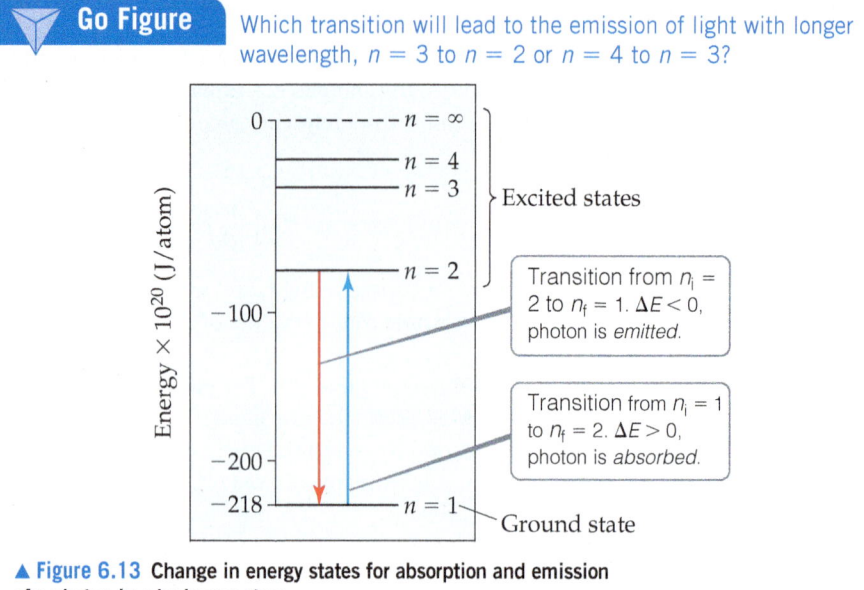

Go Figure   Which transition will lead to the emission of light with longer wavelength, $n = 3$ to $n = 2$ or $n = 4$ to $n = 3$?

▲ **Figure 6.13 Change in energy states for absorption and emission of a photon in a hydrogen atom.**

of the photon ($h\nu$) must always be positive. Thus, the sign of ΔE tells us whether the photon is absorbed or emitted:

$$\Delta E > 0 \; (n_f > n_i): \text{Photon } absorbed \text{ with } E_{photon} = h\nu = \Delta E$$

$$\Delta E < 0 \; (n_f < n_i): \text{Photon } emitted \text{ with } E_{photon} = h\nu = -\Delta E \qquad [6.7]$$

These two situations are summarized in **Figure 6.13.** We see that Bohr's model of the hydrogen atom leads to the conclusion that only the specific frequencies of light that satisfy Equation 6.7 can be absorbed or emitted by the atom.

Let's see how to apply these concepts by considering a transition in which the electron moves from $n_i = 3$ to $n_f = 1$. From Equation 6.6 we have

$$\Delta E = (-2.18 \times 10^{-18}\,\text{J})\left(\frac{1}{1^2} - \frac{1}{3^2}\right) = (-2.18 \times 10^{-18}\,\text{J})\left(\frac{8}{9}\right) = -1.94 \times 10^{-18}\,\text{J}$$

The value of ΔE is negative, as you should expect for an electron that is falling from a higher-energy orbit ($n = 3$) to a lower-energy orbit ($n = 1$). A photon is *emitted* during this transition, and the energy of the photon is equal to $E_{photon} = h\nu = -\Delta E = +1.94 \times 10^{-18}\,\text{J}$.

Knowing the energy of the emitted photon, we can calculate either its frequency or its wavelength. For the wavelength, we combine Equations 6.1 ($\lambda = c/\nu$) and 6.3 ($E_{photon} = h\nu$) to obtain

$$\lambda = \frac{c}{\nu} = \frac{hc}{E_{photon}} = \frac{hc}{-\Delta E} = \frac{(6.626 \times 10^{-34}\,\text{J-s})(2.998 \times 10^8\,\text{m/s})}{+1.94 \times 10^{-18}\,\text{J}} = 1.02 \times 10^{-7}\,\text{m}$$

Thus, a photon of wavelength 1.02×10^{-7} m (102 nm) is *emitted*.

We are now in a position to understand the remarkable simplicity of the line spectra of hydrogen, first discovered by Balmer. The line spectra are the result of emission, so $E_{photon} = h\nu = hc/\lambda = -\Delta E$ for these transitions. Combining Equations 6.5 and 6.6, we see that

$$E_{photon} = \frac{hc}{\lambda} = -\Delta E = hcR_H\left(\frac{1}{n_f^2} - \frac{1}{n_i^2}\right) \text{ (for emission)}$$

which gives us

$$\frac{1}{\lambda} = \frac{hcR_H}{hc}\left(\frac{1}{n_f^2} - \frac{1}{n_i^2}\right) = R_H\left(\frac{1}{n_f^2} - \frac{1}{n_i^2}\right), \text{ where } n_f < n_i$$

Thus, the existence of discrete spectral lines can be attributed to the quantized jumps of electrons between energy levels.

Sample Exercise 6.3

Electronic Transitions in the Hydrogen Atom

In the Bohr model of the hydrogen atom, electrons are confined to orbits with fixed radii, and those radii can be calculated. The radii of the first four orbits are 0.53, 2.12, 4.76, and 8.46 Å, respectively, as depicted below.

(**a**) If an electron makes a transition from the $n_i = 4$ level to a lower-energy level, $n_f = 3$, 2, or 1, which transition would produce a photon with the shortest wavelength? (**b**) What are the energy and wavelength of such a photon, and in which region of the electromagnetic spectrum does it lie? (**c**) The image on the right shows the output of a detector that measures the intensity of light emitted from a sample of hydrogen atoms that have been excited so that each atom begins with an electron in the $n = 4$ state. What is the final state, n_f of the transition being detected?

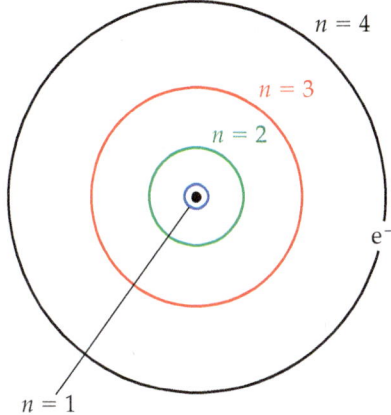

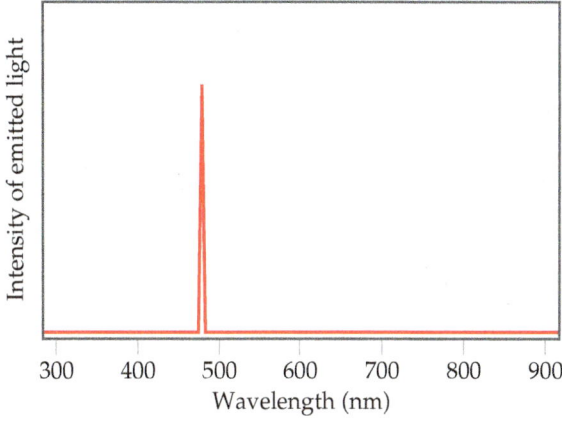

SOLUTION

Analyze We are asked to determine the energy and wavelength associated with various transitions involving an electron relaxing from the $n = 4$ state of the hydrogen atom to one of three lower-energy states.

Plan Given the integers representing the initial and final states of the electron, we can use Equation 6.6 to calculate the energy of the photon emitted and then use the relationships $E = h\nu$ and $c = \nu\lambda$ to convert energy to wavelength. The photon with the highest energy will have the shortest wavelength because photon energy is inversely proportional to wavelength.

Solve

(**a**) The wavelength of a photon is related to its energy through the relationship $E = h\nu = hc/\lambda$. Hence, the photon with the shortest wavelength will have the largest energy. The energy levels of the electron orbits decrease as n decreases.

The electron loses the most energy on transitioning from the $n_i = 4$ state to the $n_f = 1$ state. Consequently, the photon emitted in that transition has the highest energy and the shortest wavelength.

(**b**) We first calculate the energy of the photon using Equation 6.6 with $n_i = 4$ and $n_f = 1$:

$$\Delta E = -2.18 \times 10^{-18}\,\text{J}\left(\frac{1}{1^2} - \frac{1}{4^2}\right) = -2.04 \times 10^{-18}\,\text{J}$$

$$E_{\text{photon}} = -\Delta E = 2.04 \times 10^{-18}\,\text{J}$$

Next, we rearrange Planck's relationship to calculate the frequency of the emitted photon.

$$\nu = E/h = (2.04 \times 10^{-18}\,\text{J})/(6.626 \times 10^{-34}\,\text{J-s}) = 3.08 \times 10^{15}\,\text{s}^{-1}$$

Finally, we use the frequency to determine the wavelength.

Light with this wavelength falls in the ultraviolet region of the electromagnetic spectrum.

$$\lambda = c/\nu = (2.998 \times 10^8\,\text{m/s})/(3.08 \times 10^{15}\,\text{s}^{-1})$$
$$= 9.73 \times 10^{-8}\,\text{m} = 97.3\,\text{nm}$$

(**c**) From the graph, we estimate the wavelength of the photon to be approximately 480 nm. Starting from the wavelength, it is easiest to estimate n_f using Equation 6.4:

$$\frac{1}{\lambda} = R_\text{H}\left(\frac{1}{n_f^2} - \frac{1}{n_i^2}\right)$$

Continued

Rearranging:

So $n_f = 2$, and the photons seen by the detector are those emitted when an electron transitions from the $n_i = 4$ to the $n_f = 2$ state.

$$\frac{1}{n_f^2} = \frac{1}{n_i^2} + \frac{1}{R_H \lambda} = \frac{1}{4^2} + \frac{1}{(1.097 \times 10^7 \text{ m}^{-1})(480 \times 10^{-9} \text{ m})}$$

$$\frac{1}{n_f^2} = 0.25$$

Check Referring back to Figure 6.12, we confirm that the $n = 4$ to $n = 1$ transition should have the largest energy of the three possible transitions and thus the shortest wavelength.

▶ **Practice Exercise**

For each of the following transitions, give the sign of ΔE and indicate whether a photon is emitted or absorbed. **(a)** $n = 3$ to $n = 1$ **(b)** $n = 2$ to $n = 4$

Limitations of the Bohr Model

Although the Bohr model explains the line spectrum of the hydrogen atom, it cannot explain the spectra of other atoms, except in a crude way. Bohr also avoided the problem of why the negatively charged electron would not just fall into the positively charged nucleus, by simply assuming it would not happen. Furthermore, we will see that Bohr's model of an electron orbiting the nucleus at a fixed distance is not a realistic picture. As we explain in Section 6.4, the electron exhibits wave-like properties, a fact that any acceptable model of electronic structure must accommodate.

Despite its shortcomings, the Bohr model suggested an entirely different way to understand the electronic structure of atoms. Two key concepts from the Bohr model are retained in the modern picture of the hydrogen atom:

1. *Electrons exist only in certain discrete energy levels, which are described by quantum numbers.*

2. *A discrete quantity of energy is either absorbed or emitted when an electron transitions from one level to another.*

We now start to develop the successor to the Bohr model, but to do so we must first take a closer look at how the emergence of quantum theory altered our understanding of how matter behaves at the atomic and subatomic scale.

Self-Assessment Exercises

SAE 6.6 In the Bohr model of the hydrogen atom, which of the following quantities *decreases* as the value of the principal quantum number of the electron *increases*? **(a)** the distance from the electron to the nucleus **(b)** the energy of the electron **(c)** the energy required to ionize (remove an electron) the atom **(d)** the frequency of the photon emitted when the electron relaxes back to the ground state

SAE 6.7 What is the wavelength of the photon emitted when an electron in a hydrogen atom makes a transition from the $n = 3$ to the $n = 2$ state? **(a)** 821 nm **(b)** 3.03×10^{-19} m **(c)** 365 nm **(d)** 486 nm **(e)** 656 nm

6.4 | The Wave Behavior of Matter

Learning Objectives

When you finish Section 6.4, you should be able to:

▶ Describe the wave-like properties of matter, including the approximate scales where such effects are observable.

▶ Explain how the uncertainty principle limits the precision with which we can simultaneously determine the position and momentum of small particles, such as electrons.

In the early decades of the twentieth century, scientists were still mapping out the structure of the atom (Section 2.2). Once the existence of subatomic particles was accepted, scientists began to study their properties and found that they behave in surprising ways that cannot be inferred from studying larger objects. In this section we learn that, just like light, matter can behave as a particle and a wave, a key step on the path to developing a better understanding of the electronic structure of atoms.

In the years following the development of Bohr's model for the hydrogen atom, the dual nature of radiant energy became a familiar concept. Depending on the experimental circumstances, radiation appears to have either a wave-like or a particle-like (photon) character. Louis de Broglie (1892–1987), who was working on his Ph.D. thesis in physics at the Sorbonne in Paris, boldly extended this idea: If radiant energy could, under appropriate conditions, behave as though it were a stream of particles (photons), could matter, under appropriate conditions, possibly show the properties of a wave?

De Broglie suggested that an electron moving about the nucleus of an atom behaves like a wave and therefore has a wavelength. He proposed that the wavelength of the electron, or of any other particle, depends on its mass, m, and on its velocity, v:

$$\lambda = \frac{h}{mv} \qquad [6.8]$$

where h is the Planck constant. The quantity mv for any object is called its **momentum**. De Broglie used the term **matter waves** to describe the wave characteristics of material particles.

Because de Broglie's hypothesis is applicable to all matter, any object of mass m and velocity v would give rise to a characteristic matter wave. However, Equation 6.8 indicates that the wavelength associated with an object of ordinary size, such as a golf ball, is so tiny as to be completely unobservable. This is not so for an electron because its mass is so small, as we see in Sample Exercise 6.4.

A few years after de Broglie published his theory, the wave properties of the electron were demonstrated experimentally. When X rays pass through a crystal, an interference pattern results that is characteristic of the wave-like properties of electromagnetic radiation, a phenomenon called *X-ray diffraction*. As electrons pass through a crystal, they are similarly diffracted. Thus, a stream of moving electrons exhibits the same kinds of wave behavior as X rays and all other types of electromagnetic radiation. In the electron microscope, the wave characteristics of electrons are used to obtain images at the atomic scale. This technique has become an important tool for studying surface phenomena at very high magnifications (**Figure 6.14**). Electron microscopes can magnify objects by 3,000,000 times, far more than can be done with visible light (1000×), because the wavelength of the electrons is so much smaller than the wavelength of visible light.

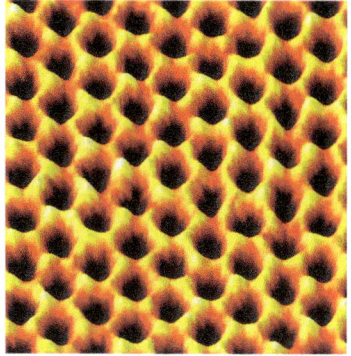

▲ **Figure 6.14 Electrons as waves.** Transmission electron micrograph of *graphene*, which has a hexagonal honeycomb arrangement of carbon atoms. Each of the bright yellow "mountains" indicates a carbon atom.

Sample Exercise 6.4
Matter Waves

What is the wavelength of an electron moving with a speed of 5.97×10^6 m/s? The mass of the electron is 9.11×10^{-31} kg.

SOLUTION

Analyze We are given the mass, m, and velocity, v, of the electron, and we must calculate its de Broglie wavelength.

Plan The wavelength of a moving particle is given by Equation 6.8, so λ is calculated by inserting the known quantities h, m, and v. In doing so, however, we must pay attention to units.

Solve
Using the value of the Planck constant, $h = 6.626 \times 10^{-34}$ J-s, we have the following:

$$\lambda = \frac{h}{mv} = \frac{(6.626 \times 10^{-34}\,\text{J-s})}{(9.11 \times 10^{-31}\,\text{kg})(5.97 \times 10^6\,\text{m/s})}\left(\frac{1\,\text{kg-m}^2/\text{s}^2}{1\,\text{J}}\right)$$

$$\lambda = 1.22 \times 10^{-10}\,\text{m} = 0.122\,\text{nm}$$

Comment By comparing this value with the wavelengths of electromagnetic radiation shown in Figure 6.4, we see that the wavelength of this electron is about the same as that of X rays.

▶ **Practice Exercise**
Calculate the velocity of a neutron whose de Broglie wavelength is 505 pm. The mass of a neutron is 1.675×10^{-27} kg.

The Uncertainty Principle

The discovery of the wave properties of matter raised some new and interesting questions. Consider, for example, a ball rolling down a ramp. Using the equations of classical physics, we can calculate, with great accuracy, the ball's position, direction of motion, and speed at any instant. Can we do the same for an electron, which exhibits wave properties? A wave extends in space, and its location is not precisely defined. We might therefore anticipate that it is impossible to determine exactly where an electron is located at a specific instant.

The German physicist Werner Heisenberg (1901–1976) proposed that the dual nature of matter places a fundamental limitation on how precisely we can know both the location and the momentum of an object at a given instant. The limitation becomes important only when we deal with matter at the subatomic level (that is, with masses as small as subatomic particles like the electron). Heisenberg's principle is called the **uncertainty principle**. When applied to the electrons in an atom, this principle states that it is impossible for us to know simultaneously both the exact momentum of an electron and its exact location in space.

Heisenberg mathematically related the uncertainty in position, Δx, and the uncertainty in momentum, $\Delta(mv)$, to a quantity involving the Planck constant:

$$\Delta x \cdot \Delta(mv) \geq \frac{h}{4\pi} \qquad [6.9]$$

A brief calculation illustrates the dramatic implications of the uncertainty principle. The electron has a mass of 9.11×10^{-31} kg and moves at an average speed of about 5×10^6 m/s in a hydrogen atom. Let's assume that we know the speed to an uncertainty of 1% [that is, an uncertainty of $(0.01)(5 \times 10^6 \text{ m/s}) = 5 \times 10^4$ m/s] and that this is the only important source of uncertainty in the momentum, so that $\Delta(mv) = m\Delta v$. We can use Equation 6.9 to calculate the uncertainty in the position of the electron:

$$\Delta x \geq \frac{h}{4\pi m \Delta v} = \left(\frac{6.626 \times 10^{-34}\text{ J-s}}{4\pi(9.11 \times 10^{-31}\text{ kg})(5 \times 10^4\text{ m/s})}\right) = 1 \times 10^{-9}\text{ m}$$

Because the diameter of a hydrogen atom is about 1×10^{-10} m, the uncertainty in the position of the electron in the atom is an order of magnitude greater than the size of the atom. Thus, we have essentially no idea where the electron is located in the atom. On the other hand, if we were to repeat the calculation with an object of ordinary mass, such as a tennis ball, the uncertainty would be so small that it would be inconsequential. In that case, m is large and Δx is out of the realm of measurement and therefore of no practical consequence.

De Broglie's hypothesis and Heisenberg's uncertainty principle set the stage for a new and more broadly applicable theory of atomic structure. In this approach, any attempt to define precisely the instantaneous location and momentum of the electron is abandoned. The wave nature of the electron is recognized, and its behavior is described in terms appropriate to waves. The result is a model that precisely describes the energy of the electron while describing its location not precisely but rather in terms of probabilities.

A CLOSER LOOK | Measurement and the Uncertainty Principle

Whenever any measurement is made, some uncertainty exists. Our experience with objects of ordinary dimensions, such as balls or trains or laboratory equipment, indicates that using more precise instruments can decrease the uncertainty of a measurement. In fact, we might expect that the uncertainty in a measurement can be made infinitesimally small. However, the uncertainty principle states that there is an actual limit to the accuracy of measurements. This limit is not a restriction on how well instruments can be made; rather, it is inherent in nature. This limit has no practical consequences when dealing with ordinary-sized objects, but its implications are enormous when dealing with subatomic particles, such as electrons.

To measure an object, we must disturb it, at least a little, with our measuring device. Imagine using a flashlight to locate a large rubber ball in a dark room. You see the ball when the light from the flashlight bounces off the ball and strikes your eyes. When a beam of photons strikes an object of this size, it does not alter its position or momentum to any practical extent. Imagine, however, that you wish to locate an electron by similarly bouncing light off it into some detector. Objects can be located to an accuracy no greater than the wavelength of the radiation used. Thus, if we want an accurate position measurement for an electron, we must use a short wavelength. This means that photons of high energy must be employed.

The more energy the photons have, the more momentum they impart to the electron when they strike it, which changes the electron's motion in an unpredictable way. The attempt to measure accurately the electron's position introduces considerable uncertainty in its momentum; the act of measuring the electron's position at one moment makes our knowledge of its future position inaccurate.

Suppose, then, that we use photons of longer wavelength. Because these photons have lower energy, the momentum of the electron is not so appreciably changed during measurement, but at the same time the longer wavelength limits the accuracy with which the electron's position can be determined. This is the essence of the uncertainty principle: *There is an uncertainty in simultaneously knowing both the position and the momentum of the electron that cannot be reduced beyond a certain minimum level.* The more accurately one is known, the less accurately the other is known.

Although we can never know the exact position and momentum of the electron, we can talk about the probability of its being at certain locations in space. In Section 6.5, we introduce a model of the atom that provides the probability of finding electrons of specific energies at certain positions in atoms.

Related Exercises: 6.51, 6.52, 6.96

Self-Assessment Exercises

SAE 6.8 The resolution of a microscope is limited to approximately one-half the wavelength of the light source. Because of this limitation, microscopes capable of obtaining atomic-scale images use electrons instead of visible light to illuminate samples. To distinguish atoms separated by a few hundred picometers, as they are in most solids, the electron wavelength should be less than or equal to approximately 200 pm. To what speed must the electrons be accelerated to attain this wavelength? (**a**) 3.6×10^{-6} m/s (**b**) 3.6×10^{6} m/s (**c**) 3.6×10^{3} m/s (**d**) 3.6×10^{9} m/s

SAE 6.9 Which of the following three statements is or are true?

(**i**) The uncertainty principle states that there is a limit to how precisely the position and momentum of a particle can simultaneously be measured.

(**ii**) The uncertainty principle helps to explain how particles can have wave-like properties.

(**iii**) The uncertainty principle is more relevant to large objects, such as a car, than to small objects, such as an electron.

(**a**) only i (**b**) only ii (**c**) both i and ii (**d**) both i and iii (**e**) All three statements are true.

6.5 | Quantum Mechanics and Atomic Orbitals

The idea that electrons have wave-like properties opens up a new way of thinking about the electronic structures of atoms. In 1926 the Austrian physicist Erwin Schrödinger (1887–1961) proposed an equation, now known as *Schrödinger's wave equation*, that incorporates both the wave-like and particle-like behaviors of the electron. His work led to a new approach to describing the behavior of electrons, an approach known as *quantum mechanics* or *wave mechanics*. The application of Schrödinger's equation requires advanced calculus that is beyond the scope of this text and is not covered here. We do qualitatively consider the results of Schrödinger's analysis, however, because they give us a powerful new way to view electronic structure.

We begin by examining the electronic structure of the simplest atom, hydrogen. Schrödinger treated the electron in a hydrogen atom like the wave on a plucked guitar string (**Figure 6.15**). Because these kinds of waves do not travel in space, they are called *standing waves*. Just as the plucked guitar string produces a standing wave that has a fundamental frequency and higher overtones (harmonics), the electron exhibits a

 Learning Objectives

When you finish Section 6.5, you should be able to:

▶ Compare and contrast the atomic orbitals used in Schrödinger's model of the atom with the orbits used in Bohr's model of the atom.

▶ State the names and symbols of the quantum numbers and qualitatively describe how they determine the size, shape, and orientation of an atomic orbital.

▶ Use the rules governing the allowed combinations of quantum numbers to identify the atomic orbitals of the hydrogen atom and arrange those orbitals in order of increasing energy.

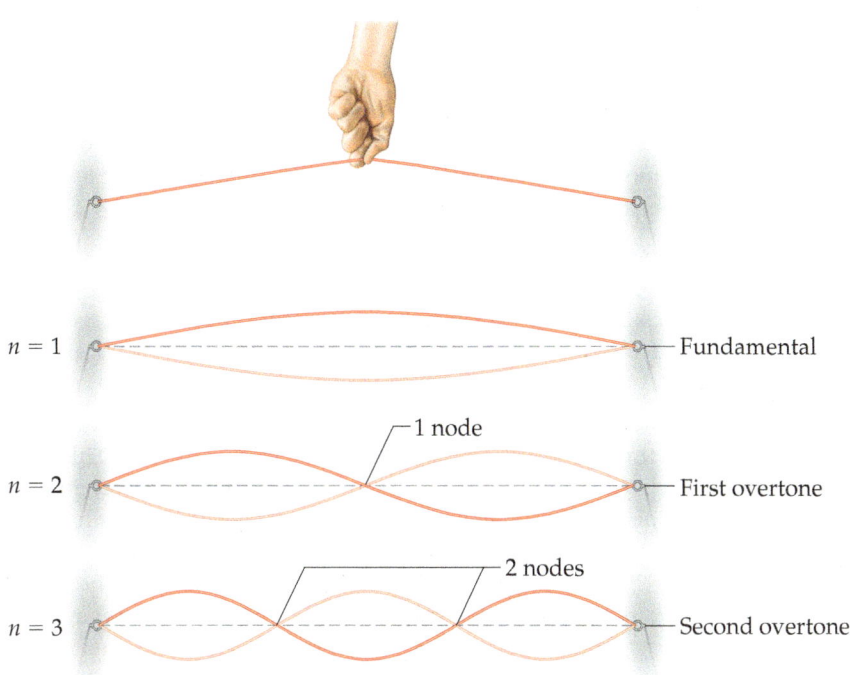

◀ Figure 6.15 Standing waves in a vibrating string.

Go Figure

Where in the figure is the region of highest electron density?

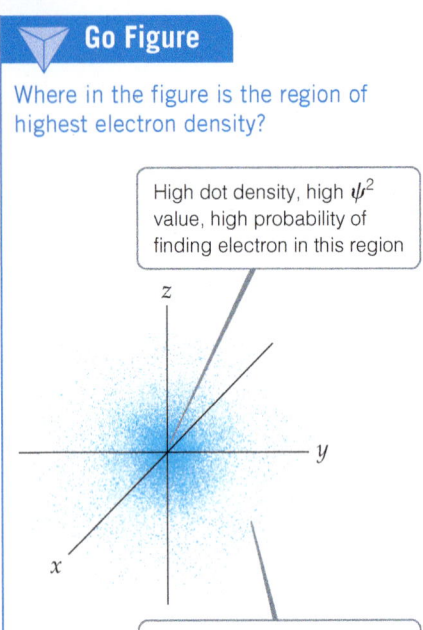

High dot density, high ψ^2 value, high probability of finding electron in this region

Low dot density, low ψ^2 value, low probability of finding electron in this region

▲ **Figure 6.16 Electron-density distribution.** This rendering represents the probability, ψ^2, of finding the electron in a hydrogen atom in its ground state. The origin of the coordinate system is at the nucleus.

lowest-energy standing wave and higher-energy ones. Furthermore, just as the overtones of the guitar string have *nodes*, points where the magnitude of the wave is zero, so do the waves that we use to describe the electron.

Solving Schrödinger's equation for the hydrogen atom leads to a series of mathematical functions called **wave functions** that describe the electron in the atom. These wave functions are usually represented by the symbol ψ (lowercase Greek letter *psi*). Although the wave function has no direct physical meaning, the square of the wave function, ψ^2, provides information about the electron's location when it is in an allowed energy state.

For the hydrogen atom, the allowed energies are the same as those predicted by the Bohr model. However, the Bohr model assumes that the electron is in a circular orbit of some particular radius about the nucleus. In the quantum-mechanical model, the electron's location cannot be described so simply.

According to the uncertainty principle, if we know the momentum of the electron with a high degree of accuracy, then our simultaneous knowledge of its location is very uncertain. Thus, we cannot hope to specify the exact location of an individual electron with respect to the nucleus, as was assumed in the Bohr model. Rather, we must be content with a kind of statistical knowledge. We therefore speak of the *probability* that the electron will be in a certain region of space at a given instant. As it turns out, the square of the wave function, ψ^2, at a given point in space represents the probability that the electron will be found at that location. For this reason, ψ^2 is called either the **probability density** or the **electron density**.

One way of representing the probability of finding the electron in various regions of an atom is shown in **Figure 6.16**, where the density of the dots represents the probability of finding the electron. The regions with a high density of dots correspond to relatively large values for ψ^2 and are therefore regions where there is a high probability of finding the electron. Based on this representation, we often describe atoms as consisting of a nucleus surrounded by an electron cloud.

Orbitals and Quantum Numbers

The solution to Schrödinger's equation for the hydrogen atom yields a set of wave functions called **orbitals**. Each orbital has a characteristic shape and energy. For example, the lowest-energy orbital in the hydrogen atom has the spherical shape illustrated in Figure 6.16 and an energy of -2.18×10^{-18} J. Note that an *orbital* (quantum-mechanical model, which describes electrons in terms of probabilities, visualized as "electron clouds") is not the same as an *orbit* (the Bohr model, which visualizes the electron moving in a physical orbit, like a planet around a star). The quantum-mechanical model does not refer to orbits because the motion of the electron in an atom cannot be precisely determined (Heisenberg's uncertainty principle).

The Bohr model introduced a single quantum number, n, to describe an orbit. The quantum-mechanical model uses three quantum numbers, n, l, and m_l, which result naturally from the mathematics used to describe an orbital.

1. The principal quantum number, n, can have positive integral values 1, 2, 3, ... As n increases, the orbital becomes larger, and the electron spends more time farther from the nucleus. An increase in n also means that the electron has a higher energy and is therefore less tightly bound to the nucleus. For the hydrogen atom, $E_n = -(2.18 \times 10^{-18}\,\text{J})(1/n^2)$, as in the Bohr model.

2. The second quantum number—the **angular momentum quantum number**, l—can have integral values from 0 to $(n-1)$ for each value of n. This quantum number is related to the fact that, even in the quantum-mechanical model, the electron can be viewed as having certain values of *angular momentum*—the momentum of a particle moving in a curved orbit. The amount of angular momentum is limited by the energy of the electron, which explains why the allowed values of l are limited by the value of n. Not all orbitals have the same shape, and this quantum number defines the shape of the orbital. The value of l for a

particular orbital is generally designated by the letters s, p, d, and f,* corresponding to l values of 0, 1, 2, and 3:

Value of l	0	1	2	3
Letter used	s	p	d	f

3. The **magnetic quantum number**, m_l, can have integral values between $-l$ and l,
including zero. When $l \neq 0$, the shape of the orbital is nonspherical and the value
of m_l defines the orientation of the orbital in space, as we discuss in Section 6.6.

A CLOSER LOOK | Schrödinger's Cat and Quantum Computing

The revolutions in scientific thinking caused by the theory of relativity and quantum theory changed far more than just science; it
also caused deep changes in how we understand the world around
us. Before relativity and quantum theory, the prevailing physical
theories were inherently *deterministic*: Once the specific conditions
of an object were given (position, velocity, forces acting on the
object), we could determine exactly the position and motion of the
object at any time in the future. These theories, from Newton's laws
to Maxwell's theory of electromagnetism, successfully described
physical phenomena such as motion of the planets, the trajectories
of projectiles, and the diffraction of light.

In this chapter we have touched on two tenets of quantum theory that lead to a nondeterministic description of matter. First, we
have seen that the descriptions of light and matter have become less
distinct—light has particle-like properties and matter has wave-like
properties. The description of matter that results—in which we can
talk only about the probability of an electron being at a certain place
as opposed to knowing exactly where it is—was very bothersome to
many. Einstein, for example, famously said that "God doesn't play
dice with the world"* about this probabilistic description. Heisenberg's uncertainty principle, which asserts that we can't know the
position and momentum of a particle exactly, also raised many phil-
osophical questions—so many, in fact, that Heisenberg wrote a
book entitled *Physics and Philosophy* in 1958.

One of the common methods scientists used to test these new
theories was through so-called thought experiments. Thought
experiments are hypothetical scenarios that can lead to paradoxes
within a given theory. One of the most famous thought experiments put forward in the early days of the quantum theory was formulated by Schrödinger and is now known as "Schrödinger's cat"
(Figure 6.17). This experiment called into question whether a system
could have multiple acceptable wave functions prior to observation
of the system. In other words, if we don't actually observe a system,
can we know anything about the state it is in? In this paradox, a
hypothetical cat is placed in a sealed box with an apparatus that will
randomly trigger a lethal dose of poison to the cat (as morbid as that
sounds). According to some interpretations of quantum theory,
until the box is opened and the cat is observed, the cat must be considered simultaneously alive and dead.

Schrödinger posed this paradox to point out the weaknesses
in some interpretations of quantum results, but the paradox has

led instead to a continuing and lively debate about the fate and
meaning of Schrödinger's cat. In 2012, the Nobel Prize in Physics
was awarded to Serge Haroche of France and David Wineland of
the United States for their ingenious methods for observing the
quantum states of photons or particles without having the act of
observation destroy the states. In so doing, they observed what is
generally called the "cat state" of the system, in which the photon or particle exists simultaneously in two different quantum
states.

As counterintuitive as this concept is at first blush, many
believe it can be harnessed to create so-called quantum computers.
In a conventional computer, each bit of information has a specific
value, either 0 or 1, and these bits can only be processed one after
the other. In a quantum computer, information is stored in a quantum bit, or *qubit*, which is a superposition of two different states.
Just as Schrödinger's cat can be simultaneously dead and alive, a
qubit can be a combination of 0 and 1 at the same time. Provided
you can keep different qubits in coherence with each other, the
computing power of a quantum computer increases rapidly as the
number of qubits increases, much more so than you can achieve by
increasing the number of binary bits in a conventional computer.
Scientists have done calculations with 20 coherent qubits, and even
with this modest number of qubits, they can create over one million superimposed states. Unfortunately, these "cat states" are very
sensitive to interactions with their environment, including thermal fluctuations, which can collapse the superposition down to
the simple 0 and 1 of a conventional bit. The main challenge for
harnessing the power of quantum computing is to find ways to
make this superposition of states more stable.

Related Exercise: 6.96

▲ **Figure 6.17** In Schrödinger's thought experiment, the hypothetical
cat in the box is a mixture or superposition of alive and dead states
until we make an observation by opening the box.

* William Hermanns, *Einstein and the Poet: In Search of the Cosmic
Man*, 1st edition, Branden Books, 1983.

* The letters come from the words *sharp*, *principal*, *diffuse*, and *fundamental*, which were used to
describe certain features of spectra before quantum mechanics was developed.

TABLE 6.2   Relationship among Values of n, l, and m_l through $n = 4$

n	Possible Values of l	Subshell Designation	Possible Values of m_l	Number of Orbitals in Subshell	Total Number of Orbitals in Shell
1	0	$1s$	0	1	1
2	0	$2s$	0	1	
	1	$2p$	$1, 0, -1$	3	4
3	0	$3s$	0	1	
	1	$3p$	$1, 0, -1$	3	
	2	$3d$	$2, 1, 0, -1, -2$	5	9
4	0	$4s$	0	1	
	1	$4p$	$1, 0, -1$	3	
	2	$4d$	$2, 1, 0, -1, -2$	5	
	3	$4f$	$3, 2, 1, 0, -1, -2, -3$	7	16

Because the value of n can be any positive integer, there are an infinite number of orbitals for the hydrogen atom. At any given instant, however, the electron in a hydrogen atom is described by only one of these orbitals—we say that the electron *occupies* a certain orbital. The remaining orbitals are *unoccupied* for that particular state of the hydrogen atom. We focus mainly on orbitals that have small values of n.

The collection of orbitals with the same value of n is called an **electron shell**. For example, all orbitals that have $n = 3$ are said to be in the third shell. The set of orbitals that have the same n and l values is called a **subshell**. Each subshell is designated by a number (the value of n) and a letter ($s, p, d,$ or f, corresponding to the value of l). For example, the orbitals that have $n = 3$ and $l = 2$ are called $3d$ orbitals and are said to be in the $3d$ subshell.

Table 6.2 summarizes the possible values of l and m_l for values of n through $n = 4$. The restrictions on possible values give rise to the following very important observations:

1. *The shell with principal quantum number* n *consists of exactly* n *subshells.* Each subshell corresponds to a different allowed value of l from 0 to $(n - 1)$ Thus, the first shell ($n = 1$) consists of only one subshell, the $1s$ ($l = 0$); the second shell ($n = 2$) consists of two subshells, the $2s$ ($l = 0$) and $2p$ ($l = 1$); the third shell consists of three subshells, $3s, 3p,$ and $3d$; and so forth.

2. *Each subshell consists of a specific number of orbitals.* Each orbital corresponds to a different allowed value of m_l. For a given value of l, there are $(2l + 1)$ allowed values of m_l, ranging from $-l$ to $+l$. Thus, each s ($l = 0$) subshell consists of one orbital; each p ($l = 1$) subshell consists of three orbitals; each d ($l = 2$) subshell consists of five orbitals, and so forth.

3. *The total number of orbitals in a shell is* n^2, *where* n *is the principal quantum number of the shell.* The resulting number of orbitals for the shells—1, 4, 9, 16—is related to a pattern seen in the periodic table: We see that the number of elements in the rows of the periodic table—2, 8, 18, and 32—equals twice these numbers. We discuss this relationship further in Section 6.9.

Figure 6.18 shows the relative energies of the hydrogen atom orbitals through $n = 3$. Each box represents an orbital, and orbitals of the same subshell, such as the three $2p$ orbitals, are grouped together. When the electron occupies the lowest-energy orbital ($1s$), the hydrogen atom is said to be in its *ground state*. When the electron occupies any other orbital, the atom is in an *excited state*. (The electron can be excited to a higher-energy orbital by absorption of a photon of appropriate energy.) At ordinary temperatures, essentially all hydrogen atoms are in the ground state.

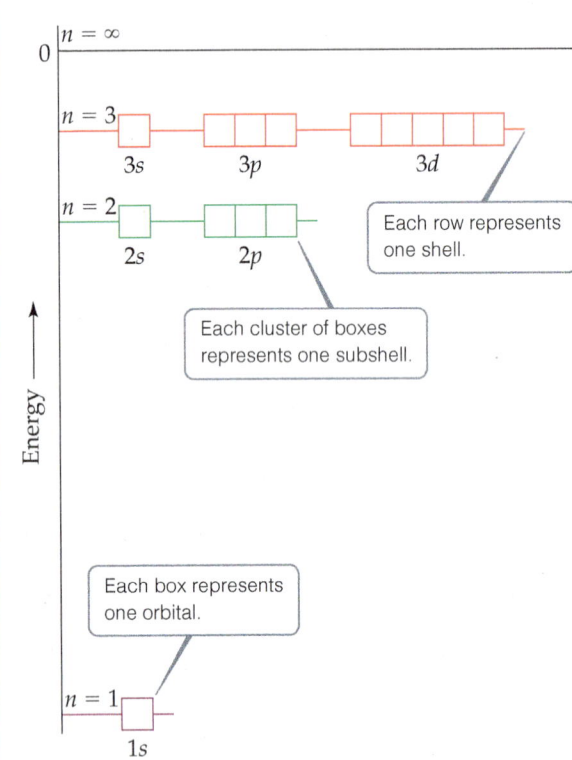

Go Figure

Are the relative energies of the $n = 1, 2,$ and 3 shells shown here the same as or different from those of the Bohr model of the atom shown in Figure 6.12?

$n = 1$ shell has one orbital
$n = 2$ shell has two subshells composed of four orbitals
$n = 3$ shell has three subshells composed of nine orbitals

▲ **Figure 6.18** Energy levels in the hydrogen atom.

Sample Exercise 6.5
Subshells of the Hydrogen Atom

(a) Without referring to Table 6.2, predict the number of subshells in the fourth shell—that is, for $n = 4$. **(b)** Give the label for each of these subshells. **(c)** How many orbitals are in each of these subshells?

SOLUTION

Analyze and Plan We are given the value of the principal quantum number, n. We need to determine the allowed values of l and m_l for this given value of n and then count the number of orbitals in each subshell.

Solve There are four subshells in the fourth shell, corresponding to the four possible values of l (0, 1, 2, and 3).

These subshells are labeled $4s$, $4p$, $4d$, and $4f$. The number given in the designation of a subshell is the principal quantum number, n; the letter designates the value of the angular momentum quantum number, l: the label is s for $l = 0$; p for $l = 1$; d for $l = 2$; and f for $l = 3$.

There is one $4s$ orbital (when $l = 0$, the only possible value of m_l is 0). There are three $4p$ orbitals (when $l = 1$, there are three possible values of m_l: 1, 0, −1). There are five $4d$ orbitals (when $l = 2$, there are five allowed values of m_l: 2, 1, 0, −1, −2). There are seven $4f$ orbitals (when $l = 3$, there are seven permitted values of m_l: 3, 2, 1, 0, −1, −2, −3).

▶ **Practice Exercise**
What is the designation for the subshell with $n = 5$ and $l = 1$? **(b)** How many orbitals are in this subshell? **(c)** Indicate the values of m_l for each of these orbitals.

Self-Assessment Exercises

SAE 6.10 In the hydrogen atom, the _____ quantum number determines the energy of an atomic orbital, and the _____ quantum number determines its shape. **(a)** principal (n), angular momentum (l) **(b)** principal (n), magnetic (m_l) **(c)** angular momentum (l), magnetic (m_l) **(d)** angular momentum (l), principal (n)

SAE 6.11 As the value of the principal quantum number, n, for an orbital increases, which of the following statements is or are true of an electron that occupies that orbital?

(i) The attraction between the electron and the nucleus increases.

(ii) The probability of finding the electron farther away from the nucleus increases.

(iii) The energy of the electron increases.

(a) only i **(b)** only iii **(c)** both i and ii **(d)** both ii and iii **(e)** All three statements are true.

SAE 6.12 Which of the following statements about the $3d$ subshell of a hydrogen atom is *false*? **(a)** All orbitals in this subshell have $n = 3$. **(b)** All orbitals in this subshell have $l = 2$. **(c)** There are nine orbitals in this subshell. **(d)** The orbitals in this subshell are at the same energy as those in the $3p$ subshell.

SAE 6.13 How many orbitals are in the fourth electron shell? **(a)** 4 **(b)** 8 **(c)** 16 **(d)** 32

6.6 | Representations of Orbitals

Thus far we have emphasized the energies of orbitals, but the wave function also provides information about an electron's probable location in space. In this section we examine the ways in which we can picture orbitals, because their shapes help us visualize how the electron density is distributed around the nucleus. We use the shapes of orbitals extensively in subsequent chapters of the text, especially when we talk about the formation of chemical bonds.

Learning Objectives

When you finish Section 6.6, you should be able to:

▶ Interpret the radial probability function of an atomic orbital.

▶ Compare and contrast the three-dimensional shapes of s, p, and d orbitals.

The s Orbitals

We have already seen one representation of the lowest-energy orbital of the hydrogen atom, the $1s$ (Figure 6.16). The electron density for the $1s$ orbital is *spherically symmetric*. In other words, the electron density at a given distance from the nucleus is the same regardless of the direction in which we proceed from the nucleus. All of the other s orbitals ($2s$, $3s$, $4s$, and so forth) are also spherically symmetric and centered on the nucleus.

The l quantum number for the s orbitals is 0, so the m_l quantum number must also be 0. Thus, for each value of n, there is only one s orbital. So how do s orbitals differ as the value of n changes? For example, how does the electron-density distribution of the hydrogen atom change when the electron is excited from the $1s$ orbital to the $2s$ orbital?

According to quantum mechanics, we must describe the position of the electron in the hydrogen atom in terms of probabilities rather than exact locations. The information about the probability is contained in the wave functions, ψ, obtained from Schrödinger's equation. Because s orbitals are spherically symmetric, the value of ψ for an electron in an s orbital depends only on its distance from the nucleus, r. The wave functions for the 1s, 2s, and 3s orbitals are shown at the top of **Figure 6.19**. All three wave functions drop off exponentially upon moving away from the nucleus, but for the 1s orbital the value of the wave function remains a positive number regardless of the distance from the nucleus. In contrast, the value of the 2s wave function crosses over from positive to negative at a distance approximately 1 Å from the nucleus, while the 3s wave function changes sign twice. Analogous to our descriptions of the standing waves of the guitar string (Figure 6.15), we refer to the points where the wave function changes sign as **nodes**.

The square of the wave function, ψ^2 (the probability density), gives the probability that the electron is located at any *point* in space. The probability density plots for the 1s, 2s, and 3s orbitals as a function of the distance r from the nucleus are shown immediately below the wave function plots in Figure 6.19. Notice that both $\psi(r)$ and $[\psi(r)]^2$ attain their largest values at the nucleus. This is true of all s orbitals, but not of the p, d, and f orbitals we discuss later.

To determine the *total* probability of finding the electron at a given distance from the nucleus, we need to "add up" the probability densities $[\psi(r)]^2$ over all points that are a distance r from the nucleus. This gives a function called the **radial probability function**. To add up all of the individual probability densities requires calculus and is therefore beyond the scope of this text, but the result of that calculation tells us that the radial probability function is equal to the probability density, $[\psi(r)]^2$, multiplied by the surface area of a sphere, $4\pi r^2$:

$$\text{Radial probability function (at a distance } r \text{ from the nucleus)} = 4\pi r^2 [\psi(r)]^2$$

The radial probability functions for the 1s, 2s, and 3s orbitals are shown in the bottom set of panels of Figure 6.19. The fact that $4\pi r^2$ increases rapidly as we move away from the nucleus makes the radial probability function and the probability density look distinctly different from each other. For example, the plot of $[\psi(r)]^2$ for the 3s orbital in Figure 6.19 generally gets smaller the farther we move away from the nucleus. But when we multiply by $4\pi r^2$, we see peaks that get larger and larger (up to a certain point) as r increases. The cutaway images shown at the bottom of Figure 6.19 give a three-dimensional picture of the radial probability function. The values of r where the radial probability density goes to 0, represented by white shells in these cutaway images, are the nodes where the wave function changes sign.

For the 1s orbital, we see that the radial probability function rises rapidly as we move away from the nucleus, maximizing at about 0.5 Å. Thus, when the electron occupies the 1s orbital, it is *most likely* to be found this distance from the nucleus.* Notice that we use the probabilistic description, consistent with the uncertainty principle. We can also see for the 1s orbital that the probability of finding the electron at a distance greater than about 3 Å from the nucleus is very small.

Turning next to the radial probability functions for the 2s and 3s orbitals, we see three trends as the value of the principal quantum number n changes:

1. For an ns orbital, the number of peaks is equal to n, with the outermost peak being larger than inner ones.

2. For an ns orbital, the number of nodes is equal to n − 1. Because these types of nodes are located at specific distances from the nucleus, we refer to them as **radial nodes**.

3. As n increases, the electron density becomes more spread out—that is, there is a greater probability of finding the electron further from the nucleus.

---

*In the quantum-mechanical model, the most probable distance at which to find the electron in the 1s orbital is actually 0.529 Å, the same as the radius of the orbit predicted by Bohr for n = 1. The distance 0.529 Å is often called the *Bohr radius*.

▼ **Go Figure**   How many maxima would you expect to find in the radial probability function for the 4s orbital of the hydrogen atom? How many nodes would you expect in this function?

▼ **Figure 6.19 The wave function** $\psi(r)$, **probability density** $[\psi(r)]^2$, **and radial probability function** $4\pi r^2[\psi(r)]^2$ **for the 1s, 2s, and 3s orbitals of hydrogen.** As n increases, the number of nodes (marked with dashed lines) increases and the most probable distance to find the electron (the highest peak in the radial probability function) moves farther from the nucleus. The cutaway views at the bottom provide a three-dimensional representation of the radial probability function.

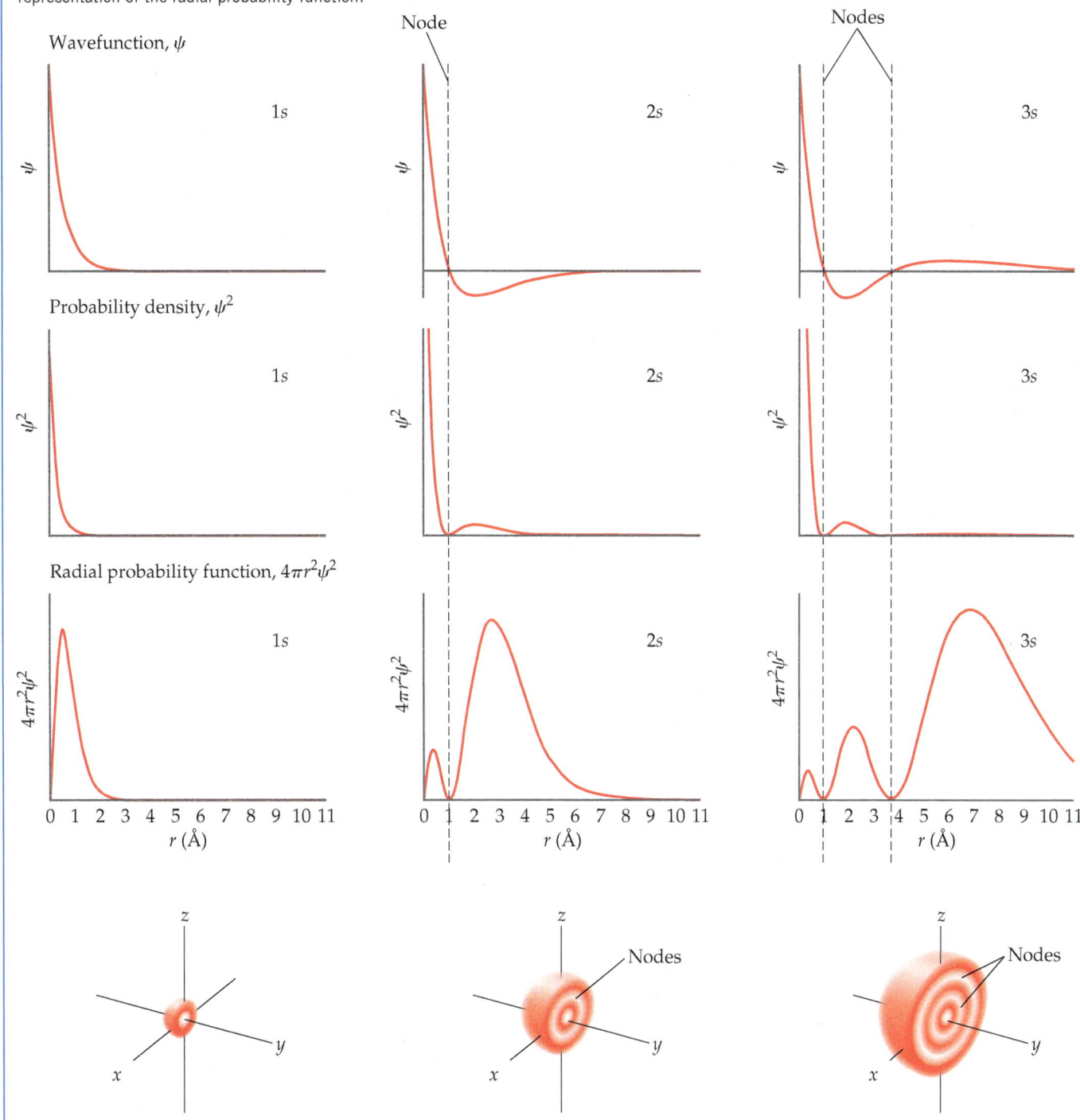

One widely used method of representing orbital *shape* is to draw a boundary surface that encloses some substantial portion, say 90%, of the electron density for the orbital. This type of drawing is called a *contour representation*, and the contour representations for the s orbitals are spheres (**Figure 6.20**). All the orbitals have the same shape, but they differ in size, becoming larger as n increases, reflecting the fact that the electron density becomes more spread out as n increases. Although the details of how electron density

▶ **Figure 6.20 Comparison of the 1s, 2s, and 3s orbitals.** (a) Electron-density distribution of a 1s orbital. (b) Contour representations of the 1s, 2s, and 3s orbitals. Each sphere is centered on the atom's nucleus and encloses the volume in which there is a 90% probability of finding the electron.

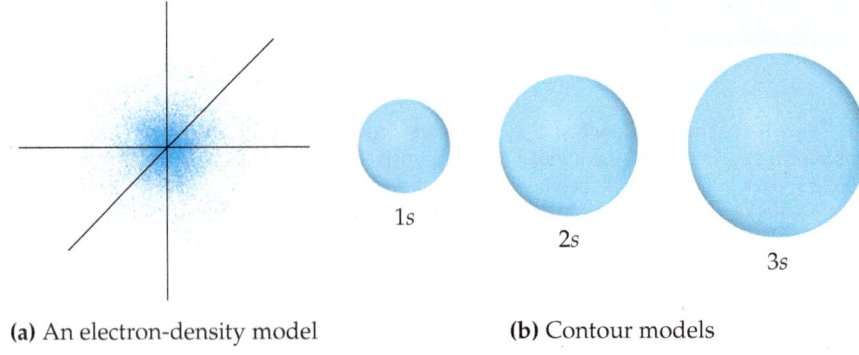

1s

2s

3s

(a) An electron-density model   **(b)** Contour models

varies within a given contour representation are lost in these representations, this is not a serious disadvantage. For qualitative discussions, the most important features of orbitals are shape and relative size, which are adequately displayed by contour representations.

The p Orbitals

The orbitals for which $l = 1$ are the p orbitals. Each p subshell has three orbitals, corresponding to the three allowed values of $m_l = -1, 0$, and 1. The distribution of electron density for one of the 2p orbitals is shown in **Figure 6.21(a)**. The electron density is not distributed spherically as in an s orbital. Instead, the density is concentrated in two regions on either side of the nucleus, separated by a node at the nucleus. We say that this dumbbell-shaped orbital has two *lobes*; for this particular 2p orbital, the two lobes are centered along the +z and −z axes of a coordinate system. Recall that we are making no statement about how the electron is moving within the orbital. Figure 6.21(a) portrays only the *averaged* distribution of the electron density in a 2p orbital. Notice that the wave function ψ and probability density $[\psi]^2$ go to zero at all points in the xy plane. When the wave function has a node that extends over an entire plane, we can refer to this type of node as an **angular node** or **nodal plane**.* For a given orbital, the number of nodal planes is equal to the value of angular momentum quantum number l.

Beginning with the $n = 2$ shell, each shell has three p orbitals (Table 6.2), one for each allowed value of m_l. Thus, there are three 2p orbitals, three 3p orbitals, and so forth. Each set of p orbitals has the dumbbell shapes shown in Figure 6.21(a). For each value of n, the three p orbitals have the same size and shape but differ from one another in spatial orientation. We usually represent p orbitals by drawing the shape and orientation of their wave functions, as shown in the contour representations in Figure 6.21(b). These orbitals can be labeled as p_x, p_y, and p_z, where the letter subscript indicates the Cartesian axis along which the orbital is oriented. Just as the $2p_z$ orbital has the xy plane as its nodal plane, the $2p_x$ and $2p_y$ orbitals have the yz and xz planes as nodal planes, respectively. Thus, we see that two orbitals with the same value of n and l, but different values of m_l, differ from each other in the way they are oriented in space.† Like s orbitals, p orbitals increase in size as we move from 2p to 3p to 4p, and so forth.

The d and f Orbitals

When n is 3 or greater, we encounter the d orbitals (for which $l = 2$). There are five 3d orbitals, five 4d orbitals, and so forth, because in each shell there are five possible values for the m_l quantum number: $-2, -1, 0, 1$, and 2. The different d orbitals in a given shell have different shapes and orientations in space, as shown in **Figure 6.22**. Four of the d-orbital

---

* For some d and f orbitals, such as the d_{z^2} orbital, the surface over which the angular node extends is a cone rather than a plane. For this reason, the term *nodal surface* is sometimes used instead of nodal plane.

† We cannot make a simple correspondence between the subscripts (x, y, and z) and the allowed m_l values (1, 0, and −1).

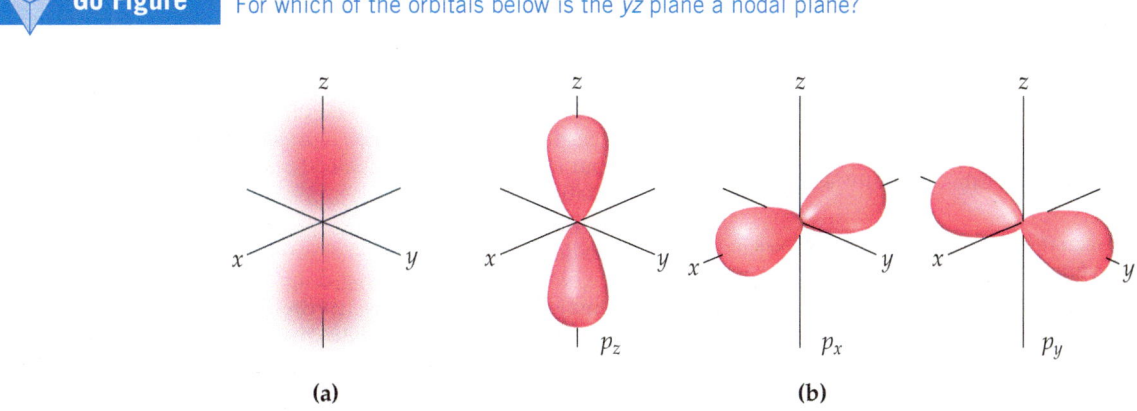

Go Figure   For which of the orbitals below is the *yz* plane a nodal plane?

(a)   (b)

▲ **Figure 6.21 The *p* orbitals. (a)** Electron-density distribution of a 2*p* orbital. **(b)** Contour representations of the three *p* orbitals. The subscript on the orbital label indicates the axis along which the orbital lies.

contour representations have a "four-leaf clover" shape, with four lobes, and each lies primarily in a plane. The d_{xy}, d_{xz}, and d_{yz} orbitals lie in the *xy*, *xz*, and *yz* planes, respectively, with the lobes oriented *between* the axes. The lobes of the $d_{x^2-y^2}$ orbital also lie in the *xy* plane, but the lobes lie *along* the *x* and *y* axes. The d_{z^2} orbital looks very different from the other four: It has two lobes along the *z* axis and a "doughnut" in the *xy* plane. Even though the d_{z^2} orbital looks different from the other *d* orbitals, it has the same energy as the other four *d* orbitals. The contour representations in Figure 6.22 are used for all *d* orbitals, regardless of the principal quantum number, but just as is the case for the *s* and *p* orbitals the size of the orbital increases as *n* increases.

When *n* is 4 or greater, there are seven equivalent *f* orbitals (for which $l = 3$). The shapes of the *f* orbitals are even more complicated than those of the *d* orbitals and are not presented here. As we discuss in Section 6.7, however, you must be aware of *f* orbitals when we consider the electronic structure of atoms in the lower part of the periodic table.

In many instances later in the text, you will find that knowing the number and shapes of atomic orbitals will help you understand chemistry at the molecular level. You will therefore find it useful to memorize the shapes of the *s*, *p*, and *d* orbitals shown in Figures 6.20, 6.21, and 6.22.

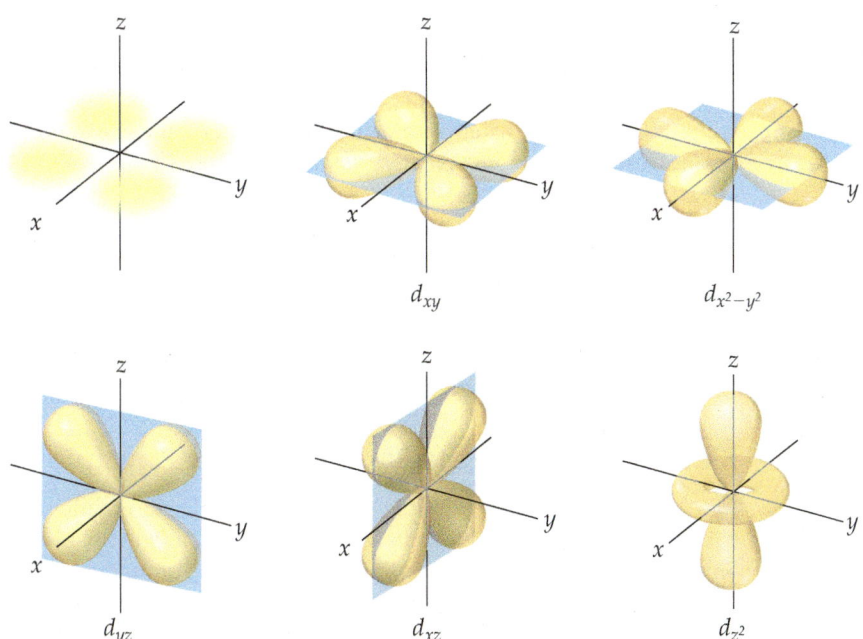

◀ **Figure 6.22 The *d* orbitals. (a)** Electron-density distribution of a 3*d* orbital. **(b)** Contour representations of the five *d* orbitals.

Self-Assessment Exercises

SAE 6.14 Which of the following statements is or are true for the wave functions describing the *s* orbitals as the value of the principal quantum number, *n*, increases?

 (i) The number of radial nodes increases.

 (ii) The number of times the probability density $[\psi(r)]^2$ changes sign increases.

 (iii) The distance between the nucleus and the largest peak in the radial distribution function increases

(a) only i **(b)** only iii **(c)** both i and ii **(d)** both i and iii **(e)** all three are true.

SAE 6.15 Which of the following statements is *true*? **(a)** The number of radial nodes in an orbital can be determined from inspection of its contour representation, but not the number of nodal planes. **(b)** The number of nodal planes in an orbital can

be determined from inspection of its contour representation, but not the number of radial nodes. **(c)** The number of radial nodes and the number of nodal planes in an orbital can be determined by inspection of its contour representation. **(d)** No information about nodes can be ascertained by inspecting the contour representation of an orbital.

SAE 6.16 Given the contour representation for an atomic orbital shown here, which of the following statements is *false*? **(a)** This orbital has two nodal planes. **(b)** For this orbital, the angular momentum quantum number *l* must be 2. **(c)** For this orbital, the principal quantum number *n* must be 3. **(d)** This is a *d* orbital.

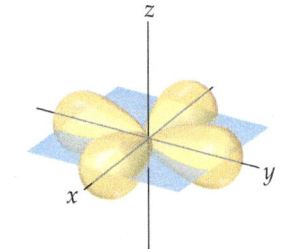

6.7 | Many-Electron Atoms

We have seen that quantum mechanics provides an elegant description of the hydrogen atom. If we want to describe atoms with more than one electron (a *many-electron* atom) the situation becomes more complicated due to interactions between the electrons. Fortunately, the basic characteristics of the wave functions are retained in many-electron atoms. In this section we consider the relative energies of the orbitals in many-electron atoms as well as how the electrons populate the available orbitals.

Orbitals and Their Energies

We can describe the electronic structure of a many-electron atom by using the orbitals we described for the hydrogen atom in Table 6.2. Thus, the orbitals of a many-electron atom also have designations such as 1*s* and 2*p*ₓ, and they have the same general shapes as the corresponding hydrogen orbitals.

Although the shapes of the orbitals of a many-electron atom are the same as those for hydrogen, the presence of more than one electron alters the energies of the orbitals. In hydrogen, the energy of an orbital depends only on its principal quantum number, *n* (Figure 6.18). For instance, in a hydrogen atom, the 3*s*, 3*p*, and 3*d* subshells all have the same energy. In a many-electron atom, however, the energies of the various subshells in a given shell are *different* because of electron–electron repulsions. To explain why this happens, we must consider the forces between the electrons and how these forces are affected by the shapes of the orbitals.

The details of this analysis are given in Chapter 7, but the important idea is this: *In a many-electron atom, for a given value of* n, *the energy of an orbital increases as the value of* l *increases*, as illustrated in **Figure 6.23**. For example, the *n* = 3 orbitals increase in energy in the order 3*s* < 3*p* < 3*d*. Nevertheless, all orbitals of a given subshell (such as the five 3*d* orbitals) retain the same energy in a many-electron atom, just as they do in the hydrogen atom. Orbitals with the same energy are said to be **degenerate**.

Figure 6.23 is a *qualitative* energy-level diagram; the exact energies of the orbitals and their spacings differ from one atom to another.

Electron Spin and the Pauli Exclusion Principle

In the hydrogen atom there is only a single electron. In the ground state it occupies the 1*s* orbital, while in an excited state it can, in principle, occupy any orbital. How do the electrons of a many-electron atom populate the available orbitals? To answer this question, we must consider an additional property of the electron.

Learning Objectives

When you finish **Section 6.7**, you should be able to:

▶ Compare the energy levels of atomic orbitals in a many-electron atom to those in a hydrogen atom.

▶ Describe the concept of electron spin.

▶ Describe the Pauli exclusion principle and explain how it limits the number of electrons each atomic orbital can hold.

 Go Figure

Not all orbitals in the *n* = 4 shell are shown in this figure. Which subshells are missing?

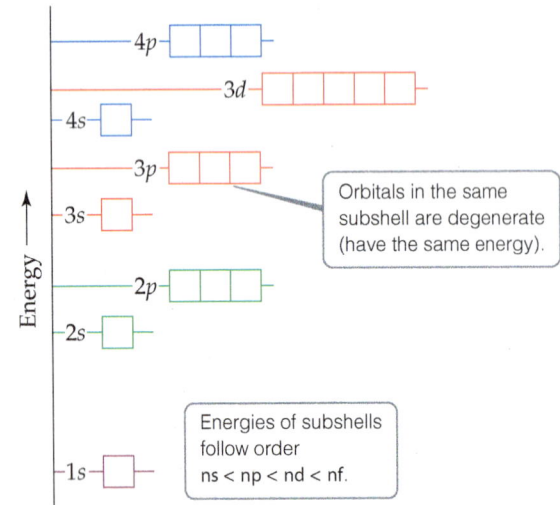

▲ **Figure 6.23** General energy ordering of orbitals for a many-electron atom.

When scientists studied the line spectra of many-electron atoms in great detail, they noticed a very puzzling feature: Lines that were originally thought to be single were actually closely spaced pairs. This meant, in essence, that there were twice as many energy levels as there were "supposed" to be. In 1925, the Dutch physicists George Uhlenbeck (1900–1988) and Samuel Goudsmit (1902–1978) proposed a solution to this dilemma. They postulated that electrons have an intrinsic property, called **electron spin**, that causes each electron to behave as if it were a tiny sphere spinning on its own axis.

By now, it may not surprise you to learn that electron spin is quantized. This observation led to the assignment of a new quantum number for the electron, in addition to n, l, and m_l, which we have already discussed. This new quantum number, the **spin magnetic quantum number**, is denoted m_s (the subscript s stands for *spin*). Two possible values are allowed for m_s, $+\frac{1}{2}$ or $-\frac{1}{2}$, or which were first interpreted as indicating the two opposite directions in which the electron can spin. The link between magnetism and spin comes from the fact that a spinning charge produces a magnetic field. The two opposite directions of spin therefore produce oppositely directed magnetic fields (**Figure 6.24**).* These two opposite magnetic fields lead to the splitting of spectral lines into closely spaced pairs.

The concept of electron spin is crucial for understanding the electronic structures of atoms. In 1925, the Austrian-born physicist Wolfgang Pauli (1900–1958) discovered the principle that governs the arrangement of electrons in many-electron atoms. The **Pauli exclusion principle** states that *no two electrons in an atom can have the same set of four quantum numbers* n, l, m_l *and* m_s. For a given orbital, the values of n, l, and m_l are fixed. Thus, if we want to put more than one electron in an orbital *and* satisfy the Pauli exclusion principle, our only choice is to assign different m_s values to the electrons. Because there are only two such values, we conclude that *an orbital can hold a maximum of two electrons and they must have opposite spins*. This restriction allows us to index the electrons in an atom, giving their quantum numbers and thereby defining the region in space where each electron is most likely to be found. It also provides the key to understanding the remarkable structure of the periodic table of the elements.

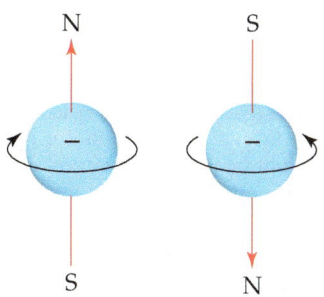

▲ **Figure 6.24 Electron spin.** The electron behaves as if it were spinning about an axis, thereby generating a magnetic field whose direction depends on the direction of spin. The two directions for the magnetic field correspond to the two possible values for the spin quantum number, m_s. The magnetic fields that emanate from materials, like iron, arise because there are more electrons with one spin direction than the other.

▲ Self-Assessment Exercises

SAE 6.17   If two electrons in a multielectron atom are in degenerate orbitals, which quantum number(s) must be the same? (**a**) n (**b**) l (**c**) m_l (**d**) n and l (**e**) n, l, and m_l

SAE 6.18   Which of the following statements is or are true?

(**i**) The value of the spin magnetic quantum number m_s depends on the value of the magnetic quantum number m_l.

(**ii**) There are two possible values of the spin magnetic quantum number m_s.

(**a**) only i (**b**) only ii (**c**) Both i and ii are true. (**d**) Neither i nor ii is true.

SAE 6.19   An oxygen atom has eight electrons. In the ground state, which orbitals will be populated with electrons? (**a**) $1s$ (**b**) $1s$ and $2s$ (**c**) $1s$, $2s$, and $2p$ (**d**) $1s$, $2s$, $2p$, $3s$, and $3p$

6.8 | Electron Configurations

Now that we know the relative energies of orbitals and the Pauli exclusion principle, we are in a position to understand the arrangements of electrons in atoms. The way electrons are distributed among the various orbitals of an atom is called the **electron configuration** of the atom. Electron configurations are central to many chemical and physical properties of the elements. The tendency for atoms to form bonds with other atoms or to participate in oxidation–reduction reactions is determined in part by the electron configuration, as are the magnetic and optical properties of many substances.

The most stable electron configuration—the ground state—is that in which the electrons are in the lowest possible energy states. If there were no restrictions on the possible values for the quantum numbers of the electrons, all the electrons would crowd into the

▲ Learning Objectives

When you finish Section 6.8, you should be able to:

▶ Determine the expected ground-state electron configuration of any atom and use various representations to express it.

▶ Use Hund's rule and an orbital diagram to determine the number of unpaired electrons for any atom or ion.

---

* As we discussed previously, the electron has both particle-like and wave-like properties. Thus, the picture of an electron as a spinning charged sphere is, strictly speaking, just a useful pictorial representation that helps us understand the two directions of magnetic field that an electron can possess.

CHEMISTRY AND LIFE | Nuclear Spin and Magnetic Resonance Imaging

A major challenge facing medical diagnosis is seeing inside the human body. For many years this was accomplished primarily by X-ray technology. X rays, however, do not give well-resolved images of overlapping physiological structures, and sometimes they fail to discern diseased or injured tissue. Moreover, because X rays are high-energy radiation, they potentially can cause physiological harm, even in low doses. An imaging technique developed in the 1980s called *magnetic resonance imaging (MRI)* does not have these disadvantages.

The foundation of MRI is a phenomenon called *nuclear magnetic resonance (NMR)*, which was discovered in the mid-1940s. Today NMR has become one of the most important spectroscopic methods used in chemistry. NMR is based on the observation that, like electrons, the nuclei of many elements possess an intrinsic spin. Like electron spin, nuclear spin is quantized. For example, the nucleus of ^{1}H has two possible magnetic nuclear spin quantum numbers, $+\frac{1}{2}$ and $-\frac{1}{2}$.

A spinning hydrogen nucleus acts like a tiny magnet. In the absence of external effects, the two spin states have the same energy. However, when the nuclei are placed in an external magnetic field, they can align either parallel or opposed (antiparallel) to the field, depending on their spin. The parallel alignment is lower in energy than the antiparallel one by a certain amount, ΔE (Figure 6.25). If the nuclei are irradiated with photons having energy equal to ΔE, the spin of the nuclei can be "flipped"—that is, excited from the parallel to the antiparallel alignment. Detection of the flipping of nuclei between the two spin states leads to an NMR spectrum. The radiation used in an NMR experiment is in the radiofrequency range, typically 100 to 900 MHz, which is far less energetic per photon than X rays.

Because hydrogen is a major constituent of aqueous body fluids and fatty tissue, the hydrogen nucleus is the most convenient one for study by MRI. In MRI a person's body is placed in a strong magnetic field. By irradiating the body with pulses of radiofrequency radiation and using sophisticated detection techniques, medical technicians can image tissue at specific depths in the body, giving pictures with spectacular detail (Figure 6.26). The ability to sample at different depths allows the technicians to construct a three-dimensional picture of the body.

MRI has had such a profound influence on the modern practice of medicine that Paul Lauterbur, a chemist, and Peter Mansfield, a physicist, were awarded the 2003 Nobel Prize in Physiology or Medicine for their discoveries concerning MRI. The major drawback of this technique is expense: The cost of a new standard MRI instrument for clinical applications can easily run in excess of a million dollars. Improvements in MRI technology now allow for millimeter resolution of biological features small enough to detect tumors before they become metastatic.

Related Exercise: 6.98

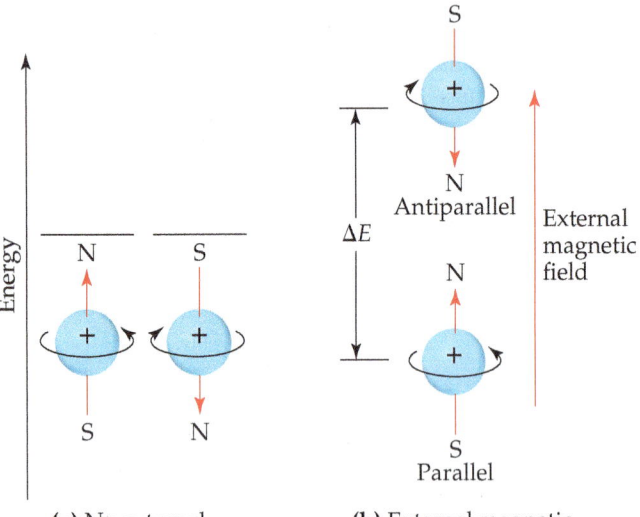

(a) No external magnetic field

(b) External magnetic field applied

▲ **Figure 6.25 Nuclear spin.** Like electron spin, nuclear spin generates a small magnetic field and its values are quantized. **(a)** In the absence of an external magnetic field, the two spin states of a hydrogen nucleus have the same energy. **(b)** When an external magnetic field is applied, the spin state in which the spin direction is parallel to the direction of the external field is lower in energy than the spin state in which the spin direction is antiparallel to the field direction. The energy difference, ΔE, between the two states is in the radio frequency region of the electromagnetic spectrum.

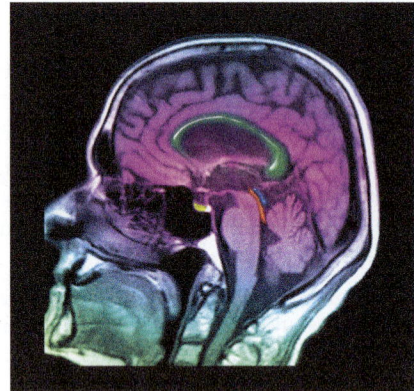

▲ **Figure 6.26 MRI image.** This image of a human head, obtained using magnetic resonance imaging, shows a normal brain, airways, and facial tissues.

1s orbital because it is the lowest in energy (Figure 6.23). According to the Pauli exclusion principle, however, there can be at most two electrons in any single orbital. Thus, *the orbitals are filled in order of increasing energy, with no more than two electrons per orbital.* The lithium atom, for example, has three electrons. (Recall that the number of electrons in a neutral atom equals its atomic number.) The 1s orbital can accommodate two of the electrons. The third one goes into the next lowest-energy orbital, the 2s.

We can represent any electron configuration by writing the symbol for the occupied subshell and adding a superscript to indicate the number of electrons in that subshell. For example, for lithium we write $1s^2 2s^1$ (read "1s two, 2s one"). We can also show the arrangement of the electrons as

Li

1s   2s

In this representation, which we call an **orbital diagram**, each orbital is denoted by a box and each electron by a half arrow. A half arrow pointing up (↑) represents an electron with a positive spin magnetic quantum number ($m_s = +\frac{1}{2}$), whereas a half arrow pointing down (↓) represents an electron with a negative spin magnetic quantum number ($m_s = -\frac{1}{2}$). This pictorial representation of electron spin, which corresponds to the directions of the magnetic fields in Figure 6.24, is quite convenient. Chemists refer to the two possible spin states as "spin-up" and "spin-down" corresponding to the directions of the half arrows.

Electrons having opposite spins are said to be *paired* when they are in the same orbital (↑↓). An *unpaired electron* is one not accompanied by a partner of opposite spin. In the lithium atom the two electrons in the 1s orbital are paired, whereas the electron in the 2s orbital is unpaired.

Hund's Rule

How do the electron configurations of the elements change as we move from element to element across the periodic table?

TABLE 6.3   Electron Configurations of Several Lighter Elements

Element	Total Electrons	Orbital Diagram				Electron Configuration
		1s	2s	2p	3s	
Li	3	↑↓	↑	☐☐☐	☐	$1s^2 2s^1$
Be	4	↑↓	↑↓	☐☐☐	☐	$1s^2 2s^2$
B	5	↑↓	↑↓	↑ ☐☐	☐	$1s^2 2s^2 2p^1$
C	6	↑↓	↑↓	↑ ↑ ☐	☐	$1s^2 2s^2 2p^2$
N	7	↑↓	↑↓	↑ ↑ ↑	☐	$1s^2 2s^2 2p^3$
Ne	10	↑↓	↑↓	↑↓ ↑↓ ↑↓	☐	$1s^2 2s^2 2p^6$
Na	11	↑↓	↑↓	↑↓ ↑↓ ↑↓	↑	$1s^2 2s^2 2p^6 3s^1$

Hydrogen Hydrogen has one electron, which occupies the 1s orbital in its ground state:

$$\text{H} \quad \boxed{↑} \quad : 1s^1$$
$$1s$$

The choice of a spin-up electron here is arbitrary; we could equally well show the ground state with one spin-down electron. It is customary, however, to show unpaired electrons with their spins up.

Helium The next element, helium, has two electrons. Because two electrons with opposite spins can occupy the same orbital, both of helium's electrons are in the 1s orbital:

$$\text{He} \quad \boxed{↑↓} \quad : 1s^2$$
$$1s$$

The two electrons present in helium complete the filling of the first shell. This arrangement represents a very stable configuration, which is consistent with the chemical inertness of helium.

Lithium The electron configurations of lithium and several elements that follow it in the periodic table are shown in **Table 6.3**. For the third electron of lithium, the change in the principal quantum number from $n = 1$ for the first two electrons to $n = 2$ for the third electron represents a large jump in energy and a corresponding jump in the average distance of the electron from the nucleus. In other words, it represents the start of a new shell occupied with electrons. As you can see by examining the periodic table, lithium starts a new row of the table. It is the first member of the alkali metals (group 1A).

Beryllium and Boron The element that follows lithium is beryllium; its electron configuration is $1s^2 2s^2$ (Table 6.3). Boron, atomic number 5, has the electron configuration $1s^2 2s^2 2p^1$. The fifth electron must be placed in a $2p$ orbital because the $2s$ orbital is filled. Because all three of the $2p$ orbitals are of equal energy, it does not matter which $2p$ orbital we place this fifth electron in.

Carbon With the next element, carbon, we encounter a new situation. We know that the sixth electron must go into a $2p$ orbital, but does it go into the $2p$ orbital that already has one electron or into one of the other two $2p$ orbitals?

> *Hund's rule* states that when filling degenerate orbitals, the lowest energy is attained when the number of electrons having the same spin is maximized.

This means that electrons occupy orbitals singly to the maximum extent possible and that these single electrons in a given subshell all have the same spin magnetic quantum number. Electrons arranged in this way are said to have *parallel spins*. For a carbon atom to achieve its lowest energy, therefore, the two $2p$ electrons must have the same spin. For this to happen, the electrons must be in different $2p$ orbitals, as shown in Table 6.3. Thus, a carbon atom in its ground state has two unpaired electrons.

Nitrogen, Oxygen, Fluorine Similarly, for nitrogen in its ground state, Hund's rule requires that the three $2p$ electrons singly occupy each of the three $2p$ orbitals. This is the only way that all three electrons can have the same spin. For oxygen and fluorine, we place four and five electrons, respectively, in the $2p$ orbitals. To achieve this, we pair up electrons in the $2p$ orbitals, as we demonstrate in Sample Exercise 6.6.

The rationale behind Hund's rule is based in large part on the fact that electrons repel one another. By occupying different orbitals, the electrons remain as far as possible from one another, thus minimizing electron–electron repulsions.

Sample Exercise 6.6
Orbital Diagrams and Electron Configurations

Draw the orbital diagram for the electron configuration of oxygen, atomic number 8. How many unpaired electrons does an oxygen atom possess?

SOLUTION

Analyze and Plan Because oxygen has an atomic number of 8, each oxygen atom has eight electrons. Figure 6.23 shows the ordering of orbitals. The electrons (represented as half arrows) are placed in the orbitals (represented as boxes) beginning with the lowest-energy orbital, the $1s$. Each orbital can hold a maximum of two electrons (the Pauli exclusion principle). Because the $2p$ orbitals are degenerate, we place one electron in each of these orbitals (spin-up) before pairing any electrons (Hund's rule).

Solve Two electrons each go into the $1s$ and $2s$ orbitals with their spins paired. This leaves four electrons for the three degenerate $2p$ orbitals. Following Hund's rule, we put one electron into each $2p$ orbital until all three orbitals have one electron each. The fourth electron is then paired up with one of the three electrons already in a $2p$ orbital, so that the orbital diagram is

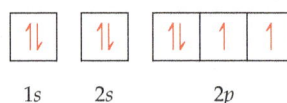

$$1s \quad 2s \quad 2p$$

The corresponding electron configuration is written $1s^2 2s^2 2p^4$. The atom has two unpaired electrons.

▶ **Practice Exercise**
(**a**) Write the electron configuration for silicon, element 14, in its ground state. (**b**) How many unpaired electrons does a ground-state silicon atom possess?

Condensed Electron Configurations

The filling of the $2p$ subshell is complete at neon (Table 6.3), which has a stable configuration with eight electrons (an *octet*) in the outermost occupied shell. The next element, sodium, atomic number 11, marks the beginning of a new row of the periodic table. Sodium has a single $3s$ electron beyond the stable configuration of neon. We can therefore abbreviate the electron configuration of sodium as

$$\text{Na:} \quad [\text{Ne}]3s^1$$

The symbol [Ne] represents the electron configuration of the ten electrons of neon, $1s^2 2s^2 2p^6$. Writing the electron configuration as $[Ne]3s^1$ focuses attention on the outer-most electron of the atom, which is the one largely responsible for how sodium behaves chemically.

We can generalize what we have just done for the electron configuration of sodium. In writing the *condensed electron configuration* of an element, the electron configuration of the nearest noble-gas element of lower atomic number is represented by its chemical symbol in brackets. For lithium, for example, we can write

$$\text{Li:} \quad [He]2s^1$$

We refer to the electrons represented by the bracketed symbol as the *noble-gas core* of the atom. These inner-shell electrons are often referred to as the **core electrons**. The electrons given after the noble-gas core are called the *outer-shell electrons*. The outer-shell electrons include the electrons involved in chemical bonding, which are called the **valence electrons**. For the elements with atomic number of 30 or less, all of the outer-shell electrons are valence electrons. By comparing the condensed electron configurations of lithium and sodium, we can appreciate why these two elements are so similar chemically. They have the same type of electron configuration in the outermost occupied shell. Indeed, all the members of the alkali metal group (1A) have a single *s* valence electron beyond a noble-gas configuration.

Transition Metals

The noble-gas element argon ($1s^2 2s^2 2p^6 3s^2 3p^6$) marks the end of the row started by sodium. The element following argon in the periodic table is potassium (K), atomic number 19. In all its chemical properties, potassium is clearly a member of the alkali metal group. The experimental facts about the properties of potassium leave no doubt that the outermost electron of this element occupies an *s* orbital. But this means that the electron with the highest energy has *not* gone into a 3*d* orbital, which we might expect it to do. Because the 4*s* orbital is lower in energy than the 3*d* orbital (Figure 6.23), the condensed electron configuration of potassium is

$$\text{K:} \quad [Ar]4s^1$$

Following the complete filling of the 4*s* orbital (this occurs in the calcium atom), the next set of orbitals to be filled is the 3*d*. (You will find it helpful as we go along to refer often to the periodic table on the front-inside cover.) Beginning with scandium and extending through zinc, electrons are added to the five 3*d* orbitals until they are completely filled. Thus, the fourth row of the periodic table is ten elements wider than the two previous rows. These ten elements are known as either **transition elements** or **transition metals**. Note the position of these elements in the periodic table.

In writing the electron configurations of the transition elements, we fill orbitals in accordance with Hund's rule; we add them to the 3*d* orbitals singly until all five orbitals have one electron each and then place additional electrons in the 3*d* orbitals with spin pairing until the shell is completely filled. The condensed electron configurations and the corresponding orbital diagram representations of two transition elements are as follows:

Once all the 3*d* orbitals have been filled with two electrons each, the 4*p* orbitals begin to be occupied until the completed octet of outer electrons ($4s^2 4p^6$) is reached with krypton (Kr), atomic number 36, another of the noble gases. Rubidium (Rb) marks the beginning of the fifth row. Refer again to the periodic table on the front-inside cover. Notice that this row is in every respect like the preceding one, except that the value for *n* is greater by 1.

The Lanthanides and Actinides

The sixth row of the periodic table begins with Cs and Ba, which have $[Xe]6s^1$ and $[Xe]6s^2$ configurations, respectively. Notice, however, that the periodic table then has a break, with elements 57–70 placed below the main portion of the table. This break point is where we begin to encounter a new set of orbitals, the $4f$ orbitals.

There are seven degenerate $4f$ orbitals, corresponding to the seven allowed values of m_l, ranging from 3 to -3. Thus, it takes 14 electrons to fill the $4f$ orbitals completely. The 14 elements corresponding to the filling of the $4f$ orbitals are known as either the **lanthanide elements** or the **rare earth elements**. These elements are set below the other elements to avoid making the periodic table unduly wide. The properties of the lanthanide elements are all quite similar, and these elements occur together in nature. For many years it was virtually impossible to separate them from one another.

Because the energies of the $4f$ and $5d$ orbitals are very close to each other, the electron configurations of some of the lanthanides involve $5d$ electrons. For example, the elements lanthanum (La), cerium (Ce), and praseodymium (Pr) have the following electron configurations:

$$[Xe]6s^25d^1 \quad [Xe]6s^25d^14f^1 \quad [Xe]6s^24f^3$$
$$\text{Lanthanum} \qquad \text{Cerium} \qquad \text{Praseodymium}$$

Because La has a single $5d$ electron, it is sometimes placed below yttrium (Y) as the first member of the third series of transition elements; Ce is then placed as the first member of the lanthanides. Based on its chemical properties, however, La can be considered the first element in the lanthanide series. Arranged this way, there are fewer apparent exceptions to the regular filling of the $4f$ orbitals among the subsequent members of the series.

After the lanthanide series, the third transition element series is completed by the filling of the $5d$ orbitals, followed by the filling of the $6p$ orbitals. This brings us to radon (Rn), the heaviest naturally occurring noble-gas element.

The final row of the periodic table begins by filling the $7s$ orbitals. The **actinide elements**, of which uranium (U, element 92) and plutonium (Pu, element 94) are the best known, are then built up by completing the $5f$ orbitals. All of the actinide elements are radioactive, and most of them are *not* found in nature.

Self-Assessment Exercises

SAE 6.20 What is the full ground-state electron configuration for Ti?
(a) $1s^22s^22p^63s^23p^64s^23d^2$
(b) $1s^22s^22p^63s^23p^64s^24p^2$
(c) $1s^22s^22p^63s^23p^63d^4$
(d) $1s^22s^22p^63s^23p^64s^4$

SAE 6.21 A ground-state Sn atom has a condensed electron configuration of _____ and possesses _____ unpaired electrons.
(a) $[Kr]5s^25p^2$, two (b) $[Kr]5s^24d^{10}5p^2$, two (c) $[Kr]5s^25p^2$, zero
(d) $[Kr]5s^24d^{10}5p^2$, zero (e) $[Kr]5s^25d^{10}5p^2$, two

6.9 | Electron Configurations and the Periodic Table

Learning Objectives

When you finish Section 6.9, you should be able to:

▶ Explain how the organization of the periodic table arises from the electron configurations of the elements.

▶ Identify and name the various blocks of elements that make up the periodic table.

Now that we've learned how to assign electron configurations, the structure of the periodic table should start to make more sense to you. In this section we examine the connection between the periodic table and the electron configurations of the elements. We also discuss how to use the periodic table as a guide to remember the relative energy levels of the different atomic orbitals in multi-electron atoms.

We just saw that the electron configurations of the elements correspond to their locations in the periodic table. Thus, elements in the same column of the table have related outer-shell (valence) electron configurations. As **Table 6.4** shows, for example, all 2A elements have an ns^2 outer configuration, and all 3A elements have an ns^2np^1 outer configuration, with the value of n increasing as we move down each column.

In Table 6.2 we saw that the total number of orbitals in each shell equals n^2: 1, 4, 9, or 16. Because we can place two electrons in each orbital, each shell accommodates up to $2n^2$ electrons: 2, 8, 18, or 32. The overall structure of the periodic table reflects these electron numbers: Each row of the table has 2, 8, 18, or 32 elements in it. As shown in **Figure 6.27**, the periodic table can be further divided into four blocks based on the filling order of orbitals. On the left are *two* blue columns of elements. These elements, known as the alkali

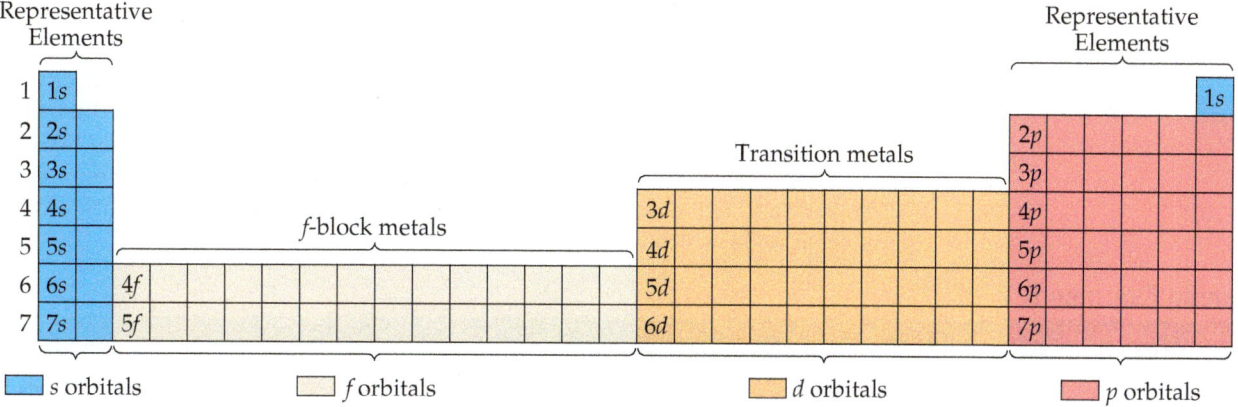

▲ **Figure 6.27 Regions of the periodic table.** The order in which electrons are added to orbitals is read left to right beginning in the top-left corner.

metals (group 1A) and alkaline earth metals (group 2A), are those in which the valence *s* orbitals are being filled. These two columns make up the *s* block of the periodic table.

On the right is a block of *six* pink columns that comprises the *p* block, where the valence *p* orbitals are being filled. The *s* block and the *p* block elements together are the **representative elements**, sometimes called the **main-group elements**. The orange block in Figure 6.27 has *ten* columns containing the **transition metals**. These are the elements in which the valence *d* orbitals are being filled and make up the *d* block. The elements in the two tan rows containing *14* columns are the ones in which the valence *f* orbitals are being filled and make up the *f* block. Consequently, these elements are often referred to as the **f-block metals**. In most instances, the *f* block is positioned below the periodic table to save space.

The number of columns in each block corresponds to the maximum number of electrons that can occupy each kind of subshell. Recall that 2, 6, 10, and 14 are the numbers of electrons that can fill the *s*, *p*, *d*, and *f* subshells, respectively. Thus, the *s* block has 2 columns, the *p* block has 6, the *d* block has 10, and the *f* block has 14. Recall also that 1*s* is the first *s* subshell, 2*p* is the first *p* subshell, 3*d* is the first *d* subshell, and 4*f* is the first *f* subshell, as Figure 6.27 shows. Using these facts, you can write the electron configuration of an element based merely on its position in the periodic table. Remember: *The periodic table is your best guide to the order in which orbitals are filled.*

Let's use the periodic table to write the electron configuration of selenium (Se, element 34). We first locate Se in the table and then move backward from it through the table, from element 34 to 33 to 32, and so forth, until we come to the noble gas that precedes Se. In this case, the noble gas is argon, Ar, element 18. Thus, the noble-gas core for Se is [Ar]. Our next step is to write symbols for the outer electrons. We do this by moving across period 4 from K, the element following Ar, to Se:

TABLE 6.4 Electron Configurations of Group 2A and 3A Elements

Group 2A	
Be	$[He]2s^2$
Mg	$[Ne]3s^2$
Ca	$[Ar]4s^2$
Sr	$[Kr]5s^2$
Ba	$[Xe]6s^2$
Ra	$[Rn]7s^2$
Group 3A	
B	$[He]2s^22p^1$
Al	$[Ne]3s^23p^1$
Ga	$[Ar]3d^{10}4s^24p^1$
In	$[Kr]4d^{10}5s^25p^1$
Tl	$[Xe]4f^{14}5d^{10}6s^26p^1$

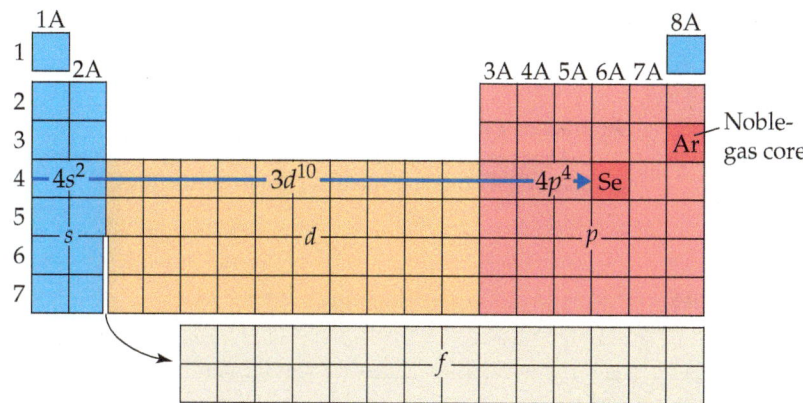

Because K is in the fourth period and the *s* block, we begin with the 4*s* electrons, meaning our first two outer electrons are written $4s^2$. We then move into the *d* block, which begins with the 3*d* electrons. The principal quantum number in the *d* block is always one less than that of the preceding elements in the *s* block, as seen in Figure 6.27. Traversing the d block adds ten electrons, $3d^{10}$. Finally, we move into the p block, whose principal quantum number is always the same as that of the s block. Counting the squares as we move across the p block to Se indicates that we need four electrons, $4p^4$. The electron configuration for Se is therefore $[Ar]4s^23d^{10}4p^4$. This configuration can also be written with the subshells arranged in order of increasing principal quantum number: $[Ar]3d^{10}4s^24p^4$.

As a check, we add the number of electrons in the [Ar] core, 18, to the number of electrons we added to the $4s$, $3d$, and $4p$ subshells. This sum should equal the atomic number of Se, 34, which it does: $18 + 2 + 10 + 4 = 34$.

Sample Exercise 6.7
Electron Configurations from the Periodic Table

(a) Based on its position in the periodic table, write the condensed electron configuration for bismuth, element 83. (b) How many unpaired electrons does a bismuth atom have?

SOLUTION

(a) Our first step is to write the noble-gas core. We do this by locating bismuth, element 83, in the periodic table. We then move backward to the nearest noble gas, which is Xe, element 54. Thus, the noble-gas core is [Xe].

Next, we trace the path in order of increasing atomic numbers from Xe to Bi. Moving from Xe to Cs, element 55, we find ourselves in period 6 of the s block. Knowing the block and the period identifies the subshell in which we begin placing outer electrons, $6s$. As we move through the s block, we add two electrons: $6s^2$.

As we move beyond the s block, from element 56 to element 57, the curved arrow below the periodic table reminds us that we are entering the f block. The first row of the f block corresponds to the $4f$ subshell. As we move across this block, we add 14 electrons: $4f^{14}$.

With element 71, we move into the third row of the d block. Because the first row of the d block is $3d$, the second

row is $4d$ and the third row is $5d$. Thus, as we move through the ten elements of the d block, from element 71 to element 80, we fill the $5d$ subshell with ten electrons: $5d^{10}$.

Moving from element 80 to element 81 puts us into the p block in the $6p$ subshell. (Remember that the principal quantum number in the p block is the same as that in the s block.) Moving across to Bi requires three electrons: $6p^3$. Putting the parts together, we obtain the condensed electron configuration: $[Xe]6s^24f^{14}5d^{10}6p^3$. This configuration can also be written with the subshells arranged in order of increasing principal quantum number: $[Xe]4f^{14}5d^{10}6s^26p^3$.

Finally, we check our result to see if the number of electrons equals the atomic number of Bi, 83: Because Xe has 54 electrons (its atomic number), we have $54 + 2 + 14 + 10 + 3 = 83$. (If we had come up with 14 electrons too few, it would be a clue that we had forgotten the f block.)

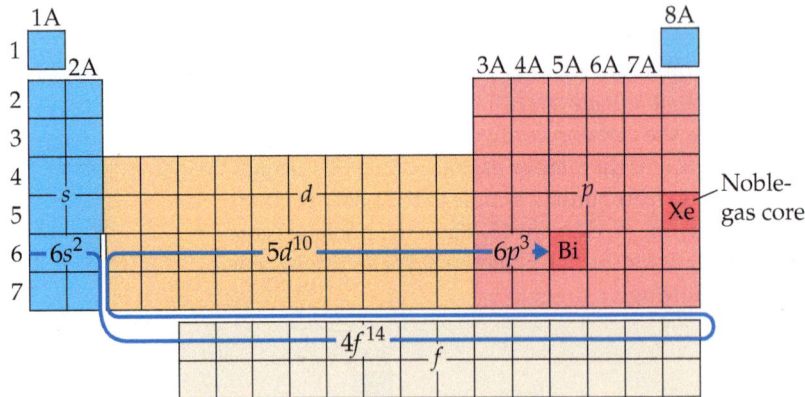

(b) We see from the condensed electron configuration that the only partially occupied subshell is $6p$. The orbital diagram representation for this subshell is

In accordance with Hund's rule, the three $6p$ electrons occupy the three $6p$ orbitals singly, with their spins parallel. Thus, there are three unpaired electrons in the bismuth atom.

▶ **Practice Exercise**
Use the periodic table to write the condensed electron configuration for (a) Co (element 27), (b) In (element 49).

Figure 6.28 gives, for all the elements, the ground-state electron configurations for the outer-shell electrons. You can use this figure to check your answers as you practice writing electron configurations. We have written these configurations with orbitals listed in order of increasing principal quantum number. As we saw in Sample Exercise 6.7, the orbitals can also be listed in order of filling, as they would be read off of the periodic table.

Go Figure A friend tells you that her favorite element has an electron configuration of [noble gas]$6s^24f^{14}5d^6$. Which element is it?

1A 1												3A 13	4A 14	5A 15	6A 16	7A 17	8A 18
1 **H** $1s^1$	2A 2																2 **He** $1s^2$
Core [He] 3 **Li** $2s^1$	4 **Be** $2s^2$	3B 3	4B 4	5B 5	6B 6	7B 7	8	8B 9	10	1B 11	2B 12	5 **B** $2s^22p^1$	6 **C** $2s^22p^2$	7 **N** $2s^22p^3$	8 **O** $2s^22p^4$	9 **F** $2s^22p^5$	10 **Ne** $2s^22p^6$
[Ne] 11 **Na** $3s^1$	12 **Mg** $3s^2$											13 **Al** $3s^23p^1$	14 **Si** $3s^23p^2$	15 **P** $3s^23p^3$	16 **S** $3s^23p^4$	17 **Cl** $3s^23p^5$	18 **Ar** $3s^23p^6$
[Ar] 19 **K** $4s^1$	20 **Ca** $4s^2$	21 **Sc** $4s^23d^1$	22 **Ti** $4s^23d^2$	23 **V** $4s^23d^3$	24 **Cr** $4s^13d^5$	25 **Mn** $4s^23d^5$	26 **Fe** $4s^23d^6$	27 **Co** $4s^23d^7$	28 **Ni** $4s^23d^8$	29 **Cu** $4s^13d^{10}$	30 **Zn** $4s^23d^{10}$	31 **Ga** $4s^23d^{10}4p^1$	32 **Ge** $4s^23d^{10}4p^2$	33 **As** $4s^23d^{10}4p^3$	34 **Se** $4s^23d^{10}4p^4$	35 **Br** $4s^23d^{10}4p^5$	36 **Kr** $4s^23d^{10}4p^6$
[Kr] 37 **Rb** $5s^1$	38 **Sr** $5s^2$	39 **Y** $5s^24d^1$	40 **Zr** $5s^24d^2$	41 **Nb** $5s^14d^3$	42 **Mo** $5s^14d^5$	43 **Tc** $5s^24d^5$	44 **Ru** $5s^14d^7$	45 **Rh** $5s^14d^8$	46 **Pd** $4d^{10}$	47 **Ag** $5s^14d^{10}$	48 **Cd** $5s^24d^{10}$	49 **In** $5s^24d^{10}5p^1$	50 **Sn** $5s^24d^{10}5p^2$	51 **Sb** $5s^24d^{10}5p^3$	52 **Te** $5s^24d^{10}5p^4$	53 **I** $5s^24d^{10}5p^5$	54 **Xe** $5s^24d^{10}5p^6$
[Xe] 55 **Cs** $6s^1$	56 **Ba** $6s^2$	71 **Lu** $6s^24f^{14}5d^1$	72 **Hf** $6s^24f^{14}5d^2$	73 **Ta** $6s^24f^{14}5d^3$	74 **W** $6s^24f^{14}5d^4$	75 **Re** $6s^24f^{14}5d^5$	76 **Os** $6s^24f^{14}5d^6$	77 **Ir** $6s^24f^{14}5d^7$	78 **Pt** $6s^14f^{14}5d^9$	79 **Au** $6s^14f^{14}5d^{10}$	80 **Hg** $6s^24f^{14}5d^{10}$	81 **Tl** $6s^24f^{14}5d^{10}6p^1$	82 **Pb** $6s^24f^{14}5d^{10}6p^2$	83 **Bi** $6s^24f^{14}5d^{10}6p^3$	84 **Po** $6s^24f^{14}5d^{10}6p^4$	85 **At** $6s^24f^{14}5d^{10}6p^5$	86 **Rn** $6s^24f^{14}5d^{10}6p^6$
[Rn] 87 **Fr** $7s^1$	88 **Ra** $7s^2$	103 **Lr** $7s^25f^{14}6d^1$	104 **Rf** $7s^25f^{14}6d^2$	105 **Db** $7s^25f^{14}6d^3$	106 **Sg** $7s^25f^{14}6d^4$	107 **Bh** $7s^25f^{14}6d^5$	108 **Hs** $7s^25f^{14}6d^6$	109 **Mt** $7s^25f^{14}6d^7$	110 **Ds** $7s^25f^{14}6d^8$	111 **Rg** $7s^25f^{14}6d^9$	112 **Cn** $7s^25f^{14}6d^{10}$	113 **Nh** $7s^25f^{14}6d^{10}7p^1$	114 **Fl** $7s^25f^{14}6d^{10}7p^2$	115 **Mc** $7s^25f^{14}6d^{10}7p^3$	116 **Lv** $7s^25f^{14}6d^{10}7p^4$	117 **Ts** $7s^25f^{14}6d^{10}7p^5$	118 **Og** $7s^25f^{14}6d^{10}7p^6$

Lanthanide series [Xe]

57 **La** $6s^25d^1$	58 **Ce** $6s^24f^15d^1$	59 **Pr** $6s^24f^3$	60 **Nd** $6s^24f^4$	61 **Pm** $6s^24f^5$	62 **Sm** $6s^24f^6$	63 **Eu** $6s^24f^7$	64 **Gd** $6s^24f^75d^1$	65 **Tb** $6s^24f^9$	66 **Dy** $6s^24f^{10}$	67 **Ho** $6s^24f^{11}$	68 **Er** $6s^24f^{12}$	69 **Tm** $6s^24f^{13}$	70 **Yb** $6s^24f^{14}$

Actinide series [Rn]

89 **Ac** $7s^26d^1$	90 **Th** $7s^26d^2$	91 **Pa** $7s^25f^26d^1$	92 **U** $7s^25f^36d^1$	93 **Np** $7s^25f^46d^1$	94 **Pu** $7s^25f^6$	95 **Am** $7s^25f^7$	96 **Cm** $7s^25f^76d^1$	97 **Bk** $7s^25f^9$	98 **Cf** $7s^25f^{10}$	99 **Es** $7s^25f^{11}$	100 **Fm** $7s^25f^{12}$	101 **Md** $7s^25f^{13}$	102 **No** $7s^25f^{14}$

☐ Metals   ☐ Metalloids   ☐ Nonmetals

▲ **Figure 6.28 Outer-shell electron configurations of the elements.**

Figure 6.28 allows us to reexamine the concept of *valence electrons*. Notice, for example, that as we proceed from Cl ([Ne]$3s^23p^5$) to Br ([Ar]$3d^{10}4s^24p^5$), we add a complete subshell of $3d$ electrons to the electrons beyond the [Ar] core. Although the $3d$ electrons are outer-shell electrons, they are not involved in chemical bonding and are therefore not considered valence electrons. Thus, we consider only the $4s$ and $4p$ electrons of Br to be valence electrons. Similarly, if we compare the electron configurations of Ag (element 47) and Au (element 79), we see that Au has a completely full $4f^{14}$ subshell beyond its noble-gas core, but those $4f$ electrons are not involved in bonding. In general, *for representative elements we do not consider the electrons in completely filled* d *or* f *subshells to be valence electrons,* and *for transition elements we do not consider the electrons in a completely filled* f *subshell to be valence electrons.*

Anomalous Electron Configurations

The electron configurations of certain elements appear to violate the rules we have just discussed. For example, Figure 6.28 shows that the electron configuration of chromium (element 24) is [Ar]$3d^54s^1$ rather than the [Ar]$3d^44s^2$ configuration we might expect. Similarly, the configuration of copper (element 29) is [Ar]$3d^{10}4s^1$ instead of [Ar]$3d^94s^2$.

This anomalous behavior is largely a consequence of the closeness of the 3*d* and 4*s* orbital energies. It frequently occurs when there are enough electrons to form precisely half-filled sets of degenerate orbitals (as in chromium) or a completely filled *d* subshell (as in copper). There are a few similar cases among the heavier transition metals (those with partially filled 4*d* or 5*d* orbitals) and among the *f*-block metals. Although these minor departures from the expected are interesting, they are not of great chemical significance.

 Self-Assessment Exercises

SAE 6.22 In which column of the periodic table would you find an atom with a valence electron configuration of ns^2np^3? (**a**) group 2A (**b**) group 3A (**c**) group 5A (**d**) group 7A

SAE 6.23 In which part of the periodic table would you expect to find the element(s) with the largest number of unpaired electrons? (**a**) the representative elements (**b**) the transition elements (**c**) the rare earth elements

 Putting Concepts Together

Boron, atomic number 5, occurs naturally as two isotopes, ^{10}B and ^{11}B, with natural abundances of 19.9% and 80.1%, respectively. (**a**) In what ways do the two isotopes differ from each other? Does the electronic configuration of ^{10}B differ from that of ^{11}B? (**b**) Draw the orbital diagram for an atom of ^{11}B. Which electrons are the valence electrons? (**c**) Indicate three ways in which the 1*s* electrons in boron differ from its 2*s* electrons. (**d**) Elemental boron reacts with fluorine to form BF_3, a gas. Write a balanced chemical equation for the reaction of solid boron with fluorine gas. (**e**) ΔH_f° for $BF_3(g)$ is -1135.6 kJ/mol. Calculate the standard enthalpy change in the reaction of boron with fluorine. (**f**) Will the mass percentage of F be the same in $^{10}BF_3$ and $^{11}BF_3$? If not, why is that the case?

SOLUTION

(**a**) The two isotopes of boron differ in the number of neutrons in the nucleus. (Sections 2.3 and 2.4) Each of the isotopes contains five protons, but ^{10}B contains five neutrons, whereas ^{11}B contains six neutrons. The two isotopes of boron have identical electron configurations, $1s^22s^22p^1$, because each has five electrons.

(**b**) The complete orbital diagram is

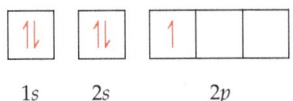

$$1s \qquad 2s \qquad 2p$$

The valence electrons are the ones in the outermost occupied shell, the $2s^2$ and $2p^1$ electrons. The $1s^2$ electrons constitute the core electrons, which we represent as [He] when we write the condensed electron configuration, $[He]2s^22p^1$.

(**c**) The 1*s* and 2*s* orbitals are both spherical, but they differ in three important respects: First, the 1*s* orbital is lower in energy than the 2*s* orbital. Second, the average distance of the 2*s* electrons from the nucleus is greater than that of the 1*s* electrons, so the 1*s* orbital is smaller than the 2*s*. Third, the 2*s* orbital has one node, whereas the 1*s* orbital has no nodes (Figure 6.19).

(**d**) The balanced chemical equation is

$$2\,B(s) + 3\,F_2(g) \longrightarrow 2\,BF_3(g)$$

(**e**) $\Delta H^\circ = 2(-1135.6) - [0 + 0] = -2271.2$ kJ. The reaction is strongly exothermic.

(**f**) As we saw in Equation 3.13 (Section 3.3), the mass percentage of an element in a substance depends on the formula weight of the substance. The formula weights of $^{10}BF_3$ and $^{11}BF_3$ are different because of the difference in the masses of the two isotopes (the isotope masses of ^{10}B and ^{11}B are 10.01294 and 11.00931 amu, respectively). The denominators in Equation 3.13 would therefore be different for the two isotopes, whereas the numerators would remain the same.

Chapter Summary and Key Terms

WAVELENGTHS AND FREQUENCIES OF LIGHT (INTRODUCTION AND SECTION 6.1) The **electronic structure** of an atom describes the energies and arrangement of electrons around the atom. Much of what is known about the electronic structure of atoms was obtained by observing the interaction of light with matter.

Visible light and other forms of **electromagnetic radiation** (also known as radiant energy) move through a vacuum at the speed of light, $c = 2.998 \times 10^8$ m/s. Electromagnetic radiation has both electric and magnetic components that vary periodically in wave-like fashion. The wave characteristics of radiant energy allow it to be described in terms of **wavelength**, λ, and **frequency**, ν, which are interrelated: $\lambda\nu = c$.

QUANTIZED ENERGY AND PHOTONS (SECTION 6.2) To explain blackbody radiation, Planck proposed that the minimum amount of radiant energy that an object can gain or lose is related to the frequency of the radiation: $E = h\nu$. This smallest quantity is called a **quantum** of energy. The constant h is called the **Planck constant**: $h = 6.626 \times 10^{-34}$ J-s.

In the quantum theory, energy is quantized, meaning that it can have only certain allowed values. Einstein used quantum theory to explain the **photoelectric effect**, the emission of electrons from metal surfaces when exposed to light. He proposed that light behaves as if it consists of quantized energy packets called **photons**. Each photon carries energy, $E = h\nu$.

BOHR MODEL OF THE HYDROGEN ATOM (SECTION 6.3) Dispersion of radiation into its component wavelengths produces a **spectrum**. If the spectrum contains all wavelengths, it is called a **continuous spectrum**; if it contains only certain specific wavelengths, the spectrum is called a **line spectrum**. The radiation emitted by excited hydrogen atoms forms a line spectrum.

Bohr proposed a model of the hydrogen atom that explains its line spectrum. In this model, the energy of the electron in the hydrogen atom depends on the value of a quantum number, n, called the **principal quantum number**. The value of n must be a positive integer (1, 2, 3, . . .), and each value of n corresponds to a different specific energy, E_n. The energy of the atom increases as n increases. The lowest energy is achieved for $n = 1$, and this is called the **ground state** of the hydrogen atom. Other values of n correspond to **excited states**. Light is emitted when the electron drops from a higher-energy state to a lower-energy state; light is absorbed to excite the electron from a lower-energy state to a higher-energy state. The frequency of light emitted or absorbed is such that $h\nu$ equals the difference in energy between two allowed states.

WAVE BEHAVIOR OF MATTER (SECTION 6.4) De Broglie proposed that matter, such as electrons, should exhibit wave-like properties. This hypothesis of **matter waves** was proved experimentally by observing the diffraction of electrons. An object has a characteristic wavelength that depends on its **momentum**, mv: $\lambda = h/mv$.

Discovery of the wave properties of the electron led to Heisenberg's **uncertainty principle**, which states that there is an inherent limit to the accuracy with which the position and momentum of a particle can be measured simultaneously.

QUANTUM MECHANICS AND ORBITALS (SECTION 6.5) In the quantum-mechanical model of the hydrogen atom, the behavior of the electron is described by mathematical functions called **wave functions**, denoted with the Greek letter ψ. Each allowed wave function has a precisely known energy, but the location of the electron cannot be determined exactly; rather, the probability of it being at a particular point in space is given by the **probability density**, ψ^2. The **electron-density** distribution is a map of the probability of finding the electron at all points in space.

The allowed wave functions of the hydrogen atom are called **orbitals**. An orbital is uniquely identified by the values of three quantum numbers. The *principal quantum number*, n, is indicated by the integers 1, 2, 3, ... This quantum number relates most directly to the size and energy of the orbital. The **angular momentum quantum number**, l, is indicated by the letters s, p, d, f, and so on, corresponding to the values of 0, 1, 2, 3, . . . , respectively. The l quantum number defines the shape of the orbital. For a given value of n, l can have integer values ranging from 0 to $(n - 1)$. The **magnetic quantum number**, m_l, relates to the orientation of the orbital in space. For a given value of l, m_l can have integral values ranging from $-l$ to l, including 0. Subscripts can be used to label the orientations of the orbitals. For example, the three $3p$ orbitals are designated $3p_x$, $3p_y$, and $3p_z$, with the subscripts indicating the axis along which the orbital is oriented.

An electron shell is the set of all orbitals with the same value of n, such as $3s$, $3p$, and $3d$. In the hydrogen atom all the orbitals in an electron shell have the same energy. A **subshell** is the set of one or more orbitals with the same n and l values; for example, $3s$, $3p$, and $3d$ are each subshells of the $n = 3$ shell. There is one orbital in an s subshell, three in a p subshell, five in a d subshell, and seven in an f subshell.

REPRESENTATIONS OF ORBITALS (SECTION 6.6) The **radial probability function** $4\pi r^2[\psi(r)]^2$ tells us the probability that the electron will be found at a certain distance from the nucleus. Points where the wave function changes sign are called **nodes**. There is zero probability that the electron will be found at a node because the wave function equals zero and so the radial probability function is

also zero at a node. Nodes that occur at a given distance from the nucleus are called **radial nodes**. They are spherical in shape, and their number increases as the value of the principal quantum number, n, increases. Contour representations are surfaces that are useful for visualizing the shapes of the orbitals. Represented in this way, s orbitals appear as spheres that increase in size as n increases.

The wave function for each p orbital has two lobes on opposite sides of the nucleus. They are oriented along either the x, y, or z axis. The two nodes are separated by a **nodal plane**, where the probability of finding the electron goes to zero. Four of the five d orbitals have contour representations with four lobes around the nucleus, due to the presence of two perpendicular nodal planes, whereas the angular nodes for the d_{z^2} orbital are cones, which produce a contour surface with two lobes along the z axis and a "doughnut" in the xy plane.

MANY-ELECTRON ATOMS (SECTION 6.7) In many-electron atoms, different subshells of the same electron shell have different energies. For a given value of n, the energy of the subshells increases as the value of l increases: $ns < np < nd < nf$. Orbitals within the same subshell are **degenerate**, meaning they have the same energy.

Electrons have an intrinsic property called **electron spin**, which is quantized. The **spin magnetic quantum number**, m_s, can have two possible values, $+\frac{1}{2}$ and $-\frac{1}{2}$, which can be envisioned as the two directions of an electron spinning about an axis. The **Pauli exclusion principle** states that no two electrons in an atom can have the same values for n, l, m_l, and m_s. This principle places a limit of two on the number of electrons that can occupy any one atomic orbital. These two electrons differ in their value of m_s.

ELECTRON CONFIGURATIONS AND THE PERIODIC TABLE (SECTIONS 6.8 AND 6.9) The **electron configuration** of an atom describes how the electrons are distributed among the orbitals of the atom. The ground-state electron configurations are generally obtained by placing the electrons in the atomic orbitals of lowest possible energy, with the restriction that each orbital can hold no more than two electrons. We depict the arrangement of the electrons pictorially using an **orbital diagram**. When electrons occupy a subshell with more than one degenerate orbital, such as the $2p$ subshell, **Hund's rule** states that the lowest energy is attained by maximizing the number of electrons with the same electron spin. For example, in the ground-state electron configuration of carbon, the two $2p$ electrons have the same spin and must occupy two different $2p$ orbitals.

Elements in any given group in the periodic table have the same type of electron arrangements in their outermost shells. For example, the electron configurations of the halogens fluorine and chlorine are $[He]2s^22p^5$ and $[Ne]3s^23p^5$, respectively. The outer-shell electrons are those that lie outside the orbitals occupied in the next lowest noble-gas element. The outer-shell electrons that are involved in chemical bonding are the **valence electrons** of an atom; for the elements with atomic number 30 or less, all of the outer-shell electrons are valence electrons. The electrons that are not valence electrons are called **core electrons**.

The periodic table is partitioned into different types of elements, based on their electron configurations. Those elements in which the outermost subshell is an s or p subshell are called the **representative** (or **main-group**) **elements**. Those elements in which a d subshell is being filled are called the **transition elements** (or **transition metals**). The elements in which the $4f$ subshell is being filled are called the **lanthanide** (or **rare earth**) **elements**. The actinide elements are those in which the $5f$ subshell is being filled. The lanthanide and **actinide elements** are collectively referred to as the **f-block metals**. These elements are shown as two rows of 14 elements below the main part of the periodic table. The structure of the periodic table, summarized in Figure 6.28, allows us to write the electron configuration of an element from its position in the periodic table.

Key Equations

- $\lambda\nu = c$ [6.1]   light as a wave: λ = wavelength in meters, ν = frequency in s^{-1}, c = speed of light (2.998×10^8 m/s)

- $E = h\nu$ [6.2]   light as a particle (photon): E = energy of photon in joules, h = Planck constant (6.626×10^{-34} J-s), ν = frequency in s^{-1}

- $\dfrac{1}{\lambda} = (R_H)\left(\dfrac{1}{n_1^2} - \dfrac{1}{n_2^2}\right)$ [6.4]   The Rydberg equation which gives the wavelengths of light λ (in m) in the line spectrum of the hydrogen atom: R_H = Rydberg constant ($1.096776 \times 10^7 \text{ m}^{-1}$); $n = 1, 2, 3, \ldots$ (any positive integer).

- $E = (-hcR_H)\left(\dfrac{1}{n^2}\right) = (-2.18 \times 10^{-18} \text{J})\left(\dfrac{1}{n^2}\right)$ [6.5]   energies of the allowed states of the hydrogen atom: h = Planck constant; c = speed of light; R_H = Rydberg constant ($1.096776 \times 10^7 \text{ m}^{-1}$); $n = 1, 2, 3, \ldots$ (any positive integer)

- $\lambda = h/mv$ [6.8]   matter as a wave: λ = wavelength, h = Planck constant, m = mass of object in kg, v = speed of object in m/s

- $\Delta x \cdot \Delta(mv) \geq \dfrac{h}{4\pi}$ [6.9]   Heisenberg's uncertainty principle. The uncertainty in position (Δx) and momentum [$\Delta(mv)$] of an object cannot be zero; the smallest value of their product is $h/4\pi$.

Exam Prep

EP 6.1 A source of electromagnetic radiation produces infrared light. Which of the following could be the wavelength of the light? (**a**) 3.0 nm (**b**) 4.7 cm (**c**) 66.8 cm (**d**) 34.5 μm (**e**) 16.5 Å

EP 6.2 The _____ of ultraviolet radiation is greater than that of microwave radiation, but the _____ of the two types of electromagnetic radiation are the same. (**a**) wavelength, speed (**b**) speed, wavelength (**c**) frequency, speed (**d**) speed, frequency (**e**) frequency, wavelength

EP 6.3 Cellular phones use radio waves to transmit information. If your cell phone uses a frequency of 1900 MHz, what is the wavelength of the electromagnetic radiation emitted by your phone? (**a**) 5.7×10^{17} m (**b**) 0.16 m (**c**) 1.6×10^5 m (**d**) 6.3 m

EP 6.4 If electrons are ejected from a given metal when irradiated with a red laser pointer, what will happen when the same metal is irradiated with a green laser pointer of comparable intensity? (**a**) Electrons will be emitted, and the maximum kinetic energies of those electrons will be similar to those emitted when irradiated with the red laser pointer. (**b**) Electrons will be emitted, and the maximum kinetic energies of those electrons will be greater than those emitted when irradiated with the red laser pointer. (**c**) Electrons will be emitted, and the maximum kinetic energies of those electrons will be less than those emitted when irradiated with the red laser pointer. (**d**) Electrons will not be emitted.

EP 6.5 Which of the following expressions correctly gives the energy of a mole of photons with wavelength λ?

(**a**) $E = \dfrac{h}{\lambda}$

(**b**) $E = N_A\dfrac{\lambda}{h}$

(**c**) $E = \dfrac{hc}{\lambda}$

(**d**) $E = N_A\dfrac{h}{\lambda}$

(**e**) $E = N_A\dfrac{hc}{\lambda}$

EP 6.6 Consider the electronic transitions in the Bohr model that give rise to the line spectrum of hydrogen. What characteristic feature do the four visible colors of light that fall in the visible portion of the spectrum (λ = 410 nm, 434 nm, 486 nm, and 656 nm) have in common? (**a**) They all have $n_f = 1$. (**b**) They all have $n_f = 2$. (**c**) They all have $n_f = 3$. (**d**) They all have $n_i = 2$. (**e**) They all have $n_i = 3$.

EP 6.7 In the Bohr model of the hydrogen atom, the electron orbits the nucleus at a fixed radius of 0.53 Å when in the ground state. This model is a violation of which scientific principle that followed? (**a**) the Pauli exclusion principle (**b**) the uncertainty principle (**c**) Hund's rule (**d**) both the Pauli exclusion principle and the

uncertainty principle (**e**) both the Pauli exclusion principle and Hund's rule

EP 6.8 Arrange the following objects in order of increasing de Broglie wavelength:

(**i**) a golf ball with a mass of 45.9 g moving at a speed of 50 m/s,

(**ii**) an electron with a mass of 9.109×10^{-31} kg moving at a speed of 3.50×10^5 m/s, and

(**iii**) a neutron with a mass of 1.675×10^{-27} kg moving at a speed of 2.3×10^2 m/s.

(**a**) i < iii < ii (**b**) ii < iii < i (**c**) iii < ii < i (**d**) i < ii < iii (**e**) iii < i < ii

EP 6.9 The average speed of an N_2 gas molecule in a container with a temperature of 100 °C is 500 m/s. If the uncertainty in our measurement of the molecule's speed is 1% (5 m/s), what is the *lower limit* on the uncertainty with which we can measure its position?

(**a**) 1.2×10^{-5} m   (**c**) 2.3×10^{-13} m

(**b**) 2.3×10^{-10} m   (**d**) 3.8×10^{-37} m

EP 6.10 Which of the following descriptions applies to Schrödinger's model *but not* Bohr's model of the hydrogen atom?

(**i**) The angular momentum of the electron is quantized and represented by the quantum number l.

(**ii**) The energy level of the electron depends only on the principal quantum number, n.

(**iii**) The distance between the electron and the nucleus cannot be specified because of the wave-like properties of the electron.

(**a**) only i (**b**) only ii (**c**) only iii (**d**) both i and iii (**e**) All three statements are true.

EP 6.11 Which aspect of the Bohr model is *not* consistent with our current understanding of how the electron in a hydrogen atom behaves? (**a**) The allowed energy levels of the electron are quantized. (**b**) A photon is emitted when the electron transitions from a higher-energy state to a lower-energy state. (**c**) The electron moves in a circular orbit around the nucleus. (**d**) As the energy of the electron increases, it experiences a weaker attraction to the nucleus.

EP 6.12 What are the principal and angular momentum quantum numbers for the lowest energy f orbital subshell? (**a**) $n = 1, l = 3$ (**b**) $n = 3, l = 3$ (**c**) $n = 4, l = 1$ (**d**) $n = 4, l = 3$ (**e**) $n = 4, l = 4$

EP 6.13 Consider two atomic orbitals, the first defined by the quantum numbers $n = 2, l = 1, m_l = 0$, while the second orbital has the quantum numbers $n = 2$, $l = 1$, $m_l = 1$. Which of the following statements comparing the two orbitals is or are true?

(i) The two orbitals have the same energy.

(ii) The two orbitals have the same shape.

(iii) The two orbitals have the same orientation.

(a) only i (b) both i and ii (c) both i and iii (d) both ii and iii (e) All three statements are true.

EP 6.14 An orbital has $n = 4$ and $m_l = -1$. What are the possible values of l for this orbital? (a) 0, 1, 2, 3 (b) −3, −2, −1, 0, 1, 2, 3 (c) 1, 2, 3 (d) −3, −2 (e) 1, 2, 3, 4

EP 6.15 The plots shown here map the probability of finding an electron in either a $3s$, $3p_x$, or $3d_{x^2-y^2}$ orbital of the hydrogen atom, each plotted as a function of distance along the x axis. Which of the following statements is or are true?

(i) The $3s$ orbital has more radial nodes than the $3p_x$ orbital.

(ii) The most probable location of the $3s$ orbital is closer to the nucleus than the most probable location of the $3d_{x^2-y^2}$ orbital.

(iii) An electron in the $3s$ orbital has a higher probability of being found very close to the nucleus ($r < 1$ Å) than an electron in the $d_{x^2-y^2}$ orbital.

(a) only i (b) both i and ii (c) both i and iii (d) both ii and iii (e) All three statements are true

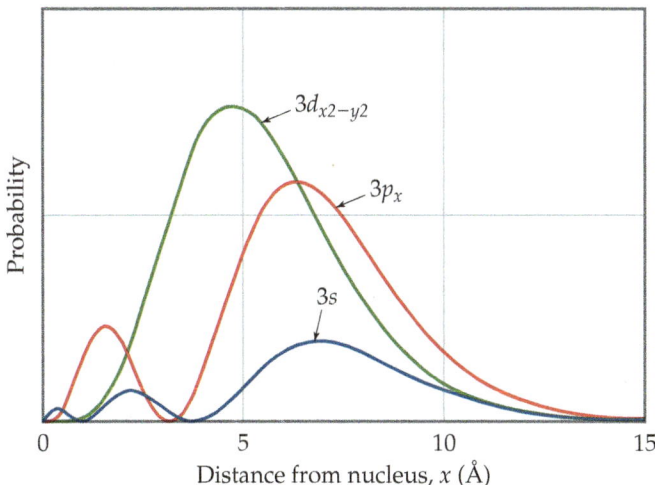

EP 6.16 Which type of atomic orbital can be described as having two lobes of electron density separated by a nodal plane? (a) s (b) p (c) d (d) f

EP 6.17 Consider three electrons in a sodium atom, with the following quantum numbers:

electron 1: $n = 3$, $l = 0$, $m_l = 0$, $m_s = +\frac{1}{2}$

electron 2: $n = 2$, $l = 1$, $m_l = 1$, $m_s = -\frac{1}{2}$

electron 3: $n = 2$, $l = 1$, $m_l = 0$, $m_s = +\frac{1}{2}$

Which electron has the lowest energy? (a) electron 1 (b) electron 2 (c) electron 3 (d) electrons 2 and 3, which are degenerate and lower in energy than electron 1 (e) electrons 1 and 3, which are degenerate and lower in energy than electron 2

EP 6.18 Given the constraints of the Pauli exclusion principle, what is the maximum number of electrons that can occupy the $4d$ subshell? (a) 4 (b) 5 (c) 6 (d) 10 (e) 14

EP 6.19 In which of the following atoms—Ti, Ca, or C—do the spin magnetic quantum numbers of all the electrons sum to zero and therefore does not possess a magnetic moment? Assume each atom is in its ground-state electron configuration. (a) only Ti (b) only Ca (c) only C (d) Ti and Ca (e) all three atoms

EP 6.20 What is the condensed electron configuration for the element osmium ($Z = 76$)?

(a) $[\text{Kr}]5s^24f^{14}4d^6$

(b) $[\text{Kr}]6s^24f^{14}5d^6$

(c) $[\text{Xe}]6s^25d^6$

(d) $[\text{Xe}]6s^24f^{14}5d^6$

(e) $[\text{Xe}]5s^24f^{14}5d^6$

EP 6.21 How many of the elements in the second row of the periodic table (Li through Ne) will have at least one unpaired electron in their electron configurations? (a) 3 (b) 4 (c) 5 (d) 6 (e) 7

EP 6.22 Which element has a [noble-gas]$5s^24d^{10}5p^4$ electron configuration? (a) Cd (b) Te (c) Sm (d) Hg (e) Se

EP 6.23 A certain atom has an ns^2np^6 electron configuration in its outermost occupied shell. Which of the following elements could it be? (a) Be (b) Si (c) I (d) Ar (e) Rb

EP 6.24 The organization of the periodic table tells us that the $5f$ subshell is filled after the _____ subshell and before the _____ subshell. (a) $4f$, $6f$ (b) $5d$, $5g$ (c) $6s$, $5d$ (d) $7s$, $7d$ (e) $7s$, $6d$

EP 6.25 The element samarium ($Z = 62$) is _____ with a partially filled _____ subshell. (a) an f-block metal, $4f$ (b) an f-block metal, $5f$ (c) a transition metal, $4d$ (d) a transition metal, $5d$ (e) a representative element, $4f$

EP 6.26 An element has a [noble gas]$ns^2(n-2)f^{14}(n-1)d^{10}np^2$ electron configuration. In which region of the periodic table will this element be found? (a) representative elements (b) f-block metals (c) transition metals (d) noble gases

Exercises

Visualizing Concepts

6.1 The speed of sound in dry air at 20 °C is 343 m/s, and the middle C on a piano keyboard has a frequency of 261 Hz. (a) What is the wavelength of the sound wave corresponding to a middle C? (b) What would be the frequency of electromagnetic radiation with the same wavelength? (c) What type of electromagnetic radiation would that correspond to? (d) Do sound waves travel faster, slower, or at the same speed as electromagnetic radiation? [Section 6.1]

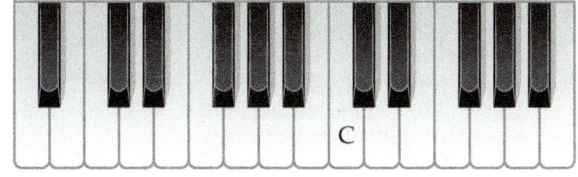

6.2 A popular kitchen appliance produces electromagnetic radiation with a frequency of 2450 MHz. With reference to Figure 6.4, answer the following: (a) Estimate the wavelength of this radiation. (b) Would the radiation produced by the appliance be visible to the human eye? (c) If the radiation is not visible, do photons of this radiation have more or less energy than photons of visible light? (d) Which of the following is the appliance likely to be? (i) a toaster oven, (ii) a microwave oven, or (iii) an electric hotplate. [Section 6.1]

6.3 The following diagrams represent two electromagnetic waves, drawn on the same scale. (a) Which wave has a longer wavelength? (**b**) Which wave has a higher frequency? (**c**) Which wave has a higher energy? [Section 6.2]

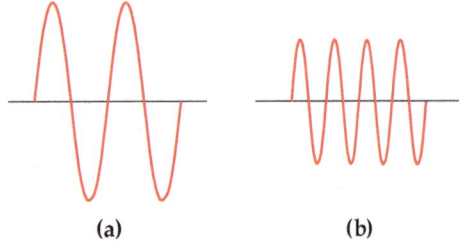

(a)   (b)

6.4 Stars do not all have the same temperature. The color of light emitted by stars is characteristic of the light emitted by hot objects. Telescopic photos of three stars are shown below: (i) the Sun, which is classified as a *yellow* star, (ii) *Rigel*, in the constellation Orion, which is classified as a *blue-white* star, and (iii) *Betelgeuse*, also in Orion, which is classified as a *red* star. (**a**) Place these three stars in order of increasing temperature. (**b**) Which of the following principles is relevant to your choice of answer for part (a): the uncertainty principle, the photoelectric effect, blackbody radiation, or line spectra? [Section 6.2]

(i) Sun   (ii) Rigel   (iii) Betelgeuse

6.5 The familiar phenomenon of a rainbow results from the diffraction of sunlight through raindrops. (**a**) Does the wavelength of light increase or decrease as we proceed outward from the innermost band of the rainbow? (**b**) Does the frequency of light increase or decrease as we proceed outward? [Section 6.3]

6.6 A certain quantum-mechanical system has the energy levels shown in the accompanying diagram. The energy levels are indexed by a single quantum number n that is an integer. (**a**) As drawn, which quantum numbers are involved in the transition that requires the most energy? (**b**) Which quantum numbers are involved in the transition that requires the least energy? (**c**) Based on the drawing, put the following in order of increasing wavelength of the light absorbed during the transition: (i) $n = 1$ to $n = 2$, (ii) $n = 2$ to $n = 3$, (iii) $n = 2$ to $n = 4$. (iv) $n = 1$ to $n = 3$. [Section 6.3]

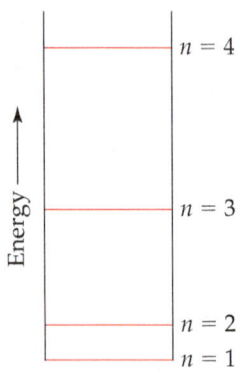

6.7 Consider the three electronic transitions in a hydrogen atom shown here, labeled A, B, and C.

(**a**) Three electromagnetic waves, all drawn on the same scale, are also shown. Each corresponds to one of the transitions. Which electromagnetic wave—(i), (ii), or (iii)—is associated with electronic transition C?

(**b**) Calculate the energy of the photon emitted for each transition.

(**c**) Calculate the wavelength of the photon emitted for each transition. Do any of these transitions lead to the emission of visible light? If so, which one(s)? [Section 6.3]

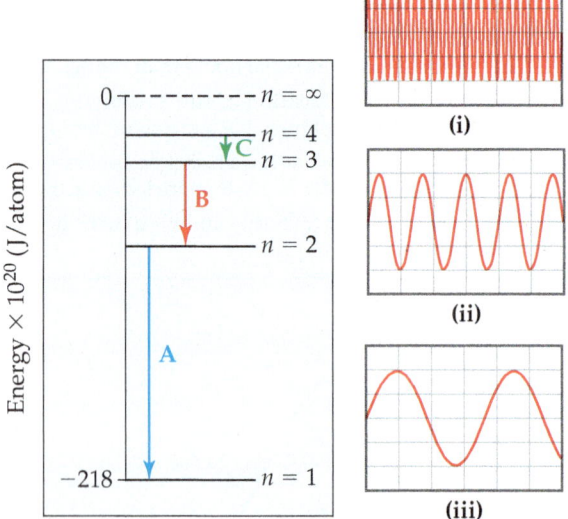

6.8 Consider a fictitious one-dimensional system with one electron. The wave function for the electron, drawn below, is $\psi(x) = \sin x$ from $x = 0$ to $x = 2\pi$. (**a**) Sketch the probability density, $\psi^2(x)$, from $x = 0$ to $x = 2\pi$. (**b**) At what value or values of x will there be the greatest probability of finding the electron? (**c**) What is the probability that the electron will be found at $x = \pi$? What is such a point in a wave function called? [Section 6.5]

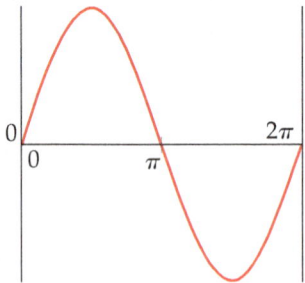

6.9 The contour representation of one of the orbitals for the $n = 3$ shell of a hydrogen atom is shown here. (**a**) What is the quantum number l for this orbital? (**b**) How do we label this orbital? (**c**) In which of the following ways would you modify this sketch if the value of the magnetic quantum number, m_l, were to change? (i) It would be drawn larger, (ii) the number of lobes would change, (iii) the lobes of the orbital would point in a different direction, (iv) there would be no change in the sketch. [Section 6.6]

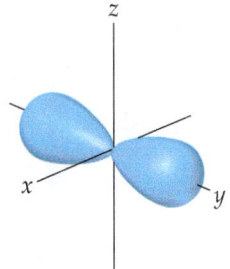

6.10 The accompanying drawing shows a contour plot for a d_{yz} orbital. Consider the quantum numbers that could potentially correspond to this orbital. (**a**) What is the smallest possible value of the principal quantum number, n? (**b**) What is the value of the angular momentum quantum number, l? (**c**) What is the largest possible value of the magnetic quantum number, m_l? (**d**) The probability density goes to zero along which of the following planes: xy, xz, or yz?

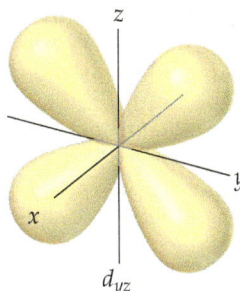

d_{yz}

6.11 Four possible electron configurations for a nitrogen atom are shown below, but only one schematic represents the correct configuration for a nitrogen atom in its ground state. Which one is the correct electron configuration? Which configurations violate the Pauli exclusion principle? Which configurations violate Hund's rule? [Section 6.8]

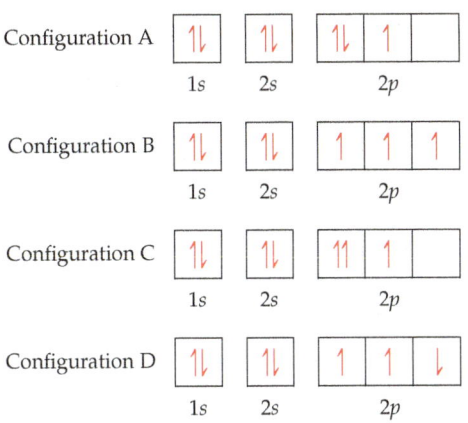

6.12 State where in the periodic table these elements appear.

(**a**) elements with the valence-shell electron configuration ns^2np^5

(**b**) elements that have three unpaired p electrons

(**c**) an element whose valence electrons are $4s^24p^1$

(**d**) the d-block elements [Section 6.9]

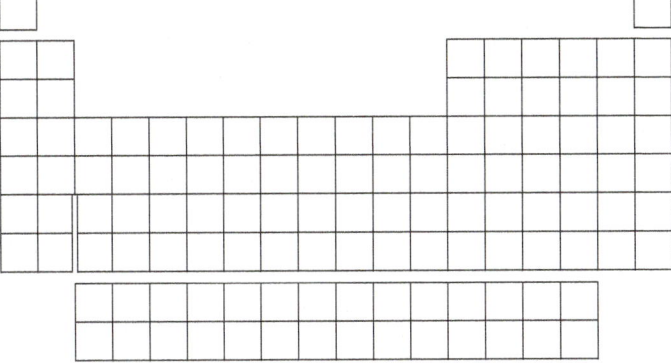

The Wave Nature of Light (Section 6.1)

6.13 What are the basic SI units for (**a**) the wavelength of light, (**b**) the frequency of light, (**c**) the speed of light?

6.14 (**a**) What is the relationship between the wavelength and the frequency of radiant energy? (**b**) Ozone in the upper atmosphere absorbs energy in the 210–230-nm range of the spectrum. In what region of the electromagnetic spectrum does this radiation occur?

6.15 Label each of the following statements as true or false. (**a**) Visible light is a form of electromagnetic radiation. (**b**) Ultraviolet light has longer wavelengths than visible light. (**c**) X rays travel faster than microwaves. (**d**) Electromagnetic radiation and sound waves travel at the same speed.

6.16 Label each of the following statements as true or false. (**a**) The frequency of radiation increases as the wavelength increases. (**b**) Electromagnetic radiation travels through a vacuum at a constant speed, regardless of wavelength. (**c**) Infrared light has higher frequencies than visible light. (**d**) The glow from a fireplace, the energy within a microwave oven, and a foghorn blast are all forms of electromagnetic radiation.

6.17 Arrange the following kinds of electromagnetic radiation in order of increasing wavelength: infrared light, green light, red light, radio waves, X rays, ultraviolet light.

6.18 List the following types of electromagnetic radiation in order of increasing wavelength: (**a**) the gamma rays produced by a radioactive nuclide used in medical imaging; (**b**) radiation from an FM radio station at 93.1 MHz on the dial; (**c**) a radio signal from an AM radio station at 680 kHz on the dial; (**d**) the yellow light from sodium vapor streetlights; (**e**) the red light of a light-emitting diode, such as in a calculator display.

6.19 (**a**) What is the frequency of radiation that has a wavelength of 10 μm, about the size of a bacterium? (**b**) What is the wavelength of radiation that has a frequency of $5.50 \times 10^{14}\,\text{s}^{-1}$? (**c**) Would the radiations in part (a) or part (b) be visible to the human eye? (**d**) What distance does electromagnetic radiation travel in 50.0 μs?

6.20 (**a**) What is the frequency of radiation whose wavelength is 0.86 nm? (**b**) What is the wavelength of radiation that has a frequency of $6.4 \times 10^{11}\,\text{s}^{-1}$? (**c**) Would the radiations in part (a) or part (b) be detected by an X-ray detector? (**d**) What distance does electromagnetic radiation travel in 0.38 ps?

6.21 A laser pointer used in a lecture hall emits light at 650 nm. What is the frequency of this radiation? Using Figure 6.4, predict the color associated with this wavelength.

6.22 It is possible to convert radiant energy into electrical energy using photovoltaic cells. Assuming equal efficiency of conversion, would infrared or ultraviolet radiation yield more electrical energy on a per-photon basis?

Quantized Energy and Photons (Section 6.2)

6.23 If human height were quantized in 1-foot increments, what would happen to the height of a child as she grew up? (**a**) The child's height would never change. (**b**) The child's height would continuously get greater. (**c**) The child's height would increase in "jumps" of 1 foot at a time. (**d**) The child's height would increase in jumps of 6 inches.

6.24 When an object like a piece of graphite is heated to 1000 K, it glows red. When the temperature is increased to around 3000 K, the color of emitted light becomes orange. What is the reason for this change in color? (**a**) The speed of the photons (quantized packets of radiant energy) increases with increasing temperature. (**b**) The average wavelength of the photons emitted increases with increasing temperature. (**c**) The average energy of the photons emitted increases as the temperature increases.

6.25 (a) Calculate the energy of a photon of electromagnetic radiation whose frequency is 2.94×10^{14} s^{-1}. (b) Calculate the energy of a photon of radiation whose wavelength is 413 nm. (c) What wavelength of radiation has photons of energy 6.06×10^{-19} J?

6.26 (a) A green laser pointer emits light with a wavelength of 532 nm. What is the frequency of this light? (b) What is the energy of one of these photons? (c) The laser pointer emits light because electrons in the material are excited (by a battery) from their ground state to an upper excited state. When the electrons return to the ground state, they lose the excess energy in the form of 532-nm photons. What is the energy gap between the ground state and excited state in the laser material?

6.27 (a) Calculate and compare the energy of a photon of wavelength 3.3 μm with that of wavelength 0.154 nm. (b) Use Figure 6.4 to identify the region of the electromagnetic spectrum to which each belongs.

6.28 An AM radio station broadcasts at 1010 kHz, and its FM partner broadcasts at 98.3 MHz. Calculate and compare the energy of the photons emitted by these two radio stations.

6.29 One type of sunburn occurs on exposure to UV light of wavelength in the vicinity of 325 nm. (a) What is the energy of a photon of this wavelength? (b) What is the energy of a mole of these photons? (c) How many photons are in a 1.00-mJ burst of this radiation? (d) These UV photons can break chemical bonds in your skin to cause sunburn—a form of radiation damage. If the 325-nm radiation provides exactly the energy to break an average chemical bond in the skin, estimate the average energy of these bonds in kJ/mol.

6.30 The energy from radiation can be used to cause the rupture of chemical bonds. A minimum energy of 242 kJ/mol is required to break the chlorine–chlorine bond in Cl_2. What is the longest wavelength of radiation that possesses the necessary energy to break the bond? What type of electromagnetic radiation is this?

6.31 A diode laser emits at a wavelength of 987 nm. (a) In what portion of the electromagnetic spectrum is this radiation found? (b) All of its output energy is absorbed in a detector that measures a total energy of 0.52 J over a period of 32 s. How many photons per second are being emitted by the laser?

6.32 A stellar object is emitting radiation at 3.55 mm. (a) What type of electromagnetic spectrum is this radiation? (b) If a detector is capturing 3.2×10^8 photons per second at this wavelength, what is the total energy of the photons detected in 1.0 hour?

6.33 Molybdenum metal must absorb radiation with a minimum frequency of 1.09×10^{15} s^{-1} before it can eject an electron from its surface via the photoelectric effect. (a) What is the minimum energy needed to eject an electron? (b) What wavelength of radiation will provide a photon of this energy? (c) If molybdenum is irradiated with light of wavelength of 120 nm, what is the maximum possible kinetic energy of the emitted electrons?

6.34 Titanium metal requires a photon with a minimum energy of 6.94×10^{-19} J to emit electrons. (a) What is the minimum frequency of light necessary to emit electrons from titanium via the photoelectric effect? (b) What is the wavelength of this light? (c) Is it possible to eject electrons from titanium metal using visible light? (d) If titanium is irradiated with light of wavelength 233 nm, what is the maximum possible kinetic energy of the emitted electrons?

Bohr's Model; Matter Waves (Sections 6.3 and 6.4)

6.35 Does the hydrogen atom "expand" or "contract" when an electron is excited from the $n = 1$ state to the $n = 3$ state?

6.36 Classify each of the following statements as either true or false. (a) A hydrogen atom in the $n = 3$ state can emit light at only two specific wavelengths, (b) a hydrogen atom in the $n = 2$ state is at a lower energy than one in the $n = 1$ state, and (c) the energy of an emitted photon equals the energy difference of the two states involved in the emission.

6.37 Is energy emitted or absorbed when the following electronic transitions occur in hydrogen: (a) from $n = 4$ to $n = 2$, (b) from an orbit of radius 2.12 Å to one of radius 8.46 Å, (c) an electron adds to the H$^+$ ion and ends up in the $n = 3$ shell?

6.38 Indicate whether energy is emitted or absorbed when the following electronic transitions occur in hydrogen: (a) from $n = 3$ to $n = 6$, (b) from an orbit of radius 4.76 Å to one of radius 0.529 Å, (c) from the $n = 6$ to the $n = 9$ state.

6.39 (a) Using Equation 6.5, calculate the energy of an electron in the hydrogen atom when $n = 2$ and when $n = 6$. Calculate the wavelength of the radiation released when an electron moves from $n = 6$ to $n = 2$. (b) Is this line in the visible region of the electromagnetic spectrum? If so, what color is it?

6.40 Consider a transition of the electron in the hydrogen atom from $n = 4$ to $n = 9$. (a) Is ΔE for this process positive or negative? (b) Determine the wavelength of light that is associated with this transition. Will the light be absorbed or emitted? (c) In which portion of the electromagnetic spectrum is the light in part (b)?

6.41 The visible emission lines observed by Balmer all involved $n_f = 2$. (a) Which of the following is the best explanation of why the lines with $n_f = 3$ are not observed in the visible portion of the spectrum: (i) Transitions to $n_f = 3$ are not allowed to happen, (ii) transitions to $n_f = 3$ emit photons in the infrared portion of the spectrum, (iii) transitions to $n_f = 3$ emit photons in the ultraviolet portion of the spectrum, or (iv) transitions to $n_f = 3$ emit photons that are at exactly the same wavelengths as those to $n_f = 2$. (b) Calculate the wavelengths of the first three lines in the Balmer series—those for which $n_i = 3$, 4, and 5—and identify these lines in the emission spectrum shown in Figure 6.11.

6.42 The Lyman series of emission lines of the hydrogen atom are those for which $n_f = 1$. (a) Determine the region of the electromagnetic spectrum in which the lines of the Lyman series are observed. (b) Calculate the wavelengths of the first three lines in the Lyman series—those for which $n_i = 2$, 3, and 4.

6.43 One of the emission lines of the hydrogen atom has a wavelength of 93.07 nm. (a) In what region of the electromagnetic spectrum is this emission found? (b) Determine the initial and final values of n associated with this emission.

6.44 The hydrogen atom can absorb light of wavelength 1094 nm. (a) In what region of the electromagnetic spectrum is this absorption found? (b) Determine the initial and final values of n associated with this absorption.

6.45 Order the following transitions in the hydrogen atom from smallest to largest frequency of light absorbed: $n = 3$ to $n = 6$, $n = 4$ to $n = 9$, $n = 2$ to $n = 3$, and $n = 1$ to $n = 2$.

6.46 Place the following transitions of the hydrogen atom in order from shortest to longest wavelength of the photon emitted: $n = 5$ to $n = 3$, $n = 4$ to $n = 2$, $n = 7$ to $n = 4$, and $n = 3$ to $n = 2$.

6.47 Use the de Broglie relationship to determine the wavelengths of the following objects: (**a**) an 85-kg person skiing at 50 km/hr, (**b**) a 10.0-g bullet fired at 250 m/s, (**c**) a lithium atom moving at 2.5×10^5 m/s, (**d**) an ozone (O_3) molecule in the upper atmosphere moving at 550 m/s.

6.48 Among the elementary subatomic particles of physics is the muon, which decays within a few microseconds after formation. The muon has a rest mass 206.8 times that of an electron. Calculate the de Broglie wavelength associated with a muon traveling at 8.85×10^5 cm/s.

6.49 Neutron diffraction is an important technique for determining the structures of molecules. Calculate the velocity of a neutron needed to achieve a wavelength of 1.25 Å. The mass of a neutron is 1.675×10^{-27} kg.

6.50 The electron microscope has been widely used to obtain highly magnified images of biological and other types of materials. When an electron is accelerated through a particular potential field, it attains a speed of 9.47×10^6 m/s. What is the characteristic wavelength of this electron? Is the wavelength comparable to the size of atoms?

6.51 Using Heisenberg's uncertainty principle, calculate the uncertainty in the position of (**a**) a 1.50-mg mosquito moving at a speed of 1.40 m/s if the speed is known to within ± 0.01 m/s; (**b**) a proton moving at a speed of $(5.00 \pm 0.01) \times 10^4$ m/s. The mass of a proton is 1.673×10^{-27} kg.

6.52 Calculate the uncertainty in the position of (**a**) an electron moving at a speed of $(3.00 \pm 0.01) \times 10^5$ m/s, (**b**) a neutron moving at this same speed. (The masses of the electron and neutron are 9.109×10^{-31} kg and 1.675×10^{-27} kg, respectively.) (**c**) Based on your answers to parts (a) and (b), which can we know with greater precision, the position of the electron or the position of the neutron?

Quantum Mechanics and Atomic Orbitals (Sections 6.5 and 6.6)

6.53 Classify the following statements as either true or false: (**a**) In a contour representation of an orbital, such as the one shown here for a $2p$ orbital, the electron is confined to move about the nucleus on the outer surface of the shape. (**b**) The probability density $[\psi(r)]^2$ gives the probability of finding the electron at a specific distance from the nucleus.

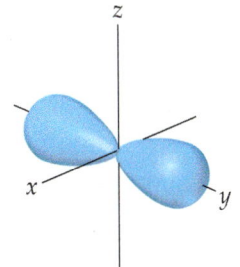

6.54 The radial probability function for a $2s$ orbital is shown here.

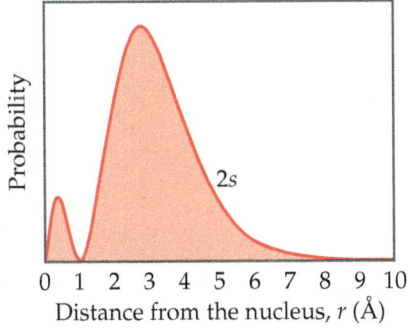

Classify the following statements as either true or false. (**a**) There are two maxima in this function because one electron spends most of its time at an approximate distance of 0.5 Å from the nucleus and the other electron spends most of its time at an approximate distance of 3 Å from the nucleus. (**b**) The radial probability function shown here and the probability density $[\psi(r)]^2$ both go to zero at the same distance from the nucleus, approximately 1 Å. (**c**) For an s orbital, the number of radial nodes is equal to the principal quantum number, n.

6.55 (**a**) For $n = 4$, what are the possible values of l? (**b**) For $l = 2$, what are the possible values of m_l? (**c**) If m_l is 2, what are the possible values for l?

6.56 How many unique combinations of the quantum numbers l and m_l are there when (**a**) $n = 3$, (**b**) $n = 4$?

6.57 Give the numerical values of n and l corresponding to each of the following orbital designations: (**a**) $3p$, (**b**) $2s$, (**c**) $4f$, (**d**) $5d$.

6.58 Give the values for n, l, and m_l for (**a**) each orbital in the $2p$ subshell and (**b**) each orbital in the $5d$ subshell.

6.59 A certain orbital of the hydrogen atom has $n = 4$ and $l = 2$. (**a**) What are the possible values of m_l for this orbital? (**b**) What are the possible values of m_s for the orbital?

6.60 A hydrogen atom orbital has $n = 5$ and $m_l = -2$. (**a**) What are the possible values of l for this orbital? (**b**) What are the possible values of m_s for the orbital?

6.61 Which of the following represent impossible combinations of n and l: (**a**) $1p$, (**b**) $4s$, (**c**) $5f$, (**d**) $2d$?

6.62 For the table that follows, write which orbital goes with the quantum numbers. Don't worry about x, y, z subscripts. If the quantum numbers are not allowed, write "not allowed."

n	l	m_l	Orbital
2	1	−1	2p (example)
1	0	0	
3	−3	2	
3	2	−2	
2	0	−1	
0	0	0	
4	2	1	
5	3	0	

6.63 Sketch the shape and orientation of the following types of orbitals: (**a**) s, (**b**) p_z, (**c**) d_{xy}.

6.64 Sketch the shape and orientation of the following types of orbitals: (**a**) p_x, (**b**) d_{z^2}, (**c**) $d_{x^2-y^2}$.

6.65 (**a**) How many radial nodes are there in the $4s$ orbital of the hydrogen atom? (**b**) How many nodal planes are there in a $2p_x$ orbital of the hydrogen atom? (**c**) Is the most probable distance from the nucleus of an electron in a $2s$ orbital larger or smaller than it would be for an electron in a $3s$ orbital? (**d**) For the hydrogen atom, list the following orbitals in order of increasing energy (that is, most stable ones first): $4f$, $6s$, $3d$, $1s$, $2p$.

6.66 (**a**) With reference to Figure 6.19, what is the relationship between the number of nodes in an s orbital and the value of the principal quantum number? (**b**) If you were to plot all the points that lie on the node of the $2s$ orbital wave function, what shape would you obtain: a set of isolated points, a plane, or a sphere? (**c**) Do any atomic orbitals have a maximum in their radial probability function at the nucleus? If so which orbital or orbitals?

Many-Electron Atoms and Electron Configurations (Sections 6.7–6.9)

6.67 (a) For an He^+ ion, do the $2s$ and $2p$ orbitals have the same energy? If not, which orbital has a lower energy? (b) If we add one electron to form the He atom, would your answer to part (a) change?

6.68 (a) The average distance from the nucleus of a $3s$ electron in a chlorine atom is smaller than that for a $3p$ electron. In light of this fact, which orbital is higher in energy? (b) Would you expect it to require more or less energy to remove a $3s$ electron from the chlorine atom as compared with a $2p$ electron?

6.69 Two possible electron configurations for an Li atom are shown here. (a) Does either configuration violate the Pauli exclusion principle? (b) Does either configuration violate Hund's rule? (c) In the absence of an external magnetic field, can we say that one electron configuration has a lower energy than the other? If so, which one has the lowest energy?

Configuration A

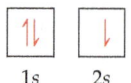

$1s$ $2s$

Configuration B

$1s$ $2s$

6.70 An experiment called the Stern–Gerlach experiment helped establish the existence of electron spin. In this experiment, a beam of silver atoms is passed through a magnetic field, which deflects half of the silver atoms in one direction and half in the opposite direction. The separation between the two beams increases as the strength of the magnetic field increases. (a) What is the electron configuration for a silver atom? (b) Would this experiment work for a beam of cadmium (Cd) atoms? (c) Would this experiment work for a beam of fluorine (F) atoms?

6.71 What is the maximum number of electrons that can occupy each of the following subshells? (a) $3p$, (b) $5d$, (c) $2s$, (d) $4f$.

6.72 What is the maximum number of electrons in an atom that can have the following quantum numbers? (a) $n = 3$, $m_l = -2$ (b) $n = 4$, $l = 3$ (c) $n = 5$, $l = 3$, $m_l = 2$ (d) $n = 4$, $l = 1$, $m_l = 0$

6.73 (a) What are "valence electrons"? (b) What are "core electrons"? (c) What does each box in an orbital diagram represent? (d) What object is represented by the half arrows in an orbital diagram? What does the direction of the arrow signify?

6.74 For each element, indicate the number of valence electrons, core electrons, and unpaired electrons in the ground state: (a) nitrogen, (b) silicon, (c) chlorine.

6.75 Write the condensed electron configurations for the following atoms, using the appropriate noble-gas core abbreviations: (a) Cs (b) Ni (c) Se (d) Cd (e) U (f) Pb

6.76 Write the condensed electron configurations for the following atoms and indicate how many unpaired electrons each has: (a) Mg (b) Ge (c) Br (d) V (e) Y (f) Lu

6.77 Identify the specific element that corresponds to each of the following electron configurations and indicate the number of unpaired electrons for each: (a) $1s^2 2s^2$ (b) $1s^2 2s^2 2p^4$ (c) $[Ar]4s^1 3d^5$ (d) $[Kr]5s^2 4d^{10} 5p^4$

6.78 Identify the group of elements that corresponds to each of the following generalized electron configurations and indicate the number of unpaired electrons for each:

(a) [noble gas]$ns^2 np^5$

(b) [noble gas]$ns^2(n-1)d^2$

(c) [noble gas]$ns^2(n-1)d^{10}np^1$

(d) [noble gas]$ns^2(n-2)f^6$

6.79 The following do not represent valid ground-state electron configurations for an atom either because they violate the Pauli exclusion principle or because orbitals are not filled in order of increasing energy. Indicate which of these two principles is violated in each example. (a) $1s^2 2s^2 3s^1$ (b) $[Xe]6s^2 5d^4$ (c) $[Ne]3s^2 3d^5$

6.80 The following electron configurations represent excited states. Identify the element and write its ground-state condensed electron configuration. (a) $1s^2 2s^2 2p^4 3s^1$ (b) $[Ar]4s^1 3d^{10} 4p^2 5p^1$ (c) $[Kr]5s^2 4d^2 5p^1$

Additional Exercises

6.81 Consider the two waves shown here, which we will consider to represent two electromagnetic radiations:

(a) What is the wavelength of wave A? Of wave B?

(b) What is the frequency of wave A? Of wave B?

(c) Identify the regions of the electromagnetic spectrum to which waves A and B belong.

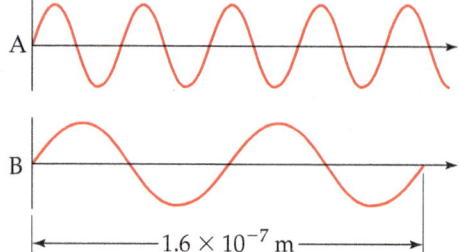

A

B

— 1.6×10^{-7} m —

6.82 If you put 120 volts of electricity through a pickle, the pickle will smoke and start glowing orange-yellow. The light is emitted because sodium ions in the pickle become excited;

their return to the ground state results in light emission. (a) The wavelength of this emitted light is 589 nm. Calculate its frequency. (b) What is the energy of 1.00 mol of these photons? (A mole of photons is called an Einstein.) (c) Calculate the energy gap between the excited and ground states for the sodium ion. (d) If you soaked the pickle for a long time in a different salt solution, such as strontium chloride, would you still observe 589-nm light emission?

6.83 Certain elements emit light of a specific wavelength when they are burned. Historically, chemists used such emission wavelengths to determine whether specific elements were present in a sample. Characteristic wavelengths for some of the elements are given in the following table:

Ag	328.1 nm	Fe	372.0 nm
Au	267.6 nm	K	404.7 nm
Ba	455.4 nm	Mg	285.2 nm
Ca	422.7 nm	Na	589.6 nm
Cu	324.8 nm	Ni	341.5 nm

(a) Determine which elements emit radiation in the visible part of the spectrum. (b) Which element emits photons of highest energy? Of lowest energy? (c) When burned, a sample of an unknown substance is found to emit light of frequency $9.23 \times 10^{14}\,s^{-1}$. Which of these elements is probably in the sample?

6.84 In August 2011, the Juno spacecraft was launched from Earth with the mission of orbiting Jupiter, arriving nearly five years later in July of 2016. The distance between the two planets varies depending on where each planet is in its orbit, but at the closest, the distance between Jupiter and Earth is 391 million miles. What is the minimum amount of time it takes for a transmitted signal from Juno to reach the Earth?

6.85 The rays of the Sun that cause tanning and burning are in the ultraviolet portion of the electromagnetic spectrum. These rays are categorized by wavelength. So-called UV-A radiation has wavelengths in the range of 320–380 nm, whereas UV-B radiation has wavelengths in the range of 290–320 nm. (a) Calculate the frequency of light that has a wavelength of 320 nm. (b) Calculate the energy of a mole of 320-nm photons. (c) Which are more energetic, photons of UV-A radiation or photons of UV-B radiation? (d) The UV-B radiation from the Sun is considered a greater cause of sunburn in humans than is UV-A radiation. Is this observation consistent with your answer to part (c)?

6.86 The watt is the derived SI unit of power, the measure of energy per unit time: $1\,W = 1\,J/s$. A semiconductor laser in a CD player has an output wavelength of 780 nm and a power level of 0.10 mW. How many photons strike the CD surface during the playing of a CD 69 minutes in length?

6.87 Carotenoids are yellow, orange, and red pigments synthesized by plants. The observed color of an object is not the color of light it absorbs but rather the complementary color, as described by a color wheel such as the one shown here. On this wheel, complementary colors are across from each other. (a) Based on this wheel, what color is absorbed most strongly if a plant is orange? (b) If a particular carotenoid absorbs photons at 455 nm, what is the energy of the photon?

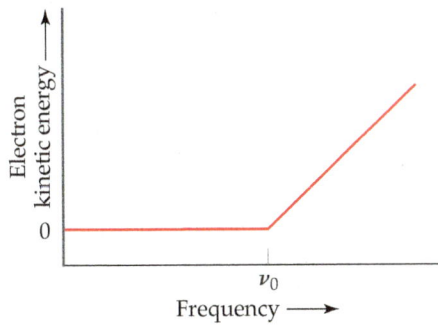

6.88 In an experiment to study the photoelectric effect, a scientist measures the kinetic energy of ejected electrons as a function of the frequency of radiation hitting a metal surface. She obtains the following plot. The point labeled "ν_0" corresponds to light with a wavelength of 542 nm. (a) What is the value of ν_0 in s^{-1}? (b) What is the value of the work function of the metal in units of kJ/mol of ejected electrons? (c) Note that when the frequency of the light is greater than ν_0, the plot shows a straight line with a nonzero slope. What is the slope of this line segment?

6.89 Consider a transition in which the electron of a hydrogen atom is excited from $n = 1$ to $n = \infty$. (a) What is the end result of this transition? (b) What wavelength of light must be absorbed to accomplish this process? (c) What will occur if light with a shorter wavelength than that in part (b) is used

to excite the hydrogen atom? (d) How are the results of parts (b) and (c) related to the plot shown in Exercise 6.88?

6.90 The human retina has three types of receptor cones, each sensitive to a different range of wavelengths of visible light, as shown in this figure (the colors are merely to differentiate the three curves from one another; they do not indicate the actual colors represented by each curve):

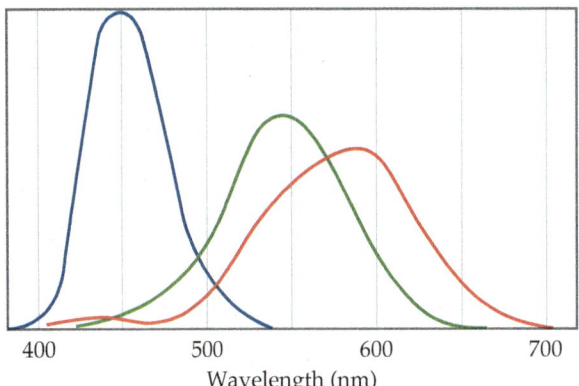

(a) Estimate the energies of photons with wavelengths at the maximum for each type of cone. (b) The color of the sky is due to scattering of solar light by the molecules of the atmosphere. Lord Rayleigh was one of the first to study scattering of this kind. He showed that the amount of scattering for very small particles such as molecules is inversely proportional to the fourth power of the wavelength. Estimate the ratio of the scattering efficiency of light at the wavelength of the maximum for the "blue" cones, as compared with that for the "green" cones. (c) Explain why the sky appears blue even though all wavelengths of solar light are scattered by the atmosphere.

6.91 The series of emission lines of the hydrogen atom for which $n_f = 3$ is called the *Paschen series*. (a) Determine the region of the electromagnetic spectrum in which the lines of the Paschen series are observed. (b) Calculate the wavelengths of the first three lines in the Paschen series—those for which $n_i = 4, 5,$ and 6.

6.92 Determine whether each of the following sets of quantum numbers for the hydrogen atom is valid. If a set is *not* valid, then indicate which of the quantum numbers has a value that is not valid:

(a) $n = 4, l = 1, m_l = 2, m_s = -\frac{1}{2}$

(b) $n = 4, l = 3, m_l = -3, m_s = +\frac{1}{2}$

(c) $n = 3, l = 2, m_l = -1, m_s = +\frac{1}{2}$

(d) $n = 5, l = 0, m_l = 0, m_s = 0$

(e) $n = 2, l = 2, m_l = 1, m_s = +\frac{1}{2}$

6.93 Which of the following explains why the Bohr model is applicable to a He^+ ion but not a neutral He atom? (i) The nuclear charge in He^+ is the same as that in the H atom. (ii) One electron in the He atom is attracted to the nucleus while the other is repelled from the nucleus, which complicates the model. (iii) The two electrons in the He atom repel one another, which complicates the model.

6.94 Bohr's model can be used for hydrogen-like ions—ions that have only one electron, such as He^+ and Li^{2+}. The ground-state energies of H, He^+, and Li^{2+} are tabulated as follows:

Atom or ion	H	He$^+$	Li^{2+}
Ground-state energy	$-2.18 \times 10^{-18}\,J$	$-8.72 \times 10^{-18}\,J$	$-1.96 \times 10^{-17}\,J$

(a) Does the ground-state energy of the electron in these atoms/ions increase or decrease as the charge of the nucleus increases? (**b**) By examining these numbers, propose a relationship between the ground-state energy of any hydrogen-like atom or ion and the nuclear charge, Z, and use the relationship to predict the ground-state energy of the C^{5+} ion.

6.95 An electron is accelerated through an electric potential to a kinetic energy of 2.15×10^{-15} J. What is its characteristic wavelength? [*Hint:* Recall that the kinetic energy of a moving object is $E = \frac{1}{2}mv^2$, where m is the mass of the object and v is the speed of the object.]

6.96 As discussed in the A Closer Look box on "Measurement and the Uncertainty Principle," the essence of the uncertainty principle is that we can't make a measurement without disturbing the system that we are measuring. (**a**) Why can't we measure the position of a subatomic particle without disturbing it? (**b**) How is this concept related to the paradox discussed in the Closer Look box on "Schrödinger's Cat and Quantum Computing"?

6.97 For orbitals that are symmetric but not spherical, the contour representations (as in Figures 6.21 and 6.22) suggest where nodal planes exist (that is, where the electron density is zero). For example, the p_x orbital has a node wherever $x = 0$. This equation is satisfied by all points on the yz plane, so this plane is called a nodal plane of the p_x orbital. (**a**) Determine the nodal plane of the p_z orbital. (**b**) What are the two nodal planes of the d_{yz} orbital? (**c**) What are the two nodal planes of the $d_{x^2-y^2}$ orbital?

6.98 The Chemistry and Life box in Section 6.7 described the techniques called NMR and MRI. (**a**) Instruments for obtaining MRI data are typically labeled with a frequency, such as 600 MHz. In what region of the electromagnetic spectrum does a photon with this frequency belong? (**b**) What is the value of ΔE in Figure 6.25 that would correspond to the absorption of a photon of radiation with frequency 450 MHz? (**c**) When the 450 MHz photon is absorbed, does it change the spin of the electron or the proton on a hydrogen atom?

6.99 Suppose that the spin quantum number, m_s, could have *three* allowed values instead of two. How would this affect the number of elements in the first four rows of the periodic table?

6.100 Using the periodic table as a guide, write the condensed electron configuration and determine the number of unpaired electrons for the ground state of (**a**) Br, (**b**) Ga, (**c**) Hf, (**d**) Sb, (**e**) Bi, (**f**) Sg.

6.101 Scientists have speculated that element 126 might have a moderate stability, allowing it to be synthesized and characterized. Predict what the condensed electron configuration of this element might be.

6.102 In the experiment shown schematically below, a beam of neutral atoms is passed through a magnetic field. Atoms that

have unpaired electrons are deflected in different directions in the magnetic field depending on the value of the electron spin quantum number. In the experiment illustrated, we envision that a beam of hydrogen atoms splits into two beams. (**a**) What is the significance of the observation that the single beam splits into two beams? (**b**) What do you think would happen if the strength of the magnet were increased? (**c**) What do you think would happen if the beam of hydrogen atoms were replaced with a beam of helium atoms? Why? (**d**) The relevant experiment was first performed by Otto Stern and Walter Gerlach in 1921. They used a beam of Ag atoms in the experiment. By considering the electron configuration of a silver atom, explain why the single beam splits into two beams.

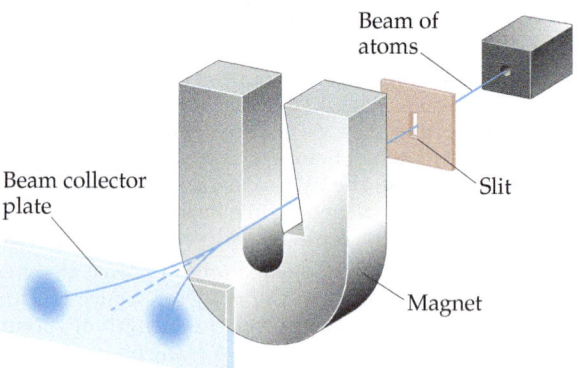

6.103 Microwave ovens use microwave radiation to heat food. The energy of the microwaves is absorbed by water molecules in food and then transferred to other components of the food. (**a**) Suppose that the microwave radiation has a wavelength of 11.2 cm. How many photons are required to heat 200 mL of coffee from 23 to 60 °C? (**b**) Suppose the microwave's power is 900 W (1 watt = 1 joule/s). How long would you have to heat the coffee in part (a)?

6.104 The stratospheric ozone (O_3) layer helps to protect us from harmful ultraviolet radiation. It does so by absorbing ultraviolet light and falling apart into an O_2 molecule and an oxygen atom, a process known as photodissociation.

$$O_3(g) \longrightarrow O_2(g) + O(g)$$

Use the data in Appendix C to calculate the enthalpy change for this reaction. What is the maximum wavelength a photon can have if it is to possess sufficient energy to cause this dissociation? In what portion of the spectrum does this wavelength occur?

Design an Experiment

In this chapter, we have learned about the *photoelectric effect* and its impact on the formulation of light as photons. We have also seen that some anomalous electron configurations of the elements are particularly favorable if each atom has one or more half-filled shell, such as the case for the Cr atom with its $[Ar]4s^1 3d^5$ electron configuration. Let's suppose it is hypothesized that more energy is required to remove an electron from a metal that has atoms with one or more half-filled shells than from those that do not. (**a**) Design a series of experiments involving the photoelectric effect that would test the

hypothesis. (**b**) What experimental apparatus would be needed to test the hypothesis? It's not necessary that you name actual equipment but rather that you imagine how the apparatus would work—think in terms of the types of measurements that would be needed and what capability you would need in your apparatus. (**c**) Describe the type of data you would collect and how you would analyze the data to see whether the hypothesis were correct. (**d**) Could your experiments be extended to test the hypothesis for other parts of the periodic table, such as the lanthanide or actinide elements?

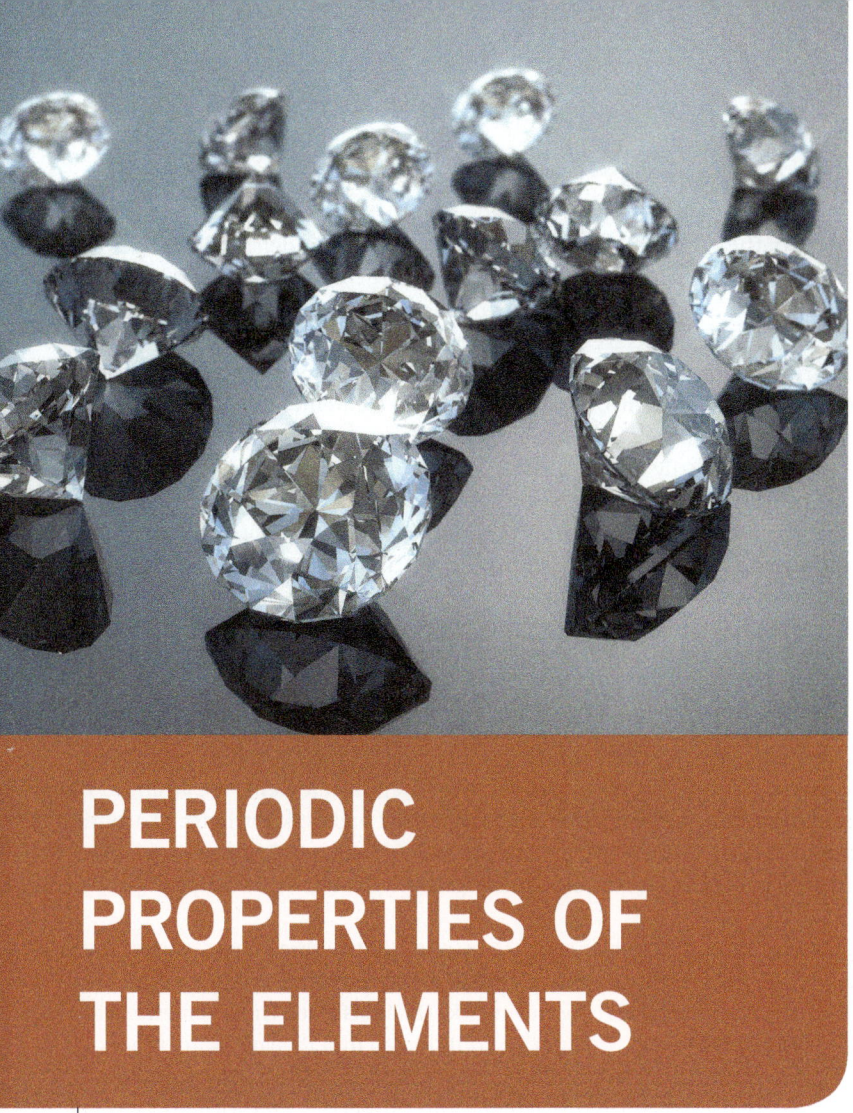

PERIODIC PROPERTIES OF THE ELEMENTS

▲ One form of the element carbon is diamond.

The periodic table is a powerful tool for understanding and predicting the physical and chemical properties of elements. Typically, elements share common characteristics with neighbors that occupy the same column of the periodic table. Sodium, potassium, and rubidium are soft metals that react violently when they come in contact with water. Neon, argon, and krypton are colorless, chemically nonreactive gases. Copper, silver, and gold are highly conductive metals that react slowly, if at all, with air and water.

In Chapter 6 we learned that elements in the same column of the periodic table contain the same number of electrons in their **valence orbitals**—the occupied orbitals that hold the electrons involved in bonding. In this chapter, we explore several fundamental characteristics of elements. We see how these characteristics change as we move across a row or down a column of the periodic table, which in turn helps us rationalize and predict the physical and chemical properties of the elements.

7.1 | Development of the Periodic Table

Learning Objective

When you finish **Section 7.1**, you should be able to:

▶ Describe how the periodic table was developed and is used to organize the chemical elements.

The periodic table of the elements is the defining framework for understanding the properties and chemistry of the elements. Elements in the same column in the periodic table have similar properties; and elements in the same row have trends that can be understood by considering the concepts we explore in this chapter.

The discovery of chemical elements has been ongoing since ancient times (**Figure 7.1**). Certain elements, such as gold (Au), appear in nature in elemental form and were thus discovered thousands of years ago. In contrast, some elements, such as technetium (Tc), are radioactive and intrinsically unstable. We know about them only because of technology developed during the twentieth century.

The majority of elements readily form compounds, so they are *not* found in nature in their elemental form. For centuries, therefore, scientists were unaware of their existence. During the early nineteenth century, advances in chemistry made it easier to isolate elements from their compounds. As a result, the number of known elements more than doubled from 31 in 1800 to 63 by 1865.

As the number of known elements increased, scientists began classifying them. In 1869, Dmitri Mendeleev (1834–1907) in Russia and Lothar Meyer (1830–1895) in Germany published nearly identical classification schemes. Both noted that similar chemical and physical properties recur periodically when the elements are arranged in order of increasing atomic weight. Scientists at that time had no knowledge of atomic numbers. Atomic weights, however, generally increase with increasing atomic number, so both Mendeleev and Meyer fortuitously arranged the elements in nearly the proper sequence.

Although Mendeleev and Meyer came to essentially the same conclusion about the periodicity of elemental properties, Mendeleev is given credit for advancing his ideas more vigorously and stimulating new work. His insistence that elements with similar characteristics be listed in the same column forced him to leave blank spaces in his table. For example, both gallium (Ga) and germanium (Ge) were unknown to Mendeleev. He boldly predicted their existence and properties, referring to them as *eka-aluminum* ("under" aluminum) and *eka-silicon* ("under" silicon), respectively, after the elements under which they appeared in his table. When these elements were discovered, their properties closely matched those predicted by Mendeleev, as shown in **Table 7.1**.

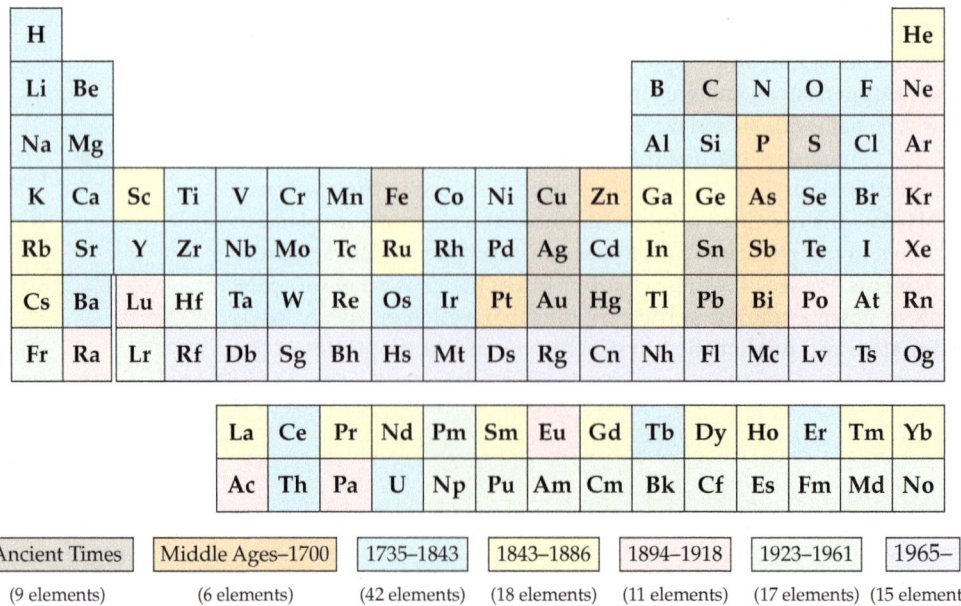

▲ **Figure 7.1** Timeline of the discovery of the elements.

TABLE 7.1   Comparison of the Properties of Eka-Silicon Predicted by Mendeleev with the Observed Properties of Germanium

Property	Mendeleev's Predictions for Eka-Silicon (made in 1871)	Observed Properties of Germanium (discovered in 1886)
Atomic weight	72	72.59
Density (g/cm³)	5.5	5.35
Specific heat (J/g-K)	0.305	0.309
Melting point (°C)	High	947
Color	Dark gray	Grayish white
Formula of oxide	XO_2	GeO_2
Density of oxide (g/cm³)	4.7	4.70
Formula of chloride	XCl_4	$GeCl_4$
Boiling point of chloride (°C)	A little under 100	84

In 1913, two years after Ernest Rutherford (1871–1937) proposed the nuclear model of the atom (Section 2.2), English physicist Henry Moseley (1887–1915) developed the concept of atomic numbers. Bombarding different elements with high-energy electrons, Moseley found that each element produced X rays of a unique frequency and that the frequency generally increased as the atomic mass increased. He arranged the X-ray frequencies in order by assigning a unique whole number, called an *atomic number*, to each element. Moseley correctly identified the atomic number as the number of protons in the nucleus of the atom. (Section 2.3)

The concept of atomic number clarified some problems in the periodic table of Moseley's day, which was based on atomic weights. For example, the atomic weight of Ar (atomic number 18) is greater than that of K (atomic number 19), yet the chemical and physical properties of Ar are much more like those of Ne and Kr than like those of Na and Rb. When the elements are arranged in order of increasing atomic number, Ar and K appear in their correct places in the table. Moseley's studies also made it possible to identify "holes" in the periodic table, which led to the discovery of new elements.

 Self-Assessment Exercise

SAE 7.1   Which of the following statements is/are true?
  (**i**) Elements that are gases were generally discovered before ones that are metals.
 (**ii**) Metallic elements that form oxides were discovered before ones that are found as pure metals.
(**iii**) The heaviest elements (largest atomic numbers) were discovered last.
(**a**) i only (**b**) ii only (**c**) iii only (**d**) ii and iii

7.2 | Effective Nuclear Charge

Many properties of atoms depend on electron configuration and on how strongly the outer electrons in the atoms are attracted to the nucleus. Coulomb's law tells us that the strength of the interaction between two electrical charges depends on the magnitudes of the charges and on the distance between them. (Section 2.3) Thus, the attractive force between an electron and the nucleus depends on the magnitude of the nuclear charge and on the average distance between the nucleus and the electron. The force increases as the nuclear charge increases and decreases as the electron moves farther from the nucleus.

Understanding the attraction between the electron and the nucleus in a hydrogen atom is straightforward because we have only one electron and one proton. In a many-electron atom, however, the situation is more complicated. In addition to the attraction of each electron to the nucleus, each electron experiences repulsion due to the

 Learning Objectives

When you finish Section 7.2, you should be able to:
▶ Predict how, and understand why, effective nuclear charge varies across the periodic table.
▶ Calculate the effective nuclear charge for a valence electron in an atom, assuming complete screening by core electrons.

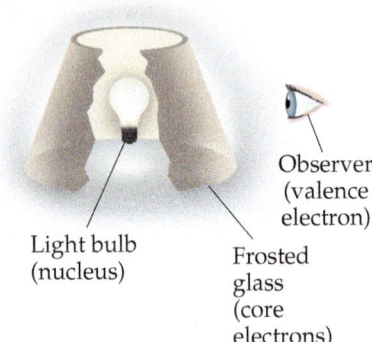

▲ Figure 7.2 An analogy for effective nuclear charge. We envision the nucleus as a light bulb, the core electrons as a frosted glass lampshade, and a valence electron as an observer. The amount of light seen by the observer depends on the intensity of the light bulb and the screening by the frosted glass lampshade.

other electrons. These electron–electron repulsions cancel some of the attraction of the electron to the nucleus so that the electron experiences less attraction than it would if the other electrons weren't there. In essence, each electron in a many-electron atom is *screened* from the nucleus by the other electrons. It therefore experiences a net attraction that is *smaller* than it would experience in the absence of other electrons.

How can we account for the combination of nuclear attraction and electron repulsions for our electron of interest? The simplest way to do so is to imagine that the electron experiences a net attraction to the nucleus that is diminished (*screened*) by repulsions from the intervening electrons. We call this partially screened nuclear charge the **effective nuclear charge**, Z_{eff}. Because the full attractive force of the nucleus has been decreased by the electron repulsions, the effective nuclear charge is always *less* than the *actual* nuclear charge ($Z_{eff} < Z$). We can define the amount of screening of the nuclear charge quantitatively by using a *screening constant*, S, such that

$$Z_{eff} = Z - S \qquad [7.1]$$

where S is a positive number. For a valence electron, most of the shielding is due to the core electrons, which are much closer to the nucleus. For the valence electrons in an atom, then, *the value of S is usually close to the number of core electrons in the atom.* (Electrons in the same valence shell do not screen one another very effectively, but they do affect the value of S slightly; see "A Closer Look: Estimating Effective Nuclear Charge.")

To better understand the notion of effective nuclear charge, we can use an analogy of a light bulb with a frosted glass shade (**Figure 7.2**). The light bulb represents the nucleus, and the observer is the electron of interest, which is usually a valence electron. The amount of light that the electron "sees" is analogous to the amount of net nuclear attraction experienced by the electron. The other electrons in the atom, especially the core electrons, act like a frosted glass lampshade, decreasing the amount of light that gets to the observer. If the light bulb gets brighter (if Z increases) while the lampshade stays the same (while S does not change), then more light is observed. Likewise, if the lampshade gets thicker (if S increases), then less light is observed. Keep this analogy in mind as we discuss trends in effective nuclear charge.

What would you expect for the magnitude of Z_{eff} for the sodium atom? Sodium has the electron configuration $[Ne]3s^1$. The nuclear charge is $Z = 11+$, and there are 10 core electrons ($1s^2 2s^2 2p^6$), which serve as a "lampshade" to screen the nuclear charge "seen" by the 3s electron. Therefore, in the simplest approach, we expect S to equal 10 and the 3s electron to experience an effective nuclear charge of $Z_{eff} = 11 - 10 = 1+$ (**Figure 7.3**). The situation is more complicated, however, because the 3s electron has a small probability of being very close to the nucleus, in the region occupied by the core electrons. (Section 6.6) Thus, this electron experiences a greater net attraction than our simple

▶ Figure 7.3 Effective nuclear charge. The effective nuclear charge experienced by the 3s electron in a sodium atom depends on the 11+ charge of the nucleus and the 10− charge of the core electrons.

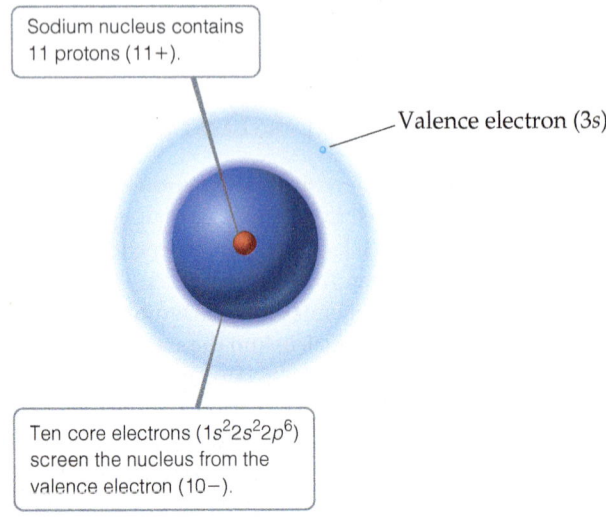

Sodium nucleus contains 11 protons (11+).

Valence electron (3s)

Ten core electrons ($1s^2 2s^2 2p^6$) screen the nucleus from the valence electron (10−).

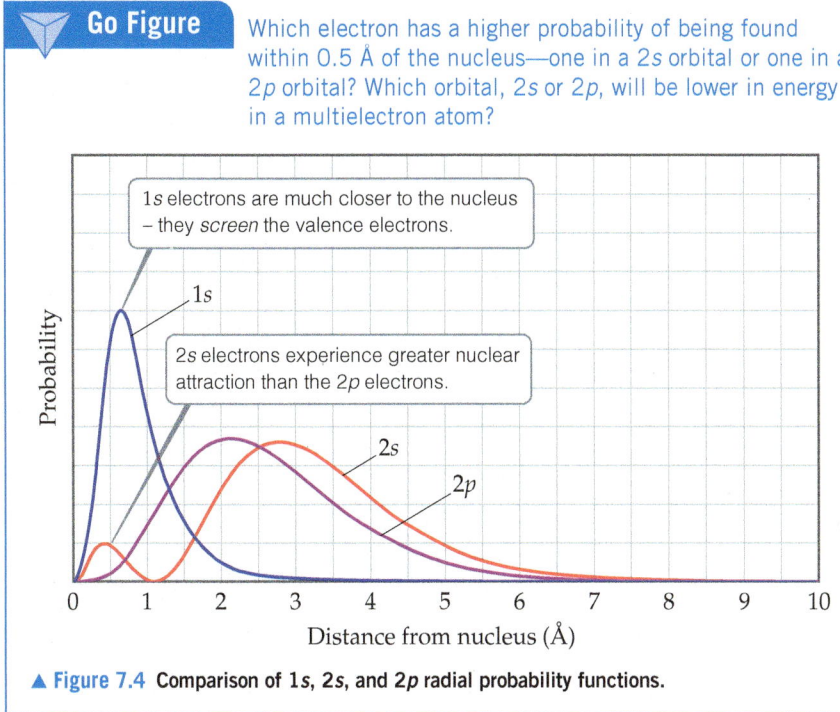

Which electron has a higher probability of being found within 0.5 Å of the nucleus—one in a 2s orbital or one in a 2p orbital? Which orbital, 2s or 2p, will be lower in energy in a multielectron atom?

1s electrons are much closer to the nucleus – they *screen* the valence electrons.

1s

2s electrons experience greater nuclear attraction than the 2p electrons.

2s

2p

Distance from nucleus (Å)

Probability

▲ **Figure 7.4 Comparison of 1s, 2s, and 2p radial probability functions.**

$S = 10$ model suggests: The actual value of Z_{eff} for the 3s electron in Na is $Z_{eff} = 2.5+$. In other words, because there is a small probability that the 3s electron is close to the nucleus, the value of S in Equation 7.1 changes from 10 to 8.5.

The notion of effective nuclear charge also explains an important effect we noted in Section 6.7: For a many-electron atom, the energies of orbitals with the same n value increase with increasing l value. For example, in the carbon atom, whose electron configuration is $1s^2 2s^2 2p^2$, the energy of the 2p orbital ($l = 1$) is higher than that of the 2s orbital ($l = 0$), even though both orbitals are in the $n = 2$ shell (Figure 6.23). This difference in energies is due to the radial probability functions for the orbitals (**Figure 7.4**). We see first that the 1s electrons are much closer to the nucleus—they serve as an effective "lampshade" for the 2s and 2p electrons. Notice next that the 2s probability function has a small peak fairly close to the nucleus, whereas the 2p probability function does not. As a result, a 2s electron is not screened as much by the core orbitals as is a 2p electron. The greater attraction between the 2s electron and the nucleus leads to a lower energy for the 2s orbital than for the 2p orbital. The same reasoning explains the general trend in orbital energies ($ns < np < nd$) in many-electron atoms.

The following very important point explains many of the properties of the elements:

The effective nuclear charge increases from left to right across a period of the periodic table.

Although the number of core electrons stays the same across the period, the number of protons increases—in our analogy, we are increasing the brightness of the light bulb while keeping the shade the same. The valence electrons added to counterbalance the increasing nuclear charge screen one another ineffectively. Thus, Z_{eff} increases steadily in going from left to right across a row in the periodic table. For example, the two core electrons of lithium ($1s^2 2s^1$) screen the 2s valence electron from the 3+ nucleus fairly efficiently. Consequently, the valence electron experiences an effective nuclear charge of roughly $3 - 2 = 1+$. For beryllium ($1s^2 2s^2$), the effective nuclear charge experienced by each valence electron is larger because here the two core 1s electrons screen a 4+ nucleus, and each 2s electron only partially screens the other. Consequently, the effective nuclear charge experienced by each 2s electron is about $4 - 2 = 2+$.

A CLOSER LOOK | Estimating Effective Nuclear Charge

To get a sense of how effective nuclear charge varies as both nuclear charge and number of electrons increase, consider **Figure 7.5**. Although the details of how the Z_{eff} values in the graph were calculated are beyond the scope of our discussion, the trends are instructive.

The effective nuclear charge felt by the outermost electrons is smaller than that felt by inner electrons because of screening by the inner electrons. In addition, the effective nuclear charge felt by the outermost electrons does not increase as steeply with increasing atomic number because the valence electrons make a small but non-negligible contribution to the screening constant S. The most striking feature associated with the Z_{eff} value for the outermost electrons is the sharp drop between the last period 2 element (Ne) and the first period 3 element (Na). This drop reflects the fact that the core electrons are much more effective than the valence electrons at screening the nuclear charge.

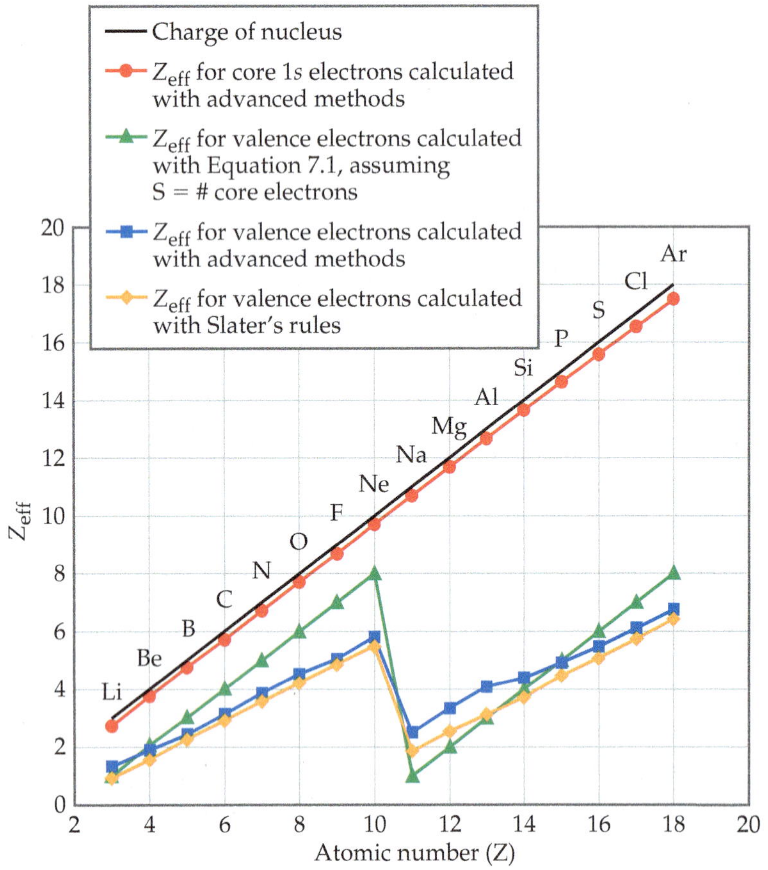

Because Z_{eff} can be used to understand many physically measurable quantities, it is desirable to have a simple method for estimating it. The value of Z in Equation 7.1 is known exactly, so the challenge boils down to estimating the value of S. In the text, we estimated S very simply by assuming that each core electron contributes 1.00 to S and the outer electrons contribute nothing. A more accurate approach was developed by John Slater (1900–1976), and we can use his approach if we limit ourselves to elements that do not have electrons in d or f subshells.

Slater's rules are as follows: Electrons for which the principal quantum number n is larger than the value of n for the electron of interest contribute 0 to the value of S. Electrons with the same value of n as the electron of interest contribute 0.35 to the value of S. Electrons that have principal quantum number $n - 1$ contribute 0.85, while those with even smaller values of n contribute 1.00. Consider fluorine, for example, which has the ground-state electron configuration $1s^2 2s^2 2p^5$. For a valence electron in fluorine, Slater's rules tell us that $S = (0.35 \times 6) + (0.85 \times 2) = 3.8$. (Slater's rules ignore the contribution of an electron to itself in screening; therefore, we consider only six $n = 2$ electrons, not all seven). Thus, $Z_{eff} = Z - S = 9 - 3.8 = 5.2+$, which is a little lower than the simple estimate of $9 - 2 = 7+$.

Values of Z_{eff} estimated using the simple method outlined in the text, as well as those estimated with Slater's rules, are plotted in Figure 7.5. While neither of these methods exactly replicates the values of Z_{eff} obtained from more sophisticated calculations, both methods effectively capture the periodic variation in Z_{eff}. While Slater's approach is more accurate, the method outlined in the text does a reasonably good job of estimating Z_{eff} despite its simplicity. For our purposes, therefore, we can assume that the screening constant S in Equation 7.1 is roughly equal to the number of core electrons.

▲ **Figure 7.5** **Variations in effective nuclear charge for period 2 and period 3 elements.** Moving from one element to the next in the periodic table, the increase in Z_{eff} felt by the innermost (1s) electrons (red circles) closely tracks the increase in nuclear charge Z (black line) because these electrons are not effectively screened from the nucleus by the remaining electrons. The results of several methods used to calculate Z_{eff} for valence electrons are shown in other colors.

Related Exercises: 7.15, 7.16, 7.31, 7.32, 7.80, 7.81

Going down a column, the effective nuclear charge experienced by valence electrons changes far less than it does across a period. For example, using our simple estimate for S, we would expect the effective nuclear charge experienced by the valence electrons in lithium and sodium to be about the same, roughly $3 - 2 = 1+$ for lithium and $11 - 10 = 1+$ for sodium. In fact, however, *effective nuclear charge increases slightly as we go down a column* because the more diffuse core electron cloud is less able to screen the valence electrons from the nuclear charge. In the case of the alkali metals, Z_{eff} increases from 1.3+ for lithium, to 2.5+ for sodium, to 3.5+ for potassium.

Self-Assessment Exercises

SAE 7.2 In going from K to Ca in the periodic table, the nuclear charge _____, and the effective nuclear charge experienced by a 3s electron _____. (**a**) increases, increases (**b**) increases, decreases (**c**) increases, stays the same (**d**) decreases, decreases

SAE 7.3 Use the simple method to calculate the effective nuclear charge of an $n = 3$ electron in silicon. (**a**) 0 (**b**) 1+ (**c**) 2+ (**d**) 4+ (**e**) 14+

SAE 7.4 Predict the relative order of Z_{eff} for the valence electrons of Na, P, S. (**a**) Na > P > S (**b**) Na < P < S (**c**) Na < S < P (**d**) S < Na < P

7.3 | Sizes of Atoms and Ions

It is tempting to think of atoms as hard, spherical objects. According to the quantum-mechanical model, however, atoms do not have sharply defined boundaries at which the electron distribution becomes zero. (Section 6.5) Nevertheless, we can define atomic size in several ways, based on the distances between atoms in various situations.

Imagine a collection of argon atoms in the gas phase. When two of these atoms collide, they ricochet apart like colliding billiard balls. This ricocheting happens because the electron clouds of the colliding atoms cannot penetrate each other to any significant extent. The shortest distance separating the two nuclei during such collisions is twice the radii of the atoms. We call this radius the *nonbonding atomic radius* or the *van der Waals* radius (Figure 7.6).

In molecules, the attractive interaction between any two adjacent atoms is what we recognize as a chemical bond. We discuss bonding in Chapters 8 and 9. For now, we need to realize that two bonded atoms are closer together than they would be in a nonbonding collision where the atoms ricochet apart. We can therefore define an atomic radius based on the distance between the nuclei when two atoms are bonded to each other, shown as distance *d* in Figure 7.6. The **bonding atomic radius** for any atom in a molecule is equal to half of this bond distance *d*. Note from Figure 7.6 that the bonding atomic radius (also known as the *covalent radius*) is smaller than the nonbonding atomic radius. Unless otherwise noted, we mean the bonding atomic radius when we speak of the "size" of an atom.

Although it is very difficult to measure the nonbonding atomic radius of an atom, scientists have developed a variety of techniques for measuring the distances separating nuclei in molecules. From observations of these distances in many molecules, each element can be assigned a bonding atomic radius. In the I_2 molecule, for example, the distance separating the nuclei is observed to be 2.66 Å, which means the bonding atomic radius of an iodine atom in I_2 is (2.66 Å)/2 = 1.33 Å.* Similarly, the distance separating adjacent carbon nuclei in diamond (a three-dimensional solid network of carbon atoms) is 1.54 Å; thus, the bonding atomic radius of carbon in diamond is 0.77 Å. By using structural information on more than 30,000 substances, a consistent set of bonding atomic radii of the elements can be defined (Figure 7.7). For the lighter noble gases, the bonding atomic radii must be estimated because there are no known compounds of these elements.

The atomic radii in Figure 7.7 allow us to estimate bond lengths in molecules. For example, the bonding atomic radii for C and Cl are 0.76 Å and 1.02 Å, respectively. In CCl_4 the measured length of the C—Cl bond is 1.77 Å, very close to the sum (0.76 + 1.02 Å = 1.78 Å) of the bonding atomic radii of Cl and C.

* Remember: The angstrom (1 Å = 10^{-10} m) is a convenient metric unit for atomic measurements of length. It is *not* an SI unit. The most commonly used SI unit for atomic measurements is the picometer (1 pm = 10^{-12} m; 1 Å = 100 pm).

Learning Objectives

When you finish Section 7.3, you should be able to:

▶ Calculate bond lengths in molecules using bonding atomic radii.

▶ Predict how atomic size varies across the periodic table.

▶ Explain how the radius of an atom changes upon losing electrons to form a cation or gaining electrons to form an anion.

▶ Write the electron configurations of ions.

▶ Predict relative sizes of ions in an isoelectronic series.

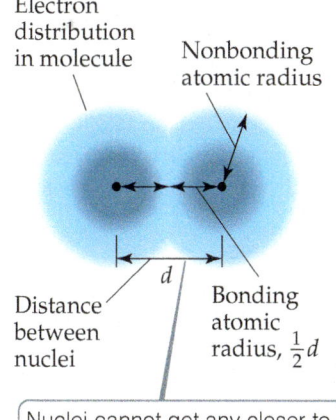

▲ Figure 7.6 **Distinction between nonbonding and bonding atomic radii within a molecule.**

Go Figure In which part of the periodic table (top/bottom, left/right) do you find the elements with the largest radii?

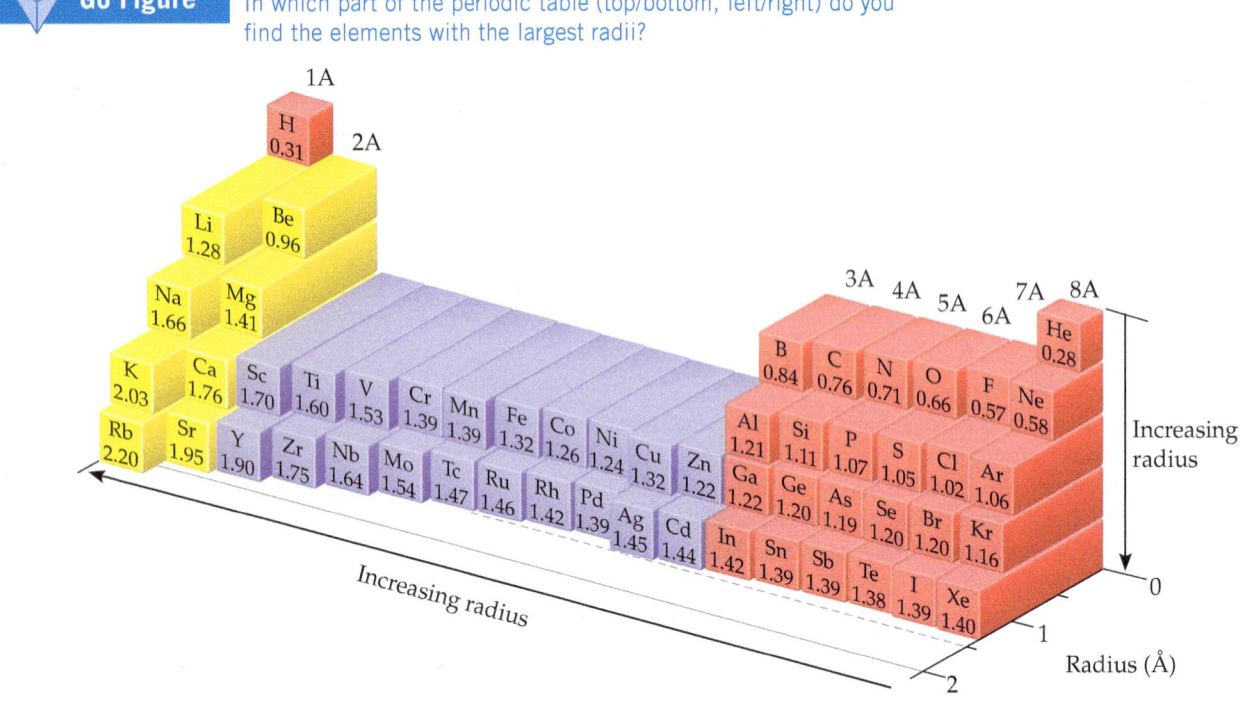

▲ **Figure 7.7 Trends in bonding atomic radii for periods 1 through 5.**

Sample Exercise 7.1
Bond Lengths in a Molecule

Natural gas used in home heating and cooking is odorless. Because natural gas leaks pose the danger of explosion or suffocation, various smelly substances are added to the gas to allow detection of a leak. One such substance is methyl mercaptan, CH_3SH. Use Figure 7.7 to predict the lengths of the C—S, C—H, and S—H bonds in this molecule.

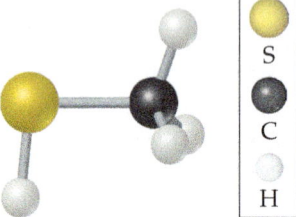

Methyl mercaptan

SOLUTION

Analyze and Plan We are given three bonds and are told to use Figure 7.7 for bonding atomic radii. We will assume that each bond length is the sum of the bonding atomic radii of the two atoms involved.

Solve

C—S bond length = bonding atomic radius of C
　　　　　　　　　+ bonding atomic radius of S
　　　　　　　　= 0.76 Å + 1.05 Å = 1.81 Å

C—H bond length = 0.76 Å + 0.31 Å = 1.07 Å

S—H bond length = 1.05 Å + 0.31 Å = 1.36 Å

Check You can look up experimentally determined bond lengths: C—S = 1.82 Å, C—H = 1.10 Å, and S—H = 1.33 Å. (In general, the lengths of bonds involving hydrogen show larger deviations from the values predicted from bonding atomic radii than do bonds involving larger atoms.)

Comment Our estimated bond lengths are close but not exact matches to the measured bond lengths. Because they are determined by averaging over many substances, bonding atomic radii must be used with some caution in estimating bond lengths.

▶ **Practice Exercise**
Using Figure 7.7, predict which is longer, the P—Br bond in PBr_3 or the As—Cl bond in $AsCl_3$.

Periodic Trends in Atomic Radii

Figure 7.7 shows two interesting trends:

1. **Within each group, bonding atomic radius tends to increase from top to bottom**. This trend results primarily from the increase in the principal quantum number (n) of the outer electrons. As we go down a column, the outer electrons have a greater probability of being farther from the nucleus, causing the atomic radius to increase.

2. **Within each period, bonding atomic radius tends to decrease from left to right** (although there are some minor exceptions, such as for Cl to Ar or As to Se). The major factor influencing this trend is the increase in Z_{eff} across a period. The increasing effective nuclear charge steadily draws the valence electrons closer to the nucleus, causing the bonding atomic radius to decrease.

Sample Exercise 7.2
Predicting Relative Sizes of Atomic Radii

Referring to the periodic table, arrange (as much as possible) the atoms B, C, Al, and Si in order of increasing size.

SOLUTION

Analyze and Plan We are given the chemical symbols for four elements and told to use their relative positions in the periodic table to predict the relative size of their atomic radii. We can use the two periodic trends described in the text to help with this problem.

Solve

B and C are in the same period, with C to the right of B. Therefore, we expect the radius of C to be smaller than that of B because radii usually decrease as we move left to right across a period.	radius C < radius B
Al and Si are in the same period, with Si to the right of Al.	radius Si < radius Al
The radius increases as we move down a group with Al and B belonging to the same group, as do C and Si.	radius B < radius Al radius C < radius Si
Combining these comparisons, we can conclude that C has the smallest radius and Al the largest. The two periodic trends available to us do not supply enough information to determine the relative sizes of B and Si.	radius C < radius B ~ radius Si < radius Al

Check Referring back to Figure 7.7, we can obtain numerical values for each atomic radius that allows us to say that the radius of Si is greater than that of B.

$$C(0.76 \text{ Å}) < B (0.84 \text{ Å}) < Si (1.11 \text{ Å}) < Al (1.21 \text{ Å})$$

If you examine Figure 7.7 carefully, you will discover that for the *s*- and *p*-block elements the increase in radius moving one element down a column tends to be greater than the increase moving one element left across a row. There are exceptions, however.

Comment The trends we have just discussed are for the *s*- and *p*-block elements. As seen in Figure 7.7, the transition elements do not show a regular decrease moving across a period.

▶ **Practice Exercise**
Arrange Be, C, K, and Ca in order of increasing atomic radius.

Periodic Trends in Ionic Radii

Just as bonding atomic radii can be determined from interatomic distances in molecules, ionic radii can be determined from interatomic distances in ionic compounds. Like the size of an atom, the size of an ion depends on its nuclear charge, the number of electrons it possesses, and the orbitals in which the valence electrons reside. When a cation is formed from a neutral atom, electrons are removed from the occupied atomic orbitals that are the most spatially extended from the nucleus. Also, when a cation is formed, the number of electron–electron repulsions is reduced. Therefore, *cations are smaller than*

Go Figure How do cations of the same charge change in radius as you move down a column in the periodic table?

	Group 1A	Group 2A	Group 3A	Group 6A	Group 7A
	Li⁺ 0.90	Be²⁺ 0.59	B³⁺ 0.41	O²⁻ 1.26	F⁻ 1.19
	Li 1.28	Be 0.96	B 0.84	O 0.66	F 0.57
	Na⁺ 1.16	Mg²⁺ 0.86	Al³⁺ 0.68	S²⁻ 1.70	Cl⁻ 1.67
	Na 1.66	Mg 1.41	Al 1.21	S 1.05	Cl 1.02
	K⁺ 1.52	Ca²⁺ 1.14	Ga³⁺ 0.76	Se²⁻ 1.84	Br⁻ 1.82
	K 2.03	Ca 1.76	Ga 1.22	Se 1.20	Br 1.20
	Rb⁺ 1.66	Sr²⁺ 1.32	In³⁺ 0.94	Te²⁻ 2.07	I⁻ 2.06
	Rb 2.20	Sr 1.95	In 1.42	Te 1.38	I 1.39

⬤ = cation   ⬤ = anion   ⬤ = neutral atom

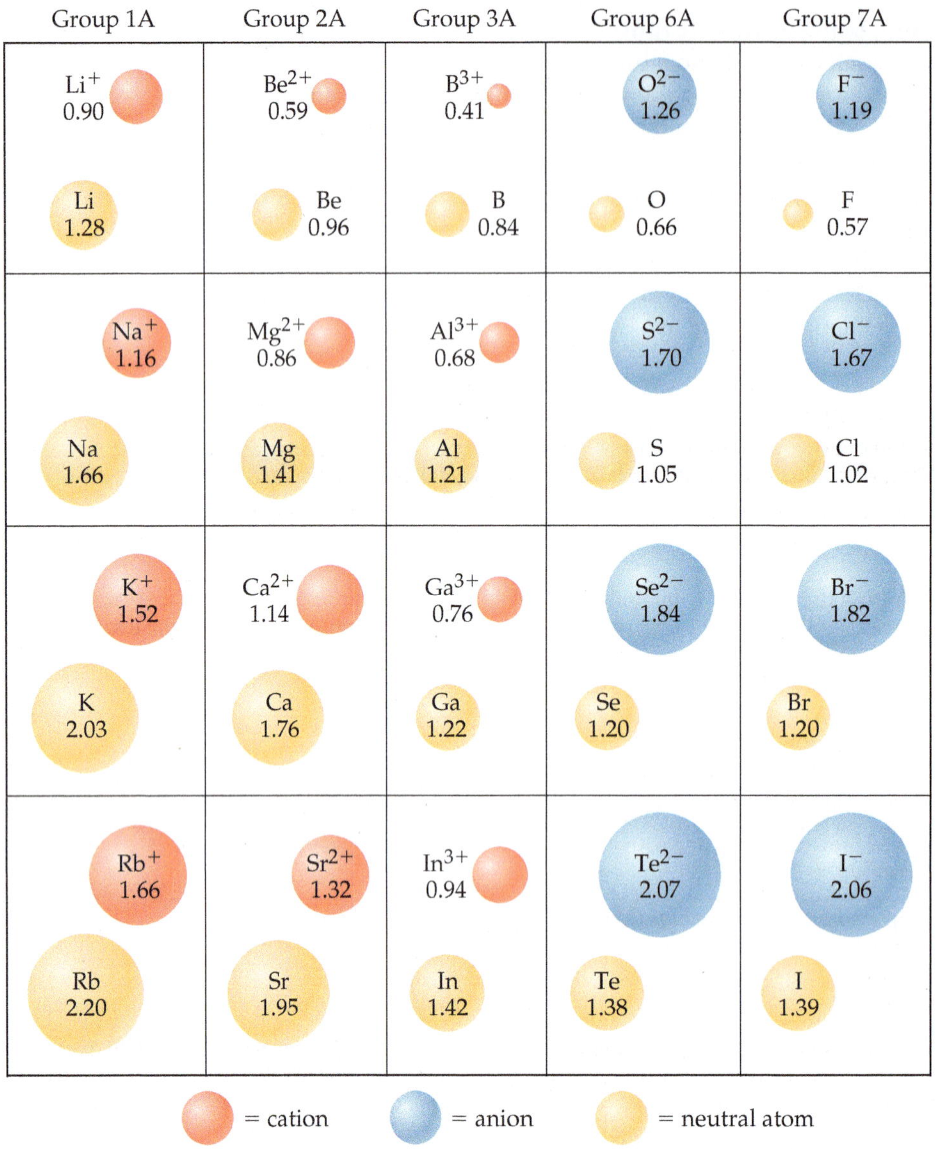

▲ **Figure 7.8 Cation and anion size.** Radii, in angstroms, of atoms and their ions for five groups of representative elements.

their parent atoms (**Figure 7.8**). The opposite is true of anions. When electrons are added to an atom to form an anion, the increased electron–electron repulsions cause the electrons to spread out more in space. Thus, *anions are larger than their parent atoms.*

For ions carrying the same charge, ionic radius increases as we move down a column in the periodic table (Figure 7.8). In other words, as the principal quantum number of the outermost occupied orbital of an ion increases, the radius of the ion increases.

Electron Configurations of Ions

When electrons are removed from an atom to form a cation, they are always removed first from the occupied orbitals having the largest principal quantum number, n. For example, when one electron is removed from a lithium atom ($1s^2 2s^1$), it is the $2s^1$ electron:

$$\text{Li}(1s^2 2s^1) \rightarrow \text{Li}^+(1s^2) + e^-$$

Likewise, when two electrons are removed from $Fe([Ar]4s^23d^6)$, the $4s^2$ electrons are the ones removed:

$$Fe([Ar]4s^23d^6) \Rightarrow Fe^{2+}([Ar]3d^6) + 2e^-$$

If an additional electron is removed, forming Fe^{3+}, it comes from a $3d$ orbital because all the orbitals with $n = 4$ are empty:

$$Fe^{2+}([Ar]3d^6) \Rightarrow Fe^{3+}([Ar]3d^5) + e^-$$

It may seem odd that $4s$ electrons are removed before $3d$ electrons in forming transition-metal cations. After all, in writing electron configurations, we added the $4s$ electrons before the $3d$ electrons. In writing electron configurations for atoms, however, we are going through an imaginary process in which we move through the periodic table from one element to another. In doing so, we are adding both an electron to an orbital and a proton to the nucleus to change the identity of the element. In ionization, we *do not* reverse this process because no protons are being removed. For example, both Ca and Ti^{2+} have 20 electrons, but a Ti^{2+} ion has more protons than a Ca atom (22 vs. 20). That changes the relative energy levels of the orbitals enough that the two species have different electron configurations: $Ca([Ar]4s^2)$ and $Ti^{2+}([Ar]3d^2)$.

If there is more than one occupied subshell for a given value of n, the electrons are first removed from the orbital with the highest value of l. For example, a tin atom loses its $5p$ electrons before it loses its $5s$ electrons:

$$Sn([Kr]5s^24d^{10}5p^2) \Rightarrow Sn^{2+}([Kr]5s^24d^{10}) + 2e^- \Rightarrow Sn^{4+}([Kr]4d^{10}) + 4e^-$$

Electrons added to an atom to form an anion are added to the empty or partially filled orbital having the lowest value of n. For example, an electron added to a fluorine atom to form the F^- ion goes into the one remaining vacancy in the $2p$ subshell:

$$F(1s^22s^22p^5) + e^- \Rightarrow F^-(1s^22s^22p^6)$$

An **isoelectronic series** is a group of ions that all contain the same number of electrons. For example, each ion in the isoelectronic series O^{2-}, F^-, Na^+, Mg^{2+}, and Al^{3+} has 10 electrons. In any isoelectronic series, we can list the members in order of increasing atomic number; therefore, the nuclear charge increases as we move through the series. Because the number of electrons remains constant, ionic radius decreases with increasing nuclear charge as the electrons are more strongly attracted to the nucleus:

Sample Exercise 7.3
Electron Configurations of Ions

Write the electron configurations for (**a**) Ca^{2+}, (**b**) Co^{3+}, and (**c**) S^{2-}.

SOLUTION

Analyze and Plan We are asked to write electron configurations for three ions. To do so, we first write the electron configuration of each parent atom and then remove or add electrons to form the ions. Electrons are first removed from the orbitals having the highest value of n. They are added to the empty or partially filled orbitals having the lowest value of n.

Solve

(**a**) Calcium (atomic number 20) has the electron configuration $[Ar]4s^2$. To form a 2+ ion, the two $4s$ electrons must be removed, giving an ion that is isoelectronic with Ar:

$$Ca^{2+}: [Ar]$$

(**b**) Cobalt (atomic number 27) has the electron configuration $[Ar]4s^23d^7$. To form a 3+ ion, three electrons must be removed. As discussed in the text, the $4s$ electrons are removed before the $3d$ electrons. Consequently, we remove the two $4s$ electrons and one of the $3d$ electrons, so the electron configuration for Co^{3+} is

$$Co^{3+}: [Ar]3d^6$$

(**c**) Sulfur (atomic number 16) has the electron configuration $[Ne]3s^23p^4$. To form a 2– ion, two electrons must be added. There is room for two additional electrons in the $3p$ orbitals. Thus, the S^{2-} electron configuration is

$$S^{2-}: [Ne]3s^23p^6 = [Ar]$$

Comment Remember that many of the common ions of the *s*- and *p*-block elements, such as Ca^{2+} and S^{2-}, have the same number of electrons as the closest noble gas. (Section 2.7)

▶ **Practice Exercise**
Write the electron configurations for (**a**) Ga^{3+}, (**b**) Cr^{3+}, and (**c**) Br^-.

Increasing nuclear charge ⟶				
8 protons	9 protons	11 protons	12 protons	13 protons
10 electrons	10 electrons	10 electrons	10 electrons	10 electrons
O^{2-}	F^-	Na^+	Mg^{2+}	Al^{3+}
1.26 Å	1.19 Å	1.16 Å	0.86 Å	0.68 Å

Decreasing ionic radius ⟶

Notice the positions and atomic numbers of these elements in the periodic table. The nonmetal anions precede the noble gas Ne in the table. The metal cations follow Ne. Oxygen, the largest ion in this isoelectronic series, has the lowest atomic number, 8. Aluminum, the smallest of these ions, has the highest atomic number, 13.

CHEMISTRY AND SUSTAINABILITY | Ionic Size and Lithium-Ion Batteries

Sustainable energy sources are those that do not rely on the combustion of fossil fuels; as we know, fossil fuels will not last forever, and the CO_2 emissions they produce lead to climate change. Solar energy is in many ways the perfect energy source for built structures: It arrives on Earth every day and it is free. Harnessing solar energy while the Sun shines is possible with solar panels, but there needs to be a way to store the energy for later use. Likewise, cars, trains, and buses at present mostly use the internal combustion engine to power their motion; electric vehicles would be one solution to avoid fossil fuel combustion. Solutions for the solar energy storage problem and for powering electric vehicles include batteries. Ionic size plays a major role in determining the properties of devices, such as batteries, that rely on the movement of ions. Lithium-ion batteries, which have become common energy sources for electronic devices, such as cell phones, iPads, laptop computers, and electric vehicles, rely in part on the small size of the lithium ion for their operation.

A fully charged battery spontaneously produces an electric current and, therefore, power when its positive and negative electrodes are connected to an electrical load, such as a device to be powered. The positive electrode is called the anode, and the negative electrode is called the cathode. The materials used for the electrodes in lithium-ion batteries are under intense development. Currently, the anode material is graphite, a form of carbon, and the cathode is a transition metal oxide, often lithium cobalt oxide, $LiCoO_2$ (**Figure 7.9**). Between anode and cathode is a *separator*, a porous solid material that allows lithium ions to pass through but not electrons.

When the battery is being charged by an external source, lithium ions migrate through the separator from the cathode to the anode where they insert between the layers of carbon atoms. The ability of an ion to move through a solid increases as the size of the ion decreases and as the charge on the ion decreases. Lithium ions are smaller than most other cations, and they carry only a 1+ charge, which allows them to migrate more readily than other ions can. As an added bonus, lithium is one of the lightest elements, which is attractive for use in electric vehicles. When the battery discharges (which is when it is delivering power), the lithium ions move from anode to cathode. To maintain charge balance, electrons simultaneously migrate from anode to cathode through an external circuit, thereby producing electricity.

At the cathode, lithium ions then insert in the oxide material. Again, the small size of lithium ions is an advantage. For every lithium ion that inserts into the lithium cobalt oxide cathode, a Co^{4+} ion is reduced to a Co^{3+} ion by an electron that has traveled through the external circuit.

The ion migration and the changes in structure that result when lithium ions enter and leave the electrode materials are complicated. Furthermore, the operation of all batteries generates heat because they are not perfectly efficient. In the case of Li-ion batteries, the heating of the separator material (typically a polymer) has led to problems as the size of the batteries has been scaled larger to increase energy capacity. In a very small number of cases, overheating of Li-ion batteries has caused the batteries to catch fire.

Teams worldwide are trying to discover new cathode and anode materials that will easily accept and release lithium ions without falling apart over many repeated cycles, with an eye toward Earth-abundant elements. New separator materials that allow for faster passage of lithium ions with less heat generation are also under development. Some research groups are looking at using sodium ions instead of lithium ions because sodium is far more abundant than lithium. This makes sodium a more sustainable choice of ion, although the larger size of sodium ions poses additional challenges. In the coming years, look for continued advances in battery technology based on alkali metal ions and improvements in lithium-ion battery recycling.

Related Exercise: 7.87

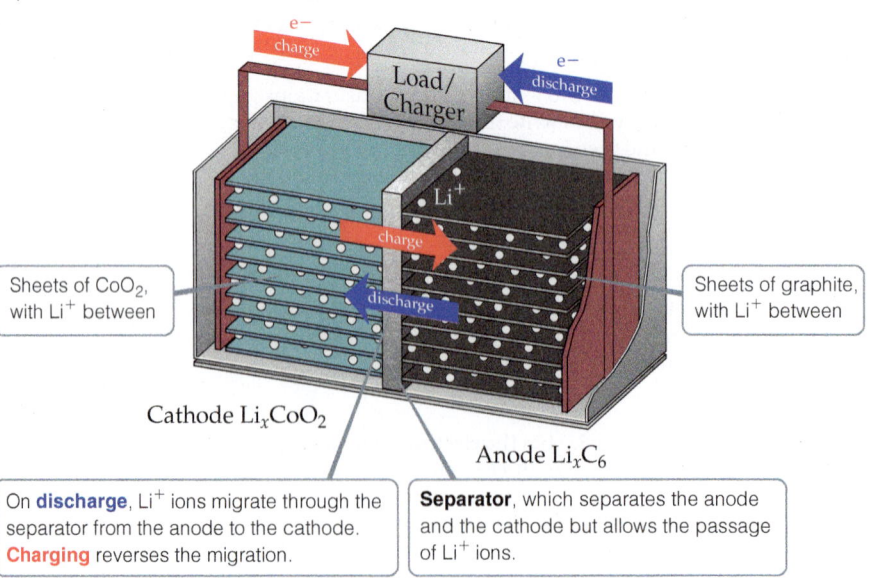

Sheets of CoO_2, with Li^+ between

Sheets of graphite, with Li^+ between

Cathode Li_xCoO_2

Anode Li_xC_6

On **discharge**, Li^+ ions migrate through the separator from the anode to the cathode. **Charging** reverses the migration.

Separator, which separates the anode and the cathode but allows the passage of Li^+ ions.

▲ Figure 7.9 Schematic of a lithium-ion battery.

Sample Exercise 7.4

Ionic Radii in an Isoelectronic Series

Arrange the ions K^+, Cl^-, Ca^{2+}, and S^{2-} in order of decreasing size.

SOLUTION

This is an isoelectronic series, with all ions having 18 electrons. In such a series, size decreases as nuclear charge (atomic number) increases. The atomic numbers of the ions are S 16, Cl 17, K 19, and Ca 20. Thus, the ions decrease in size in the order $S^{2-} > Cl^- > K^+ > Ca^{2+}$.

▶ **Practice Exercise**
In the isoelectronic series Ca^{2+}, Cs^+, Y^{3+}, which ion is largest?

Self-Assessment Exercises

SAE 7.5 Use Figure 7.7 to predict which chemical bond is the longest. (**a**) H—Cl (**b**) C—C (**c**) Ge—S (**d**) B—Br

SAE 7.6 Which corner of the periodic table has the atoms with the largest bonding atomic radii? (**a**) upper left (**b**) upper right (**c**) bottom left (**d**) bottom right

SAE 7.7 Which of these ions has a different electron configuration than the others? (**a**) S^{2-} (**b**) K^+ (**c**) Ga^{3+} (**d**) Ca^{2+}

SAE 7.8 Arrange Mg^{2+}, Ca^{2+}, and Ca in order of decreasing radius.

(**a**) $Mg^{2+} > Ca^{2+} > Ca$

(**b**) $Ca > Ca^{2+} > Mg^{2+}$

(**c**) $Ca^{2+} > Ca > Mg^{2+}$

(**d**) $Ca > Mg^{2+} > Ca^{2+}$

SAE 7.9 Which ionic compound consists of ions with the largest difference in ionic size? (**a**) NaF (**b**) KCl (**c**) RbBr (**d**) CsF (**e**) CsI

SAE 7.10 Which statement about isoelectronic ions is *false*? (**a**) Na^+ is isoelectronic to Li^+. (**b**) S^{2-} is isoelectronic to Cl^-. (**c**) Anions can be isoelectronic to cations. (**d**) Trends in ionic size for isoelectronic species follow trends in effective nuclear charge.

7.4 | Ionization Energy and Electron Affinity

The ease with which electrons can be removed from an atom or ion has a major impact on chemical behavior. The **ionization energy** of an atom or ion is the minimum energy required to remove an electron from the ground state of the isolated gaseous atom or ion. We first encountered ionization in our discussion of the Bohr model of the hydrogen atom. (Section 6.3) If the electron in an H atom is excited from $n = 1$ (the ground state) to $n = \infty$, the electron is completely removed from the atom; the atom is *ionized*. The atoms of elements with low ionization energies are likely to form cations when they make compounds. Conversely, most atoms can accept electrons in the gas phase to become anions; the energy change for this process is called the **electron affinity** of the atom. Atoms of elements with low (even negative) electron affinities are likely to form anions when they make compounds.

In general, the *first ionization energy*, I_1, is the energy needed to remove the first electron from a neutral gas-phase atom (**Figure 7.10**). As we saw in Section 6.3 for the H atom, this corresponds to exciting the electron to $n = \infty$. For example, the first ionization energy for the sodium atom is the energy required for the process

$$Na(g) \longrightarrow Na^+(g) + e^- \qquad [7.2]$$

Learning Objectives

When you finish Section 7.4, you should be able to:

▶ Write the chemical equations that relate to ionization energies.

▶ Predict periodic trends in ionization energy.

▶ Write the chemical equations that relate to electron affinities.

▶ Recognize trends in electron affinity.

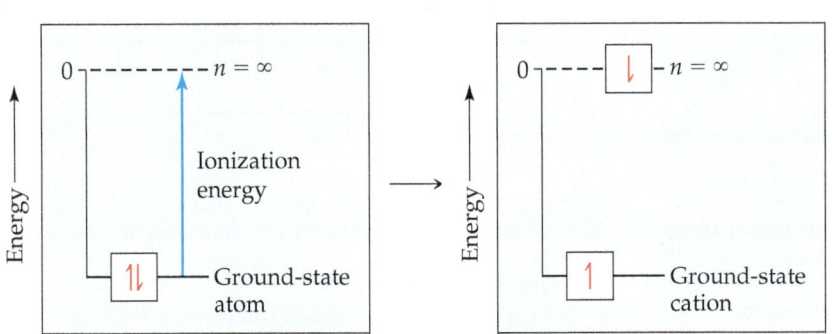

◀ Figure 7.10 **The first ionization energy (I_1) of a generic atom.** I_1 is the minimum energy required to eject an electron from the indicated orbital, leaving a cation behind.

TABLE 7.2   Successive Values of Ionization Energies, I, for the Elements Sodium through Argon (kJ/mol)

Element	I_1	I_2	I_3	I_4	I_5	I_6	I_7
Na	496	4562			(inner-shell electrons)		
Mg	738	1451	7733				
Al	578	1817	2745	11,577			
Si	786	1577	3232	4356	16,091		
P	1012	1907	2914	4964	6274	21,267	
S	1000	2252	3357	4556	7004	8496	27,107
Cl	1251	2298	3822	5159	6542	9362	11,018
Ar	1521	2666	3931	5771	7238	8781	11,995

The *second ionization energy*, I_2, is the energy needed to remove the second electron, and so forth, for successive removals of additional electrons. Thus, I_2 for the sodium atom is the energy associated with the process

$$Na^+(g) \longrightarrow Na^{2+}(g) + e^- \qquad [7.3]$$

Variations in Successive Ionization Energies

The magnitude of the ionization energy indicates how much energy is required to remove an electron; the greater the ionization energy, the more difficult it is to remove an electron. Table 7.2 presents the successive values for the ionization energies of Na through Ar. Notice that ionization energies for a given element increase as successive electrons are removed: $I_1 < I_2 < I_3$, and so forth. This trend should make sense because with each successive removal, an electron is being pulled away from an increasingly positive ion, which requires increasingly more energy.

A second important feature shown in Table 7.2 is the sharp increase in ionization energy that occurs when an inner-shell electron is removed. For silicon ($1s^2 2s^2 2p^6 3s^2 3p^2$), for example, the ionization energies increase steadily from 786 to 4356 kJ/mol for the four electrons in the 3s and 3p subshells. Removal of the fifth electron, which comes from the 2p subshell, requires a great deal more energy: 16,091 kJ/mol. The large increase occurs because the 2p electron is much more likely to be found close to the nucleus than are the four $n = 3$ electrons, and, therefore, the 2p electron experiences a much greater effective nuclear charge than do the 3s and 3p electrons.

Every element exhibits a large increase in ionization energy when the first of its inner-shell electrons is removed. This observation supports the idea that only the outermost electrons are involved in the sharing and transfer of electrons that give rise to chemical bonding and reactions. As we show when we talk about chemical bonds in Chapters 8 and 9, the inner electrons are too tightly bound to the nucleus to be lost from the atom or even shared with another atom.

Sample Exercise 7.5
Trends in Ionization Energy

Three elements are indicated in this periodic table. Which one has the largest *second* ionization energy?

SOLUTION

Analyze and Plan The locations of the elements in the periodic table allow us to predict the electron configurations. The greatest ionization energies involve removal of core electrons. Thus, we should look first for an element with only one electron in the outermost occupied shell.

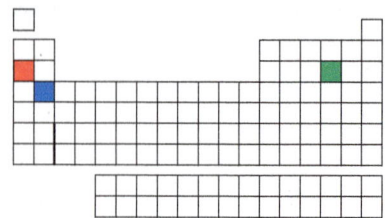

Solve The red box represents Na, which has one valence electron. The second ionization energy of this element is associated, therefore, with the removal of a core electron. The other elements indicated, S (green) and Ca (blue), have two or more valence electrons. Thus, Na should have the largest second ionization energy.

Check A chemistry handbook gives the following I_2 values: Ca, 1145 kJ/mol; S, 2252 kJ/mol; Na, 4562 kJ/mol.

▶ **Practice Exercise**
Which has the greater third ionization energy, Ca or S?

Periodic Trends in First Ionization Energies

Figure 7.11 shows, for the first 54 elements, the trends we observe in first ionization energies as we move from one element to another in the periodic table. The important trends are as follows:

1. **I_1 generally increases as we move left to right across a period.** The alkali metals show the lowest ionization energy in each period, and the noble gases show the highest. There are slight irregularities in this trend that we discuss shortly.

2. **I_1 generally decreases as we move down any column in the periodic table.** For example, the ionization energies of the noble gases follow the order He > Ne > Ar > Kr > Xe.

3. **The s- and p-block elements show a larger range of I_1 values than do the transition-metal elements.** Generally, the ionization energies of the transition metals increase slowly from left to right in a period. The f-block metals (not shown in Figure 7.10) also show only a small variation in the values of I_1.

In general, smaller atoms have higher ionization energies. The same factors that influence atomic size also influence ionization energies. The energy needed to remove an electron from the outermost occupied shell depends on both the effective nuclear charge and the average distance of the electron from the nucleus. Either increasing the effective nuclear charge or decreasing the distance from the nucleus increases the attraction between the electron and the nucleus. As this attraction increases, it becomes more difficult to remove the electron, and, thus, the ionization energy increases. As we move left to right across a period, there is both an increase in effective nuclear charge and a decrease in atomic radius, causing the ionization energy to increase. As we move down a

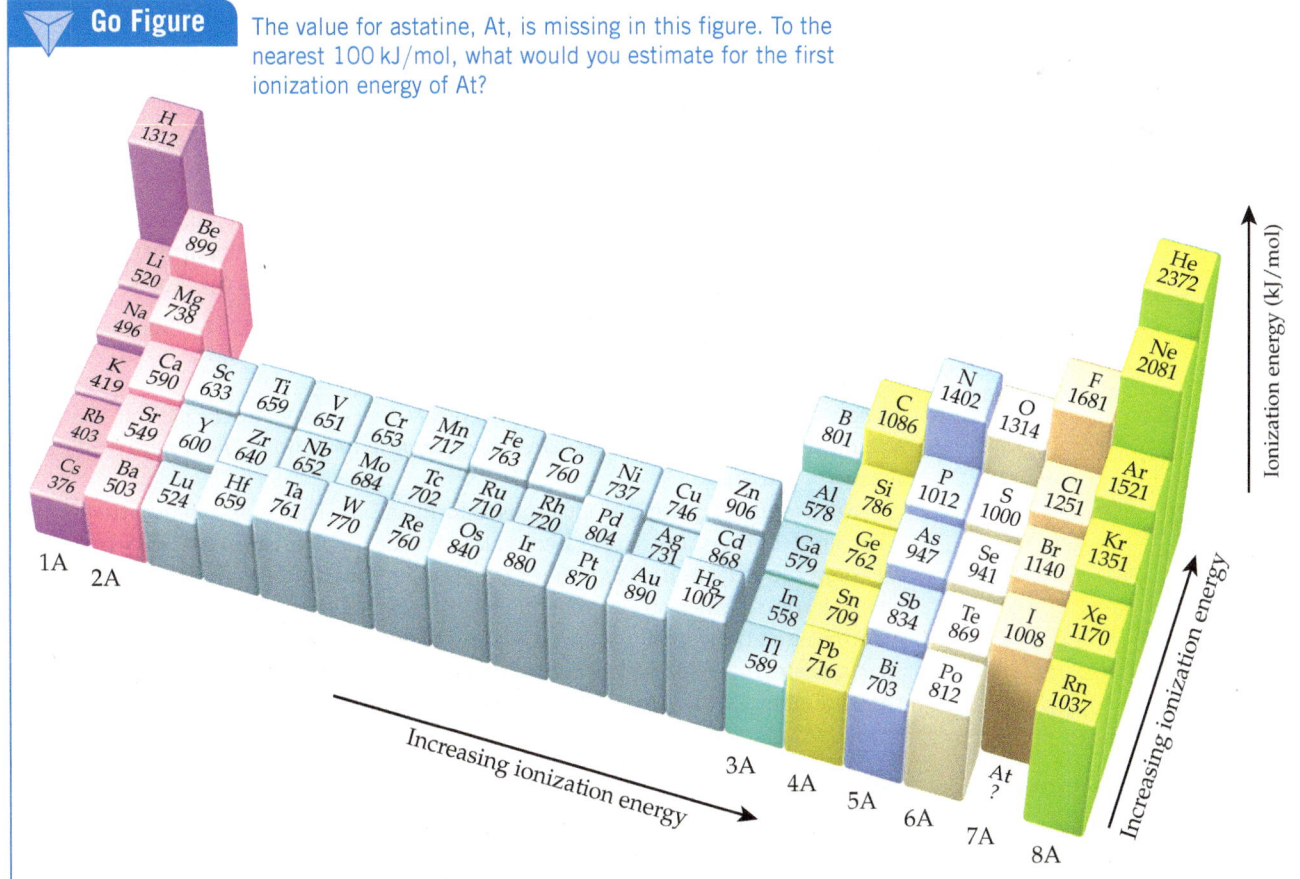

▼ **Go Figure** The value for astatine, At, is missing in this figure. To the nearest 100 kJ/mol, what would you estimate for the first ionization energy of At?

▲ **Figure 7.11** The first ionization energies of the elements in kJ/mol.

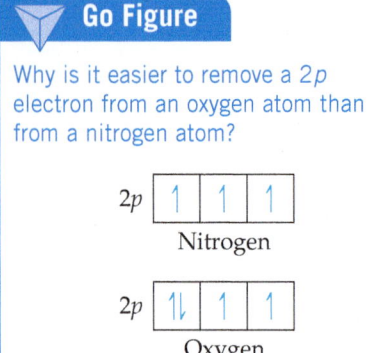

▽ **Go Figure**

Why is it easier to remove a 2*p* electron from an oxygen atom than from a nitrogen atom?

2*p* | ↑ | ↑ | ↑ |
Nitrogen

2*p* | ↑↓ | ↑ | ↑ |
Oxygen

▲ **Figure 7.12 2*p* orbital filling in nitrogen and oxygen.**

column, the atomic radius increases, while the effective nuclear charge increases only gradually. The increase in radius dominates, so the attraction between the nucleus and the electron decreases, causing the ionization energy to decrease.

The irregularities in a given period are subtle but still readily explained. For example, the decrease in ionization energy from beryllium ([He]$2s^2$) to boron ([He]$2s^22p^1$), shown in Figure 7.11, occurs because the third valence electron of B must occupy the 2*p* subshell, which is empty for Be. Recall that the 2*p* subshell is at a higher energy than the 2*s* subshell (Figure 6.23). The slight decrease in ionization energy when moving from nitrogen ([He]$2s^22p^3$) to oxygen ([He]$2s^22p^4$) is the result of the repulsion of paired electrons in the p^4 configuration (**Figure 7.12**). Remember that according to Hund's rule, each electron in the p^3 configuration resides in a different *p* orbital, which minimizes the electron–electron repulsion among the three 2*p* electrons. (Section 6.8)

Sample Exercise 7.6
Periodic Trends in Ionization Energy

Referring to the periodic table, arrange the atoms Ne, Na, P, Ar, K in order of increasing first ionization energy.

SOLUTION

Analyze and Plan We are given the chemical symbols for five elements. To rank them according to increasing first ionization energy, we need to locate each element in the periodic table. We can then use their relative positions and the trends in first ionization energies to predict their order.

Solve Ionization energy increases as we move from left to right across a period and decreases as we move down a group. Because Na, P, and Ar are in the same period, we expect I_1 to vary in the order Na $<$ P $<$ Ar. Because Ne is above Ar in group 8A, we expect Ar $<$ Ne. Similarly, K is directly below Na in group 1A, and so we expect K $<$ Na.

From these observations, we conclude that the ionization energies follow the order

$$\text{K} < \text{Na} < \text{P} < \text{Ar} < \text{Ne}$$

Check The values shown in Figure 7.11 confirm this prediction.

▶ **Practice Exercise**
Which has the lowest first ionization energy—B, Al, C, or Si? Which has the highest?

Electron Affinity

The first ionization energy of an atom is a measure of the energy change associated with removing an electron from the atom to form a cation. For example, the first ionization energy of Cl(*g*), 1251 kJ/mol, is the energy change associated with the process

$$\textit{Ionization energy:} \quad \text{Cl}(g) \longrightarrow \text{Cl}^+(g) + \text{e}^- \qquad \Delta E = 1251 \text{ kJ/mol} \qquad [7.4]$$
$$[\text{Ne}]3s^23p^5 \qquad [\text{Ne}]3s^23p^4$$

The positive value of the ionization energy means that energy must be put into the atom to remove the electron. *All ionization energies for atoms are positive: Energy must be absorbed to remove an electron.*

Most atoms can also gain electrons to form anions. The energy change that occurs when an electron is added to a gaseous atom is called the **electron affinity** because it measures the attraction, or *affinity*, of the atom for the added electron. For most atoms, energy is *released* when an electron is added. For example, the addition of an electron to a chlorine atom is accompanied by an energy change of −349 kJ/mol, where the negative sign indicates that energy is released during the process and that the anion is more stable than the atom. We therefore say that the electron affinity of Cl is −349 kJ/mol.*

---

*Two sign conventions are used for electron affinity. In most introductory texts, including this one, the thermodynamic sign convention is used: A negative sign indicates that addition of an electron is an exothermic process, as in the electron affinity for chlorine, −349 kJ/mol. Historically, however, electron affinity has been defined as the energy released when an electron is added to a gaseous atom or ion. Because 349 kJ/mol is released when an electron is added to Cl(*g*), the electron affinity by this convention would be +349 kJ/mol.

Electron affinity: $Cl(g) + e^- \longrightarrow Cl^-(g)$ $EA = -349 \text{ kJ/mol}$ [7.5]

$[Ne]3s^23p^5$ $[Ne]3s^23p^6$

Be sure you understand the difference between ionization energy and electron affinity:

- Ionization energy measures the energy change when an atom *loses* an electron.
- Electron affinity measures the energy change when an atom *gains* an electron.

The greater the attraction between an atom and an added electron, the more negative the atom's electron affinity. For some elements, such as the noble gases, the electron affinity has a positive value, meaning that the anion is higher in energy than are the separated atom and electron:

$$Ar(g) + e^- \longrightarrow Ar^-(g)   EA > 0   [7.6]$$

$[Ne]3s^23p^6$ $[Ne]3s^23p^64s^1$

The fact that the electron affinity is positive means that an electron will not attach itself to an Ar atom; in other words, the Ar^- ion is unstable and does not form.

Periodic Trends in Electron Affinity

The electron affinities for the *s*- and *p*-block elements of the first five periods are shown in **Figure 7.13**. Notice that the trends are not as evident as they are for ionization energy. The halogens, which are one electron shy of a filled *p* subshell, have the most negative electron affinities. By gaining an electron, a halogen atom forms a stable anion that has a noble-gas configuration (Equation 7.5). The addition of an electron to a noble gas, however, requires that the electron reside in a higher-energy subshell that is empty in the atom (Equation 7.6). Because occupying a higher-energy subshell is energetically unfavorable, the electron affinity is highly positive. The electron affinities of Be and Mg are positive for the same reason; the added electron would reside in a previously empty *p* subshell that is higher in energy.

The electron affinities of the group 5A elements also stand out. Because these elements have half-filled *p* subshells, the added electron must be put in an orbital that is already occupied, resulting in larger electron–electron repulsions. Consequently, these elements have electron affinities that are either positive (N) or less negative than those of their neighbors to the left (P, As, Sb). Recall from Section 7.4 that there was a discontinuity in the trends in first ionization energy for the same reason.

Electron affinities do not change greatly as we move down a group (Figure 7.13). With F, for instance, the added electron goes into a 2*p* orbital, for Cl a 3*p* orbital, for Br a 4*p* orbital, and so forth. As we proceed from F to I, therefore, the average distance between the added electron and the nucleus steadily increases, causing the electron–nucleus attraction to decrease. However, the orbital that holds the outermost electron is increasingly spread out, so that as we proceed from F to I, the electron–electron repulsions are also reduced. As a result, the reduction in the electron–nucleus attraction is counterbalanced by the reduction in electron–electron repulsions.

Go Figure

Why are the electron affinities of the Group 4A elements more negative than those of the Group 5A elements?

1A							8A
H -73	2A	3A	4A	5A	6A	7A	He >0
Li -60	Be >0	B -27	C -122	N >0	O -141	F -328	Ne >0
Na -53	Mg >0	Al -43	Si -134	P -72	S -200	Cl -349	Ar >0
K -48	Ca -2	Ga -30	Ge -119	As -78	Se -195	Br -325	Kr >0
Rb -7	Sr -5	In -30	Sn -107	Sb -103	Te -190	I -295	Xe >0

▲ **Figure 7.13** Electron affinity in kJ/mol for selected *s*- and *p*-block elements.

Self-Assessment Exercises

SAE 7.11 Which reaction corresponds to the *second* ionization energy of generic element A?

(a) $A(g) + e^- \longrightarrow A^-(g)$ **(d)** $A^+(g) \longrightarrow A^{2+}(g) + e^-$

(b) $2A(g) \longrightarrow A_2^{2+}(g) + 2e^-$ **(e)** $A(g) \longrightarrow A^{2+}(g) + 2e^-$

(c) $A(g) \longrightarrow A^+(g) + e^-$

SAE 7.12 Element X has successive ionization energies of 738 kJ/mol, 1451 kJ/mol, and 7733 kJ/mol, whereas element Z has successive ionization energies of 1000 kJ/mol, 2252 kJ/mol, and 3357 kJ/mol.

What are the possible identities of X and Z? **(a)** X could be P because the third ionization energy is so large. Z could be Na because its ionization energies keep going up in regular intervals. **(b)** X could be Mg because its third ionization energy is so large. Z could be S because its ionization energies are large and keep going up in regular intervals. **(c)** X could be Fe because its third ionization energy is so large. Z could be H because its ionization energies keep going up in regular intervals. **(d)** X could be Na because the second and third ionization energies are so much larger than the first one. Z could be a noble gas because its ionization energies are large and keep going up in regular intervals.

Continued

SAE 7.13   A certain element X has an electron affinity that is negative. The equation for the electron affinity of X is _____ and the process is _____.

(**a**) $X(g) + e^- \longrightarrow X^-(g)$, exothermic

(**b**) $X(g) + e^- \longrightarrow X^-(g)$, endothermic

(**c**) $X(g) \longrightarrow X^+(g) + e^-$, exothermic

(**d**) $X(g) \longrightarrow X^+(g) + e^-$, endothermic

SAE 7.14   According to Figure 7.13, the electron affinities for C, N, and O are -122 kJ/mol, > 0 kJ/mol, and -141 kJ/mol, respectively. Which one of these statements about these data is correct? (**a**) The first ionization energy of N must be smaller than that of C or O. (**b**) The most stable gas-phase anion of these three elements is O$^-$. (**c**) The value for N is anomalous because its ground-state electron configuration is $1s^2 2s^2 2p^3$, so adding an electron to the $3s$ orbital costs a lot of energy. (**d**) The fact that the electron affinity of N is positive means that a N atom will spontaneously lose an electron to form a N$^+$ ion.

7.5 | Metals, Nonmetals, and Metalloids

Learning Objectives

When you finish Section 7.5, you should be able to:

▶ Predict whether elements are metals, nonmetals, or metalloids based on their position in the periodic table.

▶ Differentiate the physical and chemical properties of metals from those of nonmetals.

▶ Predict the products of reactions of metal and nonmetal oxides with water or acid.

Atomic radii, ionization energies, and electron affinities are properties of individual atoms. With the exception of the noble gases, however, none of the elements exist in nature as individual atoms. To get a broader understanding of the properties of elements, we must also examine periodic trends in properties of samples that involve large collections of atoms.

The elements can be broadly grouped as metals, nonmetals, and metalloids (**Figure 7.14**). (Section 2.5) Some of the distinguishing properties of metals and nonmetals are summarized in **Table 7.3**. The more an element exhibits the physical and chemical properties of metals, the greater its **metallic character**. As indicated in Figure 7.14, metallic character generally increases as we proceed down a group of the periodic table and decreases as we proceed from left to right across a period. Let's now examine the close relationships that exist between electron configurations and the properties of metals, nonmetals, and metalloids.

Go Figure   How do the periodic trends in metallic character compare to those for ionization energy?

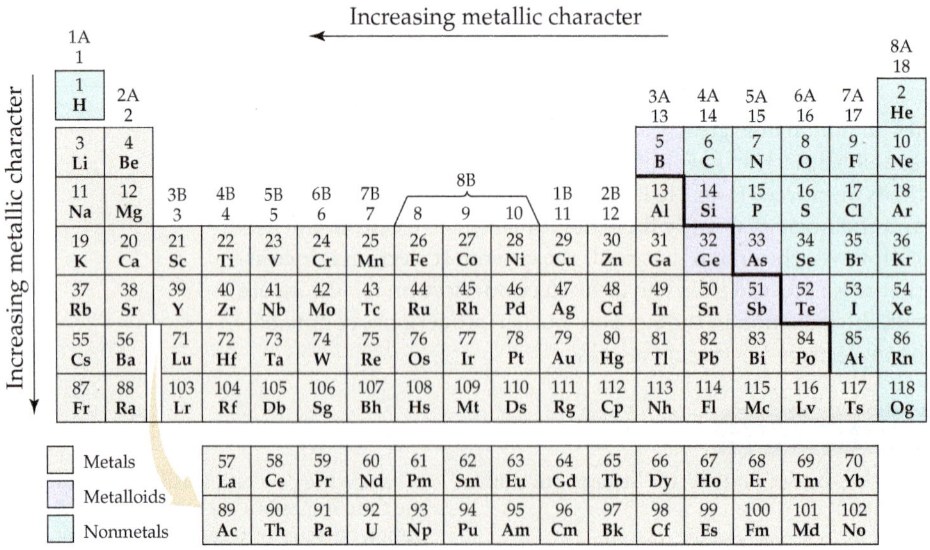

▲ **Figure 7.14 Metals, metalloids, and nonmetals.**

TABLE 7.3 Characteristic Properties of Metals and Nonmetals

Metals	Nonmetals
Have a shiny luster; various colors, although most are silvery	Do not have a luster; various colors
Solids are malleable and ductile	Solids are usually brittle; some are hard, and some are soft
Good conductors of heat and electricity	Poor conductors of heat and electricity
Most metal oxides are ionic solids that are basic	Most nonmetal oxides are molecular substances that form acidic solutions
Tend to form cations in aqueous solution	Tend to form anions or oxyanions in aqueous solution

Metals

Most metallic elements exhibit the shiny luster we associate with metals (**Figure 7.15**). Metals conduct heat and electricity. In general, they are malleable (can be pounded into thin sheets) and ductile (can be drawn into wires). All are solids at room temperature except mercury (melting point = −39 °C), which is a liquid. Cesium (28.4 °C) and gallium (29.8 °C) melt at slightly above room temperature. At the other extreme, many metals melt at very high temperatures. For example, tungsten, which is used for the filaments of incandescent light bulbs, melts at 3422 °C.

Metals tend to have low ionization energies (Figure 7.11), *so they tend to form cations relatively easily.* As a result, metals tend to be oxidized (lose electrons) when they undergo chemical reactions. Among the fundamental atomic properties (radius, electron configuration, electron affinity, and so forth), first ionization energy is the best indicator of whether an element behaves as a metal or a nonmetal.

Figure 7.16 shows the typical oxidation states of ions of some of the metals and nonmetals. As noted in Section 2.7, the charge on any alkali metal ion in a compound is always 1+, and that on any alkaline earth metal is always 2+. For atoms belonging to either of these groups, the outer *s* electrons are easily lost, yielding a noble-gas electron configuration. For metals belonging to groups with partially occupied *p* orbitals (groups 3A–7A), cations are formed either by losing only the outer *p* electrons (such as Sn^{2+}) or the outer *s* and *p* electrons (such as Sn^{4+}). The charge on transition-metal ions does not follow an obvious pattern. One characteristic of the transition metals is their ability to form more than one cation. For example, compounds of Fe^{2+} and Fe^{3+} are both very common.

Compounds made up of a metal and a nonmetal tend to be ionic substances. For example, most metal oxides and halides are ionic solids. To illustrate, the reaction between nickel metal and oxygen produces nickel oxide, an ionic solid containing Ni^{2+} and O^{2-} ions:

$$2\,Ni(s) + O_2(g) \longrightarrow 2\,NiO(s) \qquad [7.7]$$

▲ **Figure 7.15 Metals, such as gold, are shiny, malleable, and ductile.** Gold is so malleable that one troy ounce (31.1 g) of it can be hammered into a sheet that covers more than 9 m².

 **Go Figure** The red stepped line divides metals from nonmetals. How are common oxidation states divided by this line?

1A																	7A	8A
H^+	2A											3A	4A	5A	6A		H^-	N O B L E G A S E S
Li^+														N^{3-}	O^{2-}	F^-		
Na^+	Mg^{2+}			Transition metals								Al^{3+}		P^{3-}	S^{2-}	Cl^-		
K^+	Ca^{2+}	Sc^{3+}	Ti^{4+}	V^{5+} V^{4+}	Cr^{3+}	Mn^{2+} Mn^{4+}	Fe^{2+} Fe^{3+}	Co^{2+} Co^{3+}	Ni^{2+}	Cu^+ Cu^{2+}	Zn^{2+}				Se^{2-}	Br^-		
Rb^+	Sr^{2+}	Y^{3+}	Zr^{4+}						Pd^{2+}	Ag^+	Cd^{2+}	Sn^{2+} Sn^{4+}	Sb^{3+} Sb^{5+}	Te^{2-}	I^-			
Cs^+	Ba^{2+}	Lu^{3+}	Hf^{4+}						Pt^{2+}	Au^+ Au^{3+}	Hg_2^{2+} Hg^{2+}	Pb^{2+} Pb^{4+}	Bi^{3+} Bi^{5+}					

▲ **Figure 7.16 Common oxidation states of the elements.** Note that hydrogen exhibits both positive and negative oxidation numbers, +1 and −1.

Go Figure   Would you expect NiO to dissolve in an aqueous solution of NaNO₃?

Insoluble NiO

Nickel oxide (NiO), nitric acid (HNO₃), and water

NiO is insoluble in water but reacts with HNO₃ to give a green solution of the salt Ni(NO₃)₂.

▲ **Figure 7.17 Metal oxides react with acids.** NiO does not dissolve in water but does react with nitric acid (HNO₃) to give a green solution of Ni(NO₃)₂.

The oxides are particularly important because of the great abundance of oxygen in our environment.

Most metal oxides are basic. Those that dissolve in water react to form metal hydroxides, as in the following examples:

$$\text{Metal oxide + water} \longrightarrow \text{metal hydroxide}$$

$$\text{Na}_2\text{O}(s) + \text{H}_2\text{O}(l) \longrightarrow 2\,\text{NaOH}(aq) \qquad [7.8]$$

$$\text{CaO}(s) + \text{H}_2\text{O}(l) \longrightarrow \text{Ca(OH)}_2(aq) \qquad [7.9]$$

The basicity of metal oxides is due to the oxide ion, which reacts with water to form hydroxide ions:

$$\text{O}^{2-}(aq) + \text{H}_2\text{O}(l) \longrightarrow 2\,\text{OH}^-(aq) \qquad [7.10]$$

Even metal oxides that are insoluble in water demonstrate their basicity by reacting with acids to form a salt plus water, as illustrated in **Figure 7.17**:

$$\text{Metal oxide + acid} \longrightarrow \text{salt + water}$$
$$\text{NiO}(s) + 2\,\text{HNO}_3(aq) \longrightarrow \text{Ni(NO}_3)_2(aq) + \text{H}_2\text{O}(l) \qquad [7.11]$$

 Sample Exercise 7.7

Properties of Metal Oxides

(a) Would you expect scandium oxide to be a solid, liquid, or gas at room temperature?

(b) Write the balanced chemical equation for the reaction of scandium oxide with nitric acid.

SOLUTION

Analyze and Plan We are asked about one physical property of scandium oxide—its state at room temperature—and one chemical property—how it reacts with nitric acid.

Solve

(a) Because scandium oxide is the oxide of a metal, we expect it to be an ionic solid. Indeed it is, with the very high melting point of 2485 °C.

(b) In compounds, scandium has a 3+ charge, Sc^{3+}, and the oxide ion is O^{2-}. Consequently, the formula of scandium oxide is

Sc_2O_3. Metal oxides tend to be basic, so they tend to react with acids to form a salt plus water. In this case, the salt is scandium nitrate, $\text{Sc(NO}_3)_3$:

$$\text{Sc}_2\text{O}_3(s) + 6\,\text{HNO}_3(aq) \longrightarrow 2\,\text{Sc(NO}_3)_3(aq) + 3\,\text{H}_2\text{O}(l)$$

▶ **Practice Exercise**

Write the balanced chemical equation for the reaction between copper(II) oxide and sulfuric acid.

Nonmetals

Nonmetals can be solid, liquid, or gas. They are not lustrous and generally are poor conductors of heat and electricity. Their melting points are generally lower than those of metals (although diamond, a form of carbon, is an exception; it melts at 3570 °C). Under ordinary conditions, seven nonmetals exist as diatomic molecules. Five of these are gases (H_2, N_2, O_2, F_2, and Cl_2), one is a liquid (Br_2), and one is a volatile solid (I_2). Excluding the noble gases, the remaining nonmetals are solids that can be either hard, such as diamond, or soft, such as sulfur (**Figure 7.18**).

Because of their relatively large, negative electron affinities, nonmetals tend to gain electrons when they react with metals. For example, the reaction of aluminum with bromine produces the ionic compound aluminum bromide:

$$2Al(s) + 3 Br_2(l) \longrightarrow 2 AlBr_3(s) \qquad [7.12]$$

A nonmetal will typically gain enough electrons to fill its outermost occupied p subshell, thereby achieving a noble-gas electron configuration. For example, the bromine atom gains one electron to fill its $4p$ subshell:

$$Br([Ar]4s^2 3d^{10} 4p^5) + e^- \Rightarrow Br^-([Ar]4s^2 3d^{10} 4p^6)$$

Compounds composed entirely of nonmetals are typically molecular substances that tend to be gases, liquids, or low melting point solids at room temperature. Examples include the common hydrocarbons we use for fuel (methane, CH_4; propane, C_3H_8; octane, C_8H_{18}) and the gases HCl, NH_3, and H_2S. Many pharmaceuticals are molecules composed of C, H, N, O, and other nonmetals. For example, the molecular formula for the drug Lipitor®, which is used to treat high cholesterol, is $C_{33}H_{35}FN_2O_5$.

Most nonmetal oxides are acidic, which means that those that dissolve in water form acids:

$$\text{Nonmetal oxide + water} \longrightarrow \text{acid}$$

$$CO_2(g) + H_2O(l) \longrightarrow H_2CO_3(aq) \qquad [7.13]$$

$$P_4O_{10}(s) + 6 H_2O(l) \longrightarrow 4 H_3PO_4(aq) \qquad [7.14]$$

The reaction of carbon dioxide with water (**Figure 7.19**) accounts for the acidity of carbonated water and, to some extent, rainwater. Because sulfur is present in oil and coal, combustion of these common fuels produces sulfur dioxide and sulfur trioxide. These substances dissolve in water to produce acid rain, a major pollutant in many parts of the world. Like acids, most nonmetal oxides dissolve in basic solutions to form a salt plus water:

$$\text{Nonmetal oxide + base} \longrightarrow \text{salt + water}$$

$$CO_2(g) + 2 NaOH(aq) \longrightarrow Na_2CO_3(aq) + H_2O(l) \qquad [7.15]$$

▲ Figure 7.18 **Sulfur, known to the medieval world as "brimstone," is a nonmetal.**

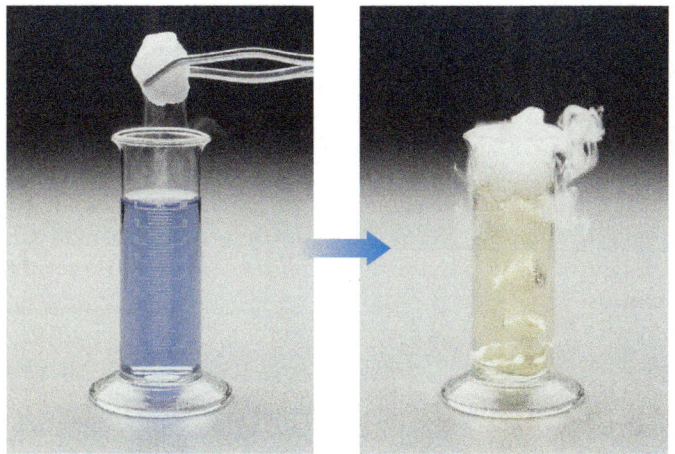

◀ Figure 7.19 **The reaction of CO_2 with water containing a bromthymol blue indicator.** Initially, the blue color tells us the water is slightly basic. When a piece of solid carbon dioxide ("dry ice") is added, the color changes to yellow, indicating an acidic solution. The mist is water droplets condensed from the air by cold CO_2 gas that sublimes before reacting.

Sample Exercise 7.8
Reactions of Nonmetal Oxides

Write a balanced chemical equation for the reaction of solid selenium dioxide, $SeO_2(s)$, with (**a**) water, (**b**) aqueous sodium hydroxide.

SOLUTION

Analyze and Plan Selenium is a nonmetal, so we need to write chemical equations for the reaction of a nonmetal oxide with water and with a base, NaOH. Nonmetal oxides are acidic, reacting with water to form an acid and with bases to form a salt and water.

Solve

(**a**) The reaction between selenium dioxide and water is like that between carbon dioxide and water (Equation 7.13):

$$SeO_2(s) + H_2O(l) \longrightarrow H_2SeO_3(aq)$$

(It does not matter that SeO_2 is a solid and that CO_2 is a gas under ambient conditions; the point is that both are water-soluble nonmetal oxides.)

(**b**) The reaction with sodium hydroxide is like that in Equation 7.15:

$$SeO_2(s) + 2\,NaOH(aq) \longrightarrow Na_2SeO_3(aq) + H_2O(l)$$

▶ **Practice Exercise**
Write a balanced chemical equation for the reaction of solid tetraphosphorus hexoxide with water.

Metalloids

Metalloids have properties intermediate between those of metals and those of nonmetals. They may have some characteristic metallic properties but lack others. For example, the metalloid silicon *looks* like a metal (**Figure 7.20**), but it is brittle rather than malleable and does not conduct heat or electricity nearly as well as metals do.

Several metalloids, most notably silicon, are electrical semiconductors and are the principal elements used in integrated circuits and computer chips. One of the reasons metalloids can be used for integrated circuits is that their electrical conductivity is intermediate between that of metals and that of nonmetals. Very pure silicon is an electrical insulator, but its conductivity can be dramatically increased with the addition of specific impurities called *dopants*. This modification provides a mechanism for controlling the electrical conductivity by controlling the chemical composition. We return to this point in Chapter 12.

▲ **Figure 7.20 Elemental silicon.**

Self-Assessment Exercises

SAE 7.15 Based on its position in the periodic table, is arsenic a (**a**) metal (**b**) nonmetal or (**c**) metalloid?

SAE 7.16 Which of these statements is true about metals? (**a**) The elements on the right side of the periodic table are metals. (**b**) Metals are brittle and fracture easily. (**c**) Metals have high electron affinities and readily form anions in aqueous solution. (**d**) Metals are good at conducting electricity but are not good at conducting heat. (**e**) None of these statements is true.

SAE 7.17 Which of the following is the expected equation for the reaction of $Fe_2O_3(s)$ with hydrochloric acid?

(**a**) $Fe_2O_3(s) + 6\,HCl(aq) \longrightarrow 2\,Fe(OH)_3(s) + 3\,Cl_2(g)$

(**b**) $Fe_2O_3(s) + 6\,HCl(aq) \longrightarrow 2\,FeCl_3(aq) + 3\,H_2O(l)$

(**c**) $Fe_2O_3(s) + 6\,HCl(aq) \longrightarrow 2\,FeH_3(s) + 3\,Cl_2O(aq)$

(**d**) $Fe_2O_3(s) + 6\,HCl(aq) \longrightarrow 2\,FeCl_2(aq) + 3\,H_2O(l)$

(**e**) $Fe_2O_3(s) + 6\,HCl(aq) \longrightarrow 2\,H_3FeCl_3(aq) + \frac{3}{2}\,O_2(g)$

7.6 | Trends for Group 1A and Group 2A Metals

As we have seen, elements in a given group possess general similarities. However, trends also exist within each group. In this section, we use the periodic table and our knowledge of electron configurations to examine the chemistry of the **alkali metals** and **alkaline earth metals**.

Group 1A: The Alkali Metals

The alkali metals are soft metallic solids (**Figure 7.21**). All have characteristic metallic properties, such as a silvery, metallic luster, and high thermal and electrical conductivity.

Learning Objectives

When you finish Section 7.6, you should be able to:

▶ Predict the chemical reactivity of Group 1A metals.

▶ Predict the chemical reactivity of Group 2A metals.

TABLE 7.4 Some Properties of the Alkali Metals

Element	Electron Configuration	Melting Point (°C)	Density (g/cm³)	Atomic Radius (Å)	I_1 (kJ/mol)
Lithium	$[He]2s^1$	181	0.53	1.28	520
Sodium	$[Ne]3s^1$	98	0.97	1.66	496
Potassium	$[Ar]4s^1$	63	0.86	2.03	419
Rubidium	$[Kr]5s^1$	39	1.53	2.20	403
Cesium	$[Xe]6s^1$	28	1.88	2.44	376

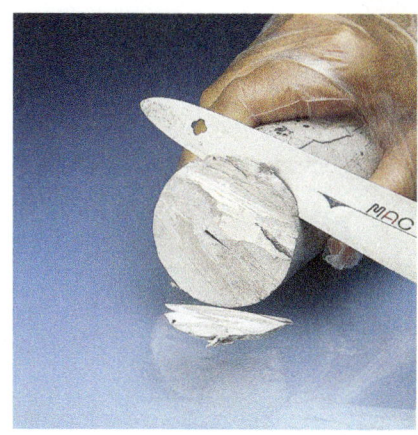

▲ Figure 7.21 Sodium, like the other alkali metals, is soft enough to be cut with a knife.

The name *alkali* comes from an Arabic word meaning "ashes." Early chemists isolated many compounds of sodium and potassium, two alkali metals, from wood ashes.

As Table 7.4 shows, the alkali metals have low densities and melting points, and these properties vary in a fairly regular way with increasing atomic number. We see the usual trends as we move down the group, such as increasing atomic radius and decreasing first ionization energy. The alkali metal of any given period has the lowest I_1 value in the period (Figure 7.11), which reflects the relative ease with which its outer *s* electron can be removed. As a result, the alkali metals are all very reactive, readily losing one electron to form ions carrying a 1+ charge. (Section 2.7)

The alkali metals exist in nature only as compounds. Sodium and potassium are relatively abundant in Earth's crust, in seawater, and in biological systems, always as the cations of ionic compounds. All alkali metals combine directly with most nonmetals. For example, they react with hydrogen to form hydrides and with sulfur to form sulfides:

$$2\,M(s) + H_2(g) \longrightarrow 2\,MH(s) \qquad [7.16]$$

$$2\,M(s) + S(s) \longrightarrow M_2S(s) \qquad [7.17]$$

where M represents any alkali metal. In hydrides of the alkali metals (LiH, NaH, and so forth), hydrogen is present as H^-, the **hydride ion**. A hydrogen atom that has *gained* an electron, the hydride ion is distinct from the hydrogen ion, H^+, formed when a hydrogen atom *loses* its electron.

The alkali metals react vigorously with water, producing hydrogen gas and a solution of an alkali metal hydroxide:

$$2\,M(s) + 2\,H_2O(l) \longrightarrow 2\,MOH(aq) + H_2(g) \qquad [7.18]$$

These reactions are very exothermic (Figure 7.22). In many cases, enough heat is generated to ignite the H_2, producing a fire or sometimes even an explosion, as in the case of

 **Go Figure** Would you expect rubidium metal to be more or less reactive with water than potassium metal?

| Li | Na | K |

▲ Figure 7.22 The alkali metals react vigorously with water.

K reacting with water. The reaction is even more violent for Rb and, especially, Cs, because their ionization energies are even lower than those of K.

Recall that the most common ion of oxygen is the oxide ion, O^{2-}. We would therefore expect that the reaction of an alkali metal with oxygen would produce the corresponding metal oxide. Indeed, reaction of Li metal with oxygen does form lithium oxide:

$$4\,Li(s) + O_2(g) \longrightarrow \underset{\text{lithium oxide}}{2\,Li_2O(s)} \tag{7.19}$$

When dissolved in water, Li_2O and other soluble metal oxides form hydroxide ions from the reaction of O^{2-} ions with H_2O (Equation 7.10).

The reactions of the other alkali metals with oxygen are more complex than we would anticipate. For example, when sodium reacts with oxygen, the main product is sodium *peroxide*, which contains the O_2^{2-} ion:

$$2\,Na(s) + O_2(g) \longrightarrow \underset{\text{sodium peroxide}}{Na_2O_2(s)} \tag{7.20}$$

Potassium, rubidium, and cesium react with oxygen to form compounds that contain the O_2^- ion, which is called the *superoxide ion*. For example, potassium forms potassium superoxide, KO_2:

$$K(s) + O_2(g) \longrightarrow \underset{\text{potassium superoxide}}{KO_2(s)} \tag{7.21}$$

The reactions in Equations 7.20 and 7.21 are somewhat unexpected; in most cases, the reaction of oxygen with a metal forms the metal oxide.

As seen from Equations 7.18–7.21, the alkali metals are extremely reactive toward water and oxygen. Because of this reactivity, the metals are usually stored submerged in a liquid hydrocarbon, such as mineral oil or kerosene.

Although alkali metal ions are colorless, each emits a characteristic color when placed in a flame (**Figure 7.23**). The ions are reduced to gaseous metal atoms in the flame. The high temperature excites the valence electron from the ground state to a higher-energy orbital, causing the atom to be in an excited state. The atom then emits energy in the form of visible light as the electron falls back into the lower-energy orbital and the atom returns to its ground state. The light emitted is at a specific wavelength for each element, just as we saw previously for the line spectra of hydrogen and neon. (Section 6.3) The characteristic yellow emission of sodium at 589 nm is the basis for sodium vapor lamps (**Figure 7.24**).

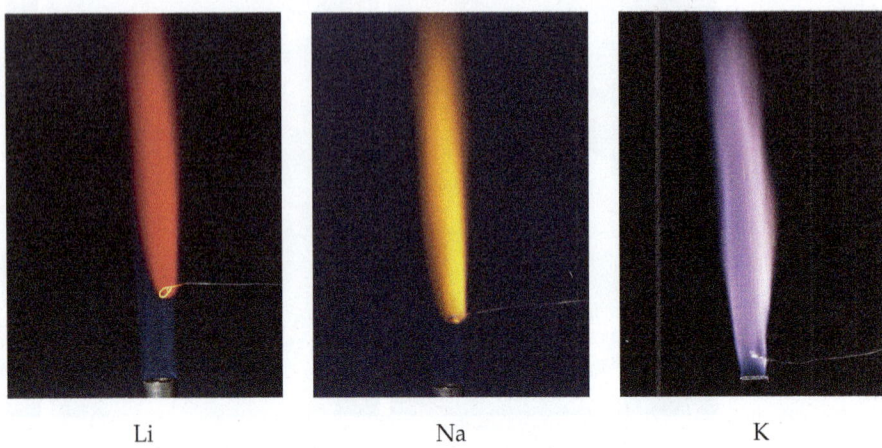

| Li | Na | K |

▲ **Figure 7.23** **Placed in a flame, ions of each alkali metal emit light of a characteristic wavelength.**

Go Figure If we had potassium vapor lamps, what color would they be?

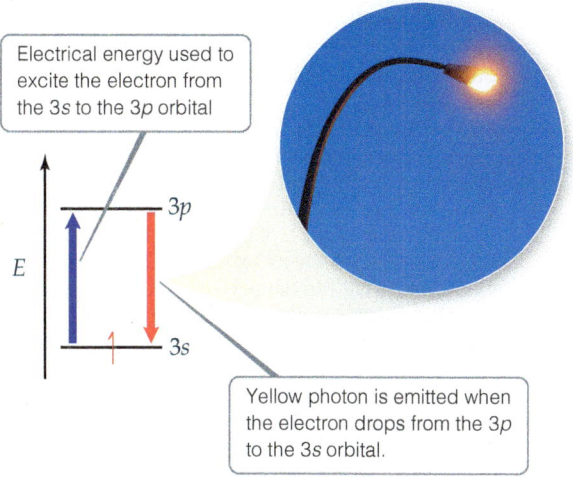

Electrical energy used to excite the electron from the 3s to the 3p orbital

E

$3p$

$3s$

Yellow photon is emitted when the electron drops from the 3p to the 3s orbital.

▲ **Figure 7.24** **The characteristic yellow light of a sodium lamp.** It results from excited electrons in the high-energy 3p orbital falling back to the lower-energy 3s orbital and releasing energy in the form of light.

 Sample Exercise 7.9

Reactions of an Alkali Metal

Write a balanced equation for the reaction of cesium metal with (a) $Cl_2(g)$, (b) $H_2O(l)$, (c) $H_2(g)$.

SOLUTION

Analyze and Plan Because cesium is an alkali metal, we expect its chemistry to be dominated by oxidation of the metal to Cs^+ ions. Further, we recognize that Cs is far down the periodic table (period 6), which means it is among the most active of all metals and probably reacts with all three substances.

Solve The reaction between Cs and Cl_2 is a simple combination reaction between a metal and a nonmetal, forming the ionic compound CsCl:

$$2\,Cs(s) + Cl_2(g) \longrightarrow 2\,CsCl(s)$$

From Equations 7.18 and 7.16, we predict the reactions of cesium with water and hydrogen to proceed as follows:

$$2\,Cs(s) + 2\,H_2O(l) \longrightarrow 2\,CsOH(aq) + H_2(g)$$
$$2\,Cs(s) + H_2(g) \longrightarrow 2\,CsH(s)$$

All three reactions are redox reactions where cesium forms a Cs^+ ion. The Cl^-, OH^-, and H^- are all $1-$ ions, which means the products have 1:1 stoichiometry with Cs^+.

▶ **Practice Exercise**

Write a balanced equation for the expected reaction between potassium metal and elemental sulfur, $S(s)$.

CHEMISTRY AND LIFE   The Improbable Development of Lithium Drugs

Alkali metal ions tend to play an unexciting role in most chemical reactions. As noted in Section 4.2, all salts of the alkali metal ions are soluble in water, and the ions are spectators in most aqueous reactions (except for those involving the alkali metals in their elemental form, such as those in Equations 7.16–7.21). However, these ions play an important role in human physiology. Sodium and potassium ions, for example, are major components of blood plasma and intracellular fluid, respectively, with average concentrations of 0.1 M. These electrolytes serve as vital charge carriers in normal cellular function. In contrast, lithium ion has no known function in normal human physiology. Since the discovery of lithium in 1817, however, people have believed that salts of the element possessed almost mystical healing powers. There were even claims

that lithium ions were an ingredient in ancient "fountain of youth" formulas. In 1927, C. L. Grigg began marketing a soft drink that contained lithium. The original unwieldy name of the beverage was "Bib-Label Lithiated Lemon-Lime Soda," which was soon changed to the simpler and more familiar name 7UP® (**Figure 7.25**).

In response to the Food and Drug Administration's concerns, lithium was removed from 7UP® during the early 1950s. At nearly the same time, psychiatrists discovered that lithium ions have a remarkable therapeutic effect on the mental condition called *bipolar disorder*. More than 5 million American adults each year suffer from this psychosis, undergoing severe mood swings ranging from deep depression to a manic euphoria. The lithium ion smooths these mood swings, allowing the bipolar patient to function more effectively in daily life.

Continued

▲ **Figure 7.25 Lithium no more.** The soft drink 7UP® originally contained a lithium salt that was claimed to give the beverage healthful benefits, including "an abundance of energy, enthusiasm, a clear complexion, lustrous hair, and shining eyes!" The lithium was removed from the beverage in the early 1950s, about the time that the antipsychotic action of Li^+ was discovered.

The antipsychotic action of Li^+ was discovered by accident in the 1940s by Australian psychiatrist John Cade (1912–1980) as he was researching the use of uric acid—a component of urine—to treat manic-depressive illness. He administered the acid to manic laboratory animals in the form of its most soluble salt, lithium urate, and found that many of the manic symptoms seemed to disappear. Later studies showed that uric acid had no role in the therapeutic effects observed; rather, the Li^+ ions were responsible. Because lithium overdose can cause severe side effects in humans, including kidney failure and death, lithium salts were not approved as antipsychotic drugs for humans until 1970. Today Li^+ is usually administered orally in the form of Li_2CO_3, which is the active ingredient in prescription drugs such as Eskalith®. Lithium drugs are effective for about 70% of the bipolar patients who take it.

In this age of sophisticated drug design and biotechnology, the simple lithium ion is still the most effective treatment of this destructive psychiatric disorder. Remarkably, despite intensive research, scientists still do not fully understand the biochemical action of lithium that leads to its therapeutic effects. Because of its similarity to Na^+, Li^+ is incorporated into blood plasma, where it can affect the behavior of nerve and muscle cells. Because Li^+ has a smaller radius than Na^+ (Figure 7.8), the way Li^+ interacts with molecules in human cells is different from the way Na^+ interacts with those molecules. Other studies indicate that Li^+ alters the function of certain neurotransmitters, which might lead to its effectiveness as an antipsychotic drug.

Group 2A: The Alkaline Earth Metals

Like the alkali metals, the alkaline earth metals are all solids at room temperature and have typical metallic properties (Table 7.5). Compared with the alkali metals, the alkaline earth metals are harder and denser, and melt at higher temperatures.

The first ionization energies of the alkaline earth metals are low but not as low as those of the alkali metals. Consequently, the alkaline earth metals are less reactive than their alkali metal neighbors. As noted in Section 7.4, the ease with which the elements lose electrons decreases as we move left to right across a period and increases as we move down a group. Thus, beryllium and magnesium, the lightest alkaline earth metals, are the least reactive.

The trend of increasing reactivity within the group is shown by the way the alkaline earth metals behave in the presence of water. Beryllium does not react with either water or steam, even when heated red-hot. Magnesium reacts slowly with liquid water and more readily with steam:

$$Mg(s) + H_2O\,(g) \longrightarrow MgO(s) + H_2(g) \qquad [7.22]$$

Calcium and the elements below it react readily with water at room temperature (although more slowly than the alkali metals adjacent to them in the periodic table). The reaction between calcium and water (Figure 7.26), for example, is

$$Ca(s) + 2\,H_2O(l) \longrightarrow Ca(OH)_2(aq) + H_2(g) \qquad [7.23]$$

TABLE 7.5   Some Properties of the Alkaline Earth Metals

Element	Electron Configuration	Melting Point (°C)	Density (g/cm³)	Atomic Radius (Å)	I_1 (kJ/mol)
Beryllium	$[He]2s^2$	1287	1.85	0.96	899
Magnesium	$[Ne]3s^2$	650	1.74	1.41	738
Calcium	$[Ar]4s^2$	842	1.55	1.76	590
Strontium	$[Kr]5s^2$	777	2.63	1.95	549
Barium	$[Xe]6s^2$	727	3.51	2.15	503

Equations 7.22 and 7.23 illustrate the dominant pattern in the reactivity of the alkaline earth elements: They tend to lose their two outer s electrons and form 2+ ions. For example, magnesium reacts with chlorine at room temperature to form $MgCl_2$ and burns with dazzling brilliance in air to give MgO:

$$Mg(s) + Cl_2(g) \longrightarrow MgCl_2(s) \qquad [7.24]$$

$$2\,Mg(s) + O_2(g) \longrightarrow 2\,MgO(s) \qquad [7.25]$$

In the presence of O_2, magnesium metal is protected by a thin coating of water-insoluble MgO. Thus, even though Mg is high in the activity series (Section 4.4), it can be incorporated into lightweight structural alloys used in, for example, automobile wheels. The heavier alkaline earth metals (Ca, Sr, and Ba) are even more reactive toward nonmetals than is magnesium.

The heavier alkaline earth ions give off characteristic colors when heated in a hot flame. Strontium salts produce the brilliant red color in fireworks, whereas barium salts produce the green color.

Like their neighbors sodium and potassium, magnesium and calcium are relatively abundant on Earth and in seawater and are essential for living organisms as cations in ionic compounds. Calcium is particularly important for the growth and maintenance of bones and teeth.

Go Figure

What is the cause of the bubbles that are formed? How could you test your answer?

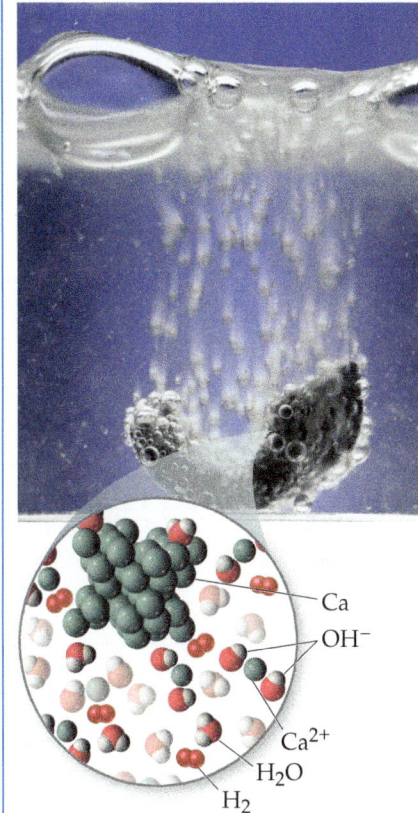

▲ **Figure 7.26** Elemental calcium reacts with water.

 Self-Assessment Exercises

SAE 7.18   Which reaction(s) is/are likely to occur?
(a) $2\,Li(s) + H_2O(l) \longrightarrow 2\,LiH(s) + \frac{1}{2}O_2(g)$ **(d)** a and c
(b) $2\,Li(s) + 2\,H_2O(l) \longrightarrow 2\,LiOH(aq) + H_2(g)$ **(e)** b and c
(c) $2\,Li(s) + H_2(g) \longrightarrow 2\,LiH(s)$

SAE 7.19   As you go from top to bottom in group 2A of the periodic table, the elements become more reactive with water. What is the explanation for this trend? **(a)** As you go down this column in the periodic table, atomic size increases, and they are more able to insert themselves into the O—H bonds of water. **(b)** As you go down this column in the periodic table, the density of the elements increases, and so they sink faster to the bottom of a flask of water and therefore have more time to react with water. **(c)** As you go down this column in the periodic table, ionization energies decrease, so it is easier for the elements to be oxidized by water. **(d)** As you go down this column in the periodic table, ionization energies decrease, so it is easier for the elements to be reduced by water.

7.7 | Trends for Selected Nonmetals

The elements on the left side of the periodic table are metals. Progressing from the right edge of the periodic table, elements can switch from nonmetals to metalloids to metals, depending on what column and which row we are considering. Here we highlight hydrogen, which is quite unusual compared to all other elements, and groups 6A (the **chalcogens**), 7A (the **halogens**), and 8A (**the noble gases**).

Hydrogen

We have seen that the chemistry of the alkali metals is dominated by the loss of their outer ns^1 electron to form cations. The $1s^1$ electron configuration of hydrogen suggests that its chemistry should have some resemblance to that of the alkali metals. The chemistry of hydrogen is much richer and more complex than that of the alkali metals, however, mainly because the ionization energy of hydrogen, 1312 kJ/mol, is higher than that of any metal and comparable to that of oxygen. As a result, hydrogen is a nonmetal that occurs as a colorless diatomic gas, $H_2(g)$, under most conditions.

Learning Objectives

When you finish Section 7.7, you should be able to:
▶ Describe the properties and chemical reactivity of hydrogen.
▶ Describe the properties and chemical reactivity of the group 6A elements.
▶ Describe the properties and chemical reactivity of the group 7A elements.
▶ Describe the properties and chemical reactivity of the group 8A elements.

The reactivity of hydrogen with nonmetals reflects its much greater tendency to hold on to its electron relative to the alkali metals. Unlike the alkali metals, hydrogen reacts with most nonmetals to form molecular compounds in which its electron is shared with, rather than completely transferred to, the other nonmetal. For example, we have seen that sodium metal reacts vigorously with chlorine gas to produce the ionic compound sodium chloride, in which the outermost sodium electron is completely transferred to a chlorine atom (Figure 2.19):

$$\text{Na}(s) + \tfrac{1}{2}\,\text{Cl}_2(g) \longrightarrow \underset{\text{ionic}}{\text{NaCl}(s)} \qquad \Delta H° = -410.9 \text{ kJ} \qquad \text{[7.26]}$$

By contrast, molecular hydrogen reacts with chlorine gas to form hydrogen chloride gas, which consists of HCl molecules:

$$\tfrac{1}{2}\text{H}_2(g) + \tfrac{1}{2}\text{Cl}_2(g) \longrightarrow \underset{\text{molecular}}{\text{HCl}(g)} \qquad \Delta H° = -92.3 \text{ kJ} \qquad \text{[7.27]}$$

Hydrogen readily forms molecular compounds with other nonmetals, such as the formation of water, $H_2O(l)$; ammonia, $NH_3(g)$; and methane, $CH_4(g)$. The ability of hydrogen to form bonds with carbon is one of the most important aspects of organic chemistry, as described in later chapters.

We have seen that, particularly in the presence of water, hydrogen does readily form H^+ ions in which the hydrogen atom has lost its electron. (Section 4.3) For example, $HCl(g)$ dissolves in H_2O to form a solution of hydrochloric acid, $HCl(aq)$, in which the electron of the hydrogen atom is transferred to the chlorine atom; a solution of hydrochloric acid consists largely of $H^+(aq)$ and $Cl^-(aq)$ ions stabilized by the H_2O solvent.* Indeed, the ability of molecular compounds of hydrogen with nonmetals to form acids in water is one of the most important aspects of aqueous chemistry. We discuss the chemistry of acids and bases in detail later in the text, particularly in Chapter 16.

Finally, as is typical for nonmetals, hydrogen also has the ability to gain an electron from a metal with a low ionization energy. For example, we saw in Equation 7.16 that hydrogen reacts with active metals to form solid metal hydrides that contain the hydride ion, H^-. The fact that hydrogen can gain an electron further illustrates that it behaves much more like a nonmetal than an alkali metal.

Group 6A: The Chalcogens

As we proceed down group 6A, the **chalcogens**, there is a change from nonmetallic to metallic character (Figure 7.14). Oxygen, sulfur, and selenium are typical nonmetals. Tellurium is a metalloid, and polonium, which is radioactive and quite rare, is a metal. While oxygen is a colorless gas at room temperature, all of the other members of group 6A are solids. Some of the physical properties of the group 6A elements are listed in Table 7.6.

As we saw in Section 2.6, oxygen exists in two molecular forms, O_2 and O_3. Because O_2 is the more common form, people generally mean O_2 when they say "oxygen," although the name *dioxygen* is more descriptive. The O_3 form is **ozone**. The two forms of oxygen are examples of *allotropes,* defined as different forms of the same element. About 21% of dry air consists of O_2 molecules. Ozone is present in very small amounts in the

TABLE 7.6   Some Properties of the Group 6A Elements

Element	Electron Configuration	Melting Point (°C)	Density	Atomic Radius (Å)	I_1 (kJ/mol)
Oxygen	$[\text{He}]2s^2 2p^4$	-218	1.43 g/L	0.66	1314
Sulfur	$[\text{Ne}]3s^2 3p^4$	115	1.96 g/cm³	1.05	1000
Selenium	$[\text{Ar}]3d^{10}4s^2 4p^4$	221	4.82 g/cm³	1.20	941
Tellurium	$[\text{Kr}]4d^{10}5s^2 5p^4$	450	6.24 g/cm³	1.38	869
Polonium	$[\text{Xe}]4f^{14}5d^{10}6s^2 6p^4$	254	9.20 g/cm³	1.40	812

*A more realistic description is one where an H^+ ion is transferred from HCl to H_2O, thereby forming Cl^- and H_3O^+. We explore this chemistry in detail in Chapter 16.

upper atmosphere and in polluted air. It is also formed from O_2 in electrical discharges, such as in lightning storms:

$$3 O_2(g) \longrightarrow 2 O_3(g) \qquad \Delta H° = +284.6 \text{ kJ} \qquad [7.28]$$

This reaction is strongly endothermic, telling us that O_3 is less stable than O_2.

Although both O_2 and O_3 are colorless and therefore do not absorb visible light, O_3 absorbs certain wavelengths of ultraviolet light that O_2 does not. Because of this difference, the presence of ozone in the upper atmosphere is beneficial, filtering out harmful UV light. Ozone and oxygen also have different chemical properties. Ozone, which has a pungent odor, is a powerful oxidizing agent. Because of this property, ozone is sometimes added to water to kill bacteria or used in low levels to help to purify air. However, the reactivity of ozone also makes its presence in polluted air near Earth's surface detrimental to human health.

Oxygen has a great tendency to attract electrons from other elements (that is, to *oxidize* them). Oxygen in combination with a metal is almost always present as the oxide ion, O^{2-}. This ion has a noble-gas configuration and is particularly stable. As shown in Figure 5.14, the formation of nonmetal oxides often is also very exothermic and thus energetically favorable.

In our discussion of the alkali metals, we noted two less common oxygen anions— the peroxide (O_2^{2-}) ion and the superoxide (O_2^-) ion. Compounds of these ions often react to produce an oxide and O_2:

$$2 H_2O_2(aq) \longrightarrow 2 H_2O(l) + O_2(g) \qquad \Delta H° = -196.1 \text{ kJ} \qquad [7.29]$$

For this reason, bottles of aqueous hydrogen peroxide are brown (to protect H_2O_2 from light, which can initiate its decomposition to water and oxygen) and are topped with caps that are able to release the $O_2(g)$ produced before the pressure inside becomes too great.

After oxygen, the most important member of group 6A is sulfur. This element exists in several allotropic forms, the most common and stable of which is the yellow solid having the molecular formula S_8. This molecule consists of an eight-membered ring of sulfur atoms (**Figure 7.27**). Even though solid sulfur consists of S_8 rings, we usually write it simply as S(s) in chemical equations to simplify the stoichiometric coefficients.

Like oxygen, sulfur has a tendency to gain electrons from other elements to form sulfides, which contain the S^{2-} ion. In fact, most sulfur in nature is present as metal sulfides. Sulfur is below oxygen in the periodic table, and the tendency of sulfur to form sulfide anions is not as great as that of oxygen to form oxide ions. As a result, the chemistry of sulfur is more complex than that of oxygen. Sulfur and its compounds (including those in coal and petroleum) can be burned in oxygen. The main product is sulfur dioxide, a major air pollutant:

$$S(s) + O_2(g) \longrightarrow SO_2(g) \qquad [7.30]$$

Below sulfur in group 6A is selenium, Se. This relatively rare element is essential for life in trace quantities, although it is toxic at high doses. Many allotropes of Se, including several eight-membered ring structures, resemble the S_8 ring.

The next element in the group is tellurium, Te. Its elemental structure is even more complex than that of Se, consisting of long, twisted chains of Te—Te bonds. Both Se and Te favor the −2 oxidation state, as do O and S.

From O to S to Se to Te, the elements form larger and larger molecules and become increasingly metallic. The thermal stability of group 6A compounds with hydrogen decreases down the column: $H_2O > H_2S > H_2Se > H_2Te$, with H_2O, water, being the most stable of the series.

Group 7A: The Halogens

Some of the properties of the group 7A elements, the **halogens**, are listed in **Table 7.7**. Astatine, which is both extremely rare and radioactive, is omitted because many of its properties are not yet known. Even less is known about the recently discovered Tennessine (element 117).

Go Figure

Suppose it were possible to flatten the S_8 ring. What shape would you expect the flattened ring to have?

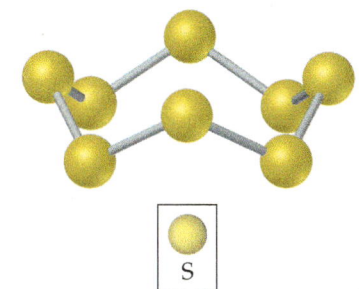

S

▲ **Figure 7.27 Elemental sulfur exists as the S_8 molecule.** At room temperature, this is the most common allotropic form of sulfur.

TABLE 7.7   Some Properties of the Halogens

Element	Electron Configuration	Melting Point (°C)	Density	Atomic Radius (Å)	I_1 (kJ/mol)
Fluorine	$[He]2s^22p^5$	-220	1.69 g/L	0.57	1681
Chlorine	$[Ne]3s^23p^5$	-102	3.12 g/L	1.02	1251
Bromine	$[Ar]4s^23d^{10}4p^5$	-7.3	3.12 g/cm^3	1.20	1140
Iodine	$[Kr]5s^24d^{10}5p^5$	114	4.94 g/cm^3	1.39	1008

Unlike the group 6A elements, all of the halogens that have been characterized are nonmetals. Their melting and boiling points increase with increasing atomic number. Fluorine and chlorine are gases at room temperature, bromine is a liquid, and iodine is a solid. Each element consists of diatomic molecules: F_2, Cl_2, Br_2, and I_2 (**Figure 7.28**).

The halogens have highly negative electron affinities (Figure 7.13). Thus, it is not surprising that the chemistry of the halogens is dominated by their tendency to gain electrons from other elements to form halide ions, X^-. (In many equations, X is used to indicate any one of the halogen elements.) Fluorine and chlorine are more reactive than bromine and iodine. In fact, fluorine removes electrons from almost any substance with which it comes into contact, including water, and usually does so very exothermically, as in the following examples:

$$2\,H_2O(l) + 2\,F_2(g) \longrightarrow 4\,HF(aq) + O_2(g) \quad \Delta H = -758.9\text{ kJ} \qquad [7.31]$$

$$SiO_2(s) + 2\,F_2(g) \longrightarrow SiF_4(g) + O_2(g) \qquad \Delta H = -704.0\text{ kJ} \qquad [7.32]$$

As a result, fluorine gas is difficult and dangerous to use in the laboratory, requiring specialized equipment.

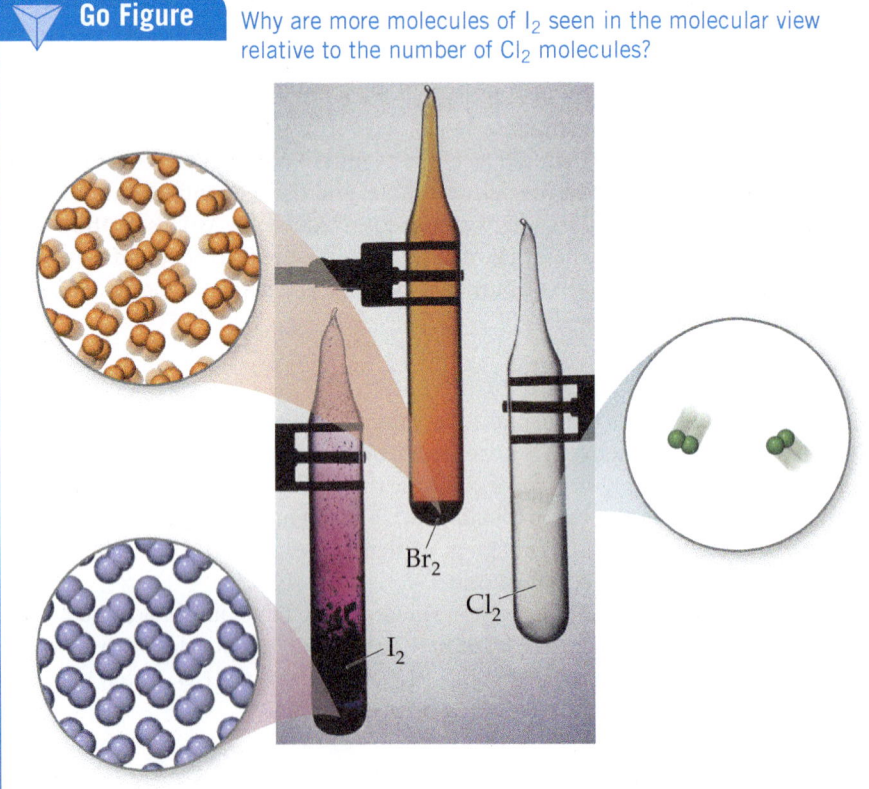

Go Figure Why are more molecules of I_2 seen in the molecular view relative to the number of Cl_2 molecules?

Br_2

Cl_2

I_2

▲ Figure 7.28 The elemental halogens exist as diatomic molecules.

Chlorine is the most industrially useful of the halogens. It is produced by a process called electrolysis, where an electrical current is used to oxidize chloride anions to molecular chlorine, Cl_2. Unlike fluorine, chlorine reacts slowly with water to form relatively stable aqueous solutions of HCl and HOCl (hypochlorous acid):

$$Cl_2(g) + H_2O(l) \longrightarrow HCl(aq) + HOCl(aq) \qquad [7.33]$$

Chlorine is often added to drinking water and swimming pools because the HOCl(aq) that is generated serves as a disinfectant.

The halogens react directly with most metals to form ionic halides. The halogens also react with hydrogen to form gaseous hydrogen halide compounds:

$$H_2(g) + X_2 \longrightarrow 2\,HX(g) \qquad [7.34]$$

All these compounds are very soluble in water and dissolve to form the hydrohalic acids. As we discussed in Section 4.3, HCl(aq), HBr(aq), and HI(aq) are strong acids, whereas HF(aq) is a weak acid.

Group 8A: The Noble Gases

The group 8A elements, known as the **noble gases**, are all nonmetals that are gases at room temperature. They are all *monatomic* (that is, they consist of single atoms rather than molecules). Some physical properties of the first six noble-gas elements are listed in Table 7.8. The high radioactivity of radon (Rn, atomic number 86) has limited the study of its reaction chemistry and some of its properties, and very little is known about oganesson (Og, atomic number 118), for which only a few atoms have ever been synthesized.

The noble gases have completely filled *s* and *p* subshells. All elements of group 8A have large first ionization energies, and we see the expected decrease as we move down the column. Because the noble gases possess such stable electron configurations, they are exceptionally unreactive. In fact, until the early 1960s the elements were called the *inert gases* because they were thought to be incapable of forming chemical compounds. In 1962, though, Neil Bartlett (1932–2008) at the University of British Columbia reasoned that the ionization energy of Xe might be low enough to allow it to form compounds. For this to happen, Xe would have to react with a substance that had an extremely high ability to remove electrons from other substances, such as fluorine. Bartlett synthesized the first noble-gas compound by combining Xe with the fluorine-containing compound PtF_6. Xenon also reacts directly with $F_2(g)$ to form the molecular compounds XeF_2, XeF_4, and XeF_6. Krypton has a higher I_1 value than xenon and is therefore less reactive. In fact, only a single stable compound of krypton is known, KrF_2. In 2000, Finnish scientists reported the first neutral molecule that contains argon, the HArF molecule, which is stable only at low temperatures.

TABLE 7.8   Some Properties of the Noble Gases

Element	Electron Configuration	Boiling Point (K)	Density (g/L)	Atomic Radius* (Å)	I_1 (kJ/mol)
Helium	$1s^2$	4.2	0.18	0.28	2372
Neon	$[He]2s^22p^6$	27.1	0.90	0.58	2081
Argon	$[Ne]3s^23p^6$	87.3	1.78	1.06	1521
Krypton	$[Ar]4s^23d^{10}4p^6$	120	3.75	1.16	1351
Xenon	$[Kr]5s^24d^{10}5p^6$	165	5.90	1.40	1170
Radon	$[Xe]6s^24f^{14}5d^{10}6p^6$	211	9.73	1.50	1037

*Only the heaviest of the noble-gas elements form chemical compounds. Thus, the atomic radii for the lighter noble-gas elements are estimated values.

Self-Assessment Exercises

SAE 7.20 Which statement is correct? (**a**) Group 6A elements are likely to form −1 ions. (**b**) Group 7A elements are very unreactive. (**c**) The halogens are likely to form −2 ions. (**d**) Noble gases form diatomic molecules. (**e**) None of these is correct.

SAE 7.21 Which statements is/are *true*?

(**i**) Hydrogen reacts with lithium to make lithium hydride, LiH, where H is in the −1 oxidation state.

(**ii**) Chlorine reacts with water to make hydrochloric acid and hypochlorous acid.

(**iii**) The element fluorine will react with glass (SiO_2).

(**a**) Only i is true. (**b**) Only ii is true. (**c**) Only iii is true. (**d**) Two of i, ii, and iii are true. (**e**) All of i, ii, and iii are true.

SAE 7.22 Which of the following statements about oxygen is *false*? (**a**) The most common form of oxygen is O_2. (**b**) Ozone is an allotrope of oxygen. (**c**) The most common ion formed by oxygen is the peroxide ion. (**d**) The main product when oxygen reacts with sulfur is SO_2.

 Putting Concepts Together

The element bismuth (Bi, atomic number 83) is the heaviest member of group 5A. A salt of the element, bismuth subsalicylate, is the active ingredient in Pepto-Bismol®, an over-the-counter medication for gastric distress.

(**a**) Based on values presented in Figure 7.7 and Tables 7.5 and 7.6, what might you expect for the bonding atomic radius of bismuth?

(**b**) What accounts for the general increase in atomic radius going down the group 5A elements?

(**c**) Another major use of bismuth has been as an ingredient in low-melting metal alloys, such as those used in fire sprinkler systems and in typesetting. The element itself is a brittle white crystalline solid. How do these characteristics fit with the fact that bismuth is in the same periodic group with such nonmetallic elements as nitrogen and phosphorus?

(**d**) Bi_2O_3 is a basic oxide. Write a balanced chemical equation for its reaction with dilute nitric acid. If 6.77 g of Bi_2O_3 is dissolved in dilute acidic solution to make 0.500 L of solution, what is the molarity of the solution of Bi^{3+} ion?

(**e**) ^{209}Bi is the heaviest stable isotope of any element. How many protons and neutrons are present in this nucleus?

(**f**) The density of Bi at 25 °C is 9.808 g/cm³. How many Bi atoms are present in a cube of the element that is 5.00 cm on each edge? How many moles of the element are present?

SOLUTION

(**a**) Bismuth is directly below antimony, Sb, in group 5A. Based on the observation that atomic radii increase as we go down a column, we would expect the radius of Bi to be greater than that of Sb, which is 1.39 Å. We also know that atomic radii generally decrease as we proceed from left to right in a period. Tables 7.5 and 7.6 each give an element in the same period—namely, Ba and Po. We would therefore expect that the radius of Bi is smaller than that of Ba (2.15 Å) and larger than that of Po (1.40 Å). We also see that in other periods, the difference in radius between the neighboring group 5A and group 6A elements is relatively small. We might therefore expect that the radius of Bi is slightly larger than that of Po—much closer to the radius of Po than to the radius of Ba. The tabulated value for the atomic radius on Bi is 1.48 Å, in accord with our expectations.

(**b**) The general increase in radius with increasing atomic number in the group 5A elements occurs because additional shells of electrons are being added, with corresponding increases in nuclear charge. The core electrons in each case largely screen the outermost electrons from the nucleus, so the effective nuclear charge does not vary greatly as we go to higher atomic numbers. However, the principal quantum number, n, of the outermost electrons steadily increases, with a corresponding increase in orbital radius.

(**c**) The contrast between the properties of bismuth and those of nitrogen and phosphorus illustrates the general rule that there is a trend toward increased metallic character as we move down in a given group. Bismuth, in fact, is a metal. The increased metallic character occurs because the outermost electrons are more readily lost in bonding, a trend that is consistent with its lower ionization energy.

(**d**) Following the procedures described in Section 4.2 for writing molecular and net ionic equations, we have the following:

Molecular equation:

$$Bi_2O_3(s) + 6\,HNO_3(aq) \longrightarrow 2\,Bi(NO_3)_3(aq) + 3\,H_2O(l)$$

Net ionic equation:

$$Bi_2O_3(s) + 6\,H^+(aq) \longrightarrow 2\,Bi^{3+}(aq) + 3\,H_2O(l)$$

In the net ionic equation, nitric acid is a strong acid and $Bi(NO_3)_3$ is a soluble salt, so we need to show only the reaction of the solid with the hydrogen ion forming the $Bi^{3+}(aq)$ ion and water. To calculate the concentration of the solution, we proceed as follows (Section 4.5):

$$\frac{6.77\text{ g }Bi_2O_3}{0.500\text{ L soln}} \times \frac{1\text{ mol }Bi_2O_3}{466.0\text{ g }Bi_2O_3} \times \frac{2\text{ mol }Bi^{3+}}{1\text{ mol }Bi_2O_3}$$

$$= \frac{0.0581\text{ mol }Bi^{3+}}{\text{L soln}} = 0.0581\ M$$

(**e**) Recall that the atomic number of any element is the number of protons and electrons in a neutral atom of the element. (Section 2.3) Bismuth is element 83, so there are 83 protons in the nucleus. Because the atomic mass number is 209, there are 209 − 83 = 126 neutrons in the nucleus.

(**f**) We can use the density and the atomic weight to determine the number of moles of Bi, and then we can use Avogadro's number to convert the result to the number of atoms. (Sections 1.4 and 3.4) The volume of the cube is $(5.00)^3$ cm³ = 125 cm³. Then we have

$$125\text{ cm}^3\,Bi \times \frac{9.808\text{ g }Bi}{1\text{ cm}^3} \times \frac{1\text{ mol }Bi}{209.0\text{ g }Bi} = 5.87\text{ mol }Bi$$

$$5.87\text{ mol }Bi \times \frac{6.022 \times 10^{23}\text{ atom }Bi}{1\text{ mol }Bi} - 3.53 \times 10^{24}\text{ atoms }Bi$$

Chapter Summary and Key Terms

DEVELOPMENT OF THE PERIODIC TABLE AND EFFECTIVE NUCLEAR CHARGE (SECTIONS 7.1 AND 7.2) The periodic table was first developed by Mendeleev and Meyer on the basis of the similarity in chemical and physical properties exhibited by certain elements. Moseley established that each element has a unique atomic number, which added more order to the periodic table.

We now know that elements in the same column of the periodic table have the same number of electrons in their **valence orbitals**. This similarity in valence electronic structure leads to the similarities among elements in the same group. Differences among elements in the same group arise because their valence orbitals are in different shells.

Many properties of atoms depend on the **effective nuclear charge**, which is the portion of the nuclear charge that an outer electron experiences after accounting for repulsions by other electrons in the atom. The core electrons are very effective in screening the outer electrons from the full charge of the nucleus, whereas electrons in the same shell do not screen each other very effectively. Because the actual nuclear charge increases as we progress through a period, the effective nuclear charge experienced by valence electrons increases as we move left to right across a period.

SIZES OF ATOMS AND IONS (SECTION 7.3) The size of an atom can be gauged by its **bonding atomic radius**, which is based on measurements of the distances separating atoms in their chemical compounds. In general, atomic radii increase down a column in the periodic table and decrease left to right across a row.

Cations are smaller than their parent atoms, whereas anions are larger than their parent atoms. For ions of the same charge, size increases going down a column of the periodic table. We can write electron configurations for ions by first writing the electron configuration of the neutral atom and then removing or adding the appropriate number of electrons. For cations, electrons are removed first from the orbitals of the neutral atom with the largest value of n. If there are two valence orbitals with the same value of n (such as $4s$ and $4p$), then the electrons are lost first from the orbital with a higher value of l (in this case, $4p$). For anions, electrons are added to orbitals in the reverse order.

An **isoelectronic series** is a series of ions that has the same number of electrons. For such a series, size decreases with increasing atomic number as the electrons are attracted more strongly to the nucleus as its positive charge increases.

IONIZATION ENERGY AND ELECTRON AFFINITY (SECTION 7.4) The first **ionization energy** of an atom is the minimum energy needed to remove an electron from the atom in the gas phase, forming a cation. The second ionization energy is the energy needed to remove a second electron from the cation, and so forth. Ionization energies show a sharp increase after all the valence electrons have been removed because of the much higher effective nuclear charge experienced by the core electrons. The first ionization energies of the elements show periodic trends that are opposite those seen for atomic radii, with smaller atoms having higher first ionization energies. Thus, first ionization energies decrease as we go down a column and increase as we proceed left to right across a row.

The **electron affinity** of an element is the energy change upon adding an electron to an atom in the gas phase, forming an anion. A negative electron affinity means that energy is released when the electron is added; hence, when the electron affinity is negative, the anion is stable. By contrast, a positive electron affinity means that the anion is not stable relative to the separated atom and electron. In general, electron affinities become more negative as we proceed from left to right across the periodic table. The halogens have the most negative electron affinities. The electron affinities of the noble gases are positive because the added electron would have to occupy a new, higher-energy subshell.

METALS, NONMETALS, AND METALLOIDS (SECTION 7.5) The elements can be categorized as metals, nonmetals, and metalloids. Most elements are metals; they occupy the left side and the middle of the periodic table. Nonmetals appear in the upper-right section of the table. Metalloids occupy a narrow band between the metals and nonmetals. The tendency of an element to exhibit the properties of metals, called the **metallic character**, increases as we proceed down a column and decreases as we proceed from left to right across a row.

Metals have a characteristic luster, and they are good conductors of heat and electricity. When metals react with nonmetals, the metal atoms are oxidized to cations and ionic substances are generally formed. Most metal oxides are basic; they react with acids to form salts and water.

Nonmetals lack metallic luster and are generally poor conductors of heat and electricity. Several are gases at room temperature. Compounds composed entirely of nonmetals are generally molecular. Nonmetals usually form anions in their reactions with metals. Nonmetal oxides are acidic; they react with bases to form salts and water. Metalloids have properties that are intermediate between those of metals and nonmetals.

TRENDS FOR GROUP 1A AND 2A METALS (SECTION 7.6) The periodic properties of the elements can help us understand the properties of groups of the representative elements. The **alkali metals** (group 1A) are soft metals with low densities and low melting points. They have the lowest ionization energies of the elements. As a result, they are very reactive toward nonmetals, easily losing their outer s electron to form 1+ ions.

The **alkaline earth metals** (group 2A) are harder and denser, and have higher melting points than the alkali metals. They are also very reactive toward nonmetals, although not as reactive as the alkali metals. The alkaline earth metals readily lose their two outer s electrons to form 2+ ions. Both alkali and alkaline earth metals react with hydrogen to form ionic substances that contain the **hydride ion**, H^-.

TRENDS FOR SELECTED NONMETALS (SECTION 7.7) Hydrogen is a nonmetal with properties that are distinct from any of the groups of the periodic table. It forms molecular compounds with other nonmetals, such as oxygen and the halogens.

Oxygen and sulfur are the most important elements in group 6A, the **chalcogens**. Oxygen is usually found as a diatomic molecule, O_2. **Ozone**, O_3, is an important allotrope of oxygen. Oxygen has a strong tendency to gain electrons from other elements, thus oxidizing them. In combination with metals, oxygen is usually found as the oxide ion, O^{2-}, although salts of the peroxide ion, O_2^{2-}, and superoxide ion, O_2^-, are sometimes formed. Elemental sulfur is most commonly found as S_8 molecules. In combination with metals, it is most often found as the sulfide ion, S^{2-}.

The **halogens** (group 7A) exist as diatomic molecules. The halogens have the most negative electron affinities of the elements. Thus, their chemistry is dominated by a tendency to form 1− ions, especially in reactions with metals.

The **noble gases** (group 8A) exist as monatomic gases. They are very unreactive because they have completely filled s and p subshells. Only the heaviest noble gases are known to form compounds, and they do so only with very active nonmetals, such as fluorine.

Key Equations

- $Z_{\text{eff}} = Z - S$ [7.1]   Estimating effective nuclear charge

Exam Prep

EP 7.1 Estimate the effective nuclear charge experienced by the $4s$ electron of potassium. (**a**) 1+ (**b**) 5+ (**c**) 18+ (**d**) 19+

EP 7.2 Hypothetical elements X and Y form a linear molecule XY_2, in which both Y atoms are bonded to atom X: Y—X—Y. X_2 and Y_2 are diatomic molecules with an X—X distance of 2.04 Å and a Y—Y distance of 1.68 Å, respectively. What would you predict for the X—Y distance in the XY_2 molecule? (**a**) 0.84 Å (**b**) 1.02 Å (**c**) 1.86 Å (**d**) 2.70 Å (**e**) 3.72 Å

EP 7.3 By referring only to the periodic table, place the following atoms in order of increasing atomic size: N, O, P, Ge.

(**a**) N < O < P < Ge
(**b**) P < N < O < Ge
(**c**) O < N < Ge < P
(**d**) O < N < P < Ge
(**e**) N < P < Ge < O

EP 7.4 Arrange the following atoms and ions in order of increasing ionic radius: F, S^{2-}, Cl, and Se^{2-}.

(**a**) F < S^{2-} < Cl < Se^{2-}
(**b**) F < Cl < S^{2-} < Se^{2-}
(**c**) F < S^{2-} < Se^{2-} < Cl
(**d**) Cl < F < Se^{2-} < S^{2-}
(**e**) S^{2-} < F < Se^{2-} < Cl

EP 7.5 Arrange the following ions in order of increasing ionic radius: Br^-, Rb^+, Se^{2-}, Sr^{2+}, Te^{2-}.

(**a**) Sr^{2+} < Rb^+ < Br^- < Se^{2-} < Te^{2-}
(**b**) Br^- < Sr^{2+} < Se^{2-} < Te^{2-} < Rb^+
(**c**) Rb^+ < Sr^{2+} < Se^{2-} < Te^{2-} < Br^-
(**d**) Rb^+ < Br^- < Sr^{2+} < Se^{2-} < Te^{2-}
(**e**) Sr^{2+} < Rb^+ < Br^- < Te^{2-} < Se^{2-}

EP 7.6 The third ionization energy of bromine is the energy required for which of the following processes?

(**a**) $Br(g) \longrightarrow Br^+(g) + e^-$
(**b**) $Br^+(g) \longrightarrow Br^{2+}(g) + e^-$
(**c**) $Br(g) \longrightarrow Br^{2+}(g) + 2e^-$
(**d**) $Br(g) \longrightarrow Br^{3+}(g) + 3e^-$
(**e**) $Br^{2+}(g) \longrightarrow Br^{3+}(g) + e^-$

EP 7.7 Consider the following statements about first ionization energies:

(**i**) Because the effective nuclear charge for Mg is greater than that for Be, the first ionization energy of Mg is greater than that for Be.

(**ii**) The first ionization energy of O is less than that of N because in O we must pair electrons in one of the $2p$ orbitals.

(**iii**) The first ionization energy of Ar is less than that of Ne because a $3p$ electron in Ar is farther from the nucleus than a $2p$ electron in Ne.

Which of the statements i, ii, and iii is/are true? (**a**) Only one of the statements is true. (**b**) Statements i and ii are true. (**c**) Statements i and iii are true. (**d**) Statements ii and iii are true. (**e**) All three statements are true.

EP 7.8 The ground-state electron configuration of a Tc atom is $[Kr]5s^24d^5$. What is the electron configuration of a Tc^{3+} ion? (**a**) $[Kr]4d^4$ (**b**) $[Kr]5s^24d^2$ (**c**) $[Kr]5s^14d^3$ (**d**) $[Kr]5s^24d^8$ (**e**) $[Kr]4d^{10}$

EP 7.9 If an element has a large negative value for electron affinity, (**a**) it is likely to be a noble gas. (**b**) it is likely to have a large negative first ionization energy. (**c**) it is likely to be a halogen. (**d**) it is likely to have a negative value for effective nuclear charge.

EP 7.10 As you go down a column in the periodic table, first ionization energy _____ and atomic size _____. (**a**) increases, increases (**b**) increases, decreases (**c**) decreases, decreases (**d**) decreases, increases

EP 7.11 Which of the following statements about effective nuclear charge is *false*? (**a**) Effective nuclear charge generally increases as we traverse across a row of the periodic table from left to right. (**b**) In the simple model of effective nuclear charge, the core electrons completely shield the valence electrons. (**c**) In the simple model of effective nuclear charge, valence electrons do not shield one another at all. (**d**) If the nuclear charge of atom A is greater than that of atom B, then the effective nuclear charge of A is greater than that of B.

EP 7.12 As you go across a row in the periodic table from left to right, first ionization energy _____ and atomic size _____. (**a**) increases, increases (**b**) increases, decreases (**c**) decreases, decreases (**d**) decreases, increases

EP 7.13 An element conducts electricity to some extent, and it tends to form molecular compounds. What is the most likely identity of the element? (**a**) O (**b**) Cl (**c**) Al (**d**) Si (**e**) Bi

EP 7.14 An element conducts electricity, and its oxidation state in ionic compounds is usually 1+. What is the most likely identity of the element? (**a**) F (**b**) Zn (**c**) Ca (**d**) Ag (**e**) Cr

EP 7.15 Suppose that a metal oxide of formula M_2O_3 were soluble in water. What would be the major product or products of dissolving the substance in water?

(**a**) $MH_3(aq) + O_2(g)$
(**b**) $M(s) + H_2(g) + O_2(g)$
(**c**) $M^{3+}(aq) + H_2O_2(aq)$
(**d**) $M(OH)_2(aq)$
(**e**) $M(OH)_3(aq)$

EP 7.16 Consider the following oxides: SO_2, Y_2O_3, MgO, Cl_2O, N_2O_5. How many are expected to form acidic solutions in water? (**a**) 1 (**b**) 2 (**c**) 3 (**d**) 4 (**e**) 5

EP 7.17 Consider the following three statements about the reactivity of an alkali metal M with oxygen gas:

(**i**) Based on their positions in the periodic table, the expected product is the ionic oxide M_2O.

(**ii**) Some of the alkali metals produce metal peroxides or metal superoxides when they react with oxygen.

(**iii**) When dissolved in water, an alkali metal oxide produces a basic solution.

Which of the statements i, ii, and iii is/are true? (**a**) Only one of the statements is true. (**b**) Statements i and ii are true. (**c**) Statements i and iii are true. (**d**) Statements ii and iii are true. (**e**) All three statements are true.

EP 7.18 Consider the mystery element "Z." Elemental Z is a powdery solid that does not conduct electricity at all. It reacts with metals to form MZ compounds, where Z is in the −2 oxidation state. What is the likely identity of Z? (**a**) O (**b**) S (**c**) F (**d**) Cl (**e**) None of these is correct

EP 7.19 Elemental cesium reacts more violently with water than does elemental sodium. Which of the following best explains this difference in reactivity?

(**a**) Sodium has more metallic character than does cesium.
(**b**) The first ionization energy of cesium is less than that of sodium.
(**c**) The electron affinity of sodium is less than that of cesium.
(**d**) The effective nuclear charge for cesium is less than that of sodium.
(**e**) The atomic radius of cesium is less than that of sodium.

EP 7.20 Which of the following are the expected products when $Sr(s)$ is added to $H_2O(l)$?

(**a**) $Sr^{2+}(aq) + OH^-(aq) + O_2(g)$
(**b**) $Sr^{3+}(aq) + OH^-(aq) + H_2(g)$
(**c**) $Sr^+(aq) + OH^-(aq) + H_2(g)$
(**d**) $Sr^{2+}(aq) + H_2(g) + O_2(g)$
(**e**) $Sr^{2+}(aq) + OH^-(aq) + H_2(g)$

Exercises

Visualizing Concepts

7.1 As discussed in the text, we can draw an analogy between the attraction of an electron to a nucleus and the act of perceiving light from a light bulb through a frosted glass shade, as shown in the illustration.

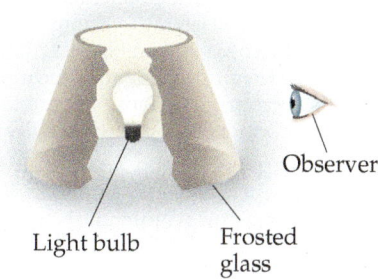

Observer

Light bulb

Frosted glass

Using the simple method of estimating effective nuclear charge, Equation 7.1, how does the intensity of the light bulb and/or the thickness of the frosting change in the following cases. **(a)** Moving from boron to carbon? **(b)** Moving from boron to aluminum? [Section 7.2]

7.2 Which of these spheres represents F, which represents Br, and which represents Br⁻? [Section 7.3]

7.3 Consider the Mg^{2+}, Cl^-, K^+, and Se^{2-} ions. The four spheres below represent these four ions, scaled according to ionic size. **(a)** Without referring to Figure 7.8, match each ion to its appropriate sphere. **(b)** In terms of size, between which of the spheres would you find the (i) Ca^{2+} and (ii) S^{2-} ions? [Section 7.3]

7.4 In the following reaction

Reactants   Products

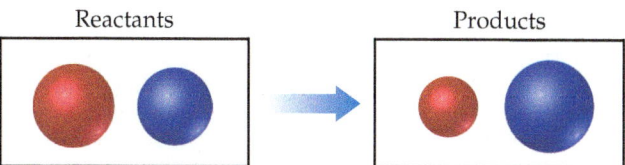

which sphere represents a metal and which represents a nonmetal? [Sections 7.3, 7.5, 7.6]

7.5 Consider the A_2X_4 molecule depicted here, where A and X are elements. The A—A bond length in this molecule is d_1, and the four A—X bond lengths are each d_2. **(a)** In terms of d_1 and d_2, how could you define the bonding atomic radii of atoms A

and X? **(b)** In terms of d_1 and d_2, what would you predict for the X—X bond length of an X_2 molecule? [Section 7.2]

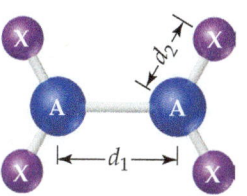

7.6 The graph below shows the ionization energies for a particular element. In which group is the element most likely a member of? [Section 7.3]

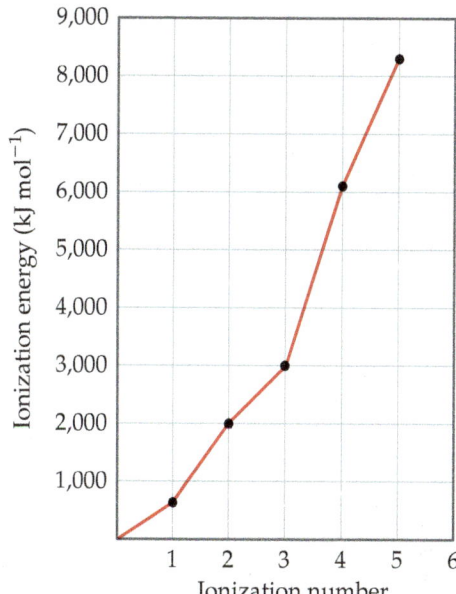

7.7 Which of the charts below shows the general periodic trends for each of the following properties of the main-group elements? (You can neglect small deviations going either left to right across a row or down a column of the periodic table.) **(1)** Bonding atomic radius, **(2)** first ionization energy, **(3)** effective nuclear charge. [Sections 7.1–7.3]

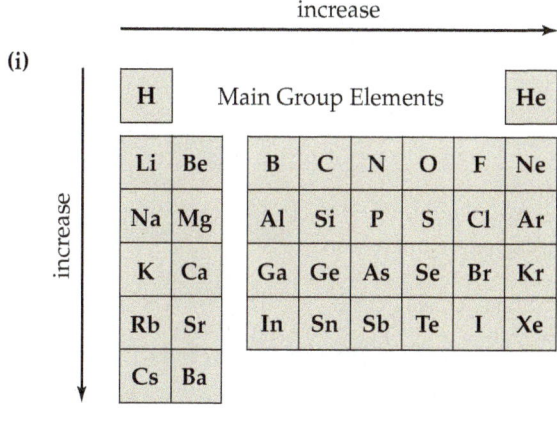

(ii)

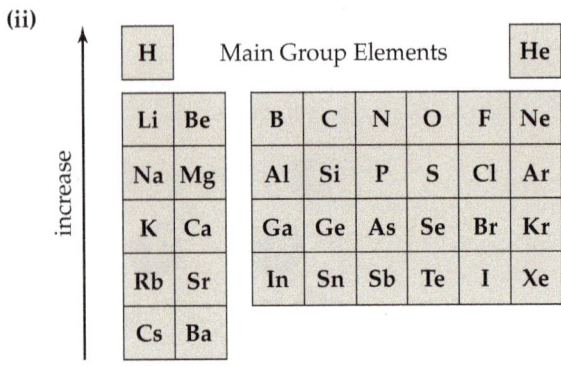

(iii)

(iv)

7.8 An element X reacts with $F_2(g)$ to form the molecular product shown here. (**a**) Write a balanced equation for this reaction (do not worry about the phases for X and the product). (**b**) If X were a metal, what oxidation state must it have? (**c**) Suggest an identity for X if X is a nonmetal. [Sections 7.4–7.6]

Periodic Table; Effective Nuclear Charge (Sections 7.1 and 7.2)

7.9 Group 1A and 2A elements are sometimes called, collectively, "the s-block." Therefore, what is an analogous name for the entire collection of the group 3A, 4A, 5A, 6A, 7A, and 8A elements?

7.10 The prefix *eka-* comes from the Sanskrit word for "one." Mendeleev used this prefix to indicate that the unknown element was one place away from the known element that followed the prefix. For example, *eka-silicon*, which we now call germanium, is one element below silicon. Mendeleev also predicted the existence of *eka-manganese*, which was not experimentally confirmed until 1937 because this element is radioactive and does not occur in nature. Based on the periodic table shown in Figure 7.1, what do we now call the element Mendeleev called *eka-manganese*?

7.11 (**a**) The five most abundant elements in Earth's crust are O, Si, Al, Fe, and Ca. Referring to Figure 7.1, are any of these elements among those known before 1700? If so, which ones? (**b**) Seven of the nine elements known since ancient times are metals. Referring to Table 4.5, are these metals mostly found at the bottom or top of the activity series?

7.12 Moseley's experiments on X rays emitted from atoms led to the concept of atomic numbers. (**a**) If arranged in order of increasing atomic mass, which element would come after chlorine? (**b**) Describe two ways in which the properties of this element differ from the other elements in group 8A.

7.13 Among elements 1–18, which element or elements have the smallest effective nuclear charge if we use Equation 7.1 to calculate Z_{eff}? Which element or elements have the largest effective nuclear charge?

7.14 Which of the following statements about effective nuclear charge for the outermost valence electron of an atom is *incorrect*?

(**a**) The effective nuclear charge can be thought of as the true nuclear charge minus a screening constant due to the other electrons in the atom.

(**b**) Effective nuclear charge increases going left to right across a row of the periodic table.

(**c**) Valence electrons screen the nuclear charge more effectively than do core electrons.

(**d**) The effective nuclear charge shows a sudden decrease when we go from the end of one row to the beginning of the next row of the periodic table.

(**e**) The change in effective nuclear charge going down a column of the periodic table is generally less than that going across a row of the periodic table.

7.15 Detailed calculations show that the value of Z_{eff} for the outermost electrons in Na and K atoms is 2.51+ and 3.49+, respectively. (**a**) What value do you estimate for Z_{eff} experienced by the outermost electron in both Na and K by assuming core electrons contribute 1.00 and valence electrons contribute 0.00 to the screening constant? (**b**) What values do you estimate for Z_{eff} using Slater's rules? (**c**) Which approach gives a more accurate estimate of Z_{eff}? (**d**) Does either method of approximation account for the gradual increase in Z_{eff} that occurs upon moving down a group? (**e**) Predict Z_{eff} for the outermost electrons in the Rb atom based on the calculations for Na and K.

7.16 Detailed calculations show that the value of Z_{eff} for the outermost electrons in Si and Cl atoms is 4.29+ and 6.12+, respectively. (**a**) What value do you estimate for Z_{eff} experienced by the outermost electron in both Si and Cl by assuming core electrons contribute 1.00 and valence electrons contribute 0.00 to the screening constant? (**b**) What values do you estimate for Z_{eff} using Slater's rules? (**c**) Which approach gives a more accurate estimate of Z_{eff}? (**d**) Which method of approximation more accurately accounts for the steady increase in Z_{eff} that occurs upon moving left to right across a period? (**e**) Predict Z_{eff} for a valence electron in P, phosphorus, based on the calculations for Si and Cl.

7.17 Which will experience the greater effective nuclear charge, the electrons in the $n = 3$ shell in Ar or the $n = 3$ shell in Kr? Which is more likely to be closer to the nucleus?

7.18 Arrange the following atoms in order of increasing effective nuclear charge experienced by the electrons in the $n = 3$ electron shell: K, Mg, P, Rh, Ti.

Atomic and Ionic Radii; Electronic Configuration of Ions (Section 7.3)

7.19 If you look up the radius of the Cl atom in various online sources, you will find values ranging from 79 to 182 pm (picometers; 1 pm = 1×10^{-12} m). The most common values are 100 pm and 182 pm. What can you conclude from these data? (**a**) Chemists must be doing something wrong if they cannot agree on a simple number like this. (**b**) The 100 pm number is probably the van der Waals radius, whereas the 182 pm number is probably the bonding atomic radius. (**c**) The 100 pm number is probably the bonding atomic radius, whereas the 182 pm number is probably the van der Waals radius. (**d**) The databases are probably including Cl ions in with atoms; so the 100 pm value is probably the chlorine anion radius, whereas the 182 pm value is probably the Cl atomic radius.

7.20 With the exception of helium, the noble gases condense to form solids when they are cooled sufficiently. At temperatures below 83 K, argon forms a close-packed solid whose structure is shown below. (**a**) What is the apparent radius of an argon atom in solid argon, assuming the atoms touch as shown in this figure? (**b**) Is this value larger or smaller than the bonding atomic radius estimated for argon in Figure 7.7? (**c**) Based on this comparison, would you say that the atoms are held together by chemical bonds in solid argon?

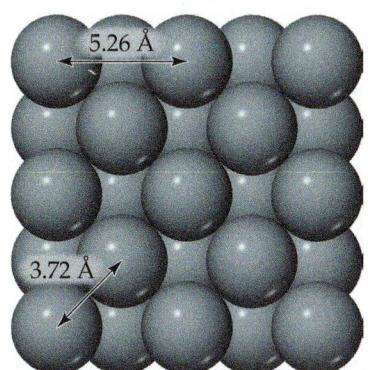

7.21 Tungsten has the highest melting point of any metal in the periodic table: 3422 °C. The distance between the centers of W atoms in tungsten metal is 2.74 Å. (**a**) What is the atomic radius of a tungsten atom in this environment? (This radius is called the *metallic radius*.) (**b**) If you put tungsten metal under high pressure, predict what would happen to the distance between W atoms.

7.22 Which of the following statements about the bonding atomic radii in Figure 7.7 is *incorrect*?

(**a**) For a given period, the radii of the representative elements generally decrease from left to right across a period.

(**b**) The radii of the representative elements for the $n = 3$ period are all larger than those of the corresponding elements in the $n = 2$ period.

(**c**) For most of the representative elements, the change in radius from the $n = 2$ to the $n = 3$ period is greater than the change in radius from $n = 3$ to $n = 4$.

(**d**) The radii of the transition elements generally increase moving from left to right within a period. (**v**) The large radii of the group 1A elements are due to their relatively small effective nuclear charges.

7.23 Estimate the As—I bond length from the data in Figure 7.7 and compare your value to the experimental As—I bond length in arsenic triiodide, AsI_3, 2.55 Å.

7.24 The experimental Bi—I bond length in bismuth triiodide, BiI_3, is 2.81 Å. Based on this value and the data in Figure 7.7, predict the atomic radius of Bi.

7.25 Using only the periodic table, arrange each set of atoms in order from largest to smallest. (**a**) K, Li, Cs, (**b**) Pb, Sn, Si, (**c**) F, O, N

7.26 Using only the periodic table, arrange each set of atoms in order of increasing radius. (**a**) Ba, Ca, Na, (**b**) In, Sn, As, (**c**) Al, Be, Si.

7.27 Identify each statement as true or false. (**a**) Cations are larger than their corresponding neutral atoms. (**b**) Li^+ is smaller than Li. (**c**) Cl^- is bigger than I^-.

7.28 For each set of atoms and ions, pick the smallest one.

(**a**) I, I^+, I^-

(**b**) Be^{2+}, Ca^{2+}, Mg^{2+}

(**c**) Fe, Fe^{2+}, Fe^{3+}

7.29 For each ion, identify the neutral atom that is isoelectronic with it. (**a**) Ga^{3+} (**b**) Zr^{4+} (**c**) Mn^{7+} (**d**) I^- (**e**) Pb^{2+}

7.30 For each ion, identify the neutral atom that is isoelectronic with it. (**a**) Cl^- (**b**) Sc^{3+} (**c**) Fe^{2+} (**d**) Zn^{2+} (**e**) Sn^{4+}

7.31 Consider the isoelectronic ions F^- and Na^+. (**a**) Which ion is smaller? (**b**) Using Equation 7.1 and assuming that core electrons contribute 1.00 and valence electrons contribute 0.00 to the screening constant, S, calculate Z_{eff} for the $2p$ electrons in both ions. (**c**) Repeat this calculation using Slater's rules to estimate the screening constant, S. (**d**) For isoelectronic ions, how are effective nuclear charge and ionic radius related?

7.32 Consider the isoelectronic ions Cl^- and K^+. (**a**) Which ion is smaller? (**b**) Using Equation 7.1 and assuming that core electrons contribute 1.00 and valence electrons contribute nothing to the screening constant, S, calculate Z_{eff} for these two ions. (**c**) Repeat this calculation using Slater's rules to estimate the screening constant, S. (**d**) For isoelectronic ions, how are effective nuclear charge and ionic radius related?

7.33 Consider S, Cl, and K and their most common ions. (**a**) List the atoms in order of increasing size. (**b**) List the ions in order of increasing size. (**c**) Explain any differences in the orders of the atomic and ionic sizes.

7.34 Arrange each of the following sets of atoms and ions in order of increasing size. (**a**) Se^{2-}, Te^{2-}, Se (**b**) Co^{3+}, Fe^{2+}, Fe^{3+} (**c**) Ca, Ti^{4+}, Sc^{3+} (**d**) Be^{2+}, Na^+, Ne

7.35 True or false? (**a**) O^{2-} is smaller than O. (**b**) S^{2-} is smaller than O^{2-}. (**c**) S^{2-} is larger than K^+. (**d**) K^+ is larger than Ca^{2+}.

7.36 In the ionic compounds LiF, NaCl, KBr, and RbI, the measured cation–anion distances are 2.01 Å (Li–F), 2.82 Å (Na–Cl), 3.30 Å (K–Br), and 3.67 Å (Rb–I), respectively. (**a**) Predict the cation–anion distance using the values of ionic radii given in Figure 7.8. (**b**) Calculate the difference between the experimentally measured ion–ion distances and the ones predicted from Figure 7.8. (**c**) What estimates of the cation–anion distance would you obtain for these four compounds using neutral atom *bonding atomic radii*? Are these estimates as accurate as the estimates using ionic radii?

7.37 Write the electron configurations for the following ions, and determine which have noble-gas configurations. (**a**) Co^{2+} (**b**) Sn^{2+} (**c**) Zr^{4+} (**d**) Ag^+ (**e**) S^{2-}

7.38 Write the electron configurations for the following ions, and determine which have noble-gas configurations. (**a**) Ru^{3+} (**b**) As^{3-} (**c**) Y^{3+} (**d**) Pd^{2+} (**e**) Pb^{2+} (**f**) Au^{3+}

7.39 Which of the ions Ni^{2+}, Fe^{2+}, Co^{3+}, and Pt^{2+} has an electron configuration of nd^8 ($n = 3, 4, 5, \ldots$)? (**a**) Ni^{2+} (**b**) Fe^{2+} (**c**) Co^{3+} (**d**) Pt^{2+} (**e**) More than one of these

7.40 Which of the ions Ni^{2+}, Fe^{2+}, Co^{3+}, and Pt^{2+} has an electron configuration of nd^6 ($n = 3, 4, 5, \ldots$)? (**a**) Ni^{2+} (**b**) Fe^{2+} (**c**) Co^{3+} (**d**) Pt^{2+} (**e**) More than one of these

Ionization Energy and Electron Affinity (Section 7.4)

7.41 (**a**) Write an equation for the second electron affinity of chlorine. (**b**) Would you predict a positive or a negative quantity for this process?

7.42 True or false: If the electron affinity for an element is a negative number, then the anion of the element is more stable than the neutral atom.

7.43 Write equations that show the processes that describe the first, second, and third ionization energies of an aluminum atom. Which process would require the least amount of energy?

7.44 Write equations that show the process for (**a**) the first two ionization energies of lead and (**b**) the fourth ionization energy of zirconium.

7.45 Which element has the highest second ionization energy: Li, K, or Be?

7.46 Identify each statement as true or false. (**a**) Ionization energies are always negative quantities. (**b**) Oxygen has a larger first ionization energy than fluorine. (**c**) The second ionization energy of an atom is always greater than its first ionization energy. (**d**) The third ionization energy is the energy needed to ionize three electrons from a neutral atom.

7.47 (**a**) Pick the correct word to complete the sentence: The larger the atom, the (smaller/larger) its first ionization energy. (**b**) Which element in the periodic table has the largest first ionization energy? (**c**) Which element has the smallest first ionization energy?

7.48 (**a**) What is the trend in first ionization energies as one proceeds down the group 7A elements? (**b**) Do the atomic radii of the group 7A elements show the same trend as first ionization energies? (**c**) What is the trend in first ionization energies as one moves across the fourth period from K to Kr? (**d**) Do the atomic radii of period 4 elements show the same trend as first ionization energies?

7.49 Based on their positions in the periodic table, predict which atom of the following pairs will have the smaller first ionization energy. (**a**) Cl, Ar, (**b**) Be, Ca, (**c**) K, Co, (**d**) S, Ge, (**e**) Sn, Te

7.50 For each of the following pairs, predict which element has the smaller first ionization energy. (**a**) Ti, Ba, (**b**) Ag, Cu, (**c**) Ge, Cl, (**d**) Pb, Sb

7.51 Would a neutral K atom or a K^+ ion have a more negative value of electron affinity?

7.52 What is the relationship between the ionization energy of an anion with a $1-$ charge such as F^- and the electron affinity of the neutral atom, F?

7.53 Consider the first ionization energy of neon and the electron affinity of fluorine. (**a**) Write equations, including electron configurations, for each process. (**b**) These two quantities have opposite signs. Which will be positive, and which will be negative? (**c**) Predict which of these quantities will be larger in magnitude.

7.54 Consider the following equation:

$$Ca^+(g) + e^- \longrightarrow Ca(g)$$

Which of the following statements are true?

(**i**) The energy change for this process is the electron affinity of the Ca^+ ion.

(**ii**) The energy change for this process is the negative of the first ionization energy of the Ca atom.

(**iii**) The energy change for this process is the negative of the electron affinity of the Ca atom.

(**a**) Only statement i is true. (**b**) Only statement ii is true. (**c**) Only statement iii is true. (**d**) Only statements i and ii are true. (**e**) All three statements are true.

Properties of Metals and Nonmetals (Section 7.5)

7.55 (**a**) Does metallic character increase, decrease, or remain unchanged as you go from left to right across a row of the periodic table? (**b**) Does metallic character increase, decrease, or remain unchanged as you go down a column of the periodic table? (**c**) Are the periodic trends in (a) and (b) the same as or different from those for first ionization energy?

7.56 You read the following statement about two solid elements X and Y: Experiments show that the first ionization energy of X is twice as great as that of Y. Which element has the greater metallic character?

7.57 True or false: An element that commonly forms a cation is a metal.

7.58 True or false: Because elements that form cations are metals, and elements that form anions are nonmetals, elements that do not form ions are metalloids.

7.59 Predict whether each of the following oxides is ionic or molecular: SnO_2, Al_2O_3, CO_2, Li_2O, Fe_2O_3, H_2O.

7.60 Some metal oxides, such as Sc_2O_3, do not react with pure water, but they do react when the solution becomes either acidic or basic. Do you expect Sc_2O_3 to react when the solution becomes acidic or when it becomes basic? Write a balanced chemical equation to support your answer.

7.61 Would you expect manganese(II) oxide, MnO, to react more readily with $HCl(aq)$ or $NaOH(aq)$?

7.62 Arrange the following oxides in order of increasing acidity: CO_2, CaO, Al_2O_3, SO_3, SiO_2, P_2O_5.

7.63 Chlorine reacts with oxygen to form Cl_2O_7. (**a**) What is the name of this product (see Table 2.6)? (**b**) Write a balanced equation for the formation of $Cl_2O_7(l)$ from the elements. (**c**) Would you expect Cl_2O_7 to be more reactive toward $H^+(aq)$ or $OH^-(aq)$? (**d**) If the oxygen in Cl_2O_7 is considered to have the -2 oxidation state, what is the oxidation state of the Cl? What is the electron configuration of Cl in this oxidation state?

7.64 An element X reacts with oxygen to form XO_2 and with chlorine to form XCl_4. XO_2 is a white solid that melts at high temperatures (above 1000 °C). Under usual conditions, XCl_4 is a colorless liquid with a boiling point of 58 °C. (**a**) XCl_4 reacts with water to form XO_2 and another product. What is the likely identity of the other product? (**b**) Do you think that element X is a metal, nonmetal, or metalloid? (**c**) Which is the most likely identity of X: Zr, C, Si, P, or S?

7.65 Write balanced equations for the following reactions. (**a**) barium oxide with water, (**b**) iron(II) oxide with perchloric acid, (**c**) sulfur trioxide with water, (**d**) carbon dioxide with aqueous sodium hydroxide.

7.66 Write balanced equations for the following reactions. (**a**) potassium oxide with water, (**b**) diphosphorus trioxide with water, (**c**) chromium(III) oxide with dilute hydrochloric acid, (**d**) selenium dioxide with aqueous potassium hydroxide.

Group Trends in Metals and Nonmetals (Sections 7.6 and 7.7)

7.67 Arrange these elements in order of increasing reactivity: Ca, Mg, K.

7.68 Silver and rubidium both form +1 ions. Predict which of these elements is more reactive to form ionic compounds.

7.69 Write a balanced equation for the reaction that occurs in each of the following cases. (**a**) Potassium metal is exposed to an atmosphere of chlorine gas. (**b**) Strontium oxide is added to water. (**c**) A fresh surface of lithium metal is exposed to oxygen gas. (**d**) Sodium metal reacts with molten sulfur.

7.70 Write a balanced equation for the reaction that occurs in each of the following cases. (**a**) Cesium is added to water. (**b**) Strontium is added to water. (**c**) Sodium reacts with oxygen. (**d**) Calcium reacts with iodine.

7.71 (**a**) As described in Section 7.7, the alkali metals react with hydrogen to form hydrides and react with halogens to form halides. Compare the roles of hydrogen and halogens in these reactions. Write balanced equations for the reaction of fluorine with calcium and for the reaction of hydrogen with calcium. (**b**) What is the oxidation number and electron configuration of calcium in each product?

7.72 Potassium and hydrogen react to form the ionic compound potassium hydride. (**a**) Write a balanced equation for this reaction. (**b**) Use the data in Figures 7.11 and 7.13 to determine the energy change in kJ/mol for the following two reactions:

$$K(g) + H(g) \longrightarrow K^+(g) + H^-(g)$$
$$K(g) + H(g) \longrightarrow K^-(g) + H^+(g)$$

(**c**) Based on your calculated energy changes in (b), which of these reactions is energetically more favorable (or less unfavorable)? (**d**) Is your answer to (c) consistent with the description of potassium hydride as containing hydride ions?

7.73 Consider the elements bromine and chlorine. For each element, write its: (**a**) electron configuration, (**b**) most common ionic charge, (**c**) first ionization energy, (**d**) primary reaction toward water, (**e**) electron affinity, (**f**) atomic radius.

7.74 Little is known about the properties of astatine, At, because of its rarity and high radioactivity. Nevertheless, it is possible for us to make many predictions about its properties. (**a**) Do you expect the element to be a gas, liquid, or solid at room temperature? (**b**) Would you expect At to be a metal, nonmetal, or metalloid? (**c**) What is the chemical formula of the compound it forms with Na?

7.75 Until the early 1960s, the group 8A elements were called the inert gases. (**a**) Why was the term *inert gases* dropped? (**b**) What discovery triggered this change in name? (**c**) What name is applied to the group now?

7.76 (**a**) Why does xenon react with fluorine, whereas neon does not? (**b**) Using appropriate reference sources, look up the bond lengths of Xe—F bonds in several molecules. How do these numbers compare to the bond lengths calculated from the atomic radii of the elements?

7.77 Write a balanced equation for the reaction that occurs in each of the following cases. (**a**) Ozone decomposes to dioxygen. (**b**) Xenon reacts with fluorine. (Write three different equations.) (**c**) Sulfur reacts with hydrogen gas. (**d**) Fluorine reacts with water.

7.78 Write a balanced equation for the reaction that occurs in each of the following cases. (**a**) Chlorine reacts with water. (**b**) Barium metal is heated in an atmosphere of hydrogen gas. (**c**) Lithium reacts with sulfur. (**d**) Fluorine reacts with magnesium metal.

Additional Exercises

7.79 Consider the stable elements through lead ($Z = 82$). In how many instances are the atomic weights of the elements out of order relative to the atomic numbers of the elements?

7.80 Figure 7.4 shows the radial probability distribution functions for the 2s orbitals and 2p orbitals. (**a**) Which orbital, 2s or 2p, has more electron density close to the nucleus? (**b**) How would you modify Slater's rules to adjust for the difference in electronic penetration of the nucleus for the 2s and 2p orbitals?

7.81 (**a**) If the core electrons were totally effective at screening the valence electrons and the valence electrons provided no screening for each other, what would be the effective nuclear charge acting on the 3s and 3p valence electrons in P? (**b**) Repeat these calculations using Slater's rules. (**c**) Detailed calculations indicate that the effective nuclear charge is 5.6+ for the 3s electrons and 4.9+ for the 3p electrons. Why are the values for the 3s and 3p electrons different? (**d**) If you remove a single electron from a P atom, which orbital will it come from?

7.82 In the series of group 5A hydrides, of general formula MH_3, the measured bond distances are P—H, 1.419 Å; As—H, 1.519 Å; Sb—H, 1.707 Å. (**a**) Compare these values with those estimated by use of the atomic radii in Figure 7.7. (**b**) Explain the steady increase in M—H bond distance in this series in terms of the electron configurations of the M atoms.

7.83 In Table 7.8, the bonding atomic radius of neon is listed as 0.58 Å, whereas that for xenon is listed as 1.40 Å. A classmate of yours states that the value for Xe is more realistic than the one for Ne. Is she correct? If so, what is the basis for her statement?

7.84 The As—As bond length in elemental arsenic is 2.48 Å. The Cl—Cl bond length in Cl_2 is 1.99 Å. (**a**) Based on these data, what is the predicted As—Cl bond length in arsenic trichloride, $AsCl_3$, in which each of the three Cl atoms is bonded to the As atom? (**b**) What bond length is predicted for $AsCl_3$, using the atomic radii in Figure 7.7?

7.85 The following observations are made about two hypothetical elements A and B: The A—A and B—B bond lengths in the elemental forms of A and B are 2.36 and 1.94 Å, respectively. A and B react to form the binary compound AB_2, which has a *linear* structure (that is, $\angle B-A-B = 180°$). Based on these statements, predict the separation between the two B nuclei in a molecule of AB_2.

7.86 Elements in group 7A in the periodic table are called the halogens; elements in group 6A are called the chalcogens. (**a**) What is the most common oxidation state of the chalcogens compared to the halogens? (**b**) For each of the following periodic properties, state whether the halogens or the chalcogens have larger values: atomic radii, ionic radii of the most common oxidation state, first ionization energy, second ionization energy.

7.87 (**a**) Which ion is smaller, Co^{3+} or Co^{4+}? (**b**) In a lithium-ion battery that is discharging to power a device, for every Li^+ that inserts into the lithium cobalt oxide electrode, a Co^{4+} ion must be reduced to a Co^{3+} ion to balance charge. Using the *CRC Handbook of Chemistry and Physics* or other standard reference, find the ionic radii of Li^+, Co^{3+}, and Co^{4+}. Order these ions from smallest to largest. (**c**) Will the lithium cobalt oxide cathode expand or contract as lithium ions are inserted? (**d**) Lithium is not nearly as abundant as sodium. If sodium-ion batteries were developed that function in the same manner as lithium-ion batteries, do you think "sodium cobalt oxide" would still work as the electrode material? Explain.

(e) If you don't think cobalt would work as the redox-active partner ion in the sodium version of the electrode, suggest an alternative metal ion and explain your reasoning.

7.88 The ionic substance strontium oxide, SrO, forms from the reaction of strontium metal with molecular oxygen. The arrangement of the ions in solid SrO is analogous to that in solid NaCl:

(a) Write a balanced equation for the formation of SrO(s) from its elements. **(b)** Based on the ionic radii in Figure 7.8, predict the length of the side of the cube in the figure (the distance from the center of an atom at one corner to the center of an atom at a neighboring corner). **(c)** The density of SrO is 5.10 g/cm³. Given your answer to part (b), how many formula units of SrO are contained in the cube shown here?

7.89 Explain the variation in the ionization energies of carbon, as displayed in this graph:

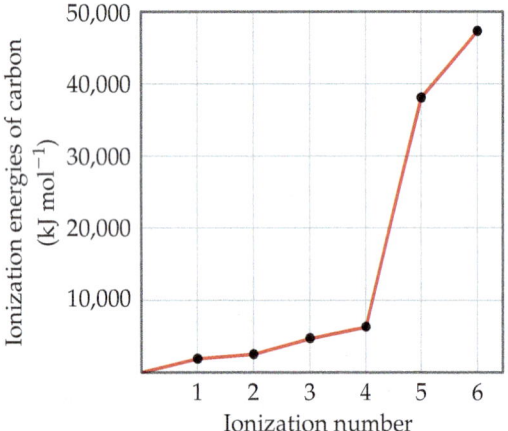

7.90 Group 4A elements have much more negative electron affinities than their neighbors in groups 3A and 5A (see Figure 7.13). Which of the following statements best explains this observation?

(a) The group 4A elements have much higher first ionization energies than their neighbors in groups 3A and 5A.

(b) The addition of an electron to a group 4A element leads to a half-filled np^3 outer electron configuration.

(c) The group 4A elements have unusually large atomic radii.

(d) The group 4A elements are easier to vaporize than are the group 3A and 5A elements.

7.91 In the chemical process called *electron transfer*, an electron is transferred from one atom or molecule to another. (We discuss electron transfer extensively in Chapter 20.) A simple electron transfer reaction is

$$A(g) + A(g) \longrightarrow A^+(g) + A^-(g)$$

In terms of the ionization energy and electron affinity of atom A, what is the energy change for this reaction? For a representative nonmetal such as chlorine, is this process

exothermic? For a representative metal such as sodium, is this process exothermic?

7.92 **(a)** Use orbital diagrams to illustrate what happens when an oxygen atom gains two electrons. **(b)** Why does O^{3-} not exist?

7.93 Use electron configurations to explain the following observations: **(a)** The first ionization energy of phosphorus is greater than that of sulfur. **(b)** The electron affinity of nitrogen is lower (less negative) than those of both carbon and oxygen. **(c)** The second ionization energy of oxygen is greater than the first ionization energy of fluorine. **(d)** The third ionization energy of manganese is greater than those of both chromium and iron.

7.94 Identify two ions that have the following ground-state electron configurations: **(a)** [Ar] **(b)** $[Ar]3d^5$ **(c)** $[Kr]5s^24d^{10}$

7.95 Which of the following chemical equations is connected to the definitions of **(a)** the first ionization energy of oxygen, **(b)** the second ionization energy of oxygen, and **(c)** the electron affinity of oxygen?

(i) $O(g) + e^- \longrightarrow O^-(g)$

(ii) $O(g) \longrightarrow O^+(g) + e^-$

(iii) $O(g) + 2e^- \longrightarrow O^{2-}(g)$

(iv) $O(g) \longrightarrow O^{2+}(g) + 2e^-$

(v) $O^+(g) \longrightarrow O^{2+}(g) + e^-$

7.96 Hydrogen is an unusual element because it behaves in some ways like the alkali metal elements and in other ways like nonmetals. Its properties can be explained in part by its electron configuration and by the values for its ionization energy and electron affinity. **(a)** Explain why the electron affinity of hydrogen is much closer to the values for the alkali elements than for the halogens. **(b)** Is the following statement true? "Hydrogen has the smallest bonding atomic radius of any element that forms chemical compounds." If not, correct it. If it is correct, explain in terms of electron configurations. **(c)** Explain why the ionization energy of hydrogen is closer to the values for the halogens than for the alkali metals. **(d)** The hydride ion is H^-. Write out the process corresponding to the first ionization energy of the hydride ion. **(e)** How does the process in part (d) compare to the process for the electron affinity of a neutral hydrogen atom?

7.97 The first ionization energy of the oxygen molecule is the energy required for the following process:

$$O_2(g) \longrightarrow O_2^+(g) + e^-$$

The energy needed for this process is 1175 kJ/mol, very similar to the first ionization energy of Xe. Would you expect O_2 to react with F_2? If so, suggest a product or products of this reaction.

7.98 It is possible to define *metallic character* as we do in this book and base it on the reactivity of the element and the ease with which it loses electrons. Alternatively, you could measure how well electricity is conducted by each of the elements to determine how "metallic" the elements are. On the basis of conductivity, there is not much of a trend in the periodic table: Silver is the most conductive metal, and manganese the least. Look up the first ionization energies of silver and manganese; which of these two elements would you call more metallic based on the way we define it in this book?

7.99 Which of the following is the expected product of the reaction of K(s) and $H_2(g)$? **(a)** KH(s), **(b)** $K_2H(s)$, **(c)** $KH_2(s)$, **(d)** $K_2H_2(s)$, **(e)** K(s) and $H_2(g)$ will not react with one another.

7.100 Elemental cesium reacts more violently with water than does elemental sodium. Which of the following best explains this difference in reactivity? **(a)** Sodium has greater metallic character than does cesium. **(b)** The first ionization energy of cesium is less than that of sodium. (**c**) The electron affinity of sodium is smaller than that of cesium. (**d**) The effective nuclear charge for cesium is less than that of sodium. (**e**) The atomic radius of cesium is smaller than that of sodium.

7.101 (**a**) One of the alkali metals reacts with oxygen to form a solid white substance. When this substance is dissolved in water, the solution gives a positive test for hydrogen peroxide, H_2O_2. When the solution is tested in a burner flame, a lilac-purple flame is produced. What is the likely identity of the metal? (**b**) Write a balanced chemical equation for the reaction of the white substance with water.

7.102 A historian discovers a nineteenth-century notebook in which some observations, dated 1822, were recorded on a substance thought to be a new element. Here are some of the data recorded in the notebook: "Ductile, silver-white, metallic looking. Softer than lead. Unaffected by water. Stable in air. Melting point: 153 °C. Density: 7.3 g/cm³. Electrical conductivity: 20% that of copper. Hardness: About 1% as hard as iron. When 4.20 g of the unknown is heated in an excess of oxygen, 5.08 g of a white solid is formed. The solid could be sublimed by heating to over 800 °C." (**a**) Using information in the text and the *CRC Handbook of Chemistry and Physics*, and making allowances for possible variations in numbers from current values, identify the element reported. (**b**) Write a balanced chemical equation for the reaction with oxygen. (**c**) Judging from Figure 7.1, might this nineteenth-century investigator have been the first to discover a new element?

7.103 In April 2010, a research team reported that it had made Element 117. This discovery was confirmed in 2012 by additional experiments. Write the ground-state electron configuration for Element 117 and estimate values for its first ionization energy, electron affinity, atomic size, and common oxidation state based on its position in the periodic table.

7.104 We show in Chapter 12 that semiconductors are materials that conduct electricity better than nonmetals but not as well as metals. The only two elements in the periodic table that are technologically useful semiconductors are silicon and germanium. Integrated circuits in computer chips today are based on silicon. Compound semiconductors are also used in the electronics industry. Examples are gallium arsenide, GaAs; gallium phosphide, GaP; cadmium sulfide, CdS; and cadmium selenide, CdSe. (**a**) What is the relationship between the compound semiconductors' compositions and the positions of their elements on the periodic table relative to Si and Ge? (**b**) Workers in the semiconductor industry refer to "II–VI" and "III–V" materials, using Roman numerals. Can you identify which compound semiconductors are II–VI and which are III–V? (**c**) Suggest other compositions of compound semiconductors based on the positions of their elements in the periodic table.

7.105 Moseley established the concept of atomic number by studying X rays emitted by the elements. The X rays emitted by some of the elements have the following wavelengths:

Element	Wavelength (Å)
Ne	14.610
Ca	3.358
Zn	1.435
Zr	0.786
Sn	0.491

(**a**) Calculate the frequency, ν, of the X rays emitted by each of the elements, in Hz. (**b**) Plot the square root of ν versus the atomic number of the element. What do you observe about the plot? (**c**) Explain how the plot in part (b) allowed Moseley

to predict the existence of undiscovered elements. (**d**) Use the result from part (b) to predict the X-ray wavelength emitted by iron. (**e**) A particular element emits X rays with a wavelength of 0.980 Å. What element do you think it is?

7.106 (**a**) Write the electron configuration for Li and estimate the effective nuclear charge experienced by the valence electron. (**b**) The energy of an electron in a one-electron atom or ion equals $\left(-2.18 \times 10^{-18} \text{J}\right)\left(\dfrac{Z^2}{n^2}\right)$, where Z is the nuclear charge and n is the principal quantum number of the electron. Estimate the first ionization energy of Li. (**c**) Compare the result of your calculation with the value reported in Table 7.4 and explain the difference. (**d**) What value of the effective nuclear charge gives the proper value for the ionization energy? Does this agree with your explanation in part (c)?

7.107 One way to measure ionization energies is ultraviolet photoelectron spectroscopy (PES), a technique based on the photoelectric effect. (Section 6.2) In PES, monochromatic light is directed onto a sample, causing electrons to be emitted. The kinetic energy of the emitted electrons is measured. The difference between the energy of the photons and the kinetic energy of the electrons corresponds to the energy needed to remove the electrons (that is, the ionization energy). Suppose that a PES experiment is performed in which mercury vapor is irradiated with ultraviolet light of wavelength 58.4 nm. (**a**) What is the energy of a photon of this light, in joules? (**b**) Write an equation that shows the process corresponding to the first ionization energy of Hg. (**c**) The kinetic energy of the emitted electrons is measured to be 1.72×10^{-18} J. What is the first ionization energy of Hg, in kJ/mol? (**d**) Using Figure 7.11, determine which of the halogen elements has a first ionization energy closest to that of mercury.

7.108 When magnesium metal is burned in air (Figure 3.5), two products are produced. One is magnesium oxide, MgO. The other is the product of the reaction of Mg with molecular nitrogen, magnesium nitride. When water is added to magnesium nitride, it reacts to form magnesium oxide and ammonia gas. (**a**) Based on the charge of the nitride ion (Table 2.5), predict the formula of magnesium nitride. (**b**) Write a balanced equation for the reaction of magnesium nitride with water. What is the driving force for this reaction? (**c**) In an experiment, a piece of magnesium ribbon is burned in air in a crucible. The mass of the mixture of MgO and magnesium nitride after burning is 0.470 g. Water is added to the crucible, further reaction occurs, and the crucible is heated to dryness until the final product is 0.486 g of MgO. What was the mass percentage of magnesium nitride in the mixture obtained after the initial burning? (**d**) Magnesium nitride can also be formed by reaction of the metal with ammonia at high temperature. Write a balanced equation for this reaction. If a 6.3-g Mg ribbon reacts with 2.57 g $NH_3(g)$ and the reaction goes to completion, which component is the limiting reactant? What mass of $H_2(g)$ is formed in the reaction? (**e**) The standard enthalpy of formation of solid magnesium nitride is -461.08 kJ/mol. Calculate the standard enthalpy change for the reaction between magnesium metal and ammonia gas.

7.109 Potassium superoxide, KO_2, is often used in oxygen masks (such as those used by firefighters) because KO_2 reacts with CO_2 to release molecular oxygen. Experiments indicate that 2 mol of $KO_2(s)$ react with each mole of $CO_2(g)$. (**a**) The products of the reaction are $K_2CO_3(s)$ and $O_2(g)$. Write a balanced equation for the reaction between $KO_2(s)$ and $CO_2(g)$. (**b**) Indicate the oxidation number for each atom involved in the reaction in part (a). What elements are being oxidized and reduced? (**c**) What mass of $KO_2(s)$ is needed to consume 18.0 g $CO_2(g)$? What mass of $O_2(g)$ is produced during this reaction?

Design an Experiment

In this chapter we have seen that the reaction of potassium metal with oxygen leads to a product that we might not expect—namely, potassium superoxide, $KO_2(s)$. Let's design some experiments to learn more about this unusual product.

(a) One of your team members proposes that the capacity to form a superoxide such as KO_2 is related to a low value for the first ionization energy. How would you go about testing this hypothesis for the metals of group 1A? What other periodic property of the alkali metals might be considered as a factor favoring superoxide formation?

(b) $KO_2(s)$ is the active ingredient in many breathing masks used by firefighters because it can be used as a source of $O_2(g)$. In principle, $KO_2(s)$ can react with both major components of human breath, $H_2O(g)$ and $CO_2(g)$, to produce $O_2(g)$ and other products (all of which follow the expected patterns of reactivity we have seen). Predict the other products in these reactions and design experiments to determine whether $KO_2(s)$ does actually react with both $H_2O(g)$ and $CO_2(g)$.

(c) Propose an experiment to determine whether either of the reactions in part (b) is more important in the operation of a firefighter's breathing mask.

(d) The reaction of $K(s)$ and $O_2(g)$ leads to a mixture of $KO_2(s)$ and $K_2O(s)$. Use ideas presented in this exercise to design an experiment to determine the percentages of $KO_2(s)$ and $K_2O(s)$ in the product mixture that results from the reaction of $K(s)$ with excess $O_2(g)$.

8

BASIC CONCEPTS OF CHEMICAL BONDING

▲ **GIANT CRYSTALS.** These gypsum crystals, large enough for people to walk on, are composed of $CaSO_4 \cdot 2H_2O$. The ionic bonding between calcium and sulfate ions at the atomic scale leads to the characteristic crystal shape at the human scale.

WHAT'S AHEAD

8.1 ▶ Lewis Symbols and the Octet Rule Learn about the three main types of chemical bonds: *ionic*, *covalent*, and *metallic*. In discussing bonding, *Lewis symbols* provide a useful shorthand for keeping track of valence electrons. We observe that atoms usually follow the *octet rule*.

8.2 ▶ Ionic Bonding Explore ionic substances, in which atoms are held together by the electrostatic attractions between ions of opposite charge. Analyze the energetics of forming ionic substances and describe the *lattice energy* of these substances.

8.3 ▶ Covalent Bonding Examine the bonding in molecular substances, in which atoms bond by sharing one or more electron pairs. In general, the electrons are shared in such a way that each atom attains an octet of electrons.

8.4 ▶ Bond Polarity and Electronegativity Learn that *electronegativity* is the ability of an atom in a compound to attract electrons to itself. In general, electron pairs are shared unequally between atoms with different electronegativities, leading to *polar covalent bonds*.

8.5 ▶ Drawing Lewis Structures Learn that *Lewis structures* are a simple yet powerful way of predicting covalent bonding patterns in molecules. In addition to the octet rule, see how the concept of *formal charge* can be used to identify the dominant Lewis structure.

8.6 ▶ Resonance Structures We see that, in some cases, more than one equivalent Lewis structure can be drawn for a molecule or polyatomic ion. The bonding description in such cases is a blend of two or more *resonance structures*.

8.7 ▶ Exceptions to the Octet Rule Recognize that the octet rule is more of a guideline than an absolute rule. Exceptions to the rule include molecules with an odd number of electrons, molecules where large differences in electronegativity prevent an atom from completing its octet, and molecules where an element from period 3 or below in the periodic table attains more than an octet of electrons.

8.8 ▶ Strengths and Lengths of Single and Multiple Bonds Observe that bond strengths and lengths vary with the number of shared electron pairs between a given atom pair.

Whenever two atoms or ions are strongly held together, we say there is a **chemical bond** between them. There are three general types of chemical bonds: *ionic, covalent,* and *metallic*. These three types of bonds are in the substances illustrated in **Figure 8.1**: table salt, water, and stainless steel.

Table salt is sodium chloride, NaCl, which consists of sodium ions, Na^+, and chloride ions, Cl^-. The structure is held together by **ionic bonds**, which are due to the electrostatic attractions between oppositely charged ions. Water consists mainly of H_2O molecules. The hydrogen and oxygen atoms are bonded to one another through **covalent bonds**, in which molecules are formed by the sharing of electrons between atoms. The spoon consists mainly of iron metal, in which Fe atoms are connected to one another by **metallic bonds**, which are formed by electrons that are relatively free to move from one atom to another. These different substances—NaCl, H_2O, and Fe metal—behave as they do because of the ways in which their constituent atoms are connected to one another. For example, NaCl readily dissolves in water, but Fe metal does not.

In this chapter and the next, we examine the relationship between the electronic structure of atoms and the ionic and covalent chemical bonds they form. We discuss metallic bonding in Chapter 12.

Learning Objective

When you finish Section 8.1, you should be able to:

▶ Draw Lewis symbols for atoms and use them to determine how many electrons an atom must lose or gain to obey the octet rule.

Go Figure

If the white powder were sugar, $C_{12}H_{22}O_{11}$, how would we have to change this picture?

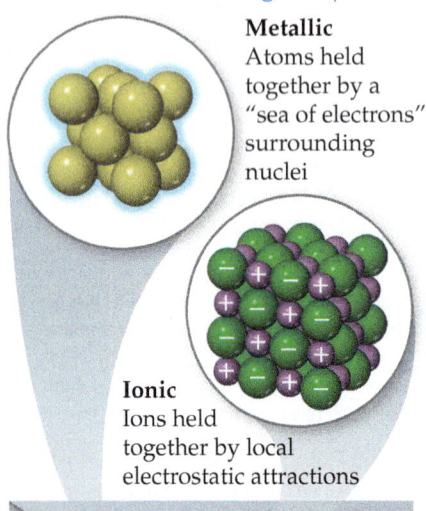

Metallic
Atoms held together by a "sea of electrons" surrounding nuclei

Ionic
Ions held together by local electrostatic attractions

Covalent
Atoms held together by sharing electrons in localized bonds

▲ **Figure 8.1 Ionic, covalent, and metallic bonds.** The three different substances shown here are held together by different types of chemical bonds.

8.1 | Lewis Symbols and the Octet Rule

The electrons involved in chemical bonding are the *valence electrons*, which, for most atoms, are those in the outermost occupied shell. (Section 6.8) The American chemist G. N. Lewis (1875–1946) suggested a simple way of showing the valence electrons in an atom and tracking them during bond formation, using what are now known as either *Lewis electron-dot symbols* or simply Lewis symbols. The **Lewis symbol** for an element consists of the element's chemical symbol plus a dot for each valence electron. Sulfur, for example, has the electron configuration $[Ne]3s^2 3p^4$ and therefore six valence electrons. Its Lewis symbol is

The dots are placed on the four sides of the symbol—top, bottom, left, and right—and each side can accommodate up to two electrons. All four sides are equivalent, which means that the choice of sides for placement of two electrons rather than one electron is arbitrary. In general, we spread out the dots as much as possible. In the Lewis symbol for S, for instance, we prefer the dot arrangement shown rather than the arrangement having two electrons on three of the sides and none on the fourth.

The electron configurations and Lewis symbols for the main-group elements of periods 2 and 3 are shown in **Figure 8.2**. Notice that the number of valence electrons in any representative element is the same as the element's group number. For example, the Lewis symbols for oxygen and sulfur, members of group 6A, both show six dots.

The Octet Rule

Atoms often gain, lose, or share electrons to achieve the same number of electrons as the noble gas closest to them in the periodic table. The noble gases have very stable electron arrangements, as evidenced by their high ionization energies, low affinity for additional electrons, and general lack of chemical reactivity. (Section 7.8) Because all the noble gases except He have eight valence electrons, many atoms undergoing reactions end up with eight valence electrons. This observation has led to a guideline known as the **octet rule**: *Atoms tend to gain, lose, or share electrons until they are surrounded by eight valence electrons.*

An octet of electrons consists of full *s* and *p* subshells in an atom. In a Lewis symbol, an octet is shown as four pairs of valence electrons arranged around the element symbol, as in the Lewis symbols for Ne and Ar in Figure 8.2. There are exceptions to the octet rule, as we explain in Section 8.7, but it provides a useful framework for introducing many important concepts of bonding. The octet rule mostly applies to atoms that have *s* and *p* valence electrons; compounds of the transition metals, having valence *d* electrons, are examined in Chapter 23.

Self-Assessment Exercises

SAE 8.1 Which of the following is the most correct Lewis symbol for an oxygen atom?

(a) $\ddot{O}\!:$ **(b)** $\because\!\ddot{O}\!:$ **(c)** $\because\!\dot{O}\!:$ **(d)** $\cdot\ddot{O}$

SAE 8.2 For each of the following Lewis symbols, indicate the group in the periodic table in which element Q belongs:

(i) $\cdot\dot{Q}$ **(ii)** $\because\!\dot{Q}\cdot$ **(iii)** $\cdot\dot{Q}\!:$

(a) i Group 1A, ii Group 5A, iii Group 3A **(b)** i Group 5A, ii Group 1A, iii Group 3A **(c)** i Group 0A, ii Group 3A, iii Group 1A **(d)** i Group 3A, ii Group 7A, iii Group 5A

Group	1A	2A	3A	4A	5A	6A	7A	8A
Element	**Li**	**Be**	**B**	**C**	**N**	**O**	**F**	**Ne**
Electron Configuration	$[He]2s^1$	$[He]2s^2$	$[He]2s^2 2p^1$	$[He]2s^2 2p^2$	$[He]2s^2 2p^3$	$[He]2s^2 2p^4$	$[He]2s^2 2p^5$	$[He]2s^2 2p^6$
Lewis Symbol	$Li\cdot$	$\cdot Be\cdot$	$\cdot\dot{B}\cdot$	$\cdot\dot{C}\cdot$	$\cdot\ddot{N}\cdot$	$:\ddot{O}:$	$:\ddot{F}:$	$:\ddot{Ne}:$
	Na	**Mg**	**Al**	**Si**	**P**	**S**	**Cl**	**Ar**
	$[Ne]3s^1$	$[Ne]3s^2$	$[Ne]3s^2 3p^1$	$[Ne]3s^2 3p^2$	$[Ne]3s^2 3p^3$	$[Ne]3s^2 3p^4$	$[Ne]3s^2 3p^5$	$[Ne]3s^2 3p^6$
	$Na\cdot$	$\cdot Mg\cdot$	$\cdot\dot{Al}\cdot$	$\cdot\dot{Si}\cdot$	$\cdot\ddot{P}\cdot$	$:\ddot{S}:$	$:\ddot{Cl}:$	$:\ddot{Ar}:$

▲ **Figure 8.2 Lewis symbols.**

8.2 | Ionic Bonding

Ionic substances generally result from the interaction of metals on the left side of the periodic table with nonmetals on the right side (excluding the noble gases, group 8A). In this section we look more closely at several aspects of the formation of ions and of ionic bonding.

When sodium metal, Na(*s*), is brought into contact with chlorine gas, $Cl_2(g)$, a violent reaction ensues (**Figure 8.3**). The product of this very exothermic reaction is sodium chloride, NaCl(*s*):

$$Na(s) + \tfrac{1}{2} Cl_2(g) \longrightarrow NaCl(s) \qquad \Delta H_f^{\circ} = -410.9 \text{ kJ} \qquad [8.1]$$

Sodium chloride is composed of Na^+ and Cl^- ions arranged in a three-dimensional array (**Figure 8.4**).

The formation of Na^+ from Na and Cl^- from Cl_2 indicates that an electron has been lost by a sodium atom and gained by a chlorine atom—we say there has been an *electron transfer* from the Na atom to the Cl atom. Two of the atomic properties discussed in Chapter 7 give us an indication of how readily electron transfer occurs: ionization energy, which shows how easily an electron can be removed from an atom; and electron affinity, which measures how much an atom wants to gain an electron. (Section 7.4) Electron transfer to form oppositely charged ions occurs when one atom readily gives up an electron (low ionization energy) and another atom readily gains an electron (high electron affinity). Thus, NaCl is a typical ionic compound because it consists of a metal of low ionization energy and a nonmetal of high electron affinity. Using Lewis electron-dot symbols (and showing a chlorine atom rather than the Cl_2 molecule), we can represent this reaction as

$$Na\cdot + \cdot\overset{\cdot\cdot}{\underset{\cdot\cdot}{Cl}}: \longrightarrow Na^+ + [:\overset{\cdot\cdot}{\underset{\cdot\cdot}{Cl}}:]^- \qquad [8.2]$$

The arrow indicates the transfer of an electron from the Na atom to the Cl atom. Each ion has an octet of electrons, the Na^+ octet being the $2s^2 2p^6$ electrons that lie below the single $3s$ valence electron of the Na atom. We have put a bracket around the chloride ion to emphasize that all eight electrons are located on it.

Ionic substances possess several characteristic properties. They are usually brittle substances with high melting points. They are usually crystalline. Furthermore, ionic crystals often can be cleaved; that is, they break apart along smooth, flat surfaces. These

Learning Objectives

When you finish **Section 8.2**, you should be able to:

▶ Analyze the factors that lead to the stability of ionic compounds.

▶ Order a series of ionic compounds from lowest to highest lattice energy.

▶ Use the octet rule to predict the preferred charges on ions.

Go Figure Do you expect a similar reaction between potassium metal and elemental bromine?

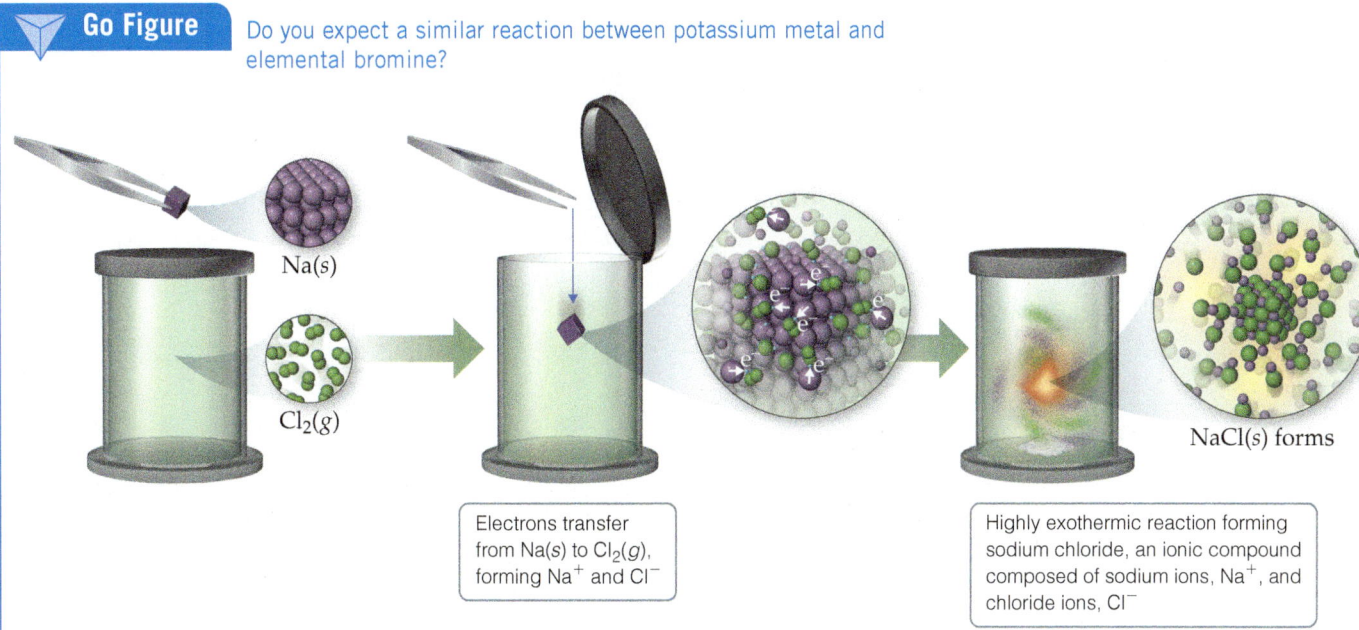

Na(*s*)

$Cl_2(g)$

Electrons transfer from Na(*s*) to $Cl_2(g)$, forming Na^+ and Cl^-

NaCl(*s*) forms

Highly exothermic reaction forming sodium chloride, an ionic compound composed of sodium ions, Na^+, and chloride ions, Cl^-

▲ **Figure 8.3** **Reaction of sodium metal with chlorine gas to form the ionic compound sodium chloride.**

Go Figure

If no color key were provided, how would you know which color ball represented Na$^+$ and which represented Cl$^-$?

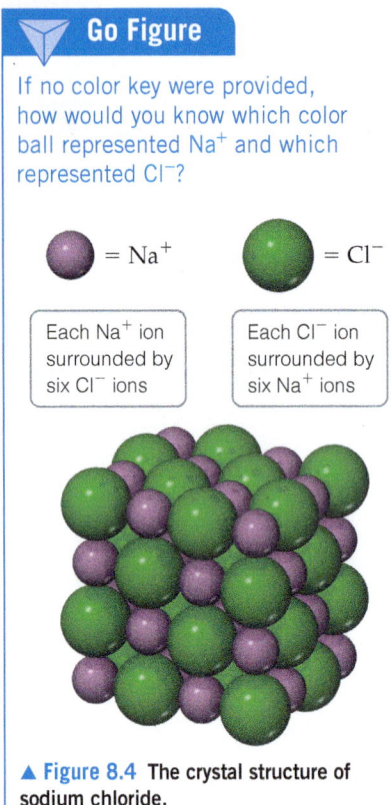

● = Na$^+$ ● = Cl$^-$

Each Na$^+$ ion surrounded by six Cl$^-$ ions

Each Cl$^-$ ion surrounded by six Na$^+$ ions

▲ **Figure 8.4** **The crystal structure of sodium chloride.**

characteristics result from electrostatic forces that maintain the ions in a rigid, well-defined, three-dimensional arrangement such as that shown in Figure 8.4.

Energetics of Ionic Bond Formation

The formation of sodium chloride from sodium and chlorine is *very* exothermic, as indicated by the large negative enthalpy of formation value given in Equation 8.1, $\Delta H_f^\circ = -410.9$ kJ. Appendix C shows that the heat of formation of other ionic substances is also quite negative. What factors make the formation of ionic compounds so exothermic?

In Equation 8.2, we represented the formation of NaCl as the transfer of an electron from Na to Cl. Recall from Section 7.4 that the loss of electrons from an atom is always an endothermic process. The energy needed to remove an electron from Na(g) to form Na$^+$(g) is the first ionization energy of Na, 496 kJ/mol. Recall that when a nonmetal gains an electron, the process is generally exothermic: The energy change when an electron is added to Cl(g) is the electron affinity of Cl, -349 kJ/mol., where the negative sign indicates that energy is released in the process. (Section 7.4) From these energies, we can see that the transfer of an electron from an Na atom to a Cl atom would not be exothermic—the overall process would be an endothermic process that requires $496 - 349 = 147$ kJ/mol. This endothermic process corresponds to the formation of gaseous sodium and chloride ions that are infinitely far apart; in other words, the positive energy change assumes that the ions do not interact with each other, which is quite different from the situation in ionic solids.

The principal reason ionic compounds are stable is the electrostatic attraction between ions of opposite charge. This attraction draws the ions together, releasing energy and leads to the formation of a crystalline solid with a repeating pattern of cations and anions, such as that shown in Figure 8.4. A measure of how much stabilization results from arranging oppositely charged ions in an ionic solid is given by the **lattice energy**, which is *the energy required to completely separate one mole of a solid ionic compound into its gaseous ions.*

To envision this process for NaCl, imagine that the structure in Figure 8.4 expands from within, so that the distances between the ions increase until the ions are very far apart. This process requires 788 kJ/mol, which is the value of the lattice energy:

$$NaCl(s) \longrightarrow Na^+(g) + Cl^-(g) \qquad \Delta H_{lattice} = +788 \text{ kJ/mol} \qquad [8.3]$$

This process is highly endothermic, so the reverse process—the coming together of Na$^+$(g) and Cl$^-$(g) to form NaCl(s)—is highly exothermic ($\Delta H = -788$ kJ/mol).

Table 8.1 lists the lattice energies for a number of ionic compounds. The large positive values indicate that the ions are strongly attracted to one another in ionic solids. The energy released by the attraction between ions of unlike charge more than makes

TABLE 8.1   Lattice Energies for Some Ionic Compounds

Compound	Lattice Energy (kJ/mol)	Compound	Lattice Energy (kJ/mol)
LiF	1030	MgCl$_2$	2326
LiCl	834	SrCl$_2$	2127
LiI	730		
NaF	910	MgO	3795
NaCl	788	CaO	3414
NaBr	732	SrO	3217
NaI	682		
KF	808	ScN	7547
KCl	701		
KBr	671		
CsCl	657		
CsI	600		

up for the endothermic nature of ionization energies, making the formation of ionic compounds an exothermic process. The strong attractions also cause most ionic materials to be hard and brittle with high melting points; NaCl, for example, melts at 801 °C.

The magnitude of the lattice energy of an ionic solid depends on the charges of the ions, their sizes, and their arrangement in the solid. We saw in Section 5.1 that the electrostatic potential energy of two interacting charged particles is given by:

$$E_{el} = \frac{\kappa Q_1 Q_2}{d} \qquad [8.4]$$

In this equation Q_1 and Q_2 are the charges on the particles in coulombs, including their signs; d is the distance between their centers in meters; and κ is a constant, 8.99×10^9 J-m/C^2. From Equation 8.4 we can see that the attractive interaction between two oppositely charged ions increases as the magnitudes of their charges increase and as the distance between their centers decreases. Thus, *for a given arrangement of ions, the lattice energy increases as the charges on the ions increase and as their radii decrease.* The variation in the magnitude of lattice energies depends more on ionic charge than on ionic radius because ionic radii vary over only a limited range compared to charges.

Sample Exercise 8.1
Magnitudes of Lattice Energies

Without consulting Table 8.1, arrange the ionic compounds NaF, CsI, and CaO in order of increasing lattice energy.

SOLUTION

Analyze From the formulas for three ionic compounds, we must determine their relative lattice energies.

Plan We need to determine the charges and relative sizes of the ions in the compounds. We then use Equation 8.4 qualitatively to determine the relative energies, knowing that (**a**) the larger the ionic charges, the greater the energy and (**b**) the farther apart the ions are, the lower the energy.

Solve NaF consists of Na$^+$ and F$^-$ ions, CsI of Cs$^+$ and I$^-$ ions, and CaO of Ca^{2+} and O^{2-} ions. Because the product Q_1Q_2 appears in the numerator of Equation 8.4, the lattice energy increases dramatically when the magnitudes of the charges increase. Thus, we expect the lattice energy of CaO, which has 2+ and 2− ions, to be the greatest of the three.

The ionic charges are the same in NaF and CsI. The difference in their lattice energies thus depends on the difference in the distance between ions in the lattice. Because ionic size increases as we go down a group in the periodic table (Section 7.3), we know that Cs$^+$ is larger than Na$^+$ and I$^-$ is larger than F$^-$. Therefore, the distance between Na$^+$ and F$^-$ ions in NaF is less than the distance between the Cs$^+$ and I$^-$ ions in CsI. As a result, the lattice energy of NaF should be greater than that of CsI. In order of increasing energy, therefore, we have CsI < NaF < CaO.

Check Table 8.1 confirms our predicted order is correct.

▶ **Practice Exercise**
Which substance do you expect to have the greatest lattice energy: MgF$_2$, CaF$_2$, or ZrO$_2$?

Because lattice energy decreases as distance between ions increases, lattice energies follow trends that parallel those in ionic radius shown in Figure 7.8. In particular, because ionic radius increases as we go down a group of the periodic table, we find that, for a given type of ionic compound, lattice energy decreases as we go down a group. **Figure 8.5** illustrates this trend for the alkali chlorides MCl (M = Li, Na, K, Rb, Cs) and the sodium halides NaX(X = F, Cl, Br, I).

Electron Configurations of Ions of the s- and p-Block Elements

The energetics of ionic bond formation helps explain why many ions tend to have noble-gas electron configurations. For example, sodium readily loses one electron to form Na$^+$, which has the same electron configuration as Ne:

$$\text{Na} \qquad 1s^2 2s^2 2p^6 3s^1 = [\text{Ne}]3s^1$$

$$\text{Na}^+ \qquad 1s^2 2s^2 2p^6 \quad = [\text{Ne}]$$

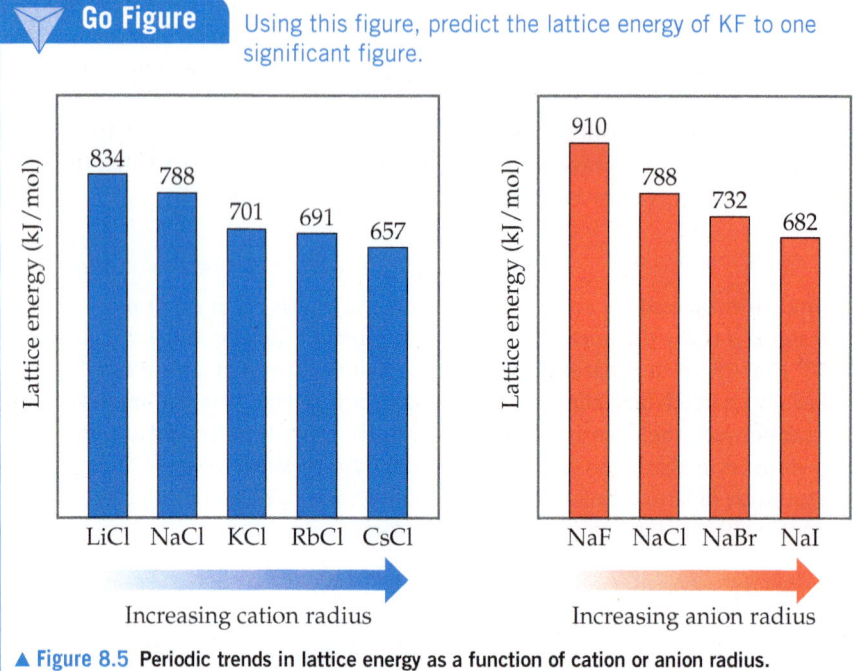

▲ **Figure 8.5 Periodic trends in lattice energy as a function of cation or anion radius.**

Even though lattice energy increases with increasing ionic charge, we never find ionic compounds that contain Na^{2+} ions. The second electron removed would have to come from an inner shell of the sodium atom, and removing electrons from an inner shell requires a very large amount of energy. (Section 7.4) The increase in lattice energy is not enough to compensate for the energy needed to remove an inner-shell electron. Thus, sodium and the other group 1A metals are found in ionic substances only as 1+ ions.

Similarly, adding electrons to nonmetals is either exothermic or only slightly endothermic as long as the electrons are added to the valence shell. Thus, a Cl atom readily adds an electron to form Cl^-, which has the same electron configuration as Ar:

$$Cl \qquad 1s^22s^22p^63s^23p^5 = [Ne]3s^23p^5$$

$$Cl^- \qquad 1s^22s^22p^63s^23p^6 = [Ne]3s^23p^6 = [Ar]$$

To form a Cl^{2-} ion, the second electron would have to be added to the next higher shell of the Cl atom, an addition that is energetically very unfavorable. Therefore, we never observe Cl^{2-} ions in ionic compounds. We thus expect ionic compounds of the representative metals from groups 1A, 2A, and 3A to contain 1+, 2+, and 3+ cations, respectively, and we usually expect ionic compounds of the representative nonmetals of groups 5A, 6A, and 7A to contain 3−, 2−, and 1− anions, respectively.

Transition-Metal Ions

We have seen that the octet rule is a useful tool for predicting the preferred ionic charges of many ions of the representative elements. It is less useful as a guide to predict the charges on ions formed by atoms of the transition elements. Consider an Fe atom, for example, which has an electron configuration of $[Ar]4s^23d^6$. (Section 6.9) Fe forms two common cations, Fe^{2+} and Fe^{3+}, which have electron configurations of $[Ar]3d^6$ and $[Ar]3d^5$, respectively. (Sections 2.8 and 7.4) Neither Fe^{2+} nor Fe^{3+} has an octet of electrons, nor is there a simple analogy to allow us to predict these as the preferred ionic charges.

The octet rule, though useful, is clearly limited in scope and is generally not useful for transition metals. That said, once we know the charge on a transition metal ion, we can predict the way it will combine with nonmetals to make ionic compounds. For example, both Fe^{2+} and Fe^{3+} form oxides—namely, FeO and Fe_2O_3, respectively. Consistent with our expectations for ionic substances, both FeO and Fe_2O_3 are brittle solids with high melting points.

A CLOSER LOOK | Calculation of Lattice Energies: The Born–Haber Cycle

Lattice energies cannot be determined directly by experiment. They can, however, be calculated by envisioning the formation of an ionic compound as occurring in a series of well-defined steps. We can then use Hess's law (Section 5.6) to combine the steps in a way that gives the lattice energy for the compound. By so doing, we construct a **Born–Haber cycle**, a thermochemical cycle named after the German scientists Max Born (1882–1970) and Fritz Haber (1868–1934), who introduced it to analyze the factors contributing to the stability of ionic compounds.

Let's use NaCl as an example. In Equation 8.3, which defines lattice energy, NaCl(s) is the reactant, and the gas-phase ions $Na^+(g)$ and $Cl^-(g)$ are the products. This equation is our target as we apply Hess's law.

In seeking a set of other equations that can be added up to give our target equation, we can use the heat of formation for NaCl (Section 5.7):

$$Na(s) + \tfrac{1}{2}Cl_2(g) \longrightarrow NaCl(s) \quad \Delta H_f^\circ[NaCl(s)] = -411 \text{ kJ} \quad [8.5]$$

This equation must be turned around, however, so that we have NaCl(s) as the reactant as we do in the equation for the lattice energy. We can use two other equations to arrive at our target, as shown below:

1. $NaCl(s) \longrightarrow Na(s) + \tfrac{1}{2}Cl_2(g)$ $\Delta H_1 = -\Delta H_f^\circ[NaCl(s)]$
   $= +411$ kJ
2. $Na(s) \longrightarrow Na^+(g) + e^-$ $\Delta H_2 = ?$
3. $e^- + \tfrac{1}{2}Cl_2(g) \longrightarrow Cl^-(g)$ $\Delta H_3 = ?$

4. $NaCl(s) \longrightarrow Na^+(g) + Cl^-(g)$ $\Delta H_4 = \Delta H_1 + \Delta H_2 + \Delta H_3$
   $= \Delta H_{lattice}$

Step 2 involves the formation of sodium ion from solid sodium, which is just the heat of formation for sodium gas and the first ionization energy for sodium (Appendix C and Figure 7.11 list numbers for these processes):

$Na(s) \longrightarrow Na(g)$ $\Delta H = \Delta H_f^\circ[Na(g)] = 108$ kJ [8.6]
$Na(g) \longrightarrow Na^+(g) + e^-$ $\Delta H = I_1(Na) = 496$ kJ [8.7]

The sum of these two processes gives us the required energy for Step 2 (above), which is 604 kJ.

Similarly, for Step 3, we have to create chlorine atoms, and then anions, from the Cl_2 molecule, in two steps. The enthalpy changes for these two steps are the sum of the enthalpy of formation of Cl(g) and the electron affinity of chlorine, EA(Cl):

$\tfrac{1}{2}Cl_2(g) \longrightarrow Cl(g)$ $\Delta H = \Delta H_f^\circ[Cl(g)] = 122$ kJ [8.8]
$e^- + Cl(g) \longrightarrow Cl^-(g)$ $\Delta H = EA(Cl) = -349$ kJ [8.9]

The sum of these two processes gives us the required energy for Step 3 (above), which is −227 kJ.

Finally, when we put it all together, we have:

1. $NaCl(s) \longrightarrow Na(s) + \tfrac{1}{2}Cl_2(g)$ $\Delta H_1 = -\Delta H_f^\circ[NaCl(s)]$
   $= +411$ kJ
2. $Na(s) \longrightarrow Na^+(g) + e^-$ $\Delta H_2 = 604$ kJ
3. $e^- + \tfrac{1}{2}Cl_2(g) \longrightarrow Cl^-(g)$ $\Delta H_3 = -227$ kJ

4. $NaCl(s) \longrightarrow Na^+(g) + Cl^-(g)$ $\Delta H_4 = 788$ kJ $= \Delta H_{lattice}$

This process is described as a "cycle" because it corresponds to the scheme in **Figure 8.6**, which shows how all the quantities we have just calculated are related. The sum of all the blue "up" arrow energies has to be equal to the sum of all the red "down" arrow energies in this cycle. Born and Haber recognized that if we know the value of every quantity in the cycle except the lattice energy, we can calculate it from this cycle.

Related Exercises: 8.30–8.32, 8.85

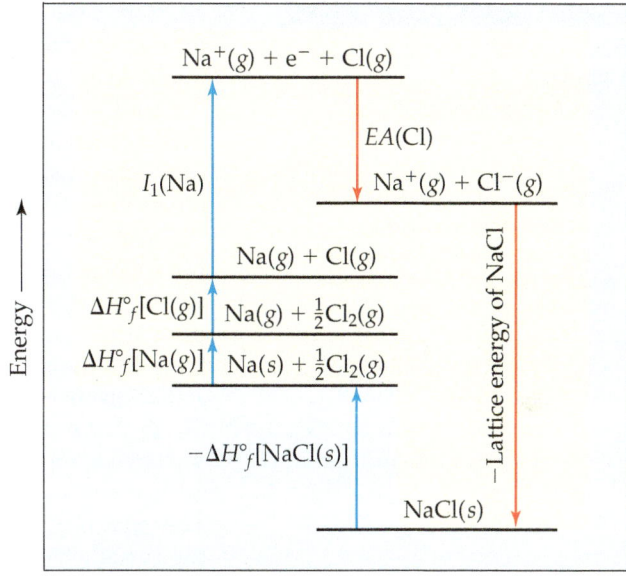

▲ **Figure 8.6 Born–Haber cycle for formation of NaCl.** This Hess's law representation shows the energetic relationships in the formation of the ionic solid from its elements.

Sample Exercise 8.2
Charges on Ions

Predict the ion generally formed by **(a)** Sr, **(b)** S, **(c)** Al.

SOLUTION

Analyze We must decide how many electrons are most likely to be gained or lost by atoms of Sr, S, and Al.

Plan In each case, we can use the element's position in the periodic table to predict whether the element forms a cation or an anion. We can then use its electron configuration to determine the most likely ion formed.

Solve
(a) Strontium is a metal in group 2A, so it forms a cation. Its electron configuration is $[Kr]5s^2$, and so we expect that the two valence electrons will be lost to give an Sr^{2+} ion.

(b) Sulfur is a nonmetal in group 6A, so it will tend to be found as an anion. Its electron configuration ($[Ne]3s^23p^4$) is two electrons short of a noble-gas configuration. Thus, we expect that sulfur will form S^{2-} ions.

(c) Aluminum is a metal in group 3A. We therefore expect it to form Al^{3+} ions.

Check The ionic charges we predict here are confirmed in Tables 2.4 and 2.5.

▶ **Practice Exercise**
Predict the charges on the ions formed when magnesium reacts with nitrogen.

Self-Assessment Exercises

SAE 8.3 Which of the following statements about ions and ionic bonding is or are *true*?

   (i) An atom with a low first ionization energy and low electron affinity is more likely to form a cation than an anion.

   (ii) The most common ion for an atom of a representative element generally adopts a noble-gas electron configuration.

   (iii) The stability of KBr is due to the electrostatic attraction between the potassium ions and the bromide ions.

(a) Only one of the statements is true. **(b)** i and ii are true. **(c)** i and iii are true. **(d)** ii and iii are true. **(e)** All three statements are true.

SAE 8.4 Which of the following statements about ionic bonding is *false*? **(a)** Ionic bonds are found in compounds formed from metals combined with nonmetals. **(b)** The attraction between cations and anions increases as the charges on the ions increase. **(c)** The attraction between cations and anions increases as the radii of the ions increase. **(d)** Lattice energy is the energy needed to break an ionic compound into gaseous ions.

SAE 8.5 Place the following ionic substances in order from smallest to largest lattice energy: CaO, CsI, KF, KI, MgO. **(a)** CsI < KI < KF < CaO < MgO **(b)** CaO < MgO < CsI < KI < KF **(c)** KF < KI < CsI < MgO < CaO **(d)** CsI < CaO < KF < KI < MgO **(e)** KI < CsI < CaO < MgO < KF

SAE 8.6 Which of the following is the electron configuration of the most common ion of a P atom? **(a)** $[Ne]3s^23p^3$ **(b)** $[Ne]3s^2$ **(c)** $1s^22s^22p^6$ **(d)** $[Ne]3s^23p^6$ **(e)** $[Ne]3s^23p^33d^3$

8.3 | Covalent Bonding

Learning Objectives

When you finish Section 8.3, you should be able to:

▶ Describe covalent bonding and differentiate it from ionic bonding.

▶ Draw Lewis structures to represent the distribution of electrons in molecules.

▶ Use Lewis structures and the octet rule to predict the presence and location of multiple bonds in molecules.

The vast majority of chemical substances do not have the characteristics of ionic materials. Most of the substances with which we come into daily contact—such as water—tend to be gases, liquids, or solids with low melting points. Many, such as gasoline, vaporize readily. Many are pliable in their solid forms—for example, plastic bags and wax.

For the very large class of substances that do not behave like ionic substances, we need a different model to describe the bonding between atoms. G. N. Lewis reasoned that atoms might acquire a noble-gas electron configuration by sharing electrons with other atoms. A chemical bond formed by sharing a pair of electrons is a *covalent bond*. The hydrogen molecule, H_2, provides the simplest example of a covalent bond. When two hydrogen atoms are close to each other, the two positively charged nuclei repel each other, the two negatively charged electrons repel each other, and the nuclei and electrons attract each other, as shown in **Figure 8.7(a)**. Because the molecule is stable, we know that the attractive forces must overcome the repulsive ones. Let's take a closer look at the attractive forces that hold H_2 together.

By using quantum-mechanical methods analogous to those applied to atoms in **Section 6.5**, we can calculate the distribution of electron density in molecules. Such a calculation for H_2 shows that the attractions between the nuclei and the electrons cause electron density to concentrate between the nuclei, as shown in Figure 8.7(b). As a result, the overall electrostatic interactions are attractive. Thus, the atoms in H_2 are held together principally because the two positive nuclei are attracted to the concentration of negative charge between them. In essence, the shared pair of electrons in any covalent bond acts as a kind of "glue" to bind atoms together.

Lewis Structures

The formation of covalent bonds can be represented with Lewis symbols. The formation of the H_2 molecule from two H atoms, for example, can be represented as

$$H\cdot + \cdot H \longrightarrow H:H$$

In forming the covalent bond, each hydrogen atom acquires a second electron, achieving the stable, two-electron, noble-gas electron configuration of helium.

Formation of a covalent bond between two Cl atoms to give a Cl_2 molecule can be represented in a similar way:

$$:\ddot{Cl}\cdot + \cdot\ddot{Cl}: \longrightarrow :\ddot{Cl}:\ddot{Cl}:$$

By sharing the bonding electron pair, each chlorine atom has eight electrons (an octet) in its valence shell, thus achieving the noble-gas electron configuration of argon.

The structures shown here for H_2 and Cl_2 are **Lewis structures**, or *Lewis dot structures*. While these structures show circles to indicate electron sharing, the more common convention is to show each shared electron pair or **bonding pair**, as a line and any unshared electron pairs (also called **lone pairs** or **nonbonding pairs**) as dots. Written this way, the Lewis structures for H_2 and Cl_2 are

$$\text{H—H} \qquad \ddot{\underset{..}{\text{Cl}}}—\ddot{\underset{..}{\text{Cl}}}:$$

For nonmetals, the number of valence electrons in a neutral atom is the same as the group number. Therefore, one might predict that 7A elements, such as F, would form one covalent bond to achieve an octet; 6A elements, such as O, would form two covalent bonds; 5A elements, such as N, would form three; and 4A elements, such as C, would form four. These predictions are borne out in many compounds, as in, for example, the compounds with hydrogen of the nonmetals of the second row of the periodic table:

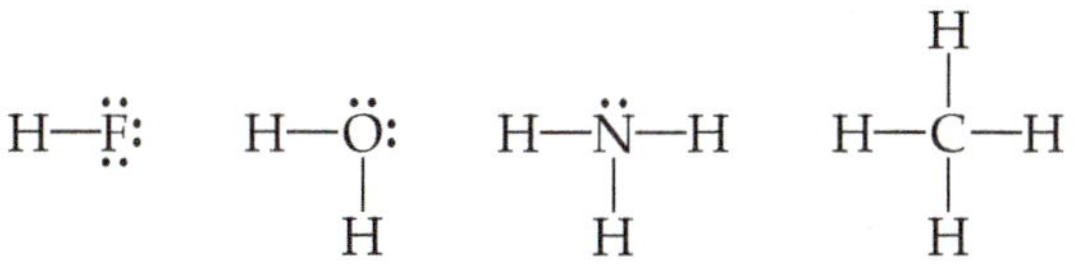

$$\text{H}—\ddot{\underset{..}{\text{F}}}: \qquad \text{H}—\underset{\underset{\text{H}}{|}}{\ddot{\text{O}}}: \qquad \text{H}—\underset{\underset{\text{H}}{|}}{\overset{..}{\text{N}}}—\text{H} \qquad \text{H}—\underset{\underset{\text{H}}{|}}{\overset{\overset{\text{H}}{|}}{\text{C}}}—\text{H}$$

Multiple Bonds

A shared electron pair constitutes a single covalent bond, generally referred to simply as a **single bond**. In many molecules, atoms attain complete octets by sharing more than one pair of electrons. When two electron pairs are shared by two atoms, two lines are drawn in the Lewis structure, representing a **double bond**. In carbon dioxide, for example, bonding occurs between carbon, with four valence electrons, and oxygen, with six:

$$:\ddot{\text{O}}: + \cdot\dot{\text{C}}\cdot + :\ddot{\text{O}}: \longrightarrow \ddot{\text{O}}::\text{C}::\ddot{\text{O}} \qquad (\text{or } \ddot{\text{O}}=\text{C}=\ddot{\text{O}})$$

As the diagram shows, each oxygen atom acquires an octet by sharing two electron pairs with carbon. In the case of CO_2, carbon acquires an octet by sharing two electron pairs with each of the two oxygen atoms; each double bond involves four electrons.

A **triple bond** corresponds to the sharing of three pairs of electrons, such as in the N_2 molecule:

$$:\dot{\text{N}}\cdot + \cdot\dot{\text{N}}: \longrightarrow :\text{N}:::\text{N}: \qquad (\text{or } :\text{N}\equiv\text{N}:)$$

Because each nitrogen atom has five valence electrons, three electron pairs must be shared to achieve the octet configuration.

The properties of N_2 are in complete accord with its Lewis structure. Nitrogen is a diatomic gas with exceptionally low reactivity that results from the very stable nitrogen–nitrogen bond. The nitrogen atoms are separated by only 1.10 Å. The short separation distance between the two N atoms is a result of the triple bond between the atoms. From studies of the structures of many different substances in which nitrogen atoms share one or two electron pairs, we have learned that the average distance between bonded nitrogen atoms varies with the number of shared electron pairs:

$$\begin{array}{ccc} \text{N—N} & \text{N=N} & \text{N}\equiv\text{N} \\ 1.47\ \text{Å} & 1.24\ \text{Å} & 1.10\ \text{Å} \end{array}$$

As a general rule, the length of the bond between two atoms decreases as the number of shared electron pairs increases. We explore this point in greater detail in Section 8.8.

Go Figure

What would happen to the concentration of electron density between the nuclei in (b) if you pulled the nuclei farther apart?

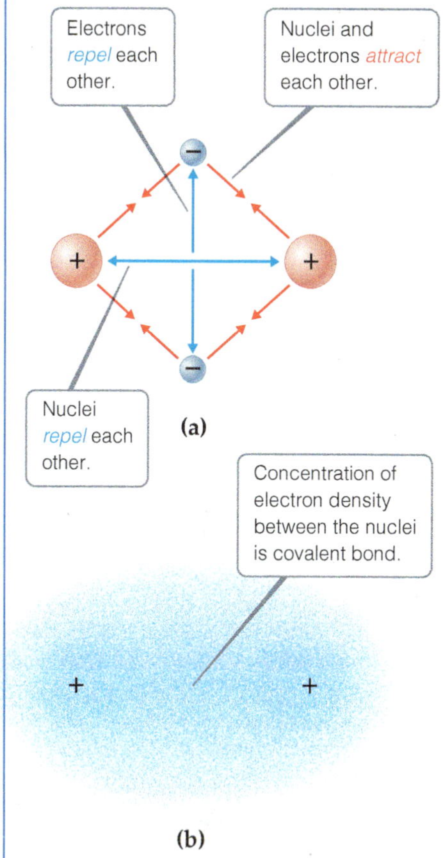

Electrons *repel* each other.

Nuclei and electrons *attract* each other.

Nuclei *repel* each other.

(a)

Concentration of electron density between the nuclei is covalent bond.

(b)

▲ **Figure 8.7 The covalent bond in H_2.** (a) The attractions and repulsions among electrons and nuclei in the hydrogen molecule. (b) Electron distribution in the H_2 molecule.

Sample Exercise 8.3
Lewis Structure of a Compound

Given the Lewis symbols for nitrogen and fluorine in Figure 8.2, predict the formula of the stable binary compound (a compound composed of two elements) formed when nitrogen reacts with fluorine and draw its Lewis structure.

SOLUTION

Analyze According to the Lewis symbols for nitrogen and fluorine, nitrogen has five valence electrons and fluorine has seven.

Plan We need to find a combination of the two elements that results in an octet of electrons around each atom. Nitrogen requires three additional electrons to complete its octet, and fluorine requires one. Sharing a pair of electrons between one N atom and one F atom will result in an octet of electrons for fluorine but not for nitrogen. We therefore need to figure out a way to get two more electrons for the N atom.

Solve Nitrogen must share a pair of electrons with three fluorine atoms to complete its octet. Thus, the binary compound these two elements form must be NF₃:

$$\cdot \ddot{N} \cdot \ + \ 3 \ \cdot \ddot{\ddot{F}} : \ \longrightarrow \ :\ddot{\ddot{F}} : \ddot{N} : \ddot{\ddot{F}} : \ \longrightarrow \ :\ddot{\ddot{F}} - \ddot{N} - \ddot{\ddot{F}} :$$
$$\qquad\qquad\qquad\qquad :\ddot{\ddot{F}}: \qquad\qquad\qquad |$$
$$\qquad\qquad\qquad\qquad\qquad\qquad\qquad :\ddot{\ddot{F}}:$$

Check The Lewis structure in the center shows that each atom is surrounded by an octet of electrons. Once you are accustomed to thinking of each line in a Lewis structure as representing *two* electrons, you can just as easily use the structure on the right to check for octets.

▶ **Practice Exercise**
Compare the Lewis structure for water, H₂O, with that for methane, CH₄. How many valence electrons are in each structure? How many bonding pairs and how many nonbonding pairs does each structure have?

Self-Assessment Exercises

SAE 8.7 The CF₄ molecule has a central C atom bonded to four F atoms, as shown in the figure. Which of the following statements about the CF₄ molecule is or are *true*?

   (**i**) The electrons in the C—F bonds are attracted to both the C and F nuclei.

   (**ii**) The correct Lewis structure of CF₄ has four bonding pairs of electrons and no nonbonding pairs.

   (**iii**) Every atom in the CF₄ molecule satisfies the octet rule.

(**a**) Only one of the statements is true. (**b**) i and ii are true. (**c**) i and iii are true. (**d**) ii and iii are true. (**e**) All three statements are true.

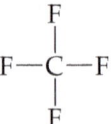

SAE 8.8 Which of the following statements about multiple bonds is *false*? (**a**) A double bond is shorter than a single bond. (**b**) A triple bond involves the sharing of three electrons between the two atoms. (**c**) The formation of multiple bonds allows atoms to satisfy the octet rule. (**d**) A molecule can have more than one multiple bond in it.

Learning Objectives

When you finish Section 8.4, you should be able to:

▶ Describe the concept of electronegativity and its periodic trends.

▶ Categorize different types of bonds based on the electronegativity difference of the two bonded atoms.

▶ Relate the dipole moment of a diatomic molecule to the partial charges on its atoms.

▶ Differentiate compounds that possess ionic bonds from those that possess covalent bonds based on their physical properties.

8.4 | Bond Polarity and Electronegativity

When two identical atoms bond, as in Cl₂ or H₂, the electron pairs must be shared equally. When two atoms from opposite sides of the periodic table bond, such as NaCl, there is relatively little sharing of electrons, which means that NaCl is best described as an ionic compound composed of Na⁺ and Cl⁻ ions. The 3s electron of the Na atom is, in effect, transferred completely to chlorine. The bonds that are found in most substances fall somewhere between these extremes. In this section, we look more closely at bonds in which electrons are shared unequally between the two atoms.

When discussing the degree of sharing of electrons between two atoms in a bond, it is helpful to introduce some new terminology. **Bond polarity** is a measure of how equally or unequally the electrons in any covalent bond are shared. A **nonpolar covalent bond** is one in which the electrons are shared equally, as in Cl₂ and N₂. In a **polar covalent bond**, one of the atoms exerts a greater attraction for the bonding electrons than the other. If the difference in relative ability to attract electrons is large enough, an ionic bond is formed.

Electronegativity

We use a quantity called *electronegativity* to estimate whether a given bond is nonpolar covalent, polar covalent, or ionic:

> **Electronegativity** is defined as the ability of an atom *in a molecule* to attract electrons to itself.

The greater an atom's electronegativity, the greater its ability to attract electrons to itself. The electronegativity of an atom in a molecule is related to the atom's ionization energy and electron affinity, which are properties of isolated atoms. An atom with a very negative electron affinity and a high ionization energy both attracts electrons from other atoms and resists having its electrons attracted away; therefore, it is highly electronegative.

Electronegativity values can be based on a variety of properties, not just ionization energy and electron affinity. The American chemist Linus Pauling (1901–1994) developed the first and most widely used electronegativity scale, which is based on thermochemical data. In the Pauling scale, electronegativity values range from 0.7 to 4.0, with no units. As **Figure 8.8** shows, there is generally an increase in electronegativity from left to right across a period—that is, from the most metallic to the most nonmetallic elements. With some exceptions (especially in the transition metals), electronegativity decreases from top to bottom in a group. This is what we expect because ionization energies decrease as we go down a group and electron affinities do not change very much.

You do not need to memorize electronegativity values. Instead, you should know the periodic trends so that you can predict which of two elements is more electronegative.

Electronegativity and Bond Polarity

We can use the difference in electronegativity between two atoms to gauge the polarity of the bond the atoms form. Consider these three fluorine-containing compounds:

	F_2	HF	LiF
Electronegativity difference	$4.0 - 4.0 = 0$	$4.0 - 2.1 = 1.9$	$4.0 - 1.0 = 3.0$
Type of bond	Nonpolar covalent	Polar covalent	Ionic

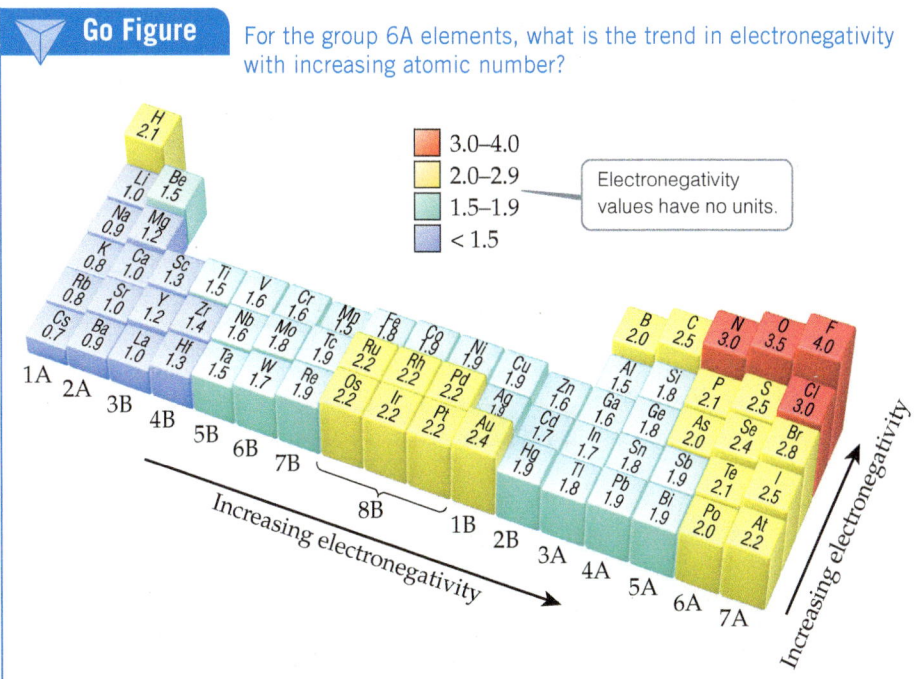

Go Figure For the group 6A elements, what is the trend in electronegativity with increasing atomic number?

▲ **Figure 8.8 Electronegativity values based on Pauling's thermochemical data.** Notice the general increase in values from left to right across a period, and the general decrease from top to bottom down a group.

In F_2 the electrons are shared equally between the fluorine atoms and, thus, the covalent bond is *nonpolar*. A nonpolar covalent bond results when the electronegativities of the bonded atoms are equal.

In HF the fluorine atom has a greater electronegativity than the hydrogen atom, with the result that the electrons are shared unequally—the bond is *polar*. In general, a polar covalent bond results when the atoms differ in electronegativity. In HF the more electronegative fluorine atom attracts electron density away from the less electronegative hydrogen atom, leaving a partial positive charge on the hydrogen atom and a partial negative charge on the fluorine atom. We can represent this charge distribution as

$$\overset{\delta+}{H}\overset{\delta-}{-F}$$

The $\delta+$ and $\delta-$ (read "delta plus" and "delta minus") symbolize the partial positive and negative charges, respectively. In a polar bond, these values are less than the full charges of the ions.

In LiF the electronegativity difference is very large, meaning that the electron density is shifted far toward F. The resulting bond is therefore most accurately described as *ionic*. Thus, if we considered the bond in LiF to be fully ionic, we could say $\delta+$ for Li is 1+ and $\delta-$ for F is 1−. If the electronegativity difference between two atoms is greater than 2.0, we generally consider a bond between these two atoms to be an ionic bond.

The shift of electron density toward the more electronegative atom in a bond can be seen from the results of calculations of electron-density distributions. For the three species in our example, the calculated electron-density distributions are shown in **Figure 8.9**. Notice that the distribution in F_2 is symmetrical, whereas the electron density in HF is clearly shifted toward fluorine, and in LiF the shift is even greater. These examples illustrate, therefore, that *the greater the difference in electronegativity between two atoms, the more polar their bond*.

Dipole Moments

The difference in electronegativity between H and F leads to a polar covalent bond in the HF molecule. As a consequence, there is a concentration of negative charge on the more electronegative F atom, leaving the less electronegative H atom at the positive end of the molecule. A molecule such as HF, in which the centers of positive and negative charge do

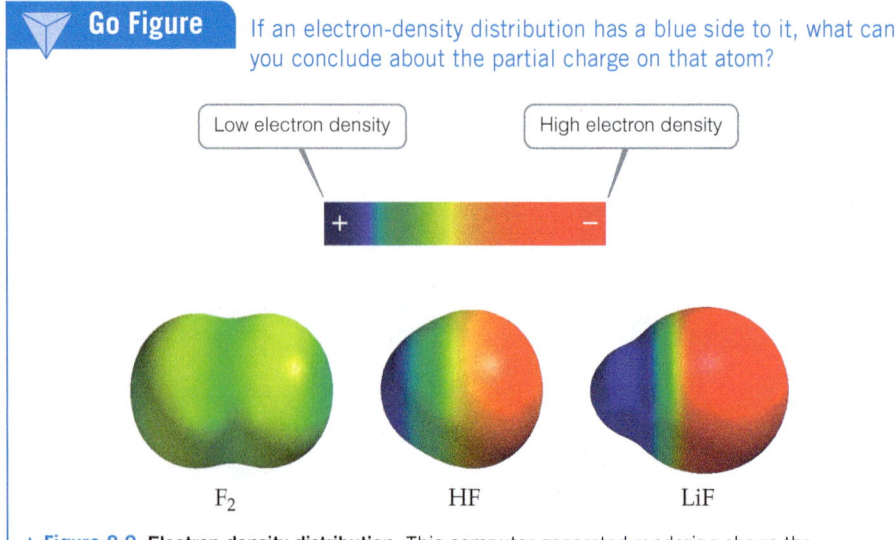

▲ **Figure 8.9 Electron-density distribution.** This computer-generated rendering shows the calculated electron-density distribution on the surface of the F_2, HF, and LiF molecules.

Sample Exercise 8.4
Bond Polarity

In each case, which bond is more polar? (**a**) B—Cl or C—Cl, (**b**) P—F or P—Cl. Indicate in each case which atom has the partial negative charge.

SOLUTION

Analyze We are asked to determine relative bond polarities, given nothing but the atoms involved in the bonds.

Plan Because we are not asked for quantitative answers, we can use the periodic table and our knowledge of electronegativity trends to answer the question.

Solve

(**a**) The chlorine atom is common to both bonds. Therefore, we just need to compare the electronegativities of B and C. Because boron is to the left of carbon in the periodic table, we predict that boron has the lower electronegativity. Chlorine, being on the right side of the table, has high electronegativity. The more polar bond will be the one between the atoms with the biggest differences in electronegativity. Consequently, the B—Cl bond is more polar; the chlorine atom carries the partial negative charge because it has a higher electronegativity.

(**b**) In this example, phosphorus is common to both bonds, and so we just need to compare the electronegativities of F and Cl. Because fluorine is above chlorine in the periodic table, it

should be more electronegative and will form the more polar bond with P. The higher electronegativity of fluorine means that it will carry the partial negative charge.

Check

(**a**) Using Figure 8.8: The difference in the electronegativities of chlorine and boron is $3.0 - 2.0 = 1.0$; the difference between the electronegativities of chlorine and carbon is $3.0 - 2.5 = 0.5$. Hence, the B—Cl bond is more polar, as we had predicted.

(**b**) Using Figure 8.8: The difference in the electronegativities of chlorine and phosphorus is $3.0 - 2.1 = 0.9$; the difference between the electronegativities of fluorine and phosphorus is $4.0 - 2.1 = 1.9$. Hence, the P—F bond is more polar, as we had predicted.

▶ **Practice Exercise**
Which of the following bonds is most polar: S—Cl, S—Br, Se—Cl, or Se—Br?

---

not coincide, is a **polar molecule**. Thus, we describe both bonds and entire molecules as being polar and nonpolar.

We can indicate the polarity of the HF molecule in two ways:

$$\overset{\delta+}{H}-\overset{\delta-}{F} \quad or \quad \overset{\longleftarrow}{H-F}$$

In the notation on the right, the arrow denotes the shift in electron density toward the fluorine atom. The crossed end of the arrow can be thought of as a plus sign designating the positive end of the molecule.

Polarity helps determine many properties we observe at the macroscopic level in the laboratory and in everyday life. Polar molecules align themselves with respect to one another, with the negative end of one molecule and the positive end of another attracting each other. Polar molecules are likewise attracted to ions. The negative end of a polar molecule is attracted to a positive ion, whereas the positive end is attracted to a negative ion. These interactions account for many properties of liquids, solids, and solutions, as we explore in Chapters 11, 12, and 13. Charge separation within molecules plays an important role in energy conversion processes such as photosynthesis and in solar cells.

How can we quantify the polarity of a molecule? Whenever two electrical charges of equal magnitude but opposite sign are separated by a distance, a **dipole** is established. The quantitative measure of the magnitude of a dipole is called its **dipole moment**, denoted with the Greek letter mu, μ. The dipole moment is what we call a *vector* quantity, meaning that it has both a magnitude and a direction. The direction of a dipole moment is given by the "crossed arrow" that we used previously to describe the polarity of HF. The magnitude of the dipole moment depends on how much charge is separated and the distance of the separation. If two equal and opposite charges $Q+$ and $Q-$ are separated by a distance r, as in **Figure 8.10**, the magnitude of the dipole moment is the product of Q and r:

$$\mu = Qr \qquad [8.10]$$

This expression tells us that the dipole moment increases as the magnitude of Q increases and as r increases. The larger the dipole moment, the more polar the bond. For a nonpolar

Go Figure

If the charged particles are moved closer together, does μ increase, decrease, or stay the same?

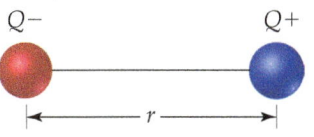

Dipole moment $\mu = Qr$

▲ **Figure 8.10 Dipole and dipole moment.** When charges of equal magnitude and opposite sign $Q+$ and $Q-$ are separated by a distance r, a dipole is produced.

molecule, such as F_2, the dipole moment is zero because the atoms are identical and the electrons in the bond between them are shared equally.

Dipole moments are experimentally measurable and are usually reported in *debyes* (D), a unit that equals 3.34×10^{-30} coulomb-meters (C-m). For molecules, we usually measure charge in units of the electronic charge e, 1.60×10^{-19} C, and distance in angstroms. This means we need to convert units whenever we want to report a dipole moment in debyes. Suppose that two charges 1+ and 1− (in units of e) are separated by 1.00 Å. The dipole moment produced is

$$\mu = Qr = (1.60 \times 10^{-19} \text{ C})(1.00 \text{ Å})\left(\frac{10^{-10} \text{ m}}{1 \text{ Å}}\right)\left(\frac{1 \text{ D}}{3.34 \times 10^{-30} \text{ C-m}}\right) = 4.79 \text{ D}$$

Measurement of the dipole moments can provide us with valuable information about the charge distributions in molecules, as illustrated in Sample Exercise 8.5.

Sample Exercise 8.5
Dipole Moments of Diatomic Molecules

The bond length in the HCl molecule is 1.27 Å. **(a)** Calculate the dipole moment, in debyes, that results if the charges on the H and Cl atoms were 1+ and 1− respectively. **(b)** The experimentally measured dipole moment of HCl(g) is 1.08 D. What magnitude of charge, in units of e, on the H and Cl atoms leads to this dipole moment?

SOLUTION

Analyze and Plan We are asked in part (a) to calculate the dipole moment of HCl that would result if there were a full charge transferred from H to Cl. We can use Equation 8.10 to obtain this result.

In part (b), we are given the actual dipole moment for the molecule and will use that value to calculate the actual partial charges on the H and Cl atoms.

Solve

(a) Because Cl is more electronegative than H, we expect that the Cl atom has a negative charge and the H atom has a positive charge. The magnitude of the charge on each atom is the electronic charge, $e = 1.60 \times 10^{-19}$ C. The separation is 1.27 Å. The dipole moment is therefore:

$$\mu = Qr = (1.60 \times 10^{-19} \text{ C})(1.27 \text{ Å})\left(\frac{10^{-10} \text{ m}}{1 \text{ Å}}\right)\left(\frac{1 \text{ D}}{3.34 \times 10^{-30} \text{ C-m}}\right) = 6.08 \text{ D}$$

(b) We know the value of μ, 1.08 D, and the value of r, 1.27 Å. We want to calculate the value of Q:

$$Q = \frac{\mu}{r} = \frac{(1.08 \text{ D})\left(\dfrac{3.34 \times 10^{-30} \text{ C-m}}{1 \text{ D}}\right)}{(1.27 \text{ Å})\left(\dfrac{10^{-10} \text{ m}}{1 \text{ Å}}\right)} = 2.84 \times 10^{-20} \text{ C}$$

We can readily convert this charge to units of e:

$$\text{Charge in } e = (2.84 \times 10^{-20} \text{ C})\left(\frac{1e}{1.60 \times 10^{-19} \text{ C}}\right) = 0.178e$$

Thus, the experimental dipole moment indicates that the charge separation in the HCl molecule is:

$$\overset{0.178+}{\text{H}} \!-\! \overset{0.178-}{\text{Cl}}$$

Because the experimental dipole moment is less than that calculated in part (a), the charges on the atoms are much less than a full electronic charge. We could have anticipated this because the H—Cl bond is polar covalent rather than ionic.

▶ **Practice Exercise**

The dipole moment of chlorine monofluoride, ClF(g), is 0.88 D. The bond length of the molecule is 1.63 Å. **(a)** Which atom is expected to have the partial negative charge? **(b)** What is the charge on that atom in units of e?

Table 8.2 presents the bond lengths and dipole moments of the hydrogen halides. Notice that as we proceed from HF to HI, the electronegativity difference decreases and the bond length increases. The first effect decreases the amount of charge separated and causes the dipole moment to decrease from HF to HI, even though the bond length is increasing. Calculations identical to those used in Sample Exercise 8.5 show that the charges on the atoms decrease from 0.41+ and 0.41− in HF to 0.057+ and 0.057− in HI. We can visualize the varying degree of electronic charge shift in these substances from computer-generated renderings based on calculations of electron distribution, as shown

TABLE 8.2 Bond Lengths, Electronegativity Differences, and Dipole Moments of the Hydrogen Halides

Compound	Bond Length (Å)	Electronegativity Difference	Dipole Moment (D)
HF	0.92	1.9	1.82
HCl	1.27	0.9	1.08
HBr	1.41	0.7	0.82
HI	1.61	0.4	0.44

in Figure 8.11. For these molecules, the change in the electronegativity difference has a greater effect on the dipole moment than does the change in bond length.

Before leaving this section, let's return to the LiF molecule in Figure 8.9. Under standard conditions, LiF exists as an ionic solid with an arrangement of atoms analogous to the sodium chloride structure shown in Figure 8.4. However, it is possible to generate LiF *molecules* by vaporizing the ionic solid at high temperature. The molecules have a dipole moment of 6.28 D and a bond distance of 1.53 Å. From these values we can calculate the charge on lithium and fluorine to be 0.857+ and 0.857−, respectively. This bond is extremely polar, and the presence of such large charges strongly favors the formation of an extended ionic lattice in which each lithium ion is surrounded by fluoride ions and vice versa. But even here, the experimentally determined charges on the ions are still not 1+ and 1−. This tells us that even in ionic compounds, there is still some covalent contribution to the bonding.

Comparing Ionic and Covalent Bonding

To understand the interactions responsible for chemical bonding, it is advantageous to treat ionic and covalent bonding separately. That is the approach taken in this chapter, as well as in most other undergraduate-level chemistry texts. In reality, however, there is a continuum between the extremes of ionic and covalent bonding. This lack of a well-defined separation between the two types of bonding may seem unsettling or confusing at first.

The simple models of ionic and covalent bonding presented here go a long way toward understanding and predicting the structures and properties of chemical compounds. When covalent bonding is dominant, we expect compounds to exist as molecules,* having all the properties we associate with molecular substances, such as relatively low melting and boiling points and nonelectrolyte behavior when dissolved in water. When ionic bonding is dominant, we expect the compounds to be brittle, high-melting solids with extended lattice structures, exhibiting strong electrolyte behavior when dissolved in water.

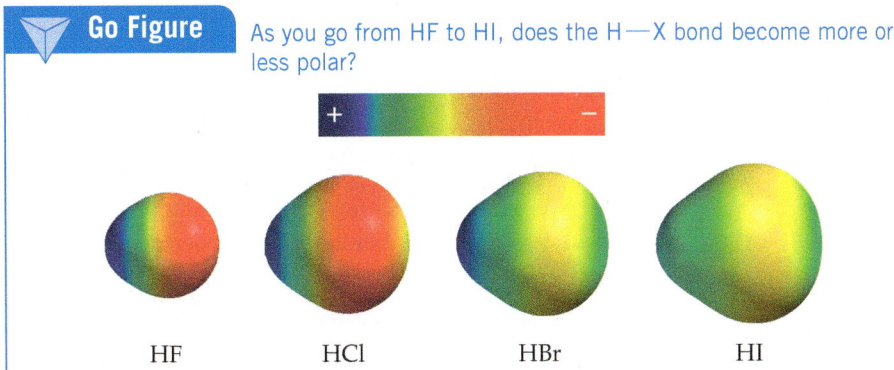

Go Figure As you go from HF to HI, does the H—X bond become more or less polar?

HF   HCl   HBr   HI

▲ Figure 8.11 **Charge separation in the hydrogen halides.** In HF, the strongly electronegative F pulls much of the electron density away from H. In HI, the I, being much less electronegative than F, does not attract the shared electrons as strongly, so the bond is far less polar.

*There are some exceptions to this statement, such as network solids, including diamond, silicon, and germanium, where an extended structure is formed even though the bonding is clearly covalent. These substances are discussed in Section 12.7.

Not surprisingly, there are exceptions to these general characterizations, some of which we examine later in the book. Nonetheless, the ability to quickly categorize the predominant bonding interactions in a substance as covalent or ionic imparts considerable insight into the properties of that substance. The question then becomes what is the best way to recognize which type of bonding dominates?

The simplest approach is to assume that the interaction between a metal and a nonmetal is ionic and that the interaction between two nonmetals is covalent. While this classification scheme is reasonably predictive, there are far too many exceptions to allow us to use it blindly. For example, tin is a metal and chlorine is a nonmetal, but $SnCl_4$ is a molecular substance that exists as a colorless liquid at room temperature. It freezes at $-33\,°C$ and boils at $114\,°C$. The characteristics of $SnCl_4$ are not those typical of an ionic substance. Is there a more predictable way of determining what type of bonding is prevalent in a compound? A more sophisticated approach is to use the difference in electronegativity as the main criterion for determining whether ionic or covalent bonding will be dominant. This approach correctly predicts the bonding in $SnCl_4$ to be polar covalent based on an electronegativity difference of 1.2 and at the same time correctly predicts the bonding in NaCl to be predominantly ionic based on an electronegativity difference of 2.1.

Evaluating bonding based on electronegativity difference is a useful system, but it has one shortcoming. The electronegativity values given in Figure 8.8 do not take into account changes in bonding that accompany changes in the oxidation state of the metal. (Section 4.4) For example, Figure 8.8 gives the electronegativity difference between manganese and oxygen as $3.5 - 1.5 = 2.0$, which falls in the range where the bonding is normally considered ionic (the electronegativity difference for NaCl is $3.0 - 0.9 = 2.1$). Therefore, it is not surprising to learn that manganese(II) oxide, MnO, is a green solid that melts at $1842\,°C$ and has the same crystal structure as NaCl.

However, the bonding between manganese and oxygen is not always ionic. Manganese(VII) oxide, Mn_2O_7, is a green liquid that freezes at $5.9\,°C$, an indication that covalent rather than ionic bonding dominates. The change in the oxidation state of manganese is responsible for the change in bonding. In general, as the oxidation state of a metal increases, so does the degree of covalent bonding. When the oxidation state of the metal is highly positive (roughly speaking, $+4$ or larger), we can expect significant covalency in the bonds it forms with nonmetals. Thus, metals in high oxidation states form molecular substances, such as Mn_2O_7, or polyatomic ions, such as MnO_4^- and CrO_4^{2-}, rather than ionic compounds.

Self-Assessment Exercises

SAE 8.9   Which of the following statements about electronegativity is *false*? (**a**) Electronegativity is a measure of the ability of an atom in a molecule to attract electrons to itself. (**b**) Fluorine is the most electronegative element. (**c**) Elements that have a high electronegativity generally have a high first ionization energy and a high-electron affinity. (**d**) Electronegativity generally increases from top to bottom in the periodic table. (**e**) The alkali metals are the family that has the lowest electronegativity values.

SAE 8.10   Arrange the following elements in order of lowest to highest electronegativity: S, Cl, Ca, As, Rb.

(**a**) Rb < Ca < As < S < Cl
(**b**) Ca < As < Rb < Cl < S
(**c**) S < Cl < Ca < As < Rb
(**d**) Rb < Ca < Cl < As < S

SAE 8.11   Arrange the following bonds in order of increasing polarity: B—Cl, P—S, Al—F, Br—Br.

(**a**) P—S < Br—Br < Al—F < B—Cl
(**b**) Br—Br < P—S < Al—F < B—Cl
(**c**) B—Cl < Br—Br < P—S < Al—F
(**d**) Br—Br < P—S < B—Cl < Al—F

SAE 8.12   The ICl molecule has a dipole moment of 1.24 D and a bond length of 2.32 Å. In units of the electronic charge, what is the partial charge on the I atom in the ICl molecule? (**a**) $0.11+$ (**b**) $0.11-$ (**c**) $0.26+$ (**d**) $0.26-$ (**e**) 0

SAE 8.13   In each of the following pairs, one compound is a molecular substance and one is an ionic substance: PCl_3 and $ScCl_3$; CrF_2 and MoF_6; OsO_4 and La_2O_3. Which of the compounds are molecular substances? (**a**) PCl_3, CrF_2, OsO_4 (**b**) $ScCl_3$, CrF_2, La_2O_3 (**c**) PCl_3, MoF_6, OsO_4 (**d**) $ScCl_3$, MoF_6, La_2O_3

8.5 | Drawing Lewis Structures

Lewis structures can help us understand the bonding in many compounds and are frequently used when discussing the properties of molecules. For this reason, drawing Lewis structures is an important skill that you should practice.

We use the following procedure to draw Lewis structures for molecules. We will need to know the atoms involved and which atoms are bonded to one another.

How to Draw Lewis Structures

1. *Sum the valence electrons from all atoms, taking into account overall charge.* Use the periodic table to help you determine the number of valence electrons in each atom. For an anion, add one electron to the total for each negative charge. For a cation, subtract one electron from the total for each positive charge. Do not worry about keeping track of which electrons come from which atoms. Only the total number is important.

2. *Write the symbols for the atoms, show which atoms are attached to which, and connect them with a single bond (a line, representing two electrons).* Chemical formulas are often written in the order in which the atoms are connected in the molecule or ion. The formula HCN, for example, tells you that the carbon atom is bonded to the H and to the N. In many polyatomic molecules and ions, the central atom is usually written first, as in $CO_3{}^{2-}$ and SF_4. Remember that the central atom is generally less electronegative than the atoms surrounding it. In other cases, you may need more information before you can draw the Lewis structure.

3. *Complete the octets around all the atoms bonded to the central atom.* Keep in mind that a hydrogen atom has only a single pair of electrons around it.

4. *Place any remaining electrons on the central atom,* even if doing so results in more than an octet of electrons around the atom.

5. *If there are not enough electrons to give the central atom an octet, try multiple bonds.* Use one or more of the unshared pairs of electrons on the atoms bonded to the central atom to form double or triple bonds.

The following examples of this procedure will help you put it into practice.

Learning Objectives

When you finish Section 8.5, you should be able to:

▶ Construct Lewis structures for molecules and polyatomic ions.

▶ Determine the formal charge for each atom in a molecule from its Lewis structure.

▶ Use formal charges to determine the dominant Lewis structure for a molecule or ion.

Sample Exercise 8.6
Drawing a Lewis Structure

Draw the Lewis structure for phosphorus trichloride, PCl_3.

SOLUTION

Analyze and Plan We are asked to draw a Lewis structure from a molecular formula. Our plan is to follow the five-step procedure just described.

Solve First, we sum the valence electrons. Phosphorus (group 5A) has five valence electrons, and each chlorine (group 7A) has seven. The total number of valence electrons is therefore

$$5 + (3 \times 7) = 26$$

Second, we arrange the atoms to show which atom is connected to which, and we draw a single bond between them. There are various ways the atoms might be arranged. It helps to know, though, that in binary compounds the first element in the chemical formula is generally surrounded by the remaining atoms. So, we proceed to draw a skeleton structure in which a single bond connects the P atom to each Cl atom:

$$\text{Cl} - \text{P} - \text{Cl}$$
$$|$$
$$\text{Cl}$$

(It is not crucial that the Cl atoms be to the left of, right of, and below the P atom—any structure that shows each of the three Cl atoms bonded to P will work.)

Third, we add Lewis electron dots to complete the octets on the atoms bonded to the central atom. Completing the octets around each Cl atom accounts for 24 electrons (remember, each line in our structure represents *two* electrons):

$$:\!\ddot{\text{C}}\text{l} - \text{P} - \ddot{\text{C}}\text{l}\!:$$
$$|$$
$$:\!\ddot{\text{C}}\text{l}\!:$$

Fourth, recalling that our total number of electrons is 26, place the remaining two electrons on the central atom, P, which completes its octet:

$$:\!\ddot{\text{C}}\text{l} - \ddot{\text{P}} - \ddot{\text{C}}\text{l}\!:$$
$$|$$
$$:\!\ddot{\text{C}}\text{l}\!:$$

This structure gives each atom an octet, so we stop at this point. (In checking for octets, remember to count a single bond as two electrons.)

▶ **Practice Exercise**
(a) How many valence electrons should appear in the Lewis structure for CH_2Cl_2?
(b) Draw the Lewis structure.

Sample Exercise 8.7
Lewis Structure with a Multiple Bond

Draw the Lewis structure for HCN.

SOLUTION

Hydrogen has one valence electron, carbon (group 4A) has four, and nitrogen (group 5A) has five. The total number of valence electrons is, therefore, $1 + 4 + 5 = 10$. In principle, there are different ways in which we might choose to arrange the atoms. Because hydrogen can accommodate only one electron pair, it always has only one single bond associated with it. Therefore, C—H—N is an impossible arrangement. The remaining two possibilities are H—C—N and H—N—C. The first is the arrangement found experimentally. You might have guessed this because the formula is written with the atoms in this order, and carbon is less electronegative than nitrogen. Thus, we begin with the skeleton structure

$$H—C—N$$

The two bonds account for four electrons. The H atom can have only two electrons associated with it, and so we will not add any more electrons to it. If we place the remaining six electrons around N to give it an octet, we do not achieve an octet on C:

$$H—C—\ddot{N}:$$

We therefore try a double bond between C and N, using one of the unshared pairs we placed on N. Again, we end up with fewer than eight electrons on C, and so we next try a triple bond. This structure gives an octet around both C and N:

$$H—C \overset{\frown}{:} \ddot{N}: \longrightarrow H—C≡N:$$

The octet rule is satisfied for the C and N atoms, and the H atom has two electrons around it. This is a correct Lewis structure.

▶ **Practice Exercise**
Draw the Lewis structure for (**a**) CO, (**b**) C_2H_4.

Sample Exercise 8.8
Lewis Structure for a Polyatomic Ion

Draw the Lewis structure for the BrO_3^- ion.

SOLUTION

Bromine (group 7A) has seven valence electrons, and oxygen (group 6A) has six. We must add one more electron to our sum to account for the 1− charge of the ion. The total number of valence electrons is, therefore, $7 + (3 \times 6) + 1 = 26$. For oxyanions—$SO_4^{2-}$, NO_3^-, CO_3^{2-}, and so forth—the oxygen atoms surround the central nonmetal atom. After arranging the O atoms around the Br atom, drawing single bonds, and distributing the unshared electron pairs, we have

$$\left[:\ddot{O}—\overset{..}{\underset{..}{Br}}—\ddot{O}: \right]^-$$
$$\underset{:\ddot{O}:}{|}$$

Notice that the Lewis structure for an ion is written in brackets and the charge is shown outside the brackets at the upper right.

▶ **Practice Exercise**
Draw the Lewis structure for (**a**) ClO_2^-, (**b**) PO_4^{3-}.

Formal Charge and Alternative Lewis Structures

When we draw a Lewis structure, we are describing how the electrons are distributed in a molecule or polyatomic ion. In some instances, we can draw two or more valid Lewis structures for a molecule that all obey the octet rule. All of these structures can be thought of as contributing to the *actual* arrangement of the electrons in the molecule, but not all of them will contribute to the same extent. How do we decide which one of several Lewis structures is the most important? One approach is to do some "bookkeeping" of the valence electrons to determine the *formal charge* of each atom in each Lewis structure. The **formal charge** of any atom in a molecule is the charge the atom would have if each bonding electron pair in the molecule were shared equally between its two atoms.

How to Calculate the Formal Charges of Atoms in Lewis Structures

1. *All* unshared (nonbonding) electrons are assigned to the atom on which they are found.

2. For any bond—single, double, or triple—*half* of the bonding electrons are assigned to each atom in the bond.

3. The formal charge of each atom is calculated by subtracting the number of electrons assigned to the atom from the number of valence electrons in the neutral atom:

Formal charge =
$$\text{valence electrons} - \left[\tfrac{1}{2}(\text{bonding electrons}) + \text{nonbonding electrons}\right] \qquad [8.11]$$

Let's practice by calculating the formal charges for the atoms in the cyanide ion, CN^-, which has the Lewis structure

$$[:C\equiv N:]^-$$

The neutral C atom has four valence electrons. There are six electrons in the cyanide triple bond and two nonbonding electrons on C. We calculate the formal charge on C as $4 - \left[\tfrac{1}{2}(6) + 2\right] = -1$. For N, the valence electron count is five; there are six electrons in the cyanide triple bond, and two nonbonding electrons on the N. The formal charge on N is $5 - \left[\tfrac{1}{2}(6) + 2\right] = 0$. We can draw the whole ion with its formal charges as

$$\overset{-1 \quad 0}{[:C\equiv N:]^-}$$

Notice that the sum of the formal charges equals the overall charge on the ion, 1−. The formal charges on a neutral molecule must add to zero, whereas those on an ion must add to give the charge on the ion.

If we can draw several Lewis structures for a molecule, the concept of formal charge can help us decide which is the most important, which we shall call the *dominant* Lewis structure. One Lewis structure for CO_2, for instance, has two double bonds. However, we can also satisfy the octet rule by drawing a Lewis structure having one single bond and one triple bond. Calculating formal charges in these structures, we have

	$\ddot{O}=C=\ddot{O}$			$:\ddot{O}-C\equiv O:$		
Valence electrons:	6	4	6	6	4	6
−(Electrons assigned to atom):	6	4	6	7	4	5
Formal charge:	0	0	0	−1	0	+1

Note that in both cases the formal charges add up to zero, as they must because CO_2 is a neutral molecule. So, which is the dominant structure? As a general rule, when more than one Lewis structure is possible, we use the following guidelines to choose the dominant one:

How to Identify the Dominant Lewis Structure

1. The dominant Lewis structure is generally the one in which the atoms bear formal charges closest to zero.

2. A Lewis structure in which any negative charges reside on the more electronegative atoms is generally more dominant than one that has negative charges on the less electronegative atoms.

Thus, the first Lewis structure of CO_2 is the dominant one because the atoms carry no formal charges and so satisfy the first guideline. The other Lewis structure shown (and the similar one that has a triple bond to the left O and a single bond to the right O) contribute to the actual structure to a much smaller extent.

Although the concept of formal charge helps us to arrange alternative Lewis structures in order of importance, it is important to remember that *formal charges do not represent real charges on atoms*. These charges are just a bookkeeping convention. The actual charge distributions in molecules and ions are determined not by formal charges but by a number of other factors, including electronegativity differences between atoms.

Sample Exercise 8.9
Lewis Structures and Formal Charges

Three possible Lewis structures for the thiocyanate ion, NCS⁻, are

$$[:\ddot{N}—C≡S:]^- \qquad [\ddot{N}=C=\ddot{S}:]^- \qquad [:N≡C—\ddot{\ddot{S}}:]^-$$

(a) Determine the formal charges in each structure.

(b) Based on the formal charges, which Lewis structure is the dominant one?

SOLUTION

(a) Neutral N, C, and S atoms have five, four, and six valence electrons, respectively. We can determine the formal charges in the three structures by using the rules we just discussed:

$$\begin{matrix} -2 & 0 & +1 \\ [:\ddot{N}— & C≡ & S:]^- \end{matrix} \qquad \begin{matrix} -1 & 0 & 0 \\ [\ddot{N}= & C= & \ddot{S}:]^- \end{matrix} \qquad \begin{matrix} 0 & 0 & -1 \\ [:N≡ & C— & \ddot{\ddot{S}}:]^- \end{matrix}$$

As they must, the formal charges in all three structures sum to 1−, the overall charge of the ion.

(b) The dominant Lewis structure generally produces formal charges of the smallest magnitude (guideline 1). That

eliminates the left structure as the dominant one. Further, as discussed in Section 8.4, N is more electronegative than C or S, so we expect any negative formal charge to reside on the N atom (guideline 2). For these two reasons, the middle Lewis structure is the dominant one for NCS⁻.

▶ **Practice Exercise**

The cyanate ion, NCO⁻, has three possible Lewis structures. **(a)** Draw these three structures and assign formal charges in each. **(b)** Which Lewis structure is dominant?

A CLOSER LOOK | Oxidation Numbers, Formal Charges, and Actual Partial Charges

In Chapter 4 we introduced the rules for assigning *oxidation numbers* to atoms. The concept of electronegativity is the basis of these numbers. An atom's oxidation number is the charge the atom would have if its bonds were completely ionic. That is, in determining oxidation numbers, all shared electrons are counted with the more electronegative atom. For example, consider the Lewis structure of HCl in Figure 8.12(a). To assign oxidation numbers, both electrons in the covalent bond between the atoms are assigned to the more electronegative Cl atom. This procedure gives Cl eight valence electrons, one more than in the neutral atom. Thus, its oxidation number is −1. Hydrogen has no valence electrons when they are counted this way, giving it an oxidation number of +1.

In assigning formal charges to the atoms in HCl [Figure 8.12(b)], we ignore electronegativity; the electrons in bonds are assigned equally to the two bonded atoms. In this case Cl has seven assigned electrons, the same as that of the neutral Cl atom, and H has one assigned electron. Thus, the formal charges of both Cl and H in this compound are 0.

Neither oxidation number nor formal charge gives an accurate depiction of the actual charges on atoms because oxidation numbers overstate the role of electronegativity and formal charges ignore it. It seems reasonable that electrons in covalent bonds should be apportioned according to the relative electronegativities of the bonded atoms. Figure 8.8 shows that Cl has an electronegativity of 3.0, while the electronegativity value of H is 2.1. The more

electronegative Cl atom might therefore be expected to have roughly $3.0/(3.0 + 2.1) = 0.59$ of the electrical charge in the bonding pair, whereas the H atom would have $2.1/(3.0 + 2.1) = 0.41$ of the charge. Because the bond consists of two electrons, the Cl atom's share is $0.59 \times 2e = 1.18e$, or $0.18e$ more than the neutral Cl atom. This gives rise to a partial negative charge of 0.18− on Cl and therefore a partial positive charge of 0.18+ on H. (Notice again that we place the plus and minus signs *before* the magnitude in writing oxidation numbers and formal charges but *after* the magnitude in writing actual charges.)

The dipole moment of HCl gives an experimental measure of the partial charge on each atom. In Sample Exercise 8.5 we saw that the dipole moment of HCl corresponds to a partial charge of 0.178+ on H and 0.178− on Cl, in remarkably good agreement with our simple approximation based on electronegativities. Although our approximation method provides ballpark numbers for the magnitude of charge on atoms, the relationship between electronegativities and charge separation is generally more complicated. As we have already seen, computer programs employing quantum-mechanical principles have been developed to obtain more accurate estimates of the partial charges on atoms, even in complex molecules. A computer-graphical representation of the calculated charge distribution in HCl is shown in Figure 8.12(c).

Related Exercises: 8.10, 8.51–8.54, 8.94

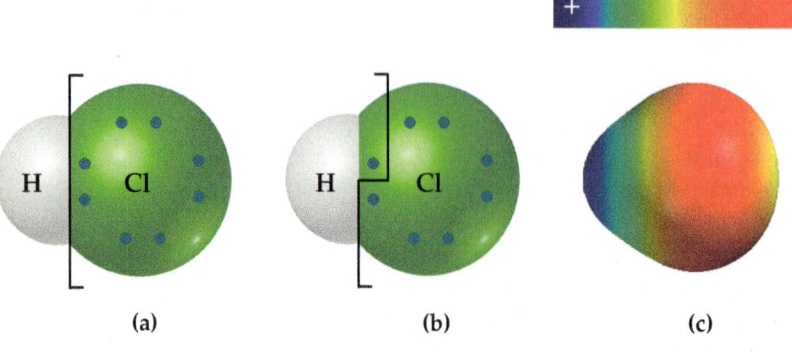

◀ **Figure 8.12** (a) Oxidation number, (b) formal charge, and (c) electron-density distribution for the HCl molecule.

Self-Assessment Exercises

SAE 8.14 The substance called *malonic acid* has the molecular formula $C_3H_4O_4$ and the connectivity shown in the figure. When the Lewis structure of malonic acid is completed, the molecule has _____ double bonds and _____ nonbonding pairs of electrons. (**a**) 0, 4 (**b**) 1, 4 (**c**) 2, 4 (**d**) 2, 8 (**e**) 4, 8

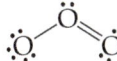

SAE 8.15 The SO_3 molecule has a central S atom with the three O atoms bonded to the S, as shown in the figure. When a single Lewis structure is written for SO_3 that obeys the octet rule for all the atoms, which of the following statements is or are true?

   (**i**) The Lewis structure has one double bond.
  (**ii**) The three oxygen atoms do not all have the same formal charge.
 (**iii**) The formal charge on the S atom is +1.

(**a**) Only one of the statements is true. (**b**) i and ii are true. (**c**) i and iii are true. (**d**) ii and iii are true. (**e**) All three statements are true.

SAE 8.16 Which of the three Lewis structures shown here for carbamic acid, NH_2COOH, is the dominant Lewis structure?

8.6 | Resonance Structures

We sometimes encounter molecules and ions in which the experimentally determined arrangement of atoms is not adequately described by a single dominant Lewis structure. In such instances, we use multiple equivalent Lewis structures to describe the molecule or ion, a phenomenon we call *resonance*.

Let's consider ozone, O_3, which is a bent molecule with two equal oxygen–oxygen bond lengths (**Figure 8.13**). Because each oxygen atom contributes 6 valence electrons, the ozone molecule has 18 valence electrons. This means the Lewis structure must have one O—O single bond and one O=O double bond to attain an octet about each atom:

However, this single structure cannot by itself be dominant because it requires that one O—O bond be different from the other, contrary to the observed structure. We would expect the O=O double bond to be shorter than the O—O single bond. In drawing the Lewis structure, however, we could just as easily have put the O=O bond on the left:

There is no reason for one of these Lewis structures to be dominant because they are equally valid representations of the molecule. The placement of the atoms in these two alternative but completely equivalent Lewis structures is the same, but the placement of the electrons is different; we call Lewis structures of this sort **resonance structures**. To describe the structure of ozone properly, we write both resonance structures and use a double-headed arrow to indicate that the real molecule is described by an average of the two:

To understand why certain molecules require more than one resonance structure, we can draw an analogy to mixing paint (**Figure 8.14**). Blue and yellow are both primary colors of paint pigment. An equal blend of blue and yellow pigments produces green pigment. We cannot describe green paint in terms of a single primary color, yet it still has its own identity. Green paint does not oscillate between its two primary colors: It is not blue

▲ **Learning Objective**

When you finish **Section 8.6**, you should be able to:

▶ Use resonance structures to describe the electron distribution in a molecule or ion that does not have a single dominant Lewis structure.

▼ **Go Figure**

What feature of this structure suggests that the two outer O atoms are in some way equivalent to each other?

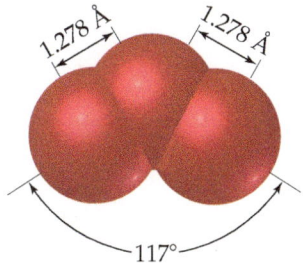

▲ **Figure 8.13 Molecular structure of ozone.**

Go Figure

Is the electron density consistent with equal contributions from the two resonance structures for O_3? Explain.

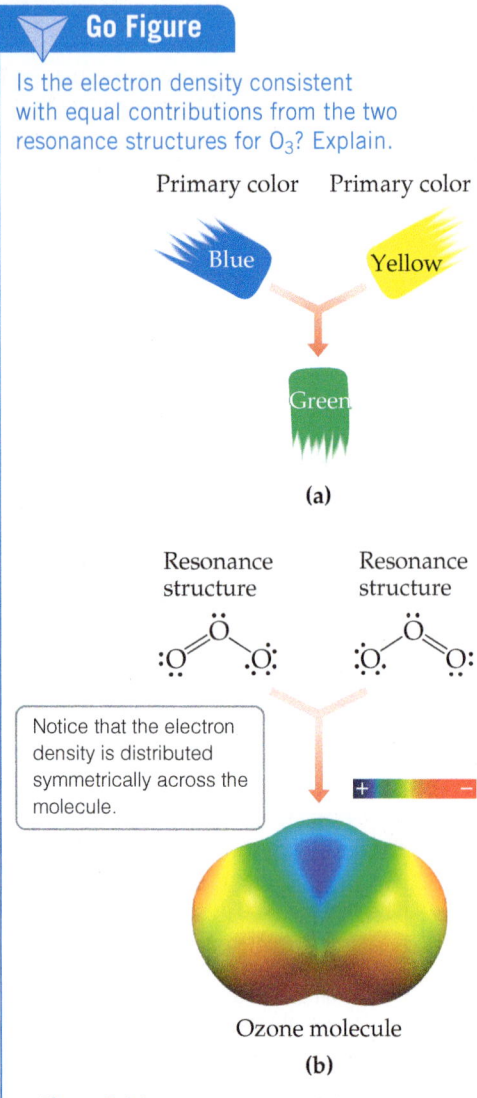

Primary color   Primary color

Blue   Yellow

Green

(a)

Resonance   Resonance
structure   structure

Notice that the electron density is distributed symmetrically across the molecule.

Ozone molecule

(b)

▲ **Figure 8.14 Resonance.** Describing a molecule as a blend of different resonance structures is similar to describing a paint color as a blend of primary colors. **(a)** Green paint is a blend of blue and yellow. We cannot describe green as a single primary color. **(b)** The ozone molecule is a blend of two resonance structures. We cannot describe the ozone molecule in terms of a single Lewis structure.

part of the time and yellow the rest of the time. Similarly, molecules such as ozone cannot be described as oscillating between the two individual Lewis structures shown previously—there are two equivalent dominant Lewis structures that contribute equally to the actual structure of the molecule.

The actual arrangement of the electrons in molecules such as O_3 must be considered as a blend of two (or more) Lewis structures. By analogy to the green paint, the molecule has its own identity separate from the individual resonance structures. For example, the ozone molecule always has two equivalent O—O bonds whose lengths are intermediate between the lengths of an oxygen–oxygen single bond and an oxygen–oxygen double bond. Another way of looking at it is to say that the rules for drawing Lewis structures do not allow us to have a single dominant structure for the ozone molecule. For example, there are no rules for drawing half-bonds. We can get around this limitation by drawing two equivalent Lewis structures that, when averaged, amount to something very much like what is observed experimentally.

As an additional example of resonance structures, consider the nitrate ion, NO_3^-, for which three equivalent Lewis structures can be drawn:

Notice that the arrangement of atoms is the same in each structure; only the placement of electrons differs. In writing resonance structures, the same atoms must be bonded to each other in all structures, so that the only differences are in the arrangements of electrons. All three NO_3^- Lewis structures are equally dominant and taken together adequately describe the ion, in which all three N=O bond lengths are the same.

For some molecules or ions, all possible Lewis structures may not be equivalent; in other words, one or more resonance structures are more dominant than others. We will encounter examples of this as we proceed.

Resonance in Benzene

Resonance is an important concept in describing the bonding in organic molecules, particularly *aromatic* organic molecules, a category that includes the hydrocarbon *benzene,* C_6H_6. The six C atoms are bonded in a hexagonal ring, and one H atom is bonded to each C atom. We can write two equivalent dominant Lewis structures for benzene, each of which satisfies the octet rule. These two structures are in resonance:

Note that the double bonds are in different places in the two structures. Each of these resonance structures shows three carbon–carbon single bonds and three carbon–carbon double bonds. However, experimental data show that all six C—C bonds are of equal length, 1.40 Å, intermediate between the typical bond lengths for a C—C single bond (1.54 Å) and a C=C double bond (1.34 Å). Each of the C—C bonds in benzene can be thought of as a blend of a single bond and a double bond (**Figure 8.15**).

Benzene is commonly represented by omitting the hydrogen atoms and showing only the carbon–carbon framework with the vertices unlabeled. In this convention, the resonance in the molecule is represented either by two structures separated by a double-headed arrow or by a shorthand notation in which we draw a hexagon with a circle inside:

The shorthand notation reminds us that benzene is a blend of two resonance structures—it emphasizes that the C=C double bonds cannot be assigned to specific edges of the hexagon. Chemists use both representations of benzene interchangeably.

The bonding arrangement in benzene confers special stability to the molecule. As a result, millions of organic compounds contain the six-membered ring characteristic of benzene. Many of these compounds are important in biochemistry, in pharmaceuticals, and in the production of modern materials.

What is the significance of the dashed bonds in this ball-and-stick model?

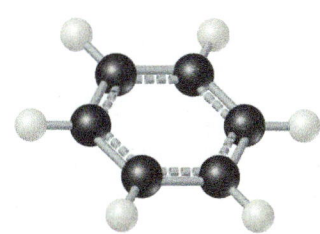

▲ **Figure 8.15 Benzene, an "aromatic" organic compound.** The benzene molecule is a regular hexagon of carbon atoms with a hydrogen atom bonded to each one. The dashed lines represent the blending of two equivalent resonance structures, leading to C—C bonds that are intermediate between single and double bonds.

Sample Exercise 8.10
Resonance Structures

Which is predicted to have the shorter sulfur–oxygen bonds, SO_3 or SO_3^{2-}?

SOLUTION

The sulfur atom has six valence electrons, as does oxygen. Thus, SO_3 contains 24 valence electrons. In writing the Lewis structure, we see that three equivalent resonance structures can be drawn:

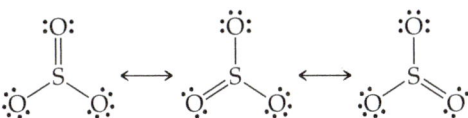

As with NO_3^- the actual structure of SO_3 is an equal blend of all three. Thus, each S—O bond length should be about one-third of the way between the length of a single bond and the length of a double bond. That is, S—O should be shorter than single bonds but not as short as double bonds.

The SO_3^{2-} ion has 26 electrons, which leads to a dominant Lewis structure in which all the S—O bonds are single:

$$\left[\ddot{\underset{..}{O}}-\underset{\underset{\displaystyle :\ddot{O}:}{|}}{S}-\ddot{\underset{..}{O}}: \right]^{2-}$$

Our analysis of the Lewis structures thus far leads us to conclude that SO_3 should have the shorter S—O bonds and SO_3^{2-} the longer ones. This conclusion is correct: The experimentally measured S—O bond lengths are 1.42 Å in SO_3 and 1.51 Å in SO_3^{2-}

▶ **Practice Exercise**
Draw two equivalent resonance structures for the formate ion, HCO_2^-.

Self-Assessment Exercises

SAE 8.17 The SO_2 molecule, which has a central S atom bonded to two O atoms, has _____ equivalent resonance structures, and the two S—O bonds are _____. (**a**) one, the same length (**b**) two, the same length (**c**) one, different lengths (**d**) two, different lengths

SAE 8.18 Based on their Lewis structures, predict the ordering of the O—O bond lengths from shortest to longest in O_2, O_2^{2-} (the peroxide ion) and O_3 (ozone).

(**a**) $O_3 < O_2 < O_2^{2-}$
(**b**) $O_3 < O_2^{2-} < O_2$
(**c**) $O_2 < O_3 < O_2^{2-}$
(**d**) $O_2^{2-} < O_2 < O_3$
(**e**) $O_2^{2-} < O_3 < O_2$

SAE 8.19 The structure of benzene, C_6H_6, is often denoted as shown in the following figure. Which of the follow statements is *false*? (**a**) This notation is used because there are no Lewis structures for benzene that satisfy the octet rule. (**b**) This notation represents a blend of two equivalent resonance structures. (**c**) The circle indicates that there are three double bonds in each individual resonance structure for the molecule. (**d**) Each vertex of the hexagon represents a carbon atom, and the hydrogen atoms have been omitted.

8.7 | Exceptions to the Octet Rule

The octet rule is so simple and useful in introducing the basic concepts of bonding that you might assume it is always obeyed. In Section 8.2, however, we noted its limitation in dealing with ionic compounds of the transition metals. The rule also fails in many situations involving covalent bonding. In this section, we examine molecules that do not conform to the octet rule.

Three main categories of molecules and ions are exceptions to the octet rule:

1. Molecules and polyatomic ions containing an odd number of electrons

2. Molecules and polyatomic ions in which an atom has fewer than an octet of valence electrons

3. Molecules and polyatomic ions in which an atom has more than an octet of valence electrons

Learning Objectives

When you finish Section 8.7, you should be able to:

▶ Determine the Lewis structures of molecules or ions that have an odd number of electrons or less than an octet of electrons around an atom.

▶ Determine the Lewis structures of molecules or ions with more than an octet of electrons around an atom.

Odd Number of Electrons

In the vast majority of molecules and polyatomic ions, the total number of valence electrons is even, and complete pairing of electrons occurs. However, in some relatively common molecules and polyatomic ions, such as ClO_2, NO, NO_2, and O_2^-, the number of valence electrons is odd. Because the number of electrons is odd, complete pairing of these electrons is impossible, so an octet around each atom cannot be achieved. For example, NO contains $5 + 6 = 11$ valence electrons. The two most important Lewis structures for this molecule are

$$\ddot{\overset{\textstyle\cdot}{N}}{=}\ddot{\overset{\textstyle\cdot}{O}} \quad \text{and} \quad \overset{\textstyle\cdot}{\ddot{N}}{=}\ddot{\overset{\textstyle\cdot}{O}}$$

In any molecule or ion with an odd number of electrons, the Lewis structure will have one unpaired electron somewhere in the structure, as we see for the NO molecule. We can still use the same tools we have already developed to determine the dominant Lewis structure. In the case of NO, we expect the structure on the left, with the unpaired electron on the N atom, to be dominant due to more favorable formal charges:

$$\overset{0 \quad\;\; 0}{\ddot{\overset{\textstyle\cdot}{N}}{=}\ddot{\overset{\textstyle\cdot}{O}}} \quad \text{and} \quad \overset{-1 \;\; +1}{\overset{\textstyle\cdot}{\ddot{N}}{=}\ddot{\overset{\textstyle\cdot}{O}}}$$

Less Than an Octet of Valence Electrons

A second type of exception occurs when there are fewer than eight valence electrons around an atom in a molecule or polyatomic ion. This situation is relatively rare (with the exception of hydrogen and helium as we have already discussed), and is most often encountered in compounds of boron and beryllium. For example, if we apply the first steps of our procedure for drawing Lewis structures to boron trifluoride, BF_3, we obtain the structure

which has only six electrons around the boron atom. The formal charge is zero on both B and F, and we could complete the octet around boron by forming a double bond (recall that if there are not enough electrons to give the central atom an octet, a multiple bond may be the answer). In so doing, we see that there are three equivalent resonance structures (the formal charges are shown in red):

Each of these structures forces a fluorine atom to share additional electrons with the boron atom, which is inconsistent with the high electronegativity of fluorine. In fact, the formal charges indicate that this is an unfavorable situation. In each structure, the F atom involved in the B=F double bond has a formal charge of +1, while the less electronegative B atom has a formal charge of −1. Thus, the resonance structures containing a B=F double bond are less important than the one in which there are fewer than an octet of valence electrons around boron:

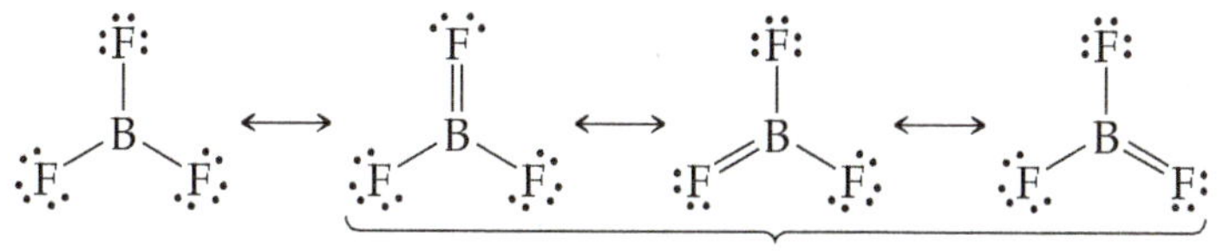

Dominant   Less important

We usually represent BF_3 solely by the dominant resonance structure, in which there are only six valence electrons around boron. The chemical behavior of BF_3 is consistent with this representation. In particular, BF_3 reacts energetically with molecules having an unshared pair of electrons that can be used to form a bond with boron, as, for example, in the reaction

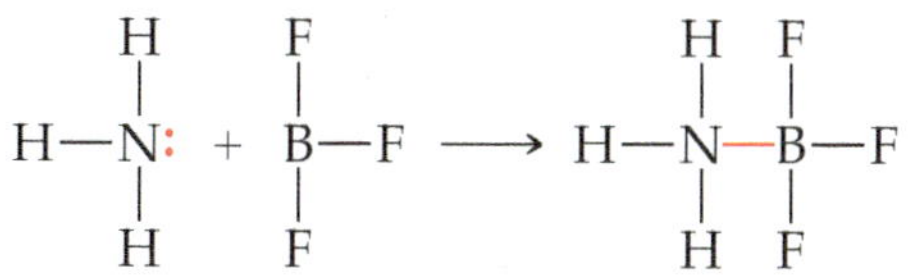

In the stable compound NH_3BF_3, boron has an octet of valence electrons. This type of behavior can be characterized as a somewhat different type of acid–base reaction—one that we discuss in Section 16.1.

More Than an Octet of Valence Electrons

The third and largest class of exceptions consists of molecules or polyatomic ions in which there are more than eight electrons in the valence shell of an atom. When we draw the Lewis structure for PF_5, for example, we are forced to place 10 electrons around the central phosphorus atom:

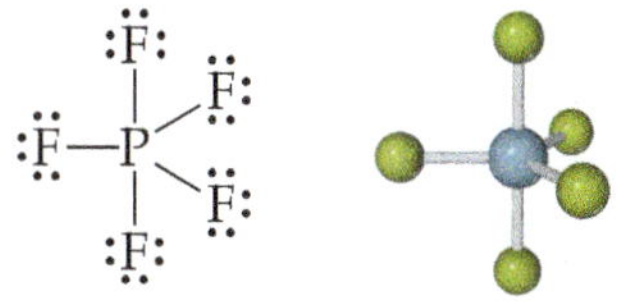

Molecules and ions with more than an octet of electrons around the central atom are often called *hypervalent*. Other examples of hypervalent species are SF_4, AsF_6^-, and ICl_4^-. The corresponding molecules with a second-period atom as the central atom, such as NCl_5 and OF_4, do *not* exist.

Hypervalent molecules are formed only for central atoms from period 3 and below in the periodic table. The principal reason for their formation is the relatively larger size of the central atom. For example, a P atom is large enough that five F (or even five Cl) atoms can be bonded to it without being too crowded. By contrast, a N atom is too small to accommodate five atoms bonded to it. Because size is a factor, hypervalent molecules occur most often when the central atom is bonded to the smallest and most electronegative atoms—F, Cl, and O.

The notion that a valence shell can contain more than eight electrons is also consistent with the presence of unfilled nd orbitals in atoms from period 3 and below. (Section 6.8) By comparison, in elements of the second period, only the $2s$ and $2p$ valence orbitals are available for bonding. However, theoretical work on the bonding in molecules such as PF_5 and SF_6 suggests that the presence of unfilled $3d$ orbitals in P and S has a relatively minor impact on the formation of hypervalent molecules. Most chemists now believe that the larger size of the atoms from periods 3 through 6 is more important to explain hypervalency than is the presence of unfilled d orbitals.

Sample Exercise 8.11

Lewis Structure for an Ion with More Than an Octet of Electrons

Draw the Lewis structure for ICl_4^-.

SOLUTION

Iodine (group 7A) has seven valence electrons. Each chlorine atom (group 7A) also has seven. An extra electron is added to account for the 1− charge of the ion. Therefore, the total number of valence electrons is $7 + (4 \times 7) + 1 = 36$.

The I atom is the central atom in the ion. Putting eight electrons around each Cl atom (including a pair of electrons between I and each Cl to represent the single bond between these atoms) requires $8 \times 4 = 32$ electrons.

We are thus left with $36 - 32 = 4$ electrons to be placed on the larger iodine:

$$\left[\begin{array}{c} :\ddot{Cl} \quad \ddot{Cl}: \\ \quad \diagdown \cdot\cdot \diagup \\ \quad \diagup I \diagdown \\ :\ddot{Cl} \quad \ddot{Cl}: \end{array}\right]^-$$

Iodine has 12 valence electrons around it, four more than needed for an octet.

▶ **Practice Exercise**
(a) Which of the following atoms is never found with more than an octet of valence electrons around it? S, C, P, Br, I.
(b) Draw the Lewis structure for XeF_2.

Finally, there are Lewis structures where you might have to choose between satisfying the octet rule and obtaining the most favorable formal charges by using more than an octet of electrons. For example, consider these Lewis structures for the phosphate ion, PO_4^{3-}:

$$\left[\begin{array}{c} \overset{-1}{:\ddot{O}:} \\ | \\ :\ddot{O}-\overset{+1}{P}-\overset{-1}{\ddot{O}:} \\ {}^{-1}\quad | \\ :\underset{-1}{\ddot{O}}: \end{array}\right]^{3-} \qquad \left[\begin{array}{c} \overset{-1}{:\ddot{O}:} \\ || \\ \overset{0}{\ddot{O}}=\overset{0}{P}-\overset{-1}{\ddot{O}:} \\ | \\ :\underset{-1}{\ddot{O}}: \end{array}\right]^{3-}$$

The formal charges on the atoms are shown in red. In the left structure, the P atom obeys the octet rule. In the right structure, however, the P atom has five electron pairs, leading to smaller formal charges on the atoms. (You should be able to visualize that there are three additional resonance structures for the Lewis structure on the right.)

Chemists are still debating which of these two structures is dominant for PO_4^{3-}. Theoretical calculations based on quantum mechanics suggest to some researchers that the left structure is the dominant one. Other researchers claim that the bond lengths in the ion are more consistent with the right structure being dominant. This disagreement is a convenient reminder that, in general, multiple Lewis structures can contribute to the actual electron distribution in an atom or molecule.

 Self-Assessment Exercises

SAE 8.20 For how many of the following molecules and ions is it *not* possible to write a Lewis structure that satisfies the octet rule: NH_4^+, NO, NO_2, N_2O? **(a)** 0 **(b)** 1 **(c)** 2 **(d)** 3 **(e)** 4

SAE 8.21 How many of the following molecules are hypervalent: PF_3, PF_5, XeF_2, XeF_4? **(a)** 0 **(b)** 1 **(c)** 2 **(d)** 3 **(e)** 4

SAE 8.22 The BrF_5 molecule has a central Br atom with the five F atoms bonded to the Br, as shown in the figure. When a Lewis structure is written for BrF_5 with five Br—F single bonds, which of the following statements is or are *true*?

(i) There are no nonbonding electron pairs on the Br atom.
(ii) The Br atom is hypervalent.
(iii) The formal charge on the Br atom is 0.

(a) Only one of the statements is true. **(b)** i and ii are true. **(c)** i and iii are true. **(d)** ii and iii are true. **(e)** All three statements are.

$$\begin{array}{c} F \\ | \\ F\diagdown \overset{|}{Br}\diagup F \\ \diagup \quad \diagdown \\ F \qquad F \end{array}$$

8.8 | Strengths and Lengths of Single and Multiple Bonds

Given how small atoms and molecules are, you may be wondering how chemists explore the effectiveness of covalent chemical bonds in holding atoms together. In this final section of Chapter 8, we explore two of the common ways that chemical bonds can be examined experimentally—namely, by looking at how strong they are and how long they are. In particular, we explore the relationship between the number of bonds between a pair of atoms and the strength and length of the bond.

The stability of a molecule is related to the strengths of its covalent bonds. We saw in Chapter 5 that we can measure the average bond enthalpy of many single bonds (Table 5.4). The information in Table 5.4 is repeated in **Table 8.3**, which also includes average bond enthalpies for multiple bonds. As you might expect, the data in Table 8.3 show that multiple bonds are generally stronger than single bonds.

> **△ Learning Objective**
>
> **When you finish Section 8.8, you should be able to:**
>
> ▶ Correlate the strengths and lengths of covalent bonds with the number of bonds between an atom pair.

TABLE 8.3   Average Bond Enthalpies (kJ/mol)

Single Bonds

C—H	413	N—H	391	O—H	463	F—F	155
C—C	348	N—N	163	O—O	146		
C—N	293	N—O	201	O—F	190	Cl—F	253
C—O	358	N—F	272	O—Cl	203	Cl—Cl	242
C—F	485	N—Cl	200	O—I	234		
C—Cl	328	N—Br	243			Br—F	237
C—Br	276			S—H	339	Br—Cl	218
C—I	240	H—H	436	S—F	327	Br—Br	193
C—S	259	H—F	567	S—Cl	253		
		H—Cl	431	S—Br	218	I—Cl	208
Si—H	323	H—Br	366	S—S	266	I—Br	175
Si—Si	226	H—I	299			I—I	151
Si—C	301						
Si—O	368						
Si—Cl	464						

Multiple Bonds

C=C	614	N=N	418	O=O	495
C≡C	839	N≡N	941		
C=N	615	N=O	607	S=O	523
C≡N	891			S=S	418
C=O	799				
C≡O	1072				

Just as we can define an average bond enthalpy, we can also define an average bond length for a number of common single and multiple bonds (**Table 8.4**). For bonds between the same atoms, notice that the bonds get shorter as the number of bonds between the atoms increases. Of particular interest is the relationship, in any atom pair, among bond enthalpy, bond length, and number of bonds between the atoms. For example, we can use data in Tables 8.3 and 8.4 to compare the bond lengths and bond enthalpies of carbon–carbon single, double, and triple bonds:

C—C	C=C	C≡C
1.54 Å	1.34 Å	1.20 Å
348 kJ/mol	614 kJ/mol	839 kJ/mol

As the number of bonds between the carbon atoms increases, the bond length decreases and the average bond enthalpy increases. That is, the carbon atoms are held more closely and more tightly together. In general, *as the number of bonds between two atoms increases, the bond grows shorter and stronger*. This trend is illustrated in **Figure 8.16** for N—N single, double, and triple bonds.

TABLE 8.4   Average Bond Lengths for Some Single, Double, and Triple Bonds

Bond	Bond Length (Å)	Bond	Bond Length (Å)
C—C	1.54	N—N	1.47
C=C	1.34	N=N	1.24
C≡C	1.20	N≡N	1.10
C—N	1.43	N—O	1.36
C=N	1.38	N=O	1.22
C≡N	1.16		
		O—O	1.48
C—O	1.43	O=O	1.21
C=O	1.23		
C≡O	1.13		

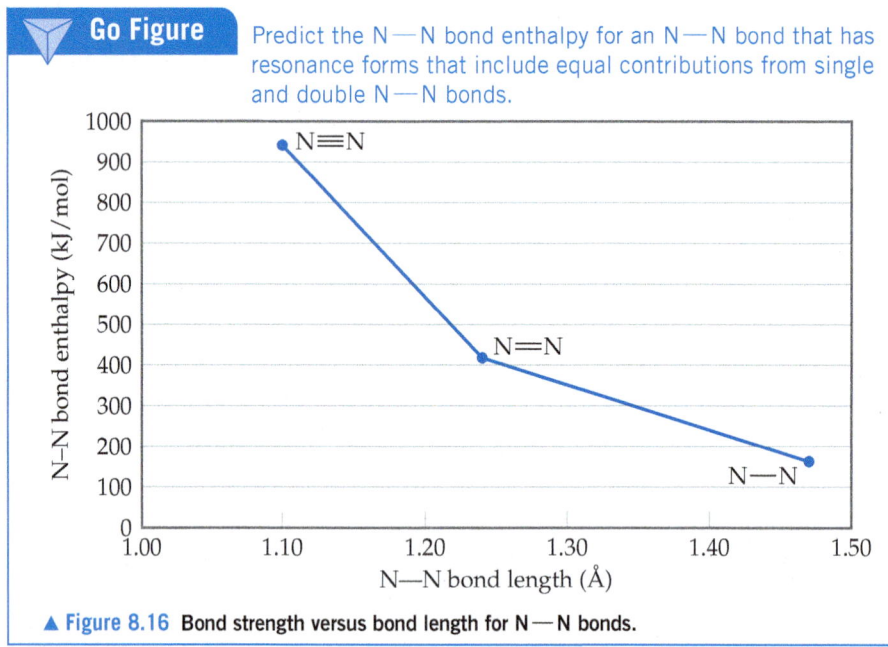

Go Figure Predict the N—N bond enthalpy for an N—N bond that has resonance forms that include equal contributions from single and double N—N bonds.

▲ **Figure 8.16** Bond strength versus bond length for N—N bonds.

You should make one final observation from the data in Table 8.3: The average bond enthalpy of a double bond between two atoms is not exactly two times the bond enthalpy for a single bond between the same atoms. Likewise, the bond enthalpy of triple bonds is not exactly three times that for single bonds. Notice that the average bond enthalpy of a C=C double bond (614 kJ/mol) is *less* than twice that of a C—C single bond (348 kJ/mol). By comparison, the bond enthalpy for a N=N double bond (418 kJ/mol) is *more* than twice that for a N—N single bond (163 kJ/mol). As we discuss in Chapter 9, a number of factors affect the types of bonds we make between atoms. The important take-away from the data here is that not all bonds between a pair of atoms have the same strength.

 Self-Assessment Exercises

SAE 8.23 Which of the following statements about the correlation of bond order with bond strength is *false*? (**a**) A double bond between two atoms is stronger than a single bond between the same two atoms. (**b**) A triple bond between two atoms is stronger than a double bond between the same two atoms. (**c**) All triple bonds have the same average bond enthalpy regardless of which atoms are bonded to one another. (**d**) The reaction enthalpy can be estimated by determining how much energy needs to be added to break bonds and how much energy is released in forming new bonds.

SAE 8.24 The average lengths for C—O, C=O, and C≡O bonds are 1.43, 1.23, and 1.13 Å, respectively. The carbonate ion, CO_3^{2-}, has a central carbon atom and three O atoms bonded to the C. All three C—O bond lengths in the carbonate ion are equal in length. Based on the Lewis structure of CO_3^{2-}, what is the most likely value for the C—O bond length in CO_3^{2-}? (**a**) less than 1.13 Å (**b**) between 1.13 and 1.23 Å (**c**) between 1.23 and 1.43 Å (**d**) greater than 1.43 Å

Putting Concepts Together

Phosgene, a substance used in poisonous gas warfare during World War I, is so named because it was first prepared by the action of sunlight on a mixture of carbon monoxide and chlorine gases. Its name comes from the Greek words *phos* (light) and *genes* (born of). Phosgene has the following elemental composition: 12.14% C, 16.17% O, and 71.69% Cl by mass. Its molar mass is 98.9 g/mol. **(a)** Determine the molecular formula of this compound. **(b)** Draw three Lewis structures for the molecule that satisfy the octet rule for each atom. (The Cl and O atoms bond to C.) **(c)** Using formal charges, determine which Lewis structure is the dominant one. **(d)** Using average bond enthalpies, estimate ΔH for the formation of gaseous phosgene from $CO(g)$ and $Cl_2(g)$.

SOLUTION

(a) The empirical formula of phosgene can be determined from its elemental composition. (Section 3.5) Assuming 100 g of the compound and calculating the number of moles of C, O, and Cl in this sample, we have:

$$(12.14 \text{ g C})\left(\frac{1 \text{ mol C}}{12.01 \text{ g C}}\right) = 1.011 \text{ mol C}$$

The ratio of the number of moles of each element, obtained by dividing each number of moles by the smallest quantity, indicates that there is one C and one O for each two Cl in the empirical formula, $COCl_2$.

$$(16.17 \text{ g O})\left(\frac{1 \text{ mol O}}{16.00 \text{ g O}}\right) = 1.011 \text{ mol O}$$

$$(71.69 \text{ g Cl})\left(\frac{1 \text{ mol Cl}}{35.45 \text{ g Cl}}\right) = 2.022 \text{ mol Cl}$$

The molar mass of the empirical formula is $12.01 + 16.00 + 2(35.45) = 98.91$ g/mol, the same as the molar mass of the molecule. Thus, $COCl_2$ is the molecular formula.

(b) Carbon has four valence electrons, oxygen has six, and chlorine has seven, giving $4 + 6 + 2(7) = 24$ electrons for the Lewis structures. Drawing a Lewis structure with all single bonds does not give the central carbon atom an octet. Using multiple bonds, we find that three structures satisfy the octet rule:

(c) Calculating the formal charges on each atom gives:

The first structure is expected to be the dominant one because it has the lowest formal charges on each atom. Indeed, the molecule is usually represented by this single Lewis structure.

(d) Writing the chemical equation in terms of the Lewis structures of the molecules, we have:

Thus, the reaction involves breaking a $C\equiv O$ bond and a $Cl-Cl$ bond and forming a $C=O$ bond and two $C-Cl$ bonds. Using bond enthalpies from Table 8.3, we have:

$$\Delta H = [D(C\equiv O) + D(Cl-Cl)] - [D(C=O) + 2D(C-Cl)]$$

$$= [1072 \text{ kJ} + 242 \text{ kJ}] - [799 \text{ kJ} + 2(328 \text{ kJ})] = -141 \text{ kJ}$$

The reaction is exothermic, but energy is still needed from sunlight or another source for the reaction to begin, as is the case for the combustion of $H_2(g)$ and $O_2(g)$ to form $H_2O(g)$ (see Figure 5.14).

Chapter Summary and Key Terms

CHEMICAL BONDS, LEWIS SYMBOLS, AND THE OCTET RULE (INTRODUCTION AND SECTION 8.1) In this chapter, we have focused on the interactions that lead to the formation of **chemical bonds**. We classify these bonds into three broad groups: **ionic bonds**, which result from the electrostatic forces that exist between ions of opposite charge; **covalent bonds**, which result from the sharing of electrons by two atoms; and **metallic bonds**, which result from a delocalized sharing of electrons in metals. The formation of bonds involves interactions of the outermost electrons of atoms, their valence electrons. The valence electrons of an atom can be represented by electron-dot symbols, called **Lewis symbols**. The tendencies of atoms to gain, lose, or share their valence electrons often follow the **octet rule**, which says that the atoms in molecules or ions (usually) have eight valence electrons.

IONIC BONDING (SECTION 8.2) Ionic bonding results from the transfer of electrons from one atom to another, leading to the formation of a three-dimensional lattice of charged particles. The stabilities of ionic substances result from the strong electrostatic attractions between an ion and the surrounding ions of opposite charge. The magnitude of these interactions is measured by the **lattice energy**, which is the energy needed to separate an ionic lattice into gaseous ions. Lattice energy increases with increasing charge on the ions and with decreasing distance between the ions. The **Born–Haber cycle** is a useful thermochemical cycle in which we use Hess's law to calculate the lattice energy as the sum of several steps in the formation of an ionic compound.

COVALENT BONDING (SECTION 8.3) A covalent bond results from the sharing of valence electrons between atoms. We can represent the electron distribution in molecules by means of **Lewis structures**, which indicate how many valence electrons are involved in forming bonds and how many remain as **nonbonding electron pairs** (or **lone pairs**). The octet rule helps determine how many bonds will be formed between two atoms. The sharing of one pair of electrons produces a **single bond**; the sharing of two or three pairs of electrons between two atoms produces **double** or **triple bonds**, respectively. Double and triple bonds are examples of multiple bonding between atoms. The bond length decreases as the number of bonds between the atoms increases.

BOND POLARITY AND ELECTRONEGATIVITY (SECTION 8.4) In covalent bonds, the electrons may not necessarily be shared equally between two atoms. **Bond polarity** helps describe unequal sharing of electrons in a bond. In a **nonpolar covalent bond**, the electrons in the bond are shared equally by the two atoms; in a **polar covalent bond**, one of the atoms exerts a greater attraction for the electrons than the other.

Electronegativity is a numerical measure of an atom's ability to compete with other atoms for the electrons shared between them. Fluorine is the most electronegative element, meaning it has the greatest ability to attract electrons from other atoms. Electronegativity values range from 0.7 for Cs to 4.0 for F. Electronegativity generally increases from left to right in a row of the periodic table and decreases going down a column. The difference in the electronegativities of bonded atoms can be used to determine the polarity of a bond. The greater the electronegativity difference, the more polar the bond.

A **polar molecule** is one whose centers of positive and negative charge do not coincide. Thus, a polar molecule has a positive side and a negative side. This separation of charge produces a **dipole**, the magnitude of which is given by the **dipole moment**, which is measured in debyes (D). Dipole moments increase with increasing amount of charge separated and increasing distance of separation. Any diatomic molecule X—Y in which X and Y have different electronegativities is a polar molecule.

Most bonding interactions lie between the extremes of covalent and ionic bonding. While it is generally true that the bonding between a metal and a nonmetal is predominantly ionic, exceptions to this guideline are not uncommon when the difference in electronegativity of the atoms is relatively small or when the oxidation state of the metal becomes large.

DRAWING LEWIS STRUCTURES AND RESONANCE STRUCTURES (SECTIONS 8.5 AND 8.6) If we know which atoms are connected to one another, we can draw Lewis structures for molecules and ions using a five-step procedure. Once we do so, we can determine the **formal charge** of each atom in a Lewis structure, which is the charge that the atom would have if all atoms had the same electronegativity. In general, the dominant Lewis structure will have low formal charges, with any negative formal charges residing on more electronegative atoms.

Sometimes a single dominant Lewis structure is inadequate to represent a particular molecule (or ion). In such situations, we describe the molecule by using two or more **resonance structures** for the molecule. The molecule is envisioned as a blend of these multiple resonance structures. Resonance structures are important in describing the bonding in molecules such as ozone, O_3, and the organic molecule benzene, C_6H_6.

EXCEPTIONS TO THE OCTET RULE (SECTION 8.7) The octet rule is not obeyed in all cases. Exceptions occur when (**a**) a molecule has an odd number of electrons, (**b**) it is not possible to complete an octet around an atom without forcing an unfavorable distribution of electrons, or (**c**) a large atom is surrounded by a sufficiently large number of small electronegative atoms that it has more than an octet of electrons around it. Lewis structures with more than an octet of electrons are observed for atoms in the third row and beyond in the periodic table.

STRENGTHS AND LENGTHS OF SINGLE AND MULTIPLE BONDS (SECTION 8.8) The average strengths and lengths of many common covalent bonds can be measured. Average bond enthalpies for multiple bonds are generally larger than those of single bonds. The average bond length between two atoms decreases as the number of bonds between the atoms increases, consistent with the bond being stronger as the number of bonds increases.

Key Equations

- $E_{el} = \dfrac{\kappa Q_1 Q_2}{d}$ [8.4]   The potential energy of two interacting charges

- $\mu = Qr$ [8.10]   The dipole moment of two charges of equal magnitude but opposite sign, separated by a distance r

- Formal charge =   The definition of formal charge

  valence electrons $- \left[\frac{1}{2}(\text{bonding electrons}) + \text{nonbonding electrons} \right]$ [8.11]

Exam Prep

EP 8.1 If the Lewis symbol of a certain element has one dot on each of the four sides of the symbol for the element, which of the following elements could it be? (**a**) P (**b**) S (**c**) Si (**d**) B (**e**) N

EP 8.2 Which of the following statements about ionic bonding is *incorrect*? (**a**) Ionic bonds form because of the electrostatic attraction between cations and anions. (**b**) Ionic bonds are formed from the elements via the transfer of one or more electrons from one atom to another. (**c**) Energy is required to break an ionic compound into gaseous ions. (**d**) The attraction between cations and anions increases as the charges on the ions increase. (**e**) The attraction between cations and anions increases as the radii of the ions increase.

EP 8.3 Which of the following orderings of lattice energy is correct for the ionic compounds CsI, MgO, NaCl, and ScN?
(**a**) NaCl > MgO > CsI > ScN   (**d**) MgO > NaCl > ScN > CsI,
(**b**) ScN > MgO > NaCl > CsI   (**e**) ScN > CsI > NaCl > MgO
(**c**) NaCl > CsI > ScN > MgO

EP 8.4 Which of these elements is most likely to form ions with a 2+ charge? (**a**) Li (**b**) Ca (**c**) O (**d**) P (**e**) Cl

EP 8.5 Which of the following statements about the bonding of two Cl atoms to form Cl_2 is *incorrect*? (**a**) Cl_2 has a single bond. (**b**) The electrons in the Cl—Cl bond are simultaneously attracted to both Cl nuclei. (**c**) In forming the Cl—Cl bond, an electron is completely transferred from one Cl atom to the other. (**d**) Each of the Cl atoms in Cl_2 has three nonbonding pairs of electrons around it. (**e**) Both Cl atoms in Cl_2 satisfy the octet rule.

EP 8.6 Which of these molecules has the same number of shared electron pairs as unshared electron pairs? (**a**) HCl (**b**) H_2S (**c**) PF_3 (**d**) CCl_2F_2 (**e**) Br_2

EP 8.7 Consider a Lewis structure for the P_2 molecule that satisfies the octet rule. Which of the following statements is or are *true*?

(**i**) Each of the P atoms has one nonbonding electron pair.

(**ii**) The molecule has a P—P triple bond.

(**iii**) Six electrons are shared between the two P atoms.

(**a**) Only one of the statements is true. (**b**) i and ii are true. (**c**) i and iii are true. (**d**) ii and iii are true. (**e**) All three statements are true.

EP 8.8 Which of the following is the best explanation of why fluorine is the most electronegative element? (**a**) A fluorine atom easily transfers an electron to other atoms in forming ionic compounds. (**b**) Fluorine has a large first ionization energy and a large electron affinity. (**c**) The ionic radius of fluorine is larger than its bonding atomic radius. (**d**) The F_2 molecule is nonpolar. (**e**) The F^{2-} ion does not exist.

EP 8.9 Which of the following bonds is the most polar? (**a**) H—F (**b**) H—I (**c**) Se—F (**d**) N—P (**e**) Ga—Cl

EP 8.10 What would be the dipole moment for HF (bond length 0.917 Å) if it is assumed that the bond is completely ionic? (**a**) 0.917 D (**b**) 1.91 D (**c**) 2.75 D (**d**) 4.39 D (**e**) 7.37 D

EP 8.11 Tungsten dioxide, WO_2, is a solid with a melting point of about 1700 °C, whereas tungsten hexafluoride, WF_6, is a colorless gas at room temperature. Which of the following is the best explanation for these observations? (**a**) Covalent bonding dominates in WO_2 whereas ionic bonding dominates in WF_6. (**b**) O_2 has a double bond whereas F_2 has a single bond. (**c**) Fluorine does not readily form ionic compounds. (**d**) W is in a higher oxidation state in WF_6, which increases the degree of covalent bonding. (**e**) The difference in electronegativity between W and O is greater than that between W and F.

EP 8.12 Which of these molecules has a Lewis structure with a central atom having no nonbonding electron pairs? (**a**) CO_2 (**b**) H_2S (**c**) PF_3 (**d**) SiF_4 (**e**) more than one of a, b, c, d

EP 8.13 Draw the Lewis structure(s) for the molecule with the chemical formula C_2H_3N, where the N is connected to only one other atom. How many double bonds are there in the correct Lewis structure? (**a**) 0 (**b**) 1 (**c**) 2 (**d**) 3 (**e**) 4

EP 8.14 How many nonbonding electron pairs are there in the Lewis structure of the peroxide ion, O_2^{2-}? (**a**) 7 (**b**) 6 (**c**) 5 (**d**) 4 (**e**) 3

EP 8.15 The sulfate ion, SO_4^{2-}, can be drawn in many ways. If you minimize formal charge on the sulfur, how many S=O double bonds should you draw in the Lewis structure? (**a**) 0 (**b**) 1 (**c**) 2 (**d**) 3 (**e**) 4

EP 8.16 Which of these statements about resonance is *true*? (**a**) When you draw resonance structures, it is permissible to alter the way atoms are connected. (**b**) Because of multiple resonance structures, the six carbon–carbon bonds in benzene, C_6H_6, all have the same length. (**c**) "Resonance" refers to the idea that molecules are resonating rapidly between different bonding patterns. (**d**) The ethylene molecule, C_2H_4, has multiple resonance structures. (**e**) All of the above are true.

EP 8.17 Consider the following ions: NO_2^-, CO_3^{2-}, and SO_4^{2-}. For each of these, you should use a Lewis structure that satisfies the octet rule. Which of these ions will exhibit resonance among multiple equivalent Lewis structures that satisfy the octet rule? (**a**) only one ion (**b**) NO_2^- and CO_3^{2-} (**c**) NO_2^-, and SO_4^{2-} (**d**) CO_3^{2-} and SO_4^{2-} (**e**) all three ions

EP 8.18 Which of the following is expected to be the dominant Lewis structure for the CF_2 molecule?

(a)   (b)   (c)   (d)

EP 8.19 In which of these molecules or ions is there only one lone pair of electrons on the central sulfur atom? (**a**) SF_4 (**b**) SF_6 (**c**) SOF_4 (**d**) SF_2 (**e**) SO_4^{2-}

EP 8.20 Of the three numbered bonds in the structure shown here, bond number _____ is the strongest and bond number _____ is the longest. (**a**) 1, 3 (**b**) 2, 3 (**c**) 2, 1 (**d**) 3, 2 (**e**) 1, 2

Exercises

Visualizing Concepts

8.1 For each of these Lewis symbols, indicate the group in the periodic table in which the element X belongs: [Section 8.1]

(**a**) ·Ẋ·   (**b**) ·X·   (**c**) :Ẋ·

8.2 Illustrated are four ions—A, B, X, and Y—showing their relative ionic radii. The ions shown in red carry positive charges: a 2+ charge for A and a 1+ charge for B. Ions shown in blue carry negative charges: a 1− charge for X and a 2− charge for Y.

(a) Which combinations of these ions produce ionic compounds where there is a 1:1 ratio of cations and anions? (b) Among the combinations in part (a), which leads to the ionic compound having the largest lattice energy? [Section 8.2]

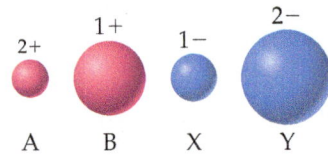

8.3 A portion of a two-dimensional "slab" of NaCl(s) is shown here (see Figure 8.4) in which the ions are numbered. (a) Which colored balls must represent sodium ions? (b) Which colored balls must represent chloride ions? (c) Consider ion 5. How many attractive electrostatic interactions are shown for it? (d) Consider ion 5. How many repulsive interactions are shown for it? (e) Is the sum of the attractive interactions in part (c) larger or smaller than the sum of the repulsive interactions in part (d)? (f) If this pattern of ions were extended indefinitely in two dimensions, would the lattice energy be positive or negative? [Section 8.2]

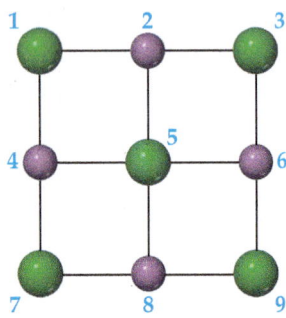

8.4 The orbital diagram that follows shows the valence electrons for a 3+ ion of an element. (a) What is the element? (b) What is the electron configuration of an atom of this element? [Section 8.2]

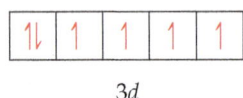

3d

8.5 Which of the following charts shows the general periodic trends for the electronegativities of the representative elements? [Section 8.4]

(a)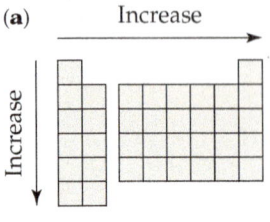

(c)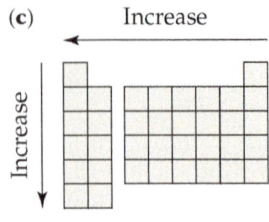

8.6 A molecule with formula C_4H_3NO has the connectivity shown in the figure. After the Lewis structure of the molecule is completed, how many of each of the following are there in the molecule: (a) single bonds, (b) double bonds, (c) triple bonds, (d) nonbonding pairs? [Sections 8.3 and 8.5]

$$\begin{array}{cccccc} & \text{O} & & & \text{H} & \\ & | & & & | & \\ \text{H}-\text{C}-&\text{C}-&\text{C}-&\text{C}-&\text{N}-&\text{H} \end{array}$$

8.7 In the Lewis structure shown here, A, D, E, Q, X, and Z represent elements in the first two rows of the periodic table. Identify all six elements so that the formal charges of all atoms are zero. [Section 8.5]

$$\begin{array}{ccc} & :\!\ddot{\text{E}}: & \text{X} \\ & \| & | \\ :\!\ddot{\text{A}}-&\text{D}-\ddot{\text{Q}}-&\text{Z} \end{array}$$

8.8 Incomplete Lewis structures for the nitrous acid molecule, HNO_2, and the nitrite ion, NO_2^-, are shown here. (a) Complete each Lewis structure by adding electron pairs as needed. (b) Is the formal charge on N the same or different in these two species? (c) Would either HNO_2 or NO_2^- be expected to exhibit resonance? (d) Would you expect the N=O bond in HNO_2 to be longer, shorter, or the same length as the N—O bonds in NO_2^-? [Sections 8.5, 8.6, and 8.8]

$$\text{H}-\text{O}-\text{N}=\text{O} \qquad \text{O}-\text{N}=\text{O}$$

8.9 The molecule shown here is *styrene*, C_8H_8, a benzene derivative that is used to make a number of polymers, including polystyrene. The shorthand notation for the benzene ring (described in Section 8.6) is used. Three of the carbon–carbon bonds are numbered in the structure. (a) Which of the three bonds is the strongest? (b) Which of the three bonds is the longest? (c) Which of the three bonds is best described as halfway between a single and a double bond? [Sections 8.6 and 8.8]

8.10 Consider the Lewis structure for the polyatomic oxyanion shown here, where X is an element from the third period (Na—Ar). By changing the overall charge, n, from 1− to 2− to 3− we get three different polyatomic ions. For each of these ions (a) identify the central atom, X; (b) determine the formal charge of the central atom, X; (c) draw a Lewis structure that makes the formal charge on the central atom equal to zero. [Sections 8.5, 8.6, and 8.7]

$$\left[\begin{array}{ccc} & :\!\ddot{\text{O}}: & \\ & | & \\ :\!\ddot{\text{O}}-&\text{X}-&\ddot{\text{O}}: \\ & | & \\ & :\!\ddot{\text{O}}: & \end{array}\right]^{n-}$$

Lewis Symbols (Section 8.1)

8.11 (a) True or false: An element's number of valence electrons is the same as its atomic number. (b) How many valence electrons does a nitrogen atom possess? (c) An atom has the electron configuration $1s^2 2s^2 2p^6 3s^2 3p^2$. How many valence electrons does the atom have?

(b)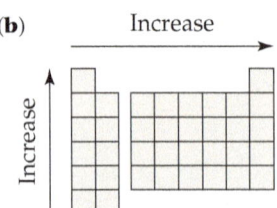

(d)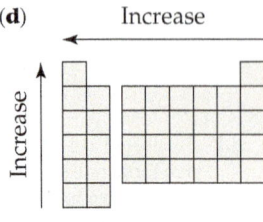

8.12 (a) True or false: The hydrogen atom is most stable when it has a full octet of electrons. (b) How many electrons must a sulfur atom gain to achieve an octet in its valence shell? (c) If an atom has the electron configuration $1s^2 2s^2 2p^3$, how many electrons must it gain to achieve an octet?

8.13 Consider the element silicon, Si. (a) Write its electron configuration. (b) How many valence electrons does a silicon atom have? (c) Which subshells hold the valence electrons?

8.14 (a) Write the electron configuration for the element titanium, Ti. How many valence electrons does this atom possess? (b) Hafnium, Hf, is also found in group 4B. Write the electron configuration for Hf. (c) Ti and Hf behave as though they possess the same number of valence electrons. Which of the subshells in the electron configuration of Hf behave as valence orbitals? Which behave as core orbitals?

8.15 Write the Lewis symbol for atoms of each of the following elements: (a) Al, (b) Br, (c) Ar, (d) Sr.

8.16 What is the Lewis symbol for each of the following atoms or ions: (a) K, (b) As, (c) Sn^{2+}, (d) N^{3-}?

Ionic Bonding (Section 8.2)

8.17 (a) Using Lewis symbols, diagram the reaction between magnesium and oxygen atoms to give the ionic substance MgO. (b) How many electrons are transferred? (c) Which atom loses electrons in the reaction?

8.18 (a) Use Lewis symbols to represent the reaction that occurs between Ca and F atoms. (b) What is the chemical formula of the most likely product? (c) How many electrons are transferred? (d) Which atom loses electrons in the reaction?

8.19 Predict the chemical formula of the ionic compound formed between the following pairs of elements: (a) Al and F, (b) K and S, (c) Y and O, (d) Mg and N.

8.20 Which ionic compound is expected to form from combining the following pairs of elements: (a) barium and fluorine, (b) cesium and chlorine, (c) lithium and nitrogen, (d) aluminum and oxygen?

8.21 Write the electron configuration for each of the following ions, and determine which ones possess noble-gas configurations: (a) Rb^+ (b) Rh^{3+} (c) P^{3-} (d) Sc^{3+} (e) S^{2-} (f) V^{2+}.

8.22 Write electron configurations for the following ions, and determine which have noble-gas configurations: (a) Sn^{2+} (b) I^- (c) Ta^{3+} (d) Se^{2-} (e) Pd^{2+} (f) Cu^{2+}.

8.23 (a) Is the process that defines lattice energy endothermic or exothermic? (b) Write the chemical equation that represents the process of lattice energy for the case of NaCl. (c) Would you expect salts like NaCl, which have singly charged ions, to have larger or smaller lattice energies compared to salts like CaO, which are composed of doubly charged ions?

8.24 NaCl and KF have the same crystal structure. The only difference between the two is the distance that separates cations and anions. (a) The lattice energies of NaCl and KF are given in Table 8.1. Based on the lattice energies, would you expect the Na—Cl or the K—F distance to be longer? (b) Use the ionic radii given in Figure 7.8 to estimate the Na—Cl and K—F distances.

8.25 The substances NaF and CaO are isoelectronic (have the same number of valence electrons). (a) What are the charges on each of the cations in each compound? (b) What are the charges of each of the anions in each compound? (c) Without looking up lattice energies, which compound is predicted to have the larger lattice energy? (d) Using the lattice energies in Table 8.1, predict the lattice energy of ScN.

8.26 (a) Does the lattice energy of an ionic solid increase or decrease (i) as the charges of the ions increase, (ii) as the sizes of the ions increase? (b) Arrange the following substances not listed in Table 8.1 according to their expected lattice energies, listing them from lowest lattice energy to the highest: MgS, KI, GaN, LiBr.

8.27 Consider the ionic compounds KF, NaCl, NaBr, and LiCl. (a) Use ionic radii (Figure 7.8) to estimate the cation–anion distance for each compound. (b) Based on your answer to part (a), arrange these four compounds in order of decreasing lattice energy. (c) Check your predictions in part (b) with the experimental values of lattice energy from Table 8.1. Are the predictions from ionic radii correct?

8.28 Which of the following trends in lattice energy is due to differences in ionic radii: (a) NaCl > RbBr > CsBr, (b) BaO > KF, (c) SrO > SrCl$_2$?

8.29 Energy is required to remove two electrons from Ca to form Ca^{2+}, and energy is required to add two electrons to O to form O^{2-}. Yet CaO is stable relative to the free elements. Which statement is the best explanation? (a) The lattice energy of CaO is large enough to overcome these processes. (b) CaO is a covalent compound, and these processes are irrelevant. (c) CaO has a higher molar mass than either Ca or O. (d) The enthalpy of formation of CaO is small. (e) CaO is stable to atmospheric conditions.

8.30 List the individual steps used in constructing a Born–Haber cycle for the formation of BaI_2 from the elements. Which of the steps would you expect to be exothermic?

8.31 Use data from Appendix C, Figure 7.11, and Figure 7.13 to calculate the lattice energy of KI.

8.32 (a) Based on the lattice energies of $MgCl_2$ and $SrCl_2$ given in Table 8.1, what is the range of values that you would expect for the lattice energy of $CaCl_2$? (b) Using data from Appendix C, Figure 7.11, Figure 7.13, and the value of the second ionization energy for Ca, 1145 kJ/mol, calculate the lattice energy of $CaCl_2$.

Covalent Bonding, Electronegativity, and Bond Polarity (Sections 8.3 and 8.4)

8.33 (a) State whether or not the bonding in each substance is likely to be covalent: (i) iron, (ii) sodium chloride, (iii) water, (iv) oxygen, (v) argon. (b) A substance XY, formed from two different elements, boils at $-33\,°C$. Is XY likely to be a covalent or an ionic substance?

8.34 Which of these elements are unlikely to form covalent bonds: S, H, K, Ar, Si?

8.35 Using Lewis symbols and Lewis structures, diagram the formation of $SiCl_4$ from Si and Cl atoms, showing valence-shell electrons. (a) How many valence electrons does Si have initially? (b) How many valence electrons does each Cl have initially? (c) How many valence electrons surround the Si in the $SiCl_4$ molecule? (d) How many valence electrons surround each Cl in the $SiCl_4$ molecule? (e) How many bonding pairs of electrons are in the $SiCl_4$ molecule?

8.36 Use Lewis symbols and Lewis structures to diagram the formation of PF_3 from P and F atoms, showing valence-shell electrons. (a) How many valence electrons does P have initially? (b) How many valence electrons does each F have initially? (c) How many valence electrons surround the P in the PF_3 molecule? (d) How many valence electrons surround each F in the PF_3 molecule? (e) How many bonding pairs of electrons are in the PF_3 molecule?

8.37 (a) Construct a Lewis structure for O_2 in which each atom achieves an octet of electrons. (b) How many bonding electrons are in the structure? (c) Would you expect the O—O bond in O_2 to be shorter or longer than the O—O bond in compounds that contain an O—O single bond? Explain.

8.38 (a) Construct a Lewis structure for hydrogen peroxide, H_2O_2, in which each atom achieves an octet of electrons. (b) How many bonding electrons are between the two oxygen atoms? (c) Do you expect the O—O bond in H_2O_2 to be longer or shorter than the O—O bond in O_2? Explain.

8.39 Which of the following statements about electronegativity is or are *true*?

   (i) The alkali metals are the family with the largest electronegativity values.

   (ii) The numerical values for electronegativity have no units.

   (iii) Electronegativity is the ability of an atom in a molecule to attract electron density toward itself.

8.40 Place the following pairs of elements in order from smallest to largest difference in electronegativity: K and F, S and O, Br and I, Ca and Se, Li and Cl.

8.41 Using only the periodic table as your guide, select the most electronegative atom in each of the following sets: (a) Na, Mg, K, Ca (b) P, S, As, Se (c) Be, B, C, Si (d) Zn, Ge, Ga, As.

8.42 By referring only to the periodic table, select (a) the most electronegative element in group 6A; (b) the least electronegative element in the group Al, Si, P; (c) the most electronegative element in the group Ga, P, Cl, Na; (d) the element in the group K, C, Zn, F that is most likely to form an ionic compound with Ba.

8.43 Which of the following bonds are polar: (a) B—F, (b) Cl—Cl, (c) Se—O, (d) H—I? Which is the more electronegative atom in each polar bond?

8.44 Arrange the bonds in each of the following sets in order of increasing polarity: (a) C—F, O—F, B—F (b) O—Cl, S—Br, C—P (c) C—S, B—F, N—O.

8.45 (a) From the data in Table 8.2, calculate the effective charges on the H and Br atoms of the HBr molecule in units of the electronic charge, e. (b) If you were to put HBr under very high pressure, so that its bond length decreased significantly, would its dipole moment increase, decrease, or stay the same, if you assume that the effective charges on the atoms do not change?

8.46 The bromine monofluoride molecule, BrF, has a bond length of 1.76 Å and a dipole moment of 1.29 D. (a) Which atom of the molecule is expected to have a negative charge? (b) In units of the electronic charge, e, what is the magnitude of the charge on the Br atom in BrF?

8.47 In the following pairs of binary compounds, determine which one is a molecular substance and which one is an ionic substance. Use the appropriate naming convention (for ionic or molecular substances) to assign a name to each compound: (a) SiF_4 and LaF_3 (b) $FeCl_2$ and $ReCl_6$ (c) $PbCl_4$ and RbCl.

8.48 In the following pairs of binary compounds, determine which one is a molecular substance and which one is an ionic substance. Use the appropriate naming convention (for ionic or molecular substances) to assign a name to each compound: (a) $TiCl_4$ and CaF_2 (b) ClF_3 and VF_3 (c) $SbCl_5$ and AlF_3.

Lewis Structures; Resonance Structures (Sections 8.5 and 8.6)

8.49 Draw Lewis structures that satisfy the octet rule for the following molecules and ions: (a) CF_4, (b) NO^+, (c) SO_3^{2-} (d) HCN (H and N are both bonded to C), (e) BF_4^-, (f) HOCl.

8.50 Write Lewis structures that satisfy the octet rule for the following molecules and ions: (a) NH_4^+, (b) C_2F_4 (the two C atoms are bonded to one another), (c) $COCl_2$ (the Cl atoms

are bonded to C), (d) HSO_3^- (H is bonded to one of the O atoms), (e) HNC (H and C are both bonded to N), (f) ClO_3^-.

8.51 Which one of these statements about formal charge is *true*? (a) Formal charge is the same as oxidation number. (b) To determine the dominant Lewis structure, you should minimize formal charge. (c) Formal charge takes into account the different electronegativities of the atoms in a molecule (d) Formal charge is most useful for ionic compounds. (e) Formal charge is used in calculating the dipole moment of a diatomic molecule.

8.52 (a) Draw the dominant Lewis structure for the phosphorus trifluoride molecule, PF_3. (b) Determine the oxidation numbers of the P and F atoms. (c) Determine the formal charges of the P and F atoms.

8.53 Write Lewis structures that obey the octet rule for each of the following, and assign oxidation numbers and formal charges to each atom: (a) OCS, (b) $SOCl_2$ (S is the central atom), (c) BrO_3^-, (d) $HClO_2$ (H is bonded to O).

8.54 For each of the following ions of nitrogen and oxygen, write a single Lewis structure that obeys the octet rule, and calculate the oxidation numbers and formal charges on all the atoms: (a) NO^+, (b) NO_2^-, (c) NO_2^+. (d) Arrange these ions in order of increasing N—O bond length.

8.55 (a) Draw a single Lewis structure that satisfies the octet rule for sulfur dioxide, SO_2. (b) With what allotrope of oxygen is it isoelectronic? (c) Are there multiple equivalent resonance structures for this molecule? (d) What would you predict for the lengths of the bonds in SO_2 relative to S—O single bonds and double bonds?

8.56 Consider the formate ion, HCO_2^-, which is the anion formed when formic acid loses an H^+ ion. The H and the two O atoms are bonded to the central C atom. (a) Draw a single Lewis structure that satisfies the octet rule for this ion. (b) Are resonance structures needed to describe the structure? (c) Would you predict that the C—O bond lengths in the formate ion would be longer or shorter relative to those in CO_2?

8.57 Predict the ordering, from shortest to longest, of the bond lengths in CO, CO_2, and CO_3^{2-}.

8.58 Based on Lewis structures, predict the ordering, from shortest to longest, of N—O bond lengths in NO^+, NO_2^-, and NO_3^-.

8.59 Consider a Lewis structure for SO_3 that satisfies the octet rule. Which of the following statements is or are *true*?

   (i) SO_3 has three equivalent resonance structures.

   (ii) There are one shorter and two longer S—O bond lengths in SO_3.

   (iii) The S atom in SO_3 has a nonzero formal charge.

8.60 Which of the following statements about benzene, C_6H_6, is or are *true*?

   (i) Benzene has two equivalent resonance structures.

   (ii) There are no nonbonding pairs in the Lewis structure for benzene.

   (iii) Benzene has three short and three long C—C bonds.

Exceptions to the Octet Rule (Section 8.7)

8.61 (a) Which of these compounds is an exception to the octet rule: carbon dioxide, water, ammonia, phosphorus trifluoride, or arsenic pentafluoride? (b) Which of these compounds or ions is an exception to the octet rule: nitrogen dioxide, borohydride (BH_4^-), borazine ($B_3N_3H_6$, which is analogous to benzene with alternating B and N in the ring), or boron trichloride?

8.62 Fill in the blank with the appropriate numbers for both electrons and bonds (considering that single bonds are counted as one, double bonds as two, and triple bonds as three).

(a) Fluorine has _____ valence electrons and makes _____ bond(s) in compounds.

(b) Oxygen has _____ valence electrons and makes _____ bond(s) in compounds.

(c) Nitrogen has _____ valence electrons and makes _____ bond(s) in compounds.

(d) Carbon has _____ valence electrons and makes _____ bond(s) in compounds.

8.63 Draw the dominant Lewis structures for these chlorine–oxygen molecules/ions: ClO, ClO^-, ClO_2^-, ClO_3^-, ClO_4^-. Which of these do not obey the octet rule?

8.64 Which of the following statements is or are *true*?

   (i) Any molecule that doesn't satisfy the octet rule is hypervalent.

   (ii) Elements in the third row of the periodic table can form hypervalent molecules, whereas those in the second row cannot.

   (iii) Any molecule in which there are four bonds to the central atom satisfies the octet rule.

8.65 Draw the Lewis structures for each of the following ions or molecules. Identify those in which the octet rule is not obeyed; state which atom in each compound does not follow the octet rule; and state, for those atoms, how many electrons surround these atoms: (a) PH_3, (b) AlH_3, (c) N_3^-, (d) CH_2Cl_2, (e) SnF_6^{2-}.

8.66 Draw the Lewis structures for each of the following molecules or ions. Identify instances where the octet rule is not obeyed; state which atom in each compound does not follow the octet rule; and state how many electrons surround these atoms: (a) NO, (b) BF_3, (c) ICl_2^-, (d) $OPBr_3$ (the P is the central atom), (e) XeF_4.

8.67 In the vapor phase, $BeCl_2$ exists as a discrete molecule. (a) Draw the Lewis structure of this molecule, using only single bonds. Does this Lewis structure satisfy the octet rule? (b) What other resonance structures are possible that satisfy the octet rule? (c) On the basis of the formal charges, which Lewis structure is expected to be dominant for $BeCl_2$?

8.68 (a) Describe the molecule xenon trioxide, XeO_3, using four possible Lewis structures, one each with zero, one, two, or three Xe—O double bonds. (b) Do any of these resonance structures satisfy the octet rule for every atom in the molecule? (c) Do any of the four Lewis structures have multiple resonance structures? If so, how many resonance structures do you find? (d) Which of the Lewis structures in part (a) yields the most favorable formal charges for the molecule?

8.69 There are many Lewis structures you could draw for sulfuric acid, H_2SO_4 (each H is bonded to an O). (a) What Lewis structure(s) would you draw to satisfy the octet rule? (b) What Lewis structure(s) would you draw to minimize formal charge?

8.70 Some chemists believe that satisfaction of the octet rule should be the top criterion for choosing the dominant Lewis structure of a molecule or ion. Other chemists believe that achieving the best formal charges should be the top criterion. Consider the dihydrogen phosphate ion, $H_2PO_4^-$, in which the H atoms are bonded to O atoms. (a) What is the predicted dominant Lewis structure if satisfying the octet rule is the top criterion? (b) What is the predicted dominant Lewis structure if achieving the best formal charges is the top criterion?

Strengths and Lengths of Covalent Bonds (Section 8.8)

8.71 Using Table 8.3, estimate ΔH for each of the following gas-phase reactions (note that lone pairs on atoms are not shown):

(a)

(b)

(c) $2\ Cl-N(Cl)-Cl \longrightarrow N\equiv N + 3\ Cl-Cl$

8.72 Using Table 8.3, estimate ΔH for the following gas-phase reactions:

(a)

(b)

(c)

8.73 State whether each of these statements is true or false. (a) The longer the bond, the larger the bond enthalpy. (b) C—C bonds are stronger than C—H bonds. (c) A typical single-bond length is in the 5–10 Å range. (d) Energy is released when a chemical bond is broken. (e) Energy is stored in chemical bonds.

8.74 State whether each of these statements is true or false. (a) A carbon–carbon triple bond is shorter than a carbon–carbon single bond. (b) There are exactly six bonding electrons in the O_2 molecule. (c) The C—O bond in carbon monoxide is shorter than the C—O bond in carbon dioxide. (d) The O—O bond in ozone is shorter than the O—O bond in O_2. (e) The average bond enthalpies of all triple bonds are the same value regardless of what atoms are involved.

8.75 We can define average bond enthalpies and bond lengths for ionic bonds, just like we have for covalent bonds. Which ionic bond is predicted to be stronger, Na—Cl or Ca—O?

8.76 We can define average bond enthalpies and bond lengths for ionic bonds, just like we have for covalent bonds. Which ionic bond is predicted to have the smaller bond enthalpy, Li—F or Cs—F?

8.77 A new compound is made that has a C—C bond length of 1.15 Å. Is this bond likely to be a single, double, or triple C—C bond?

8.78 A new compound is made that has an N—N bond length of 1.26 Å. Is this bond likely to be a single, double, or triple N—N bond?

8.79 The molecule shown is *propylene*, C_3H_6, which is an important feedstock for the polymer industry. The two carbon–carbon bonds in the molecule are numbered. (**a**) Does propylene have multiple equivalent resonance structures? (**b**) Which carbon–carbon bond is the stronger? (**c**) Which carbon–carbon bond is longer?

$$
\begin{array}{ccccc}
 & H & H & H & \\
 & | & | & | & \\
H- & C & -\!\!\overset{1}{}\!\!- & C & =\!\!\overset{2}{}\!\!= C \\
 & | & & | & \\
 & H & & H &
\end{array}
$$

8.80 The molecule shown is *methyl formate*, $C_2H_4O_2$, the simplest example of an organic *ester*. The three carbon–oxygen bonds in the molecule are numbered. Which of the following statements is or are *true*?

  (**i**) Bond 1 is the strongest carbon–oxygen bond.

  (**ii**) Bond 1 is longer than bond 2.

  (**iii**) The length of bond 3 is closer in value to that of bond 2 than to that of bond 1.

$$
\begin{array}{ccccc}
 & O & & & H \\
 & \| 1 & & & | \\
H-C & -\!\!\overset{3}{}\!\!O\overset{2}{}\!\!- & C & -H \\
 & & & & | \\
 & & & & H
\end{array}
$$

Additional Exercises

8.81 Consider the lattice energies of the following group 2A compounds: BeH_2, 3205 kJ/mol; MgH_2, 2791 kJ/mol; CaH_2, 2410 kJ/mol; SrH_2, 2250 kJ/mol; BaH_2, 2121 kJ/mol. (**a**) What is the oxidation number of H in these compounds? (**b**) Assuming that all of these compounds have the same three-dimensional arrangement of ions in the solid, which of these compounds has the shortest cation-anion distance? (**c**) Consider BeH_2. Does it require 3205 kJ of energy to break one mole of the solid into its ions, or does breaking up one mole of solid into its ions release 3205 kJ of energy? (**d**) The lattice energy of ZnH_2 is 2870 kJ/mol. Considering the trend in lattice enthalpies in the group 2A compounds, predict which group 2A element is most similar in ionic radius to the Zn^{2+} ion.

8.82 Based on data in Table 8.1, estimate (within 30 kJ/mol) the lattice energy for (**a**) LiBr, (**b**) CsBr, (**c**) $CaCl_2$.

8.83 An ionic substance of formula MX has a lattice energy of 3000 kJ/mol (to one significant figure). Based on the data in Table 8.1, is the charge on the ion M likely to be 1+, 2+, or 3+?

8.84 The ionic compound CaO crystallizes with the same structure as sodium chloride (Figure 8.3). (**a**) In this structure, how many O^{2-} are in contact with each Ca^{2+} ion (*Hint*: Remember the pattern of ions shown in Figure 8.3 repeats over and over again in all three directions.) (**b**) Would energy be consumed or released if a crystal of CaO was converted to a collection of widely separated Ca—O ion pairs? (**c**) From the ionic radii given in Figure 7.8, calculate the potential energy of a single Ca—O ion pair that is just touching (the magnitude of electronic charge is given on the inside back cover). (**d**) Calculate the energy of a mole of such pairs. How does this compare to the lattice energy of CaO? (**e**) What factor do you think accounts for most of the discrepancy between the energies in part (d)—the bonding in CaO is more covalent than ionic, or the electrostatic interactions in a crystal lattice are more complicated than those in a single ion pair?

8.85 Construct a Born–Haber cycle for the formation of the hypothetical compound $NaCl_2$, where the sodium ion has a 2+ charge (the second ionization energy for sodium is given in Table 7.2). (**a**) How large would the lattice energy need to be for the formation of $NaCl_2$ to be exothermic? (**b**) If we were to estimate the lattice energy of $NaCl_2$ to be roughly equal to that of $MgCl_2$ (2326 kJ/mol from Table 8.1), what value would you obtain for the standard enthalpy of formation, ΔH_f°, of $NaCl_2$?

8.86 A classmate of yours is convinced that they know everything about electronegativity. (**a**) In the case of atoms X and Y having different electronegativities, they say, the diatomic molecule X—Y must be polar. Is your classmate correct? (**b**) Your classmate says that the farther the two atoms are apart in a bond, the larger the dipole moment will be. Is your classmate correct?

8.87 Consider the collection of nonmetallic elements O, P, Te, I, and B. (**a**) Which two would form the most polar single bond? (**b**) Which two would form the longest single bond? (**c**) Which two would be likely to form a compound of formula XY_2? (**d**) Which combinations of elements would likely yield a compound of empirical formula X_2Y_3?

8.88 The substance chlorine monoxide, ClO(g), is important in atmospheric processes that lead to depletion of the ozone layer. The ClO molecule has an experimental dipole moment of 1.24 D, and the Cl—O bond length is 1.60 Å. (**a**) Determine the magnitude of the charges on the Cl and O atoms in units of the electronic charge, e. (**b**) Based on the electronegativities of the elements, which atom would you expect to have a partial negative charge in the ClO molecule? (**c**) Using formal charges as a guide, propose the dominant Lewis structure for the molecule. (**d**) The anion ClO^- exists. What is the formal charge on the Cl for the dominant Lewis structure for ClO^-?

8.89 (**a**) Using the electronegativities of Br and Cl, estimate the partial charges on the atoms in the Br—Cl molecule using the procedure discussed in the *A Closer Look* box in Section 8.5. (**b**) Using these partial charges and the atomic radii given in Figure 7.8, estimate the dipole moment of the molecule. (**c**) The measured dipole moment of BrCl is 0.57 D. If you assume the bond length in BrCl is the sum of the atomic radii, what are the partial charges on the atoms in BrCl using the experimental dipole moment?

8.90 A major challenge in implementing the "hydrogen economy" is finding a safe, lightweight, and compact way of storing hydrogen for use as a fuel. The hydrides of light metals are attractive for hydrogen storage because they can store a high weight percentage of hydrogen in a small volume. For example, $NaAlH_4$ can release 5.6% of its mass as H_2 upon decomposing to NaH(s), Al(s), and $H_2(g)$. $NaAlH_4$ possesses both covalent bonds, which hold polyatomic anions together, and ionic bonds. (**a**) Write a balanced equation for the decomposition of $NaAlH_4$. (**b**) Which element in $NaAlH_4$ is the most electronegative? Which one is the least electronegative? (**c**) Based on electronegativity differences, predict the identity of the polyatomic anion. Draw a Lewis structure for this ion. (**d**) What is the formal charge on hydrogen in the polyatomic ion?

8.91 Structures A, B, and C show the connectivity of the atoms in three different molecules that are isomers of C_3H_4O. By completing the Lewis structures of these molecules, complete the information in the following table:

	Isomer A	Isomer B	Isomer C
Number of single bonds			
Number of double bonds			
Number of triple bonds			
Number of nonbonding pairs			

A   H—C—C—C—O (with H H H on top carbons)

B   H—C—C—C—O—H (with H on top and bottom of first carbon)

C   H—C—C—C—O (with H H on top, H on bottom)

8.92 The *triiodide* ion, I_3^-, exists, whereas the corresponding ion with fluorine, F_3^-, does not. The I_3^- ion has a linear structure in which two outer I atoms are each bonded to a central I atom. Although I_3^- is a known ion, F_3^- is not. **(a)** Draw the Lewis structure for I_3^-, assuming that there are two I—I single bonds in the ion. **(b)** Does I_3^- satisfy the octet rule? **(c)** Which of the following statements about the existence of I_3^- versus the nonexistence of F_3^- is or are true? (i) The Lewis structure of I_3^- shows 12 electrons around the central I atom. (ii) Elements from the second row of the periodic table generally do not form hypervalent molecules and ions. (iii) An I atom can form a hypervalent molecule or ion more readily than an F atom because of the larger size of the I atom.

8.93 Calculate the formal charge on the indicated atom in each of the following molecules or ions: **(a)** the central oxygen atom in O_3, **(b)** phosphorus in PF_6^-, **(c)** nitrogen in NO_2, **(d)** iodine in ICl_3, **(e)** chlorine in $HClO_4$ (hydrogen is bonded to O).

8.94 The hypochlorite ion, ClO^-, is the active ingredient in bleach. The perchlorate ion, ClO_4^-, is a main component of rocket propellants. Draw Lewis structures for both ions. **(a)** What is the formal charge of Cl in the hypochlorite ion? **(b)** What is the formal charge of Cl in the perchlorate ion, assuming the Cl—O bonds are all single bonds? **(c)** What is the oxidation number of Cl in the hypochlorite ion? **(d)** What is the oxidation number of Cl in the perchlorate ion, assuming the Cl—O bonds are all single bonds? **(e)** In a redox reaction, which ion would you expect to be more easily reduced?

8.95 The following three Lewis structures can be drawn for N_2O:

:N≡N—Ö: ⟷ :Ṅ—N≡O: ⟷ :Ṅ=N=Ö:

(a) Using formal charges, which of these three resonance forms is likely to be the most important? **(b)** The N—N bond length in N_2O is 1.12 Å, slightly longer than a typical N≡N bond; and the N—O bond length is 1.19 Å, slightly shorter than a typical N=O bond (see Table 8.4). Based on these data, which resonance structure best represents N_2O?

8.96 Mothballs are composed of naphthalene, $C_{10}H_8$, a molecule that consists of two six-membered rings of carbon fused along an edge, as shown in this incomplete Lewis structure:

(a) Draw all of the resonance structures of naphthalene. How many are there? **(b)** Do you expect the C—C bond lengths in the molecule to be similar to those of C—C single bonds, C=C double bonds, or intermediate between C—C single and C=C double bonds? **(c)** Not all of the C—C bond lengths in naphthalene are equivalent. Based on your resonance structures, how many C—C bonds in the molecule do you expect to be shorter than the others?

8.97 **(a)** Triazine, $C_3H_3N_3$, is like benzene except that in triazine every other C—H group is replaced by a nitrogen atom. Draw the Lewis structure(s) for the triazine molecule. **(b)** Estimate the carbon–nitrogen bond distances in the ring.

8.98 *Ortho*-Dichlorobenzene, $C_6H_4Cl_2$, is obtained when two of the adjacent hydrogen atoms in benzene are replaced with Cl atoms. A skeleton of the molecule is shown here. **(a)** Complete a Lewis structure for the molecule using bonds and electron pairs as needed. **(b)** Are there any resonance structures for the molecule? If so, sketch them. **(c)** Are the resonance structures in (a) and (b) equivalent to one another as they are in benzene?

8.99 Two compounds are isomers if they have the same chemical formula but different arrangements of atoms. Use Table 8.3 to estimate ΔH for each of the following gas-phase isomerization reactions and indicate which isomer has the lower enthalpy.

(a) Ethanol → Dimethyl ether

(b) Ethylene oxide → Acetaldehyde

(c) Cyclopentene → Pentadiene

(d) $H-\underset{\underset{H}{|}}{\overset{\overset{H}{|}}{C}}-N\equiv C \longrightarrow H-\underset{\underset{H}{|}}{\overset{\overset{H}{|}}{C}}-C\equiv N$

Methyl isocyanide   Acetonitrile

8.100 (a) Draw the Lewis structure for hydrogen peroxide, H_2O_2. (b) From the data presented in Table 8.3, what is the weakest bond in hydrogen peroxide? (c) Hydrogen peroxide is sold commercially as an aqueous solution in brown bottles to protect it from light. Calculate the longest wavelength of light that has sufficient energy to break the weakest bond in hydrogen peroxide.

8.101 The electron affinity of oxygen is -141 kJ/mol, corresponding to the reaction

$$O(g) + e^- \longrightarrow O^-(g)$$

The lattice energy of $K_2O(s)$ is 2238 kJ/mol. Use these data along with data in Appendix C and Figure 7.11 to calculate the "second electron affinity" of oxygen, corresponding to the reaction

$$O^-(g) + e^- \longrightarrow O^{2-}(g)$$

8.102 You and a partner are asked to complete a lab entitled "Oxides of Ruthenium" that is scheduled to extend over two lab periods. The first lab, which is to be completed by your partner, is devoted to carrying out compositional analysis. In the second lab, you are to determine melting points. Upon going to lab you find two unlabeled vials, one containing a soft yellow substance and the other a black powder. You also find the following notes in your partner's notebook—*Compound 1*: 76.0% Ru and 24.0% O (by mass), *Compound 2*: 61.2% Ru and 38.8% O (by mass). (a) What is the empirical formula for Compound 1? (b) What is the empirical formula for Compound 2? Upon determining the melting points of these two compounds, you find that the yellow compound melts at 25 °C, while the black powder does not melt up to the maximum temperature of your apparatus, 1200 °C. (c) What is the identity of the yellow compound? (d) What is the identity of the black compound? (e) Which compound is molecular? (f) Which compound is ionic?

8.103 The compound chloral hydrate, known in detective stories as knockout drops, is composed of 14.52% C, 1.83% H, 64.30% Cl, and 13.35% O by mass, and has a molar mass of 165.4 g/mol. (a) What is the empirical formula of this substance? (b) What is the molecular formula of this substance? (c) Draw the Lewis structure of the molecule, assuming that the Cl atoms bond to a single C atom and that there are a C—C bond and two C—O bonds in the compound.

8.104 Under special conditions, sulfur reacts with anhydrous liquid ammonia to form a binary compound of sulfur and nitrogen. The compound is found to consist of 69.6% S and 30.4% N. Measurements of its molecular mass yield a value of 184.3 g/mol. The compound occasionally detonates on being struck or when heated rapidly. The sulfur and nitrogen atoms of the molecule are joined in a ring. All the bonds in the ring are of the same length. (a) Calculate the empirical and molecular formulas for the substance. (b) Write Lewis structures for the molecule, based on the information you are given. (*Hint:* You should find a relatively small number of dominant Lewis structures.) (c) Predict the bond distances between the atoms in the ring. (*Note:* The S—S distance in the S_8 ring is 2.05 Å.) (d) The enthalpy of formation of the compound is estimated to be 480 kJ/mol^{-1}. ΔH_f° of $S(g)$ is 222.8 kJ/mol. Estimate the average bond enthalpy in the compound.

8.105 A common form of elemental phosphorus is the tetrahedral P_4 molecule, where all four phosphorus atoms are equivalent:

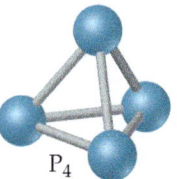

P_4

At room temperature phosphorus is a solid. (a) Are there any lone pairs of electrons in the P_4 molecule? (b) How many P—P bonds are there in the molecule? (c) Draw a Lewis structure for a linear P_4 molecule that satisfies the octet rule. Does this molecule have resonance structures? (d) On the basis of formal charges, which is more stable, the linear molecule or the tetrahedral molecule?

Design an Experiment

You have learned that the resonance of benzene, C_6H_6, gives the compound special stability.

(a) By using data in Appendix C, compare the heat of combustion of 1.0 mol $C_6H_6(g)$ to the heat of combustion of 3.0 mol acetylene, $C_2H_2(g)$, Which has the greater fuel value, 1.0 mol $C_6H_6(g)$ or 3.0 mol $C_2H_2(g)$? Are your calculations consistent with benzene being especially stable? (b) Repeat part (a), with the appropriate molecules, for toluene ($C_6H_5CH_3$), a derivative of benzene that has a —CH_3 group in place of one H. (c) Another reaction you can use to compare molecules is *hydrogenation*, the reaction of a carbon–carbon double bond with H_2 to make a C—C single bond and two C—H single bonds. The experimental heat of hydrogenation of benzene to make cyclohexane (C_6H_{10}, a six-membered ring with 6 C—C single bonds and 12 C—H bonds) is 208 kJ/mol. The experimental heat of hydrogenation of cyclohexene (C_6H_{10}, a six-membered ring with one C=C double bond, 5 C—C single bonds, and 10 C—H bonds) to make cyclohexane is 120 kJ/mol. Show how these data can provide you with an estimate of the *resonance stabilization energy* of benzene. (d) Are the bond lengths or angles in benzene, compared to other hydrocarbons, sufficient to decide if benzene exhibits resonance and is especially stable? Discuss. (e) Consider cyclooctatetraene, C_8H_8, which has the octagonal structure shown below.

Cyclooctatetraene

What experiments or calculations could you perform to determine whether cyclooctatetraene exhibits resonance?

9

MOLECULAR GEOMETRY AND BONDING THEORIES

▲ **COLOR OF TOMATOES.** The hydrocarbon molecule *lycopene*, $C_{40}H_{56}$, gives tomatoes their red color as well as some of their nutrient properties. The color of lycopene is due to its arrangement of alternating single and double bonds, which gives the molecule the ability to absorb visible light.

Lewis structures help us understand the compositions of molecules and their covalent bonds. However, Lewis structures do *not* show one of the most important aspects of molecules—their overall shapes. The shape and size of molecules—sometimes referred to as *molecular architecture*—are defined by the angles and distances between the nuclei of the component atoms.

In this chapter, our first goal is to understand the relationship between two-dimensional Lewis structures and three-dimensional molecular shapes. We show the intimate relationship between the number of electrons in a molecule and the overall shape it adopts. We then examine more closely the nature of covalent bonds. The lines used to depict bonds in Lewis structures provide important clues about the orbitals that molecules use in bonding. By examining these orbitals, we can gain a greater understanding of the behavior of molecules. Mastering the material in this chapter will help you in later discussions of the physical and chemical properties of substances.

9.1 | Molecular Shapes

Learning Objectives

When you finish **Section 9.1**, you should be able to:

▶ Describe, sketch, and name the three-dimensional shapes of AB_n molecules.

▶ Identify the characteristic bond angles associated with the shapes of AB_n molecules.

▶ Demonstrate that trigonal-pyramidal and bent geometries are obtained by removing B atoms from a tetrahedral AB_4 molecule.

In Chapter 8 we used Lewis structures to account for the formulas of covalent compounds. (Section 8.5) Lewis structures, however, do *not* indicate the shapes of molecules; they simply show the number and types of bonds. For example, the Lewis structure of CCl_4 tells us only that four Cl atoms are bonded to a central C atom:

The Lewis structure is drawn with the atoms all in the same plane. As shown in **Figure 9.1**, however, the actual three-dimensional arrangement has the Cl atoms at the corners of a *tetrahedron*, a geometric object with four corners and four faces, each an equilateral triangle.

The shape of a molecule is determined by its **bond angles**, the angles made by the lines joining the nuclei of the atoms in the molecule. The bond angles of a molecule, together with the bond lengths (Section 8.8), define the shape and size of the molecule. In Figure 9.1, you should be able to see that there are six Cl—C—Cl bond angles in CCl_4, all of which have the same value. That bond angle, 109.5°, is characteristic of a tetrahedron. In addition, all four C—Cl bonds have the same length (1.78 Å). Thus, the shape and size of CCl_4 are completely described by stating that the molecule is tetrahedral with C—Cl bonds of length 1.78 Å. To draw the three-dimensional structure of a molecule on paper, chemists use a convention shown in Figure 9.1: Regular lines imply the bond is in the plane of the paper, a heavy wedge is used to show that the bond is coming out of the paper toward you, and the dashed wedge is used to show that the bond is pointing away from you, through the back of the paper.

We begin our discussion of molecular shapes with molecules (and ions) that, like CCl_4, have a single central atom bonded to two or more atoms of the same type. Such molecules have the general formula AB_n in which the central atom A is bonded to n B atoms. Both CO_2 and H_2O are AB_2 molecules, for example, whereas SO_3 and NH_3 are AB_3 molecules, and so on.

The number of shapes possible for AB_n molecules depends on the value of n. Those commonly found for AB_2 and AB_3 molecules are shown in **Figure 9.2**. An AB_2 molecule must be either *linear* (bond angle = 180°) or *bent* (bond angle ≠ 180°). For AB_3 molecules, the two most common shapes place the B atoms at the vertices of an equilateral triangle. If the A atom lies in the same plane as the B atoms, the shape is called *trigonal planar*. If the A atom lies above the plane of the B atoms, the shape is called *trigonal*

 **Go Figure** In the space-filling model, what determines the relative sizes of the spheres?

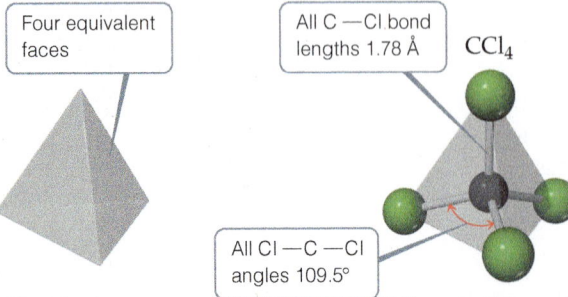

Four equivalent faces

All C—Cl bond lengths 1.78 Å

CCl_4

All Cl—C—Cl angles 109.5°

Tetrahedron   Ball-and-stick model

CCl_4

Space-filling model

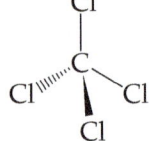

Three-dimensional atomic model

▲ **Figure 9.1 Tetrahedral shape of CCl_4.**

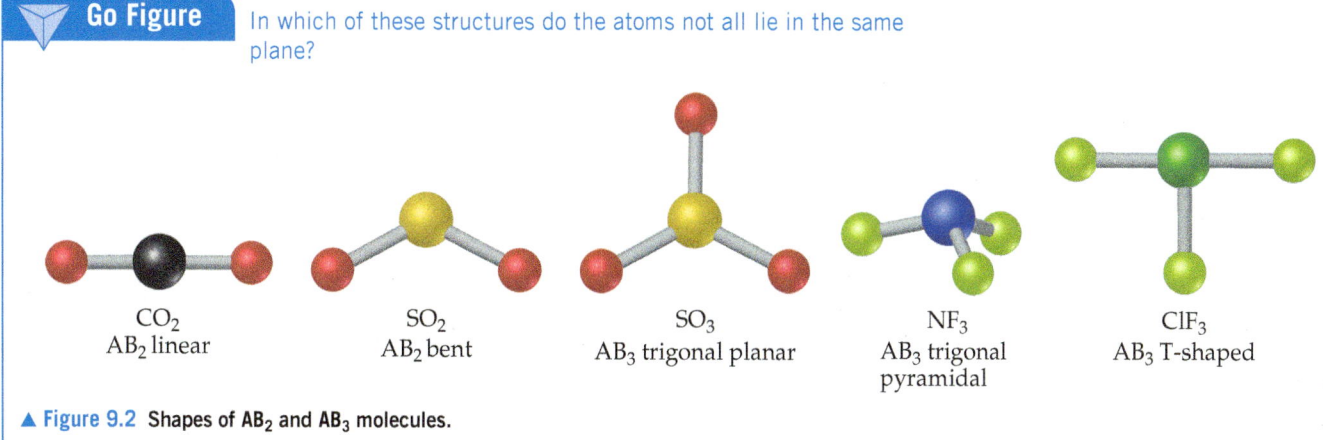

Go Figure In which of these structures do the atoms not all lie in the same plane?

CO_2	SO_2	SO_3	NF_3	ClF_3
AB_2 linear	AB_2 bent	AB_3 trigonal planar	AB_3 trigonal pyramidal	AB_3 T-shaped

▲ **Figure 9.2 Shapes of AB$_2$ and AB$_3$ molecules.**

pyramidal (a pyramid with an equilateral triangle as its base). Some AB_3 molecules, such as ClF_3, are *T-shaped*, a relatively unusual shape also shown in Figure 9.2. The atoms lie in one plane with two B—A—B angles of about 90°, and a third angle close to 180°.

Quite remarkably, the shapes of most AB_n molecules can be derived from just five basic geometric arrangements, shown in **Figure 9.3**. All of these are highly symmetric arrangements of the *n* B atoms around the central A atom. We have already seen the first three shapes: linear, trigonal planar, and tetrahedral. The trigonal-bipyramid shape for AB_5 can be thought of as a trigonal planar AB_3 arrangement with two additional atoms, one above and one below the equilateral triangle. The octahedral shape for AB_6 has all six B atoms at the same distance from atom A, with 90° B—A—B angles between all neighboring B atoms. Its symmetric shape (and its name) are derived from the *octahedron*, a geometric solid with eight faces, all of which are equilateral triangles.

Some of the shapes we have already discussed are *not* among the five shapes in Figure 9.3. In Figure 9.2, for example, neither the bent shape of the SO_2 molecule nor the trigonal-pyramidal shape of the NF_3 molecule is among the shapes in Figure 9.3. However, we can derive additional shapes, such as bent and trigonal pyramidal, by starting with one

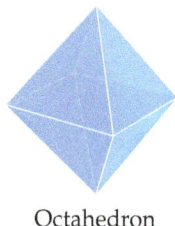

Octahedron

Go Figure Which of these molecular shapes do you expect for the SF_6 molecule?

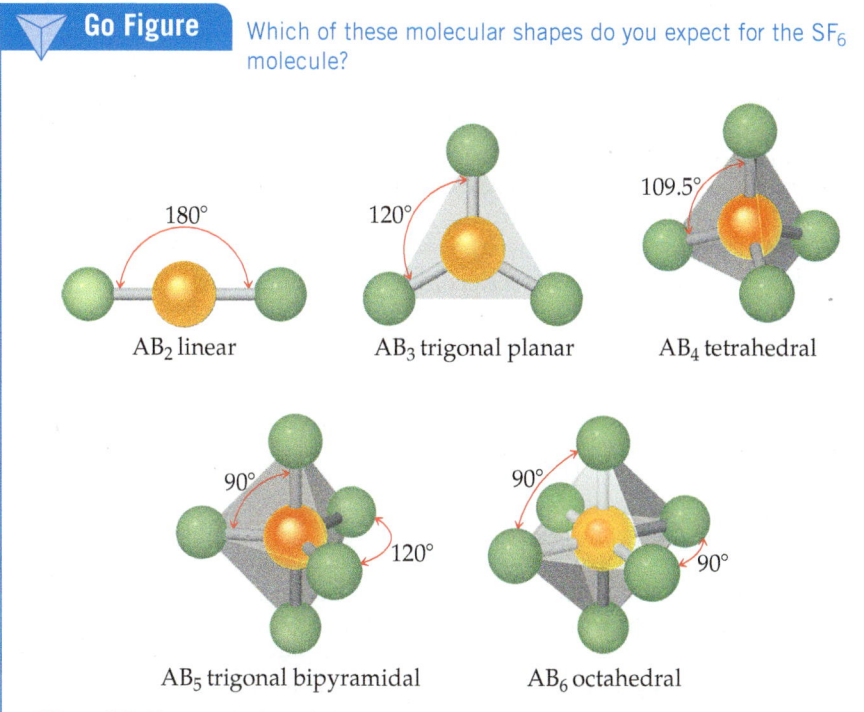

180°	120°	109.5°
AB_2 linear	AB_3 trigonal planar	AB_4 tetrahedral

90° 120°	90° 90°
AB_5 trigonal bipyramidal	AB_6 octahedral

▲ **Figure 9.3 Shapes allowing maximum distances between B atoms in AB$_n$ molecules.**

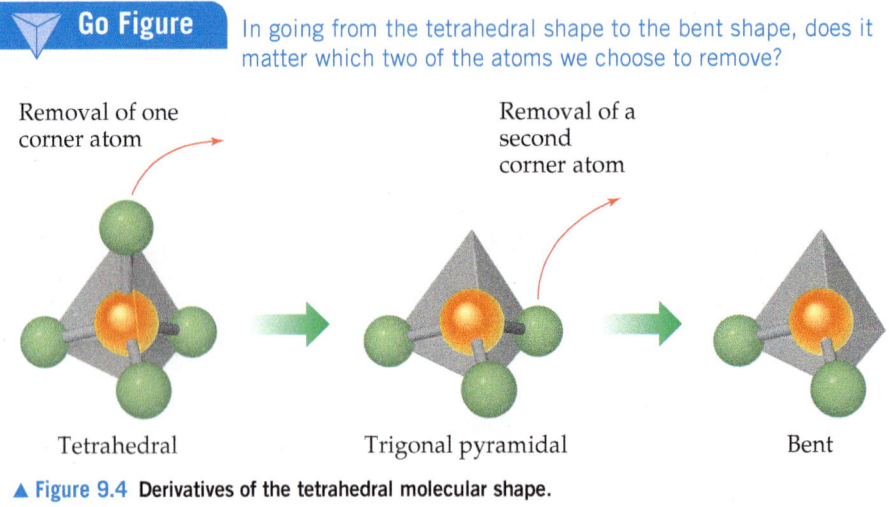

▲ **Figure 9.4** Derivatives of the tetrahedral molecular shape.

of our five basic arrangements. Starting with a tetrahedron, for example, we can remove atoms successively from the vertices, as shown in **Figure 9.4**. When an atom is removed from one vertex of a tetrahedron, the remaining AB_3 fragment has a trigonal-pyramidal geometry. When a second atom is removed, the remaining AB_2 fragment has a bent geometry.

Why do most AB_n molecules have shapes related to those shown in Figure 9.3? Can we predict these shapes? When A is a representative element (one from the *s* block or *p* block of the periodic table), we can answer these questions by using the **valence-shell electron-pair repulsion (VSEPR) model**. Although the name is rather imposing, the model is quite simple and has useful predictive capabilities.

Self-Assessment Exercises

SAE 9.1 Water, H_2O, and ammonia, NH_3, have the molecular shapes shown here. What are the names of the shapes of H_2O and NH_3? (**a**) linear, trigonal planar (**b**) linear, trigonal pyramidal (**c**) bent, trigonal planar (**d**) bent, trigonal pyramidal

SAE 9.2 The nitrate ion, NO_3^-, has a trigonal-planar shape. Which of the following statements about the nitrate ion is *false*? (**a**) All four atoms in the nitrate ion lie in the same plane. (**b**) The O—N—O bond angles in the nitrate ion are 109.5°. (**c**) There are AB_3 molecules that have a different shape than that of the nitrate ion. (**d**) The three O atoms in the nitrate ion lie on the vertices of an equilateral triangle.

SAE 9.3 Which of the following statements about AB_n molecules is or are *true*?

  (**i**) The two characteristic shapes for an AB_2 molecule are linear and bent.
  (**ii**) In a tetrahedral AB_4 molecule, all of the B—A—B angles are 109.5°.
  (**iii**) Removing one B atom from a tetrahedral AB_4 molecule leads to a trigonal pyramidal AB_3 molecule.

(**a**) Only one of the statements is true. (**b**) Statements i and ii are true. (**c**) Statements i and iii are true. (**d**) Statements ii and iii are true. (**e**) All three statements are true.

Learning Objectives

When you finish Section 9.2, you should be able to:

▶ Determine the electron-domain geometry around the central atom for an AB_n molecule or ion ($n = 2 - 6$).

▶ Predict the geometries of molecules with two, three, or four electron domains around the central atom using the VSEPR model.

▶ Predict the geometries of molecules with five or six electron domains around the central atom using the VSEPR model.

▶ Use the VSEPR model to predict the bond angles in molecules with more than one central atom.

9.2 | The VSEPR Model

Imagine tying two identical balloons together at their ends. As shown in **Figure 9.5**, the two balloons naturally orient themselves to point away from each other; that is, they try to "get out of each other's way" as much as possible. If we add a third balloon, the balloons orient themselves toward the vertices of an equilateral triangle, and if we add a fourth balloon, they adopt a tetrahedral shape. Thus, an optimum geometry exists for each number of balloons.

In some ways, the electrons in molecules behave like these balloons. We have seen that a single covalent bond is formed between two atoms when a pair of electrons occupies the space between the atoms. (Section 8.3) A *bonding pair* of electrons thus defines a region in which the electrons are most likely to be found. We refer to such a region as an **electron domain**. Likewise, a *nonbonding pair* (or *lone pair*) of electrons, which was also discussed in Section 8.3, defines an electron domain that is located principally on one atom. For example, the Lewis structure of NH_3 has four electron domains around the central nitrogen atom (three bonding pairs, represented as usual by short lines, and one nonbonding pair, represented by dots):

Nonbonding pair

$$H-\overset{\cdot\cdot}{N}-H$$
$$|$$
$$H$$

Bonding pairs

Each multiple bond in a molecule also constitutes a single electron domain. Thus, the following resonance structure for O_3 has three electron domains around the central oxygen atom (a single bond, a double bond, and a nonbonding pair of electrons):

$$:\overset{\cdot\cdot}{O}-\overset{\cdot\cdot}{O}=\overset{\cdot\cdot}{O}$$

In general, *each nonbonding pair, single bond, or multiple bond produces a single electron domain around the central atom in a molecule.*

The VSEPR model is based on the idea that electron domains are negatively charged and therefore repel one another. Like the balloons in Figure 9.5, electron domains try to stay out of one another's way:

The best arrangement of a given number of electron domains is the one that minimizes the repulsions among them.

In fact, the analogy between electron domains and balloons is so close that the same preferred geometries are found in both cases. Like the balloons in Figure 9.5, two electron domains orient *linearly*, three domains orient in a *trigonal-planar* fashion, and four orient *tetrahedrally*. These arrangements, together with those for five- and six-electron domains, are summarized in Table 9.1. Notice that the geometries in Table 9.1 are the same as those in Figure 9.3. The observation that there is an optimum shape for each specific number of electron domains around the central atom leads to the most important assumption of the VSEPR model:

The shapes of different AB_n molecules or ions depend on the number of electron domains surrounding the central atom.

The arrangement of the electron domains about the central atom of an AB_n molecule or ion is called its **electron-domain geometry**. In contrast, the **molecular geometry** is the arrangement of *only the atoms* in a molecule or ion—any nonbonding pairs in the molecule are *not* part of the description of the molecular geometry.

Applying the VSEPR Model to Determine Molecular Shapes

In determining the shape of any molecule, we first use the VSEPR model to predict the electron-domain geometry. From knowing how many of the domains are due to nonbonding pairs, we can then predict the molecular geometry. When all the electron domains in a molecule arise from bonds, the molecular geometry is identical to the electron-domain geometry. When one or more domains involve nonbonding pairs of electrons, however, we must remember that *the molecular geometry involves only electron domains due to bonds* even though the nonbonding pairs contribute to the electron-domain geometry.

▼ **Go Figure**

What arrangement do you think six balloons would adopt?

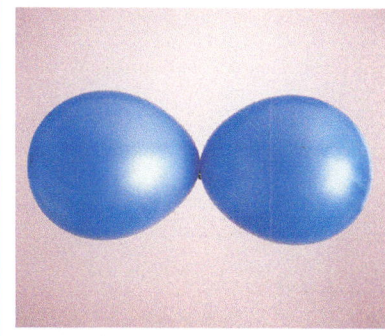
Two balloons: Linear

Three balloons: Trigonal planar

Four balloons: Tetrahedral

▲ **Figure 9.5** A balloon analogy for electron domains.

TABLE 9.1   Electron-Domain Geometries as a Function of Number of Electron Domains

Number of Electron Domains	Arrangement of Electron Domains	Electron-Domain Geometry	Predicted Bond Angles
2	180°	Linear	180°
3	120°	Trigonal planar	120°
4	109.5°	Tetrahedral	109.5°
5	90° 120°	Trigonal bipyramidal	120° 90°
6	90° 90°	Octahedral	90°

How to Predict the Shapes of Molecules and Ions Using the VSEPR Model

1. Draw the Lewis structure of the molecule or ion (Section 8.5), then count the number of electron domains around the central atom. Each nonbonding electron pair, each single bond, each double bond, and each triple bond counts as one electron domain.

2. From the number of electron domains around the central atom, use Table 9.1 to determine the *electron-domain geometry* of the molecule or ion.

3. Use the arrangement of the bonded atoms to determine the *molecular geometry*.

Table 9.2 summarizes the possible molecular geometries when an AB_n molecule has four or fewer electron domains about A. These geometries are important because they include all the shapes usually seen in molecules or ions that obey the octet rule.

TABLE 9.2   Electron-Domain and Molecular Geometries for Two, Three, and Four Electron Domains around a Central Atom

Number of Electron Domains	Electron-Domain Geometry	Bonding Domains	Nonbonding Domains	Molecular Geometry	Example
2	Linear	2	0	Linear	$\ddot{O}\!\!=\!\!C\!\!=\!\!\ddot{O}$
3	Trigonal planar	3	0	Trigonal planar	(BF₃ Lewis structure)
		2	1	Bent	$\left[\ddot{O}\!\!=\!\!N\!\!-\!\!\ddot{O}\right]^{-}$
4	Tetrahedral	4	0	Tetrahedral	(CH₄ Lewis structure)
		3	1	Trigonal pyramidal	(NH₃ Lewis structure)
		2	2	Bent	(H₂O Lewis structure)

Figure 9.6 shows how to use the three steps in the VSEPR model to predict the geometry of the NH_3 molecule. The three bonds and one nonbonding pair in the Lewis structure tell us we have four electron domains. Thus, from Table 9.1, the electron-domain geometry of NH_3 is tetrahedral. We know from the Lewis structure that one electron domain is due to a nonbonding pair, which occupies one of the four vertices of the tetrahedron. In determining the molecular geometry, we consider only the three N—H bond domains, which leads to a trigonal-pyramidal geometry. The situation is just like the middle drawing in Figure 9.4 in which removing one atom from a tetrahedral molecule results in a trigonal-pyramidal molecule. As a result, the trigonal-pyramidal molecular geometry is a direct consequence of the tetrahedral electron-domain geometry.

Because the trigonal-pyramidal molecular geometry is based on a tetrahedral electron-domain geometry, the ideal bond angles are 109.5°. As we will soon see, bond angles deviate from ideal values when the surrounding atoms and electron domains are not identical.

As one more example, let's determine the shape of the CO_2 molecule. Its Lewis structure reveals two electron domains (each one a double bond) around the central carbon:

$$\ddot{O}\!\!=\!\!C\!\!=\!\!\ddot{O}$$

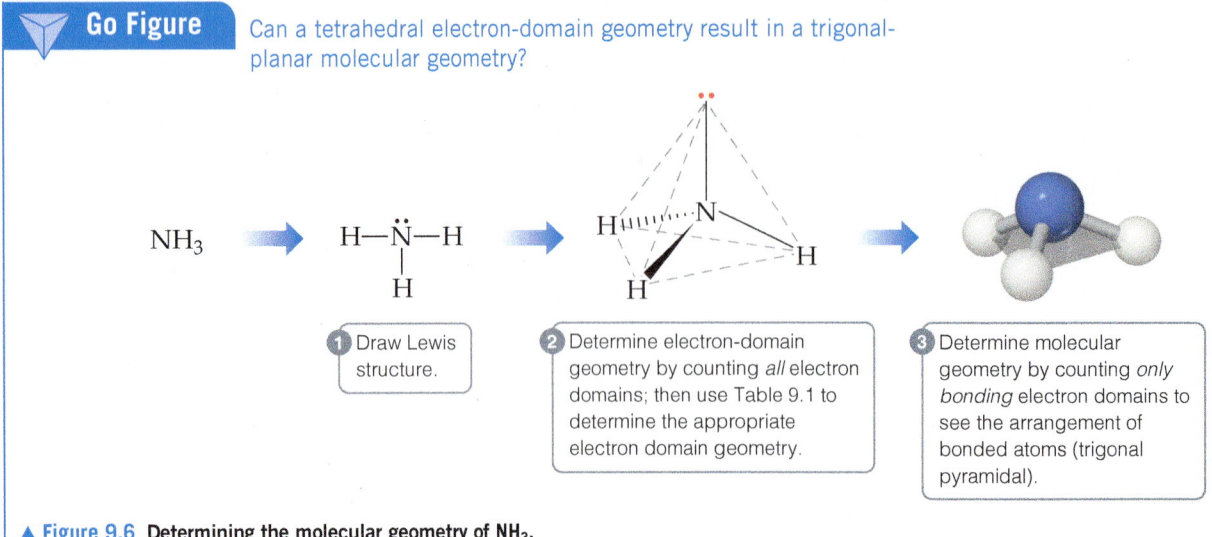

Go Figure Can a tetrahedral electron-domain geometry result in a trigonal-planar molecular geometry?

1 Draw Lewis structure.

2 Determine electron-domain geometry by counting *all* electron domains; then use Table 9.1 to determine the appropriate electron domain geometry.

3 Determine molecular geometry by counting *only* bonding electron domains to see the arrangement of bonded atoms (trigonal pyramidal).

▲ **Figure 9.6 Determining the molecular geometry of NH₃.**

Two electron domains orient in a linear electron-domain geometry (Table 9.1). Because neither domain is a nonbonding pair of electrons, the molecular geometry is also linear, and the O—C—O bond angle is 180°.

Sample Exercise 9.1
Using the VSEPR Model

Use the VSEPR model to predict the molecular geometry of (**a**) O₃, (**b**) SnCl₃⁻.

SOLUTION

Analyze We are given the molecular formulas of a molecule and a polyatomic ion, both conforming to the general formula AB_n and both having a central atom from the p block of the periodic table. (For O₃, the A and B atoms are all oxygen atoms.)

Plan To predict the molecular geometries, we draw their Lewis structures and count electron domains around the central atom to get the electron-domain geometry. We then obtain the molecular geometry from the arrangement of the domains that are due to bonds.

Solve

(**a**) We can draw two resonance structures for O₃:

$$\ddot{\text{O}}\!-\!\ddot{\text{O}}\!=\!\ddot{\text{O}} \longleftrightarrow \ddot{\text{O}}\!=\!\ddot{\text{O}}\!-\!\ddot{\text{O}}$$

Because of resonance, the bonds between the central O atom and the outer O atoms are of equal length. In both resonance structures, the central O atom is bonded to the two outer O atoms and has one nonbonding pair. Thus, there are three electron domains about the central O atoms. (Remember that a double bond counts as a single electron domain.) The arrangement of three electron domains is trigonal planar (Table 9.1). Two of the domains are from bonds, and one is due to a nonbonding pair. So, the molecular geometry is bent with an ideal bond angle of 120° (Table 9.2).

Comment As this example illustrates, when a molecule exhibits resonance, any one of the resonance structures can be used to predict the molecular geometry.

(**b**) The Lewis structure for SnCl₃⁻ is:

$$\left[\ddot{\text{Cl}}\!-\!\ddot{\text{Sn}}\!-\!\ddot{\text{Cl}}\right]^{-}$$

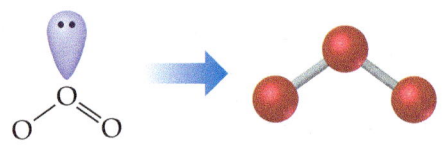

The central Sn atom is bonded to the three Cl atoms and has one nonbonding pair; thus, we have four electron domains, meaning a tetrahedral electron-domain geometry (Table 9.1) with one vertex occupied by a nonbonding pair of electrons. A tetrahedral electron-domain geometry with three bonding and one nonbonding domains leads to a trigonal-pyramidal molecular geometry (Table 9.2).

▶ **Practice Exercise**
Predict the electron-domain and molecular geometries for
(**a**) SeCl₂, (**b**) CO₃²⁻.

Effect of Nonbonding Electrons and Multiple Bonds on Bond Angles

We can refine the VSEPR model to explain slight distortions from the ideal geometries summarized in Table 9.2. For example, methane (CH_4), ammonia (NH_3), and water (H_2O) all have four electron domains around the central atom. As a result, each has a tetrahedral electron-domain geometry, but each has a different distribution of bonding and nonbonding electron domains—namely, four bonding and zero nonbonding for CH_4, three bonding and one nonbonding for NH_3, and two bonding and two nonbonding for H_2O. Experiments show that their bond angles differ slightly:

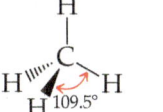  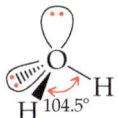

Methane, which has a tetrahedral molecular geometry, has the usual tetrahedral angles of 109.5°. For NH_3 and H_2O, however, the bond angles decrease as the number of nonbonding electron pairs increases. A bonding pair of electrons is attracted by both nuclei of the bonded atoms, but a nonbonding pair is attracted primarily by only one nucleus. Because a nonbonding pair experiences less nuclear attraction, its electron domain is spread out more in space than is the electron domain for a bonding pair (**Figure 9.7**). Nonbonding electron pairs therefore take up more space than bonding pairs; in essence, they act as larger and fatter balloons in our analogy of Figure 9.5. As a result, *electron domains for nonbonding electron pairs exert greater repulsive forces on adjacent electron domains and tend to compress bond angles*.

We face a similar situation when we compare the electron domains that result from single and multiple bonds. Because multiple bonds contain a higher electronic-charge density than single bonds, multiple bonds also represent enlarged electron domains. Consider the Lewis structure of *phosgene*, Cl_2CO:

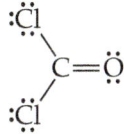

Because three electron domains surround the central atom, we might expect a trigonal-planar geometry with 120° bond angles. The double bond, however, repels other domains more strongly than does a single bond, reducing the Cl—C—Cl bond angle to 111.4°:

Cl
← 124.3°
111.4° C ═ O
← 124.3°
Cl

In general, *electron domains for multiple bonds exert a greater repulsive force on adjacent electron domains than do electron domains for single bonds*.

Molecules with Expanded Valence Shells

Recall that atoms from period 3 and beyond may be surrounded by more than four electron pairs—they can have more than an octet of electrons around the central atom. (Section 8.7) We consider here the electron-domain and molecular geometries that result when the central atom has five or six electron domains around it. The results are summarized in **Table 9.3**.

The most stable electron-domain geometry for five electron domains is the trigonal bipyramid (two trigonal pyramids sharing a base). Unlike the other arrangements we have seen, the electron domains in a trigonal bipyramid can point toward two geometrically distinct types of positions. The three positions that lie at the vertices of an equilateral triangle are called *equatorial positions*. The remaining two positions, which are above

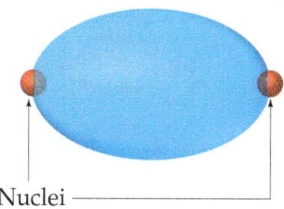

Bonding electron pair

Nuclei

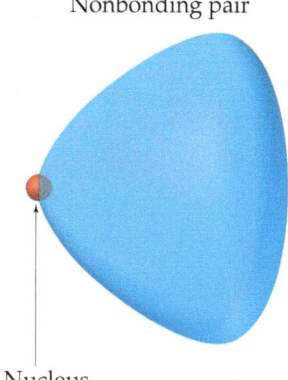

Nonbonding pair

Nucleus

▲ **Figure 9.7 Relative sizes of bonding and nonbonding electron domains.** Nonbonding electron domains are more spread out in space than are bonding domains.

TABLE 9.3   **Electron-Domain and Molecular Geometries for Five and Six Electron Domains around a Central Atom**

Number of Electron Domains	Electron-Domain Geometry	Bonding Domains	Nonbonding Domains	Molecular Geometry	Example
5	Trigonal bipyramidal	5	0	Trigonal bipyramidal	PCl_5
		4	1	Seesaw	SF_4
		3	2	T-shaped	ClF_3
		2	3	Linear	XeF_2
6	Octahedral	6	0	Octahedral	SF_6
		5	1	Square pyramidal	BrF_5
		4	2	Square planar	XeF_4

and below the plane of the equilateral triangle, are called *axial positions* (**Figure 9.8**). Each domain in an axial position (an *axial domain*) makes a 90° angle with any domain in an equatorial position (an *equatorial domain*). Each equatorial domain makes a 120° angle with either of the other two equatorial domains and a 90° angle with either axial domain.

Suppose a molecule has five electron domains, and there are one or more nonbonding pairs. Will the domains from the nonbonding pairs occupy axial or equatorial positions? To answer this question, we must determine which location minimizes the total repulsions between domains. Repulsion between two domains is much greater when they are situated 90° from each other than when they are at 120°. An equatorial domain is 90° from only two other domains (the axial domains), but an axial domain is 90° from *three* other domains (the equatorial domains). Hence, an equatorial domain experiences less repulsion than an axial domain. Because the domains from nonbonding pairs exert larger repulsions than those from bonding pairs, nonbonding domains *always* occupy the equatorial positions in a trigonal bipyramid.

The most stable electron-domain geometry for six electron domains is the *octahedron*. An octahedron is a polyhedron with six vertices and eight faces, each an equilateral triangle. An atom with six electron domains around it can be visualized as being at the center of the octahedron with the electron domains pointing toward the six vertices, as shown in Table 9.3. All the bond angles are 90°, and all six vertices are equivalent. Therefore, if an atom has five bonding electron domains and one nonbonding domain, we can put the nonbonding domain at any of the six vertices of the octahedron. The result is always a *square-pyramidal* molecular geometry—four of the outer atoms lie at the vertices of a square and the fifth outer atom occupies a position above the plane of the square. When there are two nonbonding electron domains, their repulsions are minimized by pointing them toward opposite sides of the octahedron, producing a *square-planar* molecular geometry, as shown in Table 9.3.

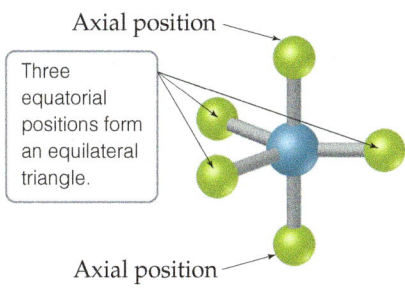

Go Figure

What is the bond angle formed by an axial position, the central atom, and any equatorial position?

Axial position

Three equatorial positions form an equilateral triangle.

Axial position

▲ **Figure 9.8 Trigonal-bipyramidal geometry.** The outer atoms occupy two types of positions.

 Sample Exercise 9.2

Molecular Geometries of Molecules with Expanded Valence Shells

Use the VSEPR model to predict the molecular geometry of (**a**) SF_4, (**b**) IF_5.

SOLUTION

Analyze The molecules are of the AB_n type with a central *p*-block atom.

Plan We first draw Lewis structures and then use the VSEPR model to determine the electron-domain geometry and molecular geometry.

Solve

(**a**) The Lewis structure for SF_4 is:

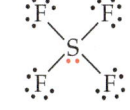

The sulfur has five electron domains around it: four from the S—F bonds and one from the nonbonding pair. Each domain points toward a vertex of a trigonal bipyramid. The domain from the nonbonding pair will point toward an equatorial position. The four bonds point toward the remaining four positions, resulting in a molecular geometry that is described as seesaw-shaped:

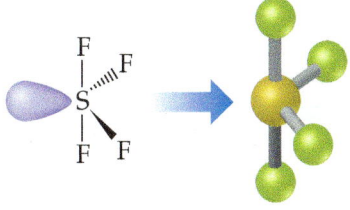

Comment The experimentally observed structure is shown on the right. We can infer that the nonbonding electron domain occupies an equatorial position, as predicted. The axial and equatorial S—F bonds are slightly bent away from the nonbonding domain, suggesting that the bonding domains are "pushed" by the nonbonding domain, which exerts a greater repulsion (Figure 9.7).

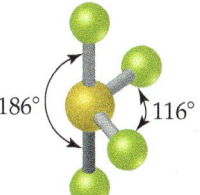

186°   116°

Solve

(**b**) The Lewis structure of IF_5 is:

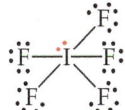

Continued

The iodine has six electron domains around it, one of which is nonbonding. The electron-domain geometry is therefore octahedral, with one position occupied by the nonbonding pair, and the molecular geometry is *square pyramidal* (Table 9.3):

Comment Because the nonbonding domain is larger than the bonding domains, we predict that the four F atoms in the base of the pyramid will be tipped up slightly toward the top F atom. Experimentally, we find that the angle between the base atoms and the top F atom is 82°, smaller than the ideal 90° angle of an octahedron.

▶ **Practice Exercise**
Predict the electron-domain and molecular geometries of **(a)** BrF_3, **(b)** SF_5^+.

Shapes of Larger Molecules

Although the molecules and ions we have considered contain only a single central atom, the VSEPR model can be extended to more complex molecules, such as acetic acid:

$$
\begin{array}{ccc}
H & :O: & \\
| & || & \\
H-C-C-\ddot{O}-H & & \\
| & & \\
H & &
\end{array}
$$

We can use the VSEPR model to predict the geometry about each atom that is bonded to two or more other atoms:

Number of electron domains	4	3	4
Electron-domain geometry	Tetrahedral	Trigonal planar	Tetrahedral
Predicted bond angles	109.5°	120°	109.5°

The C on the left has four electron domains (all bonding), so the electron-domain and molecular geometries around that atom are both tetrahedral. The central C has three electron domains (counting the double bond as one domain), making both the electron-domain and the molecular geometries trigonal planar. The O on the right has four electron domains (two bonding, two nonbonding), so its electron-domain geometry is tetrahedral and its molecular geometry is bent. The bond angles about the central C atom and the O atom are expected to deviate slightly from the ideal values of 120° and 109.5° because of the spatial demands of multiple bonds and nonbonding electron pairs.

Our analysis of the acetic acid molecule is summarized in **Figure 9.9**.

Go Figure In the actual structure of acetic acid, which bond angle is expected to be the smallest?

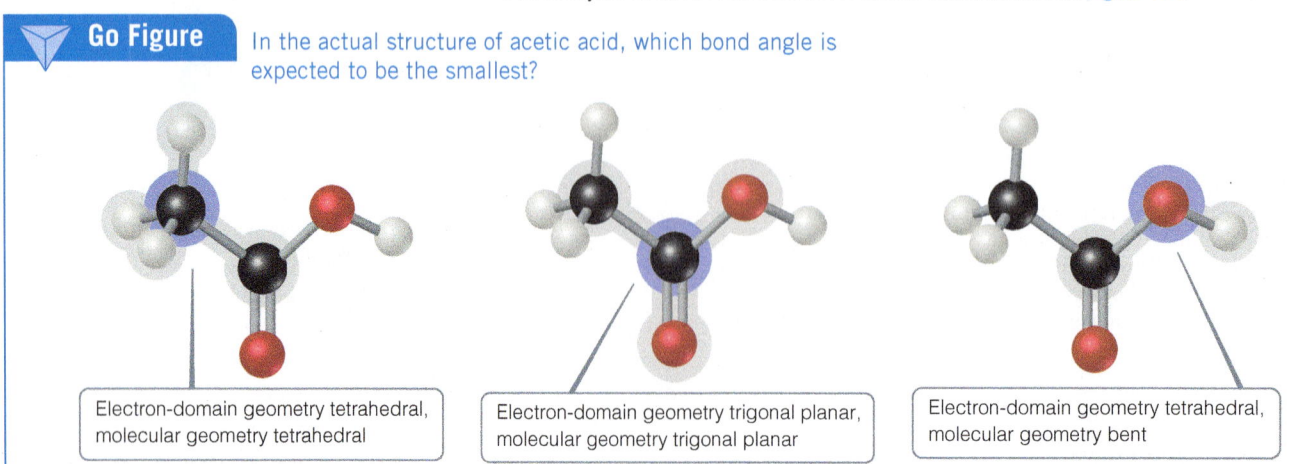

Electron-domain geometry tetrahedral, molecular geometry tetrahedral

Electron-domain geometry trigonal planar, molecular geometry trigonal planar

Electron-domain geometry tetrahedral, molecular geometry bent

▲ **Figure 9.9** The electron-domain and molecular geometries around the three central atoms of acetic acid, CH_3COOH.

Sample Exercise 9.3
Predicting Bond Angles

Eyedrops for dry eyes usually contain a water-soluble polymer called *poly(vinyl alcohol)*, which is based on the unstable organic molecule *vinyl alcohol*:

$$H-\ddot{O}-\overset{\overset{\displaystyle H}{|}}{C}=\overset{\overset{\displaystyle H}{|}}{C}-H$$

Predict the approximate values for the H—O—C and O—C—C bond angles in vinyl alcohol.

SOLUTION

Analyze We are given a Lewis structure and asked to determine two bond angles.

Plan To predict a bond angle, we determine the number of electron domains surrounding the middle atom in the angle. The ideal angle corresponds to the electron-domain geometry around the atom. The angle will be compressed somewhat by nonbonding electrons or multiple bonds.

Solve In H—O—C, the O atom has four electron domains (two bonding, two nonbonding). The electron-domain geometry around O is therefore tetrahedral, which gives an ideal angle of 109.5°. The H—O—C angle is compressed somewhat by the nonbonding pairs, so we expect this angle to be slightly less than 109.5°.

To predict the O—C—C bond angle, we examine the middle atom in the angle. In the molecule, there are three atoms bonded to this C atom and no nonbonding pairs, and so it has three electron domains about it. The predicted electron-domain geometry is trigonal planar, resulting in an ideal bond angle of 120°. Because of the larger size of the C=C domain, the actual bond angle is 126°, slightly greater than 120°.

▶ **Practice Exercise**
Predict the H—C—H and C—C—C bond angles in *propyne*:

$$H-\overset{\overset{\displaystyle H}{|}}{\underset{\underset{\displaystyle H}{|}}{C}}-C\equiv C-H$$

Self-Assessment Exercises

SAE 9.4 Consider the following four XF_4 ions:

   (**i**) BrF_4^+
   (**ii**) BF_4^-
   (**iii**) AsF_4^-
   (**iv**) ClF_4^-

Which of these ions has an octahedral electron-domain geometry? (**a**) ion i (**b**) ion ii (**c**) ion iii (**d**) ion iv (**e**) more than one of the ions

SAE 9.5 How many of the following AB_3 molecules and ions have a trigonal-pyramidal molecular geometry: NF_3, BCl_3, CH_3^-, and SF_3^+? (**a**) 0 (**b**) 1 (**c**) 2 (**d**) 3 (**e**) 4

SAE 9.6 Which of the following XF_4 ions will have a seesaw-shaped molecular geometry?

   (**i**) BrF_4^+
   (**ii**) BF_4^-
   (**iii**) AsF_4^-
   (**iv**) ClF_4^-

(**a**) ion i (**b**) ion ii (**c**) ion iii (**d**) ion iv (**e**) more than one of the ions

SAE 9.7 For each bond angle 1, 2, and 3 in the molecule shown here, predict whether it will be closer to 109.5° or 120°. (**a**) angle 1: 120°; angle 2: 109.5°; angle 3: 120° (**b**) angle 1: 109.5°; angle 2: 109.5°; angle 3: 109.5° (**c**) angle 1: 120°; angle 2: 120°; angle 3: 109.5° (**d**) angle 1: 109.5°; angle 2: 120°; angle 3: 120°

9.3 | Molecular Shape and Molecular Polarity

Now that we have a sense of the shapes that molecules adopt and why they do so, we can further explore some topics that we first discussed in Section 8.4—namely, *bond polarity* and *dipole moments*. Recall that bond polarity is a measure of how equally the electrons in a bond are shared between the two atoms of the bond. As the difference in electronegativity between the two atoms increases, so does the bond polarity. (Section 8.4) We saw that the dipole moment of a diatomic molecule is a measure of the amount of charge separation in the molecule.

▲ Learning Objectives

When you finish Section 9.3, you should be able to:

▶ Explain how the overall polarity of a molecule depends on its shape and the polarities of its bonds.

▶ Predict whether a molecule is polar or nonpolar based on its molecular geometry.

For a molecule consisting of more than two atoms, *the dipole moment depends on both the polarities of the individual bonds and the geometry of the molecule*. For each bond in the molecule, we consider the **bond dipole**, which is the dipole moment due only to the two atoms in that bond. In the linear CO_2 molecule, for example, shown in **Figure 9.10(a)**, each $C=O$ bond is polar. And because the $C=O$ bonds are identical, the bond dipoles are equal in magnitude. A plot of the molecule's electron density clearly shows that the individual bonds are polar, but what can we say about the *overall* dipole moment of the molecule?

Bond dipoles and dipole moments are *vector quantities*; that is, they have both a magnitude and a direction. (Section 8.4) The dipole moment of a polyatomic molecule is the vector sum of its bond dipoles. Both the magnitudes *and* the directions of the bond dipoles must be considered when summing vectors. The two bond dipoles in CO_2, although equal in magnitude, are opposite in direction. Adding them is the same as adding two numbers that are equal in magnitude but opposite in sign, such as $100 + (-100)$. The bond dipoles, like the numbers, "cancel" each other. Therefore, the dipole moment of CO_2 is zero, even though the individual bonds are polar. The geometry of the molecule dictates that the overall dipole moment be zero, making CO_2 a *nonpolar* molecule.

Now let's consider H_2O, a bent molecule with two polar bonds [**Figure 9.10(b)**]. Again, the two bonds are identical, and the bond dipoles are equal in magnitude. Because the molecule is bent, however, the bond dipoles do *not* directly oppose each other and therefore do *not* completely cancel. Hence, the H_2O molecule has an overall nonzero dipole moment (measured dipole moment, $\mu = 1.85$ D) and is therefore a *polar* molecule. The oxygen atom carries a partial negative charge, whereas the hydrogen atoms each have a partial positive charge, as shown in Figure 9.10(b).

CO_2 and H_2O are both AB_2 molecules, yet one is nonpolar whereas the other is polar. The shape of CO_2 dictates that the individual bond dipoles *must* cancel completely. We can generalize this observation: For AB_n molecules in which the B atoms are the same, the molecular geometry will determine whether the molecule is polar or nonpolar. **Figure 9.11** shows some examples of nonpolar and polar AB_n molecules—whether the bond dipoles completely cancel depends entirely on the molecular geometry. For example, the bond

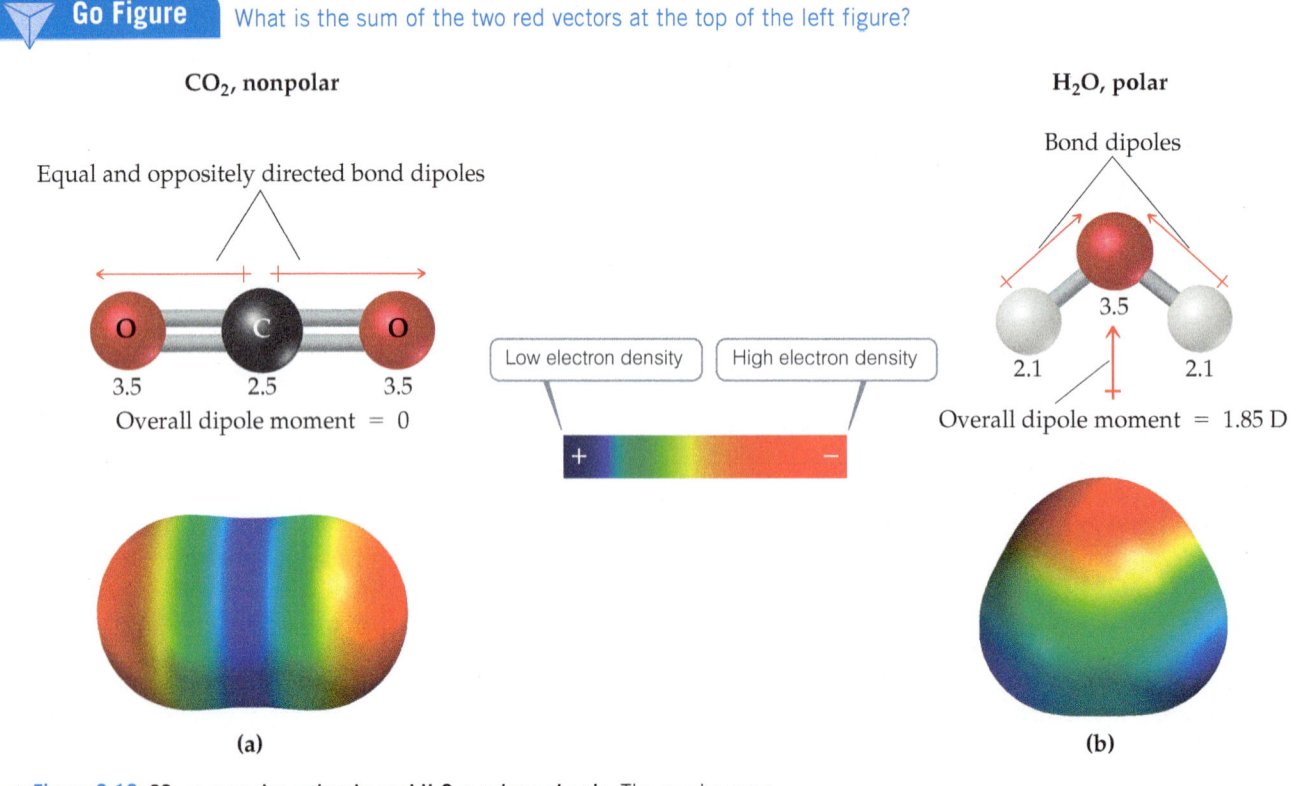

Go Figure What is the sum of the two red vectors at the top of the left figure?

CO_2, nonpolar

Equal and oppositely directed bond dipoles

O   C   O

3.5   2.5   3.5

Overall dipole moment = 0

Low electron density   High electron density

H_2O, polar

Bond dipoles

3.5

2.1   2.1

Overall dipole moment = 1.85 D

(a)   (b)

▲ **Figure 9.10** CO_2, a nonpolar molecule, and H_2O, a polar molecule. The numbers are electronegativity values for these two atoms.

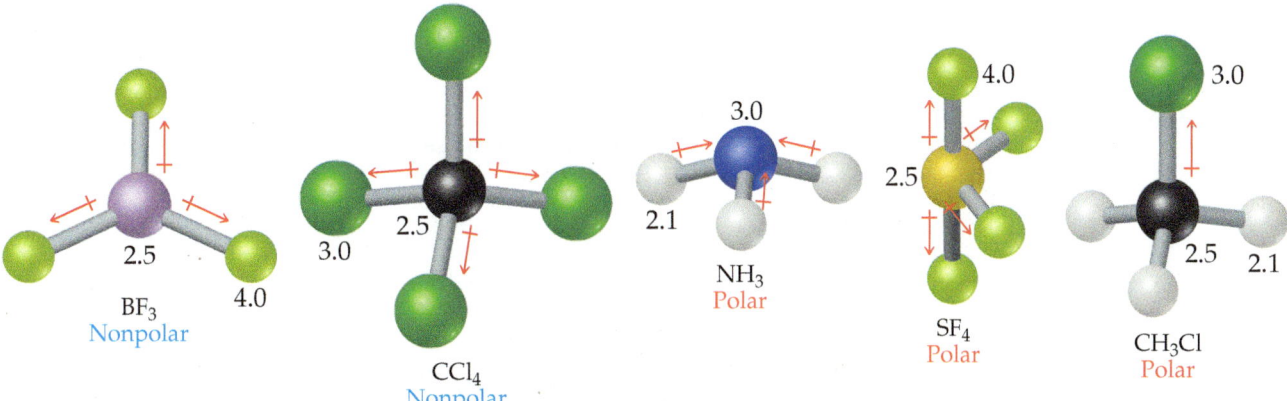

▲ **Figure 9.11 Nonpolar and polar molecules containing polar bonds.** The numbers are electronegativity values.

dipoles of trigonal planar BF_3 cancel completely, whereas in trigonal pyramidal NH_3 the bond dipoles do not completely cancel. **Table 9.4** summarizes the geometries that lead to nonpolar and polar AB_n molecules.

Figure 9.11 also shows the CH_3Cl molecule, which is polar in spite of having a tetrahedral geometry. In this instance, one of the outer atoms is different from the others, so the bond dipoles don't cancel completely. In general, we can use Table 9.4 to determine whether a molecule is nonpolar only if all of the outer atoms are the same.

TABLE 9.4   Relationship Between Molecular Shape and Molecular Polarity for AB_n Molecules

AB_n	Nonpolar	Polar
AB_2	Linear	Bent
AB_3	Trigonal planar	Trigonal pyramidal
		T-shaped
AB_4	Tetrahedral	Seesaw-shaped
	Square planar	
AB_5	Trigonal bipyramidal	Square pyramidal
AB_6	Octahedral	

Sample Exercise 9.4
Polarity of Molecules

Predict whether these molecules are polar or nonpolar: (**a**) BrCl, (**b**) SO_2, (**c**) SF_6.

SOLUTION

Analyze We are given three molecular formulas and asked to predict whether the molecules are polar.

Plan A molecule containing only two atoms is polar if the atoms differ in electronegativity. The polarity of a molecule containing three or more atoms depends on both the molecular geometry and the individual bond polarities. Thus, we must draw a Lewis structure for each molecule containing three or more atoms and determine its molecular geometry. We then use electronegativity values to determine the direction of the bond dipoles. Finally, we see whether the bond dipoles cancel to give a nonpolar molecule or reinforce each other to give a polar one.

Solve

(**a**) Chlorine is more electronegative than bromine. All diatomic molecules with polar bonds are polar molecules. Consequently, BrCl is polar, with chlorine carrying the partial negative charge:

Br—Cl

The measured dipole moment μ of BrCl is 0.57 D.

(**b**) Because oxygen is more electronegative than sulfur, SO_2 has polar bonds. Three resonance forms can be written:

$$:\ddot{O}—\ddot{S}=\ddot{O}: \longleftrightarrow :\ddot{O}=\ddot{S}—\ddot{O}: \longleftrightarrow :\ddot{O}=\ddot{S}=\ddot{O}:$$

For each of these, the VSEPR model predicts a bent molecular geometry. Because the molecule is bent, the bond dipoles do not cancel, and the molecule is polar:

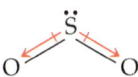

Experimentally, the dipole moment μ of SO_2 is 1.63 D.

Continued

(c) Fluorine is more electronegative than sulfur, so the bond dipoles point toward fluorine. For clarity, only one S—F dipole is shown. The six S—F bonds are arranged octahedrally around the central sulfur:

Because the octahedral molecular geometry is symmetrical, the bond dipoles cancel, and the molecule is nonpolar, meaning that $\mu = 0$.

▶ **Practice Exercise**
Determine whether the following molecules are polar or nonpolar: **(a)** SF_4, **(b)** $SiCl_4$.

⚠ Self-Assessment Exercises

SAE 9.8 Which of the following statements about bond dipoles and the polarity of molecules is *false*? **(a)** An AB_n molecule will have n bond dipoles. **(b)** If all the bond dipoles in a molecule are nonzero, then the molecule must be polar. **(c)** The overall dipole moment of a molecule is the vector sum of its individual bond dipoles. **(d)** Bond dipoles have both a magnitude and a direction. **(e)** The magnitude of a bond dipole depends on the electronegativity difference of the two atoms involved.

SAE 9.9 Which of the following molecules is polar? **(a)** CF_4 **(b)** XeF_4 **(c)** PF_5 **(d)** IF_5

9.4 | Covalent Bonding and Orbital Overlap

⚠ Learning Objective

When you finish Section 9.4, you should be able to:

▶ Describe the principles of valence-bond theory, in which bonds form between two atoms via the overlap of their atomic orbitals.

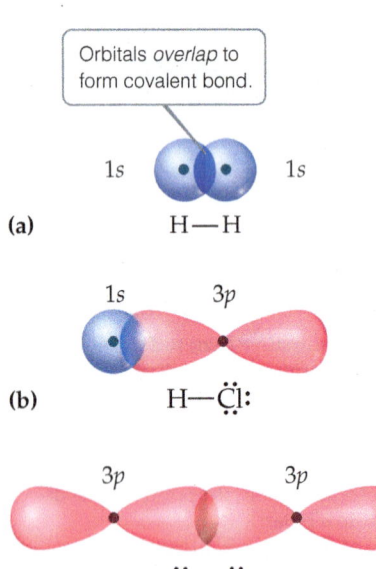

▲ **Figure 9.12** Covalent bonds in H_2, HCl, and Cl_2 result from overlap of atomic orbitals.

The VSEPR model provides a simple means for predicting molecular geometries but does not explain *why* bonds exist between atoms. In developing theories of covalent bonding, chemists have approached the problem from another direction, using quantum mechanics. How can we use atomic orbitals to explain bonding and to account for molecular geometries? The marriage of Lewis's notion of electron-pair bonds and the idea of atomic orbitals leads to a model of chemical bonding, called **valence-bond theory**, in which bonding electron pairs are concentrated in the regions between atoms, and nonbonding electron pairs lie in directed regions of space away from the bonds. By extending this approach to include the ways in which atomic orbitals can mix with one another, we obtain an explanatory picture that corresponds to the VSEPR model.

In Lewis theory, covalent bonding occurs when atoms share electrons because the sharing concentrates electron density between the nuclei. In valence-bond theory, we visualize the buildup of electron density between two nuclei as occurring when a valence atomic orbital of one atom shares space, or *overlaps*, with a valence atomic orbital of another atom. The overlap of orbitals allows two electrons of opposite spin to share the space between the nuclei, forming a covalent bond.

Figure 9.12 shows three examples of how valence-bond theory describes the coming together of two atoms to form a molecule. In the formation of H_2 [Figure 9.12(a)], each hydrogen atom has a single electron in a $1s$ orbital. As the orbitals overlap, electron density is concentrated between the nuclei. Because the electrons in the overlap region are simultaneously attracted to both nuclei, they hold the atoms together, forming a covalent bond.

The idea of orbital overlap producing a covalent bond applies equally well to other molecules. In HCl, for example, chlorine has the electron configuration [Ne]$3s^2 3p^5$. All the valence orbitals of chlorine are full except one $3p$ orbital, which contains a single electron. This $3p$ orbital overlaps with the $1s$ orbital of H to form the covalent bond that holds HCl together [Figure 9.12(b)]. Because the other two chlorine $3p$ orbitals are already each filled with a pair of electrons, they do not participate in the bonding to hydrogen. Likewise, we can explain the covalent bond in Cl_2 [Figure 9.12(c)] in terms of the overlap of the singly occupied $3p$ orbital of one Cl atom with the singly occupied $3p$ orbital of another.

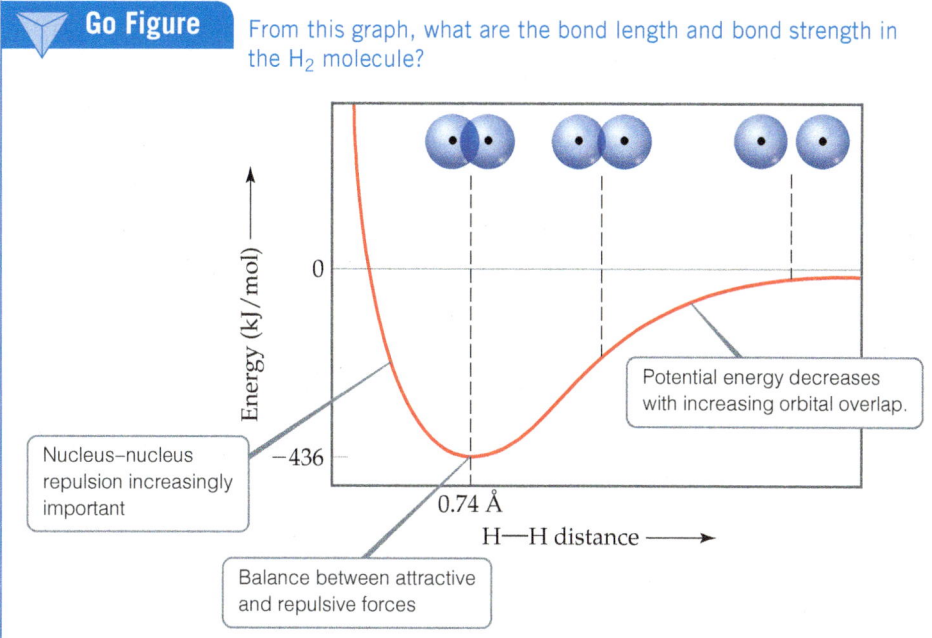

From this graph, what are the bond length and bond strength in the H_2 molecule?

Potential energy decreases with increasing orbital overlap.

Nucleus–nucleus repulsion increasingly important

Balance between attractive and repulsive forces

▲ **Figure 9.13 Formation of the H_2 molecule as atomic orbitals overlap.** The red curve shows the potential energy of the molecule as a function of H—H distance. The minimum in the curve corresponds to the bond length of the molecule, and the energy at the minimum corresponds to the bond strength of the molecule.

There is always an optimum distance between the two nuclei in any covalent bond. **Figure 9.13** shows how the potential energy of a system consisting of two H atoms changes as the atoms come together to form an H_2 molecule. When the atoms are infinitely far apart, they do not "feel" each other and so the energy approaches zero. As the distance between the atoms decreases, the overlap between their $1s$ orbitals increases. Because of the stabilizing effect of increasing the electron density between the nuclei, the potential energy of the system decreases. That is, the strength of the bond increases, as shown by the decrease in the potential energy of the two-atom system. However, Figure 9.13 also shows that the energy increases sharply when the distance between the two hydrogen nuclei is less than 0.74 Å. The increase in potential energy of the system, which becomes significant at short internuclear distances, is due mainly to the electrostatic repulsion between the positively charged nuclei. The internuclear distance at the minimum of the potential-energy curve (in this example, at 0.74 Å) corresponds to the bond *length* of the molecule. The potential energy at this minimum corresponds to the bond *strength*. Thus, the observed bond length is the distance at which the sum of the attractive forces between unlike charges (electrons and nuclei) and the repulsive forces between like charges (electron–electron and nucleus–nucleus) is most negative.

Valence-bond theory explains the energetic reasons that bonds form via the overlap of orbitals on different atoms. In order to connect the theory more directly to the molecular geometries we obtain from the VSEPR model, we need to develop one more concept—how atomic orbitals can mix in a certain way to explain the electron-domain geometries that are so key to using VSEPR. We address this issue in the next section.

▲ Self-Assessment Exercise

SAE 9.10 Which of the following statements about orbital overlap and covalent bonding is *false*? (**a**) In valence-bond theory, bonds are formed by the overlap of atomic orbitals on different atoms. (**b**) p orbitals can overlap only with other p orbitals. (**c**) As two atoms are pulled apart, the overlap between their atomic orbitals decreases. (**d**) The energy required to separate the atoms in an H_2 molecule reaches its maximum value at the observed bond length of 0.74 Å.

9.5 | Hybrid Orbitals

The VSEPR model, simple as it is, does a surprisingly good job at predicting molecular shape, despite the fact that it has no obvious relationship to the filling and shapes of atomic orbitals. For example, we would like to understand how to account for the tetrahedral arrangement of C—H bonds in methane in terms of the $2s$ and $2p$ orbitals of the central carbon atom, which are not directed toward the apices of a tetrahedron. How can we reconcile the notion that covalent bonds are formed from the overlap of atomic orbitals with the molecular geometries that come from the VSEPR model?

To begin with, recall that atomic orbitals are mathematical functions that come from the quantum-mechanical model for atomic structure. (Section 6.5) To explain molecular geometries, we often assume that the atomic orbitals on an atom (usually the central atom) mix to form new orbitals called **hybrid orbitals**. The shape of any hybrid orbital is different from the shapes of the original atomic orbitals. The process of mixing atomic orbitals is a mathematical operation called **hybridization**. The total number of atomic orbitals on an atom remains constant, so the number of hybrid orbitals on an atom equals the number of atomic orbitals that are mixed.

As we examine the common types of hybridization, notice the connection between the type of hybridization and some of the molecular geometries predicted by the VSEPR model: linear, bent, trigonal planar, and tetrahedral.

Learning Objectives

When you finish Section 9.5, you should be able to:

▶ Contrast atomic orbitals and hybrid orbitals.

▶ Demonstrate the formation of sp hybrid orbitals from an s and a p orbital on the same atom.

▶ Illustrate the formation and spatial orientation of sp^2 and sp^3 hybrid orbitals.

sp Hybrid Orbitals

To illustrate the process of hybridization, consider the BeF_2 molecule, which has the Lewis structure

$$:\!\ddot{F}\!-\!Be\!-\!\ddot{F}\!:$$

The VSEPR model predicts that BeF_2 is linear with two identical Be—F bonds, in accord with experiment. How can we use valence-bond theory to describe the bonding? The electron configuration of F ($1s^2 2s^2 2p^5$) indicates an unpaired electron in a $2p$ orbital. This electron can be paired with an unpaired Be electron to form a polar covalent bond. Which orbitals on the Be atom, however, overlap with those on the F atoms to form the Be—F bonds?

The orbital diagram for a ground-state Be atom is

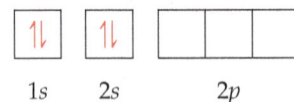

Because it has no unpaired electrons, the Be atom in its ground state cannot bond with the fluorine atoms. The Be atom could form two bonds, however, by envisioning that we "promote" one of the $2s$ electrons to a $2p$ orbital:

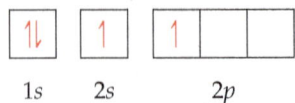

Promoting an electron requires energy. However, the Be atom now has two unpaired electrons and can therefore form two polar covalent bonds with F atoms, which would release more energy than was expended in promotion. The two bonds would not be identical, however, because a Be $2s$ orbital would be used to form one of the bonds and a $2p$ orbital would be used to form the other. Therefore, although the promotion of an electron allows two Be—F bonds to form, we still have not explained the structure of BeF_2.

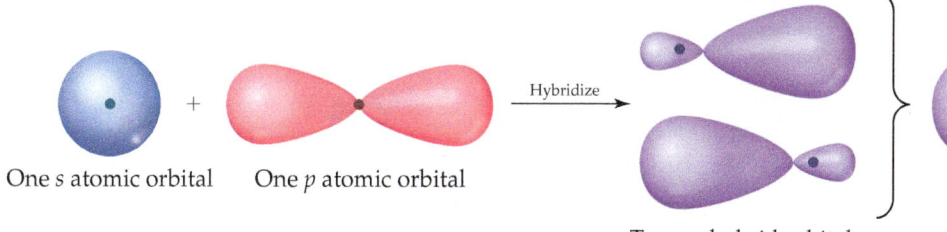

One s atomic orbital   One p atomic orbital   Hybridize   Two sp hybrid orbitals sp hybrid orbitals superimposed (large lobes only)

▲ **Figure 9.14 Formation of sp hybrid orbitals.**

We can solve this dilemma by "mixing" the $2s$ orbital with one $2p$ orbital to generate two new orbitals, as shown in **Figure 9.14**.* Like p orbitals, each new orbital has two lobes. Unlike p orbitals, however, one lobe is much larger than the other. The two new orbitals are identical in shape, but their large lobes point in opposite directions. These two new orbitals, which are shown in purple in Figure 9.14, are hybrid orbitals. Because we have hybridized one s and one p orbital, we call each hybrid an sp hybrid orbital. *According to the valence-bond model, a linear arrangement of electron domains implies sp hybridization.*

For the Be atom of BeF_2, we write the orbital diagram for the formation of two sp hybrid orbitals as

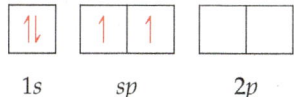

$1s$ sp $2p$

The electrons in the sp hybrid orbitals can form bonds with the two fluorine atoms (**Figure 9.15**). Because the sp hybrid orbitals are equivalent but point in opposite directions, BeF_2 has two identical bonds and a linear geometry. Furthermore, the large lobes of the hybrid orbitals allow them to overlap very effectively with the orbitals of other atoms.

We have used one of the three Be $2p$ orbitals in making the two sp hybrids; the remaining two $2p$ atomic orbitals of Be remain unhybridized and are vacant. Remember also that each fluorine atom has two other valence $2p$ atomic orbitals, each containing one nonbonding electron pair. Those atomic orbitals are omitted from Figure 9.15 to simplify the illustration.

sp^2 and sp^3 Hybrid Orbitals

Whenever we mix a certain number of atomic orbitals, we get the same number of hybrid orbitals. Each hybrid orbital is equivalent to the others but points in a different direction. Thus, mixing one $2s$ and one $2p$ atomic orbital yields two equivalent sp hybrid orbitals that point in opposite directions (Figure 9.14). Other combinations of atomic orbitals can be hybridized to obtain different geometries, such as sp^2 and sp^3.

In BF_3, for example, mixing the $2s$ and two of the $2p$ atomic orbitals yields three equivalent sp^2 (pronounced "s-p-two") hybrid orbitals (**Figure 9.16**). The three sp^2 hybrid orbitals lie in the same plane, 120° apart from one another. They are used to make three equivalent bonds with the three fluorine atoms, leading to the trigonal-planar molecular geometry of BF_3. Notice that an unfilled $2p$ atomic orbital remains unhybridized; it is oriented perpendicular to the plane defined by the three sp^2 hybrid orbitals, with one lobe

Large lobes from two Be sp hybrid orbitals

F   **Be**   **F**

F $2p$ atomic orbital   └ Overlap regions ┘   F $2p$ atomic orbital

◀ **Figure 9.15 Formation of two equivalent Be—F bonds in BeF_2.**

---

*By mixing, we mean mathematically taking two linear combinations of the $2s$ and $2p$ atomic orbital wave functions. We do not discuss the mathematical details of hybridization because they are beyond the scope of an introductory textbook.

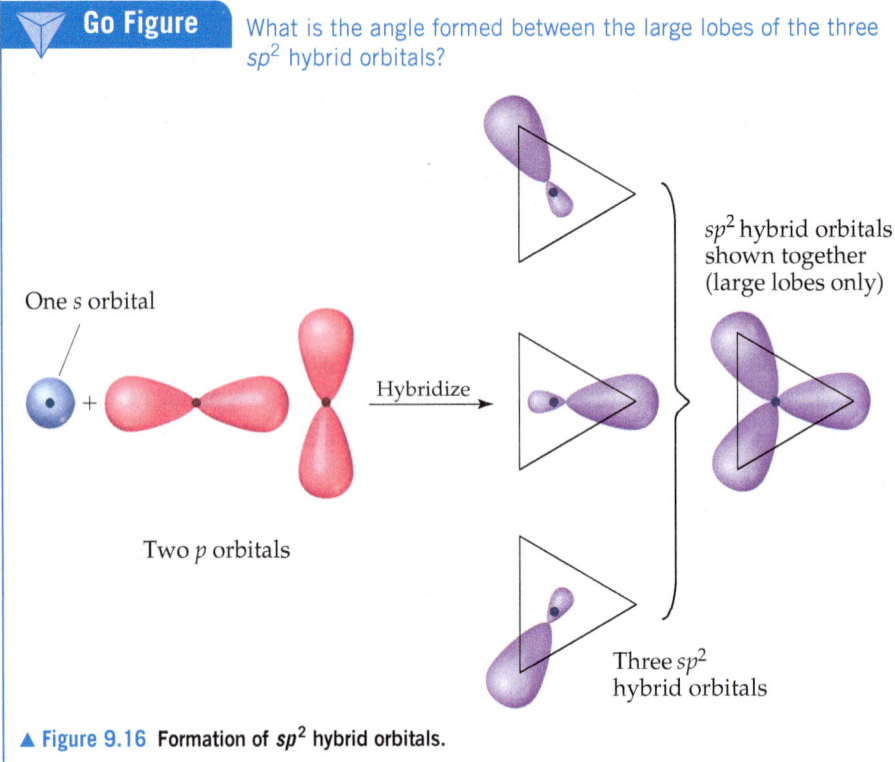

Go Figure   What is the angle formed between the large lobes of the three *sp*² hybrid orbitals?

One *s* orbital

Two *p* orbitals

Hybridize →

*sp*² hybrid orbitals shown together (large lobes only)

Three *sp*² hybrid orbitals

▲ **Figure 9.16 Formation of *sp*² hybrid orbitals.**

above and one below the plane. This unhybridized orbital becomes important when we discuss double bonds in Section 9.6.

An *s* atomic orbital can mix with all three *p* atomic orbitals in the same subshell. For example, the carbon atom in CH_4 forms four equivalent bonds with the four hydrogen atoms. We envision this process as resulting from the mixing of the 2*s* and all three 2*p* atomic orbitals of carbon to create four equivalent *sp*³ (pronounced "s-p-three") hybrid orbitals. Each *sp*³ hybrid orbital has a large lobe that points toward one vertex of a tetrahedron (**Figure 9.17**). These hybrid orbitals can be used to form two-electron bonds by overlap with the atomic orbitals of another atom, such as H. Using valence-bond theory, we can describe the bonding in CH_4 as the overlap of four equivalent *sp*³ hybrid orbitals on C with the 1*s* orbitals of the four H atoms to form four equivalent bonds.

The idea of hybridization is also used to describe the bonding in molecules containing nonbonding pairs of electrons. In H_2O, for example, the O atom has four electron domains about it—two bonding domains and two nonbonding domains. From VSEPR, we therefore expect a tetrahedral electron-domain geometry around the O atom, which we can explain using *sp*³ hybridization (**Figure 9.18**). Two of the hybrid orbitals contain nonbonding pairs of electrons, and the other two form valence bonds via overlap with the H 1*s* orbitals.

Hypervalent Molecules

So far our discussion of hybridization has extended only to period 2 elements—specifically carbon, nitrogen, and oxygen. The elements of period 3 and beyond introduce a new consideration because in many of their compounds these elements are **hypervalent**—they have more than an octet of electrons around the central atom. (Section 8.7) We saw in Section 9.2 that the VSEPR model works well to predict the geometries of hypervalent molecules such as PCl_5, SF_6, or BrF_5. But can we extend the use of hybrid orbitals to describe the bonding in these molecules? In short, the answer is no; hybrid orbitals should not be used for hypervalent molecules. Let's examine the reasons.

The valence-bond model we developed for period 2 elements works well for compounds of period 3 elements as long as we have no more than an octet of electrons in the valence-shell orbitals. Thus, for example, it is appropriate to discuss the bonding in PF_3 or H_2Se in terms of hybridized *s* and *p* orbitals on the central atom.

For compounds with more than an octet, we could imagine increasing the number of hybrid orbitals formed by including valence-shell *d* orbitals. For example, for SF_6 we could

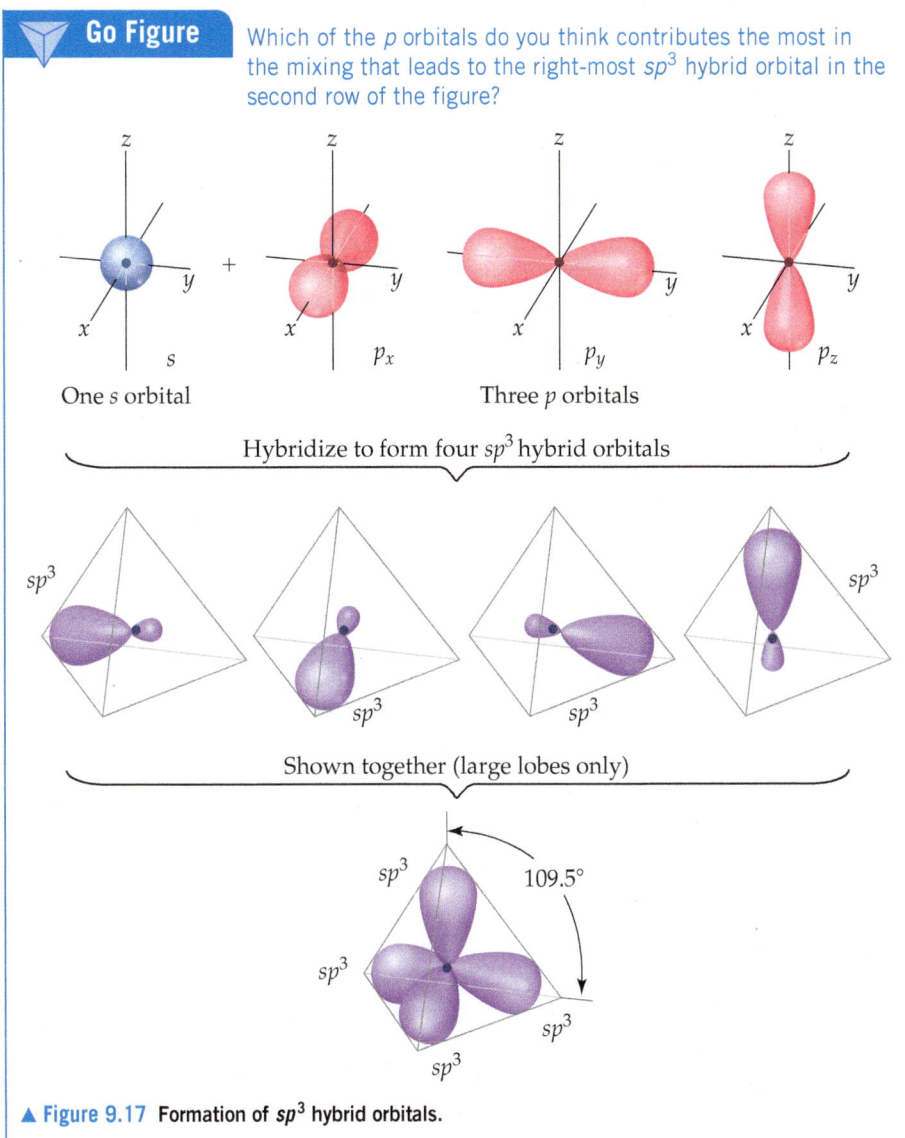

Go Figure Which of the *p* orbitals do you think contributes the most in the mixing that leads to the right-most sp^3 hybrid orbital in the second row of the figure?

▲ **Figure 9.17 Formation of sp^3 hybrid orbitals.**

envision mixing in two sulfur $3d$ orbitals in addition to the $3s$ and three $3p$ orbitals to make a total of six hybrid orbitals. However, the sulfur $3d$ orbitals are substantially higher in energy than the $3s$ and $3p$ orbitals, so the amount of energy needed to form the six hybrid orbitals is greater than the amount returned by forming bonds with the six

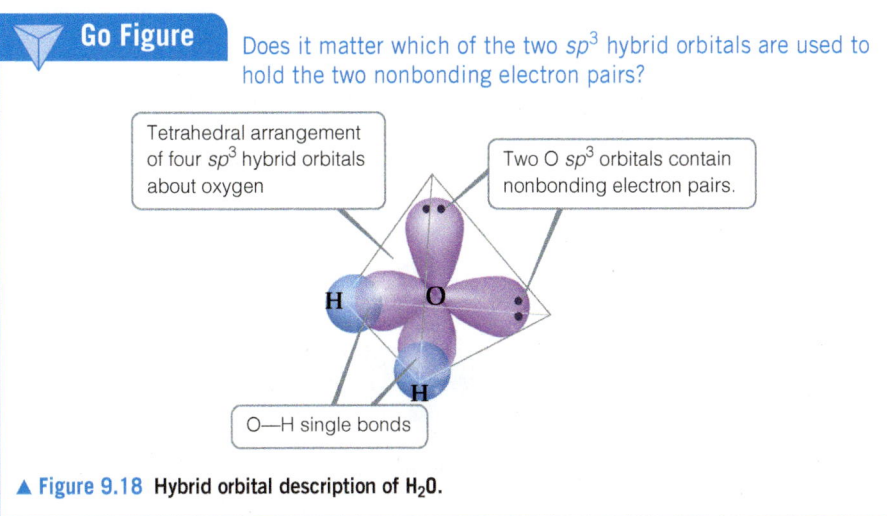

Go Figure Does it matter which of the two sp^3 hybrid orbitals are used to hold the two nonbonding electron pairs?

Tetrahedral arrangement of four sp^3 hybrid orbitals about oxygen

Two O sp^3 orbitals contain nonbonding electron pairs.

O—H single bonds

▲ **Figure 9.18 Hybrid orbital description of H_2O.**

fluorine atoms. Theoretical calculations suggest that the sulfur $3d$ orbitals do *not* participate to a significant degree in the bonding between sulfur and the six fluorine atoms, and that it would not be valid to describe the bonding in SF_6 in terms of six hybrid orbitals. The more detailed bonding model needed to discuss the bonding in SF_6 and other hypervalent molecules requires a treatment beyond the scope of a general chemistry text. Fortunately, the VSEPR model, which explains the geometrical properties of such molecules in terms of electrostatic repulsions, does a good job of predicting their geometries.

This discussion reminds us that models in science are not reality but rather they are our attempts to describe aspects of reality that we have been able to measure, such as bond distances, bond energies, molecular geometries, and so on. A model may work well up to a certain point but not beyond it, as is the case for hybrid orbitals. The hybrid orbital model for period 2 elements has proven very useful and is an essential part of any modern discussion of bonding and molecular geometry in organic chemistry. When it comes to molecules such as SF_6, however, we encounter the limitations of the model.

Hybrid Orbital Summary

Overall, hybrid orbitals provide a convenient model for using valence-bond theory to describe covalent bonds in molecules that have an octet or less of electrons around the central atom and in which the molecular geometry conforms to the electron-domain geometry predicted by the VSEPR model. While the concept of hybrid orbitals has limited predictive value, *when we know the electron-domain geometry, we can employ hybridization to describe the atomic orbitals used by the central atom in bonding.*

How to Describe the Hybrid Orbitals Used by an Atom in Bonding

1. Draw the *Lewis structure* for the molecule or ion.

2. Use the VSEPR model to determine the electron-domain geometry around the central atom.

3. Specify the *hybrid orbitals* needed to accommodate the electron pairs based on their geometric arrangement (Table 9.5).

These steps are illustrated in **Figure 9.19**, which shows how the hybridization at N in NH_3 is determined.

TABLE 9.5 Geometric Arrangements Characteristic of Hybrid Orbital Sets

Atomic Orbital Set	Hybrid Orbital Set	Geometry	Examples
s,p	Two sp	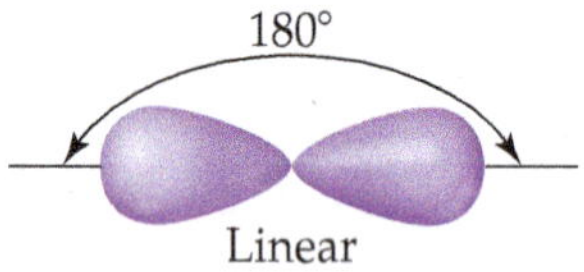 180° / Linear	BeF_2, $HgCl_2$
s,p,p	Three sp^2	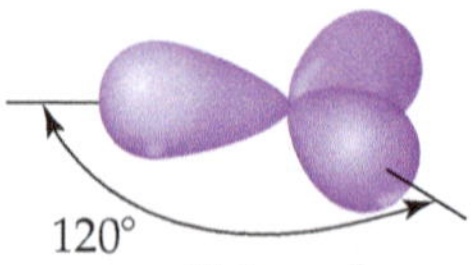 120° / Trigonal planar	BF_3, SO_3
s,p,p,p	Four sp^3	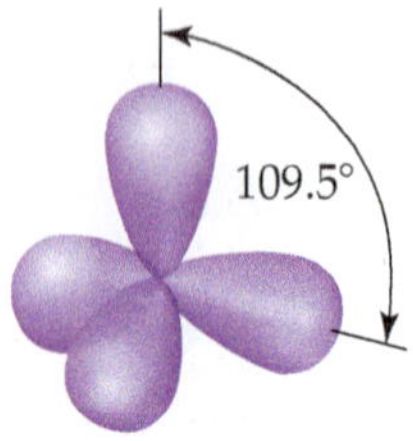 109.5° / Tetrahedral	CH_4, NH_3, H_2O, NH_4^+

Go Figure How would we modify the figure if we were looking at PH_3 rather than NH_3?

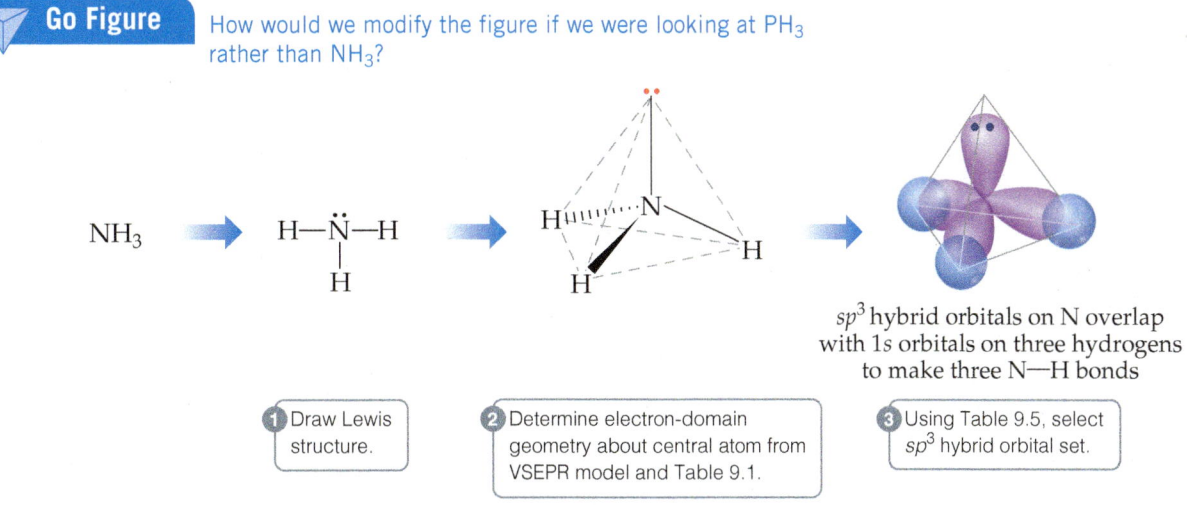

▲ **Figure 9.19 Hybrid orbital description of bonding in NH_3.** Note the similarity to Figure 9.6. Here we focus on the hybrid orbitals used to make bonds and to hold nonbonding electron pairs.

Sample Exercise 9.5

Describing the Hybridization of a Central Atom

Describe the orbital hybridization around the central atom in NH_2^-.

SOLUTION

Analyze We are given the chemical formula for a polyatomic anion and asked to describe the type of hybrid orbitals surrounding the central atom.

Plan To determine the central atom hybrid orbitals, we must know the electron-domain geometry around the atom. Thus, we draw the Lewis structure to determine the number of electron domains around the central atom. The hybridization conforms to the number and geometry of electron domains around the central atom as predicted by the VSEPR model.

Solve The Lewis structure is

$$\left[H \overset{..}{\underset{..}{N}} H \right]^-$$

Because there are four electron domains around N, the electron-domain geometry is tetrahedral. The hybridization that gives a tetrahedral electron-domain geometry is sp^3 (Table 9.5). Two of the sp^3 hybrid orbitals contain nonbonding pairs of electrons, and the other two are used to make bonds with the hydrogen atoms.

▶ **Practice Exercise**
Predict the electron-domain geometry and hybridization of the central atom in SO_3^{2-}.

Self-Assessment Exercises

SAE 9.11 Which of the following statements about the formation of hybrid orbitals is *false*? (**a**) Hybrid orbitals are a means to explain the formation of bonds consistent with the molecular shapes predicted by the VSEPR model. (**b**) The shapes of hybrid orbitals are different from the shapes of atomic orbitals. (**c**) Hybrid orbitals are created by mixing an orbital on one atom with an orbital on a different atom. (**d**) The formation of hybrid orbitals may require the promotion of an electron from an s orbital to a p orbital.

SAE 9.12 Fill in the blanks in the following statement: The formation of sp hybrid orbitals involves the mixing of an s orbital with ____ p orbital(s) to create ____ hybrids. The angle between the large lobes of the hybrids is ____ degrees. (**a**) one, two, 180 (**b**) one, two, 90 (**c**) two, three, 180 (**d**) two, three, 120 (**e**) three, four, 109.5

SAE 9.13 Which of the following statements about hybridization is or are *true*?

    (**i**) When an atom undergoes sp hybridization, there is one unhybridized p orbital remaining on the atom.
    (**ii**) Under sp^2 hybridization, the large lobes point to the vertices of an equilateral triangle.
    (**iii**) The angle between the large lobes of sp^3 hybrids is 109.5°.

(**a**) Only one of the statements is true. (**b**) Statements i and ii are true. (**c**) Statements i and iii are true. (**d**) Statements ii and iii are true. (**e**) All three statements are true.

SAE 9.14 How many of the following AB_n molecules and ions exhibit sp^3 hybridization at the central atom: H_2O, CH_3^+, BF_3, PCl_3, NO_3^-? (**a**) 0 (**b**) 1 (**c**) 2 (**d**) 3 (**e**) 4

Learning Objectives

When you finish Section 9.6, you should be able to:

▶ Define and contrast σ (sigma) and π (pi) bonds.

▶ For each σ and π bond in a molecule, identify the hybridized and/or unhybridized atomic orbitals from which it is formed.

▶ Determine whether a molecule or ion exhibits delocalized π bonding and, if so, how many electrons are in the π system.

9.6 | Multiple Bonds

In the covalent bonds we have considered thus far, the electron density is concentrated along the line connecting the nuclei (the *internuclear axis*). The line joining the two nuclei passes through the middle of the overlap region, forming a type of covalent bond called a **sigma (σ) bond**. Examples of sigma bond formation include:

- the overlap of two s orbitals; one from each H in H_2 [Figure 9.12(a)],
- the overlap of an H s orbital and a Cl p orbital in HCl [Figure 9.12(b)],
- the overlap of two p orbitals; one from each Cl in Cl_2 [Figure 9.12(c)],
- the overlap of an F p orbital and a Be sp hybrid orbital in BeF_2 (Figure 9.15),
- the overlap of a H s orbital and a N sp^3 hybrid orbital in NH_3 (Figure 9.19).

To describe multiple bonding, we must consider a second kind of bond—one that results from the overlap between two p orbitals oriented perpendicularly to the internuclear axis (**Figure 9.20**). The sideways overlap of p orbitals produces what is called a **pi (π) bond**. A π bond is one in which the overlap regions lie above and below the internuclear axis. Unlike a σ bond, the electron density in a π bond is not concentrated on the internuclear axis. Because the two p orbitals that overlap sideways in a π bond are not pointed directly at one another, they do not overlap as strongly as do the p orbitals in a σ bond. As a result, π bonds are generally weaker than σ bonds.

In almost all cases, single bonds are σ bonds. A double bond consists of one σ bond and one π bond, and a triple bond consists of one σ bond and two π bonds:

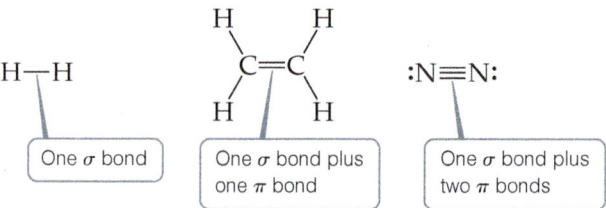

| One σ bond | One σ bond plus one π bond | One σ bond plus two π bonds |

Ethylene (C_2H_4), for example, has a C=C double bond (**Figure 9.21**). The three bond angles about each carbon are all nearly 120°, suggesting that each carbon atom uses sp^2 hybrid orbitals (Figure 9.16) to form σ bonds with the other carbon and with two hydrogens. Each carbon atom also has an *unhybridized 2p* orbital, which is directed perpendicular to the plane that contains the three sp^2 hybrid orbitals. Let's analyze more closely how the orbitals involved can be used to build the bonds of the ethylene molecule.

> **Go Figure** How many bonds are represented in each of the two parts of the figure?

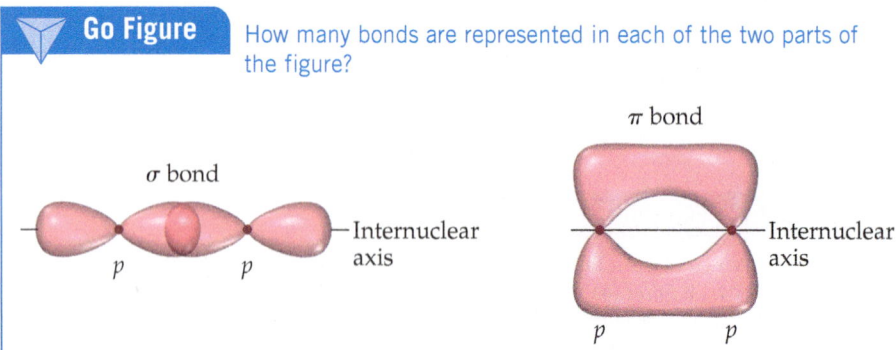

▲ **Figure 9.20 Comparison of σ and π bonds.** Note that the *two* regions of overlap in the π bond—one above and one below the internuclear axis—constitute a single π bond.

Figure 9.22 shows how we can first envision forming the C—C σ bond in ethylene by the overlap of two sp^2 hybrid orbitals, one on each carbon atom. Two electrons are used in forming the C—C σ bond. Next, the C—H σ bonds are formed by overlap of the remaining sp^2 hybrid orbitals on the C atoms with the $1s$ orbitals on each H atom. We use eight more electrons to form these four C—H bonds. Thus, 10 of the 12 valence electrons in the C_2H_4 molecule are used to form five σ bonds.

▲ **Figure 9.21 Trigonal-planar molecular geometry of ethylene.** The double bond is made up of one C—C σ bond and one C—C π bond.

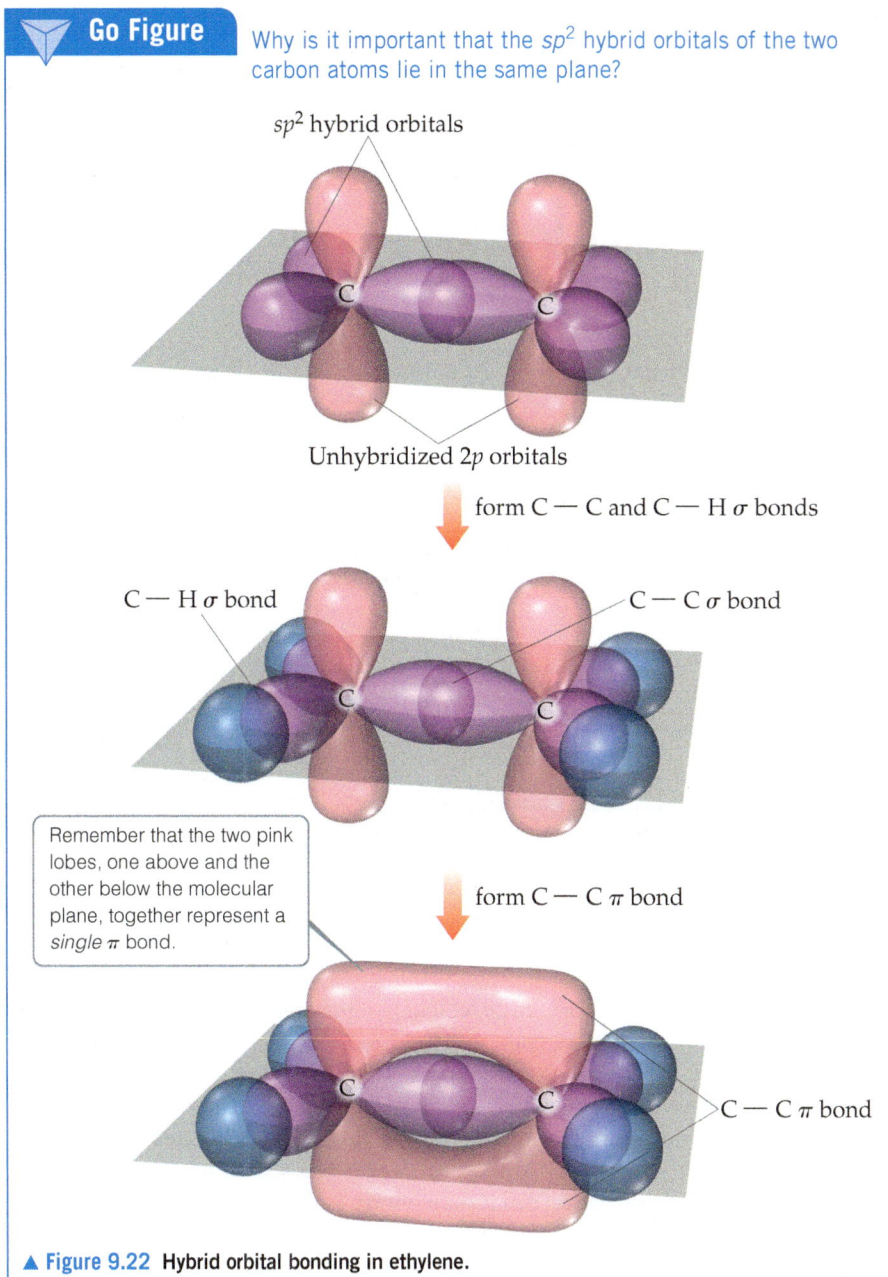

Go Figure

Why is it important that the sp^2 hybrid orbitals of the two carbon atoms lie in the same plane?

sp^2 hybrid orbitals

Unhybridized $2p$ orbitals

form C — C and C — H σ bonds

C — H σ bond

C — C σ bond

Remember that the two pink lobes, one above and the other below the molecular plane, together represent a *single* π bond.

form C — C π bond

C — C π bond

▲ **Figure 9.22 Hybrid orbital bonding in ethylene.**

The remaining two valence electrons reside in the unhybridized $2p$ orbitals, one electron on each carbon. These two orbitals can overlap sideways with each other to form a π bond, as shown in Figure 9.22. Thus, the C=C double bond in ethylene consists of one σ bond and one π bond.

Although we cannot experimentally observe a π bond directly (all we can observe are the positions of the atoms), the structure of ethylene provides strong support for its presence. First, the C—C bond length in ethylene (1.34 Å) is much shorter than in compounds with C—C single bonds (1.54 Å), consistent with the presence of a stronger C=C double bond. (Section 8.8) Second, all six atoms in C_2H_4 lie in the same plane. The p orbitals on each C atom that make up the π bond can achieve a good overlap only when the two CH_2 fragments lie in the same plane. Because π bonds require that portions of a molecule be planar, they can introduce rigidity into molecules.

Triple bonds can also be explained using hybrid orbitals. Acetylene (C_2H_2), for example, is a linear molecule containing a triple bond: H—C≡C—H. The linear geometry suggests that each carbon atom uses sp hybrid orbitals to form σ bonds with the other carbon and one hydrogen. Because only one p orbital is used in making sp hybrids on an atom,

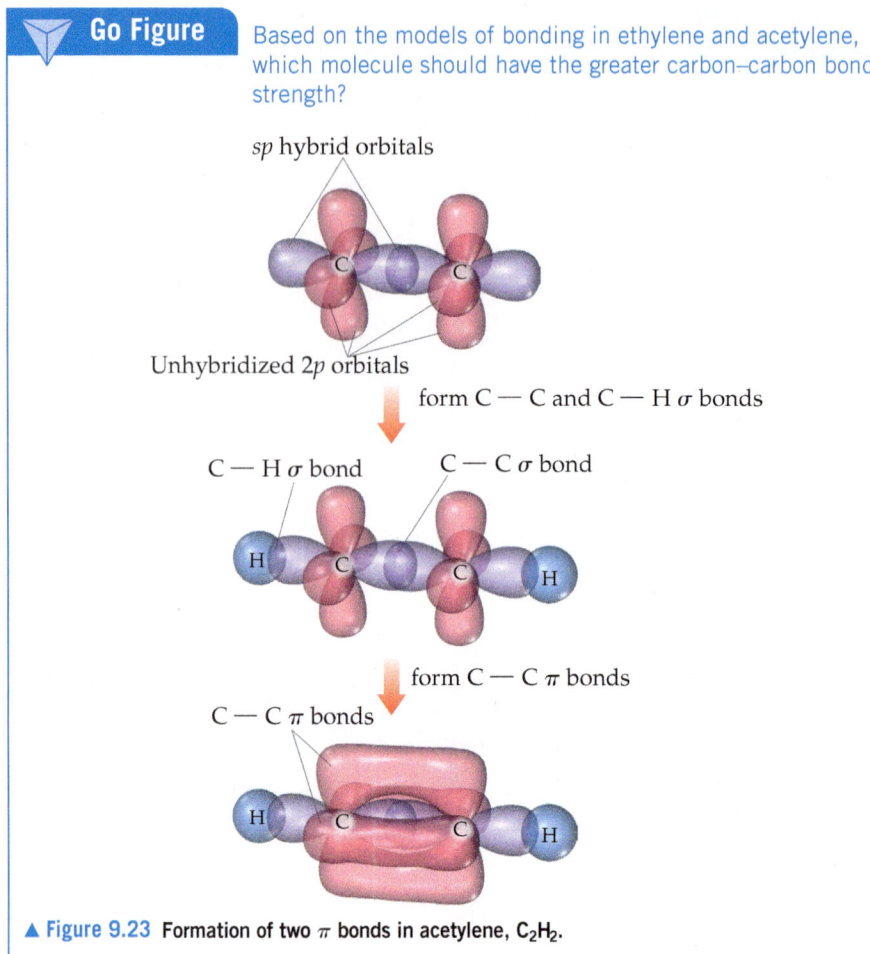

Go Figure Based on the models of bonding in ethylene and acetylene, which molecule should have the greater carbon–carbon bond strength?

sp hybrid orbitals

Unhybridized 2*p* orbitals

form C — C and C — H σ bonds

C — H σ bond   C — C σ bond

form C — C π bonds

C — C π bonds

▲ Figure 9.23 **Formation of two π bonds in acetylene, C₂H₂.**

each carbon atom has two unhybridized 2*p* orbitals at right angles to each other and to the axis of the *sp* hybrid set (**Figure 9.23**). These unhybridized *p* orbitals overlap to form a pair of π bonds. The triple bond in acetylene therefore consists of one σ bond and two π bonds.

Although it is possible to make π bonds from *d* orbitals, the only π bonds that we consider in this chapter are those formed by the overlap of *p* orbitals. These π bonds can form only if unhybridized *p* orbitals are present on the bonded atoms. Therefore, only atoms having *sp* or *sp²* hybridization can form π bonds. Further, double and triple bonds (and hence π bonds) are more common in molecules made up of period 2 atoms, especially C, N, and O. Larger atoms, such as S, P, and Si, form π bonds less readily.

Sample Exercise 9.6
Describing σ and π Bonds in a Molecule

Formaldehyde, H₂CO, has the Lewis structure

$$\begin{array}{c} H \\ \diagdown \\ C = \ddot{O}\!: \\ \diagup \\ H \end{array}$$

Describe how the bonds in formaldehyde are formed in terms of overlaps of hybrid and unhybridized orbitals.

SOLUTION

Analyze We are asked to describe the bonding in formaldehyde in terms of hybrid orbitals.

Plan Single bonds are σ bonds, whereas double bonds consist of one σ bond and one π bond. The ways in which these bonds form can be deduced from the molecular geometry, which we predict using the VSEPR model.

Solve The C atom has three electron domains around it, which suggests a trigonal-planar geometry with bond angles of about 120°. This geometry implies *sp²* hybrid orbitals on C (Table 9.5). These hybrids are used to make the two C—H and one C—O σ bonds to C. There remains an unhybridized 2*p* orbital on carbon, perpendicular to the plane of the three *sp²* hybrids.

The O atom also has three electron domains around it, and so we assume it has sp^2 hybridization as well. One of these hybrid orbitals participates in the C—O σ bond, while the other two hold the two nonbonding electron pairs of the O atom. Like the C atom, therefore, the O atom has a p orbital that is perpendicular to the plane of the molecule. The two p orbitals overlap to form a C—O π bond (**Figure 9.24**).

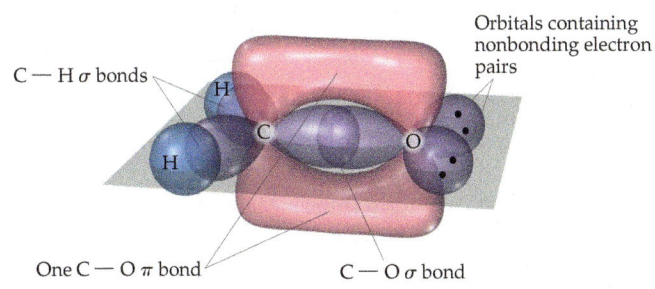

C—H σ bonds

H

Orbitals containing nonbonding electron pairs

One C—O π bond

C—O σ bond

▲ **Figure 9.24** Formation of σ and π bonds in formaldehyde, H_2CO.

► **Practice Exercise**
(a) Predict the bond angles around each carbon atom in acetonitrile:

$$H—C—C\equiv N:$$

with H atoms above, below, and to the left of the first carbon.

(b) Describe the hybridization at each carbon atom.
(c) Determine the number of σ and π bonds in the molecule.

Resonance Structures, Delocalization, and π Bonding

In the molecules we have discussed thus far in this section, the bonding electrons are *localized*. By this we mean that the σ and π electrons are associated totally with the two atoms that form the bond. In many molecules, however, we cannot adequately describe the bonding as being entirely localized. This situation arises particularly in molecules that have two or more resonance structures involving π bonds.

One molecule that cannot be described with localized π bonds is benzene (C_6H_6), which has two resonance structures: (Section 8.6)

(two hexagon resonance structures with alternating double bonds) ⟷ (hexagon) or (hexagon with inscribed circle)

Benzene has a total of 30 valence electrons. To describe the bonding in benzene using hybrid orbitals, we first choose a hybridization scheme consistent with the geometry of the molecule. Because each carbon is surrounded by three atoms at 120° angles, the appropriate hybrid set is sp^2. Six localized C—C σ bonds and six localized C—H σ bonds are formed from the sp^2 hybrid orbitals, as shown in **Figure 9.25(a)**. Thus, 24 of the valence electrons are used to form the σ bonds in the molecule.

Because the hybridization at each C atom is sp^2, there is one p orbital on each C atom, each oriented perpendicular to the plane of the molecule. The situation is very much like that in ethylene except we now have six p orbitals arranged in a ring [**Figure 9.25(b)**]. The remaining six valence electrons occupy these six p orbitals, one per orbital.

Go Figure   What are the two kinds of σ bonds found in benzene?

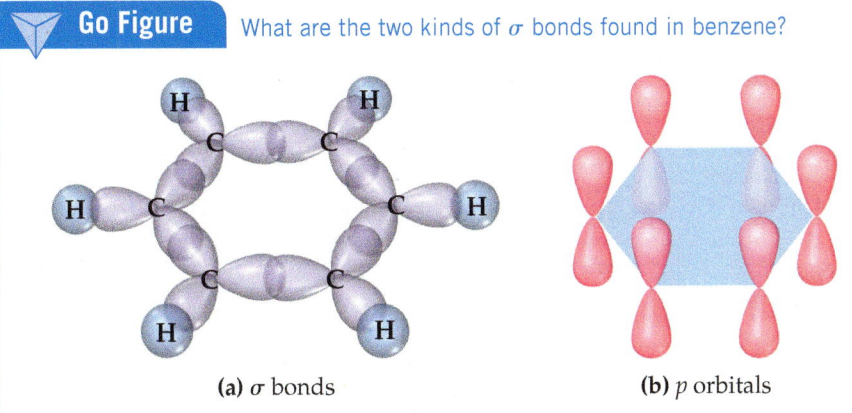

(a) σ bonds

(b) p orbitals

▲ **Figure 9.25** σ and π bond networks in benzene, C_6H_6. (a) The σ bond framework. (b) The π bonds are formed from overlap of the unhybridized $2p$ orbitals on the six carbon atoms.

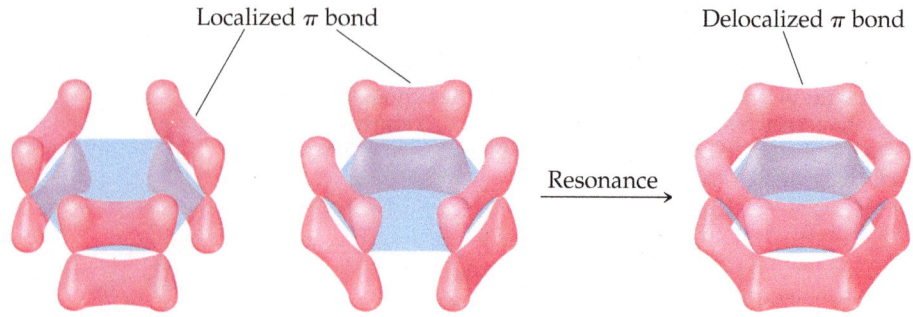

Localized π bond

Delocalized π bond

Resonance

▲ **Figure 9.26** Delocalized π bonds in benzene.

We could envision using the p orbitals to form three localized π bonds. As shown in Figure 9.26, there are two equivalent ways to make these localized bonds, each corresponding to one resonance structure. However, a representation that reflects *both* resonance structures has the six π electrons "smeared out" among all six carbon atoms, as shown on the right in Figure 9.26. Notice how this smeared representation corresponds to the circle-in-a-hexagon drawing we often use to represent benzene. This model leads us to predict that all the carbon–carbon bond lengths will be identical, with a bond length between that of a C—C single bond (1.54 Å) and that of a C=C double bond (1.34 Å). This prediction is consistent with the observed carbon–carbon bond length in benzene (1.40 Å).

Because we cannot describe the π bonds in benzene as individual bonds between neighboring atoms, we say that benzene has a six-electron π system **delocalized** among the six carbon atoms. Delocalization of the electrons in its π bonds gives benzene a special stability. Electron delocalization in π bonds is also responsible for the color of many organic molecules. A final important point to remember about delocalized π bonds is the constraint they place on the geometry of a molecule. For optimal overlap of the p orbitals, all the atoms involved in a delocalized π bonding network should lie in the same plane. This restriction imparts a certain rigidity to the molecule that is absent in molecules containing only σ bonds (see the box "Chemistry and Life: The Chemistry of Vision").

If you take a course in organic chemistry, you will see many examples of how electron delocalization influences the properties of organic molecules.

Sample Exercise 9.7

Delocalized Bonding

Describe the bonding in the nitrate ion, NO_3^-. Does this ion have delocalized π bonds?

SOLUTION

Analyze Given the chemical formula for a polyatomic anion, we are asked to describe the bonding and determine whether the ion has delocalized π bonds.

Plan Our first step is to draw Lewis structures. Multiple resonance structures involving the placement of the double bonds in different locations would suggest that the π component of the double bonds is delocalized.

Solve In Section 8.6 we saw that NO_3^- has three resonance structures:

$$\left[\overset{\displaystyle :\!\ddot{O}\!:}{\underset{:\ddot{O}: \qquad :\ddot{O}:}{N}} \right]^- \longleftrightarrow \left[\overset{\displaystyle :\!\ddot{O}\!:}{\underset{:\ddot{O}: \qquad :\ddot{O}:}{N}} \right]^- \longleftrightarrow \left[\overset{\displaystyle :\!\ddot{O}\!:}{\underset{:\ddot{O}: \qquad :\ddot{O}:}{N}} \right]^-$$

As we have discussed previously, the three N—O bonds in the ion are equivalent and are between a normal N—O single bond and a N=O double bond in length. (Section 8.6) In each resonance structure, the electron-domain geometry at nitrogen is trigonal planar, which implies sp^2 hybridization of the N atom. It is helpful when considering delocalized π bonding to consider atoms with lone pairs that are bonded to the central atom to be sp^2 hybridized as well. Thus, we can envision that each of the O atoms in the anion has three sp^2 hybrid orbitals in the plane of the ion. Each of the four atoms has an unhybridized p orbital oriented perpendicular to the plane of the ion.

The NO_3^- ion has 24 valence electrons. We can first use the sp^2 hybrid orbitals on the four atoms to construct the three N—O σ bonds. That uses all of the sp^2 hybrids on the N atom and one sp^2 hybrid on each O atom. Each of the two remaining sp^2 hybrids on each O atom is used to hold a nonbonding pair of electrons. Thus, for any of the resonance structures, we have the following arrangement in the plane of the ion:

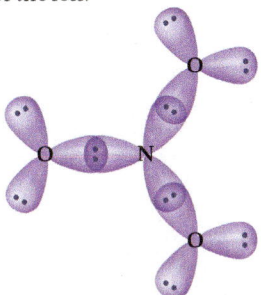

We have accounted for a total of 18 electrons—6 in the three N—O σ bonds, and 12 as nonbonding pairs on the O atoms. The remaining six electrons will reside in the π system of the ion.

The four p orbitals—one on each of the four atoms—are used to build the π system. For any one of the three resonance structures shown, we might imagine a single localized N—O π bond formed by the overlap of the p orbital on N and a p orbital on one of the O atoms. The remaining two O atoms have nonbonding pairs in

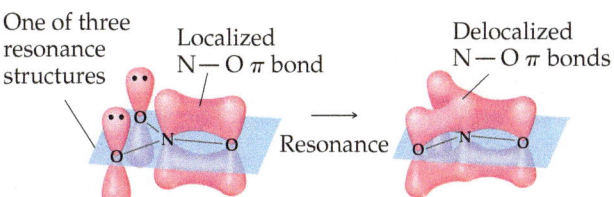

▲ **Figure 9.27** Localized and delocalized representations of the six-electron π system in NO_3^-.

their p orbitals. Thus, for each of the resonance structures, we have the situation shown in **Figure 9.27**. Because each resonance structure contributes equally to the observed structure of NO_3^-, however, we represent the π bonding as delocalized over the three N—O bonds, as shown in the figure. Thus, the NO_3^- ion has a *six-electron π system delocalized among the four atoms in the ion.*

▶ **Practice Exercise**
Which of these species have delocalized π bonding: SO_2, SO_3, SO_3^{2-}, H_2CO, NH_4^+?

General Conclusions about σ and π Bonding

On the basis of the examples we have seen, we can draw a few helpful conclusions for using hybrid orbitals to describe molecular structures:

- Every pair of bonded atoms shares one or more pairs of electrons. Each bond line we draw in a Lewis structure represents two shared electrons. In every σ bond, one pair of electrons is localized in the space between the atoms. The appropriate set of hybrid orbitals used to form the σ bonds between an atom and its neighbors is determined by the observed geometry of the molecule. The correlation between the set of hybrid orbitals and the geometry about an atom is given in Table 9.5.

- Because the electrons in σ bonds are localized in the region between two bonded atoms, they do not make a significant contribution to the bonding between any other two atoms.

- When atoms share more than one pair of electrons, one pair is used to form a σ bond; the additional pairs form π bonds. The centers of charge density in a π bond lie above and below the internuclear axis.

- Molecules can have π systems that extend over more than two bonded atoms. Electrons in extended π systems are said to be "delocalized." We can determine the number of electrons in the π system of a molecule using the procedures we discussed in this section.

CHEMISTRY AND LIFE | The Chemistry of Vision

Vision begins when light is focused by the lens of the eye onto the retina, the layer of cells lining the interior of the eyeball. The retina contains *photoreceptor* cells called rods and cones (**Figure 9.28**). The rods are sensitive to dim light and are used in night vision. The cones are sensitive to colors. The tops of the rods and cones contain a molecule called *rhodopsin,* which consists of a protein, *opsin,* bonded to a reddish purple pigment called *retinal.* Structural changes around a double bond in the retinal portion of the molecule trigger a series of chemical reactions that result in vision.

We know that a double bond between two atoms is stronger than a single bond between those same two atoms (Table 8.3). We are now in a position to appreciate another aspect of double bonds: They introduce rigidity into molecules.

Consider the C—C double bond in ethylene. Imagine rotating one —CH₂ group in ethylene relative to the other —CH₂ group, as

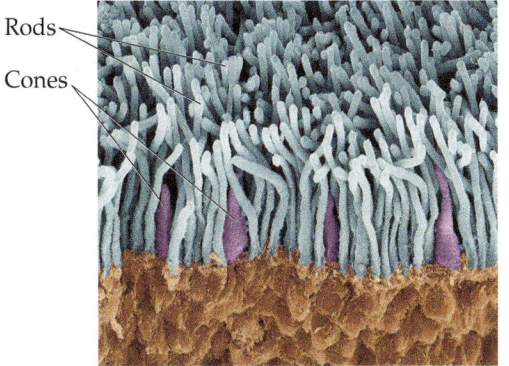

▲ **Figure 9.28** **Inside the eye.** A color-enhanced scanning electron micrograph of the rods and cones in the retina of the human eye.

Continued

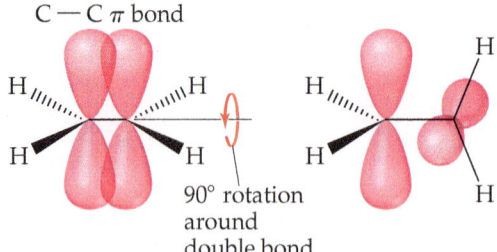

C—C π bond

90° rotation around double bond

▲ **Figure 9.29** **Rotation about the carbon—carbon double bond in ethylene breaks the π bond.**

pictured in **Figure 9.29**. This rotation destroys the overlap of the *p* orbitals, breaking the π bond, a process that requires considerable energy. Thus, the presence of a double bond restricts bond rotation in a molecule. In contrast, molecules can rotate almost freely around the bond axis in single (σ) bonds because this motion has

no effect on the orbital overlap for a σ bond. Rotation allows molecules with single bonds to twist and fold almost as if their atoms were attached by hinges.

Our vision depends on the rigidity of double bonds in retinal. In its normal form, retinal is held rigid by its double bonds. Light entering the eye is absorbed by the retinal portion of rhodopsin, and the energy of that light is used to break the π-bond portion of the double bond shown in red in **Figure 9.30**. Breaking the double bond allows rotation around the bond axis, changing the geometry of the retinal molecule. The retinal then separates from the opsin, triggering the reactions that produce a nerve impulse that the brain interprets as the sensation of vision. It takes as few as five closely spaced molecules reacting in this fashion to produce the sensation of vision. Thus, only five photons of light are necessary to stimulate the eye.

The retinal slowly reverts to its original form and reattaches to the opsin. The slowness of this process helps explain why intense bright light causes temporary blindness. The light causes all the retinal to separate from opsin, leaving no molecules to absorb light.
Related Exercises: 9.109, 9.112, Design an Experiment

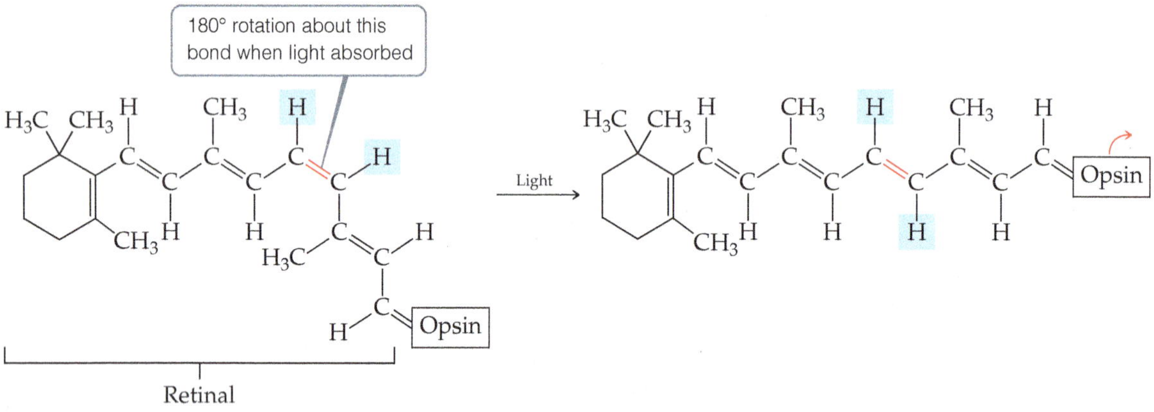

180° rotation about this bond when light absorbed

Light →

Retinal

▲ **Figure 9.30** **The rhodopsin molecule, the chemical basis of vision.** When rhodopsin absorbs visible light, the π component of the double bond shown in red breaks, allowing rotation that produces a change in molecular geometry before the π bond re-forms. The hexagonal part of the structure is a shorthand way of drawing C—C bonds with hydrogen atoms attached—each vertex represents a carbon atom with a total of four bonds.

◢ Self-Assessment Exercises

SAE 9.15 Which of the following statements about σ and π bonds is *false*? (**a**) A σ bond concentrates electron density along the internuclear axis. (**b**) A π bond consists of two lobes on either side of the internuclear axis. (**c**) A triple bond consists of one σ bond and two π bonds. (**d**) π bonds are formed from unhybridized *p* orbitals. (**e**) The formation of a σ bond must involve the overlap of two hybrid orbitals.

SAE 9.16 Acrylonitrile, C_3H_3N, has the Lewis structure shown here. The molecule has _____ σ bonds and _____ π bonds. (**a**) six and two (**b**) six and three (**c**) three and three (**d**) four and five

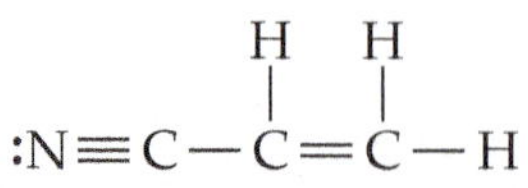

$$:N\equiv C-\overset{\overset{\displaystyle H}{|}}{C}=\overset{\overset{\displaystyle H}{|}}{C}-H$$

SAE 9.17 Which of the following statements about the acrylonitrile molecule, C_3H_3N, is or are true?

    (**i**) The C—C single bond results from overlap between an *sp* hybrid orbital and an sp^2 hybrid orbital.
    (**ii**) There are two unhybridized *p* orbitals on the N atom.
  (**iii**) All of the atoms in the molecule lie in the same plane.

(**a**) Only one of the statements is true. (**b**) Statements i and ii are true. (**c**) Statements i and iii are true. (**d**) Statements ii and iii are true. (**e**) All three statements are true.

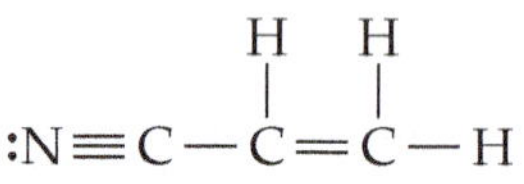

$$:N\equiv C-\overset{\overset{\displaystyle H}{|}}{C}=\overset{\overset{\displaystyle H}{|}}{C}-H$$

SAE 9.18 How many electrons are in the π system of the ozone molecule, O_3? (**a**) 2 (**b**) 4 (**c**) 6 (**d**) 14 (**e**) 18

SAE 9.19 A Lewis structure of the oxalate ion, $C_2O_4{}^{2-}$, is shown in the figure. Which of the following statements about the ion is *false*? (**a**) The ion has multiple equivalent resonance structures. (**b**) There is one unhybridized *p* orbital on each of the C atoms. (**c**) The experimentally observed structure of the ion will show two short and two long C—O bonds. (**d**) The π system of the ion is delocalized among all six atoms.

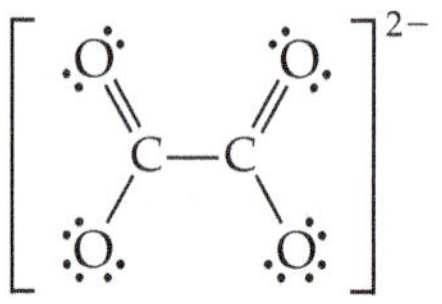

9.7 | Molecular Orbitals

While valence-bond theory helps explain some of the relationships among Lewis structures, atomic orbitals, and molecular geometries, it does not explain all aspects of bonding. It is not successful, for example, in describing the excited states of molecules, which we must understand to explain how molecules absorb light, giving them color.

Some aspects of bonding are better explained by a more sophisticated model called **molecular-orbital theory**. In Chapter 6 we saw that electrons in atoms can be described by wave functions, which we call atomic orbitals. In a similar way, molecular-orbital theory describes the electrons in molecules by using specific wave functions, each of which is called a **molecular orbital (MO)**.

Molecular orbitals have many of the same characteristics as atomic orbitals. For example, an MO can hold a maximum of two electrons (with opposite spins), it has a definite energy, and we can visualize its electron-density distribution by using a contour representation, as we did with atomic orbitals. Unlike atomic orbitals, however, MOs are associated with an entire molecule, not with a single atom.

Molecular Orbitals of the Hydrogen Molecule

We begin our study of MO theory with the hydrogen molecule, H_2. The two $1s$ atomic orbitals (one on each H atom) are used to construct molecular orbitals for H_2. *Whenever two atomic orbitals overlap, two molecular orbitals form*. Thus, the overlap of the $1s$ orbitals of two hydrogen atoms to form H_2 produces two MOs. The first MO, which is shown at the bottom right of **Figure 9.31**, is formed by adding the wave functions for the two $1s$ orbitals. We refer to this as *constructive combination*. The energy of the resulting MO is lower in energy than the two atomic orbitals from which it was made. It is called the **bonding molecular orbital**.

The second MO is formed by what is called *destructive combination*: combining the two atomic orbitals in a way that causes the electron density to be canceled in the central region where the two overlap. The process is discussed more fully in the "Closer Look" box

> ## Learning Objectives
>
> When you finish Section 9.7, you should be able to:
>
> ▶ Describe the characteristics of the bonding and antibonding molecular orbitals of the H_2 molecule.
>
> ▶ Interpret the energy-level diagram for the H_2 molecule and related molecules and molecular ions.

Go Figure What is the probability of finding an electron at the nodal plane in the antibonding orbital?

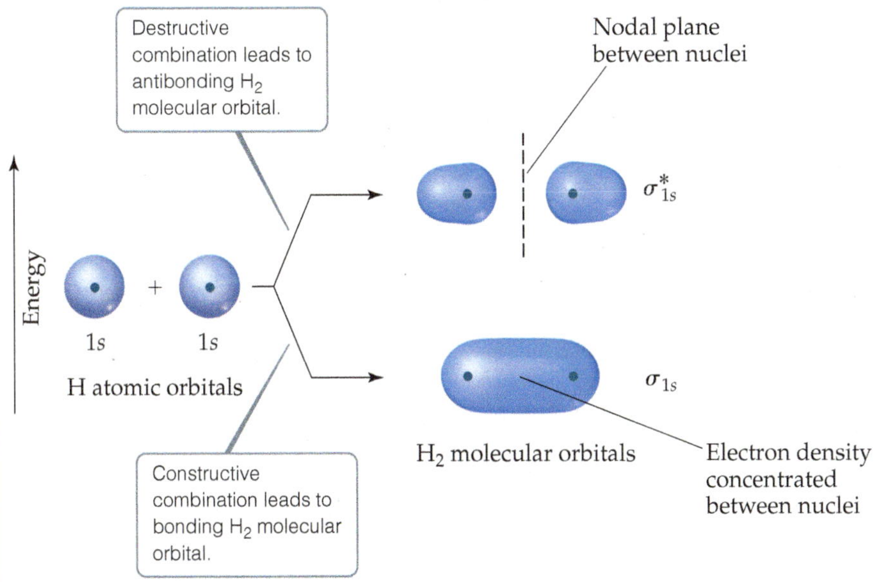

Destructive combination leads to antibonding H_2 molecular orbital.

Nodal plane between nuclei

σ^*_{1s}

Energy

$1s$ $1s$

H atomic orbitals

σ_{1s}

H_2 molecular orbitals

Electron density concentrated between nuclei

Constructive combination leads to bonding H_2 molecular orbital.

▲ **Figure 9.31** **The two molecular orbitals of H_2.** The one that is lower in energy is a bonding MO, whereas the other is an antibonding MO.

later in the chapter. The energy of the resulting MO, referred to as the **antibonding molecular orbital**, is higher than the energy of the atomic orbitals. The antibonding MO of H_2 is shown at the top right in Figure 9.31.

As illustrated in Figure 9.31, in the bonding MO electron density is concentrated in the region between the two nuclei. This sausage-shaped MO results from summing the two atomic orbitals so that the atomic orbital wave functions combine in the region between the two nuclei. Because an electron in this MO is attracted to both nuclei, the electron is more stable (it has lower energy) than it is in the $1s$ atomic orbital of an isolated hydrogen atom. Further, because this bonding MO concentrates electron density between the nuclei, it holds the atoms together in a covalent bond.

By contrast, the antibonding MO does the very opposite of a bonding MO: It actually excludes electron density between the nuclei. Instead of combining in the region between the nuclei, the atomic orbital wave functions cancel each other in this region, leaving the greatest electron density on opposite sides of the two nuclei. Instead of promoting the formation of a bond, an electron in an antibonding MO is actively discouraged from being involved in bonding. Antibonding orbitals invariably have a *plane* in the region between the nuclei where the electron density is zero. This plane is called a **nodal plane** of the MO. (Section 6.6) (The nodal plane is shown as a dashed line in Figure 9.31 and subsequent figures.) An electron in an antibonding MO is repelled from the bonding region and is therefore less stable (it has higher energy) than it is in the $1s$ atomic orbital of a hydrogen atom.

Notice from Figure 9.31 that the electron density in both the bonding MO and the antibonding MO of H_2 is centered about the internuclear axis. MOs of this type are called **sigma (σ) molecular orbitals** (by analogy to σ bonds). The bonding sigma MO of H_2 is labeled σ_{1s}, where the subscript indicates that the MO is formed from two $1s$ orbitals. The antibonding sigma MO of H_2 is labeled σ_{1s}^* (read "sigma-star-one-s"), where the asterisk denotes that the MO is antibonding.

The relative energies of two $1s$ atomic orbitals and the molecular orbitals formed from them are represented by an **energy-level diagram** (also called a **molecular-orbital diagram**). Such diagrams show the interacting atomic orbitals on the left and right and the MOs in the middle, as shown in **Figure 9.32**. Like atomic orbitals, each MO can accommodate two electrons with their spins paired (Pauli exclusion principle). (Section 6.7)

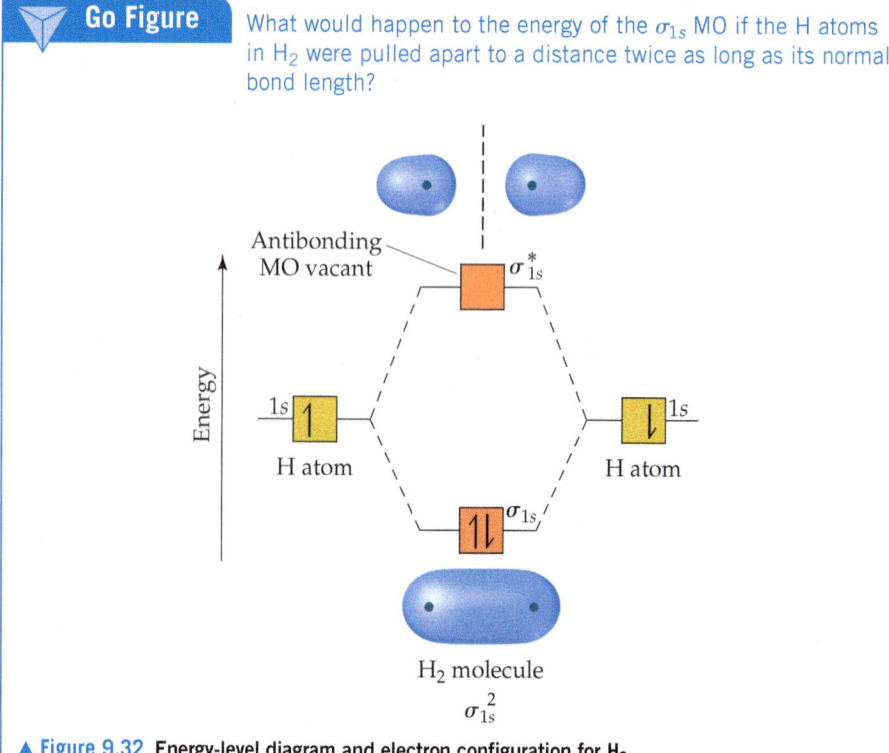

Go Figure

What would happen to the energy of the σ_{1s} MO if the H atoms in H_2 were pulled apart to a distance twice as long as its normal bond length?

▲ **Figure 9.32** Energy-level diagram and electron configuration for H_2.

As the MO diagram for H_2 in Figure 9.32 shows, each H atom has one electron, so there are two electrons in H_2. These two electrons occupy the lower-energy bonding (σ_{1s}) MO, and their spins are paired. Electrons occupying a bonding molecular orbital are called *bonding electrons*. Because the σ_{1s} MO is lower in energy than the H $1s$ atomic orbitals, the H_2 molecule is more stable than the two separate H atoms. By analogy with atomic electron configurations, the electron configurations for molecules can be written with superscripts to indicate electron occupancy. In the MO model, we therefore write the electron configuration for H_2 as σ_{1s}^2.

In constructing hybrid orbitals, which we discussed earlier in this chapter, and in constructing molecular orbitals, we are mathematically mixing atomic orbitals to create new orbitals. It is important to recognize that while both hybrids and MOs are models used to discuss bonding, they are distinctly different: In creating hybrids, we mix atomic orbitals that reside on the *same* atom. In creating MOs, we mix atomic orbitals from *different* atoms.

Bond Order

In molecular-orbital theory, the stability of a covalent bond is related to its **bond order**, defined as half the difference between the number of bonding electrons and the number of antibonding electrons:

$$\frac{\text{Bond}}{\text{order}} = \tfrac{1}{2}\,(\text{number of bonding electrons} - \text{number of antibonding electrons}) \quad [9.1]$$

We take half the difference because we are used to thinking of bonds as pairs of electrons. *A bond order of 1 represents a single bond, a bond order of 2 represents a double bond, and a bond order of 3 represents a triple bond*. Because MO theory can also treat molecules containing an odd number of electrons, bond orders of $1/2$, $3/2$, or $5/2$ are possible.

According to Figure 9.32, H_2 has two bonding electrons and zero antibonding electrons, so Equation 9.1 gives it a bond order of 1: $\tfrac{1}{2}(2 - 0) = 1$. How does this value compare to the bond order of He_2?

Figure 9.33 shows the energy-level diagram for the hypothetical He_2 molecule, which requires four electrons to fill its molecular orbitals. Because only two electrons can go in the σ_{1s} MO, the other two electrons must go in the σ_{1s}^* MO. The electron configuration of He_2 is thus $\sigma_{1s}^2\sigma_{1s}^{*2}$. The energy decrease realized in going from He atomic orbitals to the He bonding MO is offset by the energy increase realized in going from the atomic orbitals to the He antibonding MO.* Because He_2 has two bonding electrons and two antibonding electrons, it has a bond order of 0: $\tfrac{1}{2}(2 - 2) = 0$. A bond order of 0 means that no bond exists. Molecular-orbital theory correctly predicts that hydrogen forms diatomic molecules but helium does not.

◀ **Figure 9.33 Energy-level diagram and electron configuration for He_2.**

Antibonding MO occupied

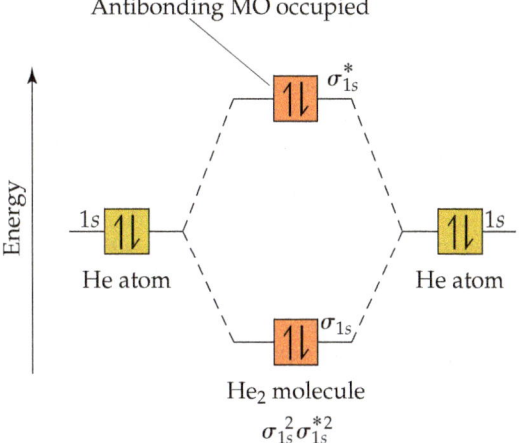

He atom   He atom

He$_2$ molecule
$\sigma_{1s}^2\sigma_{1s}^{*2}$

*Antibonding MOs are slightly more energetically unfavorable than bonding MOs are energetically favorable. Thus, whenever there is an equal number of electrons in bonding and antibonding orbitals, the energy of the molecule is slightly higher than that for the separated atoms. As a result, no bond is formed.

Sample Exercise 9.8
Bond Order

What is the bond order of the He_2^+ ion? Would you expect this ion to be stable relative to the separated He atom and He^+ ion?

SOLUTION

Analyze We will determine the bond order for the He_2^+ ion and use it to predict whether the ion is stable.

Plan To determine the bond order, we must determine the number of electrons in the molecule and how these electrons populate the available MOs. The valence electrons of He are in the $1s$ orbital, and the $1s$ orbitals combine to give an MO diagram like that for H_2 or He_2 (Figure 9.33). If the bond order is greater than 0, we expect a bond to exist, and the ion is stable.

Solve The energy-level diagram for the He_2^+ ion is shown in **Figure 9.34**. This ion has three electrons. Two are placed in the

bonding orbital and the third in the antibonding orbital. Thus, the bond order is

$$\text{Bond order} = \tfrac{1}{2}(2-1) = \tfrac{1}{2}$$

Because the bond order is greater than 0, we predict the He_2^+ ion to be stable relative to the separated He and He_2^+. Formation of He_2^+ in the gas phase has been demonstrated in laboratory experiments.

▶ **Practice Exercise**
What are the electron configuration and the bond order of the H_2^- ion?

▼ **Go Figure**

What is the electron configuration of the He_2^+ ion?

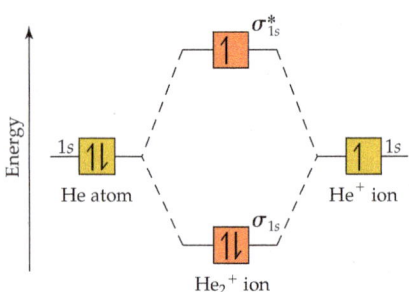

▲ **Figure 9.34** Energy-level diagram for the He_2^+ ion.

 Self-Assessment Exercises

SAE 9.20 Which of the following statements about molecular orbitals is *false*? **(a)** A molecular orbital can extend over multiple atoms. **(b)** Molecular-orbital theory describes the distribution of electrons in molecules using wave functions. **(c)** The energy of an antibonding molecular orbital is higher than the energy of the atomic orbitals used to make it. **(d)** A single molecular orbital can be used to hold all of the electrons in any molecule. **(e)** A bonding molecular orbital concentrates electron density between the two nuclei of the atoms involved.

SAE 9.21 In the energy-level diagram for the H_2^- ion, there is/are _____ electron(s) in the σ_{1s}^* molecular orbital, and the bond order of the ion is _____. **(a)** 0, 1 **(b)** 1, ½ **(c)** 1, 1 **(d)** 2, 0 **(e)** 3, ½

A CLOSER LOOK   **Phases in Atomic and Molecular Orbitals**

Our discussion of atomic orbitals in Chapter 6 and molecular orbitals here in Chapter 9 highlights some of the most important applications of quantum mechanics in chemistry. In the quantum-mechanical treatment of electrons in atoms and molecules, we are mainly interested in determining two characteristics of the electrons—their energies and their distribution in space. Recall that solving Schrödinger's wave equation yields the electron's energy, E, and wave function, ψ, but that ψ does not have a direct physical meaning. (Section 6.5) The contour representations of atomic and molecular orbitals we have presented thus far are based on ψ^2 (the *probability density*), which gives the probability of finding the electron at a given point in space.

Because probability densities are squares of functions, their values must be nonnegative (zero or positive) at all points in space. However, the functions themselves can have negative values. The situation is like that of the sine function plotted in **Figure 9.35**. In the top graph, the sine function is negative for x between 0 and $-\pi$ and positive for x between 0 and $+\pi$. We say that the *phase* of the sine function is negative between 0 and $-\pi$ and positive between 0 and $+\pi$. If we square the sine function (bottom graph), we get two

peaks that are symmetrical about the origin. Both peaks are positive because squaring a negative number produces a positive number. In other words, *we lose the phase information of the function upon squaring it.*

Like the sine function, the more complicated wave functions for atomic orbitals can also have phases. Consider, for example, the representations of the $1s$ orbital in **Figure 9.36**. Note that here we plot this orbital a bit differently from what is shown in Section 6.6. The origin is the point where the nucleus resides, and the wave function for the $1s$ orbital extends from the origin out into space. The plot shows the value of ψ for a slice taken along the z-axis. Below the plot is a contour representation of the $1s$ orbital. Notice that the value of the $1s$ wave function is always a positive number (we show positive values in red in Figure 9.36). Thus, it has only one phase. Notice also that the wave function approaches zero only at a long distance from the nucleus. It therefore has no nodes, as we saw in Figure 6.19.

In the graph for the $2p_z$ orbital in Figure 9.36, the wave function changes sign when it passes through $z = 0$. Notice that the two halves of the wave have the same shape except that one has positive

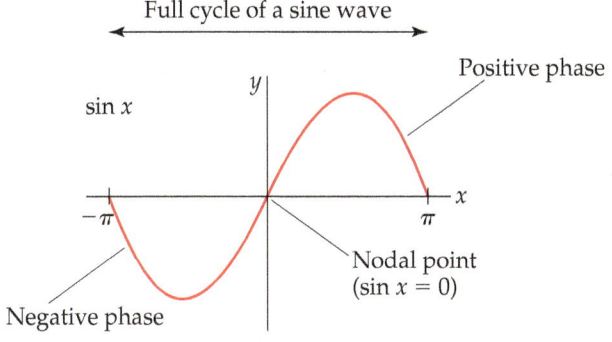

Full cycle of a sine wave

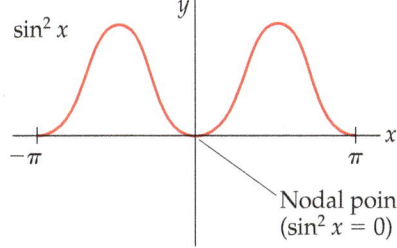

▲ **Figure 9.35 Graphs for a sine function and the same function squared.**

(red) values and the other negative (blue) values. Analogously to the sine function, the wave function changes phase when it passes through the origin. Mathematically, the $2p_z$ wave function is equal to zero whenever $z = 0$. This corresponds to any point on the xy plane, so we say that the xy plane is a *nodal plane* of the $2p_z$ orbital. The wave function for a p orbital is much like a sine function because

it has two equal parts that have opposite phases. Figure 9.36 gives a typical representation used by chemists of the wave function for a p_z orbital.* The red and blue lobes indicate the different phases of the orbital. (Note: The colors do *not* represent charge, as they did in the plots in Figures 9.10 and 9.11.) As with the sine function, the origin is a node.

The third graph in Figure 9.36 shows that when we square the wave function of the $2p_z$ orbital, we get two peaks that are symmetrical about the origin. Both peaks are positive because squaring a negative number produces a positive number. Thus, *we lose the phase information of the function upon squaring it* just as we did for the sine function. When we square the wave function for the p_z orbital, we get the probability density for the orbital, which is given as a contour representation in Figure 9.36. This is what we saw in the earlier presentation of p orbitals. (Section 6.6) For this squared wave function, both lobes have the same phase and therefore the same sign. We use this representation throughout most of this book because it has a simple physical interpretation: The square of the wave function at any point in space represents the electron density at that point.

The lobes of the wave functions for the d orbitals also have different phases. For example, the wave function for a d_{xy} orbital has four lobes, with the phase of each lobe opposite the phase of its nearest neighbors (**Figure 9.37**). The wave functions for the other d orbitals likewise have lobes in which the phase in one lobe is opposite that in an adjacent lobe.

Why do we need to consider the complexity introduced by considering the phase of the wave function? While it is true that the phase is not necessary to visualize the shape of an atomic orbital in an isolated atom, it does become important when we consider the

---

*The mathematical development of this three-dimensional function (and its square) is beyond the scope of this book, and, as is typically done by chemists, we have used lobes that are the same shape as in Figure 6.21.

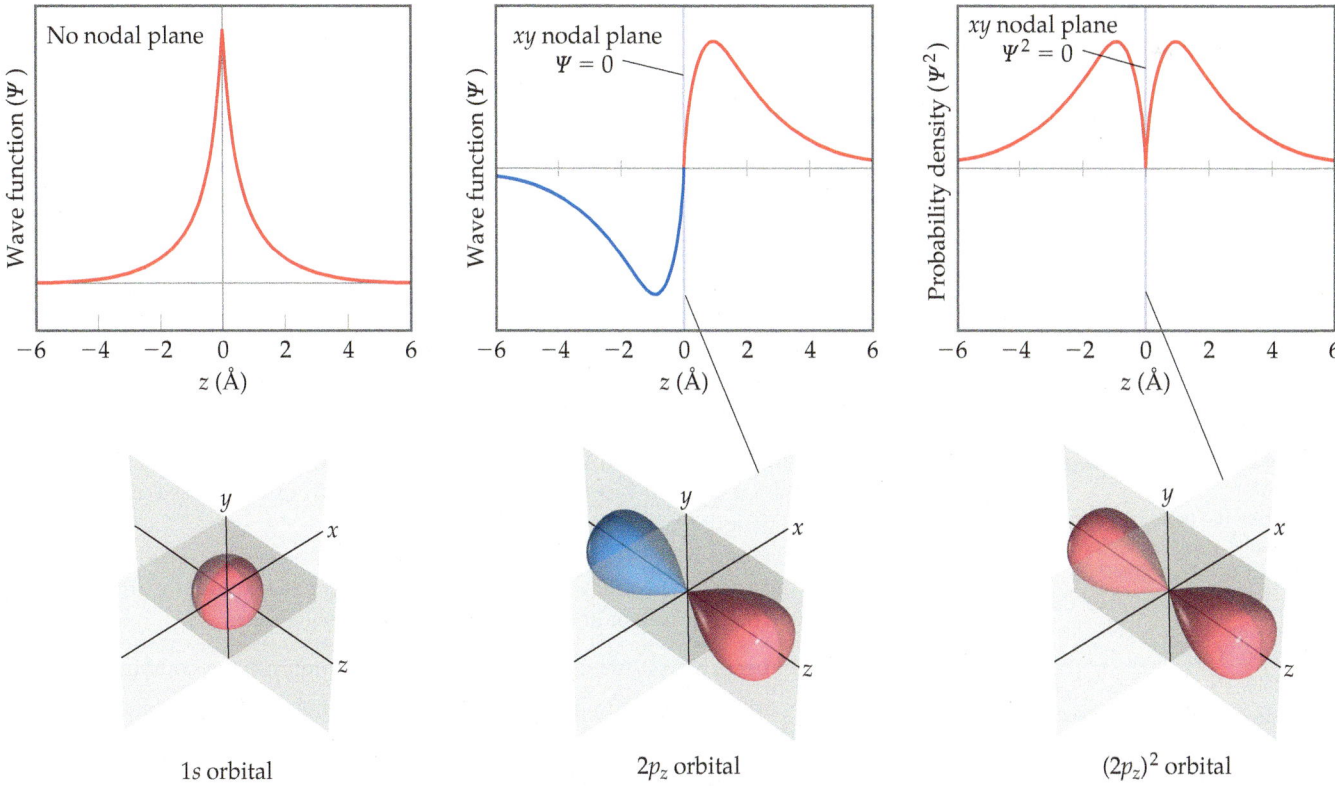

▲ **Figure 9.36 Phases in wave functions of *s* and *p* atomic orbitals.** Red shading means a positive value for the wave function, whereas blue shading means a negative value.

Continued

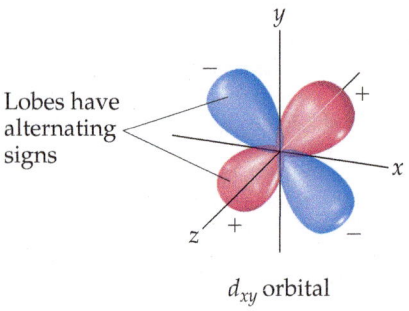

▲ **Figure 9.37** **Phases in *d* orbitals.**

overlap of orbitals in molecular-orbital theory. Let's use the sine function as an example again. If you add two sine functions having the same phase, they add *constructively,* resulting in increased amplitude:

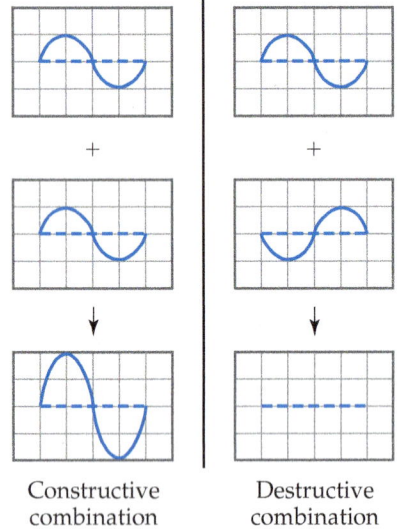

Constructive   Destructive
combination   combination

but if you add two sine functions having opposite phases, they add *destructively* and cancel each other.

The idea of constructive and destructive interactions of wave functions is key to understanding the origin of bonding and anti-bonding molecular orbitals. For example, the wave function of the σ_{1s}

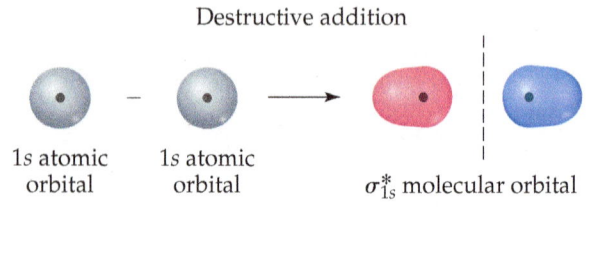

▲ **Figure 9.38** **Molecular orbitals from atomic orbital wave functions.**

MO of H_2 is generated by adding the wave function for the $1s$ orbital on one atom to the wave function for the $1s$ orbital on the other atom, with both orbitals having the same phase. The atomic wave functions overlap *constructively* in this case to increase the electron density between the two atoms (**Figure 9.38**). The wave function of the σ_{1s}^* MO of H_2 is generated by subtracting the wave function for a $1s$ orbital on one atom from the wave function for a $1s$ orbital on the other atom. The result is that the atomic orbital wave functions overlap *destructively* to create a region of zero electron density between the two atoms—a node. Notice the similarity between this figure and Figure 9.32. In Figure 9.38, we use red and blue shading to denote positive and negative phases in the H atomic orbitals. However, chemists may alternatively draw contour representations in different colors, or with one phase shaded and one unshaded, to denote the two phases.

When we square the wave function of the σ_{1s}^* MO, we get the electron density representation which we saw earlier, in Figure 9.32. Notice once again that we lose the phase information by squaring the wave function when we look at the electron density.

The wave functions of atomic and molecular orbitals are used by chemists to understand many aspects of chemical bonding, spectroscopy, and reactivity. If you take a course in organic chemistry, you will probably see orbitals drawn to show the phases as in this box.

Related Exercises: 9.13, 9.14, 9.106

9.8 | Molecular-Orbital Description of Period 2 Diatomic Molecules

 Learning Objectives

When you finish Section 9.8, you should be able to:

▶ Describe the principles for using molecular orbitals for atoms with more than $1s$ atomic orbitals.

▶ Construct the energy-level diagram for molecules in which both the $1s$ and $2s$ orbitals contribute to the molecular orbitals.

▶ Construct molecular orbitals from $2p$ atomic orbitals.

▶ Use MO theory to determine the electron configurations and properties for period 2 diatomic molecules.

In considering the MO description of diatomic molecules other than H_2, we initially restrict our discussion to *homonuclear* diatomic molecules (those composed of two identical atoms) of period 2 elements.

Period 2 atoms have valence $2s$ and $2p$ orbitals, and we need to consider how they interact to form MOs. The following rules summarize some of the guiding principles for the formation of MOs and for how they are populated by electrons:

1. The number of MOs formed equals the number of atomic orbitals combined.

2. Atomic orbitals combine most effectively with other atomic orbitals of similar energy.

3. The effectiveness with which two atomic orbitals combine is proportional to their overlap. That is, as the overlap increases, the energy of the bonding MO is lowered and the energy of the antibonding MO is raised.

4. Each MO can accommodate, at most, two electrons, with their spins paired (Pauli exclusion principle). (Section 6.7)

5. When MOs of the same energy are populated, one electron enters each orbital (with the same spin) before spin pairing occurs (Hund's rule). (Section 6.8)

Molecular Orbitals for Li_2 and Be_2

Lithium has the electron configuration $1s^2 2s^1$. When lithium metal is heated above its boiling point (1342 °C), Li_2 molecules are found in the vapor phase. The Lewis structure for Li_2 indicates a Li—Li single bond. We now use MOs to describe the bonding in Li_2.

Figure 9.39 shows that the Li $1s$ and $2s$ atomic orbitals have substantially different energy levels. From this, we can assume that the interaction of a $1s$ orbital on one Li atom with a $2s$ orbital on the other Li atom is negligible (rule 2). More generally, because of the large energy separation between core and valence orbitals, we usually assume that core orbitals interact only with core orbitals and valence orbitals interact only with valence orbitals.

Combining the two atomic orbitals on each of two Li atoms produces four MOs (rule 1). The Li $1s$ orbitals combine to form σ_{1s} and σ_{1s}^* bonding and antibonding MOs, as they did for H_2. The $2s$ orbitals interact with one another in exactly the same way, producing bonding (σ_{2s}) and antibonding (σ_{2s}^*) MOs. In general, the separation between bonding and antibonding MOs depends on the extent to which the interacting atomic orbitals overlap. Because the Li $2s$ orbitals extend farther from the nucleus than do the $1s$ orbitals, the $2s$ orbitals overlap more effectively. (Figure 6.19) As a result, the energy difference between the σ_{2s} and σ_{2s}^* orbitals is greater than the energy difference between the σ_{1s} and σ_{1s}^* orbitals. The $1s$ orbitals of Li are much lower in energy than the $2s$ orbitals, so the energy of the σ_{1s}^* antibonding MO is much lower than the energy of the σ_{2s} bonding MO.

Each Li atom has three electrons, so six electrons must be placed in Li_2 MOs. As shown in Figure 9.39, these electrons occupy the $\sigma_{1s}, \sigma_{1s}^*,$ and σ_{2s} MOs, each with two electrons. There are four electrons in bonding orbitals and two in antibonding orbitals, so the bond order is $\frac{1}{2}(4 - 2) = 1$. The molecule has a single bond, in agreement with its Lewis structure.

Because both the σ_{1s} and σ_{1s}^* MOs of Li_2 are completely filled, the $1s$ orbitals contribute virtually nothing to the bonding. The single bond in Li_2 is due essentially to the interaction of the valence $2s$ orbitals on the Li atoms. This example illustrates the general rule that *core electrons usually do not contribute significantly to bonding in molecules.* The rule is equivalent to using only the valence electrons when drawing Lewis structures. Thus, we need not consider further the $1s$ orbitals while discussing the other period 2 diatomic molecules.

The MO description of Be_2 follows readily from the energy-level diagram for Li_2. Each Be atom has four electrons ($1s^2 2s^2$), so we must place eight electrons in molecular orbitals. Therefore, we completely fill the $\sigma_{1s}, \sigma_{1s}^*, \sigma_{2s},$ and σ_{2s}^* MOs. With equal numbers of bonding and antibonding electrons, the bond order is zero; thus, Be_2 does not exist.

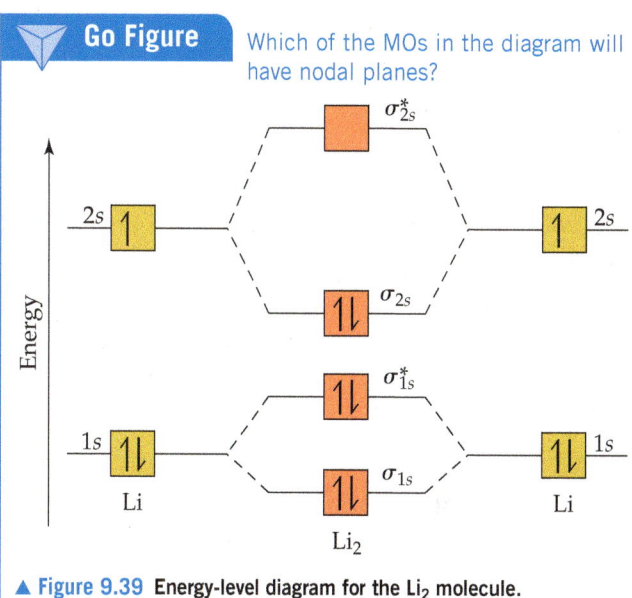

Go Figure

Which of the MOs in the diagram will have nodal planes?

▲ **Figure 9.39** Energy-level diagram for the Li_2 molecule.

Molecular Orbitals from $2p$ Atomic Orbitals

Before we can consider the remaining period 2 diatomic molecules, we must look at the MOs that result from combining $2p$ atomic orbitals. The interactions between p orbitals are shown in **Figure 9.40**, where we have arbitrarily chosen the internuclear axis to be the z axis. The $2p_z$ orbitals face each other head to head. Just as with s orbitals, we can combine $2p_z$ orbitals in two ways. One combination concentrates electron density between the nuclei and is, therefore, a bonding molecular orbital. The other combination

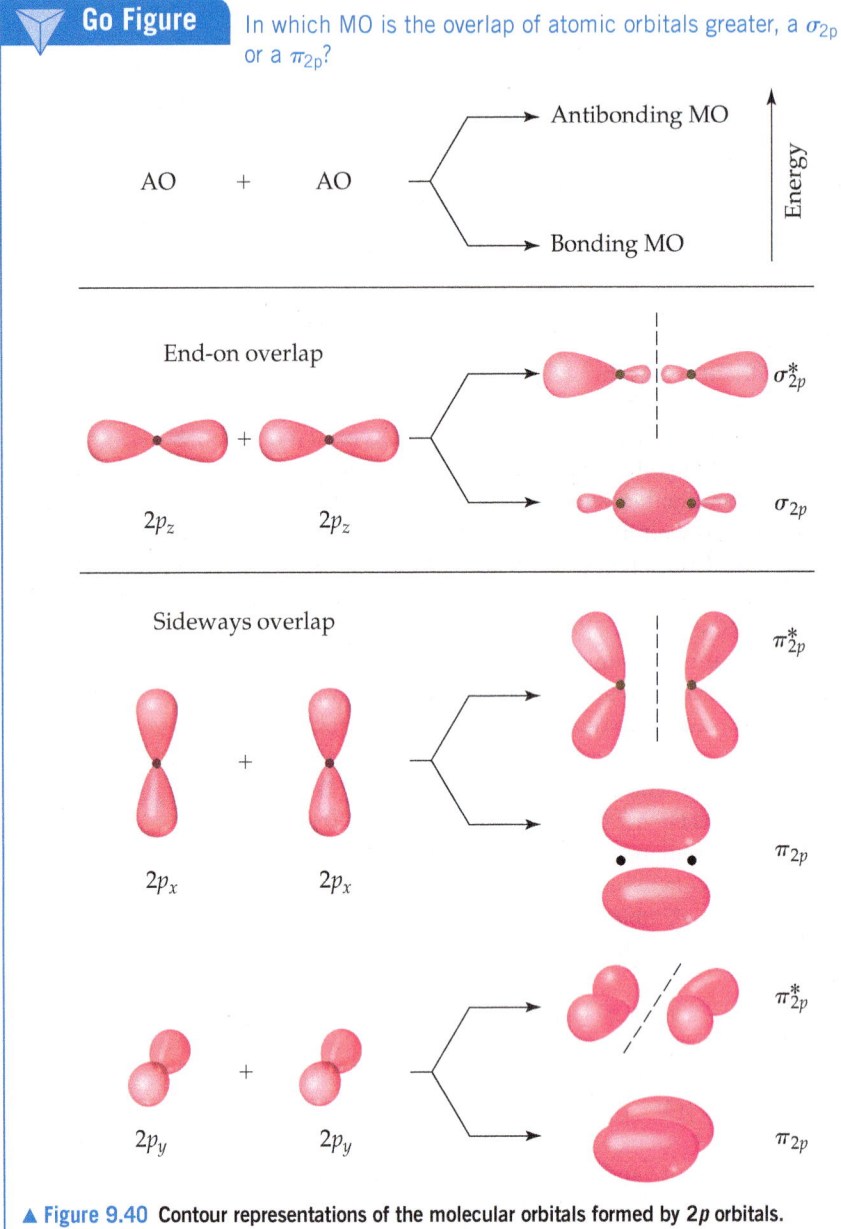

Go Figure In which MO is the overlap of atomic orbitals greater, a σ_{2p} or a π_{2p}?

▲ Figure 9.40 Contour representations of the molecular orbitals formed by **2p** orbitals.

excludes electron density from the bonding region and so is an antibonding molecular orbital. In both MOs, the electron density lies along the internuclear axis, so they are σ molecular orbitals: σ_{2p} and σ_{2p}^*. Notice that the σ_{2p}^* MO has a nodal plane halfway between the nuclei.

The other $2p$ orbitals overlap sideways and thus concentrate electron density above and below the internuclear axis. MOs of this type are called **pi (π) molecular orbitals** by analogy to π bonds. We get one π bonding MO by combining the $2p_x$ atomic orbitals and another from the $2p_y$ atomic orbitals. These two π_{2p} molecular orbitals have the same energy—they are degenerate. Likewise, we get two degenerate π_{2p}^* antibonding MOs that are perpendicular to each other like the $2p$ orbitals from which they were made. Each of these π_{2p}^* orbitals has four lobes, pointing away from the two nuclei, and a nodal plane, as shown in Figure 9.40.

The $2p_z$ orbitals on the two atoms point directly at each other. Hence, overlap of the two $2p_z$ orbitals is greater than that of the two $2p_x$ or $2p_y$ orbitals. We therefore expect the σ_{2p} MO to be lower in energy (more stable) than the π_{2p} MOs. Similarly, the σ_{2p}^* MO should be higher in energy (less stable) than the π_{2p}^* MOs.

Electron Configurations for B_2 through Ne_2

We can combine our analyses of MOs formed from s orbitals (Figure 9.32) and from p orbitals (Figure 9.40) to construct an energy-level diagram (**Figure 9.41**) for homonuclear diatomic molecules of the elements boron through neon, all of which have valence $2s$ and $2p$ atomic orbitals. The following features of the diagram are notable:

- The $2s$ atomic orbitals are substantially lower in energy than the $2p$ atomic orbitals. (Section 6.7) Consequently, both MOs formed from the $2s$ orbitals are lower in energy than the lowest-energy MO derived from the $2p$ atomic orbitals.
- The overlap of the two $2p_z$ orbitals is greater than that of the two $2p_x$ or $2p_y$ orbitals. As a result, the bonding σ_{2p} MO is lower in energy than the π_{2p} MOs, and the anti-bonding σ_{2p}^* MO is higher in energy than the π_{2p}^* MOs.
- Both the π_{2p} and π_{2p}^* MOs are *doubly degenerate*; that is, there are two degenerate MOs of each type.

Before we can add electrons to Figure 9.41, we must consider one more effect. We have constructed the diagram assuming no interaction between the $2s$ orbital on one atom and the $2p$ orbitals on the other. In fact, such interactions can and do take place. **Figure 9.42** shows the overlap of a $2s$ orbital on one of the atoms with a $2p$ orbital on the other. These interactions increase the energy difference between the σ_{2s} and σ_{2p} MOs, with the σ_{2s} energy decreasing and the σ_{2p} energy increasing (Figure 9.42). These $2s$–$2p$ interactions can be strong enough that the energetic ordering of the MOs can be altered: For B_2, C_2, and N_2, the σ_{2p} MO is *above* the π_{2p} MOs in energy. For O_2, F_2, and Ne_2, however, the σ_{2p} MO is *below* the π_{2p} MOs.

Given the energy ordering of the molecular orbitals, it is a simple matter to determine the electron configurations for the diatomic molecules B_2 through Ne_2. For example, a boron atom has three valence electrons. (Remember that we are ignoring the core $1s$ electrons.) Thus, for B_2 we must place six electrons in MOs. Four of them fill the σ_{2s} and σ_{2s}^* MOs, leading to no net bonding. The fifth electron goes in one π_{2p} MO, and the

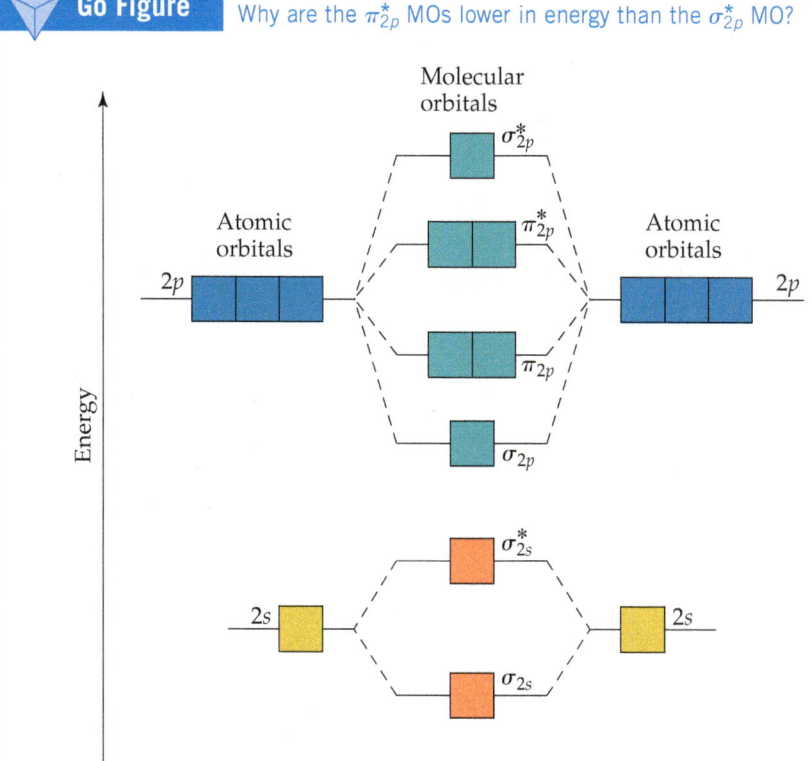

Go Figure   Why are the π_{2p}^* MOs lower in energy than the σ_{2p}^* MO?

▲ **Figure 9.41 Energy-level diagram for MOs of period 2 homonuclear diatomic molecules.** The diagram assumes no interaction between the $2s$ atomic orbital on one atom and the $2p$ atomic orbitals on the other atom, and experiment shows that it fits only for O_2, F_2, and Ne_2.

Go Figure   Which molecular orbitals have switched relative energy in the group on the right as compared with the group on the left?

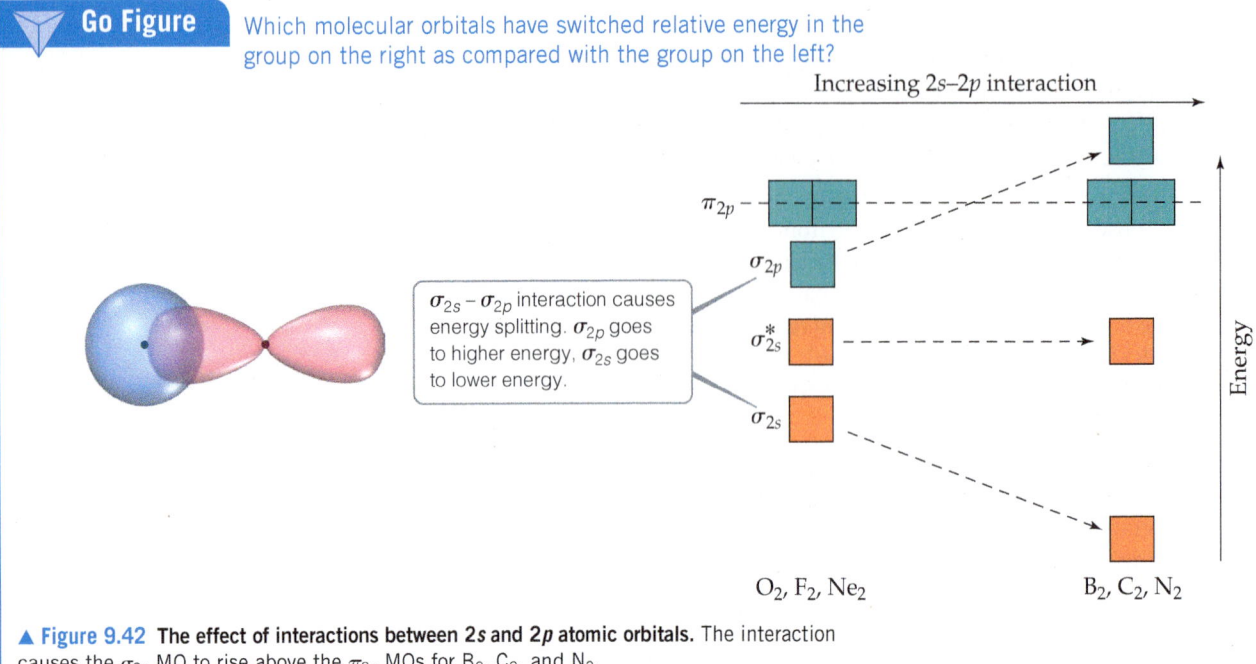

▲ **Figure 9.42 The effect of interactions between 2s and 2p atomic orbitals.** The interaction causes the σ_{2p} MO to rise above the π_{2p} MOs for B_2, C_2, and N_2.

sixth goes in the other π_{2p} MO, with the two electrons having the same spin. Therefore, B_2 has a bond order of 1.

Each time we move one element to the right in period 2, two more electrons must be placed in the diagram of Figure 9.41. For example, on moving to C_2, we have two more electrons than in B_2, and these electrons are placed in the π_{2p} MOs, completely filling them. Because it has six electrons in bonding MOs and two electrons in antibonding MOs, the C_2 molecule has a bond order of 2. The electron configurations and bond orders for B_2 through Ne_2 are given in **Figure 9.43**.

Go Figure   Which stable molecules have their highest-energy electrons in antibonding orbitals?

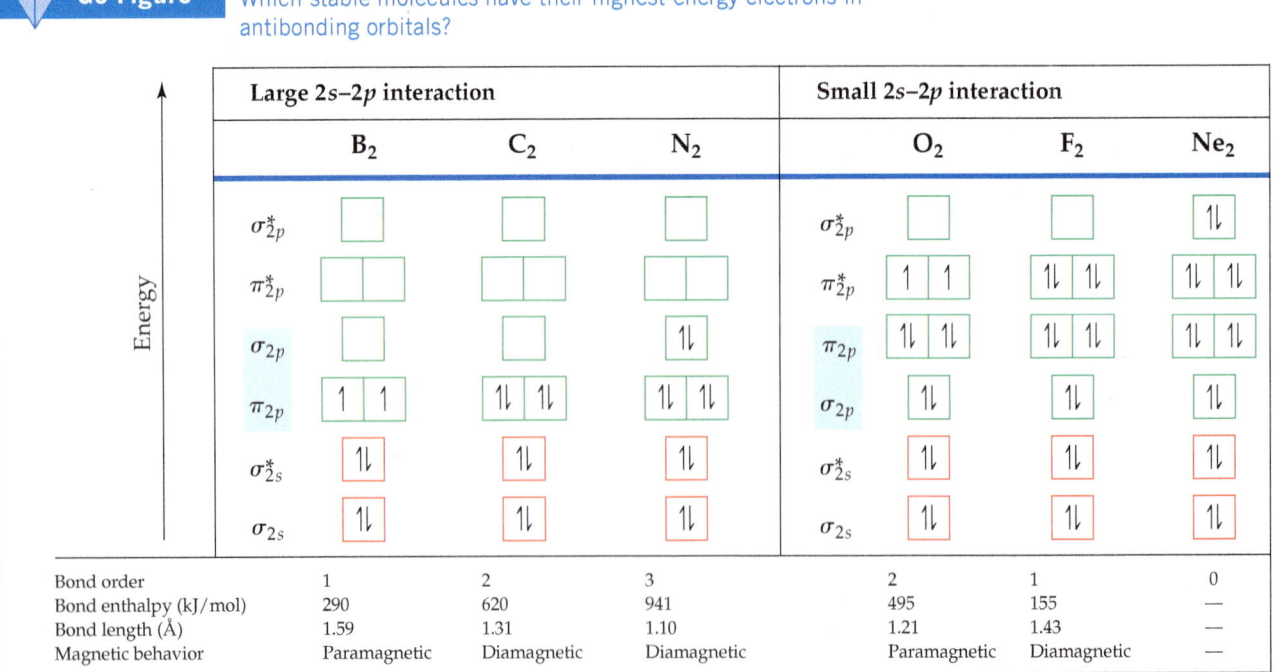

▲ **Figure 9.43** Molecular-orbital electron configurations and some experimental data for period 2 diatomic molecules.

Electron Configurations and Molecular Properties

The way a substance behaves in a magnetic field can in some cases provide insight into the arrangements of its electrons. Molecules with one or more unpaired electrons are attracted to a magnetic field. The more unpaired electrons in a species, the stronger the attractive force. This type of magnetic behavior is called **paramagnetism**.

Substances with no unpaired electrons are weakly repelled by a magnetic field. This property is called **diamagnetism**. The distinction between paramagnetism and diamagnetism is nicely illustrated in an older method for measuring magnetic properties (**Figure 9.44**). It involves weighing the substance in the presence and absence of a magnetic field. A paramagnetic substance appears to weigh more in the magnetic field, whereas a diamagnetic substance appears to weigh less. The magnetic behaviors observed for the period 2 diatomic molecules agree with the electron configurations shown in Figure 9.43.

Electron configurations in molecules can also be related to bond distances and bond enthalpies. (Sections 5.8 and 8.8) As the bond order increases, bond distances decrease and bond enthalpies increase. N_2, for example, whose bond order is 3, has a short bond distance and a large bond enthalpy. The N_2 molecule does not react readily with other substances to form nitrogen compounds. The high bond order of the molecule helps explain its exceptional stability. Note, however, that molecules with the same bond orders do *not* have the same bond distances and bond enthalpies. Bond order is only one factor influencing these properties. Other factors include nuclear charge and extent of orbital overlap.

The bonding in O_2 provides an interesting test case for molecular-orbital theory. The Lewis structure for this molecule shows a double bond and complete pairing of electrons:

$$\ddot{\text{O}}{=}\ddot{\text{O}}$$

The short oxygen–oxygen bond distance (1.21 Å) and relatively high bond enthalpy (495 kJ/mol) are in agreement with the presence of a double bond. However, Figure 9.43 tells us that the molecule contains two unpaired electrons and should therefore be paramagnetic, a detail not discernible in the Lewis structure. The paramagnetism of O_2 is demonstrated in **Figure 9.45**, which confirms the prediction from MO theory. The MO description also correctly predicts a bond order of 2 as did the Lewis structure.

Going from O_2 to F_2, we add two electrons, completely filling the π_{2p}^* MOs. Thus, F_2 is expected to be diamagnetic and have an F—F single bond, in accord with its Lewis structure. Finally, the addition of two more electrons to make Ne_2 fills all the bonding and antibonding MOs. Therefore, the bond order of Ne_2 is zero, and the molecule is not expected to exist.

Weigh sample in absence of a magnetic field

A diamagnetic sample appears to weigh less in magnetic field (weak effect)

A paramagnetic sample appears to weigh more in magnetic field

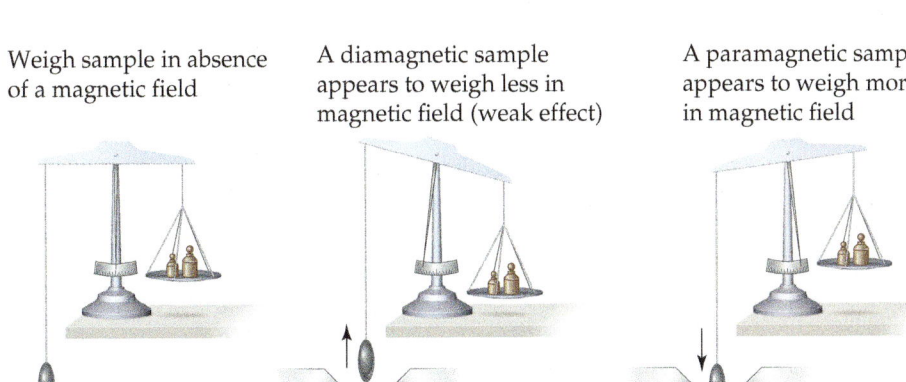

Sample

▲ **Figure 9.44 Determining the magnetic properties of a sample.**

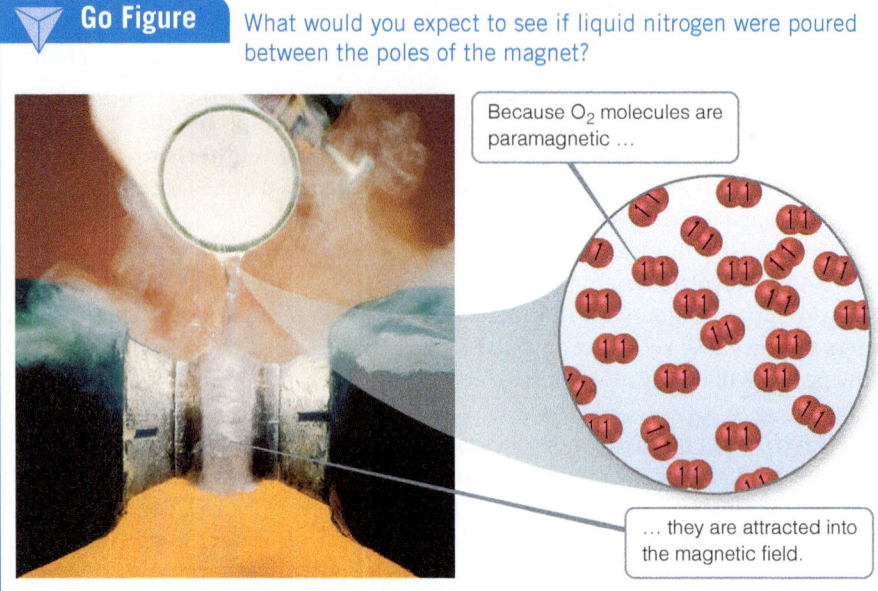

Go Figure What would you expect to see if liquid nitrogen were poured between the poles of the magnet?

Because O_2 molecules are paramagnetic …

… they are attracted into the magnetic field.

▲ **Figure 9.45 Paramagnetism of** O_2. When liquid oxygen is poured through a magnet, it "sticks" to the poles.

Sample Exercise 9.9
Molecular Orbitals of a Period 2 Diatomic Ion

For the O_2^+ ion predict (**a**) the number of unpaired electrons, (**b**) the bond order, and (**c**) the bond enthalpy and bond length.

SOLUTION

Analyze Our task is to predict several properties of the cation O_2^+.

Plan We will use the MO description of O_2^+ to determine the desired properties. We must first determine the number of electrons in O_2^+ and then draw its MO energy diagram. The unpaired electrons are those without a partner of opposite spin. The bond order is one-half the difference between the number of bonding and antibonding electrons. After calculating the bond order, we can use Figure 9.43 to estimate the bond enthalpy and bond length.

Solve

(**a**) The O_2^+ ion has 11 valence electrons, one fewer than O_2. The electron removed from O_2 to form O_2^+ is one of the two unpaired π_{2p}^* electrons (see Figure 9.43). Therefore, O_2^+ has one unpaired electron.

(**b**) The molecule has eight bonding electrons (the same as O_2) and three antibonding electrons (one fewer than O_2). Thus, its bond order is

$$\tfrac{1}{2}(8 - 3) = 2\tfrac{1}{2}$$

(**c**) The bond order of O_2^+ is between that for O_2 (bond order 2) and N_2 (bond order 3). Thus, the bond enthalpy and bond length should be about midway between those for O_2 and N_2, approximately 700 kJ/mol and 1.15 Å. (The experimentally measured values are 625 kJ/mol and 1.123 Å.)

▶ **Practice Exercise**

Predict the magnetic properties and bond orders of (**a**) the peroxide ion, O_2^{2-}; (**b**) the acetylide ion, C_2^{2-}.

Heteronuclear Diatomic Molecules

The principles we have used in developing an MO description of homonuclear diatomic molecules can be extended to *heteronuclear* diatomic molecules—those in which the two atoms in the molecule are not the same. We conclude this section with a fascinating heteronuclear diatomic molecule—nitric oxide, NO.

The NO molecule controls several important human physiological functions. Our bodies use it, for example, to relax muscles, kill foreign cells, and reinforce memory. The 1998 Nobel Prize in Physiology or Medicine was awarded to three scientists for their research that uncovered the importance of NO as a "signaling" molecule in the cardiovascular system. NO also functions as a neurotransmitter and is implicated in many other biological pathways. That NO plays such an important role in human metabolism was unsuspected before 1987 because NO has an odd number of electrons and is highly reactive. The molecule has 11 valence electrons, and two possible Lewis structures can be drawn. The Lewis structure with the lower formal charges places the odd electron on the N atom:

$$\overset{0}{\ddot{\text{N}}}=\overset{0}{\ddot{\text{O}}} \longleftrightarrow \overset{-1}{\ddot{\text{N}}}=\overset{+1}{\dot{\text{O}}}$$

Both structures indicate the presence of a double bond, but when compared with the molecules in Figure 9.43, the experimental bond length of NO (1.15 Å) suggests a bond order greater than 2. How do we treat NO using the MO model?

If the atoms in a heteronuclear diatomic molecule do not differ too greatly in electronegativities, their MOs resemble those in homonuclear diatomics, with one important modification: The energy of the atomic orbitals of the more electronegative atom is lower than that of the atomic orbitals of the less electronegative element. In **Figure 9.46**, the 2s and 2p atomic orbitals of oxygen are slightly lower than those of nitrogen because oxygen is more electronegative than nitrogen. The MO energy-level diagram for NO is much like that of a homonuclear diatomic molecule—because the 2s and 2p orbitals on the two atoms interact, the same types of MOs are produced.

There is one other important difference in the MOs of heteronuclear molecules. The MOs are still a mix of atomic orbitals from both atoms, but in general an *MO in a heteronuclear diatomic molecule has a greater contribution from the atomic orbital to which it is closer in energy*. In the case of NO, for example, the σ_{2s} bonding MO is closer in energy to the O 2s atomic orbital than to the N 2s atomic orbital. As a result, the σ_{2s} MO has a slightly greater contribution from O than from N—the orbital is no longer an equal mixture of the two atoms, as was the case for the homonuclear diatomic molecules. Similarly, the σ_{2s}^* antibonding MO is weighted more heavily toward the N atom because that MO is closest in energy to the N 2s atomic orbital.

We complete the MO diagram for NO by filling the MOs in Figure 9.46 with the 11 valence electrons. Eight bonding and three antibonding electrons give a bond order

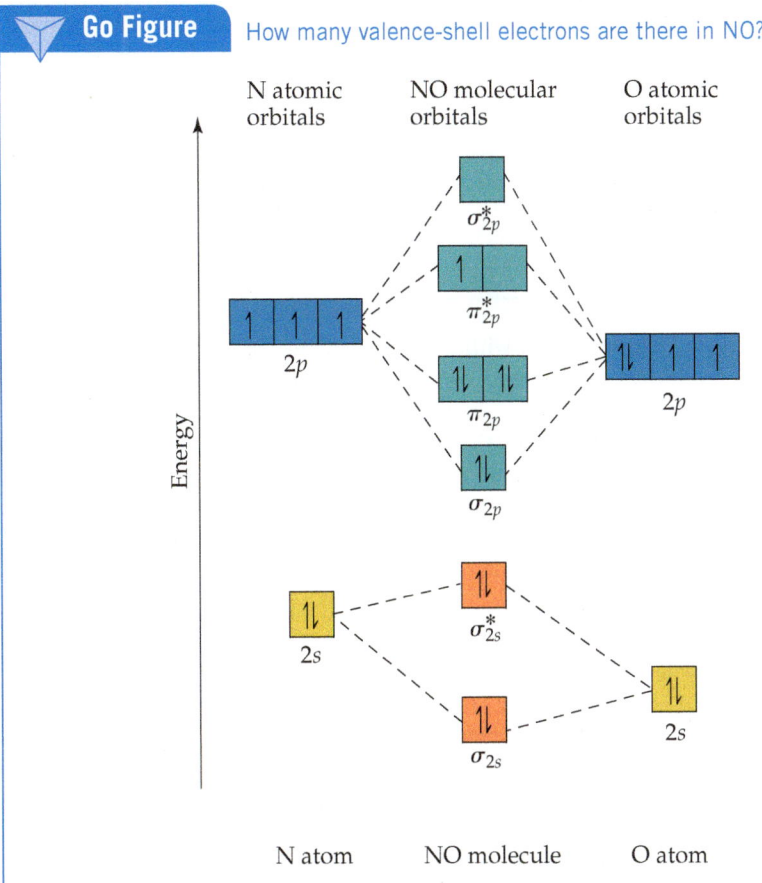

Go Figure   How many valence-shell electrons are there in NO?

▲ **Figure 9.46** The energy-level diagram for atomic and molecular orbitals in NO.

CHEMISTRY AND SUSTAINABILITY | Orbitals and Solar Energy

One of the United Nations Sustainable Development Goals is "Affordable and Clean Energy" (Figure 1.16). Indeed, one of our major technological challenges for the twenty-first century is the development of sustainable energy sources to meet the energy needs of future generations of people on our planet. One of the most remarkable sources of clean energy is the Sun, which sends enough energy to power the world for millions of years. Our challenge is to capture enough of this energy in a form that allows us to use it as needed. *Photovoltaic solar cells* convert the light from the Sun into usable electricity, and the development of more efficient solar cells is one way to address Earth's future energy needs.

How does solar energy conversion work? Fundamentally, we need to be able to use photons from the Sun, especially from the visible portion of the spectrum, to excite electrons in molecules and materials to different energy levels. The brilliant colors around you—those of your clothes, the photographs in this book, the foods you eat—are due to the selective absorption of visible light by chemicals. It is helpful to think of this process in the context of molecular-orbital theory: Light excites an electron from a filled molecular orbital to an empty one at higher energy. Because MOs have definite energies, only light of the proper wavelengths can excite electrons, similar to what we saw in our discussion of atomic line spectra. (Section 6.3)

In discussing light absorption by molecules, we can focus on the two MOs shown in **Figure 9.47**. The *highest occupied molecular orbital* (HOMO) is the MO of highest energy that has electrons in it. The *lowest unoccupied molecular orbital* (LUMO) is the MO of lowest energy that does not have electrons in it. In N_2, for example, the HOMO is the σ_{2p} MO and the LUMO is the π_{2p}^* MO (Figure 9.43).

The energy difference between the HOMO and the LUMO—known as the HOMO–LUMO gap—is related to the minimum energy needed to excite an electron in the molecule. Colorless or white substances usually have such a large HOMO–LUMO gap that visible light is not energetic enough to excite an electron to the higher level. The minimum energy needed to excite an electron from the HOMO to the LUMO in N_2 corresponds to light with a wavelength of less than 200 nm, which is far into the ultraviolet part of the spectrum. (Figure 6.4) As a result, N_2 cannot absorb visible light and is therefore colorless.

The magnitude of the energy gap between filled and empty electronic states is critical for solar energy conversion. Ideally, we want a substance that absorbs as many solar photons as possible and

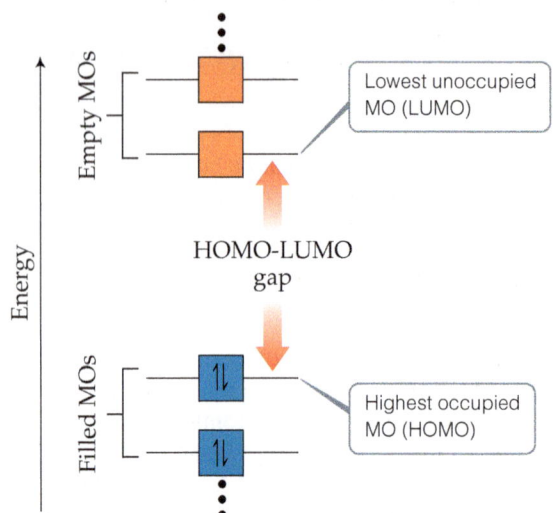

▲ **Figure 9.47 Definitions of the highest occupied and lowest unoccupied molecular orbitals** The energy difference between these is the HOMO–LUMO gap.

then converts the energy of those photons into a useful form of energy. Titanium dioxide is a readily available material that can be reasonably efficient at converting light directly into electricity. However, TiO_2 is white and absorbs only a small amount of the Sun's radiant energy. Scientists are working to make solar cells in which TiO_2 is mixed with highly colored molecules, whose HOMO–LUMO gaps correspond to visible and near-infrared light to absorb more of the solar spectrum. If the HOMO of these molecules is higher in energy than the HOMO of TiO_2, the excited electrons will flow from the molecules into the TiO_2. When this system is connected to an external circuit, the electron can flow to another cell where it adds back to the dye, restoring it to its original state. Thus, the energy of light can be used to cause a current of electricity to flow.

Efficient solar energy conversion promises to be one of the most interesting and important areas of both scientific and technological development in our future. Many of you may ultimately end up working in fields that have an impact on the world's energy portfolio.

Related Exercises: 9.109, 9.112, Design an Experiment

of $\frac{1}{2}(8-3) = 2\frac{1}{2}$, which agrees better with experiment than the Lewis structures do. The unpaired electron resides in one of the π_{2p}^* MOs, which have a greater contribution from the N atom. (We could have placed this electron in either the left or right π_{2p}^* MO.) Thus, the Lewis structure that places the unpaired electron on nitrogen (the one preferred on the basis of formal charge) is the more accurate description of the true electron distribution in the molecule.

▲ Self-Assessment Exercises

SAE 9.22 For the Li_2 molecule, rank-order the following orbitals from lowest to highest energy: $1s$, $2s$, σ_{2s}, σ_{2s}^*.
(a) $1s < 2s < \sigma_{2s} < \sigma_{2s}^*$ **(c)** $1s < \sigma_{2s}^* < 2s < \sigma_{2s}$
(b) $\sigma_{2s} < 1s < 2s < \sigma_{2s}^*$ **(d)** $1s < \sigma_{2s} < 2s < \sigma_{2s}^*$

SAE 9.23 Which of the following statements about the molecular orbitals for a diatomic molecule with $2s$ and $2p$ orbitals is or are *true*?

   (i) The σ_{2p}^* molecular orbital is higher in energy than a π_{2p}^* molecular orbital because the overlap is greater for two $2p$ orbitals that point directly at one another.
   (ii) The molecule has one σ_{2p} molecular orbital and one π_{2p} molecular orbital.

   (iii) The σ_{2p} and π_{2p} molecular orbitals both have a lower energy than the $2p$ orbitals from which they were constructed.

(a) Only one of the statements is true. **(b)** Statements i and ii are true. **(c)** Statements i and iii are true. **(d)** Statements ii and iii are true. **(e)** All three statements are true.

SAE 9.24 For the N_2^- molecular ion, the number of electrons in the π_{2p}^* molecular orbitals is _____ and the bond order of the ion is _____. **(a)** 0, 3 **(b)** 3, 1.5 **(c)** 0, 2.5 **(d)** 1, 1.5 **(e)** 1, 2.5

Putting Concepts Together

Elemental sulfur is a yellow solid that consists of S_8 molecules. The structure of the S_8 molecule is a puckered, eight-membered ring (see Figure 7.27). Heating elemental sulfur to high temperatures produces gaseous S_2 molecules:

$$S_8(s) \longrightarrow 4\,S_2(g)$$

(a) The electron configuration of which period 2 element is most similar to that of sulfur? (b) Use the VSEPR model to predict the S—S—S bond angles in S_8 and the hybridization at S in S_8. (c) Use MO theory to predict the sulfur–sulfur bond order in S_2. Do you expect this molecule to be diamagnetic or paramagnetic? (d) Use average bond enthalpies (Table 8.3) to estimate the enthalpy change for this reaction. Is the reaction exothermic or endothermic?

SOLUTION

(a) Sulfur is a group 6A element with an $[Ne]3s^23p^4$ electron configuration. It is expected to be most similar electronically to oxygen (electron configuration, $[He]2s^22p^4$), which is immediately above it in the periodic table.

(b) The Lewis structure of S_8 is

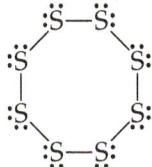

There is a single bond between each pair of S atoms and two nonbonding electron pairs on each S atom. Thus, we see four electron domains around each S atom and expect a tetrahedral electron-domain geometry corresponding to sp^3 hybridization. Because of the nonbonding pairs, we expect the S—S—S angles to be somewhat less than 109.5°, the tetrahedral angle. Experimentally, the S—S—S angle in S_8 is 108°, in good agreement with this prediction. Interestingly, if S_8 were a planar ring, it would have S—S—S angles of 135°. Instead, the S_8 ring puckers to accommodate the smaller angles dictated by sp^3 hybridization.

(c) The MOs of S_2 are analogous to those of O_2, although the MOs for S_2 are constructed from the $3s$ and $3p$ atomic orbitals of sulfur. Further, S_2 has the same number of valence electrons as O_2. Thus, by analogy with O_2, we expect S_2 to have a bond order of 2 (a double bond) and to be paramagnetic with two unpaired electrons in the π^*_{3p} molecular orbitals of S_2.

(d) We are considering the reaction in which an S_8 molecule falls apart into four S_2 molecules. From parts (b) and (c), we see that S_8 has S—S single bonds and S_2 has S=S double bonds. During the reaction, therefore, we are breaking eight S—S single bonds and forming four S=S double bonds. We can estimate the enthalpy of the reaction by using Equation 5.32 and the average bond enthalpies in Table 8.3:

$$\Delta H_{rxn} = \Sigma(\text{bond enthalpies of bonds broken}) -$$
$$\Sigma(\text{bond enthalpies of bonds formed})$$
$$= 8D\,(S—S) - 4D(S=S)$$
$$= 8(266\,\text{kJ}) - 4(418\,\text{kJ}) = +456\,\text{kJ}$$

Recall that $D(X—Y)$ represents the X—Y bond enthalpy. Because $\Delta H_{rxn} > 0$, the reaction is endothermic (Section 5.3). The very positive value of ΔH_{rxn} suggests that high temperatures are required to cause the reaction to occur.

Chapter Summary and Key Terms

MOLECULAR SHAPES (INTRODUCTION AND SECTION 9.1) The three-dimensional shapes and sizes of molecules are determined by their **bond angles** and bond lengths. Molecules with a central atom A surrounded by n atoms B, denoted AB_n, adopt a number of different geometric shapes, depending on the value of n and on the particular atoms involved. In the overwhelming majority of cases, these geometries are related to five basic shapes (linear, trigonal pyramidal, tetrahedral, trigonal bipyramidal, and octahedral).

THE VSEPR MODEL (SECTION 9.2) The **valence-shell electron-pair repulsion (VSEPR) model** rationalizes molecular geometries based on the repulsions between **electron domains**, which are regions about a central atom in which electrons are likely to be found. **Bonding pairs** of electrons, which are those involved in making bonds, and **nonbonding pairs** of electrons, also called **lone pairs**, both create electron domains around an atom. According to the VSEPR model, electron domains orient themselves to minimize electrostatic repulsions; that is, they remain as far apart as possible.

Electron domains from nonbonding pairs exert slightly greater repulsions than those from bonding pairs, which leads to certain preferred positions for nonbonding pairs and to the departure of bond angles from idealized values. Electron domains from multiple bonds exert slightly greater repulsions than those from single bonds. The arrangement of electron domains around a central atom is called the **electron-domain geometry**; the arrangement of atoms is called the **molecular geometry**.

MOLECULAR POLARITY (SECTION 9.3) The dipole moment of a polyatomic molecule depends on the vector sum of the dipole moments associated with the individual bonds, called the **bond dipoles**. Certain molecular shapes, such as linear AB_2 and trigonal-planar AB_3, lead to cancellation of the bond dipoles, producing a nonpolar molecule (one whose overall dipole moment is zero). In other shapes, such as bent AB_2 and trigonal pyramidal AB_3, the bond dipoles do *not* cancel and the molecule will be polar (that is, it will have a nonzero dipole moment).

COVALENT BONDING AND VALENCE-BOND THEORY (SECTION 9.4) **Valence-bond theory** is an extension of Lewis's notion of electron-pair bonds. In valence-bond theory, covalent bonds are formed when atomic orbitals on neighboring atoms overlap one another. The overlap region is one of greater stability for the two electrons because of their simultaneous attraction to two nuclei. The greater the overlap between two orbitals, the stronger the bond that is formed.

HYBRID ORBITALS (SECTION 9.5) To extend the ideas of valence-bond theory to polyatomic molecules, we must envision mixing *s* and *p* orbitals to form **hybrid orbitals**. The process of **hybridization** leads to hybrid atomic orbitals that have a large lobe directed to overlap with orbitals on another atom to make a bond. Hybrid orbitals can also accommodate nonbonding pairs. A particular mode of hybridization can be associated with each of three common electron-domain geometries (linear = sp; trigonal planar = sp^2; tetrahedral = sp^3). The bonding in **hypervalent** molecules—those with more than an octet of electrons—is not as readily discussed in terms of hybrid orbitals.

MULTIPLE BONDS (SECTION 9.6) Covalent bonds in which the electron density lies along the line connecting the atoms (the internuclear axis) are called **sigma (σ) bonds**. Bonds can also be formed from the sideways overlap of *p* orbitals. Such a bond is called a **pi (π) bond**. A double bond, such as that in C_2H_4, consists of one σ bond and one π bond; each carbon atom has an unhybridized *p* orbital, and these are the orbitals that overlap to form π bonds. A triple bond, such as that in C_2H_2, consists of one σ and two π bonds. The formation of a π bond requires that molecules adopt a specific orientation; the two CH_2 groups in C_2H_4, for example, must lie in the same plane. As a result, the presence of π bonds introduces rigidity into molecules. In molecules that have multiple bonds and more than one resonance structure, such as C_6H_6, the π system is **delocalized**; that is, the π electrons are spread among several atoms.

MOLECULAR ORBITALS (SECTION 9.7) **Molecular-orbital theory** is another model used to describe the bonding in molecules. In this model, the electrons exist in allowed energy states called **molecular orbitals (MOs)**. An MO can extend over all the atoms of a molecule. Like an atomic orbital, a molecular orbital has a definite energy and can hold two electrons of opposite spin. We can build molecular orbitals by combining atomic orbitals on different atomic centers. In the simplest case, the combination of two atomic orbitals leads to the formation of two MOs, one at lower energy and one at higher energy relative to the energy of the atomic orbitals. The lower-energy MO concentrates charge density in the region between the nuclei and is called a **bonding molecular orbital**. The higher-energy MO excludes electrons from the region between the nuclei and is called an **antibonding molecular orbital**. Antibonding MOs have a **nodal plane**—a place at which the electron density is zero—between the nuclei. Occupation of bonding MOs favors bond formation, whereas occupation of antibonding MOs is unfavorable. The bonding and antibonding MOs formed by the combination of *s* orbitals are **sigma (σ) molecular orbitals**; they lie on the internuclear axis.

The combination of atomic orbitals and the relative energies of the molecular orbitals are shown by an **energy-level (or molecular-orbital) diagram**. When the appropriate number of electrons is put into the MOs, we can calculate the **bond order** of a bond, which is half the difference between the number of electrons in bonding MOs and the number of electrons in antibonding MOs. A bond order of 1 corresponds to a single bond, and so forth. Bond orders can be fractional numbers.

MOLECULAR-ORBITAL DESCRIPTION OF PERIOD 2 DIATOMIC MOLECULES (SECTION 9.8) Electrons in core orbitals do not contribute to the bonding between atoms, so a molecular-orbital description usually needs to consider only the valence electrons. To describe the MOs of period 2 homonuclear diatomic molecules, we need to consider the MOs that can form by the combination of *p* orbitals. The *p* orbitals that point directly at one another can form σ bonding and σ^* antibonding MOs. The *p* orbitals that are oriented perpendicular to the internuclear axis combine to form **pi (π) molecular orbitals**. In diatomic molecules, the π molecular orbitals occur as a pair of degenerate (same energy) bonding MOs and a pair of degenerate antibonding MOs. The σ_{2p} bonding MO is expected to be lower in energy than the π_{2p} bonding MOs because of larger orbital overlap of the *p* orbitals directed along the internuclear axis. However, this ordering is reversed in B_2, C_2, and N_2 because of interaction between the $2s$ and $2p$ atomic orbitals of different atoms.

The molecular-orbital description of period 2 diatomic molecules leads to bond orders in accord with the Lewis structures of these molecules. Further, the model predicts correctly that O_2 should exhibit **paramagnetism**; a paramagnetic molecule is attracted into a magnetic field due to the influence of unpaired electrons. Molecules in which all the electrons are paired exhibit **diamagnetism**; a diamagnetic molecule is weakly repelled from a magnetic field. The molecular orbitals of heteronuclear diatomic molecules are often closely related to those of homonuclear diatomic molecules.

Key Equations

- Bond order = $\frac{1}{2}$ (number of bonding electrons − number of antibonding electrons)   [9.1]

Exam Prep

EP 9.1 BF_3 is a trigonal-planar molecule, whereas PF_3 is a trigonal-pyramidal molecule. Which of the following statements about these two molecules is *false*? (**a**) All the F—B—F angles in BF_3 are 120°. (**b**) The F atoms in BF_3 sit on the vertices of an equilateral triangle. (**c**) The shape of PF_3 can be obtained by removing one atom from a tetrahedral arrangement of atoms. (**d**) The four atoms in BF_3 all lie in the same plane. (**e**) The four atoms in PF_3 all lie in the same plane.

EP 9.2 Consider the following AB_3 molecules and ions: PCl_3, SO_3, $AlCl_3$, SO_3^{2-}, and CH_3^+. How many of these molecules and ions do you predict to have a trigonal-planar molecular geometry? (**a**) 1 (**b**) 2 (**c**) 3 (**d**) 4 (**e**) 5

EP 9.3 Which of the following statements about the structure of H_2S is *false*? (**a**) H_2S has a trigonal-planar electron-domain geometry. (**b**) The Lewis structure of H_2S has eight valence electrons. (**c**) There are two nonbonding electron domains on the S atom in H_2S. (**d**) The H—S—H angle in H_2S is less than 109.5°.

EP 9.4 A certain AB_4 molecule has a square-planar molecular geometry. Which of the following statements about the molecule is or are *true*?

   (i) The molecule has four electron domains about the central atom A.

   (ii) The B—A—B angles between neighboring B atoms are 90°.

   (iii) The molecule has two nonbonding pairs of electrons on atom A.

(**a**) Only one of the statements is true. (**b**) Statements i and ii are true. (**c**) Statements i and iii are true. (**d**) Statements ii and iii are true. (**e**) All three statements are true.

EP 9.5 The atoms of the compound methylhydrazine, CH_6N_2, which is used as a rocket propellant, are connected as shown (note that the lone pairs are not shown):

$$
\begin{array}{c}
H \\
| \\
H-C-H \\
| \\
N-N-H \\
| \quad | \\
H \quad H
\end{array}
$$

What do you predict for the ideal values (ignoring any effects from lone pairs) of the C—N—N and H—N—H angles, respectively? (a) 109.5° and 109.5° (b) 109.5° and 120° (c) 120° and 109.5° (d) 120° and 120° (e) None of the answers is correct.

EP 9.6 Which of the following statements about molecular polarity is or are *true*?

(i) The bond dipole is a vector quantity that has both a magnitude and a direction.

(ii) If a molecule has a nonzero bond dipole, then it will be a polar molecule.

(iii) The bond dipoles in a tetrahedral AB_4 molecule cancel one another, so the molecule is nonpolar.

(a) Only one of the statements is true. (b) Statements i and ii are true. (c) Statements i and iii are true. (d) Statements ii and iii are true. (e) All three statements are true.

EP 9.7 Consider an AB_3 molecule in which A and B differ in electronegativity. You are told that the molecule has an overall dipole moment of zero. Which of the following could be the molecular geometry of the molecule? (a) trigonal pyramidal (b) trigonal planar (c) T-shaped (d) tetrahedral (e) More than one of these geometries is possible.

EP 9.8 The bond length in Cl_2 is 1.99 Å. Which of the following statements about the valence-bond description of the Cl—Cl bond in Cl_2 is *false*? (a) The bond is formed from the overlap of a $3p$ orbital on one Cl atom with a $3s$ orbital on the other Cl atom. (b) The orbital overlap decreases as the Cl atoms are pulled apart. (c) At Cl—Cl distances less than 1.99 Å, the potential energy of the Cl_2 molecule increases because of the electrostatic repulsion between the two Cl atoms. (d) Each of the overlapping orbitals has one electron in it. (e) The potential energy of the Cl_2 molecule is at a minimum at a Cl—Cl distance of 1.99 Å.

EP 9.9 Which of the following statements about atomic orbitals and hybrid orbitals is *true*?

(a) To make hybrid orbitals, we mix atomic orbitals on two different atoms. (b) Hybrid orbitals can allow us to make multiple equivalent bonds when atomic orbitals can't do so. (c) Each hybrid orbital has two lobes that are the same size. (d) The number of hybrid orbitals produced is greater than the number of atomic orbitals that are mixed. (e) Each hybrid orbital can hold a maximum of one electron.

EP 9.10 Fill in the blanks in the following sentences about the CH_2^{2+} ion: "The valence-bond description of CH_2^{2+} involves the overlap of the H $1s$ orbitals with _____ hybrid orbitals on the C atom. The predicted H—C—H angle in CH_2^{2+} is _____ degrees." (a) sp, 120 (b) sp^2, 120 (c) sp, 180 (d) sp^2, 180 (e) sp^3, 109.5

EP 9.11 For which of the following molecules or ions does the following description apply? "The σ bonding can be explained using a set of sp^2 hybrid orbitals on the central atom, with one of the hybrid orbitals holding a nonbonding pair of electrons." (a) CO_2 (b) H_2S (c) O_3 (d) CO_3^{2-} (e) more than one of the molecules or ions listed

EP 9.12 Three types of orbital overlaps are illustrated in the figure. Which of these represent the formation of a σ bond? (a) only one of them (b) i and ii (c) i and iii (d) ii and iii (e) all three of them

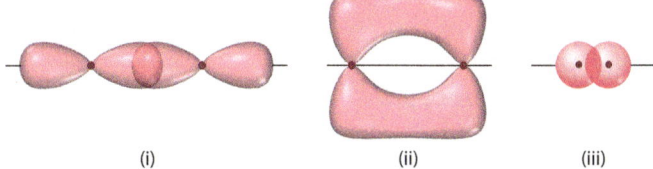

(i)   (ii)   (iii)

EP 9.13 Fill in the blanks in the following sentence about the hydrogen cyanide molecule, HCN: "In HCN, _____ electrons are used to make σ bonds and _____ electrons are used to make π bonds." (a) 2, 2 (b) 2, 4 (c) 2, 6 (d) 4, 2 (e) 4, 4

EP 9.14 A Lewis structure of the oxalate ion, $C_2O_4^{2-}$, is shown here. How many electrons are in the π system of the ion? (a) 2 (b) 4 (c) 6 (d) 8

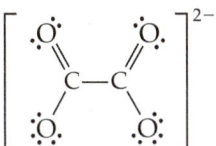

EP 9.15 One of the molecular orbitals of the H_2 molecule is shown here. The label for this orbital is _____, and it is _____. (a) σ_{1s}, bonding (b) σ_{1s}, antibonding (c) σ_{1s}^*, bonding (d) σ_{1s}^*, antibonding

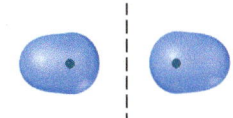

EP 9.16 Which of the following molecules and ions have a bond order of 1/2: H_2, H_2^+, H_2^-, and He_2^{2+}? (a) H_2 and H_2^+ (b) H_2^+ and H_2^- (c) H_2^- and He_2^{2+} (d) H_2, H_2^+, and H_2^- (e) H_2^+, H_2^-, and He_2^{2+}

EP 9.17 Which of the following statements about molecular-orbital theory is *false*? (a) The number of molecular orbitals for a molecule is less than the number of atomic orbitals that are used to make them. (b) Two atomic orbitals combine most effectively to make molecular orbitals when they have similar energies. (c) As the overlap of two atomic orbitals increases, the antibonding molecular orbital that arises from them becomes higher in energy. (d) A molecular orbital for a molecule can hold zero, one, or two electrons, depending on its energy and the number of electrons in the molecule. (e) The energy of a bonding molecular orbital is lower than the energy of the atomic orbitals from which it is made.

EP 9.18 In the molecular-orbital description of the Be_2^+ ion, the σ_{2s}^* MO has _____ electron(s) in it, and the bond order of the ion is _____. (a) 1, 0 (b) 2, 0 (c) 2, 1 (d) 2, ½ (e) 1, ½

EP 9.19 One of the molecular orbitals of a homonuclear diatomic molecule is shown here. Which of the following statements about this MO is or are *true*?

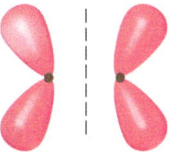

(i) This is a π_{2p}^* MO.

(ii) There is another molecular orbital for the molecule that has exactly the same energy as the one shown.

(iii) The energy of this molecular orbital is lower than the energy of the atomic orbitals from which it is formed.

(a) Only one of the statements is true. (b) Statements i and ii are true. (c) Statements i and iii are true. (d) Statements ii and iii are true. (e) All three statements are true.

EP 9.20 A homonuclear diatomic molecule has the following valence molecular-orbital electron configuration: $\sigma_{2s}^2 \sigma_{2s}^{*2} \sigma_{2p}^2 \pi_{2p}^4 \pi_{2p}^{*2}$. What is the molecule, and is it diamagnetic or paramagnetic? (a) B_2, paramagnetic (b) N_2, diamagnetic (c) N_2, paramagnetic (d) O_2, diamagnetic (e) O_2, paramagnetic

EP 9.21 Place the following molecular ions in order from smallest to largest bond order: C_2^{2+}, N_2^-, O_2^-, F_2^-. (a) $C_2^{2+} < N_2^- < O_2^- < F_2^-$ (b) $F_2^- < O_2^- < N_2^- < C_2^{2+}$ (c) $O_2^- < C_2^{2+} < F_2^- < N_2^-$ (d) $C_2^{2+} < F_2^- < O_2^- < N_2^-$ (e) $F_2^- < C_2^{2+} < O_2^- < N_2^-$

Exercises

Visualizing Concepts

9.1 A certain AB_4 molecule has a "seesaw" shape.

From which of the fundamental geometries shown in Figure 9.3 could you remove one or more atoms to create a molecule having this seesaw shape? [Section 9.1]

9.2 (a) If these three balloons are all the same size, what angle is formed between the red one and the green one? (b) If additional air is added to the blue balloon so that it gets larger, will the angle between the red and green balloons increase, decrease, or stay the same? (c) Which of the following aspects of the VSEPR model is illustrated by part (b): (i) The electron-domain geometry for four electron domains is tetrahedral. (ii) The electron domains for nonbonding pairs are larger than those for bonding pairs. (iii) The hybridization that corresponds to a trigonal-planar electron-domain geometry is sp^2. [Section 9.2]

9.3 For each molecule (a)–(f), indicate how many different electron-domain geometries are consistent with the molecular geometry shown. [Section 9.2]

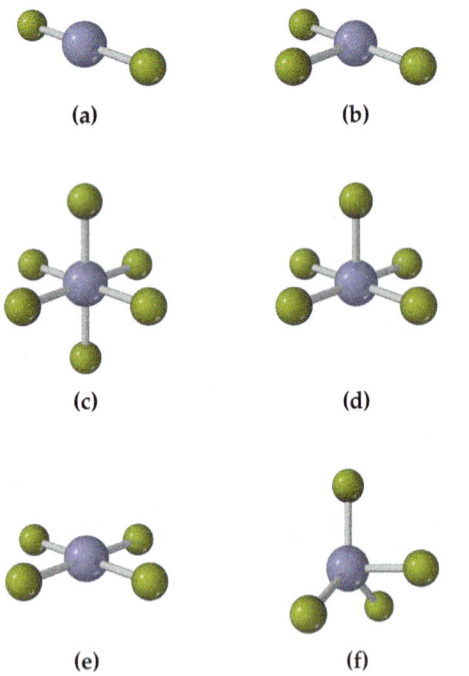

(a)   (b)

(c)   (d)

(e)   (f)

9.4 The molecule shown here is *difluoromethane* (CH_2F_2), which is used as a refrigerant called R-32. (a) Based on the structure, how many electron domains surround the C atom in this molecule? (b) Would the molecule have a nonzero dipole moment? (c) If the molecule is polar, which of the following describes the direction of the overall dipole moment vector in the molecule: (i) from the carbon atom toward a fluorine atom, (ii) from the carbon atom to a point midway between the fluorine atoms, (iii) from the carbon atom to a point midway between the hydrogen atoms, or (iv) from the carbon atom toward a hydrogen atom? [Sections 9.2 and 9.3]

9.5 The following plot shows the potential energy of two Cl atoms as a function of the distance between them. (a) If the

two atoms are very far away from each other, what is their potential energy of interaction? (b) We know that the Cl_2 molecule exists. What is the approximate bond length and bond strength for the Cl—Cl bond in Cl_2 from this graph? (c) If the Cl_2 molecule is compressed under higher and higher pressure, does the Cl—Cl bond become stronger or weaker?

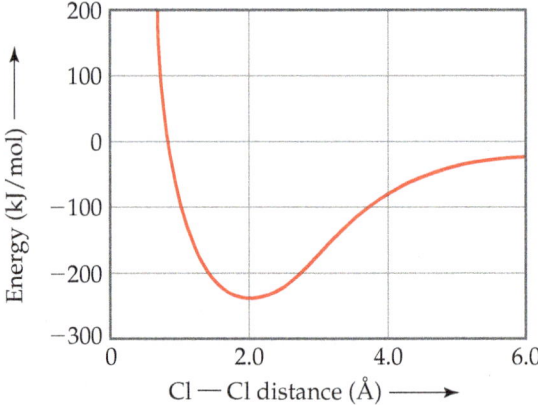

9.6 The orbital diagram that follows presents the final step in the formation of hybrid orbitals by a silicon atom. (a) Which of the following best describes what took place before the step pictured in the diagram: (i) Two $3p$ electrons became unpaired, (ii) An electron was promoted from the $2p$ orbital to the $3s$ orbital, or (iii) An electron was promoted from the $3s$ orbital to the $3p$ orbital? (b) What type of hybrid orbital is produced in this hybridization? [Section 9.5]

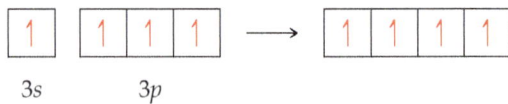

$3s$ $3p$

9.7 Consider the following hydrocarbon:

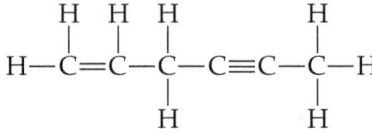

(a) What is the hybridization at each carbon atom in the molecule? (b) How many σ bonds are there in the molecule? (c) How many π bonds? (d) Identify all the 120° bond angles in the molecule. [Section 9.6]

9.8 The drawing provided shows the overlap of two hybrid orbitals to form a bond in a hydrocarbon. (a) Which of the following types of bonds is being formed: (i) C—C σ, (ii) C—C π, or (iii) C—H σ? (b) Which of the following could be the identity of the hydrocarbon: (i) CH_4 (ii) C_2H_6 (iii) C_2H_4 or (iv) C_2H_2? [Section 9.6]

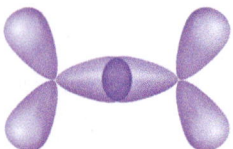

9.9 The molecule shown here is called *furan*. It is represented in the typical shorthand way for organic molecules, with hydrogen atoms not shown, and each of the four vertices representing a carbon atom.

(a) What is the molecular formula for furan? (b) How many valence electrons are there in the molecule? (c) What is the hybridization at each of the carbon atoms? (**d**) How many electrons are in the π system of the molecule? (**e**) The C—C—C bond angles in furan are much smaller than those in benzene. The likely reason is which of the following? (i) The hybridization of the carbon atoms in furan is different from that in benzene. (ii) Furan does not have another resonance structure equivalent to the one shown here. (iii) The atoms are forced to adopt smaller angles in a five-membered ring than in a six-membered ring. [Section 9.5]

9.10 The following is part of a molecular-orbital energy-level diagram for MOs constructed from $1s$ atomic orbitals.

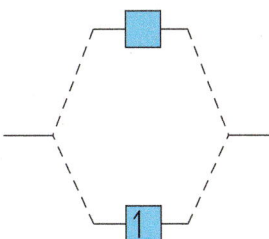

(**a**) What labels do we use for the two MOs shown? (**b**) For which of the following molecules or ions could this be the energy-level diagram: H_2, He_2, H_2^+, He_2^+, or H_2^-? (**c**) What is the bond order of the molecule or ion? (**d**) If an electron is added to the system, into which of the MOs will it be added? [Section 9.7]

9.11 For each of these contour representations of molecular orbitals, identify (**a**) the atomic orbitals (s or p) used to construct the MO (**b**) the type of MO (σ or π) (**c**) whether the MO is bonding or antibonding and (**d**) the locations of nodal planes. [Sections 9.7 and 9.8]

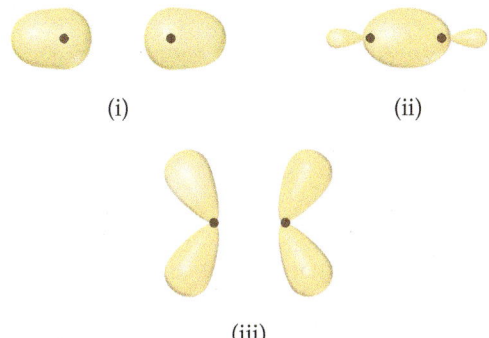

(i)   (ii)

(iii)

9.12 The diagram that follows shows the highest-energy occupied MOs of a neutral molecule CX, where element X is in the same row of the periodic table as C. (**a**) Based on the number of electrons, can you determine the identity of X? (**b**) Would the molecule be diamagnetic or paramagnetic? (**c**) Consider the π_{2p} MOs of the molecule. Would you expect them to have a greater atomic orbital contribution from C, have a greater atomic orbital contribution from X, or be an equal mixture of atomic orbitals from the two atoms? [Section 9.8]

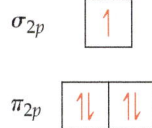

9.13 One of the $2p$ orbitals with phases is shown in the figure. (**a**) Which $2p$ orbital is it? (**b**) Which of the statements about the orbital is or are *true*? (i) The two lobes are equal in size but opposite in sign. (ii) The probability of finding an electron is positive in one direction and negative in the opposite direction. (iii) The phase information for the orbital is lost when the wave function is squared. [Section 9.8]

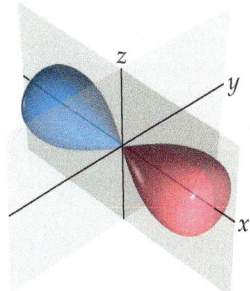

9.14 The figures show two different possible combinations of two $2p$ atomic orbitals with phases to form a molecular orbital. For each of these figures (**a**) identify whether it represents constructive or destructive addition of the two atomic orbitals, and (**b**) identify whether the resulting molecular orbital will be a σ, σ^*, π, or π^* MO. [Section 9.8]

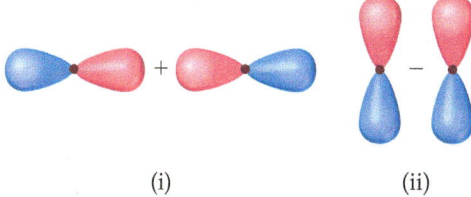

(i)   (ii)

Molecular Shapes; the VSEPR Model (Sections 9.1 and 9.2)

9.15 An AB_2 molecule that satisfies the octet rule has only A—B single bonds. (**a**) What is the electron-domain geometry of the molecule? (**b**) What is the molecular geometry of the molecule? (**c**) Do all AB_2 molecules that satisfy the octet rule have this same molecular geometry?

9.16 (**a**) Methane (CH_4) and the perchlorate ion (ClO_4^-) are both described as tetrahedral. What does this indicate about their bond angles? (**b**) The NH_3 molecule is trigonal pyramidal, while BF_3 is trigonal planar. Which of these molecules is flat?

9.17 For a particular AB_3 molecule, all the B—A—B bond angles are roughly 109°. (**a**) What is the molecular geometry of the molecule? (**b**) What is the electron-domain geometry of the molecule? (**c**) How many nonbonding electron domains are on the atom A?

9.18 Describe the bond angles to be found in each of the following molecular structures. (**a**) trigonal planar (**b**) tetrahedral (**c**) octahedral (**d**) linear

9.19 (**a**) An AB_6 molecule has no lone pairs of electrons on the A atom. What is its molecular geometry? (**b**) An AB_4 molecule has two lone pairs of electrons on the A atom (in addition to the four B atoms). What is the electron-domain geometry around the A atom? (**c**) For the AB_4 molecule in part (b), predict the molecular geometry.

9.20 Would you expect the nonbonding electron-pair domain in NH_3 to be larger or smaller in size than the corresponding one in PH_3?

9.21 In which of these molecules or ions does the presence of nonbonding electron pairs affect the molecular shape? (**a**) SiH_4 (**b**) PF_3 (**c**) HBr (**d**) HCN (**e**) SO_2

9.22 In which of the following molecules can you confidently predict the bond angles about the central atom, and for which would you be a bit uncertain? Explain in each case. (**a**) H_2S (**b**) BCl_3 (**c**) CH_3I (**d**) CBr_4 (**e**) $TeBr_4$

9.23 Give the electron-domain and molecular geometries of a molecule that has the following electron domains on its central atom: (**a**) four bonding domains and no nonbonding domains

(b) three bonding domains and two nonbonding domains
(c) five bonding domains and one nonbonding domain
(d) four bonding domains and two nonbonding domains

9.24 What are the electron-domain and molecular geometries of a molecule that has the following electron domains on its central atom? (a) three bonding domains and no nonbonding domains (b) three bonding domains and one nonbonding domain (c) two bonding domains and two nonbonding domains

9.25 Give the electron-domain and molecular geometries for the following molecules and ions. (a) HCN (b) SO_3^{2-} (c) SF_4 (d) PF_6^- (e) NH_3Cl^+ (f) N_3^-

9.26 Draw the Lewis structure for each of the following molecules or ions, and predict their electron-domain and molecular geometries. (a) AsF_3 (b) CH_3^+ (c) BrF_3 (d) ClO_3^- (e) XeF_2 (f) BrO_2^-

9.27 The figure that follows shows ball-and-stick drawings of three possible shapes of an AF_3 molecule. (a) For each shape, give the electron-domain geometry on which the molecular geometry is based. (b) For each shape, how many nonbonding electron domains are there on atom A? (c) Which of the following elements will lead to an AF_3 molecule with the shape in (ii): Li, B, N, Al, P, Cl? (d) Name an element A that is expected to lead to the AF_3 structure shown in (iii).

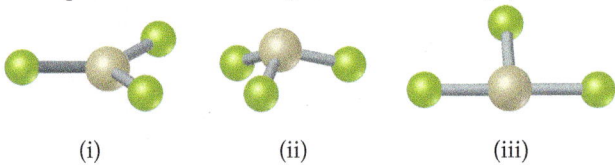

(i)   (ii)   (iii)

9.28 The figure that follows contains ball-and-stick drawings of three possible shapes of an AF_4 molecule. (a) For each shape, give the electron-domain geometry on which the molecular geometry is based. (b) For each shape, how many nonbonding electron domains are there on atom A? (c) Which of the following elements will lead to an AF_4 molecule with the shape in (iii): Be, C, S, Se, Si, Xe? (d) Name an element A that is expected to lead to the AF_4 structure shown in (i).

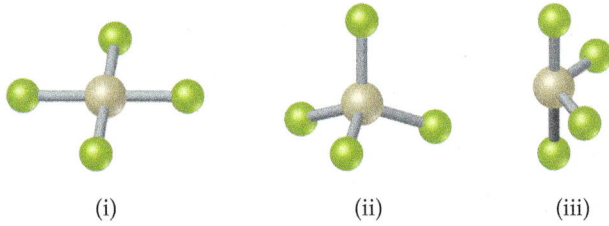

(i)   (ii)   (iii)

9.29 Give the approximate values for the indicated bond angles in the following molecules:

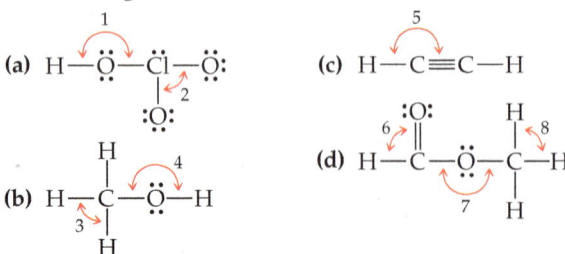

9.30 Give approximate values for the indicated bond angles in the following molecules:

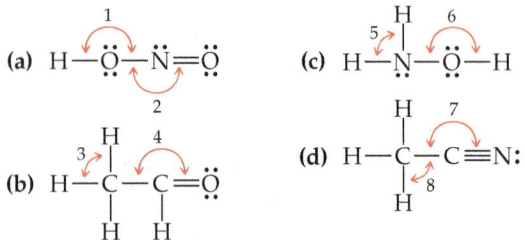

9.31 Ammonia, NH_3, reacts with incredibly strong bases to produce the amide ion, NH_2^-. Ammonia can also react with acids to produce the ammonium ion, NH_4^+. (a) Which species (amide ion, ammonia, or ammonium ion) has the largest H—N—H bond angle? (b) Which species has the smallest H—N—H bond angle?

9.32 In which of the following AF_n molecules or ions is there more than one F—A—F bond angle: SiF_4, F_5, SF_4, F_3?

9.33 Consider the following XF_4 ions: PF_4^-, BrF_4^-, ClF_4^+, and AlF_4^-. (a) Which of the ions have more than an octet of electrons around the central atom? (b) For which of the ions will the electron-domain and molecular geometries be the same? (c) Which of the ions will have an octahedral electron-domain geometry? (d) Which of the ions will exhibit a see-saw molecular geometry?

9.34 Name the proper three-dimensional molecular shapes for each of the following molecules or ions, showing lone pairs as needed: (a) ClO_2^- (b) SO_4^{2-} (c) NF_3 (d) CCl_2Br_2 (e) SF_4^{2+}

Shapes and Polarity of Polyatomic Molecules (Section 9.3)

9.35 (a) Which of the following statements are *true*?

(i) The overall dipole moment of a molecule is the vector sum of its individual bond dipoles.

(ii) If atoms A and B in an AB_n molecule have different electronegativities, then the AB_n molecule must have a nonzero dipole moment.

(iii) The bond dipoles in a tetrahedral AB_4 molecule cancel one another.

(b) An AB_2 molecule has a nonzero overall dipole moment. Is its molecular geometry linear or bent?

9.36 Consider a molecule with formula AX_3. Supposing the A—X bond is polar, how would you expect the dipole moment of the AX_3 molecule to change as the X—A—X bond angle increases from 100° to 120°?

9.37 (a) Does SCl_2 have a nonzero dipole moment? (b) If the answer to part (a) is yes, then which of the following best describes the direction of the net dipole moment? (i) It points along one of the S—Cl bonds with the positive end at the S atom. (ii) It points along one of the S—Cl bonds with the positive end at the Cl atom. (iii) It bisects the Cl—S—Cl angle with the positive end pointing toward the S atom. (iv) It bisects the Cl—S—Cl angle with the positive end pointing away from the S atom.

9.38 (a) The PH_3 molecule is polar. Does this offer experimental proof that the molecule cannot be planar? Explain. (b) It turns out that ozone, O_3, has a small dipole moment. How is this possible, given that all the atoms are the same?

9.39 (a) Is the molecule BF_3 polar or nonpolar? (b) If two electrons are added to BF_3 to make the ion BF_3^{2-}, is this ion planar? (c) Does the molecule BF_2Cl have a dipole moment?

9.40 (a) Which of the AF_3 molecules in Exercise 9.27 will have a nonzero dipole moment? (b) Which of the AF_4 molecules in Exercise 9.28 will have a zero dipole moment?

9.41 Predict whether each of the following molecules is polar or nonpolar: (a) IF (b) CS_2 (c) SO_3 (d) PCl_3 (e) SF_6 (f) IF_5

9.42 Predict whether each of the following molecules is polar or nonpolar: (a) CCl_4 (b) NH_3 (c) SF_4 (d) XeF_4 (e) CH_3Br (f) GaH_3

9.43 Dichloroethylene ($C_2H_2Cl_2$) has three forms (isomers), each of which is a different substance. (a) Draw Lewis structures of the three isomers, all of which have a carbon–carbon double bond. (b) Which of these isomers has a zero dipole moment? (c) How many isomeric forms can chloroethylene, C_2H_3Cl, have? Would they be expected to have dipole moments?

9.44 Dichlorobenzene, $C_6H_4Cl_2$, exists in three forms (isomers) called *ortho*, *meta*, and *para*:

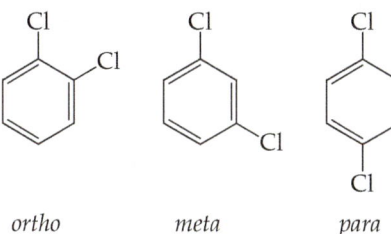

ortho   *meta*   *para*

Which of these has a nonzero dipole moment?

Orbital Overlap; Hybrid Orbitals (Sections 9.4 and 9.5)

9.45 Determine whether each of the following statements is true or false. **(a)** According to valence-bond theory, covalent bonds are made by overlapping orbitals on each atom in the bond. **(b)** A *p* orbital on one atom cannot overlap an *s* orbital on another atom. **(c)** The overlap between the orbitals on two atoms increases as the atoms are pulled apart. **(d)** The potential energy of a molecule increases when the atoms get too close together because of repulsion between the two nuclei. **(e)** The bond distance of a diatomic molecule corresponds to a minimum in the potential energy of the molecule.

9.46 Draw sketches illustrating the overlap between the following orbitals on two atoms: **(a)** the 2*s* orbital on each atom **(b)** the $2p_z$ orbital on each atom (assume both atoms are on the *z* axis) **(c)** the 2*s* orbital on one atom and the $2p_z$ orbital on the other atom.

9.47 For each statement, indicate whether it is true or false. **(a)** The greater the orbital overlap in a bond, the weaker the bond. **(b)** The greater the orbital overlap in a bond, the shorter the bond. **(c)** To create a hybrid orbital, you could use the *s* orbital on one atom with a *p* orbital on another atom. **(d)** Nonbonding electron pairs cannot occupy a hybrid orbital.

9.48 **(a)** According to valence-bond theory, which of the following correctly describes the formation of the Br—Br bond in the Br_2 molecule?

 (i) The 4*s* orbital on one atom overlaps with the 4*s* orbital on the other atom.

 (ii) The 4*s* orbital on one atom overlaps with a 4*p* orbital on the other atom.

 (iii) A 4*p* orbital on one atom overlaps with a 4*p* orbital on the other atom.

 (iv) A 4*p* orbital on one atom overlaps with a lone pair on the other atom.

 (b) According to valence-bond theory, which of the following correctly describes the formation of the H—F bond in the HF molecule?

 (i) The electron from the H atom is completely transferred to the F atom.

 (ii) The 1*s* orbital of the H atom overlaps with the 1*s* orbital on the F atom.

 (iii) The 1*s* orbital on the H atom overlaps with the 2*s* orbital on the F atom.

 (iv) The 1*s* orbital on the H atom overlaps with a 2*p* orbital on the F atom.

9.49 Consider the molecule BF_3. **(a)** What is the electron configuration of an isolated B atom? **(b)** What is the electron configuration of an isolated F atom? **(c)** What hybrid orbitals should be constructed on the B atom to make the B—F

bonds in BF_3? **(d)** What valence orbitals, if any, remain unhybridized on the B atom in BF_3?

9.50 Consider the SCl_2 molecule. **(a)** What is the electron configuration of an isolated S atom? **(b)** What is the electron configuration of an isolated Cl atom? **(c)** What hybrid orbitals should be constructed on the S atom to make the S—Cl bonds in SCl_2? **(d)** What valence orbitals, if any, remain unhybridized on the S atom in SCl_2?

9.51 Indicate the hybridization of the central atom in **(a)** BCl_3, **(b)** $AlCl_4^-$, **(c)** CS_2, **(d)** GeH_4.

9.52 What is the hybridization of the central atom in **(a)** $SiCl_4$, **(b)** HCN, **(c)** SO_3, **(d)** $TeCl_2$?

9.53 Shown here are three pairs of hybrid orbitals, with each set at a characteristic angle. For each pair, determine the type of hybridization, if any, that could lead to hybrid orbitals at the specified angle.

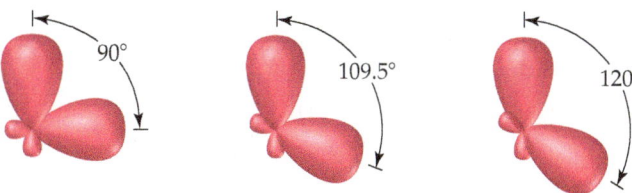

9.54 Determine whether each of the following statements about CH_3^+, CH_3^-, and CH_4 is true or false: **(a)** All three have the same hybridization at the C atom. **(b)** All three have an unhybridized orbital on the C atom. **(c)** The H—C—H angles in CH_3^+ are larger than those in the other two. **(d)** CH_3^- and CH_4 have the same electron-domain geometry.

Multiple Bonds (Section 9.6)

9.55 Which of the following statements are *true*? **(a)** A σ bond is generally stronger than a π bond. **(b)** Two *s* orbitals can make a π bond. **(c)** A π bond is made from the sideways overlap of two *p* orbitals. **(d)** A π bond has two regions of overlap on opposite sides of the internuclear axis.

9.56 **(a)** If the valence atomic orbitals of an atom are *sp* hybridized, how many unhybridized *p* orbitals remain in the valence shell? How many π bonds can the atom form? **(b)** Imagine that you could hold two atoms that are bonded together, twist them, and not change the bond length. Would it be easier to twist (rotate) around a single σ bond or around a double (σ plus π) bond, or would they be the same?

9.57 **(a)** Draw Lewis structures for ethane (C_2H_6), ethylene (C_2H_4), and acetylene (C_2H_2). **(b)** What is the hybridization of the carbon atoms in each molecule? **(c)** Predict which molecules, if any, are planar. **(d)** How many σ and π bonds are there in each molecule?

9.58 The nitrogen atoms in N_2 participate in multiple bonding, whereas those in hydrazine, N_2H_4, do not. **(a)** Draw Lewis structures for both molecules. **(b)** What is the hybridization of the nitrogen atoms in each molecule? **(c)** Which molecule has the stronger N—N bond?

9.59 Propylene, C_3H_6, is a gas that is used to form the important polymer called polypropylene. Its Lewis structure is

$$H-\overset{\displaystyle H}{\underset{}{C}}=\overset{\displaystyle H}{\underset{}{C}}-\overset{\displaystyle H}{\underset{\displaystyle H}{C}}-H$$

(a) What is the total number of valence electrons in the propylene molecule? **(b)** How many valence electrons are used to make σ bonds in the molecule? **(c)** How many valence electrons are used to make π bonds in the molecule? **(d)** How many valence electrons remain in nonbonding pairs in the molecule? (**e**) What is the hybridization at each carbon atom in the molecule?

9.60 Ethyl acetate, $C_4H_8O_2$, is a fragrant substance used both as a solvent and as an aroma enhancer. Its Lewis structure is

(**a**) What is the hybridization at each of the carbon atoms of the molecule? (**b**) What is the total number of valence electrons in ethyl acetate? (**c**) How many of the valence electrons are used to make σ bonds in the molecule? (**d**) How many valence electrons are used to make π bonds? (**e**) How many valence electrons remain in nonbonding pairs in the molecule?

9.61 Glycine, the simplest amino acid, has the following Lewis structure:

(**a**) What are the approximate bond angles about each of the two carbon atoms, and what are the hybridizations of the orbitals on each of them? (**b**) What are the hybridizations of the orbitals on the two oxygens and the nitrogen atom, and what are the approximate bond angles at the nitrogen? (**c**) What is the total number of σ bonds in the entire molecule, and what is the total number of π bonds?

9.62 Acetylsalicylic acid, better known as aspirin, has the following Lewis structure:

(**a**) What are the approximate values of the bond angles labeled 1, 2, and 3? (**b**) What hybrid orbitals are used about the central atom of each of these angles? (**c**) How many σ bonds are in the molecule?

9.63 Which of the following statements about localized and delocalized π bonding are *true*? (**a**) In a localized π bond, the two electrons in the bond are associated totally with two atoms. (**b**) A molecule or ion that has multiple resonance structures with double bonds exhibits delocalized π bonding. (**c**) In a delocalized π bond, the electrons are spread among three or more atoms. (**d**) Benzene is best described using localized π bonding.

9.64 (**a**) Write a single Lewis structure that satisfies the octet rule for SO_3. (**b**) What is the hybridization at the S atom? (**c**) Are there other equivalent Lewis structures for the molecule? (**d**) Would you expect SO_3 to exhibit delocalized π bonding? (**e**) How many electrons are in the π system of SO_3?

9.65 In the formate ion, HCO_2^-, the carbon atom is the central atom with the other three atoms attached to it. (**a**) Draw a Lewis structure for the formate ion. (**b**) What hybridization is exhibited by the C atom? (**c**) Are there multiple equivalent resonance structures for the ion? (**d**) How many electrons are in the π system of the ion?

9.66 Consider the following Lewis structure:

(**a**) Does the Lewis structure depict a neutral molecule or an ion? If it is an ion, what is the charge on the ion? (**b**) What hybridization is exhibited by each of the carbon atoms? (**c**) Are there multiple equivalent resonance structures for the species? (**d**) How many electrons are in the π system of the species?

9.67 Predict the molecular geometry of each of the following molecules:

(**a**) $H-C{\equiv}C-C{\equiv}C-C{\equiv}N$

(**b**)

(**c**) $H-N{=}N-H$

9.68 What hybridization do you expect for the atom indicated in red in each of the following species? (**a**) $CH_3CO_2^-$ (**b**) PH_4^+ (**c**) AlF_3 (**d**) $H_2C{=}CH-CH_2^+$

Molecular Orbitals and Period 2 Diatomic Molecules (Sections 9.7 and 9.8)

9.69 Which of the following statements about molecular orbitals are *true*? (**a**) A molecular orbital can hold a maximum of two electrons. (**b**) The number of molecular orbitals formed equals the number of atomic orbitals that are mixed to form the MOs. (**c**) A bonding molecular orbital will have a lower energy than that of the two atomic orbitals that are mixed to make the MO. (**d**) Electrons cannot be put in antibonding molecular orbitals. (**e**) A bonding molecular orbital concentrates electron density in the region between two atoms.

9.70 Which of the following statements about molecular orbitals and hybrid orbitals are *true*? (**a**) Molecular orbitals and hybrid orbitals are both created by mathematically mixing the wave functions of atomic orbitals. (**b**) Molecular orbitals are made by combining atomic orbitals from different atoms. (**c**) Hybrid orbitals and molecular orbitals are different names for the same things. (**d**) A hybrid orbital can hold more electrons than can a molecular orbital.

9.71 Consider the H_2^+ ion. (**a**) Sketch the molecular orbitals of the ion and draw its energy-level diagram. (**b**) How many electrons are there in the H_2^+ ion? (**c**) Write the electron configuration of the ion in terms of its MOs. (**d**) What is the bond order in H_2^+? (**e**) Suppose that the ion is excited by light so that an electron moves from a lower-energy to a higher-energy MO. Would you expect the excited-state H_2^+ ion to be stable or to fall apart? (**f**) Which of the following statements about part (e) is correct: (i) The light excites an electron from a bonding orbital to an antibonding orbital, (ii) The bond order of the ion does not change when an electron is excited, or (iii) In the excited state there are more bonding electrons than antibonding electrons?

9.72 (**a**) Sketch the molecular orbitals of the H_2^- ion and draw its energy-level diagram. (**b**) Write the electron configuration of the ion in terms of its MOs. (**c**) Calculate the bond order in H_2^-. (**d**) Suppose that the ion is excited by light, so that an electron moves from a lower-energy to a higher-energy molecular orbital. Would you expect the excited-state H_2^- ion to be stable? (**e**) Which of the following statements about part (d) is correct: (i) The light excites an electron from a bonding orbital to an antibonding orbital, (ii) The bond order of the ion does not change when an electron is excited, or (iii) In the excited state there are more bonding electrons than antibonding electrons?

9.73 Draw a picture that shows all three $2p$ orbitals on one atom and all three $2p$ orbitals on another atom. (a) Imagine the atoms coming close together to bond. How many σ bonds can the two sets of $2p$ orbitals make with each other? (b) How many π bonds can the two sets of $2p$ orbitals make with each other? (c) How many antibonding orbitals, and of what type, can be made from the two sets of $2p$ orbitals?

9.74 Indicate whether each statement is true or false. (a) s orbitals can only make σ or σ^* molecular orbitals. (b) The probability is 100% for finding an electron at the nucleus in a π^* orbital. (c) Antibonding orbitals are higher in energy than bonding orbitals (if all orbitals are created from the same atomic orbitals). (d) Electrons cannot occupy an antibonding orbital.

9.75 Consider the valence molecular orbitals for the Na_2 molecule that result from mixing the $3s$ valence atomic orbitals on each of the Na atoms. (a) How many molecular orbitals will be formed? (b) What is the bond order of the Na_2 molecule? (c) What is the bond order of the Na_2^+ molecular ion? (d) Which is expected to have the longer Na—Na bond, Na_2 or Na_2^+? (e) Which is expected to have the stronger Na—Na bond, Na_2 or Na_2^+?

9.76 (a) Based on its molecular-orbital diagram, what is the bond order of the O_2 molecule? (b) What is the expected bond order for the *peroxide* ion, O_2^{2-}? (c) What is the expected bond order for the *superoxide* ion, O_2^-? (d) From shortest to longest, predict the ordering of the bond lengths for O_2, O_2^{2-}, and O_2^-. (e) From weakest to strongest, predict the ordering of the bond strengths for O_2, O_2^{2-}, and O_2^-.

9.77 Determine whether each of the following statements about diamagnetism and paramagnetism is true or false: (a) A diamagnetic substance is weakly repelled from a magnetic field. (b) A substance with unpaired electrons will be diamagnetic. (c) A paramagnetic substance is attracted to a magnetic field. (d) The O_2 molecule is paramagnetic.

9.78 (a) Which of the following is expected to be paramagnetic: Ne, Li_2, Li_2^+, N_2, N_2^+, N_2^{2-}? (b) For each of the substances in part (a) that is paramagnetic, determine the number of unpaired electrons it has.

9.79 Using Figures 9.39 and 9.43 as guides, draw the molecular-orbital electron configuration for (a) B_2^+, (b) Li_2^+, (c) N_2^+, (d) Ne_2^{2+}. In each case indicate whether the addition of an electron to the ion would increase or decrease the bond order of the species.

9.80 If we assume that the energy-level diagrams for homonuclear diatomic molecules shown in Figure 9.43 can be applied to heteronuclear diatomic molecules and ions, predict the bond order and magnetic behavior of (a) CO^+, (b) NO^-, (c) OF^+, (d) NeF^+.

9.81 Determine the electron configurations for CN^+, CN, and CN^-. (a) Which species has the strongest C—N bond? (b) Which species, if any, is paramagnetic?

9.82 (a) The nitric oxide molecule, NO, readily loses one electron to form the NO^+ ion. Which of the following is the best explanation of why this happens? (i) Oxygen is more electronegative than nitrogen. (ii) The highest energy electron in NO lies in a π^*_{2p} molecular orbital. (iii) The π^*_{2p} MO in NO is completely filled. (b) Predict the order of the N—O bond strengths in NO, NO^+, and NO^-, and determine whether each is diamagnetic or paramagnetic. (c) With what neutral homonuclear diatomic molecules are the NO^+ and NO^- ions isoelectronic (same number of electrons)?

9.83 Assume that the MOs of diatomics from the third row of the periodic table, such as P_2, are analogous to those from the second row. (a) Which valence atomic orbitals of P are used to construct the MOs of P_2? (b) The figure that follows shows a sketch of one of the MOs for P_2. What is the label for this MO? (c) For the P_2 molecule, how many electrons occupy the MO in the figure? (d) Is P_2 expected to be diamagnetic or paramagnetic?

9.84 The iodine bromide molecule, IBr, is an *interhalogen compound*. Assume that the molecular orbitals of IBr are analogous to the homonuclear diatomic molecule F_2. (a) Which valence atomic orbitals of I and of Br are used to construct the MOs of IBr? (b) What is the bond order of the IBr molecule? (c) One of the valence MOs of IBr is sketched here. Determine whether each of the following statements about this orbital is *true*: (i) This is an antibonding orbital. (ii) The larger contribution is from the I atom. (iii) The energy of the molecular orbital is closer in energy to the valence atomic orbitals of Br than to those of I. (d) What is the label for the MO sketched here? (e) For the IBr molecule, how many electrons occupy the MO sketched here?

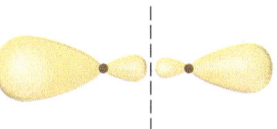

Additional Exercises

9.85 (a) Determine whether each of the following statements about the VSEPR model is *true*:

(i) The model is based on the arrangement of electron domains that minimizes the overall repulsion among them.

(ii) A double bond is considered to be two electron domains.

(iii) A nonbonding electron domain is more spread out in space than a bonding domain.

(b) Which of the following statements about the application of the VSEPR model are *true*?

(i) The optimum arrangement of four electron domains is square planar.

(ii) For a trigonal-bipyramidal electron-domain geometry, it is preferable to put nonbonding electron domains in an equatorial position.

(iii) An AB$_3$ molecule that exhibits a tetrahedral electron-domain geometry has no nonbonding electron domains on atom A.

9.86 An AB$_3$ molecule is described as having a trigonal-bipyramidal electron-domain geometry. (a) How many nonbonding domains are on atom A? (b) Based on the information given, which of the following is the molecular geometry of the molecule: (i) trigonal planar (ii) trigonal pyramidal (iii) T-shaped or (iv) tetrahedral?

9.87 Which of the following statements about the PF$_4$Cl molecule are *true*? (a) The molecule has an octahedral electron-domain geometry. (b) The P—Cl electron domain is larger than the P—F electron domains. (c) The Cl atom will occupy an equatorial site. (d) The molecule is hypervalent.

9.88 The vertices of a tetrahedron correspond to four alternating corners of a cube. By using analytical geometry, demonstrate that the angle made by connecting two of the vertices to a point at the center of the cube is 109.5°, the characteristic angle for tetrahedral molecules.

9.89 Fill in the blank spaces in the following chart. If the molecule column is blank, find an example that fulfills the conditions of the rest of the row.

Molecule	Electron-Domain Geometry	Hybridization of Central Atom	Dipole Moment? Yes or No
CO_2			
		sp^3	Yes
		sp^3	No
	Trigonal planar		No
SF_4			
	Octahedral		No
		sp^2	Yes
	Trigonal bipyramidal		No
XeF_2			

9.90 From their Lewis structures, determine the number of σ and π bonds in each of the following molecules or ions. (**a**) CO_2 (**b**) cyanogen, $(CN)_2$ (**c**) formaldehyde, H_2CO (**d**) formic acid, HCOOH, which has one H and two O atoms attached to C

9.91 The lactic acid molecule, $CH_3CH(OH)COOH$, gives sour milk its unpleasant, sour taste. (**a**) Draw the Lewis structure for the molecule, assuming that carbon always forms four bonds in its stable compounds. (**b**) How many π and how many σ bonds are in the molecule? (**c**) Which CO bond is shortest in the molecule? (**d**) What is the hybridization of atomic orbitals around the carbon atom associated with that short bond? (**e**) What are the approximate bond angles around each carbon atom in the molecule?

9.92 An AB_5 molecule adopts the geometry shown here. (**a**) What is the name of this molecular geometry? (**b**) What is the electron-domain geometry for the molecule? (**c**) Suppose the B atoms are halogen atoms. Of which group in the periodic table is atom A a member: (i) group 5A (ii) group 6A (iii) group 7A (iv) group 8A, or (v) is more information needed?

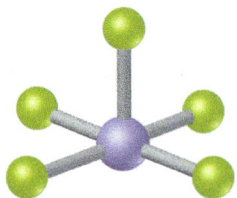

9.93 There are two compounds of the formula $Pt(NH_3)_2Cl_2$:

$$\begin{array}{cc} NH_3 & Cl \\ | & | \\ Cl-Pt-Cl & Cl-Pt-NH_3 \\ | & | \\ NH_3 & NH_3 \end{array}$$

The compound on the right is called *cisplatin*, and the compound on the left is called *transplatin*. (**a**) Which compound has a nonzero dipole moment? (**b**) One of these compounds is an anticancer drug, and one is inactive. The anticancer drug works by its chloride ions undergoing a substitution reaction with nitrogen atoms in DNA that are close together, forming a N—Pt—N angle of about 90°. Which compound would you predict to be the anticancer drug?

9.94 The O—H bond lengths in the water molecule (H_2O) are 0.96 Å, and the H—O—H angle is 104.5°. The overall dipole moment of the water molecule is 1.85 D. (**a**) Determine whether each of the following statements about the bond

dipoles and the overall dipole moment of H_2O is true: (i) The negative ends of the bond dipoles both point at the O atom. (ii) The two bond dipoles point in different directions. (iii) The overall dipole moment points in the same direction as one of the bond dipoles. (**b**) Calculate the magnitude of the bond dipole of the O—H bonds. (Note: You will need to use vector addition to do this.) (**c**) Compare your answer from part (b) to the dipole moments of the hydrogen halides (Table 8.3). Is your answer in accord with the relative electronegativity of oxygen?

9.95 (**a**) Predict the electron-domain geometry around the central Xe atom in XeF_2, XeF_4, and XeF_6. (**b**) The molecule IF_7 has a pentagonal bipyramid structure: five fluorines are equatorial, forming a flat pentagon around the central iodine atom, and the other two fluorines are axial. Predict the molecular geometry of IF_6^-.

9.96 Which of the following statements about hybrid orbitals is or are *true*? (**a**) After an atom undergoes sp hybridization, there is one unhybridized p orbital on the atom, (**b**) Under sp^2 hybridization, the large lobes point to the vertices of an equilateral triangle, and (**c**) The angle between the large lobes of sp^3 hybrids is 109.5°.

9.97 The Lewis structure for allene is

$$\begin{array}{ccc} H & & H \\ \diagdown & & \diagup \\ & C=C=C & \\ \diagup & & \diagdown \\ H & & H \end{array}$$

Make a sketch of the structure of this molecule that is analogous to Figure 9.22. In addition, answer the following questions: (**a**) What is the hybridization at the central C atom? (**b**) Is the molecule planar? (**c**) Does it have a nonzero dipole moment? (**d**) Would the π bonding in allene be described as delocalized?

9.98 The molecule C_4H_5N has the connectivity shown here. (**a**) After the Lewis structure for the molecule is completed, how many σ and how many π bonds are there in this molecule? (**b**) How many atoms in the molecule exhibit (i) sp hybridization (ii) sp^2 hybridization and (iii) sp^3 hybridization?

$$\begin{array}{ccccc} & H & & & H \\ & | & & & | \\ H-C & -C-C-N-C & -H \\ & | & & & \\ & H & & & \end{array}$$

9.99 Sodium azide is a shock-sensitive compound that releases N_2 upon physical impact. The compound is used in automobile airbags. The azide ion is N_3^-. (**a**) Draw the Lewis structure of the azide ion that minimizes formal charge (it does not form a triangle). Is it linear or bent? (**b**) State the hybridization of the central N atom in the azide ion. (**c**) How many σ bonds and how many π bonds does the central nitrogen atom make in the azide ion?

9.100 In ozone, O_3, the two oxygen atoms on the ends of the molecule are equivalent to one another. (**a**) What is the best choice of hybridization scheme for the atoms of ozone? (**b**) For one of the resonance forms of ozone, which of the orbitals are used to make bonds and which are used to hold nonbonding pairs of electrons? (**c**) Which of the orbitals can be used to delocalize the π electrons? (**d**) How many electrons are delocalized in the π system of ozone?

9.101 Butadiene, C_4H_6, is a planar molecule that has the following carbon–carbon bond lengths:

$$H_2C\!=\!\!=\!\!=\!\!CH\underset{1.48\,\text{Å}}{-\!\!-\!\!-}CH\!=\!\!=\!\!=\!\!CH_2$$
$$\overset{1.34\,\text{Å}}{}\qquad\qquad\overset{1.34\,\text{Å}}{}$$

(**a**) Predict the bond angles around each of the carbon atoms and sketch the molecule. (**b**) From left to right, what is the hybridization of each carbon atom in butadiene? (**c**) The middle C—C bond length in butadiene (1.48 Å) is a little shorter than the average C—C single bond length (1.54 Å). Does this imply that the middle C—C bond in butadiene is weaker or stronger than the average C—C single bond? (**d**) Based on your answer for part (c), discuss what additional aspects of bonding in butadiene might support the shorter middle C—C bond.

9.102 The structure of *borazine*, $B_3N_3H_6$, is a six-membered ring of alternating B and N atoms. There is one H atom bonded to each B and to each N atom. The molecule is planar. (**a**) Write a Lewis structure for borazine in which the formal charge on every atom is zero. (**b**) Write a Lewis structure for borazine in which the octet rule is satisfied for every atom. (**c**) What are the formal charges on the atoms in the Lewis structure from part (b)? Given the electronegativities of B and N, do the formal charges seem favorable or unfavorable? (**d**) Do either of the Lewis structures in parts (a) and (b) have multiple resonance structures? (**e**) What are the hybridizations at the B and N atoms in the Lewis structures from parts (a) and (b)? Would you expect the molecule to be planar for both Lewis structures? (**f**) The six B—N bonds in the borazine molecule are all identical in length at 1.44 Å. Typical values for the bond lengths of B—N single and double bonds are 1.51 Å and 1.31 Å, respectively. Does the value of the B—N bond length seem to favor one Lewis structure over the other? (**g**) How many electrons are in the π system of borazine?

9.103 The highest occupied molecular orbital of a molecule is abbreviated as the HOMO. The lowest unoccupied molecular orbital in a molecule is called the LUMO. Experimentally, you can measure the difference in energy between the HOMO and LUMO by taking the electronic absorption (UV–visible) spectrum of the molecule. Peaks in the electronic absorption spectrum can be labeled as $\pi_{2p}-\pi_{2p}^*$, $\sigma_{2s}-\sigma_{2s}^*$, and so on, corresponding to electrons being promoted from one orbital to another. The HOMO–LUMO transition corresponds to molecules going from their ground state to their first excited state. (**a**) Write out the molecular-orbital valence electron configurations for the ground state and first excited state for N_2. (**b**) Is N_2 paramagnetic or diamagnetic in its first excited state? (**c**) The electronic absorption spectrum of the N_2 molecule has the lowest energy peak at 170 nm. To what orbital transition does this correspond? (**d**) Calculate the energy of the HOMO–LUMO transition in part (a) in terms of kJ/mol. (**e**) Is the N—N bond in the first excited state stronger or weaker compared to that in the ground state?

9.104 One of the molecular orbitals of the H_2^- ion can be sketched as follows:

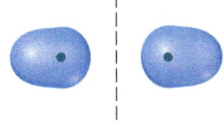

(**a**) Is the molecular orbital a σ or π MO? Is it bonding or antibonding? (**b**) In H_2^-, how many electrons occupy the MO sketched here? (**c**) What is the bond order in the H_2^- ion? (**d**) Compared to the H—H bond in H_2, the H—H bond in H_2^- is expected to be which of the following? (i) shorter and stronger (ii) longer and stronger (iii) shorter and weaker (iv) longer and weaker or (v) the same length and strength

9.105 Place the following molecules and ions in order from smallest to largest bond order: H_2^+, B_2, N_2^+, F_2^+, and Ne_2.

9.106 The following sketches show the atomic orbital wave functions (with phases) used to construct some of the MOs of a homonuclear diatomic molecule. For each sketch, determine the type of MO that will result from mixing the atomic orbital wave functions as drawn. Use the same labels for the MOs as in the "Closer Look" box on phases.

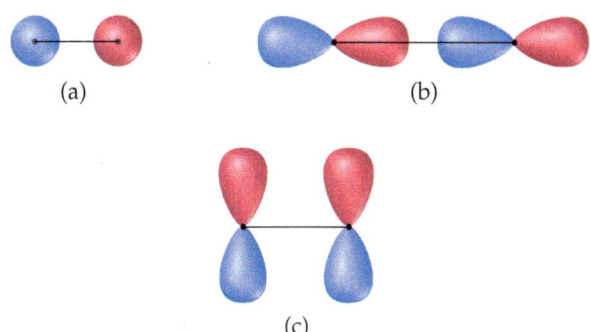

(a)　　　　　(b)

(c)

9.107 *Azo dyes* are organic dyes that are used for many applications, such as the coloring of fabrics. Many azo dyes are derivatives of the organic substance *azobenzene*, $C_{12}H_{10}N_2$. A closely related substance is *hydrazobenzene*, $C_{12}H_{12}N_2$. The Lewis structures of these two substances are

Azobenzene　　　　　Hydrazobenzene

(Recall the shorthand notation used for benzene.) (**a**) What is the hybridization at the N atom in each of the substances? (**b**) How many unhybridized atomic orbitals are there on the N and the C atoms in each of the substances? (**c**) Predict the N—N—C angles in each of the substances. (**d**) Azobenzene is said to have greater delocalization of its π electrons than hydrazobenzene. Discuss this statement in light of your answers to (a) and (b). (**e**) All the atoms of azobenzene lie in one plane, whereas those of hydrazobenzene do not. Is this observation consistent with the statement in part (d)? (**f**) Azobenzene is an intense red-orange color, whereas hydrazobenzene is nearly colorless. Which molecule would be a better one to use in a solar energy conversion device? (See the "Chemistry and Sustainability" box for more information about solar cells.)

9.108 Carbon monoxide, CO, is isoelectronic to N_2. (**a**) Draw a Lewis structure for CO that satisfies the octet rule. (**b**) Assume that the diagram in Figure 9.46 can be used to describe the MOs of CO. What is the predicted bond order for CO? Is this answer in accord with the Lewis structure you drew in part (a)? (**c**) Experimentally, it is found that the highest-energy electrons in CO reside in a σ-type MO. Which of the following statements best explains this observation? (i) The observation is consistent with the ordering of MOs in Figure 9.46. (ii) The observation suggests that the ordering of the π_{2p} and π_{2p}^* MOs needs to be modified compared to Figure 9.46. (iii) The observation suggests that the σ_{2p} MO in Figure 9.46 is pushed above the π_{2p} MO by mixing similar to that in Figure 9.42. (**d**) Would you expect the π_{2p} MOs of CO to have equal atomic orbital contributions from the C and O atoms? If not, which atom would have the greater contribution?

9.109 The energy-level diagram in Figure 9.40 shows that the sideways overlap of a pair of p orbitals produces two molecular orbitals, one bonding and one antibonding. In ethylene there is a pair of electrons in the bonding π orbital between the two carbons. Absorption of a photon of the appropriate wavelength can result in promotion of one of the bonding electrons from the π_{2p} to the π_{2p}^* molecular orbital. (**a**) Assuming this electronic transition corresponds to the HOMO–LUMO transition, what is the HOMO in ethylene? (**b**) Assuming this electronic transition corresponds to the HOMO–LUMO transition, what is the LUMO in ethylene? (**c**) Is the C—C bond in ethylene stronger or weaker in the excited state than in the ground state? Why? (**d**) Is the C—C bond in ethylene easier to twist in the ground state or in the excited state?

9.110 A compound composed of 2.1% H, 29.8% N, and 68.1% O has a molar mass of approximately 50 g/mol. (**a**) What is the molecular formula of the compound? (**b**) What is its Lewis structure if H is bonded to O? (**c**) What is the geometry of the molecule? (**d**) What is the hybridization of the orbitals around the N atom? (**e**) How many σ and how many π bonds are there in the molecule?

9.111 Sulfur tetrafluoride (SF_4) reacts slowly with O_2 to form sulfur tetrafluoride monoxide (OSF_4) according to the following unbalanced reaction:

$$SF_4(g) + O_2(g) \longrightarrow OSF_4(g)$$

The O atom and the four F atoms in OSF_4 are bonded to a central S atom. (**a**) Balance the equation. (**b**) Write a Lewis structure of OSF_4 in which the formal charges of all atoms are zero. (**c**) Use average bond enthalpies (Table 8.3) to estimate the enthalpy of the reaction. Is it endothermic or exothermic? (**d**) Determine the electron-domain geometry of OSF_4, and write two possible molecular geometries for the molecule based on this electron-domain geometry. (**e**) For each of the molecules you drew in part (d), state how many fluorine atoms are equatorial and how many are axial.

9.112 The molecule 2-butene, C_4H_8, can undergo a geometric change called *cis–trans isomerization*:

cis-2-butene   *trans*-2-butene

As discussed in the "Chemistry and Life" box on the chemistry of vision, such transformations can be induced by light and are the key to human vision. (**a**) What is the hybridization at the two central carbon atoms of 2-butene? (**b**) The isomerization occurs by rotation about the central C—C bond. With reference to Figure 9.29, which of the following statements best characterizes what happens halfway through the rotation from *cis*- to *trans*-2-butene? (i) The σ bond between the two C atoms is destroyed. (ii) The π bond between the two C atoms is destroyed. (iii) Neither the σ bond nor the π bond between the two C atoms is destroyed. (**c**) Based on average bond enthalpies (Table 8.3), how much energy per molecule must be supplied to break the C—C π bond? (**d**) What is the longest wavelength of light that will provide photons of sufficient energy to break the C—C π bond and cause the isomerization? (**e**) Is the wavelength in your answer to part (d) in the visible portion of the electromagnetic spectrum? (**f**) Based on your answer to part (e), could this isomerization reaction be the basis for human vision?

9.113 Methyl isocyanate, CH_3NCO, was made infamous in 1984 when an accidental leakage of this compound from a storage tank in Bhopal, India, resulted in the deaths of about 3800 people and severe and lasting injury to many thousands more. (**a**) Draw a Lewis structure for methyl isocyanate. (**b**) Draw a ball-and-stick model of the structure, including estimates of all the bond angles in the compound. (**c**) Predict all the bond distances in the molecule. (**d**) Do you predict that the molecule will have a dipole moment?

Design an Experiment

In this chapter we have been introduced to a number of new concepts, including the delocalization of π systems of molecules and the molecular-orbital description of molecular bonding. A connection between these concepts is provided by the field of *organic dyes*, molecules with delocalized π systems that have color. The color is due to the excitation of an electron from the *highest occupied molecular orbital* (HOMO) to the *lowest unoccupied molecular orbital* (LUMO). It is hypothesized that the energy gap between the HOMO and the LUMO depends on the length of the π system. Imagine that you are given samples of the following substances to test this hypothesis:

or, in shorthand notation for organic molecules

butadiene

hexatriene

β-carotene

β-Carotene is the substance chiefly responsible for the bright orange color of carrots. It is also an important nutrient for the body's production of retinal (see the "Chemistry and Life" box in Section 9.6). (**a**) What experiments could you design to determine the amount of energy needed to excite an electron from the HOMO to the LUMO in each of these molecules? (**b**) How might you graph your data to determine whether a relationship exists between the length of the π system and the excitation energy? (**c**) What additional molecules might you want to procure to further test the ideas developed here? (**d**) How could you design an experiment to determine whether the delocalized π systems and not some other molecular features, such as molecular length or the presence of π bonds, are important in making the excitations occur in the visible portion of the spectrum? (*Hint:* You might want to test some additional molecules not shown here.)

10

▲ Hot air is less dense than cool air, so balloons filled with hot air will rise.

GASES

In this chapter, we examine the physical properties of gases. Because gases are the simplest state of matter, they serve as an excellent starting point for probing the behavior of large collections of atoms or molecules. Indeed, it is fairly straightforward to formulate a simple model for gases that explains their behavior under most common conditions. We gain further insight by then comparing how real gases differ from this ideal model as conditions change. Thus, while we start with a description that treats all gases the same regardless of their chemical identity, we end by gaining an understanding of important aspects of the physical behavior of molecules.

WHAT'S AHEAD

10.1 ▶ Physical Characteristics of Gases Compare the distinguishing characteristics of gases with those of liquids and solids. Describe gas *pressure* and the units used to express it, and consider Earth's atmosphere and the pressure it exerts.

10.2 ▶ The Gas Laws Express the state of a gas in terms of its volume, pressure, temperature, and quantity, and examine several *gas laws*, which are empirical relationships among these four variables.

10.3 ▶ The Ideal-Gas Equation Examine the assumptions behind the ideal-gas equation, which combines the gas laws of the previous section into a single equation, $PV = nRT$, and use this equation to analyze the properties of gases.

10.4 ▶ Gas Mixtures and Partial Pressures Recognize and use the fact that in a mixture of gases, each gas exerts a pressure that is part of the total pressure. This *partial pressure* is the pressure the gas would exert if it were by itself.

10.5 ▶ The Kinetic-Molecular Theory of Gases Relate the macroscopic properties of gases to their behavior at the molecular level. The atoms or molecules that make up a gas are in constant random motion and move with an average kinetic energy that is proportional to the temperature of the gas.

10.6 ▶ Molecular Speeds, Effusion, and Diffusion Use kinetic-molecular theory to calculate the speed of gas molecules and predict relative rates of *effusion* and *diffusion*.

10.7 ▶ Real Gases: Deviations from Ideal Behavior Examine the reasons why real gases deviate from ideal behavior and the conditions where those deviations are significant. The *van der Waals equation* accounts for real gas behavior at high pressures and low temperatures.

10.1 | Physical Characteristics of Gases

Learning Objectives

When you have finished Section 10.1, you should be able to:

▶ Describe how the properties of gases differ from those of solids and liquids.

▶ Define gas pressure and determine it from experimental data.

▶ Interconvert between units of pressure.

Of the few elements that exist as gases at ordinary temperatures and pressures, He, Ne, Ar, Kr, and Xe are monatomic and H_2, N_2, O_2, F_2, and Cl_2 are diatomic. Many molecular compounds are gases under ordinary conditions, and Table 10.1 lists a few of them. All of these gases are composed entirely of nonmetallic elements. Furthermore, all have simple molecular formulas and, therefore, low molar masses.

Substances that are liquids or solids under ordinary conditions can also exist in the gaseous state, where they are often referred to as **vapors**. The substance H_2O, for example, can exist as liquid water, solid ice, or water vapor.

Even though different gaseous substances may have very different *chemical* properties, they behave quite similarly as far as their *physical* properties are concerned. For example, the N_2 and O_2 that account for approximately 99% of our atmosphere have very different chemical properties—O_2 supports human life but N_2 does not, to name just one difference—but these two components of air behave physically as one gaseous material because their physical properties are essentially identical.

The physical properties of gases differ significantly from those of solids and liquids. For example, a gas expands spontaneously to fill its container. Consequently, the volume of a gas equals the volume of its container. Gases also are highly compressible: When pressure is applied to a gas, its volume readily decreases. Solids and liquids, on the other hand, do not expand to fill their containers and are not readily compressible.

Two or more gases form a homogeneous mixture regardless of the identities or relative proportions of the gases; the atmosphere serves as an excellent example. Two or more liquids or two or more solids may or may not form homogeneous mixtures, depending on their chemical nature. For example, when water and gasoline are mixed, the two liquids remain as separate layers. In contrast, the water vapor and gasoline vapors above the liquids form a homogeneous gas mixture.

The characteristic properties of gases—expanding to fill a container, being highly compressible, forming homogeneous mixtures—arise because the molecules are relatively far apart. In any given volume of air, for example, the molecules take up only about 0.1% of the total volume, with the rest being empty space. Thus, each molecule behaves largely as though the others were not present. As a result, different gases behave similarly even though they are made up of different molecules.

Pressure

One characteristic of gases is that they all exert a *pressure* on any surface with which they are in contact. We can interpret this observation in terms of the atoms or molecules in the gas. The molecules of a gas move chaotically, colliding with each other and with the walls of their container. The impacts with the container walls exert a force—an

TABLE 10.1   Some Common Compounds That Are Gases at Room Temperature

Formula	Name	Characteristics
HCN	Hydrogen cyanide	Very toxic, slight odor of bitter almonds
H_2S	Hydrogen sulfide	Very toxic, odor of rotten eggs
CO	Carbon monoxide	Toxic, colorless, odorless
CO_2	Carbon dioxide	Colorless, odorless
CH_4	Methane	Colorless, odorless, flammable
C_2H_4	Ethene (ethylene)	Colorless, ripens fruit
C_3H_8	Propane	Colorless, odorless, bottled gas
N_2O	Nitrous oxide	Colorless, sweet odor, laughing gas
NO_2	Nitrogen dioxide	Toxic, red-brown, irritating odor
NH_3	Ammonia	Colorless, pungent odor
SO_2	Sulfur dioxide	Colorless, irritating odor

outward push against the walls. The **pressure**, P, that a gas exerts is defined as the force, F, divided by the area, A, of the surface on which the force is acting:

$$P = \frac{F}{A} \qquad [10.1]$$

Atmospheric Pressure and the Barometer

People, apples, and nitrogen molecules all experience an attractive gravitational force that pulls them toward the center of Earth. When an apple comes loose from a tree, for example, this force causes the apple to be accelerated toward Earth, its speed increasing as its potential energy is converted into kinetic energy. (Section 1.4) The gas atoms and molecules of the atmosphere also experience a gravitational acceleration. Because these particles have such tiny masses, however, their thermal energies of motion (their kinetic energies) override the gravitational forces, so the particles that make up the atmosphere don't pile up at Earth's surface. Nevertheless, the gravitational force does operate, and it causes the atmosphere as a whole to press down on Earth's surface, creating *atmospheric pressure*, defined as the force exerted by the atmosphere on a given surface area.

You can demonstrate the existence of atmospheric pressure with an empty plastic water bottle. If you suck on the mouth of the empty bottle, chances are you can cause the bottle to partially cave in. When you break the partial vacuum you have created, the bottle pops out to its original shape. The bottle caves in because, once you've sucked out some of the air molecules, the air molecules in the atmosphere exert a force on the outside of the bottle that is greater than the force exerted by the lesser number of air molecules inside the bottle.

We can calculate the magnitude of the atmospheric pressure using Equation 10.1. The force, F, exerted by any object is the product of its mass, m, and its acceleration, a: $F = ma$. When applied to our atmosphere, the force is the gravitational force, which is also commonly called weight. The acceleration is that due to gravity, $g = 9.8 \text{ m/s}^2$. Thus, the gravitational force of the atmosphere on Earth's surface (the weight of the atmosphere) is given by $F = mg$.

Now imagine a column of air, 1 m^2 in cross section, extending through the entire atmosphere (**Figure 10.1**). That column has a mass of roughly 10,000 kg. The downward gravitational force exerted on this column is

$$F = (10{,}000 \text{ kg})(9.8 \text{ m/s}^2) = 1 \times 10^5 \text{ kg-m/s}^2 = 1 \times 10^5 \text{ N}$$

where N is the abbreviation for *newton*, the SI unit for force: $1 \text{ N} = 1 \text{ kg-m/s}^2$.

The pressure exerted by the column is this force divided by the cross-sectional area, A, over which the force is applied. Because our air column has a cross-sectional area of 1 m^2, we have for the magnitude of atmospheric pressure at sea level

$$P = \frac{F}{A} = \frac{1 \times 10^5 \text{ N}}{1 \text{ m}^2} = 1 \times 10^5 \text{ N/m}^2 = 1 \times 10^5 \text{ Pa} = 1 \times 10^2 \text{ kPa}$$

The SI unit of pressure is the **pascal** (Pa), named for Blaise Pascal (1623–1662), a French scientist who studied pressure: $1 \text{ Pa} = 1 \text{ N/m}^2$. A related pressure unit is the **bar**: $1 \text{ bar} = 10^5 \text{ Pa} = 10^5 \text{ N/m}^2$. Thus, the atmospheric pressure at sea level we just calculated, 100 kPa, can be reported as 1 bar. (The actual atmospheric pressure at any location depends on weather conditions and altitude.) Another pressure unit is pounds per square inch (psi, lbs/in.^{2}). At sea level, atmospheric pressure is 14.7 psi.

In the seventeenth century, many scientists and philosophers believed that the atmosphere had no weight. Evangelista Torricelli (1608–1647), a student of Galileo's, proved this untrue. He invented the *barometer* (**Figure 10.2**), which is made from a glass tube more than 760 mm long that is closed at one end, completely filled with mercury, and inverted into a dish of mercury. (Care must be taken so that no air gets into the tube.) When the tube is inverted into the dish, some of the mercury flows out of the tube, but a column of mercury remains in the tube. Torricelli argued that the mercury surface in the dish experiences the full force of Earth's atmosphere, which pushes the mercury up the tube until the pressure exerted by the mercury column downward, due to gravity, equals the atmospheric pressure at the base of the tube. Therefore, *the height, h, of the mercury column is a measure of atmospheric pressure and changes as atmospheric pressure changes.*

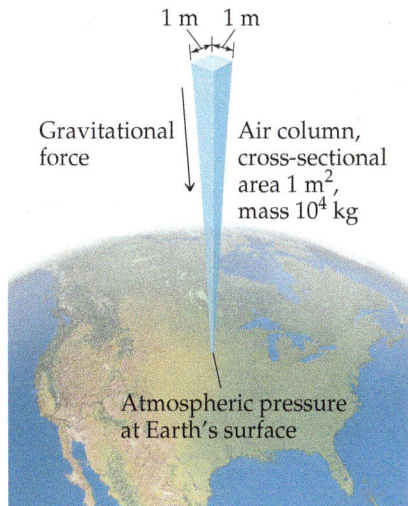

1 m   1 m

Gravitational force

Air column, cross-sectional area 1 m^2, mass 10^4 kg

Atmospheric pressure at Earth's surface

▲ **Figure 10.1** A calculation of atmospheric pressure.

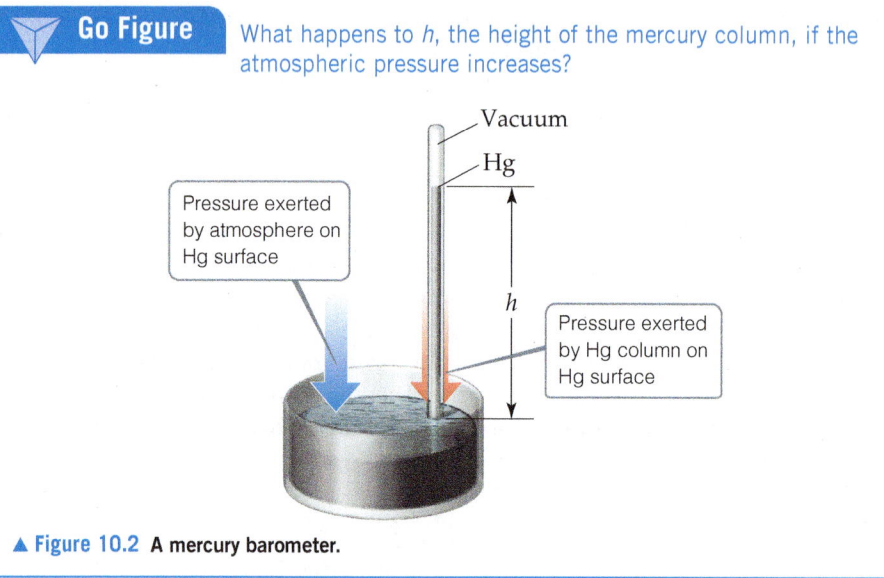

Go Figure   What happens to *h*, the height of the mercury column, if the atmospheric pressure increases?

Vacuum

Hg

Pressure exerted by atmosphere on Hg surface

h

Pressure exerted by Hg column on Hg surface

▲ **Figure 10.2 A mercury barometer.**

Standard atmospheric pressure, which corresponds to the typical pressure at sea level, is the pressure sufficient to support a column of mercury 760 mm high. In SI units, this pressure is 1.01325×10^5 Pa. Standard atmospheric pressure defines some common non-SI units used to express gas pressure, such as the **atmosphere** (atm) and the *millimeter of mercury* (mm Hg). The latter unit is also called the **torr**, after Torricelli: 1 torr = 1 mm Hg. Thus, we have

$$1 \text{ atm} = 760. \text{ mm Hg} = 760. \text{ torr} = 1.01325 \times 10^5 \text{ Pa} = 101.325 \text{ kPa} = 1.01325 \text{ bar}$$

We express gas pressure in a variety of units throughout this chapter, so you should become comfortable converting pressures from one unit to another.

Sample Exercise 10.1

Calculating Pressure

What is the pressure, in kilopascals, on the body of a diver if she is 31.0 m below the surface of the water when the atmospheric pressure on the surface is 98 kPa? Assume that the density of the water is $1.00 \text{ g/cm}^3 = 1.00 \times 10^3 \text{ kg/m}^3$. The gravitational constant is 9.81 m/s^2, and $1 \text{ Pa} = 1 \text{ kg/m-s}^2$.

SOLUTION

Analyze We are asked to calculate the pressure on the diver given the atmospheric pressure (98 KPa) and the depth of the water (31.0 m).

Plan The total pressure on the diver equals that of the atmosphere plus that of the water. The pressure of the water can be calculated starting with Equation 10.1, $P = F/A$. The force, F, due to the water above the diver is given by its mass times the acceleration due to gravity, $F = mg$, where $g = 9.81 \text{ m/s}^2$.

Solve The pressure caused by the water is

$$P = \frac{F}{A} = \frac{mg}{A}$$

The mass of the water is related to its density ($d = m/V$, so $m = d \times V$). We can treat the water as a column whose volume equals its cross-sectional area times its height: $V = A \times h$. When we make these substitutions for mass ($m = d \times V$) and volume ($V = A \times h$), we have

$$P = \frac{mg}{A} = \frac{dVg}{A} = \frac{d(Ah)g}{A} = dhg$$

Inserting SI quantities, we have

$$P = dhg = (1.00 \times 10^3 \text{ kg/m}^3)(31.0 \text{ m})(9.81 \text{ m/s}^2)$$

$$= 3.00 \times 10^5 \frac{\text{kg}}{\text{m-s}^2} = 3.00 \times 10^5 \text{ Pa}$$

Thus, the total pressure on the diver is

$$P_{\text{total}} = 98 \text{ kPa} + 300 \text{ kPa} = 398 \text{ kPa}$$

This corresponds to a pressure of 3.94 atm.

▶ **Practice Exercise**

Gallium melts just above room temperature and is liquid over a very wide temperature range (30–2204 °C), which means it would be a suitable fluid for a high-temperature barometer. Given its density, $d_{\text{Ga}} = 6.0 \text{ g/cm}^3$, what would be the height of the column if gallium is used as the barometer fluid and the external pressure is 9.5×10^4 Pa?

We use various devices to measure the pressures of enclosed gases. Tire gauges, for example, measure the pressure of air in automobile and bicycle tires. In laboratories, we sometimes use a *manometer*, which operates on a principle similar to that of a barometer, as shown in Sample Exercise 10.2.

Sample Exercise 10.2
Using a Manometer to Measure Gas Pressure

On a certain day, a laboratory barometer indicates that the atmospheric pressure is 764.7 torr. A sample of gas is placed in a flask attached to an open-end mercury manometer (**Figure 10.3**), and a meter stick is used to measure the height of the mercury in the two arms of the U tube. The height of the mercury in the open-ended arm is 136.4 mm, and the height in the arm in contact with the gas in the flask is 103.8 mm. What is the pressure of the gas in the flask (**a**) in atmospheres, (**b**) in kilopascals?

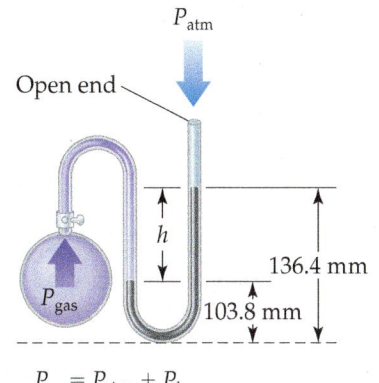

$$P_{gas} = P_{atm} + P_h$$

▲ **Figure 10.3** A mercury manometer.

SOLUTION

Analyze We are given the atmospheric pressure (764.7 torr) and the mercury heights in the two arms of the manometer, and we are asked to determine the gas pressure in the flask. Recall that millimeters of mercury is a pressure unit. We know that the gas pressure from the flask must be greater than atmospheric pressure because the mercury level in the arm on the flask side (103.8 mm) is lower than the level in the arm open to the atmosphere (136.4 mm). Therefore, the gas from the flask is pushing mercury from the arm in contact with the flask into the arm open to the atmosphere.

Plan We will use the difference in height between the two arms (*h* in Figure 10.3) to obtain the amount by which the pressure of the gas exceeds atmospheric pressure. Because an open-end mercury manometer is used, the height difference directly measures the pressure difference in mm Hg or torr between the gas and the atmosphere.

Solve

(**a**) The pressure of the gas equals the atmospheric pressure plus *h*:

$$P_{gas} = P_{atm} + h$$
$$= 764.7 \text{ torr} + (136.4 \text{ torr} - 103.8 \text{ torr})$$
$$= 797.3 \text{ torr}$$

We convert the pressure of the gas to atmospheres:

$$P_{gas} = (797.3 \text{ torr})\left(\frac{1 \text{ atm}}{760 \text{ torr}}\right) = 1.049 \text{ atm}$$

(**b**) To calculate the pressure in kPa, we employ the conversion factor between atmospheres and kPa:

$$1.049 \text{ atm}\left(\frac{101.3 \text{ kPa}}{1 \text{ atm}}\right) = 106.3 \text{ kPa}$$

Check The calculated pressure is a bit more than 1 atm, which is about 101 kPa. This makes sense because we anticipated that the pressure in the flask would be greater than the atmospheric pressure (764.7 torr = 1.01 atm) acting on the manometer.

▶ **Practice Exercise**
If the pressure of the gas inside the flask were increased and the height of the column in the open-ended arm went up by 5.0 mm in Figure 10.3, what would be the new pressure of the gas in the flask, in torr?

Self-Assessment Exercises

SAE 10.1   Which of the following statements is *false*? (**a**) Many physical properties of gases are similar to one another. (**b**) Gases are much more compressible than liquids and solids. (**c**) Given enough time, any mixture of gases will form a homogeneous mixture. (**d**) Many chemical properties of gases are similar to one another.

SAE 10.2   The atmosphere of Venus is made up mostly of CO_2. The temperature and pressure at the surface of the planet are estimated to be 740 K and 93 bar, respectively. What is the mass of a 1.0 m^2 cross-sectional column of Venus's atmosphere that extends from the surface to the upper reaches of the atmosphere? The acceleration due to gravity on Venus is 8.9 m/s^2. (**a**) 1.0×10^1 kg (**b**) 1.0×10^4 kg (**c**) 1.1×10^6 kg (**d**) 9.3×10^6 kg

SAE 10.3   A vacuum pump is used to remove most of the gas in a container, ultimately producing a pressure of 10.5 torr inside the vessel. What is the pressure in the vessel in units of atm? (**a**) 1.38×10^{-2} atm (**b**) 10.6 atm (**c**) 7.98×10^2 atm (**d**) 1.06×10^5 atm

Learning Objectives

When you have finished Section 10.2, you should be able to:

▶ Use Boyle's law to predict how changes in the pressure of a fixed quantity of a gas held at constant temperature affect the volume of the gas and vice versa.

▶ Use Charles's law to predict how changes in the temperature of a fixed quantity of a gas held at constant pressure affect the volume of the gas and vice versa.

▶ Use Avogadro's law to predict how changes in the number of moles of a gas held at constant temperature and pressure are related to the volume the gas occupies.

Go Figure

Does atmospheric pressure increase or decrease as altitude increases? (Neglect changes in temperature.)

Balloon rises up through atmosphere

▲ **Figure 10.4** As a balloon rises in the atmosphere, its volume increases.

10.2 | The Gas Laws

Four variables are needed to define the physical condition, or *state*, of a gas: temperature, pressure, volume, and amount of gas, usually expressed as number of moles. The equations that express the relationships among these four variables are known as the *gas laws*. Because volume is easily measured, the first gas laws to be studied expressed the effect of one of the variables on volume, with the remaining two variables held constant.

The Pressure–Volume Relationship: Boyle's Law

Gas volume increases as the pressure exerted on the gas decreases. Thus, an inflated weather balloon released at Earth's surface expands as it rises (**Figure 10.4**) because the pressure of the atmosphere decreases with increasing elevation.

The British chemist Robert Boyle (1627–1691) was the first person to investigate the quantitative relationship between the pressure of a gas and its volume. He found, for example, that decreasing the pressure of a gas to half its original value causes the volume to double. Conversely, doubling the pressure causes the volume to decrease to half its original value.

Boyle's law, which summarizes these observations, states that:

The volume of a fixed quantity of gas maintained at constant temperature is inversely proportional to the pressure.

When two measurements are inversely proportional, one gets smaller as the other gets larger. Boyle's law can be expressed mathematically as

$$V = \text{constant} \times \frac{1}{p} \quad \text{or} \quad PV = \text{constant} \qquad [10.2]$$

The value of the constant depends on temperature and on the amount of gas in the sample.

The graph of V versus P in **Figure 10.5** shows the curve obtained for a given quantity of gas at a fixed temperature. A linear relationship is obtained when V is plotted versus $1/P$ as shown on the right in Figure 10.5.

Boyle's law occupies a special place in the history of science because Boyle was the first to carry out experiments in which one variable was systematically changed to determine the effect on another variable. The data from the experiments were then employed to establish an empirical relationship—a "law."

We apply Boyle's law every time we breathe. The rib cage, which can expand and contract, and the diaphragm, a muscle beneath the lungs, govern the volume of the lungs. Inhalation occurs when the rib cage expands and the diaphragm moves downward. Both actions increase the volume of the lungs, thus decreasing the gas pressure inside the lungs. Atmospheric pressure then forces air into the lungs until the pressure in the lungs equals atmospheric pressure. Exhalation reverses the process—the rib cage contracts and the diaphragm moves up, decreasing the volume of the lungs. Air is forced out of the lungs by the resulting increase in pressure.

Go Figure

What would a plot of P versus $1/V$ look like for a fixed quantity of gas at a fixed temperature?

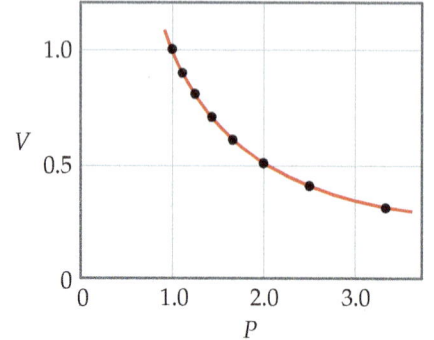
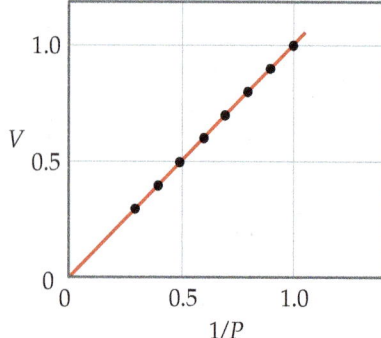

▲ **Figure 10.5** **Boyle's law.** For a fixed quantity of gas at constant temperature, the volume of the gas is inversely proportional to its pressure.

The Temperature–Volume Relationship: Charles's Law

Figure 10.6 shows that the volume of an inflated balloon increases when the tempera-ture of the gas inside the balloon increases and decreases when the temperature of the gas decreases. This relationship between gas volume and temperature was discovered in 1787 by French scientist Jacques Charles (1746–1823). Some typical volume–temperature data are shown in **Figure 10.7**. Notice that the extrapolated (dashed) line passes through −273 °C. Note also that the gas is predicted to have zero volume at this temperature. This condition is never realized, however, because all gases liquefy or solidify before reaching this temperature.

In 1848, William Thomson (1824–1907), a British physicist whose title was Lord Kel-vin, proposed an absolute-temperature scale, now known as the Kelvin scale. On this scale, 0 K, called *absolute zero*, equals −273.15 °C. (Section 1.5) In terms of the Kelvin scale, **Charles's law** states the following:

> *The volume of a fixed amount of gas maintained at constant pressure is directly proportional to its absolute temperature.*

Thus, doubling the absolute temperature causes the gas volume to double. Mathemati-cally, Charles's law takes the form

$$V = \text{constant} \times T \quad \text{or} \quad \frac{V}{T} = \text{constant} \qquad [10.3]$$

with the value of the constant depending on the pressure and on the amount of gas.

The Quantity–Volume Relationship: Avogadro's Law

The relationship between the quantity of a gas and its volume follows from the work of Joseph Louis Gay-Lussac (1778–1823) and Amedeo Avogadro (1776–1856).

Gay-Lussac was one of those extraordinary figures in the history of science who could truly be called an adventurer. In 1804, he ascended to 23,000 feet in a hot-air balloon—an exploit that held the altitude record for several decades. To better control the balloon, Gay-Lussac studied the properties of gases. In 1808, he observed the *law of combining vol-umes*: At a given pressure and temperature, the volumes of gases that react with one another are in the ratios of small whole numbers. For example, two volumes of hydrogen gas react with one volume of oxygen gas to form two volumes of water vapor. (Section 3.1)

Three years later, Amedeo Avogadro interpreted Gay-Lussac's observation by propos-ing what is now known as **Avogadro's hypothesis**:

> *Equal volumes of gases at the same temperature and pressure contain equal numbers of molecules.*

For example, 22.4 L of any gas at 0 °C and 1 atm contain 6.02×10^{23} gas molecules (that is, 1 mol), as depicted in **Figure 10.8**.

▲ **Figure 10.6** The effect of temperature on volume.

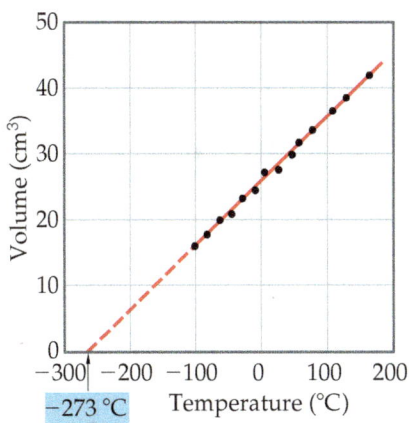

▲ **Figure 10.7** **Charles's law.** For a fixed quantity of gas at constant pressure, the volume of the gas is proportional to its temperature.

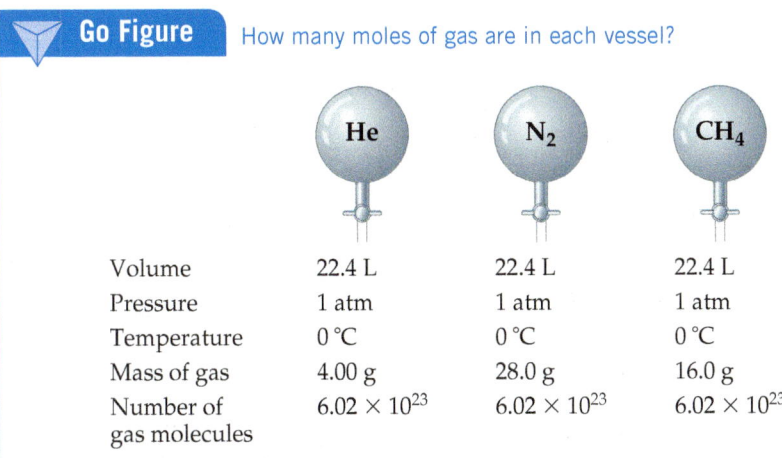

Go Figure How many moles of gas are in each vessel?

	He	N₂	CH₄
Volume	22.4 L	22.4 L	22.4 L
Pressure	1 atm	1 atm	1 atm
Temperature	0 °C	0 °C	0 °C
Mass of gas	4.00 g	28.0 g	16.0 g
Number of gas molecules	6.02×10^{23}	6.02×10^{23}	6.02×10^{23}

▲ **Figure 10.8** **Avogadro's hypothesis.** At the same volume, pressure, and temperature, samples of different gases have the same number of molecules but different masses.

Avogadro's law follows from Avogadro's hypothesis:

The volume of a gas maintained at constant temperature and pressure is directly proportional to the number of moles of the gas.

That is,

$$V = \text{constant} \times n \quad \text{or} \quad \frac{V}{n} = \text{constant} \qquad [10.4]$$

where n is number of moles. Thus, doubling the number of moles of gas, for instance, causes the volume to double if T and P remain constant.

Sample Exercise 10.3

Evaluating the Effects of Changes in *P*, *V*, *n*, and *T* on a Gas

Suppose we have a gas confined to a cylinder with a movable piston that is sealed so there are no leaks. (Sections 5.2, 5.3) How will each of the following changes affect (i) the pressure of the gas, (ii) the number of moles of gas in the cylinder, (iii) the average distance between molecules? (**a**) Heating the gas while maintaining a constant pressure; (**b**) Reducing the volume while maintaining a constant temperature; (**c**) Injecting additional gas while keeping the temperature and volume constant.

SOLUTION

Analyze We need to think about how each change affects (1) the pressure of the gas, (2) the number of moles of gas in the cylinder, and (3) the average distance between molecules.

Plan We can use the gas laws to evaluate the changes in pressure. The number of moles of gas in the cylinder will not change unless gas is either added or removed. Assessing the average distance between molecules is not quite as straightforward. For a given number of gas molecules, the average distance between molecules increases as the volume increases. Conversely, for constant volume, the average distance between molecules decreases as the number of moles increases. Thus, the average distance between molecules will be proportional to V/n.

Solve

(**a**) Because it is stipulated that the pressure remains constant, pressure is not a variable in this problem, and the total number of moles of gas will also remain constant. We know from Charles's law, however, that heating the gas while maintaining constant pressure will cause the piston to move and the volume to increase. Thus, the distance between molecules will increase.

(**b**) The reduction in volume causes the pressure to increase (Boyle's law). Compressing the gas into a smaller volume does not change the total number of gas molecules; thus, the total

number of moles remains the same. The average distance between molecules, however, must decrease because of the smaller volume.

(**c**) Injecting more gas into the cylinder means that more molecules are present and there will be an increase in the number of moles of gas in the cylinder. Because we have added more molecules while keeping the volume constant, the average distance between molecules must decrease. Avogadro's law tells us that the volume of the cylinder should have increased when we added more gas, provided the pressure and temperature were held constant. Here the volume is held constant, as is the temperature, which means the pressure must change. Knowing from Boyle's law that there is an inverse relationship between volume and pressure ($PV = \text{constant}$), we conclude that if the volume does not increase on injecting more gas, then the pressure must increase.

▶ **Practice Exercise**

An oxygen cylinder used in a hospital contains 35.4 L of oxygen gas at a pressure of 149.6 atm. How much volume would the oxygen occupy if it were transferred to a container that maintained a pressure of 1.00 atm if the temperature remained constant?

Self-Assessment Exercises

SAE 10.4 If the pressure of a particular sample of a gas at constant temperature is doubled, what will happen to its volume? (**a**) It will double. (**b**) It will remain the same. (**c**) It will be cut in half. (**d**) It will be cut to a quarter of the original volume. (**e**) Not enough information is given to answer this question.

SAE 10.5 A gas is contained inside a flexible but impermeable membrane with a volume of 1.00 L and a temperature of 25 °C. If the gas is heated to 50 °C while keeping the pressure constant, what volume will it occupy? (**a**) 0.923 L (**b**) 1.00 L (**c**) 1.08 L (**d**) 2.00 L

SAE 10.6 Three balloons, each held at room temperature and each containing a different gas (methane, CH_4; nitrogen, N_2; and oxygen, O_2), have identical volumes. Which of the following does this imply? (**a**) The same number of atoms is present in each balloon. (**b**) The same number of molecules is present in each balloon. (**c**) The same mass of gas is present in each balloon. (**d**) More than one of these statements are true.

10.3 | The Ideal-Gas Equation

All three gas laws we just examined were obtained by holding two of the four variables *P, V, T,* and *n* constant and seeing how the remaining two variables affect each other. We can express each law as a proportionality relationship. Using the symbol $\propto$ for "is proportional to," we have

$$\text{Boyle's law:} \quad V \propto \frac{1}{P} \quad (\text{constant } n, T)$$

$$\text{Charles's law:} \quad V \propto T \quad (\text{constant } n, P)$$

$$\text{Avogadro's law:} \quad V \propto n \quad (\text{constant } P, T)$$

We can combine these relationships into a general gas law:

$$V \propto \frac{nT}{P}$$

If we call the proportionality constant *R*, we obtain an equality:

$$V = R\left(\frac{nT}{P}\right)$$

which we can rearrange to

$$PV = nRT \qquad [10.5]$$

Equation 10.5 is the **ideal-gas equation** (also called the **ideal-gas law**). An **ideal gas** is a hypothetical gas whose pressure, volume, and temperature relationships are described completely by the ideal-gas equation.

In deriving the ideal-gas equation, we make two assumptions:

* the molecules of an ideal gas do not interact with one another, and
* the combined volume of the molecules is much smaller than the volume the gas occupies.

For these reasons, we consider the molecules as taking up no space in the container. In many cases, the small error introduced by these assumptions is acceptable. If more accurate calculations are needed, we can correct for the assumptions if we know something about the attraction molecules have for one another and the size of the molecules, as we explain in Section 10.7.

The term *R* in the ideal-gas equation is called the **gas constant**. The value and units of *R* depend on the units of *P, V, n,* and *T.* The value for *T* in the ideal-gas equation must *always* be the absolute temperature (in kelvins instead of degrees Celsius). The quantity of gas, *n,* is normally expressed in moles. The units chosen for pressure and volume are most often atmospheres and liters, respectively. However, other units can be used. In countries other than the United States, the pascal is the most commonly used unit for pressure. Table 10.2 lists the numerical value for *R* in various units. In working with the ideal-gas equation, you must choose the form of *R* in which the units agree with the units of *P, V, n,* and *T* given in the problem. In this chapter, we most often use $R = 0.08206$ L-atm/mol-K because pressure is most often given in atmospheres.

Suppose we have 1.000 mol of an ideal gas at 1.000 atm and 0.00 °C (273.15 K). According to the ideal-gas equation, the volume of the gas is

$$V = \frac{nRT}{P}$$

$$= \frac{(1.000 \text{ mol})(0.08206 \text{ L-atm/mol-K})(273.15 \text{ K})}{1.000 \text{ atm}} = 22.41 \text{ L}$$

The conditions 0 °C and 1 atm are referred to as **standard temperature and pressure (STP)**. The volume occupied by 1 mol of ideal gas at STP, 22.41 L, is known as the *molar volume* of an ideal gas at STP.

The ideal-gas equation accounts adequately for the properties of most gases under a variety of circumstances. The equation is not exactly correct, however, for any real gas. Thus, the measured volume for given values of *P, n,* and *T* might differ from the volume

Learning Objectives

When you have finished Section 10.3, you should be able to:

▶ Calculate the pressure, volume, moles, or temperature of a gas using the ideal-gas equation.

▶ Calculate the density or molar mass of a gas using the ideal-gas equation.

▶ Use the ideal-gas equation and the concept of reaction stoichiometry to calculate the volume of gas produced or consumed in a reaction.

TABLE 10.2 Numerical Values of the Gas Constant R in Various Units

Units	Numerical Value
L-atm/mol-K	0.08206
J/mol-K *	8.314
cal/mol-K	1.987
m³-Pa/mol-K *	8.314
L-torr/mol-K	62.36

*SI unit

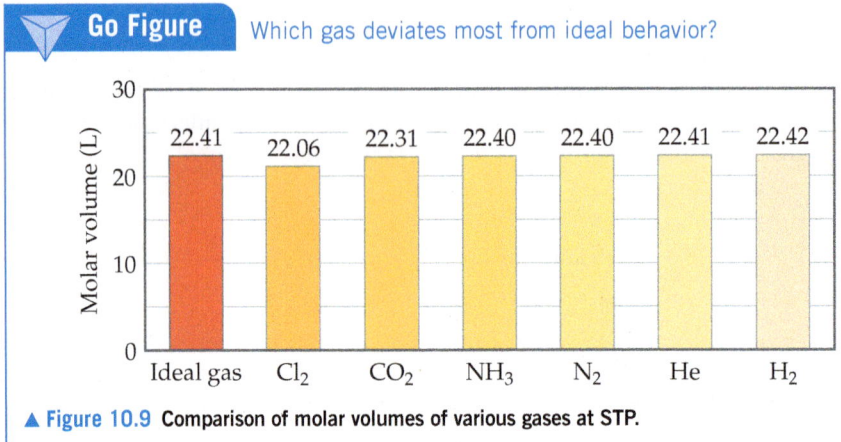

▲ **Figure 10.9 Comparison of molar volumes of various gases at STP.**

calculated from $PV = nRT$ (**Figure 10.9**). Although real gases do not always behave ideally, their behavior differs so little from ideal behavior that we can ignore any deviations for all but the most accurate work.

Sample Exercise 10.4
Using the Ideal-Gas Equation

Calcium carbonate, $CaCO_3(s)$, the principal compound in limestone, decomposes upon heating to $CaO(s)$ and $CO_2(g)$. A sample of $CaCO_3$ is decomposed, and the carbon dioxide is collected in a 250-mL flask. After decomposition is complete, the gas has a pressure of 1.3 atm at a temperature of 31 °C. How many moles of CO_2 gas were generated?

SOLUTION

Analyze We are given the volume (250 mL), pressure (1.3 atm), and temperature (31 °C) of a sample of CO_2 gas and asked to calculate the number of moles of CO_2 in the sample.

Plan Because we are given V, P, and T, we can use the ideal-gas equation to solve for the unknown quantity, n.

Solve In analyzing and solving gas law problems, it is helpful to tabulate the information given in the problems and then to convert the values to units that are consistent with those for $R(0.08206$ L-atm/mol-K). In this case, the given values are

$$V = 250 \, mL = 0.250 \, L$$
$$P = 1.3 \, atm$$
$$T = 31 \, °C = (31 + 273) \, K = 304 \, K$$

Remember: *Absolute temperature must always be used when the ideal-gas equation is solved.*

We now rearrange the ideal-gas equation (Equation 10.5) to solve for n

$$n = \frac{PV}{RT}$$

$$n = \frac{(1.3 \, atm)(0.250 \, L)}{(0.08206 \, L\text{-}atm/mol\text{-}K)(304 \, K)}$$

$$= 0.013 \, mol \, CO_2$$

Check Appropriate units cancel, thus ensuring that we have properly rearranged the ideal-gas equation and have converted to the correct units.

▶ **Practice Exercise**
Tennis balls are usually filled with either air or N_2 gas to a pressure above atmospheric pressure to increase their bounce. If a tennis ball has a volume of 144 cm³ and contains 0.33 g of N_2 gas, what is the pressure inside the ball at 24 °C?

STRATEGIES FOR SUCCESS | Calculations Involving Many Variables

In this chapter, we encounter a variety of problems based on the ideal-gas equation, which contains four variables—P, V, n, and T—and one constant, R. Depending on the type of problem, we might need to solve for any of the four variables.

To extract the necessary information from problems involving more than one variable, we suggest the following steps:

1. **Tabulate information.** Read the problems carefully to determine which variable is the unknown and which variables have numeric values given. Every time you encounter a numerical

value, jot it down. In many cases, constructing a table of the given information will be useful.

2. **Convert to consistent units.** Make certain that quantities are converted to the proper units. In using the ideal-gas equation, for example, we usually use the value of R that has units of L-atm/mol-K. If you are given a pressure in torr, you will need to convert it to atmospheres before using this value of R in your calculations.

3. **If a single equation relates the variables, solve the equation for the unknown.** For the ideal-gas equation, these algebraic rearrangements will all be used at one time or another:

$$P = \frac{nRT}{V}, \quad V = \frac{nRT}{P}, \quad n = \frac{PV}{RT}, \quad T = \frac{PV}{nR}$$

4. **Use dimensional analysis.** Carry the units through your calculation. Using dimensional analysis enables you to check that you have solved an equation correctly. If the units in the equation cancel to give the units of the desired variable, you have probably used the equation correctly.

Sometimes you will not be given explicit values for several variables, making it look like a problem that cannot be solved. In these cases, however, you should look for information that can be used to determine the needed variables. For example, suppose you are using the ideal-gas equation to calculate a pressure in a problem that gives a value for T but not for n or V. However, the problem states that "the sample contains 0.15 mol of gas per liter." You can turn this statement into the expression

$$\frac{n}{V} = 0.15 \text{ mol/L}$$

Solving the ideal-gas equation for pressure yields

$$P = \frac{nRT}{V}$$

which we can rewrite as follows:

$$P = \left(\frac{n}{V}\right)RT$$

Thus, we can solve the equation even though we are not given explicit values for n and V.

As we have continuously stressed, the most important thing you can do to become proficient at solving chemistry problems is to do the Practice Exercises at the end of each Sample Exercise, the Self-Assessment Exercises at the end of each section, and the Exam Prep questions and the Exercises at the end of each chapter. By using systematic procedures, such as those described here, you should be able to minimize difficulties in solving problems involving many variables.

Relating the Ideal-Gas Equation and the Gas Laws

The gas laws we discussed in Section 10.2 are special cases of the ideal-gas equation. For example, when n and T are held constant, the product nRT contains three constants and so must itself be a constant:

$$PV = nRT = \text{constant} \quad \text{or} \quad PV = \text{constant} \qquad [10.6]$$

Note that this rearrangement gives Boyle's law. We see that if n and T are constant, the values of P and V can change, but the product PV must remain constant.

We can use Boyle's law to determine how the volume of a gas changes when its pressure changes. For example, if a cylinder fitted with a movable piston holds 50.0 L of O_2 gas at 18.5 atm and 21 °C, what volume will the gas occupy if the temperature is maintained at 21 °C while the pressure is reduced to 1.00 atm? Because the product PV is a constant when a gas is held at constant n and T, we know that

$$P_1V_1 = P_2V_2 \qquad [10.7]$$

where P_1 and V_1 are initial values and P_2 and V_2 are final values. Dividing both sides of this equation by P_2 gives the final volume, V_2:

$$V_2 = V_1 \times \frac{P_1}{P_2} = (50.0 \text{ L})\left(\frac{18.5 \text{ atm}}{1.00 \text{ atm}}\right) = 925 \text{ L}$$

The answer is reasonable because a gas expands as its pressure decreases.

In a similar way, we can start with the ideal-gas equation and derive relationships between any other two variables, V and T (Charles's law), n and V (Avogadro's law), or P and T.

We are often faced with the situation in which P, V, and T all change for a fixed number of moles of gas. Because n is constant in this situation, the ideal-gas equation gives

$$\frac{PV}{T} = nR = \text{constant}$$

If we represent the initial and final conditions by subscripts 1 and 2, respectively, we can write an equation that is often called the *combined gas law*:

$$\frac{P_1V_1}{T_1} = \frac{P_2V_2}{T_2} \qquad [10.8]$$

Sample Exercise 10.5
Calculating the Effect of Temperature Changes on Pressure

The gas pressure in an aerosol can is 1.5 atm at 25 °C. Assuming that the gas obeys the ideal-gas equation, what is the pressure when the can is heated to 450 °C?

SOLUTION

Analyze We are given the initial pressure (1.5 atm) and temperature (25 °C) of the gas and asked for the pressure at a higher temperature (450 °C).

Plan The volume and number of moles of gas do not change, so we must use a relationship connecting pressure and temperature. Converting temperature to the Kelvin scale and tabulating the given information, we have

	P	T
Initial	1.5 atm	298 K
Final	P_2	723 K

Solve To determine how P and T are related, we start with the ideal-gas equation and isolate the quantities that do not change (n, V, and R) on one side and the variables (P and T) on the other side.

$$\frac{P}{T} = \frac{nR}{V} = \text{constant}$$

Because the quotient P/T is a constant, we can write

$$\frac{P_1}{T_1} = \frac{P_2}{T_2}$$

(where the subscripts 1 and 2 represent the initial and final states, respectively). Rearranging to solve for P_2 and substituting, we have that the given data give

$$P_2 = (1.5\,\text{atm})\left(\frac{723\,\text{K}}{298\,\text{K}}\right) = 3.6\,\text{atm}$$

Check This answer is intuitively reasonable—increasing the temperature of a gas increases its pressure.

Comment It is evident from this example why aerosol cans carry a warning not to incinerate.

▶ **Practice Exercise**
The pressure in a natural-gas tank is maintained at 2.20 atm. On a day when the temperature is −15 °C, the volume of gas in the tank is $3.25 \times 10^3\,\text{m}^3$. What is the volume of the same quantity of gas on a day when the temperature is 31 °C?

Sample Exercise 10.6
Using the Combined Gas Laws

An inflated balloon has a volume of 6.0 L at sea level (1.0 atm) and is allowed to ascend until the pressure is 0.45 atm. During ascent, the temperature of the gas falls from 22 °C to −21 °C. Calculate the volume of the balloon at its final altitude.

SOLUTION

Analyze We need to determine a new volume for a gas sample when both pressure and temperature change.

Plan Let's again proceed by converting temperatures to kelvins and tabulating our information.

	P	V	T
Initial	1.0 atm	6.0 L	295 K
Final	0.45 atm	V_2	252 K

Because n is constant, we can use Equation 10.8.

Solve Rearranging Equation 10.8 to solve for V_2 gives

$$V_2 = V_1 \times \frac{P_1}{P_2} \times \frac{T_2}{T_1}$$

$$= (6.0\,\text{L})\left(\frac{1.0\,\text{atm}}{0.45\,\text{atm}}\right)\left(\frac{252\,\text{K}}{295\,\text{K}}\right) = 11\,\text{L}$$

Check The result appears reasonable. Notice that the calculation involves multiplying the initial volume by a ratio of pressures and a ratio of temperatures. Intuitively, we expect decreasing pressure to cause the volume to increase, while decreasing the temperature should have the opposite effect. Because the change in pressure is more dramatic than the change in temperature, we expect the effect of the pressure change to predominate in determining the final volume, as it does.

▶ **Practice Exercise**
A 0.50-mol sample of oxygen gas is confined at 0 °C and 1.0 atm in a cylinder with a movable piston. The piston compresses the gas so that the final volume is half the initial volume and the final pressure is 2.2 atm. What is the final temperature of the gas in degrees Celsius?

Gas Densities and Molar Mass

Recall that density has units of mass per unit volume ($d = m/V$). (Section 1.5) We can arrange the ideal-gas equation to obtain similar units of moles per unit volume:

$$\frac{n}{V} = \frac{P}{RT}$$

If we multiply both sides of this equation by the molar mass, $\mathcal{M}$, which is the number of grams in 1 mol of a substance (Section 3.4), we obtain

$$\frac{n\mathcal{M}}{V} = \frac{P\mathcal{M}}{RT} \qquad\qquad [10.9]$$

The term on the left equals the density in grams per liter:

$$\frac{\text{moles}}{\text{liter}} \times \frac{\text{grams}}{\text{mole}} = \frac{\text{grams}}{\text{liter}}$$

Thus, the density of the gas is also given by the expression on the right in Equation 10.9:

$$d = \frac{n\mathcal{M}}{V} = \frac{P\mathcal{M}}{RT} \qquad\qquad [10.10]$$

Equation 10.10 tells us that the density of a gas depends on its pressure, molar mass, and temperature. The higher the molar mass and pressure, the denser the gas. The higher the temperature, the less dense the gas. Although gases form homogeneous mixtures, a less dense gas will lie above a denser gas in the absence of mixing. For example, CO_2 has a higher molar mass than N_2 or O_2 and is therefore denser than air. For this reason, CO_2 released from a CO_2 fire extinguisher blankets a fire, preventing O_2 from reaching the combustible material. "Dry ice," which is solid CO_2, converts directly to CO_2 gas at room temperature, and the resulting "fog" (which is actually condensed water droplets cooled by the CO_2) is carried downward by the heavier, but colorless, CO_2 (Figure 10.10).

▲ **Figure 10.10** Carbon dioxide gas flows downward because it is denser than air.

Sample Exercise 10.7
Calculating Gas Density

What is the density of carbon tetrachloride vapor at 714 torr and 125 °C?

SOLUTION

Analyze We are asked to calculate the density of a gas given its name, its pressure, and its temperature. From the name, we can write the chemical formula of the substance and determine its molar mass.

Plan We can use Equation 10.10 to calculate the density. Before we can do that, however, we must convert the given quantities to the appropriate units—namely, degrees Celsius to kelvins and pressure to atmospheres. We must also calculate the molar mass of CCl_4.

Solve The absolute temperature is $125 + 273 = 398$ K. The pressure is $(714 \text{ torr})(1 \text{ atm}/760 \text{ torr}) = 0.939$ atm. The molar mass of CCl_4 is $12.01 + (4)(35.45) = 153.8$ g/mol. Therefore,

$$d = \frac{P\mathcal{M}}{RT} = \frac{(0.939 \text{ atm})(153.8 \text{ g/mol})}{(0.08206 \text{ L-atm/mol-K})(398 \text{ K})} = 4.42 \text{ g/L}$$

Check If we divide molar mass (g/mol) by density (g/L), we end up with L/mol. The numerical value is roughly $154/4.4 = 35$, which is in the right ballpark for the molar volume of a gas heated to 125 °C at near atmospheric pressure. We may thus conclude our answer is reasonable.

▶ **Practice Exercise**
The mean molar mass of the gases in the atmosphere at the surface of Titan, Saturn's largest moon, is 28.6 g/mol. The surface temperature is 95 K, and the pressure is 1.6 atm. Assuming ideal behavior, calculate the density of Titan's atmosphere.

When we have equal molar masses of two gases at the same pressure but different temperatures, the hotter gas is less dense than the cooler one, so the hotter gas rises. The difference between the densities of hot and cold air is responsible for the lift of hot-air balloons. It is also responsible for many phenomena in weather, such as the formation of large thunderheads during thunderstorms.

Equation 10.10 can be rearranged to solve for the molar mass of a gas:

$$\mathcal{M} = \frac{dRT}{P} \qquad\qquad [10.11]$$

Thus, we can use the experimentally measured density of a gas to determine the molar mass of the gas molecules, as shown in Sample Exercise 10.8.

Sample Exercise 10.8
Calculating the Molar Mass of a Gas

A large evacuated flask initially has a mass of 134.567 g. When the flask is filled with a gas of unknown molar mass to a pressure of 735 torr at 31 °C, its mass is 137.328 g. When the flask is evacuated again and then filled with water at 31 °C, its mass is 1067.9 g. (The density of water at this temperature is 0.997 g/mL.) Assuming the ideal-gas equation applies, calculate the molar mass of the gas.

SOLUTION

Analyze We are given the temperature (31 °C) and pressure (735 torr) for a gas, together with information to determine its volume and mass, and we are asked to calculate its molar mass.

Plan The data obtained when the flask is filled with water can be used to calculate the volume of the container. The mass of the empty flask and of the flask when filled with gas can be used to calculate the mass of the gas. From these quantities we calculate the gas density and then apply Equation 10.11 to calculate the molar mass of the gas.

Solve The gas volume equals the volume of water the flask can hold, calculated from the mass and density of the water. The mass of the water is the difference between the masses of the full and evacuated flask:

$$1067.9 \text{ g} - 134.567 \text{ g} = 933.3 \text{ g}$$

Rearranging the equation for density ($d = m/V$), we have

$$V = \frac{m}{d} = \frac{(933.3 \text{ g})}{(0.997 \text{ g/mL})} = 936 \text{ mL} = 0.936 \text{ L}$$

The gas mass is the difference between the mass of the flask filled with gas and the mass of the evacuated flask:

$$137.328 \text{ g} - 134.567 \text{ g} = 2.761 \text{ g}$$

Knowing the mass of the gas (2.761 g) and its volume (0.936 L), we can calculate the density of the gas:

$$d = 2.761 \text{ g}/0.936 \text{ L} = 2.95 \text{ g/L}$$

After converting pressure to atmospheres and temperature to kelvins, we can use Equation 10.11 to calculate the molar mass:

$$
\begin{aligned}
\mathcal{M} &= \frac{dRT}{P} \\
&= \frac{(2.95 \text{ g/L})(0.08206 \text{ L-atm}/\text{mol-K})(304 \text{ K})}{(0.967 \text{ atm})} \\
&= 76.1 \text{ g/mol}
\end{aligned}
$$

Check The units work out appropriately, and the value of molar mass obtained is reasonable for a substance that is gaseous near room temperature.

▶ **Practice Exercise**
Calculate the average molar mass of dry air if it has a density of 1.17 g/L at 21 °C and 740.0 torr.

Volumes of Gases in Chemical Reactions

We are often concerned with knowing the identity and/or quantity of a gas involved in a chemical reaction. Thus, it is useful to be able to calculate the volumes of gases consumed or produced in reactions. Such calculations are based on the mole concept and balanced chemical equations. (Section 3.6) The coefficients in a balanced chemical equation tell us the relative amounts (in moles) of reactants and products in a reaction. The ideal-gas equation relates the number of moles of a gas to P, V, and T.

Sample Exercise 10.9
Relating Gas Variables and Reaction Stoichiometry

Automobile air bags are inflated by nitrogen gas generated by the rapid decomposition of sodium azide, NaN_3:

$$2 \text{ NaN}_3(s) \longrightarrow 2 \text{ Na}(s) + 3 \text{ N}_2(g)$$

If an air bag has a volume of 36 L and is to be filled with nitrogen gas at 1.15 atm and 26 °C, how many grams of NaN_3 must be decomposed?

SOLUTION

Analyze This is a multistep problem. We are given the volume, pressure, and temperature of the N_2 gas and the chemical equation for the reaction by which the N_2 is generated. We must use this information to calculate the number of grams of NaN_3 needed to obtain the necessary N_2.

Plan We need to use the gas data (P, V, and T) and the ideal-gas equation to calculate the number of moles of N_2 gas that should be formed for the air bag to operate correctly. We can then use the balanced equation to determine the number of moles of NaN_3 needed. Finally, we can convert moles of NaN_3 to grams.

Solve

The conversion sequence is:

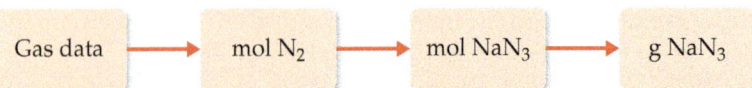

The number of moles of N_2 is determined using the ideal-gas equation:

$$n = \frac{PV}{RT} = \frac{(1.15\ \text{atm})(36\ \text{L})}{(0.08206\ \text{L-atm/mol-K})(299\ \text{K})}$$

$$= 1.7\ \text{mol}\ N_2$$

We use the coefficients in the balanced equation to calculate the number of moles of NaN_3:

$$(1.7\ \text{mol}\ N_2)\frac{2\ \text{mol}\ NaN_3}{(3\ \text{mol}\ N_2)} = 1.1\ \text{mol}\ NaN_3$$

Finally, using the molar mass of NaN_3, we convert moles of NaN_3 to grams:

$$(1.1\ \text{mol}\ NaN_3)\frac{65.0\ \text{g}\ NaN_3}{(1\ \text{mol}\ NaN_3)} = 72\ \text{g}\ NaN_3$$

Check The units cancel properly at each step in the calculation, leaving us with the correct units in the answer, g NaN_3.

▶ **Practice Exercise**

In the first step of the industrial process for making nitric acid, ammonia reacts with oxygen in the presence of a suitable catalyst to form nitric oxide and water vapor:

$$4\,NH_3(g) + 5\,O_2(g) \longrightarrow 4\,NO(g) + 6\,H_2O(g)$$

How many liters of $NH_3(g)$ at 850 °C and 5.00 atm are required to react with 1.00 mol of $O_2(g)$ in this reaction?

 Self-Assessment Exercises

SAE 10.7 Which situation will result in an increased temperature of a gaseous sample? (**a**) increasing the pressure of a fixed quantity of gas in a rigid vessel (**b**) decreasing the volume of a balloon for a fixed quantity of gas at constant pressure (**c**) increasing the number of moles of gas in a rigid vessel at constant pressure (**d**) simultaneously doubling the pressure and halving the volume of a balloon that holds a fixed quantity of gas

SAE 10.8 A sample of oxygen at a temperature of 20 °C occupies a volume of 335 mL. How many moles of gas are present if the pressure is 0.998 atm? (**a**) 1.37×10^{-4} mol (**b**) 0.0139 mol (**c**) 0.204 mol (**d**) 13.9 mol

SAE 10.9 A sample of N_2 gas at a temperature of 25 °C and a pressure of 750 torr is sealed inside a glass container of unknown volume. If the container is heated to 250 °C, what is the pressure of

the gas inside the container? (**a**) 75.0 torr (**b**) 427 torr (**c**) 1320 torr (**d**) 7500 torr

SAE 10.10 What is the density of neon in a vessel where the temperature is 35 °C and the pressure is 15.0 torr? (**a**) 0.00684 g/L (**b**) 0.0158 g/L (**c**) 0.128 g/L (**d**) 12.0 g/L

SAE 10.11 If sodium hydrogen carbonate is added to a hydrochloric acid solution, it will produce gaseous carbon dioxide according to the following reaction:

$$NaHCO_3(s) + HCl(aq) \longrightarrow NaCl(aq) + H_2O(l) + CO_2(g)$$

How many grams of $NaHCO_3$ must be added to an excess of $HCl(aq)$ to produce 50.0 mL of CO_2 at 25 °C and 0.995 atm? (**a**) 0.00203 g (**b**) 2.04 g (**c**) 171 g (**d**) 0.170 g

10.4 | Gas Mixtures and Partial Pressures

Thus far we have considered mainly pure gases—those that consist of only one substance in the gaseous state. How do we deal with mixtures of two or more different, nonreacting gases? While studying the properties of air, John Dalton (Section 2.1) made an important observation:

> *The total pressure of a mixture of gases equals the sum of the pressures that each would exert if it were present alone.*

The pressure exerted by a particular component of a mixture of gases is called the **partial pressure** of that component. Dalton's observation is known as **Dalton's law of partial pressures**.

If we let P_t be the total pressure of a mixture of gases and P_1, P_2, P_3, and so forth be the partial pressures of the individual gases, we can write Dalton's law of partial pressures as

$$P_t = P_1 + P_2 + P_3 + \ldots \qquad [10.12]$$

This equation implies that each gas behaves independently of the others, as we can see by the following analysis. Let n_1, n_2, n_3, and so forth be the number of moles of each of the gases in the mixture and let n_t be the total number of moles of gas. If each gas obeys the ideal-gas equation, we can write

$$P_1 = n_1\!\left(\frac{RT}{V}\right);\quad P_2 = n_2\!\left(\frac{RT}{V}\right);\quad P_3 = n_3\!\left(\frac{RT}{V}\right);\quad \text{and so forth}$$

 Learning Objectives

When you finish **Section 10.4**, you should be able to:

▶ Calculate the partial pressure of each gas, and the total pressure of all gases, in a mixture of gases.

▶ Interconvert between the mole fraction of a gas in a mixture and its partial pressure.

All of the gases in a container must occupy the same volume and will come to the same temperature in a relatively short period of time. Using these facts to simplify Equation 10.12, we obtain

$$P_t = (n_1 + n_2 + n_3 + \dots)\left(\frac{RT}{V}\right) = n_t\left(\frac{RT}{V}\right) \qquad [10.13]$$

That is, at constant temperature and constant volume, the total pressure of a gas sample is determined by the total number of moles of gas present, whether that total represents just one gas or a mixture of gases.

Sample Exercise 10.10
Applying Dalton's Law of Partial Pressures

A mixture of 6.00 g of $O_2(g)$ and 9.00 g of $CH_4(g)$ is placed in a 15.0-L vessel at 0 °C. What is the partial pressure of each gas, and what is the total pressure in the vessel?

SOLUTION

Analyze We need to calculate the pressure for two gases in the same volume and at the same temperature.

Plan Because each gas behaves independently, we can use the ideal-gas equation to calculate the pressure each would exert if the other were not present. Per Dalton's law, the total pressure is the sum of these two partial pressures.

Solve We first convert the mass of each gas to moles:

$$n_{O_2} = (6.00\ \text{g}\ O_2)\left(\frac{1\ \text{mol}\ O_2}{32.0\ \text{g}\ O_2}\right) = 0.188\ \text{mol}\ O_2$$

$$n_{CH_4} = (9.00\ \text{g}\ CH_4)\left(\frac{1\ \text{mol}\ CH_4}{16.0\ \text{g}\ CH_4}\right) = 0.563\ \text{mol}\ CH_4$$

We use the ideal-gas equation to calculate the partial pressure of each gas:

$$P_{O_2} = \frac{n_{O_2}RT}{V} = \frac{(0.188\ \text{mol})(0.08206\ \text{L-atm/mol-K})(273\ \text{K})}{15.0\ \text{L}}$$

$$= 0.281\ \text{atm}$$

$$P_{CH_4} = \frac{n_{CH_4}RT}{V} = \frac{(0.563\ \text{mol})(0.08206\ \text{L-atm/mol-K})(273\ \text{K})}{15.0\ \text{L}}$$

$$= 0.841\ \text{atm}$$

According to Dalton's law of partial pressures (Equation 10.12), the total pressure in the vessel is the sum of the partial pressures:

$$P_t = P_{O_2} + P_{CH_4}$$

$$= 0.281\ \text{atm} + 0.841\ \text{atm} = 1.122\ \text{atm}$$

Check A pressure of roughly 1 atm seems right for a mixture of about 0.2 mol O_2 and a bit more than 0.5 mol CH_4, together in a 15-L volume, because 1 mol of an ideal gas at 1 atm pressure and 0 °C occupies about 22 L.

▶ **Practice Exercise**
What is the total pressure exerted by a mixture of 2.00 g of $H_2(g)$ and 8.00 g of $N_2(g)$ at 273 K in a 10.0-L vessel?

Partial Pressures and Mole Fractions

Because each gas in a mixture behaves independently, we can relate the amount of a given gas in a mixture to its partial pressure. For an ideal gas, we can write

$$\frac{P_1}{P_t} = \frac{n_1RT/V}{n_tRT/V} = \frac{n_1}{n_t} \qquad [10.14]$$

The ratio n_1/n_t is called the *mole fraction of gas 1*, which we denote X_1. The **mole fraction**, X, is a dimensionless number that expresses the ratio of the number of moles of one component in a mixture to the total number of moles in the mixture. Thus, for gas 1 we have

$$X_1 = \frac{\text{moles of compound 1}}{\text{total moles}} = \frac{n_1}{n_t} \qquad [10.15]$$

We can combine Equations 10.14 and 10.15 to give

$$P_1 = \left(\frac{n_1}{n_t}\right)P_t = X_1 P_t \qquad [10.16]$$

The mole fraction of N_2 in air is 0.78—that is, 78% of the molecules in air are N_2. This means that if the barometric pressure is 760 torr, the partial pressure of N_2 is

$$P_{N_2} = (0.78)(760\ \text{torr}) = 590\ \text{torr}$$

This result makes intuitive sense: Because N_2 makes up 78% of the mixture, it contributes 78% of the total pressure.

Sample Exercise 10.11
Relating Mole Fractions and Partial Pressures

A study of the effects of certain gases on plant growth requires a synthetic atmosphere composed of 1.5 mol % CO_2, 18.0 mol % O_2 and 80.5 mol % Ar. **(a)** Calculate the partial pressure of O_2 in the mixture if the total pressure of the atmosphere is 745 torr. **(b)** If this atmosphere is held in a 121-L space at 295 K, how many moles of O_2 are needed?

SOLUTION

Analyze For **(a)** we need to calculate the partial pressure of O_2 given its mole percent and the total pressure of the mixture. For **(b)** we need to calculate the number of moles of O_2 in the mixture given its volume (121 L), temperature (295 K), and partial pressure from part (a).

Plan We calculate the partial pressures using Equation 10.16, and then use P_{O_2}, V, and T in the ideal-gas equation to calculate the number of moles of O_2.

Solve

(a) The mole percent is the mole fraction times 100. Therefore, the mole fraction of O_2 is 0.180. Equation 10.16 gives:

$$P_{O_2} = (0.180)(745\ \text{torr}) = 134\ \text{torr}$$

(b) Tabulating the given variables and converting to appropriate units, we have:

$$P_{O_2} = (134\ \text{torr})\left(\frac{1\ \text{atm}}{760\ \text{torr}}\right) = 0.176\ \text{atm}$$

$$V = 121\ \text{L}$$

$$n_{O_2} = ?$$

$$R = 0.08206\frac{\text{L-atm}}{\text{mol-K}}$$

$$T = 295\ \text{K}$$

Solving the ideal-gas equation for n_{O_2}, we have:

$$n_{O_2} = P_{O_2}\left(\frac{V}{RT}\right)$$

$$= (0.176\ \text{atm})\frac{121\ \text{L}}{(0.08206\ \text{L-atm}/\text{mol-K})(295\ \text{K})} = 0.880\ \text{mol}$$

Check The units check out, and the answer seems to be the right order of magnitude.

▶ **Practice Exercise**
From data gathered by the *Voyager 1* space probe, scientists have estimated the composition of the atmosphere of Titan, Saturn's largest moon. The pressure on the surface of Titan is 1220 torr. The atmosphere consists of 82 mol% N_2, 12 mol% Ar, and 6.0 mol% CH_4. Calculate the partial pressure of each gas.

Self-Assessment Exercises

SAE 10.12   A cylinder with a volume of 15 L contains 95 mol% N_2 and 5 mol% H_2, and the total pressure is 2.0×10^6 Pa. If 12 g of H_2 is injected into the cylinder while the temperature is held constant, the total pressure will _____ and the partial pressure of N_2 will _____. **(a)** increase, increase **(b)** increase, not change **(c)** not change, increase **(d)** not change, not change **(e)** increase, decrease

SAE 10.13   Water can be decomposed into its constituent elements, $2\ H_2O(l) \longrightarrow 2\ H_2(g)$ $+ O_2(g)$, with an electrical current in a process called electrolysis. If 1.2 g of H_2O is subjected to electrolysis and the products are collected in a cylinder with a volume of 0.20 L and a temperature of 30 °C, what is the partial pressure of H_2? **(a)** 0.55 atm **(b)** 2.8 atm **(c)** 5.5 atm **(d)** 8.3 atm

Pressure inside container comes from collisions of gas molecules with container walls.

10.5 | The Kinetic-Molecular Theory of Gases

Learning Objectives

When you finish Section 10.5, you should be able to:

▶ Describe the five assumptions of the kinetic-molecular theory of gases.

▶ Use the kinetic-molecular theory to relate macroscopic properties of gases to their microscopic behavior.

▶ Describe qualitatively how the distribution of molecular speeds in a sample of gas changes with temperature and molecular weight.

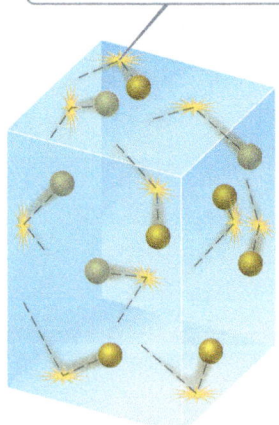

▲ **Figure 10.11** The molecular origin of gas pressure.

The ideal-gas equation describes *how* gases behave but not *why* they behave as they do. Why does a gas expand when heated at constant pressure? Or why does its pressure increase when the gas is compressed at constant temperature? To understand the physical properties of gases, we need a model that helps us picture what happens to gas particles when conditions such as pressure or temperature change. Such a model, known as the **kinetic-molecular theory of gases**, was developed over a period of about 100 years, culminating in 1857 when Rudolf Clausius (1822–1888) published a complete and satisfactory form of the theory.

The kinetic-molecular theory (the theory of moving molecules) is summarized by the following statements:

1. **Random motion** Gases consist of large numbers of molecules that are in continuous, random motion. (The word *molecule* is used here to designate the smallest particle of any gas, even though some gases, such as the noble gases, consist of individual atoms. All we learn about gas behavior from the kinetic-molecular theory applies equally to atomic gases.)

2. **Negligible molecular volume** The combined volume of all the molecules of the gas is negligible relative to the total volume in which the gas is contained.

3. **Negligible forces** Attractive and repulsive forces between gas molecules are negligible.

4. **Constant average kinetic energy** Energy can be transferred between molecules during collisions, but, as long as temperature remains constant, the *average* kinetic energy of the molecules does not change with time.

5. **Average kinetic energy proportional to temperature** The average kinetic energy of the molecules is proportional to the absolute temperature. At any given temperature, the molecules of all gases have the same average kinetic energy.

The kinetic-molecular theory explains both pressure and temperature at the molecular level. The pressure of a gas is caused by collisions of the molecules with the walls of the container (**Figure 10.11**). The magnitude of the pressure is determined by how often and how forcefully the molecules strike the walls.

The absolute temperature of a gas is a measure of the *average* kinetic energy of its molecules. If two gases are at the same temperature, their molecules have the same average kinetic energy (statement 5 of the kinetic-molecular theory). If the absolute temperature of a gas is doubled, the average kinetic energy of its molecules doubles. Thus, molecular motion increases with increasing temperature.

Distributions of Molecular Speed

Although collectively the molecules in a sample of gas have an *average* kinetic energy and hence an average speed, the individual molecules are moving at different speeds. Each molecule collides frequently with other molecules. Momentum is conserved in each collision, but one of the colliding molecules might be deflected off at high speed while the other is nearly stopped. The result is that, at any instant, the molecules in the sample have a wide range of speeds. In **Figure 10.12**(a), which shows the distribution of molecular speeds for nitrogen gas at 273 K and 373 K, notice that a larger fraction of the molecules at 373 K moves at the higher speeds. This means that the sample at 373 K has the higher average kinetic energy.

In any graph of the distribution of molecular speeds in a gas sample, the peak of the curve represents the most probable speed, u_{mp} [Figure 10.12(b)]. The most probable speeds in Figure 10.12(a), for instance, are 4×10^2 m/s for the sample at 273 K and 5×10^2 m/s for the sample at 373 K. Figure 10.12(b) also shows the **root-mean-square (rms) speed**, u_{rms}, of the molecules. This is the speed of a molecule possessing a kinetic energy identical to the average kinetic energy of the sample. The rms speed is not quite

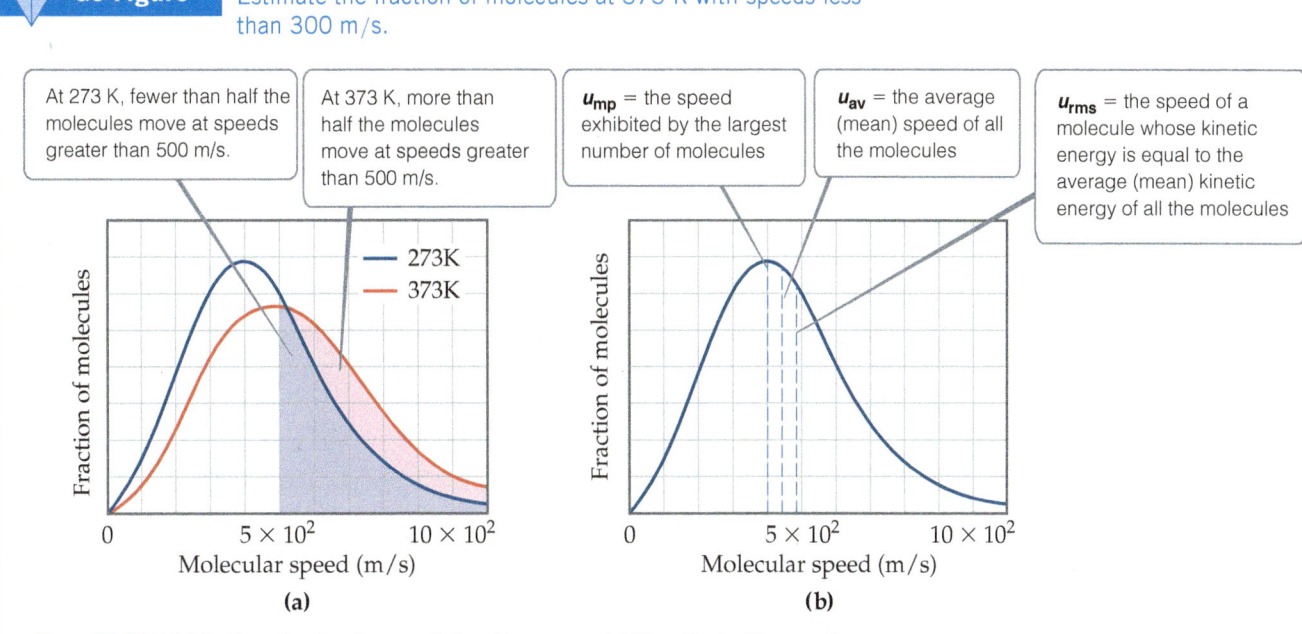

Go Figure   Estimate the fraction of molecules at 373 K with speeds less than 300 m/s.

At 273 K, fewer than half the molecules move at speeds greater than 500 m/s.

At 373 K, more than half the molecules move at speeds greater than 500 m/s.

u_{mp} = the speed exhibited by the largest number of molecules

u_{av} = the average (mean) speed of all the molecules

u_{rms} = the speed of a molecule whose kinetic energy is equal to the average (mean) kinetic energy of all the molecules

▲ **Figure 10.12 Distribution of molecular speeds for nitrogen gas. (a)** The effect of temperature on molecular speed. The relative area under the curve for a range of speeds gives the relative fraction of molecules that have those speeds. **(b)** Position of most probable (u_{mp}), average (u_{av}), and root-mean-square (u_{rms}) speeds of gas molecules. The data shown here are for nitrogen gas at 273 K.

the same as the average (mean) speed, u_{av}, because the shape of the distribution curve is not symmetric; there is a longer tail toward larger molecular speeds. The difference between the two is small, however. In Figure 10.12(b), for example, the root-mean-square speed is about 4.9×10^2 m/s, whereas the average speed is about 4.5×10^2 m/s.

If you calculate the rms speeds (as we will later), you will find that the rms speed is almost 6×10^2 m/s for the sample at 373 K, but slightly less than 5×10^2 m/s for the sample at 273 K. Notice that the distribution curve broadens and flattens as we go to a higher temperature, which tells us that the range of molecular speeds increases with temperature.

The rms speed is important because the average kinetic energy of the gas molecules in a sample is equal to $\frac{1}{2}m(u_{rms})^2$. (Section 1.4) Because mass does not change with temperature, the increase in the average kinetic energy $\frac{1}{2}m(u_{rms})^2$ as the temperature increases implies that the rms speed of the molecules (as well as u_{av} and u_{mp}) increases as temperature increases.

Application of the Kinetic-Molecular Theory to the Gas Laws

The empirical observations of gas properties as expressed by the various gas laws are readily understood in terms of the kinetic-molecular theory. The following examples illustrate this point:

1. **An increase in volume at constant temperature causes pressure to decrease.** A constant temperature means that the average kinetic energy of the gas molecules remains unchanged. This means that the rms speed of the molecules remains unchanged. When the volume is increased, the molecules must move a longer distance between collisions. Consequently, there are fewer collisions per unit time with the container walls, which means the pressure decreases. Thus, kinetic-molecular theory explains Boyle's law.

2. **A temperature increase at constant volume causes pressure to increase.** An increase in temperature means an increase in the average kinetic energy of the molecules and in u_{rms}. Because there is no change in volume, the temperature

increase causes more collisions with the walls per unit time because the molecules are all moving faster. Furthermore, the momentum in each collision increases (the molecules strike the walls more forcefully). A greater number of more forceful collisions means the pressure increases, and the theory explains this increase.

A CLOSER LOOK | The Ideal-Gas Equation

The ideal-gas equation can be derived from the five statements given in the text for the kinetic-molecular theory. Rather than perform the derivation, however, let's consider in qualitative terms how the ideal-gas equation might follow from these statements. The total force of the molecular collisions on the walls and hence the pressure (force per unit area, Section 10.1) produced by these collisions depend both on how strongly the molecules strike the walls (impulse imparted per collision) and on the rate at which the collisions occur:

$$P \propto \text{impulse imparted per collision} \times \text{collision rate}$$

For a molecule traveling at the rms speed, the impulse imparted by a collision with a wall depends on the momentum of the molecule; that is, it depends on the product of the molecule's mass and speed: mu_{rms}. The collision rate is proportional to the number of molecules per unit volume, n/V, and to their speed, which is u_{rms}

because we are talking about only molecules traveling at this speed. Thus, we have

$$P \propto mu_{rms} \times \frac{n}{V} \times u_{rms} \propto \frac{nm(u_{rms})^2}{V} \quad [10.17]$$

Because the average kinetic energy, $\frac{1}{2}m(u_{rms})^2$, is proportional to temperature, we have $m(u_{rms})^2 \propto T$. Making this substitution in Equation 10.17 gives

$$P \propto \frac{nm(u_{rms})^2}{V} \propto \frac{nT}{V} \quad [10.18]$$

If we put in a proportionality constant, calling it R, the gas constant, you can see that we obtain the ideal-gas equation:

$$P = \frac{nRT}{V} \quad [10.19]$$

Related Exercises: 10.75, 10.76

Sample Exercise 10.12
Applying the Kinetic-Molecular Theory

A sample of O_2 gas initially at STP is compressed to a smaller volume at constant temperature. What effect does this change have on (**a**) the average kinetic energy of the molecules, (**b**) their average speed, (**c**) the number of collisions they make with the container walls per unit time, (**d**) the number of collisions they make with a unit area of container wall per unit time, and (**e**) the pressure?

SOLUTION

Analyze We need to apply the concepts of the kinetic-molecular theory of gases to a gas compressed at constant temperature.

Plan We will determine how each of the quantities in (a)–(e) is affected by the change in volume at constant temperature.

Solve (**a**) Because the average kinetic energy of the O_2 molecules is determined only by temperature, this energy is unchanged by the compression. (**b**) Because the average kinetic energy of the molecules does not change, their average speed remains constant. (**c**) The number of collisions with the walls per unit time increases because the molecules are moving in a smaller volume but with the same average speed as before. Under these conditions they will strike the walls of the container more frequently. (**d**) The number of collisions with a unit area of wall per unit time increases because the total number of collisions with the walls per unit

time increases and the area of the walls decreases. (**e**) Although the average force with which the molecules collide with the walls remains constant, the pressure increases because there are more collisions per unit area of wall per unit time.

Check In a conceptual exercise of this kind, there is no numerical answer to check. All we can check in such cases is our reasoning in the course of solving the problem. The increase in pressure seen in part (**e**) is consistent with Boyle's law.

▶ **Practice Exercise**
How is the rms speed of N_2 molecules in a gas sample changed by (**a**) an increase in temperature, (**b**) an increase in volume, and (**c**) mixing with a sample of Ar at the same temperature?

Self-Assessment Exercises

SAE 10.14 Which of these statements is an assumption of the kinetic-molecular theory of gases? (**a**) Larger gas molecules take up more volume than smaller gas molecules. (**b**) Gas molecules lose energy every time they collide with something, so over time the molecules will stop moving, no matter what the conditions. (**c**) At a given temperature, all gas molecules move with the same speed. (**d**) Gas molecules neither attract nor repel each other.

SAE 10.15 A gas is contained in a closed piston–cylinder apparatus. If the cylinder is raised so that the enclosed volume of the gas doubles but the temperature is kept constant, which of the following statements are *true*?

(**i**) The pressure of the gas will decrease.
(**ii**) The average kinetic energy of the gas molecules will decrease.
(**iii**) The rate of collisions with the cylinder walls will decrease.

(**a**) only i (**b**) both i and ii (**c**) both i and iii (**d**) both ii and iii (**e**) All three statements are true.

SAE 10.16 Two flasks of equal volume, one containing argon (Ar) and the other containing krypton (Kr), are kept at the same temperature. The pressure inside both flasks is the same, and both gases are assumed to behave as ideal gases. According to the kinetic-molecular theory of gases, the average kinetic energy of the argon atoms is _____ the average kinetic energy of the krypton atoms, and the average speed of the argon atoms is _____ than the average speed of the krypton atoms.

(**a**) less than, the same speed (**b**) equal to, faster than (**c**) greater than, faster than (**d**) equal to, the same speed as (**e**) less than, slower than

SAE 10.17 Which of the following statements best describes what a graph of "number of molecules" versus "molecular speed" for a particular gas at a particular temperature would look like? (**a**) a vertical line, because all gas molecules have the same speed at the same temperature (**b**) a straight line with a negative slope, indicating that the largest number of molecules have the smallest speeds (**c**) a symmetric bell-shaped curve, indicating that the most probable speed of gas molecules is the same as the average speed (**d**) an asymmetric bell-shaped curve, indicating that the most probable speed of gas molecules is smaller than the average speed (**e**) an asymmetric bell-shaped curve, indicating that the most probable speed of gas molecules is larger than the average speed

10.6 | Molecular Speeds, Effusion, and Diffusion

According to the kinetic-molecular theory of gases, the average kinetic energy of *any* collection of gas molecules, $\frac{1}{2}m(u_{rms})^2$, has a specific value at a given temperature. Thus, for two gases at the same temperature, a gas composed of low-mass particles, such as He, has the same average kinetic energy as one composed of more massive particles, such as Xe. The mass of the particles in the He sample is smaller than that in the Xe sample. Consequently, the He particles must have a higher rms speed than the Xe particles. Equation 10.20 expresses this fact quantitatively:

$$u_{rms} = \sqrt{\frac{3RT}{\mathcal{M}}} \qquad [10.20]$$

where $\mathcal{M}$ is the molar mass of the particles. Equation 10.20 can be derived from the kinetic-molecular theory. Because $\mathcal{M}$ appears in the denominator, the rms speed increases as the molar mass of the gas particles decreases.

Figure 10.13 shows the distribution of molecular speeds for several gases at 298 K. Notice how the distributions are shifted toward higher speeds for gases of lower molar masses.

The most probable speed of a gas molecule can also be derived:

$$u_{mp} = \sqrt{\frac{2RT}{\mathcal{M}}} \qquad [10.21]$$

Finally, the average speed of a gas molecule can be derived:

$$u_{av} = \sqrt{\frac{8RT}{\pi \mathcal{M}}} \qquad [10.22]$$

Learning Objectives

When you finish **Section 10.6**, you should be able to:

▶ Calculate the root-mean-square speed, average speed, and the most probable speed of gas molecules, for a given gas at constant temperature.

▶ Use Graham's law of effusion to predict the behavior of effusing gases.

▶ Predict how the mean free path of a diffusing gas molecule will be affected by changes in pressure.

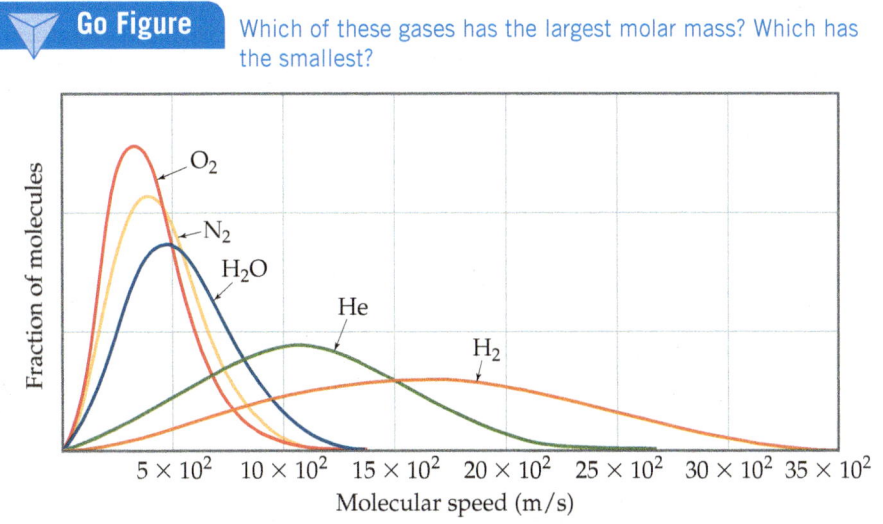

▽ **Go Figure** Which of these gases has the largest molar mass? Which has the smallest?

▲ **Figure 10.13 The effect of molar mass on molecular speed at 298 K.**

Notice that the formulas for u_{rms}, u_{mp}, and u_{av} all show that the speed is proportional to the square root of $T/\mathcal{M}$. Hence, the speed will increase as T increases and will decrease as $\mathcal{M}$ increases.

The dependence of molecular speed on mass has two interesting consequences. The first is **effusion**, which is the escape of gas molecules through a tiny hole (**Figure 10.14**). The second is **diffusion**, which is the spread of one substance throughout a space or throughout a second substance. For example, the molecules of a perfume diffuse throughout a room once the bottle is opened.

Sample Exercise 10.13
Calculating a Root-Mean-Square Speed

Calculate the rms speed of the molecules in a sample of N_2 gas at 25 °C.

SOLUTION

Analyze We are given the identity of a gas and the temperature, the two quantities we need to calculate the rms speed.

Plan We calculate the rms speed using Equation 10.20.

Solve We must convert each quantity in our equation to SI units. We will also use R in units of J/mol-K (Table 10.2) to make the units cancel correctly.

$T = 25 + 273 = 298$ K

$\mathcal{M} = 28.0\,g/mol = 28.0 \times 10^{-3}\,kg/mol$

$R = 8.314\,J/mol\text{-}K = 8.314\,kg\text{-}m^2/s^2\text{-}mol\text{-}K$ (Since 1 J $= 1$ kg-m^2/s^2)

$$u_{rms} = \sqrt{\frac{3RT}{\mathcal{M}}}$$

$$= \sqrt{\frac{3(8.314\,kg\text{-}m^2/s^2\text{-}mol\text{-}K)(298\,K)}{28.0 \times 10^{-3}\,kg/mol}} = 5.15 \times 10^2\,m/s$$

Comment This corresponds to a speed of 1150 mi/hr. Because the molecular weight of O_2 molecules is slightly greater than that of N_2, the rms speed of O_2 molecules is a little less than that for N_2 at the same temperature.

▶ **Practice Exercise**
What is the rms speed of an atom in a sample of He gas at 25 °C?

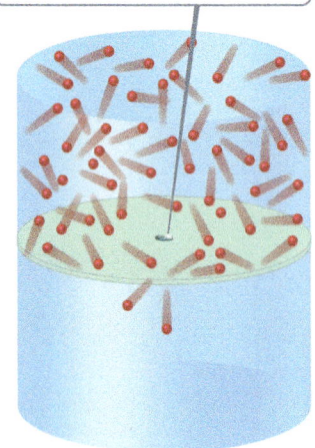

Gas molecules in top half effuse through pinhole only when they happen to hit the pinhole.

▲ **Figure 10.14** Effusion.

Graham's Law of Effusion

In 1846, Thomas Graham (1805–1869) discovered that the effusion rate of a gas is inversely proportional to the square root of its molar mass. Assume we have two gases at the same temperature and pressure in two containers with identical pinholes. If the rates of effusion of the two gases are r_1 and r_2 and their molar masses are $\mathcal{M}_1$ and $\mathcal{M}_2$, **Graham's law** states that

$$\frac{r_1}{r_2} = \sqrt{\frac{\mathcal{M}_2}{\mathcal{M}_1}} \qquad [10.23]$$

a relationship that indicates that the lighter gas has the higher effusion rate.

The only way for a molecule to escape from its container is for it to "hit" the hole in the partitioning wall of Figure 10.14. The faster the molecules are moving, the more often they hit the partition wall and the greater the likelihood that a molecule will hit the hole and effuse. This implies that the rate of effusion is directly proportional to the rms speed of the molecules. Because R and T are constant, we have, from Equation 10.22

$$\frac{r_1}{r_2} = \frac{u_{rms1}}{u_{rms2}} = \sqrt{\frac{3RT/\mathcal{M}_1}{3RT/\mathcal{M}_2}} = \sqrt{\frac{\mathcal{M}_2}{\mathcal{M}_1}} \qquad [10.24]$$

As expected from Graham's law, helium escapes from containers through tiny pinhole leaks more rapidly than other gases of higher molecular weight (**Figure 10.15**).

 **Go Figure** Because pressure and temperature are constant in this figure but volume changes, which other quantity in the ideal-gas equation must also change?

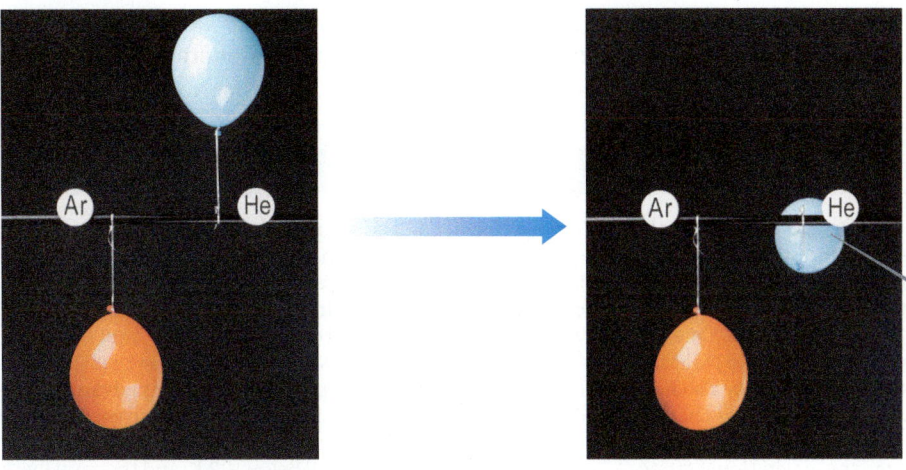

Both gases effuse through pores in balloon, but lighter helium gas effuses faster than heavier argon gas.

▲ **Figure 10.15** An illustration of Graham's law of effusion.

 ## Sample Exercise 10.14

Applying Graham's Law

An unknown gas composed of homonuclear diatomic molecules effuses at a rate that is 0.355 times the rate at which O_2 gas effuses at the same temperature. Calculate the molar mass of the unknown and identify it.

SOLUTION

Analyze We are given the rate of effusion of an unknown gas relative to that of O_2 and asked to find the molar mass and identity of the unknown. Thus, we need to connect relative rates of effusion to relative molar masses.

Plan We use Equation 10.23 to determine the molar mass of the unknown gas. If we let r_x and M_x represent the rate of effusion and molar mass of the gas, we can write

$$\frac{r_x}{r_{O_2}} = \sqrt{\frac{M_{O_2}}{M_x}}$$

Solve From the information given,

$$r_x = 0.355 \times r_{O_2}$$

Thus,

$$\frac{r_x}{r_{O_2}} = 0.355 = \sqrt{\frac{32.0 \, \text{g/mol}}{M_x}}$$

$$\frac{32.0 \, \text{g/mol}}{M_x} = (0.355)^2 = 0.126$$

$$M_x = \frac{32.0 \, \text{g/mol}}{0.126} = 254 \, \text{g/mol}$$

Because we are told that the unknown gas is composed of homonuclear diatomic molecules, it must be an element. The molar mass must represent twice the atomic weight of the atoms in the unknown gas. We conclude that the unknown gas must have an atomic weight of 127 g/mol, in which case the molecule is I_2.

▶ **Practice Exercise**
Calculate the ratio of the effusion rates of N_2 gas and O_2 gas.

Diffusion and Mean Free Path

Although diffusion, like effusion, is faster for lower-mass molecules than for higher-mass ones, molecular collisions make diffusion more complicated than effusion.

Graham's law, Equation 10.23, approximates the ratio of the diffusion rates of two gases under identical conditions. We can see from the horizontal axis in Figure 10.12 that the speeds of molecules are quite high. For example, the rms speed of molecules of N_2 gas at room temperature is 515 m/s. In spite of this high speed, if someone opens a vial of perfume at one end of a room, some time elapses—perhaps a few minutes—before the scent is detected at the other end of the room. This tells us that the diffusion rate of gases throughout a volume of space is much slower than molecular speeds.* This difference is

---

*The rate at which the perfume moves across the room also depends on how well stirred the air is from temperature gradients and the movement of people. Nevertheless, even with the aid of these factors, it still takes much longer for the molecules to traverse the room than one would expect from their rms speed.

due to molecular collisions, which occur frequently for a gas at atmospheric pressure—about 10^{10} times per second for each molecule. Collisions occur because real gas molecules have finite volumes.

Because of molecular collisions, a gas molecule's direction of motion is constantly changing. A good mathematical approximation to this motion is the "random walk,"

CHEMISTRY AND SUSTAINABILITY | HYDROGEN AND HELIUM

Hydrogen and helium are by far the most abundant elements in the universe. Together they account for 75% of the mass of the universe, yet they are only present in trace amounts in Earth's atmosphere. The concentration of He is 5 parts per million (ppm) and that of H_2 is roughly 10 times smaller, ~0.5 ppm. Why is this the case? One reason is that given their very light masses, H_2 molecules and He atoms not only have a faster average speed than other gases, they also have a broader distribution, which produces some molecules that are moving very fast (Figure 10.12). Consequently, they are more likely to achieve sufficient velocities to overcome the gravitational attraction to Earth and escape into space. It is estimated that Earth loses about 3 kg of $H_2(g)$ and about 50 g of He per second in this way.

Hydrogen is a feedstock for many important chemical reactions and forms the basis for a potential "hydrogen economy." The combination reaction between hydrogen and oxygen to create water and release energy is central to the idea of the hydrogen economy.

$$2\,H_2(g) + O_2(g) \longrightarrow 2\,H_2O(g) \quad \Delta H_{rxn} = -572\,kJ/mol$$

This is the same process (although with liquids, not gases) that is used to boost rockets into space, illustrating the significant amount of chemical energy this reaction releases (Figure 10.16). Unlike fossil fuel combustion, this reaction does not produce CO_2, making it a highly attractive prospect for combating climate change. Unfortunately, the main way to generate elemental hydrogen is through steam reforming of the fossil fuel methane, represented by the following reaction:

$$CH_4(g) + H_2O(g) \longrightarrow CO(g) + 3H_2(g)$$

Thus, a frontier area of scientific research is to develop methods to safely and efficiently obtain hydrogen that do not rely on fossil fuels like methane (and produce a toxic gas like CO). For instance, you could convert water to hydrogen and oxygen using a process called electrolysis; but that would cost at least 572 kJ/mol to run.

Helium is probably best known as the inert gas that makes balloons and airships float, a property that takes advantage of its low density, but that is far from its only use. It is also used in welding, in the manufacture of electronics, and in the detection of leaks. Helium liquifies when cooled below 4.2 K, and this very low boiling point leads to its most important commercial application. Liquid helium is used to cool many high-tech instruments, including the superconducting magnets of magnetic resonance imaging machines found in hospitals.

Helium is found, surprisingly, in subterranean natural gas deposits where it is formed when much heavier radioactive elements, like uranium and thorium, decay. After the natural gas is pumped out of the ground, helium can be isolated from the other components by distillation. Over the course of the twentieth century, the U.S. government took on a major role in providing a stable and affordable supply of helium. The National Helium Reserve, located near Amarillo, Texas, is a strategic reserve that held over 1 billion cubic meters of the gas at its peak. Starting in the mid-1990s, the federal government started to slowly but steadily privatize the collection, storage, and sale of liquid helium. This has led to concerns about the supply and price of this valuable commodity. As of this writing (spring 2021), liquid helium costs approximately $38/L, more than many good wines.

Related Exercise: 10.86

▲ **Figure 10.16** Hydrogen and oxygen combine to form water, releasing enough energy to power this rocket into space.

the details of which are beyond the scope of this book. For our purposes, it is sufficient to know that the diffusion of a molecule from one point to another consists of many short, straight-line segments as collisions buffet it around in random directions (Figure 10.17).

The average distance traveled by a molecule between collisions, called the molecule's **mean free path**, varies with pressure as the following analogy illustrates. Imagine walking through a shopping mall. When the mall is crowded (high pressure), the average distance you can walk before bumping into someone is short (short mean free path). When the mall is empty (low pressure), you can walk a long way (long mean free path) before bumping into someone. The mean free path for air molecules at sea level is about 60 nm. At about 100 km in altitude, where the air pressure is much lower, the mean free path is about 10 cm, over 1 million times longer than at Earth's surface. For a given volume and temperature, the distance between collisions will decrease as the number of molecules increases. An increase in the number of molecules at constant T and V corresponds to an increase in P, so it makes sense that the mean free path is inversely proportional to the pressure of the gas.

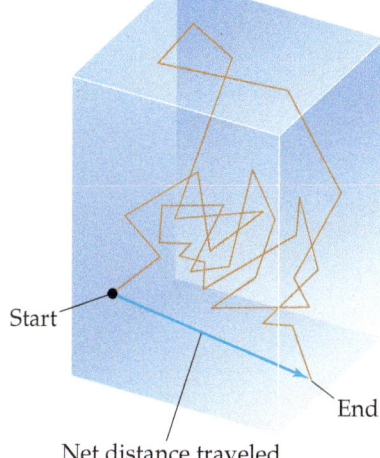

▲ **Figure 10.17 Diffusion of a gas molecule.** For clarity, no other gas molecules in the container are shown.

 Self-Assessment Exercises

SAE 10.18 What is the most probable speed of $H_2(g)$ molecules at 5500 °C, the surface temperature of the Sun? (**a**) 2.20×10^2 m/s (**b**) 6.8×10^3 m/s (**c**) 6.9×10^3 m/s (**d**) 8.5×10^3 m/s

SAE 10.19 N_2 gas is allowed to effuse through a porous membrane under constant pressure conditions. It takes 212 s for 1.0 L of N_2 gas to pass through the membrane. If it takes 337 s for 1.0 L of an unknown gas to effuse through the membrane under identical conditions, what is the molar mass of the unknown gas? (**a**) 11 g/mol (**b**) 19 g/mol (**c**) 44 g/mol (**d**) 71 g/mol

SAE 10.20 Which of the following is the correct rank order of the diffusion rates of either Cl_2 or H_2S through air at a pressure of 1.0 atm?

(**a**) H_2S at 300 K < H_2S at 250 K < Cl_2 at 300 K < Cl_2 at 250 K
(**b**) H_2S at 250 K < H_2S at 300 K < Cl_2 at 250 K < Cl_2 at 300 K
(**c**) Cl_2 at 300 K < Cl_2 at 250 K < H_2S at 300 K < H_2S at 250 K
(**d**) Cl_2 at 250 K < Cl_2 at 300 K < H_2S at 250 K < H_2S at 300 K

SAE 10.21 Which statement is *true* about the following plot?

(**a**) If both curves represent the same gas, then A is at high temperature and B is at low temperature. (**b**) The average speed of the gas

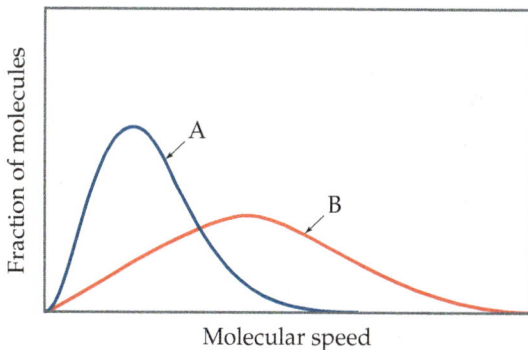

molecules in the B sample is about half that in the A sample. (**c**) If these curves represent two different gases at the same temperature, then A has a higher molar mass than B. (**d**) For most gases at temperatures near room temperature, the horizontal axis would span 1–100 m/s.

10.7 | Real Gases: Deviations from Ideal Behavior

Using the ideal-gas law for many situations is a good first step. In the real world, however, gases may not always behave ideally. We must therefore examine cases where the ideal-gas law must be modified.

The extent to which a real gas departs from ideal behavior can be seen by rearranging the ideal-gas equation to solve for n:

$$\frac{PV}{RT} = n \qquad [10.25]$$

This form of the equation tells us that for 1 mol of ideal gas, the quantity PV/RT equals 1 at all pressures. In Figure 10.18, PV/RT is plotted as a function of P for 1 mol of several real gases. At high pressures (generally above 10 atm), the deviation from ideal behavior ($PV/RT = 1$) is large and different for each gas. *Real gases, in other words, do not behave ideally at high pressure.* At lower pressures (usually below 10 atm), however, the deviation from ideal behavior is small, and we can use the ideal-gas equation without generating serious error.

 Learning Objectives

When you have finished Section 10.7, you should be able to:

▶ Recognize conditions that lead to deviations from the ideal-gas law and explain why those deviations occur.

▶ Use the van der Waals equation to calculate the characteristic properties of real gases.

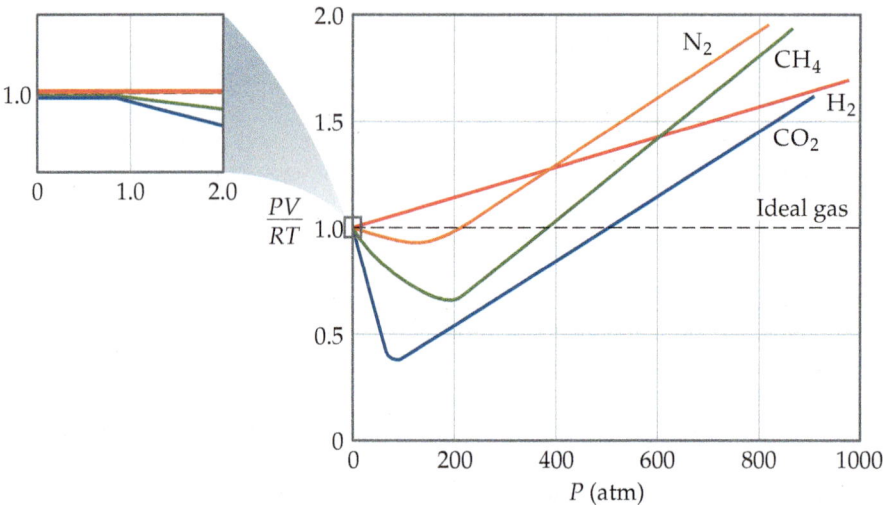

▲ **Figure 10.18 The effect of pressure on the behavior of several real gases.** Data for 1 mol of gas in all cases. The data were taken at 300 K for N_2, CH_4, and H_2, whereas they were taken at 313 K for CO_2 because under high pressure CO_2 liquefies at 300 K.

Deviation from ideal behavior also depends on temperature. As temperature increases, the behavior of a real gas more nearly approaches that of the ideal gas (**Figure 10.19**). In general, *the deviation from ideal behavior increases as temperature decreases,* becoming significant near the temperature at which the gas liquefies.

The basic assumptions of the kinetic-molecular theory of gases give us insight into why real gases deviate from ideal behavior. The molecules of an ideal gas are assumed to occupy no space and have no attraction for one another. *Real molecules, however, do have finite volumes and do attract one another.* As **Figure 10.20** shows, the unoccupied space in which real molecules can move is less than the container volume. At low pressures, the combined volume of the gas molecules is negligible relative to the container volume. Thus, the unoccupied volume available to the molecules is essentially the container volume. At high pressures, the combined volume of the gas molecules is *not* negligible relative to the container volume. Now the unoccupied volume available to the molecules is less than the container volume. At high pressures, therefore, gas volumes tend to be slightly greater than those predicted by the ideal-gas equation.

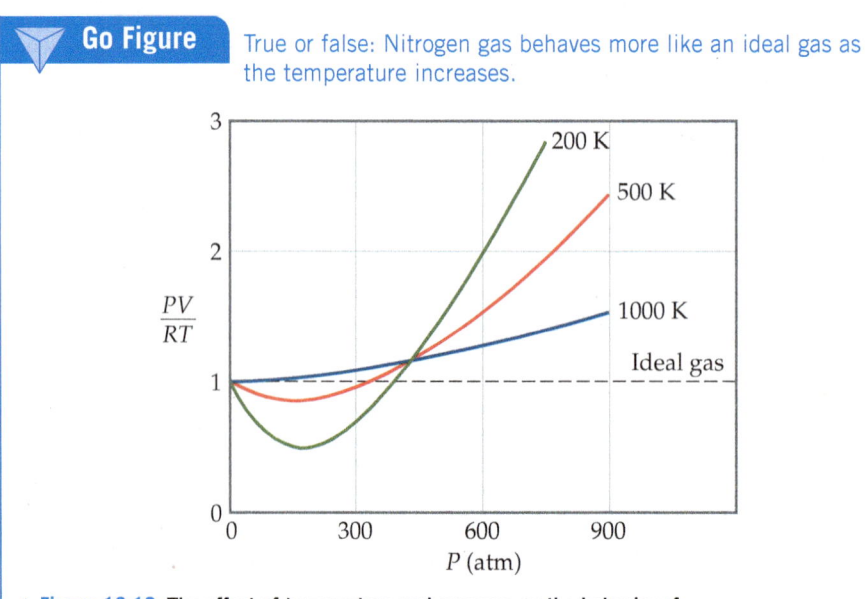

Go Figure   True or false: Nitrogen gas behaves more like an ideal gas as the temperature increases.

▲ **Figure 10.19 The effect of temperature and pressure on the behavior of nitrogen gas.**

Gas molecules occupy a small fraction of the total volume.

Gas molecules occupy a larger fraction of the total volume.

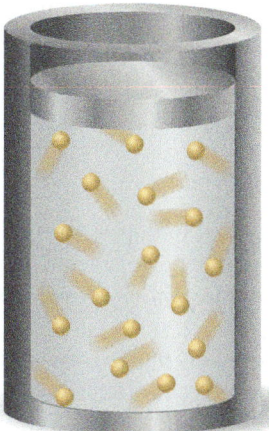 

Low pressure   High pressure

Another reason for nonideal behavior at high pressures is that the attractive forces between molecules come into play at the short intermolecular distances found when molecules are crowded together at high pressures. Because of these attractive forces, the impact of a given molecule with the container wall is lessened. If we could stop the motion in a gas, as illustrated in **Figure 10.21**, we would see that a molecule about to collide with the wall experiences the attractive forces of nearby molecules. These attractions lessen the force with which the molecule hits the wall. As a result, the gas pressure is less than that of an ideal gas. This effect decreases PV/RT to below its ideal value, as seen at the lower pressures in Figures 10.18 and 10.19. When the pressure is sufficiently high, however, the volume effects dominate and PV/RT increases to above the ideal value.

Temperature determines how effective attractive forces between gas molecules are in causing deviations from ideal behavior at lower pressures. Figure 10.19 shows that, at pressures below about 400 atm, cooling increases the extent to which a gas deviates from

Go Figure How would you expect the pressure of a gas to change if suddenly the intermolecular forces were repulsive rather than attractive?

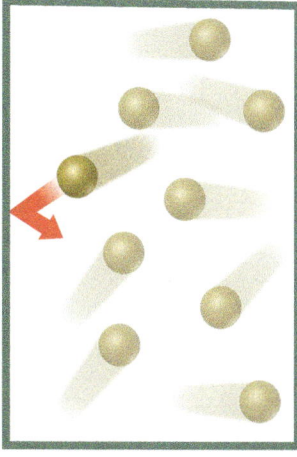 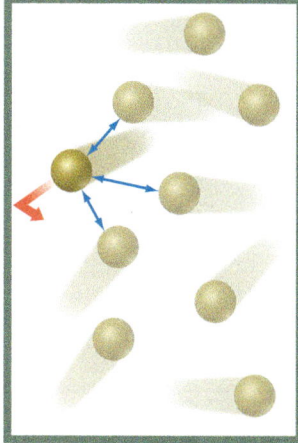

Ideal gas   Real gas

▲ **Figure 10.21** **In any real gas, attractive intermolecular forces reduce pressure to values lower than in an ideal gas.**

ideal behavior. As the gas cools, the average kinetic energy of the molecules decreases. This drop in kinetic energy means the molecules do not have the energy needed to overcome intermolecular attraction, and the molecules will be more likely to stick to each other than bounce off each other.

As the temperature of a gas increases—as, say, from 200 to 1000 K in Figure 10.19—the negative deviation of PV/RT from the ideal value of 1 disappears. As noted earlier, the deviations seen at high temperatures stem mainly from the effect of the finite volumes of the molecules.

The van der Waals Equation

Engineers and scientists who work with gases at high pressures often cannot use the ideal-gas equation because departures from ideal behavior are too large. One useful equation developed to predict the behavior of real gases was proposed by the Dutch scientist Johannes van der Waals (1837–1923).

As we have seen, a real gas has a lower pressure due to intermolecular forces, and a larger volume due to the finite volume of the molecules, relative to an ideal gas. Van der Waals recognized that it would be possible to retain the form of the ideal-gas equation, $PV = nRT$, if corrections were made to the pressure and the volume. He introduced two constants for these corrections: a, a measure of how strongly the gas molecules attract one another, and b, a measure of the finite volume occupied by the molecules. His description of gas behavior is known as the **van der Waals equation**:

$$\left(P + \frac{n^2a}{V^2}\right)(V - nb) = nRT \qquad [10.26]$$

The term n^2a/V^2 accounts for the attractive forces. The equation adjusts the pressure upward by adding n^2a/V^2 because attractive forces between molecules tend to reduce the pressure (Figure 10.21). The added term has the form n^2a/V^2 because the attractive force between pairs of molecules increases as the square of the number of molecules per unit volume, $(n/V)^2$.

The term nb accounts for the small but finite volume occupied by the gas molecules (Figure 10.20). The van der Waals equation subtracts nb to adjust the volume downward to give the volume that would be available to the molecules in the ideal case. The constants a and b, called *van der Waals constants*, are experimentally determined positive quantities that differ from one gas to another. Notice in Table 10.3 that a and b generally increase with increasing molecular mass. Larger, more massive molecules have larger volumes and tend to have greater intermolecular attractive forces.

TABLE 10.3   Van der Waals Constants for Gas Molecules

Substance	a (L²-atm/mol²)	b (L/mol)
He	0.0341	0.02370
Ne	0.211	0.0171
Ar	1.34	0.0322
Kr	2.32	0.0398
Xe	4.19	0.0510
H_2	0.244	0.0266
N_2	1.39	0.0391
O_2	1.36	0.0318
F_2	1.06	0.0290
Cl_2	6.49	0.0562
H_2O	5.46	0.0305
NH_3	4.17	0.0371
CH_4	2.25	0.0428
CO_2	3.59	0.0427
CCl_4	20.4	0.1383

Sample Exercise 10.15

Using the van der Waals Equation

If 10.00 mol of an ideal gas were confined to 22.41 L at 0.0 °C, it would exert a pressure of 10.00 atm. Use the van der Waals equation and Table 10.3 to estimate the pressure exerted by 10.00 mol of $Cl_2(g)$ in 22.41 L at 0.0 °C.

SOLUTION

Analyze We need to determine a pressure. Because we will use the van der Waals equation, we must identify the appropriate values for the constants in the equation.

Plan Rearrange Equation 10.26 to isolate P.

Solve Substituting $n = 10.00$ mol, $R = 0.08206$ L-atm/mol-K, $T = 273.2$ K, $V = 22.41$ L, $a = 6.49$ L^2-atm/mol^2, and $b = 0.0562$ L/mol:

$$P = nRT/(V - nb) - n^2a/V^2$$

$$= \frac{(10.00 \text{ mol})(0.08206 \text{ L-atm/mol-K})(273.2 \text{ K})}{22.41 \text{ L} - (10.00 \text{ mol})(0.0562 \text{ L/mol})}$$

$$- \frac{(10.00 \text{ mol})^2(6.49 \text{ L}^2\text{-atm/mol}^2)}{(22.41 \text{ L})^2}$$

$$= 10.26 \text{ atm} - 1.29 \text{ atm} = 8.97 \text{ atm}$$

Comment Notice that the term 10.26 atm is the pressure corrected for molecular volume. This value is higher than the ideal value, 10.00 atm, because the volume in which the molecules are free to move is smaller than the container volume, 22.41 L. Thus, the molecules collide more frequently with the container walls, and the pressure is higher than that of a real gas. The term 1.29 atm makes a correction in the opposite direction for intermolecular forces. The correction for intermolecular forces is the larger of the two, and thus the pressure 8.97 atm is smaller than would be observed for an ideal gas.

▶ **Practice Exercise**
A sample of 1.000 mol of $CO_2(g)$ is confined to a 3.000-L container at 0.000 °C. Calculate the pressure of the gas using (**a**) the ideal-gas equation and (**b**) the van der Waals equation.

Self-Assessment Exercises

SAE 10.22 Which of the following statements regarding the behavior of real gases is *false*? (**a**) As the attractions between molecules increase, the value of the constant b in the van der Waals equation increases. (**b**) Deviations from ideal-gas behavior are more pronounced at high pressures because the molecules experience greater attractions to one another. (**c**) Deviations from ideal-gas behavior are more pronounced at high pressures because the fraction of space occupied by molecules becomes nonnegligible. (**d**) Deviations from ideal-gas behavior are more pronounced at low temperatures because the molecules experience greater attractions to one another.

SAE 10.23 Based on their respective van der Waals constants, is Ar ($a = 1.34, b = 0.0322$) or CO_2 ($a = 3.59, b = 0.0427$) expected to behave more nearly like an ideal gas at high pressures? (**a**) It is Ar, because both the a and b parameters are smaller than that of CO_2. (**b**) It is CO_2, because both the a and b parameters are larger than that of Ar. (**c**) The deviation from ideal behavior depends only on T—the pressure has no effect. (**d**) Both gases would behave quite similarly at high pressure.

Putting Concepts Together

Cyanogen, a highly toxic gas, is 46.2% C and 53.8% N by mass. At 25 °C and 751 torr, 1.05 g of cyanogen occupies 0.500 L. (**a**) What is the molecular formula of cyanogen? Predict (**b**) its molecular structure and (**c**) its polarity.

SOLUTION

Analyze We need to determine the molecular formula of a gas from elemental analysis data and data on its properties. Then we need to predict the structure of the molecule and from that, its polarity.

(**a**) **Plan** We can use the percentage composition of the compound to calculate its empirical formula. (Section 3.5) Then we can determine the molecular formula by comparing the mass of the empirical formula with the molar mass. (Section 3.5)

Solve To determine the empirical formula, we assume we have a 100-g sample and calculate the number of moles of each element in the sample:

$$\text{Moles C} = (46.2 \text{ g C})\left(\frac{1 \text{ mol C}}{12.01 \text{ g C}}\right) = 3.85 \text{ mol C}$$

$$\text{Moles N} = (53.8 \text{ g N})\left(\frac{1 \text{ mol N}}{14.01 \text{ g N}}\right) = 3.84 \text{ mol N}$$

Continued

Because the ratio of the moles of the two elements is essentially 1:1, the empirical formula is CN. To determine the molar mass, we use Equation 10.11.

$$\mathcal{M} = \frac{dRT}{P} = \frac{(1.05\ \text{g}/0.500\ \text{L})(0.08206\ \text{L-atm}/\text{mol-K})(298\ \text{K})}{(751/760)\ \text{atm}}$$

$$= 52.0\ \text{g/mol}$$

The molar mass associated with the empirical formula CN is $12.0 + 14.0 = 26.0$ g/mol. Dividing the molar mass by that of its empirical formula gives $(52.0\ \text{g/mol})/(26.0\ \text{g/mol}) = 2.00$. Thus, the molecule has twice as many atoms of each element as the empirical formula, giving the molecular formula C_2N_2.

(b) Plan To determine the molecular structure, we must first determine the Lewis structure. (Section 8.5) We can then use the VSEPR model to predict the structure. (Section 9.2)

Solve The molecule has $2(4) + 2(5) = 18$ valence-shell electrons. By trial and error, we seek a Lewis structure with 18 valence

electrons in which each atom has an octet and the formal charges are as low as possible. The structure

$$:N\equiv C - C \equiv N:$$

meets these criteria. (This structure has zero formal charge on each atom.)

The Lewis structure shows that each atom has two electron domains. (Each nitrogen has a nonbonding pair of electrons and a triple bond, whereas each carbon has a triple bond and a single bond.) Thus, the electron-domain geometry around each atom is linear, causing the overall molecule to be linear.

(c) Plan To determine the polarity of the molecule, we must examine the polarity of the individual bonds and the overall geometry of the molecule.

Solve Because the molecule is linear, we expect the two dipoles created by the polarity in the carbon–nitrogen bond to cancel each other, leaving the molecule with no dipole moment.

Chapter Summary and Key Terms

PHYSICAL CHARACTERISTICS OF GASES (SECTION 10.1) Substances that are gases at room temperature tend to be molecular substances with low molar masses. Air, a mixture composed mainly of N_2 and O_2, is the most common gas we encounter. Some liquids and solids can also exist in the gaseous state, where they are known as **vapors**. Gases are compressible; they mix in all proportions because their component molecules are far apart from each other. To describe the state or condition of a gas, we must specify four variables: pressure (P), volume (V), temperature (T), and quantity (n). Volume is usually measured in liters, temperature in kelvins, and quantity of gas in moles. **Pressure** is the force per unit area and is expressed in SI units as **pascals**, Pa (1 Pa = 1 N/m^2). A related unit, the **bar**, equals 10^5 Pa. In chemistry, **standard atmospheric pressure** is used to define the **atmosphere** (atm) and the **torr** (also called the millimeter of mercury). One atmosphere of pressure equals 101.325 kPa, or 760 torr. A barometer is often used to measure the atmospheric pressure. A manometer can be used to measure the pressure of enclosed gases.

THE GAS LAWS (SECTION 10.2) Studies have revealed several simple gas laws: For a constant quantity of gas at constant temperature, the volume of the gas is inversely proportional to the pressure (**Boyle's law**). For a fixed quantity of gas at constant pressure, the volume is directly proportional to its absolute temperature (**Charles's law**). Equal volumes of gases at the same temperature and pressure contain equal numbers of molecules (**Avogadro's hypothesis**). For a gas at constant temperature and pressure, the volume of the gas is directly proportional to the number of moles of gas (**Avogadro's law**). Each of these gas laws is a special case of the ideal-gas equation.

THE IDEAL-GAS EQUATION (SECTION 10.3) The **ideal-gas equation**, $PV = nRT$, is the equation of state for an **ideal gas**. The term R in this equation is the **gas constant**. We can use the ideal-gas equation to calculate variations in one variable when one or more of the others are changed. Most gases at pressures less than 10 atm and temperatures near 273 K and above obey the ideal-gas equation reasonably well. The conditions of 273 K (0 °C) and 1 atm are known as the **standard temperature and pressure (STP)**. In all applications of the ideal-gas equation, we must remember to convert temperatures to the absolute-temperature scale (the Kelvin scale).

Using the ideal-gas equation, we can relate the density of a gas to its molar mass: $\mathcal{M} = dRT/P$. We can also use the ideal-gas equation to solve problems involving gases as reactants or products in chemical reactions.

GAS MIXTURES AND PARTIAL PRESSURES (SECTION 10.4) In gas mixtures, the total pressure is the sum of the **partial pressures** that each gas would exert if it were present alone under the same conditions (**Dalton's law of partial pressures**). The partial pressure of a component of a mixture is equal to its mole fraction times the total pressure: $P_1 = X_1 P_t$. The **mole fraction** X is the ratio of the moles of one component of a mixture to the total moles of all components.

THE KINETIC-MOLECULAR THEORY OF GASES (SECTION 10.5) The **kinetic-molecular theory of gases** accounts for the properties of an ideal gas in terms of a set of statements about the nature of gases. Briefly, these statements are as follows: (1) Molecules are in continuous chaotic motion; (2) The volume of gas molecules is negligible compared to the volume of their container; (3) The gas molecules neither attract nor repel each other; (4) The average kinetic energy of the gas molecules is proportional to the absolute temperature and does not change if the temperature remains constant.

The individual molecules of a gas do not all have the same kinetic energy at a given instant. Their speeds are distributed over a wide range; the distribution varies with the molar mass of the gas and with temperature.

MOLECULAR SPEEDS, EFFUSION, AND DIFFUSION (SECTION 10.6) Gas molecules are in constant motion. Their **root-mean-square (rms) speed**, u_{rms}, varies in proportion to the square root of the absolute temperature and inversely with the square root of the molar mass: $u_{rms} = \sqrt{(3RT/\mathcal{M})}$. The most probable speed of a gas molecule is given by $u_{mp} = \sqrt{(2RT/\mathcal{M})}$. The average speed of a gas molecule is given by $u_{av} = \sqrt{(8RT/\pi\mathcal{M})}$.

It follows from kinetic-molecular theory that the rate at which a gas undergoes **effusion** (escapes through a tiny hole) is inversely proportional to the square root of its molar mass (**Graham's law**). The **diffusion** of gas molecules through space is another phenomenon related to the speeds at which molecules move. Because moving molecules undergo frequent collisions with one another, the **mean free path**—the mean distance traveled between collisions—is short. Collisions between molecules limit the rate at which a gas molecule can diffuse.

REAL GASES: DEVIATIONS FROM IDEAL BEHAVIOR (SECTION 10.7) Departures from ideal behavior increase in magnitude as pressure increases and as temperature decreases. Real gases depart from ideal behavior because (1) the molecules possess finite volume and (2) the molecules experience attractive forces for one another. These two effects make the volumes of real gases larger and their pressures smaller than those of an ideal gas. The **van der Waals equation** is an equation of state for gases, which modifies the ideal-gas equation to account for intrinsic molecular volume and intermolecular forces.

Key Equations

- $PV = nRT$ [10.5]   Ideal-gas equation

- $\dfrac{P_1V_1}{T_1} = \dfrac{P_2V_2}{T_2}$ [10.8]   The combined gas law, showing how P, V, and T are related for a constant n

- $d = \dfrac{P\mathcal{M}}{RT}$ [10.10]   Density or molar mass of an ideal gas

- $P_t = P_1 + P_2 + P_3 + \ldots$ [10.12]   Relating the total pressure of a gas mixture to the partial pressures of its components (Dalton's law of partial pressures)

- $P_1 = \left(\dfrac{n_1}{n_t}\right)P_t = X_1P_t$ [10.16]   Relating partial pressure to mole fraction

- $u_{rms} = \sqrt{\dfrac{3RT}{\mathcal{M}}}$ [10.20]   Definition of the root-mean-square (rms) speed of gas molecules

- $u_{mp} = \sqrt{\dfrac{2RT}{\mathcal{M}}}$ [10.21]   Definition of the most probably (mp) speed of gas molecules

- $u_{av} = \sqrt{\dfrac{8RT}{\pi\mathcal{M}}}$ [10.22]   Definition of the average (av) speed of gas molecules

- $\dfrac{r_1}{r_2} = \sqrt{\dfrac{\mathcal{M}_2}{\mathcal{M}_1}}$ [10.24]   Relating the relative rates of effusion of two gases to their molar masses

- $\left(P + \dfrac{n^2a}{V^2}\right)(V - nb) = nRT$ [10.26]   The van der Waals equation

Exam Prep

EP 10.1 Which statement about gases is false? (a) A given gas can mix in all proportions with any other gas. (b) Gases are incompressible. (c) Gaseous substances can be atomic or molecular. (d) Substances that are gases at room temperature generally have molar masses lower than 100 g/mol.

EP 10.2 What would be the height of the column in a barometer if the external pressure was 101 kPa and water ($d = 1.00\,\text{g/cm}^3$) was used in place of mercury ($d = 13.6\,\text{g/cm}^3$)? (a) 0.0558 m (b) 0.760 m (c) 1.03×10^4 m (d) 10.3 m (e) 0.103 m

EP 10.3 If the gas inside the flask in Figure 10.3 is cooled so that its pressure is reduced to a value of 715.7 torr, what will be the height of the mercury in the open-ended arm? (*Hint:* The sum of the heights in both arms must remain constant regardless of the change in pressure.) (a) 49.0 mm (b) 95.6 mm (c) 144.6 mm (d) 120.1 mm

EP 10.4 The deepest place on Earth is 7 miles below sea level in the Pacific Ocean's Mariana Trench. The water pressure at the bottom of the Trench is reported to be a bone-crushing 8 tons per square inch. Convert this value to atmospheres (1 atm = 14.70 lb/in^2; 1 ton = 2000 lbs). (a) 0.004 atm (b) 0.06 atm (c) 100 atm (d) 1000 atm (e) 2×10^4 atm

EP 10.5 Which of the following statements is or are *true* for an ideal gas?

  (i) The volume occupied by the molecules is negligible.

  (ii) The molecules are all moving at the same speed.

  (iii) The molecules are neither attracted to nor repelled from each other.

(a) Only one of these statements is true. (b) i and ii (c) i and iii (d) ii and iii (e) All three statements are true.

EP 10.6 Which statement about gas behavior is *true*? (a) If you increase the pressure of a fixed amount of gas while keeping the temperature constant, the volume must increase. (b) If you increase the amount of a gas while holding the pressure and temperature constant, the volume must decrease. (c) If you plot V versus $1/P$ for a fixed quantity of a gas at constant temperature, you will get a straight line with a positive slope. (d) If you plot V versus $1/n$ for a gas at constant T and P, you will get a straight line with a positive slope.

EP 10.7 If you double the number of moles of an ideal gas in a fixed volume while keeping the temperature constant, what happens to the pressure? (a) It doubles. (b) It remains the same. (c) It is decreased by a factor of 2. (d) It is increased by a factor of 4. (e) It is decreased by a factor of 4.

EP 10.8 A sample of oxygen gas with a mass of 8.0 g is added to a cylinder equipped with massless and frictionless piston. If the apparatus is placed in a chamber where the temperature is 0 °C and the pressure is 1.0 atm, the gas will occupy a volume of 5.6 L. If the pressure inside the chamber is increased to 3.0 atm, the piston will shift until the gas occupies a volume of _____. (a) 17 L (b) 1.9 L (c) 22.4 L (d) 0.62 L

EP 10.9 A helium balloon is filled to a volume of 5.60 L at 25 °C. What will the volume of the balloon become if it is put into liquid nitrogen to lower the temperature of the helium to 77 K? (a) 17 L (b) 22 L (c) 1.4 L (d) 0.046 L (e) 3.7 L

EP 10.10 The Goodyear blimp contains 5.74×10^6 L of helium at 25.0 °C and 1.00 atm. What is the mass in grams of the helium inside the blimp?

(a) 2.30×10^7 g   (d) 2.34×10^5 g
(b) 2.80×10^6 g   (e) 9.39×10^5 g
(c) 1.12×10^7 g

EP 10.11 If you fill your car tire to a pressure of 32 psi (pounds per square inch) on a hot summer day when the temperature is 35 °C (95 °F), what is the pressure (in psi) on a cold winter day when the temperature is −15 °C (5 °F)? Assume no gas leaks out between measurements and the volume of the tire does not change. (a) 38 psi (b) 27 psi (c) 1.8 psi (d) 13.7 psi

EP 10.12 A gas occupies a volume of 0.75 L at 20 °C at 720 torr. What volume would the gas occupy at 41 °C and 760 torr? (a) 1.45 L (b) 0.85 L (c) 0.76 L (d) 0.66 L (e) 0.35 L

EP 10.13 What is the density of methane, CH_4, in a vessel where the pressure is 910 torr and the temperature is 255 K? (**a**) 0.92 g/L (**b**) 697 g/L (**c**) 0.057 g/L (**d**) 16 g/L (**e**) 0.72 g/L

EP 10.14 What is the molar mass of an unknown hydrocarbon whose density is measured to be 1.97 g/L at STP? (**a**) 4.04 g/mol (**b**) 30.7 g/mol (**c**) 44.1 g/mol (**d**) 48.2 g/mol

EP 10.15 Silver oxide decomposes when heated:

$$2\,Ag_2O(s) \xrightarrow{\Delta} 4\,Ag(s) + O_2(g)$$

If 5.76 g of Ag_2O is heated and the O_2 gas produced by the reaction is collected in an evacuated flask, what is the pressure of the O_2 gas if the volume of the flask is 0.65 L and the gas temperature is 25 °C? (**a**) 0.94 atm (**b**) 0.039 atm (**c**) 0.012 atm (**d**) 0.47 atm (**e**) 3.2 atm

EP 10.16 A 15-L cylinder contains 4.0 g of hydrogen and 28 g of nitrogen. If the temperature is 27 °C, what is the total pressure of the mixture? (**a**) 0.44 atm (**b**) 1.6 atm (**c**) 3.3 atm (**d**) 4.9 atm (**e**) 9.8 atm

EP 10.17 A 4.0-L vessel containing N_2 at STP and a 2.0-L vessel containing H_2 at STP are connected by a valve. If the valve is opened allowing the two gases to mix, what is the mole fraction of hydrogen in the mixture? (**a**) 0.034 (**b**) 0.33 (**c**) 0.50 (**d**) 0.67 (**e**) 0.96

EP 10.18 Consider two gas cylinders of the same volume and temperature, one containing 1.0 mol of propane, C_3H_8, and the other 2.0 mol of methane, CH_4. Which of the following statements is *true*? (**a**) The C_3H_8 and CH_4 molecules have the same u_{rms}. (**b**) The C_3H_8 and CH_4 molecules have the same average kinetic energy. (**c**) The rate at which the molecules collide with the cylinder walls is the same for both cylinders. (**d**) The gas pressure is the same in both cylinders.

EP 10.19 Which of these is *not* an assumption of the kinetic-molecular theory of gases? (**a**) Gas molecules take up no space. (**b**) Gas molecules move in circles. (**c**) The average kinetic energy of the gas molecules is proportional to temperature. (**d**) Gas molecules neither attract nor repel each other. (**e**) Gas molecules are in constant motion.

EP 10.20 Fill in the blanks for the following statement: The rms speed of the molecules in a sample of H_2 gas at 300 K will be ____ times larger than the rms speed of O_2 molecules at the same temperature, and the ratio _____ with increasing temperature. (**a**) four, will not change (**b**) four, will increase (**c**) sixteen, will not change (**d**) sixteen, will decrease (**e**) Not enough information is given to answer this question.

EP 10.21 Calculate the average speed of CO_2 molecules at −40.0 °C, the typical temperature of an airplane flying at 36,000 ft. (**a**) 10.6 m/s (**b**) 33.3 m/s (**c**) 297 m/s (**d**) 335 m/s (**e**) 364 m/s

EP 10.22 Which of the following changes will lead to an increase in the mean free path of the molecules of gas in a closed container?

(**i**) Lower the temperature without changing the volume of the container.

(**ii**) Increase the volume of the container without changing the temperature.

(**iii**) Open the valve on the container to release half of the molecules without changing the temperature or volume of the container.

(**a**) Only i (**b**) only ii (**c**) only iii (**d**) both ii and iii (**e**) all three

EP 10.23 In a system for separating gases, a tank containing a mixture of hydrogen and carbon dioxide is connected to a much larger tank where the pressure is kept very low. The two tanks are separated by a porous membrane through which the molecules must effuse. If the initial partial pressure of each gas is 5.00 atm, what will be the mole fraction of hydrogen in the tank after the partial pressure of carbon dioxide has declined to 4.50 atm? (**a**) 52.1% (**b**) 37.2% (**c**) 32.1% (**d**) 4.68% (**e**) 27.4%

EP 10.24 Calculate the pressure of a 2.975-mol sample of N_2 in a 0.7500-L flask at 300.0 °C using the van der Waals equation and then repeat the calculation using the ideal-gas equation. Within the limits of the significant figures justified by these parameters, will the ideal-gas equation overestimate or underestimate the pressure, and if so by how much?

(**a**) Underestimate by 12.38 atm

(**b**) Overestimate by 12.38 atm

(**c**) Underestimate by 3.85 atm

(**d**) Overestimate by 3.85 atm

EP 10.25 Which of the following statements regarding deviations from ideal-gas behavior are *true*?

(**i**) Gases behave nonideally because most molecules aren't spherical.

(**ii**) Gases behave nonideally because atoms and molecules occupy finite volumes.

(**iii**) Gases behave nonideally because atoms and molecules attract one another.

(**iv**) Gases will behave most ideally under conditions of high temperature and low pressure.

(**a**) i and ii (**b**) ii and iii (**c**) iii and iv (**d**) i, ii, and iii (**e**) ii, iii, and iv

Exercises

Visualizing Concepts

10.1 Mars has an average atmospheric pressure of 0.007 atm. Would it be easier or harder to drink from a straw on Mars than on Earth? Explain. [Section 10.1]

10.2 You have a sample of gas in a container with a movable piston, such as the one in the drawing. (**a**) Redraw the container to show what it might look like if the temperature of the gas is increased from 300 to 500 K while the pressure is kept constant. (**b**) Redraw the container to show what it might look like if the external pressure on the piston is increased from 1.0 atm to 2.0 atm while the temperature is kept constant. (**c**) Redraw the container to show what it might look like if the temperature of the gas decreases from 300 to 200 K while

the pressure is kept constant (assume the gas does not liquefy). [Section 10.2]

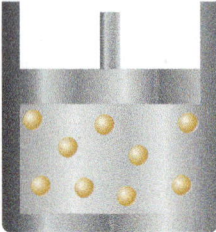

10.3 Consider the sample of gas depicted here. What would the drawing look like if the volume and temperature remained constant while you removed enough of the gas to decrease the pressure by a factor of 2? [Section 10.2]

(**a**) It would contain the same number of molecules.

(**b**) It would contain half as many molecules.

(**c**) It would contain twice as many molecules.

(**d**) There is insufficient data to say.

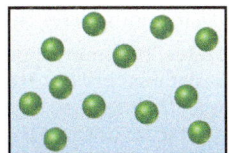

10.4 Imagine that the reaction $2\,CO(g) + O_2(g) \longrightarrow 2\,CO_2(g)$ occurs in a container that has a piston that moves to maintain a constant pressure when the reaction occurs at constant temperature. Which of the following statements describes how the volume of the container changes due to the reaction: (**a**) The volume increases by 50%. (**b**) The volume increases by 33%. (**c**) The volume remains constant. (**d**) The volume decreases by 33%. (**e**) The volume decreases by 50%. [Section 10.3]

10.5 Consider a fixed quantity of gas at constant pressure. If the volume of the gas is plotted as the temperature is changed, which of the following plots will be obtained? [Section 10.3]

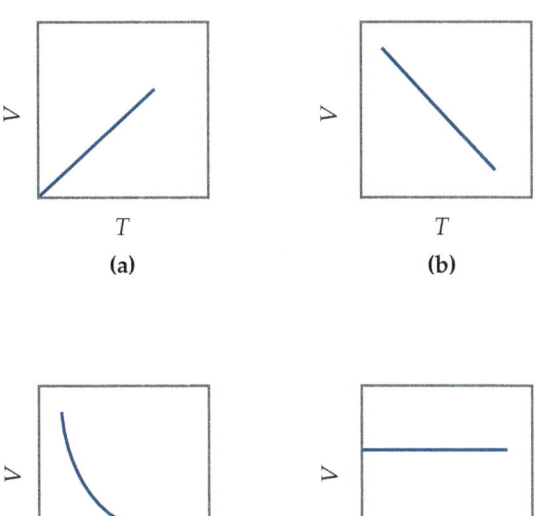

(**e**) None of the above

10.6 The apparatus shown here has two gas-filled containers and one empty container, all attached to a hollow horizontal tube closed at both ends. (**a**) How many blue gas molecules are in the left container? (**b**) How many red gas molecules are in the middle container? (**c**) When the valves are opened and the gases are allowed to mix at constant temperature, how many atoms of each type of gas end up in the originally empty container? Assume that the containers are of

equal volume and ignore the volume of the connecting tube. [Section 10.4]

10.7 The accompanying drawing represents a mixture of three different gases. (**a**) Rank the three components (red, yellow, blue) in order of increasing partial pressure. (**b**) If the total pressure of the mixture is 1.40 atm, calculate the partial pressure of each gas. [Section 10.4]

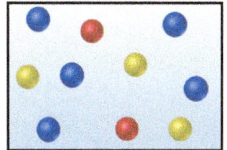

10.8 On a single plot, qualitatively sketch the distribution of molecular speeds for (**a**) Kr(g) at $-50\,°C$, (**b**) Kr(g) at $0\,°C$, (**c**) Ar(g) at $0\,°C$. [Section 10.6]

10.9 Consider the following graph. (**a**) If curves A and B refer to two different gases, He and O_2, at the same temperature, which curve corresponds to He? (**b**) If A and B refer to the same gas at two different temperatures, which represents the higher temperature? (**c**) For each curve, which speed is highest: the most probable speed, the root-mean-square speed, or the average speed? [Section 10.6]

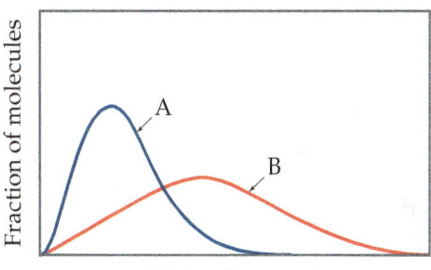

10.10 Consider the following samples of gases:

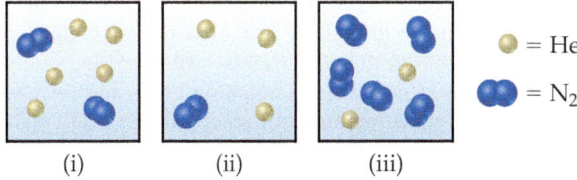

If the three samples are all at the same temperature, rank them with respect to (**a**) total pressure, (**b**) partial pressure of helium, (**c**) density, (**d**) average kinetic energy of particles. [Sections 10.5 and 10.6]

10.11 A thin glass tube 1 m long is filled with Ar gas at 1 atm, and the ends are stoppered with cotton plugs as shown in the drawing. HCl gas is introduced at one end of the tube, and simultaneously NH_3 gas is introduced at the other end. When the two gases diffuse through the cotton plugs down the tube and meet, a white ring appears due to the formation of $NH_4Cl(s)$. At which location—a, b, or c—do you expect the ring to form? [Section 10.6]

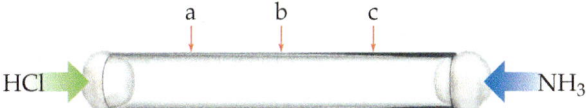

10.12 The graph provided shows the change in pressure as the temperature increases for a 1-mol sample of a gas confined to a 1-L container. The four plots correspond to an ideal gas and three real gases: CO_2, N_2, and Cl_2. (**a**) At room temperature, all three real gases have a pressure less than the ideal gas. Which van der Waals constant, a or b, accounts for the influence intermolecular forces have in lowering the pressure of a real gas? (**b**) Use the van der Waals constants in Table 10.3 to match the labels in the plot (A, B, and C) with the respective gases (CO_2, N_2, and Cl_2). [Section 10.7]

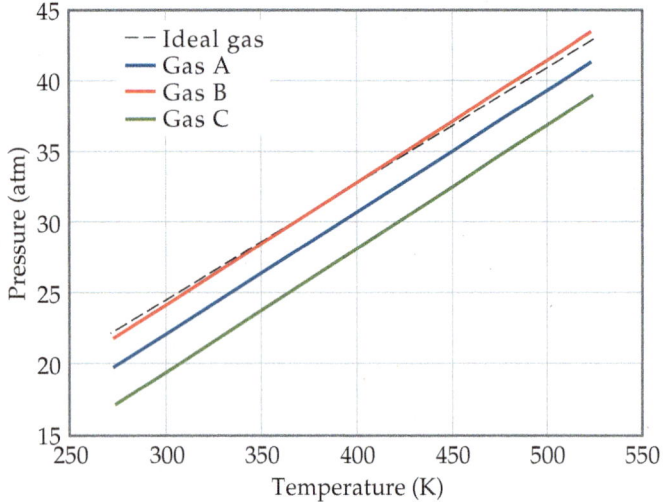

Physical Characteristics of Gases (Section 10.1)

10.13 Which of the following statements is *false*?

(**a**) Gases are far less dense than liquids.

(**b**) Gases are far more compressible than liquids.

(**c**) Because liquid water and liquid carbon tetrachloride do not mix, neither do their vapors.

(**d**) The volume occupied by a gas is determined by the volume of its container.

10.14 (**a**) Are you more likely to see the density of a gas reported in g/mL, g/L, or kg/cm³? (**b**) Which units are appropriate for expressing atmospheric pressures: N, Pa, atm, kg/m²? (**c**) Which is most likely to be a gas at room temperature and ordinary atmospheric pressure: F_2, Br_2, K_2O.

10.15 Suppose that a woman weighing 130 lb and wearing high-heeled shoes momentarily places all her weight on the heel of one foot. If the area of the heel is 0.50 in.², calculate the pressure exerted on the underlying surface in (**a**) pounds per square inch, (**b**) kilopascals, and (**c**) atmospheres.

10.16 A set of bookshelves rests on a hard floor surface on four legs, each having a cross-sectional dimension of 3.0 × 4.1 cm in contact with the floor. The total mass of the shelves plus the books stacked on them is 262 kg. Calculate the pressure in pascals exerted by the shelf footings on the surface.

10.17 (**a**) How high in meters must a column of glycerol be to exert a pressure equal to that of a 760-mm column of mercury? The density of glycerol is 1.26 g/mL, whereas that of mercury is 13.6 g/mL. (**b**) What pressure, in atmospheres, is exerted on the body of a diver if they are 15 ft below the surface of the water when the atmospheric pressure is 750 torr? Assume that the density of the water is $1.00 \text{ g/cm}^3 = 1.00 \times 10^3 \text{ kg/m}^3$. The gravitational constant is 9.81 m/s², and 1 Pa = 1 kg/m-s².

10.18 (**a**) The compound 1-iodododecane is a nonvolatile liquid with a density of 1.20 g/mL. The density of mercury is 13.6 g/mL. What do you predict for the height of a barometer column based on 1-iodododecane, when the atmospheric pressure is 749 torr? (**b**) What is the pressure, in atmospheres, on the body of a diver if they are 21 ft below the surface of the water when the atmospheric *pressure* is 742 torr?

10.19 The typical atmospheric pressure on top of Mount Everest (29,032 ft) is about 265 torr. Convert this pressure to (**a**) atm, (**b**) mm Hg, (**c**) pascals, (**d**) bars, (**e**) psi.

10.20 Perform the following conversions: (**a**) 0.912 atm to torr, (**b**) 0.685 bar to kilopascals, (**c**) 655 mm Hg to atmospheres, (**d**) 1.323×10^5 Pa to atmospheres, (**e**) 2.50 atm to psi.

10.21 In the United States, barometric pressures are generally reported in inches of mercury (in. Hg). On a beautiful summer day in Chicago, the barometric pressure is 30.45 in. Hg. (**a**) Convert this pressure to torr. (**b**) Convert this pressure to atm.

10.22 Hurricane Wilma of 2005 is the most intense hurricane on record in the Atlantic basin, with a low-pressure reading of 882 mbar (millibars). Convert this reading into (**a**) atmospheres, (**b**) torr, and (**c**) inches of Hg.

10.23 If the atmospheric pressure is 0.995 atm, what is the pressure of the enclosed gas in each of the three cases depicted in the drawing? Assume that the gray liquid is mercury.

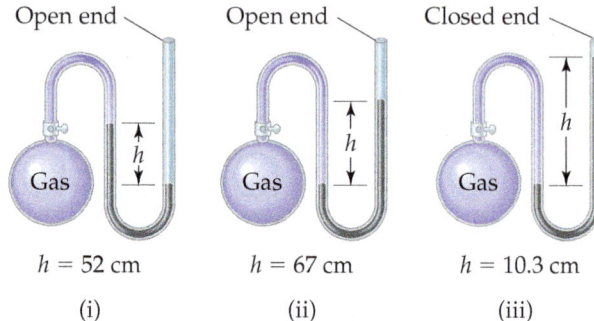

10.24 An open-end manometer containing mercury is connected to a container of gas, as depicted in Sample Exercise 10.2. What is the pressure of the enclosed gas in torr in each of the following situations? (**a**) The mercury in the arm attached to the gas is 15.4 mm higher than in the one open to the atmosphere; atmospheric pressure is 0.985 atm. (**b**) The mercury in the arm attached to the gas is 12.3 mm lower than in the one open to the atmosphere; atmospheric pressure is 0.99 atm.

The Gas Laws (Section 10.2)

10.25 You have a gas at 25 °C confined to a cylinder with a movable piston. Which of the following actions would double the gas pressure? (**a**) Lifting up on the piston to double the volume while keeping the temperature constant; (**b**) Heating the gas so that its temperature rises from 25 °C to 50 °C, while keeping the volume constant; (**c**) Pushing down on the piston to halve the volume while keeping the temperature constant.

10.26 A fixed quantity of gas at 21 °C exhibits a pressure of 752 torr and occupies a volume of 5.12 L. (**a**) Calculate the volume the gas will occupy if the pressure is increased to 1.88 atm while the temperature is held constant. (**b**) Calculate the volume the gas will occupy if the temperature is increased to 175 °C while the pressure is held constant.

10.27 (**a**) Amonton's law expresses the relationship between pressure and temperature. Use Charles's law and Boyle's law to derive the proportionality relationship between P and T. (**b**) If a car tire is filled to a pressure of 32.0 lb/in.² (psi) measured at 75 °F, what will be the tire pressure if the tires heat up to 120 °F during driving?

10.28 Nitrogen and hydrogen gases react to form ammonia gas as follows:

$$N_2(g) + 3\,H_2(g) \longrightarrow 2\,NH_3(g)$$

At a certain temperature and pressure, 1.2 L of N_2 reacts with 3.6 L of H_2. If all the N_2 and H_2 are consumed, what volume of NH_3, at the same temperature and pressure, will be produced?

The Ideal-Gas Equation (Section 10.3)

10.29 (a) What conditions are represented by the abbreviation STP? (b) What is the molar volume of an ideal gas at STP? (c) Room temperature is often assumed to be 25 °C. Calculate the molar volume of an ideal gas at 25 °C and 1 atm pressure. (d) If you measure pressure in bars instead of atmospheres, calculate the corresponding value of R in L-bar/mol-K.

10.30 To derive the ideal-gas equation, we assume that the volume of the gas atoms/molecules can be neglected. Given the atomic radius of neon, 0.69 Å, and knowing that a sphere has a volume of $4\pi r^3/3$, calculate the fraction of space that Ne atoms occupy in a sample of neon at STP.

10.31 Suppose you are given two 1-L flasks and told that one contains a gas of molar mass 30 and the other a gas of molar mass 60, both at the same temperature. The pressure in flask A is x atm, and the mass of gas in the flask is 1.2 g. The pressure in flask B is $0.5x$ atm, and the mass of gas in that flask is 1.2 g. Which flask contains the gas of molar mass 30, and which contains the gas of molar mass 60?

10.32 Suppose you are given two flasks at the same temperature, one of volume 2 L and the other of volume 3 L. The 2-L flask contains 4.8 g of gas, and the gas pressure is x atm. The 3-L flask contains 0.36 g of gas, and the gas pressure is $0.1x$. Do the two gases have the same molar mass? If not, which contains the gas of higher molar mass?

10.33 Complete the following table for an ideal gas:

P	V	n	T
2.00 atm	1.00 L	0.500 mol	? K
0.300 atm	0.250 L	? mol	27 °C
650 torr	? L	0.333 mol	350 K
? atm	585 mL	0.250 mol	295 K

10.34 Calculate each of the following quantities for an ideal gas: (a) the volume of the gas, in liters, if 1.50 mol has a pressure of 1.25 atm at a temperature of −6 °C; (b) the absolute temperature of the gas at which 3.33×10^{-3} mol occupies 478 mL at 750 torr; (c) the pressure, in atmospheres, if 0.00245 mol occupies 413 mL at 138 °C; (d) the quantity of gas, in moles, if 126.5 L at 54 °C has a pressure of 11.25 kPa.

10.35 The Goodyear blimps, which frequently fly over sporting events, hold approximately 175,000 ft³ of helium. If the gas is at 23 °C and 1.0 atm, what mass of helium is in a blimp?

10.36 A neon sign is made of glass tubing whose inside diameter is 2.5 cm and whose length is 5.5 m. If the sign contains neon at a pressure of 1.78 torr at 35 °C, how many grams of neon are in the sign? (The volume of a cylinder is $\pi r^2 h$.)

10.37 (a) Calculate the number of molecules in a deep breath of air whose volume is 2.25 L at body temperature, 37 °C, and a pressure of 735 torr. (b) The adult blue whale has a lung capacity of 5.0×10^3 L. Calculate the mass of air (assume an average molar mass of 28.98 g/mol) contained in an adult blue whale's lungs at 0.0 °C and 1.00 atm, assuming the air behaves ideally.

10.38 (a) If the pressure exerted by ozone, O_3, in the stratosphere is 3.0×10^{-3} atm and the temperature is 250 K, how many ozone molecules are in a liter? (b) Carbon dioxide makes up approximately 0.04% of Earth's atmosphere. If you collect a 2.0-L sample from the atmosphere at sea level (1.00 atm) on a warm day (27 °C), how many CO_2 molecules are in your sample?

10.39 A scuba diver's tank contains 0.29 kg of O_2 compressed into a volume of 2.3 L. (a) Calculate the gas pressure inside the tank at 9 °C. (b) What volume would this oxygen occupy at 26 °C and 0.95 atm?

10.40 An aerosol spray can with a volume of 250 mL contains 2.30 g of propane gas (C_3H_8) as a propellant. (a) If the can is at 23 °C, what is the pressure in the can? (b) What volume would the propane occupy at STP? (c) The can's label says that exposure to temperatures above 130 °F may cause the can to burst. What is the pressure in the can at this temperature?

10.41 A 35.1 g sample of solid CO_2 (dry ice) is added to a container at a temperature of 100 K with a volume of 4.0 L. If the container is evacuated (all of the gas is removed), sealed, and then allowed to warm to room temperature ($T = 298$ K) so that all of the solid CO_2 is converted to a gas, what is the pressure inside the container?

10.42 A 334-mL cylinder for use in chemistry lectures contains 5.225 g of helium at 23 °C. How many grams of helium must be released to reduce the pressure to 75 atm assuming ideal-gas behavior?

10.43 Chlorine is widely used to purify municipal water supplies and to treat swimming pool waters. Suppose that the volume of a particular sample of Cl_2 gas is 8.70 L at 895 torr and 24 °C. (a) How many grams of Cl_2 are in the sample? (b) What volume will the Cl_2 occupy at STP? (c) At what temperature will the volume be 15.00 L if the pressure is 8.76×10^2 torr? (d) At what pressure will the volume equal 5.00 L if the temperature is 58 °C?

10.44 Many gases are shipped in high-pressure containers. Consider a steel tank whose volume is 55.0 gallons that contains O_2 gas at a pressure of 16,500 kPa at 23 °C. (a) What mass of O_2 does the tank contain? (b) What volume would the gas occupy at STP? (c) At what temperature would the pressure in the tank equal 150.0 atm? (d) What would be the pressure of the gas, in kPa, if it were transferred to a container at 24 °C whose volume is 55.0 L?

10.45 In an experiment reported in the scientific literature, male cockroaches were made to run at different speeds on a miniature treadmill while their oxygen consumption was measured. In 1 h the average cockroach running at 0.08 km/h consumed 0.8 mL of O_2 at 1 atm pressure and 24 °C per gram of insect mass. (a) How many moles of O_2 would be consumed in 1 h by a 5.2-g cockroach moving at this speed? (b) This same cockroach is caught by a child and placed in a 1-qt fruit jar with a tight lid. Assuming the same level of continuous activity as in the research, will the cockroach consume more than 20% of the available O_2 in a 48-h period? (Air is 21 mol% O_2.)

10.46 The physical fitness of athletes is measured by "V_{O_2} max," which is the maximum volume of oxygen consumed by an individual during incremental exercise (for example, on a treadmill). An average male has a V_{O_2} max of 45 mL O_2/kg body mass/min, but a world-class male athlete can have a V_{O_2} max reading of 88.0 mL O_2/kg body mass/min. (a) Calculate the volume of oxygen, in mL, consumed in 1 h by an average man who weighs 185 lb and has a V_{O_2} max reading of 47.5 mL O_2/kg body mass/min. (b) If this man lost 20 lb, exercised, and increased his V_{O_2} max to 65.0 mL O_2/kg body mass/min, how many mL of oxygen would he consume in 1 h?

10.47 Rank the following gases from least dense to most dense at 1.00 atm and 298 K: CO, N_2O, Cl_2, HF.

10.48 Rank the following gases from least dense to most dense at 1.00 atm and 298 K: SO_2, HBr, CO_2.

10.49 Which of the following statements best explains why a closed balloon filled with helium gas rises in air?

(a) Helium is a monatomic gas, whereas nearly all the molecules that make up air, such as nitrogen and oxygen, are diatomic.

(b) The average speed of helium atoms is greater than the average speed of air molecules, and the greater speed of collisions with the balloon walls propels the balloon upward.

(c) Because the helium atoms are of lower mass than the average air molecule, the helium gas is less dense than air. The mass of the balloon is thus less than the mass of the air displaced by its volume.

(d) Because helium has a lower molar mass than the average air molecule, the helium atoms are in faster motion. This means that the temperature of the helium is greater than the air temperature. Hot gases tend to rise.

10.50 Which of the following statements best explains why nitrogen gas at STP is less dense than Xe gas at STP?

(a) Because Xe is a noble gas, there is less tendency for the Xe atoms to repel one another, so they pack more densely in the gaseous state.

(b) Xe atoms have a higher mass than N_2 molecules. Because both gases at STP have the same number of molecules per unit volume, the Xe gas must be denser.

(c) The Xe atoms are larger than N_2 molecules and thus take up a larger fraction of the space occupied by the gas.

(d) Because the Xe atoms are much more massive than the N_2 molecules, they move more slowly and thus exert less upward force on the gas container and make the gas appear denser.

10.51 (a) Calculate the density of NO_2 gas at 0.970 atm and 35 °C. (b) Calculate the molar mass of a gas if 2.50 g occupies 0.875 L at 685 torr and 35 °C.

10.52 (a) Calculate the density of sulfur hexafluoride gas at 707 torr and 21 °C. (b) Calculate the molar mass of a vapor that has a density of 7.135 g/L at 12 °C and 743 torr.

10.53 In the Dumas-bulb technique for determining the molar mass of an unknown liquid, you vaporize the sample of a liquid that boils below 100 °C in a boiling-water bath and determine the mass of vapor required to fill the bulb. From the following data, calculate the molar mass of the unknown liquid: mass of unknown vapor, 1.012 g; volume of bulb, 354 cm³; pressure, 742 torr; temperature, 99 °C.

Dumas bulb filled with vaporized unknown substance — Boiling water

10.54 The molar mass of a volatile substance was determined by the Dumas-bulb method described in Exercise 10.53. The unknown vapor had a mass of 0.846 g; the volume of the bulb was 354 cm³, pressure 752 torr, and temperature 100 °C. Calculate the molar mass of the unknown vapor.

10.55 Magnesium can be used as a "getter" in evacuated enclosures to react with the last traces of oxygen. (The magnesium is usually heated by passing an electric current through a wire or ribbon of the metal.) If an enclosure of 0.452 L has a partial pressure of O_2 of 3.5×10^{-6} torr at 27 °C, what mass of magnesium will react according to the following equation?

$$2\,Mg(s) + O_2(g) \longrightarrow 2\,MgO(s)$$

10.56 Calcium hydride, CaH_2, reacts with water to form hydrogen gas:

$$CaH_2(s) + 2\,H_2O(l) \longrightarrow Ca(OH)_2(aq) + 2\,H_2(g)$$

This reaction is sometimes used to inflate life rafts, weather balloons, and the like, when a simple, compact means of generating H_2 is desired. How many grams of CaH_2 are needed to generate 145 L of H_2 gas if the pressure of H_2 is 825 torr at 21 °C?

10.57 The metabolic oxidation of glucose, $C_6H_{12}O_6$, in our bodies produces CO_2, which is expelled from our lungs as a gas:

$$C_6H_{12}O_6(aq) + 6\,O_2(g) \longrightarrow 6\,CO_2(g) + 6\,H_2O(l)$$

(a) Calculate the volume of dry CO_2 produced at body temperature (37 °C) and 0.970 atm when 24.5 g of glucose is consumed in this reaction. (b) Calculate the volume of oxygen you would need, at 1.00 atm and 298 K, to completely oxidize 50.0 g of glucose.

10.58 Both Jacques Charles and Joseph Louis Guy-Lussac were avid balloonists. In his original flight in 1783, Jacques Charles used a balloon that contained approximately 31,150 L of H_2. He generated the H_2 using the reaction between iron and hydrochloric acid:

$$Fe(s) + 2\,HCl(aq) \longrightarrow FeCl_2(aq) + H_2(g)$$

How many kilograms of iron were needed to produce this volume of H_2 if the temperature was 22 °C?

10.59 Hydrogen gas is produced when zinc reacts with sulfuric acid:

$$Zn(s) + H_2SO_4(aq) \longrightarrow ZnSO_4(aq) + H_2(g)$$

If 159 mL of wet H_2 is collected over water at 24 °C and a barometric pressure of 738 torr, how many grams of Zn have been consumed? (The vapor pressure of water is tabulated in Appendix B.)

10.60 Acetylene gas, $C_2H_2(g)$, can be prepared by the reaction of calcium carbide with water:

$$CaC_2(s) + 2\,H_2O(l) \longrightarrow Ca(OH)_2(aq) + C_2H_2(g)$$

Calculate the volume of C_2H_2 that is collected over water at 23 °C by reaction of 1.524 g of CaC_2 if the total pressure of the gas is 753 torr. (The vapor pressure of water is tabulated in Appendix B.)

Gas Mixtures and Partial Pressures (Section 10.4)

10.61 Consider the apparatus shown in the following drawing. (a) When the valve between the two containers is opened and the gases are allowed to mix, how does the volume occupied by the N_2 gas change? What is the partial pressure of N_2 after mixing? (b) How does the volume of the O_2 gas change when the gases mix? What is the partial pressure of O_2 in the mixture? (**c**) What is the total pressure in the container after the gases mix?

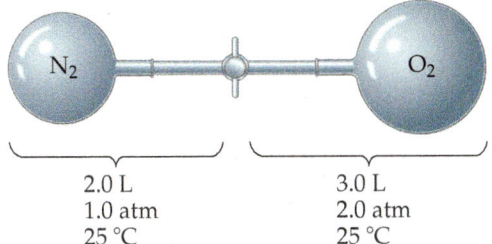

2.0 L
1.0 atm
25 °C

3.0 L
2.0 atm
25 °C

10.62 Consider a mixture of two gases, A and B, confined in a closed vessel. A quantity of a third gas, C, is added to the same vessel at the same temperature. How does the addition of gas C affect the following: (**a**) the partial pressure of gas A, (**b**) the total pressure in the vessel, (**c**) the mole fraction of gas B?

10.63 A mixture containing 0.765 mol He(g), 0.330 mol Ne(g), and 0.110 mol Ar(g) is confined in a 10.00-L vessel at 25 °C. (**a**) Calculate the partial pressure of each of the gases in the mixture. (**b**) Calculate the total pressure of the mixture.

10.64 A deep-sea diver uses a gas cylinder with a volume of 10.0 L and a content of 51.2 g of O_2 and 32.6 g of He. Calculate the partial pressure of each gas and the total pressure if the temperature of the gas is 19 °C.

10.65 The atmospheric concentration of CO_2 gas is presently 407 ppm (parts per million, by volume; that is, 407 L of every 10^6 L of the atmosphere are CO_2). What is the mole fraction of CO_2 in the atmosphere?

10.66 A plasma-screen TV contains thousands of tiny cells filled with a mixture of Xe, Ne, and He gases that emits light of specific wavelengths when a voltage is applied. A particular plasma cell, 0.900 mm × 0.300 mm × 10.0 mm, contains Xe, Ne, and He atoms in a ratio of 1:12:12, respectively, at a total pressure of 500 torr at 298 K. Calculate the number of Xe, Ne, and He atoms in the cell.

10.67 A piece of dry ice (solid carbon dioxide) with a mass of 5.50 g is placed in a 10.0-L vessel that already contains air at 705 torr and 24 °C. After the carbon dioxide has totally sublimed, what is the partial pressure of the resultant CO_2 gas, and the total pressure in the container, at 24 °C?

10.68 A sample of 5.00 mL of diethyl ether ($C_2H_5OC_2H_5$, density = 0.7134 g/mL) is introduced into a 6.00-L vessel that already contains a mixture of N_2 and O_2, whose partial pressures are $P_{N_2} = 0.751$ atm and $P_{O_2} = 0.208$ atm. The temperature is held at 35.0 °C, and the diethyl ether totally evaporates. (**a**) Calculate the partial pressure of the diethyl ether. (**b**) Calculate the total pressure in the container.

10.69 A rigid vessel containing a 3:1 mol ratio of carbon dioxide and water vapor is held at 200 °C where it has a total pressure of 2.00 atm. If the vessel is cooled to 10 °C so that all of the water vapor condenses, what is the pressure of carbon dioxide? Neglect the volume of the liquid water that forms on cooling.

10.70 If 5.15 g of Ag_2O is sealed in a 75.0-mL tube filled with 760 torr of N_2 gas at 32 °C, and the tube is heated to 320 °C, the Ag_2O decomposes to form oxygen and silver. What is the total pressure inside the tube, assuming the volume of the tube remains constant?

10.71 At an underwater depth of 250 ft, the pressure is 8.38 atm. What should the mole percent of oxygen be in the diving gas for the partial pressure of oxygen in the mixture to be 0.21 atm, the same as in air at 1 atm?

10.72 (**a**) What are the mole fractions of each component in a mixture of 15.08 g of O_2, 8.17 g of N_2, and 2.64 g of H_2? (**b**) What is the partial pressure in atm of each component of this mixture if it is held in a 15.50-L vessel at 15 °C?

10.73 A quantity of N_2 gas originally held at 5.25 atm pressure in a 1.00-L container at 26 °C is transferred to a 12.5-L container at 20 °C. A quantity of O_2 gas originally at 5.25 atm and 26 °C in a 5.00-L container is transferred to this same container. What is the total pressure in the new container?

10.74 A sample of 3.00 g of $SO_2(g)$ originally in a 5.00-L vessel at 21 °C is transferred to a 10.0-L vessel at 26 °C. A sample of 2.35 g of $N_2(g)$ originally in a 2.50-L vessel at 20 °C is transferred to this same 10.0-L vessel. (**a**) What is the partial pressure of $SO_2(g)$ in the larger container? (**b**) What is the partial pressure of $N_2(g)$ in this vessel? (**c**) What is the total pressure in the vessel?

Kinetic-Molecular Theory of Gases; Molecular Speeds, Effusion, and Diffusion (Sections 10.5 and 10.6)

10.75 Determine whether each of the following changes will increase, decrease, or not affect the rate with which gas molecules collide with the walls of their container: (**a**) increasing the volume of the container, (**b**) increasing the temperature, (**c**) increasing the molar mass of the gas.

10.76 Indicate which of the following statements regarding the kinetic-molecular theory of gases are correct. (**a**) The average kinetic energy of a collection of gas molecules at a given temperature is proportional to the square root of their mass, $m^{1/2}$. (**b**) The gas molecules are assumed to exert no forces on each other, aside from random elastic collisions. (**c**) All the molecules of a gas at a given temperature have the same kinetic energy. (**d**) The volume of the gas molecules is negligible in comparison to the total volume in which the gas is contained. (**e**) All gas molecules move with the same speed if they are at the same temperature.

10.77 WF_6 is one of the heaviest known gases. How much slower is the root-mean-square speed of WF_6 than He at 300 K?

10.78 You have an evacuated container of fixed volume and known mass and introduce a known mass of a gas sample. Measuring the pressure at constant temperature over time, you are surprised to see it slowly dropping. You measure the mass of the gas-filled container and find that the mass is what it should be—gas plus container—and the mass does not change over time, so you do not have a leak. Suggest an explanation for your observations.

10.79 The temperature of a 5.00-L container of N_2 gas is increased from 20 °C to 250 °C. If the volume is held constant, predict qualitatively how this change affects the following: (**a**) the average kinetic energy of the molecules; (**b**) the root-mean-square speed of the molecules; (**c**) the strength of the impact of an average molecule with the container walls; (**d**) the total number of collisions of molecules with walls per second.

10.80 Suppose you have two 1-L flasks, one containing N_2 at STP, the other containing CH_4 at STP. How do these systems compare with respect to (**a**) number of molecules, (**b**) density, (**c**) average kinetic energy of the molecules, (**d**) rate of effusion through a pinhole leak?

10.81 (**a**) Place the following gases in order of increasing average molecular speed at 25 °C: Ne, HBr, SO_2, NF_3, CO. (**b**) Calculate the rms speed of NF_3 molecules at 25 °C. (**c**) Calculate the most probable speed of an ozone molecule in the stratosphere, where the temperature is 270 K.

10.82 (**a**) Place the following gases in order of increasing average molecular speed at 300 K: CO, SF_6, H_2S, Cl_2, HBr. (**b**) Calculate the rms speeds of CO and Cl_2 molecules at 300 K. (**c**) Calculate the average speeds of CO and Cl_2 molecules at 300 K.

10.83 Which one or more of the following statements are *true*?

(a) O_2 will effuse faster than Cl_2.

(b) Effusion and diffusion are different names for the same process.

(c) Perfume molecules travel to your nose by the process of effusion.

(d) The higher the density of a gas, the shorter the mean free path.

10.84 At constant pressure, the mean free path (λ) of a gas molecule is directly proportional to temperature. At constant temperature, λ is inversely proportional to pressure. If you compare two different gas molecules at the same temperature and pressure, λ is inversely proportional to the square of the diameter of the gas molecules. Put these facts together to create a formula for the mean free path of a gas molecule with a proportionality constant (call it R_{mfp}, like the ideal-gas constant) and define units for R_{mfp}.

10.85 Hydrogen has two naturally occurring isotopes, ^1H and ^2H. Chlorine also has two naturally occurring isotopes, ^{35}Cl and ^{37}Cl. Thus, hydrogen chloride gas consists of four distinct types of molecules: $^1H^{35}Cl$, $^1H^{37}Cl$, $^2H^{35}Cl$, and $^2H^{37}Cl$. Place these four molecules in order of increasing rate of effusion.

10.86 The "escape velocity" for an object to escape from the top of Earth's atmosphere is 7.1 km/s. (a) At what temperature would the rms speed of a He atom equal this value? (b) The actual temperature at the top of the atmosphere is approximately 200 K. Thus, only a small proportion of He atoms at this temperature are moving fast enough to potentially escape from Earth. Calculate the rms speed of a He atom at 200 K. (c) Calculate the rms speed of a hydrogen molecule at 200 K. (d) Which gas—hydrogen or helium—is more likely to escape Earth's atmosphere?

10.87 Arsenic(III) sulfide sublimes readily, even below its melting point of 320 °C. The molecules of the vapor phase are found to effuse through a tiny hole at 0.28 times the rate of effusion of Ar atoms under the same conditions of temperature and pressure. What is the molecular formula of arsenic(III) sulfide in the gas phase?

10.88 A gas of unknown molecular mass was allowed to effuse through a small opening under constant-pressure conditions. It required 105 s for 1.0 L of the gas to effuse. Under identical experimental conditions it required 31 s for 1.0 L of O_2 gas to effuse. Calculate the molar mass of the unknown

gas. (Remember that the faster the rate of effusion, the shorter the time required for effusion of 1.0 L; in other words, rate is the amount that diffuses over the time it takes to diffuse.)

Real Gases: Deviations from Ideal Behavior (Section 10.7)

10.89 (a) List two experimental conditions under which gases deviate from ideal behavior. (b) List two reasons why the gases deviate from ideal behavior.

10.90 The planet Jupiter has a surface temperature of 140 K and a mass 318 times that of Earth. Mercury (the planet) has a surface temperature between 600 K and 700 K and a mass 0.05 times that of Earth. On which planet is the atmosphere more likely to obey the ideal-gas law?

10.91 Which statement concerning the van der Waals constants a and b is *true*?

(a) The magnitude of a relates to molecular volume, whereas b relates to attractions between molecules.

(b) The magnitude of a relates to attractions between molecules, whereas b relates to molecular volume.

(c) The magnitudes of a and b depend on pressure.

(d) The magnitudes of a and b depend on temperature.

10.92 Calculate the pressure that CCl_4 will exert at 80 °C if 1.00 mol occupies 33.3 L, assuming that (a) CCl_4 obeys the ideal-gas equation; (b) CCl_4 obeys the van der Waals equation. (Values for the van der Waals constants are given in Table 10.3.) (c) Which would you expect to deviate more from ideal behavior under these conditions, Cl_2 or CCl_4? Explain.

10.93 Table 10.3 shows that the van der Waals b parameter has units of L/mol. This implies that we can calculate the size of atoms or molecules from b. Using the value of b for Xe, calculate the radius of a Xe atom and compare it to the value found in Figure 7.7—that is, 1.40 Å. Recall that the volume of a sphere is $(4/3)\pi r^3$.

10.94 Table 10.3 shows that the van der Waals b parameter has units of L/mol. This means that we can calculate the sizes of atoms or molecules from the b parameter. Refer back to the discussion in Section 7.3. Is the van der Waals radius we calculate from the b parameter of Table 10.3 more closely associated with the bonding or nonbonding atomic radius discussed there? Explain.

Additional Exercises

10.95 A gas bubble with a volume of 1.0 mm³ originates at the bottom of a lake where the pressure is 3.0 atm. Calculate its volume when the bubble reaches the surface of the lake where the pressure is 730 torr, assuming that the temperature doesn't change.

10.96 A 15.0-L tank is filled with helium gas at a pressure of 1.00×10^2 atm. How many balloons (each 2.00 L) can be inflated to a pressure of 1.00 atm, assuming that the temperature remains constant and that the tank cannot be emptied below 1.00 atm?

10.97 To minimize the rate of evaporation of the tungsten filament, 1.4×10^{-5} mol of argon is placed in a 600-cm³ light bulb. What is the pressure of argon in the light bulb at 23 °C?

10.98 Carbon dioxide, which is recognized as the major contributor to global warming as a "greenhouse gas," is formed when fossil fuels are combusted, as in electrical power plants fueled by coal, oil, or natural gas. One potential way to reduce the amount of CO_2 added to the atmosphere is to store it as a

compressed gas in underground formations. Consider a 1000-megawatt coal-fired power plant that produces about 6×10^6 tons of CO_2 per year. (a) Assuming ideal-gas behavior, 1.00 atm, and 27 °C, calculate the volume of CO_2 produced by this power plant. (b) If the CO_2 is stored underground as a liquid at 10 °C and 120 atm and a density of 1.2 g/cm³, what volume does it possess? (c) If it is stored underground as a gas at 30 °C and 70 atm, what volume does it occupy?

10.99 Nickel carbonyl, $Ni(CO)_4$, is one of the most toxic substances known. The present maximum allowable concentration in laboratory air during an 8-h workday is 1 ppb (parts per billion) by volume, which means that there is one mole of $Ni(CO)_4$ for every 10^9 moles of gas. Assume 24 °C and 1.00 atm pressure. What mass of $Ni(CO)_4$ is allowable in a laboratory room that is 12 ft × 20 ft × 9 ft?

10.100 When a large evacuated flask is filled with argon gas, its mass increases by 3.224 g. When the same flask is again evacuated and then filled with a gas of unknown molar mass, the mass increase is 8.102 g. (**a**) Based on the molar mass of argon, estimate the molar mass of the unknown gas. (**b**) What assumptions did you make in arriving at your answer?

10.101 Consider the arrangement of bulbs shown in the drawing. Each of the bulbs contains a gas at the pressure shown. What is the pressure of the system when all the stopcocks are opened, assuming that the temperature remains constant? (We can neglect the volume of the capillary tubing connecting the bulbs.)

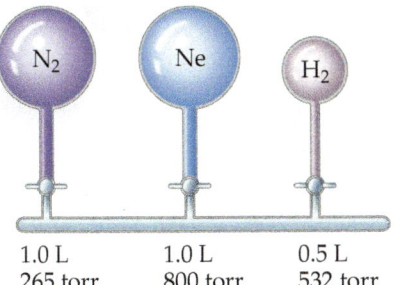

1.0 L	1.0 L	0.5 L
265 torr	800 torr	532 torr

10.102 Assume that a single cylinder of an automobile engine has a volume of 524 cm³. (**a**) If the cylinder is full of air at 74 °C and 0.980 atm, how many moles of O_2 are present? (The mole fraction of O_2 in dry air is 0.2095.) (**b**) How many grams of C_8H_{18} could be combusted by this quantity of O_2, assuming complete combustion with formation of CO_2 and H_2O?

10.103 Assume that an exhaled breath of air consists of 74.8% N_2, 15.3% O_2, 3.7% CO_2, and 6.2% water vapor. (**a**) If the total pressure of the gases is 0.985 atm, calculate the partial pressure of each component of the mixture. (**b**) If the volume of the exhaled gas is 455 mL and its temperature is 37 °C, calculate the number of moles of CO_2 exhaled. (**c**) How many grams of glucose ($C_6H_{12}O_6$) would need to be metabolized to produce this quantity of CO_2? (The chemical reaction is the same as that for combustion of $C_6H_{12}O_6$. See Section 3.2 and Problem 10.57.)

10.104 A 1.42-g sample of helium and an unknown mass of O_2 are mixed in a flask at room temperature. The partial pressure of the helium is 42.5 torr, and that of the oxygen is 158 torr. What is the mass of the oxygen?

10.105 An ideal gas at a pressure of 1.50 atm is contained in a bulb of unknown volume. A stopcock is used to connect this bulb with a previously evacuated bulb that has a volume of 0.800 L as shown here. When the stopcock is opened, the gas expands into the empty bulb. If the temperature is held constant during this process and the final pressure is 695 torr, what is the volume of the bulb that was originally filled with gas?

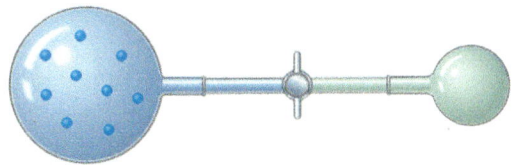

10.106 You have a sample of gas at −33 °C. You wish to increase the rms speed by a factor of 2. To what temperature should the gas be heated?

10.107 Consider the following gases, all at STP: Ne, SF_6, N_2, CH_4. (**a**) Which gas is most likely to depart from the assumption of the kinetic-molecular theory that says there are no attractive or repulsive forces between molecules? (**b**) Which one is closest to an ideal gas in its behavior? (**c**) Which one has the highest root-mean-square molecular speed at a given temperature? (**d**) Which one has the highest total molecular volume relative to the space occupied by the gas? (**e**) Which has the highest average kinetic-molecular energy? (**f**) Which one would effuse more rapidly than N_2? (**g**) Which one would have the largest van der Waals *b* parameter?

10.108 Does the effect of intermolecular attraction on the properties of a gas become more significant or less significant if (**a**) the gas is compressed to a smaller volume at constant temperature; (**b**) the temperature of the gas is increased at constant volume?

10.109 Large amounts of nitrogen gas are used in the manufacture of ammonia, principally for use in fertilizers. Suppose 120.00 kg of $N_2(g)$ is stored in a 1100.0-L metal cylinder at 280 °C. (**a**) Calculate the pressure of the gas, assuming ideal-gas behavior. (**b**) By using the data in Table 10.3, calculate the pressure of the gas according to the van der Waals equation. (**c**) Under the conditions of this problem, which correction dominates, the one for finite volume of gas molecules or the one for attractive interactions?

10.110 Cyclopropane, a gas used with oxygen as a general anesthetic, is composed of 85.7% C and 14.3% H by mass. (**a**) If 1.56 g of cyclopropane has a volume of 1.00 L at 0.984 atm and 50.0 °C, what is the molecular formula of cyclopropane? (**b**) Judging from its molecular formula, would you expect cyclopropane to deviate more or less than Ar from ideal-gas behavior at moderately high pressures and room temperature? Explain. (**c**) Would cyclopropane effuse through a pinhole faster or more slowly than methane, CH_4?

10.111 Consider the combustion reaction between 25.0 mL of liquid methanol (density = 0.850 g/mL) and 12.5 L of oxygen gas measured at STP. The products of the reaction are $CO_2(g)$ and $H_2O(g)$. Calculate the volume of liquid H_2O formed if the reaction goes to completion and you condense the water vapor.

10.112 An herbicide is found to contain only C, H, N, and Cl. The complete combustion of a 100.0-mg sample of the herbicide in excess oxygen produces 83.16 mL of CO_2 and 73.30 mL of H_2O vapor expressed at STP. A separate analysis shows that the sample also contains 16.44 mg of Cl. (**a**) Determine the percentage of the composition of the substance. (**b**) Calculate its empirical formula. (**c**) What other information would you need to know about this compound to calculate its true molecular formula?

10.113 A 4.00-g sample of a mixture of CaO and BaO is placed in a 1.00-L vessel containing CO_2 gas at a pressure of 730 torr and a temperature of 25 °C. The CO_2 reacts with the CaO and BaO, forming $CaCO_3$ and $BaCO_3$. When the reaction is complete, the pressure of the remaining CO_2 is 150 torr. (**a**) Calculate the number of moles of CO_2 that have reacted. (**b**) Calculate the mass percentage of CaO in the mixture.

10.114 Ammonia and hydrogen chloride react to form solid ammonium chloride:

$$NH_3(g) + HCl(g) \longrightarrow NH_4Cl(s)$$

Two 2.00-L flasks at 25 °C are connected by a valve, as shown in the drawing. One flask contains 5.00 g of $NH_3(g)$, and the other contains 5.00 g of HCl(g). When the valve is opened, the gases react until one is completely consumed. (**a**) Which gas will remain in the system after the reaction is complete? (**b**) What will be the final pressure of the system after the reaction is complete? (Neglect the volume of the ammonium chloride formed.) (**c**) What mass of ammonium chloride will be formed?

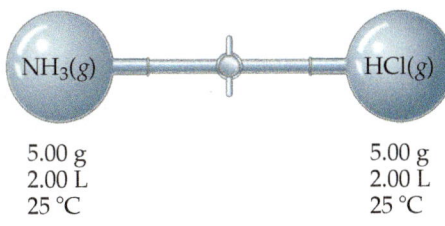

$NH_3(g)$	$HCl(g)$
5.00 g	5.00 g
2.00 L	2.00 L
25 °C	25 °C

10.115 Gas pipelines are used to deliver natural gas (methane, CH_4) to the various regions of the United States. The total volume of natural gas that is delivered is on the order of 2.7×10^{12} L per day, measured at STP. Calculate the total enthalpy change for combustion of this quantity of methane. (*Note:* Less than this amount of methane is actually combusted daily. Some of the delivered gas is passed through to other regions.)

10.116 Natural gas is very abundant in many Middle Eastern oil fields. However, the costs of shipping the gas to markets in other parts of the world are high because it is necessary to liquefy the gas, which is mainly methane and has a boiling point at atmospheric pressure of $-164\,°C$. One possible strategy is to oxidize the methane to methanol, CH_3OH, which has a boiling point of $65\,°C$ and can therefore be shipped more readily. Suppose that $10.7 \times 10^9\,ft^3$ of methane at atmospheric pressure and $25\,°C$ is oxidized to methanol. **(a)** What volume of methanol is formed if the density of CH_3OH is 0.791 g/mL? **(b)** Write balanced chemical equations for the oxidations of methane and methanol to $CO_2(g)$ and $H_2O(l)$. Calculate the total enthalpy change for complete combustion of the $10.7 \times 10^9\,ft^3$ of methane just described and for complete combustion of the equivalent amount of methanol, as calculated in part **(a)**. **(c)** Methane, when liquefied, has a density of 0.466 g/mL; the density of methanol at $25\,°C$ is 0.791 g/mL. Compare the enthalpy change upon combustion of a unit volume of liquid methane and liquid methanol. From the standpoint of energy production, which substance has the higher enthalpy of combustion per unit volume?

10.117 Gaseous iodine pentafluoride, IF_5, can be prepared by the reaction of solid iodine and gaseous fluorine:

$$I_2(s) + 5\,F_2(g) \longrightarrow 2\,IF_5(g)$$

A 5.00-L flask containing 10.0 g of I_2 is charged with 10.0 g of F_2, and the reaction proceeds until one of the reagents is completely consumed. After the reaction is complete, the temperature in the flask is $125\,°C$. **(a)** What is the partial pressure of IF_5 in the flask? **(b)** What is the mole fraction of IF_5 in the flask **(c)** Draw the Lewis structure of IF_5. **(d)** What is the total mass of reactants and products in the flask?

10.118 A 6.53-g sample of a mixture of magnesium carbonate and calcium carbonate is treated with excess hydrochloric acid. The resulting reaction produces 1.72 L of carbon dioxide gas at $28\,°C$ and 743 torr pressure. **(a)** Write balanced chemical equations for the reactions that occur between hydrochloric acid and each component of the mixture. **(b)** Calculate the total number of moles of carbon dioxide that forms from these reactions. **(c)** Assuming that the reactions are complete, calculate the percentage by mass of magnesium carbonate in the mixture.

Design an Experiment

You are given a cylinder of an unknown, nonradioactive, noble gas and tasked to determine its molar mass and use that value to identify the gas. The tools available to you are several empty mylar balloons that are about the size of a grapefruit when inflated (gases diffuse through mylar much more slowly than conventional latex balloons), an analytical balance, and three graduated glass beakers of different sizes (100 mL, 500 mL, and 2 L). **(a)** To how many significant figures would you need to determine the molar mass to identify the gas? **(b)** Propose an experiment or series of experiments that would allow you to determine the molar mass of the unknown gas. Describe the tools, calculations, and assumptions you would need to use. **(c)** If you had access to a broader range of analytical instruments, describe an alternative way you could identify the gas using any experimental methods that you have learned about in the previous chapters.

11

▲ **SOME LIQUIDS FLOW EASILY FROM ONE CONTAINER TO ANOTHER, WHILE OTHERS LIKE HONEY FLOW VERY SLOWLY.** This property, called viscosity, depends on the strength of the attractive forces between neighboring molecules.

When studying gases, we were able to largely ignore attractive forces between molecules, but this is not the case for liquids and solids, phases whose very existence depends on these forces. To understand the behavior of liquids and solids, we must first understand **intermolecular forces**, the forces that exist *between* molecules. Only by understanding the nature and strength of these forces can we understand how the composition and structure of a substance are related to its physical properties in the liquid and solid states. In this chapter we consider intermolecular forces and how they are related to the properties of liquids, with a focus that is largely on molecular substances. The structures and properties of solids are covered in Chapter 12, where we examine substances with extended bonding networks (ionic, metallic, and network-covalent solids) in addition to molecular substances.

11.1 | A Molecular Comparison of Gases, Liquids, and Solids

Learning Objectives

When you finish Section 11.1, you should be able to:

▶ Compare and contrast the behavior of gases, liquids, and solids.

▶ Explain how the balance between the kinetic energy of atoms, ions, or molecules and the strength of the intermolecular attractions determine the physical state (solid, liquid, gas) of a substance.

As we learned in Chapter 10, the molecules in a gas are widely separated and in a state of constant, chaotic motion. A key tenet of the kinetic-molecular theory of gases is the assumption that we can neglect the interactions between molecules. (Section 10.7) The properties of liquids and solids are quite different from those of gases largely because the intermolecular forces in liquids and solids are much stronger. The properties of gases, liquids, and solids are compared in Table 11.1.

In liquids, the intermolecular attractive forces are strong enough to hold particles close together. Thus, liquids are much denser and far less compressible than gases. Unlike gases, liquids have a definite volume, independent of the size and shape of their container. The attractive forces in liquids are not strong enough, however, to keep the particles from moving past one another. Thus, any liquid can be poured and assumes the shape of the container it occupies.

Increasing pressure or decreasing temperature will eventually lock particles in a crystalline arrangement.* This makes solids rigid, in which case their shape and volume are independent of their container. Solids, like liquids, are not very compressible because the particles have little free space between them. Because the particles in a solid or liquid are fairly close together compared with those of a gas, we often refer to solids and liquids as *condensed phases*.

Figure 11.1 compares the three states of matter. *The state of a substance depends largely on the balance between the kinetic energies of the particles (atoms, molecules, or ions) and the interparticle energies of attraction.* The kinetic energies, which depend on temperature, tend to keep the particles apart and moving. The interparticle attractions tend to draw the particles together. Substances that are gases at room temperature have much weaker interparticle attractions than those that are liquids; substances that are liquids have weaker interparticle attractions than those that are solids. The different states of matter adopted by the halogens at room temperature—iodine is a solid, bromine is a liquid, and chlorine is a gas—are a direct consequence of a decrease in the strength of the intermolecular forces as we move from I_2 to Br_2 to Cl_2.

We can change a substance from one state to another by heating or cooling, which changes the average kinetic energy of the particles. NaCl, for example, a solid at room temperature, melts at 1074 K and boils at 1686 K under 1 atm pressure, and Cl_2, a gas at room temperature, liquefies at 239 K and solidifies at 172 K under 1 atm pressure. As the temperature of a gas decreases, the average kinetic energy of its particles decreases, allowing the attractions between the particles to draw the particles close together, forming a liquid, and then to virtually lock them in place, forming a solid. Increasing the pressure on a gas can also drive transformations from gas to liquid to solid because the increased pressure brings the molecules closer together, thus making intermolecular forces more effective. For example, propane (C_3H_8) is a gas at room temperature and 1 atm pressure, whereas liquefied propane (LP) is a liquid at room temperature because it is stored under much higher pressure.

TABLE 11.1 Characteristic Properties of the States of Matter

Gas	Liquid	Solid
Assumes both volume and shape of its container	Assumes shape of a portion of the container it occupies	Retains its own shape and volume
Expands to fill its container	Does not expand to fill its container	Does not expand to fill its container
Is compressible	Is virtually incompressible	Is virtually incompressible
Flows readily	Flows readily	Does not flow
Diffusion within a gas occurs rapidly	Diffusion within a liquid occurs slowly	Diffusion within a solid occurs extremely slowly

*The atoms in a solid are able to vibrate in place, which are small motions compared to the molecular motion in gases and liquids. As the temperature of the solid increases, the vibrational motion increases.

Increasing intermolecular attractions

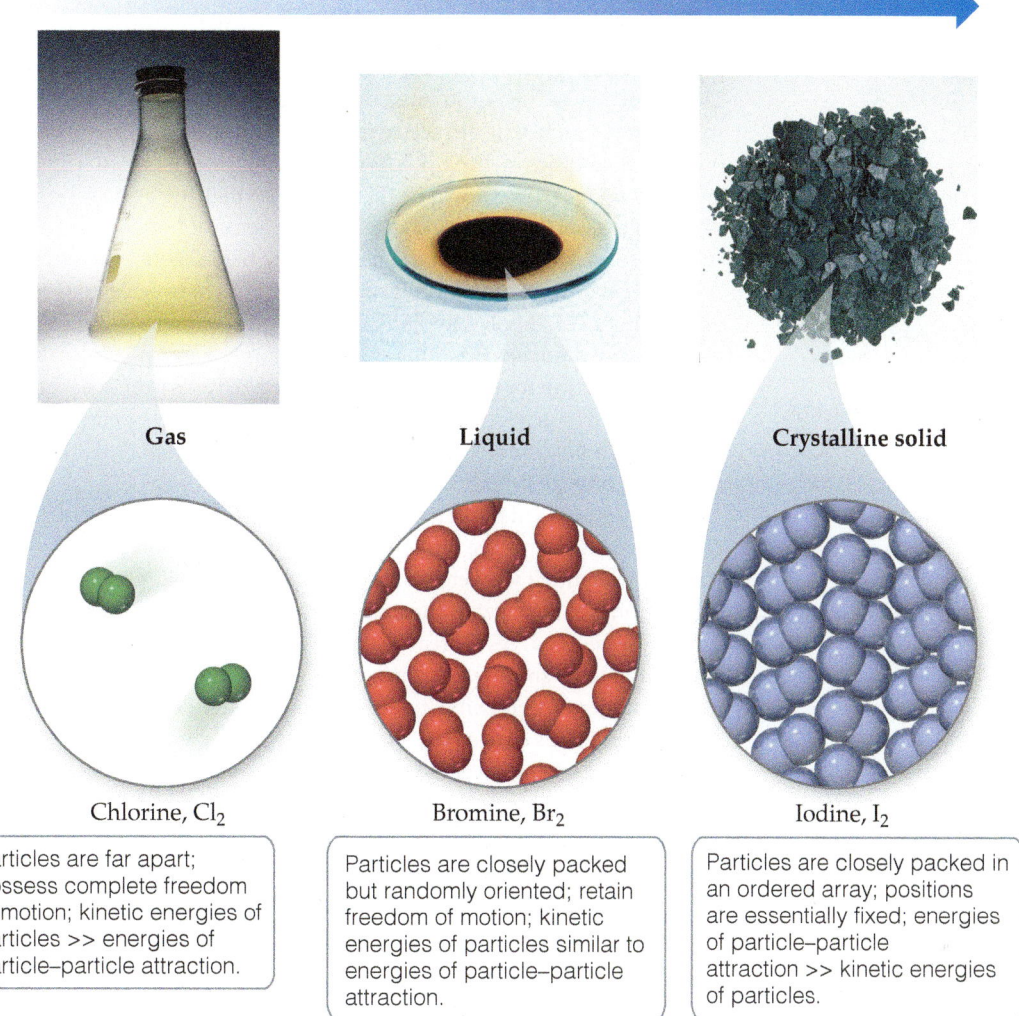

Gas

Liquid

Crystalline solid

Chlorine, Cl_2

Bromine, Br_2

Iodine, I_2

Particles are far apart; possess complete freedom of motion; kinetic energies of particles >> energies of particle–particle attraction.

Particles are closely packed but randomly oriented; retain freedom of motion; kinetic energies of particles similar to energies of particle–particle attraction.

Particles are closely packed in an ordered array; positions are essentially fixed; energies of particle–particle attraction >> kinetic energies of particles.

▲ **Figure 11.1 Gases, liquids, and solids.** Chlorine, bromine, and iodine are all diatomic molecules as a result of covalent bonding. However, due to differences in the strength of the intermolecular forces, they exist in three different states at room temperature and standard pressure: Cl_2 is a gas, Br_2 is a liquid, and I_2 is a solid.

 Self-Assessment Exercises

SAE 11.1 Which characteristic of a liquid is more like a gas than a solid? (**a**) its compressibility (**b**) its density (**c**) its ability to flow (**d**) its molar volume at standard temperature and pressure

SAE 11.2 Substance A is a liquid at room temperature and pressure, while substance B is a gas under the same conditions. Both are molecular substances. Based on this observation, we can say that the intermolecular attractions in substance A are _____ those in substance B. (**a**) stronger than (**b**) weaker than (**c**) approximately the same strength as (**d**) There is insufficient information to decide.

 Learning Objectives

When you finish Section 11.2, you should be able to:

▶ Describe what is meant by the polarizability of a molecule and how it leads to intermolecular dispersion forces.

▶ Compare the relative strengths of dispersion forces in pure substances.

▶ Compare the relative strengths of dipole–dipole and ion–dipole interactions in pure substances and mixtures.

▶ Identify hydrogen-bonding interactions and describe how they affect the physical properties of those substances and mixtures in which they occur.

▶ Compare the relative strength of the intermolecular forces in pure substances and/or mixtures that feature differing intermolecular forces.

11.2 | Intermolecular Forces

The strengths of *intermolecular* forces vary over a wide range but are generally much weaker than *intramolecular* forces—ionic, metallic, or covalent bonds (**Figure 11.2**). Less energy, therefore, is required to vaporize a liquid or melt a solid than to break covalent bonds. For example, only 16 kJ/mol is required to overcome the intermolecular attractions in liquid HCl to vaporize it. In contrast, the energy required to break the covalent bond in HCl is

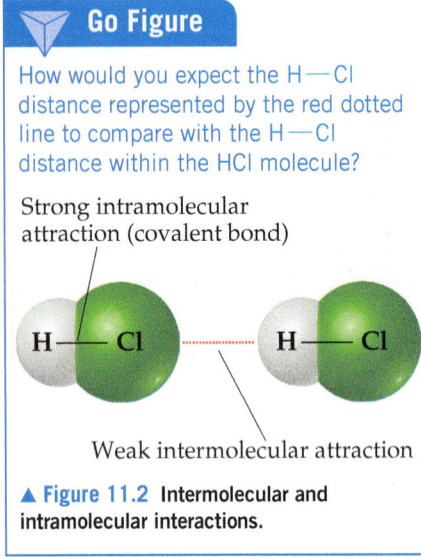

Go Figure

How would you expect the H — Cl distance represented by the red dotted line to compare with the H — Cl distance within the HCl molecule?

Strong intramolecular attraction (covalent bond)

Weak intermolecular attraction

▲ **Figure 11.2** Intermolecular and intramolecular interactions.

431 kJ/mol. Thus, when a molecular substance such as HCl changes from solid to liquid to gas, the molecules remain intact.

Many properties of liquids, including boiling points, reflect the strength of the intermolecular forces. A liquid boils when bubbles of its vapor form within the liquid. The molecules of the liquid must overcome their attractive forces to separate and form a vapor. The stronger the attractive forces, the higher the temperature at which the liquid boils. Similarly, the melting points of solids increase as the strengths of the intermolecular forces increase. As shown in **Table 11.2**, the melting and boiling points of substances in which the particles are held together by chemical bonds tend to be much higher than those of substances in which the particles are held together by intermolecular forces.

Three types of intermolecular attractions exist between electrically neutral molecules: dispersion forces, dipole–dipole attractions, and hydrogen bonding. The first two are collectively called *van der Waals forces* after Johannes van der Waals (1837–1923), who developed the equation for predicting the deviation of gases from ideal behavior. (Section 10.9) Another kind of attractive force, the ion–dipole force, is operative when ions are in close proximity to polar molecules. It plays an important role in the formation and behavior solutions of ionic compounds.

All intermolecular interactions are electrostatic, involving attractions between positive and negative species, much like ionic bonds. (Section 8.2) Why then are intermolecular forces so much weaker than ionic bonds? Recall from Equation 8.4 that electrostatic interactions get stronger as the magnitude of the charges increases and weaker as the distance between charges increases. The charges responsible for intermolecular forces are generally much smaller than the charges in ionic compounds. For example, from its dipole moment it is possible to estimate charges of +0.178 and −0.178 for the hydrogen and chlorine ends of the HCl molecule, respectively (see Sample Exercise 8.5). Furthermore, the distances between molecules are often larger than the distances between atoms held together by chemical bonds.

Dispersion Forces

You might think there would be no electrostatic interactions between electrically neutral, nonpolar atoms and/or molecules. Yet some kind of attractive interactions must exist because nonpolar gases like helium, argon, and nitrogen can be liquefied. Fritz London, a German-American physicist, first proposed the origin of this attraction in 1930. London recognized that the motion of electrons in an atom or molecule can create an *instantaneous*, or momentary, dipole moment.

In a collection of helium atoms, for example, the *average* distribution of the electrons about each nucleus is spherically symmetrical, as shown in **Figure 11.3**(**a**). The atoms are nonpolar and so possess no permanent dipole moment. The *instantaneous* distribution of the electrons, however, can be different from the average distribution. If we could freeze the motion of the electrons at any given instant, both electrons could be on one side of the nucleus. At just that instant, the atom has an instantaneous dipole moment as shown in Figure 11.3(**b**). The motions of electrons in one atom influence the motions of electrons in its neighbors. The instantaneous dipole on one atom can induce

TABLE 11.2   Melting and Boiling Points of Representative Substances

Force Holding Particles Together	Substance	Melting Point (K)	Boiling Point (K)
Chemical bonds			
Ionic bonds	Lithium fluoride (LiF)	1118	1949
Metallic bonds	Beryllium (Be)	1560	2742
Covalent bonds	Diamond (C)	3800	4300
Intermolecular forces			
Dispersion forces	Nitrogen (N_2)	63	77
Dipole–dipole interactions	Hydrogen chloride (HCl)	158	188
Hydrogen bonding	Hydrogen fluoride (HF)	190	293

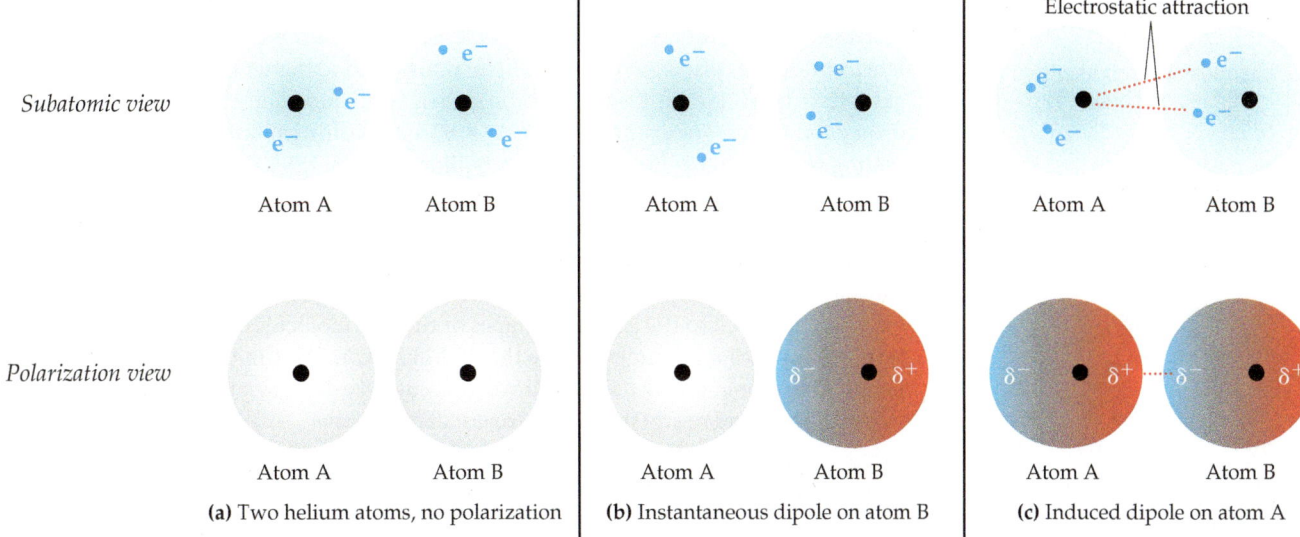

Subatomic view

Atom A   Atom B   Atom A   Atom B

Electrostatic attraction

Atom A   Atom B

Polarization view

Atom A   Atom B   Atom A   Atom B   Atom A   Atom B

(a) Two helium atoms, no polarization | **(b)** Instantaneous dipole on atom B | **(c)** Induced dipole on atom A

▲ **Figure 11.3 Dispersion forces.** "Snapshots" of the charge distribution for a pair of helium atoms at three instants.

an instantaneous dipole on an adjacent atom, causing the atoms to be attracted to each other as shown in Figure 11.3(**c**). This attractive interaction is called the **dispersion force** (also called *London dispersion forces* or *induced dipole–induced dipole interactions*). It is significant only when molecules are very close together.

The strength of the dispersion force depends on the ease with which the charge distribution in a molecule can be distorted to induce an instantaneous dipole. The ease with which the charge distribution is distorted is called the molecule's **polarizability**. We can think of the polarizability of a molecule as a measure of the "squashiness" of its electron cloud: The greater the polarizability, the more easily the electron cloud can be distorted to give an instantaneous dipole. Therefore, more polarizable molecules have larger dispersion forces.

In general, polarizability increases as the number of electrons in an atom or molecule increases and as the volume over which the electrons are distributed increases. Because the number of electrons in a molecule and molecular weight generally parallel each other, *dispersion forces tend to increase in strength with increasing molecular weight.* We can see this in the boiling points of the halogens and noble gases (**Figure 11.4**), where dispersion forces

Go Figure Why is the boiling point of the halogen always greater than the corresponding noble gas from the same period?

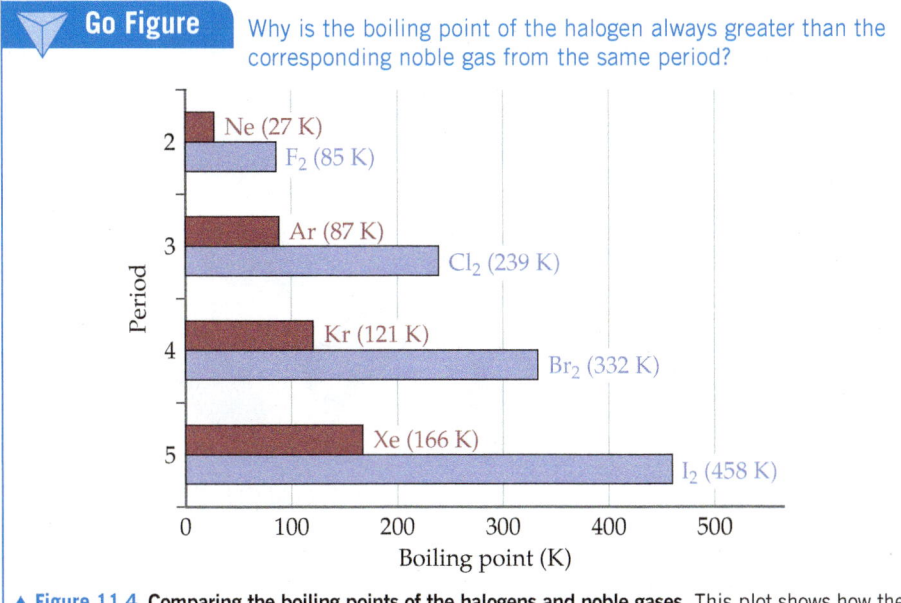

▲ **Figure 11.4 Comparing the boiling points of the halogens and noble gases.** This plot shows how the boiling points increase due to stronger dispersion forces as the atomic/molecular weight increases.

Linear molecule—larger surface area enhances intermolecular contact and increases dispersion force.

n-Pentane (C_5H_{12})
bp = 309.4 K

Spherical molecule—smaller surface area diminishes intermolecular contact and decreases dispersion force.

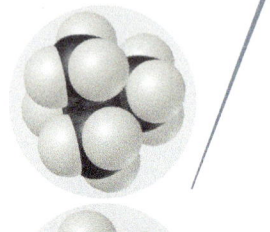

Neopentane (C_5H_{12})
bp = 282.7 K

▲ **Figure 11.5 Molecular shape affects intermolecular attraction.** Molecules of *n*-pentane make more contact with each other than do neopentane molecules. Therefore, *n*-pentane has stronger intermolecular attractive forces and a higher boiling point.

are the only intermolecular forces at work. In both families, the atomic/molecular weight increases on moving down the periodic table. The higher atomic/molecular weights translate into stronger dispersion forces, which in turn lead to higher boiling points. If molecules have a similar number of electrons, then larger molecules tend to have stronger dispersion forces. For example, Kr (AW = 83 amu, boiling point = 121 K) has a lower boiling point than Cl_2 (MW = 71 amu, boiling point = 239 K), which has a lower boiling point than *n*-pentane,* C_5H_{12} (MW = 72 amu, boiling point = 309 K).

Molecular shape also influences the magnitude of dispersion forces. For example, *n*-pentane and neopentane (**Figure 11.5**) have the same molecular formula (C_5H_{12}), yet the boiling point of *n*-pentane is about 27 K higher than that of neopentane. The difference can be traced to the different shapes of the two molecules. Intermolecular attraction is greater for *n*-pentane because the molecules can come in contact over the entire length of the long, somewhat cylindrical molecules. Less contact is possible between the more compact and nearly spherical neopentane molecules.

Dipole–Dipole Interactions

The presence of a permanent dipole moment in polar molecules gives rise to **dipole–dipole interactions**. These interactions originate from electrostatic attractions between the partially positive end of one molecule and the partially negative end of a neighboring molecule. Repulsions can also occur when the positive (or negative) ends of two molecules are in close proximity. Dipole–dipole interactions are effective only when molecules are very close together.

To see the effect of dipole–dipole interactions, we compare the boiling points of two compounds of similar molecular weight: acetonitrile (CH_3CN, MW = 41 amu, boiling point = 355 K) and propane $CH_3CH_2CH_3$, MW = 44 amu, boiling point = 231 K). On the one hand, acetonitrile is a polar molecule, with a dipole moment of 3.9 D, so dipole–dipole interactions are present. Propane, on the other hand, is essentially nonpolar, so dipole–dipole interactions are absent. Because acetonitrile and propane have similar molecular weights, both have similar dispersion forces. Therefore, the higher boiling point of acetonitrile can be attributed to dipole–dipole interactions.

To better understand these forces, compare how CH_3CN molecules pack together in the solid and liquid states. In the solid [**Figure 11.6(a)**], the molecules are arranged with the negatively charged nitrogen end of each molecule close to the positively charged —CH_3 end of its neighbors. In the liquid [Figure 11.6(**b**)], the molecules are free to move with respect to one another, so their arrangement becomes more disordered. At any given instant, then, both attractive and repulsive dipole–dipole interactions are

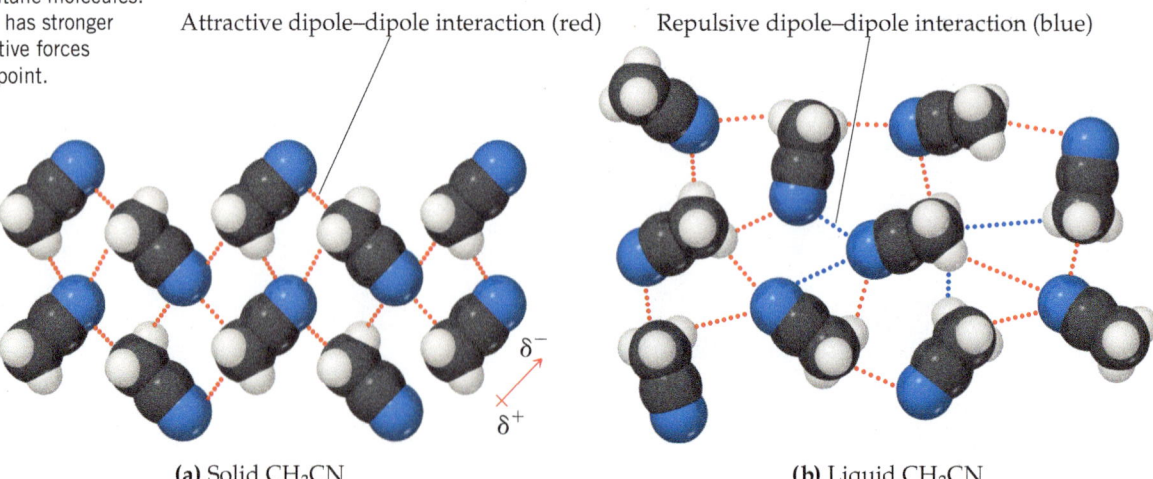

Attractive dipole–dipole interaction (red)   Repulsive dipole–dipole interaction (blue)

δ^-
δ^+

(a) Solid CH_3CN **(b)** Liquid CH_3CN

▲ **Figure 11.6 Dipole–dipole interactions.** The dipole–dipole interactions in (**a**) crystalline CH_3CN and (**b**) liquid CH_3CN.

---

* The *n* in *n*-pentane is an abbreviation for the word *normal*. A normal hydrocarbon is one in which the carbon atoms are arranged in a straight chain. (Section 2.9)

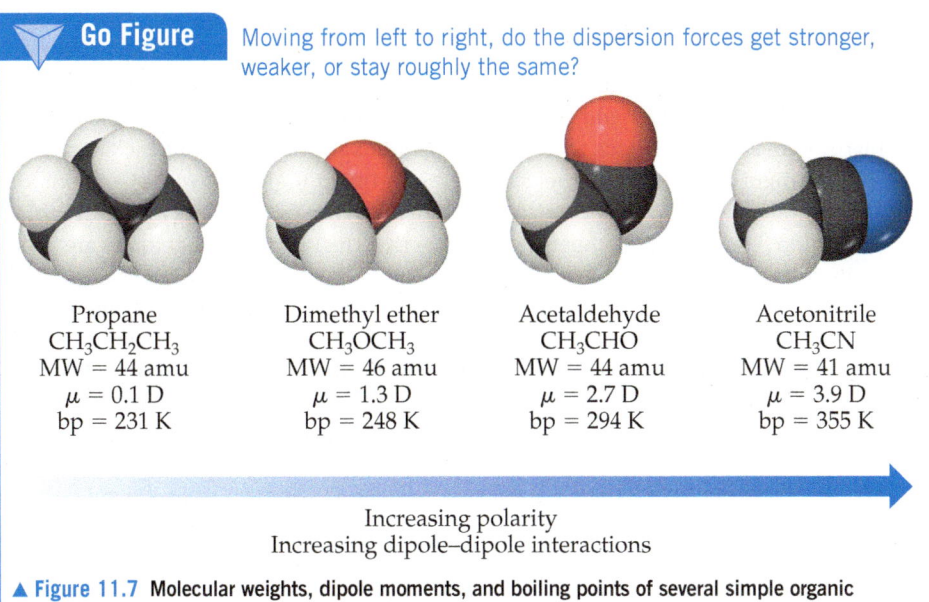

Go Figure Moving from left to right, do the dispersion forces get stronger, weaker, or stay roughly the same?

Propane	Dimethyl ether	Acetaldehyde	Acetonitrile
$CH_3CH_2CH_3$	CH_3OCH_3	CH_3CHO	CH_3CN
MW = 44 amu	MW = 46 amu	MW = 44 amu	MW = 41 amu
μ = 0.1 D	μ = 1.3 D	μ = 2.7 D	μ = 3.9 D
bp = 231 K	bp = 248 K	bp = 294 K	bp = 355 K

Increasing polarity
Increasing dipole–dipole interactions

▲ **Figure 11.7 Molecular weights, dipole moments, and boiling points of several simple organic substances.**

present. Nevertheless, the molecular orientations that lead to attractions are greater in number than those that lead to repulsions, so the overall effect is a net attraction strong enough to keep the molecules in liquid CH_3CN from moving apart to form a gas.

For molecules of approximately equal mass and size, the strength of intermolecular attractions increases with increasing polarity, a trend we see in **Figure 11.7**. Notice how the boiling point increases as the dipole moment increases.

Hydrogen Bonding

Figure 11.8 shows the boiling points of the binary compounds that form between hydrogen and the elements in groups 4A through 7A. The boiling points of the compounds containing

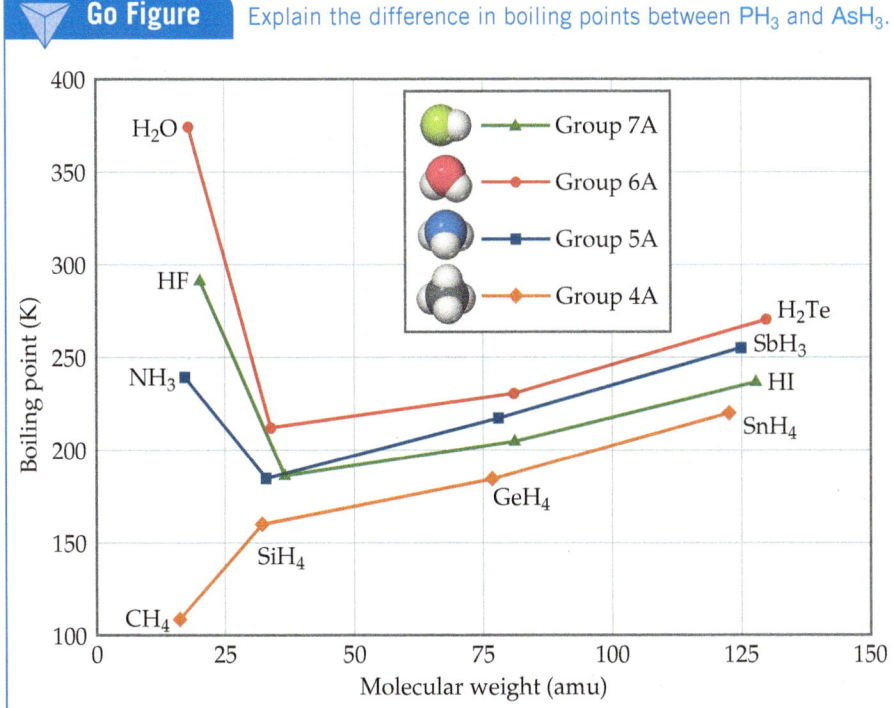

Go Figure Explain the difference in boiling points between PH_3 and AsH_3.

▲ **Figure 11.8 Boiling points of the covalent hydrides of the elements in groups 4A–7A as a function of molecular weight.**

Go Figure

To form a hydrogen bond, what must the nonhydrogen atom (N, O, or F) involved in the bond possess?

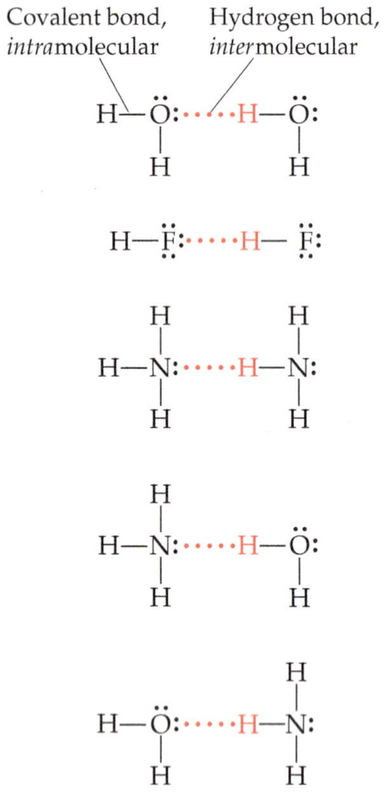

Covalent bond,
*intra*molecular

Hydrogen bond,
*inter*molecular

H—Ö:·····H—Ö:
    | |
    H   H

H—F̈:·····H—F̈:

    H   H
    | |
H—N:·····H—N:
    | |
    H   H

    H
    |
H—N:·····H—Ö:
    | |
    H   H

                 H
                 |
H—Ö:·····H—N:
    | |
    H   H

▲ **Figure 11.9 Hydrogen bonding.** Hydrogen bonding occurs between an H atom that is bonded to N, O, or F in one molecule and an N, O, or F atom in another molecule.

group 4A elements (CH$_4$ through SnH$_4$, all nonpolar) increase systematically moving down the group. This is the expected trend because polarizability and, hence, dispersion forces generally increase as molecular weight increases. The heavier members of groups 5A, 6A, and 7A follow the same trend, but NH$_3$, H$_2$O, and HF have boiling points that are much higher than expected. In fact, these three compounds also have many other characteristics that distinguish them from other substances of similar molecular weight and polarity. For example, water has a high melting point, a high specific heat, and a high heat of vaporization. Each of these properties indicates that the intermolecular forces are abnormally strong.

The strong intermolecular attractions in HF, H$_2$O, and NH$_3$ result from *hydrogen bonding. A* **hydrogen bond** *is an attraction between a hydrogen atom attached to a highly electronegative atom (usually F, O, or N) and a nearby small electronegative atom in another molecule or chemical group.* Thus, H—F, H—O, or H—N bonds in one molecule can form hydrogen bonds with an F, O, or N atom in another molecule. Several examples of hydrogen bonds are shown in **Figure 11.9**, including the hydrogen bond that exists between the H atom in an H$_2$O molecule and the O atom of an adjacent H$_2$O molecule. Notice in each case that the H atom in the hydrogen bond interacts with a nonbonding electron pair.

Hydrogen bonds can be considered a special type of dipole–dipole attraction. Because N, O, and F are so electronegative, a bond between hydrogen and any of these elements is quite polar, with hydrogen at the positive end (remember the + on the right-hand side of the dipole symbol represents the positive end and the arrowhead represents the negative end of the dipole):

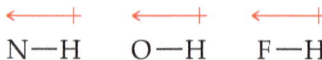

The hydrogen atom has no inner electrons. Thus, the positive side of the dipole has the concentrated charge of the nearly bare hydrogen nucleus. This positive charge is attracted to the negative charge of an electronegative atom in a nearby molecule. Because the electron-poor hydrogen is so small, it can approach an electronegative atom very closely and, thus, interact strongly with it. The strength of hydrogen bonding generally increases as the electronegativity of the atom covalently bonded to hydrogen increases. As a result, hydrogen bonding in compounds with an O—H bond tends to be stronger than that in comparable compounds with an N—H bond. Hence, methylamine, CH$_3$NH$_2$, is a gas under normal conditions, whereas methanol, CH$_3$OH, is a liquid. Hydrogen bonding is present in both compounds but is stronger in methanol.

Hydrogen bonding plays important roles in many chemical systems, especially those of biological significance. For example, hydrogen bonding helps stabilize the three-dimensional structure of proteins, which is critical to their function. Hydrogen bonding is also responsible for the double-helical structure of DNA, which is key to its genetic function.

Sample Exercise 11.1
Identifying Substances That Can Form Hydrogen Bonds

In which of these substances is hydrogen bonding likely to play an important role in determining physical properties: methane (CH$_4$), hydrazine (H$_2$NNH$_2$), methyl fluoride (CH$_3$F), hydrogen sulfide (H$_2$S)?

SOLUTION

Analyze We are given the chemical formulas of four compounds and asked to predict whether they can participate in hydrogen bonding. All the compounds contain H, but hydrogen bonding usually occurs only when the hydrogen is covalently bonded to N, O, or F.

Plan We analyze each formula to see if it contains N, O, or F directly bonded to H. There also needs to be a nonbonding pair of electrons on an electronegative atom (usually N, O, or F) in a nearby molecule, which can be revealed by drawing the Lewis structure for the molecule.

Solve The foregoing criteria eliminate CH$_4$ and H$_2$S, which do not contain H bonded to N, O, or F. They also eliminate CH$_3$F, whose Lewis structure shows a central C atom surrounded by three H atoms and an F atom. (Carbon always forms four bonds, whereas hydrogen and fluorine form one each.) Because the molecule contains a C—F bond and not an H—F bond, it does not form hydrogen bonds. In H$_2$NNH$_2$, however, we find N—H bonds, and the Lewis structure shows a nonbonding pair of electrons on each N atom, telling us hydrogen bonds can exist between the molecules:

SECTION 11.2 Intermolecular Forces

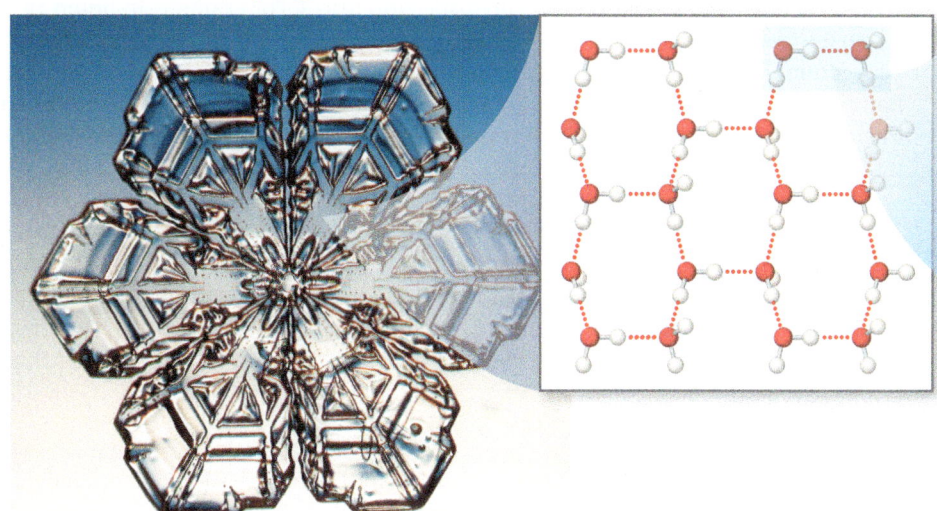

covalently bonded to H, drawing the Lewis structure for the interaction provides a way to check the prediction.

Check Although we can generally identify substances that participate in hydrogen bonding based on their containing N, O, or F

▶ **Practice Exercise**
In which of these substances is significant hydrogen bonding possible: trifluoroethane (CF_3CH_3), phosphine (PH_3), chloramine (NH_2Cl), acetone (CH_3COCH_3)?

One remarkable consequence of hydrogen bonding is seen in the densities of ice and liquid water. In most substances, the molecules in the solid are more densely packed than those in the liquid, making the solid phase denser than the liquid phase. By contrast, the density of ice at 0 °C (0.917 g/mL) is less than that of liquid water at 0 °C (1.00 g/mL), so ice floats on liquid water.

Go Figure What is the approximate H—O · · · H bond angle in ice, where H—O is the covalent bond and O · · · H is the hydrogen bond?

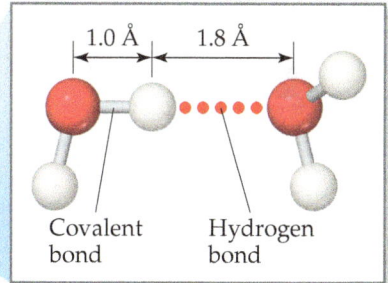

▲ **Figure 11.10 Hydrogen bonding in ice.** The empty channels in the structure of ice make water less dense as a solid than as a liquid.

The lower density of ice can be understood in terms of hydrogen bonding. In ice, the H_2O molecules assume the ordered, open arrangement shown in **Figure 11.10**. This arrangement optimizes hydrogen bonding between molecules, with each H_2O molecule forming hydrogen bonds to four neighboring H_2O molecules. These hydrogen bonds, however, create the cavities seen in the middle image of Figure 11.10. When ice melts, the motions of the molecules cause the structure to collapse. The hydrogen bonding in the liquid is more random than that in the solid but is strong enough to hold the molecules close together. Consequently, liquid water has a denser structure than ice, meaning that a given mass of water occupies a smaller volume than the same mass of ice.

The expansion of water upon freezing (**Figure 11.11**) is responsible for many phenomena we take for granted. It causes icebergs to float and water pipes to burst in cold weather. The lower density of ice compared to liquid water also profoundly affects life on Earth. Because ice floats, it covers the top of the water when a lake freezes, thereby insulating the water. If ice were denser than water, ice forming at the top of a lake would sink

▲ **Figure 11.11 Expansion of water upon freezing.**

 **Go Figure**

Why does the O side of H_2O point toward the Na^+ ion?

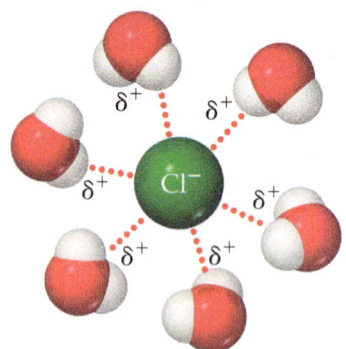

Positive ends of polar molecules are oriented toward negatively charged anion.

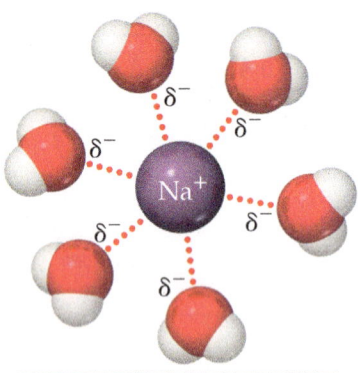

Negative ends of polar molecules are oriented toward positively charged cation.

 **Figure 11.12 Ion–dipole forces.**

to the bottom, and the lake could freeze solid. Most aquatic life could not survive under these conditions.

Ion–Dipole Forces

An **ion–dipole force** exists between an ion and a polar molecule (**Figure 11.12**). Cations are attracted to the negative end of a dipole, and anions are attracted to the positive end. The magnitude of the attraction increases as either the ionic charge or the magnitude of the dipole moment increases. Ion–dipole forces are especially important for solutions of ionic substances in polar liquids, such as a solution of NaCl in water. (Section 4.1)

Comparing Intermolecular Forces

We can identify the intermolecular forces operative in a substance by considering its composition and structure (**Figure 11.13**). *Dispersion forces are found in all substances.* The strength of these attractive forces increases with increasing molecular weight and depends on molecular shapes. With polar molecules dipole–dipole interactions are also operative, but these interactions often make a smaller contribution to the total intermolecular attraction than do dispersion forces. For example, in liquid HCl, dispersion forces are estimated to account for more than 80% of the total attraction between molecules, while dipole–dipole attractions account for the rest. Hydrogen bonds, when present, can make a greater contribution to the total intermolecular interaction—for example, in H_2O, the dispersion forces account for less than 25% of the total intermolecular forces.*

In general, the energies associated with dispersion forces are 0.1–30 kJ/mol. The wide range reflects the wide range in polarizabilities of molecules. By comparison, the energies associated with dipole–dipole interactions and hydrogen bonds are approximately 2–15 kJ/mol and 10–40 kJ/mol, respectively. Ion–dipole forces tend to be stronger than the aforementioned intermolecular forces, with energies typically exceeding 50 kJ/mol. All these interactions are considerably weaker than

_____

* Jacob Israelachvili (1992), *Intermolecular and Surface Forces* (2nd ed.), Academic Press, London.

 **Go Figure** Can the energies of dispersion forces between two molecules be larger than the energy of hydrogen bonding between the two molecules?

Type of intermolecular interaction	Atoms Examples: Ne, Ar	Nonpolar molecules Examples: BF_3, CH_4	Polar molecules without OH, NH, or HF groups Examples: HCl, CH_3CN	Polar molecules containing OH, NH, or HF groups Examples: H_2O, NH_3	Ionic solids dissolved in polar liquids Examples: NaCl in H_2O
Dispersion forces (0.1–30 kJ/mol)	✓	✓	✓	✓	✓
Dipole–dipole interactions (2–15 kJ/mol)			✓	✓	
Hydrogen bonding (10–40 kJ/mol)				✓	
Ion–dipole interactions (>50 kJ/mol)					✓

▲ **Figure 11.13 Checklist for determining intermolecular forces.** Multiple types of intermolecular forces can be operating in a given substance or mixture. Note that dispersion forces occur in all substances.

covalent and ionic bonds, which have energies that are hundreds of kilojoules per mole.

It is important to realize that the effects of all these attractions are additive. For example, acetic acid, CH_3COOH, and 1-propanol, $CH_3CH_2CH_2OH$, have the same molecular weight, 60 g/mol, and both are capable of forming hydrogen bonds. However, a pair of acetic acid molecules can form two hydrogen bonds, whereas a pair of 1-propanol molecules can form only one (**Figure 11.14**). Hence, the boiling point of acetic acid is higher. These effects can be important, especially for very large polar molecules such as proteins, which have multiple dipoles and hydrogen-bonding groups over their surfaces. These molecules can be held together in solution to a surprisingly high degree due to the presence of multiple attractive interactions.

When comparing the relative strengths of intermolecular attractions, consider these generalizations:

1. **When the molecules of two substances have comparable molecular weights and shapes, dispersion forces are approximately equal in the two substances.** Differences in the magnitudes of the intermolecular forces are due to differences in the strengths of dipole–dipole attractions. The intermolecular forces get stronger as molecule polarity increases, with those molecules capable of hydrogen bonding having the strongest interactions.

2. **When the molecules of two substances differ widely in molecular weights, and there is no hydrogen bonding, dispersion forces tend to determine which substance has the stronger intermolecular attractions.** Intermolecular attractive forces are generally higher in the substance with higher molecular weight.

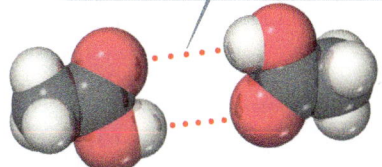

Each molecule can form two hydrogen bonds with a neighbor.

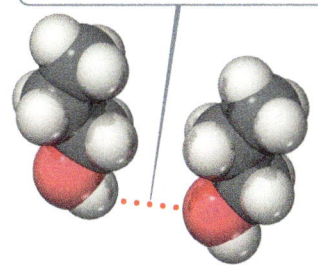

Acetic acid, CH_3COOH
MW = 60 amu
bp = 391 K

Each molecule can form one hydrogen bond with a neighbor.

1-Propanol, $CH_3CH_2CH_2OH$
MW = 60 amu
bp = 370 K

▲ **Figure 11.14 Hydrogen bonding in acetic acid and 1-propanol.** The greater the number of hydrogen bonds, the more tightly the molecules are held together and, therefore, the higher the boiling point.

 Sample Exercise 11.2

Predicting Types and Relative Strengths of Intermolecular Attractions

Arrange the substances $BaCl_2$, H_2, CO, HF, and Ne in order of increasing boiling point.

SOLUTION

Analyze We need to assess the intermolecular forces in these substances and use that information to determine the relative boiling points.

Plan The boiling point depends in part on the attractive forces in each substance. We need to order these substances according to the relative strengths of the different kinds of intermolecular attractions.

Solve The attractive forces are stronger for ionic substances than for molecular ones, so $BaCl_2$ should have the highest boiling point. The intermolecular forces of the remaining substances depend on molecular weight, polarity, and hydrogen bonding. The molecular weights are H_2, 2 amu; CO, 28 amu; HF, 20 amu; and Ne, 20 amu. The boiling point of H_2 should be the lowest because it is nonpolar and has the lowest molecular weight. The molecular weights of CO, HF, and Ne are similar. Because HF can

hydrogen-bond, it should have the highest boiling point of the three. Next is CO, which is slightly polar and has the highest molecular weight. Finally, Ne, which is nonpolar, should have the lowest boiling point of these three. The predicted order of boiling points is, therefore,

$$H_2 < Ne < CO < HF < BaCl_2$$

Check The boiling points reported in the literature are H_2, 20 K; Ne, 27 K; CO, 83 K, HF, 293 K; and $BaCl_2$, 1813 K —in agreement with our predictions.

▶ **Practice Exercise**
(a) Identify the intermolecular attractions present in the following substances and (b) select the substance with the highest boiling point: CH_3CH_3, CH_3OH, and CH_3CH_2OH.

 Self-Assessment Exercises

SAE 11.3 Which of the following statements about polarizability and dispersion forces is *false*? (a) The polarizability of an atom is a measure of how easily its electron cloud can be distorted. (b) The polarizability of an atom generally decreases as the atomic number increases. (c) Dispersion forces can be attributed to the formation of instantaneous dipoles. (d) If two molecules have a similar number of electrons, the polarizability should be larger for the molecule that occupies the larger volume.

SAE 11.4 Based on the strengths of the dispersion forces, predict the relative order of the boiling points of *n*-butane, $CH_3CH_2CH_2CH_3$; *n*-pentane, $CH_3CH_2CH_2CH_2CH_3$; and isobutane, $CH_3CH(CH_3)_2$.
(a) *n*-butane < *n*-pentane < isobutene
(b) isobutane < *n*-butane < *n*-pentane
(c) *n*-pentane < *n*-butane < isobutene
(d) *n*-butane < isobutane < *n*-pentane

Continued

SAE 11.5 In which of the following substances are dipole–dipole forces *present*, but hydrogen-bonding interactions are *absent*: CH_3F, SF_6, SF_4, HCl, NH_3? **(a)** CH_3F, SF_6, SF_4, and HCl **(b)** CH_3F, HCl, and NH_3 **(c)** CH_3F, SF_6, and SF_4 **(d)** CH_3F, SF_4, and HCl **(e)** SF_4 and HCl

SAE 11.6 Ion–dipole interactions are generally _____ than dipole–dipole interactions and _____ than ionic bonds. **(a)** weaker, weaker **(b)** weaker, stronger **(c)** stronger, weaker **(d)** stronger, stronger

SAE 11.7 Methanethiol, CH_3SH, is a major contributor to bad breath odor and is a by-product of metabolizing asparagus. What intermolecular forces must be overcome to convert methanethiol from a liquid to a gas? **(a)** only dispersion forces **(b)** only dipole–dipole interactions **(c)** dispersion forces and dipole–dipole interactions **(d)** dispersion forces and hydrogen bonding **(e)** dispersion forces, dipole–dipole interactions, and hydrogen bonding

SAE 11.8 Arrange the following substances from the lowest to highest normal boiling point: ethylenediamine, $NH_2CH_2CH_2NH_2$; butane, $CH_3CH_2CH_2CH_3$; propylamine, $CH_3CH_2CH_2NH_2$; and ethylene glycol, $HOCH_2CH_2OH$.

(a) $HOCH_2CH_2OH < NH_2CH_2CH_2NH_2 < CH_3CH_2CH_2NH_2 < CH_3CH_2CH_2CH_3$

(b) $CH_3CH_2CH_2CH_3 < CH_3CH_2CH_2NH_2 < HOCH_2CH_2OH < NH_2CH_2CH_2NH_2$

(c) $CH_3CH_2CH_2CH_3 < NH_2CH_2CH_2NH_2 < CH_3CH_2CH_2NH_2 < HOCH_2CH_2OH$

(d) $CH_3CH_2CH_2CH_3 < CH_3CH_2CH_2NH_2 < NH_2CH_2CH_2NH_2 < HOCH_2CH_2OH$

11.3 | Select Properties of Liquids

The intermolecular attractions we have just discussed can help us understand many familiar properties of liquids. In this section, we examine three: viscosity, surface tension, and capillary action.

Viscosity

Some liquids, such as honey, molasses, and motor oil, flow very slowly; others, such as water and gasoline, flow easily. The resistance of a liquid to flow is called **viscosity**. The greater a liquid's viscosity, the more slowly it flows. Viscosity can be measured by timing how long it takes a certain amount of the liquid to flow through a thin vertical tube (**Figure 11.15**). Viscosity can also be determined by measuring the rate at which steel balls fall through the liquid. The balls fall more slowly as the viscosity increases. The SI unit for viscosity is kg/m-s.

The viscosity of a liquid is related to how easily its molecules flow past one another. It depends on the attractive forces between molecules as well as their shape and flexibility. Substances made of long flexible molecules generally have a higher viscosity because the molecules tend to become entangled, much like cooked spaghetti noodles, which makes it more difficult for them to slide past one another. For a series of related compounds, viscosity increases with molecular weight, as illustrated in **Table 11.3**. By comparison, the viscosity of H_2O at 20 °C is 1.0×10^{-3} kg/m-s.

The viscosity of a substance decreases as the temperature increases. The viscosity of octane, for example, is:

$$7.06 \times 10^{-4} \text{ kg/m-s at 0 °C}$$
$$4.33 \times 10^{-4} \text{ kg/m-s at 40 °C}$$

At higher temperatures, the greater average kinetic energy of the molecules overcomes the attractive forces between molecules, allowing the molecules to slide by each other more easily.

Surface Tension

The surface of water behaves almost as if it had an elastic skin, as evidenced by the ability of certain insects to "walk" on water. This behavior is due to an imbalance of

Learning Objectives

When you finish **Section 11.3**, you should be able to:

▶ Compare the expected viscosities of liquids based on differences in their composition and/or molecular structure.

▶ Relate changes in the surface tension of a liquid to changes in the strength of the intermolecular forces that hold the molecules of the liquid together.

▶ Describe capillary action and predict the shape of a meniscus from the relative strengths of the cohesive and adhesive forces.

SAE 40
higher number
higher viscosity
slower pouring

SAE 10
lower number
lower viscosity
faster pouring

▲ **Figure 11.15 Comparing viscosities.** The Society of Automotive Engineers (SAE) has established a numeric scale to indicate motor-oil viscosity.

TABLE 11.3 Viscosities of a Series of Hydrocarbons at 20 °C

Substance	Formula	Viscosity (kg/m-s)
Hexane	$CH_3CH_2CH_2CH_2CH_2CH_3$	3.26×10^{-4}
Heptane	$CH_3CH_2CH_2CH_2CH_2CH_2CH_3$	4.09×10^{-4}
Octane	$CH_3CH_2CH_2CH_2CH_2CH_2CH_2CH_3$	5.42×10^{-4}
Nonane	$CH_3CH_2CH_2CH_2CH_2CH_2CH_2CH_2CH_3$	7.11×10^{-4}
Decane	$CH_3CH_2CH_2CH_2CH_2CH_2CH_2CH_2CH_2CH_3$	1.42×10^{-3}

intermolecular forces at the surface of the liquid. As shown in **Figure 11.16**, molecules in the interior are attracted equally in all directions, but those at the surface experience a net inward force. This net force tends to pull surface molecules toward the interior, thereby reducing the surface area and making the molecules at the surface pack closely together.

Because spheres have the smallest surface area for their volume, water droplets assume an almost spherical shape. This explains the tendency of water to "bead up" when it contacts a surface made of nonpolar molecules, like a lotus leaf or a newly waxed car.

A measure of the net inward force that must be overcome to expand the surface area of a liquid is given by its surface tension. **Surface tension** is the energy required to increase the surface area of a liquid by a unit amount. For example, the surface tension of water at 20 °C is 7.29×10^{-2} J/m^2, which means that an energy of 7.29×10^{-2} J must be supplied to increase the surface area of a given amount of water by 1 m^2. Water has a high surface tension because of its strong hydrogen bonds. The surface tension of mercury is even higher (4.6×10^{-1} J/m^2) because of even stronger metallic bonds between the atoms of mercury.

Capillary Action

Intermolecular forces that bind similar molecules to one another, such as the hydrogen bonding in water, are called *cohesive forces*. Intermolecular forces that bind a substance to a surface are called *adhesive forces*. Water placed in a glass tube adheres to the glass because the adhesive forces between the water and the glass are greater than the cohesive forces between water molecules; glass is principally SiO_2, which has a very polar surface. The curved surface, or *meniscus*, of the water is therefore concave or U-shaped (**Figure 11.17**). For mercury, however, the situation is different.

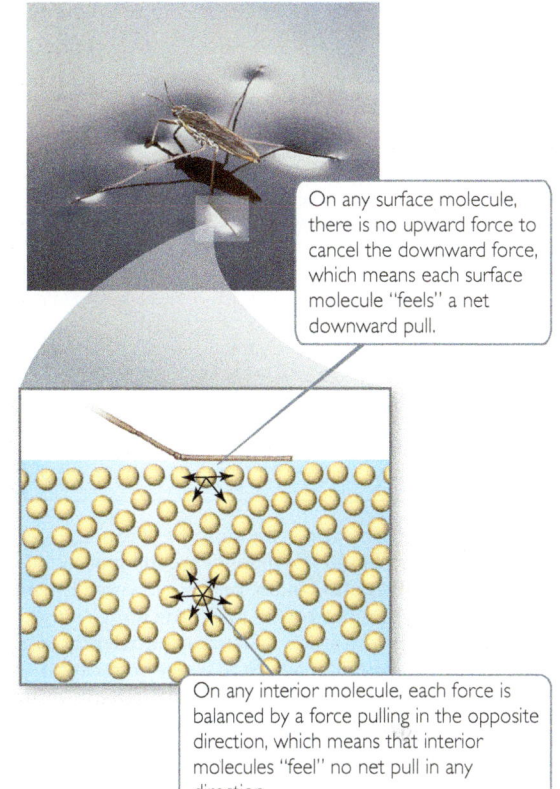

On any surface molecule, there is no upward force to cancel the downward force, which means each surface molecule "feels" a net downward pull.

On any interior molecule, each force is balanced by a force pulling in the opposite direction, which means that interior molecules "feel" no net pull in any direction.

▲ Figure 11.16 **Molecular-level view of surface tension.** The high surface tension of water keeps the water strider from sinking.

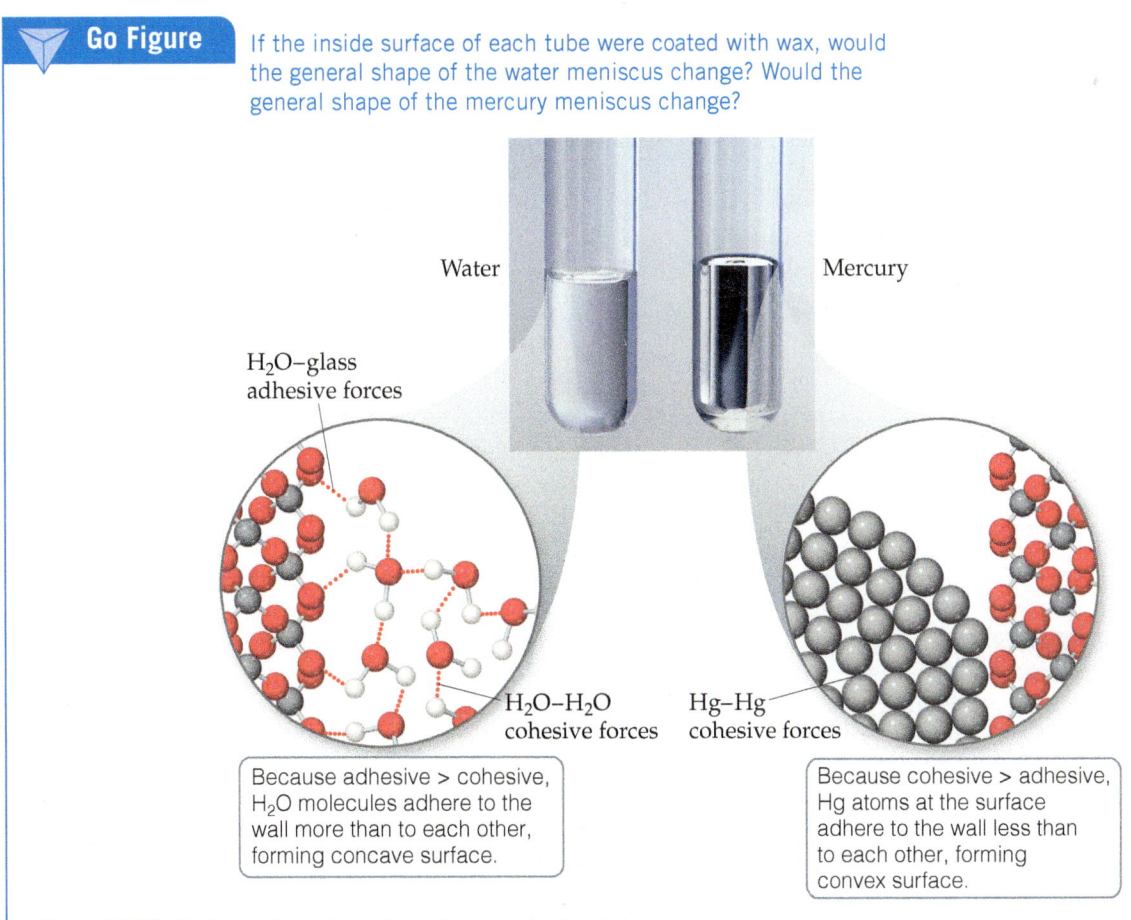

Go Figure If the inside surface of each tube were coated with wax, would the general shape of the water meniscus change? Would the general shape of the mercury meniscus change?

Water   Mercury

H$_2$O–glass adhesive forces

H$_2$O–H$_2$O cohesive forces

Hg–Hg cohesive forces

Because adhesive > cohesive, H$_2$O molecules adhere to the wall more than to each other, forming concave surface.

Because cohesive > adhesive, Hg atoms at the surface adhere to the wall less than to each other, forming convex surface.

▲ Figure 11.17 **Meniscus shapes for water and mercury in glass tubes.**

Mercury atoms can form bonds with one another but not with the glass. As a result, the cohesive forces are much greater than the adhesive forces and the meniscus is convex—that is, shaped like an inverted U.

When a small-diameter glass tube, or capillary, is placed in water, water rises in the tube. The rise of liquids up very narrow tubes is called **capillary action**. The adhesive forces between the liquid and the walls of the tube tend to increase the surface area of the liquid. The surface tension of the liquid tends to reduce the area, thereby pulling the liquid up the tube. The liquid climbs until the force of gravity on the liquid balances the adhesive and cohesive forces.

Capillary action is widespread. For example, towels absorb liquid, and "stay-dry" synthetic fabrics move sweat away from the skin by capillary action. Capillary action also plays a role in enabling plants to move water and dissolved nutrients upward against the force of gravity.

CHEMISTRY AND SUSTAINABILITY | Ionic Liquids

The strong electrostatic attractions between cations and anions explain why most ionic compounds are solids at room temperature, with high melting and boiling points. However, the melting point of an ionic compound can be low if the ionic charges are not too high and the cation–anion distance is sufficiently large. For example, the melting point of NH_4NO_3, where both cation and anion are larger polyatomic ions, is 170 °C. If the ammonium cation is replaced by the even larger ethylammonium cation, $CH_3CH_2NH_3^+$, the melting point drops to 12 °C, making ethylammonium nitrate a liquid at room temperature. Ethylammonium nitrate is an example of an *ionic liquid*: a salt that is a liquid at room temperature.

$CH_3CH_2NH_3^+$ is not only larger than NH_4^+, it is less symmetric. In general, the larger and more irregularly shaped the ions in an ionic substance, the better the chances of forming an ionic liquid. Among the cations that form ionic liquids, one of the most widely used is the 1-butyl-3-methylimidazolium cation (abbreviated bmim⁺; Figure 11.18 and Table 11.4), which has two arms of different lengths coming off a five-atom central ring. This feature gives bmim⁺ an irregular shape, which makes it difficult for the molecules to pack together in a solid. Common anions found in ionic liquids include polyatomic anions like PF_6^- and BF_4^-, as well as the halide ions.

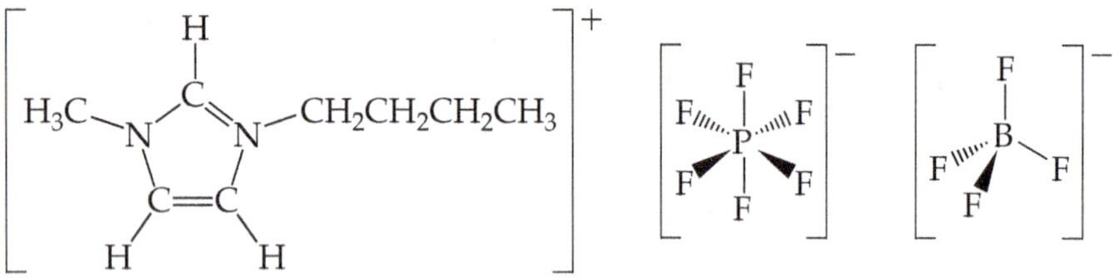

TABLE 11.4 Melting Point and Decomposition Temperature of Four 1-Butyl-3-Methylimidazoliu (bmin⁺) Salts

Cation	Anion	Melting Point (°C)	Decomposition Temperature (°C)
bmim⁺	Cl⁻	41	254
bmim⁺	I⁻	−72	265
bmim⁺	PF_6^-	10	349
bmim⁺	BF_4^-	−81	403

Ionic liquids have many useful properties. Unlike most molecular liquids, they are nonvolatile (that is, they don't evaporate readily) and nonflammable. They tend to remain in the liquid state at temperatures up to about 400 °C. Most molecular substances are liquids only at much lower temperatures, 100 °C or less in most cases. Because ionic liquids are good solvents for a wide range of substances, ionic liquids can be used for a variety of reactions and separations. These properties make them attractive replacements for volatile organic solvents in many industrial processes. Relative to traditional organic solvents, ionic liquids offer the promise of reduced volumes, safer handling, and easier reuse, all of which have the potential to reduce the environmental impact of industrial chemical processes. However, the extent to which a given ionic liquid enhances sustainability must be evaluated on a case-by-case basis. The chemicals and process needed to prepare the ionic liquid as well as its toxicity must also be taken into account.

Related Exercises: 11.33, 11.34, 11.38, 11.84

▲ **Figure 11.18** Representative ions found in ionic liquids.

1-Butyl-3-methylimidazolium (bmim⁺) cation

PF_6^- anion

BF_4^- anion

Self-Assessment Exercises

SAE 11.9 Arrange the following substances from the lowest to highest viscosity: *n*-propyl alcohol, $CH_3CH_2CH_2OH$; ethylene glycol, $HOCH_2CH_2OH$; and acetone, CH_3COCH_3.

(a) $CH_3CH_2CH_2OH < HOCH_2CH_2OH < CH_3COCH_3$

(b) $HOCH_2CH_2OH < CH_3CH_2CH_2OH < CH_3COCH_3$

(c) $CH_3COCH_3 < HOCH_2CH_2OH < CH_3CH_2CH_2OH$

(d) $CH_3COCH_3 < CH_3CH_2CH_2OH < HOCH_2CH_2OH$

SAE 11.10 When water forms beads on a surface, as it does on the hood of a freshly waxed car, we can say that the water is _____ its contact with the surface because the cohesive forces are _____ than the adhesive forces.

(a) minimizing, greater than

(b) minimizing, less than

(c) minimizing, equal to

(d) maximizing, less than

(e) maximizing, greater than

SAE 11.11 The surface tension of ethylene glycol ($HOCH_2CH_2OH$), $4.7 \times 10^{-2}\,J/m^2$ at 20 °C, is approximately double that of propanol ($CH_3CH_2CH_2OH$), $2.4 \times 10^{-2}\,J/m^2$, and ethanol ($CH_3CH_2OH$), $2.3 \times 10^{-2}\,J/m^2$. Which of the following statements best explains this observation? **(a)** Ethylene glycol can form hydrogen bonds, but propanol and ethanol cannot. **(b)** Ethylene glycol has a larger molecular weight and therefore experiences much stronger dispersion forces. **(c)** Ethylene glycol is more flexible than propanol and ethanol, and this leads the molecules to become more entangled. **(d)** An ethylene glycol molecule can hydrogen-bond with two neighboring molecules, whereas propanol and ethanol molecules can only hydrogen-bond with one neighbor.

11.4 | Phase Changes

Liquid water left uncovered in a glass eventually evaporates. An ice cube left in a warm room quickly melts. Solid CO_2 (sold as a product called dry ice) *sublimes* at room temperature; that is, it changes directly from solid to gas. In general, each state of matter—solid, liquid, gas—can transform into either of the other two states. **Figure 11.19** shows the names associated with these transformations, which are called either **phase changes** or *changes of state*.

Energy Changes Accompany Phase Changes

Every phase change is accompanied by a change in the energy of the system. In a solid, for example, the particles (molecules, ions, or atoms) are in more or less fixed positions with respect to one another and closely packed to minimize the energy of the system. As the temperature of the solid increases, the particles vibrate about their equilibrium positions with increasing energetic motion. When the average kinetic energy increases enough to overcome some of the intermolecular forces, the particles begin moving freely relative to one another, and the solid becomes a liquid—it melts. Melting is called (somewhat confusingly) *fusion*. The increased freedom of motion of the particles requires energy, measured by the **heat of fusion** or *enthalpy of fusion*, ΔH_{fus}. The heat of fusion of ice, for example, is 6.01 kJ/mol:

$$H_2O(s) \longrightarrow H_2O(l) \qquad \Delta H_{fus} = 6.01\ kJ$$

As the temperature of the liquid increases, the particles move about more vigorously. The increased motion allows some particles to escape into the gas phase. As a result, the concentration of gas-phase particles above the liquid surface increases with temperature. These gas-phase particles exert a pressure called *vapor pressure*. We explore vapor pressure in **Section 11.5**. For now we just need to understand that vapor pressure increases with increasing temperature until it equals the external pressure above the liquid, typically atmospheric pressure. At this point the liquid boils—bubbles of the vapor form within the liquid. The energy required to cause the transition of a given quantity of the liquid to the vapor is called either the **heat of vaporization** or the *enthalpy of vaporization*, ΔH_{vap}. For water, the heat of vaporization is 40.7 kJ/mol.

$$H_2O(l) \longrightarrow H_2O(g) \qquad \Delta H_{vap} = 40.7\ kJ$$

Figure 11.20 shows ΔH_{fus} and ΔH_{vap} values for four substances. The values of ΔH_{vap} tend to be larger than the values of ΔH_{fus} because in the transition from liquid to gas, particles must essentially sever all their intermolecular attractions, whereas in the transition from solid to liquid, many of these attractive interactions are still present.

Learning Objectives

When you finish **Section 11.4**, you should be able to:

▶ Analyze the energy changes that accompany phase changes.

▶ Use heating curves to determine the relationship between heat added to or removed from a substance and its change in temperature.

▶ Define the critical temperature and pressure of a substance and compare the properties of supercritical fluids with those of gases and liquids.

Go Figure How is energy evolved in deposition related to those for condensation and freezing?

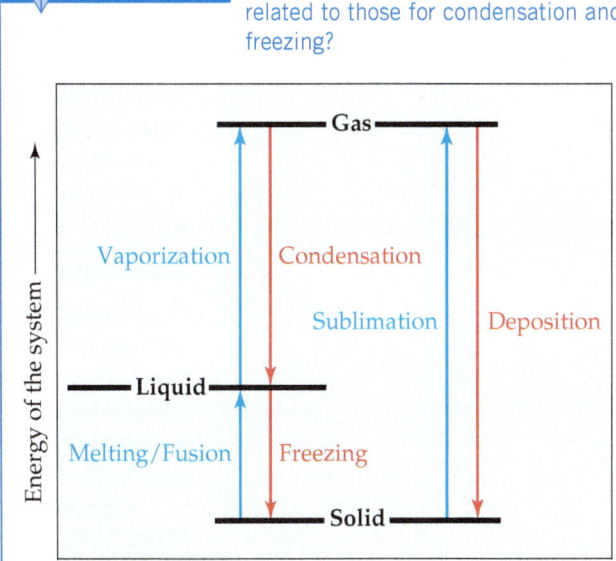

— Endothermic process (energy added to substance)
— Exothermic process (energy released from substance)

▲ **Figure 11.19 Phase changes and the names associated with them.**

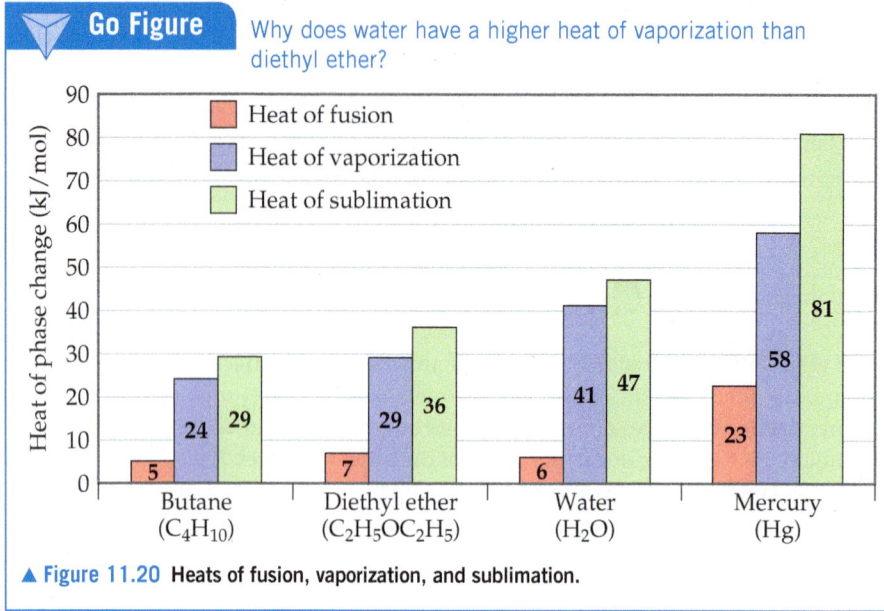

▲ **Figure 11.20 Heats of fusion, vaporization, and sublimation.**

The particles of a solid can move directly into the gaseous state. The enthalpy change required for this transition is called the **heat of sublimation**, denoted ΔH_{sub}. As illustrated in Figure 11.19, ΔH_{sub} is the sum of ΔH_{fus} and ΔH_{vap}. Thus, ΔH_{sub} for water is approximately 47 kJ/mol.

Phase changes show up in important ways in our everyday experiences. When we use ice cubes to cool a drink, for instance, the heat of fusion of the ice cools the liquid. We feel cool when we step out of a swimming pool or a warm shower because the liquid water's heat of vaporization is drawn from our bodies as the water evaporates from our skin. Our bodies use this mechanism to regulate body temperature, especially when we exercise vigorously in warm weather. (Section 5.5) A refrigerator also relies on the cooling effects of vaporization. Its mechanism contains an enclosed gas that can be liquefied under pressure. The liquid absorbs heat as it subsequently evaporates, thereby cooling the interior of the refrigerator.

What happens to the heat absorbed when the liquid refrigerant vaporizes? According to the first law of thermodynamics (Section 5.2), this absorbed heat must be released when the gas condenses to liquid. As this phase change occurs, the heat released is dissipated through cooling coils in the back of the refrigerator. Just as the heat of condensation for a given substance is equal in magnitude to the heat of vaporization and has the opposite sign, so too the *heat of deposition* for a given substance is exothermic to the same degree that the heat of sublimation is endothermic; the *heat of freezing* is exothermic to the same degree that the heat of fusion is endothermic (see Figure 11.19).

Heating Curves

When we heat an ice cube initially at −25 °C and 1 atm pressure, the temperature of the ice increases. As long as the temperature is below 0 °C, the ice cube remains in the solid state. When the temperature reaches 0 °C, the ice begins to melt. Because melting is an endothermic process, the heat we add at 0 °C is used to convert ice to liquid water, and *the temperature remains constant until all the ice has melted*. Once all the ice has melted, adding more heat causes the temperature of the liquid water to increase.

A graph of temperature versus amount of heat added is called a *heating curve*. **Figure 11.21** shows the heating curve for transforming ice, $H_2O(s)$, initially at −25 °C to steam, $H_2O(g)$, at 125 °C. As heat is added at a constant rate, the heating curve forms distinct regions:

- Line *AB:* Heating increases the temperature of $H_2O(s)$ from −25 to 0 °C.
- Line *BC:* Heating converts $H_2O(s)$ to $H_2O(l)$ as the ice melts at a constant temperature of 0 °C.

- Line *CD*: Heating increases the temperature of the $H_2O(l)$ from 0 °C to 100 °C.
- Line *DE*: Heating converts $H_2O(l)$ to $H_2O(g)$ as the water boils at a constant temperature of 100 °C.
- Line *EF*: Heating increases the temperature of the $H_2O(g)$ to 125 °C.

We can calculate the enthalpy change of the system for each region of the heating curve. Lines *AB, CD,* and *EF* show the heating of a single phase from one temperature to another. As we saw in Section 5.5, the amount of heat needed to raise the temperature of a substance is given by the product of the specific heat, mass, and temperature change (Equation 5.21). The greater the specific heat of a substance, the more heat we must add to accomplish a certain temperature increase. Because the specific heat of water is greater than that of ice, the slope of line *CD* is less than that of line *AB*. This lesser slope means the amount of heat we must add to a given mass of liquid water to achieve a 1 °C temperature change is greater than the amount we must add to achieve a 1 °C temperature change in the same mass of ice.

Lines *BC* and *DE* show the conversion of one phase to another at a constant temperature. The temperature remains constant during these phase changes because the added energy is used to overcome the attractive forces between molecules rather than to increase their average kinetic energy. For line *BC*, the enthalpy change can be calculated by using ΔH_{fus}, and for line *DE* we can use ΔH_{vap}.

If we start with 1 mol of steam at 125 °C and cool it, we move right to left across Figure 11.21. We first lower the temperature of the $H_2O(g)$ ($F \longrightarrow E$), then condense it ($E \longrightarrow D$) to $H_2O(l)$, and so forth.

Sometimes as we remove heat from a liquid, we can temporarily cool it below its freezing point without forming a solid. This phenomenon, called *supercooling*, occurs when the heat is removed so rapidly that the molecules have no time to assume the ordered structure of a solid. A supercooled liquid is unstable; particles of dust entering the solution or gentle stirring is often sufficient to cause the substance to solidify quickly.

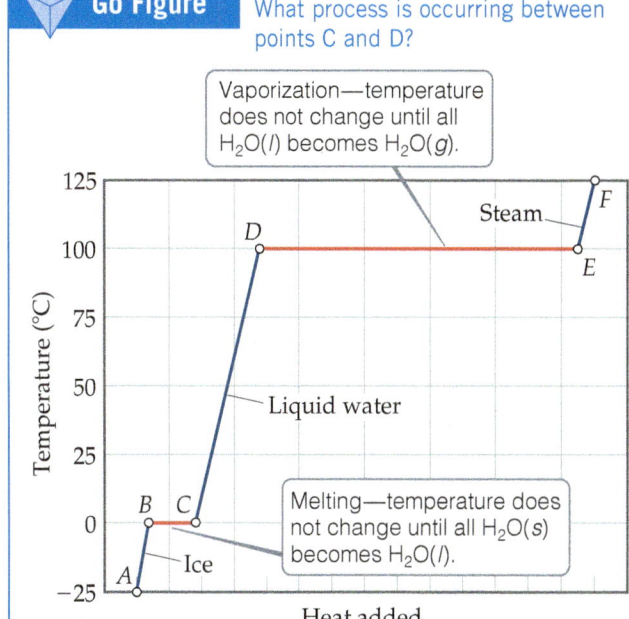

Go Figure What process is occurring between points C and D?

Vaporization—temperature does not change until all $H_2O(l)$ becomes $H_2O(g)$.

Melting—temperature does not change until all $H_2O(s)$ becomes $H_2O(l)$.

▲ **Figure 11.21 Heating curve for water.** Changes that occur when 1.00 mol of H_2O is heated from $H_2O(s)$ at −25 °C to $H_2O(g)$ at 125 °C at a constant pressure of 1 atm. Even though heat is being added continuously, the system temperature does not change during the two phase changes (red lines).

Sample Exercise 11.3

Calculating ΔH for Temperature and Phase Changes

Calculate the enthalpy change upon converting 1.00 mol of ice at −25 °C to steam at 125 °C under a constant pressure of 1 atm. The specific heats of ice, liquid water, and steam are 2.03, 4.18 and 1.84 J/g-K, respectively. For H_2O, ΔH_{fus} = 6.01 kJ/mol and ΔH_{vap} = 40.67 kJ/mol.

SOLUTION

Analyze Our goal is to calculate the total heat required to convert 1 mol of ice at −25 °C to steam at 125 °C.

Plan We can calculate the enthalpy change for each segment and then sum them to get the total enthalpy change (Hess's law, Section 5.6).

Solve

For line *AB* in Figure 11.21, we are adding enough heat to ice to increase its temperature by 25 °C. A temperature change of 25 °C is the same as a temperature change of 25 K, so we can use the specific heat of ice to calculate the enthalpy change during this process:

$$AB: \Delta H = (1.00\,\text{mol})(18.0\,\text{g/mol})(2.03\,\text{J/g-K})(25\,\text{K})$$
$$= 914\,\text{J} = 0.91\,\text{kJ}$$

For line *BC* in Figure 11.21, in which we convert ice to water at 0 °C, we can use the molar enthalpy of fusion directly:

$$BC: \Delta H = (1.00\,\text{mol})(6.01\,\text{kJ/mol}) = 6.01\,\text{kJ}$$

The enthalpy changes for lines *CD, DE,* and *EF* can be calculated in similar fashion:

$$CD: \Delta H = (1.00\,\text{mol})(18.0\,\text{g/mol})(4.18\,\text{J/g-K})(100\,\text{K})$$
$$= 7520\,\text{J} = 7.52\,\text{kJ}$$
$$DE: \Delta H = (1.00\,\text{mol})(40.67\,\text{kJ/mol}) = 40.7\,\text{kJ}$$
$$EF: \Delta H = (1.00\,\text{mol})(18.0\,\text{g/mol})(1.84\,\text{J/g-K})(25\,\text{K})$$
$$= 830\,\text{J} = 0.83\,\text{kJ}$$

Continued

The total enthalpy change is the sum of the changes of the individual steps:

$$\Delta H = 0.91 \text{ kJ} + 6.01 \text{ kJ} + 7.52 \text{ kJ} + 40.7 \text{ kJ} + 0.83 \text{ kJ} = 56.0 \text{ kJ}$$

Check The components of the total enthalpy change are reasonable relative to the horizontal lengths (heat added) of the lines in Figure 11.21. Notice that the largest component is the heat of vaporization.

▶ **Practice Exercise**

What is the enthalpy change during the process in which 100.0 g of water at 50.0 °C is cooled to ice at −30 °C? (Use the specific heats and enthalpies for phase changes given in Sample Exercise 11.3.)

Critical Temperature and Pressure

A gas normally liquefies at some point when pressure is applied. Suppose we have a cylinder fitted with a piston and the cylinder contains water vapor at 100 °C. If we increase the pressure on the water vapor, liquid water will form when the pressure is 760 torr (1 atm). However, if the temperature is 110 °C, the liquid phase does not form until the pressure is 1075 torr (1.414 atm). At 374 °C the liquid phase forms only at 1.655×10^5 torr (217.7 atm). Above this temperature, no amount of pressure causes a distinct liquid phase to form. Instead, as pressure increases, the gas becomes steadily more compressed. The highest temperature at which a distinct liquid phase can form is called the **critical temperature**. The **critical pressure** is the pressure required to bring about liquefaction at this critical temperature.

The critical temperature is the highest temperature at which a liquid can exist. Above the critical temperature, the kinetic energies of the molecules are greater than the attractive forces that lead to the liquid state regardless of how much the substance is compressed to bring the molecules closer together. *The greater the intermolecular forces, the higher the critical temperature of a substance.*

Several experimentally determined critical temperatures and pressures are listed in Table 11.5. Notice that nonpolar, low-molecular-weight substances, which have weak intermolecular attractions, have lower critical temperatures and pressures than substances that are polar or of higher molecular weight. Notice also that water and ammonia have exceptionally high critical temperatures and pressures as a consequence of strong intermolecular hydrogen-bonding forces.

Because they provide information about the conditions under which gases liquefy, critical temperatures and pressures are often of considerable importance to engineers and other people working with gases. Sometimes we want to liquefy a gas; other times we want to avoid liquefying it. It is useless to try to liquefy a gas by applying pressure if the gas is above its critical temperature. For example, O_2 has a critical temperature of 154.4 K. It must be cooled below this temperature before it can be liquefied by pressure. In contrast, ammonia has a critical temperature of 405.6 K. Thus, it can be liquefied at room temperature (approximately 295 K) by applying sufficient pressure.

TABLE 11.5 **Critical Temperatures and Pressures of Selected Substances**

Substance	Critical Temperature (K)	Critical Pressure (atm)
Nitrogen, N_2	126.1	33.5
Argon, Ar	150.9	48.0
Oxygen, O_2	154.4	49.7
Methane, CH_4	190.0	45.4
Carbon dioxide, CO_2	304.3	73.0
Phosphine, PH_3	324.4	64.5
Propane, $CH_3CH_2CH_3$	370.0	42.0
Hydrogen sulfide, H_2S	373.5	88.9
Ammonia, NH_3	405.6	111.5
Water, H_2O	647.6	217.7

When the temperature exceeds the critical temperature and the pressure exceeds the critical pressure, the liquid and gas phases are indistinguishable from each other, and the substance is in a state called a **supercritical fluid**. A supercritical fluid expands to fill its container (like a gas), but the molecules are still quite closely spaced (like a liquid).

Like liquids, supercritical fluids can behave as solvents, dissolving a wide range of substances. Using *supercritical fluid extraction*, the components of mixtures can be separated from one another. Supercritical fluid extraction has been successfully used to separate complex mixtures in the chemical, food, pharmaceutical, and energy industries. Supercritical CO_2 is a popular choice because it is relatively inexpensive and, if used in closed-loop systems, may be a way to reduce CO_2 emissions into the atmosphere.

 Self-Assessment Exercises

SAE 11.12 When solid CO_2 is heated at atmospheric pressure, it transforms directly to a gas. This is an example of _____ , which is an _____ phase transition. (**a**) fusion, exothermic (**b**) fusion, endothermic (**c**) sublimation, exothermic (**d**) sublimation, endothermic (**e**) deposition, exothermic

SAE 11.13 If you add 4.00×10^2 kJ of heat to 1.00 L of liquid water at 20.0 °C, what is the final temperature of the water? The initial density of water is 1.00 g/mL. The specific heats of liquid and gaseous

water are 4.18 and 1.84 J/g-K, and the enthalpy of vaporization is 40.67 kJ/mol. (**a**) 20.1 °C (**b**) 95.7 °C (**c**) 100 °C (**d**) 115.7 °C (**e**) 136.4 °C

SAE 11.14 Methane has a critical temperature of 190 K and a critical pressure of 45.4 atm. If you were to cool a container of methane at a pressure of 100 atm from 200 K to 180 K, you would observe a transition from a _____ to a _____ . (**a**) gas, liquid (**b**) supercritical fluid, liquid (**c**) supercritical fluid, gas (**d**) liquid, supercritical fluid

11.5 | Vapor Pressure

Molecules can escape from the surface of a liquid into the gas phase by evaporation. Suppose we place a quantity of ethanol (CH_3CH_2OH) in an evacuated, closed container, as in **Figure 11.22**. The ethanol quickly begins to evaporate. As a result, the pressure exerted by the vapor in the space above the liquid increases. After a short time, the pressure of the vapor attains a constant value, which we call the **vapor pressure**.

At any instant, some of the ethanol molecules at the liquid surface possess sufficient kinetic energy to overcome the attractive forces of their neighbors and, therefore, escape into the gas phase. As the number of gas-phase molecules increases, however, the probability increases that a molecule in the gas phase will strike the liquid surface and be recaptured by the liquid, as shown in the flask on the right in Figure 11.22. Eventually, the rate at which molecules return to the liquid equals the rate at which they escape. The number of molecules in the gas phase then reaches a steady value, and the pressure exerted by the vapor becomes constant.

 Learning Objectives

When you finish Section 11.5, you should be able to:

▶ Describe the concepts of vapor pressure and boiling point.

▶ Qualitatively predict how changes in temperature and/or strength of intermolecular forces will impact vapor pressure.

▶ Rank-order the vapor pressures of different substances at a given temperature based on differences in their composition and/or molecular structure.

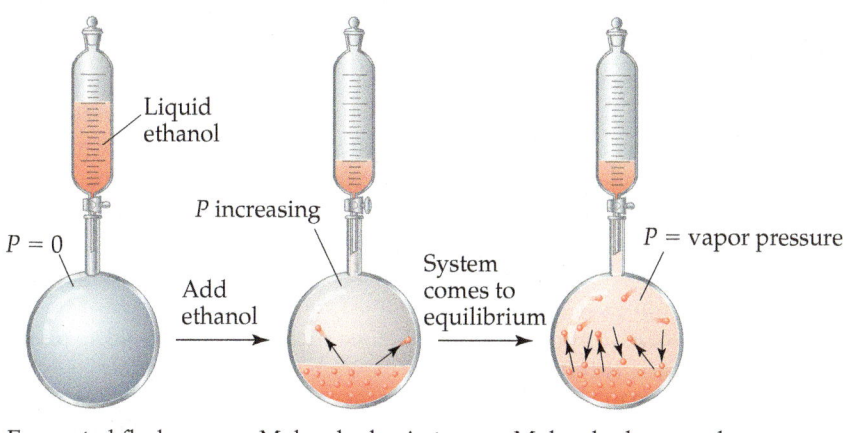

◀ Figure 11.22 Vapor pressure over a liquid.

Go Figure

As the temperature increases, does the rate of molecules escaping into the gas phase increase or decrease?

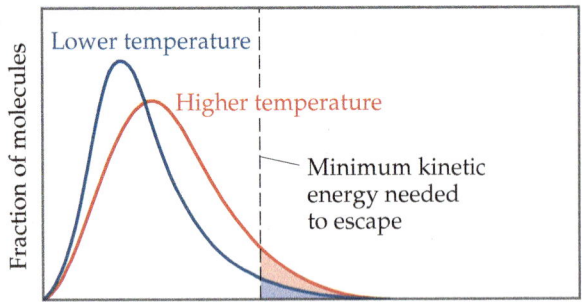

Blue area = fraction of molecules having enough energy to evaporate at lower temperature

Red + blue areas = fraction of molecules having enough energy to evaporate at higher temperature

▲ **Figure 11.23** The effect of temperature on the distribution of kinetic energies in a liquid.

Go Figure

What is the vapor pressure of ethylene glycol at its normal boiling point?

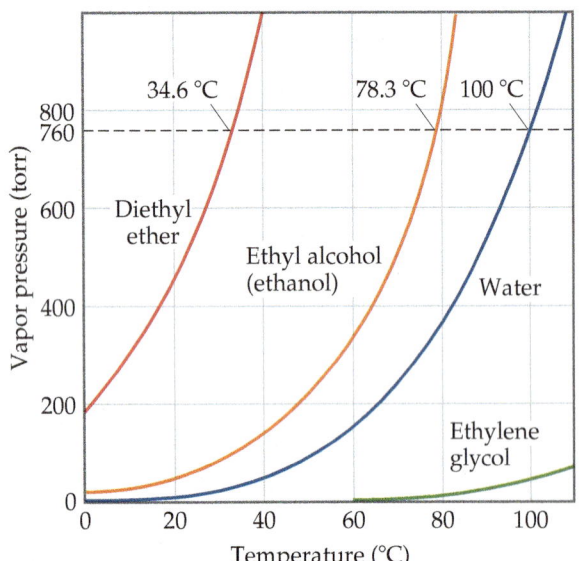

▲ **Figure 11.24 Vapor pressure for four liquids as a function of temperature.** The normal boiling point is the temperature where the vapor pressure equals standard atmospheric pressure, 760 torr, marked by the dashed line.

The condition in which two opposing processes occur simultaneously at equal rates is called **dynamic equilibrium** (or simply *equilibrium*). Chemical equilibrium (see Section 4.1) is a kind of dynamic equilibrium in which the opposing processes are chemical reactions.

A liquid and its vapor are in dynamic equilibrium when evaporation and condensation occur at equal rates. It may appear that nothing is occurring at equilibrium because there is no net change in the system. In fact, though, a great deal is happening as molecules continuously pass from liquid state to gas state and from gas state to liquid state. *The vapor pressure of a liquid is the pressure exerted by its vapor when the liquid and vapor are in dynamic equilibrium.*

Volatility, Vapor Pressure, and Temperature

When vaporization occurs in an open container, as when water evaporates from a bowl, the vapor moves away from the liquid. Little, if any, is recaptured at the surface of the liquid. Equilibrium never occurs, and the vapor continues to form until the liquid evaporates to dryness. Substances with high vapor pressure (such as gasoline) evaporate more quickly than substances with low vapor pressure (such as motor oil). Liquids that evaporate readily are said to be **volatile**.

Hot water evaporates more quickly than cold water because vapor pressure increases with increasing temperature. To see why this statement is true, we begin with the fact that the molecules of a liquid move at various speeds. Figure 11.23 shows the distribution of kinetic energies of the molecules at the surface of a liquid at two temperatures. These distributions are like those we encountered with gases in Chapter 10. (Section 10.7) As the temperature is increased, the molecules move more energetically and more of them can break free from their neighbors and enter the gas phase, increasing the vapor pressure.

Figure 11.24 depicts the variation in vapor pressure with temperature for four common substances that differ greatly in volatility. Note that the vapor pressure in all cases increases nonlinearly with increasing temperature. The weaker the intermolecular forces in the liquid, the more easily molecules can escape and, therefore, the higher the vapor pressure at a given temperature.

Vapor Pressure and Boiling Point

The **boiling point** of a liquid is the temperature at which its vapor pressure equals the external pressure, acting on the liquid surface. At this temperature, the thermal energy of the molecules is great enough for the molecules in the interior of the liquid to break free from their neighbors and enter the gas phase. As a result, bubbles of vapor form within the liquid. The boiling point increases as the external pressure increases. The boiling point of a liquid at 1 atm (760 torr) pressure is called its **normal boiling point**. From Figure 11.24 we see that the normal boiling point of water is 100 °C.

The time required to cook food in boiling water depends on the water temperature. In an open container, that temperature is 100 °C,

A CLOSER LOOK | The Clausius–Clapeyron Equation

The plots in Figure 11.24 have a distinct shape: For each substance, the vapor pressure curves sharply upward with increasing temperature. The relationship between vapor pressure and temperature is given by the *Clausius–Clapeyron equation*:

$$\ln P = \frac{-\Delta H_{vap}}{RT} + C \qquad [11.1]$$

where P is the vapor pressure, T is the absolute temperature, R is the gas constant (8.314 J/mol-K), ΔH_{vap} is the molar enthalpy of vaporization, and C is a constant. This equation predicts that a graph of ln P versus $1/T$ should give a straight line with a slope equal to $\Delta H_{vap}/R$. Using this plot, we can determine the enthalpy of vaporization of a substance:

$$\Delta H_{vap} = -\text{slope} \times R$$

As an example of how we use the Clausius–Clapeyron equation, the vapor–pressure data for ethanol shown in Figure 11.24 are graphed as ln P versus $1/T$ in **Figure 11.26**. The data lie on a straight line with a negative slope. We can use the slope to determine ΔH_{vap} for ethanol, 38.56 kJ/mol. We can also extrapolate the line to obtain the vapor pressure of ethanol at temperatures above and below the temperature range for which we have data.

Related Exercises: 11.86, 11.87

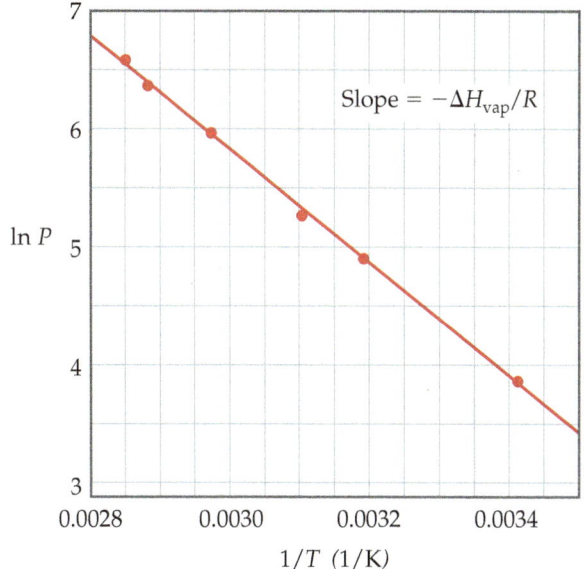

▲ **Figure 11.26** The natural logarithm of vapor pressure versus $1/T$ for ethanol.

but it is possible to boil at higher temperatures. Pressure cookers work by allowing steam to escape only when it exceeds a predetermined pressure; the pressure above the water can therefore increase above atmospheric pressure (**Figure 11.25**). The higher pressure causes the water to boil at a higher temperature, thereby allowing the food to get hotter and to cook more rapidly.

The effect of pressure on boiling point also explains why it takes longer to cook food at high elevations than it does at sea level. The atmospheric pressure is lower at higher altitudes, so water boils at a temperature lower than 100 °C, and foods generally take longer to cook.

▲ **Figure 11.25** **Pressure cooker.** Cooking food at elevated pressures elevates the boiling point of water that surrounds the food, thereby reducing the cooking time.

Sample Exercise 11.4
Relating Boiling Point to Vapor Pressure

Use Figure 11.24 to estimate the boiling point of diethyl ether under an external pressure of 0.80 atm.

SOLUTION

Analyze We are asked to read a graph of vapor pressure versus temperature to determine the boiling point of a substance at a particular pressure. The boiling point is the temperature at which the vapor pressure is equal to the external pressure.

Plan We need to convert 0.80 atm to torr because that is the pressure scale on the graph. We estimate the location of that pressure on the graph, move horizontally to the vapor-pressure curve, and then drop vertically from the curve to estimate the temperature.

Solve The pressure equals (0.80 atm)(760 torr/atm) = 610 torr (to two significant digits). From Figure 11.24 we see that the

boiling point at this pressure is about 27 °C, which is close to room temperature.

Comment We can make a flask of diethyl ether boil at room temperature by using a vacuum pump to lower the pressure above the liquid to about 0.8 atm.

▶ **Practice Exercise**
Use Figure 11.24 to determine the external pressure at which ethanol will boil at 60 °C.

Self-Assessment Exercises

SAE 11.15   The container on the left contains liquid ethanol in equilibrium with its vapor. The temperature is 293 K and the vapor pressure is 45 torr. The container on the right, which has the same volume as the one on the left, is evacuated. What will happen to the vapor pressure when the valve connecting the two containers is opened and new equilibrium is established between the liquid and the gas? Assume the temperature remains constant and that some liquid ethanol remains once the new equilibrium is established.

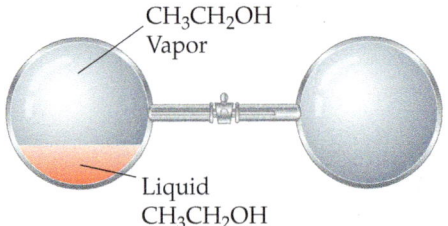

(**a**) The new vapor pressure will be halved, 22.5 torr. (**b**) The new vapor pressure will be unchanged, 45 torr. (**c**) The new vapor pressure will be doubled, 90 torr. (**d**) The new vapor pressure will decrease but not enough information is given to determine its value.

SAE 11.16   At room temperature, toluene ($C_6H_5CH_3$) is a colorless liquid. If the temperature is increased from 25 °C to 50 °C which of the following quantities will increase?

   (**i**) vapor pressure
   (**ii**) normal boiling point

(**a**) only i (**b**) only ii (**c**) both i and ii (**d**) neither i nor ii

SAE 11.17   All of the following alcohols are liquids at room temperature: *n*-propanol, $CH_3CH_2CH_2OH$; *n*-butanol, $CH_3CH_2CH_2CH_2OH$; and *n*-hexanol, $CH_3CH_2CH_2CH_2CH_2CH_2OH$. As the length of the hydrocarbon chain increases (propanol → butanol → hexanol), which of the following will *decrease*? (**a**) normal boiling point (**b**) viscosity (**c**) average strength of the intermolecular attractions between molecules (**d**) vapor pressure at room temperature

11.6 | Phase Diagrams

Learning Objective

When you finish Section 11.6, you should be able to:

▶ Interpret phase diagrams, which depict the state of a pure substance as a function of temperature and pressure.

The equilibrium between a liquid and its vapor is not the only dynamic equilibrium that can exist between states of matter. Under appropriate conditions, a solid can be in equilibrium with its liquid or even with its vapor. The temperature at which solid and liquid phases coexist at equilibrium is the *melting point* of the solid or the *freezing point* of the liquid. Solids can also undergo evaporation and therefore possess a vapor pressure.

A **phase diagram** is a graphic way to summarize the conditions under which equilibria exist between the different states of matter. Such a diagram also allows us to predict which phase of a substance is present at any given temperature and pressure.

The phase diagram for any substance that can exist in all three phases of matter is shown in **Figure 11.27**. The diagram contains three important curves, each of which represents the temperature and pressure at which the various phases can coexist at equilibrium. The only substance present in the system is the one whose phase diagram is under

Go Figure   Imagine that the pressure on the solid phase in the figure is decreased at constant temperature. If the solid eventually sublimes, what must be true about the temperature?

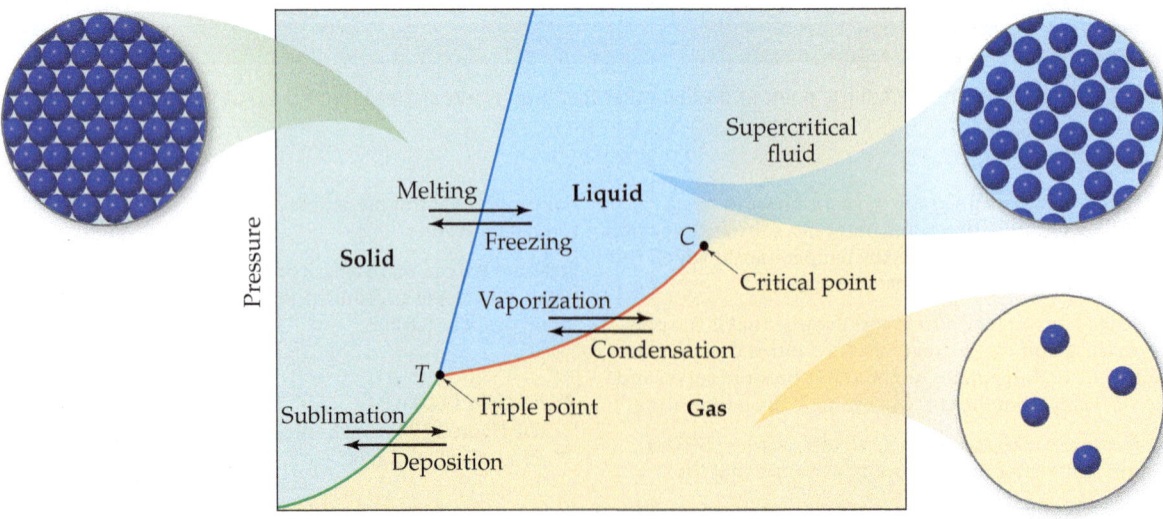

▲ **Figure 11.27 Generic phase diagram for a pure substance.** The green line is the sublimation curve, the blue line is the melting curve, and the red line is the vapor-pressure curve.

consideration. The pressure shown in the diagram is either the pressure applied to the system or the pressure generated by the substance. The curves may be described as follows:

1. The red curve is the *vapor-pressure curve* of the liquid, representing equilibrium between the liquid and gas phases. The point on this curve where the vapor pressure is 1 atm is the normal boiling point of the substance. The vapor-pressure curve ends at the **critical point** (*C*), which corresponds to the critical temperature and critical pressure of the substance. At temperatures and pressures above those at the critical point, the liquid and gas phases are indistinguishable from each other, and the substance is a *supercritical fluid*.

2. The green curve, the *sublimation curve*, separates the solid phase from the gas phase and represents the change in the vapor pressure of the solid as it sublimes at different temperatures. Each point on this curve represents a condition of equilibrium between the solid and the gas.

3. The blue curve, the *melting curve*, separates the solid phase from the liquid phase and represents the change in melting point of the solid with increasing pressure. Each point on this curve represents an equilibrium between the solid and the liquid. This curve usually slopes slightly to the right as pressure increases for most substances because the solid form is denser than the liquid form. An increase in pressure usually favors the more compact solid phase; thus, higher temperatures are required to melt the solid at higher pressures. The melting point at 1 atm is the **normal melting point**.

Point *T*, where the three curves intersect, is the **triple point**, and here all three phases are in equilibrium. Any other point on any of the three curves represents equilibrium between two phases. Any point on the diagram that does not fall on one of the curves corresponds to conditions under which only one phase is present. The gas phase, for example, is stable at low pressures and high temperatures, whereas the solid phase is stable at low temperatures and high pressures. Liquids are stable in the region between gases and solids.

The Phase Diagrams of H_2O and CO_2

Figure 11.28 shows the phase diagram of H_2O. Because of the large range of pressures covered in the diagram, a logarithmic scale is used to represent pressure. The melting curve (blue line) of H_2O is atypical, slanting slightly to the left with increasing pressure, indicating that for water the melting point *decreases* with increasing pressure. This unusual behavior occurs because water is among the very few substances whose liquid form is more compact than its solid form, as we learned in Section 11.2.

If the pressure is held constant at 1 atm, it is possible to move from the solid to liquid to gaseous regions of the phase diagram by changing the temperature, as we expect from our everyday encounters with water. The triple point of H_2O falls at a relatively low pressure, 0.00603 atm. Below this pressure, liquid water is not stable and ice sublimes to water vapor on heating. This property of water is used to "freeze-dry" foods and beverages. The food or beverage is frozen to a temperature below 0 °C. Next it is placed in a low-pressure chamber (below 0.00603 atm) and then warmed so that the water sublimes, leaving behind dehydrated food or beverage.

The phase diagram for CO_2 is shown in **Figure 11.29**. The melting curve (blue line) behaves typically, slanting to the right with increasing pressure, telling us that the melting point of CO_2 increases with increasing pressure. Because the pressure at the triple point is relatively high, 5.11 atm, CO_2 does not exist as a liquid at 1 atm, which means that solid CO_2 does not melt when heated; instead it sublimes. Thus, CO_2 does not have a normal melting point; instead, it has a normal sublimation point, −78.5 °C. Because CO_2 sublimes rather than melts as it absorbs energy at ordinary pressures, solid CO_2 (dry ice) is a convenient coolant.

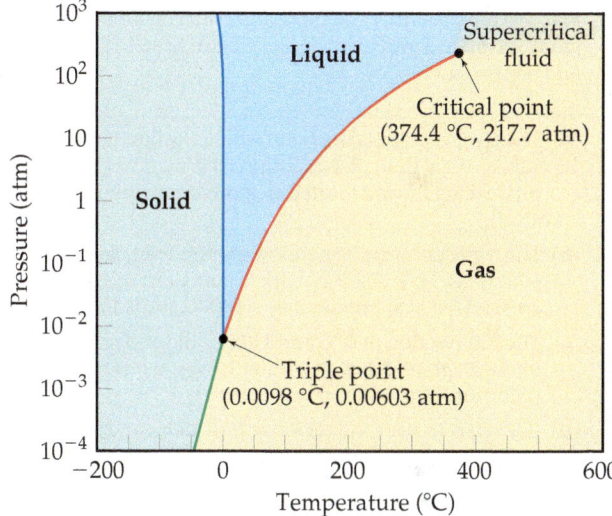

▲ **Figure 11.28 Phase diagram of H_2O.** Note that a linear scale is used to represent temperature and a logarithmic scale to represent pressure.

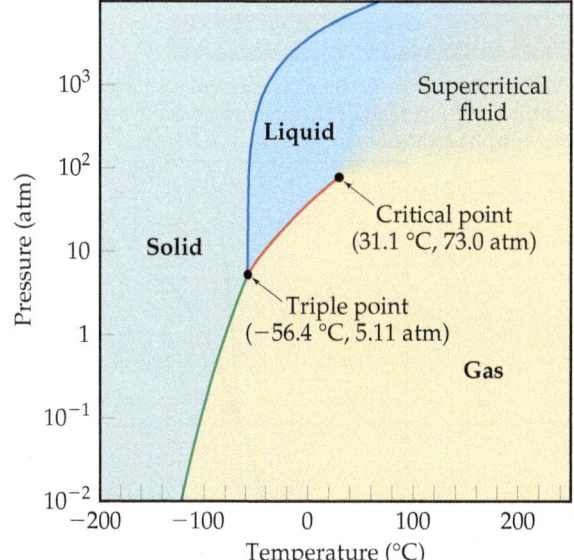

▲ **Figure 11.29 Phase diagram of CO_2.** Note that a linear scale is used to represent temperature and a logarithmic scale to represent pressure.

Sample Exercise 11.5
Interpreting a Phase Diagram

Use the phase diagram for methane, CH$_4$, shown in **Figure 11.30** to answer the following questions. (**a**) What are the approximate temperature and pressure of the critical point? (**b**) What are the approximate temperature and pressure of the triple point? (**c**) Is methane a solid, liquid, or gas at 1 atm and 0 °C? (**d**) If solid methane at 1 atm is heated while the pressure is held constant, will it melt or sublime? (**e**) If methane at 1 atm and 0 °C is compressed until a phase change occurs, in which state is the methane when the compression is complete?

SOLUTION

Analyze We are asked to identify key features of the phase diagram and to use it to deduce what phase changes occur when specific pressure and temperature changes take place.

Plan We must identify the triple and critical points on the diagram and also identify which phase exists at specific temperatures and pressures.

Solve
(**a**) The critical point is the point where the liquid, gaseous, and supercritical fluid phases coexist. It is marked point 3 in the phase diagram and located at approximately −80 °C and 45 atm.

(**b**) The triple point is the point where the solid, liquid, and gaseous phases coexist. It is marked point 1 in the phase diagram and located at approximately −180 °C and 0.1 atm.

(**c**) The intersection of 0 °C and 1 atm is marked point 2 in the phase diagram. It is well within the gaseous region of the phase diagram.

(**d**) If we start in the solid region at $P = 1$ atm and move horizontally (this means we hold the pressure constant), we cross first into the liquid region, at $T \approx -180$ °C, and then into the gaseous region, at $T \approx -160$ °C. Therefore, solid methane melts when the pressure is 1 atm. (For methane to sublime, the pressure must be below the triple point pressure.)

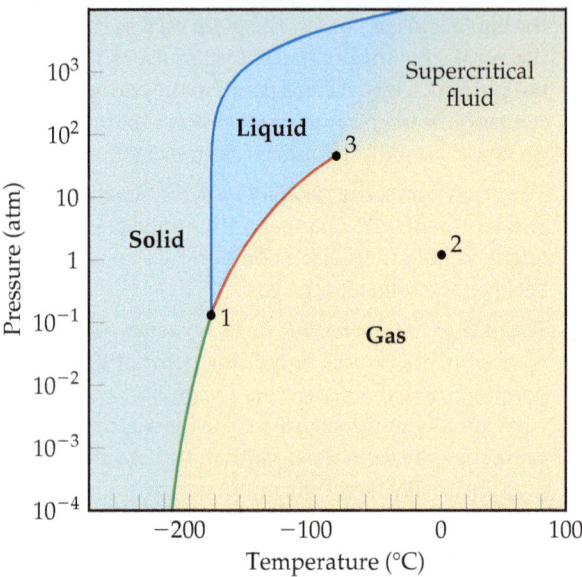

▲ **Figure 11.30 Phase diagram of CH$_4$.** Note that a linear scale is used to represent temperature and a logarithmic scale to represent pressure.

(**e**) Moving vertically up from point 2, which is 1 atm and 0 °C, the first phase change we come to is from gas to supercritical fluid. This phase change happens when we exceed the critical pressure (~50 atm).

Check The pressure and temperature at the critical point are higher than those at the triple point, which is expected. Methane is the principal component of natural gas. So it seems reasonable that it exists as a gas at 1 atm and 0 °C.

▶ **Practice Exercise**
Use the phase diagram of methane (Figure 11.30) to answer the following questions. (**a**) What is the normal boiling point of methane? (**b**) Over what pressure range does solid methane sublime? (**c**) Above what temperature does liquid methane not exist?

 ## Self-Assessment Exercises

SAE 11.18 Identify the phase present (solid, gas, liquid, or supercritical fluid) in each of the regions labeled with a letter in the following phase diagram for substance X.

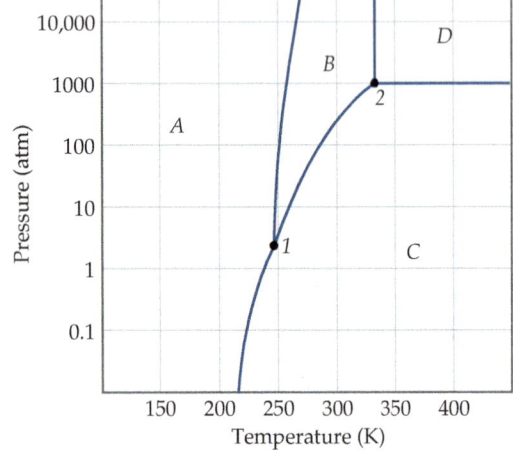

(**a**) A = liquid, B = solid, C = gas, D = supercritical fluid
(**b**) A = solid, B = liquid, C = gas, D = supercritical fluid
(**c**) A = solid, B = liquid, C = supercritical fluid, D = gas
(**d**) A = gas, B = liquid, C = solid, D = supercritical fluid

SAE 11.19 Use the phase diagram of CO$_2$ (Figure 11.29) to determine which of the following statements is *false*. (**a**) Liquid CO$_2$ does not exist at pressures lower than 5.11 atm. (**b**) Liquid CO$_2$ does not exist at pressures higher than 73.0 atm. (**c**) Liquid CO$_2$ does not exist at temperatures higher than 31.1 °C. (**d**) Liquid CO$_2$ does not exist at temperatures lower than −60 °C.

11.7 | Liquid Crystals

In 1888 Friedrich Reinitzer, an Austrian botanist, discovered that the organic compound cholesteryl benzoate has an interesting and unusual property, shown in **Figure 11.31**. Solid cholesteryl benzoate melts at 145 °C, forming a viscous milky liquid; then at 179 °C the milky liquid becomes clear and remains that way at temperatures above 179 °C. When cooled, the clear liquid turns viscous and milky at 179 °C, and the milky liquid solidifies at 145 °C.

Reinitzer's work represents the first systematic report of what we call a **liquid crystal**, the term we use today for the viscous, milky state that some substances exhibit between the liquid and solid states.

This intermediate phase has some of the structure of solids and some of the freedom of motion of liquids. Because of the partial ordering, liquid crystals may be viscous and possess properties intermediate between those of solids and those of liquids. The region in which they exhibit these properties is marked by sharp transition temperatures, as in Reinitzer's sample. Just like solid, liquid, and gas phases, the liquid crystalline phase is represented by a distinct region on a phase diagram.

Today liquid crystals are used as pressure and temperature sensors and as liquid crystals displays (LCDs) in such devices as digital watches, televisions, and computers. They can be used for these applications because the weak intermolecular forces that hold the molecules together in the liquid crystalline phase are easily affected by changes in temperature, pressure, and electric fields.

Learning Objectives

When you finish Section 11.7, you should be able to:

▶ Describe how the physical properties of liquid crystals differ from those of ordinary liquids.

▶ Describe the differences between nematic, smectic, and cholesteric liquid crystals.

▶ Identify substances likely to form liquid crystals from their molecular structure.

Types of Liquid Crystals

Substances that form liquid crystals are often composed of rod-shaped molecules that are somewhat rigid. In the liquid phase, these molecules are oriented randomly. In the liquid crystalline phase, by contrast, the molecules are arranged in specific patterns as illustrated in **Figure 11.32**. Depending on the nature of the ordering, liquid crystals are classified as nematic, smectic A, smectic C, or cholesteric.

In a **nematic liquid crystal**, the molecules are aligned so that their long axes tend to point in the same direction, but the ends are not aligned with one another. In **smectic A** and **smectic C liquid crystals**, the molecules maintain the long-axis alignment seen in nematic crystals, but in addition they pack into layers.

Two molecules that exhibit liquid crystalline phases are shown in **Figure 11.33**. The lengths of these molecules are much greater than their widths. The double bonds, including those in the benzene rings, add rigidity to the molecules, and the rings, because they are flat, help the molecules stack with one another. The polar $-OCH_3$ and $-COOH$ groups give rise to dipole–dipole interactions and promote alignment of the molecules. Thus, the molecules order themselves quite naturally along their long axes. They can, however, rotate around their axes and slide parallel to one another. In smectic liquid crystals, the intermolecular forces (dispersion forces, dipole–dipole attractions, and hydrogen bonding) limit the ability of the molecules to slide past one another.

In a **cholesteric liquid crystal**, the molecules are arranged in layers, with their long axes parallel to the other molecules within the same layer.* Upon moving from one layer to the next, the orientation of the molecules rotates by a fixed angle, resulting in a spiral pattern. These liquid crystals are so named because many derivatives of cholesterol adopt this structure.

145 °C < T < 179 °C
Liquid crystalline phase

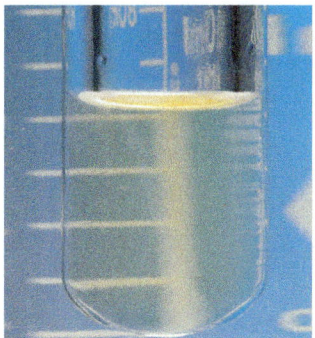

T > 179 °C
Liquid phase

▶ **Figure 11.31 Cholesteryl benzoate in its liquid and liquid crystalline states.**

---

* Cholesteric liquid crystals are sometimes called chiral nematic phases because the molecules within each plane adopt an arrangement similar to a nematic liquid crystal.

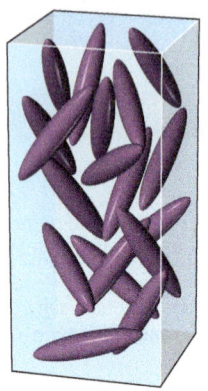 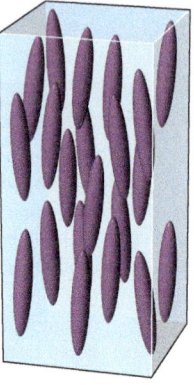 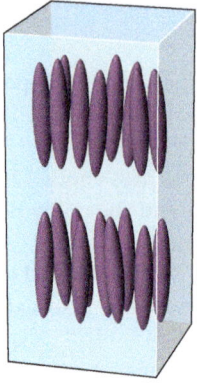 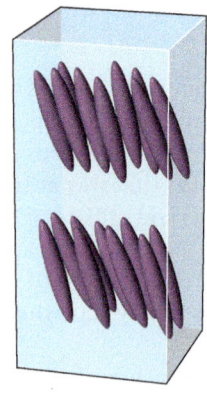 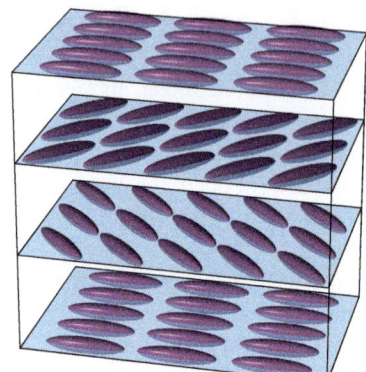

Liquid phase	**Nematic liquid crystalline phase**	**Smectic A liquid crystalline phase**	**Smectic C liquid crystalline phase**	**Cholesteric liquid crystalline phase**
Molecules arranged randomly	Long axes of molecules aligned, but ends are not aligned	Molecules aligned in layers, long axes of molecules perpendicular to layer planes	Molecules aligned in layers, long axes of molecules inclined with respect to layer planes	Molecules pack into layers, long axes of molecules in one layer rotated relative to the long axes in the layer above it

▲ **Figure 11.32 Molecular order in nematic, smectic, and cholesteric liquid crystals.** In the liquid phase of any substance, the molecules are arranged randomly, whereas in the liquid crystalline phases the molecules are arranged in a partially ordered way.

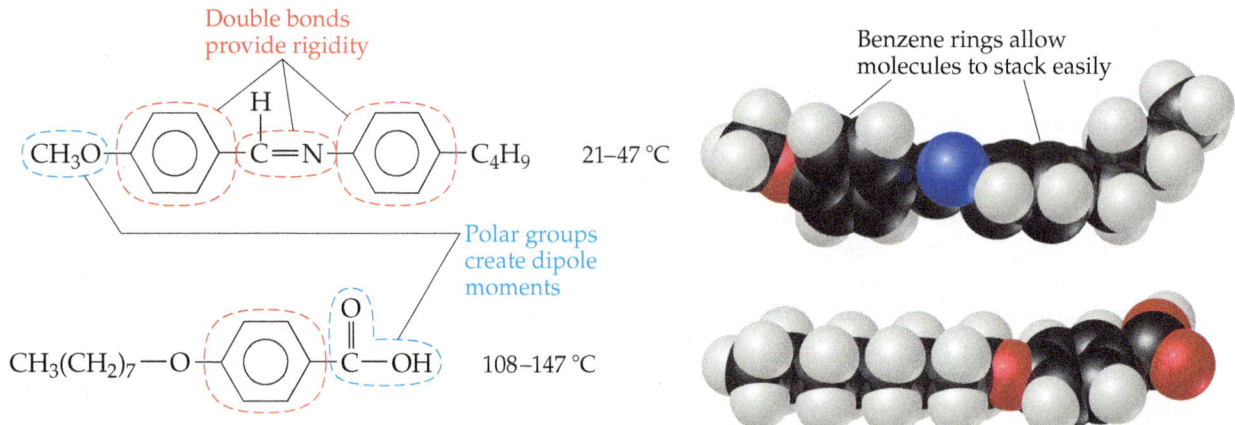

▲ **Figure 11.33 Molecular structure and liquid crystal temperature range for two typical liquid crystalline materials.**

The molecular arrangement in cholesteric liquid crystals produces unusual coloring patterns with visible light. Changes in temperature and pressure change the order and, hence, the color. Cholesteric liquid crystals are used to monitor temperature changes in situations where conventional methods are not feasible. For example, they can detect hot spots in microelectronic circuits, which may signal the presence of flaws. They can also be fashioned into thermometers for measuring the skin temperature of infants. Liquid crystals are used in the displays of devices such as calculators, computer monitors, televisions, and video projectors. Liquid crystal displays draw much less power than light-emitting diode (LED) technologies. In such a display, an applied electric field is used to change the orientation of nematic liquid crystals, and by changing the orientation one can control whether light can pass through a given pixel in the display.

Sample Exercise 11.6
Properties of Liquid Crystals

Which of these substances is most likely to exhibit liquid crystalline behavior?

$$\underset{(i)}{CH_3-CH_2-\overset{\overset{\displaystyle CH_3}{|}}{\underset{\underset{\displaystyle CH_3}{|}}{C}}-CH_2-CH_3}$$

(ii) CH₃CH₂—⟨benzene ring⟩—N=N—⟨benzene ring⟩—C(=O)—OCH₃

(iii) ⟨benzene ring⟩—CH₂—C(=O)—O⁻Na⁺

SOLUTION

Analyze We have three molecules with different structures, and we are asked to determine which one is most likely to be a liquid crystalline substance.

Plan We need to identify all structural features that might induce liquid crystalline behavior.

Solve Molecule (i) is not likely to be liquid crystalline because the absence of double and/or triple bonds makes this molecule flexible rather than rigid. Molecule (iii) is ionic, and the generally high melting points of ionic materials make it unlikely that this substance is liquid crystalline. Molecule (ii) possesses the characteristic long axis and the kinds of structural features often seen in liquid crystals: The molecule has a rod-like shape, the double bonds and benzene rings provide rigidity, and the polar COOCH₃ group creates a dipole moment.

▶ **Practice Exercise**

Suggest a reason why decane
(CH₃CH₂CH₂CH₂CH₂CH₂CH₂CH₂CH₂CH₃) does not exhibit liquid crystalline behavior.

Self-Assessment Exercises

SAE 11.20   Arrange the liquid, solid, and liquid crystalline phases in terms of the freedom of motion that molecules have to move past one another.

(**a**) liquid crystal < solid < liquid

(**b**) liquid < liquid crystal < solid

(**c**) solid < liquid < liquid crystal

(**d**) solid < liquid crystal < liquid

(**e**) solid < liquid crystal ≈ liquid

SAE 11.21   Substances that form liquid crystalline phases tend to be made up of _____ molecules that are _____ in shape. (**a**) flexible, rod-like (**b**) rigid, rod-like (**c**) flexible, spherical (**d**) rigid, spherical

SAE 11.22   Of the various liquid crystal phases shown in Figure 11.32, which phase shows the least amount of long-range intermolecular order? (**a**) nematic (**b**) smectic A (**c**) smectic C (**d**) cholesteric

Putting Concepts Together

The substance CS₂ has a melting point of −110.8 °C and a boiling point of 46.3 °C. Its density at 20 °C is 1.26 g/cm³. It is highly flammable. (**a**) What is the name of this compound? (**b**) List the intermolecular forces that CS₂ molecules exert on one another. (**c**) Write a balanced equation for the combustion of this compound in air. (You will have to decide on the most likely oxidation products.) (**d**) The critical temperature and pressure for CS₂ are 552 K and 78 atm, respectively. Compare these values with those for CO₂ in Table 11.5 and discuss the possible origins of the differences.

SOLUTION

(**a**) The compound is named carbon disulfide, in analogy with the naming of other binary molecular compounds such as carbon dioxide. (Section 2.8)

(**b**) Because there is no H atom, there can be no hydrogen bonding. If we draw the Lewis structure, we see that carbon forms double bonds with each sulfur:

$$\ddot{\underset{..}{S}}=C=\ddot{\underset{..}{S}}$$

Using the VSEPR model (Section 9.2), we conclude that the molecule is linear and therefore has no dipole moment. (Section 9.3) Thus, there are no dipole–dipole forces. Only dispersion forces operate between the CS₂ molecules.

(**c**) The most likely products of the combustion will be CO₂ and SO₂. (Section 3.2) Under some conditions, SO₃ might be formed, but this would be the less likely outcome. Thus, we have the following equation for combustion:

$$CS_2(l) + 3\,O_2(g) \longrightarrow CO_2(g) + 2\,SO_2(g)$$

(**d**) The critical temperature and pressure of CS₂ (552 K and 78 atm, respectively) are both higher than those given for CO₂ in Table 11.5 (304 K and 73 atm, respectively). The difference in critical temperatures is especially notable. The higher values for CS₂ arise from the greater dispersion attractions between the CS₂ molecules compared with CO₂. These greater attractions are due to the larger size of the sulfur compared to oxygen and, therefore, its greater polarizability.

Chapter Summary and Key Terms

A MOLECULAR COMPARISON OF GASES, LIQUIDS, AND SOLIDS (INTRODUCTION AND SECTION 11.1) Substances that are gases or liquids at room temperature are usually composed of molecules. In gases, the intermolecular attractive forces are negligible compared to the kinetic energies of the molecules; thus, the molecules are widely separated and undergo constant, chaotic motion. In liquids, the **intermolecular forces** keep the molecules in close proximity; nevertheless, the molecules are free to move with respect to one another. In solids, the intermolecular attractive forces restrain molecular motion and to force the molecules to occupy specific locations in a three-dimensional arrangement.

INTERMOLECULAR FORCES (SECTION 11.2) Three types of intermolecular forces exist between neutral molecules: **dispersion forces, dipole–dipole interactions**, and **hydrogen bonding**. Dispersion forces operate between all molecules (and atoms, for atomic substances such as He, Ne, Ar, and so forth). As molecular weight increases, the **polarizability** of a molecule increases, which results in stronger dispersion forces. Molecular shape is also an important factor. Dipole–dipole forces increase in strength as the polarity of the molecule increases. Hydrogen bonding occurs in compounds containing O—H, N—H, and F—H bonds. Hydrogen bonds are generally stronger than other dipole–dipole interactions. **Ion–dipole forces** occur when polar molecules are in close proximity to ions. They are particularly important in solutions of ionic compounds dissolved in polar solvents.

SELECT PROPERTIES OF LIQUIDS (SECTION 11.3) The stronger the intermolecular forces, the greater the **viscosity**, or resistance to flow, of a liquid. **Surface tension** is a measure of the tendency of a liquid to maintain a minimum surface area. The surface tension of a liquid also increases as intermolecular forces increase in strength. The adhesion of a liquid to the walls of a narrow tube and the cohesion of the liquid account for **capillary action**, a term used to describe the rise of liquids up narrow tubes.

PHASE CHANGES (SECTION 11.4) A substance may exist in more than one state of matter, or phase. **Phase changes** are transformations from one phase to another. Changes of a solid to liquid (melting), solid to gas (sublimation), and liquid to gas (vaporization) are all endothermic processes. Thus, the **heat of fusion** (melting), the **heat of sublimation**, and the **heat of vaporization** are all positive quantities. The reverse processes (freezing, deposition, and condensation) are exothermic.

A gas cannot be liquefied by application of pressure if the temperature is above its **critical temperature**. The pressure required to liquefy a gas at its critical temperature is called the **critical**

pressure. When the temperature exceeds the critical temperature and the pressure exceeds the critical pressure, the liquid and gas phases coalesce to form a **supercritical fluid**, a phase that expands to fill its container like a gas but has a density and compressibility that are more like a liquid.

VAPOR PRESSURE (SECTION 11.5) The **vapor pressure** of a liquid is the partial pressure of the vapor when it is in **dynamic equilibrium** with the liquid. At equilibrium the rate of transfer of molecules from the liquid to the vapor equals the rate of transfer from the vapor to the liquid. The higher the vapor pressure of a liquid, the more readily it evaporates and the more **volatile** it is. Vapor pressure increases with temperature. Boiling occurs when the vapor pressure equals the external pressure. Thus, the **boiling point** of a liquid depends on pressure. The **normal boiling point** is the temperature at which the vapor pressure equals 1 atm.

PHASE DIAGRAMS (SECTION 11.6) The equilibria between the various phases of a substance as a function of temperature and pressure are displayed on a **phase diagram**. A line indicates equilibria between any two phases. The line through the melting point usually slopes slightly to the right as pressure increases, because the solid is usually more dense than the liquid. The boundary between the solid and liquid phases at a pressure of 1 atm is the **normal melting point,** and the boundary between liquid and gas phases at 1 atm corresponds to the normal boiling point. The point on the diagram at which all three phases coexist in equilibrium is called the **triple point**. The **critical point** corresponds to the critical temperature and critical pressure. Beyond the critical point, the substance is a supercritical fluid.

LIQUID CRYSTALS (SECTION 11.7) A **liquid crystal** is a substance that exhibits one or more ordered phases at a temperature above the melting point of the solid. In a **nematic liquid crystal** the molecules are aligned along a common direction, but the ends of the molecules are not lined up. In a **smectic liquid crystal** the ends of the molecules are lined up so that the molecules form layers. In **smectic A liquid crystals** the long axes of the molecules line up perpendicular to the layers. In **smectic C liquid crystals** the long axes of molecules are inclined with respect to the layers. A **cholesteric liquid crystal** is composed of molecules that align parallel to each other within a layer, as they do in nematic liquid crystalline phases, but the direction along which the long axes of the molecules align rotates from one layer to the next to form a helical structure. Substances that form liquid crystals are generally composed of molecules that are fairly rigid, with cylindrical or rod-like shapes. These molecules also tend to have polar groups to help align molecules through dipole–dipole interactions.

Exam Prep

EP 11.1 Which of the following properties is *not* a characteristic of a liquid? (**a**) flows readily (**b**) very difficult to compress (**c**) assumes the shape of the container it occupies (**d**) fills the entire volume of the container it occupies

EP 11.2 Which of the following explains why a substance transforms from a liquid to a gas when the temperature is raised above its boiling point, T_b?

  (**i**) Because the strength of the intermolecular attractions decreases as the temperature increases.

  (**ii**) Because the average kinetic energy of the molecules increases as the temperature increases.

(**a**) only i (**b**) only ii (**c**) both i and ii (**d**) neither i nor ii

EP 11.3 Under 1 atm of pressure, carbon tetrachloride is a liquid between the temperatures of $-23\,°C$ and $77\,°C$, while carbon

tetrabromide is a liquid between the temperatures of $95\,°C$ and $190\,°C$. Which of the following statements explains why CBr_4 has higher melting and boiling points than CCl_4?

  (**i**) The attractions between CBr_4 molecules are stronger than those between CCl_4 molecules.

  (**ii**) Because it has a higher molecular weight, the average kinetic energy of the CBr_4 molecules at a given temperature is less than the average kinetic energy of the CCl_4 molecules at the same temperature.

  (**iii**) Because chlorine is more electronegative than bromine, CCl_4 will have a larger dipole moment than CBr_4, and that will lead to greater repulsions between CCl_4 molecules.

(**a**) only i (**b**) only ii (**c**) only iii (**d**) both i and ii (**e**) both i and iii

EP 11.4 Which factor or factors contribute to the higher boiling point of dibromomethane CH_2Br_2 (97 °C) versus that of dichloromethane CH_2Cl_2 (40 °C)?

(**i**) The dispersion forces are stronger in CH_2Br_2.

(**ii**) The dipole–dipole interactions are stronger in CH_2Br_2.

(**a**) only i (**b**) only ii (**c**) both i and ii (**d**) neither i nor ii

EP 11.5 The molecular structures and formulas for four substances that are liquids at room temperature and pressure are shown here. In which pure substance are hydrogen bonds *not* present?

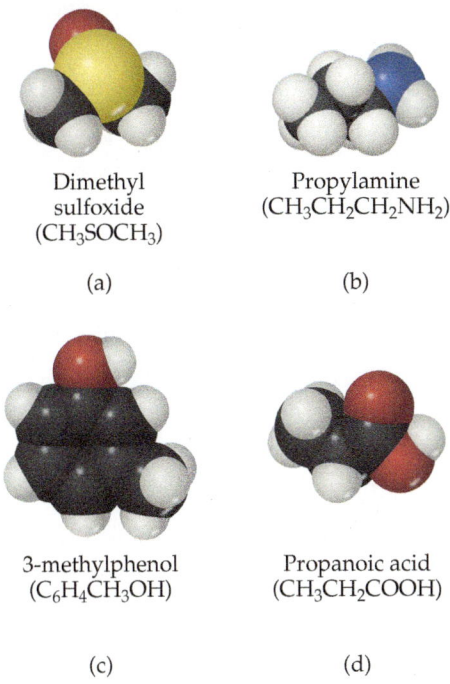

Dimethyl sulfoxide (CH_3SOCH_3)

(a)

Propylamine ($CH_3CH_2CH_2NH_2$)

(b)

3-methylphenol ($C_6H_4CH_3OH$)

(c)

Propanoic acid (CH_3CH_2COOH)

(d)

EP 11.6 Only one of the following substances is a liquid at room temperature; the others are gases. Which substance is most likely to be a liquid at room temperature? (**a**) formaldehyde, H_2CO (**b**) fluoromethane, CH_3F (**c**) hydrogen cyanide, HCN (**d**) hydrogen peroxide, H_2O_2 (**e**) hydrogen sulfide, H_2S

EP 11.7 The mineral gypsum ($CaSO_4 \cdot 2H_2O$) is a white crystalline solid that is the principal component of building materials like plaster and sheetrock. The calcium and sulfate ions are attracted to each other predominantly by _____ , while the calcium ions and water molecules are held together by _____ . (**a**) ion–dipole interactions, hydrogen bonding (**b**) dispersion forces, hydrogen bonding (**c**) ionic bonding, hydrogen bonding (**d**) ionic bonding, ion–dipole interactions (**e**) ionic bonding, dipole–dipole interactions

EP 11.8 Arrange the following substances from the lowest to highest normal boiling point: calcium oxide, CaO; oxygen, O_2; hydrogen, H_2; formaldehyde, H_2CO; and water, H_2O.

(**a**) $CaO < H_2 < O_2 < H_2CO < H_2O$

(**b**) $H_2 < O_2 < H_2CO < CaO < H_2O$

(**c**) $H_2 < O_2 < H_2CO < H_2O < CaO$

(**d**) $O_2 < H_2 < H_2CO < H_2O < CaO$

(**e**) $H_2 < O_2 < H_2O < H_2CO < CaO$

EP 11.9 List the substances Cl_2, CH_4, and CH_3CH_2COOH in order of increasing strength of intermolecular attractions.

(**a**) $CH_4 < CH_3CH_2COOH < Cl_2$

(**b**) $Cl_2 < CH_3CH_2COOH < CH_4$

(**c**) $CH_4 < Cl_2 < CH_3CH_2COOH$

(**d**) $CH_3CH_2COOH < Cl_2 < CH_4$

(**e**) $Cl_2 < CH_4 < CH_3CH_2COOH$

EP 11.10 The viscosity of ethylene glycol ($HOCH_2CH_2OH$) is approximately 8 times larger than that of 1-propanol ($CH_3CH_2CH_2OH$) (1.6×10^{-2} kg/m-s vs. 2.0×10^{-3} kg/m-s). Which of the following best explains the higher viscosity of ethylene glycol? (**a**) The molecules in ethylene glycol are held together by much stronger dispersion forces than 1-propanol. (**b**) The ethylene glycol molecule can form more hydrogen bonds per molecule than 1-propanol. (**c**) The ethylene glycol molecules are more flexible than the 1-propanol molecules. (**d**) Molecules of ethylene glycol are nonpolar, while those of 1-propanol are polar.

EP 11.11 Which of the following properties *does not* increase as the strength of the intermolecular forces increases. (**a**) viscosity (**b**) surface tension (**c**) vapor pressure (**d**) enthalpy of vaporization (**e**) normal boiling point

EP 11.12 When water is placed in a narrow glass capillary a _____ meniscus will form due to the fact that the interactions between the water molecules and the glass of the capillary are _____ than the interactions between water molecules and other water molecules. (**a**) inverted U-shaped, weaker (**b**) U-shaped, weaker (**c**) inverted U-shaped, stronger (**d**) U-shaped, stronger

EP 11.13 For a given substance, the enthalpy of sublimation will be _____ and its magnitude will be _____ than the heat of vaporization. (**a**) exothermic, smaller (**b**) exothermic, larger (**c**) endothermic, smaller (**d**) endothermic, larger

EP 11.14 What information about water is needed to calculate the enthalpy change associated with converting 1 mol $H_2O(g)$ at 100 °C to 1 mol $H_2O(l)$ at 80 °C? (**a**) heat of fusion (**b**) heat of vaporization (**c**) heat of vaporization and specific heat of $H_2O(g)$ (**d**) heat of vaporization and specific heat of $H_2O(l)$ (**e**) heat of fusion and specific heat of $H_2O(l)$

EP 11.15 What characteristic of a supercritical fluid is similar to a gas but unlike a liquid? (**a**) the spacing of the molecules (**b**) its ability to flow (**c**) its ability to assume the shape and volume of its container (**d**) its ability to dissolve other substances

EP 11.16 In the mountains where the atmospheric pressure is less than it is at sea level, an open container of water heated over a campfire will begin to boil when _____ . (**a**) its temperature exceeds the critical temperature (**b**) its vapor pressure equals the pressure of the surrounding atmosphere (**c**) its temperature reaches 100 °C (**d**) enough energy is supplied to break covalent bonds between oxygen and hydrogen

EP 11.17 A liquid and its vapor are in dynamic equilibrium in a cylinder fitted with a weightless, frictionless piston as shown here. If the temperature of the apparatus is 56.2 °C and the external pressure is 1.00 atm, which of the following statements about the system is or are true?

(**i**) Molecules are condensing and vaporizing at the same rate.

(**ii**) The normal boiling point of the substance is 56.2 °C.

(**iii**) Because the temperature is not changing, all molecules in the gas phase have the same kinetic energy.

vapor

liquid

(**a**) only i (**b**) only ii (**c**) only iii (**d**) both i and ii (**e**) All three statements are true.

EP 11.18 A liquid and its vapor are in dynamic equilibrium in a cylinder fitted with a weightless, frictionless piston as shown in the previous problem, where the temperature was 56.2 °C and the external pressure was 1.00 atm. If enough heat is added to the system to vaporize half of the remaining liquid, and equilibrium is reestablished, which of the following statements is *false*?

(a) The temperature of the liquid is the same as it was before the heat was added.

(b) The vapor pressure is the same as it was before the heat was added.

(c) The number of molecules in the gas phase is greater than it was before the heat was added.

(d) The piston is at a higher position than it was before the heat was added.

(e) The average kinetic energy of the molecules in the gas phase is greater than it was before the heat was added.

EP 11.19 At 50 °C, substances A and B (both liquids) have vapor pressures of 270 torr and 41 torr, respectively. Given this information, which of the following statements is *false*? **(a)** The normal boiling points of both substances are greater than 50 °C. **(b)** The intermolecular forces in substance B are stronger than those in substance A. **(c)** The normal boiling point of substance A is higher than that of substance B. **(d)** Substance A is more volatile than substance B.

EP 11.20 Which of the following substances would you predict will have the lowest vapor pressure at a temperature of 300 K?

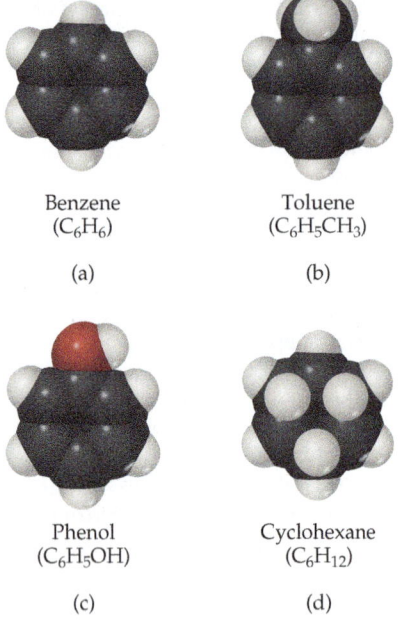

Benzene
(C_6H_6)

(a)

Toluene
($C_6H_5CH_3$)

(b)

Phenol
(C_6H_5OH)

(c)

Cyclohexane
(C_6H_{12})

(d)

EP 11.21 In which quadrant of a phase diagram would you be most likely to find a supercritical fluid? **(a)** high temperature, high pressure **(b)** high temperature, low pressure **(c)** low temperature, low pressure **(d)** low temperature, high pressure

EP 11.22 Based on the phase diagram for methane, what happens to methane as it is heated from −250 to 0 °C at a pressure of 10^{-2} atm?

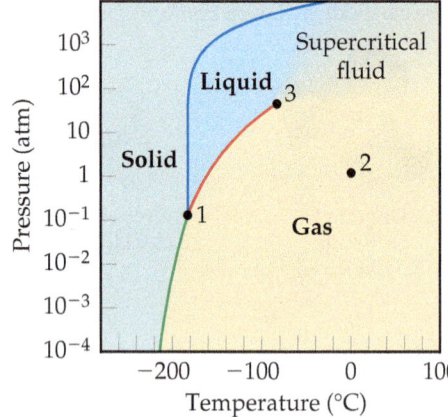

(a) It sublimes at about −200 °C. **(b)** It melts at about −200 °C. **(c)** It boils at about −200 °C. **(d)** It condenses at about −200 °C. **(e)** It reaches the triple point at about −200 °C.

EP 11.23 Arrange the following phases from the least to the greatest amount of freedom of motion the molecules possess: solid, liquid, nematic liquid crystal, smectic A liquid crystal.

(a) liquid < nematic liquid crystal < smectic A liquid crystal < solid

(b) solid < smectic A liquid crystal < nematic liquid crystal < liquid

(c) solid < liquid < nematic liquid crystal < smectic A liquid crystal

(d) solid < nematic liquid crystal < smectic A liquid crystal < liquid

(e) nematic liquid crystal < smectic A liquid crystal < solid < liquid

EP 11.24 Liquid crystalline phases tend to be found for substances with molecules that are _____ and _____ . **(a)** polar, flexible **(b)** nonpolar, flexible **(c)** polar, rigid **(d)** polar, flexible

Exercises

Visualizing Concepts

11.1 **(a)** Does the diagram best describe a crystalline solid, a liquid, or a gas? **(b)** Explain. [Section 11.1]

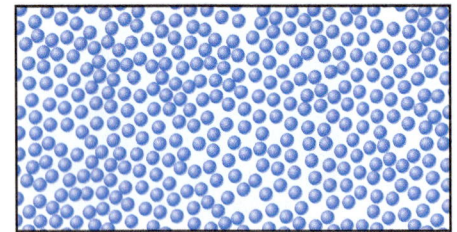

11.2 **(a)** Which kind of intermolecular attractive force is shown in each case here?

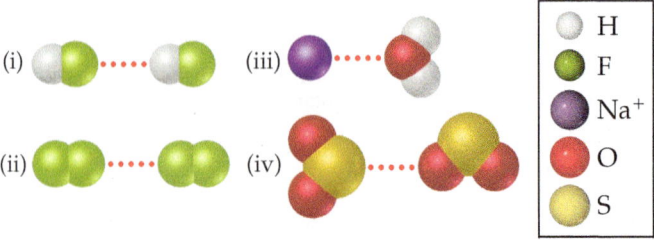

(b) Predict which of the four interactions is the weakest. [Section 11.2]

11.3 (**a**) Which of the molecules shown here can form dipole–dipole interactions with other molecules of the same type? (**b**) Which are capable of forming hydrogen bonds with other molecules of the same type? [Section 11.2]

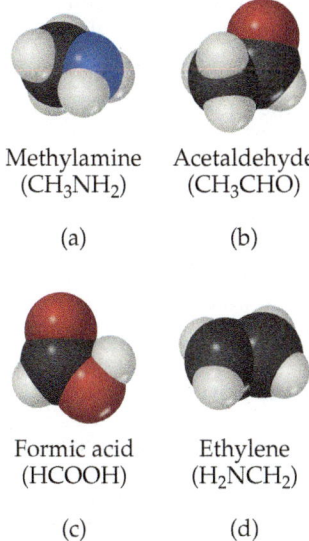

Methylamine
(CH_3NH_2)

(a)

Acetaldehyde
(CH_3CHO)

(b)

Formic acid
(HCOOH)

(c)

Ethylene
(H_2NCH_2)

(d)

11.4 (**a**) Do you expect the viscosity of glycerol, $C_3H_5(OH)_3$, to be larger or smaller than that of 1-propanol, C_3H_7OH?

(**b**) Explain. [Section 11.3]

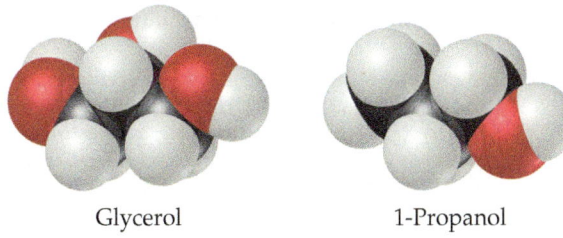

Glycerol   1-Propanol

11.5 If 42.0 kJ of heat is added to a 32.0-g sample of liquid methane under 1 atm of pressure at a temperature of −170 °C, what are the final state and temperature of the methane once the system equilibrates? Assume no heat is lost to the surroundings. The normal boiling point of methane is −161.5 °C. The specific heats of liquid and gaseous methane are 3.48 and 2.22 J/g-K, respectively. [Section 11.4]

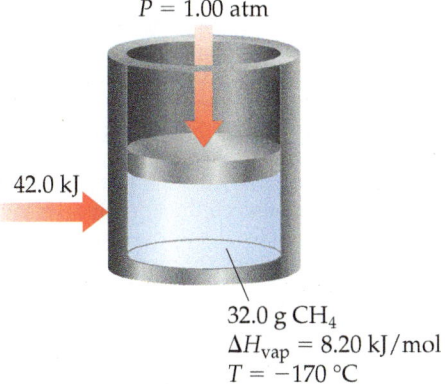

$P = 1.00$ atm

42.0 kJ

32.0 g CH_4
$\Delta H_{vap} = 8.20$ kJ/mol
$T = -170$ °C

11.6 If 54.0 kJ of heat is gradually added to a 1.00 mol sample of liquid ethanol with an initial temperature of 298 K, the heating curve shown here is obtained. Use this graph to answer the following questions. (**a**) What is the boiling point? (**b**) What is the enthalpy of vaporization in kJ/mol? (**c**) Is the specific heat of liquid ethanol greater than or less than the specific heat of gaseous ethanol? [Section 11.4]

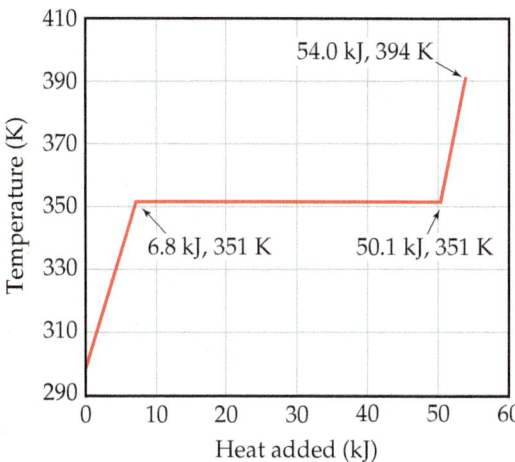

11.7 Using this graph of CS_2 data, determine (**a**) the approximate vapor pressure of CS_2 at 30 °C, (**b**) the temperature at which the vapor pressure equals 300 torr, (**c**) the normal boiling point of CS_2. [Section 11.5]

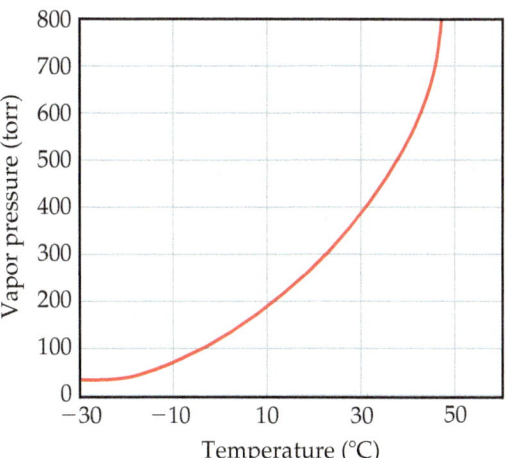

11.8 The molecules

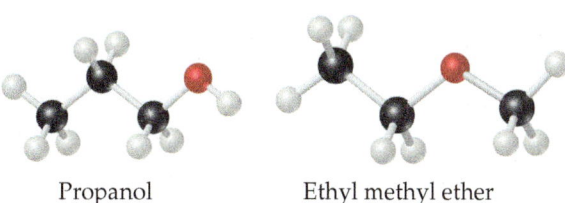

Propanol   Ethyl methyl ether

have the same molecular formula (C_3H_8O) but different chemical structures. (**a**) Which molecule(s), if any, can engage in hydrogen bonding? (**b**) Which molecule do you expect to have a larger dipole moment? (**c**) One of these molecules has a normal boiling point of 97.2 °C, while the other one has a normal boiling point of 10.8 °C. Assign each molecule to its normal boiling point. [Sections 11.2 and 11.5]

11.9 The phase diagram of a hypothetical substance is

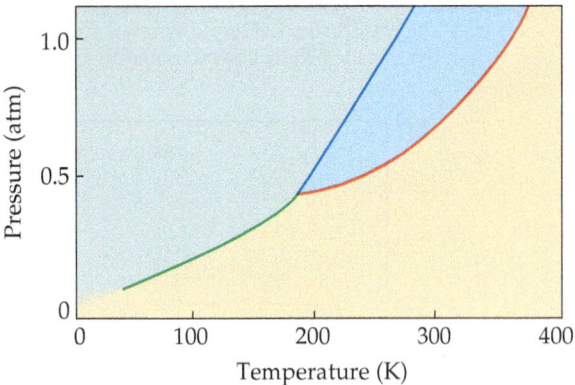

(a) Estimate the normal boiling point and freezing point of the substance. (b) What is the physical state of the substance under the following conditions? (i) $T = 150\,K, P = 0.2\,atm$; (ii) $T = 100\,K, P = 0.8\,atm$; (iii) $T = 300\,K, P = 1.0\,atm$. (c) What is the triple point of the substance? [Section 11.6]

11.10 At three different temperatures, T_1, T_2, and T_3, the molecules in a liquid crystal align in these ways:

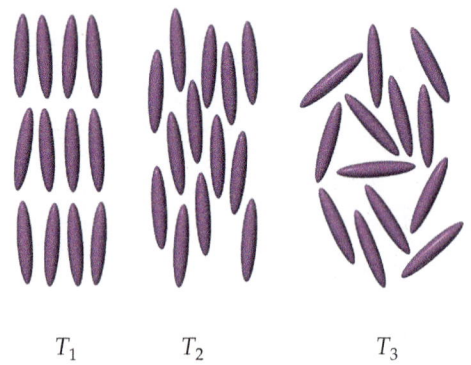

(a) At which temperature or temperatures is the substance in a liquid crystalline state? At those temperatures, which type of liquid crystalline phase is depicted? (b) Based on your answer to part (a) which temperature T_1, T_2, or T_3 is the highest? [Section 11.7]

Molecular Comparisons of Gases, Liquids, and Solids (Section 11.1)

11.11 List the three states of matter in order of (a) increasing molecular disorder and (b) increasing intermolecular attraction. (c) Which state of matter is most easily compressed?

11.12 (a) How does the average kinetic energy of molecules compare with the average energy of attraction between molecules in solids, liquids, and gases? (b) Why does increasing the temperature cause a solid substance to change in succession from a solid to a liquid to a gas? (c) What happens to a gas if you put it under extremely high pressure?

11.13 As a metal such as lead melts, what happens to (a) the average kinetic energy of the atoms and (b) the average distance between the atoms?

11.14 At room temperature, Si is a solid, CCl_4 is a liquid, and Ar is a gas. List these substances in order of (a) increasing intermolecular energy of attraction and (b) increasing boiling point.

11.15 At standard temperature and pressure, the molar volumes of Cl_2 and NH_3 gases are 22.06 and 22.40 L, respectively.

(a) Given the different molecular weights, dipole moments, and molecular shapes, why are their molar volumes nearly the same? (b) On cooling to 160 K, both substances form crystalline solids. Do you expect the molar volumes to decrease or increase on cooling the gases to 160 K? (c) The densities of crystalline Cl_2 and NH_3 at 160 K are 2.02 and 0.84 g/cm^3, respectively. Calculate their molar volumes. (d) Are the molar volumes in the solid state as similar as they are in the gaseous state? (e) Would you expect the molar volumes in the liquid state to be closer to those in the solid or gaseous state?

11.16 Benzoic acid, C_6H_5COOH, melts at 122 °C. The density in the liquid state at 130 °C is 1.08 g/cm^3. The density of solid benzoic acid at 15 °C is 1.266 g/cm^3. (a) In which of these two states is the average distance between molecules greater? (b) If you converted a cubic centimeter of liquid benzoic acid into a solid, would the solid take up more, or less, volume than the original cubic centimeter of liquid?

Intermolecular Forces (Section 11.2)

11.17 (a) Which type of intermolecular attractive force operates between all molecules? (b) Which type of intermolecular attractive force operates only between polar molecules? (c) Which type of intermolecular attractive force operates only between the hydrogen atom of a polar bond and a nearby small electronegative atom?

11.18 (a) Which is generally stronger, intermolecular interactions or intramolecular interactions? (b) Which of these kinds of interactions are broken when a liquid is converted to a gas?

11.19 Describe the intermolecular forces that must be overcome to convert these substances from a liquid to a gas: (a) SO_2, (b) CH_3COOH, (c) H_2S.

11.20 Which type of intermolecular force accounts for each of these differences? (a) CH_3OH boils at 65 °C; CH_3SH boils at 6 °C. (b) Xe is a liquid at atmospheric pressure and 120 K, whereas Ar is a gas under the same conditions. (c) Kr, atomic weight 84 amu, boils at 120.9 K, whereas Cl_2, molecular weight 71 amu, boils at 238 K. (d) Acetone boils at 56 °C, whereas 2-methylpropane boils at −12 °C.

$$\underset{\text{Acetone}}{CH_3\!-\!\overset{\displaystyle O}{\overset{\displaystyle \|}{C}}\!-\!CH_3} \qquad \underset{\text{2-Methylpropane}}{CH_3\!-\!\overset{\displaystyle CH_3}{\overset{\displaystyle |}{CH}}\!-\!CH_3}$$

11.21 (a) List the following molecules in order of increasing polarizability: $GeCl_4$, CH_4, $SiCl_4$, SiH_4, and $GeBr_4$. (b) Predict the order of boiling points of the substances in part (a).

11.22 True or false: (a) Dispersion forces become stronger as molecules become more polarizable. (b) For the noble gases the dispersion forces decrease while the boiling points increase as you go down the column in the periodic table. (c) In terms of the total attractive forces for a given substance, dipole–dipole interactions, when present, are always greater than dispersion forces. (d) All other factors being the same, dispersion forces between linear molecules are greater than those between molecules whose shapes are nearly spherical. (e) The larger the atom, the more polarizable it is.

11.23 Which member in each pair has the greater dispersion forces? (a) H_2O or H_2S (b) CO_2 or CO (c) SiH_4 or GeH_4

11.24 Which member in each pair has the stronger intermolecular dispersion forces? (a) Br_2 or O_2 (b) $CH_3CH_2CH_2CH_2SH$ or $CH_3CH_2CH_2CH_2CH_2SH$ (c) $CH_3CH_2CH_2Cl$ or $(CH_3)_2CHCl$

11.25 Given the molecular structures of butane ($CH_3CH_2CH_2CH_3$) and 2-methylpropane [$CH_3CH(CH_3)_2$] shown here, predict which substance will have the higher boiling point. What factor is primarily responsible for the difference in boiling point? Is it the difference in polarity, shape, or mass of the two molecules?

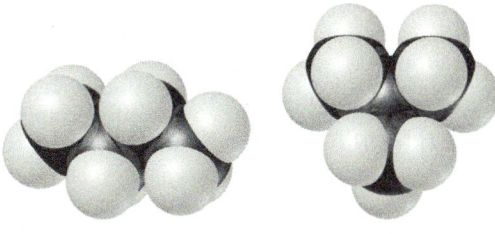

Butane   2-Methylpropane

11.26 Given the molecular structures of propyl alcohol (CH_3CH_2 CH_2OH) and isopropyl alcohol [($CH_3)_2CHOH$)] shown here, predict which substance will have the higher boiling point. What factor is primarily responsible for the difference in boiling point? Is it the difference in polarity, shape, or mass of the two molecules?

Propyl alcohol   Isopropyl alcohol

11.27 (**a**) What atoms must a molecule contain to participate in hydrogen bonding with other molecules of the same kind? (**b**) Which of the following molecules can form hydrogen bonds with other molecules of the same kind: CH_3F, CH_3NH_2, CH_3OH, CH_3Br?

11.28 Rationalize the difference in boiling points in each pair: (**a**) HF (20 °C) and HCl (−85 °C), (**b**) $CHCl_3$ (61 °C) and $CHBr_3$ (150 °C), (**c**) Br_2 (59 °C) and ICl (97 °C).

11.29 Ethylene glycol ($HOCH_2CH_2OH$), the major substance in antifreeze, has a normal boiling point of 198 °C. By comparison, ethyl alcohol (CH_3CH_2OH) boils at 78 °C at atmospheric pressure. Ethylene glycol dimethyl ether ($CH_3OCH_2CH_2OCH_3$) has a normal boiling point of 83 °C, and ethyl methyl ether ($CH_3CH_2OCH_3$) has a normal boiling point of 11 °C. (**a**) Explain why replacement of a hydrogen on the oxygen by a CH_3 group generally results in a lower boiling point. (**b**) What are the major factors responsible for the difference in boiling points of the two ethers?

11.30 Based on the type or types of intermolecular forces, predict the substance in each pair that has the higher boiling point: (**a**) propane (C_3H_8) or *n*-butane (C_4H_{10}), (**b**) diethyl ether ($CH_3CH_2OCH_2CH_3$) or 1-butanol ($CH_3CH_2CH_2CH_2OH$), (**c**) sulfur dioxide (SO_2) or sulfur trioxide (SO_3), (**d**) phosgene (Cl_2CO) or formaldehyde (H_2CO).

11.31 Look up and compare the normal boiling points and normal melting points of H_2O and H_2S. Based on these physical properties, which substance has stronger intermolecular forces? What kinds of intermolecular forces exist for each molecule?

11.32 Carbon tetrachloride, CCl_4, and chloroform, $CHCl_3$, are common organic liquids. Carbon tetrachloride's normal boiling point is 77 °C, whereas chloroform's normal boiling point is 61 °C. Which statement is the best explanation of these data? (**a**) Chloroform can hydrogen-bond, but carbon tetrachloride cannot. (**b**) Carbon tetrachloride has a larger

dipole moment than chloroform. (**c**) Carbon tetrachloride is more polarizable than chloroform.

11.33 A number of salts containing the tetrahedral polyatomic anion, BF_4^-, are ionic liquids, whereas salts containing the somewhat larger tetrahedral ion SO_4^{2-} do not form ionic liquids. Explain this observation.

11.34 The generic structural formula for a 1-alkyl-3-methylimidazolium cation is

$$\left[\begin{array}{c} H \\ H_3C-N \overset{C}{\underset{C=C}{\overset{\|}{\diagdown}}} N-R \\ H \quad H \end{array} \right]^+$$

where R is a $—CH_2(CH_2)_nCH_3$ alkyl group. The melting points of the salts that form between the 1-alkyl-3-methylimidazolium cation and the PF_6^- anion are as follows: R = CH_2CH_3 (m.p. = 60 °C), R = $CH_2CH_2CH_3$ (m.p. = 40 °C), R = $CH_2CH_2CH_2CH_3$ (m.p. = 10 °C), and R = $CH_2CH_2CH_2CH_2CH_2CH_3$ (m.p. = −61 °C). Why does the melting point decrease as the length of alkyl group increases?

Select Properties of Liquids (Section 11.3)

11.35 (**a**) What is the relationship between surface tension and temperature? (**b**) What is the relationship between viscosity and temperature? (**c**) Why do substances with high surface tension also tend to have high viscosities?

11.36 Based on their composition and structure, list CH_2Cl_2, $CH_3CH_2CH_3$, and CH_3CH_2OH in order of (**a**) increasing intermolecular forces, (**b**) increasing viscosity, (**c**) increasing surface tension.

11.37 Liquids can interact with flat surfaces just as they can with capillary tubes; the cohesive forces within the liquid can be stronger or weaker than the adhesive forces between liquid and surface:

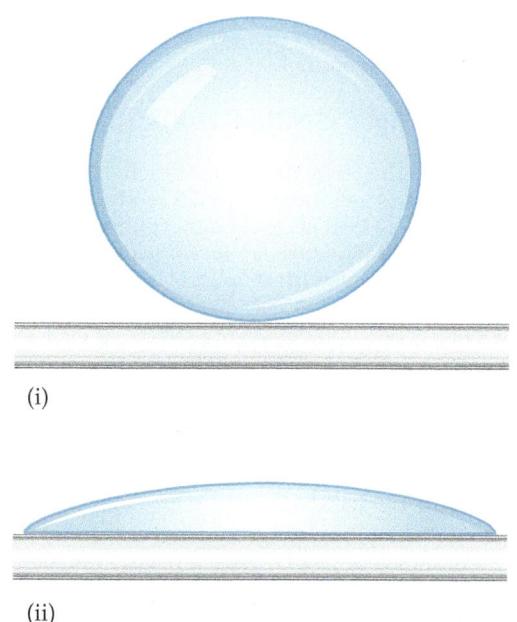

(i)

(ii)

(**a**) In which of these diagrams, i or ii, do the adhesive forces between surface and liquid exceed the cohesive forces within the liquid? (**b**) Which of these diagrams, i or ii, represents what happens when water is on a nonpolar surface? (c) Which of these diagrams, i or ii, represents what happens when water is on a polar surface?

11.38 The ionic liquid $(bmim^+)(PF_6^-)$ has a melting point of 10 °C and a surface tension of approximately $4.5 \times 10^{-2} J/m^2$ at room temperature. By comparison, the melting point of H_2O is 0 °C and its surface tension is $7.3 \times 10^{-2} J/m^2$ at room temperature. What is the most plausible explanation for the higher surface tension of water? (i) hydrogen bonding interactions are stronger than ion–ion interactions. (ii) Surface tension increases as molecules become smaller and more symmetric because the number of neighboring molecules in the liquid state increases. (iii) As the molecules become smaller, they can escape into the gas phases more easily, leading to an increase in surface tension.

11.39 The boiling points, surface tensions, and viscosities of water and several alcohols are listed in the following table:

	Boiling Point (°C)	Surface Tension (J/m^2)	Viscosity (kg/m-s)
Water, H_2O	100	7.3×10^{-2}	0.9×10^{-3}
Ethanol, CH_3CH_2OH	78	2.3×10^{-2}	1.1×10^{-3}
Propanol, $CH_3CH_2CH_2OH$	97	2.4×10^{-2}	2.2×10^{-3}
n-Butanol, $CH_3CH_2CH_2CH_2OH$	117	2.6×10^{-2}	2.6×10^{-3}
Ethylene glycol, $HOCH_2CH_2OH$	197	4.8×10^{-2}	26×10^{-3}

(a) From ethanol to propanol to n-butanol the boiling points, surface tensions, and viscosities all increase. What is the reason for this increase? (b) How do you explain the fact that propanol and ethylene glycol have similar molecular weights (60 vs. 62 amu), yet the viscosity of ethylene glycol is more than 10 times larger than propanol? (c) How do you explain the fact that water has the highest surface tension but the lowest viscosity?

11.40 (a) Would you expect the viscosity of n-pentane, $CH_3CH_2CH_2CH_2CH_3$, to be larger or smaller than the viscosity of n-hexane, $CH_3CH_2CH_2CH_2CH_2CH_3$? (b) Would you expect the viscosity of neopentane, $(CH_3)_4C$, to be smaller or larger than that of n-pentane? (See Figure 11.5 to see the shapes of these molecules.)

Phase Changes (Section 11.4)

11.41 Name the phase transition in each of the following situations and indicate whether it is exothermic or endothermic: (a) When ice is heated, it turns to water. (b) Wet clothes dry on a warm summer day. (c) Frost appears on a window on a cold winter day. (d) Droplets of water appear on a cold glass of lemonade.

11.42 Name the phase transition in each of the following situations and indicate whether it is exothermic or endothermic: (a) Bromine vapor turns to bromine liquid as it is cooled. (b) Crystals of iodine disappear from an evaporating dish as they stand in a fume hood. (c) Rubbing alcohol in an open container slowly disappears. (d) Molten lava from a volcano turns into solid rock.

11.43 (a) What phase change is represented by the "heat of fusion" of a substance? (b) Is the heat of fusion endothermic or exothermic? (c) If you compare a substance's heat of fusion to its heat of vaporization, which one is generally larger?

11.44 Ethyl chloride (C_2H_5Cl) boils at 12 °C. When liquid C_2H_5Cl under pressure is sprayed on a room-temperature (25 °C) surface in air, the surface is cooled considerably. (a) What does this observation tell us about the specific heat of $C_2H_5Cl(g)$

as compared with that of $C_2H_5Cl(l)$? (b) Assume that the heat lost by the surface is gained by the ethyl chloride. What enthalpies must you consider if you were to calculate the final temperature of the surface?

11.45 For many years, drinking water has been cooled in hot climates by evaporating it from the surfaces of canvas bags or porous clay pots. How many grams of water can be cooled from 35 to 20 °C by the evaporation of 60 g of water? (The heat of vaporization of water in this temperature range is 2.4 kJ/g. The specific heat of water is 4.18 J/g-K.)

11.46 Compounds like CCl_2F_2 are known as chlorofluorocarbons, or CFCs. These compounds were once widely used as refrigerants but have largely been replaced by compounds that are believed to be less harmful to the environment. The heat of vaporization of CCl_2F_2 is 289 J/g. What mass of this substance must evaporate to freeze 200 g of water initially at 15 °C? (The heat of fusion of water is 334 J/g; the specific heat of water is 4.18 J/g-K.)

11.47 Ethanol (C_2H_5OH) melts at −114 °C and boils at 78 °C. The enthalpy of fusion of ethanol is 5.02 kJ/mol, and its enthalpy of vaporization is 38.56 kJ/mol. The specific heats of solid and liquid ethanol are 0.97 and 2.3 J/g-K, respectively. (a) How much heat is required to convert 42.0 g of ethanol at 35 °C to the vapor phase at 78 °C? (b) How much heat is required to convert the same amount of ethanol at −155 °C to the vapor phase at 78 °C?

11.48 The fluorocarbon compound $C_2Cl_3F_3$ has a normal boiling point of 47.6 °C. The specific heats of $C_2Cl_3F_3(l)$ and $C_2Cl_3F_3(g)$ are 0.91 and 0.67 J/g-K, respectively. The heat of vaporization for the compound is 27.49 kJ/mol. Calculate the heat required to convert 35.0 g of $C_2Cl_3F_3$ from a liquid at 10.00 °C to a gas at 105.00 °C.

11.49 Indicate whether each statement is true or false: (a) The critical pressure of a substance is the pressure at which it turns into a solid at room temperature. (b) The critical temperature of a substance is the highest temperature at which the liquid phase can form. (c) Generally speaking, the higher the critical temperature of a substance, the lower its critical pressure. (d) In general, the stronger the intermolecular forces in a substance, the higher its critical temperature and pressure.

11.50 The critical temperatures and pressures of a series of halogenated methanes are as follows:

Compound	CCl_3F	CCl_2F_2	$CClF_3$	CF_4
Critical temperature (K)	471	385	302	227
Critical pressure (atm)	43.5	40.6	38.2	37.0

(a) List the intermolecular forces that occur for each compound. (b) Predict the order of increasing intermolecular attraction, from least to most, for this series of compounds. (c) Predict the critical temperature and pressure for CCl_4 based on the trends in this table. Look up the experimentally determined critical temperatures and pressures for CCl_4, using a source such as the *CRC Handbook of Chemistry and Physics*, and suggest a reason for any discrepancies.

Vapor Pressure (Section 11.5)

11.51 Which of the following affects the vapor pressure of a liquid? (a) Volume of the liquid (b) surface area (c) intermolecular attractive forces (d) temperature (e) density of the liquid

11.52 Acetone (H_3CCOCH_3) has a boiling point of 56 °C. Based on the data given in Figure 11.24, would you expect acetone to have a higher or lower vapor pressure than ethanol at 25 °C?

11.53 (a) Arrange the following substances in order of increasing volatility: CH_4, CBr_4, CH_2Cl_2, CH_3Cl, $CHBr_3$, and CH_2Br_2.

(b) How do the boiling points vary through this series? **(c)** Explain your answer to part (b) in terms of intermolecular forces.

11.54 True or false: **(a)** CBr_4 is more volatile than CCl_4. **(b)** CBr_4 has a higher boiling point than CCl_4. **(c)** CBr_4 has weaker intermolecular forces than CCl_4. **(d)** CBr_4 has a higher vapor pressure at the same temperature than CCl_4.

11.55 **(a)** Two pans of water are on different burners of a stove. One pan of water is boiling vigorously, while the other is boiling gently. What can be said about the temperature of the water in the two pans? **(b)** Two containers of water, one large and the other small, are at the same temperature. What can be said about the relative vapor pressures of the water in the two containers?

11.56 You are high up in the mountains and boil water to make some tea. However, when you drink your tea, it is not as hot as it should be. You try again and again, but the water is just not hot enough to make a hot cup of tea. Which is the best explanation for this result? **(a)** High in the mountains, it is probably very dry, and so the water is rapidly evaporating from your cup and cooling it. **(b)** High in the mountains, it is probably very windy, and so the water is rapidly evaporating from your cup and cooling it. **(c)** High in the mountains, the air pressure is significantly less than 1 atm, so the boiling point of water is much lower than at sea level. **(d)** High in the mountains, the air pressure is significantly less than 1 atm, so the boiling point of water is much higher than at sea level.

11.57 Using the vapor–pressure curves in Figure 11.24, **(a)** estimate the boiling point of ethanol at an external pressure of 200 torr, **(b)** estimate the external pressure at which ethanol will boil at 60 °C, **(c)** estimate the boiling point of diethyl ether at 400 torr, **(d)** estimate the external pressure at which diethyl ether will boil at 40 °C.

11.58 Appendix B lists the vapor pressure of water at various external pressures. **(a)** Plot the data in Appendix B, vapor pressure (torr) versus temperature (°C). From your plot, estimate the vapor pressure of water at body temperature, 37 °C. **(b)** Explain the significance of the data point at 760.0 torr and 100 °C. **(c)** A city at an altitude of 5000 ft above sea level has a barometric pressure of 633 torr. To what temperature would you have to heat water to boil it in this city? **(d)** A city at an altitude of 500 ft below sea level would have a barometric pressure of 774 torr. To what temperature would you have to heat water to boil it in this city?

Phase Diagrams (Section 11.6)

11.59 In each of the following regions of a generic phase diagram is it possible for the liquid to be the most stable phase? **(a)** pressures greater than the critical pressure **(b)** temperatures greater than the critical temperature **(c)** pressures less than the pressure of the triple point **(d)** temperatures less than the temperature of the triple point

11.60 **(a)** Of the three conventional states of matter—solid, liquid and gas—which two have the most similar densities? **(b)** In a typical phase diagram, which boundary is the least affected by changes in pressure, the vapor pressure curve, the melting curve, or the sublimation curve? **(c)** Is there a correlation between your answers to parts (a) and (b)?

11.61 Referring to Figure 11.28, describe all the phase changes that would occur in each of the following cases: **(a)** Water vapor originally at 0.005 atm and −0.5 °C is slowly compressed at constant temperature until the final pressure is 20 atm. **(b)** Water originally at 100.0 °C and 0.50 atm is cooled at constant pressure until the temperature is −10 °C.

11.62 Referring to Figure 11.29, describe the phase changes (and the temperatures at which they occur) when CO_2 is heated

from −80 to −20 °C at **(a)** a constant pressure of 3 atm and **(b)** a constant pressure of 6 atm.

11.63 The phase diagram for neon is

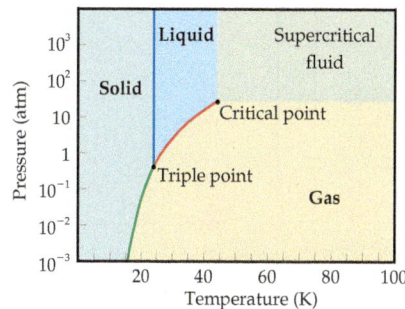

Use the phase diagram to answer the following questions. **(a)** What is the approximate value of the normal melting point? **(b)** Over what pressure range will solid neon sublime? **(c)** At room temperature ($T = 25$ °C) can neon be liquefied by compressing it?

11.64 Use the phase diagram of neon to answer the following questions. **(a)** What is the approximate value of the normal boiling point? **(b)** What can you say about the strength of the intermolecular forces in neon and argon based on the critical points of Ne and Ar (see Table 11.5.)?

11.65 The fact that water on Earth can readily be found in all three states (solid, liquid, and gas) is in part a consequence of the fact that the triple point of water ($T = 0.01$ °C, $P = 0.006$ atm) falls within a range of temperatures and pressures found on Earth. Saturn's largest moon Titan has a considerable amount of methane in its atmosphere. The conditions on the surface of Titan are estimated to be $P = 1.6$ atm and $T = -178$ °C. As seen from the phase diagram of methane (Figure 11.30), these conditions are not far from the triple point of methane, raising the tantalizing possibility that solid, liquid, and gaseous methane can be found on Titan. **(a)** In what state would you expect to find methane on the surface of Titan? **(b)** On moving upward through the atmosphere, the pressure will decrease. If we assume that the temperature does not change, what phase change would you expect to see as we move away from the surface?

11.66 At 25 °C gallium is a solid with a density of 5.91 g/cm³ and a melting point, 29.8 °C, just slightly above room temperature. The density of liquid gallium just above the melting point is 6.1 g/cm³. Based on this information, what unusual feature would you expect to find in the phase diagram of gallium?

Liquid Crystals (Section 11.7)

11.67 In terms of the arrangement and freedom of motion of the molecules, how are the nematic liquid crystalline phase and an ordinary liquid phase similar? How are they different?

11.68 What observations made by Reinitzer on cholesteryl benzoate suggested that this substance possesses a liquid crystalline phase?

11.69 Indicate whether each statement is true or false: **(a)** The liquid crystal state is another phase of matter, just like solid, liquid, and gas. **(b)** Liquid crystalline molecules are generally spherical in shape. **(c)** Molecules that exhibit a liquid crystalline phase do so at well-defined temperatures and pressures. **(d)** Molecules that exhibit a liquid crystalline phase show weaker-than-expected intermolecular forces. **(e)** Molecules containing only carbon and hydrogen are likely to form liquid crystalline phases. **(f)** Molecules can exhibit more than one liquid crystalline phase.

11.70 Two heating curves, A and B, are shown. In both cases, point 1 corresponds to the crystalline solid phase.

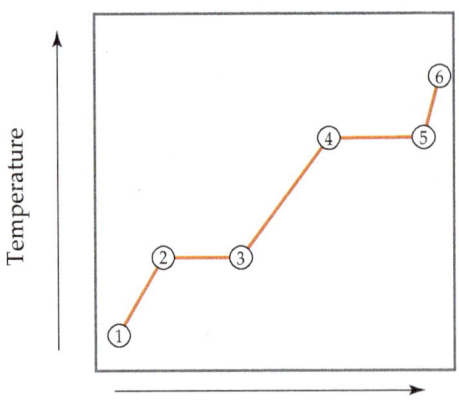

Heat added

A

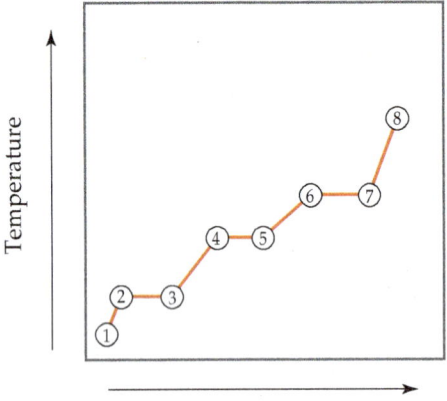

Heat added

B

(a) One of these graphs shows data for a liquid crystalline material. Which one? (b) In graph A, what process does the 2–3 line segment correspond to? (c) In graph B, what process does the 2–3 line segment correspond to? (d) In graph A, what process does the 3–4 line segment correspond to? (e) In graph B, what process does the 3–4 line segment correspond to?

11.71 The molecule *p*-azoxyanisole shown here transitions from a solid to a nematic liquid crystalline phase at 118 °C and to a liquid at 135 °C. (a) Will the phase transition that occurs upon heating from 130 °C to 140 °C be exothermic or endothermic? (b) Do you expect the viscosity to increase or decrease on heating from 130 °C to 140 °C?

11.72 It is not uncommon for molecules to adopt more than one type of liquid crystalline phase, with each phase appearing over a different temperature range. The molecule shown below, 4′-octyl-4-biphenylcarbonitrile, is one such example. The solid melts at 24 °C and undergoes further phase transitions at 34 °C and 43 °C. (a) Based on the amount of molecular order found in the various types of liquid crystals, would you expect the phase that is stable between 24 and 34 °C to be a nematic liquid crystal or a smectic A liquid crystal? (b) Could you determine if the phase transition at 43 °C is to the liquid phase or yet another liquid crystal phase from the appearance of the sample?

$CH_3(CH_2)_6CH_2$—⬡—⬡—CN

11.73 In all four liquid crystalline phases shown in Figure 11.32, the long axis of the molecule preferentially orders along one or more specific directions. In three of the four phases the molecules also lose some freedom of translational motion. In which of the four liquid crystalline phases do the molecules retain the freedom to move in all three directions that they possess in the liquid phase: nematic, smectic A, smectic C, or cholesteric?

11.74 In which type of liquid crystal do the molecules show the least amount of order: nematic, smectic A, or cholesteric?

Additional Exercises

11.75 As the intermolecular attractive forces between molecules increase in magnitude, do you expect each of the following to increase or decrease in magnitude? (a) Vapor pressure (b) heat of vaporization (c) boiling point (d) freezing point (e) viscosity (f) surface tension (g) critical temperature

11.76 The table below lists the density of O_2 at various temperatures and at 1 atm. The normal melting point of O_2 is 54 K.

Temperature (K)	Density (mol/L)
60	40.1
70	38.6
80	37.2
90	35.6
100	0.123
120	0.102
140	0.087

(a) Over what temperature range is O_2 a solid? (b) Over what temperature range is O_2 a liquid? (c) Over what temperature range in the table is O_2 a gas? (d) Estimate the normal boiling point of O_2. (e) What intermolecular forces are operative in O_2?

11.77 Suppose you have two colorless molecular liquids, one boiling at −84 °C, the other at 34 °C, and both at atmospheric pressure. Which of the following statements are true? (a) The higher-boiling liquid has stronger intermolecular forces than the lower-boiling liquid. (b) The lower-boiling liquid must consist of nonpolar molecules. (c) The lower-boiling liquid has a lower molecular weight than the higher-boiling liquid. (d) The two liquids have identical vapor pressures at their normal boiling points. (e) At −84 °C both liquids have vapor pressures of 760 mm Hg.

11.78 Two isomers of the planar compound 1,2-dichloroethylene are shown here.

cis isomer   trans isomer

(a) Which of the two isomers will have the stronger dipole–dipole forces? (b) One isomer has a boiling point of 60.3 °C and the other 47.5 °C. Which isomer has which boiling point?

11.79 The following table lists some physical properties of halogenated liquids.

Liquid	Experimental Dipole Moment (D)	Normal Boiling Point (°C)
CH_2F_2	1.93	−52
CH_2Cl_2	1.60	40
CH_2Br_2	1.43	97

Which of the following statements best explains these data? (a) The larger the dipole moment, the stronger the intermolecular forces, in which case the boiling point is lowest for the molecule with the largest dipole moment. (b) The dispersion forces increase from F to Cl to Br; because the boiling point also increases in this order, the dispersion forces must make a far greater contribution to intermolecular interactions than dipole–dipole interactions. (c) The trend in electronegativity is F > Cl > Br; therefore, the most ionic compound (CH_2F_2) has the lowest boiling point, and the most covalent compound (CH_2Br_2) has the highest boiling point. (d) Boiling point increases with molecular weight for these nonpolar compounds.

11.80 The following table lists the normal boiling points of benzene and benzene derivatives.

Compound	Structure	Normal Boiling Point (°C)
C_6H_6 (benzene)		80
C_6H_5Cl (chlorobenzene)	Cl	132
C_6H_5Br (bromobenzene)	Br	156
C_6H_5OH (phenol, or hydroxybenzene)	OH	182

(a) How many of these compounds exhibit dispersion interactions? (b) How many of these compounds exhibit dipole–dipole interactions? (c) How many of these compounds exhibit hydrogen bonding? (d) Why is the boiling point of bromobenzene higher than that of chlorobenzene? (e) Why is the boiling point of phenol the highest of all?

11.81 The DNA double helix (Figure 24.27) at the atomic level looks like a twisted ladder, where the "rungs" of the ladder consist of molecules that are hydrogen-bonded together, while sugar and phosphate groups make up the sides of the ladder. Shown are the structures of the adenine–thymine (AT) "base pair" and the guanine–cytosine (GC) base pair:

Thymine   Adenine

Cytosine   Guanine

You can see that AT base pairs are held together by *two* hydrogen bonds, whereas GC base pairs are held together by *three* hydrogen bonds. Which base pair is more stable to heating? Why?

11.82 Ethylene glycol ($HOCH_2CH_2OH$) and pentane (C_5H_{12}) are both liquids at room temperature and room pressure, and have about the same molecular weight. (a) One of these liquids is much more viscous than the other. Which one do you predict is more viscous? (b) One of these liquids has a much lower normal boiling point (36.1 °C) compared to the other one (198 °C). Which liquid has the lower normal boiling point? (c) One of these liquids is the major component in antifreeze in automobile engines. Which liquid would you expect to be used as antifreeze? (d) One of these liquids is used as a "blowing agent" in the manufacture of polystyrene foam because it is so volatile. Which liquid would you expect to be used as a blowing agent?

11.83 Use the normal boiling points

propane (C_3H_8)	−42.1 °C
butane (C_4H_{10})	−0.5 °C
pentane (C_5H_{12})	36.1 °C
hexane (C_6H_{14})	68.7 °C
heptane (C_7H_{16})	98.4 °C

to estimate the normal boiling point of octane (C_8H_{18}). Explain the trend in the boiling points.

11.84 One of the attractive features of ionic liquids is their low vapor pressure, which in turn tends to make them nonflammable. Why do you think ionic liquids have lower vapor pressures than most room-temperature molecular liquids?

11.85 (a) When you exercise vigorously, you sweat. How does this help your body cool? (b) A flask of water is connected to a vacuum pump. A few moments after the pump is turned on, the water begins to boil. After a few minutes, the water begins to freeze. Explain why these processes occur.

11.86 The following table lists the vapor pressure of hexafluorobenzene (C_6F_6) as a function of temperature:

Temperature (K)	Vapor Pressure (torr)
280.0	32.42
300.0	92.47
320.0	225.1
330.0	334.4
340.0	482.9

(**a**) By plotting these data in a suitable fashion, determine whether the Clausius–Clapeyron equation (Equation 11.1) is obeyed. If it is obeyed, use your plot to determine ΔH_{vap} for C_6F_6. (**b**) Use these data to determine the boiling point of the compound.

11.87 Suppose the vapor pressure of a substance is measured at two different temperatures. (**a**) By using the Clausius–Clapeyron equation (Equation 11.1), derive the following relationship between the vapor pressures, P_1 and P_2, and the absolute temperatures at which they were measured, T_1 and T_2:

$$\ln\frac{P_1}{P_2} = -\frac{\Delta H_{vap}}{R}\left(\frac{1}{T_1} - \frac{1}{T_2}\right)$$

(**b**) Gasoline is a mixture of hydrocarbons, a component of which is octane ($CH_3CH_2CH_2CH_2CH_2CH_2CH_2CH_3$). Octane has a vapor pressure of 13.95 torr at 25 °C and a vapor pressure of 144.78 torr at 75 °C. Use these data and the equation in part (a) to calculate the heat of vaporization of octane. (**c**) By using the equation in part (a) and the data given in part (b), calculate the normal boiling point of octane. Compare your answer to the one you obtained from Exercise 11.83. (**d**) Calculate the vapor pressure of octane at −30 °C.

11.88 Naphthalene ($C_{10}H_8$) is the main ingredient in traditional mothballs. Its normal melting point is 81 °C, its normal boiling point is 218 °C, and its triple point is 80 °C at 1000 Pa. Using these data, construct a phase diagram for naphthalene, labeling all the regions of your diagram.

11.89 A watch with a liquid crystal display (LCD) does not function properly when it is exposed to low temperatures during a trip to Antarctica. Explain why the LCD might not function well at low temperature.

11.90 A particular liquid crystalline substance has the phase diagram shown in the figure. By analogy with the phase diagram for a nonliquid crystalline substance, identify the phase present in each area.

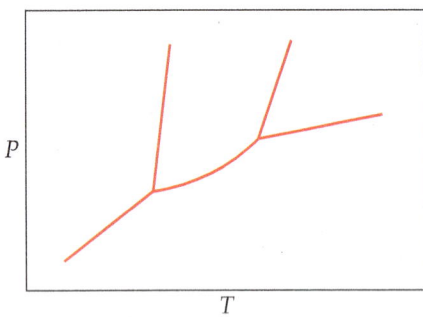

11.91 In Table 11.3, we saw that the viscosity of a series of hydrocarbons increased with molecular weight, doubling from the six-carbon molecule to the ten-carbon molecule. (**a**) The eight-carbon hydrocarbon, octane, has an isomer, isooctane. Would you predict that isooctane would have a larger or smaller viscosity than octane? (**b**) Predict the relative order of boiling points of the hydrocarbons in Table 11.4, from lowest to highest. (**c**) The surface tension of the hydrocarbon liquids in Table 11.4 increases from hexane to decane, but only by a rather small amount (20% overall, compared to the doubling of viscosity). Which of the statements (i)–(iii) is the most likely explanation for this phenomenon? (i) The flexibility of the molecules has a much larger effect on viscosity than on surface tension. (ii) Viscosity only depends on molecular weight, but surface tension depends on molecular weight and on intermolecular forces. (iii) Larger molecules can make larger liquid droplets and therefore have lower surface tension.

2,2,4-Trimethylpentane or isooctane

11.92 Acetone [$(CH_3)_2CO$] is widely used as an industrial solvent. (**a**) Draw the Lewis structure for the acetone molecule and predict the geometry around each carbon atom. (**b**) Is the acetone molecule polar or nonpolar? (**c**) What kinds of intermolecular attractive forces exist between acetone molecules? (**d**) 1-Propanol ($CH_3CH_2CH_2OH$) has a molecular weight that is very similar to that of acetone, yet acetone boils at 56.5 °C and 1-propanol boils at 97.2 °C. Explain the difference.

11.93 The table shown here lists the molar heats of vaporization for several organic compounds. Use specific examples from this list to illustrate how the heat of vaporization varies with (**a**) molar mass, (**b**) molecular shape, (**c**) molecular polarity, and (**d**) hydrogen-bonding interactions. Explain these comparisons in terms of the nature of the intermolecular forces at work. (You may find it helpful to draw out the structural formula for each compound.)

Compound	Heat of Vaporization (kJ/mol)
$CH_3CH_2CH_3$	19.0
$CH_3CH_2CH_2CH_2CH_3$	27.6
$CH_3CHBrCH_3$	31.8
CH_3COCH_3	32.0
$CH_3CH_2CH_2Br$	33.6
$CH_3CH_2CH_2OH$	47.3

11.94 The vapor pressure of ethanol (C_2H_5OH) at 19 °C is 40.0 torr. A 1.00-g sample of ethanol is placed in a 2.00 L container at 19 °C. If the container is closed and the ethanol is allowed to reach equilibrium with its vapor, how many grams of liquid ethanol remain?

11.95 Using information in Appendices B and C, calculate the minimum grams of propane, $C_3H_8(g)$, that must be combusted to provide the energy necessary to convert 5.50 kg of ice at −20 °C to liquid water at 75 °C.

Design an Experiment

Intermolecular forces are very important for predicting the physical properties of molecular substances. Sometimes, however, it is difficult to explain or predict trends in these properties because all the possible intermolecular forces can be operating at the same time, and there is a wide range of energies for these interactions. Ammonia (NH_3), for example, is a gas at $P = 1$ atm, $T = 25$ °C. Ammonia can be liquefied at −33.5 °C ($P = 1$ atm). Methylamine, CH_3NH_2, is a derivative of ammonia and is also a gas at $P = 1$ atm, $T = 25$ °C. Methylamine can be liquefied at −6.4 °C ($P = 1$ atm).

Use these experimental data to determine which molecule—ammonia or methylamine—has the stronger intermolecular attractive interactions, and suggest reasons for your conclusion. Based on your reasons, what other experiments could you do with these molecules, or related molecules, to test your hypothesis?

12

SOLIDS AND MODERN MATERIALS

▲ **LIGHT-EMITTING DIODES LEDS** convert much more of their input energy into light than heat, making them more efficient than standard incandescent light bulbs.

Modern devices like computers and cell phones are built from solids with very specific physical properties. For example, the integrated circuit that is at the heart of many electronic devices is built from semiconductors like silicon, metals like copper, and insulators like hafnium oxide.

Scientists and engineers turn almost exclusively to solids for materials used in many other technologies: *alloys* for magnets and airplane turbines, *semiconductors* for solar cells and light-emitting diodes, and *polymers* for packaging and biomedical applications. Chemists have contributed to the discovery and development of new materials either by inventing new substances or by developing the means to process natural materials to form substances that have specific electrical, magnetic, optical, catalytic, or mechanical properties. In this chapter, we explore the structures and properties of solids. As we do so, we examine some of the solid materials used in modern technology.

12.1 | Classification and Structures of Solids

⚠ **Learning Objectives**

When you finish Section 12.1, you should be able to:

▶ Classify solids based on the nature of the chemical bonding within the solid.

▶ Classify solids based on their chemical formula or physical properties.

▶ Describe the difference between crystalline and amorphous solids.

▶ Given an array of lattice points, identify the lattice vectors and primitive unit cell.

▶ Describe the position of lattice points for primitive cubic, body-centered cubic, and face-centered cubic lattices.

Solids can be as hard as diamond or as soft as wax. Some readily conduct electricity, whereas others do not. The shapes of some solids can easily be manipulated, while others are brittle and resistant to any change in shape. The physical properties as well as the structures of solids are dictated by the types of bonds that hold the atoms in place. We can classify solids according to those bonds (**Figure 12.1**).

Metallic solids consist of metal atoms held together by a delocalized "sea" of collectively shared valence electrons. This form of bonding allows metals to conduct electricity. It is also responsible for the fact that most metals are relatively strong without being brittle. **Ionic solids** consist of anions and cations held together by mutual electrostatic attraction. Differences between ionic and metallic bonding make the electrical and mechanical properties of ionic solids very different from those of metals: Ionic solids do not conduct electricity well and are brittle. The atoms in **covalent-network solids** are held together by an extended network of covalent bonds. This type of bonding can result in materials that are extremely hard, like diamond, and it is also responsible for the unique properties of semiconductors. **Molecular solids** consist of discrete molecules held together by the intermolecular forces we studied in Chapter 11: dispersion forces, dipole–dipole interactions, and hydrogen bonds. Because these forces are relatively weak, molecular solids tend to be soft and have low melting points.

We also discuss two classes of solids that do not fall neatly into the preceding categories: polymers and nanomaterials. **Polymers** contain long chains of atoms (usually carbon), where the atoms within a given chain are connected by covalent bonds and adjacent chains are held to one another largely by weaker intermolecular forces. Polymers are normally stronger and have higher melting points than molecular solids, and they are more flexible than metallic, ionic, or covalent-network solids. **Nanomaterials** are solids in which the dimensions of individual crystals have been reduced to the order of 1–100 nm. As we explain, the properties of conventional materials change when their crystals become this small.

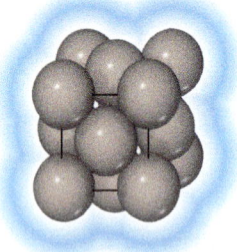

Metallic solids

Extended networks of atoms held together by metallic bonding (Cu, Fe)

Ionic solids

Extended networks of ions held together by cation–anion interactions (NaCl, MgO)

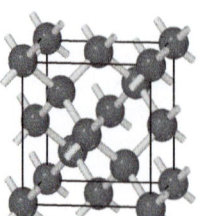

Covalent-network solids

Extended networks of atoms held together by covalent bonds (C, Si)

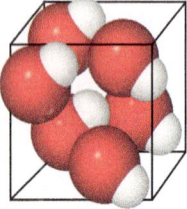

Molecular solids

Discrete molecules held together by intermolecular forces (HBr, H_2O)

▲ **Figure 12.1** Classification and examples of solids according to predominant bonding type.

Crystalline and Amorphous Solids

Solids contain large numbers of atoms. For example, a 1-carat diamond has a volume of 57 mm^3 and contains 1.0×10^{22} carbon atoms. How can we hope to describe such a large collection of atoms? Fortunately, the structures of many solids have patterns that repeat over and over in three dimensions. We can visualize the solid as being formed by stacking a large number of small, identical structural units, much like a wall can be built by stacking identical bricks.

Solids in which atoms are arranged in an orderly repeating pattern are called **crystalline solids**. These solids usually have flat surfaces, or *faces*, that make definite angles with one another. The orderly arrangements of atoms that produce these faces also cause the solids to have highly regular shapes (**Figure 12.2**). Examples of crystalline solids include sodium chloride, quartz, and diamond.

Amorphous solids (from the Greek words for "without form") lack the order found in crystalline solids. At the atomic level the structures of amorphous solids are similar to the structures of liquids, but the molecules, atoms, and/or ions lack the freedom of motion they have in liquids. Amorphous solids do not have the well-defined faces and shapes of a crystal. Familiar amorphous solids are rubber, glass, and obsidian (volcanic glass).

Unit Cells and Crystal Lattices

In a crystalline solid, there is a relatively small repeating unit, called a **unit cell**, that is made up of a unique arrangement of atoms and embodies the structure of the solid. The structure of the crystal can be built by stacking this unit over and over in all three dimensions. Thus, the structure of a crystalline solid is defined by (a) the size and shape of the unit cell and (b) the locations of atoms within the unit cell.

The geometrical pattern of points on which the unit cells are arranged is called a **crystal lattice**. The crystal lattice is, in effect, an abstract (that is, not real) scaffolding for the crystal structure. We can imagine forming the entire crystal structure by first building the scaffolding and then filling in each unit cell with the same atom or group of atoms.

Before describing the structures of solids, we need to understand the properties of crystal lattices. It is useful to begin with two-dimensional lattices because they are simpler to visualize than three-dimensional ones. **Figure 12.3** shows a two-dimensional array of **lattice points**. Each lattice point has an identical environment. The positions of the lattice points are defined by the **lattice vectors** *a* and *b*. Beginning from any lattice point, it is possible to move to any other lattice point by adding together whole-number multiples of the two lattice vectors.*

The parallelogram formed by the lattice vectors, the shaded region in Figure 12.3, defines the unit cell. In two dimensions the unit cells must *tile*, or fit together in space, in such a way that they completely cover the area of the lattice with no gaps. In three dimensions, the unit cells must stack together to fill all space.

In a two-dimensional lattice, the unit cells can take only one of the five shapes shown in **Figure 12.4**. The most general type of lattice is the *oblique lattice*. In this lattice, the lattice vectors are of different lengths and the angle γ between them is of arbitrary size, which makes the unit cell an arbitrarily shaped parallelogram. The *square lattice, rectangular lattice, hexagonal lattice,*** and *rhombic lattice* have a unique combination of γ angle and relationship between the lengths of lattice vectors *a* and *b* (shown in Figure 12.4). For a rhombic lattice, an alternative unit cell can be drawn—namely, a

Iron pyrite (FeS_2), a crystalline solid

Obsidian (*ca.* 70% SiO_2), an amorphous solid

▲ **Figure 12.2 Examples of crystalline and amorphous solids.** The atoms in crystalline solids repeat in an orderly, periodic fashion that leads to well-defined faces at the macroscopic level, as observed in iron pyrite. This order is lacking in amorphous solids like obsidian (volcanic glass).

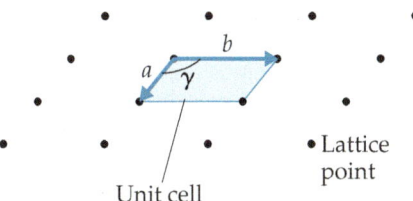

▲ **Figure 12.3 A crystalline lattice in two dimensions.** An infinite array of lattice points is generated by adding together the lattice vectors *a* and *b*. The unit cell is a parallelogram defined by the lattice vectors.

---

* A vector is a quantity involving both a direction and a magnitude. The magnitudes of the vectors in Figure 12.3 are indicated by their lengths, and their directions are indicated by the arrowheads.

** You may wonder why the hexagonal unit cell is not shaped like a hexagon. Remember that the unit cell is by definition a *parallelogram* whose size and shape are defined by the lattice vectors *a* and *b*.

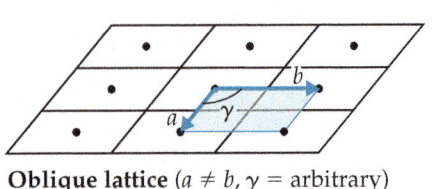

Oblique lattice ($a \neq b$, γ = arbitrary)

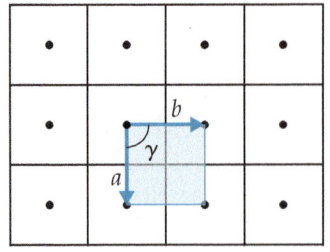

Square lattice ($a = b$, $\gamma = 90°$)

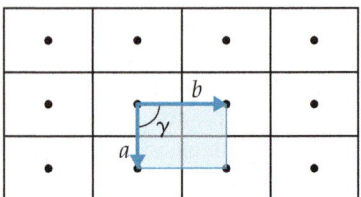

Rectangular lattice ($a \neq b$, $\gamma = 90°$)

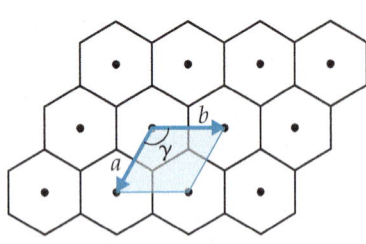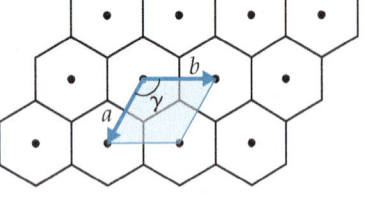

Hexagonal lattice ($a = b$, $\gamma = 120°$)

Rhombic lattice ($a = b$, γ = arbitrary)
Centered rectangular lattice

▲ **Figure 12.4 The five two-dimensional lattices.** The primitive unit cell for each lattice is shaded in blue. For the rhombic lattice, the centered rectangular unit cell is shaded in green. Unlike the primitive rhombic unit cell, the centered cell has two lattice points per unit cell.

 Go Figure

Why is there a centered rectangular lattice but not a centered square lattice?

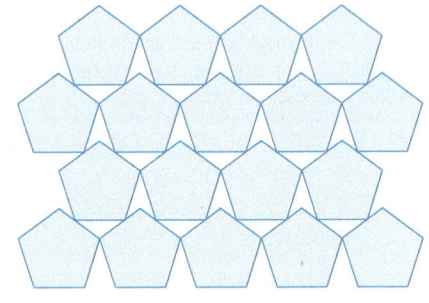

▲ **Figure 12.5 Not all shapes tile space.** Tiling means covering a surface entirely, which is impossible for some geometric shapes, as shown here for pentagons.

rectangle with lattice points on its corners *and* its center (shown in green in Figure 12.4). Because of this, the rhombic lattice is commonly referred to as a *centered rectangular lattice*. The lattices in Figure 12.4 represent five basic shapes: squares, rectangles, hexagons, rhombuses (diamonds), and arbitrary parallelograms. Other polygons, such as pentagons, cannot cover space without leaving gaps, as **Figure 12.5** shows.

To understand real crystals, we must consider three dimensions. A three-dimensional lattice is defined by *three* lattice vectors *a*, *b*, and *c* (**Figure 12.6**). These lattice vectors define a unit cell that is a parallelepiped (a six-sided figure whose faces are all parallelograms) and is described by the lengths *a*, *b*, *c* of the cell edges and the angles α, β, γ between these edges. There are seven possible shapes for a three-dimensional unit cell, as shown in Figure 12.6.

If we place a lattice point at each corner of a unit cell, we get a **primitive lattice**. All seven lattices in Figure 12.6 are primitive lattices. It is also possible to generate what are called *centered lattices* by placing additional lattice points in specific locations

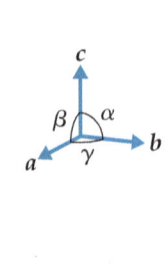

Cubic
$a = b = c$
$\alpha = \beta = \gamma = 90°$

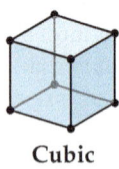

Tetragonal
$a = b \neq c$
$\alpha = \beta = \gamma = 90°$

Orthorhombic
$a \neq b \neq c$
$\alpha = \beta = \gamma = 90°$

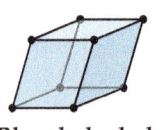

Rhombohedral
$a = b = c$
$\alpha = \beta = \gamma \neq 90°$

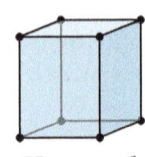

Hexagonal
$a = b \neq c$
$\alpha = \beta = 90°, \gamma = 120°$

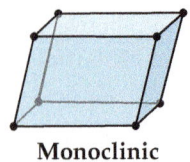

Monoclinic
$a \neq b \neq c$
$\alpha = \gamma = 90°, \beta \neq 90°$

Triclinic
$a \neq b \neq c$
$\alpha \neq \beta \neq \gamma$

▲ **Figure 12.6 The seven three-dimensional primitive lattices.**

in the unit cell. This is illustrated for a cubic lattice in **Figure 12.7**. A **body-centered cubic lattice** has one lattice point at the center of the unit cell in addition to the lattice points at the eight corners. A **face-centered cubic lattice** has one lattice point at the center of each of the six faces of the unit cell in addition to the lattice points at the eight corners. Centered lattices exist for other types of unit cells as well. For the crystals discussed in this chapter, we need consider only the lattices shown in Figures 12.6 and 12.7.

Filling the Unit Cell

The lattice by itself does not define a crystal structure. To generate a crystal structure, we need to associate an atom or group of atoms with each lattice point. In the simplest case, the crystal structure consists of identical atoms, and each atom lies directly on a lattice point. When this happens, the crystal structure and the lattice points have identical patterns. Many metallic elements adopt such structures, as we show in Section 12.3. Only for solids in which all the atoms are identical can this occur; in other words, *only elements* can form structures of this type. For compounds, even if we were to put an atom on every lattice point, the points would not be identical because the atoms are not all the same.

In most crystals, the atoms are not exactly coincident with the lattice points. Instead, a group of atoms, called a **motif**, is associated with each lattice point. The unit cell contains a specific motif of atoms, and the crystal structure is built up by repeating the unit cell over and over. This process is illustrated in **Figure 12.8** for a two-dimensional crystal based on a hexagonal unit cell and a two-carbon-atom motif. The resulting infinite two-dimensional honeycomb structure is a two-dimensional crystal called *graphene*, a material that has so many interesting properties that its modern discoverers won the Nobel Prize in Physics in 2010. Each carbon atom is covalently bonded to three neighboring carbon atoms in what amounts to an infinite sheet of interconnected hexagonal rings.

The crystal structure of graphene illustrates two important characteristics of crystals. First, we see that no atoms lie on the lattice points. While most of the structures we discuss in this chapter do have atoms on the lattice points, there are many examples, like graphene, where this is not the case. Thus, to build up a structure you must know the location and orientation of the atoms in the motif with respect to the lattice points. Second, we see that bonds can be formed between atoms in neighboring unit cells and the bonds between atoms need not be parallel to the lattice vectors.

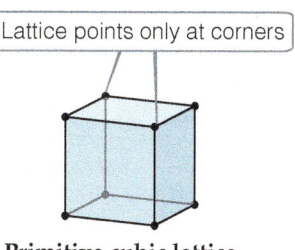

Lattice points only at corners

Primitive cubic lattice

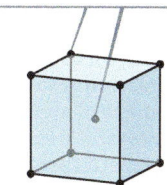

Lattice points at corners plus one lattice point in the center of unit cell

Body-centered cubic lattice

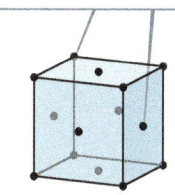

Lattice points at corners plus one lattice point at the center of each face

Face-centered cubic lattice

▲ **Figure 12.7** The three types of cubic lattices.

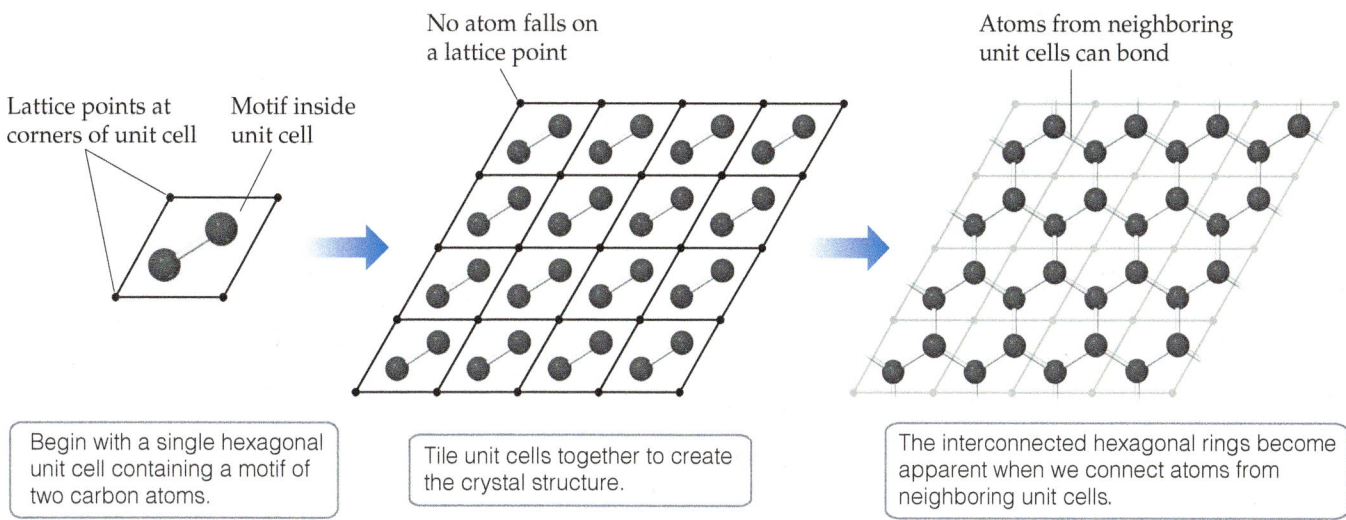

Lattice points at corners of unit cell   Motif inside unit cell

No atom falls on a lattice point

Atoms from neighboring unit cells can bond

Begin with a single hexagonal unit cell containing a motif of two carbon atoms.

Tile unit cells together to create the crystal structure.

The interconnected hexagonal rings become apparent when we connect atoms from neighboring unit cells.

▲ **Figure 12.8** Two-dimensional structure of graphene built up from a single unit cell.

A CLOSER LOOK   X-ray Diffraction

When light waves pass through a narrow slit, they are scattered in such a way that the wave seems to spread out. This physical phenomenon is called *diffraction*. When light passes through many evenly spaced narrow slits (a *diffraction grating*), the scattered waves interact to form a series of bright and dark bands, known as a diffraction pattern. The bright bands correspond to constructive overlapping of the light waves, and the dark bands correspond to destructive overlapping of the light waves. (See the "A Closer Look" box on "Phases in Atomic and Molecular Orbitals" that is located between Sections 9.7 and 9.8.) The most effective diffraction of light occurs when the wavelength of the light and the width of the slits are similar in magnitude.

The spacing of the layers of atoms in solid crystals is usually about 2–20 Å. The wavelengths of X rays are also in this range. Thus, a crystal can serve as an effective diffraction grating for X rays. X-ray diffraction results from the scattering of X rays by a regular arrangement of atoms, molecules, or ions. Much of what we know about crystal structures has been obtained by looking at the diffraction patterns that result when X rays pass through a crystal, a technique known as *X-ray crystallography*. As shown in Figure 12.9, a monochromatic beam of X rays is passed through a crystal. The diffraction pattern that results is recorded. For many years the diffracted X rays were detected by photographic film. Today, crystallographers use an *array detector*, a device analogous to that used in digital cameras, to capture and measure the intensities of the diffracted rays.

The pattern of spots on the detector in Figure 12.9 depends on the particular arrangement of atoms in the crystal. The spacing and symmetry of the bright spots, where constructive interference occurs, provide information about the size and shape of the unit cell. The intensities of the spots provide information that can be used to determine the locations of the atoms within the unit cell. When combined, these two pieces of information give the atomic structure that defines the crystal.

X-ray crystallography is used extensively to determine the structures of molecules in crystals. The instruments used to measure X-ray diffraction, known as *X-ray diffractometers*, are now computer controlled, making the collection of diffraction data highly automated. The diffraction pattern of a crystal can be determined very accurately and quickly (sometimes in a matter of hours), even though thousands of diffraction spots are measured. Computer programs are then used to analyze the diffraction data and determine the arrangement and structure of the molecules in the crystal. X-ray diffraction is an important technique in industries ranging from steel and cement manufacture to pharmaceuticals.

Related Exercises: 12.117, 12.118

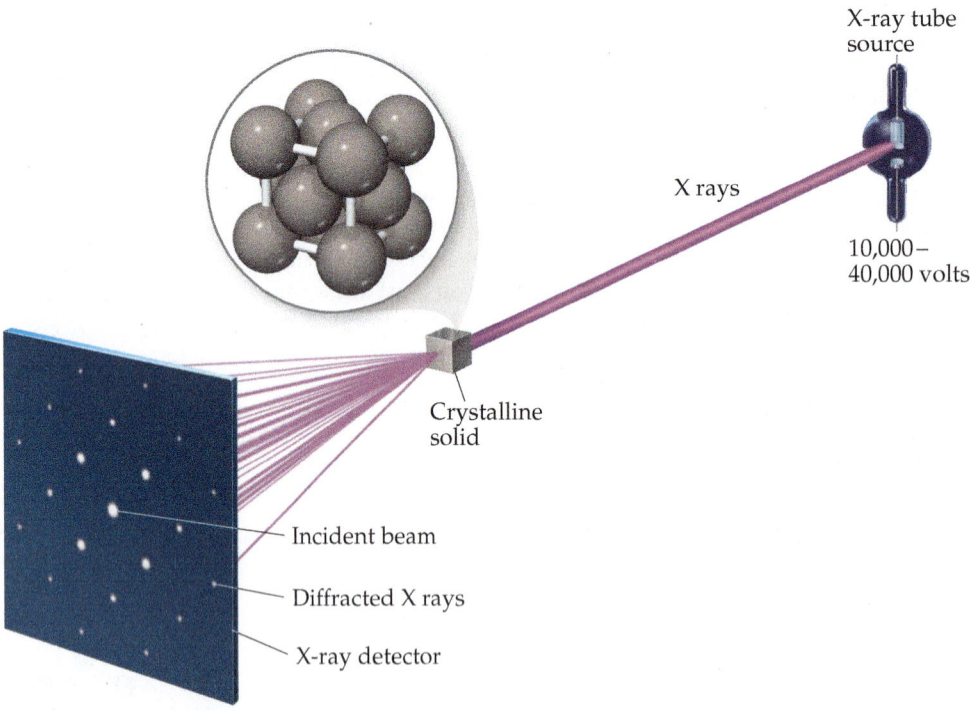

X-ray tube source

X rays

10,000– 40,000 volts

Crystalline solid

Incident beam

Diffracted X rays

X-ray detector

▲ **Figure 12.9 Diffraction of X rays by a crystal.** A monochromatic X-ray beam is passed through a crystal. The X rays are diffracted, and the resulting interference pattern is recorded. The crystal is rotated and another diffraction pattern recorded. Analysis of many diffraction patterns gives the positions of the atoms in the crystal.

▲ **Self-Assessment Exercises**

SAE 12.1 Which of the following solids are held together by the weakest interactions? (**a**) metallic solids (**b**) ionic solids (**c**) covalent-network solids (**d**) molecular solids

SAE 12.2 Select the structure shown below that represents an amorphous solid.

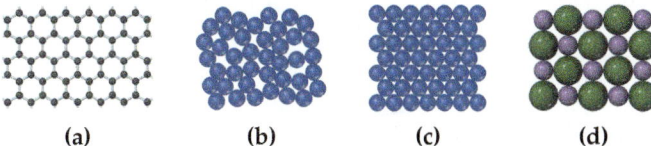

(a)   (b)   (c)   (d)

SAE 12.3 Which type of lattice contains a lattice point completely within the unit cell? (**a**) primitive cubic lattice (**b**) body-centered cubic (bcc) lattice (**c**) face-centered cubic (fcc) lattice (**d**) face-centered tetragonal lattice (**e**) More than one of a, b, c, d.

SAE 12.4 In which of the following three-dimensional primitive lattices are no two lattice vectors the same length? (**a**) cubic (**b**) tetragonal (**c**) hexagonal (**d**) orthorhombic

12.2 | Metallic Solids

Metallic solids, also simply called *metals*, consist entirely of metal atoms. The bonding in metals is unlike all other types of chemical bonding: It is too strong to be due to dispersion forces, and yet there are not enough valence electrons to form covalent bonds between atoms. The bonding, called *metallic bonding*, happens because the valence electrons are *delocalized* throughout the entire solid. That is, the valence electrons are not associated with specific atoms or bonds but are spread throughout the solid. In fact, we can visualize a metal as an array of positive ions immersed in a "sea" of delocalized valence electrons.

The chemical bonding in metals is reflected in their properties. You have probably held a length of copper wire or an iron bolt. Perhaps you have even seen the surface of a freshly cut piece of sodium metal. These substances, though distinct from one another, share certain similarities that enable us to classify them as metallic. A clean metal surface has a characteristic luster. Metals have a characteristic cold feeling when you touch them, related to their high thermal conductivity (ability to conduct heat). Metals also have high electrical conductivity, which means that electrically charged particles flow easily through them. The thermal conductivity of a metal usually parallels its electrical conductivity. Silver and copper, for example, which possess the highest electrical conductivities among the elements, also possess the highest thermal conductivities.

Most metals are *malleable*, which means that they can be hammered into thin sheets, and *ductile*, which means that they can be drawn into wires (**Figure 12.10**). These properties indicate that the atoms are capable of slipping past one another. Ionic and covalent-network solids do not exhibit such behavior; they are typically brittle.

The Structures of Metallic Solids

The crystal structures of many metals are simple and can be generated by placing a single atom on each lattice point. The structures corresponding to the three cubic lattices are shown in **Figure 12.11**. Metals with a primitive cubic structure are rare; one of the few examples is the radioactive element polonium. Body-centered cubic metals include iron, chromium, sodium, and tungsten. Examples of face-centered cubic metals include aluminum, lead, copper, silver, and gold.

Notice in the bottom row of Figure 12.11 that the atoms on the corners and faces of a unit cell do not lie wholly within the unit cell. These corner and face atoms are shared by neighboring unit cells. An atom that sits at the corner of a unit cell is shared among eight unit cells, and only 1/8 of the atom is in one particular unit cell. Because a cube has eight corners, each primitive cubic unit cell contains $(1/8) \times 8 = 1$ atom, as shown in

▲ **Learning Objectives**

When you have finished Section 12.2, you should be able to:

▶ Identify metallic solids and describe their physical properties.

▶ Determine the number of atoms and their coordination number, and derive the length of a unit cell edge for primitive, body-centered, and face-centered cubic metals.

▶ Calculate the density of a metal from its crystal structure, atomic radius, and molar mass.

▶ Distinguish between the atomic arrangements in hexagonal close-packed and cubic close-packed structures.

▶ Calculate the packing efficiencies for face-centered and body-centered cubic metals.

▶ Differentiate between elemental metals, alloys, and intermetallic compounds.

▶ Distinguish between interstitial, substitutional, and heterogeneous alloys.

▲ **Figure 12.10 Malleability and ductility.** Gold leaf demonstrates the characteristic malleability of metals, whereas copper wire demonstrates their ductility.

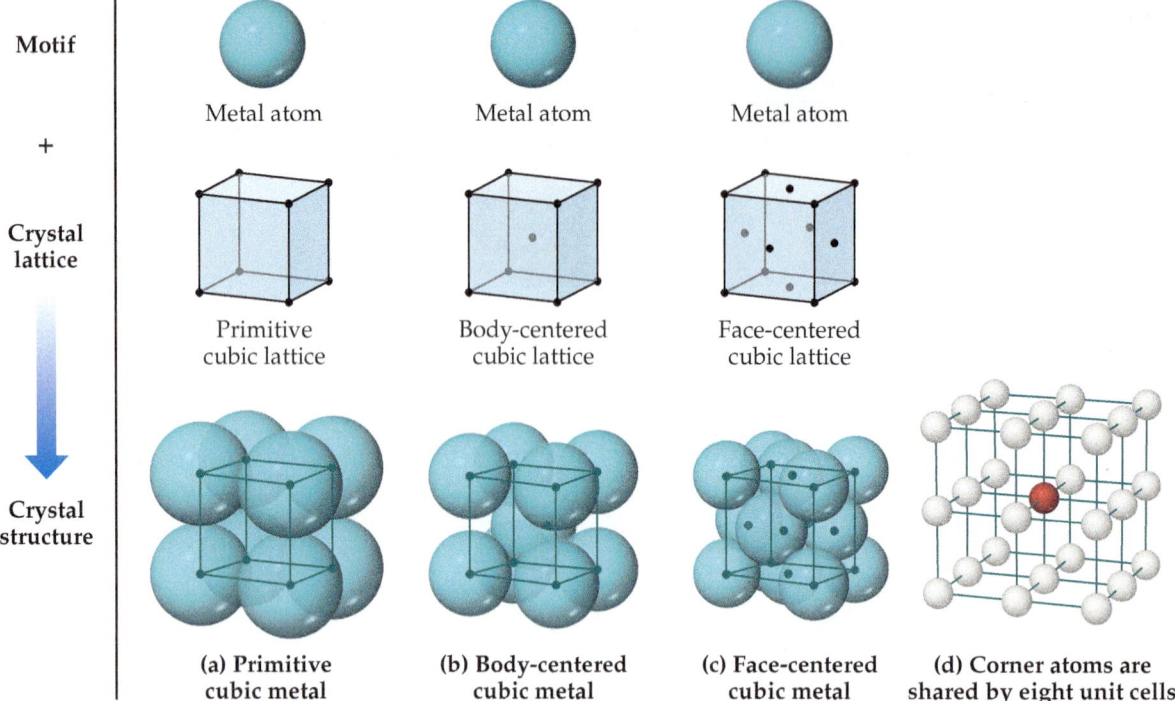

▲ Figure 12.11 The structures of (a) primitive cubic, (b) body-centered cubic, and (c) face-centered cubic metals. Each structure can be generated by the combination of a single-atom motif and the appropriate lattice. (**d**) Corner atoms (one shown in red) are shared among eight neighboring cubic unit cells.

Figure 12.12(**a**). Similarly, each body-centered cubic unit cell [Figure 12.12(**b**)] contains two atoms, $(1/8) \times 8 = 1$ from the corners and 1 at the center of the unit cell. Atoms that lie on the face of a unit cell, as they do in a face-centered cubic metal, are shared by two unit cells, so only one-half of the atom belongs to each unit cell. Therefore, a face-centered cubic unit cell [Figure 12.12(**c**)] contains four atoms, $(1/8) \times 8 = 1$ atom from the corners and $(1/2) \times 6 = 3$ atoms from the faces.

Table 12.1 summarizes how the fractional part of each atom that resides within a unit cell depends on the atom's location within the cell.

Go Figure

Which one of these unit cells would you expect to represent the densest packing of spheres?

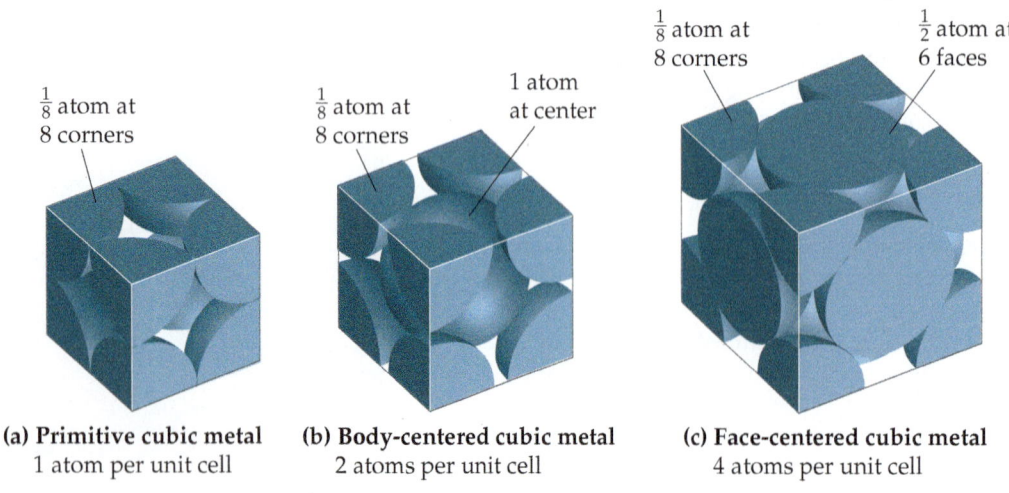

$\frac{1}{8}$ atom at 8 corners

$\frac{1}{8}$ atom at 8 corners

1 atom at center

$\frac{1}{8}$ atom at 8 corners

$\frac{1}{2}$ atom at 6 faces

(a) Primitive cubic metal
1 atom per unit cell

(b) Body-centered cubic metal
2 atoms per unit cell

(c) Face-centered cubic metal
4 atoms per unit cell

▲ Figure 12.12 A space-filling view of unit cells for metals with a cubic structure. Only the portion of each atom that falls within the unit cell is shown.

TABLE 12.1   Fraction of Any Atom as a Function of Location within the Unit Cell*

Atom Location	Number of Unit Cells Sharing Atom	Fraction of Atom within Unit Cell
Corner	8	1/8 or 12.5%
Edge	4	1/4 or 25%
Face	2	1/2 or 50%
Anywhere else	1	1 or 100%

*It is only the position of the center of the atom that matters. Atoms that reside near the boundary of the unit cell but not on a corner, edge, or face are counted as residing 100% within the unit cell.

Close Packing

The shortage of valence electrons and the fact that they are collectively shared make it favorable for the atoms in a metal to pack together closely. Because we can consider the shape of an atom to be a sphere, we can understand the structures of metals by considering how spheres pack. The most efficient way to pack one layer of equal-sized spheres is to surround each sphere by six neighbors, as shown at the top of Figure 12.13. To form a three-dimensional structure, we need to stack additional layers on top of this base layer. To maximize packing efficiency, the second layer of spheres must sit in the depressions formed by the spheres in the first layer. We can either put the next layer of atoms into the depressions marked by the yellow dot or into the depressions marked by the red dot (the spheres are too large to simultaneously fill both sets of depressions). For the sake of discussion, we arbitrarily put the second layer in the yellow depressions.

For the third layer, we have two choices for where to place the spheres. One possibility is to put the third layer in the depressions that lie directly over the spheres in the first layer. This is done on the left-hand side of Figure 12.13, as shown by the dashed red lines in the side view. Continuing with this pattern, the fourth layer would lie directly over the spheres in the second layer, leading to the ABAB stacking pattern seen on the left, which is called **hexagonal close packing** (hcp). Alternatively, the third-layer spheres could lie directly over the depressions that were marked with red dots in the first layer. In this arrangement, the spheres in the third layer do not sit directly above the spheres in either of the first two layers, as shown by the dashed red lines on the lower right-hand side of Figure 12.13. If this sequence is repeated in subsequent layers, we derive an ABCABC stacking pattern shown on the right known as **cubic close packing** (ccp). In both hexagonal close packing and cubic close packing, each sphere has 12 equidistant nearest neighbors: six neighbors in the same layer, three from the layer above, and three from the layer below. We say that each sphere has a **coordination number** of 12. The coordination number is the number of atoms immediately surrounding a given atom in a crystal structure.

The extended structure of a hexagonal close-packed metal is shown in Figure 12.14(**a**). There are two atoms in the primitive hexagonal unit cell, one from each layer. Neither atom sits directly on the lattice points, which are located at the corners of the unit cell. The presence of two atoms in the unit cell is consistent with the two-layer ABAB stacking sequence associated with hcp packing.

Although it is not immediately obvious, the structure that results from cubic close packing possesses a unit cell that is identical to the face-centered cubic unit cell we encountered in Figure 12.11(**c**). The relationship between the ABC layer stacking and the face-centered cubic unit cell is shown in Figure 12.14(**b**). In this figure, we see that the layers stack perpendicular to the body diagonal of the cubic unit cell.

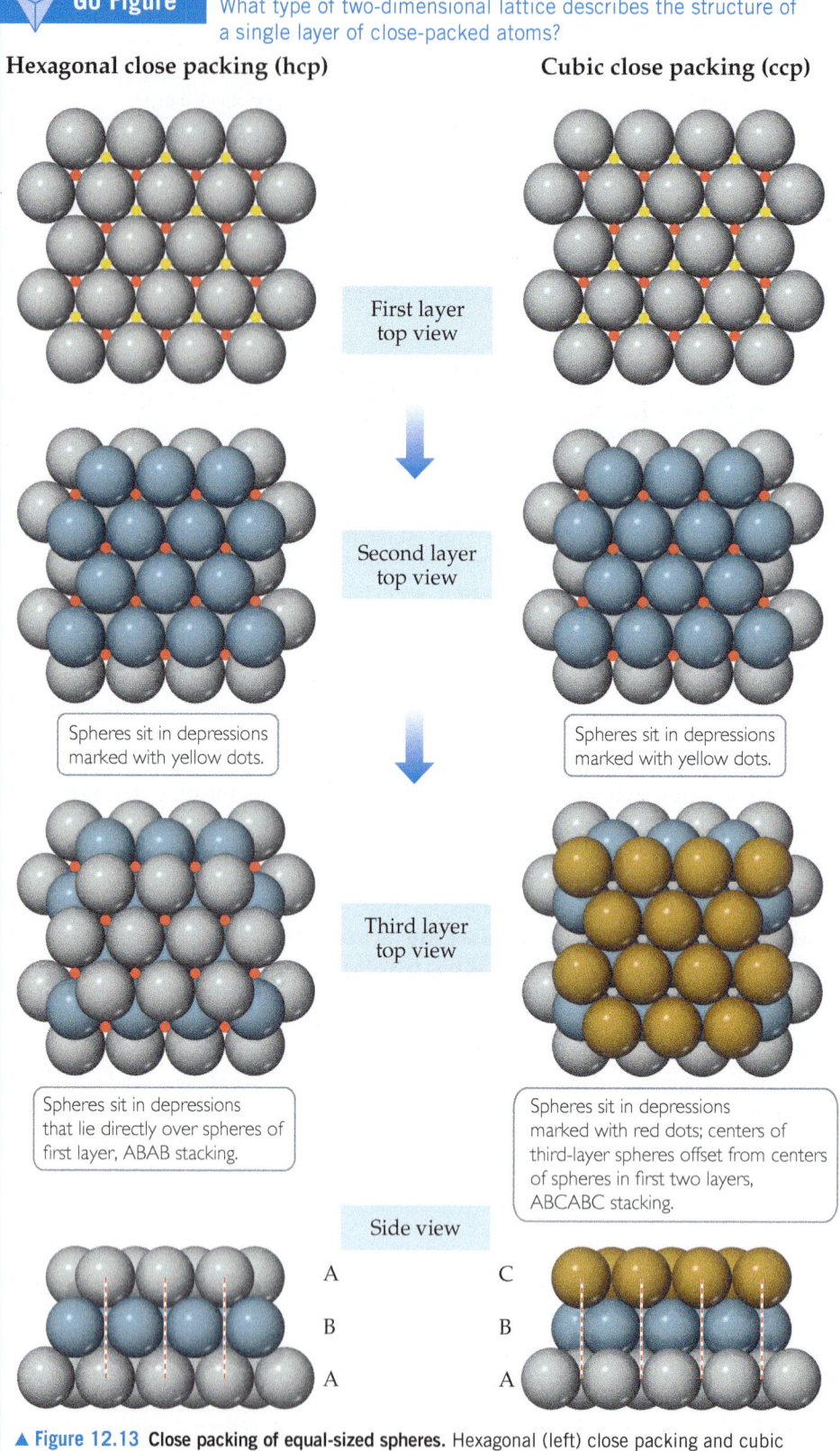

Go Figure What type of two-dimensional lattice describes the structure of a single layer of close-packed atoms?

Hexagonal close packing (hcp)

Cubic close packing (ccp)

First layer top view

Second layer top view

Spheres sit in depressions marked with yellow dots.

Spheres sit in depressions marked with yellow dots.

Third layer top view

Spheres sit in depressions that lie directly over spheres of first layer, ABAB stacking.

Spheres sit in depressions marked with red dots; centers of third-layer spheres offset from centers of spheres in first two layers, ABCABC stacking.

Side view

A   C

B   B

A   A

▲ **Figure 12.13 Close packing of equal-sized spheres.** Hexagonal (left) close packing and cubic (right) close packing are equally efficient ways of packing spheres. The red and yellow dots indicate the positions of depressions between atoms.

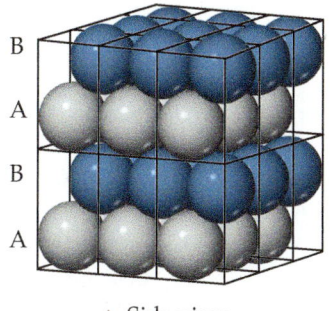

Side view   Side view

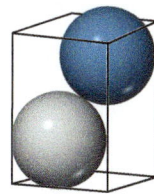   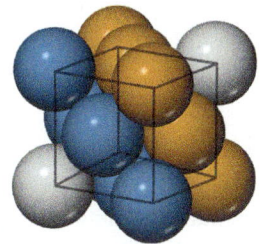

Unit cell view   Unit cell view

(a) Hexagonal close-packed metal   **(b) Cubic close-packed metal**

▲ **Figure 12.14 The unit cells for (a) a hexagonal close-packed metal and (b) a cubic close-packed metal.** The solid lines indicate the unit cell boundaries. Colors are used to distinguish one layer of atoms from another.

 Sample Exercise 12.1

Calculating Packing Efficiency

It is not possible to pack spheres together without leaving some void spaces between the spheres. *Packing efficiency* is the fraction of space in a crystal that is actually occupied by atoms. Determine the packing efficiency of a face-centered cubic metal.

SOLUTION

Analyze We must determine the volume taken up by the atoms that reside in the unit cell and divide this number by the volume of the unit cell.

Plan We can calculate the volume taken up by atoms by multiplying the number of atoms per unit cell by the volume of a sphere, $4\pi r^3/3$. To determine the volume of the unit cell, we must first identify the direction along which the atoms touch each other. We can then use geometry to express the length of the cubic unit cell edge, a, in terms of the radius of the atoms. Once we know the edge length, the cell volume is simply a^3.

Solve As shown in Figure 12.12, a face-centered cubic metal has four atoms per unit cell. Therefore, the volume occupied by the atoms is

$$\text{Occupied volume} = 4 \times \left(\frac{4\pi r^3}{3}\right) = \frac{16\pi r^3}{3}$$

For a face-centered cubic metal, the atoms touch along the diagonal of a face of the unit cell:

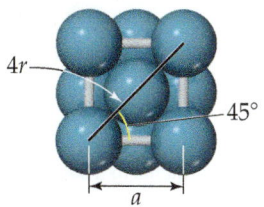

Therefore, a diagonal across a face of the unit cell is equal to four times the atomic radius, r. Using simple trigonometry, and the identity $\cos(45°) = \sqrt{2}/2$, we can show that

$$a = 4r\cos(45°) = 4r(\sqrt{2}/2) = (2\sqrt{2})r$$

Finally, we calculate the packing efficiency by dividing the volume occupied by atoms by the volume of the cubic unit cell, a^3:

$$\text{Packing efficiency} = \frac{\text{volume of atoms}}{\text{volume of unit cell}} = \frac{\left(\frac{16}{3}\right)\pi r^3}{(2\sqrt{2})^3 r^3}$$

$$= 0.74 \text{ or } 74\%$$

▶ **Practice Exercise**

Determine the packing efficiency by calculating the fraction of space occupied by atoms in a body-centered cubic metal.

TABLE 12.2 Some Common Alloys

Name	Primary Element	Typical Composition (by Mass)	Properties	Uses
Wood's metal	Bismuth	50% Bi, 25% Pb, 12.5% Sn, 12.5% Cd	Low melting point (70 °C)	Fuse plugs, automatic sprinklers
Yellow brass	Copper	67% Cu, 33% Zn	Ductile, takes polish	Hardware items
Bronze	Copper	88% Cu, 12% Sn	Tough and chemically stable in dry air	Important alloy for early civilizations
Stainless steel	Iron	80.6% Fe, 0.4% C, 18% Cr, 1% Ni	Resists corrosion	Cookware, surgical instruments
Plumber's solder	Lead	67% Pb, 33% Sn	Low melting point (275 °C)	Soldering joints
Sterling silver	Silver	92.5% Ag, 7.5% Cu	Bright surface	Tableware
Dental amalgam	Silver	70% Ag, 18% Sn, 10% Cu, 2% Hg	Easily worked	Dental fillings
Pewter	Tin	92% Sn, 6% Sb, 2% Cu	Low melting point (230 °C)	Dishes, jewelry

Alloys

An **alloy** is a material that contains more than one element and has the characteristic properties of a metal. The alloying of metals is of great importance because it is one of the primary ways of modifying the properties of pure metallic elements. Nearly all the common uses of iron, for example, involve alloy compositions (for example, stainless steel). Bronze is formed by alloying copper and tin, while brass is an alloy of copper and zinc. Pure gold is too soft to be used in jewelry, but alloys of gold with copper or silver are much harder. Other common alloys are described in **Table 12.2**.

Alloys can be divided into four categories: substitutional alloys, interstitial alloys, heterogeneous alloys, and intermetallic compounds. Substitutional and interstitial alloys are both homogeneous mixtures in which components are dispersed randomly and uniformly (**Figure 12.15**). (Section 1.2) Solids that form homogeneous mixtures are called solid solutions. When atoms of the solute in a solid solution occupy positions normally occupied by a solvent atom, we have a **substitutional alloy**. When the solute atoms occupy interstitial positions in the "holes" between solvent atoms, we have an **interstitial alloy** (Figure 12.15).

Substitutional alloys are formed when the two metallic components have similar atomic radii and chemical-bonding characteristics. For example, silver and gold form such an alloy over the entire range of possible compositions. When two metals differ in radii by more than about 15%, solubility is generally more limited.

For an interstitial alloy to form, the solute atoms must have a much smaller bonding atomic radius than the solvent atoms. Typically, the interstitial element is a nonmetal that makes covalent bonds to the neighboring metal atoms. The presence of the extra bonds provided by the interstitial component causes the metal lattice to become harder,

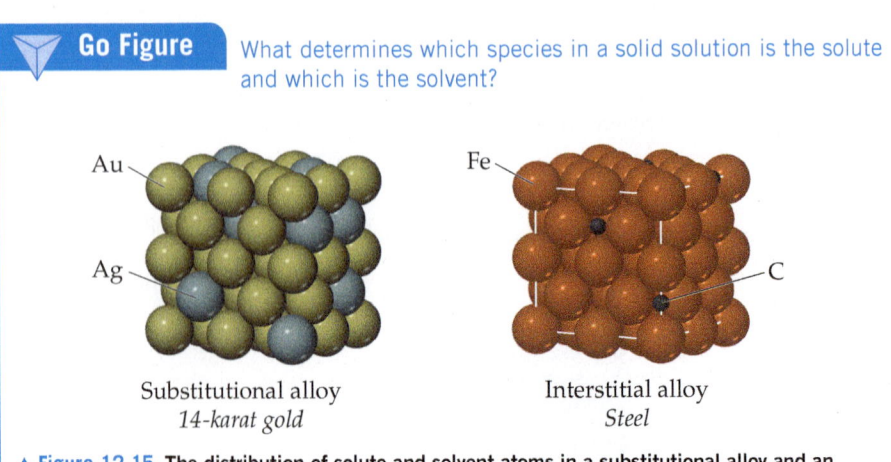

Go Figure What determines which species in a solid solution is the solute and which is the solvent?

Au
Ag

Substitutional alloy
14-karat gold

Fe
C

Interstitial alloy
Steel

▲ **Figure 12.15** The distribution of solute and solvent atoms in a substitutional alloy and an interstitial alloy. Both types of alloys are solid solutions and, therefore, homogeneous mixtures.

stronger, and less ductile. For example, steel, which is much harder and stronger than pure iron, is an alloy of iron that contains up to 3% carbon. Other elements may be added to form *alloy steels*. Vanadium and chromium may be added to impart strength, for instance, and to increase resistance to fatigue and corrosion.

One of the most important iron alloys is stainless steel, which contains about 0.4% carbon, 18% chromium, and 1% nickel. The ratio of elements present in the steel may vary over a wide range, imparting a variety of specific physical and chemical properties to the materials.

In a **heterogeneous alloy**, the components are not dispersed uniformly. For example, the heterogeneous alloy pearlite contains two phases (**Figure 12.16**). One phase is essentially pure body-centered cubic iron, whereas the other is the compound Fe_3C, known as cementite. In general, the properties of heterogeneous alloys depend on both the composition and the manner in which the solid is formed from the molten mixture. The properties of a heterogeneous alloy formed by rapid cooling of a molten mixture, for example, are distinctly different from the properties of an alloy formed by slow cooling of the same mixture.

Intermetallic compounds are compounds rather than mixtures. Because they are compounds, they have definite properties and their composition cannot be varied. Furthermore, the different types of atoms in an intermetallic compound are ordered rather than randomly distributed. The ordering of atoms in an intermetallic compound generally leads to better structural stability and higher melting points than what is observed in the constituent metals. These features can be attractive for high-temperature applications. On the negative side, intermetallic compounds are often more brittle than substitutional alloys.

Intermetallic compounds play many important roles in modern society. The intermetallic compound Ni_3Al is a major component of jet aircraft engines because of its strength at high temperature and its low density. Razor blades are often coated with Cr_3Pt, which adds hardness, allowing the blade to stay sharp longer. Both compounds have the structure shown on the left-hand side of **Figure 12.17**. The compound Nb_3Sn also shown in Figure 12.17, is a superconductor, a substance that, when cooled below a critical temperature, conducts electricity with no resistance. In the case of Nb_3Sn, superconductivity is observed only when the temperature falls below 18 K. Superconductors are used in the magnets in MRI scanners widely employed for medical imaging. See the "Chemistry and Life" box on "Nuclear Spin and Magnetic Resonance Imaging." (Section 6.7) The need to keep the magnets cooled to such a low temperature is part of the reason why MRI devices are expensive to operate. The hexagonal intermetallic compound $SmCo_5$, shown on the right-hand side of Figure 12.17, is used to make the permanent magnets found in lightweight headsets and high-fidelity speakers. A related compound with the same structure, $LaNi_5$, is used as the anode in nickel–metal hydride batteries.

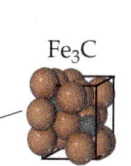

▲ **Figure 12.16 Microscopic view of the structure of the heterogeneous alloy pearlite.** The dark regions are body-centered cubic iron metal, and the lighter regions are cementite, Fe_3C.

Go Figure   In the unit cell drawing on the right, why do we see eight Sm atoms and nine Co atoms if the empirical formula is $SmCo_5$?

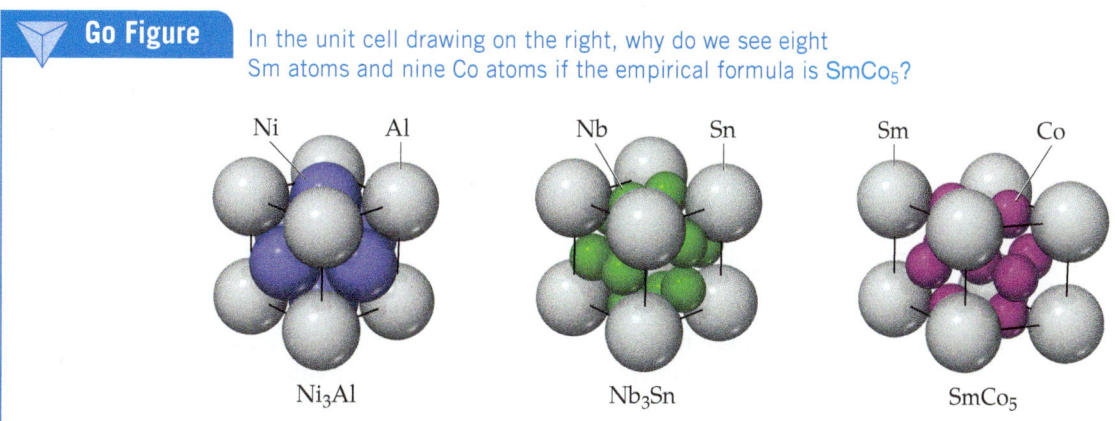

▲ **Figure 12.17 Three examples of intermetallic compounds.**

Self-Assessment Exercises

SAE 12.5 Which of the following solids would you expect to possess metallic properties?

   (i) $BaCrO_4$
   (ii) bronze alloy
   (iii) Sb

(a) only one **(b)** i and ii **(c)** i and iii **(d)** ii and iii **(e)** all three

SAE 12.6 Consider the primitive cubic unit cell with 8 atoms at the eight corners shown here. In a real solid, these unit cells are stacked on top of and next to each other in all three dimensions. How many atoms are there in this unit cell?

(a) 1 **(b)** 2 **(c)** 4 **(d)** 8

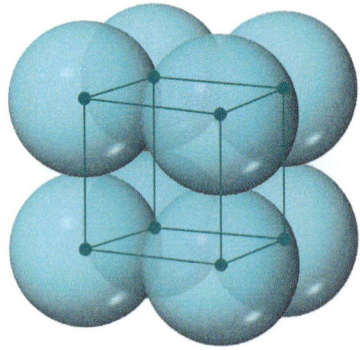

SAE 12.7 Nickel crystallizes in a face-centered cubic unit cell that has a density of 8.90 g/cm^3. Calculate the atomic radius of a nickel atom.

(a) 0.79 Å **(b)** 1.25 Å **(c)** 1.83 Å **(d)** 3.52 Å

SAE 12.8 Which of the following statements is/are *true*?

   (i) Substitutional alloys are formed when two metallic components have similar atomic radii and chemical-bonding characteristics.
   (ii) Intermetallic compounds have variable composition.
   (iii) It is not possible to form a solution from two metals.

(a) Only one of the statements is true. **(b)** Statements i and ii are true. **(c)** Statements i and iii are true. **(d)** Statements ii and iii are true. **(e)** All three statements are true.

SAE 12.9 Which of the following statements is/are *true*?

   (i) Cubic-close- packed and hexagonal-close-packed solids have the same packing efficiency.
   (ii) In both cubic-close-packed and hexagonal-close-packed solids, the coordination number of each atom is 12.
   (iii) The stacking arrangement in a hexagonal-close-packed solid of identical atoms is ABABAB.

(a) Only one of these statements is true. **(b)** Both i and ii are true. **(c)** Both ii and iii are true. **(d)** Both i and iii are true. **(e)** All three statements are true.

SAE 12.10 Gold crystallizes in the fcc lattice with an edge length of 4.08 Å. Calculate the density of gold to one decimal place. **(a)** 4.8 g/cm^3 **(b)** 9.6 g/cm^3 **(c)** 10.6 g/cm^3 **(d)** 19.3 g/cm^3

12.3 | Metallic Bonding

Learning Objectives

When you finish Section 12.3, you should be able to:

▶ Describe the electron-sea model of metallic bonding.

▶ Describe how the electronic band structure of metals arises from the overlap of atomic orbitals.

Consider the elements of the third period of the periodic table (Na–Ar). Argon, with eight valence electrons, has a complete octet; as a result, it does not form any bonds. Chlorine, sulfur, and phosphorus form molecules (Cl_2, S_8, and P_4) in which the atoms make one, two, and three bonds, respectively (**Figure 12.18**). Silicon forms an extended network solid in which each atom is bonded to four equidistant neighbors. Each of these elements forms $8 - N$ bonds, where N is the number of valence electrons. This behavior can easily be understood through the application of the octet rule.

If the $8 - N$ trend continued as we move left across the periodic table, we would expect aluminum (three valence electrons) to form five bonds. Like many other metals, however, aluminum adopts a close-packed structure with 12 nearest neighbors. Magnesium and sodium also adopt metallic structures. What is responsible for this abrupt change in the preferred bonding mechanism? The answer, as noted previously, is that metals do not have enough valence-shell electrons to satisfy their bonding requirements by forming localized electron-pair bonds. In response to this deficiency, the valence electrons are collectively shared. A structure in which the atoms are close-packed facilitates this delocalized sharing of electrons.

Electron-Sea Model

A simple model that accounts for some of the most important characteristics of metals is the **electron-sea model**, which pictures the metal as an array of metal cations in a "sea" of valence electrons (**Figure 12.19**). The electrons are confined to the metal by electrostatic attractions to the cations, and they are uniformly distributed throughout the structure. The electrons are mobile, however, and no individual electron is confined to any particular metal ion. When a voltage is applied to a metal wire, the electrons, being negatively charged, flow through the metal toward the positively charged end of the wire.

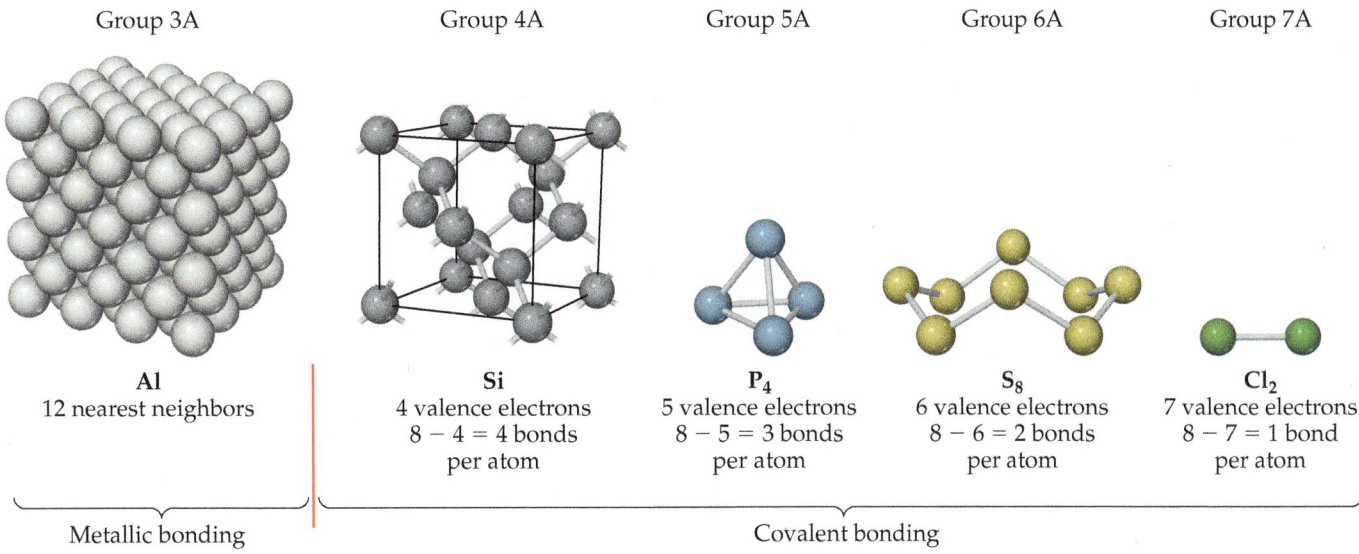

Group 3A	Group 4A	Group 5A	Group 6A	Group 7A
Al	**Si**	**P₄**	**S₈**	**Cl₂**
12 nearest neighbors	4 valence electrons $8 - 4 = 4$ bonds per atom	5 valence electrons $8 - 5 = 3$ bonds per atom	6 valence electrons $8 - 6 = 2$ bonds per atom	7 valence electrons $8 - 7 = 1$ bond per atom

Metallic bonding   Covalent bonding

▲ **Figure 12.18 Bonding in period 3 elements.**

The high thermal conductivity of metals is also accounted for by the presence of mobile electrons. The movement of electrons in response to temperature gradients permits ready transfer of kinetic energy throughout the solid.

The ability of metals to deform (their malleability and ductility) can be explained by the fact that metal atoms form bonds to many neighbors. Changes in the positions of the atoms brought about in reshaping the metal are partly accommodated by a redistribution of electrons.

The Molecular-Orbital Model and Electronic Band Structure

Although the electron-sea model works surprisingly well given its simplicity, it does not adequately explain many properties of metals. According to the model, for example, the strength of bonding between metal atoms should steadily increase as the number of valence electrons increases, resulting in a corresponding increase in the melting points. However, elements near the middle of the transition metal series, rather than those at the end, have the highest melting points in their respective periods (**Figure 12.20**). This trend implies that the strength of metallic bonding first increases with increasing number of electrons and then decreases. Similar trends are seen in other physical properties of the metals, such as the boiling point, heat of fusion, and hardness.

To obtain a more accurate picture of the bonding in metals, we must turn to molecular-orbital theory. In Sections 9.7 and 9.8, we learned how molecular orbitals are created from the overlap of atomic orbitals. Let's briefly review some of the rules of molecular-orbital theory:

1. Atomic orbitals combine to make molecular orbitals that can extend over the entire molecule.
2. A molecular orbital can contain zero, one, or two electrons.
3. The number of molecular orbitals in a molecule equals the number of atomic orbitals that combine to form molecular orbitals.
4. Adding electrons to a bonding molecular orbital strengthens bonding, while adding electrons to antibonding molecular orbitals weakens bonding.

The electronic structures of crystalline solids and small molecules have similarities as well as differences. To illustrate, consider how the molecular-orbital diagram for a chain of lithium atoms changes as we increase the length of the chain (**Figure 12.21**). Each lithium atom contains a half-filled 2s orbital in its valence shell. The molecular-orbital

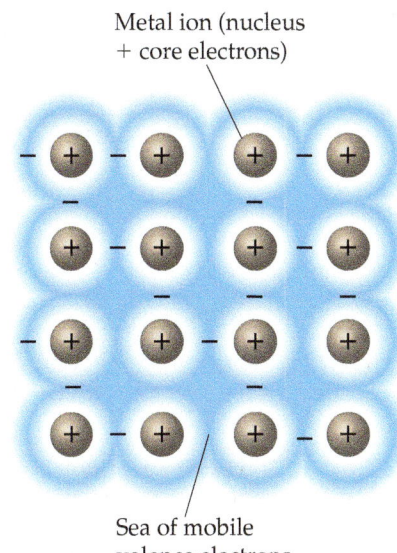

Metal ion (nucleus + core electrons)

Sea of mobile valence electrons

▲ Figure 12.19 **Electron-sea model of metallic bonding.** The valence electrons delocalize to form a sea of mobile electrons that surrounds and binds together an extended array of metal ions.

Go Figure

Which element in each period has the highest melting point? In each case, is the element you named at the beginning, middle, or end of its period?

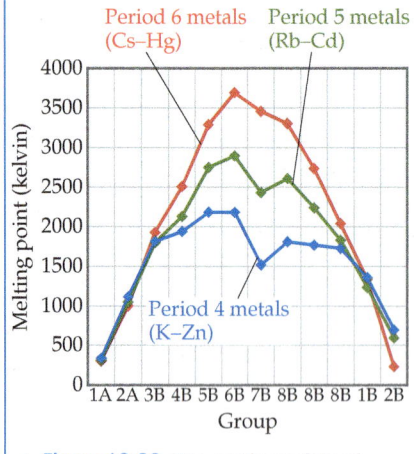

▲ **Figure 12.20** The melting points of metals from periods 4, 5, and 6.

diagram for Li_2 is analogous to that of an H_2 molecule: one filled bonding molecular orbital and one empty antibonding molecular orbital with a nodal plane between the atoms. (Section 9.7) For Li_4, there are four molecular orbitals, ranging from the lowest-energy orbital, where the orbital interactions are completely bonding (zero nodal planes), to the highest-energy orbital, where all interactions are antibonding (three nodal planes).

As the length of the chain increases, the number of molecular orbitals increases. Regardless of chain length, the lowest-energy orbitals are always the most bonding and the highest-energy orbitals always the most antibonding. Furthermore, because each lithium atom has only one valence-shell atomic orbital, the number of molecular orbitals is equal to the number of lithium atoms in the chain. Because each lithium atom has one valence electron, half of the molecular orbitals are fully occupied and the other half are empty, regardless of chain length.[*]

If the chain becomes very long, there are so many molecular orbitals that the energy separation between them becomes vanishingly small. As the chain length goes to infinity, the allowed energy states become a continuous **band**. For a crystal large enough to see with the eye (or even an optical microscope), the number of atoms is extremely large. Consequently, the electronic structure of the crystal is like that of the infinite chain, consisting of bands, as shown on the right-hand side of Figure 12.21.

The electronic structures of most metals are more complicated than those shown in Figure 12.21 because we have to consider more than one type of atomic orbital on each atom. Because each type of orbital can give rise to its own band, the electronic structure of a solid usually consists of a series of bands. The electronic structure of a bulk solid is referred to as a **band structure**.

The band structure of a typical metal is shown schematically in **Figure 12.22**. The electron filling depicted corresponds to nickel metal, but the basic features of other

Go Figure

How does the energy spacing between molecular orbitals change as the number of atoms in the chain increases?

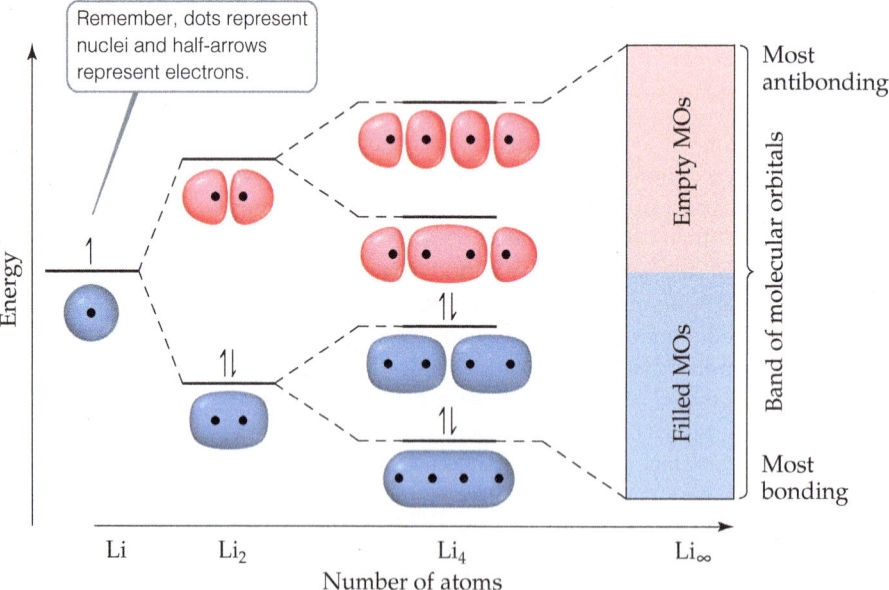

▲ **Figure 12.21** Discrete energy levels in individual molecules become continuous energy bands in a solid. Occupied orbitals are shaded blue, whereas empty orbitals are pink.

[*] This is strictly true only for chains with an even number of atoms.

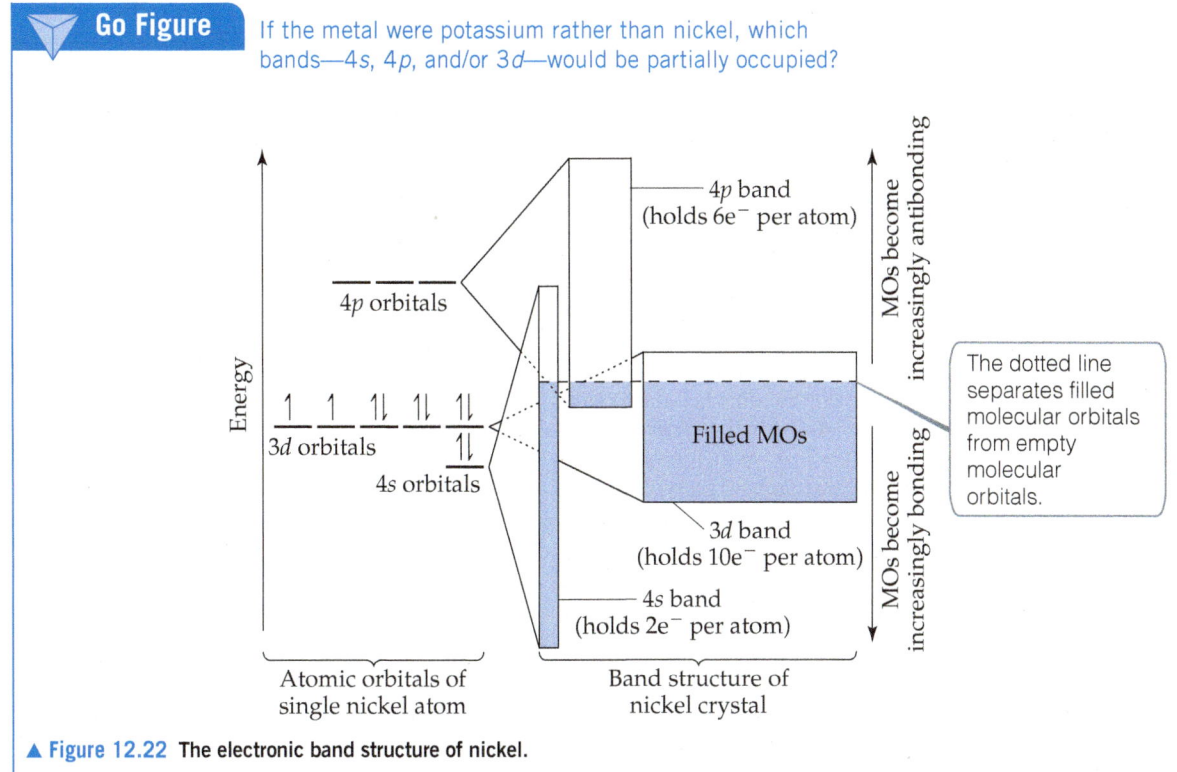

Go Figure   If the metal were potassium rather than nickel, which bands—4s, 4p, and/or 3d—would be partially occupied?

▲ **Figure 12.22** **The electronic band structure of nickel.**

metals are similar. The electron configuration of a nickel atom is $[Ar]4s^2 3d^8$, as shown on the left side of the figure. The energy bands that form from each of these orbitals are shown on the right side. The 4s, 4p, and 3d orbitals are treated independently, each giving rise to a band of molecular orbitals. In practice, these overlapping bands are not completely independent of each other, but for our purposes this simplification is reasonable.

The 4s, 4p, and 3d bands differ from one another in the energy range they span (represented by the heights of the rectangles on the right side of Figure 12.22) and in the number of electrons they can hold (represented by the area of the rectangles). The 4s, 4p, and 3d bands can hold 2, 6, and 10 electrons per atom, respectively, corresponding to two per orbital, as dictated by the Pauli exclusion principle. (Section 6.7) The energy range spanned by the 3d band is smaller than the range spanned by the 4s and 4p bands because the 3d orbitals are smaller and, therefore, overlap with orbitals on neighboring atoms less effectively.

Many properties of metals can be understood from Figure 12.22. We can think of the energy band as a partially filled container for electrons. The incomplete filling of the energy band gives rise to characteristic metallic properties. The electrons in orbitals near the top of the occupied levels require very little energy input to be "promoted" to higher-energy orbitals that are unoccupied. Under the influence of any source of excitation, such as an applied electrical potential or an input of thermal energy, electrons move into previously vacant levels and are thus freed to move through the lattice, giving rise to electrical and thermal conductivity.

Without the overlap of energy bands, the periodic properties of metals could not be explained. In the absence of the d- and p-bands, we would expect the s-band to be half-filled for the alkali metals (group 1A) and completely filled for the alkaline-earth metals (group 2A). If that were true, metals like magnesium, calcium, and strontium would not be good electrical and thermal conductors, in disagreement with experimental observations.

While the conductivity of metals can be qualitatively understood using either the electron-sea model or the molecular-orbital model, many physical properties of transition metals, such as the melting points plotted in Figure 12.20, can be explained only with the latter model. The molecular-orbital model predicts that bonding first becomes stronger as the number of valence electrons increases and the bonding orbitals are increasingly populated. Upon moving past the middle elements of the transition metal series, the bonds grow weaker as electrons populate antibonding orbitals. Strong bonds between atoms lead to metals with higher melting and boiling points, higher heats of fusion, higher hardness, and so forth.

 Self-Assessment Exercises

SAE 12.11   The electron-sea model does not account for which of the following? (**a**) The high thermal conductivity of metals. (**b**) The malleability of metals. (**c**) The strength of metallic bonding first increasing with the increasing number of electrons and then decreasing. (**d**) The high electrical conductivity of metals.

SAE 12.12   Which metal would you expect to have the highest melting point? (**a**) calcium (**b**) vanadium (**c**) nickel (**d**) zinc

12.4 | Ionic Solids

Learning Objectives

When you finish Section 12.4, you should be able to:

▶ Describe the physical properties of ionic solids.

▶ Determine the empirical formula of an ionic solid, and the coordination numbers of its cation and anion, from the structure of the unit cell.

▶ Calculate the density of an ionic solid from its structure, ionic radii, and molar masses of the ions.

Ionic solids are held together by the electrostatic attraction between cations and anions: ionic bonds. (Section 8.2) The high melting and boiling points of ionic compounds are a testament to the strength of the ionic bonds. The strength of an ionic bond depends on the charges and sizes of the ions. As discussed in Chapters 8 and 11, the attractions between cations and anions increase as the charges of the ions increase. Thus, NaCl, where the ions have charges of 1+ and 1−, melts at 801 °C, whereas MgO, where the ions have charges of 2+ and 2−, melts at 2852 °C. The interactions between cations and anions also increase as the ions get smaller, as we see from the melting points of the alkali metal halides listed in Table 12.3. These trends mirror the trends in lattice energy discussed in Section 8.2.

Although both ionic and metallic solids have high melting and boiling points, the differences between ionic and metallic bonding are responsible for important differences in their properties. Because the valence electrons in ionic compounds are confined to the anions, rather than being delocalized, ionic compounds are typically electrical insulators. They tend to be brittle, a property explained by repulsive interactions between ions of like charge. When stress is applied to an ionic solid, as in Figure 12.23, the planes of atoms, which before the stress were arranged with cations next to anions, shift so that the alignment becomes cation–cation, anion–anion. The resulting repulsive interaction causes the planes to split away from each other, a property that lends itself to the carving of certain gemstones (such as ruby, composed principally of Al_2O_3).

Structures of Ionic Solids

Like metallic solids, ionic solids tend to adopt structures with symmetric, close-packed arrangements of atoms. However, important differences arise because we now have to pack together spheres that have different radii and opposite charges. Because cations

TABLE 12.3   Properties of the Alkali Metal Halides

Compound	Cation–Anion Distance (Å)	Lattice Energy (kJ/mol)	Melting Point (°C)
LiF	2.01	1030	845
NaCl	2.83	788	801
KBr	3.30	671	734
RbI	3.67	632	674

are often considerably smaller than anions (Section 7.3), the coordination numbers in ionic compounds are smaller than those in close-packed metals. Even if the anions and cations were the same size, the close-packed arrangements seen in metals cannot be replicated without letting ions of like charge come in contact with each other. The repulsions between ions of the same type make such arrangements unfavorable. The most favorable structures are those where the cation–anion distances are as close as those permitted by ionic radii, but the anion–anion and cation–cation distances are maximized.

Three common ionic structure types are shown in **Figure 12.24**. The cesium chloride (CsCl) structure is based on a primitive cubic lattice. Anions sit on the lattice points at the corners of the unit cell, and a cation sits at the center of each cell. (Remember, there is no lattice point inside a primitive unit cell.) With this arrangement, both cations and anions are surrounded by a cube of eight ions of the opposite type.

The sodium chloride (NaCl; also called the rock salt structure) and zinc blende (ZnS) structures are based on a face-centered cubic lattice. In both structures, the anions sit on the lattice points that lie on the corners and faces of the unit cell, but the two-atom motif is slightly different for the two structures. In NaCl the Na^+ ions are displaced from the Cl^- ions along the edge of the unit cell, whereas in ZnS the Zn^{2+} ions are displaced from the S^{2-} ions along the body diagonal of the unit cell. This difference leads to different coordination numbers. In sodium chloride, each cation and each anion are surrounded by six ions of the opposite type, leading to an octahedral coordination environment. In zinc blende, each cation and each anion are surrounded by four ions of the opposite type, leading to a tetrahedral coordination geometry. The cation coordination environments can be seen in **Figure 12.25**.

For a given ionic compound, we might ask which type of structure is most favorable. A number of factors come into play, but two of the most important are the relative sizes of the ions and the stoichiometry. Consider first ion size. Notice in Figure 12.25 that the coordination number changes from 8 to 6 to 4 on moving from CsCl to NaCl to ZnS. This trend is driven in part by the fact that for these three compounds the ionic radius of the cation gets smaller while the ionic radius of the anion changes very little. When the cation and anion are similar in size, a large coordination number is favored and the CsCl structure is often realized. As the relative size of the cation gets smaller, eventually it is no longer possible to maintain the cation–anion contacts and simultaneously keep the anions from touching each other. When this occurs, the coordination number drops from 8 to 6, and the sodium chloride structure becomes more favorable. As the cation size decreases further, eventually the coordination number must be reduced again, this time from 6 to 4, and the zinc blende structure becomes favored. Remember that, in ionic crystals, ions of opposite charge touch each other but ions of the same charge should not touch.

The relative number of cations and anions also helps determine the most stable structure type. All the structures in Figure 12.25 have equal numbers of cations and anions. These structure types (cesium chloride, sodium chloride, zinc blende) can be realized only for ionic compounds in which the number of cations and anions is equal. When this is not the case, other crystal structures must result. As an example, consider NaF, MgF_2, and ScF_3 (**Figure 12.26**). Sodium fluoride has the sodium chloride structure with a coordination number of 6 for both cation and anion, as you might expect because NaF and NaCl are quite similar. Magnesium fluoride, however, has two anions for every cation, resulting in a tetragonal crystal structure called the *rutile structure*. The cation coordination number is still 6, but the fluoride coordination number is now only 3. In the scandium fluoride structure, there are three anions for every cation; the cation coordination number is still 6, but the fluoride coordination number has dropped to 2. As the cation/anion ratio goes down, there are fewer cations to surround each anion, and so the anion coordination number must decrease. The empirical formula of an ionic compound can be described quantitatively by the relationship

$$\frac{\text{Number of cations per formula unit}}{\text{Number of anions per formula unit}} = \frac{\text{anion coordination number}}{\text{cation coordination number}} \quad [12.1]$$

Go Figure

Why don't metals cleave in the way depicted here for ionic substances?

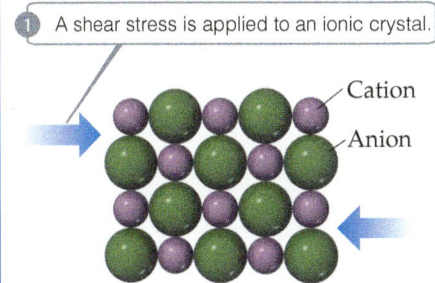

1. A shear stress is applied to an ionic crystal.

Cation
Anion

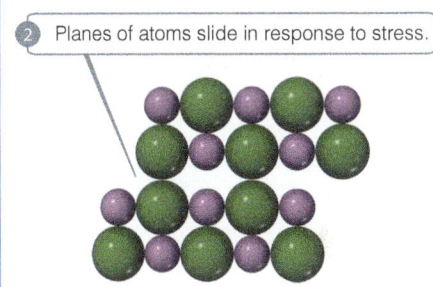

2. Planes of atoms slide in response to stress.

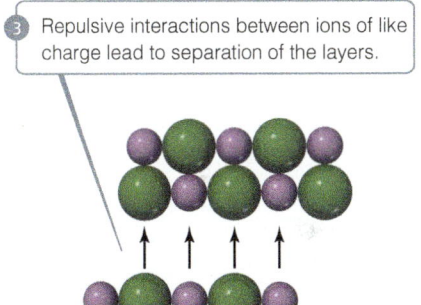

3. Repulsive interactions between ions of like charge lead to separation of the layers.

(a)

(b)

▲ **Figure 12.23 Brittleness and faceting in ionic crystals. (a)** When a shear stress (blue arrows) is applied to an ionic solid, the crystal separates along a plane of atoms as shown. **(b)** This property of ionic crystals is used to facet gemstones, such as rubies.

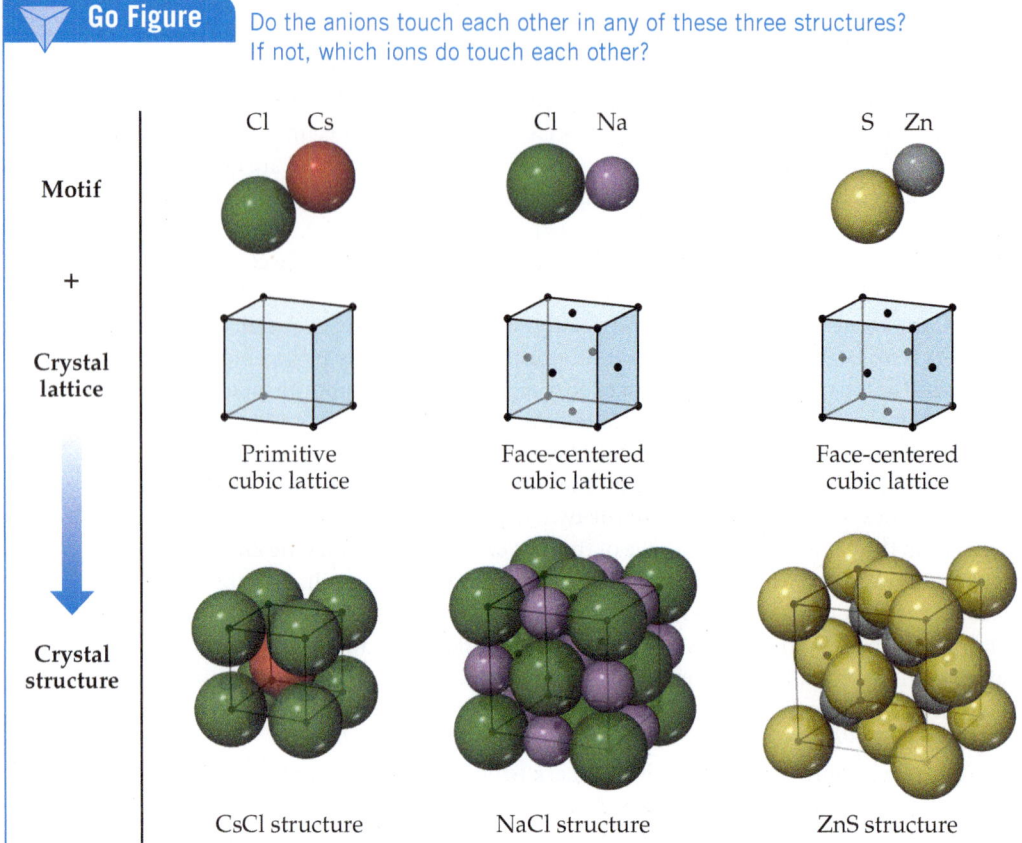

Go Figure   Do the anions touch each other in any of these three structures? If not, which ions do touch each other?

▲ **Figure 12.24 The structures of CsCl, NaCl, and ZnS.** Each structure type can be generated by the combination of a two-atom motif and the appropriate lattice.

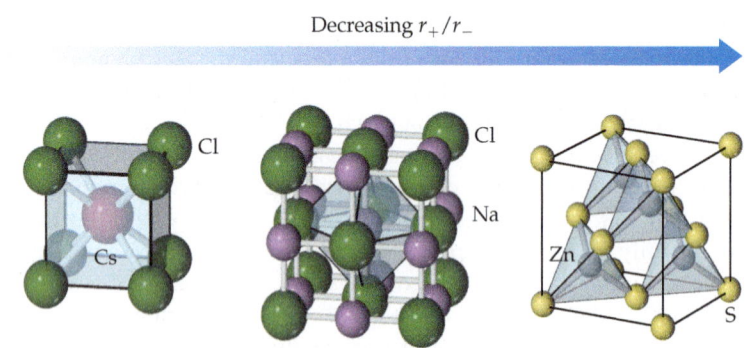

Decreasing r_+/r_-

	CsCl	NaCl	ZnS
Cation radius, r_+ (Å)	1.81	1.16	0.88
Anion radius, r_- (Å)	1.67	1.67	1.70
r_+/r_-	1.08	0.69	0.52
Cation coordination number	8	6	4
Anion coordination number	8	6	4

▲ **Figure 12.25 Coordination environments in CsCl, NaCl, and ZnS.** The sizes of the ions have been reduced to show the coordination environments clearly.

Go Figure How many cations are there per unit cell for each of these structures? How many anions per unit cell?

Increasing anion-to-cation ratio

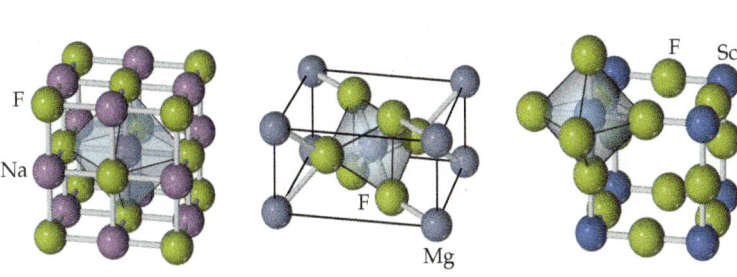

	NaF	MgF$_2$	ScF$_3$
Cation coordination number	6	6	6
Cation coordination geometry	Octahedral	Octahedral	Octahedral
Anion coordination number	6	3	2
Anion coordination geometry	Octahedral	Trigonal planar	Linear

▲ **Figure 12.26 Coordination numbers depend on stoichiometry.** The sizes of the ions have been reduced to show the coordination environments clearly.

Sample Exercise 12.2

Calculating the Density of an Ionic Solid

Rubidium iodide crystallizes with the same structure as sodium chloride. (**a**) How many iodide ions are there per unit cell? (**b**) How many rubidium ions are there per unit cell? (**c**) Use the ionic radii and molar masses of Rb$^+$ (1.66 Å, 85.47 g/mol) and I$^-$ (2.06 Å, 126.90 g/mol) to estimate the density of rubidium iodide in g/cm^3.

SOLUTION

Analyze and Plan

(**a**) We need to count the number of anions in the unit cell of the sodium chloride structure, remembering that ions on the corners, edges, and faces of the unit cell are only partially inside the unit cell.

(**b**) We can apply the same approach to determine the number of cations in the unit cell. We can double-check our answer by writing the empirical formula to make sure the charges of the cations and anions are balanced.

(**c**) Because density is an intensive property, the density of the unit cell is the same as the density of a bulk crystal. To calculate the density, we must divide the mass of the atoms per unit cell by the volume of the unit cell. To determine the volume of the unit cell, we need to estimate the length of the unit cell edge by first identifying the direction along which the ions touch and then using ionic radii to estimate the length. Once we have the length of the unit cell edge, we can cube it to determine its volume.

Solve

(**a**) The crystal structure of rubidium iodide looks just like NaCl with Rb$^+$ ions replacing Na$^+$ and I$^-$ ions replacing Cl$^-$. From the views of the NaCl structure in Figures 12.24 and 12.25, we see that there is an anion at each corner of the unit cell and at the center of each face. From Table 12.1 we see that the ions sitting on the corners are equally shared by eight unit cells ($\frac{1}{8}$ ion per unit cell), while those ions sitting on the faces are equally shared by two unit cells ($\frac{1}{2}$ ion per unit cell). A cube has eight corners and six faces, so the total number of I$^-$ ions is $8(\frac{1}{8}) + 6(\frac{1}{2}) = 4$ per unit cell.

(**b**) Using the same approach for the rubidium cations, we see that there is a rubidium ion on each edge and one at the center of the unit cell. Using Table 12.1 again, we see that the ions sitting on the edges are equally shared by four unit cells (1/4 ion per unit cell), whereas the cation at the center of the unit cell is not shared. A cube has 12 edges, so the total number of rubidium ions is $12(\frac{1}{4}) + 1 = 4$. This answer makes sense because the number of Rb$^+$ ions must be the same as the number of I$^-$ ions to maintain charge balance.

(**c**) In ionic compounds, cations and anions touch each other. In RbI the cations and anions touch along the edge of the unit cell as shown in the following figure.

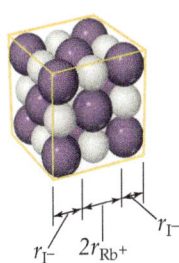

Continued

The length of the unit cell edge is equal to $r(I^-) + 2r(Rb^+) + r(I^-) = 2r(I^-) + 2r(Rb^+)$. Plugging in the ionic radii we get $2(2.06\ \text{Å}) + 2(1.66\ \text{Å}) = 7.44\ \text{Å}$. The volume of a cubic unit cell is just the edge length cubed. Converting from Å to cm and cubing we get

$$\text{Volume} = (7.44 \times 10^{-8}\ \text{cm})^3 = 4.12 \times 10^{-22}\ \text{cm}^3$$

From parts (a) and (b) we know that there are four rubidium and four iodide ions per unit cell. Using this result and the molar masses, we can calculate the mass per unit cell

$$\text{Mass} = \frac{4(85.47\ \text{g/mol}) + 4(126.90\ \text{g/mol})}{6.022 \times 10^{23}\ \text{mol}^{-1}} = 1.411 \times 10^{-21}\ \text{g}$$

The density is the mass per unit cell divided by the volume of a unit cell

$$\text{Density} = \frac{\text{mass}}{\text{volume}} = \frac{1.411 \times 10^{-21}\ \text{g}}{4.12 \times 10^{-22}\ \text{cm}^3} = 3.43\ \text{g/cm}^3$$

Check The densities of most solids fall between the density of lithium ($0.5\ \text{g/cm}^3$) and that of iridium ($22.6\ \text{g/cm}^3$), so this value is reasonable.

▶ **Practice Exercise**
Estimate the length of the cubic unit cell edge and the density of CsCl (Figure 12.24) from the ionic radii of cesium, 1.81 Å, and chloride, 1.67 Å. (*Hint:* Ions in CsCl touch along the body diagonal, a vector running from one corner of a cube through the body center to the opposite corner. Using trigonometry, we can show that the body diagonal of a cube is $\sqrt{3}$ times longer than the edge.)

 Self-Assessment Exercises

SAE 12.13   The unit cell of rubidium iodide is as follows:

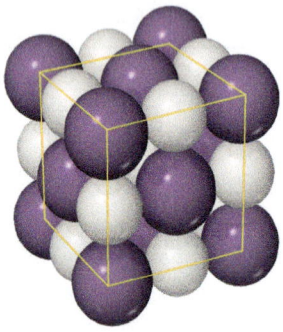

What is the coordination number of I^-, which is represented by the purple spheres?

(**a**) 1 (**b**) 4 (**c**) 6 (**d**) 8

SAE 12.14   The structure shown here has cations, shown in blue, at each corner and at the center of each face of the unit cell. All of the anions, shown in green, are located inside the unit cell.

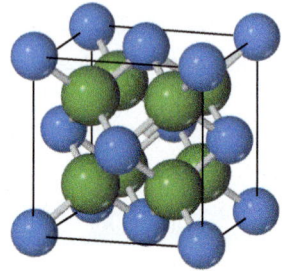

If we use A to represent the cations (in blue) and X to represent the anions (in green), what is the empirical formula of this ionic compound? (**a**) AX_2 (**b**) A_2X (**c**) A_4X_8 (**d**) $A_{14}X_8$

SAE 12.15   Magnesium oxide crystallizes with the same structure as sodium chloride and has a unit-cell edge length of 4.213 Å. What is the density of magnesium oxide? (**a**) $0.895\ \text{g/cm}^3$ (**b**) $1.79\ \text{g/cm}^3$ (**c**) $3.58\ \text{g/cm}^3$ (**d**) $7.16\ \text{g/cm}^3$

12.5 | Molecular and Covalent-Network Solids

 Learning Objectives

When you finish Section 12.5, you should be able to:

▶ Predict the relative melting points of molecular and covalent-network solids.

▶ Describe the electronic band structure of semiconductors.

▶ Predict what type of impurity atom is required to produce an n-type or p-type semiconductor for a given host material.

Molecular solids consist of atoms or neutral molecules held together by dipole–dipole forces, dispersion forces, and/or hydrogen bonds. Because these intermolecular forces are weak, molecular solids are soft and have relatively low melting points (usually below 200 °C). Most substances that are gases or liquids at room temperature form molecular solids at low temperature. Examples include Ar, H_2O, and CO_2.

The properties of molecular solids depend in large part on the strengths of the forces between molecules. Consider, for example, the properties of sucrose (table sugar, $C_{12}H_{22}O_{11}$). Each sucrose molecule has eight —OH groups, which allow for the formation of multiple hydrogen bonds. Consequently, sucrose exists as a crystalline solid at room temperature, and its melting point, 184 °C, is relatively high for a molecular solid.

Molecular shape is also important because it dictates how efficiently molecules pack together in three dimensions. Benzene (C_6H_6), for example, is a highly symmetrical planar molecule. (Section 8.6) It has a higher melting point than toluene, a

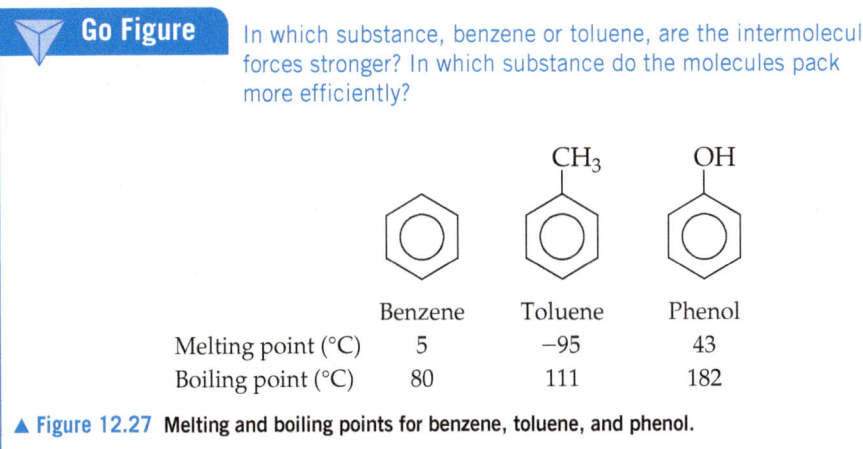

Go Figure In which substance, benzene or toluene, are the intermolecular forces stronger? In which substance do the molecules pack more efficiently?

	Benzene	Toluene	Phenol
Melting point (°C)	5	−95	43
Boiling point (°C)	80	111	182

▲ **Figure 12.27** Melting and boiling points for benzene, toluene, and phenol.

compound in which one of the hydrogen atoms of benzene has been replaced by a CH_3 group (**Figure 12.27**). The lower symmetry of toluene molecules prevents them from packing in a crystal as efficiently as benzene molecules. As a result, the intermolecular forces that depend on close contact are not as effective and the melting point is lower. In contrast, the boiling point of toluene is *higher* than that of benzene, indicating that the intermolecular attractive forces are larger in liquid toluene than in liquid benzene. The melting and boiling points of phenol, another substituted benzene shown in Figure 12.27, are higher than those of benzene because the OH group of phenol can form hydrogen bonds.

Covalent-network solids consist of atoms held together in large networks by covalent bonds. Because covalent bonds are much stronger than intermolecular forces, these solids are much harder and have higher melting points than molecular solids. Diamond and graphite, two allotropes of carbon, are two of the most familiar covalent-network solids. Other examples are silicon, germanium, quartz (SiO_2), silicon carbide (SiC), and boron nitride (BN). In all cases, the bonding between atoms is either completely covalent or more covalent than ionic.

In diamond, each carbon atom is bonded tetrahedrally to four other carbon atoms (**Figure 12.28**). The structure of diamond can be derived from the zinc blende structure (Figure 12.25) if carbon atoms replace both the zinc and sulfide ions. The carbon atoms are sp^3-hybridized and are held together by strong carbon—carbon single covalent bonds. The strength and directionality of these bonds make diamond the hardest known material. For this reason, industrial-grade diamonds are employed in saw blades used for the most demanding cutting jobs. The stiff, interconnected bond

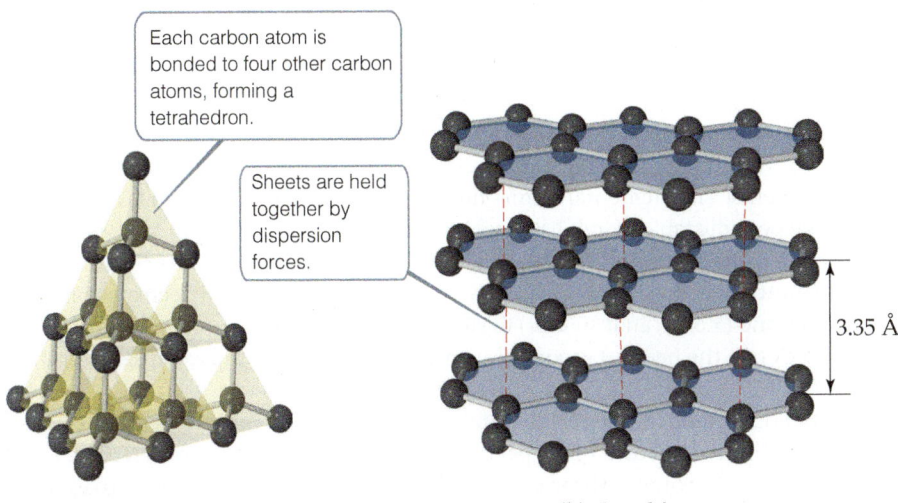

Each carbon atom is bonded to four other carbon atoms, forming a tetrahedron.

Sheets are held together by dispersion forces.

3.35 Å

(a) Diamond

(b) Graphite

◀ **Figure 12.28** The structures of (a) diamond and (b) graphite.

network also explains why diamond is one of the best-known thermal conductors, and yet is not electrically conductive. Diamond has a high melting point, 3550 °C.

In graphite, [Figure 12.28(b)], the carbon atoms form covalently bonded layers that are held together by intermolecular forces. The layers in graphite are the same as those in the graphene sheet shown in Figure 12.8. Graphite has a hexagonal unit cell containing two layers offset so that the carbon atoms in a given layer sit over the middle of the hexagons of the layer below. Each carbon is covalently bonded to three other carbons in the same layer to form interconnected hexagonal rings. The distance between adjacent carbon atoms in the plane, 1.42 Å, is very close to the C—C distance in benzene, 1.395 Å. In fact, the bonding resembles that of benzene, with delocalized π bonds extending over the layers. (Section 9.6) Electrons move freely through the delocalized orbitals, making graphite a good electrical conductor along the layers. (In fact, graphite is used as a conducting electrode in batteries.) These sp^2-hybridized sheets of carbon atoms are separated by 3.35 Å from one another, and the sheets are held together only by dispersion forces. Thus, the layers readily slide past one another when rubbed, giving graphite a greasy feel. This tendency is enhanced when impurity atoms are trapped between the layers, as is typically the case in commercial forms of the material.

Graphite is used as a lubricant and as the "lead" in pencils. The enormous differences in physical properties of graphite and diamond—both of which are pure carbon—arise from differences in their three-dimensional structure and bonding.

Semiconductors

Metals conduct electricity extremely well. Many solids, however, conduct electricity somewhat, but nowhere near as well as metals, which is why such materials are called **semiconductors**. Two examples of semiconductors are silicon and germanium, which lie immediately below carbon in the periodic table. Like carbon, each of these elements has four valence electrons, just the right number to satisfy the octet rule by forming single covalent bonds with four neighbors. Hence, silicon and germanium, as well as the gray form of tin, crystallize with the same infinite network of covalent bonds as diamond.

When atomic s and p orbitals overlap, they form bonding molecular orbitals and antibonding molecular orbitals. Each pair of s orbitals overlaps to give one bonding and one antibonding molecular orbital, whereas the p orbitals overlap to give three bonding and three antibonding molecular orbitals. (Section 9.8) The extended network of bonds leads to the formation of the same type of bands we saw for metals in Section 12.3. However, unlike metals, in semiconductors an energy gap develops between the filled and empty states, much like the energy gap between bonding and antibonding orbitals. (Section 9.7) The band that forms from bonding molecular orbitals is called the **valence band**, and the band that forms the antibonding orbitals is called the **conduction band** (**Figure 12.29**). In a semiconductor, the valence band is filled with electrons and the conduction band is empty. These two bands are separated by the energy **band gap** E_g. In the semiconductor community, energies are given in electron volts (eV); 1 eV $= 1.602 \times 10^{-19}$ J. Band gaps greater than ~3.5 eV are so large that the material is not a semiconductor; it is an **insulator** and does not conduct electricity.

Semiconductors can be divided into two classes: elemental semiconductors, which contain only one type of atom, and compound semiconductors, which contain two or more elements. The elemental semiconductors all come from group 4A. As we move down the periodic table, bond distances increase, which decreases orbital overlap. This decrease in overlap reduces the energy difference between the top of the valence band and the bottom of the conduction band. As a result, the band gap decreases on going from diamond (5.5 eV, an insulator) to silicon (1.11 eV) to germanium (0.67 eV) to gray tin (0.08 eV). In the heaviest group 4A element, lead, the band gap collapses altogether. As a result, lead has the structure and properties of a metal.

Compound semiconductors maintain the same *average* valence electron count as elemental semiconductors—four per atom. For example, in gallium arsenide, GaAs, each Ga atom contributes three electrons and each As atom contributes five, which averages out to four per atom—the same number as in silicon or germanium. Hence,

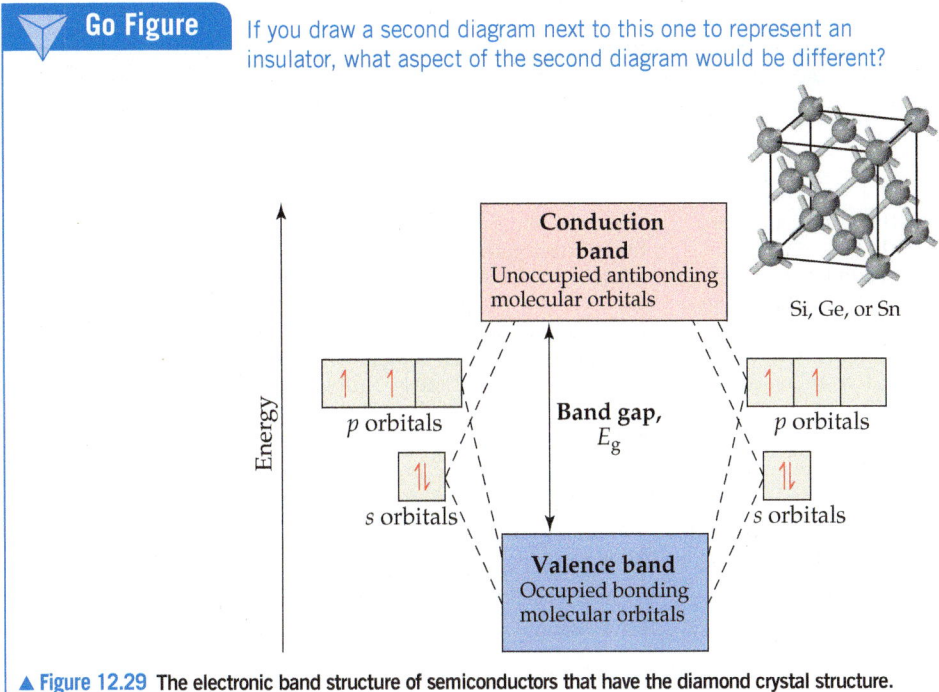

Go Figure If you draw a second diagram next to this one to represent an insulator, what aspect of the second diagram would be different?

▲ **Figure 12.29** The electronic band structure of semiconductors that have the diamond crystal structure.

GaAs is a semiconductor. Other examples are InP, where indium contributes three valence electrons and phosphorus contributes five, and CdTe, where cadmium provides two valence electrons and tellurium contributes six. In both cases, the average is again four valence electrons per atom. GaAs, InP, and CdTe all crystallize with a zinc blende structure.

The band gap of a compound semiconductor tends to increase as the difference in group numbers increases. For example, $E_g = 0.67$ eV in Ge, but $E_g = 1.43$ eV in GaAs. If we increase the difference in group number to four, as in ZnSe (groups 2B and 6A), the band gap increases to 2.70 eV. This progression is a result of the transition from pure covalent bonding in elemental semiconductors to polar covalent bonding in compound semiconductors. As the difference in electronegativity of the elements increases, the bonding becomes more polar and the band gap increases.

Electrical engineers manipulate both the orbital overlap and the bond polarity to control the band gaps of compound semiconductors for use in a wide range of electrical and optical devices. The band gaps of several elemental and compound semiconductors are listed in **Table 12.4**.

TABLE 12.4   Band Gaps of Select Elemental and Compound Semiconductors

Material	Structure Type	E_g, eV[†]
Si	Diamond	1.11
AlP	Zinc blende	2.43
Ge	Diamond	0.67
GaAs	Zinc blende	1.43
ZnSe	Zinc blende	2.58
Sn[‡]	Diamond	0.08
InSb	Zinc blende	0.18
CdTe	Zinc blende	1.50

[†] Band gap energies are room temperature values, 1 eV = 1.602×10^{-19} J.
[‡] These data are for gray tin, the semiconducting allotrope of tin. The other allotrope, white tin, is a metal.

Sample Exercise 12.3
Qualitative Comparison of Semiconductor Band Gaps

Will GaP have a larger or smaller band gap than ZnS? Will it have a larger or smaller band gap than GaN?

SOLUTION

Analyze The size of the band gap depends on the vertical and horizontal positions of the elements in the periodic table. The band gap will increase when either of the following conditions is met: (1) The elements are located higher up in the periodic table, where enhanced orbital overlap leads to a larger splitting between bonding and antibonding orbital energies, or (2) the horizontal separation between the elements increases, which leads to an increase in the electronegativity difference and bond polarity.

Plan We must look at the periodic table and compare the relative positions of the elements in each case.

Solve Gallium is in the fourth period and group 3A. Phosphorus is in the third period and group 5A. Zinc and sulfur are in the same periods as gallium and phosphorus, respectively. However, zinc, in group 2B, is one element to the left of gallium; sulfur, in group 6A,

is one element to the right of phosphorus. Thus, we would expect the electronegativity difference to be larger for ZnS, which should result in ZnS having a larger band gap than GaP.

For both GaP and GaN, the more electropositive element is gallium. So, we need only compare the positions of the more electronegative elements, P and N. Nitrogen is located above phosphorus in group 5A. Therefore, based on increased orbital overlap, we would expect GaN to have a larger band gap than GaP.

Check External references show that the band gap of GaP is 2.26 eV, ZnS is 3.6 eV, and GaN is 3.4 eV.

▶ **Practice Exercise**
Will ZnSe have a larger or smaller band gap than ZnS?

Semiconductor Doping

The electrical conductivity of a semiconductor is influenced by the presence of small numbers of impurity atoms. The process of adding controlled amounts of impurity atoms to a material is known as **doping**. Consider what happens when a few phosphorus atoms (known as dopants) replace silicon atoms in a silicon crystal. In pure Si, all of the valence-band molecular orbitals are filled and all of the conduction-band molecular orbitals are empty, as **Figure 12.30**(**a**) shows. Because phosphorus has five valence electrons but silicon has only four, the "extra" electrons that come with the dopant phosphorus atoms are forced to occupy the conduction band [Figure 12.30(**b**)]. The doped material is called an *n-type* semiconductor, where *n* signifies that the number of *n*egatively charged

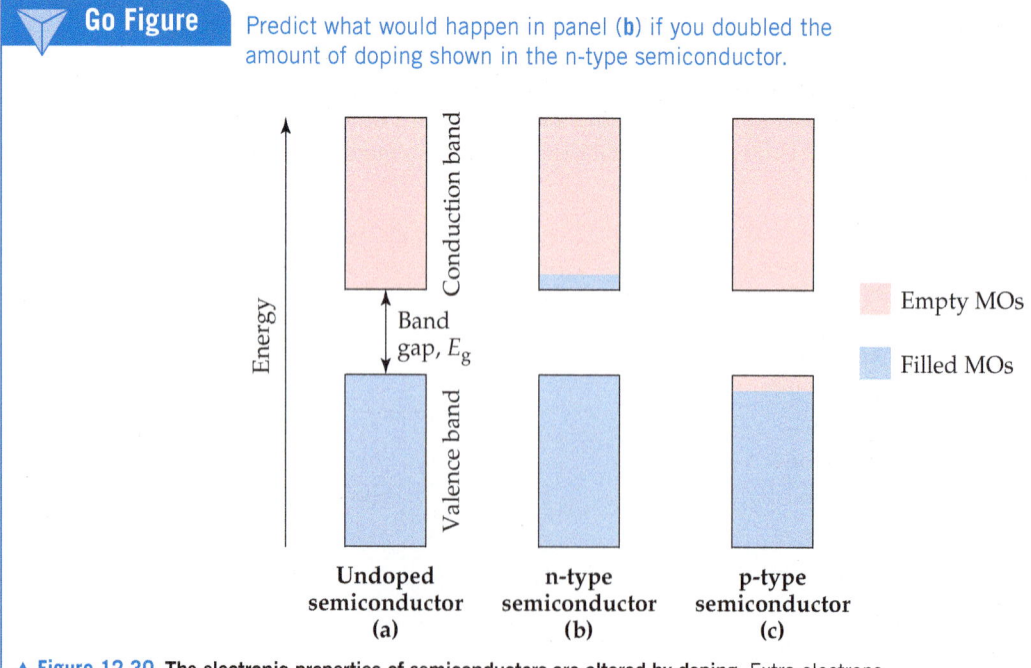

Go Figure Predict what would happen in panel (**b**) if you doubled the amount of doping shown in the n-type semiconductor.

▲ **Figure 12.30 The electronic properties of semiconductors are altered by doping.** Extra electrons in conduction band produce an n-type material, while extra holes in the valence band produce a p-type material.

CHEMISTRY AND SUSTAINABILITY | Solid-State Lighting

Artificial lighting is so widespread that we take it for granted. Major savings in energy would be realized if all incandescent lights could be replaced by light-emitting diodes (LEDs). Because LEDs are made from semiconductors, this is an appropriate place to take a closer look at the operation of an LED.

The heart of an LED is a p–n diode, which is formed by bringing an n-type semiconductor into contact with a p-type semiconductor. In the junction where they meet, there are very few electrons or holes to carry the charge across the interface between them, and the conductivity decreases. When an appropriate voltage is applied, electrons are driven from the conduction band of the n-doped side into the junction, where they meet holes that have been driven from

the valence band of the p-doped side. The electrons fall into the empty holes, and their energy is converted into light whose photons have energy equal to the band gap (Figure 12.31). In this way electrical energy is converted into optical energy.

Because the wavelength of light that is emitted depends on the band gap of the semiconductor, the color of light produced by the LED can be controlled by appropriate choice of semiconductor. Most red LEDs are made of a mixture of GaP and GaAs. The band gap of GaP is 2.26 eV (3.62×10^{-19} J), which corresponds to a green photon with a wavelength of 549 nm, while GaAs has a band gap of 1.43 eV (2.29×10^{-19} J), which corresponds to an infrared photon with a wavelength of 867 nm. (Sections 6.1 and 6.2) By forming solid solutions of these two compounds, with stoichiometries of $GaP_{1-x}As_x$, the band gap can be adjusted to any intermediate value. Thus, $GaP_{1-x}As_x$ is the solid solution of choice for red, orange, and yellow LEDs. Green LEDs are made from mixtures of GaP and AlP ($E_g = 2.43$ eV, $\lambda = 510$ nm).

Red LEDs have been in the market for decades, but to make white light, an efficient blue LED was needed. The first prototype bright blue LED was demonstrated in a Japanese laboratory in 1993. In 2010, less than 20 years later, over $10 billion worth of blue LEDs were sold worldwide. The blue LEDs are based on combinations of GaN ($E_g = 3.4$ eV, $\lambda = 365$ nm) and InN ($E_g = 2.4$ eV, $\lambda = 517$ nm). Many colors of LEDs are now available and are used in everything from barcode scanners to traffic lights. Because the light emission results from semiconductor structures that can be made extremely small and because they emit little heat, LEDs are replacing standard incandescent and fluorescence light bulbs in many applications.

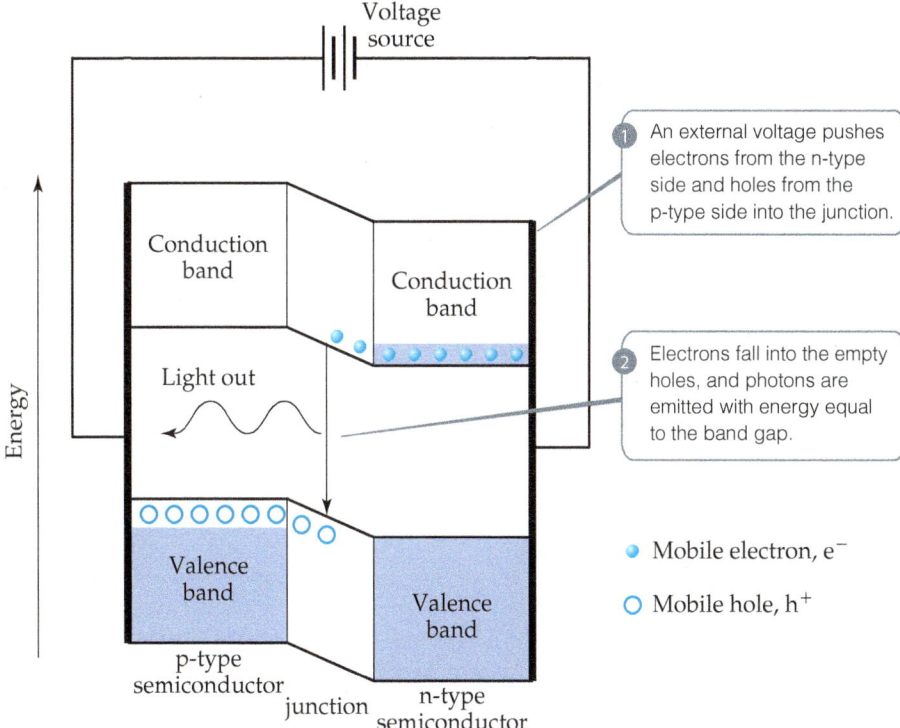

Voltage source

An external voltage pushes electrons from the n-type side and holes from the p-type side into the junction.

Electrons fall into the empty holes, and photons are emitted with energy equal to the band gap.

● Mobile electron, e⁻

○ Mobile hole, h⁺

▲ **Figure 12.31 Light-emitting diodes.** The heart of a light-emitting diode is a p–n junction in which an applied voltage drives electrons and holes together, where they combine and give off light.

Related Exercises: 12.77–12.80

electrons in the conduction band has increased. These extra electrons can move very easily in the conduction band. Thus, just a few parts per million (ppm) of phosphorus in silicon can increase silicon's intrinsic conductivity by a factor of a million.

The dramatic change in conductivity in response to the addition of a trace amount of a dopant means that extreme care must be taken to control the impurities in semiconductors. The semiconductor industry uses "nine-nines" silicon to make integrated circuits; what this means is that Si must be 99.999999999% pure (nine nines after the decimal place) to be technologically useful. Doping provides an opportunity for controlling the electrical conductivity through precise control of the type and concentration of dopants.

It is also possible to dope semiconductors with atoms that have fewer valence electrons than the host material. Consider what happens when a few aluminum atoms replace silicon atoms in a silicon crystal. Aluminum has only three valence electrons compared to silicon's four. Thus, there are electron vacancies, known as **holes**, in the valence band when silicon is doped with aluminum [Figure 12.30(c)]. Because the negatively charged electron is not there, the hole can be thought of as having a positive charge. Any adjacent electron that jumps into the hole leaves behind a new hole. Thus, the positive hole moves about in the lattice like a particle.* A material like this is called a *p-type* semiconductor, where *p* signifies that the number of *p*ositive holes in the material has increased.

As with n-type conductivity, p-type dopant levels of only parts per million can lead to a millionfold increase in conductivity—but in this case, the holes in the valence band are doing the conduction [Figure 12.30(c)].

The control of the electrical properties of silicon by doping forms the basis for the computers we have today. The junction of an n-type semiconductor with a p-type semiconductor leads to devices such as diodes (see the "Chemistry and Sustainability" box on "Solid-State Lighting"), transistors, and solar cells.

Sample Exercise 12.4
Identifying Types of Semiconductors

Which of the following elements, if doped into silicon, would yield an n-type semiconductor: Ga, As, or C?

SOLUTION

Analyze An n-type semiconductor means that the dopant atoms must have more valence electrons than the host material. Silicon is the host material in this case.

Plan We must look at the periodic table and determine the number of valence electrons associated with Si, Ga, As, and C. The elements with more valence electrons than silicon are the ones that will produce an n-type material upon doping.

Solve Si is in group 4A, so it has four valence electrons. Ga is in group 3A, so it has three valence electrons. As is in group 5A, so it has five valence electrons; C is in group 4A, so it has four valence electrons. Therefore, As, if doped into silicon, would yield an n-type semiconductor.

▶ **Practice Exercise**

Compound semiconductors can be doped to make n-type and p-type materials, but the scientist has to make sure that the proper atoms are substituted. For example, if Ge were doped into GaAs, Ge could substitute for Ga, making an n-type semiconductor; but if Ge substituted for As, the material would be p-type. Suggest a way to dope CdSe to create a p-type material.

Self-Assessment Exercises

SAE 12.16 Which of the following statements is/are *true*?

  (i) The melting points of molecular solids are lower than those of ionic solids because the bonds in molecular solids are weaker than those in ionic solids.

  (ii) The shape of a molecule does not influence its melting point or boiling point.

(a) i only **(b)** ii only **(c)** both i and ii **(d)** neither i nor ii

SAE 12.17 Which of the following statements associated with covalent-network solids is/are *true*?

  (i) Covalent-network solids generally have lower electrical conductivities than metals.

  (ii) The presence of impurities can increase the conductivity of covalent-network solids.

  (iii) Covalent-network solids generally have similar melting points to molecular solids.

(a) Only one of the statements is true. **(b)** Statements i and ii are true. **(c)** Statements i and iii are true. **(d)** Statements ii and iii are true. **(e)** All three statements are true.

SAE 12.18 Which of the following will have the largest band gap? **(a)** GaSe **(b)** InSe **(c)** ZnSe **(d)** CdSe

SAE 12.19 Which of the following diagrams accurately portrays the electronic structure of germanium doped with phosphorus?

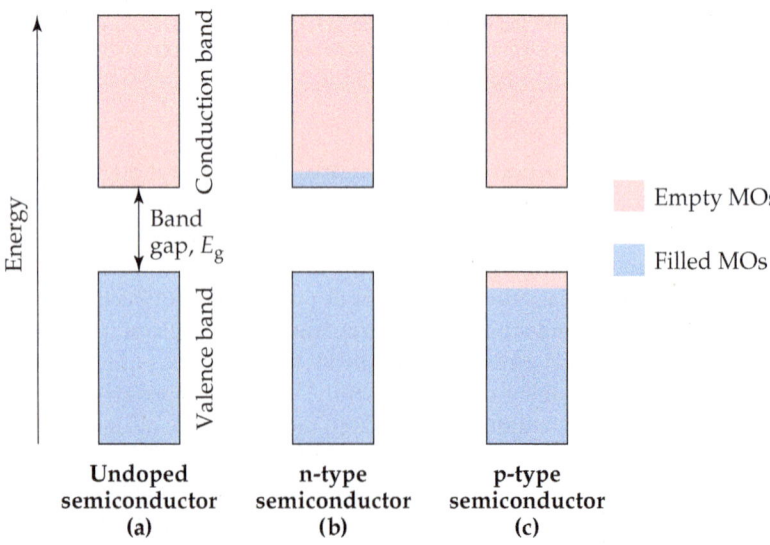

* This movement is analogous to watching people changing seats in a classroom; you can watch the people (electrons) move about the seats (atoms), or you can watch the empty seats (holes) "move."

12.6 | Polymers

In nature we find many substances of very high molecular weight, running up to millions of amu, that make up much of the structure of living organisms and tissues. Some examples are starch and cellulose, which abound in plants, as well as proteins, which are found in both plants and animals. In 1827 Jöns Jakob Berzelius coined the word **polymer** (from the Greek *polys*, "many," and *meros*, "parts") to denote molecular substances of high molecular weight formed by the *polymerization* (joining together) of **monomers**, molecules with low molecular weight.

Historically, natural polymers, such as wool, leather, silk, and natural rubber, were processed into usable materials. During the past 70 years or so, chemists have learned to form synthetic polymers by polymerizing monomers through controlled chemical reactions. A great many of these synthetic polymers have a backbone of carbon–carbon bonds because carbon atoms have an exceptional ability to form strong stable bonds with one another.

Plastics are polymeric solids that can be formed into various shapes, usually by the application of heat and pressure. There are several types of plastics.

Thermoplastics can be reshaped. For example, plastic milk containers are made from the thermoplastic polymer *polyethylene*. These containers can be melted down and the polymer recycled for some other use. Over two million tons of plastic are recycled every year in the United States. If you look at the bottom of a plastic container, you are likely to see a recycle symbol containing a number, as shown in Figure 12.32. The number and the letter abbreviations below it indicate the kind of polymer from which the container is made, as summarized in Table 12.5. These symbols make it possible to sort containers by composition. In general, the lower the number, the greater the ease with which the material can be recycled.

Unlike a thermoplastic, a **thermosetting plastic** (also called a *thermoset*) is shaped through irreversible chemical processes and, therefore, cannot be reshaped readily. Vulcanized rubber and polyurethanes found in commercial products, including insulating foams and mattresses, are familiar examples of thermosetting plastics.

Another type of plastic is the **elastomer**, which is a material that exhibits rubbery or elastic behavior. When subjected to stretching or bending, an elastomer regains its original shape upon removal of the distorting force, if it has not been distorted beyond some elastic limit. Rubber is the most familiar example of an elastomer.

Some polymers, such as nylon and polyesters, both of which are thermosetting plastics, can be formed into fibers that, like hair, are very long relative to their cross-sectional area. These fibers can be woven into fabrics and cords and fashioned into clothing, tire cord, and other useful objects.

Making Polymers

A good example of a polymerization reaction is the formation of polyethylene from ethylene molecules (Figure 12.33). In this reaction, the double bond in each ethylene molecule "opens up," and two of the electrons originally in this bond are used to form

▲ Figure 12.32 **Recycling symbols.** Most plastic containers manufactured today carry a recycling symbol that indicates the type of polymer used to make the container and the polymer's suitability for recycling.

Learning Objectives

When you finish Section 12.6, you should be able to:

▶ Describe how polymers are formed from monomers.

▶ Differentiate polymers from other types of solids.

▶ Describe the relationships between the structure, bonding, and physical properties of polymers.

TABLE 12.5   Categories Used for Recycling Polymeric Materials in the United States

Number	Abbreviation	Polymer
1	PET or PETE	Polyethylene terephthalate
2	HDPE	High-density polyethylene
3	PVC	Polyvinyl chloride
4	LDPE	Low-density polyethylene
5	PP	Polypropylene
6	PS	Polystyrene
7	None	Other

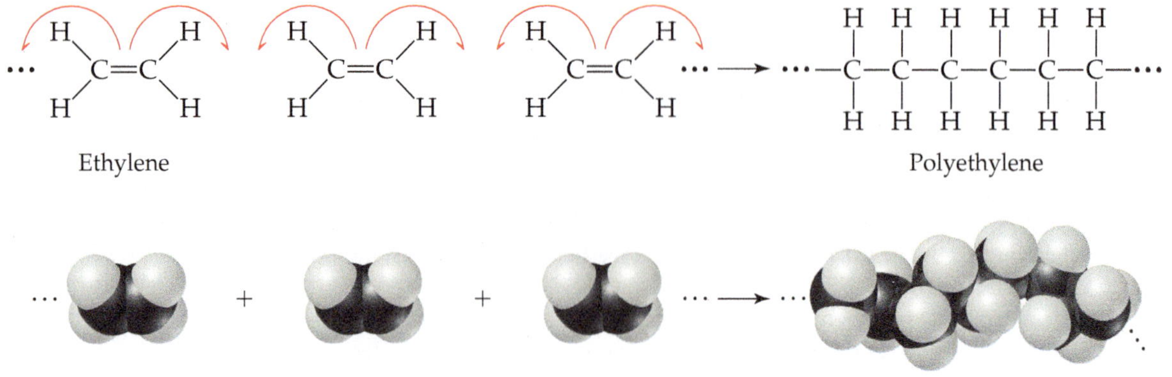

▲ Figure 12.33 **The polymerization of ethylene monomers to make the polymer polyethylene.**

new C—C single bonds with two other ethylene molecules. This type of polymerization, in which monomers are coupled through their multiple bonds, is called **addition polymerization**.

We can write the equation for the polymerization reaction as follows:

$$n\ \text{CH}_2\!=\!\text{CH}_2 \longrightarrow \left[\begin{array}{cc} \text{H} & \text{H} \\ | & | \\ \text{C}\!-\!\text{C} \\ | & | \\ \text{H} & \text{H} \end{array}\right]_n$$

Here n represents the large number—ranging from hundreds to many thousands—of monomer molecules (ethylene in this case) that react to form one polymer molecule. Within the polymer, a repeat unit (the partial structure shown in brackets for the polymerization of ethylene) appears over and over along the entire chain. The ends of the chain are capped by carbon–hydrogen bonds or by some other bond, so that the end carbons have four bonds.

Polyethylene is an important material; its annual production exceeds 190 billion pounds each year. Although its composition is simple, the polymer is not easy to make. The right manufacturing conditions were identified only after many years of research. Today many forms of polyethylene, varying widely in physical properties, are known.

Polymers of other chemical compositions provide still greater variety in physical and chemical properties. **Table 12.6** lists several other common polymers obtained by addition polymerization.

A second general reaction used to synthesize commercially important polymers is **condensation polymerization**. In a condensation reaction, two molecules are joined to form a larger molecule by elimination of a small molecule, such as H_2O. For example, an amine (a compound containing —NH_2) reacts with a carboxylic acid (a compound containing —COOH) to form a bond between N and C plus an H_2O molecule (**Figure 12.34**).

Polymers formed from two different monomers are called **copolymers**. In the formation of many nylons, a *diamine*, a compound with the —NH_2 group at each end, is reacted with a *diacid*, a compound with the —COOH group at each end. For example, the copolymer nylon 6,6 is formed when a diamine that has six carbon atoms and an amino group on each end is reacted with adipic acid, which also has six carbon atoms (**Figure 12.35**). A condensation reaction occurs on each end of the diamine and the acid. Water is released, and N—C bonds are formed between molecules.

TABLE 12.6 Polymers of Commercial Importance

Polymer	Structure	Uses
Addition Polymers		
Polyethylene	$\left[CH_2-CH_2\right]_n$	Films, packaging, bottles
Polypropylene	$\left[CH_2-CH(CH_3)\right]_n$	Kitchenware, fibers, appliances
Polystyrene	$\left[CH_2-CH(C_6H_5)\right]_n$	Packaging, disposable food containers, insulation
Polyvinyl chloride (PVC)	$\left[CH_2-CH(Cl)\right]_n$	Pipe fittings, plumbing
Condensation Polymers		
Polyurethane	$\left[NH-R-NH-\underset{O}{C}-O-R'-O-\underset{O}{C}\right]_n$ R, R' = $-CH_2-CH_2-$ (for example)	"Foam" furniture stuffing, spray-on insulation, automotive parts, footwear, water-protective coatings
Polyethylene terephthalate (a polyester)	$\left[O-CH_2-CH_2-O-\underset{O}{C}-C_6H_4-\underset{O}{C}\right]_n$	Tire cord, magnetic tape, apparel, soft-drink bottles
Nylon 6,6	$\left[NH-(CH_2)_6-NH-\underset{O}{C}-(CH_2)_4-\underset{O}{C}\right]_n$	Home furnishings, apparel, carpet, fishing line, toothbrush bristles
Polycarbonate	$\left[O-C_6H_4-\underset{CH_3}{\overset{CH_3}{C}}-C_6H_4-O-\underset{O}{C}\right]_n$	Shatterproof eyeglass lenses, CDs, DVDs, bulletproof windows, greenhouses

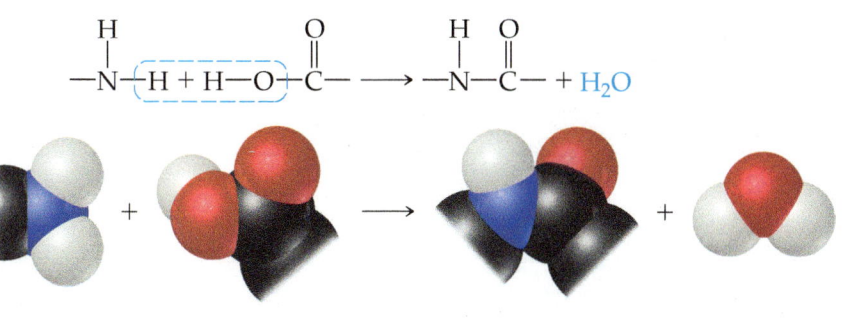

◀ Figure 12.34 A condensation polymerization.

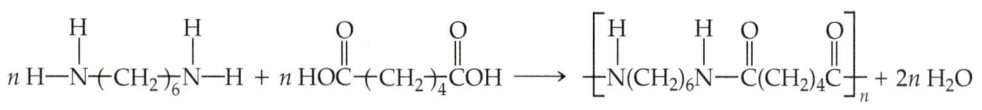

◀ Figure 12.35 The formation of the copolymer nylon 6,6.

Diamine   Adipic acid   Nylon 6,6

Table 12.6 lists nylon 6,6 and some other common polymers obtained by condensation polymerization. Notice that these polymers have backbones containing N or O atoms as well as C atoms.

CHEMISTRY AND SUSTAINABILITY | Modern Materials in the Automobile

There are over a billion motor vehicles in the world. Improvement in the fundamental understanding of the structure and properties of materials has enabled the development of motor vehicles that are safer, more powerful, more comfortable, and more fuel efficient. Let's take a look at some of these modern materials (Figure 12.36).

Metals and metal alloys are incorporated into many parts of an automobile. For example, aluminum is the primary component of the radiator, intake manifold, and engine block. Steel is typically the material used for the frame and body. Stainless steel is utilized in mufflers, exhaust silencers, and catalytic converters. The inside of the catalytic converter contains small particles of platinum-group metals that are deposited onto a honeycomb-shaped structural ceramic. The ceramic is composed of the ionic solids alumina (Al_2O_3) coated onto cordierite ($Mg_2Al_4Si_5O_{18}$). Semiconducting oxides are used for the oxygen sensor to monitor the air/fuel ratio in the exhaust gases, which is controlled by the engine control computer that is based on silicon, a covalent network solid. Polymers are found in a motor vehicle, too, with polyesters in seat covers and carpets, polycarbonate optical reflectors, and polypropylene in bumpers and car batteries. As electric vehicles become more common worldwide, the internal combustion engine of conventional motor vehicles will be replaced by batteries; we discuss more about the materials in various types of batteries in Chapter 20.

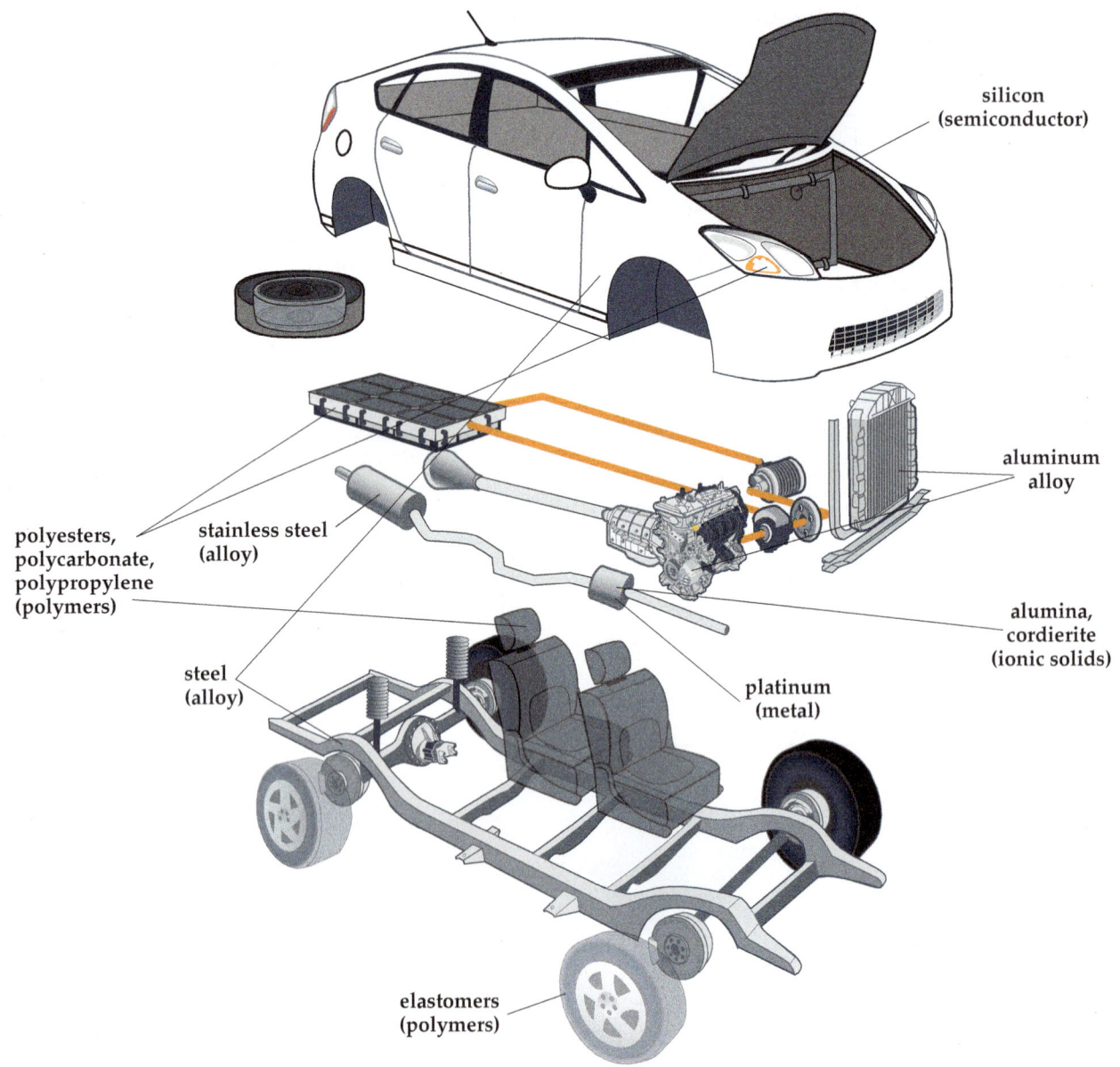

silicon (semiconductor)

aluminum alloy

polyesters, polycarbonate, polypropylene (polymers)

stainless steel (alloy)

alumina, cordierite (ionic solids)

steel (alloy)

platinum (metal)

elastomers (polymers)

▲ **Figure 12.36** Selected modern material components of an automobile.

Structure and Physical Properties of Polymers

The simple structural formulas given for polyethylene and other polymers are deceptive. Because four bonds surround each carbon atom in polyethylene, the atoms are arranged in a tetrahedral fashion, in which case the chain is not straight as we have depicted it. Furthermore, the atoms are relatively free to rotate around the C—C single bonds. Rather than being straight and rigid, therefore, the chains are flexible, folding readily (Figure 12.37). The flexibility in the molecular chains causes any material made of this polymer to be very flexible.

Both synthetic and natural polymers commonly consist of a collection of *macromolecules* (large molecules) of different molecular weights. Depending on the conditions of formation, the molecular weights may be distributed over a wide range or may be closely clustered around an average value. In part because of this distribution in molecular weights, polymers are largely amorphous (noncrystalline) materials. Rather than exhibiting a well-defined crystalline phase with a sharp melting point, polymers soften over a range of temperatures. They may, however, possess short-range order in some regions of the solid, with chains lined up in regular arrays as shown in Figure 12.38. The extent of such ordering is indicated by the degree of **crystallinity** of the polymer. Mechanical stretching or pulling to align the chains as the molten polymer is drawn through small holes can frequently enhance the crystallinity of a polymer. Intermolecular forces between the polymer chains hold the chains together in the ordered crystalline regions, making the polymer denser, harder, less soluble, and more resistant to heat. Table 12.7 shows how the properties of polyethylene change as the degree of crystallinity increases.

The linear structure of polyethylene is conducive to intermolecular interactions that lead to crystallinity. However, the degree of crystallinity in polyethylene strongly depends on the average molecular weight. Polymerization results in a mixture of macromolecules with varying values of *n* (numbers of monomer molecules) and, hence, varying molecular weights. Low-density polyethylene (LDPE), used in forming films and sheets, has an average molecular weight in the range of 10^4 amu; a density of less than 0.94 g/cm³; and substantial chain branching. That is, there are side chains off the main chain of the polymer. These side chains inhibit the formation of crystalline regions, reducing the density of the material. High-density polyethylene (HDPE), used to form bottles, drums, and pipes, has an average molecular weight in the range of 10^6 amu and a density of 0.94 g/cm³ or higher. This form has fewer side chains and thus a higher degree of crystallinity.

Polymers can be made stiffer by introducing chemical bonds between chains. Forming bonds between chains is called **cross-linking** (Figure 12.39). The greater the number of cross-links, the more rigid the polymer. Whereas thermoplastic materials consist of independent polymer chains, thermosetting plastics become cross-linked when heated; the cross-links allow them to hold their shapes.

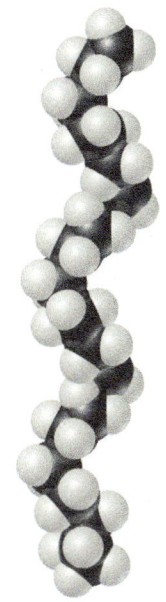

▲ Figure 12.37 **A segment of a polyethylene chain.** The segment shown here consists of 28 carbon atoms. In commercial polyethylenes, the number of CH_2 units ranges from approximately 10^3 to 10^5.

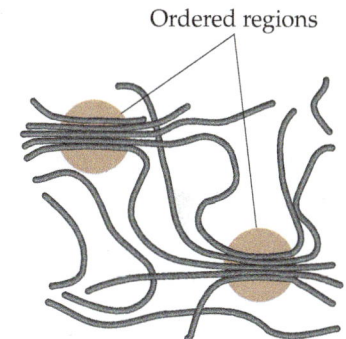

▲ Figure 12.38 **Interactions between polymer chains.** In the circled regions, the forces that operate between adjacent segments of the chains lead to ordering analogous to the ordering in crystals, though less regular.

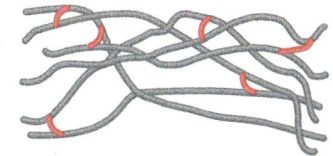

▲ Figure 12.39 **Cross-linking of polymer chains.** The cross-linking groups (red) constrain the relative motions of the polymer chains, making the material harder and less flexible than when the cross-links are not present.

TABLE 12.7   Properties of Polyethylene as a Function of Crystallinity

Properties	Crystallinity				
	55%	62%	70%	77%	85%
Melting point (°C)	109	116	125	130	133
Density (g/cm³)	0.92	0.93	0.94	0.95	0.96
Stiffness*	25	47	75	120	165
Yield stress*	1700	2500	3300	4200	5100

*These test results show that the mechanical strength of the polymer increases with increased crystallinity. The physical units for the stiffness test are psi $\times$ 10^{-3} (psi = pounds per square inch); those for the yield stress test are psi. Discussion of the exact meaning and significance of these tests is beyond the scope of this text.

An important example of cross-linking is the **vulcanization** of natural rubber, a process discovered by Charles Goodyear in 1839. Natural rubber is formed from a liquid resin derived from the inner bark of the *Hevea brasiliensis* tree. Chemically, it is a polymer of isoprene, C_5H_8 (**Figure 12.40**). Because rotation about the carbon–carbon double bond does not readily occur, the orientation of the groups bound to the carbons is rigid. In natural rubber, the chain extensions are on the same side of the double bond, as shown in Figure 12.40(**a**).

Natural rubber is not a useful polymer because it is too soft and too chemically reactive. Goodyear accidentally discovered that adding sulfur and then heating the mixture make the rubber harder and reduce its susceptibility to oxidation and other chemical degradation reactions. The sulfur changes rubber into a thermosetting polymer by cross-linking the polymer chains through reactions at some of the double bonds, as shown schematically in Figure 12.40(**b**). Cross-linking of about 5% of the double bonds creates a flexible, resilient rubber. When the rubber is stretched, the cross-links help prevent the chains from slipping, so that the rubber retains its elasticity. Because heating was an important step in his process, Goodyear named it after Vulcan, the Roman god of fire.

Most polymers contain sp^3-hybridized carbon atoms lacking delocalized π electrons (Section 9.6), so they are usually electrical insulators and are colorless (which implies a large band gap). However, if the backbone of the polymer has resonance (Sections 8.6 and 9.6), the electrons can become delocalized over long distances, which can lead to semiconducting behavior in the polymer. Such "plastic electronics" or "organic electronics" are of great current interest for lightweight and flexible organic solar cells, light-emitting diodes, wearable electronics, and other devices that are based on carbon rather than inorganic semiconductors like silicon.

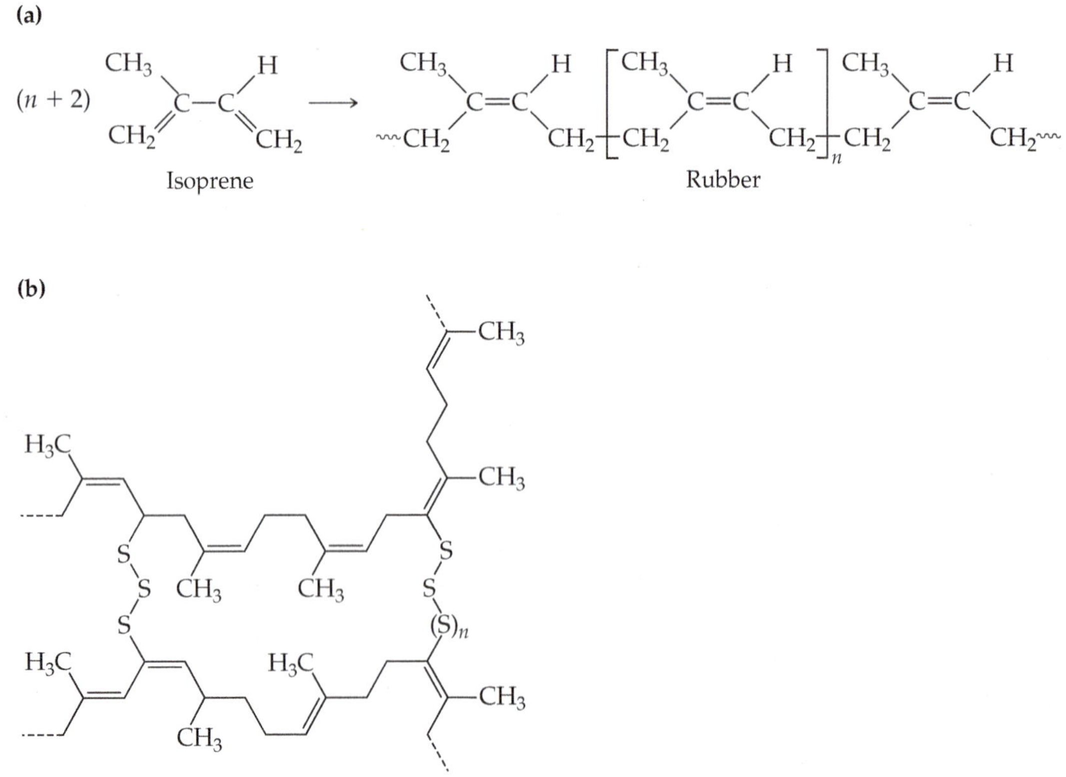

(a)

Isoprene

Rubber

(b)

▲ **Figure 12.40 Vulcanization of natural rubber.** (**a**) Formation of polymeric natural rubber from the monomer isoprene. (**b**) Adding sulfur to rubber creates carbon–sulfur bonds and sulfur–atom links between chains.

Self-Assessment Exercises

SAE 12.20 Which of the following can be used as a monomer?

    (**i**) Vinyl chloride
    (**ii**) Glucose
    (**iii**) Amino acids

(**a**) only i (**b**) i and ii (**c**) i and iii (**d**) ii and iii (**e**) all three

SAE 12.21 Polymers that are electrical insulators and colorless lack delocalized π electrons. What is the hybridization of the carbon atoms of these polymers? (**a**) sp^2 (**b**) sp^3 (**c**) both sp^2 and sp^3 (**d**) neither sp^2 nor sp^3

SAE 12.22 Which of the following statements is/are *true*?

    (**i**) Polystyrene is a condensation polymer.
    (**ii**) Crosslinked polymers are generally stiffer than their non-crosslinked counterparts.
    (**iii**) The more crystalline a polymer, the lower its melting point.

(**a**) Only one of these statements is true. (**b**) Statements i and ii are true. (**c**) Statements ii and iii are true. (**d**) Statements i and iii are true. (**e**) All three statements are true.

12.7 | Nanomaterials

The prefix *nano* means 10^{-9}. (Section 1.5) When people speak of "nanotechnology," they usually mean making devices that are on the 1–100-nm scale. It turns out that the properties of semiconductors and metals change in this size range. **Nanomaterials**—materials that have dimensions on the 1–100-nm scale—are under intense investigation in research laboratories around the world, and chemistry plays a central role in this investigation.

Semiconductors on the Nanoscale

Figure 12.21 shows that, in small molecules, electrons occupy discrete molecular orbitals, whereas in macroscale solids the electrons occupy delocalized bands. At what point does a molecule get so large that it starts behaving as though it has delocalized bands rather than localized molecular orbitals? For semiconductors, both theory and experiment tell us that the answer is roughly at 1 to 10 nm (about 10–100 atoms across). The exact number depends on the specific semiconductor material. The equations of quantum mechanics that were used for electrons in atoms can be applied to electrons (and holes) in semiconductors to estimate the size where materials undergo a crossover from molecular orbitals to bands. Because these effects become important at 1 to 10 nm, semiconductor particles with diameters in this size range are called *quantum dots*.

One of the most spectacular effects of reducing the size of a semiconductor crystal is that the band gap changes substantially with size in the 1–10-nm range. As the particle gets smaller, the band gap gets larger, an effect observable by the naked eye, as shown in Figure 12.41. On the macro level, the semiconductor cadmium phosphide looks black because its band gap is small ($E_g = 0.5$ eV), and it absorbs all wavelengths of visible light. As the crystals are made smaller, the material progressively changes color until it looks white. It looks white because now no visible light is absorbed. The band gap is so large that only high-energy ultraviolet light can excite electrons into the conduction band ($E_g > 3.0$ eV).

> ### Learning Objectives
>
> When you finish Section 12.7, you should be able to:
>
> ▶ Describe how the bulk properties of metals and semiconductors change as the size of the crystals decrease to the nanometer-length scale.
>
> ▶ Describe the structures and unique properties of fullerenes, carbon nanotubes, and graphene.

Bulk   Semiconductor particles ~1–10 nm diameter are "quantum dots."   Molecular limit

Decreasing size   Increasing band gap energy

◀ **Figure 12.41 Cd_3P_2 powders with different particle sizes.** The arrow indicates decreasing particle size and a corresponding increase in the band gap energy, resulting in different colors.

As the size of the quantum dots decreases, does the wavelength of the emitted light increase or decrease?

Size of CdSe quantum dots

2 nm ⟶ 7 nm

2.7 eV ⟶ 2.0 eV

Band gap energy, E_g

▲ **Figure 12.42 Photoluminescence depends on particle size at the nanoscale.** When illuminated with ultraviolet light, these solutions, each containing nanoparticles of the semiconductor CdSe, emit light that corresponds to their respective band gap energies. The wavelength of the light emitted depends on the size of the CdSe nanoparticles.

Making quantum dots is most easily accomplished using chemical reactions in solution. For example, to make CdS, you can mix $Cd(NO_3)_2$ and Na_2S in water. If you do not do anything else, you will precipitate large crystals of CdS. However, if you first add a negatively charged polymer to the water (such as polyphosphate, $-(OPO_2^-)_n-$), the Cd^{2+} associates with the polymer, like tiny "meatballs" in the polymer "spaghetti." When sulfide is added, CdS particles grow, but the polymer keeps them from forming large crystals. A great deal of fine-tuning of reaction conditions is necessary to produce nanocrystals that are of uniform size and shape.

As we learned in Section 12.5, some semiconductor devices can emit light when a voltage is applied. Another way to make semiconductors emit light is to illuminate them with light whose photons have energies larger than the energy of the band gap of the semiconductor, a process called *photoluminescence*. A valence-band electron absorbs a photon and is promoted to the conduction band. If the excited electron then falls back down into the hole it left in the valence band, it emits a photon having energy equal to the band gap energy. In the case of quantum dots, the band gap is tunable with the crystal size, and thus all the colors of the rainbow can be obtained from just one material, as shown for CdSe in **Figure 12.42**.

Quantum dots are being explored for applications ranging from electronics to lasers to medical imaging because they are very bright, very stable, and small enough to be taken up by living cells even after being coated with a biocompatible surface layer.

Semiconductors do not have to be shrunk to the nanoscale in all three dimensions to show new properties. They can be laid down in relatively large two-dimensional areas on a substrate but can be only a few nanometers thick to make *quantum wells*. *Quantum wires*, in which the semiconductor wire diameter is only a few nanometers but its length is very long, have also been made by various chemical routes. In both quantum wells and quantum wires, measurements along the nanoscale dimension(s) show quantum behavior, but in the long dimension, the properties seem to be just like those of the bulk material.

Metals on the Nanoscale

Metals also have unusual properties on the 1–100-nm-length scale. Fundamentally, this is because the mean free path (Section 10.6) of an electron in a metal at room temperature is typically about 1–100 nm. So, when the particle size of a metal is 100 nm or less, you might expect unusual effects because the "sea of electrons" encounters a "shore" (the surface of the particle).

Although it was not fully understood, people have known for hundreds of years that metals are different when they are very finely divided. Dating back to the Middle Ages, the makers of stained-glass windows knew that gold dispersed in molten glass made the glass a beautiful deep red (**Figure 12.43**). Much later, in 1857, Michael Faraday reported that dispersions of small gold particles could be made stable and were deeply colored—some of the original colloidal solutions that he made are still in the Royal Institution of Great Britain's Faraday Museum in London.

Other physical and chemical properties of metallic nanoparticles are also different from the properties of the bulk materials. Gold particles less than 20 nm in diameter melt at a far lower temperature than bulk gold, for instance, and when the particles are between 2 and 3 nm in diameter, gold is no longer a "noble," unreactive metal. In this size range, it becomes chemically reactive and is under active exploration as a catalyst.

At nanoscale dimensions, silver has properties analogous to those of gold in its beautiful colors, although it is more reactive than gold. Currently, research laboratories around the world are showing great interest in taking advantage of the unusual optical properties of metal nanoparticles for applications in biomedical imaging and chemical detection.

▲ **Figure 12.43 Stained glass window from the Chartres Cathedral in France.** Gold nanoparticles are responsible for the red color in this window, which dates back to the twelfth century.

CHEMISTRY AND SUSTAINABILITY | Microporous and Mesoporous Materials

Macroporous materials have pores that are visible to the naked eye. Examples include the everyday synthetic sponge [Figure 12.44(**a**)] and the honeycomb-like cordierite core of an automobile catalytic converter [Figure 12.44(**b**)], with pore sizes in the mm size range and the tens of μm size range, respectively. *Microporous* and *mesoporous* materials have much smaller pores that are not visible to the naked eye. **Microporous** solids have pores up to 2 nm in size, whereas **mesoporous** solids have pore sizes in the 2–50 nm range.

Microporous and mesoporous materials have a large surface area relative to their volume because of their numerous pores and cavities. **Nanomaterials**, on the other hand, have a large surface area relative to their volume because of their small particle size. The size-dependent properties of these materials have led researchers to investigate their fundamental science and application.

Zeolites, which occur naturally and can also be synthesized, are a class of aluminosilicates that have been known since 1756. There are several hundred types of microporous and mesoporous zeolites. These substances adopt a variety of structures with polyhedral cavities connected by tunnels that often resemble a honeycomb [Figure 12.44(**c**)]. The interior surfaces attract ions and molecules that weakly interact with the rigid framework of aluminum, silicon, and oxygen atoms. Various pore and cavity sizes can be prepared by varying the chemical composition and synthesis method.

Zeolites may be synthesized with weakly interacting ions occupying the cavities. Upon exposure to ions that interact more strongly with the interior surfaces, there is a preferential exchange of weakly interacting ions for more strongly interacting ions. This effectively creates what we might think of as an ionic sponge. An example that illustrates this behavior is the use of a sodium zeolite to remove radioactive cesium (^{134}Cs and ^{137}Cs) from contaminated areas around the Fukushima Daiichi nuclear power plants in Japan, which were damaged by an earthquake and tsunami in 2011. The cesium ions are attracted into the cavities of the zeolite where an ion exchange occurs (Cs^+ for Na^+). Another example is treatment of water from wells that contain relatively high concentrations of calcium, magnesium, and iron ions—so-called hard water. Heated hard water can cause problems by forming deposits inside of pipes, reducing flow over time. Hard water can be "softened" by passing it through zeolites containing sodium ions that are replaced or exchanged for calcium ions in the hard water. Periodic flushing of the zeolite with an aqueous solution containing a high concentration of sodium ions removes the calcium ions and renews the zeolite for further use. More recently, chemists have developed "metal-organic frameworks," or MOFs, which are the most porous materials on Earth; for example, a MOF the size of a pea has the surface area of a football field. Such highly porous materials are being examined for many applications, ranging from environmental remediation to new catalysts.

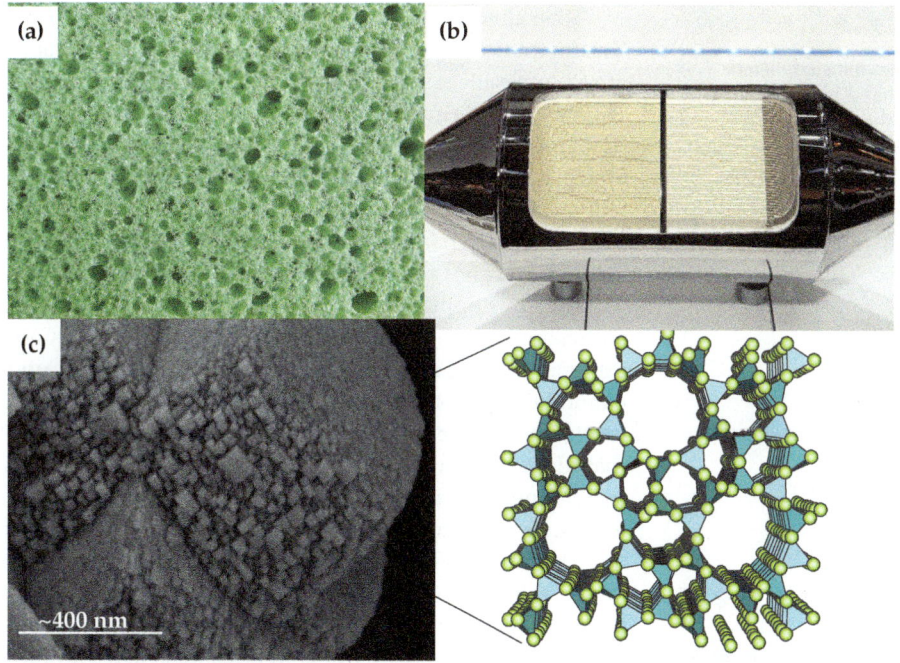

▲ **Figure 12.44 Porous materials. (a)** Sponges and **(b)** the core of an automobile catalytic converter are macroporous. **(c)** Zeolite ZSM-5 is a microporous material with pore sizes near 0.5 nm.

Carbon on the Nanoscale

We have seen that elemental carbon is quite versatile. In its bulk sp^3-hybridized solid-state form, it is diamond; in its bulk sp^2-hybridized solid-state form, it is graphite. Over the past three decades, scientists have discovered that sp^2-hybridized carbon can also form discrete molecules, one-dimensional nanoscale tubes, and two-dimensional nanoscale sheets. Each of these forms of carbon shows very interesting properties.

Go Figure

How many bonds does each carbon atom in C_{60} make? Based on this observation, would you expect the bonding in C_{60} to be more like that in diamond or that in graphite?

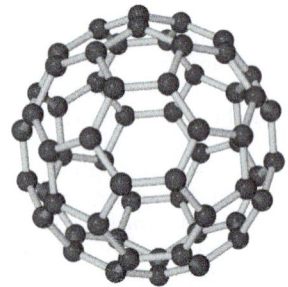

▲ **Figure 12.45 Buckminsterfullerene, C_{60}.** The molecule has a highly symmetric structure in which the 60 carbon atoms sit at the vertices of a truncated icosahedron. The bottom view shows only the bonds between carbon atoms.

Until the mid-1980s, pure solid carbon was thought to exist in only two forms: the covalent-network solids diamond and graphite. In 1985, however, a group of researchers led by Richard Smalley and Robert Curl of Rice University and Harry Kroto of the University of Sussex, England, vaporized a sample of graphite with an intense pulse of laser light and used a stream of helium gas to carry the vaporized carbon into a mass spectrometer. [See the "A Closer Look" box on "The Mass Spectrometer." (Section 2.4)] The mass spectrum showed peaks corresponding to clusters of carbon atoms, with a particularly strong peak corresponding to molecules composed of 60 carbon atoms, C_{60}.

Because C_{60} clusters were so preferentially formed, the group proposed a radically different form of carbon—namely, nearly spherical C_{60} *molecules*. They proposed that the carbon atoms of C_{60} form a "ball" with 32 faces, 12 of them pentagons and 20 hexagons (Figure 12.45), exactly like a soccer ball. The shape of this molecule is reminiscent of the geodesic dome invented by the U.S. engineer and philosopher R. Buckminster Fuller (1895–1983), so C_{60} was whimsically named "buckminsterfullerene," or "buckyball" for short. Since the discovery of C_{60}, other related molecules made of pure carbon have been discovered. These molecules are now known as fullerenes.

Appreciable amounts of buckyball can be prepared by electrically evaporating graphite in an atmosphere of helium gas. About 14% of the resulting soot consists of C_{60} and a related molecule, C_{70}, which has a more elongated structure. The carbon-rich gases from which C_{60} and C_{70} condense also contain other fullerenes, mostly containing more carbon atoms, such as C_{76} and C_{84}. The smallest possible fullerene, C_{20}, was first detected in 2000. This small, ball-shaped molecule is much more reactive than the larger fullerenes. Because fullerenes are molecules, they dissolve in various organic solvents, whereas diamond and graphite do not. This solubility permits fullerenes to be separated from the other components of soot and even from one another. It also allows the study of their reactions in solution.

Soon after the discovery of C_{60}, chemists discovered carbon nanotubes (Figure 12.46). You can think of these as sheets of graphite rolled up and capped at one or both ends by half of a C_{60} molecule. Carbon nanotubes are made in a manner similar to that used to make C_{60}. They can be made in either *multiwall* or *single-walled* forms. Multiwall carbon nanotubes consist of tubes within tubes, nested together, whereas single-walled carbon nanotubes consist of single tubes. Single-walled carbon nanotubes can be 1000 nm long or even longer but are only about 1 nm in diameter. Depending on the diameter of the graphite sheet and how it is rolled up, carbon nanotubes can behave as either semiconductors or metals.

The fact that carbon nanotubes can be made either semiconducting or metallic without any doping is unique among solid-state materials, and laboratories worldwide are making and testing carbon-based electronic devices. Carbon nanotubes are also being explored for their mechanical properties. The carbon–carbon bonded framework of the nanotubes means that the imperfections that might appear in a metal nanowire of similar dimensions are nearly absent. Experiments on individual carbon nanotubes suggest that they are stronger than steel, if steel were the dimensions of a carbon nanotube. Carbon nanotubes have been spun into fibers with polymers, adding great strength and toughness to the composite material.

The two-dimensional form of carbon, graphene, is the most recent low-dimensional form of carbon to be experimentally isolated and studied. Although its properties had been the subject of theoretical predictions for over 60 years, it was not until 2004 that researchers at the University of Manchester in England isolated and identified individual sheets of carbon atoms with the honeycomb structure shown in Figure 12.47. Amazingly, the technique they used to isolate single-layer graphene was to successively peel away thin layers of graphite using adhesive tape. Individual layers of graphene were then transferred to a silicon wafer having a precisely defined overcoat of SiO_2. When a single layer of graphene is left on the wafer, an interference-like contrast pattern results that can be seen with an optical microscope. If not for this simple yet effective way to scan for individual graphene crystals, they would probably still remain undiscovered. Subsequently, it has been shown that graphene can be deposited on clean surfaces of other types of crystals. The scientists who led the effort at the University of Manchester, Andre Geim and Konstantin Novoselov, were awarded the 2010 Nobel Prize in Physics for their work.

The properties of graphene are remarkable. It is very strong and has a record thermal conductivity, topping carbon nanotubes in both categories. Graphene is a semimetal,

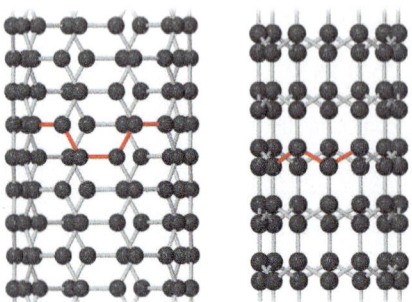

▲ **Figure 12.46 Atomic models of carbon nanotubes.** Left: "Armchair" nanotube, which shows metallic behavior. Right: "Zigzag" nanotube, which can be either semiconducting or metallic, depending on tube diameter.

which means its electronic structure is like that of a semiconductor in which the energy gap is exactly zero. The combination of graphene's two-dimensional character and the fact that it is a semimetal allows the electrons to travel very long distances, up to 0.3 μm, without scattering from another electron, atom, or impurity. Graphene can sustain electrical current densities six orders of magnitude higher than those sustainable in copper. Even though it is only one atom thick, graphene can absorb 2.3% of sunlight that strikes it. Scientists are currently exploring ways to incorporate graphene in various technologies, including electronics, sensors, batteries, and solar cells.

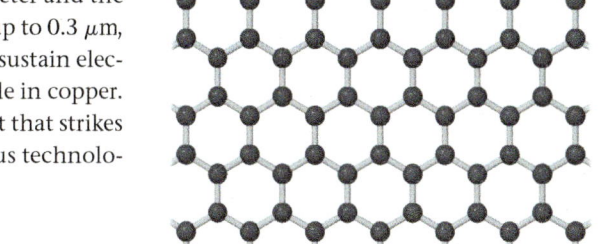

▲ Figure 12.47 A portion of a two-dimensional graphene sheet.

 Self-Assessment Exercises

SAE 12.23   Which of the following statements is *false*?

(**a**) Nanomaterials have dimensions on the 1–100 nm scale. (**b**) The band gap of a semiconductor crystal does not depend on particle size. (**c**) The physical and chemical properties of metallic nanoparticles are different from the properties of the bulk materials. (**d**) The wavelength of light emitted in quantum dots is influenced by particle size.

SAE 12.24   Which of the following statements is/are *true*?

   (**i**) Fullerenes are soluble in organic solvents.
   (**ii**) In order to be conducting, carbon nanotubes must be doped with an impurity.

(**a**) i only (**b**) ii only (**c**) both i and ii (**d**) neither i nor ii

---

 Putting Concepts Together

Polymers that can conduct electricity are called *conducting polymers*. Some polymers can be made semiconducting; others can be nearly metallic. Polyacetylene is an example of a polymer that is a semiconductor. It can also be doped to increase its conductivity.

    Polyacetylene is made from acetylene in a reaction that looks simple but is actually tricky to do:

$$H-C\equiv C-H \qquad \text{┤}CH=CH\text{├}_n$$

$$\text{Acetylene} \qquad\qquad \text{Polyacetylene}$$

(**a**)   What is the hybridization of the carbon atoms, and the geometry around those atoms, in acetylene and in polyacetylene?

(**b**)   Write a balanced equation to make polyacetylene from acetylene.

(**c**)   Acetylene is a gas at room temperature and pressure (298 K, 1.00 atm). How many grams of polyacetylene can you make from a 5.00-L vessel of acetylene gas at room temperature and room pressure? Assume that acetylene behaves ideally and that the polymerization reaction occurs with 100% yield.

(**d**)   Using the average bond enthalpies in Table 8.3, predict whether the formation of polyacetylene from acetylene is endothermic or exothermic.

(**e**)   A sample of polyacetylene absorbs light from 300 nm down to 650 nm. What is its band gap, in electron volts?

SOLUTION

Analyze For part (a), we need to recall what we have learned about sp, sp^2, and sp^3 hybridization and geometry. (Section 9.5) For part (b), we need to write a balanced equation. For part (c), we need to use the ideal-gas equation. (Section 10.3) For part (d), we need to recall the definitions of endothermic and exothermic and how bond enthalpies can be used to predict overall reaction enthalpies. (Section 8.8) For part (e), we need to relate the absorption of light to the differences in energy levels between filled and empty states in a material. (Section 6.3)

Plan For part (a), we should draw out the chemical structures of the reactant and product. For part (b), we need to make sure the

equation is properly balanced. For part (c), we need to convert from liters of gas to moles of gas, using the ideal-gas equation ($PV = nRT$); then we need to convert from moles of acetylene gas to moles of polyacetylene using the answer from part (b); finally, we can convert to grams of polyacetylene. For part (d), we need to recall that $\Delta H_{rxn} = \sum(\text{bond enthalpies of bonds broken}) - \sum(\text{bond enthalpies of bonds formed})$. (Section 8.8) For part (e), we need to realize that the lowest energy absorbed by a material will tell us its band gap E_g (for a semiconductor or insulator) and combine $E = h\nu$ and $c = \lambda\nu$ together ($E = hc/\lambda$) to solve for E_g.

Solve

(a) Carbon usually forms four bonds. Thus, each C atom must have a single bond to H and a triple bond to the other C atom in acetylene. As a result, each C atom has two electron domains and must be *sp* hybridized. This *sp* hybridization also means that the H—C—C angles in acetylene are 180° and the molecule is linear. We can write out the partial structure of polyacetylene as follows:

Each carbon is identical but now has three bonding electron domains that surround it. Therefore, the hybridization of each carbon atom is *sp*², and each carbon has local trigonal planar geometry with 120° angles.

(b) We can write:

Note that all atoms originally present in acetylene end up in the polyacetylene product.

$$n C_2H_2(g) \longrightarrow \text{—}[CH\text{—}CH]_n$$

(c) We can use the ideal-gas equation as follows:

$$PV = nRT$$
$$(1.00 \text{ atm})(5.00 \text{ L}) = n(0.08206 \text{ L-atm/K-mol})(298 \text{ K})$$
$$n = 0.204 \text{ mol}$$

Acetylene has a molar mass of 26.0 g/mol; therefore, the mass of 0.204 mol is:

$$(0.204 \text{ mol})(26.0 \text{ g/mol}) = 5.32 \text{ g acetylene}$$

Note that from the answer to part (b), all the atoms in acetylene go into polyacetylene. Due to conservation of mass, then, the mass of polyacetylene produced must also be 5.32 g, if we assume 100% yield.

(d) Let's consider the case for $n = 1$. We note that the reactant side of the equation in part (b) has one C≡C triple bond and two C—H single bonds. The product side of the equation in part (b) has one C=C double bond, one C—C single bond (to link to the adjacent monomer), and two C—H single bonds. Therefore, we are breaking one C≡C triple bond and are forming one C=C double bond and one C—C single bond. Accordingly, the enthalpy change for polyacetylene formation is:

$$\Delta H_{rxn} = (C≡C \text{ enthalpy}) - (C=C \text{ enthalpy}) - (C—C \text{ enthalpy})$$
$$= (839 \text{ kJ/mol}) - (614 \text{ kJ/mol}) - (348 \text{ kJ/mol})$$
$$= -123 \text{ kJ/mol}$$

Because ΔH is a negative number, the reaction releases heat and is exothermic.

(e) The sample of polyacetylene absorbs many wavelengths of light, but the one we care about is the longest one, which corresponds to the lowest energy.

We recognize that this energy corresponds to the energy difference between the bottom of the conduction band and the top of the valence band, so it is equivalent to the band gap E_g. Now we have to convert the number to electron volts. Because $1.602 \times 10^{-19} \text{ J} = 1 \text{ eV}$, we find that:

$$E = hc/\lambda$$
$$= (6.626 \times 10^{-34} \text{ J-s})(3.00 \times 10^8 \text{ m-s}^{-1})/(650 \times 10^{-9} \text{ m})$$
$$= 3.06 \times 10^{-19} \text{ J}$$

$$E_g = 1.91 \text{ eV}$$

Chapter Summary and Key Terms

CLASSIFICATION AND STRUCTURES OF SOLIDS (SECTION 12.1) The structures and properties of solids can be classified according to the forces that hold the atoms together. **Metallic solids** are held together by a delocalized sea of collectively shared valence electrons. **Ionic solids** are held together by the mutual attraction between cations and anions. **Covalent-network solids** are held together by an extended network of covalent bonds. **Molecular solids** are held together by weak intermolecular forces. **Polymers** contain very long chains of atoms held together by covalent bonds. These chains are usually held to one another by weaker intermolecular forces. **Nanomaterials** are solids where the dimensions of individual crystals are on the order of 1–100 nm. In **crystalline solids**, particles are arranged in a regularly repeating pattern. In **amorphous solids**, however, particles show no long-range order. In a crystalline solid, the smallest repeating unit is called a **unit cell**. All unit cells in a crystal contain an identical arrangement of atoms. The geometrical pattern of points on which the unit cells are arranged is called a **crystal lattice**. To generate a crystal structure, a **motif**, which is an atom or group of atoms, is associated with each and every **lattice point**.

In two dimensions, the unit cell is a parallelogram whose size and shape are defined by two **lattice vectors** (*a* and *b*). There are five **primitive lattices**, lattices in which the lattice points are located only at the corners of the unit cell: square, hexagonal, rectangular, rhombic, and oblique. In three dimensions, the unit cell is a parallelepiped whose size and shape are defined by three lattice vectors (*a*, *b*, and *c*), and there are seven primitive lattices: cubic, tetragonal, hexagonal, rhombohedral, orthorhombic, monoclinic, and triclinic. Placing an additional lattice point at the center of a cubic unit cell leads to a **body-centered cubic lattice**, while placing an additional point at the center of each face of the unit cell leads to a **face-centered cubic lattice**.

METALLIC SOLIDS (SECTION 12.2) **Metallic solids** are typically good conductors of electricity and heat, *malleable*, which means that they can be hammered into thin sheets, and *ductile*, which means that they can be drawn into wires. Metals tend to form structures where the atoms are closely packed. Two related forms of close packing, **cubic close packing** and **hexagonal close packing**, are possible. In both, each atom has a **coordination number** of 12.

Alloys are materials that possess characteristic metallic properties and are composed of more than one element. The elements in an alloy can be distributed either homogeneously or heterogeneously. Alloys, which contain homogeneous mixtures of elements, can be either substitutional or interstitial alloys. In a **substitutional alloy**, the atoms of the minority element(s) occupy positions normally occupied by atoms of the majority element. In an **interstitial alloy**, atoms of the minority element(s), often smaller nonmetallic atoms, occupy interstitial positions that lie in the "holes" between atoms of the majority element. In a **heterogeneous alloy**, the elements are not distributed uniformly; instead, two or more distinct phases with characteristic compositions are present. **Intermetallic compounds** are alloys that have a fixed composition and definite properties.

METALLIC BONDING (SECTION 12.3) The properties of metals can be accounted for in a qualitative way by the **electron-sea model**, in which the electrons are visualized as being free to move throughout the metal. In the molecular-orbital model, the valence atomic orbitals of the metal atoms interact to form energy **bands** that are incompletely filled by valence electrons. Consequently, the electronic structure of a bulk solid is referred to as a **band structure**. The orbitals that constitute the energy band are delocalized over the atoms of the metal, and their energies are closely spaced. In a metal the valence shell *s*, *p*, and *d* orbitals form bands, and these bands overlap, resulting in one or more partially filled bands. Because the energy differences between orbitals *within a band* are extremely small, promoting electrons to higher-energy orbitals requires very little energy. This gives rise to high electrical and thermal conductivity, as well as other characteristic metallic properties.

IONIC SOLIDS (SECTION 12.4) **Ionic solids** consist of cations and anions held together by electrostatic attractions. Because these interactions are quite strong, ionic compounds tend to have high melting points. The attractions become stronger as the charges of the ions increase and/or the sizes of the ions decrease. The presence of both attractive (cation–anion) and repulsive (cation–cation and anion–anion) interactions helps to explain why ionic compounds are brittle. Like metals, the structures of ionic compounds tend to be symmetric, but to minimize direct contact between ions of like charge, the coordination numbers (typically 4 to 8) are necessarily smaller than those seen in close-packed metals. The exact structure depends on the relative sizes of the ions and the cation-to-anion ratio in the empirical formula.

MOLECULAR AND COVALENT-NETWORK SOLIDS (SECTION 12.5) **Molecular solids** consist of atoms or molecules held together by intermolecular forces. Because these forces are relatively weak, molecular solids tend to be soft and possess low melting points. The melting point depends on the strength of the intermolecular forces, as well as the efficiency with which the molecules can pack together.

Covalent-network solids consist of atoms held together in large networks by covalent bonds. These solids are much harder and have higher melting points than molecular solids. Important examples include diamond, where the carbons are tetrahedrally coordinated to each other, and graphite, where the sp^2-hybridized carbon atoms form hexagonal layers. **Semiconductors** are solids that do conduct electricity, but to a far lesser extent than metals. **Insulators** do not conduct electricity at all.

Elemental semiconductors, like Si and Ge, as well as compound semiconductors, like GaAs, InP, and CdTe, are important examples of covalent-network solids. In a semiconductor, the filled bonding molecular orbitals make up the **valence band**, while the empty anti-bonding molecular orbitals make up the **conduction band**. The valence and conduction bands are separated by an energy that is referred to as the **band gap**, E_g. The size of the band gap increases as the bond length decreases and as the difference in electronegativity between the two elements increases.

Doping semiconductors changes their ability to conduct electricity by orders of magnitude. An n-type semiconductor is one that is doped so that there are excess electrons in the conduction band; a p-type semiconductor is one that is doped so that there are missing electrons, which are called **holes**, in the valence band.

POLYMERS (SECTION 12.6) **Polymers** are molecules of high molecular weight formed by joining large numbers of small molecules called **monomers**. **Plastics** are materials that can be formed into various shapes, usually by the application of heat and pressure. **Thermoplastic** polymers can be reshaped, typically through heating, in contrast to **thermosetting plastics**, which are formed into objects through an irreversible chemical process and cannot readily be reshaped. An **elastomer** is a material that exhibits elastic behavior; that is, it returns to its original shape following stretching or bending.

In an **addition polymerization** reaction, the molecules form new linkages by opening existing π bonds. Polyethylene forms, for example, when the carbon–carbon double bonds of ethylene open up. In a **condensation polymerization** reaction, the monomers are joined by eliminating a small molecule between them. The various kinds of nylon are formed, for example, by removing a water molecule between an amine and a carboxylic acid. A polymer formed from two different monomers is called a **copolymer**.

Polymers are largely amorphous, but some materials possess a degree of **crystallinity**. For a given chemical composition, the crystallinity depends on the molecular weight and the degree of branching along the main polymer chain. Polymer properties are also strongly affected by **cross-linking**, in which short chains of atoms connect the long polymer chains. Rubber is cross-linked by short chains of sulfur atoms in a process called **vulcanization**.

NANOMATERIALS (SECTION 12.7) When one or more dimensions of a material become sufficiently small, generally smaller than 100 nm, the properties of the material change. Materials with dimensions on this length scale are called **nanomaterials**. Quantum dots are semiconductor particles with diameters of 1–10 nm. In this size range the material's band gap energy becomes size-dependent. Metal nanoparticles have different chemical and physical properties in the 1–100-nm size range. Nanoparticles of gold, for example, are more reactive than bulk gold and no longer have a golden color. Nanoscience has produced a number of previously unknown forms of sp^2-hybridized carbon. Fullerenes, like C_{60}, are large molecules containing only carbon atoms. Carbon nanotubes are sheets of graphite rolled up. They can behave as either semiconductors or metals, depending on how the sheet was rolled. Graphene, which is an isolated layer from graphite, is a two-dimensional form of carbon. These nanomaterials are being developed now for many applications in electronics, batteries, solar cells, and medicine.

Key Equations

$$\frac{\text{Number of cations per formula unit}}{\text{Number of anions per formula unit}} = \frac{\text{anion coordination number}}{\text{cation coordination number}}$$ [12.1] Relationship between cation and anion coordination numbers and the empirical formula of an ionic compound

Exam Prep

EP 12.1 Ruthenium(IV) oxide is a black crystalline solid. It has a melting point of 1300 °C, is a conductor of electricity, and is insoluble in water. What type of solid is RuO_2? **(a)** molecular **(b)** metallic **(c)** ionic **(d)** covalent-network

EP 12.2 Germanium dioxide can be prepared in both crystalline and amorphous forms. What would allow you to differentiate between them?

   (i) density

  (ii) empirical formula

 (iii) percent composition

(a) only i **(b)** i and ii **(c)** i and iii **(d)** ii and iii **(e)** all three

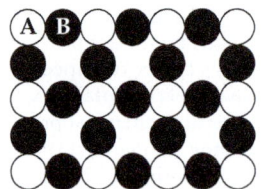

EP 12.3 For the two-dimensional crystal structure shown here, what is the lattice type and how many atoms are there per unit cell?

(a) lattice = square, atoms per unit cell = 1 A + 1 B

(b) lattice = square, atoms per unit cell = 1 A + 2 B

(c) lattice = centered rectangular, atoms per unit cell = 1 A + 2 B

(d) lattice = square, atoms per unit cell = 4 A + 4 B

(e) lattice = diamond, atoms per unit cell = 1 A + 2 B

EP 12.4 The structure of an intermetallic compound is shown here:

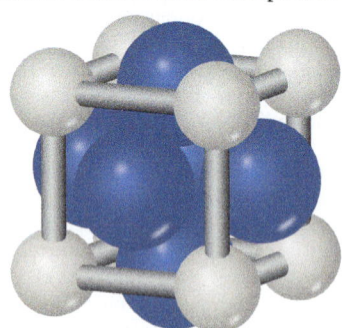

The lattice vectors a, b, and c are equal, and the angles between these vectors are 90°. What lattice corresponds to this structure?

(a) primitive cubic   **(d)** primitive tetragonal

(b) face-centered cubic   **(e)** primitive hexagonal

(c) body-centered cubic

EP 12.5 Of the seven three-dimensional primitive lattices, which ones have a unit cell where the lattice vectors (a, b, and c) are all the same length?

(a) tetragonal   **(d)** rhombohedral

(b) orthorhombic   **(e)** hexagonal

(c) monoclinic

EP 12.6 Which solid would you expect to have metallic properties?

(a) Ge   **(d)** NiAs

(b) Ir   **(e)** both (b) and (d)

(c) $SnCl_4$

EP 12.7 What is the coordination number of the atoms in the polonium structure?

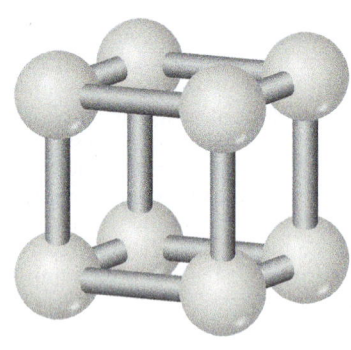

Polonium (PO)

(a) 2 **(b)** 3 **(c)** 6 **(d)** 7 **(e)** 8

EP 12.8 Silver crystallizes with a face-centered cubic structure. Given the molar mass (107.87 g/mol) and the density (10.5 g/cm³ at 20 °C) of silver, what value would you calculate for the radius of a silver atom?

(a) 1.29 Å **(b)** 1.35 Å **(c)** 1.45 Å **(d)** 1.60 **(e)** 1.72 Å

EP 12.9 Consider the two-dimensional square lattice shown here (see also Figure 12.4). The "packing efficiency" for a two-dimensional structure would be the area of the atoms divided by the area of the unit cell times 100%. What is the packing efficiency for a square lattice for atoms of radius $a/2$ that are centered at the lattice points?

(a) 3.14% **(b)** 15.7% **(c)** 31.8% **(d)** 74.0% **(e)** 78.5%

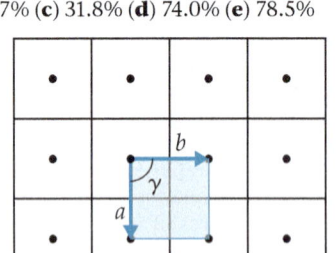

Square lattice ($a = b$, $\gamma = 90°$)

EP 12.10 Which of the following statements is *false*?

(a) Interstitial alloys and substitutional alloys both have variable compositions.

(b) Nonmetallic elements are typically present in both interstitial and substitutional alloys.

(c) Substitutional alloys are usually more malleable and ductile than interstitial alloys.

(d) Interstitial alloys and substitutional alloys are both examples of homogeneous solid solutions.

(e) Both bronze and brass are alloys that contain copper.

EP 12.11 Which of the following properties of metals *cannot* be adequately explained by the electron-sea model?

(a) the high electrical conductivity of metals

(b) the high thermal conductivity of metals

(c) the close-packed structures of most metals

(d) the high melting points of metals in the middle of the transition series, like rhenium (Re) and tungsten (W)

(e) the low melting points of metals at the very end of the transition series, like cadmium (Cd) and mercury (Hg)

EP 12.12 Which metal would you expect to have the lowest melting point?

(a) Cs **(b)** Sr **(c)** Nb **(d)** Mo

EP 12.13 What is the empirical formula of the following intermetallic structure if blue spheres are Ni and white spheres are Sn?

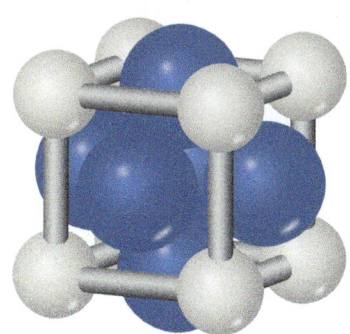

(a) Ni_6Sn_8

(b) Ni_3Sn_4

(c) Ni_3Sn

(d) Ni_3Sn_2

EP 12.14 Given the ionic radii and molar masses of Sc^{3+} (0.88 Å, 45.0 g/mol) and F^- (1.19 Å, 19.0 g/mol), what value do you estimate for the density of ScF_3, whose structure is shown here (see also Figure 12.26)?

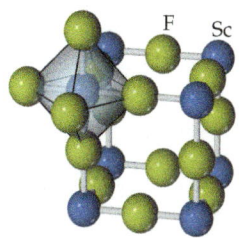

(a) 5.99 g/cm^3

(b) $1.44 \times 10^{24} \text{ g/mol}$

(c) 19.1 g/cm^3

(d) 2.39 g/cm^3

(e) 5.72 g/cm^3

EP 12.15 Which of the following statements is/are *true*?

(i) Molecular solids are hard and brittle.

(ii) Molecular solids are held together by the electrostatic interaction between a cation and anion.

(a) i only **(b)** ii only **(c)** both i and ii **(d)** neither i nor ii

EP 12.16 Which of the following properties is *not* characteristic of a covalent-network solid?

(a) They have high melting points.

(b) They are brittle.

(c) They are insoluble in polar solvents like water.

(d) Their electrical conductivity can range from insulating to semiconducting to metallic.

(e) The atoms in the crystal have a large number (8–12) of nearest neighbors.

EP 12.17 Which of these statements is *false*?

(a) As you go down group 4A in the periodic table, the elemental solids become more electrically conducting. **(b)** As you go down group 4A in the periodic table, the band gaps of the elemental solids decrease. **(c)** The valence electron count for a compound semiconductor averages out to four per atom. **(d)** Band gap energies of semiconductors range from ~0.1 to 3.5 eV. **(e)** In general, the more polar the bonds are in compound semiconductors, the smaller the band gap.

EP 12.18 Which of these doped semiconductors would yield a p-type material? (These choices are written as host atom:dopant atom.)

(a) Ge:P **(b)** Si:Ge **(c)** Si:Al **(d)** Ge:S **(e)** Si:N

Exercises

Visualizing Concepts

12.1 Two solids are shown in the photographs. One is a semiconductor and one is an insulator. Which one is which? Explain your reasoning. [Sections 12.1, 12.5]

12.2 For each of the two-dimensional structures shown here, **(a)** draw the unit cell, **(b)** determine the type of two-dimensional lattice (from Figure 12.4), and **(c)** determine how many of each type of circle (white or black) there are per unit cell. [Section 12.1]

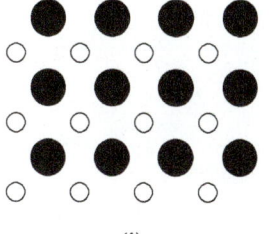

(i)   (ii)

12.3 Shown here are sketches of two processes. Which of the processes refers to the ductility of metals and which refers to the malleability of metals? [Section 12.2]

(a)

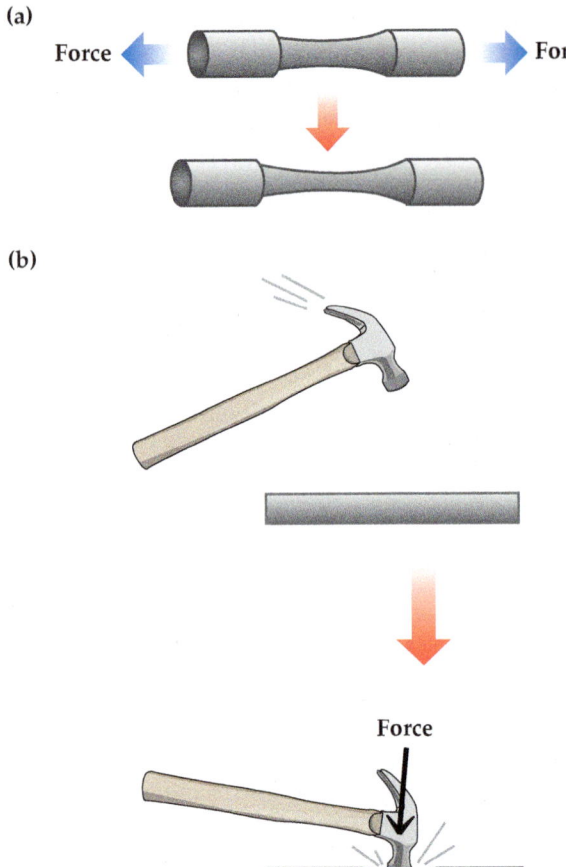

(b)

12.4 Which arrangement of atoms in a lattice represents close-packing? [Section 12.2]

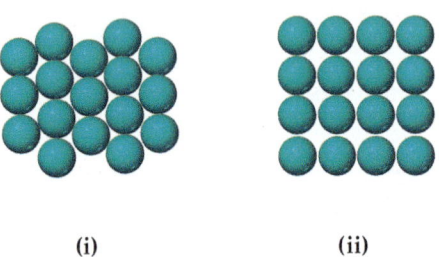

(i)   (ii)

12.5 **(a)** What kind of packing arrangement is seen in the accompanying photo? **(b)** What is the coordination number of each cannonball in the interior of the stack? **(c)** What are the coordination numbers for the numbered cannonballs on the visible side of the stack? [Section 12.2]

12.6 Which arrangement of cations (yellow) and anions (blue) in a lattice is the more stable? Explain your reasoning. [Section 12.4]

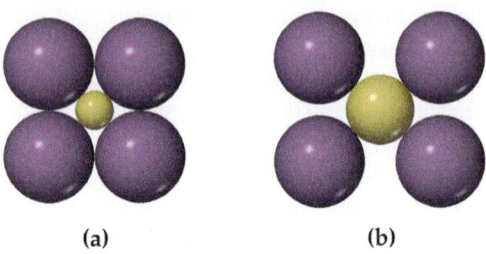

(a)   (b)

12.7 Which of these molecular fragments would you expect to be more likely to give rise to electrical conductivity? Explain your reasoning. [Sections 12.4, 12.5]

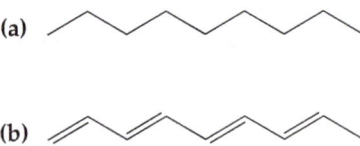

12.8 The electronic structure of a doped semiconductor is shown here. **(a)** Which band, A or B, is the valence band? **(b)** Which band is the conduction band? **(c)** Which region of the diagram represents the band gap? **(d)** Which band consists of bonding molecular orbitals? **(e)** Is this an example of an n-type or p-type semiconductor? **(f)** If the semiconductor is germanium, which of the following elements could be the dopant: Ga, Si, or P? [Section 12.5]

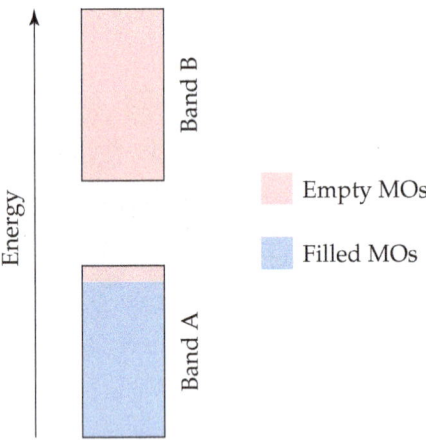

Empty MOs

Filled MOs

12.9 Shown here are cartoons of two different polymers. Which of these polymers would you expect to be more crystalline? Which one would have the higher melting point? [Section 12.6]

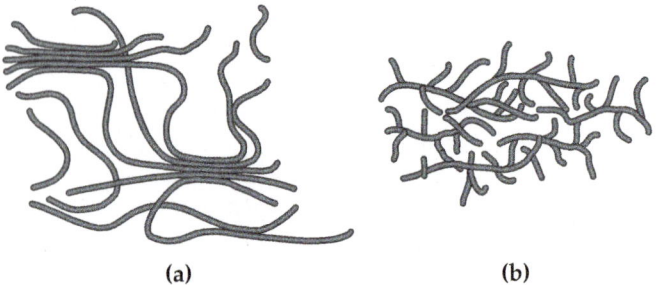

(a)   (b)

12.10 The accompanying image shows photoluminescence from four different samples of CdTe nanocrystals, each embedded in a polymer matrix. The photoluminescence occurs because the samples are being irradiated by a UV light source. The nanocrystals in each vial have different average sizes. The sizes are 4.0, 3.5, 3.2, and 2.8 nm. **(a)** Which vial contains the 4.0-nm nanocrystals? **(b)** Which vial contains the 2.8-nm nanocrystals? **(c)** Crystals of CdTe that have sizes that are larger than approximately 100 nm have a band gap of 1.5 eV. What would be the wavelength and frequency of light emitted from these crystals? What type of light is this? [Sections 12.5 and 12.7]

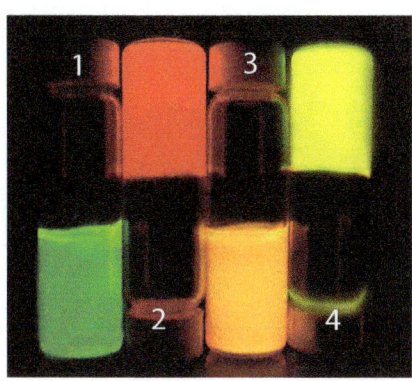

Classification and Structures of Solids (Section 12.1)

12.11 Covalent bonding occurs in both molecular and covalent-network solids. Which of the following statements best explains why these two kinds of solids differ so greatly in their hardness and melting points? **(a)** The molecules in molecular solids have stronger covalent bonding than covalent-network solids do. **(b)** The molecules in molecular solids are held together by weak intermolecular interactions. **(c)** The atoms in covalent-network solids are more polarizable than those in molecular solids. **(d)** Molecular solids are denser than covalent-network solids.

12.12 Silicon is the fundamental component of integrated circuits. Si has the same structure as diamond. **(a)** Is Si a molecular, metallic, ionic, or covalent-network solid? **(b)** Silicon readily reacts to form silicon dioxide, SiO_2, which is quite hard and is insoluble in water. Is SiO_2 most likely a molecular, metallic, ionic, or covalent-network solid?

12.13 What kinds of attractive forces exist between particles (atoms, molecules, or ions) in **(a)** molecular crystals, **(b)** covalent-network crystals, **(c)** ionic crystals, **(d)** and metallic crystals?

12.14 Which type (or types) of crystalline solid is characterized by each of the following? **(a)** high mobility of electrons throughout the solid **(b)** softness, relatively low melting point **(c)** high melting point and poor electrical conductivity **(d)** network of covalent bonds

12.15 Indicate the type of solid (molecular, metallic, ionic, or covalent-network) for each compound: **(a)** $CaSO_4$, **(b)** Pd, **(c)** Ta_2O_5 (melting point, 1872 °C), **(d)** caffeine ($C_8H_{10}N_4O_2$), **(e)** toluene (C_7H_8), **(f)** P_4.

12.16 Indicate the type of solid (molecular, metallic, ionic, or covalent-network) for each compound: **(a)** InAs, **(b)** MgO, **(c)** HgS, **(d)** In, **(e)** HBr.

12.17 You are given a gray substance that melts at 700 °C; the solid is a conductor of electricity and is insoluble in water. Which type of solid (molecular, metallic, covalent-network, or ionic) might this substance be?

12.18 You are given a white substance that melts at 100 °C. The substance is soluble in water. Neither the solid nor the solution is a conductor of electricity. Which type of solid (molecular, metallic, covalent-network, or ionic) might this substance be?

12.19 **(a)** Draw a picture that represents a crystalline solid at the atomic level. **(b)** Now draw a picture that represents an amorphous solid at the atomic level.

12.20 Amorphous silica, SiO_2, has a density of about 2.2 g/cm³, whereas the density of crystalline quartz, another form of SiO_2, is 2.65 g/cm³. Which of the following statements is the best explanation for the difference in density? **(a)** Amorphous silica is a network-covalent solid, but quartz is metallic. **(b)** Amorphous silica crystallizes in a primitive cubic lattice. **(c)** Quartz is harder than amorphous silica. **(d)** Quartz must have a larger unit cell than amorphous silica. **(e)** The atoms in amorphous silica do not pack as efficiently in three dimensions as compared to the atoms in quartz.

12.21 Two patterns of packing for two different circles of the same size are shown here. For each structure **(a)** draw the two-dimensional unit cell; **(b)** determine the angle between the lattice vectors, γ, and determine whether the lattice vectors are of the same length or of different lengths; and **(c)** determine the type of two-dimensional lattice (from Figure 12.4).

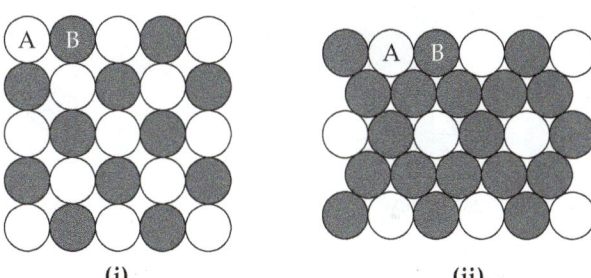

(i)   (ii)

12.22 Two patterns of packing two different circles of the same size are shown here. For each structure **(a)** draw the two-dimensional unit cell; **(b)** determine the angle between the lattice vectors, γ, and determine whether the lattice vectors are of the same length or of different lengths; **(c)** determine the type of two-dimensional lattice (from Figure 12.4).

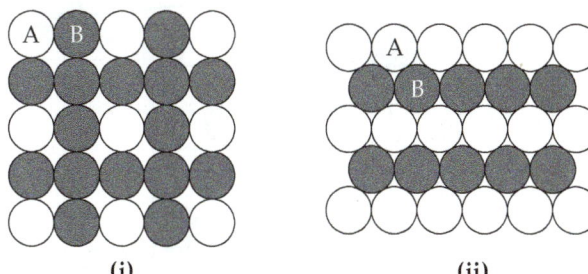

(i)   (ii)

12.23 Imagine the primitive cubic lattice. Now imagine grabbing the top of it and stretching it straight up. All angles remain 90°. What kind of primitive lattice have you made?

12.24 Imagine the primitive cubic lattice. Now imagine grabbing opposite corners and stretching it along the body diagonal while keeping the edge lengths equal. The three angles between the lattice vectors remain equal but are no longer 90°. What kind of primitive lattice have you made?

12.25 Which of the three-dimensional primitive lattices has a unit cell where none of the internal angles is 90°? (**a**) orthorhombic (**b**) hexagonal (**c**) rhombohedral (**d**) triclinic (**e**) both rhombohedral and triclinic

12.26 Besides the cubic unit cell, which other unit cell(s) has edge lengths that are all equal to each other? (**a**) orthorhombic (**b**) hexagonal (**c**) rhombohedral (**d**) triclinic (**e**) both rhombohedral and triclinic

12.27 What is the minimum number of atoms that could be contained in the unit cell of an element with a body-centered cubic lattice? (**a**) 1 (**b**) 2 (**c**) 3 (**d**) 4 (**e**) 5

12.28 What is the minimum number of atoms that could be contained in the unit cell of an element with a face-centered cubic lattice? (**a**) 1 (**b**) 2 (**c**) 3 (**d**) 4 (**e**) 5

12.29 The unit cell of nickel arsenide is shown here. (**a**) What type of lattice does this crystal possess? (**b**) What is the empirical formula?

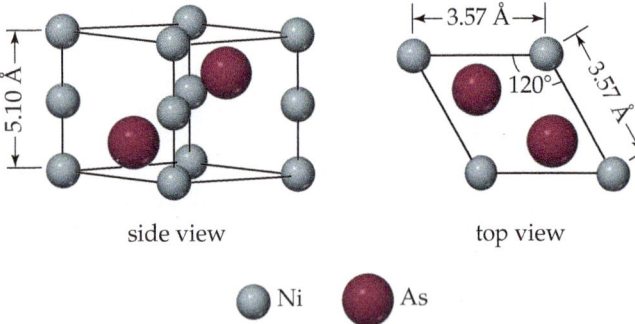

side view   top view

Ni   As

12.30 The unit cell of a compound containing potassium, aluminum, and fluorine is shown here. (**a**) What type of lattice does this crystal possess (all three lattice vectors are mutually perpendicular)? (**b**) What is the empirical formula?

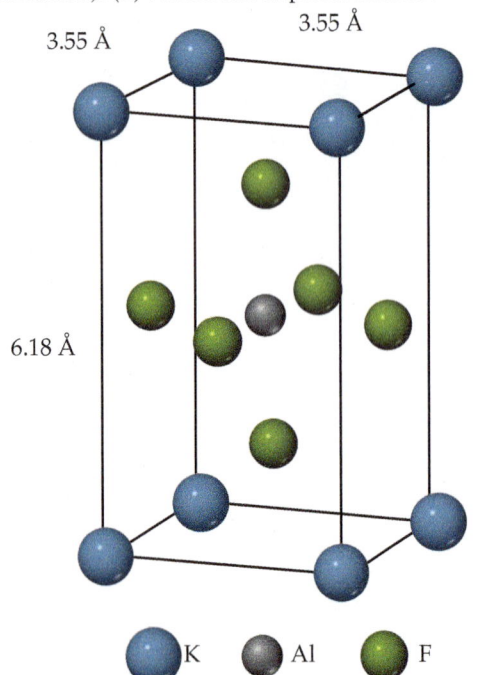

K   Al   F

Metallic Solids (Section 12.2)

12.31 The densities of the elements K, Ca, Sc, and Ti are 0.86, 1.5, 3.2, and 4.5 g/cm³, respectively. One of these elements crystallizes in a body-centered cubic structure; the other three crystallize in a face-centered cubic structure. Which one crystallizes in the body-centered cubic structure?

12.32 For each of these solids, state whether you would expect it to possess metallic properties: (**a**) $TiCl_4$, (**b**) NiCo alloy, (**c**) W, (**d**) Ge, (**e**) ScN.

12.33 Consider the unit cells shown here for three different structures that are commonly observed for metallic elements. (**a**) Which structure(s) corresponds to the densest packing of atoms? (**b**) Which structure(s) corresponds to the least dense packing of atoms?

      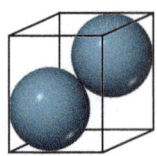

Structure type A   Structure type B   Structure type C

12.34 Sodium metal (atomic weight 22.99 g/mol) adopts a body-centered cubic structure with a density of 0.97 g/cm³. (**a**) Use this information and Avogadro's number ($N_A = 6.022 \times 10^{23}$/mol) to estimate the atomic radius of sodium. (**b**) If sodium didn't react so vigorously, it could float on water. Use the answer from part (a) to estimate the density of Na if its structure were that of a cubic close-packed metal. Would it still float on water?

12.35 Iridium crystallizes in a face-centered cubic unit cell that has an edge length of 3.833 Å. (**a**) Calculate the atomic radius of an iridium atom. (**b**) Calculate the density of iridium metal.

12.36 Calcium crystallizes in a body-centered cubic structure at 467 °C. (**a**) How many Ca atoms are contained in each unit cell? (**b**) How many nearest neighbors does each Ca atom possess? (**c**) Estimate the length of the unit cell edge, a, from the atomic radius of calcium (1.97 Å). (**d**) Estimate the density of Ca metal at this temperature.

12.37 Calcium crystallizes in a face-centered cubic unit cell at room temperature that has an edge length of 5.588 Å. (**a**) Calculate the atomic radius of a calcium atom. (**b**) Calculate the density of Ca metal at this temperature.

12.38 Calculate the volume in Å³ of each of the following types of cubic unit cells if it is composed of atoms with an atomic radius of 1.82 Å. (**a**) primitive (**b**) face-centered cubic

12.39 Aluminum metal crystallizes in a face-centered cubic unit cell. (**a**) How many aluminum atoms are in a unit cell? (**b**) What is the coordination number of each aluminum atom? (**c**) Estimate the length of the unit cell edge, a, from the atomic radius of aluminum (1.43 Å). (**d**) Calculate the density of aluminum metal.

12.40 An element crystallizes in a face-centered cubic lattice. The edge of the unit cell is 4.078 Å, and the density of the crystal is 19.30 g/cm³. Calculate the atomic weight of the element and identify the element.

12.41 Which of these statements about alloys and intermetallic compounds is *false*? (**a**) Bronze is an example of an alloy. (**b**) "Alloy" is just another word for "a chemical compound of fixed composition that is made of two or more metals." (**c**) Intermetallics are compounds of two or more metals that have a definite composition and are not considered alloys. (**d**) If you mix two metals together and, at the atomic level, they separate into two or more different compositional phases, you have created a heterogeneous alloy. (**e**) Alloys can be formed even if the atoms that comprise them are rather different in size.

12.42 Determine if each statement is *true* or *false*: (**a**) Substitutional alloys are solid solutions, but interstitial alloys are heterogenous alloys. (**b**) Substitutional alloys have "solute" atoms that replace "solvent" atoms in a lattice, but interstitial alloys have "solute" atoms that are in between the "solvent" atoms in a lattice. (**c**) The atomic radii of the atoms in a substitutional alloy are similar to each other, but in an interstitial alloy, the interstitial atoms are a lot smaller than the host lattice atoms.

12.43 For each of the following alloy compositions, indicate whether you would expect it to be a substitutional alloy, an interstitial alloy, or an intermetallic compound:

(**a**) $Fe_{0.97}Si_{0.03}$, (**b**) $Fe_{0.60}Ni_{0.40}$, (**c**) $SmCo_5$.

12.44 For each of the following alloy compositions, indicate whether you would expect it to be a substitutional alloy, an interstitial alloy, or an intermetallic compound:

(**a**) $Cu_{0.66}Zn_{0.34}$, (**b**) Ag_3Sn, (**c**) $Ti_{0.99}O_{0.01}$.

12.45 Indicate whether each statement is *true* or *false*:

(**a**) Substitutional alloys tend to be more ductile than interstitial alloys.

(**b**) Interstitial alloys tend to form between elements with similar ionic radii.

(**c**) Nonmetallic elements are never found in alloys.

12.46 Indicate whether each statement is *true* or *false*:

(**a**) Intermetallic compounds have a fixed composition.

(**b**) Copper is the majority component in both brass and bronze.

(**c**) In stainless steel, the chromium atoms occupy interstitial positions.

12.47 Pure gold crystallizes in a face-centered cubic unit cell with an edge length of 4.08 Å. The alloy called 18-karat gold consists of 75% Au, 15% Ag, and 10% Cu by mass. What is the empirical formula of 18-karat gold, to two significant digits?

12.48 An increase in temperature causes most metals to undergo *thermal expansion*, which means the volume of the metal increases upon heating. (**a**) How does thermal expansion affect the unit cell length? (**b**) What is the effect of an increase in temperature on the density of a metal?

Metallic Bonding (Section 12.3)

12.49 State whether each sentence is *true or false*:

(**a**) Metals have high electrical conductivities because the electrons in the metal are delocalized.

(**b**) Metals have high electrical conductivities because they are denser than other solids.

(**c**) Metals have large thermal conductivities because they expand when heated.

(**d**) Metals have small thermal conductivities because the delocalized electrons cannot easily transfer the kinetic energy imparted to the metal from heat.

12.50 Imagine that you have a metal bar sitting half in the Sun and half in the dark. On a sunny day, the part of the metal that has been sitting in the Sun feels hot. If you touch the part of the metal bar that has been sitting in the dark, will it feel hot or cold? Justify your answer in terms of thermal conductivity.

12.51 The molecular-orbital diagrams for two- and four-atom linear chains of lithium atoms are shown in Figure 12.21. Construct a molecular-orbital diagram for a chain containing six lithium atoms and use it to answer the following questions: (**a**) How many molecular orbitals are there in the diagram? (**b**) How many nodes are in the lowest-energy molecular

orbital? (**c**) How many nodes are in the highest-energy molecular orbital? (**d**) How many nodes are in the highest-energy occupied molecular orbital (HOMO)? (**e**) How many nodes are in the lowest-energy unoccupied molecular orbital (LUMO)? (**f**) How does the HOMO–LUMO energy gap for this case compare to that of the four-atom case?

12.52 Repeat Exercise 12.51 for a linear chain of eight lithium atoms.

12.53 Which would you expect to be the more ductile element: (**a**) Ag or Mo or (**b**) Zn or Si?

12.54 Which of the following statements does not follow from the fact that the alkali metals have relatively weak metal–metal bonding?

(**a**) The alkali metals are less dense than other metals.

(**b**) The alkali metals are soft enough to be cut with a knife.

(**c**) The alkali metals are more reactive than other metals.

(**d**) The alkali metals have higher melting points than other metals.

(**e**) The alkali metals have low ionization energies.

12.55 Arrange the following metals in increasing order of expected melting point: Mo, Zr, Y, Nb.

12.56 For each of the following groups, which metal would you expect to have the highest melting point: (**a**) gold, rhenium, or cesium; (**b**) rubidium, molybdenum, or indium; (**c**) ruthenium, strontium, or cadmium?

Ionic Solids (Section 12.4)

12.57 Tausonite, a mineral composed of Sr, O, and Ti, has the cubic unit cell shown in the drawing. (**a**) What is the empirical formula of this mineral? (**b**) How many oxygens are coordinated to titanium? (**c**) To see the full coordination environment of the other ions, we have to consider neighboring unit cells. How many oxygens are coordinated to strontium?

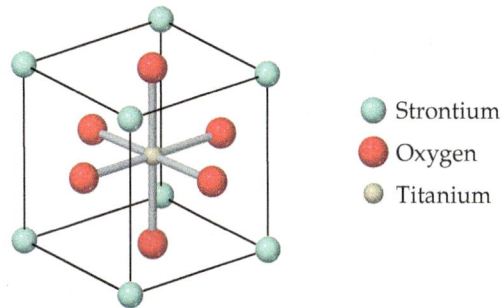

● Strontium
● Oxygen
● Titanium

12.58 The unit cell of a compound containing Co and O has a unit cell shown in the diagram. The Co atoms are on the corners, and the O atoms are completely within the unit cell. (**a**) What is the empirical formula of this compound? (**b**) What is the oxidation state of the metal?

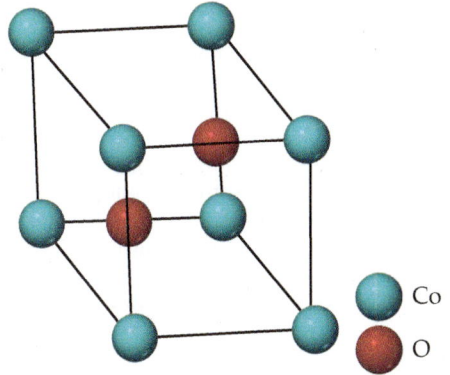

● Co
● O

12.59 Alabandite is a mineral composed of manganese(II) sulfide (MnS). The mineral adopts the rock salt structure. The length of an edge of the MnS unit cell is 5.223 Å at 25 °C. Determine the density of MnS in g/cm³.

12.60 Clausthalite is a mineral composed of lead selenide (PbSe). The mineral adopts the rock salt structure. The density of PbSe at 25 °C is 8.27 g/cm³. Calculate the length of an edge of the PbSe unit cell.

12.61 A particular form of cinnabar (HgS) adopts the zinc blende structure. The length of the unit cell edge is 5.852 Å. (**a**) Calculate the density of HgS in this form. (**b**) The mineral tiemannite (HgSe) also forms a solid phase with the zinc blende structure. The length of the unit cell edge in this mineral is 6.085 Å. What accounts for the larger unit cell length in tiemannite? (**c**) Which of the two substances has the higher density? How do you account for the difference in densities?

12.62 At room temperature and pressure, RbI crystallizes with the NaCl-type structure. (**a**) Use ionic radii to predict the length of the cubic unit cell edge. (**b**) Use this value to estimate the density. (**c**) At high pressure, the structure transforms to one with a CsCl-type structure. Use ionic radii to predict the length of the cubic unit cell edge for the high-pressure form of RbI. (**d**) Use this value to estimate the density. How does this density compare with the density you calculated in part (b)?

12.63 CuI, CsI, and NaI each adopt a different type of structure. The three different structures to consider are those shown in Figure 12.25 for CsCl, NaCl, and ZnS. (**a**) Use ionic radii, Cs^+ ($r = 1.81$ Å), Na^+ ($r = 1.16$ Å), Cu^+ ($r = 0.74$ Å), and, I^- ($r = 2.06$ Å), to predict which compound will crystallize with which structure. (**b**) What is the coordination number of iodide in each of these structures?

12.64 The rutile and fluorite structures, shown here (anions are colored green), are two of the most common structure types of ionic compounds where the cation to anion ratio is 1 : 2. (**a**) For CaF_2 and ZnF_2 use ionic radii, Ca^{2+} ($r = 1.14$ Å), Zn^{2+} ($r = 0.88$ Å), and F^- ($r = 1.19$ Å), to predict which compound is more likely to crystallize with the fluorite structure and which with the rutile structure. (**b**) What are the coordination numbers of the cations and anions in each of these structures?

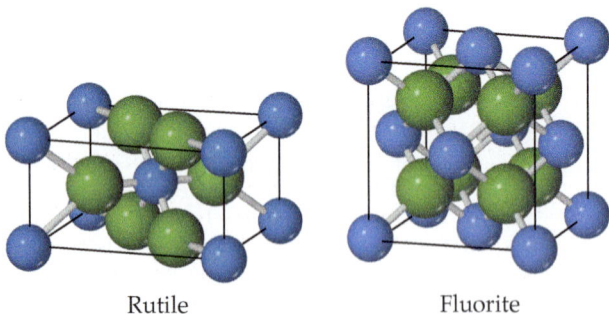

Rutile   Fluorite

12.65 The coordination number for Mg^{2+} ion is usually six. Assuming this assumption holds, determine the anion coordination number in the following compounds: (**a**) MgS, (**b**) MgF_2, (**c**) MgO.

12.66 The coordination number for the Al^{3+} ion is typically between four and six. Use the anion coordination number to determine the Al^{3+} coordination number in the following compounds: (**a**) AlF_3, where the fluoride ions are two coordinate; (**b**) Al_2O_3, where the oxygen ions are six coordinate; (**c**) AlN, where the nitride ions are four coordinate.

Molecular and Covalent-Network Solids (Section 12.5)

12.67 Indicate whether each of the following statements is *true* or *false*:

(**a**) Although both molecular solids and covalent-network solids have covalent bonds, the melting points of molecular solids are much lower because their covalent bonds are much weaker.

(**b**) Other factors being equal, highly symmetric molecules tend to form solids with higher melting points than asymmetrically shaped molecules.

12.68 Indicate whether each of the following statements is *true* or *false*:

(**a**) For molecular solids, the melting point generally increases as the strengths of the covalent bonds increase.

(**b**) For molecular solids, the melting point generally increases as the strengths of the intermolecular forces increase.

12.69 Both covalent-network solids and ionic solids can have melting points well in excess of room temperature, and both can be poor conductors of electricity in their pure form. In other ways, however, their properties are quite different.

(**a**) Which type of solid is more likely to dissolve in water?

(**b**) Which type of solid can become a considerably better conductor of electricity via chemical substitution?

12.70 Which of the following properties are typical characteristics of a covalent-network solid, a metallic solid, or both? (**a**) ductility (**b**) hardness (**c**) high melting point

12.71 For each of the following pairs of semiconductors, which one will have the larger band gap? (**a**) CdS or CdTe (**b**) GaN or InP (**c**) GaAs or InAs

12.72 For each of the following pairs of semiconductors, which one will have the larger band gap (**a**) InP or InAs, (**b**) Ge or AlP (**c**) AgI or CdT?

12.73 If you want to dope GaAs to make an n-type semiconductor with an element to replace Ga, which element would you pick? (**a**) Zn (**b**) Al (**c**) In or (**d**) Si

12.74 If you want to dope GaAs to make a p-type semiconductor with an element to replace As, which group of elements would you pick? (**a**) group 1A (**b**) group 2B (**c**) group 4A (**d**) group 5A or (**e**) group 6A

12.75 Silicon has a band gap of 1.1 eV at room temperature. (**a**) What wavelength of light would a photon of this energy correspond to? (**b**) Draw a vertical line at this wavelength in the figure shown, which shows the light output of the Sun as a function of wavelength. Does silicon absorb all, none, or a portion of the visible light that comes from the Sun? (**c**) You can estimate the portion of the overall solar spectrum that silicon absorbs by considering the area under the curve. If you call the area under the entire curve "100%," what approximate percentage of the area under the curve is absorbed by silicon?

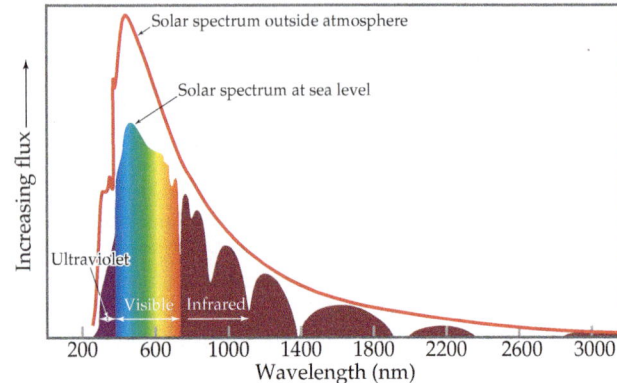

12.76 Cadmium telluride is an important material for solar cells. (**a**) What is the band gap of CdTe? (**b**) What wavelength of light would a photon of this energy correspond to? (**c**) Draw a vertical line at this wavelength in the figure shown in Exercise 12.75, which shows the light output of the Sun as a function of wavelength. (**d**) With respect to silicon, does CdTe absorb a larger or smaller portion of the solar spectrum?

12.77 The semiconductor CdSe has a band gap of 1.74 eV. What wavelength of light would be emitted from an LED made from CdSe? What region of the electromagnetic spectrum is this?

12.78 The first LEDs were made from GaAs, which has a band gap of 1.43 eV. What wavelength of light would be emitted from an LED made from GaAs? What region of the electromagnetic spectrum does this light correspond to: ultraviolet, visible, or infrared?

12.79 GaAs and GaP make solid solutions that have the same crystal structure as the parent materials, with As and P randomly distributed throughout the crystal. GaP_xAs_{1-x} exists for any value of x. If we assume that the band gap varies linearly with composition between $x = 0$ and $x = 1$, estimate the band gap for $GaP_{0.5}As_{0.5}$. (GaAs and GaP band gaps are 1.43 eV and 2.26 eV, respectively.) What wavelength of light does this correspond to?

12.80 Red light-emitting diodes are made from GaAs and GaP solid solutions, GaP_xAs_{1-x} (see Exercise 12.79). The original red LEDs emitted light with a wavelength of 660 nm. If we assume that the band gap varies linearly with composition between $x = 0$ and $x = 1$, estimate the composition (the value of x) that is used in these LEDs.

Polymers (Section 12.6)

12.81 (**a**) What is a monomer? (**b**) Which of these molecules can be used as a monomer: ethanol, ethene (also called ethylene), or methane?

12.82 The molecular formula of n-decane is $CH_3(CH_2)_8CH_3$. Decane is not considered a polymer, whereas polyethylene is. What is the distinction?

12.83 State whether each of these numbers is a reasonable value for a polymer's molecular weight: 100 amu, 10,000 amu, 100,000 amu, 1,000,000 amu?

12.84 Indicate whether the following statement is *true* or *false*: For an addition polymerization, there are no by-products of the reaction (assuming 100% yield).

12.85 An ester is a compound formed by a condensation reaction between a carboxylic acid and an alcohol that eliminates a water molecule. Which of the following, then, would be the repeat unit for a polyester?

(a)

(b)

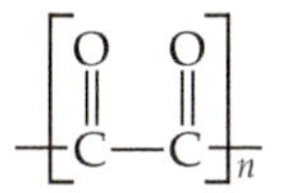

(c)

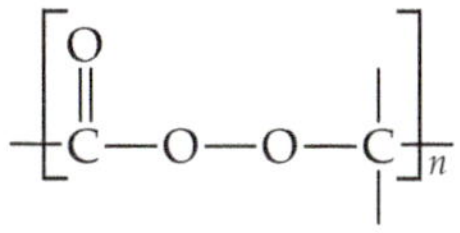

(d)

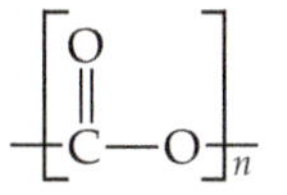

12.86 Write a balanced chemical equation for the formation of a polymer via a condensation reaction from the monomers succinic acid ($HOOCCH_2CH_2COOH$) and ethylenediamine ($H_2NCH_2CH_2NH_2$).

12.87 An addition polymerization forms the polymer originally used as Saran™ wrap. It has the following structure $-[CCl_2-CH_2]_n-$. Draw the structure of the monomer.

12.88 Write the chemical equation that represents the formation of

(**a**) polychloroprene from chloroprene (polychloroprene is used in highway-pavement seals, expansion joints, conveyor belts, and wire and cable jackets)

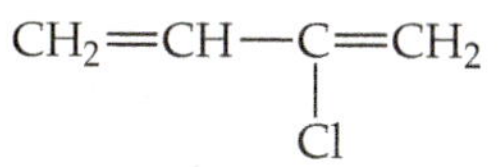

$$CH_2=CH-C=CH_2$$
$$|$$
$$Cl$$

Chloroprene

(**b**) polyacrylonitrile from acrylonitrile (polyacrylonitrile is used in home furnishings, craft yarns, clothing, and many other items).

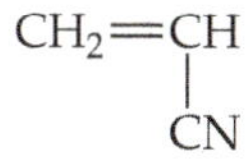

$$CH_2=CH$$
$$|$$
$$CN$$

Acrylonitrile

12.89 The polymer Kevlar, a condensation polymer, is used as reinforcement in car tires and in strings of archery bows and as a component of bulletproof vests.

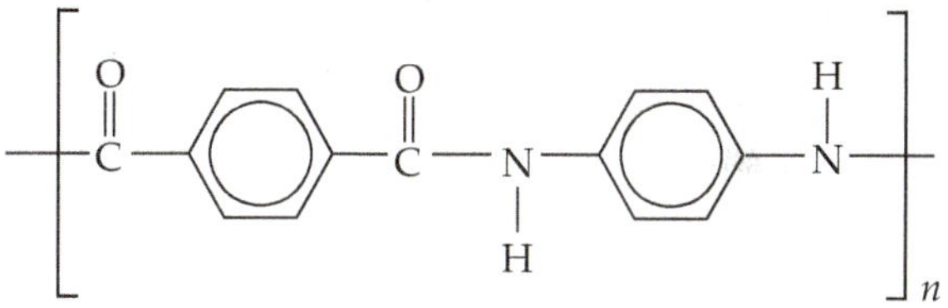

Repeat unit

Draw the structures of the two monomers that yield Kevlar.

12.90 Proteins are naturally occurring polymers formed by condensation reactions of amino acids, which have the general structure

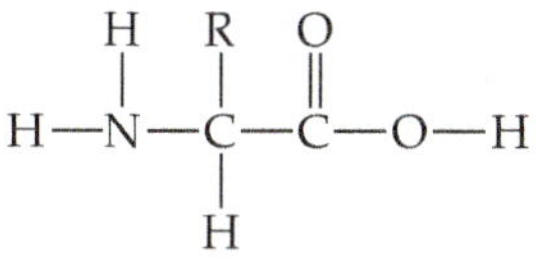

In this structure, $-R$ represents $-H$, $-CH_3$, or another group of atoms; there are 20 different natural amino acids, and each has one of 20 different R groups. (**a**) Draw the general structure of a protein formed by condensation polymerization of the generic amino acid shown here. (**b**) When only a few amino acids react to make a chain, the product is called a "peptide" rather than a protein; only when there are 50 amino acids or more in the chain would the molecule be called a protein. For three amino acids (distinguished by having three different R groups, R1, R2, and R3), draw the peptide that results from their condensation reactions. (**c**) The order in which the R groups exist in a peptide or protein has a huge influence on its biological activity. To distinguish different peptides and proteins, chemists call the first amino acid the one at the "N terminus" and the last one the one at the "C terminus." From your drawing in part (b) you should be able to figure out what "N terminus" and "C terminus" mean. How many different peptides can be made from your three different amino acids?

12.91 (**a**) What molecular features make a polymer flexible? (**b**) If you cross-link a polymer, is it more flexible or less flexible than it was before?

12.92 What molecular structural features cause high-density polyethylene to be denser than low-density polyethylene?

12.93 If you want to make a polymer for plastic wrap, should you strive to make a polymer that has a high or low degree of crystallinity?

12.94 Indicate whether each statement is *true* or *false*:

(a) Elastomers are rubbery solids.

(b) Thermosets cannot be reshaped.

(c) Thermoplastic polymers can be recycled.

Nanomaterials (Section 12.7)

12.95 In what size range are materials considered to be "nanomaterials"?

12.96 CdS has a band gap of 2.4 eV. If large crystals of CdS are illuminated with ultraviolet light, they emit light equal to the band gap energy. (a) What color is the emitted light? (b) Would appropriately sized CdS quantum dots be able to emit blue light? (c) What about red light?

12.97 Indicate whether each statement is *true* or *false*:

(a) The band gap of a semiconductor decreases as the particle size decreases in the 1–10-nm range.

(b) The light that is emitted from a semiconductor, upon external stimulation, becomes longer in wavelength as the particle size of the semiconductor decreases.

12.98 Indicate whether this statement is *true* or *false*:

If you want a semiconductor that emits blue light, you could either use a material that has a band gap corresponding to the energy of a blue photon or you could use a material that has a smaller band gap but make an appropriately sized nanoparticle of the same material.

12.99 Gold adopts a face-centered cubic structure with a unit cell edge of 4.08 Å. How many gold atoms are there in a sphere that is 20 nm in diameter? Recall that the volume of a sphere is $\frac{4}{3}\pi r^3$.

12.100 An ideal quantum dot for use in TVs does not contain any cadmium due to concerns about disposal. One potential material for this purpose is InP, which adopts the zinc blende (ZnS) structure (face-centered cubic). The unit cell edge length is 5.869 Å. (a) If the quantum dot is shaped like a cube, how many of each type of atom are there in a cubic crystal with an edge length of 3.00 nm? 5.00 nm? (b) If one of the nanoparticles in part (a) emits blue light and the other emits orange light, which color is emitted by the crystal with the 3.00-nm edge length? With the 5.00-nm edge length?

12.101 Which statement correctly describes a difference between graphene and graphite?

(a) Graphene is a molecule but graphite is not. (b) Graphene is a single sheet of carbon atoms and graphite contains many, and larger, sheets of carbon atoms. (c) Graphene is an insulator but graphite is a metal. (d) Graphite is pure carbon but graphene is not. (e) The carbons are sp^2 hybridized in graphene but sp^3 hybridized in graphite.

12.102 What evidence supports the notion that buckyballs are actual molecules and not extended materials?

(a) Buckyballs are made of carbon.

(b) Buckyballs have a well-defined atomic structure and molecular weight.

(c) Buckyballs have a well-defined melting point.

(d) Buckyballs are semiconductors.

(e) More than one of the previous choices.

Additional Exercises

12.103 Selected chlorides have the following melting points: NaCl (801 °C), MgCl$_2$ (714 °C), PCl$_3$ (−94 °C), SCl$_2$ (−121 °C) (a) For each compound, indicate what type its solid form is (molecular, metallic, ionic, or covalent-network). (b) Predict which of the following compounds has a higher melting point: CaCl$_2$ or SiCl$_4$.

12.104 A face-centered tetragonal lattice is not one of the 14 three-dimensional lattices. Show that a face-centered tetragonal unit cell can be redefined as a body-centered tetragonal lattice with a smaller unit cell.

12.105 Imagine the primitive cubic lattice. Now imagine pushing on top of it, straight down. Next, stretch another face by pulling it to the right. All angles remain 90°. What kind of primitive lattice have you made?

12.106 Pure iron crystallizes in a body-centered cubic structure, but small amounts of impurities can stabilize a face-centered cubic structure. Which form of iron has a higher density?

12.107 Ni$_3$Al is used in the turbines of aircraft engines because of its strength and low density. Nickel metal has a cubic close-packed structure with a face-centered cubic unit cell, while Ni$_3$Al has the ordered cubic structure shown in Figure 12.17. The length of the cubic unit cell edge is 3.53 Å for nickel and 3.56 Å for Ni$_3$Al. Use these data to calculate and compare the densities of these two materials.

12.108 What type of lattice—namely, primitive cubic, body-centered cubic, or face-centered cubic—does each of the following structure types possess? (a) CsCl (b) Au (c) NaCl (d) Po (e) ZnS

12.109 Cinnabar (HgS) was utilized as a pigment known as vermillion. It has a band gap of 2.20 eV near room temperature for the bulk solid. What wavelength of light (in nm) would a photon of this energy correspond to?

12.110 The electrical conductivity of aluminum is approximately 10^9 times greater than that of its neighbor in the periodic table, silicon. Aluminum has a face-centered cubic structure, and silicon has the diamond structure. A classmate of yours tells you that density is the reason aluminum is a metal but silicon is not; therefore, if you were to put silicon under high pressure, it too would act like a metal. Discuss this idea with your classmates, looking up data about Al and Si as needed.

12.111 Silicon carbide, SiC, has the three-dimensional structure shown in the figure.

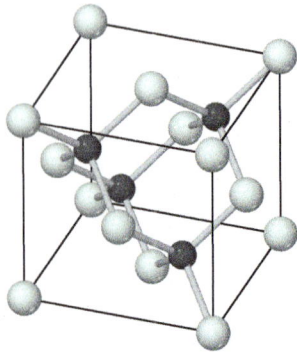

(a) Name another compound that has the same structure. (b) Would you expect the bonding in SiC to be predominantly ionic, metallic, or covalent? (c) How do the bonding and structure of SiC lead to its high thermal stability (to 2700 °C) and exceptional hardness?

12.112 Energy bands are considered continuous due to the large number of closely spaced energy levels. The range of energy levels in a crystal of copper is approximately 1×10^{-19} J. Assuming equal spacing between levels, one can approximate the spacing between energy levels by dividing the range of energies by the number of atoms in the crystal. (**a**) How many copper atoms are in a piece of copper metal in the shape of a cube with edge length 0.5 mm? The density of copper is 8.96 g/cm^3. (**b**) Determine the average spacing in J between energy levels in the copper metal in part (a). (**c**) Is this spacing larger, substantially smaller, or about the same as the 1×10^{-18} J separation between energy levels in a hydrogen atom?

12.113 The electrical conductivity of a metal generally decreases as the temperature is raised. Unlike metals, semiconductors increase their conductivity as you heat them (up to a point). What is the most reasonable explanation for the behavior of semiconductors? (**a**) Semiconductors, unlike metals, become more dense as they are heated, improving atomic orbital overlap. (**b**) As the temperature is raised, electrons are more likely to populate the conduction band in semiconductors, leading to increased conductivity. (**c**) Orbital overlap extends over more and more atoms as the temperature is raised in semiconductors but not in metals. (**d**) Atomic defects that enable conductivity crystallize in semiconductors as the temperature is raised.

12.114 Sodium oxide (Na_2O) adopts a cubic structure, with Na atoms represented by green spheres and O atoms by red spheres.

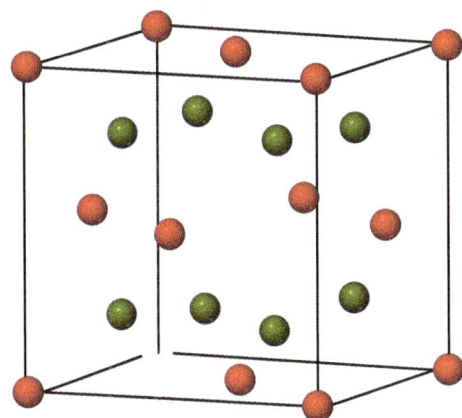

(**a**) How many atoms of each type are there in the unit cell?

(**b**) Determine the coordination number and describe the shape of the coordination environment for the sodium ion.

(**c**) The unit cell edge length is 5.550 Å. Determine the density of Na_2O.

12.115 Teflon is a polymer formed by the polymerization of $F_2C{=}CF_2$. (**a**) Draw the structure of a section of this polymer. (**b**) What type of polymerization reaction is required to form Teflon?

12.116 Hydrogen bonding between polyamide chains plays an important role in determining the properties of a nylon such as nylon 6,6 (Table 12.6). Draw the structural formulas for two adjacent chains of nylon 6,6 and show where hydrogen-bonding interactions could occur between them.

12.117 In their study of X-ray diffraction, William and Lawrence Bragg determined that the relationship among the wavelength of the radiation (λ), the angle at which the radiation is diffracted (θ), and the distance between planes of atoms in the crystal that cause the diffraction (d) is given by $n\lambda = 2d \sin\theta$. X rays from a copper X-ray tube that have a wavelength of 1.54 Å are diffracted at an angle of 14.22 degrees by crystalline silicon. Using the Bragg equation, calculate the distance between the planes of atoms responsible for diffraction in this crystal, assuming $n = 1$ (first-order diffraction).

12.118 Germanium has the same structure as silicon, but the unit cell size is different because Ge and Si atoms are not the same size. If you were to repeat the experiment described in Additional Exercise 12.117, but replace the Si crystal with a Ge crystal, would you expect the X rays to be diffracted at a larger or smaller angle θ?

12.119 (**a**) The density of diamond is 3.5 g/cm^3, whereas that of graphite is 2.3 g/cm^3. Based on the structure of buckminsterfullerene, what would you expect its density to be relative to these other forms of carbon? (**b**) X-ray diffraction studies of buckminsterfullerene show that it has a face-centered cubic lattice of C_{60} molecules. The length of an edge of the unit cell is 14.2 Å. Calculate the density of buckminsterfullerene.

12.120 When you shine light of band gap energy or higher on a semiconductor and promote electrons from the valence band to the conduction band, do you expect the conductivity of the semiconductor to (**a**) remain unchanged, (**b**) increase, or (**c**) decrease?

12.121 Spinel is a mineral that contains 37.9% Al, 17.1% Mg, and 45.0% O, by mass, and has a density of 3.57 g/cm^3. The unit cell is cubic with an edge length of 8.09 Å. How many atoms of each type are in the unit cell?

12.122 (**a**) What are the C—C—C bond angles in diamond? (**b**) What are they in graphite (in one sheet)? (**c**) What atomic orbitals are involved in the stacking of graphite sheets with each other?

12.123 Employing the bond enthalpy values listed in Table 8.3 estimate the molar enthalpy change occurring upon (**a**) polymerization of ethylene, (**b**) formation of nylon 6,6, (**c**) formation of polyethylene terephthalate (PET).

12.124 Although polyethylene can twist and turn in random ways, the most stable form is a linear one, with the carbon backbone oriented as shown in the following figure:

The solid wedges in the figure indicate bonds from carbon that come out of the plane of the page, whereas the dashed wedges indicate bonds that lie behind the plane of the page.

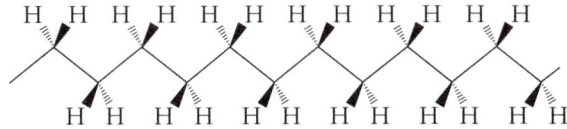

(**a**) What is the hybridization of orbitals at each carbon atom? What angles do you expect between the bonds?

(**b**) Now imagine that the polymer is polypropylene rather than polyethylene. Draw structures for polypropylene in which (i) the CH_3 groups all lie on the same side of the plane of the paper (this form is called isotactic polypropylene), (ii) the CH_3 groups lie on alternating sides of the plane (syndiotactic polypropylene), or (iii) the CH_3 groups are randomly distributed on either side (atactic polypropylene). Which of these forms would you expect to have the highest crystallinity and melting point, and which the lowest? Explain in terms of intermolecular interactions and molecular shapes.

(**c**) Polypropylene fibers have been employed in athletic wear. The product is said to be superior to cotton or polyester clothing in wicking moisture away from the body through the fabric to the outside. Explain the difference between polypropylene and polyester or cotton (which has many —OH groups along the molecular chain) in terms of intermolecular interactions with water.

12.125 (a) In polyvinyl chloride shown in Table 12.6, which bonds have the lowest average bond enthalpy? (b) When subjected to high pressure and heated, polyvinyl chloride converts to diamond. During this transformation, which bonds are most likely to break first? (c) Employing the values of average bond enthalpy in Table 8.3, estimate the overall enthalpy change for converting PVC to diamond.

12.126 Silicon has the diamond structure with a unit cell edge length of 5.43 Å and eight atoms per unit cell. (a) How many silicon atoms are there in 1 cm^3 of material? (b) Suppose you dope that 1 cm^3 sample of silicon with 1 ppm of phosphorus that will increase the conductivity by a factor of a million. How many milligrams of phosphorus are required?

12.127 One method to synthesize ionic solids is by heating two reactants at high temperatures. Consider the reaction of FeO with TiO_2 to form $FeTiO_3$. Determine the amount of each of the two reactants to prepare 2.500 g $FeTiO_3$, assuming the reaction goes to completion.

(a) Write a balanced chemical reaction.

(b) Calculate the formula weight of $FeTiO_3$.

(c) Determine the moles of $FeTiO_3$.

(d) Determine the moles and mass (g) of FeO required.

(e) Determine the moles and mass (g) of TiO_2 required.

12.128 Look up the diameter of a silicon atom, in Å. The latest semiconductor chips have fabricated lines as small as 7 nm. How many silicon atoms does this correspond to?

Design an Experiment

Polymers were commercially made by the DuPont Company starting in the late 1920s. At that time, some chemists still could not believe that polymers were molecules; they thought covalent bonding would not "last" for millions of atoms and that polymers were really clumps of molecules held together by weak intermolecular forces. Design an experiment to demonstrate that polymers really are large molecules and not little clumps of small molecules that are held together by weak intermolecular forces.

13

PROPERTIES OF SOLUTIONS

▲ **THE DEAD SEA.** The Dead Sea is one of the saltiest bodies of water on Earth. The concentration of dissolved salts, like $MgCl_2$ and NaCl, is nearly 10 times higher than ocean water—so high that neither plants nor animals can survive in the solution that is the Dead Sea. Finding ways to remove dissolved salts from the ocean and other "salty" bodies of water is becoming an increasingly important pursuit as the population in arid regions grows.

In Chapters 10, 11, and 12, we explored the properties of pure gases, liquids, and solids. However, the matter that we encounter in our daily lives, such as air, tap water, and glass, are frequently mixtures. Here in Chapter 13, we examine homogeneous mixtures, which as we learned in previous chapters, are called *solutions*. (Sections 1.2 and 4.1)

When we think of solutions, liquids immediately come to mind, but solutions can also be solids or gases. For example, sterling silver is a homogeneous mixture of approximately 7% copper in silver and is considered a solid solution. The air we breathe is a homogeneous mixture of several gases, making air a gaseous solution. Nevertheless, liquid solutions play a particularly important role in chemistry and biology and are the focus of this chapter.

Each substance in a solution is a *component* of the solution. As we saw in Chapter 4, the *solvent* is normally the component present in the greatest amount, and all the other components are called *solutes*. In this chapter, we compare the physical properties of solutions with the properties of the components in their pure form. We are particularly concerned with aqueous solutions, which contain water as the solvent and either a gas, liquid, or solid as the solute.

13.1 | The Solution Process

Learning Objectives

When you finish Section 13.1, you should be able to:

▶ Qualitatively describe the entropy and enthalpy changes associated with solution formation.

▶ Identify the intermolecular forces that hold molecules and/or ions together in pure substances and solutions.

▶ Evaluate the various enthalpy changes associated with forming a solution and use those values to predict the likelihood of solution formation.

Solutions are all around us, yet not all substances form solutions. Whether two or more substances mix to form a solution depends on two factors: (1) the natural tendency of substances to mix and spread into larger volumes when not restrained in some way and (2) the intermolecular forces that hold molecules together in both pure substances and solutions. In this section we consider the thermodynamics behind solution formation.

The Natural Tendency toward Mixing

Suppose we have $O_2(g)$ and $Ar(g)$ separated by a barrier, as in **Figure 13.1**. If the barrier is removed, the gases mix to form a solution. The molecules experience very little intermolecular interactions and behave like ideal gas particles. As a result, their molecular motion causes them to spread through the larger volume, resulting in a gaseous solution.

The mixing of gases is a *spontaneous* process, meaning it occurs of its own accord without any input of energy from outside the system. When the molecules mix and become more randomly distributed, there is an increase in a thermodynamic quantity called *entropy*. Entropy is defined more rigorously in Chapter 19; but for now, we can consider that the entropy of a system increases if its degree of disorder increases, if the atoms or molecules gain more freedom of motion, or if its energy becomes dispersed over a greater number of particles. The *formation of solutions is favored by the increase in entropy that accompanies mixing.* However, to fully evaluate the energetics of forming a solution, we must also consider the enthalpy change that accompanies mixing. (Section 5.3) Chemical bonds as well as intermolecular forces (dispersion forces, dipole–dipole interactions, hydrogen bonding, ion–dipole interactions) and their corresponding energies contribute to the enthalpy of the system. If mixing optimizes the intermolecular attractions between atoms, it leads to a *decrease* in the enthalpy of the system, which is energetically *favorable*. If it *raises* the enthalpy of the system by disrupting the most favorable intermolecular interactions, then the enthalpy *increases* upon forming a solution.

When two gases mix, the change in enthalpy is very small because the gas molecules are far apart and the interactions between them are negligible. Thus, mixing of gases depends only on entropy and is always thermodynamically favored. However, when the solvent and/or solute is a solid or a liquid, intermolecular forces become important in determining whether a solution forms. For example, although ionic bonds hold sodium and chloride ions together in solid sodium chloride (Section 8.2), the solid dissolves in water because of the compensating strength of the attractive forces between the ions and water molecules. (Section 11.2) Sodium chloride does not dissolve in gasoline, however, because the intermolecular forces between the ions and the hydrocarbon molecules that make up gasoline are too weak.

 **Go Figure**   What aspect of the kinetic-molecular theory of gases tells us that the gases will mix?

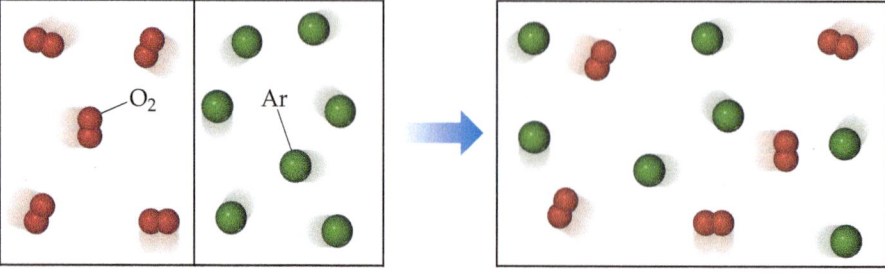

▲ **Figure 13.1** Spontaneous mixing of two gases to form a homogeneous mixture (a solution).

The Effect of Intermolecular Forces on Solution Formation

Any of the intermolecular forces discussed in Chapter 11 can operate between solute and solvent particles in a solution. These forces are summarized in **Figure 13.2**. Dispersion forces, for example, dominate when one nonpolar substance, such as C_7H_{16}, dissolves in another, such as C_5H_{12}, and ion–dipole forces dominate in solutions of ionic substances in water.

Three kinds of intermolecular interactions are involved in solution formation:

1. *Solute–solute* interactions between solute particles must be overcome to disperse the solute particles through the solvent.

2. *Solvent–solvent* interactions between solvent particles must be overcome to make room for the solute particles in the solvent.

3. *Solvent–solute* interactions between the solvent and solute particles occur as the particles mix.

The extent to which one substance is able to dissolve in another depends on the relative magnitudes of these three types of interactions. Solutions form when the magnitudes of the solvent–solute interactions are either comparable to or greater than the solute–solute and solvent–solvent interactions. For example, heptane (C_7H_{16}) and pentane (C_5H_{12}) dissolve in each other in all proportions. For this discussion, we can arbitrarily call heptane the solvent and pentane the solute. Both substances are nonpolar, and the magnitudes of the solvent–solute interactions (attractive dispersion forces) are comparable to the solute–solute and solvent–solvent interactions. Thus, no forces impede mixing, and the tendency to mix (increase entropy) causes the solution to form spontaneously.

An everyday example of solution formation occurs when you dissolve salt in water. As noted previously, solid NaCl dissolves readily in water because the attractive solvent–solute interactions between the polar H_2O molecules and the ions are strong enough to largely compensate for the loss of attractive solute–solute interactions between ions in the NaCl(*s*) and attractive solvent–solvent interactions between H_2O molecules when the substances mix. When NaCl is added to water (**Figure 13.3**), the water molecules orient themselves on the surface of the NaCl crystals, with the positive end of the water dipole oriented toward the Cl^- ions and the negative end oriented toward the Na^+ ions. These ion–dipole attractions are strong enough to pull the surface ions away from the solid, thus overcoming the solute–solute interactions. For the solid to dissolve, some solvent–solvent interactions must also be overcome to create room for the ions to "fit" among all the water molecules.

Once separated from the solid, the Na^+ and Cl^- ions are surrounded by water molecules. Interactions such as this between solute and solvent molecules are known as **solvation**. When the solvent is water, the interactions are referred to as **hydration**.

Energetics of Solution Formation

Solution processes are typically accompanied by changes in enthalpy. For example, when NaCl dissolves in water, the process is slightly endothermic, $\Delta H_{soln} = 3.9$ kJ/mol. We

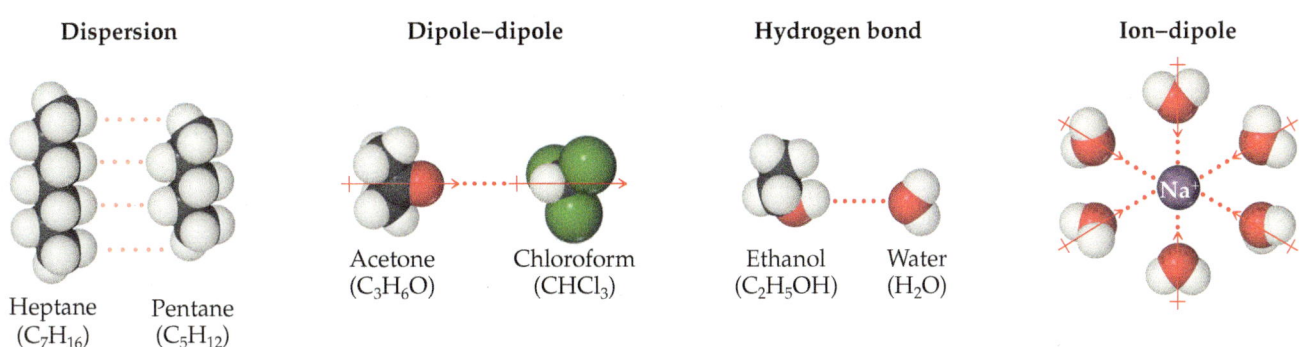

| Dispersion | Dipole–dipole | Hydrogen bond | Ion–dipole |

Acetone   Chloroform
(C_3H_6O)   ($CHCl_3$)

Ethanol   Water
(C_2H_5OH)   (H_2O)

Heptane   Pentane
(C_7H_{16})   (C_5H_{12})

▲ **Figure 13.2 Intermolecular interactions involved in solutions.**

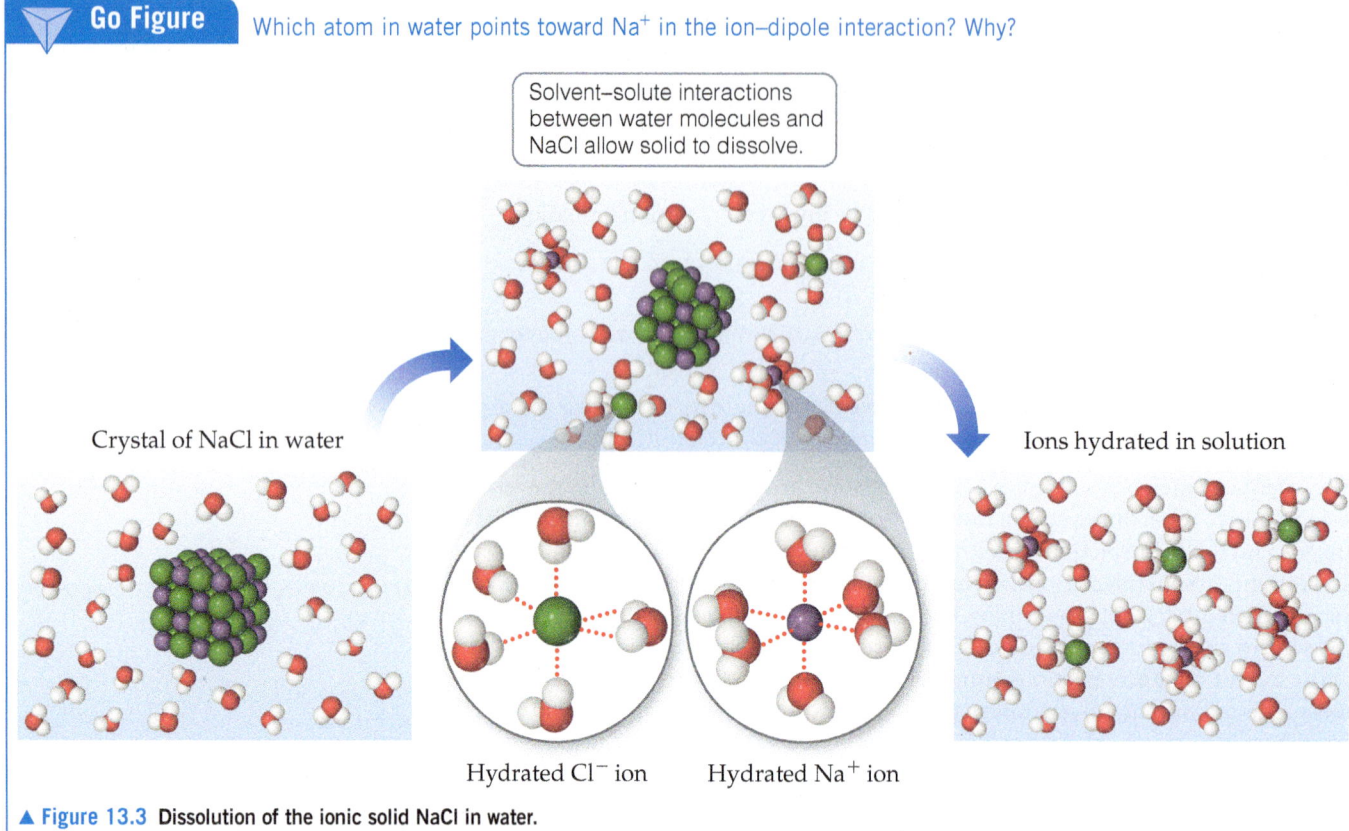

Go Figure Which atom in water points toward Na$^+$ in the ion–dipole interaction? Why?

Solvent–solute interactions between water molecules and NaCl allow solid to dissolve.

Crystal of NaCl in water

Ions hydrated in solution

Hydrated Cl$^-$ ion   Hydrated Na$^+$ ion

▲ **Figure 13.3** Dissolution of the ionic solid NaCl in water.

can use Hess's law to analyze how the solute–solute, solvent–solvent, and solute–solvent interactions influence the enthalpy of solution. (Section 5.6)

We can imagine the solution process as having three components, each with an associated enthalpy change: A cluster of n solute particles must separate from one another (ΔH_{solute}), a cluster of m solvent particles separate from one another ($\Delta H_{solvent}$), and these solute and solvent particles mix (ΔH_{mix}).

1. (solute)$_n$ $\rightleftharpoons$ n solute ΔH_{solute}

2. (solvent)$_m$ $\rightleftharpoons$ m solvent $\Delta H_{solvent}$

3. n solute + m solvent $\rightleftharpoons$ solution ΔH_{mix}

4. (solute)$_n$ + (solvent)$_m$ $\rightleftharpoons$ solution $\Delta H_{soln} = \Delta H_{solute} + \Delta H_{solvent} + \Delta H_{mix}$

Thus, the overall enthalpy change, ΔH_{soln}, is the sum of the three steps:

$$\Delta H_{soln} = \Delta H_{solute} + \Delta H_{solvent} + \Delta H_{mix} \qquad [13.1]$$

Separation of the solute particles from one another always requires an input of energy to overcome their attractive interactions. The process is therefore endothermic ($\Delta H_{solute} > 0$). Likewise, separation of solvent molecules to accommodate the solute always requires energy ($\Delta H_{solvent} > 0$). The third component, which arises from the attractive interactions between solute particles and solvent particles, is always exothermic ($\Delta H_{mix} < 0$).

The three enthalpy terms in Equation 13.1 can be added together to give either a negative or a positive sum, depending on the actual numbers for the system being considered (**Figure 13.4**). Thus, the formation of a solution can be either exothermic or endothermic. For example, when magnesium sulfate (MgSO$_4$) is added to water, the solution process is exothermic: $\Delta H_{soln} = -91.2$ kJ/mol. In contrast, the dissolution of ammonium nitrate (NH$_4$NO$_3$) is endothermic: $\Delta H_{soln} = +26.4$ kJ/mol. These

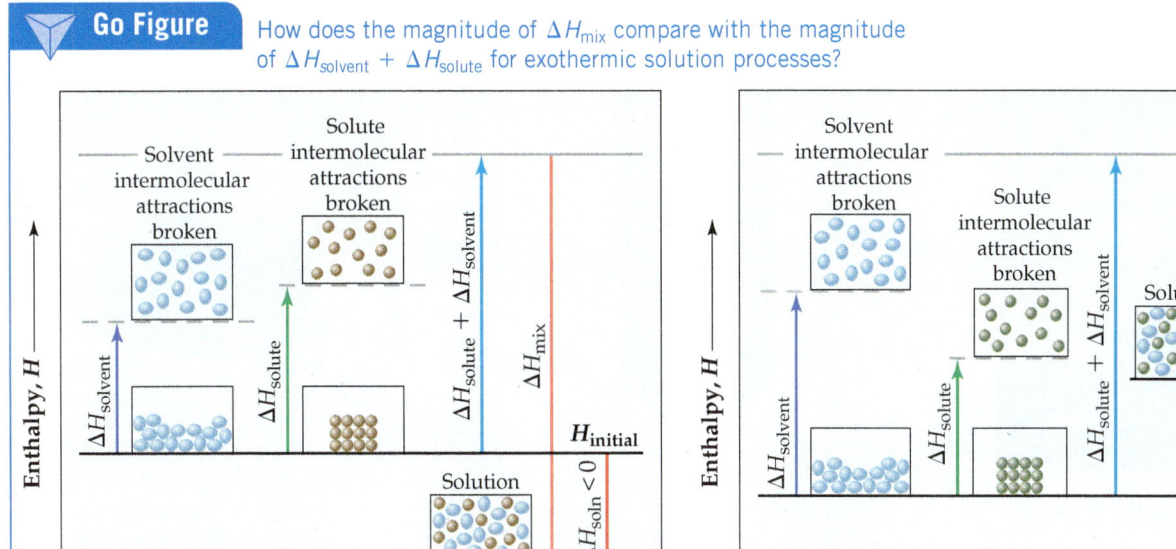

Go Figure How does the magnitude of ΔH_{mix} compare with the magnitude of $\Delta H_{solvent} + \Delta H_{solute}$ for exothermic solution processes?

Exothermic solution process

Endothermic solution process

▲ **Figure 13.4** Enthalpy changes accompanying the solution process.

particular salts are the main components in the instant heat packs and ice packs used to treat athletic injuries (**Figure 13.5**). The packs consist of a pouch of water and the solid salt sealed off from the water—$MgSO_4(s)$ for hot packs and $NH_4NO_3(s)$ for cold packs. When the pack is squeezed, the seal separating the solid from the water is broken and a solution forms, either increasing or decreasing the temperature. This means that a freshly made solution of $MgSO_4$ in water feels warm and a freshly made solution of NH_4NO_3 feels cold.

The enthalpy change for a process can provide insight into the extent to which the process occurs. (Section 5.4) Exothermic processes tend to proceed spontaneously. On the other hand, if ΔH_{soln} is too endothermic, the solute might not dissolve to any significant extent in the chosen solvent. Thus, for solutions to form, the solvent–solute interaction must be strong enough to make ΔH_{mix} comparable in magnitude to $\Delta H_{solute} + \Delta H_{solvent}$. This fact further explains why ionic solutes do not dissolve in nonpolar solvents. The nonpolar solvent molecules experience only weak attractive interactions with the ions, and these interactions do not compensate for the energies required to separate the ions from one another.

By similar reasoning, a polar liquid solute, such as water, does not dissolve in a nonpolar liquid solvent, such as octane (C_8H_{18}). The water molecules experience strong hydrogen-bonding interactions with one another (Section 11.2)—attractive forces that must be overcome if the water molecules are to be dispersed throughout the octane solvent. The energy required to separate the H_2O molecules from one another is not recovered in the form of attractive interactions between the H_2O and C_8H_{18} molecules.

Solution Formation and Chemical Reactions

In discussing solutions, we must be careful to distinguish the physical process of solution formation from chemical reactions that lead to a solution. For example, nickel metal dissolves on contact with an aqueous hydrochloric acid solution because the following reaction occurs:

$$Ni(s) + 2\,HCl(aq) \longrightarrow NiCl_2(aq) + H_2(g) \qquad [13.2]$$

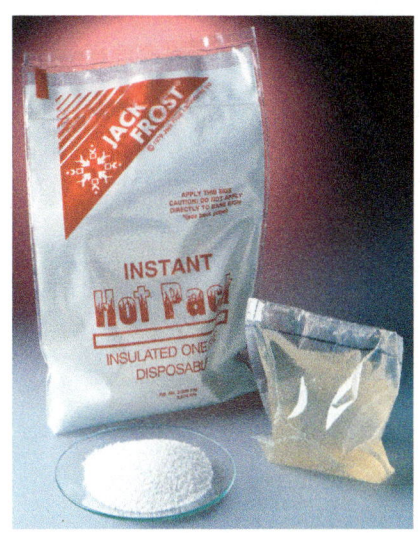

▲ **Figure 13.5 Magnesium sulfate instant hot pack.** When $MgSO_4$ dissolves in H_2O, heat is released, making the pack feel hot.

Nickel metal and hydrochloric acid

Nickel reacts with hydrochloric acid, forming $NiCl_2(aq)$ and $H_2(g)$. The solution is of $NiCl_2$, not Ni metal

$NiCl_2 \cdot 6 H_2O(s)$ remains when solvent evaporated

▲ **Figure 13.6 The reaction between nickel metal and hydrochloric acid is *not* a simple dissolution.** The product is nickel(II) chloride hexahydrate, $NiCl_2 \cdot 6H_2O(s)$, which has exactly six waters of hydration in the crystal for every nickel ion.

In this instance, one of the resulting solutes is not Ni metal but rather its salt $NiCl_2$. If the solution is evaporated to dryness, $NiCl_2 \cdot 6H_2O(s)$ is recovered (**Figure 13.6**). Compounds such as $NiCl_2 \cdot 6H_2O(s)$ that have a defined number of water molecules in the crystal lattice are known as *hydrates*. When $NaCl(s)$ is dissolved in water, on the other hand, no chemical reaction occurs. If the solution is evaporated to dryness, NaCl is recovered. Our focus throughout this chapter is on solutions from which the solute can be recovered unchanged from the solution.

 Self-Assessment Exercises

SAE 13.1   In which of these situations is entropy increasing? (**a**) 200 mL of water is poured from a 250-mL beaker into a 500-mL beaker. (**b**) A pinch of sugar is dissolved into 200 mL of water. (**c**) 200 mL of olive oil is poured on top of 200 mL of water. (**d**) 200 mL of water is frozen into ice.

SAE 13.2   Lithium bromide (LiBr) dissolves in methanol (CH_3OH) to form a solution. Which of the following images most accurately portrays the arrangement of methanol molecules around the Li^+ cations (oxygen = red, carbon = black, hydrogen = white, lithium = blue)?

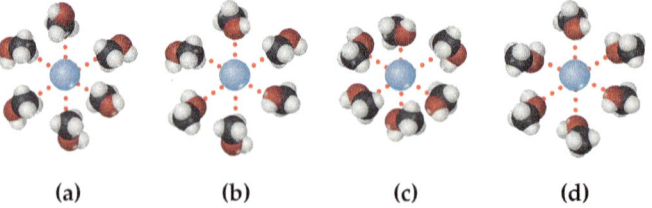

(a)   (b)   (c)   (d)

SAE 13.3   Consider a solution where lithium bromide (LiBr) is the solute and methanol (CH_3OH) is the solvent. The molecules of methanol will be _____ the lithium cations predominantly by _____ interactions. (**a**) attracted to, hydrogen-bonding (**b**) attracted to, dipole–dipole (**c**) attracted to, ion–dipole (**d**) repelled from, dipole–dipole (**e**) repelled from, ion–dipole

SAE 13.4   When 0.10 mol of the ionic solid AX_2 is dissolved in 500 mL water in a beaker, the temperature of the beaker increases. When 0.10 mol of the ionic solid DZ_2 is dissolved in 500 mL of water in a separate beaker, the temperature of the beaker decreases. What do these observations tell us about the signs of ΔH_{soln} for forming aqueous solutions of each solid? (**a**) $\Delta H_{soln} > 0$ when the solute is AX_2 and $\Delta H_{soln} > 0$ when the solute is DZ_2 (**b**) $\Delta H_{soln} > 0$ when the solute is AX_2 and $\Delta H_{soln} < 0$ when the solute is DZ_2 (**c**) $\Delta H_{soln} < 0$ when the solute is AX_2 and $\Delta H_{soln} < 0$ when the solute is DZ_2 (**d**) $\Delta H_{soln} < 0$ when the solute is AX_2 and $\Delta H_{soln} > 0$ when the solute is DZ_2

13.2 | Saturated Solutions and Solubility

As a solid solute begins to dissolve in a solvent, the concentration of solute particles in solution increases, increasing the chances that some solute particles will collide with the surface of the solid and reattach. This process, which is the opposite of the solution process, is called **crystallization**. Thus, two opposing processes occur in a solution in contact with undissolved solute. This situation is represented in the following chemical equation:

$$\text{Solute} + \text{solvent} \underset{\text{crystallize}}{\overset{\text{dissolve}}{\rightleftharpoons}} \text{solution} \qquad [13.3]$$

When the rates of these opposing processes become equal, a *dynamic equilibrium* is established, and there is no further net increase in the amount of solute in solution. (Section 4.1)

A solution that is in equilibrium with undissolved solute is referred to as **saturated**. Additional solute will not dissolve if it is added to a saturated solution. The amount of solute needed to form a saturated solution in a given quantity of solvent is known as the **solubility** of that solute. That is,

the solubility of a particular solute in a particular solvent is the maximum amount of the solute that can dissolve in a given amount of the solvent at a specified temperature.

For example, the solubility of NaCl in water at 0 °C is 35.7 g per 100 mL of water. This is the maximum amount of NaCl that can be dissolved in water to give a stable equilibrium solution at that temperature.

If we dissolve less solute than the amount needed to form a saturated solution, the solution is said to be **unsaturated**. Thus, a solution containing 10.0 g of NaCl per 100 mL of water at 0 °C is unsaturated because it has the capacity to dissolve more solute.

Under suitable conditions, it is possible to form solutions that contain a greater amount of solute than needed to form a saturated solution. Such solutions are called **supersaturated**. For example, when a saturated solution of sodium acetate is made at a high temperature and then slowly cooled, all of the solute may remain dissolved even though its solubility decreases as the temperature decreases. Because the solute in a supersaturated solution is present in a concentration higher than the equilibrium concentration, supersaturated solutions are unstable with respect to crystallization of the solute. For crystallization to occur, however, the solute particles must arrange themselves properly to form crystals. The addition of a small crystal of the solute (a seed crystal) provides a template for crystallization of the excess solute, leading to a saturated solution in contact with excess solid (**Figure 13.7**).

Learning Objectives

When you finish Section 13.2, you should be able to:

▶ Describe the dynamic equilibrium between solid, solute, and solvent.

▶ Differentiate saturated, unsaturated, and supersaturated solutions.

Go Figure How many grams of sodium acetate are in solution in the third beaker?

Amount of sodium acetate dissolved is greater than its solubility at this temperature.

1 Seed crystal of sodium acetate is added to supersaturated solution.

2 Excess sodium acetate starts to crystallize from solution.

3 Solution is saturated and excess solute has crystallized.

▲ **Figure 13.7 Precipitation from a supersaturated sodium acetate solution.** The solution on the left was formed by dissolving 170 g of the salt in 100 mL of water at 100 °C and then slowly cooling it to 20 °C. Because the solubility of sodium acetate in water at 20 °C is 46 g per 100 mL of water, the solution is supersaturated. Addition of a sodium acetate crystal causes the excess solute to crystallize from solution.

Self-Assessment Exercises

SAE 13.5 A 10-g sample of $Ba(OH)_2$ is added to 100 mL of water and left to sit out with intermittent stirring. Initially, the $Ba(OH)_2$ starts to dissolve, but after a while the amount of solid barium hydroxide at the bottom of the beaker stops changing. Which graph of $Ba(OH)_2$ concentration versus time best represents this process?

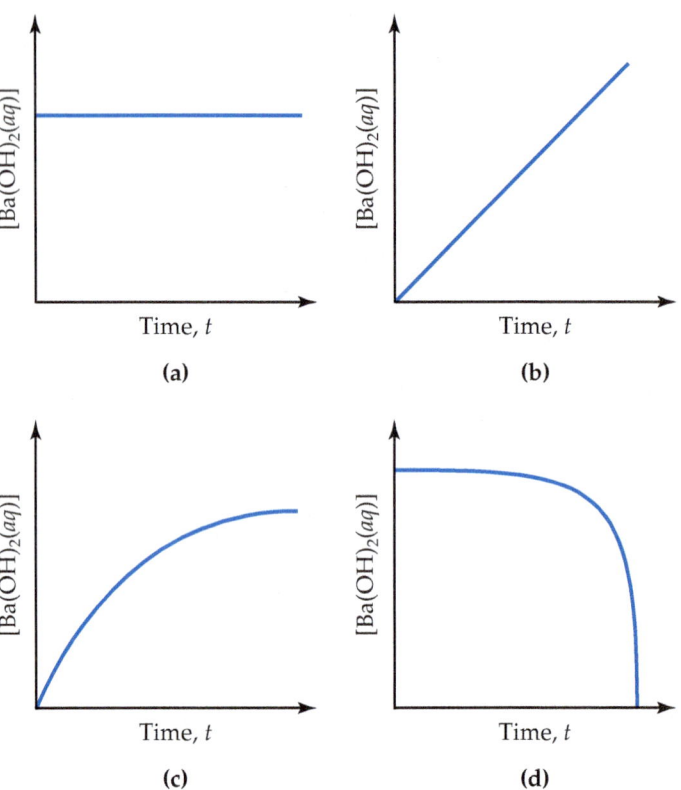

SAE 13.6 Two aqueous solutions of zinc chloride are prepared in beakers labeled A and B and given sufficient time to reach a dynamic equilibrium. Solution A is colorless and has no apparent solid at the bottom of the beaker. Solution B is colorless and has a white solid at the bottom of the beaker. An aliquot of each solution is collected, being careful not to disturb the solid at the bottom of solution B. The two aliquots are analyzed and found to have the same concentration of $Zn^{2+}(aq)$. From this we can conclude that the solution in beaker A is _____ and that in beaker B is _____. (**a**) saturated, supersaturated (**b**) unsaturated, saturated (**c**) saturated, saturated (**d**) saturated, unsaturated (**e**) supersaturated, unsaturated

13.3 | Factors Affecting Solubility

The extent to which one substance dissolves in another depends on the nature of both substances. The intermolecular forces operative in the pure solvent, the pure solute, and the solution play a large role in determining the solubility of a solute in a particular solvent. (Section 13.1) Solubility also depends on temperature, and for gaseous solutes, on pressure.

Solute–Solvent Interactions

The natural tendency of substances to mix and the various interactions among solute and solvent particles are all involved in determining solubilities. Nevertheless, insight into variations in solubility can often be gained by focusing on the interaction between the solute and solvent. The data in **Table 13.1** show that the solubilities of various gases in water increase with increasing molar mass. Because the attractive forces between the gas molecules and solvent molecules are mainly dispersion forces, which increase with increasing size and mass of the gas molecules (Section 11.2), the solubilities of gases in

▲ **Learning Objectives**

When you finish Section 13.3, you should be able to:

▶ Identify solutes that will be soluble in a given solvent based on the "like dissolves like" rule.

▶ Use Henry's law to determine the solubility of a gas from the partial pressure of the gas in contact with the solution.

▶ Explain how the solubility of gases and solids in aqueous solutions typically respond to changes in temperature.

water increase as the attraction between solute (gas) and solvent (water) increases. In general, when other factors are comparable, *the stronger the attractions between solute and solvent molecules, the greater the solubility of the solute in that solvent.*

Because of favorable dipole–dipole attractions between polar solvent molecules and polar solute molecules, *polar liquids tend to dissolve in polar solvents.* Water is both polar and able to form hydrogen bonds. (Section 11.2) Thus, polar molecules, especially those that can form hydrogen bonds with water molecules, tend to be soluble in water. For example, acetone, a polar molecule with the structural formula shown below, mixes in all proportions with water. Acetone has a strongly polar C=O bond, and the pairs of nonbonding electrons on the O atom can form hydrogen bonds with water.

TABLE 13.1 Solubilities of Gases in Water at 20°C and 1 atm Gas Pressure

Gas	Molar Mass (g/mol)	Solubility (M)
N_2	28.0	0.69×10^{-3}
O_2	32.0	1.38×10^{-3}
Ar	39.9	1.50×10^{-3}
Kr	83.8	2.79×10^{-3}

$$:O:$$
$$\|$$
$$CH_3CCH_3$$

Acetone

Liquids that mix in all proportions, such as acetone and water, are **miscible**, whereas those that do not dissolve in one another are **immiscible**. Gasoline, which is a mixture of hydrocarbons, is immiscible with water. Hydrocarbons are nonpolar substances because of several factors: The C—C bonds are nonpolar, the C—H bonds are nearly nonpolar, and the molecules are symmetrical enough to cancel much of the weak C—H bond dipoles. The attraction between the polar water molecules and the nonpolar hydrocarbon molecules is not sufficiently strong to allow the formation of a solution. *Nonpolar liquids tend to be insoluble in polar liquids,* as **Figure 13.8** shows for hexane (C_6H_{14}) and water.

Many organic compounds have polar groups attached to a nonpolar framework of carbon and hydrogen atoms. For example, the series of organic compounds in **Table 13.2** all contain the polar OH group. Organic compounds with this molecular feature are called *alcohols.* (Section 2.9) The presence of the O—H group leads to the formation of hydrogen bonds. For example, ethanol (CH_3CH_2OH) molecules can form hydrogen bonds with water molecules as well as with each other (**Figure 13.9**). As a result, the solute–solute, solvent–solvent, and solute–solvent forces are not greatly different in a mixture of CH_3CH_2OH and H_2O. No major change occurs in the environments of the molecules as they are mixed. Therefore, the increase in entropy that occurs when the components mix favors solution formation, and ethanol is completely miscible with water.

Notice in Table 13.2 that the number of carbon atoms in an alcohol affects its solubility in water. As this number increases, the polar OH group becomes an even smaller part of the molecule, and the molecule behaves more like a hydrocarbon. The solubility of the alcohol in water decreases correspondingly. On the other hand, the solubility of alcohols in a nonpolar solvent like hexane (C_6H_{14}) increases as the nonpolar hydrocarbon chain lengthens.

Hexane

Water

▲ **Figure 13.8 Hexane, a hydrocarbon, is immiscible with water.** Hexane is the top layer because it is less dense than water.

TABLE 13.2 Solubilities of Some Alcohols in Water and in Hexane*

Alcohol	Solubility in H_2O	Solubility in C_6H_{14}
CH_3OH (methanol)	∞	0.12
CH_3CH_2OH (ethanol)	∞	∞
$CH_3CH_2CH_2OH$ (propanol)	∞	∞
$CH_3CH_2CH_2CH_2OH$ (butanol)	0.11	∞
$CH_3CH_2CH_2CH_2CH_2OH$ (pentanol)	0.030	∞
$CH_3CH_2CH_2CH_2CH_2CH_2OH$ (hexanol)	0.0058	∞

*Expressed in mol alcohol/100 g solvent at 20 °C. The infinity symbol (∞) indicates that the alcohol is completely miscible with the solvent.

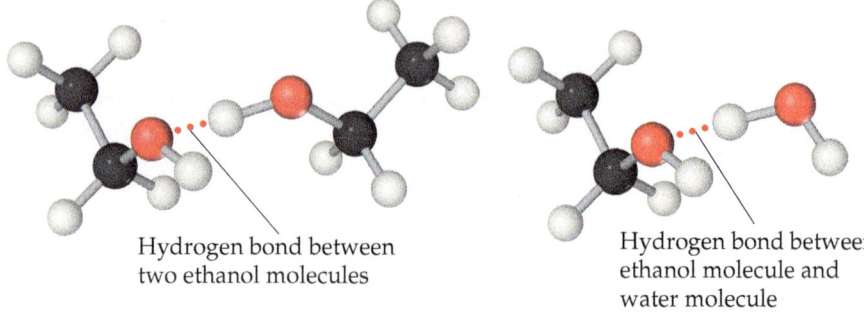

Hydrogen bond between
two ethanol molecules

Hydrogen bond between
ethanol molecule and
water molecule

▲ **Figure 13.9 Hydrogen bonding involving OH groups.**

Cyclohexane, C_6H_{12}, which
has no polar OH groups, is
essentially insoluble in water.

In glucose, the OH groups enhance
the aqueous solubility because of
their ability to hydrogen-bond
with H_2O.

Hydrogen-bonding
sites

▲ **Figure 13.10** The correlation of molecular
structure with solubility.

One way to enhance the solubility of a substance in water is to increase the number of polar groups the substance contains. For example, increasing the number of OH groups in a solute increases the extent of hydrogen bonding between that solute and water, thereby increasing solubility. Glucose ($C_6H_{12}O_6$, **Figure 13.10**) has five OH groups on a six-carbon framework, which makes the molecule very soluble in water: 830 g dissolves in 1.00 L of water at 17.5 °C. In contrast, cyclohexane (C_6H_{12}), which has a similar structure to glucose but with all of the OH groups replaced by H, is essentially insoluble in water (only 55 mg of cyclohexane can dissolve in 1.00 L of water at 25 °C).

Over years of study, examination of different solvent–solute combinations has led to an important generalization:

*Substances with similar intermolecular attractive forces tend to be
soluble in one another.*

This generalization is often simply stated as *"like dissolves like."* Thus, nonpolar substances are more likely to be soluble in nonpolar solvents, whereas ionic and polar solutes are more likely to be soluble in polar solvents. Network solids such as diamond and quartz are not soluble in either polar or nonpolar solvents because of the strong bonding within the solid.

Sample Exercise 13.1
Predicting Solubility Patterns

Predict whether each of the following substances is more likely to dissolve in the nonpolar solvent carbon tetrachloride or in water: C_7H_{16}, Na_2SO_4, HCl, and I_2.

SOLUTION

Analyze We are given two solvents, one that is nonpolar (CCl_4) and the other that is polar (H_2O) and asked to determine which will be the better solvent for each solute listed.

Plan By examining the formulas of the solutes, we can predict whether they are ionic or molecular. For those that are molecular, we can predict whether they are polar or nonpolar. We can then apply the idea that the nonpolar solvent will be better for the nonpolar solutes, whereas the polar solvent will be better for the ionic and polar solutes.

Solve C_7H_{16} is a hydrocarbon, so it is molecular and nonpolar. Na_2SO_4, a compound containing a metal and nonmetals, is ionic. HCl, a diatomic molecule containing two nonmetals that differ in electronegativity, is polar. I_2, a diatomic molecule with atoms of

equal electronegativity, is nonpolar. We would therefore predict that C_7H_{16} and I_2 (the nonpolar solutes) would be more soluble in the nonpolar CCl_4 than in polar H_2O, whereas water would be the better solvent for Na_2SO_4 and HCl (the ionic and polar covalent solutes).

▶ **Practice Exercise**
Arrange the following substances in order of increasing solubility in water:
(**a**) $CH_3CH_2CH_2CH_2CH_3$
(**b**) $CH_3CH_2CH_2CH_2CH_2OH$
(**c**) $HOCH_2CH_2CH_2CH_2CH_2OH$
(**d**) $CH_3CH_2CH_2CH_2CH_2Cl$

CHEMISTRY AND LIFE | Fat-Soluble and Water-Soluble Vitamins

Vitamins have unique chemical structures that affect their solubilities in different parts of the human body. Vitamin C and the B vitamins are soluble in water, for example, whereas vitamins A, D, E, and K are soluble in nonpolar solvents and in fatty tissue (which is nonpolar). Because of their water solubility, vitamins B and C are not stored to any appreciable extent in the body, and so foods containing these vitamins should be included in the daily diet. In contrast, the fat-soluble vitamins are stored in sufficient quantities to keep vitamin-deficiency diseases from appearing even after a person has subsisted for a long period on a vitamin-deficient diet.

The structures of the vitamins explain why some are soluble in water and others are not. Notice in Figure 13.11 that vitamin A (retinol) is an alcohol with a very long hydrocarbon chain. Because the OH group is such a small part of the molecule, the molecule resembles the long-chain alcohols listed in Table 13.2. This vitamin is nearly nonpolar. In contrast, the vitamin C molecule is smaller and has several OH groups that can form hydrogen bonds with water, somewhat like glucose.

Related Exercises: 13.7, 13.48

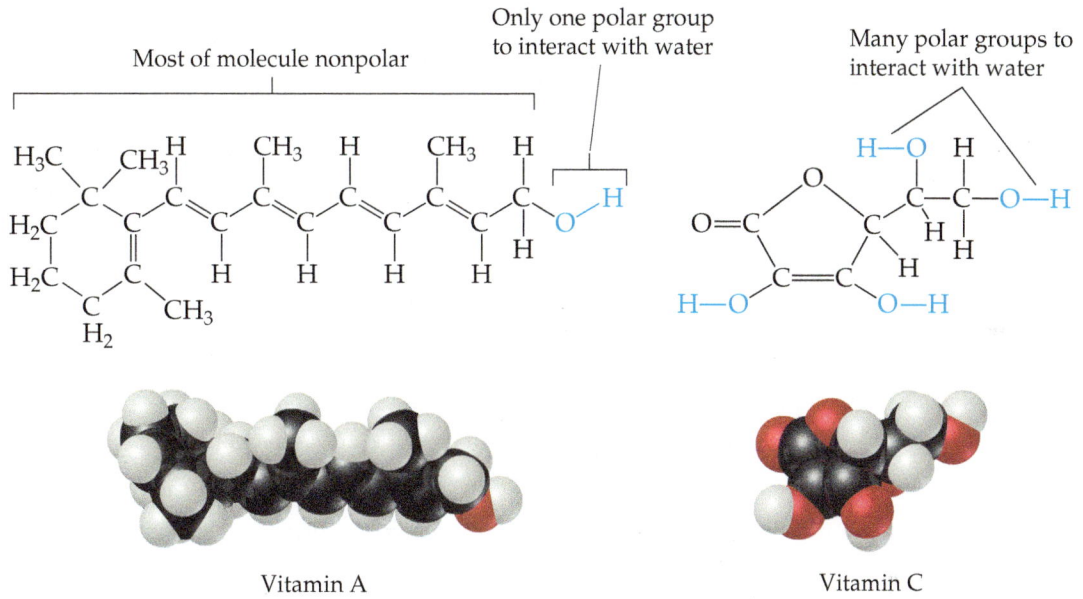

Vitamin A   Vitamin C

▲ **Figure 13.11 The molecular structures of vitamins A and C.**

Pressure Effects

The solubilities of solids and liquids are not appreciably affected by pressure, whereas *the solubility of a gas in any solvent is increased as the partial pressure of the gas above the solvent increases.* We can understand the effect of pressure on gas solubility by considering Figure 13.12, which shows carbon dioxide gas distributed between the gas and solution phases. When equilibrium is established, the rate at which gas molecules enter the solution equals the rate at which solute molecules escape from the solution to enter the gas phase. The equal number of up and down arrows in the left container in Figure 13.12 represents these opposing processes.

Now suppose we exert greater pressure on the piston and compress the gas above the solution, as shown in the middle container in Figure 13.12. If we reduce the gas volume to half its original value, the pressure of the gas increases to twice its original value. As a result of this pressure increase, the rate at which gas molecules strike the liquid surface and enter the solution phase increases, but the rate at which solute molecules escape from the solution remains unchanged. Thus, the solubility of the gas in the solution increases until equilibrium is again established; that is, solubility increases until the rate at which gas molecules enter the solution equals the rate at which they escape from the

Go Figure   If the partial pressure of a gas over a solution is doubled, how has the concentration of gas in the solution changed after equilibrium is restored?

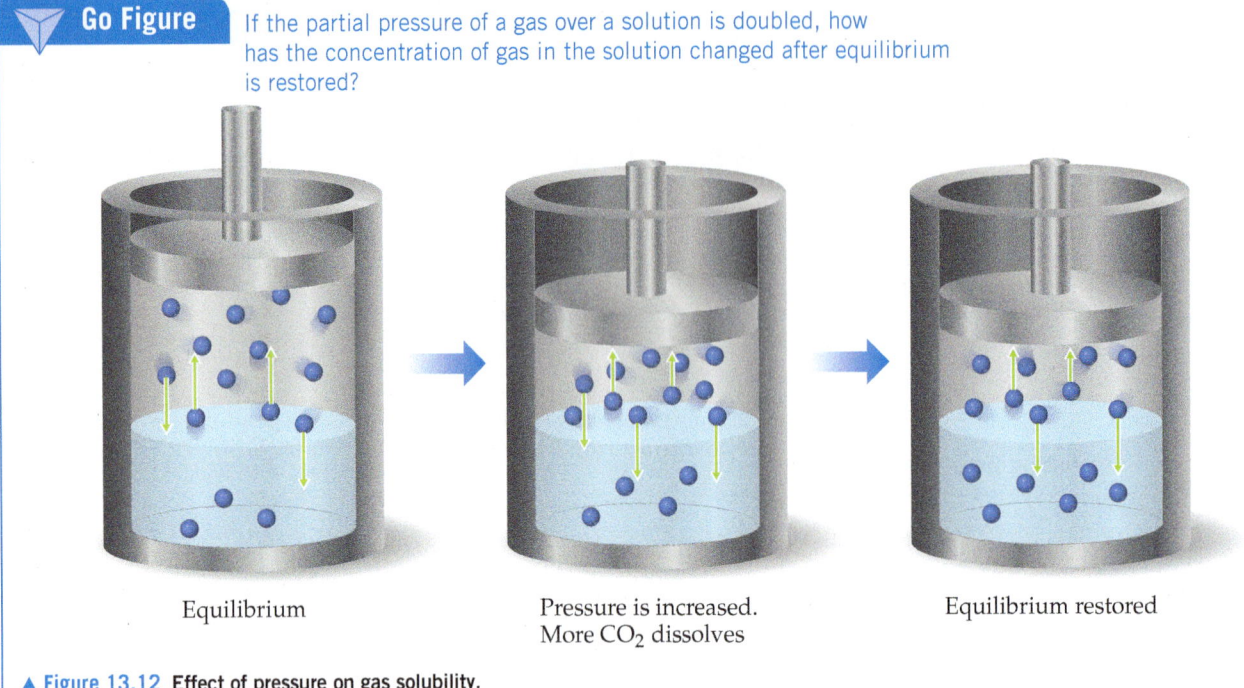

Equilibrium   Pressure is increased.   Equilibrium restored
                     More CO_2 dissolves

▲ **Figure 13.12** **Effect of pressure on gas solubility.**

solution. Thus, *the solubility of a gas in a liquid solvent increases in direct proportion to the partial pressure of the gas above the solution* (**Figure 13.13**).

The relationship between pressure and gas solubility is expressed by **Henry's law**:

$$S_g = kP_g \qquad [13.4]$$

Here, S_g is the solubility of the gas in the solvent (usually expressed as molarity), P_g is the partial pressure of the gas over the solution, and k is a proportionality constant known as the *Henry's law constant*. The value of this constant depends on the solute, solvent, and temperature. As an example, the solubility of N_2 gas in water at 25 °C and 0.78 atm pressure is $4.75 \times 10^{-4} M$. The Henry's law constant for N_2 in 25 °C water is thus $(4.75 \times 10^{-4} \, \text{mol/L})/0.78 \, \text{atm} = 6.1 \times 10^{-4} \, \text{mol/L-atm}$. If the partial pressure of N_2 is doubled, Henry's law predicts that the solubility in water at 25 °C also doubles to $9.50 \times 10^{-4} M$.

Bottlers use the effect of pressure on solubility in producing carbonated beverages, which are bottled under a carbon dioxide pressure greater than 1 atm. When the bottles are opened to the air, the partial pressure of CO_2 above the solution decreases. Hence, the solubility of CO_2 decreases, and $CO_2(g)$ escapes from the solution as bubbles (**Figure 13.14**).

Go Figure

Does the gas with the largest molar mass have the highest or lowest solubility in water?

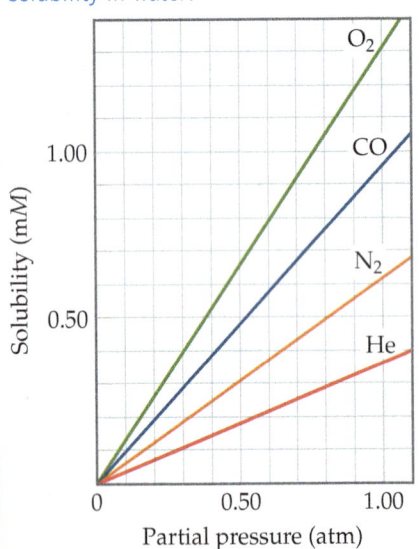

▲ **Figure 13.13** **The solubility of a gas in water is directly proportional to the partial pressure of the gas.** The solubilities are in millimoles of gas per liter of solution.

▲ **Figure 13.14** **Gas solubility decreases as pressure decreases.** CO_2 bubbles out of solution when a carbonated beverage is opened because the CO_2 partial pressure above the solution is reduced.

Sample Exercise 13.2
Using Henry's Law to Calculate the Solubility of a Gas

Calculate the concentration of CO_2 in a soft drink that is bottled with a partial pressure of CO_2 of 4.0 atm over the liquid at 25 °C. The Henry's law constant for CO_2 in water at this temperature is 3.4×10^{-2} mol/L-atm.

SOLUTION

Analyze We are given the partial pressure of CO_2, P_{CO_2}, and the Henry's law constant, k, and asked to calculate the concentration of CO_2 in the solution.

Plan With the information given, we can use Henry's law, Equation 13.4, to calculate the solubility, S_{CO_2}.

Solve $S_{CO_2} = kP_{CO_2} = (3.4 \times 10^{-2} \text{ mol/L-atm})(4.0 \text{ atm})$

$$= 0.14 \text{ mol/L} = 0.14\, M$$

Check The units are correct for solubility, and the answer has two significant figures, consistent with both the partial pressure of CO_2 and the value of the Henry's law constant.

▶ **Practice Exercise**

Calculate the concentration of CO_2 in a soft drink after the bottle is opened and the solution equilibrates at 25 °C under a CO_2 partial pressure of 3.0×10^{-4} atm.

Temperature Effects

Many aspects of cleaning, such as washing the dishes, doing a load of laundry, or mopping the floor, involve dissolving unwanted substances in an aqueous solution. More often than not we use hot water as the solvent for these tasks, under the assumption that the substances we are trying to dissolve become more soluble as the temperature of the solution increases. The graphs in **Figure 13.15** show, indeed, that *the solubility of most solid solutes in water increases as the solution temperature increases*. There are exceptions to this rule, however, as seen for $Ce_2(SO_4)_3$, whose solubility curve slopes downward with increasing temperature.

In contrast to solid solutes, *the solubility of gases in water decreases with increasing temperature* (**Figure 13.16**). If a glass of cold tap water is warmed, you can see bubbles on the inside of the glass because some of the dissolved air comes out of solution. Similarly, as carbonated beverages are allowed to warm, the solubility of carbon dioxide decreases, and $CO_2(g)$ escapes from the solution.

Why do the solubilities of solids and gases generally trend in opposite directions as the temperature increases? The changes in entropy that accompany solution formation can help us understand this behavior. When a solid dissolves to form a solution, the molecules or ions that make up the solid usually experience an increase in their freedom of motion, and in this way the dissolution usually increases the entropy of the system. As we explain in Chapter 19, an increase in entropy favors a process occurring, and this effect becomes more pronounced as the temperature increases. Thus, the solubilities of most solids increase as the temperature rises, increasing the entropy of the system. Entropy changes in the opposite sense for gaseous solutes. A gas molecule has less freedom of motion in solution, where it is surrounded by solvent molecules, than it does in the gaseous state, where it is well separated from other molecules. So, entropy *decreases* as the concentration of gas molecules dissolved in the solvent *increases*, and the solubility *decreases* as the temperature *increases*.

The solution process is a complicated one, and enthalpy as well as entropy must be considered to fully understand how temperature affects solubility. As a general rule, when the enthalpy of solution is endothermic ($\Delta H_{soln} > 0$), solubility increases as the temperature increases, whereas solutes that dissolve exothermically in solution ($\Delta H_{soln} < 0$), like gases dissolving in water, tend to show the opposite temperature dependence. We explore some of these effects in greater detail in Chapter 19. In the meantime, the change in entropy on dissolution provides a straightforward explanation of why the solubility of solids generally increases with increasing temperature, whereas the solubility of gases decreases.

Go Figure

How does the solubility of KCl at 80 °C compare with that of NaCl at the same temperature?

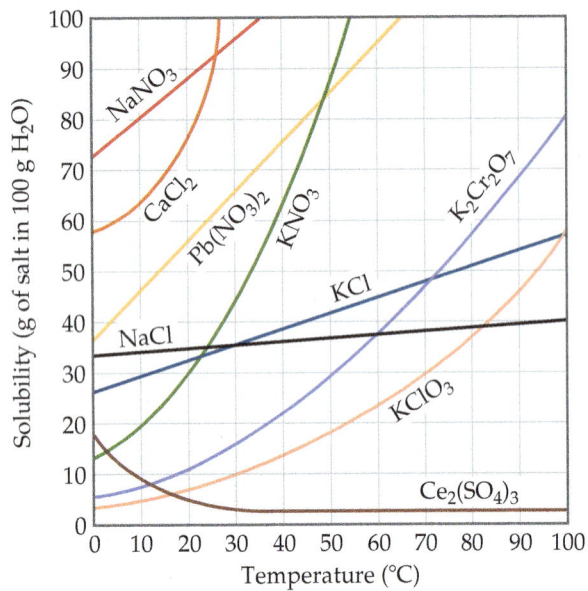

▲ **Figure 13.15** Solubilities of some ionic compounds in water as a function of temperature.

Go Figure

Between which two gases would you expect N_2 to fit on this graph?

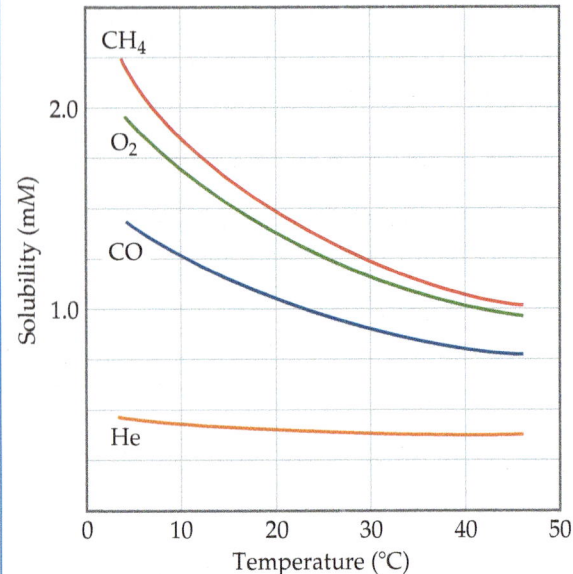

▲ **Figure 13.16** Solubilities of four gases in water as a function of temperature. The solubilities are in millimoles per liter of solution, for a constant total pressure of 1 atm in the gas phase.

Self-Assessment Exercises

SAE 13.7 Acetonitrile (CH_3CN) is a colorless liquid used as a solvent for numerous chemical reactions. Which of the following liquids is *least likely* to be miscible with acetonitrile? (**a**) water, H_2O (**b**) acetone, CH_3COCH_3 (**c**) hexane, C_6H_{14} (**d**) chloroform, CH_2Cl_2 (**e**) methanol, CH_3OH

SAE 13.8 A sealed aluminum can contains 330 mL of soda that is at 20 °C and in equilibrium with CO_2 gas at a pressure of 2.0 atm. If the Henry's law constant for CO_2 is 3.4×10^{-2} mol/L-atm at 20 °C, how many moles of CO_2 are dissolved in the soda? (**a**) 0.068 mol (**b**) 0.034 mol (**c**) 0.022 mol (**d**) 0.21 mol

SAE 13.9 Which of the following statements about solubility is or are *true*?

(**i**) The solubility of a gas in contact with water generally increases as the temperature of the water increases.

(**ii**) The solubility of a gas in contact with water generally increases as the partial pressure of the gas increases.

(**iii**) The solubility of an ionic solid in contact with water generally increases as pressure exerted on the solution increases.

(**a**) only i (**b**) only ii (**c**) only iii (**d**) both ii and iii (**e**) All three statements are true.

SAE 13.10 Which of the following would increase the concentration of CO_2 dissolved in the ocean?

(**i**) increasing the concentration of CO_2 in the atmosphere
(**ii**) increasing the temperature of the ocean
(**iii**) increasing the concentration of O_2 in the atmosphere

(**a**) only i (**b**) only ii (**c**) both i and ii (**d**) both i and iii (**e**) all three

13.4 | Expressing Solution Concentration

The concentration of a solution can be expressed either qualitatively or quantitatively. The terms *dilute* and *concentrated* are used to describe a solution qualitatively. A solution with a relatively small concentration of solute is said to be dilute, whereas one with a large concentration is said to be concentrated. In many instances, however, we need to express the concentration of a solution quantitatively. In this section we examine several different ways to do so.

Learning Objectives

When you finish Section 13.4, you should be able to:

▶ Express the concentration of a solute either in parts per million (ppm), parts per billion (ppb), or mass percentage.

▶ Express the concentration of each component of a solution as a mole fraction when given the amounts of each component of a solution.

▶ Express the concentration of a solute either in molarity or molality and convert between the two.

Mass Percentage, ppm, and ppb

One of the simplest quantitative expressions of concentration is the **mass percentage** of a component in a solution, given by

$$\text{Mass \% of component} = \frac{\text{mass of component in soln}}{\text{total mass of soln}} \times 100\% \qquad [13.5]$$

Because *percent* means "per hundred," a solution of hydrochloric acid that is 36% HCl by mass contains 36 g of HCl for each 100 g of solution.

We often express the concentration of very dilute solutions in **parts per million (ppm)** or **parts per billion (ppb)**. These quantities are similar to mass percentage but use 10^6 (a million) or 10^9 (a billion), respectively, in place of 100, as a multiplier for the ratio of the mass of solute to the mass of solution. Thus, parts per million is defined as

$$\text{ppm of component} = \frac{\text{mass of component in soln}}{\text{total mass of soln}} \times 10^6 \qquad [13.6]$$

A solution whose solute concentration is 1 ppm contains 1 g of solute for each million (10^6) grams of solution or, equivalently, 1 mg of solute per kilogram of solution. Because the density of water is 1 g/mL, 1 kg of a dilute aqueous solution has a volume very close to 1 L. Thus, 1 ppm also corresponds to 1 mg of solute per liter of aqueous solution.

The acceptable maximum concentrations of toxic or carcinogenic substances in the environment are often expressed in ppm or ppb. For example, the maximum allowable concentration of arsenic in drinking water in the United States is 0.010 ppm; that is, 0.010 mg of arsenic per liter of water. This concentration corresponds to 10 ppb.

Sample Exercise 13.3

Calculation of Mass-Related Concentrations

(a) A solution is made by dissolving 13.5 g of glucose ($C_6H_{12}O_6$) in 0.100 kg of water. What is the mass percentage of solute in this solution?

(b) A 2.5-g sample of groundwater was found to contain 5.4 μg of Zn^{2+}. What is the concentration of Zn^{2+} in parts per million?

SOLUTION

(a) **Analyze** We are given the number of grams of solute (13.5 g) and the number of grams of solvent (0.100 kg = 100 g). From this, we must calculate the mass percentage of solute.

Plan We can calculate the mass percentage by using Equation 13.5. The mass of the solution is the sum of the mass of solute (glucose) and the mass of solvent (water).

Solve Mass % of glucose $= \dfrac{\text{mass glucose}}{\text{mass soln}} \times 100\%$

$$= \frac{13.5\,\text{g}}{(13.5 + 100)\,\text{g}} \times 100\% = 11.9\%$$

Comment The mass percentage of water in this solution is $(100 - 11.9)\% = 88.1\%$.

(b) **Analyze** In this case we are given the number of micrograms of solute. Because 1 μg is 1×10^{-6} g, 5.4 μg = 5.4×10^{-6} g.

Plan We calculate the parts per million using Equation 13.6.

Solve ppm $= \dfrac{\text{mass of solute}}{\text{mass of soln}} \times 10^6$

$$= \frac{5.4 \times 10^{-6}\,\text{g}}{2.5\,\text{g}} \times 10^6 = 2.2\,\text{ppm}$$

▶ **Practice Exercise**

A commercial bleaching solution contains 3.62% by mass of sodium hypochlorite, NaOCl. What is the mass of NaOCl in a bottle containing 2.50 kg of bleaching solution?

Mole Fraction, Molarity, and Molality

Concentration expressions are often based on the number of moles of one or more components of the solution. Recall from Section 10.4 that the *mole fraction* of a component of a solution is given by

$$\text{Mole fraction of component} = \frac{\text{moles of component}}{\text{total moles of all components}} \qquad [13.7]$$

The symbol X is commonly used for mole fraction, with a subscript to indicate the component of interest. For example, the mole fraction of HCl in a hydrochloric acid solution is represented as X_{HCl}. Thus, if a solution contains 1.00 mol of HCl (36.5 g) and 8.00 mol of water (144 g), the mole fraction of HCl is $X_{HCl} = (1.00\ \text{mol})/(1.00\ \text{mol} + 8.00\ \text{mol}) = 0.111$. Mole fractions have no units because the units in the numerator and the denominator cancel. The sum of the mole fractions of all components of a solution must equal 1. Thus, in the aqueous HCl solution, $X_{H_2O} = 1.000 - 0.111 = 0.889$. Mole fractions are very useful when dealing with gases, as we saw in Section 10.4, but have limited use when dealing with liquid solutions.

Recall from Section 4.5 that the *molarity* (M) of a solute in a solution is defined as

$$\text{Molarity} = \frac{\text{moles of solute}}{\text{liters of soln}} \qquad [13.8]$$

For example, if you dissolve 0.500 mol of Na_2CO_3 in enough water to form 0.250 L of solution, the molarity of Na_2CO_3 in the solution is $(0.500\ \text{mol})/(0.250\ \text{L}) = 2.00\ M$. Molarity is especially useful for relating the volume of a solution to the quantity of solute contained in that volume, as we saw in our discussions of titrations. (Section 4.6)

The **molality** of a solution, denoted m, is a concentration unit that is also based on moles of solute. Molality equals the number of moles of solute per kilogram of solvent:

$$\text{Molality} = \frac{\text{moles of solute}}{\text{kilograms of solvent}} \qquad [13.9]$$

Thus, if you form a solution by mixing 0.200 mol of NaOH (8.00 g) and 0.500 kg of water (500 g), the concentration of the solution is $(0.200\ \text{mol})/(0.500\ \text{kg}) = 0.400\ m$ (that is, 0.400 molal) in NaOH.

The definitions of molarity and molality are similar enough that they can be easily confused.

- Molarity depends on the *volume of solution*.
- Molality depends on the *mass of solvent*.

When water is the solvent, the molality and molarity of dilute solutions are numerically similar because 1 kg of solvent is nearly the same as 1 kg of solution, and 1 kg of the solution has a volume of about 1 L.

The molality of a given solution does not vary with temperature because masses do not vary with temperature. The molarity of the solution does change with temperature, however, because the volume of the solution expands or contracts with temperature. Thus, molality is often the concentration unit of choice when a solution is to be used over a range of temperatures.

Sample Exercise 13.4
Calculation of Molality

A solution is made by dissolving 4.35 g of glucose ($C_6H_{12}O_6$) in 25.0 mL of water at 25 °C. Calculate the molality of glucose in the solution. At this temperature water has a density of 0.997 g/mL.

SOLUTION

Analyze We are asked to calculate a solution concentration in units of molality. To do this, we must determine the number of moles of solute (glucose) and the number of kilograms of solvent (water).

Plan We use the molar mass of $C_6H_{12}O_6$ to convert grams of glucose to moles. We use the density of water to convert milliliters of water to kilograms. The molality equals the number of moles of solute (glucose) divided by the number of kilograms of solvent (water).

Solve

Use the molar mass of glucose, 180.2 g/mol, to convert grams to moles:

$$\text{Mol } C_6H_{12}O_6 = (4.35 \text{ g } C_6H_{12}O_6)\left(\frac{1 \text{ mol } C_6H_{12}O_6}{180.2 \text{ g } C_6H_{12}O_6}\right) = 0.0241 \text{ mol } C_6H_{12}O_6$$

Because water has a density of 0.997 g/mL the mass of the solvent is:

$$(25.0 \text{ mL})(0.997 \text{ g/mL}) = 24.9 \text{ g} = 0.0249 \text{ kg}$$

Finally, use Equation 13.9 to obtain the molality:

$$\text{Molality of } C_6H_{12}O_6 = \frac{0.0241 \text{ mol } C_6H_{12}O_6}{0.0249 \text{ kg } H_2O} = 0.968 \; m$$

▶ **Practice Exercise**

What is the molality of a solution made by dissolving 36.5 g of naphthalene ($C_{10}H_8$) in 425 g of toluene (C_7H_8)?

Converting Concentration Units

Often, we need to convert from one method of expressing solution concentration to another. **Figure 13.17** summarizes the six different ways of expressing solution concentration we have encountered. In all cases the amount of solute is found in the numerator, expressed in terms of either its mass or number of moles. If we need to convert from a concentration found on the left side of Figure 13.17 to one on the right, the molar mass of the solute can be used to convert the numerator to the proper units.

The denominators of the different expressions in Figure 13.17 are more diverse. First, we must determine if the denominator is the amount of solution (mass percentage, ppm, ppb, molarity), the amount of solvent (molality), or the total number of moles of each component in the solution (mole fraction). Some of the ways in which we can convert between different concentration units are as follows:

- If we know the mass of the solution, we can convert to volume or vice versa using its density. This would be necessary to convert from mass percentage, ppm, or ppb to molarity.
- Converting between molality, molarity, and mole fraction requires an additional conversion. For example, if we know molarity and want to calculate molality, we must first use density to convert the volume of solution to its mass, and once we know the mass of solution, we can subtract the mass of the solute to get the mass of solvent.
- If we need to convert to mole fraction, we can use the molar mass of the solvent to convert its mass to moles.

Sample Exercise 13.5 illustrates some of these conversions.

$$\text{Mass \%} = \frac{\text{mass of solute}}{\text{total mass of solution}} \times 100\%$$

$$\text{Mole Fraction} = \frac{\text{moles of solute}}{\text{total moles of all components}}$$

$$\text{ppm} = \frac{\text{mass of solute}}{\text{total mass of solution}} \times 10^6$$

$$\text{Molarity} = \frac{\text{moles of solute}}{\text{liters of solution}}$$

$$\text{ppb} = \frac{\text{mass of solute}}{\text{total mass of solution}} \times 10^9$$

$$\text{Molarity} = \frac{\text{moles of solute}}{\text{kilograms of solvent}}$$

Concentrations based on **mass** of solute

Concentrations based on **moles** of solute

▲ **Figure 13.17 Various ways of expressing solution concentration.**

Sample Exercise 13.5

Converting between mass percentage, molality, and molarity

A solution is made by mixing 5.0 g of toluene (C_7H_8) and 225 g of benzene (C_6H_6). The density of the solution is measured and found to be 0.876 g/mL. Calculate the (**a**) mass percentage, (**b**) molality, and (**c**) molarity of toluene in this solution.

SOLUTION

Analyze We are asked to express the concentration of toluene in three different ways shown in Figure 3.17, given the mass of solute and the mass of solvent.

Plan (**a**) We can calculate the total mass of the solution by summing the mass of the solute (toluene) and the solvent (benzene). We can then determine the mass percentage (Equation 13.5) of toluene by dividing the mass of toluene (5.0 g) by the total mass of the solution, then multiplying by 100%.

(**b**) Using the molar mass of toluene, we can calculate the number of moles toluene from its mass. Dividing this quantity by the mass (in kg) of benzene, we can determine the molality of the solution (Equation 13.9). (**c**) The molarity of a solution is the number of moles of solute divided by the number of liters of solution (Equation 13.8). We determined the number of moles solute when calculating molality, and the volume of the solution can be determined from the mass of the solution and its density.

Solve

(**a**) The mass percentage toluene can be calculated from the masses given in the problem:

$$\text{Mass \% toluene } C_7H_8 = \frac{5.0 \text{ g } C_7H_8}{(5.0 \text{ g } C_7H_8 + 225 \text{ g } C_6H_6)} \times 100\% = 2.2\%$$

(**b**) The number of moles of solute is:

$$\text{Moles } C_7H_8 = (5.0 \text{ g } C_7H_8)\left(\frac{1 \text{ mol } C_7H_8}{92 \text{ g } C_7H_8}\right) = 0.054 \text{ mol}$$

The molality is the moles solute (toluene) divided by the mass (in kg) of solvent (benzene):

$$\text{Molality} = \left(\frac{\text{moles } C_7H_8}{\text{kg } C_6H_6}\right) = \left(\frac{0.054 \text{ mol } C_7H_8}{0.225 \text{ kg } C_6H_6}\right) = 0.24 \text{ } m$$

(**c**) To determine the molarity, we need to know the volume of the solution. The density of the solution is used to convert the mass of the solution to its volume:

$$\text{Milliliters soln} = (230 \text{ g})\left(\frac{1 \text{ mL}}{0.876 \text{ g}}\right) = 263 \text{ mL}$$

Molarity is moles of solute per liter of solution:

$$\text{Molarity} = \left(\frac{\text{moles } C_7H_8}{\text{liter soln}}\right) = \left(\frac{0.054 \text{ mol } C_7H_8}{263 \text{ mL soln}}\right)\left(\frac{1000 \text{ mL soln}}{1 \text{ L soln}}\right) = 0.21 \text{ } M$$

Check The magnitudes of our answers are reasonable. Rounding moles to 0.05 and liters to 0.25 gives a molarity of $(0.05 \text{ mol})/(0.25 \text{ L}) = 0.2 \text{ } M$. The molality and molarity are similar in magnitude, as we would expect for a solution with a density not too different from 1 g/mL. Each answer has two significant figures, corresponding to the number of significant figures in the mass of the solute (2).

▶ **Practice Exercise**

A solution containing equal masses of glycerol ($C_3H_8O_3$) and water has a density of 1.10 g/mL. Calculate (**a**) the molality of glycerol, (**b**) the mole fraction of glycerol, (**c**) the molarity of glycerol in the solution.

Self-Assessment Exercises

SAE 13.11   The drinking water standards in the United States mandate that the maximum safe amount of arsenic in water is 10 ppb. The level of arsenic in natural waters is generally less than 2 μg/L. What is this concentration in ppb? To two significant figures, the density of water is 1.0 g/mL. (**a**) 2000 ppb (**b**) 2×10^{-9} ppb (**c**) 20 ppb (**d**) 0.2 ppb (**e**) 2 ppb

SAE 13.12   Calculate the molality (m) and molarity (M) of a solution made from mixing 802 g of glucose, $C_6H_{12}O_6$, with 1.00 L of water (density = 0.997 g/mL at 25 °C) to form 1.20 L of solution. (**a**) 2.47 m, 4.47 M (**b**) 4.47 m, 4.46 M (**c**) 4.47 m, 3.71 M (**d**) 0.668 m, 668 M

SAE 13.13   An aqueous solution of hydrochloric acid contains 36% HCl by mass. What is the mole fraction of HCl in this solution? (**a**) 0.22 (**b**) 15 (**c**) 0.018 (**d**) 0.36 (**e**) 0.28

SAE 13.14   Vinegar is an aqueous solution of acetic acid (CH_3COOH). If the mass percent acetic acid in a bottle of vinegar is 5.0% and the density of the solution is 1.006 g/mL, what is the molarity of acetic acid in the vinegar? Assume that the vinegar contains only acetic acid and water. (**a**) 5.0 M (**b**) 0.88 M (**c**) 50. M (**d**) 0.84 M (**e**) 2.5 M

13.5 | Colligative Properties

Some physical properties of solutions differ in important ways from those of the pure solvent. For example, pure water freezes at 0 °C, but aqueous solutions freeze at lower temperatures. We use this behavior to our advantage when we add ethylene glycol antifreeze to a car's radiator to lower the freezing point of the solution. The added solute also raises the boiling point of the solution above that of pure water, making it possible to operate the engine at a higher temperature.

Lowering of the freezing point and raising of the boiling point are physical properties of solutions that depend on the *quantity* (concentration) but not on the *identity* of the solute particles. Such properties are called **colligative properties**. (*Colligative* means "depending on the collection"; colligative properties depend on the collective effect of the number of solute particles.)

In addition to lowering the freezing point and raising the boiling point, other important colligative properties include vapor–pressure lowering and osmotic pressure. As we examine each one, notice how solute concentration quantitatively affects the property.

Vapor–Pressure Lowering

A liquid in a closed container establishes an equilibrium with its vapor. (Section 11.5) The *vapor pressure* is the pressure exerted by the vapor when it is at equilibrium with the liquid (i.e., when the rate of vaporization equals the rate of condensation). A substance that has no measurable vapor pressure is said to be *nonvolatile,* whereas one that exhibits a vapor pressure is *volatile*.

A solution consisting of a *volatile* liquid solvent and a *nonvolatile* solute forms spontaneously because of the increase in entropy that accompanies their mixing. In effect, this process stabilizes the solvent molecules in their liquid state, thus causing them to have a lower tendency to escape into the vapor state. Therefore, when a *nonvolatile* solute is present, the vapor pressure of the solvent is lower than the vapor pressure of the pure solvent, as illustrated in Figure 13.18.

Ideally, the vapor pressure of a *volatile* solvent above a solution containing a *nonvolatile* solute is proportional to the solvent's concentration in the solution. This relationship is expressed quantitatively by **Raoult's law**, which states that the partial pressure exerted by solvent vapor above the solution, $P_{solution}$, equals the product of the mole fraction of the solvent, $X_{solvent}$, times the vapor pressure of the pure solvent, $P^{\circ}_{solvent}$:

$$P_{solution} = X_{solvent} P^{\circ}_{solvent} \qquad [13.10]$$

For example, the vapor pressure of pure water at 20 °C is $P^{\circ}_{H_2O} = 17.5$ torr. Imagine holding the temperature constant while adding glucose ($C_6H_{12}O_6$) to the water so that the mole fractions in the resulting solution are $X_{H_2O} = 0.800$ and $X_{C_6H_{12}O_6} = 0.200$.

Learning Objectives

When you finish **Section 13.5**, you should be able to:

▶ Use Raoult's law to calculate the vapor pressure of the solvent above a solution containing a nonvolatile solute.

▶ Define what is meant by an ideal solution and explain how deviations from ideal behavior affect the vapor pressure of a solution.

▶ Calculate the normal boiling and freezing points of a solution and describe how they depend on the nature and concentration of the solute.

▶ Describe osmotic pressure and calculate it for two solutions separated by a semipermeable membrane.

▶ Explain why the colligative properties of solutions containing electrolytes differ from those containing nonelectrolytes and calculate the size of this effect.

▶ Use one or more colligative properties of a solution containing an unknown solute to determine the molar mass of the solute.

● Volatile solvent particles

● Nonvolatile solute particles

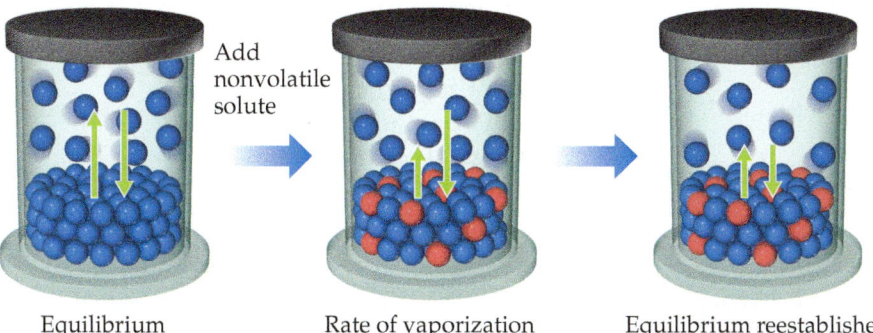

| Equilibrium | Rate of vaporization reduced by presence of nonvolatile solute | Equilibrium reestablished with fewer molecules in gas phase |

◀ **Figure 13.18 Vapor–pressure lowering.** The presence of nonvolatile solute particles in a liquid solvent results in a reduction of the vapor pressure above the liquid.

According to Equation 13.10, the vapor pressure of the water above this solution is 80.0% that of pure water:

$$P_{solution} = (0.800)(17.5 \text{ torr}) = 14.0 \text{ torr}$$

The presence of the *nonvolatile* solute lowers the vapor pressure of the *volatile* solvent by 17.5 torr − 14.0 torr = 3.5 torr.

The vapor–pressure lowering, ΔP, is directly proportional to the mole fraction of the solute, X_{solute}:

$$\Delta P = X_{solute}P^{\circ}_{solvent} \qquad [13.11]$$

Thus, for the example of the solution of glucose in water, we have

$$\Delta P = X_{C_6H_{12}O_6} P^{\circ}_{H_2O} = (0.200)(17.5 \text{ torr}) = 3.50 \text{ torr}$$

The vapor–pressure lowering caused by adding a *nonvolatile* solute depends on the total concentration of solute particles, regardless of whether they are molecules or ions. Remember that vapor-pressure lowering is a colligative property, so its value for any solution depends on the concentration of solute particles and not on their identity. This becomes important for solutions of electrolytes like NaCl or HNO_3, where we must consider the concentration of cations and anions as separate particles.

An ideal gas is defined as one that obeys the ideal-gas equation (Section 10.3), and an **ideal solution** is defined as one that obeys Raoult's law. Whereas ideality for a gas arises from a complete lack of intermolecular interaction, ideality for a solution implies total uniformity of interaction. The molecules in an ideal solution all influence one another in the same way; in other words, solute–solute, solvent–solvent, and solute–solvent interactions are indistinguishable from one another. Real solutions best approximate ideal behavior when the solute concentration is low and solute and solvent have similar molecular sizes and take part in similar types of intermolecular attractions.

Many solutions do not obey Raoult's law exactly and so are not ideal. If, for instance, the solvent–solute interactions in a solution are weaker than either the solvent–solvent or solute–solute interactions, the vapor pressure tends to be greater than that predicted by Raoult's law. When the solute–solvent interactions in a solution are exceptionally strong, as might be the case when hydrogen bonding exists, the vapor pressure is lower than that predicted by Raoult's law. Although you should be aware that these departures from ideality occur, we will largely ignore them for the remainder of this chapter.

Sample Exercise 13.6

Calculation of Vapor Pressure of a Solution

Glycerin ($C_3H_8O_3$) is a nonvolatile nonelectrolyte with a density of 1.26 g/mL at 25 °C. Calculate the vapor pressure at 25 °C of a solution made by adding 50.0 mL of glycerin to 500.0 mL of water. The vapor pressure of pure water at 25 °C is 23.8 torr (Appendix B), and its density is 1.00 g/mL.

SOLUTION

Analyze Our goal is to calculate the vapor pressure of a solution, given the volume and density of both the solute and the solvent.

Plan We can use Raoult's law (Equation 13.10) to calculate the vapor pressure of a solution from the mole fraction of the

solvent, $X_{solvent}$. To calculate the mole fraction of the solvent (Equation 13.7), we use the volume, density, and molar mass of both the solvent and the solute to determine the number of moles of each that are present in the solution.

Solve

We begin by determining the number of moles of $C_3H_8O_3$ and H_2O from their volume, density, and molar mass:

$$\text{Moles } C_3H_8O_3 = (50.0 \text{ mL } C_3H_8O_3)\left(\frac{1.26 \text{ g } C_3H_8O_3}{1 \text{ mL } C_3H_8O_3}\right)\left(\frac{1 \text{ mol } C_3H_8O_3}{92.1 \text{ g } C_3H_8O_3}\right) = 0.684 \text{ mol}$$

$$\text{Moles } H_2O = (500.0 \text{ mL } H_2O)\left(\frac{1.00 \text{ g } H_2O}{1 \text{ mL } H_2O}\right)\left(\frac{1 \text{ mol } H_2O}{18.0 \text{ g } H_2O}\right) = 27.8 \text{ mol}$$

We use these values to calculate the mole fraction of water in the solution:

$$X_{H_2O} = \frac{\text{mol } H_2O}{\text{mol } H_2O + \text{mol } C_3H_8O_3} = \frac{27.8}{27.8 + 0.684} = 0.976$$

Finally, we use Raoult's law to calculate the vapor pressure of water for the solution:

$$P_{H_2O} = X_{H_2O} P^{\circ}_{H_2O} = (0.976)(23.8 \text{ torr}) = 23.2 \text{ torr}$$

Comment The vapor pressure of the solution has been lowered by 23.8 torr − 23.2 torr = 0.6 torr relative to that of pure water. The vapor–pressure lowering can be calculated directly using Equation 13.11 together with the mole fraction of the solute, $C_3H_8O_3$: $\Delta P = X_{C_3H_8O_3} P^{\circ}_{H_2O} = (0.024)(23.8\ \text{torr}) = 0.57\ \text{torr}$. Notice that the use of Equation 13.11 gives one more significant figure than the number obtained by subtracting the vapor pressure of the solution from that of the pure solvent.

▶ **Practice Exercise**
The vapor pressure of pure water at 110 °C is 1070 torr. A solution of ethylene glycol and water has a vapor pressure of 1.00 atm at 110 °C. Assuming that Raoult's law is obeyed, what is the mole fraction of ethylene glycol in the solution?

A CLOSER LOOK | Ideal Solutions with Two or More Volatile Components

Solutions sometimes have two or more volatile components. Gasoline, for example, is a solution of several volatile liquids. To gain some understanding of such mixtures, consider an ideal solution of two volatile liquids, A and B. (For our purposes here, it does not matter which we call the solute and which the solvent.) The partial pressures above the solution are given by Raoult's law:

$$P_A = X_A P^{\circ}_A \quad \text{and} \quad P_B = X_B P^{\circ}_B$$

and the total vapor pressure above the solution is

$$P_{total} = P_A + P_B = X_A P^{\circ}_A + X_B P^{\circ}_B$$

▲ **Figure 13.19** The volatile components of organic mixtures can be separated on an industrial scale in these distillation towers.

Consider a mixture of 1.0 mol of benzene (C_6H_6) and 2.0 mol of toluene (C_7H_8) ($X_{ben} = 0.33$, $X_{tol} = 0.67$). At 20 °C, the vapor pressures of the pure substances are $P^{\circ}_{ben} = 75$ torr and $P^{\circ}_{tol} = 22$ torr. Thus, the partial pressures above the solution are

$$P_{ben} = (0.33)(75\ \text{torr}) = 25\ \text{torr}$$

$$P_{tol} = (0.67)(22\ \text{torr}) = 15\ \text{torr}$$

and the total vapor pressure above the liquid is

$$P_{total} = P_{ben} + P_{tol} = 25\ \text{torr} + 15\ \text{torr} = 40\ \text{torr}$$

Note that the vapor is richer in benzene, the more volatile component.

The mole fraction of benzene in the vapor is given by the ratio of its vapor pressure to the total pressure (Equations 10.14 and 10.15):

$$X_{ben}\ \text{in vapor} = \frac{P_{ben}}{P_{tol}} = \frac{25\ \text{torr}}{40\ \text{torr}} = 0.63$$

Although benzene constitutes only 33% of the molecules in the solution, it makes up 63% of the molecules in the vapor.

When an ideal liquid solution containing two volatile components is in equilibrium with its vapor, the more volatile component will be relatively richer in the vapor. This fact forms the basis of *distillation*, a technique used to separate (or partially separate) mixtures containing volatile components. (Section 1.3) Distillation is a way of purifying liquids and is the procedure by which petrochemical plants achieve the separation of crude petroleum into gasoline, diesel fuel, lubricating oil, and other products (Figure 13.19). Distillation is also used routinely on a small scale in the laboratory.
Related Exercises: 13.69, 13.70

Boiling-Point Elevation

In Sections 11.5 and 11.6, we examined the vapor pressures of pure substances and how to use them to construct phase diagrams. How does the phase diagram of a solution and, hence, its boiling and freezing points differ from that of the pure solvent? The addition of a nonvolatile solute lowers the vapor pressure of the solution. Thus, in Figure 13.20 the vapor–pressure curve of the solution is shifted downward relative to the vapor–pressure curve of the pure solvent.

Recall from Section 11.5 that the normal boiling point of a liquid is the temperature at which its vapor pressure equals 1 atm. Because the solution has a lower vapor pressure than the pure solvent, a higher temperature is required for the solution to achieve a vapor pressure of 1 atm. As a result, *the boiling point of the solution is higher than that of the pure solvent*. This effect is seen in Figure 13.20.

The increase in the boiling point of a solution, relative to the pure solvent, depends on the molality of the solute. Remember, though, that boiling-point elevation is proportional to the *total* concentration of solute particles, regardless of whether the particles are molecules or ions. When NaCl dissolves in water, 2 mol of solute particles (1 mol of Na^+ and 1 mol of Cl^-) are formed for each mole of NaCl that dissolves. We take this into

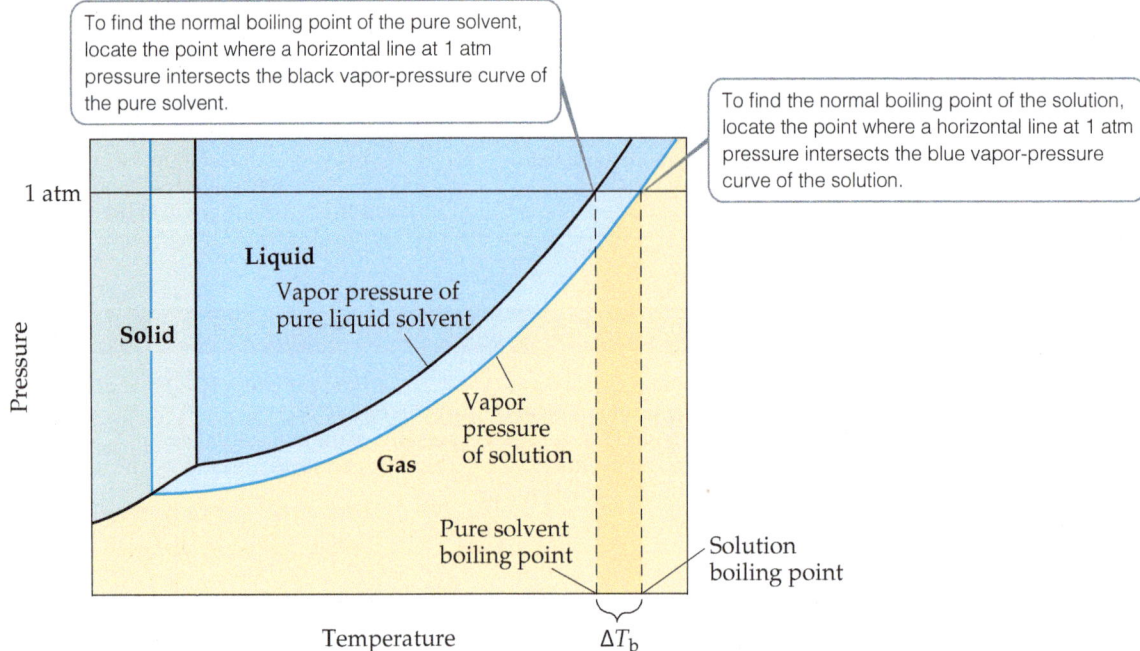

To find the normal boiling point of the pure solvent, locate the point where a horizontal line at 1 atm pressure intersects the black vapor-pressure curve of the pure solvent.

To find the normal boiling point of the solution, locate the point where a horizontal line at 1 atm pressure intersects the blue vapor-pressure curve of the solution.

▲ **Figure 13.20 Phase diagram illustrating boiling-point elevation.** The black lines show the pure solvent's phase equilibria curves, whereas the blue lines show the solution's phase equilibria curves.

account by defining i, the **van't Hoff factor**, as the number of fragments that a solute breaks up into when it dissolves in a particular solvent. The change in boiling point for a solution compared to the pure solvent is:

$$\Delta T_b = T_b(\text{solution}) - T_b(\text{solvent}) = iK_b m \qquad [13.12]$$

In this equation, $T_b(\text{solution})$ is the boiling point of the solution, $T_b(\text{solvent})$ is the boiling point of the pure solvent, m is the molality of the solute, K_b is the **molal boiling-point-elevation constant** for the solvent (which is a proportionality constant that is experimentally determined for each solvent), and i is the van't Hoff factor. For a nonelectrolyte, we can always assume $i = 1$; for an electrolyte, i will depend on how the substance ionizes in that solvent. For instance, $i = 2$ for NaCl in water, assuming complete dissociation of ions. As a result, we expect the boiling-point elevation of a $1\ m$ aqueous solution of NaCl to be twice as large as the boiling-point elevation of a $1\ m$ solution of a nonelectrolyte such as sucrose. Thus, to properly predict the effect of a particular solute on boiling-point elevation (or any other colligative property), it is important to know whether the solute is an electrolyte or a nonelectrolyte. (Sections 4.1 and 4.3)

Freezing-Point Depression

The vapor–pressure curves for the liquid and solid phases meet at the triple point. (Section 11.6) In **Figure 13.21** the triple-point temperature of the solution is lower than the triple-point temperature of pure liquid because the solution has a lower vapor pressure than the pure liquid.

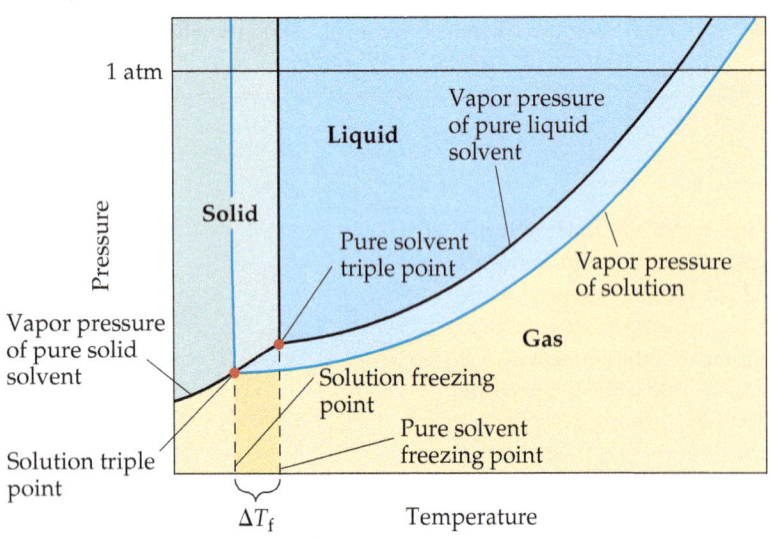

▲ **Figure 13.21 Phase diagram illustrating freezing-point depression.** The black lines show the pure solvent's phase equilibria curves, whereas the blue lines show the solution's phase equilibria curves.

TABLE 13.3 Molal Boiling-Point-Elevation and Freezing-Point-Depression Constants

Solvent	Normal Boiling Point (°C)	K_b (°C/m)	Normal Freezing Point (°C)	K_f (°C/m)
Water, H_2O	100.0	0.51	0.0	1.86
Benzene, C_6H_6	80.1	2.53	5.5	5.12
Ethanol, C_2H_5OH	78.4	1.22	−114.6	1.99
Carbon tetrachloride, CCl_4	76.8	5.02	−22.3	29.8
Chloroform, $CHCl_3$	61.2	3.63	−63.5	4.68

The freezing point of a solution is the temperature at which the first crystals of pure solvent form in equilibrium with the solution. Recall from Section 11.6 that the line representing the solid–liquid equilibrium rises nearly vertically from the triple point. In Figure 13.21, the triple-point temperature of the solution is lower than that of the pure liquid, but it is also true for all points along the solid–liquid equilibrium curve. Thus, *the freezing point of the solution is lower than that of the pure liquid.*

Like the boiling-point elevation, the change in freezing point ΔT_f is directly proportional to solute molality, taking into account the van't Hoff factor *i*:

$$\Delta T_f = T_f(\text{solution}) - T_f(\text{solvent}) = -iK_f m \qquad [13.13]$$

The proportionality constant K_f is the **molal freezing-point-depression constant**, analogous to K_b for boiling-point elevation. Note that because the solution freezes at a *lower* temperature than does the pure solvent, the value of ΔT_f is *negative*.

Values of K_b and K_f for several common solvents are listed in **Table 13.3**. For water, $K_b = 0.51\,°C/m$, which means that the boiling point of any aqueous solution that is 1 *m* in nonvolatile solute particles is 0.51 °C higher than the boiling point of pure water. Because solutions generally do not behave ideally, the constants listed in Table 13.3 work best for solutions that are rather dilute. For water, $K_f = 1.86\,°C/m$. Therefore, any aqueous solution that is 1 *m* in nonvolatile solute particles (such as 1 *m* $C_6H_{12}O_6$ or 0.5 *m* NaCl) freezes at the temperature that is 1.86 °C lower than the freezing point of pure water.

The freezing-point depression caused by solutes has useful applications: It is why antifreeze works in car cooling systems and why calcium chloride ($CaCl_2$) promotes the melting of ice on roads and sidewalks during winter.

Sample Exercise 13.7

Calculation of Boiling-Point Elevation and Freezing-Point Depression

Automotive antifreeze is an aqueous solution of ethylene glycol, $HOCH_2CH_2OH$, a nonvolatile nonelectrolyte. Calculate the boiling point and freezing point of an aqueous solution that is 25.0% ethylene glycol by mass.

SOLUTION

Analyze We are told that the solution contains 25.0% by mass of a nonvolatile, nonelectrolyte solute, and we are asked to calculate the boiling and freezing points of the solution. To do this, we need to calculate the boiling-point elevation and freezing-point depression.

Plan To calculate the boiling-point elevation and the freezing-point depression using Equations 13.12 and 13.13, we must express the concentration of the solution as molality. Let's assume for convenience that we have 1000 g of solution. Because the solution is 25.0% by mass ethylene glycol, the masses of ethylene glycol and water in the solution are 250 and 750 g, respectively. Using these quantities, we can calculate the molality of the solution, which we use with the molal boiling-point-elevation and freezing-point-depression constants (Table 13.3) to calculate ΔT_b and ΔT_f. We add ΔT_b to the boiling point and ΔT_f to the freezing point of the solvent to obtain the boiling point and freezing point of the solution.

Continued

Solve

The molality of the solution is calculated as follows:

$$\text{Molality} = \frac{\text{moles } C_2H_6O_2}{\text{kilograms } H_2O}$$

$$= \left(\frac{250 \text{ g } C_2H_6O_2}{750 \text{ g } H_2O}\right)\left(\frac{1 \text{ mol } C_2H_6O_2}{62.1 \text{ g } C_2H_6O_2}\right)\left(\frac{1000 \text{ g } H_2O}{1 \text{ kg } H_2O}\right) = 5.37 \text{ } m$$

We can now use Equations 13.12 and 13.13 to calculate the changes in the boiling and freezing points:

$$\Delta T_b = iK_b m = (1)(0.51°C/m)(5.37 \text{ } m) = 2.7 °C$$

$$\Delta T_f = -iK_f m = -(1)(1.86°C/m)(5.37 \text{ } m) = -10.0 °C$$

Hence, the boiling and freezing points of the solution are readily calculated:

$$\Delta T_b = T_b(\text{solution}) - T_b(\text{solvent})$$

$$2.7°C = T_b(\text{solution}) - 100.0°C$$

$$T_b(\text{solution}) = 102.7°C$$

$$\Delta T_f = T_f(\text{solution}) - T_f(\text{solvent})$$

$$-10.0 °C = T_f(\text{solution}) - 0.0°C$$

$$T_f(\text{solution}) = -10.0°C$$

Comment Notice that the solution of water and ethylene glycol is a liquid over a larger temperature range than pure water.

▶ **Practice Exercise**

Referring to Table 13.3, calculate the freezing point of a solution containing 0.600 kg of $CHCl_3$ and 42.0 g of eucalyptol ($C_{10}H_{18}O$), a fragrant substance found in the leaves of eucalyptus trees.

A CLOSER LOOK   The van't Hoff Factor

The colligative properties of solutions depend on the *total* concentration of solute particles, regardless of whether the particles are ions or molecules. Thus, we expect a 0.100 m solution of NaCl to have a freezing-point depression of $(2)(0.100 \text{ } m)(1.86 °C/m) = 0.372 °C$, because it is 0.100 m in $Na^+(aq)$ and 0.100 m in $Cl^-(aq)$. The measured freezing-point depression is only 0.348 °C, however, and the situation is similar for other strong electrolytes. A 0.100 m solution of KCl, for example, freezes at −0.344 °C.

The difference between expected and observed colligative properties for strong electrolytes is due to electrostatic attractions between ions. As the ions move about in solution, ions of opposite charge collide and "stick together" for brief moments. While they are together, they behave as a single particle called an *ion pair* (**Figure 13.22**). The number of independent particles is thereby reduced, causing a reduction in the freezing-point depression (as well as in boiling-point elevation, vapor–pressure reduction, and osmotic pressure).

We have been assuming that the van't Hoff factor, i, is equal to the number of ions per formula unit of the electrolyte. The true (measured) value of this factor, however, is given by the ratio of the measured value of a colligative property to the value calculated when the substance is assumed to be a nonelectrolyte. Using the freezing-point depression, for example, we have

$$i = \frac{\Delta T_f(\text{measured})}{\Delta T_f(\text{calculated for nonelectrolyte})} \qquad [13.14]$$

The limiting value of i can be determined for a salt from the number of ions per formula unit. For NaCl, for example, the limiting van't Hoff factor is 2 because NaCl consists of one Na^+ and one Cl^- per formula unit; for K_2SO_4 it is 3 because K_2SO_4 consists of two

K^+ and one SO_4^{2-} per formula unit. In the absence of any information about the actual value of i for a solution, we will use the limiting value in calculations.

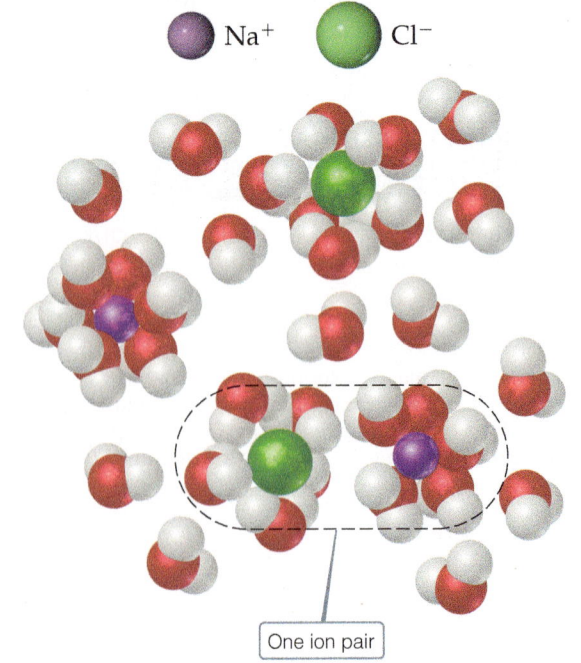

▲ **Figure 13.22 Ion pairing and colligative properties.** A solution of NaCl contains not only separated $Na^+(aq)$ and $Cl^-(aq)$ ions but ion pairs as well.

Two trends are evident in Table 13.4, which gives measured van't Hoff factors for several substances at different dilutions. First, dilution affects the value of i for electrolytes; that is, the more dilute the solution, the more closely i approaches the ideal value based on the number of ions in the formula unit. Thus, we conclude that the extent of ion pairing in electrolyte solutions decreases upon dilution. Second, the lower the charges on the ions, the less i departs from the ideal value because the extent of ion pairing decreases as the ionic charges decrease. Both trends are consistent with simple electrostatics: The force of interaction between charged particles decreases as their separation increases and as their charges decrease.

Related Exercises: 13.71, 13.85, 13.86, 13.105

TABLE 13.4   Measured and Expected van't Hoff Factors for Several Substances at 25 °C

| Compound | Concentration | | | Expected Value |
	0.100 m	0.0100 m	0.00100 m	
Sucrose	1.00	1.00	1.00	1.00
NaCl	1.87	1.94	1.97	2.00
K_2SO_4	2.32	2.70	2.84	3.00
$MgSO_4$	1.21	1.53	1.82	2.00

Osmosis

Certain materials, including many membranes in biological systems and synthetic substances such as cellophane, are *semipermeable*. When in contact with a solution, these materials allow only ions or small molecules—water molecules, for instance—to pass through their network of tiny pores.

Imagine a situation in which only solvent molecules are able to pass through a semipermeable membrane placed between two solutions of different concentrations. The rate at which the solvent molecules pass from the less concentrated solution (lower solute concentration but higher solvent concentration) to the more concentrated solution (higher solute concentration but lower solvent concentration) is greater than the rate in the opposite direction. Thus, there is a net movement of solvent molecules from the solution with a lower solute concentration into the one with a higher solute concentration. In this process, called **osmosis**, *the net movement of solvent is always toward the solution with the lower solvent (higher solute) concentration*, as if the solutions were driven to attain equal concentrations.

Figure 13.23 shows the osmosis that occurs between an aqueous solution and pure water, separated by a semipermeable membrane that allows the passage of water

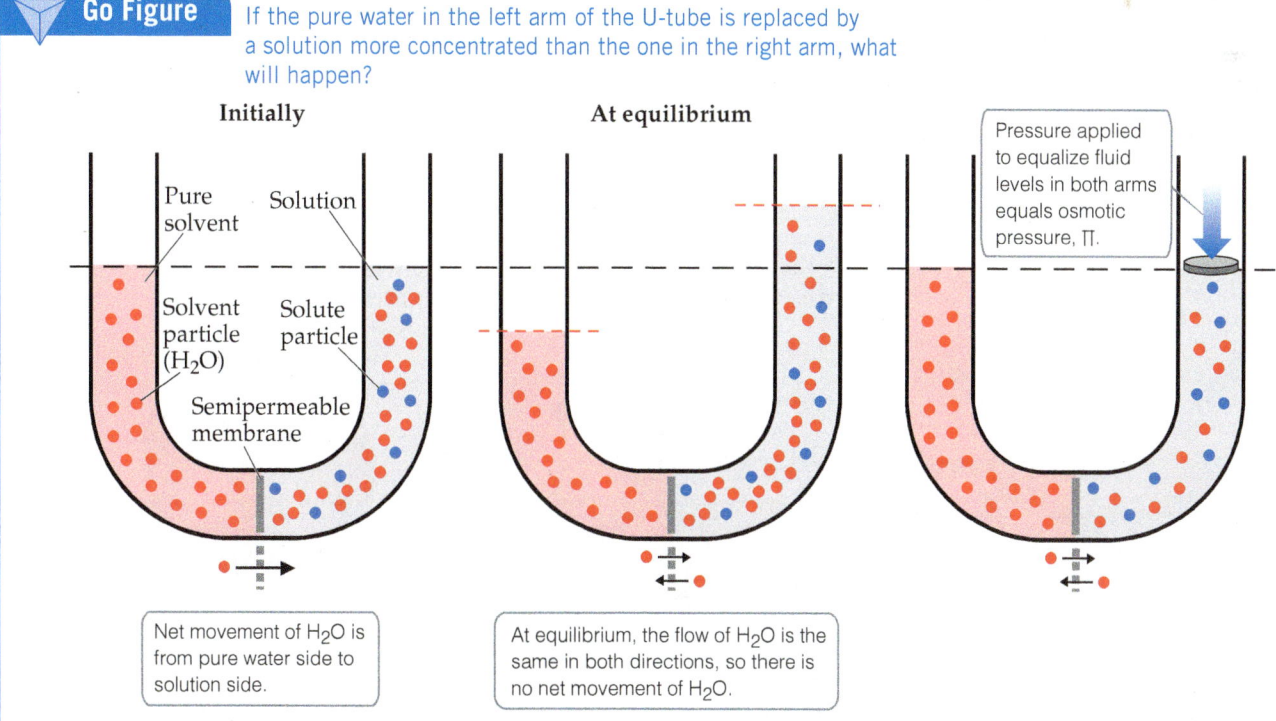

Go Figure If the pure water in the left arm of the U-tube is replaced by a solution more concentrated than the one in the right arm, what will happen?

Initially   At equilibrium

Pure solvent   Solution

Solvent particle (H_2O)   Solute particle

Semipermeable membrane

Pressure applied to equalize fluid levels in both arms equals osmotic pressure, Π.

Net movement of H_2O is from pure water side to solution side.

At equilibrium, the flow of H_2O is the same in both directions, so there is no net movement of H_2O.

▲ **Figure 13.23** Osmosis is the process of a solvent moving from one compartment to another, across a semipermeable membrane, toward higher solute concentration. Osmotic pressure is generated at equilibrium due to the different heights of liquid on either side of the membrane and is equivalent to the pressure needed to equalize the fluid levels across the membrane.

molecules in both directions. The U-tube contains water on the left and an aqueous solution on the right. Initially, there is a net movement of water through the membrane from left to right, leading to unequal liquid levels in the two arms of the U-tube. Eventually, at equilibrium (middle panel of Figure 13.23), the pressure difference resulting from the unequal liquid heights becomes so large that the net flow of water ceases. This pressure, which stops osmosis, is the **osmotic pressure**, Π, of the solution. If an external pressure equal to the osmotic pressure is applied to the solution, the liquid levels in the two arms can be equalized, as shown in the right panel of Figure 13.23.

The osmotic pressure obeys a law similar in form to the ideal-gas law, $\Pi V = inRT$ where Π is the osmotic pressure, V is the volume of the solution, i is the van't Hoff factor, n is the number of moles of solute, R is the ideal-gas constant, and T is the absolute temperature. From this equation, we can write

$$\Pi = i\left(\frac{n}{V}\right)RT = iMRT \qquad [13.15]$$

where M is the molarity of the solution. Because the osmotic pressure for any solution depends on the solution concentration, osmotic pressure is a colligative property.

If two solutions of identical osmotic pressure are separated by a semipermeable membrane, no osmosis will occur. The two solutions are *isotonic* with respect to each other. If one solution is of lower osmotic pressure, it is *hypotonic* with respect to the more concentrated solution. The more concentrated solution is *hypertonic* with respect to the dilute solution.

Osmosis plays an important role in living systems. The membranes of red blood cells, for example, are semipermeable. Placing a red blood cell in a solution that is *hyper*tonic relative to the intracellular solution (the solution inside the cells) causes water to move out of the cell (**Figure 13.24**). This causes the cell to shrivel, a process called *crenation*. Placing the cell in a solution that is *hypo*tonic relative to the intracellular fluid causes water to move into the cell, which may cause the cell to rupture, a process called *hemolysis*. People who need body fluids or nutrients replaced but cannot be fed orally are given solutions by intravenous (IV) infusion, which

Go Figure If the fluid surrounding a patient's red blood cells is depleted in electrolytes, is crenation or hemolysis more likely to occur?

The arrows represent the net movement of water molecules.

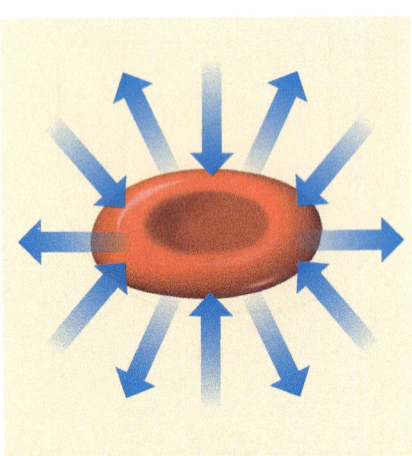

Red blood cell in isotonic medium neither swells nor shrinks.

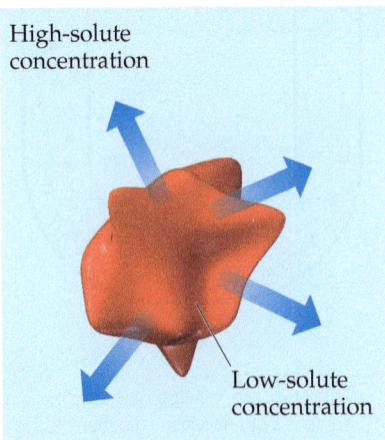

High-solute concentration

Low-solute concentration

Crenation of red blood cell placed in hypertonic environment

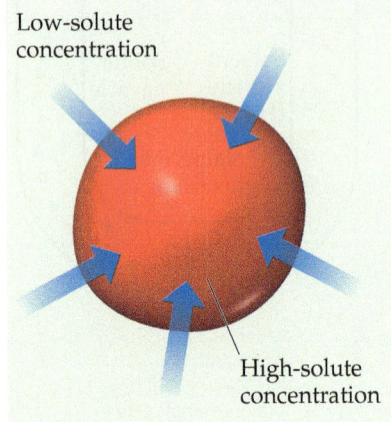

Low-solute concentration

High-solute concentration

Hemolysis of red blood cell placed in hypotonic environment

▲ **Figure 13.24 Osmosis through red blood cell walls.** If water moves out of the red blood cell, it shrivels (crenation); if water moves into the red blood cell, it will swell and may burst (hemolysis).

feeds nutrients directly into the veins. To prevent crenation or hemolysis of red blood cells, the IV solutions must be isotonic with the intracellular fluids of the blood cells.

There are many interesting biological examples of osmosis. A cucumber placed in concentrated brine loses water via osmosis and shrivels into a pickle. People who eat a lot of salty food retain water in tissue cells and intercellular space because of osmosis. The resultant swelling or puffiness is called *edema*. Water moves from soil into plant roots partly because of osmosis. Bacteria on salted meat or candied fruit lose water through osmosis, shrivel, and die—thus preserving the food.

Sample Exercise 13.8
Osmotic Pressure Calculations

The average osmotic pressure of blood is 7.7 atm at 25 °C. What molarity of glucose ($C_6H_{12}O_6$) will be isotonic with blood?

SOLUTION

Analyze We are asked to calculate the concentration of glucose in water that would be isotonic with blood, given that the osmotic pressure of blood at 25 °C is 7.7 atm.

Plan Because we are given the osmotic pressure and temperature, we can solve for the concentration, using Equation 13.15. Because glucose is a nonelectrolyte, $i = 1$.

Solve

$$\Pi = iMRT$$

$$M = \frac{\Pi}{iRT} = \frac{(7.7 \text{ atm})}{(1)\left(0.0821\dfrac{\text{L-atm}}{\text{mol-K}}\right)(298 \text{ K})} = 0.31\ M$$

Comment In clinical situations, the concentrations of solutions are generally expressed as mass percentages. The mass percentage of a 0.31 M solution of glucose is 5.3%. The concentration of NaCl that is isotonic with blood is 0.16 M because $i = 2$ for NaCl in water (a 0.155 M solution of NaCl is 0.310 M in particles). A 0.16 M solution of NaCl is 0.9% mass in NaCl. This kind of solution is known as a physiological saline solution.

▶ **Practice Exercise**
What is the osmotic pressure, in atm, of a 0.0020 M sucrose ($C_{12}H_{22}O_{11}$) solution at 20 °C?

Determination of Molar Mass from Colligative Properties

In Chapter 2 we learned that an instrument called a mass spectrometer can be used to determine the molecular weight of a substance. (Section 2.4) In many cases, it is possible to obtain good estimates of molecular and formula weights by preparing solutions and measuring one or more colligative properties, an approach that can be carried out in the lab with much less sophisticated instrumentation. The procedure for determining molar mass from a colligative property is outlined in **Figure 13.25**. To do so, we first weigh out a known mass of the solute and dissolve it into an appropriate solvent to prepare a solution. Next, we measure at least one of the colligative properties of the solution. Because each colligative property depends only on the concentration of the solute, we can use its value to determine the concentration of the solution—mole fraction from vapor pressure, molality from the freezing or boiling point, molarity from osmotic pressure. Once the concentration is known, the number of moles of the solute can be calculated from the volume of the solution. At this stage we are in a position to calculate the molar mass of the solute:

$$\text{Molar mass of solute} = \frac{\text{mass of solute}}{\text{moles of solute}} \qquad [13.16]$$

Sample Exercise 13.9 illustrates this approach by showing how the molar mass of a protein can be estimated from the osmotic pressure of a solution containing the protein.

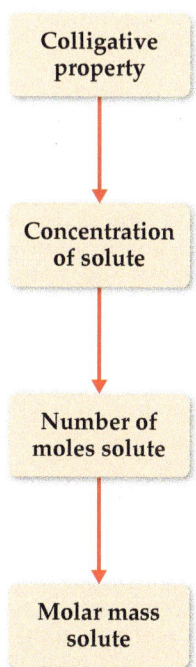

▲ **Figure 13.25** Procedure for estimating the molar mass of a substance from the colligative properties of a solution containing the substance.

Sample Exercise 13.9
Molar Mass from Osmotic Pressure

The osmotic pressure of an aqueous solution of a certain protein was measured to determine the protein's molar mass. The solution contained 3.50 mg of protein dissolved in sufficient water to form 5.00 mL of solution. The osmotic pressure of the solution at 25 °C was found to be 1.54 torr. Treating the protein as a nonelectrolyte, calculate its molar mass.

SOLUTION

Analyze Our goal is to calculate the molar mass of a large protein molecule, based on the osmotic pressure of a solution containing the protein. The protein will be considered as a nonelectrolyte, so $i = 1$.

Plan The temperature ($T = 25 °C$) and osmotic pressure ($\Pi = 1.54$ torr) are given, and we know the value of R, so we can use

Equation 13.15 to calculate the molarity of the solution, M. In doing so, we must convert temperature from °C to K and the osmotic pressure from torr to atm. We then use the molarity and the volume of the solution (5.00 mL) to determine the number of moles of solute. Finally, we use Equation 13.16 to obtain the molar mass of the protein.

Solve

Solving Equation 13.15 for molarity gives:

$$\text{Molarity} = \frac{\Pi}{iRT} = \frac{(1.54 \text{ torr})\left(\dfrac{1 \text{ atm}}{760 \text{ torr}}\right)}{(1)\left(0.0821\dfrac{\text{L-atm}}{\text{mol-K}}\right)(298 \text{ K})} = 8.28 \times 10^{-5}\frac{\text{mol}}{\text{L}}$$

Because the volume of the solution is $5.00 \text{ mL} = 5.00 \times 10^{-3}$ L, the number of moles of protein must be:

$$\text{Moles} = (8.28 \times 10^{-5} \text{ mol/L})(5.00 \times 10^{-3} \text{ L}) = 4.14 \times 10^{-7} \text{ mol}$$

The molar mass is the number of grams per mole of the substance. Because we know the sample has a mass of $3.50 \text{ mg} = 3.50 \times 10^{-3}$ g, we can calculate the molar mass by dividing the number of grams in the sample by the number of moles we just calculated:

$$\text{Molar mass} = \frac{\text{mass of solute}}{\text{moles of solute}} = \frac{3.50 \times 10^{-3} \text{ g}}{4.14 \times 10^{-7} \text{ mol}} = 8.45 \times 10^{3} \text{ g/mol}$$

Comment Because small pressures can be measured easily and accurately, osmotic pressure measurements provide a useful way to determine the molar masses of large molecules.

▶ **Practice Exercise**
A sample of 2.05 g of polystyrene of uniform polymer chain length was dissolved in enough toluene to form 0.100 L of solution. The osmotic pressure of this solution was found to be 1.21 kPa at 25 °C. Calculate the molar mass of the polystyrene.

Self-Assessment Exercises

SAE 13.15 Which aqueous solution would you predict has the lowest freezing point? (**a**) 0.015 m NaCl (**b**) 0.025 m sucrose (**c**) 0.012 m MgCl$_2$ (**d**) 0.010 m glucose

SAE 13.16 Calculate the expected vapor pressure at 25 °C for a solution prepared by dissolving 20.0 g of NaCl in 100.0 g of water. The vapor pressure of water at this temperature is 23.8 torr. Neglect the effects of ion pairing. (*Hint:* Don't forget to account for the fact that each mole of NaCl dissolves to give one mole of Na$^+$ and one mole of Cl$^-$.) (**a**) 23.8 torr (**b**) 22.4 torr (**c**) 21.2 torr (**d**) 19.8 torr

SAE 13.17 Benzene, C$_6$H$_6$, which boils at 80.1 °C, has a boiling-point elevation constant of 2.53 °C/m and a density of 0.8765 g/mL. Predict the boiling point of a solution made up of 150.0 g of benzene and 12.6 g of a hydrophobic, nonelectrolyte, experimental drug with a molar mass of 256 g/mol. (**a**) 0.83 °C (**b**) 79.3 °C (**c**) 80.2 °C (**d**) 80.9 °C (**e**) 81.1 °C

SAE 13.18 A solution is made by dissolving 0.010 g of polystyrene, a nonelectrolyte polymer of unknown molar mass, in 1.00 mL cyclohexane (density = 0.779 g/mol). When this solution is separated from pure cyclohexane by a semipermeable membrane, an osmotic pressure of 68 Pa is measured at 317 K.

What is the molar mass of the polystyrene? (**a**) 3.9×10^2 g/mol (**b**) 8.2×10^3 g/mol (**c**) 3.9×10^4 g/mol (**d**) 1.8×10^5 g/mol (**e**) 3.9×10^5 g/mol

SAE 13.19 The vapor pressure of an aqueous solution is predicted by Raoult's law to be 14.7 torr at 20 °C, assuming the solute is nonvolatile. When the vapor pressure of this solution is measured, it is found to be 14.2 torr. What is the most likely reason for this deviation from Raoult's law? (**a**) The solute must also be volatile, and when the solute is volatile, it lowers the total vapor pressure of the solution. (**b**) The interactions between solute and solvent molecules are weaker than the solute–solute and solvent–solvent interactions. (**c**) The interactions between solute and solvent molecules are stronger than the solute–solute and solvent–solvent interactions. (**d**) Being less dense than water, the solute molecules rise to the top of the solution, thus lowering the mole fraction of water close to the surface.

SAE 13.20 At 20 °C a saturated solution of NaCl can be prepared by dissolving 35.9 g of NaCl in 100.0 g of water to form a solution with a density of 1.20 g/mL. What osmotic pressure would be predicted for this solution if we neglect the effects of ion pairing? (**a**) 17.8 atm (**b**) 130 atm (**c**) 261 atm (**d**) 295 atm

CHEMISTRY AND SUSTAINABILITY | Desalination and Reverse Osmosis

Because of its high salt content, seawater is unfit for human consumption, irrigation, and most other practical uses. In the United States, the salt content of municipal water supplies is restricted by health codes to no more than about 0.05% by mass. This amount is much lower than the 3.5% dissolved salts present in seawater and the 0.5% or so present in brackish water found underground in some regions. The removal of salts from seawater or brackish water to make the water usable is called desalination.

Because water is a volatile substance and the salts are nonvolatile, distillation can be used for desalination. (Section 1.3) The principle of distillation (see Figure 1.12) is simple enough, but carrying out the process on a large scale presents many problems. Distillation is an energy-intensive process, and as the distillation process proceeds, the salts become more and more concentrated and eventually precipitate out. So, while distillation can be used to desalinate water on a small scale, less energy-intensive approaches are favored for industrial-scale operations.

As we've learned, osmosis is the net movement of solvent molecules, but not solute particles, through a semipermeable membrane. In normal osmosis, the solvent passes from the more dilute solution into the more concentrated one, but if sufficient external pressure is applied, the flow of water molecules into the saltier solution can be stopped and, at still higher pressures, reversed. At such pressures, solvent molecules pass from the more concentrated solution across the membrane to the more dilute solution. This process, called *reverse osmosis*, is the preferred method for large-scale desalination.

The first step at a reverse osmosis desalination plant is to remove algae, organic materials, and other particles from the incoming seawater by passing it through a filter of sand and gravel. This water is then further filtered to remove microscopic sediments that can clog the membrane. Finally, the treated salt water is forced under pressure into the hollow fibers made from a polymer composite that acts as the semipermeable membrane (**Figure 13.26**). The osmotic pressure of seawater is on the order of 27 atm, which means that external pressures higher than that must be applied to push water molecules through the membrane. In a typical desalination plant, pressures ranging from 50 to 70 atm are used to drive the process. A significant amount of electrical energy is needed to raise the pressure of the salt water to such levels.

The world's largest desalination plants are in the Middle East: Saudi Arabia, the United Arab Emirates, and Israel all get a substantial portion of their freshwater from desalination plants. The largest plant, Ras Al-Khair in Saudi Arabia, generates over $1 \times 10^6 \, \text{m}^3$ freshwater per day ($1 \times 10^9 \, \text{L/day}$). Such plants are becoming more common in the United States. The largest, which opened in late 2015 near Carlsbad, California, produces approximately $1.9 \times 10^8 \, \text{L/day}$. To operate at full capacity, it uses approximately 38 MW of electricity, enough to power 38,500 homes. Two gallons of seawater are needed to produce one gallon of freshwater.

Disposing of the very salty "brine" that is a by-product of desalination poses another challenge. The most common approach is to discharge the brine back into the ocean, but the high salt content can have negative environmental impacts if it is not managed properly. Dispersing the brine over a large area and/or diluting it with cooling water from a co-located power plant are two approaches to managing this issue.

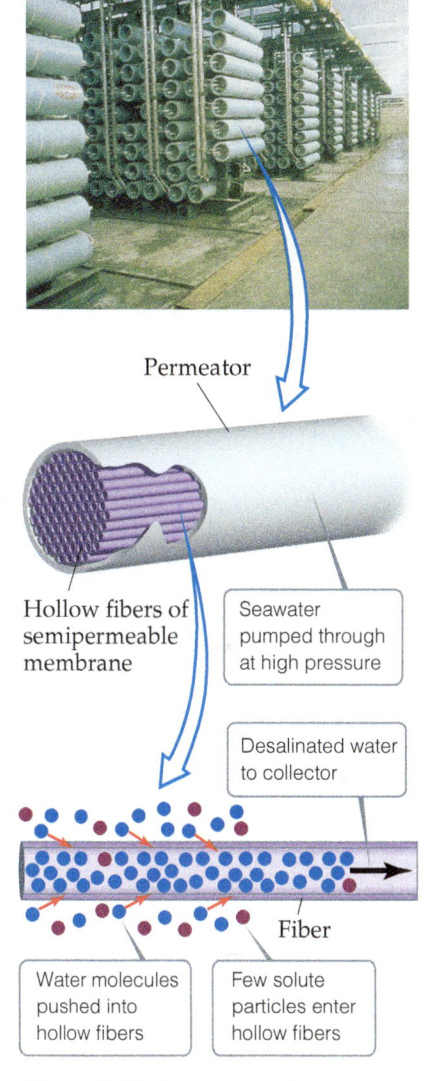

▲ **Figure 13.26 Reverse osmosis.**

Related Exercises: 13.61, 13.62, 13.100

13.6 | Colloids

Some substances appear to initially dissolve in a solvent, but over time, the substance separates from the pure solvent. For example, finely divided clay particles dispersed in water eventually settle out because of gravity. Gravity affects the clay particles because they are much larger than most molecules, consisting of tens of thousands or even millions of atoms. In contrast, the dispersed particles in a true solution (ions in a salt solution or glucose molecules in a sugar solution) are small. Between these extremes lie dispersed particles that are larger than typical molecules but not so large that the components of the mixture separate under the influence of gravity. These intermediate types of dispersions are called either **colloidal dispersions** or simply **colloids**. Colloids form the dividing line between solutions and heterogeneous mixtures. Like solutions, colloids can be made up of gases, liquids, or solids. Examples of each are listed in **Table 13.5**.

Particle size can be used to classify a mixture as colloid or solution. Colloid particles range in diameter from approximately 5 to 1000 nm, whereas in solutions the solute particles are generally smaller than 5 nm in diameter. The nanomaterials we saw in Chapter 12 (Section 12.7), when dispersed in a liquid, are colloids. A colloid particle may even

▲ Learning Objectives

When you finish Section 13.6, you should be able to:

▶ Differentiate solutions, colloidal dispersions, and heterogeneous mixtures.

▶ Use the phase (solid, liquid, or gas) of the dispersed particle and that of the dispersing medium to classify different types of colloids.

▶ Classify colloidal dispersions in aqueous solutions as either hydrophilic or hydrophobic based on the composition of the colloidal particle and its interaction with the solvent.

TABLE 13.5 **Types of Colloids**

Phase of Colloid	Dispersing (solvent-like) Substance	Dispersed (solute-like) Substance	Colloid Type	Example
Gas	Gas	Gas	—	None (all are solutions)
Gas	Gas	Liquid	Aerosol	Fog
Gas	Gas	Solid	Aerosol	Smoke
Liquid	Liquid	Gas	Foam	Whipped cream
Liquid	Liquid	Liquid	Emulsion	Milk
Liquid	Liquid	Solid	Sol	Paint
Solid	Solid	Gas	Solid foam	Marshmallow
Solid	Solid	Liquid	Solid emulsion	Butter
Solid	Solid	Solid	Solid sol	Ruby glass

consist of a single giant molecule. The hemoglobin molecule, for example, which carries oxygen in your blood, has molecular dimensions of $6.5 \times 5.5 \times 5.0$ nm and a molar mass of 64,500 g/mol.

Although colloid particles may be so small that the dispersion appears uniform even under a microscope, they are large enough to scatter light. Consequently, most colloids appear cloudy or opaque unless they are very dilute. (For example, homogenized milk is a colloid of fat and protein molecules dispersed in water.) Furthermore, because they scatter light, a light beam can be seen as it passes through a colloidal dispersion (**Figure 13.27**). This scattering of light by colloidal particles, known as the **Tyndall effect**, makes it possible to see the light beam of an automobile on a dusty dirt road, or the sunlight streaming through trees or clouds. Not all wavelengths are scattered to the same extent. Colors at the blue end of the visible spectrum are scattered more than those at the red end by the molecules and small dust particles in the atmosphere. As a result, our sky appears blue. At sunset, light from the sun travels through more of the atmosphere; blue light is scattered even more, allowing the reds and yellows to pass through and be seen.

Hydrophilic and Hydrophobic Colloids

The most important colloids are arguably those in which the dispersing medium is water. These colloids may be **hydrophilic** ("water loving") or **hydrophobic**

▶ **Figure 13.27 Tyndall effect in the laboratory.** The glass on the right contains a colloidal dispersion; that on the left contains a solution.

("water fearing"). Hydrophilic colloids are most like the solutions that we have previously examined. In the human body, the extremely large protein molecules such as enzymes and antibodies are kept in suspension by interaction with surrounding water molecules. A hydrophilic molecule folds in such a way that its nonpolar hydrophobic groups are kept away from the solvent molecules, on the inside of the folded molecule, while its hydrophilic, polar groups are on the surface, interacting with the water molecules. The hydrophilic groups generally contain oxygen or nitrogen and often carry a charge (**Figure 13.28**).

Hydrophobic colloids can be dispersed in water only if they are stabilized in some way. Otherwise, their natural lack of affinity for water causes them to separate from the water. One method of stabilization involves adsorbing ions on the surface of the hydrophobic particles (**Figure 13.29**). (*Adsorption* means to adhere to a surface. It differs from *absorption*, which means to pass into the interior, as when a sponge absorbs water.) The adsorbed ions can interact with water, thereby stabilizing the colloid. At the same time, the electrostatic repulsion between adsorbed ions on neighboring colloid particles keeps the particles from sticking together rather than dispersing in the water.

Hydrophobic colloids can also be stabilized by molecules that are hydrophobic on one end and hydrophilic on the other. Oil drops are hydrophobic, for example, and they do not remain suspended in water. Instead, they aggregate, forming an oil slick on the surface of the water. Sodium stearate (**Figure 13.30**), or any similar substance having one end that is hydrophilic (either polar or charged) and one end that is hydrophobic (nonpolar), will stabilize a suspension of oil in water. Stabilization results from the interaction of the hydrophobic ends of the stearate ions with the oil drops and the hydrophilic ends with the water molecules.

Colloid stabilization has an interesting application in the human digestive system. When fats in our diet reach the small intestine, a hormone causes the gallbladder to excrete a fluid called bile. Among the components of bile are compounds that have chemical structures similar to sodium stearate; that is, they have a hydrophilic (polar) end and a hydrophobic (nonpolar) end. These compounds emulsify the fats in the intestine and thus permit digestion and absorption of fat-soluble vitamins through the intestinal wall. The term *emulsify* means "to form an emulsion," a suspension of one liquid in another, with milk being one example (Table 13.5). A substance that aids in forming an emulsion is called an emulsifying agent. If you read the labels on foods and other materials, you will find that a variety of chemicals are used as emulsifying agents. These chemicals typically have a hydrophilic end and a hydrophobic end.

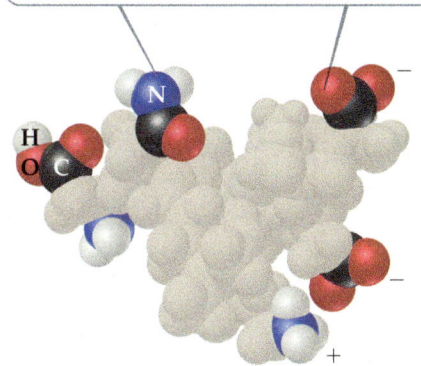

Hydrophilic polar and charged groups on a molecule's surface help the molecule remain dispersed in water and other polar solvents.

▲ **Figure 13.28 Hydrophilic colloidal particle.** Examples of the hydrophilic groups that help to keep a giant molecule (macromolecule) suspended in water.

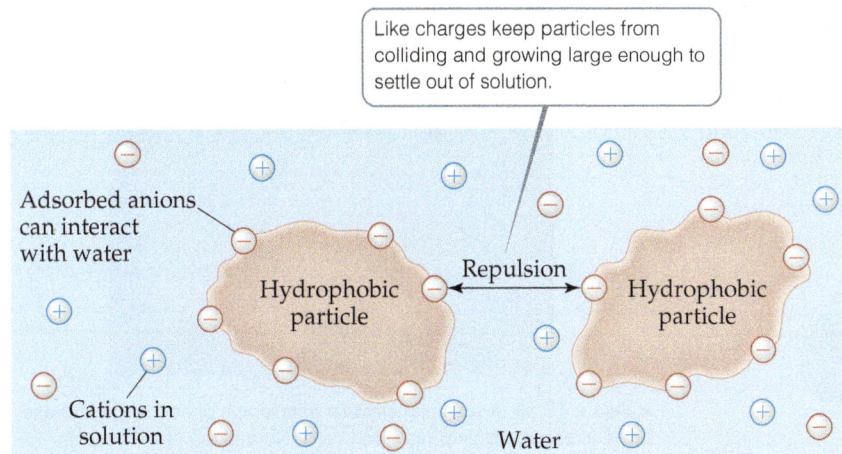

Like charges keep particles from colliding and growing large enough to settle out of solution.

Adsorbed anions can interact with water

Hydrophobic particle   Repulsion   Hydrophobic particle

Cations in solution

Water

▲ **Figure 13.29 Hydrophobic colloids stabilized in water by adsorbed anions.**

Go Figure Which kind of intermolecular force attracts the stearate ion to the oil drop?

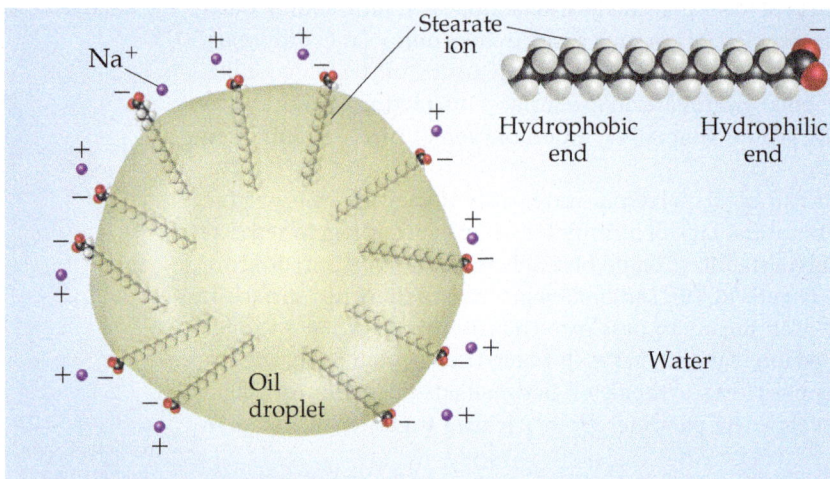

▲ **Figure 13.30** Stabilization of an emulsion of oil in water by stearate ions.

CHEMISTRY AND LIFE | Sickle-Cell Anemia

Our blood contains the complex protein hemoglobin, which carries oxygen from the lungs to other parts of the body. In the genetic disease sickle-cell anemia, hemoglobin molecules are abnormal and have a lower solubility in water, especially in their unoxygenated form. Consequently, as much as 85% of the hemoglobin in red blood cells crystallizes out of solution.

The cause of the insolubility is a structural change in one part of an amino acid. Normal hemoglobin molecules contain an amino acid that has a —CH_2CH_2COOH group:

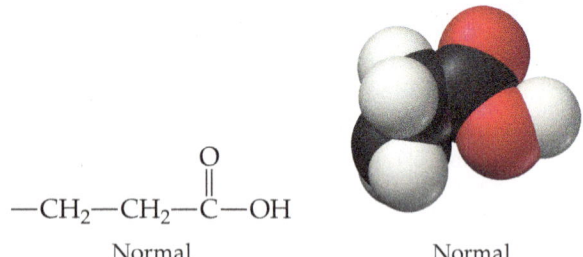

Normal   Normal

The polarity of the —COOH group contributes to the solubility of the hemoglobin molecule in water. In the hemoglobin molecules of sickle-cell anemia patients, the —CH_2CH_2COOH chain is absent and in its place is the nonpolar (hydrophobic) —$CH(CH_3)_2$ group:

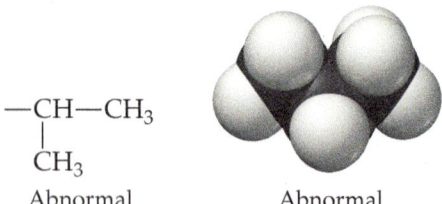

Abnormal   Abnormal

This change leads to the aggregation of the defective form of hemoglobin into particles too large to remain suspended in biological fluids. It also causes the cells to distort into the sickle shape shown in **Figure 13.31**. The sickled cells tend to clog capillaries, causing severe pain, weakness, and the gradual deterioration of vital organs. The disease is hereditary, and if both parents carry the defective genes, it is likely that their children will possess only abnormal hemoglobin.

Why has a life-threatening disease such as sickle-cell anemia persisted in humans through evolutionary time? The answer in part is that people with the gene are far less susceptible to malaria. Thus, in tropical climates rife with malaria, those with sickle-cell genes have lower incidence of this debilitating disease.

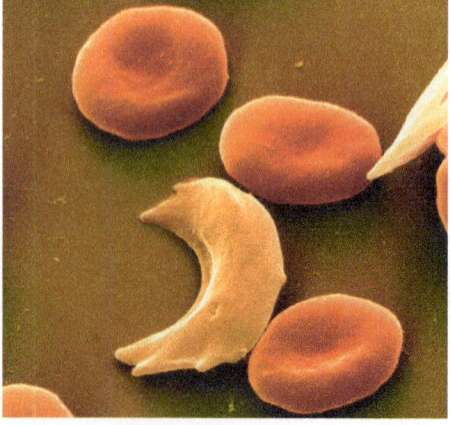

▲ **Figure 13.31** A scanning electron micrograph of normal (round) and sickle (crescent-shaped) red blood cells. Normal red blood cells are about 6×10^{-3} mm in diameter.

Colloidal Motion in Liquids

Gas molecules move at some average speed that depends inversely on their molar mass, in a straight line, until they collide with something. The *mean free path* is the average distance molecules travel between collisions. (Section 10.6) The kinetic-molecular theory of gases assumes, moreover, that gas molecules are in continuous, random motion. (Section 10.5)

Colloidal particles in a solution undergo random motion as a result of collisions with solvent molecules. Because the colloidal particles are massive in comparison to solvent molecules, their movements from any one collision are very tiny. There are many such collisions, however, and they cause a random motion of the entire colloidal particle. When viewed with a microscope, the random motions of the colloids can be seen, but the solvent molecules cannot. This effect was described by Scottish botanist Robert Brown in 1827 while studying pollen under a microscope. This type of random, solvent driven movement is called **Brownian motion**. In 1905, Albert Einstein developed an equation for the average square of the displacement of a colloidal particle, an important development in the history of science because it indirectly confirmed the existence of molecules in the solvent and gave an estimate of their size. Today, the understanding of Brownian motion is applied to diverse problems in everything from cheese-making to medical imaging.

Self-Assessment Exercises

SAE 13.21 Which of the following systems represents a sol? **(a)** CO_2 dissolved in whipped cream **(b)** gold nanoparticles suspended in water **(c)** oil droplets suspended in water **(d)** nanodiamonds dispersed in glass

SAE 13.22 Several grams of a protein are added to 50 mL of water, and the mixture is stirred vigorously until the protein dissolves. After several days, the mixture appears homogeneous and shows no evidence of a solid. It is transparent to visible light and does not show the Tyndall effect. Based on these observations, the mixture can be classified as a _____. **(a)** solution **(b)** colloidal dispersion **(c)** heterogeneous mixture **(d)** Not enough information is given to answer this question.

SAE 13.23 Which compound is most likely to stabilize an emulsion of oil droplets in water? **(a)** NaCl **(b)** hexane, $CH_3(CH_2)_4CH_3$ **(c)** ammonium chloride, NH_4Cl **(d)** sodium dodecylsulfate, $CH_3(CH_2)_{11}OSO_3^-Na^+$

Putting Concepts Together

A 0.100-L solution is made by dissolving 0.441 g of $CaCl_2(s)$ in water. **(a)** Calculate the osmotic pressure of this solution at 27 °C, assuming that it is completely dissociated into its component ions. **(b)** The measured osmotic pressure of this solution is 2.56 atm at 27 °C. Explain why it is less than the value calculated in (a), and calculate the van't Hoff factor, i, for the solute in this solution. **(c)** The enthalpy of solution for $CaCl_2$ is $\Delta H = -81.3$ kJ/mol. If the final temperature of the solution is 27 °C, what was its initial temperature? (Assume that the density of the solution is 1.00 g/mL, that its specific heat is 4.18 J/g-K, and that the solution loses no heat to its surroundings.)

SOLUTION

(a) We can calculate the molarity of the solution from the mass of $CaCl_2$ and the volume of the solution:

$$\text{Molarity} = \left(\frac{0.441 \text{ g CaCl}_2}{0.100 \text{ L}}\right)\left(\frac{1 \text{ mol CaCl}_2}{110 \text{ g CaCl}_2}\right)$$

$$= 0.0397 \text{ mol CaCl}_2/\text{L}$$

Soluble ionic compounds are strong electrolytes. (Sections 4.1 and 4.3) Thus, $CaCl_2$ consists of metal cations (Ca^{2+}) and nonmetal anions (Cl^-). When completely dissociated, each $CaCl_2$ unit forms three ions (one Ca^{2+} and two Cl^-). Given the molarity, van't Hoff factor, and temperature (300 K), we can calculate the osmotic pressure using Equation 13.15:

$$\Pi = iMRT = (3)(0.0397 \text{ mol/L})(0.0821 \text{ L-atm/mol-K})(300 \text{ K})$$

$$= 2.93 \text{ atm}$$

Continued

(b) The actual values of colligative properties of electrolytes are less than those calculated because the electrostatic interactions between ions limit their independent movements. In this case, the van't Hoff factor, which measures the extent to which electrolytes actually dissociate into ions, is given by:

$$i = \frac{\Pi(\text{measured})}{\Pi(\text{calculated for nonelectrolyte})}$$

Thus, the solution behaves as if the $CaCl_2$ has dissociated into 2.62 particles instead of the ideal 3.

$$= \frac{2.56 \text{ atm}}{(0.0397 \text{ mol/L})(0.0821 \text{ L-atm/mol-K})(300 \text{ K})} = 2.62$$

(c) If the solution is 0.0397 M in $CaCl_2$ and has a total volume of 0.100 L, the number of moles of solute is:

$$(0.100 \text{ L})(0.0397 \text{ mol/L}) = 0.00397 \text{ mol}$$

Hence, the quantity of heat generated in forming the solution is:

$$(0.00397 \text{ mol})(-81.3 \text{ kJ/mol}) = -0.323 \text{ kJ}$$

The solution absorbs this heat, causing its temperature to increase. The relationship between temperature change and heat is given by Equation 5.22:

$$q = (\text{specific heat of solution})(\text{grams of solution})(\Delta T)$$

The heat absorbed by the solution is $q = +0.323 \text{ kJ} = 323 \text{ J}$. The mass of the 0.100 L of solution is $(100 \text{ mL})(1.00 \text{ g/mL}) = 100 \text{ g}$ (to three significant figures). Thus, the temperature change is:

$$\Delta T = \frac{q}{(\text{specific heat of solution})(\text{grams of solution})}$$
$$= \frac{323 \text{ J}}{(4.18 \text{ J/g-K})(100 \text{ g})} = 0.773 \text{ K}$$

A kelvin has the same size as a degree Celsius. (Section 1.4) Because the solution temperature increases by 0.773 °C, the initial temperature was:

$$27.0 \,°C - 0.773 \,°C = 26.2 \,°C$$

Chapter Summary and Key Terms

THE SOLUTION PROCESS (SECTION 13.1) Solutions form when one substance disperses uniformly throughout another. The attractive interaction of solvent molecules with solute is called **solvation**. When the solvent is water, the interaction is called **hydration**. The dissolution of ionic substances in water is promoted by hydration of the separated ions by the polar water molecules. The overall enthalpy change upon solution formation may be either positive or negative. Solution formation is favored both by a positive entropy change, corresponding to an increased dispersal of the components of the solution, and by a negative enthalpy change, indicating an exothermic process.

SATURATED SOLUTIONS AND SOLUBILITY (SECTION 13.2) The equilibrium between a saturated solution and undissolved solute is dynamic; the process of solution and the reverse process, **crystallization,** occur simultaneously. In a solution in equilibrium with undissolved solute, the two processes occur at equal rates, giving a **saturated** solution. If there is less solute present than is needed to saturate the solution, the solution is **unsaturated**. When solute concentration is greater than the equilibrium concentration value, the solution is **supersaturated**. This is an unstable condition, and crystallization of some solute from the solution will occur if the process is initiated with a solute seed crystal. The amount of solute needed to form a saturated solution at any particular temperature is the **solubility** of that solute at that temperature.

FACTORS AFFECTING SOLUBILITY (SECTION 13.3) The solubility of one substance in another depends on the tendency of systems to become more random, by becoming more dispersed in space, and on the relative strength of intermolecular solute–solute and solvent–solvent interactions compared with solute–solvent interactions. Polar and ionic solutes tend to dissolve in polar solvents, and nonpolar solutes tend to dissolve in nonpolar solvents ("like dissolves like"). Liquids that mix in all proportions are **miscible**; those that do not dissolve significantly in one another are **immiscible**. Hydrogen-bonding interactions between solute and solvent often play an important role in determining solubility; for example, ethanol and water, whose molecules form hydrogen bonds with each other, are miscible.

The solubilities of gases in a liquid are generally proportional to the pressure of the gas over the solution, as expressed by **Henry's law**: $S_g = kP_g$. The solubilities of most solid solutes in water increase as the temperature of the solution increases. In contrast, the solubilities of gases in water generally decrease with increasing temperature.

EXPRESSING SOLUTION CONCENTRATIONS (SECTION 13.4) Concentrations of solutions can be expressed quantitatively by several different measures, including **mass percentage, parts per million (ppm), parts per billion (ppb),** and mole fraction. Molarity, M, is defined as moles of solute per liter of solution; **molality**, m, is defined as moles of solute per kilogram of solvent. Conversions between different ways of expressing solution concentration involve use of the molar mass of the solute to convert from mass-based concentrations (mass percentage, ppm, ppb) and mole-based concentrations (mole fraction, molarity, molality). The density of the solution must be known to convert between molarity and other concentration units.

COLLIGATIVE PROPERTIES (SECTION 13.5) A physical property of a solution that depends on the concentration of solute particles present, but not their identity, is a **colligative property**. Colligative properties include vapor–pressure lowering, freezing-point depression, boiling-point elevation, and osmotic pressure. **Raoult's law** can be used to quantify the extent of vapor–pressure lowering. An **ideal solution** is one where the solute–solute, solute–solvent, and solute–solvent interactions are all the same strength. Deviations from Raoult's law are observed when solutions depart from ideal behavior.

A solution containing a nonvolatile solute possesses a higher boiling point than the pure solvent. The **molal boiling-point-elevation constant**, K_b, represents the increase in boiling point for a 1 m solution of solute particles as compared with the pure solvent. Similarly, the **molal freezing-point-depression constant**, K_f, measures the lowering of the freezing point of a solution for a 1 m solution of solute particles. The temperature changes are given by the equations $\Delta T_b = iK_b m$ and $\Delta T_f = -iK_f m$ where i is the **van't Hoff factor**, which represents how many particles the solute breaks up into upon dissolution. When

NaCl dissolves in water, two moles of solute particles are formed for each mole of dissolved salt. The boiling point or freezing point is thus elevated or depressed, respectively, approximately twice as much as that of a nonelectrolyte solution of the same concentration. Similar considerations apply to other strong electrolytes.

Osmosis is the movement of solvent molecules through a semipermeable membrane from a less concentrated to a more concentrated solution. This net movement of solvent generates an **osmotic pressure**, Π, which can be measured in units of gas pressure, such as atm. The osmotic pressure of a solution is proportional to the solution molarity: $\Pi = iMRT$. Osmosis is a very important process in living systems, in which cell walls act as semipermeable membranes, permitting the passage of water but restricting the passage of ionic and macromolecular components.

COLLOIDS (SECTION 13.6) Particles that are large on the molecular scale but still small enough to remain suspended indefinitely in a solvent form **colloids**, or **colloidal dispersions**. Colloids, which are intermediate between solutions and heterogeneous mixtures, have many practical applications. One useful physical property of colloids, the scattering of visible light, is referred to as the **Tyndall effect**. Aqueous colloids are classified as **hydrophilic** or **hydrophobic**. Hydrophilic colloids are common in living organisms, in which large molecular aggregates (enzymes, antibodies) remain suspended because they have many polar, or charged, atomic groups on their surfaces that interact with water. Hydrophobic colloids, such as small droplets of oil, may remain in suspension through adsorption of charged particles on their surfaces or through the use of emulsifying agents, molecules with nonpolar, hydrophobic tails and polar head groups. The hydrophobic tails penetrate into the hydrophobic colloid and the polar head groups, which are often charged, remain on the surface stabilizing the colloidal particle.

Colloids undergo **Brownian motion** in liquids, analogous to the random three-dimensional motion of gas molecules.

Key Equations

$\bullet$ $S_g = kP_g$	[13.4]	Henry's law, which relates gas solubility to partial pressure
$\bullet$ Mass % of component $= \dfrac{\text{mass of component in soln}}{\text{total mass of soln}} \times 100\%$	[13.5]	Concentration in terms of mass percent
$\bullet$ ppm of component $= \dfrac{\text{mass of component in soln}}{\text{total mass of soln}} \times 10^6$	[13.6]	Concentration in terms of parts per million (ppm)
$\bullet$ Mole fraction of component $= \dfrac{\text{moles of component}}{\text{total moles of all components}}$	[13.7]	Concentration in terms of mole fraction
$\bullet$ Molarity $= \dfrac{\text{moles of solute}}{\text{liters of soln}}$	[13.8]	Concentration in terms of molarity
$\bullet$ Molality $= \dfrac{\text{moles of solute}}{\text{kilograms of solvent}}$	[13.9]	Concentration in terms of molality
$\bullet$ $P_{\text{solution}} = X_{\text{solvent}} P^{\circ}_{\text{solvent}}$	[13.10]	Raoult's law, calculating vapor pressure of solvent above a solution
$\bullet$ $\Delta T_b = iK_b m$	[13.12]	The boiling-point elevation of a solution
$\bullet$ $\Delta T_f = -iK_f m$	[13.13]	The freezing-point depression of a solution
$\bullet$ $\Pi = i\left(\dfrac{n}{V}\right)RT = iMRT$	[13.15]	The osmotic pressure of a solution

Exam Prep

EP 13.1 A solution is made by dissolving 10.0 g KBr in 150 mL of H_2O. Once the mixing is complete, no solid KBr remains, and the temperature of the solution drops by 2 °C from the temperature before mixing. If we define the system as the KBr plus the H_2O, the enthalpy of the system _____ and the entropy of the system _____ upon forming the solution. (**a**) increases, increases (**b**) increases, decreases (**c**) decreases, increases (**d**) decreases, decreases

EP 13.2 In which of the following solutions are the solute and solvent attracted to each other through hydrogen bonding that is absent in the pure solute? (**a**) cesium iodide (CsI) dissolved in methanol (CH_3OH) (**b**) acetone (CH_3COCH_3) dissolved in H_2O (**c**) ammonia (NH_3) dissolved in H_2O (**d**) acetic acid (CH_3COOH) dissolved in H_2O (**e**) diethyl ether ($C_2H_5OC_2H_5$) dissolved in hexane (C_6H_{14})

EP 13.3 The dissolution of sodium nitrate in water is endothermic, $\Delta H_{\text{soln}} = +20.5$ kJ/mol, yet $NaNO_3$ is highly soluble in water. Based on these facts, which of the following statements is or are *true*?

   (**i**) The entropy of the system must increase when $NaNO_3$ dissolves in water.

   (**ii**) The enthalpy of mixing must be endothermic, $\Delta H_{\text{mix}} > 0$.

   (**iii**) The temperature of the solution will rise when $NaNO_3(s)$ dissolves in water.

(**a**) only i (**b**) only ii (**c**) only iii (**d**) both i and ii (**e**) both i and iii

EP 13.4 When $CO_2(g)$ dissolves in water to form a solution (carbonated water), the overall process is exothermic. Based on this observation, what can be said about the magnitudes of $\Delta H_{\text{solvent}}$, ΔH_{mix}, and ΔH_{solute} for this solution?

   (**a**) $|\Delta H_{\text{mix}}| < |\Delta H_{\text{solvent}}| < |\Delta H_{\text{solute}}|$

   (**b**) $|\Delta H_{\text{solute}}| < |\Delta H_{\text{solvent}}| < |\Delta H_{\text{mix}}|$

   (**c**) $|\Delta H_{\text{solute}}| = |\Delta H_{\text{solvent}}| < |\Delta H_{\text{mix}}|$

   (**d**) $|\Delta H_{\text{solute}}| < |\Delta H_{\text{mix}}| < |\Delta H_{\text{solvent}}|$

   (**e**) Not enough information is given to answer this question.

EP 13.5 A 10.0-g sample of $KClO_3$ is added to 100 mL of water, and the solution is stirred periodically until it comes to equilibrium, at which point 1.4 g of undissolved $KClO_3$ settles out on the bottom of the beaker. Once the system reaches this equilibrium, the rate of dissolution will be _____ the rate of crystallization, and the solution will be _____. (**a**) less than, unsaturated (**b**) equal to, unsaturated (**c**) less than, saturated (**d**) equal to, saturated (**e**) greater than, saturated

EP 13.6   If additional solute is added to a saturated solution, it will _____. (**a**) trigger a precipitation of the solute that has already dissolved, lowering the concentration of the solution (**b**) dissolve, increasing the concentration of the solution (**c**) settle to the bottom without dissolving, leaving the concentration of the solution unchanged (**d**) raise the temperature of the solution, increasing the concentration of the solution

EP 13.7   Which of the following solvents will most effectively dissolve paraffin wax, which is a mixture of large hydrocarbon molecules (alkanes) containing between 20 and 40 carbon atoms? (**a**) benzene (C_6H_6) (**b**) acetone (CH_3COCH_3) (**c**) acetonitrile (CH_3CH) (**d**) water (H_2O)

EP 13.8   A rigid, metal cylinder is filled to the 75% level with water, and N_2 gas is injected into the head space over the water to a pressure of 2 atm. The N_2 and H_2O are then given time to equilibrate. Which of the following changes will further increase the concentration of N_2 molecules dissolved in the water?

(**i**)   Lowering the temperature of the cylinder from 20 °C to 10 °C

(**ii**)   Injecting additional N_2 into the head space of the cylinder

(**iii**) Removing one-quarter of the water in the cylinder

(**a**) only i (**b**) only ii (**c**) only iii (**d**) both i and ii (**e**) both ii and iii

EP 13.9   The Henry's law constant for argon dissolved in water at 25 °C is 1.4×10^{-3} mol/L-atm, whereas that of helium is 3.7×10^{-4} mol/L-atm. From these values we can determine that _____ is more soluble in water, and a partial pressure of _____ argon would yield the same gas concentration in water as 1.00 atm of helium. (**a**) argon, 0.26 atm (**b**) argon, 3.8 atm (**c**) helium, 0.26 atm (**d**) helium, 3.8 atm

EP 13.10   According to a standard reference, the Henry's law constant for methane, CH_4, dissolved in H_2O at 25 °C is 1.4×10^{-5} mol/m^{3}Pa. What is the concentration of methane, in units of mol/L, in an aqueous solution at 25 °C in equilibrium with methane at a pressure of 40 atm? (**a**) 57 mol/L (**b**) 18 mol/L (**c**) 0.057 mol/L (**d**) 5.6×10^{-7} mol/L

EP 13.11   An aqueous solution of SO_2 contains 2.3×10^{-4} g of SO_2 per liter of solution. If the density of the solution is 1.00 g/mL, what is the concentration of SO_2 in ppb? (**a**) 2.3×10^5 ppb (**b**) 230 ppb (**c**) 3.6 ppb (**d**) 0.23 ppb (**e**) 3.6×10^{-6} ppb

EP 13.12   Calculate the mass percentage of NaCl in a solution containing 1.50 g of NaCl in 50.0 g of water. (**a**) 0.0291% (**b**) 0.0300% (**c**) 0.0513% (**d**) 2.91% (**e**) 3.00%

EP 13.13   At 40 °C the density of water is 0.992 g/mL and the solubility of oxygen gas is 1.00 mmol per liter of water. At this temperature, what is the mole fraction of O_2 in a saturated solution? (**a**) 1.00×10^{-6} (**b**) 1.82×10^{-5} (**c**) 1.00×10^{-2} (**d**) 1.82×10^{-2} (**e**) 5.55×10^{-2}

EP 13.14   Maple syrup has a density of 1.325 g/mL, and 100.00 g of maple syrup contains 67 mg of calcium in the form of Ca^{2+} ions. What is the molarity of Ca^{2+} in maple syrup? (**a**) 0.017 M (**b**) 0.022 M (**c**) 0.89 M (**d**) 12.6 M (**e**) 45.4 M

EP 13.15   Concentrated nitric acid is an aqueous solution that is 68% HNO_3 by mass. If the density of this solution is 1.40 g/mL, what is its molality? (**a**) 0.034 m (**b**) 0.38 m (**c**) 11 m (**d**) 15 m (**e**) 34 m

EP 13.16   In one container 0.2 mol NaCl is added to 500 mL of water, while in a second container 0.2 mol glucose ($C_6H_{12}O_6$) is added to the same quantity of water. Both containers are then sealed, and the vapor pressure is monitored. In which container will the vapor pressure be higher? (*Hint:* You can neglect any volume changes associated with adding the solute and assume that both NaCl and glucose are nonvolatile.) (**a**) The vapor pressure will be higher in the container with the NaCl solution. (**b**) The vapor pressure will be higher in the container with the glucose solution. (**c**) The vapor

pressure will be the same in both beakers. (**d**) Not enough information is given to answer this question.

EP 13.17   The vapor pressure of benzene, C_6H_6, is 100.0 torr at 26.1 °C. Assuming Raoult's law is obeyed, how many moles of a nonvolatile solute must be added to 100.0 mL of benzene to decrease its vapor pressure to 90.0 torr at 26.1 °C? The density of benzene is 0.8765 g/cm^3. (**a**) 0.0112 mol (**b**) 0.112 mol (**c**) 0.125 mol (**d**) 8.77 mol (**e**) 0.142 mol

EP 13.18   Which aqueous solution will have the lowest freezing point? (**a**) 0.075 m $CaCl_2$ (**b**) 0.15 m NaCl (**c**) 0.10 m HCl (**d**) 0.050 m CH_3COOH (**e**) 0.20 m $C_{12}H_{22}O_{11}$

EP 13.19   An ideal solution is defined as a solution where _____. (**a**) the interactions between solute and solvent molecules are assumed to be so weak they can be neglected (**b**) the interactions between solute and solvent molecules are assumed to be so strong that other intermolecular interactions can be neglected (**c**) the effects of entropy are neglected (**d**) the solute–solute, solvent–solvent, and solute–solvent interactions are all assumed to be the same strength

EP 13.20   If you were to administer an intravenous drip of pure water to a patient's bloodstream, the "solution" would be classified as _____ with respect to the blood and it would lead to _____ of red blood cells. (**a**) hypotonic, crenation (**b**) hypotonic, hemolysis (**c**) hypertonic, crenation (**d**) hypertonic, hemolysis (**e**) isotonic, hemolysis

EP 13.21   Which of the following actions will raise the osmotic pressure of a solution?

(**i**)   Diluting the solution by adding more solvent

(**ii**)   Increasing the temperature

(**iii**) Increasing the volume of the solution while maintaining the same concentration

(**a**) only i (**b**) only ii (**c**) only iii (**d**) both i and ii (**e**) both ii and iii

EP 13.22   When 80 mg of an unknown white powder found at a crime scene is dissolved in 1.50 mL of ethanol (density = 0.789 g/mL, normal freezing point −114.6 °C, K_f = 1.99 °C/m), the freezing point is lowered to −115.5 °C. Which of the following substances could be responsible for this freezing-point depression? (**a**) sucrose ($C_{12}H_{22}O_{11}$) (**b**) cocaine ($C_{17}H_{21}NO_4$) (**c**) codeine ($C_{18}H_{21}NO_3$) (**d**) norfenefrine ($C_8H_{11}NO_2$) (**e**) fructose($C_6H_{12}O_6$)

EP 13.23   Proteins frequently form complexes in which two, three, four, or even more individual proteins ("monomers") interact specifically with each other via hydrogen bonds or electrostatic interactions. The entire assembly of proteins can act as one unit in solution, and this assembly is called the "quaternary structure" of the protein. Suppose you discover a new protein whose monomer molar mass is 25,000 g/mol. You measure an osmotic pressure of 0.0916 atm at 37 °C for 7.20 g of the protein in 10.00 mL of an aqueous solution. How many protein monomers form the quaternary protein structure in solution? Treat the protein as a nonelectrolyte. (**a**) one (**b**) two (**c**) three (**d**) four (**e**) eight

EP 13.24   In cold climates, road crews "salt" the roads to lower the freezing point of water. How many grams of $CaCl_2$ should be added to 1000 L of water to lower its freezing point from 0.0 °C to −10.0 °C? The freezing-point depression constant for water is 1.86 °C/m, and the density of water is 1.00 g/mL. (**a**) 0.200 kg (**b**) 1.79 kg (**c**) 71.7 kg (**d**) 199 kg (**e**) 596 kg

EP 13.25   In which of the following are the molecules likely to have *both* hydrophobic (nonpolar) and hydrophilic (polar) regions?

(**i**)   emulsifying agent

(**ii**) hydrophobic colloid

(**iii**) hydrophilic colloid

(**a**) only i (**b**) only ii (**c**) only iii (**d**) both i and ii (**e**) both i and iii

EP 13.26 Gelatin-based desserts like the one shown here are made from protein-rich, animal-based collagen that surrounds and entraps tiny drops of aqueous solution. What type of colloid is this? (**a**) emulsion (**b**) solid foam (**c**) foam (**d**) solid emulsion (**e**) sol

EP 13.27 Which of the following is a property of a colloidal dispersion?

(**i**) inhomogeneous on a microscopic length scale

(**ii**) scatters light

(**iii**) given enough time will separate into a heterogeneous mixture

(**a**) only i (**b**) only ii (**c**) both i and ii (**d**) both ii and iii (**e**) all three i, ii, and iii

Exercises

Visualizing Concepts

13.1 Rank the contents of the following containers in order of increasing entropy: [Section 13.1]

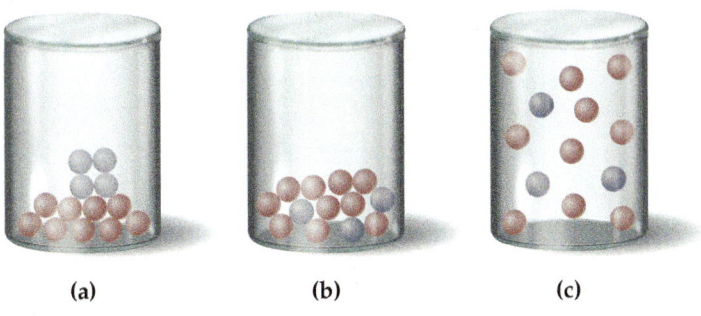

(a)   (b)   (c)

13.2 This figure shows the interaction of a cation with surrounding water molecules.

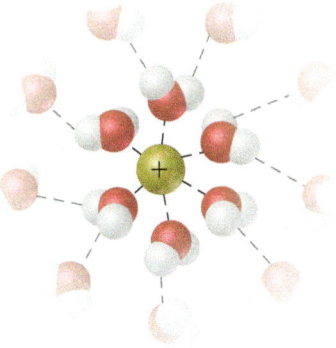

(**a**) Which atom of water is associated with the cation? Explain.

(**b**) Which of the following explanations accounts for the fact that the ion–solvent interaction is greater for Li^+ than for K^+?

a. Li^+ is of lower mass than K^+.

b. The ionization energy of Li is higher than that for K.

c. Li^+ has a smaller ionic radius than K^+.

d. Li has a lower density than K.

e. Li reacts with water more slowly than K. [Section 13.1]

13.3 Consider two ionic solids, both composed of singly charged ions, that have different lattice energies. (**a**) Will the solids have the same solubility in water? (**b**) If not, which solid will be more soluble in water, the one with the larger lattice energy or the one with the smaller lattice energy? Assume that solute–solvent interactions are the same for both solids. [Section 13.1]

13.4 Which *two* statements about gas mixtures are *true*? [Section 13.1]

(**a**) Gases always mix with other gases because the gas particles are too far apart to feel significant intermolecular attractions or repulsions.

(**b**) Just like water and oil don't mix in the liquid phase, two gases can be immiscible and not mix in the gas phase.

(**c**) If you cool a gaseous mixture, you will liquefy all the gases at the same temperature.

(**d**) Gases mix in all proportions in part because the entropy of the system increases upon doing so.

13.5 Which of the following is the best representation of a saturated solution? [Section 13.2]

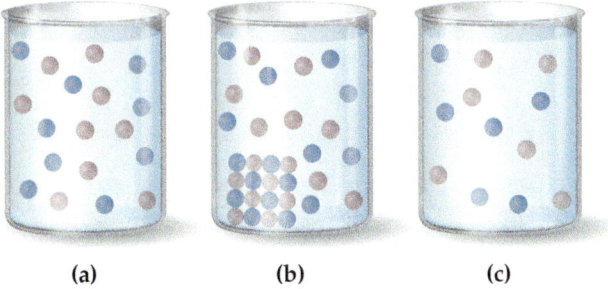

(a)   (b)   (c)

13.6 If you compare the solubilities of the noble gases in water, you find that solubility increases from smallest atomic weight to largest, Ar < Kr < Xe. Which of the following statements is the best explanation? [Section 13.3]

(**a**) The heavier the gas, the more it sinks to the bottom of the water and leaves room for more gas molecules at the top of the water.

(**b**) The heavier the gas, the more dispersion forces it has, and therefore the more attractive interactions it has with water molecules.

(**c**) The heavier the gas, the more likely it is to hydrogen-bond with water.

(**d**) The heavier the gas, the more likely it is to make a saturated solution in water.

13.7 Using the structures of vitamins B$_6$ and E provided, predict which is more water-soluble and which is more fat-soluble. [Section 13.3]

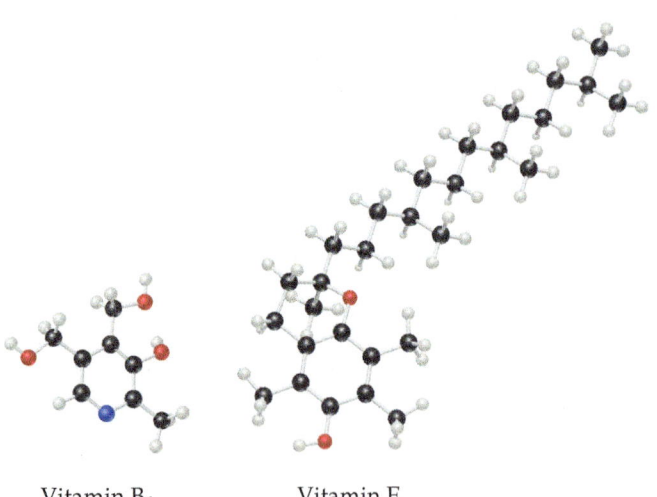

Vitamin B$_6$ Vitamin E

13.8 You take a sample of water that is at room temperature and in contact with air and put it under a vacuum. Right away, you see bubbles leave the water, but after a little while, the bubbles stop. As you keep applying the vacuum, more bubbles appear. A friend tells you that the first bubbles were water vapor and that the low pressure had reduced the boiling point of water, causing the water to boil. Another friend tells you that the first bubbles were gas molecules from the air (oxygen, nitrogen, and so forth) that were dissolved in the water. Which friend is most likely to be correct? What, then, is responsible for the second batch of bubbles? [Section 13.4]

13.9 The figure shows two identical volumetric flasks containing the same solution at two temperatures.

(a) Does the molarity of the solution change with the change in temperature?

(b) Does the molality of the solution change with the change in temperature? [Section 13.4]

25 °C   55 °C

13.10 This portion of a phase diagram shows the vapor–pressure curves of a volatile solvent and of a solution of that solvent containing a nonvolatile solute. (a) Which line represents the solution? (b) What are the normal boiling points of the solvent and the solution? [Section 13.5]

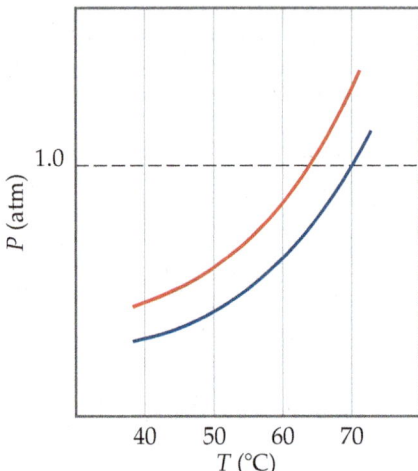

13.11 Suppose you had a balloon made of some highly flexible semipermeable membrane. The balloon is filled completely with a 0.2 M solution of some solute and is submerged in a 0.1 M solution of the same solute:

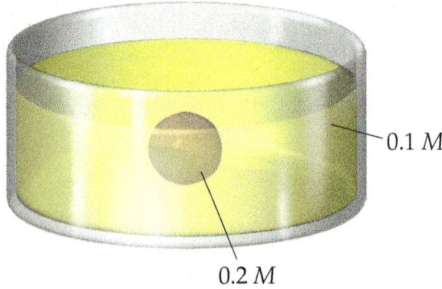

0.1 M

0.2 M

Initially, the volume of solution in the balloon is 0.25 L. Assuming the volume outside the semipermeable membrane is large, as the illustration shows, what would you expect for the solution volume inside the balloon once the system has come to equilibrium through osmosis? [Section 13.5]

13.12 The diagrams shown represent an emulsion, a true solution, and a liquid crystal. The colored balls represent different liquid molecules. Which diagram corresponds to which type of mixture? [Section 13.6]

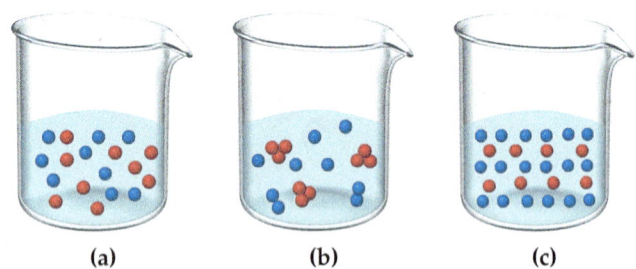

(a)   (b)   (c)

The Solution Process (Section 13.1)

13.13 Indicate whether each statement is *true* or *false*: (a) A solute will dissolve in a solvent if solute–solute interactions are stronger than solute–solvent interactions. (b) In making a solution, the enthalpy of mixing is always a positive number. (c) An increase in entropy favors mixing.

13.14 Indicate whether each statement is *true* or *false*: (a) NaCl dissolves in water but not in benzene (C$_6$H$_6$) because benzene is denser than water. (b) NaCl dissolves in water but not in benzene because water has a large dipole moment and benzene has a zero dipole moment. (c) NaCl dissolves in water but not in benzene because the water–ion interactions are stronger than the benzene–ion interactions.

13.15 Indicate the type of solute–solvent interaction that should be most important in each of the following solutions: (**a**) CCl_4 in benzene (C_6H_6), (**b**) methanol (CH_3OH) in water, (**c**) KBr in water, (**d**) HCl in acetonitrile (CH_3CN).

13.16 Indicate the principal type of solute–solvent interaction in each of the following solutions, and rank the solutions from weakest to strongest solute–solvent interaction: (**a**) KCl in water, (**b**) CH_2Cl_2 in benzene (C_6H_6), (**c**) methanol (CH_3OH) in water.

13.17 An ionic compound has a very negative ΔH_{soln} in water. (**a**) Would you expect it to be very soluble or nearly insoluble in water? (**b**) Which term would you expect to be the largest negative number: $\Delta H_{solvent}$, ΔH_{solute}, or ΔH_{mix}?

13.18 When ammonium chloride dissolves in water, the solution becomes colder. (**a**) Is the solution process exothermic or endothermic? (**b**) Why does the solution form?

13.19 (**a**) In Equation 13.1, which of the enthalpy terms for dissolving an ionic solid would correspond to the lattice energy? (**b**) Which energy term in this equation is always exothermic?

13.20 For the dissolution of LiCl in water, $\Delta H_{soln} = -37\ kJ/mol$. Which term would you expect to be the largest negative number: $\Delta H_{solvent}$, ΔH_{solute}, or ΔH_{mix}?

13.21 Two nonpolar organic liquids, hexane (C_6H_{14}) and heptane (C_7H_{16}), are mixed. (**a**) Do you expect ΔH_{soln} to be a large positive number, a large negative number, or close to zero? Explain. (**b**) Hexane and heptane are miscible with each other in all proportions. In making a solution of them, is the entropy of the system increased, decreased, or close to zero, compared to the separate pure liquids?

13.22 KBr is relatively soluble in water, yet its enthalpy of solution is $+19.8\ kJ/mol$. Which of the following statements provides the best explanation for this behavior? (**a**) Potassium salts are always soluble in water. (**b**) The entropy of mixing must be unfavorable. (**c**) The enthalpy of mixing must be less than the enthalpies for breaking up water–water interactions and K–Br ionic interactions. (**d**) KBr has a high molar mass compared to other salts like NaCl.

Saturated Solutions; Factors Affecting Solubility (Sections 13.2 and 13.3)

13.23 The solubility of $Cr(NO_3)_3 \cdot 9\ H_2O$ in water is 208 g per 100 g of water at 15 °C. A solution of $Cr(NO_3)_3 \cdot 9\ H_2O$ in water at 35 °C is formed by dissolving 324 g in 100 g of water. When this solution is slowly cooled to 15 °C, no precipitate forms. (**a**) Is the solution that has cooled down to 15 °C unsaturated, saturated, or supersaturated? (**b**) You take a metal spatula and scratch the side of the glass vessel that contains this cooled solution, and crystals start to appear. What has just happened? (**c**) At equilibrium, what mass of crystals do you expect to form?

13.24 The solubility of $MnSO_4 \cdot H_2O$ in water at 20 °C is 70 g per 100 mL of water. (**a**) Is a 1.22 M solution of $MnSO_4 \cdot H_2O$ in water at 20 °C saturated, supersaturated, or unsaturated? (**b**) Given a solution of $MnSO_4 \cdot H_2O$ of unknown concentration, what experiment could you perform to determine whether the new solution is saturated, supersaturated, or unsaturated?

13.25 By referring to Figure 13.15, determine whether the addition of 40.0 g of each of the following ionic solids to 100 g of water at 40 °C will lead to a saturated solution: (**a**) $NaNO_3$, (**b**) KCl, (**c**) $K_2Cr_2O_7$, (**d**) $Pb(NO_3)_2$.

13.26 Use Figure 13.15 to determine the mass of each of the following salts required to form a saturated solution in 250 g of water at 30 °C: (**a**) $KClO_3$, (**b**) $Pb(NO_3)_2$, (**c**) $Ce_2(SO_4)_3$.

13.27 (**a**) Would you expect water and glycerol, $CH_2(OH)CH(OH)CH_2OH$, to be miscible in all proportions? (**b**) List the intermolecular attractions that occur between a water molecule and a glycerol molecule.

13.28 Oil and water are immiscible. Which is the most likely reason? (**a**) Oil molecules are denser than water. (**b**) Oil molecules are composed mostly of carbon and hydrogen. (**c**) Oil molecules have higher molar masses than water. (**d**) Oil molecules have higher vapor pressures than water. (**e**) Oil molecules have higher boiling points than water.

13.29 Common laboratory solvents include acetone (CH_3COCH_3), methanol (CH_3OH), toluene ($C_6H_5CH_3$), and water. Which of these is the best solvent for nonpolar solutes?

13.30 Would you expect alanine (an amino acid) to be more soluble in water or in hexane?

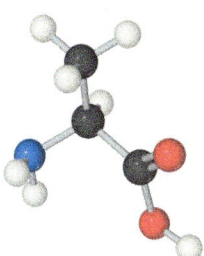

Alanine

13.31 (**a**) Would you expect stearic acid, $CH_3(CH_2)_{16}COOH$, to be more soluble in water or in carbon tetrachloride?

(**b**) Which would you expect to be more soluble in water, cyclohexane or dioxane?

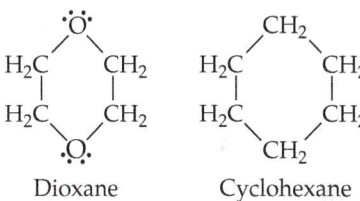

Dioxane   Cyclohexane

13.32 Ibuprofen, widely used as a pain reliever, has a limited solubility in water—less than 1 mg/mL. Which part of the molecule's structure (gray, white, red) contributes to its water solubility? Which part of the molecule (gray, white, red) contributes to its water insolubility?

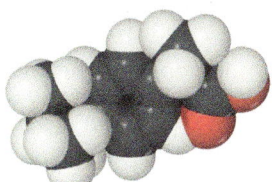

Ibuprofen

13.33 Which of the following in each pair is likely to be more soluble in hexane, C_6H_{14}: (**a**) CCl_4 or $CaCl_2$, (**b**) benzene (C_6H_6) or glycerol, $CH_2(OH)CH(OH)CH_2OH$, (**c**) octanoic acid, $CH_3(CH_2)_6COOH$ or acetic acid, CH_3COOH? Explain your answer in each case.

13.34 Which of the following in each pair is likely to be more soluble in water: (**a**) cyclohexane (C_6H_{12}) or glucose ($C_6H_{12}O_6$), (**b**) propionic acid (CH_3CH_2COOH) or sodium propionate (CH_3CH_2COONa), (**c**) HCl or ethyl chloride (CH_3CH_2Cl)? Explain in each case.

13.35 Indicate whether each statement is *true* or *false*: (**a**) The higher the temperature, the more soluble most gases are in water. (**b**) The higher the temperature, the more soluble most ionic solids are in water. (**c**) As you cool a saturated solution from high temperature to low temperature, solids start to crystallize out of solution if you achieve a supersaturated solution. (**d**) If you take a saturated solution and raise its temperature, you can (usually) add more solute and make the solution even more concentrated.

13.36 Indicate whether each statement is *true* or *false*: (**a**) When Henry's law is obeyed, the solubility of a gas in water is directly proportional to the absolute temperature. (**b**) In most cases, entropy increases when an ionic solid dissolves in water to form an aqueous solution. (**c**) When a gas dissolves in water to form a saturated solution, the process is endothermic ($\Delta H_{soln} < 0$). (**d**) As the molecular weight of a gaseous substance increases, its Henry's law constant, k, should also increase.

13.37 The Henry's law constant for helium gas in water at 30 °C is $3.7 \times 10^{-4}\ M/atm$, and the constant for N_2 at 30 °C is $6.0 \times 10^{-4}\ M/atm$. If the two gases are each present at 1.5 atm pressure, calculate the solubility of each gas.

13.38 Some beers like Guinness Irish Stout are pressurized not with pure CO_2, but with a mixture called "beer gas" that is 25 mole percent CO_2 and 75 mole percent N_2. (**a**) If the total pressure of the beer gas is 1.0 atm, what is the concentration of CO_2 dissolved in the beer at 5 °C? The Henry's law constants for CO_2 and N_2 at 5 °C are 6.1×10^{-2} and $8.3 \times 10^{-4}\ mol/L\text{-}atm$, respectively. (**b**) What is the concentration of N_2 dissolved in the beer under the same conditions? (**c**) What is the ratio of CO_2 to N_2 in the beer?

Concentrations of Solutions (Section 13.4)

13.39 (**a**) Calculate the mass percentage of Na_2SO_4 in a solution containing 10.6 g of Na_2SO_4 in 483 g of water. (**b**) An ore contains 2.86 g of silver per ton of ore. What is the concentration of silver in ppm?

13.40 (**a**) What is the mass percentage of iodine in a solution containing 0.035 mol I_2 in 125 g of CCl_4? (**b**) Seawater contains 0.0079 g of Sr^{2+} per kilogram of water. What is the concentration of Sr^{2+} in ppm?

13.41 A solution is made containing 14.6 g of CH_3OH in 184 g of H_2O. Calculate (**a**) the mole fraction of CH_3OH, (**b**) the mass percent of CH_3OH, (**c**) the molality of CH_3OH.

13.42 A solution is made containing 20.8 g of phenol (C_6H_5OH) in 425 g of ethanol (CH_3CH_2OH). Calculate (**a**) the mole fraction of phenol, (**b**) the mass percent of phenol, (**c**) the molality of phenol.

13.43 Calculate the molarity of the following aqueous solutions: (**a**) 0.540 g of $Mg(NO_3)_2$ in 250.0 mL of solution, (**b**) 22.4 g of $LiClO_4 \cdot 3\ H_2O$ in 125 mL of solution, (**c**) 25.0 mL of 3.50 M HNO_3 diluted to 0.250 L.

13.44 What is the molarity of each of the following solutions: (**a**) 15.0 g of $Al_2(SO_4)_3$ in 0.250 mL solution, (**b**) 5.25 g of $Mn(NO_3)_2 \cdot 2\ H_2O$ in 175 mL of solution, (**c**) 35.0 mL of 9.00 M H_2SO_4 diluted to 0.500 L?

13.45 Calculate the molality of each of the following solutions: (**a**) 8.66 g of benzene (C_6H_6) dissolved in 23.6 g of carbon tetrachloride (CCl_4), (**b**) 4.80 g of NaCl dissolved in 0.350 L of water.

13.46 (**a**) What is the molality of a solution formed by dissolving 1.12 mol of KCl in 16.0 mol of water? (**b**) How many grams of sulfur (S_8) must be dissolved in 100.0 g of naphthalene ($C_{10}H_8$) to make a 0.12 m solution?

13.47 A sulfuric acid solution containing 571.6 g of H_2SO_4 per liter of solution has a density of 1.329 g/cm^3. Calculate (**a**) the mass percentage, (**b**) the mole fraction, (**c**) the molality, (**d**) the molarity of H_2SO_4 in this solution.

13.48 Ascorbic acid (vitamin C, $C_6H_8O_6$) is a water-soluble vitamin. A solution containing 80.5 g of ascorbic acid dissolved in 210 g of water has a density of 1.22 g/mL at 55 °C. Calculate (**a**) the mass percentage, (**b**) the mole fraction, (**c**) the molality, (**d**) the molarity of ascorbic acid in this solution.

13.49 The density of acetonitrile (CH_3CN) is 0.786 g/mL and the density of methanol (CH_3OH) is 0.791 g/mL. A solution is made by dissolving 22.5 mL of CH_3OH in 98.7 mL of CH_3CN. (**a**) What is the mole fraction of methanol in the solution? (**b**) What is the molality of the solution? (**c**) Assuming that the volumes are additive, what is the molarity of CH_3OH in the solution?

13.50 The density of toluene (C_7H_8) is 0.867 g/mL, and the density of thiophene (C_4H_4S) is 1.065 g/mL. A solution is made by dissolving 8.10 g of thiophene in 250.0 mL of toluene. (**a**) Calculate the mole fraction of thiophene in the solution. (**b**) Calculate the molality of thiophene in the solution. (**c**) Assuming that the volumes of the solute and solvent are additive, what is the molarity of thiophene in the solution?

13.51 Calculate the number of moles of solute present in each of the following aqueous solutions: (**a**) 600 mL of 0.250 M $SrBr_2$, (**b**) 86.4 g of 0.180 m KCl, (**c**) 124.0 g of a solution that is 6.45% glucose ($C_6H_{12}O_6$) by mass.

13.52 Calculate the number of moles of solute present in each of the following solutions: (**a**) 255 mL of 1.50 M $HNO_3(aq)$, (**b**) 50.0 mg of an aqueous solution that is 1.50 m NaCl, (**c**) 75.0 g of an aqueous solution that is 1.50% sucrose ($C_{12}H_{22}O_{11}$) by mass.

13.53 Describe how you would prepare each of the following aqueous solutions from water and solid KBr: (**a**) 0.75 L of $1.5 \times 10^{-2}\ M$ KBr, (**b**) 125 g of 0.180 m KBr.

13.54 Describe how you would prepare each of the following aqueous solutions from water and the solid solute: (**a**) 1.50 L of 0.110 M $(NH_4)_2SO_4$ solution; (**b**) 225 g of a solution that is 0.65 m in Na_2CO_3.

13.55 Commercial aqueous nitric acid has a density of 1.42 g/mL and is 16 M. Calculate the percent HNO_3 by mass in the solution.

13.56 Commercial concentrated aqueous ammonia is 28% NH_3 by mass and has a density of 0.90 g/mL. What is the molarity of this solution?

13.57 Brass is a substitutional alloy consisting of a solution of copper and zinc. A particular sample of red brass consisting of 80.0% Cu and 20.0% Zn by mass has a density of 8750 kg/m^3. (**a**) What is the molality of Zn in the solid solution? (**b**) What is the molarity of Zn in the solution?

13.58 Caffeine ($C_8H_{10}N_4O_2$) is a stimulant found in coffee and tea. If a solution of caffeine in the solvent chloroform ($CHCl_3$) has a concentration of 0.0500 m, calculate (**a**) the percentage of caffeine by mass, (**b**) the mole fraction of caffeine in the solution.

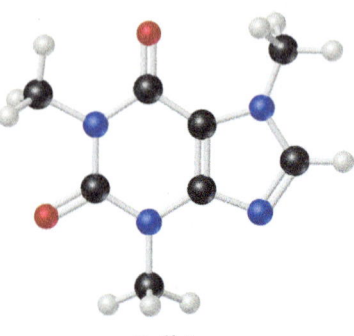

Caffeine

13.59 Describe how you would prepare the following solutions from water and the solid solute: (**a**) 1.85 L of an aqueous solution that is 12.0% KBr by mass and has a density of 1.10 g/mL; (**b**) 1.20 L of a solution that is 15.0% $Pb(NO_3)_2$ by mass and has a density of 1.19 g/mL.

13.60 (a) What volume of a 0.150 *M* solution of KBr is needed to precipitate 16.0 g of AgBr from a solution containing 0.480 mol of $AgNO_3$. (b) What volume of a 0.50 *M* solution of HCl would just neutralize 250 mL of a solution containing 5.5 g of $Ba(OH)_2$?

Colligative Properties (Section 13.5)

13.61 Suppose that you want to use reverse osmosis to reduce the salt content of brackish water containing 0.22 *M* total salt concentration to a value of 0.01 *M*, thus rendering it usable for human consumption. What is the minimum pressure that needs to be applied in the permeators (Figure 13.26) to achieve this goal, assuming that the operation occurs at 298 K?

13.62 Assume that a portable reverse-osmosis apparatus operates on seawater, whose effective concentration (the concentration of dissolved ions) is 1.12 *M*, and that the desalinated water output has an effective molarity of about 0.02 *M*. What minimum pressure must be applied by hand pumping at 297 K to cause reverse osmosis to occur?

13.63 You make a solution of a nonvolatile solute with a liquid solvent. Indicate whether each of the following statements is *true* or *false*. (a) The freezing point of the solution is *higher* than that of the pure solvent. (b) The freezing point of the solution is *lower* than that of the pure solvent. (c) The boiling point of the solution is *higher* than that of the pure solvent. (d) The boiling point of the solution is *lower* than that of the pure solvent.

13.64 You make a solution of a nonvolatile solute with a liquid solvent. Indicate if each of the following statements is *true* or *false*. (a) The solid that forms as the solution freezes is nearly pure solute. (b) The freezing point of the solution is independent of the concentration of the solute. (c) The boiling point of the solution increases in proportion to the concentration of the solute. (d) At any temperature, the vapor pressure of the solvent over the solution is lower than what it would be for the pure solvent.

13.65 Consider two solutions, one formed by adding 10 g of glucose ($C_6H_{12}O_6$) to 1 L of water and the other formed by adding 10 g of sucrose ($C_{12}H_{22}O_{11}$) to 1 L of water. Calculate the vapor pressure for each solution at 20 °C; the vapor pressure of pure water at this temperature is 17.5 torr.

13.66 (a) The vapor pressure of pure water at 60 °C is 149 torr. What vapor pressure is predicted by Raoult's law for a solution at 60 °C that is 50 mol% water and 50 mol% ethylene glycol (a nonvolatile solute)? (b) If the vapor pressure is measured to be 67 torr, does that suggest the interactions between ethylene glycol molecules and water are stronger, weaker, or the same strength as the water–water and ethylene glycol–ethylene glycol interactions?

13.67 (a) Calculate the vapor pressure of water above a solution prepared by adding 22.5 g of lactose ($C_{12}H_{22}O_{11}$) to 200.0 g of water at 338 K. (Vapor–pressure data for water are given in Appendix B.) (b) Calculate the mass of propylene glycol ($C_3H_8O_2$) that must be added to 0.340 kg of water to reduce the vapor pressure by 2.88 torr at 40 °C.

13.68 (a) Calculate the vapor pressure of water above a solution prepared by dissolving 28.5 g of glycerin ($C_3H_8O_3$) in 125 g of water at 343 K. (The vapor pressure of water is given in Appendix B.) (b) Calculate the mass of ethylene glycol ($C_2H_6O_2$) that must be added to 1.00 kg of ethanol (C_2H_5OH) to reduce its vapor pressure by 10.0 torr at 35 °C. The vapor pressure of pure ethanol at 35 °C is 1.00×10^2 torr.

13.69 At 63.5 °C, the vapor pressure of H_2O is 175 torr, and that of ethanol (C_2H_5OH) is 400 torr. A solution is made by mixing equal masses of H_2O and C_2H_5OH. (a) What is the mole fraction of ethanol in the solution? (b) Assuming ideal-solution behavior, what is the vapor pressure of the solution at 63.5 °C? (c) What is the mole fraction of ethanol in the vapor above the solution?

13.70 At 20 °C, the vapor pressure of benzene (C_6H_6) is 75 torr, and that of toluene (C_7H_8) is 22 torr. Assume that benzene and toluene form an ideal solution. (a) What is the composition in mole fraction of a solution that has a vapor pressure of 35 torr at 20 °C? (b) What is the mole fraction of benzene in the vapor above the solution described in part (a)?

13.71 (a) What boiling point would you predict for an aqueous sodium hydroxide solution that is 10 mole percent NaOH, assuming no ion pairing in solution? (b) If the boiling point is measured to be 105.0 °C, what is the true value of the van't Hoff factor for this solution?

13.72 Arrange the following aqueous solutions, each 10% by mass in solute, in order of increasing boiling point: glucose ($C_6H_{12}O_6$), sucrose ($C_{12}H_{22}O_{11}$), sodium nitrate ($NaNO_3$).

13.73 List the following aqueous solutions in order of increasing boiling point: 0.120 *m* glucose, 0.050 *m* LiBr, 0.050 *m* $Zn(NO_3)_2$.

13.74 List the following aqueous solutions in order of decreasing freezing point: 0.040 *m* glycerin ($C_6H_8O_3$), 0.020 *m* KBr, 0.030 *m* phenol (C_6H_5OH).

13.75 Use data from Table 13.3 to calculate the freezing and boiling points of each of the following solutions: (a) 0.22 *m* glycerol ($C_3H_8O_3$) in ethanol, (b) 0.240 mol of naphthalene ($C_{10}H_8$) in 2.45 mol of chloroform, (c) 1.50 g NaCl in 0.250 kg of water, (d) 2.04 g KBr and 4.82 g glucose ($C_6H_{12}O_6$) in 188 g of water.

13.76 Use data from Table 13.3 to calculate the freezing and boiling points of each of the following solutions: (a) 0.25 *m* glucose in ethanol; (b) 20.0 g of decane, $C_{10}H_{22}$, in 50.0 g $CHCl_3$; (c) 3.50 g NaOH in 175 g of water, (d) 0.45 mol ethylene glycol and 0.15 mol KBr in 150 g H_2O.

13.77 How many grams of ethylene glycol ($C_2H_6O_2$) must be added to 1.00 kg of water to produce a solution that freezes at −5.00 °C?

13.78 What is the freezing point of an aqueous solution that boils at 105.0 °C?

13.79 What is the osmotic pressure formed by dissolving 44.2 mg of aspirin ($C_9H_8O_4$) in 0.358 L of water at 25 °C?

13.80 Seawater contains 34 g of salts for every liter of solution. Assuming that the solute consists entirely of NaCl (in fact, over 90% of the salt is indeed NaCl), calculate the osmotic pressure of seawater at 20 °C.

13.81 Adrenaline is the hormone that triggers the release of extra glucose molecules in times of stress or emergency. A solution of 0.64 g of adrenaline in 36.0 g of CCl_4 elevates the boiling point by 0.49 °C. Calculate the approximate molar mass of adrenaline from these data.

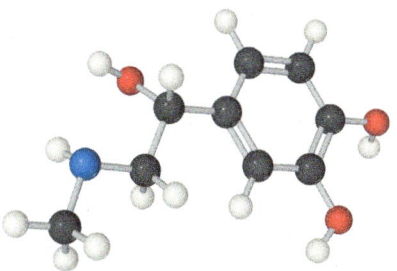

Adrenaline

13.82 Lauryl alcohol is obtained from coconut oil and is used to make detergents. A solution of 5.00 g of lauryl alcohol in 0.100 kg of benzene freezes at 4.1 °C. What is the molar mass of lauryl alcohol from these data? See Table 13.3 for the normal freezing point and K_f of benzene.

13.83 Lysozyme is an enzyme that breaks bacterial cell walls. A solution containing 0.150 g of this enzyme in 210 mL of solution has an osmotic pressure of 0.953 torr at 25 °C. What is the molar mass of lysozyme?

13.84 A dilute aqueous solution of an organic compound soluble in water is formed by dissolving 2.35 g of the compound in water to form 0.250 L of solution. The resulting solution has an osmotic pressure of 0.605 atm at 25 °C. Assuming that the organic compound is a nonelectrolyte, what is its molar mass?

13.85 The osmotic pressure of a 0.010 M aqueous solution of $CaCl_2$ is found to be 0.674 atm at 25 °C. Calculate the van't Hoff factor, i, for the solution.

13.86 Based on the data given in Table 13.4, which solution would give the larger freezing-point lowering, a 0.030 m solution of NaCl or a 0.020 m solution of K_2SO_4?

Colloids (Section 13.6)

13.87 In each of the following examples, identify the type of colloid: (**a**) mayonnaise, which is a dispersion of oil droplets in vinegar with proteins from the egg yolks acting as an emulsifying agent; (**b**) an ink produced by dispersing pigment particles in an organic solvent; (**c**) an inverse opal, made by first forming a close packed array of polymer spheres, then depositing an inorganic matrix of SiO_2 around the spheres, and finally heating to remove the polymer spheres via a combustion reaction with oxygen.

13.88 In each of the following examples, identify the type of colloid: (**a**) the meringue topping on a pie; (**b**) a cloud; (**c**) a stained glass window in a medieval church that gets its red coloration from gold nanocrystals embedded in the glass.

13.89 An "emulsifying agent" is a compound that helps stabilize a hydrophobic colloid in a hydrophilic solvent (or a hydrophilic colloid in a hydrophobic solvent). Which of the following choices is the best emulsifying agent? (**a**) CH_3COOH, (**b**) $CH_3CH_2CH_2COOH$, (**c**) $CH_3(CH_2)_{11}COOH$, (**d**) $CH_3(CH_2)_{11}COONa$.

13.90 Aerosols are important components of the atmosphere. Does the presence of aerosols in the atmosphere increase or decrease the amount of sunlight that arrives at Earth's surface, compared to an "aerosol-free" atmosphere? Explain your reasoning.

13.91 Proteins can be precipitated out of aqueous solution by the addition of an electrolyte; this process is called "salting out" the protein. (**a**) Do you think that all proteins would be precipitated out to the same extent by the same concentration of the same electrolyte? (**b**) If a protein has been salted out, are the protein–protein interactions stronger or weaker than they were before the electrolyte was added? (**c**) A friend of yours who is taking a biochemistry class says that salting out works because the waters of hydration that surround the protein prefer to surround the electrolyte as the electrolyte is added; therefore, the protein's hydration shell is stripped away, leading to protein precipitation. Another friend of yours in the same biochemistry class says that salting out works because the incoming ions adsorb tightly to the protein, making ion pairs on the protein surface, which end up giving the protein a zero net charge in water and therefore leading to precipitation. Discuss these two hypotheses. What kind of measurements would you need to make to distinguish between these two hypotheses?

13.92 Soaps consist of compounds such as sodium stearate, $CH_3(CH_2)_{16}COO^-Na^+$, that have both hydrophobic and hydrophilic parts. Consider the hydrocarbon part of sodium stearate to be the "tail" and the charged part to be the "head."

(**a**) Which part of sodium stearate, head or tail, is more likely to be solvated by water?

(**b**) Grease is a complex mixture of (mostly) hydrophobic compounds. Which part of sodium stearate, head or tail, is most likely to bind to grease?

(**c**) If you have large deposits of grease that you want to wash away with water, you can see that adding sodium stearate will help you produce an emulsion. What intermolecular interactions are responsible for this?

Additional Exercises

13.93 The "free-base" form of cocaine ($C_{17}H_{21}NO_4$) and its protonated hydrochloride form ($C_{17}H_{22}ClO_4$) are shown below; the free-base form can be converted to the hydrochloride form with one equivalent of HCl. For clarity, not all the carbon and hydrogen atoms are shown; each vertex represents a carbon atom with the appropriate number of hydrogen atoms so that each carbon makes four bonds to other atoms.

(**a**) Which form of cocaine, the free base or the hydrochloride, is relatively water-soluble?

(**b**) Which form, the free base or the hydrochloride, is relatively insoluble in water?

(**c**) The free-base form of cocaine has a solubility of 1.00 g in 6.70 mL ethanol (CH_3CH_2OH). Calculate the molarity of a saturated solution of the free-base form of cocaine in ethanol.

(**d**) The hydrochloride form of cocaine has a solubility of 1.00 g in 0.400 mL water. Calculate the molarity of a saturated solution of the hydrochloride form of cocaine in water.

(**e**) How many mL of a concentrated 18.0 M HCl aqueous solution would it take to convert 1.00 kg (a "kilo") of the free-base form of cocaine into its hydrochloride form?

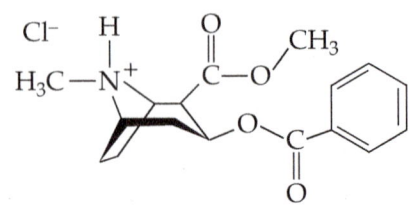

Cocaine   Cocaine hydrochloride
(Free base)

13.94 A supersaturated solution of sucrose ($C_{12}H_{22}O_{11}$) is made by dissolving sucrose in hot water and slowly letting the solution cool to room temperature. After a long time, the excess sucrose crystallizes out of the solution. Indicate whether each of the following statements is *true* or *false*: (**a**) After the excess sucrose has crystallized out, the remaining solution is saturated. (**b**) After the excess sucrose has crystallized out, the system is now unstable and is not in equilibrium. (**c**) After the excess sucrose has crystallized out, the rate of sucrose molecules leaving the surface of the crystals to be hydrated by water is equal to the rate of sucrose molecules in water attaching to the surface of the crystals.

13.95 Most fish need at least 4 ppm dissolved O_2 in water for survival. (**a**) What is this concentration in mol/L? (**b**) What partial pressure of O_2 above water is needed to obtain 4 ppm O_2 in water at 10 °C? (The Henry's law constant for O_2 at this temperature is 1.71×10^{-3} mol/L-atm.)

13.96 The presence of the radioactive gas radon (Rn) in well water presents a possible health hazard in parts of the United States. (**a**) Assuming that the solubility of radon in water with 1 atm pressure of the gas over the water at 30 °C is $7.27 \times 10^{-3} M$, what is the Henry's law constant for radon in water at this temperature? (**b**) A sample consisting of various gases contains 3.5×10^{-6} mole fraction of radon. This gas at a total pressure of 32 atm is shaken with water at 30 °C. Calculate the molar concentration of radon in the water.

13.97 Glucose makes up about 0.10% by mass of human blood. Calculate this concentration in (**a**) ppm, (**b**) molality.

13.98 The strength of alcoholic beverages like beer, wine, and spirits is typically expressed in units of alcohol by volume or abv. The abv is calculated by dividing the volume of ethanol (C_2H_5OH) by the total volume of solution. If the abv of a strong beer like a doppelbock is 8.0%, we can express the concentration in other units from the density of H_2O and C_2H_5OH, which are 1.0 g/mL and 0.79 g/mL at 20 °C, respectively. What is the concentration of this beer in (**a**) mass percent (also known as alcohol by weight or abw), (**b**) mole percent, (**c**) molarity, (**d**) molality?

13.99 The maximum allowable concentration of lead in drinking water is 9.0 ppb. (**a**) Calculate the molarity of lead in a 9.0-ppb solution. (**b**) How many grams of lead are in a swimming pool containing 9.0 ppb lead in 60 m³ of water?

13.100 The first stage of treatment at a reverse osmosis plant is to flow the water through rock, sand, and gravel as shown here. Would this step remove particulate matter? Would this step remove dissolved salts?

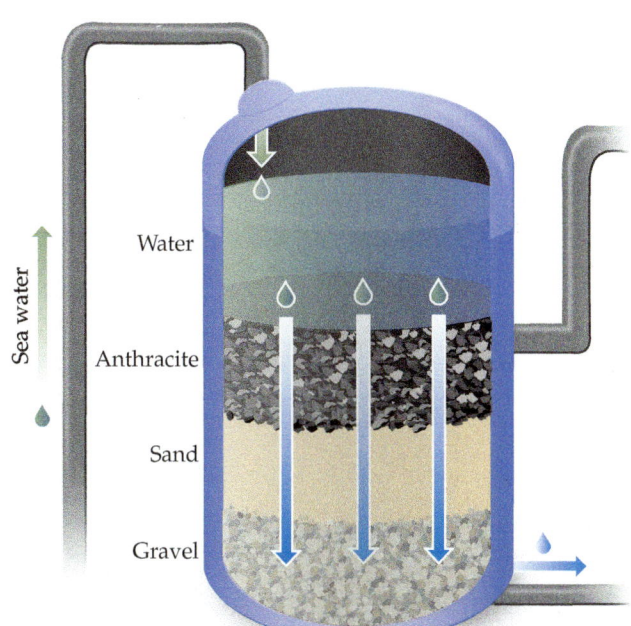

13.101 Acetonitrile (CH_3CN) is a polar organic solvent that dissolves a wide range of solutes, including many salts. The density of a 1.80 M LiBr solution in acetonitrile is 0.826 g/cm³. Calculate the concentration of the solution in (**a**) molality, (**b**) mole fraction of LiBr, (**c**) mass percentage of CH_3CN.

13.102 A solution contains 0.115 mol H_2O and an unknown number of moles of sodium chloride. The vapor pressure of the solution at 30 °C is 25.7 torr. The vapor pressure of pure water at this temperature is 31.8 torr. Calculate the number of grams of sodium chloride in the solution. (*Hint:* Remember that sodium chloride is a strong electrolyte.)

13.103 Two beakers are placed in a sealed box at 25 °C. One beaker contains 30.0 mL of a 0.050 M aqueous solution of a nonvolatile nonelectrolyte. The other beaker contains 30.0 mL of a 0.035 M aqueous solution of NaCl. The water vapor from the two solutions reaches equilibrium. (**a**) In which beaker does the solution level rise, and in which one does it fall? (**b**) What are the volumes in the two beakers when equilibrium is attained, assuming ideal behavior?

13.104 The normal boiling point of ethanol, CH_3CH_2OH, is 78.4 °C. When 9.15 g of a soluble nonelectrolyte is dissolved in 100.0 g of ethanol at that temperature, the vapor pressure of the solution is 7.40×10^2 torr. What is the molar mass of the solute?

13.105 Calculate the freezing point of a 0.100 m aqueous solution of K_2SO_4, (**a**) ignoring interionic attractions, and (**b**) taking interionic attractions into consideration by using the van't Hoff factor (Table 13.4).

13.106 Carbon disulfide (CS_2) boils at 46.30 °C and has a density of 1.261 g/mL. (**a**) When 0.250 mol of a nonelectrolyte solute is dissolved in 400.0 mL of CS_2, the solution boils at 47.46 °C. What is the molal boiling-point-elevation constant for CS_2? (**b**) When 5.39 g of a unknown nonelectrolyte solute is dissolved in 50.0 mL of CS_2, the solution boils at 47.08 °C. What is the molar mass of the unknown solute?

13.107 Fluorocarbons (compounds that contain both carbon and fluorine) were, until recently, used as refrigerants. The compounds listed in the following table are all gases at 25 °C, and their solubilities in water at 25 °C and 1 atm fluorocarbon pressure are given as mass percentages. (**a**) For each fluorocarbon, calculate the molality of a saturated solution. (**b**) Which molecular property best predicts the solubility of these gases in water: molar mass, dipole moment, or ability to hydrogen-bond to water? (**c**) Infants born with severe respiratory problems are sometimes given *liquid ventilation*: They breathe a liquid that can dissolve more oxygen than air can hold. One of these liquids is a fluorinated compound, $CF_3(CF_2)_7Br$. The solubility of oxygen in this liquid is 66 mL O_2 per 100 mL liquid. In contrast, air is 21% oxygen by volume. Calculate the moles of O_2 present in an infant's lungs (volume: 15 mL) if the infant takes a full breath of air compared to taking a full "breath" of a saturated solution of O_2 in the fluorinated liquid. Assume a pressure of 1 atm in the lungs.

Fluorocarbon	Solubility (mass %)
CF_4	0.0015
$CClF_3$	0.009
CCl_2F_2	0.028
$CHClF_2$	0.30

13.108 A lithium salt used in lubricating grease has the formula $LiC_nH_{2n+1}O_2$. The salt is soluble in water to the extent of 0.036 g per 100 g of water at 25 °C. The osmotic pressure of this solution is found to be 57.1 torr. Assuming that molality and molarity in such a dilute solution are the same and that the lithium salt is completely dissociated in the solution, determine an appropriate value of n in the formula for the salt.

13.109 At ordinary body temperature (37 °C), the solubility of N_2 in water at ordinary atmospheric pressure (1.0 atm) is 0.015 g/L. Air is approximately 78 mol % N_2. (**a**) Calculate the number of moles of N_2 dissolved per liter of blood, assuming blood is a simple aqueous solution. (**b**) At a depth of 100 ft in water, the external pressure is 4.0 atm. What is the solubility of N_2 from air in blood at this pressure? (**c**) If a scuba diver suddenly surfaces from this depth, how many milliliters of N_2 gas, in the form of tiny bubbles, are released into the bloodstream from each liter of blood?

13.110 The following table presents the solubilities of several gases in water at 25 °C under a total pressure of gas and water vapor of 1 atm. (**a**) What volume of $CH_4(g)$ under standard conditions of temperature and pressure is contained in 4.0 L of a saturated solution at 25 °C? (**b**) The solubilities (in water) of the hydrocarbons are as follows: methane < ethane < ethylene. Is this because ethylene is the most polar molecule? (**c**) What intermolecular interactions can these hydrocarbons have with water? (**d**) Draw the Lewis dot structures for the three hydrocarbons. Which of these hydrocarbons possess π bonds? Based on their solubilities, would you say π bonds are more or less polarizable than σ bonds? (**e**) Explain why NO is more soluble in water than either N_2 or O_2. (**f**) H_2 is more water-soluble than almost all the other gases in the table. What intermolecular forces is H_2S likely to have with water? (**g**) SO_2 is by far the most water-soluble gas in the table. What intermolecular forces is SO_2 likely to have with water?

Gas	Solubility (mM)
CH_4(methane)	1.3
C_2H_6(ethane)	1.8
C_2H_4(ethylene)	4.7
N_2	0.6
O_2	1.2
NO	1.9
H_2S	99
SO_2	1476

13.111 At 35 °C the vapor pressure of acetone, $(CH_3)_2CO$, is 360 torr, and that of chloroform, $CHCl_3$, is 300 torr. Acetone and chloroform can form very weak hydrogen bonds between one another; the chlorines on the carbon give the carbon a sufficient partial positive charge to enable this behavior:

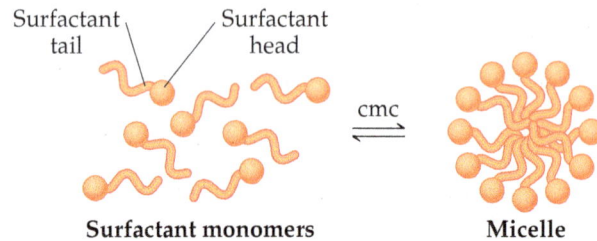

A solution composed of an equal number of moles of acetone and chloroform has a vapor pressure of 250 torr at 35 °C. (**a**) What would be the vapor pressure of the solution if it exhibited ideal behavior? (**b**) Based on the behavior of the solution, predict whether the mixing of acetone and chloroform is an exothermic ($\Delta H_{soln} < 0$) or endothermic ($\Delta H_{soln} > 0$) process.

13.112 Compounds like sodium stearate, called "surfactants" in general, can form structures known as micelles in water, once the solution concentration reaches the value known as the critical micelle concentration (cmc). Micelles contain dozens to hundreds of molecules. The cmc depends on the substance, the solvent, and the temperature.

Surfactant tail   Surfactant head

cmc

Surfactant monomers   **Micelle**

At and above the cmc, the properties of the solution vary drastically.

(**a**) The turbidity (the amount of light scattering) of solutions increases dramatically at the cmc. Suggest an explanation. (**b**) The ionic conductivity of the solution dramatically changes at the cmc. Suggest an explanation. (**c**) Chemists have developed fluorescent dyes that glow brightly only when the dye molecules are in a hydrophobic environment. Predict how the intensity of such fluorescence would relate to the concentration of sodium stearate as the sodium stearate concentration approaches and then increases past the cmc.

Design an Experiment

Based on Figure 13.18, you might think that volatile solvent molecules in a solution are less likely to escape to the gas phase, compared to the pure solvent, because the solute molecules are physically blocking the solvent molecules from leaving at the surface. This is a common misconception. Design an experiment to test the hypothesis that solute blocking of solvent vaporization is *not* the reason that solutions have lower vapor pressures than pure solvents.

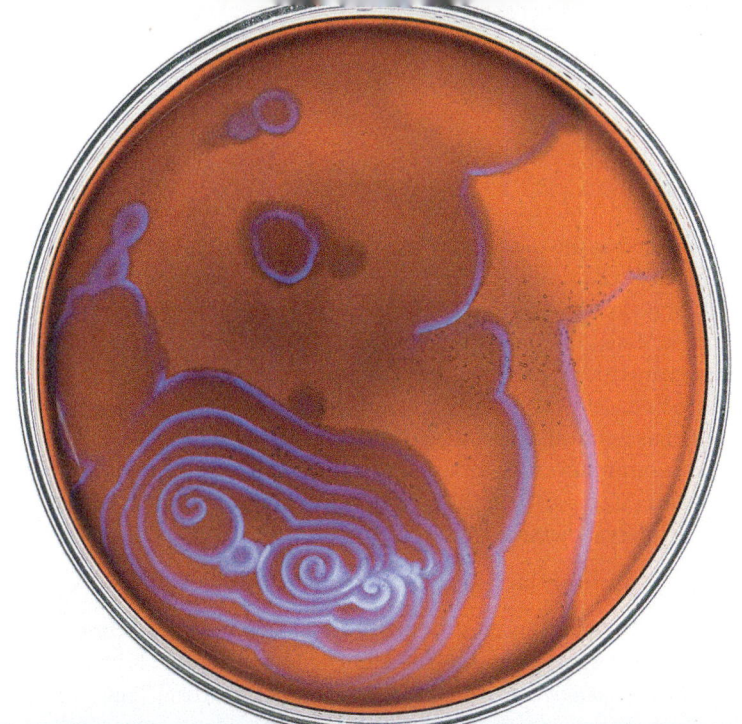

14

CHEMICAL KINETICS

▲ "Oscillating" chemical reactions in a dish show both reactants and products changing in time and space.

Chemical reactions take time to occur. Some reactions, such as the rusting of iron, occur relatively slowly, requiring days, months, or years to complete. Others, such as the decomposition of sodium azide, the reaction used to inflate automobile air bags, occur so quickly they are difficult to measure. As chemists, we need to be concerned about the *speed* of chemical reactions as well as the products of those reactions.

The speed at which a chemical reaction occurs is called the **reaction rate**. Many variables affect how fast a reaction proceeds, such as temperature and the concentration of the reactants. The part of chemistry that deals with reaction rates is known as **chemical kinetics**. It plays an important role in processes as diverse as the production of chemicals on an industrial scale, the design of drug delivery systems, and the decay of radioactive isotopes used in medicine.

Chemical kinetics is also useful in providing information about how reactions occur—the order in which chemical bonds are broken and formed during the course of a reaction. To understand how reactions happen, we must examine the reaction rates and the factors that influence them. Experimental information on the rate of a given reaction provides important evidence that helps us formulate a **reaction mechanism**, which is a step-by-step, molecular-level view of the pathway from reactants to products.

Our goal in this chapter is to understand how to determine reaction rates and to consider the factors that control these rates. What factors determine how rapidly food spoils, for instance? What determines the rate at which steel rusts? How can we remove hazardous pollutants in automobile exhaust before the exhaust leaves the tailpipe? Although we do not address all of these specific questions, we show that the rates of all chemical reactions are subject to the same principles.

WHAT'S AHEAD

14.1 ▶ Reaction Rates Describe the relationships between reaction rates and the factors that affect those rates: concentration, physical states of reactants, temperature, and presence of catalysts. Learn the difference between average and instantaneous reaction rates, and use reaction stoichiometry to determine the relative rates of reactant disappearance and product appearance.

14.2 ▶ Rate Laws and Rate Constants: The Method of Initial Rates Learn how to determine *rate laws*, expressions that quantitatively express reaction rates in terms of reactant concentrations, and experimentally determined *rate constants*. One method to determine rate laws and associated rate constants involves running reactions at different initial concentrations and measuring initial rates.

14.3 ▶ Integrated Rate Laws Learn that rate equations can be written to express how concentrations change over time, and look at several classifications of rate equations: *zero-order, first-order*, and *second-order* reactions.

14.4 ▶ Temperature and Rate: Activation Energy and the Arrhenius Equation Learn that reaction rate depends in part on the *activation energy*, a minimum energy that must be supplied for the reaction to occur. Thermal energy is needed to overcome this barrier, and the Arrhenius equation quantitatively connects the rate constant, temperature, and activation energy.

14.5 ▶ Reaction Mechanisms Understand how rate laws are derived from *reaction mechanisms*, the step-by-step molecular pathways leading from reactants to products.

14.6 ▶ Catalysis Examine the role of catalysts, substances that increase the rates of reactions but do not appear in the overall reaction equation. Learn how biological catalysts, called *enzymes*, work.

△ **Learning Objectives**

△ **Learning Objectives**

When you finish **Section 14.1**, you should be able to:

▶ Identify the factors that affect the rate of a reaction.

▶ Calculate the average and instantaneous rates of a reaction from the appropriate data.

▶ Use the stoichiometry of a reaction to compare the rate at which each product is appearing and each reactant is disappearing.

▼ **Go Figure**

If a heated steel nail were placed in pure O_2, would you expect it to burn as readily as the steel wool does?

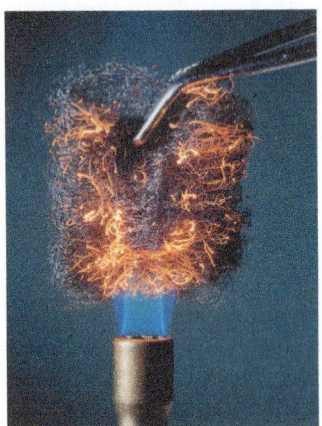

Steel wool heated in air (about 20% O_2) glows red-hot but oxidizes to Fe_2O_3 slowly.

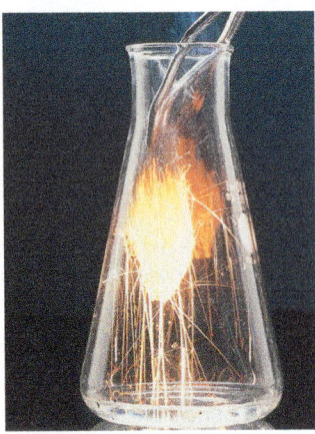

Red-hot steel wool in 100% O_2 burns vigorously, forming Fe_2O_3 quickly.

▲ **Figure 14.1 Effect of concentration on reaction rate.** The difference in behavior is due to the different concentrations of O_2 in the two environments.

14.1 | Reaction Rates

The *speed* of an event is defined as the *change* that occurs in a given *time* interval, which means that whenever we talk about speed, we necessarily bring in the notion of time. For example, the speed of a car is expressed as the change in the car's position over a certain time interval. In the United States, the speed of cars is usually measured in units of miles per hour—that is, the quantity that is changing (position measured in miles) divided by a time interval (measured in hours).

Similarly, the speed of a chemical reaction—its reaction rate—is the change in the concentration of reactants or products per unit of time. The units for reaction rate are usually molarity per second (M/s)—that is, the change in concentration measured in molarity divided by a time interval measured in seconds.

Four factors affect the rate at which any particular reaction occurs:

1. *Physical state of the reactants.* Reactants must come together to react. The more readily reactant molecules collide with one another, the more rapidly they react. Reactions may broadly be classified as *homogeneous*, which involve either all gases or all liquids, or as *heterogeneous*, in which reactants are in different phases. Under heterogeneous conditions, a reaction is limited by the area of contact of the reactants. Thus, heterogeneous reactions that involve solids tend to proceed more rapidly if the surface area of the solid is increased. For example, a medicine in the form of a fine powder dissolves in the stomach and enters the blood more quickly than the same medicine in the form of a tablet.

2. *Reactant concentrations.* Most chemical reactions proceed more quickly if the concentration of one or more reactants is increased. For example, steel wool burns only slowly in air, which contains 20% O_2, but bursts into flame in pure oxygen (**Figure 14.1**). As reactant concentration increases, the frequency with which the reactant molecules collide increases, leading to increased rates.

3. *Reaction temperature.* Reaction rates generally increase as temperature is increased. The bacterial reactions that spoil milk, for instance, proceed more rapidly at room temperature than at the lower temperature of a refrigerator. Increasing temperature increases the kinetic energies of molecules. (Section 10.5) As molecules move more rapidly, they collide more frequently and with higher energy, leading to increased reaction rates.

4. *The presence of a catalyst.* Catalysts are agents that increase reaction rates without themselves being used up. They affect the kinds of collisions (and therefore alter the mechanism) that lead to reaction. Catalysts play many crucial roles in living organisms.

On a molecular level, reaction rates depend on the frequency of collisions between molecules. *The greater the frequency of collisions, the higher the reaction rate.* For a collision to lead to a reaction, however, it must occur with sufficient energy to break bonds and with suitable orientation for new bonds to form in the proper locations. We will consider these factors as we proceed through this chapter.

Let's consider the hypothetical reaction A ⟶ B, depicted in **Figure 14.2**. Each red sphere represents 0.01 mol of A, each blue sphere represents 0.01 mol of B, and the container has a volume of 1.00 L. At the beginning of the reaction, there is 1.00 mol A, so the concentration is $1.00 \, mol/L = 1.00 \, M$. After 20 s, the concentration of A has fallen to $0.54 \, M$ and the concentration of B has risen to $0.46 \, M$. The sum of the concentrations is still $1.00 \, M$ because 1 mol of B is produced for each mole of A that reacts. After 40 s, the concentration of A is $0.30 \, M$ and that of B is $0.70 \, M$.

The rate of this reaction can be expressed either as the rate of disappearance of reactant A or as the rate of appearance of product B. The *average* rate of appearance of B over a particular time interval is given by the change in concentration of B divided by the change in time:

$$\text{Average rate of appearance of B} = \frac{\text{change in concentration of B}}{\text{change in time}}$$

$$= \frac{[B] \text{ at } t_2 - [B] \text{ at } t_1}{t_2 - t_1} = \frac{\Delta[B]}{\Delta t} \quad [14.1]$$

Go Figure   Estimate the number of moles of A in the mixture after 30 s.

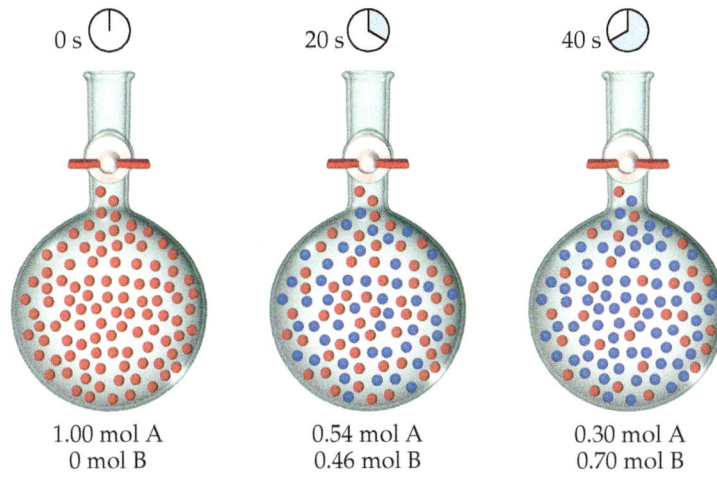

0 s   20 s   40 s

1.00 mol A   0.54 mol A   0.30 mol A
0 mol B   0.46 mol B   0.70 mol B

▲ **Figure 14.2** **Progress of a hypothetical reaction A ⟶ B.** The volume of the flask is 1.0 L.

We use brackets around a chemical formula, as in [B], to indicate molarity. The Greek letter delta, Δ, is read "change in" and is always equal to a final value minus an initial value. (Equation 5.3, Section 5.2) The average rate of appearance of B over the 20 s interval from the beginning of the reaction ($t_1 = 0$ s to $t_2 = 20$ s) is

$$\text{Average rate} = \frac{0.46\,M - 0.00\,M}{20\,\text{s} - 0\,\text{s}} = 2.3 \times 10^{-2}\,M/\text{s}$$

We could equally well express the reaction rate in terms of the reactant, A. In this case, we would be describing the rate of disappearance of A, which we express as

$$\text{Average rate of disappearance of A} = -\frac{\text{change in concentration of A}}{\text{change in time}}$$

$$= -\frac{\Delta[A]}{\Delta t} \qquad\qquad [14.2]$$

Because [A] decreases, $\Delta[A]$ is a negative number. The minus sign we put in the equation converts the negative $\Delta[A]$ to a positive rate of disappearance. *By convention, reaction rates are always expressed as positive quantities.*

Because one molecule of A is consumed for every molecule of B that forms, the average rate of disappearance of A equals the average rate of appearance of B:

$$\text{Average rate} = -\frac{\Delta[A]}{\Delta t} = -\frac{0.54\,M - 1.00\,M}{20\,\text{s} - 0\,\text{s}} = 2.3 \times 10^{-2}\,M/\text{s}$$

Sample Exercise 14.1
Calculating an Average Rate of Reaction

From the data in Figure 14.2, calculate the average rate at which A disappears over the time interval from 20 s to 40 s.

SOLUTION

Analyze We are given the concentration of A at 20 s (0.54 M) and at 40 s (0.30 M), and we are asked to calculate the average rate of reaction over this time interval.

Plan The average rate is given by the change in concentration, $\Delta[A]$, divided by the change in time, Δt. Because A is a reactant, a minus sign is used in the calculation to make the rate a positive quantity.

Solve

$$\text{Average rate} = -\frac{\Delta[A]}{\Delta t} = -\frac{0.30\,M - 0.54\,M}{40\,\text{s} - 20\,\text{s}}$$

$$= 1.2 \times 10^{-2}\,M/\text{s}$$

▶ **Practice Exercise**

Use the data in Figure 14.2 to calculate the average rate of appearance of B over the time interval from 0 s to 40 s.

TABLE 14.1   Rate Data for Reaction of C_4H_9Cl with Water

Time, t(s)	$[C_4H_9Cl]$ (M)	Average Rate (M/s)
0.0	0.1000	
		1.9×10^{-4}
50.0	0.0905	
		1.7×10^{-4}
100.0	0.0820	
		1.6×10^{-4}
150.0	0.0741	
		1.4×10^{-4}
200.0	0.0671	
		1.22×10^{-4}
300.0	0.0549	
		1.01×10^{-4}
400.0	0.0448	
		0.80×10^{-4}
500.0	0.0368	
		0.560×10^{-4}
800.0	0.0200	
10,000	0	

Change of Rate with Time

Butyl chloride (C_4H_9Cl) reacts with water to form butyl alcohol (C_4H_9OH) and hydro-chloric acid:

$$C_4H_9Cl(aq) + H_2O(l) \longrightarrow C_4H_9OH(aq) + HCl(aq) \qquad [14.3]$$

Suppose we prepare a 0.1000-M aqueous solution of C_4H_9Cl and then measure the concentration of C_4H_9Cl at various times after time zero (which is the instant at which the reactants are mixed, thereby initiating the reaction). We can use the resulting data, shown in the first two columns of **Table 14.1**, to calculate the average rate of disappearance of C_4H_9Cl over various time intervals; these rates are given in the third column. Notice that the average rate decreases over each 50-s interval for the first several measurements and continues to decrease over even larger intervals through the remaining measurements. *Reaction rates usually decrease as a reaction proceeds because the concentration of reactants decreases.* The change in rate as the reaction proceeds is also seen in a graph of $[C_4H_9Cl]$ versus time (**Figure 14.3**). Notice how the steepness of the curve decreases with time, indicating a decreasing reaction rate.

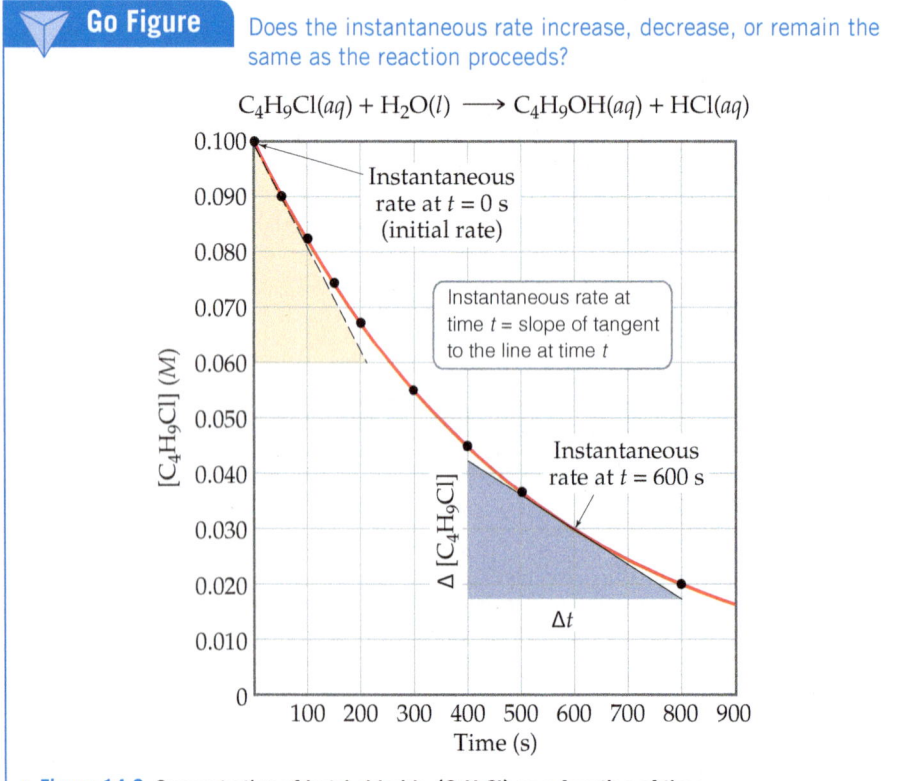

Go Figure   Does the instantaneous rate increase, decrease, or remain the same as the reaction proceeds?

$$C_4H_9Cl(aq) + H_2O(l) \longrightarrow C_4H_9OH(aq) + HCl(aq)$$

Instantaneous rate at $t = 0$ s (initial rate)

Instantaneous rate at time t = slope of tangent to the line at time t

Instantaneous rate at $t = 600$ s

$\Delta [C_4H_9Cl]$

Δt

▲ **Figure 14.3** Concentration of butyl chloride (C_4H_9Cl) as a function of time.

Instantaneous Rate

Graphs such as Figure 14.3 that show how the concentration of a reactant or product changes with time allow us to evaluate the **instantaneous rate** of a reaction, which is the rate at a particular instant during the reaction. The instantaneous rate is determined from the slope of the curve at a particular point in time. We have drawn two tangent lines in Figure 14.3, a dashed line running through the point at $t = 0$ s and a solid line running through the point at $t = 600$ s. The slopes of these tangent lines give the instantaneous rates at these two time points.* To determine the instantaneous rate at 600 s, for example, we construct horizontal and vertical lines to form the blue right triangle in Figure 14.3. The slope of the tangent line is the ratio of the height of the vertical side to the length of the horizontal side:

$$\text{Instantaneous rate} = -\frac{\Delta[C_4H_9Cl]}{\Delta t} = -\frac{(0.017 - 0.042)\,M}{(800 - 400)\,s}$$

$$= 6.3 \times 10^{-5}\,M/s$$

In the discussions that follow, the term *rate* means instantaneous rate unless indicated otherwise. The instantaneous rate at $t = 0$ is called the *initial rate* of the reaction. To understand the difference between average and instantaneous rates, imagine you have just driven 98 mi in 2.0 h. Your average speed for the trip is 49 mi/h, but your instantaneous speed at any moment during the trip is the speedometer reading at that moment.

Sample Exercise 14.2

Calculating an Instantaneous Rate of Reaction

Using Figure 14.3, calculate the instantaneous rate of disappearance of C_4H_9Cl at $t = 0$ s (the initial rate).

SOLUTION

Analyze We are asked to determine an instantaneous rate from a graph of reactant concentration versus time.

Plan To obtain the instantaneous rate at $t = 0$ s, we must determine the slope of the curve at $t = 0$. The tangent is drawn on the graph as the hypotenuse of the tan triangle. The slope of this straight line equals the change in the vertical axis divided by the corresponding change in the horizontal axis (which, in the case of this example, is the change in molarity over change in time).

Solve The tangent line falls from $[C_4H_9Cl] = 0.100\,M$ to $0.060\,M$ in the time change from 0 s to 210 s. Thus, the initial rate is

$$\text{Rate} = -\frac{\Delta[C_4H_9Cl]}{\Delta t} = -\frac{(0.060 - 0.100)\,M}{(210 - 0)\,s}$$

$$= 1.9 \times 10^{-4}\,M/s$$

▶ **Practice Exercise**
Using Figure 14.3, determine the instantaneous rate of disappearance of C_4H_9Cl at $t = 300$ s.

Reaction Rates and Stoichiometry

During our discussion of the hypothetical reaction in Figure 14.2, we saw that the stoichiometry requires that the rate of disappearance of A equal the rate of appearance of B. Likewise, the stoichiometry of Equation 14.3 indicates that 1 mol of C_4H_9OH is produced for each mole of C_4H_9Cl consumed. Therefore, the rate of appearance of C_4H_9OH equals the rate of disappearance of C_4H_9Cl:

$$\text{Rate} = -\frac{\Delta[C_4H_9Cl]}{\Delta t} = \frac{\Delta[C_4H_9OH]}{\Delta t}$$

---

*You may want to review graphical determination of slopes in Appendix A. If you are familiar with calculus, you may recognize that the average rate approaches the instantaneous rate as the time interval approaches zero. This limit, in the notation of calculus, is the negative of the derivative of the curve at time t, $-d[C_4H_9Cl]/dt$.

What happens when the stoichiometric relationships are not one-to-one, such as in the reaction $2 HI(g) \longrightarrow H_2(g) + I_2(g)$? We can measure either the rate of disappearance of HI or the rate of appearance of either H_2 or I_2. Because 2 mol of HI disappears for each mole of H_2 or I_2 that forms, the rate of disappearance of HI is *twice* the rate of appearance of either H_2 or I_2. How do we decide which number to use for the rate of the reaction? Depending on whether we monitor HI, I_2 or H_2, the rates can differ by a factor of 2. To fix this problem, we need to take into account the reaction stoichiometry. To arrive at a number for the reaction rate that does not depend on which component we measured, we must divide the rate of disappearance of HI by 2 (its coefficient in the balanced chemical equation):

$$\text{Rate} = -\frac{1}{2}\frac{\Delta[\text{HI}]}{\Delta t} = \frac{\Delta[\text{H}_2]}{\Delta t} = \frac{\Delta[\text{I}_2]}{\Delta t}$$

In general, for the reaction

$$a\,\text{A} + b\,\text{B} \longrightarrow c\,\text{C} + d\,\text{D}$$

the rate is given by

$$\text{Rate} = -\frac{1}{a}\frac{\Delta[\text{A}]}{\Delta t} = -\frac{1}{b}\frac{\Delta[\text{B}]}{\Delta t} = \frac{1}{c}\frac{\Delta[\text{C}]}{\Delta t} = \frac{1}{d}\frac{\Delta[\text{D}]}{\Delta t} \qquad [14.4]$$

When we speak of the rate of a reaction without specifying a particular reactant or product, we utilize the definition in Equation 14.4.*

Sample Exercise 14.3

Relating Rates at Which Products Appear and Reactants Disappear

(a) How is the rate at which ozone disappears related to the rate at which oxygen appears in the reaction $2 O_3(g) \longrightarrow 3 O_2(g)$?

(b) If the rate at which O_2 appears, $\Delta[O_2]/\Delta t$, is 6.0×10^{-5} *M*/s at a particular instant, at what rate is O_3 disappearing at this same time, $-\Delta[O_3]/\Delta t$?

SOLUTION

Analyze We are given a balanced chemical equation and asked to relate the rate of appearance of the product to the rate of disappearance of the reactant.

Plan We can use the coefficients in the chemical equation as shown in Equation 14.4 to express the relative rates of reactions.

Solve

(a) Using the coefficients in the balanced equation and the relationship given by Equation 14.4, we have:

$$\text{Rate} = -\frac{1}{2}\frac{\Delta[\text{O}_3]}{\Delta t} = \frac{1}{3}\frac{\Delta[\text{O}_2]}{\Delta t}$$

(b) Solving the equation from part (a) for the rate at which O_3 disappears, $-\Delta[O_3]/\Delta t$, we have:

$$-\frac{\Delta[\text{O}_3]}{\Delta t} = \frac{2}{3}\frac{\Delta[\text{O}_2]}{\Delta t} = \frac{2}{3}(6.0 \times 10^{-5}\,M/s) = 4.0 \times 10^{-5}\,M/s$$

Check We can apply a stoichiometric factor to convert the O_2 formation rate to the O_3 disappearance rate:

$$-\frac{\Delta[\text{O}_3]}{\Delta t} = \left(6.0 \times 10^{-5}\,\frac{\text{mol O}_2/\text{L}}{\text{s}}\right)\left(\frac{2\text{ mol O}_3}{3\text{ mol O}_2}\right) = 4.0 \times 10^{-5}\,\frac{\text{mol O}_3/\text{L}}{\text{s}}$$

$$= 4.0 \times 10^{-5}\,M/s$$

▶ **Practice Exercise**

If the rate of decomposition of N_2O_5 in the reaction $2 N_2O_5(g) \longrightarrow 4 NO_2(g) + O_2(g)$ at a particular instant is

4.2×10^{-7} *M*/s, what is the rate of appearance of (**a**) NO_2 and (**b**) O_2 at that instant?

---

*Equation 14.4 does not hold true if substances other than C and D are formed in significant amounts. For example, sometimes intermediate substances build in concentration before forming the final products. In that case, the relationship between the rate of disappearance of reactants and the rate of appearance of products is not given by Equation 14.4. All reactions whose rates we consider in this chapter obey Equation 14.4.

Self-Assessment Exercises

SAE 14.1 Which of the following statements is or are *true*?

(**i**) As a reaction proceeds, the rate is expected to increase.
(**ii**) Reaction rates generally increase as the temperature is increased.
(**iii**) The units for reaction rates are typically expressed as moles/s.

(**a**) Only one of the statements is true. (**b**) Statements i and ii are true. (**c**) Statements i and iii are true. (**d**) Statements ii and iii are true. (**e**) All three statements are true.

SAE 14.2 The concentrations of A and B are plotted as a function of time in the figure shown here:

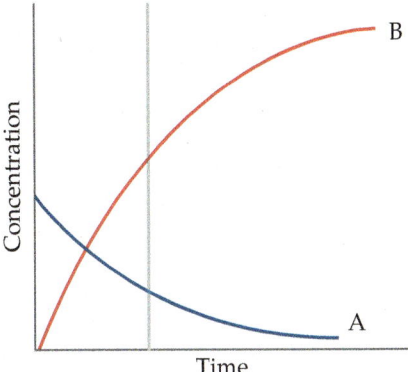

Which of the following statements is or are *true*?

(**i**) The stoichiometry of the overall reaction is A $\longrightarrow$ B.
(**ii**) Assuming the vertical line on the plot indicates time t, the instantaneous rate of appearance of B is greater than the instantaneous rate of disappearance of A at time $= t$.

(**a**) only i (**b**) only ii (**c**) both i and ii (**d**) neither i nor ii

SAE 14.3 Nitrogen monoxide (NO; also called nitric oxide) combines with oxygen to form nitrogen dioxide according to the following reaction: $2\,NO(g) + O_2(g) \longrightarrow 2\,NO_2(g)$. When the rate of formation of $NO_2(g)$ is $5.5 \times 10^{-4}\,M/s$, what is the rate of disappearance of $O_2(g)$? (**a**) $3.0 \times 10^{-7}\,M/s$ (**b**) $2.8 \times 10^{-4}\,M/s$ (**c**) $5.5 \times 10^{-4}\,M/s$ (**d**) $1.1 \times 10^{-3}\,M/s$

14.2 | Rate Laws and Rate Constants: The Method of Initial Rates

 Learning Objectives

When you have finished Section 14.2, you should be able to:

▶ Determine the reaction order, the magnitude of the rate constant, and the units of the rate constant, given the rate law and the initial concentrations of the reactants.

▶ Use the method of initial rates to determine the rate law for a reaction.

One way of studying the effect of concentration on reaction rate is to determine the way in which the initial rate of a reaction depends on the initial concentrations. For example, we might study the rate of the reaction

$$NH_4^+(aq) + NO_2^-(aq) \longrightarrow N_2(g) + 2\,H_2O(l)$$

by measuring the concentration of NH_4^+ or NO_2^- as a function of time or by measuring the volume of N_2 collected as a function of time. Because the stoichiometric coefficients on NH_4^+, NO_2^-, and N_2 are the same, all of these rates are the same.

Table 14.2 shows that changing the initial concentration of either reactant changes the initial reaction rate. If we double $[NH_4^+]$ while holding $[NO_2^-]$ constant, the rate

TABLE 14.2   Rate Data for the Reaction of Ammonium and Nitrite Ions in Water at 25 °C

Experiment Number	Initial $[NH_4^+]$ Concentration (M)	Initial $[NO_2^-]$ Concentration (M)	Observed Initial Rate (M/s)
1	0.0100	0.200	5.4×10^{-7}
2	0.0200	0.200	10.8×10^{-7}
3	0.0400	0.200	21.5×10^{-7}
4	0.200	0.0202	10.8×10^{-7}
5	0.200	0.0404	21.6×10^{-7}
6	0.200	0.0808	43.3×10^{-7}

A CLOSER LOOK | Using Spectroscopic Methods to Measure Reaction Rates: Beer's Law

A variety of techniques can be used to monitor reactant and product concentration during a reaction, including spectroscopic methods, which rely on the ability of substances to absorb (or emit) light. Spectroscopic kinetic studies are often performed with the reaction mixture in the sample compartment of a *spectrometer*, an instrument that measures the amount of light transmitted or absorbed by a sample at different wavelengths. For kinetic studies, the spectrometer is set to measure the light absorbed at a wavelength characteristic of one of the reactants or products. In the decomposition of $HI(g)$ into $H_2(g)$ and $I_2(g)$, for example, both HI and H_2 are colorless, whereas I_2 is violet. During the reaction, the violet color of the reaction mixture gets more intense as I_2 forms. Thus, visible light of appropriate wavelength can be used to monitor the reaction (Figure 14.4).

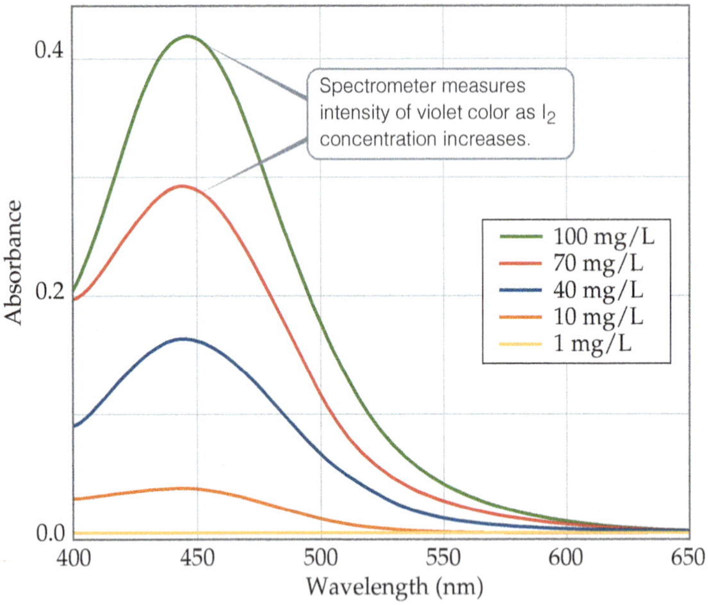

Figure 14.5 shows the components of a spectrometer. The spectrometer measures the amount of light absorbed by the sample by comparing the intensity of the light emitted from the light source with the intensity of the light transmitted through the sample, for various wavelengths. As the concentration of I_2 increases and its color becomes more intense, the amount of light absorbed by the reaction mixture increases, as Figure 14.4 shows, causing less light to reach the detector.

How can we relate the amount of light detected by the spectrometer to the concentration of a species? A relationship called *Beer's law* connects the amount of light absorbed to the concentration of the absorbing substance:

$$A = \varepsilon bc \qquad [14.5]$$

In this equation, A is the measured absorbance, ε is the extinction coefficient (a characteristic of the substance being monitored at a given wavelength of light), b is the path length through which the light passes, and c is the molar concentration of the absorbing substance. Thus, the concentration is directly proportional to absorbance. Many chemical and pharmaceutical companies routinely use Beer's law to calculate the concentration of purified solutions of the compounds that they make. In the laboratory portion of your course, you may very well perform one or more experiments in which you use Beer's law to relate the absorption of light to concentration.

Related Exercises: 14.101, 14.102, Design an Experiment

Spectrometer measures intensity of violet color as I_2 concentration increases.

Legend:
- 100 mg/L
- 70 mg/L
- 40 mg/L
- 10 mg/L
- 1 mg/L

▲ **Figure 14.4** Visible spectra of I_2 at different concentrations.

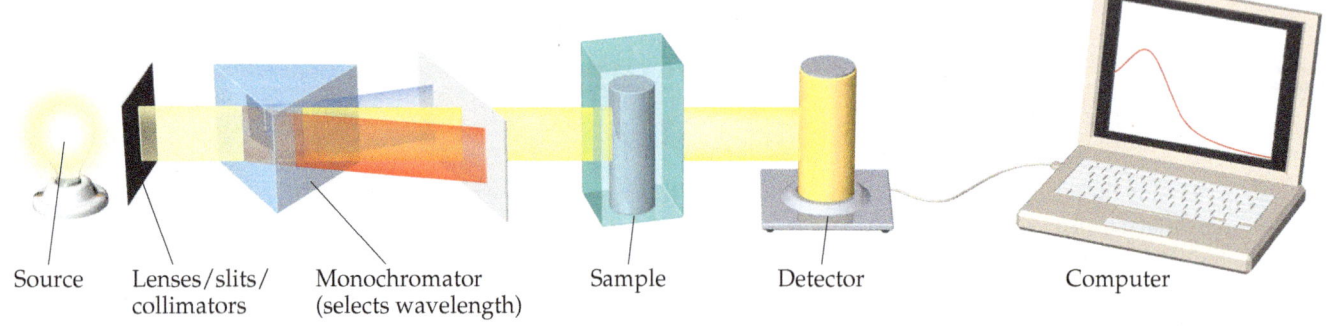

Source　Lenses/slits/collimators　Monochromator (selects wavelength)　Sample　Detector　Computer

▲ **Figure 14.5** Components of a spectrometer.

doubles (compare experiments 1 and 2). If we increase $[NH_4^+]$ by a factor of 4 but leave $[NO_2^-]$ unchanged (experiments 1 and 3), the rate changes by a factor of 4, and so forth. These results indicate that the initial reaction rate is proportional to $[NH_4^+]$. When $[NO_2^-]$ is similarly varied while $[NH_4^+]$ is held constant, the rate is affected in the same manner. Thus, the rate is also directly proportional to the concentration of $[NO_2^-]$.

We express the way in which the rate depends on the reactant concentrations by the equation

$$\text{Rate} = k[NH_4^+][NO_2^-] \qquad [14.6]$$

An equation such as Equation 14.6, which shows how the rate depends on reactant concentrations, is called a **rate law**. For the general reaction

$$a\,A + b\,B \longrightarrow c\,C + d\,D$$

the rate law generally has the form

$$\text{Rate} = k[A]^m[B]^n \qquad [14.7]$$

Notice that only the concentrations of the reactants generally appear in the rate law. The constant k is called the **rate constant**. The magnitude of k changes with temperature and therefore determines how temperature affects rate, as we explain in Section 14.4. The exponents m and n are typically small whole numbers, whose values are not necessarily equal to the coefficients a and b from the balanced equation. As we describe shortly, if we know m and n for a reaction, we can gain great insight into the individual steps that occur during the reaction.

Once we know the rate law for a reaction and the reaction rate for a set of reactant concentrations, we can calculate the value of k. For example, using the values for experiment 1 in Table 14.2, we can substitute into Equation 14.6:

$$5.4 \times 10^{-7}\,M/s = k(0.0100\,M)(0.200\,M)$$

$$k = \frac{5.4 \times 10^{-7}\,M/s}{(0.0100\,M)(0.200\,M)} = 2.7 \times 10^{-4}\,M^{-1}s^{-1}$$

You should verify that this same value of k is obtained using any of the other experimental results in Table 14.2.

Once we have both the rate law and the k value for a reaction, we can calculate the reaction rate for any set of concentrations. For example, using Equation 14.7 with $k = 2.7 \times 10^{-4}\,M^{-1}s^{-1}$, $m = 1$, and $n = 1$, we can calculate the rate for $[NH_4^+] = 0.100\,M$ and $[NO_2^-] = 0.100\,M$:

$$\text{Rate} = (2.7 \times 10^{-4}\,M^{-1}s^{-1})(0.100\,M)(0.100\,M) = 2.7 \times 10^{-6}\,M/s$$

Reaction Orders: The Exponents in the Rate Law

The rate law for most reactions has the form

$$\text{Rate} = k[\text{reactant 1}]^m[\text{reactant 2}]^n \ldots \qquad [14.8]$$

The exponents m and n are called **reaction orders**. In the rate law for the reaction of NH_4^+ with NO_2^-,

$$\text{Rate} = k[NH_4^+][NO_2^-]$$

the exponent of $[NH_4^+]$ is 1, so the rate is *first order* in NH_4^+. The rate is also first order in NO_2^-. (The exponent 1 is not shown in rate laws.) The **overall reaction order** is the sum of the orders with respect to each reactant represented in the rate law. Thus, for the reaction between NH_4^+ and NO_2^- the rate law has an overall reaction order of $1 + 1 = 2$, and the reaction is *second order overall*.

The exponents in a rate law indicate how the rate is affected by each reactant concentration. Because the rate at which NH_4^+ reacts with NO_2^- depends on $[NH_4^+]$ raised to the first power, the rate doubles when $[NH_4^+]$ doubles, triples when $[NH_4^+]$ triples, and so forth. Doubling or tripling $[NO_2^-]$ likewise doubles or triples the rate. If a rate law is second order with respect to a reactant, $[A]^2$, then doubling the concentration of that substance causes the reaction rate to quadruple because $[2]^2 = 4$, whereas tripling the concentration causes the rate to increase ninefold: $[3]^2 = 9$.

The following are some additional examples of experimentally determined rate laws:

$$2\,N_2O_5(g) \longrightarrow 4\,NO_2(g) + O_2(g) \quad \text{Rate} = k[N_2O_5] \qquad [14.9]$$

$$H_2(g) + I_2(g) \longrightarrow 2\,HI(g) \qquad \qquad \text{Rate} = k[H_2][I_2] \qquad [14.10]$$

$$CHCl_3(g) + Cl_2(g) \longrightarrow CCl_4(g) + HCl(g) \quad \text{Rate} = k[CHCl_3][Cl_2]^{1/2} \quad [14.11]$$

Although the exponents in a rate law are sometimes the same as the coefficients in the balanced equation, this is not necessarily the case, as Equations 14.9 and 14.11 show.

For any reaction, the rate law must be determined experimentally.

In most rate laws, reaction orders are 0, 1, or 2. However, we also occasionally encounter rate laws in which the reaction order is fractional (as is the case with Equation 14.11) or even negative.

Sample Exercise 14.4
Relating a Rate Law to the Effect of Concentration on Rate

Consider a reaction A + B ⟶ C for which rate = $k[A][B]^2$. Each of the following boxes represents a reaction mixture in which A is shown as red spheres and B as purple ones. Rank these mixtures in order of increasing rate of reaction.

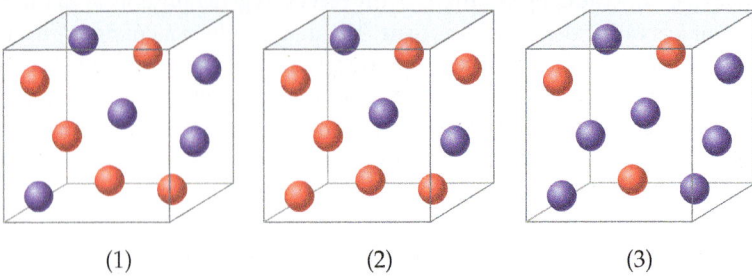

(1)   (2)   (3)

SOLUTION

Analyze We are given three boxes containing different numbers of spheres representing mixtures containing different reactant concentrations. We are asked to use the given rate law and the compositions of the boxes to rank the mixtures in order of increasing reaction rates.

Plan Because all three boxes have the same volume, we can put the number of spheres of each kind into the rate law and calculate the rate for each box.

Solve Box 1 contains five red spheres and five purple spheres, giving the following rate:

$$\text{Box 1: Rate} = k(5)(5)^2 = 125k$$

Box 2 contains seven red spheres and three purple spheres:

$$\text{Box 2: Rate} = k(7)(3)^2 = 63k$$

Box 3 contains three red spheres and seven purple spheres:

$$\text{Box 3: Rate} = k(3)(7)^2 = 147k$$

The slowest rate is $63k$ (Box 2), and the fastest is $147k$ (Box 3). Thus, the rates vary in the order $2 < 1 < 3$.

Check Each box contains 10 spheres. The rate law indicates that in this case [B] has a greater influence on rate than [A] because B has a larger reaction order. Hence, the mixture with the highest concentration of B (most purple spheres) should react fastest. This analysis confirms the order $2 < 1 < 3$.

▶ **Practice Exercise**
Assuming that rate = $k[A][B]$, rank the mixtures represented in this Sample Exercise in order of increasing rate.

Magnitudes and Units of Rate Constants

If chemists want to compare reactions to evaluate which ones are relatively fast and which ones are relatively slow, the quantity of interest is the rate constant. A good general rule is that a large value of k (~10^9 or higher) means a fast reaction, whereas a small value of k (10 or lower) means a slow reaction.

The units of the rate constant depend on the overall reaction order of the rate law. In a reaction that is second order overall, for example, the units of the rate constant must satisfy the equation:

$$\text{Units of rate} = (\text{units of rate constant})(\text{units of concentration})^2$$

Hence, in our usual units of molarity for concentration and seconds for time, we have

$$\text{Units of rate constant} = \frac{\text{units of rate}}{(\text{units of concentration})^2} = \frac{M/s}{M^2} = M^{-1}s^{-1}$$

Sample Exercise 14.5
Determining Reaction Orders and Units for Rate Constants

(a) What are the overall reaction orders for the reactions described in Equations 14.9 and 14.11?

(b) What are the units of the rate constant for the rate law in Equation 14.9?

SOLUTION

Analyze We are given two rate laws and asked to express (a) the overall reaction order for each and (b) the units for the rate constant for the first reaction.

Plan The overall reaction order is the sum of the exponents in the rate law. The units for the rate constant, k, are found by using the normal units for rate (M/s) and concentration (M) in the rate law and applying algebra to solve for k.

Solve

(a) The rate of the reaction in Equation 14.9 is first order in N_2O_5 and first order overall. The reaction in Equation 14.11 is first order in $CHCl_3$ and one-half order in Cl_2. The overall reaction order is three halves.

(b) For the rate law for Equation 14.9, we have

Units of rate $=$ (units of rate constant)(units of concentration)

so

$$\text{Units of rate constant} = \frac{\text{units of rate}}{\text{units of concentration}} = \frac{M/\text{s}}{M} = \text{s}^{-1}$$

Thus, if the reaction order changes, then the units of the rate constant change.

▶ **Practice Exercise**
(a) What is the reaction order of the reactant H_2 in Equation 14.10?
(b) What are the units of the rate constant for Equation 14.10?

Using Initial Rates to Determine Rate Laws

We have seen that the rate law for most reactions has the general form

$$\text{Rate} = k[\text{reactant 1}]^m[\text{reactant 2}]^n \ldots$$

Thus, the task of determining the rate law becomes one of determining the reaction orders, m and n. In most reactions, the reaction orders are 0, 1, or 2. As noted previously in this section, we can use the response of the reaction rate to a change in initial concentration to determine the reaction order.

In working with rate laws, it is important to realize that the *rate* of a reaction depends on concentration but the *rate constant* does not. As we explain later in this chapter, the rate constants (and hence the reaction rate) are affected by temperature and by the presence of a catalyst.

Sample Exercise 14.6
Determining a Rate Law from Initial Rate Data

The initial rate of a reaction A + B $\longrightarrow$ C was measured for three different starting concentrations of A and B, and the results are as follows:

Experiment Number	[A](M)	[B](M)	Initial Rate (M/s)
1	0.100	0.100	4.0×10^{-5}
2	0.100	0.200	4.0×10^{-5}
3	0.200	0.100	16.0×10^{-5}

Using these data, determine **(a)** the rate law for the reaction, **(b)** the rate constant, **(c)** the rate of the reaction when [A] = 0.050 M and [B] = 0.100 M.

SOLUTION

Analyze We are given a table of data that relates concentrations of reactants with initial rates of reaction and asked to determine **(a)** the rate law, **(b)** the rate constant, and **(c)** the rate of reaction for a set of concentrations not listed in the table.

Plan (a) We assume that the rate law has the following form: Rate $= k[A]^m[B]^n$. We will use the given data to deduce the reaction orders m and n by determining how changes in the concentration change the rate. **(b)** Once we know m and n, we can use the rate law and one of the sets of data to determine the rate constant k. **(c)** Upon determining both the rate constant and the reaction orders, we can use the rate law with the given concentrations to calculate rate.

Solve

(a) If we compare experiments 1 and 2, we see that [A] is held constant and [B] is doubled. Thus, this pair of experiments shows how [B] affects the rate, allowing us to deduce the order of the rate law with respect to B.

$$\frac{\text{Rate 1}}{\text{Rate 2}} = \frac{k[A_1]^m[B_1]^n}{k[A_2]^m[B_2]^n}$$

Inserting values of rate and concentration from the experiments gives:

$$\frac{4.0 \times 10^{-5}\, M/\text{s}}{4.0 \times 10^{-5}\, M/\text{s}} = \frac{k[0.100\, M]^m[0.100\, M]^n}{k[0.100\, M]^m[0.200\, M]^n}$$

The only way this equation can be true is if $n = 0$. Therefore, the rate law is zero order in B, which means that the rate is independent of [B].

$$1 = (1/2)^n$$

Continued

In experiments 1 and 3, [B] is held constant, so these data allow us to determine the order of the rate law with respect to [A].

$$\frac{\text{Rate 1}}{\text{Rate 3}} = \frac{k[A_1]^m[B_1]^n}{k[A_3]^m[B_3]^n}$$

Inserting values of rate and concentration from the experiments gives:

$$\frac{16.0 \times 10^{-5}\,M/s}{4.0 \times 10^{-5}\,M/s} = \frac{k[0.200\,M]^m[0.100\,M]^n}{k[0.100\,M]^m[0.100\,M]^n}$$

$$4 = (2)^m$$

Because the rate increases by a factor of 4 when [A] is doubled, we can conclude that $m = 2$ and the rate law is second order in A.

Combining these results, we arrive at the rate law:

$$\text{Rate} = k[A]^2[B]^0 = k[A]^2$$

(b) Using the rate law and the data from experiment 1, we have:

$$k = \frac{\text{rate}}{[A]^2} = \frac{4.0 \times 10^{-5}\,M/s}{(0.100\,M)^2} = 4.0 \times 10^{-3}\,M^{-1}s^{-1}$$

(c) Using the rate law from part (a) and the rate constant from part (b), we have:

$$\text{Rate} = k[A]^2$$
$$= (4.0 \times 10^{-3}\,M^{-1}s^{-1})(0.050\,M)^2$$
$$= 1.0 \times 10^{-5}\,M/s$$

Because [B] is not part of the rate law, it is irrelevant to the rate if there is at least some B present to react with A.

Check A good way to check our rate law is to use the concentrations in experiment 2 or 3 and see if we can correctly calculate the rate. Using data from experiment 3, we have

$$\text{Rate} = k[A]^2 = (4.0 \times 10^{-3}\,M^{-1}s^{-1})(0.200\,M)^2$$
$$= 1.6 \times 10^{-4}\,M/s$$

Thus, the rate law correctly reproduces the data, giving both the correct number and the correct units for the rate.

▶ **Practice Exercise**
The following data were measured for the reaction of nitric oxide with hydrogen:

$$2\,NO(g) + 2\,H_2(g) \longrightarrow N_2(g) + 2\,H_2O(g)$$

Experiment Number	[NO](M)	[H₂](M)	Initial Rate (M/s)
1	0.10	0.10	1.23×10^{-3}
2	0.10	0.20	2.46×10^{-3}
3	0.20	0.10	4.92×10^{-3}

(a) Determine the rate law for this reaction.
(b) Calculate the rate constant.
(c) Calculate the rate when [NO] = 0.050 M and [H₂] = 0.150 M.

 Self-Assessment Exercises

SAE 14.4 The rate law for the reaction, $2\,NO(g) + 2H_2(g) \longrightarrow N_2(g) + 2\,H_2O(g)$, is first order in H₂ and second order in NO. What happens to the rate when the concentrations of H₂ and NO are both doubled? **(a)** The rate doubles. **(b)** The rate increases by a factor of 4. **(c)** The rate increases by a factor of 8. **(d)** The rate increases by a factor of 16.

SAE 14.5 The following data were collected for the reaction $CH_3Br(aq) + OH^-(aq) \longrightarrow CH_3OH(aq) + Br^-(aq)$.

Experiment	[CH₃Br] (M)	[OH⁻] (M)	Initial Rate (M/s)
1	0.010	0.015	0.0415
2	0.010	0.030	0.0830
3	0.030	0.015	0.125

What is the rate law for the reaction?

(a) Rate = $k[CH_3Br][OH^-]$ **(c)** Rate = $k[CH_3Br][OH^-]^2$
(b) Rate = $k[CH_3Br]^2[OH^-]$ **(d)** Rate = $k[CH_3Br]^2[OH^-]^2$.

SAE 14.6 Which of the following set of statements is or are *true*?
 (i) All rate constants have the units s⁻¹.
 (ii) The rate law for the reaction $A + B \longrightarrow C + D$ must be first order in A and first order in B.
 (iii) For two reactions that are first order overall, the one with the larger rate constant will be faster.

(a) Only i is true. **(b)** Only ii is true. **(c)** Only iii is true. **(d)** Only two of i, ii, and iii are true. **(e)** None of i, ii, and iii is true.

SAE 14.7 The following data were collected for the reaction $2\,NO(g) + Cl_2(g) \longrightarrow 2\,NOCl(g)$.

Experiment	[NO] (M)	[Cl₂] (M)	Initial Rate (M/s)
1	0.150	0.200	0.00475
2	0.150	0.300	0.00713
3	0.300	0.200	0.0190

Predict the initial rate when the initial [NO] = 0.450 M and the initial [Cl₂] = 0.450 M. **(a)** 0.00475 M/s **(b)** 0.0321 M/s **(c)** 0.0721 M/s **(d)** 0.0962 M/s

14.3 | Integrated Rate Laws

The rate laws we have examined so far enable us to calculate the initial rate of a reaction from the rate constant and initial reactant concentrations. In this section, we show that rate laws can also be converted into equations that describe how the concentrations of reactants or products change as a function of time. The mathematics required to accomplish this conversion involve calculus. We do not expect you to be able to perform the calculus operations, but you should be able to use the resulting equations. We apply this conversion to three of the simplest rate laws—namely, those that are first order overall, those that are second order overall, and those that are zero order overall.

First-Order Reactions

A **first-order reaction** is one whose rate depends on the concentration of a single reactant raised to the first power. If a reaction of the type A $\longrightarrow$ products is first order, the rate law is:

$$\text{Rate} = -\frac{\Delta[A]}{\Delta t} = k[A]$$

This form of a rate law, which expresses how rate depends on concentration, is called the *differential rate law*. Using the operation from calculus called integration, we can transform this relationship into an equation known as the *integrated rate law* for a first-order reaction that relates the initial concentration of A, $[A]_0$, to its concentration at any other time t, $[A]_t$:

$$\ln[A]_t - \ln[A]_0 = -kt \quad \text{or} \quad \ln\frac{[A]_t}{[A]_0} = -kt \qquad [14.12]$$

The function "ln" in Equation 14.12 is the natural logarithm (Appendix A.2). Equation 14.12 can also be rearranged to

$$\ln[A]_t = -kt + \ln[A]_0 \qquad [14.13]$$

Equations 14.12 and 14.13 can be used with any concentration units as long as the units are the same for both $[A]_t$ and $[A]_0$. In solution, the usual units are molarity. When dealing with gases, we can use pressure as a concentration in Equations 14.12 and 14.13. This substitution is allowed because the ideal gas law (see Section 10.3) dictates that at constant temperature the pressure is directly proportional to the concentration (n/V).

For a first-order reaction, Equation 14.12 or 14.13 can be used in several ways. Given any three of the following quantities, we can solve for the fourth: k, t, $[A]_0$, and $[A]_t$. Thus, you can use these equations to determine (1) the concentration of a reactant remaining at any time after the reaction has started, (2) the time interval required for a given fraction of a sample to react, or (3) the time interval required for a reactant concentration to fall to a certain level.

Equation 14.13 can be used to verify whether a reaction is first order and to determine its rate constant. This equation has the form of the general equation for a straight line, $y = mx + b$, in which m is the slope and b is the y-intercept of the line (Appendix A.4):

$$\ln [A]_t = -kt + \ln [A]_0$$
$$y = mx + b$$

For a first-order reaction, therefore, a graph of $\ln[A]_t$ versus time gives a straight line with a slope of $-k$ and a y-intercept of $\ln[A]_0$. A reaction that is not first order will not yield a straight line.

As an example, consider the conversion of methyl isonitrile (CH_3NC) to its isomer acetonitrile (CH_3CN) (Figure 14.6). Because experiments show that the reaction is first order, the concentration of CH_3NC will vary with time according to the integrated rate law in Equation 14.13:

$$\ln[CH_3NC]_t = -kt + \ln[CH_3NC]_0$$

Learning Objectives

When you have finished **Section 14.3**, you should be able to:

▶ Distinguish between zero-, first-, and second-order reactions based on how concentrations change as a function of time.

▶ Use integrated rate law expressions to calculate the concentrations of reactants and products at any point in time.

▶ Analyze integrated rate laws to obtain the half-life for a reaction.

Methyl isonitrile

Acetonitrile

▲ **Figure 14.6** The first-order reaction of CH_3NC conversion into CH_3CN.

Go Figure   Why is the slope of the line in part (b) negative?

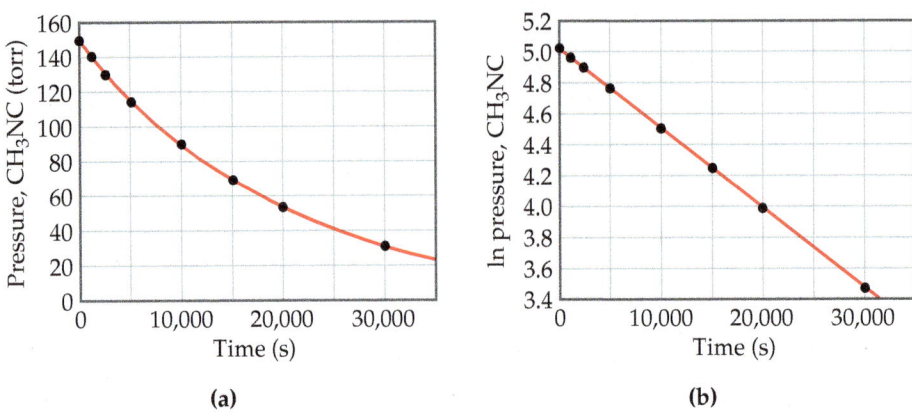

▲ **Figure 14.7 Kinetic data for conversion of methyl isonitrile into acetonitrile.**

We can run the reaction at a temperature at which both substances are gaseous (199 °C). **Figure 14.7(a)** shows how the pressure of $CH_3NC(g)$ varies with time. Figure 14.7(b) shows that a plot of the natural logarithm of the pressure versus time is a straight line. The slope of this line is $-5.1 \times 10^{-5} s^{-1}$. (Verify this for yourself, but remember that your result may vary slightly from ours because of inaccuracies associated with reading the graph.) Because the slope of the line equals $-k$, the rate constant for this reaction equals $5.1 \times 10^{-5} s^{-1}$.

Sample Exercise 14.7

Using the Integrated First-Order Rate Law

The decomposition of a certain insecticide in water at 12 °C follows first-order kinetics with a rate constant of $1.45 \, yr^{-1}$. A quantity of this insecticide is washed into a lake on June 1, leading to a concentration of $5.0 \times 10^{-7} \, g/cm^3$. Assume that the temperature of the lake is constant (so that there are no effects of temperature variation on the rate). **(a)** What is the concentration of the insecticide on June 1 of the following year? **(b)** How long will it take for the insecticide concentration to decrease to $3.0 \times 10^{-7} \, g/cm^3$?

SOLUTION

Analyze We are given the rate constant for a reaction that obeys first-order kinetics, as well as information about concentrations and times, and we are asked to calculate how much reactant (insecticide) remains after one year. We must also determine the time interval needed to reach a particular insecticide concentration.

Plan In part (a) we are given the rate constant, a period of time, and the initial concentration of the reactant, so we can use Equation 14.13 to determine the concentration of the reactant after one year has passed. In part (b) we are given the initial and final concentrations and the rate constant. In this case, we can use Equation 14.13 to calculate the time that must pass to reach the desired concentration.

Solve

(a) Substituting the known quantities into Equation 14.13, we have:

$$\ln[\text{insecticide}]_{t=1\,yr} = -(1.45 \, yr^{-1})(1.00 \, yr) + \ln(5.0 \times 10^{-7})$$

We use the ln function on a calculator to evaluate the second term on the right [that is, $\ln(5.0 \times 10^{-7})$], giving:

$$\ln[\text{insecticide}]_{t=1\,yr} = -1.45 + (-14.51) = -15.96$$

To obtain $[\text{insecticide}]_{t=1\,yr}$, we use the inverse natural logarithm, or e^x, function on the calculator:

$$[\text{insecticide}]_{t=1\,yr} = e^{-15.96} = 1.2 \times 10^{-7} \, g/cm^3$$

Note that the concentration units for $[A]_t$ and $[A]_0$ must be the same.

(b) Again, substituting into Equation 14.13, with $[\text{insecticide}]_t = 3.0 \times 10^{-7} \, g/cm^3$, gives:

$$\ln(3.0 \times 10^{-7}) = -(1.45 \, yr^{-1})(t) + \ln(5.0 \times 10^{-7})$$

Solving for t gives:

$$t = -[\ln(3.0 \times 10^{-7}) - \ln(5.0 \times 10^{-7})]/1.45 \, \text{yr}^{-1}$$

$$= -(-15.02 + 14.51)/1.45 \, \text{yr}^{-1} = 0.35 \, \text{yr}$$

Check In part (a), the concentration remaining after 1.00 yr (that is, 1.2×10^{-7} g/cm^3) is less than the original concentration (5.0×10^{-7} g/cm^3), as it should be. In part (b), the given concentration (3.0×10^{-7} g/cm^3) is greater than that remaining after 1.00 yr, indicating that the time must be less than a year. Thus, $t = 0.35$ yr is a reasonable answer.

▶ **Practice Exercise**

The decomposition of dimethyl ether, $(CH_3)_2O$, at 510 °C is a first-order process with a rate constant of 6.8×10^{-4} s^{-1}:

$$(CH_3)_2O(g) \longrightarrow CH_4(g) + H_2(g) + CO(g)$$

If the initial pressure of $(CH_3)_2O$ is 135 torr, what is its pressure after 1420 s?

Second-Order Reactions

A **second-order reaction** is one for which the rate depends either on a reactant concentration raised to the second power or on the concentrations of two reactants each raised to the first power. For simplicity, let's consider reactions of the type A $\longrightarrow$ products or A + B $\longrightarrow$ products that are second order in just one reactant, A:

$$\text{Rate} = -\frac{\Delta[A]}{\Delta t} = k[A]^2$$

With the use of calculus, this differential rate law can be used to derive the integrated rate law for second-order reactions:

$$\frac{1}{[A]_t} = kt + \frac{1}{[A]_0} \qquad\qquad [14.14]$$

This equation, like Equation 14.13, has four variables, k, t, $[A]_0$, and $[A]_t$, and any one of these can be calculated knowing the other three. Equation 14.14 also has the form of a straight line ($y = mx + b$). If the reaction is second order, a plot of $1/[A]_t$ versus t yields a straight line with slope k and y-intercept $1/[A]_0$. One way to distinguish between first- and second-order rate laws is to graph both $\ln[A]_t$ and $1/[A]_t$ as a function of t. If the $\ln[A]_t$ plot is linear, the reaction is *first* order; if the $1/[A]_t$ plot is linear, the reaction is *second* order.

Sample Exercise 14.8

Determining Reaction Order from the Integrated Rate Law

The following data were obtained for the gas-phase decomposition of nitrogen dioxide at 300 °C, $NO_2(g) \longrightarrow NO(g) + \frac{1}{2}O_2(g)$. Is the reaction first or second order in NO_2?

Time (s)	$[NO_2]$ (M)
0.0	0.01000
50.0	0.00787
100.0	0.00649
200.0	0.00481
300.0	0.00380

SOLUTION

Analyze We are given the concentrations of a reactant at various times during a reaction, and we are asked to determine whether the reaction is first or second order.

Plan We can plot $\ln[NO_2]$ and $1/[NO_2]$ against time. If one plot or the other is linear, we will know the reaction is either first or second order.

Solve

To graph $\ln[NO_2]$ and $1/[NO_2]$ against time, we first make the following calculations from the data given:

Time (s)	$[NO_2]$ (M)	$\ln[NO_2]$	$1/[NO_2]$ (1/M)
0.0	0.01000	−4.605	100
50.0	0.00787	−4.845	127
100.0	0.00649	−5.037	154
200.0	0.00481	−5.337	208
300.0	0.00380	−5.573	263

Continued

As Figure 14.8 shows, only the plot of $1/[NO_2]$ versus time is linear. Thus, the reaction obeys a second-order rate law: Rate = $k[NO_2]^2$. From the slope of this straight-line graph, we determine that $k = 0.543\ M^{-1}s^{-1}$ for the disappearance of NO_2.

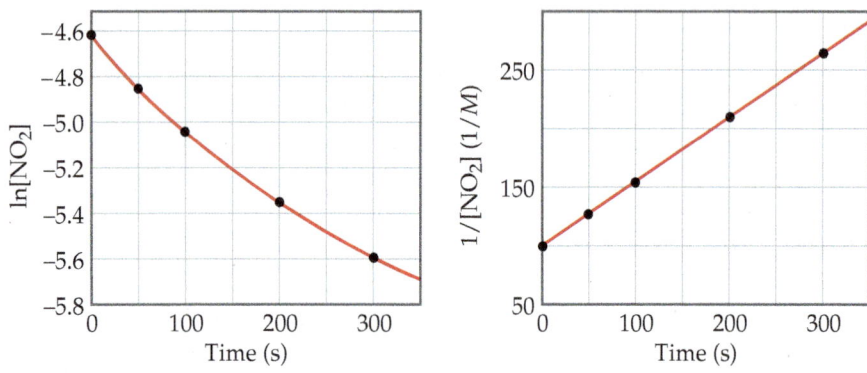

▲ **Figure 14.8** Kinetic data for decomposition of NO_2.

▶ **Practice Exercise**

The decomposition of NO_2 discussed in this Sample Exercise is second order in NO_2 with $k = 0.543\ M^{-1}s^{-1}$. If the initial concentration of NO_2 in a closed vessel is 0.0500 M, what is the concentration of this reactant after 0.500 h?

Zero-Order Reactions

The concentration of a reactant A decreases nonlinearly in a first-order reaction, as shown by the red curve in **Figure 14.9**. As [A] declines, the *rate* at which it disappears declines in proportion. A **zero-order reaction** is one in which the rate of disappearance of A is *independent* of [A]. The rate law for a zero-order reaction is

$$\text{Rate} = \frac{-\Delta[A]}{\Delta t} = k \qquad [14.15]$$

The integrated rate law for a zero-order reaction is

$$[A]_t = -kt + [A]_0 \qquad [14.16]$$

where $[A]_t$ is the concentration of A at time t and $[A]_0$ is the initial concentration. This is the equation for a straight line with vertical intercept $[A]_0$ and slope $-kt$, as shown in the blue curve in Figure 14.9.

Go Figure At which times during the reaction would you have trouble distinguishing a zero-order reaction from a first-order reaction?

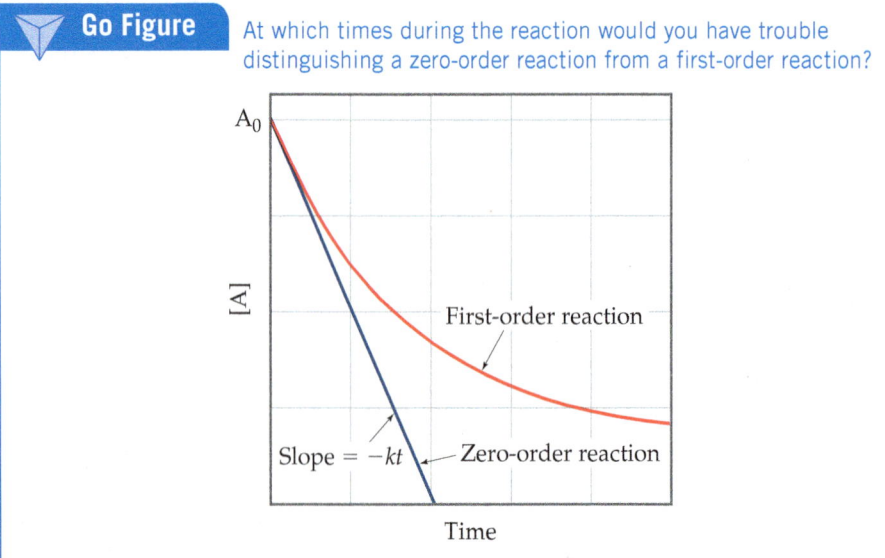

▲ **Figure 14.9** Comparison of first-order and zero-order reactions for the disappearance of reactant A with time.

The most common type of zero-order reaction occurs when a gas undergoes decomposition on the surface of a solid. If the surface is completely covered by decomposing molecules, the rate of reaction is constant because the number of reacting surface molecules is constant, as long as there is some gas-phase substance left.

Half-Life

The **half-life** of a reaction, $t_{1/2}$, is the time required for the concentration of a reactant to reach half its initial value, $[A]_{t_{1/2}} = \frac{1}{2}[A]_0$. Half-life is a convenient way to describe how fast a reaction occurs, especially if it is a first-order process. A fast reaction has a short half-life.

We can determine the half-life of a first-order reaction by substituting $[A]_{t_{1/2}} = \frac{1}{2}[A]_0$ for $[A]_t$ and $t_{1/2}$ for t in Equation 14.12:

$$\ln\frac{\frac{1}{2}[A]_0}{[A]_0} = -kt_{1/2}$$

$$\ln\tfrac{1}{2} = -kt_{1/2}$$

$$t_{1/2} = -\frac{\ln\tfrac{1}{2}}{k} = \frac{0.693}{k} \qquad [14.17]$$

For a first-order rate law, then, Equation 14.17 shows that $t_{1/2}$ does *not* depend on the initial concentration of any reactant. Consequently, the half-life remains constant throughout the reaction. If, for example, the initial concentration of a reactant is 0.120 M, then it will be $\frac{1}{2}(0.120\ M) = 0.060\ M$ after one half-life. After one more half-life passes, the concentration will drop to 0.030 M, and so on. Equation 14.17 also indicates that, for a first-order reaction, we can calculate $t_{1/2}$ if we know k, and we can calculate k if we know $t_{1/2}$.

The change in concentration over time for the first-order rearrangement of gaseous methyl isonitrile at 199 °C is graphed in **Figure 14.10**. Because the concentration of this gas is directly proportional to its pressure during the reaction, we have chosen to plot pressure rather than concentration in this graph. The first half-life occurs at 13,600 s (3.78 h). At a time 13,600 s later, the methyl isonitrile pressure (and therefore, the concentration) has decreased to half of one-half, or one-fourth, of the initial value.

In a first-order reaction, the concentration of the reactant decreases by one-half *in each of a series of regularly spaced time intervals, where each interval is equal to $t_{1/2}$.*

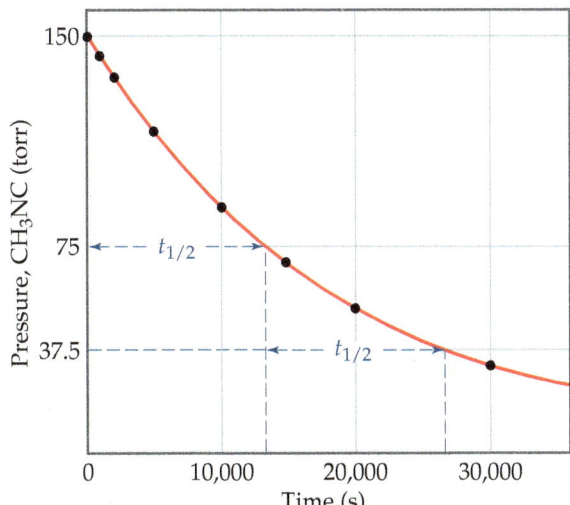

▲ **Figure 14.10** Kinetic data for the gas-phase rearrangement of methyl isonitrile to acetonitrile at 199 °C, showing the half-life of the reaction.

CHEMISTRY AND SUSTAINABILITY   Methyl Bromide in the Atmosphere

The compounds known as chlorofluorocarbons (CFCs) are responsible for the destruction of Earth's ozone layer, which, as we discuss in Chapter 18, protects us from harmful ultraviolet radiation. Another simple molecule that has the potential to destroy the stratospheric ozone layer is methyl bromide, CH_3Br (Figure 14.11). Because it has a wide range of uses, including antifungal treatment of plant seeds, CH_3Br has been produced in large quantities in the past (about 150 million pounds per year worldwide in 1997, at the height of its production). In the stratosphere, the C—Br bond is broken through absorption of short-wavelength (that is,

high-energy) radiation. The resultant Br atoms then catalyze the decomposition of O_3.

Methyl bromide is removed from the lower atmosphere by a variety of mechanisms, including a slow reaction with ocean water:

$$CH_3Br(g) + H_2O(l) \longrightarrow CH_3OH(aq) + HBr(aq) \quad [14.18]$$

To determine the potential importance of CH_3Br in destruction of the ozone layer, it is important to know how rapidly the reaction in Equation 14.18 and all other reactions remove CH_3Br from the lower atmosphere before it can diffuse into the stratosphere.

The average lifetime of CH_3Br in Earth's lower atmosphere is difficult to measure because the conditions that exist in the atmosphere are too complex to be simulated in the laboratory. Instead, scientists analyzed nearly 4000 atmospheric samples collected above the Pacific Ocean for the presence of several trace organic substances, including methyl bromide. From these measurements, it was possible to estimate the *atmospheric residence time* for CH_3Br.

The atmospheric residence time is related to the half-life for CH_3Br in the lower atmosphere, assuming CH_3Br decomposes by a first-order process. From the experimental data, the half-life for methyl bromide in the lower atmosphere is estimated to be 0.8 ± 0.1 yr. That is, a collection of CH_3Br molecules present at any given time will, on average, be 50% decomposed after 0.8 yr, 75% decomposed after 1.6 yr, and so on. A half-life of 0.8 yr, while comparatively short, is still sufficiently long so that CH_3Br contributes significantly to the destruction of the ozone layer.

In 1997 an international agreement was reached to phase out use of methyl bromide in developed countries by 2005. Although exemptions for critical agricultural use have been granted, global consumption since the phase-out is only a small fraction of the levels in the early 1990s.

Related Exercise: 14.122

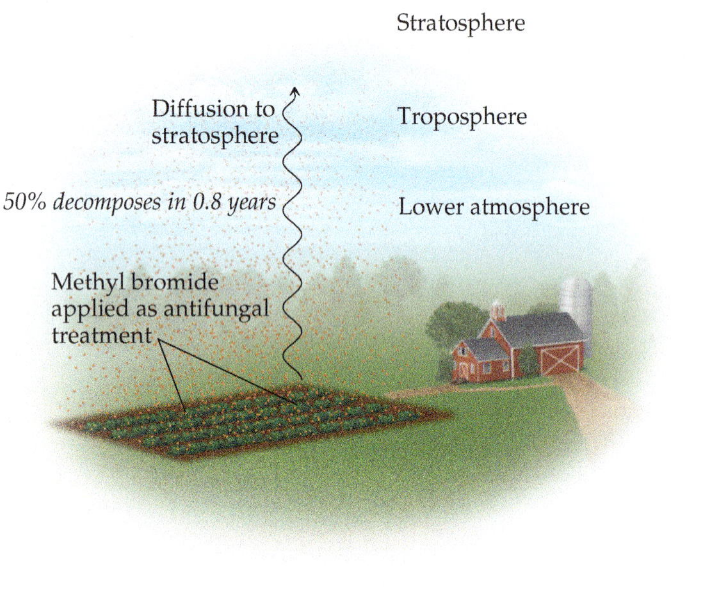

Stratosphere

Troposphere

Diffusion to stratosphere

50% decomposes in 0.8 years

Lower atmosphere

Methyl bromide applied as antifungal treatment

▲ **Figure 14.11** **Distribution and fate of methyl bromide in Earth's atmosphere.**

Sample Exercise 14.9

Determining the Half-Life of a First-Order Reaction

The reaction of C_4H_9Cl with water is a first-order reaction. (**a**) Use Figure 14.3 to estimate the half-life for this reaction. (**b**) Use the half-life from (a) to calculate the rate constant.

SOLUTION

Analyze We are asked to estimate the half-life of a reaction from a graph of concentration versus time and then to use the half-life to calculate the rate constant for the reaction.

Plan

(**a**) To estimate a half-life, we can select a concentration and then determine the time required for the concentration to decrease to half of that value.

(**b**) Equation 14.17 is used to calculate the rate constant from the half-life.

Solve

(**a**) From the graph, we see that the initial value of $[C_4H_9Cl]$ is 0.100 M. The half-life for this first-order reaction is the time required for $[C_4H_9Cl]$ to decrease to 0.050 M, which we can read off the graph. This point occurs at approximately 340 s.

(**b**) Solving Equation 14.17 for k, we have

$$k = \frac{0.693}{t_{1/2}} = \frac{0.693}{340 \text{ s}} = 2.0 \times 10^{-3} \text{ s}^{-1}$$

Check At the end of the second half-life, which should occur at 680 s, the concentration should have decreased by yet another factor of 2, to 0.025 M. Inspection of the graph shows that this is indeed the case.

▶ **Practice Exercise**

(**a**) Using Equation 14.17, calculate $t_{1/2}$ for the decomposition of the insecticide described in Sample Exercise 14.7.

(**b**) How long does it take for the concentration of the insecticide to reach one-quarter of the initial value?

The half-life for second-order and other reactions depends on reactant concentrations and therefore changes as the reaction progresses. We obtained Equation 14.17 for the half-life for a first-order reaction by substituting $[A]_{t_{1/2}} = \frac{1}{2}[A]_0$ for $[A]_t$ and $t_{1/2}$ for t in Equation 14.12. We find the half-life of a second-order reaction by making the same substitutions into Equation 14.14:

$$\frac{1}{\frac{1}{2}[A]_0} = kt_{1/2} + \frac{1}{[A]_0}$$

$$\frac{2}{[A]_0} - \frac{1}{[A]_0} = kt_{1/2}$$

$$t_{1/2} = \frac{1}{k[A]_0} \qquad [14.19]$$

In this case, the half-life depends on the initial concentration of reactant—the lower the initial concentration, the longer the half-life.

 ## Self-Assessment Exercises

SAE 14.8 Consider the hypothetical reaction A $\longrightarrow$ B. If the rate law is second order in [A], which of the following linear plots is expected?

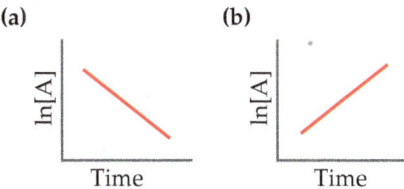

(a)   (b)

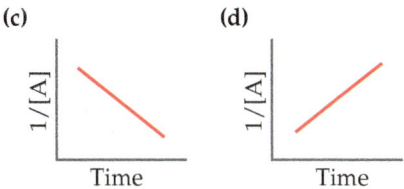

(c)   (d)

SAE 14.9 The decomposition of N_2O_5 is given by the reaction $2\,N_2O_5 \longrightarrow 4\,NO_2 + O_2$. The rate law is first order in N_2O_5. At 64 °C, the rate constant is $4.82 \times 10^{-3}\,s^{-1}$. If you start with $0.100\,M\,N_2O_5$ at this temperature, what is the concentration of N_2O_5 after 120 s? (**a**) 0.0013 *M* (**b**) 0.056 *M* (**c**) 0.095 *M* (**d**) 0.18 *M*

SAE 14.10 In 6 *M* HCl, the complex ion $[Ru(NH_3)_6]^{3+}$ decomposes to a variety of products. The reaction is first order in $[Ru(NH_3)_6]^{3+}$ and has a rate constant of 0.0495 h^{-1} at 250 °C. If the initial concentration of $[Ru(NH_3)_6]^{3+}$ is 0.100 *M*, what is the half-life of $[Ru(NH_3)_6]^{3+}$ at 250 °C? (**a**) 0.0714 h (**b**) 14.0 h (**c**) 20.2 h (**d**) 202 h

14.4 | Temperature and Rate: Activation Energy and the Arrhenius Equation

 ## Learning Objectives

When you finish Section 14.4, you should be able to:

▶ Use the collision model to explain how temperature influences the rate of a reaction.

▶ Interpret reaction profiles that illustrate the relationship between potential energy and reaction progress.

▶ Use the Arrhenius equation to calculate activation energy and rate constants at different temperatures.

The rates of most chemical reactions increase as the temperature rises. For example, dough rises faster at room temperature than when refrigerated, and plants grow more rapidly in warm weather than in cold. We can see the effect of temperature on reaction rate by observing a chemiluminescence reaction (one that produces light), such as that in Cyalume® light sticks (**Figure 14.12**).

How is this experimentally observed temperature effect reflected in the rate law? The faster rate at higher temperature is due to an increase in the rate constant with increasing temperature. For example, let's reconsider the first-order reaction we saw in Figure 14.6—namely,

Hot water   Cold water

▲ **Figure 14.12 Temperature affects the rate of the chemiluminescence reaction in light sticks:** The chemiluminescent reaction occurs more rapidly in hot water, and more light is produced.

Go Figure Would you expect this curve to eventually go back down to lower values? Why or why not?

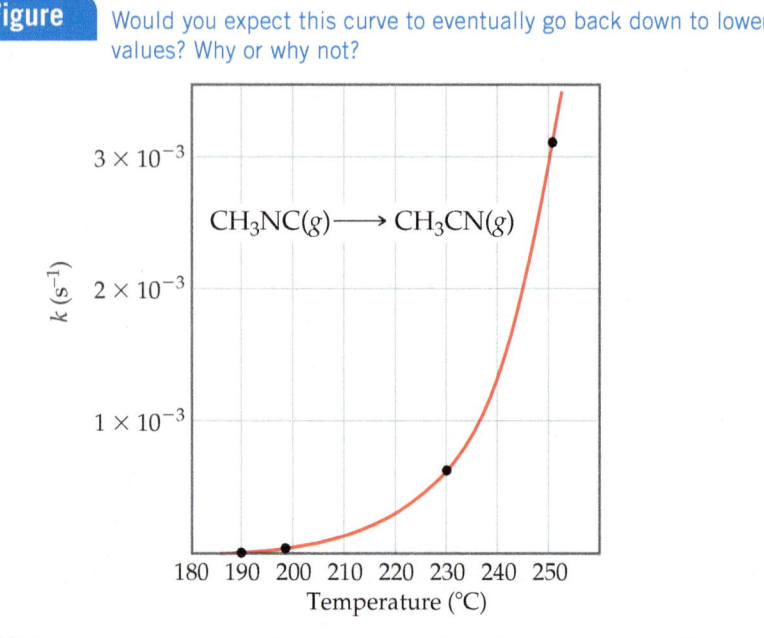

▲ **Figure 14.13 Temperature dependence of the rate constant for methyl isonitrile conversion to acetonitrile.** The four points indicated are used in Sample Exercise 14.11.

$CH_3NC \longrightarrow CH_3CN$. **Figure 14.13** shows the rate constant for this reaction as a function of temperature. The rate constant and, hence, the rate of the reaction, increase rapidly with temperature, approximately doubling for each 10 °C rise.

The Collision Model

Reaction rates are affected by both reactant concentrations and temperature. The **collision model**, based on the kinetic-molecular theory (Section 10.5), accounts for both of these effects at the molecular level. The central idea of the collision model is that molecules must collide to react. The greater the number of collisions per second, the greater the reaction rate. As reactant concentration increases, therefore, the number of collisions increases, leading to an increase in reaction rate. According to the kinetic-molecular theory of gases, increasing the temperature increases molecular speeds. As molecules move faster, they collide more forcefully (with more energy) and more frequently, both of which increase the reaction rate.

For a reaction to occur, though, more is required than simply a collision—it must be the right kind of collision. For most reactions, in fact, only a tiny fraction of collisions leads to a reaction. For example, in a mixture of H_2 and I_2 at ordinary temperatures and pressures, each molecule undergoes about 10^{10} collisions per second. If every collision between H_2 and I_2 resulted in the formation of HI, the reaction would be over in much less than a second. Instead, at room temperature the reaction proceeds very slowly because only about one in every 10^{13} collisions produces a reaction. What keeps the reaction from occurring more rapidly?

The Orientation Factor

In most reactions, collisions between molecules result in a chemical reaction only if the molecules are oriented in a certain way when they collide. The relative orientations of the molecules during collision determine whether the atoms are suitably positioned to form new bonds. For example, the reaction

$$Cl + NOCl \longrightarrow NO + Cl_2$$

takes place if the collision brings Cl atoms together to form Cl_2, as shown in the top panel of **Figure 14.14**. In contrast, the two Cl atoms are *not* colliding directly with one another in the lower panel, so no products are formed.

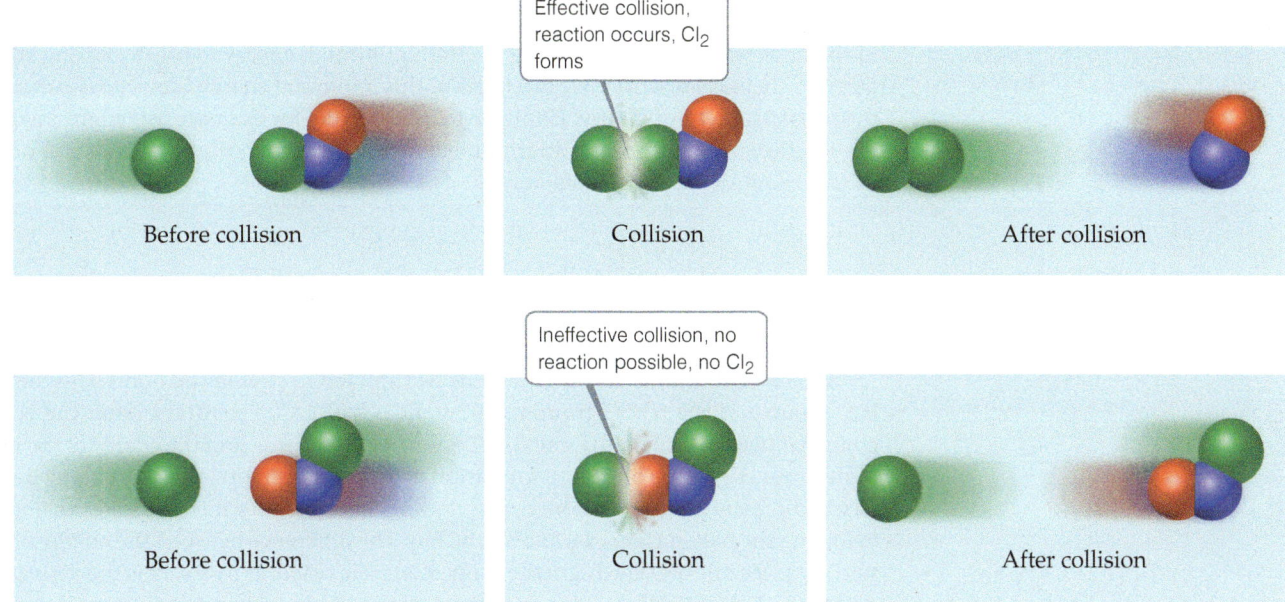

▲ **Figure 14.14** **Molecular collisions may or may not lead to a chemical reaction between Cl and NOCl.**

Activation Energy

Molecular orientation is not the only factor influencing whether a molecular colli-sion will produce a reaction. In 1888 the Swedish chemist Svante Arrhenius suggested that molecules must possess a certain minimum amount of energy to react. According to the collision model, this energy comes from the kinetic energies of the colliding molecules. Upon collision, the kinetic energy of the molecules can be used to stretch, bend, and ultimately break bonds, leading to chemical reactions. That is, the kinetic energy is used to change the potential energy of the molecule. If molecules are moving too slowly—in other words, with too little kinetic energy—then they merely bounce off one another without changing. The minimum energy required to initiate a chem-ical reaction is called the **activation energy**, E_a, and its value varies from reaction to reaction.

The situation during reactions is analogous to that shown in **Figure 14.15**, where the golfer hits the ball to make it roll over the hill in the direction of the cup. The hill is a *barrier* between ball and cup. To reach the cup, the player must impart enough kinetic

Go Figure Which of the following determines how hard the golfer must strike the ball: the difference in elevation between the ball and the hole or between the ball and the top of the barrier?

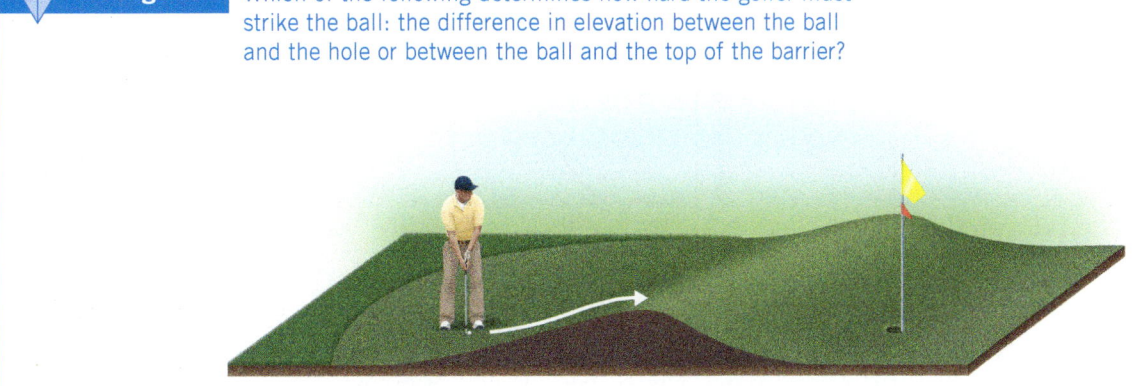

▲ **Figure 14.15** **Energy is needed to overcome a barrier between initial and final states.**

energy with the putter to move the ball to the top of the barrier. If they do not impart enough energy, the ball will roll partway up the hill and then back down toward them. In the same way, molecules require a certain minimum energy to break existing bonds during a chemical reaction. We can think of this minimum energy as an *energy barrier*. In the rearrangement of methyl isonitrile to acetonitrile, for example, we might imagine the reaction passing through an intermediate state in which the N≡C portion of the methyl isonitrile molecule is sideways:

$$H_3C-N\equiv C: \longrightarrow \left[H_3C\cdots \overset{\ddot{C}}{\underset{\ddot{N}}{|||}} \right] \longrightarrow H_3C-C\equiv N:$$

Figure 14.16 shows that energy must be supplied to stretch the bond between the H_3C group and the N≡C group to allow the N≡C group to rotate. After the N≡C group has twisted sufficiently, the C—C bond begins to form, and the energy of the molecule drops. Thus, the barrier to formation of acetonitrile represents the energy necessary to force the molecule through the relatively unstable intermediate state, analogous to forcing the ball in Figure 14.15 over the hill. The difference between the energy of the starting molecule and the highest energy along the reaction pathway is the activation energy, E_a. The molecule having the arrangement of atoms shown at the top of the barrier is called either the **activated complex** or the **transition state**.

The conversion of $H_3C-N\equiv C$ to $H_3C-C\equiv N$ is exothermic. Figure 14.16 therefore shows the product as having a lower energy than the reactant. The overall energy change for the reaction, ΔE, has no effect on reaction rate, however. In general:

The rate constant depends on the magnitude of E_a; usually, the lower the value of E_a is, the larger the rate constant and the faster the reaction.

Notice that the reverse reaction is endothermic. The activation energy for the reverse reaction is equal to the energy that must be overcome if approaching the barrier from the right: $\Delta E + E_a$. Thus, to reach the activated complex for the reverse reaction requires more energy than for the forward reaction—for this reaction, there is a larger barrier to overcome going from right to left than from left to right.

Any particular methyl isonitrile molecule acquires sufficient energy to overcome the energy barrier through collisions with other molecules. Recall from the kinetic-molecular

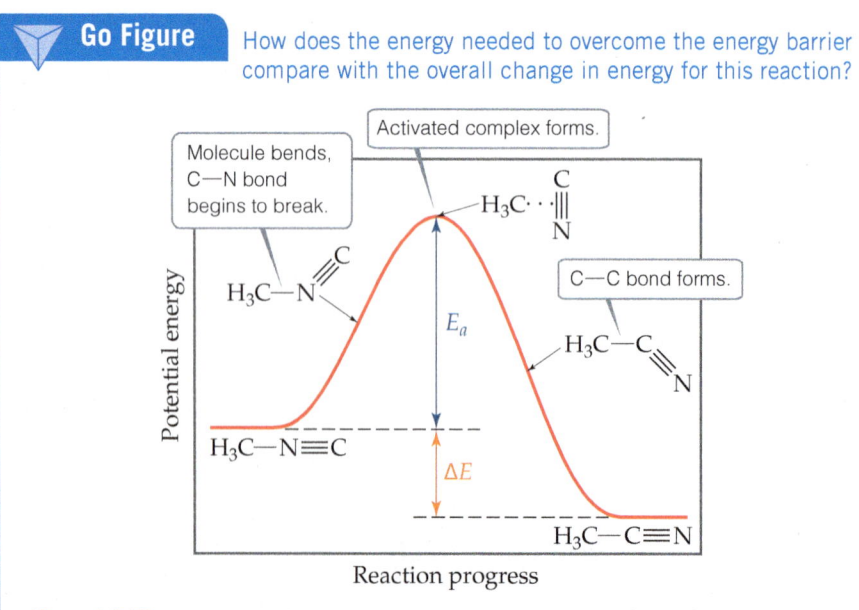

Go Figure How does the energy needed to overcome the energy barrier compare with the overall change in energy for this reaction?

▲ **Figure 14.16** Energy profile for the conversion of methyl isonitrile (H_3CNC) to its isomer acetonitrile (H_3CCN).

What would the curve look like for a temperature higher than that for the red curve in the figure?

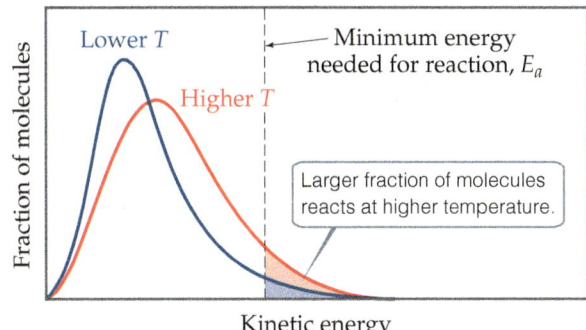

▲ **Figure 14.17** The effect of temperature on the distribution of kinetic energies of molecules in a sample.

theory of gases that, at any instant, gas molecules are distributed in energy over a wide range. (Section 10.5) **Figure 14.17** shows the distribution of kinetic energies for two temperatures, comparing them with the minimum energy needed for reaction, E_a. At the higher temperature, a much greater fraction of the molecules has kinetic energy greater than E_a, which leads to a greater rate of reaction.

For a collection of molecules in the gas phase, the fraction of molecules that have kinetic energy equal to or greater than E_a is given by the expression

$$f = e^{-E_a/RT} \qquad [14.20]$$

In this equation, R is the gas constant (8.314 J/mol-K) and T is the absolute temperature. To get an idea of the magnitude of f, let's suppose that E_a is 100 kJ/mol, a value typical of many reactions, and that T is 300 K. The calculated value of f is 3.9×10^{-18}, an extremely small number! At 320 K, $f = 4.7 \times 10^{-17}$. Thus, only a 20° increase in temperature produces a more than tenfold increase in the fraction of molecules possessing at least 100 kJ/mol of energy.

The Arrhenius Equation

Svante Arrhenius (1859–1927) noted that for most reactions the increase in rate with increasing temperature is nonlinear (Figure 14.13). He found that most reaction-rate data obeyed an equation based on (a) the fraction of molecules possessing energy E_a or greater, (b) the number of collisions per second, and (c) the fraction of collisions that have the appropriate orientation. These three factors are incorporated into the **Arrhenius equation**:

$$k = Ae^{-E_a/RT} \qquad [14.21]$$

In this equation, k is the rate constant, E_a is the activation energy, R is the gas constant (8.314 J/mol-K), and T is the absolute temperature. The **frequency factor**, A, is constant, or nearly so, as temperature is varied. This factor (also known as the *pre-exponential factor*) is related to the frequency of collisions and the probability that the collisions are favorably oriented for reaction.* As the magnitude of E_a increases, k decreases because the fraction of molecules that possess the required energy is smaller. Thus, at fixed values of T and A, *the rate constant decreases as E_a increases*.

---

*Because collision frequency increases with temperature, A also has some temperature dependence, but this dependence is much smaller than the exponential term. Therefore, A is considered to be approximately constant.

Sample Exercise 14.10
Activation Energies and Speeds of Reaction

Consider a series of reactions having the following energy profiles:

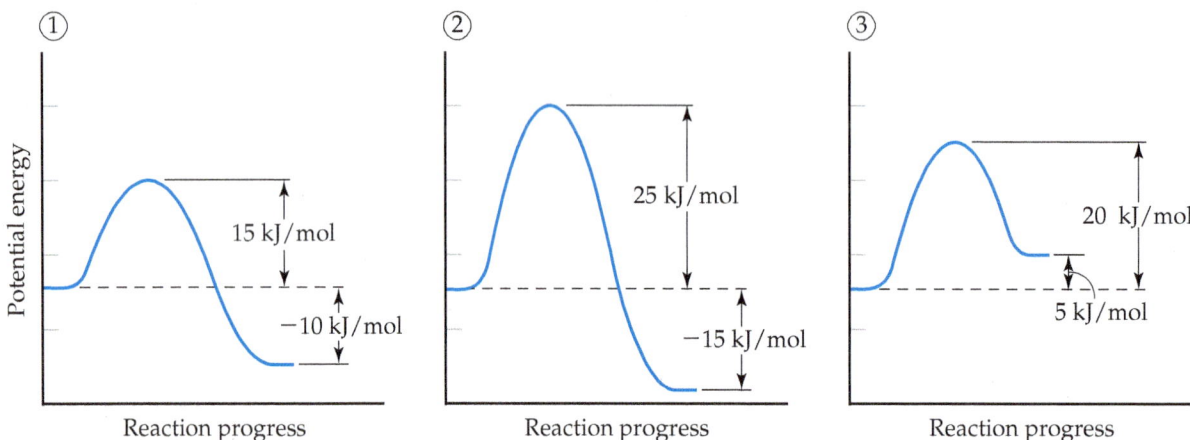

Rank the forward rate constants from smallest to largest, assuming all three reactions have nearly the same value for the frequency factor A.

SOLUTION

The lower the activation energy, the larger the rate constant and the faster the reaction. The value of ΔE does not affect the value of the rate constant. Hence, the order of the rate constants is $2 < 3 < 1$.

▶ **Practice Exercise**

Rank the rate constants of the reverse reactions from slowest to fastest.

Determining the Activation Energy

We can calculate the activation energy for a reaction by manipulating the Arrhenius equation. Taking the natural log of both sides of Equation 14.21, we obtain

$$\ln k = \ln A e^{-E_a/RT}$$

$$\ln k = \ln e^{-E_a/RT} + \ln A$$

$$\ln k = -\frac{E_a}{RT} + \ln A$$

$$\underset{y}{\downarrow} \quad = \quad \underset{mx}{\downarrow} \quad + \quad \underset{b}{\downarrow}$$

[14.22]

which has the form of the equation for a straight line. A graph of $\ln k$ versus $1/T$ is a line with a slope equal to $-E_a/R$ and a y-intercept equal to $\ln A$. Thus, the activation energy can be determined by measuring k at a series of temperatures, graphing $\ln k$ versus $1/T$, and calculating E_a from the slope of the resultant line.

We can also use Equation 14.22 to evaluate E_a in a nongraphical way if we know the rate constant of a reaction at two or more temperatures. For example, suppose that at two different temperatures T_1 and T_2 a reaction has rate constants k_1 and k_2. For each condition, we have

$$\ln k_1 = -\frac{E_a}{RT_1} + \ln A \quad \text{and} \quad \ln k_2 = -\frac{E_a}{RT_2} + \ln A$$

Subtracting $\ln k_2$ from $\ln k_1$ gives

$$\ln k_1 - \ln k_2 = \left(-\frac{E_a}{RT_1} + \ln A\right) - \left(-\frac{E_a}{RT_2} + \ln A\right)$$

Assuming temperature-dependent changes in the frequency factor A are negligible, we can simplify to obtain

$$\ln\frac{k_1}{k_2} = \frac{E_a}{R}\left(\frac{1}{T_2} - \frac{1}{T_1}\right)$$

[14.23]

Equation 14.23 provides a convenient way to calculate a rate constant k_1 at some temperature T_1 when we know the activation energy and the rate constant k_2 at some other temperature T_2.

Sample Exercise 14.11
Determining the Activation Energy

The table at right shows the rate constants for the rearrangement of methyl isonitrile at various temperatures (these are the data points in Figure 14.13):

(a) From these data, calculate the activation energy for the reaction.

(b) What is the value of the rate constant at 430.0 K?

Temperature (°C)	$k\,(s^{-1})$
189.7	2.52×10^{-5}
198.9	5.25×10^{-5}
230.3	6.30×10^{-4}
251.2	3.16×10^{-3}

SOLUTION

Analyze We are given rate constants, k, measured at several temperatures, and we are asked to determine the activation energy, E_a, and the rate constant, k, at a particular temperature.

Plan We can obtain E_a from the slope of a graph of $\ln k$ versus $1/T$. Once we know E_a, we can use Equation 14.23 together with the given rate data to calculate the rate constant at 430.0 K.

Solve

(a) We must first convert the temperatures from degrees Celsius to kelvins. We then take the inverse of each temperature, $1/T$, and the natural log of each rate constant, $\ln k$. This gives us the following table:

$T\,(K)$	$1/T\,(K^{-1})$	$\ln k$
462.9	2.160×10^{-3}	-10.589
472.1	2.118×10^{-3}	-9.855
503.5	1.986×10^{-3}	-7.370
524.4	1.907×10^{-3}	-5.757

A graph of $\ln k$ versus $1/T$ is a straight line (Figure 14.18).

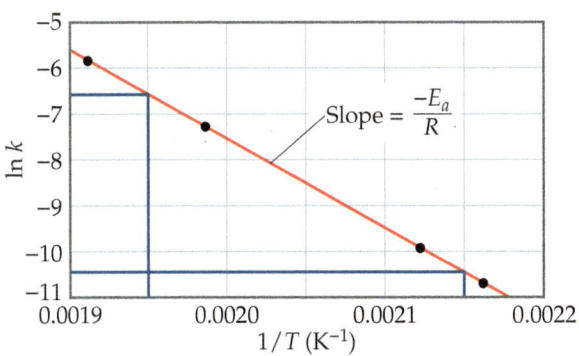

▲ **Figure 14.18** Graphical determination of activation energy E_a.

The slope of the line is obtained by choosing any two well-separated points and using the coordinates of each:

$$\text{Slope} = \frac{\Delta y}{\Delta x} = \frac{-6.6 - (-10.4)}{0.00195 - 0.00215} = -1.9 \times 10^4$$

Because logarithms have no units, the numerator in this equation is dimensionless. The denominator has the units of $1/T$—namely, K^{-1}. Thus, the overall units for the slope are K. The slope equals $-E_a/R$. We use the value for the gas constant R in units of J/mol-K (Table 10.2). We thus obtain

$$\text{Slope} = -\frac{E_a}{R}$$

$$E_a = -(\text{slope})(R) = -(-1.9 \times 10^4\,K)\left(8.314\frac{J}{\text{mol-K}}\right)\left(\frac{1\,kJ}{1000\,J}\right)$$

$$= 1.6 \times 10^2\,kJ/mol = 160\,kJ/mol$$

We report the activation energy to only two significant figures because we are limited by the precision with which we can read the graph in Figure 14.18.

(b) To determine the rate constant, k_1, at $T_1 = 430.0$ K, we can use Equation 14.23 with $E_a = 160$ kJ/mol and one of the rate constants and temperatures from the given data, such as $k_2 = 2.52 \times 10^{-5}\,s^{-1}$ and $T_2 = 462.9$ K:

$$\ln\left(\frac{k_1}{2.52 \times 10^{-5}\,s^{-1}}\right) =$$

$$\left(\frac{160\,kJ/mol}{8.314\,J/\text{mol-K}}\right)\left(\frac{1}{462.9\,K} - \frac{1}{430.0\,K}\right)\left(\frac{1000\,J}{1\,kJ}\right) = -3.18$$

Thus,

$$\frac{k_1}{2.52 \times 10^{-5}\,s^{-1}} = e^{-3.18} = 4.15 \times 10^{-2}$$

$$k_1 = (4.15 \times 10^{-2})(2.52 \times 10^{-5}\,s^{-1}) = 1.0 \times 10^{-6}\,s^{-1}$$

Note that the units of k_1 are the same as those of k_2 and the rate constant at 430.0 K is smaller than it is at 462.9 K, as it should be.

▶ **Practice Exercise**
To one significant figure, what is the value for the frequency factor A for the data presented in this Sample Exercise?

▲ **Self-Assessment Exercises**

SAE 14.11   The rate of a reaction typically doubles for each 10 °C rise in temperature. Which statement best describes *why* the rate increases? (**a**) The number of collisions per second increases. (**b**) More reactant molecules have kinetic energy greater than the activation energy. (**c**) The activation energy of the reaction decreases. (**d**) The frequency factor, *A*, increases.

SAE 14.12   Consider plots 1 and 2 from Sample Exercise 14.10:

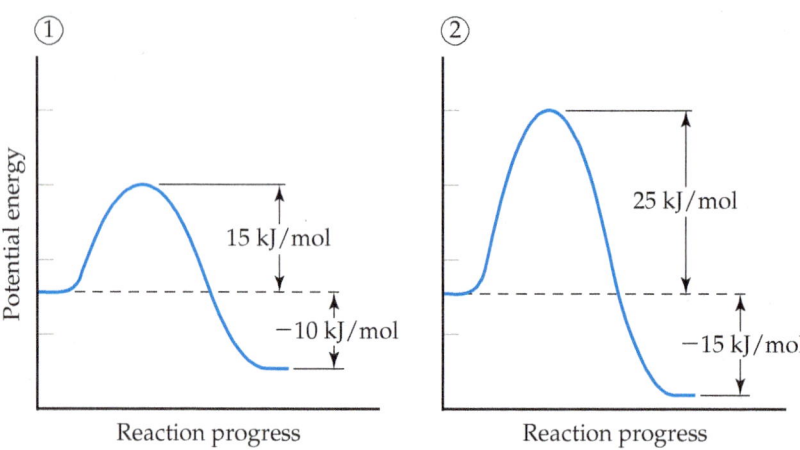

Assuming the collision factors are comparable, which reaction will have the smallest rate constant?

(**a**) forward reaction 1   (**c**) forward reaction 2
(**b**) reverse reaction 1   (**d**) reverse reaction 2

SAE 14.13   The activation energy of a first-order reaction is 83.5 kJ/mol. The rate constant is $3.54 \times 10^{-5}\,\text{s}^{-1}$ at 45 °C. What is the rate constant at 65 °C?

(**a**) $5.46 \times 10^{-6}\,\text{s}^{-1}$ (**c**) $2.29 \times 10^{-4}\,\text{s}^{-1}$
(**b**) $3.55 \times 10^{-5}\,\text{s}^{-1}$ (**d**) $2.36 \times 10^{25}\,\text{s}^{-1}$

14.5 | Reaction Mechanisms

A balanced equation for a chemical reaction indicates the substances present at the start of the reaction and those present at the end of the reaction. It provides no information, however, about the detailed steps that occur at the molecular level as the reactants are turned into products. The steps by which a reaction occurs is called the **reaction mechanism**. At the most sophisticated level, a reaction mechanism describes the order in which bonds are broken and formed and the changes in relative positions of the atoms in the course of the reaction.

Elementary Reactions

We have seen that reactions take place because of collisions between reacting molecules. For example, the collisions between molecules of methyl isonitrile (CH_3NC) can provide the energy that allows CH_3NC to rearrange to acetonitrile:

Similarly, the reaction of NO and O_3 to form NO_2 and O_2 appears to occur as a result of a single collision involving suitably oriented and sufficiently energetic NO and O_3 molecules:

$$NO(g) + O_3(g) \longrightarrow NO_2(g) + O_2(g) \qquad [14.24]$$

Both reactions occur in a single event or step and are called **elementary reactions**.

▲ **Learning Objectives**

When you have finished Section 14.5, **you should be able to:**

▶ Describe the molecularity and rate laws of elementary reactions.

▶ Derive the rate law and overall reaction equation for a multistep reaction from the reaction mechanism and the relative rates of the elementary steps.

▶ Identify any intermediates and predict the rate law based on the relative rates of the elementary reactions when given a reaction mechanism.

The number of molecules that participate as reactants in an elementary reaction defines the **molecularity** of the reaction. If a single molecule is involved, the reaction is **unimolecular**. The rearrangement of methyl isonitrile is a unimolecular process. Elementary reactions involving the collision of two reactant molecules are **bimolecular**. The reaction between NO and O_3 is bimolecular. Elementary reactions involving the simultaneous collision of three molecules are **termolecular**. Termolecular reactions are far less probable than unimolecular or bimolecular processes and are extremely rare. The chance that four or more molecules will collide simultaneously with any regularity is even more remote. Consequently, such collisions are never proposed as part of a reaction mechanism. Thus, nearly all reaction mechanisms contain only unimolecular and/or bimolecular elementary reactions.

Multistep Mechanisms

The net change represented by a balanced chemical equation often occurs by a *multistep mechanism* consisting of a sequence of elementary reactions. For example, below 225 °C, the reaction

$$NO_2(g) + CO(g) \longrightarrow NO(g) + CO_2(g) \qquad [14.25]$$

appears to proceed in two elementary reactions (or two *elementary steps*), each of which is bimolecular. First, two NO_2 molecules collide, and an oxygen atom is transferred from one to the other. The resultant NO_3 then collides with a CO molecule and transfers an oxygen atom to it:

$$NO_2(g) + NO_2(g) \longrightarrow NO_3(g) + NO(g)$$
$$NO_3(g) + CO(g) \longrightarrow NO_2(g) + CO_2(g)$$

Thus, we say that the reaction occurs by a two-step mechanism. In general:

> *The chemical equations for the elementary reactions in a multistep mechanism must always add to give the chemical equation of the overall process.*

In the present example, the sum of the two elementary reactions is

$$2\,NO_2(g) + NO_3(g) + CO(g) \longrightarrow NO_2(g) + NO_3(g) + NO(g) + CO_2(g)$$

Simplifying this equation by eliminating substances that appear on both sides gives Equation 14.25, the net equation for the process.

Because NO_3 is neither a reactant nor a product of the reaction—it is formed in one elementary reaction and consumed in the next—it is called an **intermediate**. Multistep mechanisms involve one or more intermediates. Intermediates are not the same as transition states, as shown in **Figure 14.19**. Intermediates can be stable and can therefore

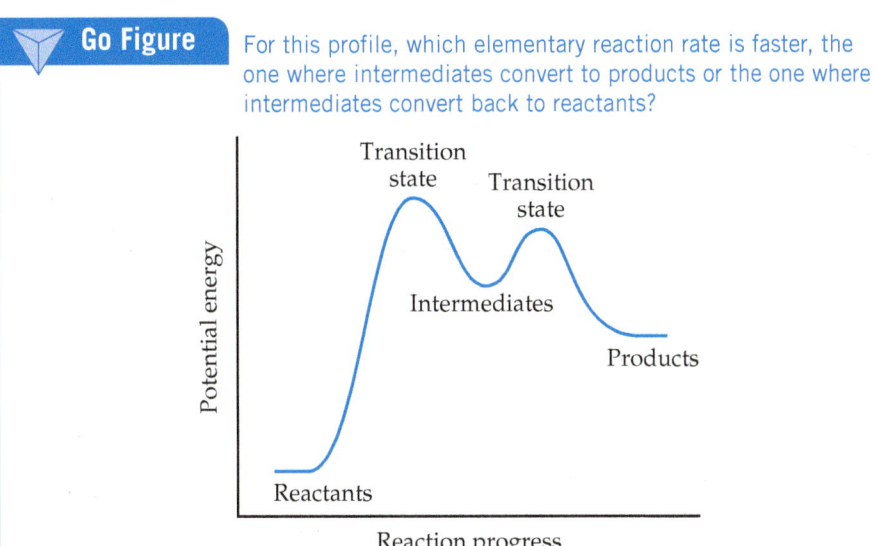

Go Figure For this profile, which elementary reaction rate is faster, the one where intermediates convert to products or the one where intermediates convert back to reactants?

▲ Figure 14.19 **The energy profile of a reaction, showing transition states and intermediates.**

sometimes be identified and even isolated. Transition states, on the other hand, are always inherently unstable and as such can never be isolated. Nevertheless, the use of advanced "ultrafast" techniques sometimes allows us to characterize them.

Sample Exercise 14.12
Determining Molecularity and Identifying Intermediates

It has been proposed that the conversion of ozone into O_2 proceeds by a two-step mechanism:

$$O_3(g) \longrightarrow O_2(g) + O(g)$$
$$O_3(g) + O(g) \longrightarrow 2\,O_2(g)$$

(a) Describe the molecularity of each elementary reaction in this mechanism.
(b) Write the equation for the overall reaction.
(c) Identify the intermediate(s).

SOLUTION

Analyze We are given a two-step mechanism and asked for (a) the molecularities of each of the two elementary reactions, (b) the equation for the overall process, and (c) the intermediate.

Plan The molecularity of each elementary reaction depends on the number of reactant molecules in the equation for that reaction. The overall equation is the sum of the equations for the elementary reactions. The intermediate is a substance formed in one step of the mechanism and used in another and therefore not part of the equation for the overall reaction.

Solve

(a) The first elementary reaction involves a single reactant and is consequently unimolecular. The second reaction, which involves two reactant molecules, is bimolecular.

(b) Adding the two elementary reactions gives

$$2\,O_3(g) + O(g) \longrightarrow 3\,O_2(g) + O(g)$$

Because $O(g)$ appears in equal amounts on both sides of the equation, it can be eliminated to give the net equation for the chemical process:

$$2\,O_3(g) \longrightarrow 3\,O_2(g)$$

(c) The intermediate is $O(g)$. It is neither an original reactant nor a final product but is formed in the first step of the mechanism and consumed in the second.

▶ **Practice Exercise**
For the reaction

$$Mo(CO)_6 + P(CH_3)_3 \longrightarrow Mo(CO)_5 P(CH_3)_3 + CO$$

the proposed mechanism is

$$Mo(CO)_6 \longrightarrow Mo(CO)_5 + CO$$
$$Mo(CO)_5 + P(CH_3)_3 \longrightarrow Mo(CO)_5 P(CH_3)_3$$

(a) Is the proposed mechanism consistent with the equation for the overall reaction? (b) What is the molecularity of each step of the mechanism? (c) Identify the intermediate(s).

Rate Laws for Elementary Reactions

In Section 14.2, we stressed that rate laws must be determined experimentally; they cannot be predicted from the coefficients of balanced chemical equations. We are now in a position to understand why this is so. Every reaction is made up of a series of one or more elementary steps, and the rate laws and relative speeds of these steps dictate the overall rate law for the reaction. Indeed, the rate law for a reaction can be determined from its mechanism, as we explain shortly, and it can be compared with the experimental rate law. Thus, our next challenge in kinetics is to arrive at reaction mechanisms that lead to rate laws consistent with those observed experimentally. We start by examining the rate laws of elementary reactions.

Elementary reactions are significant in a very important way: *If a reaction is elementary, its rate law is based directly on its molecularity.* For example, consider the unimolecular reaction

$$A \longrightarrow products$$

As the number of A molecules increases, the number that reacts in a given time interval increases proportionally. Thus, the rate of a unimolecular process is first order:

$$Rate = k[A]$$

TABLE 14.3   Elementary Reactions and Their Rate Laws

Molecularity	Elementary Reaction	Rate Law
*Uni*molecular	A $\longrightarrow$ products	Rate = $k[A]$
*Bi*molecular	A + A $\longrightarrow$ products	Rate = $k[A]^2$
*Bi*molecular	A + B $\longrightarrow$ products	Rate = $k[A][B]$
*Ter*molecular	A + A + A $\longrightarrow$ products	Rate = $k[A]^3$
*Ter*molecular	A + A + B $\longrightarrow$ products	Rate = $k[A]^2[B]$
*Ter*molecular	A + B + C $\longrightarrow$ products	Rate = $k[A][B][C]$

For bimolecular elementary steps, the rate law is second order, as in the reaction

$$A + B \longrightarrow \text{products} \quad \text{Rate} = k[A][B]$$

The second-order rate law follows directly from collision theory. If we double the concentration of A, the number of collisions between the molecules of A and B doubles; likewise, if we double [B], the number of collisions between A and B doubles. Therefore, the rate law is first order in both [A] and [B] and second order overall.

The rate laws for all feasible elementary reactions are listed in **Table 14.3**. Notice how each rate law follows directly from the molecularity of the reaction. It is important to remember, however, that we cannot tell by merely looking at a balanced, overall chemical equation whether the reaction involves one or several elementary steps.

Sample Exercise 14.13
Predicting the Rate Law for an Elementary Reaction

If the following reaction occurs in a single elementary reaction, predict its rate law:

$$H_2(g) + Br_2(g) \longrightarrow 2\,HBr(g)$$

SOLUTION

Analyze We are given the equation and are asked for its rate law, assuming that it is an elementary process.

Plan Because we are assuming that the reaction occurs as a single elementary reaction, we are able to write the rate law using the coefficients for the reactants in the equation as the reaction orders.

Solve The reaction is bimolecular, involving one molecule of H_2 and one molecule of Br_2. Thus, the rate law is first order in each reactant and second order overall:

$$\text{Rate} = k[H_2][Br_2]$$

Comment Experimental studies of this reaction show that the reaction actually has a different rate law:

$$\text{Rate} = k[H_2][Br_2]^{1/2}$$

Because the experimental rate law differs from the one obtained by assuming a single elementary reaction, we can conclude that the mechanism cannot occur by a single elementary step. It must, therefore, involve two or more elementary steps.

▶ **Practice Exercise**
Consider the following reaction: $2\,NO(g) + Br_2(g) \longrightarrow 2\,NOBr(g)$. **(a)** Write the rate law for the reaction, assuming it involves a single elementary reaction. **(b)** Is a single-step mechanism likely for this reaction?

The Rate-Determining Step for a Multistep Mechanism

As with the reaction in Sample Exercise 14.13, most reactions occur by mechanisms that involve two or more elementary reactions. Each step of the mechanism has its own rate constant and activation energy. Often one step is much slower than the others, and the overall rate of a reaction cannot exceed the rate of the slowest elementary step. Because

Go Figure   In scenario (a), would increasing the rate at which cars pass through toll plaza B increase the rate at which cars go from point 1 to point 3?

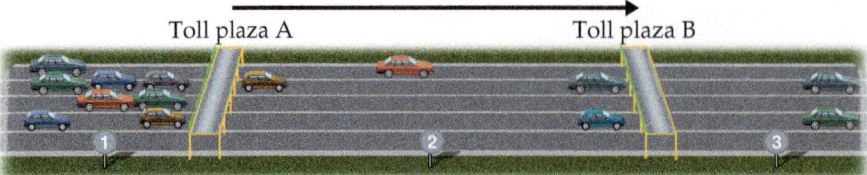

(a) Cars slowed at toll plaza A, but not at B; rate-determining step is passage through A

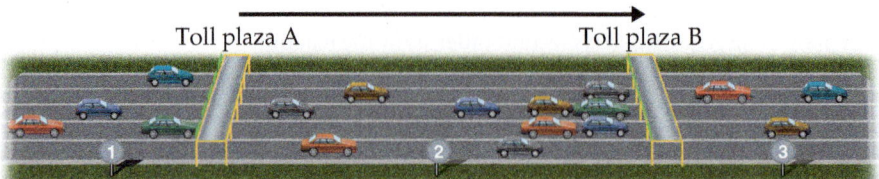

(b) Cars slowed at toll plaza B, but not at A; rate-determining step is passage through B

▲ **Figure 14.20 Rate-determining steps in traffic flow on a toll road.**

the slow step limits the overall reaction rate, it is called the **rate-determining step** (or *rate-limiting step*).

To understand the concept of the rate-determining step for a reaction, consider a toll road with two toll plazas (**Figure 14.20**). Cars enter the toll road at point 1 and pass through toll plaza A. They then pass an intermediate point 2 before passing through toll plaza B and arriving at point 3. We can envision this trip along the toll road as occurring in two elementary steps:

Step 1:	Point 1 ⟶ Point 2	(through toll plaza A)
Step 2:	Point 2 ⟶ Point 3	(through toll plaza B)
Overall:	Point 1 ⟶ Point 3	(through both toll plazas)

Now suppose that one or more gates at toll plaza A are malfunctioning, so that traffic backs up behind the gates, as depicted in Figure 14.20(a). The rate at which cars can get to point 3 is limited by the rate at which they can get through the traffic jam at plaza A. Thus, step 1 is the rate-determining step of the journey along the toll road. If, however, all gates at A are functioning but one or more at B are not, traffic flows quickly through A but gets backed up at B, as depicted in Figure 14.20(b). In this case, step 2 is the rate-determining step.

In the same way, *the slowest step in a multistep reaction determines the overall rate.* By analogy to Figure 14.20(a), the rate of a fast step following the rate-determining step does not speed up the overall rate. If the slow step is not the first one, as is the case in Figure 14.20(b), the faster preceding steps produce intermediate products that accumulate before being consumed in the slow step. In either case, *the rate-determining step governs the rate law for the overall reaction.*

Mechanisms with a Slow Initial Step

We can most easily see the relationship between the slow step in a mechanism and the rate law for the overall reaction by studying an example in which the first step in a multistep mechanism is the rate-determining step. Consider the reaction of NO_2 and CO to produce NO and CO_2 (Equation 14.25). Below 225 °C, it is found experimentally that the rate law for this reaction is second order in NO_2 and zero order in CO: Rate = $k[NO_2]^2$.

Can we propose a reaction mechanism consistent with this rate law? Consider the two-step mechanism:*

$$\text{Step 1:} \quad NO_2(g) + NO_2(g) \xrightarrow{k_1} NO_3(g) + NO(g) \quad \text{(slow)}$$
$$\text{Step 2:} \quad NO_3(g) + CO(g) \xrightarrow{k_2} NO_2(g) + CO_2(g) \quad \text{(fast)}$$
$$\text{Overall:} \quad NO_2(g) + CO(g) \longrightarrow NO(g) + CO_2(g)$$

Step 2 is much faster than step 1; that is, $k_2 \gg k_1$, telling us that the intermediate $NO_3(g)$ is slowly produced in step 1 and immediately consumed in step 2.

Because step 1 is slow and step 2 is fast, step 1 is the rate-determining step. Thus, the rate of the overall reaction depends on the rate of step 1, and the rate law of the overall reaction equals the rate law of step 1. Step 1 is a bimolecular process that has the rate law

$$\text{Rate} = k_1[NO_2]^2$$

Thus, the rate law predicted by this mechanism agrees with the one observed experimentally. The reactant CO is absent from the rate law because it reacts in a step that follows the rate-determining step.

A scientist would not, at this point, say that we have "proved" that this mechanism is correct. All we can say is that the rate law predicted by the mechanism is *consistent with experiment*. We can often envision a different sequence of steps that leads to the same rate law. If, however, the predicted rate law of the proposed mechanism *disagrees* with the experiment, we know for certain that the mechanism cannot be correct.

Sample Exercise 14.14
Determining the Rate Law for a Multistep Mechanism

The decomposition of nitrous oxide, N_2O, is believed to occur by a two-step mechanism:

$$N_2O(g) \longrightarrow N_2(g) + O(g) \quad \text{(slow)}$$
$$N_2O(g) + O(g) \longrightarrow N_2(g) + O_2(g) \quad \text{(fast)}$$

(**a**) Write the equation for the overall reaction. (**b**) Write the rate law for the overall reaction.

SOLUTION

Analyze Given a multistep mechanism with the relative speeds of the steps, we are asked to write the overall reaction and the rate law for that overall reaction.

Plan (**a**) Find the overall reaction by adding the elementary steps and eliminating the intermediates. (**b**) The rate law for the overall reaction will be that of the slow, rate-determining step.

Solve

(**a**) Adding the two elementary reactions gives

$$2 N_2O(g) + O(g) \longrightarrow 2 N_2(g) + O_2(g) + O(g)$$

Omitting the intermediate, $O(g)$, which occurs on both sides of the equation, gives the overall reaction:

$$2 N_2O(g) \longrightarrow 2 N_2(g) + O_2(g)$$

(**b**) The rate law for the overall reaction is just the rate law for the slow, rate-determining elementary reaction. Because that slow step is a unimolecular elementary reaction, the rate law is first order:

$$\text{Rate} = k[N_2O]$$

▶ **Practice Exercise**

Ozone reacts with nitrogen dioxide to produce dinitrogen pentoxide and oxygen:

$$O_3(g) + 2 NO_2(g) \longrightarrow N_2O_5(g) + O_2(g)$$

The reaction is believed to occur in two steps:

$$O_3(g) + NO_2(g) \longrightarrow NO_3(g) + O_2(g)$$
$$NO_3(g) + NO_2(g) \longrightarrow N_2O_5(g)$$

The experimental rate law is rate $= k[O_3][NO_2]$. What can you say about the relative rates of the two steps of the mechanism?

---

*Note the rate constants k_1 and k_2 written above the reaction arrows. The subscript on each rate constant identifies the elementary step involved. Thus, k_1 is the rate constant for step 1, and k_2 is the rate constant for step 2. A negative subscript refers to the rate constant for the reverse of an elementary step. For example, k_{-1} is the rate constant for the reverse of the first step.

Mechanisms with a Fast Initial Step

It is possible, but not particularly straightforward, to derive the rate law for a mechanism in which an intermediate is a reactant in the rate-determining step. This situation arises in multistep mechanisms when the first step is fast and therefore *not* the rate-determining step. The gas-phase reaction of nitric oxide (NO) with bromine (Br_2) is one example:

$$2\,NO(g) + Br_2(g) \longrightarrow 2\,NOBr(g) \tag{14.26}$$

The experimentally determined rate law for this reaction is second order in NO and first order in Br_2:

$$Rate = k[NO]^2[Br_2] \tag{14.27}$$

We seek a reaction mechanism that is consistent with this rate law. One possibility is that the reaction occurs in a single termolecular step:

$$NO(g) + NO(g) + Br_2(g) \longrightarrow 2\,NOBr(g) \quad Rate = k[NO]^2[Br_2] \tag{14.28}$$

This seems unlikely, however, because termolecular processes are so rare.

Consider, instead, the following alternative mechanism, which does not involve a termolecular step:

$$\text{Step 1:} \qquad NO(g) + Br_2(g) \underset{k_{-1}}{\overset{k_1}{\rightleftharpoons}} NOBr_2(g) \qquad \text{(fast)}$$

$$\text{Step 2:} \quad NOBr_2(g) + NO(g) \xrightarrow{k_2} 2\,NOBr(g) \quad \text{(slow)} \tag{14.29}$$

In this mechanism, step 1 involves two processes: a forward reaction and its reverse.

Because step 2 is the rate-determining step, the rate law for that step governs the rate of the overall reaction:

$$Rate = k_2[NOBr_2][NO] \tag{14.30}$$

Note that $NOBr_2$ is an intermediate generated in the forward reaction of step 1. Intermediates are usually unstable and have a low, unknown concentration. Thus, the rate law of Equation 14.30 depends on the unknown concentration of an intermediate, which isn't desirable. We want instead to express the rate law for a reaction in terms of the reactants, or the products if necessary, of the reaction.

With the aid of some assumptions, we can express the concentration of the intermediate $NOBr_2$ in terms of the concentrations of the starting reactants NO and Br_2. We first assume that $NOBr_2$ is unstable and does not accumulate to any significant extent in the reaction mixture. Once formed, $NOBr_2$ can be consumed either by reacting with NO to form NOBr or by falling back apart into NO and Br_2. The first of these possibilities is step 2 of our alternative mechanism, a slow process. The second is the reverse of step 1, a unimolecular process:

$$NOBr_2(g) \xrightarrow{k_{-1}} NO(g) + Br_2(g)$$

Because step 2 is slow, we assume that most of the $NOBr_2$ falls apart according to the reverse of step 1. Thus, we have both the forward and reverse reactions of step 1 occurring much faster than step 2. Because they occur rapidly relative to step 2, the forward and reverse reactions of step 1 establish an equilibrium. As in any other dynamic equilibrium, the rate of the forward reaction equals that of the reverse reaction:

$$\underset{\text{Rate of forward reaction}}{k_1[NO][Br_2]} = \underset{\text{Rate of reverse reaction}}{k_{-1}[NOBr_2]}$$

Solving for $[NOBr_2]$, we have

$$[NOBr_2] = \frac{k_1}{k_{-1}}[NO][Br_2]$$

Substituting this relationship into Equation 14.30, we have

$$\text{Rate} = k_2 \frac{k_1}{k_{-1}}[\text{NO}][\text{Br}_2][\text{NO}] = k[\text{NO}]^2[\text{Br}_2]$$

where the experimental rate constant k equals $k_2 k_1 / k_{-1}$. This expression is consistent with the experimental rate law (Equation 14.27). Thus, our alternative mechanism (Equation 14.29), which involves two steps but only unimolecular and bimolecular processes, is far more probable than the single-step termolecular mechanism of Equation 14.28. This assumption, that an intermediate has a relatively constant concentration during most of the reaction, is also called the *steady-state assumption* or the *steady-state approximation*:

> *In general, whenever a fast step precedes a slow one, we can solve for the concentration of an intermediate by assuming that an equilibrium is established in the fast step.*

Sample Exercise 14.15

Deriving the Rate Law for a Mechanism with a Fast Initial Step

Show that the following mechanism for Equation 14.26 also produces a rate law consistent with the experimentally observed one:

Step 1: $\text{NO}(g) + \text{NO}(g) \underset{k_{-1}}{\overset{k_1}{\rightleftharpoons}} \text{N}_2\text{O}_2(g)$ (fast, equilibrium)

Step 2: $\text{N}_2\text{O}_2(g) + \text{Br}_2(g) \overset{k_2}{\longrightarrow} 2\,\text{NOBr}(g)$ (slow)

SOLUTION

Analyze We are given a mechanism with a fast initial step, and we are asked to write the rate law for the overall reaction.

Plan The rate law of the slow elementary step in a mechanism determines the rate law for the overall reaction. Thus, we first write the rate law based on the molecularity of the slow step. In this case, the slow step involves the intermediate N_2O_2 as a reactant. Experimental rate laws, however, do not contain the concentrations of intermediates; instead, they are expressed in terms of the concentrations of reactants, and in some cases products. Thus, we must relate the concentration of N_2O_2 to the concentration of NO by assuming that an equilibrium is established in the first step.

Solve The second step is rate determining, so the overall rate is

$$\text{Rate} = k_2[\text{N}_2\text{O}_2][\text{Br}_2]$$

We solve for the concentration of the intermediate N_2O_2 by assuming that an equilibrium is established in step 1; thus, the rates of the forward and reverse reactions in step 1 are equal:

$$k_1[\text{NO}]^2 = k_{-1}[\text{N}_2\text{O}_2]$$

Solving for the concentration of the intermediate, N_2O_2, gives

$$[\text{N}_2\text{O}_2] = \frac{k_1}{k_{-1}}[\text{NO}]^2$$

Substituting this expression into the rate expression gives

$$\text{Rate} = k_2 \frac{k_1}{k_{-1}}[\text{NO}]^2[\text{Br}_2] = k[\text{NO}]^2[\text{Br}_2]$$

Notice that we have combined the three rate constants into one generic final rate constant. Thus, this mechanism also yields a rate law consistent with the experimental one. Remember: There may be more than one mechanism that leads to an observed experimental rate law!

▶ **Practice Exercise**
The first step of a mechanism involving the reaction of bromine is

$$\text{Br}_2(g) \underset{k_{-1}}{\overset{k_1}{\rightleftharpoons}} 2\,\text{Br}(g) \quad \text{(fast, equilibrium)}$$

What is the expression relating the concentration of Br(g) to that of $\text{Br}_2(g)$?

So far we have considered only three reaction mechanisms: one for a reaction that occurs in a single elementary step and two for simple multistep reactions where there is one rate-determining step. There are other more complex mechanisms. If you take a biochemistry class, for example, you will learn about cases in which the concentration of an intermediate cannot be neglected in deriving the rate law. Furthermore, some mechanisms require a large number of steps, sometimes 35 or more, to arrive at a rate law that agrees with experimental data!

A CLOSER LOOK | Diffusion-Controlled Reactions and Activation-Controlled Reactions

Imagine that an overall reaction $A + B \longrightarrow C$ occurs via an activated complex AB:

$$A + B \underset{k_{-1}}{\overset{k_1}{\rightleftharpoons}} AB \overset{k_2}{\longrightarrow} C$$

Because a dynamic equilibrium leads to the activated complex AB, the rate constant for forming AB is given by k_1, whereas the rate constant for AB falling apart back into A and B is k_{-1}. The rate constant for AB forming product C is given by k_2. The values for k_1, k_{-1}, and k_2 will be governed by many factors, including how tightly the activated complex AB is held together, and the rate at which A and B collide to form AB.

In solution, the rate at which A and B collide depends not only on their concentrations, but also on *diffusion*, the process in which molecules move from a position of high concentration to one of lower concentration, leading ultimately to the same equilibrium concentration throughout the sample. Diffusion has its own rate and activation energy for molecules making their way through the solvent. The *diffusion coefficient* is a quantitative measure of the ability of a solute to diffuse in a solvent. It is somewhat analogous to a rate constant in the sense that a larger diffusion coefficient means the molecule is moving faster (just like a larger rate constant means a reaction is faster). Thus, large objects

in high-viscosity (that is, molasses-like) solvents have very low diffusion coefficients. Conversely, small objects in low-viscosity liquids, such as small molecules in most organic solvents, have large diffusion coefficients and are thus able to move relatively large distances per time. In water, small molecules such as oxygen and methanol have diffusion coefficients of $\sim 10^{-5} \text{ cm}^2/\text{s}$ at room temperature. Aluminum atoms in copper metal, in contrast, have a diffusion coefficient of $10^{-30} \text{ cm}^2/\text{s}$ at room temperature.

There are two limiting cases for reactions: *diffusion-controlled* and *activation-controlled* (also called *reaction-controlled*). In a diffusion-controlled reaction, $k_2 \gg k_{-1}$, so the reaction rate is limited by k_1—that is, by how fast molecules A and B diffuse through the solution and collide to form AB. A classic test to determine if a reaction is diffusion-controlled is to change the stirring rate of the reaction: If the rate increases as the stirring rate is increased, then the reaction is likely diffusion-controlled. In an activation-controlled reaction, E_a is large (more than 20 kJ/mol), in which case $k_2 \ll k_{-1}$. Thus, the rate of reaction is governed by the activation energy. Most of the reactions we discuss in this book are activation-controlled. In your future studies, however, you may encounter reactions, such as certain biomolecular processes that involve enzymes, which are diffusion-controlled.

Self-Assessment Exercises

SAE 14.14 Consider the following energy profile.

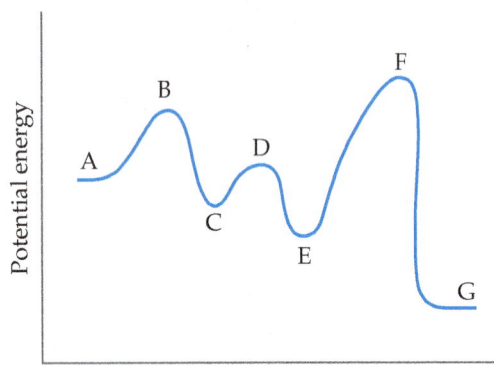

Which of the following statements is or are *true*?

  (i) C and E are intermediates.
 (ii) B, D, and F are transition states.
(iii) This is a three-step mechanism.

(a) Only one of the statements is true. **(b)** Statements i and ii are true. **(c)** Statements i and iii are true. **(d)** Statements ii and iii are true. **(e)** All three statements are true.

SAE 14.15 The formation of $2 NO_2(g)$ from $NO(g)$ and $O_2(g)$ is believed to occur by a two-step mechanism:

$$NO(g) + O_2(g) \rightleftharpoons NO_3(g) \text{ (fast)}$$
$$NO_3(g) + NO(g) \longrightarrow 2 NO_2(g) \text{ (slow)}$$

Which of the following rate laws is consistent with this mechanism?
(a) Rate $= k[NO]^2[O_2]$ **(c)** Rate $= k[NO][O_2]$
(b) Rate $= k[NO][O_2]^2$ **(d)** Rate $= k[NO_3][NO]$

SAE 14.16 When the reaction $2 NO(g) + Cl_2(g) \longrightarrow 2 NOCl(g)$ was run, the following data were obtained under conditions of constant $[Cl_2]$:

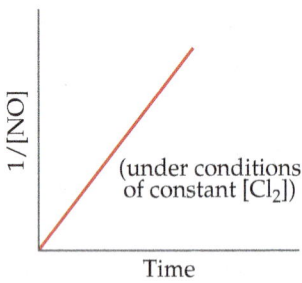

Which of the following mechanisms is or are consistent with the data?

 (i) $2 NO(g) \rightleftharpoons N_2O_2(g)$ (fast)

      $N_2O_2(g) + Cl_2(g) \longrightarrow 2 NOCl(g)$ (slow)

(ii) $NO(g) + Cl_2(g) \longrightarrow NOCl(g) + Cl(g)$ (slow)

      $NO(g) + Cl(g) \longrightarrow NOCl(g)$ (fast)

(a) only i **(b)** only ii **(c)** both i and ii **(d)** neither i nor ii

14.6 | Catalysis

A **catalyst** is a substance that changes the speed of a chemical reaction without undergoing a permanent chemical change itself. Most reactions in an organism (its metabolism), the atmosphere, and the oceans occur with the help of catalysts. Much industrial chemical research is devoted to the search for more effective catalysts for reactions of commercial importance. Extensive research efforts also are devoted to finding means of inhibiting or removing certain catalysts that promote undesirable reactions, such as those that corrode metals, age our bodies, and cause tooth decay.

Homogeneous Catalysis

A catalyst that is present in the same phase as the reactants in a reaction mixture is called a **homogeneous catalyst**. Examples abound both in solution and in the gas phase. Consider, for example, the decomposition of aqueous hydrogen peroxide, $H_2O_2(aq)$, into water and oxygen:

$$2 H_2O_2(aq) \longrightarrow 2 H_2O(l) + O_2(g) \qquad [14.31]$$

In the absence of a catalyst, this reaction occurs extremely slowly. Many substances are capable of catalyzing the reaction, however, including bromide ion, which reacts with hydrogen peroxide in acidic solution, forming aqueous bromine and water (**Figure 14.21**).

$$2 Br^-(aq) + H_2O_2(aq) + 2 H^+(aq) \longrightarrow Br_2(aq) + 2 H_2O(l) \qquad [14.32]$$

Learning Objectives

When you have finished Section 14.6, you should be able to:

▶ Identify species that act as catalysts from inspection of the reaction mechanism.

▶ Distinguish between homogeneous and heterogeneous catalysts.

▶ Describe the role of enzymes in living organisms.

Go Figure Which species is responsible for the brownish color in the middle cylinder: H_2O_2, Br_2, Na^+, Br^-, or O_2? Is the brown substance a catalyst or an intermediate?

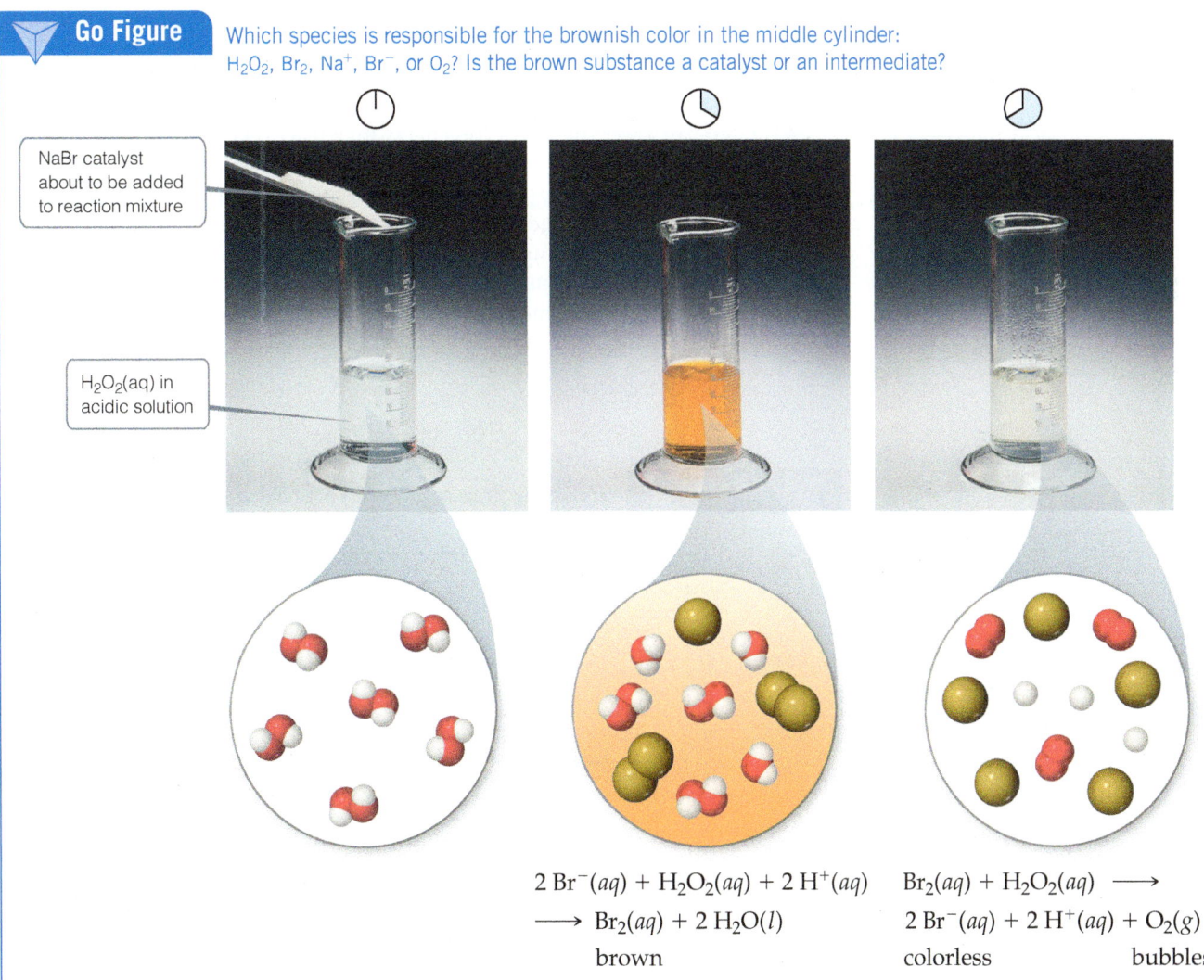

NaBr catalyst about to be added to reaction mixture

$H_2O_2(aq)$ in acidic solution

$2 Br^-(aq) + H_2O_2(aq) + 2 H^+(aq)$
$\longrightarrow Br_2(aq) + 2 H_2O(l)$
brown

$Br_2(aq) + H_2O_2(aq) \longrightarrow$
$2 Br^-(aq) + 2 H^+(aq) + O_2(g)$
colorless   bubbles

▲ **Figure 14.21 Homogeneous catalysis.** The effect of a catalyst on the speed of hydrogen peroxide decomposition to water and oxygen gas.

If this were the complete reaction, bromide ion would not be a catalyst because it undergoes chemical change during the reaction. However, hydrogen peroxide also reacts with the $Br_2(aq)$ generated in Equation 14.32:

$$Br_2(aq) + H_2O_2(aq) \longrightarrow 2\,Br^-(aq) + 2\,H^+(aq) + O_2(g) \qquad [14.33]$$

The sum of Equations 14.32 and 14.33 is just Equation 14.31, a result that you should check for yourself.

When the H_2O_2 has been completely decomposed, we are left with a colorless solution of $Br^-(aq)$, which means that this ion is indeed a catalyst of the reaction because it speeds up the reaction without itself undergoing any net change. In contrast, Br_2 is an intermediate because it is first formed (Equation 14.32) and then consumed (Equation 14.33). The color change we see in Figure 14.22 illustrates that the presence of intermediates can in some cases be readily detected. Neither the catalyst nor the intermediate appears in the equation for the overall reaction. Notice, however, that *the catalyst is present at the start of the reaction, whereas the intermediate is formed during the course of the reaction.*

How does a catalyst work? Based on the general form of rate laws (Equation 14.7, rate $= k[A]^m[B]^n$), the catalyst must affect the numerical value of k, the rate constant. On the basis of the Arrhenius equation (Equation 14.21, $k = Ae^{-E_a/RT}$), k is determined by the activation energy (E_a) and the frequency factor (A). A catalyst may affect the rate of reaction by altering the value of either E_a or A. We can envision this happening in two ways: The catalyst could provide a new mechanism for the reaction that has an E_a value lower than the E_a value for the uncatalyzed reaction, or the catalyst could assist in the orientation of reactants and so increase A. The most dramatic catalytic effects come from lowering E_a. As a general rule, *a catalyst lowers the overall activation energy for a chemical reaction.*

A catalyst can lower the activation energy for a reaction by providing a different mechanism for the reaction. In the decomposition of hydrogen peroxide, for example, two successive reactions of H_2O_2, first with bromide and then with bromine, take place. Because these two reactions together serve as a catalytic pathway for hydrogen peroxide decomposition, *both* of them must have significantly lower activation energies than the uncatalyzed decomposition (**Figure 14.22**). And because the catalyzed reaction can follow a different series of steps in its mechanism, the rate law for the catalyzed reaction might be different than that of the uncatalyzed reaction.

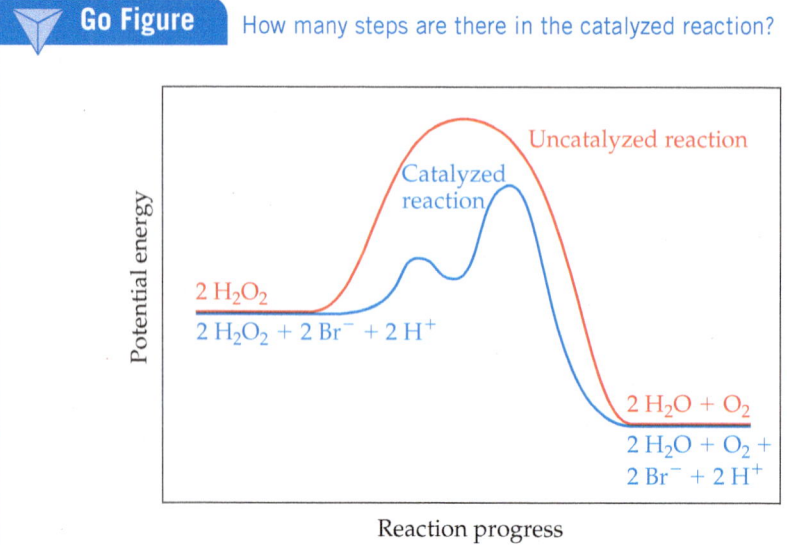

▲ **Figure 14.22** Energy profiles for the uncatalyzed and bromide-catalyzed decomposition of H_2O_2.

Heterogeneous Catalysis

A **heterogeneous catalyst** is one that exists in a phase different from the phase of the reactant molecules, usually as a solid in contact with either gaseous reactants or reactants in a liquid solution. Many industrially important reactions are catalyzed by the surfaces of solids. For example, raw petroleum is transformed into smaller hydrocarbon molecules by using what are called "cracking" catalysts. Heterogeneous catalysts are often composed of metals or metal oxides.

The initial step in heterogeneous catalysis is usually **adsorption** of reactants. **Ad**sorption refers to the binding of molecules to a surface, whereas **ab**sorption refers to the uptake of molecules into the interior of a substance. (Section 13.6) Adsorption occurs because the atoms or ions at the surface of a solid are extremely reactive. Because the catalyzed reaction occurs on the surface, special methods are often used to prepare catalysts so that they have very large surface areas. Unlike their counterparts in the interior of the substance, surface atoms and ions have unused bonding capacity that can be used to bond molecules from the gas or solution phase to the surface of the solid.

The reaction of hydrogen gas with ethylene gas to form ethane gas provides an example of heterogeneous catalysis:

$$\underset{\text{Ethylene}}{C_2H_4(g)} + H_2(g) \longrightarrow \underset{\text{Ethane}}{C_2H_6(g)} \qquad \Delta H° = -137 \text{ kJ/mol} \qquad [14.34]$$

Even though this reaction is exothermic, it occurs very slowly in the absence of a catalyst. In the presence of a finely powdered metal, however, such as nickel, palladium, or platinum, the reaction occurs easily at room temperature via the mechanism diagrammed in **Figure 14.23**. Both ethylene and hydrogen are adsorbed on the metal surface. Upon adsorption, the H—H bond of H_2 breaks, leaving two H atoms initially bonded to the metal surface but relatively free to move. When a hydrogen encounters an adsorbed ethylene molecule, it can form a σ bond to one of the carbon atoms, effectively destroying the C—C π bond and leaving an *ethyl group* (C_2H_5) bonded to the surface via a metal-to-carbon σ bond. This σ bond is relatively weak, so when the other carbon atom also encounters a hydrogen atom, a sixth C—H σ bond is readily formed, and an ethane molecule (C_2H_6) is released from the metal surface.

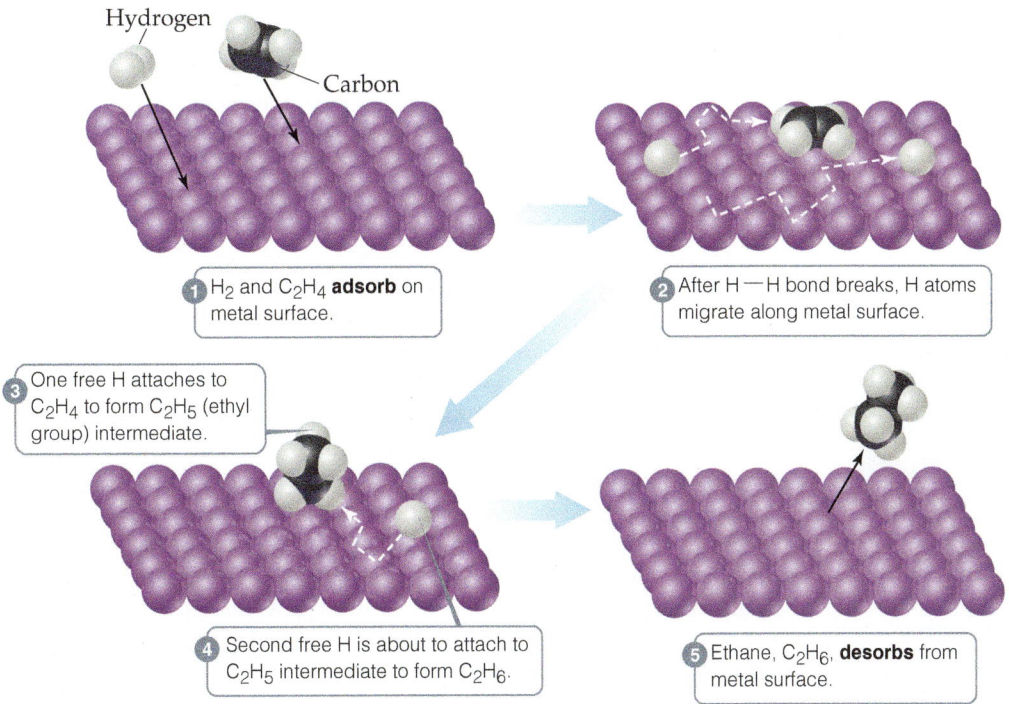

① H_2 and C_2H_4 **adsorb** on metal surface.

② After H—H bond breaks, H atoms migrate along metal surface.

③ One free H attaches to C_2H_4 to form C_2H_5 (ethyl group) intermediate.

④ Second free H is about to attach to C_2H_5 intermediate to form C_2H_6.

⑤ Ethane, C_2H_6, **desorbs** from metal surface.

▲ **Figure 14.23 Heterogeneous catalysis.** The mechanism for the reaction of ethylene with hydrogen on a catalytic surface.

CHEMISTRY AND SUSTAINABILITY   Catalytic Converters

Heterogeneous catalysis plays a major role in the fight against urban air pollution. Two components of automobile exhausts that help form photochemical smog are nitrogen oxides and unburned hydrocarbons. In addition, automobile exhaust may contain considerable quantities of carbon monoxide. Even with the most careful attention to engine design, it is impossible under normal driving conditions to reduce the quantity of these pollutants to an acceptable level in the exhaust gases. It is therefore necessary to remove them from the exhaust before they are vented to the air. This removal is accomplished in the *catalytic converter*.

The catalytic converter, which is part of an automobile's exhaust system, must perform two functions: (1) oxidation of CO and unburned hydrocarbons (C_xH_y) to carbon dioxide and water, and (2) reduction of nitrogen oxides to nitrogen gas:

$$CO, C_xH_y \xrightarrow{O_2} CO_2 + H_2O$$

$$NO, NO_2 \longrightarrow N_2 + O_2$$

These two functions require different catalysts, so the development of a successful catalyst system is a difficult challenge. The catalysts must be effective over a wide range of operating temperatures. They must continue to be active despite the fact that various components of the exhaust can block the active sites of the catalyst. And the catalysts must be sufficiently rugged to withstand exhaust gas turbulence and the mechanical shocks of driving under various conditions for thousands of miles.

Catalysts that promote the combustion of CO and hydrocarbons are, in general, the transition-metal oxides and the noble metals. These materials are supported on a structure (**Figure 14.24**) that allows the best possible contact between the flowing exhaust gas and the catalyst surface. A honeycomb structure is made from alumina (Al_2O_3) and impregnated with the catalyst that is employed. Such catalysts operate by first adsorbing oxygen gas present in the exhaust gas. This adsorption weakens the O—O bond in O_2, so that oxygen atoms are available for reaction with adsorbed CO to form CO_2. Hydrocarbon oxidation probably proceeds somewhat similarly, with the hydrocarbons first being adsorbed followed by rupture of a C—H bond.

Transition-metal oxides and noble metals are also the most effective catalysts for the reduction of NO to N_2 and O_2. The catalysts that are most effective in one reaction, however, are usually much less effective in the other. It is therefore necessary to have two catalytic components.

Catalytic converters contain remarkably efficient heterogeneous catalysts. The automotive exhaust gases are in contact with the catalyst for only 100 to 400 ms, but in this very short time, 96% of the hydrocarbons and CO are converted to CO_2 and H_2O, and the emission of nitrogen oxides is reduced by 76%.

Although the exact combination of catalysts used varies from one catalytic converter to another, precious metals are an essential component of any catalytic converter. Platinum is very good at catalyzing the oxidation reactions and has good resistance to impurities such as lead, sulfur, and phosphorus that can poison or disable the catalyst. Palladium is a slightly less expensive alternative to platinum, but it is more sensitive to poisoning from impurities in the exhaust stream. Rhodium is the metal of choice for the reduction of nitrogen oxides and has reasonable activity for the oxidation reactions. Unfortunately, it is even rarer and more expensive than platinum. Catalytic converters currently account for approximately 35% of the world's use of platinum, 65% of palladium, and 95% of rhodium. Deposits of these metals tend to be concentrated in South Africa and Russia, which makes their supply subject to complex global political issues.

Related Exercises: 14.62, 14.81, 14.82, 14.124

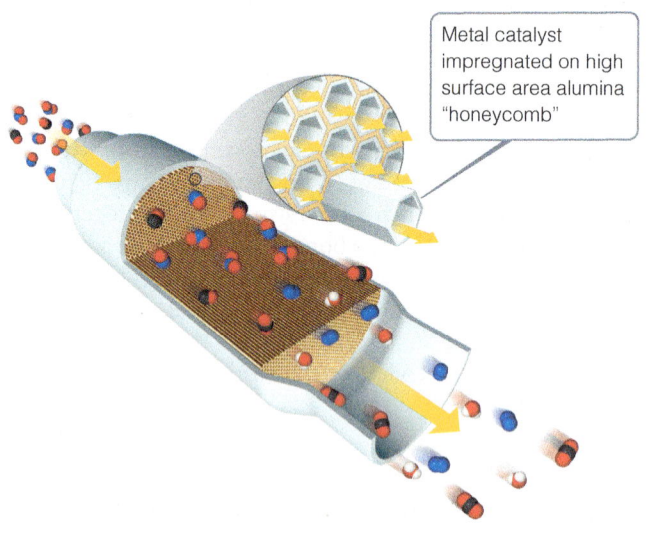

Metal catalyst impregnated on high surface area alumina "honeycomb"

▲ **Figure 14.24** **Cross section of a catalytic converter.**

▼ **Go Figure**

Why is the reaction faster when the liver is ground up?

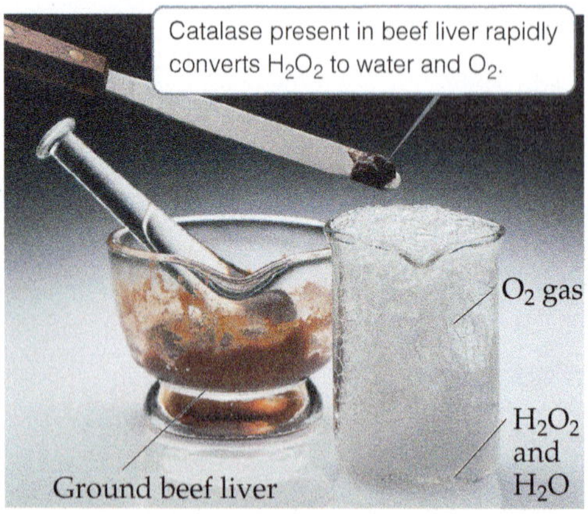

Catalase present in beef liver rapidly converts H_2O_2 to water and O_2.

O_2 gas

H_2O_2 and H_2O

Ground beef liver

▲ **Figure 14.25** **Enzymes speed up reactions.**

Enzymes

The human body is characterized by an extremely complex system of interrelated chemical reactions, all of which must occur at carefully controlled rates to maintain life. A large number of marvelously efficient biological catalysts known as **enzymes** are necessary for many of these reactions to occur at suitable rates. Most enzymes are large protein molecules with molecular weights ranging from about 10,000 to about 1 million amu. They are very selective in the reactions they catalyze, and some are absolutely specific, operating for only one substance in only one reaction. The decomposition of hydrogen peroxide, for example, is an important biological process. Because hydrogen peroxide is strongly oxidizing, it can be physiologically harmful. For this reason, the blood and liver of mammals contain an enzyme, *catalase*, that catalyzes the decomposition of hydrogen peroxide into water and oxygen (Equation 14.31). **Figure 14.25** shows the dramatic acceleration of this chemical reaction by the catalase in beef liver.

The reaction any given enzyme catalyzes takes place at a specific location in the enzyme called the **active site**. The substances that react at this site are

Go Figure Which molecules must bind more tightly to the active site, substrates or products?

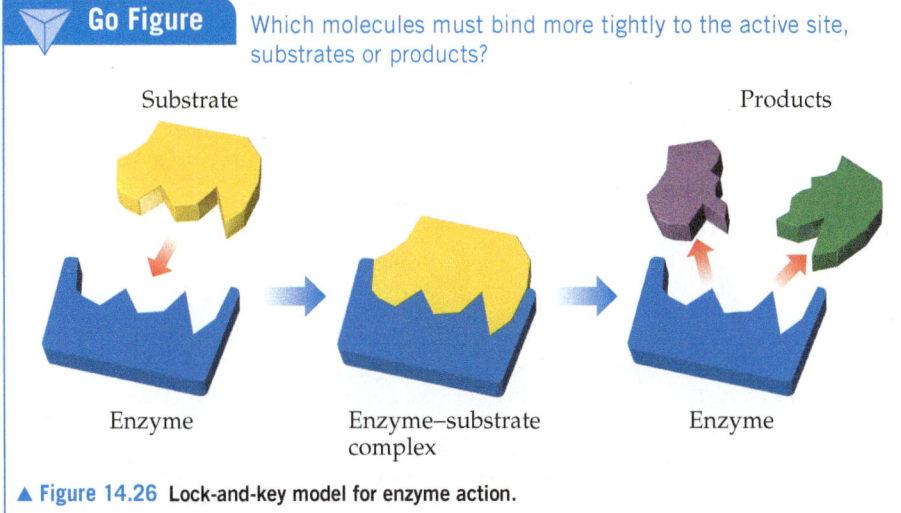

Substrate

Products

Enzyme

Enzyme–substrate complex

Enzyme

▲ **Figure 14.26 Lock-and-key model for enzyme action.**

called **substrates**. The **lock-and-key model** provides a simple explanation for the specificity of an enzyme (**Figure 14.26**). The substrate is pictured as fitting neatly into the active site, much like a key fits into a lock.

Lysozyme is an enzyme that is important to the functioning of our immune system because it accelerates reactions that damage (or "lyse") bacterial cell walls. **Figure 14.27** shows a model of the enzyme lysozyme first without and then with a bound substrate molecule.

The combination of enzyme and substrate is called the *enzyme–substrate complex*. Although Figure 14.26 shows both the active site and its substrate as having a fixed shape, the active site is often fairly flexible and so may change shape as it binds the substrate. The binding between substrate and active site involves dipole–dipole attractions, hydrogen bonds, and dispersion forces. (Section 11.2)

As substrate molecules enter the active site, they are somehow activated so that they are capable of reacting rapidly. This activation process may occur, for example, by the withdrawal or donation of electron density from a particular bond or group of atoms in the enzyme's active site. In addition, the substrate may become distorted in the process of fitting into the active site and made more reactive. Once the reaction occurs, the products depart from the active site, allowing another substrate molecule to enter.

The activity of an enzyme is destroyed if some molecule other than the substrate specific to that enzyme binds to the active site and blocks entry of the substrate. Such substances are called *enzyme inhibitors*. Nerve poisons and certain toxic metal ions, such as lead and mercury, are believed to act in this way to inhibit enzyme activity. Some other poisons act by attaching elsewhere on the enzyme, thereby distorting the active site so that the substrate no longer fits.

Enzymes are enormously more efficient than nonbiochemical catalysts. The number of individual catalyzed reaction events occurring at a particular active site, called the *turnover number*, is generally in the range of 10^3 to 10^7 per second. Such large turnover numbers correspond to very low activation energies. Compared with a simple chemical catalyst, enzymes can increase the rate constant for a given reaction by a millionfold or

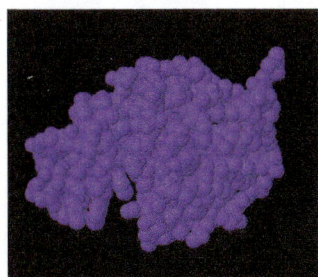
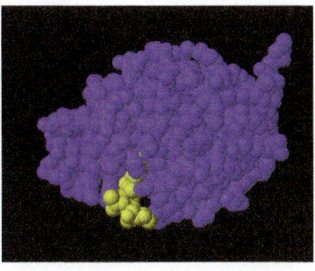

▲ **Figure 14.27 Lysozyme was one of the first enzymes for which a structure–function relationship was described.** This model shows how the substrate (yellow) "fits" into the active site of the enzyme.

CHEMISTRY AND LIFE | Nitrogen Fixation and Nitrogenase

Nitrogen, one of the most essential elements in living organisms, is found in many compounds vital to life, including proteins, nucleic acids, vitamins, and hormones. Nitrogen is continually cycling through the biosphere in various forms, as shown in Figure 14.28. For example, certain microorganisms convert the nitrogen in animal waste and dead plants and animals into $N_2(g)$, which then returns to the atmosphere. For the food chain to be sustained, there must be a means of converting atmospheric $N_2(g)$ into a form plants can use. For this reason, if a chemist were asked to name the most important chemical reaction in the world, she might say *nitrogen fixation*, the process by which atmospheric $N_2(g)$ is converted into compounds suitable for plant use. Some fixed nitrogen results from the action of lightning on the atmosphere, and some is produced industrially using a process we discuss in Chapter 15. About 60% of fixed nitrogen, however, is a consequence of the action of the remarkable and complex enzyme *nitrogenase*. This enzyme is *not* present in humans or other animals; rather, it is found in bacteria that live in the root nodules of certain plants, such as the legumes clover and alfalfa.

Nitrogenase converts N_2 into NH_3, a process that, in the absence of a catalyst, has a very large activation energy. This process is a *reduction* reaction in which the oxidation state of N is reduced from 0 in N_2 to -3 in NH_3. The mechanism by which nitrogenase reduces N_2 is not fully understood. Like many other enzymes, including catalase (Figure 14.25), the active site of nitrogenase contains transition-metal atoms; such enzymes are called *metalloenzymes*. Because transition metals can readily change oxidation state, metalloenzymes are especially useful for effecting transformations in which substrates are either oxidized or reduced.

It has been known for nearly 40 years that a portion of nitrogenase contains iron and molybdenum atoms. This portion, called the *FeMo-cofactor*, is thought to serve as the active site of the enzyme. The FeMo-cofactor of nitrogenase is a cluster of seven Fe atoms and one Mo atom, all linked by sulfur atoms (Figure 14.29).

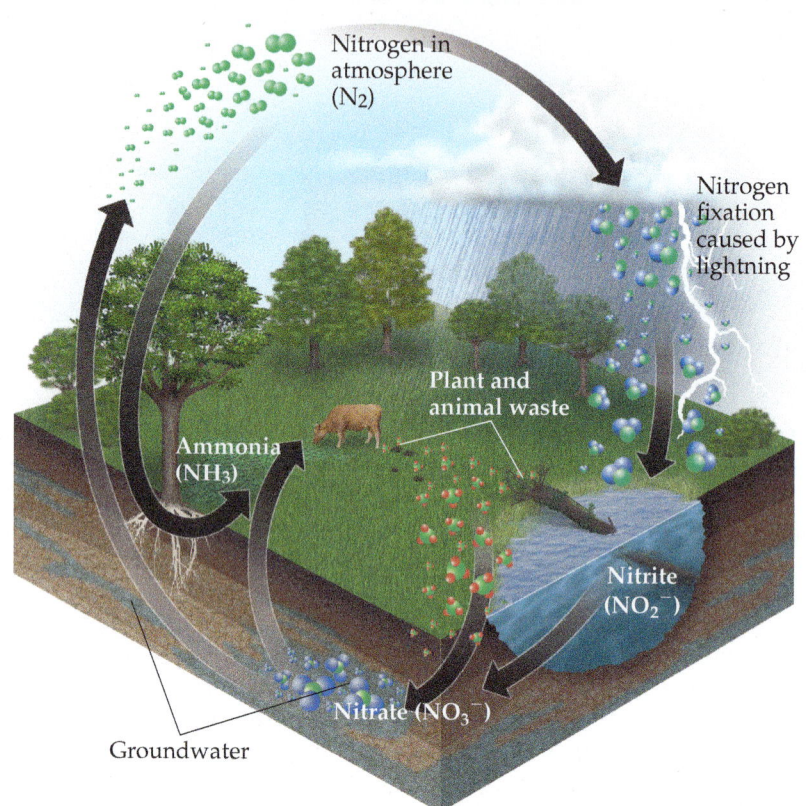

▲ Figure 14.28 Simplified picture of the nitrogen cycle.

It is one of the wonders of life that simple bacteria can contain beautifully complex and vitally important enzymes such as nitrogenase. Because of this enzyme, nitrogen is continually cycled between its comparatively inert role in the atmosphere and its critical role in living organisms. Without nitrogenase, life as we know it could not exist on Earth.

Related Exercises: 14.86, 14.115, 14.116

▶ Figure 14.29 **The FeMo-cofactor of nitrogenase.** Nitrogenase is found in nodules in the roots of certain plants, such as the white clover roots shown at the left. The cofactor, which is thought to be the active site of the enzyme, contains seven Fe atoms and one Mo atom, linked by sulfur atoms. The molecules on the outside of the cofactor connect it to the rest of the protein.

◀ **Figure 14.30 Frances Arnold.** She shared the 2018 Nobel Prize in Chemistry for her work in engineering enzymes to perform new functions.

more. Some enzymes are nearly "perfect" in the sense that the rate-limiting step is the diffusion of the substrate to the active site, and thus the overall reactions are diffusion-controlled (see A Closer Look box on diffusion: "Diffusion-Controlled Reactions and Activation-Controlled Reactions").

While enzymes have evolved to work over millions of years in aqueous solutions, scientists can use the techniques of molecular biology to "direct" the evolution of enzymes to catalyze unnatural reactions or to catalyze reactions in nonaqueous solutions. Chemical engineer Frances Arnold (**Figure 14.30**) shared the 2018 Nobel Prize in Chemistry for her work in this area.

 Self-Assessment Exercises

SAE 14.17 Which of the following statements is or are *true*?

(**i**) A catalyst influences the rate of a reaction by altering the value of either E_a or A.
(**ii**) A catalyst alters the ΔE of a reaction.
(**iii**) Catalysts are often prepared in such a way as to have very large surface areas.

(**a**) Only one of the statements is true. (**b**) Statements i and ii are true. (**c**) Statements i and iii are true. (**d**) Statements ii and iii are true. (**e**) All three statements are true.

SAE 14.18 Which of the following is *true* for the reaction sequence shown?

Step 1: $Cl(g) + O_3(g) \longrightarrow ClO(g) + O_2(g)$
Step 2: $ClO(g) + O(g) \longrightarrow Cl(g) + O_2(g)$
Overall: $O_3(g) + O(g) \longrightarrow 2\,O_2(g)$

(**a**) Cl is an intermediate; ClO is a catalyst.(**b**) Cl is a catalyst; ClO is an intermediate.(**c**) Both Cl and ClO are catalysts. (**d**) Both Cl and ClO are intermediates,

SAE 14.19 There are two main ways to convert nitrogen to ammonia: Nature uses nitrogenase enzymes, whereas humans use the Haber–Bosch process, in which gaseous N_2 and H_2 are heated to high temperatures and flowed over an iron-containing catalyst bed to produce ammonia. Which statements are *true* about these systems?

(**i**) Nitrogenase is a heterogeneous catalyst.
(**ii**) The energy change for the overall reaction is the same, no matter which catalyst system is used.
(**iii**) The activation energy for the overall reaction is the same, no matter which catalyst system is used.

(**a**) Only i is true. (**b**) Only ii is true. (**c**) Only iii is true. (**d**) Statements i and ii are true. (**e**) Statements ii and iii are true.

 Putting Concepts Together

Formic acid (HCOOH) decomposes in the gas phase at elevated temperatures as follows:

$$HCOOH(g) \longrightarrow CO_2(g) + H_2(g)$$

The uncatalyzed decomposition reaction is determined to be first order. A graph of the partial pressure of HCOOH versus time for decomposition at 838 K is shown as the red curve in **Figure 14.31**. When a small amount of solid ZnO is added to the reaction chamber, the partial pressure of acid versus time varies as shown by the blue curve in Figure 14.31.

(**a**) Estimate the half-life and first-order rate constant for formic acid decomposition.

(**b**) What can you conclude from the effect of added ZnO on the decomposition of formic acid?

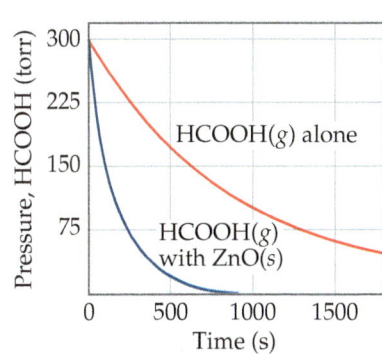

▲ **Figure 14.31** Variation in pressure of HCOOH(*g*) as a function of time at 838 K.

Continued

(c) The progress of the reaction was followed by measuring the partial pressure of formic acid vapor at selected times. Suppose that, instead, we had plotted the concentration of formic acid in units of mol/L. What effect would this have had on the calculated value of k?

(d) The pressure of formic acid vapor at the start of the reaction is 3.00×10^2 torr. Assuming constant temperature and ideal-gas behavior, what is the pressure in the system at the end of the reaction? If the volume of the reaction chamber is $436\ \text{cm}^3$, how many moles of gas occupy the reaction chamber at the end of the reaction?

(e) The standard heat of formation of formic acid vapor is $\Delta H_f^\circ = -378.6\ \text{kJ/mol}$. Calculate ΔH° for the overall reaction. If the activation energy (E_a) for the reaction is $184\ \text{kJ/mol}$, sketch an approximate energy profile for the reaction, and label E_a, ΔH°, and the transition state.

SOLUTION

(a) The initial pressure of HCOOH is 3.00×10^2 torr. On the graph we move to the level at which the partial pressure of HCOOH is 1.50×10^2 torr, half the initial value. This corresponds to a time of about 6.60×10^2 s, which is therefore the half-life. The first-order rate constant is given by Equation 14.17: $k = 0.693/t_{1/2} = 0.693/660\ \text{s} = 1.05 \times 10^{-3}\ \text{s}^{-1}$.

(b) The reaction proceeds much more rapidly in the presence of solid ZnO, so the surface of the oxide must be acting as a catalyst for the decomposition of the acid. This is an example of heterogeneous catalysis.

(c) If we had graphed the concentration of formic acid in units of moles per liter, we would still have determined that the half-life for decomposition is 660 s, and we would have computed the same value for k. Because the units for k are s^{-1}, the value for k is independent of the units used for concentration.

(d) According to the stoichiometry of the reaction, two moles of product are formed for each mole of reactant. When reaction is completed, therefore, the pressure will be 600 torr, just twice the initial pressure, assuming ideal-gas behavior. (Because we are working at quite high temperature and fairly low gas pressure, we assume ideal-gas behavior is reasonable.) The number of moles of gas present can be calculated using the ideal-gas equation (Section 10.3):

$$n = \frac{PV}{RT} = \frac{(600/760\ \text{atm})(0.436\ \text{L})}{(0.08206\ \text{L-atm/mol-K})(838\ \text{K})} = 5.00 \times 10^{-3}\ \text{mol}$$

(e) We first calculate the overall change in energy, ΔH° (Section 5.7 and Appendix C), as in

$$\Delta H^\circ = \Delta H_f^\circ(CO_2(g)) + \Delta H_f^\circ(H_2(g)) - \Delta H_f^\circ(HCOOH(g))$$
$$= -393.5\ \text{kJ/mol} + 0 - (-378.6\ \text{kJ/mol})$$
$$= -14.9\ \text{kJ/mol}$$

From this and the given value for E_a, we can draw an approximate energy profile for the reaction, in analogy to Figure 14.16.

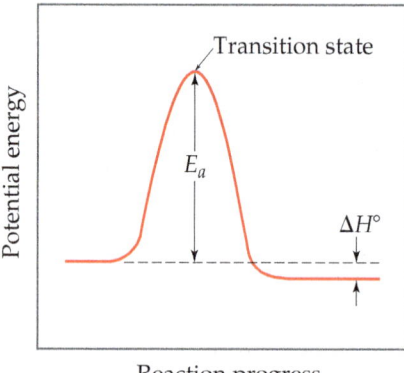

Chapter Summary and Key Terms

REACTION RATES (SECTION 14.1) **Chemical kinetics** is the area of chemistry in which **reaction rates** are studied. Factors that affect reaction rate are the physical state of the reactants, concentration, temperature, and the presence of catalysts. Reaction rates are usually expressed as changes in concentration per unit time: Typically, for reactions in solution, rates are given in units of molarity per second (M/s). For most reactions, a plot of molarity versus time shows that the rate slows down as the reaction proceeds. The **instantaneous rate** is the slope of a line drawn tangent to the concentration-versus-time curve at a specific time. Rates can be written in terms of the appearance of products or the disappearance of reactants; the stoichiometry of the reaction dictates the relationship between rates of appearance and disappearance.

REACTION LAWS AND RATE CONSTANTS: THE METHOD OF INITIAL RATES (SECTION 14.2) The quantitative relationship between rate and concentration is expressed by a **rate law**, which usually has the following form:

$$\text{Rate} = k[\text{reactant 1}]^m[\text{reactant 2}]^n \ldots$$

The constant k in the rate law is called the **rate constant**; the exponents m, n, and so forth are called **reaction orders** for the reactants. The sum of the reaction orders gives the **overall reaction order**. Reaction orders must be determined experimentally; one common way to do so is to run the reaction with different initial reactant concentrations and measure initial rates. The units of the rate constant depend on the overall reaction order. For a reaction in which the overall reaction order is 1, k has units of s^{-1}; for one in which the overall reaction order is 2, k has units of $M^{-1}\,\text{s}^{-1}$.

Spectroscopy is one technique that can be used to monitor the course of a reaction. According to Beer's law, the absorption of electromagnetic radiation by a substance at a particular wavelength is directly proportional to its concentration.

INTEGRATIVE RATE LAWS (SECTION 14.3) Rate laws can be used to determine the concentrations of reactants or products at any time during a reaction. In a **first-order reaction**, the rate is proportional to the concentration of a single reactant raised to the first power: $\text{Rate} = k[A]$. In such cases, the integrated form of the rate law is $\ln[A]_t = -kt + \ln[A]_0$, where $[A]_t$ is the concentration of reactant A at time t, k is the rate constant, and $[A]_0$ is the initial concentration of A. Thus, for a first-order reaction, a graph of $\ln[A]$ versus time yields a straight line of slope $-k$.

A **second-order reaction** is one for which the overall reaction order is 2. If a second-order rate law depends on the concentration of only one reactant, then rate $= k[A]^2$, and the time dependence of $[A]$ is given by the integrated form of the rate law: $1/[A]_t = 1/[A]_0 + kt$. In this case, a graph of $1/[A]_t$ versus time yields a straight line. A **zero-order reaction** is one for which the overall reaction order is 0. Rate $= k$ if the reaction is zero order.

The **half-life** of a reaction, $t_{1/2}$, is the time required for the concentration of a reactant to drop to one-half of its original value. For a first-order reaction, the half-life depends only on the rate constant and not on the initial concentration: $t_{1/2} = 0.693/k$. The half-life of a second-order reaction depends on both the rate constant and the initial concentration of A: $t_{1/2} = 1/(k[A]_0)$.

TEMPERATURE AND RATE: ACTIVATION ENERGY AND THE ARRHENIUS EQUATION (SECTION 14.4)

The **collision model**, which assumes that reactions occur as a result of collisions between molecules, helps explain why the magnitudes of rate constants increase with increasing temperature. The greater the kinetic energy of the colliding molecules, the greater is the energy of collision. The minimum energy required for a reaction to occur is called the **activation energy**, E_a. A collision with energy E_a or greater can cause the atoms of the colliding molecules to reach the **activated complex** (or **transition state**), which is the highest energy arrangement in the pathway from reactants to products. Even if a collision is energetic enough, it may not lead to reaction; the reactants must also be correctly oriented relative to one another in order for a collision to be effective.

Because the kinetic energy of molecules depends on temperature, the rate constant of a reaction is very dependent on temperature. The relationship between k and temperature is given by the **Arrhenius equation**: $k = Ae^{-E_a/RT}$, where A is called the **frequency factor**; it relates to the number of collisions that are favorably oriented for reaction. The Arrhenius equation is often used in logarithmic form: $\ln k = \ln A - E_a/RT$. Thus, a graph of $\ln k$ versus $1/T$ yields a straight line with slope $-E_a/R$. Reactions in solution may be diffusion-controlled or activation-controlled; in a diffusion-controlled reaction, the rate is limited by the diffusion of the reactants to form the activated complex. In an activation-controlled reaction, the rate is limited by the activation energy of the reaction.

REACTION MECHANISMS (SECTION 14.5)

A **reaction mechanism** details the individual steps that occur in the course of a reaction. Each of these steps, called **elementary reactions**, has a well-defined rate law that depends on the number of molecules (the **molecularity**) of the step. Elementary reactions are defined as either **unimolecular**, **bimolecular**, or **termolecular**, depending on whether one, two, or three reactant molecules are involved, respectively. Termolecular elementary reactions are very rare. Unimolecular, bimolecular, and termolecular reactions follow rate laws that are first order overall, second order overall, and third order overall, respectively.

Many reactions occur by a multistep mechanism, involving two or more elementary reactions, or steps. An **intermediate** is produced in one elementary step and is consumed in a later elementary step, so it does not appear in the overall equation for the reaction. When a mechanism has several elementary steps, the overall rate is limited by the slowest elementary step, called the **rate-determining step**. A fast elementary step that follows the rate-determining step will have no effect on the rate law of the reaction. A fast step that precedes the rate-determining step often creates an equilibrium that involves an intermediate. For a mechanism to be valid, the rate law predicted by the mechanism must be the same as that observed experimentally.

CATALYSIS (SECTION 14.6)

A **catalyst** is a substance that increases the rate of a reaction without undergoing a net chemical change itself. It does so by providing a different mechanism for the reaction—namely, one that has a lower activation energy. A **homogeneous catalyst** is one that is in the same phase as the reactants, whereas a **heterogeneous catalyst** has a different phase from the reactants. Finely divided metals are often used as heterogeneous catalysts for solution- and gas-phase reactions. Reacting molecules can undergo binding, or **adsorption**, at the surface of the catalyst. The adsorption of a reactant at specific sites on the surface makes bond breaking easier by lowering the activation energy. Catalysis in living organisms is achieved by **enzymes**, large protein molecules that usually catalyze a very specific reaction. The specific reactant molecules involved in an enzymatic reaction are called **substrates**. The site of the enzyme where the catalysis occurs is called the **active site**. In the **lock-and-key model** for enzyme catalysis, substrate molecules bind very specifically to the active site of the enzyme, after which they can undergo reaction.

Key Equations

- $\text{Rate} = -\dfrac{1}{a}\dfrac{\Delta[A]}{\Delta t} = -\dfrac{1}{b}\dfrac{\Delta[B]}{\Delta t} = \dfrac{1}{c}\dfrac{\Delta[C]}{\Delta t} = \dfrac{1}{d}\dfrac{\Delta[D]}{\Delta t}$ [14.4]

  Definition of reaction rate in terms of the components of the balanced chemical equation $a\,A + b\,B \longrightarrow c\,C + d\,D$

- $\text{Rate} = k[A]^m[B]^n$ [14.7]

  General form of a rate law for the reaction $A + B \longrightarrow$ products

- $\ln[A]_t - \ln[A]_0 = -kt \quad \text{or} \quad \ln\dfrac{[A]_t}{[A]_0} = -kt$ [14.12]

  The integrated form of a first-order rate law for the reaction $A \longrightarrow$ products

- $\dfrac{1}{[A]_t} = kt + \dfrac{1}{[A]_0}$ [14.14]

  The integrated form of a second-order rate law for the reaction $A \longrightarrow$ products

- $[A]_t = -kt + [A]_0$ [14.16]

  The integrated form of a zero-order rate law for the reaction $A \longrightarrow$ products

- $t_{1/2} = \dfrac{0.693}{k}$ [14.17]

  Relating the half-life and rate constant for a first-order reaction

- $k = Ae^{-E_a/RT}$ [14.21]

  The Arrhenius equation, which expresses how the rate constant depends on temperature

- $\ln k = -\dfrac{E_a}{RT} + \ln A$ [14.22]

  Logarithmic form of the Arrhenius equation

Exam Prep

EP 14.1 Which factor does *not* affect reaction rate? (**a**) temperature (**b**) reactant concentration (**c**) collision frequency of reactants (**d**) size of reaction vessel (**e**) whether the reactants are in the gas phase or in solution

EP 14.2 If the experiment in Figure 14.2 (included here) is run for 60 s, 0.16 mol A remain. Which of the following statements is or are *true*?

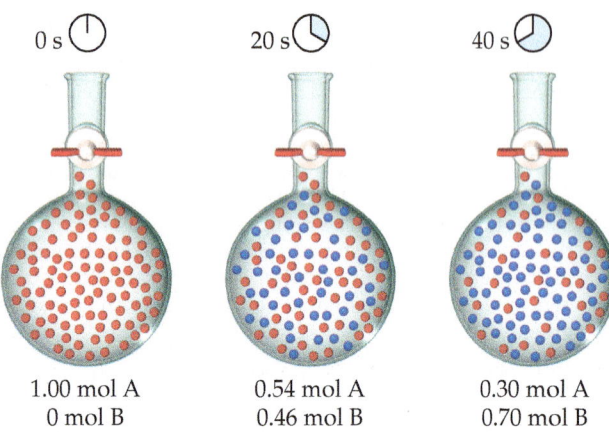

0 s

20 s

40 s

1.00 mol A
0 mol B

0.54 mol A
0.46 mol B

0.30 mol A
0.70 mol B

  (**i**) There are 0.84 mol B in the flask after 60 s.

 (**ii**) The decrease in the number of moles of A from $t_1 = 0$ s to $t_2 = 20$ s is greater than that from $t_1 = 40$ to $t_2 = 60$ s.

(**iii**) The average rate for the reaction from $t_1 = 40$ s to $t_2 = 60$ s is 7.0×10^{-3} M/s.

(**a**) Only statement (i) is true. (**b**) Statements (i) and (ii) are true. (**c**) Statements (i) and (iii) are true. (**d**) Statements (ii) and (iii) are true. (**e**) All three statements are true.

EP 14.3 Which of the following could be the instantaneous rate of the reaction in Figure 14.3 (included here) at $t = 1000$ s?

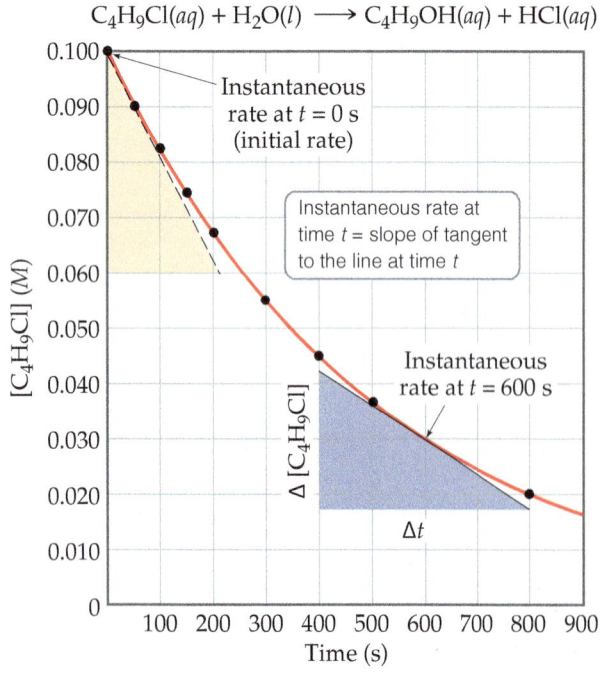

$$C_4H_9Cl(aq) + H_2O(l) \longrightarrow C_4H_9OH(aq) + HCl(aq)$$

Instantaneous rate at $t = 0$ s (initial rate)

Instantaneous rate at time t = slope of tangent to the line at time t

Instantaneous rate at $t = 600$ s

$\Delta[C_4H_9Cl]$

Δt

$[C_4H_9Cl]$ (M)

Time (s)

(**a**) 1.2×10^{-4} M/s   (**c**) 6.3×10^{-5} M/s

(**b**) 8.8×10^{-5} M/s   (**d**) 2.7×10^{-5} M/s

EP 14.4 At a certain time in a reaction, substance A is disappearing at a rate of 4.2×10^{-2} M/s, substance B is appearing at a rate of 2.0×10^{-2} M/s, and substance C is appearing at a rate of 6.0×10^{-2} M/s. Which of the following could be the stoichiometry for the reaction being studied?

(**a**) $2A + B \longrightarrow 3C$ (**d**) $4A \longrightarrow 2B + 3C$

(**b**) $A \longrightarrow 2B + 3C$ (**e**) $A + 2B \longrightarrow 3C$

(**c**) $2A \longrightarrow B + 3C$

EP 14.5 The rate law for a reaction between A and B is found to be rate $= k[A]^2[B]$. Which of the following statements is or are *true*?

  (**i**) The reaction is second-order overall.

 (**ii**) If the initial concentration of A is doubled, the rate doubles.

(**iii**) The rate constant has units of $M^{-1}s^{-1}$.

(**a**) Only statement iii is true. (**b**) Statements i and ii are true. (**c**) Statements i and iii are true. (**d**) Statements ii and iii are true. (**e**) None of statements i, ii, and iii is true.

EP 14.6 Given the following reaction and its rate law, what are the units of the rate constant?

$$CHCl_3(g) + Cl_2(g) \longrightarrow CCl_4(g) + HCl(g) \quad \text{Rate} = k[CHCl_3][Cl_2]^{1/2}$$

(**a**) $M^{-1/2}s^{-1}$ (**b**) $M^{-1/2}s^{-1/2}$ (**c**) $M^{1/2}s^{-1}$ (**d**) $M^{-3/2}s^{-1}$ (**e**) $M^{-3/2}s^{-1/2}$

EP 14.7 The rate law of the reaction $A + B \longrightarrow C$ is rate $= k[A]^2$. If the concentration of B is doubled, the rate of disappearance of B _____, whereas if the concentration of A is doubled, the rate of disappearance of B _____. (**a**) does not change; increases by a factor of 2 (**b**) increases by a factor of 2; increases by a factor of 2 (**c**) increases by a factor of 4; increases by a factor of 2 (**d**) does not change; increases by a factor of 4 (**e**) increases by a factor of 4; does not change

EP 14.8 At 25 °C, the decomposition of dinitrogen pentoxide, $N_2O_5(g)$, into $NO_2(g)$ and $O_2(g)$ follows first-order kinetics with $k = 3.4 \times 10^{-5}$ s^{-1}. A sample of N_2O_5 with an initial pressure of 760 torr decomposes at 25 °C until its partial pressure is 650 torr. How much time (in seconds) has elapsed? (**a**) 5.3×10^{-6} (**b**) 2000 (**c**) 4600 (**d**) 34,000 (**e**) 190,000

EP 14.9 For a certain reaction $A \longrightarrow$ products, a plot of ln[A] versus time produces a straight line with a slope of -3.0×10^{-2} s^{-1}. Which of the following statements is or are *true*?

  (**i**) The reaction follows first-order kinetics.

 (**ii**) The rate constant for the reaction is 3.0×10^{-2} s^{-1}.

(**iii**) The initial concentration of [A] was 1.0 M.

(**a**) Only one of the statements is true. (**b**) Statements (i) and (ii) are true. (**c**) Statements (i) and (iii) are true. (**d**) Statements (ii) and (iii) are true. (**e**) All three statements are true.

EP 14.10 At 25 °C the decomposition of $N_2O_5(g)$ into $NO_2(g)$ and $O_2(g)$ follows first-order kinetics with $k = 3.4 \times 10^{-5}$ s^{-1}. How long will it take for a sample originally containing 2.0 atm of N_2O_5 to reach a partial pressure of 380 torr? (**a**) 5.7 h (**b**) 8.2 h (**c**) 11 h (**d**) 16 h (**e**) 32 h

EP 14.11 Which of the following changes *always* leads to an increase in the rate constant for a reaction:

  (**i**) Decreasing the temperature

 (**ii**) Decreasing the activation energy

(**iii**) Making the value of the overall energy change for the reaction ΔE more negative

(**a**) Only (ii) (**b**) (i) and (ii) (**c**) (i) and (iii) (**d**) (ii) and (iii) (**e**) All three—(i), (ii), and (iii)—increase the reaction rate.

EP 14.12 The rate constant for the rearrangement of methyl isonitrile is $2.52 \times 10^{-5}\,\text{s}^{-1}$ at 189.7 °C. If the activation energy is 160 kJ/mol, what is the rate constant at 320 °C?

(a) $8.1 \times 10^{-15}\,\text{s}^{-1}$
(d) $2.3 \times 10^{-1}\,\text{s}^{-1}$
(b) $2.2 \times 10^{-13}\,\text{s}^{-1}$
(e) $9.2 \times 10^{3}\,\text{s}^{-1}$
(c) $2.7 \times 10^{-9}\,\text{s}^{-1}$

EP 14.13 Consider the following two-step reaction mechanism:

$$A(g) + B(g) \longrightarrow X(g) + Y(g)$$
$$X(g) + C(g) \longrightarrow Y(g) + Z(g)$$

Which of the following statements about this mechanism is or are *true*?

(i) Both of the steps in this mechanism are bimolecular.

(ii) The overall reaction is $A(g) + B(g) + C(g) \longrightarrow Y(g) + Z(g)$.

(iii) The substance $X(g)$ is an intermediate in this mechanism.

(a) Only statement (i) is true. (b) Statements (i) and (ii) are true. (c) Statements (i) and (iii) are true. (d) Statements (ii) and (iii) are true. (e) All three statements are true.

EP 14.14 Consider the following reaction: $2\,A + B \longrightarrow X + 2\,Y$. The rate law of the first step in the mechanism of this reaction is rate $= k[A][B]$. Which of the following could be the first step in the reaction mechanism (note that substance Z is an intermediate)?

(a) $A + A \longrightarrow Y + Z$
(d) $B \longrightarrow X + Y$
(b) $A \longrightarrow X + Z$
(e) $A + B \longrightarrow X + Z$
(c) $A + A + B \longrightarrow X + Y + Y$

EP 14.15 The rate of the reaction $2\,C + D \longrightarrow J + 2\,K$ is second order overall and second order in [C]. Could any of the following be a rate-determining first step in a reaction mechanism that is consistent with the observed rate law for the reaction (note that substance Z is an intermediate)?

(a) $C + C \longrightarrow K + Z$
(b) $C + D \longrightarrow J + Z$
(c) $C \longrightarrow J + Z$
(d) $D \longrightarrow J + K$
(e) None of these are consistent with the observed rate law.

EP 14.16 Consider the following hypothetical reaction: $2\,P + Q \longrightarrow 2\,R + S$. The following mechanism is proposed for this reaction:

$$P + P \rightleftharpoons T \quad (\text{fast})$$
$$Q + T \longrightarrow R + U \quad (\text{slow})$$
$$U \longrightarrow R + S \quad (\text{fast})$$

Substances T and U are unstable intermediates. What rate law is predicted by this mechanism?

(a) Rate $= k[P]^2$
(d) Rate $= k[P][Q]^2$
(b) Rate $= k[P][Q]$
(e) Rate $= k[U]$
(c) Rate $= k[P]^2[Q]$

EP 14.17 Which compound in the following reaction mechanism is an intermediate?

Step 1: $A + B \longrightarrow C$

Step 2: $C + D \longrightarrow E$

(a) A (b) B (c) C (d) D (e) E

EP 14.18 Which of the following statements about catalysts is *false*? (a) A catalyst does not appear in the overall stoichiometry of the reaction. (b) A catalyst increases the rate of a chemical reaction. (c) The addition of a catalyst to a mixture at equilibrium will have no effect. (d) A catalyst cannot affect the mechanism of a reaction. (e) Enzymes are proteins that act as catalysts.

EP 14.19 As noted earlier, Frances Arnold shared the 2018 Nobel Prize in Chemistry for her work in the directed evolution of enzymes—that is, mutating enzymes (altering their original amino acids) to make them work in unnatural environments, to make them catalyze their particular reaction even faster, or to catalyze new reactions they could not before. Her lab recently evolved an enzyme to form a C—Si bond, a reaction that no enzyme can do in nature. This new enzyme had a 15-fold increase in the turnover number for C—Si bond formation compared to the standard homogeneous catalyst used in the chemical industry. What does this mean? (a) Her enzyme lowered the activation energy by a factor of 15 compared to the standard catalyst. (b) Her enzyme could repeat its reaction a total of 15 times compared to just one time for the standard catalyst. (c) It would take 1/15 of the enzyme concentration to do the job compared to the standard catalyst. (d) The number of reactions per second done by her enzyme was 15 times that of the standard catalyst.

Exercises

Visualizing Concepts

14.1 An automotive fuel injector dispenses a fine spray of gasoline into the automobile cylinder, as shown in the bottom drawing here. When an injector gets clogged, as shown in the top drawing, the spray is not as fine or even and the performance of the car declines. How is this observation related to chemical kinetics? [Section 14.1]

14.2 Consider the following graph of the concentration of a substance X over time. Is each of the following statements *true* or *false*? (a) X is a product of the reaction. (b) The rate of the reaction remains the same as time progresses. (c) The average rate between points 1 and 2 is greater than the average rate between points 1 and 3. (d) As time progresses, the curve will eventually turn downward toward the *x*-axis. [Section 14.1]

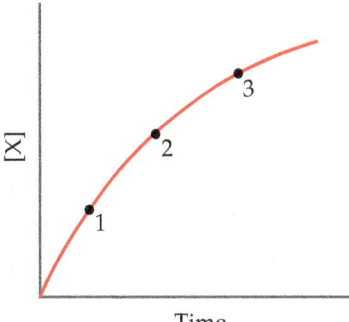

14.3 You study the rate of a reaction, measuring both the concentration of the reactant and the concentration of the product as a function of time, and obtain the following results:

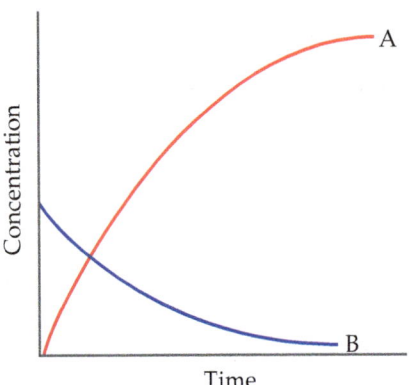

(a) Which chemical equation is consistent with these data?
(i) A $\longrightarrow$ B (ii) B $\longrightarrow$ A (iii) A $\longrightarrow$ 2 B
(iv) B $\longrightarrow$ 2 A?

(b) Write equivalent expressions for the rate of the reaction in terms of the appearance or disappearance of the two substances. [Section 14.1]

14.4 Suppose that for the reaction K + L $\longrightarrow$ M, you monitor the production of M over time, and then plot the following graph from your data:

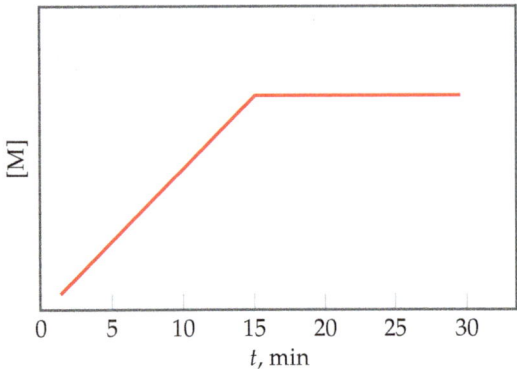

(a) Is the reaction occurring at a constant rate from $t = 0$ to $t = 15$ min? (b) Is the reaction completed at $t = 15$ min? (c) Suppose the reaction as plotted here were started with 0.20 mol K and 0.40 mol L. After 30 min, an additional 0.20 mol K are added to the reaction mixture. Which of the following correctly describes how the plot would look from $t = 30$ min to $t = 60$ min? (i) [M] would remain at the same constant value it has at $t = 30$ min, (ii) [M] would increase with the same slope as $t = 0$ to 15 min, until $t = 45$ min, at which point the plot becomes horizontal again, or (iii) [M] decreases and reaches 0 at $t = 45$ min. [Section 14.2]

14.5 The following diagrams represent mixtures of NO(g) and O$_2$(g). These two substances react as follows:

$$2\,NO(g) + O_2(g) \longrightarrow 2\,NO_2(g)$$

It has been determined experimentally that the rate is second order in NO and first order in O$_2$. Based on this fact, which of the following mixtures will have the fastest initial rate? [Section 14.2]

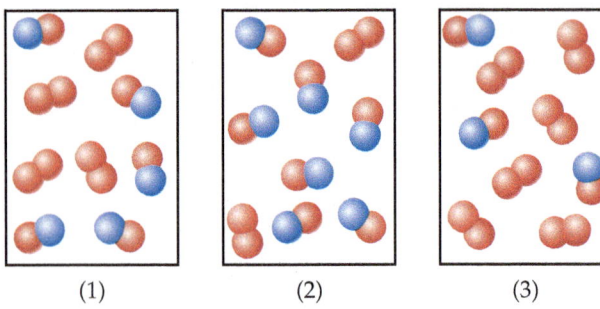

(1)   (2)   (3)

14.6 A friend studies a first-order reaction and obtains the following three graphs for experiments done at two different temperatures. (a) Which two graphs represent experiments done at the same temperature? What accounts for the difference in these two graphs? In what way are they the same? (b) Which two graphs represent experiments done with the same starting concentration but at different temperatures? Which graph probably represents the lower temperature? How do you know? [Section 14.3]

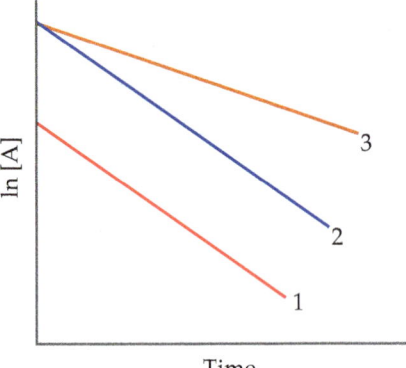

14.7 (a) Given the following diagrams at $t = 0$ min and $t = 30$ min, what is the half-life of the reaction if it follows first-order kinetics?

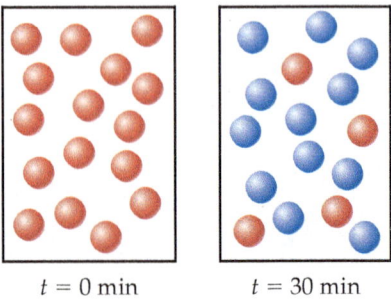

$t = 0$ min $t = 30$ min

(b) After four half-life periods for a first-order reaction, what fraction of reactant remains? [Section 14.3]

14.8 Which of the following linear plots do you expect for a reaction A $\longrightarrow$ products if the kinetics are (a) zero order, (b) first order, or (c) second order? [Section 14.3]

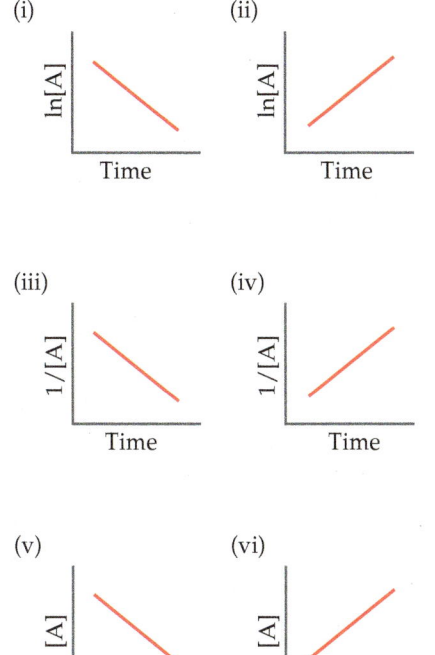

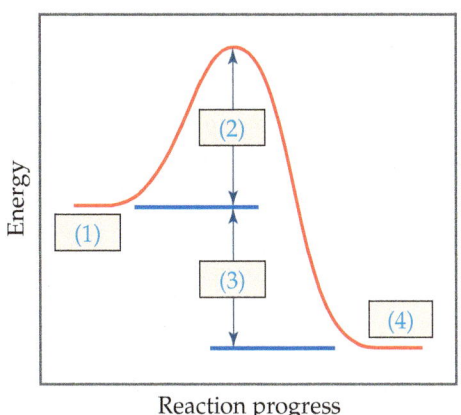

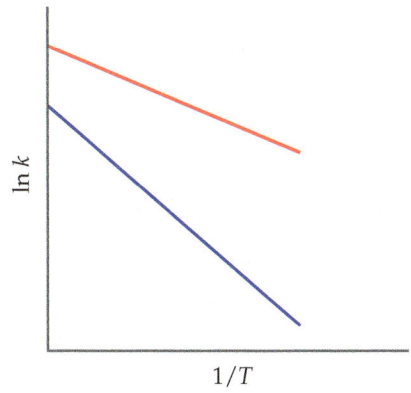

14.9 The following diagram shows a reaction profile. Label the components indicated by the boxes. [Section 14.4]

14.10 The accompanying graph shows plots of ln *k* versus $1/T$ for two different reactions. The plots have been extrapolated to the y-intercepts. Which reaction (red or blue) has (**a**) the larger value for E_a, and (**b**) the larger value for the frequency factor, *A*? [Section 14.4]

14.11 The following graph shows two different reaction pathways for the same overall reaction at the same temperature. Is each of the following statements *true* or *false*? (**a**) The rate is faster for the red path than for the blue path. (**b**) For both paths, the rate of the reverse reaction is slower than the rate of the forward reaction. (**c**) The energy change ΔE is the same for both paths. [Section 14.5]

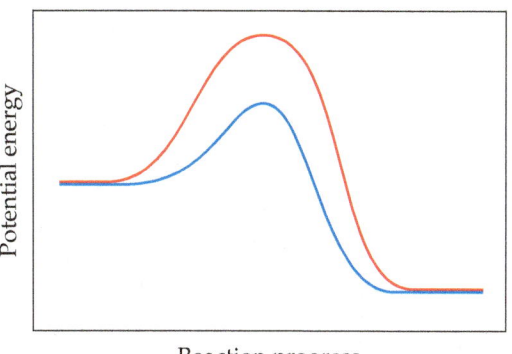

14.12 Consider the diagram that follows, which represents two steps in an overall reaction. The red spheres are oxygen, the blue ones are nitrogen, and the green ones are fluorine. (**a**) Write the chemical equation for each step in the reaction. (**b**) Write the equation for the overall reaction. (**c**) Identify the intermediate in the mechanism. (**d**) Write the rate law for the overall reaction if the first step is the slow, rate-determining step. [Section 14.5]

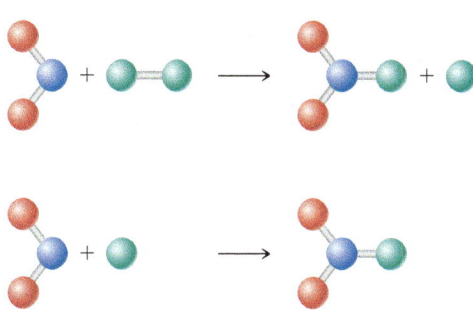

14.13 Based on the following reaction profile, how many intermediates are formed in the reaction A ⟶ C? How many transition states are there? Which step, A ⟶ B or B ⟶ C, is the faster? For the reaction A ⟶ C, is ΔE positive, negative, or zero? [Section 14.5]

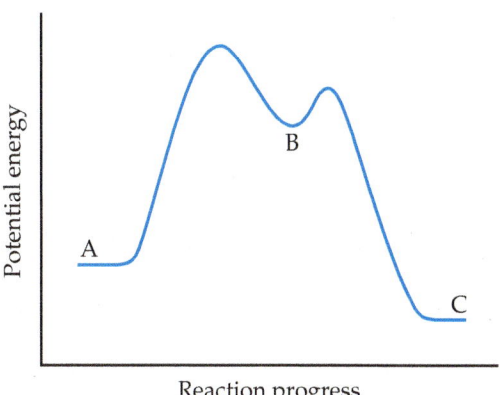

14.14 Draw a possible transition state for the bimolecular reaction depicted here. (The blue spheres are nitrogen atoms, whereas the red ones are oxygen atoms.) Use dashed lines to represent the bonds that are in the process of being broken or made in the transition state. [Sections 14.4 and 14.5]

14.15 The following diagram represents an imaginary two-step mechanism. Let the red spheres represent element A, the green ones element B, and the blue ones element C. (**a**) Write the equation for the net reaction that is occurring. (**b**) Identify the intermediate. (**c**) Identify the catalyst. [Sections 14.5 and 14.6]

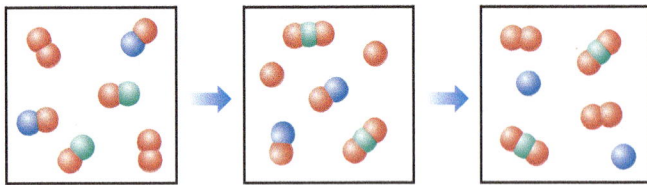

14.16 Draw a graph showing the reaction pathway for an overall exothermic reaction with two intermediates that are produced at different rates. On your graph indicate the reactants, products, intermediates, transition states, and activation energies. [Sections 14.4 and 14.5]

Reaction Rates (Section 14.1)

14.17 (**a**) What is meant by the term *reaction rate*? (**b**) Name three factors that can affect the rate of a chemical reaction. (**c**) Is the rate of disappearance of reactants always the same as the rate of appearance of products?

14.18 (**a**) What are the units usually used to express the rates of reactions occurring in solution? (**b**) As the temperature increases, does the reaction rate usually increase or decrease? (**c**) As a reaction proceeds, does the instantaneous reaction rate increase or decrease?

14.19 Consider the following hypothetical aqueous reaction: $A(aq) \longrightarrow B(aq)$. A flask is charged with 0.065 mol of A in a total volume of 100.0 mL. The following data are collected:

Time (min)	0	10	20	30	40
Moles of A	0.065	0.051	0.042	0.036	0.031

(**a**) Calculate the number of moles of B at each time in the table, assuming that there are no molecules of B at time zero and that A cleanly converts to B with no intermediates. (**b**) Calculate the average rate of disappearance of A for each 10-min interval in units of M/s. (**c**) Between $t = 10$ min and $t = 30$ min, what is the average rate of appearance of B in units of M/s? Assume that the volume of the solution is constant.

14.20 A flask is charged with 0.100 mol of A and allowed to react to form B according to the hypothetical gas-phase reaction $A(g) \longrightarrow B(g)$. The following data are collected:

Time (s)	0	40	80	120	160
Moles of A	0.100	0.067	0.045	0.030	0.020

(**a**) Calculate the number of moles of B at each time in the table, assuming that A is cleanly converted to B with no

intermediates. (**b**) Calculate the average rate of disappearance of A for each 40-s interval in units of mol/s. (**c**) Which of the following would be needed to calculate the rate in units of concentration per time? (i) the pressure of the gas at each time (ii) the volume of the reaction flask (iii) the temperature or (iv) the molecular weight of A.

14.21 The isomerization of methyl isonitrile (CH_3NC) to acetonitrile (CH_3CN) was studied in the gas phase at 215 °C, and the following data were obtained:

Time (s)	[CH₃NC] (*M*)
0	0.0165
2000	0.0110
5000	0.00591
8000	0.00314
12,000	0.00137
15,000	0.00074

(**a**) Calculate the average rate of reaction, in M/s, for the time interval between each measurement. (**b**) Calculate the average rate of reaction over the entire time of the data from $t = 0$ to $t = 15,000$ s. (**c**) Which is greater, the average rate between $t = 2000$ and $t = 12,000$ s, or that between $t = 8000$ and $t = 15,000$ s? (**d**) Graph [CH_3NC] versus time and determine the instantaneous rates in M/s at $t = 5000$ s and $t = 8000$ s.

14.22 The rate of disappearance of HCl was measured for the following reaction:

$$CH_3OH(aq) + HCl(aq) \longrightarrow CH_3Cl(aq) + H_2O(l)$$

The following data were collected:

Time (min)	[HCl] (*M*)
0.0	1.85
54.0	1.58
107.0	1.36
215.0	1.02
430.0	0.580

(**a**) Calculate the average rate of reaction, in M/s, for the time interval between each measurement. (**b**) Calculate the average rate of reaction for the entire time for the data from $t = 0.0$ min to $t = 430.0$ min. (**c**) Which is greater, the average rate between $t = 54.0$ and $t = 215.0$ min, or that between $t = 107.0$ and $t = 430.0$ min? (**d**) Graph [HCl] versus time and determine the instantaneous rates in M/min and M/s at $t = 75.0$ min and $t = 250$ min.

14.23 For each of the following gas-phase reactions, indicate how the rate of disappearance of each reactant is related to the rate of appearance of each product:
(**a**) $H_2O_2(g) \longrightarrow H_2(g) + O_2(g)$
(**b**) $2 N_2O(g) \longrightarrow 2 N_2(g) + O_2(g)$
(**c**) $N_2(g) + 3 H_2(g) \longrightarrow 2 NH_3(g)$
(**d**) $C_2H_5NH_2(g) \longrightarrow C_2H_4(g) + NH_3(g)$

14.24 For each of the following gas-phase reactions, write the rate expression in terms of the appearance of each product and disappearance of each reactant:
(**a**) $2 H_2O(g) \longrightarrow 2 H_2(g) + O_2(g)$
(**b**) $2 SO_2(g) + O_2(g) \longrightarrow 2 SO_3(g)$
(**c**) $2 NO(g) + 2 H_2(g) \longrightarrow N_2(g) + 2 H_2O(g)$
(**d**) $N_2(g) + 2 H_2(g) \longrightarrow N_2H_4(g)$

14.25 **(a)** Consider the combustion of hydrogen, $2\,H_2(g) + O_2(g) \longrightarrow 2\,H_2O(g)$. If hydrogen is burning at the rate of 0.48 mol/s, what is the rate of consumption of oxygen? What is the rate of formation of water vapor? **(b)** The reaction $2\,NO(g) + Cl_2(g) \longrightarrow 2\,NOCl(g)$ is carried out in a closed vessel. If the partial pressure of NO is decreasing at the rate of 56 torr/min, what is the rate of change of the total pressure of the vessel?

14.26 **(a)** Consider the combustion of ethylene, $C_2H_4(g) + 3\,O_2(g) \longrightarrow 2\,CO_2(g) + 2\,H_2O(g)$. If the concentration of C_2H_4 is decreasing at the rate of 0.036 M/s, what are the rates of change in the concentrations of CO_2 and H_2O? **(b)** The rate of decrease in N_2H_4 partial pressure in a closed reaction vessel from the reaction $N_2H_4(g) + H_2(g) \longrightarrow 2\,NH_3(g)$ is 74 torr per hour. What are the rates of change of NH_3 partial pressure and total pressure in the vessel?

Rate Laws and Rate Constants: The Method of Initial Rates (Section 14.2)

14.27 A reaction $A + B \longrightarrow C$ obeys the following rate law: Rate $= k[B]^2$. **(a)** If [A] is doubled, how will the rate change? Will the rate constant change? **(b)** What are the reaction orders for A and B? What is the overall reaction order? **(c)** What are the units of the rate constant?

14.28 Consider a hypothetical reaction between A, B, and C that is first order in A, zero order in B, and second order in C. **(a)** Write the rate law for the reaction. **(b)** How does the rate change when [A] is doubled and the other reactant concentrations are held constant? **(c)** How does the rate change when [B] is tripled and the other reactant concentrations are held constant? **(d)** How does the rate change when [C] is tripled and the other reactant concentrations are held constant? **(e)** By what factor does the rate change when the concentrations of all three reactants are tripled? **(f)** By what factor does the rate change when the concentrations of all three reactants are cut in half?

14.29 The decomposition reaction of N_2O_5 in carbon tetrachloride is $2\,N_2O_5 \longrightarrow 4\,NO_2 + O_2$. The rate law is first order in N_2O_5. At 64 °C the rate constant is $4.82 \times 10^{-3}\,s^{-1}$. **(a)** Write the rate law for the reaction. **(b)** What is the rate of reaction when $[N_2O_5] = 0.0240\,M$? **(c)** What happens to the rate when the concentration of N_2O_5 is doubled to $0.0480\,M$? **(d)** What happens to the rate when the concentration of N_2O_5 is halved to $0.0120\,M$?

14.30 Consider the following reaction:

$$2\,NO(g) + 2\,H_2(g) \longrightarrow N_2(g) + 2\,H_2O(g)$$

(a) The rate law for this reaction is first order in H_2 and second order in NO. Write the rate law. **(b)** If the rate constant for this reaction at 1000 K is $6.0 \times 10^4\,M^{-2}s^{-1}$, what is the reaction rate when $[NO] = 0.035\,M$ and $[H_2] = 0.015\,M$? **(c)** What is the reaction rate at 1000 K when the concentration of NO is increased to 0.10 M, while the concentration of H_2 is 0.010 M? **(d)** What is the reaction rate at 1000 K if [NO] is decreased to 0.010 M and $[H_2]$ is increased to 0.030 M?

14.31 Consider the following reaction:

$$CH_3Br(aq) + OH^-(aq) \longrightarrow CH_3OH(aq) + Br^-(aq)$$

The rate law for this reaction is first order in CH_3Br and first order in OH^-. When $[CH_3Br]$ is $5.0 \times 10^{-3}\,M$ and $[OH^-]$ is 0.050 M, the reaction rate at 298 K is 0.0432 M/s. **(a)** What is the value of the rate constant? **(b)** What are the units of the rate constant? **(c)** What would happen to the rate if the concentration of OH^- were tripled? **(d)** What would happen to the rate if the concentration of both reactants were tripled?

14.32 The reaction between ethyl bromide (C_2H_5Br) and hydroxide ion in ethyl alcohol at 330K, $C_2H_5Br(alc) + OH^-(alc) \longrightarrow C_2H_5OH(l) + Br^-(alc)$, is first order each in ethyl bromide and hydroxide ion. When $[C_2H_5Br]$ is 0.0477 M and $[OH^-]$ is 0.100 M, the rate of disappearance of ethyl bromide is $1.7 \times 10^{-7}\,M/s$. **(a)** What is the value of the rate constant? **(b)** What are the units of the rate constant? **(c)** How would the rate of disappearance of ethyl bromide change if the solution were diluted by adding an equal volume of pure ethyl alcohol to the solution?

14.33 The iodide ion reacts with hypochlorite ion (the active ingredient in chlorine bleaches) in the following way: $OCl^- + I^- \longrightarrow OI^- + Cl^-$. This rapid reaction gives the following rate data:

$[OCl^-]\,(M)$	$[I^-]\,(M)$	Initial Rate (M/s)
1.5×10^{-3}	1.5×10^{-3}	1.36×10^{-4}
3.0×10^{-3}	1.5×10^{-3}	2.72×10^{-4}
1.5×10^{-3}	3.0×10^{-3}	2.72×10^{-4}

(a) Write the rate law for this reaction. **(b)** Calculate the rate constant with proper units. **(c)** Calculate the rate when $[OCl^-] = 2.0 \times 10^{-3}\,M$ and $[I^-] = 5.0 \times 10^{-4}\,M$.

14.34 The reaction $2\,ClO_2(aq) + 2\,OH^-(aq) \longrightarrow ClO_3^-(aq) + ClO_2^-(aq) + H_2O(l)$ was studied with the following results:

Experiment	$[ClO_2](M)$	$[OH^-](M)$	Initial Rate (M/s)
1	0.060	0.030	0.0248
2	0.020	0.030	0.00276
3	0.020	0.090	0.00828

(a) Determine the rate law for the reaction. **(b)** Calculate the rate constant with proper units. **(c)** Calculate the rate when $[ClO_2] = 0.100\,M$ and $[OH^-] = 0.050\,M$.

14.35 The following data were measured for the reaction $BF_3(g) + NH_3(g) \longrightarrow F_3BNH_3(g)$:

Experiment	$[BF_3](M)$	$[NH_3]\,(M)$	Initial Rate (M/s)
1	0.250	0.250	0.2130
2	0.250	0.125	0.1065
3	0.200	0.100	0.0682
4	0.350	0.100	0.1193
5	0.175	0.100	0.0596

(a) What is the rate law for the reaction? **(b)** What is the overall order of the reaction? **(c)** Calculate the rate constant with proper units? **(d)** What is the rate when $[BF_3] = 0.100\,M$ and $[NH_3] = 0.500\,M$?

14.36 The following data were collected for the rate of disappearance of NO in the reaction $2\,NO(g) + O_2(g) \longrightarrow 2\,NO_2(g)$:

Experiment	$[NO]\,(M)$	$[O_2]\,(M)$	Initial Rate (M/s)
1	0.0126	0.0125	1.41×10^{-2}
2	0.0252	0.0125	5.64×10^{-2}
3	0.0252	0.0250	1.13×10^{-1}

(a) What is the rate law for the reaction? **(b)** What are the units of the rate constant? **(c)** What is the average value of the rate constant calculated from the three data sets? **(d)** What is the rate of disappearance of NO when $[NO] = 0.0750\,M$ and $[O_2] = 0.0100\,M$? **(e)** What is the rate of disappearance of O_2 at the concentrations given in part (d)?

14.37 Consider the gas-phase reaction between nitric oxide and bromine at $273°$ C: $2 NO(g) + Br_2(g) \longrightarrow 2 NOBr(g)$. The following data for the initial rate of appearance of NOBr were obtained:

Experiment	[NO](M)	[Br$_2$](M)	Initial Rate (M/s)
1	0.10	0.20	24
2	0.25	0.20	150
3	0.10	0.50	60
4	0.35	0.50	735

(a) Determine the rate law. (b) Calculate the average value of the rate constant for the appearance of NOBr from the four data sets. (c) How is the rate of appearance of NOBr related to the rate of disappearance of Br_2? (d) What is the rate of disappearance of Br_2 when $[NO] = 0.075\,M$ and $[Br_2] = 0.25\,M$?

14.38 Consider the reaction of peroxydisulfate ion ($S_2O_8^{2-}$) with iodide ion (I^-) in aqueous solution:

$$S_2O_8^{2-}(aq) + 3I^-(aq) \longrightarrow 2SO_4^{2-}(aq) + I_3^-(aq)$$

At a particular temperature, the initial rate of disappearance of $S_2O_8^{2-}$ varies with reactant concentrations in the following manner:

Experiment	[S$_2$O$_8$$^{2-}$] (M)	[I$^-$](M)	Initial Rate (M/s)
1	0.018	0.036	2.6×10^{-6}
2	0.027	0.036	3.9×10^{-6}
3	0.036	0.054	7.8×10^{-6}
4	0.050	0.072	1.4×10^{-5}

(a) Determine the rate law for the reaction and state the units of the rate constant. (b) What is the average value of the rate constant for the disappearance of $S_2O_8^{2-}$ based on the four sets of data? (c) How is the rate of disappearance of $S_2O_8^{2-}$ related to the rate of disappearance of I^-? (d) What is the rate of disappearance of I^- when $[S_2O_8^{2-}] = 0.025\,M$ and $[I^-] = 0.050\,M$?

Integrated Rate Laws (Section 14.3)

14.39 (a) For the generic reaction A $\longrightarrow$ B what quantity, when graphed versus time, will yield a straight line for a first-order reaction? (b) How can you calculate the rate constant for a first-order reaction from the graph you made in part (a)?

14.40 (a) For a generic second-order reaction A $\longrightarrow$ B, what quantity, when graphed versus time, will yield a straight line? (b) What is the slope of the straight line from part (a)? (c) Does the half-life of a second-order reaction increase, decrease, or remain the same as the reaction proceeds?

14.41 (a) The gas-phase decomposition of SO_2Cl_2, $SO_2Cl_2(g)$ $\longrightarrow SO_2(g) + Cl_2(g)$, is first order in SO_2Cl_2. At 600 K the half-life for this process is 2.3×10^5 s. What is the rate constant at this temperature? (b) At $320°$C the rate constant is $2.2 \times 10^{-5}\,s^{-1}$. What is the half-life at this temperature?

14.42 Molecular iodine, $I_2(g)$, dissociates into iodine atoms at 625 K with a first-order rate constant of $0.271\,s^{-1}$. (a) What is the half-life for this reaction? (b) If you start with $0.050\,M\,I_2$ at this temperature, how much will remain after 5.12 s, assuming that the iodine atoms do not recombine to form I_2?

14.43 As described in Exercise 14.41, the decomposition of sulfuryl chloride (SO_2Cl_2) is a first-order process. The rate constant for the decomposition at 660 K is $4.5 \times 10^{-2}\,s^{-1}$. (a) If we begin with an initial SO_2Cl_2 pressure of 450 torr, what is the partial pressure of this substance after 60 s? (b) At what time will the partial pressure of SO_2Cl_2 decline to one-tenth its initial value?

14.44 The first-order rate constant for the decomposition of N_2O_5, $2 N_2O_5(g) \longrightarrow 4 NO_2(g) + O_2(g)$, at $70°$C is $6.82 \times 10^{-3}\,s^{-1}$. Suppose we start with 0.0250 mol of $N_2O_5(g)$ in a volume of 2.0 L. (a) How many moles of N_2O_5 will remain after 5.0 min? (b) How many minutes will it take for the quantity of N_2O_5 to drop to 0.010 mol? (c) What is the half-life of N_2O_5 at $70°$C?

14.45 The reaction $SO_2Cl_2(g) \longrightarrow SO_2(g) + Cl_2(g)$ is first order in SO_2Cl_2. Using the following kinetic data, determine the magnitude and units of the first-order rate constant:

Time (s)	Pressure SO$_2$Cl$_2$ (atm)
0	1.000
2500	0.947
5000	0.895
7500	0.848
10,000	0.803

14.46 From the following data for the first-order gas-phase isomerization of CH_3NC to CH_3CN at $215°$C, calculate the first-order rate constant and half-life for the reaction:

Time (s)	Pressure CH$_3$NC (torr)
0	502
2000	335
5000	180
8000	95.5
12,000	41.7
15,000	22.4

14.47 Consider the data presented in Exercise 14.19. (a) By using appropriate graphs, determine whether the reaction is first order or second order. (b) What is the rate constant for the reaction? (c) What is the half-life for the reaction?

14.48 Consider the data presented in Exercise 14.20. (a) Determine whether the reaction is first order or second order. (b) What is the rate constant? (c) What is the half-life?

14.49 The gas-phase decomposition of NO_2, $2 NO_2(g) \longrightarrow 2 NO(g) + O_2(g)$, was studied at $383°$C, giving the following data:

Time (s)	[NO$_2$] (M)
0.0	0.100
5.0	0.017
10.0	0.0090
15.0	0.0062
20.0	0.0047

(a) Is the reaction first order or second order with respect to the concentration of NO_2? (b) What is the rate constant? (c) Predict the reaction rates at the beginning of the reaction for initial concentrations of $0.200\,M$, $0.100\,M$, and $0.050\,M\,NO_2$.

14.50 Sucrose ($C_{12}H_{22}O_{11}$), commonly known as table sugar, reacts in dilute acid solutions to form two simpler sugars, glucose and fructose, both of which have the formula $C_6H_{12}O_6$. At $23°$C and in 0.5 M HCl, the following data were obtained for the disappearance of sucrose:

Time (min)	[C$_{12}$H$_{22}$O$_{11}$] (M)
0	0.316
39	0.274
80	0.238
140	0.190
210	0.146

(a) Is the reaction first order or second order with respect to $[C_{12}H_{22}O_{11}]$? **(b)** What is the rate constant? **(c)** Using this rate constant, calculate the concentration of sucrose at 39, 80, 140, and 210 min if the initial sucrose concentration was 0.316 M and the reaction were zero order in sucrose.

Activation Energy and the Arrhenius Equation (Section 14.4)

14.51 **(a)** What factors determine whether a collision between two molecules will lead to a chemical reaction? **(b)** Does the rate constant for a reaction generally increase or decrease with an increase in reaction temperature? **(c)** Which factor is most sensitive to changes in temperature—the frequency of collisions, the orientation factor, or the fraction of molecules with energy greater than the activation energy?

14.52 **(a)** In which of the following reactions would you expect the orientation factor to be least important in leading to reaction: $NO + O \longrightarrow NO_2$ or $H + Cl \longrightarrow HCl$? **(b)** Does the orientation factor depend on temperature?

14.53 Calculate the fraction of atoms in a sample of argon gas at 400 K that has an energy of 10.0 kJ or greater.

14.54 **(a)** The activation energy for the isomerization of methyl isonitrile (Figure 14.6) is 160 kJ/mol. Calculate the fraction of methyl isonitrile molecules that has an energy equal to or greater than the activation energy at 500 K. **(b)** Calculate this fraction for a temperature of 520 K. What is the ratio of the fraction at 520 K to that at 500 K?

14.55 The gas-phase reaction $Cl(g) + HBr(g) \longrightarrow HCl(g) + Br(g)$ has an overall energy change of -66 kJ. The activation energy for the reaction is 7 kJ. **(a)** Sketch the energy profile for the reaction, and label E_a and ΔE. **(b)** What is the activation energy for the reverse reaction?

14.56 For the elementary process $N_2O_5(g) \longrightarrow NO_2(g) + NO_3(g)$, the activation energy (E_a) and overall ΔE are 154 kJ/mol and 136 kJ/mol, respectively. **(a)** Sketch the energy profile for this reaction, and label E_a and ΔE. **(b)** What is the activation energy for the reverse reaction?

14.57 Indicate whether each statement is *true* or *false*.

(a) If you compare two reactions with similar collision factors, the one with the larger activation energy will be faster.

(b) A reaction that has a small rate constant must have a small frequency factor.

(c) Increasing the reaction temperature increases the fraction of successful collisions between reactants.

14.58 Indicate whether each statement is *true* or *false*.

(a) If you measure the rate constant for a reaction at different temperatures, you can calculate the overall enthalpy change for the reaction.

(b) Exothermic reactions are faster than endothermic reactions.

(c) If you double the temperature for a reaction, you cut the activation energy in half.

14.59 Based on their activation energies and energy changes and assuming that all collision factors are the same, rank the following reactions from slowest to fastest.

(a) $E_a = 45$ kJ/mol; $\Delta E = -25$ kJ/mol

(b) $E_a = 35$ kJ/mol; $\Delta E = -10$ kJ/mol

(c) $E_a = 55$ kJ/mol; $\Delta E = 10$ kJ/mol

14.60 Which of the reactions in Exercise 14.59 will be fastest in the reverse direction? Which will be slowest?

14.61 **(a)** A certain first-order reaction has a rate constant of 2.75×10^{-2} s^{-1} at 20 °C. What is the value of k at 60 °C if $E_a = 75.5$ kJ/mol? **(b)** Another first-order reaction also has a rate constant of 2.75×10^{-2} s^{-1} at 20 °C. What is the value of k at 60 °C if $E_a = 125$ kJ/mol? **(c)** What assumptions do you need to make in order to calculate answers for parts (a) and (b)?

14.62 Understanding the high-temperature behavior of nitrogen oxides is essential for controlling pollution generated in automobile engines. The decomposition of nitric oxide (NO) to N_2 and O_2 is second order with a rate constant of 0.0796 $M^{-1}s^{-1}$ at 737 °C and 0.0815 $M^{-1}s^{-1}$ at 947 °C. Calculate the activation energy for the reaction.

14.63 The rate of the reaction

$$CH_3COOC_2H_5(aq) + OH^-(aq) \longrightarrow CH_3COO^-(aq) + C_2H_5OH(aq)$$

was measured at several temperatures, and the following data were collected:

Temperature (°C)	$k(M^{-1}s^{-1})$
15	0.0521
25	0.101
35	0.184
45	0.332

Calculate the value of E_a by constructing an appropriate graph.

14.64 The temperature dependence of the rate constant for a reaction is tabulated as follows:

Temperature (K)	$k(M^{-1}s^{-1})$
600	0.028
650	0.22
700	1.3
750	6.0
800	23

Calculate E_a and A.

Reaction Mechanisms (Section 14.5)

14.65 **(a)** What is meant by the term *elementary reaction*? **(b)** What is the difference between a *unimolecular* and a *bimolecular* elementary reaction? **(c)** What is a *reaction mechanism*? **(d)** What is meant by the term *rate-determining step*?

14.66 **(a)** Can an intermediate appear as a reactant in the first step of a reaction mechanism? **(b)** On a reaction energy profile diagram, is an intermediate represented as a peak or a valley? **(c)** If a molecule like Cl_2 falls apart in an elementary reaction, what is the molecularity of the reaction?

14.67 What is the molecularity of each of the following elementary reactions? Write the rate law for each.

(a) $Cl_2(g) \longrightarrow 2\,Cl(g)$

(b) $OCl^-(aq) + H_2O(l) \longrightarrow HOCl(aq) + OH^-(aq)$

(c) $NO(g) + Cl_2(g) \longrightarrow NOCl_2(g)$

14.68 What is the molecularity of each of the following elementary reactions? Write the rate law for each.

(a) $2\,NO(g) \longrightarrow N_2O_2(g)$

(b) $H_2C \overset{CH_2}{\overbrace{}} CH_2(g) \longrightarrow CH_2{=}CH{-}CH_3(g)$

(c) $SO_3(g) \longrightarrow SO_2(g) + O(g)$

14.69 **(a)** Based on the following reaction profile, how many intermediates are formed in the reaction $A \longrightarrow D$? **(b)** How many transition states are there? **(c)** Which step is the fastest? (**d**) For the reaction A ⟶ D, is ΔE positive, negative, or zero?

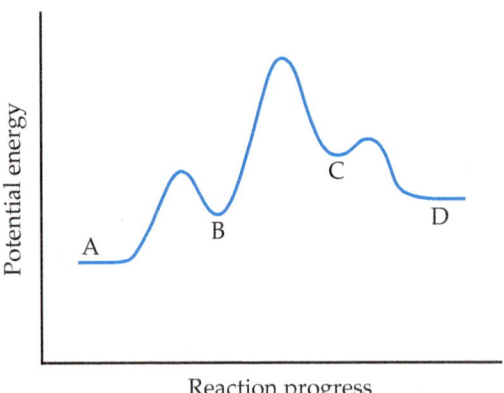

Reaction progress

14.70 Consider the following energy profile.

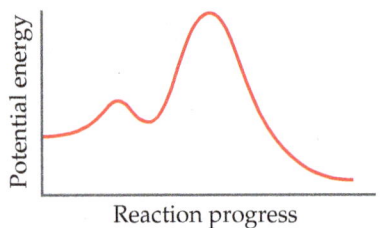

Reaction progress

(**a**) How many elementary reactions are in the reaction mechanism? (**b**) How many intermediates are formed in the reaction? (**c**) Which step is rate limiting? (**d**) For the overall reaction, is ΔE positive, negative, or zero?

14.71 The following mechanism has been proposed for the gas-phase reaction of H_2 with ICl:

$$H_2(g) + ICl(g) \longrightarrow HI(g) + HCl(g)$$
$$HI(g) + ICl(g) \longrightarrow I_2(g) + HCl(g)$$

(**a**) Write the balanced equation for the overall reaction. (**b**) Identify any intermediates in the mechanism. (**c**) If the first step is slow and the second one is fast, which rate law do you expect to be observed for the overall reaction?

14.72 The decomposition of hydrogen peroxide is catalyzed by iodide ion. The catalyzed reaction is thought to proceed by a two-step mechanism:

$$H_2O_2(aq) + I^-(aq) \longrightarrow H_2O(l) + IO^-(aq) \quad \text{(slow)}$$
$$IO^-(aq) + H_2O_2(aq) \longrightarrow H_2O(l) + O_2(g) + I^-(aq) \quad \text{(fast)}$$

(**a**) Write the chemical equation for the overall process. (**b**) Identify the intermediate, if any, in the mechanism. (**c**) Assuming that the first step of the mechanism is rate determining, predict the rate law for the overall process.

14.73 The reaction $2 NO(g) + Cl_2(g) \longrightarrow 2 NOCl(g)$ was performed and the following data were obtained under conditions of constant $[Cl_2]$:

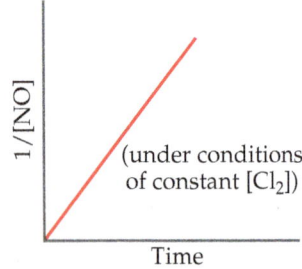

Time

(**a**) Is the following mechanism consistent with the data?

$$NO(g) + Cl_2(g) \rightleftharpoons NOCl_2(g) \quad \text{(fast)}$$
$$NOCl_2(g) + NO(g) \longrightarrow 2 NOCl(g) \quad \text{(slow)}$$

(**b**) Does the linear plot guarantee that the overall rate law is second order?

14.74 You have studied the gas-phase oxidation of HBr by O_2:

$$4 HBr(g) + O_2(g) \longrightarrow 2 H_2O(g) + 2 Br_2(g)$$

You find the reaction to be first order with respect to HBr and first order with respect to O_2. You propose the following mechanism:

$$HBr(g) + O_2(g) \longrightarrow HOOBr(g)$$
$$HOOBr(g) + HBr(g) \longrightarrow 2 HOBr(g)$$
$$HOBr(g) + HBr(g) \longrightarrow H_2O(g) + Br_2(g)$$

(**a**) Confirm that the elementary reactions add to give the overall reaction. (**b**) Based on the experimentally determined rate law, which step is rate determining? (**c**) What are the intermediates in this mechanism? (**d**) If you are unable to detect HOBr or HOOBr among the products, does this disprove your mechanism?

Catalysis (Section 14.6)

14.75 (**a**) What is a catalyst? (**b**) What is the difference between a homogeneous and a heterogeneous catalyst? (**c**) Do catalysts affect the overall enthalpy change for a reaction, the activation energy, or both?

14.76 (**a**) Most commercial heterogeneous catalysts are extremely finely divided solid materials. Why is particle size important? (**b**) What role does adsorption play in the action of a heterogeneous catalyst?

14.77 In Figure 14.21, we saw that $Br^-(aq)$ catalyzes the decomposition of $H_2O_2(aq)$ into $H_2O(l)$ and $O_2(g)$. Suppose that some KBr(s) is added to an aqueous solution of hydrogen peroxide. Make a sketch of $[Br^-(aq)]$ versus time from the addition of the solid to the end of the reaction.

14.78 In solution, chemical species as simple as H^+ and OH^- can serve as catalysts for reactions. Imagine you could measure the $[H^+]$ of a solution containing an acid-catalyzed reaction as it occurs. Assume the reactants and products themselves are neither acids nor bases. Sketch the $[H^+]$ concentration profile you would measure as a function of time for the reaction, assuming $t = 0$ is when you add a drop of acid to the reaction.

14.79 The oxidation of SO_2 to SO_3 is accelerated by NO_2. The reaction proceeds according to:

$$NO_2(g) + SO_2(g) \longrightarrow NO(g) + SO_3(g)$$
$$2 NO(g) + O_2(g) \longrightarrow 2 NO_2(g)$$

(**a**) Show that, with appropriate coefficients, the two reactions can be summed to give the overall oxidation of SO_2 by O_2 to give SO_3. (**b**) Do we consider NO_2 a catalyst or an intermediate in this reaction? (**c**) Would you classify NO as a catalyst or as an intermediate? (**d**) Is this an example of homogeneous catalysis or heterogeneous catalysis?

14.80 The addition of NO accelerates the decomposition of N_2O, possibly by the following mechanism:

$$NO(g) + N_2O(g) \longrightarrow N_2(g) + NO_2(g)$$
$$2 NO_2(g) \longrightarrow 2 NO(g) + O_2(g)$$

(a) What is the chemical equation for the overall reaction? Show how the two steps can be added to give the overall equation. **(b)** Is NO serving as a catalyst or an intermediate in this reaction? **(c)** If experiments show that during the decomposition of N_2O, NO_2 does *not* accumulate in measurable quantities, does this rule out the proposed mechanism?

14.81 Many metallic catalysts, particularly the precious-metal ones, are often deposited as very thin films on a substance of high surface area per unit mass, such as alumina (Al_2O_3) or silica (SiO_2). **(a)** Why is this an effective way of utilizing the catalyst material compared to having powdered metals? **(b)** How does the surface area affect the rate of reaction?

14.82 **(a)** If you were going to build a system to check the effectiveness of automobile catalytic converters on cars, what substances would you want to look for in the car exhaust? **(b)** Automobile catalytic converters have to work at high temperatures, as hot exhaust gases stream through them. In what ways could this be an advantage? In what ways a disadvantage? **(c)** Why is the rate of flow of exhaust gases over a catalytic converter important?

14.83 When D_2 reacts with ethylene (C_2H_4) in the presence of a finely divided catalyst, ethane with two deuterium atoms, CH_2D-CH_2D, is formed. (Deuterium, D, is an isotope of hydrogen of mass 2.) Very little ethane forms in which two deuterium atoms are bound to one carbon (for example, CH_3-CHD_2). Use the sequence of steps involved in the reaction (Figure 14.23) to explain why this is so.

14.84 Heterogeneous catalysts that perform hydrogenation reactions, as illustrated in Figure 14.23, are subject to "poisoning," which shuts down their catalytic ability. Compounds of sulfur are often poisons. Suggest a mechanism by which such compounds might act as poisons.

14.85 The enzyme carbonic anhydrase catalyzes the reaction $CO_2(g) + H_2O(l) \longrightarrow HCO_3^-(aq) + H^+(aq)$. In water, without the enzyme, the reaction proceeds with a rate constant of $0.039 \, s^{-1}$ at 25 °C. In the presence of the enzyme in water, the reaction proceeds with a rate constant of $1.0 \times 10^6 \, s^{-1}$ at 25 °C. Assuming the collision factor is the same for both situations, calculate the difference in activation energies for the uncatalyzed versus enzyme-catalyzed reaction.

14.86 The enzyme urease catalyzes the reaction of urea, (NH_2CONH_2), with water to produce carbon dioxide and ammonia. In water, without the enzyme, the reaction proceeds with a first-order rate constant of $4.15 \times 10^{-5} \, s^{-1}$ at 100 °C. In the presence of the enzyme in water, the reaction proceeds with a rate constant of $3.4 \times 10^4 \, s^{-1}$ at 21 °C. **(a)** Write out the balanced equation for the reaction catalyzed by urease. **(b)** If the rate of the catalyzed reaction were the same at 100 °C as it is at 21 °C, what would be the difference in the activation energy between the catalyzed and uncatalyzed reactions? **(c)** In actuality, what would you expect for the rate of the catalyzed reaction at 100 °C as compared to that at 21 °C? **(d)** On the basis of parts (b) and (c), what can you conclude about the difference in activation energies for the catalyzed and uncatalyzed reactions?

14.87 The activation energy of an uncatalyzed reaction is 95 kJ/mol. The addition of a catalyst lowers the activation energy to 55 kJ/mol. Assuming that the collision factor remains the same, by what factor will the catalyst increase the rate of the reaction at **(a)** 25 °C, **(b)** 125 °C?

14.88 Suppose that a certain biologically important reaction is quite slow at physiological temperature (37 °C) in the absence of a catalyst. Assuming that the collision factor remains the same, by how much must an enzyme lower the activation energy of the reaction to achieve a 1×10^5-fold increase in the reaction rate?

Additional Exercises

14.89 Consider the reaction $A + B \longrightarrow C + D$. Is each of the following statements *true* or *false*? **(a)** The rate law for the reaction must be rate = $k[A][B]$. **(b)** If the reaction is an elementary reaction, the rate law is second order. **(c)** If the reaction is an elementary reaction, the rate law of the reverse reaction is first order. **(d)** The activation energy for the reverse reaction must be greater than that for the forward reaction.

14.90 Hydrogen sulfide (H_2S) is a common and troublesome pollutant in industrial wastewaters. One way to remove H_2S is to treat the water with chlorine, in which case the following reaction occurs:

$$H_2S(aq) + Cl_2(aq) \longrightarrow S(s) + 2 \, H^+(aq) + 2 \, Cl^-(aq)$$

The rate of this reaction is first order in each reactant. The rate constant for the disappearance of H_2S at 28 °C is $3.5 \times 10^{-2} \, M^{-1}s^{-1}$. If at a given time the concentration of H_2S is $2.0 \times 10^{-4} \, M$ and that of Cl_2 is 0.025 M, what is the rate of formation of Cl^-?

14.91 The reaction $2 \, NO(g) + O_2(g) \longrightarrow 2 \, NO_2(g)$ is second order in NO and first order in O_2. When [NO] = 0.040 M, and [O_2] = 0.035 M, the observed rate of disappearance of NO is $9.3 \times 10^{-5} \, M/s$. **(a)** What is the rate of disappearance of O_2 at this moment? **(b)** What is the value of the rate constant? **(c)** What are the units of the rate constant? **(d)** What would happen to the rate if the concentration of NO were increased by a factor of 1.8?

14.92 You perform a series of experiments for the reaction $A \longrightarrow B + C$ and find that the rate law has the form rate = $k[A]^x$. Determine the value of x in each of the following cases: **(a)** There is no rate change when [A_0] is tripled. **(b)** The rate increases by a factor of 9 when [A]$_0$ is tripled. **(c)** When [A]$_0$ is doubled, the rate increases by a factor of 8.

14.93 Consider the following reaction between mercury(II) chloride and oxalate ion:

$$2 \, HgCl_2(aq) + C_2O_4^{2-}(aq) \longrightarrow 2 \, Cl^-(aq) + 2 \, CO_2(g) + Hg_2Cl_2(s)$$

The initial rate of this reaction was determined for several concentrations of $HgCl_2$ and $C_2O_4^{2-}$, and the following rate data were obtained for the rate of disappearance of $C_2O_4^{2-}$:

Experiment	$[HgCl_2] \, (M)$	$[C_2O_4^{2-}] \, (M)$	Rate (M/s)
1	0.164	0.15	3.2×10^{-5}
2	0.164	0.45	2.9×10^{-4}
3	0.082	0.45	1.4×10^{-4}
4	0.246	0.15	4.8×10^{-5}

(a) What is the rate law for this reaction? **(b)** What is the value of the rate constant with proper units? **(c)** What is the reaction rate when the initial concentration of $HgCl_2$ is 0.100 M and that of $C_2O_4^{2-}$ is 0.25 M if the temperature is the same as that used to obtain the data shown?

14.94 The following kinetic data are collected for the initial rates of a reaction $2 X + Z \longrightarrow$ products:

Experiment	$[X]_0(M)$	$[Z]_0(M)$	Rate (M/s)
1	0.25	0.25	4.0×10^1
2	0.50	0.50	3.2×10^2
3	0.50	0.75	7.2×10^2

(a) What is the rate law for this reaction? (b) What is the value of the rate constant with proper units? (c) What is the reaction rate when the initial concentration of X is 0.75 M and that of Z is 1.25 M?

14.95 The reaction $2 NO_2 \longrightarrow 2 NO + O_2$ has the rate constant $k = 0.63 \, M^{-1}s^{-1}$. (a) Based on the units for k, is the reaction first or second order in NO_2? (b) If the initial concentration of NO_2 is 0.100 M, how would you determine how long it would take for the concentration to decrease to 0.025 M?

14.96 Consider two reactions. Reaction (1) has a constant half-life, whereas reaction (2) has a half-life that gets longer as the reaction proceeds. What can you conclude about the rate laws of these reactions from these observations?

14.97 A first-order reaction $A \longrightarrow B$ has the rate constant $k = 3.2 \times 10^{-3} \, s^{-1}$. If the initial concentration of A is $2.5 \times 10^{-2} \, M$, what is the rate of the reaction at $t = 660 \, s$?

14.98 (a) The reaction $H_2O_2(aq) \longrightarrow H_2O(l) + \frac{1}{2}O_2(g)$ is first order. At 300 K the rate constant equals $7.0 \times 10^{-4} \, s^{-1}$. Calculate the half-life at this temperature. (b) If the activation energy for this reaction is 75 kJ/mol, at what temperature would the reaction rate be doubled?

14.99 Americium-241 is used in smoke detectors. It has a first-order rate constant for radioactive decay of $k = 1.6 \times 10^{-3} \, yr^{-1}$. By contrast, iodine-125, which is used to test for thyroid functioning, has a rate constant for radioactive decay of $k = 0.011 \, day^{-1}$. (a) What are the half-lives of these two isotopes? (b) Which one decays at a faster rate? (c) How much of a 1.00-mg sample of each isotope remains after three half-lives? (d) How much of a 1.00-mg sample of each isotope remains after 4 days?

14.100 Urea (NH_2CONH_2) is the end product in protein metabolism in animals. The decomposition of urea in 0.1 M HCl occurs according to the reaction

$$NH_2CONH_2(aq) + H^+(aq) + 2 H_2O(l) \longrightarrow 2 NH_4^+(aq) + HCO_3^-(aq)$$

The reaction is first order in urea and first order overall. When $[NH_2CONH_2] = 0.200 \, M$, the rate at 61.05 °C is $8.56 \times 10^{-5} \, M/s$. (a) What is the rate constant, k? (b) What is the concentration of urea in this solution after $4.00 \times 10^3 \, s$ if the starting concentration is 0.500 M? (c) What is the half-life for this reaction at 61.05 °C?

14.101 The rate of a first-order reaction is followed by spectroscopy, monitoring the absorbance of a colored reactant at 520 nm. The reaction occurs in a 1.00-cm sample cell, and the only colored species in the reaction has an extinction coefficient of $5.60 \times 10^3 \, M^{-1} \, cm^{-1}$ at 520 nm. (a) Calculate the initial concentration of the colored reactant if the absorbance is 0.605 at the beginning of the reaction. (b) The absorbance falls to 0.250 at 30.0 min. Calculate the rate constant in units of s^{-1}. (c) Calculate the half-life of the reaction. (d) How long does it take for the absorbance to fall to 0.100?

14.102 A colored dye compound decomposes to give a colorless product. The original dye absorbs at 608 nm and has an extinction coefficient of $4.7 \times 10^4 \, M^{-1}cm^{-1}$ at that wavelength. You perform the decomposition reaction in a 1-cm cuvette in a spectrometer and obtain the following data:

Time (min)	Absorbance at 608 nm
0	1.254
30	0.941
60	0.752
90	0.672
120	0.545

From these data, determine the rate law for the reaction "dye $\longrightarrow$ product" and determine the rate constant.

14.103 Cyclopentadiene (C_5H_6) reacts with itself to form dicyclopentadiene ($C_{10}H_{12}$). A 0.0400 M solution of C_5H_6 was monitored as a function of time as the reaction $2 \, C_5H_6 \longrightarrow C_{10}H_{12}$ proceeded. The following data were collected:

Time (s)	$[C_5H_6] \, (M)$
0.0	0.0400
50.0	0.0300
100.0	0.0240
150.0	0.0200
200.0	0.0174

Plot $[C_5H_6]$ versus time, $\ln[C_5H_6]$ versus time, and $1/[C_5H_6]$ versus time. (a) What is the order of the reaction? (b) What is the value of the rate constant?

14.104 The first-order rate constant for reaction of a particular organic compound with water varies with temperature as follows:

Temperature (K)	Rate Constant (s^{-1})
300	3.2×10^{-11}
320	1.0×10^{-9}
340	3.0×10^{-8}
355	2.4×10^{-7}

From these data, calculate the activation energy in units of kJ/mol.

14.105 At 28 °C, raw milk sours in 4.0 h but takes 48 h to sour in a refrigerator at 5 °C. Estimate the activation energy in kJ/mol for the reaction that leads to the souring of milk.

14.106 The following is a quote from an article in the August 18, 1998, issue of *The New York Times* about the breakdown of cellulose and starch: "A drop of 18 degrees Fahrenheit [from 77 °F to 59 °F] lowers the reaction rate six times; a 36-degree drop [from 77 °F to 41 °F] produces a fortyfold decrease in the rate." (a) Calculate activation energies for the breakdown process based on the two estimates of the effect of temperature on rate. Are the values consistent? (b) Assuming the value of E_a calculated from the 36° drop and that the rate of breakdown is first order with a half-life at 25 °C of 2.7 yr, calculate the half-life for breakdown at a temperature of -15 °C.

14.107 The following mechanism has been proposed for the reaction of NO with H_2 to form N_2O and H_2O:

$$NO(g) + NO(g) \longrightarrow N_2O_2(g)$$
$$N_2O_2(g) + H_2(g) \longrightarrow N_2O(g) + H_2O(g)$$

(a) Show that the elementary reactions of the proposed mechanism add to provide a balanced equation for the reaction. (**b**) Write a rate law for each elementary reaction in the mechanism. (**c**) Identify any intermediates in the mechanism. (**d**) The observed rate law is rate $= k[NO]^2[H_2]$. If the proposed mechanism is correct, what can we conclude about the relative speeds of the first and second reactions?

14.108 Ozone in the upper atmosphere can be destroyed by the following two-step mechanism:

$$Cl(g) + O_3(g) \longrightarrow ClO(g) + O_2(g)$$
$$ClO(g) + O(g) \longrightarrow Cl(g) + O_2(g)$$

(**a**) What is the overall equation for this process? (**b**) What is the catalyst in the reaction? (**c**) What is the intermediate in the reaction?

14.109 The gas-phase decomposition of ozone is thought to occur by the following two-step mechanism.

Step 1: $O_3(g) \rightleftharpoons O_2(g) + O(g)$ (fast)

Step 2: $O(g) + O_3(g) \longrightarrow 2\,O_2(g)$ (slow)

(**a**) Write the balanced equation for the overall reaction. (**b**) Derive the rate law that is consistent with this mechanism. (*Hint:* The product appears in the rate law.) (**c**) Is O a catalyst or an intermediate? (**d**) If instead the reaction occurred in a single step, would the rate law change? If so, what would it be?

14.110 The following mechanism has been proposed for the gas-phase reaction of chloroform ($CHCl_3$) and chlorine:

Step 1: $Cl_2(g) \underset{k_{-1}}{\overset{k_1}{\rightleftharpoons}} 2\,Cl(g)$ (fast)

Step 2: $Cl(g) + CHCl_3(g) \xrightarrow{k_2} HCl(g) + CCl_3(g)$ (slow)

Step 3: $Cl(g) + CCl_3(g) \xrightarrow{k_3} CCl_4$ (fast)

(**a**) What is the overall reaction? (**b**) What are the intermediates in the mechanism? (**c**) What is the molecularity of each of the elementary reactions? (**d**) What is the rate-determining step? (**e**) What is the rate law predicted by this mechanism? (*Hint:* The overall reaction order is not an integer.)

14.111 Consider the hypothetical reaction $2\,A + B \longrightarrow 2\,C + D$. The following two-step mechanism is proposed for the reaction:

Step 1: $A + B \longrightarrow C + X$

Step 2: $A + X \longrightarrow C + D$

X is an unstable intermediate. (**a**) What is the predicted rate law expression if Step 1 is rate determining? (**b**) What is the predicted rate law expression if Step 2 is rate determining? (**c**) Your result for part (b) might be considered surprising for which of the following reasons: (i) The concentration of a product is in the rate law. (ii) There is a negative reaction order in the rate law. (iii) Both reasons (i) and (ii). (iv) Neither reason (i) nor (ii).

14.112 In a hydrocarbon solution, the gold compound $(CH_3)_3AuPH_3$ decomposes into ethane (C_2H_6) and a different gold compound, $(CH_3)AuPH_3$. The following mechanism has been proposed for the decomposition of $(CH_3)_3AuPH_3$:

Step 1: $(CH_3)_3AuPH_3 \underset{k_{-1}}{\overset{k_1}{\rightleftharpoons}} (CH_3)_3Au + PH_3$ (fast)

Step 2: $(CH_3)_3Au \xrightarrow{k_2} C_2H_6 + (CH_3)Au$ (slow)

Step 3: $(CH_3)Au + PH_3 \xrightarrow{k_3} (CH_3)AuPH_3$ (fast)

(**a**) What is the overall reaction? (**b**) What are the intermediates in the mechanism? (**c**) What is the molecularity of each of the elementary steps? (**d**) What is the rate-determining step? (**e**) What is the rate law predicted by this mechanism?

(**f**) What would be the effect on the reaction rate of adding PH_3 to the solution of $(CH_3)_3AuPH_3$?

14.113 Platinum nanoparticles of diameter ~2 nm are important catalysts in carbon monoxide oxidation to carbon dioxide. Platinum crystallizes in a face-centered cubic arrangement with an edge length of 3.924 Å. (**a**) Estimate how many platinum atoms would fit into a 2.0-nm sphere; the volume of a sphere is $(4/3)\pi r^3$. Recall that $1\,\text{Å} = 1 \times 10^{-10}\,\text{m}$ and $1\,\text{nm} = 1 \times 10^{-9}\,\text{m}$. (**b**) Estimate how many platinum atoms are on the surface of a 2.0-nm Pt sphere, using the surface area of a sphere ($4\pi r^2$) and assuming that the "footprint" of one Pt atom can be estimated from its atomic diameter of 2.8 Å. (**c**) Using your results from (a) and (b), calculate the percentage of Pt atoms that are on the surface of a 2.0-nm nanoparticle. (**d**) Repeat these calculations for a 5.0-nm platinum nanoparticle. (**e**) Which size of nanoparticle would you expect to be more catalytically active and why?

14.114 One of the many remarkable enzymes in the human body is carbonic anhydrase, which catalyzes the interconversion of carbon dioxide and water with bicarbonate ion and protons. If it were not for this enzyme, the body could not rid itself rapidly enough of the CO_2 accumulated by cell metabolism. The enzyme catalyzes the dehydration (release to air) of up to 10^7 CO_2 molecules per second. Which components of this description correspond to the terms *enzyme*, *substrate*, and *turnover number*?

14.115 Suppose that, in the absence of a catalyst, a certain biochemical reaction occurs x times per second at normal body temperature (37 °C). In order to be physiologically useful, the reaction needs to occur 5000 times faster than when it is uncatalyzed. By how many kJ/mol must an enzyme lower the activation energy of the reaction to make it useful?

14.116 Enzymes are often described as following the two-step mechanism:

$$E + S \rightleftharpoons ES \quad (\text{fast})$$
$$ES \longrightarrow E + P \quad (\text{slow})$$

where E = enzyme, S = substrate,

ES = enzyme−substrate complex, and P = product.

(**a**) If an enzyme follows this mechanism, what rate law is expected for the reaction? (**b**) Molecules that can bind to the active site of an enzyme but are not converted into product are called *enzyme inhibitors*. Write an additional elementary step to add into the preceding mechanism to account for the reaction of E with I, an inhibitor.

14.117 Dinitrogen pentoxide (N_2O_5) decomposes in chloroform as a solvent to yield NO_2 and O_2. The decomposition is first order with a rate constant at 45 °C of $1.0 \times 10^{-5}\,\text{s}^{-1}$. Calculate the partial pressure of O_2 produced from 1.00 L of $0.600\,M$ N_2O_5 solution at 45 °C over a period of 20.0 h if the gas is collected in a 10.0-L container. (Assume that the products do not dissolve in chloroform.)

14.118 The reaction between ethyl iodide and hydroxide ion in ethanol (C_2H_5OH) solution, $C_2H_5I(alc) + OH^-(alc) \longrightarrow C_2H_5OH(l) + I^-(alc)$, has an activation energy of 86.8 kJ/mol and a frequency factor of $2.10 \times 10^{11}\,M^{-1}\,\text{s}^{-1}$. (**a**) Predict the rate constant for the reaction at 35 °C. (**b**) A solution of KOH in ethanol is made up by dissolving 0.335 g KOH in ethanol to form 250.0 mL of solution. Similarly, 1.453 g of C_2H_5I is dissolved in ethanol to form 250.0 mL of solution. Equal volumes of the two solutions are mixed. Assuming the reaction is first order in each reactant, what is the initial rate at 35 °C? (**c**) Which reagent in the reaction is limiting, assuming the reaction proceeds to completion? (**d**) Assuming the frequency factor and activation energy do not change as a function of temperature, calculate the rate constant for the reaction at 50 °C.

14.119 You obtain kinetic data for a reaction at a set of different temperatures. You plot ln k versus $1/T$ and obtain the following graph:

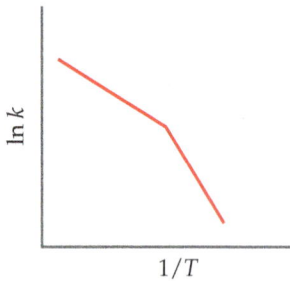

Suggest a molecular-level interpretation of these unusual data.

14.120 The gas-phase reaction of NO with F_2 to form NOF and F has an activation energy of $E_a = 6.3$ kJ/mol. and a frequency factor of $A = 6.0 \times 10^8 \, M^{-1}s^{-1}$. The reaction is believed to be bimolecular:

$$NO(g) + F_2(g) \longrightarrow NOF(g) + F(g)$$

(a) Calculate the rate constant at 100 °C. **(b)** Draw the Lewis structures for the NO and the NOF molecules, given that the chemical formula for NOF is misleading because the nitrogen atom is actually the central atom in the molecule. **(c)** Predict the shape for the NOF molecule. **(d)** Draw a possible transition state for the formation of NOF, using dashed lines to indicate the weak bonds that are beginning to form. **(e)** Suggest a reason for the low activation energy for the reaction.

14.121 The mechanism for the oxidation of HBr by O_2 to form 2 H_2O and Br_2 is shown in Exercise 14.74. **(a)** Calculate the overall standard enthalpy change for the reaction process. **(b)** HBr does not react with O_2 at a measurable rate at room temperature under ordinary conditions. What can you infer from this about the magnitude of the activation energy for the rate-determining step? **(c)** Draw a plausible Lewis structure for the intermediate HOOBr. To what familiar compound of hydrogen and oxygen does it appear similar?

14.122 The rates of many atmospheric reactions are accelerated by the absorption of light by one of the reactants. For example, consider the reaction between methane and chlorine to produce methyl chloride and hydrogen chloride:

$$\text{Reaction 1: } CH_4(g) + Cl_2(g) \longrightarrow CH_3Cl(g) + HCl(g)$$

This reaction is very slow in the absence of light. However, $Cl_2(g)$ can absorb light to form Cl atoms:

$$\text{Reaction 2: } Cl_2(g) + h\nu \longrightarrow 2\,Cl(g)$$

Once the Cl atoms are generated, they can catalyze the reaction of CH_4 and Cl_2, according to the following proposed mechanism:

$$\text{Reaction 3: } CH_4(g) + Cl(g) \longrightarrow CH_3(g) + HCl(g)$$
$$\text{Reaction 4: } CH_3(g) + Cl_2(g) \longrightarrow CH_3Cl(g) + Cl(g)$$

The enthalpy changes and activation energies for these two reactions are tabulated as follows:

Reaction	$\Delta H°$ (kJ/mol)	E_a (kJ/mol)
3	+4	17
4	−109	4

(a) By using the bond enthalpy for Cl_2 (Table 8.3), determine the longest wavelength of light that is energetic enough to cause reaction 2 to occur. In which portion of the electromagnetic spectrum is this light found? **(b)** By using the data tabulated here, sketch a quantitative energy profile for the catalyzed reaction represented by reactions 3 and 4. **(c)** By using bond enthalpies, estimate where the reactants, $CH_4(g) + Cl_2(g)$, should be placed on your diagram in part (b). Use this result to estimate the value of E_a for the reaction $CH_4(g) + Cl_2(g) \longrightarrow CH_3(g) + HCl(g) + Cl(g)$. **(d)** The species $Cl(g)$ and $CH_3(g)$ in reactions 3 and 4 are radicals—that is, atoms or molecules with unpaired electrons. Draw a Lewis structure of CH_3, and verify that it is a radical. **(e)** The sequence of reactions 3 and 4 comprises a radical chain mechanism. Why do you think this is called a "chain reaction"? Propose a reaction that will terminate the chain reaction.

14.123 Many primary amines, RNH_2, where R is a carbon-containing fragment such as CH_3, CH_3CH_2, and so on, undergo reactions where the transition state is tetrahedral. (a) Draw a hybrid orbital picture to visualize the bonding at the nitrogen in a primary amine (just use a C atom for "R"). (b) What kind of reactant with a primary amine can produce a tetrahedral intermediate?

14.124 The NO_x waste stream from automobile exhaust includes species such as NO and NO_2. Catalysts that convert these species to N_2 are desirable to reduce air pollution. **(a)** Draw the Lewis dot and VSEPR structures of NO, NO_2, and N_2. **(b)** Using a resource such as Table 8.3, look up the energies of the bonds in these molecules. In what region of the electromagnetic spectrum are these energies? **(c)** Design a spectroscopic experiment to monitor the conversion of NO_x into N_2, describing what wavelengths of light need to be monitored as a function of time.

Design an Experiment

Let's explore the chemical kinetics of our favorite hypothetical reaction: $a\,A + b\,B \longrightarrow c\,C + d\,D$. We shall assume that all the substances are soluble in water and that we carry out the reaction in aqueous solution. Substances A and C both absorb visible light, and the absorption maxima are 510 nm for A and 640 nm for C. Substances B and D are colorless. You are provided with pure samples of all four substances and you know their chemical formulas. You are also provided appropriate instrumentation to obtain visible absorption spectra (see the Closer Look box in Section 14.2: "Using Spectroscopic Methods to Measure Reaction Rates"). Let's design an experiment to ascertain the kinetics of our reaction. **(a)** What experiments could you design to determine the rate law and the rate constant for the reaction at room temperature? Would you need to know the values of the stoichiometric constants a and c in order to find the rate law? **(b)** Design an experiment to determine the activation energy for the reaction. What challenges might you face in actually carrying out this experiment? **(c)** You now want to test whether a particular water-soluble substance X is a homogeneous catalyst for the reaction. What experiments can you carry out to test this notion? **(d)** If X does indeed catalyze the reaction, what follow-up experiments might you undertake to learn more about the reaction profile for the reaction?

15

CHEMICAL EQUILIBRIUM

▲ **EQUILIBRIUM IN A FOUNTAIN.** In a decorative fountain, the amount of water in the basin remains constant because water leaves and returns to the basin at the same rate. Like a fountain, chemical equilibrium is characterized by two opposing processes that occur at the same rate.

WHAT'S AHEAD

15.1 ▶ **The Concept of Chemical Equilibrium** Relate the concept of equilibrium to reversible chemical reactions.

15.2 ▶ **The Equilibrium Constant** Define the *equilibrium constant* based on rates of forward and reverse reactions, and develop *equilibrium-constant expressions* for homogeneous reactions.

15.3 ▶ **Using Equilibrium Constants** Interpret the magnitude of an equilibrium constant and how its value depends on the way the corresponding chemical equation is expressed.

15.4 ▶ **Heterogeneous Equilibria** Write equilibrium-constant expressions for heterogeneous reactions.

15.5 ▶ **Calculating Equilibrium Constants** Calculate equilibrium constants from the equilibrium concentrations of reactants and products.

15.6 ▶ **Some Applications of Equilibrium Constants** Use equilibrium constants to predict equilibrium concentrations of reactants and products and to determine the direction in which a reaction mixture must proceed to achieve equilibrium.

15.7 ▶ **Le Châtelier's Principle** Use *Le Châtelier's principle* to make qualitative predictions of how a system at equilibrium will respond to changes in concentration, volume, pressure, and temperature.

To be in equilibrium is to be in a state of balance. A tug of war in which the two sides pull with equal force so that the rope does not move is an example of a *static* equilibrium, one in which an object is at rest. Equilibria can also be *dynamic*, whereby a forward process and the reverse process take place at the same rate so that no net change occurs.

Over the past few chapters we've encountered several examples of dynamic equilibrium involving physical changes of matter, including vapor pressure (Section 11.5), the formation of saturated solutions (Section 13.2), and Henry's law (Section 13.3). Consider vapor pressure as an illustrative example. In a closed container, the pressure of a vapor above a liquid stops changing when the rate at which molecules escaping from the liquid into the gas phase equals the rate at which molecules from the gas phase are captured by and reenter the liquid.

Just like evaporation and condensation, or dissolution and crystallization, chemical reactions can run in the forward and reverse directions. In the forward reaction, reactants are converted to products, while in the reverse reaction, the products are converted back to products. We can eventually reach a state in which these two processes occur at the same rate, which is a dynamic equilibrium we call **chemical equilibrium**.

In this and the next two chapters, we explore chemical equilibrium in detail. Later, in Chapter 19, we explain how to relate chemical equilibria to thermodynamics. Here, we learn how to express the equilibrium state of a reaction in quantitative terms and study the factors that determine the relative concentrations of reactants and products in equilibrium mixtures.

15.1 | The Concept of Chemical Equilibrium

Learning Objective

When you finish Section 15.1, you should be able to:

▶ Relate the notion of chemical equilibrium to the rates of a chemical reaction in the forward and reverse directions.

The notion of chemical equilibrium rests on the fact that reactions can occur in both the forward and reverse directions. When a reaction is at equilibrium, the rate at which the products form from the reactants equals the rate at which the reactants form from the products. As a result, concentrations cease to change, and the reaction appears to stop, even though reactants and products continue to convert into one another. To summarize:

Chemical equilibrium occurs when the forward and reverse reactions proceed at equal rates.

Let's examine a simple chemical reaction to see how it reaches an *equilibrium state*—a mixture of reactants and products whose concentrations no longer change with time. We begin with N_2O_4, a colorless substance that dissociates to form brown NO_2. **Figure 15.1** shows a sample of frozen N_2O_4 inside a sealed tube. The solid N_2O_4 becomes a gas as it is warmed above its boiling point (21.2 °C) and the gas mixture turns darker as the colorless N_2O_4 gas dissociates into brown NO_2 gas. Eventually, even though there is still N_2O_4 in the tube, the color stops getting darker because the system reaches equilibrium. We are left with an *equilibrium mixture* of N_2O_4 and NO_2 in which the concentrations of the gases no longer change as time passes. Because the reaction is in a closed system, no gases escape, and the equilibrium mixture can be maintained.

Go Figure If you were to let the tube on the right sit overnight and then take another picture, would the brown color look darker, lighter, or the same?

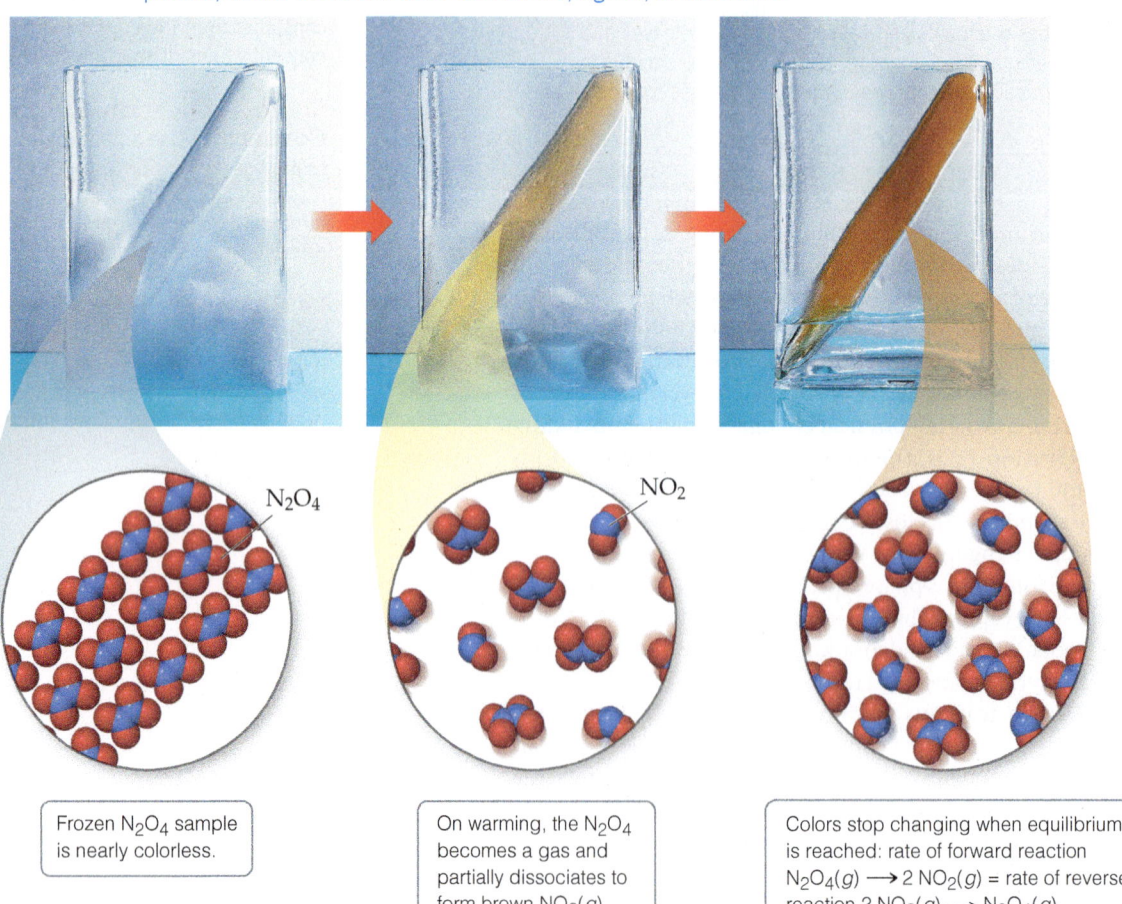

Frozen N_2O_4 sample is nearly colorless.	On warming, the N_2O_4 becomes a gas and partially dissociates to form brown $NO_2(g)$.	Colors stop changing when equilibrium is reached: rate of forward reaction $N_2O_4(g) \longrightarrow 2\,NO_2(g)$ = rate of reverse reaction $2\,NO_2(g) \longrightarrow N_2O_4(g)$.

▲ **Figure 15.1 The equilibrium between colorless N_2O_4 and brown NO_2.**

The equilibrium mixture results because the reaction is *reversible*: N_2O_4 can form NO_2 and NO_2 can form N_2O_4. Dynamic equilibrium is represented by writing the equation for the reaction with two half arrows pointing in opposite directions (Section 4.1):

$$N_2O_4(g) \rightleftharpoons 2\,NO_2(g) \qquad\qquad\qquad [15.1]$$
Colorless $\qquad$ Brown

We can analyze this equilibrium using our knowledge of kinetics. Let's call the decomposition of N_2O_4 into NO_2 the forward reaction and the formation of N_2O_4 from NO_2 the reverse reaction. In this case, both the forward reaction and the reverse reaction are *elementary reactions*. (Section 14.5) As we learned in Section 14.5, the rate laws for elementary reactions can be written from their chemical equations:

$$\text{Forward reaction: } N_2O_4(g) \longrightarrow 2\,NO_2(g) \quad \text{Rate}_f = k_f[N_2O_4] \qquad [15.2]$$

$$\text{Reverse reaction: } 2\,NO_2(g) \longrightarrow N_2O_4(g) \quad \text{Rate}_r = k_r[NO_2]^2 \qquad [15.3]$$

Here, k_f and Rate$_f$ are the rate constant and reaction rate in the *forward* direction, whereas k_r and Rate$_r$ are the rate constant and reaction rate in the *reverse* direction.

At equilibrium, the rate at which NO_2 forms in the forward reaction equals the rate at which N_2O_4 forms in the reverse reaction:

$$\underbrace{k_f[N_2O_4]}_{\text{Forward reaction}} = \underbrace{k_r[NO_2]^2}_{\text{Reverse reaction}} \qquad\qquad [15.4]$$

Rearranging this equation gives

$$\frac{[NO_2]^2}{[N_2O_4]} = \frac{k_f}{k_r} = \text{a constant} \qquad\qquad [15.5]$$

From Equation 15.5, we see that the quotient of two rate constants is another constant, which is called the *equilibrium constant*. We also see that, at equilibrium, the ratio of the concentration terms equals this same constant. It makes no difference whether we start with N_2O_4 or with NO_2, or even with some mixture of the two. At equilibrium, at a given temperature, the ratio equals a specific value. Thus, there is an important constraint on the proportions of N_2O_4 and NO_2 at equilibrium.

Once equilibrium is established, the concentrations of N_2O_4 and NO_2 no longer change, as shown in **Figure 15.2 (a)**. However, the fact that the composition of the equilibrium mixture remains constant with time does not mean that N_2O_4 and NO_2 stop reacting. On the contrary, the equilibrium is *dynamic*, so some N_2O_4 is always converting to NO_2 and some NO_2 is always converting to N_2O_4. At equilibrium, however, the two processes occur at the same rate, as shown in Figure 15.2 (**b**). In the next section we show that the general form of Equation 15.5 of having the concentrations of products divided by the concentrations of reactants can be generalized even if we don't know the mechanism of a reaction.

This example demonstrates several important lessons about equilibrium:

- At equilibrium, the concentrations of reactants and products no longer change with time.
- For equilibrium to occur, neither reactants nor products can escape from the system.
- At equilibrium, a particular ratio of concentration terms equals a constant.

 **Go Figure** Why does the rate of the forward reaction slow down as the reaction proceeds?

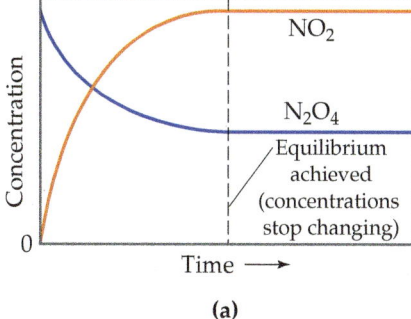

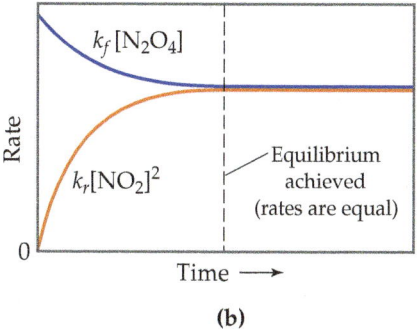

(a) $\qquad\qquad\qquad\qquad\qquad\qquad\qquad\qquad$ (b)

▲ **Figure 15.2 Achieving chemical equilibrium in the reaction $N_2O_4(g) \longrightarrow 2\,NO_2(g)$.** Equilibrium occurs when the rate of the forward reaction equals the rate of the reverse reaction.

Self-Assessment Exercises

SAE 15.1 A certain reaction A $\rightleftharpoons$ B reaches equilibrium. Which of the following statements about the equilibrium is or are *true*?

(**i**) At equilibrium, the concentrations of A and B in the reaction mixture must be equal.
(**ii**) At equilibrium, the processes A $\longrightarrow$ B and B $\longrightarrow$ A have the same reaction rates.
(**iii**) At equilibrium, molecules of A are no longer converted into molecules of B.

(**a**) Only statement i is true. (**b**) Only ii is true. (**c**) Only iii is true. (**d**) Both i and ii are true. (**e**) Both ii and iii are true.

SAE 15.2 Consider the following two elementary reactions, which are the reverse of one another:

$$C + D \longrightarrow E + E \quad \text{Rate} = k_1[C][D]$$
$$E + E \longrightarrow C + D \quad \text{Rate} = k_2[E]^2$$

Which of the following statements about an equilibrium mixture of C, D, and E is *false*? (**a**) At equilibrium, $k_1[C][D] = k_2[E]^2$ (**b**) At equilibrium, C and D still react to form E. (**c**) At equilibrium, $[C][D]/[E]^2$ is a constant. (**d**) At equilibrium, the concentrations of C and D must be the same. (**e**) k_1 and k_2 can have different values.

15.2 | The Equilibrium Constant

 Learning Objectives

When you finish Section 15.2, you should be able to:

▶ Construct equilibrium-constant expressions using the law of mass action.

▶ For a given equilibrium reaction, evaluate K_c from the equilibrium concentrations of reactants and products.

▶ Convert between K_c and K_p as equilibrium constants.

A reaction in which reactants convert to products and products convert to reactants in the same reaction vessel naturally leads to an equilibrium, regardless of how complicated the reaction is and regardless of the nature of the kinetic processes for the forward and reverse reactions. Consider the synthesis of ammonia from nitrogen and hydrogen:

$$N_2(g) + 3\,H_2(g) \rightleftharpoons 2\,NH_3(g) \qquad [15.6]$$

This reaction is the basis for the *Haber–Bosch process*, more commonly known as the **Haber process**. The Haber process is critical for the production of fertilizers and therefore critical to the world's food supply (see the "Chemistry and Sustainability" box on "The Haber Process: Feeding the World"). In this process, N_2 and H_2 react at high pressure and temperature in the presence of a catalyst to form ammonia. In a closed system, however, the reaction does not lead to complete consumption of the N_2 and H_2. Rather, at some point the reaction appears to stop, with all three components of the reaction mixture present at the same time.

How the concentrations of H_2, N_2, and NH_3 vary with time is shown in **Figure 15.3**. Notice that an equilibrium mixture is obtained regardless of whether we begin with N_2 and H_2 or with NH_3. *The equilibrium condition is reached from either direction.*

An expression similar to Equation 15.5 governs the concentrations of N_2, H_2, and NH_3 at equilibrium. If we were to systematically change the relative amounts of the three gases in the starting mixture and then analyze each equilibrium mixture, we could determine the relationship among the equilibrium concentrations.

▼ **Go Figure** Is the rate of disappearance of H_2 related to the rate of disappearance of N_2? If so, how are they related?

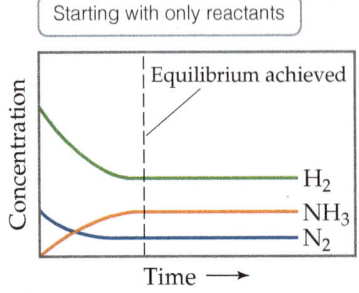
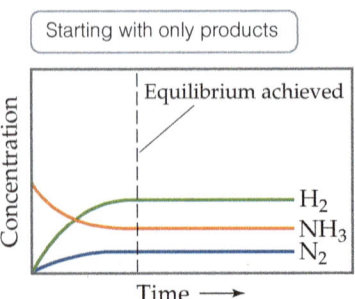

▲ **Figure 15.3 Equilibrium in the reaction $N_2(g) + 3\,H_2(g) \rightleftharpoons 2\,NH_3(g)$**. The same equilibrium is reached whether we start with only reactants (N_2 and H_2) or with only product (NH_3).

Chemists carried out studies of this kind on other chemical systems in the nineteenth century before Haber's work. In 1864, Cato Maximilian Guldberg (1836–1902) and Peter Waage (1833–1900) postulated their **law of mass action**, which expresses, for any reaction, the relationship between the concentrations of the reactants and products present at equilibrium. Suppose we have the general equilibrium equation

$$a\,A + b\,B \rightleftharpoons d\,D + e\,E \qquad [15.7]$$

where A, B, D, and E are the chemical species involved and a, b, d, and e are their coefficients in the balanced chemical equation. According to the law of mass action, the equilibrium condition is described by the expression

$$K_c = \frac{[\text{D}]^d[\text{E}]^e}{[\text{A}]^a[\text{B}]^b} \quad \begin{array}{l} \longleftarrow \text{ products} \\ \longleftarrow \text{ reactants} \end{array} \qquad [15.8]$$

We call this relationship the **equilibrium-constant expression** (or merely the *equilibrium expression*) for the reaction. The constant K_c, the **equilibrium constant**, is the numerical value obtained when we substitute molar equilibrium concentrations into the equilibrium-constant expression. The subscript c on the K indicates that concentrations expressed in molarity are used to evaluate the constant.

The numerator of the equilibrium-constant expression is the product of the concentrations of all substances on the product side of the equilibrium equation, each raised to a power equal to its coefficient in the balanced equation. The denominator is similarly

CHEMISTRY AND SUSTAINABILITY The Haber Process: Feeding the World

Addressing world hunger is one of the UN Sustainable Development Goals (Figure 1.16). The quantity of food required to feed the ever-increasing human population far exceeds that provided by nitrogen-fixing plants (see the "Chemistry and Life" box on "Nitrogen Fixation and Nitrogenase"). (Section 14.6) Therefore, human agriculture requires substantial amounts of ammonia-based fertilizers for croplands. Of all the chemical reactions that humans have learned to control for their own purposes, the synthesis of ammonia from hydrogen and atmospheric nitrogen is one of the most important.

In 1912, the German chemist Fritz Haber (1868–1934) developed the Haber process (Equation 15.6). The process is sometimes also called the *Haber–Bosch process* to honor Karl Bosch, the engineer who developed the industrial process on a large scale. The engineering needed to implement the Haber process requires the use of temperatures (approximately 500 °C) and pressures (200 to 600 atm) that were very challenging to achieve at that time.

The Haber process provides a historically interesting example of the complex impact of chemistry on our lives. At the start of World War I, in 1914, Germany depended on nitrate deposits in Chile for the nitrogen-containing compounds needed to manufacture explosives. During the war, the Allied naval blockade of South America cut off this supply. However, by using the Haber reaction to fix nitrogen from air, Germany was able to continue to produce explosives. Experts have estimated that World War I would have ended perhaps a year earlier had it not been for the Haber process.

From these unhappy beginnings as a major factor in international warfare, the Haber process has become the world's principal source of fixed nitrogen. The same process that prolonged World War I has enabled the manufacture of fertilizers that have increased crop yields, thereby saving millions of people from starvation. About 40 billion pounds of ammonia are manufactured annually in the United States, mostly by the Haber process. The ammonia can be applied directly to the soil (Figure 15.4), or it can be converted into ammonium salts that are also used as fertilizers. The importance of the Haber process in feeding the world is paramount—without it, it is estimated that more than half of the world's population would go hungry.

Although the Haber process has been critical in sustaining the global food supply, it also has severe negative impacts with respect to sustainability. The growth in the use nitrogen-based fertilizers has led to ecological problems of nitrogen runoff into lakes, streams, and watersheds, as is discussed in Chapter 18. Most of the hydrogen gas used in the Haber process is produced using methane, $CH_4(g)$, as a feedstock. The main by-product of H_2 production is CO_2, which contributes to climate change. The Haber process also requires enormous amounts of energy to achieve the temperatures and pressures needed, and most of that energy is provided by combustion of fossil fuels. Between the production of H_2 and the energy needs for the process, the production of ammonia via the Haber process is responsible for more than 1% of global human-made CO_2 emissions—in the United States, about 2.1 tons of CO_2 are emitted for every ton of NH_3 produced. Much current research is exploring alternative ways to make H_2 and the use of renewable energy sources to produce ammonia via the Haber process in a way that leaves a smaller carbon footprint.

Haber was a patriotic German who gave enthusiastic support to his nation's war effort. He served as chief of Germany's Chemical Warfare Service during World War I and developed the use of chlorine as a poison-gas weapon. Consequently, the decision to award him the Nobel Prize in Chemistry in 1918 was the subject of considerable controversy and criticism. The ultimate irony, however, came in 1933 when Haber was expelled from Germany because he was Jewish.

Related Exercises: 15.44, 15.75, 15.78

▲ **Figure 15.4 Ammonia and agriculture.** Ammonia produced using the Haber process is applied directly to crop fields as a nitrogen fertilizer.

derived from the reactant side of the equilibrium equation. Thus, for the Haber process, $N_2(g) + 3\,H_2(g) \rightleftharpoons 2\,NH_3(g)$, the equilibrium-constant expression is

$$K_c = \frac{[NH_3]^2}{[N_2][H_2]^3} \qquad\qquad [15.9]$$

Once we know the balanced chemical equation for a reaction that reaches equilibrium, we can write the equilibrium-constant expression even if we do not know the reaction mechanism.

> *The equilibrium-constant expression depends only on the stoichiometry of the reaction, not on its mechanism.*

In this way equilibrium-constant expressions are different from rate laws.

The value of the equilibrium constant at any given temperature does not depend on the initial amounts of reactants and products. It also does not matter whether other substances are present, as long as they do not react with a reactant or a product. The value of K_c depends only on the particular reaction and on the temperature.

Sample Exercise 15.1
Writing Equilibrium-Constant Expressions

Write the equilibrium expression for K_c for the following reactions:

(a) $2\,O_3(g) \rightleftharpoons 3\,O_2(g)$

(b) $2\,NO(g) + Cl_2(g) \rightleftharpoons 2\,NOCl(g)$

(c) $Ag^+(aq) + 2\,NH_3(aq) \rightleftharpoons Ag(NH_3)_2^+(aq)$

SOLUTION

Analyze We are given three equations and are asked to write an equilibrium-constant expression for each.

Plan Using the law of mass action, we write each expression as a quotient having the product concentration terms in the numerator and the reactant concentration terms in the denominator. Each concentration term is then raised to the power of its coefficient in the balanced chemical equation.

Solve

(a) $K_c = \dfrac{[O_2]^3}{[O_3]^2}$ (b) $K_c = \dfrac{[NOCl]^2}{[NO]^2[Cl_2]}$ (c) $K_c = \dfrac{[Ag(NH_3)_2^+]}{[Ag^+][NH_3]^2}$

▶ **Practice Exercise**

Write the equilibrium-constant expression K_c for

(a) $H_2(g) + I_2(g) \rightleftharpoons 2\,HI(g)$,

(b) $Cd^{2+}(aq) + 4\,Br^-(aq) \rightleftharpoons CdBr_4^{2-}(aq)$.

Evaluating K_c

We can illustrate how the law of mass action was discovered empirically and demonstrate that the equilibrium constant is independent of starting concentrations by examining a series of experiments involving dinitrogen tetroxide and nitrogen dioxide:

$$N_2O_4(g) \rightleftharpoons 2\,NO_2(g) \qquad K_c = \frac{[NO_2]^2}{[N_2O_4]} \qquad\qquad [15.10]$$

We start with several sealed tubes containing different concentrations of NO_2 and N_2O_4. The tubes are kept at 100 °C until equilibrium is reached. We then analyze the mixtures and determine the equilibrium concentrations of NO_2 and N_2O_4, which are listed in Table 15.1.

To evaluate K_c, we insert the equilibrium concentrations into the equilibrium-constant expression. For example, using Experiment 1 data, $[NO_2] = 0.0172\,M$ and $[N_2O_4] = 0.00140\,M$, we find

$$K_c = \frac{[NO_2]^2}{[N_2O_4]} = \frac{[0.0172]^2}{0.00140} = 0.211$$

TABLE 15.1 Initial and Equilibrium Concentrations of $N_2O_4(g)$ and $NO_2(g)$ at 100 °C

Experiment	Initial $[N_2O_4]$ (M)	Initial $[NO_2]$ (M)	Equilibrium $[N_2O_4]$ (M)	Equilibrium $[NO_2]$ (M)	K_c
1	0.0	0.0200	0.00140	0.0172	0.211
2	0.0	0.0300	0.00280	0.0243	0.211
3	0.0	0.0400	0.00452	0.0310	0.213
4	0.0200	0.0	0.00452	0.0310	0.213

Proceeding in the same way, we calculate the values of K_c for the other samples. Note from Table 15.1 that the value for K_c is constant (within the limits of experimental error), even though the initial concentrations vary, as do the final concentrations. Furthermore, Experiment 4 shows that equilibrium can be achieved beginning with N_2O_4 rather than with NO_2. That is, equilibrium can be approached from either direction. Figure 15.5 shows how Experiments 3 and 4 result in the same equilibrium mixture even though the two experiments start with very different NO_2 concentrations.

Notice that no units are given for K_c either in Table 15.1 or in the calculation we just did using Experiment 1 data. It is common practice to write equilibrium constants without units for reasons that we address later in this section.

Equilibrium Constants in Terms of Pressure, K_p

When the reactants and products in a chemical reaction are gases, we can formulate the equilibrium-constant expression in terms of partial pressures. When partial pressures in atmospheres are used in the expression, we denote the equilibrium constant K_p (where the subscript p stands for pressure). For the general reaction in Equation 15.7, we have

$$K_p = \frac{(P_D)^d(P_E)^e}{(P_A)^a(P_B)^b} \qquad [15.11]$$

where P_A is the partial pressure of A in atmospheres, P_B is the partial pressure of B in atmospheres, and so forth. For example, for our N_2O_4/NO_2 reaction, we have

$$K_p = \frac{(P_{NO_2})^2}{P_{N_2O_4}}$$

For a given reaction, the numerical value of K_c is generally different from the numerical value of K_p. We must therefore take care to indicate, via subscript c or p, which constant we are using. It is possible, however, to calculate one from the other using the ideal-gas equation (Section 10.3):

$$PV = nRT, \text{ so } P = \frac{n}{V}RT \qquad [15.12]$$

The usual units for n/V are mol/L, which equals molarity, M. For substance A in our generic reaction, we therefore see that

$$P_A = \frac{n_A}{V}RT = [A]RT \qquad [15.13]$$

Go Figure

In which experiment, 3 or 4, does the concentration of N_2O_4 decrease to reach equilibrium?

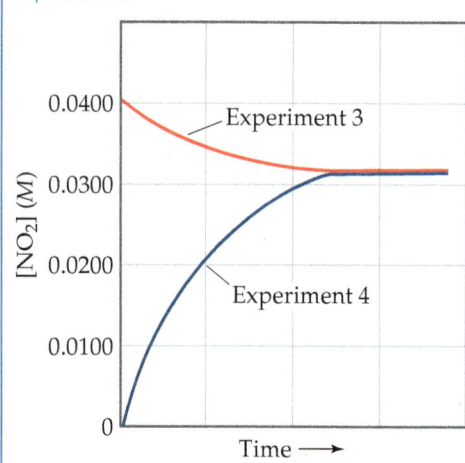

▲ **Figure 15.5** The same equilibrium mixture is produced regardless of the initial NO_2 concentration. The concentration of NO_2 either increases or decreases until equilibrium is reached.

When we substitute Equation 15.13 and like expressions for the other gaseous components of the reaction into Equation 15.11, we obtain a general expression relating K_p and K_c:

$$K_p = \frac{([D]RT)^d([E]RT)^e}{([A]RT)^a([B]RT)^b} = \left(\frac{[D]^d[E]^e}{[A]^a[B]^b}\right)\frac{(RT)^{d+e}}{(RT)^{a+b}} \qquad [15.14]$$

The term in brackets is equal to K_c, so we can simplify the expression:

$$K_p = K_c(RT)^{(d+e)-(a+b)} = K_c(RT)^{\Delta n} \qquad [15.15]$$

The quantity Δn is the change in the number of moles of gas in the balanced chemical equation. It equals the sum of the coefficients of the gaseous products minus the sum of the coefficients of the gaseous reactants:

$$\Delta n = (\text{moles of gaseous product}) - (\text{moles of gaseous reactant}) \qquad [15.16]$$

For example, in the $N_2O_4(g) \rightleftharpoons 2\,NO_2(g)$ reaction, there are 2 mol of product NO_2 and 1 mol of reactant N_2O_4 Therefore, $\Delta n = 2 - 1 = 1$, and $K_p = K_c(RT)$ for this reaction. In our derivation, gas pressures are expressed in atmospheres and concentrations in moles per liter, so the appropriate form of the gas constant is $R = 0.08206$ L-atm/mol-K.

Sample Exercise 15.2
Converting between K_c and K_p

For the Haber process,

$$N_2(g) + 3\,H_2(g) \rightleftharpoons 2\,NH_3(g)$$

$K_c = 9.60$ at 300 °C. Calculate K_p for this reaction at this temperature.

SOLUTION

Analyze We are given K_c for a reaction and asked to calculate K_p.

Plan The relationship between K_c and K_p is given by Equation 15.15. To apply that equation, we must determine Δn by comparing the number of moles of product with the number of moles of reactants (Equation 15.16).

Solve With 2 mol of gaseous products (2 NH_3) and 4 mol of gaseous reactants (1 N_2 + 3 H_2), $\Delta n = 2 - 4 = -2$. (Remember that Δ functions are always based on *products minus reactants*.) The temperature is $300 + 273 = 573$ K. The value for the ideal-gas

constant, R, is 0.08206 L-atm/mol-K. Using $K_c = 9.60$, we therefore have

$$K_p = K_c(RT)^{\Delta n} = (9.60)(0.08206 \times 573)^{-2}$$

$$= \frac{(9.60)}{(0.08206 \times 573)^2} = 4.34 \times 10^{-3}$$

▶ **Practice Exercise**
For the equilibrium $2\,SO_3(g) \rightleftharpoons 2\,SO_2(g) + O_2(g)$, K_c is 4.08×10^{-3} at 1000 K. Calculate the value for K_p.

Equilibrium Constants and Units

Why are equilibrium constants reported without units? The equilibrium constant is related to the kinetics of a reaction as well as to the thermodynamics. (We explore this latter connection in Chapter 19.) Equilibrium constants derived from thermodynamic measurements are defined in terms of *activities* rather than concentrations or partial pressures.

The activity of any substance in an *ideal* mixture is the ratio of the concentration or pressure of the substance either to a reference concentration (1 *M*) or to a reference pressure (1 atm). For example, if the concentration of a substance in an equilibrium mixture is 0.010 *M*, its activity is 0.010 *M*/1 *M* = 0.010. The units of such ratios always cancel and, consequently, activities have no units. Furthermore, the numerical value of the activity equals the concentration. For pure solids and pure liquids, the situation is even simpler because the activities then merely equal 1 (again with no units).

In real systems, activities are also ratios that have no units. Even though these activities may not be exactly numerically equal to concentrations, we assume those differences are small enough that we can neglect them. All we need to know at this point is that activities have no units. As a result, the *thermodynamic equilibrium constants* derived from them also have no units. It is therefore common practice to write all types of equilibrium constants without units, a practice that we adhere to in this text. In more advanced chemistry courses, you may make more rigorous distinctions between concentrations and activities.

Self-Assessment Exercises

SAE 15.3 For the reaction $C_2H_2(g) + 2H_2(g) \rightleftharpoons C_2H_6(g)$, which of the following is the correct expression for K_c?

(a) $K_c = \dfrac{[C_2H_6]}{[C_2H_2][H_2]^2}$

(d) $K_c = \dfrac{[C_2H_6]}{[C_2H_2][H_2]}$

(b) $K_c = \dfrac{2[C_2H_2][H_2]}{[C_2H_6]}$

(e) $K_c = \dfrac{[C_2H_6]}{2[C_2H_2][H_2]}$.

(c) $K_c = \dfrac{[C_2H_2][H_2]^2}{[C_2H_6]}$

SAE 15.4 When the reaction $A(aq) + B(aq) \rightleftharpoons 2C(aq)$ achieves equilibrium in a particular experiment, the equilibrium concentrations of A, B, and C are $[A] = 0.150\,M$, $[B] = 0.350\,M$, and $[C] = 0.600\,M$. What is the value of K_c for this reaction? **(a)** 0.146 **(b)** 6.86 **(c)** 11.4 **(d)** 22.9 **(e)** 131

SAE 15.5 Consider the equilibrium $C_2H_2(g) + 2H_2(g) \rightleftharpoons C_2H_6(g)$. Which of the following is the correct relationship between K_c and K_p for this equilibrium? **(a)** $K_p = K_c(RT)^2$ **(b)** $K_p = K_c(RT)$ **(c)** $K_p = K_c$ **(d)** $K_p = K_c/(RT)$ **(e)** $K_p = K_c/(RT)^2$

SAE 15.6 At 900 K, $K_p = 2.90$ for the following reaction:

$$2SO_3(g) \rightleftharpoons 2SO_2(g) + O_2(g)$$

What is the value of K_c for this equilibrium at 900 K? **(a)** 7.20×10^{-6} **(b)** 3.22×10^{-3} **(c)** 3.93×10^{-2} **(d)** 2.90 **(e)** 214

15.3 | Using Equilibrium Constants

The numerical values of equilibrium constants are powerful tools for chemists—the values of K often allow chemists to "tune" reactions to maximize the amount of the product they seek. Later in this chapter, and in Chapters 16 and 17, we describe many ways in which we can use the numerical values of equilibrium constants quantitatively. In this section, we show how we can glean important *qualitative* information from the *magnitude* of the equilibrium constant. We also show how the values of equilibrium constants change depending on how the chemical equation is written, and how to combine equilibrium constants when two or more equilibria are added together.

 Learning Objectives

When you finish Section 15.3, you should be able to:

▶ Deduce qualitatively the extent to which an equilibrium favors either products or reactants from the magnitude of the equilibrium constant.

▶ Determine how equilibrium constants change when a reaction is reversed, multiplied by a constant, or when two or more equilibria are added together.

The Magnitude of Equilibrium Constants

The magnitude of the equilibrium constant for a reaction gives us important information about the composition of the equilibrium mixture. For example, consider the experimental data for the reaction of carbon monoxide gas and chlorine gas at 100 °C to form phosgene ($COCl_2$), a toxic gas used in the manufacture of certain polymers and insecticides:

$$CO(g) + Cl_2(g) \rightleftharpoons COCl_2(g) \qquad K_c = \frac{[COCl_2]}{[CO][Cl_2]} = 4.56 \times 10^9$$

For the equilibrium constant to be so large, the numerator of the equilibrium-constant expression must be approximately a billion (10^9) times larger than the denominator. Thus, the equilibrium concentration of $COCl_2$ must be much greater than that of CO or Cl_2; in fact, this is just what we find experimentally. We say that this equilibrium *lies to the right* (that is, toward the product side). Likewise, a very small equilibrium constant indicates that the equilibrium mixture contains mostly reactants. We then say that the equilibrium *lies to the left*. In general,

   If $K \gg 1$ (large K): Equilibrium lies to right, products predominate

   If $K \ll 1$ (small K): Equilibrium lies to left, reactants predominate

These situations are summarized in **Figure 15.6**. *Remember, it is the forward and reverse reaction rates, not the reactant and product concentrations, that are equal at equilibrium.*

▼ **Go Figure**

What would this figure look like for a reaction in which $K \approx 1$?

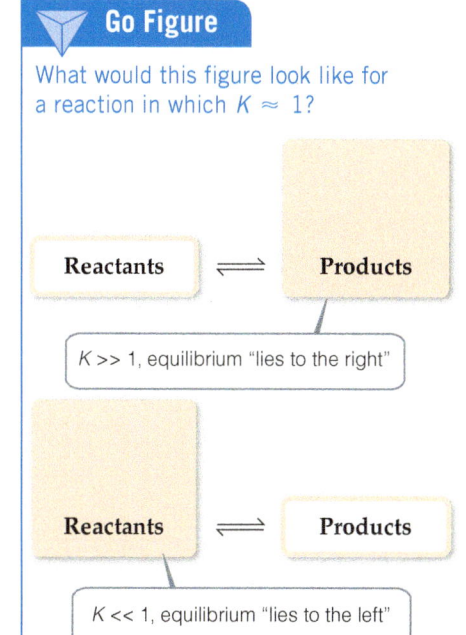

▲ **Figure 15.6** Relationship between the magnitude of *K* and the composition of an equilibrium mixture.

Sample Exercise 15.3
Interpreting the Magnitude of an Equilibrium Constant

The following diagrams represent three systems at equilibrium, all in the same-size containers. (**a**) Without doing any calculations, rank the systems in order of increasing K_c. (**b**) If the volume of the containers is 1.0 L and each sphere represents 0.10 mol, calculate K_c for each system.

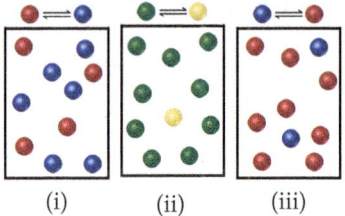

(i)   (ii)   (iii)

SOLUTION

Analyze We are asked to judge the relative magnitudes of three equilibrium constants and then to calculate them.

Plan (**a**) The more product present at equilibrium, relative to reactant, the larger the equilibrium constant. (**b**) The equilibrium constant is given by Equation 15.8.

Solve

(**a**) Each box contains 10 spheres. The amount of product in each varies as follows: (i) 6, (ii) 1, (iii) 8. Therefore, the equilibrium constant varies in the order (ii) < (i) < (iii), from smallest (most reactants) to largest (most products).

(**b**) In (i), we have 0.60 mol/L product and 0.40 mol/L reactant, giving K_c = 0.60/0.40 = 1.5. (You will get the same result by merely dividing the number of spheres of each kind: 6 spheres/4 spheres = 1.5.) In (ii), we have 0.10 mol/L product and 0.90 mol/L reactant, giving K_c = 0.10/0.90 = 0.11

(or 1 sphere/9 spheres = 0.11). In (iii), we have 0.80 mol/L product and 0.20 mol/L reactant, giving K_c = 0.80/0.20 = 4.0 (or 8 spheres/2 spheres = 4.0). These calculations verify the order in (a).

Comment Imagine a drawing that represents a reaction with a very small or very large value of K_c. For example, what would the drawing look like if $K_c = 1 \times 10^{-5}$? In that case, there would need to be 100,000 reactant molecules for only 1 product molecule. But then, that would be impractical to draw.

▶ **Practice Exercise**
For the reaction $H_2(g) + I_2(g) \rightleftharpoons 2\,HI(g)$, $K_p = 794$ at 298 K and $K_p = 55$ at 700 K. Is the formation of HI favored more at the higher or lower temperature?

The Direction of the Chemical Equation and K

We have seen in Table 15.1 that we can represent the N_2O_4/NO_2 equilibrium as

$$N_2O_4(g) \rightleftharpoons 2\,NO_2(g) \quad K_c = \frac{[NO_2]^2}{[N_2O_4]} = 0.212 \quad \text{(at 100 °C)} \qquad [15.17]$$

We could equally well consider this equilibrium in terms of the reverse reaction:

$$2\,NO_2(g) \rightleftharpoons N_2O_4(g)$$

The equilibrium expression is then

$$K_c = \frac{[N_2O_4]}{[NO_2]^2} = \frac{1}{0.212} = 4.72 \quad \text{(at 100 °C)} \qquad [15.18]$$

Equation 15.18 is the reciprocal of the expression in Equation 15.17. *The equilibrium-constant expression for a reaction written in one direction is the reciprocal of the expression for the reaction written in the reverse direction.* Consequently, the numerical value of the equilibrium constant for the reaction written in one direction is the reciprocal of that for the reverse reaction. Both expressions are equally valid, but it is meaningless to say that the equilibrium constant for the equilibrium between NO_2 and N_2O_4 is "0.212" or "4.72" unless we indicate how the equilibrium reaction is written and specify the temperature. Therefore, whenever you are using an equilibrium constant, you should always write the associated balanced chemical equation.

Relating Chemical Equation Stoichiometry and Equilibrium Constants

There are many ways to write a balanced chemical equation for a given reaction. For example, if we multiply Equation 15.1, $N_2O_4(g) \rightleftharpoons 2\,NO_2(g)$ by 2, we have

$$2\,N_2O_4(g) \rightleftharpoons 4\,NO_2(g)$$

This chemical equation is balanced and might be written this way in some contexts. Therefore, the equilibrium-constant expression for this equation is

$$K_c = \frac{[NO_2]^4}{[N_2O_4]^2}$$

which is the square of the equilibrium-constant expression given in Equation 15.10 for the reaction as written in Equation 15.1: $K_c = [NO_2]^2/[N_2O_4]$. Because the new equilibrium-constant expression equals the original expression squared, the new equilibrium constant K_c equals the original constant squared: $0.212^2 = 0.0449$ (at 100 °C). Once again, you must relate each equilibrium constant you work with to a *specific* balanced chemical equation. *The concentrations of the substances in the equilibrium mixture will be the same no matter how you write the chemical equation, but the value of K_c you calculate depends on how you write the reaction.*

It is also possible to calculate the equilibrium constant for a reaction if we know the equilibrium constants for other reactions that add up to give us the one we want, similar to the manner in which the reaction enthalpy of an unknown reaction can be determined from known reaction enthalpies using Hess's law. (Section 5.6) For example, consider the following two reactions, their equilibrium-constant expressions, and their equilibrium constants at 100 °C:

1. $2\,NOBr(g) \rightleftharpoons 2\,NO(g) + Br_2(g)$ $K_{c1} = \dfrac{[NO]^2[Br_2]}{[NOBr]^2} = 0.014$

2. $Br_2(g) + Cl_2(g) \rightleftharpoons 2\,BrCl(g)$ $K_{c2} = \dfrac{[BrCl]^2}{[Br_2][Cl_2]} = 7.2$

The net sum of these two equations is:

3. $2\,NOBr(g) + Cl_2(g) \rightleftharpoons 2\,NO(g) + 2\,BrCl(g)$

You can prove algebraically that the equilibrium-constant expression for the net reaction is the product of the expressions for individual reactions:

$$K_{c3} = \frac{[NO]^2[BrCl]^2}{[NOBr]^2[Cl_2]} = \frac{[NO]^2[Br_2]}{[NOBr]^2} \times \frac{[BrCl]^2}{[Br_2][Cl_2]}$$

Thus,

$$K_{c3} = (K_{c1})(K_{c2}) = (0.014)(7.2) = 0.10$$

To summarize:

1. The equilibrium constant of a reaction in the *reverse* direction is the *inverse* (or *reciprocal*) of the equilibrium constant of the reaction in the forward direction:

$$A + B \rightleftharpoons C + D \quad K_1$$
$$C + D \rightleftharpoons A + B \quad K = 1/K_1$$

2. The equilibrium constant of a reaction that has been *multiplied* by a number is equal to the original equilibrium constant raised to a *power* equal to that number.

$$A + B \quad\rightleftharpoons C + D \quad K_1$$
$$n\,A + n\,B \rightleftharpoons n\,C + n\,D \quad K = K_1{}^n$$

3. The equilibrium constant for a net reaction made up by adding *two or more reactions* is the *product* of the equilibrium constants for the individual reactions:

$$A + B \rightleftharpoons C + D \quad K_1$$
$$\underline{C + F \rightleftharpoons G + A \quad K_2}$$
$$B + F \rightleftharpoons D + G \quad K_3 = (K_1)(K_2)$$

Sample Exercise 15.4

Combining Equilibrium Expressions

Given the reactions

$$HF(aq) \rightleftharpoons H^+(aq) + F^-(aq) \qquad K_c = 6.8 \times 10^{-4}$$
$$H_2C_2O_4(aq) \rightleftharpoons 2\,H^+(aq) + C_2O_4^{2-}(aq) \quad K_c = 3.8 \times 10^{-6}$$

determine the value of K_c for the reaction

$$2\,HF(aq) + C_2O_4^{2-}(aq) \rightleftharpoons 2\,F^-(aq) + H_2C_2O_4(aq)$$

Continued

SOLUTION

Analyze We are given two equilibrium equations and the corresponding equilibrium constants and are asked to determine the equilibrium constant for a third equation, which is related to the first two.

Plan We cannot simply add the first two equations to get the third. Instead, we need to determine how to manipulate these equations to come up with equations that we can add to give us the desired equation.

Solve

If we multiply the first equation by 2 and make the corresponding change to its equilibrium constant (raising to the power 2), we get

$$2\,HF(aq) \rightleftharpoons 2\,H^+(aq) + 2\,F^-(aq)$$

$$K_c = (6.8 \times 10^{-4})^2 = 4.6 \times 10^{-7}$$

Reversing the second equation and again making the corresponding change to its equilibrium constant (taking the reciprocal) gives

$$2\,H^+(aq) + C_2O_4{}^{2-}(aq) \rightleftharpoons H_2C_2O_4(aq)$$

$$K_c = \frac{1}{3.8 \times 10^{-6}} = 2.6 \times 10^5$$

Now, we have two equations that sum to give the net equation, and we can multiply the individual K_c values to get the desired equilibrium constant.

$$2\,HF(aq) \rightleftharpoons 2\,H^+(aq) + 2\,F^-(aq) \qquad K_c = 4.6 \times 10^{-7}$$
$$2\,H^+(aq) + C_2O_4{}^{2-}(aq) \rightleftharpoons H_2C_2O_4(aq) \qquad K_c = 2.5 \times 10^5$$
$$\overline{2\,HF(aq) + C_2O_4{}^{2-}(aq) \rightleftharpoons 2\,F^-(aq) + H_2C_2O_4(aq) \qquad K_c = (4.6 \times 10^{-7})(2.6 \times 10^5) = 0.12}$$

▶ **Practice Exercise**

Given that, at 700 K, $K_p = 54.0$ for the reaction $H_2(g) + I_2(g) \rightleftharpoons 2\,HI(g)$ and $K_p = 1.04 \times 10^{-4}$ for the reaction $N_2(g) + 3\,H_2(g) \rightleftharpoons 2\,NH_3(g)$, determine the value of K_p for the reaction $2\,NH_3(g) + 3\,I_2(g) \rightleftharpoons 6\,HI(g) + N_2(g)$ at 700 K.

▲ Self-Assessment Exercises

SAE 15.7   The following reaction has an equilibrium constant $K_p < 1$:

$$A(g) + B(g) \rightleftharpoons C(g).$$

Which of the following charts could describe the approach to equilibrium starting from a mixture of B(g) and C(g)?

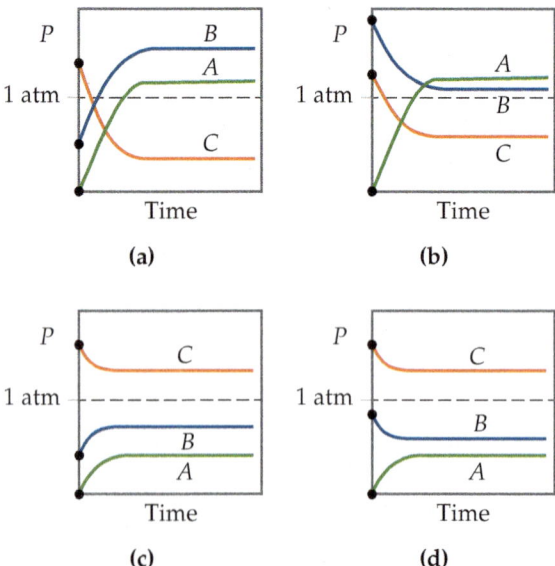

(a)   (b)

(c)   (d)

SAE 15.8   Consider the following two equilibria at 300 K:

$$2\,SO_3(g) \rightleftharpoons 2\,SO_2(g) + O_2(g) \quad K_c = 2.3 \times 10^{-7}$$

$$2\,NO_3(g) \rightleftharpoons 2\,NO_2(g) + O_2(g) \quad K_c = 1.4 \times 10^{-3}$$

What is the value of K_c for the following equilibrium at 300 K?

$$SO_2(g) + NO_3(g) \rightleftharpoons SO_3(g) + NO_2(g)$$

(a) 1.8×10^{-5} **(b)** 7.8×10^1 **(c)** 3.0×10^3 **(d)** 6.1×10^3

15.4 | Heterogeneous Equilibria

Many equilibria involve substances that are all in the same phase. Such equilibria are called **homogeneous equilibria**. The equilibrium between $N_2O_4(g)$ and $NO_2(g)$ shown in Figure 15.1 is one such example. In some cases, however, the substances in equilibrium are in different phases, giving rise to **heterogeneous equilibria**. An example occurs when solid lead(II) chloride dissolves in water to form a saturated solution:

$$PbCl_2(s) \rightleftharpoons Pb^{2+}(aq) + 2\,Cl^-(aq) \qquad [15.19]$$

This system consists of a solid in equilibrium with two aqueous species. If we want to write the equilibrium-constant expression for this process, we encounter a problem we have not encountered previously: How do we express the concentration of a solid? As it turns out, we don't need to worry about it! Experiments show that the amount of a solid present does not affect the equilibrium (as long as there is some of the solid present). For the reaction in Equation 15.19, the equilibrium-constant expression is

$$K_c = [Pb^{2+}][Cl^-]^2 \qquad [15.20]$$

Thus, our problem of how to express the concentration of a solid is not relevant in the end because $PbCl_2(s)$ does not show up in the equilibrium-constant expression. More generally, we can state that *whenever a pure solid or a pure liquid is involved in a heterogeneous equilibrium, its concentration is not included in the equilibrium-constant expression.*

The fact that pure solids and pure liquids are excluded from equilibrium-constant expressions can be explained in two ways. First, the concentration of a pure solid or liquid has a constant value. If the mass of a solid is doubled, its volume also doubles. Thus, its concentration, which relates to the ratio of mass to volume, stays the same. Because equilibrium-constant expressions include terms only for reactants and products whose concentrations can change during a chemical reaction, the concentrations of pure solids and pure liquids are omitted.

Second, recall from Section 15.2 that what is substituted into a thermodynamic equilibrium expression is the *activity* of each substance, which is a ratio of the concentration to a reference value. For a pure substance, the reference value is the concentration of the pure substance, so the activity of any pure solid or liquid is always 1.

The decomposition of calcium carbonate is another example of a heterogeneous reaction:

$$CaCO_3(s) \rightleftharpoons CaO(s) + CO_2(g)$$

Omitting the concentrations of the solids from the equilibrium-constant expression gives

$$K_c = [CO_2] \quad \text{and} \quad K_p = P_{CO_2}$$

These equations tell us that at a given temperature, an equilibrium among $CaCO_3$, CaO, and CO_2 always leads to the same CO_2 partial pressure as long as all three components are present. As shown in **Figure 15.7**, we have the same CO_2 pressure regardless of the relative amounts of CaO and $CaCO_3$.

When a solvent is a reactant or product in an equilibrium, its concentration is omitted from the equilibrium-constant expression, provided the concentrations of reactants and products are low, so that the solvent is essentially a pure substance. Applying this guideline to an equilibrium involving water as a solvent,

$$H_2O(l) + CO_3^{2-}(aq) \rightleftharpoons OH^-(aq) + HCO_3^-(aq) \qquad [15.21]$$

gives an equilibrium-constant expression that does not contain $[H_2O]$:

$$K_c = \frac{[OH^-][HCO_3^-]}{[CO_3^{2-}]} \qquad [15.22]$$

Learning Objective

When you finish Section 15.4, you should be able to:

▶ Write equilibrium-constant expressions for reactions that involve a pure solid or pure liquid.

Go Figure If some of the $CO_2(g)$ were released from the bell jar on the left, the seal then restored, and the system allowed to return to equilibrium, would the amount of $CaCO_3(s)$ increase, decrease, or remain the same?

$$CaCO_3(s) \rightleftharpoons CaO(s) + CO_2(g)$$

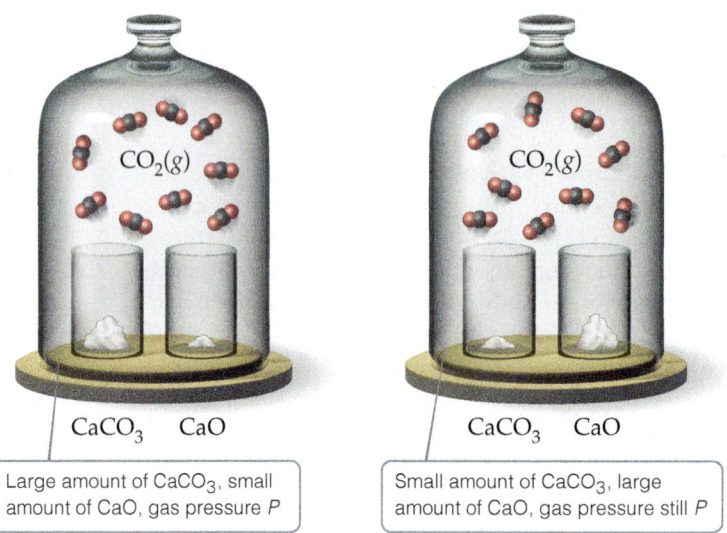

CaCO₃   CaO

Large amount of $CaCO_3$, small amount of CaO, gas pressure P

CaCO₃   CaO

Small amount of $CaCO_3$, large amount of CaO, gas pressure still P

▲ **Figure 15.7 At a given temperature, the equilibrium pressure of CO₂ in the bell jars is the same no matter how much of each solid is present.**

Sample Exercise 15.5

Writing Equilibrium-Constant Expressions for Heterogeneous Reactions

Write the equilibrium-constant expression K_c for

(a) $CO_2(g) + H_2(g) \rightleftharpoons CO(g) + H_2O(l)$

(b) $SnO_2(s) + 2\,CO(g) \rightleftharpoons Sn(s) + 2\,CO_2(g)$

SOLUTION

Analyze We are given two chemical equations, both for heterogeneous equilibria, and asked to write the corresponding equilibrium-constant expressions.

Plan We use the law of mass action, remembering to omit any pure solids and pure liquids from the expressions.

Solve

(a) The equilibrium-constant expression is

$$K_c = \frac{[CO]}{[CO_2][H_2]}$$

Because H_2O appears in the reaction as a liquid, its concentration does not appear in the equilibrium-constant expression.

(b) The equilibrium-constant expression is

$$K_c = \frac{[CO_2]^2}{[CO]^2}$$

Because SnO_2 and Sn are pure solids, their concentrations do not appear in the equilibrium-constant expression.

▶ **Practice Exercise**

Write the following equilibrium-constant expressions:
(a) K_c for $Cr(s) + 3\,Ag^+(aq) \rightleftharpoons Cr^{3+}(aq) + 3\,Ag(s)$,
(b) K_p for $3\,Fe(s) + 4\,H_2O(g) \rightleftharpoons Fe_3O_4(s) + 4\,H_2(g)$.

Sample Exercise 15.6
Analyzing a Heterogeneous Equilibrium

Each of these mixtures was placed in a closed container and allowed to stand:

(a) $CaCO_3(s)$

(b) $CaO(s)$ and $CO_2(g)$ at a pressure greater than the value of K_p

(c) $CaCO_3(s)$ and $CO_2(g)$ at a pressure greater than the value of K_p

(d) $CaCO_3(s)$ and $CaO(s)$

Determine whether or not each mixture can attain the equilibrium

$$CaCO_3(s) \rightleftharpoons CaO(s) + CO_2(g)$$

SOLUTION

Analyze We are asked which of several combinations of species can establish an equilibrium between calcium carbonate and its decomposition products, calcium oxide and carbon dioxide.

Plan For equilibrium to be achieved, it must be possible for both the forward process and the reverse process to occur. For the forward process to occur, some calcium carbonate must be present. For the reverse process to occur, both calcium oxide and carbon dioxide must be present. In both cases, either the necessary compounds may be present initially or they may be formed by reaction of the other species.

Solve Equilibrium can be reached in all cases except (c) as long as sufficient quantities of solids are present. (a) $CaCO_3$ simply decomposes, forming $CaO(s)$ and $CO_2(g)$ until the equilibrium pressure of CO_2 is attained. There must be enough $CaCO_3$,

however, to allow the CO_2 pressure to reach equilibrium. (b) CO_2 continues to combine with CaO until the partial pressure of the CO_2 decreases to the equilibrium value. (c) Because there is no CaO present, equilibrium cannot be attained; there is no way the CO_2 pressure can decrease to its equilibrium value (which would require some CO_2 to react with CaO). (d) The situation is essentially the same as in (a): $CaCO_3$ decomposes until equilibrium is attained. The presence of CaO initially makes no difference.

▶ **Practice Exercise**
When added to $Fe_3O_4(s)$ in a closed container, which one of the following substances—$H_2(g), H_2O(g), O_2(g)$— allows equilibrium to be established in the reaction $3\,Fe(s) + 4\,H_2O(g) \rightleftharpoons Fe_3O_4(s) + 4\,H_2(g)$?

Self-Assessment Exercises

SAE 15.9 Sodium azide, $NaN_3(s)$, is used to inflate automotive air bags using the following equilibrium:

$$2\,NaN_3(s) \rightleftharpoons 2\,Na(s) + 3\,N_2(g)$$

Which of the following statements about this equilibrium reaction is or are *true*?

(i) In writing the equilibrium-constant expression for the reaction, we do not include NaN_3 and Na because they are solids.

(ii) For an equilibrium mixture of the three substances, the equilibrium constant K_p will equal the equilibrium pressure of $N_2(g)$.

(iii) An equilibrium mixture can be reached starting with pure $NaN_3(s)$.

(a) Only statement i is true. (b) Statements i and ii are true. (c) Statements i and iii are true. (d) Statements ii and iii are true. (e) All three statements are true.

SAE 15.10 The following equilibrium is established in a saturated solution of lead bromide:

$$PbBr_2(s) \rightleftharpoons Pb^{2+}(aq) + 2\,Br^-(aq).$$

If additional solid $PbBr_2$ is added to this solution, which of the following statements is *true*? (a) The total mass of undissolved $PbBr_2$ will return to the value before more $PbBr_2$ was added. (b) The value of K_c will increase. (c) The concentration of Br^- will increase more than the concentration of Pb^{2+} does. (d) The concentrations of Pb^{2+} and Br^- will not change.

Learning Objectives

When you finish Section 15.5, you should be able to:

▶ Determine the value of K from known equilibrium concentrations of the reactants and products.

▶ Deduce the value of K when the initial concentrations of reactants and products are known and the equilibrium concentration of at least one species is known.

15.5 | Calculating Equilibrium Constants

The value of an equilibrium constant is a powerful quantitative tool in chemistry. By using the equilibrium-constant expression and the equilibrium concentrations of at least some of the reactants and products, we can often determine the numerical value of the equilibrium constant at a specific temperature (remember that the value of an equilibrium constant changes when the temperature changes). In this section, we look at some of these techniques more closely.

Let's start with the simplest way to determine the value of an equilibrium constant: If we can measure the equilibrium concentrations of all the reactants and products in a chemical reaction, as we did with the data in Table 15.1, then calculating the value of the equilibrium constant is straightforward. We simply insert all the equilibrium concentrations into the equilibrium-constant expression for the reaction, as shown in Sample Exercise 15.7.

Sample Exercise 15.7
Calculating *K* When All Equilibrium Concentrations Are Known

After a mixture of hydrogen and nitrogen gases in a reaction vessel is allowed to attain equilibrium at 472 °C, it is found to contain 7.38 atm H_2, 2.46 atm N_2, and 0.166 atm NH_3. From these data, calculate the equilibrium constant K_p for the reaction

$$N_2(g) + 3\,H_2(g) \rightleftharpoons 2\,NH_3(g)$$

SOLUTION

Analyze We are given a balanced equation and equilibrium partial pressures and are asked to calculate the value of the equilibrium constant.

Plan Using the balanced equation, we write the equilibrium-constant expression for K_p (Equation 15.11). We then substitute the equilibrium partial pressures into the expression and solve for K_p.

Solve

$$K_p = \frac{(P_{NH_3})^2}{(P_{N_2})(P_{H_2})^3} = \frac{(0.166)^2}{(2.46)(7.38)^3} = 2.79 \times 10^{-5}$$

Comment Whenever you calculate the value of an equilibrium constant, you should look at the value and ask yourself what

information the value conveys. In this case, the very small value of K_p tells us that an equilibrium mixture contains much more N_2 and H_2 than NH_3—this equilibrium lies significantly to the left at this temperature.

▶ **Practice Exercise**
An aqueous solution of acetic acid is found to have the following equilibrium concentrations at 25 °C: $[CH_3COOH] = 1.65 \times 10^{-2}\,M$; $[H^+] = 5.44 \times 10^{-4}\,M$; and $[CH_3COO^-] = 5.44 \times 10^{-4}\,M$. Calculate the equilibrium constant K_c for the ionization of acetic acid at 25 °C. The equilibrium reaction is

$$CH_3COOH(aq) \rightleftharpoons H^+(aq) + CH_3COO^-(aq)$$

Often, we do not know the equilibrium concentrations of all species in an equilibrium mixture. If we know the initial concentrations and the equilibrium concentration of at least one species, however, we can generally use the stoichiometry of the reaction to deduce the equilibrium concentrations of the others. The following steps outline the procedure:

How to Determine Concentrations of Unknown Species in an Equilibrium Mixture

1. Tabulate all known initial and equilibrium concentrations of the species that appear in the equilibrium-constant expression.

2. For those species for which initial and equilibrium concentrations are known, calculate the change in concentration that occurs as the system reaches equilibrium.

3. Use the stoichiometry of the reaction (that is, the coefficients in the balanced chemical equation) to calculate the changes in concentration for all other species in the equilibrium-constant expression.

4. Use initial concentrations from Step 1 and changes in concentration from Step 3 to calculate any equilibrium concentrations not tabulated in Step 1.

5. Determine the value of the equilibrium constant.

The best way to illustrate this procedure is by example, as we do in Sample Exercise 15.8.

Sample Exercise 15.8
Calculating *K* from Initial and Equilibrium Concentrations

A reaction vessel containing $1.000 \times 10^{-3}\,M\,H_2(g)$ and $2.000 \times 10^{-3}\,M\,I_2(g)$ is heated to 448 °C, where the following reaction takes place

$$H_2(g) + I_2(g) \rightleftharpoons 2\,HI(g)$$

What is the value of the equilibrium constant K_c if once the system comes to equilibrium at 448 °C the concentration of HI is $1.87 \times 10^{-3}\,M$?

SOLUTION

Analyze We are given the initial concentrations of H_2 and I_2 and the equilibrium concentration of HI. We are asked to calculate the equilibrium constant K_c for $H_2(g) + I_2(g) \rightleftharpoons 2\,HI(g)$.

Plan We construct a table to find equilibrium concentrations of all species and then use the equilibrium concentrations to calculate the equilibrium constant.

Solve

(1) We tabulate the initial and equilibrium concentrations of as many species as we can. We also provide space in our table for listing the changes in concentrations. As shown, it is convenient to use the chemical equation as the heading for the table.

	$H_2(g) + I_2(g) \rightleftharpoons 2\,HI(g)$		
Initial concentration (M)	1.000×10^{-3}	2.000×10^{-3}	0
Change in concentration (M)			
Equilibrium concentration (M)			1.87×10^{-3}

(2) We calculate the change in HI concentration, which is the difference between the equilibrium and initial values:

Change in $[HI] = 1.87 \times 10^{-3}\,M - 0 = 1.87 \times 10^{-3}M$

(3) We use the coefficients in the balanced equation to relate the change in $[HI]$ to the changes in $[H_2]$ and $[I_2]$:

$$\left(1.87 \times 10^{-3}\frac{\text{mol HI}}{L}\right)\left(\frac{1\,\text{mol } H_2}{2\,\text{mol HI}}\right) = 0.935 \times 10^{-3}\frac{\text{mol } H_2}{L}$$

$$\left(1.87 \times 10^{-3}\frac{\text{mol HI}}{L}\right)\left(\frac{1\,\text{mol } I_2}{2\,\text{mol HI}}\right) = 0.935 \times 10^{-3}\frac{\text{mol } I_2}{L}$$

(4) We calculate the equilibrium concentrations of H_2 and I_2, using initial concentrations and changes in concentration. The equilibrium concentration equals the initial concentration minus that consumed:

$$[H_2] = (1.000 \times 10^{-3}\,M) - (0.935 \times 10^{-3}\,M) = 0.065 \times 10^{-3}\,M$$

$$[I_2] = (2.000 \times 10^{-3}\,M) - (0.935 \times 10^{-3}\,M) = 1.065 \times 10^{-3}\,M$$

(5) Our table now is complete (with equilibrium concentrations in blue for emphasis):

	$H_2(g)$	$+$	$I_2(g)$	$\rightleftharpoons$	$2\,HI(g)$
Initial concentration (M)	1.000×10^{-3}		2.000×10^{-3}		0
Change in concentration (M)	-0.935×10^{-3}		-0.935×10^{-3}		$+1.87 \times 10^{-3}$
Equilibrium concentration (M)	0.065×10^{-3}		1.065×10^{-3}		1.87×10^{-3}

Notice that the entries for the changes are negative when a reactant is consumed and positive when a product is formed.

Finally, we use the equilibrium-constant expression to calculate the equilibrium constant:

$$K_c = \frac{[HI]^2}{[H_2][I_2]} = \frac{(1.87 \times 10^{-3})^2}{(0.065 \times 10^{-3})(1.065 \times 10^{-3})} = 51$$

Comment The same method can be applied to gaseous equilibrium problems to calculate K_p, in which case partial pressures are used as table entries in place of molar concentrations. Your instructor may refer to this kind of table as an ICE chart, where ICE stands for Initial – Change – Equilibrium.

▶ **Practice Exercise**
The gaseous compound BrCl decomposes at high temperature in a sealed container: $2\,BrCl(g) \rightleftharpoons Br_2(g) + Cl_2(g)$. Initially, the vessel is charged at 500 K with $BrCl(g)$ at a partial pressure of 0.500 atm. At equilibrium, the $BrCl(g)$ partial pressure is 0.040 atm. Calculate the value of K_p at 500 K.

 Self-Assessment Exercises

SAE 15.11 A mixture of $CH_4(g)$ and $H_2O(g)$ is heated and allowed to equilibrate at a certain temperature according to the following reaction:

$$CH_4(g) + H_2O(g) \rightleftharpoons CO(g) + 3\,H_2(g)$$

The equilibrium mixture of the gases has the following partial pressures: $P(CH_4) = 0.310$ atm, $P(H_2O) = 0.830$ atm, $P(CO) = 0.570$ atm, and $P(H_2) = 2.26$ atm. What is the value of K_p for this equilibrium? **(a)** 3.91×10^{-2} **(b)** 5.01 **(c)** 15.0 **(d)** 25.6

SAE 15.12 A 2.00-L flask is charged with 2.00 mol pure $NOBr(g)$, which equilibrates at constant temperature according to the following reaction:

$$2\,NOBr(g) \rightleftharpoons 2\,NO(g) + Br_2(g)$$

The equilibrium mixture of the three gases contains 0.86 mol $NOBr(g)$. What is the value of K_c for this equilibrium at this temperature? **(a)** 0.13 **(b)** 0.38 **(c)** 0.44 **(d)** 0.50 **(e)** 1.0

SAE 15.13 Consider the following equilibrium: $3\,A(g) \rightleftharpoons B(g)$.

A flask is charged with 1.00 atm of pure $A(g)$ and allowed to equilibrate at a fixed temperature. When equilibrium is established, the partial pressure of $A(g)$ in the flask is 0.40 atm. What is the value of K_p for this equilibrium at this temperature? **(a)** 0.50 **(b)** 3.1 **(c)** 6.3 **(d)** 9.4

15.6 | Some Applications of Equilibrium Constants

In this section, we continue our quantitative treatment of equilibria by looking at two important applications of equilibrium constants: (1) Determining whether a system is at equilibrium and, if not, how it must respond to achieve equilibrium, and (2) using the value of the equilibrium constant to determine the equilibrium concentrations of the reactants and products. The techniques we learn here will be particularly useful in the material presented on acid–base equilibria and aqueous equilibria in Chapters 16 and 17.

⚠ Learning Objectives

When you finish Section 15.6, you should be able to:

▶ Use the reaction quotient Q to predict the direction of a reaction that is not at equilibrium.

▶ Calculate equilibrium concentrations from initial concentrations, the equilibrium-constant expression, and the value of K.

Predicting the Direction of Reaction

Let's look again at the formation of NH_3 from N_2 and H_2 (Equation 15.6), for which $K_c = 0.105$ at 472 °C. Suppose we place 2.00 mol of H_2, 1.00 mol of N_2, and 2.00 mol of NH_3 in a 1.00-L container at 472 °C. How will the mixture react to reach equilibrium? Will N_2 and H_2 react to form more NH_3, or will NH_3 decompose to N_2, and H_2?

To answer this question, we substitute the starting concentrations of N_2, H_2, and NH_3 into the equilibrium-constant expression and compare its value to that of the equilibrium constant:

$$\frac{[NH_3]^2}{[N_2][H_2]^3} = \frac{(2.00)^2}{(1.00)(2.00)^3} = 0.500 \quad \text{whereas} \quad K_c = 0.105 \qquad [15.23]$$

To reach equilibrium, the quotient $[NH_3]^2/[N_2][H_2]^3$ must decrease from the starting value of 0.500 to the equilibrium value of 0.105. Because the system is closed, this change can happen only if $[NH_3]$ decreases and $[N_2]$ and $[H_2]$ increase. Thus, the reaction proceeds toward equilibrium by forming N_2 and H_2 from NH_3; that is, the reaction as written in Equation 15.6 proceeds from *right to left*, forming more reactants from products.

This approach can be formalized by defining a quantity called the reaction quotient:

> The **reaction quotient**, Q, is a number obtained by substituting reactant and product concentrations or partial pressures at any point during a reaction into an equilibrium-constant expression.

Therefore, for the general reaction

$$a\,A + b\,B \rightleftharpoons d\,D + e\,E$$

the reaction quotient in terms of molar concentrations is

$$Q_c = \frac{[D]^d[E]^e}{[A]^a[B]^b} \qquad [15.24]$$

A related quantity Q_p can be written for any reaction that involves gases by using partial pressures instead of concentrations.

Although we use what looks like the equilibrium-constant expression to calculate the reaction quotient, the concentrations we use may or may not be the equilibrium concentrations. For example, when we substituted the starting concentrations into the equilibrium-constant expression of Equation 15.23, we obtained $Q_c = 0.500$ whereas $K_c = 0.105$. The equilibrium constant has only one value at each temperature. The reaction quotient, however, varies as the reaction proceeds.

Of what use is Q? One practical thing we can do with Q is tell whether our reaction really is at equilibrium, which is an especially valuable option when a reaction is very slow. We can take samples of our reaction mixture as the reaction proceeds, separate the components, and measure their concentrations. Then we insert these numbers into Equation 15.24 for our reaction. To determine whether we are at equilibrium, or in which direction the reaction proceeds to achieve equilibrium, we compare the values of Q_c and K_c or Q_p and K_p. There are three possible scenarios (Figure 15.8), which can be summarized as follows:

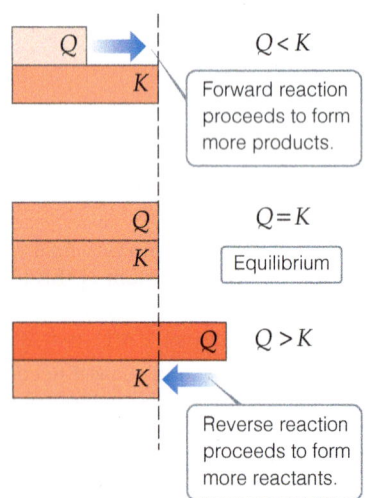

At equilibrium

$Q < K$ — Forward reaction proceeds to form more products.

$Q = K$ — Equilibrium

$Q > K$ — Reverse reaction proceeds to form more reactants.

▲ **Figure 15.8 Predicting the direction of a reaction by comparing Q and K at a given temperature.**

How to Use Q to Analyze Reaction Progress

- $Q < K$: The concentration of products is too small and that of reactants is too large. The reaction achieves equilibrium by forming more products; it proceeds from left to right.
- $Q = K$: The reaction quotient equals the equilibrium constant only if the system is at equilibrium.
- $Q > K$: The concentration of products is too large and that of reactants is too small. The reaction achieves equilibrium by forming more reactants; it proceeds from right to left.

 Sample Exercise 15.9

Predicting the Direction of Approach to Equilibrium

At 448 °C, the equilibrium constant K_c for the reaction

$$H_2(g) + I_2(g) \rightleftharpoons 2\,HI(g)$$

is 50.5. Predict in which direction the reaction proceeds to reach equilibrium if we start with 2.0×10^{-2} mol of HI, 1.0×10^{-2} mol of H_2, and 3.0×10^{-2} mol of I_2 in a 2.00-L container.

SOLUTION

Analyze We are given a volume and initial molar amounts of the species in a reaction and asked to determine in which direction the reaction must proceed to achieve equilibrium.

Plan We can determine the starting concentration of each species in the reaction mixture. We can then substitute the starting concentrations into the equilibrium-constant expression to calculate the reaction quotient, Q_c. Comparing the magnitudes of the equilibrium constant, which is given, and the reaction quotient will tell us in which direction the reaction will proceed.

Solve The initial concentrations are

$$[HI] = 2.0 \times 10^{-2}\,\text{mol}/2.00\,\text{L} = 1.0 \times 10^{-2}\,M$$

$$[H_2] = 1.0 \times 10^{-2}\,\text{mol}/2.00\,\text{L} = 5.0 \times 10^{-3}\,M$$

$$[I_2] = 3.0 \times 10^{-2}\,\text{mol}/2.00\,\text{L} = 1.5 \times 10^{-2}\,M$$

The reaction quotient Q_c is therefore

$$Q_c = \frac{[HI]^2}{[H_2][I_2]} = \frac{(1.0 \times 10^{-2})^2}{(5.0 \times 10^{-3})(1.5 \times 10^{-2})} = 1.3$$

Because $K_c = 50.5$, $Q_c < K_c$ and therefore the concentration of HI must increase and the concentrations of H_2 and I_2 must decrease to reach equilibrium; the reaction as written proceeds left to right to attain equilibrium.

▶ **Practice Exercise**

At 1000 K, the value of K_p for the reaction $2\,SO_3(g) \rightleftharpoons 2\,SO_2(g) + O_2(g)$ is 0.338. Calculate the value for Q_p, and predict the direction in which the reaction proceeds toward equilibrium if the initial partial pressures are $P_{SO_3} = 0.16$ atm; $P_{SO_2} = 0.41$ atm; $P_{O_2} = 2.5$ atm.

Calculating Equilibrium Concentrations

Chemists frequently need to calculate the amounts of reactants and products present at equilibrium in a reaction for which they know the equilibrium constant. The approach in solving problems of this type is similar to the one we used for evaluating equilibrium constants: We tabulate initial concentrations or partial pressures, changes in those concentrations or pressures, and final equilibrium concentrations or partial pressures. Usually, we end up using the equilibrium-constant expression to derive an equation that must be solved for an unknown quantity, as demonstrated in Sample Exercise 15.10.

 Sample Exercise 15.10

Calculating Equilibrium Concentrations

For the Haber process, $N_2(g) + 3\,H_2(g) \rightleftharpoons 2\,NH_3(g)$, $K_p = 1.45 \times 10^{-5}$, at 500 °C. In an equilibrium mixture of the three gases at 500 °C, the partial pressure of H_2 is 0.928 atm and that of N_2 is 0.432 atm. What is the partial pressure of NH_3 in this equilibrium mixture?

SOLUTION

Analyze We are given an equilibrium constant, K_p, and the equilibrium partial pressures of two of the three substances in the equation (N_2 and H_2), and we are asked to calculate the equilibrium partial pressure for the third substance (NH_3).

Plan We can set K_p equal to the equilibrium-constant expression and substitute in the partial pressures that we know. Then we can solve for the only unknown in the equation.

Solve

We tabulate the equilibrium pressures:

$$N_2(g) + 3\,H_2(g) \rightleftharpoons 2\,NH_3(g)$$

Equilibrium pressure (atm)	0.432 0.928	x

Continued

Because we do not know the equilibrium pressure of NH_3, we represent it with x. At equilibrium, the pressures must satisfy the equilibrium-constant expression:

$$K_p = \frac{(P_{NH_3})^2}{P_{N_2}(P_{H_2})^3} = \frac{x^2}{(0.432)(0.928)^3} = 1.45 \times 10^{-5}$$

We now rearrange the equation to solve for x:

$$x^2 = (1.45 \times 10^{-5})(0.432)(0.928)^3 = 5.01 \times 10^{-6}$$

$$x = \sqrt{5.01 \times 10^{-6}} = 2.24 \times 10^{-3}\,\text{atm} = P_{NH_3}$$

Check We can always check our answer by using it to recalculate the value of the equilibrium constant:

$$K_p = \frac{(2.24 \times 10^{-3})^2}{(0.432)(0.928)^3} = 1.45 \times 10^{-5}$$

▶ **Practice Exercise**

At 500 K, the reaction $PCl_5(g) \rightleftharpoons PCl_3(g) + Cl_2(g)$ has $K_p = 0.497$. In an equilibrium mixture at 500 K, the partial pressure of PCl_5 is 0.860 atm and that of PCl_3 is 0.350 atm. What is the partial pressure of Cl_2 in the equilibrium mixture?

In many situations, we know the value of the equilibrium constant and the initial amounts of all species. We must then solve for the equilibrium amounts. Solving this type of problem usually entails treating the change in concentration as a variable. The stoichiometry of the reaction gives us the relationship between the changes in the amounts of all the reactants and products, and the calculations frequently involve the quadratic formula, as shown in Sample Exercise 15.11.

Sample Exercise 15.11

Calculating Equilibrium Concentrations from Initial Concentrations

A 1.000-L flask is filled with 1.000 mol of $H_2(g)$ and 2.000 mol of $I_2(g)$ at 448 °C. The value of the equilibrium constant K_c for the reaction

$$H_2(g) + I_2(g) \rightleftharpoons 2\,HI(g)$$

at 448 °C is 50.5. What are the equilibrium concentrations of H_2, I_2, and HI in moles per liter?

SOLUTION

Analyze We are given the volume of a container, an equilibrium constant, and starting amounts of reactants in the container and are asked to calculate the equilibrium concentrations of all species.

Plan In this case, we are not given any of the equilibrium concentrations. We must develop some relationships that relate the initial concentrations to those at equilibrium. The procedure is similar in many regards to that outlined in Sample Exercise 15.8, where we calculated an equilibrium constant using initial concentrations.

Solve

(1) The initial concentrations of H_2 and I_2 are:

$$[H_2] = 1.000\,M \quad \text{and} \quad [I_2] = 2.000\,M$$

(2) Construct a table that includes the initial concentrations:

	$H_2(g)$ +	$I_2(g)$ $\rightleftharpoons$	2 HI(g)
Initial concentration (M)	1.000	2.000	0
Change in concentration (M)			
Equilibrium concentration (M)			

(3) Use the stoichiometry of the reaction to determine the changes in concentration that occur as the reaction proceeds to equilibrium. The H_2 and I_2 concentrations will decrease as equilibrium is established and that of HI will increase. Let's represent the change in concentration of H_2 by x. The balanced chemical equation tells us that for each x mol of H_2 that reacts, x mol of I_2 are consumed and $2x$ mol of HI are produced:

	$H_2(g)$ +	$I_2(g)$ $\rightleftharpoons$	2 HI(g)
Initial concentration (M)	1.000	2.000	0
Change in concentration (M)	$-x$	$-x$	$+2x$
Equilibrium concentration (M)			

(4) Use initial concentrations and changes in concentrations, as dictated by stoichiometry, to express the equilibrium concentrations. With all our entries, our table now looks like this:

	$H_2(g)$ +	$I_2(g)$ $\rightleftharpoons$	2 HI(g)
Initial concentration (M)	1.000	2.000	0
Change in concentration (M)	$-x$	$-x$	$+2x$
Equilibrium concentration (M)	$1.000 - x$	$2.000 - x$	$2x$

(**5**) Substitute the equilibrium concentrations into the equilibrium-constant expression and solve for x:

$$K_c = \frac{[HI]^2}{[H_2][I_2]} = \frac{(2x)^2}{(1.000 - x)(2.000 - x)} = 50.5$$

If you have an equation-solving calculator, you can solve this equation directly for x. If not, expand this expression to obtain a quadratic equation in x:

$$4x^2 = 50.5(x^2 - 3.000x + 2.000)$$

$$46.5x^2 - 151.5x + 101.0 = 0$$

Solving the quadratic equation (Appendix A.3) leads to two solutions for x:

$$x = \frac{-(-151.5) \pm \sqrt{(-151.5)^2 - 4(46.5)(101.0)}}{2(46.5)} = 2.323 \text{ or } 0.935$$

When we substitute $x = 2.323$ into the expressions for the equilibrium concentrations, we find *negative* concentrations of H_2 and I_2. Because a negative concentration is not chemically meaningful, we reject this solution. We then use $x = 0.935$ to find the equilibrium concentrations:

$$[H_2] = 1.000 - x = 0.065 \, M$$

$$[I_2] = 2.000 - x = 1.065 \, M$$

$$[HI] = 2x = 1.87 \, M$$

Check We can check our solution by putting these numbers into the equilibrium-constant expression to assure that we correctly calculate the equilibrium constant:

$$K_c = \frac{[HI]^2}{[H_2][I_2]} = \frac{(1.87)^2}{(0.065)(1.065)} = 51$$

Comment Whenever you use a quadratic equation to solve an equilibrium problem, one of the solutions to the equation will give you a value that leads to negative concentrations and thus is not chemically meaningful. Reject this solution to the quadratic equation.

▶ **Practice Exercise**

For the equilibrium $PCl_5(g) \rightleftharpoons PCl_3(g) + Cl_2(g)$, the equilibrium constant K_p is 0.497 at 500 K. A gas cylinder at 500 K is charged with $PCl_5(g)$ at an initial pressure of 1.66 atm. What are the equilibrium pressures of PCl_5, PCl_3, and Cl_2 at this temperature?

 Self-Assessment Exercises

SAE 15.14 At 950 K, $K_p = 1.75$ for the following equilibrium:

$$C(s) + CO_2(g) \rightleftharpoons 2\,CO(g).$$

An empty vessel at 950 K is charged with 0.50 atm $CO_2(g)$ and 3.0 atm $CO(g)$. What happens as this system attempts to reach equilibrium? (**a**) The system is already at equilibrium. (**b**) $CO_2(g)$ decomposes to form more $CO(g)$. (**c**) $CO(g)$ decomposes to form $C(s)$ and $CO_2(g)$. (**d**) The system is unable to reach equilibrium until $C(s)$ is added to the vessel.

SAE 15.15 At a certain temperature T, the equilibrium constant for the following reaction is $K_c = 8.0$:

$$A(aq) \rightleftharpoons B(aq)$$

A flask at temperature T is charged with 400.0 mL of 1.0 M A(aq). What is [B(aq)] when the reaction reaches equilibrium? (**a**) 0.36 M (**b**) 0.89 M (**c**) 2.8 M (**d**) 8.0 M (**e**) More information is needed.

SAE 15.16 At $T = 700$ K, the equilibrium constant for the following reaction is $K_p = 0.76$:

$$CCl_4(g) \rightleftharpoons C(s) + 2\,Cl_2(g)$$

A flask is charged with 3.0 atm of $CCl_4(g)$ at 700 K, which is then allowed to reach equilibrium. What is the partial pressure of $Cl_2(g)$ in the flask at equilibrium? (**a**) 0.666 atm (**b**) 0.826 atm (**c**) 1.12 atm (**d**) 1.18 atm (**e**) 1.33 atm

15.7 | Le Châtelier's Principle

Many of the products we use in everyday life are obtained from the chemical industry. Chemists and chemical engineers in industry spend a great deal of time and effort to maximize the yield of valuable products and minimize waste. For example, when Haber developed his process for making ammonia from N_2 and H_2, he examined how reaction conditions might be varied to increase the yield of NH_3. Using the values of the equilibrium constant at various temperatures, he calculated the equilibrium amounts of NH_3 formed under a variety of conditions. Some of Haber's results are shown in **Figure 15.9**. Notice that the percent of NH_3 present at equilibrium decreases with increasing temperature and increases with increasing pressure.

We can understand these effects in terms of a principle first put forward by Henri-Louis Le Châtelier* (1850–1936), a French industrial chemist:

 Learning Objectives

When you finish Section 15.7, you should be able to:

▶ Use Le Châtelier's principle to predict how a system at equilibrium will respond to a change that takes the system out of equilibrium.

▶ Describe the effect of a catalyst on chemical equilibrium.

---

*Pronounced "le-SHOT-lee-ay."

If a system at equilibrium is disturbed by a change in temperature, pressure, or a component concentration, the system will shift its equilibrium position so as to counteract the effect of the disturbance.

In this section, we use Le Châtelier's principle to make qualitative predictions about how a system at equilibrium responds to various changes in external conditions. We consider three ways in which a chemical equilibrium can be disturbed: (1) adding or removing a reactant or product, (2) changing the pressure by changing the volume of the reaction vessel, and (3) changing the temperature. The impact of Le Châtelier's principle under these changes is summarized in the chart in **Figure 15.10**. You will find it helpful to refer to this chart as we go through the discussion.

Go Figure At what combination of pressure and temperature should you run the reaction to maximize NH₃ yield?

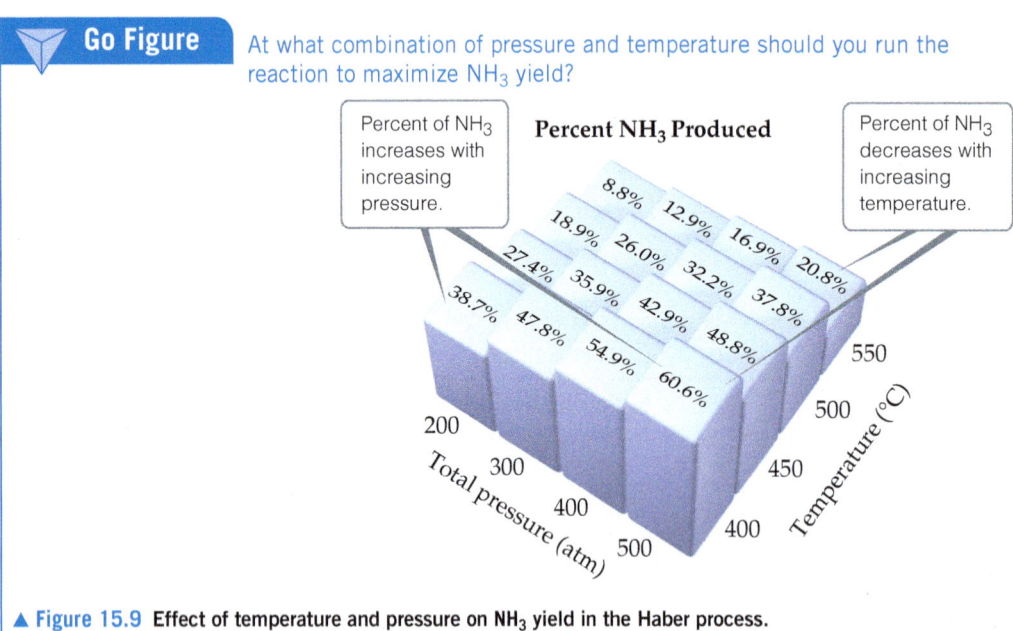

Percent NH₃ Produced

Percent of NH₃ increases with increasing pressure.

Percent of NH₃ decreases with increasing temperature.

▲ **Figure 15.9 Effect of temperature and pressure on NH₃ yield in the Haber process.** Each mixture was produced by starting with a 3:1 molar mixture of H₂ and N₂.

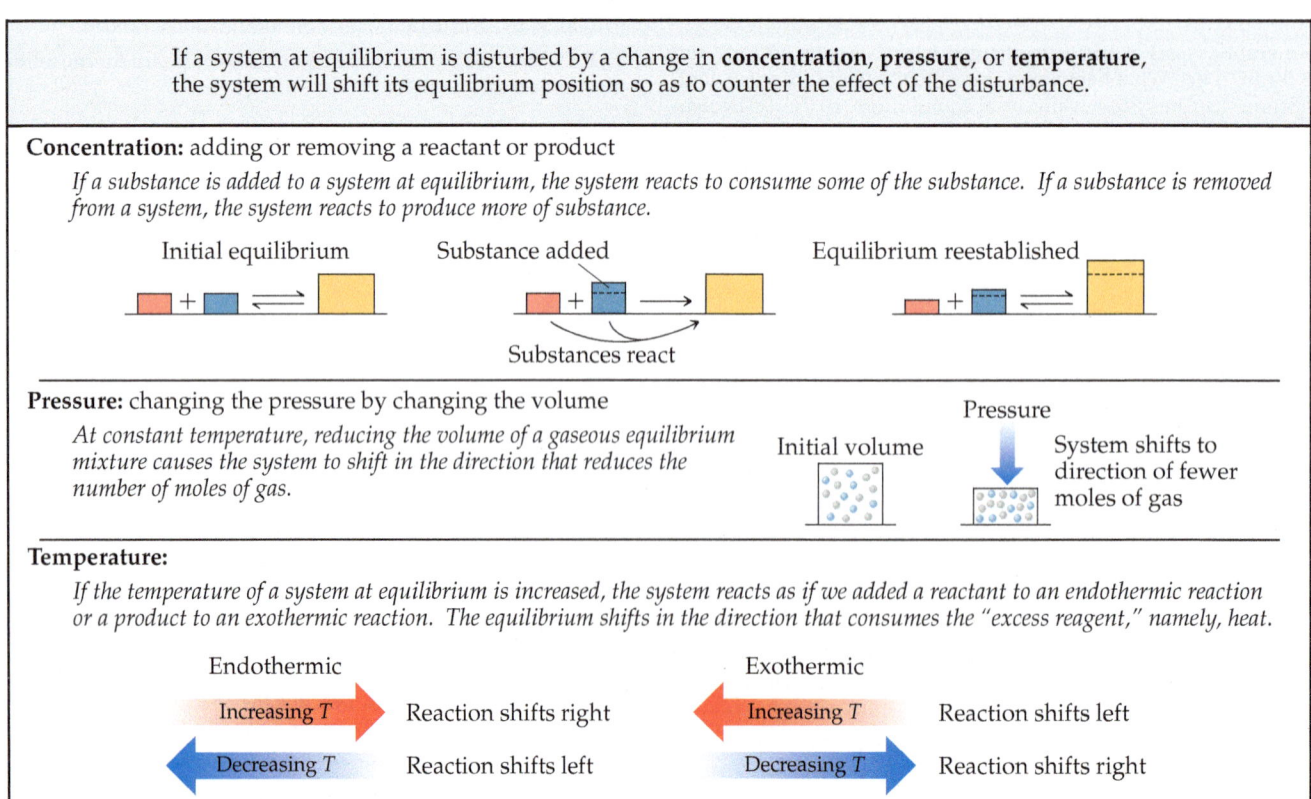

If a system at equilibrium is disturbed by a change in **concentration**, **pressure**, or **temperature**, the system will shift its equilibrium position so as to counter the effect of the disturbance.

Concentration: adding or removing a reactant or product

If a substance is added to a system at equilibrium, the system reacts to consume some of the substance. If a substance is removed from a system, the system reacts to produce more of substance.

Initial equilibrium   Substance added   Equilibrium reestablished

Substances react

Pressure: changing the pressure by changing the volume

At constant temperature, reducing the volume of a gaseous equilibrium mixture causes the system to shift in the direction that reduces the number of moles of gas.

Pressure

Initial volume   System shifts to direction of fewer moles of gas

Temperature:

If the temperature of a system at equilibrium is increased, the system reacts as if we added a reactant to an endothermic reaction or a product to an exothermic reaction. The equilibrium shifts in the direction that consumes the "excess reagent," namely, heat.

Endothermic   Exothermic

Increasing *T*   Reaction shifts right   Increasing *T*   Reaction shifts left

Decreasing *T*   Reaction shifts left   Decreasing *T*   Reaction shifts right

▲ **Figure 15.10 Summary of Le Châtelier's principle.**

Change in Reactant or Product Concentration

A system at dynamic equilibrium is in a state of balance. When the concentrations of species in the reaction are altered, the equilibrium shifts until a new state of balance is attained. What does *shift* mean? It means that reactant and product concentrations change over time to accommodate the new situation. *Shift* does *not* mean that the equilibrium constant itself is altered; the equilibrium constant remains the same as long as the temperature remains the same. Le Châtelier's principle states that the shift is in the direction that minimizes or reduces the effect of the change:

> *If a chemical system is already at equilibrium and the concentration of any substance in the mixture is increased (either reactant or product), the system reacts to consume some of that substance. Conversely, if the concentration of a substance is decreased, the system reacts to produce some of that substance.*

As an example, consider our familiar equilibrium mixture of N_2, H_2, and NH_3:

$$N_2(g) + 3\,H_2(g) \rightleftharpoons 2\,NH_3(g)$$

Adding H_2 takes the system out of equilibrium, and as a result, the system will shift so as to reduce the concentration of H_2 **(Figure 15.11)**. This change can occur only if the reaction consumes H_2 and simultaneously consumes N_2 to form more NH_3. Adding N_2 to the equilibrium mixture likewise causes the reaction to shift toward forming more NH_3. Removing NH_3 also causes a shift toward producing more NH_3, whereas *adding* NH_3 to the system at equilibrium causes the reaction to shift in the direction that reduces the increased NH_3 concentration: Some of the added ammonia decomposes to form N_2 and H_2. All of these "shifts" are entirely consistent with predictions that we would make by comparing the reaction quotient Q after the addition or removal of a reactant or product with the equilibrium constant K.

In the Haber reaction, therefore, removing NH_3 from an equilibrium mixture of N_2, H_2, and NH_3 causes the reaction to shift right to form more NH_3. If the NH_3 can be removed continuously as it is produced, the yield can be increased dramatically. In the industrial production of ammonia, the NH_3 is continuously removed by selectively liquefying it **(Figure 15.12)**. (The boiling point of NH_3, $-33\,°C$, is much higher than those of N_2, $-196\,°C$, and H_2, $-253\,°C$.) The liquid NH_3 is removed, and the N_2 and H_2 are

 **Go Figure**   Why does the nitrogen concentration decrease after hydrogen is added?

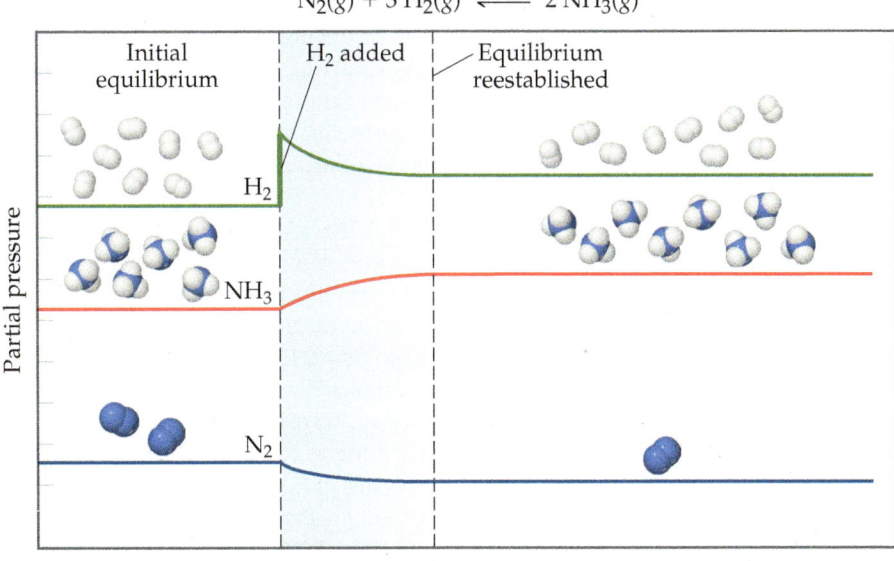

▲ **Figure 15.11 Effect of adding H_2 to an equilibrium mixture of N_2, H_2, and NH_3.** Adding H_2 causes the reaction as written to shift to the right, consuming some N_2 to produce more NH_3.

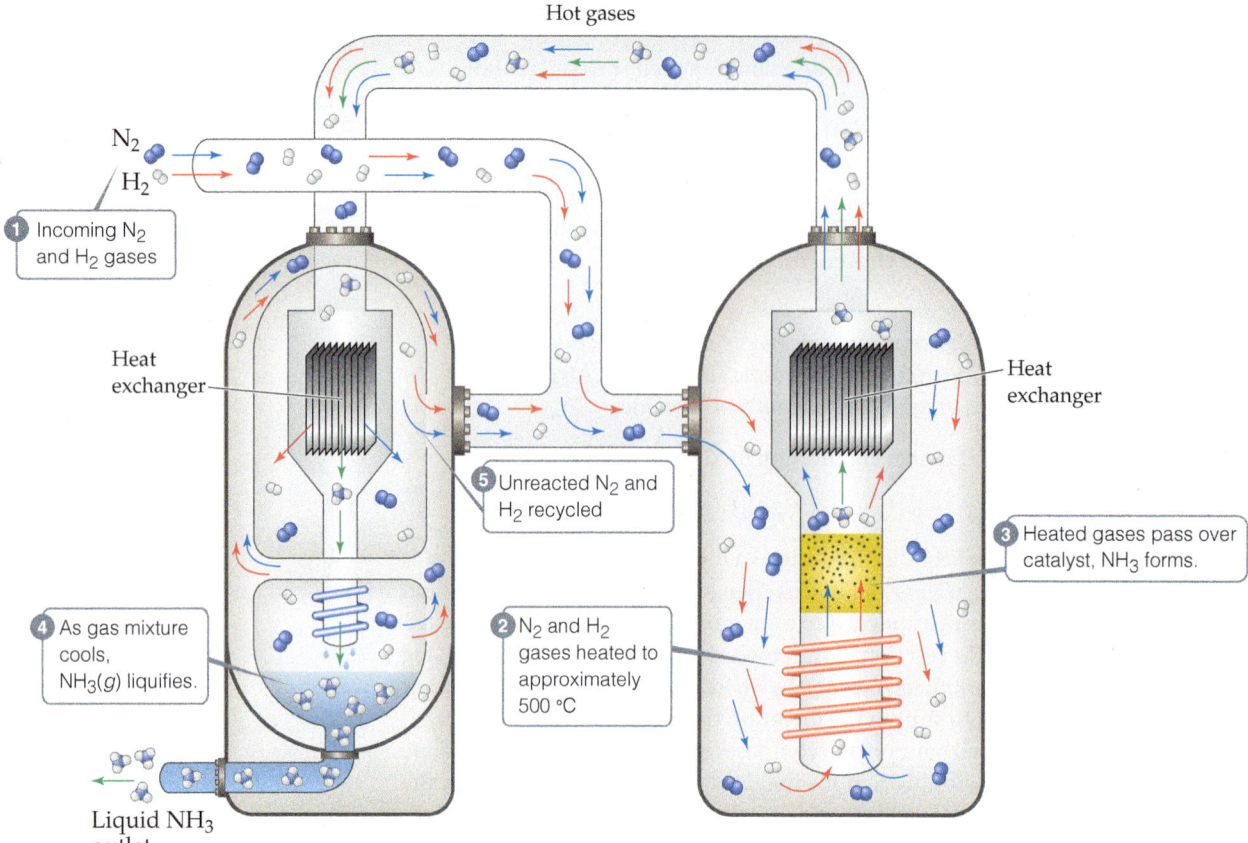

▲ Figure 15.12 Diagram of the industrial production of ammonia using the Haber process. Incoming $N_2(g)$ and $H_2(g)$ are heated to approximately 500 °C and passed over a catalyst. When the resultant N_2, H_2, and NH_3 mixture is cooled, the NH_3 liquefies and is removed from the mixture, shifting the reaction to produce more NH_3.

1 Incoming N_2 and H_2 gases

2 N_2 and H_2 gases heated to approximately 500 °C

3 Heated gases pass over catalyst, NH_3 forms.

4 As gas mixture cools, $NH_3(g)$ liquifies.

5 Unreacted N_2 and H_2 recycled

Hot gases

N_2

H_2

Heat exchanger

Heat exchanger

Liquid NH_3 outlet

recycled to form more NH_3. As a result of the product being continuously removed, the reaction is driven essentially to completion.

Effects of Volume and Pressure Changes

If a system containing one or more gases is at equilibrium and its volume is decreased, thereby increasing its total pressure, Le Châtelier's principle indicates that the system responds by shifting its equilibrium position to reduce the pressure. A system can reduce its pressure by reducing the total number of gas molecules (fewer molecules of gas exert a lower pressure). Thus, at constant temperature, *reducing the volume of a gaseous equilibrium mixture causes the system to shift in the direction that reduces the number of moles of gas.* Increasing the volume causes a shift in the direction that produces more gas molecules **(Figure 15.13)**.

In the reaction $N_2(g) + 3 H_2(g) \rightleftharpoons 2 NH_3(g)$, four molecules of reactant are consumed for every two molecules of product produced. Consequently, an increase in pressure (caused by a decrease in volume) shifts the reaction in the direction that produces fewer gas molecules, which leads to the formation of more NH_3, as indicated in Figure 15.9. In the reaction $H_2(g) + I_2(g) \rightleftharpoons 2 HI(g)$, the number of molecules of gaseous products (two) equals the number of molecules of gaseous reactants; therefore, changing the pressure does not influence the position of equilibrium.

Keep in mind that, as long as temperature remains constant, pressure–volume changes do *not* change the value of K. Rather, these changes alter the partial pressures of the gaseous substances. In Sample Exercise 15.7, we calculated $K_p = 2.79 \times 10^{-5}$ for the Haber reaction, $N_2(g) + 3 H_2(g) \rightleftharpoons 2 NH_3(g)$, in an equilibrium mixture at 472 °C containing 7.38 atm H_2, 2.46 atm N_2, and 0.166 atm NH_3. Consider what happens when we suddenly reduce the volume of the system by one-half. If there were no shift in equilibrium, this volume change would cause the partial pressures of all substances to

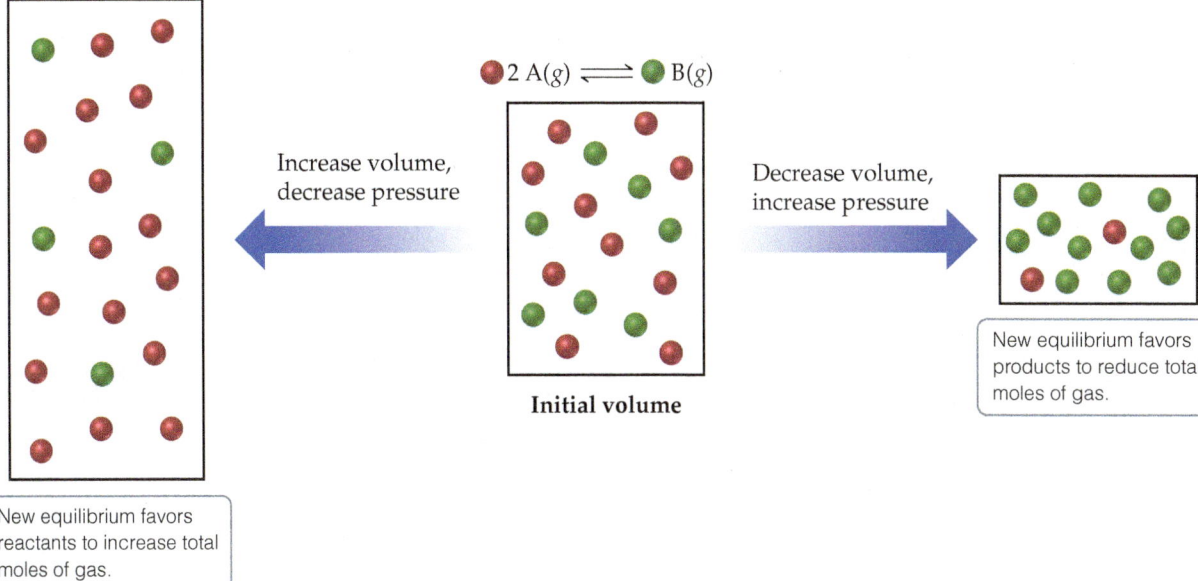

2 A(g) ⇌ B(g)

Increase volume, decrease pressure

Decrease volume, increase pressure

Initial volume

New equilibrium favors products to reduce total moles of gas.

New equilibrium favors reactants to increase total moles of gas.

▲ **Figure 15.13 Pressure and Le Châtelier's principle.**

double, giving $P_{H_2} = 14.76\,atm$, $P_{N_2} = 4.92\,atm$, and $P_{NH_3} = 0.332\,atm$. The reaction quotient would then no longer equal the equilibrium constant:

$$Q_p = \frac{(P_{NH_3})^2}{P_{N_2}(P_{H_2})^3} = \frac{(0.332)^2}{(4.92)(14.76)^3} = 6.97 \times 10^{-6} < K_p$$

Because $Q_p < K_p$, the system would no longer be at equilibrium. Equilibrium would be reestablished by increasing P_{NH_3} and decreasing both P_{N_2} and P_{H_2} until $Q_p = K_p = 2.79 \times 10^{-5}$. Therefore, the equilibrium shifts to the right in the reaction as written, as Le Châtelier's principle predicts.

It is possible to change the pressure of a system in which a chemical reaction is running without changing its volume. For example, pressure increases if additional amounts of any reacting components are added to the system. We have already seen how to deal with a change in concentration of a reactant or product. However, the *total* pressure in the reaction vessel might also be increased by adding a gas that is not involved in the equilibrium. For example, argon might be added to the ammonia equilibrium system. The argon would not alter the *partial* pressures of any of the reacting components and therefore would not cause a shift in equilibrium.

Effect of Temperature Changes

Changes in concentrations or partial pressures shift equilibria without changing the value of the equilibrium constant. In contrast, almost every equilibrium constant varies as the temperature changes. For example, consider the equilibrium established when aqueous chloride ions, $Cl^-(aq)$, react with hydrated cobalt(II), $Co(H_2O)_6^{2+}(aq)$, a reaction that is endothermic:

$$\underset{\text{Pale pink}}{Co(H_2O)_6^{2+}(aq)} + 4\,Cl^-(aq) \rightleftharpoons \underset{\text{Deep blue}}{CoCl_4^{2-}(aq)} + 6\,H_2O(l) \qquad \Delta H > 0 \qquad [15.25]$$

Because $Co(H_2O)_6^{2+}$ is pink and $CoCl_4^{2-}$ is blue, the position of this equilibrium is readily apparent from the color of the solution **(Figure 15.14)**. When the solution is heated it turns blue, indicating that the equilibrium has shifted to form more $CoCl_4^{2-}$. Cooling the solution leads to a pink solution, indicating that the equilibrium has shifted to produce more $Co(H_2O)_6^{2+}$. We can monitor this reaction by spectroscopic methods, measuring the concentration of all species at the different temperatures. (Section 14.2) We can then calculate the equilibrium constant at each temperature. How do we explain why the equilibrium constants and therefore the position of equilibrium both depend on temperature?

$\Delta H > 0$, endothermic reaction

$$\text{Heat} + \text{Co(H}_2\text{O)}_6{}^{2+}(aq) + 4\,\text{Cl}^-(aq) \;\rightleftharpoons\; \text{CoCl}_4{}^{2-}(aq) + 6\,\text{H}_2\text{O}(l)$$

Pink   Blue

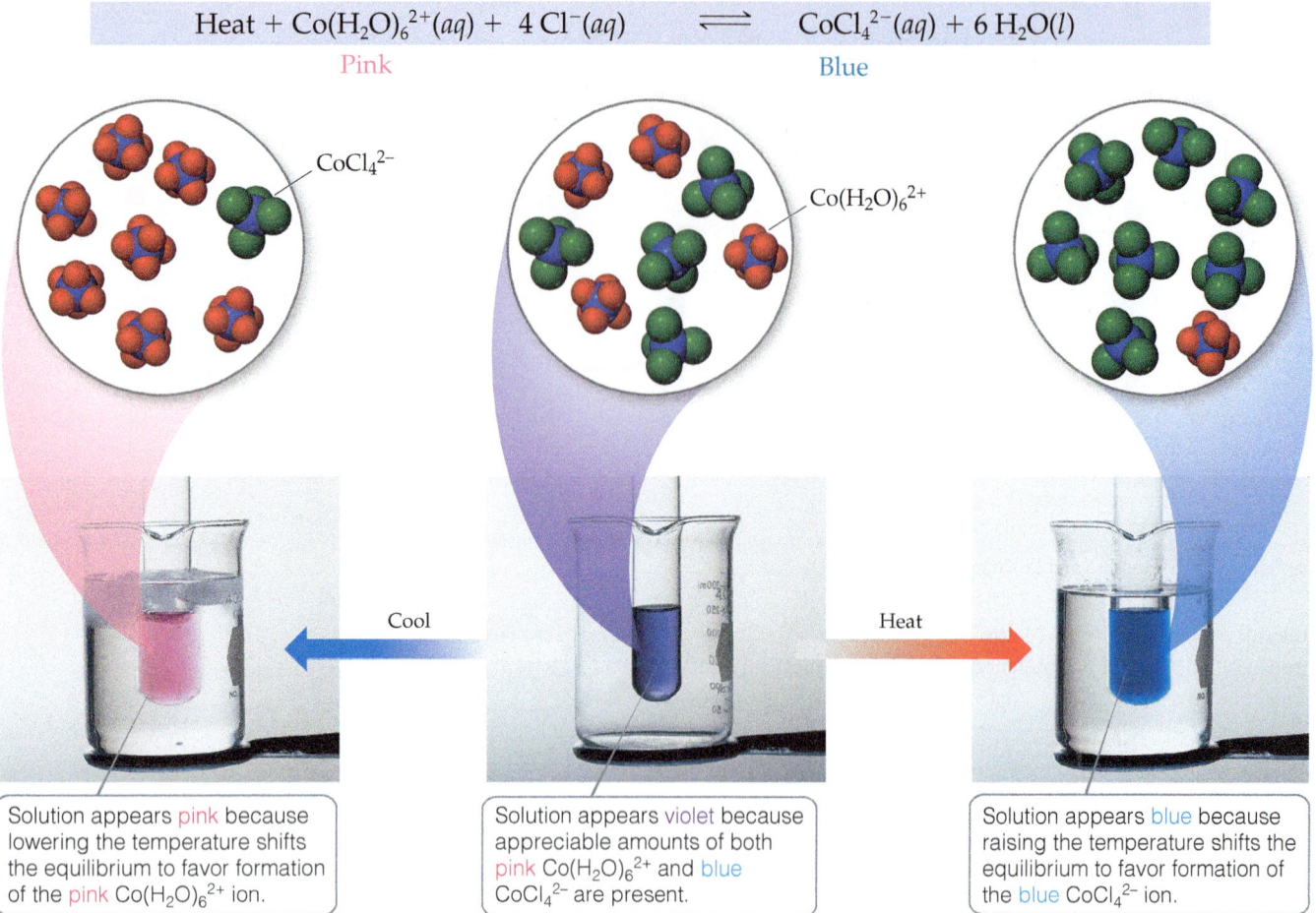

CoCl$_4{}^{2-}$

Co(H$_2$O)$_6{}^{2+}$

Cool   Heat

| Solution appears pink because lowering the temperature shifts the equilibrium to favor formation of the pink Co(H$_2$O)$_6{}^{2+}$ ion. | Solution appears violet because appreciable amounts of both pink Co(H$_2$O)$_6{}^{2+}$ and blue CoCl$_4{}^{2-}$ are present. | Solution appears blue because raising the temperature shifts the equilibrium to favor formation of the blue CoCl$_4{}^{2-}$ ion. |

▲ **Figure 15.14 Temperature and Le Châtelier's principle.** In the molecular level views, only the CoCl$_4{}^{2-}$ and Co(H$_2$O)$_6{}^{2+}$ ions are shown for clarity.

We can deduce the rules for the relationship between K and temperature from Le Châtelier's principle. We do this by treating heat as a chemical reagent. In an *endothermic* (heat-absorbing) reaction, we consider heat a *reactant*, and in an *exothermic* (heat-releasing) reaction, we consider heat a *product*:

Endothermic: Reactants + *heat* $\rightleftharpoons$ products
Exothermic: Reactants $\rightleftharpoons$ products + *heat*

When the temperature of a system at equilibrium is increased, the system reacts as if we added a reactant to an endothermic reaction or a product to an exothermic reaction. The equilibrium shifts in the direction that consumes the excess reactant (or product)—namely, heat.

In an endothermic reaction, such as Equation 15.25, heat is absorbed as reactants are converted to products. Thus, increasing the temperature causes the equilibrium to shift to the right, in the direction of making more products, and K increases. In an exothermic reaction, the opposite occurs: Heat is produced as reactants are converted to products. Thus, increasing the temperature in this case causes the equilibrium to shift to the left, in the direction of making more reactants, and K decreases.

Endothermic: Increasing T results in higher K value
Exothermic: Increasing T results in lower K value

Cooling a reaction has the opposite effect. As we lower the temperature, the equilibrium shifts in the direction that produces heat. Thus, cooling an endothermic reaction shifts the equilibrium to the left, decreasing K, as shown in Figure 15.14, and cooling an exothermic reaction shifts the equilibrium to the right, increasing K.

A CLOSER LOOK | Temperature Changes and Le Châtelier's Principle

By thinking of heat as a chemical reagent, we can use Le Châtelier's principle to predict how an equilibrium mixture of reactants and products will respond to a change in temperature. For an endothermic reaction, the equilibrium constant K increases as the temperature rises and decreases as the temperature is lowered, while exothermic reactions respond in the opposite manner. The underlying reason for this behavior can be understood by taking a closer look at the relationships among the equilibrium constant, the forward and reverse reaction rates, and their activation energies.

To illustrate the underlying relationships, consider an elementary reaction A $\rightleftharpoons$ B. At equilibrium, the rates of the forward and reverse reactions are equal:

$$k_f[A] = k_r[B] \qquad [15.26]$$

The equilibrium constant, $K = [B]/[A]$ can be expressed in terms of reaction rates by rearranging Equation 15.26

$$K = \frac{[B]}{[A]} = \frac{k_f}{k_r} \qquad [15.27]$$

It's instructive to consider shifts in equilibrium through the lens of Equation 15.27. If the reaction is endothermic, a decrease in temperature will lead to a decrease in K, thereby shifting the equilibrium to the left. Equation 15.27 tells us that in order for K to decrease, the rate constant of the forward reaction k_f must decrease by a larger amount than the rate constant of the reverse reaction k_r.

To understand why k_f decreases faster than k_r, consider the relationship between reaction rate and activation energy, E_a. For the sake of illustration, consider the effect of decreasing the temperature from $T_1 = 398$ K to $T_2 = 298$ K. We can calculate the change in the forward rate constant using Equation 14.23:

$$\ln\frac{k_{f1}}{k_{f2}} = \frac{E_a(\text{forward})}{R}\left(\frac{1}{T_2} - \frac{1}{T_1}\right) \qquad [15.28]$$

$$\ln\frac{k_{f1}}{k_{f2}} = \frac{E_a(\text{forward})}{R}\left(\frac{1}{298\text{ K}} - \frac{1}{398\text{ K}}\right) = (8.43 \times 10^{-4})\frac{E_a(\text{forward})}{R} \qquad [15.29]$$

Using the same approach, we can write an equation that gives the change in the reverse rate constant:

$$\ln\frac{k_{r1}}{k_{r2}} = \frac{E_a(\text{reverse})}{R}\left(\frac{1}{298\text{ K}} - \frac{1}{398\text{ K}}\right) = (8.43 \times 10^{-4})\frac{E_a(\text{reverse})}{R} \qquad [15.30]$$

These equations are identical with one exception: The activation energies for the forward and reverse reactions are not the same, as shown on the left-hand side of **Figure 15.15**. For an endothermic reaction, the activation energy in the forward direction is always larger than that of the reverse reaction, $E_a(\text{forward}) > E_a(\text{reverse})$. Consequently, the decrease in the forward rate constant (Equation 15.29) will be larger than the decrease in the reverse rate constant (Equation 15.30), and the overall value of the equilibrium constant K must decrease.

For an exothermic reaction, $E_a(\text{reverse}) > E_a(\text{forward})$, and the opposite relationships apply as shown on the right-hand side of Figure 15.15. As temperature is lowered, the reverse rate constant will decrease more rapidly than the forward rate constant and the equilibrium constant K will increase; that is, the equilibrium will shift to the right.

Related Exercises: 15.69, 15.70, 15.93, 15.95

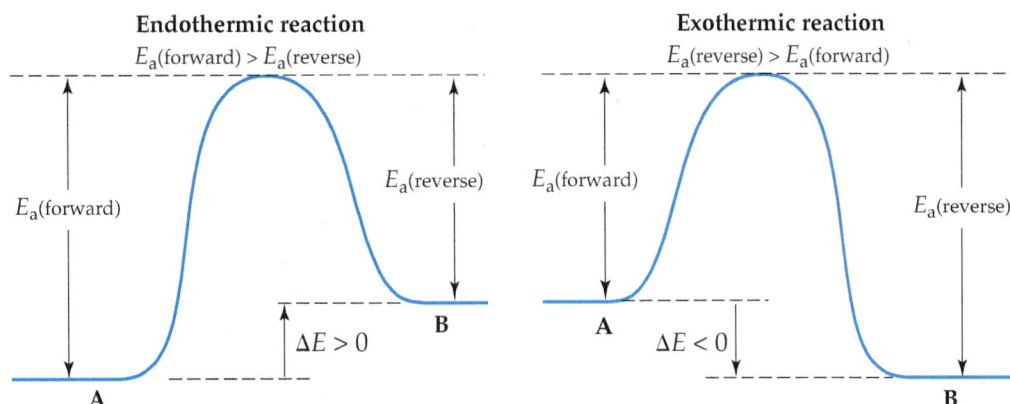

▲ **Figure 15.15** The energy profile for an endothermic reaction (left) and an exothermic reaction (right).

Sample Exercise 15.12

Using Le Châtelier's Principle to Predict Shifts in Equilibrium

Consider the equilibrium

$$N_2O_4(g) \rightleftharpoons 2\,NO_2(g) \qquad \Delta H° = 58.0\text{ kJ}$$

In which direction will the equilibrium shift when (**a**) N_2O_4 is added, (**b**) NO_2 is removed, (**c**) the pressure is increased by addition of $N_2(g)$, (**d**) the volume is increased, (**e**) the temperature is decreased?

SOLUTION

Analyze We are given a series of changes to be made to a system at equilibrium and are asked to predict what effect each change will have on the position of the equilibrium.

Plan Le Châtelier's principle can be used to determine the effects of each of these changes.

Continued

Solve

(a) The system will adjust to decrease the concentration of the added N_2O_4, so the equilibrium shifts to the right, in the direction of the product.

(b) The system will adjust to the removal of NO_2 by shifting to the side that produces more NO_2; thus, the equilibrium shifts to the right.

(c) Adding N_2 will increase the total pressure of the system, but N_2 is not involved in the reaction. The partial pressures of NO_2 and N_2O_4 are therefore unchanged, and there is no shift in the position of the equilibrium.

(d) If the volume is increased, the system will shift in the direction that occupies a larger volume (more gas molecules); thus, the equilibrium shifts to the right.

(e) The reaction is endothermic, so we can imagine heat as a reagent on the reactant side of the equation. Decreasing the

temperature will shift the equilibrium in the direction that produces heat, so the equilibrium shifts to the left, toward the formation of more N_2O_4.

Of all the changes proposed in this problem, only the last one—the change in temperature—affects the value of the equilibrium constant, K.

▶ **Practice Exercise**

For the reaction

$$PCl_5(g) \rightleftharpoons PCl_3(g) + Cl_2(g) \qquad \Delta H° = 87.9 \text{ kJ}$$

in which direction will the equilibrium shift when **(a)** $Cl_2(g)$ is removed, **(b)** the temperature is decreased, **(c)** the volume of the reaction system is increased, **(d)** $PCl_3(g)$ is added?

The Effect of Catalysts

What happens if we add a catalyst to a chemical system that is at equilibrium? As shown in **Figure 15.16**, a catalyst lowers the activation barrier between reactants and products. The activation energies for both the forward and reverse reactions are lowered. The catalyst thereby increases the rates of both forward and reverse reactions. Because K is the ratio of the forward and reverse rate constants for a reaction, you can predict, correctly, that the presence of a catalyst, even though it changes the reaction *rate,* does not affect the numeric value of K (Figure 15.16). As a result, *a catalyst increases the rate at which equilibrium is achieved but does not change the composition of the equilibrium mixture.*

The rate at which a reaction approaches equilibrium is an important practical consideration. As an example, let's again consider the synthesis of ammonia from N_2 and H_2. In designing his process, Haber had to deal with a rapid decrease in the equilibrium

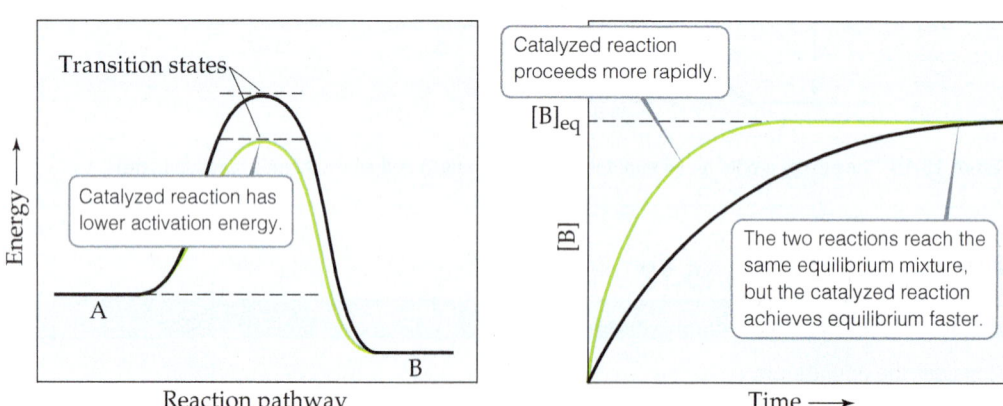

Go Figure What quantity dictates the speed of a reaction: **(a)** the energy difference between the initial state and the transition state or **(b)** the energy difference between the initial state and the final state?

▲ **Figure 15.16 An energy profile for the reaction A ⇌ B (left), and the change in concentration of B as a function of time (right), with and without a catalyst.** Green curves show the reaction with a catalyst; black curves show the reaction without a catalyst.

constant with increasing temperature, as you can see from Table 15.2. At temperatures sufficiently high to give a satisfactory reaction rate, the amount of ammonia formed was too small. The solution to this dilemma was to develop a catalyst that would produce a reasonably rapid approach to equilibrium at a sufficiently low temperature, so that the equilibrium constant remained reasonably large. The development of a suitable catalyst thus became the focus of Haber's research efforts.

After trying different substances to see which would be most effective, Carl Bosch settled on iron mixed with metal oxides. Variants of this catalyst formulation are still used today [see the "Chemistry and Sustainability" box on "The Haber Process: Feeding the World" (Section 15.2)]. These catalysts make it possible to obtain a reasonably rapid approach to equilibrium at around 400 to 500 °C and 200 to 600 atm. The high pressures are needed to obtain a satisfactory equilibrium amount of NH_3. If a catalyst could be found that leads to sufficiently rapid reaction at lower temperatures, then it would be possible to obtain the same extent of equilibrium conversion at pressures much lower than 200 to 600 atm. This would result in great savings in both the cost of the high-pressure equipment and the energy consumed in the production of ammonia. As noted in the box on the Haber process, it is estimated that the Haber process consumes approximately 1% of the energy generated in the world each year. Not surprisingly, chemists and chemical engineers are actively searching for improved catalysts for the Haber process. A breakthrough in this field would not only increase the supply of ammonia for fertilizers, it would also reduce the global consumption of fossil fuels in a significant way.

TABLE 15.2 Variation in K_p with Temperature for $N_2 + 3 H_2 \rightleftharpoons 2 NH_3$

Temperature (°C)	K_p
300	4.34×10^{-3}
400	1.64×10^{-4}
450	4.51×10^{-5}
500	1.45×10^{-5}
550	5.38×10^{-6}
600	2.25×10^{-6}

CHEMISTRY AND SUSTAINABILITY   Controlling Nitric Oxide Emissions

As we explore further in Chapter 18, the oxides of nitrogen are among the most important contributors to urban air pollution. One of the most significant ways in which nitrogen oxides are produced is the formation of nitric oxide, NO, in automobile engines via the reaction of N_2 and O_2 from air:

$$\tfrac{1}{2} N_2(g) + \tfrac{1}{2} O_2(g) \rightleftharpoons NO(g) \qquad \Delta H° = 90.4 \text{ kJ} \qquad [15.31]$$

This reaction provides an interesting example of the practical importance of the fact that equilibrium constants and reaction rates change with temperature. By applying Le Châtelier's principle to this endothermic reaction and treating heat as a reactant, we deduce that an increase in temperature shifts the equilibrium in the direction of more NO. The equilibrium constant K_p for the formation of 1 mol of NO from its elements at 300 K is only about 1×10^{-15} (Figure 15.17). At 2400 K, however, the equilibrium constant is about 0.05, which is 10^{13} times larger than the 300 K value.

Figure 15.17 helps explain why NO is a pollution problem. In the cylinder of a modern high-compression automobile engine, the temperature during the fuel-burning part of the cycle is approximately 2400 K. Also, there is a fairly large excess of air in the cylinder. These conditions favor the formation of NO. After combustion, however, the gases cool quickly. As the temperature drops, the equilibrium in Equation 15.31 shifts to the left (because the reactant heat is being removed). However, the lower temperature also means that the reaction rate decreases, so the NO formed at 2400 K is essentially "trapped" in that form as the gas cools.

The gases exhausting from the cylinder are still quite hot, perhaps 1200 K. At this temperature, Figure 15.17 shows that the equilibrium constant for formation of NO is about 5×10^{-4}, much smaller than the value at 2400 K. However, the rate of conversion of NO to N_2 and O_2 is too slow to permit much loss of NO before the gases are cooled further.

As discussed in the "Chemistry and Sustainability" box on "Catalytic Converters," (Section 14.6) one of the goals of automotive catalytic converters is to achieve rapid conversion of NO to N_2 and O_2 at the temperature of the exhaust gas. Some catalysts

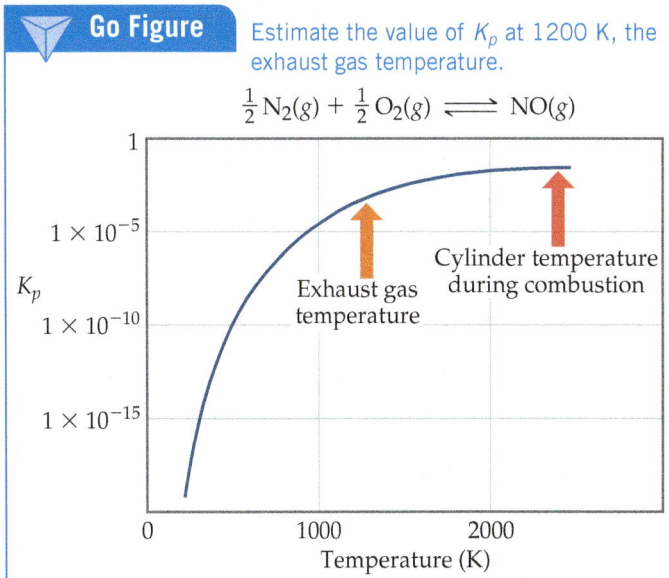

Go Figure Estimate the value of K_p at 1200 K, the exhaust gas temperature.

$$\tfrac{1}{2} N_2(g) + \tfrac{1}{2} O_2(g) \rightleftharpoons NO(g)$$

▲ **Figure 15.17 Equilibrium and temperature.** The equilibrium constant increases with increasing temperature because the reaction is endothermic. It is necessary to use a log scale for K_p because the values vary over such a large range.

developed for this reaction are reasonably effective under the grueling conditions in automotive exhaust systems. Nevertheless, scientists and engineers are continuously searching for new materials that provide even more effective catalysis of the decomposition of nitrogen oxides. The shift from internal combustion to electric vehicles will serve to decrease this source of air pollution as well as having its obvious positive impact on the generation of greenhouse gases.

Self-Assessment Exercises

SAE 15.17   A mixture of substances A, B, and C at a fixed temperature is in equilibrium in a closed vessel according to the following equation:

$$A(g) \rightleftharpoons B(g) + C(g)$$

Without changing the temperature or the volume of the vessel, more B(g) is added to the mixture. Which of the following graphs shows the return of the system to a new equilibrium?

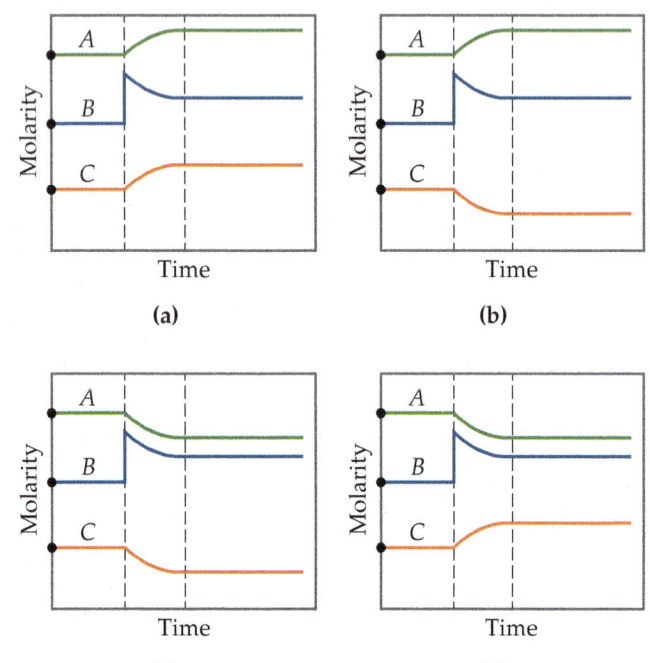

(a)   (b)

(c)   (d)

SAE 15.18   For which of the following equilibria will the addition of more $O_2(g)$ cause the equilibrium to shift to the *right*?

(i) $N_2(g) + O_2(g) \rightleftharpoons 2\,NO(g)$

(ii) $2\,CO(g) + O_2(g) \rightleftharpoons 2\,CO_2(g)$

(iii) $2\,O_3(g) \rightleftharpoons 3\,O_2(g)$

(iv) $N_2(g) + 2\,O_2(g) \rightleftharpoons 2\,NO_2(g)$

(a) equilibria i and ii (b) equilibria ii and iii (c) equilibria iii and iv (d) equilibria i, ii, and iii (e) equilibria i, ii, and iv

SAE 15.19   The following reaction is endothermic ($\Delta H > 0$):

$$CCl_4(g) \rightleftharpoons C(s) + 2\,Cl_2(g)$$

Suppose that the reaction is carried out in a vessel that allows for variable temperature (T) and variable volume (V). To maximize the fraction of Cl_2 in an equilibrium mixture, which of the following conditions would you choose? (a) high T, high V (b) high T, low V (c) low T, high V (d) low T, low V (e) The fraction of $Cl_2(g)$ at equilibrium will not depend on T or V.

SAE 15.20   Which of the following statements about equilibrium and catalysts at a fixed temperature is or are *true*?

(i) A catalyst allows a system to reach equilibrium faster.

(ii) A catalyst increases the rate of both the forward and reverse reactions of an equilibrium.

(iii) A catalyst does not change the value of the equilibrium constant K.

(a) Only one statement is true. (b) Statements i and ii are true. (c) Statements i and iii are true. (d) Statements ii and iii are true. (e) All three statements are true.

 Putting Concepts Together

At temperatures near 800 °C, steam passed over hot coke (a form of carbon obtained from coal) reacts to form CO and H_2:

$$C(s) + H_2O(g) \rightleftharpoons CO(g) + H_2(g)$$

The mixture of gases that results is an important industrial fuel called *water gas*. (a) At 800 °C the equilibrium constant for this reaction is $K_p = 14.1$. What are the equilibrium partial pressures of H_2O, CO, and H_2 in the equilibrium mixture at this temperature if we start with solid carbon and 0.100 mol of H_2O in a 1.00-L vessel? (b) What is the minimum amount of carbon required to achieve equilibrium under these conditions? (c) What is the total pressure in the vessel at equilibrium? (d) At 25 °C the value of K_p for this reaction is 1.7×10^{-21}. Is the reaction exothermic or endothermic? (e) To produce the maximum amount of CO and H_2 at equilibrium, should the pressure of the system be increased or decreased?

SOLUTION

(a) To determine the equilibrium partial pressures, we use the ideal-gas equation, first determining the starting partial pressure of water.

$$P_{H_2O} = \frac{n_{H_2O}RT}{V} = \frac{(0.100\ \text{mol})(0.08206\ \text{L-atm/mol-K})(1073\ \text{K})}{1.00\ \text{L}} = 8.81\ \text{atm}$$

We then construct a table of initial partial pressures and their changes as equilibrium is achieved:

		$C(s)$ +	$H_2O(g)$ $\rightleftharpoons$	$CO(g)$ +	$H_2(g)$
Initial partial pressure (atm)			8.81	0	0
Change in partial pressure (atm)			$-x$	$+x$	$+x$
Equilibrium partial pressure (atm)			$8.81 - x$	x	x

There are no entries in the table under C(*s*) because the reactant, being a solid, does not appear in the equilibrium-constant expression. Substituting the equilibrium partial pressures of the other species into the equilibrium-constant expression for the reaction gives:

$$K_p = \frac{P_{CO}P_{H_2}}{P_{H_2O}} = \frac{(x)(x)}{(8.81 - x)} = 14.1$$

Multiplying through by the denominator gives a quadratic equation in *x*:

$$x^2 = (14.1)(8.81 - x)$$

Solving this equation for *x* using the quadratic formula yields *x* = 6.14 atm. Hence, the equilibrium partial pressures are $P_{CO} = x = 6.14$ atm, $P_{H_2} = x = 6.14$ atm, and $P_{H_2O} = (8.81 - x) = 2.67$ atm.

$$x^2 + 14.1x - 124.22 = 0$$

(b) Part (a) shows that *x* = 6.14 atm of H_2O must react for the system to achieve equilibrium. We can use the ideal-gas equation to convert this partial pressure into a mole amount.

$$n = \frac{PV}{RT}$$

Thus, 0.0697 mol of H_2O and the same amount of C must react to achieve equilibrium. As a result, at least 0.0697 mol of C (0.836 g C) must be present among the reactants at the start of the reaction.

$$= \frac{(6.14 \text{ atm})(1.00 \text{ L})}{(0.08206 \text{ L-atm/mol-K})(1073 \text{ K})} = 0.0697 \text{ mol}$$

(c) The total pressure in the vessel at equilibrium is simply the sum of the equilibrium partial pressures:

$$P_{total} = P_{H_2O} + P_{CO} + P_{H_2}$$
$$= 2.67 \text{ atm} + 6.14 \text{ atm} + 6.14 \text{ atm} = 14.95 \text{ atm}$$

(d) In discussing Le Châtelier's principle, we saw that endothermic reactions exhibit an increase in K_p with increasing temperature. Because the equilibrium constant for this reaction increases as temperature increases, the reaction must be endothermic. From the enthalpies of formation given in Appendix C, we can verify our prediction by calculating the enthalpy change for the reaction:

$$\Delta H° = \Delta H_f°(CO(g)) + \Delta H_f°(H_2(g)) - \Delta H_f°[C(s, \text{graphite})]$$
$$- \Delta H_f°(H_2O(g)) = +131.3 \text{ kJ}$$

The positive sign for $\Delta H°$ indicates that the reaction is endothermic.

(e) According to Le Châtelier's principle, a decrease in the pressure causes a gaseous equilibrium to shift toward the side of the equation with the greater number of moles of gas. In this case, there are 2 mol of gas on the product side and only one on the reactant side. Therefore, the pressure should be decreased to maximize the yield of the CO and H_2.

Chapter Summary and Key Terms

THE CONCEPT OF CHEMICAL EQUILIBRIUM (SECTION 15.1) A chemical reaction can achieve a state in which the forward and reverse processes are occurring at the same rate. This condition is called **chemical equilibrium**, and it results in the formation of an equilibrium mixture of reactants and products. The composition of an equilibrium mixture does not change with time if temperature is held constant.

THE EQUILIBRIUM CONSTANT (SECTION 15.2) An equilibrium that is used throughout this chapter is the reaction $N_2(g) + 3 H_2(g) \rightleftharpoons 2 NH_3(g)$. This reaction is the basis of the **Haber process** for the production of ammonia. The relationship between the concentrations of the reactants and products of a system at equilibrium is given by the **law of mass action**. For an equilibrium equation of the form $a A + b B \rightleftharpoons d D + e E$, the **equilibrium-constant expression** is written as

$$K_c = \frac{[D]^d[E]^e}{[A]^a[B]^b}$$

where K_c is a dimensionless constant called the **equilibrium constant**. When the equilibrium system of interest consists of gases,

it is often convenient to express the concentrations of reactants and products in terms of gas pressures:

$$K_p = \frac{(P_D)^d(P_E)^e}{(P_A)^a(P_B)^b}$$

K_c and K_p are related by the expression $K_p = K_c(RT)^{\Delta n}$, where Δn is the number of moles of gas in the products minus those in the reactants.

USING EQUILIBRIUM CONSTANTS (SECTION 15.3) The value of the equilibrium constant changes with temperature. A large value of K_c indicates that the equilibrium mixture contains more products than reactants and therefore lies toward the product side of the equation ("lies to the right"). A small value for the equilibrium constant means that the equilibrium mixture contains less products than reactants and therefore lies toward the reactant side ("lies to the left"). The equilibrium-constant expression and the equilibrium constant of the reverse of a reaction are the reciprocals of those of the forward reaction. If a reaction is the sum of two or more reactions, its equilibrium constant will be the product of the equilibrium constants for the individual reactions.

HETEROGENEOUS EQUILIBRIA (SECTION 15.4) Equilibria can be divided into two types: **homogeneous equilibria**, where all of the reactants and products are in the same phase, and **heterogeneous equilibria**, where more than one phase is present. Because their activities are exactly 1, the concentrations of pure solids and liquids are left out of the equilibrium-constant expression for a heterogeneous equilibrium.

CALCULATING EQUILIBRIUM CONSTANTS (SECTION 15.5) If the concentrations of all species in an equilibrium are known, the equilibrium-constant expression can be used to calculate the equilibrium constant. The concentrations of reactants and products at equilibrium can often be determined by using the stoichiometry of the reaction.

SOME APPLICATIONS OF EQUILIBRIUM CONSTANTS (SECTION 15.6) The **reaction quotient**, Q, is found by substituting reactant and product concentrations or partial pressures at any point during a reaction into the equilibrium-constant expression. If the system is at equilibrium, $Q = K$. If $Q \neq K$, however, the system is not at equilibrium. When $Q < K$, the reaction will move toward equilibrium by forming more products (the reaction proceeds from left to right); when $Q > K$, the reaction will move toward equilibrium by forming more reactants (the reaction proceeds from right to left). Knowing the value of K makes it possible to calculate the equilibrium amounts of reactants and products, often by the solution of an equation in which the unknown is the change in a partial pressure or concentration.

LE CHÂTELIER'S PRINCIPLE (SECTION 15.7) Le Châtelier's principle states that if a system at equilibrium is disturbed, the equilibrium will shift to minimize the disturbing influence. Therefore, if a reactant or product is added to a system at equilibrium, the equilibrium will shift to consume the added substance. The effects of removing reactants or products and of changing the pressure or volume of a reaction can be similarly deduced. For example, if the volume of the system is reduced, the equilibrium will shift in the direction that decreases the number of gas molecules. While changes in concentration or pressure lead to shifts in the equilibrium concentrations, they do not change the value of the equilibrium constant, K.

Changes in temperature affect both the equilibrium concentrations and the equilibrium constant. We can use the enthalpy change for a reaction to determine how an increase in temperature affects the equilibrium: For an endothermic reaction, an increase in temperature shifts the equilibrium to the right; for an exothermic reaction, a temperature increase shifts the equilibrium to the left. Catalysts affect the speed at which equilibrium is reached but do not affect the magnitude of K.

Key Equations

- $K_c = \dfrac{[D]^d[E]^e}{[A]^a[B]^b}$ [15.8]

  The equilibrium-constant expression for a general reaction of the type $a\,A + b\,B \rightleftharpoons d\,D + e\,E$, the concentrations are equilibrium concentrations only

- $K_p = \dfrac{(P_D)^d(P_E)^e}{(P_A)^a(P_B)^b}$ [15.11]

  The equilibrium-constant expression in terms of equilibrium partial pressures

- $K_p = K_c(RT)^{\Delta n}$ [15.15]

  Relating the equilibrium constant based on pressures to the equilibrium constant based on concentrations

- $Q_c = \dfrac{[D]^d[E]^e}{[A]^a[B]^b}$ [15.24]

  The reaction quotient. The concentrations are for any time during a reaction. If the concentrations are equilibrium concentrations, then $Q_c = K_c$.

Exam Prep

EP 15.1 When the reaction $N_2O_4(g) \rightleftharpoons 2\,NO_2(g)$ reaches equilibrium, which of the following must be equal?

   (i) the forward and reverse rate constants k_f and k_r

   (ii) the concentrations $[N_2O_4]$ and $[NO_2]$

   (iii) the forward and reverse reaction rates

(a) i only **(b)** ii only **(c)** iii only **(d)** i and ii **(e)** i and iii

EP 15.2 For the reaction $2\,SO_2(g) + O_2(g) \rightleftharpoons 2\,SO_3(g)$, which of the following is the correct equilibrium-constant expression?

(a) $K_c = \dfrac{[SO_2]^2[O_2]}{[SO_3]^2}$ **(c)** $K_c = \dfrac{[SO_3]^2}{[SO_2]^2[O_2]}$

(b) $K_c = \dfrac{2[SO_2][O_2]}{2[SO_3]}$ **(d)** $K_c = \dfrac{2[SO_3]}{2[SO_2][O_2]}$

EP 15.3 Consider the following equilibrium:

$$2\,SO_2(g) + O_2(g) \rightleftharpoons 2\,SO_3(g)$$

At a certain temperature, the following equilibrium concentrations are found: $[SO_2] = 1.83 \times 10^{-3}\,M$, $[O_2] = 6.16 \times 10^{-3}\,M$, and $[SO_3] = 6.23 \times 10^{-3}\,M$. What is the value of K_c for this reaction at this temperature?

(a) 5.32×10^{-4} **(b)** 3.44 **(c)** 5.52×10^2 **(d)** 1.88×10^3 **(e)** 3.02×10^5

EP 15.4 The equilibrium constant for the reaction $N_2O_4(g) \rightleftharpoons 2\,NO_2(g)$ at 2 °C is $K_c = 2.0$. If each yellow sphere represents 1 mol of N_2O_4 and each brown sphere 1 mol of NO_2, which of the following 1.0-L containers represents the equilibrium mixture at 2 °C?

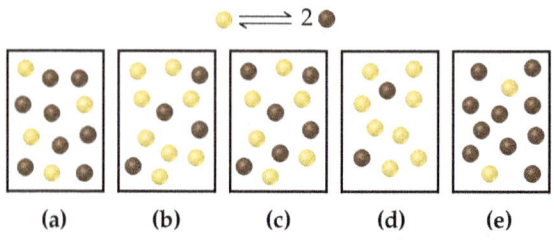

   (a)   **(b)**   **(c)**   **(d)**   **(e)**

EP 15.5 For which of the following reactions is the ratio K_p/K_c largest at 300 K?

(a) $N_2(g) + O_2(g) \rightleftharpoons 2\,NO(g)$

(b) $CaCO_3(s) \rightleftharpoons CaO(s) + CO_2(g)$

(c) $C(s) + 2\,H_2(g) \rightleftharpoons CH_4(g)$

(d) $Ni(CO)_4(g) \rightleftharpoons Ni(s) + 4\,CO(g)$

EP 15.6 A sample of pure A(g) is placed in a vessel at a pressure of 1.0 atm and is allowed to equilibrate according to the reaction: A(g) $\rightleftharpoons$ B(g) + C(g). The equilibrium constant for this reaction is K_p = 1.0 × 10^{-6}. Which of the following statements about the equilibrium mixture of A, B, and C is or are *true*?

(**i**) The equilibrium mixture contains more B(g) and C(g) than A(g).

(**ii**) At equilibrium, the partial pressures of B(g) and C(g) are the same.

(**iii**) The equilibrium constant for the reverse reaction is K_p = 1.0 × 10^6.

(**a**) Only one of the statements is true. (**b**) Statements i and ii are true. (**c**) Statements i and iii are true. (**d**) Statements ii and iii are true. (**e**) All three statements are true.

EP 15.7 Given the equilibrium constants for the following two reactions in aqueous solution at 25 °C,

$HNO_2(aq) \rightleftharpoons H^+(aq) + NO_2^-(aq)$ $K_c = 4.5 \times 10^{-4}$

$H_2SO_3(aq) \rightleftharpoons 2H^+(aq) + SO_3^-(aq)$ $K_c = 1.1 \times 10^{-9}$

what is the value of K_c for the reaction?

$2\,HNO_2(aq) + SO_3^{2-}(aq) \rightleftharpoons H_2SO_3(aq) + 2\,NO_2^-(aq)$

(**a**) 4.9 × 10^{-13} (**b**) 4.1 × 10^5 (**c**) 8.2 × 10^5 (**d**) 1.8 × 10^2 (**e**) 5.4 × 10^{-3}

EP 15.8 Consider the equilibrium that is established in a saturated solution of silver chloride, Ag$^+$(aq) + Cl$^-$(aq) $\rightleftharpoons$ AgCl(s). If solid AgCl is added to this solution, what will happen to the concentrations of the Ag$^+$ and Cl$^-$ ions in solution?

(**a**) [Ag$^+$] and [Cl$^-$] will both increase. (**b**) [Ag$^+$] and [Cl$^-$] will both decrease. (**c**) [Ag$^+$] will increase and [Cl$^-$] will decrease. (**d**) [Ag$^+$] will decrease and [Cl$^-$] will increase. (**e**) Neither [Ag$^+$] nor [Cl$^-$] will change.

EP 15.9 If 8.0 g of NH$_4$HS(s) is placed in a sealed vessel with a volume of 1.0 L and heated to 200 °C, the reaction NH$_4$HS(s) $\rightleftharpoons$ NH$_3$(g) + H$_2$S(g) will occur. When the system comes to equilibrium, some NH$_4$HS(s) is still present. Which of the following changes will lead to a reduction in the amount of NH$_4$HS(s) that is present, assuming in all cases that equilibrium is re-established following the change? (**a**) adding more NH$_3$(g) to the vessel (**b**) adding more H$_2$S(g) to the vessel (**c**) adding more NH$_4$HS(s) to the vessel (**d**) increasing the volume of the vessel

EP 15.10 A mixture of gaseous sulfur dioxide and oxygen are added to a reaction vessel and heated to 1000 K, where they react to form SO$_3$(g). If the vessel contains 0.669 atm SO$_2$(g), 0.395 atm O$_2$(g), and 0.0851 atm SO$_3$(g) after the system has reached equilibrium, what is the value of the equilibrium constant K_p for the reaction 2 SO$_2$(g) + O$_2$(g) $\rightleftharpoons$ 2 SO$_3$(g) at 1000 K? (**a**) 0.0410 (**b**) 0.322 (**c**) 24.4 (**d**) 3.36 (**e**) 3.11

EP 15.11 When 9.20 g of N$_2$O$_4$ is added to a 0.500-L reaction vessel that is heated to 400 K and allowed to equilibrate, the equilibrium concentration of N$_2$O$_4$ is determined to be 0.057 M. Given this information, what is the value of K_c for the reaction N$_2$O$_4$(g) $\rightleftharpoons$ 2 NO$_2$(g) at 400 K? (**a**) 0.23 (**b**) 0.36 (**c**) 0.13 (**d**) 1.4 (**e**) 2.5

EP 15.12 Chlorine, Cl$_2$(g), and bromine, Br$_2$(g), react in the gas phase to form BrCl(g) according to the following equilibrium:

$$Cl_2(g) + Br_2(g) \rightleftharpoons 2\,BrCl(g)$$

A cylinder is charged with 0.500 atm Cl$_2$(g) and 0.300 atm Br$_2$(g) at 298 K, and they react until equilibrium is reached. The equilibrium mixture contains 0.0754 atm BrCl(g). What is the value of K_p for the reaction at 298 K?

(**a**) 2.7 × 10^{-2} (**b**) 4.7 × 10^{-2} (**c**) 5.9 × 10^{-2} (**d**) 6.2 × 10^{-1}

EP 15.13 Which of the following statements about the reaction quotient, Q, is or are *true*?

(**i**) The expression used to evaluate Q looks the same as that used to evaluate K.

(**ii**) When the value of Q < K, the reaction must form more products to achieve equilibrium.

(**iii**) A large value of Q indicates that a reaction will proceed more quickly to equilibrium.

(**a**) Only statement i is true. (**b**) Only statement ii is true. (**c**) Statements i and ii are true. (**d**) Statements ii and iii are true. (**e**) All three statements are true.

EP 15.14 At 500 K, the reaction 2 NO(g) + Cl$_2$(g) $\rightleftharpoons$ 2 NOCl(g) has K_p = 51. In an equilibrium mixture at 500 K, the partial pressure of NO is 0.125 atm, and that of Cl$_2$ is 0.165 atm. What is the partial pressure of NOCl in the equilibrium mixture? (**a**) 0.13 atm (**b**) 0.36 atm (**c**) 1.0 atm (**d**) 5.1 × 10^{-5} atm (**e**) 0.017 atm

EP 15.15 For the equilibrium Br$_2$(g) + Cl$_2$(g) $\rightleftharpoons$ 2 BrCl(g), the equilibrium constant K_p is 7.0 at 400 K. If a cylinder is charged with BrCl(g) at an initial pressure of 1.00 atm and the system is allowed to come to equilibrium, what is the final (equilibrium) pressure of BrCl? (**a**)0.57 atm (**b**) 0.22 atm (**c**) 0.45 atm (**d**) 0.15 atm (**e**) 0.31 atm

EP 15.16 Consider the reaction N$_2$(g) + 2 O$_2$(g) $\rightleftharpoons$ 2 NO$_2$(g).

How does each of the following changes impact the partial pressure of NO$_2$ at equilibrium?

(**i**) Addition of more N$_2$ to the reaction vessel will _____ the partial pressure of NO$_2$ at equilibrium.

(**ii**) Removing O$_2$ from the reaction vessel will _____ the partial pressure of NO$_2$ at equilibrium.

(**iii**) Adding a catalyst to the reaction vessel will _____ the partial pressure of NO$_2$ at equilibrium.

(**a**) increase, decrease, increase (**b**) leave unchanged, increase, decrease (**c**) decrease, increase, leave unchanged (**d**) increase, decrease, leave unchanged (**e**) decrease, leave unchanged, increase

EP 15.17 For the reaction

$$4\,NH_3(g) + 5\,O_2(g) \rightleftharpoons 4\,NO(g) + 6\,H_2O(g) \quad \Delta H° = -904\,kJ$$

which of the following changes will shift the equilibrium to the right, toward the formation of more products? (**a**) increasing the volume of the reaction vessel (**b**) increasing the temperature (**c**) adding more water vapor (**d**) removing O$_2$(g) (**e**) The total pressure of the system is increased by adding 1 atm of Ne(g) to the reaction vessel.

EP 15.18 A mixture of CO$_2$(g), C(s), and CO(g) is in equilibrium according to the following reaction:

$$CO_2(g) + C(s) \rightleftharpoons 2\,CO(g)$$

What is the effect of adding more C(s) to the equilibrium mixture? (**a**) The equilibrium will shift to the left. (**b**) The equilibrium will not shift. (**c**) The equilibrium will shift to the right. (**d**) More information is needed.

EP 15.19 We have seen that the use of catalysts is important in a number of industrial and sustainable processes. Which of the following statements about the use of catalysts is *false*?

(**a**) A catalyst can be used to allow a reaction to reach an equilibrium more quickly than if it were not catalyzed.

(**b**) A catalyst can be used to achieve an equilibrium at a lower temperature, where the equilibrium constant may be more favorable for the desired reaction.

(**c**) At a fixed temperature, a catalyst can shift the equilibrium toward the creation of more products.

(**d**) The development of a suitable catalyst was one of the most important challenges of using the Haber process to synthesize ammonia on a large scale.

Exercises

Visualizing Concepts

15.1 **(a)** Based on the following energy profile, predict whether $k_f > k_r$ or $k_f < k_r$. **(b)** Using Equation 15.5, predict whether the equilibrium constant for the process is greater than 1 or less than 1. [Section 15.1]

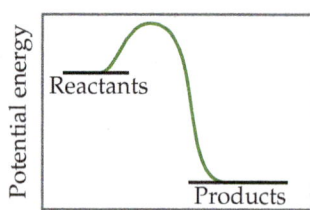

15.2 The following diagrams represent a hypothetical reaction A $\longrightarrow$ B, with A represented by red spheres and B represented by blue spheres. The sequence from left to right represents the system as time passes. Does the system reach equilibrium? If so, in which diagram(s) is the system in equilibrium? [Sections 15.1 and 15.2]

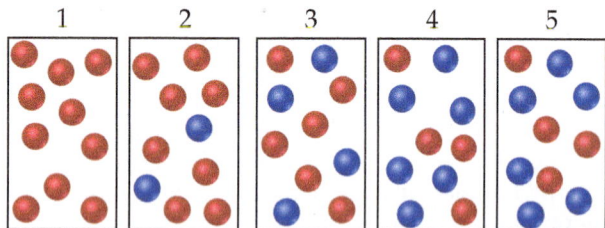

15.3 The following diagram represents an equilibrium mixture produced for a reaction of the type A + X $\rightleftharpoons$ AX. If the volume is 1 L and each atom/molecule in the diagram represents 1 mol, is K greater than, equal to, or less than 1? [Section 15.2]

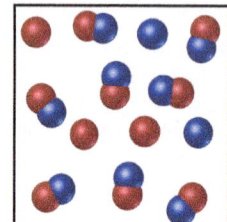

15.4 The following diagram represents a reaction shown reaching equilibrium. Each molecule in the diagram represents 0.1 mol, and the volume of the box is 1.0 L. **(a)** Letting atom A = red spheres and atom B = blue spheres, write a balanced equation for the reaction. **(b)** Write the equilibrium-constant expression for the reaction. **(c)** Calculate the value of K_c. **(d)** Assuming that all of the molecules are in the gas phase, calculate Δn, the change in the number of gas molecules that accompanies the reaction. **(e)** Calculate the value of K_p. [Section 15.2]

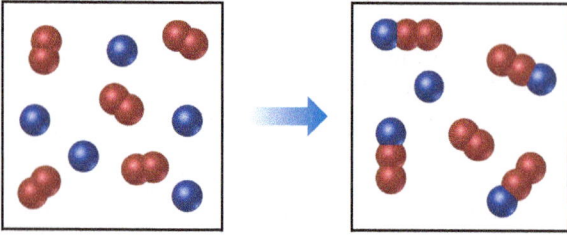

15.5 Snapshots of two hypothetical reactions, A(g) + B(g) $\rightleftharpoons$ AB(g) and X(g) + Y(g) $\rightleftharpoons$ XY(g) at five different times are shown here. Which reaction has a larger equilibrium constant? [Sections 15.1 and 15.2]

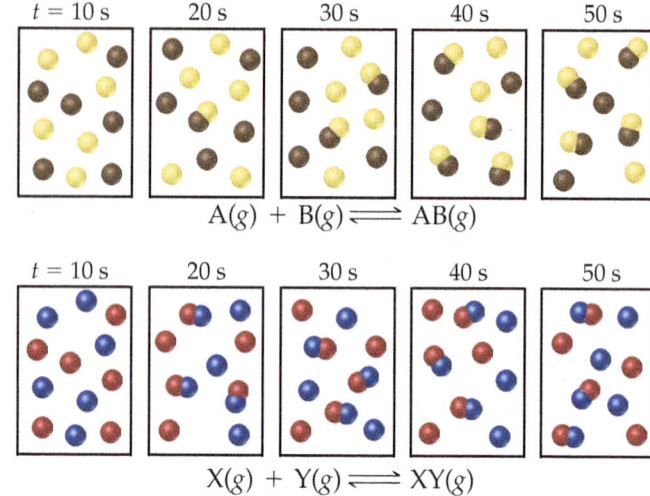

A(g) + B(g) $\rightleftharpoons$ AB(g)

X(g) + Y(g) $\rightleftharpoons$ XY(g)

15.6 Ethene (C$_2$H$_4$) reacts with halogens (X$_2$) by the following reaction:

$$C_2H_4(g) + X_2(g) \rightleftharpoons C_2H_4X_2(g)$$

The following figures represent the concentrations at equilibrium at the same temperature when X$_2$ is Cl$_2$ (green), Br$_2$ (brown), and I$_2$ (purple). List the equilibria from smallest to largest equilibrium constant. [Section 15.3]

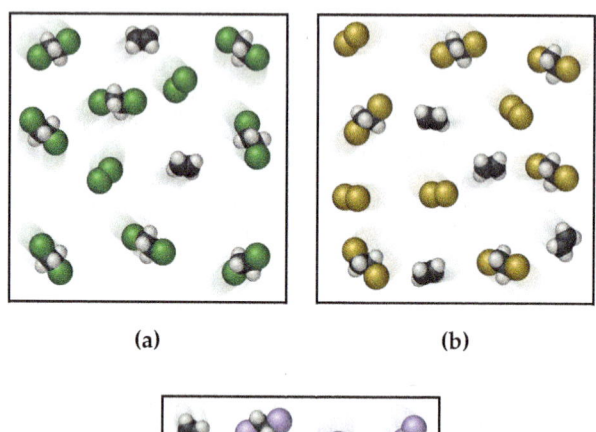

(a)   (b)

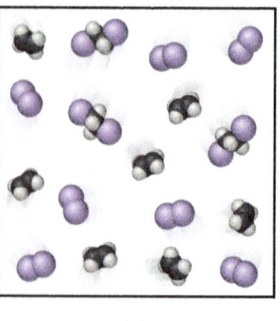

(c)

15.7 The chart presented here shows the progress of the reaction A(g) $\rightleftharpoons$ B(g) + C(g) to equilibrium in a closed vessel at fixed temperature. Which of the following statements about the progress of this system to equilibrium is or are *true*? **(a)** The vessel originally contained a mixture of B(g) and C(g). **(b)** At equilibrium, the vessel contains the same number of moles of B(g) and C(g). **(c)** The value of K_p for this reaction is > 1. [Section 15.3]

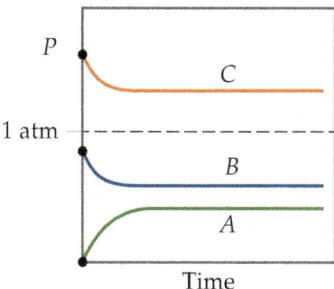

15.8 When lead(IV) oxide is heated above 300 °C, it decomposes according to the reaction, $2\,PbO_2(s) \rightleftharpoons 2\,PbO(s) + O_2(g)$. Consider the two sealed vessels of PbO_2 shown here. If both vessels are heated to 400 °C and allowed to come to equilibrium, which of the following statements is or are *true*? **(a)** There will be less PbO_2 remaining in vessel A than in vessel B. **(b)** The solid left at the bottom of each vessel will be a mixture of $PbO_2(s)$ and $PbO(s)$. **(c)** The partial pressure of $O_2(g)$ will be the same in vessels A and B. [Section 15.4]

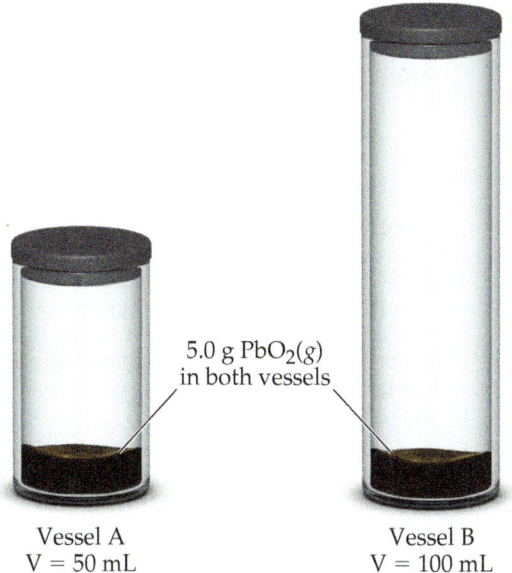

5.0 g $PbO_2(g)$ in both vessels

Vessel A
V = 50 mL

Vessel B
V = 100 mL

15.9 The reaction $A_2 + B_2 \rightleftharpoons 2\,AB$ has an equilibrium constant $K_c = 1.5$. The following diagrams represent reaction mixtures containing A_2 molecules (red), B_2 molecules (blue), and AB molecules. **(a)** Which reaction mixture is at equilibrium? **(b)** For those mixtures that are not at equilibrium, how will the reaction proceed to reach equilibrium? [Sections 15.5 and 15.6]

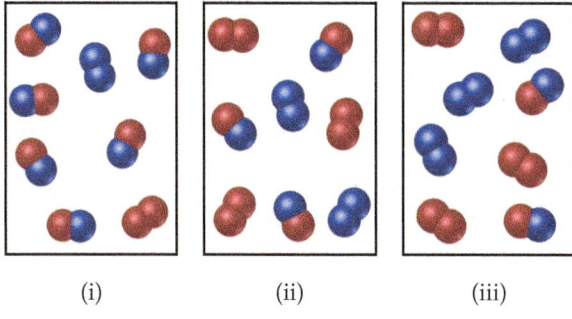

(i)   (ii)   (iii)

15.10 The diagram shown here represents the equilibrium state for the reaction $A_2(g) + 2\,B(g) \rightleftharpoons 2\,AB(g)$. **(a)** Assuming the volume is 2 L, calculate the equilibrium constant K_c for the reaction. **(b)** If the volume of the equilibrium mixture is

decreased, will the number of AB molecules increase or decrease? [Sections 15.5 and 15.7]

15.11 The following diagrams represent equilibrium mixtures for the reaction $A_2 + B \rightleftharpoons A + AB$ at 300 K and 500 K. The A atoms are red, and the B atoms are blue. **(a)** Is the value of K_p at 500 K greater than, equal to, or less than the value of K_p at 300 K? **(b)** Is the reaction exothermic or endothermic? [Section 15.7]

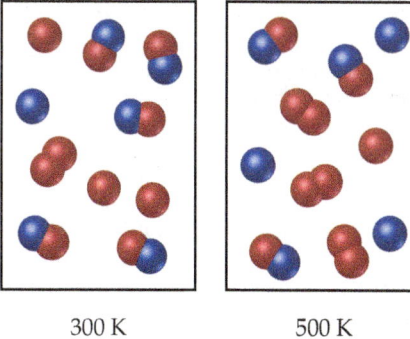

300 K   500 K

15.12 The following graph represents the yield of the compound AB at equilibrium in the reaction $A(g) + B(g) \longrightarrow AB(g)$ at two different pressures, x and y, as a function of temperature.

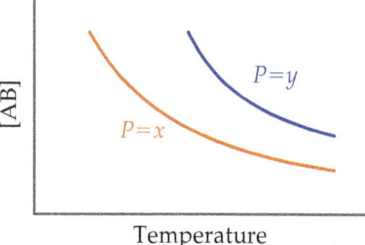

(a) Is this reaction exothermic or endothermic? **(b)** Is $P = x$ greater or smaller than $P = y$? [Section 15.7]

Equilibrium; Using the Equilibrium Constant (Sections 15.1–15.4)

15.13 Suppose that the gas-phase reactions $A \longrightarrow B$ and $B \longrightarrow A$ are both elementary reactions with rate constants of $4.7 \times 10^{-3}\,s^{-1}$ and $5.8 \times 10^{-1}\,s^{-1}$, respectively. **(a)** What is the value of the equilibrium constant for the equilibrium $A(g) \rightleftharpoons B(g)$? **(b)** Which is greater at equilibrium, the partial pressure of A or the partial pressure of B?

15.14 The equilibrium constant for the dissociation of molecular iodine, $I_2(g) \rightleftharpoons 2\,I(g)$, at 800 K is $K_c = 3.1 \times 10^{-5}$. **(a)** Which species predominates at equilibrium I_2 or I? **(b)** Assuming both forward and reverse reactions are elementary reactions, which reaction has the larger rate constant, the forward or the reverse reaction?

15.15 Write the expression for K_c for the following reactions. In each case indicate whether the reaction is homogeneous or heterogeneous.

(a) $3 NO(g) \rightleftharpoons N_2O(g) + NO_2(g)$

(b) $CH_4(g) + 2 H_2S(g) \rightleftharpoons CS_2(g) + 4 H_2(g)$

(c) $Ni(CO)_4(g) \rightleftharpoons Ni(s) + 4 CO(g)$

(d) $HF(aq) \rightleftharpoons H^+(aq) + F^-(aq)$

(e) $2 Ag(s) + Zn^{2+}(aq) \rightleftharpoons 2 Ag^+(aq) + Zn(s)$

(f) $H_2O(l) \rightleftharpoons H^+(aq) + OH^-(aq)$

(g) $2 H_2O_2(l) \rightleftharpoons 2 H_2O(l) + O_2(g)$

15.16 Write the expressions for K_c for the following reactions. In each case indicate whether the reaction is homogeneous or heterogeneous.

(a) $2 O_3(g) \rightleftharpoons 3 O_2(g)$

(b) $Ti(s) + 2 Cl_2(g) \rightleftharpoons TiCl_4(l)$

(c) $2 C_2H_4(g) + 2 H_2O(g) \rightleftharpoons 2 C_2H_6(g) + O_2(g)$

(d) $C(s) + 2 H_2(g) \rightleftharpoons CH_4(g)$

(e) $4 HCl(aq) + O_2(g) \rightleftharpoons 2 H_2O(l) + 2 Cl_2(g)$

(f) $2 C_8H_{18}(l) + 25 O_2(g) \rightleftharpoons 16 CO_2(g) + 18 H_2O(g)$

(g) $2 C_8H_{18}(l) + 25 O_2(g) \rightleftharpoons 16 CO_2(g) + 18 H_2O(l)$

15.17 When the following reactions come to equilibrium, does the equilibrium mixture contain mostly reactants or mostly products?

(a) $N_2(g) + O_2(g) \rightleftharpoons 2 NO(g)$ $K_c = 1.5 \times 10^{-10}$

(b) $2 SO_2(g) + O_2(g) \rightleftharpoons 2 SO_3(g)$ $K_p = 2.5 \times 10^9$

15.18 Which of the following reactions lies to the right, favoring the formation of products, and which lies to the left, favoring the formation of reactants?

(a) $2 NO(g) + O_2(g) \rightleftharpoons 2 NO_2(g)$ $K_p = 5.0 \times 10^{12}$

(b) $2 HBr(g) \rightleftharpoons H_2(g) + Br_2(g)$ $K_c = 5.8 \times 10^{-18}$

15.19 Which of the following statements are true and which are false? (a) The equilibrium constant can never be a negative number. (b) In reactions that we draw with a single-headed arrow, the equilibrium constant has a value that is very large. (c) As the value of the equilibrium constant increases, the speed at which a reaction reaches equilibrium must increase.

15.20 Which of the following statements about the equilibrium $2 NO_2(g) \rightleftharpoons 2 NO(g) + O_2(g)$ at 300 K are true and which are false? (a) The values of K_c and K_p are different. (b) If the equilibrium is reached by starting with pure $NO_2(g)$, then the equilibrium partial pressures of $NO_2(g)$ and $NO(g)$ must be the same. (c) If the equilibrium is reached by starting with pure $NO_2(g)$, then the equilibrium partial pressures of $NO(g)$ and $O_2(g)$ must be different.

15.21 If $K_c = 0.042$ for $PCl_3(g) + Cl_2(g) \rightleftharpoons PCl_5(g)$ at 500 K, what is the value of K_p for this reaction at this temperature?

15.22 Calculate K_c at 303 K for $SO_2(g) + Cl_2(g) \rightleftharpoons SO_2Cl_2(g)$ if $K_p = 34.5$ at this temperature.

15.23 The equilibrium constant for the reaction

$$2 NO(g) + Br_2(g) \rightleftharpoons 2 NOBr(g)$$

is $K_c = 1.3 \times 10^{-2}$ at 1000 K. (a) At this temperature does the equilibrium favor NO and Br_2, or does it favor NOBr? (b) Calculate K_c for $2 NOBr(g) \rightleftharpoons 2 NO(g) + Br_2(g)$. (c) Calculate K_c for $NOBr(g) \rightleftharpoons NO(g) + \frac{1}{2}Br_2(g)$.

15.24 Consider the following equilibrium:

$$2 H_2(g) + S_2(g) \rightleftharpoons 2 H_2S(g) \quad K_c = 1.08 \times 10^7 \text{ at } 700\,°C$$

(a) Calculate K_p. (b) Does the equilibrium mixture contain mostly H_2 and S_2 or mostly H_2S? (c) Calculate the value of K_c if you rewrote the equation $H_2(g) + \frac{1}{2}S_2(g) \rightleftharpoons H_2S(g)$.

15.25 At 1000 K, $K_p = 1.85$ for the reaction

$$SO_2(g) + \tfrac{1}{2}O_2(g) \rightleftharpoons SO_3(g)$$

(a) What is the value of K_p for this reaction? (b) What is the value of K_p for the reaction $2 SO_2(g) + O_2(g) \rightleftharpoons 2 SO_3(g)$? (c) What is the value of K_c for the reaction in part (b)?

15.26 Consider the following equilibrium, for which at $K_p = 0.0752$ at 480 °C:

$$2 Cl_2(g) + 2 H_2O(g) \rightleftharpoons 4 HCl(g) + O_2(g)$$

(a) What is the value of K_p for the reaction $4 HCl(g) + O_2(g) \rightleftharpoons 2 Cl_2(g) + 2 H_2O(g)$?

(b) What is the value of K_p for the reaction $Cl_2(g) + H_2O(g) \rightleftharpoons 2 HCl(g) + \frac{1}{2}O_2(g)$?

(c) What is the value of K_c for the reaction in part (b)?

15.27 The following equilibria were attained at 823 K:

$$CoO(s) + H_2(g) \rightleftharpoons Co(s) + H_2O(g) \quad K_c = 67$$
$$CoO(s) + CO(g) \rightleftharpoons Co(s) + CO_2(g) \quad K_c = 490$$

Based on these equilibria, calculate the value of K_c for $H_2(g) + CO_2(g) \rightleftharpoons CO(g) + H_2O(g)$ at 823 K.

15.28 Consider the equilibrium

$$N_2(g) + O_2(g) + Br_2(g) \rightleftharpoons 2 NOBr(g)$$

Calculate the equilibrium constant K_p for this reaction, given the following information at 298 K:

$$2 NO(g) + Br_2(g) \rightleftharpoons 2 NOBr(g) \quad K_c = 2.0$$
$$2 NO(g) \rightleftharpoons N_2(g) + O_2(g) \quad K_c = 2.1 \times 10^{30}$$

15.29 Upon heating, mercury(I) oxide decomposes into elemental mercury and elemental oxygen: $2 Hg_2O(s) \rightleftharpoons 4 Hg(l) + O_2(g)$. (a) Write the equilibrium-constant expression for K_p for this reaction. (b) Suppose you run this reaction in a solvent that dissolves elemental mercury and elemental oxygen. Rewrite the equilibrium-constant expression in terms of molarities for the reaction, using (solv) to indicate solvation.

15.30 Consider the equilibrium $Na_2O(s) + SO_2(g) \rightleftharpoons Na_2SO_3(s)$. (a) Write the equilibrium-constant expression for this reaction. (b) All the compounds in this reaction are soluble in water. Write the equilibrium-constant expression for K_c for the aqueous reaction.

Calculating Equilibrium Constants (Section 15.5)

15.31 Methanol (CH_3OH) is produced commercially by the catalyzed reaction of carbon monoxide and hydrogen: $CO(g) + 2 H_2(g) \rightleftharpoons CH_3OH(g)$. An equilibrium mixture in a 2.00-L vessel is found to contain 0.0406 mol CH_3OH, 0.170 mol CO, and 0.302 mol H_2 at 500 K. Calculate K_c at this temperature.

15.32 Gaseous hydrogen iodide is placed in a closed container at 425 °C, where it partially decomposes to hydrogen and iodine: $2 HI(g) \rightleftharpoons H_2(g) + I_2(g)$. At equilibrium it is found that $[HI] = 3.53 \times 10^{-3}\,M$, $[H_2] = 4.79 \times 10^{-4}\,M$, and $[I_2] = 4.79 \times 10^{-4}\,M$. What is the value of K_c at this temperature?

15.33 The equilibrium $2 NO(g) + Cl_2(g) \rightleftharpoons 2 NOCl(g)$ is established at 500.0 K. An equilibrium mixture of the three gases has partial pressures of 0.095 atm, 0.171 atm, and 0.28 atm for NO, Cl_2, and NOCl, respectively. (a) Calculate K_p for this reaction at 500.0 K. (b) If the vessel has a volume of 5.00 L, calculate K_c at this temperature.

15.34 Phosphorus trichloride gas and chlorine gas react to form phosphorus pentachloride gas: $PCl_3(g) + Cl_2(g) \rightleftharpoons PCl_5(g)$. A 7.5-L gas vessel is charged with a mixture of $PCl_3(g)$ and $Cl_2(g)$, which is allowed to equilibrate at 450 K. At equilibrium the partial pressures of the three gases are $P_{PCl_3} = 0.124$ atm, $P_{Cl_2} = 0.157$ atm, and $P_{PCl_5} = 1.30$ atm. (a) What is the value of K_p at this temperature? (b) Does the equilibrium favor reactants or products? (c) Calculate K_c for this reaction at 450 K.

15.35 A mixture of 0.10 mol of NO, 0.050 mol of H_2, and 0.10 mol of H_2O is placed in a 1.0-L vessel at 300 K. The following equilibrium is established:

$$2\,NO(g) + 2\,H_2(g) \rightleftharpoons N_2(g) + 2\,H_2O(g)$$

At equilibrium $[NO] = 0.062\,M$. (a) Calculate the equilibrium concentrations of H_2, N_2, and H_2O. (b) Calculate the value of K_c at 300 K.

15.36 A mixture of 1.374 g of H_2 and 70.31 g of Br_2 is heated in a 2.00-L vessel at 700 K. These substances react according to

$$H_2(g) + Br_2(g) \rightleftharpoons 2\,HBr(g)$$

At equilibrium, the vessel is found to contain 0.566 g of H_2. (a) Calculate the equilibrium concentrations of H_2, Br_2, and HBr. (b) Calculate the value of K_c at 700 K.

15.37 A mixture of 0.2000 mol of CO_2, 0.1000 mol of H_2, and 0.1600 mol of H_2O is placed in a 2.000-L vessel. The following equilibrium is established at 500 K:

$$CO_2(g) + H_2(g) \rightleftharpoons CO(g) + H_2O(g)$$

(a) Calculate the initial partial pressures of CO_2, H_2, and H_2O. (b) At equilibrium $P_{H_2O} = 3.51$ atm. Calculate the equilibrium partial pressures of CO_2, H_2, and CO. (c) Calculate K_p for the reaction. (d) Calculate K_c for the reaction.

15.38 A flask is charged with 1.500 atm of $N_2O_4(g)$ and 1.00 atm $NO_2(g)$ at 250 °C, and the following equilibrium is achieved:

$$N_2O_4(g) \rightleftharpoons 2\,NO_2(g)$$

After equilibrium is reached, the partial pressure of NO_2 is 0.512 atm. (a) What is the equilibrium partial pressure of N_2O_4? (b) Calculate the value of K_p for the reaction. (c) Calculate K_c for the reaction.

15.39 Two different proteins X and Y are dissolved in aqueous solution at 37 °C. The proteins bind in a 1:1 ratio to form the complex XY. A solution that is initially 1.00 mM in each protein is allowed to reach equilibrium. At equilibrium, 0.20 mM of free X and 0.20 mM of free Y remain. What is K_c for the reaction?

15.40 A chemist at a pharmaceutical company is measuring equilibrium constants for reactions in which drug candidate molecules bind to a protein involved in cancer. The drug molecules bind the protein in a 1:1 ratio to form a drug–protein complex. The protein concentration in aqueous solution at 25 °C is $1.50 \times 10^{-6}\,M$. Drug A is introduced into the protein solution at an initial concentration of $2.00 \times 10^{-6}\,M$. Drug B is introduced into a separate, identical protein solution at an initial concentration of $2.00 \times 10^{-6}\,M$. At equilibrium, the drug A–protein solution has an A–protein complex concentration of $1.00 \times 10^{-6}\,M$, and the drug B solution has a B–protein complex concentration of $1.40 \times 10^{-6}\,M$. Calculate the K_c values for the A–protein-binding reaction and for the B–protein-binding reaction. Assuming that the drug that binds more strongly will be more effective, which drug is the better choice for further research?

Some Applications of Equilibrium Constants (Section 15.6)

15.41 Indicate whether each of the following statements about the reaction quotient Q is true or false: (a) The expression for Q_c looks the same as the expression for K_c. (b) If $Q_c < K_c$, the reaction needs to proceed to the right to reach equilibrium. (c) If $Q_p > K_p$, the partial pressures of the products need to

increase compared to those of the reactants in order to reach equilibrium.

15.42 A certain equilibrium $A(g) + B(g) \rightleftharpoons C(g)$ has $K_c = 2.0$. For each of the following concentrations of A, B, and C, determine whether the reaction mixture is at equilibrium, or whether it needs to proceed to the left or to the right in order to reach equilibrium: (a) $[A] = 1.0\,M$, $[B] = 1.0\,M$, $[C] = 1.0\,M$, (b) $[A] = 1.0\,M$, $[B] = 0.50\,M$, $[C] = 1.0\,M$, (c) $[A] = 0.5\,M$, $[B] = 0.50\,M$, $[C] = 1.0\,M$, (d) $[A] = 1.0\,M$, $[B] = 2.0\,M$, $[C] = 2.0\,M$.

15.43 At 100 °C, the equilibrium constant for the reaction $COCl_2(g) \rightleftharpoons CO(g) + Cl_2(g)$ has the value $K_c = 2.19 \times 10^{-10}$. Are the following mixtures of $COCl_2$, CO, and Cl_2 at 100 °C at equilibrium? If not, indicate the direction that the reaction must proceed to achieve equilibrium.

(a) $[COCl_2] = 2.00 \times 10^{-3}\,M$, $[CO] = 3.3 \times 10^{-6}\,M$, $[Cl_2] = 6.62 \times 10^{-6}\,M$

(b) $[COCl_2] = 4.50 \times 10^{-2}\,M$, $[CO] = 1.1 \times 10^{-7}\,M$, $[Cl_2] = 2.25 \times 10^{-6}\,M$

(c) $[COCl_2] = 0.0100\,M$, $[CO] = [Cl_2] = 1.48 \times 10^{-6}\,M$

15.44 As shown in Table 15.2, K_p for the equilibrium

$$N_2(g) + 3\,H_2(g) \rightleftharpoons 2\,NH_3(g)$$

is 4.51×10^{-5} at 450 °C. For each of the mixtures listed here, indicate whether the mixture is at equilibrium at 450 °C. If it is not at equilibrium, indicate the direction (toward product or toward reactants) in which the mixture must shift to achieve equilibrium.

(a) 98 atm NH_3, 45 atm N_2, 55 atm H_2

(b) 57 atm NH_3, 143 atm N_2, no H_2

(c) 13 atm NH_3, 27 atm N_2, 82 atm H_2

15.45 At 100 °C, $K_c = 0.078$ for the reaction

$$SO_2Cl_2(g) \rightleftharpoons SO_2(g) + Cl_2(g)$$

In an equilibrium mixture of the three gases, the concentrations of SO_2Cl_2 and SO_2 are 0.108 M and 0.052 M, respectively. What is the partial pressure of Cl_2 in the equilibrium mixture?

15.46 At 900 K, the following reaction has $K_p = 0.345$:

$$2\,SO_2(g) + O_2(g) \rightleftharpoons 2\,SO_3(g)$$

In an equilibrium mixture the partial pressures of SO_2 and O_2 are 0.135 atm and 0.455 atm, respectively. What is the equilibrium partial pressure of SO_3 in the mixture?

15.47 At 1285 °C, the equilibrium constant for the reaction $Br_2(g) \rightleftharpoons 2\,Br(g)$ is $K_c = 1.04 \times 10^{-3}$. A 0.200-L vessel containing an equilibrium mixture of the gases has 0.245 g $Br_2(g)$ in it. What is the mass of $Br(g)$ in the vessel?

15.48 For the reaction $H_2(g) + I_2(g) \rightleftharpoons 2\,HI(g)$, $K_c = 55.3$ at 700 K. In a 2.00-L flask containing an equilibrium mixture of the three gases, there are 0.056 g H_2 and 4.36 g I_2. What is the mass of HI in the flask?

15.49 At 800 K, the equilibrium constant for $I_2(g) \rightleftharpoons 2\,I(g)$ is $K_c = 3.1 \times 10^{-5}$. If an equilibrium mixture in a 10.0-L vessel contains 2.67×10^{-2} g of $I(g)$, how many grams of I_2 are in the mixture?

15.50 For $2\,SO_2(g) + O_2(g) \rightleftharpoons 2\,SO_3(g)$, $K_p = 3.0 \times 10^4$ at 700 K. In a 2.00-L vessel, the equilibrium mixture contains 1.17 g of SO_3 and 0.105 g of O_2. How many grams of SO_2 are in the vessel?

15.51 At 218 °C, $K_c = 1.2 \times 10^{-4}$ for the equilibrium

$$NH_4SH(s) \rightleftharpoons NH_3(g) + H_2S(g)$$

(a) Calculate the equilibrium concentrations of NH_3 and H_2S if a sample of solid NH_4SH is placed in a closed vessel at 218 °C and decomposes until equilibrium is reached. (b) If the volume of the flask is 1.00 L, what is the minimum mass of $NH_4SH(s)$ that must be added to the flask to reach equilibrium?

15.52 At 80 °C, $K_c = 1.87 \times 10^{-3}$ for the reaction

$$PH_3BCl_3(s) \rightleftharpoons PH_3(g) + BCl_3(g)$$

(a) Calculate the equilibrium concentrations of PH_3 and BCl_3 if a solid sample of PH_3BCl_3 is placed in a closed vessel at 80 °C and decomposes until equilibrium is reached. (b) If the flask has a volume of 0.250 L, what is the minimum mass of $PH_3BCl_3(s)$ that must be added to the flask to achieve equilibrium?

15.53 At 25 °C, the reaction

$$CaCrO_4(s) \rightleftharpoons Ca^{2+}(aq) + CrO_4^{2-}(aq)$$

has an equilibrium constant $K_c = 7.1 \times 10^{-4}$. What are the equilibrium concentrations of Ca^{2+} and CrO_4^{2-} in a saturated solution of $CaCrO_4$?

15.54 Consider the reaction

$$CaSO_4(s) \rightleftharpoons Ca^{2+}(aq) + SO_4^{2-}(aq)$$

At 25 °C, the equilibrium constant is $K_c = 2.4 \times 10^{-5}$ for this reaction. (a) If excess $CaSO_4(s)$ is mixed with water at 25 °C to produce a saturated solution of $CaSO_4$, what are the equilibrium concentrations of Ca^{2+} and SO_4^{2-}? (b) If the resulting solution has a volume of 1.4 L, what is the minimum mass of $CaSO_4(s)$ needed to achieve equilibrium?

15.55 At 2000 °C, the equilibrium constant for the reaction

$$2 NO(g) \rightleftharpoons N_2(g) + O_2(g)$$

is $K_c = 2.4 \times 10^3$. If the initial concentration of NO is 0.175 M, what are the equilibrium concentrations of NO, N_2, and O_2?

15.56 For the reaction $I_2(g) + Br_2(g) \rightleftharpoons 2 IBr(g)$, $K_c = 280$ at 150 °C. Suppose that 0.500 mol IBr in a 2.00-L flask is allowed to reach equilibrium at 150 °C. What are the equilibrium concentrations of IBr, I_2, and Br_2?

15.57 For the equilibrium

$$Br_2(g) + Cl_2(g) \rightleftharpoons 2 BrCl(g)$$

at 400 K, $K_c = 7.0$. If 0.25 mol of Br_2 and 0.55 mol of Cl_2 are introduced into a 3.0-L container at 400 K, what will be the equilibrium concentrations of Br_2, Cl_2, and BrCl?

15.58 At 373 K, $K_p = 0.416$ for the equilibrium

$$2 NOBr(g) \rightleftharpoons 2 NO(g) + Br_2(g)$$

If the equilibrium partial pressures of NOBr(g) and $Br_2(g)$ are both 0.100 atm at 373 K, what is the equilibrium partial pressure of NO(g)?

15.59 Methane, CH_4, reacts with I_2 according to the reaction $CH_4(g) + I_2(g) \rightleftharpoons CH_3I(g) + HI(g)$. At 630 K, K_p for this reaction is 2.26×10^{-4}. A reaction was set up at 630 K with initial partial pressures of 105.1 torr for CH_4 and 7.96 torr for I_2. Calculate the pressures, in torr, of all reactants and products at equilibrium.

15.60 The reaction of an organic acid with an alcohol, in organic solvent, to produce an ester and water is commonly done in the pharmaceutical industry. This reaction is catalyzed by strong acid (usually H_2SO_4). A simple example is the reaction of acetic acid with ethyl alcohol to produce ethyl acetate and water:

$$CH_3COOH(solv) + CH_3CH_2OH(solv) \rightleftharpoons$$
$$CH_3COOCH_2CH_3(solv) + H_2O(solv)$$

where (*solv*) indicates that all reactants and products are in solution but not an aqueous solution. The equilibrium constant for this reaction at 55 °C is 6.68. A pharmaceutical chemist makes up 15.0 L of a solution that is initially 0.275 M in acetic acid and 3.85 M in ethanol. At equilibrium, how many grams of ethyl acetate are formed?

Le Châtelier's Principle (Section 15.7)

15.61 Consider the following equilibrium, for which $\Delta H < 0$

$$2 SO_2(g) + O_2(g) \rightleftharpoons 2 SO_3(g)$$

How will each of the following changes affect an equilibrium mixture of the three gases: (a) $O_2(g)$ is added to the system; (b) the reaction mixture is heated; (c) the volume of the reaction vessel is doubled; (d) a catalyst is added to the mixture; (e) the total pressure of the system is increased by adding a noble gas; (f) $SO_3(g)$ is removed from the system?

15.62 Consider the reaction

$$4 NH_3(g) + 5 O_2(g) \rightleftharpoons$$
$$4 NO(g) + 6 H_2O(g), \Delta H = -904.4 \text{ kJ}$$

Does each of the following increase, decrease, or leave unchanged the yield of NO at equilibrium? (a) increase[NH_3] (b) increase [H_2O] (c) decrease [O_2] (d) decrease the volume of the container in which the reaction occurs (e) add a catalyst (f) increase temperature

15.63 How do the following changes affect the value of the equilibrium constant for a gas-phase exothermic reaction? (a) removal of a reactant (b) removal of a product (c) decrease in the volume (d) decrease in the temperature (e) addition of a catalyst

15.64 For a certain gas-phase reaction, the fraction of products in an equilibrium mixture is increased by either increasing the temperature or by increasing the volume of the reaction vessel. (a) Is the reaction exothermic or endothermic? (b) Does the balanced chemical equation have more molecules on the reactant side or product side?

15.65 Consider the following equilibrium between oxides of nitrogen

$$3 NO(g) \rightleftharpoons NO_2(g) + N_2O(g)$$

(a) At constant temperature, would a change in the volume of the container affect the fraction of products in the equilibrium mixture? (b) Use data in Appendix C to calculate $\Delta H°$ for this reaction. (c) Based on your answer for part (b), will the equilibrium constant for the reaction increase or decrease with increasing temperature?

15.66 Methanol (CH_3OH) can be made by the reaction of CO with H_2:

$$CO(g) + 2 H_2(g) \rightleftharpoons CH_3OH(g)$$

(a) Use thermochemical data in Appendix C to calculate $\Delta H°$ for this reaction. (b) To maximize the equilibrium yield of methanol, would you use a high or low temperature? (c) To maximize the equilibrium yield of methanol, would you use a high or low pressure?

15.67 Ozone, O_3, decomposes to molecular oxygen in the stratosphere according to the reaction $2 O_3(g) \longrightarrow 3 O_2(g)$. Would increasing the pressure by decreasing the size of the reaction vessel favor the formation of ozone or of oxygen?

15.68 The water–gas shift reaction $CO(g) + H_2O(g) \rightleftharpoons CO_2(g) + H_2(g)$ is used industrially to produce hydrogen. The reaction enthalpy is $\Delta H° = -41$ kJ. (a) To increase the equilibrium yield of hydrogen, would you use high or low temperature? (b) Could you increase the equilibrium yield of hydrogen by controlling the pressure of this reaction? If so, would high or low pressure favor formation of $H_2(g)$?

15.69 (a) Is the dissociation of fluorine molecules into atomic fluorine, $F_2(g) \rightleftharpoons 2 F(g)$, an exothermic or endothermic process? (b) If the temperature is raised by 100 K, does the equilibrium constant for this reaction increase or decrease? (c) If the temperature is raised by 100 K, does the forward rate constant k_f increase by a larger or smaller amount than the reverse rate constant k_r?

15.70 True or false: When the temperature of an exothermic reaction increases, the rate constant of the forward reaction decreases, which leads to a decrease in the equilibrium constant, K_c.

Additional Exercises

15.71 Both the forward reaction and the reverse reaction in the following equilibrium are believed to be elementary reactions:

$$CO(g) + Cl_2(g) \rightleftharpoons COCl(g) + Cl(g)$$

At 25 °C, the rate constants for the forward and reverse reactions are $1.4 \times 10^{-28}\,M^{-1}\,s^{-1}$ and $9.3 \times 10^{10}\,M^{-1}\,s^{-1}$, respectively. **(a)** What is the value for the equilibrium constant at 25 °C? **(b)** Are reactants or products more plentiful at equilibrium?

15.72 Suppose the equilibrium $2\,A(g) \rightleftharpoons B(g)$ has $K_c = 1$. **(a)** What is the relationship between [A] and [B] at equilibrium? **(b)** If one starts with $[A] = 1.0\,M$ and $[B] = 0\,M$, what is [B] when the system reaches equilibrium? **(c)** If the volume of the reaction vessel were now decreased at constant T, would the equilibrium shift to the right, shift to the left, or remain unchanged?

15.73 A mixture of CH_4 and H_2O is passed over a nickel catalyst at 1000 K. The emerging gas is collected in a 5.00-L flask and is found to contain 8.62 g of CO, 2.60 g of H_2, 43.0 g of CH_4, and 48.4 g of H_2O. Assuming that equilibrium has been reached, calculate K_c and K_p for the reaction $CH_4(g) + H_2O(g) \rightleftharpoons CO(g) + 3\,H_2(g)$.

15.74 When 2.00 mol of SO_2Cl_2 is placed in a 2.00-L flask at 303 K, 56% of the SO_2Cl_2 decomposes to SO_2 and Cl_2:

$$SO_2Cl_2(g) \rightleftharpoons SO_2(g) + Cl_2(g)$$

(a) Calculate K_c for this reaction at this temperature. **(b)** Calculate K_p for this reaction at 303 K. **(c)** According to Le Châtelier's principle, would the percent of SO_2Cl_2 that decomposes increase, decrease, or stay the same if the mixture were transferred to a 15.00-L vessel? **(d)** Use the equilibrium constant you calculated above to determine the percentage of SO_2Cl_2 that decomposes when 2.00 mol of SO_2Cl_2 is placed in a 15.00-L vessel at 303 K.

15.75 The value of the equilibrium constant K_c for the reaction $N_2(g) + 3\,H_2(g) \rightleftharpoons 2\,NH_3(g)$ changes in the following manner as a function of temperature

Temperature (°C)	K
300	9.6
400	0.50
500	0.058

(a) Based on the changes in K_c is this reaction exothermic or endothermic? **(b)** Use the standard enthalpies of formation given in Appendix C to determine the ΔH for this reaction at standard conditions. Does this value agree with your prediction from part (a)? **(c)** If 0.025 mol of gaseous NH_3 is added to a 1.00-L container and heated to 500 °C, what is the concentration of NH_3 once the sample reaches equilibrium?

15.76 A sample of nitrosyl bromide (NOBr) decomposes according to the equation

$$2\,NOBr(g) \rightleftharpoons 2\,NO(g) + Br_2(g)$$

An equilibrium mixture in a 5.00-L vessel at 100 °C contains 3.22 g of NOBr, 2.46 g of NO, and 6.55 g of Br_2 **(a)** Calculate K_c. **(b)** What is the total pressure exerted by the mixture of gases? **(c)** What was the mass of the original sample of NOBr?

15.77 Consider the hypothetical reaction $A(g) \rightleftharpoons 2\,B(g)$. A flask is charged with 0.75 atm of pure A, after which it is allowed to reach equilibrium at 0 °C. At equilibrium, the partial pressure of A is 0.36 atm. **(a)** What is the total pressure in the flask at equilibrium? **(b)** What is the value of K_p? **(c)** To maximize the yield of product B, would you make the reaction flask larger or smaller?

15.78 As shown in Table 15.2, the equilibrium constant for the reaction $N_2(g) + 3\,H_2(g) \rightleftharpoons 2\,NH_3(g)$ is $K_p = 4.34 \times 10^{-3}$ at 300 °C. Pure NH_3 is placed in a 1.00-L flask and allowed to reach equilibrium at this temperature. There are 1.05 g NH_3 in the equilibrium mixture. **(a)** What are the masses of N_2 and H_2 in the equilibrium mixture? **(b)** What was the initial mass of ammonia placed in the vessel? **(c)** What is the total pressure in the vessel?

15.79 For the equilibrium

$$2\,IBr(g) \rightleftharpoons I_2(g) + Br_2(g)$$

$K_p = 8.5 \times 10^{-3}$ at 150 °C. If 0.025 atm of IBr is placed in a 2.0-L container, what is the partial pressure of all substances after equilibrium is reached?

15.80 For the equilibrium

$$PH_3BCl_3(s) \rightleftharpoons PH_3(g) + BCl_3(g)$$

$K_p = 0.052$ at 60 °C. **(a)** Calculate K_c. **(b)** A closed 1.500-L vessel at 60 °C is charged with 0.0500 g of $BCl_3(g)$; 3.00 g of solid PH_3BCl_3 is then added to the flask, and the system is allowed to equilibrate. What is the equilibrium concentration of PH_3?

15.81 Solid NH_4SH is introduced into an evacuated flask at 24 °C. The following reaction takes place:

$$NH_4SH(s) \rightleftharpoons NH_3(g) + H_2S(g)$$

At equilibrium, the total pressure (for NH_3 and H_2S taken together) is 0.614 atm. What is K_p for this equilibrium at 24 °C?

15.82 A 0.831-g sample of SO_3 is placed in a 1.00-L container and heated to 1100 K. The SO_3 decomposes to SO_2 and O_2:

$$2\,SO_3(g) \rightleftharpoons 2\,SO_2(g) + O_2(g)$$

At equilibrium, the total pressure in the container is 1.300 atm. Find the values of K_p and K_c for this reaction at 1100 K.

15.83 At 100 °C, $K_c = 0.212$ for the following equilibrium:

$$N_2O_4(g) \rightleftharpoons 2\,NO_2(g)$$

(a) What is the value of K_p for this equilibrium? **(b)** At 100 °C, a 2.00-L flask is filled with 1.00 mol $N_2O_4(g)$. After the system reaches equilibrium, what is the concentration of $NO_2(g)$?

15.84 Nitric oxide (NO) reacts readily with chlorine gas as follows:

$$2\,NO(g) + Cl_2(g) \rightleftharpoons 2\,NOCl(g)$$

At 700 K, the equilibrium constant K_p for this reaction is 0.26. For each of the following mixtures at this temperature, indicate whether the mixture is at equilibrium, or, if not, whether it needs to produce more products or reactants to reach equilibrium.

(a) $P_{NO} = 0.15$ atm, $P_{Cl_2} = 0.31$ atm, $P_{NOCl} = 0.11$ atm

(b) $P_{NO} = 0.12$ atm, $P_{Cl_2} = 0.10$ atm, $P_{NOCl} = 0.050$ atm

(c) $P_{NO} = 0.15$ atm, $P_{Cl_2} = 0.20$ atm, $P_{NOCl} = 5.10 \times 10^{-3}$ atm

15.85 At 900 °C, $K_c = 0.0108$ for the reaction

$$CaCO_3(s) \rightleftharpoons CaO(s) + CO_2(g)$$

A mixture of $CaCO_3$, CaO, and CO_2 is placed in a 10.0-L vessel at 900 °C. For the following mixtures, will the amount of $CaCO_3$ increase, decrease, or remain the same as the system approaches equilibrium?

(a) 15.0 g $CaCO_3$, 15.0 g CaO, and 4.25 g CO_2

(b) 2.50 g $CaCO_3$, 25.0 g CaO, and 5.66 g CO_2

(c) 30.5 g $CaCO_3$, 25.5 g CaO, and 6.48 g CO_2

15.86 When 1.50 mol CO_2 and 1.50 mol H_2 are placed in a 3.00-L container at 395 °C, the following reaction occurs: $CO_2(g) + H_2(g) \rightleftharpoons CO(g) + H_2O(g)$. If $K_c = 0.802$, what are the concentrations of each substance in the equilibrium mixture?

15.87 The equilibrium constant K_c for C(s) + $CO_2(g) \rightleftharpoons 2\,CO(g)$ is 1.9 at 1000 K and 0.133 at 298 K. (a) If excess C is allowed to react with 25.0 g of CO_2 in a 3.00-L vessel at 1000 K, how many grams of CO are produced? (b) How many grams of C are consumed? (c) If a smaller vessel is used for the reaction, will the yield of CO be greater or smaller? (d) Is the reaction endothermic or exothermic?

15.88 NiO is to be reduced to nickel metal in an industrial process by use of the reaction

$$NiO(s) + CO(g) \rightleftharpoons Ni(s) + CO_2(g)$$

At 1600 K, the equilibrium constant for the reaction is $K_p = 6.0 \times 10^2$. (a) A 50.0-L reactor at 1600 K is charged with 50.0 g of NiO(s) and 1.00 atm of CO(g). After equilibrium is reached, what is the partial pressure of $CO_2(g)$ in the reactor? (b) What mass of Ni(s) is produced in the reaction in part (a)?

15.89 At 700 K, the equilibrium constant for the reaction

$$CCl_4(g) \rightleftharpoons C(s) + 2\,Cl_2(g)$$

is $K_p = 0.76$. A flask is charged with 2.00 atm of CCl_4, which then reaches equilibrium at 700 K. (a) What fraction of the CCl_4 is converted into C and Cl_2? (b) What are the partial pressures of CCl_4 and Cl_2 at equilibrium?

15.90 The reaction $PCl_3(g) + Cl_2(g) \rightleftharpoons PCl_5(g)$ has $K_p = 0.0870$ at 300 °C. A flask is charged with 0.50 atm PCl_3, 0.50 atm Cl_2, and 0.20 atm PCl_5 at this temperature. (a) Use the reaction quotient to determine the direction the reaction must proceed to reach equilibrium. (b) Calculate the equilibrium partial pressures of the gases. (c) What effect will increasing the volume of the system have on the mole fraction of Cl_2 in the equilibrium mixture? (d) The reaction is exothermic. What effect will increasing the temperature of the system have on the mole fraction of Cl_2 in the equilibrium mixture?

15.91 An equilibrium mixture of H_2, I_2, and HI at 458 °C contains 0.112 mol H_2, 0.112 mol I_2, and 0.775 mol HI in a 5.00-L vessel. What are the equilibrium partial pressures when equilibrium is reestablished following the addition of 0.200 mol of HI?

15.92 Consider the hypothetical reaction A(g) + 2 B(g) $\rightleftharpoons$ 2 C(g), for which $K_c = 0.25$ at a certain temperature. A 1.00-L reaction vessel is loaded with 1.00 mol of compound C, which is allowed to reach equilibrium. Let the variable x represent the number of mol/L of compound A present at equilibrium. (a) In terms of x, what are the equilibrium concentrations of compounds B and C? (b) What limits must be placed on the value of x so that all concentrations are positive? (c) By

putting the equilibrium concentrations (in terms of x) into the equilibrium-constant expression, derive an equation that can be solved for x. (d) The equation from part (c) is a cubic equation (one that has the form $ax^3 + bx^2 + cx + d = 0$). In general, cubic equations cannot be solved in closed form. However, you can estimate the solution by plotting the cubic equation in the allowed range of x that you specified in part (b). The point at which the cubic equation crosses the x-axis is the solution. (e) From the plot in part (d), estimate the equilibrium concentrations of A, B, and C. (*Hint:* You can check the accuracy of your answer by substituting these concentrations into the equilibrium expression.)

15.93 At a temperature of 700 K, the forward and reverse rate constants for the reaction 2 HI(g) $\rightleftharpoons$ $H_2(g) + I_2(g)$ are $k_f = 1.8 \times 10^{-3} M^{-1} s^{-1}$ and $k_r = 0.063 M^{-1} s^{-1}$. (a) What is the value of the equilibrium constant K_c at 700 K? (b) Is the forward reaction endothermic or exothermic if the rate constants for the same reaction have values of $k_f = 0.097 M^{-1} s^{-1}$ and $k_r = 2.6 M^{-1} s^{-1}$ at 800 K?

15.94 Consider the reaction $IO_4^-(aq) + 2\,H_2O(l) \rightleftharpoons H_4IO_6^-(aq)$; $K_c = 3.5 \times 10^{-2}$. If you start with 25.0 mL of a 0.905 M solution of $NaIO_4$, and then dilute it with water to 500.0 mL, what is the concentration of $H_4IO_6^-$ at equilibrium?

15.95 At 800 K, the equilibrium constant for the reaction $A_2g \rightleftharpoons 2\,A(g)$ is $K_c = 3.1 \times 10^{-4}$. (a) Assuming both forward and reverse reactions are elementary reactions, which rate constant do you expect to be larger, k_f or k_r? (b) If the value of $k_f = 0.27\,s^{-1}$, what is the value of k_r at 800 K? (c) Based on the nature of the reaction, do you expect the forward reaction to be endothermic or exothermic? (d) If the temperature is raised to 1000 K, will the reverse rate constant k_r increase or decrease? Will the change in k_r be larger or smaller than the change in k_f?

15.96 Water molecules in the atmosphere can form hydrogen-bonded dimers, $(H_2O)_2$. The presence of these dimers is thought to be important in the nucleation of ice crystals in the atmosphere and in the formation of acid rain. (a) Using valence-shell electron-pair repulsion (VSEPR) theory, draw the structure of a water dimer, using dashed lines to indicate intermolecular interactions. (b) What kind of intermolecular forces are involved in water dimer formation? (c) The K_p for water dimer formation in the gas phase is 0.050 at 300 K and 0.020 at 350 K. Is water dimer formation endothermic or exothermic?

15.97 The protein hemoglobin (Hb) transports O_2 in mammalian blood. Each Hb can bind four O_2 molecules. The equilibrium constant for the O_2 binding reaction is higher in fetal hemoglobin than in adult hemoglobin. In discussing protein oxygen-binding capacity, biochemists use a measure called the *P50 value*, defined as the partial pressure of oxygen at which 50% of the protein is saturated. Fetal hemoglobin has a P50 value of 19 torr, whereas adult hemoglobin has a P50 value of 26.8 torr. Use these data to estimate how much larger K_c is for the aqueous reaction 4 $O_2(g)$ + Hb(aq) $\rightleftharpoons$ [Hb(O_2)$_4$(aq)] in a fetus, compared to K_c for the same reaction in an adult.

Design an Experiment

The reaction between hydrogen and iodine to form hydrogen iodide was used to illustrate Beer's law in Chapter 14 (Figure 14.4). The reaction can be monitored using visible-light spectroscopy because I_2 has a violet color, while H_2 and HI are colorless. At 300 K, the equilibrium constant for the reaction $H_2(g) + I_2(g) \rightleftharpoons 2\,HI(g)$ is $K_c = 794$. To answer the following questions, assume you have access to hydrogen, iodine, hydrogen iodide, a transparent reaction vessel, a visible-light spectrometer, and a means for changing the temperature. (a) Which

gas or gases concentration could you readily monitor with the spectrometer? (b) To use Beer's law (Equation 14.5) you need to determine the extinction coefficient, ε, for the substance in question. How would you determine ε? (c) Describe an experiment for determining the equilibrium constant at 600 K. (d) Use the bond enthalpies in Table 8.3 to estimate the enthalpy of this reaction. (e) Based on your answer to part (d), would you expect K_c to be larger or smaller at 600 K than at 300 K?

16

ACID–BASE EQUILIBRIA

▲ CITRUS FRUITS. Most fruits contain a variety of acids that contribute to their characteristic flavors.

Acids and bases are among the most important substances in chemistry, and they affect our daily lives in innumerable ways. For example, citric acid and ascorbic acid (also known as vitamin C) are what give citrus fruits their characteristic tangy, sour tastes (**Figure 16.1**). Not only are they present in our foods, but acids and bases are also crucial components of living systems, such as the amino acids that are used to synthesize proteins and the nucleic acids that code genetic information. Both citric and malic acids are among several acids involved in the Krebs cycle (also called the citric acid cycle) that is used to generate energy in aerobic organisms. The application of acid–base chemistry has also had a critical role in shaping modern society, including such human-driven activities as industrial manufacturing, the creation of advanced pharmaceuticals, and many aspects of the environment.

We first encountered acids and bases in Sections 2.8 and 4.3, in which we discussed the naming of acids and some simple acid–base reactions, respectively. In this chapter, we take a closer look at how acids and bases are identified and characterized. In doing so, we consider their structure and bonding as well as the chemical equilibria in which they participate.

▶ Figure 16.1 **Two organic acids: Citric acid, C$_6$H$_8$O$_7$, and ascorbic acid, C$_6$H$_8$O$_6$.**

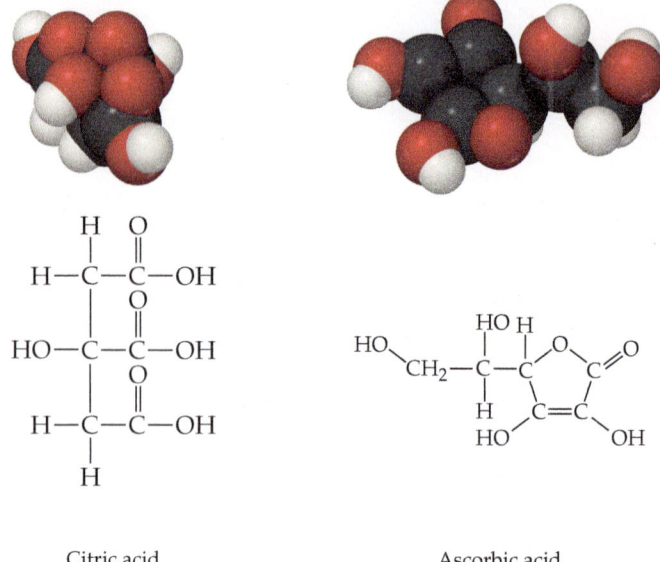

Citric acid   Ascorbic acid

16.1 | Classifications of Acids and Bases

Learning Objectives

When you finish Section 16.1, you should be able to:

▶ Define Arrhenius acids and bases and identify substances that act as Arrhenius acids or bases.

▶ Define Brønsted–Lowry acids and bases and identify molecules and ions that act as Brønsted–Lowry acids or bases.

▶ Write balanced chemical equations for proton-transfer reactions.

▶ Define Lewis acids and bases and identify molecules and ions that can act as Lewis acids or bases.

From the earliest days of experimental chemistry, scientists have recognized acids and bases by their characteristic properties. Acids have a sour taste and cause certain dyes to change color, whereas bases have a bitter taste and feel slippery (soap is a good example). Use of the term *base* comes from the old English meaning of the word, "to bring low." (We still use the word *debase* in this sense, meaning to lower the value of something.) When a base is added to an acid, the base "lowers" the amount of acid. Indeed, when acids and bases are mixed in the right proportions, their characteristic properties seem to disappear altogether. (Section 4.3)

Not surprisingly, our understanding of acids and bases has evolved over the centuries. Along the way, the definition of what makes a substance an acid or a base has changed. In general, the criteria have become more encompassing with each successive definition. For this reason, the most appropriate definition often depends on the context, making it important to be familiar with the different ways of classifying acids and bases.

Arrhenius Acids and Bases

By 1830, it was evident that all acids contain hydrogen but not all hydrogen-containing substances are acids. During the 1880s, the Swedish chemist Svante Arrhenius (1859–1927) defined acids as substances that produce H$^+$ ions in water and bases as substances that produce OH$^-$ ions in water. Over time, the Arrhenius concept of acids and bases came to be stated in the following way:

• An **Arrhenius acid** is a substance that, when dissolved in water, increases the concentration of H$^+$ ions.

• An **Arrhenius** *base* is a substance that, when dissolved in water, increases the concentration of OH$^-$ ions.

Hydrogen chloride gas, which is highly soluble in water, is an Arrhenius acid. When it dissolves in water, HCl(*g*) produces hydrated H$^+$ and Cl$^-$ ions:

$$HCl(g) \xrightarrow{H_2O} H^+(aq) + Cl^-(aq) \qquad [16.1]$$

The aqueous solution of HCl is known as *hydrochloric acid*. Concentrated hydrochloric acid is about 37% HCl by mass and is 12 *M* in HCl.

Sodium hydroxide is an Arrhenius base. Because NaOH is a soluble ionic compound, it dissociates into Na$^+$ and OH$^-$ ions when it dissolves in water, thereby increasing the concentration of OH$^-$ ions in the solution.

Brønsted–Lowry Acids and Bases

The Arrhenius concept of acids and bases, while useful, is rather limited. For one thing, it is restricted to aqueous solutions. In 1923 the Danish chemist Johannes Brønsted (1879–1947) and the English chemist Thomas Lowry (1874–1936) independently proposed a more general definition of acids and bases. Their concept is based on the fact that *acid–base reactions involve the transfer of H⁺ ions from one substance to another*. To understand this definition better, we need to examine the behavior of the H⁺ ion in water more closely.

We might at first imagine that ionization of HCl in water produces just H⁺ and Cl⁻. A hydrogen ion is no more than a bare proton—a very small particle with a positive charge. As such, an H⁺ ion interacts strongly with any source of electron density, such as the nonbonding electron pairs on the oxygen atoms of water molecules. When a proton interacts with water in this way, the **hydronium ion**, $H_3O^+(aq)$, is formed:

$$H^+ \;+\; :\!\overset{\displaystyle .}{\underset{\displaystyle H}{O}}\!-\!H \longrightarrow \left[H\!-\!\overset{\displaystyle .}{\underset{\displaystyle H}{O}}\!-\!H \right]^+$$

[16.2]

The behavior of H⁺ ions in liquid water is complex because hydronium ions interact with additional water molecules via the formation of hydrogen bonds. (Section 11.2) For example, the H_3O^+ ion bonds to additional H_2O molecules to generate such ions as $H_5O_2^+$ and $H_9O_4^+$ (**Figure 16.2**).

Chemists use the notations $H^+(aq)$ and $H_3O^+(aq)$ interchangeably to represent the hydrated proton responsible for the characteristic properties of aqueous solutions of acids. We often use the notation $H^+(aq)$ for simplicity and convenience, as we did in Chapter 4 and Equation 16.1. The notation $H_3O^+(aq)$, however, more closely represents reality.

In the reaction that occurs when HCl dissolves in water, the HCl molecule transfers an H⁺ ion (a proton) to a water molecule. Thus, we can represent the reaction as occurring between an HCl molecule and a water molecule to form hydronium and chloride ions:

$$HCl(g) \;+\; H_2O(l) \longrightarrow Cl^-(aq) \;+\; H_3O^+(aq)$$

[16.3]

$$:\!\overset{\displaystyle ..}{\underset{\displaystyle ..}{Cl}}\!-\!H \;+\; :\!\overset{\displaystyle .}{\underset{\displaystyle H}{O}}\!-\!H \longrightarrow :\!\overset{\displaystyle ..}{\underset{\displaystyle ..}{Cl}}\!:^- \;+\; \left[H\!-\!\overset{\displaystyle .}{\underset{\displaystyle H}{O}}\!-\!H \right]^+$$

Acid   Base

Notice that the reaction in Equation 16.3 involves a *proton donor* (HCl) and a *proton acceptor* (H₂O). The notion of transfer from a proton donor to a proton acceptor is the key idea in the Brønsted–Lowry definition of acids and bases:

- A **Brønsted–Lowry acid** is a substance (molecule or ion) that *donates* a proton to another substance.
- A **Brønsted–Lowry base** is a substance (molecule or ion) that *accepts* a proton.

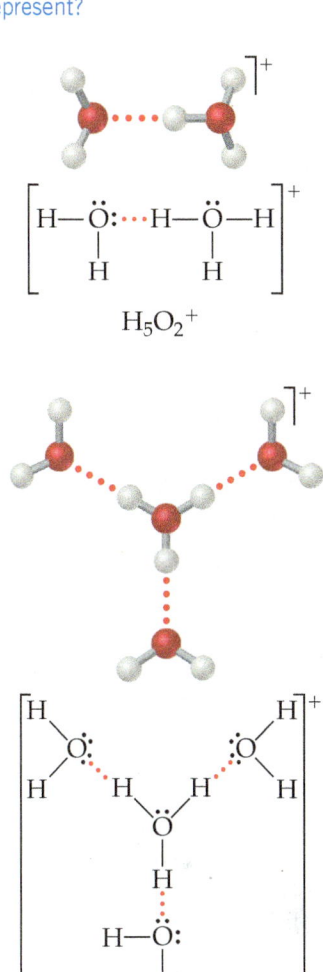

Go Figure

Which type of intermolecular force do the dotted lines in this figure represent?

$H_5O_2^+$

$H_9O_4^+$

▲ **Figure 16.2 Ball-and-stick models and Lewis structures for two hydrated hydronium ions.**

Thus, when HCl dissolves in water (Equation 16.3), HCl acts as a Brønsted–Lowry acid (it donates a proton to H_2O), and H_2O acts as a Brønsted–Lowry base (it accepts a proton from HCl). The H_2O molecule serves as a proton acceptor by using one of the nonbonding pairs of electrons on the O atom to "attach" the proton.

Because the emphasis in the Brønsted–Lowry concept is on proton transfer, the concept also applies to reactions that do not occur in aqueous solution. In the reaction between gas phase HCl and NH_3, for example, a proton is transferred from the acid HCl to the base NH_3:

[16.4]

Acid   Base

The hazy film that forms on the windows of general chemistry laboratories and on glassware in the laboratory (**Figure 16.3**) is largely solid NH_4Cl formed by the gas-phase reaction between HCl and NH_3.

To better understand the relationship between the Arrhenius and Brønsted–Lowry definitions of acids and bases, consider an aqueous solution of ammonia, in which we have the following equilibrium:

$$NH_3(aq) + H_2O(l) \rightleftharpoons NH_4^+(aq) + OH^-(aq)$$

Base   Acid

[16.5]

Ammonia is a Brønsted–Lowry base because it accepts a proton from H_2O. It is also an Arrhenius base because adding it to water leads to an increase in the concentration of $OH^-(aq)$. A key aspect of the Brønsted–Lowry approach to classifying acids and bases is summarized in the following sentence:

The transfer of a proton always involves both an acid (donor) and a base (acceptor).

In other words, a substance can function as an acid only if another substance simultaneously behaves as a base. To be a Brønsted–Lowry acid, a molecule or ion must have a hydrogen atom it can lose as an H^+ ion. To be a Brønsted–Lowry base, a molecule or ion must have a nonbonding pair of electrons it can use to bind the H^+ ion.

Some substances can act as an acid in one reaction and as a base in another. For example, H_2O is a Brønsted–Lowry base in Equation 16.3 and a Brønsted–Lowry acid in Equation 16.5. A substance capable of acting as either an acid or a base is called **amphiprotic**. An amphiprotic substance acts as a base when combined with something more strongly acidic than itself and as an acid when combined with something more strongly basic than itself.

▲ **Figure 16.3** Vapors of HCl(g) and NH_3(g) from the reagent bottles below react to create a fog of NH_4Cl(s).

 ## Sample Exercise 16.1
Writing Equations for Brønsted–Lowry Acid–Base Reactions

The hydrogen sulfite ion (HSO_3^-) is amphiprotic. Write an equation for the reaction of HSO_3^- with water (**a**) in which the ion acts as an acid and (**b**) in which the ion acts as a base.

SOLUTION

Analyze and Plan We are asked to write two equations representing reactions between HSO_3^- and water, one in which HSO_3^- should donate a proton to water, thereby acting as a Brønsted–Lowry acid, and one in which HSO_3^- should accept a proton from water, thereby acting as a base.

Solve

(**a**) In order for HSO_3^- to act as an acid, it must donate a proton to H_2O. After donating the proton, it will become SO_3^{2-}, and H_2O will be transformed to H_3O^+ upon accepting the proton.

$$HSO_3^-(aq) + H_2O(l) \rightleftharpoons SO_3^{2-}(aq) + H_3O^+(aq)$$

(**b**) In order for HSO_3^- to act as a base, it must accept a proton from H_2O. This proton transfer will create sulfurous acid (H_2SO_3) and hydroxide ions.

$$HSO_3^-(aq) + H_2O(l) \rightleftharpoons H_2SO_3(aq) + OH^-(aq)$$

▶ **Practice Exercise**

When lithium oxide (Li_2O) is dissolved in water, the solution turns basic from the reaction of the oxide ion (O^{2-}) with water. Write the balanced chemical equation for this reaction.

Lewis Acids and Bases

For a substance to be a proton acceptor (a Brønsted–Lowry base), it must have an unshared pair of electrons for binding the proton, as illustrated by ammonia. Using Lewis structures, we can write the reaction between H^+ and NH_3 as

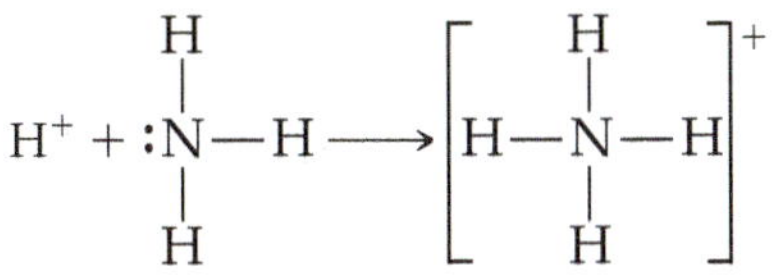

$$\text{H}^+ + \text{:N—H} \longrightarrow \left[\text{H—N—H}\right]^+$$

[16.6]

G. N. Lewis was the first to notice this aspect of acid–base reactions. He proposed a more general definition of acids and bases that emphasizes the shared electron pair:

- A **Lewis acid** is an electron-pair acceptor.
- A **Lewis base** is an electron-pair donor.

Everything that is a base in the Brønsted–Lowry sense (a proton acceptor) is also a base in the Lewis sense (an electron-pair donor). In the Lewis theory, however, a base can donate its electron pair to something other than H^+. The Lewis definition therefore greatly increases the number of species that can be considered acids. For example, the reaction between NH_3 and BF_3 occurs because BF_3 has a vacant orbital in its valence shell. (Section 8.7) It therefore acts as an electron-pair acceptor (a Lewis acid) toward NH_3, which donates the electron pair:

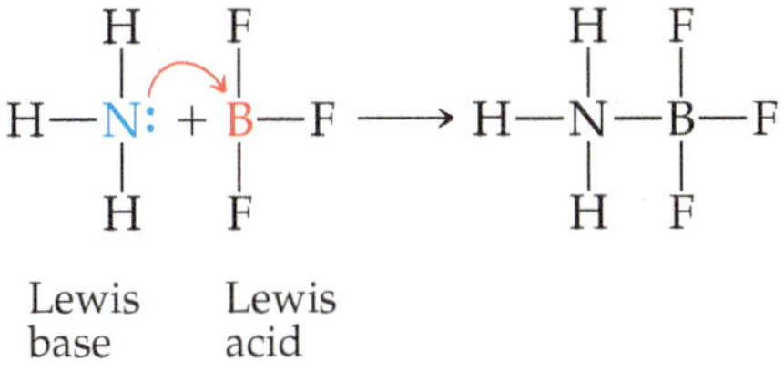

$$\text{H—N: + B—F} \longrightarrow \text{H—N—B—F}$$

| Lewis | Lewis |
| base | acid |

[16.7]

Lewis acids include molecules that have an incomplete octet of electrons, like BF_3. In addition, many simple cations can function as Lewis acids. For example, Fe^{3+} interacts strongly with cyanide ions to form the *ferricyanide ion* $[Fe(CN)_6]^{3-}$:

$$Fe^{3+} + 6[\text{:C}\equiv\text{N:}]^- \longrightarrow [Fe(\text{:C}\equiv\text{N:})_6]^{3-}$$

[16.8]

The Fe^{3+} ion has vacant orbitals that accept the electron pairs donated by the cyanide ions. (In Chapter 23, we explain in more detail which orbitals are used by the Fe^{3+} ion.) The metal ion is highly charged, too, which contributes to the interaction with the negatively charged CN^- ions. We show in Section 16.9 that this property of metal cations is important for understanding the acid–base properties of salt solutions.

Some compounds containing multiple bonds can behave as Lewis acids. For example, the reaction of carbon dioxide with water to form carbonic acid (H_2CO_3) can be pictured as an attack by a water molecule on CO_2, in which the water acts as an electron-pair donor and the CO_2 acts as an electron-pair acceptor:

$$\text{H—O: C} \longrightarrow \text{H—O—C} \longrightarrow \text{H—O—C}$$

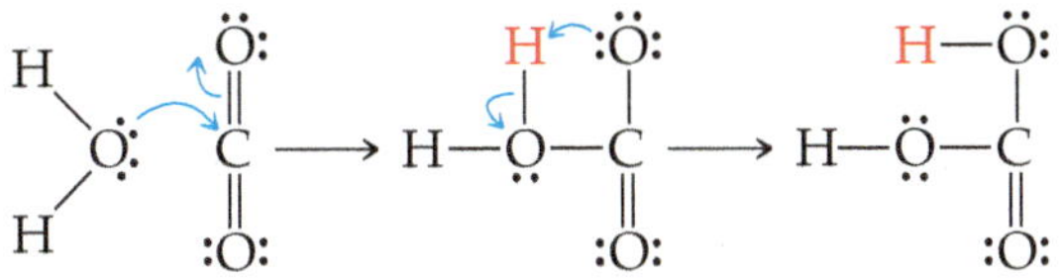

[16.9]

One electron pair of one of the carbon–oxygen double bonds is moved onto the oxygen, leaving a vacant orbital on the carbon, which means the carbon can accept an electron pair donated by H_2O. The initial acid–base product rearranges by transferring a proton from the water oxygen to a carbon dioxide oxygen, forming carbonic acid.

Our emphasis throughout this chapter is on water as the solvent and on the proton as the source of acidic properties. In such cases we find the Brønsted–Lowry definition of acids and bases to be the most useful. In fact, when we speak of a substance as being acidic or basic, we are usually thinking of aqueous solutions and using these terms in the Arrhenius or Brønsted–Lowry sense. To avoid confusion, a substance such as BF_3 is rarely called an acid, unless it is clear from the context that we are using the term in the sense of the Lewis definition. Instead, substances that function as electron-pair acceptors are referred to explicitly as "Lewis acids."

The Lewis acid–base concept allows many ideas developed in this chapter to be used more broadly in chemistry, including reactions in solvents other than water. If you take a course in organic chemistry, you will see a number of important reactions that require the presence of a Lewis acid in order to proceed. The interaction of lone pairs on one molecule or ion with vacant orbitals on another molecule or ion is one of the most important concepts in chemistry, as you will see throughout your studies.

 Self-Assessment Exercises

SAE 16.1 When sodium fluoride is added to water, it dissolves, and the resulting fluoride ions react with water according to the following reaction: $F^-(aq) + H_2O(l) \rightleftharpoons HF(aq) + OH^-(aq)$. How would you classify the fluoride ions in this reaction? (**a**) a Brønsted–Lowry base but not an Arrhenius base (**b**) a Brønsted–Lowry base and an Arrhenius base (**c**) a Brønsted–Lowry but not an Arrhenius acid (**d**) a Brønsted–Lowry acid and an Arrhenius acid

SAE 16.2 Which of the following molecules can act as a Lewis acid but not a Brønsted–Lowry acid? (**a**) $AlCl_3$ (**b**) CH_3NH_2 (**c**) $HClO_4$ (**d**) H_2O

SAE 16.3 The hydrogen phosphate ion, HPO_4^{2-}, acts as a Brønsted–Lowry base when it reacts with water. What are the products of this reaction? (**a**) H_3O^+ and PO_4^{3-} (**b**) H_3O^+ and $H_2PO_4^-$ (**c**) OH^- and PO_4^{3-} (**d**) OH^- and $H_2PO_4^-$

SAE 16.4 When $NH_4Cl(s)$ is dissolved in water, the solution turns acidic due to the reaction of ammonium cations and water: $NH_4^+(aq) + H_2O(l) \rightleftharpoons NH_3(aq) + H_3O^+(aq)$. In this reaction the ammonium cation acts as ____. (**a**) an Arrhenius acid (**b**) a Brønsted–Lowry acid (**c**) a Lewis acid (**d**) a Brønsted–Lowry acid and a Lewis acid (**e**) an acid by all three definitions

16.2 | Conjugate Acid–Base Pairs

 Learning Objectives

When you finish Section 16.2, you should be able to:

▶ Given a Brønsted–Lowry acid or base, determine the identity of its conjugate.

▶ Correlate the strength of an acid to the strength of its conjugate base.

▶ Use the relative strengths of acids and bases in a reaction to predict whether the equilibrium in a Brønsted–Lowry acid–base reaction favors the reactants or the products.

In any Brønsted–Lowry acid–base equilibrium, both the forward reaction (to the right) and the reverse reaction (to the left) involve proton transfer. For example, consider the reaction of an acid HA with water:

$$HA(aq) + H_2O(l) \rightleftharpoons A^-(aq) + H_3O^+(aq) \qquad [16.10]$$

In the forward reaction, HA donates a proton to H_2O. Therefore, HA is the Brønsted–Lowry acid and H_2O is the Brønsted–Lowry base. In the reverse reaction, the H_3O^+ ion donates a proton to the A^- ion, so H_3O^+ is the acid and A^- is the base. When the acid HA donates a proton, it leaves behind a substance, A^-, that can act as a base. Likewise, when H_2O acts as a base, it generates H_3O^+, which can act as an acid.

An acid and a base such as HA and A^- that differ only in the presence or absence of a proton are called a **conjugate acid–base pair**. Every acid has a **conjugate base**, formed by removing a proton from the acid. For example, OH^- is the conjugate base of H_2O, and A^- is the conjugate base of HA. Every base has a **conjugate acid**, formed by adding a proton to the base. Thus, H_3O^+ is the conjugate acid of H_2O, and HA is the conjugate acid of A^-.

In any Brønsted–Lowry acid–base reaction, we can identify two sets of conjugate acid–base pairs. For example, consider the reaction between nitrous acid and water:

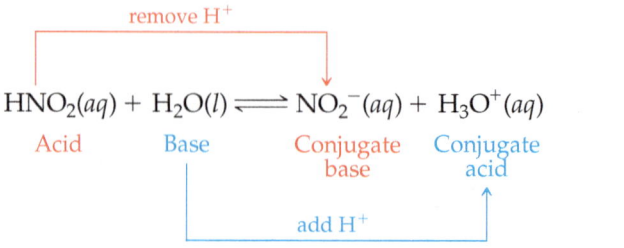

$$[16.11]$$

Likewise, for the reaction between NH_3 and H_2O (Equation 16.12), we have

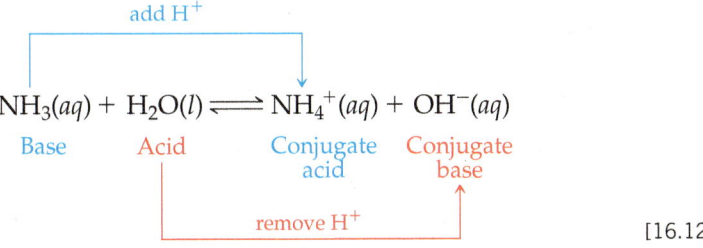

$$NH_3(aq) + H_2O(l) \rightleftharpoons NH_4^+(aq) + OH^-(aq)$$

Base   Acid   Conjugate   Conjugate
                acid   base

[16.12]

Once you become proficient at identifying conjugate acid–base pairs, it is not difficult to write equations for reactions involving Brønsted–Lowry acids and bases (*proton-transfer reactions*).

Sample Exercise 16.2
Identifying Conjugate Acids and Bases

(**a**) What is the conjugate base of $HClO_4$, H_2S, PH_4^+, HCO_3^-?

(**b**) What is the conjugate acid of CN^-, SO_4^{2-}, H_2O, HCO_3^-?

SOLUTION

Analyze We are asked to give the conjugate base for several acids and the conjugate acid for several bases.

Plan The conjugate base of a substance is simply the parent substance minus one proton, whereas the conjugate acid of a substance is the parent substance plus one proton.

Solve

(**a**) If we remove a proton from $HClO_4$, we obtain ClO_4^-, which is its conjugate base. The other conjugate bases are HS^-, PH_3, and CO_3^{2-}.

(**b**) If we add a proton to CN^-, we get HCN, its conjugate acid. The other conjugate acids are HSO_4^-, H_3O^+, and H_2CO_3. Notice that the hydrogen carbonate ion (HCO_3^-) is amphiprotic. It can act as either an acid or a base.

▶ **Practice Exercise**
Write the formula for the conjugate acid of HSO_3^-, F^-, CH_3NH_2, and PO_4^{3-}.

Relative Strengths of Acids and Bases

Some acids are better proton donors than others, and some bases are better proton acceptors than others. If we arrange acids in the order of their ability to donate a proton, we find that the more easily a substance gives up a proton, the less easily its conjugate base accepts a proton. Similarly, the more easily a base accepts a proton, the less easily its conjugate acid gives up a proton. In other words, *the stronger an acid, the weaker its conjugate base*, and *the stronger a base, the weaker its conjugate acid*. Thus, if we know how readily an acid donates protons, we also know something about how readily its conjugate base accepts protons.

The inverse relationship between the strengths of acids and their conjugate bases is illustrated in **Figure 16.4**. Here we have grouped acids and bases into three broad categories based on their behavior in water:

1. A *strong acid* completely transfers its protons to water, leaving essentially no undissociated molecules in solution. (Section 4.3) Its conjugate base has a negligible tendency to accept protons in aqueous solution. (*The conjugate base of a strong acid shows negligible basicity.*)

2. A *weak acid* only partially dissociates in aqueous solution and therefore exists in the solution as a mixture of the undissociated acid and its conjugate base. Therefore, the conjugate base of a weak acid shows a slight ability to remove protons from water. (*The conjugate base of a weak acid is a weak base.*)

3. A substance with *negligible acidity* contains hydrogen but does not demonstrate any acidic behavior in water. Its conjugate base is a strong base, reacting completely with water to form OH^- ions. (*The conjugate base of a substance with negligible acidity is a strong base.*)

Go Figure   If O^{2-} ions are added to water, what reaction, if any, occurs?

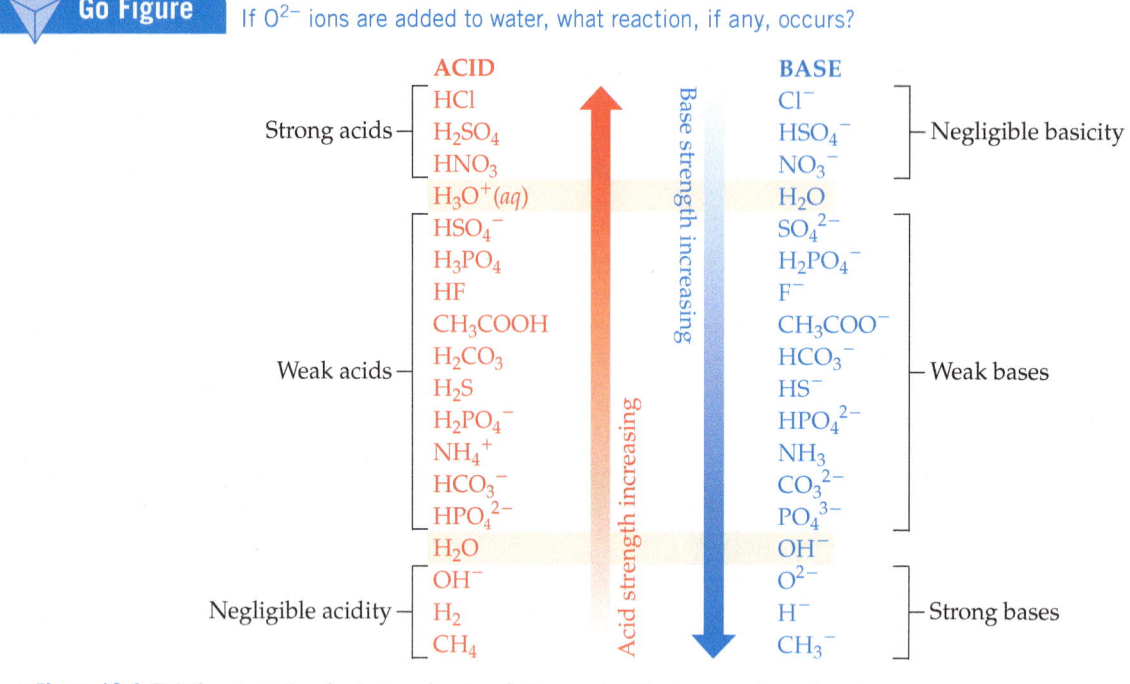

▲ **Figure 16.4 Relative strengths of select conjugate acid–base pairs.** The two members of each pair are listed opposite each other in the two columns.

The ions $H_3O^+(aq)$ and $OH^-(aq)$ are, respectively, the strongest possible acid and strongest possible base that can exist at equilibrium in aqueous solution. Stronger acids react with water to produce $H_3O^+(aq)$ ions, and stronger bases react with water to produce $OH^-(aq)$ ions, a phenomenon known as the *leveling effect*.

We can think of proton-transfer reactions as being governed by the relative abilities of two bases to abstract protons. For example, consider the proton transfer that occurs when an acid HA dissolves in water:

$$HA(aq) + H_2O(l) \rightleftharpoons H_3O^+(aq) + A^-(aq) \qquad [16.13]$$

If H_2O (the base in the forward reaction) is a stronger base than A^- (the conjugate base of HA), it is favorable to transfer the proton from HA to H_2O, producing H_3O^+ and A^-. As a result, the equilibrium lies to the right. Taken to the limit, this describes the behavior of a strong acid in water. For example, when HCl dissolves in water, the solution consists almost entirely of H_3O^+ and Cl^- ions with a negligible concentration of HCl molecules:

$$HCl(g) + H_2O(l) \longrightarrow H_3O^+(aq) + Cl^-(aq) \qquad [16.14]$$

H_2O is a stronger base than Cl^- (Figure 16.4), so H_2O acquires the proton to become the hydronium ion. Because the reaction lies completely to the right, we write Equation 16.14 with only an arrow to the right rather than using the double arrows for an equilibrium.

When A^- is a stronger base than H_2O, the equilibrium lies to the left. This situation occurs when HA is a weak acid. For example, an aqueous solution of acetic acid consists mainly of CH_3COOH molecules with relatively few H_3O^+ and CH_3COO^- ions:

$$CH_3COOH(aq) + H_2O(l) \rightleftharpoons H_3O^+(aq) + CH_3COO^-(aq) \qquad [16.15]$$

The CH_3COO^- ion is a stronger base than H_2O (Figure 16.4), so the reverse reaction is favored more than the forward reaction.

From these examples, we conclude that *in every acid–base reaction, equilibrium favors transfer of the proton from the stronger acid to the stronger base to form the weaker acid and the weaker base.*

Sample Exercise 16.3

Predicting the Position of a Proton-Transfer Equilibrium

For the following proton-transfer reaction, use Figure 16.4 to predict whether the equilibrium lies to the left ($K_c < 1$) or to the right ($K_c > 1$):

$$HSO_4^-(aq) + CO_3^{2-}(aq) \rightleftharpoons SO_4^{2-}(aq) + HCO_3^-(aq)$$

SOLUTION

Analyze We are asked to predict whether an equilibrium lies to the right, favoring products, or to the left, favoring reactants.

Plan This is a proton-transfer reaction, and the position of the equilibrium will favor the proton going to the stronger of two bases. The two bases in the equation are CO_3^{2-}, the base in the forward reaction, and SO_4^{2-}, the conjugate base of HSO_4^-. We can find the relative positions of these two bases in Figure 16.4 to determine which is the stronger base.

Solve The CO_3^{2-} ion appears lower in the right-hand column in Figure 16.4 and is therefore a stronger base than SO_4^{2-}. Therefore, CO_3^{2-} will get the proton preferentially to become HCO_3^-, while SO_4^{2-} will remain mostly unprotonated. The resulting equilibrium lies to the right, favoring products (that is, $K_c > 1$):

$$\underset{\text{Acid}}{HSO_4^-(aq)} + \underset{\text{Base}}{CO_3^{2-}(aq)} \rightleftharpoons \underset{\text{Conjugate base}}{SO_4^{2-}(aq)} + \underset{\text{Conjugate acid}}{HCO_3^-(aq)} \quad K_c > 1$$

Comment Of the two acids, HSO_4^- and HCO_3^-, the stronger one (HSO_4^-) gives up a proton more readily, and the weaker one (HCO_3^-) tends to retain its proton. Thus, the equilibrium favors the direction in which the proton moves from the stronger acid and becomes bonded to the stronger base.

▶ **Practice Exercise**
For each reaction, use Figure 16.4 to predict whether the equilibrium lies to the left or to the right:
(a) $HPO_4^{2-}(aq) + H_2O(l) \rightleftharpoons H_2PO_4^-(aq) + OH^-(aq)$
(b) $NH_4^+(aq) + OH^-(aq) \rightleftharpoons NH_3(aq) + H_2O(l)$

Self-Assessment Exercises

SAE 16.5 The conjugate base of $HPO_4^{2-}(aq)$ is _____, and the conjugate acid of $NH_3(aq)$ is _____. (a) $H_2PO_4^-$, NH_4^+ (b) PO_4^{3-}, NH_4^+ (c) $H_2PO_4^-$, NH_2^- (d) PO_4^{3-}, NH_2^- (e) PO_4, NH_4

SAE 16.6 Arrange the following molecules/ions from the weakest to the strongest base: Cl^-, OH^-, O^{2-}, NH_3. (a) weakest $Cl^- < NH_3 < O^{2-} < OH^-$ strongest (b) weakest $O^{2-} < Cl^- < NH_3 < OH^-$ strongest (c) weakest $Cl^- < NH_3 < OH^- < O^{2-}$ strongest (d) weakest $Cl^- < O^{2-} < NH_3 < OH^-$ strongest (e) weakest $NH_3 < Cl^- < OH^- < O^{2-}$ strongest

SAE 16.7 Use Figure 16.4 to determine in which of the following acid–base reactions the equilibrium lies to the right ($K_c > 1$)?

(i) $NH_3(aq) + H_2PO_4^-(aq) \rightleftharpoons NH_4^+(aq) + HPO_4^{2-}(aq)$
(ii) $HF(aq) + SO_4^{2-}(aq) \rightleftharpoons F^-(aq) + HSO_4^-(aq)$
(iii) $CH_3COOH(aq) + HCO_3^-(aq) \rightleftharpoons CH_3COO^-(aq) + H_2CO_3(aq)$

(a) reaction (i) (b) reaction (iii) (c) both reactions (i) and (ii) (d) both reactions (i) and (iii) (e) all three reactions

16.3 | The Autoionization of Water

One of the most important chemical properties of water is its ability to act as either a Brønsted–Lowry acid or a Brønsted–Lowry base. In the presence of an acid, it acts as a proton acceptor; in the presence of a base, it acts as a proton donor. In fact, one water molecule can donate a proton to another water molecule:

$$H_2O(l) + H_2O(l) \rightleftharpoons OH^-(aq) + H_3O^+(aq) \qquad [16.16]$$

Learning Objective

When you finish **Section 16.3**, you should be able to:

▶ Determine the concentration of H_3O^+ or OH^- ions in an aqueous solution from the ion-product constant of water K_w and the concentration of the other ion.

We call this process the **autoionization** of water.

Because the forward and reverse reactions in Equation 16.16 are extremely rapid, no water molecule remains ionized for long. At room temperature, only about two out of every 10^9 water molecules are ionized at any given instant. Thus, pure water consists almost entirely of H_2O molecules and is an extremely poor conductor of electricity. Nevertheless, the autoionization of water is very important, as we soon explain.

The Ion Product of Water

The equilibrium-constant expression for the autoionization of water is

$$K_c = [H_3O^+][OH^-] \qquad [16.17]$$

The term $[H_2O]$ is excluded from the equilibrium-constant expression because we exclude the concentrations of pure solids and liquids. (Section 15.4) Because this expression refers specifically to the autoionization of water, we use the symbol K_w to denote the equilibrium constant, which we call the **ion-product constant** for water. At 25 °C, K_w equals 1.0×10^{-14}. Thus, we have

$$K_w = [H_3O^+][OH^-] = 1.0 \times 10^{-14} \quad \text{(at 25 °C)} \qquad [16.18]$$

Because we use $H^+(aq)$ and $H_3O^+(aq)$ interchangeably to represent the hydrated proton, the autoionization reaction for water can also be written as

$$H_2O(l) \rightleftharpoons H^+(aq) + OH^-(aq) \qquad [16.19]$$

Likewise, the expression for K_w can be written in terms of either H_3O^+ or H^+, and K_w has the same value in either case:

$$K_w = [H_3O^+][OH^-] = [H^+][OH^-] = 1.0 \times 10^{-14} \quad \text{(at 25 °C)} \qquad [16.20]$$

This equilibrium-constant expression and the value of K_w at 25 °C are extremely important, and you should commit them to memory.

A solution in which $[H^+] = [OH^-]$ is said to be *neutral*. In most solutions, however, the H^+ and OH^- concentrations are not equal. As the concentration of one of these ions increases, the concentration of the other must decrease, so that the product of their concentrations always equals 1.0×10^{-14} (**Figure 16.5**). Aqueous solutions in which $[H^+] > [OH^-]$ are called *acidic*, whereas aqueous solutions in which $[OH^-] > [H^+]$ are called *basic*.

What makes Equation 16.20 particularly useful is that it is applicable both to pure water and to any aqueous solution. Although the equilibrium between $H^+(aq)$ and $OH^-(aq)$ as well as other ionic equilibria are affected somewhat by the presence of additional ions in solution, it is customary to ignore these ionic effects except in work requiring exceptional accuracy. Thus, Equation 16.20 is taken to be valid for any dilute aqueous solution and can be used to calculate either $[H^+]$ (if $[OH^-]$ is known) or $[OH^-]$ (if $[H^+]$ is known).

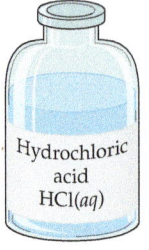

Acidic solution
$[H^+] > [OH^-]$
$[H^+][OH^-] = 1.0 \times 10^{-14}$

Neutral solution
$[H^+] = [OH^-]$
$[H^+][OH^-] = 1.0 \times 10^{-14}$

Basic solution
$[H^+] < [OH^-]$
$[H^+][OH^-] = 1.0 \times 10^{-14}$

▶ **Figure 16.5 Relative concentrations of H^+ and OH^- in aqueous solutions at 25 °C.**

Sample Exercise 16.4
Calculating [H⁺] from [OH⁻]

Calculate the concentration of $H^+(aq)$ in (**a**) a solution in which [OH⁻] is 0.010 M, (**b**) a solution in which [OH⁻] is 1.8×10^{-9} M. *Note:* In this problem and all the problems that follow, we assume, unless stated otherwise, that the temperature is 25 °C.

SOLUTION

Analyze We are asked to calculate the [H⁺] concentration in an aqueous solution where the hydroxide concentration is known.

Plan We can use the equilibrium-constant expression for the auto-ionization of water and the value of K_w to solve for each unknown concentration.

Solve

(**a**) Using Equation 16.20, we have

$$[H^+][OH^-] = 1.0 \times 10^{-14}$$
$$[H^+] = \frac{(1.0 \times 10^{-14})}{[OH^-]} = \frac{1.0 \times 10^{-14}}{0.010} = 1.0 \times 10^{-12}\,M$$

This solution is basic because

$$[OH^-] > [H^+]$$

(**b**) In this instance

$$[H^+] = \frac{(1.0 \times 10^{-14})}{[OH^-]} = \frac{1.0 \times 10^{-14}}{1.8 \times 10^{-9}} = 5.6 \times 10^{-6}\,M$$

This solution is acidic because

$$[H^+] > [OH^-]$$

▶ **Practice Exercise**
Calculate the concentration of $OH^-(aq)$ in a solution in which (**a**) $[H^+] = 2 \times 10^{-6}\,M$; (**b**) $[H^+] = [OH^-]$; (**c**) $[H^+] = 200 \times [OH^-]$.

Self-Assessment Exercises

SAE 16.8   At a temperature of 75 °C the ion-product constant for water is $K_w = 2.0 \times 10^{-13}$. For pure water at this temperature $[H_3O^+]$ is _____ $1 \times 10^{-7}\,M$ and is _____ [OH⁻]. (**a**) greater than, greater than (**b**) less than, less than (**c**) equal to, equal to (**d**) greater than, equal to (**e**) less than, equal to

SAE 16.9   When an unknown substance is dissolved in water at 25 °C, the concentration of OH⁻ ions is determined to be $2.5 \times 10^{-9}\,M$. From this we can conclude that $[H_3O^+] =$ _____ and that the solution is _____. (**a**) $2.5 \times 10^{-9}\,M$, neutral (**b**) $1.0 \times 10^{-7}\,M$, acidic (**c**) $4.0 \times 10^{-6}\,M$, basic (**d**) $4.0 \times 10^{-6}\,M$, acidic

---

16.4 | The pH Scale

The molar concentration of $H^+(aq)$ in an aqueous solution is usually very small. For convenience, we therefore usually express [H⁺] in terms of **pH**, which is the negative logarithm in base 10 of [H⁺]:*

$$pH = -\log[H^+] \qquad [16.21]$$

If you need to review the use of logarithms, see Appendix A.

We can use Equation 16.21 to calculate the pH of a neutral solution at 25 °C, where $[H^+] = 1.0 \times 10^{-7}\,M$:

$$pH = -\log(1.0 \times 10^{-7}) = -(-7.00) = 7.00$$

Notice that the pH is reported with two decimal places. We do so because *only the numbers to the right of the decimal point are the significant figures in a logarithm.* Because our original value for the concentration (1.0×10^{-7}) has two significant figures, the corresponding pH has two decimal places (7.00).

Learning Objectives

When you finish Section 16.4, you should be able to:

▶ Calculate the pH of a solution from the concentration of either H⁺ or OH⁻ and use that result to classify the solution as acidic, basic, or neutral at 25 °C.

▶ Calculate the pOH of a solution from either the OH⁻ concentration or the pH.

▶ Describe methods for measuring pH experimentally.

---

*Because [H⁺] and [H₃O⁺] are used interchangeably, you might see pH defined as $-\log[H_3O^+]$.

TABLE 16.1   Relationships between $[H^+]$, $[OH^-]$, and pH at 25 °C for Acidic, Neutral, and Basic Aqueous Solutions

	Acidic	Neutral	Basic
pH	<7.00	7.00	>7.00
$[H^+]$ (M)	>1.0×10^{-7}	1.0×10^{-7}	<1.0×10^{-7}
$[OH^-]$ (M)	<1.0×10^{-7}	1.0×10^{-7}	>1.0×10^{-7}

What happens to the pH of a solution as we make the solution more acidic, so that $[H^+]$ increases? Because of the negative sign in the logarithm term of Equation 16.21, *the pH decreases as $[H^+]$ increases*. For example, when we add sufficient acid to make $[H^+] = 1.0 \times 10^{-3} M$, the pH is

$$pH = -\log(1.0 \times 10^{-3}) = -(-3.00) = 3.00$$

At 25 °C the pH of an acidic solution is less than 7.00.

We can also calculate the pH of a basic solution, one in which $[OH^-] > 1.0 \times 10^{-7} M$. For example, suppose $[OH^-] = 2.0 \times 10^{-3} M$. We can use Equation 16.20 to calculate $[H^+]$ for this solution and Equation 16.21 to calculate the pH:

$$[H^+] = \frac{K_w}{[OH^-]} = \frac{1.0 \times 10^{-14}}{2.0 \times 10^{-3}} = 5.0 \times 10^{-12} M$$

$$pH = -\log(5.0 \times 10^{-12}) = 11.30$$

At 25 °C the pH of a basic solution is greater than 7.00. The relationships among $[H^+]$, $[OH^-]$, and pH are summarized in **Table 16.1**.

Even when $[H^+]$ is very small, as is often the case, its value is still important. Remember that many chemical processes depend on the ratio of changes in concentration. For example, if a kinetic rate law is first order in $[H^+]$, doubling the H^+ concentration doubles the rate even if the change is merely from $1 \times 10^{-7} M$ to $2 \times 10^{-7} M$. (Section 14.3) In biological systems, many reactions involve proton transfers and have rates that depend on $[H^+]$. Because the speeds of these reactions are crucial, the pH of biological fluids must be maintained within narrow limits. For example, human blood has a normal pH range of 7.35 to 7.45. Illness and even death can result if the pH varies much from this narrow range.

Sample Exercise 16.5
Calculating pH from $[H^+]$

Calculate the pH values for the two solutions of Sample Exercise 16.4.

SOLUTION

Analyze We are asked to determine the pH of aqueous solutions for which we have already calculated $[H^+]$.

Plan We can calculate pH using its defining equation, Equation 16.21.

Solve

(a) In the first instance we found $[H^+]$ to be $1.0 \times 10^{-12} M$, so

$$pH = -\log(1.0 \times 10^{-12}) = -(-12.00) = 12.00$$

Because 1.0×10^{-12} has two significant figures, the pH has two decimal places, 12.00.

(b) For the second solution, $[H^+] = 5.6 \times 10^{-6} M$. Before performing the calculation, it is helpful to estimate the pH. To do so, we note that $[H^+]$ lies between 1×10^{-6} and 1×10^{-5}.

Thus, we expect the pH to lie between 6.0 and 5.0. We use Equation 16.21 to calculate the pH:

$$pH = -\log(5.6 \times 10^{-6}) = 5.25$$

Check After calculating a pH, it is useful to compare it to your estimate. In this case the pH, as we predicted, falls between 6 and 5. Had the calculated pH and the estimate not agreed, we should have reconsidered our calculation or estimate or both.

▶ **Practice Exercise**
(a) In a sample of lemon juice, $[H^+] = 3.8 \times 10^{-4} M$. What is the pH?
(b) A commonly available window-cleaning solution has $[OH^-] = 1.9 \times 10^{-6} M$. What is the pH?

Go Figure Which is more acidic, black coffee or lemon juice?

	$[H^+]$ (M)	pH	pOH	$[OH^-]$ (M)	
	$1 (1 \times 10^0)$	0.0	14.0	1×10^{-14}	
Stomach acid	1×10^{-1}	1.0	13.0	1×10^{-13}	
Lemon juice	1×10^{-2}	2.0	12.0	1×10^{-12}	
Cola, vinegar	1×10^{-3}	3.0	11.0	1×10^{-11}	
Wine / Tomatoes	1×10^{-4}	4.0	10.0	1×10^{-10}	
Black coffee	1×10^{-5}	5.0	9.0	1×10^{-9}	
Rain / Saliva / Milk	1×10^{-6}	6.0	8.0	1×10^{-8}	
	1×10^{-7}	7.0	7.0	1×10^{-7}	Human blood
	1×10^{-8}	8.0	6.0	1×10^{-6}	Seawater
	1×10^{-9}	9.0	5.0	1×10^{-5}	Borax
	1×10^{-10}	10.0	4.0	1×10^{-4}	Lime water
	1×10^{-11}	11.0	3.0	1×10^{-3}	
	1×10^{-12}	12.0	2.0	1×10^{-2}	Household ammonia / Household bleach
	1×10^{-13}	13.0	1.0	1×10^{-1}	
	1×10^{-14}	14.0	0.0	$1 (1 \times 10^0)$	

Increasing acidity

Increasing basicity

$$pH + pOH = 14$$

$$[H^+][OH^-] = 1 \times 10^{-14}$$

▲ **Figure 16.6** The values of $[H^+]$, $[OH^-]$, pH, and pOH of some common solutions at 25 °C.

pOH and Other "p" Scales

The negative logarithm is a convenient way of expressing the magnitudes of other small quantities. We use the convention that the negative logarithm of a quantity is labeled "p" (quantity). Thus, we can express the concentration of $[OH^-]$ as pOH:

$$pOH = -\log[OH^-] \qquad [16.22]$$

Likewise, pK_w equals $-\log K_w$.

By taking the negative logarithm of both sides of the equilibrium-constant expression for water, $K_w = [H^+][OH^-]$, we obtain

$$-\log[H^+] + (-\log[OH^-]) = -\log K_w \qquad [16.23]$$

from which we obtain the useful expression

$$pH + pOH = 14.00 \quad \text{(at 25 °C)} \qquad [16.24]$$

The pH and pOH values characteristic of a number of familiar solutions are shown in **Figure 16.6**. Notice that a change in $[H^+]$ by a factor of 10 causes the pH to change by 1. Thus, the concentration of $H^+(aq)$ in a solution of pH 5 is ten times larger than the $H^+(aq)$ concentration in a solution of pH 6.

Measuring pH

The pH of a solution can be measured with a *pH meter* (**Figure 16.7**). A complete understanding of how this important device works requires knowledge of electrochemistry, a subject we take up in Chapter 20. In brief, a pH meter consists of a pair of electrodes

▲ **Figure 16.7 A digital pH meter.** The electrodes immersed in a solution produce a voltage that depends on the pH of the solution.

Sample Exercise 16.6
Calculating [H⁺] from pOH

A sample of freshly pressed apple juice has a pOH of 10.24. Calculate $[H^+]$.

SOLUTION

Analyze We need to calculate $[H^+]$ from pOH.

Plan We will first use Equation 16.24, pH + pOH = 14.00, to calculate pH from pOH. Then we will use Equation 16.21 to determine the concentration of H^+.

Solve From Equation 16.24, we have

$$pH = 14.00 - pOH$$
$$pH = 14.00 - 10.24 = 3.76$$

Next we use Equation 16.21:

$$pH = -\log[H^+] = 3.76$$

Thus,

$$\log[H^+] = -3.76$$

To find $[H^+]$, we need to determine the *antilogarithm* of −3.76. Your calculator will show this command as 10^x or INV log (these functions are usually above the log key). We use this function to perform the calculation:

$$[H^+] = \text{antilog}\,(-3.76) = 10^{-3.76} = 1.7 \times 10^{-4}\,M$$

Comment The number of significant figures in $[H^+]$ is two because the number of decimal places in the pH is two.

Check Because the pH is between 3.0 and 4.0, we know that $[H^+]$ will be between $1.0 \times 10^{-3}\,M$ and $1.0 \times 10^{-4}\,M$. Our calculated $[H^+]$ falls within this estimated range.

▶ **Practice Exercise**
A solution formed by dissolving an antacid tablet has a pOH of 4.82. What is $[H^+]$ in this solution?

connected to a meter capable of measuring small voltages, on the order of millivolts. A voltage, which varies with pH, is generated when the electrodes are placed in a solution. This voltage is read by the meter, which is calibrated to give pH.

Although less precise, acid–base indicators can be used to measure pH. An acid–base indicator is a colored substance that can exist in either an acid or a base form. The two forms have different colors. Thus, the indicator has one color at lower pH and another at higher pH. If you know the pH at which the indicator turns from one form to the other, you can determine whether a solution has a higher or lower pH than this value. Litmus, for example, changes color in the vicinity of pH 7. The color change, however, is not very sharp. Red litmus indicates a pH of about 5 or lower, and blue litmus indicates a pH of about 8 or higher.

Some common indicators are listed in **Figure 16.8**. The chart tells us, for instance, that methyl red changes color over the pH interval from about 4.5 to 6.0. Below pH 4.5 it

 Go Figure If a colorless solution turns pink when we add phenolphthalein, what can we conclude about the pH of the solution?

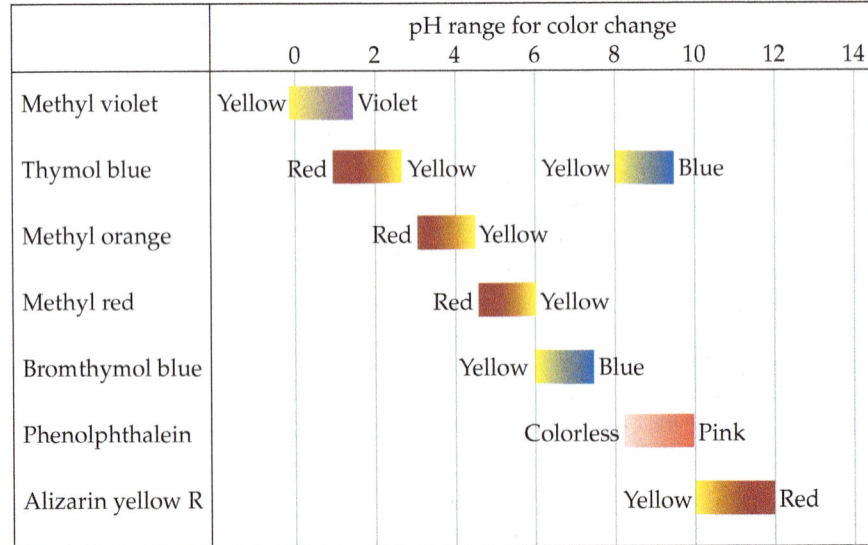

▲ **Figure 16.8 pH ranges for common acid–base indicators.** Most indicators change color over a range of about 2 pH units.

Go Figure Which of these indicators is best suited to distinguish between a solution that is slightly acidic and one that is slightly basic?

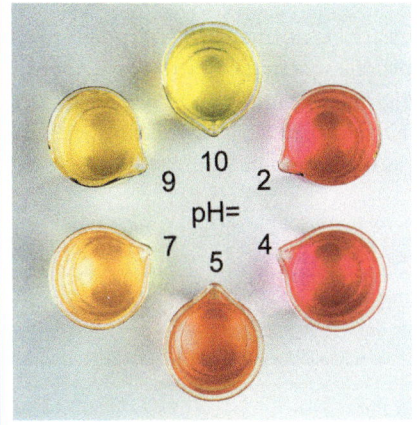

Methyl red

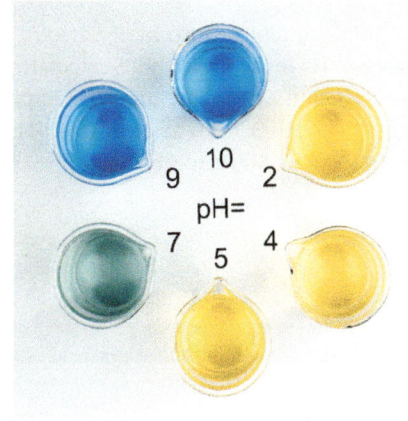

Bromthymol blue

Phenolphthalein

▲ **Figure 16.9** Solutions containing three common acid–base indicators at various pH values.

is in the acid form, which is red. In the interval between 4.5 and 6.0, it is gradually converted to its basic form, which is yellow. Once the pH rises above 6, the conversion is complete and the solution is yellow. This color change, along with that of the indicators bromthymol blue and phenolphthalein, is shown in **Figure 16.9**. Paper tape impregnated with several indicators is widely used for determining approximate pH values.

Self-Assessment Exercises

SAE 16.10 A particular nitric acid solution has an H^+ concentration of 5.0 M. What is the pH of this solution? **(a)** 5.0 **(b)** −0.70 **(c)** 0.70 **(d)** 14.70

SAE 16.11 The concentration of OH^- in an aqueous solution is $8.0 \times 10^{-5}\ M$ at room temperature. What is the pH of this solution? **(a)** 1.2×10^{-10} **(b)** 8.00 **(c)** 9.90 **(d)** 4.10

SAE 16.12 If you wanted to determine if a solution is acidic or basic by adding an acid–base indicator, which of the indicators in Figure 16.8 would be best suited to the task? **(a)** methyl orange **(b)** thymol blue **(c)** bromthymol blue **(d)** phenolphthalein

16.5 | Strong Acids and Bases

The chemistry of an aqueous solution often depends critically on pH. It is therefore important to examine how pH relates to acid and base concentrations. The simplest cases are those involving strong acids and strong bases. Strong acids and bases are *strong electrolytes*, existing in aqueous solution entirely as ions. We encountered strong acids and bases in Chapter 4. (Section 4.3) While the list of strong acids and bases is not extensive (see Table 4.2), several, including sulfuric acid and sodium hydroxide, are made in very large quantities for use in numerous applications.

Learning Objective

When you finish Section 16.5, you should be able to:

▶ Calculate the pH of a solution of a strong acid or a strong base from its concentration.

Strong Acids

The seven most common strong acids include six monoprotic acids (HCl, HBr, HI, HNO_3, $HClO_3$, $HClO_4$), and one diprotic acid (H_2SO_4). Nitric acid (HNO_3) exemplifies the behavior of the monoprotic strong acids. For all practical purposes, an aqueous solution of HNO_3 consists entirely of H_3O^+ and NO_3^- ions:

$$HNO_3(aq) + H_2O(l) \longrightarrow H_3O^+(aq) + NO_3^-(aq) \quad \text{(complete ionization)} \quad [16.25]$$

We have not used equilibrium arrows for this equation because the reaction lies entirely to the right. (Section 4.1) As noted in Section 16.3, we use $H_3O^+(aq)$ and $H^+(aq)$

interchangeably to represent the hydrated proton in water. Thus, we can simplify this acid ionization equation to

$$HNO_3(aq) \longrightarrow H^+(aq) + NO_3^-(aq)$$

In an aqueous solution of a strong acid, the acid is normally the only significant source of H^+ ions.* As a result, calculating the pH of a solution of a strong monoprotic acid is straightforward because $[H^+]$ equals the original concentration of acid. In a 0.20 M solution of $HNO_3(aq)$, for example, $[H^+] = [NO_3^-] = 0.20\, M$. The situation with the diprotic acid H_2SO_4 is somewhat more complex, as we explain in Section 16.6.

Sample Exercise 16.7
Calculating the pH of a Strong Acid Solution

What is the pH of a 0.040 M solution of $HClO_4$?

SOLUTION
Analyze and Plan Because $HClO_4$ is a strong acid, it is completely ionized, giving $[H^+] = [ClO_4^-] = 0.040\, M$.

Solve

$$pH = -\log(0.040) = 1.40$$

Check Because $[H^+]$ lies between 1×10^{-2} and 1×10^{-1}, the pH will be between 2.0 and 1.0. Our calculated pH falls within the estimated range. Furthermore, because the concentration has two significant figures, the pH has two decimal places.

▶ **Practice Exercise**
An aqueous solution of HNO_3 has a pH of 2.34. What is the concentration of the acid?

Strong Bases

The most common soluble strong bases are the ionic hydroxides of the alkali metals, such as NaOH, KOH, and the ionic hydroxides of heavier alkaline earth metals, such as $Sr(OH)_2$. These compounds completely dissociate into ions in aqueous solution. Thus, a solution labeled 0.30 M NaOH consists of 0.30 M $Na^+(aq)$ and 0.30 M $OH^-(aq)$; there is essentially no undissociated NaOH.

Although all of the alkali metal hydroxides are strong electrolytes, LiOH, RbOH, and CsOH are not commonly encountered in the laboratory. The hydroxides of the heavier alkaline earth metals—$Ca(OH)_2$, $Sr(OH)_2$, and $Ba(OH)_2$—are also strong electrolytes. They have limited solubility, however, so they are used only when high solubility is not critical.

Strongly basic solutions are also created by certain substances that react with water to form $OH^-(aq)$. The most common of these contain the oxide ion. Ionic metal oxides, especially Na_2O and CaO, are often used in industry when a strong base is needed. The O^{2-} ion reacts exothermically with water to form two OH^- ions, leaving virtually no O^{2-} in the solution:

$$O^{2-}(aq) + H_2O(l) \longrightarrow 2\, OH^-(aq) \tag{16.26}$$

Thus, a solution formed by dissolving 0.010 mol of $Na_2O(s)$ in enough water to form 1.0 L of solution has $[OH^-] = 0.020\, M$ and a pH of 12.30.

---

*Normally, the concentration of H^+ from H_2O is so small that it can be neglected, but if the concentration of the acid is $10^{-6}\, M$ or less, we also need to consider H^+ ions that result from H_2O autoionization.

Sample Exercise 16.8
Calculating the pH of a Strong Base

What is the pH of (**a**) a 0.028 *M* solution of NaOH, (**b**) a 0.0011 *M* solution of Ca(OH)$_2$?

SOLUTION

Analyze We are asked to calculate the pH of two solutions of strong bases.

Plan We can calculate each pH by either of two equivalent methods. First, we could use Equation 16.20 to calculate [H$^+$], and then we could use Equation 16.21 to calculate the pH. Alternatively, we could use [OH$^-$] to calculate pOH and then use Equation 16.24 to calculate the pH.

Solve

(**a**) NaOH dissociates in water to give one OH$^-$ ion per formula unit. Therefore, the OH$^-$ concentration for the solution in (a) equals the stated concentration of NaOH—namely, 0.028 *M*.

Method 1:

$$[H^+] = \frac{1.0 \times 10^{-14}}{0.028} = 3.57 \times 10^{-13} M$$
$$pH = -\log(3.57 \times 10^{-13}) = 12.45$$

Method 2:

$$pOH = -\log(0.028) = 1.55$$
$$pH = 14.00 - pOH = 12.45$$

(**b**) Ca(OH)$_2$ is a strong base that dissociates in water to give *two* OH$^-$ ions per formula unit. Thus, the concentration of OH$^-$(*aq*) for the solution in part (b) is

Method 1:

$$[H^+] = \frac{1.0 \times 10^{-14}}{0.022} = 4.55 \times 10^{-12} M$$
$$pH = -\log(4.55 \times 10^{-12}) = 11.34$$

Method 2:

$$pOH = -\log(0.0022) = 2.66$$
$$pH = 14.00 - pOH = 11.34$$

▶ **Practice Exercise**
What is the concentration of a solution of (**a**) KOH for which the pH is 11.89, (**b**) Ca(OH)$_2$ for which the pH is 11.68?

Self-Assessment Exercises

SAE 16.13 What is the pH of a solution made by diluting 25 mL of 6.0 *M* HCl(*aq*) to a total volume of 0.200 L? (**a**) −0.78 (**b**) 0.12 (**c**) 0.75 (**d**) 0.90

SAE 16.14 What is the pH of a 0.044 *M* Sr(OH)$_2$ solution? (**a**) 1.06 (**b**) 1.36 (**c**) 12.64 (**d**) 12.94

SAE 16.15 The pH of a hydrobromic acid solution is measured to be 3.28. What is the concentration of HBr in the solution? (**a**) 3.28 *M* (**b**) 1.9 × 10^3 *M* (**c**) 5.2 × 10^{-4} *M* (**d**) 2.6 × 10^{-4} *M*

16.6 | Weak Acids

Most acidic substances are weak acids and therefore are only partially ionized in aqueous solution (**Figure 16.10**). We can use the equilibrium constant for the ionization reaction to express the extent to which a weak acid ionizes. If we represent a general weak acid as HA, we can write the equation for its ionization in either of the following ways, depending on whether the hydrated proton is represented as H$_3$O$^+$(*aq*) or H$^+$(*aq*):

$$HA(aq) + H_2O(l) \rightleftharpoons H_3O^+(aq) + A^-(aq) \qquad [16.27]$$

or

$$HA(aq) \rightleftharpoons H^+(aq) + A^-(aq) \qquad [16.28]$$

These equilibria are in aqueous solution, so we will use equilibrium-constant expressions based on concentrations. Because H$_2$O is the solvent, it is omitted from the equilibrium-constant expression. (Section 15.4) Further, we add the subscript *a* on the equilibrium constant to indicate that it is an equilibrium constant for the ionization of an *acid*. Thus, we can write the equilibrium-constant expression as either:

$$K_a = \frac{[H_3O^+][A^-]}{[HA]} \quad \text{or} \quad K_a = \frac{[H^+][A^-]}{[HA]} \qquad [16.29]$$

K_a is called the **acid-dissociation constant** for acid HA.

Learning Objectives

When you finish Section 16.6, you should be able to:

▶ Define the acid-dissociation constant K_a of a weak acid and calculate its value from the pH of an aqueous solution of the acid.

▶ Calculate the pH of an aqueous solution of a weak acid from its concentration and its K_a value.

▶ Calculate the percent ionization of a weak acid in a solution from the initial concentration of the acid and its K_a value.

▶ Write equations and the corresponding equilibrium constant expressions for each successive ionization of a polyprotic acid.

▶ Calculate the pH and the concentrations of various ions in an aqueous solution of a polyprotic acid.

▶ **Figure 16.10** Species present in solutions of a strong acid and a weak acid.

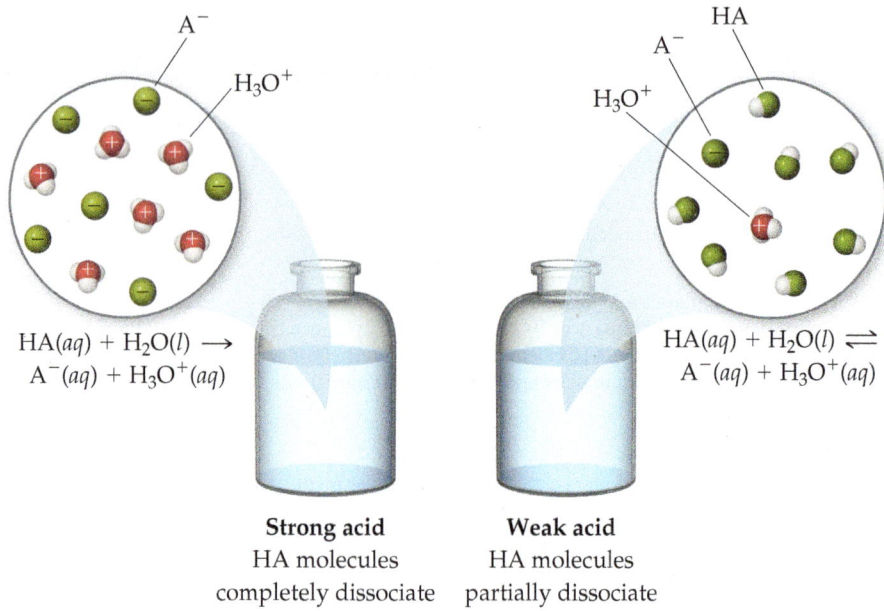

$$HA(aq) + H_2O(l) \longrightarrow$$
$$A^-(aq) + H_3O^+(aq)$$

$$HA(aq) + H_2O(l) \rightleftharpoons$$
$$A^-(aq) + H_3O^+(aq)$$

Strong acid
HA molecules
completely dissociate

Weak acid
HA molecules
partially dissociate

Table 16.2 shows the structural formulas, conjugate bases, and K_a values for a number of weak acids. Appendix D provides a more complete list. Many weak acids are organic compounds composed entirely of carbon, hydrogen, and oxygen. These compounds usually contain some hydrogen atoms bonded to carbon atoms and others bonded to oxygen atoms. In almost all cases, the hydrogen atoms bonded to carbon do not ionize in water; instead, the acidic behavior of these compounds is due to the hydrogen atoms attached to oxygen atoms.

The magnitude of K_a indicates the tendency of the acid to ionize in water: *The larger the value of K_a, the stronger the acid*. Chlorous acid ($HClO_2$), for example, is the strongest acid in Table 16.2, and phenol (HOC_6H_5) is the weakest. For most weak acids K_a values range from 10^{-2} to 10^{-10}.

TABLE 16.2   Characteristics of Selected Weak Acids in Water at 25 °C

Acid	Structural Formula*	Conjugate Base	K_a
Chlorous acid ($HClO_2$)	H—O—Cl—O	ClO_2^-	1.0×10^{-2}
Hydrofluoric acid (HF)	H—F	F^-	6.8×10^{-4}
Nitrous acid (HNO_2)	H—O—N=O	NO_2^-	4.5×10^{-4}
Benzoic acid (C_6H_5COOH)	H—O—C(=O)—⟨C₆H₅⟩	$C_6H_5COO^-$	6.3×10^{-5}
Acetic acid (CH_3COOH)	H—O—C(=O)—C(H)(H)—H	CH_3COO^-	1.8×10^{-5}
Hypochlorous acid (HOCl)	H—O—Cl	OCl^-	3.0×10^{-8}
Hydrocyanic acid (HCN)	H—C≡N	CN^-	4.9×10^{-10}
Phenol (HOC_6H_5)	H—O—⟨C₆H₅⟩	$C_6H_5O^-$	1.3×10^{-10}

*The proton that ionizes is shown in red.

Calculating K_a from pH

In order to calculate either the K_a value for a weak acid or the pH of its solutions, you will need to use many of the skills for solving equilibrium problems developed in Section 15.5. In many cases, the small magnitude of K_a means you can use approximations to simplify the problem. In doing these calculations, it is important to realize that proton-transfer reactions are generally very rapid. As a result, the measured or calculated pH for a weak acid always represents an equilibrium condition. We show the procedure for obtaining the K_a from pH in Sample Exercise 16.9.

Sample Exercise 16.9
Calculating K_a from Measured pH

A student prepared a 0.10 M solution of formic acid (HCOOH) and found its pH at 25 °C to be 2.38. Calculate K_a for formic acid at this temperature.

SOLUTION

Analyze We are given the molar concentration of an aqueous solution of weak acid and the pH of the solution, and we are asked to determine the value of K_a for the acid.

Plan Although we are dealing specifically with the ionization of a weak acid, this problem is similar to the equilibrium problems in Chapter 15. We can solve this problem using the method outlined in Sample Exercise 15.8, starting with the chemical reaction and a tabulation of initial and equilibrium concentrations.

Solve

The first step in solving any equilibrium problem is to write the equation for the equilibrium reaction. The ionization of formic acid can be written as:

$$\text{HCOOH}(aq) \rightleftharpoons \text{H}^+(aq) + \text{HCOO}^-(aq)$$

The equilibrium-constant expression is:

$$K_a = \frac{[\text{H}^+][\text{HCOO}^-]}{[\text{HCOOH}]}$$

From the measured pH, we can calculate $[\text{H}^+]$:

$$\text{pH} = -\log[\text{H}^+] = 2.38$$
$$\log[\text{H}^+] = -2.38$$
$$[\text{H}^+] = 10^{-2.38} = 4.2 \times 10^{-3}\, M$$

To determine the concentrations of the species involved in the equilibrium, we consider first that the solution is initially 0.10 M in HCOOH molecules. We then consider the ionization of the acid into H^+ and HCOO^-. For each HCOOH molecule that ionizes, one H^+ ion and one HCOO^- ion are produced in solution. Because the pH measurement indicates that $[\text{H}^+] = 4.2 \times 10^{-3}\,M$ at equilibrium, we can construct the following table:

	$\text{HCOOH}(aq)$	$\rightleftharpoons$ $\text{H}^+(aq)$	$+$ $\text{HCOO}^-(aq)$
Initial concentration (M)	0.10	0	0
Change in concentration (M)	-4.2×10^{-3}	$+4.2 \times 10^{-3}$	$+4.2 \times 10^{-3}$
Equilibrium concentration (M)	$(0.10 - 4.2 \times 10^{-3})$	4.2×10^{-3}	4.2×10^{-3}

Notice that we have neglected the very small concentration of $\text{H}^+(aq)$ due to H_2O autoionization. Notice also that the amount of HCOOH that ionizes is very small compared with the initial concentration of the acid. To the number of significant figures we are using, the subtraction yields 0.10 M:

$$(0.10 - 4.2 \times 10^{-3})\, M \approx 0.10\, M$$

We can now insert the equilibrium concentrations into the expression for K_a:

$$K_a = \frac{(4.2 \times 10^{-3})(4.2 \times 10^{-3})}{0.10} = 1.8 \times 10^{-4}$$

Check The magnitude of our answer is reasonable because K_a for a weak acid is usually between 10^{-2} and 10^{-10}.

▶ **Practice Exercise**

Niacin, one of the B vitamins, has the molecular structure shown here. A 0.020 M solution of niacin has a pH of 3.26. What is the acid-dissociation constant for niacin?

Using K_a to Calculate pH

Knowing the value of K_a and the initial concentration of a weak acid, we can calculate the concentration of $H^+(aq)$ in a solution of the acid. The procedure can be broken into the following four steps:

How to Calculate pH Using K_a

1. Write the ionization equilibrium.
2. Write the equilibrium-constant expression, including the appropriate value of the equilibrium constant, K_a.
3. Express the concentrations involved in the equilibrium reaction.
4. Substitute the equilibrium concentrations into the equilibrium-constant expression and solve for x.

Let's use this process to calculate the pH at 25 °C of a 0.30 M solution of acetic acid (CH_3COOH).

1. The ionization equilibrium is:

$$CH_3COOH(aq) \rightleftharpoons H^+(aq) + CH_3COO^-(aq) \qquad [16.30]$$

Notice that the hydrogen that ionizes is the one attached to an oxygen atom.

2. Taking $K_a = 1.8 \times 10^{-5}$ from Table 16.2, we write the equilibrium-constant expression and its value.

$$K_a = \frac{[H^+][CH_3COO^-]}{[CH_3COOH]} = 1.8 \times 10^{-5} \qquad [16.31]$$

3. Next we express the concentrations involved in the equilibrium reaction. This can be done with a little accounting, as described in Sample Exercise 16.9. Because we want to find the equilibrium value for $[H^+]$, let's call this quantity x. The concentration of acetic acid before any of it ionizes is 0.30 M. The chemical equation tells us that for each molecule of CH_3COOH that ionizes, one $H^+(aq)$ and one $CH_3COO^-(aq)$ are formed. Consequently, if x moles per liter of $H^+(aq)$ form at equilibrium, x moles per liter of $CH_3COO^-(aq)$ must also form and x moles per liter of CH_3COOH must be ionized:

	$CH_3COOH(aq)$	$\rightleftharpoons$	$H^+(aq)$	+	$CH_3COO^-(aq)$
Initial concentration (M)	0.30		0		0
Change in concentration (M)	$-x$		$+x$		$+x$
Equilibrium concentration (M)	$(0.30 - x)$		x		x

4. Finally, we substitute the equilibrium concentrations into the equilibrium-constant expression and solve for x:

$$K_a = \frac{[H^+][CH_3COO^-]}{[CH_3COOH]} = \frac{(x)(x)}{0.30 - x} = 1.8 \times 10^{-5} \qquad [16.32]$$

This expression leads to a quadratic equation in x, which we can solve by using either an equation-solving calculator or the quadratic formula. We can simplify the problem, however, by noting that the value of K_a is quite small. As a result, the equilibrium lies far to the left, and x is much smaller than the initial concentration of acetic acid. Thus, we *assume* that x is negligible relative to 0.30, so that $0.30 - x$ is essentially equal to 0.30. We can (and should!) check the validity of this assumption when we finish the problem. By using this assumption, Equation 16.32 becomes

$$K_a = \frac{x^2}{0.30} = 1.8 \times 10^{-5}$$

Solving for x, we have

$$x^2 = (0.30)(1.8 \times 10^{-5}) = 5.4 \times 10^{-6}$$

$$x = \sqrt{5.4 \times 10^{-6}} = 2.3 \times 10^{-3}$$

$$[H^+] = x = 2.3 \times 10^{-3}M$$

$$pH = -\log(2.3 \times 10^{-3}) = 2.64$$

Now we check the validity of our simplifying assumption that $0.30 - x \approx 0.30$. The value of x we determined is so small that, for this number of significant figures, the assumption is entirely valid. We are thus satisfied that the assumption was a reasonable one to make. *As a general rule, if x is more than about 5% of the initial concentration value, it is better to use the quadratic formula.* You should always check the validity of any simplifying assumptions after you have finished solving a problem.

We have also made one other assumption—namely, that all of the H^+ in the solution comes from ionization of CH_3COOH. Are we justified in neglecting the autoionization of H_2O? The answer is yes—the additional $[H^+]$ due to water, which would be on the order of $10^{-7}\,M$, is negligible compared to the $[H^+]$ from the acid (which in this case is on the order of $10^{-3}\,M$). In extremely precise work, or in cases involving very dilute solutions of acids, we would need to consider the autoionization of water more fully.

Finally, we can compare the pH value of this weak acid with the pH of a solution of a strong acid of the same concentration. The pH of the 0.30 M acetic acid is 2.64, but the pH of a 0.30 M solution of a strong acid such as HCl is $-\log(0.30) = 0.52$. As expected, the pH of a solution of a weak acid is higher than that of a solution of a strong acid of the same molarity. (Remember, the higher the pH value, the *less* acidic the solution.)

Sample Exercise 16.10
Using K_a to Calculate pH

Calculate the pH of a 0.20 M solution of HCN. (Refer to Table 16.2 or Appendix D for the value of K_a.)

SOLUTION

Analyze We are given the molarity of a weak acid and are asked for the pH. From Table 16.2, K_a for HCN is 4.9×10^{-10}.

Plan We proceed as in the example just worked in the text, writing the chemical equation and constructing a table of initial and equilibrium concentrations in which the equilibrium concentration of H^+ is our unknown.

Solve Writing both the chemical equation for the ionization reaction that forms $H^+(aq)$ and the equilibrium-constant (K_a) expression for the reaction:

$$HCN(aq) \rightleftharpoons H^+(aq) + CN^-(aq)$$

$$K_a = \frac{[H^+][CN^-]}{[HCN]} = 4.9 \times 10^{-10}$$

Next, we tabulate the concentrations of the species involved in the equilibrium reaction, letting $x = [H^+]$ at equilibrium:

	HCN(aq)	$\rightleftharpoons$	H⁺(aq)	+	CN⁻(aq)
Initial concentration (M)	0.20		0		0
Change in concentration (M)	$-x$		$+x$		$+x$
Equilibrium concentration (M)	$(0.20 - x)$		x		x

Substituting the equilibrium concentrations into the equilibrium-constant expression yields:

$$K_a = \frac{(x)(x)}{0.20 - x} = 4.9 \times 10^{-10}$$

We next make the simplifying approximation that x, the amount of acid that dissociates, is small compared with the initial concentration of acid: $0.20 - x \approx 0.20$. Thus,

$$\frac{x^2}{0.20} = 4.9 \times 10^{-10}$$

Solving for x, we have:

$$x^2 = (0.20)(4.9 \times 10^{-10}) = 0.98 \times 10^{-10}$$

$$x = \sqrt{0.98 \times 10^{-10}} = 9.9 \times 10^{-6}\,M = [H^+]$$

Continued

A concentration of $9.9 \times 10^{-6} M$ is much smaller than 5% of $0.20 M$, the initial HCN concentration. Our simplifying approximation is therefore appropriate. We now calculate the pH of the solution:

$$pH = -\log[H^+] = -\log(9.9 \times 10^{-6}) = 5.00$$

Comment The concentration of H^+ in this example is quite small ($\sim 1 \times 10^{-5}$), but it is still approximately 100 times higher than the concentration from the autoionization of water (1×10^{-7}). Thus, our decision to neglect the autoionization of water is still a reasonable assumption. But if the $[H^+]$ concentration from ionization of the weak acid gets much smaller, then that would no longer be the case.

▶ **Practice Exercise**
The K_a for niacin (Sample Exercise 16.9) is 1.5×10^{-5}. What is the pH of a 0.010 M solution of niacin?

Percent Ionization

We have seen that the magnitude of K_a indicates the strength of a weak acid. Another measure of acid strength is **percent ionization**, defined as

$$\text{Percent ionization} = \frac{\text{concentration of ionized HA}}{\text{original concentration of HA}} \times 100\% \qquad [16.33]$$

The stronger the acid, the greater the percent ionization.

If we assume that the autoionization of H_2O is negligible, the concentration of acid that ionizes equals the concentration of $H^+(aq)$ that forms. Thus, the percent ionization for an acid HA can be expressed as

$$\text{Percent ionization} = \frac{[H^+]_{\text{equilibrium}}}{[HA]_{\text{initial}}} \times 100\% \qquad [16.34]$$

For example, a 0.035 M solution of HNO_2 contains $3.7 \times 10^{-3} M H^+(aq)$, and its percent ionization is

$$\text{Percent ionization} = \frac{[H^+]_{\text{equilibrium}}}{[HNO_2]_{\text{initial}}} \times 100\% = \frac{3.7 \times 10^{-3} M}{0.035 M} \times 100\% = 11\%$$

As the concentration of a weak acid increases, the equilibrium concentration of $H^+(aq)$ increases, as you should expect. However, as shown in **Figure 16.11**, *the percent ionization decreases as the concentration increases.* Thus, the concentration of $H^+(aq)$ is not directly proportional to the concentration of the weak acid. For example, doubling the concentration of a weak acid does not double the concentration of $H^+(aq)$.

 **Go Figure** Is the trend observed in this graph consistent with Le Châtelier's principle? Explain.

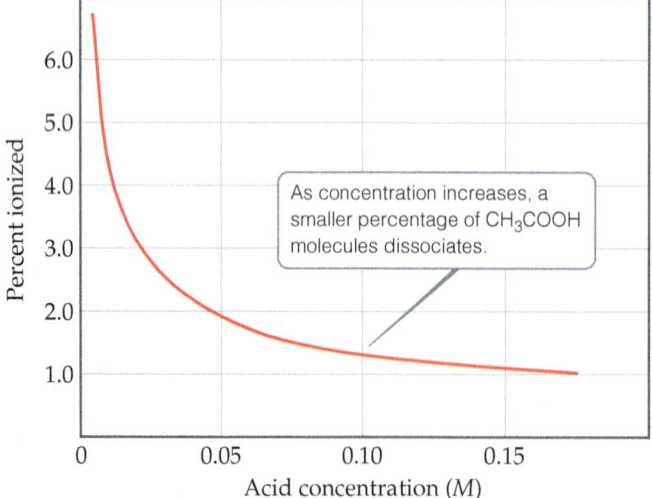

As concentration increases, a smaller percentage of CH_3COOH molecules dissociates.

▲ **Figure 16.11 Effect of concentration on percent ionization in an acetic acid solution.**

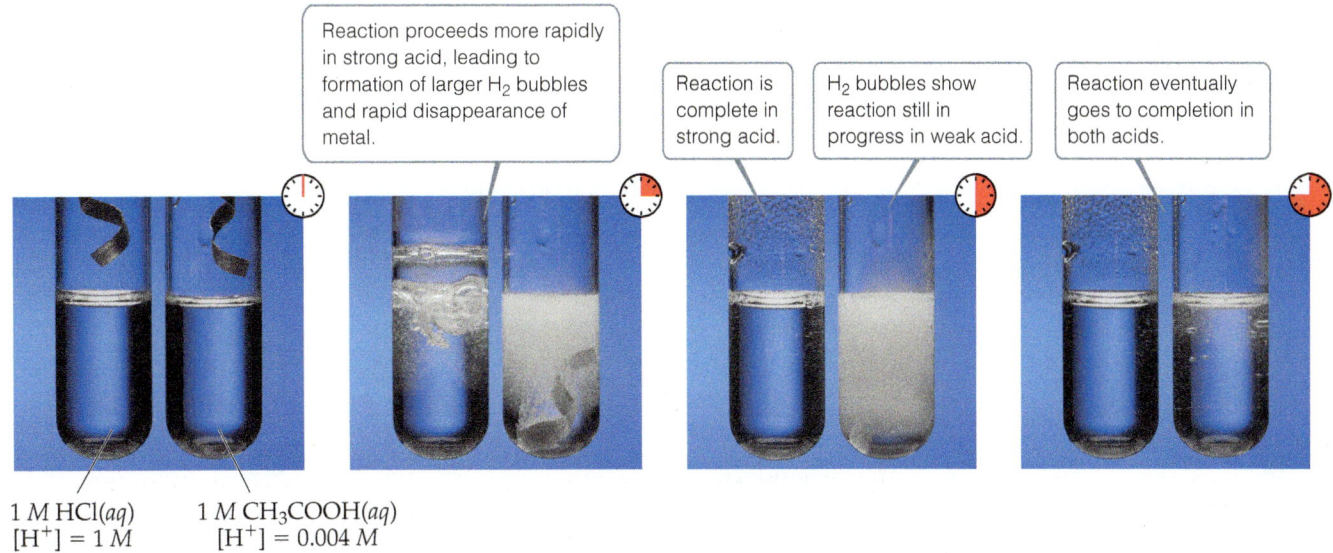

Reaction proceeds more rapidly in strong acid, leading to formation of larger H₂ bubbles and rapid disappearance of metal.

Reaction is complete in strong acid.

H₂ bubbles show reaction still in progress in weak acid.

Reaction eventually goes to completion in both acids.

$1\ M\ \text{HCl}(aq)$
$[\text{H}^+] = 1\ M$

$1\ M\ \text{CH}_3\text{COOH}(aq)$
$[\text{H}^+] = 0.004\ M$

▲ **Figure 16.12 The same reaction has different rates in a weak acid solution compared to a strong acid solution.** The bubbles are H₂ gas, which along with metal cations, is produced when a metal is oxidized by an acid. (Section 4.4)

The properties of an acid solution that relate directly to the concentration of $\text{H}^+(aq)$, such as electrical conductivity and rate of reaction with an active metal, are less evident for a solution of a weak acid than for a solution of a strong acid of the same concentration. **Figure 16.12** presents an experiment that demonstrates this difference with $1\ M$ CH₃COOH and $1\ M$ HCl. The concentration of $\text{H}^+(aq)$ in $1\ M$ CH₃COOH is only $0.004\ M$, whereas the $1\ M$ HCl solution contains $1\ M\ \text{H}^+(aq)$. As a result, the reaction rate with the metal is much faster in the HCl solution.

Sample Exercise 16.11

Using the Quadratic Equation to Calculate pH and Percent Ionization

Calculate the pH and percentage of HF molecules ionized in a 0.10 *M* HF solution.

SOLUTION

Analyze We are asked to calculate the percent ionization of a solution of HF. From Appendix D, we find $K_a = 6.8 \times 10^{-4}$.

Plan We approach this problem as for previous equilibrium problems: We write the chemical equation for the equilibrium and tabulate the known and unknown concentrations of all species. We then substitute the equilibrium concentrations into the equilibrium-constant expression and solve for the unknown concentration of H^+. The pH can then be calculated using Equation 16.21 and the percent ionization from Equation 16.33.

Solve

The equilibrium reaction and equilibrium concentrations are as follows:

	HF(aq) $\rightleftharpoons$	H⁺(aq) +	F⁻(aq)
Initial concentration (*M*)	0.10	0	0
Change in concentration (*M*)	−x	+x	+x
Equilibrium concentration (*M*)	(0.10 − x)	x	x

The equilibrium-constant expression is:

$$K_a = \frac{[\text{H}^+][\text{F}^-]}{[\text{HF}]} = \frac{(x)(x)}{0.10 - x} = 6.8 \times 10^{-4}$$

Continued

When we try solving this equation using the approximation $0.10 - x \approx 0.10$ (that is, by neglecting the concentration of acid that ionizes), we obtain:

$$x = 8.2 \times 10^{-3} M$$

Because this approximation is greater than 5% of $0.10\ M$, however, we should work the problem in standard quadratic form. Rearranging, we have:

$$x^2 = (0.10 - x)(6.8 \times 10^{-4})$$
$$= 6.8 \times 10^{-5} - (6.8 \times 10^{-4})x$$
$$x^2 + (6.8 \times 10^{-4})x - 6.8 \times 10^{-5} = 0$$

Substituting these values in the standard quadratic formula gives:

$$x = \frac{-6.8 \times 10^{-4} \pm \sqrt{(6.8 \times 10^{-4})^2 - 4(-6.8 \times 10^{-5})}}{2}$$
$$= \frac{-6.8 \times 10^{-4} \pm 1.65 \times 10^{-2}}{2}$$

Of the two solutions (7.9×10^{-3}, -8.6×10^{-2}), only the positive value for x is chemically reasonable. From that value, we can determine $[H^+]$ and hence the pH:

$$x = [H^+] = [F^-] = 7.9 \times 10^{-3} M$$
$$pH = -\log[H^+] = -\log(7.9 \times 10^{-3}) = 2.10$$

From our result, we can calculate the percent of molecules ionized:

$$\text{Percent ionization of HF} = \frac{\text{concentration ionized}}{\text{original concentration}} \times 100\%$$
$$= \frac{7.9 \times 10^{-3}\ M}{0.10\ M} \times 100\% = 7.9\%$$

▶ **Practice Exercise**

Calculate the percentage of niacin molecules ($K_a = 1.5 \times 10^{-5}$) ionized in a solution that is (**a**) 0.010 M, (**b**) 1.0×10^{-3} M.

Polyprotic Acids

Acids that have more than one ionizable H atom are known as **polyprotic acids**. Sulfurous acid (H_2SO_3), for example, can undergo two successive ionizations:

$$H_2SO_3(aq) \rightleftharpoons H^+(aq) + HSO_3^-(aq) \qquad K_{a1} = 1.7 \times 10^{-2} \qquad [16.35]$$
$$HSO_3^-(aq) \rightleftharpoons H^+(aq) + SO_3^{2-}(aq) \qquad K_{a2} = 6.4 \times 10^{-8} \qquad [16.36]$$

Note that the acid-dissociation constants are labeled K_{a1} and K_{a2}. The numbers on the constants refer to the particular proton of the acid that is ionizing. Thus, K_{a2} always refers to the equilibrium involving removal of the second proton of a polyprotic acid.

K_{a2} for sulfurous acid is much smaller than K_{a1}. Because of electrostatic attractions, we would expect a positively charged proton to be lost more readily from the neutral H_2SO_3 molecule than from the negatively charged HSO_3^- ion. This observation is general: *It is always easier to remove the first proton from a polyprotic acid than to remove the second*. Similarly, for an acid with three ionizable protons, it is easier to remove the second proton than the third. Thus, the K_a values become smaller as successive protons are removed. The acid-dissociation constants for common polyprotic acids are listed in Table 16.3, and Appendix D provides a more complete list. Citric acid (Figure 16.13), for example, has multiple ionizable protons.

Notice in Table 16.3 that in most cases the K_a values for successive losses of protons differ by a factor of at least 10^3. Notice also that the value of K_{a1} for sulfuric acid is listed simply as "large." Sulfuric acid is a strong acid with respect to the removal of the first proton. Thus, the reaction for the first ionization step lies completely to the right:

$$H_2SO_4(aq) \longrightarrow H^+(aq) + HSO_4^-(aq) \quad \text{(complete ionization)}$$

However, HSO_4^- is a weak acid for which $K_{a2} = 1.2 \times 10^{-2}$.

▼ **Go Figure**

How many protons per citric acid molecule are ionizable?

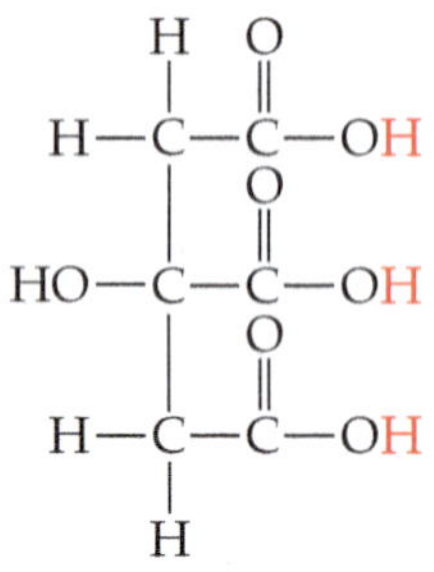

Citric acid

▲ **Figure 16.13** The structure of citric acid.

TABLE 16.3 Acid-Dissociation Constants of Some Common Polyprotic Acids

Name	Formula	K_{a1}	K_{a2}	K_{a3}
Ascorbic	$H_2C_6H_6O_6$	8.0×10^{-5}	1.6×10^{-12}	
Carbonic	H_2CO_3	4.3×10^{-7}	5.6×10^{-11}	
Citric	$H_3C_6H_5O_7$	7.4×10^{-4}	1.7×10^{-5}	4.0×10^{-7}
Oxalic	$HOOC—COOH$	5.9×10^{-2}	6.4×10^{-5}	
Phosphoric	H_3PO_4	7.5×10^{-3}	6.2×10^{-8}	4.2×10^{-13}
Sulfurous	H_2SO_3	1.7×10^{-2}	6.4×10^{-8}	
Sulfuric	H_2SO_4	Large	1.2×10^{-2}	
Tartaric	$C_2H_2O_2(COOH)_2$	1.0×10^{-3}	4.6×10^{-5}	

For many polyprotic acids, K_{a1} is much larger than subsequent dissociation constants, in which case the $H^+(aq)$ in the solution comes almost entirely from the first ionization reaction. As long as successive K_a values differ by a factor of 10^3 or more, it is usually possible to obtain a satisfactory estimate of the pH of polyprotic acid solutions by treating the acids as if they were monoprotic, considering only K_{a1}, as we illustrate in Sample Exercise 16.12.

Sample Exercise 16.12
Calculating the pH of a Solution of a Polyprotic Acid

The solubility of CO_2 in water at 25 °C and a CO_2 partial pressure of 0.1 atm is 0.0037 M. The common practice is to assume that all the dissolved CO_2 is in the form of carbonic acid (H_2CO_3), which is produced in the reaction

$$CO_2(aq) + H_2O(l) \rightleftharpoons H_2CO_3(aq)$$

What is the pH of a 0.0037 M solution of H_2CO_3?

SOLUTION

Analyze We are asked to determine the pH of a 0.0037 M solution of a polyprotic acid.

Plan H_2CO_3 is a diprotic acid; the two acid-dissociation constants, K_{a1} and K_{a2} (Table 16.3), differ by more than a factor of 10^3. Consequently, the pH can be determined by considering only K_{a1}, which means we can approach this problem essentially as though H_2CO_3 was a monoprotic acid.

Solve

Proceeding as in Sample Exercises 16.10 and 16.11, we can write the equilibrium reaction and equilibrium concentrations as:

$$H_2CO_3(aq) \rightleftharpoons H^+(aq) + HCO_3^-(aq)$$

	H_2CO_3	H^+	HCO_3^-
Initial concentration (M)	0.0037	0	0
Change in concentration (M)	$-x$	$+x$	$+x$
Equilibrium concentration (M)	$(0.0037 - x)$	x	x

The equilibrium-constant expression is:

$$K_{a1} = \frac{[H^+][HCO_3^-]}{[H_2CO_3]} = \frac{(x)(x)}{0.0037 - x} = 4.3 \times 10^{-7}$$

Solving this quadratic equation, we get:

$$x = 4.0 \times 10^{-5} \, M$$

Alternatively, because K_{a1} is small, we can make the simplifying approximation that x is small, so that:

$$0.0037 - x \simeq 0.0037$$

Thus,

$$\frac{(x)(x)}{0.0037} = 4.3 \times 10^{-7}$$

Continued

Solving for x, we have:

$$x^2 = (0.0037)(4.3 \times 10^{-7}) = 1.6 \times 10^{-9}$$

$$x = [H^+] = [HCO_3^-] = \sqrt{1.6 \times 10^{-9}} = 4.0 \times 10^{-5} M$$

Because we get the same value (to two significant figures), our simplifying assumption was justified. The pH is therefore:

$$pH = -\log[H^+] = -\log(4.0 \times 10^{-5}) = 4.40$$

Comment If we were asked for $[CO_3^{2-}]$, we would need to use K_{a2}. Let's illustrate that calculation. Using our calculated values of $[HCO_3^-]$ and $[H^+]$ and setting $[CO_3^{2-}] = y$, we have:

	$HCO_3^-(aq)$ $\rightleftharpoons$	$H^+(aq)$ +	$CO_3^{2-}(aq)$
Initial concentration (M)	4.0×10^{-5}	4.0×10^{-5}	0
Change in concentration (M)	$-y$	$+y$	$+y$
Equilibrium concentration (M)	$(4.0 \times 10^{-5} - y)$	$(4.0 \times 10^{-5} + y)$	y

Assuming that y is small relative to 4.0×10^{-5}, we have:

$$K_{a2} = \frac{[H^+][CO_3^{2-}]}{[HCO_3^-]} = \frac{(4.0 \times 10^{-5})(y)}{4.0 \times 10^{-5}} = 5.6 \times 10^{-11}$$

$$y = 5.6 \times 10^{-11} M = [CO_3^{2-}]$$

The value for y is indeed very small compared with 4.0×10^{-5}, showing that our assumption was justified. It also shows that the ionization of HCO_3^- is negligible relative to that of H_2CO_3, as far as production of H^+ is concerned. However, it is the *only* source of CO_3^{2-}, which has a very low concentration in the solution. Our calculations thus tell us that in a solution of carbon dioxide in water, most of the CO_2 is in the form of CO_2 or H_2CO_3; only a small fraction ionizes to form H^+ and HCO_3^-, and an even smaller fraction ionizes to give CO_3^{2-}. Notice also that $[CO_3^{2-}]$ is numerically equal to K_{a2}.

▶ **Practice Exercise**
(a) Calculate the pH of a 0.020 M solution of oxalic acid $(H_2C_2O_4)$. (See Table 16.3 for K_{a1} and K_{a2}.)
(b) Calculate the concentration of oxalate ion, $[C_2O_4^{2-}]$, in this solution.

A CLOSER LOOK   Polyprotic Acids and pH

Different protonation states of polyprotic acids exist at different pH values. Shown in **Figure 16.14** are the relative equilibrium concentrations of phosphoric acid and its successive conjugate bases as a function of pH.

We can learn many things from Figure 16.14. For instance, if the pH is greater than 4, there is essentially no H_3PO_4 left in solution. At pH > 12, the major component is the phosphate ion, PO_4^{3-}. We can also see that when the pH is equal to one of the pK_a values (just as $pK_w = -\log K_w$, $pK_a = -\log K_a$), the concentrations of the relevant conjugate acid–base pairs are equal. For instance, if you look at pH = 7.21, $[H_2PO_4^-] = [HPO_4^-]$. This makes sense if we consider that the equilibrium reaction at that pH is:

$$H_2PO_4^-(aq) + H_2O(l) \rightleftharpoons HPO_4^{2-}(aq) + H_3O^+(aq)$$

Therefore, K_{a2} corresponds to

$$K_{a2} = [HPO_4^{2-}][H_3O]^+/[H_2PO_4^-]$$

Using our log rules,

$$pK_{a2} = \log([HPO_4^{2-}]/[H_2PO_4^-]) + pH$$

Therefore, if pK_{a2} has the same value as the pH, then the ratio of $[H_2PO_4^-]$ to $[HPO_4^{2-}]$ must be 1 (since the log of 1 is equal to zero).

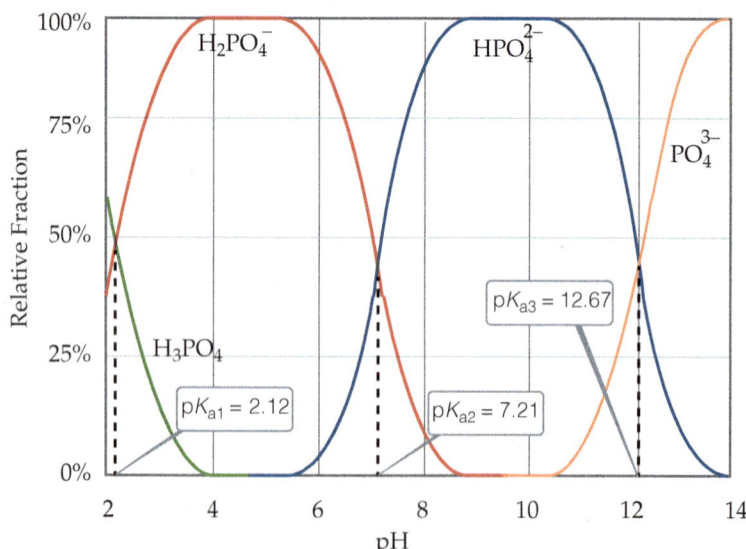

▲ **Figure 16.14** Relative fraction of phosphate species in water as a function of pH for phosphoric acid.

Self-Assessment Exercises

SAE 16.16 A 5.5×10^{-3} M solution of fluoroacetic acid, CH_2FCOOH, has a pH of 2.57. What is the acid-dissociation constant, K_a, of this acid? (**a**) 2.5×10^{-3} (**b**) 1.3×10^{-3} (**c**) 0.95 (**d**) 1.8×10^{-5}

SAE 16.17 What is the percent ionization of fluoroacetic acid in the previous problem (SAE 16.16)? (**a**) 51% (**b**) 49% (**c**) 5.1% (**d**) 0.049%.

SAE 16.18 Chlorous acid ($HClO_2$) has an acid-dissociation constant $K_a = 0.010$ at 25 °C. What is the pH of a 5.0×10^{-3} M solution of chlorous acid? (**a**) 2.30 (**b**) 2.15 (**c**) 2.87 (**d**) 2.44

SAE 16.19 What is the percent ionization of a 2.0 M solution of nitrous acid, HNO_2 ($K_a = 4.5 \times 10^{-4}$)? (**a**) 0.045% (**b**) 1.1% (**c**) 1.5% (**d**) 3.0%

SAE 16.20 Malic acid, $H_2C_4H_4O_5$, is a diprotic acid found in fruits and wine. K_{a2} is the equilibrium constant for the reaction _____, and the magnitude of K_{a2} will be _____ K_{a1}.

(**a**) $H_2C_4H_4O_5(aq) \rightleftharpoons 2\,H^+(aq) + C_4H_4O_5^{2-}(aq)$, less than

(**b**) $H_2C_4H_4O_5(aq) \rightleftharpoons 2\,H^+(aq) + C_4H_4O_5^{2-}(aq)$, greater than

(**c**) $HC_4H_4O_5^-(aq) \rightleftharpoons H^+(aq) + C_4H_4O_5^{2-}(aq)$, less than

(**d**) $HC_4H_4O_5^-(aq) \rightleftharpoons H^+(aq) + C_4H_4O_5^{2-}(aq)$, greater than

SAE 16.21 Rank-order the species present in a 1 M sulfurous acid solution from highest to lowest concentration. Sulfurous acid, H_2SO_3, is a diprotic acid with acid-dissociation constants $K_{a1} = 1.7 \times 10^{-2}$ and $K_{a2} = 6.4 \times 10^{-8}$.

(**a**) $[H_2SO_3] > [HSO_3^-] > [SO_3^{2-}] > [H_3O^+]$

(**b**) $[H_2SO_3] > [HSO_3^-] \approx [H_3O^+] > [SO_3^{2-}]$

(**c**) $[H_3O^+] > [HSO_3^-] \approx [SO_3^{2-}] > [H_2SO_3]$

(**d**) $[H_3O^+] > [H_2SO_3] > [HSO_3^-] > [SO_3^{2-}]$

(**e**) $[H_2SO_3] > [HSO_3^-] > [H_3O^+] \approx [SO_3^{2-}]$

16.7 | Weak Bases

In every acid–base reaction, there must be both an acid and a base. In the previous section, we explored the properties of molecules that act as weak acids in aqueous solutions. Now we turn our attention to substances that behave as weak bases when they dissolve in water by abstracting protons from H_2O. The products of this reaction are the conjugate acid of the base and OH^- ions:

$$B(aq) + H_2O(l) \rightleftharpoons HB^+(aq) + OH^-(aq) \qquad [16.37]$$

The equilibrium-constant expression for this reaction can be written as

$$K_b = \frac{[BH^+][OH^-]}{[B]} \qquad [16.38]$$

Water is the solvent, so it is omitted from the equilibrium-constant expression. These relationships are illustrated with ammonia, one of the most common weak bases:

$$NH_3(aq) + H_2O(l) \rightleftharpoons NH_4^+(aq) + OH^-(aq) \quad K_b = \frac{[NH_4^+][OH^-]}{[NH_3]} \quad [16.39]$$

As with K_w and K_a, the subscript b in K_b denotes that the equilibrium constant refers to a particular type of reaction—namely, the ionization of a weak base in water. The **base-dissociation constant**, K_b, *always refers to the equilibrium in which a base reacts with H_2O to form the corresponding conjugate acid and OH^-.*

Table 16.4 lists the Lewis structures, conjugate acids, and K_b values for a number of weak bases in water. Appendix D includes a more extensive list. These bases contain one or more lone pairs of electrons because a lone pair is necessary to form the bond with H^+, which has no electrons of its own. The other bases listed are anions derived from weak acids.

Weak bases fall into two general categories. The first category is neutral substances that have an atom with a nonbonding pair of electrons that can accept a proton. Most of

 Learning Objectives

When you finish Section 16.7, you should be able to:

▶ Describe the characteristic features of molecules and ions that can act as weak bases.

▶ Calculate the base-dissociation constant K_b of a weak base from its concentration and the pH of an aqueous solution of the base.

▶ Calculate the pH of an aqueous solution of a weak base from its concentration and its K_b value.

Go Figure

When hydroxylamine acts as a base, which atom accepts the proton?

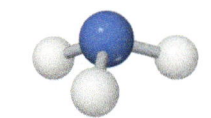

Ammonia
NH_3

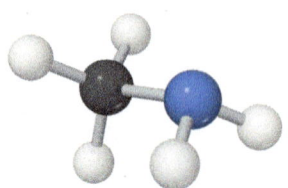

Methylamine
CH_3NH_2

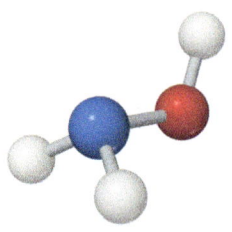

Hydroxylamine
NH_2OH

▲ **Figure 16.15** Structures of ammonia and two simple amines.

TABLE 16.4   Some Weak Bases in Water at 25 °C

Base	Structural Formula*	Conjugate Acid	K_b
Ammonia (NH_3)	H—N̈—H with H below	NH_4^+	1.8×10^{-5}
Pyridine (C_5H_5N)	(ring) N:	$C_5H_5NH^+$	1.7×10^{-9}
Hydroxylamine ($HONH_2$)	H—N̈—ÖH with H below	$HONH_3^+$	1.1×10^{-8}
Methylamine (CH_3NH_2)	H—N̈—CH_3 with H below	$CH_3NH_3^+$	4.4×10^{-4}
Hydrosulfide ion (HS^-)	[H—S̈:]$^-$	H_2S	1.8×10^{-7}
Carbonate ion (CO_3^{2-})	[:Ö: / C / :O—C—O:]$^{2-}$	HCO_3^-	1.8×10^{-4}
Hypochlorite ion (ClO^-)	[:C̈l—Ö:]$^-$	$HClO$	3.3×10^{-7}

*The atom that accepts the proton is shown in blue.

these bases, including all uncharged bases in Table 16.4, contain a nitrogen atom. These substances include ammonia and a related class of compounds called **amines** (Figure 16.15). In organic amines, at least one N—H bond in NH_3 is replaced with an N—C bond. Like NH_3, amines can abstract a proton from a water molecule by forming an N—H bond, as shown here for methylamine:

$$H—\overset{..}{N}—CH_3(aq) + H_2O(l) \rightleftharpoons \left[H—\overset{H}{\underset{H}{N}}—CH_3 \right]^+(aq) + OH^-(aq)$$

[16.40]

Anions of weak acids make up the second general category of weak bases. In an aqueous solution of sodium hypochlorite (NaClO), for example, NaClO dissociates to Na^+ and ClO^- ions. The Na^+ ion is always a spectator ion in acid–base reactions. (Section 4.3) The ClO^- ion, however, is the conjugate base of a weak acid, hypochlorous acid. Consequently, the ClO^- ion acts as a weak base in water:

$$ClO^-(aq) + H_2O(l) \rightleftharpoons HClO(aq) + OH^-(aq) \qquad K_b = 3.3 \times 10^{-7} \quad [16.41]$$

In Figure 16.6 we saw that bleach is quite basic (pH values of 12–13). Common chlorine bleach is typically a 5% NaOCl solution.

Calculations Involving Weak Bases

For all of the equilibrium calculations we encountered with weak acids, a counterpart exists for weak bases. Given the concentration and pH of a weak base solution, we can calculate the base–dissociation constant, K_b. Once we know K_b, we can calculate the pH of a weak base solution from its concentration or vice versa. The reactions and the equilibrium constant expressions will look different, as we saw in Equations 16.37 and 16.38, but our approach to solving the problem is basically the same. Because these reactions produce OH^- rather than H^+, it is necessary to convert between pH and pOH (remember pH + pOH = 14). Sample Exericses 16.13 and 16.14 illustrate the types of calculations you will be expected to perform when working with weak bases.

Sample Exercise 16.13
Using K_b to Calculate $[OH^-]$ and pH

Calculate the concentration of OH^- in a 0.15 M solution of NH_3, and the pH of this solution.

SOLUTION

Analyze We are given the concentration of a weak base and asked to determine the concentration of OH^- and the pH of the solution.

Plan We will use essentially the same procedure here as we used in solving problems involving the ionization of weak acids. We begin by writing the chemical equation and the corresponding equilibrium constant expression. Next we tabulate the initial and equilibrium concentrations and plug those values back into the equilibrium constant expression. Finally, we solve for the unknown quantity, the concentration of OH^-. Once we know $[OH^-]$ we can determine pOH and subtract that value from 14 to get the pH.

Solve

The ionization reaction and equilibrium-constant expression are:

$$NH_3(aq) + H_2O(l) \rightleftharpoons NH_4^+(aq) + OH^-(aq)$$

$$K_b = \frac{[NH_4^+][OH^-]}{[NH_3]} = 1.8 \times 10^{-5}$$

Ignoring the concentration of H_2O, because it is not involved in the equilibrium-constant expression, the equilibrium concentrations are:

	$NH_3(aq)$	+ $H_2O(l)$	$\rightleftharpoons$ $NH_4^+(aq)$	+ OH^-
Initial concentration (M)	0.15	—	0	0
Change in concentration (M)	$-x$	—	$+x$	$+x$
Equilibrium concentration (M)	$(0.15 - x)$	—	x	x

Inserting these quantities into the equilibrium-constant expression gives:

$$K_b = \frac{[NH_4^+][OH^-]}{[NH_3]} = \frac{(x)(x)}{0.15 - x} = 1.8 \times 10^{-5}$$

Because K_b is small, the amount of NH_3 that reacts with water is much smaller than the NH_3 concentration, and so we can neglect x relative to 0.15 M. Then we have:

$$\frac{x^2}{0.15} = 1.8 \times 10^{-5}$$

$$x^2 = (0.15)(1.8 \times 10^{-5}) = 2.7 \times 10^{-6}$$

$$x = [NH_4^+] = [OH^-] = \sqrt{2.7 \times 10^{-6}} = 1.6 \times 10^{-3} M$$

Once we have determined $[OH^-]$ we can follow the procedure outlined in Sample Exercise 16.8 to calculate the pH of the solution.

$$pOH = -\log(1.6 \times 10^{-3}) = 2.80$$

$$pH = 14.00 - pOH = 11.20$$

Check The value obtained for x is only about 1% of the NH_3 concentration, 0.15 M. Therefore, neglecting x relative to 0.15 was justified. The pH is greater than 7, as we would expect for a basic solution.

▶ **Practice Exercise**

Which of the following compounds should produce the highest pH as a 0.05 M solution: pyridine, methylamine, or nitrous acid?

Sample Exercise 16.14

Using pH to Determine the Concentration of a Salt

A solution made by adding solid sodium hypochlorite (NaClO) to enough water to make 2.00 L of solution has a pH of 10.50. Using the reaction given in Equation 16.41 and the appropriate base dissociation constant K_b from Table 16.4, calculate the number of moles of NaClO added to the water.

SOLUTION

Analyze NaClO is an ionic compound consisting of Na^+ and ClO^- ions. As such, it is a strong electrolyte that completely dissociates in solution into Na^+, a spectator ion, and ClO^-, a weak base with $K_b = 3.3 \times 10^{-7}$. Given this information, we must calculate the number of moles of NaClO needed to increase the pH of 2.00 L of water to 10.50.

Plan From the pH, we can determine the equilibrium concentration of OH^-. We can then construct a table of initial and equilibrium concentrations in which the initial concentration of ClO^- is our unknown. We can calculate $[ClO^-]$ using the expression for K_b.

Solve

We can calculate $[OH^-]$ by using either Equation 16.20 or Equation 16.24; we use the latter method here:

$$pOH = 14.00 - pH = 14.00 - 10.50 = 3.50$$
$$[OH^-] = 10^{-3.50} = 3.2 \times 10^{-4}\,M$$

This concentration is high enough that we can assume that Equation 16.41 is the only source of OH^-; that is, we can neglect any OH^- produced by the autoionization of H_2O. We now assume a value of x for the initial concentration of ClO^- and solve the equilibrium problem in the usual way.

	$ClO^-(aq)$	$+ H_2O(l) \rightleftharpoons$	$HClO(aq)$	$+ OH^-(aq)$
Initial concentration (M)	x	—	0	0
Change in concentration (M)	-3.2×10^{-4}	—	$+3.2 \times 10^{-4}$	$+3.2 \times 10^{-4}$
Equilibrium concentration (M)	$(x - 3.2 \times 10^{-4})$	—	3.2×10^{-4}	3.2×10^{-4}

We now use the expression for the base-dissociation constant to solve for x:

$$K_b = \frac{[HClO][OH^-]}{[ClO^-]} = \frac{(3.2 \times 10^{-4})^2}{x - 3.2 \times 10^{-4}} = 3.3 \times 10^{-7}$$

$$x = \frac{(3.2 \times 10^{-4})^2}{3.3 \times 10^{-7}} + (3.2 \times 10^{-4}) = 0.31\,M$$

We say that the solution is 0.31 M in NaClO, even though some of the ClO^- ions have reacted with water. Because the solution is 0.31 M in NaClO and the total volume of solution is 2.00 L, 0.62 mol of NaClO is the amount of the salt that was added to the water.

▶ **Practice Exercise**

What is the molarity of an aqueous NH_3 solution that has a pH of 11.17?

Self-Assessment Exercises

SAE 16.22 Which of the following molecules will act as a weak base?

(i) $C_3N_2H_4$ (ii) CH_3CH_2COOH (iii) $CH_3CH_2CH_2CH_2OH$

(**a**) only i (**b**) only ii (**c**) only iii (**d**) i and iii (**e**) all three molecules

SAE 16.23 A 0.15 M solution of imidazole, $C_3N_2H_4$, a weak organic base, has a pH of 10.11. What is the K_b of imidazole? (**a**) 4.0×10^{-20} (**b**) 1.3×10^{-4} (**c**) 2.5×10^{-9} (**d**) 1.1×10^{-7}

SAE 16.24 The base-dissociation constant of pyridine (C_5H_5N) is $K_b = 1.7 \times 10^{-9}$. What is the pH of a 0.069 M solution of pyridine? (**a**) 4.97 (**b**) 9.03 (**c**) 9.61 (**d**) 9.93

16.8 | Relationship Between K_a and K_b

We have seen in a qualitative way that the stronger an acid, the weaker its conjugate base. In this section we quantify this relationship. Let's begin by considering the NH_4^+ and NH_3 conjugate acid–base pair. Each species reacts with water. For the acid, NH_4^+, the equilibrium is

$$NH_4^+(aq) + H_2O(l) \rightleftharpoons NH_3(aq) + H_3O^+(aq)$$

or written in its simpler form:

$$NH_4^+(aq) \rightleftharpoons NH_3(aq) + H^+(aq) \qquad [16.42]$$

For the base, NH_3, the equilibrium is

$$NH_3(aq) + H_2O(l) \rightleftharpoons NH_4^+(aq) + OH^-(aq) \qquad [16.43]$$

Each equilibrium is expressed by a dissociation constant:

$$K_a = \frac{[NH_3][H^+]}{[NH_4^+]} \qquad K_b = \frac{[NH_4^+][OH^-]}{[NH_3]}$$

When we add Equations 16.42 and 16.43, the NH_4^+ and NH_3 species cancel, and we are left with the autoionization of water:

$$\begin{aligned} NH_4^+(aq) &\rightleftharpoons NH_3(aq) + H^+(aq) \\ NH_3(aq) + H_2O(l) &\rightleftharpoons NH_4^+(aq) + OH^-(aq) \\ \hline H_2O(l) &\rightleftharpoons H^+(aq) + OH^-(aq) \end{aligned}$$

Recall that when two equations are added to give a third, the equilibrium constant associated with the third equation equals the product of the equilibrium constants of the first two equations. (Section 15.3)

When we multiply K_a and K_b for our current example, we obtain

$$K_a \times K_b = \left(\frac{[NH_3][H^+]}{[NH_4^+]} \right)\left(\frac{[NH_4^+][OH^-]}{[NH_3]} \right)$$

$$= [H^+][OH^-] = K_w$$

Thus, the product of K_a and K_b is the ion-product constant for water, K_w (Equation 16.20). We expect this result because adding Equations 16.42 and 16.43 gave us the autoionization equilibrium for water, for which the equilibrium constant is K_w.

This result holds for any conjugate acid–base pair. In general, *the product of the acid-dissociation constant for an acid and the base-dissociation constant for its conjugate base equals the ion-product constant for water:*

$$K_a \times K_b = K_w \quad \text{(for a conjugate acid–base pair)} \qquad [16.44]$$

Note that this relationship only holds true for conjugate acid–base pairs. Do not use Equation 16.44 for any random pair of acids and bases! As the strength of an acid increases (that is, as K_a gets larger), the strength of its conjugate base must decrease (that is, K_b must get smaller), so that the product $K_a \times K_b$ remains 1.0×10^{-14} at 25 °C. Table 16.5 demonstrates this relationship.

By using Equation 16.44, we can calculate K_b for any weak base if we know K_a for its conjugate acid. Similarly, we can calculate K_a for a weak acid if we know K_b for its conjugate base. As a practical consequence, ionization constants are often listed for only one member of a conjugate acid–base pair. For example, Appendix D does not contain K_b values for the anions of weak acids because they can be readily calculated from the tabulated K_a values for their conjugate acids.

> **⚠ Learning Objective**
>
> When you finish Section 16.8, you should be able to:
>
> ▶ Given the acid-dissociation constant K_a of a weak acid, calculate the base-dissociation constant K_b of its conjugate base, and vice versa.

TABLE 16.5   Some Conjugate Acid–Base Pairs

Acid	K_a	Base	K_b
HNO_3	(Strong acid)	NO_3^-	(Negligible basicity)
HF	6.8×10^{-4}	F^-	1.5×10^{-11}
CH_3COOH	1.8×10^{-5}	CH_3COO^-	5.6×10^{-10}
H_2CO_3	4.3×10^{-7}	HCO_3^-	2.3×10^{-8}
NH_4^+	5.6×10^{-10}	NH_3	1.8×10^{-5}
HCO_3^-	5.6×10^{-11}	CO_3^{2-}	1.8×10^{-4}
OH^-	(Negligible acidity)	O^{2-}	(Strong base)

Recall that we often express $[H^+]$ as pH: $pH = -\log[H^+]$. (Section 16.4) This "p" nomenclature is often used for other very small numbers. For example, if you look up the values for acid- or base-dissociation constants in a chemistry handbook, you may find them expressed as pK_a or pK_b:

$$pK_a = -\log K_a \quad \text{and} \quad pK_b = -\log K_b \qquad [16.45]$$

Using this nomenclature, we can write Equation 16.44 in terms of pK_a and pK_b by taking the negative logarithm of both sides:

$$pK_a + pK_b = pK_w = 14.00 \quad \text{at 25 °C (conjugate acid–base pair)} \qquad [16.46]$$

CHEMISTRY AND LIFE | Amines and Amine Hydrochlorides

Many low-molecular-weight amines have a fishy odor. Amines and NH_3 are produced by the anaerobic (absence of O_2) decomposition of dead animal or plant matter. Two such amines with very disagreeable aromas are *putrescine*, $H_2N(CH_2)_4NH_2$, and *cadaverine*, $H_2N(CH_2)_5NH_2$. The names of these substances reflect their repugnant odors!

Many drugs, including quinine, codeine, caffeine, and amphetamine, are amines. Like other amines, these substances are weak bases; the amine nitrogen is readily protonated upon treatment with an acid. The resulting products are called *acid salts*. If we use A as the abbreviation for an amine, the acid salt formed by reaction with hydrochloric acid can be written AH^+Cl^-. It can also be written as $A \cdot HCl$ and referred to as a hydrochloride. Amphetamine hydrochloride, for example, is the acid salt formed by treating amphetamine with HCl:

Amphetamine

Amphetamine hydrochloride

Acid salts are much less volatile, more stable, and generally more water-soluble than the corresponding amines. For this reason, many drugs that are amines are sold and administered as acid salts. Some examples of over-the-counter medications that contain amine hydrochlorides as active ingredients are shown in Figure 16.16.

Related Exercises: 16.10, 16.77, 16.78

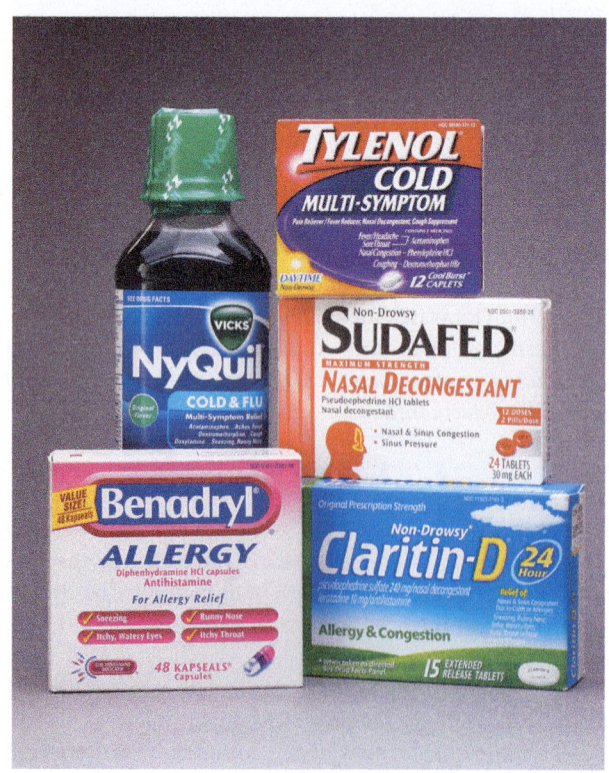

▲ **Figure 16.16** Some over-the-counter medications in which an amine hydrochloride is a major active ingredient.

Sample Exercise 16.15

Calculating K_a or K_b for a Conjugate Acid–Base Pair

Calculate (**a**) K_b for the fluoride ion from the acid–dissociation constant of HF, (**b**) K_a for the ammonium ion from the base–dissociation constant of ammonia.

SOLUTION

Analyze We are asked to determine dissociation constants for F^-, the conjugate base of HF, and NH_4^+, the conjugate acid of NH_3.

Plan We can use the tabulated K values for HF and NH_3 and the relationship between K_a and K_b to calculate the dissociation constants for their conjugates, F^- and NH_4^+.

Solve

(**a**) For the weak acid HF, Table 16.2 and Appendix D give $K_a = 6.8 \times 10^{-4}$. We can use Equation 16.44 to calculate K_b for the conjugate base, F^-:

$$K_b = \frac{K_w}{K_a} = \frac{1.0 \times 10^{-14}}{6.8 \times 10^{-4}} = 1.5 \times 10^{-11}$$

(**b**) For NH_3, Table 16.4 and Appendix D give $K_b = 1.8 \times 10^{-5}$, and this value in Equation 16.44 gives us K_a for the conjugate acid, NH_4^+:

$$K_a = \frac{K_w}{K_b} = \frac{1.0 \times 10^{-14}}{1.8 \times 10^{-5}} = 5.6 \times 10^{-10}$$

Check The respective K values for F^- and NH_4^+ are listed in Table 16.5, where we see that the values calculated here agree with those presented in Table 16.5.

▶ **Practice Exercise**

(**a**) Based on information in Appendix D, which of these anions has the largest base-dissociation constant: NO_2^-, PO_4^{3-}, or N_3^-?

(**b**) The base quinoline has the structure

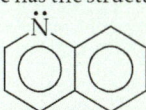

Its conjugate acid is listed in handbooks as having a pK_a of 4.90. What is the base-dissociation constant for quinoline?

 Self-Assessment Exercises

SAE 16.25 The acid-dissociation constant for hydrocyanic acid, HCN, is $K_a = 4.9 \times 10^{-10}$. What is the identity and the pK_b value of its conjugate base? (**a**) CN^-, 4.69 (**b**) CN^-, 2.0×10^{-5} (**c**) H_2CN^+, 4.69 (**d**) H_2CN^+, 9.31 (**e**) CN^-, 9.31

SAE 16.26 Given the following acid- and base-dissociation constants: NH_3 ($K_b = 1.8 \times 10^{-5}$), H_2CO_3 ($K_{a1} = 4.3 \times 10^{-7}, K_{a2} = 5.6 \times 10^{-11}$), and H_2SO_3 ($K_{a1} = 1.7 \times 10^{-2}, K_{a2} = 6.4 \times 10^{-8}$),

arrange the following species from the weakest to the strongest acid: NH_4^+, HCO_3^-, HSO_3^-.

(**a**) weakest acid $NH_4^+ < HCO_3^- < HSO_3^-$ strongest acid
(**b**) weakest acid $HSO_3^- < HCO_3^- < NH_4^+$ strongest acid
(**c**) weakest acid $HCO_3^- < NH_4^+ < HSO_3^-$ strongest acid
(**d**) weakest acid $HSO_3^- < NH_4^+ < HCO_3^-$ strongest acid

16.9 | Acid–Base Properties of Salt Solutions

Even before you began this chapter, you were undoubtedly aware of neutral molecules that form acidic solutions when dissolved in water, like HNO_3, HCl, and H_2SO_4. Likewise, we know many substances that form basic solutions when dissolved in water, such as NaOH and NH_3. It may surprise you to learn that salts like $AlCl_3$ and Na_2CO_3 can also have a profound effect on the pH when they form aqueous solutions. A clue to this behavior comes from the observation that ions can exhibit acidic or basic properties. For example, we calculated K_a for NH_4^+ and K_b for F^- in Sample Exercise 16.15. Because salts are made up of cations and anions, it follows that salt solutions can be acidic or basic. Before proceeding with further discussions of acids and bases, let's examine how dissolved salts affect pH.

Because nearly all salts are strong electrolytes, we can assume that any salt dissolved in water is completely dissociated. Consequently, the acid–base properties of salt solutions are due to the behavior of the cations and anions. Many ions react with water to generate $H^+(aq)$ or $OH^-(aq)$ ions. This type of reaction, called **hydrolysis**, is responsible for the acid–base characteristics of a salt.

 Learning Objectives

When you finish Section 16.9, you should be able to:

▶ Predict whether an aqueous solution of a salt will be acidic, basic, or neutral.

▶ Calculate the pH of a salt solution from the concentration and identity of the salt and the relevant aqueous equilibrium constants.

Anion Hydrolysis

In general, an anion A^- in solution can be considered the conjugate base of an acid. For example, Cl^- is the conjugate base of HCl, and CH_3COO^- is the conjugate base of CH_3COOH. Whether an anion reacts with water to produce hydroxide ions depends on the strength of the anion's conjugate acid. To identify the acid and assess its strength, we add a proton to the anion's formula. If the acid HA determined in this way is one of the seven strong acids listed at the beginning of Section 16.5, the anion has a negligible tendency to produce OH^- ions from water and does not affect the pH of the solution. For example, because Cl^- is the conjugate base of hydrochloric acid (a strong acid), its presence in an aqueous solution does not result in the production of OH^- and does not affect the pH. Thus, Cl^- is always a spectator ion in acid–base chemistry.

If HA is *not* one of the seven common strong acids, it is a weak acid. In this case, the conjugate base A^- is a weak base, and it reacts to a small extent with water to produce the weak acid and hydroxide ions:

$$A^-(aq) + H_2O(l) \rightleftharpoons HA(aq) + OH^-(aq) \qquad [16.47]$$

The OH^- ion generated in this way increases the pH of the solution, making it basic. Acetate ion, for example, being the conjugate base of a weak acid, reacts with water to produce acetic acid and hydroxide ions, thereby increasing the pH of the solution:

$$CH_3COO^-(aq) + H_2O(l) \rightleftharpoons CH_3COOH(aq) + OH^-(aq) \qquad [16.48]$$

The situation is more complicated for salts containing anions that have ionizable protons, such as HSO_3^-. These salts are amphiprotic (Section 16.1), and how they behave in water is determined by the relative magnitudes of K_a and K_b for the ion, as shown in Sample Exercise 16.17. If $K_a > K_b$, the ion causes the solution to be acidic. If $K_b > K_a$, the solution is made basic by the ion.

Cation Hydrolysis

Polyatomic cations containing one or more protons can be considered the conjugate acids of weak bases. The NH_4^+ ion, for example, is the conjugate acid of the weak base NH_3. Thus, NH_4^+ is a weak acid and will donate a proton to water, producing hydronium ions and thereby lowering the pH:

$$NH_4^+(aq) + H_2O(l) \rightleftharpoons NH_3(aq) + H_3O^+(aq) \qquad [16.49]$$

Other neutral bases that become cations upon accepting a proton behave in a similar way. For example, methylamine, CH_3NH_2, becomes the methylammonium cation, $CH_3NH_3^+$, upon protonation. Just like ammonium salts, those containing $CH_3NH_3^+$ will act as weak acids upon dissolving in water.

A more unusual situation is encountered with some metal salts. For instance, if you dissolve $Fe(NO_3)_3$ in water, the solution becomes quite acidic. Why is this the case? You might think that the nitrate ion is somehow producing nitric acid, but you would be wrong. Recall that nitrate is the conjugate base of a strong acid, and therefore its reaction with water is negligible. If the nitrate ion is not playing a role, the change in pH must be due in some way to the presence of Fe^{3+} in solution. It turns out that small, highly charged metal cations like Fe^{3+} can create surprisingly acidic solutions in water (Table 16.6). A comparison of Fe^{2+} and Fe^{3+} values in the table illustrates how acidity increases as ionic charge increases.

Notice that K_a values for the 3+ ions in Table 16.6 are comparable to the values for familiar weak acids, such as acetic acid ($K_a = 1.8 \times 10^{-5}$). In contrast, the ions of alkali and alkaline earth metals, being relatively large and not highly charged, do not react with water to any appreciable extent and therefore do not affect pH. Note that these are the same cations found in the strong bases (Section 16.5). The different tendencies of four cations to lower the pH of a solution are illustrated in Figure 16.17.

Let's take a closer look at the reactions that occur when a metal cation acts as an acid. Because metal ions are positively charged, they attract the unshared electron pairs of water molecules and become hydrated. (Section 13.1) We can describe the hydration of a metal cation as a Lewis acid–base reaction. The Fe^{3+} ion, being electron deficient,

TABLE 16.6 Acid-Dissociation Constants for the Hydrolysis of Metal Cations in Aqueous Solution at 25 °C

Cation	$K_a{}^*$
Fe^{3+}	6.3×10^{-3}
Cr^{3+}	1.6×10^{-4}
Al^{3+}	1.4×10^{-5}
Fe^{2+}	3.2×10^{-10}
Zn^{2+}	2.5×10^{-10}
Ni^{2+}	2.5×10^{-11}

*K_a is the equilibrium constant for the reaction shown in Figure 16.18.

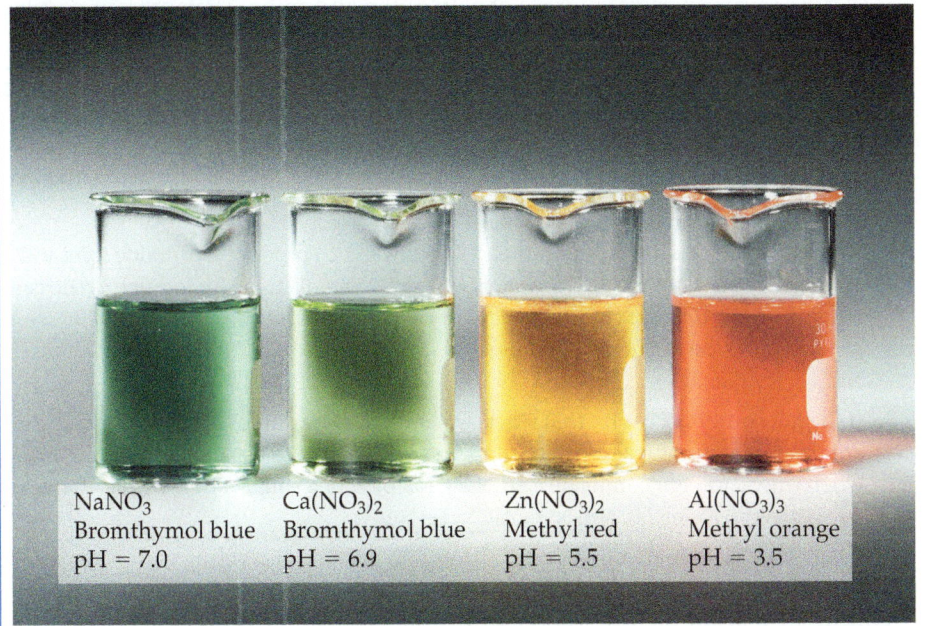

Go Figure Why do we need to use multiple acid–base indicators in this figure?

| NaNO₃ | Ca(NO₃)₂ | Zn(NO₃)₂ | Al(NO₃)₃ |

NaNO$_3$
Bromthymol blue
pH = 7.0

Ca(NO$_3$)$_2$
Bromthymol blue
pH = 6.9

Zn(NO$_3$)$_2$
Methyl red
pH = 5.5

Al(NO$_3$)$_3$
Methyl orange
pH = 3.5

▲ **Figure 16.17** **Effect of cations on solution pH.** The pH values of 1.0 *M* solutions of four nitrate salts are estimated using acid–base indicators.

acts as a Lewis acid, while the water molecules, having nonbonding electron pairs, act as the Lewis base. When a water molecule interacts with the positively charged metal ion, electron density is drawn from the oxygen. This shift of electron density toward the metal cation causes the O—H bond to become more polarized; as a result, water molecules bound to the metal ion are more acidic than those in the bulk solvent. The mechanism by which hydrated metal cations react with water to produce $H_3O^+(aq)$ is shown in **Figure 16.18**. As the cation charge increases, it becomes a stronger Lewis acid, and this in turn leads to more polarization of the O—H bonds of the hydrating water molecules. This explains why 3+ cations are much more acidic than 2+ cations, and why we can normally neglect the acidity of 1+ cations.

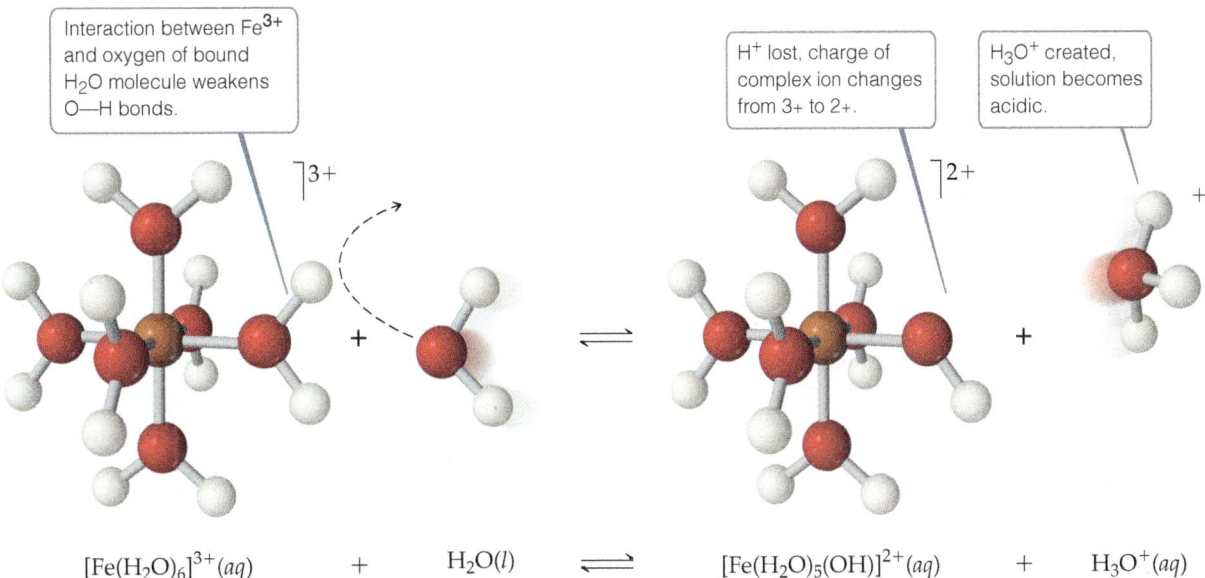

Interaction between Fe^{3+} and oxygen of bound H$_2$O molecule weakens O—H bonds.

H$^+$ lost, charge of complex ion changes from 3+ to 2+.

H$_3$O$^+$ created, solution becomes acidic.

$$[Fe(H_2O)_6]^{3+}(aq) + H_2O(l) \rightleftharpoons [Fe(H_2O)_5(OH)]^{2+}(aq) + H_3O^+(aq)$$

▲ **Figure 16.18** A hydrated **Fe$^+$** ion acts as an acid by donating an H$^+$ from a bound water molecule to a free H$_2$O molecule, forming H$_3$O$^+$.

Salts Where Cations and/or Anions Undergo Hydrolysis

To determine whether a salt forms an acidic, a basic, or a neutral solution when dissolved in water, we must consider the action of both cation and anion. There are four possible combinations.

1. If the salt contains an anion that does not react with water and a cation that does not react with water, we expect the pH to be neutral. Such is the case when the anion is a conjugate base of a strong acid and the cation is either from group 1A or one of the heavier members of group 2A (Ca^{2+}, Sr^{2+}, Ba^{2+}). *Examples*: NaCl, $Ba(NO_3)_2$, $RbClO_4$.

2. If the salt contains an anion that reacts with water to produce hydroxide ions and a cation that does not react with water, we expect the pH to be basic. Such is the case when the anion is the conjugate base of a weak acid and the cation is either from group 1A or one of the heavier members of group 2A (Ca^{2+}, Sr^{2+}, Ba^{2+}). *Examples*: NaClO, RbF, $BaSO_3$.

3. If the salt contains a cation that reacts with water to produce hydronium ions and an anion that does not react with water, we expect the pH to be acidic. Such is the case when the cation is a conjugate acid of a weak base or a small cation with a charge of 2+ or greater. *Examples*: NH_4NO_3, $AlCl_3$, $Fe(NO_3)_3$.

4. If the salt contains an anion and a cation *both* capable of reacting with water, both hydroxide ions and hydronium ions are produced. Whether the solution is basic, neutral, or acidic depends on the relative abilities of the ions to react with water. *Examples*: NH_4ClO, $Al(CH_3COO)_3$, CrF_3.

Sample Exercise 16.16
Determining Whether Salt Solutions Are Acidic, Basic, or Neutral

Determine whether the aqueous solutions of each of these salts are acidic, basic, or neutral: **(a)** $Ba(CH_3COO)_2$, **(b)** NH_4Cl, **(c)** CH_3NH_3Br, **(d)** KNO_3, **(e)** $Al(ClO_4)_3$.

SOLUTION

Analyze We are given the chemical formulas of five ionic compounds (salts) and asked whether their aqueous solutions will be acidic, basic, or neutral.

Plan We can determine whether a solution of a salt is acidic, basic, or neutral by identifying the ions in solution and by assessing how each ion will affect the pH.

Solve

(a) This solution contains barium ions and acetate ions. The cation, Ba^{2+}, is an ion of a heavy alkaline earth metal and will therefore not affect the pH. The anion, CH_3COO^-, is the conjugate base of the weak acid CH_3COOH and will hydrolyze to produce OH^- ions, thereby making the solution basic.

(b) In this solution, NH_4^+ is the conjugate acid of a weak base (NH_3) and is therefore acidic. Cl^- is the conjugate base of a strong acid (HCl) and therefore has no influence on the pH of the solution. Because the solution contains an ion that is acidic (NH_4^+) and one that has no influence on pH (Cl^-), the solution will be acidic.

(c) Here $CH_3NH_3^+$ is the conjugate acid of a weak base (CH_3NH_2, an amine) and is therefore acidic, and Br^- is the conjugate base of a strong acid (HBr) and therefore pH neutral. Because the solution contains one ion that is acidic and one that has no influence on pH, the solution will be acidic.

(d) This solution contains the K^+ ion, which is a cation of group 1A, and the NO_3^- ion, which is the conjugate base of the strong acid HNO_3. Neither of the ions will react with water to any appreciable extent, making the solution neutral.

(e) This solution contains Al^{3+} and ClO_4^- ions. Cations, such as Al^{3+}, that have a charge of 3+ or higher are acidic. The ClO_4^- ion is the conjugate base of a strong acid ($HClO_4$) and therefore does not affect pH. Thus, the solution of $Al(ClO_4)_3$ will be acidic.

▶ **Practice Exercise**
Indicate which salt in each of the following pairs forms the more acidic (or less basic) 0.010 *M* solution: **(a)** $NaNO_3$ or $Fe(NO_3)_3$, **(b)** KBr or KBrO, **(c)** CH_3NH_3Cl or $BaCl_2$, **(d)** NH_4NO_2 or NH_4NO_3.

Sample Exercise 16.17

Predicting Whether the Solution of an Amphiprotic Anion Is Acidic or Basic

Predict whether the salt Na_2HPO_4 forms an acidic solution or a basic solution when dissolved in water.

SOLUTION

Analyze We are asked to predict whether a solution of Na_2HPO_4 is acidic or basic. This substance is an ionic compound composed of Na^+ and HPO_4^{2-} ions.

Plan We need to evaluate each ion, predicting whether it is acidic or basic. Because Na^+ is a cation of group 1A, it has no influence on pH. Thus, our analysis of whether the solution is acidic or basic must focus on the behavior of the HPO_4^{2-} ion. We need to realize that HPO_4^{2-} is amphiprotic: it can act as either an acid or a base:

As acid $HPO_4^{2-}(aq) \rightleftharpoons H^+(aq) + PO_4^{3-}(aq)$

As base $HPO_4^{2-}(aq) + H_2O \rightleftharpoons H_2PO_4^-(aq) + OH^-(aq)$

Of these two reactions, the one with the larger equilibrium constant determines whether the solution is acidic or basic.

Solve The value of K_a for HPO_4^{2-} acting as an acid is equivalent to K_{a3} for H_3PO_4: 4.2×10^{-13} (Table 16.3). For the second equation, where HPO_4^{2-} acts as a base, we must calculate K_b for the base

HPO_4^{2-} from the value of K_a for its conjugate acid, $H_2PO_4^-$, and the relationship $K_a \times K_b = K_w$ (Equation 16.44). The relevant value of K_a for $H_2PO_4^-$ is K_{a2} for H_3PO_4: 6.2×10^{-8} (from Table 16.3). We therefore have

$$K_b(HPO_4^{2-}) \times K_a(H_2PO_4^-) = K_w = 1.0 \times 10^{-14}$$

$$K_b(HPO_4^{2-}) = \frac{1.0 \times 10^{-14}}{6.2 \times 10^{-8}} = 1.6 \times 10^{-7}.$$

This K_b value is more than 10^5 times larger than K_a for HPO_4^{2-}; thus, the tendency for HPO_4^{2-} to act as a base and abstract a proton from water dominates, and the solution is basic.

▶ **Practice Exercise**
Predict whether the dipotassium salt of citric acid $(K_2HC_6H_5O_7)$ forms an acidic or a basic solution in water (see Table 16.3 for data).

Self-Assessment Exercises

SAE 16.27 Which of the following salts will form acidic solutions when dissolved in water? $AlCl_3$, Na_2SO_3, $CH_3NH_3NO_3$ (**a**) only $AlCl_3$ (**b**) only Na_2SO_3 (**c**) only $CH_3NH_3NO_3$ (**d**) $AlCl_3$ and $CH_3NH_3NO_3$ (**e**) $AlCl_3$, Na_2SO_3, and $CH_3NH_3NO_3$

SAE 16.28 What is the pH of a 0.50 M solution of NaBrO? The acid-dissociation constant of HBrO is $K_a = 2.5 \times 10^{-9}$. (**a**) 2.85 (**b**) 4.45 (**c**) 9.67 (**d**) 11.15

16.10 | Acid–Base Behavior and Chemical Structure

As we have seen throughout this chapter when a substance is dissolved in water, it may behave as an acid or a base, or it may exhibit no acid–base properties whatsoever. How does the chemical structure of a substance determine which of these behaviors is exhibited by the substance? For example, why do some substances that contain OH groups behave as bases, releasing OH^- into solution (like NaOH), whereas others behave as acids, ionizing to release H^+ (like HOCl), while others do neither (like CH_3OH)? In this section, we examine the effects of chemical structure on acid–base behavior.

Factors That Affect Acid Strength

The strength of an acid is affected by three separate factors.

- **H—A Bond Polarity** A molecule containing H will act as a proton donor (an acid) only if the H—A bond is polarized such that the H atom has a partial positive charge. (Section 8.4) Recall that we indicate such polarization in this way:

$$H—A$$

In ionic hydrides, such as NaH, the bond is polarized in the opposite way: The H atom possesses a negative charge and behaves as a proton acceptor (a base). Nonpolar H—A bonds, such as the H—C bond in CH_4 produce neither acidic nor basic aqueous solutions. *In general, as the H—A bond polarity increases, thereby drawing more electron density from H, the stronger the acid.*

Learning Objectives

When you finish Section 16.10, you should be able to:

▶ Explain how the polarity and strength of the bonds in a molecule correlate to its tendency to act as either an acid or a base.

▶ Rank-order the strengths of chemically related acids based on their composition and molecular structure.

4A	5A	6A	7A
CH$_4$ Neither acid nor base	**NH$_3$** Weak base $K_b = 1.8 \times 10^{-5}$	**H$_2$O**	**HF** Weak acid $K_a = 6.8 \times 10^{-4}$
SiH$_4$ Neither acid nor base	**PH$_3$** Very weak base $K_b = 4 \times 10^{-28}$	**H$_2$S** Weak acid $K_a = 9.5 \times 10^{-8}$	**HCl** Strong acid
		H$_2$Se Weak acid $K_a = 1.3 \times 10^{-4}$	**HBr** Strong acid

Increasing acid strength →

Increasing acid strength ↓

▲ **Figure 16.19 Trends in acid strength for the binary hydrides of periods 2–4.**

- **H—A Bond Strength** The strength of the bond (Section 8.8) also helps determine whether a molecule containing an H—A bond donates a proton. Strong bonds are less easily broken than weaker ones. This factor is important, for example, in the hydrogen halides. The H—F bond is the most polar H—A bond. You therefore might expect HF to be a strong acid if bond polarity were all that mattered. However, the H—A bond strength increases as you move up the group: 299 kJ/mol in HI, 366 kJ/mol in HBr, 431 kJ/mol in HCl, and 567 kJ/mol in HF. Because HF has the highest bond strength among the hydrogen halides, it is a weak acid, whereas all the other hydrogen halides are strong acids in water. *In general, the strength of an acid increases as the H—A bond strength decreases.*

- **Conjugate Base Stability** A third factor that affects the ease with which a hydrogen atom ionizes from HA is the stability of the conjugate base, A$^-$. *In general, the more stable the conjugate base, the stronger the acid.*

Binary Acids

For a series of binary acids HA in which A represents members of the same *group* in the periodic table, the strength of the H—A bond is generally the most important factor determining acid strength. The strength of an H—A bond tends to decrease as the element A increases in size. As a result, the bond strength decreases and acidity increases down a group. Thus, HCl is a stronger acid than HF, and H$_2$S is a stronger acid than H$_2$O.

Bond polarity is the major factor determining acidity for binary acids HA when A represents members of the same *period*. Thus, acidity increases as the electronegativity of the element A increases, as it generally does moving from left to right across a period. (Section 8.4) For example, the difference in acidity of the period 2 elements is CH$_4$ < NH$_3$ $\ll$ H$_2$O < HF. Because the C—H bond is essentially nonpolar, CH$_4$ shows no tendency to form H$^+$ and CH$_3^-$ ions. Although the N—H bond is polar, NH$_3$ has a nonbonding pair of electrons on the nitrogen atom that dominates its chemistry, so NH$_3$ acts as a base rather than an acid.

The periodic trends in the acid strengths of binary compounds of hydrogen and the nonmetals of periods 2 and 3 are summarized in **Figure 16.19**.

Oxyacids

Many common acids, such as sulfuric acid, contain one or more O—H bonds:

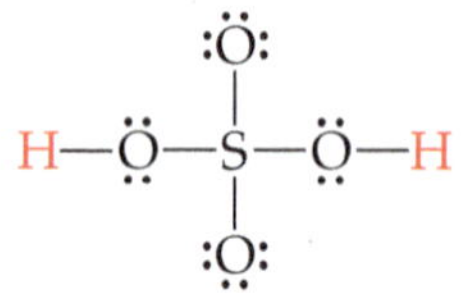

$$
\begin{array}{c}
\ddot{\text{O}}{:} \\
| \\
\text{H}-\ddot{\text{O}}-\text{S}-\ddot{\text{O}}-\text{H} \\
| \\
{:}\ddot{\text{O}}{:}
\end{array}
$$

Acids in which OH groups and possibly additional oxygen atoms are bound to a central atom are called **oxyacids**. At first it may seem confusing that the OH group, which we know behaves as a base, is also present in some acids. Let's take a closer look at what factors determine whether a given OH group behaves as a base or as an acid.

Consider an OH group bound to some atom Y, which might in turn have other groups attached to it:

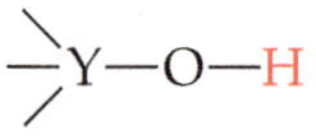

$$\text{Y}-\text{O}-\text{H}$$

 **Go Figure**   At equilibrium, which of the two species with a halogen atom (green) is present in greater concentration?

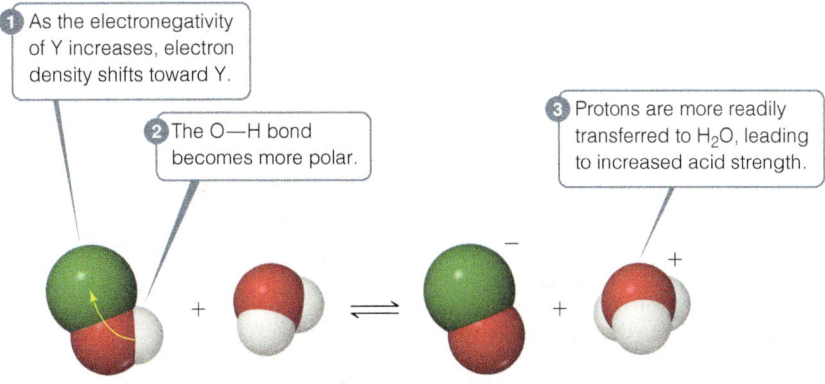

1 As the electronegativity of Y increases, electron density shifts toward Y.

2 The O—H bond becomes more polar.

3 Protons are more readily transferred to H_2O, leading to increased acid strength.

Substance	Y—OH	Electronegativity of Y	Dissociation constant
Hypochlorous acid	Cl—OH	3.0	$K_a = 3.0 \times 10^{-8}$
Hypobromous acid	Br—OH	2.8	$K_a = 2.5 \times 10^{-9}$
Hypoiodous acid	I—OH	2.5	$K_a = 2.3 \times 10^{-11}$
Water	H—OH	2.1	$K_w = 1.0 \times 10^{-14}$

▲ **Figure 16.20** Acidity of the hypohalous oxyacids (YOH) and water as a function of electronegativity of Y.

At one extreme, Y might be a metal, such as Na or Mg. Because of the low electronegativity of metals, the pair of electrons shared between Y and O is completely transferred to oxygen, and an ionic compound containing OH^- is formed. Such compounds are therefore sources of OH^- ions and behave as bases; examples include NaOH and $Mg(OH)_2$.

When Y is a nonmetal, the bond to O is covalent and the substance does not readily lose OH^-. Instead, these compounds are either acidic or neutral. *Generally, as the electronegativity of Y increases, so does the acidity of the substance.* This happens for two reasons: First, as electron density is drawn toward Y, the O—H bond becomes weaker and more polar, thereby favoring loss of H^+. Second, because the conjugate base of any acid YOH is usually an anion, its stability generally increases as the electronegativity of Y increases. This trend is illustrated by the K_a values of the hypohalous acids (YOH acids where Y is a halide ion), which decrease as the electronegativity of the halogen atom decreases (**Figure 16.20**).

Many oxyacids contain additional oxygen atoms bonded to the central atom Y. These atoms pull electron density from the O—H bond, further increasing its polarity. Increasing the number of oxygen atoms also helps stabilize the conjugate base by increasing its ability to "spread out" its negative charge. Thus, *the strength of an acid increases as additional electronegative atoms bond to the central atom Y.* For example, the strength of the chlorine oxyacids (Y = Cl) steadily increases as O atoms are added:

Hypochlorous	Chlorous	Chloric	Perchloric
H—Ö—Cl:	H—Ö—Cl—Ö:	H—Ö—Cl—Ö:	H—Ö—Cl—Ö:
$K_a = 3.0 \times 10^{-8}$	$K_a = 1.1 \times 10^{-2}$	Strong acid	Strong acid

Increasing acid strength →

Because the oxidation number of Y increases as the number of attached O atoms increases, this correlation can be stated in an equivalent way: In a series of oxyacids, the acidity increases as the oxidation number of the central atom increases.

Sample Exercise 16.18

Predicting Relative Acidities from Composition and Structure

Arrange the compounds in each series in order of increasing acid strength: **(a)** AsH_3, HBr, KH, H_2Se; **(b)** H_2SO_4, H_2SeO_3, H_2SeO_4.

SOLUTION

Analyze We are asked to arrange two sets of compounds in order from weakest acid to strongest acid. In (a), the substances are binary compounds containing H, and in (b), the substances are oxyacids.

Plan For the binary compounds, all four elements bound to hydrogen are in the fourth period, so we will consider the electronegativities of As, Br, K, and Se relative to the electronegativity of H. The higher the electronegativity of these atoms, the higher the partial positive charge on H and so the more acidic the compound. For the oxyacids, we will consider both the electronegativities of the central atom and the number of oxygen atoms bonded to the central atom.

Solve

(a) Because K is on the left side of the periodic table, it has a very low electronegativity (0.8, from Figure 8.8). As a result, the hydrogen in KH carries a negative charge. Thus, KH should be the least acidic (most basic) compound in the series.

Arsenic and hydrogen have similar electronegativities, 2.0 and 2.1, respectively. This means that the As—H bond is nonpolar, and so AsH_3 has little tendency to donate a proton in aqueous solution.

The electronegativity of Se is 2.4, and that of Br is 2.8. Consequently, the H—Br bond is more polar than the H—Se bond, giving HBr the greater tendency to donate a proton. (This expectation is confirmed by Figure 16.19, where we see that H_2Se is a weak acid and HBr a strong acid.) Thus, the order of increasing acidity is $KH < AsH_3 < H_2Se < HBr$.

(b) The acids H_2SO_4 and H_2SeO_4 have the same number of O atoms and the same number of OH groups. In such cases, the acid strength increases with increasing electronegativity of the central atom. Because S is slightly more electronegative than Se (2.5 vs. 2.4), we predict that H_2SO_4 is more acidic than H_2SeO_4.

For acids with the same central atom, the acidity increases as the number of oxygen atoms bonded to the central atom increases. Thus, H_2SeO_4 should be a stronger acid than H_2SeO_3. We predict the order of increasing acidity to be $H_2SeO_3 < H_2SeO_4 < H_2SO_4$.

▶ **Practice Exercise**
In each pair, choose the compound that gives the more acidic (or less basic) solution: **(a)** HBr, HF; **(b)** PH_3, H_2S; **(c)** HNO_2, HNO_3; **(d)** H_2SO_3, H_2SeO_3.

Carboxylic Acids

Another large group of acids is illustrated by acetic acid, a weak acid ($K_a = 1.8 \times 10^{-5}$):

The portion of the structure shown in red is called the *carboxyl group*, which is often written COOH. Thus, the chemical formula of acetic acid is written as CH_3COOH, where only the hydrogen atom in the carboxyl group can be ionized. Acids that contain a carboxyl group are called **carboxylic acids**, and they form the largest category of organic acids. Formic acid and benzoic acid are further examples of this large and important category of acids:

Formic acid   Benzoic acid

Two factors contribute to the acidic behavior of carboxylic acids. First, the additional oxygen atom attached to the carbon of the carboxyl group draws electron density from the O—H bond, increasing its polarity and helping to stabilize the conjugate base.

CHEMISTRY AND LIFE   The Amphiprotic Behavior of Amino Acids

As we discuss in greater detail in Chapter 24, *amino acids* are the building blocks of proteins. The general structure of amino acids is

where different amino acids have different R groups attached to the central carbon atom. For example, in *glycine*, the simplest amino acid, R is a hydrogen atom, and in *alanine* R is a CH_3 group:

Glycine   Alanine

Amino acids contain a carboxyl group and can therefore serve as acids. They also contain an NH_2 group, characteristic of amines (Section 16.7), and thus they can also act as bases. Amino acids, therefore, are amphiprotic. For glycine, we might expect the acid and base reactions with water to be

Acid: $H_2N-CH_2-COOH(aq) + H_2O(l) \rightleftharpoons$
$$H_2N-CH_2-COO^-(aq) + H_3O^+(aq) \quad [16.50]$$

Base: $H_2N-CH_2-COOH(aq) + H_2O(l) \rightleftharpoons$
$$^+H_3N-CH_2-COOH(aq) + OH^-(aq) \quad [16.51]$$

The pH of a solution of glycine in water is about 6.0, indicating that it is a slightly stronger acid than base.

The acid–base chemistry of amino acids is more complicated, however, than what is shown in Equations 16.50 and 16.51. Because the COOH group can act as an acid and the NH_2 group can act as a base, amino acids undergo a "self-contained" Brønsted–Lowry acid–base reaction in which the proton of the carboxyl group is transferred to the basic nitrogen atom:

Neutral molecule   Zwitterion

Although the form of the amino acid on the right in this equation is electrically neutral overall, it has a positively charged end and a negatively charged end. A molecule of this type is called a *zwitterion* (German for "hybrid ion").

Do amino acids exhibit any properties indicating that they behave as zwitterions? If so, their behavior should be similar to that of ionic substances. (Section 8.2) Crystalline amino acids have relatively high melting points, usually above 200 °C, which is characteristic of ionic solids. Amino acids are far more soluble in water than in nonpolar solvents, too. In addition, the dipole moments of amino acids are large, consistent with a large separation of charge in the molecule. Thus, the ability of amino acids to act simultaneously as acids and bases has important effects on their properties.

Related Exercise: 16.113

Second, the conjugate base of a carboxylic acid (a *carboxylate anion*) can exhibit resonance (Section 8.6), which contributes to the stability of the anion by spreading the negative charge over several atoms:

Self-Assessment Exercises

SAE 16.29 Both bromous acid ($HBrO_2$) and hypobromous acid (HBrO) are weak acids. In both molecules, the hydrogen atom is bound to _____. Yet, the bond is more polar in _____, which makes it a stronger acid. **(a)** bromine, HBrO **(b)** bromine, $HBrO_2$ **(c)** oxygen, HBrO **(d)** oxygen, $HBrO_2$

SAE 16.30 Which of the following statements are *true*?

  (i) The strength of an acid increases as the polarity of the H–X bond increases.
  (ii) The strength of an acid increases as the strength of the H–X bond increases.

(a) only i **(b)** only ii **(c)** Both i and ii are true. **(d)** Neither i nor ii is true.

Putting Concepts Together

Phosphorous acid (H_3PO_3) has the Lewis structure shown here.
(a) Explain why H_3PO_3 is diprotic and not triprotic. **(b)** A 25.0-mL sample of an H_3PO_3 solution titrated with 0.102 M NaOH requires 23.3 mL of NaOH to neutralize both acidic protons. What is the molarity of the H_3PO_3 solution? **(c)** The original solution from part (b) has a pH of 1.59. Calculate the percent ionization and K_{a1} for H_3PO_3, assuming that $K_{a1} \gg K_{a2}$. **(d)** How does the osmotic pressure of a 0.050 M solution of HCl compare qualitatively with that of a 0.050 M solution of H_3PO_3? Explain.

$$
\begin{array}{c}
\text{H} \\
| \\
\ddot{\text{O}}-\text{P}-\ddot{\text{O}}-\text{H} \\
| \\
\ddot{\text{O}}-\text{H}
\end{array}
$$

SOLUTION

We will use what we have learned about molecular structure and its impact on acidic behavior to answer part (a). We will then use stoichiometry and the relationship between pH and [H^+] to answer parts (b) and (c). Finally, we will consider percent ionization in order to compare the osmotic pressure of the two solutions in part (d).

(a) Acids have polar H—X bonds. According to Figure 8.8, the electronegativity of H is 2.1 and that of P is also 2.1. Because the two elements have the same electronegativity, the H—P bond is nonpolar. (Section 8.4) Thus, this H cannot be acidic. The other two H atoms, however, are bonded to O, which has an electronegativity of 3.5. The H—O bonds, therefore, are polar, with H having a partial positive charge. These two H atoms are consequently acidic.

(b) The chemical equation for the neutralization reaction is:

$$H_3PO_3(aq) + 2\,NaOH(aq) \longrightarrow Na_2HPO_3(aq) + 2\,H_2O(l)$$

From the definition of molarity, $M = $ mol/L, we see that moles $= M \times$ L. (Section 4.5) Thus, the number of moles of NaOH added to the solution is:

$$(0.0233\ \text{L})(0.102\ \text{mol/L}) = 2.38 \times 10^{-3}\ \text{mol NaOH}$$

The balanced equation indicates that 2 mol of NaOH is consumed for each mole of H_3PO_3. Thus, the number of moles of H_3PO_3 in the sample is:

$$(2.38 \times 10^{-3}\ \text{mol NaOH})\left(\frac{1\ \text{mol}\ H_3PO_3}{2\ \text{mol NaOH}}\right) = 1.19 \times 10^{-3}\ \text{mol}\ H_3PO_3$$

The concentration of the H_3PO_3 solution, therefore, equals $(1.19 \times 10^{-3}\ \text{mol})/(0.0250\ \text{L}) = 0.0476\ M$.

(c) From the pH of the solution, 1.59, we can calculate [H^+] at equilibrium:

$$[H^+] = \text{antilog}(-1.59) = 10^{-1.59} = 0.026\ M\ \text{(two significant figures)}$$

Because $K_{a1} \gg K_{a2}$, the vast majority of the ions in solution are from the first ionization step of the acid.

Because one $H_2PO_3^-$ ion forms for each H^+ ion formed, the equilibrium concentrations of H^+ and $H_2PO_3^-$ are equal: $[H^+] = [H_2PO_3^-] = 0.026\ M$. The equilibrium concentration of H_3PO_3 equals the initial concentration minus the amount that ionizes to form H^+ and $H_2PO_3^-$: $[H_3PO_3] = 0.0476\ M - 0.026\ M = 0.022\ M$ (two significant figures). These results can be tabulated as follows:

	$H_3PO_3(aq)$	$\rightleftharpoons$	$H^+(aq)$	$+$	$H_2PO_3^-(aq)$
Initial concentration (M)	0.0476		0		0
Change in concentration (M)	-0.026		$+0.026$		$+0.026$
Equilibrium concentration (M)	0.022		0.026		0.026

The percent ionization is:

$$\text{percent ionization} = \frac{[H^+]_{\text{equilibrium}}}{[H_3PO_3]_{\text{initial}}} \times 100\% = \frac{0.026\ M}{0.0476\ M} \times 100\% = 55\%$$

The first acid-dissociation constant is:

$$K_{a1} = \frac{[H^+][H_2PO_3^-]}{[H_3PO_3]} = \frac{(0.026)(0.026)}{0.022} = 0.031$$

(d) Osmotic pressure is a colligative property and depends on the total concentration of particles in solution. (Section 13.5) Because HCl is a strong acid, a 0.050 M solution will contain 0.050 $M\ H^+(aq)$ and 0.050 $M\ Cl^-(aq)$, or a total of 0.100 mol/L of particles. Because H_3PO_3 is a weak acid, it ionizes to a lesser extent than HCl; hence, there are fewer particles in the H_3PO_3 solution. As a result, the H_3PO_3 solution will have the lower osmotic pressure.

Chapter Summary and Key Terms

CLASSIFICATIONS OF ACIDS AND BASES (SECTION 16.1) Acids and bases were first recognized by the properties of their aqueous solutions. For example, acids turn litmus red, whereas bases turn litmus blue. Arrhenius recognized that the properties of acidic solutions are due to $H^+(aq)$ ions and those of basic solutions are due to $OH^-(aq)$ ions. **Arrhenius acids** are substances that increase the concentration of H^+ ions when dissolved in water. **Arrhenius bases** are substances that increase the concentration of OH^- ions when dissolved in water.

The Brønsted–Lowry concept of acids and bases is more general than the Arrhenius concept and emphasizes the transfer of a proton (H^+) from an acid to a base. The H^+ ion is strongly bound to water. For this reason, the **hydronium ion**, $H_3O^+(aq)$, is often used to represent the predominant form of H^+ in water instead of the simpler $H^+(aq)$. A **Brønsted–Lowry acid** is a substance that donates a proton to another substance; a **Brønsted–Lowry base** is a substance that accepts a proton from another substance. Water is an example of an **amphiprotic** substance, one that can function as either a Brønsted–Lowry acid or base, depending on the substance with which it reacts.

The Lewis concept of acids and bases emphasizes the shared electron pair rather than the proton. A **Lewis acid** is an electron-pair acceptor, whereas a **Lewis base** is an electron-pair donor. The Lewis concept is more general than the Brønsted–Lowry concept because it can apply to cases in which the acid does not contain hydrogen.

CONJUGATE ACID–BASE PAIRS (SECTION 16.2) The **conjugate base** of a Brønsted–Lowry acid is the species that remains when a proton is removed from the acid. The **conjugate acid** of a Brønsted–Lowry base is the species formed by adding a proton to the base. Together, an acid and its conjugate base (or a base and its conjugate acid) are called a **conjugate acid–base pair**.

The acid–base strengths of conjugate acid–base pairs are related: The stronger an acid, the weaker is its conjugate base; the weaker an acid, the stronger is its conjugate base. In every acid–base reaction, the position of the equilibrium favors the transfer of the proton from the stronger acid to the stronger base.

AUTOIONIZATION OF WATER (SECTION 16.3) Water ionizes to a slight degree, forming $H^+(aq)$ and $OH^-(aq)$. The extent of this **autoionization** is expressed by the **ion-product constant** for water: $K_w = [H^+][OH^-] = 1.0 \times 10^{-14}$ (25 °C). This relationship holds for both pure water and aqueous solutions. The K_w expression indicates that the product of $[H^+]$ and $[OH^-]$ is a constant. Thus, as $[H^+]$ increases, $[OH^-]$ decreases. Acidic solutions are those that contain more $H^+(aq)$ than $OH^-(aq)$, whereas basic solutions contain more $OH^-(aq)$ than $H^+(aq)$. When $[H^+] = [OH^-]$, the solution is neutral.

THE PH SCALE (SECTION 16.4) The concentration of $H^+(aq)$ can be expressed in terms of **pH**: $pH = -\log[H^+]$. At 25 °C the pH of a neutral solution is 7.00, whereas the pH of an acidic solution is below 7.00, and the pH of a basic solution is above 7.00. This p notation is also used to represent the negative logarithm of other small quantities, as in pOH and pK_w. The pH of a solution can be measured using a pH meter, or it can be estimated using acid–base indicators.

STRONG ACIDS AND BASES (SECTION 16.5) Strong acids are strong electrolytes, ionizing completely in aqueous solution. The common strong acids are HCl, HBr, HI, HNO_3, $HClO_3$, $HClO_4$ and H_2SO_4. The conjugate bases of strong acids have negligible basicity. Common strong bases are the ionic hydroxides of the alkali metals and the heavy alkaline earth metals.

WEAK ACIDS (SECTION 16.6) Weak acids are weak electrolytes, so only a small fraction of the molecules exist in solution in ionized form. The extent of ionization is expressed by the **acid-dissociation constant**, K_a, which is the equilibrium constant for the reaction $HA(aq) \rightleftharpoons H^+(aq) + A^-(aq)$, which can also be written as $HA(aq) + H_2O(l) \rightleftharpoons H_3O^+(aq) + A^-(aq)$. The larger the value of K_a, the stronger is the acid. For solutions of the same concentration, a stronger acid also has a larger **percent ionization**. The concentration of a weak acid and its K_a value can be used to calculate the pH of a solution.

Polyprotic acids, such as H_3PO_4, have more than one ionizable proton. These acids have acid-dissociation constants that decrease in magnitude in the order $K_{a1} > K_{a2} > K_{a3}$. Because nearly all the $H^+(aq)$ in a polyprotic acid solution comes from the first dissociation step, the pH can usually be estimated satisfactorily by considering only K_{a1}.

WEAK BASES (SECTION 16.7) Weak bases include NH_3, **amines**, and those anions that are conjugate bases of weak acids. The extent to which a weak base reacts with water to generate the corresponding conjugate acid and OH^- is measured by the **base-dissociation constant**, K_b. K_b is the equilibrium constant for the reaction $B(aq) + H_2O(l) \rightleftharpoons HB^+(aq) + OH^-(aq)$, where B is the base. The larger the value of K_b, the stronger base.

RELATIONSHIP BETWEEN K_a AND K_b (SECTION 16.8) The relationship between the strength of an acid and the strength of its conjugate base is expressed quantitatively by the equation $K_a \times K_b = K_w$, where K_a and K_b are dissociation constants for conjugate acid–base pairs. This equation explains the inverse relationship between the strength of an acid and the strength of its conjugate base.

ACID–BASE PROPERTIES OF SALT SOLUTIONS (SECTION 16.9) The acid–base properties of salts can be ascribed to the behavior of their respective cations and anions. The reaction of ions with water, with a resulting change in pH, is called **hydrolysis**. The cations of the alkali metals and the alkaline earth metals as well as the anions of strong acids, such as Cl^-, Br^-, I^- and NO_3^-, do not undergo hydrolysis. They are always spectator ions in acid–base chemistry. A cation that is the conjugate acid of a weak base produces H^+ upon hydrolysis. An anion that is the conjugate base of a weak acid produces OH^- upon hydrolysis. Highly charged metal cations, such as Fe^{3+}, are hydrated in water; the metal-bound water molecules undergo reaction with free water to make H_3O^+ and therefore are acidic.

ACID–BASE BEHAVIOR AND CHEMICAL STRUCTURE (SECTION 16.10) The tendency of a substance to show acidic or basic characteristics in water can be correlated with its chemical structure. Acid character requires the presence of a highly polar H—X bond. Acidity is also favored when the H—X bond is weak and when the X^- ion is very stable.

For **oxyacids** with the same number of OH groups and the same number of O atoms, acid strength increases with increasing electronegativity of the central atom. For oxyacids with the same central atom, acid strength increases as the number of oxygen atoms attached to the central atom increases. **Carboxylic acids**, which are organic acids containing the COOH group, are the most important class of organic acids. The presence of delocalized π bonding in the conjugate base is a major factor responsible for the acidity of these compounds.

Key Equations

- $K_w = [H_3O^+][OH^-] = [H^+][OH^-] = 1.0 \times 10^{-14}$ [16.20]   Ion-product constant of water at 25 °C

- $pH = -\log[H^+]$ [16.21]   Definition of pH

- $pOH = -\log[OH^-]$ [16.22]   Definition of pOH

- $pH + pOH = 14.00$ [16.24]   Relationship between pH and pOH

- $K_a = \dfrac{[H_3O^+][A^-]}{[HA]}$ or $K_a = \dfrac{[H^+][A^-]}{[HA]}$ [16.29]   Acid-dissociation constant for a weak acid, HA

- Percent ionization $= \dfrac{[H^+]_{equilibrium}}{[HA]_{initial}} \times 100\%$ [16.34]   Percent ionization of a weak acid

- $K_b = \dfrac{[BH^+][OH^-]}{[B]}$ [16.38]   Base-dissociation constant for a weak base, B

- $K_a \times K_b = K_w$ [16.44]   Relationship between acid- and base-dissociation constants of a conjugate acid–base pair

- $pK_a = -\log K_a$ and $pK_b = -\log K_b$ [16.45]   Definitions of pK_a and pK_b

Exam Prep

EP 16.1 A Brønsted–Lowry base is defined as a proton _____; a Lewis base is defined as an electron pair _____. (a) acceptor, acceptor (b) acceptor, donor (c) donor, acceptor (d) donor, donor

EP 16.2 Consider the following equilibrium reaction:

$$HSO_4^-(aq) + OH^-(aq) \rightleftharpoons SO_4^{2-}(aq) + H_2O(l)$$

Which substances are acting as acids in the reaction? (a) HSO_4^- and OH^- (b) HSO_4^- and H_2O (c) OH^- and SO_4^{2-} (d) SO_4^{2-} and H_2O (e) OH^- and H_2O

EP 16.3 A reaction between two gas-phase molecules *cannot* be classified as an acid–base reaction under which classification scheme? (a) Arrhenius (b) Brønsted–Lowry (c) Lewis (d) Arrhenius and Lewis (e) A gas-phase reaction could be an acid–base reaction under all three classification schemes.

EP 16.4 When sodium hydride, NaH, dissolves in water, the hydride ion reacts with water as described by the following net ionic reaction:

$$H^-(aq) + H_2O(l) \longrightarrow H_2(g) + OH^-(aq)$$

Which of the following statements pertaining to this reaction are *true*? (i) NaH is an Arrhenius base. (ii) H_2 is the conjugate acid of H^-. (iii) H_2O acts as a Brønsted–Lowry acid.

(a) both i and ii (b) both i and iii (c) both ii and iii (d) All three statements are true.

EP 16.5 The dihydrogen phosphate ion, $H_2PO_4^-$, is amphiprotic. In which of the following reactions is this ion acting as a base?

(i) $H_3O^+(aq) + H_2PO_4^-(aq) \rightleftharpoons H_3PO_4(aq) + H_2O(l)$

(ii) $H_3O^+(aq) + HPO_4^{2-}(aq) \rightleftharpoons H_2PO_4^-(aq) + H_2O(l)$

(iii) $H_3PO_4(aq) + HPO_4^{2-}(aq) \rightleftharpoons 2\,H_2PO_4^-(aq)$

(a) i only (b) i and ii (c) i and iii (d) ii and iii (e) i, ii, and iii

EP 16.6 What species is the conjugate base of the hydroxide ion, OH^-? (a) H_2O (b) O^{2-} (c) H_3O^+ (d) H_2O^- (e) OH^- has no conjugate base.

EP 16.7 Based on information in Figure 16.4, place the following equilibria in order from smallest to largest value of K_c:

(i) $CH_3COOH(aq) + HS^-(aq) \rightleftharpoons CH_3COO^-(aq) + H_2S(aq)$

(ii) $F^-(aq) + NH_4^+(aq) \rightleftharpoons HF(aq) + NH_3(aq)$

(iii) $H_2CO_3(aq) + Cl^-(aq) \rightleftharpoons HCO_3^-(aq) + HCl(aq)$

(a) i < ii < iii (b) ii < i < iii (c) iii < i < ii (d) ii < iii < i (e) iii < ii < i

EP 16.8 A solution has $[OH^-] = 4.0 \times 10^{-8} M$ at 25 °C. What is the value of $[H^+]$ for this solution? (a) $2.5 \times 10^{-8} M$ (b) $4.0 \times 10^{-8} M$ (c) $2.5 \times 10^{-7} M$ (d) $2.5 \times 10^{-6} M$ (e) $4.0 \times 10^{-6} M$

EP 16.9 If $[H^+]$ is 100 times greater than $[OH^-]$, in an aqueous solution at 25 °C, what is the concentration of OH^-? (a) $1.0 \times 10^{-8} M$ (b) $1.0 \times 10^{-7} M$ (c) $1.0 \times 10^{-6} M$ (d) $1.0 \times 10^{-2} M$ (e) $1.0 \times 10^{-9} M$

EP 16.10 As the temperature increases, the equilibrium constant for the autoionization of water, K_w, increases. Based on this fact, a sample of pure water at 50 °C will have a pH _____, and $[H^+]$ will be _____ $[OH^-]$. (a) greater than 7, equal to (b) greater than 7, greater than (c) equal to 7, equal to (d) less than 7, equal to (e) less than 7, less than

EP 16.11 A solution at 25 °C has $[OH^-] = 6.7 \times 10^{-3} M$. What is the pH of the solution? (a) 0.83 (b) 2.2 (c) 2.17 (d) 11.83 (e) 12

EP 16.12 A solution at 25 °C has pOH = 10.53. In this solution $[OH^-]$ is _____ $[H^+]$, and the pH = _____. (a) greater than, 10.53 (b) less than, 10.53 (c) greater than, 3.47 (d) less than, 3.47

EP 16.13 Order the following three solutions from smallest to largest pH:

(i) 0.20 M $HClO_3$ (ii) 0.0030 M HNO_3 (iii) 1.50 M HCl

(a) i < ii < iii (b) ii < i < iii (c) iii < i < ii (d) ii < iii < i (e) iii < ii < i

EP 16.14 What is the pH of a 0.015 M solution of $Ba(OH)_2$? (a) 1.52 (b) 1.82 (c) 12.48 (d) 12.18 (e) 8.83

EP 16.15 Order the following three solutions from smallest to largest pH:

(i) 0.030 M $Sr(OH)_2$ (ii) 0.040 M KOH (iii) pure water

(a) i < ii < iii (b) ii < i < iii (c) iii < i < ii (d) ii < iii < i (e) iii < ii < i

EP 16.16 The indicator bromothymol blue is yellow when pH < 6.2 and turns blue at higher pH values. Phenolphthalein is colorless at pH < 8.0 and turns pink at higher pH values. Solutions A, B, and C exhibit the following colors with each of these indicators:

Solution A: blue with bromothymol blue, pink with phenolphthalein

Solution B: yellow with bromothymol blue, colorless with phenolphthalein

Solution C: blue with bromothymol blue, colorless with phenolphthalein

Order these solutions from lowest to highest pH. (**a**) A < B < C (**b**) B < C < A (**c**) A < C < B (**d**) C < A < B (**e**) C < B < A

EP 16.17 A 0.50 M solution of a weak acid HA has pH = 2.24. What is the value of K_a for the acid? (**a**) 1.7×10^{-12} (**b**) 3.3×10^{-5} (**c**) 6.7×10^{-5} (**d**) 5.8×10^{-3} (**e**) 1.2×10^{-2}

EP 16.18 A 0.077 M solution of an acid HA has pH = 2.16. What is the percentage of the acid that is ionized? (**a**) 0.090% (**b**) 0.69% (**c**) 0.90% (**d**) 3.6% (**e**) 9.0%

EP 16.19 What is the pH of a 0.40 M solution of benzoic acid, C_6H_5COOH ($K_a = 6.3 \times 10^{-5}$)? (**a**) 2.30 (**b**) 2.10 (**c**) 1.90 (**d**) 4.20 (**e**) 4.60

EP 16.20 What is the pH of a 0.010 M solution of HF ($K_a = 6.8 \times 10^{-4}$)? (**a**) 1.58 (**b**) 2.10 (**c**) 2.30 (**d**) 2.58 (**e**) 2.64

EP 16.21 Which of the following equilibrium-constant expressions corresponds to K_{a2} for carbonic acid, H_2CO_3?

(**a**) $K_{a2} = \dfrac{[H^+]^2[CO_3^{2-}]}{[H_2CO_3]}$

(**b**) $K_{a2} = \dfrac{[H^+][CO_3^{2-}]}{[H_2CO_3]}$

(**c**) $K_{a2} = \dfrac{[H^+][CO_3^{2-}]}{[HCO_3^-]}$

(**d**) $K_{a2} = \dfrac{[HCO_3^-]}{[H^+][CO_3^{2-}]}$

EP 16.22 What is the pH of a 0.28 M solution of ascorbic acid ($K_{a1} = 8.0 \times 10^{-5}$ and $K_{a2} = 1.6 \times 10^{-12}$)? (**a**) 2.04 (**b**) 2.32 (**c**) 2.82 (**d**) 4.65 (**e**) 6.17

EP 16.23 What is the concentration of SO_4^{2-} ions in a 0.80 M solution of sulfuric acid (K_{a1} = large, $K_{a2} = 1.2 \times 10^{-2}$) (**a**) 0.012 (**b**) 0.096 (**c**) 0.79 M (**d**) 0.80 M (**e**) 0.812

EP 16.24 What is the pH of a 0.65 M solution of pyridine, C_5H_5N ($K_b = 1.7 \times 10^{-9}$)? (**a**) 4.48 (**b**) 8.96 (**c**) 9.52 (**d**) 9.62 (**e**) 9.71

EP 16.25 Which of the following molecules or ions will act as a weak base in an aqueous solution: aniline, $C_6H_5NH_2$; methanol, CH_3OH; sulfite ion, SO_3^{2-}? (**a**) aniline (**b**) methanol (**c**) sulfite ion (**d**) aniline and sulfite ion (**e**) all three

EP 16.26 A 0.066 M solution of a weak base has pH = 11.31. What is the value of K_b for this base? (**a**) 3.0×10^{-2} (**b**) 2.0×10^{-3} (**c**) 6.3×10^{-5} (**d**) 4.0×10^{-6} (**e**) 3.6×10^{-22}

EP 16.27 The benzoate ion, $C_6H_5COO^-$, is a weak base with $K_b = 1.6 \times 10^{-10}$. How many moles of sodium benzoate are present in 0.50 L of a solution of NaC_6H_5COO if the pH is 9.04? (**a**) 0.38 (**b**) 0.66 (**c**) 0.76 (**d**) 1.5 (**e**) 2.9

EP 16.28 Given the following acid- and base-dissociation constants—chloroacetic acid, $CH_2ClCOOH$ ($K_a = 1.4 \times 10^{-3}$); pyridine, C_5H_5N ($K_b = 1.7 \times 10^{-9}$); and hypochlorous acid, HClO ($pK_a = 7.5$)—arrange the following species from the weakest to the strongest acid: $CH_2ClCOOH$, $C_5H_5NH^+$, HClO.

(**a**) weakest acid $CH_2ClCOOH < C_5H_5NH^+ <$ HClO strongest acid

(**b**) weakest acid HClO $< C_5H_5NH^+ < CH_2ClCOOH$ strongest acid

(**c**) weakest acid $C_5H_5NH^+ < CH_2ClCOOH <$ HClO strongest acid

(**d**) weakest acid $C_5H_5NH^+ <$ HClO $< CH_2ClCOOH$ strongest acid

(**e**) weakest acid HClO $< CH_2ClCOOH < C_5H_5NH^+$ strongest acid

EP 16.29 By using information from Appendix D, arrange the following three substances in order of weakest to strongest base: trimethylamine, $(CH_3)_3N$; formate ion, $HCOO^-$; hypobromite ion, BrO^-.

(**a**) weakest base $HCOO^- < BrO^- < (CH_3)_3N$ strongest base

(**b**) weakest base $BrO^- < (CH_3)_3N < HCOO^-$ strongest base

(**c**) weakest base $HCOO^- < (CH_3)_3N < BrO^-$ strongest base

(**d**) weakest base $(CH_3)_3N < BrO^- < HCOO^-$ strongest base

(**e**) weakest base $BrO^- < HCOO^- < (CH_3)_3N$ strongest base

EP 16.30 Order the following solutions from lowest to highest pH: (i) 0.10 M NaClO, (ii) 0.10 M KBr, (iii) 0.10 M NH_4ClO_4. (**a**) i < ii < iii (**b**) ii < i < iii (**c**) iii < i < ii (**d**) ii < iii < i (**e**) iii < ii < i

EP 16.31 Which of the following salts are expected to produce acidic solutions (see Table 16.3 for data): $NaHSO_4$, NaH_2PO_4, and $NaHCO_3$? (**a**) none (**b**) $NaHSO_4$ (**c**) NaH_2PO_4 (**d**) $NaHSO_4$ and NaH_2PO_4 (**e**) $NaHSO_4$, NaH_2PO_4, and $NaHCO_3$

EP 16.32 Arrange the following substances in order from weakest to strongest acid: $HClO_3$, HOI, $HBrO_2$, $HClO_2$, HIO_2.

(**a**) $HIO_2 <$ HOI $< HClO_3 < HBrO_2 < HClO_2$

(**b**) HOI $< HIO_2 < HBrO_2 < HClO_2 < HClO_3$

(**c**) $HBrO_2 < HIO_2 < HClO_2 <$ HOI $< HClO_3$

(**d**) $HClO_3 < HClO_2 < HBrO_2 < HIO_2 <$ HOI

(**e**) HOI $< HClO_2 < HBrO_2 < HIO_2 < HClO_3$

Exercises

Visualizing Concepts

16.1 (**a**) Identify the Brønsted–Lowry acid and base in the reaction

(b) Identify the Lewis acid and base in the reaction. [Section 16.1]

16.2 For each of these reactions, identify the acid and base among the reactants, and state if the acids and bases are Lewis, Arrhenius, and/or Brønsted–Lowry:

(**a**) $PCl_4^+ + Cl^- \longrightarrow PCl_5$

(**b**) $NH_3 + BF_3 \longrightarrow H_3NBF_3$

(**c**) $[Al(H_2O)_6]^{3+} + H_2O \longrightarrow [Al(H_2O)_5OH]^{2+} + H_3O^+$

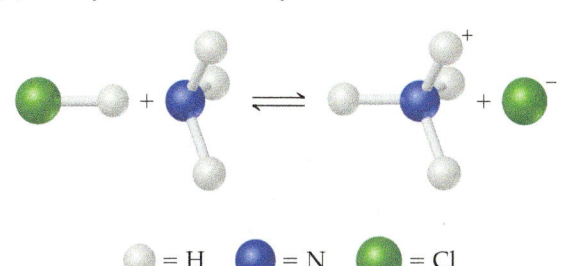

= H   = N   = Cl

16.3 The following diagrams represent the aqueous solutions of two monoprotic acids, HA (A = X or Y). The water molecules have been omitted for clarity. **(a)** Which is the stronger acid, HX or HY? **(b)** Which is the stronger base, X^- or Y^-? **(c)** If you mix equal concentrations of HX and NaY, will the equilibrium

$$HX(aq) + Y^-(aq) \rightleftharpoons HY(aq) + X^-(aq)$$

lie mostly to the right ($K_c > 1$) or to the left ($K_c < 1$)? [Section 16.2]

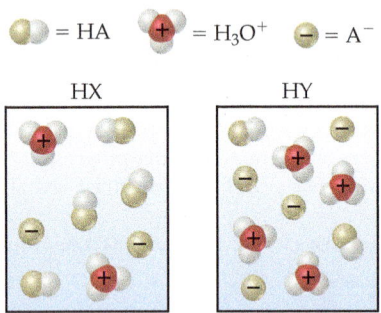

= HA   = H_3O^+ = A^-

HX   HY

16.4 The indicator methyl orange has been added to both of the following solutions. Based on the colors, classify each statement as *true* or *false*:

(a) The pH of solution A is definitely less than 7.00.

(b) The pH of solution B is definitely greater than 7.00.

(c) The pH of solution B is greater than that of solution A. [Section 16.4]

Solution A   Solution B

16.5 The probe of the pH meter shown here is sitting in a beaker that contains a clear liquid. **(a)** You are told the liquid is pure water, a solution of HCl(*aq*), or a solution of KOH(*aq*). Which one is it? **(b)** If the liquid is one of the solutions, what is its molarity? **(c)** Why is the temperature given on the pH meter? [Sections 16.4 and 16.5]

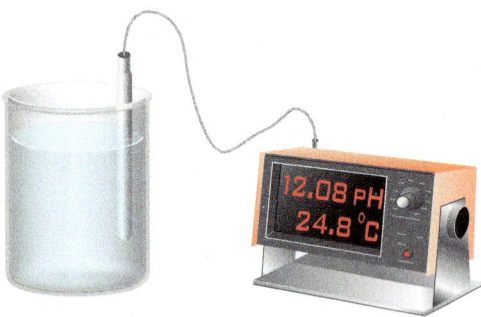

16.6 The following diagrams represent the aqueous solutions of three acids, HX, HY, and HZ. The water molecules have been omitted for clarity, and the hydrated proton is represented as H^+ rather than H_3O^+. **(a)** Which of the acids is a strong acid? Explain. **(b)** Which acid would have the smallest acid-dissociation constant, K_a? **(c)** Which solution would have the highest pH? [Sections 16.5 and 16.6]

HX   HY   HZ

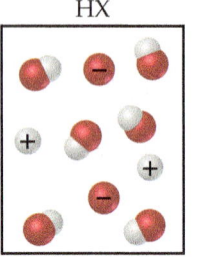   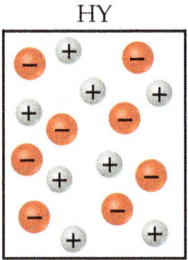   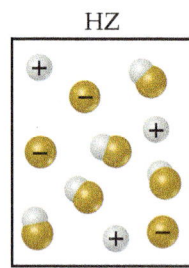

16.7 The graph given below shows $[H^+]$ versus concentration for an aqueous solution of an unknown substance. **(a)** Is the substance a strong acid, a weak acid, a strong base, or a weak base? **(b)** Based on your answer to (a), can you determine the value of the pH of the solution when the concentration is 0.18 *M*? **(c)** Would the line go exactly through the origin of the plot? [Sections 16.5 and 16.6]

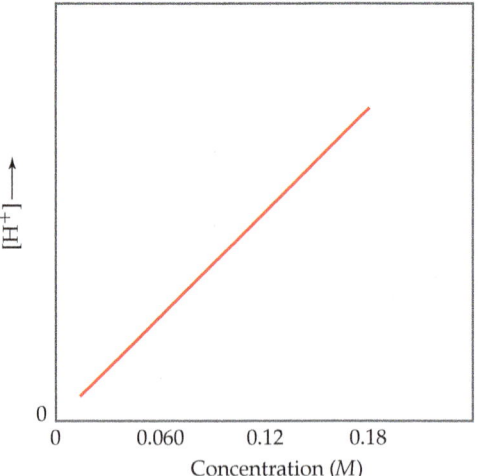

16.8 Which of these statements about how the percent ionization of a weak acid depends on acid concentration is true? [Section 16.6]

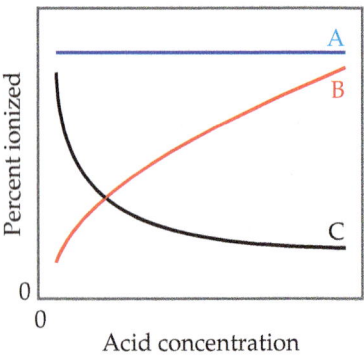

(a) Line A is most accurate because K_a does not depend on concentration.

(b) Line A is the most accurate because the percent ionization of the acid does not depend on concentration.

(c) Line B is the most accurate because as the acid concentration increases, a greater proportion of it is ionized.

(d) Line B is the most accurate because as the acid concentration increases, K_a increases.

(e) Line C is the most accurate because as the acid concentration increases, a lesser proportion of it ionized.

(f) Line C is the most accurate because as the acid concentration increases, K_a decreases.

16.9 Each of the three molecules shown here contains an OH group, but one molecule acts as a base, one as an acid, and the third is neither acid nor base. (a) Which one acts as a base? (b) Which one acts as an acid? (c) Which one is neither acidic nor basic? [Sections 16.6 and 16.7]

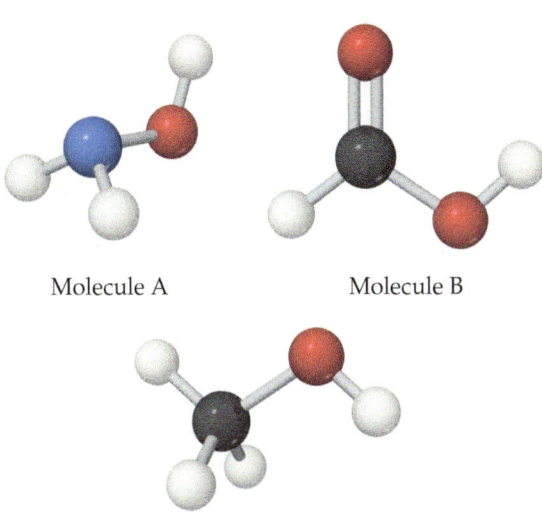

Molecule A   Molecule B

Molecule C

16.10 *Phenylephrine*, an organic substance with molecular formula $C_9H_{13}NO_2$, is used as a nasal decongestant in over-the-counter medications. The molecular structure of phenylephrine is shown using the usual shortcut organic structure. (a) Would you expect a solution of phenylephrine to be acidic, neutral, or basic? (b) One of the active ingredients in Alka-Seltzer PLUS® cold medication is phenylephrine hydrochloride. How does this ingredient differ from the structure shown here? (c) Would you expect a solution of phenylephrine hydrochloride to be acidic, neutral, or basic? [Sections 16.8 and 16.9]

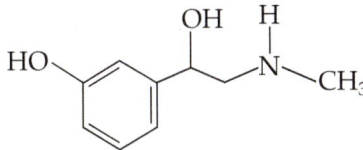

16.11 Which of the following diagrams best represents an aqueous solution of NaF? (For clarity, the water molecules are not shown.) Will this solution be acidic, neutral, or basic? [Section 16.9]

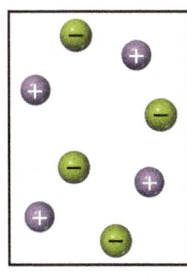 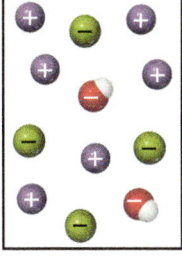 

Solution A   Solution B   Solution C

 Na⁺   F⁻   OH⁻   HF

16.12 Consider the molecular models shown here, where X represents a halogen atom. (a) If X is the same atom in both molecules, which molecule will be more acidic? (b) Does the acidity of each molecule increase or decrease as the electronegativity of the atom X increases? [Section 16.10]

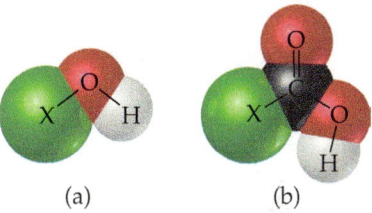

(a)   (b)

Classifications of Acids and Bases (Section 16.1)

16.13 $NH_3(g)$ and $HCl(g)$ react to form the ionic solid $NH_4Cl(s)$. Which substance is the Brønsted–Lowry acid in this reaction? Which is the Brønsted–Lowry base?

16.14 Which of the following statements is *false*?

(a) An Arrhenius base increases the concentration of OH⁻ in water.

(b) A Brønsted–Lowry base is a proton acceptor.

(c) Water can act as a Brønsted–Lowry acid.

(d) Water can act as a Brønsted–Lowry base.

(e) Any compound that contains an –OH group acts as a Brønsted–Lowry base.

16.15 Identify the reactants that act as Brønsted–Lowry acids in each of the following reactions:

(a) $NaHCO_3(s) + CH_3COOH(aq) \longrightarrow Na^+(aq) + H_2O(l) + CO_2(g) + CH_3COO^-(aq)$

(b) $F^-(aq) + H_2O(l) \rightleftharpoons HF(aq) + OH^-(aq)$

(c) $CH_3NH_3^+(aq) + H_2O(l) \rightleftharpoons CH_3NH_2(aq) + H_3O^+(aq)$

16.16 Identify the reactants that act as Brønsted–Lowry bases in each of the following reactions:

(a) $H_3PO_4(aq) + HPO_4^{2-}(aq) \rightleftharpoons 2 H_2PO_4^-(aq)$

(b) $HSO_3^-(aq) + H_2O(l) \rightleftharpoons H_2SO_3(aq) + OH^-(aq)$

(c) $CaO(s) + H_2O(l) \longrightarrow Ca^{2+}(aq) + 2 OH^-(aq)$

16.17 Identify the Lewis acid and Lewis base among the reactants in each of the following reactions:

(a) $Fe(ClO_4)_3(s) + 6 H_2O(l) \rightleftharpoons$
$[Fe(H_2O)_6]^{3+}(aq) + 3 ClO_4^-(aq)$

(b) $CN^-(aq) + H_2O(l) \rightleftharpoons HCN(aq) + OH^-(aq)$

(c) $(CH_3)_3N(g) + BF_3(g) \rightleftharpoons (CH_3)_3NBF_3(s)$

(d) $HIO(lq) + NH_2^-(lq) \rightleftharpoons NH_3(lq) + IO^-(lq)$ (lq denotes liquid ammonia as solvent)

16.18 Identify the Lewis acid and Lewis base in each of the following reactions:

(a) $HNO_2(aq) + OH^-(aq) \rightleftharpoons NO_2^-(aq) + H_2O(l)$

(b) $FeBr_3(s) + Br^-(aq) \rightleftharpoons FeBr_4^-(aq)$

(c) $Zn^{2+}(aq) + 4\,NH_3(aq) \rightleftharpoons Zn(NH_3)_4{}^{2+}(aq)$

(d) $SO_2(g) + H_2O(l) \rightleftharpoons H_2SO_3(aq)$

Conjugate Acid–Base Pairs (Section 16.2)

16.19 (a) Give the conjugate base of the following Brønsted–Lowry acids: (i) HIO_3, (ii) NH_4^+. (b) Give the conjugate acid of the following Brønsted–Lowry bases: (i) O^{2-}, (ii) $H_2PO_4^-$.

16.20 (a) Give the conjugate base of the following Brønsted–Lowry acids: (i) $HCOOH$, (ii) HPO_4^{2-}. (b) Give the conjugate acid of the following Brønsted–Lowry bases: (i) SO_4^{2-}, (ii) CH_3NH_2.

16.21 Identify the Brønsted–Lowry acid and the Brønsted–Lowry base on the left side of each of the following equations, and also identify the conjugate acid and conjugate base of each on the right side:

(a) $NH_4^+(aq) + CN^-(aq) \rightleftharpoons HCN(aq) + NH_3(aq)$

(b) $(CH_3)_3N(aq) + H_2O(l) \rightleftharpoons$
$(CH_3)_3NH^+(aq) + OH^-(aq)$

(c) $HCOOH(aq) + PO_4{}^{3-}(aq) \rightleftharpoons$
$HCOO^-(aq) + HPO_4{}^{2-}(aq)$

16.22 Identify the Brønsted–Lowry acid and the Brønsted–Lowry base on the left side of each equation, and also identify the conjugate acid and conjugate base of each on the right side.

(a) $HBrO(aq) + H_2O(l) \rightleftharpoons H_3O^+(aq) + BrO^-(aq)$

(b) $HSO_4^-(aq) + HCO_3^-(aq) \rightleftharpoons SO_4{}^{2-}(aq) + H_2CO_3(aq)$

(c) $HSO_3^-(aq) + H_3O^+(aq) \rightleftharpoons H_2SO_3(aq) + H_2O(l)$

16.23 (a) The hydrogen sulfite ion (HSO_3^-) is amphiprotic. Write a balanced chemical equation showing how it acts as an acid toward water and another equation showing how it acts as a base toward water. (b) What is the conjugate acid of HSO_3^-? What is its conjugate base?

16.24 (a) Write an equation for the reaction in which $H_2C_6H_7O_5^-(aq)$ acts as a base in $H_2O(l)$. (b) Write an equation for the reaction in which $H_2C_6H_7O_5^-(aq)$ acts as an acid in $H_2O(l)$. (c) What is the conjugate acid of $H_2C_6H_7O_5^-(aq)$? What is its conjugate base?

16.25 Label each of the following as being a strong base, a weak base, or a species with negligible basicity. In each case, write the formula of its conjugate acid, and indicate whether the conjugate acid is a strong acid, a weak acid, or a species with negligible acidity: (a) CH_3COO^-, (b) HCO_3^-, (c) O^{2-}, (d) Cl^-, (e) NH_3.

16.26 Label each of the following as being a strong acid, a weak acid, or a species with negligible acidity. In each case write the formula of its conjugate base, and indicate whether the conjugate base is a strong base, a weak base, or a species with negligible basicity: (a) $HCOOH$, (b) H_2, (c) CH_4, (d) HF, (e) NH_4^+.

16.27 (a) Which of the following is the stronger Brønsted–Lowry acid, $HBrO$ or HBr? (b) Which is the stronger Brønsted–Lowry base, F^- or Cl^-?

16.28 (a) Which of the following is the stronger Brønsted–Lowry acid, $HClO_3$ or $HClO_2$? (b) Which is the stronger Brønsted–Lowry base, HS^- or HSO_4^-?

16.29 Predict the products of the following acid–base reactions, and predict whether the equilibrium lies to the left or to the right of the reaction arrow:

(a) $O^{2-}(aq) + H_2O(l) \rightleftharpoons$

(b) $CH_3COOH(aq) + HS^-(aq) \rightleftharpoons$

(c) $NO_2^-(aq) + H_2O(l) \rightleftharpoons$

16.30 Predict the products of the following acid–base reactions, and predict whether the equilibrium lies to the left or to the right of the reaction arrow:

(a) $NH_4^+(aq) + OH^-(aq) \rightleftharpoons$

(b) $CH_3COO^-(aq) + H_3O^+(aq) \rightleftharpoons$

(c) $HCO_3^-(aq) + F^-(aq) \rightleftharpoons$

Autoionization of Water (Section 16.3)

16.31 If a neutral solution of water, with pH = 7.00, is cooled to 10 °C, the pH rises to 7.27. Which of the following three statements is correct for the cooled water: (i) $[H^+] > [OH^-]$, (ii) $[H^+] = [OH^-]$, or (iii) $[H^+] < [OH^-]$?

16.32 (a) Write a chemical equation that illustrates the autoionization of water. (b) Write the expression for the ion-product constant for water, K_w. (c) If a solution is described as basic, which of the following is true: (i) $[H^+] > [OH^-]$, (ii) $[H^+] = [OH^-]$, or (iii) $[H^+] < [OH^-]$?

16.33 Calculate $[H^+]$ for each of the following solutions, and indicate whether the solution is acidic, basic, or neutral: (a) $[OH^-] = 0.00045\ M$; (b) $[OH^-] = 8.8 \times 10^{-9}\ M$; (c) a solution in which $[OH^-]$ is 100 times greater than $[H^+]$.

16.34 Calculate $[OH^-]$ for each of the following solutions, and indicate whether the solution is acidic, basic, or neutral: (a) $[H^+] = 0.0505\ M$; (b) $[H^+] = 2.5 \times 10^{-10}\ M$; (c) a solution in which $[H^+]$ is 1000 times greater than $[OH^-]$.

16.35 At the freezing point of water (0 °C), $K_w = 1.2 \times 10^{-15}$. Calculate $[H^+]$ and $[OH^-]$ for a neutral solution at this temperature.

16.36 Deuterium oxide (D_2O, where D is deuterium, the hydrogen-2 isotope) has an ion-product constant, K_w, of 8.9×10^{-16} at 20 °C. Calculate $[D^+]$ and $[OD^-]$ for pure (neutral) D_2O at this temperature.

The pH Scale (Section 16.4)

16.37 By what factor does $[H^+]$ change for a pH change of (a) 2.00 units, (b) 0.50 units?

16.38 If $[H^+]$ in solution A is 250 times greater than $[H^+]$ in solution B, then what is the difference in the pH values of the two solutions?

16.39 Complete the following table by calculating the missing entries and indicating whether the solution is acidic or basic.

$[H^+]$	$[OH^-]$	pH	pOH	Acidic or Basic?
$7.5 \times 10^{-3}\ M$				
	$3.6 \times 10^{-10}\ M$			
		8.25		
			5.70	

16.40 Complete the following table by calculating the missing entries. In each case, indicate whether the solution is acidic or basic.

pH	pOH	$[H^+]$	$[OH^-]$	Acidic or Basic?
5.25				
	2.02			
		$4.4 \times 10^{-10}\ M$		
			$8.5 \times 10^{-2}\ M$	

16.41 The average pH of normal arterial blood is 7.40. At normal body temperature (37°C), $K_w = 2.4 \times 10^{-14}$. Calculate $[H^+], [OH^-]$, and pOH for blood at this temperature.

16.42 Carbon dioxide in the atmosphere dissolves in raindrops to produce carbonic acid (H_2CO_3), causing the pH of clean, unpolluted rain to range from about 5.2 to 5.6. What are the ranges of $[H^+]$ and $[OH^-]$ in the raindrops?

16.43 Addition of the indicator methyl orange to an unknown solution leads to a yellow color. The addition of bromthymol blue to the same solution also leads to a yellow color. (**a**) Is the solution acidic, neutral, or basic? (**b**) What is the range (in whole numbers) of possible pH values for the solution? (**c**) Is there another indicator you could use to narrow the range of possible pH values for the solution?

16.44 Addition of phenolphthalein to an unknown colorless solution does not cause a color change. The addition of bromthymol blue to the same solution leads to a yellow color. (**a**) Is the solution acidic, neutral, or basic? (**b**) Which of the following can you establish about the solution: (i) a minimum pH, (ii) a maximum pH, or (iii) a specific range of pH values? (**c**) What other indicator or indicators would you want to use to determine the pH of the solution more precisely?

Strong Acids and Bases (Section 16.5)

16.45 Is each of the following statements *true* or *false*? (**a**) All strong acids contain one or more H atoms. (**b**) A strong acid is a strong electrolyte. (**c**) A 1.0-M solution of a strong acid will have pH = 1.0.

16.46 Determine whether each of the following is *true* or *false*: (**a**) All strong bases are salts of the hydroxide ion. (**b**) The addition of a strong base to water produces a solution of pH > 7.0. (**c**) Because $Mg(OH)_2$ is not very soluble, it cannot be a strong base.

16.47 Calculate the pH of each of the following strong acid solutions: (**a**) $8.5 \times 10^{-3}\,M$ HBr, (**b**) 1.52 g of HNO_3 in 575 mL of solution, (**c**) 5.00 mL of 0.250 M $HClO_4$ diluted to 50.0 mL, (**d**) a solution formed by mixing 10.0 mL of 0.100 M HBr with 20.0 mL of 0.200 M HCl.

16.48 Calculate the pH of each of the following strong acid solutions: (**a**) 0.0167 M HNO_3 (**b**) 0.225 g of $HClO_3$ in 2.00 L of solution, (**c**) 15.00 mL of 1.00 M HCl diluted to 0.500 L, (**d**) a mixture formed by adding 50.0 mL of 0.020 M HCl to 125 mL of 0.010 M HI.

16.49 Calculate $[OH^-]$ and pH for (**a**) $1.5 \times 10^{-3}\,M$ $Sr(OH)_2$, (**b**) 2.250 g of LiOH in 250.0 mL of solution, (**c**) 1.00 mL of 0.175 M NaOH diluted to 2.00 L, (**d**) a solution formed by adding 5.00 mL of 0.105 M KOH to 15.0 mL of $9.5 \times 10^{-2}\,M$ $Ca(OH)_2$.

16.50 Calculate $[OH^-]$ and pH for each of the following strong base solutions: (**a**) 0.182 M KOH, (**b**) 3.165 g of KOH in 500.0 mL of solution, (**c**) 10.0 mL of 0.0105 M $Ca(OH)_2$ diluted to 500.0 mL, (**d**) a solution formed by mixing 20.0 mL of 0.015 M $Ba(OH)_2$ with 40.0 mL of $8.2 \times 10^{-3}\,M$ NaOH.

16.51 Calculate the concentration of an aqueous solution of NaOH that has a pH of 11.50.

16.52 Calculate the concentration of an aqueous solution of $Ca(OH)_2$ that has a pH of 10.05.

Weak Acids (Section 16.6)

16.53 Write the chemical equation and the K_a expression for the ionization of each of the following acids in aqueous solution. First show the reaction with $H^+(aq)$ as a product and then with the hydronium ion: (**a**) $HBrO_2$, (**b**) C_2H_5COOH.

16.54 Write the chemical equation and the K_a expression for the acid dissociation of each of the following acids in aqueous solution. First show the reaction with $H^+(aq)$ as a product and then with the hydronium ion: (**a**) C_6H_5COOH, (**b**) HCO_3^-.

16.55 Lactic acid ($CH_3CH(OH)COOH$) has one acidic hydrogen. A 0.10 M solution of lactic acid has a pH of 2.44. Calculate K_a.

16.56 Phenylacetic acid ($C_6H_5CH_2COOH$) is one of the substances that accumulates in the blood of people with phenylketonuria, an inherited disorder that can cause mental retardation or even death. A 0.085 M solution of $C_6H_5CH_2COOH$ has a pH of 2.68. Calculate the K_a value for this acid.

16.57 A 0.100 M solution of chloroacetic acid ($ClCH_2COOH$) is 11.0% ionized. Using this information, calculate $[ClCH_2COO^-], [H^+], [ClCH_2COOH]$, and K_a for chloroacetic acid.

16.58 A 0.100 M solution of bromoacetic acid ($BrCH_2COOH$) is 13.2% ionized. Calculate $[H^+], [BrCH_2COO^-], [BrCH_2COOH]$ and K_a for bromoacetic acid.

16.59 A particular sample of vinegar has a pH of 2.90. If acetic acid is the only acid that vinegar contains ($K_a = 1.8 \times 10^{-5}$), calculate the concentration of acetic acid in the vinegar.

16.60 If a solution of HF ($K_a = 6.8 \times 10^{-4}$) has a pH of 3.65, calculate the concentration of hydrofluoric acid.

16.61 The acid-dissociation constant for benzoic acid (C_6H_5COOH) is 6.3×10^{-5}. Calculate the equilibrium concentrations of $H_3O^+, C_6H_5COO^-$, and C_6H_5COOH in the solution if the initial concentration of C_6H_5COOH is 0.050 M.

16.62 The acid-dissociation constant for chlorous acid ($HClO_2$) is 1.1×10^{-2}. Calculate the concentrations of H_3O^+, ClO_2^-, and $HClO_2$ at equilibrium if the initial concentration of $HClO_2$ is 0.0125 M.

16.63 Calculate the pH of each of the following solutions (K_a and K_b values are given in Appendix D): (**a**) 0.095 M propionic acid (C_2H_5COOH), (**b**) 0.100 M hydrogen chromate ion ($HCrO_4^-$), (**c**) 0.120 M pyridine (C_5H_5N).

16.64 Determine the pH of each of the following solutions (K_a and K_b values are given in Appendix D): (**a**) 0.095 M hypochlorous acid, (**b**) 0.0085 M hydrazine, (**c**) 0.165 M hydroxylamine.

16.65 Saccharin, a sugar substitute, is a weak acid with $pK_a = 2.32$ at 25 °C. It ionizes in aqueous solution as follows:

$$HNC_7H_4SO_3(aq) \rightleftharpoons H^+(aq) + NC_7H_4SO_3^-(aq)$$

What is the pH of a 0.10 M solution of this substance?

16.66 The active ingredient in aspirin is acetylsalicylic acid ($HC_9H_7O_4$), a monoprotic acid with $K_a = 3.3 \times 10^{-4}$ at 25 °C. What is the pH of a solution obtained by dissolving two extra-strength aspirin tablets, containing 500 mg of acetylsalicylic acid each, in 250 mL of water?

16.67 Calculate the percent ionization of hydrazoic acid (HN_3) in solutions of each of the following concentrations (K_a is given in Appendix D): (**a**) 0.400 M, (**b**) 0.100 M, (**c**) 0.0400 M.

16.68 Calculate the percent ionization of propionic acid (C_2H_5COOH) in solutions of each of the following concentrations (K_a is given in Appendix D): (**a**) 0.250 M, (**b**) 0.0800 M, (**c**) 0.0200 M.

16.69 Citric acid, which is present in citrus fruits, is a triprotic acid (Table 16.3). (**a**) Calculate the pH of a 0.040 M solution of citric acid. (**b**) Did you have to make any approximations or assumptions in completing your calculations? (**c**) Is the concentration of citrate ion ($C_6H_5O_7^{3-}$) equal to, less than, or greater than the H^+ ion concentration?

16.70 Tartaric acid is found in many fruits, including grapes, and is partially responsible for the dry texture of certain wines. Calculate the pH and the tartrate ion ($C_4H_4O_6^{2-}$) concentration for a 0.250 M solution of tartaric acid, for which the acid-dissociation constants are listed in Table 16.3. Did you have to make any approximations or assumptions in your calculation?

Weak Bases (Section 16.7)

16.71 Consider the base hydroxylamine, NH_2OH. (a) What is the conjugate acid of hydroxylamine? (b) When it acts as a base, which atom in hydroxylamine accepts a proton? (c) There are two atoms in hydroxylamine that have nonbonding electron pairs that could act as proton acceptors. Use Lewis structures and formal charges (Section 8.5) to rationalize why one of these two atoms is a much better proton acceptor than the other.

16.72 The hypochlorite ion, ClO^-, acts as a weak base. (a) Is ClO^- a stronger or weaker base than hydroxylamine, NH_2OH? (b) When ClO^- acts as a base, which atom, Cl or O, acts as the proton acceptor? (c) Can you use formal charges (Section 8.5) to rationalize your answer to part (b)?

16.73 Write the chemical equation and the K_b expression for the reaction of each of the following bases with water: (a) dimethylamine, $(CH_3)_2NH$; (b) carbonate ion, CO_3^{2-}; (c) formate ion, CHO_2^-.

16.74 Write the chemical equation and the K_b expression for the reaction of each of the following bases with water: (a) propylamine, $C_3H_7NH_2$; (b) monohydrogen phosphate ion, HPO_4^{2-}; (c) benzoate ion, $C_6H_5CO_2^-$.

16.75 Calculate the molar concentration of OH^- in a 0.075 M solution of ethylamine ($C_2H_5NH_2$; $K_b = 6.4 \times 10^{-4}$). Calculate the pH of this solution.

16.76 Calculate the molar concentration of OH^- in a 0.724 M solution of hypobromite ion (BrO^-; $K_b = 4.0 \times 10^{-6}$). What is the pH of this solution?

16.77 Ephedrine, a central nervous system stimulant, is used in nasal sprays as a decongestant. This compound is a weak organic base:

$$C_{10}H_{15}ON(aq) + H_2O(l) \rightleftharpoons C_{10}H_{15}ONH^+(aq) + OH^-(aq)$$

A 0.035 M solution of ephedrine has a pH of 11.33. (a) What are the equilibrium concentrations of $C_{10}H_{15}ON$, $C_{10}H_{15}ONH^+$, and OH^-? (b) Calculate K_b for ephedrine.

16.78 Codeine ($C_{18}H_{21}NO_3$) is a weak organic base. A 5.0×10^{-3} M solution of codeine has a pH of 9.95. Calculate the value of K_b for this substance. What is the pK_b for this base?

The K_a – K_b Relationship; Acid–Base Properties of Salt Solutions (Sections 16.8 and 16.9)

16.79 Phenol, C_6H_5OH, has a K_a of 1.3×10^{-10}.

(a) Write out the K_a reaction for phenol.

(b) Calculate K_b for phenol's conjugate base.

(c) Is phenol a stronger or weaker acid than water?

16.80 Use the acid-dissociation constants in Table 16.3 to arrange these oxyanions from strongest to weakest base: SO_4^{2-}, CO_3^{2-}, SO_3^{2-}, and PO_4^{3-}.

16.81 (a) Given that K_a for acetic acid is 1.8×10^{-5} and that for hypochlorous acid is 3.0×10^{-8}, which is the stronger acid? (b) Which is the stronger base, the acetate ion or the hypochlorite ion? (c) Calculate K_b values for CH_3COO^- and ClO^-.

16.82 (a) Given that K_b for ammonia is 1.8×10^{-5} and that for hydroxylamine is 1.1×10^{-8}, which is the stronger base? (b) Which is the stronger acid, the ammonium ion or the hydroxylammonium ion? (c) Calculate K_a values for NH_4^+ and H_3NOH^+.

16.83 Using data from Appendix D, calculate $[OH^-]$ and pH for each of the following solutions: (a) 0.10 M NaBrO, (b) 0.080 M NaHS, (c) a mixture that is 0.10 M in $NaNO_2$ and 0.20 M in $Ca(NO_2)_2$.

16.84 Using data from Appendix D, calculate $[OH^-]$ and pH for each of the following solutions: (a) 0.105 M NaF, (b) 0.035 M Na_2S, (c) a mixture that is 0.045 M in $NaCH_3COO$ and 0.055 M in $Ba(CH_3COO)_2$.

16.85 A solution of sodium acetate ($NaCH_3COO$) has a pH of 9.70. What is the molarity of the solution?

16.86 Pyridinium bromide (C_5H_5NHBr) is a strong electrolyte that dissociates completely into $C_5H_5NH^+$ and Br^-. An aqueous solution of pyridinium bromide has a pH of 2.95.

(a) Write out the reaction that leads to this acidic pH.

(b) Using Appendix D, calculate the K_a for pyridinium bromide.

(c) A solution of pyridinium bromide has a pH of 2.95. What is the concentration of the pyridinium cation at equilibrium, in units of molarity?

16.87 Predict whether aqueous solutions of the following compounds are acidic, basic, or neutral: (a) NH_4Br, (b) $FeCl_3$, (c) Na_2CO_3, (d) $KClO_4$, (e) $NaHC_2O_4$.

16.88 Predict whether aqueous solutions of the following substances are acidic, basic, or neutral: (a) $AlCl_3$, (b) NaBr, (c) NaClO, (d) $[CH_3NH_3]NO_3$, (e) Na_2SO_3.

16.89 Predict which member of each pair produces the more acidic aqueous solution: (a) K^+ or Cu^{2+}, (b) Fe^{2+} or Fe^{3+}, (c) Al^{3+} or Ga^{3+}.

16.90 Which member of each pair produces the more acidic aqueous solution: (a) $ZnBr_2$ or $CdCl_2$ (b) CuCl or $Cu(NO_3)_2$ (c) $Ca(NO_3)_2$ or $NiBr_2$?

16.91 An unknown salt is either NaF, NaCl, or NaOCl. When 0.050 mol of the salt is dissolved in water to form 0.500 L of solution, the pH of the solution is 8.08. What is the identity of the salt?

16.92 An unknown salt is either KBr, NH_4Cl, KCN, or K_2CO_3. If a 0.100 M solution of the salt is neutral, what is the identity of the salt?

Acid–Base Character and Chemical Structure (Section 16.10)

16.93 Predict the stronger acid in each pair: (a) HNO_3 or HNO_2; (b) H_2S or H_2O; (c) H_2SO_4 or H_2SeO_4; (d) CH_3COOH or CCl_3COOH.

16.94 Predict the stronger acid in each pair: (a) HCl or HF; (b) H_3PO_4 or H_3AsO_4; (c) $HBrO_3$ or $HBrO_2$; (d) $H_2C_2O_4$ or $HC_2O_4^-$; (e) benzoic acid (C_6H_5COOH) or phenol (C_6H_5OH).

16.95 Based on their compositions and structures and on conjugate acid–base relationships, select the stronger base in each of the following pairs: (a) BrO^- or ClO^-, (b) BrO^- or BrO_2^-, (c) HPO_4^{2-} or $H_2PO_4^-$.

16.96 Based on their compositions and structures and on conjugate acid–base relationships, select the stronger base in each of the following pairs: (a) NO_3^- or NO_2^-, (b) PO_4^{3-} or AsO_4^{3-}, (c) HCO_3^- or CO_3^{2-}.

16.97 Indicate whether each of the following statements is *true* or *false*. For each statement that is false, correct the statement to make it true. **(a)** In general, the acidity of binary acids increases from left to right in a given row of the periodic table. **(b)** In a series of acids that have the same central atom, acid strength increases with the number of hydrogen atoms bonded to the central atom. **(c)** Hydrotelluric acid (H_2Te) is a stronger acid than H_2S because Te is more electronegative than S.

16.98 Indicate whether each of the following statements is *true* or *false*. For each statement that is false, correct the statement to make it true. **(a)** Acid strength in a series of H—A molecules increases with increasing size of A. **(b)** For acids of the same general structure but differing electronegativities of the central atoms, acid strength decreases with increasing electronegativity of the central atom. **(c)** The strongest acid known is HF because fluorine is the most electronegative element.

Additional Exercises

16.99 Indicate whether each of the following statements is correct or incorrect.

(a) Every Brønsted–Lowry acid is also a Lewis acid.

(b) Every Lewis acid is also a Brønsted–Lowry acid.

(c) Conjugate acids of weak bases produce more acidic solutions than conjugate acids of strong bases.

(d) The K^+ ion is acidic in water because it causes hydrating water molecules to become more acidic.

(e) The percent ionization of a weak acid in water increases as the concentration of acid decreases.

16.100 A solution is made by adding 0.300 g $Ca(OH)_2(s)$, 50.0 mL of 1.40 M HNO_3, and enough water to make a final volume of 75.0 mL. Assuming that all of the solid dissolves, what is the pH of the final solution?

16.101 Which, if any, of the following statements are *true*?

(a) The stronger the base, the smaller the pK_b.

(b) The stronger the base, the larger the pK_b.

(c) The stronger the base, the smaller the K_b.

(d) The stronger the base, the larger the K_b.

(e) The stronger the base, the smaller the pK_a of its conjugate acid.

(f) The stronger the base, the larger the pK_a of its conjugate acid.

16.102 Predict how each molecule or ion would act, in the Brønsted–Lowry sense, in aqueous solution by writing "acid," "base," "both," or "neither" on the line provided.

(a) HCO_3^-, the bicarbonate ion: _____

(b) Prozac: _____

(c) PABA (formerly in sunscreen): _____

(d) TNT, trinitrotoluene: _____

(e) *N*-Methylpyridinium: _____

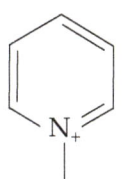

16.103 Calculate the pH of a solution made by adding 2.50 g of lithium oxide (Li_2O) to enough water to make 1.500 L of solution.

16.104 Benzoic acid (C_6H_5COOH) and aniline ($C_6H_5NH_2$) are both derivatives of benzene. Benzoic acid is an acid with $K_a = 6.3 \times 10^{-5}$ and aniline is a base with $K_b = 4.3 \times 10^{-10}$.

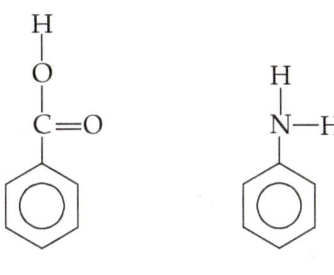

Benzoic acid   Aniline

(a) What are the conjugate base of benzoic acid and the conjugate acid of aniline? **(b)** Anilinium chloride ($C_6H_5NH_3Cl$) is a strong electrolyte that dissociates into anilinium ions ($C_6H_5NH_3^+$) and chloride ions. Which will be more acidic, a 0.10 M solution of benzoic acid or a 0.10 M solution of anilinium chloride? **(c)** What is the value of the equilibrium constant for the following equilibrium?

$$C_6H_5COOH(aq) + C_6H_5NH_2(aq) \rightleftharpoons$$
$$C_6H_5COO^-(aq) + C_6H_5NH_3^+(aq)$$

16.105 What is the pH of a solution that is $2.5 \times 10^{-9} M$ in NaOH? Does your answer make sense? What assumption do we normally make that is not valid in this case?

16.106 Oxalic acid ($H_2C_2O_4$) is a diprotic acid. By using data in Appendix D as needed, determine whether each of the following statements is true: **(a)** $H_2C_2O_4$ can serve as both a Brønsted–Lowry acid and a Brønsted–Lowry base. **(b)** $C_2O_4^{2-}$ is the conjugate base of $HC_2O_4^-$. **(c)** An aqueous solution of the strong electrolyte KHC_2O_4 will have pH < 7.

16.107 Succinic acid ($H_2C_4H_6O_4$), which we will denote H_2Suc, is a biologically relevant diprotic acid with the structure shown below. At 25 °C the acid-dissociation constants for succinic acid are $K_{a1} = 6.9 \times 10^{-5}$ and $K_{a2} = 2.5 \times 10^{-6}$. **(a)** Determine the pH of a 0.32 M solution of H_2Suc at 25 °C, assuming that only the first dissociation is relevant. **(b)** Determine the molar concentration of Suc^{2-} in the solution in part (a). (**c**) Is the assumption you made in part (a) justified by the result from part (b)? (**d**) Will a solution of the salt NaHSuc be acidic, neutral, or basic?

16.108 Butyric acid is responsible for the foul smell of rancid butter. The pK_a of butyric acid is 4.84. (**a**) Calculate the pK_b for the butyrate ion. (**b**) Calculate the pH of a 0.050 M solution of butyric acid. (**c**) Calculate the pH of a 0.050 M solution of sodium butyrate.

16.109 Arrange the following 0.10 M solutions in order of increasing acidity: (i) NH_4NO_3, (ii) $NaNO_3$, (iii) CH_3COONH_4, (iv) NaF, (v) CH_3COONa.

16.110 A 0.25 M solution of a salt NaA has pH = 9.29. What is the value of K_a for the parent acid HA?

16.111 The following observations are made about a diprotic acid H_2A: (i) A 0.10 M solution of H_2A has pH = 3.30. (ii) A 0.10 M solution of the salt NaHA is acidic. Which of the following could be the value of pK_{a2} for H_2A: (a) 3.22, (b) 5.30, (c) 7.47, or (d) 9.82?

16.112 Many moderately large organic molecules containing basic nitrogen atoms are not very soluble in water as neutral molecules, but they are frequently much more soluble as their acid salts. Assuming that pH in the stomach is 2.5, indicate whether each of the following compounds would be present in the stomach as the neutral base or in the protonated form: nicotine, $K_b = 7 \times 10^{-7}$; caffeine, $K_b = 4 \times 10^{-14}$; strychnine, $K_b = 1 \times 10^{-6}$; quinine, $K_b = 1.1 \times 10^{-6}$.

16.113 The amino acid glycine ($H_2N—CH_2—COOH$) can participate in the following equilibria in water:

$$H_2N—CH_2—COOH + H_2O \rightleftharpoons$$
$$H_2N—CH_2—COO^- + H_3O^+ \quad K_a = 4.3 \times 10^{-3}$$

$$H_2N—CH_2—COOH + H_2O \rightleftharpoons$$
$$^+H_3N—CH_2—COOH + OH^- \quad K_b = 6.0 \times 10^{-5}$$

(**a**) Use the values of K_a and K_b to estimate the equilibrium constant for the intramolecular proton transfer to form a zwitterion:

$$H_2N—CH_2—COOH \rightleftharpoons {}^+H_3N—CH_2—COO^-$$

(**b**) What is the pH of a 0.050 M aqueous solution of glycine?

(**c**) What would be the predominant form of glycine in a solution with pH 13? With pH 1?

16.114 The pK_b of water is _____.
(**a**) 1 (**b**) 7 (**c**) 14 (**d**) not defined (**e**) none of these

16.115 Calculate the number of $H^+(aq)$ ions in 1.0 mL of pure water at 25 °C.

16.116 How many milliliters of concentrated hydrochloric acid solution (36.0% HCl by mass, density = 1.18 g/mL) are required to produce 10.0 L of a solution that has a pH of 2.05?

16.117 Atmospheric CO_2 levels have risen by nearly 20% over the past 40 years from 320 ppm to over 400 ppm.

(**a**) Given that the average pH of clean, unpolluted rain today is 5.4, determine the pH of unpolluted rain 40 years ago. Assume that carbonic acid (H_2CO_3) formed by the reaction of CO_2 and water is the only factor influencing pH.

$$CO_2(g) + H_2O(l) \rightleftharpoons H_2CO_3(aq)$$

(**b**) What volume of CO_2 at 25 °C and 1.0 atm is dissolved in a 20.0-L bucket of today's rainwater?

16.119 At 50 °C, the ion-product constant for H_2O has the value $K_w = 5.48 \times 10^{-14}$. (**a**) What is the pH of pure water at 50 °C? (**b**) Based on the change in K_w with temperature, predict whether ΔH is positive, negative, or zero for the autoionization reaction of water:

$$2 H_2O(l) \rightleftharpoons H_3O^+(aq) + OH^-(aq)$$

16.120 In many reactions, the addition of $AlCl_3$ produces the same effect as the addition of H^+.

(**a**) Draw a Lewis structure for $AlCl_3$ in which no atoms carry formal charges, and determine its structure using the VSEPR method.

(**b**) What characteristic is notable about the structure in part (a) that helps us understand the acidic character of $AlCl_3$?

(**c**) Predict the result of the reaction between $AlCl_3$ and NH_3 in a solvent that does not participate as a reactant.

(**d**) Which acid–base theory is most suitable for discussing the similarities between $AlCl_3$ and H^+?

Design an Experiment

Your professor gives you a bottle that contains a clear liquid. You are told that the liquid is a pure substance that is volatile, soluble in water, and might be an acid or a base. Design experiments to obtain the following information about this unknown sample. (**a**) Determine whether the substance in the sample is an acid or a base. (**b**) Suppose the substance is an acid. How would you determine whether it is a strong acid or a weak acid? (**c**) If the substance is a weak acid, how would you determine the value of K_a for the substance? (**d**) Suppose that the substance were a weak acid and that you are also given a solution of NaOH(aq) of known molarity. What procedure would you use to isolate a pure sample of the sodium salt of the substance? (**e**) Now suppose that the substance were a base rather than an acid. How would you adjust the procedures in parts (b) and (c) to determine if the substance were a strong or weak base, and, if weak, the value of K_b?

AQUEOUS EQUILIBRIA: BUFFERS, TITRATIONS, AND SOLUBILITY

▲ **GREAT BARRIER REEF** These structures are made of calcium carbonate, $CaCO_3$.

WHAT'S AHEAD

17.1 ▶ The Common-Ion Effect Consider a specific example of Le Châtelier's principle known as the common-ion effect, and use it to calculate equilibrium concentrations of reagents in solution.

17.2 ▶ Buffers Recognize the composition of buffered solutions and learn how they resist pH change when small amounts of a strong acid or strong base are added to them.

17.3 ▶ Acid–Base Titrations Examine acid–base titrations and explore how to determine pH at any point in an acid–base titration.

17.4 ▶ Solubility Equilibria Use *solubility-product constants* to determine to what extent a sparingly soluble salt dissolves in water.

17.5 ▶ Factors That Affect Solubility Investigate some of the factors that affect solubility, including the common-ion effect and the effect of acids.

17.6 ▶ Precipitation and Separation of Ions Learn how differences in solubility can be used to separate ions through selective precipitation.

17.7 ▶ Qualitative Analysis for Metallic Elements Use the principles of solubility and complexation equilibria to identify ions in solution.

Water, the most common and most important solvent on Earth, occupies its position of importance because of its abundance and its exceptional ability to dissolve a wide variety of substances. Coral reefs are a striking example of aqueous chemistry at work in nature. Coral reefs are built by tiny animals called stony corals, which secrete a hard calcium carbonate exoskeleton. Over time, the stony corals build up large networks of calcium carbonate on which a reef is built. The size of such structures can be immense, as illustrated by the Great Barrier Reef.

To understand the chemistry that underlies coral reef formation and other processes in the ocean and in aqueous systems such as living cells, we must develop a deeper understanding of the concepts of aqueous equilibria. In this chapter, we consider not only acid–base equilibria, in which there is a single solute, but also equilibria containing a mixture of solutes. We then broaden our discussion to include two additional types of aqueous equilibria: those involving slightly soluble salts and those involving the formation of metal complexes in solution. The discussions and calculations in this chapter are extensions of those in Chapters 15 and 16.

17.1 | The Common-Ion Effect

▲ **Learning Objectives**

When you finish Section 17.1, you should be able to:

▶ Determine the equilibrium concentrations of species in solutions containing multiple solutes with a common ion.

▶ Calculate the pH of a solution involving a common ion.

In Chapter 16, we examined the equilibrium concentrations of ions in solutions containing a weak acid or a weak base. We now consider solutions that contain a weak acid, such as acetic acid (CH_3COOH), and a soluble salt of that acid, such as sodium acetate (CH_3COONa). These solutions contain two substances that share a *common ion*, CH_3COO^-. It is instructive to view these solutions from the perspective of Le Châtelier's principle. (Section 15.7)

Sodium acetate is a soluble ionic compound and therefore a strong electrolyte. (Section 4.1) Consequently, it dissociates completely in aqueous solution to form Na^+ and CH_3COO^- ions:

$$CH_3COONa(aq) \longrightarrow Na^+(aq) + CH_3COO^-(aq)$$

In contrast, CH_3COOH is a weak electrolyte that ionizes only partially, represented by the dynamic equilibrium

$$CH_3COOH(aq) \rightleftharpoons H^+(aq) + CH_3COO^-(aq) \qquad [17.1]$$

The equilibrium constant for Equation 17.1 is $K_a = 1.8 \times 10^{-5}$ at 25 °C (Table 16.2). If we add sodium acetate to a solution of acetic acid in water, the CH_3COO^- from CH_3COONa causes the equilibrium concentrations of the substances in Equation 17.1 to shift to the left, as expected from Le Châtelier's principle, thereby decreasing the equilibrium concentration of $H^+(aq)$:

$$CH_3COOH(aq) \rightleftharpoons H^+(aq) + CH_3COO^-(aq)$$

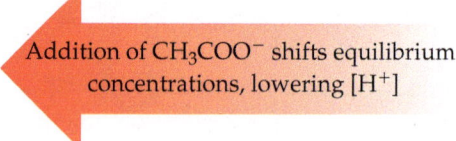

Addition of CH_3COO^- shifts equilibrium concentrations, lowering $[H^+]$

In other words, the presence of the added acetate ion causes the acetic acid to ionize less than it normally would. We call this observation the **common-ion effect**.

> *Whenever a weak electrolyte and a strong electrolyte containing a common ion are together in solution, the weak electrolyte ionizes less than it would if it were alone in solution.*

Note that the equilibrium constant itself does not change; instead, it is the relative concentrations of products and reactants in the equilibrium expression that change. We verify this prediction in Sample Exercises 17.1 and 17.2.

 Sample Exercise 17.1

Calculating the pH When a Common Ion Is Involved

What is the pH of a solution made by adding 0.30 mol of acetic acid and 0.30 mol of sodium acetate to enough water to make 1.0 L of solution?

SOLUTION

Analyze We are asked to determine the pH of a solution of a weak electrolyte (CH_3COOH) and a strong electrolyte (CH_3COONa) that share a common ion, CH_3COO^-.

Plan In any problem in which we must determine the pH of a solution containing a mixture of solutes, it is helpful to proceed by a series of logical steps:

1. Consider which solutes are strong electrolytes and which are weak electrolytes, and identify the major species in solution.

2. Identify the important equilibrium reaction that is the source of H^+ and therefore determines pH.

3. Tabulate the concentrations of ions involved in the equilibrium.

4. Use the equilibrium-constant expression to calculate $[H^+]$ and then pH.

Solve

First, because CH_3COOH is a weak electrolyte and CH_3COONa is a strong electrolyte, the major species in the solution are CH_3COOH (a weak acid), Na^+ (which is neither acidic nor basic and is therefore a spectator in the acid–base chemistry), and CH_3COO^- (which is the conjugate base of CH_3COOH).

Second, $[H^+]$ and, therefore, the pH are controlled by the dissociation equilibrium of CH_3COOH:

$$CH_3COOH(aq) \rightleftharpoons H^+(aq) + CH_3COO^-(aq)$$

[We have written the equilibrium using $H^+(aq)$ rather than $H_3O^+(aq)$, but both representations of the hydrated hydrogen ion are equally valid.]

Third, we tabulate the initial and equilibrium concentrations as we did in solving other equilibrium problems in Chapters 15 and 16:

The equilibrium concentration of CH_3COO^- (the common ion) is the initial concentration that is due to CH_3COONa (0.30 M) plus the change in concentration (x) that is due to the ionization of CH_3COOH.

$CH_3COOH(aq) \rightleftharpoons$	$H^+(aq) +$	$CH_3COO^-(aq)$
Initial (M) 0.30	0	0.30
Change (M) $-x$	$+x$	$+x$
Equilibrium (M) $(0.30 - x)$	x	$(0.30 + x)$

Now we can use the equilibrium-constant expression:

$$K_a = 1.8 \times 10^{-5} = \frac{[H^+][CH_3COO^-]}{[CH_3COOH]}$$

The dissociation constant for CH_3COOH at 25 °C is from Table 16.2, or Appendix D; addition of CH_3COONa does *not* change the value of this constant. Substituting the equilibrium-constant concentrations from our table into the equilibrium expression gives:

$$K_a = 1.8 \times 10^{-5} = \frac{x(0.30 + x)}{0.30 - x}$$

Because K_a is small, we assume that x is small compared to the original concentrations of CH_3COOH and CH_3COO^- (0.30 M each). Thus, we can ignore the very small x relative to 0.30 M, giving:

$$K_a = 1.8 \times 10^{-5} = \frac{x(0.30)}{0.30}$$

The resulting value of x is indeed small relative to 0.30, justifying the approximation made in simplifying the problem.

$$x = 1.8 \times 10^{-5} M = [H^+]$$

Finally, we calculate the pH from the equilibrium concentration of $H^+(aq)$:

$$pH = -\log(1.8 \times 10^{-5}) = 4.74$$

Comment In Section 16.6, we calculated that a 0.30 M solution of CH_3COOH has a pH of 2.64, corresponding to $[H^+] = 2.3 \times 10^{-3} M$. Thus, the addition of CH_3COONa has substantially decreased $[H^+]$ (that is, substantially increased pH), as we expect from Le Châtelier's principle.

▶ **Practice Exercise**

Calculate the pH of a solution containing 0.085 M nitrous acid (HNO_2, $K_a = 4.5 \times 10^{-4}$) and 0.10 M potassium nitrite (KNO_2).

Sample Exercise 17.2

Calculating Ion Concentrations When a Common Ion Is Involved

Calculate the fluoride ion concentration and pH of a solution that is 0.20 M in HF and 0.10 M in HCl.

SOLUTION

Analyze We are asked to determine the concentration of F^- and the pH in a solution containing the weak acid HF and the strong acid HCl. In this case the common ion is H^+.

Plan We can again use the four steps outlined in Sample Exercise 17.1.

Continued

Solve

Because HF is a weak acid and HCl is a strong acid, the major species in solution are HF, H^+, and Cl^-. The Cl^-, which is the conjugate base of a strong acid, is merely a spectator ion in any acid–base chemistry. The problem asks for $[F^-]$, which is formed by ionization of HF. Thus, the important equilibrium is

$$HF(aq) \rightleftharpoons H^+(aq) + F^-(aq)$$

The common ion in this problem is the hydrogen (or hydronium) ion. Now we can tabulate the initial and equilibrium concentrations of each species involved in this equilibrium:

	$HF(aq)$ $\rightleftharpoons$	$H^+(aq)$ +	$F^-(aq)$
Initial (M)	0.20	0.10	0
Change (M)	$-x$	$+x$	$+x$
Equilibrium (M)	$(0.20 - x)$	$(0.10 + x)$	x

The equilibrium constant for the ionization of HF, from Appendix D, is 6.8×10^{-4}. Substituting the equilibrium concentrations and equilibrium constant into the equilibrium expression gives:

$$K_a = 6.8 \times 10^{-4} = \frac{[H^+][F^-]}{[HF]} = \frac{(0.10 + x)(x)}{0.20 - x}$$

If we assume that x is small relative to 0.10 or 0.20 M, this expression simplifies to:

$$\frac{(0.10)(x)}{0.20} = 6.8 \times 10^{-4}$$

$$x = \frac{0.20}{0.10}(6.8 \times 10^{-4}) = 1.4 \times 10^{-3} M = [F^-]$$

Because 1.4×10^{-3} is much smaller than 0.10 and 0.20, our assumption about x was valid; otherwise, we would have had to solve for x using the quadratic equation. The calculated F^- concentration is substantially smaller than it would be in 0.20 M solution of HF with no added HCl. The common ion, H^+, suppresses the ionization of HF. The concentration of $H^+(aq)$ is:

Thus,

$$[H^+] = (0.10 + x) M \approx 0.10 M$$
$$pH = 1.00$$

Comment Notice that for all practical purposes, the hydrogen ion concentration is due entirely to the HCl; the HF makes a negligible contribution by comparison.

▶ **Practice Exercise**

Calculate the formate ion concentration and pH of a solution that is 0.050 M in formic acid (HCOOH, $K_a = 1.8 \times 10^{-4}$) and 0.10 M in HNO_3.

Sample Exercises 17.1 and 17.2 both involve weak acids. The ionization of a weak base is also decreased by the addition of a common ion. For example, the addition of NH_4^+ (as from the strong electrolyte NH_4Cl) causes the equilibrium concentrations of the reagents to shift to the left, decreasing the equilibrium concentration of OH^- and lowering the pH:

$$NH_3(aq) + H_2O(l) \rightleftharpoons NH_4^+(aq) + OH^-(aq)$$

Addition of $NH_4{}^+$ shifts equilibrium concentrations, lowering $[OH^-]$

▲ Self-Assessment Exercises

SAE 17.1 Consider the equilibrium of a generic weak acid in solution: $HA(aq) + H_2O(l) \rightleftharpoons A^-(aq) + H_3O^+(aq)$. If a highly soluble salt containing the anion A^- is added to the solution, which of the following statements is *true*?

   (**i**) The equilibrium constant for this reaction, K_a, will decrease.
   (**ii**) The concentration of A^- will decrease.
   (**iii**) The pH of the solution will decrease.

(**a**) only i (**b**) only ii (**c**) only iii (**d**) both ii and iii (**e**) None of the three statements is true.

SAE 17.2 What is the pH of a solution made by mixing 50.0 mL of a 0.20 M solution of HNO_2 ($K_a = 4.5 \times 10^{-4}$) with 20.0 mL of a 0.46 M $NaNO_2$ solution? (**a**) 2.02 (**b**) 2.10 (**c**) 3.31 (**d**) 3.71

17.2 | Buffers

Solutions that contain high concentrations (10^{-3} M or more) of a weak conjugate acid–base pair and that resist drastic changes in pH when small amounts of strong acid or strong base are added to them are called **buffered solutions** (or merely **buffers**). Human blood, for example, is a complex buffered solution that maintains the blood pH at about 7.4 (see the *Chemistry and Life* box, "Blood as a Buffered Solution" later in this section). Much of the chemical behavior of seawater is determined by its pH, buffered to 8.1 near the surface due to the HCO_3^-/CO_3^{2-} acid–base pair. Buffers find many important applications in the laboratory and in medicine (**Figure 17.1**). Many biological reactions occur at the optimal rates only when properly buffered. If you ever work in a biochemistry lab, you will very likely have to prepare specific buffers in which to run your biochemical reactions.

Composition and Action of Buffers

A buffer resists changes in pH because it contains both an acid to neutralize added OH^- ions and a base to neutralize added H^+ ions. The acid and base that make up the buffer, however, must not consume each other through a neutralization reaction. (Section 4.3) These requirements are fulfilled by a weak acid–base conjugate pair, such as CH_3COOH/CH_3COO^- or NH_4^+/NH_3. The key is to have roughly equal concentrations of both the weak acid and its conjugate base. There are two ways to make a buffer:

- Mix a weak acid or a weak base with a salt of that acid or base. For example, the CH_3COOH/CH_3COO^- buffer can be prepared by adding CH_3COONa to a solution of CH_3COOH. Similarly, the NH_4^+/NH_3 buffer can be prepared by adding NH_4Cl to a solution of NH_3.

- Make the conjugate acid or base from a solution of weak base or acid by the addition of a strong acid or base. For example, to make the CH_3COOH/CH_3COO^- buffer, you could start with a solution of CH_3COOH and add some NaOH to the solution—enough to neutralize about half of CH_3COOH according to the reaction

$$CH_3COOH(aq) + OH^-(aq) \longrightarrow CH_3COO^-(aq) + H_2O(l)$$

(Section 4.3) Neutralization reactions have very large equilibrium constants, and so the amount of acetate formed will only be limited by the relative amounts of the acid and strong base that are mixed. The resulting solution is the same as if you added sodium acetate to the acetic acid solution: You will have comparable quantities of both acetic acid and its conjugate base in solution.

By choosing appropriate components and adjusting their relative concentrations, we can buffer a solution at virtually any pH.

To understand how a buffer works, let's consider one composed of a weak acid HA and one of its salts MA, where M^+ could be Na^+, K^+, or any other cation that does not react with water. The acid-dissociation equilibrium in this buffered solution involves both the acid and its conjugate base:

$$HA(aq) \rightleftharpoons H^+(aq) + A^-(aq)$$

The corresponding acid-dissociation-constant expression is

$$K_a = \frac{[H^+][A^-]}{[HA]} \qquad [17.2]$$

Solving this expression for $[H^+]$, we have

$$[H^+] = K_a \frac{[HA]}{[A^-]} \qquad [17.3]$$

Learning Objectives

When you finish **Section 17.2**, you should be able to:

▶ Identify buffered solutions from their composition.

▶ Calculate the pH and the concentrations of the reagents in buffered solutions.

▶ Calculate the concentrations of reagents needed to prepare a buffer with a particular pH.

▶ Calculate the pH after acid or base is added to a buffered solution.

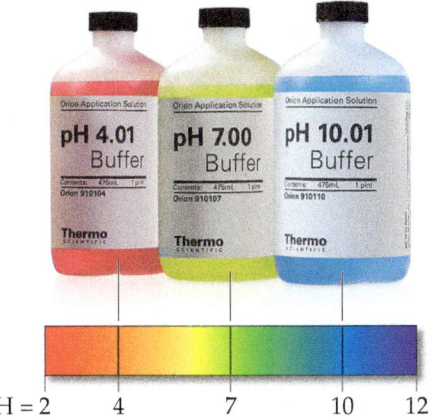

▲ **Figure 17.1** **Standard buffers.** For laboratory work, prepackaged buffers at specific pH values can be purchased.

We see from this expression that $[H^+]$ and, thus, the pH are determined by two factors:

- the value of K_a for the weak-acid component of the buffer;
- the ratio of the equilibrium concentrations of the conjugate acid–base pair, $[HA]/[A^-]$.

If OH^- ions are added to this buffered solution, they react with the buffer acid component to produce water and A^-:

$$OH^-(aq) + \underset{\text{added base}}{HA(aq)} \longrightarrow H_2O(l) + A^-(aq)$$

This neutralization reaction causes $[HA]$ to decrease and $[A^-]$ to increase. As long as the amounts of HA and A^- in the buffer are large relative to the amount of OH^- added, the ratio $[HA]/[A^-]$ does not change much and, thus, the change in pH is small.

If H^+ ions are added, they react with the base component of the buffer:

$$H^+(aq) + \underset{\text{added acid}}{A^-(aq)} \longrightarrow HA(aq)$$

This reaction can also be represented using H_3O^+:

$$H_3O^+(aq) + A^-(aq) \longrightarrow HA(aq) + H_2O(l)$$

Using either chemical equation, we see that this reaction causes $[A^-]$ to decrease and $[HA]$ to increase. As long as the change in the ratio $[HA]/[A^-]$ is small, the change in pH will be small.

An example of an HA/A^- buffer is the CH_3COOH/CH_3COO^- buffered solution shown in **Figure 17.2**. The buffer consists of equal concentrations of acetic acid, CH_3COOH, and acetate ion, CH_3COO^- (center). The addition of OH^- reduces $[CH_3COOH]$ and slightly increases $[CH_3COO^-]$, whereas the addition of H^+ reduces $[CH_3COO^-]$ and slightly increases $[CH_3COOH]$.

It is possible to overwhelm a buffer by adding too much strong acid or strong base. We examine this issue in more detail a little later in this chapter.

Calculating the pH of a Buffer

Because conjugate acid–base pairs share a common ion, we can use the same procedures to calculate the pH of a buffer that we used to treat the common-ion effect in Sample Exercise 17.1. Alternatively, we can take an approach based on an equation derived from Equation 17.3. Taking the negative logarithm of both sides of Equation 17.3, we have

$$-\log[H^+] = -\log\left(K_a\frac{[HA]}{[A^-]}\right) = -\log K_a - \log\frac{[HA]}{[A^-]}$$

▼ **Figure 17.2 Buffer action.** The pH of a buffered solution containing both a weak acid, CH_3COOH, and its conjugate base, CH_3COO^-, changes by only a small amount in response to addition of an external strong acid or base.

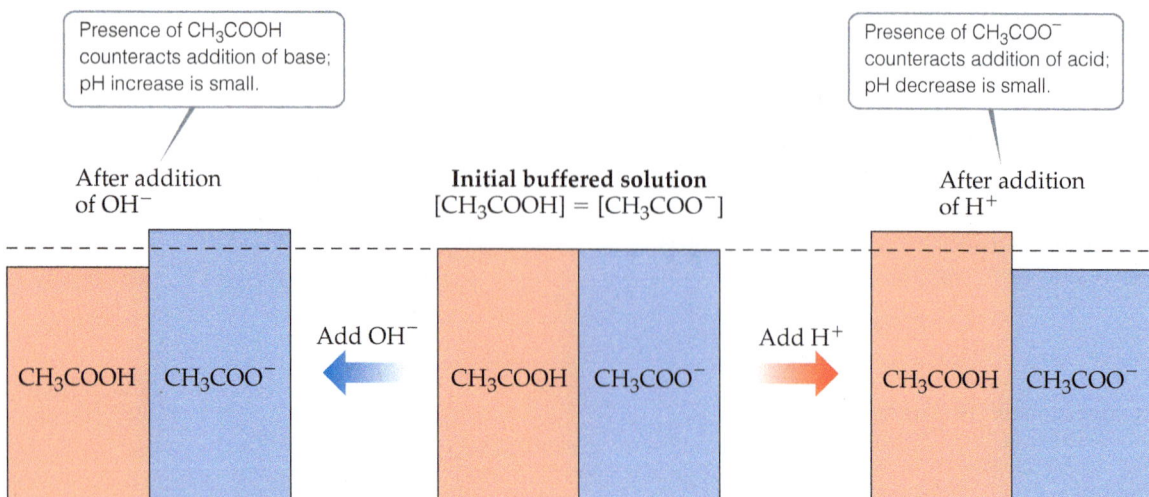

Presence of CH_3COOH counteracts addition of base; pH increase is small.

Presence of CH_3COO^- counteracts addition of acid; pH decrease is small.

After addition of OH^-

Initial buffered solution
$[CH_3COOH] = [CH_3COO^-]$

After addition of H^+

Add OH^- ⟵

Add H^+ ⟶

CH_3COOH | CH_3COO^- CH_3COOH | CH_3COO^- CH_3COOH | CH_3COO^-

$H_2O(l) + CH_3COO^-(aq) \longleftarrow CH_3COOH(aq) + OH^-(aq)$ $CH_3COO^-(aq) + H^+(aq) \longrightarrow CH_3COOH(aq)$

Because $-\log[\text{H}^+] = \text{pH}$ and $-\log K_a = \text{p}K_a$, we have

$$\text{pH} = \text{p}K_a - \log\frac{[\text{HA}]}{[\text{A}^-]} = \text{p}K_a + \log\frac{[\text{A}^-]}{[\text{HA}]} \qquad [17.4]$$

(Remember the logarithm rules in Appendix A.2, if you are not sure how this calculation works.)

In general,

$$\text{pH} = \text{p}K_a + \log\frac{[\text{base}]}{[\text{acid}]} \qquad [17.5]$$

where [acid] and [base] refer to the equilibrium concentrations of the *conjugate acid–base pair*. Note that when $[\text{base}] = [\text{acid}]$, we have $\text{pH} = \text{p}K_a$.

Equation 17.5 is known as the **Henderson–Hasselbalch equation**. Biologists, biochemists, and others who work frequently with buffers often use this equation to calculate the pH of buffers. In doing equilibrium calculations, we have seen that we can normally neglect the amounts of the acid and base of the buffer that ionize. Therefore, we can usually use the *initial* concentrations of the acid and base components of the buffer directly in Equation 17.5, as seen in Sample Exercise 17.3. However, the assumption that the initial concentrations of the acid and base components in the buffer are equal to the equilibrium concentrations is just that: an assumption. There may be times when you will need to be more careful, as seen in Sample Exercise 17.4.

Sample Exercise 17.3

Calculating the pH of a Buffer

What is the pH of a buffer that is 0.12 *M* in lactic acid [$CH_3CH(OH)COOH$, or $HC_3H_5O_3$] and 0.10 *M* in sodium lactate [$CH_3CH(OH)COONa$ or $NaC_3H_5O_3$]? For lactic acid, $K_a = 1.4 \times 10^{-4}$.

SOLUTION

Analyze We are asked to calculate the pH of a buffer containing lactic acid ($HC_3H_5O_3$) and its conjugate base, the lactate ion ($C_3H_5O_3^-$).

Plan We will first determine the pH using the method described in Section 17.1. Because $HC_3H_5O_3$ is a weak electrolyte and

$NaC_3H_5O_3$ is a strong electrolyte, the major species in solution are $HC_3H_5O_3$, Na^+, and $C_3H_5O_3^-$. The Na^+ ion is a spectator ion. The $HC_3H_5O_3/C_3H_5O_3^-$ conjugate acid–base pair determines $[\text{H}^+]$ and, thus, pH; $[\text{H}^+]$ can be determined using the acid-dissociation equilibrium of lactic acid.

Solve

The initial and equilibrium concentrations of the species involved in this equilibrium are:

	$CH_3CH(OH)COOH(aq)$	$\rightleftharpoons$	$H^+(aq)$	$+$	$CH_3CH(OH)COO^-(aq)$
Initial (*M*)	0.12		0		0.10
Change (*M*)	$-x$		$+x$		$+x$
Equilibrium (*M*)	$(0.12 - x)$		x		$(0.10 + x)$

The equilibrium concentrations are governed by the equilibrium expression:

$$K_a = 1.4 \times 10^{-4} = \frac{[\text{H}^+][\text{C}_3\text{H}_5\text{O}_3^-]}{[\text{HC}_3\text{H}_5\text{O}_3]} = \frac{x(0.10 + x)}{0.12 - x}$$

Because K_a is small and a common ion is present, we expect x to be small relative to either 0.12 or 0.10 *M*. Thus, our equation can be simplified to give:

$$K_a = 1.4 \times 10^{-4} = \frac{x(0.10)}{0.12}$$

Solving for x gives a value that justifies our approximation:

$$[\text{H}^+] = x = \left(\frac{0.12}{0.10}\right)(1.4 \times 10^{-4}) = 1.7 \times 10^{-4}\,M$$

Continued

Then, we can solve for pH:

$$pH = -\log(1.7 \times 10^{-4}) = 3.77$$

Alternatively, we can use the Henderson–Hasselbalch equation (Equation 17.5) with the initial concentrations of acid and base to calculate pH directly:

$$pH = pK_a + \log\frac{[\text{base}]}{[\text{acid}]} = 3.85 + \log\left(\frac{0.10}{0.12}\right)$$

$$= 3.85 + (-0.08) = 3.77$$

▶ **Practice Exercise**

Calculate the pH of a buffer composed of 0.12 M benzoic acid and 0.20 M sodium benzoate. (Refer to Appendix D.)

Sample Exercise 17.4

Calculating pH When the Henderson–Hasselbalch Equation May Not Be Accurate

Calculate the pH of a buffer that initially contains 1.00×10^{-3} $M\,CH_3COOH$ and 1.00×10^{-4} $M\,CH_3COONa$ in the following two ways: (i) using the Henderson–Hasselbalch equation and (ii) making no assumptions about quantities (which means you will need to use the quadratic equation). The K_a of CH_3COOH is 1.80×10^{-5}.

SOLUTION

Analyze We are asked to calculate the pH of a buffer in two different ways. We know the initial concentrations of the weak acid and its conjugate base, and the K_a of the weak acid.

Plan We will first use the Henderson–Hasselbalch equation, which relates pK_a and the ratio of acid–base concentrations to the pH. This will be straightforward. Then, we will redo the calculation making no assumptions about any quantities, which means we will need to write out the initial/change/equilibrium concentrations, as we have done before. In addition, we will need to solve for quantities using the quadratic equation (because we cannot make assumptions about unknowns being small).

Solve

(i) The Henderson–Hasselbalch equation is:

$$pH = pK_a + \log\frac{[\text{base}]}{[\text{acid}]}$$

We know the K_a of the acid (1.8×10^{-5}), so we know pK_a ($pK_a = -\log K_a = 4.74$). We know the initial concentrations of the base, sodium acetate, and the acid, acetic acid, which we will assume are the same as the equilibrium concentrations.

Therefore, we have:

$$pH = 4.74 + \log\frac{(1.00 \times 10^{-4})}{(1.00 \times 10^{-3})}$$

$$= 4.74 - 1.00 = 3.74$$

(ii) Now we will redo the calculation, without making any assumptions at all. We will solve for x, which represents the H^+ concentration at equilibrium, in order to calculate pH.

	$CH_3COOH(aq)$	$\rightleftharpoons$	$CH_3COO^-(aq)$	$+$	$H^+(aq)$
Initial (M)	1.00×10^{-3}		1.00×10^{-4}		0
Change (M)	$-x$		$+x$		$+x$
Equilibrium (M)	$(1.00 \times 10^{-3} - x)$		$(1.00 \times 10^{-4} + x)$		x

$$\frac{[CH_3COO^-][H^+]}{[CH_3COOH]} = K_a$$

$$\frac{(1.00 \times 10^{-4} + x)(x)}{(1.00 \times 10^{-3} - x)} = 1.8 \times 10^{-5}$$

$$1.00 \times 10^{-4}x + x^2 = 1.8 \times 10^{-5}(1.00 \times 10^{-3} - x)$$

$$x^2 + 1.00 \times 10^{-4}x = 1.8 \times 10^{-8} - 1.8 \times 10^{-5}x$$

$$x^2 + 1.18 \times 10^{-4}x - 1.8 \times 10^{-8} = 0$$

$$x = \frac{-1.18 \times 10^{-4} \pm \sqrt{(1.18 \times 10^{-4})^2 - 4(1)(-1.8 \times 10^{-8})}}{2(1)}$$

$$= \frac{-1.18 \times 10^{-4} \pm \sqrt{8.5924 \times 10^{-8}}}{2}$$

$$= 8.76 \times 10^{-5} = [H^+]$$

$$pH = 4.06$$

Comment In Sample Exercise 17.3, the calculated pH is the same whether we solve exactly using the quadratic equation or make the simplifying assumption that the equilibrium concentrations of acid and base are equal to their initial concentrations. The simplifying assumption works because the concentrations of the acid–base conjugate pair are both a thousand times larger than K_a. Here in Sample Exercise 17.4, however, the acid–base conjugate pair concentrations are only 10–100 as large as K_a. Therefore, we cannot assume that x is small compared to the initial concentrations (that is, that the initial concentrations are essentially equal to the equilibrium concentrations). The best answer to this Sample Exercise is pH = 4.06, obtained without assuming x is small. Therefore, we see that the assumptions behind the Henderson–Hasselbalch equation are not valid when the initial concentration of the weak acid (or base) is small compared to its K_a (or K_b).

▶ **Practice Exercise**
Calculate the final, equilibrium pH of a buffer that initially contains $6.50 \times 10^{-4}\,M$ HOCl and $7.50 \times 10^{-4}\,M$ NaOCl. The K_a of HOCl is 3.0×10^{-5}.

In Sample Exercise 17.3 we calculated the pH of a buffered solution. Often we will need to work in the opposite direction by calculating the amounts of the acid and its conjugate base needed to achieve a specific pH. This calculation is illustrated in Sample Exercise 17.5.

Sample Exercise 17.5
Preparing a Buffer

How many moles of NH_4Cl must be added to 2.0 L of 0.10 M NH_3 to form a buffer whose pH is 9.00? (Assume that the addition of NH_4Cl does not change the volume of the solution.)

SOLUTION

Analyze We are asked to determine the amount of NH_4^+ ion required to prepare a buffer of a specific pH.

Plan The major species in the solution will be NH_4^+, Cl^-, and NH_3. Of these, the Cl^- ion is a spectator (it is the conjugate base of a strong acid). Thus, the NH_4^+/NH_3 conjugate acid–base pair will determine the pH of the buffer. The equilibrium relationship between NH_4^+ and NH_3 is given by the base-dissociation reaction for NH_3:

$$NH_3(aq) + H_2O(l) \rightleftharpoons NH_4^+(aq) + OH^-(aq)$$

$$K_b = \frac{[NH_4^+][OH^-]}{[NH_3]} = 1.8 \times 10^{-5}$$

The key to this exercise is to use this K_b expression to calculate $[NH_4^+]$.

Solve

We obtain $[OH^-]$ from the given pH:

pOH = 14.00 − pH = 14.00 − 9.00 = 5.00

and so:

$[OH^-] = 1.0 \times 10^{-5}\,M$

Because K_b is small and the common ion $[NH_4^+]$ is present, the equilibrium concentration of NH_3 essentially equals its initial concentration:

$[NH_3] = 0.10\,M$

We now use the expression for K_b to calculate $[NH_4^+]$:

$$[NH_4^+] = K_b \frac{[NH_3]}{[OH^-]} = (1.8 \times 10^{-5})\frac{(0.10)}{(1.0 \times 10^{-5})} = 0.18\,M$$

Thus, for the solution to have pH = 9.00, $[NH_4^+]$ must equal 0.18 M. The number of moles of NH_4Cl needed to produce this concentration is given by the product of the volume of the solution and its molarity:

$(2.0\,L)(0.18\,mol\,NH_4Cl/L) = 0.36\,mol\,NH_4Cl$

Comment Because NH_4^+ and NH_3 are a conjugate acid–base pair, we could use the Henderson–Hasselbalch equation (Equation 17.5) to solve this problem. To do so requires first calculating pK_a for NH_4^+ from the value of pK_b for NH_3. We suggest you try this approach to convince yourself that you can use the Henderson–Hasselbalch equation for buffers for which you are given K_b for the conjugate base rather than K_a for the conjugate acid.

▶ **Practice Exercise**
Calculate the concentration of sodium benzoate that must be present in a 0.20 M solution of benzoic acid (C_6H_5COOH) to produce a pH of 4.00. Refer to Appendix D.

Buffer Capacity and pH Range

Two important characteristics of a buffer are its capacity and its effective pH range. **Buffer capacity** is the amount of acid or base the buffer can neutralize before the pH begins to change to an appreciable degree. The buffer capacity depends on the amount of acid and base used to prepare the buffer. According to Equation 17.3, for example, the pH of a 1-L solution that is 1 M in CH_3COOH and 1 M in CH_3COONa is the same as the pH of a 1-L solution that is 0.1 M in CH_3COOH and 0.1 M in CH_3COONa. The first solution has a greater buffering capacity, however, because it contains more CH_3COOH and CH_3COO^-.

The pH range of any buffer is the pH range over which the buffer acts effectively. Buffers most effectively resist a change in pH in *either* direction when the concentrations of weak acid and conjugate base are about the same. From Equation 17.5 we see that when the concentrations of weak acid and conjugate base are equal, pH = pK_a. This relationship gives the optimal pH of any buffer. Thus, we usually try to select a buffer whose acid form has a pK_a close to the desired pH. In practice, we find that if the concentration of one component of the buffer is more than 10 times the concentration of the other component, the buffering action is poor. Because log 10 = 1, *buffers usually have a usable range within* ± 1 *pH unit of* pK_a (that is, a range of pH = $pK_a \pm 1$).

How Buffers Respond to the Addition of Strong Acids or Bases

Let's now consider in a more quantitative way how a buffered solution responds to addition of a strong acid or base. In this discussion, it is important to understand that *neutralization reactions between strong acids and weak bases proceed essentially to completion, as do those between strong bases and weak acids.* This is because water is a product of the reaction, and you have an equilibrium constant of $1/K_w = 10^{14}$ in your favor when making water. (Section 16.3) Thus, as long as we do not exceed the buffering capacity of the buffer, we can assume that the strong acid or strong base is completely consumed by reaction with the buffer.

Consider a buffer that contains a weak acid HA and its conjugate base A^-. When a strong acid is added to this buffer, the added H^+ is consumed by A^- to produce HA; thus, [HA] increases and [A^-] decreases. Upon addition of a strong base, the added OH^- is consumed by HA to produce A^-; in this case [HA] decreases and [A^-] increases. These two situations are summarized in Figure 17.2, where HA = CH_3COOH and $A^- = CH_3COO^-$.

To calculate how the pH of the buffer responds to the addition of a strong acid or a strong base, we follow the strategy outlined in **Figure 17.3**:

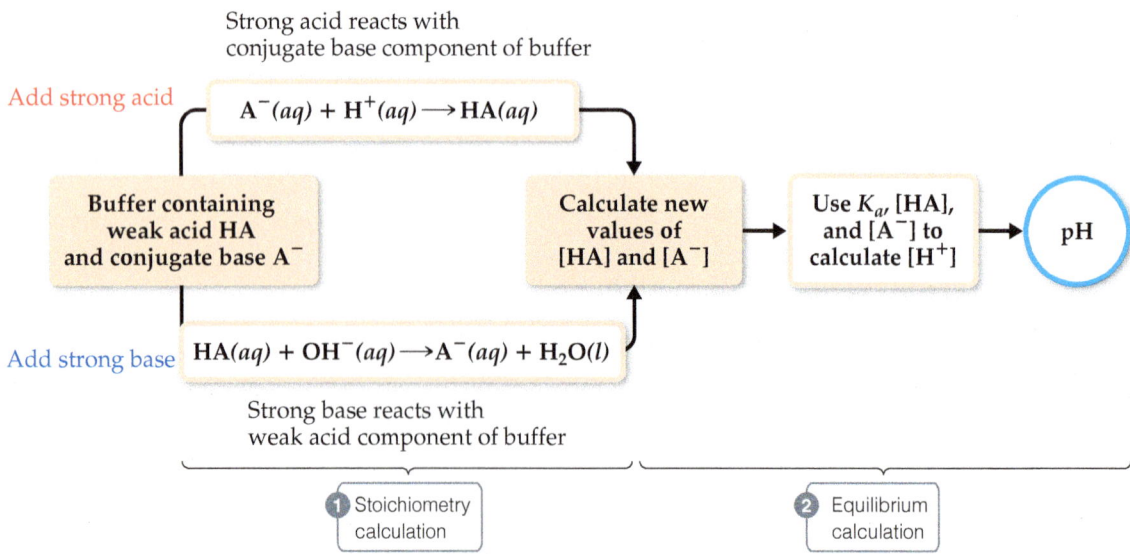

▲ **Figure 17.3 Calculating the pH of a buffer after addition of a strong acid or strong base: react, then equilibrate.**

1. Consider the acid–base neutralization reaction and determine its effect on [HA] and [A⁻]. This step is a *limiting reactant stoichiometry calculation.* (Sections 3.6 and 3.7)

2. Use the calculated values of [HA] and [A⁻] along with K_a to calculate [H⁺]. This step is an *equilibrium calculation* and is most easily done using the Henderson–Hasselbalch equation (if the concentrations of the weak acid–base pair are very large compared to K_a for the acid).

Sample Exercise 17.6
Calculating the pH of a Buffer after Strong Base Is Added

A buffer is made by adding 0.300 mol CH_3COOH and 0.300 mol CH_3COONa to enough water to make 1.000 L of solution. The pH of the buffer is 4.74 (Sample Exercise 17.1). (**a**) Calculate the pH of this solution after 5.0 mL of 4.0 *M* NaOH(*aq*) solution is added. (**b**) For comparison, calculate the pH of a solution made by adding 5.0 mL of 4.0 *M* NaOH(*aq*) solution to 1.000 L of pure water.

SOLUTION

Analyze We are asked to determine the pH of a buffer after addition of a small amount of strong base and to compare the pH change with the pH that would result if we were to add the same amount of strong base to pure water.

Plan Solving this problem involves the two steps outlined in Figure 17.3. First we do a stoichiometry calculation to determine how the added OH⁻ affects the buffer composition. Then we use the resulting buffer composition and either the Henderson–Hasselbalch equation or the equilibrium-constant expression for the buffer to determine the pH.

Solve

(**a**) *Stoichiometry Calculation:* The OH⁻ provided by NaOH reacts with CH_3COOH, the weak acid component of the buffer. Because volumes are changing, it is prudent to figure out how many moles of reactants and products would be produced, then divide by the final volume later to obtain concentrations. Prior to this neutralization reaction, there are 0.300 mol each of CH_3COOH and CH_3COO^-. The amount of base added is 0.0050 L × 4.0 mol/L = 0.020 mol. Neutralizing the 0.020 mol OH⁻ requires 0.020 mol of CH_3COOH. Consequently, the amount of CH_3COOH *decreases* by 0.020 mol, and the amount of the product of the neutralization, CH_3COO^-, *increases* by 0.020 mol. We can create a before/change/after table (Sections 3.6) to see how the composition of the buffer changes as a result of its reaction with OH⁻:

	$CH_3COOH(aq)$ +	$OH^-(aq)$ ⟶	$H_2O(l)$ +	$CH_3COO^-(aq)$
Before reaction (mol)	0.300	0.020	—	0.300
Change (limiting reactant) (mol)	−0.020	−0.020	—	+0.020
After reaction (mol)	0.280	0	—	0.320

Equilibrium Calculation: We now turn our attention to the equilibrium for the ionization of acetic acid, the relationship that determines the buffer pH:

$$CH_3COOH(aq) \rightleftharpoons H^+(aq) + CH_3COO^-(aq)$$

Using the quantities of CH_3COOH and CH_3COO^- remaining in the buffer after the reaction with strong base, we determine the pH using the Henderson–Hasselbalch equation. The volume of the solution is now 1.000 L + 0.0050 L = 1.005 L due to addition of the NaOH solution:

$$pH = 4.74 + \log\frac{0.320\ \text{mol}/1.005\ \text{L}}{0.280\ \text{mol}/1.005\ \text{L}} = 4.80$$

Continued

(b) To determine the pH of a solution made by adding 0.020 mol of NaOH to 1.000 L of pure water, we first determine the concentration of OH^- ions in solution:

$$[OH^-] = 0.020 \text{ mol}/1.005 \text{ L} = 0.020 \, M$$

We use this value in Equation 16.22 to calculate pOH and then use our calculated pOH value to obtain pH:

$$pOH = -\log[OH^-] = -\log(0.020) = +1.70$$
$$pH = 14 - (+1.70) = 12.30$$

1.000 L buffer
0.300 M CH$_3$COOH
0.300 M CH$_3$COO$^-$

Add 5.0 mL of
4.0 M NaOH(aq)

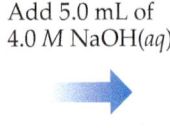

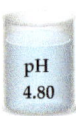

pH increases by
0.06 pH units

(a)

1.000 L H$_2$O

Add 5.0 mL of
4.0 M NaOH(aq)

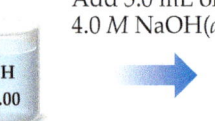

pH increases by
5.30 pH units

(b)

▲ **Figure 17.4　Effect of adding a strong base to (a) a buffered solution and to (b) water.**

Comment Note that the small amount of added NaOH changes the pH of water significantly. In contrast, the pH of the buffer changes very little when the NaOH is added, as summarized in **Figure 17.4**.

▶ **Practice Exercise**
Determine **(a)** the pH of the original buffer described in Sample Exercise 17.6 after the addition of 0.020 mol HCl and **(b)** the pH of the solution that would result from the addition of 0.020 mol HCl to 1.000 L of pure water.

CHEMISTRY AND LIFE | Blood as a Buffered Solution

Chemical reactions that occur in living systems are often extremely sensitive to pH. Many of the enzymes that catalyze important biochemical reactions are effective only within a narrow pH range. For this reason, the human body maintains a remarkably intricate system of buffers, both within cells and in the fluids that transport cells. Blood, the fluid that transports oxygen to all parts of the body, is one of the most prominent examples of the importance of buffers in living beings.

Human blood has a normal pH of 7.35 to 7.45. Any deviation from this range can have extremely disruptive effects on the stability of cell membranes, the structures of proteins, and the activities of enzymes. Death may result if the blood pH falls below 6.8 or rises above 7.8. When the pH falls below 7.35, the condition is called *acidosis*; when it rises above 7.45, the condition is called *alkalosis*. Acidosis is the more common tendency because metabolism generates several acids in the body.

The major buffer system used to control blood pH is the *carbonic acid–bicarbonate buffer system*. Carbonic acid (H_2CO_3) and bicarbonate ion (HCO_3^-) are a conjugate acid–base pair. In addition, carbonic acid decomposes into carbon dioxide gas and water. The important equilibria in this buffer system are

$$H^+(aq) + HCO_3^-(aq) \rightleftharpoons H_2CO_3(aq)$$
$$\rightleftharpoons H_2O(l) + CO_2(g) \quad [17.6]$$

Several aspects of these equilibria are notable. First, although carbonic acid is diprotic, the carbonate ion (CO_3^{2-}) is unimportant in this system. Second, one component of this equilibrium, CO_2, is a gas, which provides a mechanism for the body to adjust the equilibria. Removal of CO_2 via exhalation shifts the equilibria to the right, consuming H^+ ions. Third, the buffer system in blood operates at pH 7.4, which is fairly far removed from the pK_{a1} value of H_2CO_3 (6.1 at physiological temperatures). For the buffer to have a pH of 7.4, the

ratio [base]/[acid] must be about 20. In normal blood plasma, the concentrations of HCO_3^- and H_2CO_3 are about 0.024 M and 0.0012 M, respectively. Consequently, the buffer has a high capacity to neutralize additional acid but only a low capacity to neutralize additional base.

The principal organs that regulate the pH of the carbonic acid–bicarbonate buffer system are the lungs and kidneys. When the concentration of CO_2 rises, the equilibrium concentrations in Equation 17.6 shift to the left, which leads to the formation of more H^+ and a drop in pH. This change is detected by receptors in the brain that trigger a reflex to breathe faster and deeper, increasing the rate at which CO_2 is expelled from the lungs and thereby shifting the equilibrium concentrations in Equation 17.6 back to the right. When the blood pH becomes too high, the kidneys remove HCO_3^- from the blood. This shifts the equilibrium concentrations to the left, increasing the concentration of H^+. As a result, the pH decreases.

Regulation of blood pH relates directly to the effective transport of O_2 throughout the body. The protein hemoglobin, found in red blood cells (**Figure 17.5**), carries oxygen. Hemoglobin (Hb) reversibly binds both O_2 and H^+. These two substances compete for the Hb, which can be represented approximately by the equilibrium

$$HbH^+(aq) + O_2(aq) \rightleftharpoons HbO_2(aq) + H^+(aq) \quad [17.7]$$

Oxygen enters the blood through the lungs, where it passes into the red blood cells and binds to Hb. When the blood reaches tissue in which the concentration of O_2 is low, the equilibrium concentrations in Equation 17.7 shift to the left and O_2 is released.

During periods of strenuous exertion, three factors work together to ensure delivery of O_2 to active tissues. The role of each factor can be understood by applying Le Châtelier's principle to the hemoglobin–O_2 equilibrium:

1. O_2 is consumed, causing the equilibrium concentrations in Equation 17.7 to shift to the left, releasing more O_2.
2. Large amounts of CO_2 are produced by metabolism, which increases $[H^+]$ (Equation 17.6) and causes the equilibrium concentrations in Equation 17.7 to shift to the left, releasing O_2.
3. Body temperature rises. Because the hemoglobin–O_2 reaction is exothermic, the increase in temperature shifts the equilibrium concentrations in Equation 17.7 to the left, releasing O_2.

In addition to the factors causing release of O_2 to tissues, the decrease in pH stimulates an increase in breathing rate, which furnishes more O_2 and eliminates CO_2. Without this elaborate series of equilibrium shifts and pH changes, the O_2 in tissues would be rapidly depleted, making further activity impossible. Under such conditions, the buffering capacity of the blood and the exhalation of CO_2 through the lungs are essential to keep the pH from dropping too low, thereby triggering acidosis.

Related Exercise: 17.29

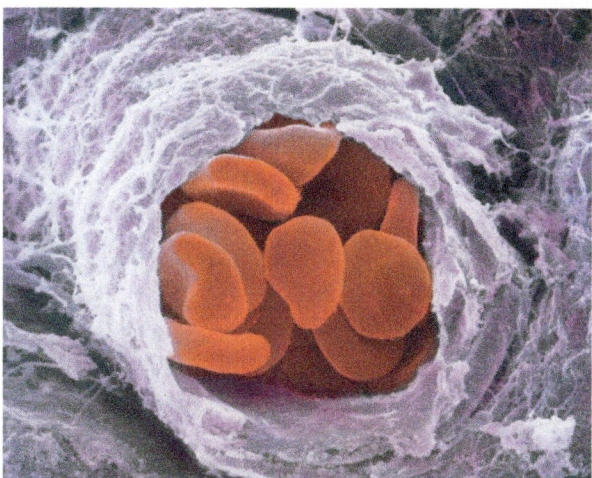

▲ **Figure 17.5 Red blood cells.** A scanning electron micrograph of red blood cells traveling through a small branch of an artery. The red blood cells are approximately 0.010 mm in diameter.

 Self-Assessment Exercises

SAE 17.3 Consider the solutions formed by adding 50 mL of a 1.00 *M* solution of NH_3 to each of the following beakers:

Beaker 1: 50 mL of 2.00 *M* HCl(*aq*)

Beaker 2: 50 mL of 0.50 *M* HCl(*aq*)

Beaker 3: 50 mL of 1.00 *M* NH_4Cl(*aq*)

Which beaker(s) will contain a buffered solution once the mixing is complete? (**a**) beaker 3 (**b**) beakers 1 and 2 (**c**) beakers 1 and 3 (**d**) beakers 2 and 3 (**e**) beakers 1, 2, and 3

SAE 17.4 What is the pH of a buffer formed by adding 0.78 g of sodium benzoate, NaC_6H_5COO (molar mass = 144.1 g/mol), to 150 mL of a 0.88 *M* solution of benzoic acid, C_6H_5COOH ($K_a = 6.3 \times 10^{-5}$)? Neglect any changes in volume that occur upon addition of the salt. (**a**) 3.81 (**b**) 4.20 (**c**) 4.58 (**d**) 5.15

SAE 17.5 Which of the following combinations of reagents would be most appropriate for forming a solution buffered at pH = 10.5? (**a**) hydrofluoric acid, HF(*aq*) ($K_a = 6.8 \times 10^{-4}$), and NaF (**b**) pyridine, C_5H_5N(*aq*) ($K_b = 1.7 \times 10^{-9}$), and C_5H_5NHCl (**c**) methylamine, CH_3NH_2 ($K_b = 4.4 \times 10^{-4}$), and CH_3NH_3Cl; (**d**) sodium hydroxide, NaOH(*aq*), and NaCl.

SAE 17.6 A buffer is made by adding 0.68 g of sodium formate, NaHCOO (molar mass = 68.0 g/mol), to 1.00 L of 0.015 *M* formic acid, HCOOH ($K_a = 1.8 \times 10^{-4}$). The pH of the buffer is 3.57. What will the pH be after 10.0 mL of 0.50 *M* HCl(*aq*) is added to the buffer? (**a**) 2.30 (**b**) 3.14 (**c**) 3.57 (**d**) 3.92 (**e**) 4.34

17.3 | Acid–Base Titrations

Titrations are procedures in which one reactant is slowly added into a solution of another reactant, while equilibrium concentrations along the way are monitored. (Section 4.6) There are two main reasons to do titrations:

- to determine the concentration of one of the reactants
- to determine the equilibrium constant for the reaction

In an acid–base titration, a solution containing a known concentration of base is slowly added to an acid (or the acid is added to the base). (Section 4.6) Acid–base indicators can be used to signal the *equivalence point* of a titration (the point at which stoichiometrically equivalent quantities of acid and base have been brought together). Alternatively, a pH meter can be used to monitor the progress of the reaction (**Figure 17.6**), producing a **pH titration curve**, a graph of the pH as a function of the volume of titrant added. The shape of the titration curve makes it possible to determine the equivalence point. The curve can also be used to select suitable indicators and to determine the K_a of the weak acid or the K_b of the weak base being titrated.

 Learning Objectives

When you finish Section 17.3, you should be able to:

► Interpret a pH titration curve to find the equivalence point.

► Calculate the pH at any point during a strong acid–strong base titration.

► Calculate the pH at any point, including the equivalence point, during a weak acid–strong base (or strong acid–weak base) titration.

► Chose a suitable indicator to estimate the equivalence point in an acid–base titration.

► Interpret a pH titration curve to estimate the pK_a of a weak acid.

Go Figure

In which direction do you expect the pH to change as NaOH is added to the HCl solution?

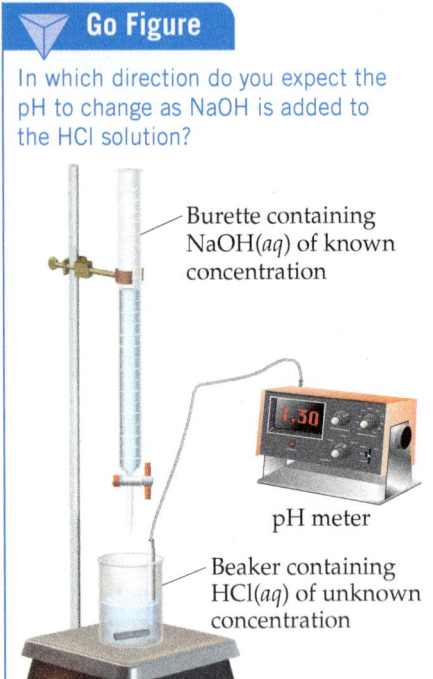

Burette containing NaOH(aq) of known concentration

pH meter

Beaker containing HCl(aq) of unknown concentration

▲ **Figure 17.6 Measuring pH during a titration.**

To understand why titration curves have certain characteristic shapes, we examine the curves for three kinds of titrations: (1) strong acid–strong base, (2) weak acid–strong base, and (3) polyprotic acid–strong base. We also briefly consider how these curves relate to those involving weak bases.

Strong Acid–Strong Base Titrations

The titration curve produced when a strong base is added to a strong acid has the general shape shown in **Figure 17.7**, which depicts the pH change that occurs as 0.100 M NaOH is added to 50.0 mL of 0.100 M HCl. The pH can be calculated at various stages of the titration. To help understand these calculations, we can divide the curve into four regions:

1. **Initial pH:** The pH of the solution before the addition of any base is determined by the initial concentration of the strong acid. For a solution of 0.100 M HCl, $[H^+] = 0.100\ M$ and pH $= -\log(0.100) = 1.000$. Thus, the initial pH is low.

2. **Between initial pH and equivalence point:** As NaOH is added, the pH increases slowly at first and then rapidly in the vicinity of the equivalence point. The pH before the equivalence point is determined by the concentration of acid not yet neutralized. This calculation is illustrated in Sample Exercise 17.7(a).

Go Figure   What volume of NaOH(aq) would be needed to reach the equivalence point if the concentration of the added base were 0.200 M?

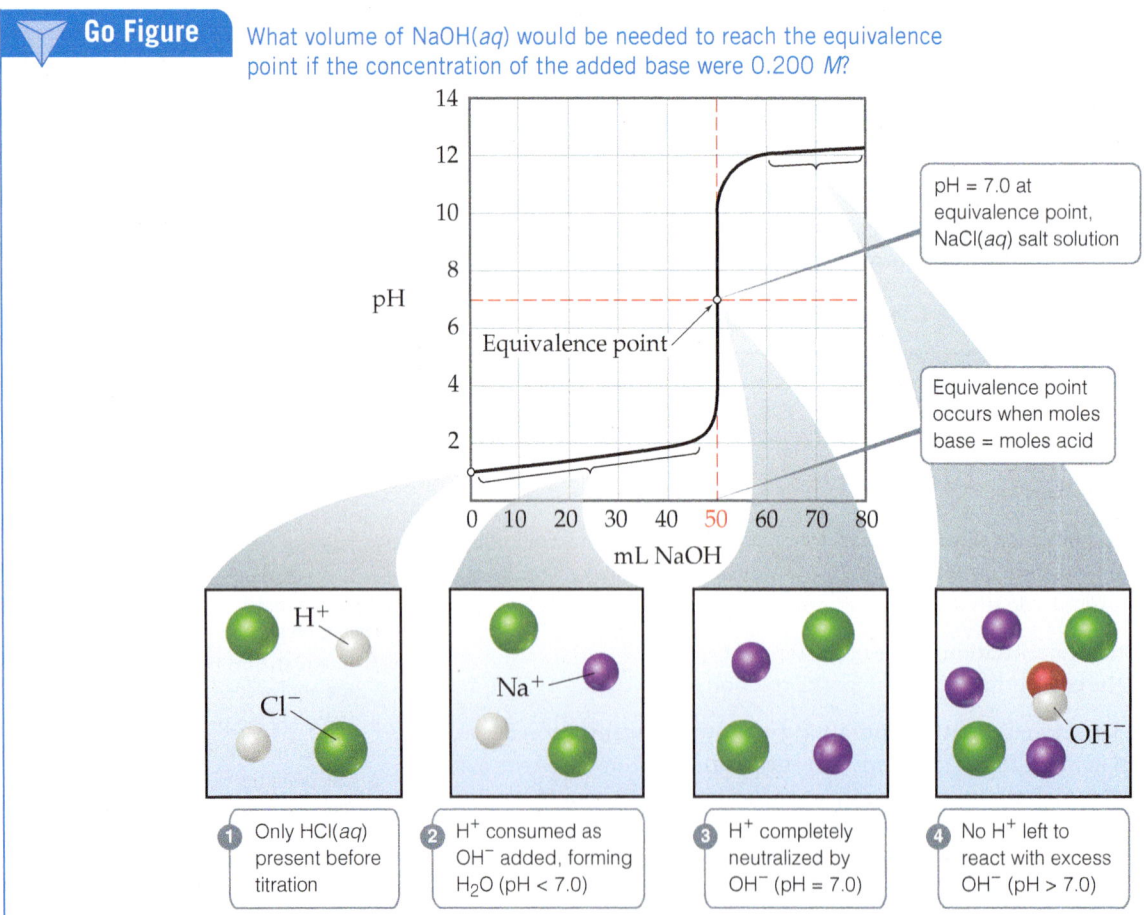

pH = 7.0 at equivalence point, NaCl(aq) salt solution

Equivalence point

Equivalence point occurs when moles base = moles acid

1. Only HCl(aq) present before titration

2. H$^+$ consumed as OH$^-$ added, forming H$_2$O (pH < 7.0)

3. H$^+$ completely neutralized by OH$^-$ (pH = 7.0)

4. No H$^+$ left to react with excess OH$^-$ (pH > 7.0)

▲ **Figure 17.7 Titration of a strong acid with a strong base.** The pH curve for titration of 50.0 mL of a 0.100 M solution of hydrochloric acid with a 0.100 M solution of NaOH(aq). For clarity, water molecules have been omitted from the molecular art.

3. **Equivalence point:** At the equivalence point an equal number of moles of NaOH and HCl have reacted, leaving only a solution of their salt, NaCl. The pH of the solution is 7.00 because the cation of a strong base (in this case Na^+) and the anion of a strong acid (in this case Cl^-) are neither acids nor bases and, therefore, have no appreciable effect on pH. (Section 16.9)

4. **After equivalence point:** The pH of the solution after the equivalence point is determined by the concentration of excess NaOH in the solution. This calculation is illustrated in Sample Exercise 17.7(b).

Sample Exercise 17.7
Calculations for a Strong Acid–Strong Base Titration

Calculate the pH when **(a)** 49.0 mL and **(b)** 51.0 mL of 0.100 M NaOH solution have been added to 50.0 mL of 0.100 M HCl solution.

SOLUTION

Analyze We are asked to calculate the pH at two points in the titration of a strong acid with a strong base. The first point is just before the equivalence point, so we expect the pH to be determined by the small amount of strong acid that has not yet been neutralized. The second point is just after the equivalence point, so we expect this pH to be determined by the small amount of excess strong base.

Plan (a) As the NaOH solution is added to the HCl solution, $H^+(aq)$ reacts with $OH^-(aq)$ to form H_2O. Both Na^+ and Cl^- are spectator ions, having negligible effect on the pH. To determine the pH of the solution, we must first determine how many moles

of H^+ were originally present and how many moles of OH^- were added. We can then calculate how many moles of each ion remain after the neutralization reaction. To calculate $[H^+]$, and hence pH, we must also remember that the volume of the solution increases as we add titrant, thus diluting the concentration of all solutes present. Therefore, it is best to deal with moles first, and then convert to molarities using total solution volumes (volume of acid plus volume of base). **(b)** We proceed in the same way as in part (a) except we are now past the equivalence point and have more OH^- in the solution than H^+.

Solve

(a) The number of moles of H^+ in the original HCl solution is given by the product of the volume of the solution and its molarity:

$$(0.0500 \text{ L soln})\left(\frac{0.100 \text{ mol } H^+}{1 \text{ L soln}}\right) = 5.00 \times 10^{-3} \text{ mol } H^+$$

Likewise, the number of moles of OH^- in 49.0 mL of 0.100 M NaOH is:

$$(0.0490 \text{ L soln})\left(\frac{0.100 \text{ mol } OH^-}{1 \text{ L soln}}\right) = 4.90 \times 10^{-3} \text{ mol } OH^-$$

Because we have not reached the equivalence point, there are more moles of H^+ present than OH^-. Therefore, OH^- is the limiting reactant. Each mole of OH^- reacts with 1 mol of H^+. Using the convention introduced in Sample Exercise 17.6, we have:

	$H^+(aq)$	$+$ $OH^-(aq)$	$\longrightarrow$ $H_2O(l)$
Before reaction (mol)	5.00×10^{-3}	4.90×10^{-3}	—
Change (limiting reactant) (mol)	-4.90×10^{-3}	-4.90×10^{-3}	—
After reaction (mol)	0.10×10^{-3}	0	—

The volume of the reaction mixture increases as the NaOH solution is added to the HCl solution. Thus, at this point in the titration, the volume in the titration flask is:

$$50.0 \text{ mL} + 49.0 \text{ mL} = 99.0 \text{ mL} = 0.0990 \text{ L}$$

Thus, the concentration of $H^+(aq)$ in the flask is:

$$[H^+] = \frac{\text{moles } H^+(aq)}{\text{liters soln}} = \frac{0.10 \times 10^{-3} \text{ mol}}{0.09900 \text{ L}} = 1.0 \times 10^{-3} M$$

The corresponding pH is:

$$-\log(1.0 \times 10^{-3}) = 3.00$$

(b) As before, the initial number of moles of each reactant is determined from their volumes and concentrations. The reactant present in a smaller stoichiometric amount (the limiting reactant) is consumed completely, leaving an excess of hydroxide ion.

	$H^+(aq)$	$+$ $OH^-(aq)$	$\longrightarrow$ $H_2O(l)$
Before reaction (mol)	5.00×10^{-3}	5.10×10^{-3}	—
Change (limiting reactant) (mol)	-5.00×10^{-3}	-5.00×10^{-3}	—
After reaction (mol)	0	0.10×10^{-3}	—

Continued

In this case, the volume in the titration flask is: Hence, the concentration of $OH^-(aq)$ in the flask is:

$$50.0 \text{ mL} + 51.0 \text{ mL} = 101.0 \text{ mL} = 0.1010 \text{ L}$$

$$[OH^-] = \frac{\text{moles } OH^-(aq)}{\text{liters soln}} = \frac{0.10 \times 10^{-3} \text{ mol}}{0.1010 \text{ L}} = 1.0 \times 10^{-3} M$$

and we have:

$$pOH = -\log(1.0 \times 10^{-3}) = 3.00$$

$$pH = 14.00 - pOH = 14.00 - 3.00 = 11.00$$

Comment Note that the pH increased by only two pH units, from 1.00 (Figure 17.7) to 3.00, after the first 49.0 mL of NaOH solution was added, but jumped by eight pH units, from 3.00 to 11.00, as 2.0 mL of base solution was added near the equivalence point. Such a rapid rise in pH near the equivalence point is a characteristic of titrations involving strong acids and strong bases.

▶ **Practice Exercise**
Calculate the pH when (**a**) 24.9 mL and (**b**) 25.1 mL of 0.100 *M* HNO_3 have been added to 25.0 mL of 0.100 *M* KOH solution.

Titration of a solution of a strong base with a solution of a strong acid yields an analogous curve of pH versus added acid. In this case, however, the pH is high at the outset of the titration and low at its completion (**Figure 17.8**). The pH at the equivalence point is still 7.0 (at 25 °C), just like the strong acid–strong base titration.

Weak Acid–Strong Base Titrations

The curve for titration of a weak acid by a strong base is similar but not identical to the curve in Figure 17.7. Consider, for example, the curve for titration of 50.0 mL of 0.100 *M* acetic acid with 0.100 *M* NaOH shown in **Figure 17.9**. We can calculate the pH at points along this curve, using principles we discussed previously, which means again dividing the curve into four regions:

1. **Initial pH:** We use K_a to calculate this pH, as shown in Section 16.6. The calculated pH of a 0.100 *M* CH_3COOH solution is 2.89.

2. **Between initial pH and equivalence point:** Prior to reaching the equivalence point, the acid is being neutralized, and its conjugate base is being formed:

$$CH_3COOH(aq) + OH^-(aq) \longrightarrow CH_3COO^-(aq) + H_2O(l)$$

Thus, the solution contains a mixture of CH_3COOH and CH_3COO^-, which makes it a buffered solution. Calculating the pH in this region involves two steps. First, we consider the neutralization reaction between CH_3COOH and OH^- to determine $[CH_3COOH]$ and $[CH_3COO^-]$. Next, we calculate the pH of this buffer pair using the equilibration procedures developed in Sections 17.1 and 17.2. The general procedure is diagrammed in **Figure 17.10** and illustrated in Sample Exercise 17.7.

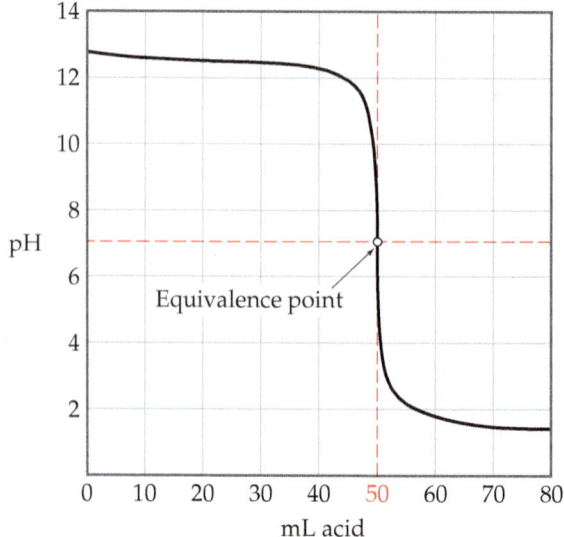

▲ **Figure 17.8 Titration of a strong base with a strong acid.** The pH curve for titration of 50.0 mL of a 0.100 *M* solution of a strong base with a 0.100 *M* solution of a strong acid.

Go Figure

If the acetic acid being titrated here were replaced by hydrochloric acid, would the amount of base needed to reach the equivalence point change? Would the pH at the equivalence point change?

pH change near the equivalence point is smaller than in strong acid–strong base titration.
14
12
10
8
6
4
2
pH
Equivalence point
0 10 20 30 40 50 60 70 80
mL NaOH
pH > 7.0 at equivalence point because CH_3COO^- is a weak base that reacts with water to make OH^-.
Equivalence point occurs when moles base = moles acid.
CH_3COOH
CH_3COO^-
Na^+
OH^-
1 $CH_3COOH(aq)$ solution before titration
2 Added OH^- converts $CH_3COOH(aq)$ into $CH_3COO^-(aq)$, forming buffer solution.
3 Acid completely neutralized by added base, $CH_3COONa(aq)$ salt solution results
4 No acid left to react with excess OH^-

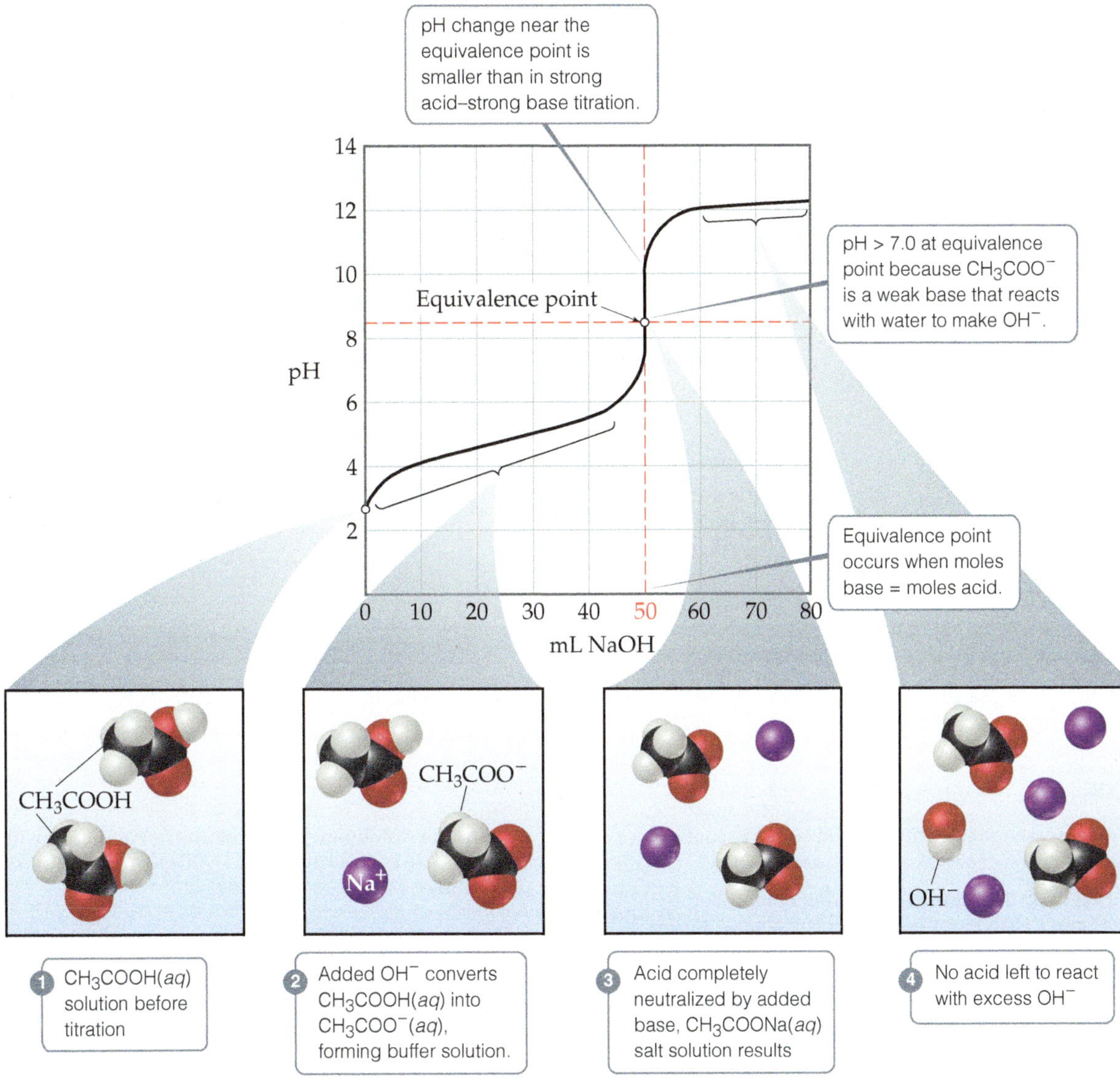

▲ **Figure 17.9 Titration of a weak acid with a strong base.** The pH curve for titration of 50.0 mL of a 0.100 *M* solution of acetic acid with a 0.100 *M* solution of NaOH(*aq*). For clarity, water molecules have been omitted from the molecular art.

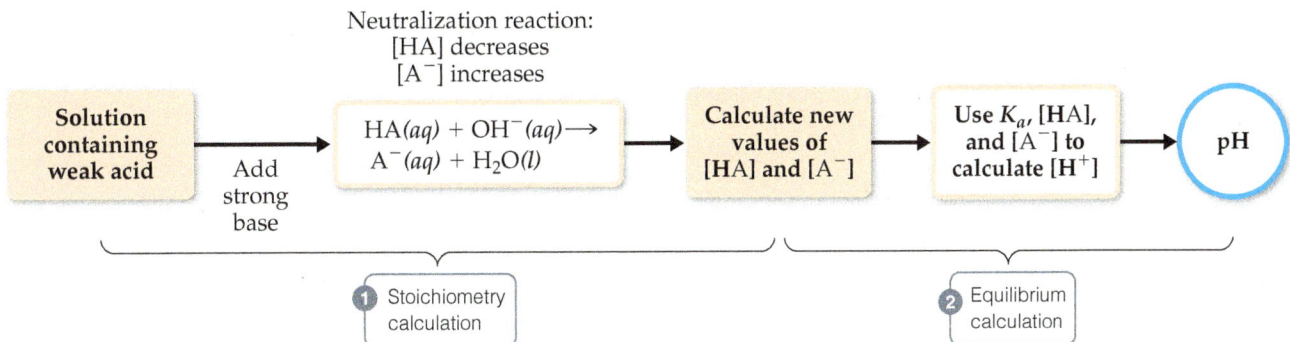

Neutralization reaction:
[HA] decreases
$[A^-]$ increases
Solution containing weak acid
Add strong base
$HA(aq) + OH^-(aq) \longrightarrow A^-(aq) + H_2O(l)$
Calculate new values of [HA] and $[A^-]$
Use K_a, [HA], and $[A^-]$ to calculate $[H^+]$
pH
1 Stoichiometry calculation
2 Equilibrium calculation

▲ **Figure 17.10 Procedure for calculating pH when a weak acid is partially neutralized by a strong base.**

3. **Equivalence point:** The equivalence point is reached when 50.0 mL of 0.100 M NaOH has been added to the 50.0 mL of 0.100 M CH$_3$COOH. At this point, the 5.00×10^{-3} mol of NaOH completely reacts with the 5.00×10^{-3} mol of CH$_3$COOH to form 5.00×10^{-3} mol of CH$_3$COONa. The Na$^+$ ion of this salt has no significant effect on the pH. The CH$_3$COO$^-$ ion, however, is a weak base whose reaction with water cannot be neglected, and the pH at the equivalence point is therefore greater than 7. In general, the pH at the equivalence point is always above 7 in a weak acid–strong base titration because the anion of the salt formed is a weak base. The procedure for calculating the pH of the solution of a weak base is described in Sections 16.7 and 16.9, and is illustrated in Sample Exercise 17.8.

4. **After equivalence point** (**excess base**): In this region, [OH$^-$] from the reaction of CH$_3$COO$^-$ with water is negligible relative to [OH$^-$] from the excess NaOH. Thus, the pH is determined by the concentration of OH$^-$ from the excess NaOH. The method for calculating pH in this region is therefore like that illustrated in Sample Exercise 17.7(b). Thus, the addition of 51.0 mL of 0.100 M NaOH to 50.0 mL of either 0.100 M HCl or 0.100 M CH$_3$COOH yields the same pH, 11.00. Notice that the titration curves for a strong acid (Figure 17.7) and a weak acid (Figure 17.9) are nearly the same after the equivalence point.

We now demonstrate how to calculate the pH for a weak acid–strong base titration prior to the equivalence point (Sample Exercise 17.8) and at the equivalence point (Sample Exercise 17.9).

Sample Exercise 17.8
Calculation for a Weak Acid–Strong Base Titration

Calculate the pH of the solution formed when 45.0 mL of 0.100 M NaOH is added to 50.0 mL of 0.100 M CH$_3$COOH ($K_a = 1.8 \times 10^{-5}$).

SOLUTION

Analyze We are asked to calculate the pH before the equivalence point of the titration of a weak acid with a strong base.

Plan We first must determine the number of moles of CH$_3$COOH and CH$_3$COO$^-$ present after the neutralization

reaction (the stoichiometry calculation). We then calculate pH using K_a, [CH$_3$COOH], and [CH$_3$COO$^-$] (the equilibrium calculation).

Solve

Stoichiometry Calculation: The product of the volume and concentration of each solution gives the number of moles of each reactant present before the neutralization:

$$(0.0500 \text{ L soln})\left(\frac{0.100 \text{ mol CH}_3\text{COOH}}{1 \text{ L soln}}\right) = 5.00 \times 10^{-3} \text{ mol CH}_3\text{COOH}$$

$$(0.0450 \text{ L soln})\left(\frac{0.100 \text{ mol NaOH}}{1 \text{ L soln}}\right) = 4.50 \times 10^{-3} \text{ mol NaOH}$$

The 4.50×10^{-3} mol of NaOH consumes 4.50×10^{-3} mol of CH$_3$COOH:

	CH$_3$COOH(aq) +	OH$^-$(aq) $\longrightarrow$	CH$_3$COO$^-$(aq) +	H$_2$O(l)
Before reaction (mol)	5.00×10^{-3}	4.50×10^{-3}	0	—
Change (limiting reactant) (mol)	-4.50×10^{-3}	-4.50×10^{-3}	$+4.50 \times 10^{-3}$	
After reaction (mol)	0.50×10^{-3}	0	4.50×10^{-3}	—

The total volume of the solution is:

$$45.0 \text{ mL} + 50.0 \text{ mL} = 95.0 \text{ mL} = 0.0950 \text{ L}$$

The resulting molarities of CH$_3$COOH and CH$_3$COO$^-$ after the reaction are therefore:

$$[\text{CH}_3\text{COOH}] = \frac{0.50 \times 10^{-3} \text{ mol}}{0.0950 \text{ L}} = 0.0053 \, M$$

$$[\text{CH}_3\text{COO}^-] = \frac{4.50 \times 10^{-3} \text{ mol}}{0.0950 \text{ L}} = 0.0474 \, M$$

Equilibrium Calculation: The equilibrium between CH_3COOH and CH_3COO^- must obey the equilibrium-constant expression for CH_3COOH:

$$K_a = \frac{[H^+][CH_3COO^-]}{[CH_3COOH]} = 1.8 \times 10^{-5}$$

Solving for $[H^+]$ gives:

$$[H^+] = K_a \times \frac{[CH_3COOH]}{[CH_3COO^-]} = (1.8 \times 10^{-5}) \times \left(\frac{0.0053}{0.0474}\right) = 2.0 \times 10^{-6}\ M$$

$$pH = -\log(2.0 \times 10^{-6}) = 5.70$$

Comment We could have solved for pH equally well using the Henderson–Hasselbalch equation in the last step.

▶ **Practice Exercise**

(a) Calculate the pH in the solution formed by adding 10.0 mL of 0.050 M NaOH to 40.0 mL of 0.0250 M benzoic acid (C_6H_5COOH, $K_a = 6.3 \times 10^{-5}$). (b) Calculate the pH in the solution formed by adding 10.0 mL of 0.100 M HCl to 20.0 mL of 0.100 M NH_3.

Sample Exercise 17.9

Calculating the pH at the Equivalence Point for a Weak Acid–Strong Base Titration

Calculate the pH at the equivalence point in the titration of 50.0 mL of 0.100 M CH_3COOH with 0.100 M NaOH.

SOLUTION

Analyze We are asked to determine the pH at the equivalence point of the titration of a weak acid with a strong base. Because the neutralization of a weak acid produces its anion, a conjugate base that can react with water, we expect the pH at the equivalence point to be greater than 7.

Plan The initial number of moles of acetic acid equals the number of moles of acetate ion at the equivalence point. We use the volume of the solution at the equivalence point to calculate the concentration of acetate ion. Because the acetate ion is a weak base, we can calculate the pH using K_b and $[CH_3COO^-]$.

Solve

The number of moles of acetic acid in the initial solution is obtained from the volume and molarity of the solution:

Moles $= M \times V = (0.100\ \text{mol/L})(0.0500\ \text{L})$

$= 5.00 \times 10^{-3}\ \text{mol } CH_3COOH$

Hence, 5.00×10^{-3} mol of CH_3COO^- is formed. It will take 50.0 mL of NaOH to reach the equivalence point (Figure 17.9). The volume of this salt solution at the equivalence point is the sum of the volumes of the acid and base, 50.0 mL + 50.0 mL = 100.0 mL = 0.1000 L. Thus, the concentration of CH_3COO^- is:

$$[CH_3COO^-] = \frac{5.00 \times 10^{-3}\ \text{mol}}{0.1000\ \text{L}} = 0.0500\ M$$

The CH_3COO^- ion is a weak base:

$$CH_3COO^-(aq) + H_2O(l) \rightleftharpoons CH_3COOH(aq) + OH^-(aq)$$

The K_b for CH_3COO^- can be calculated from the K_a value of its conjugate acid, $K_b = K_w/K_a = (1.0 \times 10^{-14})/(1.8 \times 10^{-5}) = 5.6 \times 10^{-10}$. Using the K_b expression, we have:

$$K_b = \frac{[CH_3COOH][OH^-]}{[CH_3COO^-]} = \frac{(x)(x)}{0.0500 - x} = 5.6 \times 10^{-10}$$

Making the approximation that $0.0500 - x \approx 0.0500$, and then solving for x, we have:

$x = [OH^-] = 5.3 \times 10^{-6}\ M$, which gives pOH = 5.28 and pH = 8.72

Check The pH is above 7, as expected for the salt of a weak acid and strong base.

▶ **Practice Exercise**

Calculate the pH at the equivalence point when (a) 40.0 mL of 0.025 M benzoic acid (C_6H_5COOH, $K_a = 6.3 \times 10^{-5}$) is titrated with 0.050 M NaOH and (b) 40.0 mL of 0.100 M NH_3 is titrated with 0.100 M HCl.

Go Figure

How does the pH at the equivalence point change as the acid being titrated becomes weaker? How does the volume of NaOH(aq) needed to reach the equivalence point change?

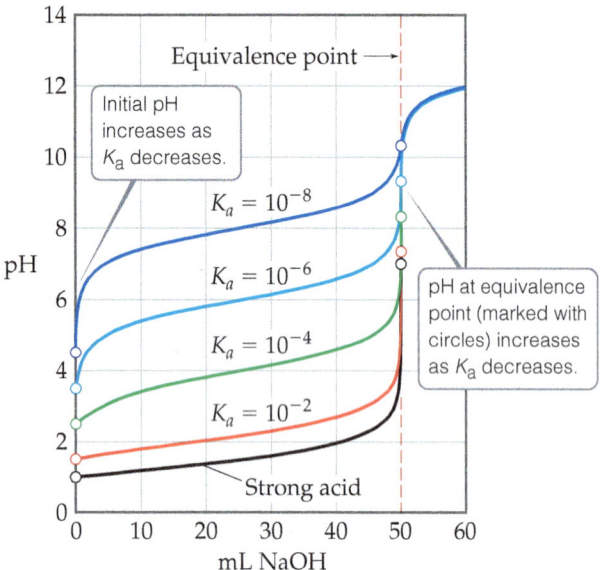

▲ **Figure 17.11** **A set of curves showing the effect of acid strength on the characteristics of the titration curve when a weak acid is titrated by a strong base.** Each curve represents titration of 50.0 mL of 0.10 *M* acid with 0.10 *M* NaOH.

The titration curve for a weak acid–strong base titration (Figure 17.9) differs from the curve for a strong acid–strong base titration (Figure 17.7) in three noteworthy ways:

1. The solution of the weak acid has a higher initial pH than a solution of a strong acid of the same concentration.

2. The pH change in the rapid-rise portion of the curve near the equivalence point is smaller for the weak acid than for the strong acid.

3. The pH at the equivalence point is above 7.00 for the weak acid titration.

The weaker the acid, the more pronounced these differences become. As an illustration, consider the family of titration curves shown in **Figure 17.11**. Notice that as the acid becomes weaker (that is, as K_a decreases), the initial pH increases and the pH change near the equivalence point becomes less marked. Furthermore, the pH at the equivalence point steadily increases as K_a decreases because the strength of the conjugate base of the weak acid increases. It is virtually impossible to determine the equivalence point when pK_a is 10 or higher because the pH change is too small and gradual.

pH titration experiments are an excellent way to measure the pK_a of a weak acid. Notice in Figure 17.11 that for each acid solution, 50 mL of strong base is required to reach the equivalence point. This means that 50 mL of base is required to convert the HA molecules into A$^-$ conjugate base anions. Notice, too, that halfway to each equivalence point (at 25 mL base added), the pH of the solution is nearly equal to the pK_a of the acid. Is this a coincidence? No! Recall that $K_a = [H^+][A^-]/[HA]$, where all concentrations are those at equilibrium. Halfway to the equivalence point, half of [HA] has been converted to [A$^-$]. In other words, [HA] = [A$^-$]. Therefore, [HA]/[A$^-$] = 1. In which case, $K_a = [H^+]$, and therefore pK_a = pH.

It becomes possible, then, to determine the pK_a of a weak acid from its pH titration curve. Once you identify the amount of base needed to reach the equivalence point, find the pH on the curve at halfway to the equivalence point. The pH you read off the graph at that point corresponds to the pK_a of the weak acid. If the acid's concentration is too low, however, the autoionization of water becomes significant, and this graphical way of measuring pK_a is not as accurate.

Titrations with an Acid–Base Indicator

An indicator is often used in an acid–base titration, rather than a pH meter. An indicator is a compound that changes color in solution over a specific pH range because its acid/base conjugate forms are different colors. Optimally, an indicator should change color abruptly at the equivalence point in a titration. In practice, however, an indicator need not precisely mark the equivalence point. The pH changes very rapidly near the equivalence point, and in this region one drop of titrant can change the pH by several units. Thus, an indicator beginning and ending its color change anywhere on the rapid-rise portion of the titration curve gives a sufficiently accurate measure of the titrant volume needed to reach the equivalence point. The point in a titration where the indicator changes color is called the *end point* to distinguish it from the equivalence point that it closely approximates.

Figure 17.12 shows the curve for titration of a strong base (NaOH) with a strong acid (HCl). We see from the vertical part of the curve that the pH changes rapidly from roughly 11 to 3 near the equivalence point. Consequently, an indicator for this titration can change color anywhere in this range. Most strong acid–strong base titrations are carried out using phenolphthalein as an indicator because it changes color in this range

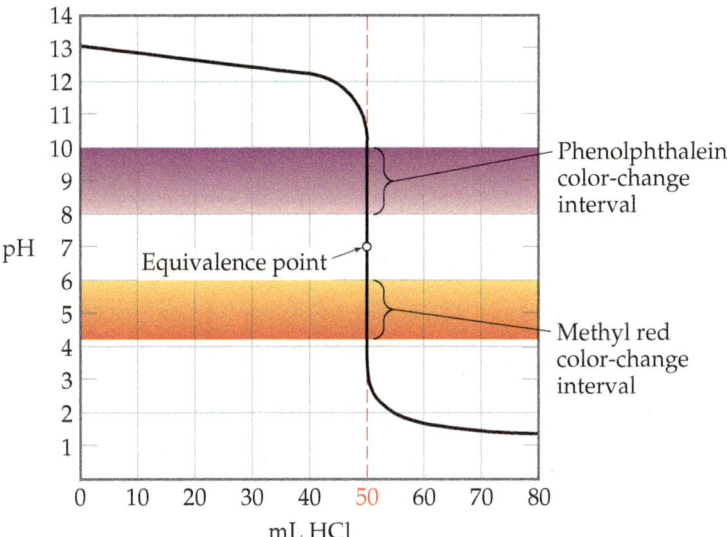

◀ Figure 17.12 Using color indicators **for titration of a strong base with a strong acid.** Both phenolphthalein and methyl red change color in the rapid-change portion of the titration curve.

(see Figure 16.8). Several other indicators would also be satisfactory, including methyl red, which, as the lower color band in Figure 17.12 shows, changes color in the pH range from about 4.2 to 6.0.

As noted in our discussion of Figure 17.11, the pH change near the equivalence point becomes smaller as K_a decreases, so the choice of indicator for a weak acid–strong base titration is more critical than it is for titrations where both acid and base are strong. When $0.100\,M\,CH_3COOH$ ($K_a = 1.8 \times 10^{-5}$) is titrated with $0.100\,M\,NaOH$, for example, the pH increases rapidly only over the pH range from about 7 to 11 (Figure 17.13). Phenolphthalein is therefore an ideal indicator because it changes color from pH 8.3 to 10.0, close to the pH at the equivalence point. Methyl red is a poor choice, however, because its color change, from 4.2 to 6.0, begins well before the equivalence point is reached.

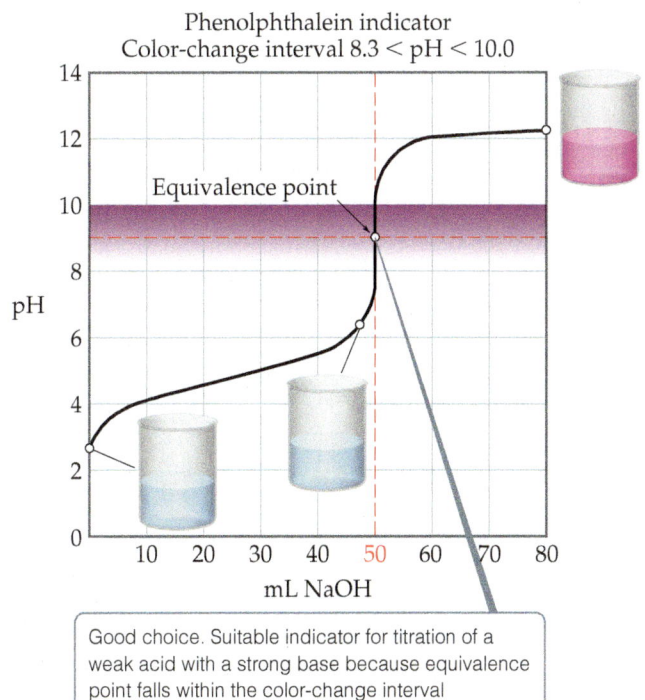

Phenolphthalein indicator
Color-change interval 8.3 < pH < 10.0

Good choice. Suitable indicator for titration of a weak acid with a strong base because equivalence point falls within the color-change interval

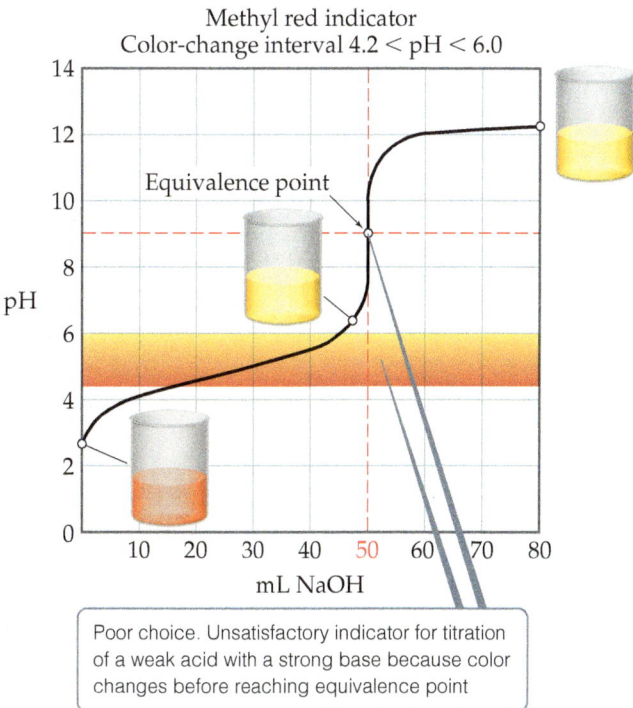

Methyl red indicator
Color-change interval 4.2 < pH < 6.0

Poor choice. Unsatisfactory indicator for titration of a weak acid with a strong base because color changes before reaching equivalence point

▲ Figure 17.13 **Good and poor indicators for the titration of a weak acid with a strong base.**

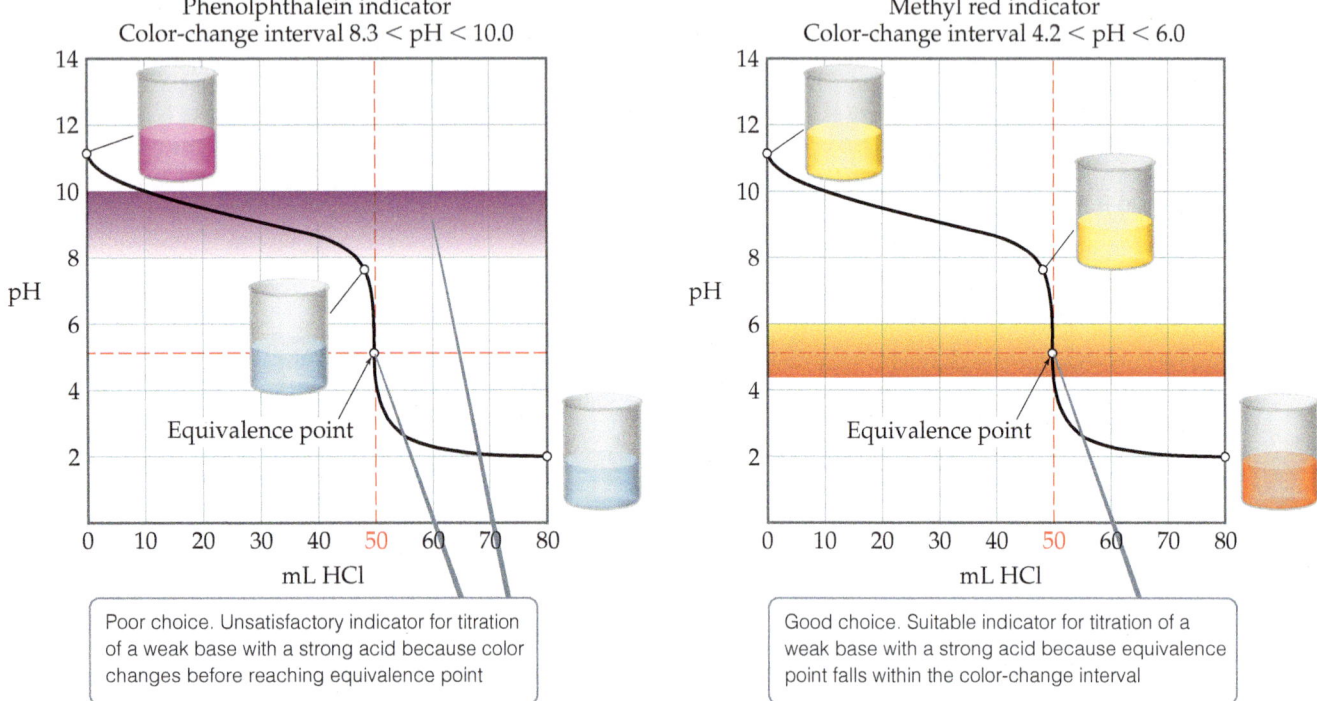

▲ **Figure 17.14 Good and poor indicators for the titration of a weak base with a strong acid.**

Titration of a weak base (such as $0.100 \, M \, NH_3$) with a strong acid solution (such as $0.100 \, M \, HCl$) leads to the titration curve shown in **Figure 17.14**. In this example, the equivalence point occurs at pH 5.28. Thus, methyl red is an ideal indicator, but phenolphthalein would be a poor choice.

Titrations of Polyprotic Acids

When weak acids contain more than one ionizable H atom, the reaction with OH^- occurs in a series of steps. Neutralization of phosphorous acid, H_3PO_3, for example, proceeds in two steps (the third H is bonded to the P and does not ionize):

Step 1: $H_3PO_3(aq) + OH^-(aq) \longrightarrow H_2PO_3^-(aq) + H_2O(l)$

Step 2: $H_2PO_3^-(aq) + OH^-(aq) \longrightarrow HPO_3^{2-}(aq) + H_2O(l)$

When the neutralization steps of a polyprotic acid or polybasic base are sufficiently separated, the titration has multiple equivalence points. **Figure 17.15** shows the two equivalence points corresponding to Steps 1 and 2.

You can use titration data such as that shown in Figure 17.15 to figure out the pK_as of the weak polyprotic acid. For example, let's write the K_{a1} and K_{a2} reactions for phosphorous acid:

$$H_3PO_3(aq) \rightleftharpoons H_2PO_3^-(aq) + H^+(aq) \quad K_{a1} = \frac{[H_2PO_3^-][H^+]}{[H_3PO_3]}$$

$$H_2PO_3^-(aq) \rightleftharpoons HPO_3^{2-}(aq) + H^+(aq) \quad K_{a2} = \frac{[HPO_3^{2-}][H^+]}{[H_2PO_3^-]}$$

If we rearrange these equilibrium expressions, you see that we obtain Henderson–Hasselbalch equations:

$$pH = pK_{a1} + \log\frac{[H_2PO_3^-]}{[H_3PO_3]}$$

$$pH = pK_{a2} + \log\frac{[HPO_3^{2-}]}{[H_2PO_3^-]}$$

Therefore, if the concentrations of each acid and base conjugate pair are identical for each equilibrium, $\log(1) = 0$ and so $pH = pK_a$. When does this happen during the titration?

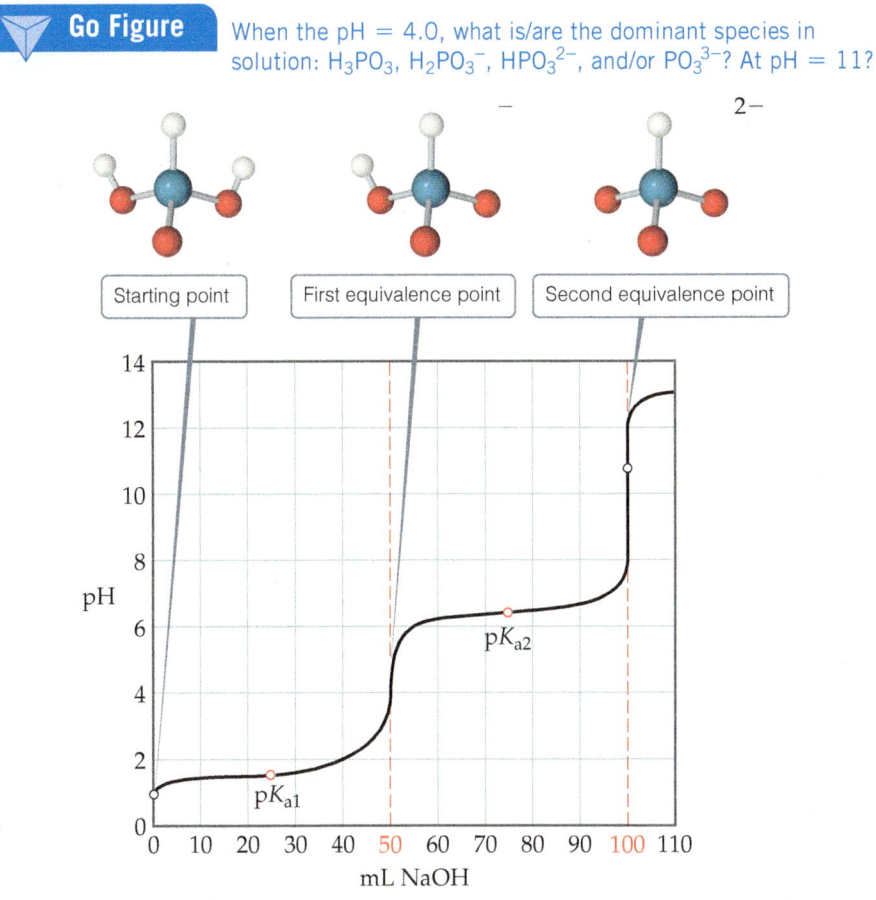

When the pH = 4.0, what is/are the dominant species in solution: H_3PO_3, $H_2PO_3^-$, HPO_3^{2-}, and/or PO_3^{3-}? At pH = 11?

Starting point | First equivalence point | Second equivalence point

▲ **Figure 17.15 Titration curve for a diprotic acid.** The curve shows the pH change when 50.0 mL of 0.10 M H_3PO_3 is titrated with 0.10 M NaOH.

At the beginning of the titration, the acid is H_3PO_3 initially; at the first equivalence point, it is all converted to $H_2PO_3^-$. Therefore, halfway to the first equivalence point, half of the H_3PO_3 is converted to $H_2PO_3^-$. Thus, halfway to the equivalence point, the concentration of H_3PO_3 is equal to that of $H_2PO_3^-$. and at that point, pH $=$ pK_{a1}. Similar logic is true for the second equilibrium reaction: halfway between the first and second equivalence points, pH $=$ pK_{a2}.

We can then just look at titration data and estimate the pK_as for the polyprotic acid directly from the titration curve. This procedure is especially useful if you are trying to identify an unknown polyprotic acid. In Figure 17.15, for instance, the first equivalence point occurs for 50 mL NaOH added. Halfway to the equivalence point corresponds to 25 mL NaOH. Because the pH at 25 mL NaOH is about 1.5, we may estimate pK_{a1} $=$ 1.5 for phosphorous acid. The second equivalence point occurs at 100 mL NaOH added; halfway between the first and second equivalence points is at 75 mL NaOH added. The graph indicates the pH at 75 mL NaOH added is about 6.5, and we therefore estimate that pK_{a2} for phosphorous acid is 6.5. The actual values for the two pK_as are pK_{a1} $=$ 1.3 and pK_{a2} $=$ 6.7 (close to our estimates).

Titrations in the Laboratory

Most of the acid–base titrations you will do in the laboratory do not involve calculation of the pH at each point; you will most likely measure the pH as either strong acid or strong base that is added to your sample (Figure 17.6), leading to data that resembles Figures 17.7, 17.8, 17.9, or 17.15. As we have just seen in Figure 17.15, inspection of a pH titration curve can give you good estimates for the pK_a(s) of a weak acid, which may allow you to identify your acid among several choices. Another common laboratory experiment is to use an acid–base titration to calculate the amount of acid or base in your sample for quantitative analysis. This type of calculation is illustrated in Sample Exercise 4.6. A combination of these concepts is illustrated in Sample Exercise 17.10.

Sample Exercise 17.10
Identification of a Weak Acid Using Acid–Base Titration Data

A 0.1004-g sample of an unknown monoprotic acid requires 22.10 mL of 0.0500 M NaOH to reach the equivalence point. **(a)** What is the molar mass of the unknown acid? **(b)** As the acid is titrated, the pH of the solution after the addition of 11.05 mL of the base is 4.89. What is the K_a of the acid? **(c)** Using Appendix D, suggest the identity of the acid.

SOLUTION

Analyze In part (a), we are given the mass of an unknown monoprotic acid (let's call it HA), and we are told what volume of a certain concentration of base is required to reach the equivalence point; from that we need to find the molar mass of HA. In part (b), we are given titration data, and we are asked to find the K_a of HA. Finally, in part (c), we are asked to suggest an identity of HA, using molar mass and K_a.

Plan In part (a), we recognize that moles base = moles acid at the equivalence point. From the information we have about the base, we can calculate moles of base and therefore will obtain moles of acid. Because we also know how many grams of HA are in the sample, we can solve for molar mass by dividing grams of HA by moles of HA. In part (b), we can use our knowledge of titration curves to find K_a. Finally, in part (c), we can use Appendix D to find an acid that is monoprotic, has a K_a similar to the one we calculate, and has a molar mass similar to the one we calculate.

Solve

(a) We can solve for the moles of base required to react the end point in the titration:

$(0.0500 \text{ mol NaOH/L}) (22.10 \times 10^{-3} \text{ L}) = 1.105 \times 10^{-3}$ mol NaOH

Because moles base = moles acid at the equivalence point, there must be 1.105×10^{-3} mol HA in the sample.

Because we know how many grams of HA are in the sample, we can solve for molar mass:

$0.1004 \text{ g HA}/1.105 \times 10^{-3} \text{ mol HA} = 94.5 \text{ g/mol}$

(b) Notice that the volume of base in the titration to reach the end point was 22.10 mL. Then we were told that the pH of the solution was 4.89 with 11.05 mL base added. You can see that 11.05 is half of 22.10. From our discussion about Figure 17.11, recall that halfway to the equivalence point, the pH of the solution is equal to the pK_a of the weak acid. Therefore, the pK_a of HA is 4.89. Then the K_a is $10^{-4.89} = 1.29 \times 10^{-5}$.

(c) In Appendix D, we need to look for an acid that is monoprotic, has a K_a of 1.3×10^{-5}, and has a molar mass of 95 g/mol (to two significant figures). Of the choices in Appendix D, butanoic acid, $CH_3CH_2CH_2COOH$, is the best one; it is monoprotic, it has a molar mass of 88 g/mol, and its K_a is 1.5×10^{-5}.

▶ **Practice Exercise**
A 325 mg tablet of aspirin (a monoprotic acid whose formal name is acetylsalicylic acid) is crushed, dissolved in 100.0 mL water, and titrated with 0.1000 M NaOH, requiring 18.1 mL of the base to reach the equivalence point. What is the molar mass of aspirin?

 ## Self-Assessment Exercises

SAE 17.7 In a weak acid–strong base titration, which of the following quantities *does not depend* on the acid dissociation constant K_a of the weak acid? **(a)** the pH at the equivalence point **(b)** the initial pH **(c)** the volume of base needed to reach the equivalence point **(d)** the rise in pH near the equivalence point

SAE 17.8 A 25.0-mL sample of 0.100 M HBr solution is titrated with 0.050 M NaOH solution. What is the pH after 25.0 mL of the NaOH solution have been added? **(a)** 1.00 **(b)** 1.30 **(c)** 1.60 **(d)** 7.00 **(e)** 12.70

SAE 17.9 A 25.0-mL sample of 0.150 M acetic acid (CH_3COOH) solution ($K_a = 1.8 \times 10^{-5}$) is titrated with 0.150 M NaOH solution. What is the pH after 15.0 mL of the NaOH solution have been added? **(a)** 3.09 **(b)** 4.74 **(c)** 4.92 **(d)** 5.25 **(e)** 6.89

SAE 17.10 A 25.0-mL sample of 0.150 M acetic acid (CH_3COOH) solution ($K_a = 1.8 \times 10^{-5}$) is titrated with 0.150 M NaOH solution. What is the pH at the equivalence point? **(a)** 5.19 **(b)** 7.00 **(c)** 8.15 **(d)** 8.81 **(e)** 8.96

SAE 17.11 The acid–base indicator thymol blue undergoes two color changes as shown in the diagram provided here (see Figure 16.8). If this indicator was used in the titration of acetic acid with a sodium hydroxide solution, what would be the color change on passing through the equivalence point? **(a)** red to yellow **(b)** yellow to red **(c)** yellow to blue **(d)** blue to yellow **(e)** red to blue

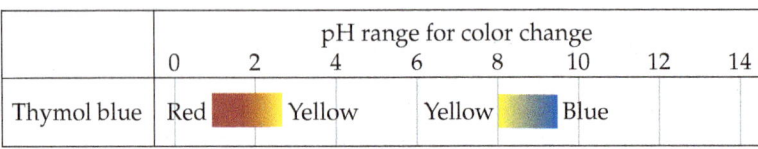

SAE 17.12 Consider the titration data presented in the following graph. Which of the statements is or are *true*?

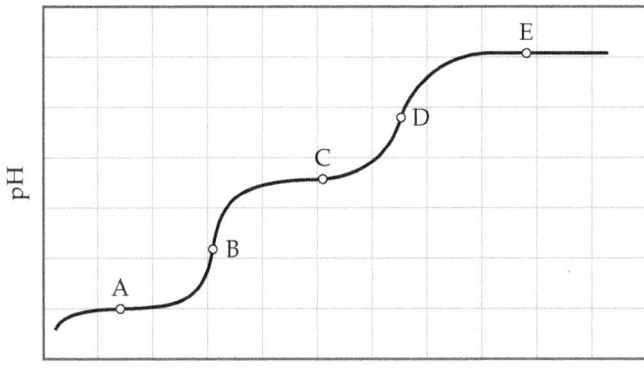

(**i**) Points A and C are regions where the solution is well buffered.
(**ii**) Point D is an equivalence point.
(**iii**) This is the titration curve of a triprotic acid.

(**a**) Only (i) is true. (**b**) Only ii is true. (**c**) Only ii is true. (**d**) Statements i and ii are true. (**e**) All three statements are true.

SAE 17.13 The monoprotic acid oleic acid, molecular weight 282.5 g/mol, is a major component of olive oil. A 25.6-mL sample containing oleic acid and other nonacidic oils is titrated with 0.150 M NaOH, requiring 15.7 mL to reach the equivalence point. What percent oleic acid by volume is in the sample? Assume all compounds are liquids with a density of 0.925 g/mL.

(**a**) 2.40% (**b**) 2.61% (**c**) 2.81% (**d**) 3.26%

17.4 | Solubility Equilibria

The equilibria we have considered thus far in this chapter have involved acids and bases. Furthermore, they have been homogeneous; that is, all the species have been in the same phase (aqueous solutions). Throughout the rest of the chapter, we consider the aqueous equilibria involved in the dissolution or precipitation of ionic compounds. These reactions are heterogeneous because they involve the equilibrium between solids and aqueous ions dissolved in solution.

Dissolution and precipitation occur both within us and around us. Tooth enamel dissolves in acidic solutions, for example, causing tooth decay, and the precipitation of certain salts in our kidneys produces kidney stones. The waters of Earth contain salts dissolved as water passes over and through the ground. Precipitation of $CaCO_3$ from groundwater is responsible for the formation of stalactites and stalagmites within limestone caves.

In our previous discussion of precipitation reactions, we considered general rules for predicting the solubility of common salts in water. (Section 4.2) These rules give us a *qual*itative sense of whether a compound has a low or high solubility in water. By considering solubility equilibria, however, we can make *quant*itative predictions about solubility.

> ⚠ **Learning Objectives**
>
> **When you finish Section 17.4, you should be able to:**
>
> ▶ Write the K_{sp} expression for an ionic solid in contact with water.
>
> ▶ Interconvert between mass solubility, molar solubility, and K_{sp}.

The Solubility-Product Constant, K_{sp}

Recall that a *saturated solution* is one in which the solution is in contact with undissolved solute. (Section 13.2) Consider, for example, a saturated aqueous solution of $BaSO_4$ in contact with solid $BaSO_4$. Because the solid is an ionic compound, it is a strong electrolyte and yields $Ba^{2+}(aq)$ and $SO_4^{2-}(aq)$ ions when dissolved in water, readily establishing the equilibrium

$$BaSO_4(s) \rightleftharpoons Ba^{2+}(aq) + SO_4^{2-}(aq)$$

As with any other equilibrium, the extent to which this dissolution reaction occurs is expressed by the magnitude of the equilibrium constant. Because this equilibrium equation describes the dissolution of a solid, the equilibrium constant indicates how soluble the solid is in water and is referred to as the **solubility-product constant** (or simply the **solubility product**). It is denoted K_{sp}, where *sp* stands for solubility product.

The equilibrium-constant expression for the equilibrium between a solid and an aqueous solution of its component ions (K_{sp}) is written according to the rules that apply to any other equilibrium-constant expression. Remember, however, that solids do *not* appear in the equilibrium-constant expressions for heterogeneous equilibrium. (Section 15.4)

Thus, the solubility-product expression for $BaSO_4$ is

$$K_{sp} = [Ba^{2+}][SO_4^{2-}]$$

The coefficient for each ion in the equilibrium equation also equals its subscript in the compound's chemical formula.

> *In general, the solubility product K_{sp} of a compound equals the product of the concentration of the ions involved in the equilibrium, each raised to the power of its coefficient in the equilibrium equation.*

Thus, for the generic ionic solid A_nX_m,

$$K_{sp} = [A^+]^n[X^-]^m \qquad [17.8]$$

The values of K_{sp} at 25 °C for many ionic solids are tabulated in Appendix D. The value of K_{sp} for $BaSO_4$ is 1.1×10^{-10}, a very small number, indicating that only a very small amount of the solid dissolves in 25 °C water.

Sample Exercise 17.11
Writing Solubility-Product (K_{sp}) Expressions

Write the expression for the solubility-product constant for CaF_2, and look up the corresponding K_{sp} value in Appendix D.

SOLUTION

Analyze We are asked to write an equilibrium-constant expression for the process by which CaF_2 dissolves in water.

Plan We apply the general rules for writing an equilibrium-constant expression, excluding the solid reactant from the expression. We assume that the compound dissociates completely into its component ions:

$$CaF_2(s) \rightleftharpoons Ca^{2+}(aq) + 2\,F^-(aq)$$

Solve The expression for K_{sp} is

$$K_{sp} = [Ca^{2+}][F^-]^2 \text{ Appendix D gives } 3.9 \times 10^{-11} \text{ for this } K_{sp}.$$

▶ **Practice Exercise**
Give the solubility-product-constant expressions and K_{sp} values (from Appendix D) for **(a)** barium carbonate and **(b)** silver sulfate.

Solubility and K_{sp}

It is important to distinguish carefully between solubility and the solubility-product constant. The solubility of a substance is the quantity that dissolves to form a saturated solution. (Section 13.2) Solubility is often expressed as grams of solute per liter of solution (g/L); this quantity is also called *mass solubility*. *Molar solubility* is the number of moles of solute that dissolve in forming 1 L of saturated solution of the solute (mol/L). The solubility-product constant (K_{sp}) is the equilibrium constant for the equilibrium between an ionic solid and its saturated solution and is a unitless number. Thus, the magnitude of K_{sp} is a measure of how much of the solid dissolves to form a saturated solution.

The solubility of a substance can change considerably in response to a number of factors. For example, the solubilities of hydroxide salts, like $Mg(OH)_2$, depend on the pH of the solution. The solubility is also affected by concentrations of other ions in solution, especially common ions. In other words, the numeric value of the solubility of a given solute does change as the other species in solution change. In contrast, the solubility-product constant, K_{sp}, has only one value for a given solute at any specific temperature.*

Figure 17.16 summarizes the relationships among various expressions of solubility and K_{sp}.

In principle, it is possible to use the K_{sp} value of a salt to calculate solubility under a variety of conditions. In practice, great care must be taken in doing so for the reasons indicated in "A Closer Look: Limitations of Solubility Products" at the end of this section. Agreement between the measured solubility and that calculated from K_{sp} is usually best for salts whose ions have low charges (1+ and 1−) and do not react with water.

---

* This is strictly true only for very dilute solutions, because K_{sp} values change somewhat when the concentration of ionic substances in water is increased. However, we ignore these effects, which are taken into consideration only for work that requires exceptional accuracy.

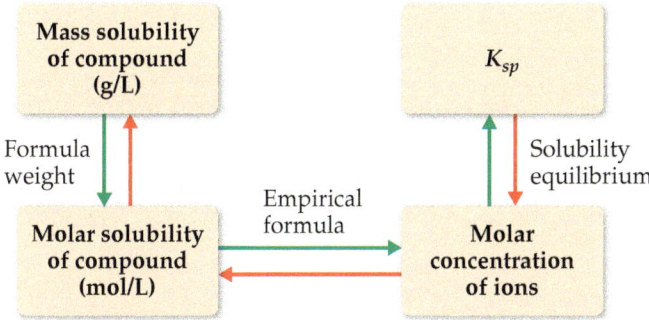

▲ **Figure 17.16** **Procedure for converting between solubility and** K_{sp}. Starting from the mass solubility, follow the green arrows to determine K_{sp}. Starting from K_{sp}, follow the red arrows to determine either molar solubility or mass solubility.

Sample Exercise 17.12
Calculating K_{sp} from Solubility

Solid silver chromate is added to pure water at 25 °C, and some of the solid remains undissolved. The mixture is stirred for several days to ensure that equilibrium is achieved between the undissolved $Ag_2CrO_4(s)$ and the solution. Analysis of the equilibrated solution shows that its silver ion concentration is 1.3×10^{-4} M. Assuming that the Ag_2CrO_4 solution is saturated and that there are no other important equilibria involving Ag^+ or CrO_4^{2-} ions in the solution, calculate K_{sp} for this compound.

SOLUTION

Analyze We are given the equilibrium concentration of Ag^+ in a saturated solution of Ag_2CrO_4, and we are asked to determine the value of K_{sp} for Ag_2CrO_4.

Plan The equilibrium equation and the expression for K_{sp} are

$$Ag_2CrO_4(s) \rightleftharpoons 2\,Ag^+(aq) + CrO_4^{2-}(aq)$$

$$K_{sp} = [Ag^+]^2[CrO_4^{2-}]$$

To calculate K_{sp}, we need the equilibrium concentrations of Ag^+ and CrO_4^{2-}. We know that at equilibrium $[Ag^+] = 1.3 \times 10^{-4}$ M. All the Ag^+ and CrO_4^{2-} ions in the solution come from the Ag_2CrO_4 that dissolves. Thus, we can use $[Ag^+]$ to calculate $[CrO_4^{2-}]$.

Solve From the chemical formula of silver chromate, we know that there must be two Ag^+ ions in solution for each CrO_4^{2-} ion in solution. Consequently, the concentration of CrO_4^{2-} is half the concentration of Ag^+:

$$[CrO_4^{2-}] = \left(\frac{1.3 \times 10^{-4}\ \text{mol Ag}^+}{L}\right)\left(\frac{1\ \text{mol CrO}_4^{2-}}{2\ \text{mol Ag}^+}\right) = 6.5 \times 10^{-5}\ M$$

and K_{sp} is

$$K_{sp} = [Ag^+]^2[CrO_4^{2-}] = (1.3 \times 10^{-4})^2(6.5 \times 10^{-5}) = 1.1 \times 10^{-12}$$

Check We obtain a small value, as expected for a slightly soluble salt. Furthermore, the calculated value agrees well with the one given in Appendix D, 1.2×10^{-12}.

▶ **Practice Exercise**
A saturated solution of $Mg(OH)_2$ in contact with undissolved $Mg(OH)_2(s)$ is prepared at 25 °C. The pH of the solution is found to be 10.17. Assuming that there are no other simultaneous equilibria involving the Mg^{2+} or OH^- ions, calculate K_{sp} for this compound.

Sample Exercise 17.13
Calculating Solubility from K_{sp}

The K_{sp} for CaF_2 is 3.9×10^{-11} at 25 C. Assuming that equilibrium is established between solid and dissolved CaF_2. and that there are no other important equilibria affecting its solubility, calculate the solubility of CaF_2 in grams per liter.

SOLUTION

Analyze We are given K_{sp} for CaF_2 and are asked to determine solubility. Recall that the solubility of a substance is the quantity that can dissolve in solvent, whereas the solubility-product constant, K_{sp}, is an equilibrium constant.

Plan To go from K_{sp} to solubility, we follow the steps indicated by the red arrows in Figure 17.16. We first write the chemical equation for the dissolution and set up a table of initial and equilibrium concentrations. We then use the equilibrium-constant expression.

In this case we know K_{sp}, and so we solve for the concentrations of the ions in solution. Once we know these concentrations, we use the formula weight to determine solubility in g/L.

Continued

Solve

Assume that initially no salt has dissolved, and then allow x mol/L of CaF_2 to dissociate completely when equilibrium is achieved:

	$CaF_2(s)$ $\rightleftharpoons$	$Ca^{2+}(aq)$	+ $2F^-(aq)$
Initial concentration (M)	—	0	0
Change (M)	—	$+x$	$+2x$
Equilibrium concentration (M)	—	x	$2x$

The stoichiometry of the equilibrium dictates that $2x$ mol/L of F^- are produced for each x mol/L of CaF_2 that dissolve. We now use the expression for K_{sp} and substitute the equilibrium concentrations to solve for the value of x:

$$K = [Ca^{2+}][F^-]^2 = (x)(2x)^2 = 4x^3 = 3.9 \times 10^{-11}$$

(Remember that $\sqrt[3]{y} = y^{1/3}$.) Thus, the molar solubility of CaF_2 is 2.1×10^{-4} mol/L.

$$x = \sqrt[3]{\frac{3.9 \times 10^{-11}}{4}} = 2.1 \times 10^{-4}$$

The mass of CaF_2 that dissolves in water to form 1 L of solution is:

$$\left(\frac{2.1 \times 10^{-4} \text{ mol } CaF_2}{1 \text{ L soln}}\right)\left(\frac{78.1 \text{ g } CaF_2}{1 \text{ mol } CaF_2}\right) = 1.6 \times 10^{-2} \text{ g } CaF_2/\text{L soln}$$

Check We expect a small number for the solubility of a slightly soluble salt. If we reverse the calculation, we should be able to recalculate the solubility product: $K_{sp} = (2.1 \times 10^{-4})(4.2 \times 10^{-4})^2 = 3.7 \times 10^{-11}$, close to the value given in the problem statement, 3.9×10^{-11}.

Comment Because F^- is the anion of a weak acid, you might expect hydrolysis of the ion to affect the solubility of CaF_2. The basicity of F^- is so small ($K_b = 1.5 \times 10^{-11}$), however, that the hydrolysis occurs to only a slight extent and does not significantly influence the solubility. The reported solubility is 0.017 g/L at 25 °C, in good agreement with our calculation.

▶ **Practice Exercise**
The K_{sp} for LaF_3 is 2×10^{-19}. What is the solubility of LaF_3 in water in moles per liter?

A CLOSER LOOK | Limitations of Solubility Products

Ion concentrations calculated from K_{sp} values sometimes deviate appreciably from those found experimentally. In part, these deviations are due to electrostatic interactions between ions in solution, which can lead to ion pairs. See the "A Closer Look" box on "The van't Hoff Factor." (Section 13.5) These interactions increase in magnitude both as the concentrations of the ions increase and as their charges increase. The solubility calculated from K_{sp} tends to be low unless corrected to account for these interactions.

As an example of the effect of these interactions, consider $CaCO_3$ (calcite), whose solubility product, 4.5×10^{-9}, gives a calculated solubility of 6.7×10^{-5} mol/L; correcting for ionic interactions in the solution yields 7.3×10^{-5} mol/L. The reported solubility, however, is 1.4×10^{-4} mol/L, indicating that additional factors must be involved.

Another common source of error in calculating ion concentrations from K_{sp} is ignoring other equilibria that occur simultaneously in the solution. It is possible, for example, that acid–base equilibria take place simultaneously with solubility equilibria. In particular, both basic anions and cations with high charge-to-size ratios undergo hydrolysis reactions that can measurably increase the solubilities of their salts. For example, $CaCO_3$ contains the basic carbonate ion ($K_b = 1.8 \times 10^{-4}$), which reacts with water:

$$CO_3^{2-}(aq) + H_2O(l) \rightleftharpoons HCO_3^-(aq) + OH^-(aq)$$

If we consider the effect of ion–ion interactions as well as simultaneous solubility and K_b equilibria, we calculate a solubility of 1.4×10^{-4} mol/L, in agreement with the measured value for calcite.

Finally, we generally assume that ionic compounds dissociate completely when they dissolve, but this assumption is not always valid. When MgF_2 dissolves, for example, it yields not only Mg^{2+} and F^- ions but also MgF^+ ions.

Self-Assessment Exercises

SAE 17.14 The ionic compounds $La(IO_3)_3$ and $PbCO_3$ both have K_{sp} values equal to 7.4×10^{-14}. What can we infer about the molar solubility from this fact? **(a)** The molar solubility of $La(IO_3)_3$ will be equal to the molar solubility of $PbCO_3$. **(b)** The molar solubility of $La(IO_3)_3$ will be greater than the molar solubility of $PbCO_3$. **(c)** The molar solubility of $La(IO_3)_3$ will be less than the molar solubility of $PbCO_3$.

SAE 17.15 The mass solubility of silver iodate, $AgIO_3$ (molar mass = 283 g/mol), is 0.098 g/L at 25 °C. What is the value of the solubility product constant, K_{sp}? **(a)** 9.6×10^{-3} **(b)** 3.5×10^{-4} **(c)** 1.2×10^{-7} **(d)** 1.7×10^{-10}

SAE 17.16 The K_{sp} of $Ni(OH)_2$ at 25 °C is 6.0×10^{-16}. What is the pH of a saturated solution of $Ni(OH)_2$ at this temperature, assuming no other equilibria need to be considered? **(a)** 5.0 **(b)** 7.6 **(c)** 8.9 **(d)** 9.0 **(e)** 9.2

17.5 | Factors That Affect Solubility

Solubility is affected by temperature and by the presence of other solutes. The presence of an acid, for example, can have a major influence on the solubility of a substance. In Section 17.4, we considered solutions made by dissolving ionic compounds in pure water. In this section, we examine three factors that affect the solubility of ionic compounds: (1) the presence of common ions, (2) the solution pH, and (3) the presence of complexing agents. We also examine the phenomenon of *amphoterism*, which is related to the effects of both pH and complexing agents.

> ### Learning Objectives
>
> **When you finish Section 17.5, you should be able to:**
>
> ▶ Calculate how the presence of a common ion will affect the solubility of an ionic compound.
>
> ▶ Calculate how the solubility of an ionic compound will respond to changes in pH.
>
> ▶ Identify complex ions and predict how their presence will affect the solubility of ionic compounds.

The Common-Ion Effect

The presence of either $Ca^{2+}(aq)$ or $F^-(aq)$ in a solution reduces the solubility of CaF_2, shifting the equilibrium concentrations to the left:

$$CaF_2(s) \rightleftharpoons Ca^{2+}(aq) + 2F^-(aq)$$

Addition of Ca^{2+} or F^- shifts equilibrium concentrations, reducing solubility

This reduction in solubility is another manifestation of the common-ion effect we discussed in Section 17.1. In general, *the solubility of a slightly soluble salt is decreased by the presence of a second solute that furnishes a common ion*, as **Figure 17.17** shows for CaF_2.

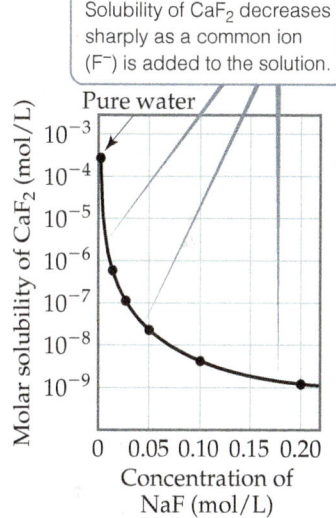

Solubility of CaF_2 decreases sharply as a common ion (F^-) is added to the solution.

▲ **Figure 17.17 Common-ion effect.** Notice that the CaF_2 solubility is on a logarithmic scale.

 Sample Exercise 17.14

Calculating the Effect of a Common Ion on Solubility

Calculate the molar solubility of CaF_2 at 25 °C in a solution that is (**a**) 0.010 *M* in $Ca(NO_3)_2$ and (**b**) 0.010 *M* in NaF.

SOLUTION

Analyze We are asked to determine the solubility of CaF_2 in the presence of two strong electrolytes, each containing an ion common to CaF_2. In (**a**) the common ion is Ca^{2+}, and NO_3^- is a spectator ion. In (**b**) the common ion is F^-, and Na^+ is a spectator ion.

Plan Because the slightly soluble compound is CaF_2, we need to use K_{sp} for this compound, which Appendix D gives as

3.9×10^{-11}. *The value of K_{sp} is unchanged by the presence of additional solutes. Because of the common-ion effect, however, the solubility of the salt decreases in the presence of common ions. We use our standard equilibrium techniques of starting with the equation for CaF_2 dissolution, setting up a table of initial and equilibrium concentrations, and using the K_{sp} expression to determine the concentration of the ion that comes only from CaF_2.*

Solve

(**a**) The initial concentration of Ca^{2+} is 0.010 *M* because of the dissolved $Ca(NO_3)_2$:

	$CaF_2(s)$ $\rightleftharpoons$	$Ca^{2+}(aq)$ +	$2F^-(aq)$
Initial Concentration (*M*)	—	0.010	0
Change (*M*)	—	+*x*	+2*x*
Equilibrium Concentration (*M*)	—	(0.010 + *x*)	2*x*

Substituting into the solubility-product expression gives:

$$K_{sp} = 3.9 \times 10^{-11} = [Ca^{2+}][F^-]^2 = (0.010 + x)(2x)^2$$

Continued

If we assume that x is small compared to 0.010, we have:

$$3.9 \times 10^{-11} = (0.010)(2x)^2$$

This very small value for x validates the simplifying assumption we made. Our calculation indicates that 3.1×10^{-5} mol of solid CaF_2 dissolves per liter of 0.010 M $Ca(NO_3)_2$ solution.

$$x^2 = \frac{3.9 \times 10^{-11}}{4(0.010)} = 9.8 \times 10^{-10}$$

$$x = \sqrt{9.8 \times 10^{-10}} = 3.1 \times 10^{-5} M$$

(b) The common ion is F^-, and at equilibrium we have:

$$[Ca^{2+}] = x \quad \text{and} \quad [F^-] = 0.010 + 2x$$

Assuming that $2x$ is much smaller than 0.010 M (that is, $0.010 + 2x \approx 0.010$), we have:

$$3.9 \times 10^{-11} = (x)(0.010 + 2x)^2 \approx x(0.010)^2$$

Thus, 3.9×10^{-7} mol of solid CaF_2 should dissolve per liter of 0.010 M NaF solution.

$$x = \frac{3.9 \times 10^{-11}}{(0.010)^2} = 3.9 \times 10^{-7} M$$

Comment The molar solubility of CaF_2 in water is $2.1 \times 10^{-4} M$ (Sample Exercise 17.13). By comparison, our calculations here give a CaF_2 solubility of $3.1 \times 10^{-5} M$ in the presence of 0.010 M Ca^{2+} and $3.9 \times 10^{-7} M$ in the presence of 0.010 M F^- ion. Thus, the addition of either Ca^{2+} or F^- to a solution of CaF_2 decreases the solubility. However, the effect of F^- on the solubility is more pronounced than that of Ca^{2+} because $[F^-]$ appears to the second power in the K_{sp} expression for CaF_2, whereas $[Ca^{2+}]$ appears to the first power.

▶ **Practice Exercise**

For manganese(II) hydroxide, $Mn(OH)_2$, $K_{sp} = 1.6 \times 10^{-13}$. Calculate the molar solubility of $Mn(OH)_2$ in a solution that contains 0.020 M $NaOH$.

Solubility and pH

The solubility of almost any ionic compound is affected if the solution is made sufficiently acidic or basic. The effects are noticeable, however, only when one (or both) ions in the compound is at least moderately acidic or basic. The metal hydroxides, such as $Mg(OH)_2$, are examples of compounds containing a strongly basic ion, the hydroxide ion. Let's look at $Mg(OH)_2$, for which the solubility equilibrium is

$$Mg(OH)_2(s) \rightleftharpoons Mg^{2+}(aq) + 2\,OH^-(aq) \quad K_{sp} = 1.8 \times 10^{-11}$$

A saturated solution of $Mg(OH)_2$ has a calculated pH of 10.52, and its Mg^{2+} concentration is $1.7 \times 10^{-4} M$. Now suppose that solid $Mg(OH)_2$ is equilibrated with a solution buffered at pH 9.0. The pOH, therefore, is 5.0, so $[OH^-] = 1.0 \times 10^{-5}$. Inserting this value for $[OH^-]$ into the solubility-product expression, we have

$$K_{sp} = [Mg^{2+}][OH^-]^2 = 1.8 \times 10^{-11}$$

$$[Mg^{2+}](1.0 \times 10^{-5})^2 = 1.8 \times 10^{-11}$$

$$[Mg^{2+}] = \frac{1.8 \times 10^{-11}}{(1.0 \times 10^{-5})^2} = 0.18\ M$$

Thus, the $Mg(OH)_2$ dissolves until $[Mg^{2+}] = 0.18\ M$. It is apparent that $Mg(OH)_2$ is much more soluble in this solution.

If $[OH^-]$ were reduced further by making the solution even more acidic, the Mg^{2+} concentration would have to increase to maintain the equilibrium condition. Thus, a sample of $Mg(OH)_2(s)$ dissolves completely if sufficient acid is added, as we saw in Figure 4.8.

As we have seen, the solubility of $Mg(OH)_2$ greatly increases as the acidity of the solution increases. Based on this observation, we can make the following generalization:

In general, the solubility of a compound containing a basic anion (that is, the anion is the conjugate base of a weak acid) increases as the solution becomes more acidic.

The solubility of PbF_2 increases as the solution becomes more acidic, too, because F^- is a base (it is the conjugate base of the weak acid HF). As a result, the solubility equilibrium of PbF_2 is shifted to the right as the concentration of F^- is reduced by protonation to form HF. Thus, the solution process can be understood in terms of two consecutive reactions:

Step 1: $$PbF_2(s) \rightleftharpoons Pb^{2+}(aq) + 2\,F^-(aq)$$

Step 2: $$F^-(aq) + H^+(aq) \rightleftharpoons HF(aq)$$

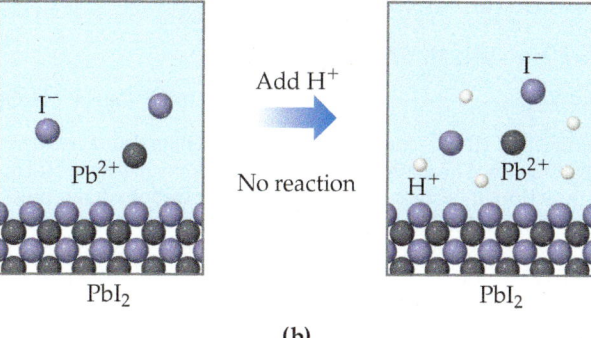

▲ **Figure 17.18 Response of two ionic compounds to addition of a strong acid.** (a) The solubility of PbF_2 increases upon addition of acid. (b) The solubility of PbI_2 is not affected by the addition of acid. The water molecules and the anion of the strong acid have been omitted for clarity.

The equation for the overall process is

$$PbF_2(s) + 2\,H^+(aq) \rightleftharpoons Pb^{2+}(aq) + 2\,HF(aq)$$

The processes responsible for the increase in solubility of PbF_2 in acidic solution are illustrated in **Figure 17.18**(a).

Other salts that contain basic anions, such as CO_3^{2-}, PO_4^{3-}, CN^-, or S^{2-}, behave similarly. These examples illustrate a general rule: *The solubility of slightly soluble salts containing basic anions increases as $[H^+]$ increases (as pH is lowered)*. The more basic the anion, the more the solubility is influenced by pH. The solubility of salts with anions of negligible basicity (anions that are conjugate bases of strong acids), such as Cl^-, Br^-, I^-, and NO_3^-, is unaffected by pH changes, as shown in Figure 17.18(**b**).

Sample Exercise 17.15

Predicting the Effect of Acid on Solubility

Which of these substances are more soluble in acidic solution than in basic solution?
(**a**) $Ni(OH)_2(s)$ (**b**) $CaCO_3(s)$ (**c**) $BaF_2(s)$ (**d**) $AgCl(s)$

SOLUTION

Analyze The problem lists four sparingly soluble salts, and we are asked to determine which are more soluble at low pH than at high pH.

Plan We will identify ionic compounds that dissociate to produce a basic anion because these are especially soluble in acid solution.

Solve

(**a**) $Ni(OH)_2(s)$ is more soluble in acidic solution because of the basicity of OH^-; the H^+ reacts with the OH^- ion, forming water:

$$Ni(OH)_2(s) \rightleftharpoons Ni^{2+}(aq) + 2\,OH^-(aq)$$
$$\underline{2\,OH^-(aq) + 2\,H^+(aq) \longrightarrow 2\,H_2O(l)}$$
Overall: $Ni(OH)_2(s) + 2\,H^+(aq) \rightleftharpoons Ni^{2+}(aq) + 2\,H_2O(l)$

(**b**) Similarly, $CaCO_3(s)$ dissolves in acid solutions because CO_3^{2-} is a basic anion:

The reaction between CO_3^{2-} and H^+ occurs in steps, with HCO_3^- forming first and H_2CO_3 forming in appreciable amounts only when $[H^+]$ is sufficiently high.

$$CaCO_3(s) \rightleftharpoons Ca^{2+}(aq) + CO_3^{2-}(aq)$$
$$CO_3^{2-}(aq) + 2\,H^+(aq) \rightleftharpoons H_2CO_3(aq)$$
$$\underline{H_2CO_3(aq) \rightleftharpoons CO_2(g) + H_2O(l)}$$
Overall: $CaCO_3(s) + 2\,H^+(aq) \rightleftharpoons Ca^{2+}(aq) + CO_2(g) + H_2O(l)$

(**c**) The solubility of BaF_2 is enhanced by lowering the pH because F^- is a basic anion:

$$BaF_2(s) \rightleftharpoons Ba^{2+}(aq) + 2\,F^-(aq)$$
$$\underline{2\,F^-(aq) + 2\,H^+(aq) \rightleftharpoons 2\,HF(aq)}$$
Overall: $BaF_2(s) + 2\,H^+(aq) \rightleftharpoons Ba^{2+}(aq) + 2\,HF(aq)$

(**d**) The solubility of AgCl is unaffected by changes in pH because Cl^- is the anion of a strong acid and therefore has negligible basicity.

▶ **Practice Exercise**

Write the net ionic equation for the reaction between a strong acid and (**a**) CuS, (**b**) $Cu(N_3)_2$.

CHEMISTRY AND LIFE | Tooth Decay and Fluoridation

Tooth enamel consists mainly of the mineral hydroxyapatite, $Ca_{10}(PO_4)_6(OH)_2$, the hardest substance in the body. Tooth cavities form when acids dissolve tooth enamel:

$$Ca_{10}(PO_4)_6(OH)_2(s) + 8\,H^+(aq) \longrightarrow$$

$$10\,Ca^{2+}(aq) + 6\,HPO_4^{2-}(aq) + 2\,H_2O(l)$$

The Ca^{2+} and HPO_4^{2-} ions diffuse out of the enamel and are washed away by saliva. The acids that attack the hydroxyapatite are formed by the action of bacteria on sugars and other carbohydrates present in the plaque adhering to the teeth.

Fluoride ion, which is added to municipal water systems and toothpastes, can react with hydroxyapatite to form fluoroapatite, $Ca_{10}(PO_4)_6F_2$. This mineral, in which F^- has replaced OH^-, is much more resistant to attack by acids because the fluoride ion is a much weaker Brønsted–Lowry base than the hydroxide ion.

The usual concentration of F^- in municipal water systems is 1 mg/L (1 ppm). The compound added may be NaF or Na_2SiF_6. The silicon–fluorine anion reacts with water to release fluoride ions:

$$SiF_6^{2-}(aq) + 2\,H_2O(l) \longrightarrow 6\,F^-(aq) + 4\,H^+(aq) + SiO_2(s)$$

About 80% of all toothpastes now sold in the United States contain fluoride compounds, usually at the level of 0.1% fluoride by mass. The most common compounds in toothpastes are sodium fluoride (NaF), sodium monofluorophosphate (Na_2PO_3F), and stannous fluoride (SnF_2).

Related Exercises: 17.100, 17.116

Formation of Complex Ions

A characteristic property of metal ions is their ability to act as Lewis acids toward water molecules, which serve as Lewis bases. (Section 16.1) Lewis bases other than water can also interact with metal ions, particularly transition-metal ions. Such interactions can dramatically affect the solubility of a metal salt. For example, AgCl ($K_{sp} = 1.8 \times 10^{-10}$) dissolves in the presence of aqueous ammonia because Ag^+ interacts with the Lewis base NH_3, as shown in **Figure 17.19**. This process can be viewed as the sum of two reactions:

Step 1: $$AgCl(s) \rightleftharpoons Ag^+(aq) + Cl^-(aq)$$

Step 2: $$Ag^+(aq) + 2\,NH_3(aq) \rightleftharpoons Ag(NH_3)_2^+(aq)$$

Overall: $$AgCl(s) + 2\,NH_3(aq) \rightleftharpoons Ag(NH_3)_2^+(aq) + Cl^-(aq)$$

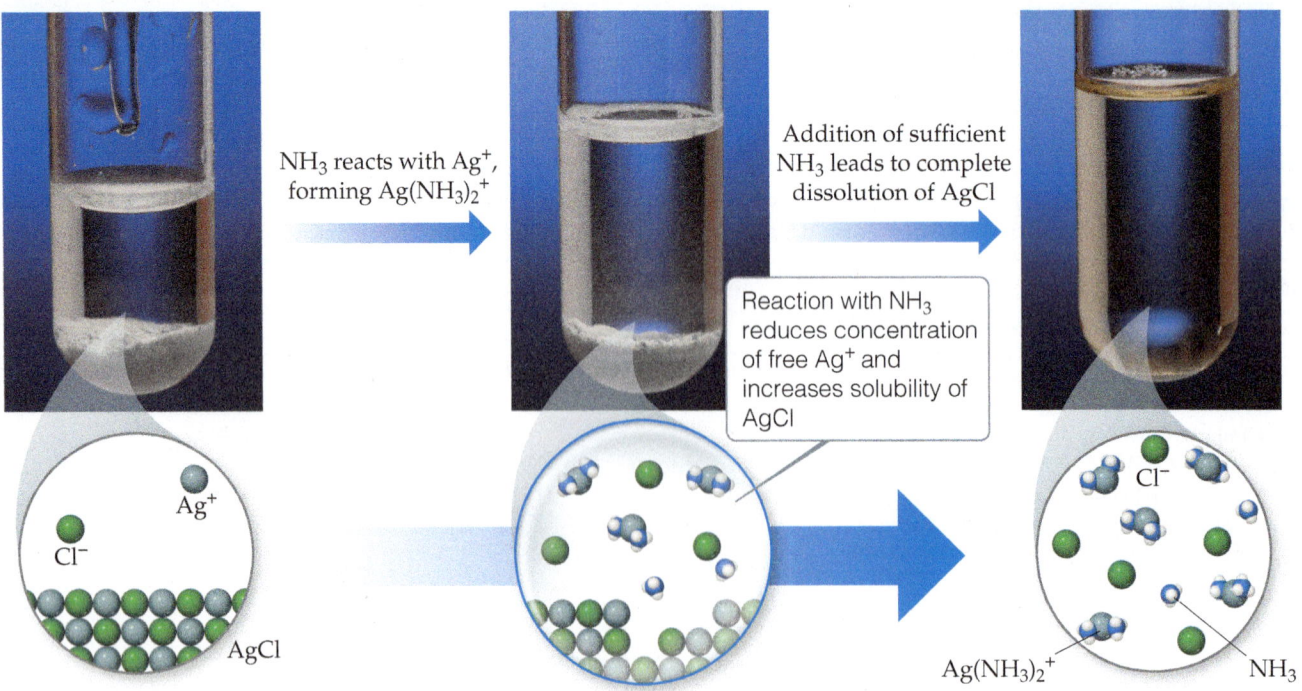

$$AgCl(s) + 2\,NH_3(aq) \rightleftharpoons Ag(NH_3)_2^+(aq) + Cl^-(aq)$$

▲ **Figure 17.19 Concentrated NH₃(aq) dissolves AgCl(s), which otherwise has very low solubility in water.**

TABLE 17.1 Formation Constants for Some Metal Complex Ions in Water at 25 °C

Complex Ion	K_f	Chemical Equation
$Ag(NH_3)_2^+$	1.7×10^7	$Ag^+(aq) + 2\,NH_3(aq) \rightleftharpoons Ag(NH_3)_2^+\,(aq)$
$Ag(CN)_2^-$	1×10^{21}	$Ag^+(aq) + 2\,CN^-(aq) \rightleftharpoons Ag(CN)_2^-\,(aq)$
$Ag(S_2O_3)_2^{3-}$	2.9×10^{13}	$Ag^+(aq) + 2\,S_2O_3^{2-}(aq) \rightleftharpoons Ag(S_2O_3)_2^{3-}\,(aq)$
$Al(OH)_4^-$	1.1×10^{33}	$Al^{3+}(aq) + 4\,OH^-(aq) \rightleftharpoons Al(OH)_4^-\,(aq)$
$CdBr_4^{2-}$	5×10^3	$Cd^{2+}(aq) + 4\,Br^-(aq) \rightleftharpoons CdBr_4^{2-}\,(aq)$
$Cr(OH)_4^-$	8×10^{29}	$Cr^{3+}(aq) + 4\,OH^-(aq) \rightleftharpoons Cr(OH)_4^-\,(aq)$
$Co(SCN)_4^{2-}$	1×10^3	$Co^{2+}(aq) + 4\,SCN^-(aq) \rightleftharpoons Co(SCN)_4^{2-}\,(aq)$
$Cu(NH_3)_4^{2+}$	5×10^{12}	$Cu^{2+}(aq) + 4\,NH_3(aq) \rightleftharpoons Cu(NH_3)_4^{2+}\,(aq)$
$Cu(CN)_4^{2-}$	1×10^{25}	$Cu^{2+}(aq) + 4\,CN^-(aq) \rightleftharpoons Cu(CN)_4^{2+}\,(aq)$
$Ni(NH_3)_6^{2+}$	1.2×10^9	$Ni^{2+}(aq) + 6\,NH_3(aq) \rightleftharpoons Ni(NH_3)_6^{2+}\,(aq)$
$Fe(CN)_6^{4-}$	1×10^{35}	$Fe^{2+}(aq) + 6\,CN^-(aq) \rightleftharpoons Fe(CN)_6^{4-}\,(aq)$
$Fe(CN)_6^{3-}$	1×10^{42}	$Fe^{3+}(aq) + 6\,CN^-(aq) \rightleftharpoons Fe(CN)_6^{3-}\,(aq)$
$Zn(OH)_4^{2-}$	4.6×10^{17}	$Zn^{2+}(aq) + 4\,OH^-(aq) \rightleftharpoons Zn(OH)_4^{2-}\,(aq)$

The presence of NH_3 drives the reaction, the dissolution of AgCl, to the right as $Ag^+(aq)$ is consumed to form $Ag(NH_3)_2^+$, which is a very soluble species.

For a Lewis base such as NH_3 to increase the solubility of a metal salt, the base must be able to interact more strongly with the metal ion than water does. In other words, the NH_3 must displace solvating H_2O molecules (Section 13.1) in order to form $[Ag(NH_3)_2]^+$:

$$Ag^+(aq) + 2\,NH_3(aq) \rightleftharpoons Ag(NH_3)_2^+(aq) \qquad [17.9]$$

An assembly of a metal ion and the Lewis bases bonded to it, such as $Ag(NH_3)_2^+$, is called a **complex ion**. Complex ions are very soluble in water. The stability of a complex ion in aqueous solution can be judged by the size of the equilibrium constant for its formation from the hydrated metal ion. For example, the equilibrium constant for the formation of $Ag(NH_3)_2^+$ is

$$K_f = \frac{[Ag(NH_3)_2^+]}{[Ag^+][NH_3]^2} = 1.7 \times 10^7 \qquad [17.10]$$

Note that the equilibrium constant for this kind of reaction is called a **formation constant**, K_f. The formation constants for several complex ions are listed in Table 17.1.

The general rule is that the solubility of metal salts increases in the presence of suitable Lewis bases, such as NH_3, CN^-, or OH^-, provided the metal forms a complex with the base. The ability of metal ions to form complexes is an extremely important aspect of their chemistry.

Sample Exercise 17.16

Evaluating an Equilibrium Involving a Complex Ion

Calculate the concentration of Ag^+ present in solution at equilibrium when concentrated ammonia is added to a 0.010 M solution of $AgNO_3$ to give an equilibrium concentration of $[NH_3] = 0.20\ M$. Neglect the small volume change that occurs when NH_3 is added.

SOLUTION

Analyze Addition of $NH_3(aq)$ to $Ag^+(aq)$ forms $Ag(NH_3)_2^+(aq)$, as shown in Equation 17.9. We are asked to determine what concentration of $Ag^+(aq)$ remains uncombined when the NH_3 concentration is brought to 0.20 M in a solution originally 0.010 M in $AgNO_3$.

Plan We assume that the $AgNO_3$ is completely dissociated, giving 0.010 M Ag^+. Because K_f for the formation of $Ag(NH_3)_2^+$ is quite large (see Equation 17.10), we assume that essentially all the Ag^+ is converted to $Ag(NH_3)_2^+$ and approach the problem as though we are concerned with the dissociation of $Ag(NH_3)_2^+$ rather than its formation. To facilitate this approach, we need to reverse the K_f

Continued

reaction (Equation 17.9) and make the corresponding change to the equilibrium constant:

$$Ag(NH_3)_2^+(aq) \rightleftharpoons Ag^+(aq) + 2\,NH_3(aq)$$

$$\frac{1}{K_f} = \frac{1}{1.7 \times 10^7} = 5.9 \times 10^{-8}$$

Solve If $[Ag^+]$ is 0.010 M initially, $[Ag(NH_3)_2^+]$ will be 0.010 M following addition of the NH_3. We construct a table to solve this equilibrium problem. Note that the NH_3 concentration given in the problem is an equilibrium concentration rather than an initial concentration.

	$Ag(NH_3)_2^+(aq) \rightleftharpoons$	$Ag^+(aq) +$	$2\,NH_3(aq)$
Initial (M)	0.010	0	—
Change (M)	$-x$	$+x$	—
Equilibrium (M)	$(0.010-x)$	x	0.20

Because $[Ag^+]$ is very small, we can assume that x is small compared to 0.010. Substituting these values into the equilibrium-constant expression for the dissociation of $Ag(NH_3)_2^+$, we obtain

$$\frac{[Ag^+][NH_3]^2}{[Ag(NH_3)_2^+]} = \frac{(x)(0.20)^2}{0.010} = 5.9 \times 10^{-8}$$

$$x = 1.5 \times 10^{-8}\,M = [Ag^+]$$

Formation of the $Ag(NH_3)_2^+$ complex drastically reduces the concentration of free Ag^+ ion in solution.

▶ **Practice Exercise**
Calculate $[Cr^{3+}]$ in equilibrium with $Cr(OH)_4^-(aq)$ when 0.010 mol of $Cr(NO_3)_3$ is dissolved in 1 L of solution buffered at pH 10.0.

Amphoterism

Some metal oxides and hydroxides that are relatively insoluble in water dissolve in strongly acidic and strongly basic solutions. These substances, called **amphoteric oxides** and **amphoteric hydroxides**,* are soluble in strong acids and bases because they themselves are capable of behaving as either an acid or base. Examples of amphoteric substances include the oxides and hydroxides of Al^{3+}, Cr^{3+}, Zn^{2+}, and Sn^{2+}.

Like other metal oxides and hydroxides, amphoteric species dissolve in acidic solutions because their anions, O^{2-} or OH^-, react with acids. What makes amphoteric oxides and hydroxides special, though, is that they also dissolve in strongly basic solutions. This behavior results from the formation of complex anions containing several (typically four) hydroxides bound to the metal ion (**Figure 17.20**):

$$Al(OH)_3(s) + OH^-(aq) \rightleftharpoons Al(OH)_4^-(aq)$$

The extent to which an insoluble metal hydroxide reacts with either acid or base varies with the particular metal ion involved. Many metal hydroxides—such as $Ca(OH)_2$, $Fe(OH)_2$, and $Fe(OH)_3$—are capable of dissolving in acidic solution but do not react with excess base. These hydroxides are not amphoteric.

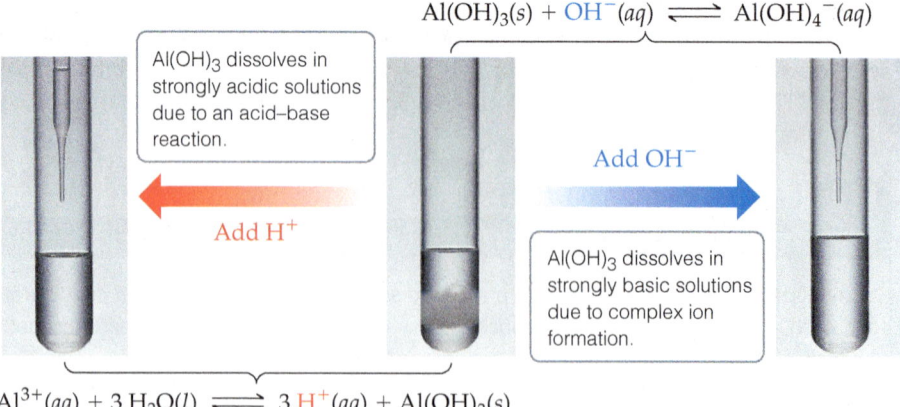

$$Al^{3+}(aq) + 3\,H_2O(l) \rightleftharpoons 3\,H^+(aq) + Al(OH)_3(s)$$

▲ **Figure 17.20 Amphoterism.** Some metal oxides and hydroxides, such as $Al(OH)_3$, are amphoteric, which means they dissolve in both strongly acidic and strongly basic solutions.

---

*Notice that the term *amphoteric* is applied to the behavior of insoluble oxides and hydroxides that dissolve in acidic or basic solutions. The similar term *amphiprotic* (Section 16.1) relates more generally to any molecule or ion that can either gain or lose a proton.

The purification of aluminum ore in the manufacture of aluminum metal provides an interesting application of amphoterism. As we have seen, $Al(OH)_3$ is amphoteric, whereas $Fe(OH)_3$ is not. Aluminum occurs in large quantities as the ore *bauxite*, which is essentially hydrated Al_2O_3 contaminated with Fe_2O_3. When bauxite is added to a strongly basic solution, the Al_2O_3 dissolves because the aluminum forms complex ions, such as $Al(OH)_4^-$. The Fe_2O_3 impurity, however, is not amphoteric and remains as a solid. The solution is filtered, getting rid of the iron impurity. Aluminum hydroxide is then precipitated by addition of acid. The purified hydroxide receives further treatment and eventually yields aluminum metal.

A CLOSER LOOK | Lead Contamination in Drinking Water

Access to clean drinking water is something most people in the industrialized world take for granted. Unfortunately, there are rare instances in which tap water is not safe to drink, as illustrated by the discovery in 2015 of elevated levels of lead in the municipal water supply of Flint, Michigan.

Lead is detrimental to many organs in the human body, but the brain and central nervous system are particularly sensitive to its presence. In the brain, Pb^{2+} ions interfere with cell communication and growth by mimicking Ca^{2+} ions. One of the most serious side effects of lead poisoning occurs in young children, where it leads to cognitive impairment. Although lead compounds were once used in a variety of applications—as a gasoline additive, in pigments, shotgun pellets, glass, and water pipes—our daily exposure to lead dropped dramatically once governmental agencies started regulating its use in the 1970s. According to the National Health and Nutrition Examination Survey, the mean concentration of lead in the blood for an average U.S. resident dropped from 150 ppb in 1976 to 16 ppb by 2002, nearly an order-of-magnitude decrease.

The regulatory limit set by the U.S. Environmental Protection Agency (EPA) for lead in drinking water is 15 parts per billion (ppb). According to EPA regulations, utilities serving more than 50,000 people must monitor the level of lead in their water and take corrective action if more than 10% of the homes sampled exceed the 15 ppb limit. Tests performed on samples collected in September 2015 by researchers from Virginia Tech found that in 10% of the 252 Flint homes tested the lead concentration exceeded 25 ppb, and in several homes the concentration exceeded 100 ppb. At the same time, a local pediatrician analyzed the results of infant blood tests and found that the percentage of children with elevated concentrations of lead in their bloodstream (>50 ppb) had doubled from 2.4% in 2013 to 4.9% in 2015.

The troubles began in April 2014 when the city began using the nearby Flint River as the natural source of its municipal water. Prior to that, Flint obtained water from Detroit, where water taken from Lake Huron was treated before piping it to Flint. The source of the lead was not the Flint River itself, but corrosion from lead pipes that are present in the underground water distribution network. When water is properly treated, a passivation layer of insoluble lead salts builds up on the inner surface of the lead pipes (Figure 17.21). This layer prevents corrosion that would otherwise allow lead to be oxidized and dissolve into the water as Pb^{2+} ions. The water treatment facility in Detroit was adding phosphate ions, PO_4^{3-}, to their water to inhibit corrosion, whereas the people managing water treatment in Flint elected not to do so. The presence of PO_4^{3-} ions promotes the formation of highly insoluble phosphate salts on the inner surface of the pipes that helps prevent corrosion.

Another factor that appears to have contributed to the problem is a drop in the pH of the water, from 8.0 in December 2014 to 7.3 in August 2015. Because the insoluble lead salts that form the passivating layer, like $Pb_3(PO_4)_2$ and $PbCO_3$, contain anions that can act as weak bases, anything that makes the water more acidic increases their solubility.

Another contributing factor was the presence of high levels of chloride ions. The treated water from Detroit had chloride levels of about 11 ppm, whereas the treated Flint water had chloride levels of 85 ppm in August 2015. While $PbCl_2$ is fairly insoluble ($K_{sp} = 1.7 \times 10^{-5}$), high concentrations of chloride ions can lead to the formation of soluble complex ions such as $PbCl_3^-$ and $PbCl_4^{2-}$. The increased chloride levels were due in part to the addition of $FeCl_3$, which was used to help coagulate and filter out unwanted organic matter that was leading to problems with *E. coli* contamination. Chloride ions are also produced when unwanted organic matter is oxidized by hypochlorite ions, which are added to kill bacteria. Runoff containing chloride salts used to treat icy roads in the winter may have also contributed.

Although the use of lead in plumbing has been banned in the United States since 1986, it is estimated that millions of miles of buried lead pipe are still in use in America's cities. Vigilance by water treatment facilities and environmental protection agencies is needed to avoid a repeat of the tragedy in Flint.

Related Exercises: 17.97, 17.101

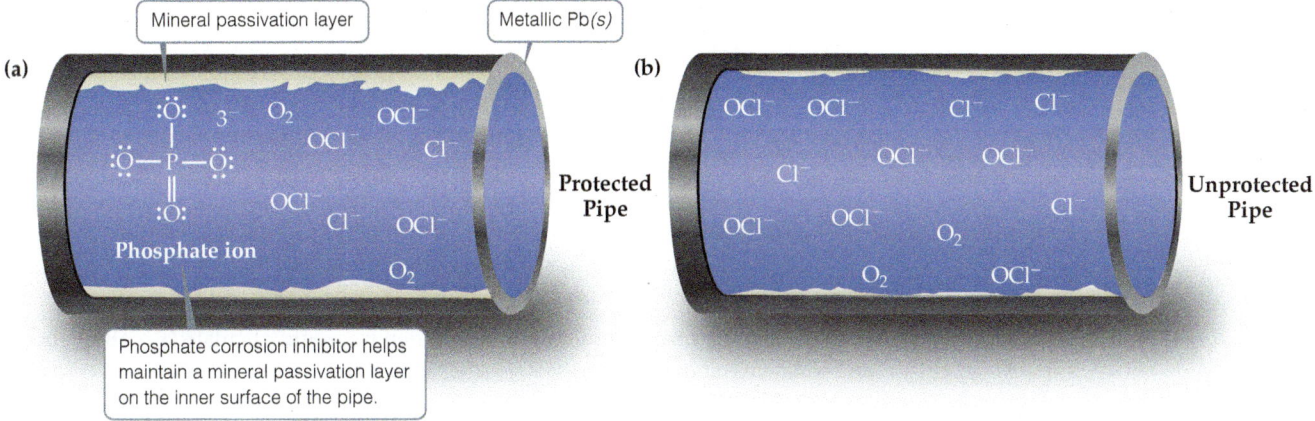

▲ **Figure 17.21 Protected and unprotected lead pipes. (a)** A lead pipe that has a protective passivation layer, and **(b)** a lead pipe where the lack of a phosphate corrosion inhibitor causes the passivation layer to dissolve and fall off, exposing the lead to oxidizing agents such as O_2 and OCl^-.

Self-Assessment Exercises

SAE 17.17 What is the molar solubility of $BaSO_4$ (molar mass — 233.4 g/mol, $K_{sp} = 1.1 \times 10^{-10}$) in a 0.050 M solution of $BaCl_2$? **(a)** 0.050 M **(b)** $1.0 \times 10^{-5} M$ **(c)** $5.1 \times 10^{-7} M$ **(d)** $2.2 \times 10^{-9} M$ **(e)** $1.1 \times 10^{-10} M$

SAE 17.18 Consider a saturated solution of $ZnCO_3$ ($K_{sp} = 1 \times 10^{-10}$) in contact with undissolved $ZnCO_3$. If a small amount of concentrated hydrochloric acid $HCl(aq)$ is added to the solution, the amount of undissolved $ZnCO_3$ will _____. If some solid $Zn(NO_3)_2$ is dissolved in the solution, the amount of undissolved $ZnCO_3$ will _____. **(a)** increase, increase **(b)** increase, decrease **(c)** decrease, increase **(d)** decrease, decrease **(e)** stay the same, increase

SAE 17.19 If 1.70 g of $AgNO_3$ (molar mass = 169.9 g/mol) is dissolved in 150 mL of 0.20 M hydrochloric acid, the complex ion $AgCl_2^-$ ($K_f = 1.1 \times 10^5$) will form. When the solution reaches equilibrium, what will be the concentrations of Ag^+ and $AgCl_2^-$? **(a)** $[Ag^+] = 1.4 \times 10^{-4} M$, $[AgCl_2^-] = 6.7 \times 10^{-2} M$ **(b)** $[Ag^+] = 1.4 \times 10^{-4} M$, $[AgCl_2^-] = 1.4 \times 10^{-4} M$ **(c)** $[Ag^+] = 9.1 \times 10^{-6} M$, $[AgCl_2^-] = 6.7 \times 10^{-2} M$ **(d)** $[Ag^+] = 3.4 \times 10^{-5} M$, $[AgCl_2^-] = 6.7 \times 10^{-2} M$

17.6 | Precipitation and Separation of Ions

Learning Objectives

When you finish Section 17.6, you should be able to:

▶ Predict whether an ionic compound will precipitate when aqueous solutions containing soluble salts are mixed.

▶ Predict the order of precipitation of ionic compounds, and thus a way to separate compounds, when soluble salts are mixed.

Equilibrium can be achieved starting with the substances on either side of a chemical equation. For example, consider the dissolution of barium sulfate again:

$$BaSO_4(s) \rightleftharpoons Ba^{2+}(aq) + SO_4^{2-}(aq)$$

The equilibrium that exists between $BaSO_4(s)$, $Ba^{2+}(aq)$, and $SO_4^{2-}(aq)$ can be achieved by starting either with $BaSO_4(s)$ or with solutions containing Ba^{2+} and SO_4^{2-}. If we mix, say, a $BaCl_2$ aqueous solution with a Na_2SO_4 aqueous solution, solid $BaSO_4$ may or may not precipitate out, depending on the concentrations of the two ions. In this section we explain how to predict whether a precipitate will form under various conditions.

Recall that we used the reaction quotient Q in Section 15.6 to determine the direction in which a reaction must proceed to reach equilibrium. The form of Q is the same as the equilibrium-constant expression for a reaction, but instead of using equilibrium concentrations, we can use the concentrations that are present at any point in time. The direction in which a reaction proceeds to reach equilibrium depends on the relationship between Q and K for the reaction. If $Q < K$, the product concentrations are too low and reactant concentrations are too high relative to the equilibrium concentrations, and so the reaction will proceed to the right (toward products) to achieve equilibrium. If $Q > K$, product concentrations are too high and reactant concentrations are too low, and so the reaction will proceed to the left (toward reactants) to achieve equilibrium. If $Q = K$, the reaction is at equilibrium.

For solubility-product equilibria, the relationship between Q and K_{sp} is exactly like that for other equilibria. For K_{sp} reactions, products are always the soluble ions, and the reactant is always the solid.

Therefore, for solubility equilibria,

- If $Q = K_{sp}$, the system is at equilibrium, which means the solution is saturated; this is the highest concentration the solution can have without precipitating.
- If $Q < K_{sp}$, the reaction will proceed to the right, toward the soluble ions; no precipitate will form.
- If $Q > K_{sp}$, the reaction will proceed to the left, toward the solid; precipitate will form.

For the case of the barium sulfate solution, then we would calculate $Q = [Ba^{2+}][SO_4^{2-}]$, and compare this quantity to the K_{sp} for barium sulfate.

Sample Exercise 17.17
Predicting Whether a Precipitate Forms

Does a precipitate form when 0.10 L of $8.0 \times 10^{-3}\,M\,Pb(NO_3)_2$ is added to 0.40 L of $5.0 \times 10^{-3}\,M\,Na_2SO_4$?

SOLUTION

Analyze The problem asks us to determine whether a precipitate forms when two salt solutions are combined.

Plan We should determine the concentrations of all ions just after the solutions are mixed and compare the value of Q with K_{sp} for any potentially insoluble product. The possible metathesis products are $PbSO_4$ and $NaNO_3$. Like all sodium salts, $NaNO_3$ is soluble, but $PbSO_4$ has a K_{sp} of 6.3×10^{-7} (Appendix D) and will precipitate if the Pb^{2+} and SO_4^{2-} concentrations are high enough for Q to exceed K_{sp}.

Solve

When the two solutions are mixed, the volume is $0.10\,L + 0.40\,L = 0.50\,L$. The number of moles of Pb^{2+} in 0.10 L of $8.0 \times 10^{-3}\,M\,Pb(NO_3)_2$ is:

$$(0.10\,L)\left(\frac{8.0 \times 10^{-3}\,mol}{L}\right) = 8.0 \times 10^{-4}\,mol$$

The concentration of Pb^{2+} in the 0.50-L mixture is therefore:

$$[Pb^{2+}] = \frac{8.0 \times 10^{-4}\,mol}{0.50\,L} = 1.6 \times 10^{-3}\,M$$

The number of moles of SO_4^{2-} in 0.40 L of $5.0 \times 10^{-3}\,M\,Na_2SO_4$ is:

$$(0.40\,L)\left(\frac{5.0 \times 10^{-3}\,mol}{L}\right) = 2.0 \times 10^{-3}\,mol$$

Therefore:

$$[SO_4^{2-}] = \frac{2.0 \times 10^{-3}\,mol}{0.50\,L} = 4.0 \times 10^{-3}\,M$$

and:

$$Q = [Pb^{2+}][SO_4^{2-}] = (1.6 \times 10^{-3})(4.0 \times 10^{-3}) = 6.4 \times 10^{-6}$$

Because $Q > K_{sp}$, $PbSO_4$ precipitates.

▶ **Practice Exercise**
Does a precipitate form when 0.050 L of $2.0 \times 10^{-2}\,M$ NaF is mixed with 0.010 L of $1.0 \times 10^{-2}\,M\,Ca(NO_3)_2$?

Selective Precipitation of Ions

Ions can be separated from each other based on the solubilities of their salts. Consider a solution containing both Ag^+ and Cu^{2+}. If HCl is added to the solution, AgCl ($K_{sp} = 1.8 \times 10^{-10}$) precipitates, while Cu^{2+} remains in solution because $CuCl_2$ is soluble. Separation of ions in an aqueous solution by using a reagent that forms a precipitate with one or more (but not all) of the ions is called *selective precipitation*.

Sulfide ion is often used to separate metal ions because the solubilities of sulfide salts span a wide range and depend greatly on solution pH. For example, Cu^{2+} and Zn^{2+} can be separated by bubbling H_2S gas through an acidified solution containing these two cations. Because CuS ($K_{sp} = 6 \times 10^{-37}$) is less soluble than ZnS ($K_{sp} = 2 \times 10^{-25}$), CuS precipitates from an acidified solution (pH $\approx$ 1) while ZnS does not (**Figure 17.22**):

$$Cu^{2+}(aq) + H_2S(aq) \rightleftharpoons CuS(s) + 2\,H^+(aq) \qquad [17.11]$$

The CuS can be separated from the Zn^{2+} solution by filtration. The separated CuS can then be dissolved by raising the concentration of H^+ even further, shifting the equilibrium concentrations of the compounds in Equation 17.11 to the left.

Sample Exercise 17.18
Selective Precipitation

A solution contains $1.0 \times 10^{-2}\,M\,Ag^+(aq)$ and $2.0 \times 10^{-2}\,M\,Pb^{2+}(aq)$. When $Cl^-(aq)$ is added, both AgCl ($K_{sp} = 1.8 \times 10^{-10}$) and $PbCl_2$ ($K_{sp} = 1.7 \times 10^{-5}$) can precipitate. What concentration of $Cl^-(aq)$ is necessary to begin the precipitation of each salt? Which salt precipitates first?

Continued

SOLUTION

Analyze We are asked to determine the concentration of $Cl^-(aq)$ necessary to begin the precipitation from a solution containing $Ag^+(aq)$ and $Pb^{2+}(aq)$ ions, and to predict which metal chloride will begin to precipitate first.

Plan We are given K_{sp} values for the two precipitates. Using these and the metal ion concentrations, we can calculate what $Cl^-(aq)$ concentration is necessary to precipitate each salt. The salt requiring the lower $Cl^-(aq)$ ion concentration precipitates first.

Solve For AgCl we have $K_{sp} = [Ag^+][Cl^-] = 1.8 \times 10^{-10}$.

Because $[Ag^+] = 1.0 \times 10^{-2} M$, the greatest concentration of $Cl^-(aq)$ that can be present without causing precipitation of AgCl can be calculated from the K_{sp} expression:

$$K_{sp} = (1.0 \times 10^{-2})[Cl^-] = 1.8 \times 10^{-10}$$

$$[Cl^-] = \frac{1.8 \times 10^{-10}}{1.0 \times 10^{-2}} = 1.8 \times 10^{-8} M$$

Comment Precipitation of AgCl will keep the $Cl^-(aq)$ concentration low until the number of moles of $Cl^-(aq)$ added exceeds the number of moles of $Ag^+(aq)$ in the solution. Once past this point, $[Cl^-]$ rises sharply and $PbCl_2$ will soon begin to precipitate.

Any $Cl^-(aq)$ in excess of this very small concentration will cause AgCl to precipitate from solution. Proceeding similarly for $PbCl_2$, we have

$$K_{sp} = [Pb^{2+}][Cl^-]^2 = 1.7 \times 10^{-5}$$

$$(2.0 \times 10^{-2})[Cl^-]^2 = 1.7 \times 10^{-5}$$

$$[Cl^-]^2 = \frac{1.7 \times 10^{-5}}{2.0 \times 10^{-2}} = 8.5 \times 10^{-4}$$

$$[Cl^-] = \sqrt{8.5 \times 10^{-4}} = 2.9 \times 10^{-2} M$$

Thus, a concentration of $Cl^-(aq)$ in excess of $2.9 \times 10^{-2} M$ causes $PbCl_2$ to precipitate.

Comparing the $Cl^-(aq)$ concentration required to precipitate each salt, we see that as $Cl^-(aq)$ is added, AgCl precipitates first because it requires a much smaller concentration of Cl^-. Thus, $Ag^+(aq)$ can be separated from $Pb^{2+}(aq)$ by slowly adding $Cl^-(aq)$ so that the chloride ion concentration remains between $1.8 \times 10^{-8} M$ and $2.9 \times 10^{-2} M$.

▶ **Practice Exercise**
A solution consists of $0.050 M Mg^{2+}(aq)$ and $Cu^{2+}(aq)$. Which ion precipitates first as $OH^-(aq)$ is added? What concentration of $OH^-(aq)$ is necessary to begin the precipitation of each cation? [$K_{sp} = 1.8 \times 10^{-11}$ for $Mg(OH)_2$, and $K_{sp} = 4.8 \times 10^{-20}$ for $Cu(OH)_2$.]

▼ **Go Figure** Why is the sulfide reagent HS^- in the third test tube and not H_2S?

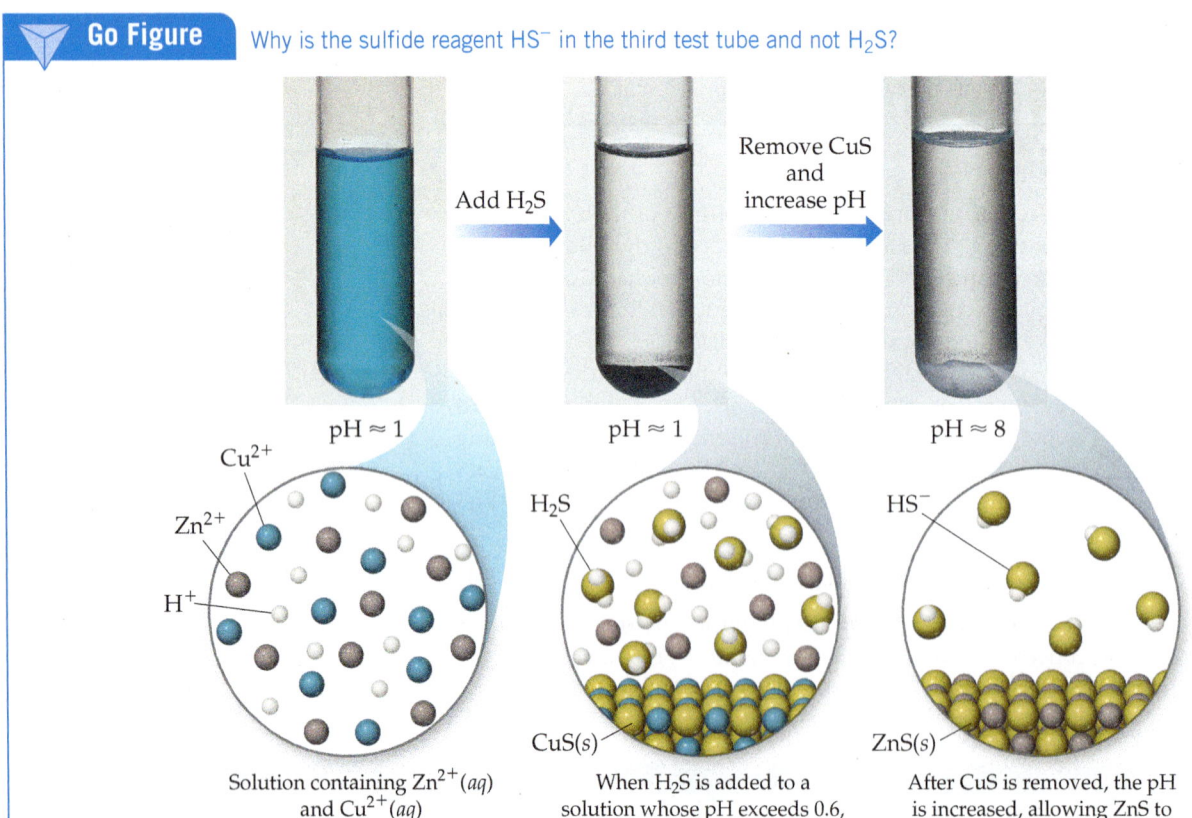

| Solution containing $Zn^{2+}(aq)$ and $Cu^{2+}(aq)$ | When H_2S is added to a solution whose pH exceeds 0.6, CuS precipitates | After CuS is removed, the pH is increased, allowing ZnS to precipitate |

▲ **Figure 17.22 Selective precipitation.** In this example, Cu^{2+} ions are separated from Zn^{2+} ions.

Self-Assessment Exercises

SAE 17.20 In the laboratory, you mix 75 mL of a 0.100 M $Pb(NO_3)_2$ solution with 25 mL of 0.044 M KCl solution, anticipating the precipitation of $PbCl_2$ ($K_{sp} = 1.7 \times 10^{-5}$). The reaction quotient for the precipitation reaction will be $Q =$ _____, which means that $PbCl_2$ _____ precipitate from solution. (a) 1.9×10^{-4}, will (b) 1.9×10^{-4}, will not (c) 9.1×10^{-6}, will (d) 9.1×10^{-6}, will not (e) 8.2×10^{-4}, will

SAE 17.21 An aqueous solution contains a mixture of Ag^+ and Pb^{2+} ions, with concentrations of $[Ag^+] = 0.32 M$ and $[Pb^{2+}] = 8.8 \times 10^{-3} M$. If a $Na_2SO_4(aq)$ solution is added dropwise, _____ will precipitate first, when $[SO_4^{2-}]$ reaches _____. The solubility product constant of Ag_2SO_4 is $K_{sp} = 1.5 \times 10^{-5}$, and that of $PbSO_4$ is $K_{sp} = 6.3 \times 10^{-7}$. (a) Ag_2SO_4, $4.7 \times 10^{-5} M$ (b) Ag_2SO_4, $1.5 \times 10^{-4} M$ (c) $PbSO_4$, $7.2 \times 10^{-5} M$ (d) $PbSO_4$, $1.2 \times 10^{-3} M$

17.7 | Qualitative Analysis for Metallic Elements

In this final section, we look at how solubility equilibria and complex-ion formation can be used to detect the presence of particular metal ions in solution. Before the development of modern analytical instrumentation, it was necessary to analyze mixtures of metals in a sample by what were called *wet chemical methods*. For example, an ore sample that might contain several metallic elements was dissolved in a concentrated acid solution that was then tested in a systematic way for the presence of various metal ions.

Qualitative analysis determines only the presence or absence of a particular metal ion relative to some threshold, whereas **quantitative analysis** determines how much of a given substance is present. Even though wet methods of qualitative analysis have become less important in the chemical industry, they are frequently used in general chemistry laboratory programs to illustrate equilibria, to teach the properties of common metal ions in solution, and to develop laboratory skills. Typically, such analyses proceed in three stages: (1) The ions are separated into broad groups on the basis of solubility properties. (2) The ions in each group are separated by selectively dissolving members in the group. (3) The ions are identified by means of specific tests.

A scheme in general use divides the common cations into five groups (**Figure 17.23**). The order in which reagents are added is important in this scheme. The most selective separations—those that involve the smallest number of ions—are carried out first. The reactions used must proceed so far toward completion that any concentration of cations remaining in the solution is too small to interfere with subsequent tests.

Let's look at each of these five groups of cations, briefly examining the logic used in this qualitative analysis scheme.

Group 1. *Insoluble chlorides:* Of the common metal ions, only $Ag^+(aq)$, $Hg_2^{2+}(aq)$, and $Pb^{2+}(aq)$ form insoluble chlorides. Therefore, when HCl is added to a mixture of cations, only $AgCl$, Hg_2Cl_2, and $PbCl_2$ precipitate, leaving the other cations in solution. The absence of a precipitate indicates that the starting solution contains no $Ag^+(aq)$, $Hg_2^{2+}(aq)$, or $Pb^{2+}(aq)$.

Group 2. *Acid-insoluble sulfides:* After any insoluble chlorides have been removed, the remaining solution, now acidic from HCl treatment, is treated with H_2S. Because H_2S is a weak acid compared to HCl, its role here is to act as a source for small amounts of sulfide. Only the most insoluble metal sulfides—CuS, Bi_2S_3, CdS, PbS, HgS, As_2S_3, Sb_2S_3, and SnS_2—precipitate. (Note the very small values of K_{sp} for some of these sulfides in Appendix D.) Those metal ions whose sulfides are somewhat more soluble—for example, ZnS or NiS—remain in solution.

Group 3. *Base-insoluble sulfides and hydroxides:* After the solution is filtered to remove any acid-insoluble sulfides, it is made slightly basic, and $(NH_4)_2S$ is added. The concentration of $S^{2-}(aq)$ is higher in basic solutions than in acidic solutions. Under these conditions, the ion products for many of the more soluble sulfides exceed their K_{sp} values and thus precipitation occurs. The metal ions precipitated at this stage are $Al^{3+}(aq)$, $Cr^{3+}(aq)$, $Fe^{3+}(aq)$, $Zn^{2+}(aq)$, $Ni^{2+}(aq)$, $Co^{2+}(aq)$, and $Mn^{2+}(aq)$. [The $Al^{3+}(aq)$, $Fe^{3+}(aq)$, and $Cr^{3+}(aq)$ ions do not form insoluble sulfides; instead they precipitate as insoluble hydroxides, as Figure 17.23 shows.]

 **Learning Objective**

When you finish Section 17.7, **you should be able to:**

▶ Use differences in solubility to identify specific groups of cations in a solution containing an unknown mixture of cations.

Go Figure   If a solution contained a mixture of $Cu^{2+}(aq)$ and $Zn^{2+}(aq)$ ions, would this separation scheme work? After which step would the first precipitate be observed?

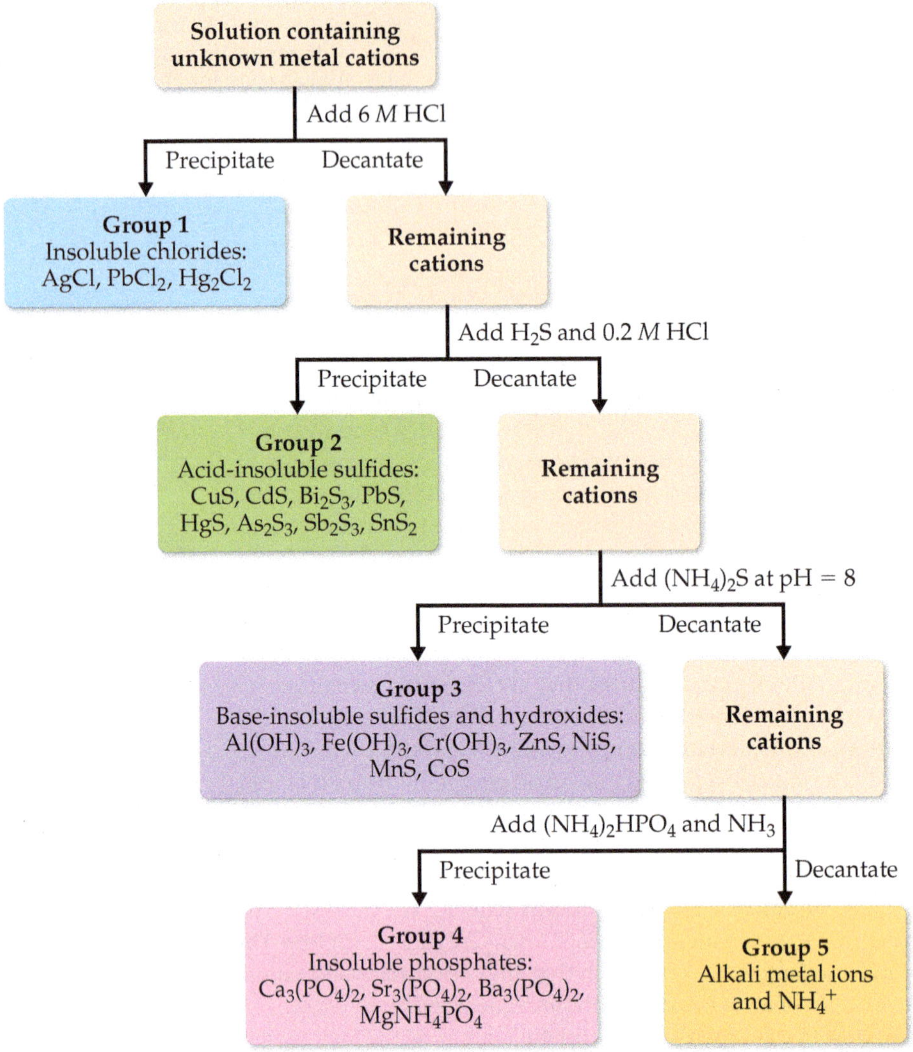

Solution containing unknown metal cations

Add 6 M HCl

Precipitate   Decantate

Group 1
Insoluble chlorides:
$AgCl$, $PbCl_2$, Hg_2Cl_2

Remaining cations

Add H_2S and 0.2 M HCl

Precipitate   Decantate

Group 2
Acid-insoluble sulfides:
CuS, CdS, Bi_2S_3, PbS,
HgS, As_2S_3, Sb_2S_3, SnS_2

Remaining cations

Add $(NH_4)_2S$ at pH = 8

Precipitate   Decantate

Group 3
Base-insoluble sulfides and hydroxides:
$Al(OH)_3$, $Fe(OH)_3$, $Cr(OH)_3$, ZnS, NiS,
MnS, CoS

Remaining cations

Add $(NH_4)_2HPO_4$ and NH_3

Precipitate   Decantate

Group 4
Insoluble phosphates:
$Ca_3(PO_4)_2$, $Sr_3(PO_4)_2$, $Ba_3(PO_4)_2$,
$MgNH_4PO_4$

Group 5
Alkali metal ions
and NH_4^+

▲ **Figure 17.23 Qualitative analysis.** A flowchart showing a common scheme for identifying cations.

Group 4. *Insoluble phosphates:* At this point, the solution contains only metal ions from groups 1A and 2A of the periodic table. Adding $(NH_4)_2HPO_4$ to a basic solution precipitates the group 2A elements $Mg^{2+}(aq)$, $Ca^{2+}(aq)$, $Sr^{2+}(aq)$, and $Ba^{2+}(aq)$ because these metals form insoluble phosphates.

Group 5. *The alkali metal ions and $NH_4^+(aq)$:* The ions that remain after removing the insoluble phosphates are tested for individually. A flame test can be used to determine the presence of $K^+(aq)$, for example, because the flame turns a characteristic violet color if $K^+(aq)$ is present (Figure 7.23).

 Self-Assessment Exercise

SAE 17.22   You are given a 25.0-mL solution and told that it may contain any combination of the following ions: Ag^+, Bi^{3+}, Zn^{2+}, and/or Ca^{2+}. Following the qualitative analysis scheme, you first add several drops of 6 M HCl(*aq*) to lower the pH to approximately 1, and you see no apparent reaction. Next, you bubble H_2S through the solution, which leads to the formation of a dark, brown–black precipitate. Next, you raise the pH to approximately 8 by adding NaOH(*aq*), before adding $(NH_4)_2S$. No apparent reaction is observed. Finally, the solution is treated with $(NH_4)_2HPO_4$, leading to the formation of a white precipitate. Which ions are present in the original solution? (**a**) Ag^+, Bi^{3+} and Ca^{2+} (**b**) Ag^+ and Ca^{2+} (**c**) Bi^{3+} and Ca^{2+} (**d**) Bi^{3+} and Zn^{2+} (**e**) Bi^{3+}, Zn^{2+} and Ca^{2+}

Putting Concepts Together

A sample of 1.25 L of HCl gas at 21 °C and 0.950 atm is bubbled through 0.500 L of 0.150 M NH_3 solution. Calculate the pH of the resulting solution assuming that all the HCl dissolves and that the volume of the solution remains 0.500 L.

SOLUTION

The number of moles of HCl gas is calculated from the ideal-gas law:

$$n = \frac{PV}{RT} = \frac{(0.950 \text{ atm})(1.25 \text{ L})}{(0.0821 \text{ L-atm/mol-K})(294 \text{ K})} = 0.0492 \text{ mol HCl}$$

The number of moles of NH_3 in the solution is given by the product of the volume of the solution and its concentration:

$$\text{Moles } NH_3 = (0.500 \text{ L})(0.150 \text{ mol } NH_3/\text{L}) = 0.0750 \text{ mol } NH_3$$

The acid HCl and base NH_3 react, transferring a proton from HCl to NH_3, producing NH_4^+ and Cl^- ions:

$$HCl(g) + NH_3(aq) \longrightarrow NH_4^+(aq) + Cl^-(aq)$$

To determine the pH of the solution, we first calculate the amount of each reactant and each product present at the completion of the reaction. Because you can assume this neutralization reaction proceeds as far toward the product side as possible, this is a limiting reactant problem.

	$HCl(g)$	$+\ NH_3(aq)$	$\longrightarrow NH_4^+(aq)$	$+\ Cl^-(aq)$
Before reaction (mol)	0.0492	0.0750	0	0
Change (limiting reactant) (mol)	−0.0492	−0.0492	+0.0492	+0.0492
After reaction (mol)	0	0.0258	0.0492	0.0492

Thus, the reaction produces a solution containing a mixture of NH_3, NH_4^+, and Cl^-. The NH_3 is a weak base ($K_b = 1.8 \times 10^{-5}$), NH_4^+ is its conjugate acid, and Cl^- is neither acidic nor basic. Consequently, the pH depends on $[NH_3]$ and $[NH_4^+]$:

$$[NH_3] = \frac{0.0258 \text{ mol } NH_3}{0.500 \text{ L soln}} = 0.0516 \, M$$

$$[NH_4^+] = \frac{0.0492 \text{ mol } NH_4^+}{0.500 \text{ L soln}} = 0.0984 \, M$$

We can calculate the pH using either K_b for NH_3 or K_a for NH_4^+. Using the K_b expression, we have:

	$NH_3(aq)$	$+\ H_2O(l) \rightleftharpoons$	$NH_4^+(aq)$	$+\ OH^-(aq)$
Initial (M)	0.0516	—	0.0984	0
Change (M)	−x	—	+x	+x
Equilibrium (M)	$(0.0516 - x)$	—	$(0.0984 + x)$	x

$$K_b = \frac{[NH_4^+][OH^-]}{[NH_3]} = \frac{(0.0984 + x)(x)}{(0.0516 - x)} \cong \frac{(0.0984)x}{0.0516} = 1.8 \times 10^{-5}$$

$$x = [OH^-] = \frac{(0.0516)(1.8 \times 10^{-5})}{0.0984} = 9.4 \times 10^{-6} \, M$$

Hence, pOH = $-\log(9.4 \times 10^{-6}) = 5.03$
and pH = $14.00 - \text{pOH} = 14.00 - 5.03 = 8.97$.

Chapter Summary and Key Terms

THE COMMON ION EFFECT (SECTION 17.1) AND BUFFERS (SECTION 17.2) The dissociation of a weak acid or weak base is repressed by the presence of a strong electrolyte that provides an ion common to the equilibrium (the **common-ion effect**). A particularly important type of acid–base mixture is that of a weak conjugate acid–base pair that functions as a **buffered solution (buffer)**. Addition of small amounts of a strong acid or a strong base to a buffered solution causes only small changes in pH because the buffer reacts with the added acid or base. (Strong acid–strong base, strong acid–weak base, and weak acid–strong base reactions proceed essentially to completion.) Buffered solutions are usually prepared from a weak acid and a salt of that acid or from a weak base and a salt of that base. Two important characteristics of a buffered solution are its **buffer capacity** and its pH range. The optimal pH of a buffer is equal to pK_a (or pK_b) of the acid (or base) used to prepare the buffer. The relationship between pH, pK_a, and the concentrations of an acid and its conjugate base can be expressed by the **Henderson–Hasselbalch equation**. It is important to realize that the Henderson–Hasselbalch equation is an approximation, and more detailed calculations may need to be performed to obtain equilibrium concentrations.

ACID–BASE TITRATIONS (SECTION 17.3) The plot of the pH of an acid (or base) as a function of the volume of added base (or acid) is called a **pH titration curve**. The titration curve of a strong acid–strong base titration exhibits a large change in pH in the immediate vicinity of the equivalence point; at the equivalence point for such a titration pH = 7. For strong acid–weak base or weak acid–strong base titrations, the pH change in the vicinity of the equivalence point is not as large as for a strong acid–strong base titration, nor will the pH equal 7 at the equivalence point in these cases. Instead, what determines the pH at the equivalence point is the acid–base characteristics of the salt solution that results from the neutralization reaction. For this reason, it is important to choose an indicator whose color change is near the pH at the equivalence point for titrations involving either weak acids or weak bases. It is possible to calculate the pH at any point of the titration curve by first considering the effects of the acid–base reaction on solution concentrations and then examining equilibria involving the remaining solute species.

SOLUBILITY EQUILIBRIA (SECTION 17.4) The equilibrium between an ionic solid and its ions in solution provides an example of heterogeneous equilibrium. The **solubility-product constant** (or simply the **solubility product**), K_{sp}, is an equilibrium constant that expresses quantitatively the extent to which the compound dissolves. The K_{sp} can be used to calculate the solubility of an ionic compound, and the solubility can be used to calculate K_{sp}.

FACTORS THAT AFFECT SOLUBILITY (SECTION 17.5) Several experimental factors, including temperature, affect the solubilities of ionic compounds in water. The solubility of a slightly soluble ionic compound is decreased by the presence of a second solute that furnishes a common ion (the common-ion effect). The solubility of compounds containing basic anions increases as the solution is made more acidic (as pH decreases). Salts with anions of negligible basicity (the anions of strong acids) are unaffected by pH changes.

The solubility of metal salts is also affected by the presence of certain Lewis bases that react with metal ions to form stable **complex ions**. Complex-ion formation in aqueous solution involves the displacement by Lewis bases (such as NH_3 and CN^-) of water molecules attached to the metal ion. The extent to which such complex formation occurs is expressed quantitatively by the **formation constant** for the complex ion. **Amphoteric oxides** and **hydroxides** are those that are only slightly soluble in water but dissolve on addition of either acid or base.

PRECIPITATION AND SEPARATION OF IONS (SECTION 17.6) Comparison of the reaction quotient, Q, with the value of K_{sp} can be used to judge whether a precipitate will form when solutions are mixed or whether a slightly soluble salt will dissolve under various conditions. Precipitates form when $Q > K_{sp}$. If two salts have sufficiently different solubilities, selective precipitation can be used to precipitate one ion while leaving the other in solution, effectively separating the two ions.

QUALITATIVE ANALYSIS FOR METALLIC ELEMENTS (SECTION 17.7) Metallic elements vary a great deal in the solubilities of their salts, in their acid–base behavior, and in their tendencies to form complex ions. These differences can be used to separate and detect the presence of metal ions in mixtures. **Qualitative analysis** determines the presence or absence of species in a sample, whereas **quantitative analysis** determines how much of each species is present. The qualitative analysis of metal ions in solution can be carried out by separating the ions into groups on the basis of precipitation reactions and then analyzing each group for individual metal ions.

Key Equations

- $$pH = pK_a + \log\frac{[\text{base}]}{[\text{acid}]} \qquad [17.5]$$

  The Henderson–Hasselbalch equation, used to estimate the pH of a buffer from the concentrations of a conjugate acid–base pair

- $$K_{sp} = [A^+]^n[X^-]^m \qquad [17.8]$$

  The definition of the solubility-product constant for a generic ionic solid A_nX_m

Exam Prep

EP 17.1 For the generic equilibrium, $HA(aq) \rightleftharpoons H^+(aq) + A^-(aq)$, which of these statements is *true*?

(a) The equilibrium constant for this reaction changes as the pH changes.

(b) If you add the soluble salt KA to a solution of HA that is at equilibrium, the concentration of HA would decrease.

(c) If you add the soluble salt KA to a solution of HA that is at equilibrium, the concentration of A^- would decrease.

(d) If you add the soluble salt KA to a solution of HA that is at equilibrium, the pH would increase.

EP 17.2 Calculate the concentration of the lactate ion in a solution that is 0.100 M in lactic acid $[CH_3CH(OH)COOH, pK_a = 3.86]$ and 0.080 M in HCl.

(a) 4.83 M

(b) 0.0800 M

(c) 7.3 $\times$ 10^{-3} M

(d) 3.65 $\times$ 10^{-3} M

(e) 1.73 $\times$ 10^{-4} M

EP 17.3 If the pH of a buffer solution is equal to the pK_a of the acid in the buffer, what does this tell you about the relative concentrations of the acid and conjugate base forms of the buffer components?

(a) The acid concentration must be zero. (b) The base concentration must be zero. (c) The acid and base concentrations must be equal. (d) The acid and base concentrations must be equal to the K_a. (e) The base concentration must be 2.3 times as large as the acid concentration.

EP 17.4 A buffer is made with sodium acetate (CH_3COONa) and acetic acid (CH_3COOH); the K_a for acetic acid is 1.80×10^{-5}. The pH of the buffer is 3.98. What is the ratio of the equilibrium concentration of sodium acetate to that of acetic acid? (a) 0.174 (b) 0.760 (c) 0.840 (d) 5.75 (e) Not enough information is given to answer this question.

EP 17.5 Calculate the number of grams of ammonium chloride that must be added to 2.00 L of a 0.500 M ammonia solution to obtain a buffer of pH = 9.20. Assume the volume of the solution does not change as the solid is added. The K_b for ammonia is 1.8×10^{-5}. (a) 60.7 g (b) 30.4 g (c) 1.52 g (d) 0.568 g (e) 1.59×10^{-5} g

EP 17.6 For the titration curve shown, at what point is the pH = pK_{a2}?

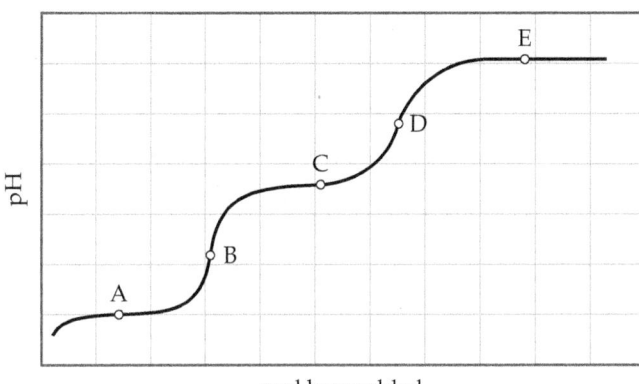

y-axis: pH; x-axis: mol base added

(a) A (b) B (c) C (d) D (e) E

EP 17.7 An acid–base titration is performed: 250.0 mL of an unknown concentration of HCl (aq) is titrated to the equivalence point with 36.7 mL of a 0.1000 M aqueous solution of NaOH. For this titration, which of the following statements is *false*? (a) The HCl solution is less concentrated than the NaOH solution. (b) The pH is less than 7 after adding 25 mL of NaOH solution. (c) The pH at the equivalence point is 7.00. (d) If an additional 1.00 mL of NaOH solution is added beyond the equivalence point, the pH of the solution will be greater than 7.00. (e) At the equivalence point, the OH⁻ concentration in the solution is $3.67 \times 10^{-3} M$.

EP 17.8 When NH_3 is titrated with HCl, the pH at the equivalence point will be _____ because of the presence of _____. (a) greater than 7, NH_3 (b) greater than 7, Cl^- (c) equal to 7, NH_4Cl (d) less than 7, NH_4^+ (e) less than 7, HCl

EP 17.9 Why is pH at the equivalence point greater than 7 when you titrate a weak acid with a strong base? (a) There is excess strong base at the equivalence point. (b) There is excess weak acid at the equivalence point. (c) The conjugate base that is formed at the equivalence point is a strong base. (d) The conjugate base that is formed at the equivalence point reacts with water. (e) This statement is false: The pH is always 7 at an equivalence point in a pH titration.

EP 17.10 A 25.0-mL sample of 0.100 M HBr solution is titrated with 0.050 M NaOH solution. What is the pH after 25.0 mL of the NaOH solution have been added? (a) 1.00 (b) 1.30 (c) 1.60 (d) 7.00 (e) 12.70

EP 17.11 Which of these expressions correctly expresses the solubility-product constant for Ag_3PO_4 in water? (a) $[Ag]^3[PO_4]$ (b) $[Ag^+][PO_4^{3-}]$ (c) $[Ag^+]^3[PO_4^{3-}]$ (d) $[Ag^+][PO_4^{3-}]^3$ (e) $[Ag^+]^3[PO_4^{3-}]^3$

EP 17.12 You add 10.0 grams of solid copper(II) phosphate, $Cu_3(PO_4)_2$, to a beaker and then add 100.0 mL of water to the beaker at $T = 298$ K. The solid does not appear to dissolve. You wait a long time, with occasional stirring, and eventually measure the equilibrium concentration of $Cu^{2+}(aq)$ in the water to be $5.01 \times 10^{-8} M$. What is the K_{sp} of copper(II) phosphate? (a) 5.01×10^{-8} (d) 3.16×10^{-37} (b) 2.50×10^{-15} (e) 1.40×10^{-37} (c) 4.20×10^{-15}

EP 17.13 Of the five salts listed below, which has the highest concentration of its cation in solution? Assume that all five aqueous solutions are saturated and that the ions do not undergo any additional reactions.
(a) lead(II) chromate, $K_{sp} = 2.8 \times 10^{-13}$
(b) cobalt(II) hydroxide, $K_{sp} = 1.3 \times 10^{-15}$
(c) cobalt(II) sulfide, $K_{sp} = 5 \times 10^{-22}$
(d) chromium(III) hydroxide, $K_{sp} = 1.6 \times 10^{-30}$
(e) silver sulfide, $K_{sp} = 6 \times 10^{-51}$

EP 17.14 Consider a saturated solution of the salt MA_3, in which M is a metal cation with a 3+ charge and A is an anion with a 1− charge, in water at 298 K. Which of the following factors will affect the K_{sp} of MA_3 in water? (a) The addition of more M^{3+} to the solution. (b) The addition of more A^- to the solution. (c) Diluting the solution. (d) Raising the temperature of the solution. (e) More than one of the factors listed.

EP 17.15 Which of the following actions will increase the solubility of AgBr in water? (a) increasing the pH (b) decreasing the pH (c) adding NaBr (d) adding $NaNO_3$ (e) none of these actions

EP 17.16 You have an aqueous solution of chromium(III) nitrate that you titrate with an aqueous solution of sodium hydroxide. After a certain amount of titrant has been added, you observe a precipitate forming. You add more sodium hydroxide solution and the precipitate dissolves, leaving a solution again. What has happened? (a) The precipitate was sodium hydroxide, which redissolved in the larger volume. (b) The precipitate was chromium hydroxide, which dissolved once more solution was added, forming $Cr^{3+}(aq)$. (c) The precipitate was chromium hydroxide, which then reacted with more hydroxide to produce a soluble complex ion, $Cr(OH)_4^-(aq)$. (d) The precipitate was sodium nitrate, which reacted with more nitrate to produce the soluble complex ion $Na(NO_3)_3^{2-}(aq)$.

EP 17.17 An insoluble salt MA has a K_{sp} of 1.0×10^{-16}. Two solutions, MNO_3 and NaA, are mixed to yield a final solution that is $1.0 \times 10^{-8} M$ in $M^+(aq)$ and $1.00 \times 10^{-7} M$ in $A^-(aq)$. For this situation, Q is _____ than K_{sp} and a precipitate _____ form. (a) larger, will (b) larger, will not (c) smaller, will (d) smaller, will not

EP 17.18 You take an aqueous solution of an unknown metal cation and add a concentrated solution of HCl to it; nothing apparently happens. You then bubble hydrogen sulfide gas under acidic conditions into the solution; apparently nothing happens. Then you add ammonium sulfide under basic conditions to the solution, and now you observe a colored precipitate forming. What is a likely identity of the unknown cation? (a) silver (b) lead (c) mercury (d) aluminum (e) strontium

EP 17.19 The K_{sp} for $Mg(OH)_2$ is 1.8×10^{-11}. What is the pH of a saturated solution of $Mg(OH)_2$? (a) 3.78 (b) 8.62 (c) 10.21 (d) 10.52 (e) 10.74

Exercises

Visualizing Concepts

17.1 The following boxes represent aqueous solutions containing a weak acid, HA, and its conjugate base, A⁻. Water molecules, hydronium ions, and cations are not shown. Which solution has the highest pH? Explain. [Section 17.1]

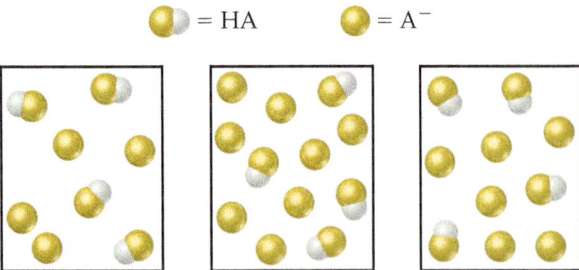

17.2 The beaker on the right contains 0.1 *M* acetic acid solution with methyl orange as an indicator. The beaker on the left contains a mixture of 0.1 *M* acetic acid and 0.1 *M* sodium acetate with methyl orange. (**a**) Using Figures 16.8 and 16.9, which solution has a higher pH? (**b**) Which solution is better able to maintain its pH when small amounts of NaOH are added? Explain. [Sections 17.1 and 17.2]

17.3 A buffer contains a weak acid, HA, and its conjugate base. The weak acid has a pK_a of 4.5, and the buffer has a pH of 4.3. Without doing a calculation, state which of these possibilities are correct at pH 4.3? (**a**) [HA] = [A⁻] (**b**) [HA] > [A⁻] or (**c**) [HA] < [A⁻] [Section 17.2]

17.4 The following diagram represents a buffer composed of equal concentrations of a weak acid, HA, and its conjugate base, A⁻. The heights of the columns are proportional to the concentrations of the components of the buffer. (**a**) Which of the three drawings—(1), (2), or (3)— represents the buffer after the addition of a strong acid? (**b**) Which of the three represents the buffer after the addition of a strong base? (**c**) Which of the three represents a situation that cannot arise from the addition of either an acid or a base? [Section 17.2]

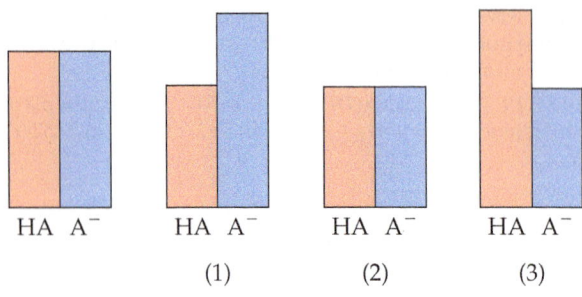

17.5 The following figure represents solutions at various stages of the titration of a weak acid, HA, with NaOH. (The Na⁺ ions and water molecules have been omitted for clarity.) To which of the following regions of the titration curve does each drawing correspond? (**a**) before addition of NaOH (**b**) after addition of NaOH but before the equivalence point (**c**) at the equivalence point (**d**) after the equivalence point [Section 17.3]

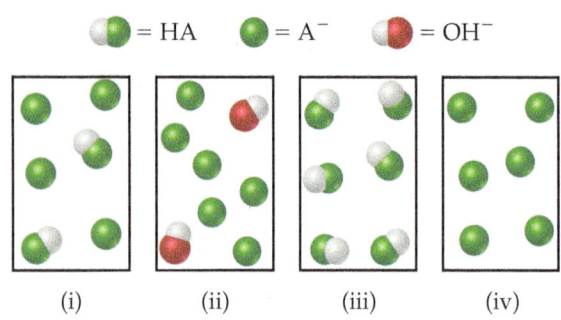

17.6 Match the following descriptions of titration curves with the diagrams: (**a**) strong acid added to strong base, (**b**) strong base added to weak acid, (**c**) strong base added to strong acid, (**d**) strong base added to polyprotic acid. [Section 17.3]

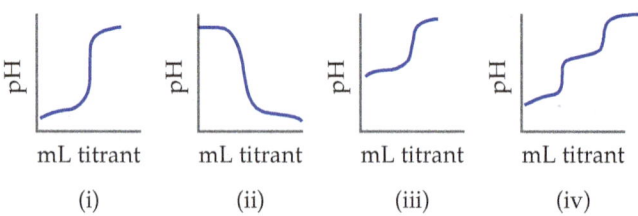

17.7 Equal volumes of two acids are titrated with 0.10 *M* NaOH, resulting in the two titration curves shown in the following figure. (**a**) Which curve corresponds to the more concentrated acid solution? (**b**) Which corresponds to the acid with the larger K_a? [Section 17.3]

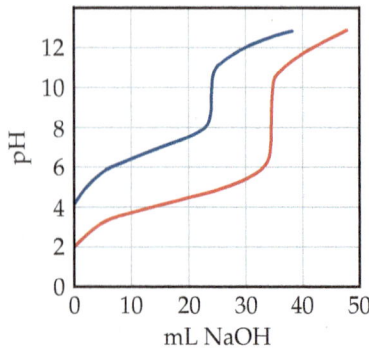

17.8 A saturated solution of Cd(OH)₂ is shown in the middle beaker. If hydrochloric acid solution is added, the solubility of Cd(OH)₂ will increase, causing additional solid to dissolve. Which of the two choices, beaker A or beaker B, accurately represents the solution after equilibrium is reestablished? (The water molecules and Cl^- ions are omitted for clarity.) [Sections 17.4 and 17.5]

Saturated solution

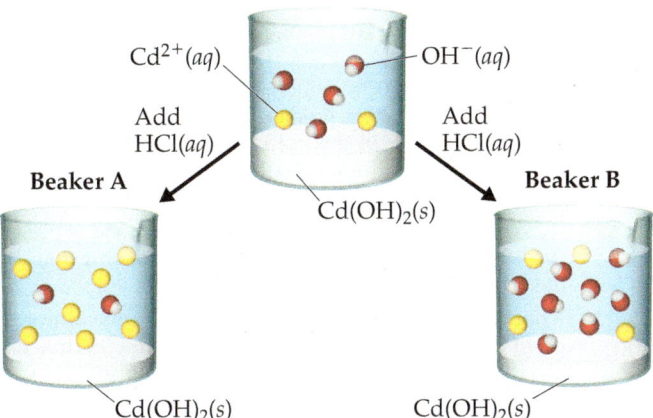

17.9 The following graphs represent the behavior of $BaCO_3$ under different circumstances. In each case, the vertical axis indicates the solubility of the $BaCO_3$ and the horizontal axis represents the concentration of some other reagent. **(a)** Which graph represents what happens to the solubility of $BaCO_3$ as HNO_3 is added? **(b)** Which graph represents what happens to the $BaCO_3$ solubility as Na_2CO_3 is added? **(c)** Which graph represents what happens to the $BaCO_3$ solubility as $NaNO_3$ is added? [Section 17.5]

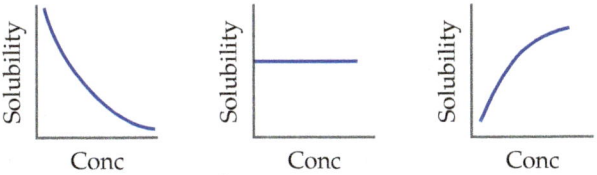

17.10 $Ca(OH)_2$ has a K_{sp} of 6.5×10^{-6}. **(a)** If 0.370 g of $Ca(OH)_2$ is added to 500 mL of water and the mixture is allowed to come to equilibrium, will the solution be saturated? **(b)** If 50 mL of the solution from part (a) is added to each of the beakers shown here, in which beakers, if any, will a precipitate form? In those cases where a precipitate forms, what is its identity? [Section 17.6]

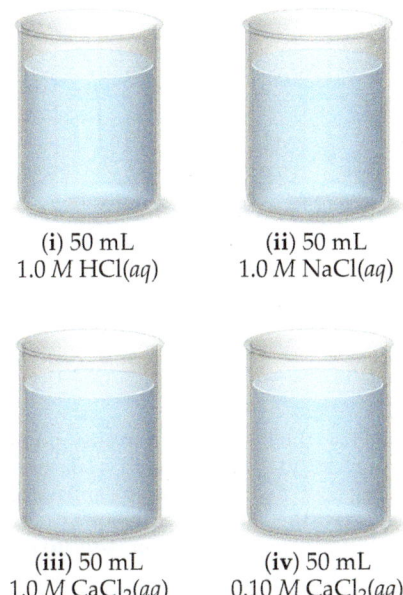

(i) 50 mL
1.0 M HCl(aq)

(ii) 50 mL
1.0 M NaCl(aq)

(iii) 50 mL
1.0 M CaCl$_2$(aq)

(iv) 50 mL
0.10 M CaCl$_2$(aq)

17.11 The graph provided shows the solubility of a salt as a function of pH. Which of the following choices explain the shape of this graph? **(a)** None; this behavior is not possible. **(b)** A soluble salt reacts with acid to form a precipitate, and additional acid reacts with this product to dissolve it. **(c)** A soluble salt forms an insoluble hydroxide, and then additional base reacts with this product to dissolve it. **(d)** The solubility of the salt increases with pH, then decreases because of the heat generated from the neutralization reactions. [Section 17.5]

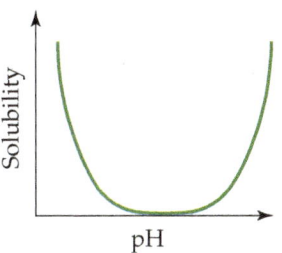

17.12 Three cations, Ni^{2+}, Cu^{2+}, and Ag^+, are separated using two different precipitating agents. Based on the qualitative analysis scheme in Figure 17.23, what two precipitating agents could be used? Using these agents, indicate which of the cations is A, which is B, and which is C. [Section 17.7]

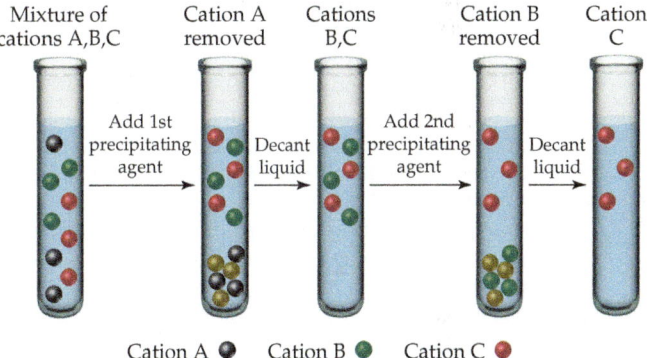

Cation A ⚫   Cation B 🟢   Cation C 🔴

The Common-Ion Effect (Section 17.1)

17.13 Which of these statements about the common-ion effect is most correct? **(a)** The solubility of a salt MA is decreased in a solution that already contains either M^+ or A^-. **(b)** Common ions alter the equilibrium constant for the reaction of an ionic solid with water. **(c)** The common-ion effect does not apply to unusual ions like SO_3^{2-}. **(d)** The solubility of a salt MA is affected equally by the addition of either A^- or a non-common ion.

17.14 Consider the equilibrium

$$B(aq) + H_2O(l) \rightleftharpoons HB^+(aq) + OH^-(aq).$$

Suppose that a salt of $HB^+(aq)$ is added to a solution of $B(aq)$ at equilibrium. **(a)** Will the equilibrium constant for the reaction increase, decrease, or stay the same? **(b)** Will the concentration of $B(aq)$ increase, decrease, or stay the same? **(c)** Will the pH of the solution increase, decrease, or stay the same?

17.15 Use information from Appendix D to calculate the pH of **(a)** a solution that is 0.060 M in potassium propionate (C_2H_5COOK or $KC_3H_5O_2$) and 0.085 M in propionic acid (C_2H_5COOH or $HC_3H_5O_2$); **(b)** a solution that is 0.075 M in trimethylamine, $(CH_3)_3N$, and 0.10 M in trimethylammonium chloride, $(CH_3)_3NHCl$; **(c)** a solution that is made by mixing 50.0 mL of 0.15 M acetic acid and 50.0 mL of 0.20 M sodium acetate.

17.16 Use information from Appendix D to calculate the pH of (a) a solution that is 0.250 M in sodium formate (HCOONa) and 0.100 M in formic acid (HCOOH), (b) a solution that is 0.510 M in pyridine (C_5H_5N) and 0.450 M in pyridinium chloride (C_5H_5NHCl), (c) a solution that is made by combining 55 mL of 0.050 M hydrofluoric acid with 125 mL of 0.10 M sodium fluoride.

17.17 (a) Calculate the percent ionization of 0.007 M butanoic acid ($K_a = 1.5 \times 10^{-5}$). (b) Calculate the percent ionization of 0.0075 M butanoic acid in a solution containing 0.085 M sodium butanoate.

17.18 (a) Calculate the percent ionization of 0.125 M lactic acid ($K_a = 1.4 \times 10^{-4}$). (b) Calculate the percent ionization of 0.125 M lactic acid in a solution containing 0.0075 M sodium lactate.

Buffers (Section 17.2)

17.19 Which of the following solutions is a buffer? (a) 0.10 M CH_3COOH and 0.10 M CH_3COONa (b) 0.10 M CH_3COOH (c) 0.10 M HCl and 0.10 M NaCl (d) both a and c (e) all of a, b, and c

17.20 Which of the following solutions is a buffer? (a) A solution made by mixing 100 mL of 0.100 M CH_3COOH and 50 mL of 0.100 M NaOH (b) A solution made by mixing 100 mL of 0.100 M CH_3COOH and 500 mL of 0.100 M NaOH (c) A solution made by mixing 100 mL of 0.100 M CH_3COOH and 50 mL of 0.100 M HCl (d) A solution made by mixing 100 mL of 0.100 M CH_3COOK and 50 mL of 0.100 M KCl

17.21 (a) Calculate the pH of a buffer that is 0.12 M in lactic acid and 0.11 M in sodium lactate. (b) Calculate the pH of a buffer formed by mixing 85 mL of 0.13 M lactic acid with 95 mL of 0.15 M sodium lactate.

17.22 (a) Calculate the pH of a buffer that is 0.105 M in $NaHCO_3$ and 0.125 M in Na_2CO_3. (b) Calculate the pH of a solution formed by mixing 65 mL of 0.20 M $NaHCO_3$ with 75 mL of 0.15 M Na_2CO_3.

17.23 A buffer is prepared by adding 20.0 g of sodium acetate (CH_3COONa) to 500 mL of a 0.150 M acetic acid (CH_3COOH) solution. (a) Determine the pH of the buffer. (b) Write the complete ionic equation for the reaction that occurs when a few drops of hydrochloric acid are added to the buffer. (c) Write the complete ionic equation for the reaction that occurs when a few drops of sodium hydroxide solution are added to the buffer.

17.24 A buffer is prepared by adding 10.0 g of ammonium chloride (NH_4Cl) to 250 mL of 1.00 M NH_3 solution. (a) What is the pH of this buffer? (b) Write the complete ionic equation for the reaction that occurs when a few drops of nitric acid are added to the buffer. (c) Write the complete ionic equation for the reaction that occurs when a few drops of potassium hydroxide solution are added to the buffer.

17.25 You are asked to prepare a pH = 3.00 buffer solution starting from 1.25 L of a 1.00 M solution of hydrofluoric acid (HF) and any amount you need of sodium fluoride (NaF). (a) What is the pH of the hydrofluoric acid solution prior to adding sodium fluoride? (b) How many grams of sodium fluoride should be added to prepare the buffer solution? Neglect the small volume change that occurs when the sodium fluoride is added.

17.26 You are asked to prepare a pH = 4.00 buffer starting from 1.50 L of 0.0200 M solution of benzoic acid (C_6H_5COOH) and any amount you need of sodium benzoate (C_6H_5COONa). (a) What is the pH of the benzoic acid solution prior to adding sodium benzoate? (b) How many grams of sodium benzoate should be added to prepare the buffer?

Neglect the small volume change that occurs when the sodium benzoate is added.

17.27 A buffer contains 0.10 mol of acetic acid and 0.13 mol of sodium acetate in 1.00 L. (a) What is the pH of this buffer? (b) What is the pH of the buffer after the addition of 0.020 mol of KOH? (c) What is the pH of the buffer after the addition of 0.020 mol of HNO_3?

17.28 A buffer contains 0.15 mol of propionic acid (C_2H_5COOH) and 0.10 mol of sodium propionate (C_2H_5COONa) in 1.20 L. (a) What is the pH of this buffer? (b) What is the pH of the buffer after the addition of 0.010 mol of NaOH? (c) What is the pH of the buffer after the addition of 0.010 mol of HI?

17.29 (a) What is the ratio of HCO_3^- to H_2CO_3 in blood of pH 7.4? (b) What is the ratio of HCO_3^- to H_2CO_3 in an exhausted marathon runner whose blood pH is 7.1?

17.30 A buffer, consisting of $H_2PO_4^-$ and HPO_4^{2-}, helps control the pH of physiological fluids. Many carbonated soft drinks also use this buffer system. What is the pH of a soft drink in which the major buffer ingredients are 6.5 g of NaH_2PO_4 and 8.0 g of Na_2HPO_4 per 355 mL of solution?

17.31 You have to prepare a pH = 3.50 buffer, and you have the following 0.10 M solutions available: HCOOH, CH_3COOH, H_3PO_4, HCOONa, CH_3COONa, and NaH_2PO_4. Which solutions would you use? How many milliliters of each solution would you use to make approximately 1 L of the buffer?

17.32 You have to prepare a pH = 5.00 buffer, and you have the following 0.10 M solutions available: HCOOH, HCOONa, CH_3COOH, CH_3COONa, HCN, and NaCN. Which solutions would you use? How many milliliters of each solution would you use to make approximately 1 L of the buffer?

Acid–Base Titrations (Section 17.3)

17.33 The accompanying graph shows the titration curves for two monoprotic acids. (a) Which curve is that of a strong acid? (b) What is the approximate pH at the equivalence point of each titration? (c) If 40.0 mL of each acid was titrated with a 0.100 M base, then which acid is more concentrated? (d) Estimate the pK_a of the weak acid.

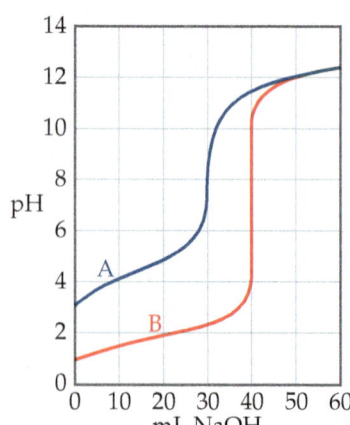

17.34 Compare the titration of a strong, monoprotic acid with a strong base to the titration of a weak, monoprotic acid with a strong base. Assume the strong and weak acid solutions initially have the same concentrations. Indicate whether the following statements are true or false. (a) More base is required to reach the equivalence point for the strong acid than the weak acid. (b) The pH at the beginning of the titration is lower for the weak acid than the strong acid. (**c**) The pH at the equivalence point is 7 no matter which acid is titrated.

17.35 The samples of nitric and acetic acids shown here are both titrated with a 0.100 *M* solution of NaOH(*aq*).

25.0 mL of 1.0 *M* HNO₃(*aq*)   25.0 mL of 1.0 *M* CH₃COOH(*aq*)

Determine whether each of the following statements concerning these titrations is *true* or *false*.

(**a**) A larger volume of NaOH(*aq*) is needed to reach the equivalence point in the titration of HNO_3.

(**b**) The pH at the equivalence point in the HNO_3 titration will be lower than the pH at the equivalence point in the CH_3COOH titration.

(**c**) Phenolphthalein would be a suitable indicator for both titrations.

17.36 Determine whether each of the following statements concerning the titrations in Problem 17.35 is *true* or *false*.

(**a**) The pH at the beginning of the two titrations will be the same.

(**b**) Both titration curves will be essentially the same after passing the equivalence point.

(**c**) Methyl red would be a suitable indicator for both titrations.

17.37 Predict whether the equivalence point of each of the following titrations is below, above, or at pH 7: (**a**) $NaHCO_3$ titrated with NaOH, (**b**) NH_3 titrated with HCl, (**c**) KOH titrated with HBr.

17.38 Predict whether the equivalence point of each of the following titrations is below, above, or at pH 7: (**a**) formic acid titrated with NaOH, (**b**) calcium hydroxide titrated with perchloric acid, (**c**) pyridine titrated with nitric acid.

17.39 As shown in Figure 16.8, the indicator thymol blue has two color changes. Which color change will generally be more suitable for the titration of a weak acid with a strong base?

17.40 Assume that 30.0 mL of a 0.10 *M* solution of a weak base B that accepts one proton is titrated with a 0.10 *M* solution of the monoprotic strong acid HA. (**a**) How many moles of HA have been added at the equivalence point? (**b**) What is the predominant form of B at the equivalence point? (**a**) Is the pH 7, less than 7, or more than 7 at the equivalence point? (**d**) Which indicator, phenolphthalein or methyl red, is likely to be the better choice for this titration?

17.41 How many milliliters of 0.0850 *M* NaOH are required to titrate each of the following solutions to the equivalence point? (**a**) 40.0 mL of 0.0900 *M* HNO_3 (**b**) 35.0 mL of 0.0850 *M* CH_3COOH (**c**) 50.0 mL of a solution that contains 1.85 g of HCl per liter

17.42 How many milliliters of 0.105 *M* HCl are needed to titrate each of the following solutions to the equivalence point? (**a**) 45.0 mL of 0.0950 *M* NaOH (**b**) 22.5 mL of 0.118 *M* NH_3 (**c**) 125.0 mL of a solution that contains 1.35 g of NaOH per liter

17.43 A 20.0-mL sample of 0.200 *M* HBr solution is titrated with 0.200 *M* NaOH solution. Calculate the pH of the solution

after the following volumes of base have been added: (**a**) 15.0 mL, (**b**) 19.9 mL, (**c**) 20.0 mL, (**d**) 20.1 mL, (**e**) 35.0 mL.

17.44 A 20.0-mL sample of 0.150 *M* KOH is titrated with 0.125 *M* $HClO_4$ solution. Calculate the pH after the following volumes of acid have been added: (**a**) 20.0 mL, (**b**) 23.0 mL, (**c**) 24.0 mL, (**d**) 25.0 mL, (**e**) 30.0 mL.

17.45 A 35.0-mL sample of 0.150 *M* acetic acid (CH_3COOH) is titrated with 0.150 *M* NaOH solution. Calculate the pH after the following volumes of base have been added: (**a**) 0 mL, (**b**) 17.5 mL, (**c**) 34.5 mL, (**d**) 35.0 mL, (**e**) 35.5 mL, (**f**) 50.0 mL.

17.46 Consider the titration of 30.0 mL of 0.050 *M* NH_3 with 0.025 *M* HCl. Calculate the pH after the following volumes of titrant have been added: (**a**) 0 mL, (**b**) 20.0 mL, (**c**) 59.0 mL, (**d**) 60.0 mL, (**e**) 61.0 mL, (**f**) 65.0 mL.

17.47 Calculate the pH at the equivalence point for titrating 0.200 *M* solutions of each of the following bases with 0.200 *M* HBr: (**a**) sodium hydroxide (NaOH), (**b**) hydroxylamine (NH_2OH), (**c**) aniline ($C_6H_5NH_2$).

17.48 Calculate the pH at the equivalence point in titrating 0.100 *M* solutions of each of the following with 0.080 *M* NaOH: (**a**) hydrobromic acid (HBr), (**b**) chlorous acid ($HClO_2$), (**c**) benzoic acid (C_6H_5COOH).

Solubility Equilibria and Factors Affecting Solubility (Sections 17.4 and 17.5)

17.49 For each statement, indicate whether it is *true* or *false*.

(**a**) The solubility of a slightly soluble salt can be expressed in units of moles per liter.

(**b**) The solubility product of a slightly soluble salt is simply the square of the solubility.

(**c**) The solubility of a slightly soluble salt is independent of the presence of a common ion.

(**d**) The solubility product of a slightly soluble salt is independent of the presence of a common ion.

17.50 The solubility of two slightly soluble salts of M^{2+}, MA and MZ_2, is the same, 4×10^{-4} mol/L. (**a**) Which has the larger numerical value for the solubility product constant? (**b**) In a saturated solution of each salt in water, which has the higher concentration of M^{2+}? (**c**) If you added an equal volume of a solution saturated in MA to one saturated in MZ_2, what would be the equilibrium concentration of the cation, M^{2+}?

17.51 Write the expression for the solubility-product constant for each of the following ionic compounds: AgI, $SrSO_4$, $Fe(OH)_2$, and Hg_2Br_2.

17.52 (**a**) *True* or *false*: "Solubility" and "solubility-product constant" are the same number for a given compound. (**b**) Write the expression for the solubility-product constant for each of the following ionic compounds: $MnCO_3$, $Hg(OH)_2$, and $Cu_3(PO_4)_2$.

17.53 (**a**) If the molar solubility of CaF_2 at 35 °C is 1.24×10^{-3} mol/L, what is K_{sp} at this temperature? (**b**) It is found that 1.1×10^{-2} g SrF_2 dissolves per 100 mL of aqueous solution at 25 °C. Calculate the solubility product for SrF_2. (**c**) The K_{sp} of $Ba(IO_3)_2$ at 25 °C is 6.0×10^{-10}. What is the molar solubility of $Ba(IO_3)_2$?

17.54 (**a**) The molar solubility of $PbBr_2$ at 25 °C is 1.0×10^{-2} mol/L. Calculate K_{sp}. (**b**) If 0.0490 g of $AgIO_3$ dissolves per liter of solution, calculate the solubility-product constant. (**c**) Using the appropriate K_{sp} value from Appendix D, calculate the pH of a saturated solution of $Ca(OH)_2$.

17.55 A 1.00-L solution saturated at 25 °C with calcium oxalate (CaC_2O_4) contains 0.0061 g of CaC_2O_4. Calculate the solubility-product constant for this salt at 25 °C.

17.56 A 1.00-L solution saturated at 25 °C with lead(II) iodide contains 0.54 g of PbI_2. Calculate the solubility-product constant for this salt at 25 °C.

17.57 Using Appendix D, calculate the molar solubility of AgBr in (a) pure water, (b) $3.0 \times 10^{-2} \, M \, AgNO_3$ solution, (c) 0.10 M NaBr solution.

17.58 Calculate the solubility of LaF_3 in grams per liter in (a) pure water, (b) 0.010 M KF solution, (c) 0.050 M $LaCl_3$ solution.

17.59 Consider a beaker containing a saturated solution of CaF_2 in equilibrium with undissolved $CaF_2(s)$. Solid $CaCl_2$ is then added to the solution. (a) Will the amount of solid CaF_2 at the bottom of the beaker increase, decrease, or remain the same? (b) Will the concentration of Ca^{2+} ions in solution increase or decrease? (c) Will the concentration of F^- ions in solution increase or decrease?

17.60 Consider a beaker containing a saturated solution of PbI_2 in equilibrium with undissolved $PbI_2(s)$. Now solid KI is added to this solution. (a) Will the amount of solid PbI_2 at the bottom of the beaker increase, decrease, or remain the same? (b) Will the concentration of Pb^{2+} ions in solution increase or decrease? (c) Will the concentration of I^- ions in solution increase or decrease?

17.61 Calculate the solubility of $Mn(OH)_2$ in grams per liter when buffered at pH (a) 7.0, (b) 9.5, (c) 11.8.

17.62 Calculate the molar solubility of $Ni(OH)_2$ when buffered at pH (a) 8.0, (b) 10.0, (c) 12.0.

17.63 Which of the following salts will be substantially more soluble in acidic solution than in pure water? (a) $ZnCO_3$ (b) ZnS (c) BiI_3 (d) AgCN (e) $Ba_3(PO_4)_2$

17.64 For each of the following slightly soluble salts, write the net ionic equation, if any, for a reaction with a strong acid: (a) MnS, (b) PbF_2, (c) $AuCl_3$, (d) $Hg_2C_2O_4$, (e) CuBr.

17.65 From the value of K_f listed in Table 17.1, calculate the concentration of $Ni^{2+}(aq)$ and $Ni(NH_3)_6{}^{2+}$ that are present at equilibrium after dissolving 1.25 g $NiCl_2$ in 100.0 mL of 1.00 M $NH_3(aq)$.

17.66 From the value of K_f listed in Table 17.1, calculate the concentration of NH_3 required to just dissolve 0.020 mol of NiC_2O_4 ($K_{sp} = 4 \times 10^{-10}$) in 1.00 L of solution? (*Hint:* You can neglect the hydrolysis of $C_2O_4{}^{2-}$ because the solution will be quite basic.)

17.67 Use values of K_{sp} for AgI and K_f for $Ag(CN)_2{}^-$ to (a) calculate the molar solubility of AgI in pure water, (b) calculate the equilibrium constant for the reaction $AgI(s) + 2 CN^-(aq) \rightleftharpoons Ag(CN)_2{}^-(aq) + I^-(aq)$, (c) determine the molar solubility of AgI in a 0.100 M NaCN solution.

17.68 Using the value of K_{sp} for Ag_2S, K_{a1} and K_{a2} for H_2S, and $K_f = 1.1 \times 10^5$ for $AgCl_2{}^-$, calculate the equilibrium constant for the following reaction:

$$Ag_2S(s) + 4 Cl^-(aq) + 2 H^+(aq) \rightleftharpoons 2 AgCl_2{}^-(aq) + H_2S(aq)$$

Precipitation and Separation of Ions (Section 17.6)

17.69 (a) Will $Ca(OH)_2$ precipitate from solution if the pH of a 0.050 M solution of $CaCl_2$ is adjusted to 8.0? (b) Will Ag_2SO_4 precipitate when 100 mL of 0.050 M $AgNO_3$ is mixed with 10 mL of $5.0 \times 10^{-2} \, M \, Na_2SO_4$ solution?

17.70 (a) Will $Co(OH)_2$ precipitate from solution if the pH of a 0.020 M solution of $Co(NO_3)_2$ is adjusted to 8.5? (b) Will $AgIO_3$ precipitate when 20 mL of 0.010 M $AgIO_3$ is mixed with 10 mL of 0.015 M $NaIO_3$? (K_{sp} of $AgIO_3$ is 3.1×10^{-8}).

17.71 Calculate the minimum pH needed to precipitate $Mn(OH)_2$ so completely that the concentration of $Mn^{2+}(aq)$ is less than 1 μg per liter [1 part per billion (ppb)].

17.72 Suppose that a 10-mL sample of a solution is to be tested for I^- ion by addition of 1 drop (0.2 mL) of 0.10 M $Pb(NO_3)_2$. What is the minimum number of grams of I^- that must be present for $PbI_2(s)$ to form?

17.73 A solution contains $2.0 \times 10^{-4} \, M \, Ag^+(aq)$ and $1.5 \times 10^{-3} M$ $Pb^{2+}(aq)$. If NaI is added, will AgI ($K_{sp} = 8.3 \times 10^{-17}$) or PbI_2 ($K_{sp} = 7.9 \times 10^{-9}$) precipitate first? Specify the concentration of $I^-(aq)$ needed to begin precipitation.

17.74 A solution of Na_2SO_4 is added dropwise to a solution that is 0.010 M in $Ba^{2+}(aq)$ and 0.010 M in $Sr^{2+}(aq)$. (a) What concentration of $SO_4{}^{2-}$ is necessary to begin precipitation? (Neglect volume changes. $BaSO_4$: $K_{sp} = 1.1 \times 10^{-10}$; $SrSO_4$: $K_{sp} = 3.2 \times 10^{-7}$.) (b) Which cation precipitates first? (c) What is the concentration of $SO_4{}^{2-}(aq)$ when the second cation begins to precipitate?

17.75 A solution contains three anions with the following concentrations: 0.20 M $CrO_4{}^{2-}$, 0.10 M $CO_3{}^{2-}$, and 0.010 M Cl^-. If a dilute $AgNO_3$ solution is slowly added to the solution, what is the first compound to precipitate: Ag_2CrO_4 ($K_{sp} = 1.2 \times 10^{-12}$), Ag_2CO_3 ($K_{sp} = 8.1 \times 10^{-12}$), or AgCl ($K_{sp} = 1.8 \times 10^{-10}$)?

17.76 A 1.0 M Na_2SO_4 solution is slowly added to 10.0 mL of a solution that is 0.20 M in Ca^{2+} and 0.30 M in Ag^+. (a) Which compound will precipitate first: $CaSO_4$ ($K_{sp} = 2.4 \times 10^{-5}$) or Ag_2SO_4 ($K_{sp} = 1.5 \times 10^{-5}$)? (b) How much Na_2SO_4 solution must be added to initiate the precipitation?

Qualitative Analysis for Metallic Elements (Section 17.7)

17.77 A solution containing several metal ions is treated with dilute HCl; no precipitate forms. The pH is adjusted to about 1, and H_2S is bubbled through. Again, no precipitate forms. The pH of the solution is then adjusted to about 8. Again, H_2S is bubbled through. This time a precipitate forms. The filtrate from this solution is treated with $(NH_4)_2HPO_4$. No precipitate forms. Which of these metal cations are either possibly present or definitely absent: Al^{3+}, Na^+, Ag^+, Mg^{2+}?

17.78 An unknown solid is entirely soluble in water. On addition of dilute HCl, a precipitate forms. After the precipitate is filtered off, the pH is adjusted to about 1 and H_2S is bubbled in; a precipitate again forms. After filtering off this precipitate, the pH is adjusted to 8 and H_2S is again added; no precipitate forms. No precipitate forms upon addition of $(NH_4)_2HPO_4$. The remaining solution shows a yellow color in a flame test (see Figure 7.22). Based on these observations, which of the following compounds might be present, which are definitely present, and which are definitely absent? CdS: $Pb(NO_3)_2$, HgO, $ZnSO_4$, $Cd(NO_3)_2$, and Na_2SO_4

17.79 In the course of various qualitative analysis procedures, the following mixtures are encountered: (a) Zn^{2+} and Cd^{2+}, (b) $Cr(OH)_3$ and $Fe(OH)_3$, (c) Mg^{2+} and K^+, (d) Ag^+ and Mn^{2+}. Suggest how each mixture might be separated.

17.80 Suggest how the cations in each of the following solution mixtures can be separated: (a) Na^+ and Cd^{2+}, (b) Cu^{2+} and Mg^{2+}, (c) Pb^{2+} and Al^{3+}, (d) Ag^+ and Hg^{2+}.

17.81 (a) Precipitation of the group 4 cations of Figure 17.23 requires a basic medium. Why is this so? (b) What is the most significant difference between the sulfides precipitated in group 2 and those precipitated in group 3? (c) Suggest a procedure that would serve to redissolve the group 3 cations following their precipitation.

17.82 A student who is in a great hurry to finish his laboratory work decides that his qualitative analysis unknown contains a metal ion from group 4 of Figure 17.23. He therefore tests his sample directly with $(NH_4)_2HPO_4$, skipping earlier tests for the metal ions in groups 1, 2, and 3. He observes a precipitate and concludes that a metal ion from group 4 is indeed present. Why is this possibly an erroneous conclusion?

Additional Exercises

17.83 Which of these equations relates the pOH of a buffer to the pK_b of its weak base, analogous to the Henderson–Hasselbalch equation for weak acids? (**a**) $pK_b = pOH + \log[\text{acid}]/[\text{base}]$ (**b**) $pK_b = pOH - \log[\text{acid}]/[\text{base}]$ (**c**) $pK_b = pOH - \log[\text{base}]/[\text{acid}]$ (**d**) $pK_b = pOH + \log[\text{base}]/[\text{acid}]$

17.84 Rainwater is acidic because $CO_2(g)$ dissolves in the water, creating carbonic acid, H_2CO_3. If the rainwater is too acidic, it will react with limestone and seashells (which are principally made of calcium carbonate, $CaCO_3$). Calculate the concentrations of carbonic acid, bicarbonate ion (HCO_3^-) and carbonate ion (CO_3^{2-}) that are in a raindrop that has a pH of 5.60, assuming that the sum of all three species in the raindrop is $1.0 \times 10^{-5}\ M$.

17.85 Furoic acid ($HC_5H_3O_3$) has a K_a value of 6.76×10^{-4} at 25 °C. Calculate the pH at 25 °C of (**a**) a solution formed by adding 25.0 g of furoic acid and 30.0 g of sodium furoate ($NaC_5H_3O_3$) to enough water to form 0.250 L of solution, (**b**) a solution formed by mixing 30.0 mL of $0.250\ M\ HC_5H_3O_3$ and 20.0 mL of $0.22\ M\ NaC_5H_3O_3$ and diluting the total volume to 125 mL, (**c**) a solution prepared by adding 50.0 mL of $1.65\ M$ NaOH solution to 0.500 L of $0.0850\ M\ HC_5H_3O_3$.

17.86 The acid–base indicator bromcresol green is a weak acid. The yellow acid and blue base forms of the indicator are present in equal concentrations in a solution when the pH is 4.68. What is the pK_a for bromcresol green?

17.87 Equal quantities of $0.010\ M$ solutions of an acid HA and a base B are mixed. The pH of the resulting solution is 9.2. (**a**) Write the chemical equation and equilibrium-constant expression for the reaction between HA and B. (**b**) If K_a for HA is 8.0×10^{-5}, what is the value of the equilibrium constant for the reaction between HA and B? (**c**) What is the value of K_b for B?

17.88 Two buffers are prepared by adding an equal number of moles of formic acid (HCOOH) and sodium formate (HCOONa) to enough water to make 1.00 L of solution. Buffer A is prepared using 1.00 mol each of formic acid and sodium formate. Buffer B is prepared by using 0.010 mol of each. (**a**) Calculate the pH of each buffer. (**b**) Which buffer will have the greater buffer capacity? (**c**) Calculate the change in pH for each buffer upon the addition of 1.0 mL of $1.00\ M$ HCl. (**d**) Calculate the change in pH for each buffer upon the addition of 10 mL of $1.00\ M$ HCl.

17.89 A biochemist needs 750 mL of an acetic acid–sodium acetate buffer with pH 4.50. Solid sodium acetate (CH_3COONa) and glacial acetic acid (CH_3COOH) are available. Glacial acetic acid is 99% CH_3COOH by mass and has a density of 1.05 g/mL. If the buffer is to be $0.15\ M$ in CH_3COOH, how many grams of CH_3COONa and how many milliliters of glacial acetic acid must be used?

17.90 A sample of 0.2140 g of an unknown monoprotic acid was dissolved in 25.0 mL of water and was titrated with $0.0950\ M$ NaOH. The acid required 30.0 mL of base to reach the equivalence point. (**a**) What is the molar mass of the acid? (**b**) After 15.0 mL of base had been added in the titration, the pH was found to be 6.50. What is the K_a for the unknown acid?

17.91 A sample of 0.1687 g of an unknown monoprotic acid was dissolved in 25.0 mL of water and titrated with $0.1150\ M$ NaOH. The acid required 15.5 mL of base to reach the equivalence point. (**a**) What is the molar mass of the acid? (**b**) After 7.25 mL of base had been added in the titration, the pH was found to be 2.85. What is the K_a for the unknown acid?

17.92 Along which of the following parts of a pH titration curve for a weak monoprotic acid/strong base is the solution buffered? (**a**) at the beginning, before any base is added (**b**) halfway to the equivalence point (**c**) at the equivalence point (**d**) well past the equivalence point

17.93 A weak monoprotic acid is titrated with $0.100\ M$ NaOH. It requires 50.0 mL of the NaOH solution to reach the equivalence point. After 25.0 mL of base is added, the pH of the solution is 3.62. Estimate the pK_a of the weak acid.

17.94 What is the pH of a solution made by mixing 0.30 mol NaOH, 0.25 mol Na_2HPO_4, and 0.20 mol H_3PO_4 with water and diluting to 1.00 L?

17.95 Suppose you want to do a physiological experiment that calls for a pH 6.50 buffer. You find that the organism with which you are working is not sensitive to the weak acid H_2A ($K_{a1} = 2 \times 10^{-2}; K_{a2} = 5.0 \times 10^{-7}$) or its sodium salts. You have available a $1.0\ M$ solution of this acid and a $1.0\ M$ solution of NaOH. How much of the NaOH solution should be added to 1.0 L of the acid to give a buffer at pH 6.50? (Ignore any volume change.)

17.96 How many microliters of $1.000\ M$ NaOH solution must be added to 25.00 mL of a $0.1000\ M$ solution of lactic acid [$CH_3CH(OH)COOH$ or $HC_3H_5O_3$] to produce a buffer with pH = 3.75?

17.97 Lead(II) carbonate, $PbCO_3$, is one of the components of the passivating layer that forms inside lead pipes. (**a**) If the K_{sp} for $PbCO_3$ is 7.4×10^{-14}, then what is the molarity of Pb^{2+} in a saturated solution of lead(II) carbonate? (**b**) What is the concentration in ppb of Pb^{2+} ions in a saturated solution? (**c**) Will the solubility of $PbCO_3$ increase or decrease as the pH is lowered? (**d**) The EPA threshold for acceptable levels of lead ions in water is 15 ppb. Does a saturated solution of lead(II) carbonate produce a solution that exceeds the EPA limit?

17.98 For each pair of compounds, use K_{sp} values to determine which has the greater molar solubility: (**a**) CdS or CuS, (**b**) $PbCO_3$ or $BaCrO_4$, (**c**) $Ni(OH)_2$ or $NiCO_3$, (**d**) AgI or Ag_2SO_4.

17.99 The solubility of $CaCO_3$ is pH dependent. (**a**) Calculate the molar solubility of $CaCO_3$ ($K_{sp} = 4.5 \times 10^{-9}$), neglecting the acid–base character of the carbonate ion. (**b**) Use the K_b expression for the CO_3^{2-} ion to determine the equilibrium constant for the reaction

$$CaCO_3(s) + H_2O(l) \rightleftharpoons$$
$$Ca^{2+}(aq) + HCO_3^-(aq) + OH^-(aq)$$

(**c**) If we assume that the only sources of Ca^{2+}, HCO_3^-, and OH^- ions are from the dissolution of $CaCO_3$, what is the molar solubility of $CaCO_3$ using the equilibrium expression from part (b)? (**d**) What is the molar solubility of $CaCO_3$ at the pH of the ocean (8.3)? (**e**) If the pH is buffered at 7.5, what is the molar solubility of $CaCO_3$?

17.100 Tooth enamel is composed of hydroxyapatite, whose simplest formula is $Ca_5(PO_4)_3OH$, and whose corresponding $K_{sp} = 6.8 \times 10^{-27}$. As discussed in the "Chemistry and Life" box on "Tooth Decay and Fluoridation" in Section 17.5, fluoride in fluorinated water or in toothpaste reacts with hydroxyapatite to form fluoroapatite, $Ca_5(PO_4)_3F$, whose $K_{sp} = 1.0 \times 10^{-60}$. (**a**) Write the expression for the solubility-constant for hydroxyapatite and for fluoroapatite. (**b**) Calculate the molar solubility of each of these compounds.

17.101 Salts containing the phosphate ion are added to municipal water supplies to prevent the corrosion of lead pipes. (**a**) Based on the pK_a values for phosphoric acid ($pK_{a1} = 7.5 \times 10^{-3}$, $pK_{a2} = 6.2 \times 10^{-8}$, $pK_{a3} = 4.2 \times 10^{-13}$), what is the K_b value for the PO_4^{3-} ion? (**b**) What is the pH of a $1 \times 10^{-3}\ M$ solution of Na_3PO_4 (you can ignore the formation of $H_2PO_4^-$ and H_3PO_4)?

17.102 Calculate the solubility of $Mg(OH)_2$ in 0.50 M NH_4Cl.

17.103 The solubility-product constant for barium permanganate, $Ba(MnO_4)_2$, is 2.5×10^{-10}. Assume that solid $Ba(MnO_4)_2$ is in equilibrium with a solution of $KMnO_4$. What concentration of $KMnO_4$ is required to establish a concentration of $2.0 \times 10^{-8} M$ for the Ba^{2+} ion in solution?

17.104 Calculate the ratio of $[Ca^{2+}]$ to $[Fe^{2+}]$ in a lake in which the water is in equilibrium with deposits of both $CaCO_3$ and $FeCO_3$. Assume that the water is slightly basic and that the hydrolysis of the carbonate ion can therefore be ignored.

17.105 The solubility product constants of $PbSO_4$ and $SrSO_4$ are 6.3×10^{-7} and 3.2×10^{-7}, respectively. What are the values of $[SO_4{}^{2-}]$, $[Pb^{2+}]$, and $[Sr^{2+}]$ in a solution at equilibrium with both substances?

17.106 A buffer of what pH is needed to give a Mg^{2+} concentration of $3.0 \times 10^{-2} M$ in equilibrium with solid magnesium oxalate?

17.107 The value of K_{sp} for $Mg_3(AsO_4)_2$ is 2.1×10^{-20}. The $AsO_4{}^{3-}$ ion is derived from the weak acid H_3AsO_4 ($pK_{a1} = 2.22$; $pK_{a2} = 6.98$; $pK_{a3} = 11.50$). (a) Calculate the molar solubility of $Mg_3(AsO_4)_2$ in water. (b) Calculate the pH of a saturated solution of $Mg_3(AsO_4)_2$ in water.

17.108 The solubility product for $Zn(OH)_2$ is 3.0×10^{-16}. The formation constant for the hydroxo complex, $Zn(OH)_4{}^{2-}$, is 4.6×10^{17}. What concentration of OH^- is required to dissolve 0.015 mol of $Zn(OH)_2$ in a liter of solution?

17.109 The value of K_{sp} for $Cd(OH)_2$ is 2.5×10^{-14}. (a) What is the molar solubility of $Cd(OH)_2$? (b) The solubility of $Cd(OH)_2$ can be increased through formation of the complex ion $CdBr_4{}^{2-}$ ($K_f = 5 \times 10^3$). If solid $Cd(OH)_2$ is added to a NaBr solution, what is the initial concentration of NaBr needed to increase the molar solubility of $Cd(OH)_2$ to 1.0×10^{-3} mol/L?

17.110 (a) Write the net ionic equation for the reaction that occurs when a solution of hydrochloric acid (HCl) is mixed with a solution of sodium formate ($NaCHO_2$). (b) Calculate the equilibrium constant for this reaction. (c) Calculate the equilibrium concentrations of Na^+, Cl^-, H^+, $CHO_2{}^-$, and $HCHO_2$ when 50.0 mL of 0.15 M HCl is mixed with 50.0 mL of 0.15 M $NaCHO_2$.

17.111 A sample of 7.5 L of NH_3 gas at 22 °C and 735 torr is bubbled into a 0.50-L solution of 0.40 M HCl. Assuming that all the NH_3 dissolves and that the volume of the solution remains 0.50 L, calculate the pH of the resulting solution.

17.112 What is the pH at 25 °C of water saturated with CO_2 at a partial pressure of 1.10 atm? The Henry's law constant for CO_2 at 25 °C is 3.1×10^{-2} mol/L-atm.

17.113 Excess $Ca(OH)_2$ is shaken with water to produce a saturated solution. The solution is filtered, and a 50.00-mL sample titrated with HCl requires 11.23 mL of 0.0983 M HCl to reach the end point. Calculate K_{sp} for $Ca(OH)_2$. Compare your result with that in Appendix D. Suggest a reason for any differences you find between your value and the one in Appendix D.

17.114 The osmotic pressure of a saturated solution of strontium sulfate at 25 °C is 21 torr. What is the solubility product of this salt at 25 °C?

17.115 A concentration of 10–100 parts per billion (by mass) of Ag^+ is an effective disinfectant in swimming pools. However, if the concentration exceeds this range, the Ag^+ can cause adverse health effects. One way to maintain an appropriate concentration of Ag^+ is to add a slightly soluble salt to the pool. Using K_{sp} values from Appendix D, calculate the equilibrium concentration of Ag^+ in parts per billion that would exist in equilibrium with (a) AgCl, (b) AgBr, (c) AgI.

17.116 Fluoridation of drinking water is employed in many places to aid in the prevention of tooth decay. Typically. the F^- ion concentration is adjusted to about 1 ppm. Some water supplies are also "hard"; that is, they contain certain cations such as Ca^{2+} that interfere with the action of soap. Consider a case where the concentration of Ca^{2+} is 8 ppm. Could a precipitate of CaF_2 form under these conditions? (Make any necessary approximations.)

17.117 Baking soda (sodium bicarbonate, $NaHCO_3$) reacts with acids in foods to form carbonic acid (H_2CO_3), which in turn decomposes to water and carbon dioxide gas. In a cake batter, the $CO_2(g)$ forms bubbles and causes the cake to rise. (a) A rule of thumb in baking is that 1/2 teaspoon of baking soda is neutralized by one cup of sour milk. The acid component in sour milk is lactic acid, $CH_3CH(OH)COOH$. Write the chemical equation for this neutralization reaction. (b) The density of baking soda is 2.16 g/cm^3. Calculate the concentration of lactic acid in one cup of sour milk (assuming the rule of thumb applies), in units of mol/L. (One cup = 236.6 mL = 48 teaspoons.) (c) If 1/2 teaspoon of baking soda is indeed completely neutralized by the lactic acid in sour milk, calculate the volume of carbon dioxide gas that would be produced at 1 atm pressure, in an oven set to 350 °F.

17.118 In nonaqueous solvents, it is possible to react HF to create H_2F^+. Which of these statements follows from this observation? (a) HF can act like a strong acid in nonaqueous solvents. (b) HF can act like a base in nonaqueous solvents. (c) HF is thermodynamically unstable. (d) There is an acid in the nonaqueous medium that is a stronger acid than HF.

Design an Experiment

You are cleaning up an old chemistry lab and find a glass bottle labeled "6.00 M NaOH." The bottle looks like it holds about 5 mL of solution. However, it is possible that sodium hydroxide, over a long period of time, has reacted with glass (SiO_2). It is also possible that the bottle was not well sealed, and some water evaporated.

Design an experiment to determine the concentration of the NaOH, using a very small amount from the bottle (less than 1 mL). Consider that you have unlimited water available, a stock solution of 2.00 M HCl, and any pH indicators you need. Also consider that your equipment only allows you to measure volumes to the nearest mL.

18

CHEMISTRY OF THE ENVIRONMENT

▲ **EARTH'S ATMOSPHERE** is something most of us take for granted, but its role in supporting life on our planet cannot be overstated. Not only does it provide the oxygen we need to breathe, it protects the surface of our planet from harmful solar radiation, and it moderates daily and seasonal swings in temperature.

The richness of life on Earth, represented in the chapter-opening photograph, is made possible by our planet's supportive atmosphere, the energy received from the Sun, and an abundance of water. These are the signature environmental features believed to be necessary for life.

We are now in a position to apply the principles we have learned in preceding chapters to an understanding of how our environment operates and how human activities affect it. To understand and protect the environment in which we live, we must understand how human-made and natural chemical compounds interact on land and in the sea and sky. Our daily actions as consumers turn on the same choices made by leading experts and governmental leaders: Each decision should reflect the costs versus the benefits of our choices. Unfortunately, the environmental impacts of our decisions are often subtle and not immediately evident.

18.1 | Earth's Atmosphere

Because most of us have never been very far from Earth's surface, we often take for granted the many ways in which the atmosphere determines the environment in which we live. In this section we examine some of the important characteristics of our planet's atmosphere.

The temperature of the atmosphere varies with altitude (**Figure 18.1**), and the atmosphere is divided into four regions based on this temperature profile. Just above the surface, in the **troposphere**, the temperature normally decreases with increasing altitude, reaching a minimum of about 215 K at about 10 km. Nearly all of us live our entire lives in the troposphere. Howling winds and soft breezes, rain, and sunny skies—all that we normally think of as "weather"—occur in this region. Commercial jet aircraft typically fly about 10 km (33,000 ft) above Earth, an altitude that defines the upper limit of the troposphere, which we call the *tropopause*.

Above the tropopause, air temperature increases with altitude, reaching a maximum of about 275 K at about 50 km. The region from 10 km to 50 km is the **stratosphere**, and above it are the *mesosphere* and *thermosphere*. Notice in Figure 18.1 that the temperature extremes that form the boundaries between adjacent regions are denoted by the suffix-*pause*. The boundaries are important because gases mix across them relatively slowly. For example, pollutant gases generated in the troposphere pass through the tropopause and find their way into the stratosphere only very slowly.

Atmospheric pressure decreases with increasing elevation (Figure 18.1), declining much more rapidly at lower elevations than at higher ones because of the atmosphere's

Learning Objectives

When you finish Section 18.1, you should be able to:

▶ Identify the four regions of Earth's atmosphere and describe how the temperature and pressure vary with altitude across each of these regions.

▶ List the major components of dry air near sea level.

▶ Express the concentrations of gases in either parts per million (ppm) or mole fraction and interconvert between the two.

▶ Explain what is meant by photodissociation and photoionization and list examples of each process that occur in Earth's atmosphere.

▶ Describe the reactions that lead to the cycle of formation and decomposition of ozone in the upper atmosphere.

Go Figure   At what altitude is the atmospheric temperature lowest?

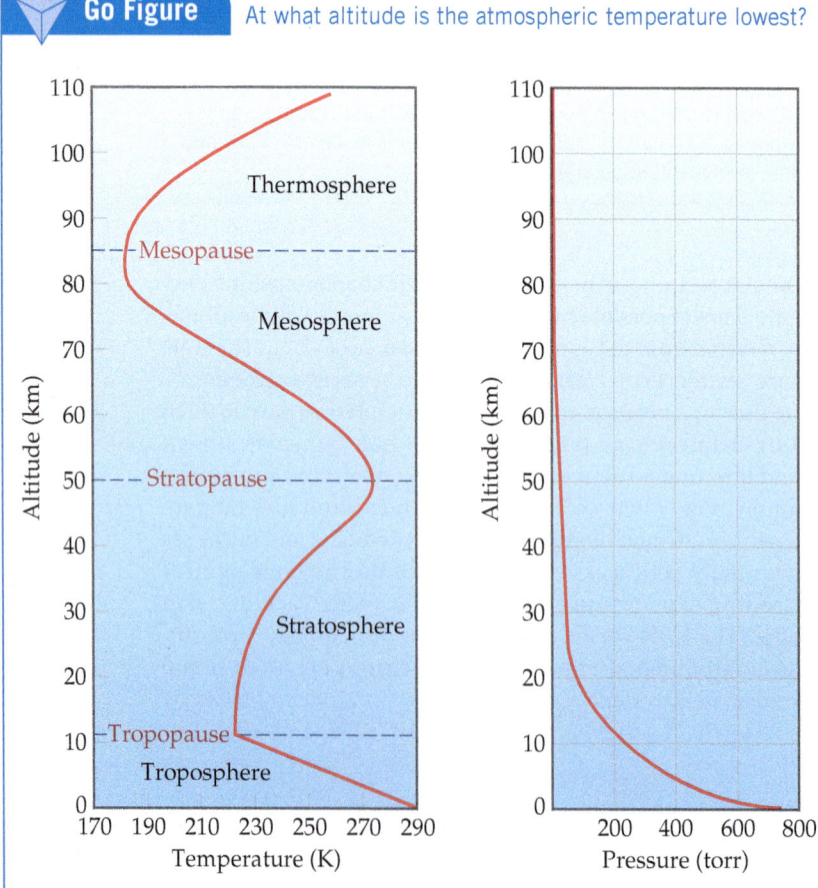

▲ **Figure 18.1** Temperature and pressure in the atmosphere vary as a function of altitude above sea level.

compressibility. Thus, the pressure decreases from an average value of 760 torr at sea level to 2.3×10^{-3} torr at 100 km, to only 1.0×10^{-6} torr at 200 km.

The troposphere and stratosphere together account for 99.9% of the mass of the atmosphere, 75% of which is in the troposphere. Nevertheless, the thin upper atmosphere plays many important roles in determining the conditions of life at the surface.

Composition of the Atmosphere

Earth's atmosphere is constantly bombarded by radiation and energetic particles from the Sun. This barrage of energy has profound chemical and physical effects, especially in the upper regions of the atmosphere, above about 80 km (Figure 18.2). In addition, because of Earth's gravitational field, heavier atoms and molecules tend to sink in the atmosphere, leaving lighter atoms and molecules at the top of the atmosphere. (This is why, as just noted, 75% of the atmosphere's mass is in the troposphere.) Because of all these factors, the composition of the atmosphere is not uniform.

Table 18.1 lists the composition of dry air near sea level. Although traces of many substances are present, N_2 and O_2 make up about 99% of sea-level air. The noble gases and CO_2 make up most of the remainder.

When applied to substances in aqueous solution, the concentration unit *parts per million* (ppm) refers to grams of substance per million grams of solution. (Section 13.4) When dealing with gases, however, 1 ppm means one part by *volume* in 1 million volumes of the whole. Because volume is proportional to number of moles of gas via the ideal-gas equation ($PV = nRT$), volume fraction and mole fraction are the same. Thus, 1 ppm of a trace constituent of the atmosphere amounts to 1 mol of that constituent in 1 million moles of air; that is, the concentration in parts per million is equal to the mole fraction times 10^6. For example, Table 18.1 lists the mole fraction of CO_2 in the atmosphere as 0.000415, which means its concentration in parts per million is $0.000415 \times 10^6 = 415$ ppm. Other minor constituents of the troposphere, in addition to CO_2, are listed in Table 18.2.

Before we consider the chemical processes that occur in the atmosphere, let's review some of the properties of the two major components, N_2 and O_2. Recall that the N_2 molecule possesses a triple bond between the nitrogen atoms. (Section 8.3) This very strong bond (bond energy 941 kJ/mol) is largely responsible for the very low reactivity of N_2. The bond energy in O_2 is only 495 kJ/mol, making O_2 much more reactive than N_2. For example, oxygen reacts with many substances to form oxides.

High-energy solar particles create excited N and O atoms; visible light results as electrons in these atoms fall from excited states to lower energy states.

▲ Figure 18.2 The aurora borealis (northern lights).

TABLE 18.1 The Major Components of Dry Air near Sea Level

Component*	Content (mole fraction)	Molar Mass (g/mol)
Nitrogen	0.78084	28.013
Oxygen	0.20946	31.998
Argon	0.00934	39.948
Carbon dioxide	0.000415	44.0099
Neon	0.00001818	20.183
Helium	0.00000524	4.003
Methane	0.000002	16.043
Krypton	0.00000114	83.80
Hydrogen	0.0000005	2.0159
Nitrous oxide	0.0000005	44.0128
Xenon	0.000000087	131.30

*Ozone, sulfur dioxide, nitrogen dioxide, ammonia, and carbon monoxide are present as trace gases in variable amounts.

TABLE 18.2 Sources and Typical Concentrations of Some Minor Atmospheric Constituents

Constituent	Sources	Typical Concentration
Carbon dioxide, CO_2	Decomposition of organic matter, release from oceans, fossil fuel combustion	415 ppm throughout troposphere
Carbon monoxide, CO	Decomposition of organic matter, industrial processes, fossil fuel combustion	0.05 ppm in unpolluted air; 1–50 ppm in urban areas
Methane, CH_4	Decomposition of organic matter, natural-gas seepage, livestock emissions	1.82 ppm throughout troposphere
Nitric oxide, NO	Atmospheric electrical discharges, internal combustion engines, combustion of organic matter	0.01 ppm in unpolluted air; 0.2 ppm in smog
Ozone, O_3	Atmospheric electrical discharges, diffusion from the stratosphere, photochemical smog	0–0.01 ppm in unpolluted air; 0.5 ppm in photochemical smog
Sulfur dioxide, SO_2	Volcanic gases, forest fires, bacterial action, fossil fuel combustion, industrial processes	0–0.01 ppm in unpolluted air; 0.1–2 ppm in polluted urban areas

Sample Exercise 18.1

Calculating Concentration from Partial Pressure

What is the concentration, in parts per million, of water vapor in a sample of air if the partial pressure of the water is 0.80 torr and the total pressure of the air is 735 torr?

SOLUTION

Analyze We are given the partial pressure of water vapor and the total pressure of an air sample, and we are asked to determine the water vapor concentration.

Plan Recall Equation 10.16, which states that the partial pressure of a component in a mixture of gases is given by the product of its mole fraction and the total pressure of the mixture (Section 10.4):

$$P_{H_2O} = X_{H_2O}P_t$$

Solve Solving for the mole fraction of water vapor in the mixture, X_{H_2O}, gives

$$X_{H_2O} = \frac{P_{H_2O}}{P_t} = \frac{0.80 \text{ torr}}{735 \text{ torr}} = 0.0011$$

The concentration in ppm is the mole fraction times 10^6:

$$0.0011 \times 10^6 = 1100 \text{ ppm}$$

▶ **Practice Exercise**
The concentration of CO in a sample of air is 4.3 ppm. What is the partial pressure of the CO if the total air pressure is 695 torr?

Photochemical Reactions in the Atmosphere

Although the atmosphere beyond the stratosphere contains only a small fraction of the atmospheric mass, it forms the outer defense against the hail of radiation and high-energy particles that continuously bombard Earth. As the bombarding radiation passes through the upper atmosphere, it causes two kinds of chemical changes: *photodissociation* and *photoionization*. These processes protect us from high-energy radiation by absorbing most of the radiation before it reaches the troposphere. If it were not for these photochemical processes, plant and animal life as we know it could not exist on Earth.

The Sun emits radiant energy over a wide range of wavelengths (**Figure 18.3**). To understand the connection between the wavelength of radiation and its effect on atoms and molecules, recall that electromagnetic radiation can be pictured as a stream of photons. (Section 6.2) The energy of each photon is given by $E = h\nu$, where h is the Planck constant and ν is the frequency of the radiation. For a chemical change to occur when radiation strikes atoms or molecules, two conditions must be met. First, the incoming photons must have sufficient energy to break a chemical bond or remove an electron from the atom or molecule. Second, the atoms or molecules being bombarded must absorb these photons. When these requirements are met, the energy of the photons is used to do the work associated with some chemical change.

The rupture of a chemical bond resulting from absorption of a photon by a molecule is called **photodissociation**. No ions are formed when the bond between two atoms is

 **Go Figure** Why doesn't the solar spectrum at sea level perfectly match the solar spectrum outside the atmosphere?

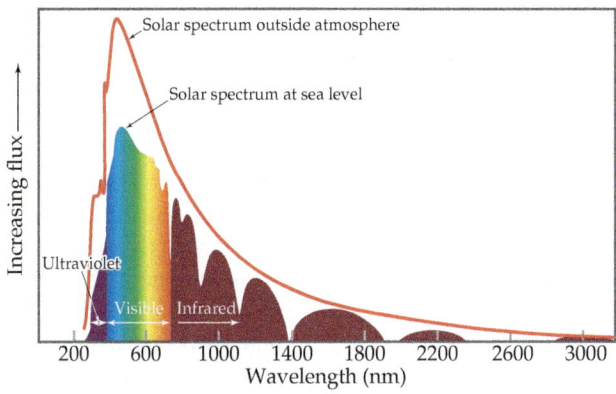

▲ **Figure 18.3 The solar spectrum above Earth's atmosphere compared to that at sea level.** The more structured curve at sea level is due to gases in the atmosphere absorbing specific wavelengths of light. "Flux," the unit on the vertical axis, is light energy per area per unit of time.

cleaved by photodissociation. Instead, half the bonding electrons stay with one atom and half stay with the other atom. The result is two electrically neutral particles.

One of the most important processes occurring above an altitude of about 120 km is photodissociation of the oxygen molecule:

$$:\ddot{O}{=}\ddot{O}: + h\nu \longrightarrow :\ddot{\ddot{O}} + \ddot{\ddot{O}}: \qquad [18.1]$$

The minimum energy required to cause this change is determined by the bond energy (or *dissociation energy*) of O_2, 495 kJ/mol.

 Sample Exercise 18.2

Calculating the Wavelength Required to Break a Bond

What is the maximum wavelength of light, in nanometers, that has enough energy per photon to dissociate the O_2 molecule?

SOLUTION

Analyze We are asked to determine the wavelength of a photon that has just enough energy to break the $O{=}O$ double bond in O_2.

Plan We first need to calculate the energy required to break the $O{=}O$ double bond in one molecule and then find the wavelength of a photon of this energy.

Solve The dissociation energy of O_2 is 495 kJ/mol. Using this value and Avogadro's number, we can calculate the amount of energy needed to break the bond in a single O_2 molecule:

$$\left(495 \times 10^3 \frac{J}{mol}\right)\left(\frac{1\ mol}{6.022 \times 10^{23}\ molecules}\right)$$
$$= 8.22 \times 10^{-19}\frac{J}{molecule}$$

We next use the Planck relationship (Equation 6.2), $E = h\nu$, (Section 6.2) to calculate the frequency ν of a photon that has this amount of energy:

$$\nu = \frac{E}{h} = \frac{8.22 \times 10^{-19}\ J}{6.626 \times 10^{-34}\ J\text{-}s} = 1.24 \times 10^{15}\ s^{-1}$$

Finally, we use the relationship between frequency and wavelength (Equation 6.1) (Section 6.1) to calculate the wavelength of the light:

$$\lambda = \frac{c}{\nu} = \left(\frac{3.00 \times 10^8\ m/s}{1.24 \times 10^{15}/s}\right)\left(\frac{10^9\ nm}{1\ m}\right) = 242\ nm$$

Comment Light of wavelength 242 nm, which is in the ultraviolet region of the electromagnetic spectrum, has sufficient energy per photon to photodissociate an O_2 molecule. Because photon energy increases as wavelength *decreases*, any photon of wavelength *shorter* than 242 nm will have sufficient energy to dissociate O_2.

▶ **Practice Exercise**
The bond energy in N_2 is 941 kJ/mol. What is the longest wavelength a photon can have and still have sufficient energy to dissociate N_2?

Fortunately for us, O_2 absorbs much of the high-energy, short-wavelength radiation from the solar spectrum before that radiation reaches the lower atmosphere. As it does, atomic oxygen, O, is formed. The dissociation of O_2 is very extensive at higher elevations. At 400 km, for example, only 1% of the oxygen is in the form of O_2 and the rest is atomic oxygen. At 130 km, O_2 and atomic oxygen are just about equally abundant. Below 130 km, O_2 is more abundant than atomic oxygen because most of the photons with energies sufficient to photodissociate O_2 have been absorbed in the upper atmosphere.

The dissociation energy of N_2 is very high, 941 kJ/mol. As you should have seen in working out the Practice Exercise associated with Sample Exercise 18.2, only photons having a wavelength shorter than 127 nm possess sufficient energy to dissociate N_2. Furthermore, N_2 does not readily absorb photons, even when they possess sufficient energy. As a result, very little atomic nitrogen is formed in the upper atmosphere by photodissociation of N_2.

Other photochemical processes besides photodissociation occur in the upper atmosphere, although their discovery has taken many twists and turns. In 1901, Guglielmo Marconi received a radio signal in St. John's, Newfoundland, that had been transmitted from Land's End, England, 2900 km away. Because people at the time thought radio waves traveled in straight lines, they assumed that the curvature of Earth's surface would make radio communication over large distances impossible. Marconi's successful experiment suggested that Earth's atmosphere in some way substantially affects radio-wave propagation. His discovery led to intensive study of the upper atmosphere. In about 1924, the existence of electrons in the upper atmosphere was established by experimental studies.

The electrons in the upper atmosphere result mainly from **photoionization**, which occurs when a molecule in the upper atmosphere absorbs solar radiation and the absorbed energy causes an electron to be ejected from the molecule. The molecule then becomes a positively charged ion. For photoionization to occur, therefore, a molecule must absorb a photon, and the photon must have enough energy to remove an electron. (Section 7.4) This is a *very* different process from photodissociation.

Four important photoionization processes occurring in the atmosphere above about 90 km are listed in **Table 18.3**. Photons of any wavelength shorter than the maximum lengths given in the table have enough energy to cause photoionization. A look back at Figure 18.3 shows you that virtually all of these high-energy photons are filtered out of the radiation reaching Earth because they are absorbed by the upper atmosphere.

Ozone in the Stratosphere

Although N_2, O_2, and atomic oxygen absorb photons having wavelengths shorter than 240 nm, ozone, O_3, is the key absorber of photons whose wavelengths range from 240 to 310 nm in the ultraviolet region of the electromagnetic spectrum (**Figure 18.4**). Ozone in the upper atmosphere protects us from these harmful high-energy photons, which would otherwise penetrate to Earth's surface. Let's consider how ozone forms in the upper atmosphere and how it absorbs photons.

By the time radiation from the Sun reaches an altitude of 90 km above Earth's surface, most of the short-wavelength radiation capable of photoionization has been absorbed. Nevertheless, radiation capable of dissociating the O_2 molecule is sufficiently intense for photodissociation of O_2 (Equation 18.1) to remain important down to an altitude of 30 km. In the region between 30 and 90 km, the concentration of O_2 is much

Go Figure

In what region of the electromagnetic spectrum does ozone most strongly absorb light: (a) IR, (b) visible, or (c) UV?

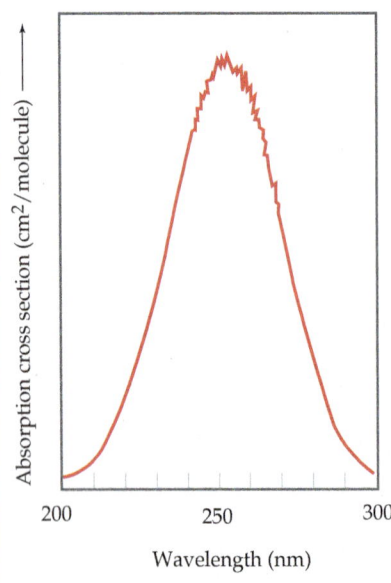

▲ **Figure 18.4** The absorption spectrum of ozone.

TABLE 18.3   Photoionization Reactions for Four Components of the Atmosphere

Process	Ionization Energy (kJ/mol)	λ_{max} (nm)
$N_2 + h\nu \longrightarrow N_2^+ + e^-$	1495	80.1
$O_2 + h\nu \longrightarrow O_2^+ + e^-$	1205	99.3
$O + h\nu \longrightarrow O^+ + e^-$	1313	91.2
$NO + h\nu \longrightarrow NO^+ + e^-$	890	134.5

greater than the concentration of atomic oxygen. Consequently, the oxygen atoms formed by photodissociation of O_2 in this region frequently collide with O_2 molecules and form ozone:

$$:\ddot{O} + O_2 \longrightarrow O_3^* \qquad [18.2]$$

The asterisk on O_3 denotes that the product contains an excess of energy, because the reaction is exothermic. The 105 kJ/mol that is released must be transferred away from the O_3^* molecule quickly or else the molecule will fly apart into O_2 and atomic O—a decomposition that is the reverse of the reaction by which O_3^* is formed.

An energy-rich O_3^* molecule can release its excess energy by colliding with another atom or molecule and transferring some of the excess energy to it. Let's use M to represent the atom or molecule with which O_3^* collides. (M is usually N_2 or O_2, because these are the most abundant molecules in the atmosphere.) The formation of O_3^* and the transfer of excess energy to M are summarized by the following equations:

$$O(g) + O_2(g) \rightleftharpoons O_3^*(g) \qquad [18.3]$$
$$\underline{O_3^*(g) + M(g) \longrightarrow O_3(g) + M^*(g)} \qquad [18.4]$$
$$O(g) + O_2(g) + M(g) \longrightarrow O_3(g) + M^*(g) \qquad [18.5]$$

The rate at which the reactions of Equations 18.3 and 18.4 proceed depends on two factors that vary in opposite directions with increasing altitude. First, the formation of O_3^* (Equation 18.3) depends on the presence of O atoms. At low altitudes, most of the radiation energetic enough to dissociate O_2 molecules into O atoms has been absorbed; thus, O atoms are not very plentiful at low altitudes. Second, Equations 18.3 and 18.4 both depend on molecular collisions. (Section 14.4) The concentration of molecules is greater at low altitudes, and so the rates of both reactions are greater at lower altitudes. Because these two effects vary with altitude in opposite directions, the highest rate of O_3 formation occurs in a band at an altitude of about 50 km, near the stratopause (Figure 18.1). Overall, roughly 90% of Earth's ozone is found in the stratosphere.

The photodissociation of ozone reverses the reaction that forms it. We thus have a cycle of ozone formation and decomposition, summarized as follows:

$$O_2(g) + h\nu \longrightarrow O(g) + O(g)$$
$$O(g) + O_2(g) + M(g) \longrightarrow O_3(g) + M^*(g) \quad \text{(heat released)}$$
$$O_3(g) + h\nu \longrightarrow O_2(g) + O(g)$$
$$O(g) + O(g) + M(g) \longrightarrow O_2(g) + M^*(g) \quad \text{(heat released)}$$

The first and third processes are photochemical; they use a solar photon to initiate a chemical reaction. The second and fourth are exothermic chemical reactions. The net result of the four reactions is a cycle in which solar radiant energy is converted into thermal energy. The ozone cycle in the stratosphere is responsible for the rise in temperature that reaches its maximum at the stratopause (Figure 18.1).

The reactions of the ozone cycle account for some, but not all, of the facts about the ozone layer. Many chemical reactions occur that involve substances other than oxygen, and we must consider the effects of turbulence and winds that mix the stratosphere. The combination of these factors leads to the upper-atmosphere ozone profile shown in **Figure 18.5**, with a maximum ozone concentration occurring at an altitude of about 25 km. This band of relatively high ozone concentration is referred to as the "ozone layer" or the "ozone shield."

Photons with wavelengths shorter than about 300 nm are energetic enough to break many kinds of single chemical bonds. Thus, the "ozone shield" is essential for our continued well-being. The ozone molecules that form this essential shield against high-energy radiation represent only a tiny fraction of the oxygen atoms present in the stratosphere, however, because these molecules are continually destroyed even as they are formed.

Go Figure

Estimate the ozone concentration in moles per liter for the peak value in this graph.

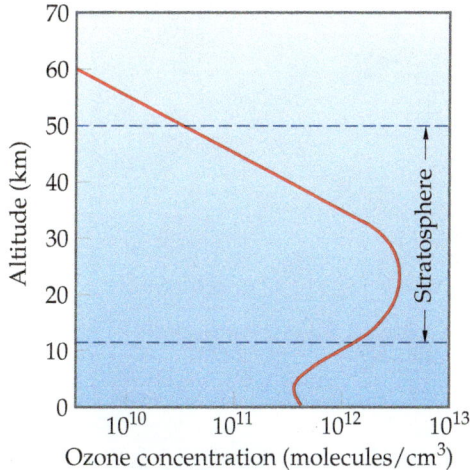

▲ **Figure 18.5** Variation in ozone concentration in the atmosphere as a function of altitude.

Self-Assessment Exercises

SAE 18.1   In the troposphere, temperature _____ with increasing altitude; in the stratosphere, temperature _____ with increasing altitude. (**a**) increases, increases (**b**) increases, decreases (**c**) decreases, increases (**d**) decreases, decreases

SAE 18.2   Which two gases account for nearly 99% of Earth's atmosphere? (**a**) O_2 and CO_2 (**b**) O_2 and H_2 (**c**) O_2 and N_2 (**d**) N_2 and CO_2 (**e**) N_2 and CH_4

SAE 18.3   The mole fraction of argon in dry air near sea level is 0.00001818. What is the concentration of argon in ppm? (**a**) 1.818×10^{-11} ppm (**b**) 18.18 ppm (**c**) 1818 ppm (**d**) 5.501×10^{10} ppm

SAE 18.4   Which of the following statements is or are *true*?

  (**i**)   The minimum energy photon needed to photoionize O_2 has a shorter wavelength than the minimum energy photon needed photodissociate O_2.

  (**ii**)   Photodissociation of N_2 readily occurs in the upper atmosphere.

  (**iii**)   Photoionization leads to the formation of a cation.

(**a**) only i (**b**) both i and ii (**c**) both ii and iii (**d**) both i and iii (**e**) All three statements are true.

SAE 18.5   Which step in the reaction cycle of ozone formation and decomposition absorbs the highest energy photons?

(**a**) $O_2(g) + h\nu \longrightarrow O(g) + O(g)$
(**b**) $O(g) + O_2(g) + M(g) \longrightarrow O_3(g) + M^*(g)$
(**c**) $O_3(g) + h\nu \longrightarrow O_2(g) + O(g)$
(**d**) $O(g) + O(g) + M(g) \longrightarrow O_2(g) + M^*(g)$

18.2 | Human Activities and Earth's Atmosphere

Learning Objectives

When you finish Section 18.2, you should be able to:

▶ Explain how chlorofluorocarbons (CFCs) catalyze depletion of ozone in the stratosphere.

▶ Describe the sources of sulfur oxides in the atmosphere and the reactions that lead to acid rain.

▶ Describe sources of nitrogen oxides in the atmosphere and the reactions that lead to photochemical smog.

▶ Explain how greenhouse gases like H_2O and CO_2 impact Earth's climate.

Both natural and *anthropogenic* (human-caused) events can modify Earth's atmosphere. In this section we focus primarily on how human activities are impacting the atmosphere, but natural events can also have a significant impact. One such event was the eruption of Mount Pinatubo in June 1991 (**Figure 18.6**). The volcano ejected approximately 10 km³ of material into the stratosphere, causing a 10% drop in the amount of sunlight reaching Earth's surface during the next 2 years. That drop in sunlight led to a temporary 0.5 °C drop in Earth's surface temperature. The volcanic particles that made it to the stratosphere remained there for approximately 3 years, *raising* the temperature of the stratosphere by several degrees due to light absorption. Measurements of the stratospheric ozone concentration showed significantly increased ozone decomposition in this 3-year period.

The Ozone Layer and Its Depletion

The ozone layer protects Earth's surface from the damaging ultraviolet (UV) radiation. Therefore, if the concentration of ozone in the stratosphere decreases substantially, more UV radiation will reach Earth's surface, causing unwanted photochemical reactions, including reactions correlated with skin cancer. Satellite monitoring of ozone, which began in 1978, has revealed a depletion of ozone in the stratosphere that is particularly severe over Antarctica, a phenomenon known as the *ozone hole* (**Figure 18.7**). The first scientific paper on this phenomenon appeared in 1985, and the National Aeronautics and Space Administration (NASA) maintains an "Ozone Hole Watch" website with daily updates and data from 1999 to the present.

In 1995, the Nobel Prize in Chemistry was awarded to F. Sherwood Rowland, Mario Molina, and Paul Crutzen for their studies of ozone depletion. In 1970, Crutzen showed that naturally occurring nitrogen oxides catalytically destroy ozone. Rowland and Molina recognized in 1974 that chlorine from **chlorofluorocarbons** (CFCs) may deplete the ozone layer. These substances, principally $CFCl_3$ and CF_2Cl_2, do not occur in nature and have been widely used as propellants in spray cans, as refrigerant and air-conditioner gases, and as foaming agents for plastics. They are virtually unreactive in the lower atmosphere. Furthermore, they are relatively insoluble in water and are therefore not removed from the atmosphere by rainfall or by dissolution in the oceans. Unfortunately, the lack of reactivity that makes them commercially useful also allows them to survive in the atmosphere and to diffuse into the stratosphere. It is estimated that several million tons of chlorofluorocarbons are now present in the atmosphere.

As CFCs diffuse into the stratosphere, they are exposed to high-energy radiation, which can cause photodissociation. Because C—Cl bonds are considerably weaker than

▲ **Figure 18.6** Mount Pinatubo erupts, June 1991.

C—F bonds, free chlorine atoms are formed readily in the presence of light with wavelengths in the range from 190 to 225 nm, as shown in the following typical reaction:

$$CF_2Cl_2(g) + h\nu \longrightarrow CF_2Cl(g) + Cl(g) \qquad [18.6]$$

Calculations suggest that chlorine atom formation occurs at the greatest rate at an altitude of about 30 km, the altitude at which ozone is at its highest concentration.

Atomic chlorine reacts rapidly with ozone to form chlorine monoxide and molecular oxygen:

$$Cl(g) + O_3(g) \longrightarrow ClO(g) + O_2(g) \qquad [18.7]$$

This reaction follows a second-order rate law with a very large rate constant:

$$\text{Rate} = k[Cl][O_3] \quad k = 7.2 \times 10^9 \, M^{-1} \, s^{-1} \text{ at 298 K} \qquad [18.8]$$

Under certain conditions, the ClO generated in Equation 18.7 can react to regenerate free Cl atoms. One way that this can happen is by photodissociation of ClO:

$$ClO(g) + h\nu \longrightarrow Cl(g) + O(g) \qquad [18.9]$$

The Cl atoms generated in Equations 18.6 and 18.9 can react with more O_3, according to Equation 18.7. The result is a sequence of reactions that accomplishes the Cl-catalyzed decomposition of O_3 to O_2:

$$
\begin{aligned}
2\,Cl(g) + 2\,O_3(g) &\longrightarrow 2\,ClO(g) + 2\,O_2(g) \\
2\,ClO(g) + h\nu &\longrightarrow 2\,Cl(g) + 2\,O(g) \\
O(g) + O(g) &\longrightarrow O_2(g) \\
\hline
2\,Cl(g) + 2\,O_3(g) + 2\,ClO(g) + 2\,O(g) &\longrightarrow 2\,Cl(g) + 2\,ClO(g) + 3\,O_2(g) + 2\,O(g)
\end{aligned}
$$

The equation can be simplified by eliminating like species from each side to give

$$2\,O_3(g) \xrightarrow{\;Cl\;} 3\,O_2(g) \qquad [18.10]$$

Because the rate of Equation 18.7 increases linearly with [Cl], the rate at which ozone is destroyed increases as the quantity of Cl atoms increases. Thus, the higher the concentration of CFCs in the stratosphere, the faster the destruction of the ozone layer. Even though troposphere-to-stratosphere diffusion rates are slow, a substantial thinning of the ozone layer over the South Pole has been observed, particularly during September and October (Figure 18.7).

Because of the environmental problems associated with CFCs, steps have been taken to limit their manufacture and use. A major step was the signing in 1987 of the Montreal Protocol on Substances That Deplete the Ozone Layer, in which participating nations agreed to reduce CFC production. More stringent limits were set in 1992, when representatives of approximately 100 nations agreed to ban the production and use of CFCs by 1996, with some exceptions for "essential uses." Since then, the production of CFCs has indeed dropped precipitously. Images such as that shown in Figure 18.7 taken annually reveal that the depth and size of the ozone hole have begun to slowly decline. Because CFCs are unreactive and diffuse so slowly into the stratosphere, scientists estimate that it will take many decades for ozone concentrations in the stratosphere to return to pre-1980 levels.

What substances have replaced CFCs? At this time, the main alternatives are hydrofluorocarbons (HFCs), compounds in which C—H bonds replace the C—Cl bonds of CFCs. One HFC in current use is CH_2FCF_3, known as HFC-134a. While the HFCs are a big improvement over the CFCs because they contain no C—Cl bonds, it turns out that they are potent greenhouse warming gases.

There are no naturally occurring CFCs, but some natural sources contribute chlorine and bromine to the atmosphere, and, just like halogens from CFCs, these naturally occurring Cl and Br atoms can participate in ozone-depleting reactions. The principal natural sources are methyl bromide and methyl chloride, which are emitted from the oceans. It is estimated that these molecules contribute less than a third of the total Cl and Br in the atmosphere; the remaining two-thirds is a result of human activities. Volcanoes are a source of HCl, but generally the HCl they release reacts with water in the troposphere and does not make it to the upper atmosphere.

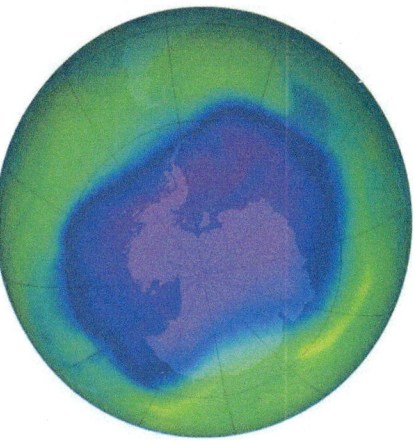

Total ozone (Dobson units)

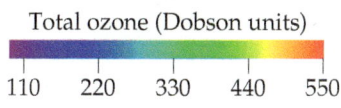

110   220   330   440   550

▲ Figure 18.7 Ozone present in the Southern Hemisphere, September 24, 2006. The data were taken from an orbiting satellite. This day had the lowest stratospheric ozone concentration yet recorded. One "Dobson unit" corresponds to 2.69×10^{16} ozone molecules in a 1 cm^2 column of atmosphere.

TABLE 18.4   Median Concentrations of Atmospheric Pollutants in a Typical Urban Atmosphere

Pollutant	Concentration (ppm)
Carbon monoxide	10
Hydrocarbons	3
Sulfur dioxide	0.08
Nitrogen oxides	0.05
Total oxidants (ozone and others)	0.02

Sulfur Compounds and Acid Rain

Sulfur-containing compounds are present to some extent in the natural, unpolluted atmosphere. They originate in the bacterial decay of organic matter, in volcanic gases, and from other sources listed in Table 18.2. Sulfur compounds, chiefly sulfur dioxide, SO_2, are among the most unpleasant and harmful of the common pollutant gases. It is estimated that approximately two-thirds of SO_2 emitted into the atmosphere can be attributed to human activity, largely from the burning of fossil fuels rich in sulfur. Table 18.4 lists the concentrations of several pollutant gases in a *typical* urban environment (where by *typical* we mean one that is not particularly affected by smog). According to these data, the level of sulfur dioxide is 0.08 ppm or higher about half the time. This concentration is considerably lower than that of other pollutants, notably carbon monoxide. Nevertheless, SO_2 is regarded as the most serious health hazard among the pollutants shown, especially for people with respiratory difficulties.

The three principal anthropogenic sources of SO_2 emissions in order of decreasing contribution are coal-fired power plants, combustion of natural gas and oil, and smelters, where sulfide-containing ores are refined to produce metals. Historically, SO_2 emissions in the United States were more problematic in the midwestern and northeastern parts of the country because coal from the eastern part of the country has a higher sulfur content, up to 6% by mass. In 2010, the U.S. Environmental Protection Agency set new standards to reduce SO_2 emissions. The old standard of 140 parts per billion, measured over 24 h, was replaced by a standard of 75 parts per billion, measured over 1 h. Over the period 1990–2020, the annual emissions of SO_2 nationwide have declined by 95%. This is in large part because of the adoption of technologies to remove most of the sulfur oxides from the gases that are exhausted at power plants. A shift toward generating electricity by burning natural gas instead of coal has also helped to reduce SO_2 emissions (Figure 18.8).

Annual sulfur dioxide emissions (1990)

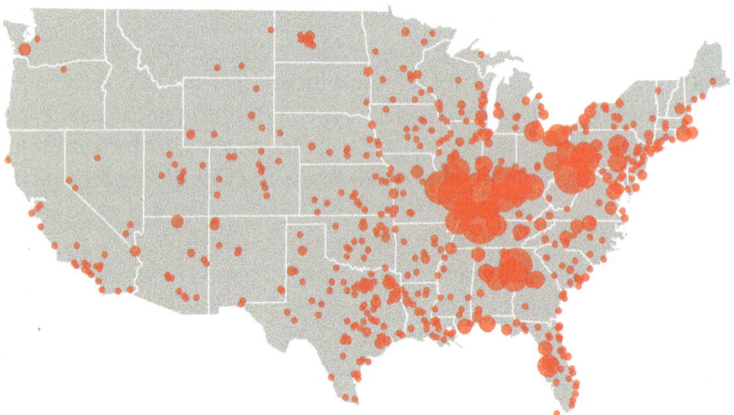

Annual sulfur dioxide emissions (2020)

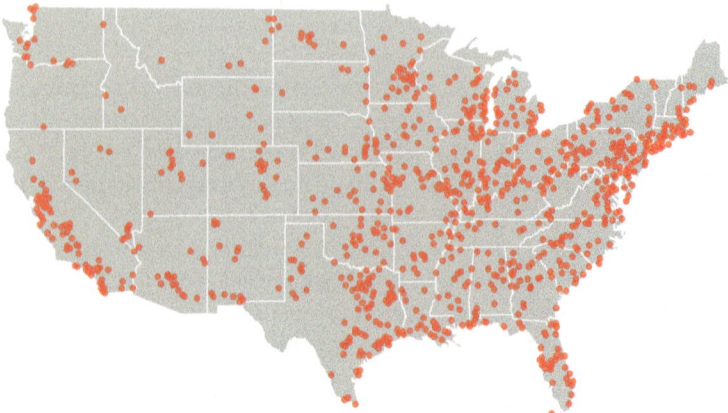

▲ **Figure 18.8   Annual SO$_2$ emissions across the United States in 1990 and 2020.** The larger the circle, the greater the SO_2 emissions from individual power plants. The largest circles represent 6×10^8 lbs and the smallest circles $< 4 \times 10^7$ lbs SO_2 emitted per year.

Sulfur dioxide is harmful to both human health and property; furthermore, atmospheric SO_2 can be oxidized to SO_3 by several pathways (such as reaction with O_2 or O_3). When SO_3 dissolves in water, it produces sulfuric acid:

$$SO_3(g) + H_2O(l) \longrightarrow H_2SO_4(aq)$$

Many of the environmental effects ascribed to SO_2 are actually due to H_2SO_4.

The presence of SO_2 in the atmosphere and the sulfuric acid it produces result in the phenomenon of **acid rain**. (Nitrogen oxides, which form nitric acid, are also major contributors to acid rain.) Uncontaminated rainwater generally has a pH value of about 5.6. The primary source of this natural acidity is CO_2, which reacts with water to form carbonic acid, H_2CO_3. Acid rain typically has a pH value of about 4. This shift toward greater acidity has affected many lakes in northern Europe, the northern United States, and Canada, reducing fish populations and affecting other parts of the ecological network in these lakes and the forests that surround them.

The pH of most natural waters containing living organisms is between 6.5 and 8.5. At pH levels below 4.0, all vertebrates, most invertebrates, and many microorganisms are destroyed. The lakes most susceptible to damage are those with low concentrations of basic ions, such as HCO_3^-, that would act as a buffer to minimize changes in pH. Many of these lakes are recovering as sulfur emissions from fossil fuel combustion have decreased, in part because of the Clean Air Act.

Because acids react with metals and with carbonates, acid rain is corrosive both to metals and to stone building materials. Marble and limestone, for example, whose major constituent is $CaCO_3$, are readily attacked by acid rain. Billions of dollars each year are lost because of corrosion due to SO_2 pollution.

One way to reduce the quantity of SO_2 released into the environment is to remove sulfur from coal and oil before these fuels are burned. Although difficult and expensive, several methods have been developed and widely implemented. Powdered limestone ($CaCO_3$), for example, can be injected into the furnace of a power plant, where it decomposes into lime (CaO) and carbon dioxide:

$$CaCO_3(s) \longrightarrow CaO(s) + CO_2(g)$$

The CaO then reacts with SO_2 to form calcium sulfite:

$$CaO(s) + SO_2(g) \longrightarrow CaSO_3(s)$$

The solid particles of $CaSO_3$, as well as much of the unreacted SO_2, can be removed from the furnace gas by passing it through an aqueous suspension of CaO (**Figure 18.9**). The $CaSO_3$ can be further oxidized to gypsum, $CaSO_4 \cdot 2H_2O$, which can be used in the building industry.

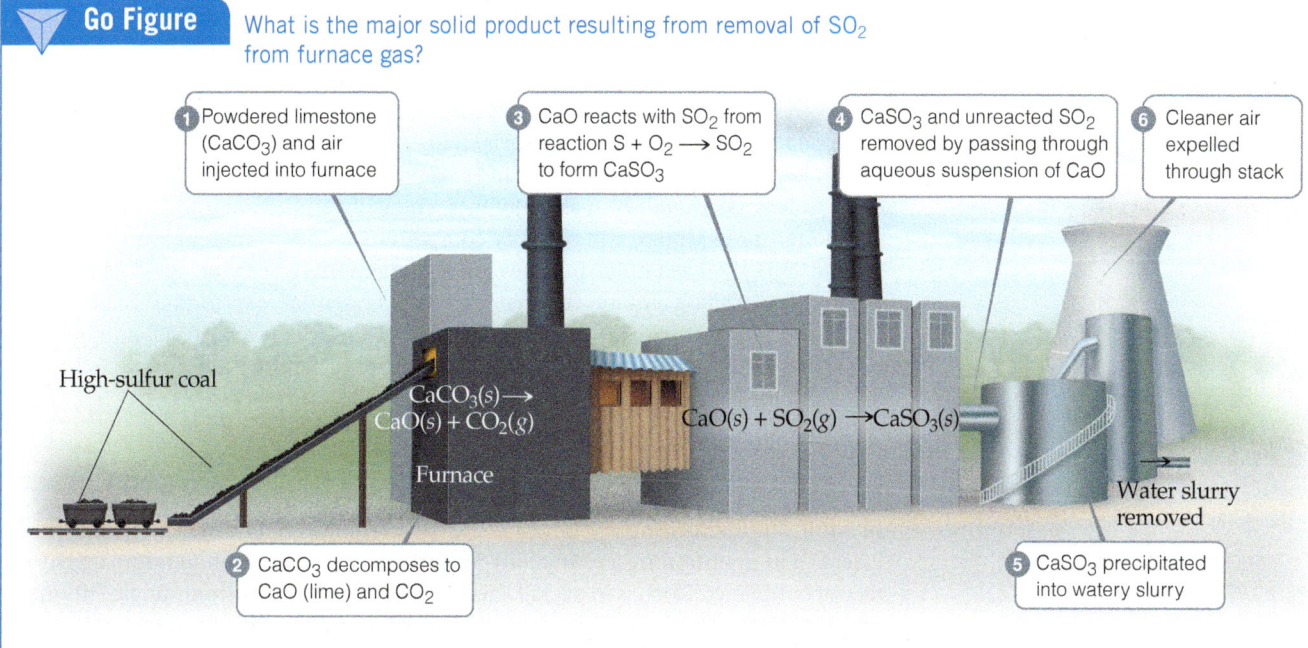

Go Figure What is the major solid product resulting from removal of SO_2 from furnace gas?

1. Powdered limestone ($CaCO_3$) and air injected into furnace

3. CaO reacts with SO_2 from reaction $S + O_2 \longrightarrow SO_2$ to form $CaSO_3$

4. $CaSO_3$ and unreacted SO_2 removed by passing through aqueous suspension of CaO

6. Cleaner air expelled through stack

High-sulfur coal

$CaCO_3(s) \longrightarrow$
$CaO(s) + CO_2(g)$

Furnace

$CaO(s) + SO_2(g) \longrightarrow CaSO_3(s)$

Water slurry removed

2. $CaCO_3$ decomposes to CaO (lime) and CO_2

5. $CaSO_3$ precipitated into watery slurry

▲ **Figure 18.9** One method for removing SO_2 from combusted fuel.

▲ **Figure 18.10 Photochemical smog.** It is produced largely by the action of sunlight on vehicle exhaust gases, as illustrated in this photo of Beijing, China.

Nitrogen Oxides and Photochemical Smog

Nitrogen oxides are the primary components of smog, a phenomenon with which city dwellers are all too familiar. The term *smog* refers to the pollution condition that occurs in certain urban environments when weather conditions produce a relatively stagnant air mass. The smog made famous by Los Angeles, but now common in many other urban areas as well, is more accurately described as **photochemical smog** because photochemical processes play a major role in its formation (**Figure 18.10**).

The majority of nitrogen oxide emissions (about 50%) comes from cars, buses, and other forms of transportation. Nitric oxide, NO, forms in small quantities in the cylinders of internal combustion engines in the reaction

$$N_2(g) + O_2(g) \rightleftharpoons 2\,NO(g) \quad \Delta H = 180.8\;kJ \qquad [18.11]$$

As noted in the "Chemistry and Sustainability" box on "Controlling Nitric Oxide Emissions" (Section 15.7), the equilibrium constant for this reaction increases from about 10^{-15} at 300 K to about 0.05 at 2400 K (approximate temperature in the cylinder of an engine during combustion). Thus, the reaction is more favorable at higher temperatures. In fact, some NO is formed in any high-temperature combustion process.

Before the installation of pollution-control devices on automobiles, typical emission levels of NO_x were 4 g/mi. (The x is either 1 or 2 because both NO and NO_2 are formed, although NO predominates.) Starting in 2004, the auto emission standards for NO_x called for a phased-in reduction to 0.07 g/mi by 2009, which was achieved.

In air, nitric oxide is rapidly oxidized to nitrogen dioxide:

$$2\,NO(g) + O_2(g) \rightleftharpoons 2\,NO_2(g) \quad \Delta H = -113.1\;kJ \qquad [18.12]$$

The equilibrium constant for this reaction decreases from about 10^{12} at 300 K to about 10^{-5} at 2400 K.

The photodissociation of NO_2 initiates the reactions associated with photochemical smog. Dissociation of NO_2 requires 304 kJ/mol, which corresponds to a photon wavelength of 393 nm. In sunlight, therefore, NO_2 dissociates to NO and O:

$$NO_2(g) + h\nu \longrightarrow NO(g) + O(g) \qquad [18.13]$$

The atomic oxygen formed undergoes several reactions, one of which gives ozone, as described previously:

$$O(g) + O_2(g) + M(g) \longrightarrow O_3(g) + M^*(g) \qquad [18.14]$$

Although it is essential to screen out harmful UV radiation in the upper atmosphere, ozone is an undesirable pollutant in the troposphere. It is extremely reactive and toxic, and breathing air that contains appreciable amounts of ozone can be especially dangerous for asthma sufferers, exercisers, and the elderly. We therefore have two ozone problems: excessive amounts in many urban environments, where it is harmful, and depletion in the stratosphere, where its presence is vital.

In addition to nitrogen oxides and carbon monoxide, an automobile engine also emits unburned *hydrocarbons* as pollutants. These organic compounds are the principal components of gasoline and of many compounds we use as fuel (propane, C_3H_8, and butane, C_4H_{10}, for example), but they are also major ingredients of smog. Reduction or elimination of smog requires that the ingredients essential to its formation be removed from automobile exhaust. Catalytic converters reduce the levels of NO_x and hydrocarbons, two of the major ingredients of smog. [See the "Chemistry and Sustainability" box on "Catalytic Converters." (Section 14.6)]

Greenhouse Gases: Water Vapor, Carbon Dioxide, and Climate

In addition to screening out harmful short-wavelength radiation, the atmosphere is essential in maintaining a reasonably uniform and moderate temperature on the surface of our planet. Earth is in overall thermal balance with its surroundings. This means that the planet radiates energy into space at a rate equal to the rate at which it absorbs energy from the Sun. **Figure 18.11** shows the flow of energy into and out of Earth, and

Go Figure What is the total amount of energy absorbed by the surface? What fraction of that energy is emitted upward as infrared radiation?

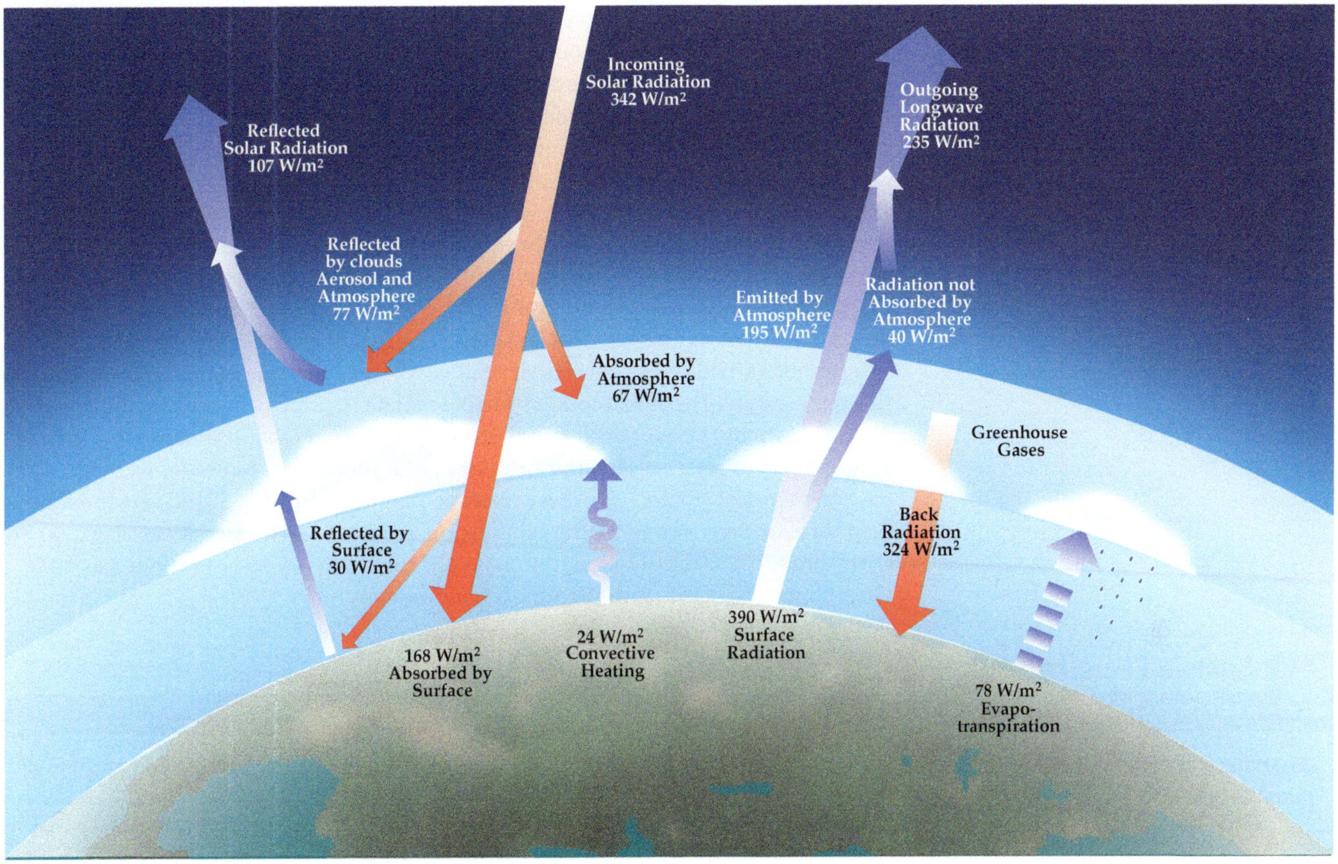

Incoming
Solar Radiation
342 W/m^2

Outgoing
Longwave
Radiation
235 W/m^2

Reflected
Solar Radiation
107 W/m^2

Reflected
by clouds
Aerosol and
Atmosphere
77 W/m^2

Absorbed by
Atmosphere
67 W/m^2

Emitted by
Atmosphere
195 W/m^2

Radiation not
Absorbed by
Atmosphere
40 W/m^2

Greenhouse
Gases

Back
Radiation
324 W/m^2

Reflected by
Surface
30 W/m^2

168 W/m^2
Absorbed by
Surface

24 W/m^2
Convective
Heating

390 W/m^2
Surface
Radiation

78 W/m^2
Evapo-
transpiration

▲ **Figure 18.11 Earth's thermal balance.** The amount of radiation reaching the surface of the planet is equal to the amount radiated back into space.

Figure 18.12 shows the fraction of infrared radiation leaving the surface that is absorbed by atmospheric water vapor and carbon dioxide. In absorbing this radiation, these two atmospheric gases help maintain a livable uniform temperature at the surface by effectively holding in the infrared radiation, which we feel as heat. As a counter example, consider the temperature swings observed on Mars, which has a very thin atmosphere.

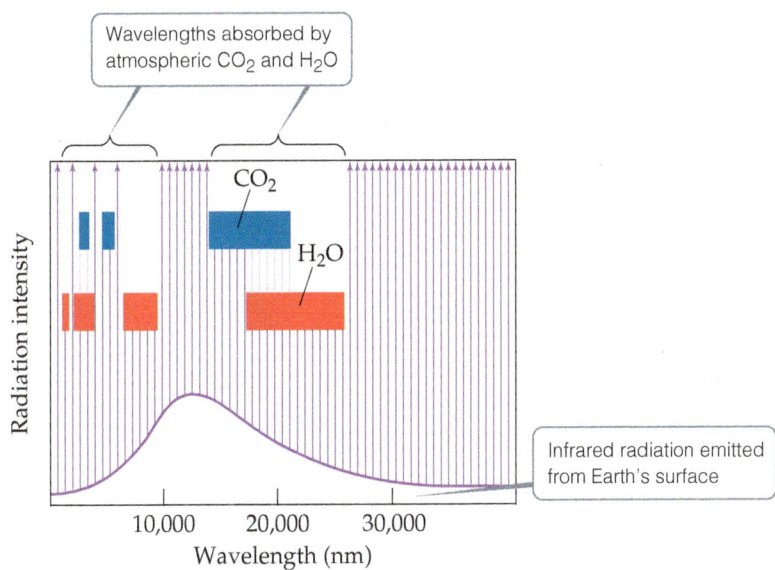

Wavelengths absorbed by atmospheric CO_2 and H_2O

CO_2

H_2O

Radiation intensity

Infrared radiation emitted from Earth's surface

10,000   20,000   30,000

Wavelength (nm)

◀ **Figure 18.12** Portions of the infrared radiation emitted by Earth's surface that are absorbed by atmospheric CO_2 and H_2O.

With little atmosphere to hold in the heat, temperature swings of 70 to 80 °C (126 to 144 °F) between the daytime high and nighttime low are commonplace on Mars.

The influence of H_2O, CO_2, and certain other atmospheric gases on Earth's temperature is called the *greenhouse effect* because in trapping infrared radiation these gases act much like the glass of a greenhouse. The gases themselves are called **greenhouse gases**.

Water vapor makes the largest contribution to the greenhouse effect. The partial pressure of water vapor in the atmosphere varies greatly from place to place and time to time but is generally highest near Earth's surface and decreases with increasing elevation. Because water vapor absorbs infrared radiation so strongly, it plays the major role in maintaining the atmospheric temperature at night, when the surface is emitting radiation into space and not receiving energy from the Sun. In very dry desert climates, where the water-vapor concentration is low, it may be extremely hot during the day but very cold at night. In the absence of a layer of water vapor to absorb and then radiate part of the infrared radiation back to Earth, the surface loses this radiation into space and cools off very rapidly.

Carbon dioxide plays a secondary but very important role in maintaining the surface temperature. The global combustion of fossil fuels, principally coal and oil, on a prodigious scale in the modern era has sharply increased carbon dioxide levels in the atmosphere. To get a sense of the amount of CO_2 produced—for example, by the combustion of hydrocarbons and other carbon-containing substances, which are the components of fossil fuels—consider the combustion of butane, C_4H_{10}. Combustion of 1.00 g of C_4H_{10} produces 3.03 g of CO_2. (Section 3.6). Human activity releases about 3.6×10^{16} g (36 billion tons) of CO_2 into the atmosphere annually.

Sample Exercise 18.3
Estimating the Quantity of CO_2 Released from Combustion of Gasoline

What mass of carbon dioxide is produced when 1.0 gallon of gasoline is combusted? The approximate density and composition of gasoline are 0.70 g/mL and C_8H_{18}, respectively.

SOLUTION

Analyze We are asked to calculate the mass of CO_2 produced when 1.0 gallon of C_8H_{18} reacts with oxygen to form carbon dioxide and water.

Plan We first must determine the mass in grams of C_8H_{18} in 1.0 gallon of gasoline. We then write a balanced chemical equation for the combustion of C_8H_{18} and use it to determine the theoretical yield of CO_2.

Solve

First, convert the volume from gallons to mL.

$$(1.0 \text{ gallon})\left(\frac{3.785 \text{ L}}{1 \text{ gallon}}\right)\left(\frac{1000 \text{ mL}}{1 \text{ L}}\right) = 3.79 \times 10^3 \text{ mL}$$

Next, use the density of gasoline to calculate its mass in grams.

$$(3.79 \times 10^3 \text{ mL})\left(\frac{0.70 \text{ g}}{\text{mL}}\right) = 2.65 \times 10^3 \text{ g}$$

To calculate the theoretical yield of carbon dioxide, we must write a balanced chemical equation for the combustion of C_8H_{18} (octane):

$$2 \, C_8H_{18}(l) + 25 \, O_2(g) \longrightarrow 16 \, CO_2(g) + 18 \, H_2O(g)$$

Using the balanced equation and the mass of C_8H_{18}, we can calculate the theoretical yield of CO_2.

$$(2.65 \times 10^3 \text{ g } C_8H_{18})\left(\frac{1 \text{ mol } C_8H_{18}}{114.2 \text{ g } C_8H_{18}}\right)\left(\frac{16 \text{ mol } CO_2}{2 \text{ mol } C_8H_{18}}\right)\left(\frac{44 \text{ g } CO_2}{1 \text{ mol } CO_2}\right)$$
$$= 8.2 \times 10^3 \text{ g } CO_2$$

So, we see that for every gallon of gasoline burned, 8.2 kg of CO_2 are produced.

▶ **Practice Exercise**

A typical propane, C_3H_8, tank for an outdoor grill contains 15 pounds of propane. What mass of CO_2 is produced by combusting the propane in such a tank?

Much CO_2 is absorbed by the oceans or used by plants. Nevertheless, we are now generating CO_2 much faster than it is being absorbed or used. Analysis of air trapped in ice cores taken from Antarctica and Greenland makes it possible to determine the atmospheric levels

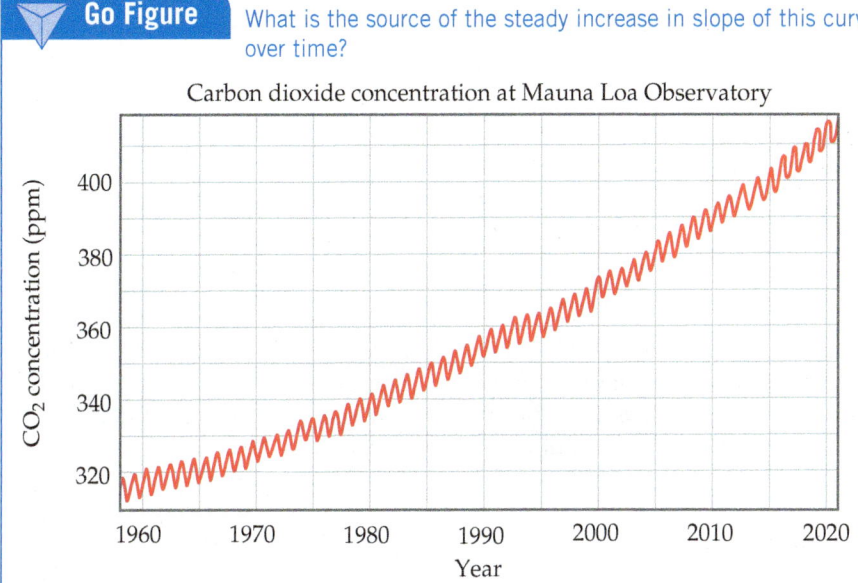

▲ **Figure 18.13 Rising CO₂ levels.** The concentration of CO_2 has been measured at the Mauna Loa Observatory in Hawaii since 1958, and over that period the concentration has increased over 100 ppm. The sawtooth shape of the graph is due to regular seasonal variations in CO_2 concentration.

of CO_2 during the past 160,000 years. These measurements reveal that the level of CO_2 remained fairly constant from the last Ice Age, some 10,000 years ago, until roughly the beginning of the Industrial Revolution, about 300 years ago. Since that time, the concentration of CO_2 has increased to a current high of approximately 415 ppm (**Figure 18.13**). Climate scientists believe that the CO_2 level has not been this high since 3 to 5 million years ago.

The consensus among climate scientists is that the increase in atmospheric CO_2 is perturbing Earth's climate and playing a role in the observed increase in the average global surface temperature (**Figure 18.14**). Scientists often use the term *climate change* instead of *global warming* to refer to this effect because as Earth's temperature increases, winds and ocean currents are affected in ways that cool some areas and warm others.

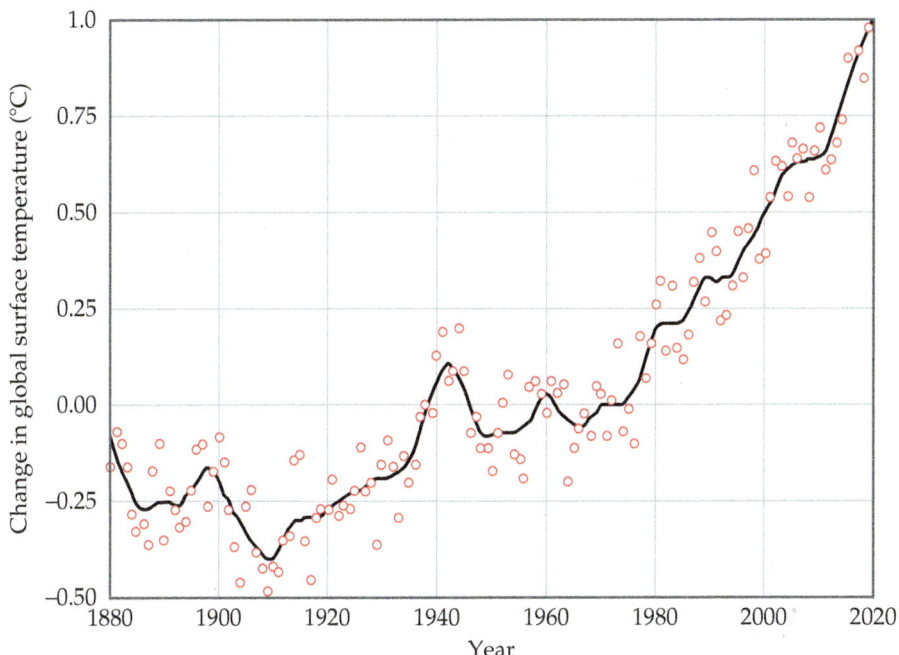

▲ **Figure 18.14 Global surface temperature over the past 140 years.** This graph illustrates the change in global surface temperature relative to the average temperature experienced between 1951 and 1980, as compiled by the NASA Goddard Institute for Space Studies.

A CLOSER LOOK | Other Greenhouse Gases

Although CO_2 receives most of the attention, other gases contribute to the greenhouse effect, including methane, CH_4, nitrous oxide, N_2O, hydrofluorocarbons (HFCs), and chlorofluorocarbons (CFCs).

Methane makes a significant contribution to the greenhouse effect. Studies of atmospheric gas trapped long ago in the Greenland and Antarctic ice sheets show that the atmospheric methane concentration has increased from preindustrial values of 0.3 to 0.7 ppm to the present value of about 1.9 ppm. The major sources of methane are associated with agriculture, fossil-fuel production, and decomposition of organic waste.

Methane is formed in biological processes that occur in low-oxygen environments. Anaerobic bacteria, which flourish in swamps and landfills, near the roots of rice plants, and in the digestive systems of cattle, sheep and other ruminant animals, produce methane (**Figure 18.15**). It also leaks into the atmosphere during natural-gas extraction and transport. It is estimated that about two-thirds of present-day methane emissions, which are increasing by about 1% per year, are related to human activities.

Methane has a half-life in the atmosphere of approximately 9–12 years, whereas CO_2 is much longer-lived. This might seem a good thing, but there are indirect effects to consider. Methane is oxidized in the stratosphere, producing water vapor, a powerful greenhouse gas that is otherwise virtually absent from the stratosphere. In the troposphere, methane is attacked by reactive species such as OH radicals or nitrogen oxides, eventually producing other greenhouse gases, such as O_3. It has been estimated that on a per mass basis, the global warming potential of CH_4 is about 25 times that of CO_2. Given this large contribution, important reductions of the greenhouse effect could be achieved by reducing methane emissions or capturing the emissions for use as a fuel.

Nitrous oxide, N_2O, is another important greenhouse gas. The majority of N_2O emissions result from the use of nitrogen-based fertilizers in agriculture. Although fertilizers play a key role in supplying food to support Earth's population, more efficient use of fertilizers would help to reduce N_2O emissions. Globally, about

▲ **Figure 18.15 Methane production.** Ruminant animals, such as cattle and sheep, produce methane in their digestive systems.

40 percent of total N_2O emissions come from human activities. It is estimated that the global warming effect of nitrous oxide on a per mass basis is 250–300 times larger than CO_2.

HFCs have replaced CFCs in a host of applications, including refrigerants and air-conditioner gases. Although they do not contribute to the depletion of the ozone layer, HFCs are nevertheless potent greenhouse gases. For example, one of the by-product molecules from production of HFCs that are used in commerce is HCF_3, which is estimated to have a global warming potential, gram for gram, more than 14,000 times that of CO_2. The total concentration of HFCs in the atmosphere has been increasing about 10% per year. Unlike CO_2, CH_4, or N_2O, all HFC emissions can be attributed to human activity because they are not naturally occurring molecules.

Related Exercise: 18.62

Because so many factors go into determining climate, it is challenging to predict with certainty how the climate will change in the future. Much of it depends on the extent to which humankind is able to mitigate the unprecedented rise in atmospheric concentrations of CO_2 and other heat-trapping gases. The effects of these changes on the climate are already being felt, and if left unchecked they have the potential to substantially alter the climate of the planet. The consequences of that change will likely have a profound effect on ecosystems and societies across the planet.

 Self-Assessment Exercises

SAE 18.6 Why do chlorofluorocarbons (CFCs), like $CFCl_3$, deplete ozone in the upper atmosphere, but hydrofluorocarbons (HFCs), like CH_2FCF_3, do not? (**a**) HFCs are more reactive than CFCs and therefore react with other molecules before reaching the upper atmosphere. (**b**) HFCs don't absorb infrared radiation. (**c**) HFCs don't undergo photodissociation in the upper atmosphere to produce atomic chlorine. (**d**) HFCs are more polar than CFCs.

SAE 18.7 Which of the following molecules contribute to acid rain: NO_2, CH_4, SO_2? (**a**) only NO_2 (**b**) only SO_2 (**c**) both CH_4 and SO_2 (**d**) both NO_2 and SO_2 (**e**) All three contribute to acid rain.

SAE 18.8 Which molecule is not a component of smog? (**a**) NO_2 (**b**) CO (**c**) CF_2Cl_2 (**d**) O_3

SAE 18.9 How does an increase in atmospheric concentration of CO_2 contribute to an increase in Earth's surface temperature? (**a**) CO_2 absorbs infrared radiation emitted from the surface and radiates a portion of it back toward the surface. (**b**) CO_2 reacts with molecules in the atmosphere like O_3 that absorb energy from the Sun, thereby reducing their concentration. (**c**) Because CO_2 doesn't absorb infrared radiation, it allows more heat from the Sun to reach the surface of Earth. (**d**) CO_2 undergoes exothermic photochemical reactions in the atmosphere, releasing heat that warms the planet.

SAE 18.10 Which of the following industrial processes is *not* a significant source of SO_2 emissions? (**a**) burning coal (**b**) anaerobic decomposition of landfill waste (**c**) smelting of ores to produce metals (**d**) combustion of oil

18.3 | Earth's Water

Water covers 72% of Earth's surface and is essential to life. Our bodies are about 65% water by mass. Because of extensive hydrogen bonding, water has unusually high melting and boiling points and a high heat capacity. (Section 11.2) Water's highly polar character is responsible for its exceptional ability to dissolve a wide range of ionic and polar-covalent substances. Many reactions occur in water, including reactions in which H_2O itself is a reactant. Recall, for example, that H_2O can participate in acid–base reactions as either a proton donor or a proton acceptor. (Section 16.3) All these properties make water an exceptional molecule and solvent. It's no surprise that water plays a prominent role in our environment.

The Global Water Cycle

All the water on Earth is connected in a global water cycle (Figure 18.16). Most of the processes depicted here rely on the phase changes of water. For instance, warmed by the Sun, liquid water in the oceans evaporates into the atmosphere as water vapor and condenses into liquid water droplets that we see as clouds. Water droplets in the clouds can crystallize to ice, which can precipitate as hail or snow. Once on the ground, the hail or snow melts to liquid water, which soaks into the ground. If conditions are right, it is also possible for ice on the ground to sublime to water vapor in the atmosphere. Because the processes occurring on the surface (evaporation, sublimation) are endothermic and those occurring in the atmosphere (condensation, crystallization) are exothermic, they act to transfer heat absorbed by the surface into the atmosphere.

Salt Water: Earth's Oceans and Seas

The vast layer of salty water that covers so much of the planet is in actuality one large connected body and is generally constant in composition. For this reason, oceanographers speak of a *world ocean* rather than of the separate oceans we learn about in geography books.

Learning Objectives

When you finish Section 18.3, you should be able to:

▶ Identify exothermic and endothermic processes in the global water cycle and explain how they facilitate the transfer of heat from the surface to the atmosphere.

▶ Describe the composition of seawater and how its properties vary with depth.

▶ Describe the sources, distribution, and characteristics of freshwater.

Go Figure Which processes shown in this figure involve the phase transition $H_2O(l) \longrightarrow H_2O(g)$?

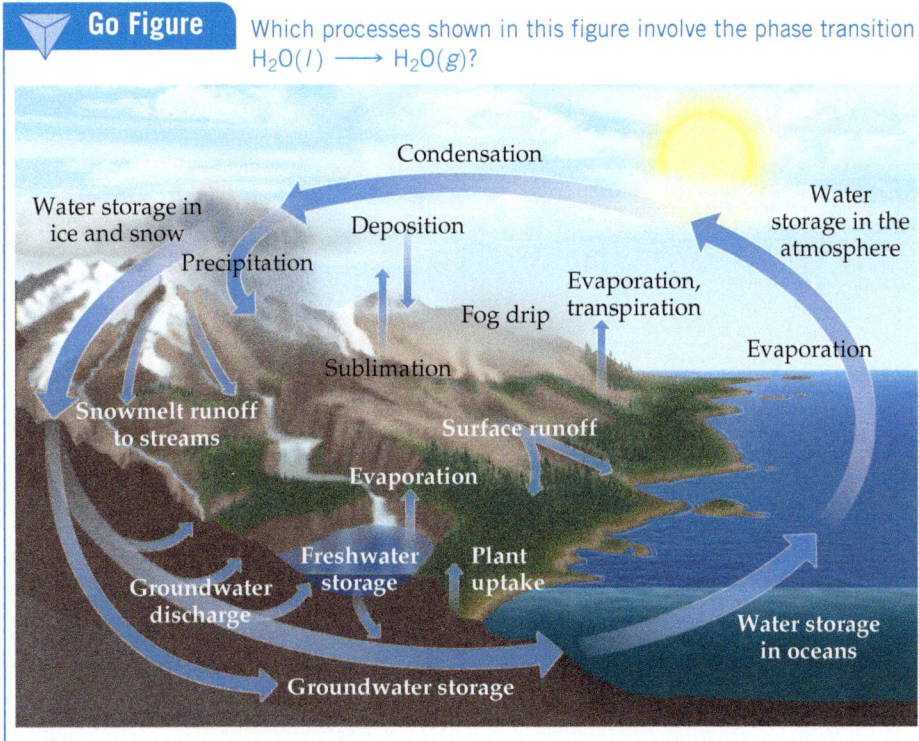

▲ Figure 18.16 The global water cycle.

TABLE 18.5   Ionic Constituents of Seawater Present in Concentrations Greater than 0.001 g/kg (1 ppm)

Ionic Constituent	Salinity	Concentration (M)
Chloride, Cl^-	19.35	0.55
Sodium, Na^+	10.76	0.47
Sulfate, SO_4^{2-}	2.71	0.028
Magnesium, Mg^{2+}	1.29	0.054
Calcium, Ca^{2+}	0.412	0.010
Potassium, K^+	0.40	0.010
Carbon dioxide*	0.106	2.3×10^{-3}
Bromide, Br^-	0.067	8.3×10^{-4}
Boric acid, H_3BO_3	0.027	4.3×10^{-4}
Strontium, Sr^{2+}	0.0079	9.1×10^{-5}
Fluoride, F^-	0.0013	7.0×10^{-5}

*CO_2 is present in seawater as HCO_3^- and CO_3^{2-}.

The world ocean is huge, having a volume of 1.35×10^9 km^3 and containing 97.2% of all the water on Earth. Of the remaining 2.8%, three-quarters is in the form of ice caps and glaciers. All the freshwater—in lakes, in rivers, and in the ground—amounts to only 0.6% of the water on Earth. Most of the remaining 0.1% is in brackish (salty) water, such as that in the Dead Sea or the Great Salt Lake.

Seawater is often referred to as saline water. The **salinity** of seawater is the mass in grams of dry salts present in 1 kg of seawater. In the world ocean, salinity averages about 35. To put it another way, seawater contains about 3.5% dissolved salts by mass. The list of elements present in seawater is very long. Most, however, are present only in very low concentrations. **Table 18.5** lists the 11 ionic species most abundant in seawater.

Seawater temperature varies as a function of depth (**Figure 18.17**), as do salinity and density. Sunlight penetrates well only 200 m into the water; the region between 200 m and 1000 m deep is the "twilight zone," where visible light is faint. Below 1000 m, the

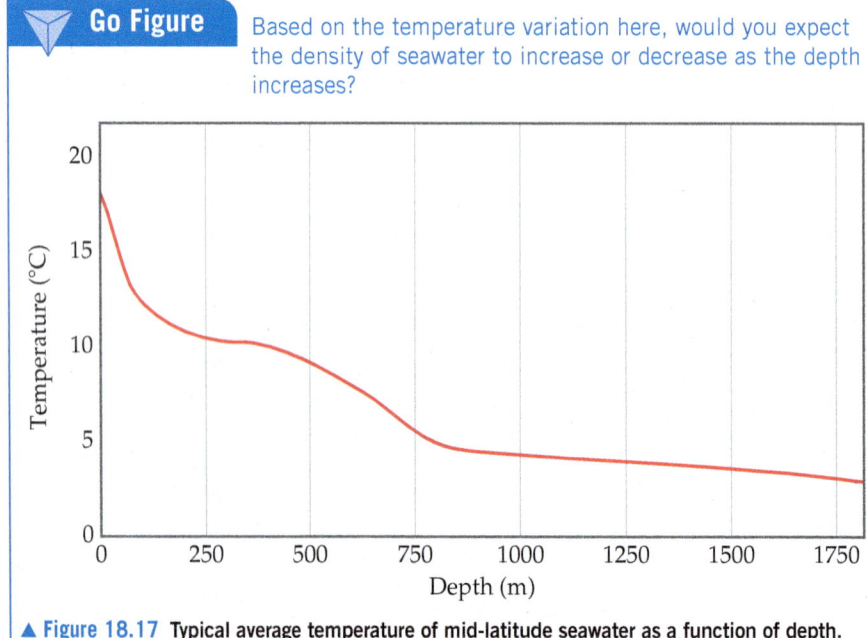

Go Figure   Based on the temperature variation here, would you expect the density of seawater to increase or decrease as the depth increases?

▲ Figure 18.17 **Typical average temperature of mid-latitude seawater as a function of depth.**

ocean is pitch-black and cold, about 4 °C. The transport of heat, salt, and other chemicals throughout the ocean is influenced by these changes in the physical properties of seawater, and in turn the changes in the way heat and substances are transported affects ocean currents and the global climate.

The sea is so vast that if the concentration of a substance in seawater is 1 part per billion (1×10^{-6} g/kg of water), there is 1×10^{12} kg of the substance in the world ocean. Nevertheless, because of high extracting costs, only three substances are obtained from seawater in commercially important amounts: sodium chloride, bromine (from bromide salts), and magnesium (from its salts).

Absorption of CO_2 by the ocean plays a large role in global climate. Because carbon dioxide and water form carbonic acid, the H_2CO_3 concentration in the ocean increases as the water absorbs atmospheric CO_2. Most of the carbon in the ocean, however, is in the form of HCO_3^{2-} and CO_3^{2-} ions, which form a buffer system that maintains the ocean's pH between 8.0 and 8.3. The pH of the ocean is predicted to decrease as the concentration of CO_2 in the atmosphere increases, as discussed in the "Chemistry and Sustainability" box on "Ocean Acidification" (Section 18.4).

Freshwater and Groundwater

Freshwater is the term used to denote natural waters that have low concentrations (less than 500 ppm) of dissolved salts and solids. Freshwater includes the waters of lakes, rivers, ponds, and streams. The United States is fortunate in its abundance of freshwater, with an estimated reserve of 1.7×10^{15} L (660 trillion gallons). Approximately 9×10^{11} L of freshwater is used every day in the United States. Most of this is used for agriculture (41%) and hydroelectric power (39%), with small amounts for industry (6%), household needs (6%), and drinking water (1%). Across the globe, the average adult drinks about 2 L of water per day. In the United States, our daily use of water per person far exceeds this subsistence level, amounting to an average of about 300 L/day for personal consumption and hygiene. We use about 8 L/person for cooking and drinking, about 120 L/person for cleaning (bathing, laundering, and housecleaning), 80 L/person for flushing toilets, and 80 L/person for watering lawns and gardens.

The total amount of freshwater on Earth is not a very large fraction of the total water present. Indeed, freshwater is one of our most precious resources. It forms by evaporation from the oceans and the land. The water vapor that accumulates in the atmosphere is transported by global atmospheric circulation, eventually returning to Earth as rain, snow, and other forms of precipitation (Figure 18.16).

As water runs off the land on its way to the oceans, it dissolves a variety of cations (mainly Na^+, K^+, Mg^{2+}, Ca^{2+}, and Fe^{2+}), anions (mainly Cl^-, SO_4^{2-}, and HCO_3^-), and gases (principally O_2, N_2, and CO_2). As we use water, it becomes laden with additional dissolved material, including the wastes of human society. As the population grows and the output of environmental pollutants increases, ever-increasing amounts of money and resources must be spent to guarantee a supply of freshwater.

Approximately 20% of the world's freshwater is under the soil, in the form of *groundwater*. Groundwater resides in *aquifers*, which are layers of porous rock that hold water. The water in aquifers can be very pure and accessible for human consumption if near the surface. Dense underground formations that do not allow water to readily penetrate can hold groundwater for years or even millennia. When their water is removed by drilling and pumping, such aquifers are slow to recharge via the diffusion of surface water.

The nature of the rock that contains the groundwater has a large influence on the water's chemical composition. If minerals in the rock are water-soluble to some extent, ions can leach out of the rock and remain dissolved in the groundwater. Water that has a relatively high concentration of dissolved ions is referred to as *hard water*. In some instances, toxic substances are leached out of the rocks in contact with groundwater. Arsenic in the form of $HAsO_4^{2-}$, $H_2AsO_4^-$, and H_3AsO_3 is found in many groundwater sources across the world, most infamously in Bangladesh, at concentrations poisonous to humans.

CHEMISTRY AND SUSTAINABILITY The Ogallala Aquifer—A Shrinking Resource

The Ogallala Aquifer, also referred to as the High Plains Aquifer, is an enormous underground body of water lying beneath the Great Plains of the United States. One of the world's largest aquifers, it covers an area of approximately 450,000 km² (170,000 mi²), encompassing portions of eight states: South Dakota, Nebraska, Wyoming, Colorado, Kansas, Oklahoma, New Mexico, and Texas. (Figure 18.18). The water-saturated depth of the underground aquifer ranges from as little as 1 m to more than 300 m. The total volume of water stored in the aquifer is more than that of Lake Huron.

Anyone who has flown over the Great Plains is familiar with the view of huge circles made by the center pivot irrigators nearly covering the land. The center post irrigation system, developed in the post–World War II era, permitted application of water onto large areas. As a result, the Great Plains became one of the most productive agricultural areas in the world. Unfortunately, the premise that the aquifer is an inexhaustible source of freshwater proved false. Recharge of the aquifer from surface water is slow, taking hundreds, perhaps thousands of years. Water levels in parts of the southern half of the aquifer have fallen by more than 50 m over the past seven decades, making the costs of bringing water to the surface prohibitive. As aquifer levels continue to drop, less water will be available for the needs of cities, residences, and businesses.

Related Exercise: 18.42

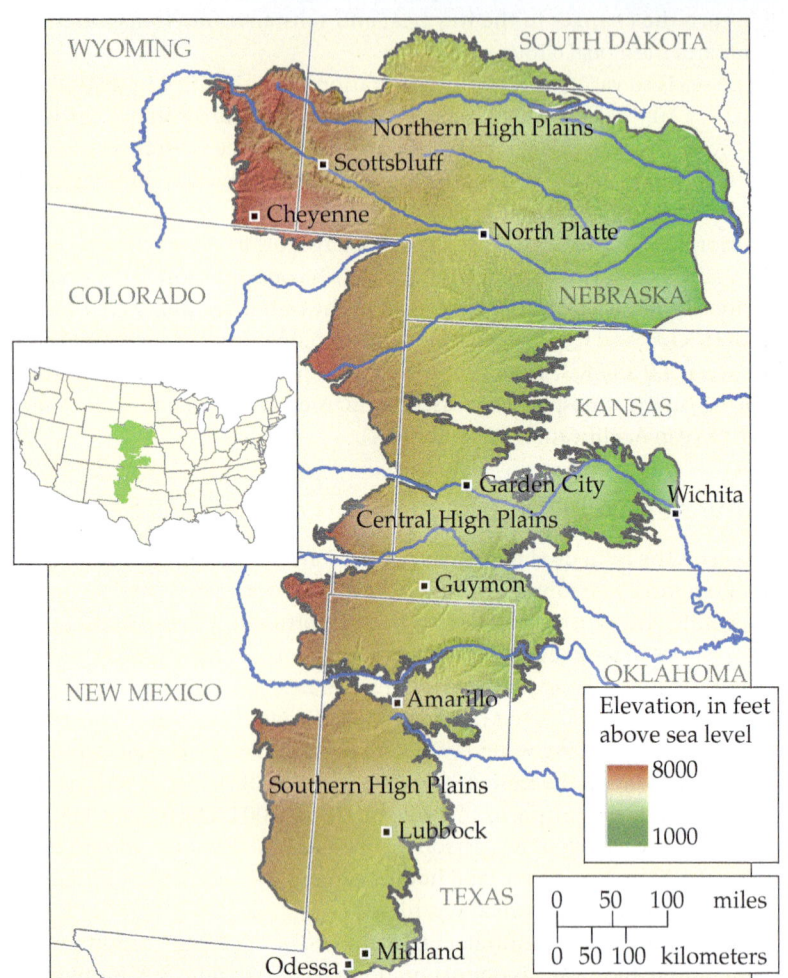

▲ Figure 18.18 **A map showing the extent of the Ogallala (High Plains) Aquifer.** Note that the elevation of the land varies greatly. The aquifer follows the topography of the formations that underlie the area.

▲ Self-Assessment Exercises

SAE 18.11 Which of the following processes of the global water cycle releases heat to the surroundings?

(**i**) evaporation
(**ii**) condensation
(**iii**) sublimation

(**a**) only i (**b**) only ii (**c**) only iii (**d**) both i and iii (**e**) all three processes

SAE 18.12 Which ion is present in the greatest concentration in seawater? (**a**) Cl^- (**b**) Na^+ (**c**) CO_3^{2-} (**d**) SO_4^{2-} (**e**) both Na^+ and Cl^-, which are present in equal amounts

SAE 18.13 Which of the following statements about Earth's water is or are *true*?

(**i**) Seawater temperature varies as a function of depth.
(**ii**) Freshwater accounts for a considerable amount of Earth's total water.
(**iii**) The pH of the ocean increases as the concentration of CO_2 in the atmosphere increases.

(**a**) only i (**b**) both i and ii (**c**) both i and iii (**d**) both ii and iii (**e**) All three statements are true.

18.4 | Human Activities and Water Quality

All life on Earth depends on the availability of suitable water. Many human activities introduce wastes into natural waters without treatment. These practices result in contaminated water that is detrimental to both plant and animal life. In this section we explore some of the consequences of human activity on the quality of freshwater in the environment, and we take a closer look at methods of treating water to make it suitable for human consumption.

Dissolved Oxygen and Water Quality

The amount of O_2 dissolved in water is an important indicator of water quality. Water fully saturated with air at 1 atm and 20 °C contains about 9 ppm of O_2. Oxygen is necessary for fish and most other aquatic life. Cold-water fish require water containing at least 5 ppm of dissolved oxygen for survival. Aerobic bacteria consume dissolved oxygen to oxidize organic matter for energy. The organic material the bacteria are able to oxidize is said to be **biodegradable**.

Excessive quantities of biodegradable organic materials in water are detrimental because they remove the oxygen necessary to sustain normal animal life. Typical sources of these biodegradable materials, which are called *oxygen-demanding wastes*, include sewage, industrial wastes from food-processing plants and paper mills, and liquid waste from meatpacking plants.

In the presence of oxygen, the carbon, hydrogen, nitrogen, sulfur, and phosphorus in biodegradable material end up mainly as CO_2, HCO_3^-, H_2O, NO_3^-, SO_4^{2-}, and phosphates. The formation of these oxidation products can reduce the amount of dissolved oxygen to the point where aerobic bacteria can no longer survive. Anaerobic bacteria then take over the decomposition process, forming CH_4, NH_3, H_2S, PH_3, and other products, several of which contribute to the offensive odors of some polluted waters.

Plant nutrients, particularly nitrogen and phosphorus, contribute to water pollution by stimulating excessive growth of aquatic plants. The most visible results of excessive plant growth are floating algae and murky water. What is more significant, however, is that as plant growth becomes excessive, the amount of dead and decaying plant matter increases rapidly, a process called *eutrophication* (**Figure 18.19**). The process by which plants decay consumes O_2, and without sufficient oxygen, the water cannot sustain animal life.

The most significant sources of nitrogen and phosphorus compounds in water are domestic sewage (phosphate-containing detergents and nitrogen-containing body wastes), runoff from agricultural land (fertilizers contain both nitrogen and phosphorus), and runoff from livestock areas (animal wastes contain nitrogen).

Water Purification: Municipal Treatment

The water needed for domestic, agricultural, and industrial use is taken either from lakes, rivers, reservoirs, or underground wells. Much of the water that finds its way into municipal water systems is "used" water, meaning it has already passed through one or more sewage systems or industrial plants. Consequently, this water must be treated before it is distributed to our faucets.

Municipal water treatment usually involves five steps (**Figure 18.20**).

1. After coarse filtration through a screen, the water is allowed to stand in large sedimentation tanks, where sand and other minute particles settle out. To aid in removing very small particles, the water may first be made slightly basic with CaO.

2. Then $Al_2(SO_4)_3$ is added and reacts with OH^- ions to form a spongy, gelatinous precipitate of $Al(OH)_3$ ($K_{sp} = 1.3 \times 10^{-33}$). This precipitate settles slowly, carrying suspended particles down with it, thereby removing nearly all finely divided matter and most bacteria.

3. The water is then filtered through a bed of sand.

4. Following filtration, the water may be sprayed into the air (aeration) to hasten oxidation of dissolved inorganic ions of iron and manganese, reduce concentrations of any H_2S or NH_3 that may be present, and reduce bacterial concentrations.

Learning Objectives

When you finish Section 18.4, you should be able to:

▶ Explain how the presence of biodegradable matter in natural water systems can lower the amount of dissolved oxygen available for aquatic life.

▶ Describe water treatment and purification processes employed to make water suitable for human consumption.

▲ **Figure 18.19 Eutrophication.** This rapid accumulation of dead and decaying plant matter in a body of water uses up the water's oxygen supply, making the water unsuitable for aquatic animals.

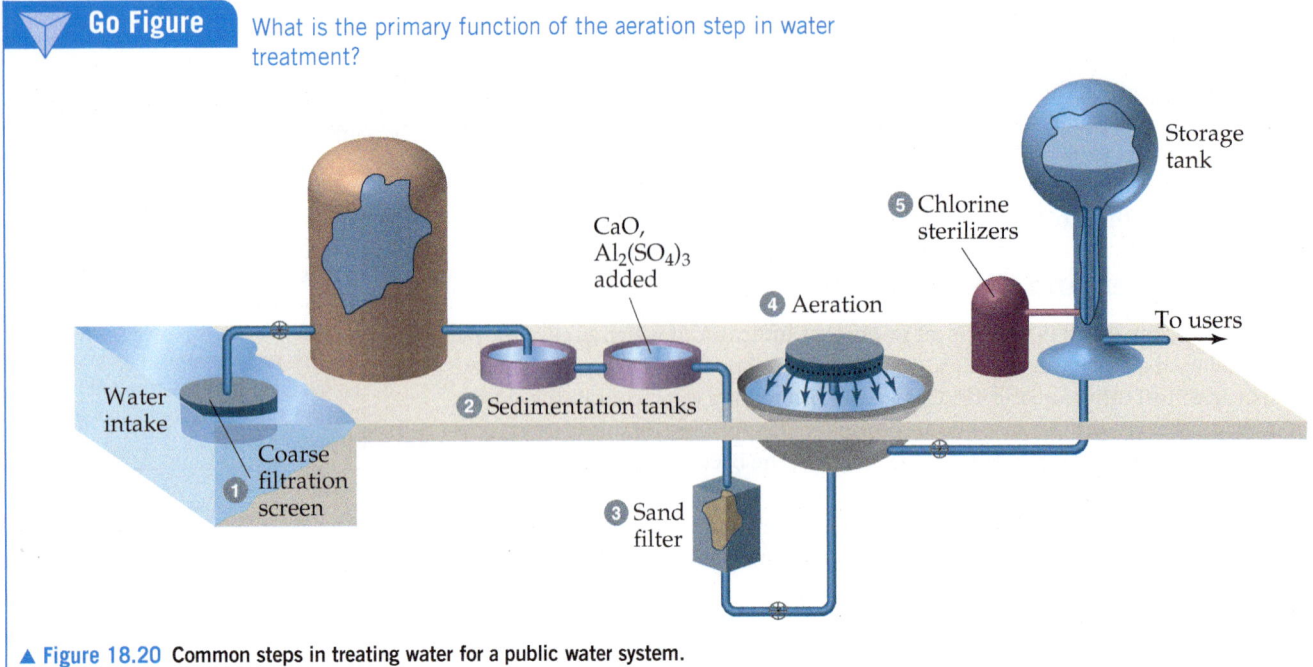

Go Figure   What is the primary function of the aeration step in water treatment?

▲ **Figure 18.20 Common steps in treating water for a public water system.**

5. The final step normally involves treating the water with a chemical agent to ensure the destruction of bacteria. Ozone is more effective, but chlorine is less expensive. Liquefied Cl_2 is dispensed from tanks through a metering device directly into the water supply. The amount used depends on the presence of other substances with which the chlorine might react and on the concentrations of bacteria and viruses to be removed.

The sterilizing action of chlorine is not due to Cl_2 itself but to hypochlorous acid, which forms when chlorine reacts with water:

$$Cl_2(aq) + H_2O(l) \longrightarrow HClO(aq) + H^+(aq) + Cl^-(aq) \qquad [18.15]$$

It is estimated that about 880 million people worldwide lack access to clean water. According to the World Health Organization and UNICEF, an estimated 2.4 billion people are still without any form of improved sanitation, and roughly 13% of the world's population have no access to sanitation facilities of any type and practice open defecation. Not surprisingly, nearly 80% of the health maladies in developing countries can be traced to waterborne diseases associated with unsanitary water.

Water disinfection is one of the greatest public health innovations in human history. It has dramatically decreased the incidence of waterborne bacterial diseases such as cholera and typhus. However, this great benefit comes at a price. In 1974 scientists in Europe and the United States discovered that chlorination of water produces a group of by-products previously undetected. These by-products are called *trihalomethanes* (THMs) because they all have a single carbon atom and three halogen atoms: $CHCl_3$, $CHCl_2Br$, $CHClBr_2$, and $CHBr_3$. These and many other chlorine- and bromine-containing organic substances are produced by the reaction of dissolved chlorine with the organic materials present in nearly all natural waters, as well as with substances that are by-products of human activity.

Recall that chlorine dissolves in water to form the oxidizing agent HClO, as shown in Equation 18.15. The HClO reacts, in turn, with organic substances to form THMs. Bromine enters the reaction sequence through the reaction of HClO with dissolved bromide ion:

$$HClO(aq) + Br^-(aq) \longrightarrow HBrO(aq) + Cl^-(aq) \qquad [18.16]$$

Then both $HBrO(aq)$ and $HClO(aq)$ can halogenate organic substances to form the THMs.

CHEMISTRY AND SUSTAINABILITY   Fracking and Water Quality

In recent years **fracking**, short for *hydraulic fracturing,* has become widely used to greatly increase the availability of petroleum reserves. In fracking, a large volume of water, typically two million gallons or more, mixed with various additives, is injected at high pressure into wellbores extended horizontally into rock formations (Figure 18.21). The water is laden with sand, ceramic materials, and other additives, including gels, foams, and compressed gases, that serve to increase the yield in the process. The high-pressure fluid finds its way into tiny faults in geological formations, releasing petroleum and natural gas. Fracking has greatly increased petroleum reserves, particularly of natural gas, in many parts of the world.

Unfortunately, the potential for environmental damage from fracking is significant. The large volume of fracking fluid required to create a well must be returned to the surface. Without purification the fluid is rendered unfit for other uses and becomes a large-scale environmental problem. Often the waste water is allowed to sit in open pits. The 2005 Energy Policy Act and other federal legislation exempt hydraulic fracturing operations from certain provisions of the Safe Drinking Water Act and other regulations. Some areas of the country that are already facing water shortages thus have one more large demand for a limited supply. Because fracturing of rock formations increases the pathways for flows of petroleum and various gases, bodies of underground water that have been serving as municipal water supplies or wells for individual homes in some locales have become contaminated with petroleum, hydrogen sulfide, and other toxic substances. The escape of a variety of gases, including methane and other hydrocarbons, from the wellheads contributes to air pollution. In a study published in 2013, methane emissions to the atmosphere during hydraulic fracking operations in Utah were estimated to be in the range of 6–12% of the amount of methane produced. As discussed in the "A Closer Look" box on "Other Greenhouse Gases" (Section 18.2), methane is a potent greenhouse gas.

The many environmental issues surrounding the practice of fracking have generated widespread concern and adverse public reaction. Fracking represents yet one more instance of the conflict between those who advocate the availability of low-cost energy and those who are more focused on sustaining the long-term quality of the environment.

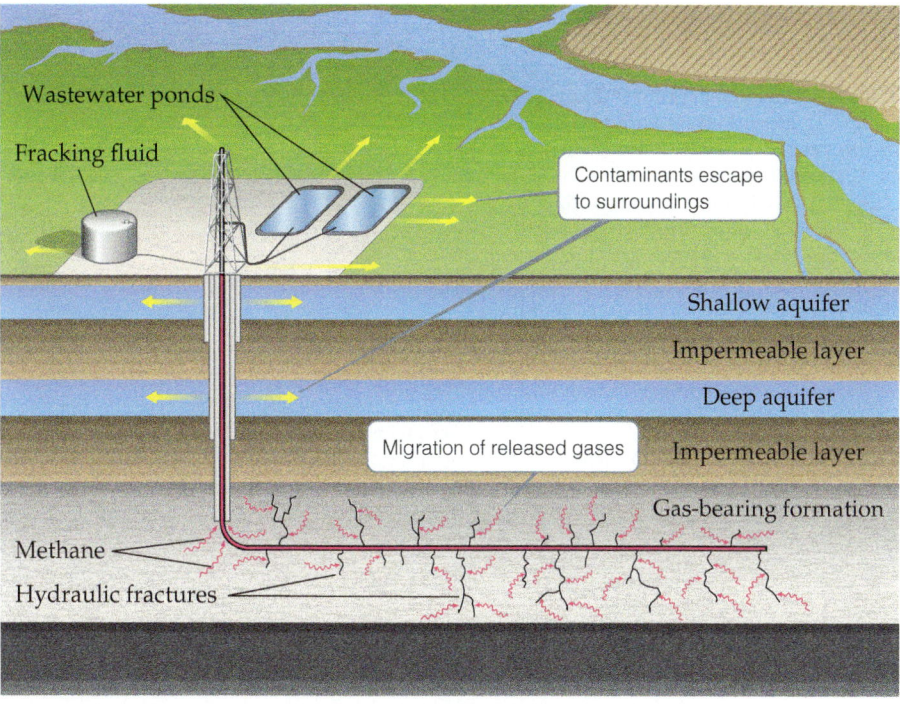

▲ Figure 18.21 **A schematic of a well site employing fracking.** The yellow arrows indicate the avenues through which contaminants can enter the environment.

Some THMs and other halogenated organic substances are suspected carcinogens; others interfere with the body's endocrine system. As a result, the World Health Organization and the U.S. Environmental Protection Agency have placed concentration limits of 80 µg/L (80 ppb) on the total quantity of THMs in drinking water. The goal is to reduce the levels of THMs and other disinfection by-products in the drinking water supply while preserving the antibacterial effectiveness of the water treatment. In some cases, lowering the concentration of chlorine may provide adequate disinfection while reducing the concentrations of THMs formed. Alternative oxidizing agents, such as ozone or chlorine dioxide, produce less of the halogenated substances but have their own disadvantages. For example, each is capable of oxidizing dissolved bromide, as shown here for ozone:

$$O_3(aq) + Br^-(aq) + H_2O(l) \longrightarrow HBrO(aq) + O_2(aq) + OH^-(aq) \qquad [18.17]$$

$$HBrO(aq) + 2\,O_3(aq) \longrightarrow BrO_3^-(aq) + 2\,O_2(aq) + H^+(aq) \qquad [18.18]$$

Bromate ion, BrO_3^-, has been shown to cause cancer in animal tests.

CHEMISTRY AND SUSTAINABILITY | Ocean Acidification

Seawater is a weakly basic solution, with pH values typically between 8.0 and 8.3. This pH range is maintained through a carbonic acid buffer system similar to the one in blood [see Equation 17.6 in the "Chemistry and Life" box on "Blood as a Buffered Solution" (Section 17.2)]. Because the pH of seawater is higher than that of blood (7.35–7.45), the second dissociation of carbonic acid cannot be neglected and CO_3^{2-} becomes an important aqueous species.

The availability of carbonate ions plays an important role in shell formation for a number of marine organisms, including stony corals, clams, mussels, and many others (Figure 18.22). These organisms, which are referred to as marine *calcifiers* and play an important role in the food chains of nearly all oceanic ecosystems, depend on dissolved Ca^{2+} and CO_3^{2-} ions to form their shells and exoskeletons. The relatively low solubility-product constant of $CaCO_3$,

$$CaCO_3(s) \rightleftharpoons Ca^{2+}(aq) + CO_3^{2-}(aq) \quad K_{sp} = 4.5 \times 10^{-9} \quad [18.19]$$

and the fact that the ocean contains saturated concentrations of Ca^{2+} and CO_3^{2-} mean that $CaCO_3$ is usually quite stable once formed. In fact, calcium carbonate skeletons of creatures that died millions of years ago are not uncommon in the fossil record.

The concentration of dissolved CO_2 in the ocean is sensitive to changes in atmospheric CO_2 levels. As discussed in Section 18.2, the atmospheric CO_2 concentration has risen by approximately 33% over the past three centuries to the present level of approximately 415 ppm. Human activity has played a prominent role in this increase. Scientists estimate that one-third to one-half of the CO_2 emissions resulting from human activity have been absorbed by Earth's oceans. While this absorption helps mitigate the greenhouse gas effects of CO_2, the extra CO_2 in the ocean produces carbonic acid, H_2CO_3, which lowers the pH. The effect on pH is mitigated to some extent by the CO_3^{2-}/HCO_3^-, buffer system. Nonetheless, the addition of carbonic acid will convert some carbonate ions into hydrogen carbonate ions:

$$CO_3^{2-}(aq) + H^+(aq) \rightleftharpoons HCO_3^-(aq) \quad [18.20]$$

This consumption of carbonate ion shifts the $CaCO_3$ dissolution equilibrium expressed in Equation 18.19 to the right, increasing the solubility of $CaCO_3$, which can lead to partial dissolution of calcium carbonate shells and exoskeletons.

For coral reefs and the ecosystems that surround them, ocean acidification is only one of many stresses they face. Warm-water corals have a symbiotic relationship with microscopic algae called zooxanthellae. The corals allow the algae to live inside them, and in return algae photosynthesis provides corals with most of their energy. When the water temperature becomes too warm, corals expel the algae and turn white, a phenomenon known as coral bleaching. Worryingly, coral bleaching events have become more common in recent years. Increased ocean temperatures, ocean acidification, pollution, and overfishing all pose threats to the health of coral reefs.

Related Exercises: 18.6, 18.8, 18.66

▲ **Figure 18.22 Marine calcifiers.** Many sea-dwelling organisms use $CaCO_3$ for their shells and exoskeletons. Coral reefs, like the one shown here, are perhaps the most striking example. Other creatures with $CaCO_3$ exoskeletons include crustaceans, sea urchins, starfish, and some types of phytoplankton.

At present, there seem to be no completely satisfactory alternatives to chlorination or ozonation, and we are faced with a consideration of benefit versus risk. In this case, the risks of cancer from THMs and related substances in municipal water are very low relative to the risks of cholera, typhus, and gastrointestinal disorders from untreated water. When the water supply is cleaner to begin with, less disinfectant is needed and thus the risk of THMs is lowered. Once THMs form, their concentrations in the water supply can be reduced by aeration because the THMs are more volatile than water. Alternatively, they can be removed by adsorption onto activated charcoal or other adsorbents.

 Self-Assessment Exercises

SAE 18.14 The Henry's law constant for $O_2(g)$ in water at 20 °C is $1.38 \times 10^{-3} M/atm$. What is the concentration, in both molarity and parts per million, of oxygen in a body of water in contact with the atmosphere at a pressure of 1.00 atm? Assume the density of the aqueous solution is 1.00 g/mL. **(a)** $2.88 \times 10^{-4} M$, 9.23 ppm **(b)** $1.38 \times 10^{-3} M$, 44.2 ppm **(c)** $2.88 \times 10^{-4} M$, 0.288 ppm **(d)** $1.38 \times 10^{-3} M$, 9.23 ppm **(e)** $1.38 \times 10^{-3} M$, 1.38 ppm

SAE 18.15 Which of the following processes reduces the concentration of dissolved oxygen in bodies of water containing biodegradable organic matter?

(i) Anaerobic bacteria consume oxygen in the process of decomposing the organic matter.
(ii) Aerobic bacteria consume oxygen in the process of decomposing the organic matter.
(iii) The decay of dead plant matter consumes oxygen.

(a) only i **(b)** only ii **(c)** only iii **(d)** both i and ii **(e)** both ii and iii

SAE 18.16 Which of the following processes, when employed at a water treatment facility, *would not* lead to oxidation of unwanted contaminants? **(a)** treatment with O_3 **(b)** treatment with Cl_2 **(c)** treatment with $Al_2(SO_4)_3$ **(d)** aeration

18.5 | Green Chemistry

The planet on which we live is, to a large extent, a *closed system*—namely, one that exchanges energy but not matter with its surroundings. If humankind is to thrive in the future, all the processes we carry out should be in balance with Earth's natural processes and physical resources. This goal requires that no toxic materials be released to the environment, that our needs be met with renewable resources, and that we consume the least possible amount of energy. Although the chemical industry is but a small part of human activity, chemical processes are involved in nearly all aspects of modern life. Chemistry is therefore at the heart of efforts to accomplish these goals.

Green chemistry is an initiative that promotes the design and application of chemical products and processes that are compatible with human health and that preserve the environment. Green chemistry rests on a set of 12 principles:

1. **Prevention** It is better to prevent waste than to clean it up after it has been created.

2. **Atom Economy** Methods to make chemical compounds should be designed to maximize the incorporation of all starting atoms into the final product.

3. **Less Hazardous Chemical Syntheses** Wherever practical, synthetic methods should be designed to use and generate substances that possess little or no toxicity to human health and the environment.

4. **Design of Safer Chemicals** Chemical products should be designed to minimize toxicity and yet maintain their desired function.

5. **Safer Solvents and Auxiliaries** Auxiliary substances (for example, solvents, separation agents) should be used as little as possible. Those that are used should be as nontoxic as possible.

6. **Design for Energy Efficiency** Energy requirements of chemical processes should be recognized for their environmental and economic impacts and should be minimized. If possible, chemical reactions should be conducted at room temperature and pressure.

7. **Use of Renewable Feedstocks** A raw material or feedstock should be renewable whenever technically and economically practical.

8. **Reduction of Derivatives** Unnecessary derivatization (intermediate compound formation, temporary modification of physical/chemical processes) should be minimized or avoided if possible because such steps require additional reagents and can generate waste.

9. **Catalysis** Catalytic reagents (as selective as possible) improve product yields within a given time and with a lower energy cost compared to noncatalytic processes and are, therefore, preferred to noncatalytic alternatives.

10. **Design for Degradation** The end products of chemical processing should break down at the end of their useful lives into innocuous degradation products that do not persist in the environment.

11. **Real-Time Analysis for Pollution Prevention** Analytical methods need to be developed that allow for real-time, in-process monitoring and control prior to the formation of hazardous substances.

12. **Inherently Safer Chemistry for Accident Prevention** Reagents and solvents used in a chemical process should be chosen to minimize the potential for chemical accidents, including releases, explosions, and fires.*

To illustrate how green chemistry works, consider the manufacture of styrene, an important building block for many polymers, including the expanded polystyrene packages used to pack eggs and restaurant takeout food. The global demand for styrene is more than 3.5×10^{10} kg per year. For many years, styrene has been produced in a two-step process: Benzene and ethylene react to form ethyl benzene, followed by the ethyl

Learning Objectives

When you finish Section 18.5, **you should be able to:**

▶ Describe the principles of green chemistry and explain how their adoption can promote sustainability and preserve the environment.

▶ Compare alternative chemical routes to a given product and determine which process is more consistent with the principles of green chemistry.

---

*Adapted from P. T. Anastas and J. C. Warner, *Green Chemistry: Theory and Practice*. New York: Oxford University Press 1998, p. 30. See also Mike Lancaster, *Green Chemistry: An Introductory Text*. Cambridge, UK: RSC Publishing, 2010, Second Edition, Chapter 1.

benzene being mixed with high-temperature steam and passed over an iron oxide catalyst to form styrene:

Benzene   Ethylene   Ethyl benzene   Styrene

This process has several shortcomings. One is that both benzene, which is formed from crude oil, and ethylene, which is formed from natural gas, are high-priced starting materials for a product that should be a low-priced commodity. Another is that benzene is a known carcinogen. In a recently developed process that bypasses some of these shortcomings, the two-step process is replaced by a one-step process in which toluene is reacted with methanol at 425 °C over a special catalyst:

Toluene   Methanol   Styrene

The one-step process saves money both because toluene and methanol are less expensive than benzene and ethylene, and because the reaction requires less energy input. Additional benefits are that the methanol can be produced from biomass and that benzene is replaced by less toxic toluene. The hydrogen formed in the reaction can be recycled as a source of energy. This example demonstrates how finding the right catalyst is often key in discovering a new process.

Let's consider some other examples in which green chemistry can operate to improve environmental quality.

Supercritical Solvents

A major area of concern in chemical processes is the use of volatile organic compounds as solvents. Generally, the solvent in which a reaction is run is not consumed in the reaction, and there are unavoidable releases of solvent into the atmosphere even in the most carefully controlled processes. Further, the solvent may be toxic or may decompose to some extent during the reaction, thus creating waste products.

The use of supercritical fluids represents a way to replace conventional solvents. Recall that a supercritical fluid is an unusual state of matter that has properties of both a gas and a liquid. (Section 11.4) Water and carbon dioxide are the two most popular choices as supercritical fluid solvents. One recently developed industrial process, for example, replaces chlorofluorocarbon solvents with liquid or supercritical CO_2 in the production of polytetrafluoroethylene ($[CF_2CF_2]_n$, sold as Teflon®). Though CO_2 is a greenhouse gas, no new CO_2 need be manufactured for use as a supercritical fluid solvent.

As a further example, *para*-xylene is oxidized to form terephthalic acid, which is used to make polyethylene terephthalate (PET) plastic and polyester fiber [see Table 12.5 (Section 12.6)]:

para-Xylene   Terephthalic acid

This commercial process requires pressurization and a relatively high temperature. Oxygen is the oxidizing agent, and acetic acid, CH_3COOH, is the solvent. An alternative route employs supercritical water as the solvent and hydrogen peroxide as the oxidant. This alternative process has several potential advantages, most particularly the elimination of acetic acid as solvent.

Greener Reagents and Processes

Let us examine two more examples of green chemistry in action.

Hydroquinone, HOC_6H_4OH, is a common intermediate used to make polymers. The standard industrial route to hydroquinone, used until recently, yields many by-products that are treated as waste:

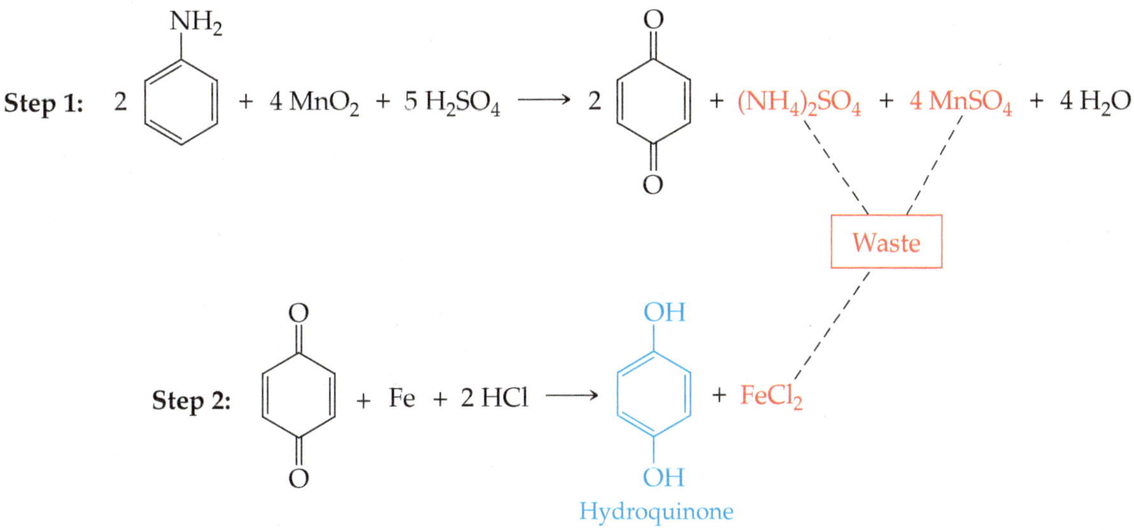

Step 1: 2 $+ 4\,MnO_2 + 5\,H_2SO_4 \longrightarrow 2$ $+ (NH_4)_2SO_4 + 4\,MnSO_4 + 4\,H_2O$

Waste

Step 2: $+ Fe + 2\,HCl \longrightarrow$ $+ FeCl_2$

Hydroquinone

Using the principles of green chemistry, researchers have improved this process. The new process for hydroquinone production uses a new starting material. Two of the by-products of the new reaction (shown in green) can be isolated and used to make the new starting material.

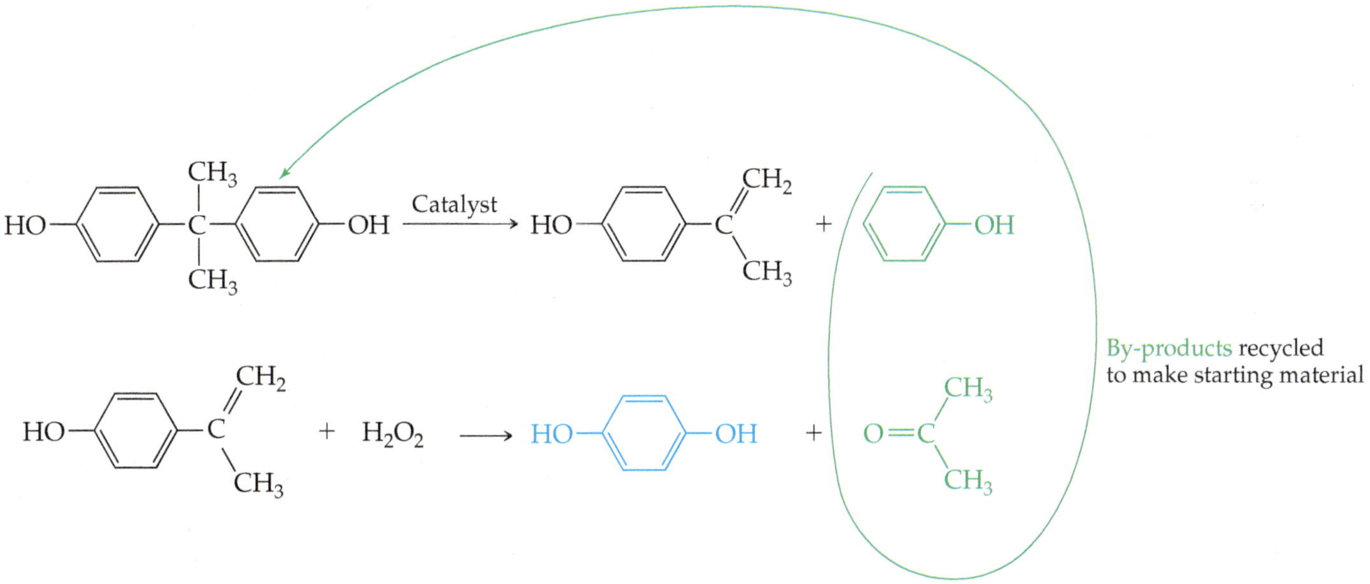

By-products recycled to make starting material

The new process is an example of excellent "atom economy" (green chemistry principle #2), because a high percentage of the atoms from the starting materials end up in the product.

Another example of excellent atom economy is the following reaction in which, at room temperature and in the presence of a copper(I) catalyst, an organic *azide* and an *alkyne* form one product molecule:

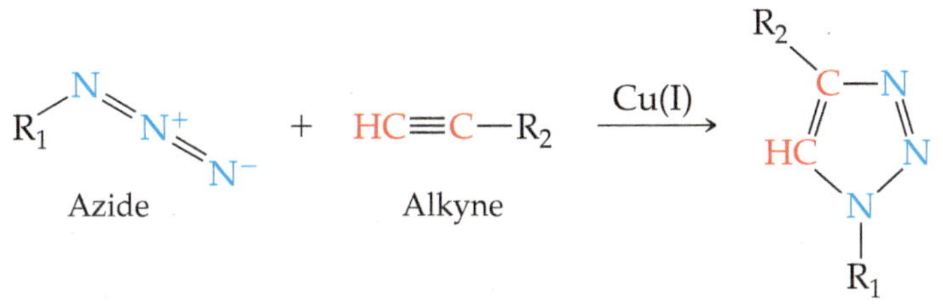

Azide   Alkyne

This reaction is informally called a *click reaction*. The yield—actual, not just theoretical—is close to 100%, and there are no by-products. Depending on the type of azide and type of alkyne we start with, this very efficient click reaction can be used to create any number of valuable product molecules.

Self-Assessment Exercises

SAE 18.17 Ionic liquids [see the "Chemistry and Sustainability" box on "Ionic Liquids" (Section 11.3)] typically have negligible vapor pressure, which makes them nonflammable and minimal risk for inhalation. Which of the following principles of green chemistry would definitely apply to a new process where a volatile organic solvent was replaced by an ionic liquid?

   (**i**) atom economy
  (**ii**) safer solvents and auxiliaries
 (**iii**) design for energy efficiency

(**a**) only i (**b**) only ii (**c**) only iii (**d**) both i and ii (**e**) both ii and iii

SAE 18.18 The Haber process for making ammonia, $N_2(g) + 3 H_2(g) \rightleftharpoons 2 NH_3(g)$, has been discussed several times in this book (Sections 15.2 and 15.7) The Haber process is an excellent example of the green chemistry principle of _____, but deviates significantly from the green chemistry principle of _____.

(**a**) use of renewable feedstocks, catalysis (**b**) atom economy, catalysis (**c**) atom economy, design for energy efficiency (**d**) use of renewable feedstocks, design for energy efficiency

SAE 18.19 Which of the following is an example of making a change that leads to a greener process?

  (**i**) Using ethyl lactate as a solvent, which is derived from processing corn, in place of a petrochemical solvent like toluene.
 (**ii**) Using an oil-based paint containing volatile organic components (VOCs) instead of a water-based paint with low VOCs.

(**a**) only i (**b**) only ii (**c**) both i and ii (**d**) neither i nor ii

Putting Concepts Together

(**a**) Acid rain is no threat to lakes in areas where the rock is limestone (calcium carbonate), which can neutralize the acid. Where the rock is granite, however, no neutralization occurs. How does limestone neutralize acid? (**b**) Acidic water can be treated with basic substances to increase the pH, although such a procedure is usually only a temporary cure. Calculate the minimum mass of lime, CaO, needed to adjust the pH of a small lake ($V = 4 \times 10^9$ L) from 5.0 to 6.5. Why might more lime be needed?

SOLUTION

Analyze We need to remember what a neutralization reaction is and calculate the amount of a substance needed to effect a certain change in pH.

Plan For (a), we need to think about how acid can react with calcium carbonate, a reaction that evidently does not happen with acid and granite. For (b), we need to think about what reaction between an acid and CaO is possible and do stoichiometric calculations. From the proposed change in pH, we can calculate the change in proton concentration needed and then figure out how much CaO is needed.

Solve

(**a**) The carbonate ion, CO_3^{2-}, which is the anion of a weak acid, is basic (Sections 16.2 and 16.7) and so reacts with $H^+(aq)$. If the concentration of $H^+(aq)$ is low, the major product is the bicarbonate ion, HCO_3^-. If the concentration of $H^+(aq)$ is high, H_2CO_3 forms and decomposes to CO_2 and H_2O. (Section 4.3)

(**b**) The initial and final concentrations of $H^+(aq)$ in the lake are obtained from their pH values:

$$[H^+]_{initial} = 10^{-5.0} = 1 \times 10^{-5}\,M$$

$$[H^+]_{final} = 10^{-6.5} = 3 \times 10^{-7}\,M$$

Using the lake volume, we can calculate the number of moles of $H^+(aq)$ at both pH values:

$$(1 \times 10^{-5}\,\text{mol/L})(4.0 \times 10^9\,\text{L}) = 4 \times 10^4\,\text{mol}$$
$$(3 \times 10^{-7}\,\text{mol/L})(4.0 \times 10^9\,\text{L}) = 1 \times 10^3\,\text{mol}$$

Hence, the change in the amount of $H^+(aq)$ is $4 \times 10^4\,\text{mol} - 1 \times 10^3\,\text{mol} \approx 4 \times 10^4\,\text{mol}$.

Let's assume that all the acid in the lake is completely ionized, so that only the free $H^+(aq)$ contributing to the pH needs to be neutralized. We need to neutralize at least that much acid, although there may be a great deal more than that amount in the lake.

The oxide ion of CaO is very basic. (Section 16.5) In the neutralization reaction, 1 mol of CaO reacts with 2 mol of H^+ to form H_2O and Ca^{2+} ions. Thus, 4×10^4 mol of H^+ requires

$$(4 \times 10^4\,\text{mol H}^+)\left(\frac{1\,\text{mol CaO}}{2\,\text{mol H}^+}\right)\left(\frac{56.1\,\text{g CaO}}{1\,\text{mol CaO}}\right) = 1 \times 10^6\,\text{g CaO}$$

This is approximately a ton of CaO. That would not be very costly because CaO is inexpensive, selling for less than $100 per ton when purchased in large quantities. This amount of CaO is the minimum amount needed, however, because there are likely to be weak acids in the water that must also be neutralized.

This liming procedure has been used to bring the pH of some small lakes into the range necessary for fish to live. The lake in our example would be about a half mile long and a half mile wide and have an average depth of 20 ft.

Chapter Summary and Key Terms

EARTH'S ATMOSPHERE (SECTION 18.1) In this section we examined the physical and chemical properties of Earth's atmosphere. The complex temperature variations in the atmosphere give rise to four regions, each with characteristic properties. The lowest of these regions, the **troposphere**, extends from Earth's surface up to an altitude of about 12 km. Above the troposphere, in order of increasing altitude, are the **stratosphere**, mesosphere, and thermosphere. In the upper reaches of the atmosphere, only the simplest chemical species can survive the bombardment of highly energetic particles and radiation from the Sun. The average molecular weight of the atmosphere at high elevations is lower than that at Earth's surface because the lightest atoms and molecules diffuse upward and also because of **photodissociation**, which is the breaking of bonds in molecules because of the absorption of light. Absorption of radiation also leads to the formation of ions via **photoionization**.

HUMAN ACTIVITIES AND EARTH'S ATMOSPHERE (SECTION 18.2) Ozone is produced in the upper atmosphere from the reaction of atomic oxygen with O_2. Ozone is itself decomposed by absorption of a photon or by reaction with a reactive species such as Cl. **Chlorofluorocarbons** can undergo photodissociation in the stratosphere, introducing atomic chlorine, which is capable of catalytically destroying ozone. A marked reduction in the ozone level in the upper atmosphere would have serious adverse consequences because the ozone layer filters out certain wavelengths of harmful ultraviolet light that are not removed by any other atmospheric component.

In the troposphere the chemistry of trace atmospheric components is of major importance. Many of these minor components are pollutants. Sulfur dioxide is one of the more noxious and prevalent examples. It is oxidized in air to form sulfur trioxide, which, upon dissolving in water, forms sulfuric acid. The oxides of sulfur are major contributors to **acid rain**. One method of preventing the escape of SO_2 from industrial operations is to react it with CaO to form calcium sulfite $CaSO_3$. **Photochemical smog** is a complex mixture in which both nitrogen oxides and ozone play important roles. Smog components are generated mainly in automobile engines, and smog control consists largely of controlling auto emissions.

Carbon dioxide and water vapor are the major components of the atmosphere that strongly absorb infrared radiation. CO_2 and H_2O are therefore critical in maintaining Earth's surface temperature. The concentrations of CO_2 and other so-called **greenhouse gases** in the atmosphere are thus important in determining global climate. Because of the extensive combustion of fossil fuels (coal, oil, and natural gas), the concentration of carbon dioxide in the atmosphere is steadily increasing, which in turn appears to be contributing to an increase in the average temperature of the Earth.

EARTH'S WATER (SECTION 18.3) Earth's water is largely in the oceans and seas; only a small fraction is freshwater. Seawater contains about 3.5% by mass of dissolved salts and is described as having a **salinity** (grams of dry salts per 1 kg seawater) of 35. Seawater's density and salinity vary with depth. The global water cycle involves continuous phase changes of water that effectively transport heat from the surface to the atmosphere.

HUMAN ACTIVITIES AND WATER QUALITY (SECTION 18.4) Freshwater contains many dissolved substances, including dissolved oxygen, which is necessary for fish and other aquatic life. Substances that are decomposed by bacteria are said to be **biodegradable**. Because the oxidation of biodegradable substances by aerobic bacteria consumes dissolved oxygen, these substances are called oxygen-demanding wastes. The presence of an excess amount of oxygen-demanding wastes in water can deplete the dissolved oxygen sufficiently to kill aquatic animals like fish. Plant nutrients can contribute to the problem by stimulating the growth of plants that become oxygen-demanding wastes when they die.

The water available from freshwater sources may require treatment before it can be used domestically. The several steps generally used in municipal water treatment include coarse filtration, sedimentation, sand filtration, aeration, and sterilization.

GREEN CHEMISTRY (SECTION 18.5) The **green chemistry** initiative promotes the design and application of chemical products and processes that are compatible with human health and that preserve the environment. The areas in which the principles of green chemistry can operate to improve environmental quality include choices of solvents and reagents for chemical reactions, development of alternative processes, and improvements in existing systems and practices.

Exam Prep

EP 18.1 Nearly 75% of the mass of the atmosphere is concentrated in which layer? (**a**) troposphere (**b**) stratosphere (**c**) mesosphere (**d**) thermosphere

EP 18.2 Photoionization reactions occur predominantly in which layer of Earth's atmosphere? (**a**) troposphere (**b**) stratosphere (**c**) mesosphere (**d**) thermosphere

EP 18.3 Carbon dioxide, oxygen, and methane are generated by various life forms on Earth. Rank-order the concentrations of these three gases in the atmosphere.

(**a**) $CO_2 > O_2 > CH_4$

(**b**) $O_2 > CO_2 > CH_4$

(**c**) $O_2 > CH_4 > CO_2$

(**d**) $CO_2 > CH_4 > O_2$

(**e**) $CH_4 > CO_2 > O_2$

EP 18.4 The concentration of ozone, O_3, in smog can reach levels of 0.5 ppm. To compare with levels of O_3 in the stratosphere (see Figure 18.5), express the concentration of ozone in smog, 0.5 ppm, in units of molecules/cm^3. You can assume the total pressure is 1.0 atm and the temperature is 298 K.

(**a**) 1×10^{16} molecules/cm^3

(**b**) 4×10^{-5} molecules/cm^3

(**c**) 2×10^{-11} molecules/cm^3

(**d**) 1×10^{13} molecules/cm^3

(**e**) 5×10^{14} molecules/cm^3

EP 18.5 The mole fraction of argon in dry air is 0.00934. What is the partial pressure of argon in dry air at an elevation where the atmospheric pressure is 668 mm Hg? (**a**) 3.12 mm Hg (**b**) 7.09 mm Hg (**c**) 6.24 mm Hg (**d**) 9.34 mm Hg (**e**) 39.9 mm Hg

EP 18.6 The bond dissociation energy of the Br—Br bond is 193 kJ/mol. What wavelength of light has just enough energy to cause Br—Br bond dissociation? (**a**) 620 nm (**b**) 310 nm (**c**) 148 nm (**d**) 6200 nm (**e**) 563 nm

EP 18.7 Use the average bond enthalpies for an oxygen–oxygen double bond (495 kJ/mol) and single bond (146 kJ/mol) to estimate ΔH for the reaction that leads to the formation of ozone in upper atmosphere:

$$O_2(g) + O(g) \longrightarrow O_3(g)$$

(**a**) +146 kJ/mol (**b**) −146 kJ/mol (**c**) +495 kJ/mol (**d**) −495 kJ/mol (**e**) −641 kJ/mol

EP 18.8 The presence of atomic chlorine in the atmosphere leads to ozone decomposition through the following three-step mechanism:

$$2\,Cl(g) + 2\,O_3(g) \longrightarrow 2\,ClO(g) + 2\,O_2(g)$$
$$2\,ClO(g) + h\nu \longrightarrow 2\,Cl(g) + 2\,O(g)$$
$$O(g) + O(g) \longrightarrow O_2(g)$$

In this reaction mechanism, Cl is as a(n) _____ , ClO is a(n) _____ , and O_2 is a(n) _____ .

(a) catalyst, product, product (b) catalyst, intermediate, product (c) intermediate, catalyst, product (d) reactant, intermediate, product (e) reactant, product, product

EP 18.9 In areas that are free of pollution, the pH of rainwater is approximately _____ , due to the reaction _____ .

(a) 7.0, $2\,H_2O(l) \rightleftharpoons H_3O^+(aq) + OH^-(aq)$
(b) 4.0, $SO_3(g) + H_2O(l) \longrightarrow H_2SO_4(aq)$
(c) 4.0, $CO_2(g) + H_2O(l) \longrightarrow H_2CO_3(aq)$
(d) 5.6, $SO_3(g) + H_2O(l) \longrightarrow H_2SO_4(aq)$
(e) 5.6, $CO_2(g) + H_2O(l) \longrightarrow H_2CO_3(aq)$

EP 18.10 Which component of urban smog *does not* form in the absence of sunlight? (a) NO_2 (b) NO (c) O_3 (d) SO_2 (e) hydrocarbons

EP 18.11 Many cars can run on a mixture of gasoline and ethanol, C_2H_5OH. What mass of CO_2 is produced by combustion of 1.0 gallon of ethanol (density = 0.789 g/mL)? (a) 2.9 kg (b) 5.7 kg (c) 6.0 kg (d) 8.6 kg (e) 9.2 kg

EP 18.12 Which of the following characteristics is or are necessary for a substance to act as a greenhouse gas?

(i) supports plant life

(ii) absorbs infrared radiation

(iii) absorbs UV radiation

(a) only i (b) only ii (c) only iii (d) both i and ii (e) all three

EP 18.13 Table 8.1 lists the principal gases that make up the atmosphere. Three of the eleven gases in this table act as greenhouse gases—carbon dioxide, methane, and nitrous oxide. What feature do these gases have in common with each other, but not with the other eight gases in Table 8.1? (a) They don't readily undergo chemical reactions in the atmosphere. (b) They contain multiple bonds. (c) They contain more than two atoms. (d) They have relatively large molar masses. (e) They can act as oxidizing agents.

EP 18.14 Of the phase changes that make up the global water cycle, _____ processes tend to occur at the surface and _____ processes tend to occur in the atmosphere. (a) exothermic, exothermic (b) exothermic, endothermic (c) endothermic, exothermic (d) endothermic, endothermic

EP 18.15 Which of the following substances is *not* extracted from seawater in commercially important quantities? (a) calcium oxide (b) magnesium (c) sodium chloride (d) bromine

EP 18.16 Which element(s) is or are largely responsible for freshwater eutrophication?

(i) calcium

(ii) nitrogen

(iii) phosphorous

(a) only Ca (b) both Ca and N (c) both N and P (d) both Ca and P (e) Ca, N, and P

EP 18.17 Which of the following statements regarding municipal water treatment is *false*? (a) Water is treated with a chemical agent, typically ozone or liquified Cl_2, to ensure the destruction of bacteria. (b) When $Al_2(SO_4)_3$ is added, it reacts with a variety of cations to form sulfate precipitates. These precipitates settle slowly, carrying suspended particles down with them. (c) Water is allowed to stand in large sedimentation tanks, where sand and other minute particles settle out. (d) An aeration step is used to hasten oxidation of unwanted contaminants.

EP 18.18 Which of the following is *not* a principle of the green chemistry initiative? (a) Methods to make chemical compounds should be designed to maximize the incorporation of starting atoms into the final product. (b) Reagents and solvents used in a chemical process should be chosen to minimize the potential for chemical accidents, including releases, explosions, and fires. (c) All waste should be carefully disposed of after it has been created. (d) Wherever practical, synthetic methods should be designed to use and generate substances that possess little or no toxicity to human health and the environment.

EP 18.19 Which of the following, if incorporated in a chemical process, can lead to a greener process.

(i) volatile organic solvent

(ii) supercritical fluid

(iii) catalyst

(a) only i (b) only ii (c) only iii (d) both ii and iii (e) all three

Exercises

Visualizing Concepts

18.1 At 273 K and 1 atm pressure, 1 mol of an ideal gas occupies 22.4 L. (Section 10.3) (a) Looking at Figure 18.1, predict whether a 1 mol sample of the atmosphere in the middle of the stratosphere would occupy a greater or smaller volume than 22.4 L (b) Looking at Figure 18.1, we see that the temperature is lower at 85 km altitude than at 50 km. Does this mean that one mole of an ideal gas would occupy less volume at 85 km than at 50 km? Explain. (c) In which parts of the atmosphere would you expect gases to behave most ideally (ignoring any photochemical reactions)? [Section 18.1]

18.2 Molecules in the upper atmosphere tend to contain double and triple bonds rather than single bonds. Suggest an explanation. [Section 18.1]

18.3 The figure shows the three lowest regions of Earth's atmosphere. (a) Name each and indicate the approximate elevations at which the boundaries occur. (b) In which region is ozone a pollutant? In which region does it filter UV solar radiation? (c) In which region is infrared radiation from Earth's surface most strongly reflected back? (d) An aurora borealis is due to excitation of atoms and molecules in the atmosphere

55–95 km above Earth's surface. Which regions in the figure are involved in an aurora borealis? (e) Compare the changes in relative concentrations of water vapor and carbon dioxide with increasing elevation in these three regions [Section 18.1].

18.4 Where does the energy come from to evaporate the estimated 425,000 km³ of water that annually leaves the oceans, as illustrated here? [Section 18.3]

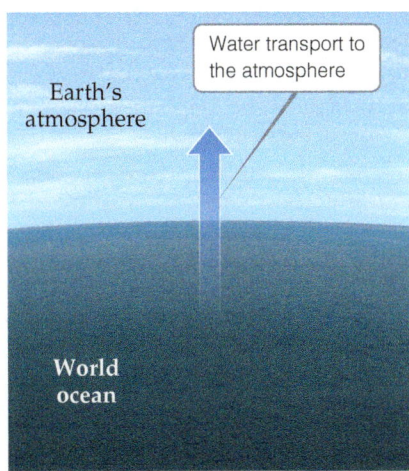

18.5 Earth's oceans have a salinity of 35. What is the concentration of dissolved salts in seawater when expressed in ppm? What percentage of salts must be removed from seawater before it

can be considered freshwater (dissolved salts < 500 ppm)? [Section 18.3]

18.6 Describe what changes occur when atmospheric CO_2 interacts with the world ocean as illustrated here. [Section 18.3]

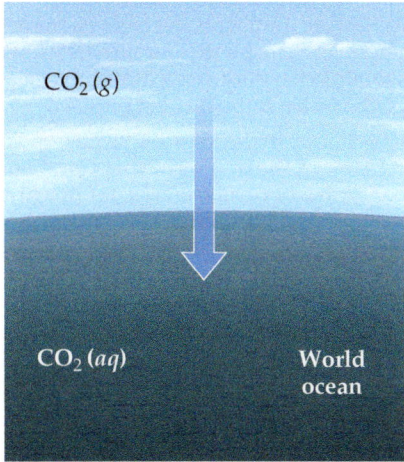

18.7 From study of Figure 18.21 describe the various ways in which operation of a fracking well site could lead to environmental contamination.

18.8 One mystery in environmental science is the imbalance in the "carbon dioxide budget." Considering only human activities, scientists have estimated that 1.6 billion metric tons of CO_2 is added to the atmosphere every year because of deforestation (plants use CO_2, and fewer plants will leave more CO_2 in the atmosphere). Another 5.5 billion tons per year is put into the atmosphere because of burning fossil fuels. It is further

estimated (again, considering only human activities) that the atmosphere actually takes up about 3.3 billion tons of this CO_2 per year, while the oceans take up 2 billion tons per year, leaving about 1.8 billion tons of CO_2 per year unaccounted for. Describe a mechanism by which CO_2 is removed from the atmosphere and ultimately ends up below the surface (*Hint:* What is the source of the fossil fuels?). [Sections 18.1–18.3]

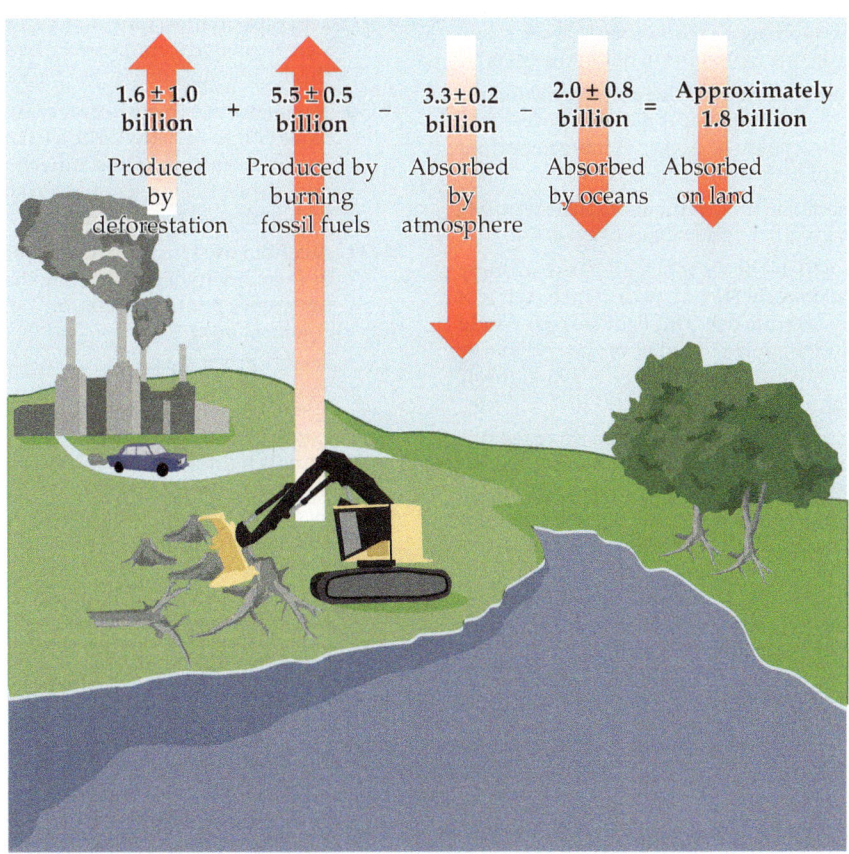

Earth's Atmosphere (Section 18.1)

18.9 (a) What is the primary basis for the division of the atmosphere into different regions? (b) Name the regions of the atmosphere, indicating the altitude interval for each one.

18.10 (a) How are the boundaries between the regions of the atmosphere determined? (b) Explain why the stratosphere, which is about 35 km thick, has a smaller total mass than the troposphere, which is about 12 km thick.

18.11 Air pollution in the Mexico City metropolitan area is among the worst in the world. The concentration of ozone in Mexico City has been measured at 441 ppb (0.441 ppm). Mexico City sits at an altitude of 7400 feet, which means its atmospheric pressure is only 0.67 atm. (a) Calculate the partial pressure of ozone at 441 ppb if the atmospheric pressure is 0.67 atm. (b) How many ozone molecules are in 1.0 L of air in Mexico City? Assume $T = 25\,°C$.

18.12 From the data in Table 18.1, calculate the partial pressures of carbon dioxide and argon when the total atmospheric pressure is 1.05 bar.

18.13 The average concentration of carbon monoxide in air in an Ohio city in 2006 was 3.5 ppm. Calculate the number of CO molecules in 1.0 L of this air at a pressure of 759 torr and a temperature of 22 $°C$.

18.14 (a) From the data in Table 18.1, what is the concentration of neon in the atmosphere in ppm? (b) What is the concentration of neon in the atmosphere in molecules per liter, assuming an atmospheric pressure of 730 torr and a temperature of 296 K?

18.15 The dissociation energy of a carbon–bromine bond is typically about 276 kJ/mol. (a) What is the maximum wavelength of photons that can cause C—Br bond dissociation? (b) Which kind of electromagnetic radiation—ultraviolet, visible, or infrared—does the wavelength you calculated in part (a) correspond to?

18.16 In CF_3Cl the C—Cl bond-dissociation energy is 339 kJ/mol. In CCl_4 the C—Cl bond-dissociation energy is 293 kJ/mol. What is the range of wavelengths of photons that can cause C—Cl bond rupture in one molecule but not in the other?

18.17 (a) Distinguish between *photodissociation* and *photoionization*. (b) Use the energy requirements of these two processes to explain why photodissociation of oxygen is more important than photoionization of oxygen at altitudes below about 90 km.

18.18 Why is the photodissociation of N_2 in the atmosphere relatively unimportant compared with the photodissociation of O_2?

18.19 The wavelength at which the O_2 molecule most strongly absorbs light is approximately 145 nm. (a) In which region of the electromagnetic spectrum does this light fall? (b) Would a photon whose wavelength is 145 nm have enough energy to photodissociate O_2 whose bond energy is 495 kJ/mol? Would it have enough energy to photoionize O_2?

18.20 The ultraviolet spectrum can be divided into three regions based on wavelength: UV-A (315–400 nm), UV-B (280–315 nm), and UV-C (100–280 nm). (a) Photons from which region have the highest energy and therefore are the most harmful to living tissue? (b) In the absence of ozone, which of these three regions, if any, are absorbed by the atmosphere? (c) When appropriate concentrations of ozone are present in the stratosphere, is all of the UV light absorbed before reaching Earth's surface? If not, which region or regions are not filtered out?

Human Activities and Earth's Atmosphere (Section 18.2)

18.21 Do the reactions involved in ozone depletion involve changes in oxidation state of the O atoms? Explain.

18.22 Which of the following reactions in the stratosphere cause an increase in temperature there?

(a) $O(g) + O_2(g) \longrightarrow O_3{}^*(g)$

(b) $O_3{}^*(g) + M(g) \longrightarrow O_3(g) + M{}^*(g)$

(c) $O_2(g) + h\nu \longrightarrow 2\,O(g)$

(d) $O(g) + N_2(g) \longrightarrow NO(g) + N(g)$

(e) All of these reactions cause an increase in the temperature of the stratosphere

18.23 (a) What is the difference between chlorofluorocarbons and hydrofluorocarbons? (b) Why are hydrofluorocarbons potentially less harmful to the ozone layer than CFCs?

18.24 Draw the Lewis structure for the chlorofluorocarbon CFC-11, $CFCl_3$. What chemical characteristics of this substance allow it to effectively deplete stratospheric ozone?

18.25 The average bond enthalpies of the C—F and C—Cl bonds are 485 kJ/mol and 328 kJ/mol, respectively. (a) What is the maximum wavelength that a photon can possess and still have sufficient energy to break the C—F and C—Cl bonds, respectively? (b) Given the fact that O_2, N_2, and O in the upper atmosphere absorb most of the light with wavelengths shorter than 240 nm, would you expect the photodissociation of C—F bonds to be significant in the lower atmosphere?

18.26 (a) When chlorine atoms react with atmospheric ozone, what are the products of the reaction? (b) Based on average bond enthalpies, would you expect a photon capable of dissociating a C—Cl bond to have sufficient energy to dissociate a C—Br bond? (c) Would you expect the substance $CFBr_3$ to accelerate depletion of the ozone layer?

18.27 Nitrogen oxides like NO_2 and NO are a significant source of acid rain. For each of these molecules write an equation that shows how an acid is formed from the reaction with water.

18.28 Why is rainwater naturally acidic, even in the absence of polluting gases such as SO_2?

18.29 (a) Write a chemical equation that describes the attack of acid rain on limestone, $CaCO_3$. (b) If a limestone sculpture were treated to form a surface layer of calcium sulfate, would this help to slow down the effects of acid rain? Explain.

18.30 The first stage in corrosion of iron upon exposure to air is oxidation to Fe^{2+}. (a) Write a balanced chemical equation to show the reaction of iron with oxygen and protons from acid rain. (b) Would you expect the same sort of reaction to occur with a silver surface? Explain.

18.31 Alcohol-based fuels for automobiles lead to the production of formaldehyde, CH_2O, in exhaust gases. Formaldehyde undergoes photodissociation, which contributes to photochemical smog:

$$CH_2O + h\nu \longrightarrow CHO + H$$

The maximum wavelength of light that can cause this reaction is 335 nm. (a) In what part of the electromagnetic spectrum is light with this wavelength found? (b) What is the maximum strength of a bond, in kJ/mol, that can be broken by absorption of a photon of 335-nm light? (c) Compare your answer from part (b) to the appropriate value from Table 8.3. What do you conclude about C—H bond energy in formaldehyde? (d) Write out the formaldehyde photodissociation reaction, showing Lewis electron-dot structures.

18.32 An important reaction in the formation of photochemical smog is the photodissociation of NO_2:

$$NO_2 + h\nu \longrightarrow NO(g) + O(g)$$

The maximum wavelength of light that can cause this reaction is 420 nm. (a) In what part of the electromagnetic spectrum is light with this wavelength found? (b) What is the maximum strength of a bond, in kJ/mol, that can be broken by absorption of a photon of 420-nm light? **(c)** Write out the photodissociation reaction showing Lewis electron-dot structures.

18.33 Consider Earth's energy balance shown in Figure 18.11. **(a)** How many different sources transfer energy to the atmosphere? Which makes the largest contribution? What is the total amount of energy transferred into the atmosphere in W/m^2? **(b)** To maintain a balance, the atmosphere must lose an equal amount of energy by emitting radiation, either into space or back toward the surface. What fraction is radiated back to the surface?

18.34 The atmosphere of Mars is 96% CO_2, with a pressure of approximately 6×10^{-3} atm at the surface. Based on measurements taken over a period of several years by the Rover Environmental Monitoring Station (REMS), the average daytime temperature at the REMS location on Mars is $-5.7\,°C$ ($22\,°F$), while the average nighttime temperature is $-79\,°C$ ($-109\,°F$). This daily variation in temperature is much larger than what we experience on Earth. What factor plays the largest role in this wide temperature variation, the composition or the density of the atmosphere?

Earth's Water (Section 18.3)

18.35 What is the molarity of Na^+ in a solution of NaCl whose salinity is 5.6 if the solution has a density of 1.03 g/mL?

18.36 Phosphorus is present in seawater to the extent of 0.07 ppm by mass. Assuming that the phosphorus is present as dihydrogenphosphate, $H_2PO_4^-$, calculate the corresponding molar concentration of $H_2PO_4^-$ in seawater.

18.37 The enthalpy of evaporation of water is 40.67 kJ/mol. Sunlight striking Earth's surface supplies $168\ W/m^2$ ($1\ W = 1\ watt = 1\ J/s$). **(a)** Assuming that evaporation of water is due only to energy input from the Sun, calculate how many grams of water could be evaporated from a $1.00\ m^2$ patch of ocean over a 12-h day. **(b)** The specific heat capacity of liquid water is $4.184\ J/g\,°C$. If the initial surface temperature of a $1.00\ m^2$ patch of ocean is $26\,°C$, what is its final temperature after being in sunlight for 12 h, assuming no phase changes and assuming that sunlight penetrates uniformly to a depth of 10.0 cm?

18.38 The enthalpy of fusion of water is 6.01 kJ/mol. Sunlight striking Earth's surface supplies $168\ W/m^2$ ($1\ W = 1\ watt = 1\ J/s$). **(a)** Assuming that melting of ice is due only to energy input from the Sun, calculate how many grams of ice could be melted from a $1.00\ m^2$ patch of ice over a 12-h day. **(b)** The specific heat capacity of ice is $2.032\ J/g\,°C$. If the initial temperature of a $1.00\ m^2$ patch of ice is $-5.0\,°C$, what is its final temperature after being in sunlight for 12 h, assuming no phase changes and assuming that sunlight penetrates uniformly to a depth of 1.00 cm?

18.39 A first-stage recovery of magnesium from seawater is precipitation of $Mg(OH)_2$ with CaO:

$$Mg^{2+}(aq) + CaO(s) + H_2O(l) \longrightarrow Mg(OH)_2(s) + Ca^{2+}(aq)$$

What mass of CaO, in grams, is needed to precipitate 1000 lb of $Mg(OH)_2$?

18.40 Gold is found in seawater at very low levels, about 0.05 ppb by mass. Assuming that gold is worth about $1300 per troy ounce, how many liters of seawater would you have to process to obtain $1,000,000 worth of gold? Assume the density of seawater is 1.03 g/mL and that your gold recovery process is 50% efficient.

18.41 Although there are many ions in seawater, the overall charges of the dissolved cations and anions must maintain charge neutrality. Consider only the six most abundant ions in seawater, as listed in Table 18.5 (Cl^-, Na^+, SO_4^{2-}, Mg^{2+},

Ca^{2+}, and K^+), and calculate the total charge in coulombs of the cations in 1.0 L of seawater. Calculate the total charge in coulombs of the anions in 1.0 L of seawater. To how many significant figures are the two numbers equal?

18.42 The Ogallala Aquifer described in the "A Closer Look" box: "The Ogallala Aquifer—A Shrinking Resource" (Section 18.3), provides 82% of the drinking water for the people who live in the region, although more than 75% of the water that is pumped from it is for irrigation. Irrigation withdrawals are approximately 18 billion gallons per day. Assuming that 2% of the rainfall that falls on an area of 600,000 km^2 recharges the aquifer, what average annual rainfall would be required to replace the water removed for irrigation?

Human Activities and Water Quality (Section 18.4)

18.43 List the common products formed when an organic material containing the elements carbon, hydrogen, oxygen, sulfur, and nitrogen decomposes **(a)** under aerobic conditions, **(b)** under anaerobic conditions.

18.44 **(a)** Explain why the concentration of dissolved oxygen in freshwater is an important indicator of the quality of the water. **(b)** Find graphical data in the text that show variations of gas solubility with temperature, and estimate to two significant figures the percent solubility of O_2 in water at $30\,°C$ as compared with $20\,°C$. How do these data relate to the quality of natural waters?

18.45 The organic anion

is found in most detergents. Assume that the anion undergoes aerobic decomposition in the following manner:

$$2\ C_{18}H_{29}SO_3^-(aq) + 51\ O_2(aq) \longrightarrow$$
$$36\ CO_2(aq) + 28\ H_2O(l) + 2\ H^+(aq) + 2\ SO_4^{2-}(aq)$$

What is the total mass of O_2 required to biodegrade 10.0 g of this substance?

18.46 The average daily mass of O_2 taken up by sewage discharged in the United States is 59 g per person. How many liters of water at 9 ppm O_2 are 50% depleted of oxygen in 1 day by a population of 1,200,000?

18.47 Magnesium ions are removed in water treatment by the addition of slaked lime, $Ca(OH)_2$. Write a balanced chemical equation to describe what occurs in this process.

18.48 In the lime soda process, once used in large-scale municipal water softening, calcium hydroxide prepared from lime and sodium carbonate are added to precipitate Ca^{2+} as $CaCO_3(s)$ and Mg^{2+} as $Mg(OH)_2(s)$:

$$Ca^{2+}(aq) + CO_3^{2-}(aq) \longrightarrow CaCO_3(s)$$
$$Mg^{2+}(aq) + 2\ OH^-(aq) \longrightarrow Mg(OH)_2(s)$$

How many moles of $Ca(OH)_2$ and Na_2CO_3 should be added to soften (remove the Ca^{2+} and Mg^{2+}) 1200 L of water in which

$$[Ca^{2+}] = 5.0 \times 10^{-4}\ M \quad \text{and}$$
$$[Mg^{2+}] = 7.0 \times 10^{-4}\ M?$$

18.49 **(a)** What are *trihalomethanes* (THMs)? **(b)** Draw the Lewis structures of two example THMs.

18.50 (a) Suppose that tests of a municipal water system reveal the presence of bromate ion, BrO_3^-. What are the likely origins of this ion? (b) Is bromate ion an oxidizing or reducing agent?

Green Chemistry (Section 18.5)

18.51 One of the principles of green chemistry is that it is better to use as few steps as possible in making new chemicals. In what ways does following this rule advance the goals of green chemistry? How does this principle relate to energy efficiency?

18.52 Discuss how catalysts can make processes more energy efficient.

18.53 A reaction for converting ketones to lactones, called the Baeyer–Villiger reaction,

is used in the manufacture of plastics and pharmaceuticals. 3-Chloroperbenzoic acid is shock-sensitive, however, and prone to explode. Also, 3-chlorobenzoic acid is a waste product. An alternative process being developed uses hydrogen peroxide and a catalyst consisting of tin deposited within a solid support. The catalyst is readily recovered from the reaction mixture. (a) What would you expect to be the other product of oxidation of the ketone to lactone by hydrogen peroxide? (b) What principles of green chemistry are addressed by use of the proposed process?

18.54 The hydrogenation reaction shown here was performed with an iridium catalyst, both in supercritical CO_2 ($scCO_2$) and in the chlorinated solvent CH_2Cl_2. The kinetic data for the reaction in both solvents are plotted in the graph. In what respects is the use of $scCO_2$ a good example of a green chemical reaction?

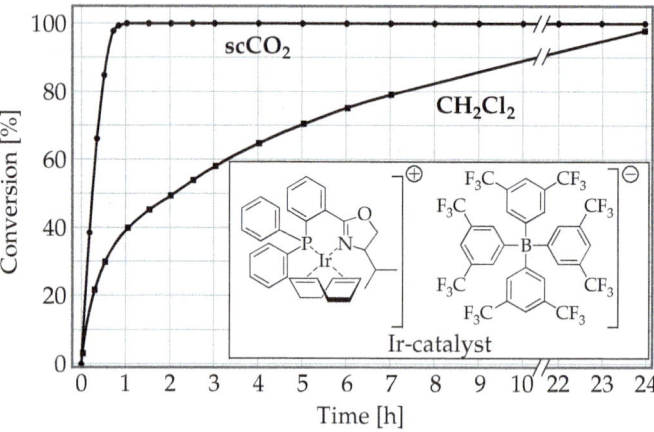

18.55 In the following three instances, which choice is greener in each situation? Explain. (a) Benzene as a solvent or water as a solvent. (b) The reaction temperature is 500 K or 1000 K. (c) Sodium chloride as a by-product or chloroform ($CHCl_3$) as a by-product.

18.56 In the following three instances, which choice is greener in a chemical process? Explain. (a) A reaction that can be run at 350 K for 12 h without a catalyst or one that can be run at 300 K for 1 h with a reusable catalyst. (b) A reagent for the reaction that can be obtained from corn husks or one that is obtained from petroleum. (c) A process that produces no by-products or one in which the by-products are recycled for another process.

Additional Exercises

18.57 A friend of yours has seen each of the following items in newspaper articles and would like an explanation: (a) acid rain, (b) greenhouse gas, (c) photochemical smog, (d) ozone depletion. Give a brief explanation of each term and identify one or two of the chemicals associated with each.

18.58 If an average O_3 molecule "lives" only 100–200 seconds in the stratosphere before undergoing dissociation, how can O_3 offer any protection from ultraviolet radiation?

18.59 Show how Equations 18.7 and 18.9, and the combination reaction that leads to the formation of molecular oxygen, $2\,O(g) \longrightarrow O_2(g)$, can be added to give Equation 18.10.

18.60 What properties of CFCs make them ideal for various commercial applications but also make them a long-term problem in the stratosphere?

18.61 *Halons* are fluorocarbons that contain bromine, such as $CBrF_3$. They are used extensively as foaming agents for fighting fires. Like CFCs, halons are very unreactive and ultimately can diffuse into the stratosphere. (a) Based on the data in Table 8.3,

would you expect photodissociation of Br atoms to occur in the stratosphere? (b) Propose a mechanism by which the presence of halons in the stratosphere could lead to the depletion of stratospheric ozone.

18.62 (a) What is the difference between a CFC and an HFC? (b) It is estimated that the lifetime for HFCs in the stratosphere is 2–7 years. Why is this number significant? (c) Why have HFCs been used to replace CFCs? (d) What is the major disadvantage of HFCs as replacements for CFCs?

18.63 Explain, using Le Châtelier's principle, why the equilibrium constant for the formation of NO from N_2 and O_2 increases with increasing temperature, whereas the equilibrium constant for the formation of NO_2 from NO and O_2 decreases with increasing temperature.

18.64 Natural gas consists primarily of methane, $CH_4(g)$. (a) Write a balanced chemical equation for the complete combustion of methane to produce $CO_2(g)$ as the only carbon-containing product. (b) Write a balanced chemical equation for the incomplete combustion of methane to produce $CO(g)$ as the only carbon-containing product. (**c**) At 25 °C and 1.0 atm pressure, what is the minimum quantity of dry air needed to combust 1.0 L of $CH_4(g)$ completely to $CO_2(g)$?

18.65 It was estimated that the eruption of the Mount Pinatubo volcano resulted in the injection of 20 million metric tons of SO_2 into the atmosphere. Most of this SO_2 underwent oxidation to SO_3, which reacts with atmospheric water to form an aerosol. (**a**) Write chemical equations for the processes leading to formation of the aerosol. (**b**) The aerosols caused a 0.5–0.6 °C drop in surface temperature in the northern hemisphere. What is the mechanism by which this occurs? (**c**) The sulfate aerosols, as they are called, also cause loss of ozone from the stratosphere. How might this occur?

18.66 One of the possible consequences of climate change is an increase in the temperature of ocean water. The oceans serve as a "sink" for CO_2 by dissolving large amounts of it. (**a**) The graph provided shows the solubility of CO_2 in water as a function of temperature. Does CO_2 behave more or less similarly to other gases in this respect? (**b**) What are the implications of this figure for the problem of climate change?

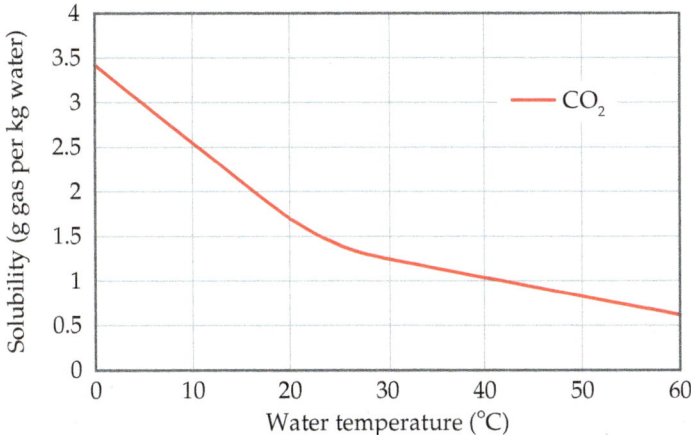

18.67 The rate of solar energy striking Earth averages 168 W/m². The rate of energy radiated from Earth's surface averages 390 W/m². Comparing these numbers, you might expect that the planet would cool quickly, yet it does not. Why not?

18.68 The solar power striking Earth every day averages 168 W/m². The highest ever recorded electrical power usage in New York City was 13,200 MW—a record established in July of 2013. Considering that the present technology for solar energy conversion is about 10% efficient, from how many square meters of land must sunlight be collected in order to provide this peak power? (For comparison, the total area of New York City is 830 km².)

18.69 Write balanced chemical equations for each of the following reactions: (**a**) The nitric oxide molecule undergoes photodissociation in the upper atmosphere. (**b**) The nitric oxide molecule undergoes photoionization in the upper atmosphere. (**c**) Nitric oxide undergoes oxidation by ozone in the stratosphere. (**d**) Nitrogen dioxide dissolves in water to form nitric acid and nitric oxide.

18.70 (**a**) Explain why $Mg(OH)_2$ precipitates when $CO_3{}^{2-}$ ions are added to a solution containing Mg^{2+}. (**b**) Will $Mg(OH)_2$ precipitate when 4.0 g of Na_2CO_3 is added to 1.00 L of a solution containing 125 ppm of Mg^{2+}?

18.71 (**a**) The EPA threshold for acceptable levels of lead ions in water is < 15 ppb. What is the molarity of an aqueous solution with a concentration of 15 ppb? (**b**) Concentrations of lead in the bloodstream are often quoted in units of $\mu g/dL$. Averaged over the entire country, the mean concentration of lead in the blood was measured to be 1.6 $\mu g/dL$ in 2008. Express this concentration in ppb.

18.72 In 1986 an electrical power plant in Taylorsville, Georgia, burned 8,376,726 tons of coal, a national record at that time. (**a**) Assuming that the coal was 83% carbon and 2.5% sulfur and that combustion was complete, calculate the number of tons of carbon dioxide and sulfur dioxide produced by the plant during the year. (**b**) If 55% of the SO_2 could be removed by reaction with powdered CaO to form $CaSO_3$, how many tons of $CaSO_3$ would be produced?

18.73 The water supply for a midwestern city contains the following impurities: coarse sand, finely divided particulates, nitrate ions, trihalomethanes, dissolved phosphorus in the form of phosphates, potentially harmful bacterial strains, and dissolved organic substances. Which of the following processes or agents, if any, is effective in removing each of these impurities: coarse sand filtration, activated carbon filtration, aeration, ozonization, and precipitation with aluminum hydroxide?

18.74 The concentration of H_2O in the stratosphere is about 5 ppm. It undergoes photodissociation according to:

$$H_2O(g) \longrightarrow H(g) + OH(g)$$

(**a**) Write out the Lewis electron-dot structures for both products and reactant.

(**b**) Given that the average bond enthalpy for an O—H bond is 463 kJ/mol, calculate the maximum wavelength for a photon that could cause this dissociation.

(**c**) The hydroxyl radical, OH, can react with ozone, giving the following reactions:

$$OH(g) + O_3(g) \longrightarrow HO_2(g) + O_2(g)$$
$$HO_2(g) + O(g) \longrightarrow OH(g) + O_2(g)$$

What overall reaction results from these two elementary reactions? What is the catalyst in the overall reaction? Explain.

18.75 The standard enthalpies of formation of ClO and ClO_2 are 101 and 102 kJ/mol, respectively. Using these data and the thermodynamic data in Appendix C, calculate the overall enthalpy change for each step in the following catalytic cycle:

$$ClO(g) + O_3(g) \longrightarrow ClO_2(g) + O_2(g)$$
$$ClO_2(g) + O(g) \longrightarrow ClO(g) + O_2(g)$$

What is the enthalpy change for the overall reaction that results from these two steps?

18.76 A reaction that contributes to the depletion of ozone in the stratosphere is the direct reaction of oxygen atoms with ozone:

$$O(g) + O_3(g) \longrightarrow 2 O_2(g)$$

At 298 K the rate constant for this reaction is $4.8 \times 10^5 \, M^{-1} \, s^{-1}$. (**a**) Based on the units of the rate constant, write the likely rate law for this reaction. (**b**) Would you expect this reaction to occur via a single elementary process? Explain why or why not. (**c**) Use $\Delta H_f°$ values from Appendix C to estimate the enthalpy change for this reaction. Would this reaction raise or lower the temperature of the stratosphere?

18.77 The following data were collected for the destruction of O_3 by H ($O_3 + H \longrightarrow O_2 + OH$) at very low concentrations:

Trial	$[O_3]$ (*M*)	$[H]$ (*M*)	Initial Rate (*M/s*)
1	5.17×10^{-33}	3.22×10^{-26}	1.88×10^{-14}
2	2.59×10^{-33}	3.25×10^{-26}	9.44×10^{-15}
3	5.19×10^{-33}	6.46×10^{-26}	3.77×10^{-14}

(a) Write the rate law for the reaction.

(b) Calculate the rate constant.

18.78 The Henry's law constant for CO_2 in water at 25 °C is $3.1 \times 10^{-2} \, M \, atm^{-1}$. (a) What is the solubility of CO_2 in water at this temperature if the solution is in contact with air at normal atmospheric pressure? (b) Assume that all of this CO_2 is in the form of H_2CO_3 produced by the reaction between CO_2 and H_2O:

$$CO_2(aq) + H_2O(l) \longrightarrow H_2CO_3(aq)$$

What is the pH of this solution?

18.79 The precipitation of $Al(OH)_3$ ($K_{sp} = 1.3 \times 10^{-33}$) is sometimes used to purify water. (a) Estimate the pH at which precipitation of $Al(OH)_3$ will begin if 5.0 lb of $Al_2(SO_4)_3$ is added to 2000 gal of water. (b) Approximately how many pounds of CaO must be added to the water to achieve this pH?

18.80 The valuable polymer polyurethane is made by a condensation reaction of alcohols (ROH) with compounds that contain an isocyanate group (RNCO). Two reactions that can generate a urethane monomer are shown here:

(i) $RNH_2 + CO_2 \longrightarrow R-N=C=O + 2 H_2O$

$$R-N=C=O + R'OH \longrightarrow R-\overset{H}{\underset{}{N}}-\overset{O}{\underset{\|}{C}}-OR'$$

(ii) $RNH_2 + \underset{Cl \quad Cl}{\overset{O}{\underset{\|}{C}}} \longrightarrow R-N=C=O + 2 HCl$

$$R-N=C=O + R'OH \longrightarrow R-\overset{H}{\underset{}{N}}-\overset{O}{\underset{\|}{C}}-OR'$$

(a) Which process, i or ii, is greener? Explain.

(b) What are the hybridization and geometry of the carbon atoms in each C-containing compound in each reaction?

(c) If you wanted to promote the formation of the isocyanate intermediate in each reaction, what could you do, using Le Châtelier's principle?

18.81 The pH of a particular raindrop is 5.6. (a) Assuming the major species in the raindrop are $H_2CO_3(aq)$, $HCO_3^-(aq)$, and $CO_3^{2-}(aq)$, calculate the concentrations of these species in the raindrop, assuming the total carbonate concentration is 1.0×10^{-5} M. The appropriate K_a values are given in Table 16.3. (b) What experiments could you do to test the hypothesis that the rain also contains sulfur-containing species that contribute to its pH? Assume you have a large sample of rain to test.

Design an Experiment

Considerable fracking of petroleum/gas wells [see the "Chemistry and Sustainability" box on "Fracking and Water Quality" (Section 18.4)] has occurred in recent years in a particular rural area. The residents have complained that the water in the residential wells serving their domestic water needs has become contaminated with chemicals associated with the fracking operations. The well operators respond that the chemicals about which complaints are lodged occur naturally and are not the result of well-drilling activities.

Describe experiments that you could conduct on the waters from residential wells to help determine whether and to what extent well contaminants are due to fracking operations. Among the chemicals that might be expected to be employed in fracking operations are hydrochloric acid, sodium chloride, ethylene glycol, borate salts, water-soluble gelling agents such as guar gum, citric acid, methanol, and other alcohols such as isopropanol and methane. Assume that you have available the techniques to measure concentrations of these substances in the residential wells. What experiments would you conduct, and what analyses of the results would you carry out in an attempt to settle the question of whether fracking operations have led to contamination of the well water? Would simply measuring the concentrations of some or all of these substances in the well waters be sufficient to settle the issue?

19

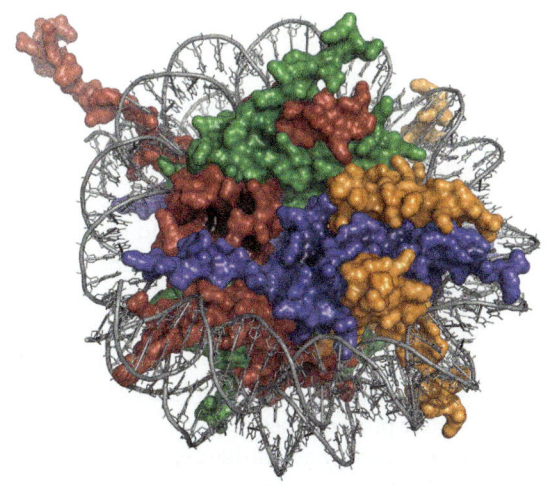

CHEMICAL THERMODYNAMICS

▲ **THE NUCLEOSOME** Inside the nucleus of a living cell, DNA (the gray outer double-helical portion) surrounds eight protein molecules (the colored molecular models). This overall DNA/protein structure, called the nucleosome, is the basic unit of chromosomes in the nuclei of our cells. These structures are highly ordered, yet they also must be unraveled for gene expression to take place. Both packaging and unpackaging of DNA in the nucleosome involve changes in the energy and the degree of order for the system.

The amazing organization of living systems, from complex molecular structures such as those of a nucleosome, to cells, to tissues, and finally to whole plants and animals, is an unending source of wonder and delight to the scientists who study them. Energy must be spent, somehow, to form and maintain these organized systems. But how is that energy channeled to accomplish these tasks?

Understanding natural processes, whether they be DNA replication, photosynthesis, or merely the rusting of a nail, relies on understanding the general laws that govern chemical reactions. There are two basic questions that chemists ask when studying reactions: "How fast is the reaction?" and "How far does it proceed?" The first question is addressed by chemical kinetics, which we discussed in Chapter 14. The second question relates to the equilibrium constant, the focus of Chapter 15.

We have already seen that the rate of a reaction is controlled largely by the activation energy of the reaction. (Section 14.4) Chemical equilibrium is reached when a given reaction and its reverse reaction occur at the same rate. (Section 15.1) In this chapter, we explore the connection between energy and the extent of a reaction. Doing so requires a deeper look at *chemical thermodynamics*, the area of chemistry that deals with energy relationships. We first considered thermodynamics in Chapter 5, where we discussed the first law of thermodynamics and the concept of *enthalpy*. We also discuss the second and third laws of thermodynamics, both of which involve the concept of *entropy*, a thermodynamic quantity that we encountered briefly in Chapter 13. (Section 13.1)

WHAT'S AHEAD

19.1 ▶ Spontaneous Processes Recognize that changes that occur in nature have a directional character. They move *spontaneously* in one direction but not in the reverse direction.

19.2 ▶ Entropy and the Second Law of Thermodynamics Develop more fully the concept of *entropy*, a thermodynamic state function important in determining whether a process is spontaneous. The *second law of thermodynamics* tells us that in any spontaneous process, the entropy of the universe (the system plus surroundings) increases.

19.3 ▶ The Molecular Interpretation of Entropy and the Third Law of Thermodynamics At the molecular level, recognize that the entropy of a system is related to the number of accessible *microstates*. The entropy of the system increases as the randomness of the system increases. The *third law of thermodynamics* states that, at 0 K, the entropy of a perfect crystalline solid is zero.

19.4 ▶ Entropy Changes in Chemical Reactions Calculate the standard entropy changes for systems undergoing reaction, using tabulated *standard molar entropies*.

19.5 ▶ Gibbs Free Energy Define another thermodynamic state function, the *free energy* (or *Gibbs free energy*), which is a measure of how far removed a system is from equilibrium. The change in free energy measures the maximum amount of useful work obtainable from a process and tells us the direction in which a chemical reaction is spontaneous.

19.6 ▶ Free Energy and Temperature Develop the relationships among free-energy change, enthalpy change, and entropy change, which provide insight into how temperature affects the spontaneity of a process.

19.7 ▶ Free Energy and the Equilibrium Constant Describe how the standard free-energy change for a chemical reaction can be used to calculate the equilibrium constant for the reaction.

19.1 | Spontaneous Processes

Learning Objectives

When you finish Section 19.1, you should be able to:

▶ Classify certain processes as thermodynamically spontaneous or nonspontaneous.

▶ Classify processes as thermodynamically reversible or irreversible.

Go Figure

Does the potential energy of the eggs change during this process?

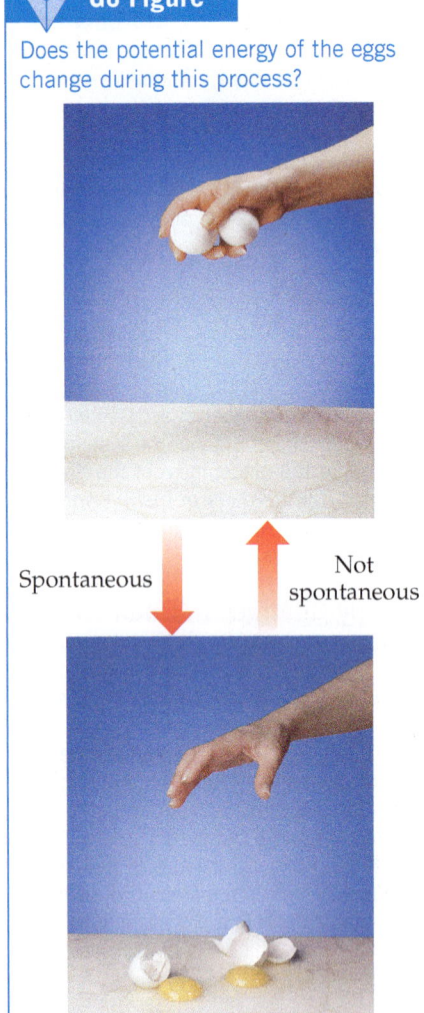

Spontaneous   Not spontaneous

▲ **Figure 19.1 A spontaneous process!**

If you release a brick from your hand, it will fall. It never jumps back into your hand no matter how long you wait. Likewise, if you let a nail sit in the rain, it will eventually rust. The rusted nail will never convert to its original condition even when the sun shines and time passes. These are but two examples illustrating that there is directionality to events. Why do these events occur in a certain direction? The first law of thermodynamics tells us that *energy is conserved* during these processes, but it does not tell us anything about their preferred direction. (Section 5.2) In order to understand why processes occur in a certain direction, we need to develop more deeply our understanding of thermodynamics.

Events such as a brick falling under the influence of gravity or a wet nail rusting occur without any outside intervention—these processes occur *spontaneously*. A **spontaneous process** is one that occurs on its own without any outside assistance.

A spontaneous process occurs in one direction only, and the reverse of any spontaneous process is always *nonspontaneous*. Drop an egg above a hard surface, for example, and it breaks on impact (**Figure 19.1**). Now, imagine seeing a video clip in which a broken egg rises from the floor, reassembles itself, and ends up in someone's hand. You would conclude that the video is running in reverse because you know that broken eggs do not magically rise and reassemble themselves! An egg falling and breaking is a spontaneous process. The reverse process, the egg rising and reassembling itself, is a nonspontaneous process, even though energy is conserved in both the forward and reverse processes.

We know of other spontaneous and nonspontaneous processes that relate more directly to our study of chemistry. For example, a gas spontaneously expands into a vacuum (**Figure 19.2**), but the reverse process, in which the gas moves back entirely into one of the flasks, does not happen. In other words, expansion of the gas is spontaneous, whereas the reverse process is nonspontaneous. In general:

Processes that are spontaneous in one direction are nonspontaneous in the opposite direction.

Experimental conditions, such as temperature and pressure, are often important in determining whether a process is spontaneous. There are situations, such as the melting of ice, in which a forward process is spontaneous at one temperature, but the reverse process is spontaneous at a different temperature. At atmospheric pressure, ice melts spontaneously when the temperature of the surroundings is above 0 °C. When the surroundings are below 0 °C, however, the reverse process—liquid water freezing—is spontaneous (**Figure 19.3**).

We must not confuse the *spontaneity* of a process with *speed*. Just because a process is spontaneous does not necessarily mean that it will occur at an observable rate. A spontaneous process can be fast, as in the case of acid–base neutralization, or slow, as in the rusting of iron. Thermodynamics tells us the *direction* and *extent* of reaction but nothing about the rate.

It is also important to understand that nonspontaneous does not mean impossible. For example, although the decomposition of table salt, NaCl, into sodium and chlorine is nonspontaneous under ordinary conditions, it is possible to decompose molten NaCl by supplying energy from an external source. In contrast, a spontaneous process occurs without such outside intervention.

Eggs dropping and breaking, or gases expanding into a vacuum, are processes that probably strike you as obviously spontaneous. By using thermodynamics we can address a more general question: Can we look at a specific chemical process and determine whether it is spontaneous or nonspontaneous? We will see that we can, and that doing so requires our learning some additional aspects of thermodynamics.

▼ **Go Figure**

No work is done when the gas initially expands from bulb A to bulb B. Why is that?

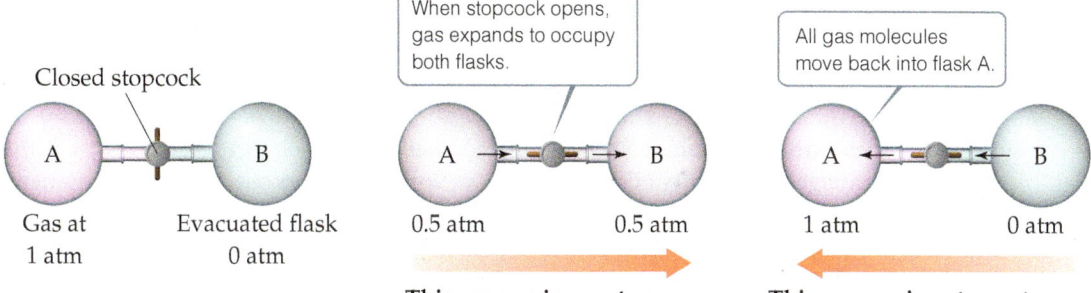

Closed stopcock

When stopcock opens, gas expands to occupy both flasks.

All gas molecules move back into flask A.

Gas at 1 atm   Evacuated flask 0 atm

0.5 atm   0.5 atm

1 atm   0 atm

This process is spontaneous   **This process is not spontaneous**

▲ **Figure 19.2 Expansion of a gas into an evacuated space is a spontaneous process.** The reverse process—gas molecules initially distributed evenly in two flasks all moving into one flask—is not spontaneous.

▼ **Go Figure**   In which direction is this process exothermic?

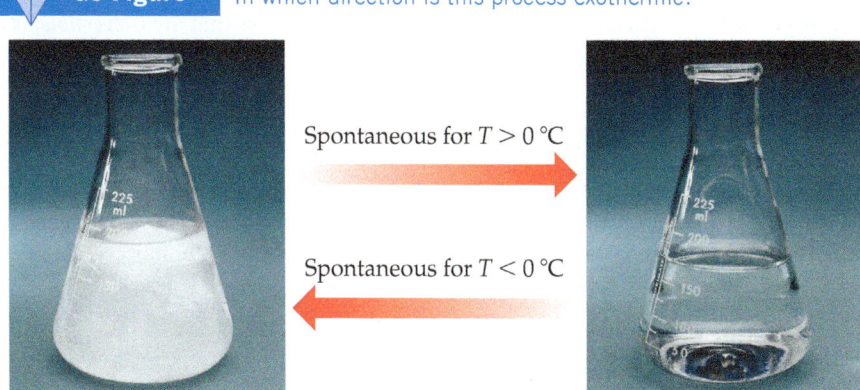

Spontaneous for $T > 0\,°C$

Spontaneous for $T < 0\,°C$

▲ **Figure 19.3 Spontaneity can depend on temperature.** At $T > 0\,°C$, ice melts spontaneously to liquid water. At $T < 0\,°C$, the reverse process, water freezing to ice, is spontaneous. At $T = 0\,°C$ the two states are in equilibrium.

Sample Exercise 19.1

Identifying Spontaneous Processes

Predict whether each process is spontaneous as described, spontaneous in the reverse direction, or at equilibrium: **(a)** Water at 40 °C gets hotter when a piece of metal heated to 150 °C is added. **(b)** Water at room temperature decomposes into $H_2(g)$ and $O_2(g)$. **(c)** Benzene vapor, $C_6H_6(g)$, at a pressure of 1 atm condenses to liquid benzene at the normal boiling point of benzene, 80.1 °C.

SOLUTION

Analyze We are asked to judge whether each process is spontaneous in the direction indicated, in the reverse direction, or in neither direction.

Plan We need to think about whether each process is consistent with our experience about the natural direction of events or whether we expect the reverse process to occur.

Solve

(a) This process is spontaneous. Whenever two objects at different temperatures are brought into contact, heat is transferred from the hotter object to the colder one. (Section 5.1) Thus, heat is transferred from the hot metal to the cooler water. The final temperature, after the metal and water achieve the same temperature (thermal equilibrium), will

be somewhere between the initial temperatures of the metal and the water.

(b) Experience tells us that this process is not spontaneous— we certainly have never seen hydrogen and oxygen gases spontaneously bubbling up out of water! Rather, the *reverse* process— the reaction of H_2 and O_2 to form H_2O—is spontaneous.

(c) The normal boiling point is the temperature at which a vapor at 1 atm is in equilibrium with its liquid. Thus, this is an equilibrium situation. If the temperature were below 80.1 °C, condensation would be spontaneous.

▶ **Practice Exercise**

At 1 atm pressure, $CO_2(s)$ sublimes at −78 °C. Is this process spontaneous at −100 °C and 1 atm pressure?

Seeking a Criterion for Spontaneity

A marble rolling down an incline or a brick falling from your hand both lose potential energy. The loss of some form of energy is a common feature of spontaneous change in mechanical systems. Does that mean that the direction of spontaneous changes in chemical systems is determined by the loss of energy? If so, does that mean that all spontaneous chemical and physical changes are exothermic? It takes only a few moments to find exceptions to this notion. For example, the melting of ice at room temperature is spontaneous and endothermic. Similarly, many spontaneous dissolution processes, such as the dissolving of NH_4NO_3, are endothermic. (Section 13.1) We can conclude that although the majority of spontaneous reactions are exothermic, there are spontaneous endothermic ones as well. As a result, some other factor must be at work in determining the natural direction of processes.

To understand why certain processes are spontaneous, we need to consider more closely the ways in which the state of a system can change. Recall from Section 5.2 that quantities such as temperature, internal energy, and enthalpy are *state functions*, properties that define a state and do not depend on how we reach that state. The heat transferred between the system and its surroundings, q, and the work done on or by the system, w, are *not* state functions—their values depend on the specific way in which the change occurs. That is, their values depend on the *path* taken between states. Understanding two kinds of paths, those that are *reversible* and those that are *irreversible*, is a key to understanding spontaneity.

Reversible and Irreversible Processes

For any process, we can imagine a hypothetical, ideal path that can be reversed to restore both the system and its surroundings to exactly their original states. This means that after the process is reversed, both the system and surroundings are unchanged. Such an ideal process is said to be reversible.

- A **reversible process** is one for which we can restore the system to its original condition with no change to the surroundings.
- An **irreversible process** is one that leaves the surroundings somehow changed when the system is restored to its original state.

Sometimes these processes are referred to as being *thermodynamically reversible* or *thermodynamically irreversible* to add greater clarity to their meanings.

As an example, let's consider a process involving the transfer of heat. When two objects at different temperatures are in contact, heat flows spontaneously from the hotter object to the colder one. Because it is impossible to make the heat flow in the opposite direction, from colder to hotter, this is an irreversible process. Given this fact, can we imagine any hypothetical conditions under which the heat transfer could be reversible? The answer is yes, but only if we consider temperature changes that are infinitesimally small.

Imagine the system and its surroundings at essentially the same temperature, with just an infinitesimal temperature difference, δT, between them (**Figure 19.4**). If the surroundings are at temperature T and the system is at an infinitesimally higher temperature $T + \delta T$, then an infinitesimal amount of heat flows from the system to surroundings. We can reverse the direction of heat flow by making an infinitesimal change of temperature in the opposite direction, lowering the system temperature to $T - \delta T$. Now the direction of heat flow is from surroundings to system. Thus, for the process to be reversible, the amounts of heat must be infinitesimally small and the transfer of heat must occur infinitely slowly. *Reversible processes are those that reverse direction whenever an infinitesimal change is made in some property of the system.**

---

*For a process to be truly reversible, the amounts of heat or work transferred must be infinitesimally small and the transfer must occur infinitely slowly. Thus, no process that we can observe is truly reversible. The notion of infinitesimal amounts is related to the infinitesimals that you may have encountered in a calculus course.

Go Figure If the flow of heat into or out of the system is to be reversible, what must be true of δT?

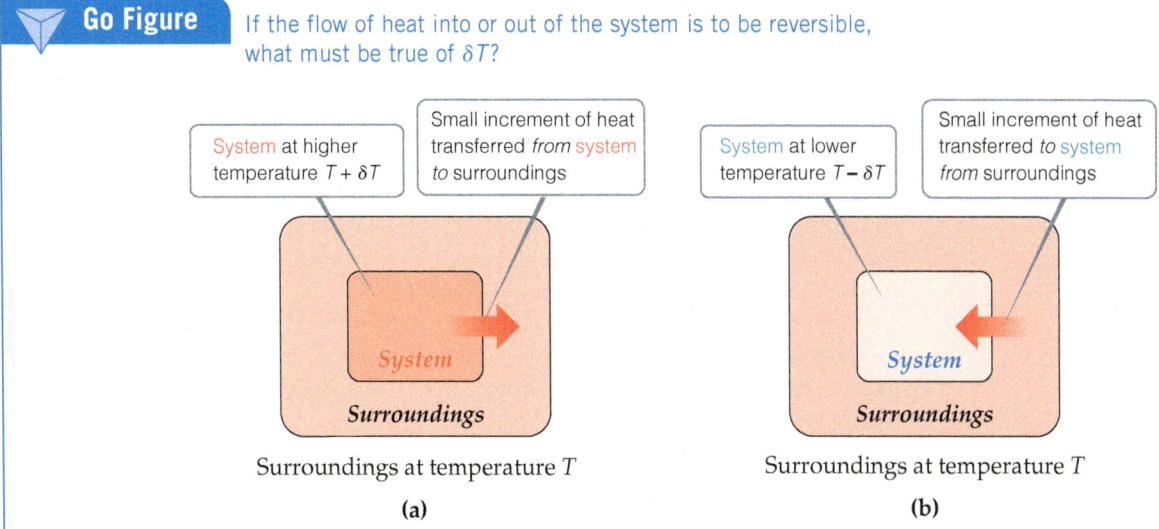

System at higher temperature $T + \delta T$

Small increment of heat transferred *from* system *to* surroundings

System at lower temperature $T - \delta T$

Small increment of heat transferred *to* system *from* surroundings

System

Surroundings

System

Surroundings

Surroundings at temperature T

Surroundings at temperature T

(a)

(b)

▲ **Figure 19.4 Reversible flow of heat.** Heat can flow reversibly between a system and its surroundings only if the two have an infinitesimally small difference in temperature δT. (a) Increasing the temperature of the system by δT causes heat to flow from the hotter system to the colder surroundings. (b) Decreasing the temperature of the system by δT causes heat to flow from the hotter surroundings to the colder system.

Now let's consider another example, the expansion of an ideal gas at constant temperature (referred to as an **isothermal** process). In the cylinder–piston arrangement of **Figure 19.5**, when the partition is removed, the gas expands spontaneously to fill the evacuated space. Can we determine whether this particular isothermal expansion is reversible or irreversible? Because the gas expands into a vacuum with no external pressure, it does no $P-V$ work on the surroundings. (Section 5.3) Thus, $w = 0$ for the expansion. We can use the piston to compress the gas back to its original state, but doing so requires that the surroundings do work on the system, meaning that $w > 0$ for the compression. In other words, the path that restores the system to its original state requires a different value of w (and, by the first law, a different value of q) than the path by which the system was first changed. The fact that the same path cannot be followed to restore the system to its original state indicates that the process is irreversible.

Go Figure After the partition is opened, why is $w = 0$ when the gas expands to fill the cylinder?

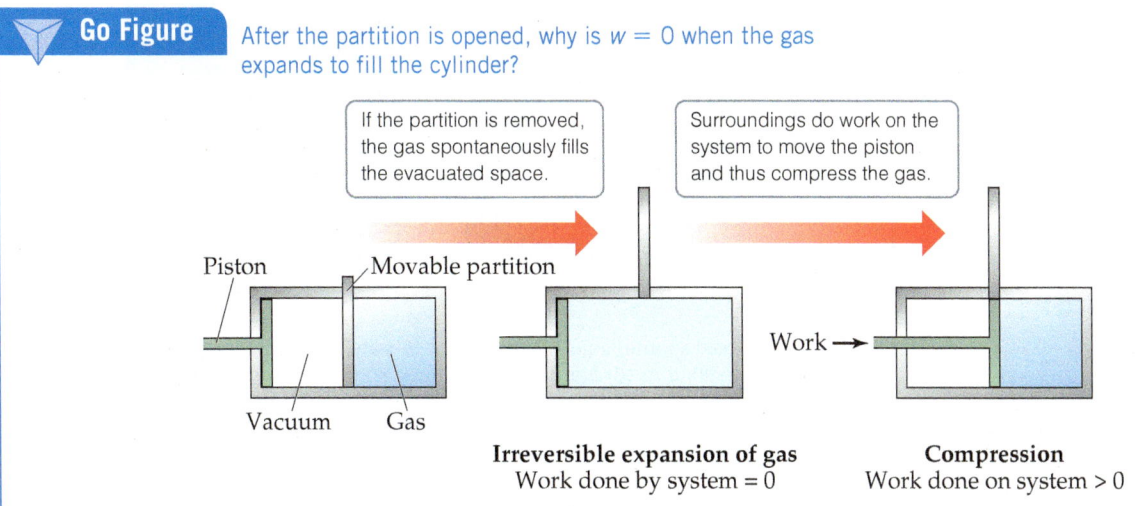

If the partition is removed, the gas spontaneously fills the evacuated space.

Surroundings do work on the system to move the piston and thus compress the gas.

Piston

Movable partition

Work →

Vacuum   Gas

Irreversible expansion of gas
Work done by system = 0

Compression
Work done on system > 0

▲ **Figure 19.5 An irreversible process.** Initially, an ideal gas is confined to the right half of a cylinder. When the partition is removed, the gas spontaneously expands to fill the whole cylinder. No work is done by the system during this expansion. Using the piston to compress the gas back to its original state requires the surroundings to do work on the system.

What might a *reversible* isothermal expansion of an ideal gas be? This process will occur only if initially, when the gas is confined to half the cylinder, the external pressure acting on the piston exactly balances the pressure exerted by the gas on the piston. If the external pressure is reduced infinitely slowly, the piston will move outward, allowing the pressure of the confined gas to readjust to maintain the pressure balance. This infinitely slow process, in which the external pressure and internal pressure are always in equilibrium, is reversible. Interestingly, *a reversible process produces the maximum amount of work that can be done by a system on its surroundings*. Hence, the work done by the system on the surroundings during the reversible expansion is greater than the work done by the system on the surroundings via *any* irreversible path.

If we reverse the process and compress the gas in the same infinitely slow manner, we can return the gas to its original volume. The complete cycle of expansion and compression in this hypothetical process, moreover, is accomplished without any net change to the surroundings.

From these and other examples we learn two important facts:

- Because real processes can at best only approximate the infinitesimal changes associated with reversible processes, *all real processes are irreversible.*
- Because spontaneous processes are real processes, *all spontaneous processes are irreversible.*

Thus, for any spontaneous change, returning the system to its original condition results in a net change in the surroundings. But what kind of change occurs? The answer to this question is given in the second law of thermodynamics, which we now discuss.

 Self-Assessment Exercises

SAE 19.1   Which of the following statements about spontaneous processes is or are *true*?

(**i**) The combustion of methane gas with oxygen gas is a spontaneous process.
(**ii**) A process that is spontaneous in one direction will be nonspontaneous in the opposite direction.
(**iii**) A spontaneous process will always occur at a fast rate.

(**a**) Only statement i is true. (**b**) Statements i and ii are true. (**c**) Statements i and iii are true. (**d**) Statements ii and iii are true. (**e**) All three statements are true.

SAE 19.2   Consider the two states of a system that consists of water in a flask at 1 atm pressure that are shown in the figure. What can you conclude about this system?

State A   State B

(**a**) It will proceed spontaneously from state A to state B. (**b**) It will proceed spontaneously from state B to state A. (**c**) Nothing will happen spontaneously. (**d**) More information is needed to draw any conclusions.

SAE 19.3   Which of the following statements about reversible and irreversible processes is *false*?

(**a**) The expansion of a gas into a vacuum is an irreversible process.
(**b**) In a reversible process, we can restore the system to its original state with no change to the surroundings.
(**c**) A reversible process will always be spontaneous.
(**d**) All real processes are irreversible.

19.2 | Entropy and the Second Law of Thermodynamics

In Section 5.2, we introduced the first law of thermodynamics, which we presented using words ("energy is conserved") and as an equation ($\Delta E = q + w$). The first law assures that in any of the processes that we have discussed, the internal energy of the universe is unchanged, whether that process is spontaneous or not. Thus, changes in internal energy can't explain the concepts of spontaneity and reversibility that we have just discussed. In order to understand why some processes are spontaneous, we need to look more closely at the thermodynamic quantity called *entropy*, which was first mentioned in Section 13.1.

Entropy is a measure of the degree of *randomness* or *disorder* associated with a system. For a system that consists of particles, such as atoms or molecules, the entropy is a measure of the number of ways that the particles can move or arrange themselves—we refer to these motions and arrangements as the *degrees of freedom* of the particles.

In this section, we consider how entropy changes are related to heat transfer and temperature. Our analysis will bring us to a profound statement about entropy changes and spontaneity known as the second law of thermodynamics.

Learning Objectives

When you finish Section 19.2, you should be able to:

▶ Define and interpret the entropy, S, as a thermodynamic state function that measures the degree of randomness or disorder of a system.

▶ State the second law of thermodynamics and its implications about changes in the entropy of the universe for both reversible and irreversible processes.

The Relationship between Entropy Change and Heat

The entropy, S, of a system is a state function just like internal energy, E, and enthalpy, H. As with these other quantities, the value of S is a characteristic of the state of a system. (Section 5.2) Thus, the change in entropy, ΔS, in a system depends only on the initial and final states of the system and not on the path taken from one state to the other:

$$\Delta S = S_{final} - S_{initial} \qquad [19.1]$$

Note that when $\Delta S > 0$, the entropy of the system has increased—its final state is *more* random or disordered than its initial state. Conversely, when $\Delta S < 0$, the entropy of the system has decreased—its final state is *less* random or more ordered than its initial state.

For the special case of an isothermal process, ΔS is equal to the heat that would be transferred if the process were reversible, q_{rev}, divided by the absolute temperature at which the process occurs:

$$\Delta S = \frac{q_{rev}}{T} \qquad \text{(constant } T) \qquad [19.2]$$

Although many possible paths can take the system from one state to another, only one path is associated with a reversible process. Thus, the value of q_{rev} is uniquely defined for any two states of the system. Because S is a state function, we can use Equation 19.2 to calculate ΔS for *any* isothermal process between states, not just the reversible one.

ΔS for Phase Changes

The melting of a substance at its melting point and the vaporization of a substance at its boiling point are isothermal processes. (Section 11.4) Consider the melting of ice. At 1 atm pressure, ice and liquid water are in equilibrium at 0 °C. Imagine melting 1 mol of ice at 0 °C, 1 atm to form 1 mol of liquid water at 0 °C, 1 atm. We can achieve this change by adding heat to the system from the surroundings: $q = \Delta H_{fusion}$, where ΔH_{fusion} is the heat of melting. Imagine adding the heat infinitely slowly, raising the temperature of the surroundings only infinitesimally above 0 °C. When we make the change in this fashion, the process is reversible because we can reverse it by infinitely slowly removing the same amount of heat, ΔH_{fusion}, from the system, using immediate surroundings that are infinitesimally below 0 °C. Thus, $q_{rev} = \Delta H_{fusion}$ for the melting of ice at $T = 0 °C = 273$ K.

The enthalpy of fusion for H_2O is $\Delta H_{fusion} = 6.01$ kJ/mol (a positive value because melting is an endothermic process). Thus, we can use Equation 19.2 to calculate ΔS_{fusion} for melting 1 mol of ice at 273 K:

$$\Delta S_{fusion} = \frac{q_{rev}}{T} = \frac{\Delta H_{fusion}}{T} = \frac{(1 \text{ mol})(6.01 \times 10^3 \text{ J/mol})}{273 \text{ K}} = 22.0 \text{ J/K}$$

Notice that (1) we must use the absolute temperature in Equation 19.2, and (2) the units for ΔS are energy divided by absolute temperature, J/K, as we expect from Equation 19.2.

Sample Exercise 19.2
Calculating ΔS for a Phase Change

Elemental mercury is a silver liquid at room temperature. Its normal freezing point is $-38.9\,°C$, and its molar enthalpy of fusion is $\Delta H_{fusion} = 2.29$ kJ/mol. What is the entropy change of the system when 50.0 g of Hg(l) freezes at the normal freezing point?

SOLUTION

Analyze We first recognize that freezing is an *exothermic* process, which means heat is transferred from the system to the surroundings and $q < 0$. Because freezing is the reverse of melting, the enthalpy change that accompanies the freezing of 1 mol of Hg is $-\Delta H_{fusion} = -2.29$ kJ/mol.

Plan We can use $-\Delta H_{fusion}$ and the atomic weight of Hg to calculate q for freezing 50.0 g of Hg. Then we use this value of q as q_{rev} in Equation 19.2 to determine ΔS for the system.

Solve

For q we have:

$$q = (50.0 \text{ g Hg})\left(\frac{1 \text{ mol Hg}}{200.59 \text{ g Hg}}\right)\left(\frac{-2.29 \text{ kJ}}{1 \text{ mol Hg}}\right)\left(\frac{1000 \text{ J}}{1 \text{ kJ}}\right) = -571 \text{ J}$$

Before using Equation 19.2, we must first convert the given Celsius temperature to kelvins:

$$-38.9\,°C = (-38.9 + 273.15) \text{ K} = 234.3 \text{ K}$$

We can now calculate ΔS_{sys}:

$$\Delta S_{sys} = \frac{q_{rev}}{T} = \frac{-571 \text{ J}}{234.3 \text{ K}} = -2.44 \text{ J/K}$$

Check The entropy change is negative because our q_{rev} value is negative, which it must be because heat flows out of the system in this exothermic process.

Comment It makes sense that the change in entropy is negative upon freezing because the atoms of mercury have less freedom to move (fewer degrees of freedom) as a solid than as a liquid, so entropy decreases upon freezing. The procedure used here can be used to calculate ΔS for other isothermal phase changes, such as the vaporization of a liquid at its boiling point.

▶ **Practice Exercise**
The normal boiling point of ethanol, C_2H_5OH, is 78.3 °C, and its molar enthalpy of vaporization is 38.56 kJ/mol. What is the change in entropy in the system when 68.3 g of $C_2H_5OH(g)$ at 1 atm condenses to liquid at the normal boiling point?

A CLOSER LOOK | The Entropy Change When a Gas Expands Isothermally

In general, the entropy of any system increases as the system becomes more random or more spread out. Thus, we expect the spontaneous expansion of a gas to result in an increase in entropy. To see how this entropy increase can be calculated, consider the expansion of an ideal gas that is initially constrained by a piston, as in the rightmost part of Figure 19.5. Imagine that we allow the gas to undergo a reversible isothermal expansion by infinitesimally decreasing the external pressure on the piston. The work done on the surroundings by the reversible expansion of the system against the piston can be calculated using calculus (we do not show the derivation):

$$w_{rev} = -nRT \ln \frac{V_2}{V_1}$$

In this equation, n is the number of moles of gas, R is the ideal-gas constant (Section 10.3), T is the absolute temperature, V_1 is the initial volume, and V_2 is the final volume. Notice that if $V_2 > V_1$, as it must be in our expansion, then $w_{rev} < 0$, meaning that the expanding gas does work on the surroundings.

One characteristic of an ideal gas is that its internal energy depends only on temperature, not on pressure. Thus, when an ideal gas expands isothermally, $\Delta E = 0$. Because $\Delta E = q_{rev} + w_{rev} = 0$, we see that $q_{rev} = -w_{rev} = nRT \ln(V_2/V_1)$. Then, using Equation 19.2, we can calculate the entropy change in the system:

$$\Delta S_{sys} = \frac{q_{rev}}{T} = \frac{nRT \ln(V_2/V_1)}{T} = nR \ln(V_2/V_1) \qquad [19.3]$$

Let's calculate the entropy change for 1.00 L of an ideal gas at 1.00 atm pressure, 0 °C temperature, expanding to 2.00 L. From the ideal-gas equation, we can calculate the number of moles in 1.00 L of an ideal gas at 1.00 atm and 0 °C as we did in Chapter 10:

$$n = \frac{PV}{RT} = \frac{(1.00 \text{ atm})(1.00 \text{ L})}{(0.08206 \text{ L-atm/mol-K})(273 \text{ K})} = 4.46 \times 10^{-2} \text{ mol}$$

The gas constant, R, can also be expressed as 8.314 J/mol-K (Table 10.2), and this is the value we must use in Equation 19.3 because we want our answer to be expressed in terms of J rather than in L-atm. Thus, for the expansion of the gas from 1.00 L to 2.00 L, we have

$$\Delta S_{sys} = (4.46 \times 10^{-2} \text{ mol})\left(8.314 \frac{J}{\text{mol-K}}\right)\left(\ln \frac{2.00 \text{ L}}{1.00 \text{ L}}\right)$$

$$= 0.26 \text{ J/K}$$

We see that ΔS_{sys} is positive; the entropy of the system increases upon expansion. That makes sense given that the gas molecules now have a greater volume in which to move—they are more disordered and have more degrees of freedom.
Related Exercises: 19.27, 19.28

The Second Law of Thermodynamics

The key idea of the first law of thermodynamics is that energy is conserved in any process. (Section 5.2) Is entropy also conserved in a spontaneous process in the same way that energy is conserved?

Let's try to answer this question by calculating the entropy change of a system and the entropy change of its surroundings when our system is 1 mol of ice (a piece roughly the size of an ice cube) melting in the palm of your hand, which is part of the surroundings. The process is not reversible because the system and surroundings are at different temperatures. Nevertheless, because ΔS is a state function, its value is the same regardless of whether the process is reversible or irreversible. Previously (just before Sample Exercise 19.2) we calculated the entropy change of this system:

$$\Delta S_{sys} = \frac{q_{rev}}{T} = \frac{(1\ \text{mol})(6.01 \times 10^3\ \text{J/mol})}{273\ \text{K}} = 22.0\ \text{J/K}$$

The portion of the surroundings immediately in contact with the ice is your palm, which we assume is at body temperature, 37 °C = 310 K. We will use this as the temperature of the surroundings. The quantity of heat lost by your palm is -6.01×10^3 J/mol, which is equal in magnitude to the quantity of heat gained by the ice but has the opposite sign. Hence, the entropy change of the surroundings is

$$\Delta S_{surr} = \frac{q_{rev}}{T} = \frac{(1\ \text{mol})(-6.01 \times 10^3\ \text{J/mol})}{310\ \text{K}} = -19.4\ \text{J/K}$$

Recall that, in thermodynamics, the universe consists of the system of interest and its surroundings. (Section 5.2) Therefore, $\Delta S_{univ} = \Delta S_{sys} + \Delta S_{surr}$. Thus, the overall entropy change of the universe is positive in our example:

$$\Delta S_{univ} = \Delta S_{sys} + \Delta S_{surr} = (22.0\ \text{J/K}) + (-19.4\ \text{J/K}) = 2.6\ \text{J/K}$$

If the temperature of the surroundings were not 310 K, but rather some temperature infinitesimally above 273 K, the melting would be reversible instead of irreversible. In that case, the entropy change of the surroundings would equal -22.0 J/K and ΔS_{univ} would be zero.

In general, any irreversible process results in an increase in the entropy of the universe, whereas any reversible process results in no change in the entropy of the universe:

$$\textit{Reversible process:}\quad \Delta S_{univ} = \Delta S_{sys} + \Delta S_{surr} = 0$$

$$\textit{Irreversible process:}\quad \Delta S_{univ} = \Delta S_{sys} + \Delta S_{surr} > 0 \qquad [19.4]$$

These equations summarize the **second law of thermodynamics**. Because spontaneous processes are irreversible, we can also state the second law as *the entropy of the universe increases for any spontaneous process.*

The second law of thermodynamics tells us the essential character of any spontaneous change—it is always accompanied by an increase in the entropy of the universe. We can use this law as a criterion to predict whether or not a given process is spontaneous. Before seeing how this is done, however, we will find it useful to explore entropy from a molecular perspective.

A word on notation before we proceed: Throughout most of the remainder of this chapter, we focus on systems rather than surroundings. To simplify the notation, we usually refer to the entropy change of the system as ΔS rather than explicitly indicating ΔS_{sys}.

 Self-Assessment Exercises

SAE 19.4 Which of the following statements about entropy is *false*? (a) The entropy change between two states of a system depends on the amount of heat transferred during the specific path taken between the two states. (b) The entropy of a system can be related to its degree of randomness or disorder. (c) Entropy is a state function. (d) The units of entropy are energy per kelvin. (e) The entropy change associated with a phase change can be determined from the enthalpy change for the phase change and the temperature at which the phase change occurs.

SAE 19.5 The melting point of tin metal, Sn(s), is 232 °C, and its molar enthalpy of fusion is $\Delta H_{fusion} = 7.03$ kJ/mol. What is the entropy change of the system when one mole of molten tin solidifies at 232 °C? (a) -30.3 J/K (b) -13.9 J/K (c) $+13.9$ J/K (d) $+30.3$ J/K (e) More information is needed.

SAE 19.6 Which of the following statements about the second law of thermodynamics is or are *true*?

(i) For a reversible process, $\Delta S_{surr} = -\Delta S_{sys}$.
(ii) The entropy of the universe is conserved in any process.
(iii) For any spontaneous process, $\Delta S_{surr} + \Delta S_{sys} > 0$.

(a) Only one statement is true. (b) Statements i and ii are true. (c) Statements i and iii are true. (d) Statements ii and iii are true. (e) All three statements are true.

19.3 | The Molecular Interpretation of Entropy and the Third Law of Thermodynamics

Learning Objectives

When you finish Section 19.3, you should be able to:

▶ Analyze the concept of entropy at the molecular level to compare qualitatively the entropies of different substances and states.

▶ Describe and quantify the relationship between the entropy of a system and the number of microstates of the system.

▶ Predict the sign of ΔS for chemical processes.

▶ Interpret the third law of thermodynamics.

As chemists, we are interested in molecules. What does entropy have to do with them and with their transformations? What molecular property does entropy reflect? Ludwig Boltzmann (1844–1906) gave another conceptual meaning to the notion of entropy, and to understand his contribution, we need to examine the ways in which we can interpret entropy at the molecular level.

Expansion of a Gas at the Molecular Level

Consider a simple spontaneous process—the expansion of a gas into a vacuum as shown previously in Figure 19.2. Though we now understand that the entropy of the universe increases during the expansion, how can we explain the spontaneity of this process at the molecular level? We can get a sense of what makes this expansion spontaneous by envisioning the gas as a collection of particles in constant motion, as we did in discussing the kinetic-molecular theory of gases. (Section 10.6) When the stopcock in Figure 19.2 is opened, we can view the expansion of the gas as the ultimate result of the gas molecules moving randomly throughout the larger volume.

Let's look at this idea more closely by tracking two of the gas molecules as they move around. Before the stopcock is opened, both molecules are confined to the left flask, as shown in **Figure 19.6**(a). After the stopcock is opened, the molecules travel randomly throughout the entire apparatus. As Figure 19.6(b) shows, there are four possible arrangements for the two molecules once both flasks are available to them. Because the molecular motion is random, all four arrangements are equally likely. Note that now only one arrangement corresponds to the situation before the stopcock was opened (that is, both molecules in the left flask).

Figure 19.6(b) shows that with both flasks available to the molecules, the probability of the red molecule being in the left flask is two in four (top right and bottom left arrangements), and the probability of the blue molecule being in the left flask is the same (top left and bottom left arrangements). Because the probability is $^2/_4 = ^1/_2$ that each molecule is in the left flask, the probability that *both* are there is $(^1/_2)^2 = ^1/_4$. If we apply the same analysis to *three* gas molecules, we find that the probability that all three are in the left flask at the same time is $(^1/_2)^3 = ^1/_8$.

Now let's consider a *mole* of gas. The probability that all the molecules are in the left flask at the same time is $(^1/_2)^N$, where $N = 6.02 \times 10^{23}$. This is a vanishingly small number! Thus, there is essentially zero likelihood that all the gas molecules will be in the left flask at the same time. This analysis of the microscopic behavior of the gas molecules leads to the expected macroscopic behavior: The gas spontaneously expands to fill both the left and right flasks, and it does not spontaneously all go back in the left flask.

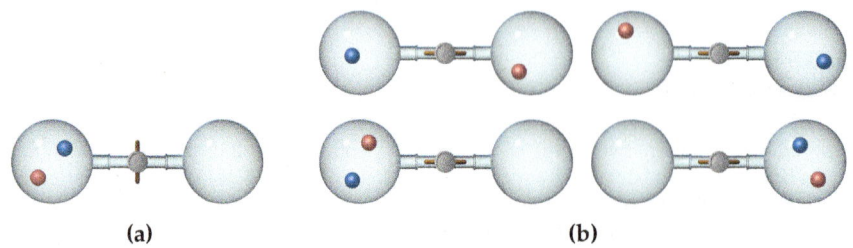

Go Figure   After the stopcock is opened, which is more probable: finding both molecules in the left flask, or one molecule in each flask?

(a)
The two molecules are colored red and blue to keep track of them.

(b)
Four arrangements (microstates) are possible once the stopcock is opened.

▲ **Figure 19.6 Possible arrangements of two gas molecules in two flasks.** (a) Before the stopcock is opened, both molecules are in the left flask. (b) After the stopcock is opened, there are four possible arrangements of the two molecules.

This molecular view of gas expansion shows the tendency of the molecules to "spread out" among the different arrangements they can take. Before the stopcock is opened, there is only one possible arrangement: all molecules in the left flask. When the stopcock is opened, the arrangement in which all the molecules are in the left flask is but one of an extremely large number of possible arrangements. The most probable arrangements by far are those in which there are essentially equal numbers of molecules in the two flasks. When the gas spreads throughout the apparatus, any given molecule can be in either flask rather than confined to the left flask. We say that with the stopcock opened, the gas molecules have more degrees of freedom and the arrangement of gas molecules is more random or disordered than when the molecules are all confined in the left flask.

We will see that this notion of increasing randomness helps us understand entropy at the molecular level.

Boltzmann's Equation and Microstates

The science of thermodynamics developed as a means of describing the properties of matter in our macroscopic world without regard to microscopic structure. In fact, thermodynamics was a well-developed field before the modern view of atomic and molecular structure was even known. The thermodynamic properties of water, for example, addressed the behavior of bulk water (or ice or water vapor) as a substance without considering any specific properties of individual H_2O molecules.

To connect the microscopic and macroscopic descriptions of matter, scientists have developed the field of *statistical thermodynamics*, which uses the tools of statistics and probability to link the microscopic and macroscopic worlds. Here we show how entropy, which is a property of bulk matter, can be connected to the behavior of atoms and molecules. Because the mathematics of statistical thermodynamics is complex, our discussion is largely conceptual.

In our discussion of two gas molecules in the two-flask system in Figure 19.6, we saw that the number of possible arrangements helped explain why the gas expands. Suppose we now consider one mole of an ideal gas in a particular thermodynamic state, which we can define by specifying the temperature, T, and volume, V, of the gas. What is happening to this gas at the microscopic level, and how does what is going on at the microscopic level relate to the entropy of the gas?

Imagine taking a snapshot of the positions and speeds of all the molecules at a given instant. The speed of each molecule relates to its kinetic energy. That particular set of 6.02×10^{23} positions and kinetic energies of the individual gas molecules is what we call a *microstate* of the system. A **microstate** is a single possible arrangement of the positions and kinetic energies of the molecules when the molecules are in a specific thermodynamic state. We could envision continuing to take snapshots of our system to see other possible microstates.

As you should see, there would be such a staggeringly large number of microstates that taking individual snapshots of all of them is not feasible. Because we are examining such a large number of particles, however, we can use the tools of statistics and probability to determine the total number of microstates for the thermodynamic state. (That is where the *statistical* part of the name *statistical thermodynamics* comes in.) Each thermodynamic state has a characteristic number of microstates associated with it, and we use the symbol W for that number.

It is helpful to distinguish between the state of a system and the microstates associated with the state.

- A *state* describes the macroscopic view of the system as characterized, for example, by the pressure and temperature of a sample of gas.
- A *microstate* is a particular microscopic arrangement of the atoms or molecules of the system that corresponds to the given state of the system.

Each of the snapshots we described is a microstate—the positions and kinetic energies of individual gas molecules will change from snapshot to snapshot, but each one is a possible arrangement of the collection of molecules corresponding to a single state. For macroscopically sized systems, such as a mole of gas, there is an extremely large number of microstates for each state—that is, W is generally an extremely large number.

▲ **Figure 19.7 Ludwig Boltzmann's gravestone.** Boltzmann's gravestone in Vienna is inscribed with his relationship between the entropy of a state, *S*, and the number of available microstates, *W*. (In Boltzmann's time, "log" was used to represent the natural logarithm.)

The connection between the number of microstates of a system, *W*, and the entropy of the system, *S*, is expressed in a beautifully simple equation developed by Boltzmann and engraved on his tombstone (**Figure 19.7**):

$$S = k \ln W \qquad [19.5]$$

In this equation, *k* is the *Boltzmann constant*, 1.38×10^{-23} J/K. We see from the equation that:

> *Entropy is a measure of how many microstates are associated with a particular macroscopic state.*

From Equation 19.5, we see that the entropy change accompanying any process is

$$\Delta S = k \ln W_{final} - k \ln W_{initial} = k \ln \frac{W_{final}}{W_{initial}} \qquad [19.6]$$

Any change in the system that leads to an increase in the number of microstates ($W_{final} > W_{initial}$) leads to a positive value of ΔS:

> *Entropy increases with the number of microstates of the system.*

Let's consider two modifications to our ideal-gas sample and see how the entropy changes in each case. First, suppose we increase the volume of the system while keeping the temperature constant, which is analogous to allowing the gas to expand isothermally. A greater volume means a greater number of positions available to the gas atoms and therefore a greater number of microstates. The entropy therefore increases as the volume increases, as we saw in the "A Closer Look" box on "The Entropy Change When a Gas Expands Isothermally" in Section 19.2.

Second, suppose we keep the volume fixed but increase the temperature. How does this change affect the entropy of the system? Recall the distribution of molecular speeds presented in Figure 10.12(a). An increase in temperature increases the most probable speed of the molecules and also broadens the distribution of speeds. Hence, the molecules have a greater number of possible kinetic energies, and the number of microstates increases. Thus, the entropy of the system increases with increasing temperature.

Molecular Motions and Energy

When a substance is heated, the motion of its molecules increases. In Section 10.5 we found that the average kinetic energy of the molecules of an ideal gas is directly proportional to the absolute temperature of the gas. That means the higher the temperature, the faster the molecules move and the more kinetic energy they possess. Moreover, hotter systems have a broader distribution of molecular speeds, as Figure 10.12(a) shows.

The particles of an ideal gas are idealized points with no volume and no bonds, however—points that we visualize as flitting around through space. Like a particle of an ideal gas or an atom, a molecule can move through space as a whole. We call such movement **translational motion**. The molecules in a gas have more freedom of translational motion than those in a liquid, which have more freedom of translational motion than the molecules of a solid.

In addition to translational motion, a molecule has two additional forms of more complex motion—each of these forms of motion is an additional *degree of freedom* of the molecule. The first of these is **vibrational motion**, in which the atoms in the molecule move periodically toward and away from one another like two weights attached to the ends of a spring. The second additional degree of freedom is **rotational motion**, in which the molecule spins about an axis. **Figure 19.8** shows the vibrational motions and one of the rotational motions possible for the water molecule. The translational, vibrational, and rotational motions of a molecule are called the *motional degrees of freedom* of the molecule.

Go Figure   Describe another possible rotational motion for this molecule.

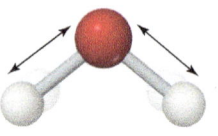 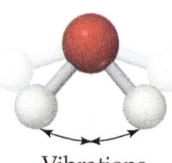 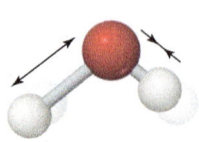 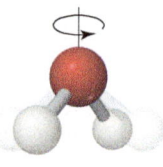

└────────── Vibrations ──────────┘   └── Rotation ──┘

▲ **Figure 19.8 Vibrational and rotational motions in a water molecule.**

The vibrational and rotational motions possible in real molecules lead to arrangements that a single atom cannot have. In general, *the number of microstates possible for a system increases with an increase in volume, an increase in temperature, or an increase in the number of molecules, because any of these changes increases the possible positions and degrees of freedom of the molecules making up the system.* A collection of molecules therefore has a greater number of possible microstates than does the same number of ideal-gas particles. Moreover, the number of microstates increases as the complexity of the molecule increases because there are more vibrational degrees of freedom available.

Chemists have several ways of describing an increase in the number of microstates possible for a system and therefore an increase in the entropy for the system. Each way seeks to capture a sense of the increased freedom of motion that causes molecules to spread out when not restrained by physical barriers or chemical bonds.

The most common way of describing an increase in entropy is the increase in the *randomness*, or *disorder*, of the system. Another way likens an entropy increase to an increased *dispersion (spreading out) of energy* because there is an increase in the number of ways the positions and energies of the molecules can be distributed throughout the system. Each description (randomness or energy dispersal) is conceptually helpful if applied correctly.

Making Qualitative Predictions about ΔS

It is usually not difficult to estimate qualitatively how the entropy of a system changes during a simple process. As noted previously, an increase in either the temperature or the volume of a system leads to an increase in the number of microstates, and hence an increase in the entropy. One more factor that correlates with number of microstates is the number of independently moving particles.

We can usually make qualitative predictions about entropy changes by focusing on these factors. For example, when water vaporizes, the molecules spread out into a larger volume. Because they occupy a larger volume, there is an increase in their freedom of motion, giving rise to a greater number of possible microstates, and hence an increase in entropy.

Now consider the phases of water. In ice, hydrogen bonding leads to the rigid structure shown in **Figure 19.9**. Each molecule in the ice is free to vibrate, but its translational

 **Go Figure**   In which phase are water molecules least able to have rotational motion?

Increasing entropy

Ice	Liquid water	Water vapor

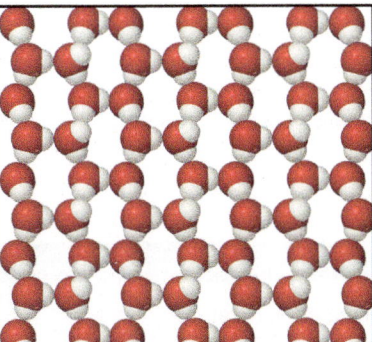 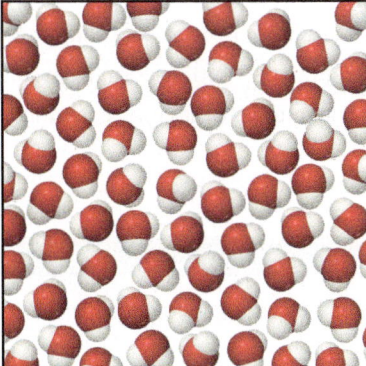 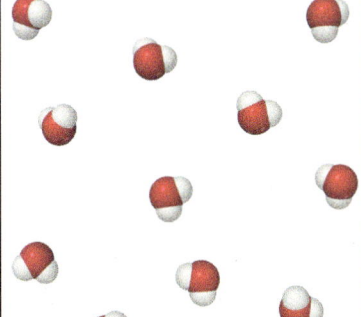

Rigid, crystalline structure

Motion restricted to **vibration** only

Smallest number of microstates

Increased freedom with respect to **translation**

Free to **vibrate** and **rotate**

Larger number of microstates

Molecules spread out, essentially independent of one another

Complete freedom for **translation**, **vibration**, and **rotation**

Largest number of microstates

▲ **Figure 19.9 Entropy and the phases of water.** The larger the number of possible microstates, the greater the entropy of the system.

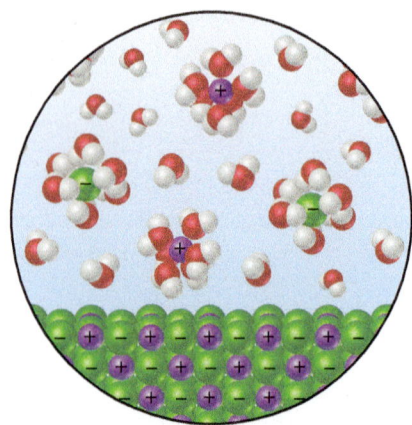

▲ **Figure 19.10 Entropy changes when an ionic solid dissolves in water.** The ions become more spread out and disordered, but the water molecules that hydrate the ions become less disordered.

  Go Figure

What major factor leads to a decrease in entropy as this reaction takes place?

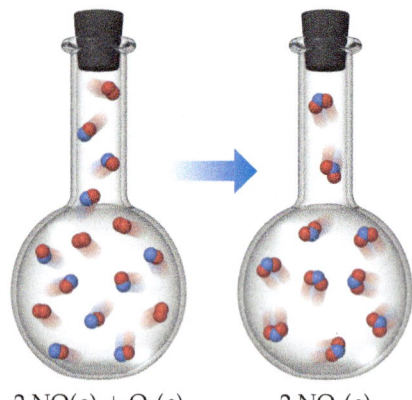

2 NO(g) + O₂(g)   2 NO₂(g)

▲ **Figure 19.11 Entropy decreases when NO(g) is oxidized by O₂(g) to NO₂(g)** A decrease in the number of gaseous molecules leads to a decrease in the entropy of the system.

and rotational motions are much more restricted than in liquid water. Although there are hydrogen bonds in liquid water, the molecules can more readily move about relative to one another (translation) and tumble around (rotation). During melting, therefore, the number of possible microstates increases, and so does the entropy. In water vapor, the molecules are essentially independent of one another and have their full range of translational, vibrational, and rotational motions. Thus, water vapor has an even greater number of possible microstates and therefore a higher entropy than liquid water or ice.

When an ionic solid dissolves in water, a mixture of water and ions replaces the pure solid and pure water, as shown for KCl in **Figure 19.10**. The ions in the liquid move in a volume that is larger than the volume in which they were able to move in the crystal lattice and so undergo more motion. This increased motion might lead us to conclude that the entropy of the system has increased. We have to be careful, however, because some water molecules have lost some freedom of motion because they are now bound to the ions as waters of hydration. (Section 13.1) These water molecules are in a *more* ordered state than before because they are now confined to the immediate environment of the ions. Therefore, the dissolving of a salt involves both a disordering process (the ions become less confined) and an ordering process (some water molecules become more confined). The disordering processes are usually dominant, and so the overall effect is an increase in the randomness of the system when most salts dissolve in water.

Now, imagine arranging biomolecules into a highly organized biochemical system, such as the nucleosome in the chapter-opening figure. We might expect that the creation of this well-ordered structure would lead to a decrease in the entropy of the system. But this is frequently not the case. Waters of hydration and counterions can be expelled from the interface as two large biomolecules interact, and so the entropy of the system can actually increase—if you consider the water and counterions to be part of the system.

The same ideas apply to chemical reactions, such as the one between nitric oxide gas and oxygen gas to form nitrogen dioxide gas:

$$2\,NO(g) + O_2(g) \longrightarrow 2\,NO_2(g) \qquad [19.7]$$

This reaction results in a decrease in the number of molecules—three molecules of gaseous reactants form two molecules of gaseous products (**Figure 19.11**). The formation of new N—O bonds reduces the motions of the atoms in the system. The formation of new bonds decreases the number of degrees of freedom, or forms of motion, available to the atoms. That is, the atoms are less free to move in random fashion because of the formation of new bonds. The decrease in the number of molecules and the resultant decrease in motion result in fewer possible microstates and therefore a decrease in the entropy of the system.

In summary, we generally expect the entropy of a system to increase for processes in which

1. Gases form from either solids or liquids.

2. Liquids or solutions form from solids.

3. The number of gas molecules increases during a chemical reaction.

  Sample Exercise 19.3

Predicting the Sign of ΔS

Predict whether ΔS is positive or negative for each process, assuming each occurs at constant temperature:

(a) $H_2O(l) \longrightarrow H_2O(g)$

(b) $Ag^+(aq) + Cl^-(aq) \longrightarrow AgCl(s)$

(c) $4\,Fe(s) + 3\,O_2(g) \longrightarrow 2\,Fe_2O_3(s)$

(d) $N_2(g) + O_2(g) \longrightarrow 2\,NO(g)$

SOLUTION

Analyze Given four reactions, we are asked to predict the sign of ΔS for each.

Plan We expect ΔS to be positive if there is an increase in temperature, an increase in volume, or an increase in the number of gas particles. The question states that the temperature is constant, so we need to concern ourselves only with changes in volume and number of particles.

Solve

(a) Evaporation involves a large increase in volume as liquid changes to gas. One mole of water (18 g) occupies about 18 mL as a liquid, and if it could exist as a gas at STP it would occupy 22.4 L. Because the molecules are distributed throughout a much larger volume in the gaseous state, an increase in motional freedom accompanies vaporization and ΔS is positive.

(b) In this process, ions, which are free to move throughout the volume of the solution, form a solid, in which they are confined to a smaller volume and restricted to more highly constrained positions. Thus, ΔS is negative.

(c) The particles of a solid are confined to specific locations and have fewer ways to move (fewer microstates) than do the molecules of a gas. Because O_2 gas is converted into part of the solid product Fe_2O_3, ΔS is negative.

(d) The number of moles of reactant gases is the same as the number of moles of product gases, and so the entropy change is expected to be small. The sign of ΔS is impossible to predict based on our discussions thus far, but we can predict that ΔS will be close to zero.

▶ **Practice Exercise**
The entropy of the universe increases for spontaneous processes. Does that mean the entropy of the universe decreases for nonspontaneous processes?

Sample Exercise 19.4
Predicting Relative Entropies

In each pair, choose the system that has greater entropy and explain your choice: **(a)** 1 mol of NaCl(s) or 1 mol of HCl(g) at 25 °C, **(b)** 2 mol of HCl(g) or 1 mol of HCl(g) at 25 °C, **(c)** 1 mol of HCl(g) or 1 mol of Ar(g) at 298 K.

SOLUTION

Analyze We need to select the system in each pair that has the greater entropy.

Plan We examine the state of each system and the complexity of the molecules it contains.

Solve (a) HCl(g) has the higher entropy because the particles in gases are more disordered and have more freedom of motion than the particles in solids. **(b)** When these two systems are at the same pressure, the sample containing 2 mol of HCl has twice the number of molecules as the sample containing 1 mol. Thus, the 2-mol

sample has twice the number of microstates and twice the entropy. **(c)** The HCl system has the higher entropy because the number of ways in which an HCl molecule can store energy is greater than the number of ways in which an Ar atom can store energy. (Molecules can rotate and vibrate; atoms cannot.)

▶ **Practice Exercise**
Choose the system with the greater entropy in each case: **(a)** 1 mol of $H_2(g)$ at STP or 1 mol of $SO_2(g)$ at STP, **(b)** 1 mol of $N_2O_4(g)$ at STP or 2 mol of $NO_2(g)$ at STP.

The Third Law of Thermodynamics

If we decrease the thermal energy of a system by lowering the temperature, the energy stored in translational, vibrational, and rotational motion decreases. As less energy is stored, the entropy of the system decreases because it has fewer and fewer microstates available. If we keep lowering the temperature, do we reach a state in which these motions are essentially shut down, a point described by a single microstate? This question is addressed by the **third law of thermodynamics**:

The entropy of a pure, perfect crystalline substance at absolute zero is zero: S (0 K) = 0.

Consider a pure, perfect crystalline solid. At absolute zero, the individual atoms or molecules in the lattice would be perfectly ordered in position. Because none of them would have thermal motion, there is only one possible microstate. As a result, Equation 19.5 becomes $S = k \ln W = k \ln 1 = 0$. As the temperature is increased from absolute zero, the atoms or molecules in the crystal gain energy in the form of vibrational motion about their lattice positions. This means that the degrees of freedom and the entropy both increase. What happens to the entropy, however, as we continue to heat the crystal? We consider this important question in the next section.

CHEMISTRY AND SUSTAINABILITY   Entropy and Human Society

The sustainability of Earth and its living organisms is highly dependent on the laws of thermodynamics. We have already seen that increasing human demands for energy have led to issues related to climate change and the depletion of natural resources. Those energy demands are used to provide heat and work, the components of the first law.

The second law of thermodynamics is also highly relevant in discussions about our existence and about our capacity to advance as a civilization. Any living organism is a complex, highly organized, well-ordered system, even at the molecular level like the nucleosome we saw at the beginning of this chapter. Our entropy content is much lower than it would be if we were completely

Continued

decomposed into carbon dioxide, water, and several other simple chemicals. Does this mean that life is a violation of the second law? No, because the thousands of chemical reactions necessary to produce and maintain life have caused a large increase in the entropy of the rest of the universe. Thus, as the second law requires, the overall entropy change during the lifetime of a human, or any other living system, is positive.

The second law applies also to the way we humans order our surroundings. In addition to being complex living systems ourselves, we are masters of producing order in the world around us. We build impressive, highly ordered structures and buildings. We manipulate and order matter at the nanoscale level in order to produce the technological breakthroughs that have become so commonplace in the twenty-first century. We use tremendous quantities of raw materials to produce highly ordered materials. In so doing, we expend a great deal of energy to, in essence, "fight" the second law of thermodynamics.

For every bit of order we produce, however, we produce an even greater amount of disorder. Petroleum, coal, and natural gas are burned to provide the energy necessary for us to achieve highly ordered structures, but their combustion increases the entropy of the universe by releasing $CO_2(g)$, $H_2O(g)$, and heat. Thus, even as we strive to create more impressive discoveries and greater order in our society, we drive the entropy of the universe higher, as the second law says we must.

We humans are, in effect, using up our storehouse of energy-rich materials to create order and advance technology. As noted in Chapter 5, we must learn to harness new energy sources, such as solar energy, and thereby reduce our dependence on nonrenewable sources. However, even as we discover new ways to use energy more sustainably, we are subject to the constraints of the second law of thermodynamics. The discovery of new materials for solar cells, more efficient wind-capturing devices, better ways to store and distribute energy, and so forth will all require the processing of raw materials into the high-performance devices that we seek. Every step of that journey—from the conversion of SiO_2 from sand into ultrapure silicon to the creation of new lightweight materials—is subject to the second law of thermodynamics. Thus, one of our major challenges in making the world more sustainable is our inevitable battle against the ever-increasing entropy of the universe.

Self-Assessment Exercises

SAE 19.7 For the isothermal expansion of an ideal gas into a vacuum, $q = 0$, $w = 0$, and $\Delta E = 0$. Which of the following is the best explanation of why this is a spontaneous process? **(a)** The entropy change of the universe is negative for the process. **(b)** The fact that $w = 0$ means that the expansion is favorable because no work is done on the system. **(c)** The particles of the ideal gas repel one another, so expansion is favorable. **(d)** $\Delta E = 0$, so the gas can expand and compress reversibly. **(e)** The number of microstates of the system increases when the volume of the system increases.

SAE 19.8 Suppose we have one mole each of $F_2(g)$, $C_2H_6(g)$, and $Ar(g)$, all at the same temperature and occupying the same volume. What is the correct order of these gases from lowest to highest entropy? **(a)** $C_2H_6 < F_2 < Ar$ **(b)** $Ar < C_2H_6 < F_2$ **(c)** $F_2 < Ar < C_2H_6$ **(d)** $Ar < F_2 < C_2H_6$

SAE 19.9 For which of the following reactions will ΔS be positive?
(i) $2\,CO(g) + O_2(g) \longrightarrow 2\,CO_2(g)$
(ii) $C_6H_6(l) \longrightarrow C_6H_6(g)$
(iii) $Cl_2(g) \longrightarrow 2\,Cl(g)$

(a) i and ii **(b)** i and iii **(c)** ii and iii **(d)** i, ii, and iii

SAE 19.10 Which of the following statements about the third law of thermodynamics is *false*? **(a)** The third law relates entropy to enthalpy. **(b)** The third law states that the entropy of a pure, perfect crystalline substance at absolute zero is zero. **(c)** When $S = 0$, only one microstate is available to the system. **(d)** The third law is consistent with Boltzmann's equation relating entropy to the number of microstates.

19.4 | Entropy Changes in Chemical Reactions

In Section 5.5 we discussed how calorimetry can be used to measure ΔH for chemical reactions. No comparable method exists for measuring ΔS for a reaction. However, because the third law establishes a zero point for entropy, we can use experimental measurements to determine the *absolute value of the entropy*, S. To see schematically how this is done, let's review in greater detail the variation in the entropy of a substance with temperature.

Temperature Variation of Entropy

We know that the entropy of a pure, perfect crystalline solid at 0 K is zero and that the entropy increases as the temperature of the crystal is increased. Figure 19.12 shows that the entropy of the solid increases steadily with increasing temperature up to the melting point of the solid. When the solid melts, the atoms or molecules are free to move about the entire volume of the sample. The added degrees of freedom increase the randomness of the substance, thereby increasing its entropy. We therefore see a sharp increase in the entropy at the melting point. After all the solid has melted, the temperature again increases and with it, the entropy.

Learning Objectives

When you finish Section 19.4, you should be able to:

► Define the standard molar entropy of a substance and predict the relative standard molar entropies of different substances.

► Use standard molar entropies to calculate standard entropy changes in chemical reactions.

At the boiling point of the liquid, another abrupt increase in entropy occurs. We can understand this increase as resulting from the increased volume available to the atoms or molecules as they enter the gaseous state. When the gas is heated further, the entropy increases steadily as more energy is stored in the translational motion of the gas atoms or molecules.

Another change occurring at higher temperatures is the skewing of molecular speeds toward higher values [Figure 10.12(a)]. The expansion of the range of speeds leads to increased kinetic energy and a greater number of degrees of freedom—and, hence, increased entropy. The conclusions we reach in examining Figure 10.12 are consistent with what we noted previously: Entropy generally increases with increasing temperature because the increased motional energy leads to a greater number of possible microstates.

Entropy versus temperature graphs such as Figure 19.12 can be obtained by carefully measuring how the heat capacity of a substance (Section 5.5) varies with temperature, and we can use the data to obtain the absolute entropies at different temperatures. (The theory and methods used for these measurements and calculations are beyond the scope of this text.) Entropies are usually tabulated as molar quantities, in units of joules per mole-kelvin (J/mol-K).

Standard Molar Entropies

Molar entropies for substances in their standard states are known as **standard molar entropies** and are denoted $S°$. The standard state for any substance is defined as the pure substance at 1 atm pressure.* Table 19.1 lists the values of $S°$ for a number of substances at 298 K; Appendix C gives a more extensive list.

We can make several observations about the $S°$ values in Table 19.1:

1. Unlike enthalpies of formation, standard molar entropies of elements at the reference temperature of 298 K are *not* zero.

2. The standard molar entropies of gases are greater than those of liquids and solids, consistent with our interpretation of experimental observations, as represented in Figure 19.12.

3. Standard molar entropies generally increase with increasing molar mass.

4. Standard molar entropies generally increase with an increasing number of atoms in the formula of a substance.

Point 4 is related to the molecular motion discussed in Section 19.3. In general, as the number of atoms increases, the number of possible microstates also increases. **Figure 19.13**

Go Figure Why does the plot show vertical jumps at the melting and boiling points?

▲ **Figure 19.12** Entropy increases with increasing temperature.

TABLE 19.1 Standard Molar Entropies of Selected Substances at 298 K

Substance	$S°$(J/mol-K)
$H_2(g)$	130.7
$N_2(g)$	191.6
$O_2(g)$	205.2
$H_2O(g)$	188.8
$NH_3(g)$	192.8
$CH_3OH(g)$	237.6
$C_6H_6(g)$	269.2
$H_2O(l)$	69.9
$CH_3OH(l)$	127.2
$C_6H_6(l)$	173.3
$Li(s)$	29.1
$Na(s)$	51.3
$K(s)$	64.7
$Fe(s)$	27.3
$FeCl_3(s)$	142.2
$NaCl(s)$	72.1

Go Figure What might you expect for the value of $S°$ for butane, C_4H_{10}?

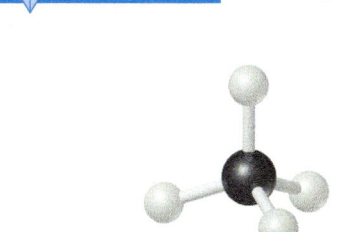

Methane, CH_4
$S° = 186.3$ J/mol-K

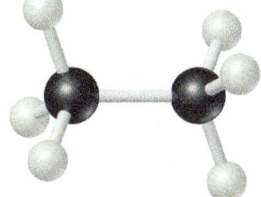

Ethane, C_2H_6
$S° = 229.5$ J/mol-K

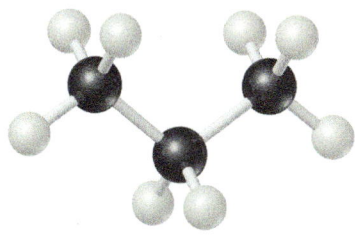

Propane, C_3H_8
$S° = 269.9$ J/mol-K

▲ **Figure 19.13** Entropy increases with increasing molecular complexity.

---

* The standard pressure used in thermodynamics is no longer 1 atm but rather is based on the SI unit for pressure, the pascal (Pa). The standard pressure is 10^5 Pa, a quantity known as a *bar*: 1 bar = 10^5 Pa = 0.987 atm. Because 1 bar differs from 1 atm by only 1.3%, we will continue to refer to the standard pressure as 1 atm.

compares the standard molar entropies of three hydrocarbons in the gas phase. Notice how the entropy increases as the number of atoms in the molecule increases.

Calculating the Standard Entropy Change for a Reaction

Because entropy is a state function, we can calculate the standard entropy change for a chemical reaction, $\Delta S°$, in a way analogous to the calculation of the standard enthalpy change for a reaction, $\Delta H°$ (Equation 5.31). (Section 5.7) The standard entropy change equals the sum of the entropies of the products minus the sum of the entropies of the reactants:

$$\Delta S° = \sum nS°(\text{products}) - \sum mS°(\text{reactants}) \tag{19.8}$$

As in Equation 5.31, the coefficients n and m are the coefficients in the balanced chemical equation for the reaction.

Sample Exercise 19.5
Calculating $\Delta S°$ from Tabulated Entropies

Calculate the change in the standard entropy of the system, $\Delta S°$, for the synthesis of ammonia from $N_2(g)$ and $H_2(g)$ at 298 K:

$$N_2(g) + 3 H_2(g) \longrightarrow 2 NH_3(g)$$

SOLUTION

Analyze We are asked to calculate the standard entropy change for the synthesis of $NH_3(g)$ from its constituent elements.

Plan We can make this calculation using Equation 19.8 and the standard molar entropy values in Table 19.1 and Appendix C.

Solve

Using Equation 19.8, we have:

$$\Delta S° = 2 S°(NH_3) - [S°(N_2) + 3 S°(H_2)]$$

Substituting the appropriate $S°$ values from Table 19.1 yields:

$$\Delta S° = (2 \text{ mol})(192.8 \text{ J/mol-K}) - [(1 \text{ mol})(191.6 \text{ J/mol-K})$$
$$+ (3 \text{ mol})(130.7 \text{ J/mol-K})]$$
$$= -198.1 \text{ J/K}$$

Check The value for $\Delta S°$ is negative, in agreement with our qualitative prediction based on the decrease in the number of molecules of gas during the reaction.

▶ **Practice Exercise**

Using the standard molar entropies in Appendix C, calculate the standard entropy change, $\Delta S°$, for the following reaction at 298 K:

$$Al_2O_3(s) + 3 H_2(g) \longrightarrow 2 Al(s) + 3 H_2O(g)$$

Entropy Changes in the Surroundings

We can use tabulated absolute entropy values to calculate the standard entropy change in a system, such as a chemical reaction, as just described. But what about the entropy change in the surroundings? We encountered this situation in Section 19.2, but it is good to revisit it now that we are examining chemical reactions.

We should recognize that the surroundings for any system serve essentially as a large, constant-temperature heat source (or heat sink if the heat flows from the system to the surroundings). The change in entropy of the surroundings depends on how much heat is absorbed or given off by the system.

For an isothermal process, the entropy change of the surroundings is given by

$$\Delta S_{surr} = \frac{q_{surr}}{T} = \frac{-q_{sys}}{T}$$

Because in a constant-pressure process, q_{sys} is simply the enthalpy change for the reaction, ΔH, we can write

$$\Delta S_{surr} = \frac{-\Delta H_{sys}}{T} \quad [\text{at constant } P] \tag{19.9}$$

For the ammonia synthesis reaction in Sample Exercise 19.5, q_{sys} is the enthalpy change for the reaction under standard conditions, $\Delta H°$, so the changes in entropy will be standard entropy changes, $\Delta S°$. Therefore, using the procedures described in Section 5.7, we have

$$\Delta H_{rxn}° = 2 \Delta H_f°[NH_3(g)] - 3 \Delta H_f°[H_2(g)] - \Delta H_f°[N_2(g)]$$
$$= 2(-45.94 \text{ kJ}) - 3(0 \text{ kJ}) - (0 \text{ kJ}) = -91.88 \text{ kJ}$$

The negative value tells us that at 298 K the formation of ammonia from $H_2(g)$ and $N_2(g)$ is exothermic. The surroundings absorb the heat given off by the system, which means an increase in the entropy of the surroundings:

$$\Delta S_{surr}° = \frac{91.88 \text{ kJ}}{298 \text{ K}} = 0.308 \text{ kJ/K} = 308 \text{ J/K}$$

Notice that the magnitude of the entropy gained by the surroundings is greater than that lost by the system, calculated as -198.1 J/K in Sample Exercise 19.5.

The overall entropy change for the reaction is

$$\Delta S_{univ}° = \Delta S_{sys}° + \Delta S_{surr}° = -198.1 \text{ J/K} + 310 \text{ J/K} = 110 \text{ J/K}$$

Because $\Delta S_{univ}°$ is positive for any spontaneous reaction, this calculation indicates that when $NH_3(g)$, $H_2(g)$, and $N_2(g)$ are together at 298 K in their standard states (each at 1 atm pressure), the reaction moves spontaneously toward formation of $NH_3(g)$.

Keep in mind that while the thermodynamic calculations indicate that formation of ammonia is spontaneous, they do not tell us anything about the rate at which ammonia is formed. Establishing equilibrium in this system within a reasonable period requires a catalyst, as discussed in Section 15.7.

 Self-Assessment Exercises

SAE 19.11 Rank the following substances from lowest to highest standard molar entropy: $H_2O(l)$, $H_2O(g)$, $C_6H_6(l)$, $C_6H_6(g)$, $CO_2(g)$.
(a) $H_2O(l) < H_2O(g) < CO_2(g) < C_6H_6(l) < C_6H_6(g)$
(b) $H_2O(g) < CO_2(g) < C_6H_6(g) < H_2O(l) < C_6H_6(l)$
(c) $C_6H_6(l) < H_2O(l) < H_2O(g) < CO_2(g) < C_6H_6(g)$
(d) $H_2O(l) < C_6H_6(l) < H_2O(g) < CO_2(g) < C_6H_6(g)$

SAE 19.12 Given the standard molar entropies of $CH_4(g)$ (186.3 J/mol-K) and $CO_2(g)$ (213.6 J/mol-K) and the data in Table 19.1, calculate $\Delta S°$ for the following reaction: $CH_4(g) + 2 O_2(g) \longrightarrow CO_2(g) + 2 H_2O(g)$. **(a)** -243.3 J/K **(b)** -5.5 J/K **(c)** 10.9 J/K **(d)** 404.9 J/K **(e)** 1187.9 J/K

19.5 | Gibbs Free Energy

We have seen examples of endothermic processes that are spontaneous, such as the dissolution of ammonium nitrate in water. (Section 13.1) We learned in our discussion of the solution process that a spontaneous process that is endothermic must be accompanied by an increase in the entropy of the system. However, we have also encountered processes that are spontaneous and yet proceed with a *decrease* in the entropy of the system, such as the highly exothermic formation of sodium chloride from its constituent elements. (Section 8.2) Spontaneous processes that result in a decrease in the system's entropy are always exothermic. Thus, the spontaneity of a reaction seems to involve two thermodynamic concepts, enthalpy and entropy.

How can we use ΔH and ΔS to predict whether a given reaction occurring at constant temperature and pressure will be spontaneous? The means for doing so was first developed by the American mathematician J. Willard Gibbs (1839–1903). Gibbs proposed a new state function, now called the **Gibbs free energy** (or just **free energy**), G, and defined as

$$G = H - TS \qquad [19.10]$$

 Learning Objectives

When you finish Section 19.5, you should be able to:

▸ Describe Gibbs free energy, G, as a thermodynamic state function.

▸ Use standard free energies of formation to calculate standard free energy changes in chemical reactions.

Go Figure

Are the processes that move a system toward equilibrium spontaneous or nonspontaneous?

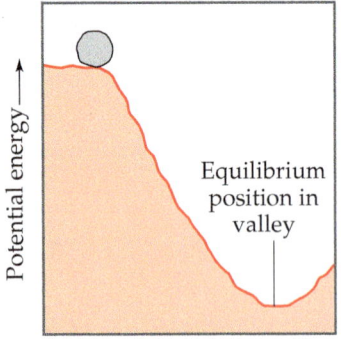

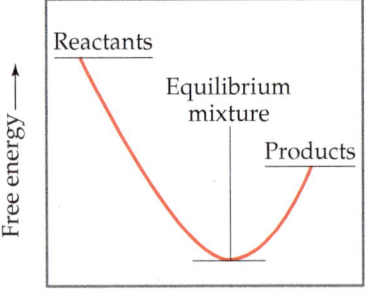

▲ **Figure 19.14 Potential energy and free energy.** We can make an analogy between the gravitational potential-energy change of a boulder rolling down a hill and the free-energy change in a spontaneous reaction. Free energy always decreases in a spontaneous process when pressure and temperature are held constant.

where T is the absolute temperature. For an isothermal process, the change in the free energy of the system, ΔG, is

$$\Delta G = \Delta H - T\Delta S \qquad [19.11]$$

Under standard conditions, this equation becomes

$$\Delta G° = \Delta H° - T\Delta S° \qquad [19.12]$$

To see how the state function G relates to reaction spontaneity, recall that for a reaction occurring at constant temperature and pressure

$$\Delta S_{univ} = \Delta S_{sys} + \Delta S_{surr} = \Delta S_{sys} + \left(\frac{-\Delta H_{sys}}{T}\right)$$

where Equation 19.9 substitutes for ΔS_{surr}. Multiplying both sides by $-T$ gives

$$-T\Delta S_{univ} = \Delta H_{sys} - T\Delta S_{sys} \qquad [19.13]$$

Comparing Equations 19.11 and 19.13, we see that in a process occurring at constant temperature and pressure, the free-energy change, ΔG, is equal to $-T\Delta S_{univ}$. We know that for spontaneous processes, ΔS_{univ} is always positive and, therefore, $-T\Delta S_{univ}$ is always negative. Thus, the sign of ΔG provides us with extremely valuable information about the spontaneity of processes that occur at constant temperature and pressure. If both T and P are constant, the relationship between the sign of ΔG and the spontaneity of a reaction is:

- If $\Delta G < 0$, the reaction is spontaneous in the forward direction.
- If $\Delta G = 0$, the reaction is at equilibrium.
- If $\Delta G > 0$, the reaction in the forward direction is nonspontaneous (work must be done to make it occur) but the reverse reaction is spontaneous.

It is more convenient to use ΔG as a criterion for spontaneity than to use ΔS_{univ} because ΔG relates to the system alone and avoids the complication of having to examine the surroundings.

An analogy is often drawn between the free-energy change during a spontaneous reaction and the potential-energy change when a boulder rolls down a hill (**Figure 19.14**). Potential energy in a gravitational field "drives" the boulder until it reaches a state of minimum potential energy in the valley. Similarly, the free energy of a chemical system decreases until it reaches a minimum value. When this minimum is reached, a state of equilibrium exists. *In any spontaneous process carried out at constant temperature and pressure, the free energy always decreases.*

To illustrate these ideas, let's return to the Haber process for the synthesis of ammonia from nitrogen and hydrogen, which we discussed extensively in Chapter 15 (Section 15.2):

$$N_2(g) + 3\,H_2(g) \rightleftharpoons 2\,NH_3(g)$$

Imagine that we have a reaction vessel that allows us to maintain a constant temperature and pressure and that we have a catalyst that allows the reaction to proceed at a reasonable rate. What happens when we load the vessel with a certain number of moles of N_2 and three times that number of moles of H_2? As we saw in Figure 15.3, the N_2 and H_2 react spontaneously to form NH_3 until equilibrium is achieved. Similarly, Figure 15.3 shows that if we load the vessel with pure NH_3, it decomposes spontaneously to N_2 and H_2 until equilibrium is reached. In each case, the free energy of the system gets progressively lower and lower as the reaction moves toward equilibrium, which represents a minimum in the free energy. We illustrate these cases in **Figure 19.15**.

This is a good time to remind ourselves of the significance of the reaction quotient, Q, for a system that is not at equilibrium. (Section 15.6) Recall that when $Q < K$, there is an excess of reactants relative to products, and the reaction proceeds spontaneously in the forward direction to reach equilibrium, as noted in Figure 19.15. When $Q > K$, the reaction proceeds spontaneously in the reverse direction. At equilibrium $Q = K$.

Go Figure Why are the spontaneous processes shown sometimes said to be "downhill" in free energy?

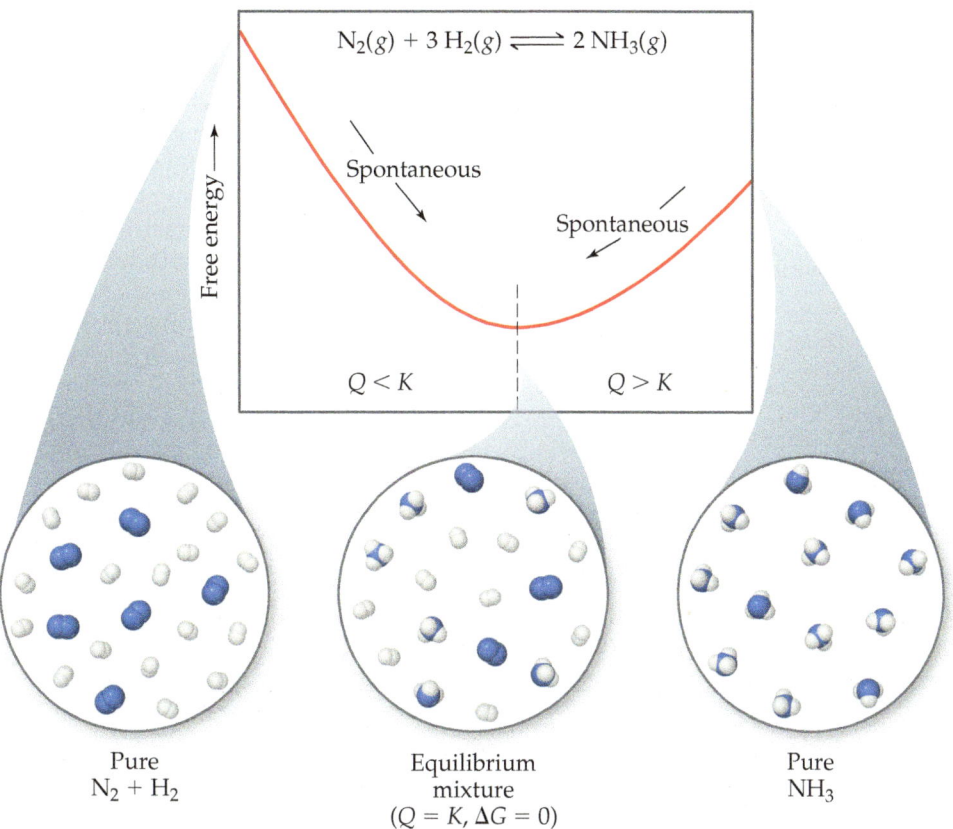

Pure
$N_2 + H_2$

Equilibrium
mixture
$(Q = K, \Delta G = 0)$

Pure
NH_3

▲ **Figure 19.15 Free energy and approaching equilibrium in the reaction** $N_2(g) + 3H_2(g) \rightleftharpoons 2NH_3(g)$. If the reaction mixture has too much N_2 and H_2 relative to NH_3 (left), $Q < K$ and NH_3 forms spontaneously. If there is more NH_3 in the mixture relative to the reactants N_2 and H_2 (right), $Q > K$ and the NH_3 decomposes spontaneously into N_2 and H_2.

Sample Exercise 19.6
Calculating Free-Energy Change from $\Delta H°$, T, and $\Delta S°$

Calculate the standard free-energy change for the formation of $NO(g)$ from $N_2(g)$ and $O_2(g)$ at 298 K:

$$N_2(g) + O_2(g) \longrightarrow 2NO(g)$$

given that $\Delta H° = 180.7$ kJ and $\Delta S° = 24.4$ J/K. Is the reaction spontaneous under these conditions?

SOLUTION

Analyze We are asked to calculate $\Delta G°$ for the indicated reaction (given $\Delta H°$, $\Delta S°$, and T) and to predict whether the reaction is spontaneous under standard conditions at 298 K.

Plan To calculate $\Delta G°$, we use Equation 19.12, $\Delta G° = \Delta H° - T\Delta S°$. To determine whether the reaction is spontaneous under standard conditions, we look at the sign of $\Delta G°$.

Solve

$$\Delta G° = \Delta H° - T\Delta S°$$

$$= 180.7 \text{ kJ} - (298 \text{ K})(24.4 \text{ J/K})\left(\frac{1 \text{ kJ}}{10^3 \text{ J}}\right)$$

$$= 180.7 \text{ kJ} - 7.3 \text{ kJ}$$

$$= 173.4 \text{ kJ}$$

Because $\Delta G°$ is positive, the reaction is not spontaneous under standard conditions at 298 K.

Comment Notice that we converted the units of the $T\Delta S°$ term to kJ so that they could be added to the $\Delta H°$ term, whose units are kJ.

▶ **Practice Exercise**
Calculate $\Delta G°$ for a reaction for which $\Delta H° = 24.6$ kJ and $\Delta S° = 132$ J/K at 298 K. Is the reaction spontaneous under these conditions?

TABLE 19.2 Conventions Used in Establishing Standard Free Energies

State of Matter	Standard State
Solid	Pure solid
Liquid	Pure liquid
Gas	1 atm pressure
Solution	1 M concentration
Element	$\Delta G_f^\circ = 0$ for most stable form of element in standard state

Standard Free Energy of Formation

Recall that we defined *standard enthalpies of formation*, ΔH_f°, as the enthalpy change when a substance is formed from its elements under defined standard conditions. (Section 5.7) We can define **standard free energies of formation**, ΔG_f°, in a similar way: ΔG_f° for a substance is the free-energy change for its formation from its elements under standard conditions. As is summarized in Table 19.2, standard state means 1 atm pressure for gases, the pure solid for solids, and the pure liquid for liquids. For substances in solution, the standard state is normally a concentration of 1 M. (In very accurate work it may be necessary to make certain corrections, but we need not worry about these.)

The temperature usually chosen for purposes of tabulating data is 25 °C, but we calculate ΔG at other temperatures as well. Just as for the standard heats of formation, the standard free energy of formation of the most stable form of an element under standard conditions is set to zero. This arbitrary choice of a reference point has no effect on the quantity in which we are interested, which is the *difference* in free energy between reactants and products. Standard free energies of formation are listed in Appendix C.

Standard free energies of formation are useful in calculating the *standard free-energy change* for chemical processes. The procedure is analogous to the calculation of ΔH° (Equation 5.31) and ΔS° (Equation 19.8):

$$\Delta G^\circ = \sum n\Delta G_f^\circ(\text{products}) - \sum m\Delta G_f^\circ(\text{reactants}) \qquad [19.14]$$

Sample Exercise 19.7
Calculating Standard Free-Energy Change from Free Energies of Formation

(a) Use data from Appendix C to calculate the standard free-energy change for the reaction $P_4(g) + 6\,Cl_2(g) \longrightarrow 4\,PCl_3(g)$ at 298 K. (b) What is ΔG° for the reverse of this reaction?

SOLUTION

Analyze We are asked to calculate the free-energy change for a reaction and then to determine the free-energy change for the reverse reaction.

Plan We look up the free-energy values for the products and reactants and use Equation 19.14. We multiply the molar quantities by the coefficients in the balanced equation and subtract the total for the reactants from that for the products.

Solve

(a) $Cl_2(g)$ is in its standard state, so ΔG_f° is zero for this reactant. $P_4(g)$, however, is not in its standard state, so ΔG_f° is not zero for this reactant. From the balanced equation and values from Appendix C, we have

$$\Delta G_{rxn}^\circ = 4\,\Delta G_f^\circ[PCl_3(g)] - \Delta G_f^\circ[P_4(g)] - 6\,\Delta G_f^\circ[Cl_2(g)]$$
$$= (4\,mol)(-269.6\,kJ/mol) - (1\,mol)(24.4\,kJ/mol) - 0$$
$$= -1102.8\,kJ$$

ΔG° is negative, so a mixture of $P_4(g)$, $Cl_2(g)$, and $PCl_3(g)$ at 25 °C, each present at a partial pressure of 1 atm, should react spontaneously in the forward direction to form more PCl_3. Remember, however, that the value of ΔG° tells us nothing about the rate at which the reaction occurs.

(b) When we reverse the reaction, we reverse the roles of the reactants and products. Thus, reversing the reaction changes the sign of ΔG in Equation 19.14, just as reversing the reaction changes the sign of ΔH. (Section 5.4) Hence, using the result from part (a), we have

$$4\,PCl_3(g) \longrightarrow P_4(g) + 6\,Cl_2(g) \quad \Delta G^\circ = +1102.8\,kJ$$

▶ **Practice Exercise**
Use data from Appendix C to ΔG° at 298 K for the combustion of methane:

$$CH_4(g) + 2\,O_2(g) \longrightarrow CO_2(g) + 2\,H_2O(g)$$

Sample Exercise 19.8
Predicting and Calculating ΔG°

In Section 5.7 we used Hess's law to calculate ΔH° for the combustion of propane gas at 298 K:

$$C_3H_8(g) + 5\,O_2(g) \longrightarrow 3\,CO_2(g) + 4\,H_2O(l) \quad \Delta H^\circ = -2220\,kJ$$

(a) *Without using data from Appendix C*, predict whether ΔG° for this reaction is more negative or less negative than ΔH°.
(b) Use data from Appendix C to calculate ΔG° for the reaction at 298 K. Is your prediction from part (a) correct?

SOLUTION

Analyze In part (a) we must predict the value for $\Delta G°$ relative to that for $\Delta H°$ on the basis of the balanced equation for the reaction. In part (b) we must calculate the value for $\Delta G°$ and compare this value with our qualitative prediction.

Plan The free-energy change incorporates both the change in enthalpy and the change in entropy for the reaction (Equation 19.11), so under standard conditions

$$\Delta G° = \Delta H° - T\Delta S°$$

To determine whether $\Delta G°$ is more negative or less negative than $\Delta H°$, we need to determine the sign of the term $T\Delta S°$. Because T is the absolute temperature, 298 K, it is always a positive number. We can predict the sign of $\Delta S°$ by looking at the reaction.

Solve

(a) The reactants are six molecules of gas, and the products are three molecules of gas and four molecules of liquid. Thus, the number of molecules of gas has decreased significantly during the reaction. By using the general rules discussed in Section 19.3, we expect a decrease in the number of gas molecules to lead to a decrease in the entropy of the system—the products have

fewer possible microstates than the reactants. We therefore expect $\Delta S°$ and $T\Delta S°$ to be negative. Because we are subtracting $T\Delta S°$, which is a negative number, we predict that $\Delta G°$ is *less negative* than $\Delta H°$.

(b) Using Equation 19.14 and values from Appendix C, we have

$$\Delta G° = 3\,\Delta G_f°[CO_2(g)] + 4\,\Delta G_f°[H_2O(l)]$$
$$\quad - \Delta G_f°[C_3H_8(g)] - 5\,\Delta G_f°[O_2(g)]$$
$$= 3\text{ mol}(-394.4\text{ kJ/mol}) + 4\text{ mol}(-237.13\text{ kJ/mol}) -$$
$$1\text{ mol}(-24.16\text{ kJ/mol}) - 5\text{ mol}(0\text{ kJ/mol}) = -2108\text{ kJ}$$

Notice that we have been careful to use the value of $\Delta G_f°$ for $H_2O(l)$. As in calculating ΔH values, the phases of the reactants and products are important. As we predicted, $\Delta G°$ is less negative than $\Delta H°$ because of the decrease in entropy during the reaction.

▶ **Practice Exercise**

For the combustion of butane at 298 K,
$2\,C_4H_{10}(g) + 13\,O_2(g) \longrightarrow 8\,CO_2(g) + 10\,H_2O(g)$, do you expect $\Delta G°$ to be more negative or less negative than $\Delta H°$?

A CLOSER LOOK | What's "Free" About Free Energy?

The Gibbs free energy is a remarkable thermodynamic quantity. Because so many chemical reactions are carried out under conditions of near-constant pressure and temperature, chemists, biochemists, and engineers consider the sign and magnitude of ΔG as exceptionally useful tools in the design of chemical and biochemical reactions. We present examples of the usefulness of ΔG throughout the remainder of this chapter and this text.

When first learning about ΔG, two common questions often arise: Why is the sign of ΔG an indicator of the spontaneity of reactions? And what is "free" about free energy?

In Section 19.2, we saw that the second law of thermodynamics governs the spontaneity of processes. In order to apply the second law (Equation 19.4), however, we must determine ΔS_{univ}, which is often difficult to evaluate. When T and P are constant, however, we can relate ΔS_{univ} to the changes in entropy and enthalpy of just the *system* by substituting the Equation 19.9 expression for ΔS_{surr} into Equation 19.4:

$$\Delta S_{univ} = \Delta S_{sys} + \Delta S_{surr} = \Delta S_{sys} + \left(\frac{-\Delta H_{sys}}{T}\right) \text{ (constant } T, P)$$

$$[19.15]$$

Thus, at constant temperature and pressure, the second law becomes

Reversible process: $\quad \Delta S_{univ} = \Delta S_{sys} - \dfrac{\Delta H_{sys}}{T} = 0$

Irreversible process: $\quad \Delta S_{univ} = \Delta S_{sys} - \dfrac{\Delta H_{sys}}{T} > 0 \quad [19.16]$

$$\text{(constant } T, P)$$

Now we can see the relationship between ΔG_{sys} (which we call simply ΔG) and the second law. From Equation 19.11 we know that $\Delta G = \Delta H_{sys} - T\Delta S_{sys}$. If we multiply Equations 19.16 by $-T$ and rearrange, we reach the following conclusion:

Reversible process: $\quad \Delta G = \Delta H_{sys} - T\Delta S_{sys} = 0$

Irreversible process: $\quad \Delta G = \Delta H_{sys} - T\Delta S_{sys} < 0 \quad [19.17]$

$$\text{(constant } T, P)$$

Equations 19.17 allow us to use the sign of ΔG to conclude whether a reaction is spontaneous, nonspontaneous, or at equilibrium. When

$\Delta G < 0$, a process is irreversible and, therefore, spontaneous. When $\Delta G = 0$, the process is reversible and, therefore, at equilibrium. If a process has $\Delta G > 0$, then the reverse process will have $\Delta G < 0$; thus, the process as written is nonspontaneous, but its reverse reaction will be irreversible and spontaneous.

The magnitude of ΔG is also significant. A reaction for which ΔG is large and negative, such as the burning of gasoline, is much more capable of doing work on the surroundings than is a reaction for which ΔG is small and negative, such as ice melting at room temperature. In fact, thermodynamics tells us that *the change in free energy for a process, ΔG, equals the maximum useful work that can be done by the system on its surroundings in a spontaneous process occurring at constant temperature and pressure*:

$$\Delta G = -w_{max} \quad [19.18]$$

(Remember our sign convention from Table 5.1: Work done *by* a system is negative, whereas work done *on* a system in positive.) In other words, ΔG gives the theoretical limit to how much work can be done by a process.

The relationship in Equation 19.18 explains why ΔG is called the *free* energy change—it is the portion of the energy change of a spontaneous reaction that is free to do useful work. The remainder of the energy enters the environment as heat. For example, the theoretical maximum work obtained for the combustion of gasoline is given by the value of ΔG for the combustion reaction. On average, standard internal combustion engines are inefficient in utilizing this potential work—more than 60% of the potential work is lost (primarily as heat) in converting the chemical energy of the gasoline to mechanical energy to move the vehicle. When other losses are considered—idling time, braking, aerodynamic drag, and so forth—only 12–30% of the potential work from the gasoline is used to move the car.

Advances in the design of internal combustion automobiles—such as hybrid technology, and new lightweight materials—have the potential to increase the percentage of useful work obtained from the gasoline. Part of the excitement about electric vehicles (EVs) is their greater utilization of the free energy available from the chemical energy of batteries (which we discuss in Chapter 20): When energy is recovered through regenerative braking, EVs are able to utilize 77–100% of the free energy in driving the wheels.

Self-Assessment Exercises

SAE 19.13 Which of the following statements about the Gibbs free energy, G, is *false*? **(a)** The Gibbs free energy for a process does not change as the temperature changes. **(b)** G is a state function. **(c)** When $\Delta G = 0$ for a reaction, the reaction is at equilibrium. **(d)** For a process that occurs at constant pressure and temperature, ΔG can be used to determine whether the process is spontaneous.

SAE 19.14 For a reaction $A(g) \longrightarrow B(g)$, $\Delta H = -300\,kJ$ and $\Delta S = -200\,J/K$ at $27\,°C$. Fill in the following blanks: ΔG for the reaction is _____ and the reaction will be _____. **(a)** $-360\,kJ$, nonspontaneous **(b)** $-240\,kJ$, nonspontaneous **(c)** $-360\,kJ$, spontaneous **(d)** $-240\,kJ$, spontaneous **(e)** More information is needed.

SAE 19.15 From data in Appendix C, calculate $\Delta G°$ for the following reaction: $4\,NO(g) \longrightarrow 2\,N_2O(g) + O_2(g)$. **(a)** $-198.3\,kJ$ **(b)** $-139.7\,kJ$ **(c)** $+16.9\,kJ$ **(d)** $+554.0\,kJ$ **(e)** More information is needed.

19.6 | Free Energy and Temperature

Learning Objective

When you finish Section 19.6, you should be able to:

▶ Predict the effect of changing temperature on the magnitude and sign of ΔG.

Tabulations of $\Delta G_f°$, such as those in Appendix C, make it possible to calculate $\Delta G°$ for reactions at the standard temperature of $25\,°C$, but we are often interested in examining reactions at other temperatures. To see how ΔG is affected by temperature, let's look again at Equation 19.11:

$$\Delta G = \Delta H - T\Delta S = \underbrace{\Delta H}_{\text{Enthalpy term}} + \underbrace{(-T\Delta S)}_{\text{Entropy term}}$$

Notice that we have written the expression for ΔG as a sum of two contributions, an enthalpy term, ΔH, and an entropy term, $-T\Delta S$. Because the value of $-T\Delta S$ depends directly on the absolute temperature T, ΔG varies with temperature. We know that the enthalpy term, ΔH, can be either positive or negative and that T is positive at all temperatures other than absolute zero. The entropy term, $-T\Delta S$, can also be positive or negative. When ΔS is positive, which means the final state has greater randomness (a greater number of microstates) than the initial state, the term $-T\Delta S$ is negative. When ΔS is negative, $-T\Delta S$ is positive.

The sign of ΔG, which tells us whether a process is spontaneous, depends on the signs and magnitudes of ΔH and $-T\Delta S$. The various combinations of ΔH and $-T\Delta S$ signs are given in Table 19.3.

Note in Table 19.3 that when ΔH and $-T\Delta S$ have opposite signs, the sign of ΔG depends on the magnitudes of these two terms. In these instances, temperature is an important consideration. Generally, we will assume that ΔH and ΔS change very little with temperature. However, the value of T directly affects the magnitude of $-T\Delta S$. As the temperature increases, the magnitude of $-T\Delta S$ increases, and this term becomes relatively more important in determining the sign and magnitude of ΔG.

As an example, let's consider once more the melting of ice to liquid water at 1 atm:

$$H_2O(s) \longrightarrow H_2O(l) \quad \Delta H > 0, \Delta S > 0$$

This process is endothermic, which means that ΔH is positive. Because the entropy increases during the process, ΔS is positive, which makes $-T\Delta S$ negative. At temperatures below $0\,°C$ ($273\,K$), the magnitude of ΔH is greater than that of $-T\Delta S$. Hence, the positive enthalpy term dominates, and ΔG is positive. This positive value of ΔG means that ice melting is not spontaneous at $T < 0\,°C$, just as our everyday experience tells us; rather, the reverse process, the freezing of liquid water into ice, is spontaneous at these temperatures.

TABLE 19.3   How Signs of ΔH and ΔS Affect Reaction Spontaneity

ΔH	ΔS	$-T\Delta S$	$\Delta G = \Delta H - T\Delta S$	Reaction Characteristics	Example
$-$	$+$	$-$	$-$	Spontaneous at all temperatures	$2\,O_3(g) \longrightarrow 3\,O_2(g)$
$+$	$-$	$+$	$+$	Nonspontaneous at all temperatures	$3\,O_2(g) \longrightarrow 2\,O_3(g)$
$-$	$-$	$+$	$+$ or $-$	Spontaneous at low T; nonspontaneous at high T	$H_2O(l) \longrightarrow H_2O(s)$
$+$	$+$	$-$	$+$ or $-$	Spontaneous at high T; nonspontaneous at low T	$H_2O(s) \longrightarrow H_2O(l)$

What happens at temperatures greater than 0 °C? As T increases, so does the magnitude of $-T\Delta S$. When $T > 0$ °C, the magnitude of $-T\Delta S$ is greater than the magnitude of ΔH, which means that the $-T\Delta S$ term dominates and ΔG is negative. The negative value of ΔG tells us that ice melting is spontaneous at $T > 0$ °C.

At the normal melting point of water, $T = 0$ °C, the two phases are in equilibrium. Recall that $\Delta G = 0$ at equilibrium; at $T = 0$ °C, ΔH and $-T\Delta S$ are equal in magnitude and opposite in sign, so they cancel and give $\Delta G = 0$.

Our discussion of the temperature dependence of ΔG is also relevant to standard free-energy changes. We can calculate the values of $\Delta H°$ and $\Delta S°$ at 298 K from the data in Appendix C. If we assume that these values do not change with temperature, we can then use Equation 19.12 to estimate ΔG at temperatures other than 298 K.

Sample Exercise 19.9
Determining the Effect of Temperature on Spontaneity

The Haber process for the production of ammonia involves the equilibrium

$$N_2(g) + 3\,H_2(g) \rightleftharpoons 2\,NH_3(g)$$

For this reaction, $\Delta H° = -91.88$ kJ and $\Delta S° = -198.1$ J/K. Assume that $\Delta H°$ and $\Delta S°$ for this reaction do not change with temperature. **(a)** Predict the direction in which ΔG for the reaction changes with increasing temperature. **(b)** Calculate ΔG at 25 °C and at 500 °C.

SOLUTION

Analyze In part (a) we are asked to predict the direction in which ΔG changes as temperature increases. In part (b) we need to determine ΔG for the reaction at two temperatures.

Plan We can answer part (a) by determining the sign of ΔS for the reaction and then using that information to analyze Equation 19.12. In part (b) we use the given $\Delta H°$ and $\Delta S°$ values for the reaction together with Equation 19.12 to calculate ΔG.

Solve

(a) The temperature dependence of ΔG comes from the entropy term in Equation 19.12, $\Delta G = \Delta H - T\Delta S$. We expect ΔS for this reaction to be negative because the number of molecules of gas is smaller in the products. Because ΔS is negative, $-T\Delta S$ is positive and increases with increasing temperature. As a result, ΔG becomes less negative (or more positive) with increasing temperature. Thus, the driving force for the production of NH_3 decreases with increasing temperature.

(b) If we assume that the values for $\Delta H°$ and $\Delta S°$ do not change with temperature, we can use Equation 19.12 to calculate ΔG at any temperature. At $T = 25$ °C $= 298$ K, we have

$$\Delta G° = -91.88 \text{ kJ} - (298 \text{ K})(-198.1 \text{ J/K})\left(\frac{1 \text{ kJ}}{1000 \text{ J}}\right)$$

$$= -91.88 \text{ kJ} + 59.0 \text{ kJ} = -32.8 \text{ kJ}$$

At $T = 500$ °C $= 773$ K, we have

$$\Delta G = -91.88 \text{ kJ} - (773 \text{ K})(-198.1 \text{ J/K})\left(\frac{1 \text{ kJ}}{1000 \text{ J}}\right)$$

$$= -91.88 \text{ kJ} + 153 \text{ kJ} = 61 \text{ kJ}$$

Notice that we had to convert the units of $-T\Delta S°$ to kJ in both calculations so that this term can be added to the $\Delta H°$ term, which has units of kJ.

Comment Increasing the temperature from 298 K to 773 K changes ΔG from -32.8 kJ to $+61$ kJ, where the result at 773 K assumes that $\Delta H°$ and $\Delta S°$ do not change with temperature. Although these values do change slightly with temperature, the result at 773 K should be a reasonable approximation.

The positive increase in ΔG with increasing T agrees with our prediction in part (a). Our result indicates that in a mixture of $N_2(g)$, $H_2(g)$, and $NH_3(g)$, each present at a partial pressure of 1 atm, the $N_2(g)$ and $H_2(g)$ react spontaneously at 298 K to form more $NH_3(g)$. At 773 K, the positive value of ΔG tells us that the reverse reaction is spontaneous. Thus, when the mixture of these gases, each at a partial pressure of 1 atm, is heated to 773 K, some of the $NH_3(g)$ spontaneously decomposes into $N_2(g)$ and $H_2(g)$.

▶ **Practice Exercise**
(a) Using standard enthalpies of formation and standard entropies in Appendix C, calculate $\Delta H°$ and $\Delta S°$ at 298 K for the reaction $2\,SO_2(g) + O_2(g) \longrightarrow 2\,SO_3(g)$. **(b)** Use your values from part (a) to estimate ΔG at 400 K.

Self-Assessment Exercises

SAE 19.16 Consider the reaction $A(g) + B(g) \longrightarrow C(g) + D(g)$ for which $\Delta H° = +85.0$ kJ and $\Delta S° = -66.0$ J/K. If we assume that $\Delta H°$ and $\Delta S°$ do not change with temperature, what can you conclude about this reaction? **(a)** It is spontaneous at all temperatures. **(b)** The reverse reaction is spontaneous at all temperatures. **(c)** It is spontaneous at low T, but nonspontaneous at high T. **(d)** It is nonspontaneous at low T but spontaneous at high T.

SAE 19.17 From data in Appendix C, calculate $\Delta G°$ at -25 °C for the following reaction: $2\,NO_2(g) \longrightarrow N_2O_4(g)$. **(a)** -102 kJ **(b)** -62.4 kJ **(c)** -40.0 kJ **(d)** -14.2 kJ **(e)** -5.3 kJ

19.7 | Free Energy and the Equilibrium Constant

In Section 19.5 we saw a special relationship between ΔG and equilibrium: For a system at equilibrium, $\Delta G = 0$. We have also seen how to use tabulated thermodynamic data to calculate values of the standard free-energy change, $\Delta G°$. In this final section, we learn two more ways in which we can use free energy to analyze chemical reactions—namely, using $\Delta G°$ to calculate ΔG under *nonstandard* conditions and relating the values of $\Delta G°$ and K for a reaction.

> **Learning Objectives**
>
> **When you finish Section 19.7, you should be able to:**
>
> ▶ Calculate the free energy change, ΔG, for a chemical reaction under nonstandard conditions.
>
> ▶ Describe the relationship between $\Delta G°$ and the equilibrium constant K and use that relationship to interconvert between the two quantities.

Free Energy under Nonstandard Conditions

The set of standard conditions for which $\Delta G°$ values pertain is given in Table 19.2. Most chemical reactions occur under nonstandard conditions. For any chemical process, the relationship between the free-energy change under standard conditions, $\Delta G°$, and the free-energy change under any other conditions, ΔG, is given by

$$\Delta G = \Delta G° + RT \ln Q \qquad [19.19]$$

In this equation R is the ideal-gas constant, 8.314 J/mol-K; T is the absolute temperature; and Q is the reaction quotient for the reaction mixture of interest. (Section 15.6) Recall that the reaction quotient Q is calculated like an equilibrium constant, except that you use the concentrations at any point of interest in the reaction; if $Q = K$, then the reaction is at equilibrium. Under standard conditions, the concentrations of all the reactants and products are equal to 1 M. Thus, under standard conditions $Q = 1$, $\ln Q = 0$, and Equation 19.19 reduces to $\Delta G = \Delta G°$ under standard conditions, as it should.

Sample Exercise 19.10

Relating ΔG to a Phase Change at Equilibrium

(a) Write the chemical equation that defines the normal boiling point of liquid carbon tetrachloride, $CCl_4(l)$. **(b)** What is the value of $\Delta G°$ for the equilibrium in part (a)? **(c)** Use data from Appendix C and Equation 19.12 to estimate the normal boiling point of CCl_4.

SOLUTION

Analyze (a) We must write a chemical equation that describes the physical equilibrium between liquid and gaseous CCl_4 at the normal boiling point. **(b)** We must determine the value of $\Delta G°$ for CCl_4, in equilibrium with its vapor at the normal boiling point. **(c)** We must estimate the normal boiling point of CCl_4, based on available thermodynamic data.

Plan (a) The chemical equation is the change of state from liquid to gas. For **(b)**, we need to analyze Equation 19.19 at equilibrium ($\Delta G = 0$), and for **(c)** we can use Equation 19.12 to calculate T when $\Delta G = 0$.

Solve

(a) The normal boiling point is the temperature at which a pure liquid is in equilibrium with its vapor at a pressure of 1 atm:

$$CCl_4(l) \rightleftharpoons CCl_4(g) \quad P = 1 \text{ atm}$$

(b) At equilibrium, $\Delta G = 0$. In any normal boiling-point equilibrium, both liquid and vapor are in their standard state of pure liquid and vapor at 1 atm (Table 19.2). Consequently, $Q = 1$, $\ln Q = 0$, and $\Delta G = \Delta G°$ for this process. We conclude that $\Delta G° = 0$ for the equilibrium representing the normal boiling point of any liquid. (We would also find that $\Delta G° = 0$ for the equilibria relevant to normal melting points and normal sublimation points.)

$$\Delta G° = 0$$

(c) Combining Equation 19.12 with the result from part (b), we see that the equality at the normal boiling point, T_b, of $CCl_4(l)$ (or any other pure liquid) is:

$$\Delta G° = \Delta H° - T_b \Delta S° = 0$$

Solving the equation for T_b, we obtain:

$$T_b = \Delta H° / \Delta S°$$

Strictly speaking, we need the values of $\Delta H°$ and $\Delta S°$ for the $CCl_4(l)/CCl_4(g)$ equilibrium at the normal boiling point to do this calculation. However, we can *estimate* the boiling point by using the values of $\Delta H°$ and $\Delta S°$ for the phases of CCl_4 at 298 K, which we obtain from Appendix C and Equations 5.31 and 19.8:

$$\Delta H° = (1\ \text{mol})(-106.7\ \text{kJ/mol}) - (1\ \text{mol})(-139.3\ \text{kJ/mol}) = +32.6\ \text{kJ}$$

$$\Delta S° = (1\ \text{mol})(309.4\ \text{J/mol-K}) - (1\ \text{mol})(214.4\ \text{J/mol-K}) = +95.0\ \text{J/K}$$

As expected, the process is endothermic ($\Delta H > 0$) and produces a gas, thus increasing the entropy ($\Delta S > 0$). We now use these values to estimate T_b for $CCl_4(l)$:

$$T_b = \frac{\Delta H°}{\Delta S°} = \left(\frac{32.6\ \text{kJ}}{95.0\ \text{J/K}}\right)\left(\frac{1000\ \text{J}}{1\ \text{kJ}}\right) = 343\ \text{K} = 70\ °\text{C}$$

Note that we have used the conversion factor between joules and kilojoules to make the units of $\Delta H°$ and $\Delta S°$ match.

Check The experimental normal boiling point of $CCl_4(l)$ is 76.5 °C. The small deviation of our estimate from the experimental value is due to the assumption that $\Delta H°$ and $\Delta S°$ do not change with temperature.

▶ **Practice Exercise**
Use data in Appendix C to estimate the normal boiling point, in K, for elemental bromine, $Br_2(l)$. (The experimental value is given in Figure 11.4.)

When the concentrations of reactants and products are nonstandard, we must calculate Q in order to determine ΔG. We illustrate how this is done in Sample Exercise 19.11. At this stage in our discussion, therefore, it becomes important to note the units used to calculate Q when using Equation 19.19. The convention used for standard states is used when applying this equation: In determining the value of Q, the concentrations of gases are always expressed as partial pressures in atmospheres and solutes are expressed as their concentrations in molarities.

Sample Exercise 19.11

Calculating the Free-Energy Change under Nonstandard Conditions

Calculate ΔG at 298 K for a mixture of 1.0 atm N_2, 3.0 atm H_2, and 0.50 atm NH_3 being used in the Haber process:

$$N_2(g) + 3\,H_2(g) \rightleftharpoons 2\,NH_3(g)$$

SOLUTION

Analyze We are asked to calculate ΔG under nonstandard conditions.

Plan We can use Equation 19.19 to calculate ΔG. Doing so requires that we calculate the value of the reaction quotient Q for the specified partial pressures, for which we use the partial-pressures form of Equation 15.24 (the equilibrium constant expression). We then use a table of standard free energies of formation to evaluate $\Delta G°$.

Solve

The partial-pressures form of Equation 15.24 gives:

$$Q = \frac{P_{NH_3}{}^2}{P_{N_2}P_{H_2}{}^3} = \frac{(0.50)^2}{(1.0)(3.0)^3} = 9.3 \times 10^{-3}$$

In Sample Exercise 19.9 we calculated $\Delta G° = -32.8$ kJ for this reaction. We will have to change the units of this quantity in applying Equation 19.19, however. For the units in Equation 19.19 to work out, we will use kJ/mol as our units for $\Delta G°$, where "per mole" means "per mole of the reaction as written." Thus, $\Delta G° = -32.8$ kJ/mol implies per 1 mol of N_2, per 3 mol of H_2, and per 2 mol of NH_3.

We now use Equation 19.19 to calculate ΔG for these nonstandard conditions:

$$\begin{aligned} \Delta G &= \Delta G° + RT \ln Q \\ &= (-32.8\ \text{kJ/mol}) \\ &\quad + (8.314\ \text{J/mol-K})(298\ \text{K})(1\ \text{kJ}/1000\ \text{J}) \ln(9.3 \times 10^{-3}) \\ &= (-32.8\ \text{kJ/mol}) + (-11.6\ \text{kJ/mol}) = -44.4\ \text{kJ/mol} \end{aligned}$$

Comment We see that ΔG becomes more negative as the pressures of N_2, H_2, and NH_3 are changed from 1.0 atm (standard conditions, $\Delta G°$) to 1.0 atm, 3.0 atm, and 0.50 atm, respectively. The larger negative value for ΔG indicates a larger "driving force" to produce NH_3.

We would make the same prediction based on Le Châtelier's principle. (Section 15.7) Relative to standard conditions, we have increased the pressure of a reactant (H_2) and decreased the pressure of the product (NH_3). Le Châtelier's principle predicts that both changes shift the reaction to the product side, thereby forming more NH_3.

▶ **Practice Exercise**
Calculate ΔG at 298 K for the Haber reaction if the reaction mixture consists of 0.50 atm N_2, 0.75 atm H_2, and 2.0 atm NH_3.

Relationship between $\Delta G°$ and K

We can now use Equation 19.19 to derive the relationship between $\Delta G°$ and the equilibrium constant, K. At equilibrium, $\Delta G = 0$ and $Q = K$. Thus, at equilibrium, Equation 19.19 transforms as follows:

$$\Delta G = \Delta G° + RT \ln Q$$

$$0 = \Delta G° + RT \ln K$$

$$\Delta G° = -RT \ln K \qquad [19.20]$$

Equation 19.20 is a very important one, with broad significance in chemistry. By relating K to $\Delta G°$, we can also relate K to entropy and enthalpy changes for a reaction.

We can also solve Equation 19.20 for K, to yield an expression that allows us to calculate K if we know the value of $\Delta G°$:

$$\ln K = \frac{\Delta G°}{-RT}$$

$$K = e^{-\Delta G°/RT} \qquad [19.21]$$

As usual, we must be careful in our choice of units. In Equations 19.20 and 19.21 we again express $\Delta G°$ in kJ/mol. In the equilibrium-constant expression, we use atmospheres for gas pressures, molarities for solutions, and solids, liquids, and solvents do not appear in the expression. (Section 15.4) Thus, the equilibrium constant is K_p for gas-phase reactions and K_c for reactions in solution. (Section 15.2)

From Equation 19.20 we see that if $\Delta G°$ is negative, $\ln K$ must be positive, which means $K > 1$. Therefore, the more negative $\Delta G°$ is, the larger K is. Conversely, if $\Delta G°$ is positive, $\ln K$ is negative, which means $K < 1$. Finally, if $\Delta G°$ is zero, $K = 1$.

Sample Exercise 19.12

Calculating an Equilibrium Constant from ΔG

The standard free-energy change for the Haber process at 25 °C was obtained in Sample Exercise 19.9 for the Haber reaction:

$$N_2(g) + 3 H_2(g) \rightleftharpoons 2 NH_3(g) \quad \Delta G° = -32.8 \text{ kJ/mol} = -32,800 \text{ J/mol}$$

Use this value of $\Delta G°$ to calculate the equilibrium constant for the process at 25 °C.

SOLUTION

Analyze We are asked to calculate K for a reaction, given $\Delta G°$.

Plan We can use Equation 19.21 to calculate K.

Solve Remembering to use the absolute temperature for T in Equation 19.21 and the form of R that matches our units, we have

$$K = e^{-\Delta G°/RT} = e^{-(-32,800 \text{ J/mol})/(8.314 \text{ J/mol-K})(298 \text{ K})} = e^{13.2} = 5 \times 10^5$$

Comment This is a large equilibrium constant, which indicates that the product, NH_3, is greatly favored in the equilibrium mixture at 25 °C. The equilibrium constants for the Haber reaction at temperatures in the range 300 °C to 600 °C, given in Table 15.2, are much smaller than the value at 25 °C. Thus, a low-temperature equilibrium

favors the production of ammonia more than a high-temperature one. Nevertheless, the Haber process is carried out at high temperatures because the reaction is extremely slow at room temperature.

Remember Thermodynamics can tell us the direction and extent of a reaction but tells us nothing about the rate at which it will occur. If a catalyst were found that would permit the reaction to proceed at a rapid rate at room temperature, high pressures would not be needed to force the equilibrium toward NH_3.

▶ **Practice Exercise**

Use data from Appendix C to calculate $\Delta G°$ and K at 298 K for the reaction $H_2(g) + Br_2(l) \rightleftharpoons 2 HBr(g)$.

CHEMISTRY AND LIFE | Driving Nonspontaneous Reactions: Coupling Reactions

Many desirable chemical reactions, including a large number that are central to living systems, are nonspontaneous as written. For example, consider the extraction of copper metal from the mineral *chalcocite*, which contains Cu_2S. The decomposition of Cu_2S to its elements is nonspontaneous:

$$Cu_2S(s) \longrightarrow 2 Cu(s) + S(s) \quad \Delta G° = +86.2 \text{ kJ}$$

Because $\Delta G°$ is very positive, we cannot obtain $Cu(s)$ directly via this reaction. Instead, we must find some way to "do work" on the reaction to force it to occur as we wish. We can do this by coupling the

reaction to another one so that the overall reaction *is* spontaneous. For example, we can envision the $S(s)$ reacting with $O_2(g)$ to form $SO_2(g)$:

$$S(s) + O_2(g) \longrightarrow SO_2(g) \quad \Delta G° = -300.4 \text{ kJ}$$

By coupling (adding together) these reactions, we can extract much of the copper metal via a spontaneous reaction:

$$Cu_2S(s) + O_2(g) \longrightarrow 2 Cu(s) + SO_2(g)$$

$$\Delta G° = (+86.2 \text{ kJ}) + (-300.4 \text{ kJ}) = -214.2 \text{ kJ}$$

In essence, we have used the spontaneous reaction of $S(s)$ with $O_2(g)$ to provide the free energy needed to extract the copper metal from the mineral.

Biological systems employ the same principle of using spontaneous reactions to drive nonspontaneous ones. Many of the biochemical reactions that are essential for the formation and maintenance of highly ordered biological structures are not spontaneous. These necessary reactions are made to occur by coupling them with spontaneous reactions that release energy. The metabolism of food is the usual source of the free energy needed to do the work of maintaining biological systems. For example, complete oxidation of the sugar glucose, $C_6H_{12}O_6$, to CO_2 and H_2O yields substantial free energy:

$$C_6H_{12}O_6(s) + 6\,O_2(g) \longrightarrow 6\,CO_2(g) + 6\,H_2O(l) \quad \Delta G° = -2880 \text{ kJ}$$

This energy can be used to drive nonspontaneous reactions in the body. However, a means is necessary to transport the energy released by glucose metabolism to the reactions that require energy. One way, shown in Figure 19.16, involves the interconversion of adenosine triphosphate (ATP) and adenosine diphosphate (ADP), molecules that are related to the building blocks of nucleic acids. The conversion of ATP to ADP releases free energy ($\Delta G° = -30.5$ kJ) that can be used to drive other reactions.

In the human body, the metabolism of glucose occurs via a complex series of reactions, most of which release free energy. The free energy released during these steps is used in part to reconvert lower-energy ADP back to higher-energy ATP. Thus, the ATP–ADP interconversions are used to store energy during metabolism and to release it as needed to drive nonspontaneous reactions in the body. If you take a course in general biology or biochemistry, you will have the opportunity to learn more about the remarkable sequence of reactions used to transport free energy throughout the human body.

Related Exercises: 19.99, 19.100

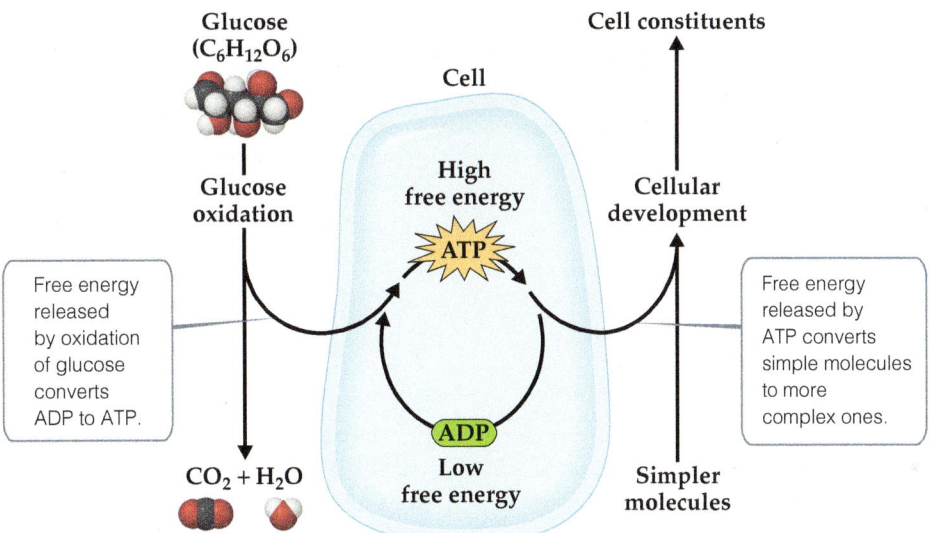

▲ **Figure 19.16 Schematic representation of free-energy changes during cell metabolism.** The oxidation of glucose to CO_2 and H_2O produces free energy that is then used to convert ADP into the more energetic ATP. The ATP is then used, as needed, as an energy source to drive nonspontaneous reactions, such as the conversion of simple molecules into more complex cell constituents.

 Self-Assessment Exercises

SAE 19.18 Consider the reaction $2\,NO_2(g) \longrightarrow N_2(g) + 2\,O_2(g)$. $\Delta G_f°$ for $NO_2(g)$ is 51.84 kJ/mol. What is ΔG at 298 K for this reaction when the partial pressures of NO_2, N_2, and O_2 are 0.100 atm, 1.00 atm, and 2.00 atm, respectively? **(a)** -104 kJ **(b)** -91.4 kJ **(c)** -39.5 kJ **(d)** 12.3 kJ **(e)** 116 kJ

SAE 19.19 The following thermodynamic data are known about the liquid and gas phases of $HSiCl_3$, which is used as a principal source of ultrapure silicon in the electronics industry:

	$\Delta H_f°$ (kJ/mol)	$S°$ (J/mol-K)
$HSiCl_3(l)$	-539.3	227.6
$HSiCl_3(g)$	-513.0	313.9

You may assume that $\Delta H_f°$ and $S°$ do not vary with temperature. What is the normal boiling point of $HSiCl_3$ in °C? **(a)** 31.8 °C **(b)** 86.3 °C **(c)** 113 °C **(d)** 199 °C **(e)** 305 °C

SAE 19.20 Consider the following equilibrium: $4\,Ag(s) + O_2(g) \rightleftharpoons 2\,Ag_2O(s)$. At 298 K, the equilibrium constant for this reaction is $K = 8.44 \times 10^3$. What is $\Delta G_f°$ for $Ag_2O(s)$? **(a)** -4.86 kJ **(b)** -8.44 kJ **(c)** -11.2 kJ **(d)** -22.4 kJ **(e)** More information is needed.

Putting Concepts Together

We will examine the equilibria in which the simple salts NaCl(s) and AgCl(s) dissolve in water to form aqueous solutions of ions:

$$NaCl(s) \rightleftharpoons Na^+(aq) + Cl^-(aq)$$
$$AgCl(s) \rightleftharpoons Ag^+(aq) + Cl^-(aq)$$

(a) Calculate the value of $\Delta G°$ at 298 K for each of the preceding reactions. (b) The two values from part (a) are very different. Is this difference primarily due to the enthalpy term or the entropy term of the standard free-energy change? (c) Use the values of $\Delta G°$ to calculate the K_{sp} values for the two salts at 298 K. (d) Sodium chloride is considered a soluble salt, whereas silver chloride is considered insoluble. Are these descriptions consistent with the answers to part (c)? (e) How will $\Delta G°$ for the solution process of these salts change with increasing T? What effect should this change have on the solubility of the salts?

SOLUTION

(a) We will use Equation 19.14 along with $\Delta G_f°$ values from Appendix C to calculate the $\Delta G_{soln}°$ values for each equilibrium. (As we did in Section 13.1, we use the subscript "soln" to indicate that these are thermodynamic quantities for the formation of a solution.) We find

$$\Delta G_{soln}°(NaCl) = (-261.9 \text{ kJ/mol}) + (-131.2 \text{ kJ/mol})$$

$$-(-384.1 \text{ kJ/mol})$$

$$= -9.0 \text{ kJ/mol}$$

$$\Delta G_{soln}°(AgCl) = (+77.11 \text{ kJ/mol}) + (-131.2 \text{ kJ/mol})$$

$$-(-109.70 \text{ kJ/mol})$$

$$= +55.6 \text{ kJ/mol}$$

(b) We can write $\Delta G_{soln}°$ as the sum of an enthalpy term, $\Delta H_{soln}°$, and an entropy term, $-T\Delta S_{soln}°$: $\Delta G_{soln}° = \Delta H_{soln}° + (-T\Delta S_{soln}°)$. We can calculate the values of $\Delta H_{soln}°$ and $\Delta S_{soln}°$ by using Equations 5.31 and 19.8. We can then calculate $-T\Delta S_{soln}°$ at $T = 298$ K. The results are summarized in the following table:

Salt	$\Delta H_{soln}°$	$\Delta S_{soln}°$	$T\Delta S_{soln}°$
NaCl	+3.8 kJ/mol	+43.4 J/mol-K	−12.9 kJ/mol
AgCl	+65.7 kJ/mol	+34.3 J/mol-K	−10.2 kJ/mol

The entropy terms for the solution of the two salts are very similar. That seems sensible because each solution process should lead to a similar increase in randomness as the salt dissolves, forming hydrated ions. (Section 13.1) In contrast, we see a very large difference in the enthalpy term for the solution of the two salts. The difference in the values of $\Delta G_{soln}°$ is dominated by the difference in the values of $\Delta H_{soln}°$.

(c) The solubility product, K_{sp}, is the equilibrium constant for the solution process. (Section 17.4) As such, we can relate K_{sp} directly to $\Delta G_{soln}°$ by using Equation 19.21:

$$K_{sp} = e^{-\Delta G_{soln}°/RT}$$

We can calculate the K_{sp} values in the same way we applied Equation 19.21 in Sample Exercise 19.12. We use the $\Delta G_{soln}°$ values we obtained in part (a), remembering to convert them from kJ/mol to J/mol:

$$NaCl: K_{sp} = [Na^+][Cl^-] = e^{-(-9100)/[(8.314)(298)]}$$

$$= e^{+3.6} = 38$$

$$AgCl: K_{sp} = [Ag^+][Cl^-] = e^{-(+55600)/[(8.314)(298)]}$$

$$= e^{-22.4}$$

$$= 1.9 \times 10^{-10}$$

The value calculated for the K_{sp} of AgCl is very close to that listed in Appendix D.

(d) A soluble salt is one that dissolves appreciably in water. (Section 4.2) The K_{sp} value for NaCl is greater than 1, indicating that NaCl dissolves to a great extent. The K_{sp} value for AgCl is very small, indicating that very little dissolves in water. Silver chloride should indeed be considered an insoluble salt.

(e) As we expect, the solution process has a positive value of ΔS for both salts (see the table in part b). As such, the entropy term of the free-energy change, $-T\Delta S_{soln}°$, is negative. If we assume that $\Delta H_{soln}°$ and $\Delta S_{soln}°$ do not change much with temperature, then an increase in T will serve to make $\Delta S_{soln}°$ more negative. Thus, the driving force for dissolution of the salts will increase with increasing T, and we therefore expect the solubility of the salts to increase with increasing T. In Figure 13.15, we see that the solubility of NaCl (and that of nearly any other salt) increase with increasing temperature. (Section 13.3)

Chapter Summary and Key Terms

SPONTANEOUS PROCESSES (SECTION 19.1) Most reactions and chemical processes have an inherent directionality: They are **spontaneous** in one direction and nonspontaneous in the reverse direction. The spontaneity of a process is related to the thermodynamic path the system takes from the initial state to the final state. In a **reversible process**, both the system and its surroundings can be restored to their original state by exactly reversing the change. In an **irreversible process**, the system cannot return to its original state without a permanent change in the surroundings. Any spontaneous process is irreversible. A process that occurs at a constant temperature is said to be **isothermal**.

ENTROPY AND THE SECOND LAW OF THERMODYNAMICS (SECTION 19.2) The spontaneous nature of processes is related to a thermodynamic state function called **entropy**, denoted S. For a process that occurs at constant temperature, the entropy change of the system is given by the heat absorbed by the system along a reversible path, divided by the temperature: $\Delta S = q_{rev}/T$. For any process, the entropy change of the universe equals the entropy change of the system plus the entropy change of the surroundings: $\Delta S_{univ} = \Delta S_{sys} + \Delta S_{surr}$. The way entropy controls the spontaneity of processes is given by the **second law of thermodynamics**, which states that in an irreversible (spontaneous) process $\Delta S_{univ} > 0$. Entropy values are usually expressed in units of joules per kelvin, J/K.

THE MOLECULAR INTERPRETATION OF ENTROPY AND THE THIRD LAW OF THERMODYNAMICS (SECTION 19.3) A particular combination of motions and locations of the atoms and molecules of a system at a particular instant is called a **microstate**. The entropy of a system is a measure of its randomness or disorder. The entropy is related to the number of microstates, W, corresponding to the state of the system: $S = k \ln W$. Molecules can undergo three kinds of motion: In **translational motion** the entire molecule moves in space. Molecules can also undergo **vibrational motion**, in which the atoms of the molecule move toward and away from one another in periodic fashion, and **rotational motion**, in which the entire molecule spins like a top. The number of available microstates, and therefore the entropy, increases with an increase in volume, temperature, or motion of molecules because any of these changes increases the possible motions and locations of the molecules. As a result, entropy generally increases when liquids or solutions are formed from solids, gases are formed from either solids or liquids, or the number of molecules of gas increases during a chemical reaction. The **third law of thermodynamics** states that the entropy of a pure crystalline solid at 0 K is zero.

ENTROPY CHANGES IN CHEMICAL REACTIONS (SECTION 19.4) The third law allows us to assign entropy values for substances at different temperatures. Under standard conditions, the entropy of a mole of a substance is called its **standard molar entropy**, denoted $S°$. From tabulated values of $S°$, we can calculate the entropy change

for any process under standard conditions. For an isothermal process, the entropy change in the surroundings is equal to $-\Delta H/T$.

GIBBS FREE ENERGY (SECTION 19.5) The **Gibbs free energy** (or just **free energy**), G, is a thermodynamic state function that combines the two state functions enthalpy and entropy: $G = H - TS$. For processes that occur at constant temperature, $\Delta G = \Delta H - T\Delta S$. For a process occurring at constant temperature and pressure, the sign of ΔG relates to the spontaneity of the process. When ΔG is negative, the process is spontaneous. When ΔG is positive, the process is nonspontaneous, but the reverse process is spontaneous. At equilibrium the process is reversible and ΔG is zero. The free energy is also a measure of the maximum useful work that can be performed by a system in a spontaneous process. The standard free-energy change, $\Delta G°$, for any process can be calculated from tabulations of **standard free energies of formation**, $\Delta G_f°$, which are defined in a fashion analogous to standard enthalpies of formation, $\Delta H_f°$. The value of $\Delta G_f°$ for a pure element in its standard state is defined to be zero.

FREE ENERGY, TEMPERATURE, AND THE EQUILIBRIUM CONSTANT (SECTIONS 19.6 AND 19.7) The values of ΔH and ΔS for a chemical process generally do not vary much with temperature. Therefore, the dependence of ΔG with temperature is governed mainly by the value of T in the expression $\Delta G = \Delta H - T\Delta S$. The entropy term $-T\Delta S$ has the greater effect on the temperature dependence of ΔG and, hence, on the spontaneity of the process. For example, a process for which $\Delta H > 0$ and $\Delta S > 0$, such as the melting of ice, can be nonspontaneous ($\Delta G > 0$) at low temperatures and spontaneous ($\Delta G < 0$) at higher temperatures. Under nonstandard conditions, ΔG is related to $\Delta G°$ and the value of the reaction quotient, Q: $\Delta G = \Delta G° + RT \ln Q$. At equilibrium ($\Delta G = 0$, $Q = K$), $\Delta G° = -RT \ln K$. Thus, the standard free-energy change is directly related to the equilibrium constant for the reaction. This relationship expresses the temperature dependence of equilibrium constants.

Key Equations

- $\Delta S = \dfrac{q_{rev}}{T}$ (constant T)   [19.2]   Relating entropy change to the heat absorbed or released in a reversible process

- *Reversible process:* $\Delta S_{univ} = \Delta S_{sys} + \Delta S_{surr} = 0$

  Irreversible process: $\Delta S_{univ} = \Delta S_{sys} + \Delta S_{surr} > 0$ [19.4]   The second law of thermodynamics

- $S = k \ln W$ [19.5]   Relating entropy to the number of microstates

- $\Delta S° = \sum n S°(\text{products}) - \sum m S°(\text{reactants})$ [19.8]   Calculating the standard entropy change from standard molar entropies

- $\Delta S_{surr} = \dfrac{-\Delta H_{sys}}{T}$ [19.9]   The entropy change of the surroundings for a process at constant temperature and pressure

- $\Delta G = \Delta H - T\Delta S$ [19.11]   Calculating the Gibbs free-energy change from enthalpy and entropy changes at constant temperature

- $\Delta G° = \sum n \Delta G_f°(\text{products}) - \sum m \Delta G_f°(\text{reactants})$ [19.14]   Calculating the standard free-energy change from standard free energies of formation

- *Reversible process:* $\Delta G = \Delta H_{sys} - T\Delta S_{sys} = 0$

- *Irreversible process:* $\Delta G = \Delta H_{sys} - T\Delta S_{sys} < 0$ [19.17]   Relating the free-energy change to the reversibility of a process at constant temperature and pressure

- $\Delta G = -w_{max}$ [19.18]   Relating the free-energy change to the maximum work a system can perform

- $\Delta G = \Delta G° + RT \ln Q$ [19.19]   Calculating the free-energy change under nonstandard conditions

- $\Delta G° = -RT \ln K$ [19.20]   Relating the standard free-energy change and the equilibrium constant

Exam Prep

EP 19.1 The process of iron being oxidized to make iron(III) oxide (rust) is spontaneous. Which of these statements about this process is *true*? (**a**) The reduction of iron(III) oxide to iron is also spontaneous. (**b**) Because the process is spontaneous, the oxidation of iron must be fast. (**c**) The oxidation of iron is endothermic. (**d**) Equilibrium is achieved in a closed system when the rate of iron oxidation is equal to the rate of iron(III) oxide reduction. (**e**) The energy of the universe is decreased when iron is oxidized to rust.

EP 19.2 A system goes from state A to state B by two different paths: Path 1 is a reversible path and path 2 is an irreversible path. Which of the following statements about this scenario is or are *true*?

(**i**) ΔE for path 1 must be the same as ΔE for path 2.

(**ii**) w for path 1 must be the same as w for path 2.

(**iii**) If the system is be returned to state A by reversing path 1, the surroundings are unchanged.

(**a**) only statement i (**b**) i and ii (**c**) i and iii (**d**) ii and iii (**e**) all three statements

EP 19.3 Which of the following statements about entropy is *false*? (**a**) ΔS can be positive, negative, or zero. (**b**) Entropy is a measure of the degree of randomness or disorder in a system. (**c**) At constant T, the magnitude of ΔS is related to the heat transferred when a system changes states along a reversible path. (**d**) The value of ΔS for a system depends on the path taken between two states of the system.

EP 19.4 Do all exothermic phase changes have a negative value for the entropy change of the system? (**a**) Yes, because the heat transferred from the system has a negative sign. (**b**) Yes, because the temperature decreases during the phase transition. (**c**) No, because the entropy change depends on the sign of the heat transferred to or from the system. (**d**) No, because the heat transferred to the system has a positive sign. (**e**) More than one of the previous answers is correct.

EP 19.5 Which of the following statements about the second law of thermodynamics is or are *true*?

(**i**) The entropy change of the universe equals the sum of the entropy change of the system and the entropy change of the surroundings.

(**ii**) For a reversible process, the entropy change of the universe is zero.

(**iii**) For an irreversible process, the entropy change of the universe is greater than zero.

(**a**) Only one statement is true. (**b**) Statements i and ii are true. (**c**) Statements i and iii are true. (**d**) Statements ii and iii are true. (**e**) All three statements are true.

EP 19.6 The possible arrangements of two atoms between two flasks is shown here. Consider two possible states of this system: state 1, in which both atoms are in the left-hand flask, and state 2, in which there is one atom in each of the two flasks. Based on this figure, complete the following sentence: "State 1 has ___ microstates, state 2 has ___ microstates, and therefore the entropy of state 1 is ___ the entropy of state 2."

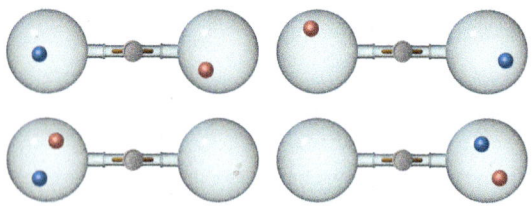

(**a**) one, one, equal to (**b**) one, two, greater than (**c**) two, one, less than (**d**) two, one, greater than (**e**) one, two, less than

EP 19.7 Which of the following processes produces a *decrease* in the entropy of the system?

(**i**) $CaO(s) + CO_2(g) \longrightarrow CaCO_3(s)$

(**ii**) $CO_2(s) \longrightarrow CO_2(g)$

(**iii**) $2 SO_2(g) + O_2(g) \longrightarrow 2 SO_3(g)$

(**a**) i only (**b**) i and ii (**c**) i and iii (**d**) ii and iii (**e**) i, ii, and iii

EP 19.8 Which system has the greatest entropy? (**a**) 1 mol of $H_2(g)$ at STP (**b**) 1 mol of $H_2(g)$ at 100 °C and 0.5 atm (**c**) 1 mol of $H_2O(s)$ at 0 °C (**d**) 1 mol of $H_2O(l)$ at 25 °C

EP 19.9 Which of the following statements about the third law of thermodynamics is *false*? (**a**) The third law describes the entropy of a pure crystalline substance at $T = 0$ K. (**b**) At $T = 0$ K, a pure crystalline solid has no translational, vibrational, or rotational motion. (**c**) At $T = 0$ K, a pure crystalline solid has zero possible microstates. (**d**) A pure crystalline solid at $T = 0$ K has $S = 0$.

EP 19.10 Which of the following statements about the standard molar entropies is *false*?

(**a**) The standard molar entropy of the gaseous form of a substance is greater than that of its liquid form.

(**b**) The standard molar entropy of a pure element in its most stable form is not equal to zero.

(**c**) Standard molar entropies generally increase with increasing molar mass.

(**d**) Standard molar entropies can have positive and negative values.

(**e**) The standard molar entropies of molecules generally increase with increasing molecular complexity.

EP 19.11 Using the standard molar entropies in Table 19.1, calculate the standard entropy change, $\Delta S°$, for the "water-splitting" reaction at 298 K:

$$2 H_2O(l) \longrightarrow 2 H_2(g) + O_2(g)$$

(**a**) 326.8 J/K (**b**) 266.0 J/K (**c**) 163.8 J/K (**d**) 88.6 J/K (**e**) −326.8 J/K

EP 19.12 Which of these statements about spontaneous reactions is *true*? (**a**) All spontaneous reactions have a negative entropy change. (**b**) All spontaneous reactions have a positive entropy change. (**c**) All spontaneous reactions have a negative free-energy change. (**d**) All spontaneous reactions have a positive free-energy change. (**e**) All spontaneous reactions have a negative enthalpy change.

EP 19.13 From data in Appendix C, calculate $\Delta G°$ for the following reaction: $N_2H_4(g) + H_2(g) \longrightarrow 2 NH_3(g)$.

(**a**) −212.4 kJ (**b**) −192.7 kJ (**c**) −175.8 kJ (**d**) +126.6 kJ (**e**) More information is needed.

EP 19.14 If a reaction is exothermic and its entropy change is positive, which statement is *true*? (**a**) The reaction is spontaneous at all temperatures. (**b**) The reaction is nonspontaneous at all temperatures. (**c**) The reaction is spontaneous only at higher temperatures. (**d**) The reaction is spontaneous only at lower temperatures.

EP 19.15 The Haber process for the production of ammonia involves the equilibrium

$$N_2(g) + 3 H_2(g) \rightleftharpoons 2 NH_3(g)$$

For this reaction, $\Delta H° = -91.88$ kJ and $\Delta S° = -198.1$ J/K. Assume that $\Delta H°$ and $\Delta S°$ for this reaction do not change with temperature. What is the temperature above which the Haber ammonia process becomes nonspontaneous?

(**a**) 25 °C (**b**) 47 °C (**c**) 61 °C (**d**) 191 °C (**e**) 500 °C

EP 19.16 If the normal boiling point of a liquid is 67 °C, and the standard molar entropy change for the boiling process is +100 J/K, estimate the standard molar enthalpy change for the boiling process.

(**a**) + 6700 J (**b**) −6700 J (**c**) + 34,000 J (**d**) −34,000 J

EP 19.17 The standard free energy change at 298 K for the aqueous reaction $A(aq) + B(aq) \longrightarrow C(aq)$ is $\Delta G° = -3.0$ kJ. What is ΔG at 298 K for this reaction when $[A] = 0.50\,M$, $[B] = 0.50\,M$, and $[C] = 2.0\,M$?

(a) -8.2 kJ (b) -3.0 kJ (c) -0.76 kJ (d) $+0.43$ kJ (e) $+2.2$ kJ

EP 19.18 The K_{sp} for a very insoluble salt is 4.2×10^{-47} at 298 K. What is $\Delta G°$ for the dissolution of the salt in water?

(a) -8.2 kJ/mol (b) -115 kJ/mol (c) -2.61 kJ/mol (d) $+115$ kJ/mol (e) $+265$ kJ/mol

EP 19.19 Consider the following equilibrium: $H_2(g) + I_2(s) \rightleftharpoons 2\,HI(g)$, for which $\Delta H°_{298} = +53.0$ kJ and $\Delta S°_{298} = +165.8$ J/K. If we assume that $\Delta H°$ and $\Delta S°$ do not change with temperature, what is the equilibrium constant for the reaction at 100 °C?

(a) $K = 9.7 \times 10^{-20}$ (b) $K = 0.059$ (c) $K = 0.31$ (d) $K = 17$

Exercises

Visualizing Concepts

19.1 Two different gases occupy the two bulbs shown here. Consider the process that occurs when the stopcock is opened, assuming the gases behave ideally. (a) Draw the final (equilibrium) state. (b) Predict the signs of ΔH and ΔS for the process. (c) Is the process that occurs when the stopcock is opened a reversible one? (d) How does the process affect the entropy of the surroundings? [Sections 19.1 and 19.2]

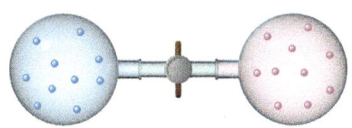

19.2 As shown here, one type of computer keyboard cleaner contains liquefied 1,1-difluoroethane ($C_2H_4F_2$), which is a gas at atmospheric pressure. When the nozzle is squeezed, the 1,1-difluoroethane vaporizes out of the nozzle at high pressure, blowing dust out of objects. (a) Based on your experience, is the vaporization a spontaneous process at room temperature? (b) Defining the 1,1-difluoroethane as the system, do you expect q_{sys} for the process to be positive or negative? (c) Predict whether ΔS is positive or negative for this process. (d) Given your answers to (a), (b), and (c), do you think the operation of this product depends more on enthalpy or entropy? [Sections 19.1 and 19.2]

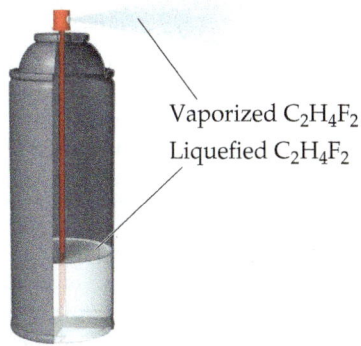

Vaporized $C_2H_4F_2$

Liquefied $C_2H_4F_2$

19.3 (a) What are the signs of ΔS and ΔH for the process depicted here? (b) If energy can flow in and out of the system to maintain a constant temperature during the process, what can you say about the entropy change of the surroundings as a result of this process? [Sections 19.2 and 19.5]

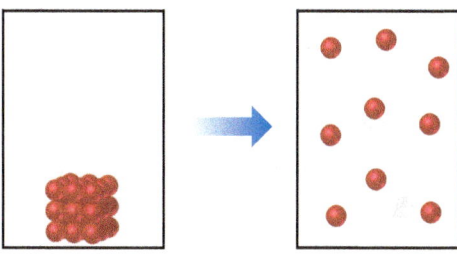

19.4 Predict the signs of ΔH and ΔS for this reaction. Explain your choice. [Section 19.3]

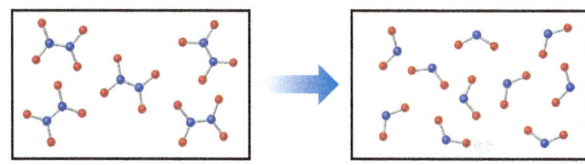

19.5 The accompanying diagram shows how entropy varies with temperature for a substance that is a gas at the highest temperature shown. (a) What processes correspond to the entropy increases along the vertical lines labeled 1 and 2? (b) Why is the entropy change for 2 larger than that for 1? (c) If this substance is a perfect crystal at $T = 0$ K, what is the value of S at this temperature? [Section 19.3]

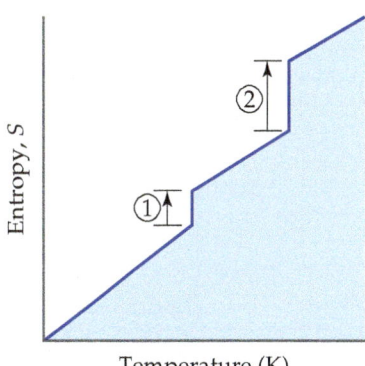

Temperature (K)

19.6 *Isomers* are molecules that have the same chemical formula but different arrangements of atoms, as shown here for two isomers of pentane, C_5H_{12}. (a) Do you expect a significant difference in the enthalpy of combustion of the two isomers? Explain. (**b**) Which isomer do you expect to have the higher standard molar entropy? Explain. [Section 19.4]

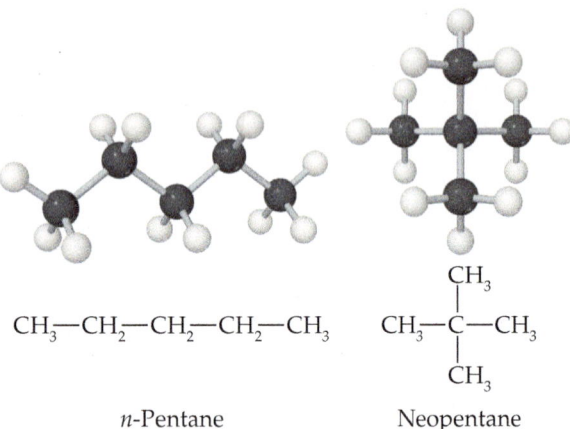

CH₃—CH₂—CH₂—CH₂—CH₃

n-Pentane

$$CH_3-\underset{\underset{CH_3}{|}}{\overset{\overset{CH_3}{|}}{C}}-CH_3$$

Neopentane

19.7 The accompanying diagram shows how ΔH (red line) and $T\Delta S$ (blue line) change with temperature for a hypothetical reaction. Which of the following statements is or are *true*? (**a**) ΔS for the reaction is less than zero. (**b**) The system is at equilibrium at $T = 300$ K. (**c**) The reaction is spontaneous at $T > 300$ K. [Section 19.6]

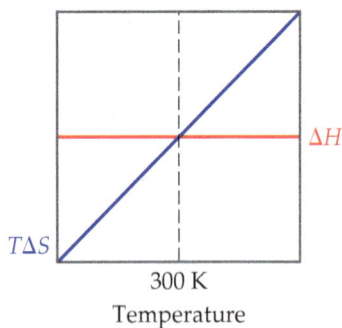

300 K

Temperature

19.8 The accompanying diagram shows how ΔG for a hypothetical reaction changes as temperature changes. (**a**) At what temperature is the system at equilibrium? (**b**) In what temperature range is the reaction spontaneous? (**c**) Is ΔH positive or negative? (**d**) Is ΔS positive or negative? [Sections 19.5 and 19.6]

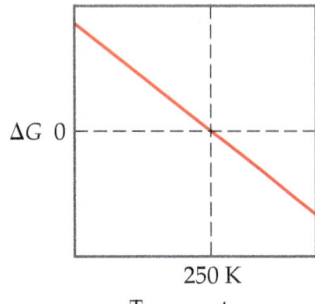

250 K

Temperature

19.9 Consider a reaction $A_2(g) + B_2(g) \rightleftharpoons 2\,AB(g)$, with atoms of A shown in red in the diagram and atoms of B shown in blue. (**a**) If $K_c = 1$, which box represents the system at equilibrium? (**b**) If $K_c = 1$, which box represents the system at

$Q < K_c$? (**c**) Rank the boxes in order of increasing magnitude of ΔG for the reaction. [Sections 19.5 and 19.7]

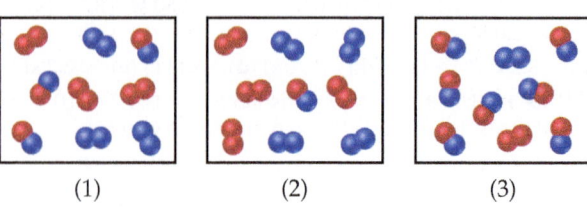

(1)   (2)   (3)

19.10 The accompanying diagram shows how the free energy, G, changes during a hypothetical reaction $A(g) + B(g) \longrightarrow C(g)$. On the left are pure reactants A and B, each at 1 atm, and on the right is the pure product, C, also at 1 atm. Indicate whether each of the following statements is true or false. (**a**) The minimum of the graph corresponds to the equilibrium mixture of reactants and products for this reaction. (**b**) At equilibrium, all of A and B have reacted to give pure C. (**c**) The entropy change for this reaction is positive. (**d**) The "*x*" on the graph corresponds to ΔG for the reaction. (**e**) ΔG for the reaction corresponds to the difference between the top left of the curve and the bottom of the curve. [Section 19.7]

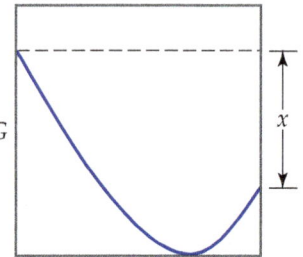

Progress of reaction

Spontaneous Processes (Section 19.1)

19.11 Which of the following processes are spontaneous and which are nonspontaneous? (**a**) the ripening of a banana (**b**) dissolution of sugar in a cup of hot coffee (**c**) the reaction of nitrogen atoms to form N_2 molecules at 25 °C and 1 atm (**d**) lightning (**e**) formation of CH_4 and O_2 molecules from CO_2 and H_2O at room temperature and 1 atm of pressure

19.12 Which of the following processes are spontaneous? (**a**) the melting of ice cubes at -10 °C and 1 atm pressure (**b**) separating a mixture of N_2 and O_2 into two separate samples, one that is pure N_2 and one that is pure O_2 (**c**) alignment of iron filings in a magnetic field (**d**) the reaction of hydrogen gas with oxygen gas to form water vapor at room temperature (**e**) the dissolution of $HCl(g)$ in water to form concentrated hydrochloric acid

19.13 Indicate whether each statement is true or false. (**a**) A reaction that is spontaneous in one direction will be nonspontaneous in the reverse direction under the same reaction conditions. (**b**) All spontaneous processes are fast. (**c**) Most spontaneous processes are reversible. (**d**) An isothermal process is one in which the system loses no heat. (**e**) The maximum amount of work can be accomplished by an irreversible process rather than a reversible one.

19.14 (**a**) Can endothermic chemical reactions be spontaneous? (**b**) Can a process be spontaneous at one temperature and nonspontaneous at a different temperature? (**c**) Water can be decomposed to form hydrogen and oxygen, and the hydrogen and oxygen can be recombined to form water. Does this mean that the processes are thermodynamically reversible? (**d**) Does the amount of work that a system can do on its surroundings depend on the path of the process?

19.15 Consider the vaporization of liquid water to steam at a pressure of 1 atm. **(a)** Is this process endothermic or exothermic? **(b)** In what temperature range is it a spontaneous process? **(c)** In what temperature range is it a nonspontaneous process? **(d)** At what temperature are the two phases in equilibrium?

19.16 The normal freezing point of *n*-octane (C_8H_{18}) is −57 °C. **(a)** Is the freezing of *n*-octane an endothermic or exothermic process? **(b)** In what temperature range is the freezing of *n*-octane a spontaneous process? **(c)** In what temperature range is it a nonspontaneous process? **(d)** Is there any temperature at which liquid *n*-octane and solid *n*-octane are in equilibrium? Explain.

19.17 A system goes from state 1 to state 2 and back to state 1 following a *reversible* path in both directions. Which of the following statements about this process is or are *true*? **(a)** The value of ΔE on going from state 1 to state 2 is equal in magnitude and opposite in sign to the value of ΔE on going from state 2 back to state 1. **(b)** The value of *q* on going from state 1 to state 2 is equal in magnitude and opposite in sign to the value of *q* on going from state 2 back to state 1. **(c)** The value of *w* on going from state 1 to state 2 is equal in magnitude and opposite in sign to the value of *w* on going from state 2 back to state 1.

19.18 A system goes from state X to state Y and back to state X following an *irreversible* path in both directions. Which of the following statements about this process is or are *true*? **(a)** The value of ΔE on going from state X to state Y is equal in magnitude and opposite in sign to the value of ΔE on going from state Y back to state X. **(b)** The value of *q* on going from state X to state Y is equal in magnitude and opposite in sign to the value of *q* on going from state Y back to state X. **(c)** The value of *w* on going from state X to state Y is equal in magnitude and opposite in sign to the value of *w* on going from state Y back to state X.

19.19 Consider a system consisting of an ice cube melting at 20 °C under constant pressure. **(a)** Is this process spontaneous or nonspontaneous? **(b)** Is this process reversible or irreversible? **(c)** What is the sign of ΔH for this process? **(d)** What is the sign of ΔS for this process?

19.20 Consider what happens when a sample of the explosive TNT is detonated under atmospheric pressure. **(a)** Is the detonation a reversible process? **(b)** What is the sign of *q* for this process? **(c)** Is *w* positive, negative, or zero for the process?

Entropy and the Second Law of Thermodynamics (Section 19.2)

19.21 Indicate whether each statement is true or false. **(a)** *S* is a state function. **(b)** If a system undergoes a reversible change, the entropy of the universe increases. **(c)** If a system undergoes a reversible process, the change in entropy of the system is exactly matched by an equal and opposite change in the entropy of the surroundings. **(d)** If a system undergoes a reversible process, the entropy change of the system must be zero.

19.22 Indicate whether each statement is true or false. **(a)** The entropy of the universe increases for any spontaneous process. **(b)** The entropy change of the system is equal and opposite that of the surroundings for any irreversible process. **(c)** The entropy of the system must increase in any spontaneous process. **(d)** The entropy change for an isothermal process depends on both the absolute temperature and the amount of heat reversibly transferred.

19.23 The normal boiling point of $Br_2(l)$ is 58.8 °C, and its molar enthalpy of vaporization is $\Delta H_{vap} = 29.6$ kJ/mol. **(a)** When $Br_2(l)$ boils at its normal boiling point, does its entropy increase or decrease? **(b)** Calculate the value of ΔS when 1.00 mol of $Br_2(l)$ is vaporized at 58.8 °C.

19.24 The element gallium (Ga) freezes at 29.8 °C, and its molar enthalpy of fusion is $\Delta H_{fus} = 5.59$ kJ/mol. **(a)** When molten gallium solidifies to Ga(*s*) at its normal melting point, is ΔS positive or negative? **(b)** Calculate the value of ΔS when 60.0 g of Ga(*l*) solidifies at 29.8 °C.

19.25 Indicate whether each statement is true or false. **(a)** The second law of thermodynamics says that entropy is conserved. **(b)** If the entropy of the system increases during a reversible process, the entropy change of the surroundings must decrease by the same amount. **(c)** In a certain spontaneous process, the system undergoes an entropy change of 4.2 J/K; therefore, the entropy change of the surroundings must be −4.2 J/K.

19.26 **(a)** Does the entropy of the surroundings increase for spontaneous processes? **(b)** In a particular spontaneous process, the entropy of the system decreases. What can you conclude about the sign and magnitude of ΔS_{surr}? **(c)** During a certain reversible process, the surroundings undergo an entropy change, $\Delta S_{surr} = -78$ J/K. What is the entropy change of the system for this process?

19.27 **(a)** What sign for ΔS do you expect when the volume of 0.200 mol of an ideal gas at 27 °C is increased isothermally from an initial volume of 10.0 L? **(b)** If the final volume is 18.5 L, calculate the entropy change for the process. **(c)** Which of the following statements about this process are *true*? (i) The entropy change you calculated will be the same for at any other constant temperature. (ii) The value of ΔS you calculated is valid only if the expansion is done reversibly. (iii) If the number of moles of gas expanding doubled, the entropy change would double.

19.28 **(a)** What sign for ΔS do you expect when the pressure on 0.600 mol of an ideal gas at 350 K is increased isothermally from an initial pressure of 0.750 atm? **(b)** If the final pressure on the gas is 1.20 atm, calculate the entropy change for the process. **(c)** Which of the following statements about this process are *true*? (i) The entropy change you calculated will be the same for at any other constant temperature. (ii) The value of ΔS you calculated is valid only if the compression is done irreversibly. (iii) If the number of moles of gas being compressed were decreased by a factor of three, the entropy change would increase by a factor of three.

The Molecular Interpretation of Entropy and the Third Law of Thermodynamics (Section 19.3)

19.29 For the isothermal expansion of a gas into a vacuum, $\Delta E = 0$, $q = 0$, and $w = 0$. **(a)** Is this a spontaneous process? **(b)** Explain why no work is done by the system during this process. **(c)** Is the "driving force" for the expansion of the gas due to enthalpy or entropy?

19.30 **(a)** What is the difference between a *state* and a *microstate* of a system? **(b)** As a system goes from state A to state B, its entropy decreases. What can you say about the number of microstates corresponding to each state? **(c)** In a particular spontaneous process, the number of microstates available to the system decreases. What can you conclude about the sign of ΔS_{surr}?

19.31 Would each of the following changes increase, decrease, or have no effect on the number of microstates available to a system? **(a)** increase in temperature **(b)** decrease in volume **(c)** change of state from liquid to gas

19.32 **(a)** Using the heat of vaporization in Appendix B, calculate the entropy change for the vaporization of water at 25 °C and at 100 °C. **(b)** From your knowledge of microstates and the structure of liquid water, explain the difference in these two values.

19.33 (a) What do you expect for the sign of ΔS in a chemical reaction in which 2 mol of gaseous reactants are converted to 3 mol of gaseous products? (b) For which of the processes in Exercise 19.11 does the entropy of the system increase?

19.34 (a) In a chemical reaction, two gases combine to form a solid. What do you expect for the sign of ΔS? (b) How does the entropy of the system change in the processes described in Exercise 19.12?

19.35 Does the entropy of the system increase, decrease, or stay the same when (a) a solid melts, (b) a gas liquefies, (c) a solid sublimes?

19.36 Does the entropy of the system increase, decrease, or stay the same when (a) the temperature of the system increases, (b) the volume of a gas increases, (c) equal volumes of ethanol and water are mixed to form a solution?

19.37 Indicate whether each statement is true or false. (a) The third law of thermodynamics says that the entropy of a perfect, pure crystal at absolute zero increases with the mass of the crystal. (b) "Translational motion" of molecules refers to their change in spatial location as a function of time. (c) "Rotational" and "vibrational" motions contribute to the entropy in atomic gases like He and Xe. (d) The larger the number of atoms in a molecule, the more degrees of freedom of rotational and vibrational motion it likely has.

19.38 Indicate whether each statement is true or false. (a) Unlike enthalpy, where we can only ever know changes in H, we can know absolute values of S. (b) If you heat a gas such as CO_2, you will increase its degrees of translational, rotational, and vibrational motions. (c) $CO_2(g)$ and $Ar(g)$ have nearly the same molar mass. At a given temperature, they will have nearly the same number of microstates.

19.39 For each of the following pairs, predict which substance has the higher entropy per mole at a given temperature. (a) $Ar(l)$ or $Ar(g)$ (b) $He(g)$ at 3 atm pressure or $He(g)$ at 1.5 atm pressure (c) 1 mol of $Ne(g)$ in 15.0 L or 1 mol of $Ne(g)$ in 1.50 L (d) $CO_2(g)$ or $CO_2(s)$

19.40 For each of the following pairs, predict which substance possesses the larger entropy per mole. (a) 1 mol of $O_2(g)$ at 300 °C, 0.01 atm, or 1 mol of $O_3(g)$ at 300 °C, 0.01 atm (b) 1 mol of $H_2O(g)$ at 100 °C, 1 atm, or 1 mol of $H_2O(l)$ at 100 °C, 1 atm (c) 0.5 mol of $N_2(g)$ at 298 K, 20-L volume, or 0.5 mol $CH_4(g)$ at 298 K, 20-L volume (d) 100 g $Na_2SO_4(s)$ at 30 °C or 100 g $Na_2SO_4(aq)$ at 30 °C

19.41 Predict the sign of the entropy change of the system for each of the following reactions:

(a) $N_2(g) + 3 H_2(g) \longrightarrow 2 NH_3(g)$

(b) $CaCO_3(s) \longrightarrow CaO(s) + CO_2(g)$

(c) $3 C_2H_2(g) \longrightarrow C_6H_6(g)$

(d) $Al_2O_3(s) + 3 H_2(g) \longrightarrow 2 Al(s) + 3 H_2O(g)$

19.42 Predict the sign of ΔS_{sys} for each of the following processes: (a) Molten gold solidifies. (b) Gaseous Cl_2 dissociates in the stratosphere to form gaseous Cl atoms. (c) Gaseous CO reacts with gaseous H_2 to form liquid methanol, CH_3OH. (d) Calcium phosphate precipitates upon mixing $Ca(NO_3)_2(aq)$ and $(NH_4)_3PO_4(aq)$.

Entropy Changes in Chemical Reactions (Section 19.4)

19.43 With reference to Figure 19.12, determine whether each of the following statements about water under 1 atm pressure is *true*? (a) The value of ΔS for $H_2O(s) \longrightarrow H_2O(l)$ at 0 °C is greater than the value of ΔS for $H_2O(l) \longrightarrow H_2O(g)$ at 100 °C. (b) When ice is melted at 0 °C, the entropy increases while the temperature remains constant until all the ice is melted. (c) At $T > 100$ °C, the entropy increases as T increases.

19.44 The normal melting and boiling points of propanol (C_3H_7OH) are −126.5 °C and 97.4 °C, respectively. With reference to Figure 19.12, determine whether each of the following statements about propanol under 1 atm pressure is *true*? (a) The value of ΔS for $C_3H_7OH(s) \longrightarrow C_3H_7OH(l)$ at −126.5 °C is smaller than the value of ΔS for $C_3H_7OH(l) \longrightarrow C_3H_7OH(g)$ at 97.4 °C. (b) When liquid propanol vaporizes at 97.4 °C, the entropy increases while the temperature remains constant until all the propanol is vaporized. (c) Between −126.5 °C and 97.4 °C, the entropy decreases as T increases.

19.45 In each of the following pairs, which would you expect to have the higher standard molar entropy? (a) $Br_2(g)$ or $Br_2(l)$ (b) $CH_3OH(l)$ or $C_2H_5OH(l)$ (c) $CCl_4(l)$ or $SiCl_4(l)$ (d) $Fe(s)$ at 25 °C or $Fe(s)$ at 500 °C?

19.46 Cyclopropane and propylene are isomers that both have the formula C_3H_6. Based on the molecular structures shown, which of these isomers would you expect to have the higher standard molar entropy at 25 °C?

Cyclopropane   Propylene

19.47 Predict which member of each of the following pairs has the greater standard entropy at 25 °C: (a) $Sc(s)$ or $Sc(g)$, (b) $NH_3(g)$ or $NH_3(aq)$, (c) $O_2(g)$ or $O_3(g)$, (d) C(graphite) or C(diamond). Use Appendix C to find the standard entropy of each substance.

19.48 Predict which member of each of the following pairs has the greater standard entropy at 25 °C: (a) $C_6H_6(l)$ or $C_6H_6(g)$, (b) $CO(g)$ or $CO_2(g)$, (c) 1 mol $N_2O_4(g)$ or 2 mol $NO_2(g)$ (d) $HCl(g)$ or $HCl(aq)$. Use Appendix C to find the standard entropy of each substance.

19.49 Ammonium nitrate dissolves spontaneously and endothermally in water at room temperature. What can you deduce about the sign of ΔS for this solution process?

19.50 The crystalline hydrate $Cd(NO_3)_2 \cdot 4H_2O(s)$ loses water when placed in a large, closed, dry vessel at room temperature:

$$Cd(NO_3)_2 \cdot 4H_2O(s) \longrightarrow Cd(NO_3)_2(s) + 4 H_2O(g)$$

This process is spontaneous and $\Delta H°$ is positive at room temperature. (a) What is the sign of $\Delta S°$ at room temperature? (b) If the hydrated compound is placed in a large, closed vessel that already contains a large amount of water vapor, does $\Delta S°$ change for this reaction at room temperature?

19.51 Using $S°$ values from Appendix C, calculate $\Delta S°$ values for the following reactions. In each case, account for the sign of $\Delta S°$.

(a) $C_2H_4(g) + H_2(g) \longrightarrow C_2H_6(g)$

(b) $N_2O_4(g) \longrightarrow 2 NO_2(g)$

(c) $Be(OH)_2(s) \longrightarrow BeO(s) + H_2O(g)$

(d) $2 CH_3OH(g) + 3 O_2(g) \longrightarrow 2 CO_2(g) + 4 H_2O(g)$

19.52 Calculate $\Delta S°$ values for the following reactions by using tabulated $S°$ values from Appendix C. In each case, explain the sign of $\Delta S°$.

(a) $HNO_3(g) + NH_3(g) \longrightarrow NH_4NO_3(s)$

(b) $2 Fe_2O_3(s) \longrightarrow 4 Fe(s) + 3 O_2(g)$

(c) $CaCO_3(s, \text{calcite}) + 2 HCl(g) \longrightarrow$
$\quad CaCl_2(s) + CO_2(g) + H_2O(l)$

(d) $3 C_2H_6(g) \longrightarrow C_6H_6(l) + 6 H_2(g)$

Gibbs Free Energy (Sections 19.5 and 19.6)

19.53 (**a**) For a process that occurs at constant temperature, does the change in Gibbs free energy depend on changes in the enthalpy and entropy of the system? (**b**) For a certain process that occurs at constant T and P, the value of ΔG is positive. Is the process spontaneous? (**c**) If ΔG for a process is large, is the rate at which it occurs fast?

19.54 (**a**) Is the standard free-energy change, $\Delta G°$, always larger than ΔG? (**b**) For any process that occurs at constant temperature and pressure, what is the significance of $\Delta G = 0$? (**c**) For a certain process, ΔG is large and negative. Does this mean that the process necessarily has a low activation barrier?

19.55 For a certain chemical reaction, $\Delta H° = -35.4$ kJ and $\Delta S° = -85.5$ J/K. (**a**) Is the reaction exothermic or endothermic? (**b**) Does the reaction lead to an increase or decrease in the randomness or disorder of the system? (**c**) Calculate $\Delta G°$ for the reaction at 298 K. (**d**) Is the reaction spontaneous at 298 K under standard conditions?

19.56 A certain reaction has $\Delta H° = +23.7$ kJ and $\Delta S° = +52.4$ J/K. (**a**) Is the reaction exothermic or endothermic? (**b**) Does the reaction lead to an increase or decrease in the randomness or disorder of the system? (**c**) Calculate $\Delta G°$ for the reaction at 298 K. (**d**) Is the reaction spontaneous at 298 K under standard conditions?

19.57 Using data in Appendix C, calculate $\Delta H°$, $\Delta S°$, and $\Delta G°$ at 298 K for each of the following reactions.

(**a**) $H_2(g) + F_2(g) \longrightarrow 2\,HF(g)$

(**b**) $C(s, \text{graphite}) + 2\,Cl_2(g) \longrightarrow CCl_4(g)$

(**c**) $2\,PCl_3(g) + O_2(g) \longrightarrow 2\,POCl_3(g)$

(**d**) $2\,CH_3OH(g) + H_2(g) \longrightarrow C_2H_6(g) + 2\,H_2O(g)$

19.58 Use data in Appendix C to calculate $\Delta H°$, $\Delta S°$, and $\Delta G°$ at 25 °C for each of the following reactions.

(**a**) $4\,Cr(s) + 3\,O_2(g) \longrightarrow 2\,Cr_2O_3(s)$

(**b**) $BaCO_3(s) \longrightarrow BaO(s) + CO_2(g)$

(**c**) $2\,P(s) + 10\,HF(g) \longrightarrow 2\,PF_5(g) + 5\,H_2(g)$

(**d**) $K(s) + O_2(g) \longrightarrow KO_2(s)$

19.59 Using data from Appendix C, calculate $\Delta G°$ for the following reactions. Indicate whether each reaction is spontaneous at 298 K under standard conditions.

(**a**) $2\,SO_2(g) + O_2(g) \longrightarrow 2\,SO_3(g)$

(**b**) $NO_2(g) + N_2O(g) \longrightarrow 3\,NO(g)$

(**c**) $6\,Cl_2(g) + 2\,Fe_2O_3(s) \longrightarrow 4\,FeCl_3(s) + 3\,O_2(g)$

(**d**) $SO_2(g) + 2\,H_2(g) \longrightarrow S(s) + 2\,H_2O(g)$

19.60 Using data from Appendix C, calculate the change in Gibbs free energy for each of the following reactions. In each case, indicate whether the reaction is spontaneous at 298 K under standard conditions.

(**a**) $2\,Ag(s) + Cl_2(g) \longrightarrow 2\,AgCl(s)$

(**b**) $P_4O_{10}(s) + 16\,H_2(g) \longrightarrow 4\,PH_3(g) + 10\,H_2O(g)$

(**c**) $CH_4(g) + 4\,F_2(g) \longrightarrow CF_4(g) + 4\,HF(g)$

(**d**) $2\,H_2O_2(l) \longrightarrow 2\,H_2O(l) + O_2(g)$

19.61 Octane (C_8H_{18}) is a liquid hydrocarbon at room temperature that is a constituent of gasoline. (**a**) Write a balanced equation for the combustion of $C_8H_{18}(l)$ to form $CO_2(g)$ and $H_2O(l)$. (**b**) Without using thermochemical data, predict whether $\Delta G°$ for this reaction is more negative or less negative than $\Delta H°$.

19.62 Sulfur dioxide reacts with strontium oxide as follows:

$$SO_2(g) + SrO(g) \longrightarrow SrSO_3(s)$$

(**a**) Without using thermochemical data, predict whether $\Delta G°$ for this reaction is more negative or less negative than

$\Delta H°$. (**b**) If you had only standard enthalpy data for this reaction, how would you estimate the value of $\Delta G°$ at 298 K, using data from Appendix C on other substances.

19.63 Classify each of the following reactions as one of the four possible types summarized in Table 19.3.

  (**i**) spontaneous at all temperatures

 (**ii**) not spontaneous at any temperature

(**iii**) spontaneous at low T but not spontaneous at high T

 (**iv**) spontaneous at high T but not spontaneous at low T

(**a**) $N_2(g) + 3\,F_2(g) \longrightarrow 2\,NF_3(g)$
$\Delta H° = -249$ kJ; $\Delta S° = -278$ J/K

(**b**) $N_2(g) + 3\,Cl_2(g) \longrightarrow 2\,NCl_3(g)$
$\Delta H° = 460$ kJ; $\Delta S° = -275$ J/K

(**c**) $N_2F_4(g) \longrightarrow 2\,NF_2(g)$
$\Delta H° = 85$ kJ; $\Delta S° = 198$ J/K

19.64 From the values given for $\Delta H°$ and $\Delta S°$, calculate $\Delta G°$ for each of the following reactions at 298 K. If the reaction is not spontaneous under standard conditions at 298 K, at what temperature (if any) would the reaction become spontaneous?

(**a**) $2\,PbS(s) + 3\,O_2(g) \longrightarrow 2\,PbO(s) + 2\,SO_2(g)$
$\Delta H° = -844$ kJ; $\Delta S° = -165$ J/K

(**b**) $2\,POCl_3(g) \longrightarrow 2\,PCl_3(g) + O_2(g)$
$\Delta H° = 572$ kJ; $\Delta S° = 179$ J/K

19.65 A particular constant-pressure reaction is barely spontaneous at 390 K. The enthalpy change for the reaction is +23.7 kJ. Estimate ΔS for the reaction.

19.66 A certain constant-pressure reaction is barely nonspontaneous at 45 °C. The entropy change for the reaction is 72 J/K. Estimate ΔH.

19.67 For a particular reaction, $\Delta H = -32$ kJ and $\Delta S = -98$ J/K. Assume that ΔH and ΔS do not vary with temperature. (**a**) At what temperature will the reaction have $\Delta G = 0$? (**b**) If T is increased from that in part (a), will the reaction be spontaneous or nonspontaneous?

19.68 Reactions in which a substance decomposes by losing CO are called *decarbonylation* reactions. The decarbonylation of acetic acid proceeds according to:

$$CH_3COOH(l) \longrightarrow CH_3OH(g) + CO(g)$$

By using data from Appendix C, calculate the minimum temperature at which this process will be spontaneous under standard conditions. Assume that $\Delta H°$ and $\Delta S°$ do not vary with temperature.

19.69 Consider the following reaction between oxides of nitrogen:

$$NO_2(g) + N_2O(g) \longrightarrow 3\,NO(g)$$

(**a**) Use data in Appendix C to predict how ΔG for the reaction varies with increasing temperature. (**b**) Calculate ΔG at 800 K, assuming that $\Delta H°$ and $\Delta S°$ do not change with temperature. Under standard conditions is the reaction spontaneous at 800 K? (**c**) Calculate ΔG at 1000 K. Is the reaction spontaneous under standard conditions at this temperature?

19.70 Methanol (CH_3OH) can be made by the controlled oxidation of methane:

$$CH_4(g) + \tfrac{1}{2}O_2(g) \longrightarrow CH_3OH(g)$$

(**a**) Use data in Appendix C to calculate $\Delta H°$ and $\Delta S°$ for this reaction. (**b**) Will ΔG for the reaction increase, decrease, or stay unchanged with increasing temperature? (**c**) Calculate $\Delta G°$ at 298 K. Under standard conditions, is the reaction spontaneous at this temperature? (**d**) Is there a temperature at which the reaction would be at equilibrium under standard conditions and that is low enough so that the compounds involved are likely to be stable?

19.71 (a) Use data in Appendix C to estimate the boiling point of benzene, $C_6H_6(l)$. (b) How does your answer to part (a) compare to the true experimental boiling point (readily obtained from on-line sources)?

19.72 (a) Using data in Appendix C, estimate the temperature at which the free-energy change for the transformation from $I_2(s)$ to $I_2(g)$ is zero. (b) Use a reference source, such as Web Elements (www.webelements.com), to find the experimental melting and boiling points of I_2. (c) Which of the values in part (b) is closer to the value you obtained in part (a)?

19.73 Acetylene gas, $C_2H_2(g)$, is used in welding. (a) Write a balanced equation for the combustion of acetylene gas to $CO_2(g)$ and $H_2O(l)$. (b) How much heat is produced in burning 1 mol of C_2H_2 under standard conditions if both reactants and products are brought to 298 K? (c) What is the maximum amount of useful work that can be accomplished under standard conditions by this reaction?

19.74 The fuel in high-efficiency natural-gas vehicles consists primarily of methane (CH_4). (a) How much heat is produced in burning 1 mol of $CH_4(g)$ under standard conditions if reactants and products are brought to 298 K and $H_2O(l)$ is formed? (b) What is the maximum amount of useful work that can be accomplished under standard conditions by this system?

Free Energy and Equilibrium (Section 19.7)

19.75 Indicate whether ΔG increases, decreases, or stays the same for each of the following reactions as the partial pressure of O_2 is increased:

(a) $2\,CO(g) + O_2(g) \longrightarrow 2\,CO_2(g)$

(b) $2\,H_2O_2(l) \longrightarrow 2\,H_2O(l) + O_2(g)$

(c) $2\,KClO_3(s) \longrightarrow 2\,KCl(s) + 3\,O_2(g)$

19.76 Indicate whether ΔG increases, decreases, or does not change when the partial pressure of H_2 is increased in each of the following reactions:

(a) $N_2(g) + 3\,H_2(g) \longrightarrow 2\,NH_3(g)$

(b) $2\,HBr(g) \longrightarrow H_2(g) + Br_2(g)$

(c) $2\,H_2(g) + C_2H_2(g) \longrightarrow C_2H_6(g)$

19.77 Consider the reaction $2\,NO_2(g) \longrightarrow N_2O_4(g)$. (a) Using data from Appendix C, calculate $\Delta G°$ at 298 K. (b) Calculate ΔG at 298 K if the partial pressures of NO_2 and N_2O_4 are 0.40 atm and 1.60 atm, respectively.

19.78 Consider the reaction $3\,CH_4(g) \longrightarrow C_3H_8(g) + 2\,H_2(g)$. (a) Using data from Appendix C, calculate $\Delta G°$ at 298 K. (b) Calculate ΔG at 298 K if the reaction mixture consists of 40.0 atm of CH_4, 0.0100 atm of $C_3H_8(g)$, and 0.0180 atm of H_2.

19.79 Use data from Appendix C to calculate the equilibrium constant, K, and $\Delta G°$ at 298 K for each of the following reactions:

(a) $H_2(g) + I_2(g) \rightleftharpoons 2\,HI(g)$

(b) $C_2H_5OH(g) \rightleftharpoons C_2H_4(g) + H_2O(g)$

(c) $3\,C_2H_2(g) \rightleftharpoons C_6H_6(g)$

19.80 Using data from Appendix C, write the equilibrium-constant expression and calculate the value of the equilibrium constant and the free-energy change for these reactions at 298 K:

(a) $NaHCO_3(s) \rightleftharpoons NaOH(s) + CO_2(g)$

(b) $2\,HBr(g) + Cl_2(g) \rightleftharpoons 2\,HCl(g) + Br_2(g)$

(c) $2\,SO_2(g) + O_2(g) \rightleftharpoons 2\,SO_3(g)$

19.81 Consider the decomposition of barium carbonate:

$$BaCO_3(s) \rightleftharpoons BaO(s) + CO_2(g)$$

Using data from Appendix C, calculate the equilibrium pressure of CO_2 at (a) 298 K and (b) 1100 K.

19.82 Consider the reaction

$$PbCO_3(s) \rightleftharpoons PbO(s) + CO_2(g)$$

Using data in Appendix C, calculate the equilibrium pressure of CO_2 in the system at (a) 400 °C and (b) 180 °C.

19.83 The value of K_a for nitrous acid (HNO_2) at 25 °C is given in Appendix D. (a) Write the chemical equation for the equilibrium that corresponds to K_a. (b) By using the value of K_a, calculate $\Delta G°$ for the dissociation of nitrous acid in aqueous solution. (c) What is the value of ΔG at equilibrium? (d) What is the value of ΔG when $[H^+] = 5.0 \times 10^{-2}\,M$, $[NO_2^-] = 6.0 \times 10^{-4}\,M$, and $[HNO_2] = 0.20\,M$?

19.84 The K_b for methylamine (CH_3NH_2) at 25 °C is given in Appendix D. (a) Write the chemical equation for the equilibrium that corresponds to K_b. (b) By using the value of K_b, calculate $\Delta G°$ for the equilibrium in part (a). (c) What is the value of ΔG at equilibrium? (d) What is the value of ΔG when $[H^+] = 6.7 \times 10^{-9}\,M$, $[CH_3NH_3^+] = 2.4 \times 10^{-3}\,M$, and $[CH_3NH_2] = 0.098\,M$?

Additional Exercises

19.85 (a) Which of the thermodynamic quantities $T, E, q, w,$ and S are state functions? (b) Which depend on the path taken from one state to another? (c) How many *reversible* paths are there between two states of a system? (d) For a reversible isothermal process, write an expression for ΔE in terms of q and w and an expression for ΔS in terms of q and T.

19.86 Indicate whether each of the following statements is true or false. (a) The feasibility of manufacturing NH_3 from N_2 and H_2 depends entirely on the value of ΔH for the process $N_2(g) + 3\,H_2(g) \longrightarrow 2\,NH_3(g)$. (b) The reaction of $Na(s)$ with $Cl_2(g)$ to form $NaCl(s)$ is a spontaneous process. (c) A spontaneous process can in principle be conducted reversibly. (d) Spontaneous processes in general require that work be done to force them to proceed. (e) Spontaneous processes are those that are exothermic and that lead to a higher degree of order in the system.

19.87 For each of the following processes, indicate whether the signs of ΔS and ΔH are expected to be positive, negative, or about zero. (a) A solid sublimes. (b) The temperature of a sample of $Co(s)$ is lowered from 60 °C to 25 °C. (c) Ethyl alcohol evaporates from a beaker. (d) A diatomic molecule dissociates into atoms. (e) A piece of charcoal is combusted to form $CO_2(g)$ and $H_2O(g)$.

19.88 Consider the following standard entropies at 298 K: $Br_2(l)$ 152.3 J/mol-K, $Br_2(g)$ 245.3 J/mol-K, $I_2(s)$ 116.7 J/mol-K, $I_2(g)$ 260.6 J/mol-K. Which of the following statements about these entropies is or are *true*? (a) The standard entropy of $I_2(g)$ is greater than that of $I_2(s)$ because gases have more degrees of freedom than do solids. (b) The standard entropy of $I_2(g)$ is greater than that of $Br_2(g)$ because I_2 has a higher molar mass than does Br_2. (c) The standard entropy of $Br_2(l)$ is greater than that of $I_2(s)$ because Br is more electronegative than I.

19.89 The reaction $2\,Mg(s) + O_2(g) \longrightarrow 2\,MgO(s)$ is highly spontaneous. A classmate calculates the entropy change for this reaction and obtains a large negative value for $\Delta S°$. Did your classmate make a mistake in the calculation? Explain.

19.90 Consider a system that consists of two standard playing dice, with the state of the system defined by the sum of the values shown on the top faces. (a) The two arrangements of top faces shown here can be viewed as two possible microstates of the system. Explain. (**b**) To which state does each micro-state correspond? (**c**) How many possible states are there for the system? (**d**) Which state or states have the highest entropy? Explain. (**e**) Which state or states have the lowest entropy? Explain. (**f**) Calculate the absolute entropy of the two-dice system.

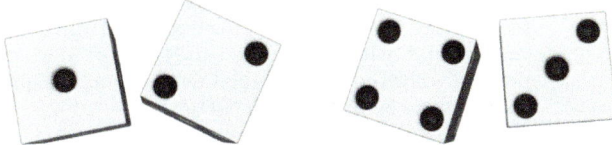

19.91 A standard air conditioner involves a *refrigerant* that is typically now a fluorinated hydrocarbon, such as CH_2F_2. An air-conditioner refrigerant has the property that it readily vaporizes at atmospheric pressure and is easily compressed to its liquid phase under increased pressure. The operation of an air conditioner can be thought of as a closed system made up of the refrigerant going through the two stages shown here (the air circulation is not shown in this diagram).

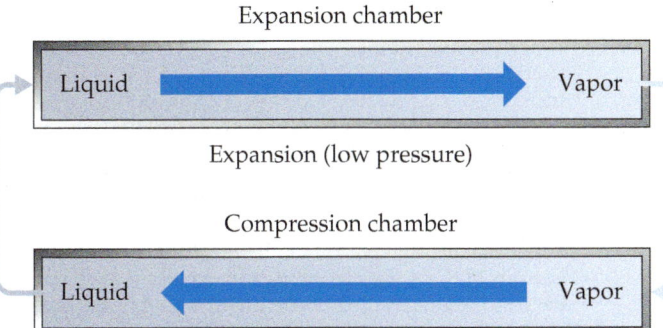

During *expansion*, the liquid refrigerant is released into an expansion chamber at low pressure, where it vaporizes. The vapor then undergoes *compression* at high pressure back to its liquid phase in a compression chamber. (**a**) What is the sign of q for the expansion? (**b**) What is the sign of q for the compression? (**c**) In a central air-conditioning system, one chamber is inside the home and the other is outside. Which chamber is where, and why? (**d**) Imagine that a sample of liquid refrigerant undergoes expansion followed by compression, so that it is back to its original state. Would you expect that to be a reversible process? (**e**) Suppose that a house and its exterior are both initially at 31 °C. Some time after the air conditioner is turned on, the house is cooled to 24 °C. Is this process spontaneous or nonspontaneous?

19.92 *Trouton's rule* states that for many liquids at their normal boiling points, the standard molar entropy of vaporization is about 88 J/mol-K. (**a**) Estimate the normal boiling point of bromine, Br_2, by determining $\Delta H°_{vap}$ for Br_2 using data from Appendix C. Assume that $\Delta H°_{vap}$ remains constant with temperature and that Trouton's rule holds. (**b**) Look up the normal boiling point of Br_2 in a chemistry handbook or at the WebElements website (www.webelements.com) and compare it to your calculation. What are the possible sources of error, or incorrect assumptions, in the calculation?

19.93 (**a**) Write the chemical equations that correspond to $\Delta G°_f$ for $NH_3(g)$ and for $CO(g)$. (**b**) For which of these formation reactions will the value of $\Delta G°_f$ be more positive (less negative) than $\Delta H°_f$? (**c**) In general, under which condition is $\Delta G°_f$ more positive (less negative) than $\Delta H°_f$ (i) When the temperature is high, (ii) when the reaction is reversible, (iii) when $\Delta S°_f$ is negative.

19.94 Consider the following three reactions:

(**i**) $Ti(s) + 2 Cl_2(g) \longrightarrow TiCl_4(g)$

(**ii**) $C_2H_6(g) + 7 Cl_2(g) \longrightarrow 2 CCl_4(g) + 6 HCl(g)$

(**iii**) $BaO(s) + CO_2(g) \longrightarrow BaCO_3(s)$

(**a**) For each of the reactions, use data in Appendix C to calculate $\Delta H°$, $\Delta G°$, K, and $\Delta S°$ at 25 °C. (**b**) Which of these reactions are spontaneous under standard conditions at 25 °C? (**c**) For each of the reactions, predict the manner in which the change in free energy varies with an increase in temperature.

19.95 Using the data in Appendix C and given the pressures listed, calculate K_p and ΔG for each of the following reactions:

(**a**) $N_2(g) + 3 H_2(g) \longrightarrow 2 NH_3(g)$
$P_{N_2} = 2.6 \text{ atm}, P_{H_2} = 5.9 \text{ atm}, P_{NH_3} = 1.2 \text{ atm}$

(**b**) $2 N_2H_4(g) + 2 NO_2(g) \longrightarrow 3 N_2(g) + 4 H_2O(g)$
$P_{N_2H_4} = P_{NO_2} = 5.0 \times 10^{-2} \text{ atm},$
$P_{N_2} = 0.5 \text{ atm}, P_{H_2O} = 0.3 \text{ atm}$

(**c**) $N_2H_4(g) \longrightarrow N_2(g) + 2 H_2(g)$
$P_{N_2H_4} = 0.5 \text{ atm}, P_{N_2} = 1.5 \text{ atm}, P_{H_2} = 2.5 \text{ atm}$

19.96 (**a**) For each of the following reactions, predict the sign of $\Delta H°$ and $\Delta S°$ without doing any calculations. (**b**) Based on your general chemical knowledge, predict which of these reactions will have $K > 1$. (**c**) In each case, indicate whether K should increase or decrease with increasing temperature.

(**i**) $2 Mg(s) + O_2(g) \rightleftharpoons 2 MgO(s)$

(**ii**) $2 KI(s) \rightleftharpoons 2 K(g) + I_2(g)$

(**iii**) $Na_2(g) \rightleftharpoons 2 Na(g)$

(**iv**) $2 V_2O_5(s) \rightleftharpoons 4 V(s) + 5 O_2(g)$

19.97 Acetic acid can be manufactured by combining methanol with carbon monoxide, an example of a *carbonylation* reaction:

$$CH_3OH(l) + CO(g) \longrightarrow CH_3COOH(l)$$

(**a**) Calculate the equilibrium constant for the reaction at 25 °C. (**b**) Industrially, this reaction is run at temperatures above 25 °C. Will an increase in temperature produce an increase or decrease in the mole fraction of acetic acid at equilibrium? (**c**) Which of the following likely explains why elevated temperatures are used? (i) The equilibrium shifts to the right at higher temperature. (ii) The reaction reaches equilibrium faster at higher temperature. (iii) Both (i) and (ii) are reasons the reaction is run at higher temperature. (**d**) At what temperature will this reaction have an equilibrium constant equal to 1? (You may assume that $\Delta H°$ and $\Delta S°$ are temperature independent, and you may ignore any phase changes that might occur.)

19.98 The oxidation of glucose ($C_6H_{12}O_6$) in body tissue produces CO_2 and H_2O, a reaction that produces much of the energy used in physiological activity:

Oxidation: $C_6H_{12}O_6(s) + 6 O_2(g) \rightleftharpoons 6 CO_2(g) + 6 H_2O(l)$

In contrast, anaerobic decomposition, which occurs during fermentation, produces ethanol (C_2H_5OH) and CO_2:

Anaerobic decomposition: $C_6H_{12}O_6(s) \rightleftharpoons$
$$2 C_2H_5OH(l) + 2 CO_2(g)$$

(**a**) Using data given in Appendix C, calculate $\Delta H°$ and $\Delta G°$ for the oxidation of glucose. (**b**) Calculate $\Delta H°$ and $\Delta G°$ for the anaerobic decomposition of glucose. (**c**) Which process, oxidation or anaerobic decomposition, releases more heat under standard conditions? (**d**) For the anaerobic decomposition of glucose, what is the value of the equilibrium constant K_p at 298 K?

19.99 The conversion of natural gas, which is mostly methane, into products that contain two or more carbon atoms, such as ethane (C_2H_6), is a very important industrial chemical process. In principle, methane can be converted into ethane and hydrogen:

$$2\,CH_4(g) \longrightarrow C_2H_6(g) + H_2(g)$$

In practice, this reaction is carried out in the presence of oxygen:

$$2\,CH_4(g) + \tfrac{1}{2}O_2(g) \longrightarrow C_2H_6(g) + H_2O(g)$$

(**a**) Using the data in Appendix C, calculate K for these reactions at 25 °C. (**b**) Is the difference in $\Delta G°$ for the two reactions due primarily to the enthalpy term (ΔH) or the entropy term ($-T\Delta S$)? (**c**) Will the equilibrium constant for the second reaction increase, decrease, or stay the same if the temperature is increased to 500 °C? (**d**) The reaction of CH_4 and O_2 to form C_2H_6 and H_2O must be carried out carefully to avoid a competing reaction. What is the most likely competing reaction?

19.100 Cells use the hydrolysis of adenosine triphosphate (ATP) as a source of energy (Figure 19.16). The conversion of ATP to ADP has a standard free-energy change of −30.5 kJ/mol. If all the free energy from the metabolism of glucose,

$$C_6H_{12}O_6(s) + 6\,O_2(g) \longrightarrow 6\,CO_2(g) + 6\,H_2O(l)$$

goes into the conversion of ADP to ATP, how many moles of ATP can be produced for each mole of glucose?

19.101 The potassium-ion concentration in blood plasma is about $5.0 \times 10^{-3}\,M$, whereas the concentration in muscle-cell fluid is much greater (0.15 M). The plasma and intracellular fluid are separated by the cell membrane, which we assume is permeable only to K^+. (**a**) What is ΔG for the transfer of 1 mol of K^+ from blood plasma to the cellular fluid at body temperature 37 °C? (**b**) What is the minimum amount of work that must be used to transfer this K^+?

19.102 At what temperatures is the following reaction, the reduction of magnetite by graphite to elemental iron, spontaneous?

$$Fe_3O_4(s) + 2\,C(s, graphite) \longrightarrow 2\,CO_2(g) + 3\,Fe(s)$$

19.103 Consider the following equilibrium:

$$N_2O_4(g) \rightleftharpoons 2\,NO_2(g)$$

Thermodynamic data on these gases are given in Appendix C. You may assume that $\Delta H°$ and $\Delta S°$ do not vary with temperature. (**a**) At what temperature will an equilibrium mixture contain equal amounts of the two gases? (**b**) At what temperature will an equilibrium mixture of 1 atm total pressure contain twice as much NO_2 as N_2O_4? (**c**) At what temperature will an equilibrium mixture of 10 atm total pressure contain twice as much NO_2 as N_2O_4?

19.104 The reaction

$$SO_2(g) + 2\,H_2S(g) \rightleftharpoons 3\,S(s) + 2\,H_2O(g)$$

is the basis of a suggested method for removal of SO_2 from power-plant stack gases. The standard free energy of each substance is given in Appendix C. (**a**) What is the equilibrium constant for the reaction at 298 K? (**b**) In principle, is this reaction a feasible method of removing SO_2? (**c**) If $P_{SO_2} = P_{H_2S}$ and the vapor pressure of water is 25 torr, calculate the equilibrium SO_2 pressure in the system at 298 K. (**d**) Would you expect the process to be more or less effective at higher temperatures?

19.105 When most elastomeric polymers (e.g., a rubber band) are stretched, the molecules become more ordered, as illustrated here:

Suppose you stretch a rubber band. (**a**) Do you expect the entropy of the system to increase or decrease? (**b**) If the rubber band were stretched isothermally, would heat need to be absorbed or emitted to maintain constant temperature? (**c**) Try this experiment: Stretch a rubber band and wait a moment. Then place the stretched rubber band on your upper lip, and let it return suddenly to its unstretched state (remember to keep holding on!). What do you observe? Are your observations consistent with your answer to part (b)?

Design an Experiment

You are measuring the equilibrium constant for a drug candidate binding to its DNA target over a series of different temperatures. You chose your drug candidate based on computer-aided molecular modeling, which indicates that the drug molecule likely would make many hydrogen bonds and favorable dipole–dipole interactions with the DNA site. You perform a set of experiments in buffer solution for the drug–DNA complex and generate a table of K's at different T's. (**a**) Derive an equation that relates equilibrium constant to standard enthalpy and entropy changes. (*Hint:* Equilibrium constant, enthalpy, and entropy are all related to free energy.) (**b**) Show how you can graph your K and T data to calculate the standard entropy and enthalpy changes for the drug candidate + DNA binding interaction. (**c**) You are surprised to learn that the enthalpy change for the binding reaction is close to zero, and the entropy change is large and positive. Suggest an explanation and design an experiment to test it. (*Hint:* Think about water and ions.) (**d**) You try another drug candidate with the DNA target and find that this drug candidate has a large negative enthalpy change upon DNA binding, and the entropy change is small and positive. Suggest an explanation, at the molecular level, and design an experiment to test your hypothesis.

20

ELECTROCHEMISTRY

▲ **BATTERIES FOR GRID-SCALE ENERGY STORAGE.** The electricity generated by renewable energy sources like solar and wind is intermittent in nature. For these technologies to become our main sources of electrical power, the energy generated at peak times must be stored for use when the demand for electricity exceeds supply. Batteries like the ones shown here offer a reliable means of storing energy.

Modern society depends on electricity for everything from lighting to computing to air conditioning. In 2020, approximately 1.4×10^{19} J of electrical energy was consumed in the United States alone—a quantity 13 times greater than electricity use in 1950! Most of the energy needed to generate electricity comes from chemical reactions, especially the combustion of hydrocarbon fuels. Given the concerns over CO_2 emissions and the effects of climate change, there is a major push to generate ever increasing amounts of electricity from renewable sources, with solar and wind power playing prominent roles. Unfortunately, the intermittent nature of these resources poses challenges. Demand for electricity does not vanish when the sun isn't shining or the wind stops blowing; energy generated during peak times must be stored, and one important way of doing that is to convert it into chemical energy. Chemical energy can readily be stored and converted back to electrical energy when needed using either a battery or a fuel cell.

Batteries are essential for technologies that go far beyond grid energy storage. The portability of batteries makes them the preferred power source for everyday necessities like laptop computers, cell phones, pacemakers, and, increasingly, automobiles. A considerable amount of effort is currently focused on research and development of new batteries. At the heart of these devices are oxidation–reduction reactions that power batteries. Redox reactions are involved not only in the operation of batteries but also in a wide variety of important natural processes, including the rusting of iron, the browning of foods, and the respiration of animals. The subject of this chapter, **electrochemistry**, is the study of the relationships between electricity and chemical reactions.

WHAT'S AHEAD

20.1 ▶ **Oxidation States and Oxidation–Reduction Reactions** Review oxidation states and *oxidation–reduction (redox) reactions*.

20.2 ▶ **Balancing Redox Equations** Learn how to balance redox equations using the method of *half-reactions*.

20.3 ▶ **Voltaic Cells** Consider *voltaic cells*, which produce electricity from spontaneous redox reactions. Solid electrodes serve as the surfaces at which oxidation and reduction take place. The electrode where oxidation occurs is the *anode*, and the electrode where reduction occurs is the *cathode*.

20.4 ▶ **Cell Potentials under Standard Conditions** Learn about *cell potential*, *E*, which is the potential difference or voltage between the two electrodes of a voltaic cell. The standard cell potential, *E°*, can be calculated from the *standard reduction potentials* of the half-reactions occurring at each electrode.

20.5 ▶ **Free Energy and Redox Reactions** Relate the Gibbs free energy, ΔG, to cell potential, *E*.

20.6 ▶ **Cell Potentials under Nonstandard Conditions** Calculate cell potentials under nonstandard conditions using standard cell potentials and the Nernst equation.

20.7 ▶ **Batteries and Fuel Cells** Learn about batteries and fuel cells, commercially important energy sources that use electrochemical reactions to convert chemical energy to electrical energy.

20.8 ▶ **Corrosion** Investigate corrosion, a spontaneous electrochemical process involving the oxidation of metals.

20.9 ▶ **Electrolysis** Examine *electrolytic cells*, which use electricity to drive chemical reactions that would otherwise be nonspontaneous.

20.1 | Oxidation States and Oxidation–Reduction Reactions

⚠ **Learning Objectives**

When you finish Section 20.1, you should be able to:

▶ Determine the oxidation numbers of the individual elements in substances.

▶ Distinguish redox reactions from other types of chemical reactions and determine the identities of the oxidizing and reducing agent in such reactions.

As discussed in Chapter 4, *oxidation* takes place when an atom loses electrons; the opposite process, gaining electrons, is called *reduction*. Thus, oxidation–reduction (redox) reactions occur when electrons are transferred from an atom that is oxidized to an atom that is reduced. (Section 4.4) We determine whether a given chemical reaction is an oxidation–reduction reaction by keeping track of the *oxidation numbers (oxidation states)* of the elements involved in the reaction.* A detailed procedure for assigning oxidation numbers is given in Section 4.4. In any redox reaction, one or more elements undergoes a change in oxidation number. For example, consider the net reaction that occurs spontaneously when zinc metal is added to a strong acid (**Figure 20.1**):

$$Zn(s) + 2\,H^+(aq) \longrightarrow Zn^{2+}(aq) + H_2(g) \qquad [20.1]$$

Assigning oxidation numbers to all species in the reaction, we have

$$Zn(s) + 2\,H^+(aq) \longrightarrow Zn^{2+}(aq) + H_2(g) \qquad [20.2]$$

H⁺ reduced

0   +1   +2   0

Zn oxidized

The oxidation numbers below the equation show that the oxidation number of Zn changes from 0 to +2, while that of H changes from +1 to 0. Because there is a change in oxidation numbers, we can identify this reaction as an oxidation–reduction reaction. Electrons are transferred from zinc atoms to hydrogen ions; Zn is oxidized and H⁺ is reduced.

Go Figure Is this reaction exothermic or endothermic?

$$Zn(s) \;+\; 2\,HCl(aq) \longrightarrow ZnCl_2(aq) \;+\; H_2(g)$$

▲ **Figure 20.1 Oxidation of zinc by hydrochloric acid.**

---

*The convention used by chemists is to write a + or − sign before the numerical value when expressing an oxidation number (for example, +6 or −1), and after the number when representing the charge of an ion (for example, 1+ or 2−).

In a reaction such as Equation 20.2, a clear transfer of electrons occurs. In some reactions, however, the oxidation numbers change, but we cannot say that any substance literally gains or loses electrons. For example, in the combustion of hydrogen gas,

$$2 \, H_2(g) + O_2(g) \longrightarrow 2 \, H_2O(g) \qquad [20.3]$$

hydrogen is oxidized from the 0 to the +1 oxidation state and oxygen is reduced from the 0 to the −2 oxidation state, indicating that Equation 20.3 is an oxidation–reduction reaction. While keeping track of oxidation states can be reliably used to identify redox reactions and offers a convenient form of "bookkeeping," you should not generally equate the oxidation state of an atom with its actual charge in a chemical compound. See the "A Closer Look" box on "Oxidation Numbers, Formal Charges, and Actual Partial Charges." (Section 8.5)

In any redox reaction, both oxidation and reduction must occur. If one substance is oxidized, another must be reduced. The substance that oxidizes another substance is called either the **oxidizing agent** or the **oxidant**. The oxidizing agent acquires electrons from the other substance and therefore is reduced. A **reducing agent**, or **reductant**, is a substance that gives up electrons, thereby causing another substance to be reduced. The reducing agent is oxidized in the process. In Equation 20.2, $H^+(aq)$, the species that is reduced, is the oxidizing agent, and $Zn(s)$, the species that is oxidized, is the reducing agent.

Sample Exercise 20.1
Identifying Oxidizing and Reducing Agents

The nickel–cadmium (NiCad) battery uses the following redox reaction to generate electricity:

$$Cd(s) + NiO_2(s) + 2 \, H_2O(l) \longrightarrow Cd(OH)_2(s) + Ni(OH)_2(s)$$

First, identify the substances that are oxidized and reduced, then indicate which reactant is the oxidizing agent and which is the reducing agent.

SOLUTION

Analyze We are given a redox equation and asked to identify the substance oxidized and the substance reduced and to label the oxidizing agent and the reducing agent.

Plan First, we use the rules outlined earlier (Section 4.4) to assign oxidation states, or numbers, to all the atoms and determine which elements change oxidation state. Second, we apply the definitions of oxidation and reduction.

Solve

$$Cd(s) + NiO_2(s) + 2 \, H_2O(l) \longrightarrow Cd(OH)_2(s) + Ni(OH)_2(s)$$

The oxidation state of Cd increases from 0 to +2, and that of Ni decreases from +4 to +2. Thus, the Cd atom is oxidized (loses electrons) and is the reducing agent. The oxidation state of Ni decreases as NiO_2 is converted into $Ni(OH)_2$. Thus, NiO_2 is reduced (gains electrons) and is the oxidizing agent.

Comment A common mnemonic for remembering oxidation and reduction is "LEO the lion says GER": *l*osing *e*lectrons is *o*xidation; *g*aining *e*lectrons is *r*eduction.

▶ **Practice Exercise**

Identify the oxidizing and reducing agents in the reaction

$$2 \, H_2O(l) + Al(s) + MnO_4^-(aq) \longrightarrow Al(OH)_4^-(aq) + MnO_2(s)$$

Self-Assessment Exercises

SAE 20.1 In which of the following compounds is the oxidation number of oxygen equal to −1?

    (**i**) Na_2O_2 (**ii**) Li_2O (**iii**) H_2SO_4

(**a**) only i (**b**) only ii (**c**) only iii (**d**) both i and ii (**e**) both i and iii

SAE 20.2 Which of the following reactions can be classified as redox reactions?

    (**i**) $2 \, HClO_4(aq) + SrO(s) \longrightarrow H_2O(l) + Sr(ClO_4)_2(aq)$
    (**ii**) $2 \, Li(s) + 2 \, H_2O(l) \longrightarrow 2 \, LiOH(aq) + H_2(g)$
    (**iii**) $SO_3(g) + H_2O(l) \longrightarrow H_2SO_4(aq)$

(**a**) only i (**b**) only ii (**c**) only iii (**d**) both i and ii (**e**) both i and iii

SAE 20.3 The following combination reaction leads to the formation of hydrogen chloride:

$$Cl_2(g) + H_2(g) \longrightarrow 2 \, HCl(g)$$

In this reaction, _____ is reduced, and _____ is the reducing agent.

(**a**) $Cl_2(g)$, $Cl_2(g)$ (**b**) $Cl_2(g)$, $H_2(g)$ (**c**) $H_2(g)$, $Cl_2(g)$ (**d**) $H_2(g)$, $H_2(g)$

SAE 20.4 The following combination reaction between hydrogen and oxygen forms water:

$$2 \, H_2(g) + O_2(g) \longrightarrow 2 \, H_2O(l)$$

Which of the following statements about this reaction is or are *true*?

    (**i**) Hydrogen is reduced in this reaction.
    (**ii**) The oxidation state of oxygen in water is +2.
    (**iii**) This is an acid–base reaction, not an oxidation–reduction reaction.

(**a**) only i (**b**) only ii (**c**) only iii (**d**) both i and iii (**e**) None of the statements is true.

20.2 | Balancing Redox Equations

Learning Objective

When you finish Section 20.2, you should be able to:

▶ Use the method of half-reactions to complete and balance redox equations in either acidic or basic conditions.

Whenever we balance a chemical equation, we must obey the law of conservation of mass: The amount of each element must be the same on both sides of the equation. (Atoms are neither created nor destroyed in any chemical reaction.) (Section 3.1) As we balance oxidation–reduction reactions, there is an additional requirement: The gains and losses of electrons must also be balanced. If a substance loses a certain number of electrons during a reaction, another substance must gain that same number of electrons. (Electrons are neither created nor destroyed in any chemical reaction.)

In many simple chemical equations, such as Equation 20.2, balancing the electrons is handled "automatically"—that is, we balance the equation without explicitly accounting for the transfer of electrons. Many redox equations are more complex than Equation 20.2, however, and cannot be balanced easily without taking into account the number of electrons lost and gained. In this section, we examine the *method of half-reactions*, a systematic procedure for balancing redox equations.

Half-Reactions

Although oxidation and reduction must take place simultaneously, it is often convenient to consider them as separate processes. For example, the oxidation of Sn^{2+} by Fe^{3+},

$$Sn^{2+}(aq) + 2\,Fe^{3+}(aq) \longrightarrow Sn^{4+}(aq) + 2\,Fe^{2+}(aq)$$

can be considered as consisting of two processes—namely, oxidation of Sn^{2+} and reduction of Fe^{3+}:

$$\textit{Oxidation:}\quad Sn^{2+}(aq) \longrightarrow Sn^{4+}(aq) + 2\,e^- \qquad [20.4]$$

$$\textit{Reduction:}\quad 2\,Fe^{3+}(aq) + 2\,e^- \longrightarrow 2\,Fe^{2+}(aq) \qquad [20.5]$$

Notice that electrons are shown as products in the oxidation process and as reactants in the reduction process.

Equations that show either oxidation or reduction alone, such as Equations 20.4 and 20.5, are called **half-reactions**. In the overall redox reaction, the number of electrons lost in the oxidation half-reaction must equal the number of electrons gained in the reduction half-reaction. When this condition is met and each half-reaction is balanced, the electrons on the two sides cancel when the two half-reactions are added to give the balanced oxidation–reduction equation.

Balancing Equations by the Method of Half-Reactions

How can we use half-reactions to balance a chemical reaction that involves oxidation–reduction? By assigning oxidation numbers, we can be assured that oxidation and reduction have occurred. We then look at two "skeleton" equations—one showing the species that is being oxidized and one showing the species that is being reduced. We then balance each half-reaction. We will find that H_2O and either H^+ (for acidic solutions) or OH^- (for basic solutions) are often involved as reactants or products in balancing the half-reactions. Unless H_2O, H^+, or OH^- is being oxidized or reduced, these species do not appear in the skeleton equations. Their presence, however, can be deduced as we balance the overall equation.

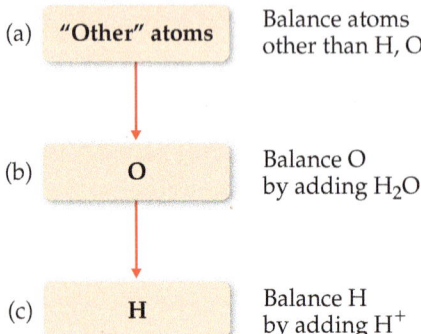

(a) "Other" atoms — Balance atoms other than H, O

(b) O — Balance O by adding H_2O

(c) H — Balance H by adding H^+

(d) e^- — Balance electrons

How to Balance Redox Reactions in Acidic Aqueous Solution

1. Divide the equation into one oxidation half-reaction and one reduction half-reaction.

2. Balance each half-reaction.
   (a) First, balance elements other than H and O.
   (b) Next, balance O atoms by adding H_2O as needed.
   (c) Then balance H atoms by adding H^+ as needed.
   (d) Finally, balance charge by adding e^- as needed.

This specific sequence (a)–(d) is important, and it is summarized in the diagram in the margin. At this point, you can check whether the number of electrons in each half-reaction corresponds to the changes in oxidation state.

3. Multiply half-reactions by integers as needed to make the number of electrons lost in the oxidation half-reaction equal the number of electrons gained in the reduction half-reaction.

4. Add half-reactions and, if possible, simplify by canceling species appearing on both sides of the combined equation.

5. Check to make sure that atoms and charges are balanced.

As an example, let's consider the reaction between permanganate ion (MnO_4^-) and oxalate ion ($C_2O_4^{2-}$) in acidic aqueous solution (**Figure 20.2**). When MnO_4^- is added to an acidified solution of $C_2O_4^{2-}$, the deep purple color of the MnO_4^- ion fades, bubbles of CO_2 form, and the solution takes on the pale pink color of Mn^{2+}. We can write the skeleton equation as

$$MnO_4^-(aq) + C_2O_4^{2-}(aq) \longrightarrow Mn^{2+}(aq) + CO_2(aq) \qquad [20.6]$$

Experiments show that H^+ is consumed and H_2O is produced in the reaction. We will see that their involvement in the reaction is deduced during the course of balancing the equation.

To complete and balance Equation 20.6, we first write the two half-reactions (Step 1). One half-reaction must have Mn on both sides of the arrow, and the other must have C on both sides of the arrow:

$$MnO_4^-(aq) \longrightarrow Mn^{2+}(aq)$$
$$C_2O_4^{2-}(aq) \longrightarrow CO_2(g)$$

We next complete and balance each half-reaction. First, we balance all the atoms except H and O (Step 2a). In the permanganate half-reaction, we have one manganese atom on each side of the equation and so need to do nothing. In the oxalate half-reaction, we add a coefficient 2 on the right to balance the two carbons on the left:

$$MnO_4^-(aq) \longrightarrow Mn^{2+}(aq)$$
$$C_2O_4^{2-}(aq) \longrightarrow 2\,CO_2(g)$$

Next we balance O (Step 2b). The permanganate half-reaction has four oxygens on the left and none on the right; to balance these four oxygen atoms, we add four H_2O molecules on the right:

$$MnO_4^-(aq) \longrightarrow Mn^{2+}(aq) + 4\,H_2O(l)$$

The eight hydrogen atoms now in the products must be balanced by adding 8 H^+ to the reactants (Step 2c):

$$8\,H^+(aq) + MnO_4^-(aq) \longrightarrow Mn^{2+}(aq) + 4\,H_2O(l)$$

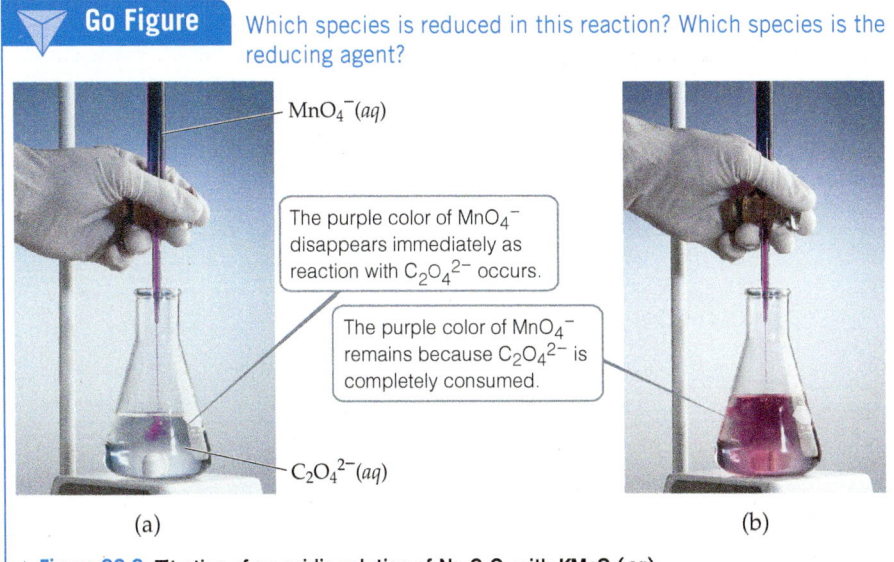

Go Figure   Which species is reduced in this reaction? Which species is the reducing agent?

$MnO_4^-(aq)$

The purple color of MnO_4^- disappears immediately as reaction with $C_2O_4^{2-}$ occurs.

The purple color of MnO_4^- remains because $C_2O_4^{2-}$ is completely consumed.

$C_2O_4^{2-}(aq)$

(a)   (b)

▲ **Figure 20.2** Titration of an acidic solution of $Na_2C_2O_4$ with $KMnO_4(aq)$.

Now there are equal numbers of each type of atom on the two sides of the equation, but the charge still needs to be balanced. The charge of the reactants is $8(1+) + 1(1-) = 7+$, whereas that of the products is $1(2+) + 4(0) = 2+$. To balance the charge, we add five electrons to the reactant side (Step 2d):

$$5\,e^- + 8\,H^+(aq) + MnO_4^-(aq) \longrightarrow Mn^{2+}(aq) + 4\,H_2O(l)$$

We can use oxidation states to check our result. In this half-reaction, Mn goes from the +7 oxidation state in MnO_4^- to the +2 oxidation state of Mn^{2+}. Therefore, each Mn atom gains five electrons, in agreement with our balanced half-reaction.

In the oxalate half-reaction, we have C and O balanced (Step 2a). We balance the charge (Step 2d) by adding two electrons to the products:

$$C_2O_4^{2-}(aq) \longrightarrow 2\,CO_2(g) + 2\,e^-$$

We can check this result using oxidation states. Carbon goes from the +3 oxidation state in $C_2O_4^{2-}$ to the +4 oxidation state in CO_2. Thus, each C atom loses one electron. There are two C atoms in $C_2O_4^{2-}$, however, so they lose two electrons, in agreement with our balanced half-reaction.

Now we multiply each half-reaction by an appropriate integer so that the number of electrons gained in one half-reaction equals the number of electrons lost in the other (Step 3). In this case, we multiply the MnO_4^- half-reaction by 2 and the $C_2O_4^{2-}$ half-reaction by 5:

$$10\,e^- + 16\,H^+(aq) + 2\,MnO_4^-(aq) \longrightarrow 2\,Mn^{2+}(aq) + 8\,H_2O(l)$$

$$\underline{\hspace{1.2cm} 5\,C_2O_4^{2-}(aq) \longrightarrow 10\,CO_2(g) + 10\,e^- \hspace{1.2cm}}$$

$$16\,H^+(aq) + 2\,MnO_4^-(aq) + 5\,C_2O_4^{2-}(aq) \longrightarrow 2\,Mn^{2+}(aq) + 8\,H_2O(l) + 10\,CO_2(g)$$

The balanced equation is the sum of the balanced half-reactions (Step 4). Note that the electrons on the reactant and product sides of the equation cancel each other.

We check the balanced equation by counting atoms and charges (Step 5). There are 16 H, 2 Mn, 28 O, 10 C, and a net charge of 4+ on each side of the equation, confirming that the equation is correctly balanced.

Sample Exercise 20.2
Balancing Redox Equations in Acidic Solution

Complete and balance this equation by the method of half-reactions:

$$Cr_2O_7^{2-}(aq) + Cl^-(aq) \longrightarrow Cr^{3+}(aq) + Cl_2(g) \text{ (acidic solution)}$$

SOLUTION

Analyze We are given an incomplete, unbalanced (skeleton) equation for a redox reaction occurring in acidic solution and asked to complete and balance it.

Plan We use the half-reaction procedure we just learned.

Solve

Step 1: We divide the equation into two half-reactions:

$$Cr_2O_7^{2-}(aq) \longrightarrow Cr^{3+}(aq)$$

$$Cl^-(aq) \longrightarrow Cl_2(g)$$

Step 2: We balance each half-reaction. In the first half-reaction the presence of one $Cr_2O_7^{2-}$ among the reactants requires two Cr^{3+} among the products. The seven oxygen atoms in $Cr_2O_7^{2-}$ are balanced by adding seven H_2O to the products. The 14 hydrogen atoms in 7 H_2O are then balanced by adding 14 H^+ to the reactants:

$$14\,H^+(aq) + Cr_2O_7^{2-}(aq) \longrightarrow 2\,Cr^{3+}(aq) + 7\,H_2O(l)$$

We then balance the charge by adding electrons to the left side of the equation so that the total charge is the same on the two sides:

$$6\,e^- + 14\,H^+(aq) + Cr_2O_7^{2-}(aq) \longrightarrow 2\,Cr^{3+}(aq) + 7\,H_2O(l)$$

We can check this result by looking at the oxidation state changes. Each chromium atom goes from +6 to +3, gaining three electrons; therefore, the two Cr atoms in $Cr_2O_7^{2-}$ gain six electrons, in agreement with our half-reaction.

In the second half-reaction, two Cl^- are required to balance one Cl_2:

$$2\,Cl^-(aq) \longrightarrow Cl_2(g)$$

We add two electrons to the right side to attain charge balance:

$$2\,Cl^-(aq) \longrightarrow Cl_2(g) + 2\,e^-$$

This result agrees with the oxidation state changes. Each chlorine atom goes from −1 to 0, losing one electron; therefore, the two chlorine atoms lose two electrons.

Step 3: We equalize the number of electrons transferred in the two half-reactions. To do so, we multiply the Cl half-reaction by 3 so that the number of electrons gained in the Cr half-reaction (6) equals the number lost in the Cl half-reaction, allowing the electrons to cancel when the half-reactions are added:

$$6\,Cl^-(aq) \longrightarrow 3\,Cl_2(g) + 6\,e^-$$

Step 4: The equations are added to give the balanced equation:

$$14\,H^+(aq) + Cr_2O_7^{2-}(aq) + 6\,Cl^-(aq) \longrightarrow 2\,Cr^{3+}(aq) + 7\,H_2O(l) + 3\,Cl_2(g)$$

Step 5: There are equal numbers of atoms of each kind on the two sides of the equation (14 H, 2 Cr, 7 O, 6 Cl). In addition, the charge is the same on the two sides (6+). Thus, the equation is balanced.

▶ **Practice Exercise**

Complete and balance the following equation in acidic solution using the method of half-reactions.

$$Cu(s) + NO_3^-(aq) \longrightarrow Cu^{2+}(aq) + NO_2(g)$$

Balancing Equations for Reactions Occurring in Basic Solution

If a redox reaction occurs in basic solution, the equation must be balanced by using OH^- and H_2O rather than H^+ and H_2O. Because the water molecule and the hydroxide ion both contain hydrogen, this approach can take more moving back and forth from one side of the equation to the other to arrive at the appropriate half-reaction. Here we use an alternate approach that is often simpler and leads to the same balanced equation.

How to Balance Redox Reactions in Basic Aqueous Solution

1. Balance the half-reactions as if they occurred in acidic solution.

2. Count the number of H^+ in each half-reaction, and then add the same number of OH^- to each side of the half-reaction.

Using this method, you see that the reaction is mass-balanced because you are adding the same thing to both sides. In essence, what you are doing is "neutralizing" the protons to form water ($H^+ + OH^- \longrightarrow H_2O$) on the side containing H^+, while the OH^- ions end up on the opposite side. The resulting water molecules can be canceled as needed.

Sample Exercise 20.3
Balancing Redox Equations in Basic Solution

Complete and balance this equation for a redox reaction that takes place in basic solution:

$$CN^-(aq) + MnO_4^-(aq) \longrightarrow CNO^-(aq) + MnO_2(s) \text{ (basic solution)}$$

SOLUTION

Analyze We are given an incomplete equation for a basic redox reaction and asked to balance it.

Plan We go through the first steps of our procedure as if the reaction were occurring in acidic solution. We then add the appropriate number of OH^- to each side of the equation, combining H^+ and OH^- to form H_2O. We complete the process by simplifying the equation.

Solve

Step 1: We write the incomplete, unbalanced half-reactions:

$$CN^-(aq) \longrightarrow CNO^-(aq)$$
$$MnO_4^-(aq) \longrightarrow MnO_2(s)$$

Step 2: We balance each half-reaction as if it took place in acidic solution:

$$CN^-(aq) + H_2O(l) \longrightarrow CNO^-(aq) + 2\,H^+(aq) + 2\,e^-$$
$$3\,e^- + 4\,H^+(aq) + MnO_4^-(aq) \longrightarrow MnO_2(s) + 2\,H_2O(l)$$

Now we must take into account that the reaction occurs in basic solution, adding OH^- to both sides of both half-reactions to neutralize H^+:

$$CN^-(aq) + H_2O(l) + 2\,OH^-(aq) \longrightarrow CNO^-(aq) + 2\,H^+(aq) + 2\,e^- + 2\,OH^-(aq)$$
$$3\,e^- + 4\,H^+(aq) + MnO_4^-(aq) + 4\,OH^-(aq) \longrightarrow MnO_2(s) + 2\,H_2O(l) + 4\,OH^-(aq)$$

We "neutralize" H^+ and OH^- by forming H_2O when they are on the same side of either half-reaction:

$$CN^-(aq) + H_2O(l) + 2\,OH^-(aq) \longrightarrow CNO^-(aq) + 2\,H_2O(l) + 2\,e^-$$
$$3\,e^- + 4\,H_2O(l) + MnO_4^-(aq) \longrightarrow MnO_2(s) + 2\,H_2O(l) + 4\,OH^-(aq)$$

Next, we cancel water molecules that appear as both reactants and products:

$$CN^-(aq) + 2\,OH^-(aq) \longrightarrow CNO^-(aq) + H_2O(l) + 2\,e^-$$
$$3\,e^- + 2\,H_2O(l) + MnO_4^-(aq) \longrightarrow MnO_2(s) + 4\,OH^-(aq)$$

Both half-reactions are now balanced. You can check the atoms and the overall charge.

Step 3: We multiply the cyanide half-reaction by 3, which gives six electrons on the product side, and multiply the permanganate half-reaction by 2, which gives six electrons on the reactant side:

$$3\,CN^-(aq) + 6\,OH^-(aq) \longrightarrow 3\,CNO^-(aq) + 3\,H_2O(l) + 6\,e^-$$
$$6\,e^- + 4\,H_2O(l) + 2\,MnO_4^-(aq) \longrightarrow 2\,MnO_2(s) + 8\,OH^-(aq)$$

Step 4: We add the two half-reactions together and simplify by canceling species that appear as both reactants and products:

$$3\,CN^-(aq) + H_2O(l) + 2\,MnO_4^-(aq) \longrightarrow 3\,CNO^-(aq) + 2\,MnO_2(s) + 2\,OH^-(aq)$$

Step 5: Check that the atoms and charges are balanced.

There are 3 C, 3 N, 2 H, 9 O, 2 Mn, and a charge of 5− on both sides of the equation.

Comment It is important to remember that this procedure does not imply that H^+ ions are involved in the chemical reaction. Recall that in aqueous solutions at 25 °C, $K_w = [H^+][OH^-] = 1.0 \times 10^{-14}$. Thus, $[H^+]$ is very small in basic solutions. (Section 16.3)

▶ **Practice Exercise**
Complete and balance the following oxidation–reduction reaction in basic solution:

$$Cr(OH)_3(s) + ClO^-(aq) \longrightarrow CrO_4^{2-}(aq) + Cl_2(g)$$

Self-Assessment Exercises

SAE 20.5 Calcium metal reacts with water to form aqueous calcium hydroxide and hydrogen gas. What is the balanced half-reaction for the oxidation that occurs during this reaction?

(a) $2\,H_2O(l) \longrightarrow 2\,H_2(g) + O_2(g)$
(b) $Ca(s) + 2\,H_2O(l) \longrightarrow Ca(OH)_2(aq) + H_2(g)$
(c) $Ca(s) \longrightarrow Ca^{2+}(aq) + 2\,e^-$
(d) $2\,H_2O(l) + 2\,e^- \longrightarrow 2\,OH^-(aq) + H_2(g)$

SAE 20.6 Use the method of half-reactions to balance the following redox reaction between copper and nitric acid in *acidic* solution:

$$Cu(s) + HNO_3(aq) \longrightarrow Cu^{2+}(aq) + NO_2(g)$$

In the balanced equation, there are _____ H_2O molecules on the _____ side. (a) 1, reactant (b) 2, reactant (c) 1, product (d) 2, product (e) There are no H_2O molecules in the balanced reaction.

SAE 20.7 Use the method of half-reactions to balance the following redox reaction between hydrogen peroxide and chlorine dioxide in *basic* solution:

$$H_2O_2(aq) + ClO_2(aq) \longrightarrow ClO_2^-(aq) + O_2(g)$$

In the balanced reaction, there are _____ OH⁻ ions on the _____ side. (**a**) 1, reactant (**b**) 2, reactant (**c**) 1, product (**d**) 2, product (**e**) There are no OH⁻ ions in the balanced reaction.

20.3 | Voltaic Cells

The energy released in a spontaneous redox reaction can be used to perform electrical work. This task is accomplished through a **voltaic** (or **galvanic**) **cell**, a device in which the transfer of electrons takes place through an external pathway rather than directly between reactants present in the same reaction vessel.

One such spontaneous reaction occurs when a strip of zinc is placed in contact with a solution containing $Cu^{2+}(aq)$. As the reaction proceeds, the blue color of $Cu^{2+}(aq)$ ions fades and copper metal deposits on the zinc. At the same time, the zinc begins to dissolve. These transformations, shown in **Figure 20.3**, are summarized by the equation

$$Zn(s) + Cu^{2+}(aq) \longrightarrow Zn^{2+}(aq) + Cu(s) \qquad [20.7]$$

Figure 20.4 shows a voltaic cell that uses the redox reaction given in Equation 20.7. Although the setup in Figure 20.4 is more complex than that in Figure 20.3, the reaction is the same in both cases. The significant difference is that in the voltaic cell the Zn metal and $Cu^{2+}(aq)$ are not in direct contact with each other. Instead, Zn metal is in contact with $Zn^{2+}(aq)$ in one compartment, and Cu metal is in contact with $Cu^{2+}(aq)$ in the other compartment. Consequently, $Cu^{2+}(aq)$ reduction can occur only by the flow of electrons through an external circuit—namely, a wire connecting the Zn and Cu strips. Electrons flowing through a wire and ions moving in solution both constitute an *electrical current*. This flow of electrical charge can be used to accomplish electrical work.

> **Learning Objectives**
>
> When you finish Section 20.3, you should be able to:
>
> ▶ Explain the operation of a voltaic cell and identify the key components (anode, cathode, salt bridge).
>
> ▶ Determine the half-reactions that occur at each electrode of a voltaic cell from the overall cell reaction.

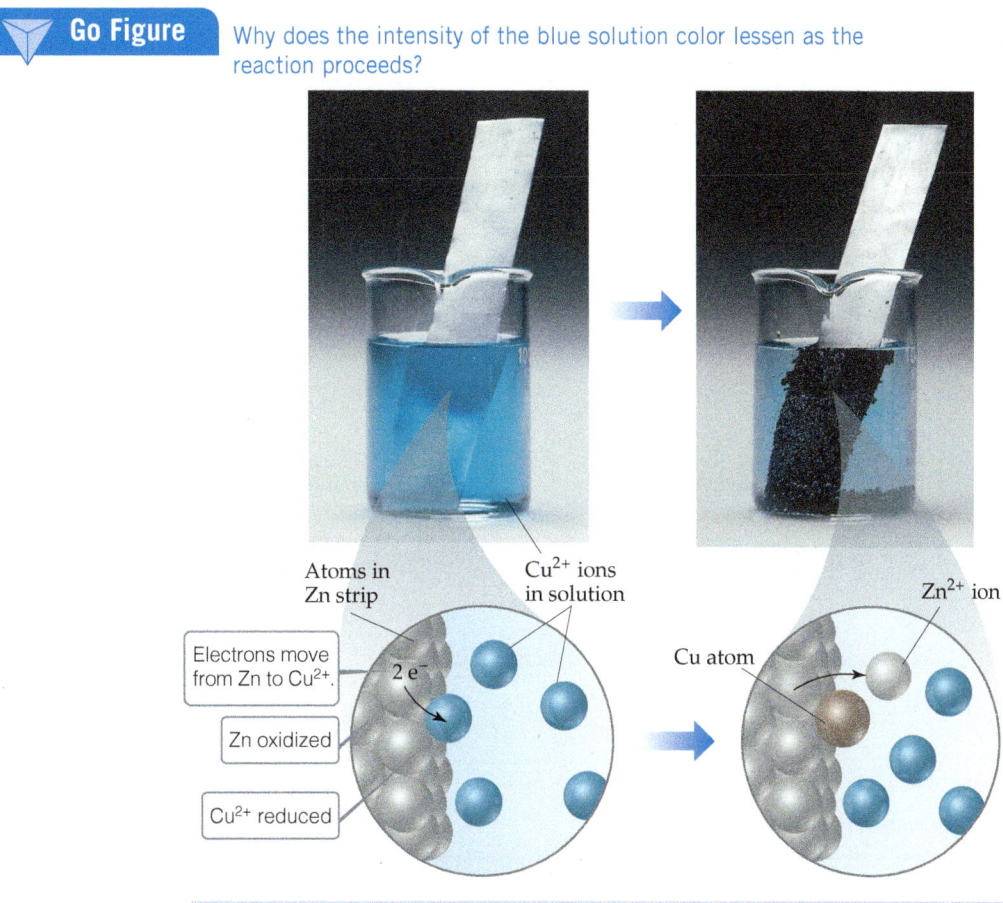

▼ **Go Figure** Why does the intensity of the blue solution color lessen as the reaction proceeds?

Atoms in Zn strip

Cu^{2+} ions in solution

Zn^{2+} ion

Electrons move from Zn to Cu^{2+}.

Zn oxidized

Cu^{2+} reduced

$2 e^-$

Cu atom

$$Zn(s) + Cu^{2+}(aq) \longrightarrow Zn^{2+}(aq) + Cu(s)$$

▲ **Figure 20.3** A spontaneous oxidation–reduction reaction involving zinc and copper.

Go Figure

Which metal, Cu or Zn, is oxidized in this voltaic cell?

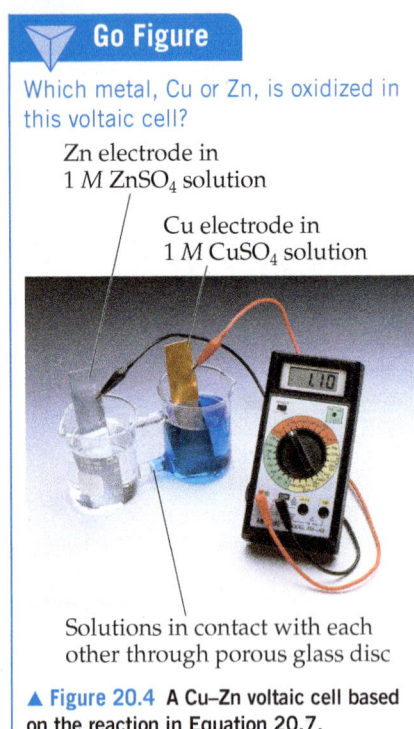

Zn electrode in 1 M $ZnSO_4$ solution

Cu electrode in 1 M $CuSO_4$ solution

Solutions in contact with each other through porous glass disc

▲ **Figure 20.4** A Cu–Zn voltaic cell based on the reaction in Equation 20.7.

The two solid metals connected by the external circuit are called *electrodes*. By definition, the electrode at which oxidation occurs is the **anode,** and the electrode at which reduction occurs is the **cathode.*** The electrodes can be made of materials that participate in the reaction, as in the present example. Over the course of the reaction, the Zn electrode gradually disappears and the copper electrode gains mass. More typically, the electrodes are made of a conducting material, such as platinum or graphite, that does not gain or lose mass during the reaction but serves as a surface at which electrons are transferred.

Each compartment of a voltaic cell is called a *half-cell*. One half-cell is the site of the oxidation half-reaction, and the other is the site of the reduction half-reaction. In our present example, Zn is oxidized and Cu^{2+} is reduced:

Anode *(oxidation half-reaction)*	$Zn(s) \longrightarrow Zn^{2+}(aq) + 2\,e^-$
Cathode *(reduction half-reaction)*	$Cu^{2+}(aq) + 2\,e^- \longrightarrow Cu(s)$

Electrons become available as zinc metal is oxidized at the anode. They flow through the external circuit to the cathode, where they are consumed as $Cu^{2+}(aq)$ is reduced. Because $Zn(s)$ is oxidized in the cell, the zinc electrode loses mass, and the concentration of the $Zn^{2+}(aq)$ solution increases as the cell operates. At the same time, the Cu electrode gains mass, and the $Cu^{2+}(aq)$ solution becomes less concentrated as $Cu^{2+}(aq)$ is reduced to $Cu(s)$.

For a voltaic cell to work, the solutions in the two half-cells must remain electrically neutral. As Zn is oxidized in the anode half-cell, $Zn^{2+}(aq)$ ions enter the solution, upsetting the initial Zn^{2+}/SO_4^{2-} charge balance. To keep the solution electrically neutral, there must be some means for $Zn^{2+}(aq)$ cations to migrate out of the anode half-cell and for anions to migrate in. Similarly, the reduction of $Cu^{2+}(aq)$ at the cathode removes these cations from the solution, leaving an excess of SO_4^{2-} anions in that half-cell. To maintain electrical neutrality, some of these anions must migrate out of the cathode half-cell, and positive ions must migrate in. In fact, no measurable electron flow occurs between electrodes unless a means is provided for ions to migrate through the solution from one half-cell to the other, thereby completing the circuit.

In Figure 20.4, a porous glass disc separating the two half-cells allows ions to migrate and maintain the electrical neutrality of the solutions. In **Figure 20.5**, a *salt bridge* serves

Go Figure

How is electrical balance maintained in the left beaker as $Zn^{2+}(aq)$ are formed at the anode?

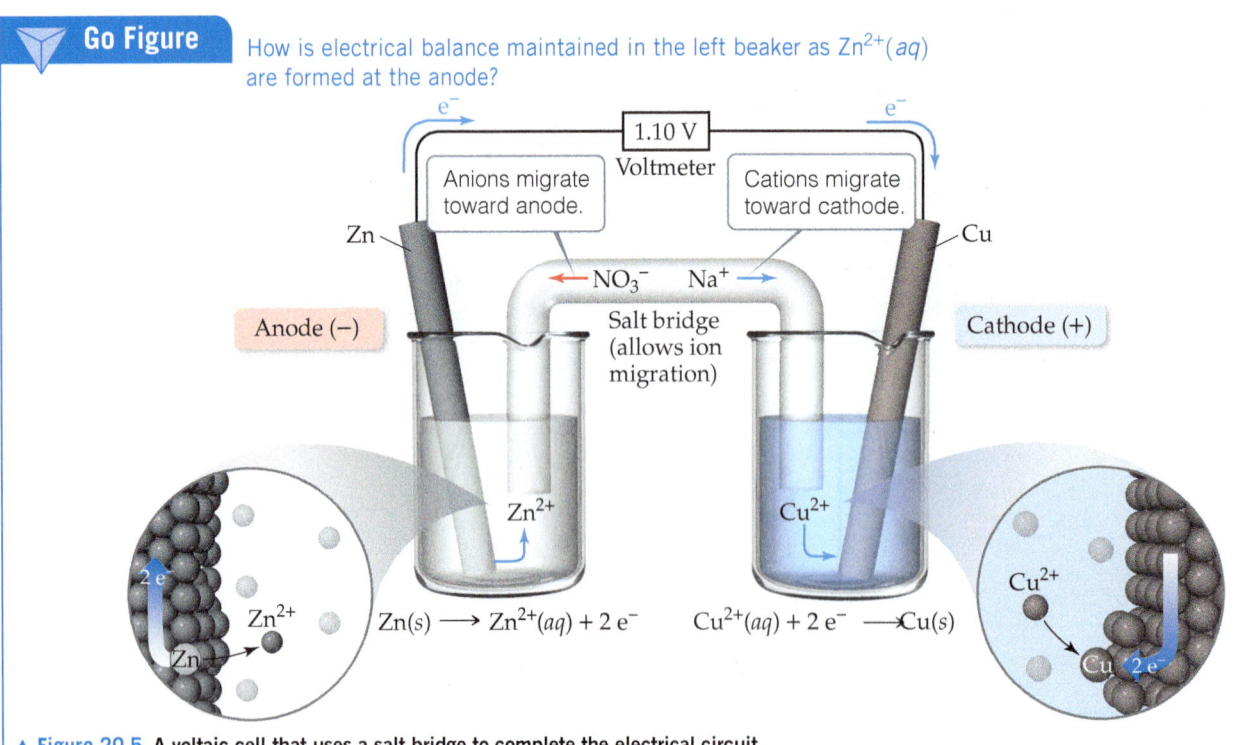

▲ **Figure 20.5** A voltaic cell that uses a salt bridge to complete the electrical circuit.

---

*To help remember these definitions, note that *anode* and *oxidation* both begin with a vowel, whereas *cathode* and *reduction* both begin with a consonant.

this purpose. The salt bridge consists of a U-shaped tube containing an electrolyte solution, such as $NaNO_3(aq)$, whose ions will not react with other ions in the voltaic cell or with the electrodes. The electrolyte is often incorporated into a paste or gel so that the electrolyte solution does not pour out when the U-tube is inverted. As oxidation and reduction proceed at the electrodes, ions from the salt bridge migrate into the two half-cells—cations migrating to the cathode half-cell and anions migrating to the anode half-cell—to neutralize charge in the half-cell solutions. Whichever device is used to allow ions to migrate between half-cells, *anions always migrate toward the anode and cations always migrate toward the cathode*.

Figure 20.6 summarizes the various relationships in a voltaic cell. Notice in particular that *electrons flow from the anode through the external circuit to the cathode*. Because of this directional flow, the anode in a voltaic cell is labeled with a negative sign and the cathode is labeled with a positive sign. We can envision the electrons as being attracted to the positive cathode from the negative anode through the external circuit.

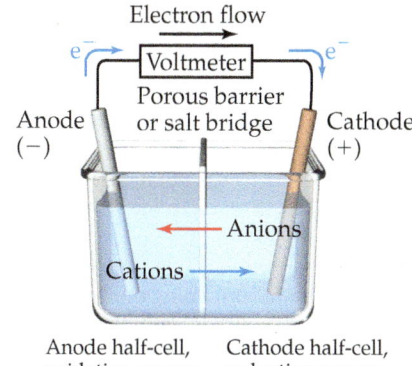

▲ **Figure 20.6 Summary of reactions occurring in a voltaic cell.** The half-cells can be separated by either a porous glass disc (as in Figure 20.4) or a salt bridge (as in Figure 20.5).

Sample Exercise 20.4
Describing a Voltaic Cell

The oxidation–reduction reaction

$$Cr_2O_7^{2-}(aq) + 14 H^+(aq) + 6 I^-(aq) \longrightarrow 2 Cr^{3+}(aq) + 3 I_2(s) + 7 H_2O(l)$$

is spontaneous. A solution containing $K_2Cr_2O_7$ and H_2SO_4 is poured into one beaker, and a solution of KI is poured into another. A salt bridge is used to join the beakers. A metallic conductor that will not react with either solution (such as platinum foil) is suspended in each solution, and the two conductors are connected with wires through a voltmeter or some other device to detect an electric current. The resulting voltaic cell generates an electric current. Indicate the reaction occurring at the anode, the reaction at the cathode, the direction of electron migration, the direction of ion migration, and the signs of the electrodes.

SOLUTION

Analyze We are given the equation for a spontaneous reaction that takes place in a voltaic cell and a description of how the cell is constructed. We are asked to write the half-reactions occurring at the anode and at the cathode, as well as the directions of electron and ion movements and the signs assigned to the electrodes.

Plan Our first step is to divide the chemical equation into half-reactions so that we can identify the oxidation and the reduction processes. We then use the definitions of anode and cathode and the other terminologies summarized in Figure 20.6.

Solve In one half-reaction, $Cr_2O_7^{2-}(aq)$ is converted into $Cr^{3+}(aq)$. Starting with these ions and then completing and balancing the half-reaction, we have

$$Cr_2O_7^{2-}(aq) + 14 H^+(aq) + 6 e^- \longrightarrow 2 Cr^{3+}(aq) + 7 H_2O(l)$$

In the other half-reaction, $I^-(aq)$ is converted to $I_2(s)$:

$$6 I^-(aq) \longrightarrow 3 I_2(s) + 6 e^-$$

Now we can use the summary in Figure 20.6 to help us describe the voltaic cell. The first half-reaction is the reduction process (electrons on the reactant side of the equation). By definition, the reduction process occurs at the cathode. The second half-reaction is the oxidation process (electrons on the product side of the equation), which occurs at the anode.

The I^- ions are the source of electrons, and the $Cr_2O_7^{2-}$ ions accept the electrons. Hence, the electrons flow through the external circuit from the electrode immersed in the KI solution (the anode) to the electrode immersed in the $K_2Cr_2O_7$–H_2SO_4 solution (the cathode). The electrodes themselves do not react in any way; they merely provide a means of transferring electrons from or to the solutions. The cations move through the solutions toward the cathode, and the anions move toward the anode. The anode (from which the electrons move) is the negative electrode, and the cathode (toward which the electrons move) is the positive electrode.

▶ **Practice Exercise**
The two half-reactions in a voltaic cell are

$$Zn(s) \longrightarrow Zn^{2+}(aq) + 2 e^- \quad (electrode = Zn)$$

$$ClO_3^-(aq) + 6 H^+(aq) + 6 e^- \longrightarrow Cl^-(aq) + 3 H_2O(l)$$

$$(electrode = Pt)$$

(a) Indicate which reaction occurs at the anode and which at the cathode. (b) Does the zinc electrode gain, lose, or retain the same mass as the reaction proceeds? (c) Does the platinum electrode gain, lose, or retain the same mass as the reaction proceeds? (d) Which electrode is positive?

Self-Assessment Exercises

SAE 20.8 Consider a voltaic cell with the overall reaction $Ni(s) + 2 Ag^+(aq) \longrightarrow Ni^{2+}(aq) + 2 Ag(s)$. Which of the following statements is or are *true* about this cell?

(**i**) The silver electrode is the anode.
(**ii**) The silver electrode's mass will increase as the reaction proceeds.
(**iii**) Anions migrate through the salt bridge to the compartment containing the Ni electrode.

(**a**) only i (**b**) only ii (**c**) only iii (**d**) both ii and iii (**e**) All three statements are true.

SAE 20.9 If the overall reaction for a voltaic cell is $Hg_2Cl_2(s) + H_2(g) \longrightarrow 2 Hg(l) + 2 Cl^-(aq) + 2 H^+(aq)$, what is the cathode half-reaction?

(**a**) $H_2(g) \longrightarrow 2 H^+(aq)$
(**b**) $H_2(g) \longrightarrow 2 H^+(aq) + 2 e^-$
(**c**) $Hg_2Cl_2(s) \longrightarrow 2 Hg(l) + 2 Cl^-(aq)$
(**d**) $Hg_2Cl_2(s) + 2 e^- \longrightarrow 2 Hg(l) + 2 Cl^-(aq)$
(**e**) $Hg_2Cl_2(s) \longrightarrow 2 Hg(l) + Cl_2(aq)$

20.4 | Cell Potentials under Standard Conditions

Learning Objectives

When you finish **Section 20.4**, you should be able to:

▶ Use tabulated values of standard reduction potentials to calculate the standard cell potential, E°_{cell}, (standard emf) of a voltaic cell.

▶ Determine the relative strengths of oxidizing and reducing agents from their standard reduction potentials.

Why do electrons transfer spontaneously from a Zn atom to a Cu^{2+} ion, either directly as in Figure 20.3 or through an external circuit as in Figure 20.4? In a simple sense, we can compare the movement of electrons to the flow of water in a waterfall (**Figure 20.7**). Water flows spontaneously over a waterfall because of a difference in potential energy between the top of the falls and the bottom. (Section 5.1) In a similar fashion, electrons flow spontaneously through an external circuit from the anode of a voltaic cell to the cathode because of a difference in potential energy. The potential energy of electrons is higher in the anode than in the cathode. Thus, electrons flow spontaneously toward the electrode with the more positive electrical potential.

The difference in potential energy per electrical charge (the *potential difference*) between two electrodes is measured in *volts*. One volt (V) is the potential difference required to impart 1 joule (J) of energy to a charge of 1 coulomb (C):

$$1 V = 1 \frac{J}{C}$$

Recall that one electron has a charge of 1.602×10^{-19} C. (Section 2.2)

The potential difference between the two electrodes of a voltaic cell is called the **cell potential**, denoted E_{cell}. Because the potential difference provides the driving force that pushes electrons through the external circuit, we also call it the **electromotive** ("causing electron motion") **force**, or **emf**. Because E_{cell} is measured in volts, it is also commonly called the *voltage* of the cell.

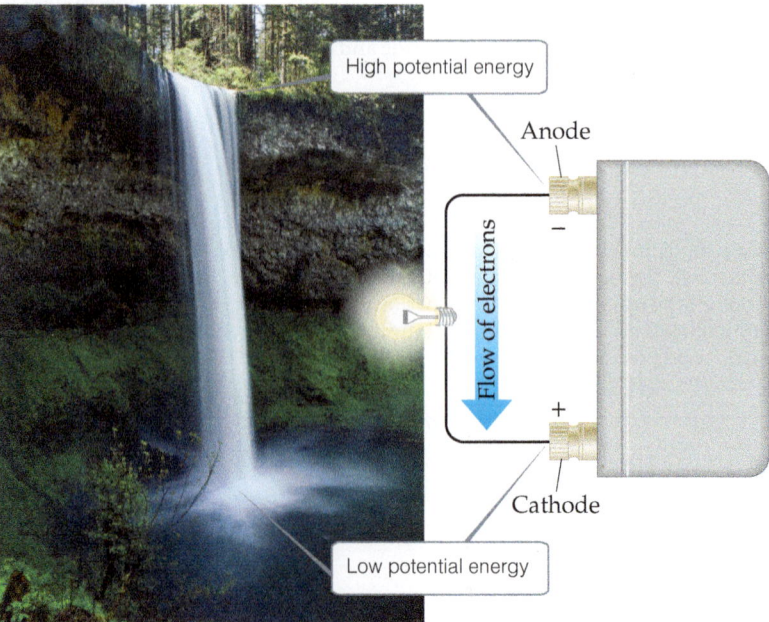

▶ **Figure 20.7 Water analogy for electron flow.** Just as water spontaneously flows downhill, electrons flow spontaneously from the anode to the cathode in a voltaic cell.

The cell potential of any voltaic cell is positive. The magnitude of the cell potential depends on the reactions that occur at the cathode and anode, the concentrations of reactants and products, and the temperature, which we will assume to be 25 °C unless otherwise noted. In this section, we focus on cells that are operated at 25 °C under *standard conditions*. Recall from Table 19.2 that standard conditions include 1 M concentration for reactants and products in solution and 1 atm pressure for gaseous reactants and products. The cell potential under standard conditions is called either the **standard cell potential** or **standard emf** and is denoted E°_{cell}. For the Zn–Cu voltaic cell in Figure 20.5, for example, the standard cell potential at 25 °C is +1.10 V:

$$Zn(s) + Cu^{2+}(aq, 1\,M) \longrightarrow Zn^{2+}(aq, 1\,M) + Cu(s) \quad E^\circ_{cell} = +1.10\,V$$

Recall that the superscript ° indicates standard-state conditions. (Section 5.7)

Standard Reduction Potentials

The standard cell potential of a voltaic cell, E°_{cell}, depends on the particular cathode and anode half-cells. We could, in principle, tabulate the standard cell potentials for all possible cathode–anode combinations. However, it is not necessary to undertake this arduous task. Rather, we can assign a standard potential to each half-cell and then use these half-cell potentials to determine E°_{cell}. The cell potential is the difference between two half-cell potentials. By convention, the potential associated with each electrode is chosen to be the potential for *reduction* at that electrode. Thus, standard half-cell potentials are tabulated for reduction reactions, which means they are **standard reduction potentials**, denoted E°_{red}. The standard cell potential, E°_{cell} is the standard reduction potential of the cathode reaction, E°_{red} (cathode), *minus* the standard reduction potential of the anode reaction, E°_{red} (anode):

$$E^\circ_{cell} = E^\circ_{red}\,(\text{cathode}) - E^\circ_{red}\,(\text{anode}) \qquad [20.8]$$

It is not possible to measure the standard reduction potential of a half-reaction directly. If we assign a standard reduction potential to a certain reference half-reaction, however, we can then determine the standard reduction potentials of other half-reactions relative to that reference value. The reference half-reaction is the reduction of $H^+(aq)$ to $H_2(g)$ under standard conditions, which is assigned a standard reduction potential of exactly 0 V:

$$2\,H^+(aq, 1\,M) + 2\,e^- \longrightarrow H_2(g, 1\,atm) \quad E^\circ_{red} = 0\,V \qquad [20.9]$$

An electrode designed to produce this half-reaction is called a **standard hydrogen electrode** (SHE). A SHE consists of a platinum wire connected to a piece of platinum foil covered with finely divided platinum that serves as an inert surface for the reaction (**Figure 20.8**). The SHE allows the platinum to be in contact with both 1 $M\,H^+(aq)$ and a

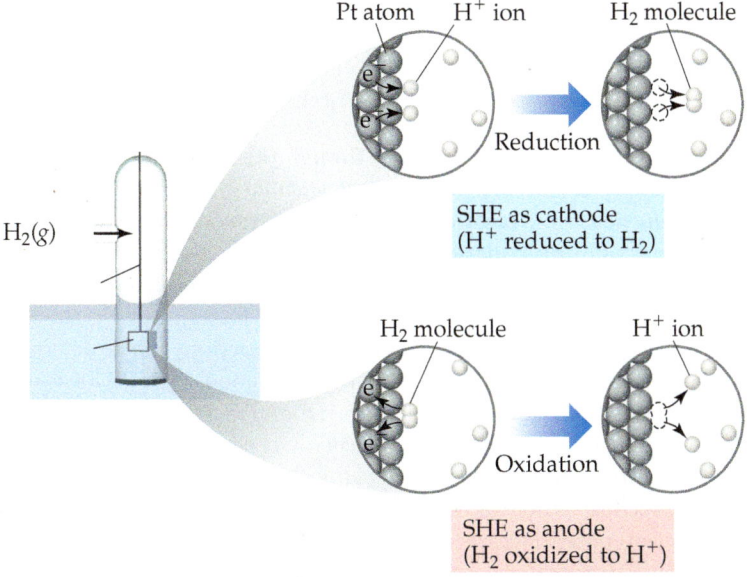

▲ **Figure 20.8** The standard hydrogen electrode (SHE) is used as a reference electrode.

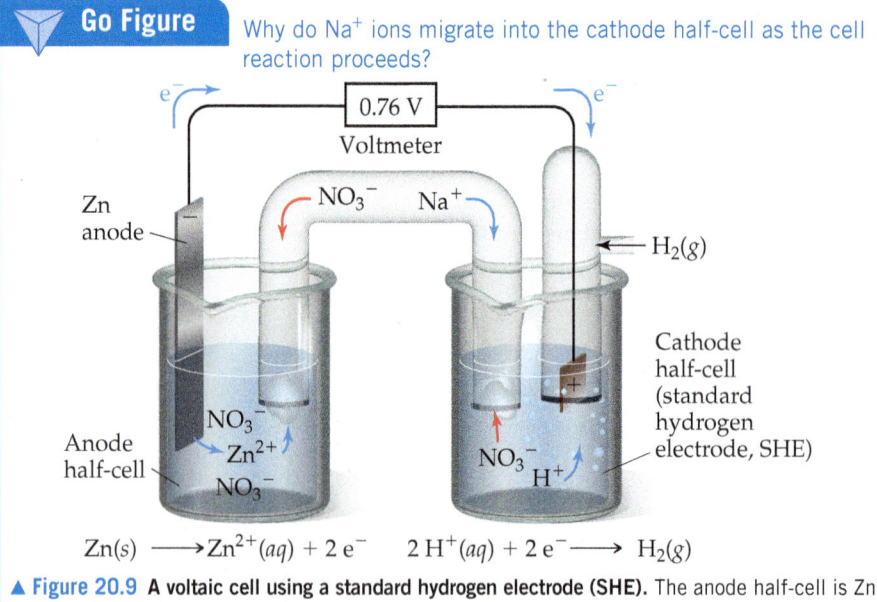

▲ **Figure 20.9 A voltaic cell using a standard hydrogen electrode (SHE).** The anode half-cell is Zn metal in a $Zn(NO_3)_2(aq)$ solution, and the cathode half-cell is the SHE in a $HNO_3(aq)$ solution.

stream of hydrogen gas at 1 atm. The SHE can operate as either the anode or cathode of a cell, depending on the nature of the other electrode.

Figure 20.9 shows a voltaic cell using a SHE. The spontaneous reaction is the one shown in Figure 20.1—namely, oxidation of Zn and reduction of H^+:

$$Zn(s) + 2\,H^+(aq) \longrightarrow Zn^{2+}(aq) + H_2(g)$$

When the cell is operated under standard conditions, the cell potential is +0.76 V. By using the standard cell potential ($E^\circ_{cell} = 0.76$ V), the defined standard reduction potential of H^+ ($E^\circ_{red} = 0$ V) and Equation 20.8, we can determine the standard reduction potential for the Zn^{2+}/Zn half-reaction:

$$E^\circ_{cell} = E^\circ_{red}\,(\text{cathode}) - E^\circ_{red}\,(\text{anode})$$

$$+0.76\,V = 0\,V - E^\circ_{red}\,(\text{anode})$$

$$E^\circ_{red}\,(\text{anode}) = -0.76\,V$$

Thus, a standard reduction potential of −0.76 V can be assigned to the reduction of Zn^{2+} to Zn:

$$Zn^{2+}(aq, 1\,M) + 2\,e^- \longrightarrow Zn(s) \quad E^\circ_{red} = -0.76\,V$$

We write the reaction as a reduction even though the Zn reaction in Figure 20.9 is an oxidation. *Whenever we assign an electrical potential to a half-reaction, we write the reaction as a reduction.* Half-reactions, however, are reversible, being able to operate as either reductions or oxidations. Consequently, half-reactions are sometimes written using two arrows (⇌) between reactants and products, as in equilibrium reactions.

The standard reduction potentials for other half-reactions can be determined in a fashion analogous to that used for the Zn^{2+}/Zn half-reaction. **Table 20.1** lists some standard reduction potentials; a more complete list is found in Appendix E. These standard reduction potentials, often called *half-cell potentials*, can be combined to calculate E°_{cell} values for a large variety of voltaic cells.

Because electrical potential measures potential energy per electrical charge, standard reduction potentials are intensive properties. (Section 1.3) In other words,

TABLE 20.1   Standard Reduction Potentials in Water at 25°C

$E°_{red}$(V)	Reduction Half-Reaction
+2.87	$F_2(g) + 2\,e^- \longrightarrow 2\,F^-(aq)$
+1.51	$MnO_4^-(aq) + 8\,H^+(aq) + 5\,e^- \longrightarrow Mn^{2+}(aq) + 4\,H_2O(l)$
+1.36	$Cl_2(g) + 2\,e^- \longrightarrow 2\,Cl^-(aq)$
+1.33	$Cr_2O_7^{2-}(aq) + 14\,H^+(aq) + 6\,e^- \longrightarrow 2\,Cr^{3+}(aq) + 7\,H_2O(l)$
+1.23	$O_2(g) + 4\,H^+(aq) + 4\,e^- \longrightarrow 2\,H_2O(l)$
+1.06	$Br_2(l) + 2\,e^- \longrightarrow 2\,Br^-(aq)$
+0.96	$NO_3^-(aq) + 4\,H^+(aq) + 3\,e^- \longrightarrow NO(g) + 2\,H_2O(l)$
+0.80	$Ag^+(aq) + e^- \longrightarrow Ag(s)$
+0.77	$Fe^{3+}(aq) + e^- \longrightarrow Fe^{2+}(aq)$
+0.68	$O_2(g) + 2\,H^+(aq) + 2\,e^- \longrightarrow H_2O_2(aq)$
+0.59	$MnO_4^-(aq) + 2\,H_2O(l) + 3\,e^- \longrightarrow MnO_2(s) + 4\,OH^-(aq)$
+0.54	$I_2(s) + 2\,e^- \longrightarrow 2\,I^-(aq)$
+0.40	$O_2(g) + 2\,H_2O(l) + 4\,e^- \longrightarrow 4\,OH^-(aq)$
+0.34	$Cu^{2+}(aq) + 2\,e^- \longrightarrow Cu(s)$
0 [defined]	$2\,H^+(aq) + 2\,e^- \longrightarrow H_2(g)$
−0.28	$Ni^{2+}(aq) + 2\,e^- \longrightarrow Ni(s)$
−0.44	$Fe^{2+}(aq) + 2\,e^- \longrightarrow Fe(s)$
−0.76	$Zn^{2+}(aq) + 2\,e^- \longrightarrow Zn(s)$
−0.83	$2\,H_2O(l) + 2\,e^- \longrightarrow H_2(g) + 2\,OH^-(aq)$
−1.66	$Al^{3+}(aq) + 3\,e^- \longrightarrow Al(s)$
−2.71	$Na^+(aq) + e^- \longrightarrow Na(s)$
−3.05	$Li^+(aq) + e^- \longrightarrow Li(s)$

if we increase the amounts of substances in a redox reaction, we increase both the energy and the charges involved, but the ratio of energy (joules) to electrical charge (coulombs) remains constant ($V = J/C$). Thus, *changing the stoichiometric coefficient in a half-reaction does not affect the value of the standard reduction potential.* For example, $E°_{red}$ for the reduction of 10 mol Zn^{2+} is the same as that for the reduction of 1 mol Zn^{2+}:

$$10\,Zn^{2+}(aq, 1\,M) + 20\,e^- \longrightarrow 10\,Zn(s) \qquad E°_{red} = -0.76\,V$$

Sample Exercise 20.5

Calculating $E°_{red}$ from $E°_{cell}$

For the Zn–Cu voltaic cell shown in Figure 20.5, we have

$$Zn(s) + Cu^{2+}(aq, 1\,M) \longrightarrow Zn^{2+}(aq, 1\,M) + Cu(s) \qquad E°_{cell} = 1.10\,V$$

Given that the standard reduction potential of Zn^{2+} to $Zn(s)$ is −0.76 V, calculate the $E°_{red}$ for the reduction of Cu^{2+} to Cu:

$$Cu^{2+}(aq, 1\,M) + 2\,e^- \longrightarrow Cu(s)$$

SOLUTION

Analyze We are given $E°_{cell}$ and $E°_{red}$ for Zn^{2+} and asked to calculate $E°_{red}$ for Cu^{2+}.

Plan In the voltaic cell, Zn is oxidized and is therefore the anode. Thus, the given $E°_{red}$ for Zn^{2+} is $E°_{red}$ (anode). Because

Cu^{2+} is reduced, it is in the cathode half-cell. Thus, the unknown reduction potential for Cu^{2+} is $E°_{red}$ (cathode). Knowing $E°_{cell}$ and $E°_{red}$ (anode), we can use Equation 20.8 to solve for $E°_{red}$ (cathode).

Continued

Solve

$$E^\circ_{cell} = E^\circ_{red} \text{(cathode)} - E^\circ_{red} \text{(anode)}$$

$$1.10 \text{ V} = E^\circ_{red} \text{(cathode)} - (-0.76 \text{ V})$$

$$E^\circ_{red} \text{(cathode)} = 1.10 \text{ V} - 0.76 \text{ V} = 0.34 \text{ V}$$

Check This standard reduction potential agrees with the one listed in Table 20.1.

Comment The standard reduction potential for Cu^{2+} can be represented as $E^\circ_{Cu^{2+}} = 0.34$ V and that for Zn^{2+} as $E^\circ_{Zn^{2+}} = -0.76$ V. The subscript identifies the ion that is reduced in the reduction half-reaction.

▶ **Practice Exercise**

The standard cell potential is 1.46 V for a voltaic cell based on the following half-reactions:

$$In^+(aq) \longrightarrow In^{3+}(aq) + 2\,e^-$$
$$Br_2(l) + 2\,e^- \longrightarrow 2\,Br^-(aq)$$

Using Table 20.1, calculate E°_{red} for the reduction of In^{3+} to In^+.

Sample Exercise 20.6

Calculating E°_{cell} from E°_{red}

Use Table 20.1 to calculate E°_{cell} for the voltaic cell described in Sample Exercise 20.4, which is based on the reaction

$$Cr_2O_7^{2-}(aq) + 14\,H^+(aq) + 6\,I^-(aq) \longrightarrow 2\,Cr^{3+}(aq) + 3\,I_2(s) + 7\,H_2O(l)$$

SOLUTION

Analyze We are given the equation for a redox reaction, and we are asked to use data in Table 20.1 to calculate the standard cell potential for the associated voltaic cell.

Plan Our first step is to identify the half-reactions that occur at the cathode and anode, which we did in Sample Exercise 20.4. Then we use Table 20.1 and Equation 20.8 to calculate the standard cell potential.

Solve

The half-reactions are:

Cathode: $Cr_2O_7^{2-}(aq) + 14\,H^+(aq) + 6\,e^- \longrightarrow 2\,Cr^{3+}(aq) + 7\,H_2O(l)$

Anode: $6\,I^-(aq) \longrightarrow 3\,I_2(s) + 6\,e^-$

According to Table 20.1, the standard reduction potential for the reduction of $Cr_2O_7^{2-}$ to Cr^{3+} is +1.33 V, and the standard reduction potential for the reduction of I_2 to I^- (the reverse of the oxidation half-reaction) is +0.54 V. We use these values in Equation 20.8:

$$E^\circ_{cell} = E^\circ_{red}\text{(cathode)} - E^\circ_{red}\text{(anode)} = 1.33 \text{ V} - 0.54 \text{ V} = 0.79 \text{ V}$$

Comment Although we must multiply the iodide half-reaction by 3 to obtain a balanced equation, we do *not* multiply the E°_{red} value by 3. As we have noted, the standard reduction potential is an intensive property, so it is independent of the stoichiometric coefficients.

Check The cell potential, 0.79 V, is a positive number. As noted previously, a voltaic cell must have a positive potential.

▶ **Practice Exercise**

Using data in Table 20.1, calculate the standard emf for a cell that employs the overall cell reaction $2\,Al(s) + 3\,I_2(s) \longrightarrow 2\,Al^{3+}(aq) + 6\,I^-(aq)$.

For each half-cell in a voltaic cell, the standard reduction potential provides a measure of the tendency for reduction to occur: *The more positive the value of E°_{red}, the greater the tendency for reduction under standard conditions.* In any voltaic cell operating under standard conditions, the E°_{red} value for the reaction at the cathode is more positive than the E°_{red} value for the reaction at the anode. Thus, electrons flow spontaneously through the external circuit from the electrode with the more negative value of E°_{red} to the electrode with the more positive value of E°_{red}. **Figure 20.10** graphically illustrates the relationship between the standard reduction potentials for the two half-reactions in the Zn–Cu voltaic cell of Figure 20.5.

Sample Exercise 20.7

Determining Half-Reactions at Electrodes and Calculating Cell Potentials

A voltaic cell is based on the two standard half-reactions

$$Cd^{2+}(aq) + 2\,e^- \longrightarrow Cd(s)$$
$$Sn^{2+}(aq) + 2\,e^- \longrightarrow Sn(s)$$

Use data in Appendix E to determine **(a)** which half-reaction occurs at the cathode and which occurs at the anode and **(b)** the standard cell potential.

SOLUTION

Analyze We have to look up $E°_{red}$ for two half-reactions. We then use these values first to determine the cathode and the anode and then to calculate the standard cell potential, $E°_{cell}$.

Plan The cathode will have the reduction with the more positive $E°_{red}$ value, and the anode will have the less positive $E°_{red}$ To write the half-reaction at the anode, we reverse the half-reaction written for the reduction, so that the half-reaction is written as an oxidation.

Solve

(a) According to Appendix E, $E°_{red}(Cd^{2+}/Cd) = -0.40\ V$ and $E°_{red}(Sn^{2+}/Sn) = -0.14\ V$. The standard reduction potential for Sn^{2+} is more positive (less negative) than that for Cd^{2+}. Hence, the reduction of Sn^{2+} is the reaction that occurs at the cathode:

Cathode: $Sn^{2+}(aq) + 2\,e^- \longrightarrow Sn(s)$

The anode reaction, therefore, is the loss of electrons by Cd:

Anode: $Cd(s) \longrightarrow Cd^{2+}(aq) + 2\,e^-$

(b) The cell potential is given by the difference in the standard reduction potentials at the cathode and anode (Equation 20.8):

$E°_{cell} = E°_{red}\ (\text{cathode}) - E°_{red}\ (\text{anode})$

$= (-0.14\ V) - (-0.40\ V) = 0.26\ V$

Comment Notice that it is unimportant that the $E°_{red}$ values of both half-reactions are negative; the negative values merely indicate how these reductions compare to the reference reaction, the reduction of $H^+(aq)$.

Check The cell potential is positive, as it must be for a voltaic cell.

▶ **Practice Exercise**
A voltaic cell is based on a Co^{2+}/Co half-cell and a $AgCl/Ag$ half-cell. **(a)** What half-reaction occurs at the anode? **(b)** What is the standard cell potential?

Strengths of Oxidizing and Reducing Agents

Table 20.1 lists half-reactions in the order of decreasing tendency to undergo reduction. For example, F_2 is located at the top of the table, having the most positive value for $E°_{red}$. This makes F_2 the most easily reduced species in Table 20.1 and therefore the strongest oxidizing agent listed.

Among the most frequently used oxidizing agents are the halogens, O_2, and oxyanions such as MnO_4^-, $Cr_2O_7^{2-}$, and NO_3^-, whose central atoms have high positive oxidation states. As seen in Table 20.1, all of these species have large positive values of $E°_{red}$ and therefore easily undergo reduction.

The lower the tendency for a half-reaction to occur in one direction, the greater the tendency for it to occur in the opposite direction. Thus, *the half-reactions with the most negative reduction potentials in Table 20.1 are the ones most easily reversed to run as an oxidation.* Being at the bottom of Table 20.1, $Li^+(aq)$ is the most difficult species in the list to reduce and is therefore the poorest oxidizing agent listed. Although $Li^+(aq)$ has little tendency to gain electrons, the reverse reaction, oxidation of $Li(s)$ to $Li^+(aq)$, is highly favorable. Thus, Li is the strongest reducing agent among the substances listed in Table 20.1. (Note that, because Table 20.1 lists half-reactions as reductions, only the substances on the reactant side of these equations can serve as oxidizing agents; only those on the product side can serve as reducing agents.)

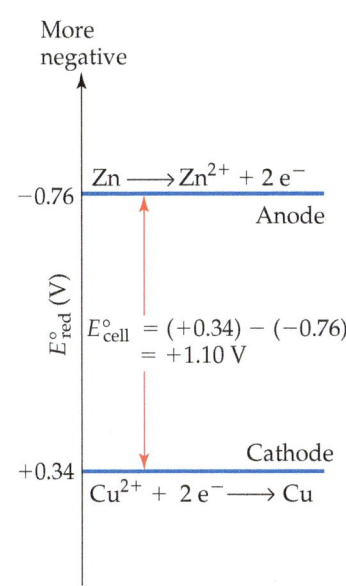

▲ Figure 20.10 Half-cell potentials and standard cell potential for the Zn–Cu voltaic cell.

Commonly used reducing agents include H_2 and the active metals, such as the alkali metals and the alkaline earth metals. Other metals whose cations have negative E_{red}° values, such as Zn and Fe, are also used as reducing agents. Solutions of reducing agents are difficult to store for extended periods because of the ubiquitous presence of O_2, a good oxidizing agent.

The information contained in Table 20.1 is summarized graphically in **Figure 20.11**. For the half-reactions at the top of Table 20.1, the substances on the reactant side of the equation are the most readily reduced species in the table and are therefore the strongest oxidizing agents. Substances on the product side of these reactions are the most difficult to oxidize and so are the weakest reducing agents in the table. Thus, Figure 20.11 shows $F_2(g)$ as the strongest oxidizing agent and $F^-(aq)$ as the weakest reducing agent. Conversely, the reactants in half-reactions at the bottom of Table 20.1, such as $Li^+(aq)$, are the most difficult to reduce and so are the weakest oxidizing agents, while the products of these reactions, such as $Li(s)$, are the most readily oxidized species in the table and so are the strongest reducing agents.

This inverse relationship between oxidizing and reducing strength is similar to the inverse relationship between the strengths of conjugate acids and bases. (Section 16.2 and Figure 16.3)

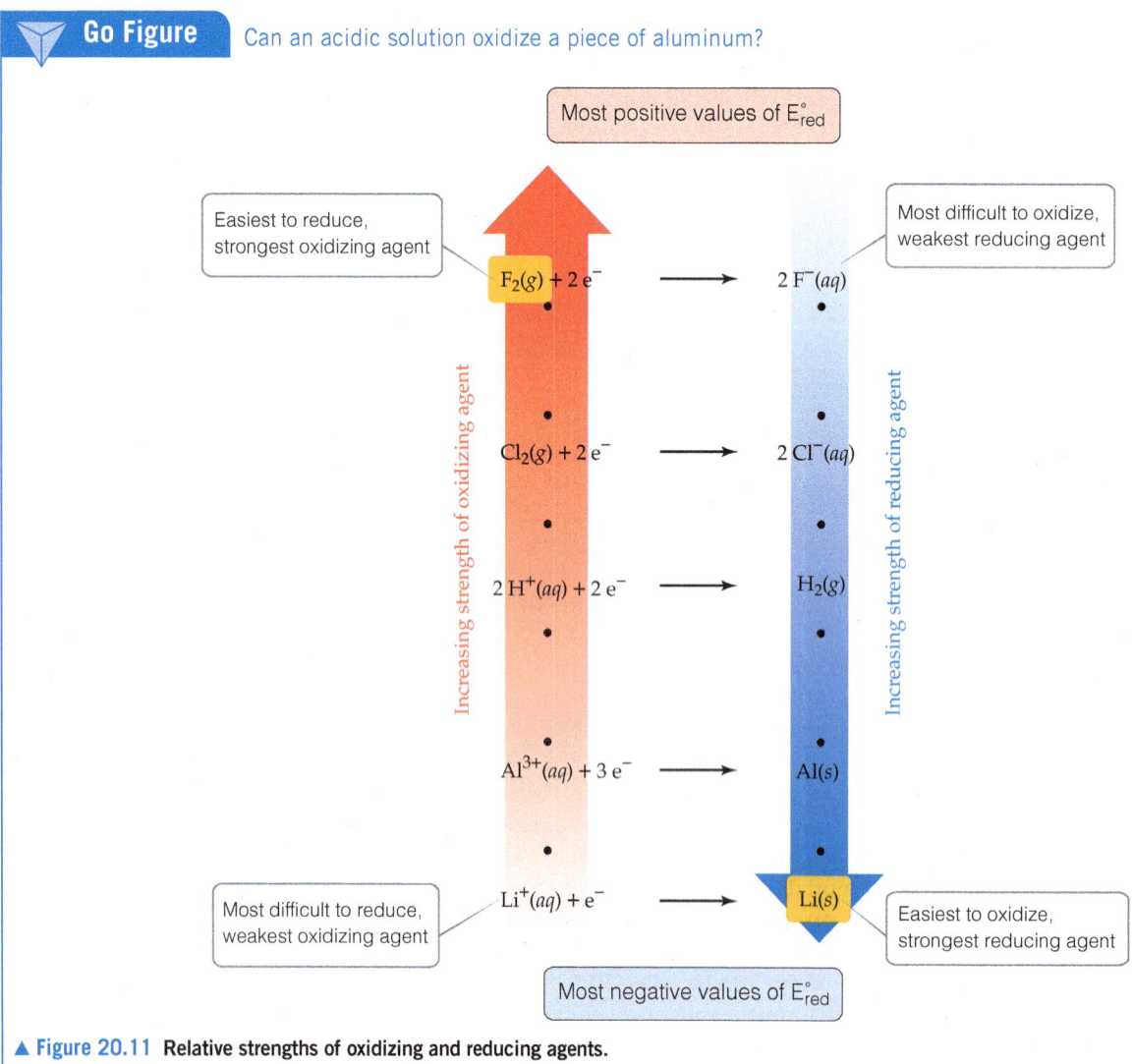

Go Figure   Can an acidic solution oxidize a piece of aluminum?

▲ **Figure 20.11** **Relative strengths of oxidizing and reducing agents.**

Sample Exercise 20.8
Determining Relative Strengths of Oxidizing Agents

Use Table 20.1 to rank the following ions in the order of increasing strength as oxidizing agents: $NO_3^-(aq)$, $Ag^+(aq)$, $Cr_2O_7^{2-}(aq)$.

SOLUTION

Analyze We are asked to rank the abilities of several ions to act as oxidizing agents.

Plan The more readily an ion is reduced (the more positive its E_{red}° value), the stronger it is as an oxidizing agent.

Solve

From Table 20.1, we have:

$$NO_3^-(aq) + 4\,H^+(aq) + 3\,e^- \longrightarrow NO(g) + 2\,H_2O(l) \qquad E_{red}^\circ = +0.96\ V$$

$$Ag^+(aq) + e^- \longrightarrow Ag(s) \qquad E_{red}^\circ = +0.80\ V$$

$$Cr_2O_7^{2-}(aq) + 14\,H^+(aq) + 6\,e^- \longrightarrow 2\,Cr^{3+}(aq) + 7\,H_2O(l) \quad E_{red}^\circ = +1.33\ V$$

Because the standard reduction potential of $Cr_2O_7^{2-}(aq)$ is the most positive, $Cr_2O_7^{2-}(aq)$ is the strongest oxidizing agent of the three. The rank order is:

$$Ag^+ < NO_3^- < Cr_2O_7^{2-}$$

▶ **Practice Exercise**

Use Table 20.1 to rank the following species from the strongest to the weakest reducing agent: $I^-(aq)$, $Fe(s)$, $Al(s)$.

Self-Assessment Exercises

SAE 20.10 Consider a voltaic cell based on one half-cell that contains a Zn electrode in a 1.0 M solution of $Zn^{2+}(aq)$ and another half-cell that contains a Sn electrode in a 1.0 M solution of $Sn^{2+}(aq)$. What is the value of E_{cell}°? See Appendix E for the relevant standard reduction potentials. **(a)** −0.90 V **(b)** −0.62 V **(c)** 0.62 V **(d)** 0.90 V **(e)** It is not possible to create a voltaic cell because both half-cell potentials have negative standard reduction potentials.

SAE 20.11 A voltaic cell is constructed by placing a nickel electrode in a 1.0 M solution of $NiCl_2(aq)$ and a gold electrode in a 1.0 M solution of $AuNO_3(aq)$. A voltmeter shows $E_{cell}^\circ = 1.97$ V. Given that Table 20.1 shows that the standard reduction potential for $Ni^{2+}(aq)$ is −0.28 V, in this cell the gold electrode acts as the _____ and the standard reduction potential for the half reaction $Au^+(aq) + e^- \longrightarrow Au(s)$ is $E_{red}^\circ =$ _____. **(a)** anode, +1.69 V **(b)** anode, +2.25 V **(c)** cathode, +1.69 V **(d)** cathode, +2.25 V

SAE 20.12 Use Table 20.1 to rank the following species from strongest to weakest oxidizing agent in acidic solution: O_2, Fe^{2+}, MnO_4^-. **(a)** $O_2 > MnO_4^- > Fe^{2+}$ **(b)** $MnO_4^- > Fe^{2+} > O_2$ **(c)** $Fe^{2+} > MnO_4^- > O_2$ **(d)** $MnO_4^- > O_2 > Fe^{2+}$ **(e)** $O_2 > Fe^{2+} > MnO_4^-$

20.5 | Free Energy and Redox Reactions

We have observed that voltaic cells use spontaneous redox reactions to produce a positive cell potential. Given half-cell potentials, we can determine whether a given redox reaction is spontaneous. In this endeavor, we can use a form of Equation 20.8 that describes redox reactions in general, not just reactions in voltaic cells:

$$E^\circ = E_{red}^\circ \text{ (reduction process)} - E_{red}^\circ \text{ (oxidation process)} \qquad [20.10]$$

In writing the equation this way, we have dropped the subscript "cell" to indicate that the calculated emf does not necessarily refer to a voltaic cell. Also, we have generalized the standard reduction potentials by using the general terms *reduction* and *oxidation* rather than the terms specific to voltaic cells, *cathode* and *anode*. We can now make a general statement about the spontaneity of a reaction and its associated emf, E: *A positive value of* E *indicates a spontaneous process; a negative value of* E *indicates a nonspontaneous process.* We use E to represent the emf under nonstandard conditions and E° to indicate the standard emf.

Learning Objectives

When you finish Section 20.5, you should be able to:

▶ Determine whether a redox reaction is spontaneous in the forward direction under standard conditions.

▶ Calculate the Gibbs free energy, ΔG, from the cell potential, E, and vice versa.

▶ Calculate the equilibrium constant, K, for a redox reaction from the standard cell potential, E°, and vice versa.

Sample Exercise 20.9

Determining Spontaneity

Use Table 20.1 to determine whether the following reactions are spontaneous under standard conditions.

(a) $Cu(s) + 2\,H^+(aq) \longrightarrow Cu^{2+}(aq) + H_2(g)$

(b) $Cl_2(g) + 2\,I^-(aq) \longrightarrow 2\,Cl^-(aq) + I_2(s)$

SOLUTION

Analyze We are given two reactions and must determine whether each is spontaneous.

Plan To determine whether a redox reaction is spontaneous under standard conditions, we first need to write its reduction and oxidation half-reactions. We can then use the standard reduction potentials and Equation 20.10 to calculate the standard emf, $E°$, for the reaction. If a reaction is spontaneous, its standard emf must be a positive number.

Solve

(a) We first must identify the oxidation and reduction half-reactions that, when combined, give the overall reaction.

Reduction:	$2\,H^+(aq) + 2\,e^- \longrightarrow H_2(g)$
Oxidation:	$Cu(s) \longrightarrow Cu^{2+}(aq) + 2\,e^-$

We look up standard reduction potentials for both half-reactions and use them to calculate $E°$ using Equation 20.10:

$E° = E°_{red}$ (reduction process) $- E°_{red}$ (oxidation process)

$= (0\,V) - (0.34\,V) = -0.34\,V$

Because $E°$ is negative, the reaction is not spontaneous in the direction written. Copper metal does not react with acids as written in Equation (a). The reverse reaction, however, *is* spontaneous and has a positive $E°$ value:

$Cu^{2+}(aq) + H_2(g) \longrightarrow Cu(s) + 2\,H^+(aq) \quad E° = +0.34\,V$

Thus, Cu^{2+} can be reduced by H_2.

(b) We follow a procedure analogous to that in (a):

Reduction:	$Cl_2(g) + 2\,e^- \longrightarrow 2\,Cl^-(aq)$
Oxidation:	$2\,I^-(aq) \longrightarrow I_2(s) + 2\,e^-$

In this case:

$E° = (1.36\,V) - (0.54\,V) = +0.82\,V$

Because the value of $E°$ is positive, this reaction is spontaneous.

▶ **Practice Exercise**

Using the standard reduction potentials listed in Appendix E, determine which of the following reactions are spontaneous under standard conditions:

(a) $I_2(s) + 5\,Cu^{2+}(aq) + 6\,H_2O(l) \longrightarrow$
$\qquad\qquad 2\,IO_3^-(aq) + 5\,Cu(s) + 12\,H^+(aq)$

(b) $Hg^{2+}(aq) + 2\,I^-(aq) \longrightarrow Hg(l) + I_2(s)$

(c) $H_2SO_3(aq) + 2\,Mn(s) + 4\,H^+(aq) \longrightarrow$
$\qquad\qquad S(s) + 2\,Mn^{2+}(aq) + 3\,H_2O(l)$

We can use standard reduction potentials to understand the activity series of metals. (Section 4.4) Recall that any metal in the activity series (Table 4.5) is oxidized by the ions of any metal below it. We can now recognize the origin of this rule based on standard reduction potentials. The activity series is based on the oxidation reactions of the metals, ordered from strongest reducing agent at the top to weakest reducing agent at the bottom. (Thus, the ordering is inverted relative to that in Table 20.1.) For example, nickel lies above silver in the activity series, making nickel the stronger reducing agent. Because a reducing agent is oxidized in any redox reaction, nickel is more easily oxidized than silver. In a mixture of nickel metal and silver cations, therefore, we expect a displacement reaction in which the silver ions are displaced in the solution by nickel ions:

$$Ni(s) + 2\,Ag^+(aq) \longrightarrow Ni^{2+}(aq) + 2\,Ag(s)$$

In this reaction, Ni is oxidized and Ag^+ is reduced. Therefore, the standard emf for the reaction is

$$E° = E°_{red}(Ag^+/Ag) - E°_{red}(Ni^{2+}/Ni)$$
$$= (+0.80\,V) - (-0.28\,V) = +1.08\,V$$

The positive value of $E°$ indicates that the displacement of silver by nickel resulting from oxidation of Ni metal and reduction of Ag^+ is a spontaneous process. Remember that although we multiply the silver half-reaction by 2, the reduction potential is not multiplied.

Emf, Free Energy, and the Equilibrium Constant

The change in the Gibbs free energy, ΔG, is a measure of the spontaneity of a process that occurs at constant temperature and pressure. (Section 19.5) The emf, E, of a redox reaction also indicates whether the reaction is spontaneous. Therefore, it should not come as a surprise to learn that these two quantities are related by the equation

$$\Delta G = -nFE \qquad [20.11]$$

In this equation, n is a positive number without units that represents the number of moles of electrons transferred according to the balanced equation for the reaction, and F is the **Faraday constant**, named after Michael Faraday (**Figure 20.12**):

$$F = 96{,}485 \text{ C/mol} = 96{,}485 \text{ J/V-mol}$$

The Faraday constant is the quantity of electrical charge on 1 mol of electrons.

The units of ΔG calculated with Equation 20.11 are J/mol. As in Equation 19.19, we use "per mole" to mean per mole of reaction as indicated by the coefficients in the balanced equation. (Section 19.7)

Because both n and F are positive numbers, a positive value of E in Equation 20.11 leads to a negative value of ΔG. Remember: *A positive value of* E *and a negative value of* ΔG *both indicate a spontaneous reaction.* When the reactants and products are all in their standard states, Equation 20.11 can be modified to relate $\Delta G°$ and $E°$:

$$\Delta G° = -nFE° \qquad [20.12]$$

Because $\Delta G°$ is related to the equilibrium constant, K, for a reaction by the expression $\Delta G° = -RT \ln K$ (Equation 19.20), we can relate $E°$ to K by solving Equation 20.12 for $E°$ and then substituting the Equation 19.20 expression for $\Delta G°$.

$$E° = \frac{\Delta G°}{-nF} = \frac{-RT \ln K}{-nF} = \frac{RT}{nF} \ln K \qquad [20.13]$$

Figure 20.13 summarizes the relationships among $E°$, $\Delta G°$, and K.

▲ **Figure 20.12 Michael Faraday.** Faraday (1791–1867) was born in England, a child of a poor blacksmith. At the age of 14, he was apprenticed to a bookbinder who gave him time to read and to attend lectures. In 1812 he became an assistant in Humphry Davy's laboratory at the Royal Institution. He succeeded Davy as the most famous and influential scientist in England, making an amazing number of important discoveries, including his formulation of the quantitative relationships between electrical current and the extent of chemical reaction in electrochemical cells.

▼ **Go Figure** | What does the variable n represent in the $\Delta G°$ and $E°$ equations?

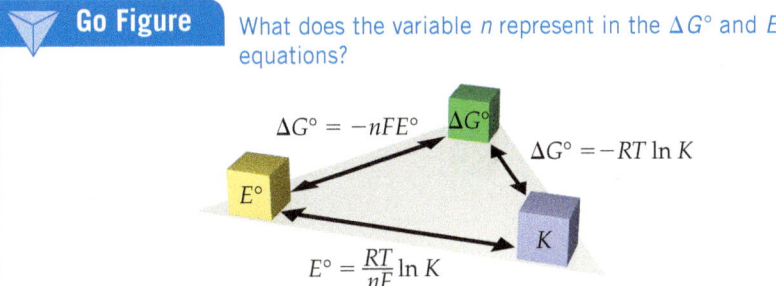

$\Delta G° = -nFE°$ $\Delta G°$

$\Delta G° = -RT \ln K$

$E°$

$E° = \dfrac{RT}{nF} \ln K$ K

▲ **Figure 20.13 Relationships between $E°$, $\Delta G°$, and K.** Any one of these important parameters can be used to calculate the other two. The signs of $E°$ and $\Delta G°$ determine the direction in which the reaction proceeds under standard conditions. The magnitude of K determines the relative amounts of reactants and products in an equilibrium mixture.

Sample Exercise 20.10
Using Standard Reduction Potentials to Calculate $\Delta G°$ and K

(a) Use the standard reduction potentials in Table 20.1 to calculate the standard free-energy change, $\Delta G°$, and the equilibrium constant, K, at 298 K for the reaction

$$4\,Ag(s) + O_2(g) + 4\,H^+(aq) \longrightarrow 4\,Ag^+(aq) + 2\,H_2O(l)$$

(b) Suppose the reaction in part (a) is written

$$2\,Ag(s) + \frac{1}{2}\,O_2(g) + 2\,H^+(aq) \longrightarrow 2\,Ag^+(aq) + H_2O(l)$$

What are the values of $E°$, $\Delta G°$, and K when the reaction is written in this way?

SOLUTION

Analyze We are asked to determine $\Delta G°$ and K for a redox reaction, using standard reduction potentials.

Plan We use the data in Table 20.1 and Equation 20.10 to determine $E°$ for the reaction, and then we use $E°$ in Equation 20.12 to calculate $\Delta G°$. We can then use either Equation 19.20 or Equation 20.13 to calculate K.

Solve

(a) We first calculate $E°$ by breaking the equation into two half-reactions and obtaining $E°_{red}$ values from Table 20.1 (or Appendix E):

Reduction:	$O_2(g) + 4\,H^+(aq) + 4\,e^- \longrightarrow 2\,H_2O(l)$	$E°_{red} = +1.23$ V
Oxidation:	$4\,Ag(s) \longrightarrow 4\,Ag^+(aq) + 4\,e^-$	$E°_{red} = +0.80$ V

Even though the second half-reaction has 4 Ag, we use the $E°_{red}$ value directly from Table 20.1 because emf is an intensive property. Using Equation 20.10, we have:

$$E° = (1.23\text{ V}) - (0.80\text{ V}) = 0.43\text{ V}$$

The half-reactions show the transfer of four electrons. Thus, for this reaction, $n = 4$. We now use Equation 20.12 to calculate $\Delta G°$:

$$\Delta G° = -nFE°$$
$$= -(4)(96,485\text{ J/V-mol})(+0.43\text{ V})$$
$$= -1.7 \times 10^5\text{ J/mol} = -170\text{ kJ/mol}$$

Now we need to calculate the equilibrium constant, K, using $\Delta G° = RT\ln K$. Because $\Delta G°$ is a large negative number, which means the reaction is thermodynamically very favorable, we expect K to be large.

$$\Delta G° = -RT\ln K$$
$$-1.7 \times 10^5\text{ J/mol} = -(8.314\text{ J/mol-K})(298\text{ K})\ln K$$
$$\ln K = \frac{-1.7 \times 10^5\text{ J/mol}}{-(8.314\text{ J/mol-K})(298\text{ K})}$$
$$\ln K = 69$$
$$K = 9 \times 10^{29}$$

(b) The overall equation is the same as that in part (a), multiplied by 1/2. The half-reactions are:

Reduction:	$\frac{1}{2}\,O_2(g) + 2\,H^+(aq) + 2\,e^- \longrightarrow H_2O(l)$	$E°_{red} = +1.23$ V
Oxidation:	$2\,Ag(s) \longrightarrow 2\,Ag^+(aq) + 2\,e^-$	$E°_{red} = +0.80$ V

The values of $E°_{red}$ are the same as they were in part (a); they are not changed by multiplying the half-reactions by $\frac{1}{2}$. Thus, $E°$ has the same value as in part (a): $E° = 0.43$ V. Notice, though, that the value of n has changed to $n = 2$, which is one-half the value in part (a). Thus, $\Delta G°$ is half as large as in part (a):

$$\Delta G° = -(2)(96,485\text{ J/V-mol})(+0.43\text{ V}) = -83\text{ kJ/mol}$$

The value of $\Delta G°$ is half that in part (a) because the coefficients in the chemical equation are half those in (a).

Now we can calculate K as before:

$$-8.3 \times 10^4\text{ J/mol} = -(8.314\text{ J/mol-K})(298\text{ K})\ln K$$
$$K = 4 \times 10^{14}$$

Comment $E°$ is an *intensive* quantity, so multiplying a chemical equation by a certain factor will not affect the value of $E°$. Multiplying an equation will change the value of n, however, and hence the value of $\Delta G°$. The change in free energy, in units of J/mol of reaction as written, is an *extensive* quantity. The equilibrium constant is also an extensive quantity.

▶ **Practice Exercise**
Consider the reaction
$2\,Ag^+(aq) + H_2(aq) \longrightarrow 2\,Ag(s) + 2\,H^+(aq)$. Calculate $\Delta G°_f$ for the $Ag^+(aq)$ ion from the standard reduction potentials in Table 20.1 and from the fact that $\Delta G°_f$ for $H_2(g)$, $Ag(s)$, and $H^+(aq)$ are all zero. Compare your answer with the value given in Appendix C.

A CLOSER LOOK | Electrical Work

For any spontaneous process, ΔG is a measure of the maximum useful work, w_{max}, that can be extracted from the process: $\Delta G = w_{max}$. (Section 19.5) Because $\Delta G = -nFE$, the maximum useful electrical work obtainable from a voltaic cell is

$$w_{max} = -nFE_{cell} \qquad [20.14]$$

Because cell emf, E_{cell}, is always positive for a voltaic cell, w_{max} is negative, indicating that work is done *by* a system *on* its surroundings, as we expect for a voltaic cell. (Section 5.2)

As Equation 20.14 shows, the more charge a voltaic cell moves through a circuit (that is, the larger nF is) and the larger the emf pushing the electrons through the circuit (that is, the larger $E°_{cell}$ is), the more work the cell can accomplish. In Sample Exercise 20.10, we calculated $\Delta G° = -170$ kJ/mol for the reaction $4\,Ag(s) + O_2(g) + 4\,H^+(aq) \longrightarrow 4\,Ag^+(aq) + 2\,H_2O(l)$. Thus, a voltaic cell utilizing this reaction could perform a maximum of 170 kJ of work in consuming 4 mol Ag, 1 mol O_2, and 4 mol H^+.

If a reaction is not spontaneous, ΔG is positive and E is negative. To force a nonspontaneous reaction to occur in an electrochemical cell, we need to apply an external potential, E_{ext}, that exceeds $|E_{cell}|$. For example, if a nonspontaneous process has $E = -0.9$ V, then the external potential E_{ext} must be greater than $+0.9$ V in order for the process to occur. We examine such nonspontaneous processes in Section 20.9.

Electrical work can be expressed in energy units of watts × time. The *watt* (W) is a unit of electrical power (that is, rate of energy expenditure):

$$1\,W = 1\,J/s$$

Thus, a watt-second is a joule. The unit employed by electric utilities is the kilowatt-hour (kWh), which equals 3.6×10^6 J:

$$1\,kWh = (1000\,W)(1\,h)\left(\frac{3600\,s}{1\,h}\right)\left(\frac{1\,J/s}{1\,W}\right) = 3.6 \times 10^6\,J$$

Related Exercises: 20.63, 20.64

 Self-Assessment Exercises

SAE 20.13 Which of the following statements is or are *true*?

 (i) All redox reactions with positive emfs are spontaneous.
 (ii) If a redox reaction is spontaneous, it must be fast.
 (iii) In a spontaneous redox reaction, the species that is being reduced will have a more negative standard reduction potential than the species that is being oxidized.

(a) only i **(b)** only ii **(c)** only iii **(d)** both i and iii **(e)** All three statements are true.

SAE 20.14 Consider the following standard reduction potentials:

$$AgI(s) + e^- \longrightarrow Ag(s) + I^-(aq) \quad E°_{red} = -0.15\,V$$
$$I_2(s) + 2\,e^- \longrightarrow 2\,I^-(aq) \quad E°_{red} = +0.54\,V$$

Based on these values, we can conclude that under standard conditions the redox reaction, $2\,Ag(s) + I_2(s) \longrightarrow 2\,AgI(s)$, is _____, and $\Delta G°$ for this reaction is _____. **(a)** nonspontaneous, 1.3×10^2 kJ **(b)** nonspontaneous, 67 kJ **(c)** spontaneous, -1.3×10^2 kJ **(d)** spontaneous, -67 kJ **(e)** spontaneous, -75 kJ

SAE 20.15 Use the standard reduction potentials given in Table 20.1 to determine the equilibrium constant under standard conditions for the reaction

$$Ag^+(aq) + Fe^{2+}(aq) \longrightarrow Ag(s) + Fe^{3+}(aq)$$

(a) 0.31 **(b)** 3.2 **(c)** 2.9×10^3 **(d)** 3.6×10^{26} **(e)** 2.8×10^{51}

20.6 | Cell Potentials under Nonstandard Conditions

 Learning Objectives

When you finish Section 20.6, you should be able to:

▶ Use the Nernst equation to relate the cell potential, E, (or emf) of a voltaic cell under nonstandard conditions to the concentrations of the reactants and products and the temperature.

▶ Analyze concentration cells to relate the cell potential, E, (or emf) and direction of electron flow to the concentrations of reactants and products in each half-cell.

We have seen how to calculate the cell potential, or emf, of a cell when the reactants and products are under standard conditions. As a voltaic cell is discharged, however, reactants are consumed and products are generated, so concentrations change. The emf progressively drops until $E = 0$, at which point we say the cell is "dead." In this section, we examine how the emf generated under nonstandard conditions can be calculated by using an equation first derived by Walther Nernst (1864–1941), a German chemist who established many of the theoretical foundations of electrochemistry.

The Nernst Equation

The effect of concentration on cell emf can be obtained from the effect of concentration on free-energy change. (Section 19.7) Recall that the free-energy change for any chemical reaction, ΔG, is related to the standard free-energy change for the reaction, $\Delta G°$:

$$\Delta G = \Delta G° + RT \ln Q \qquad [20.15]$$

The quantity Q is the reaction quotient, which has the form of the equilibrium-constant expression except that the concentrations are those that exist in the reaction mixture at a given moment. (Section 15.6)

Substituting $\Delta G = -nFE$ (Equation 20.11) into Equation 20.15 gives

$$-nFE = -nFE° + RT \ln Q$$

Solving this equation for E gives the **Nernst equation**:

$$E = E° - \frac{RT}{nF} \ln Q \qquad [20.16]$$

This equation is customarily expressed in terms of the base-10 logarithm:

$$E = E° - \frac{2.303\, RT}{nF} \log Q \qquad [20.17]$$

At $T = 298$ K, the quantity $2.303\, RT/F$ equals 0.0592, with units of volts, and so the Nernst equation simplifies to

$$E = E° - \frac{0.0592\text{ V}}{n} \log Q \qquad (T = 298\text{ K}) \qquad [20.18]$$

We can use this equation to find the emf E produced by a cell under nonstandard conditions or to determine the concentration of a reactant or product by measuring E for the cell. For example, consider the following reaction:

$$Zn(s) + Cu^{2+}(aq) \longrightarrow Zn^{2+}(aq) + Cu(s)$$

In this case $n = 2$ (two electrons are transferred from Zn to Cu^{2+}), and the standard emf is $+1.10$ V. (Section 20.4) Thus, at 298 K, the Nernst equation gives

$$E = 1.10\text{ V} - \frac{0.0592\text{ V}}{2} \log \frac{[Zn^{2+}]}{[Cu^{2+}]} \qquad [20.19]$$

Recall that pure solids are excluded from the expression for Q. (Section 15.6) According to Equation 20.19, the emf increases as $[Cu^{2+}]$ increases and as $[Zn^{2+}]$ decreases. For example, when $[Cu^{2+}]$ is 5.0 M and $[Zn^{2+}]$ is 0.050 M, we have

$$E = 1.10\text{ V} - \frac{0.0592\text{ V}}{2} \log\left(\frac{0.050}{5.0}\right)$$

$$= 1.10\text{ V} - \frac{0.0592\text{ V}}{2}(-2.00) = 1.16\text{ V}$$

Thus, increasing the concentration of reactant Cu^{2+} and decreasing the concentration of product Zn^{2+} relative to standard conditions increases the emf of the cell relative to standard conditions ($E° = +1.10$ V).

The Nernst equation helps us understand why the emf of a voltaic cell drops as the cell discharges. As reactants are converted to products, the value of Q increases, so the value of E decreases, eventually reaching $E = 0$. Because $\Delta G = -nFE$ (Equation 20.11), it follows that $\Delta G = 0$ when $E = 0$. Recall that a system is at equilibrium when $\Delta G = 0$. (Section 19.7) Thus, when $E = 0$, the cell reaction has reached equilibrium, and no net reaction occurs.

In general, increasing the concentration of reactants or decreasing the concentration of products increases the driving force for the reaction, resulting in a higher emf. Conversely, decreasing the concentration of reactants or increasing the concentration of products causes the emf to decrease from its value under standard conditions.

Sample Exercise 20.11
Cell Potential under Nonstandard Conditions

Calculate the emf at 298 K generated by a voltaic cell in which the reaction is

$$Cr_2O_7^{2-}(aq) + 14\,H^+(aq) + 6\,I^-(aq) \longrightarrow 2\,Cr^{3+}(aq) + 3\,I_2(s) + 7\,H_2O(l)$$

when

$$[Cr_2O_7^{2-}] = 2.0\ M,\ [H^+] = 1.0\ M,\ [I^-] = 1.0\ M,\ \text{and}\ [Cr^{3+}] = 1.0 \times 10^{-5}\ M$$

SOLUTION

Analyze We are given a chemical equation for a voltaic cell and the concentrations of reactants and products under which it operates. We are asked to calculate the emf of the cell under these nonstandard conditions.

Plan To calculate the emf of a cell under nonstandard conditions, we use the Nernst equation in the form of Equation 20.18.

Solve

We calculate $E°$ for the cell from standard reduction potentials (Table 20.1 or Appendix E). The standard emf for this reaction was calculated in Sample Exercise 20.6: $E° = 0.79$ V. As that exercise shows, six electrons are transferred from reducing agent to oxidizing agent, so $n = 6$. The reaction quotient, Q, is:

$$Q = \frac{[Cr^{3+}]^2}{[Cr_2O_7^{2-}][H^+]^{14}[I^-]^6}$$

$$= \frac{(1.0 \times 10^{-5})^2}{(2.0)(1.0)^{14}(1.0)^6} = 5.0 \times 10^{-11}$$

Using Equation 20.18, we have:

$$E = 0.79\,V - \left(\frac{0.0592\,V}{6}\right)\log(5.0 \times 10^{-11})$$

$$= 0.79\,V - \left(\frac{0.0592\,V}{6}\right)(-10.30)$$

$$= 0.79\,V + 0.10\,V = 0.89\,V$$

Check This result is qualitatively what we expect: Because the concentration of $Cr_2O_7^{2-}$ (a reactant) is greater than 1 M and the concentration of Cr^{3+} (a product) is less than 1 M, the emf is greater than $E°$. Because Q is about 10^{-10}, $\log Q$ is about -10. Thus, the correction to $E°$ is about $0.06 \times 10/6$, which is 0.1, in agreement with the more detailed calculation.

▶ **Practice Exercise**
For the Zn–Cu voltaic cell depicted in Figure 20.5, would the emf increase, decrease, or stay the same if you increased the $Cu^{2+}(aq)$ concentration by adding $CuSO_4 \cdot 5H_2O$ to the cathode compartment?

Sample Exercise 20.12
Calculating Concentrations in a Voltaic Cell

If the potential of a Zn–H_2 cell (like that in Figure 20.9) is 0.45 V at 25 °C when $[Zn^{2+}] = 1.0\ M$ and $P_{H_2} = 1.0$ atm, what is the pH of the cathode solution?

SOLUTION

Analyze We are given a description of a voltaic cell, its emf, the concentration of Zn^{2+}, and the partial pressure of H_2 (both products in the cell reaction). We are asked to calculate the pH of the cathode solution, which we can calculate from the concentration of H^+, a reactant.

Plan We write the equation for the cell reaction and use standard reduction potentials to calculate $E°$ for the reaction. After determining the value of n from our reaction equation, we solve the Nernst equation, Equation 20.18, for Q. We use the equation for the cell reaction to write an expression for Q that contains $[H^+]$ to determine $[H^+]$. Finally, we use $[H^+]$ to calculate pH.

Solve

The cell reaction is:	$Zn(s) + 2\,H^+(aq) \longrightarrow Zn^{2+}(aq) + H_2(g)$
The standard emf is:	$E° = E°_{red}\,(\text{reduction}) - E°_{red}\,(\text{oxidation})$
	$= 0\,V - (-0.76\,V) = +0.76\,V$
Because each Zn atom loses two electrons,	$n = 2$

Continued

Using Equation 20.18, we can solve for Q:	$0.45 \text{ V} = 0.76 \text{ V} - \dfrac{0.0592 \text{ V}}{2} \log Q$ $Q = 10^{10.5} = 3 \times 10^{10}$
Q has the form of the equilibrium constant for the reaction:	$Q = \dfrac{[\text{Zn}^{2+}]P_{\text{H}_2}}{[\text{H}^+]^2} = \dfrac{(1.0)(1.0)}{[\text{H}^+]^2} = 3 \times 10^{10}$
Solving for $[\text{H}^+]$, we have:	$[\text{H}^+]^2 = \dfrac{1.0}{3 \times 10^{10}} = 3 \times 10^{-11}$ $[\text{H}^+] = \sqrt{3 \times 10^{-11}} = 6 \times 10^{-6} M$
Finally, we use $[\text{H}^+]$ to calculate the pH of the cathode solution.	$\text{pH} = \log[\text{H}^+] = -\log(6 \times 10^{-6}) = 5.2$

Comment A voltaic cell whose cell reaction involves H^+ can be used to measure $[\text{H}^+]$ or pH. A pH meter is a specially designed voltaic cell with a voltmeter calibrated to read pH directly. (Section 16.4)

▶ **Practice Exercise**
What is the pH of the solution in the cathode half-cell in Figure 20.9 when $P_{\text{H}_2} = 1.0$ atm, $[\text{Zn}^{2+}] = 0.10 M$ in the anode half-cell, and the cell emf is 0.542 V?

Concentration Cells

In the voltaic cells we have looked at thus far, the reactive species at the anode has been different from the reactive species at the cathode. Cell emf depends on concentration, however, so a voltaic cell can be constructed using the *same* species in both half-cells as long as the concentrations are different. A cell based solely on the emf generated because of a difference in a concentration is called a **concentration cell**.

An example of a concentration cell is diagrammed in **Figure 20.14**(**a**). One half-cell consists of a strip of nickel metal immersed in a $1.00 \times 10^{-3} M$ solution of $\text{Ni}^{2+}(aq)$. The other half-cell also has an Ni(s) electrode, but it is immersed in a $1.00 M$ solution of $\text{Ni}^{2+}(aq)$. The two half-cells are connected by a salt bridge and by an external wire running through a voltmeter. The half-cell reactions are the reverse of each other:

Anode:	$\text{Ni}(s) \longrightarrow \text{Ni}^{2+}(aq) + 2\,\text{e}^-$	$E^\circ_{\text{red}} = -0.28 \text{ V}$
Cathode:	$\text{Ni}^{2+}(aq) + 2\,\text{e}^- \longrightarrow \text{Ni}(s)$	$E^\circ_{\text{red}} = -0.28 \text{ V}$

 **Go Figure** Which electrode, if any, gains mass as the reaction proceeds?

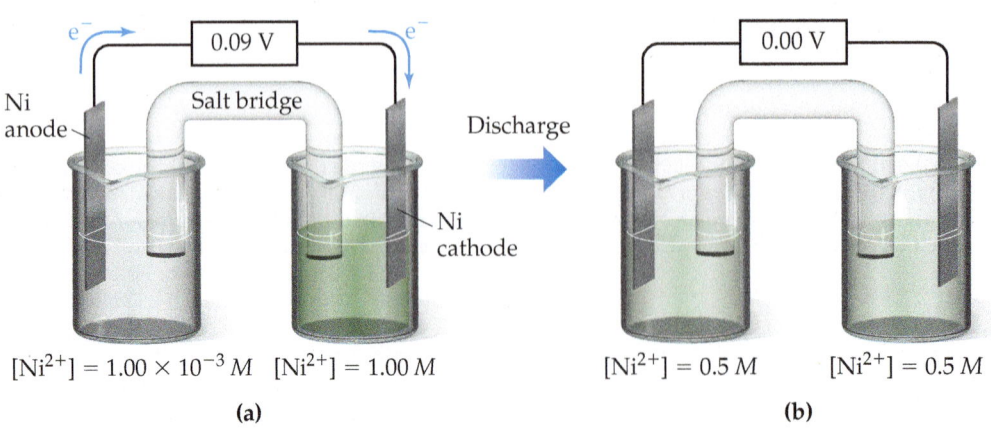

▲ **Figure 20.14 Concentration cell based on Ni^{2+}–Ni cell reaction.** (**a**) Concentrations of $\text{Ni}^{2+}(aq)$ in the two half-cells are unequal, and the cell generates an electrical current and a voltage. (**b**) The cell operates until $[\text{Ni}^{2+}]$ is the same in the two half-cells, at which point the cell has reached equilibrium and the emf goes to zero.

Although the *standard* emf for this cell is zero,

$$E^\circ_{cell} = E^\circ_{red}\,(\text{cathode}) - E^\circ_{red}\,(\text{anode}) = (-0.28\text{ V}) - (-0.28\text{ V}) = 0\text{ V}$$

the cell operates under *nonstandard* conditions because the concentration of $Ni^{2+}(aq)$ is not 1 M in both half-cells. In fact, the cell operates until $[Ni^{2+}]_{anode} = [Ni^{2+}]_{cathode}$. Oxidation of $Ni(s)$ occurs in the half-cell containing the more dilute solution, which means this is the anode of the cell. Reduction of $Ni^{2+}(aq)$ occurs in the half-cell containing the more concentrated solution, making it the cathode. The *overall* cell reaction is therefore

Anode:	$Ni(s) \longrightarrow Ni^{2+}(aq, \text{dilute}) + 2\,e^-$
Cathode:	$Ni^{2+}(aq, \text{concentrated}) + 2\,e^- \longrightarrow Ni(s)$
Overall:	$Ni^{2+}(aq, \text{concentrated}) \longrightarrow Ni^{2+}(aq, \text{dilute})$

We can calculate the emf of a concentration cell by using the Nernst equation. For this particular cell, we see that $n = 2$. The expression for the reaction quotient for the overall reaction is $Q = [Ni^{2+}]_{dilute}/[Ni^{2+}]_{concentrated}$. Thus, the emf at 298 K is

$$E = E^\circ - \frac{0.0592\text{ V}}{n}\log Q$$

$$= 0 - \frac{0.0592\text{ V}}{2}\log\frac{[Ni^{2+}]_{dilute}}{[Ni^{2+}]_{concentrated}} = -\frac{0.0592\text{ V}}{2}\log\frac{1.00\times10^{-3}\,M}{1.00\,M}$$

$$= +0.089\text{ V}$$

This concentration cell generates an emf of nearly 0.09 V, even though $E^\circ = 0$. The difference in concentration provides the driving force for the cell. When the concentrations in the two half-cells become the same, $Q = 1$ and $E = 0$.

The idea of generating a potential by a concentration difference is the basis for the operation of pH meters. It is also a critical aspect in biology. For example, nerve cells in the brain generate a potential across the cell membrane by having different concentrations of ions on the two sides of the membrane. Electric eels use cells called electrocytes that are based on a similar principle to generate short, but intense, pulses of electricity to stun prey and dissuade predators (**Figure 20.15**). The regulation of the heartbeat in mammals, as discussed in the following Chemistry and Life box, is another example of the importance of electrochemistry to living organisms.

◀ **Figure 20.15 An electric eel.** Differences in ion concentrations, mainly Na^+ and K^+, in special cells called electrocytes, produce an emf on the order of 0.1 V. By connecting thousands of these cells in series, these South American fish can generate short electric pulses as high as 500 V.

CHEMISTRY AND LIFE | Heartbeats and Electrocardiography

The human heart is a marvel of efficiency and dependability. In a typical day, an adult's heart pumps more than 7000 L of blood through the circulatory system, usually with no maintenance required beyond a sensible diet and lifestyle. We generally think of the heart as a mechanical device, a muscle that circulates blood via regularly spaced muscular contractions. However, more than two centuries ago, two pioneers in electricity, Luigi Galvani (1729–1787) and Alessandro Volta (1745–1827), discovered that the contractions of the heart are controlled by electrical phenomena, as are nerve impulses throughout the body. The pulses of electricity that cause the heart to beat result from a remarkable combination of electrochemistry and the properties of semipermeable membranes. (Section 13.5)

Cell walls are membranes with variable permeability with respect to a number of physiologically important ions (especially Na^+, K^+, and Ca^{2+}). The concentrations of these ions are different for the fluids inside the cells (the *intracellular fluid*, or ICF) and outside the cells (the *extracellular fluid*, or ECF). In cardiac muscle cells, for example, the concentrations of K^+ in the ICF and ECF are typically about 135 millimolar (mM) and 4 mM, respectively. For Na^+, however, the concentration difference between the ICF and ECF is opposite to that for K^+; typically, $[Na^+]_{ICF} = 10$ mM and $[Na^+]_{ECF} = 145$ mM.

The cell membrane is initially permeable to K^+ ions but is much less so to Na^+ and Ca^{2+}. The difference in concentration of K^+ ions between the ICF and ECF generates a concentration cell. Even though the same ions are present on both sides of the membrane, there is a potential difference between the two fluids that we can calculate using the Nernst equation with $E° = 0$. At physiological temperature (37 °C), the potential in millivolts for moving K^+ from the ECF to the ICF is

$$E = E° - \frac{2.30\,RT}{nF} \log \frac{[K^+]_{ICF}}{[K^+]_{ECF}}$$

$$= 0 - (61.5\,\text{mV}) \log\left(\frac{135\ \text{m}M}{4\ \text{m}M}\right) = -94\ \text{mV}$$

In essence, the interior of the cell and the ECF together serve as a voltaic cell. The negative sign for the potential indicates that work is required to move K^+ into the ICF.

Changes in the relative concentrations of the ions in the ECF and ICF lead to changes in the emf of the voltaic cell. The cells of the heart that govern the rate of heart contraction are called the *pacemaker cells*. The membranes of the cells regulate the concentrations of ions in the ICF, allowing them to change in a systematic way. The concentration changes cause the emf to change in a cyclic fashion, as shown in Figure 20.16. The emf cycle determines the rate at which the heart beats. If the pacemaker cells malfunction because of disease or injury, an artificial pacemaker can be surgically

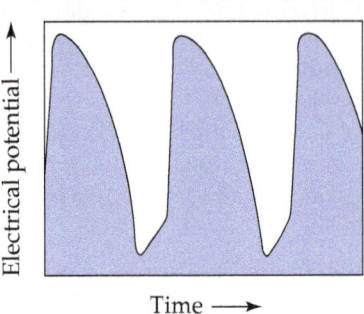

▲ **Figure 20.16 Changes in electrical potential in the human heart.** Variation of the electrical potential caused by changes of ion concentrations in the pacemaker cells of the heart.

implanted. The artificial pacemaker contains a small battery that generates the electrical pulses needed to trigger the contractions of the heart.

During the late 1800s, scientists discovered that the electrical impulses that cause the contraction of the heart muscle are strong enough to be detected at the surface of the body. This observation formed the basis for *electrocardiography*, noninvasive monitoring of the heart by using a complex array of electrodes on the skin to measure voltage changes during heartbeats. A typical electrocardiogram is shown in Figure 20.17. It is quite striking that, although the heart's major function is the *mechanical* pumping of blood, it is most easily monitored by using the *electrical* impulses generated by tiny voltaic cells.

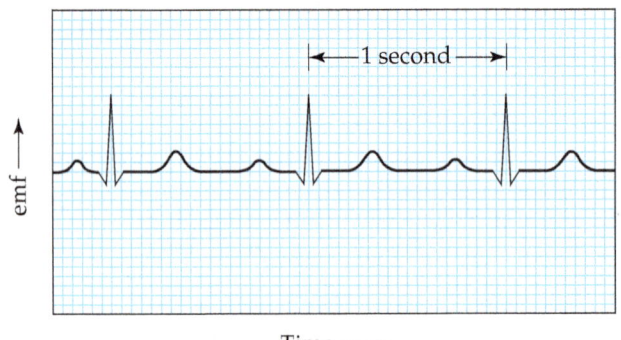

▲ **Figure 20.17 A typical electrocardiogram.** The printout records the electrical events monitored by electrodes attached to the body surface.

Sample Exercise 20.13

Determining pH Using a Concentration Cell

A voltaic cell is constructed with two hydrogen electrodes. Electrode 1 has $P_{H_2} = 1.00$ atm and an unknown concentration of $H^+(aq)$. Electrode 2 is a standard hydrogen electrode ($P_{H_2} = 1.00$ atm, $[H^+] = 1.00\ M$). At 298 K the measured cell potential is 0.211 V, and the electrical current is observed to flow from electrode 1 through the external circuit to electrode 2. What is the pH of the solution at electrode 1?

SOLUTION

Analyze We are given the potential of a concentration cell and the direction in which the current flows. We also have the concentrations or partial pressures of all reactants and products, except for $[H^+]$ in half-cell 1, which is our unknown.

Plan We can use the Nernst equation to determine Q and then use Q to calculate the unknown concentration. Because this is a concentration cell, $E°_{cell} = 0$ V.

Solve

Using the Nernst equation, we have:

$$0.211 \text{ V} = 0 - \frac{0.0592 \text{ V}}{2} \log Q$$

$$\log Q = -(0.211 \text{ V})\left(\frac{2}{0.0592 \text{ V}}\right) = -7.13$$

$$Q = 10^{-7.13} = 7.4 \times 10^{-8}$$

Because electrons flow from electrode 1 to electrode 2, electrode 1 is the anode of the cell and electrode 2 is the cathode. The electrode reactions are therefore as follows, with the concentration of $H^+(aq)$ in electrode 1 represented with the unknown x:

Electrode 1: (anode) $H_2(g, 1.00 \text{ atm}) \longrightarrow 2 \text{ H}^+(aq, x \text{ M}) + 2 \text{ e}^- \quad E^\circ_{red} = 0$

Electrode 2: (cathode) $2 \text{ H}^+(aq, 1.00 \text{ M}) + 2 \text{ e}^- \longrightarrow H_2(g, 1.00 \text{ atm}) \quad E^\circ_{red} = 0$

Overall: $2 \text{ H}^+(aq, 1.00 \text{ M}) \longrightarrow 2 \text{ H}^+(aq, x \text{ M})$

Thus,

$$Q = \frac{[\text{H}^+(aq, x \text{ M})]^2}{[\text{H}^+(aq, 1.00 \text{ M})]^2}$$

$$= \frac{x^2}{(1.00)^2} = x^2 = 7.4 \times 10^{-8}$$

$$x = [\text{H}^+] = \sqrt{7.4 \times 10^{-8}} = 2.7 \times 10^{-4}$$

At electrode 1, therefore, the pH of the solution is:

$$\text{pH} = -\log[\text{H}^+] = -\log(2.7 \times 10^{-4}) = 3.57$$

Comment The concentration of H^+ at electrode 1 is lower than that in electrode 2, which is why electrode 1 is the anode of the cell: The oxidation of H_2 to $H^+(aq)$ increases $[H^+]$ at electrode 1.

▶ **Practice Exercise**
A concentration cell is constructed with two $Zn(s)$–$Zn^{2+}(aq)$ half-cells. In one half-cell $[Zn^{2+}] = 1.35 \text{ M}$, and in the other $[Zn^{2+}] = 3.75 \times 10^{-4} \text{ M}$. **(a)** Which half-cell is the anode? **(b)** What is the emf of the cell?

 ## Self-Assessment Exercises

SAE 20.16 Silver ions oxidize copper metal to produce silver metal and Cu^{2+} ions. Calculate the emf of this reaction when $T = 40\,°C$ and the reagent concentrations are $[Ag^+(aq)] = 0.040 \text{ M}$, $[Cu^{2+}(aq)] = 0.040 \text{ M}$. **(a)** 0.37 V **(b)** 0.42 V **(c)** 0.44 V **(d)** 0.46 V **(e)** 1.10 V

SAE 20.17 For a given voltaic cell, if you double the concentrations of all soluble components from 1.0 M to 2.0 M, how does the cell potential change? **(a)** It does not change, because the cell potential does not depend on the concentration of reagents. **(b)** It doubles. **(c)** It is decreased by half. **(d)** This question cannot be answered without knowing the overall cell reaction.

SAE 20.18 The standard reduction potential for O_2 in acid is $+1.23$ V. What is the reduction potential for O_2 at pH 7, if all other conditions are standard? **(a)** 0.40 V **(b)** 0.82 V **(c)** 1.13 V **(d)** 1.23 V **(e)** 1.64 V

SAE 20.19 What is the absolute value of the voltage across a biological membrane that has $[Na^+]_{outside} = 140 \text{ mM}$ and $[Na^+]_{inside} = 12 \text{ mM}$? Assume that all other conditions (including temperature) are standard. **(a)** 0.00 mV **(b)** 63.1 mV **(c)** 65.6 mV **(d)** 300 mV

SAE 20.20 A concentration cell shown here contains a 1.0 M solution of $Zn(NO_3)_2(aq)$ in beaker 1 and an unknown concentration of $Zn^{2+}(aq)$ in beaker 2. Both half-cells contain a solid Zn electrode, and the temperature is 298 K. A voltmeter shows the cell emf to be 0.042 V and indicates that the electrons flow from beaker 2 to beaker 1. What is the concentration of Zn^{2+} in beaker 2? **(a)** 0.038 M **(b)** 26.2 M **(c)** 0.20 M **(d)** 5.1 M **(e)** 1.4 M

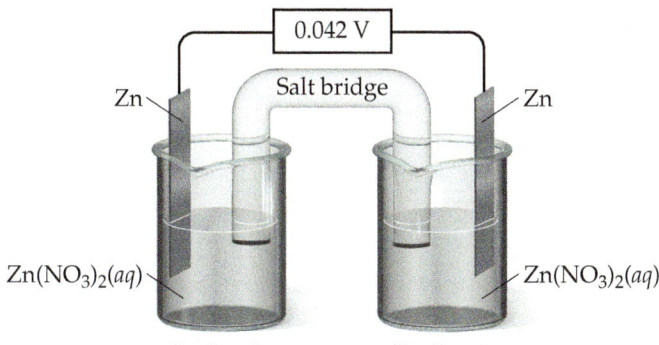

Learning Objectives

When you finish Section 20.7, you should be able to:

▶ Explain how a battery operates and distinguish between primary and secondary cells.

▶ Describe the half-reactions associated with common battery types, including alkaline, lead–acid, and lithium-ion batteries.

▶ Explain the operating principles of a fuel cell.

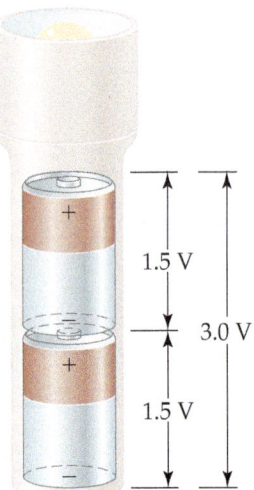

▲ **Figure 20.18 Combining batteries.** When batteries are connected in series, as in most flashlights, the total voltage is the sum of the individual voltages.

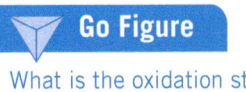

Go Figure

What is the oxidation state of lead in the cathode of this battery?

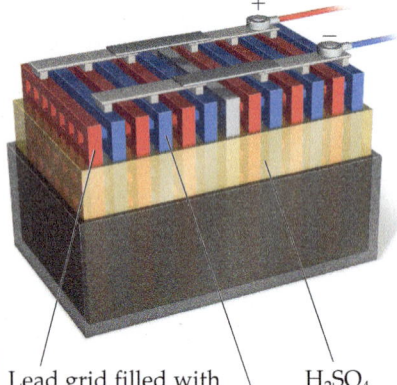

Lead grid filled with spongy lead (anode)

H_2SO_4 electrolyte

Lead grid filled with PbO_2 (cathode)

▲ **Figure 20.19 A 12-V automotive lead–acid battery.** Each anode/cathode pair in this schematic cutaway produces a voltage of about 2 V. Six pairs of electrodes are connected in series, producing 12 V.

20.7 | Batteries and Fuel Cells

A **battery** is a portable, self-contained electrochemical power source that consists of one or more voltaic cells. For example, the 1.5-V batteries used to power flashlights and many consumer electronic devices are single voltaic cells. When cells are connected in series (which means the cathode of one attached to the anode of another), the battery produces a voltage that is the sum of the voltages of the individual cells (**Figure 20.18**). Battery electrodes are marked following the convention of Figure 20.6—plus for cathode and minus for anode.

Although any spontaneous redox reaction can serve as the basis for a voltaic cell, making a commercial battery that has specific performance characteristics requires considerable ingenuity. The substances oxidized at the anode and reduced by the cathode determine the voltage, and the usable life of the battery depends on the quantities of these substances packaged in the battery. Usually, a barrier analogous to the porous barrier of Figure 20.6 separates the anode and cathode half-cells.

Different applications require batteries with different properties. The battery required to start a car, for example, must be capable of delivering a large electrical current for a short time period, whereas the battery that powers a heart pacemaker must be very small and capable of delivering a small but steady current over an extended time period. Some batteries are **primary cells**, meaning they cannot be recharged and must be either discarded or recycled after the voltage drops to zero. A **secondary cell** is powered by reversible redox reactions, so it can be recharged from an external power source after its voltage has dropped.

As we consider some common batteries, notice how the principles we have discussed so far help us understand these important sources of portable electrical energy.

Lead–Acid Battery

Lead–acid batteries are widely used in automobiles powered by combustion engines. The cathode of each cell is lead dioxide (PbO_2) and the anode is lead (**Figure 20.19**). Both electrodes are immersed in sulfuric acid.

The reactions that occur during discharge are

Cathode: $\quad PbO_2(s) + HSO_4^-(aq) + 3\,H^+(aq) + 2\,e^- \longrightarrow PbSO_4(s) + 2\,H_2O(l)$

Anode: $\qquad\qquad\qquad Pb(s) + HSO_4^-(aq) \longrightarrow PbSO_4(s) + H^+(aq) + 2\,e^-$

Overall: $PbO_2(s) + Pb(s) + 2\,HSO_4^-(aq) + 2\,H^+(aq) \longrightarrow 2\,PbSO_4(s) + 2\,H_2O(l)$ [20.20]

The standard cell potential can be obtained from the standard reduction potentials found in Appendix E:

$$E^\circ_{cell} = E^\circ_{red}\,(\text{cathode}) - E^\circ_{red}\,(\text{anode}) = (+1.69\,\text{V}) - (-0.36\,\text{V}) = +2.05\,\text{V}$$

The output voltage of approximately 12 V common to most automotive batteries is produced by electrically connecting six individual cells in series. Because these reactants are solids, there is no need to separate the cell into half-cells; the Pb and PbO_2 cannot come into contact with each other unless one electrode touches another. To keep the electrodes from touching, wood or glass-fiber spacers are placed between them (Figure 20.19). Using a reaction whose reactants and products are solids has another benefit. Because solids are excluded from the reaction quotient Q, the relative amounts of $Pb(s)$, $PbO_2(s)$, and $PbSO_4(s)$ have no effect on the voltage of the lead storage battery, helping the battery maintain a relatively constant voltage during discharge. The voltage does vary somewhat with use, though, because the concentration of H_2SO_4 varies with the extent of discharge. As Equation 20.20 indicates, H_2SO_4 is consumed during the discharge.

A major advantage of the lead–acid battery is that it can be recharged. During recharging, an external source of energy is used to reverse the direction of the cell reaction, regenerating $Pb(s)$ and $PbO_2(s)$:

$$2\,PbSO_4(s) + 2\,H_2O(l) \longrightarrow PbO_2(s) + Pb(s) + 2\,HSO_4^-(aq) + 2\,H^+(aq)$$

In an automobile, the alternator provides the energy necessary for recharging the battery. Recharging is possible because $PbSO_4$ formed during discharge adheres to the electrodes. As the external source forces electrons from one electrode to the other, the $PbSO_4$ is converted to Pb at one electrode and to PbO_2 at the other.

Alkaline Battery

The most common primary (nonrechargeable) battery is the alkaline battery (**Figure 20.20**). The anode is powdered zinc metal immobilized in a gel in contact with a concentrated solution of KOH (hence, the name *alkaline* battery). The cathode is a mixture of $MnO_2(s)$ and graphite, separated from the anode by a porous fabric. The battery is sealed in a steel can to reduce the risk of concentrated KOH escaping.

The cell reactions are complex but can be approximately represented as:

Cathode: $2 MnO_2(s) + 2 H_2O(l) + 2 e^- \longrightarrow 2 MnO(OH)(s) + 2 OH^-(aq)$

Anode: $Zn(s) + 2 OH^-(aq) \longrightarrow Zn(OH)_2(s) + 2 e^-$

A fully charged alkaline battery delivers an emf of approximately 1.5 V.

Nickel–Cadmium and Nickel–Metal Hydride Batteries

The tremendous growth in high-power-demand portable electronic devices has increased the demand for lightweight, readily rechargeable batteries. In the late 20th century, one of the most popular rechargeable batteries was the nickel–cadmium (NiCad) battery. During discharge, cadmium metal is oxidized at the anode, while nickel oxyhydroxide, $NiO(OH)(s)$, is reduced at the cathode:

Cathode: $2 NiO(OH)(s) + 2 H_2O(l) + 2 e^- \longrightarrow 2 Ni(OH)_2(s) + 2 OH^-(aq)$

Anode: $Cd(s) + 2 OH^-(aq) \longrightarrow Cd(OH)_2(s) + 2 e^-$

As in the lead–acid battery, the solid reaction products adhere to the electrodes, which permits the electrode reactions to be reversed during charging. A single nickel–cadmium voltaic cell has a voltage of 1.30 V. NiCad battery packs typically contain three or more cells in series to produce the higher voltages needed by most electronic devices.

Although nickel–cadmium batteries have a number of attractive characteristics, the use of cadmium as the anode introduces significant limitations. Because cadmium is toxic, these batteries must be recycled. The toxicity of cadmium has led to a decline in their popularity from a peak annual production level of approximately 1.5 billion batteries in the early 2000s. In 2006 the European Union outlawed the use of cadmium in portable batteries, with exceptions for a few specialized applications. Cadmium also has a relatively high density, which increases battery weight, an undesirable characteristic for use in portable devices and electric vehicles.

The shortcomings associated with the use of cadmium fueled the development of nickel–metal hydride (NiMH) batteries. The cathode reaction is the same as that for nickel–cadmium batteries, but the anode reaction is very different. The anode consists of a metal alloy, typically with AM_5 stoichiometry, where A is lanthanum (La) or a mixture of metals from the lanthanide series, and M is mostly nickel alloyed with smaller amounts of other transition metals. On charging, water is reduced at the anode to form hydroxide ions and hydrogen atoms that are absorbed into the AM_5 alloy. When the battery is operating (discharging), the hydrogen atoms are oxidized and the resulting H^+ ions react with OH^- ions to form H_2O. Nickel–metal hydride batteries output a voltage similar to a nickel–cadmium battery but can store approximately three times more energy on a per mass basis.

Lithium-Ion Batteries

Currently, most portable electronic devices, including cell phones and laptop computers, are powered by rechargeable lithium-ion (Li-ion) batteries. Because lithium is a very light element, Li-ion batteries achieve a greater *specific energy density*—the amount of energy stored per unit mass—than nickel-based batteries. Because Li^+ has a very large negative standard reduction potential (Table 20.1), Li-ion batteries produce a higher voltage per cell than other batteries. A Li-ion battery produces

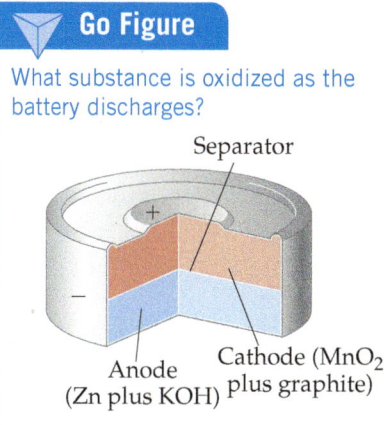

Go Figure

What substance is oxidized as the battery discharges?

Separator

Anode (Zn plus KOH)   Cathode (MnO_2 plus graphite)

▲ **Figure 20.20** Cutaway view of a miniature alkaline battery.

a maximum voltage of 3.7 V per cell, nearly three times higher than the 1.3 V per cell that nickel–cadmium and nickel–metal hydride batteries generate. As a result, a Li-ion battery can deliver more power than other batteries of comparable size, which leads to a higher *volumetric energy density*—the amount of energy stored per unit volume.

The technology of Li-ion batteries is based on the ability of Li^+ ions to be inserted into and removed from certain layered solids. In most commercial cells, the anode is made of graphite, which contains layers of sp^2 bonded carbon atoms [Figure 12.28(b)]. The cathode is made of a transition metal oxide that also has a layered structure, typically lithium cobalt oxide ($LiCoO_2$). The two electrodes are separated by an electrolyte, which functions like a salt bridge by allowing Li^+ ions to pass through it. When the cell is being charged, cobalt ions are oxidized and Li^+ ions migrate out of $LiCoO_2$ and into the graphite. During discharge, when the battery is producing electricity for use, the Li^+ ions spontaneously migrate from the graphite anode through the electrolyte to the cathode, enabling electrons to flow through the external circuit (**Figure 20.21**). John Goodenough, Stanley Whittingham, and Akira Yoshino received the 2019 Nobel Prize in Chemistry for their work that led to the development of Li-ion batteries.

Go Figure   When a Li-ion battery is fully discharged, the cathode has an empirical formula of $LiCoO_2$. What is the oxidation number of cobalt in this compound? Does the oxidation number of the cobalt increase or decrease as the battery charges?

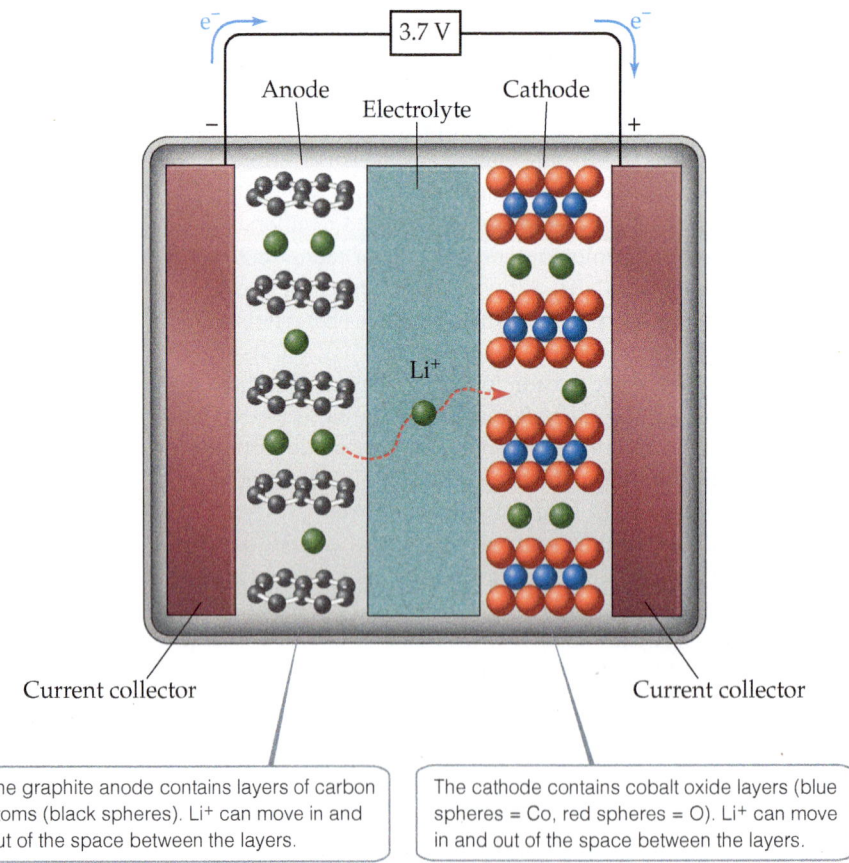

The graphite anode contains layers of carbon atoms (black spheres). Li^+ can move in and out of the space between the layers.

The cathode contains cobalt oxide layers (blue spheres = Co, red spheres = O). Li^+ can move in and out of the space between the layers.

▲ **Figure 20.21 Schematic of a Li-ion battery.** When the battery is discharging (operating), Li^+ ions move out of the anode and migrate through the electrolyte, where they enter the spaces between the cobalt oxide layers, reducing the cobalt ions. To recharge the battery, electrical energy is used to drive the Li^+ back to the anode, oxidizing the cobalt ions in the cathode.

CHEMISTRY AND SUSTAINABILITY | Batteries for Hybrid and Electric Vehicles

There has been a tremendous growth in recent years in the development of electric vehicles. Today both hybrid electric vehicles and fully electric vehicles are commercially available, with sales of fully electric vehicles topping three million worldwide in 2020. Hybrid electric vehicles can be powered either by electricity from batteries or by a conventional combustion engine, while fully electric vehicles are powered exclusively by the batteries (**Figure 20.22**). Hybrid electric vehicles can be further divided into plug-in hybrids, which require the owner to charge the battery by plugging it into a conventional outlet, or regular hybrids, which use regenerative braking and power from the combustion engine to charge the batteries.

Among the many technological advances needed to make electric vehicles practical, none is more important than advances in battery technology. Batteries for electric vehicles must have a high specific energy density, to reduce the weight of the car, as well as a high volumetric energy density, to minimize the space needed for the battery pack. A plot of energy densities for various types of rechargeable batteries is shown in **Figure 20.23**. The lead–acid batteries used in gasoline-powered automobiles are reliable and inexpensive, but their energy densities are far too low for practical use in an electric vehicle. Nickel–metal hydride batteries offer roughly three times higher energy density and until recently were the batteries of choice for commercial hybrid vehicles, such as the Toyota Prius.

Today's electric vehicles rely largely on Li-ion batteries because they offer the highest energy density of all commercially available batteries. However, the Li-ion batteries found in most electric vehicles do not use the $LiCoO_2$ anode shown in Figure 20.21. Instead, favored materials include $LiFePO_4$, $LiMn_2O_4$, and solid solutions of $LiCoO_2$, where a considerable fraction of the cobalt is replaced by

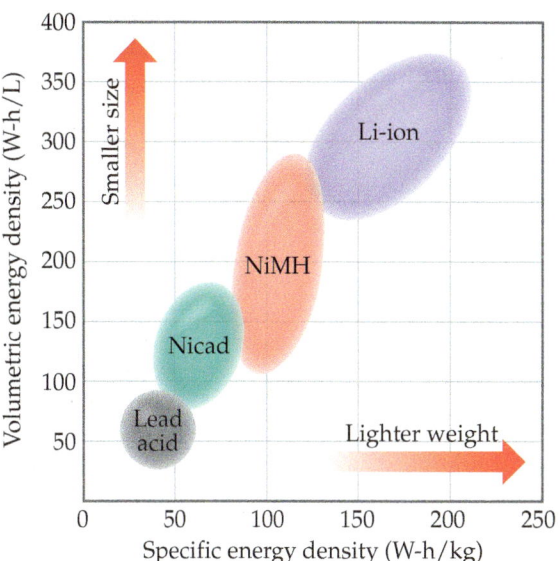

▲ **Figure 20.23 Energy densities of various types of batteries.** The higher the volumetric energy density, the smaller the amount of space needed for the batteries. The higher the specific energy density, the smaller the mass of the batteries. A Watt-hour (W-h) is equivalent to 3.6×10^3 joules.

metals such as nickel, manganese, and/or aluminum, those with nickel and manganese, $LiNi_xMn_yCo_{1-x-y}O_2$, are sometimes called Li NMC materials, for "nickel-manganese-cobalt". Batteries made with these cathodes have several advantages. They are not prone to thermal runaway events that can lead to the batteries catching fire, and they tend to have longer lifetimes than $LiCoO_2$. Concerns about the price and availability of cobalt have also driven the shift to alternative cathodes. Cobalt is a rare metal that is obtained largely as a by-product of either copper or nickel mining. Even more problematic is the concentration of cobalt reserves in a few countries; more than half of the global supply originates in the Democratic Republic of Congo. Concerns over political instability, as well as ethical issues associated with cobalt mining, provide strong incentives to minimize the use of cobalt in Li-ion batteries. However, the alternative anode materials mentioned here have one shortcoming—they have a smaller energy density than batteries made with a $LiCoO_2$ anode, which limits the range an electric vehicle can travel on a single charge. Scientists and engineers are intensively looking for new materials that will lead to further increases in energy density, lifetime, and safety of batteries, while trying to reduce the cost and environmental impact of the raw materials from which they are made.

Related Exercises: 20.10, 20.83, 20.84

▲ **Figure 20.22 Electric car.** This electric automobile can travel over 250 miles on a fully charged battery pack.

Hydrogen Fuel Cells

The thermal energy released by burning fuels can be converted to electrical energy. The thermal energy may convert water to steam, for instance, which drives a turbine that in turn drives an electrical generator. Typically, a maximum of only 40% of the free energy from combustion is converted to electricity in this manner; the remainder is lost as heat. The direct production of electricity from fuels by a voltaic cell could, in principle, yield a higher rate of conversion of chemical energy to electrical energy. Voltaic cells that perform this conversion using conventional fuels, such as H_2 and CH_4, are called **fuel cells**. Unlike batteries, fuel cells are not self-contained systems—the fuel must be continuously supplied to generate electricity.

One of the most popular fuel cells is based on the reaction of $H_2(g)$ and $O_2(g)$ to form $H_2O(l)$. These cells can generate electricity twice as efficiently as the best internal combustion engine. Further, they produce only water as a by-product and therefore do not increase CO_2 levels in the atmosphere. Under acidic conditions, the reactions of such a fuel cell are

Cathode:	$O_2(g) + 4\,H^+ + 4\,e^- \longrightarrow 2\,H_2O(l)$
Anode:	$2\,H_2(g) \longrightarrow 4\,H^+ + 4\,e^-$
Overall:	$2\,H_2(g) + O_2(g) \longrightarrow 2\,H_2O(l)$

These cells employ hydrogen gas as the fuel and oxygen gas from air as the oxidant and generate about 1.2 V.

Fuel cells are often named for either the fuel or the electrolyte used. In the hydrogen–PEM fuel cell (the acronym PEM stands for either proton-exchange membrane or polymer-electrolyte membrane), the anode and cathode are separated by a membrane that is permeable to protons but not to electrons (**Figure 20.24**). The membrane, therefore, acts as the salt bridge. The electrodes are typically made from graphite.

The hydrogen-PEM cell operates at around 80 °C. At this temperature, the electrochemical reactions would normally occur very slowly, and so small islands of platinum are deposited on each electrode to catalyze the reactions. The high cost and relative scarcity of platinum are two factors that limit wider use of hydrogen-PEM fuel cells.

In order to power a vehicle, multiple cells must be assembled into a fuel cell *stack*. The amount of power generated by a stack depends on the number and size of the fuel cells in the stack and on the surface area of the PEM.

Much fuel cell research today is directed toward improving electrolytes and catalysts and developing cells that use fuels such as hydrocarbons and alcohols, which are less difficult to handle and distribute than hydrogen gas.

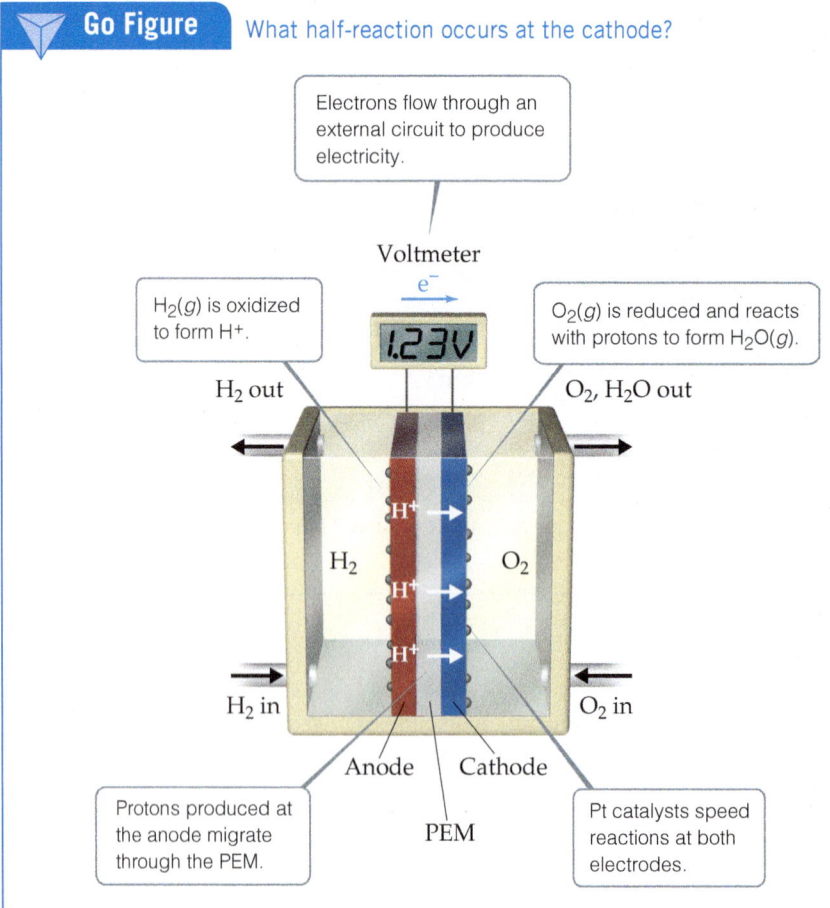

Go Figure   What half-reaction occurs at the cathode?

Electrons flow through an external circuit to produce electricity.

$H_2(g)$ is oxidized to form H^+.

$O_2(g)$ is reduced and reacts with protons to form $H_2O(g)$.

Voltmeter

e^-

1.23 V

H_2 out

O_2, H_2O out

H^+

H_2

O_2

H^+

H^+

H_2 in

O_2 in

Anode   Cathode

Protons produced at the anode migrate through the PEM.

PEM

Pt catalysts speed reactions at both electrodes.

▲ **Figure 20.24 A hydrogen-PEM fuel cell.** The proton-exchange membrane (PEM) allows H^+ ions generated by H_2 oxidation at the anode to migrate to the cathode, where H_2O is formed.

Self-Assessment Exercises

SAE 20.21   Which of the statements is or are *true* about batteries?

(**i**)   The overall reaction in a battery must have a positive emf.
(**ii**)   Batteries must have some mechanism to allow for ion transport from one half-cell to the other.
(**iii**)   Lithium-ion batteries provide higher voltages than most other batteries.

(**a**) only i (**b**) only ii (**c**) only iii (**d**) both ii and iii (**e**) All three statements are true.

SAE 20.22   The overall reaction for the lead–acid battery, $PbO_2(s) + Pb(s) + 2 HSO_4^-(aq) + 2 H^+(aq) \longrightarrow 2 PbSO_4(s) + 2 H_2O(l)$, has a standard cell potential, $E°_{cell} = +2.05$ V. When used in an automobile, the battery pack supplies approximately 12 V. What accounts for the discrepancy between $E°_{cell}$ and the output voltage? (**a**) Six individual batteries are connected in series to achieve $6 \times 2\,V \approx 12\,V$. (**b**) The concentration of sulfuric acid is much greater than 1 M, leading to an increase in $E°_{cell}$ to 12 V. (**c**) The temperature in a working automobile is not the standard temperature, so the nonstandard voltage is 12 V. (**d**) The concentration of sulfuric acid is much smaller than 1 M, leading to an increase in $E°_{cell}$ to 12 V.

SAE 20.23   Which of the following statements describing fuel cells is or are *true*?

(**i**)   A fuel cell is just another name for a battery.
(**ii**)   The reactants that power a fuel cell need to be continuously replenished.
(**iii**)   A fuel cell must have a positive emf.

(**a**) only i (**b**) only ii (**c**) only iii (**d**) both ii and iii (**e**) All three statements are true.

20.8 | Corrosion

In this section, we examine the undesirable redox reactions that lead to **corrosion** of metals. Corrosion reactions are spontaneous redox reactions in which a metal is attacked by some substance in its environment and converted to an unwanted compound.

For nearly all metals, oxidation is thermodynamically favorable in air at room temperature. When oxidation of a metal object is not inhibited, it can destroy the object. Oxidation can form an insulating protective oxide layer, however, that prevents further reaction of the underlying metal. Based on the standard reduction potential for Al^{3+}, for example, we expect aluminum metal to be readily oxidized. The many aluminum soft-drink and beer cans that litter the environment are ample evidence, however, that aluminum undergoes only very slow chemical corrosion. The exceptional stability of this active metal in air is due to the formation of a thin protective coat of oxide—a hydrated form of Al_2O_3—on the metal surface. The oxide coat is impermeable to O_2 or H_2O and so protects the underlying metal from further corrosion.

Magnesium metal is similarly protected, and some metal alloys, such as stainless steel, likewise form protective impervious oxide coats.

 Learning Objective

When you finish **Section 20.8**, you should be able to:

▶ Explain how corrosion occurs and how it is prevented by cathodic protection.

Corrosion of Iron (Rusting)

The rusting of iron is a familiar corrosion process that carries a significant economic impact. Up to 20% of the iron produced annually in the United States is used to replace iron objects that have been discarded because of rust damage.

Rusting of iron requires both oxygen and water, and the process can be accelerated by other factors, such as pH, presence of salts, contact with metals more difficult to oxidize than iron, and stress on the iron. The corrosion process involves oxidation and reduction, and the metal conducts electricity. Thus, electrons can move through the metal from a region where oxidation occurs to a region where reduction occurs, as in voltaic cells. Because the standard reduction potential for reduction of $Fe^{2+}(aq)$ is less positive than that for reduction of O_2, $Fe(s)$ can be oxidized by $O_2(g)$:

Cathode: $O_2(g) + 4 H^+(aq) + 4 e^- \longrightarrow 2 H_2O(l)$ $E°_{red} = 1.23$ V

Anode: $Fe(s) \longrightarrow Fe^{2+}(aq) + 2 e^-$ $E°_{red} = -0.44$ V

A portion of the iron, often associated with a dent or region of strain, can serve as an anode at which Fe is oxidized to Fe^{2+} (**Figure 20.25**). The electrons produced in the oxidation

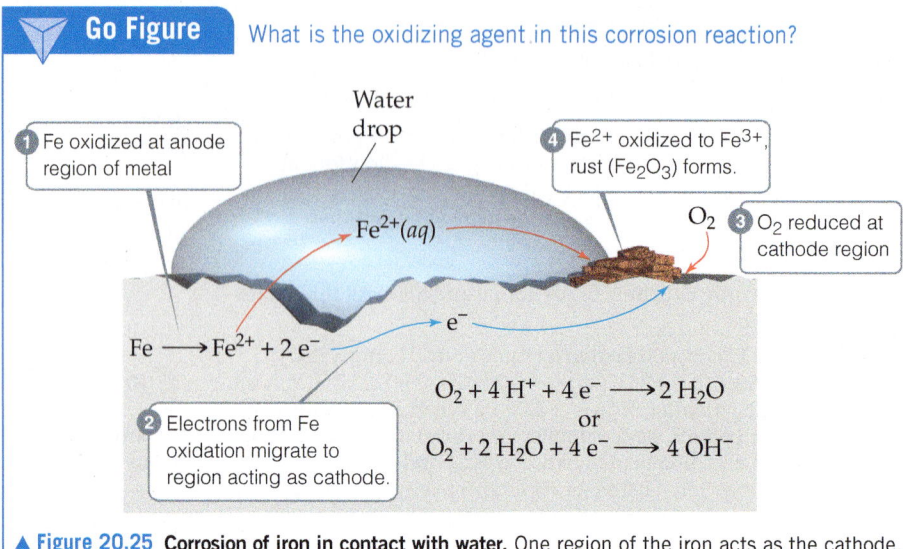

Go Figure   What is the oxidizing agent in this corrosion reaction?

1 Fe oxidized at anode region of metal

4 Fe^{2+} oxidized to Fe^{3+}, rust (Fe_2O_3) forms.

Water drop

$Fe^{2+}(aq)$

O_2 3 O_2 reduced at cathode region

$Fe \longrightarrow Fe^{2+} + 2\,e^-$

e^-

2 Electrons from Fe oxidation migrate to region acting as cathode.

$$O_2 + 4\,H^+ + 4\,e^- \longrightarrow 2\,H_2O$$
or
$$O_2 + 2\,H_2O + 4\,e^- \longrightarrow 4\,OH^-$$

▲ **Figure 20.25 Corrosion of iron in contact with water.** One region of the iron acts as the cathode, whereas another region acts as the anode.

migrate through the metal from this anodic region to another portion of the surface, which serves as the cathode where O_2 is reduced. The reduction of O_2 requires H^+, so lowering the concentration of H^+ (increasing the pH) makes O_2 reduction less favorable. Iron in contact with a solution whose pH is greater than 9 does not corrode.

The Fe^{2+} formed at the anode is eventually oxidized to Fe^{3+}, which forms the hydrated iron(III) oxide known as rust:*

$$4\,Fe^{2+}(aq) + O_2(g) + 4\,H_2O(l) + x\,H_2O(l) \longrightarrow 2\,Fe_2O_3 \cdot x\,H_2O(s) + 8\,H^+(aq)$$

Because the cathode is generally the area having the largest supply of O_2, rust often deposits there. If you look closely at a shovel after it has stood outside in the moist air with wet dirt adhered to its blade, you may notice that pitting has occurred under the dirt but that rust has formed elsewhere, where O_2 is more readily available. The enhanced corrosion caused by the presence of salts is usually evident on autos in areas where roads are heavily salted during winter. Like a salt bridge in a voltaic cell, the ions of the salt provide the electrolyte necessary to complete the electrical circuit.

Preventing Corrosion of Iron

Objects made of iron are often covered with a coat of paint or another metal to protect against corrosion. Covering the surface with paint prevents oxygen and water from reaching the iron surface. If the coating is broken, however, and the iron is exposed to oxygen and water, corrosion begins as the iron is oxidized.

With *galvanized iron*, which is iron coated with a thin layer of zinc, the iron is protected from corrosion even after the surface coat is broken. The standard reduction potentials are

$$Fe^{2+}(aq) + 2\,e^- \longrightarrow Fe(s) \quad E^\circ_{red} = -0.44\,V$$
$$Zn^{2+}(aq) + 2\,e^- \longrightarrow Zn(s) \quad E^\circ_{red} = -0.76\,V$$

Because E°_{red} for Fe^{2+} is less negative (more positive) than E°_{red} for Zn^{2+}, $Zn(s)$ is more readily oxidized than $Fe(s)$. Thus, even if the zinc coating is broken and the galvanized iron is exposed to oxygen and water, as in **Figure 20.26**, the zinc serves as the anode and is corroded (oxidized) instead of the iron. The iron serves as the cathode at which O_2 is reduced.

---

*Frequently, metal compounds obtained from aqueous solution have water associated with them. For example, copper(II) sulfate crystallizes from water with 5 mol of water per mole of $CuSO_4$. We represent this substance by the formula $CuSO_4 \cdot 5H_2O$. Such compounds are called hydrates. (Section 13.1) Rust is a hydrate of iron(III) oxide with a variable amount of water of hydration. We represent this variable water content by writing the formula $Fe_2O_3 \cdot xH_2O$.

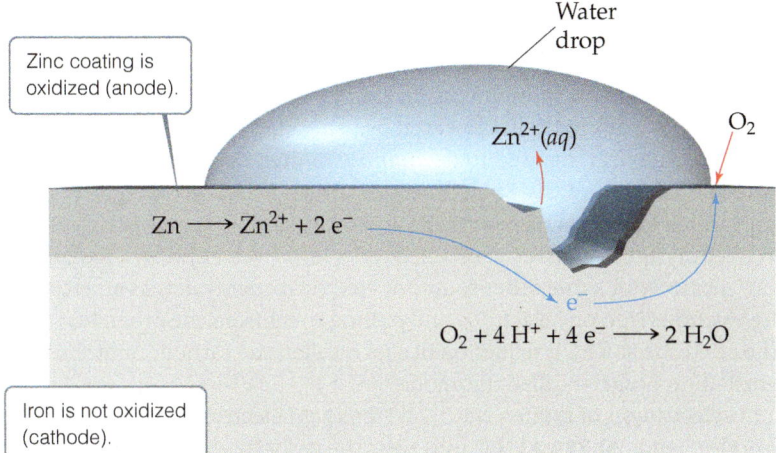

◀ **Figure 20.26 Cathodic protection of iron in contact with zinc.** The standard reduction potentials are $E^\circ_{red, Fe^{2+}} = -0.44$ V, $E^\circ_{red, Zn^{2+}} = -0.76$ V, making the zinc more readily oxidized.

Protecting a metal from corrosion by making it the cathode in an electrochemical cell is known as **cathodic protection**. The metal that is oxidized while protecting the cathode is called the *sacrificial anode*. Underground pipelines and storage tanks made of iron are often protected against corrosion by making the iron the cathode of a voltaic cell. For example, pieces of a metal that is more easily oxidized than iron, such as magnesium ($E^\circ_{red} = -2.37$ V), are buried near the pipe or storage tank and connected to it by wire (**Figure 20.27**). In moist soil, where corrosion can occur, the sacrificial metal serves as the anode, and the pipe or tank experiences cathodic protection.

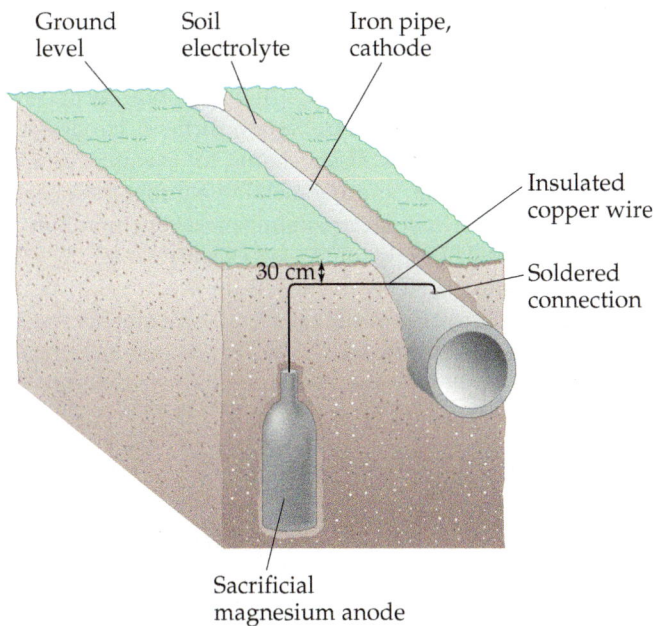

◀ **Figure 20.27 Cathodic protection of an iron pipe.** A mixture of gypsum, sodium sulfate, and clay surrounds the sacrificial magnesium anode to promote conductivity of ions.

Self-Assessment Exercises

SAE 20.24 Which of the following statements about corrosion of iron is or are *true*?

(**i**) Oxygen gas is reduced during the corrosion of iron.

(**ii**) The oxidation state of iron in rust is +2.

(**iii**) Because H$^+(aq)$ is a product of the reactions that lead to corrosion, the formation of rust is inhibited in acidic conditions.

(**a**) only i (**b**) only ii (**c**) only iii (**d**) both i and ii (**e**) both i and iii

SAE 20.25 If you want to protect a zinc pipe using cathodic protection, which metal would make a suitable sacrificial anode? (**a**) aluminum (**b**) iron (**c**) silver (**d**) tin (**e**) nickel

20.9 | Electrolysis

Voltaic cells are based on spontaneous redox reactions. It is also possible for *nonspontaneous* redox reactions to occur, however, by using electrical energy to drive them. For example, electricity can be used to decompose molten sodium chloride into its component elements, Na and Cl_2. Such processes driven by an outside source of electrical energy are called **electrolysis reactions** and take place in **electrolytic cells**.

An electrolytic cell consists of two electrodes immersed either in a molten salt or in a solution. A battery or some other source of electrical energy acts as an electron pump, pushing electrons into one electrode and pulling them from the other. Just as in voltaic cells, the electrode at which reduction occurs is called the cathode, and the electrode at which oxidation occurs is called the anode.

In the electrolysis of molten NaCl, Na^+ ions gain electrons and are reduced to Na at the cathode (**Figure 20.28**). As Na^+ ions near the cathode are depleted, additional Na^+ ions migrate in. Similarly, there is net movement of Cl^- ions to the anode, where they are oxidized. The electrode reactions for the electrolysis are

Cathode:	$2\,Na^+(l) + 2\,e^- \longrightarrow 2\,Na(l)$
Anode:	$2\,Cl^-(l) \longrightarrow Cl_2(g) + 2\,e^-$
Overall:	$2\,Na^+(l) + 2\,Cl^-(l) \longrightarrow 2\,Na(l) + Cl_2(g)$

Notice how the energy source is connected to the electrodes in Figure 20.28. The positive terminal is connected to the anode and the negative terminal is connected to the cathode, which forces electrons to move from the anode to the cathode.

Because of the high melting points of ionic substances, the electrolysis of molten salts requires very high temperatures. Do we obtain the same products if we electrolyze the aqueous solution of a salt instead of the molten salt? Frequently, the answer is no because water itself might be oxidized to form O_2 or reduced to form H_2 rather than the ions of the salt.

In our examples of the electrolysis of NaCl, the electrodes are *inert*; they do not react but merely serve as the surface where oxidation and reduction occur. Several practical applications of electrochemistry, however, are based on *active* electrodes—electrodes that participate in the electrolysis process. *Electroplating*, for example, uses electrolysis to deposit a thin layer of one metal on another metal to improve its appearance or resistance to corrosion. Examples include electroplating nickel or chromium onto steel and electroplating a precious metal like silver onto a less expensive one.

Figure 20.29 illustrates an electrolytic cell for electroplating nickel onto a piece of steel. The anode is a strip of nickel metal, and steel is the cathode. The electrodes are immersed in a solution of $NiSO_4(aq)$. When an external voltage is applied, reduction occurs at the cathode. The standard reduction potential of Ni^{2+} ($E^\circ_{red} = -0.28\,V$) is less negative than that of H_2O ($E^\circ_{red} = -0.83\,V$), so Ni^{2+} is preferentially reduced, depositing a layer of nickel metal on the steel cathode.

At the anode, the nickel metal is oxidized. To explain this behavior, we need to compare the substances in contact with the anode, H_2O and $NiSO_4(aq)$, with the anode material, Ni. For the $NiSO_4(aq)$ solution, neither Ni^{2+} nor SO_4^{2-} can be oxidized (because both already have their elements in their highest common oxidation state). Both the H_2O solvent and the Ni atoms in the anode, however, can undergo oxidation:

$$2\,H_2O(l) \longrightarrow O_2(g) + 4\,H^+(aq) + 4\,e^- \qquad E^\circ_{red} = +1.23\,V$$
$$Ni(s) \longrightarrow Ni^{2+}(aq) + 2\,e^- \qquad E^\circ_{red} = -0.28\,V$$

We saw in Section 20.4 that the half-reaction with the more negative E°_{red} undergoes oxidation more readily. (Remember Figure 20.11: The strongest reducing agents, which are the substances oxidized most readily, have the

Learning Objectives

When you finish Section 20.9, you should be able to:

▶ Explain the differences between electrolytic cells and voltaic cells.

▶ Calculate the amounts of reactants consumed and/or products produced in an electrolytic cell given the current and the length of time the cell operates.

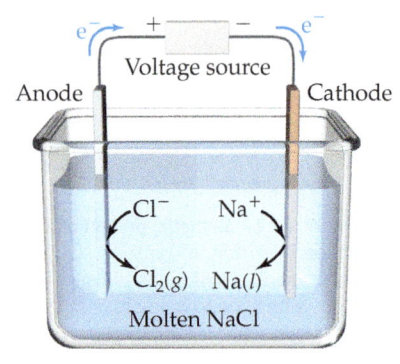

$2\,Cl^-(l) \longrightarrow Cl_2(g) + 2\,e^-$

$2\,Na^+(l) + 2\,e^- \longrightarrow 2\,Na(l)$

▲ **Figure 20.28 Electrolysis of molten sodium chloride.** Pure NaCl melts at 801 °C.

Go Figure What is E° for this cell?

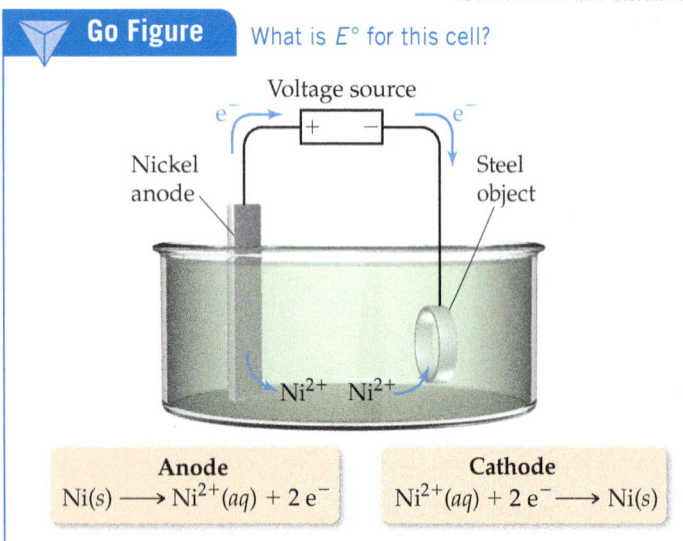

Anode	Cathode
$Ni(s) \longrightarrow Ni^{2+}(aq) + 2\,e^-$	$Ni^{2+}(aq) + 2\,e^- \longrightarrow Ni(s)$

▲ **Figure 20.29 Electrolytic cell with an active metal electrode.** Nickel dissolves from the anode to form $Ni^{2+}(aq)$. At the cathode, $Ni^{2+}(aq)$ is reduced and forms a nickel "plate" on the steel cathode.

most negative E°_{red} values.) Thus, it is the Ni(s), with its $E^\circ_{red} = -0.28$ V, that is oxidized at the anode rather than the H_2O. If we look at the overall reaction, it appears as if nothing has been accomplished. However, this is not true because Ni atoms are transferred from the Ni anode to the steel cathode, plating the steel with a thin layer of nickel atoms.

The standard emf for the overall reaction is

$$E^\circ_{cell} = E^\circ_{red}(\text{cathode}) - E^\circ_{red}(\text{anode}) = (-0.28\text{ V}) - (-0.28\text{ V}) = 0$$

Because the standard emf is zero, only a small emf is needed to cause the transfer of nickel atoms from one electrode to the other.

Quantitative Aspects of Electrolysis

The stoichiometry of a half-reaction shows how many electrons are needed to achieve an electrolytic process. For example, the reduction of Na^+ to Na is a one-electron process:

$$Na^+ + e^- \longrightarrow Na$$

Thus, 1 mol of electrons plates out 1 mol of Na metal, 2 mol of electrons plate out 2 mol of Na metal, and so forth. Similarly, 2 mol of electrons are required to produce 1 mol of Cu from Cu^{2+}, and 3 mol of electrons are required to produce 1 mol of Al from Al^{3+}:

$$Cu^{2+} + 2e^- \longrightarrow Cu$$
$$Al^{3+} + 3e^- \longrightarrow Al$$

For any half-reaction, the amount of substance reduced or oxidized in an electrolytic cell is directly proportional to the number of electrons passed into the cell.

The quantity of charge passing through an electrical circuit, such as that in an electrolytic cell, is generally measured in *coulombs*. As noted in Section 20.5, the charge on 1 mol of electrons is 96,485 C. A coulomb is the quantity of charge passing a point in a circuit in 1 s when the current is 1 ampere (A). Therefore, the number of coulombs passing through a cell can be obtained by multiplying the current in amperes by the elapsed time in seconds.

$$\text{coulombs} = \text{amperes} \times \text{seconds} \qquad [20.21]$$

Figure 20.30 shows how the quantities of substances produced or consumed in electrolysis are related to the quantity of electrical charge used. The same relationships can also be applied to voltaic cells. In other words, electrons can be thought of as "reagents" in electrolysis reactions.

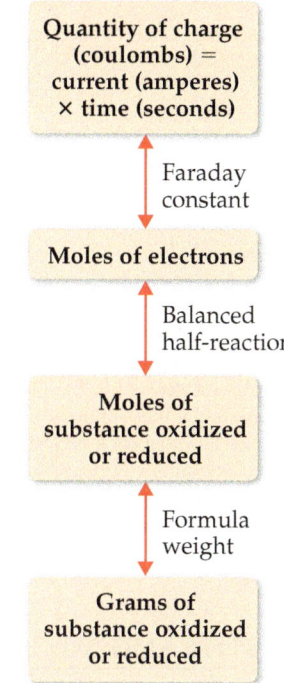

▲ Figure 20.30 Relationship between charge and amount of reactant and product in electrolysis reactions.

 Sample Exercise 20.14

Relating Electrical Charge and Quantity of Electrolysis

Calculate the number of grams of aluminum produced in 1.00 h by the electrolysis of molten $AlCl_3$ if the electrical current is 10.0 A.

SOLUTION

Analyze We are told that $AlCl_3$ is electrolyzed to form Al, and we are asked to calculate the number of grams of Al produced in 1.00 h with 10.0 A.

Plan Figure 20.30 provides a roadmap for this problem. Using the current, time, a balanced half-reaction, and the atomic weight of aluminum, we can calculate the mass of Al produced.

Solve

First, we calculate the coulombs of electrical charge passed into the electrolytic cell (note that 10.0 A = 10.0 C/s):

$$\text{Coulombs} = \text{amperes} \times \text{seconds} = (10.0\text{ C/s})(1.00\text{ h})\left(\frac{3600\text{ s}}{\text{h}}\right) = 3.60 \times 10^4\text{ C}$$

Second, we calculate the number of moles of electrons that pass into the cell:

$$\text{Moles e}^- = (3.60 \times 10^4\text{ C})\left(\frac{1\text{ mol e}^-}{96,485\text{ C}}\right) = 0.373\text{ mol e}^-$$

Third, we relate the number of moles of electrons to the number of moles of aluminum formed, using the half-reaction for the reduction of Al^{3+}:

$$Al^{3+} + 3e^- \longrightarrow Al$$

Continued

Thus, 3 mol of electrons are required to form 1 mol of Al:	Moles Al $= (0.373 \text{ mol e}^-)\left(\dfrac{1 \text{ mol Al}}{3 \text{ mol e}^-}\right) = 0.124 \text{ mol Al}$
Finally, we convert moles to grams:	Grams Al $= (0.124 \text{ mol Al})\left(\dfrac{27.0 \text{ g Al}}{1 \text{ mol Al}}\right) = 3.36 \text{ g Al}$
Or, we could have combined all of the above steps:	Grams Al $= (3.60 \times 10^4 \text{ C})\left(\dfrac{1 \text{ mol e}^-}{96{,}485 \text{ C}}\right)\left(\dfrac{1 \text{ mol Al}}{3 \text{ mol e}^-}\right)\left(\dfrac{27.0 \text{ g Al}}{1 \text{ mol Al}}\right) = 3.36 \text{ g Al}$

▶ **Practice Exercise**

(a) The half-reaction for formation of magnesium metal upon electrolysis of molten $MgCl_2$ is $Mg^{2+} + 2\,e^- \longrightarrow Mg$. Calculate the mass of magnesium formed upon passage of a current of 60.0 A for a period of 4.00×10^3 s. **(b)** How many seconds would be required to produce 50.0 g of Mg from $MgCl_2$ if the current is 100.0 A?

CHEMISTRY AND SUSTAINABILITY | Electrometallurgy of Aluminum

Many processes used to produce or refine metals are based on electrolysis. Collectively, these processes are referred to as *electrometallurgy*. Electrometallurgical procedures can be broadly differentiated according to whether they involve electrolysis of a molten salt or of an aqueous solution.

Electrolytic methods using molten salts are important for obtaining the more active metals, such as sodium, magnesium, and aluminum. These metals cannot be obtained from aqueous solution because water is more easily reduced than the metal ions. The standard reduction potentials of water under both acidic ($E^\circ_{red} = 0.00$ V) and basic ($E^\circ_{red} = -0.83$ V) conditions are more positive than those of Na^+ ($E^\circ_{red} = -2.71$ V), Mg^{2+} ($E^\circ_{red} = -2.37$ V), and Al^{3+} ($E^\circ_{red} = -1.66$ V).

Historically, obtaining aluminum metal has been a challenge. It is obtained from bauxite ore, which is chemically treated to concentrate aluminum oxide (Al_2O_3). The melting point of aluminum oxide is above 2000 °C, which is too high to permit its use as a molten medium for electrolysis.

The electrolytic process used commercially to produce aluminum is the *Hall–Héroult process*, named after its inventors, Charles M. Hall and Paul Héroult. Hall (1863–1914) began working on the problem of reducing aluminum in about 1885 after he had learned from a professor of the difficulty of reducing ores of very active metals. Before the development of an electrolytic process, aluminum was obtained by a chemical reduction using sodium or potassium as the reducing agent, a costly procedure that made aluminum metal very expensive. As late as 1852, the cost of aluminum was $545 per pound, far greater than the cost of gold. During the Paris Exposition in 1855, aluminum was exhibited as a rare metal, even though it is the third most abundant element in Earth's crust.

Hall, who was 21 years old when he began his research, utilized handmade and borrowed equipment in his studies and used a woodshed near his Ohio home as his laboratory. In about a year's time, he developed an electrolytic procedure using an ionic compound that melts to form a conducting medium that dissolves Al_2O_3 but does not interfere with the electrolysis reactions. The ionic compound he selected was the relatively rare mineral cryolite (Na_3AlF_6). Héroult, who was the same age as Hall, independently made the same discovery in France at about the same time. Because of the research of these two unknown young scientists, large-scale production of aluminum became commercially feasible, and aluminum became a common and familiar metal. Indeed, the factory that Hall subsequently built to produce aluminum evolved into the Alcoa Corporation.

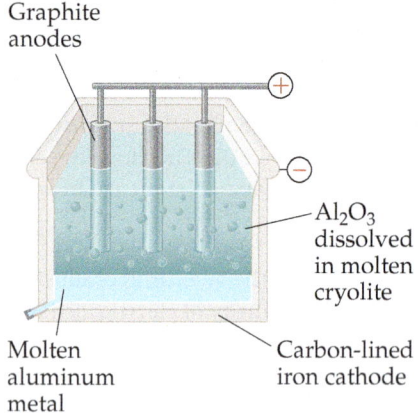

Graphite anodes

Al_2O_3 dissolved in molten cryolite

Molten aluminum metal

Carbon-lined iron cathode

▲ **Figure 20.31 The Hall–Héroult process.** Because molten aluminum is denser than the mixture of cryolite (Na_3AlF_6) and Al_2O_3, the metal collects at the bottom of the cell.

In the Hall–Héroult process, Al_2O_3 is dissolved in molten cryolite, which melts at 1012 °C and via the migration of ions is an effective electrical conductor (**Figure 20.31**). Graphite rods are employed as anodes and are consumed in the electrolysis:

Anode: $\quad C(s) + 2\,O^{2-}(l) \longrightarrow CO_2(g) + 4\,e^-$

Cathode: $\quad 3\,e^- + Al^{3+}(l) \longrightarrow Al(l)$

A large amount of electrical energy is needed in the Hall–Héroult process, which makes the aluminum industry a major consumer of electricity. Not surprisingly, most aluminum is manufactured in locations where electricity is relatively inexpensive. Approximately 75% of the electricity used to make aluminum in North America, South America, and Europe is generated at hydroelectric plants, whereas in China electricity from coal-fired plants plays a more important role. Because recycled aluminum requires less than 10% of the energy needed to produce "new" aluminum, considerable energy savings can be realized by increasing aluminum recycling rates. In the United States, approximately 50% of aluminum beverage containers are recycled. It is estimated that every 10% increase in aluminum recycling rates leads to a 15% decrease in greenhouse gas emissions from the aluminum industry.

Related Exercise: 20.109

Self-Assessment Exercises

SAE 20.26 Which of the following statements about electrolysis is *false*? (**a**) Electrolytic reactions are nonspontaneous. (**b**) Electrolytic reactions require an external source of electrical energy. (**c**) Oxidation takes place at the cathode of an electrolytic cell. (**d**) Electrolysis can be used to electroplate one metal onto another.

SAE 20.27 An electrolytic cell is set up to plate out copper metal on a flat electrode that is 3.00 cm × 3.00 cm in size. If a 10.0 A of current is applied to a cell containing 0.25 M CuCl$_2$, how long will it take to plate out a 1.00 mm thick layer of copper on the electrode (assume copper only plates out on only one side of the electrode)? The density of copper is 8.96 g/cm^3. (**a**) 19 min (**b**) 42 min (**c**) 45 min (**d**) 6.8 h (**e**) 68 h

Putting Concepts Together

The K_{sp} at 298 K for iron(II) fluoride is 2.4×10^{-6}. (**a**) Write a half-reaction that gives the likely products of the two-electron reduction of FeF$_2$(s) in water. (**b**) Use the K_{sp} value and the standard reduction potential of Fe^{2+}(aq) to calculate the standard reduction potential for the half-reaction in part (a). (**c**) Rationalize the difference between the reduction potential in part (a) and the reduction potential for Fe^{2+}(aq).

SOLUTION

Analyze We are going to combine what we know about equilibrium constants and electrochemistry to obtain reduction potentials.

Plan For (a) we need to determine which ion, Fe^{2+} or F^{-}, is more likely to be reduced by two electrons and complete the overall reaction FeF$_2$ + 2 e^{-} ⟶ ?. For (b) we need to write the chemical equation associated with the K_{sp} and see how it relates to $E°$ for the reduction half-reaction in (a). For (c) we need to compare $E°$ from (b) with the value for the reduction of Fe^{2+}.

Solve

(**a**) Iron(II) fluoride is an ionic substance that consists of Fe^{2+} and F^{-} ions. We are asked to predict where two electrons could be added to FeF$_2$. We cannot envision adding the electrons to the F^{-} ions to form F^{2-}, so it seems likely that Fe^{2+} will be reduced to Fe(s). We therefore predict the half-reaction:

$$\text{FeF}_2(s) + 2\,\text{e}^- \longrightarrow \text{Fe}(s) + 2\,\text{F}^-(aq)$$

(**b**) The K_{sp} for FeF$_2$ refers to the following equilibrium: (Section 17.4)

$$\text{FeF}_2(s) \rightleftharpoons \text{Fe}^{2+}(aq) + 2\,\text{F}^-(aq) \quad K_{sp} = [\text{Fe}^{2+}][\text{F}^-]^2 = 2.4 \times 10^{-6}$$

We were also asked to use the standard reduction potential of Fe^{2+}, whose half-reaction and standard reduction potentials are listed in Appendix E:

$$\text{Fe}^{2+}(aq) + 2\,\text{e}^- \longrightarrow \text{Fe}(s) \quad E = -0.440\,\text{V}$$

According to Hess's law, if we can add chemical equations to get a desired equation, then we can add their associated thermodynamic state functions, like ΔH or ΔG, to determine the thermodynamic quantity for the desired reaction. (Section 5.6) So we need to consider whether the three equations we are working with can be combined in a similar fashion. Notice that if we add the K_{sp} reaction to the standard reduction half-reaction for Fe^{2+}, we get the half-reaction we want:

Reaction 3 is still a half-reaction, so we do see the free electrons.

Overall:

1. $\quad \text{FeF}_2(s) \longrightarrow \text{Fe}^{2+}(aq) + 2\,\text{F}^-(aq)$

2. $\text{Fe}^{2+}(aq) + 2\,\text{e}^- \longrightarrow \text{Fe}(s)$

3. $\quad \overline{\text{FeF}_2(s) + 2\,\text{e}^- \longrightarrow \text{Fe}(s) + 2\,\text{F}^-(aq)}$

If we knew $\Delta G°$ for reactions 1 and 2, we could add them to get $\Delta G°$ for reaction 3. We can relate $\Delta G°$ to $E°$ by $\Delta G° = -nFE°$ (Equation 20.12) and to K by $\Delta G° = -RT \ln K$ (Equation 19.20; see also Figure 20.13). Furthermore, we know that K for reaction 1 is the K_{sp} of FeF$_2$, and we know $E°$ for reaction 2. Therefore, we can calculate $\Delta G°$ for reactions 1 and 2:
(Recall that 1 volt is 1 joule per coulomb.)

Reaction 1:

$$\Delta G° = -RT \ln K = -(8.314\,\text{J/K mol})(298\,\text{K}) \ln (2.4 \times 10^{-6}) = 3.21 \times 10^4\,\text{J/mol}$$

Reaction 2:

$$\Delta G° = -nFE° = -(2)(96{,}485\,\text{C/mol})(-0.440\,\text{J/C}) = 8.49 \times 10^4\,\text{J/mol}$$

Continued

Then $\Delta G°$ for reaction 3, the one we want, is the sum of the $\Delta G°$ values for reactions 1 and 2:

$$3.21 \times 10^4 \, \text{J/mol} + 8.49 \times 10^4 \, \text{J/mol} = 1.17 \times 10^5 \, \text{J/mol}$$

We can convert this to $E°$ from the relationship $\Delta G° = -nFE°$:

$$1.17 \times 10^5 \, \text{J/mol} = -(2)(96{,}485 \, \text{C/mol}) \, E°$$

$$E° = \frac{1.17 \times 10^5 \, \text{J/mol}}{-(2)(96{,}485 \, \text{C/mol})} = -0.606 \, \text{J/C} = -0.606 \, \text{V}$$

(c) The standard reduction potential for FeF_2 (-0.606 V) is more negative than that for Fe^{2+} (-0.440 V), telling us that the reduction of FeF_2 is the less favorable process. When FeF_2 is reduced, we not only reduce the Fe^{2+} but also break up the ionic solid. Because this additional energy must be overcome, the reduction of FeF_2 is less favorable than the reduction of Fe^{2+}.

Chapter Summary and Key Terms

OXIDATION STATES AND OXIDATION–REDUCTION REACTIONS (INTRODUCTION AND SECTION 20.1) In this chapter, we have focused on **electrochemistry**, the branch of chemistry that relates electricity and chemical reactions. Electrochemistry involves oxidation–reduction reactions, also called redox reactions. These reactions involve a change in the oxidation state of one or more elements. In every oxidation–reduction reaction, one substance is oxidized (its oxidation state, or number, increases) and one substance is reduced (its oxidation state, or number, decreases). The substance that is oxidized is referred to as a **reducing agent**, or **reductant**, because it causes the reduction of some other substance. Similarly, the substance that is reduced is referred to as an **oxidizing agent**, or **oxidant**, because it causes the oxidation of some other substance.

BALANCING REDOX EQUATIONS (SECTION 20.2) An oxidation–reduction reaction can be balanced by dividing the reaction into two **half-reactions**, one for oxidation and one for reduction. A half-reaction is a balanced chemical equation that includes electrons. In oxidation half-reactions, the electrons are on the product (right) side of the equation. In reduction half-reactions, the electrons are on the reactant (left) side of the equation. Each half-reaction is balanced separately, and the two are brought together with proper coefficients to balance the electrons on each side of the equation, so the electrons cancel when the half-reactions are added.

VOLTAIC CELLS (SECTION 20.3) A **voltaic (or galvanic) cell** uses a spontaneous oxidation–reduction reaction to generate electricity. In a voltaic cell, the oxidation and reduction half-reactions often occur in separate half-cells. Each half-cell has a solid surface called an electrode, where the half-reaction occurs. The electrode where oxidation occurs is called the **anode**, and the electrode where reduction occurs is called the **cathode**. The electrons released at the anode flow through an external circuit (where they do electrical work) to the cathode. Electrical neutrality in the solution is maintained by the migration of ions between the two half-cells through a device such as a salt bridge.

CELL POTENTIALS UNDER STANDARD CONDITIONS (SECTION 20.4) A voltaic cell generates an **electromotive force (emf)** that moves the electrons from the anode to the cathode through the external circuit. The origin of emf is a difference in the electrical potential energy of the two electrodes in the cell. The emf of a cell is called its **cell potential**, E_{cell} and is measured in volts ($1 \, \text{V} = 1 \, \text{J/C}$). The cell potential under standard conditions is called the **standard emf**, or the **standard cell potential**, and is denoted $E°_{cell}$.

A **standard reduction potential**, $E°_{red}$, can be assigned for an individual half-reaction. This is achieved by comparing the potential of the half-reaction to that of the **standard hydrogen electrode** (SHE), which is defined to have $E°_{red} = 0$ V.

The standard cell potential of a voltaic cell is the difference between the standard reduction potentials of the half-reactions that occur at the cathode and the anode:

$$E°_{cell} = E°_{red} \, (\text{cathode}) - E°_{red} \, (\text{anode}).$$

The value of $E°_{cell}$ is positive for a voltaic cell.

For a reduction half-reaction, $E°_{red}$ is a measure of the tendency of the reduction to occur; the more positive the value for $E°_{red}$, the greater the tendency of the substance to be reduced. Substances that are easily reduced act as strong oxidizing agents; thus, $E°_{red}$ provides a measure of the oxidizing strength of a substance. Substances that are strong oxidizing agents produce products that are weak reducing agents and vice versa.

FREE ENERGY AND REDOX REACTIONS (SECTION 20.5) The emf, E, is related to the change in the Gibbs free energy, $\Delta G = -nFE$, where n is the number of moles of electrons transferred during the redox process and F is the **Faraday constant**, defined as the quantity of electrical charge on one mole of electrons: $F = 96{,}485$ C/mol. Because E is related to ΔG, the sign of E indicates whether a redox process is spontaneous: $E > 0$ indicates a spontaneous process, and $E < 0$ indicates a nonspontaneous one. Because ΔG is also related to the equilibrium constant for a reaction ($\Delta G° = -RT \ln K$), we can relate $E°$ to K.

CELL POTENTIALS UNDER NONSTANDARD CONDITIONS (SECTION 20.6) The emf of a redox reaction varies with temperature and with the concentrations of reactants and products. The **Nernst equation** relates the emf under nonstandard conditions to the standard emf and the reaction quotient Q:

$$E = E° - (RT/nF) \ln Q = E° - (0.0592/n) \log Q$$

The factor 0.0592 is valid when $T = 298$ K. A **concentration cell** is a voltaic cell in which the same half-reaction occurs at both the anode and the cathode but with different concentrations of reactants in each half-cell. At equilibrium, $Q = K$ and $E = 0$.

BATTERIES AND FUEL CELLS (SECTION 20.7) A **battery** is a self-contained electrochemical power source that contains one or more voltaic cells. Batteries are based on a variety of different redox reactions. Batteries that cannot be recharged are called **primary cells**, while those that can be recharged are called **secondary cells**.

The common alkaline dry cell battery is an example of a primary cell battery. Lead–acid, nickel–cadmium, nickel–metal hydride, and lithium-ion batteries are examples of secondary cells. **Fuel cells** are voltaic cells that utilize redox reactions in which reactants have to be continuously supplied to the cell to generate voltage.

CORROSION (SECTION 20.8) Electrochemical principles help us understand **corrosion**, undesirable redox reactions in which a metal is attacked by some substance in its environment. The corrosion of iron into rust is caused by the presence of water and oxygen, and it is accelerated by the presence of electrolytes, such as road salt. The protection of a metal by putting it in contact with another metal that more readily undergoes oxidation is called **cathodic protection**. Galvanized iron, for example, is coated with a thin layer of zinc; because zinc is oxidized more readily than iron, the zinc serves as a sacrificial anode in the redox reaction.

ELECTROLYSIS (SECTION 20.9) An **electrolysis reaction**, which is carried out in an **electrolytic cell**, employs an external source of electricity to drive a nonspontaneous electrochemical reaction. The current-carrying medium within an electrolytic cell may be either a molten salt or an electrolyte solution. The electrodes in an electrolytic cell can be inert or active, meaning that the electrode can be involved in the electrolysis reaction. Active electrodes are important in electroplating and in metallurgical processes.

The quantity of substances formed during electrolysis can be calculated by considering the number of electrons involved in the redox reaction and the amount of electrical charge that passes into the cell. The amount of electrical charge is measured in coulombs and is related to the magnitude of the current and the time it flows $(1\ C = 1\ A\text{-s})$.

Key Equations

- $E^{\circ}_{cell} = E^{\circ}_{red}(\text{cathode}) - E^{\circ}_{red}(\text{anode})$ [20.8]   Relating standard emf to standard reduction potentials of the reduction (cathode) and oxidation (anode) half-reactions

- $\Delta G = -nFE$ [20.11]   Relating free-energy change and emf

- $E = E^{\circ} - \dfrac{0.0592\ V}{n} \log Q$ (at 298 K) [20.18]   The Nernst equation, expressing the effect of concentration on cell potential

Exam Prep

EP 20.1 In which of the following substances does nitrogen have the largest (most positive) oxidation number? (**a**) AlN (**b**) NO (**c**) HNO_3 (**d**) N_2O (**e**) N_2

EP 20.2 Which of the following can be classified as redox reactions?

  (**i**) $Ba(NO_3)_2(aq) + K_2SO_4(aq) \longrightarrow BaSO_4(s) + 2\ KNO_3(aq)$
  (**ii**) $PbS(s) + 4\ H_2O_2(aq) \longrightarrow PbSO_4(s) + 4\ H_2O(l)$
  (**iii**) $Mg(s) + O_2(g) \longrightarrow MgO(s)$

(**a**) only i (**b**) only ii (**c**) only iii (**d**) both i and ii (**e**) both ii and iii

EP 20.3 What is the reducing agent in the following reaction?

$$2\ Br^-(aq) + H_2O_2(aq) + 2\ H^+(aq) \longrightarrow Br_2(aq) + 2\ H_2O(l)$$

(**a**) $Br^-(aq)$ (**b**) $H_2O_2(aq)$ (**c**) $H^+(aq)$ (**d**) There is no reducing agent because this is not a redox reaction.

EP 20.4 In acidic solution, permanganate ions react with methanol to produce formic acid according to the following unbalanced chemical equation: $MnO_4^-(aq) + CH_3OH(aq) \longrightarrow Mn^{2+}(aq) + HCOOH(aq)$. Which statement(s) is or are *true* about this reaction?

  (**i**) After balancing the equation, there are 2 $H^+(aq)$ on the product side of the equation for every HCOOH.
  (**ii**) Manganese is oxidized during the course of the reaction.
  (**iii**) After balancing the equation, there are 11 water molecules on the product side of the equation.

(**a**) only i (**b**) only ii (**c**) only iii (**d**) Both i and iii (**e**) None of these statements is true.

EP 20.5 If you complete and balance the following equation in acidic solution,

$$Mn^{2+}(aq) + NaBiO_3(s) \longrightarrow Bi^{3+}(aq) + MnO_4^-(aq) + Na^+(aq)$$

how many water molecules are there in the balanced equation with the smallest whole-number coefficients? (**a**) four on the reactant side (**b**) three on the product side (**c**) one on the reactant side (**d**) seven on the product side (**e**) two on the product side

EP 20.6 If you complete and balance the following oxidation–reduction reaction in basic solution,

$$NO_2^-(aq) + Al(s) \longrightarrow NH_3(aq) + Al(OH)_4^-(aq)$$

how many hydroxide ions are there in the balanced equation with the smallest whole-number coefficients? (**a**) one on the reactant side (**b**) one on the product side (**c**) four on the reactant side (**d**) seven on the product side (**e**) none

EP 20.7 The following two half-reactions occur in a voltaic cell:

$$Ni(s) \longrightarrow Ni^{2+}(aq) + 2\ e^- \quad (\text{electrode} = Ni)$$
$$Cu^{2+}(aq) + 2\ e^- \longrightarrow Cu(s) \quad (\text{electrode} = Cu)$$

Which of the following descriptions most accurately describes what is occurring in the half-cell containing the Cu electrode and $Cu^{2+}(aq)$ solution?

(**a**) The electrode is losing mass, and cations from the salt bridge are flowing into the half-cell.

(**b**) The electrode is gaining mass, and cations from the salt bridge are flowing into the half-cell.

(**c**) The electrode is losing mass, and anions from the salt bridge are flowing into the half-cell.

(**d**) The electrode is gaining mass, and anions from the salt bridge are flowing into the half-cell.

EP 20.8 Consider a voltaic cell with a silver electrode and $AgNO_3(aq)$ in one cell compartment, a copper electrode and $CuCl_2(aq)$ solution in the other compartment, and an overall cell reaction of $2\ Ag^+(aq) + Cu(s) \longrightarrow 2\ Ag(s) + Cu^{2+}(aq)$. In this voltaic cell, _____ occurs at the copper electrode, which acts as the _____. (**a**) oxidation, cathode (**b**) oxidation, anode (**c**) reduction, cathode (**d**) reduction, anode

EP 20.9 A voltaic cell based on the reaction, $2\ Eu^{2+}(aq) + Ni^{2+}(aq) \longrightarrow 2\ Eu^{3+}(aq) + Ni(s)$, generates $E^{\circ}_{cell} = 0.07\ V$. Given the standard reduction potential for the reaction $Ni^{2+}(aq) + 2\ e^- \longrightarrow Ni(s)$, $E^{\circ}_{red} = -0.28\ V$, what is the standard reduction potential for the reaction $Eu^{3+}(aq) + e^- \longrightarrow Eu^{2+}(aq)$? (**a**) $-0.35\ V$ (**b**) $+0.35\ V$ (**c**) $-0.21\ V$ (**d**) $+0.21\ V$ (**e**) $-0.18\ V$

EP 20.10 Using the data in Table 20.1, what value would you calculate for the standard emf (E°_{cell}) of a voltaic cell that employs the overall cell reaction $2\ Ag^+(aq) + Ni(s) \longrightarrow 2\ Ag(s) + Ni^{2+}(aq)$? (**a**) $+0.52\ V$ (**b**) $-0.52\ V$ (**c**) $+1.08\ V$ (**d**) $-1.08\ V$ (**e**) $+0.80\ V$

EP 20.11 Consider three voltaic cells, each of which is similar to the one shown in Figure 20.5. In each voltaic cell, one half-cell contains a 1.0 M Fe(NO$_3$)$_2$(aq) solution with an Fe electrode. The contents of the other half-cells are as follows:

Cell 1: a 1.0 M CuCl$_2$(aq) solution with a Cu electrode

Cell 2: a 1.0 M NiCl$_2$(aq) solution with a Ni electrode

Cell 3: a 1.0 M ZnCl$_2$(aq) solution with a Zn electrode

In which voltaic cell(s) does iron act as the anode? (**a**) cell 1 (**b**) cell 2 (**c**) cell 3 (**d**) cells 1 and 2 (**e**) all three cells

EP 20.12 Based on the data in Table 20.1, which of the following species would you expect to be the strongest oxidizing agent? (**a**) Cl$^-$(aq) (**b**) Cl$_2$(g) (**c**) O$_2$(g) (**d**) H$^+$(aq) (**e**) Na$^+$(aq)

EP 20.13 Which of the following elements is capable of oxidizing Fe^{2+}(aq) ions to Fe^{3+}(aq) ions: chlorine, bromine, and/or iodine? (**a**) I$_2$ (**b**) Cl$_2$ (**c**) Cl$_2$ and I$_2$ (**d**) Cl$_2$ and Br$_2$ (**e**) all three elements

EP 20.14 Use the standard reduction potentials given in Appendix E to determine which of the following reactions is spontaneous in the forward direction under standard conditions.

(**i**) Pb(s) + Fe^{2+}(aq) $\longrightarrow$ Fe(s) + Pb^{2+}(aq)

(**ii**) Co^{3+}(aq) + Ce^{3+}(aq) $\longrightarrow$ Co^{2+}(aq) + Ce^{4+}(aq)

(**iii**) Ni(s) + I$_2$(s) $\longrightarrow$ NiI$_2$(aq)

(**a**) only i (**b**) only ii (**c**) only iii (**d**) both ii and iii (**e**) All three reactions are spontaneous.

EP 20.15 For the reaction,

3 Ni^{2+}(aq) + 2 Cr(OH)$_3$(s) + 10 OH$^-$(aq) $\longrightarrow$
$$3 \text{ Ni}(s) + 2 \text{ CrO}_4^{2-}(aq) + 8 \text{ H}_2\text{O}(l)$$

$\Delta G° = +87$ kJ/mol. Given the Ni^{2+}(aq) standard reduction potential $E°_{red} = -0.28$ V shown in Table 20.1, what value do you calculate for the standard reduction potential of the following half-reaction?

CrO$_4^{2-}$(aq) + 4 H$_2$O(l) + 3 e$^-$ $\longrightarrow$ Cr(OH)$_3$(s) + 5 OH$^-$(aq)

(**a**) −0.43 V (**b**) −0.28 V (**c**) 0.02 V (**d**) −0.13 V (**e**) −0.15 V

EP 20.16 Given the standard reduction potentials,

Pb^{2+}(aq) + 2 e$^-$ $\longrightarrow$ Pb(s) $E°_{red} = -0.13$ V

2 H$^+$(aq) + 2 e$^-$ $\longrightarrow$ H$_2$(g) $E°_{red} = 0.00$ V

what is the equilibrium constant for the redox reaction Pb(s) + 2 H$^+$(aq) $\longrightarrow$ Pb^{2+}(aq) + H$_2$(g) at 298 K? (**a**) 1.0 (**b**) 4.0 × 10^{-5} (**c**) 158 (**d**) 2.5 × 10^4

EP 20.17 Consider a voltaic cell whose overall reaction is Pb^{2+}(aq) + Zn(s) $\longrightarrow$ Pb(s) + Zn^{2+}(aq). The standard reduction potentials of the two half reactions are

Zn^{2+}(aq) + 2 e$^-$ $\longrightarrow$ Zn(s) $E°_{red} = -0.76$ V

Pb^{2+}(aq) + 2 e$^-$ $\longrightarrow$ Pb(s) $E°_{red} = -0.13$ V

What is the emf generated by this voltaic cell when the ion concentrations are [Pb^{2+}] = 1.5 × 10^{-3} M and [Zn^{2+}] = 0.55 M? (**a**) 0.71 V (**b**) 0.55 V (**c**) 0.49 V (**d**) 0.79 V (**e**) 0.63 V

EP 20.18 Consider a voltaic cell where the anode half-reaction is Zn(s) $\longrightarrow$ Zn^{2+}(aq) + 2 e$^-$ and the cathode half-reaction is Sn^{2+}(aq) + 2 e$^-$ $\longrightarrow$ Sn(s). The standard reduction potentials of the two half reactions are

Zn^{2+}(aq) + 2 e$^-$ $\longrightarrow$ Zn(s) $E°_{red} = -0.760$ V

Sn^{2+}(aq) + 2 e$^-$ $\longrightarrow$ Sn(s) $E°_{red} = -0.136$ V

What is the concentration of Sn^{2+} in the cathode compartment if [Zn^{2+}] = 2.5 × 10^{-3} M and the cell emf is 0.660 V? (**a**) 1.0 × 10^{-2} M (**b**) 8.4 × 10^{-3} M (**c**) 4.1 × 10^{-2} M (**d**) 1.5 × 10^{-4} M (**e**) 2.5 × 10^{-3} M

EP 20.19 A concentration cell is constructed from two hydrogen electrodes, both with $P_{H_2} = 1.00$ atm. One electrode is immersed in pure H$_2$O and the other in 6.0 M hydrochloric acid. What is the emf generated by the cell, and what is the identity of the electrode that is immersed in hydrochloric acid? (**a**) 0.23 V, cathode (**b**) 0.23 V, anode (**c**) 0.46 V, cathode (**d**) 0.46 V, anode (**e**) 0.046 V, cathode

EP 20.20 Which of the following statements about an alkaline battery is *false*? (**a**) It is a primary cell. (**b**) The reaction quotient $Q = 1$ for the overall cell reaction, so the emf should not change as the battery discharges. (**c**) As the battery discharges, K$^+$ migrate toward the anode and OH$^-$ migrate toward the cathode to maintain charge balance. (**d**) Zinc metal is oxidized at the anode. (**e**) The redox reactions occur in basic conditions.

EP 20.21 In a hydrogen fuel cell, what is the function of the polymer-electrolyte membrane (PEM)? (**a**) It allows electrons to move from the anode to the cathode. (**b**) It allows hydrogen molecules to migrate to the cathode where they react with oxygen molecules. (**c**) It allows H$^+$ ions to migrate from the anode to the cathode. (**d**) It catalyzes the reaction between protons and oxygen molecules so that the electron transfer occurs at a reasonable rate.

EP 20.22 Which of the following conditions or change in conditions will tend to slow or inhibit the corrosion of iron? (**a**) An increase in the partial pressure of oxygen (**b**) An increase in the pH (**c**) The presence of water (**d**) The presence of salts, like NaCl (**e**) Contact with metals that are less easily oxidized, like Ni or Ag

EP 20.23 Metallic sodium is produced industrially from electrolysis of molten sodium chloride. Why is it not possible to produce sodium from electrolysis of seawater, which contains a relatively high concentration of sodium ions (≈ 0.5 M)? (**a**) The electrolysis reaction releases so much energy that it evaporates the water. (**b**) Na$^+$ ions migrate through the salt bridge and react at the anode. (**c**) Water is reduced to H$_2$ gas at the cathode more easily than Na$^+$ ions are reduced to sodium metal. (**d**) Electrolysis of NaCl(aq) produces chlorine gas at the anode, which is an unwanted by-product due to its toxicity.

EP 20.24 How much time is needed to deposit 1.0 g of chromium metal from an aqueous solution of CrCl$_3$ using a current of 1.5 A? (**a**) 3.8 × 10^{-2} s (**b**) 21 min (**c**) 62 min (**d**) 139 min (**e**) 3.2 × 10^3 min

Exercises

Visualizing Concepts

20.1 In the Brønsted–Lowry concept of acids and bases, acid–base reactions are viewed as proton-transfer reactions. The stronger the acid, the weaker is its conjugate base. If we were to think of redox reactions in a similar way, what particle would be analogous to the proton? Would strong oxidizing agents be analogous to strong acids or strong bases? [Sections 20.1 and 20.2]

20.2 You may have heard that antioxidants are good for your health. Is an antioxidant an oxidizing agent or a reducing agent? [Sections 20.1 and 20.2]

20.3 The diagram that follows represents a molecular view of a process occurring at an electrode in a voltaic cell.

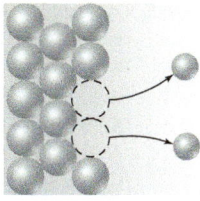

(**a**) Does the process represent oxidation or reduction? (**b**) Is the electrode the anode or cathode? (**c**) Why are the atoms in the electrode represented by larger spheres than those in the solution? [Section 20.3]

20.4 Assume that you want to construct a voltaic cell that uses the following half-reactions:

$$A^{2+}(aq) + 2\,e^- \longrightarrow A(s) \qquad E^\circ_{red} = -0.10\text{ V}$$
$$B^{2+}(aq) + 2\,e^- \longrightarrow B(s) \qquad E^\circ_{red} = -1.10\text{ V}$$

You begin with the incomplete cell pictured here in which the electrodes are immersed in water.

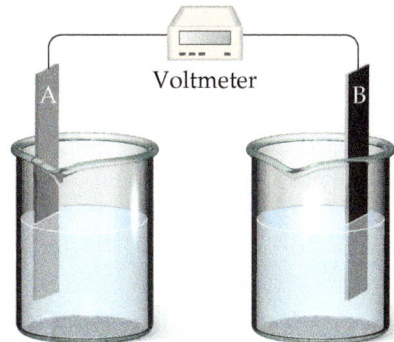

(**a**) What additions must you make to the cell for it to generate a standard emf? (**b**) Which electrode functions as the cathode? (**c**) Which direction do electrons move through the external circuit? (**d**) What voltage will the cell generate under standard conditions? [Sections 20.3 and 20.4]

20.5 For a spontaneous reaction $A(aq) + B(aq) \longrightarrow A^-(aq) + B^+(aq)$, answer the following questions:

(**a**) If you made a voltaic cell out of this reaction, what half-reaction would be occurring at the cathode, and what half-reaction would be occurring at the anode?

(**b**) Which half-reaction from (a) is higher in potential energy?

(**c**) What is the sign of E°_{red}? [Section 20.3]

20.6 Consider the following table of standard electrode potentials for a series of hypothetical reactions in aqueous solution:

Reduction Half-Reaction	E°(V)
$A^+(aq) + e^- \longrightarrow A(s)$	1.33
$B^{2+}(aq) + 2\,e^- \longrightarrow B(s)$	0.87
$C^{3+}(aq) + e^- \longrightarrow C^{2+}(aq)$	−0.12
$D^{3+}(aq) + 3\,e^- \longrightarrow D(s)$	−1.59

(**a**) Which substance is the strongest oxidizing agent? Which is weakest?

(**b**) Which substance is the strongest reducing agent? Which is weakest?

(**c**) Which substance(s) can oxidize C^{2+}? [Sections 20.4 and 20.5]

20.7 Consider a redox reaction for which E° is a negative number.

(**a**) What is the sign of ΔG° for the reaction?

(**b**) Will the equilibrium constant for the reaction be larger or smaller than 1?

(**c**) Can an electrochemical cell based on this reaction accomplish work on its surroundings? [Section 20.5]

20.8 Consider the following voltaic cell:

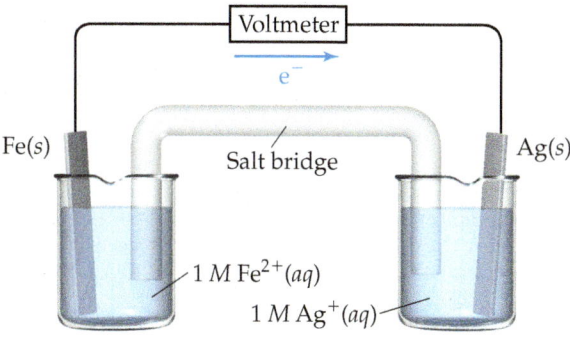

(**a**) Which electrode is the cathode?

(**b**) What is the standard emf generated by this cell?

(**c**) What is the change in the cell voltage when the ion concentrations in the cathode half-cell are increased by a factor of 10?

(**d**) What is the change in the cell voltage when the ion concentrations in the anode half-cell are increased by a factor of 10? [Sections 20.4 and 20.6]

20.9 Consider the half-reaction $Ag^+(aq) + e^- \longrightarrow Ag(s)$. (**a**) Which of the lines in the following diagram indicates how the reduction potential varies with the concentration of $Ag^+(aq)$? (**b**) What is the value of E_{red} when $\log[Ag^+] = 0$? [Section 20.6]

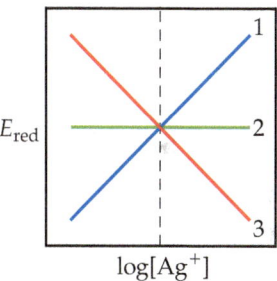

20.10 The electrodes in a silver oxide battery are silver oxide (Ag_2O) and zinc. (**a**) Which electrode acts as the anode? (**b**) Which battery do you think has an energy density most similar to the silver oxide battery: a Li-ion battery, a nickel–cadmium battery, or a lead–acid battery? [Section 20.7]

20.11 Bars of iron are put into each of the three beakers as shown here. In which beaker—A, B, or C—would you expect the iron to show the most corrosion? [Section 20.8]

Beaker A	Beaker B	Beaker C
Pure water	Dilute HCl(aq)	Dilute NaOH(aq)
pH = 7.0	solution	solution
	pH = 4.0	pH = 10.0

20.12 Magnesium, the element, is produced commercially by electrolysis from a molten salt (the "electrolyte") using a cell similar to the one shown here. (a) What is the most common oxidation number for Mg when it is part of a salt? (b) Chlorine gas is evolved as voltage is applied in the cell. Knowing this, identify the electrolyte. (c) Recall that in an electrolytic cell the anode is given the + sign and the cathode is given the − sign, which is the opposite of what we see in batteries. What half-reaction occurs at the anode in this electrolytic cell? (d) What half-reaction occurs at the cathode? (e) Assuming that the cells are 96% efficient in producing the desired products in electrolysis, what mass of Mg is formed by passing a current of 97,000 A for a 24 h period? [Section 20.9]

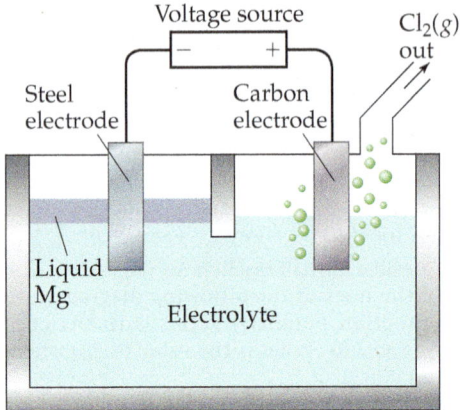

Oxidation–Reduction Reactions (Section 20.1)

20.13 (a) What is meant by the term *oxidation*? (b) On which side of an oxidation half-reaction do the electrons appear? (c) What is meant by the term *oxidant*? (d) What is meant by the term *oxidizing agent*?

20.14 (a) What is meant by the term *reduction*? (b) On which side of a reduction half-reaction do the electrons appear? (c) What is meant by the term *reductant*? (d) What is meant by the term *reducing agent*?

20.15 Indicate whether each of the following statements is *true* or *false*:

(a) If something is oxidized, it is formally losing electrons.

(b) For the reaction $Fe^{3+}(aq) + Co^{2+}(aq) \longrightarrow Fe^{2+}(aq) + Co^{3+}(aq)$, $Fe^{3+}(aq)$ is the reducing agent and $Co^{2+}(aq)$ is the oxidizing agent.

(c) If there are no changes in the oxidation state of the reactants or products of a particular reaction, that reaction is not a redox reaction.

20.16 Indicate whether each of the following statements is *true* or *false*:

(a) If something is reduced, it is formally losing electrons.

(b) A reducing agent gets oxidized as it reacts.

(c) An oxidizing agent is needed to convert CO into CO_2.

20.17 For each of the following balanced oxidation–reduction reactions, (i) identify the oxidation numbers for all the elements in the reactants and products and (ii) state the total number of electrons transferred in each reaction.

(a) $I_2O_5(s) + 5\,CO(g) \longrightarrow I_2(s) + 5\,CO_2(g)$

(b) $2\,Hg^{2+}(aq) + N_2H_4(aq) \longrightarrow 2\,Hg(l) + N_2(g) + 4\,H^+(aq)$

(c) $3\,H_2S(aq) + 2\,H^+(aq) + 2\,NO_3^-(aq) \longrightarrow 3\,S(s) + 2\,NO(g) + 4\,H_2O(l)$

20.18 For each of the following balanced oxidation–reduction reactions, (i) identify the oxidation numbers for all the elements in the reactants and products and (ii) state the total number of electrons transferred in each reaction.

(a) $2\,MnO_4^-(aq) + 3\,S^{2-}(aq) + 4\,H_2O(l) \longrightarrow 3\,S(s) + 2\,MnO_2(s) + 8\,OH^-(aq)$

(b) $4\,H_2O_2(aq) + Cl_2O_7(g) + 2\,OH^-(aq) \longrightarrow 2\,ClO_2^-(aq) + 5\,H_2O(l) + 4\,O_2(g)$

(c) $Ba^{2+}(aq) + 2\,OH^-(aq) + H_2O_2(aq) + 2\,ClO_2(aq) \longrightarrow Ba(ClO_2)_2(s) + 2\,H_2O(l) + O_2(g)$

20.19 Indicate whether the following balanced equations involve oxidation–reduction. If they do, identify the elements that undergo changes in oxidation number.

(a) $PBr_3(l) + 3\,H_2O(l) \longrightarrow H_3PO_3(aq) + 3\,HBr(aq)$

(b) $NaI(aq) + 3\,HOCl(aq) \longrightarrow NaIO_3(aq) + 3\,HCl(aq)$

(c) $3\,SO_2(g) + 2\,HNO_3(aq) + 2\,H_2O(l) \longrightarrow 3\,H_2SO_4(aq) + 2\,NO(g)$

20.20 Indicate whether the following balanced equations involve oxidation–reduction. If they do, identify the elements that undergo changes in oxidation number.

(a) $2\,AgNO_3(aq) + CoCl_2(aq) \longrightarrow 2\,AgCl(s) + Co(NO_3)_2(aq)$

(b) $2\,PbO_2(s) \longrightarrow 2\,PbO(s) + O_2(g)$

(c) $2\,H_2SO_4(aq) + 2\,NaBr(s) \longrightarrow Br_2(l) + SO_2(g) + Na_2SO_4(aq) + 2\,H_2O(l)$

Balancing Oxidation–Reduction Reactions (Section 20.2)

20.21 At 900 °C, titanium tetrachloride vapor reacts with molten magnesium metal to form solid titanium metal and molten magnesium chloride. (a) Write a balanced equation for this reaction. (b) What is being oxidized, and what is being reduced? (c) Which substance is the reductant, and which is the oxidant?

20.22 Hydrazine (N_2H_4) and dinitrogen tetroxide (N_2O_4) form a self-igniting mixture that has been used as a rocket propellant. The reaction products are N_2 and H_2O. (a) Write a balanced chemical equation for this reaction. (b) What is being oxidized, and what is being reduced? (c) Which substance serves as the reducing agent and which as the oxidizing agent?

20.23 Complete and balance the following half-reactions in acidic solution. In each case, indicate whether the half-reaction is an oxidation or a reduction.

(a) $Sn^{2+}(aq) \longrightarrow Sn^{4+}(aq)$

(b) $TiO_2(s) \longrightarrow Ti^{2+}(aq)$

(c) $ClO_3^-(aq) \longrightarrow Cl^-(aq)$

(d) $N_2(g) \longrightarrow NH_4^+(aq)$

20.24 Complete and balance the following half-reactions in basic solution. In each case, indicate whether the half-reaction is an oxidation or a reduction.

(a) $OH^-(aq) \longrightarrow O_2(g)$

(b) $SO_3^{2-}(aq) \longrightarrow SO_4^{2-}(aq)$

(c) $N_2(g) \longrightarrow NH_3(g)$

(d) $HO_2^-(aq) \longrightarrow OH^-(aq)$

20.25 Complete and balance the following half-reactions in basic solution. In each case, indicate whether the half-reaction is an oxidation or a reduction.

(a) $O_2(g) \longrightarrow H_2O(l)$

(b) $Mn^{2+}(aq) \longrightarrow MnO_2(s)$

(c) $Cr(OH)_3(s) \longrightarrow CrO_4^{2-}(aq)$

(d) $N_2H_4(aq) \longrightarrow N_2(g)$

20.26 Complete and balance the following half-reactions in acidic solution. In each case indicate whether the half-reaction is an oxidation or a reduction.

(a) $Mo^{3+}(aq) \longrightarrow Mo(s)$ (c) $NO_3^-(aq) \longrightarrow NO(g)$

(b) $H_2SO_3(aq) \longrightarrow SO_4^{2-}(aq)$ (d) $O_2(g) \longrightarrow H_2O(l)$

20.27 Complete and balance the following equations, and identify the oxidizing and reducing agents:

(a) $Cr_2O_7^{2-}(aq) + I^-(aq) \longrightarrow Cr^{3+}(aq) + IO_3^-(aq)$

(acidic solution)

(b) $I_2(s) + OCl^-(aq) \longrightarrow IO_3^-(aq) + Cl^-(aq)$

(acidic solution)

(c) $MnO_4^-(aq) + Br^-(aq) \longrightarrow MnO_2(s) + BrO_3^-(aq)$

(basic solution)

20.28 Complete and balance the following equations, and identify the oxidizing and reducing agents:

(a) $MnO_4^-(aq) + CH_3OH(aq) \longrightarrow Mn^{2+}(aq) + HCOOH(aq)$

(acidic solution)

(b) $As_2O_3(s) + NO_3^-(aq) \longrightarrow H_3AsO_4(aq) + N_2O_3(aq)$

(acidic solution)

(c) $Pb(OH)_4^{2-}(aq) + ClO^-(aq) \longrightarrow PbO_2(s) + Cl^-(aq)$

(basic solution)

20.29 Complete and balance the following equations, and identify the oxidizing and reducing agents.

(a) $NO_2^-(aq) + Cr_2O_7^{2-}(aq) \longrightarrow Cr^{3+}(aq) + NO_3^-(aq)$

(acidic solution)

(b) $Cr_2O_7^{2-}(aq) + CH_3OH(aq) \longrightarrow HCOOH(aq) + Cr^{3+}(aq)$

(acidic solution)

(c) $NO_2^-(aq) + Al(s) \longrightarrow NH_4^+(aq) + AlO_2^-(aq)$

(basic solution)

20.30 Complete and balance the following equations, and identify the oxidizing and reducing agents. (Recall that the O atoms in hydrogen peroxide, H_2O_2, have an atypical oxidation state.)

(a) $S(s) + HNO_3(aq) \longrightarrow H_2SO_3(aq) + N_2O(g)$

(acidic solution)

(b) $BrO_3^-(aq) + N_2H_4(g) \longrightarrow Br^-(aq) + N_2(g)$

(acidic solution)

(c) $H_2O_2(aq) + ClO_2(aq) \longrightarrow ClO_2^-(aq) + O_2(g)$

(basic solution)

Voltaic Cells (Section 20.3)

20.31 Indicate whether each statement is *true* or *false*: (a) The cathode is the electrode at which oxidation takes place. (b) A galvanic cell is another name for a voltaic cell. (c) Electrons flow spontaneously from anode to cathode in a voltaic cell.

20.32 Indicate whether each statement is *true* or *false*: (a) The anode is the electrode at which oxidation takes place. (b) A voltaic cell always has a positive emf. (c) A salt bridge or permeable barrier is necessary to allow a voltaic cell to operate.

20.33 A voltaic cell similar to that shown in Figure 20.5 is constructed. One electrode half-cell consists of a silver strip placed in a solution of $AgNO_3$, and the other has an iron strip placed in a solution of $FeCl_2$. The overall cell reaction is

$$Fe(s) + 2\,Ag^+(aq) \longrightarrow Fe^{2+}(aq) + 2\,Ag(s)$$

(a) What is being oxidized, and what is being reduced?
(b) Write the half-reactions that occur in the two half-cells.

(c) Which electrode is the anode, and which is the cathode? (d) Indicate the signs of the electrodes. (e) Do electrons flow from the silver electrode to the iron electrode or from the iron to the silver? (f) In which directions do the cations and anions migrate through the solution?

20.34 A voltaic cell similar to that shown in Figure 20.5 is constructed. One half-cell consists of an aluminum strip placed in a solution of $Al(NO_3)_3$, and the other has a nickel strip placed in a solution of $NiSO_4$. The overall cell reaction is

$$2\,Al(s) + 3\,Ni^{2+}(aq) \longrightarrow 2\,Al^{3+}(aq) + 3\,Ni(s)$$

(a) What is being oxidized, and what is being reduced?
(b) Write the half-reactions that occur in the two half-cells.
(c) Which electrode is the anode, and which is the cathode?
(d) Indicate the signs of the electrodes. (e) Do electrons flow from the aluminum electrode to the nickel electrode or from the nickel to the aluminum? (f) In which directions do the cations and anions migrate through the solution? Assume the Al is not coated with its oxide.

Cell Potentials under Standard Conditions (Section 20.4)

20.35 (a) What is the definition of the *volt*? (b) Do all voltaic cells produce a positive cell potential?

20.36 (a) Which electrode of a voltaic cell, the cathode or the anode, corresponds to the higher potential energy for the electrons? (b) What are the units for electrical potential? How does this unit relate to energy expressed in joules?

20.37 (a) Write the half-reaction that occurs at a hydrogen electrode in acidic aqueous solution when it serves as the cathode of a voltaic cell. (b) Write the half-reaction that occurs at a hydrogen electrode in acidic aqueous solution when it serves as the anode of a voltaic cell. (c) What is *standard* about the standard hydrogen electrode?

20.38 (a) What conditions must be met for a reduction potential to be a *standard reduction potential*? (b) What is the standard reduction potential of a standard hydrogen electrode? (c) Why is it impossible to measure the standard reduction potential of a single half-reaction?

20.39 A voltaic cell that uses the reaction

$$Tl^{3+}(aq) + 2\,Cr^{2+}(aq) \longrightarrow Tl^+(aq) + 2\,Cr^{3+}(aq)$$

has a measured standard cell potential of +1.19 V. (a) Write the two half-cell reactions. (b) Use data from Appendix E to determine E°_{red} for the reduction of $Tl^{3+}(aq)$ to $Tl^+(aq)$. (c) Sketch the voltaic cell, label the anode and cathode, and indicate the direction of electron flow.

20.40 A voltaic cell that uses the reaction

$$PdCl_4^{2-}(aq) + Cd(s) \longrightarrow Pd(s) + 4\,Cl^-(aq) + Cd^{2+}(aq)$$

has a measured standard cell potential of +1.03 V. (a) Write the two half-cell reactions. (b) Use data from Appendix E to determine E°_{red} for the reaction involving Pd. (c) Sketch the voltaic cell, label the anode and cathode, and indicate the direction of electron flow.

20.41 Use standard reduction potentials (Appendix E) to calculate the standard emf for each of the following reactions:

(a) $Cl_2(g) + 2\,I^-(aq) \longrightarrow 2\,Cl^-(aq) + I_2(s)$

(b) $Ni(s) + 2\,Ce^{4+}(aq) \longrightarrow Ni^{2+}(aq) + 2\,Ce^{3+}(aq)$

(c) $Fe(s) + 2\,Fe^{3+}(aq) \longrightarrow 3\,Fe^{2+}(aq)$

(d) $2\,NO_3^-(aq) + 8\,H^+(aq) + 3\,Cu(s) \longrightarrow 2\,NO(g) + 4\,H_2O(l) + 3\,Cu^{2+}(aq)$

20.42 Use data in Appendix E to calculate the standard emf for each of the following reactions:

(a) $H_2(g) + F_2(g) \longrightarrow 2\,H^+(aq) + 2\,F^-(aq)$

(b) $Cu^{2+}(aq) + Ca(s) \longrightarrow Cu(s) + Ca^{2+}(aq)$

(c) $3\,Fe^{2+}(aq) \longrightarrow Fe(s) + 2\,Fe^{3+}(aq)$

(d) $2\,ClO_3^-(aq) + 10\,Br^-(aq) + 12\,H^+(aq) \longrightarrow Cl_2(g) + 5\,Br_2(l) + 6\,H_2O(l)$

20.43 The standard reduction potentials of the following half-reactions are given in Appendix E:

$$Ag^+(aq) + e^- \longrightarrow Ag(s)$$
$$Cu^{2+}(aq) + 2\,e^- \longrightarrow Cu(s)$$
$$Ni^{2+}(aq) + 2\,e^- \longrightarrow Ni(s)$$
$$Cr^{3+}(aq) + 3\,e^- \longrightarrow Cr(s)$$

(a) Determine which combination of these half-cell reactions leads to the cell reaction with the largest positive cell potential and calculate the value. (b) Determine which combination of these half-cell reactions leads to the cell reaction with the smallest positive cell potential and calculate the value.

20.44 Given the following half-reactions and associated standard reduction potentials:

$$AuBr_4^-(aq) + 3\,e^- \longrightarrow Au(s) + 4\,Br^-(aq)$$
$$E_{red}^\circ = -0.86\ V$$
$$Eu^{3+}(aq) + e^- \longrightarrow Eu^{2+}(aq)$$
$$E_{red}^\circ = -0.43\ V$$
$$IO^-(aq) + H_2O(l) + 2\,e^- \longrightarrow I^-(aq) + 2\,OH^-(aq)$$
$$E_{red}^\circ = +0.49\ V$$

(a) Write the equation for the combination of these half-cell reactions that leads to the largest positive emf and calculate the value. (b) Write the equation for the combination of half-cell reactions that leads to the smallest positive emf and calculate that value.

20.45 A 1 M solution of $Cu(NO_3)_2$ is placed in a beaker with a strip of Cu metal. A 1 M solution of $SnSO_4$ is placed in a second beaker with a strip of Sn metal. A salt bridge connects the two beakers, and wires to a voltmeter link the two metal electrodes. (a) Which electrode serves as the anode, and which as the cathode? (b) Which electrode gains mass, and which loses mass as the cell reaction proceeds? (c) Write the equation for the overall cell reaction. (d) What is the emf generated by the cell under standard conditions?

20.46 A voltaic cell consists of a strip of cadmium metal in a solution of $Cd(NO_3)_2$ in one beaker, and in the other beaker a platinum electrode is immersed in a NaCl solution, with Cl_2 gas bubbled around the electrode. A salt bridge connects the two beakers. (a) Which electrode serves as the anode, and which serves as the cathode? (b) Does the Cd electrode gain or lose mass as the cell reaction proceeds? (c) Write the equation for the overall cell reaction. (d) What is the emf generated by the cell under standard conditions?

Strengths of Oxidizing and Reducing Agents (Section 20.4)

20.47 From each of the following pairs of substances, use data in Appendix E to choose the one that is the stronger reducing agent:

(a) Fe(s) or Mg(s)

(b) Ca(s) or Al(s)

(c) H_2(g, acidic solution) or $H_2S(g)$

(d) $BrO_3^-(aq)$ or $IO_3^-(aq)$

20.48 From each of the following pairs of substances, use data in Appendix E to choose the one that is the stronger oxidizing agent:

(a) $Cl_2(g)$ or $Br_2(l)$

(b) $Zn^{2+}(aq)$ or $Cd^{2+}(aq)$

(c) $Cl^-(aq)$ or $ClO_3^-(aq)$

(d) $H_2O_2(aq)$ or $O_3(g)$

20.49 Use the data in Appendix E to determine whether each of the following substances is likely to serve as an oxidant or a reductant: (a) $Cl_2(g)$, (b) MnO_4^- (aq, acidic solution), (c) Ba(s), (d) Zn(s).

20.50 Is each of the following substances likely to serve as an oxidant or a reductant: (a) $Ce^{3+}(aq)$, (b) Ca(s), (c) $ClO_3^-(aq)$, (d) $N_2O_5(g)$?

20.51 (a) Assuming standard conditions, arrange the following in order of increasing strength as oxidizing agents in acidic solution: $Cr_2O_7^{2-}$, H_2O_2, Cu^{2+}, Cl_2, O_2. (b) Arrange the following in order of increasing strength as reducing agents in acidic solution: Zn, I^-, Sn^{2+}, H_2O_2, Al.

20.52 Based on the data in Appendix E, (a) which of the following is the strongest oxidizing agent, and which is the weakest in acidic solution: Br_2, H_2O_2, Zn, $Cr_2O_7^{2-}$? (b) Which of the following is the strongest reducing agent, and which is the weakest in acidic solution: F^-, Zn, $N_2H_5^+$, I_2, NO?

20.53 The standard reduction potential for the reduction of $Eu^{3+}(aq)$ to $Eu^{2+}(aq)$ is -0.43 V. Use Appendix E to determine which of the following substances is capable of reducing $Eu^{3+}(aq)$ to $Eu^{2+}(aq)$ under standard conditions: Al, Co, H_2O_2, $N_2H_5^+$, $H_2C_2O_4$.

20.54 The standard reduction potential for the reduction of $RuO_4^-(aq)$ to $RuO_4^{2-}(aq)$ is $+0.59$ V. Use Appendix E to determine which of the following substances can oxidize $RuO_4^{2-}(aq)$ to $RuO_4^-(aq)$ under standard conditions: $Br_2(l)$, $BrO_3^-(aq)$, $Mn^{2+}(aq)$, $O_2(g)$, $Sn^{2+}(aq)$?

Free Energy and Redox Reactions (Section 20.5)

20.55 Given the following reduction half-reactions:

$$Fe^{3+}(aq) + e^- \longrightarrow Fe^{2+}(aq) \quad E_{red}^\circ = +0.77\ V$$
$$S_2O_6^{2-}(aq) + 4\,H^+(aq) + 2\,e^- \longrightarrow 2\,H_2SO_3(aq) \quad E_{red}^\circ = +0.60\ V$$
$$N_2O(g) + 2\,H^+(aq) + 2\,e^- \longrightarrow N_2(g) + H_2O(l) \quad E_{red}^\circ = -1.77\ V$$
$$VO_2^+(aq) + 2\,H^+(aq) + e^- \longrightarrow VO^{2+} + H_2O(l) \quad E_{red}^\circ = +1.00\ V$$

(a) Write balanced chemical equations for the oxidation of $Fe^{2+}(aq)$ by $S_2O_6^{2-}(aq)$, by $N_2O(aq)$, and by $VO_2^+(aq)$. (b) Calculate ΔG° for each reaction at 298 K. (c) Calculate the equilibrium constant K for each reaction at 298 K.

20.56 For each of the following reactions, write a balanced equation, calculate the standard emf, calculate ΔG° at 298 K, and calculate the equilibrium constant K at 298 K. (a) Aqueous iodide ion is oxidized to $I_2(s)$ by $Hg_2^{2+}(aq)$. (b) In acidic solution, copper(I) ion is oxidized to copper(II) ion by nitrate ion. (c) In basic solution, $Cr(OH)_3(s)$ is oxidized to $CrO_4^{2-}(aq)$ by $ClO^-(aq)$.

20.57 If the equilibrium constant for a two-electron redox reaction at 298 K is 1.5×10^{-4}, calculate the corresponding ΔG° and E°.

20.58 If the equilibrium constant for a one-electron redox reaction at 298 K is 8.7×10^4, calculate the corresponding ΔG° and E°.

20.59 Use the standard reduction potentials listed in Appendix E to calculate the equilibrium constant for each of the following reactions at 298 K:

(a) $Fe(s) + Ni^{2+}(aq) \longrightarrow Fe^{2+}(aq) + Ni(s)$

(b) $Co(s) + 2\,H^+(aq) \longrightarrow Co^{2+}(aq) + H_2(g)$

(c) $10\,Br^-(aq) + 2\,MnO_4^-(aq) + 16\,H^+(aq) \longrightarrow 2\,Mn^{2+}(aq) + 8\,H_2O(l) + 5\,Br_2(l)$

20.60 Use the standard reduction potentials listed in Appendix E to calculate the equilibrium constant for each of the following reactions at 298 K:

(a) $Cu(s) + 2 Ag^+(aq) \longrightarrow Cu^{2+}(aq) + 2 Ag(s)$

(b) $3 Ce^{4+}(aq) + Bi(s) + H_2O(l) \longrightarrow 3 Ce^{3+}(aq) +$ $BiO^+(aq) + 2 H^+(aq)$

(c) $N_2H_5^+(aq) + 4 Fe(CN)_6^{3-}(aq) \longrightarrow N_2(g) + 5 H^+(aq) +$ $4 Fe(CN)_6^{4-}(aq)$

20.61 A cell has a standard cell potential of +0.177 V at 298 K. What is the value of the equilibrium constant for the reaction (a) if $n = 1$? (b) if $n = 2$? (c) if $n = 3$?

20.62 At 298 K a cell reaction has a standard cell potential of +0.17 V. The equilibrium constant for the reaction is 5.5×10^5. What is the value of n for the reaction?

20.63 A voltaic cell is based on the reaction

$$Sn(s) + I_2(s) \longrightarrow Sn^{2+}(aq) + 2 I^-(aq)$$

Under standard conditions, what is the maximum electrical work, in joules, that the cell can accomplish if 75.0 g of Sn is consumed?

20.64 Consider the voltaic cell illustrated in Figure 20.5, which is based on the cell reaction

$$Zn(s) + Cu^{2+}(aq) \longrightarrow Zn^{2+}(aq) + Cu(s)$$

Under standard conditions, what is the maximum electrical work, in joules, that the cell can accomplish if 50.0 g of copper is formed?

Cell EMF under Nonstandard Conditions (Section 20.6)

20.65 (a) In the Nernst equation, what is the numerical value of the reaction quotient, Q, under standard conditions? (b) Can the Nernst equation be used at temperatures other than room temperature?

20.66 A voltaic cell is constructed with all reactants and products in their standard states. Will the concentration of the reactants increase, decrease, or remain the same as the cell operates?

20.67 What is the effect on the emf of the cell shown in Figure 20.9, which has the overall reaction $Zn(s) + 2 H^+(aq) \longrightarrow$ $Zn^{2+}(aq) + H_2(g)$, for each of the following changes? (a) The pressure of the H_2 gas is increased in the cathode half-cell. (b) Zinc nitrate is added to the anode half-cell. (c) Sodium hydroxide is added to the cathode half-cell, decreasing $[H^+]$. (d) The surface area of the anode is doubled.

20.68 A voltaic cell utilizes the following reaction:

$$Al(s) + 3 Ag^+(aq) \longrightarrow Al^{3+}(aq) + 3 Ag(s)$$

What is the effect on the cell emf of each of the following changes? (a) Water is added to the anode half-cell, diluting the solution. (b) The size of the aluminum electrode is increased. (c) A solution of $AgNO_3$ is added to the cathode half-cell, increasing the quantity of Ag^+ but not changing its concentration. (d) HCl is added to the $AgNO_3$ solution, precipitating some of the Ag^+ as AgCl.

20.69 A voltaic cell is constructed that uses the following reaction and operates at 298 K:

$$Zn(s) + Ni^{2+}(aq) \longrightarrow Zn^{2+}(aq) + Ni(s)$$

(a) What is the emf of this cell under standard conditions? (b) What is the emf of this cell when $[Ni^{2+}] = 3.00\,M$ and $[Zn^{2+}] = 0.100\,M$? (c) What is the emf of the cell when $[Ni^{2+}] = 0.200\,M$ and $[Zn^{2+}] = 0.900\,M$?

20.70 A voltaic cell utilizes the following reaction and operates at 298 K:

$$3 Ce^{4+}(aq) + Cr(s) \longrightarrow 3 Ce^{3+}(aq) + Cr^{3+}(aq)$$

(a) What is the emf of this cell under standard conditions? (b) What is the emf of this cell when $[Ce^{4+}] = 3.0\,M$, $[Ce^{3+}] = 0.10\,M$, and $[Cr^{3+}] = 0.010\,M$? (c) What is the emf of the cell when $[Ce^{4+}] = 0.010\,M, [Ce^{3+}] = 2.0\,M$, and $[Cr^{3+}] = 1.5\,M$?

20.71 A voltaic cell utilizes the following reaction:

$$4 Fe^{2+}(aq) + O_2(g) + 4 H^+(aq) \longrightarrow 4 Fe^{3+}(aq) + 2 H_2O(l)$$

(a) What is the emf of this cell under standard conditions?

(b) What is the emf of this cell when $[Fe^{2+}] = 1.3\,M$, $[Fe^{3+}] = 0.010\,M, P_{O_2} = 0.50$ atm, and the pH of the solution in the cathode half-cell is 3.50?

20.72 A voltaic cell utilizes the following reaction:

$$2 Fe^{3+}(aq) + H_2(g) \longrightarrow 2 Fe^{2+}(aq) + 2 H^+(aq)$$

(a) What is the emf of this cell under standard conditions? (b) What is the emf for this cell when $[Fe^{3+}] = 3.50\,M, P_{H_2} = 0.95$ atm, $[Fe^{2+}] = 0.0010\,M$, and the pH in both half-cells is 4.00?

20.73 A voltaic cell is constructed with two $Zn^{2+} - Zn$ electrodes. The two half-cells have $[Zn^{2+}] = 1.8\,M$ and $[Zn^{2+}] = 1.00 \times 10^{-2}\,M$, respectively. (a) Which electrode is the anode of the cell? (b) What is the standard emf of the cell? (c) What is the cell emf for the concentrations given? (d) For each electrode, predict whether $[Zn^{2+}]$ will increase, decrease, or stay the same as the cell operates.

20.74 A voltaic cell is constructed with two silver–silver chloride electrodes, each of which is based on the following half-reaction:

$$AgCl(s) + e^- \longrightarrow Ag(s) + Cl^-(aq)$$

The two half-cells have $[Cl^-] = 0.0150\,M$ and $[Cl^-] = 2.55\,M$, respectively. (a) Which electrode is the cathode of the cell? (b) What is the standard emf of the cell? (c) What is the cell emf for the concentrations given? (d) For each electrode, predict whether $[Cl^-]$ will increase, decrease, or stay the same as the cell operates.

20.75 The cell in Figure 20.9 could be used to provide a measure of the pH in the cathode half-cell. Calculate the pH of the cathode half-cell solution if the cell emf at 298 K is measured to be +0.684 V when $[Zn^{2+}] = 0.30\,M$ and $P_{H_2} = 0.90$ atm.

20.76 A voltaic cell is constructed that is based on the following reaction:

$$Sn^{2+}(aq) + Pb(s) \longrightarrow Sn(s) + Pb^{2+}(aq)$$

(a) If the concentration of Sn^{2+} in the cathode half-cell is 1.00 M and the cell generates an emf of +0.22 V, what is the concentration of Pb^{2+} in the anode half-cell? (b) If the anode half-cell contains $[SO_4^{2-}] = 1.00\,M$ in equilibrium with $PbSO_4(s)$, what is the K_{sp} of $PbSO_4$?

Batteries and Fuel Cells (Section 20.7)

20.77 During a period of discharge of a lead–acid battery, 402 g of Pb from the anode is converted into $PbSO_4(s)$. (a) What mass of $PbO_2(s)$ is reduced at the cathode during this same period? (b) How many coulombs of electrical charge are transferred from Pb to PbO_2?

20.78 During the discharge of an alkaline battery, 4.50 g of Zn is consumed at the anode of the battery. (a) What mass of MnO_2 is reduced at the cathode during this discharge? (b) How many coulombs of electrical charge are transferred from Zn to MnO_2?

20.79 Heart pacemakers are often powered by lithium–silver chromate "button" batteries. The overall cell reaction is

$$2 \, Li(s) + Ag_2CrO_4(s) \longrightarrow Li_2CrO_4(s) + 2 \, Ag(s)$$

(a) Lithium metal is the reactant at one of the electrodes of the battery. Is it the anode or the cathode? (b) Choose the two half-reactions from Appendix E that *most closely approximate* the reactions that occur in the battery. What standard emf would be generated by a voltaic cell based on these half-reactions? (c) The battery generates an emf of +3.5 V. How close is this value to the one calculated in part (b)? (d) Calculate the emf that would be generated at body temperature, 37 °C. How does this compare to the emf you calculated in part (b)?

20.80 Mercuric oxide dry-cell batteries are often used where a flat discharge voltage and long life are required, such as in watches and cameras. The two half-cell reactions that occur in the battery are

$$HgO(s) + H_2O(l) + 2 \, e^- \longrightarrow Hg(l) + 2 \, OH^-(aq)$$

$$Zn(s) + 2 \, OH^-(aq) \longrightarrow ZnO(s) + H_2O(l) + 2 \, e^-$$

(a) Write the overall cell reaction. (b) The value of E_{red}° for the cathode reaction is +0.098 V. The overall cell potential is +1.35 V. Assuming that both half-cells operate under standard conditions, what is the standard reduction potential for the anode reaction? (c) Why is the potential of the anode reaction different than would be expected if the reaction occurred in an acidic medium?

20.81 (a) Suppose that an alkaline battery was manufactured using cadmium metal rather than zinc. What effect would this have on the cell emf? (b) What environmental advantage is provided by the use of nickel–metal hydride batteries over nickel–cadmium batteries?

20.82 In some applications, nickel–cadmium batteries have been replaced by nickel–zinc batteries. The overall cell reaction for this battery is:

$$2 \, H_2O(l) + 2 \, NiO(OH)(s) + Zn(s)$$
$$\longrightarrow 2 \, Ni(OH)_2(s) + Zn(OH)_2(s)$$

(a) What is the cathode half-reaction? (b) What is the anode half-reaction? (c) A single nickel–cadmium cell has a voltage of 1.30 V. Based on the difference in the standard reduction potentials of Cd^{2+} and Zn^{2+}, what voltage would you estimate a nickel–zinc battery will produce? (d) Would you expect the specific energy density of a nickel–zinc battery to be higher or lower than that of a nickel–cadmium battery?

20.83 In a Li-ion battery the composition of the cathode is $LiCoO_2$ when completely discharged. On charging, approximately 50% of the Li^+ ions can be extracted from the cathode and transported to the graphite anode, where they intercalate between the layers. (a) What is the composition of the cathode when the battery is fully charged? (b) What is the oxidation number of cobalt in the cathode of a fully charged battery? (c) If the $LiCoO_2$ cathode has a mass of 10 g (when fully discharged), how many coulombs of electricity can be delivered on completely discharging a fully charged battery?

20.84 Li-ion batteries used in automobiles often use a $LiMn_2O_4$ cathode in place of the $LiCoO_2$ cathode found in most Li-ion batteries. (a) Calculate the mass percent lithium in each electrode material. (b) Which material has a higher percentage of lithium? Does this help to explain why batteries made with $LiMn_2O_4$ cathodes deliver less power on discharging? (c) In a battery that uses a $LiCoO_2$ cathode, approximately 50% of the lithium migrates from the cathode to the anode on charging. In a battery that uses a $LiMn_2O_4$ cathode, what fraction of the lithium in $LiMn_2O_4$ would need to migrate

out of the cathode to deliver the same amount of lithium to the graphite anode?

20.85 (a) Which reaction is spontaneous in the hydrogen fuel cell: hydrogen gas plus oxygen gas makes water, or water makes hydrogen gas plus oxygen gas? (b) Use the standard reduction potentials in Appendix E to calculate the standard voltage generated by the hydrogen fuel cell in acidic solution.

20.86 (a) What is the difference between a battery and a fuel cell? (b) Can the "fuel" of a fuel cell be a solid?

Corrosion (Section 20.8)

20.87 (a) Write the anode and cathode reactions that cause the corrosion of iron metal to aqueous iron(II). (b) Write the balanced half-reactions involved in the air oxidation of $Fe^{2+}(aq)$ to $Fe_2O_3 \cdot 3 \, H_2O(s)$.

20.88 (a) Based on standard reduction potentials, would you expect copper metal to oxidize under standard conditions in the presence of oxygen and hydrogen ions? (b) When the Statue of Liberty was refurbished, Teflon spacers were placed between the iron skeleton and the copper metal on the surface of the statue. What role do these spacers play?

20.89 (a) Magnesium metal is used as a sacrificial anode to protect underground pipes from corrosion. Why is the magnesium referred to as a "sacrificial anode"? (b) Looking in Appendix E, suggest what metal the underground pipes could be made from in order for magnesium to be successful as a sacrificial anode.

20.90 An iron object is plated with a coating of cobalt to protect against corrosion. Does the cobalt protect iron by cathodic protection?

20.91 Iron corrodes to produce rust, Fe_2O_3, but other corrosion products that can form are $Fe(O)(OH)$, iron oxyhydroxide, and magnetite, Fe_3O_4. (a) What is the oxidation number of Fe in iron oxyhydroxide, assuming oxygen's oxidation number is −2? (b) The oxidation number for Fe in magnetite was controversial for a long time. If we assume that oxygen's oxidation number is −2, and Fe has a unique oxidation number, what is the oxidation number for Fe in magnetite? (c) It turns out that there are two different kinds of Fe in magnetite that have different oxidation numbers. Suggest what these oxidation numbers are and what their relative stoichiometry must be, assuming oxygen's oxidation number is −2.

20.92 Copper corrodes to cuprous oxide, Cu_2O, or cupric oxide, CuO, depending on environmental conditions. (a) What is the oxidation state of copper in cuprous oxide? (b) What is the oxidation state of copper in cupric oxide? (c) Copper peroxide is another oxidation product of elemental copper. Suggest a formula for copper peroxide based on its name. (d) Copper(III) oxide is another unusual oxidation product of elemental copper. Suggest a chemical formula for copper(III) oxide.

Electrolysis (Section 20.9)

20.93 (a) What is *electrolysis*? (b) Are electrolysis reactions thermodynamically spontaneous? (c) What process occurs at the anode in the electrolysis of molten NaCl? (d) Why is sodium metal not obtained when an aqueous solution of NaCl undergoes electrolysis?

20.94 (a) What is an *electrolytic cell*? (b) The negative terminal of a voltage source is connected to an electrode of an electrolytic cell. Is the electrode the anode or the cathode of the cell? Explain. (c) The electrolysis of water is often done with a small amount of sulfuric acid added to the water. What is the role of the sulfuric acid? (d) Why are active metals such as Al obtained by electrolysis using molten salts rather than aqueous solutions?

20.95 **(a)** A $Cr^{3+}(aq)$ solution is electrolyzed, using a current of 7.60 A. What mass of $Cr(s)$ is plated out after 2.00 days? **(b)** What amperage is required to plate out 0.250 mol Cr from a Cr^{3+} solution in a period of 8.00 h?

20.96 Metallic magnesium can be made by the electrolysis of molten $MgCl_2$. **(a)** What mass of Mg is formed by passing a current of 4.55 A through molten $MgCl_2$, for 4.50 days? **(b)** How many minutes are needed to plate out 25.00 g Mg from molten $MgCl_2$ using 3.50 A of current?

20.97 **(a)** Calculate the mass of Li formed by electrolysis of molten LiCl by a current of 7.5×10^4 A flowing for a period of 24 h. Assume the electrolytic cell is 85% efficient. **(b)** What is the minimum voltage required to drive the reaction?

20.98 Elemental calcium is produced by the electrolysis of molten $CaCl_2$. **(a)** What mass of calcium can be produced by this process if a current of 7.5×10^3 A is applied for 48 h? Assume that the electrolytic cell is 68% efficient. **(b)** What is the minimum voltage needed to cause the electrolysis?

Additional Exercises

20.99 A *disproportionation* reaction is an oxidation–reduction reaction in which the same substance is oxidized and reduced. Complete and balance the following disproportionation reactions:

(a) $Ni^+(aq) \longrightarrow Ni^{2+}(aq) + Ni(s)$ (acidic solution)

(b) $MnO_4^{2-}(aq) \longrightarrow MnO_4^-(aq) + MnO_2(s)$ (acidic solution)

(c) $H_2SO_3(aq) \longrightarrow S(s) + HSO_4^-(aq)$ (acidic solution)

(d) $Cl_2(aq) \longrightarrow Cl^-(aq) + ClO^-(aq)$ (basic solution)

20.100 A common shorthand way to represent a voltaic cell is

anode | anode solution || cathode solution | cathode

A double vertical line represents a salt bridge or a porous barrier. A single vertical line represents a change in phase, such as from solid to solution. **(a)** Write the half-reactions and overall cell reaction represented by $Fe|Fe^{2+}||Ag^+|Ag$; calculate the standard cell emf using data in Appendix E. **(b)** Write the half-reactions and overall cell reaction represented by $Zn|Zn^{2+}||H^+|H_2$; calculate the standard cell emf using data in Appendix E and use Pt for the hydrogen electrode. **(c)** Using the notation just described, represent a cell based on the following reaction:

$$ClO_3{}^-(aq) + 3\,Cu(s) + 6\,H^+(aq)$$
$$\longrightarrow Cl^-(aq) + 3\,Cu^{2+}(aq) + 3\,H_2O(l)$$

Pt is used as an inert electrode in contact with the ClO_3^- and Cl^-. Calculate the standard cell emf, given: $ClO_3^-(aq) + 6\,H^+(aq) + 6\,e^- \longrightarrow Cl^-(aq) + 3\,H_2O(l)$; $E^\circ = 1.45$ V.

20.101 Predict whether the following reactions will be spontaneous in acidic solution under standard conditions: **(a)** oxidation of Sn to Sn^{2+} by I_2 (to form I^-), **(b)** reduction of Ni^{2+} to Ni by I^- (to form I_2), **(c)** reduction of Ce^{4+} to Ce^{3+} by H_2O_2, **(d)** reduction of Cu^{2+} to Cu by Sn^{2+} (to form Sn^{4+}).

20.102 Gold exists in two common positive oxidation states, +1 and +3. The standard reduction potentials for these oxidation states are

$$Au^+(aq) + e^- \longrightarrow Au(s) \quad E^\circ_{red} = +1.69 \text{ V}$$
$$Au^{3+}(aq) + 3\,e^- \longrightarrow Au(s) \quad E^\circ_{red} = +1.50 \text{ V}$$

(a) Can you use these data to explain why gold does not tarnish in the air? **(b)** Suggest several substances that should be strong enough oxidizing agents to oxidize gold metal. **(c)** Miners obtain gold by soaking gold-containing ores in an aqueous solution of sodium cyanide. A very soluble complex ion of gold forms in the aqueous solution because of the redox reaction

$$4\,Au(s) + 8\,NaCN(aq) + 2\,H_2O(l) + O_2(g)$$
$$\longrightarrow 4\,Na[Au(CN)_2](aq) + 4\,NaOH(aq)$$

What is being oxidized, and what is being reduced in this reaction? **(d)** Gold miners then react the basic aqueous

product solution from part (c) with Zn dust to get gold metal. Write a balanced redox reaction for this process. What is being oxidized, and what is being reduced?

20.103 A voltaic cell is constructed from an $Ni^{2+}(aq) - Ni(s)$ half-cell and an $Ag^+(aq) - Ag(s)$ half-cell. The initial concentration of $Ni^{2+}(aq)$ in the $Ni^{2+} - Ni$ half-cell is $[Ni^{2+}] = 0.0100$ M. The initial cell voltage is +1.12 V. **(a)** By using data in Appendix E, calculate the standard emf of this voltaic cell. **(b)** Will the concentration of $Ni^{2+}(aq)$ increase or decrease as the cell operates? **(c)** What is the initial concentration of $Ag^+(aq)$ in the Ag^+–Ag half-cell?

20.104 A voltaic cell is constructed that uses the following half-cell reactions:

$$Cu^+(aq) + e^- \longrightarrow Cu(s)$$
$$I_2(s) + 2\,e^- \longrightarrow 2\,I^-(aq)$$

The cell is operated at 298 K with $[Cu^+] = 0.25$ M and $[I^-] = 0.035$ M. **(a)** Determine E for the cell at these concentrations. **(b)** Which electrode is the anode of the cell? **(c)** Is the answer to part (b) the same as it would be if the cell were operated under standard conditions? **(d)** If $[Cu^+]$ were equal to 0.15 M, at what concentration of I^- would the cell have zero potential?

20.105 **(a)** Write the reactions for the discharge and charge of a nickel–cadmium (NiCad) rechargeable battery. **(b)** Given the following reduction potentials, calculate the standard emf of the cell:

$$Cd(OH)_2(s) + 2\,e^- \longrightarrow Cd(s) + 2\,OH^-(aq)$$
$$E^\circ_{red} = -0.76 \text{ V}$$

$$NiO(OH)(s) + H_2O(l) + e^- \longrightarrow Ni(OH)_2(s) + OH^-(aq)$$
$$E^\circ_{red} = +0.49 \text{ V}$$

(c) A typical NiCad voltaic cell generates an emf of +1.30 V. Why is there a difference between this value and the one you calculated in part (b)? **(d)** Calculate the equilibrium constant for the overall NiCad reaction based on this typical emf value.

20.106 The capacity of batteries such as the typical AA alkaline battery is expressed in units of milliamp-hours (mAh). An AA alkaline battery yields a nominal capacity of 2850 mAh. **(a)** What quantity of interest to the consumer is being expressed by the units of mAh? **(b)** The starting voltage of a fresh alkaline battery is 1.55 V. The voltage decreases during discharge and is 0.80 V when the battery has delivered its rated capacity. If we assume that the voltage declines linearly as current is withdrawn, estimate the total maximum electrical work the battery could perform during discharge.

20.107 Disulfides are compounds that have S—S bonds, and peroxides have O—O bonds. Thiols are organic compounds that have the general formula R—SH, where R is a generic hydrocarbon. The SH$^-$ ion is the sulfur counterpart of hydroxide, OH$^-$. Two thiols can react to make a disulfide, R—S—S—R. (a) What is the oxidation state of sulfur in a thiol? (b) What is the oxidation state of sulfur in a disulfide? (c) If you react two thiols to make a disulfide, are you oxidizing or reducing the thiols? (d) If you wanted to convert a disulfide to two thiols, should you add a reducing agent or an oxidizing agent to the solution? (e) Suggest what happens to the hydrogens in the thiols when they form disulfides.

20.108 (a) How many coulombs are required to plate a layer of chromium metal 0.25 mm thick on an auto bumper with a total area of 0.32 m^2 from a solution containing CrO$_4^{2-}$? The density of chromium metal is 7.20 g/cm^3. (b) What current flow is required for this electroplating if the bumper is to be plated in 10.0 s? (c) If the external source has an emf of +6.0 V and the electrolytic cell is 65% efficient, how much electrical power is expended to electroplate the bumper?

20.109 Calculate the number of kilowatt-hours of electricity required to produce 1.0×10^3 kg (1 metric ton) of aluminum by electrolysis of Al^{3+} if the applied voltage is 4.50 V and the process is 45% efficient.

20.110 Some years ago a unique proposal was made to raise the *Titanic*. The plan involved placing pontoons within the ship using a surface-controlled submarine-type vessel. The pontoons would contain cathodes and would be filled with hydrogen gas formed by the electrolysis of water. It has been estimated that it would require about 7×10^8 mol of H$_2$ to provide the buoyancy to lift the ship (*J. Chem. Educ.*, 1973, Vol. 50, 61). (a) How many coulombs of electrical charge would be required? (b) What is the minimum voltage required to generate H$_2$ and O$_2$ if the pressure on the gases at the depth of the wreckage (2 mi) is 300 atm? (c) What is the minimum electrical energy required to raise the *Titanic* by electrolysis? (d) What is the minimum cost of the electrical energy required to generate the necessary H$_2$ if the electricity costs 85 cents per kilowatt-hour to generate at the site?

20.111 Aqueous solutions of ammonia (NH$_3$) and bleach (active ingredient NaOCl) are sold as cleaning fluids, but bottles of both of them warn: "Never mix ammonia and bleach, as toxic gases may be produced." One of the toxic gases that

can be produced is chloroamine, NH$_2$Cl. (a) What is the oxidation number of chlorine in bleach? (b) What is the oxidation number of chlorine in chloramine? (c) Is Cl oxidized, reduced, or neither, upon the conversion of bleach to chloramine? (d) Another toxic gas that can be produced is nitrogen trichloride, NCl$_3$. What is the oxidation number of N in nitrogen trichloride? (e) Is N oxidized, reduced, or neither, upon the conversion of ammonia to nitrogen trichloride?

20.112 A voltaic cell is based on Ag$^+(aq)$/Ag(s) and Fe$^{3+}(aq)$/Fe$^{2+}(aq)$ half-cells. (a) What is the standard emf of the cell? (b) Which reaction occurs at the cathode and which at the anode of the cell? (c) Use $S°$ values in Appendix C and the relationship between cell potential and free-energy change to predict whether the standard cell potential increases or decreases when the temperature is raised above 25 °C.

20.113 Cytochrome, a complicated molecule that we will represent as CyFe^{2+}, reacts with the air we breathe to supply energy required to synthesize adenosine triphosphate (ATP). The body uses ATP as an energy source to drive other reactions (Section 19.7). At pH 7.0 the following reduction potentials pertain to this oxidation of CyFe^{2+}:

$$O_2(g) + 4\,H^+(aq) + 4\,e^- \longrightarrow 2\,H_2O(l) \qquad E°_{red} = +0.82\text{ V}$$

$$CyFe^{3+}(aq) + e^- \longrightarrow CyFe^{2+}(aq) \qquad E°_{red} = +0.22\text{ V}$$

(a) What is ΔG for the oxidation of CyFe^{2+} by air? (b) If the synthesis of 1.00 mol of ATP from adenosine diphosphate (ADP) requires a ΔG of 37.7 kJ, how many moles of ATP are synthesized per mole of O$_2$?

20.114 The standard potential for the reduction of AgSCN(s) is +0.09 V.

$$AgSCN(s) + e^- \longrightarrow Ag(s) + SCN^-(aq)$$

Use this value and the electrode potential for Ag$^+(aq)$ to calculate the K_{sp} for AgSCN.

20.115 A student designs an ammeter (a device that measures electrical current) that is based on the electrolysis of water into hydrogen and oxygen gases. When electrical current of unknown magnitude is run through the device for 2.00 min, 12.3 mL of water-saturated H$_2(g)$ is collected. The temperature of the system is 25.5 °C, and the atmospheric pressure is 768 torr. What is the magnitude of the current in amperes?

Design an Experiment

You are asked to construct a voltaic cell that would simulate an alkaline battery by providing an electrical output of 1.50 V at the beginning of its discharge. After you complete it, your voltaic cell will be used to power an external device that draws a constant current of 0.50 amperes for 2.0 hours. You are given the following supplies: electrodes of each transition metal from manganese to zinc, the chloride salts of the 2+ transition metal ions from Mn^{2+} to Zn^{2+} (MnCl$_2$, FeCl$_2$, CoCl$_2$, NiCl$_2$, CuCl$_2$ and ZnCl$_2$), two 100-mL beakers, a salt bridge, a voltmeter, and wires to make electrical connections between the electrodes and the voltmeter. (a) Sketch out

your voltaic cell, labeling the metal used for each electrode and the type and concentration of the solutions in which each electrode is immersed. Be sure to describe how many grams of the salt are dissolved and the total volume of solution in each beaker. (b) What will be the concentrations of the transition metal ion in each solution at the end of the 2-h discharge? (c) What voltage will the cell register at the end of the discharge? (d) How long would your cell run before it died because the reactant in one of the half-cells was completely consumed? Assume the current stays constant throughout the discharge.

21

NUCLEAR CHEMISTRY

▲ **THE SUN IS A HUGE SPHERE** so hot that nuclei and electrons move independently. It accounts for 99.86% of the mass of our solar system and is composed of 73.8% hydrogen, 24.8% helium, and 1.4% other elements. Most of the Sun's energy is generated in its core by the fusion of hydrogen nuclei to form helium nuclei. The Sun's surface releases this energy as electromagnetic radiation accompanied by a stream of charged particles called the solar wind. Bursts of radiation and particles continuously erupt from the surface, producing solar flares.

In our study of the structure and properties of matter, we have seen electrons as the major players. Chemical reactions involve the making and breaking of bonds, resulting in various changes in the electronic environments of the atoms involved. Nuclei, though certainly not unimportant, remain unchanged during chemical reactions. There is, however, another kind of reaction that we have yet to examine, one in which nuclei undergo change, thereby altering the very identities of the atoms involved.

Transformations of atomic nuclei are fittingly called **nuclear reactions**. Some nuclei change spontaneously at room temperature, emitting radiation. They are then said to be **radioactive**. As we describe in this chapter, there are also other kinds of nuclear changes.

Nuclear reactions are the source of the energy in nuclear power plants, nuclear weapons, and stars. They are involved in various radiation therapies utilized to both diagnose and treat diseases. In addition, radioactive elements are used to help determine the mechanisms of chemical reactions, to trace the movement of atoms in biological systems and in the environment, and to date historical artifacts.

Nuclear reactions can release enormous amounts of energy—far more than the amounts involved in even the most energetic chemical

WHAT'S AHEAD

21.1 ▶ Radioactivity and Nuclear Equations Describe nuclear reactions using equations analogous to chemical equations, in which the nuclear charges and masses of reactants and products are in balance. Also learn that radioactive nuclei most commonly decay by emission of *alpha*, *beta*, or *gamma* radiation.

21.2 ▶ Patterns of Nuclear Stability Recognize that nuclear stability is determined largely by the *neutron-to-proton ratio*. For stable nuclei, this ratio increases with increasing atomic number. All nuclei with 84 or more protons are radioactive. Heavy nuclei gain stability by a series of nuclear disintegrations leading to stable nuclei.

21.3 ▶ Nuclear Transmutations Describe *nuclear transmutations*, which are nuclear reactions induced by bombardment of a nucleus by a neutron or an accelerated charged particle.

21.4 ▶ Rates of Radioactive Decay Observe that radioisotope decays are first-order kinetic processes with characteristic half-lives. Decay rates can be used to determine the age of ancient artifacts and geological formations.

21.5 ▶ Detection of Radioactivity See that the radiation emitted by a radioactive substance can be detected by a variety of devices, including dosimeters, Geiger counters, and scintillation counters.

21.6 ▶ Energy Changes in Nuclear Reactions Interpret energy changes in nuclear reactions in terms of mass changes via Einstein's equation, $E = mc^2$. The *nuclear binding energy* of a nucleus is the difference between the mass of the nucleus and the sum of the masses of its nucleons.

21.7 ▶ Nuclear Power: Fission Describe *nuclear fission*, in which a heavy nucleus splits to form two or more product nuclei. Nuclear fission is the energy source for nuclear power plants.

21.8 ▶ Nuclear Power: Fusion Recognize that in *nuclear fusion* two light nuclei are fused together to form a more stable, heavier nucleus.

21.9 ▶ Radiation in the Environment and Living Systems Discover that naturally occurring radioisotopes bathe our planet— and us—with low levels of radiation. The radiation emitted in nuclear reactions can damage living cells but also has diagnostic and therapeutic applications.

reactions. Our opening photo shows the surface of the Sun. The tremendous energy that it releases is generated by nuclear reactions, principally by fusing hydrogen nuclei to form helium nuclei. Without nuclear reactions, there would be no sunlight and consequently no life on Earth.

In this chapter, we examine various common kinds of nuclear reactions and the factors that relate to the stability of the nucleus. We also consider how the rates of nuclear reactions are described and used, how radioactivity is detected, and how energy changes associated with nuclear reactions can be calculated. Finally, we discuss the effects of radiation on matter, especially living systems.

21.1 | Radioactivity and Nuclear Equations

> ### ⚠ Learning Objectives
>
> **When you finish Section 21.1, you should be able to:**
>
> ▶ Identify the reactants and products of a nuclear reaction.
>
> ▶ Compare the three primary types of radiation given off in radioactive decay.
>
> ▶ Write nuclear equations that relate the mode of radioactive decay and the changes to the mass and/or atomic number.

Nuclear reactions have some similarities to and differences with the chemical reactions we have discussed thus far. To understand nuclear reactions, we must review and develop some ideas introduced in Section 2.3:

- Two types of subatomic particles reside in the nucleus: *protons* and *neutrons*. We refer to these particles as **nucleons**.
- All atoms of a given element have the same number of protons; this number is the element's *atomic number*.
- Atoms of a given element can have different numbers of neutrons, which means they can have different mass numbers. The *mass number* is the total number of nucleons (protons + neutrons) in the nucleus.
- Atoms with the same atomic number but different mass numbers are known as *isotopes*.

The different isotopes of an element are distinguished by their mass numbers. For example, the three naturally occurring isotopes of uranium are uranium-234, uranium-235, and uranium-238, where the numerical suffixes represent the mass numbers. These isotopes are also written $^{234}_{92}U$, $^{235}_{92}U$, and $^{238}_{92}U$, where the superscript is the mass number and the subscript is the atomic number.*

Different isotopes of an element have different natural abundances. For example, 99.3% of naturally occurring uranium is uranium-238, 0.7% is uranium-235, and only a trace is uranium-234. Different isotopes of an element also have different stabilities. Indeed, the nuclear properties of any given isotope depend on the number of protons and neutrons in its nucleus.

A *nuclide* is a nucleus containing a specified number of protons and neutrons. Nuclides that are radioactive are called **radionuclides**, and atoms containing these nuclei are called **radioisotopes**.

Nuclear Equations

Most nuclei in nature are stable and remain intact indefinitely. Radionuclides, however, are unstable: They change identity while spontaneously emitting particles and electromagnetic radiation. Emission of radiation is one of the ways in which an unstable nucleus is transformed into a more stable one that has less energy. The emitted radiation is the carrier of the excess energy. Uranium-238, for example, is radioactive, undergoing a nuclear reaction emitting helium-4 nuclei. The helium-4 particles are known as **alpha (α) particles**, and a stream of them is called *alpha radiation*. When a $^{238}_{92}U$ nucleus loses an alpha particle, the remaining fragment has an atomic number of 90 and a mass

---

*As noted in Section 2.3, we often do not explicitly write the atomic number of an isotope because the element symbol is specific to the atomic number. In studying nuclear chemistry, however, it is useful to include the atomic number in order to help us keep track of changes in the nuclei.

number of 234. The element with atomic number 90 is Th, thorium. Therefore, the products of uranium-238 decomposition are an alpha particle and a thorium-234 nucleus. We represent this reaction by the *nuclear equation*

$$^{238}_{92}\text{U} \longrightarrow {}^{234}_{90}\text{Th} + {}^{4}_{2}\text{He} \qquad [21.1]$$

When a nucleus spontaneously decomposes in this way, it is said to be *radioactive* and to have decayed or to have undergone *radioactive decay*. Because an alpha particle is involved in this reaction, scientists also describe the process as *alpha decay* or **alpha emission**.

In Equation 21.1 the sum of the mass numbers is the same on both sides of the equation ($238 = 234 + 4$). Likewise, the sum of the atomic numbers on both sides of the equation is equal ($92 = 90 + 2$). Mass numbers and atomic numbers must be balanced in all nuclear equations.

The radioactive properties of the nucleus in an atom are independent of the chemical state of the atom. In writing nuclear equations, therefore, we are not concerned with the chemical form (element or compound) of the atom in which the nucleus resides.

Sample Exercise 21.1
Predicting the Product of a Nuclear Reaction

What product is formed when radium-226 undergoes alpha emission?

SOLUTION

Analyze We are asked to determine the nucleus that results when radium-226 loses an alpha particle.

Plan We can best do this by writing a balanced nuclear reaction for the process.

Solve The periodic table shows that radium has an atomic number of 88. The complete chemical symbol for radium-226 is therefore $^{226}_{88}\text{Ra}$. An alpha particle is a helium-4 nucleus, and so its symbol is $^{4}_{2}\text{He}$. The alpha particle is a product of the nuclear reaction, and so the equation is of the form

$$^{226}_{88}\text{Ra} \longrightarrow {}^{A}_{Z}\text{X} + {}^{4}_{2}\text{He}$$

where A is the mass number of the product nucleus and Z is its atomic number. Mass numbers and atomic numbers must balance, so

$$226 = A + 4$$

and

$$88 = Z + 2$$

Hence,

$$A = 222 \quad \text{and} \quad Z = 86$$

Again, from the periodic table, the element with $Z = 86$ is radon (Rn). The product, therefore, is $^{222}_{86}\text{Rn}$, and the nuclear equation is

$$^{226}_{88}\text{Ra} \longrightarrow {}^{222}_{86}\text{Rn} + {}^{4}_{2}\text{He}$$

▶ **Practice Exercise**
Which element undergoes alpha emission to form lead-208?

Types of Radioactive Decay

The three most common kinds of radiation given off when a radionuclide decays are alpha (α), beta (β), and gamma (γ) radiation. (Section 2.2) **Table 21.1** summarizes some of the important properties of these types of radiation.

Alpha Radiation As described previously, alpha radiation consists of a stream of helium-4 nuclei known as alpha particles, which we denote as $^{4}_{2}\text{He}$ or simply α.

TABLE 21.1   Properties of Alpha, Beta, and Gamma Radiation

	Type of Radiation		
Property	α	β	γ
Charge	2+	1−	0
Mass	6.64×10^{-24} g	9.11×10^{-28} g	0
Relative penetrating power	1	100	10,000
Nature of radiation	$^{4}_{2}\text{He}$ nuclei	Electrons	High-energy photons

Beta Radiation *Beta radiation* consists of streams of **beta (β) particles**, which are high-speed electrons emitted by an unstable nucleus. Beta particles are represented in nuclear equations by $_{-1}^{0}e$ or, more commonly, by β^{-}. The superscript 0 indicates that the mass of the electron is exceedingly small relative to the mass of a nucleon. The subscript -1 represents the negative charge of the beta particle, which is opposite that of the proton.

Iodine-131 is an isotope that undergoes decay by **beta emission**:

$$_{53}^{131}I \longrightarrow {}_{54}^{131}Xe + {}_{-1}^{0}e \qquad [21.2]$$

We see from this equation that beta decay causes the atomic number of the reactant to increase from 53 to 54, which means a proton was created. Therefore, beta emission is equivalent to the conversion of a neutron ($_{0}^{1}n$ or simply n) to a proton ($_{1}^{1}H$ or simply p):

$$_{0}^{1}n \longrightarrow {}_{1}^{1}H + {}_{-1}^{0}e \quad \text{or} \quad n \longrightarrow p + \beta^{-} \qquad [21.3]$$

Although an electron is emitted from a nucleus in beta decay, we should *not* think that the nucleus is composed of these particles any more than we should consider a match to be composed of sparks simply because it gives them off when struck. The beta-particle electron comes into being only when the nucleus undergoes a nuclear reaction. Furthermore, the speed of the beta particle is sufficiently high that it does not end up in an orbital of the decaying atom.

Gamma Radiation *Gamma (γ) radiation* (or **gamma rays**) consists of high-energy photons (that is, electromagnetic radiation of very short wavelength; see Figure 6.4). It changes neither the atomic number nor the mass number of a nucleus and is represented as either $_{0}^{0}\gamma$ or simply γ. Gamma radiation usually accompanies other radioactive emission because it represents the energy lost when the nucleons in a nuclear reaction reorganize into more stable arrangements. Often gamma rays are not explicitly shown when writing nuclear equations.

Positron Emission and Electron Capture Two other types of radioactive decay are positron emission and electron capture. A **positron**, represented as $_{+1}^{0}e$, or simply as β^{+}, is a particle that has the same mass as an electron (thus, we use the letter e and superscript 0 for the mass) but the opposite charge (represented by the +1 subscript).*

The isotope carbon-11 decays by **positron emission**:

$$_{6}^{11}C \longrightarrow {}_{5}^{11}B + {}_{+1}^{0}e \qquad [21.4]$$

Positron emission causes the atomic number of the reactant in this equation to decrease from 6 to 5. In general, positron emission has the effect of converting a proton to a neutron, thereby decreasing the atomic number of the nucleus by 1 while not changing the mass number:

$$_{1}^{1}p \longrightarrow {}_{0}^{1}n + {}_{+1}^{0}e \quad \text{or} \quad p \longrightarrow n + \beta^{+} \qquad [21.5]$$

Electron capture is the capture by the nucleus of an electron from the electron cloud surrounding the nucleus, as in this rubidium-81 decay:

$$_{37}^{81}Rb + {}_{-1}^{0}e \text{ (orbital electron)} \longrightarrow {}_{36}^{81}Kr \qquad [21.6]$$

Because the electron is consumed rather than formed in the process, it is shown on the reactant side of the equation. Electron capture, like positron emission, has the effect of converting a proton to a neutron:

$$_{1}^{1}p + {}_{-1}^{0}e \longrightarrow {}_{0}^{1}n \qquad [21.7]$$

Table 21.2 summarizes the symbols used to represent the particles commonly encountered in nuclear reactions. The various types of radioactive decay are summarized in Table 21.3.

TABLE 21.2 Particles Found in Nuclear Reactions

Particle	Symbol
Neutron	$_{0}^{1}n$ or n
Proton	$_{1}^{1}H$ or p
Electron	$_{-1}^{0}e$
Alpha particle	$_{2}^{4}He$ or α
Beta particle	$_{-1}^{0}e$ or β^{-}
Positron	$_{+1}^{0}e$ or β^{+}

---

*The positron has a very short life because it is annihilated when it collides with an electron, producing gamma rays: $_{+1}^{0}e + {}_{-1}^{0}e \longrightarrow 2\,{}_{0}^{0}\gamma$.

TABLE 21.3 Types of Radioactive Decay

Type	Nuclear Equation	Change in Atomic Number	Change in Mass Number
Alpha emission	$^{A}_{Z}X \longrightarrow ^{A-4}_{Z-2}Y + ^{4}_{2}He$	-2	-4
Beta emission	$^{A}_{Z}X \longrightarrow ^{A}_{Z+1}Y + ^{0}_{-1}e$	$+1$	Unchanged
Positron emission	$^{A}_{Z}X \longrightarrow ^{A}_{Z-1}Y + ^{0}_{+1}e$	-1	Unchanged
Electron capture*	$^{A}_{Z}X + ^{0}_{-1}e \longrightarrow ^{A}_{Z-1}Y$	-1	Unchanged

*The electron captured comes from the electron cloud surrounding the nucleus.

Sample Exercise 21.2
Writing Nuclear Equations

Write nuclear equations for (**a**) mercury-201 undergoing electron capture; (**b**) thorium-231 decaying to protactinium-231.

SOLUTION

Analyze We must write balanced nuclear equations in which the masses and charges of reactants and products are equal.

Plan We can begin by writing the complete chemical symbols for the nuclei and decay particles that are given in the problem statement.

Solve
(**a**) The information given in the question can be summarized as

$$^{201}_{80}Hg + ^{0}_{-1}e \longrightarrow ^{A}_{Z}X$$

The mass numbers must have the same sum on both sides of the equation:

$$201 + 0 = A$$

Thus, the product nucleus must have a mass number of 201. Similarly, balancing the atomic numbers gives

$$80 - 1 = Z$$

Thus, the atomic number of the product nucleus must be 79, which identifies it as gold (Au):

$$^{201}_{80}Hg + ^{0}_{-1}e \longrightarrow ^{201}_{79}Au$$

(**b**) In this case, we must determine what type of particle is emitted in the course of the radioactive decay:

$$^{231}_{90}Th \longrightarrow ^{231}_{91}Pa + ^{A}_{Z}X$$

From $231 = 231 + A$ and $90 = 91 + Z$, we deduce $A = 0$ and $Z = -1$. According to Table 21.2, the particle with these characteristics is the beta particle (electron). We therefore write

$$^{231}_{90}Th \longrightarrow ^{231}_{91}Pa + ^{0}_{-1}e \quad \text{or} \quad ^{231}_{90}Th \longrightarrow ^{231}_{91}Pa + \beta^{-}$$

▶ **Practice Exercise**
Write a balanced nuclear equation for the reaction in which oxygen-15 undergoes positron emission.

Self-Assessment Exercises

SAE 21.1 When a certain nuclide undergoes alpha emission, astatine-217 is produced. What is the identity of the nuclide that underwent decay? (**a**) astatine-221 (**b**) actinium-221 (**c**) francium-221 (**d**) actinium-219 (**e**) francium-217

SAE 21.2 Which of the following statements about the radiation associated with radioactive decay is or are *true*?

  (**i**) Alpha radiation consists of a stream of hydrogen atom nuclei.
 (**ii**) Beta radiation consists of a stream of high-speed electrons.
(**iii**) When a nuclide undergoes gamma ray emission, neither the atomic number nor the mass number is changed.

(**a**) Only one statement is true. (**b**) Statements i and ii are true. (**c**) Statements i and iii are true. (**d**) Statements ii and iii are true. (**e**) All three statements are true.

SAE 21.3 Which of the following statements is *false*? (**a**) A positron is a particle with the same mass as an electron but the opposite charge. (**b**) When a nuclide undergoes electron capture, its atomic number is unchanged. (**c**) When a nuclide undergoes positron emission, its mass number is unchanged. (**d**) Electron capture has the effect of converting a proton to a neutron.

SAE 21.4 What particle is produced during the following decay process? $^{56}_{25}Mn$ decays to $^{56}_{26}Fe$ (**a**) beta particle (β^{-}) (**b**) alpha particle (α) (**c**) gamma ray (γ) (**d**) positron (β^{+})

SAE 21.5 Radon-222 decays and produces an alpha particle. Which isotope is also a product of this decay? (**a**) $^{222}_{84}Po$ (**b**) $^{222}_{87}Fr$ (**c**) $^{218}_{84}Po$ (**d**) $^{218}_{86}Rn$

Learning Objectives

When you finish **Section 21.2**, you should be able to:

▶ Relate nuclear stability to the neutron-to-proton ratio in different parts of the periodic table.

▶ Predict which nuclides are radioactive and how those nuclides decay.

21.2 | Patterns of Nuclear Stability

Some nuclides, such as $^{12}_{6}C$ and $^{13}_{6}C$ are stable, whereas others, such as $^{14}_{6}C$, are unstable and undergo radioactive decay. Why does a small difference in the number of neutrons affect the stability of a nuclide? No single rule allows us to predict whether a particular nucleus is radioactive and, if it is, how it might decay. However, several empirical observations can help us predict the stability of a nucleus.

Neutron-to-Proton Ratio

Because like charges repel each other, it may seem surprising that a large number of protons can reside within the small volume of the nucleus. At close distances, however, a strong force of attraction, called the *strong nuclear force*, exists between nucleons [see the "A Closer Look" box on "Basic Forces". (Section 2.3)] Neutrons are intimately involved in this attractive force.

All nuclei other than ^1_1H contain neutrons. As the number of protons in a nucleus increases, there is an increasing need for neutrons to counteract the proton–proton repulsions. Stable nuclei with atomic numbers up to about 20 have approximately equal numbers of neutrons and protons. For nuclei with atomic number above 20, the number of neutrons exceeds the number of protons. Indeed, the number of neutrons necessary to create a stable nucleus increases more rapidly than the number of protons. Thus, the neutron-to-proton ratios of stable nuclei increase with increasing atomic number, as illustrated by the following isotopes: $^{12}_6C$ (n/p = 1), manganese, $^{55}_{25}Mn$ (n/p = 1.20), and gold, $^{197}_{79}Au$ (n/p = 1.49).

Figure 21.1 shows all known isotopes of the elements through $Z = 100$ plotted according to their numbers of protons and neutrons. Notice how the plot goes above the line for 1:1 neutron-to-proton for heavier elements. The dark blue dots in the figure represent stable (nonradioactive) isotopes. The region of the graph covered by these dark blue dots is known as the *belt of stability*. The belt of stability ends at element 83 (bismuth), which means that *all nuclei with 84 or more protons are radioactive*. For example, all isotopes of uranium, $Z = 92$, are radioactive.

Different radionuclides decay in different ways. The type of decay that occurs depends largely on a nuclide's neutron-to-proton ratio and how it compares with the

---

▼ Go Figure   Estimate the optimal number of neutrons for a nucleus of Yb (*Z* = 70).

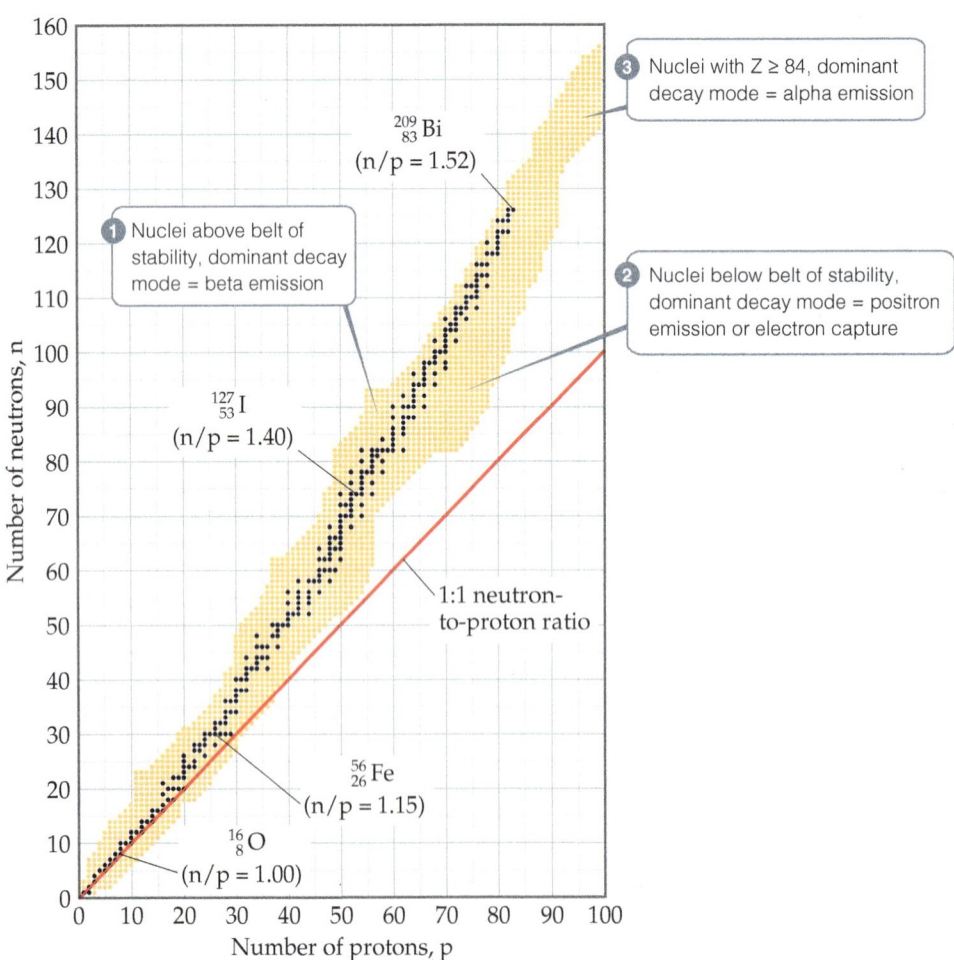

▲ Figure 21.1 Stable and radioactive isotopes as a function of numbers of neutrons and protons in a nucleus. The stable nuclei (dark blue dots) define a region known as the belt of stability.

ratio of nearby nuclei that lie within the belt of stability. We can envision three general situations, which are labeled 1, 2, and 3, in Figure 21.1.

1. **Nuclei above the belt of stability (high neutron-to-proton ratios).** These neutron-rich nuclei can lower their ratio and thereby move toward the belt of stability by emitting a beta particle because beta emission decreases the number of neutrons and increases the number of protons (Equation 21.3).

2. **Nuclei below the belt of stability (low neutron-to-proton ratios).** These proton-rich nuclei can increase their ratio and so move closer to the belt of stability by either positron emission or electron capture because both decays increase the number of neutrons and decrease the number of protons (Equations 21.5 and 21.7). Positron emission is more common among lighter nuclei. Electron capture becomes increasingly common as the nuclear charge increases.

3. **Nuclei with atomic numbers ≥ 84.** These heavy nuclei tend to undergo alpha emission, which decreases both the number of neutrons and the number of protons by two, moving the nucleus diagonally toward the belt of stability.

Sample Exercise 21.3
Predicting Modes of Nuclear Decay

Predict the mode of decay of (**a**) carbon-14, (**b**) xenon-118.

SOLUTION

Analyze We are asked to predict the modes of decay of two nuclei.

Plan We must locate the respective nuclei in Figure 21.1 and determine their positions with respect to the belt of stability in order to predict the most likely mode of decay.

Solve

(**a**) Carbon is element 6. Thus, carbon-14 has 6 protons and $14 - 6 = 8$ neutrons, giving it a neutron-to-proton ratio of 1.25. Elements with $Z < 20$ normally have stable nuclei that contain approximately equal numbers of neutrons and protons ($n/p = 1$). Thus, carbon-14 is located above the belt of stability, and we expect it to decay by emitting a beta particle to decrease the n/p ratio:

$$^{14}_{6}\text{C} \longrightarrow {}^{14}_{7}\text{N} + {}^{0}_{-1}\text{e}$$

This is indeed the mode of decay observed for carbon-14, a reaction that lowers the n/p ratio from 1.25 to 1.0.

(**b**) Xenon is element 54. Thus, xenon-118 has 54 protons and $118 - 54 = 64$ neutrons, giving it an n/p ratio of 1.18. According to Figure 21.1, stable nuclei in this region of the belt

of stability have higher neutron-to-proton ratios than xenon-118. The nucleus can increase this ratio by either positron emission or electron capture:

$$^{118}_{54}\text{Xe} \longrightarrow {}^{118}_{53}\text{I} + {}^{0}_{+1}\text{e}$$

$$^{118}_{54}\text{Xe} + {}^{0}_{-1}\text{e} \longrightarrow {}^{118}_{53}\text{I}$$

In this case, both modes of decay are observed.

Comment Keep in mind that our guidelines do not always work. For example, thorium-233 ($n/p = 143/90 = 1.59$), which we might expect to undergo alpha decay, actually undergoes beta emission. Furthermore, a few radioactive nuclei lie within the belt of stability. Both $^{146}_{60}\text{Nd}$ and $^{148}_{60}\text{Nd}$, for example, are stable ($n/p = 1.43$ and 1.47, respectively) and lie in the belt of stability. $^{147}_{60}\text{Nd}$, however, which lies between them and has $n/p = 1.45$, is radioactive.

▶ **Practice Exercise**

Predict the mode of decay of (**a**) plutonium-239, (**b**) indium-120.

Radioactive Decay Chains

Some nuclei cannot gain stability by a single emission. Consequently, a series of successive emissions occurs, as shown for uranium-238 in **Figure 21.2**. Decay continues until a stable nucleus—lead-206 in this case—is formed. A series of nuclear reactions that begins with an unstable nucleus and terminates with a stable one is known as a **radioactive decay chain** or a **nuclear disintegration series**. Three such series occur in nature: uranium-238 to lead-206, uranium-235 to lead-207, and thorium-232 to lead-208. All of the decay processes in these series are either alpha emissions or beta emissions.

Further Observations

Two further observations can help us to predict stable nuclei:

- Nuclei with the **magic numbers** of 2, 8, 20, 28, 50, or 82 protons or 2, 8, 20, 28, 50, 82, or 126 neutrons are generally more stable than nuclei that do not contain these numbers of nucleons.

▼ **Go Figure** Write the nuclear equation for the step shown for the first decay of Th.

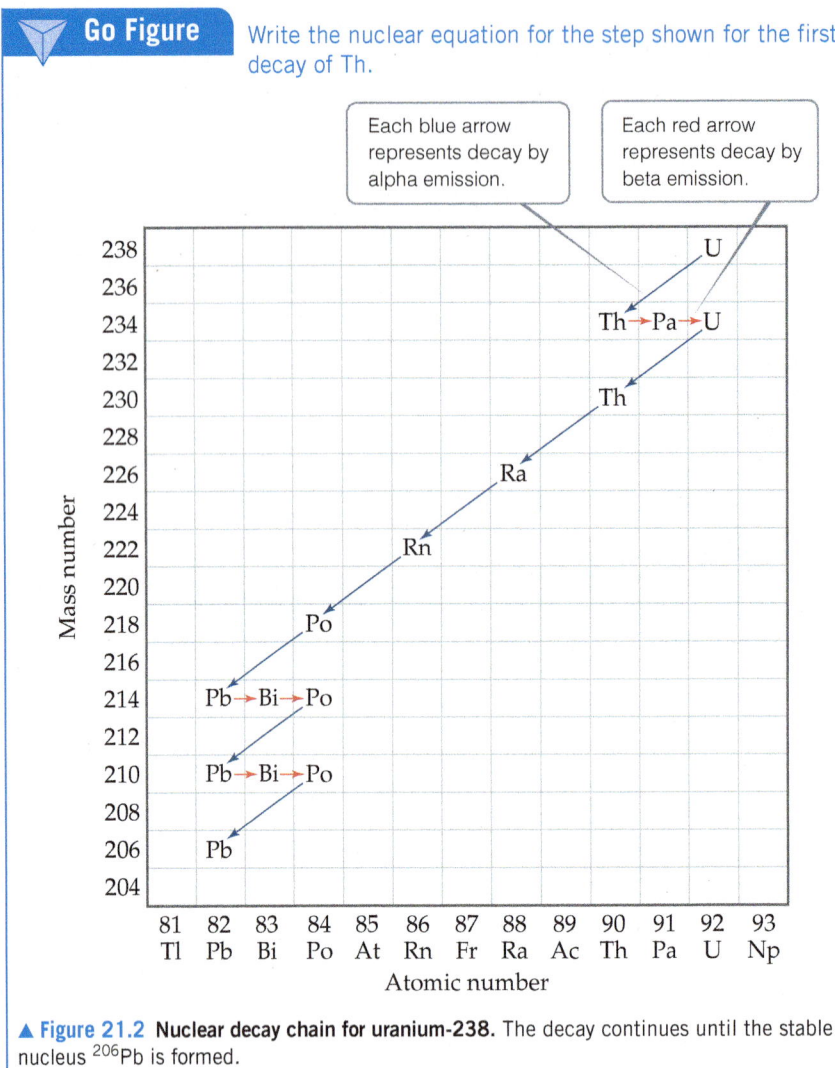

▲ **Figure 21.2 Nuclear decay chain for uranium-238.** The decay continues until the stable nucleus ^{206}Pb is formed.

- Nuclei with even numbers of protons, neutrons, or both are more likely to be stable than those with odd numbers of protons and/or neutrons. Approximately 60% of stable nuclei have an even number of both protons and neutrons, whereas less than 2% have odd numbers of both (**Table 21.4**).

These observations can be understood in terms of the *shell model of the nucleus*, in which nucleons are described as residing in shells analogous to the shell structure for electrons in atoms. Just as certain numbers of electrons correspond to stable filled-shell electron configurations, so too do certain numbers (known as magic numbers) of nucleons represent filled shells in nuclei.

There are several examples of the stability of nuclei with magic numbers of nucleons. For example, the radioactive series in Figure 21.2 ends with the stable $^{206}_{82}$Pb nucleus, which has a magic number of protons (82). Another example is the observation that tin, which has a magic number of protons (50), has 10 stable isotopes, more than any other element.

Evidence also suggests that pairs of protons and pairs of neutrons have a special stability, analogous to the pairs of electrons in molecules. This evidence accounts for the observation that stable nuclei with an even number of protons and/or neutrons are far more numerous than those with odd numbers. The preference for even numbers of protons is illustrated in **Figure 21.3**, which shows the number of stable isotopes for all elements up to Xe. Notice that once we move past nitrogen, the elements with an odd number of protons invariably have fewer stable isotopes than their neighbors with an even number of protons.

TABLE 21.4 Number of Stable Isotopes with Even and Odd Numbers of Protons and Neutrons

Number of Stable Isotopes	Proton Number	Neutron Number
157	Even	Even
53	Even	Odd
50	Odd	Even
5	Odd	Odd

Go Figure Among the elements shown here, how many have an even number of protons and fewer than three stable isotopes? How many have an odd number of protons and more than two stable isotopes?

1 H (2)																	2 He (2)
3 Li (2)	4 Be (1)											5 B (2)	6 C (2)	7 N (2)	8 O (3)	9 F (1)	10 Ne (3)
11 Na (1)	12 Mg (3)											13 Al (1)	14 Si (3)	15 P (1)	16 S (4)	17 Cl (2)	18 Ar (3)
19 K (2)	20 Ca (5)	21 Sc (1)	22 Ti (5)	23 V (2)	24 Cr (4)	25 Mn (1)	26 Fe (4)	27 Co (1)	28 Ni (5)	29 Cu (2)	30 Zn (5)	31 Ga (2)	32 Ge (4)	33 As (1)	34 Se (5)	35 Br (2)	36 Kr (6)
37 Rb (1)	38 Sr (3)	39 Y (1)	40 Zr (4)	41 Nb (1)	42 Mo (6)	43 Tc (0)	44 Ru (7)	45 Rh (1)	46 Pd (6)	47 Ag (2)	48 Cd (6)	49 In (1)	50 Sn (10)	51 Sb (2)	52 Te (6)	53 I (1)	54 Xe (9)

Number of stable isotopes

Elements with two or fewer stable isotopes

Elements with three or more stable isotopes

▲ Figure 21.3 **Number of stable isotopes for elements 1–54.**

Self-Assessment Exercises

SAE 21.6 Which of the following nuclides will likely undergo a beta emission for its dominant radioactive decay mode? (**a**) yttrium-76 (**b**) cesium-120 (**c**) barium-146 (**d**) tin-105

SAE 21.7 The final two steps in the decay chain for uranium-238 are:

bismuth-210 ⟶ polonium-210 ⟶ lead-206

Lead-206 is a stable isotope. What are the radioactive decay processes for these two steps? (**a**) two successive alpha emissions (**b**) alpha emission followed by electron capture (**c**) electron capture followed by alpha emission (**d**) alpha emission followed by beta emission (**e**) beta emission followed by alpha emission

SAE 21.8 Which of the following statements about the stability of nuclides is *false*?

(**a**) Nuclides in the belt of stability always have the same ratio of neutrons to protons. (**b**) Nuclei with greater than 84 protons tend to decay by alpha emission. (**c**) A nuclide with an even number of protons and an even number of neutrons is more likely to be stable than a nuclide with an odd number of each. (**d**) Nearly all stable nuclei have a greater number of neutrons than protons. (**e**) A nucleus that decays via electron capture has a neutron-to-proton ratio that is lower than expected for stability.

21.3 | Nuclear Transmutations

Thus far we have examined nuclear reactions in which a nucleus decays spontaneously. A nucleus can also change identity if it is struck by a neutron or by another nucleus. Nuclear reactions induced in this way are known as **nuclear transmutations**.

In 1919, Ernest Rutherford performed the first conversion of one nucleus into another, using alpha particles emitted by radium to convert nitrogen-14 into oxygen-17:

$$^{14}_{7}N + ^{4}_{2}He \longrightarrow ^{17}_{8}O + ^{1}_{1}H \quad \text{or} \quad ^{14}_{7}N + \alpha \longrightarrow ^{17}_{8}O + p \qquad [21.8]$$

Such reactions have allowed scientists to synthesize hundreds of radioisotopes in the laboratory.

A shorthand notation often used to represent nuclear transmutations lists the target nucleus, the bombarding particle, and the ejected particle in parentheses, followed by the product nucleus. Using this condensed notation, we see that Equation 21.8 becomes

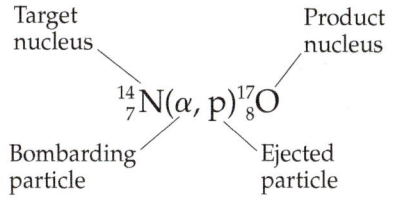

Target nucleus $\qquad$ Product nucleus

$$^{14}_{7}N(\alpha, p)^{17}_{8}O$$

Bombarding particle $\qquad$ Ejected particle

Learning Objective

When you finish **Section 21.3**, you should be able to:

▶ Describe and analyze nuclear transmutation reactions.

Sample Exercise 21.4
Writing a Balanced Nuclear Equation

Write the balanced nuclear equation for the process summarized as $^{27}_{13}\text{Al}(n, \alpha)^{24}_{11}\text{Na}$.

SOLUTION

Analyze We must go from the condensed descriptive form of the reaction to the balanced nuclear equation.

Plan We arrive at the balanced equation by writing n and α, each with its associated subscripts and superscripts.

Solve The n is the abbreviation for a neutron (^1_0n), and α represents an alpha particle (^4_2He). The neutron is the bombarding particle, and the alpha particle is a product. Therefore, the

nuclear equation is

$$^{27}_{13}\text{Al} + {}^1_0\text{n} \longrightarrow {}^{24}_{11}\text{Na} + {}^4_2\text{He} \quad \text{or} \quad {}^{27}_{13}\text{Al} + \text{n} \longrightarrow {}^{24}_{11}\text{Na} + \alpha$$

▶ **Practice Exercise**
Write the condensed version of the nuclear reaction

$$^{16}_{8}\text{O} + {}^1_1\text{H} \longrightarrow {}^{13}_{7}\text{N} + {}^4_2\text{He}$$

Accelerating Charged Particles

Alpha particles and other positively charged particles must move very fast to overcome the electrostatic repulsion between them and the target nucleus. The higher the nuclear charge on either the bombarding particle or the target nucleus, the faster the bombarding particle must move to bring about a nuclear reaction. Many methods have been devised to accelerate charged particles, using strong magnetic and electrostatic fields. These **particle accelerators**, popularly called "atom smashers," bear such names as *cyclotron* and *synchrotron*.

A common theme of all particle accelerators is the need to create charged particles so that they can be manipulated by electric and magnetic fields. In addition, the region through which the particles move must be kept at high vacuum so that they do not collide with any gas-phase molecules.

Figure 21.4(a) shows a multistage linear accelerator. A charged particle, such as a proton, is accelerated through a series of tubes of increasing length. The electrical charge on the tubes is changed from positive to negative, so that the particle is always attracted to the tube it is approaching and repelled by the one it is leaving. As a result, the particle accelerates until it has sufficient kinetic energy to smash into a target nucleus. **Figure 21.4(b)** shows the Stanford linear accelerator, which is 1.9 mi in length.

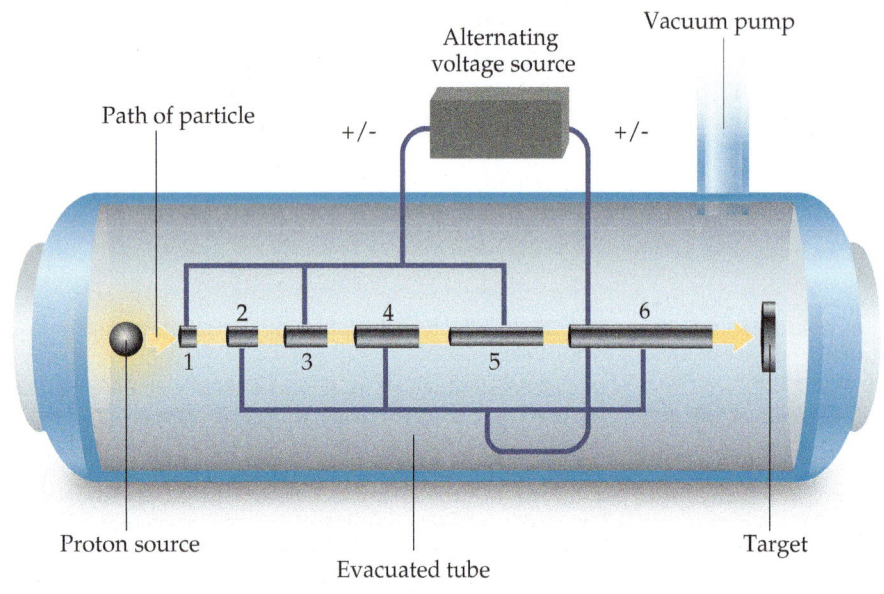

(a)

(b)

▲ **Figure 21.4 The linear accelerator.** (a) A schematic showing how it works and (b) the Stanford Linear Accelerator in California.

Figure 21.5(a) shows a *cyclotron*. In this device, charged particles move in a spiral path within two D-shaped electrodes. Alternating charges on the electrodes accelerate the particles, while magnets above and below the device constrain the particles to a spiral path of increasing radius. In a *synchrotron*, the magnetic fields are synchronized so that the particle moves in a circular rather than a spiral path. **Figure 21.5(b)** shows the Fermi National Accelerator Lab at Batavia, Illinois, which has a circumference of 3.9 mi.

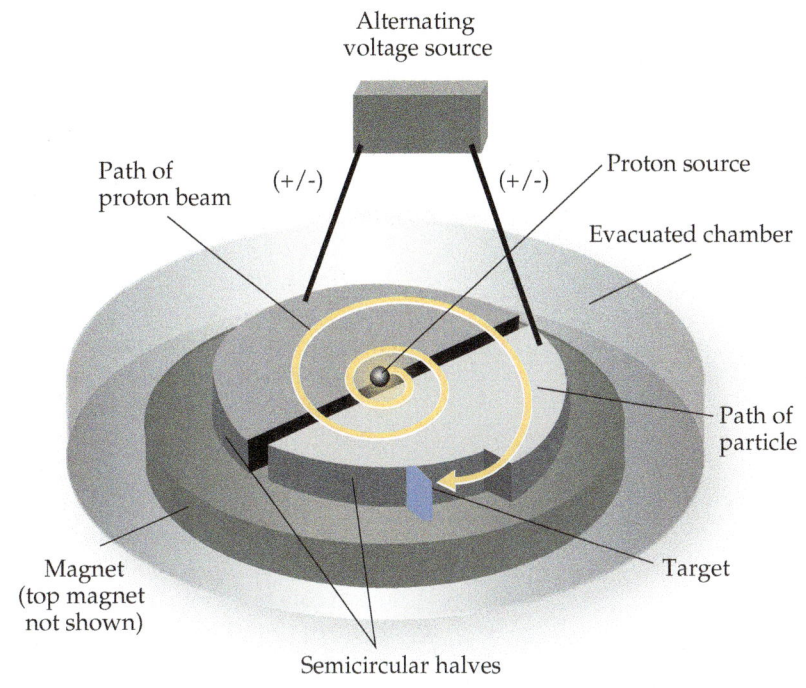

(a)

Reactions Involving Neutrons

Most synthetic isotopes used in medicine and scientific research are made using neutrons as the bombarding particles. Because neutrons are neutral, they are not repelled by the nucleus. Consequently, they do not need to be accelerated to cause nuclear reactions. The neutrons are produced in nuclear reactors (see Section 21.7). For example, cobalt-60, which is used in cancer radiation therapy, is produced by neutron capture. Iron-58 is placed in a nuclear reactor and bombarded by neutrons to trigger the following sequence of reactions:

$$\ce{^{58}_{26}Fe} + \ce{^1_0n} \longrightarrow \ce{^{59}_{26}Fe} \qquad [21.9]$$

$$\ce{^{59}_{26}Fe} \longrightarrow \ce{^{59}_{27}Co} + \ce{^0_{-1}e} \qquad [21.10]$$

$$\ce{^{59}_{27}Co} + \ce{^1_0n} \longrightarrow \ce{^{60}_{27}Co} \qquad [21.11]$$

Transuranium Elements

Nuclear transmutations have been used to produce the elements with atomic number above 92, collectively known as the **transuranium elements** because they follow uranium in the periodic table. Elements

(b)

▲ **Figure 21.5 The cyclotron.** (a) A schematic showing how it works and (b) the Fermi National Accelerator Lab in Illinois.

93 (neptunium, Np) and 94 (plutonium, Pu) were produced in 1940 by bombarding uranium-238 with neutrons:

$$\ce{^{238}_{92}U} + \ce{^1_0n} \longrightarrow \ce{^{239}_{92}U} \longrightarrow \ce{^{239}_{93}Np} + \ce{^0_{-1}e} \qquad [21.12]$$

$$\ce{^{239}_{93}Np} \longrightarrow \ce{^{239}_{94}Pu} + \ce{^0_{-1}e} \qquad [21.13]$$

Elements with still larger atomic numbers are normally formed in small quantities in particle accelerators. Curium-242, for example, is formed when a plutonium-239 target is bombarded with accelerated alpha particles:

$$\ce{^{239}_{94}Pu} + \ce{^4_2He} \longrightarrow \ce{^{242}_{96}Cm} + \ce{^1_0n} \qquad [21.14]$$

New advances in the detection of the decay patterns of single atoms have led to recent additions to the periodic table. Between 1994 and 2010, elements 110 through 118 were discovered via the nuclear reactions that occur when nuclei of much lighter elements collide with high energy. For example, in 1996 a team of European scientists based in Germany synthesized element 112, copernicium, Cn, by bombarding a lead target continuously for three weeks with a beam of zinc atoms:

$$\,^{208}_{82}\text{Pb} + \,^{70}_{30}\text{Zn} \longrightarrow \,^{277}_{112}\text{Cn} + \,^{1}_{0}\text{n} \qquad \qquad [21.15]$$

Amazingly, their discovery was based on the detection of only one atom of the new element, which decays after roughly 100 μs by alpha decay to form darmstadtium-273 (element 110). Within one minute, another five alpha decays take place producing fermium-253 (element 100). The finding has been verified in both Japan and Russia.

Because experiments to create new elements are very complicated and produce only a small number of atoms of the new elements, they need to be carefully evaluated and reproduced before the new element is made an official part of the periodic table.

The International Union for Pure and Applied Chemistry (IUPAC) is the international body that authorizes names of new elements after their experimental discovery and confirmation. According to the IUPAC, new elements can be named after a mythological concept, a mineral, a place or country, a property, or a scientist. In 2016, IUPAC approved the following names and symbols for elements 113, 115, 117, and 118, as suggested by their discoverers: nihonium, Nh, for element 113; moscovium, Mc, for element 115; tennessine, Ts, for element 117; and organesson, Og, for element 118.

Learning Objectives

When you finish Section 21.4, you should be able to:

▶ Correlate the amount of a radioisotope with its half-life.

▶ Calculate the age of an object based on radioactive decay.

Self-Assessment Exercise

SAE 21.9   Identify the nucleus X in the following transmutation:

$$\,^{239}_{94}\text{Pu}(\text{n},\beta^-)\text{X}$$

(**a**) $\,^{240}_{95}\text{Am}$ (**b**) $\,^{240}_{94}\text{Pu}$ (**c**) $\,^{240}_{96}\text{Cm}$ (**d**) $\,^{241}_{97}\text{Bk}$

21.4 | Rates of Radioactive Decay

Some radioisotopes, such as uranium-238, are found in nature even though they are not stable. Other radioisotopes do not exist in nature but can be synthesized in nuclear reactions. To understand this distinction, we must realize that different nuclei undergo radioactive decay at different rates. Many radioisotopes essentially decay completely in fractions of a second, so we do not find them in nature. Uranium-238, on the other hand, decays very slowly. Therefore, despite its instability, we can still observe what remains from its formation in the early history of the universe.

Radioactive decay is a first-order kinetic process. Recall that a first-order process has a characteristic **half-life**, which is the time required for half of any given quantity of a substance to react. (Section 14.3) Nuclear decay rates are commonly expressed in terms of half-lives. Each radioisotope has its own characteristic half-life. For example, strontium-90 has a half-life of 28.8 yr:

$$\,^{90}_{38}\text{Sr} \longrightarrow \,^{90}_{39}\text{Y} + \,^{0}_{-1}\text{e} \qquad t_{1/2} = 28.8 \text{ yr} \qquad [21.16]$$

Thus, if we start with 10.0 g of strontium-90, only 5.0 g of that isotope remains after 28.8 yr, 2.5 g remains after another 28.8 yr, and so on (**Figure 21.6**).

Half-lives as short as millionths of a second and as long as billions of years are known. The half-lives of some radioisotopes are listed in **Table 21.5**. One important feature of half-lives for nuclear decay is that they are unaffected by external conditions such as temperature, pressure, or state of chemical combination. Unlike toxic chemicals, therefore, radioactive atoms cannot be rendered harmless by chemical reaction or by any other practical treatment.

Go Figure

If we start with a 50.0-g sample, how much remains after three half-lives have passed?

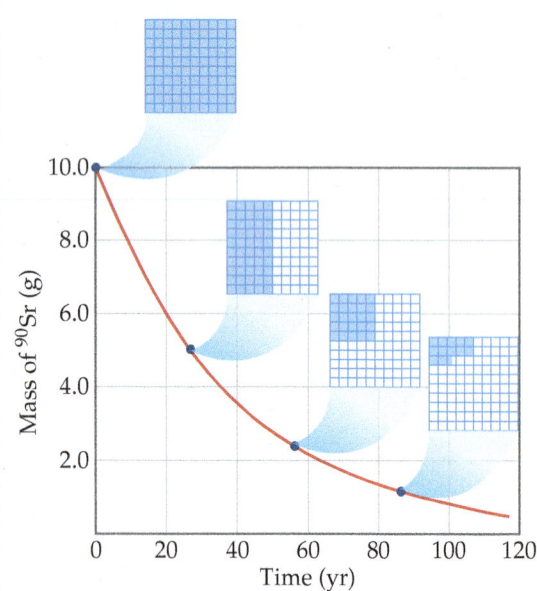

▲ **Figure 21.6 Decay of a 10.0-g sample of strontium-90** (**$t_{1/2}$ = 28.8 yr**). The 10 × 10 grids show how much of the radioactive isotope remains after various amounts of time.

TABLE 21.5   The Half-Lives and Type of Decay for Several Radioisotopes

	Isotope	Half-Life (yr)	Type of Decay
Natural radioisotopes	$^{238}_{92}U$	4.5×10^9	Alpha
	$^{235}_{92}U$	7.0×10^8	Alpha
	$^{232}_{90}Th$	1.4×10^{10}	Alpha
	$^{40}_{19}K$	1.3×10^9	Beta
	$^{14}_{6}C$	5730	Beta
Synthetic radioisotopes	$^{239}_{94}Pu$	24,000	Alpha
	$^{137}_{55}Cs$	30.2	Beta
	$^{90}_{38}Sr$	28.8	Beta
	$^{131}_{53}I$	0.022	Beta

Sample Exercise 21.5

Calculation Involving Half-Lives

The half-life of cobalt-60 is 5.27 yr. How much of a 1.000-mg sample of cobalt-60 is left after 15.81 yr?

SOLUTION

Analyze We are given the half-life for cobalt-60, and we are asked to calculate the amount of cobalt-60 remaining from an initial 1.000-mg sample after 15.81 yr.

Plan We will use the fact that the amount of a radioactive substance decreases by 50% for every half-life that passes.

Solve Because $5.27 \times 3 = 15.81$, 15.81 yr is three half-lives for cobalt-60. At the end of one half-life, 0.500 mg of cobalt-60 remains; 0.250 mg remains at the end of two half-lives, and 0.125 mg remains at the end of three half-lives.

▶ **Practice Exercise**

Carbon-11, used in medical imaging, has a half-life of 20.4 min. The carbon-11 nuclides are formed, and the carbon atoms are then incorporated into an appropriate compound. The resulting sample is injected into a patient, and the medical image is obtained. If the entire process takes five half-lives, what percentage of the original carbon-11 remains at this time?

Radiometric Dating

Because the half-life of any particular nuclide is constant, the half-life can serve as a "nuclear clock" to determine the age of objects. The method of dating objects based on their isotopes and isotope abundances is called *radiometric dating*.

When carbon-14 is used in radiometric dating, the technique is known as *radiocarbon dating*. The procedure is based on the formation of carbon-14 as neutrons created by cosmic rays in the upper atmosphere convert nitrogen-14 into carbon-14 (**Figure 21.7**). The ^{14}C reacts with oxygen to form $^{14}CO_2$ in the atmosphere, and this "labeled" CO_2 is taken up by plants and introduced into the food chain through photosynthesis. This process provides a small but reasonably constant source of carbon-14, which is radioactive and undergoes beta decay with a half-life of 5730 yr (to three significant figures):

$$^{14}_{6}C \longrightarrow ^{14}_{7}N + ^{0}_{-1}e \qquad [21.17]$$

Because a living plant or animal has a constant intake of carbon compounds, it is able to maintain a ratio of carbon-14 to carbon-12 that is nearly identical with that of the atmosphere. Once the organism dies, however, it no longer ingests carbon compounds to replenish the carbon-14 lost through radioactive decay. The ratio of carbon-14 to carbon-12 therefore decreases. By measuring this ratio and comparing it with that of the atmosphere, we can estimate the age of an object. For example, if the ratio diminishes to half that of the atmosphere, we can conclude that the age of the object is one half-life, or 5730 yr.

Go Figure How does $^{14}CO_2$ become incorporated into the mammalian food chain?

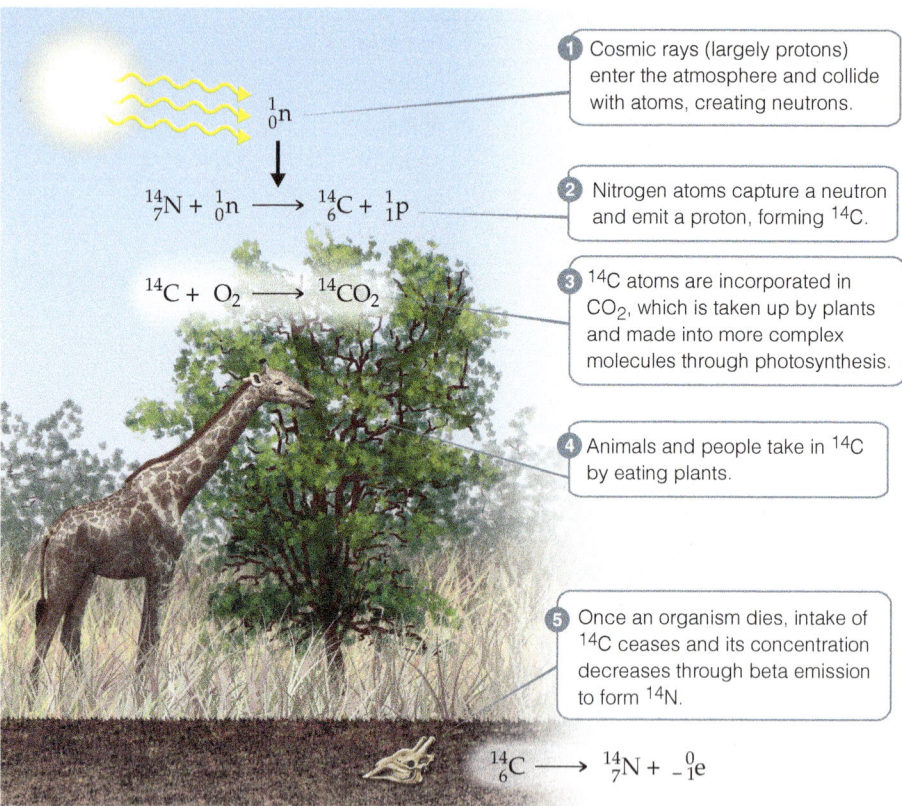

1. Cosmic rays (largely protons) enter the atmosphere and collide with atoms, creating neutrons.

$$^{1}_{0}n$$

$$^{14}_{7}N + ^{1}_{0}n \longrightarrow ^{14}_{6}C + ^{1}_{1}p$$

2. Nitrogen atoms capture a neutron and emit a proton, forming ^{14}C.

$$^{14}C + O_2 \longrightarrow ^{14}CO_2$$

3. ^{14}C atoms are incorporated in CO_2, which is taken up by plants and made into more complex molecules through photosynthesis.

4. Animals and people take in ^{14}C by eating plants.

5. Once an organism dies, intake of ^{14}C ceases and its concentration decreases through beta emission to form ^{14}N.

$$^{14}_{6}C \longrightarrow ^{14}_{7}N + ^{0}_{-1}e$$

▲ **Figure 21.7 Creation and distribution of carbon-14.** The ratio of carbon-14 to carbon-12 in a dead animal or plant is related to the time since death occurred.

This method cannot be used to date objects older than about 50,000 yr, because after this length of time the radioactivity is too low to be measured accurately.

In radiocarbon dating, a reasonable assumption is that the ratio of carbon-14 to carbon-12 in the atmosphere has been relatively constant for the past 50,000 yr. However, because variations in solar activity control the amount of carbon-14 produced in the atmosphere, that ratio can fluctuate. We can correct for this effect by using other kinds of data. Recently, scientists have compared carbon-14 data with data from tree rings, corals, lake sediments, ice cores, and other natural sources to correct variations in the carbon-14 "clock" back to 26,000 yr.

Other isotopes can be similarly used to date other types of objects. For example, it takes 4.5×10^9 yr for half of a sample of uranium-238 to decay to lead-206. The age of rocks containing uranium can therefore be determined by measuring the ratio of lead-206 to uranium-238. If the lead-206 had somehow become incorporated into the rock by normal chemical processes instead of by radioactive decay, the rock would also contain large amounts of the more abundant isotope lead-208. In the absence of large amounts of this "geonormal" isotope of lead, it is assumed that all of the lead-206 was at one time uranium-238.

The oldest rocks found on Earth are approximately 3×10^9 yr old. This age indicates that Earth's crust has been solid for at least this length of time. Scientists estimate that it required 1×10^9 to 1.5×10^9 yr for Earth to cool and its surface to become solid, making the age of Earth 4.0 to 4.5×10^9 yr.

Calculations Based on Half-Life

The rate at which a sample decays is called its **activity**, and it is often expressed as number of disintegrations per unit time. The **becquerel** (Bq) is the SI unit for expressing activity. A becquerel is defined as one nuclear disintegration per second. An older, but still widely used, unit of activity is the **curie** (Ci), defined as 3.7×10^{10} disintegrations per second, which is the rate of decay of 1 g of radium. Thus, a 4.0-mCi sample of cobalt-60 undergoes

$$4.0 \times 10^{-3} \text{ Ci} \times \frac{3.7 \times 10^{10} \text{ disintegrations/s}}{1 \text{ Ci}} = 1.5 \times 10^8 \text{ disintegrations/s}$$

and so has an activity of 1.5×10^8 Bq.

As a radioactive sample decays, the amount of radiation emanating from the sample decays as well. For example, the half-life of cobalt-60 is 5.27 yr. The 4.0-mCi sample of cobalt-60 would, after 5.27 yr, have a radiation activity of 2.0 mCi, or 7.5×10^7 Bq.

Because radioactive decay is a first-order kinetic process, its rate is proportional to the number of radioactive nuclei N in a sample:

$$\text{Rate} = kN \qquad [21.18]$$

The first-order rate constant, k, is called the *decay constant* and is related to the half-life:

$$k = \frac{0.693}{t_{1/2}} \qquad [21.19]$$

Thus, if we know the value of either the half-life or the decay constant, we can calculate the value of the other.

As we saw in Section 14.3 a first-order rate law can be expressed in the following form:

$$\ln \frac{N_t}{N_0} = -kt \qquad [21.20]$$

In this equation, t is the time interval of decay, k is the decay constant, N_0 is the initial number of nuclei (at time zero), and N_t is the number remaining after the time interval. Both the mass of a particular radioisotope and its activity are proportional to the number of radioactive nuclei. Thus, either the ratio of the mass at any time t to the mass at time $t = 0$ or the ratio of the activities at time t and $t = 0$ can be substituted for N_t/N_0 in Equation 21.20.

Sample Exercise 21.6
Calculating the Age of Objects Using Radioactive Decay

A rock contains 0.257 mg of lead-206 for every milligram of uranium-238. The half-life for the decay of uranium-238 to lead-206 is 4.5×10^9 yr. How old is the rock?

SOLUTION

Analyze We are asked to calculate the age of a rock containing uranium-238 and lead-206, given the half-life of uranium-238 and the relative amounts of uranium-238 and lead-206.

Plan Lead-206 is the product of the radioactive decay of uranium-238. We will assume that the only source of lead-206 in the rock is from the decay of uranium-238, which has a known half-life. To apply first-order kinetics expressions (Equations 21.19 and 21.20) to calculate the time elapsed since the rock was formed, we first need to calculate how much initial uranium-238 there was for every 1 mg that remains today.

Solve Let's assume that the rock currently contains 1.000 mg of uranium-238 and therefore 0.257 mg of lead-206. The amount of uranium-238 in the rock when it was first formed therefore equals 1.000 mg plus the quantity that has decayed to lead-206. Because the mass of lead atoms is not the same as the mass of uranium atoms, we cannot just add 1.000 mg and 0.257 mg. We have to multiply the present mass of lead-206 (0.257 mg) by the ratio of the mass number of uranium to that of lead, into which it has decayed. Therefore, the original mass of $^{238}_{92}\text{U}$ was

$$\text{Original } ^{238}_{92}\text{U} = 1.000 \text{ mg} + \frac{238}{206}(0.257 \text{ mg})$$

$$= 1.297 \text{ mg}$$

Using Equation 21.19, we can calculate the decay constant for the process from its half-life:

$$k = \frac{0.693}{4.5 \times 10^9 \text{ yr}} = 1.5 \times 10^{-10} \text{ yr}^{-1}$$

Rearranging Equation 21.20 to solve for time, t, and substituting known quantities gives

$$t = -\frac{1}{k} \ln \frac{N_t}{N_0} = -\frac{1}{1.5 \times 10^{-10} \text{ yr}^{-1}} \ln \frac{1.000}{1.297} = 1.7 \times 10^9 \text{ yr}$$

▶ **Practice Exercise**

A wooden object from an archeological site is subjected to radiocarbon dating. The activity due to ^{14}C is measured to be 11.6 disintegrations per second. The activity of a carbon sample of equal mass from fresh wood is 15.2 disintegrations per second. The half-life of ^{14}C is 5730 yr. What is the age of the archeological sample?

Sample Exercise 21.7
Calculations Involving Radioactive Decay and Time

If we start with 1.000 g of strontium-90, 0.953 g will remain after 2.00 yr. (**a**) What is the half-life of strontium-90? (**b**) How much strontium-90 will remain after 5.00 yr?

SOLUTION

Analyze (**a**) We are asked to calculate a half-life, $t_{1/2}$, based on data that tell us how much of a radioactive nucleus has decayed in a time interval $t = 2.00$ yr and the information $N_0 = 1.000$ g, $N_t = 0.953$ g. (**b**) We are asked to calculate the amount of a radionuclide remaining after a given period of time.

Plan (**a**) We first calculate the rate constant for the decay, k, and then we use that to compute $t_{1/2}$. (**b**) We need to calculate N_t, the amount of strontium present at time t, using the initial quantity, N_0, and the rate constant for decay, k, calculated in part (a).

Solve

(**a**) Equation 21.20 is solved for the decay constant, k, and then Equation 21.19 is used to calculate half-life, $t_{1/2}$:

$$k = -\frac{1}{t}\ln\frac{N_t}{N_0} = -\frac{1}{2.00\text{ yr}}\ln\frac{0.953\text{ g}}{1.000\text{ g}}$$

$$= -\frac{1}{2.00\text{ yr}}(-0.0481) = 0.0241\text{ yr}^{-1}$$

$$t_{1/2} = \frac{0.693}{k} = \frac{0.693}{0.0241\text{ yr}^{-1}} = 28.8\text{ yr}$$

(**b**) Again, using Equation 21.20, with $k = 0.0241$ yr^{-1}, we have:

$$\ln\frac{N_t}{N_0} = -kt = -(0.0241\text{ yr}^{-1})(5.00\text{ yr}) = -0.120$$

N_t/N_0 is calculated from $\ln(N_t/N_0) = -0.120$ using the e^x or INV LN function of a calculator:

$$\frac{N_t}{N_0} = e^{-0.120} = 0.887$$

Because $N_0 = 1.000$ g, we have:

$$N_t = (0.887)N_0 = (0.887)(1.000\text{ g}) = 0.887\text{ g}$$

▶ **Practice Exercise**
A sample to be used for medical imaging is labeled with ^{18}F, which has a half-life of 110 min. What percentage of the original activity in the sample remains after 300 min?

Self-Assessment Exercises

SAE 21.10 Iodine-131 can be used in diagnostic imaging of the thyroid gland and has a half-life of 8.02 days. If the preparation laboratory started with 224 μg, how much iodine-131 is left after 32.1 days? (**a**) 7.00 μg (**b**) 14.0 μg (**c**) 28.0 μg (**d**) 56.0 μg

SAE 21.11 It takes 45 hours for a 6.00 mg sample of sodium-24 to decay to 0.750 mg. What is the half-life of sodium-24? (**a**) 45 h (**b**) 30 h (**c**) 15 h (**d**) 65 h (**e**) 7.5 h

SAE 21.12 Cesium-137, a component of radioactive waste, has a half-life of 30.2 yr. If a sample of waste has an initial activity of

15.0 Ci due to cesium-137, how long will it take for the activity due to cesium-137 to drop to 0.250 Ci? (**a**) 0.728 yr (**b**) 60.4 yr (**c**) 78.2 yr (**d**) 124 yr (**e**) 178 yr

SAE 21.13 The half-life for the decay of radium-226 to radon-222, ^{226}Ra $\longrightarrow$ ^{222}Rn, is 1600 yr. Radon-222 is an inert gas with a short half-life. A mineral sample originally contained 30.0 mg of ^{226}Ra and now contains 28.5 mg. How long has the sample been sitting on the shelf? (**a**) 360 yr (**b**) 12 yr (**c**) 37 yr (**d**) 120 yr (**e**) 82 yr

---

21.5 | Detection of Radioactivity

Learning Objectives

When you finish Section 21.5, you should be able to:

▶ Describe common ways in which radioactivity is detected.

▶ Describe how radiotracers are used in practical applications.

A variety of methods have been devised to detect emissions from radioactive substances. Henri Becquerel discovered radioactivity because radiation caused fogging of photographic plates, and since that time photographic plates and film have been used to detect radioactivity. The radiation affects photographic film in much the same way as X rays do. The greater the extent of exposure to radiation, the darker the area of the developed negative. People who work with radioactive substances carry film badges to record the extent of their exposure to radiation (**Figure 21.8**).

Go Figure Which type of radiation—alpha, beta, or gamma—is likely to fog a film that is sensitive to X rays?

The film strip is white before exposure to radiation.

The film strip is darkened on exposure to radiation.

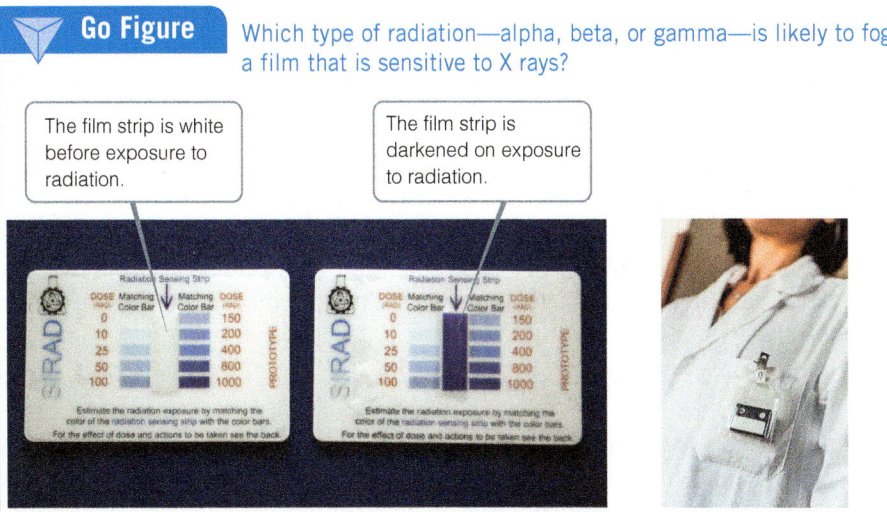

▲ **Figure 21.8 Badge dosimeters monitor the extent to which the individual has been exposed to high-energy radiation.** The radiation dose is determined from the extent of darkening of the film in the dosimeter.

Radioactivity can also be detected and measured by a Geiger counter. The operation of this device is based on the fact that radiation is able to ionize matter. The ions and electrons produced by the ionizing radiation permit conduction of an electrical current. The basic design of a Geiger counter is shown in **Figure 21.9**. A current pulse between the anode and the metal cylinder occurs whenever entering radiation produces ions. Each pulse is counted in order to estimate the amount of radiation.

Some substances, called *phosphors*, emit light when radiation strikes or passes through them. The radioactivity excites the atoms, ions, or molecules of the phosphor to a higher-energy state, and they release this energy as light as they return to their ground

Go Figure Which property of the atoms of gas inside a Geiger counter is most relevant to operation of the device?

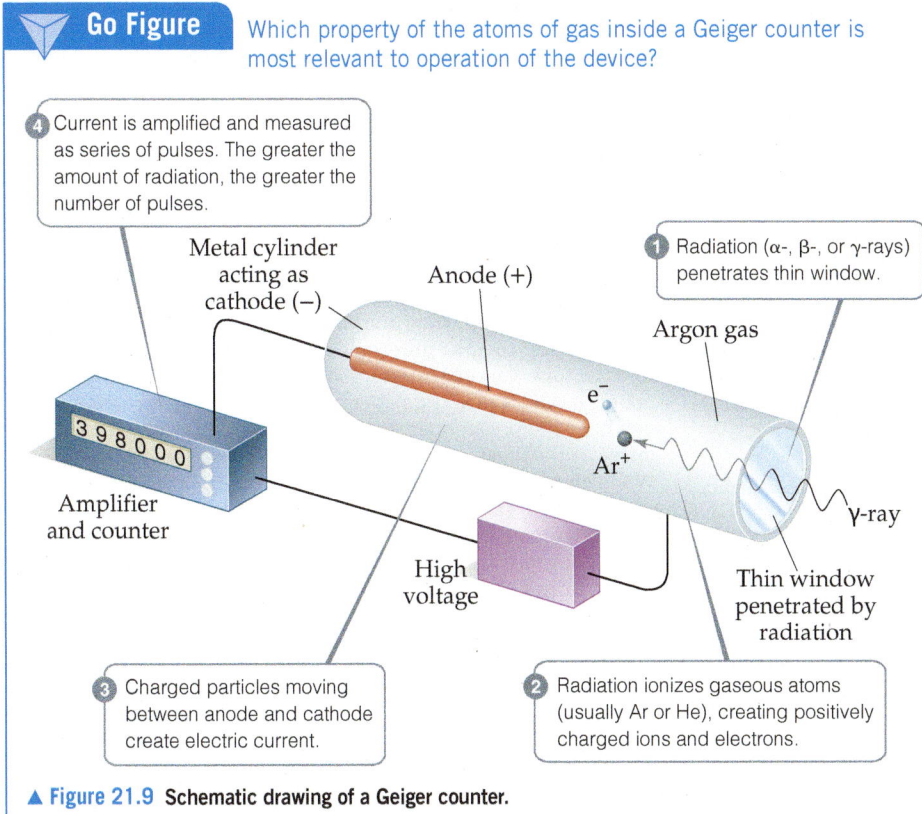

▲ **Figure 21.9 Schematic drawing of a Geiger counter.**

states. For example, ZnS responds this way to alpha radiation. An instrument called a *scintillation counter* detects and counts the flashes of light produced when radiation strikes the phosphor. The flashes of light are magnified electronically and counted to measure the amount of radiation.

Radiotracers

Because radioisotopes can be detected readily, they can be used to follow an element through its chemical reactions. The incorporation of carbon atoms from CO_2 into glucose during photosynthesis, for example, has been studied using CO_2 enriched in carbon-14:

$$6 \, ^{14}CO_2 + 6 \, H_2O \xrightarrow[\text{Chlorophyll}]{\text{Sunlight}} \, ^{14}C_6H_{12}O_6 + 6 \, O_2 \qquad [21.21]$$

Use of the carbon-14 label provides direct experimental evidence that carbon dioxide in the environment is chemically converted to glucose in plants. Analogous labeling experiments using oxygen-18 show that the O_2 produced during photosynthesis comes from water, not carbon dioxide. When it is possible to isolate and purify intermediates and products from reactions, detection devices such as scintillation counters can be used to "follow" the radioisotope as it moves from starting material through intermediates to final product. These types of experiments are useful for identifying elementary steps in a reaction mechanism. (Section 14.5)

The use of radioisotopes is possible because all isotopes of an element have essentially identical chemical properties. When a small quantity of a radioisotope is mixed with the naturally occurring stable isotopes of the same element, all the isotopes go through the same reactions together. The element's path is revealed by the radioactivity of the radioisotope. Because the radioisotope can be used to trace the path of the element, it is called a **radiotracer**.

CHEMISTRY AND LIFE | Medical Applications of Radiotracers

Radiotracers have found wide use as diagnostic tools in medicine. Table 21.6 lists some radiotracers and their uses. These radioisotopes are incorporated into a compound that is administered to the patient, usually intravenously. The diagnostic use of these isotopes is based on the ability of the radioactive compound to localize and concentrate in the organ or tissue under investigation. Iodine-131, for example, has been used to test the activity of the thyroid gland. This gland is the only place in which iodine is incorporated significantly in the body. The patient drinks a solution of NaI containing iodine-131. Only a very small amount is used so that the patient does not receive a harmful dose of radioactivity. A Geiger counter placed close to the thyroid, in the neck region, determines the ability of the thyroid to take up the iodine. A normal thyroid will absorb about 12% of the iodine within a few hours.

The medical applications of radiotracers are further illustrated by *positron emission tomography* (PET). PET is used for clinical diagnosis of many diseases. In this method, compounds containing radionuclides that decay by positron emission are injected into a patient. These compounds are chosen to enable researchers to monitor blood flow, oxygen and glucose metabolic rates, and other biological functions. Some of the most interesting work involves study of the brain, which depends on glucose for most of its energy. Changes in how this sugar is metabolized or used by the brain may signal a disease such as cancer, epilepsy, Parkinson's disease, or schizophrenia.

The radionuclides that are most widely used in PET are carbon-11 ($t_{1/2} = 20.4$ min), fluorine-18 ($t_{1/2} = 110$ min), oxygen-15 ($t_{1/2} = 2$ min), and nitrogen-13 ($t_{1/2} = 10$ min). Glucose, for example, can be labeled with carbon-11. Because the half-lives of positron emitters are so short, they must be generated on site using a cyclotron, and the chemist must quickly incorporate the radionuclide into the sugar (or other appropriate) molecule and inject the compound immediately. The patient is placed in an instrument that measures the positron emission and constructs a computer-based image of the organ in which the emitting compound is localized. When the element decays, the emitted positron quickly collides with an electron. The positron and electron are annihilated in the collision, producing two gamma rays that move in opposite directions. The gamma rays are detected by an encircling ring of scintillation counters (**Figure 21.10**). Because the rays move in opposite directions but were created in the same place at the same time, it is possible to accurately locate the point in the body where the radioactive isotope decayed. The nature of this image provides clues to the presence of disease or other abnormality and helps medical researchers understand how a particular disease affects the functioning of the brain. For example, the images shown in **Figure 21.11** reveal that levels of activity in brains of patients with Alzheimer's disease are different from the levels in those without the disease.

Related Exercises: 21.55, 21.56, 21.81

TABLE 21.6 Some Radionuclides Used as Radiotracers

Nuclide	Half-Life	Area of the Body Studied
Iodine-131	8.04 days	Thyroid
Iron-59	44.5 days	Red blood cells
Phosphorus-32	14.3 days	Eyes, liver, tumors
Technetium-99[a]	6.0 hours	Heart, bones, liver, and lungs
Thallium-201	73 hours	Heart, arteries
Sodium-24	14.8 hours	Circulatory system

[a]The isotope of technetium is actually a special isotope of Tc-99 called Tc-99*m*, where the *m* indicates a so-called *metastable* isotope.

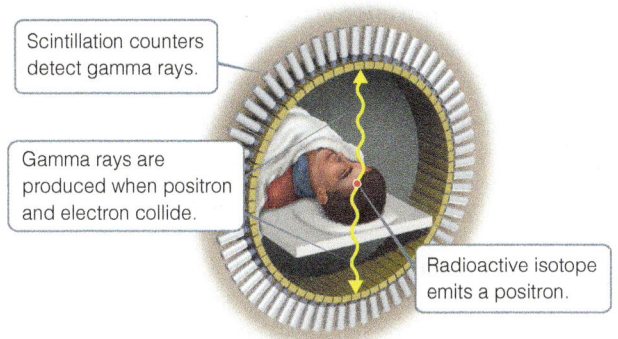

▲ **Figure 21.10 Schematic representation of a positron emission tomography (PET) scanner.**

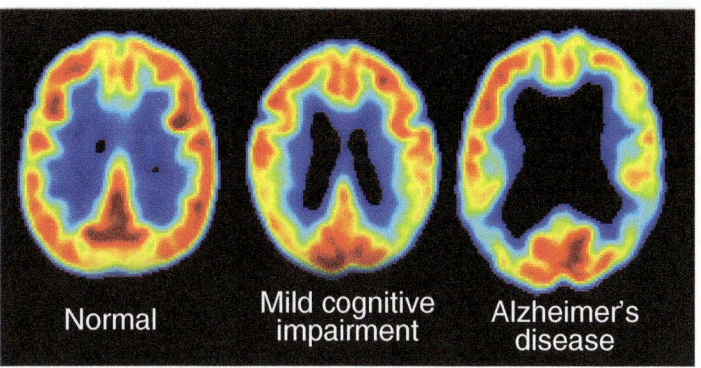

▲ **Figure 21.11 Positron emission tomography (PET) scans showing glucose metabolism levels in the brain.** Red and yellow colors show higher levels of glucose metabolism.

 Self-Assessment Exercise

SAE 21.14 What property of radioisotopes makes them useful as radiotracers to follow the path of elements in reactions? (**a**) The chemical properties of radioisotopes are identical to those of the nonradioactive isotopes. (**b**) Radioisotopes are used in high volumes to alter the element's reaction characteristics. (**c**) Radiation detectors, such as scintillation counters, only detect the nonradioactive isotope. (**d**) Radioisotopes often separate from the nonradioactive isotope and create unique reactions.

21.6 | Energy Changes in Nuclear Reactions

Why are the energies associated with nuclear reactions so large, in many cases orders of magnitude larger than those associated with nonnuclear chemical reactions? The answer to this question begins with Einstein's celebrated equation from the theory of relativity that relates mass and energy:

$$E = mc^2 \qquad [21.22]$$

In this equation E stands for energy, m for mass, and c for the speed of light, 2.9979×10^8 m/s. This equation states that mass and energy are equivalent and can be converted into one another. If a system loses mass, it loses energy; if it gains mass, it gains energy. Because the proportionality constant between energy and mass, c^2, is such a large number, even small changes in mass are accompanied by large changes in energy.

The mass changes in chemical reactions are too small to detect. For example, the mass change associated with the combustion of 1 mol of CH_4 (an exothermic process) is -9.9×10^{-9} g. Because the mass change is so small, it is possible to treat chemical reactions as though mass is conserved. (Section 2.1)

The mass changes and the associated energy changes in nuclear reactions are much greater than those in chemical reactions. The mass change accompanying the radioactive decay of 1 mol of uranium-238, for example, is 50,000 times greater than that for the combustion of 1 mol of CH_4. Let's examine the energy change for the nuclear reaction

$$^{238}_{92}\text{U} \longrightarrow {}^{234}_{90}\text{Th} + {}^4_2\text{He}$$

The masses of the nuclei are $^{238}_{92}$U, 238.0003 amu; $^{234}_{90}$Th, 233.9942 amu; and $^4_2$He, 4.0015 amu. The mass change, Δm, is the total mass of the products minus the total mass of the reactants. The mass change for the decay of 1 mol of uranium-238 can then be expressed in grams:

$$233.9942\,\text{g} + 4.0015\,\text{g} - 238.0003\,\text{g} = -0.0046\,\text{g}$$

The fact that the system has lost mass indicates that the process is exothermic. All spontaneous nuclear reactions are exothermic.

 Learning Objectives

When you finish **Section 21.6**, you should be able to:

▶ Use Einstein's equation to relate mass and energy in nuclear reactions.

▶ Determine the energy associated with separating the nucleus of an atom into smaller pieces.

The energy change per mole associated with this reaction is

$$\Delta E = \Delta(mc^2) = c^2 \Delta m$$

$$= (2.9979 \times 10^8 \text{ m/s})^2(-0.0046 \text{ g})\left(\frac{1 \text{ kg}}{1000 \text{ g}}\right)$$

$$= -4.1 \times 10^{11}\frac{\text{kg-m}^2}{\text{s}^2} = -4.1 \times 10^{11}\text{ J}$$

Notice that Δm must be converted to kilograms, the SI unit of mass, to obtain ΔE in joules, the SI unit of energy. The negative sign for the energy change indicates that energy is released in the reaction—in this case, over 400 billion joules per mole of uranium! This energy would provide the average annual household electricity for about 10,000 homes in the United States.

Sample Exercise 21.8
Calculating Mass Change in a Nuclear Reaction

How much energy is lost or gained when 1 mol of cobalt-60 undergoes beta decay, $^{60}_{27}\text{Co} \longrightarrow {}^{60}_{28}\text{Ni} + {}^{0}_{-1}\text{e}$? The mass of a $^{60}_{27}\text{Co}$ atom is 59.933819 amu, and that of a $^{60}_{28}\text{Ni}$ atom is 59.930788 amu.

SOLUTION

Analyze We are asked to calculate the energy change in a nuclear reaction.

Plan We must first calculate the mass change in the process. We are given atomic masses, but we need the masses of the nuclei in the reaction. We calculate these by taking account of the masses of the electrons that contribute to the atomic masses.

Solve

A $^{60}_{27}\text{Co}$ atom has 27 electrons. The mass of an electron is 5.4858×10^{-4} amu. (See the list of fundamental constants in the back inside cover.) We subtract the mass of the 27 electrons from the mass of the $^{60}_{27}\text{Co}$ *atom* to find the mass of the $^{60}_{27}\text{Co}$ *nucleus*:

$59.933819 \text{ amu} - (27)(5.4858 \times 10^{-4} \text{ amu})$
$= 59.919007 \text{ amu (or 59.919007 g/mol)}$

Likewise, for $^{60}_{28}\text{Ni}$, the mass of the nucleus is:

$59.930788 \text{ amu} - (28)(5.4858 \times 10^{-4} \text{ amu})$
$= 59.915428 \text{ amu (or 59.915428 g/mol)}$

The mass change in the nuclear reaction is the total mass of the products minus the mass of the reactant:

$\Delta m = \text{mass of electron} + \text{mass } {}^{60}_{28}\text{Ni nucleus} - \text{mass of } {}^{60}_{27}\text{Co nucleus}$
$= 0.00054858 \text{ amu} + 59.915428 \text{ amu} - 59.919007 \text{ amu}$
$= -0.003030 \text{ amu}$

Thus, when a mole of cobalt-60 decays,

$\Delta m = -0.003030 \text{ g}$

Because the mass decreases ($\Delta m < 0$), energy is released ($\Delta E < 0$). The quantity of energy released *per mole* of cobalt-60 is calculated using Equation 21.22:

$\Delta E = c^2 \Delta m$

$= (2.9979 \times 10^8 \text{ m/s})^2(-0.003030 \text{ g})\left(\frac{1 \text{ kg}}{1000 \text{ g}}\right)$

$= -2.723 \times 10^{11}\frac{\text{kg-m}^2}{\text{s}^2} = -2.723 \times 10^{11}\text{ J}$

▶ **Practice Exercise**
Positron emission from ^{11}C, $^{11}_{6}\text{C} \longrightarrow {}^{11}_{5}\text{B} + {}^{0}_{+1}\text{e}$, occurs with release of 2.87×10^{11} J per mole of ^{11}C. What is the mass change per mole of ^{11}C in this nuclear reaction? The masses of ^{11}B and ^{11}C are 11.009305 and 11.011434 amu, respectively.

Nuclear Binding Energies

Scientists discovered in the 1930s that the masses of nuclei are always less than the masses of the individual nucleons of which they are composed. For example, the helium-4 nucleus (an alpha particle) has a mass of 4.00150 amu. The mass of a proton is 1.00728 amu and that of a neutron is 1.00866 amu. Consequently, two protons and two neutrons have a total mass of 4.03188 amu:

$$\text{Mass of two protons} = 2(1.00728 \text{ amu}) = 2.01456 \text{ amu}$$
$$\text{Mass of two neutrons} = 2(1.00866 \text{ amu}) = \underline{2.01732 \text{ amu}}$$
$$\text{Total mass} = 4.03188 \text{ amu}$$

The mass of the individual nucleons is 0.03038 amu greater than that of the helium-4 nucleus:

$$\text{Mass of two protons and two neutrons} = 4.03188 \text{ amu}$$
$$\text{Mass of } {}^{4}_{2}\text{He nucleus} = \underline{4.00150 \text{ amu}}$$
$$\text{Mass difference } \Delta m = 0.03038 \text{ amu}$$

The mass difference between a nucleus and its constituent nucleons is called the **mass defect**. The origin of the mass defect is readily understood if we consider that energy must be added to a nucleus to break it into separated protons and neutrons:

$$\text{Energy} + {}^{4}_{2}\text{He} \longrightarrow 2\,{}^{1}_{1}\text{H} + 2\,{}^{1}_{0}\text{n} \qquad [21.23]$$

By Einstein's relation, the addition of energy to a system must be accompanied by a proportional increase in mass. The mass change we just calculated for the conversion of helium-4 into separated nucleons is $\Delta m = 0.03038$ amu. Therefore, the energy required for this process is

$$\Delta E = c^2 \Delta m$$

$$= (2.9979 \times 10^8 \text{ m/s})^2 (0.03038 \text{ amu}) \left(\frac{1 \text{ g}}{6.022 \times 10^{23} \text{ amu}} \right) \left(\frac{1 \text{ kg}}{1000 \text{ g}} \right)$$

$$= 4.534 \times 10^{-12} \text{ J}$$

The energy required to separate a nucleus into its individual nucleons is called the **nuclear binding energy**. The mass defect and nuclear binding energy for three elements are compared in Table 21.7.

Values of binding energies per nucleon can be used to compare the stabilities of different combinations of nucleons (such as two protons and two neutrons arranged either as ${}^{4}_{2}\text{He}$ or as $2\,{}^{2}_{1}\text{H}$). Figure 21.12 shows average binding energy per nucleon plotted against mass number. Binding energy per nucleon at first increases in magnitude as mass number increases, reaching about 1.4×10^{-12} J for nuclei whose mass numbers are in the vicinity of iron-56. It then decreases slowly to about 1.2×10^{-12} J for very heavy nuclei. *This trend indicates that nuclei of intermediate mass numbers are more tightly bound (and therefore more stable) than those with either smaller or larger mass numbers.*

This trend has two significant consequences: First, heavy nuclei gain stability and therefore give off energy if they are fragmented into two midsized nuclei. This process, known as **fission**, is used to generate energy in nuclear power plants. Second, because of the sharp increase in the graph for small mass numbers, even greater amounts of energy are released if very light nuclei are combined, or fused together, to give more massive nuclei. This **fusion** process is the essential energy-producing process in the Sun and other stars.

TABLE 21.7   **Mass Defects and Binding Energies for Three Nuclei**

Nucleus	Mass of Nucleus (amu)	Mass of Individual Nucleons (amu)	Mass Defect (amu)	Binding Energy (J)	Binding Energy per Nucleon (J)
${}^{4}_{2}\text{He}$	4.00150	4.03188	0.03038	4.53×10^{-12}	1.13×10^{-12}
${}^{56}_{26}\text{Fe}$	55.92068	56.44914	0.52846	7.90×10^{-11}	1.41×10^{-12}
${}^{238}_{92}\text{U}$	238.00031	239.93451	1.93420	2.89×10^{-10}	1.21×10^{-12}

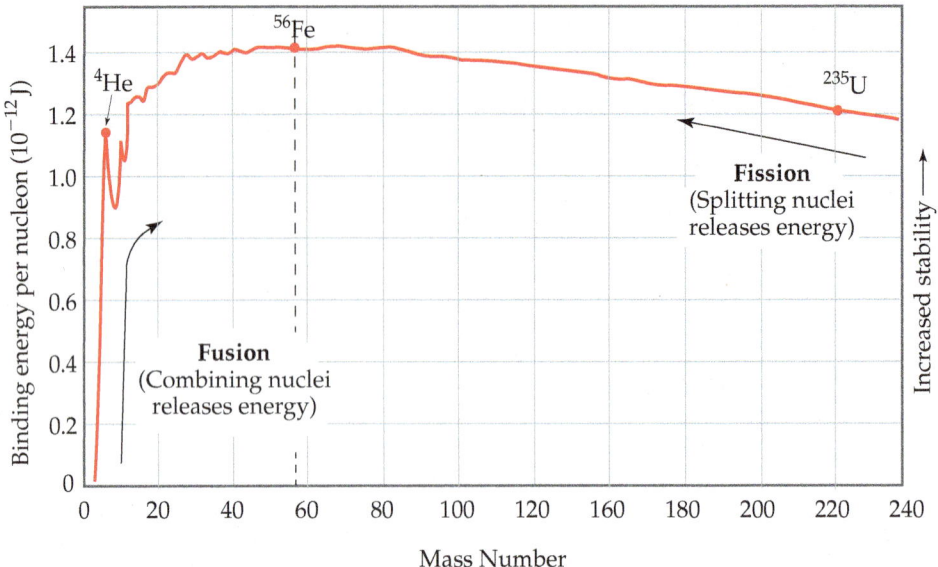

▶ **Figure 21.12 Nuclear binding energies.** The average binding energy per nucleon increases initially as the mass number increases and then decreases slowly. Because of these trends, fusion of light nuclei and fission of heavy nuclei are exothermic processes.

▲ Self-Assessment Exercises

SAE 21.15 What energy change (in J) would accompany the loss of 0.0035 mg in mass? $1 J = 1 kg\text{-}m^2/s^2$. (**a**) 1.0 J (**b**) 1.0×10^6 J (**c**) 3.1×10^8 J (**d**) 3.1×10^{14} J

SAE 21.16 The mass of a nickel-62 nucleus is 61.928345 amu. What is the binding energy for this nucleus? The mass of a neutron is 1.008664916 amu, the mass of a proton is 1.007276466 amu, the speed of light is 2.9979×10^8 m/s, and 1 amu = $1.660538921 \times 10^{-27}$ kg. (**a**) 8.506×10^{-11} J (**b**) 0.569935 J (**c**) 8.506×10^{-8} J (**d**) 5.122×10^{13} J

21.7 | Nuclear Power: Fission

▲ Learning Objectives

When you finish Section 21.7, you should be able to:

▶ Analyze a nuclear fission reaction to determine the product nuclei.

▶ Describe the principles in the operation of a nuclear power plant.

Nuclear fission is the process used to generate energy in nuclear power plants. Over 10% of the electricity generated worldwide comes from nuclear power plants, though the percentage varies from one country to the next, as **Figure 21.13** shows. There are about 440 commercial nuclear power plants in operation in 32 countries, and approximately another 50 are under construction, especially in China and India.

Most nuclear reactors rely on the fission of uranium-235. This was the first nuclear fission reaction to be discovered. This nucleus, as well as those of

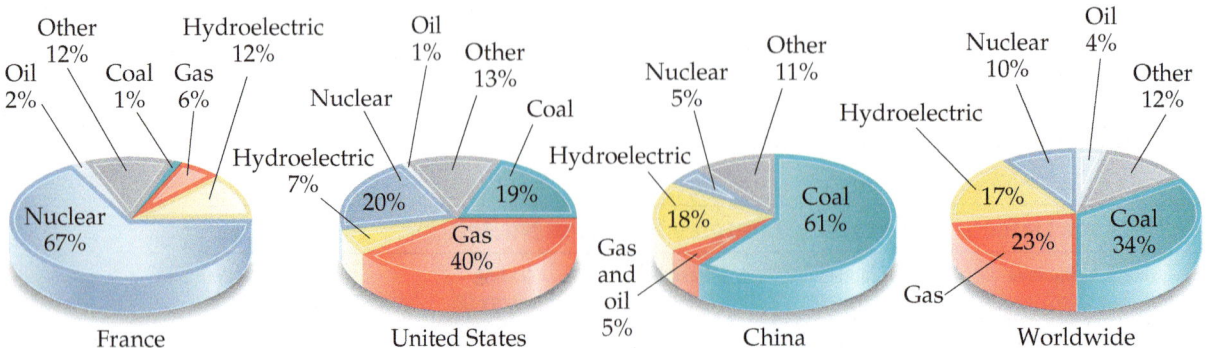

▲ **Figure 21.13 Sources of electricity generation, worldwide and for select countries.** The mix of energy sources varies widely from country to country. "Other" includes other sources of renewable energy (wind, solar, and biomass).

[**Source:** Ember Global Electricity Dashboard (ember-climate.org/data/global-electricity/), 2020 data]

Go Figure What is the relationship between the sum of the mass numbers on the two sides of this reaction?

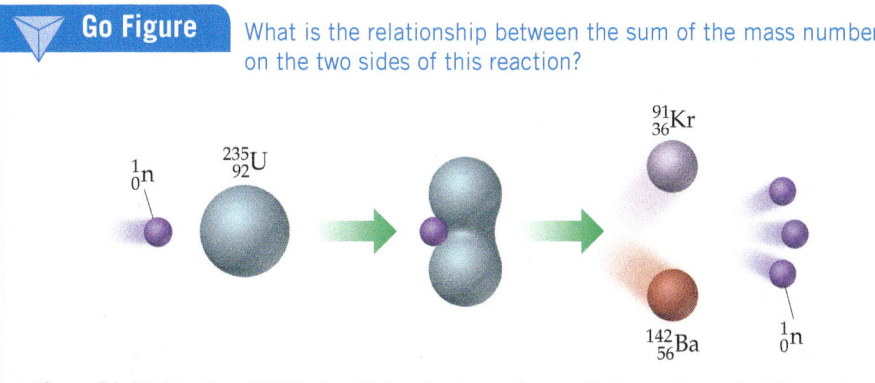

▲ **Figure 21.14 Uranium-235 fission.** This is just one of many fission patterns. In this reaction, 3.5×10^{-11} J of energy is released per ^{235}U nucleus that is split.

uranium-233 and plutonium-239, undergoes fission when struck by a slow-moving neutron (**Figure 21.14**).*

A heavy nucleus can split in many ways, giving rise to a variety of smaller nuclei. Two ways that the uranium-235 nucleus splits, for instance, are

$$\begin{array}{ll} {}^{1}_{0}\text{n} + {}^{235}_{92}\text{U} \left\{ \begin{array}{l} \longrightarrow {}^{137}_{52}\text{Te} + {}^{97}_{40}\text{Zr} + 2\,{}^{1}_{0}\text{n} \qquad [21.24] \\[2ex] \longrightarrow {}^{142}_{56}\text{Ba} + {}^{91}_{36}\text{Kr} + 3\,{}^{1}_{0}\text{n} \qquad [21.25] \end{array} \right. \end{array}$$

The nuclei produced in Equations 21.24 and 21.25—called the *fission products*—are themselves radioactive and undergo further nuclear decay. More than 200 isotopes of 35 elements have been found among the fission products of uranium-235. Most of them are radioactive.

Slow-moving neutrons are required for the fission of uranium-235 because the process involves initial absorption of the neutron by the nucleus. The resulting more massive nucleus is extremely unstable and spontaneously undergoes fission. Fast neutrons tend to bounce off the nucleus, in which case little fission occurs.

Note that the coefficients of the neutrons produced in Equations 21.24 and 21.25 are 2 and 3, respectively. On average, 2.4 neutrons are produced by every fission of a uranium-235 nucleus. If one fission produces two neutrons, then the two neutrons can cause two additional fissions, each producing two neutrons. The four neutrons thereby released can produce four fissions, and so forth, as shown in **Figure 21.15**. The number

Go Figure If this figure were extended one more "generation" down, how many neutrons would be produced?

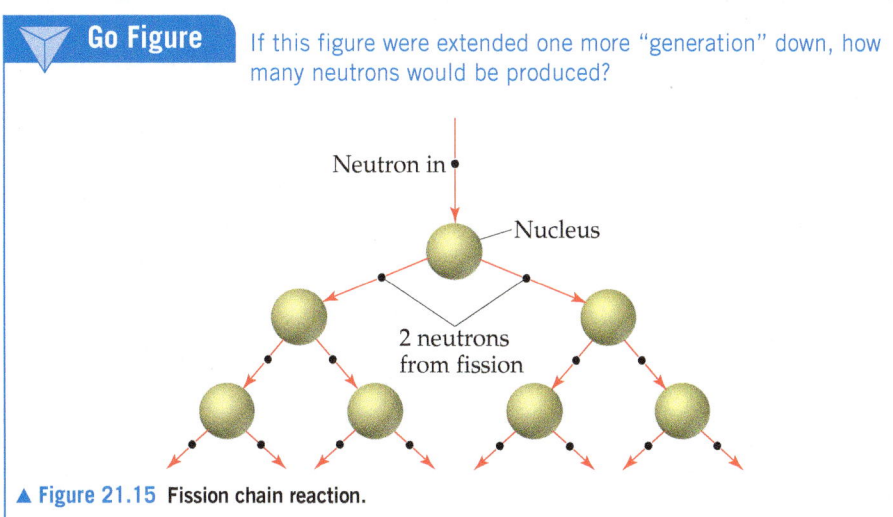

▲ **Figure 21.15 Fission chain reaction.**

---

*Other heavy nuclei can also undergo fission. However, these three are the only ones of practical importance.

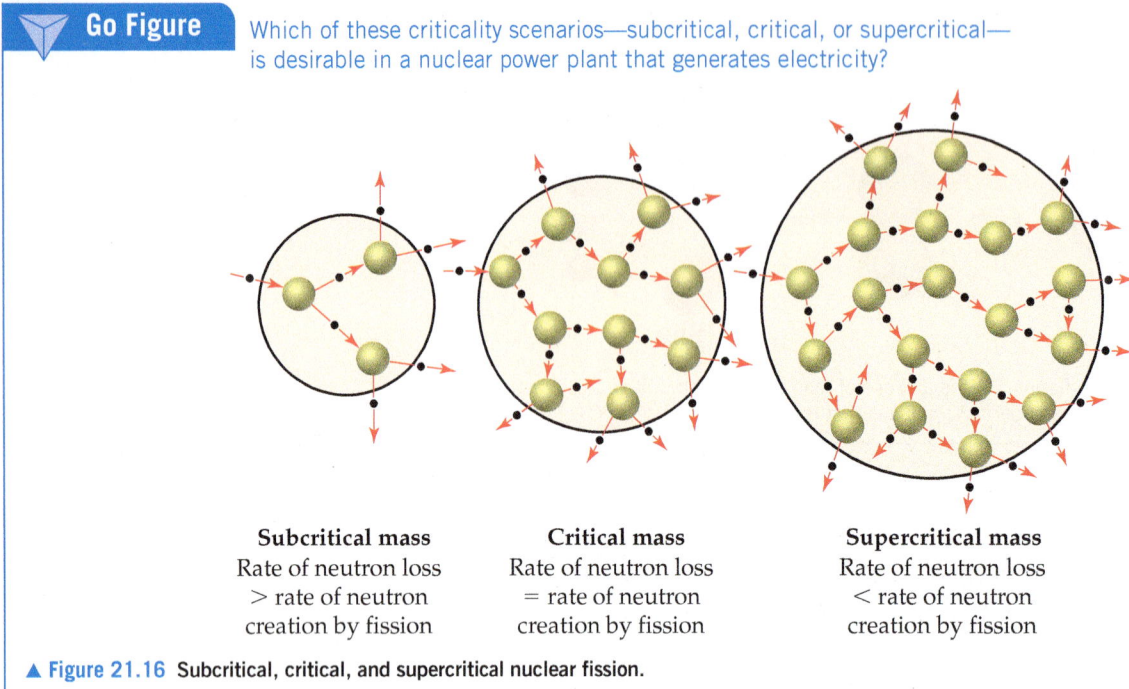

Go Figure Which of these criticality scenarios—subcritical, critical, or supercritical—is desirable in a nuclear power plant that generates electricity?

Subcritical mass
Rate of neutron loss
> rate of neutron
creation by fission

Critical mass
Rate of neutron loss
= rate of neutron
creation by fission

Supercritical mass
Rate of neutron loss
< rate of neutron
creation by fission

▲ **Figure 21.16** Subcritical, critical, and supercritical nuclear fission.

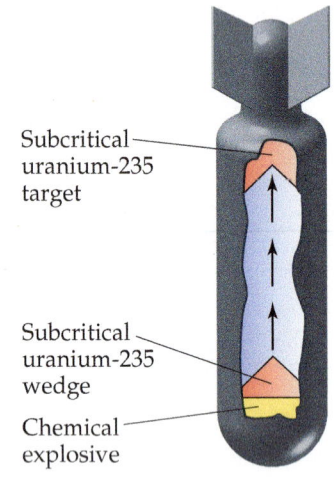

Subcritical
uranium-235
target

Subcritical
uranium-235
wedge

Chemical
explosive

▲ **Figure 21.17 Schematic drawing of an atomic bomb.** A conventional explosive is used to bring two subcritical masses together to form a supercritical mass.

of fissions and the energy released quickly escalate, and if the process is unchecked, the result is a violent explosion. Reactions that multiply in this fashion are called **chain reactions**.

For a fission chain reaction to occur, the sample of fissionable material must have a certain minimum mass. Otherwise, neutrons escape from the sample before they have the opportunity to strike other nuclei and cause additional fission. The amount of fissionable material large enough to maintain a chain reaction with a constant rate of fission is called the **critical mass**. When a critical mass of material is present, one neutron on average from each fission is subsequently effective in producing another fission and the fission continues at a constant, controllable rate. The critical mass of uranium-235 is about 50 kg for a bare sphere of the metal.*

If more than a critical mass of fissionable material is present, very few neutrons escape. The chain reaction thus multiplies the number of fissions, which can lead to a nuclear explosion. A mass in excess of a critical mass is referred to as a **supercritical mass**. The effect of mass on a fission reaction is illustrated in **Figure 21.16**.

Figure 21.17 shows a schematic diagram of the first atomic bomb used in warfare. The bomb, code-named "Little Boy," was dropped on Hiroshima, Japan, on August 6, 1945. The bomb contained about 64 kg of uranium-235, which had been separated from the nonfissionable uranium-238 primarily by gaseous diffusion of uranium hexafluoride, UF_6. (Section 10.6) To trigger the fission reaction, two subcritical masses of uranium-235 were slammed together using chemical explosives. The combined masses of the uranium formed a supercritical mass, which led to a rapid, uncontrolled chain reaction and, ultimately, a nuclear explosion. The energy released by the bomb dropped on Hiroshima was equivalent to that of 16,000 tons of TNT (it therefore is called a *16-kiloton* bomb). Unfortunately, the basic design of a fission-based atomic bomb is quite simple, and the fissionable materials are potentially available to any nation with a nuclear reactor. The combination of design simplicity and materials availability has generated international concerns about the proliferation of atomic weapons.

_____

*The exact value of the critical mass depends on the shape of the radioactive substance. The critical mass can be reduced if the radioisotope is surrounded by a material that reflects some neutrons.

A CLOSER LOOK   The Dawning of the Nuclear Age

Uranium-235 fission was first achieved during the late 1930s by Enrico Fermi and coworkers in Rome and shortly thereafter by Otto Hahn and coworkers in Berlin. Both groups were trying to produce transuranium elements. In 1938, Hahn identified barium among his reaction products. He was puzzled by this observation and questioned the identification because the presence of barium was so unexpected. He sent a letter describing his experiments to Lise Meitner, a former coworker who had been forced to leave Germany because of the anti-Semitism of the Third Reich and had settled in Sweden. She surmised that Hahn's experiment indicated a nuclear process was occurring in which the uranium-235 split. She called this process *nuclear fission*.

Meitner passed word of this discovery to her nephew, Otto Frisch, a physicist working at Niels Bohr's institute in Copenhagen. Frisch repeated the experiment, verifying Hahn's observations, and found that tremendous energies were involved. In January 1939, Meitner and Frisch published a short article describing the reaction. In March 1939, Leo Szilard and Walter Zinn at Columbia University discovered that more neutrons are produced than are used in each fission. As we have seen, this result allows a chain reaction to occur.

News of these discoveries and an awareness of their potential use in explosive devices spread rapidly within the scientific community. Several scientists finally persuaded Albert Einstein, the most famous physicist of the time, to write a letter to President Franklin D. Roosevelt explaining the implications of these discoveries. Einstein's letter, written in August 1939, outlined the possible military applications of nuclear fission and emphasized the danger that weapons based on fission would pose if they were developed by the Nazis. Roosevelt judged it imperative that the United States investigate the possibility of such weapons. Late in 1941, the decision was made to build a bomb based on the fission reaction. An enormous research project, known as the Manhattan Project, began.

On December 2, 1942, the first artificial self-sustaining nuclear fission chain reaction was achieved in an abandoned squash court at the University of Chicago. This accomplishment, achieved remarkably quickly after the initiation of the Manhattan Project, led to the development of the first atomic bomb, at Los Alamos National Laboratory in New Mexico in July 1945 (Figure 21.18). In August 1945 the United States dropped atomic bombs on two Japanese cities, Hiroshima and Nagasaki. The nuclear age had arrived, albeit in a sadly destructive fashion. Humanity has struggled with the conflict between the positive potential of nuclear energy and its terrifying potential as a weapon ever since.

▲ **Figure 21.18 The Trinity test for the atom bomb developed during World War II.** The first human-made nuclear explosion took place on July 16, 1945, on the Alamogordo test range in New Mexico.

Nuclear Reactors

Nuclear power plants use nuclear fission to generate energy. The core of a typical nuclear reactor consists of four principal components: fuel elements, control rods, a moderator, and a primary coolant (Figure 21.19). The fuel is a fissionable substance, such as uranium-235. The natural isotopic abundance of uranium-235 is only 0.7%, too low to sustain a chain reaction in most reactors. Therefore, the ^{235}U content of the fuel must be enriched to 3–5% for use in a reactor. The *fuel elements* contain enriched uranium in the form of UO_2 pellets encased in zirconium or stainless steel tubes.

The *control rods* are composed of materials that absorb neutrons, such as boron-10 or an alloy of silver, indium, and cadmium. These rods regulate the flux of neutrons to keep the reaction chain self-sustaining and to prevent the reactor core from overheating.*

The probability that a neutron will trigger fission of a ^{235}U nucleus depends on the speed of the neutron. The neutrons produced by fission have high speeds (typically in excess of 10,000 km/s). The function of the *moderator* is to slow down the neutrons (to speeds of a few kilometers per second) so that they can be captured more readily by the fissionable nuclei. The moderator is typically either water or graphite.

The *primary coolant* is a substance that transports the heat generated by the nuclear chain reaction away from the reactor core. In a *pressurized water reactor*, which is the most common commercial reactor design, water acts as both the moderator and the primary coolant.

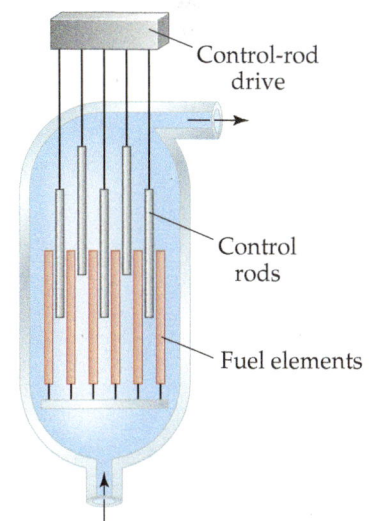

Water acts as both moderator and coolant

▲ **Figure 21.19** Schematic diagram of a pressurized water reactor core.

---

*The reactor core cannot reach supercritical levels and explode with the violence of an atomic bomb because the concentration of uranium-235 is too low. However, if the core overheats, sufficient damage can lead to release of radioactive materials into the environment, as was the case in the Chernobyl disaster in the Soviet Union in 1986.

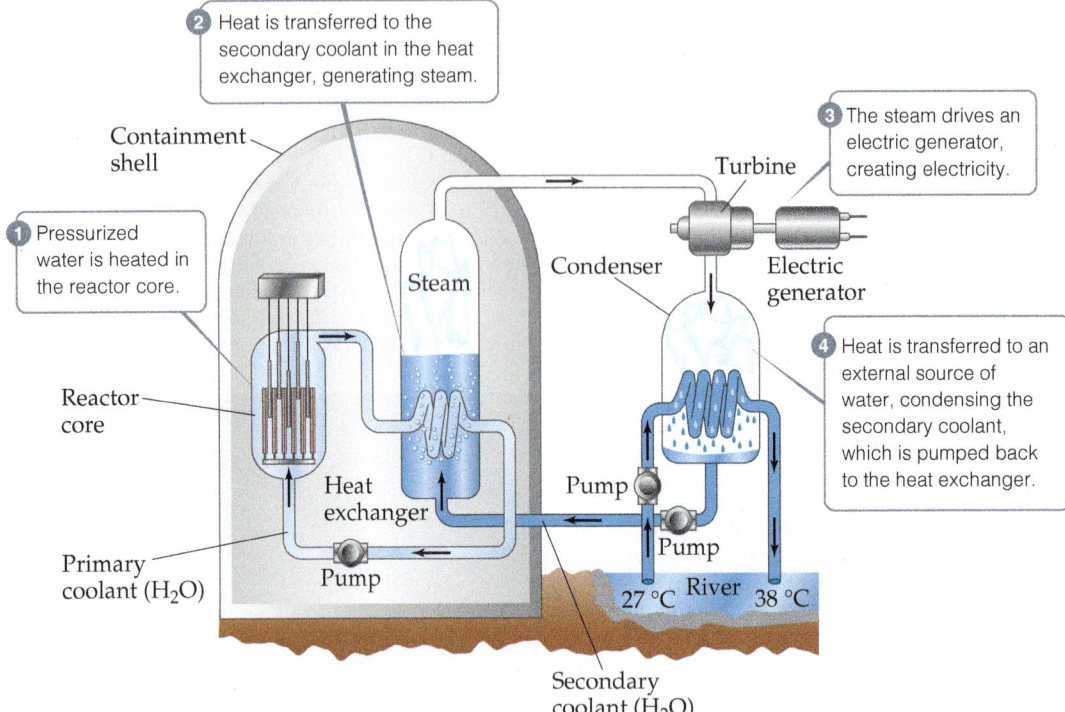

① Pressurized water is heated in the reactor core.

② Heat is transferred to the secondary coolant in the heat exchanger, generating steam.

③ The steam drives an electric generator, creating electricity.

④ Heat is transferred to an external source of water, condensing the secondary coolant, which is pumped back to the heat exchanger.

Containment shell

Turbine

Condenser

Electric generator

Steam

Reactor core

Heat exchanger

Pump

Pump

Pump

Primary coolant (H_2O)

27 °C   River   38 °C

Secondary coolant (H_2O)

▲ **Figure 21.20** Basic design of a pressurized water reactor nuclear power plant.

The design of a nuclear power plant is basically the same as that of a power plant that burns fossil fuel (except that the burner is replaced by a reactor core). The nuclear power plant design shown in **Figure 21.20**, a pressurized water reactor, is currently the most popular. The primary coolant passes through the core in a closed system, which lessens the chance that radioactive products could escape the core. As an added safety precaution, the reactor is surrounded by a reinforced concrete *containment shell* to shield personnel and nearby residents from radiation and to protect the reactor from external forces. After passing through the reactor core, the very hot primary coolant passes through a heat exchanger, where much of its heat is transferred to a *secondary coolant*, converting the latter to high-pressure steam that is used to drive a turbine. The secondary coolant is then condensed by transferring heat to an external source of water, such as a river or lake. The cooling systems of nuclear power plants are necessary for proper and safe operation. The nuclear disaster in Fukushima, Japan, in March 2011 occurred when a tsunami damaged the reactor cooling systems, resulting in a large-scale release of radioactive materials.

Although about two-thirds of all commercial reactors are pressurized water reactors, there are several variations on this basic design, each with advantages and disadvantages. A *boiling water reactor* generates steam by boiling the primary coolant; thus, no secondary coolant is needed. The reactors at Fukushima, Japan, were boiling water reactors. Pressurized water reactors and boiling water reactors are collectively referred to as *light water reactors* because they use H_2O as moderator and primary coolant. A *heavy water reactor* uses D_2O (D = deuterium, ^2H) as moderator and primary coolant, and a *gas-cooled reactor* uses a gas, typically CO_2, as primary coolant and graphite as the moderator. Use of either D_2O or graphite as the moderator has the advantage that both substances absorb fewer neutrons than H_2O. Consequently, the uranium fuel does not need to be as enriched.

Nuclear Waste

The fission products that accumulate as a reactor operates decrease the efficiency of the reactor by capturing neutrons. For this reason, commercial reactors must be stopped periodically to either replace or reprocess the nuclear fuel. When the fuel elements are removed from the reactor, they are initially very radioactive. It was originally intended that they be stored for several months in pools at the reactor site to allow decay of short-lived radioactive nuclei. They were then to be transported in shielded containers to reprocessing plants where the unspent fuel would be separated from the fission products. Reprocessing plants have been plagued with operational difficulties, however, and there is intense opposition in the United States to the transport of nuclear wastes on the nation's roads and rails.

Even if the transportation difficulties could be overcome, the high level of radioactivity of the spent fuel makes reprocessing a hazardous operation. At present in the United States, spent fuel elements are kept in storage at reactor sites. Spent fuel is reprocessed, however, in France, Russia, the United Kingdom, India, and Japan.

Storage of spent nuclear fuel poses a major problem because the fission products are extremely radioactive. It is estimated that 10 half-lives are required for their radioactivity to reach levels acceptable for biological exposure. Based on the 28.8-yr half-life of strontium-90, one of the longer-lived and most dangerous of the products, the wastes must be stored for nearly 300 yr. Plutonium-239 is one of the by-products present in spent fuel elements. It is formed by absorption of a neutron by uranium-238, followed by two successive beta emissions. (Remember that most of the uranium in the fuel elements is uranium-238.) If the elements are reprocessed, the plutonium-239 is largely recovered because it can be used as a nuclear fuel. However, if the plutonium is not removed, spent elements must be stored for a very long time because plutonium-239 has a half-life of 24,000 yr.

A *fast breeder reactor* offers one approach to getting more power out of existing uranium sources and potentially reducing radioactive waste. This type of reactor is so named because it creates ("breeds") more fissionable material than it consumes. The reactor operates without a moderator, which means the neutrons used are not slowed down. In order to capture the fast neutrons, the fuel must be highly enriched with both uranium-235 and plutonium-239. Water cannot be used as a primary coolant because it would moderate the neutrons, and so a liquid metal, usually sodium, is used. The core is surrounded by a blanket of uranium-238 that captures neutrons that escape the core, producing plutonium-239 in the process. The plutonium can later be separated by reprocessing and used as fuel in a future cycle.

Because fast neutrons are more effective at decaying many radioactive nuclides, the material separated from the uranium and plutonium during reprocessing is less radioactive than waste from other reactors. However, the generation of relatively high levels of plutonium, coupled with the need for reprocessing, is problematic in terms of nuclear nonproliferation. Thus, political factors, coupled with increased safety concerns and higher operational costs, make fast breeder reactors quite rare.

A considerable amount of research is being devoted to the disposal of radioactive wastes. At present, the most attractive possibilities appear to be formation of glass, ceramic, or synthetic rock from the wastes as a means of immobilizing them. These solid materials would then be placed in containers of high corrosion resistance and durability and buried deep underground. The process of selecting appropriate deep repositories for high-level waste and spent fuel is now under way in several countries. In the United States, the Department of Energy (DOE) had designated Yucca Mountain in Nevada as a disposal site, but in 2010 this project was discontinued because of technical and political concerns. Unfortunately, there is now no long-term solution to nuclear waste storage in the United States.

In spite of these difficulties, nuclear power is making a modest comeback as an energy source. Concerns about climate change caused by escalating atmospheric CO_2 levels (Section 18.2) have increased support for nuclear power as a major energy source in the future, although nuclear power is still a major source of political discourse and disagreement. Increasing demand for power in rapidly developing countries, particularly China, has sparked a rise in construction of new nuclear power plants in those parts of the world.

 Self-Assessment Exercises

SAE 21.17 Identify the missing product in the following nuclear fission reaction:

$$^{239}_{94}\text{Pu} + ^{1}_{0}\text{n} \longrightarrow ? + ^{94}_{36}\text{Kr} + 2\,^{1}_{0}\text{n}$$

(a) $^{144}_{31}\text{Ga}$ (b) $^{144}_{58}\text{Ce}$ (c) $^{145}_{58}\text{Ce}$ (d) $^{146}_{58}\text{Ce}$

SAE 21.18 Which of the following statements about nuclear fission power plants is or are *true*?

(i) One of the advantages of nuclear fission reactors is that they don't produce any waste.
(ii) The primary fuel used in nuclear fission power plants is ^{235}U.
(iii) The role of the control rods is to absorb neutrons as a means of controlling the rate at which fission occurs.

(a) Only one of the statements is true. (b) Statements i and ii are true. (c) Statements i and iii are true. (d) Statements ii and iii are true. (e) All three statements are true.

21.8 | Nuclear Power: Fusion

Learning Objective

When you finish Section 21.8, you should be able to:

▶ Describe the principles involved in nuclear fusion reactions.

In Sections 21.6 and 21.7, we saw that energy is released when heavy nuclei are split into smaller nuclei. Energy is also produced when very light nuclei fuse into heavier ones. Reactions of this type are responsible for the energy produced by the Sun. Spectroscopic studies indicate that the mass composition of the Sun is 74% H, 25% He, and only 1% all other elements. The following reactions are among the numerous nuclear fusion processes believed to occur in the Sun:

$$^1_1\text{H} + ^1_1\text{H} \longrightarrow ^2_1\text{H} + ^0_{+1}\text{e} \qquad [21.26]$$

$$^1_1\text{H} + ^2_1\text{H} \longrightarrow ^3_2\text{He} \qquad [21.27]$$

$$^3_2\text{He} + ^3_2\text{He} \longrightarrow ^4_2\text{He} + 2\,^1_1\text{H} \qquad [21.28]$$

$$^3_2\text{He} + ^1_1\text{H} \longrightarrow ^4_2\text{He} + ^0_{+1}\text{e} \qquad [21.29]$$

Fusion is appealing as an energy source because of the availability of light isotopes on Earth and because fusion products are generally not radioactive. Despite this fact, fusion is not presently used to generate energy. The primary problem is that extremely high temperatures and pressures are needed to overcome the electrostatic repulsion between nuclei in order to fuse them. The lowest temperature required for any fusion is about 40,000,000 K, the temperature needed to fuse deuterium and tritium:

$$^2_1\text{H} + ^3_1\text{H} \longrightarrow ^4_2\text{He} + ^1_0\text{n} \qquad [21.30]$$

Fusion reactions are therefore also known as **thermonuclear reactions**.

Such high temperatures have been achieved by using an atomic bomb to initiate fusion. This is the operating principle behind a thermonuclear, or hydrogen, bomb. This approach is obviously unacceptable for a power generation plant.*

Numerous problems must be overcome before fusion becomes a practical energy source. In addition to the high temperatures necessary to initiate the reaction, there is the problem of confining the reaction. No known structural material is able to withstand the enormous temperatures necessary for fusion. Much research has centered on the use of an apparatus called a *tokamak*, which uses strong magnetic fields to contain and to heat the reaction. Temperatures of over 100,000,000 K have been achieved in a tokamak. Unfortunately, scientists have not yet been able to generate more power than is consumed over a sustained period of time.

---

*Historically, a nuclear weapon that relies solely on a fission process to release energy is called an atomic bomb, whereas one that also releases energy via a fusion reaction is called a hydrogen bomb.

A CLOSER LOOK　Nuclear Synthesis of the Elements

The lightest elements—hydrogen and helium along with very small amounts of lithium and beryllium—were formed as the universe expanded in the moments following the Big Bang. All the heavier elements owe their existence to subsequent nuclear reactions that occur in stars. These heavier elements are not all created in equal amounts, however. In our solar system, for example, carbon and oxygen are a million times more abundant than lithium and boron, and over 100 million times more abundant than beryllium (Figure 21.21)! In fact, of the elements heavier than helium, carbon and oxygen are the most abundant. This is more than an academic curiosity given the fact that these elements, together with hydrogen, are the most important elements for life on Earth. Let's look at the factors responsible for the relatively high abundance of carbon and oxygen in the universe.

A star is born from a cloud of gas and dust called a *nebula*. When conditions are right, gravitational forces collapse the cloud, and its core density and temperature rise until nuclear fusion commences. Hydrogen nuclei fuse to form deuterium, ^2_1H, and eventually ^4_2He through the reactions shown in Equations 21.26 through 21.29. Because ^4_2He has a larger binding energy than any of its immediate neighbors (Figure 21.12), these reactions release an enormous amount of energy. This process, called *hydrogen burning*, is the dominant process for most of a star's lifetime.

Once a star's supply of hydrogen is nearly exhausted, several important changes occur as the star enters the next phase of its life and is transformed into a *red giant*. The decrease in nuclear fusion causes the core to contract, triggering an increase in core temperature and pressure. At the same time, the outer regions expand and

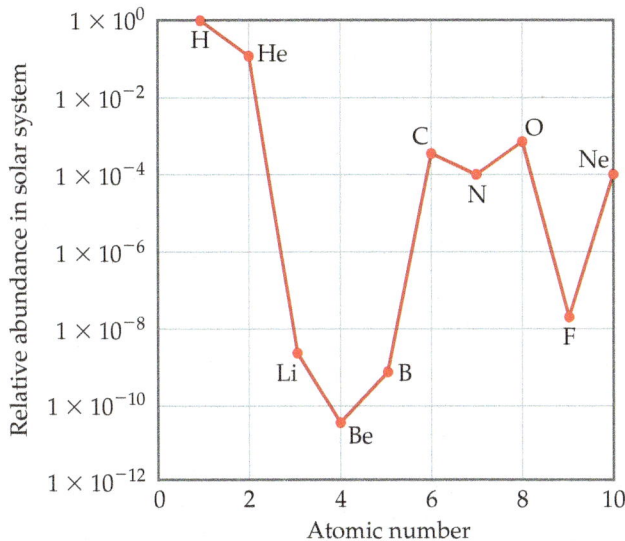

▲ **Figure 21.21** **Relative abundance of elements 1–10 in the solar system.** Note the logarithmic scale used for the *y* axis.

can be produced from carbon through a series of reactions involving proton capture and positron emission.

Most stars gradually cool and dim as their helium is converted to carbon and oxygen, ending their lives as *white dwarfs*, a phase in which stars become incredibly dense—generally about one million times denser than the Sun. The extreme density of white dwarfs is accompanied by much higher temperatures and pressures at the core, where a variety of fusion processes lead to synthesis of the elements from neon to sulfur. These fusion reactions are collectively called *advanced burning*.

Eventually, progressively heavier elements form at the core until it becomes predominantly ^{56}Fe, as shown in **Figure 21.22**. Because this is such a stable nucleus, further fusion to heavier nuclei consumes energy rather than releasing it. When this happens, the fusion reactions that power the star diminish, and immense gravitational forces lead to a dramatic collapse called a supernova *explosion*. Neutron capture coupled with subsequent radioactive decay in the dying moments of such a star is responsible for the presence of all elements heavier than iron and nickel.

Without these dramatic supernova events in the past history of the universe, heavier elements that are so familiar to us, such as silver, gold, iodine, lead, and uranium, would not exist.

Related Exercises: 21.73, 21.75

cool enough to make the star emit red light (thus, the name *red giant*). The star now must use $^4_2$He nuclei as its fuel. The simplest reaction that can occur in the He-rich core, fusion of two alpha particles to form a $^8_4$Be nucleus, does occur. The binding energy per nucleon for $^8_4$Be is very slightly smaller than that for $^4_2$He, so this fusion process is very slightly endothermic. The $^8_4$Be nucleus is highly unstable (half-life of 7×10^{-17} s)) and so falls apart almost immediately. In a tiny fraction of cases, however, a third $^4_2$He collides with a $^8_4$Be nucleus before it decays, forming carbon-12:

$$^4_2\text{He} + \,^4_2\text{He} \longrightarrow \,^8_4\text{Be}$$

$$^8_4\text{Be} + \,^4_2\text{He} \longrightarrow \,^{12}_6\text{C}$$

Some of the $^{12}_6$C nuclei go on to react with alpha particles to form oxygen-16:

$$^{12}_6\text{C} + \,^4_2\text{He} \longrightarrow \,^{16}_8\text{O}$$

This stage of nuclear fusion is called *helium burning*. Notice that carbon, element 6, is formed without prior formation of elements 3, 4, and 5, explaining in part their unusually low abundance. Nitrogen is relatively abundant because it

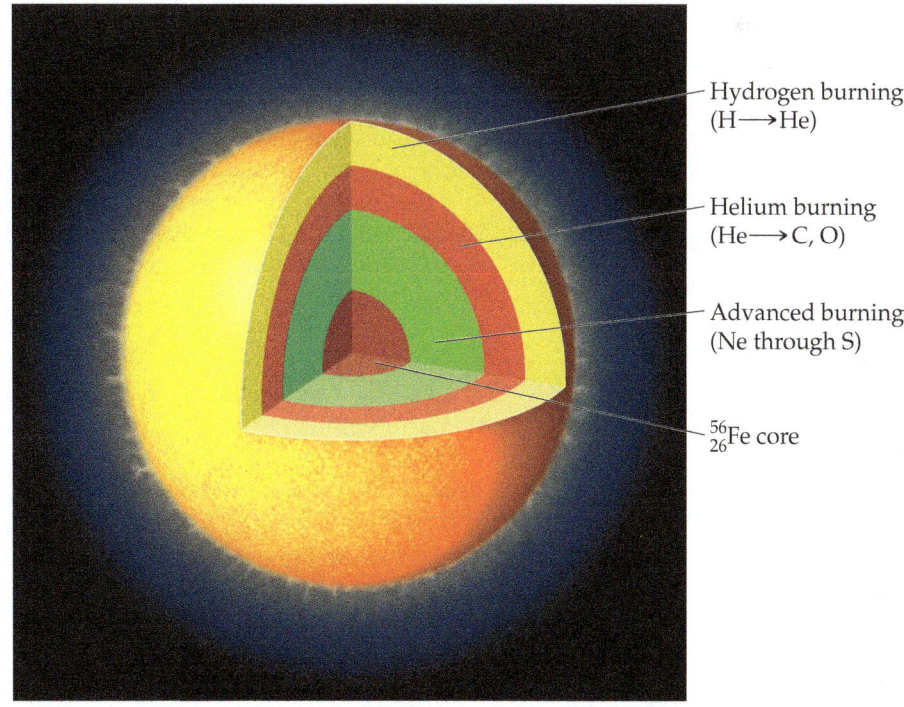

▲ **Figure 21.22** **Fusion processes that occur in a red giant just prior to a supernova explosion.**

— Hydrogen burning (H⟶He)

— Helium burning (He⟶C, O)

— Advanced burning (Ne through S)

— $^{56}_{26}$Fe core

△ **Self-Assessment Exercise**

SAE 21.19 Identify the missing product in the following nuclear fusion reaction:

$^8_4$Be + $^4_2$He ⟶ ____

(a) $^8_4$Be **(b)** $^{12}_6$C **(c)** $^{16}_8$O **(d)** $^{14}_6$C

21.9 | Radiation in the Environment and Living Systems

⚠ **Learning Objectives**

When you finish Section 21.9, you should be able to:

▶ Compare the damaging and therapeutic effects of radiation.

▶ Calculate a dose of radiation.

We are continuously bombarded by radiation from both natural and artificial sources. We are exposed to infrared, ultraviolet, and visible radiation from the Sun; radio waves from radio and television stations; microwaves from microwave ovens; X rays from medical procedures; and radioactivity from natural materials (**Table 21.8**). Understanding the different energies of these various kinds of radiation is necessary in order to understand their different effects on matter.

When matter absorbs radiation, the radiation energy can cause atoms in the matter to be either excited or ionized. In general, radiation that causes ionization, called **ionizing radiation**, is far more harmful to biological systems than radiation that does not cause ionization. The latter, called **nonionizing radiation**, is generally of lower energy, such as radiofrequency electromagnetic radiation (Section 6.1) or slow-moving neutrons.

Most living tissue contains at least 70% water by mass. When living tissue is irradiated, water molecules absorb most of the energy of the radiation. Thus, it is common to define ionizing radiation as radiation that can ionize water, a process requiring a minimum energy of 1216 kJ/mol. Alpha, beta, and gamma rays (as well as X rays and higher-energy ultraviolet radiation) possess energies in excess of this quantity and are therefore forms of ionizing radiation.

When ionizing radiation passes through living tissue, electrons are removed from water molecules, forming highly reactive H_2O^+ ions. An H_2O^+ ion can react with another water molecule to form an H_3O^+ ion and a neutral OH molecule:

$$H_2O^+ + H_2O \longrightarrow H_3O^+ + OH \qquad [21.31]$$

The unstable and highly reactive OH molecule is a **free radical**, a substance with one or more unpaired electrons, as seen in the Lewis structure shown in the margin.

The OH molecule is also called the *hydroxyl radical*, and the presence of the unpaired electron is often emphasized by writing the species with a single dot, ·OH. In cells and tissues, hydroxyl radicals can attack biomolecules to produce new free radicals, which in turn attack yet other biomolecules. Thus, the formation of a single hydroxyl radical via Equation 21.31 can initiate a large number of chemical reactions that are ultimately able to disrupt the normal operations of cells.

The damage produced by radiation depends on the activity and energy of the radiation, the length of exposure, and whether the source is inside or outside the body. Gamma rays are particularly harmful outside the body because they penetrate human tissue very effectively, just as X rays do. Consequently, their damage is not limited to the skin. In contrast, most alpha rays are stopped by skin, and beta rays are able to penetrate only about 1 cm beyond the skin surface (**Figure 21.23**). Therefore, neither alpha rays nor beta rays are as dangerous as gamma rays, *unless* the radiation source somehow enters the body. Within the body, alpha rays are particularly dangerous because they transfer their energy efficiently to the surrounding tissue, causing considerable damage.

TABLE 21.8   Average Abundances and Activities of Natural Radionuclides[†]

	Potassium-40	Rubidium-87	Thorium-232	Uranium-238
Land elemental abundance (ppm)	28,000	112	10.7	2.8
Land activity (Bq/kg)	870	102	43	35
Ocean elemental concentration (mg/L)	339	0.12	1×10^{-7}	0.0032
Ocean activity (Bq/L)	12	0.11	4×10^{-7}	0.040
Ocean sediments elemental abundance (ppm)	17,000	—	5.0	1.0
Ocean sediments activity (Bq/kg)	500	—	20	12
Human body activity (Bq)	4000	600	0.08	0.4[‡]

[†]Data from "Ionizing Radiation Exposure of the Population of the United States," Report 93, 1987, and Report 160, 2009, National Council on Radiation Protection.

[‡]Includes lead-210 and polonium-210, daughter nuclei of uranium-238.

In general, the tissues damaged most by radiation are those that reproduce rapidly, such as bone marrow, blood-forming tissues, and lymph nodes. The principal effect of extended exposure to low doses of radiation is to cause cancer. Cancer is caused by damage to the growth-regulation mechanism of cells, inducing the cells to reproduce uncontrollably. Leukemia, which is characterized by excessive growth of white blood cells, is probably the major type of radiation-caused cancer.

In light of the biological effects of radiation, it is important to determine whether any levels of exposure are safe. Unfortunately, we are hampered in our attempts to set realistic standards because we do not fully understand the effects of long-term exposure. Scientists concerned with setting health standards have used the hypothesis that the effects of radiation are proportional to exposure. *Any* amount of radiation is assumed to cause some finite risk of injury, and the effects of high-dosage rates are extrapolated to those of lower ones. Other scientists believe, however, that there is a threshold below which there are no radiation risks. Until scientific evidence enables us to settle the matter with some confidence, it is safer to assume that even low levels of radiation present some danger.

Radiation Doses

Two units are commonly used to measure exposure to radiation. The **gray** (Gy), the SI unit of absorbed dose, corresponds to the absorption of 1 J of energy per kilogram of tissue. The **rad** (radiation *a*bsorbed *d*ose) corresponds to the absorption of 1×10^{-2} J of energy per kilogram of tissue. Thus, 1 Gy = 100 rad. The rad is the unit most often used in medicine.

Not all forms of radiation harm biological materials to the same extent even at the same level of exposure. For example, 1 rad of alpha radiation can produce more damage than 1 rad of beta radiation. To correct for these differences, the radiation dose is multiplied by a factor that measures the relative damage caused by the radiation. This multiplication factor is known as the *relative biological effectiveness*, *RBE*. The RBE is approximately 1 for gamma and beta radiation, and 10 for alpha radiation.

The exact value of the RBE varies with dose rate, total dose, and type of tissue affected. The product of the radiation dose in rads and the RBE of the radiation give the *effective dosage* in **rem** (*r*oentgen *e*quivalent for *m*an):

$$\text{Number of rem} = (\text{number of rad})(\text{RBE}) \qquad [21.32]$$

The SI unit for effective dose is the *sievert* (Sv), obtained by multiplying the RBE times the SI unit for radiation dose, the gray; because a gray is 100 times larger than a rad, 1 Sv = 100 rem. The rem is the unit of radiation damage usually used in medicine.

The effects of short-term exposure to radiation appear in **Table 21.9**. An exposure of 600 rem is fatal to most humans. To put this number in perspective, a typical dental X ray entails an exposure of about 0.5 mrem (that is, 0.0005 rem). The average exposure for a person in 1 yr due to all natural sources of ionizing radiation (called *background radiation*) is about 360 mrem (**Figure 21.24**).

Radon

Radon-222 is a product of the nuclear disintegration series of uranium-238 (Figure 21.2) and is continuously generated as uranium in rocks and soil decays. As Figure 21.24 indicates, radon exposure is estimated to account for more than half the 360-mrem average annual exposure to ionizing radiation.

Go Figure

Why are alpha rays much more dangerous when the source of radiation is located inside the body?

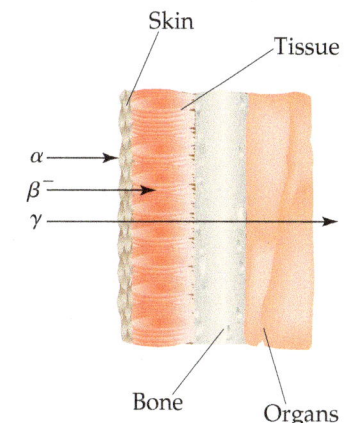

▲ **Figure 21.23 Relative penetrating abilities of alpha, beta, and gamma radiation.**

TABLE 21.9 Effects of Short-Term Exposures to Radiation

Dose (rem)	Effect
0–25	No detectable clinical effects
25–50	Slight, temporary decrease in white blood cell counts
100–200	Nausea; marked decrease in white blood cell counts
500	Death of half the exposed population within 30 days

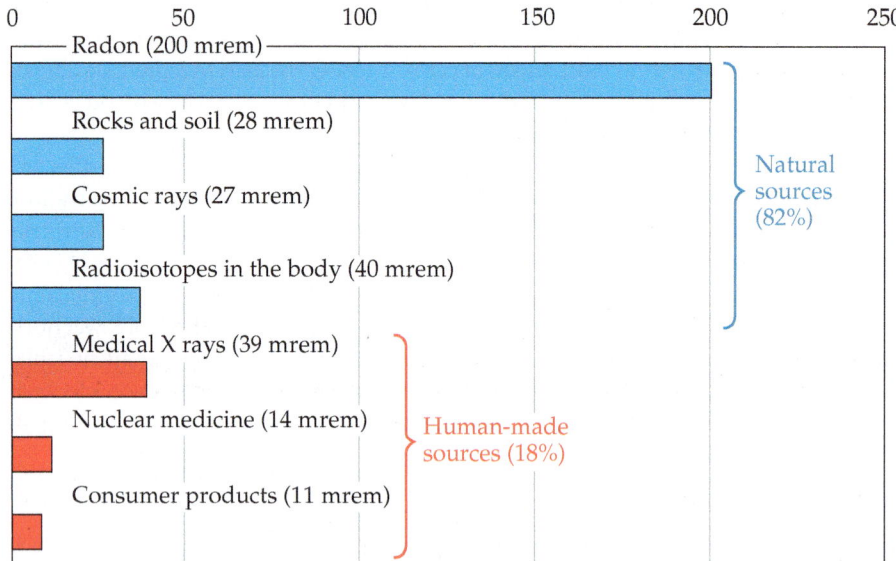

Average annual exposure (mrem)

Radon (200 mrem)
Rocks and soil (28 mrem)
Cosmic rays (27 mrem)
Radioisotopes in the body (40 mrem)
Medical X rays (39 mrem)
Nuclear medicine (14 mrem)
Consumer products (11 mrem)

Natural sources (82%)

Human-made sources (18%)

▲ **Figure 21.24 Sources of U.S. average annual exposure to high-energy radiation.** The total average annual exposure is 360 mrem.
Data from "Ionizing Radiation Exposure of the Population of the United States," Report 93, 1987 and Report 160, 2009, National Council on Radiation Protection.

The interplay between the chemical and nuclear properties of radon makes it a health hazard. Because radon is a noble gas, it is extremely unreactive and is therefore free to escape from the ground without chemically reacting along the way. It is readily inhaled and exhaled with no direct chemical effects. Its half-life, however, is only 3.82 days. It decays, by losing an alpha particle, into a radioisotope of polonium:

$$^{222}_{86}\text{Rn} \longrightarrow {}^{218}_{84}\text{Po} + {}^{4}_{2}\text{He} \qquad [21.33]$$

Because radon has such a short half-life and because alpha particles have a high RBE, inhaled radon is considered a probable cause of lung cancer. Even worse than the radon, however, is the decay product, because polonium-218 is an alpha-emitting chemically active element that has an even shorter half-life (3.11 min) than radon-222:

$$^{218}_{84}\text{Po} \longrightarrow {}^{214}_{82}\text{Pb} + {}^{4}_{2}\text{He} \qquad [21.34]$$

When a person inhales radon, therefore, atoms of polonium-218 can become trapped in the lungs, where they bathe the delicate tissue with harmful alpha radiation. The resulting damage is estimated to contribute to 10% of all lung cancer deaths in the United States.

The U.S. Environmental Protection Agency (EPA) has recommended that radon-222 levels not exceed 4 pCi per liter of air in homes. Homes located in areas where the natural uranium content of the soil is high are particularly at high risk for radon levels much greater than that, and those areas have had significantly higher incidences of radon-induced lung cancer. In 2015, the EPA, in partnership with other organizations, launched the National Radon Action Plan, with the goal of reducing the radon risk for 5 million homes, mainly by increased testing and mitigation.

CHEMISTRY AND LIFE Radiation Therapy

Healthy cells are either destroyed or damaged by high-energy radiation, leading to physiological disorders. This radiation can also destroy *unhealthy* cells, however, including cancerous cells. All cancers are characterized by runaway cell growth that can produce *malignant tumors*. These tumors can be caused by the exposure of healthy cells to high-energy radiation. Paradoxically, however, they can be destroyed by the same radiation that caused them because the rapidly reproducing cells of the tumors are very susceptible to radiation damage. Thus, cancerous cells are more susceptible to destruction by radiation than healthy ones, allowing radiation to be used effectively in the treatment of cancer. As early as 1904, physicians used the radiation emitted by radioactive substances to treat tumors by destroying the mass of unhealthy tissue. The treatment of disease by high-energy radiation is called *radiation therapy*.

TABLE 21.10 Some Radioisotopes Used in Radiation Therapy

Isotope	Half-Life	Isotope	Half-Life
^{32}P	14.3 days	^{137}Cs	30 yr
^{60}Co	5.27 yr	^{192}Ir	74.2 days
^{90}Sr	28.8 yr	^{198}Au	2.7 days
^{125}I	60.25 days	^{222}Rn	3.82 days
^{131}I	8.04 days	^{226}Ra	1600 yr

Many radionuclides are currently used in radiation therapy. Some of the more commonly used ones are listed in Table 21.10. Most of them have short half-lives, meaning that they emit a great deal of radiation in a short period of time (Figure 21.25).

The radiation source used in radiation therapy may be inside or outside the body. In almost all cases, radiation therapy uses gamma radiation emitted by radioisotopes. Any alpha or beta radiation that is emitted concurrently can be blocked by appropriate packaging. For example, ^{192}Ir is often administered as "seeds" consisting of a core of radioactive isotope coated with 0.1 mm of platinum metal. The platinum coating stops the alpha and beta rays, but the gamma rays penetrate it readily. The radioactive seeds can be surgically implanted in a tumor.

In some cases, human physiology allows a radioisotope to be ingested. For example, most of the iodine in the human body ends up in the thyroid gland, so thyroid cancer can be treated by using large doses of ^{131}I. Radiation therapy on deep organs, where a surgical implant is impractical, often uses a ^{60}Co "gun" outside the body to shoot a beam of gamma rays at the tumor. Particle accelerators are also used as an external source of high-energy radiation for radiation therapy.

Because gamma radiation is so strongly penetrating, it is nearly impossible to avoid damaging healthy cells during radiation therapy. Many cancer patients undergoing radiation treatment experience unpleasant and dangerous side effects, such as fatigue, nausea, hair loss, a weakened immune system, and occasionally

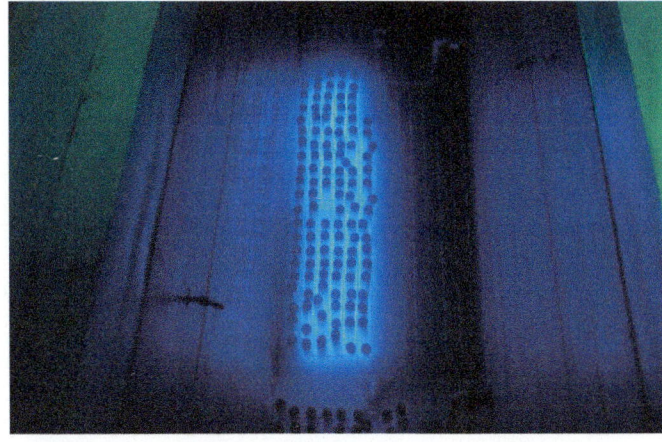
▲ **Figure 21.25 Storage of radioactive cesium.** The vials contain a salt of cesium-137, a beta emitter that is both a waste product of nuclear fission and used in radiation therapy. The blue glow is from the radioactivity of the cesium.

even death. However, if other treatments such as *chemotherapy* (the use of drugs to combat cancer) fail, radiation therapy can be a good option.

Much current research in radiation therapy is engaged in developing new drugs that specifically target tumors using a method called *neutron capture therapy*. In this technique, a nonradioactive isotope, usually boron-10, is concentrated in the tumor by using specific tumor-seeking reagents. The boron-10 is then irradiated with neutrons, where it undergoes the following nuclear reaction, producing alpha particles:

$$^{10}_{5}B + ^{1}_{0}n \longrightarrow ^{7}_{3}Li + ^{4}_{2}He$$

Tumor cells are killed or damaged by exposure to the alpha particles. Healthy tissue farther away from the tumor is unaffected because of the short-range penetrating power of alpha particles. Thus, neutron-capture therapy holds promise as a "silver bullet" that specifically targets unhealthy cells for exposure to radiation.

Related Exercises: 21.37, 21.55, 21.56

 ## Self-Assessment Exercises

SAE 21.20 Why is ionizing radiation more harmful to biological systems than nonionizing radiation? (**a**) Ionizing radiation tends not to have enough energy to have an effect on biological systems. (**b**) Ionizing radiation is generally lower in energy and does not ionize water. (**c**) Ionizing radiation removes electrons from water molecules and forms damaging free radicals. (**d**) Ionizing radiation does not cause tissue damage or cancer.

SAE 21.21 If a 75-kg man is irradiated uniformly by 0.15-J alpha radiation, what is the effective dosage in rem? (**a**) 0.00020 rem (**b**) 0.0020 rem (**c**) 0.20 rem (**d**) 2.0 rem

 ## Putting Concepts Together

Potassium ion is present in foods and is an essential nutrient in the human body. One of the naturally occurring isotopes of potassium, potassium-40, is radioactive. Potassium-40 has a natural abundance of 0.0117% and a half-life of $t_{1/2} = 1.28 \times 10^9$ yr. It undergoes radioactive decay in three ways: 98.2% is by electron capture, 1.35% is by beta emission, and 0.49% is by positron emission. (**a**) Why should we expect ^{40}K to be radioactive? (**b**) Write the nuclear equations for the three modes by which ^{40}K decays. (**c**) How many $^{40}K^+$ ions are present in 1.00 g of KCl? (**d**) How long does it take for 1.00% of the ^{40}K in a sample to undergo radioactive decay?

Continued

SOLUTION

(a) The ^{40}K nucleus contains 19 protons and 21 neutrons. There are very few stable nuclei with odd numbers of both protons and neutrons (Section 21.2).

(b) Electron capture is capture of an inner-shell electron by the nucleus:

$$^{40}_{19}\text{K} + {}^{0}_{-1}\text{e} \longrightarrow {}^{40}_{18}\text{Ar}$$

Beta emission is loss of a beta particle $({}^{0}_{-1}\text{e})$ by the nucleus:

$$^{40}_{19}\text{K} \longrightarrow {}^{40}_{20}\text{Ca} + {}^{0}_{-1}\text{e}$$

Positron emission is loss of a positron $({}^{0}_{+1}\text{e})$ by the nucleus:

$$^{40}_{19}\text{K} \longrightarrow {}^{40}_{18}\text{Ar} + {}^{0}_{+1}\text{e}$$

(c) The total number of K$^+$ ions in the sample is:

$$(1.00 \text{ g KCl})\left(\frac{1 \text{ mol KCl}}{74.55 \text{ g KCl}}\right)\left(\frac{1 \text{ mol K}^+}{1 \text{ mol KCl}}\right)\left(\frac{6.022 \times 10^{23} \text{ K}^+}{1 \text{ mol K}^+}\right) = 8.08 \times 10^{21} \text{ K}^+ \text{ ions}$$

Of these, 0.0117% are ^{40}K$^+$ ions:

$$(8.08 \times 10^{21} \text{ K}^+\text{ions})\left(\frac{0.0117 \, ^{40}\text{K}^+ \text{ ions}}{100 \text{ K}^+ \text{ ions}}\right) = 9.45 \times 10^{17} \, ^{40}\text{K}^+ \text{ ions}$$

(d) The decay constant (the rate constant) for the radioactive decay can be calculated from the half-life, using Equation 21.19:

$$k = \frac{0.693}{t_{1/2}} = \frac{0.693}{1.28 \times 10^9 \text{ yr}} = (5.41 \times 10^{-10})/\text{yr}$$

The rate equation, Equation 21.20, then allows us to calculate the time required:

$$\ln \frac{N_t}{N_0} = -kt$$

$$\ln \frac{99}{100} = -[(5.41 \times 10^{-10})/\text{yr}]t$$

$$-0.01005 = -[(5.41 \times 10^{-10})/\text{yr}]t$$

$$t = \frac{-0.01005}{(-5.41 \times 10^{-10})/\text{yr}} = 1.86 \times 10^7 \text{ yr}$$

That is, it would take 18.6 million years for just 1.00% of the ^{40}K in a sample to decay.

Chapter Summary and Key Terms

RADIOACTIVITY AND NUCLEAR EQUATIONS (SECTION 21.1) The nucleus of an atom contains protons and neutrons, both of which are called **nucleons**. Reactions that involve changes in atomic nuclei are called **nuclear reactions**. Nuclei that spontaneously change by emitting radiation are said to be **radioactive**. Radioactive nuclei are called **radionuclides**, and the atoms containing them are called **radioisotopes**. Radionuclides spontaneously change through a process called radioactive decay. The three most important types of radiation given off as a result of radioactive decay are **alpha (α) particles** ($^4_2$He or α), **beta (β) particles** ($^0_{-1}$e or β$^-$), and **gamma (γ) radiation** ($^0_0$γ or γ). **Positrons** ($^0_{+1}$e or β$^+$), which are particles with the same mass as an electron but the opposite charge, can also be produced when a radioisotope decays.

In nuclear equations, reactant and product nuclei are represented by giving their mass numbers and atomic numbers, as well as their chemical symbol. The totals of the mass numbers on both sides of the equation are equal; the totals of the atomic numbers on both sides are also equal. There are four common modes of radioactive decay: **alpha emission**, which reduces the atomic number by 2 and the mass number by 4; **beta emission**, which increases the atomic number by 1 and leaves the mass number unchanged; and **positron emission** and **electron capture**, both of which reduce the atomic number by 1 and leave the mass number unchanged.

PATTERNS OF NUCLEAR STABILITY (SECTION 21.2) The neutron-to-proton ratio is an important factor determining nuclear stability. By comparing a nuclide's neutron-to-proton ratio with those of stable nuclei, we can predict the mode of radioactive decay. In general, neutron-rich nuclei tend to emit beta particles; proton-rich nuclei tend to either emit positrons or undergo electron capture; and heavy nuclei tend to emit alpha particles. The presence of **magic numbers** of nucleons and an even number of protons and neutrons also help determine the stability of a nucleus. A nuclide may undergo a series of decay steps before a stable nuclide forms. This series of steps is called a **radioactive decay chain** or a **nuclear disintegration series**.

NUCLEAR TRANSMUTATIONS (SECTION 21.3) **Nuclear transmutations**, induced conversions of one nucleus into another, can be brought about by bombarding nuclei with either charged particles or neutrons. **Particle accelerators** increase the kinetic energies of positively charged particles, allowing these particles to overcome their electrostatic repulsion by the nucleus. Nuclear transmutations are used to produce the **transuranium elements**—those elements with atomic numbers greater than that of uranium.

RADIOACTIVE DECAY RATES AND DETECTION OF RADIOACTIVITY (SECTIONS 21.4 AND 21.5) The SI unit for the activity of a radioactive source is the **becquerel** (Bq), defined as one nuclear disintegration per second. A related unit, the **curie** (Ci), corresponds to 3.7×10^{10} disintegrations per second. Nuclear decay is a first-order process. The decay rate (**activity**) is therefore directly proportional to the number of radioactive nuclei. The **half-life** of a radionuclide, which is a constant independent of temperature, is the time needed for one-half of the nuclei to decay. Some radioisotopes can be used to date objects; ^{14}C, for example, is used to date organic objects. Geiger counters and scintillation counters count the emissions from radioactive samples. The ease of detection of radioisotopes also permits their use as **radiotracers** to follow elements through reactions.

ENERGY CHANGES IN NUCLEAR REACTIONS (SECTION 21.6) The energy produced in nuclear reactions is accompanied by measurable changes of mass in accordance with Einstein's relationship, $\Delta E = c^2 \Delta m$. The difference in mass between nuclei and the nucleons of which they are composed is known as the **mass defect**. The mass defect of a nuclide makes it possible to calculate its **nuclear binding energy**, the energy required to separate the nucleus into individual nucleons. Because of trends in the nuclear binding energy with atomic number, energy is produced when heavy nuclei split (**fission**) and when light nuclei fuse (**fusion**).

NUCLEAR FISSION AND NUCLEAR FUSION (SECTIONS 21.7 AND 21.8) Uranium-235, uranium-233, and plutonium-239 undergo fission when they capture a neutron, splitting into lighter nuclei and releasing more neutrons. The neutrons produced in one fission can cause further fission reactions, which can lead to a nuclear **chain reaction**. A reaction that maintains a constant rate is said to be critical, and the mass necessary to maintain this constant rate is called a **critical mass**. A mass in excess of the critical mass is termed a **supercritical mass**.

In nuclear reactors, the fission rate is controlled to generate a constant power. The reactor core consists of fuel elements containing fissionable nuclei, control rods, a moderator, and a primary coolant. A nuclear power plant resembles a conventional power plant except that the reactor core replaces the fuel burner. There is concern about the disposal of highly radioactive nuclear wastes that are generated in nuclear power plants.

Nuclear fusion requires high temperatures because nuclei must have large kinetic energies to overcome their mutual repulsions. Fusion reactions are therefore called **thermonuclear reactions**. It is not yet possible to generate power on Earth through a controlled fusion process.

RADIATION IN THE ENVIRONMENT AND LIVING SYSTEMS (SECTION 21.9) **Ionizing radiation** is energetic enough to remove an electron from a water molecule; radiation with less energy is called **nonionizing radiation**. Ionizing radiation generates **free radicals**, reactive substances with one or more unpaired electrons. The effects of long-term exposure to low levels of radiation are not completely understood, but there is evidence that the extent of biological damage varies in direct proportion to the level of exposure.

The amount of energy deposited in biological tissue by radiation is called the radiation dose and is measured in units of gray or rad. One **gray** (Gy) corresponds to a dose of 1 J/kg of tissue. It is the SI unit of radiation dose. The **rad** is a smaller unit; 100 rad = 1 Gy. The effective dose, which measures the biological damage created by the deposited energy, is measured in units of rem or sievert (Sv). The **rem** is obtained by multiplying the number of rad by the relative biological effectiveness (RBE); 100 rem = 1 Sv.

Key Equations

- $k = \dfrac{0.693}{t_{1/2}}$ [21.19]   Relationship between nuclear decay constant and half-life; this is derived from the following equation at $N_t = \frac{1}{2}N_0$

- $\ln\dfrac{N_t}{N_0} = -kt$ [21.20]   First-order rate law for nuclear decay

- $E = mc^2$ [21.22]   Einstein's equation that relates mass and energy

Exam Prep

EP 21.1 What product forms when plutonium-238 undergoes alpha emission? (**a**) plutonium-234 (**b**) uranium-234 (**c**) uranium-238 (**d**) thorium-236 (**e**) neptunium-237

EP 21.2 Which of the following statements is or are *true*?

  (**i**)   Gamma radiation is more penetrating than alpha radiation.

 (**ii**)   Alpha radiation consists of a stream of helium-4 nuclei.

(**iii**) When a nuclide undergoes beta emission, both the atomic number and the mass number are changed.

(**a**) Only statement i is true. (**b**) Statements i and ii are true. (**c**) Statements i and iii are true. (**d**) Statements ii and iii are true. (**e**) All three statements are true.

EP 21.3 The radioactive decay of thorium-232 occurs in multiple steps, called a *radioactive decay chain*. The second product produced in this chain is actinium-228. Which of the following processes could lead to this product starting with thorium-232?

(**a**) alpha decay followed by beta emission

(**b**) beta emission followed by electron capture

(**c**) positron emission followed by alpha decay

(**d**) electron capture followed by positron emission

(**e**) More than one of the above is consistent with the observed transformation.

EP 21.4 Iron-55 undergoes radioactive decay by electron capture. What is the product of this decay process?
(**a**) chromium-51   (**c**) iron-56
(**b**) cobalt-55   (**d**) manganese-55

EP 21.5 Which of the following nuclides would you expect to be radioactive?

(**a**) $^{90}_{40}\text{Zr}$ (**c**) $^{140}_{58}\text{Ce}$

(**b**) $^{134}_{55}\text{Cs}$ (**d**) $^{208}_{82}\text{Pb}$

EP 21.6 Which of the following radioactive nuclei is most likely to decay via emission of a β^- particle?

(**a**) nitrogen-13   (**d**) iodine-131

(**b**) magnesium-23   (**e**) neptunium-237

(**c**) rubidium-83

EP 21.7 What is the identity of nucleus X in the following nuclear transmutation? $^{238}_{92}\text{U}(n, \beta^-)\text{X}$.

(**a**) $^{238}_{93}\text{Np}$ (**d**) $^{235}_{90}\text{Th}$

(**b**) $^{239}_{92}\text{U}$ (**e**) $^{239}_{93}\text{Np}$

(**c**) $^{239}_{92}\text{U}^+$

EP 21.8 A radioisotope of technetium is useful in medical imaging techniques. A sample initially contains 80.0 mg of this isotope. After 24.0 h, only 5.0 mg of the technetium isotope remains. What is the half-life of the isotope?

(a) 3.0 h

(b) 6.0 h

(c) 12.0 h

(d) 16.0 h

(e) 24.0 h

EP 21.9 Cesium-137, which has a half-life of 30.2 yr, is a component of the radioactive waste from nuclear power plants. If the activity due to cesium-137 in a sample of radioactive waste has decreased to 35.2% of its initial value, how old is the sample?

(a) 1.04 yr

(b) 15.4 yr

(c) 31.5 yr

(d) 45.5 yr

(e) 156 yr

EP 21.10 A team of archeologists recover a plank of wood believed to be from an ancient ship. The plank has a carbon-14 activity of 35.8 Bq per gram of carbon. In comparison, a living organism has a C-14 activity of 47.5 Bq per gram of carbon. If the half-life of carbon-14 is 5730 years, what is the age of the plank?

(a) 1020 yr

(b) 1620 yr

(c) 2340 yr

(d) 11,000 yr

EP 21.11 Electricity is generated in space vehicles from the heat produced by the radioactive decay of plutonium-238: $^{238}_{94}Pu \longrightarrow ^{234}_{92}U + ^4_2He$. The atomic masses of plutonium-238 and uranium-234 are 238.049554 amu and 234.040946 amu, respectively. The mass of an alpha particle is 4.001506 amu. How much energy in kJ is released when 1.00 g of plutonium-238 decays to uranium-234?

(a) 2.27×10^6 kJ

(b) 2.68×10^6 kJ

(c) 3.10×10^6 kJ

(d) 3.15×10^6 kJ

(e) 7.37×10^8 kJ

EP 21.12 The mass of a magnesium-25 nucleus is 24.98584 amu. What is the binding energy of this nucleus? The mass of a neutron is 1.008664916 amu, the mass of a proton is 1.007276466 amu, the speed of light is 2.9979×10^8 m/s, and 1 amu $= 1.660538921 \times 10^{-27}$ kg.

(a) $3.72887048 \times 10^{-9}$ J

(b) $6.24499767 \times 10^{-9}$ J

(c) $3.19553534 \times 10^{-11}$ J

(d) $2.92616053 \times 10^{-11}$ J

EP 21.13 Identify the missing product in the following nuclear fission reaction:

$$^{235}_{92}U + ^1_0n \longrightarrow ^{90}_{38}Sr + \underline{\quad} + 2\,^1_0n$$

(a) $^{144}_{54}Xe$

(b) $^{142}_{54}Xe$

(c) $^{145}_{54}Xe$

(d) $^{144}_{52}Te$

EP 21.14 Which of the following statements about nuclear fission and fusion is or are *true*?

  (i) Nuclear fission is a favorable process for very heavy nuclei.

 (ii) Nuclear fusion occurs between very light nuclei.

(iii) In both nuclear fission and nuclear fusion, the energy released is a consequence of the variation of nuclear binding energies among the elements.

(a) Only one of the statements is true. (b) Statements i and ii are true. (c) Statements i and iii are true. (d) Statements ii and iii are true. (e) All three statements are true.

Exercises

Visualizing Concepts

21.1 Indicate whether each of the following nuclides lies within the belt of stability in Figure 21.2: (a) neon-24, (b) chlorine-32, (c) tin-108, (d) polonium-216. For any that do not, describe a nuclear decay process that would alter the neutron-to-proton ratio in the direction of increased stability. [Section 21.2]

21.2 Write the balanced nuclear equation for the reaction represented by the diagram shown here. [Section 21.2]

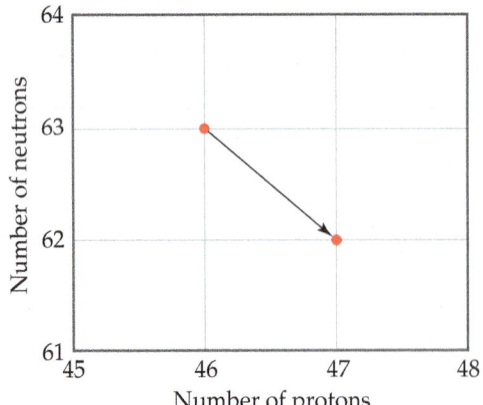

21.3 Draw a diagram similar to that shown in Exercise 21.2 that illustrates the nuclear reaction $^{211}_{83}Bi \longrightarrow ^4_2He + ^{207}_{81}Tl$. [Section 21.2]

21.4 In the sketch shown here, the red spheres represent protons and the gray spheres represent neutrons. (a) What are the identities of the four particles involved in the reaction

depicted? (b) Write the transformation represented below using condensed notation. (c) Based on its atomic number and mass number, do you think the product nucleus is stable or radioactive? [Section 21.3]

21.5 The diagram shows three of the steps in the radioactive decay chain for $^{232}_{90}Th$. The half-life of each isotope is shown below the symbol of the isotope. (a) Identify the type of radioactive decay for each of the steps (i), (ii), and (iii). (b) Which of the isotopes shown has the highest activity? (c) Which of the isotopes shown has the lowest activity? (d) The next step in the decay chain is an alpha emission. What is the next isotope in the chain? [Sections 21.2 and 21.4]

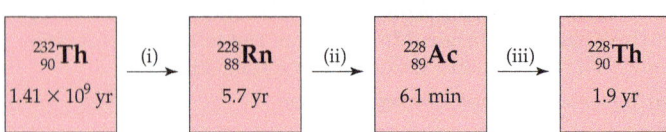

21.6 The accompanying graph illustrates the decay of $^{88}_{42}Mo$, which decays via positron emission. (a) What is the half-life of the decay? (b) What is the rate constant for the decay? (c) What fraction of the original sample of $^{88}_{42}Mo$ remains after 12 min? (**d**) What is the product of the decay process? [Section 21.4]

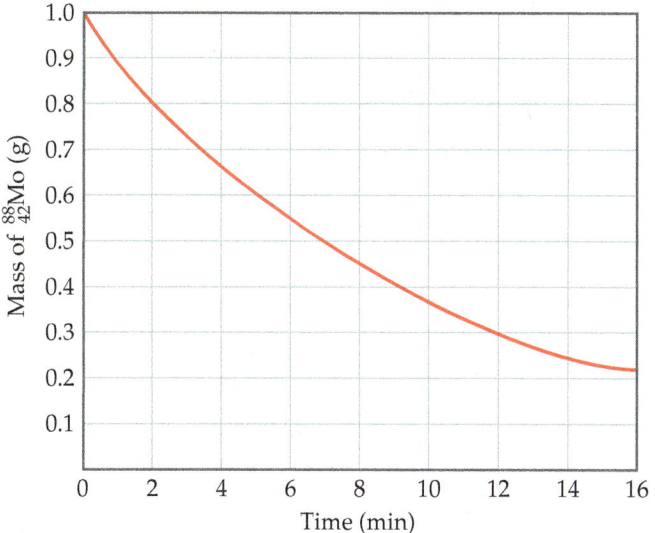

21.7 All the stable isotopes of boron, carbon, nitrogen, oxygen, and fluorine are shown in the accompanying chart (in red), along with their radioactive isotopes with $t_{1/2} > 1$ min (in blue). (**a**) Write the chemical symbols, including mass and atomic numbers, for all of the stable isotopes. (**b**) Which radioactive isotopes are most likely to decay by beta emission? (**c**) Some of the isotopes shown are used in positron emission tomography. Which ones would you expect to be most useful for this application? (**d**) Which isotope would decay to 12.5% of its original concentration after 1 hour? [Sections 21.2, 21.4, and 21.5]

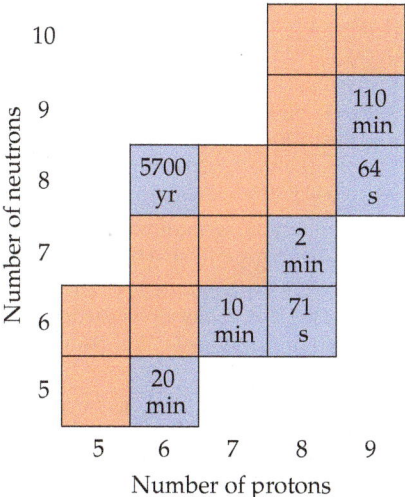

21.8 The diagram shown here illustrates a fission process. (**a**) What is the unidentified product of the fission? (**b**) Use Figure 21.2 to predict whether the nuclear products of this fission reaction are stable. [Section 21.7]

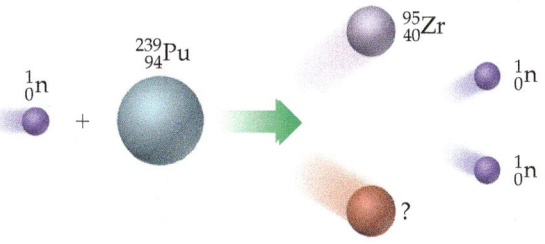

Radioactivity and Nuclear Equations (Section 21.1)

21.9 Indicate the number of protons and neutrons in the following nuclei: (**a**) $^{56}_{24}Cr$, (**b**) ^{193}Tl, (**c**) argon-38.

21.10 Indicate the number of protons and neutrons in the following nuclei: (**a**) $^{129}_{53}I$, (**b**) ^{138}Ba, (**c**) neptunium-237.

21.11 Give the symbol for (**a**) a neutron, (**b**) an alpha particle, (**c**) gamma radiation.

21.12 Give the symbol for (**a**) a proton, (**b**) a beta particle, (**c**) a positron.

21.13 Write balanced nuclear equations for the following processes: (**a**) rubidium-90 undergoes beta emission; (**b**) selenium-72 undergoes electron capture; (**c**) krypton-76 undergoes positron emission; (**d**) radium-226 emits alpha radiation.

21.14 Write balanced nuclear equations for the following transformations: (**a**) bismuth-213 undergoes alpha decay; (**b**) nitrogen-13 undergoes electron capture; (**c**) technicium-98 undergoes electron capture; (**d**) gold-188 decays by positron emission.

21.15 Decay of which nucleus will lead to the following products? (**a**) bismuth-211 by beta decay (**b**) chromium-50 by positron emission (**c**) tantalum-179 by electron capture (**d**) radium-226 by alpha decay

21.16 What particle is produced during the following decay processes? (**a**) Sodium-24 decays to magnesium-24. (**b**) Mercury-188 decays to gold-188. (**c**) Iodine-122 decays to xenon-122. (**d**) Plutonium-242 decays to uranium-238.

21.17 The naturally occurring radioactive decay series that begins with $^{235}_{92}U$ stops with formation of the stable $^{207}_{82}Pb$ nucleus. The decays proceed through a series of alpha-particle and beta-particle emissions. How many of each type of emission are involved in this series?

21.18 A radioactive decay series that begins with $^{232}_{90}Th$ ends with formation of the stable nuclide $^{208}_{82}Pb$. How many alpha-particle emissions and how many beta-particle emissions are involved in the sequence of radioactive decays?

Patterns of Nuclear Stability (Section 21.2)

21.19 Predict the type of radioactive decay process for the following radionuclides: (**a**) $^{8}_{5}B$, (**b**) $^{68}_{29}Cu$, (**c**) phosphorus-32, (**d**) chlorine-39.

21.20 Each of the following nuclei undergoes either beta decay or positron emission. Predict the type of emission for each: (**a**) tritium, $^{3}_{1}H$, (**b**) $^{89}_{38}Sr$, (**c**) iodine-120, (**d**) silver-102.

21.21 One of the nuclides in each of the following pairs is radioactive. Predict which is radioactive and which is stable: (**a**) $^{39}_{19}K$ and $^{40}_{19}K$, (**b**) ^{209}Bi and ^{208}Bi, (**c**) nickel-58 and nickel-65.

21.22 One nuclide in each of these pairs is radioactive. Predict which is radioactive and which is stable: (**a**) $^{40}_{20}Ca$ and $^{45}_{20}Ca$, (**b**) ^{12}C and ^{14}C, (**c**) lead-206 and thorium-230. Explain your choice in each case.

21.23 Which of the following nuclides have magic numbers of both protons and neutrons? (**a**) helium-4 (**b**) oxygen-18 (**c**) calcium-40 (**d**) zinc-66 (**e**) lead-208

21.24 Despite the similarities in the chemical reactivity of elements in the lanthanide series, their abundance in Earth's crust varies by two orders of magnitude. This graph shows the relative abundance as a function of atomic number. Which of the following statements best explains the saw-tooth variation across the series?

(**a**) The elements with an odd atomic number lie above the belt of stability.

(**b**) The elements with an odd atomic number lie below the belt of stability.

(c) The elements with an even atomic number have a magic number of protons.

(d) Pairs of protons have a special stability.

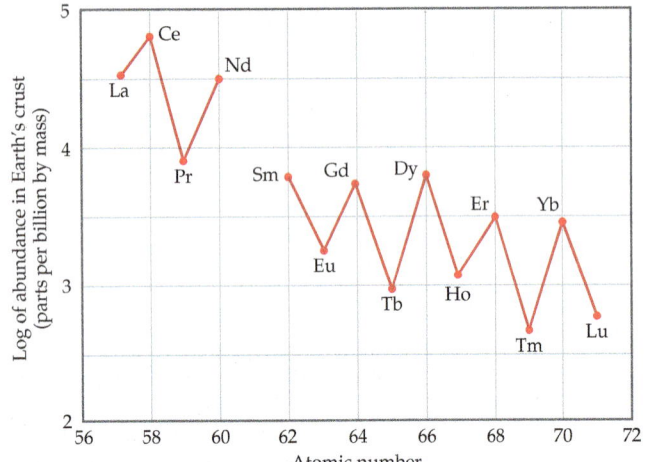

21.25 Which of the following statements best explains why alpha emission is relatively common, but proton emission is extremely rare?

(a) Alpha particles are very stable due to magic numbers of protons and neutrons.

(b) Alpha particles occur in the nucleus.

(c) Alpha particles are the nuclei of an inert gas.

(d) An alpha particle has a higher charge than a proton.

21.26 Which of the following nuclides would you expect to be radioactive: $^{58}_{26}Fe$, $^{60}_{27}Co$, $^{92}_{41}Nb$, mercury-202, radium-226? Justify your choices.

Nuclear Transmutations (Section 21.3)

21.27 Which statement best explains why nuclear transmutations involving neutrons are generally easier to accomplish than those involving protons or alpha particles?

(a) Neutrons are not a magic number particle.

(b) Neutrons do not have an electrical charge.

(c) Neutrons are smaller than protons or alpha particles.

(d) Neutrons are attracted to the nucleus even at long distances, whereas protons and alpha particles are repelled.

21.28 In 1930, the American physicist Ernest Lawrence designed the first cyclotron in Berkeley, California. In 1937 Lawrence bombarded a molybdenum target with deuterium ions, producing for the first time an element not found in nature. What was this element? Starting with molybdenum-96 as your reactant, write a nuclear equation to represent this process.

21.29 Complete and balance the following nuclear equations by supplying the missing particle:

(a) $^{252}_{98}Cf + ^{10}_{5}B \longrightarrow 3\,^{1}_{0}n + ?$

(b) $^{2}_{1}H + ^{3}_{2}He \longrightarrow ^{4}_{2}He + ?$

(c) $^{1}_{1}H + ^{11}_{5}B \longrightarrow 3\,?$

(d) $^{122}_{53}I \longrightarrow ^{122}_{54}Xe + ?$

(e) $^{59}_{26}Fe \longrightarrow ^{0}_{-1}e + ?$

21.30 Complete and balance the following nuclear equations by supplying the missing particle:

(a) $^{14}_{7}N + ^{4}_{2}He \longrightarrow ? + ^{1}_{1}H$

(b) $^{40}_{19}K + ^{0}_{-1}e$ (orbital electron) $\longrightarrow ?$

(c) $? + ^{4}_{2}He \longrightarrow ^{30}_{14}Si + ^{1}_{1}H$

(d) $^{58}_{26}Fe + 2\,^{1}_{0}n \longrightarrow ^{60}_{27}Co + ?$

(e) $^{235}_{92}U + ^{1}_{0}n \longrightarrow ^{135}_{54}Xe + 2\,^{1}_{0}n + ?$

21.31 Write balanced equations for **(a)** $^{238}_{92}U(\alpha, n)^{241}_{94}Pu$, **(b)** $^{14}_{7}N(\alpha, p)^{17}_{8}O$, **(c)** $^{56}_{26}Fe(\alpha, \beta^-)^{60}_{29}Cu$.

21.32 Write balanced equations for each of the following nuclear reactions: **(a)** $^{238}_{92}U(n, \gamma)^{239}_{92}U$, **(b)** $^{16}_{8}O(p, \alpha)^{13}_{7}N$, **(c)** $^{18}_{8}O(n, \beta^-)^{19}_{9}F$.

Rates of Radioactive Decay (Section 21.4)

21.33 Which of the following statements about the half-lives of radionuclides are *true*?

(a) As the half-life increases, the decay rate constant decreases.

(b) After two half-lives, a sample of a radionuclide will have decreased to 1/4 of its original size.

(c) The half-life is a measure of the type of radiation emitted by the radionuclide.

21.34 It has been suggested that strontium-90 (generated by nuclear testing) deposited in the hot desert will undergo radioactive decay more rapidly because it will be exposed to much higher average temperatures. **(a)** Is this a reasonable suggestion? **(b)** Does the process of radioactive decay have an activation energy, like the Arrhenius behavior of many chemical reactions? (Section 14.4)

21.35 Some watch dials are coated with a phosphor, like ZnS, and a polymer in which some of the ^{1}H atoms have been replaced by ^{3}H atoms (tritium). The phosphor emits light when struck by the beta particle from the tritium decay, causing the dials to glow in the dark. The half-life of tritium is 12.3 yr. If the light given off is assumed to be directly proportional to the amount of tritium, by how much will a dial be dimmed in a watch that is 50 yr old?

21.36 It takes 4 h 39 min for a 2.00-mg sample of radium-230 to decay to 0.25 mg. What is the half-life of radium-230?

21.37 Cobalt-60 is a strong gamma emitter that has a half-life of 5.26 yr. The cobalt-60 in a radiotherapy unit must be replaced when its radioactivity falls to 75% of the original sample. If an original sample was purchased in June 2021, when will it be necessary to replace the cobalt-60?

21.38 How much time is required for a 6.25-mg sample of ^{51}Cr to decay to 0.75 mg if it has a half-life of 27.8 days?

21.39 Radium-226, which undergoes alpha decay, has a half-life of 1600 yr. **(a)** How many alpha particles are emitted in 5.0 min by a 10.0-mg sample of ^{226}Ra? **(b)** What is the activity of the sample in mCi?

21.40 Cobalt-60, which undergoes beta decay, has a half-life of 5.26 yr. **(a)** How many beta particles are emitted in 600 s by a 3.75-mg sample of ^{60}Co? **(b)** What is the activity of the sample in Bq?

21.41 The cloth shroud from around a mummy is found to have a ^{14}C activity of 9.7 disintegrations per minute per gram of carbon as compared with living organisms that undergo 16.3 disintegrations per minute per gram of carbon. From the half-life for ^{14}C decay, 5730 yr, calculate the age of the shroud.

21.42 A wooden artifact from a Chinese temple has a ^{14}C activity of 38.0 counts per minute as compared with an activity of 58.2 counts per minute for a standard of zero age. From the half-life for ^{14}C decay, 5730 yr, determine the age of the artifact.

21.43 Potassium-40 decays to argon-40 with a half-life of 1.27×10^9 yr. What is the age of a rock in which the mass ratio of ^{40}Ar to ^{40}K is 4.2?

21.44 The half-life for the process $^{238}\text{U} \longrightarrow {}^{206}\text{Pb}$ is 4.5×10^9 yr. A mineral sample contains 75.0 mg of ^{238}U and 18.0 mg of ^{206}Pb. What is the age of the mineral?

Energy Changes in Nuclear Reactions (Section 21.6)

21.45 An analytical laboratory balance typically measures mass to the nearest 0.1 mg. What energy change would accompany the loss of 0.1 mg in mass?

21.46 The thermite reaction, $\text{Fe}_2\text{O}_3(s) + 2\,\text{Al}(s) \longrightarrow 2\,\text{Fe}(s) + \text{Al}_2\text{O}_3(s)$, $\Delta H° = -851.5$ kJ/mol, is one of the most exothermic reactions known. Because the heat released is sufficient to melt the iron product, the reaction is used to weld metal under the ocean. How much heat is released per mole of Al_2O_3 produced? How does this amount of thermal energy compare with the energy released when 2 mol of protons and 2 mol of neutrons combine to form 1 mol of alpha particles?

21.47 How much energy must be supplied to break a single aluminum-27 nucleus into separated protons and neutrons if an aluminum-27 atom has a mass of 26.9815386 amu? How much energy is required for 100.0 g of aluminum-27? (The mass of an electron is given on the inside back cover.)

21.48 How much energy must be supplied to break a single ^{21}Ne nucleus into separated protons and neutrons if the nucleus has a mass of 20.98846 amu? What is the nuclear binding energy for 1 mol of ^{21}Ne?

21.49 The atomic masses of hydrogen-2 (deuterium), helium-4, and lithium-6 are 2.014102 amu, 4.002602 amu, and 6.0151228 amu, respectively. For each isotope, calculate (**a**) the nuclear mass, (**b**) the nuclear binding energy, (**c**) the nuclear binding energy per nucleon. (**d**) Which of these three isotopes has the largest nuclear binding energy per nucleon? Does this agree with the trends plotted in Figure 21.12?

21.50 The atomic masses of nitrogen-14, titanium-48, and xenon-129 are 13.999234 amu, 47.935878 amu, and 128.904779 amu, respectively. For each isotope, calculate (**a**) the nuclear mass, (**b**) the nuclear binding energy, (**c**) the nuclear binding energy per nucleon.

21.51 The energy from solar radiation falling on Earth is 1.07×10^{16} kJ/min. (**a**) How much loss of mass from the Sun occurs in one day from just the energy falling on Earth? (**b**) If the energy released in the reaction

$$^{235}\text{U} + {}^1_0\text{n} \longrightarrow {}^{141}_{56}\text{Ba} + {}^{92}_{36}\text{Kr} + 3\,{}^1_0\text{n}$$

(^{235}U nuclear mass, 234.9935 amu; ^{141}Ba nuclear mass, 140.8833 amu; ^{92}Kr nuclear mass, 91.9021 amu) is taken to be typical of that occurring in a nuclear reactor, what mass of uranium-235 is required to equal 0.10% of the solar energy that falls on Earth in 1.0 day?

21.52 Based on the following atomic mass values—^1H, 1.00782 amu; ^2H, 2.01410 amu; ^3H, 3.01605 amu; ^3He, 3.01603 amu; ^4He, 4.00260 amu—and the mass of the neutron given in the text, calculate the energy released per mole in each of the following nuclear reactions, all of which are possibilities for a controlled fusion process:

(**a**) $^2_1\text{H} + {}^3_1\text{H} \longrightarrow {}^4_2\text{He} + {}^1_0\text{n}$

(**b**) $^2_1\text{H} + {}^2_1\text{H} \longrightarrow {}^3_2\text{He} + {}^1_0\text{n}$

(**c**) $^2_1\text{H} + {}^3_2\text{He} \longrightarrow {}^4_2\text{He} + {}^1_1\text{H}$

21.53 Using Figure 21.12, predict which of the following nuclei are likely to have the largest mass defect per nucleon: (**a**) ^{11}B, (**b**) ^{51}V, (**c**) ^{118}Sn, (**d**) ^{243}Cm.

21.54 The isotope $^{62}_{28}\text{Ni}$ has the largest binding energy per nucleon of any isotope. Calculate this value from the atomic mass of nickel-62 (61.928345 amu) and compare it with the value given for iron-56 in Table 21.7.

Nuclear Power and Radioisotopes (Sections 21.7, 21.8, and 21.9)

21.55 Iodine-131 is a convenient radioisotope to monitor thyroid activity in humans. It is a beta emitter with a half-life of 8.02 days. The thyroid is the only gland in the body that uses iodine. A person undergoing a test of thyroid activity drinks a solution of NaI, in which only a small fraction of the iodide is radioactive. (**a**) Which of the following best explains why NaI is a good choice for the source of iodine: (i) The half-life of iodine-131 is longer in an iodide ion than it is in an iodine atom. (ii) NaI is soluble in water and will therefore be readily absorbed into bodily fluids. (iii) The sodium ion absorbs most of the radiation emitted by the radioactive iodine. (**b**) A normal thyroid will take up about 12% of the ingested iodide in a few hours. How long will it take for the radioactive iodide taken up and held by the thyroid to decay to 0.01% of the original amount?

21.56 Why is it important that radioisotopes used as diagnostic tools in nuclear medicine produce gamma radiation when they decay? Why are alpha emitters not used as diagnostic tools?

21.57 (**a**) Which of the following are required characteristics of an isotope to be used as a fuel in a nuclear power reactor? (i) It must emit gamma radiation. (ii) On decay, it must release two or more neutrons. (iii) It must have a half-life less than one hour. (iv) It must undergo fission upon the absorption of a neutron. (**b**) What is the most common fissionable isotope in a commercial nuclear power reactor?

21.58 Which of the following statements about the uranium used in nuclear reactors is or are *true*? (i) Natural uranium has too little ^{235}U to be used as a fuel. (ii) ^{238}U cannot be used as a fuel because it forms a supercritical mass too easily. (iii) To be used as fuel, uranium must be enriched so that it is more than 50% ^{235}U in composition. (iv) The neutron-induced fission of ^{235}U releases more neutrons per nucleus than fission of ^{238}U.

21.59 What is the function of the control rods in a nuclear reactor? What substances are used to construct control rods? Why are these substances chosen?

21.60 (**a**) What is the function of the moderator in a nuclear reactor? (**b**) What substance acts as the moderator in a pressurized water generator? (**c**) What other substances are used as a moderator in nuclear reactor designs?

21.61 Complete and balance the nuclear equations for the following fission or fusion reactions:

(**a**) $^2_1\text{H} + {}^2_1\text{H} \longrightarrow {}^3_2\text{He} + \underline{\hspace{1em}}$

(**b**) $^{239}_{92}\text{U} + {}^1_0\text{n} \longrightarrow {}^{133}_{51}\text{Sb} + {}^{98}_{41}\text{Nb} + \underline{\hspace{1em}}\,{}^1_0\text{n}$

21.62 Complete and balance the nuclear equations for the following fission reactions:

(**a**) $^{235}_{92}\text{U} + {}^1_0\text{n} \longrightarrow {}^{160}_{62}\text{Sm} + {}^{72}_{30}\text{Zn} + \underline{\hspace{1em}}\,{}^1_0\text{n}$

(**b**) $^{239}_{94}\text{Pu} + {}^1_0\text{n} \longrightarrow {}^{144}_{58}\text{Ce} + \underline{\hspace{1em}} + 2\,{}^1_0\text{n}$

21.63 A portion of the Sun's energy comes from the following reaction:

$$4\,{}^1_1\text{H} \longrightarrow {}^4_2\text{He} + 2\,{}^0_1\text{e}$$

Use the mass of the helium-4 nucleus given in Table 21.7 to determine how much energy is released per mole of hydrogen atoms.

21.64 The spent fuel elements from a fission reactor are much more intensely radioactive than the original fuel elements. (**a**) What does this tell you about the products of the fission process in relationship to the belt of stability (see Figure 21.2)? (**b**) Given that only two or three neutrons are released per fission event and knowing that the nucleus undergoing fission has a neutron-to-proton ratio characteristic of a heavy nucleus, what sorts of decay would you expect to be dominant among the fission products?

21.65 Which type or types of nuclear reactors have the following characteristics?

(a) Does not use a secondary coolant

(b) Creates more fissionable material than it consumes

(c) Uses a gas, such as He or CO_2, as the primary coolant

21.66 Which type or types of nuclear reactors have the following characteristics?

(a) Can use natural uranium as a fuel

(b) Does not use a moderator

(c) Can be refueled without shutting down

21.67 Hydroxyl radicals can pluck hydrogen atoms from molecules ("hydrogen abstraction"), and hydroxide ions can pluck protons from molecules ("deprotonation"). Write the reaction equations and Lewis dot structures for the hydrogen abstraction and deprotonation reactions for the generic carboxylic acid R—COOH with hydroxyl radical and hydroxide ion, respectively. Why is hydroxyl radical more toxic to living systems than hydroxide ion?

21.68 Which are classified as ionizing radiation: X rays, alpha particles, microwaves from a cell phone, and gamma rays?

21.69 A laboratory rat is exposed to an alpha-radiation source whose activity is 14.3 mCi. (a) What is the activity of the radiation in disintegrations per second? In becquerels? (b) The rat has a mass of 385 g and is exposed to the radiation for 14.0 s, absorbing 35% of the emitted alpha particles, each having an energy of 9.12×10^{-13} J. Calculate the absorbed dose in millirads and grays. (c) If the RBE of the radiation is 9.5, calculate the effective absorbed dose in mrem and Sv.

21.70 A 65-kg person is accidentally exposed for 240 s to a 15-mCi source of beta radiation coming from a sample of ^{90}Sr. (a) What is the activity of the radiation source in disintegrations per second? In becquerels? (b) Each beta particle has an energy of 8.75×10^{-14} J, and 7.5% of the radiation is absorbed by the person. Assuming that the absorbed radiation is spread over the person's entire body, calculate the absorbed dose in rads and in grays. (c) If the RBE of the beta particles is 1.0, what is the effective dose in mrem and in sieverts? (d) Is the radiation dose equal to, greater than, or less than that for a typical mammogram (300 mrem)?

Additional Exercises

21.71 The table provided gives the number of protons (p) and neutrons (n) for four isotopes, identified only as (i)–(iv). (a) Write the symbol for each of the isotopes. (b) Which of the isotopes is most likely to be unstable? (c) Which of the isotopes involves a magic number of protons and/or neutrons? (d) Which isotope will yield potassium-39 following positron emission?

	(i)	(ii)	(iii)	(iv)
p	19	19	20	20
n	19	21	19	20

21.72 Radon-222 decays to a stable nucleus by a series of three alpha emissions and two beta emissions. What is the stable nucleus that is formed?

21.73 Equation 21.28 is the nuclear reaction responsible for much of the helium-4 production in the Sun. How much energy is released in this reaction?

21.74 Chlorine has two stable nuclides, ^{35}Cl and ^{37}Cl. In contrast, ^{36}Cl is a radioactive nuclide that decays by beta emission. (a) What is the product of decay of ^{36}Cl? (b) Which of the following is the most likely explanation as to why ^{36}Cl is less stable than either ^{35}Cl or ^{37}Cl: (i) The neutron-to-proton ratio for ^{36}Cl is higher than that for the other two isotopes. (ii) Nuclides with odd mass numbers are more stable than those with even mass numbers. (iii) ^{36}Cl has an odd number of protons and an odd number of neutrons.

21.75 When two protons fuse in a star, the product is ^{2}H plus a positron. Write the nuclear equation for this process.

21.76 Nuclear scientists have synthesized approximately 1600 nuclei not known in nature. More might be discovered with heavy-ion bombardment using high-energy particle accelerators. Complete and balance the following reactions, which involve heavy-ion bombardments:

(a) $^{6}_{3}\text{Li} + ^{56}_{28}\text{Ni} \longrightarrow$?

(b) $^{40}_{20}\text{Ca} + ^{248}_{96}\text{Cm} \longrightarrow ^{147}_{62}\text{Sm} + $?

(c) $^{88}_{38}\text{Sr} + ^{84}_{36}\text{Kr} \longrightarrow ^{116}_{46}\text{Pd} + $?

(d) $^{40}_{20}\text{Ca} + ^{238}_{92}\text{U} \longrightarrow ^{70}_{30}\text{Zn} + 4\,^{1}_{0}\text{n} + 2$?

21.77 In 2010, a team of scientists from Russia and the United States reported the creation of the first atom of element 117, which is named tennessine and whose symbol is Ts. The synthesis involved the collision of a target of $^{249}_{97}$Bk with accelerated ions of an isotope which we will denote Q. The product atom, which we will call Z, immediately releases neutrons and forms $^{294}_{117}$Ts:

$$^{249}_{97}\text{Bk} + \text{Q} \longrightarrow \text{Z} \longrightarrow ^{294}_{117}\text{Ts} + 3\,^{1}_{0}\text{n}$$

(a) What are the identities of isotopes Q and Z? (b) Isotope Q is unusual in that it is very long-lived (its half-life is on the order of 10^{19} yr) in spite of having an unfavorable neutron-to-proton ratio (Figure 21.1). Can you propose a reason for its unusual stability? (c) Collision of ions of isotope Q with a target was also used to produce the first atoms of livermorium, Lv. The initial product of this collision was $^{296}_{116}$Lv. What was the target isotope with which Q collided in this experiment?

21.78 The synthetic radioisotope technetium-99, which decays by beta emission, is the most widely used isotope in nuclear medicine. The following data were collected on a sample of ^{99}Tc:

Disintegrations per Minute	Time (h)
180	0
130	2.5
104	5.0
77	7.5
59	10.0
46	12.5
24	17.5

Using these data, make an appropriate graph and curve fit to determine the half-life.

21.79 According to current regulations, the maximum permissible dose of strontium-90 in the body of an adult is 1 μCi (1×10^{-6} Ci).

Using the relationship rate = kN, calculate the number of atoms of strontium-90 to which this dose corresponds. To what mass of strontium-90 does this correspond? The half-life for strontium-90 is 28.8 yr.

21.80 Methyl acetate (CH_3COOCH_3) is formed by the reaction of acetic acid with methyl alcohol. If the methyl alcohol is labeled with oxygen-18, the oxygen-18 ends up in the methyl acetate:

$$CH_3\overset{O}{\overset{\|}{C}}OH + H^{18}OCH_3 \longrightarrow CH_3\overset{O}{\overset{\|}{C}}{}^{18}OCH_3 + H_2O$$

(a) Do the C—OH bond of the acid and the O—H bond of the alcohol break in the reaction, or do the O—H bond of the acid and the C—OH bond of the alcohol break? (b) Imagine a similar experiment using the radioisotope ^3H, which is called *tritium* and is usually denoted T. Would the reaction between CH_3COOH and $TOCH_3$ provide the same information about which bond is broken as does the above experiment with $H^{18}OCH_3$?

21.81 Each of the following transmutations produces a radionuclide used in positron emission tomography (PET). (a) In equations (i) and (ii), identify the species signified as "X." (b) In equation (iii), one of the species is indicated as "d." What do you think it represents?

(i) $^{14}N(p, \alpha)X$

(ii) $^{18}O(p, X)^{18}F$

(iii) $^{14}N(d, n)^{15}O$

21.82 The nuclear masses of ^7Be, ^9Be, and ^{10}Be are 7.0147, 9.0100, and 10.0113 amu, respectively. Which of these nuclei has the largest binding energy per nucleon?

21.83 A 26.00-g sample of water containing tritium, ^3_1H, emits 1.50×10^3 beta particles per second. Tritium is a weak beta emitter with a half-life of 12.3 yr. What fraction of all the hydrogen in the water sample is tritium?

21.84 The Sun radiates energy into space at the rate of 3.9×10^{26} J/s. (a) Calculate the rate of mass loss from the Sun in kg/s. (b) It is estimated that the Sun contains 9×10^{56} free protons. How many protons per second are consumed in nuclear reactions in the Sun?

21.85 The average energy released in the fission of a single uranium-235 nucleus is about 3×10^{-11} J. If the conversion of this energy to electricity in a nuclear power plant is 40%

efficient, what mass of uranium-235 undergoes fission in a year in a plant that produces 1000 megawatts? Recall that a watt is 1 J/s.

21.86 Tests on human subjects in Boston in 1965 and 1966, following the era of atomic bomb testing, revealed average quantities of about 2 pCi of plutonium radioactivity in the average person. How many disintegrations per second does this level of activity imply? If each alpha particle deposits 8×10^{-13} J of energy and if the average person weighs 75 kg, calculate the number of rads and rems of radiation in 1 yr from such a level of plutonium.

21.87 A 53.8-mg sample of sodium perchlorate contains radioactive chlorine-36 (whose atomic mass is 36.0 amu). If 29.6% of the chlorine atoms in the sample are chlorine-36 and the remainder are naturally occurring nonradioactive chlorine atoms, how many disintegrations per second are produced by this sample? The half-life of chlorine-36 is 3.0×10^5 yr.

21.88 Naturally found uranium consists of 99.274% ^{238}U, 0.720% ^{235}U, and 0.006% ^{233}U. As we have seen, ^{235}U is the isotope that can undergo a nuclear chain reaction. Most of the ^{235}U used in the first atomic bomb was obtained by gaseous diffusion of uranium hexafluoride, $UF_6(g)$. (a) What is the mass of UF_6 in a 30.0-L vessel of UF_6 at a pressure of 695 torr at 350 K? (b) What is the mass of ^{235}U in the sample described in part (a)? (c) Now suppose that the UF_6 is diffused through a porous barrier and that the change in the ratio of ^{238}U and ^{235}U in the diffused gas can be described by Equation 10.24. What is the mass of ^{235}U in a sample of the diffused gas analogous to that in part (a)? (d) After one more cycle of gaseous diffusion, what is the percentage of $^{235}UF_6$ in the sample?

21.89 A sample of an alpha emitter having an activity of 0.18 Ci is stored in a 25.0-mL sealed container at 22 °C for 245 days. (a) How many alpha particles are formed during this time? (b) Assuming that each alpha particle is converted to a helium atom, what is the partial pressure of helium gas in the container after this 245-day period?

21.90 Charcoal samples from Stonehenge in England were burned in O_2, and the resultant CO_2 gas bubbled into a solution of $Ca(OH)_2$ (limewater), resulting in the precipitation of $CaCO_3$. The $CaCO_3$ was removed by filtration and dried. A 788-mg sample of the $CaCO_3$ had a radioactivity of 1.5×10^{-2} Bq due to carbon-14. By comparison, living organisms undergo 15.3 disintegrations per minute per gram of carbon. Using the half-life of carbon-14, 5730 yr, calculate the age of the charcoal sample.

Design an Experiment

Because radioactivity can have harmful effects on human health, stringent experimental procedures and precautions are required when undertaking experiments on radioactive materials. As such, we typically do not have experiments involving radioactive substances in general chemistry laboratories. We can nevertheless ponder the design of some hypothetical experiments that would allow us to explore some of the properties of radium, which was discovered by Marie and Pierre Curie in 1898.

(a) A key aspect of the discovery of radium was Marie Curie's observation that *pitchblende*, a natural ore of uranium, had greater radioactivity than pure uranium metal. Design an experiment to reproduce this observation and to obtain a ratio of the activity of pitchblende relative to that of pure uranium.

(b) Radium was first isolated as halide salts. Suppose you had pure samples of radium metal and radium bromide. The sample sizes are on the order of milligrams and are not amenable to the usual forms of elemental analysis. Could you use a device that measures radioactivity quantitatively to determine the empirical formula of radium bromide? What information must you use that the Curies may not have had at the time of their discovery?

(c) Suppose you had a 1-yr time period in order to measure the half-life of radium and related elements. You have some pure samples and a device that measures radioactivity quantitatively. Could you determine the half-life of the elements in the samples? Would you have different experimental constraints depending on whether the half-life were 10 yr or 1000 yr?

(**d**) Before its negative health effects were better understood, small amounts of radium salts were used in "glow in the dark" watches, such as the one shown here. The glow is not due to the radioactivity of radium directly; rather, the radium is combined with a luminescent substance, such as zinc sulfide, which glows when it is exposed to radiation. Suppose you had pure samples of radium and zinc sulfide. How could you determine whether the glow of zinc sulfide is due to alpha, beta, or gamma radiation? What type of device could you design to use the glow as a quantitative measure of the amount of radioactivity in a sample?

22

CHEMISTRY OF THE NONMETALS

▲ **A TROPICAL BEACH.** The water, sand, and air are composed of nonmetals.

Much of the world is composed of nonmetals. Water, for example, is H_2O, and sand is mostly SiO_2. Although we cannot see it, the air contains principally N_2 and O_2 with much lesser amounts of other nonmetallic substances.

In this chapter, we take a panoramic view of the descriptive chemistry of the nonmetallic elements, starting with hydrogen and progressing group by group across the periodic table. We consider how the elements occur in nature, how they are isolated from their sources, and how they are used. We emphasize hydrogen, oxygen, nitrogen, and carbon because these four nonmetals form many commercially important compounds and account for 99% of the atoms required by living cells.

As you study this *descriptive chemistry*, it is important to look for trends rather than trying to memorize all the facts presented. The periodic table is your most valuable tool in this task.

22.1 | Periodic Trends and Chemical Reactions

⚠️ **Learning Objectives**

When you finish Section 22.1, you should be able to

▶ Predict trends in the properties of the elements based on their position in the periodic table.

▶ Predict which elements in a group are most likely to undergo π bonding.

▶ Predict the products of chemical reactions based on combustion or acid–base chemistry.

Recall that we can classify elements as metals, metalloids, and nonmetals. (Section 7.5) Except for hydrogen, which is a special case, the nonmetals occupy the upper right portion of the periodic table. This division of elements relates nicely to trends in their properties as summarized in **Figure 22.1**. Electronegativity, for example, increases as we move left to right across a period and decreases as we move down a group. The nonmetals thus have higher electronegativities than the metals. This difference leads to the formation of ionic solids in reactions between metals and nonmetals. (Sections 7.5, 8.2, and 8.4) In contrast, compounds formed between two or more nonmetals are usually molecular substances. (Sections 7.7 and 8.4)

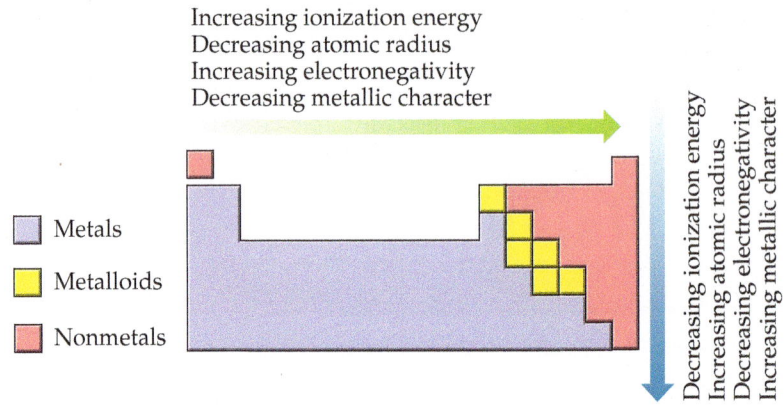

▲ Figure 22.1 **Trends in elemental properties.**

The chemistry exhibited by the first member of a nonmetal group can differ from that of subsequent members in important ways. Two differences are particularly notable:

- The first member follows the octet rule, but heavier group members can expand their octet. (Section 8.7) For example, nitrogen is able to bond to a maximum of three Cl atoms, NCl_3, whereas phosphorus can bond to five, PCl_5. The small size of nitrogen is largely responsible for this difference.
- The first member can more readily form π bonds and hence double and triple bonds. This trend is also due, in part, to size because small atoms are able to approach each other more closely. As a result, the overlap of p orbitals, which results in the formation of π bonds, is more effective for the first element in each group (**Figure 22.2**). More effective overlap means stronger π bonds, reflected in bond enthalpies. (Sections 5.8 and 8.8) For example, the difference between the enthalpies of the C—C bond and the C=C bond is about 270 kJ/mol (Table 8.3); this large value reflects the "strength" of a carbon–carbon π bond. On the other hand, the difference between Si—Si and Si=Si bonds is only about 100 kJ/mol, significantly lower than that for carbon, reflecting much weaker π bonding.

As we shall see, π bonds are particularly important in the chemistry of carbon, nitrogen, and oxygen, each the first member in its group. The heavier elements in these groups have a tendency to form only single bonds.

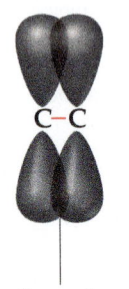
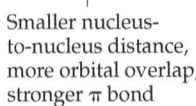
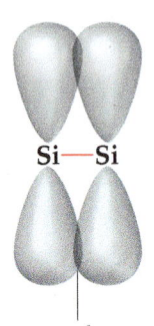

Smaller nucleus-to-nucleus distance, more orbital overlap, stronger π bond

Larger nucleus-to-nucleus distance, less orbital overlap, weaker π bond

▲ Figure 22.2 **Comparing π bonds in period 2 and period 3 elements.**

Sample Exercise 22.1
Identifying Elemental Properties

Of the elements Li, K, N, P, and Ne, (**a**) which is the most electronegative, (**b**) which has the greatest metallic character, (**c**) which can bond to more than four atoms in a molecule, and (**d**) which forms π bonds most readily?

SOLUTION

Analyze We are given a list of elements and asked to predict several properties that can be related to periodic trends.

Plan We can use Figures 22.1 and 22.2 to guide us to the answers.

Solve

(**a**) Electronegativity increases as we proceed toward the upper right portion of the periodic table, excluding the noble gases. Thus, N is the most electronegative element of our choices.

(**b**) Metallic character correlates inversely with electronegativity—the less electronegative an element, the greater its metallic character. The element with the greatest metallic character is therefore K, which is closest to the lower left corner of the periodic table.

(**c**) Nonmetals tend to form molecular compounds, so we can narrow our choice to the three nonmetals on the list: N, P, and Ne. To form more than four bonds, an element must be able to expand its valence shell to allow more than an octet

of electrons around it. Valence-shell expansion occurs for period 3 elements and below; N and Ne are both in period 2 and do not undergo valence-shell expansion. Thus, the answer is P.

(**d**) Period 2 nonmetals form π bonds more readily than elements in period 3 and below. There are no compounds known that contain covalent bonds to Ne. Thus, N is the element from the list that forms π bonds most readily.

▶ **Practice Exercise**

Of the elements Be, C, Cl, Sb, and Cs, (**a**) which has the lowest electronegativity, (**b**) which has the greatest nonmetallic character, (**c**) which is most likely to participate in extensive π bonding, and (**d**) which is most likely to be a metalloid?

The ready ability of period 2 elements to form π bonds is an important factor in determining the elemental forms of these elements. Compare, for example, carbon and silicon. Carbon has five major crystalline *allotropes*: diamond, graphite, buckminsterfullerene, graphene, and carbon nanotubes. (Sections 12.5 and 12.7) Diamond is a covalent-network solid that has C—C σ bonds but no π bonds. Other allotropes of carbon have π bonds that result from the sideways overlap of *p* orbitals. Elemental silicon, however, exists only as a diamond-like covalent-network solid with σ bonds; it has no forms analogous to graphite, buckminsterfullerene, graphene, or carbon nanotubes, probably because Si—Si π bonds are too weak.

We likewise see significant differences in the dioxides of carbon and silicon as a result of their relative abilities to form π bonds (**Figure 22.3**). CO_2 is a molecular substance containing C=O double bonds, whereas SiO_2 is a covalent-network solid in which four oxygen atoms are bonded to each silicon atom by single bonds, forming an extended structure that has the empirical formula SiO_2.

Chemical Reactions

Because O_2 and H_2O are abundant in our environment, it is particularly important to consider how these substances react with other compounds. About one-third of the reactions discussed in this chapter involve either O_2 (oxidation or combustion reactions) or H_2O (especially proton-transfer reactions).

In combustion reactions (Section 3.2), hydrogen-containing compounds produce H_2O. Carbon-containing ones produce CO_2 (unless the amount of O_2 is insufficient, in which case CO or even C can form). Nitrogen-containing compounds tend to form N_2, although NO can form in special cases or in small amounts. The following reaction illustrates these points:

$$4\,CH_3NH_2(g) + 9\,O_2(g) \longrightarrow 4\,CO_2(g) + 10\,H_2O(g) + 2\,N_2(g) \qquad [22.1]$$

The formation of H_2O, CO_2, and N_2 reflects the high thermodynamic stability of these substances, indicated by the large bond energies for the O—H, C=O, and N≡N bonds (463, 799, and 941 kJ/mol, respectively). (Sections 5.8 and 8.8)

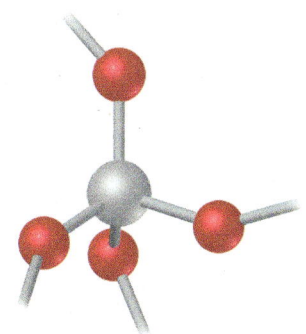

Fragment of extended SiO_2 lattice; Si forms only single bonds

CO_2; C forms double bonds

▲ **Figure 22.3** Comparison of the bonds in SiO_2 and CO_2.

When dealing with proton-transfer reactions, remember that the weaker a Brønsted–Lowry acid, the stronger its conjugate base. (Section 16.2) For example, H_2, OH^-, NH_3, and CH_4 are exceedingly weak proton donors that have *no* tendency to act as acids in water. Thus, the species formed by removing one or more protons from them are extremely strong bases. All of them react readily with water, removing protons from H_2O to form OH^-. Two representative reactions are:

$$CH_3^-(aq) + H_2O(l) \longrightarrow CH_4(g) + OH^-(aq) \qquad [22.2]$$

$$N^{3-}(aq) + 3\,H_2O(l) \longrightarrow NH_3(aq) + 3\,OH^-(aq) \qquad [22.3]$$

Sample Exercise 22.2
Predicting the Products of Chemical Reactions

Predict the products formed in each of the following reactions, and write a balanced equation:

(a) $CH_3NHNH_2(g) + O_2(g) \longrightarrow$?   **(b)** $Mg_3P_2(s) + H_2O(l) \longrightarrow$?

SOLUTION

Analyze We are given the reactants for two chemical equations, and we are asked to predict the products and then balance the equations.

Plan We need to examine the reactants to see if we might recognize a reaction type. In **(a)** the carbon compound is reacting with O_2, which suggests a combustion reaction. In **(b)** water reacts with an ionic compound. The anion P^{3-} is a strong base and H_2O is able to act as an acid, so the reactants suggest an acid–base (proton-transfer) reaction.

Solve

(a) Based on the elemental composition of the carbon compound, this combustion reaction should produce CO_2, H_2O, and N_2:

$$2\,CH_3NHNH_2(g) + 5\,O_2(g) \longrightarrow 2\,CO_2(g) + 6\,H_2O(g) + 2\,N_2(g)$$

(b) Mg_3P_2 is ionic, consisting of Mg^{2+} and P^{3-} ions. The P^{3-} ion, like N^{3-}, has a strong affinity for protons and reacts with H_2O to form OH^- and PH_3 (PH^{2-}, PH_2^-, and PH_3 are all exceedingly weak proton donors).

$$Mg_3P_2(s) + 6\,H_2O(l) \longrightarrow 2\,PH_3(g) + 3\,Mg(OH)_2(s)$$

$Mg(OH)_2$ has low solubility in water and will precipitate.

▶ **Practice Exercise**
Write a balanced equation for the reaction of solid sodium hydride with water.

Self-Assessment Exercises

SAE 22.1 Which of the elements Na, Rb, F, Si, and Ne is the most electronegative? **(a)** F **(b)** Na **(c)** Rb **(d)** Si

SAE 22.2 Which of the elements Na, Rb, F, Si, and Ne has the greatest metallic character? **(a)** Na **(b)** F **(c)** Si **(d)** Rb **(e)** Ne

SAE 22.3 Which of the elements Rb, F, Si, and Ne most readily forms a positive ion? **(a)** F **(b)** Si **(c)** Rb **(d)** Ne

SAE 22.4 Which of the elements Al, N, P, and As is most likely to form multiple π bonds with itself? **(a)** Al **(b)** N **(c)** P **(d)** As

SAE 22.5 Complete and balance the following equation:

$$C_2H_5OH(l) + O_2(g) \longrightarrow$$

(a) $C_2H_5OH(l) + 2\,O_2(g) \longrightarrow 2\,H_2O(l) + 2\,CHO_2(g)$
(b) $C_2H_5OH(l) + 3\,O_2(g) \longrightarrow 3\,H_2O(l) + 2\,CO_2(g)$
(c) $3\,C_2H_5OH(l) + O_2(g) \longrightarrow H_2O(l) + 6\,CO_2(g)$
(d) $C_2H_5OH(l) + 3\,H_2O(l) \longrightarrow 6\,H_2(g) + 2\,CO_2(g)$

Learning Objectives

When you finish Section 22.2, you should be able to:

▶ Compare and contrast the isotopes of hydrogen.

▶ Classify a given hydride as ionic, metallic, or molecular.

▶ Distinguish the properties and reactivity of hydrogen compared to other elements.

22.2 | Hydrogen

Because hydrogen produces water when burned in air, the French chemist Antoine Lavoisier (1734–1794) gave it the name *hydrogen*, which means "water producer" (Greek: *hydro*, water; *gennao*, to produce).

Hydrogen is the most abundant element in the universe. [See the "Chemistry and Sustainability" box on "Hydrogen and Helium." (Section 10.6)] It is the nuclear fuel consumed by our Sun and other stars to produce energy. (Section 21.8) Although about 75% of the known mass of the universe is hydrogen, it constitutes only 0.87% of Earth's mass. Most of the hydrogen on our planet is found associated with oxygen. Water, which is 11% hydrogen by mass, is the most abundant hydrogen compound.

Isotopes of Hydrogen

The most common isotope of hydrogen, $_1^1H$, has a nucleus consisting of a single proton. This isotope, sometimes referred to as **protium**,* makes up 99.9844% of naturally occurring hydrogen.

Two other isotopes are known: $_1^2H$, whose nucleus contains a proton and a neutron, and $_1^3H$, whose nucleus contains a proton and two neutrons. The $_1^2H$ isotope, **deuterium**, makes up 0.0156% of naturally occurring hydrogen. It is not radioactive, and it is often given the symbol D in chemical formulas, as in D_2O (deuterium oxide), which is known as *heavy water*.

Because an atom of deuterium is about twice as massive as an atom of protium, the properties of deuterium-containing substances vary somewhat from those of the protium-containing analogs. For example, the normal melting and boiling points of D_2O are 3.81 °C and 101.42 °C, respectively, versus 0.00 °C and 100.00 °C for H_2O. Not surprisingly, the density of D_2O at 25 °C (1.104 g/mL) is greater than that of H_2O (0.997 g/mL). Replacing protium with deuterium (a process called *deuteration*) can also have a profound effect on reaction rates, a phenomenon called a *kinetic-isotope effect*. For example, heavy water can be obtained from the electrolysis $[2 H_2O(l) \longrightarrow 2 H_2(g) + O_2(g)]$ of ordinary water because the small amount of naturally occurring D_2O in the sample undergoes electrolysis more slowly than H_2O and, therefore, becomes concentrated during the reaction.

The third isotope, $_1^3H$, **tritium**, is radioactive, with a half-life of 12.3 yr:

$$_1^3H \longrightarrow {}_2^3He + {}_{-1}^0e \quad t_{1/2} = 12.3 \text{ yr} \qquad [22.4]$$

Because of its short half-life, only trace quantities of tritium exist naturally. The isotope can be synthesized in nuclear reactors by neutron bombardment of lithium-6:

$$_3^6Li + {}_0^1n \longrightarrow {}_1^3H + {}_2^4He \qquad [22.5]$$

Deuterium and tritium are useful in studying reactions of compounds containing hydrogen. A compound is "labeled" by replacing one or more ordinary hydrogen atoms with deuterium or tritium at specific locations in a molecule. By comparing the locations of the label atoms in reactants and products, the reaction mechanism can often be inferred. When methyl alcohol (CH_3OH) is placed in D_2O, for example, the H atom of the O—H bond exchanges rapidly with the D atoms, forming CH_3OD. The H atoms of the CH_3 group do not exchange. This experiment demonstrates the kinetic stability of C—H bonds and reveals the speed at which the O—H bond in the molecule breaks and re-forms.

Properties of Hydrogen

Hydrogen is the only element that is not a member of any family in the periodic table. Because of its $1s^1$ electron configuration, it is generally placed above lithium in the table. However, it is definitely *not* an alkali metal. It forms a positive ion much less readily than any alkali metal. The ionization energy of the hydrogen atom is 1312 kJ/mol, whereas that of lithium is 520 kJ/mol.

Hydrogen is sometimes placed above the halogens in the periodic table because the hydrogen atom can pick up one electron to form the *hydride ion*, H^-, which has the same electron configuration as helium. However, the electron affinity of hydrogen, $EA = -73$ kJ/mol, is not as large as that of any halogen. In general, hydrogen shows no closer resemblance to the halogens than it does to the alkali metals.

Elemental hydrogen exists at room temperature as a colorless, odorless, tasteless gas composed of diatomic molecules. We can call H_2 *dihydrogen*, but it is more commonly referred to as either *molecular hydrogen* or simply *hydrogen*. Because H_2 is nonpolar and has only two electrons, attractive forces between molecules are extremely weak. As a result, its melting point (-259 °C) and boiling point (-253 °C) are very low.

The H—H bond enthalpy (436 kJ/mol) is high for a single bond. (Table 8.3) By comparison, the Cl—Cl bond enthalpy is only 242 kJ/mol. Because H_2 has a strong bond, most reactions involving H_2 are slow at room temperature. However, the molecule is readily activated by heat, irradiation, or catalysis. The activation generally produces hydrogen atoms, which are very reactive. Once H_2 is activated, it reacts rapidly and exothermically with a wide variety of substances.

---

*Giving unique names to isotopes is limited to hydrogen. Because of the proportionally large differences in their masses, the isotopes of H show appreciably more differences in their properties than isotopes of heavier elements.

Hydrogen forms strong covalent bonds with many other elements, including oxygen; the O—H bond enthalpy is 463 kJ/mol. The formation of the strong O—H bond makes hydrogen an effective reducing agent for many metal oxides. When H_2 is passed over heated CuO, for example, copper is produced:

$$CuO(s) + H_2(g) \longrightarrow Cu(s) + H_2O(g) \qquad [22.6]$$

When H_2 is ignited in air, a vigorous reaction occurs, forming H_2O. Air containing as little as 4% H_2 by volume is potentially explosive. Combustion of hydrogen–oxygen mixtures is used in liquid-fuel rocket engines to launch space vehicles. The hydrogen and oxygen are stored at low temperatures in liquid form.

Production of Hydrogen

When a small quantity of H_2 is needed in the laboratory, it is usually obtained by the reaction between an active metal such as zinc and a dilute strong acid such as HCl or H_2SO_4:

$$Zn(s) + 2\,H^+(aq) \longrightarrow Zn^{2+}(aq) + H_2(g) \qquad [22.7]$$

Large quantities of H_2 are produced by reacting methane, principally from natural gas, with steam at 1100 °C. We can view this process as involving two reactions:

$$CH_4(g) + H_2O(g) \longrightarrow CO(g) + 3\,H_2(g) \qquad [22.8]$$

$$CO(g) + H_2O(g) \longrightarrow CO_2(g) + H_2(g) \qquad [22.9]$$

Equation 22.8 is known as *steam reforming* and is an endothermic reaction. Equation 22.9 shows that one product of steam reforming, CO, can react with water to produce CO_2 and H_2. Equation 22.9 is known as the *water-gas shift reaction* and is slightly exothermic. Taken together, the products of Equations 22.8 and 22.9 (CO, CO_2, and H_2) are referred to as *syngas*. Syngas is used as industrial fuel in electricity generation. Both Equations 22.8 and 22.9 require catalysts to proceed.

Carbon heated with water to about 1000 °C is another source of H_2:

$$C(s) + H_2O(g) \longrightarrow H_2(g) + CO(g) \qquad [22.10]$$

This mixture, known as *water gas*, is also used as an industrial fuel. Water gas is a product of Equation 22.8 as well.

Electrolysis of water consumes too much energy and is consequently too costly to be used commercially to produce H_2. However, H_2 is produced as a by-product in the electrolysis of brine (NaCl) solutions in the course of commercial Cl_2 and NaOH manufacture:

$$2\,NaCl(aq) + 2\,H_2O(l) \xrightarrow{\text{electrolysis}} H_2(g) + Cl_2(g) + 2\,NaOH(aq) \qquad [22.11]$$

CHEMISTRY AND SUSTAINABILITY   The Hydrogen Economy

The reaction of hydrogen with oxygen is highly exothermic:

$$2\,H_2(g) + O_2(g) \longrightarrow 2\,H_2O(g) \quad \Delta H = -483.6\ \text{kJ} \qquad [22.12]$$

Because H_2 has a low molar mass and a high enthalpy of combustion, it has a high-energy density by mass. (That is, its combustion produces high energy per gram.) The only product of the reaction is water vapor, which means that hydrogen is environmentally cleaner than fossil fuels. Thus, the prospect of using hydrogen widely as a fuel is attractive. Furthermore, the hydrogen can be utilized as a fuel in hydrogen fuel cells, enhancing the efficiency of its use. (Section 20.7)

The term *hydrogen economy* is used to describe the concept of delivering and using hydrogen as a fuel in place of fossil fuels. [This concept was also mentioned in the "Chemistry and Sustainability" box on "Hydrogen and Helium." (Section 10.6)] In order to develop a hydrogen economy, it would be necessary to generate elemental hydrogen on a large scale and arrange for its transport and storage. These matters provide significant technical challenges.

Figure 22.4 illustrates various sources and uses of H_2 fuel. The generation of H_2 through electrolysis of water is in principle the cleanest route, because this process—the reverse of Equation 22.12—produces only hydrogen and oxygen. (Figure 1.6 and Section 20.9) However, the energy required to electrolyze water must come from somewhere. If we burn fossil fuels to generate this energy, we have not advanced very far toward a true hydrogen economy. If the energy for electrolysis came instead from a hydroelectric or nuclear power plant, solar cells, or wind generators, consumption of nonrenewable energy sources and undesired production of CO_2 could be avoided.

The storage of hydrogen is another technical obstacle that must be overcome in developing a hydrogen economy. Although $H_2(g)$ has a high-energy density by mass, it has a low-energy density by volume. Thus, storing hydrogen as a gas requires a large volume compared to the energy it delivers. There are also safety issues associated with handling and storing the gas because its combustion can be explosive. Storing hydrogen in the form of various hydride compounds, such as $LiAlH_4$, is being investigated as a means of reducing the volume and increasing the safety. One problem with this approach, however, is that such compounds have high-energy density by volume but low-energy density by mass.

Related Exercises: 22.27, 22.28, 22.93

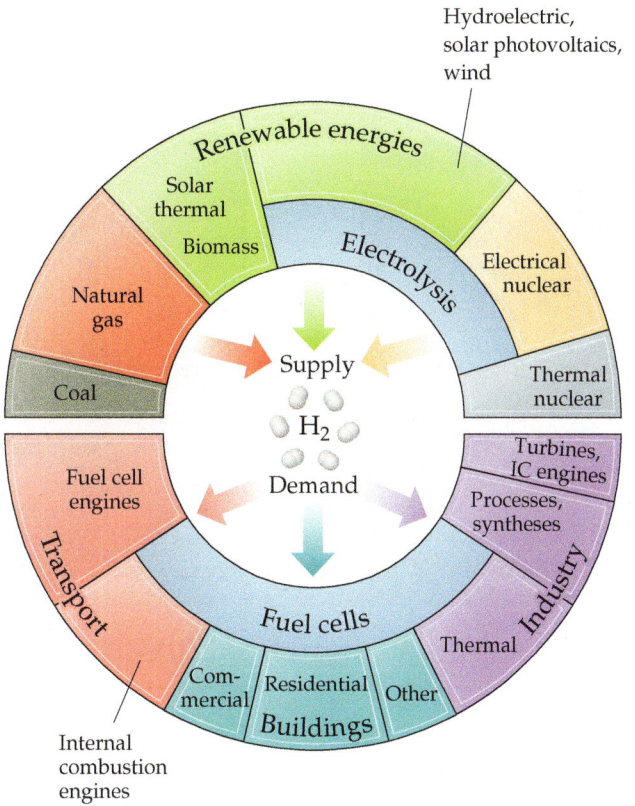

▲ **Figure 22.4 Supply and demand in the hydrogen economy.** Hydrogen would need to be produced from various sources, and it would be used in energy-related applications.

Uses of Hydrogen

Hydrogen is commercially important. About half of the H_2 produced is used to synthesize ammonia by the Haber process. (Section 15.2) Much of the remaining hydrogen is used to convert high-molecular-weight hydrocarbons from petroleum into lower-molecular-weight hydrocarbons suitable for fuel (gasoline, diesel, and others) in a process known as *cracking*. Hydrogen is also used to manufacture methanol via the catalytic reaction of CO and H_2 at high pressure and temperature:

$$CO(g) + 2\,H_2(g) \longrightarrow CH_3OH(g) \qquad [22.13]$$

Binary Hydrogen Compounds

Hydrogen reacts with other elements to form three types of compounds: (1) ionic hydrides, (2) metallic hydrides, and (3) molecular hydrides.

The **ionic hydrides** are formed by the alkali metals and by the heavier alkaline earths (Ca, Sr, and Ba). These active metals are much less electronegative than hydrogen. (Figure 8.8) Consequently, hydrogen acquires electrons from them to form hydride ions (H^-):

$$Ca(s) + H_2(g) \longrightarrow CaH_2(s) \qquad [22.14]$$

The hydride ion is very basic and reacts readily with compounds having even weakly acidic protons to form H_2 as shown in **Figure 22.5** and summarized by the following equation:

$$H^-(aq) + H_2O(l) \longrightarrow H_2(g) + OH^-(aq) \qquad [22.15]$$

Ionic hydrides can therefore be used as convenient (although expensive) sources of H_2.

Calcium hydride (CaH_2) is used to inflate life rafts, weather balloons, and the like where a simple, compact means of generating H_2 is desired (Figure 22.5).

Metallic hydrides are formed when hydrogen reacts with transition metals. These compounds are so named because they retain their metallic properties. In many metallic hydrides, the ratio of metal atoms to hydrogen atoms is not fixed or in small whole numbers. The composition can vary within a range, depending on reaction conditions. TiH_2

Go Figure

This reaction is exothermic. Is the beaker on the right warmer or colder than the beaker on the left?

CaH$_2$

H$_2$O with pH indicator

H$_2$ gas

Color change indicates presence of OH$^-$.

$$CaH_2(s) + 2\,H_2O(l) \longrightarrow Ca(OH)_2(aq) + 2\,H_2(g)$$

▲ **Figure 22.5** The reaction of CaH$_2$ with water.

Go Figure

Which is the most thermodynamically stable hydride? Which is the least thermodynamically stable?

4A	5A	6A	7A
CH$_4$(g)	NH$_3$(g)	H$_2$O(l)	HF(g)
−50.8	−16.7	−237	−271
SiH$_4$(g)	PH$_3$(g)	H$_2$S(g)	HCl(g)
+56.9	+18.2	−33.0	−95.3
GeH$_4$(g)	AsH$_3$(g)	H$_2$Se(g)	HBr(g)
+117	+111	+71	−53.2
	SbH$_3$(g)	H$_2$Te(g)	HI(g)
	+187	+138	+1.30

▲ **Figure 22.6** Standard free energies of formation of molecular hydrides. All values are kilojoules per mole of hydride.

can be produced, for example, but preparations usually yield TiH$_{1.8}$. These nonstoichiometric metallic hydrides are sometimes called *interstitial hydrides*. Because hydrogen atoms are small enough to fit between the sites occupied by the metal atoms, many metal hydrides behave like interstitial alloys. (Section 12.2)

The **molecular hydrides**, formed by nonmetals and metalloids, are either gases or liquids under standard conditions. The simple molecular hydrides are listed in **Figure 22.6**, together with their standard free energies of formation, ΔG_f°. (Section 19.5) In each family, the thermal stability (measured as ΔG_f°) decreases as we move down the family. (Recall that the more stable a compound is with respect to its elements under standard conditions, the more negative ΔG_f° is.)

Self-Assessment Exercises

SAE 22.6 Which of the following isotopes of hydrogen does not occur in nature? (**a**) $^5_1$H (**b**) $^3_1$H (**c**) $^2_1$H (**d**) $^1_1$H

SAE 22.7 Which element would you predict reacts with hydrogen to form an ionic hydride? (**a**) C (**b**) Fe (**c**) F (**d**) Sr

SAE 22.8 Which element would you predict reacts with hydrogen to form a metallic hydride? (**a**) C (**b**) K (**c**) Fe (**d**) F

SAE 22.9 Which of the following statements is/are *true*?
(**i**) The H—H bond in hydrogen is quite strong (greater than 400 kJ/mol).
(**ii**) Atomic hydrogen, electron configuration $1s^1$, has a large ionization energy relative to lithium, electron configuration $1s^2 2s^1$.
(**iii**) The common oxidation states of hydrogen are −1, 0, and +1.

(**a**) Only one of the statements is true. (**b**) Only statements i and ii are true. (**c**) Only statements i and iii are true. (**d**) Only statements ii and iii are true. (**e**) All three statements are true.

Learning Objectives

When you finish **Section 22.3**, you should be able to:

▶ Predict what elements are most likely to react with noble gases.

▶ Determine the oxidation states of the elements in noble gas compounds.

▶ Predict the structures of noble-gas compounds using the VSPER model.

22.3 | Group 8A: The Noble Gases

The elements of group 8A are chemically unreactive. Indeed, most of our references to these elements have been in relation to their physical properties, such as when we discussed intermolecular forces. (Section 11.2) The relative inertness of these elements is due to the presence of a completed octet of valence-shell electrons (except He, which only has a filled 1s shell). The stability of such an arrangement is reflected in the high ionization energies of the group 8A elements. (Section 7.4)

The group 8A elements are all gases at room temperature. They are components of Earth's atmosphere, except for radon, which exists only as a short-lived radioisotope. (Section 21.9) Only argon is relatively abundant. (Table 18.1)

Neon, argon, krypton, and xenon are used in lighting, display, and laser applications in which the atoms are excited electrically and electrons that are in a higher-energy state emit light as they return to the ground state. (Section 6.2) Neon signs are probably the most familiar example of noble gases used in lighting. Argon is also used as a protective atmosphere to prevent oxidation in welding and certain high-temperature metallurgical processes.

Helium is in many ways the most important noble gas. Liquid helium is used as a coolant to conduct experiments at very low temperatures. [See the "Chemistry and Sustainability" box on "Hydrogen and Helium." (Section 10.6)] Helium boils at 4.2 K and 1 atm, the lowest boiling point of any substance. It is found in relatively high concentrations in many natural-gas wells from which it is isolated.

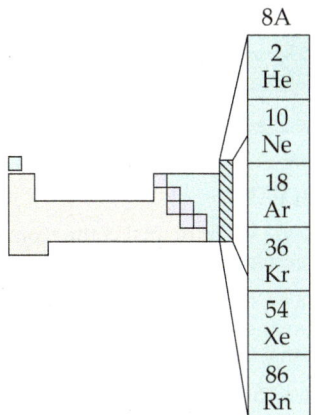

Noble-Gas Compounds

Because the noble gases are exceedingly stable, they react only under rigorous conditions. We expect the heavier ones to be most likely to form compounds because their ionization energies are lower. (Figure 7.11) A lower ionization energy suggests the possibility of sharing an electron with another atom, leading to a chemical bond. In addition, the group 8A elements (except helium) already contain eight electrons in their valence shell, so formation of covalent bonds will require an expanded valence shell. Valence-shell expansion occurs most readily with larger atoms. (Section 8.7)

The first noble-gas compound was reported in 1962. This discovery caused a sensation because it undercut the belief that the noble-gas elements were chemically inert. The initial study involved xenon in combination with fluorine, the most electronegative element and thus the element we would expect to be most able to pull electron density from another (very polarizable) atom. Since that time, chemists have prepared several xenon compounds of fluorine and oxygen (**Table 22.1**). The fluorides XeF_2, XeF_4, and XeF_6 are made by direct reaction of the elements. By varying the ratio of reactants and altering reaction conditions, each of the three compounds can be obtained. The oxygen-containing compounds are formed when the fluorides react with water as, for example,

$$XeF_6(s) + 3 H_2O(l) \longrightarrow XeO_3(aq) + 6 HF(aq) \qquad [22.16]$$

The other noble-gas elements form compounds much less readily than xenon. For many years, only one binary krypton compound, KrF_2, was known with certainty, and it decomposes to its elements at $-10\,°C$. Other compounds of krypton have been isolated at very low temperatures (40 K). In 2000, a compound of argon, HArF, was discovered, but it can exist only in argon matrices at very low temperatures. In 2017, the unusual compound Na_2He, disodium helide, was reported via reaction of the elements under high pressure, causing another sensation in the field of chemistry. In this compound the sodium atoms are partially charged cations, the He atoms bear slight negative charges, and the bulk of the negative charge to balance the Na positive charges comes from free electrons in the crystal lattice. Helium's ionization energy, 2372 kJ/mol (Figure 7.11), is the largest of the elements and thus makes He oxidation very unfavorable; thus, the usual trick of using a highly electronegative element to react with a noble gas does not work for helium.

TABLE 22.1   Properties of Xenon Compounds

Compound	Oxidation State of Xe	Melting Point (°C)	$\Delta H°_f$ (kJ/mol)[a]
XeF_2	+2	129	$-109(g)$
XeF_4	+4	117	$-218(g)$
XeF_6	+6	49	$-298(g)$
$XeOF_4$	+6	-41 to -28	$+146(l)$
XeO_3	+6	$-$[b]	$+402(s)$
XeO_2F_2	+6	31	$+145(s)$
XeO_4	+8	$-$[c]	$-$

[a]At 25 °C, for the compound in the state indicated.
[b]A solid; decomposes at 40 °C.
[c]A solid; decomposes at $-40\,°C$.

Sample Exercise 22.3
Predicting a Molecular Structure

Use the VSEPR model to predict the structure of XeF_4.

SOLUTION

Analyze We must predict the geometrical structure given only the molecular formula.

Plan We must first write the Lewis structure for the molecule. We then count the number of electron pairs (domains) around the Xe atom and use that number and the number of bonds to predict the geometry.

Solve There are 36 valence-shell electrons (8 from xenon and 7 from each fluorine). If we make four single Xe—F bonds, each fluorine has its octet satisfied. Xe then has 12 electrons in its valence shell, so we expect an octahedral arrangement of six electron pairs. Two of these are nonbonded pairs. Because nonbonded pairs require more volume than bonded pairs (Section 9.2), it is reasonable to expect these nonbonded pairs to be opposite each other. The expected structure is square planar, as shown in **Figure 22.7**.

Comment The experimentally determined structure agrees with this prediction.

▲ **Figure 22.7 Xenon tetrafluoride.**

▶ **Practice Exercise**
Describe the electron-domain geometry and molecular geometry of KrF_2.

Self-Assessment Exercises

SAE 22.10 Radon, below Xe in the periodic table, has no stable isotopes and thus is radioactive and difficult to study. Predict what type of reactions Rn is most likely to undergo. (**a**) Redox chemistry with fluorine to make radon fluorides (**b**) Redox chemistry with chlorine to make radon chlorides (**c**) Redox chemistry with calcium to make calcium radide salts (**d**) Alloys with argon at high pressure to make Ar_xRn_y compounds

SAE 22.11 What is the oxidation state of the noble gas in the compound krypton difluoride, KrF_2? (**a**) +1 (**b**) −1 (**c**) +2 (**d**) 0

SAE 22.12 The oxidation state of Xe in XeO_3 is _____ and its molecular geometry is _____. (**a**) +3, trigonal planar (**b**) +3, trigonal pyramidal (**c**) +6, trigonal planar (**d**) +6, trigonal pyramidal (**e**) +6, T-shaped

---

22.4 | Group 7A: The Halogens

The elements of group 7A, the halogens, have the outer-electron configuration ns^2np^5, where n ranges from 2 through 6. The halogens have large negative electron affinities (Section 7.4), and they most often achieve a noble-gas configuration by gaining an electron, which results in a −1 oxidation state. Fluorine, being the most electronegative element, exists in compounds only in the −1 state. The other halogens exhibit positive oxidation states up to +7 in combination with more electronegative atoms such as O. In the positive oxidation states, the halogens tend to be good oxidizing agents, readily accepting electrons.

Chlorine, bromine, and iodine are found as the halides in seawater and in salt deposits. Fluorine occurs in the minerals fluorspar (CaF_2), cryolite (Na_3AlF_6), and fluorapatite [$Ca_5(PO_4)_3F$].* Only fluorspar is an important commercial source of fluorine.

All isotopes of astatine are radioactive. The longest-lived isotope is astatine-210, which has a half-life of 8.1 h and decays mainly by electron capture. Because astatine is so unstable, very little is known about its chemistry.

Properties and Production of the Halogens

Most properties of the halogens vary in a regular fashion as we go from fluorine to iodine (**Table 22.2**).

Under ordinary conditions the halogens exist as diatomic molecules. The molecules are held together in the solid and liquid states by dispersion forces. (Section 11.2) Because

When you are finished with **Section 22.4**, you should be able to:

▶ Compare and contrast the physical and chemical properties of the halogens and their compounds.

▶ Predict the structure of compounds of group 7A using the VSEPR model and assign oxidation numbers to the elements within them.

Learning Objectives

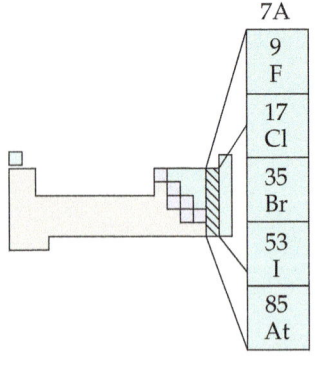

---

*Minerals are solid substances that occur in nature. They are usually known by their common names rather than by their chemical names. What we know as rock is merely an aggregate of different minerals.

TABLE 22.2   Some Properties of the Halogens

Property	F	Cl	Br	I
Atomic radius (Å)	0.57	1.02	1.20	1.39
Ionic radius, X^- (Å)	1.33	1.81	1.96	2.20
First ionization energy (kJ/mol)	1681	1251	1140	1008
Electron affinity (kJ/mol)	−328	−349	−325	−295
Electronegativity	4.0	3.0	2.8	2.5
X—X single-bond enthalpy (kJ/mol)	155	242	193	151
Reduction potential (V):				
$\frac{1}{2}X_2(aq) + e^- \longrightarrow X^-(aq)$	2.87	1.36	1.07	0.54

I_2 is the largest and most polarizable halogen molecule, the intermolecular forces between I_2 molecules are the strongest. Thus, I_2 has the highest melting point and boiling point. At room temperature and 1 atm, I_2 is a purple solid, Br_2 is a red-brown liquid, and Cl_2 and F_2 are gases. (Figure 7.28) Chlorine readily liquefies upon compression at room temperature and is normally stored and handled in liquid form under pressure in steel containers.

The comparatively low bond enthalpy of F_2 (155 kJ/mol) accounts in part for the extreme reactivity of elemental fluorine. Because of its high reactivity, F_2 is difficult to work with. Certain metals, such as copper and nickel, can be used to contain F_2 because their surfaces form a protective coating of metal fluoride. Chlorine and the heavier halogens are also reactive, though less so than fluorine.

Because of their high electronegativities, the halogens tend to gain electrons from other substances and thereby serve as oxidizing agents. The oxidizing ability of the halogens, indicated by their standard reduction potentials, decreases going down the group. As a result, a given halogen is able to oxidize the halide anions below it. For example, Cl_2 oxidizes Br^- and I^- but not F^-, as seen in **Figure 22.8**.

Notice in Table 22.2 that the standard reduction potential of F_2 is exceptionally high. As a result, fluorine gas readily oxidizes water:

$$F_2(aq) + H_2O(l) \longrightarrow 2\,HF(aq) + \tfrac{1}{2}O_2(g) \quad E° = 1.80\text{ V} \qquad [22.17]$$

Fluorine cannot be prepared by electrolytic oxidation of aqueous solutions of fluoride salts because water is oxidized more readily than F^-. (Section 20.9) In practice, the element is formed by electrolytic oxidation of a solution of KF in anhydrous HF.

Chlorine is produced mainly by electrolysis of either molten or aqueous sodium chloride. Both bromine and iodine are obtained commercially from brines containing the halide ions; the reaction used is oxidation with Cl_2.

Go Figure

Do Br_5 and I_2 appear to be more or less soluble in CCl_4 than in H_2O?

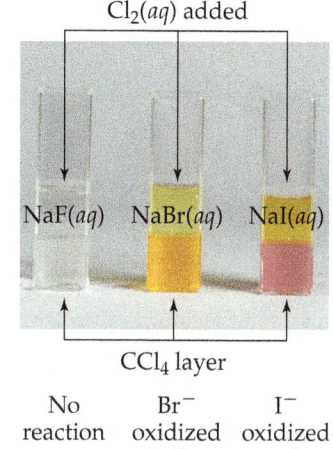

$Cl_2(aq)$ added

NaF(aq)   NaBr(aq)   NaI(aq)

CCl_4 layer

| No reaction | Br^- oxidized to Br_2 | I^- oxidized to I_2 |

▲ **Figure 22.8 Reaction of Cl_2 with aqueous solutions of NaF, NaBr, and NaI in the presence of carbon tetrachloride.** The top liquid layer in each vial is water; the bottom liquid layer is carbon tetrachloride. The $Cl_2(aq)$, which has been added to each vial, is colorless. The brown color in the carbon tetrachloride layer indicates the presence of Br_2, whereas purple indicates the presence of I_2.

Sample Exercise 22.4

Predicting Chemical Reactions among the Halogens

Write the balanced equation for the reaction, if any, between **(a)** $I^-(aq)$ and $Br_2(l)$, **(b)** $Cl^-(aq)$ and $I_2(s)$.

SOLUTION

Analyze We are asked to determine whether a reaction occurs when a particular halide and halogen are combined.

Plan A given halogen is able to oxidize anions of the halogens below it in the periodic table. Thus, in each pair the halogen having the smaller atomic number ends up as the halide ion. If the halogen with the smaller atomic number is already the halide ion, there is no reaction. Thus, the key to determining whether a reaction occurs is locating the elements in the periodic table.

Solve

(a) Br_2 can oxidize (remove electrons from) the anions of the halogens below it in the periodic table. Thus, it oxidizes I^-:

$$2\,I^-(aq) + Br_2(aq) \longrightarrow I_2(s) + 2\,Br^-(aq)$$

(b) Cl^- is the anion of a halogen above iodine in the periodic table. Thus, I_2 cannot oxidize Cl^-; there is no reaction.

▶ **Practice Exercise**
Write the balanced chemical equation for the reaction between $Br^-(aq)$ and $Cl_2(aq)$.

Go Figure

What is the repeating unit in this polymer?

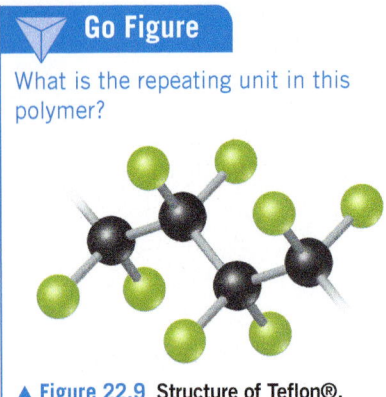

▲ **Figure 22.9** Structure of Teflon®, a fluorocarbon polymer.

Uses of the Halogens

Fluorine is used to prepare fluorocarbons—very stable carbon–fluorine compounds used as refrigerants, lubricants, and plastics. Teflon® (**Figure 22.9**) is a polymeric fluorocarbon noted for its high thermal stability and lack of chemical reactivity.

Chlorine is by far the most commercially important halogen. About half of the chlorine is used to manufacture chlorine-containing organic compounds, such as the vinyl chloride (C_2H_3Cl) used in making polyvinyl chloride (PVC) plastics. (Section 12.6) Much of the remainder is used as a bleaching agent in the paper and textile industries.

When Cl_2 dissolves in cold dilute base, it converts into Cl^- and hypochlorite, ClO^-:

$$Cl_2(aq) + 2\,OH^-(aq) \rightleftharpoons Cl^-(aq) + ClO^-(aq) + H_2O(l) \qquad [22.18]$$

Sodium hypochlorite (NaClO) is the active ingredient in many liquid bleaches. Chlorine is also used in water treatment to oxidize and thereby destroy bacteria. (Section 18.4)

A common use of iodine is as KI in table salt. Iodized salt provides the small amount of iodine necessary in our diets; it is essential for the formation of thyroxin, a hormone secreted by the thyroid gland. Lack of iodine in the diet results in an enlarged thyroid gland, a condition called *goiter*.

The Hydrogen Halides

All the halogens form stable diatomic molecules with hydrogen. The hydrogen halides can be formed by direct reaction of the elements.

The hydrogen halides form hydrohalic acid solutions when dissolved in water. These solutions have the characteristic properties of acids, such as reactions with active metals to produce hydrogen gas. (Section 4.4) Hydrofluoric acid also reacts readily with **silica** (SiO_2) and with silicates to form hexafluorosilicic acid (H_2SiF_6):

$$SiO_2(s) + 6\,HF(aq) \longrightarrow H_2SiF_6(aq) + 2\,H_2O(l) \qquad [22.19]$$

Interhalogen Compounds

Because the halogens exist as diatomic molecules, diatomic molecules made up of two different halogen atoms exist. These compounds are the simplest examples of **interhalogens**, compounds, such as ClF and IF_5, formed between two halogen elements.

The vast majority of the higher interhalogen compounds have a central Cl, Br, or I atom surrounded by fluorine atoms. The large size of the iodine atom allows the formation of IF_3, IF_5, and IF_7, in which the oxidation state of I is +3, +5, and +7, respectively. With the smaller bromine and chlorine atoms, only compounds with three or five fluorines form. The only higher interhalogen compounds that do not have outer F atoms are ICl_3 and ICl_5; the large size of the I atom can accommodate five Cl atoms, whereas Br is not large enough to allow even $BrCl_3$ to form. All of the interhalogen compounds are powerful oxidizing agents.

Oxyacids and Oxyanions

Table 22.3 summarizes the formulas of the known oxyacids of the halogens and the way they are named.* (Section 2.8) The acid strengths of the oxyacids increase with the increasing oxidation state of the central halogen atom. (Section 16.10) All the oxyacids are strong oxidizing agents. The oxyanions, formed on removal of H^+ from the oxyacids, are generally more stable than the oxyacids. Hypochlorite salts are used as bleaches and disinfectants because of the powerful oxidizing capabilities of the ClO^- ion. Chlorate salts are similarly very reactive. For example, potassium chlorate is used to make matches and fireworks.

---

*Fluorine forms one oxyacid, HOF. Because the electronegativity of fluorine is greater than that of oxygen, we must consider fluorine to be in a −1 oxidation state and oxygen to be in the 0 oxidation state in this compound.

TABLE 22.3   The Stable Oxyacids of the Halogens

Oxidation State of Halogen	Formula of Acid			Acid Name
	Cl	Br	I	
+1	HClO	HBrO	HIO	*Hypo*halous acid
+3	HClO₂	—	—	Hal*ous* acid
+5	HClO₃	HBrO₃	HIO₃	Hal*ic* acid
+7	HClO₄	HBrO₄	HIO₄	*Per*halic acid

$$10 \, Al(s) + 6 \, NH_4ClO_4(s) \longrightarrow$$
$$4 \, Al_2O_3(s) + 2 \, AlCl_3(s)$$
$$+ 12 \, H_2O(g) + 3 \, N_2(g)$$

The large volume of gases produced provides thrust for the booster rockets.

▲ **Figure 22.10 Launch of the Space Shuttle *Columbia*** from the Kennedy Space Center.

Perchloric acid and its salts are the most stable oxyacids and oxyanions. Dilute solutions of perchloric acid are quite safe, and many perchlorate salts are stable except when heated with organic materials. When heated, however, perchlorates can become vigorous, even violent, oxidizers. Considerable caution should be exercised, therefore, when handling these substances, and it is crucial to avoid contact between perchlorates and readily oxidized material. The use of ammonium perchlorate (NH_4ClO_4) as the oxidizer in the solid booster rockets for space vehicle launches demonstrates the oxidizing power of perchlorates. The solid propellant contains a mixture of NH_4ClO_4 and powdered aluminum, the reducing agent. Each launch requires about 6×10^5 kg (700 tons) of NH_4ClO_4 (**Figure 22.10**).

 Self-Assessment Exercises

SAE 22.13   The smallest halogen is _____, and the easiest halogen to oxidize is _____. (**a**) fluorine, fluorine (**b**) fluorine, iodine (**c**) iodine, fluorine (**d**) iodine, iodine

SAE 22.14   The oxidation number of iodine in IF₃ is _____ and its molecular geometry is _____. (**a**) +3, trigonal (**b**) +3, T-shaped (**c**) +3, trigonal pyramidal (**d**) +3, trigonal bipyramidal (**e**) −3, trigonal pyramidal

22.5 | Oxygen

Oxygen is found in combination with other elements in a great variety of compounds; water (H_2O), silica (SiO_2), alumina (Al_2O_3), and the iron oxides (Fe_2O_3, Fe_3O_4) are common examples. Indeed, oxygen is the most abundant element by mass both in Earth's crust and in the human body. (Section 1.2) It is the oxidizing agent for the metabolism of our foods and is crucial to human life.

Properties of Oxygen

Oxygen has two allotropes, O_2 and O_3. When we speak of molecular oxygen or simply oxygen, it is usually understood that we are speaking of *dioxygen* (O_2), the normal form of the element; O_3 is ozone.

At room temperature, dioxygen is a colorless and odorless gas. Dioxygen is only slightly soluble in water (0.04 g/L, or 0.001 M at 25 °C), but its presence in water is essential to marine life.

The electron configuration of the oxygen atom is [He]$2s^2 2p^4$. Thus, oxygen can complete its octet of valence electrons either by picking up two electrons to form the oxide ion (O^{2-}) or by sharing two electrons. In its covalent compounds, it tends to form either two single bonds, as in H_2O, or a double bond, as in formaldehyde ($H_2C{=}O$). The O_2 molecule contains a double bond. The bond in O_2 is very strong (bond enthalpy 495 kJ/mol). Oxygen also forms strong bonds with many other elements. Consequently, many oxygen-containing compounds are thermodynamically more stable than O_2. In the absence of a catalyst, however, most reactions of O_2 have high activation energies

 Learning Objectives

When you finish Section 22.5, you should be able to:

▶ Write balanced chemical equations for reactions involving oxygen.

▶ Compare the acid–base properties of metal oxides and nonmetal oxides.

▶ Assign oxidation states to oxygen atoms in reactive oxygen species.

$$2\,C_2H_2(g) + 5\,O_2(g) \longrightarrow$$
$$4\,CO_2(g) + 2\,H_2O(g);$$
$$\Delta H^\circ = -2510 \text{ kJ}$$

▲ **Figure 22.11 Welding with an oxyacetylene torch.**

and thus require high temperatures to proceed at a suitable rate. Once a sufficiently exothermic reaction begins, it may accelerate rapidly, producing a reaction of explosive violence.

Production of Oxygen

Nearly all commercial oxygen is obtained from air. The normal boiling point of O_2 is $-183\,^\circ C$, whereas that of N_2, the other principal component of air, is $-196\,^\circ C$. Thus, when air is liquefied and then allowed to warm, the N_2 boils off, leaving liquid O_2 contaminated mainly by small amounts of N_2 and Ar.

Much of the O_2 in the atmosphere is replenished through photosynthesis, in which green plants use the energy of sunlight to generate O_2 (along with glucose, $C_6H_{12}O_6$) from atmospheric CO_2:

$$6\,CO_2(g) + 6\,H_2O(l) \longrightarrow C_6H_{12}O_6(aq) + 6\,O_2(g) \qquad [22.20]$$

Uses of Oxygen

In industrial use, oxygen ranks behind only sulfuric acid (H_2SO_4) and nitrogen (N_2). Oxygen is by far the most widely used oxidizing agent in industry. Over half of the O_2 produced is used in the steel industry, mainly to remove impurities from steel. It is also used to bleach pulp and paper. (Oxidation of colored compounds often gives colorless products.) Oxygen is used together with acetylene (C_2H_2) in oxyacetylene welding (**Figure 22.11**). The reaction between C_2H_2 and O_2 is highly exothermic, producing temperatures in excess of $3000\,^\circ C$.

Ozone

Ozone is a pale blue, poisonous gas with a sharp, irritating odor. Many people can detect as little as 0.01 ppm in air. Exposure to 0.1 to 1 ppm produces headaches, burning eyes, and irritation to the respiratory passages.

The O_3 molecule possesses π electrons that are delocalized over the three oxygen atoms. (Section 8.6) The molecule dissociates readily, forming reactive oxygen atoms:

$$O_3(g) \longrightarrow O_2(g) + O(g) \quad \Delta H^\circ = 105 \text{ kJ} \qquad [22.21]$$

Ozone is a stronger oxidizing agent than dioxygen. Ozone forms oxides with many elements under conditions where O_2 will not react; indeed, it oxidizes all the common metals except gold and platinum.

Ozone can be prepared by passing electricity through dry O_2. During thunderstorms, ozone is generated (and can be smelled, if you are too close) from lightning strikes:

$$3\,O_2(g) \xrightarrow{\text{electricity}} 2\,O_3(g) \quad \Delta H^\circ = 284.6 \text{ kJ} \qquad [22.22]$$

Ozone is sometimes used to treat drinking water. Like Cl_2, ozone kills bacteria and oxidizes organic compounds. The largest use of ozone, however, is in the preparation of pharmaceuticals, synthetic lubricants, and other commercially useful organic compounds, where O_3 is used to sever carbon–carbon double bonds.

Ozone is an important component of the upper atmosphere, where it screens out ultraviolet radiation and so protects us from the effects of these high-energy rays. For this reason, depletion of stratospheric ozone is a major scientific concern. (Section 18.2) In the lower atmosphere, ozone is considered an air pollutant and is a major constituent of smog. (Section 18.2) Because of its oxidizing power, ozone damages living systems and structural materials, especially rubber.

Oxides

The electronegativity of oxygen is second only to that of fluorine. As a result, oxygen has negative oxidation states in all compounds except OF_2 and O_2F_2. The -2 oxidation state is by far the most common. Compounds that contain oxygen in this oxidation state are called *oxides*.

Nonmetals form covalent oxides, most of which are simple molecules with low melting and boiling points. Both SiO_2 and B_2O_3, however, have extended structures. Most nonmetal oxides combine with water to give oxyacids. Sulfur dioxide (SO_2), for example, dissolves in water to give sulfurous acid (H_2SO_3):

$$SO_2(g) + H_2O(l) \longrightarrow H_2SO_3(aq) \qquad [22.23]$$

This reaction and that of SO_3 with H_2O to form H_2SO_4 are largely responsible for acid rain. (Section 18.2) The analogous reaction of CO_2 with H_2O to form carbonic acid (H_2CO_3) causes the acidity of carbonated water.

Oxides that form acids when they react with water are called either **acidic anhydrides** (anhydride means "without water") or **acidic oxides**. A few nonmetal oxides, especially ones with the nonmetal in a low oxidation state—such as N_2O, NO, and CO— do not react with water and are not acidic anhydrides.

Most metal oxides are ionic compounds. The ionic oxides that dissolve in water form hydroxides and, consequently, are called either **basic anhydrides** or **basic oxides**. Barium oxide, for example, reacts with water to form barium hydroxide (**Figure 22.12**). These kinds of reactions are due to the high basicity of the O^{2-} ion and its virtually complete hydrolysis in water:

$$O^{2-}(aq) + H_2O(l) \longrightarrow 2\,OH^-(aq) \qquad [22.24]$$

Even those ionic oxides that are insoluble in water tend to dissolve in strong acids. Iron(III) oxide, for example, dissolves in acids:

$$Fe_2O_3(s) + 6\,H^+(aq) \longrightarrow 2\,Fe^{3+}(aq) + 3\,H_2O(l) \qquad [22.25]$$

This reaction is used to remove rust ($Fe_2O_3 \cdot nH_2O$) from iron or steel before a protective coat of zinc or tin is applied.

Oxides that can exhibit both acidic and basic characters are said to be *amphoteric*. (Section 17.5) If a metal forms more than one oxide, the basic character of the oxide decreases as the oxidation state of the metal increases (**Table 22.4**); or said a different way, the higher the oxidation state of the metal, the more acidic the oxide. The reasons for this are complex, but it comes down to M—O bonding in the oxide.

Go Figure   Is this reaction a redox reaction?

H$_2$O with indicator

Pink color indicates basic solution

$$BaO(s) \quad + \quad H_2O(l) \quad \longrightarrow \quad Ba(OH)_2(aq)$$

▲ **Figure 22.12** Reaction of a basic oxide with water.

$$4 \, KO_2(s) + 2 \, H_2O(l, \text{from breath}) \longrightarrow$$
$$4 \, K^+(aq) + 4 \, OH^-(aq) + 3 \, O_2(g)$$

$$2 \, OH^-(aq) + CO_2(g, \text{from breath}) \longrightarrow$$
$$H_2O(l) + CO_3^{2-}(aq)$$

▲ **Figure 22.13** A self-contained breathing apparatus.

TABLE 22.4   Acid–Base Character of Chromium Oxides

Oxide	Oxidation State of Cr	Nature of Oxide
CrO	+2	Basic
Cr_2O_3	+3	Amphoteric
CrO_3	+6	Acidic

Lower-oxidation-state metals (CrO in Table 22.4) have more ionic M—O bonds, and thus react with water to produce hydroxide (Figure 22.12). Higher-oxidation-state metals in metal oxides (CrO_3 in Table 22.4) have more covalent M—O bonding, are more acidic and thus react with hydroxide to produce water:

$$CrO_3(s) + 2 \, OH^-(aq) \longrightarrow CrO_4^{2-}(aq) + H_2O(l) \qquad [22.26]$$

Peroxides, Superoxides and Reactive Oxygen Species

Compounds containing O—O bonds and oxygen in the −1 oxidation state are *peroxides*. Oxygen has an oxidation state of $-\frac{1}{2}$ in O_2^-, which is called the *superoxide* ion. The most active (easily oxidized) metals (K, Rb, and Cs) react with O_2 to give superoxides (KO_2, RbO_2, and CsO_2). Their active neighbors in the periodic table (Na, Ca, Sr, and Ba) react with O_2, producing peroxides (Na_2O_2, CaO_2, SrO_2, and BaO_2). Less active metals and nonmetals produce normal oxides. (Section 7.5)

When superoxides dissolve in water, O_2 is produced:

$$4 \, KO_2(s) + 2 \, H_2O(l) \longrightarrow 4 \, K^+(aq) + 4 \, OH^-(aq) + 3 \, O_2(g) \qquad [22.27]$$

Because of this reaction, potassium superoxide is used as an oxygen source in masks worn by rescue workers (**Figure 22.13**). For proper breathing in toxic environments, oxygen must be generated in the mask and exhaled carbon dioxide in the mask must be eliminated. Moisture in the breath causes the KO_2 to decompose to O_2 and KOH, and the KOH removes CO_2 from the exhaled breath:

$$2 \, OH^-(aq) + CO_2(g) \longrightarrow H_2O(l) + CO_3^{2-}(aq) \qquad [22.28]$$

Hydrogen peroxide (**Figure 22.14**) is the most familiar and commercially important peroxide. Pure hydrogen peroxide is a clear, syrupy liquid that melts at −0.4 °C. Concentrated hydrogen peroxide is dangerously reactive because the decomposition to water and oxygen is very exothermic:

$$2 \, H_2O_2(l) \longrightarrow 2 \, H_2O(l) + O_2(g) \quad \Delta H° = -196.1 \, kJ \qquad [22.29]$$

This is an example of a **disproportionation** reaction, in which an element is simultaneously oxidized and reduced. The oxidation number of oxygen changes from −1 to −2 and 0.

Hydrogen peroxide is marketed as a chemical reagent in aqueous solutions of up to about 30% by mass. A solution containing about 3% H_2O_2 by mass is sold in drugstores and used as a mild antiseptic. Somewhat more concentrated solutions are used to bleach fabrics.

The peroxide ion is a byproduct of metabolism that results from the reduction of O_2. Peroxide, superoxide, and similar O-containing compounds that are byproducts of metabolism are called **reactive oxygen species** (ROS). Some ROS is always present in living systems. Elevated levels of ROS in biological systems are associated with stress, and can lead to oxidation of key compounds such as proteins and DNA, leading to irreversible oxidative damage and potentially disease. To counteract these reactions, the body has numerous enzymes that catalyze the destruction of ROS. *Superoxide dismutase*, for example, catalyzes the conversion of superoxide into oxygen and hydrogen peroxide. *Catalase* catalyzes the decomposition of hydrogen peroxide into water and oxygen, the same reaction as Equation 22.29.

Go Figure

Does H_2O_2 have a dipole moment?

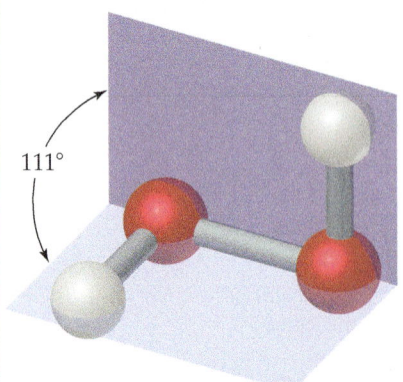

111°

▲ **Figure 22.14** Molecular structure of hydrogen peroxide in the gas phase. The repulsive interaction of the O—H bonds with the lone pairs of electrons on each O atom restricts the free rotation around the O—O single bond.

Self-Assessment Exercises

SAE 22.15   Complete and balance the following reaction:

$$I_2O_5 + H_2O \longrightarrow ?$$

(a) $I_2O_5 + H_2O \longrightarrow 2\,HIO_3$
(b) $I_2O_5 + H_2O \longrightarrow 6\,H_2O + I_2$
(c) $I_2O_5 + H_2O \longrightarrow HIO_3$
(d) $I_2O_5 + H_2O \longrightarrow 2\,HI + 3\,O_2$

SAE 22.16   Predict which compound in each pair forms the more acidic oxide when it reacts with water.

$$CO_2 \text{ and } GeO_2$$

$$MnO \text{ and } MnO_2$$

(a) CO_2 and MnO_2 (b) CO_2 and MnO (c) GeO_2 and MnO (d) GeO_2 and MnO_2

SAE 22.17   The hydroxyl radical, OH·, called "O–H dot," is another example of a reactive oxygen species. Two hydroxy radicals would be formed if the O—O bond in hydrogen peroxide was cut in half. How many valence electrons does one hydroxyl radical contain? (a) 6 (b) 7 (c) 8 (d) 9 (e) 10

22.6 | The Other Group 6A Elements: S, Se, Te, and Po

The other group 6A elements are sulfur, selenium, tellurium, and polonium. Of these, sulfur is the most important, and polonium, which has no stable isotopes and is found only in minute quantities in radium-containing minerals, is least important.

The group 6A elements possess the general outer-electron configuration ns^2np^4 with n ranging from 2 to 6. Thus, these elements attain a noble-gas electron configuration by adding two electrons, which results in a -2 oxidation state. Except for oxygen, the group 6A elements are also commonly found in positive oxidation states up to $+6$, and they can have expanded valence shells. Thus, we have such compounds as SF_6, SeF_6, and TeF_6 with the central atom in the $+6$ oxidation state.

Table 22.5 summarizes some properties of the group 6A elements.

Occurrence and Production of S, Se, and Te

Sulfur, selenium, and tellurium can all be mined from the earth. Large underground deposits are the principal source of elemental sulfur (**Figure 22.15**). Sulfur also occurs widely as sulfide (S^{2-}) and sulfate (SO_4^{2-}) minerals. Its presence as a minor component of coal and petroleum poses a major problem. Combustion of these "unclean" fuels leads to serious pollution by sulfur oxides. (Section 18.2) Much effort has been directed at removing this sulfur, and these efforts have increased the availability of sulfur.

Selenium and tellurium occur in rare minerals, such as Cu_2Se, $PbSe$, Cu_2Te, and $PbTe$, and as minor constituents in sulfide ores of copper, iron, nickel, and lead.

Properties and Uses of Sulfur, Selenium, and Tellurium

Elemental sulfur is yellow, tasteless, and nearly odorless. It is insoluble in water and exists in several allotropic forms. The thermodynamically stable form at room temperature is

Learning Objectives

When you are finished with Section 22.6, you should be able to:

▶ Predict the physical and chemical properties of the group 6A elements beyond oxygen.

▶ Predict the structure of compounds of group 6A using the VSEPR model and assign oxidation numbers to the elements within them.

TABLE 22.5   Some Properties of the Group 6A Elements

Property	O	S	Se	Te
Atomic radius (Å)	0.66	1.05	1.21	1.38
X^{2-} ionic radius (Å)	1.40	1.84	1.98	2.21
First ionization energy (kJ/mol)	1314	1000	941	869
Electron affinity (kJ/mol)	−141	−200	−195	−190
Electronegativity	3.5	2.5	2.4	2.1
X—X single-bond enthalpy (kJ/mol)	146*	266	172	126
Reduction potential to H_2X in acidic solution (V)	1.23	0.14	−0.40	−0.72

*Based on O—O bond enthalpy in H_2O_2.

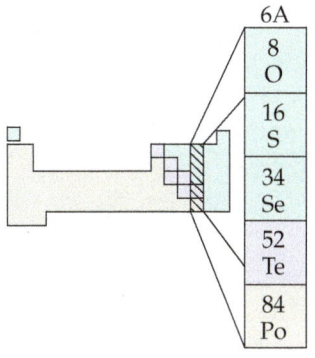

6A
8 O
16 S
34 Se
52 Te
84 Po

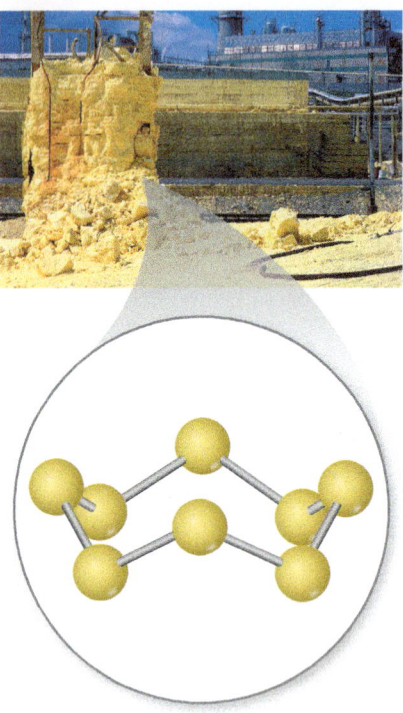

▲ **Figure 22.15** Massive amounts of sulfur are extracted every year from the earth.

▲ **Figure 22.16** Iron pyrite (FeS$_2$, on the right) with gold for comparison.

▲ **Figure 22.17** Food label warning of sulfites.

rhombic sulfur, which consists of puckered S$_8$ rings with each sulfur atom forming two bonds (Figure 22.15).

Most of the sulfur produced in the United States each year is used to manufacture sulfuric acid. Sulfur is also used to vulcanize rubber, a process that toughens rubber by introducing cross-linking between polymer chains. (Section 12.6)

Selenium is used in photoelectric cells and light meters because its electrical conductivity increases greatly upon exposure to light. Photocopiers also depend on the photoconductivity of selenium. Photocopy machines contain a belt or drum coated with a film of selenium. This drum is electrostatically charged and then exposed to light reflected from the image being photocopied. The charge drains from the regions where the selenium film has been made conductive by exposure to light. A black powder (the toner) sticks only to the areas that remain charged. The photocopy is made when the toner is transferred to a sheet of plain paper.

Sulfides

When an element is less electronegative than sulfur, *sulfides* that contain S^{2-} form. Many metallic elements are found in the form of sulfide ores, such as PbS (galena) and HgS (cinnabar). A series of related ores containing the disulfide ion, S$_2^{2-}$ (analogous to the peroxide ion), are known as *pyrites*. Iron pyrite, FeS$_2$, occurs as golden yellow cubic crystals (**Figure 22.16**). Because it has been occasionally mistaken for gold by miners, iron pyrite is often called fool's gold.

One of the most important sulfides is hydrogen sulfide (H$_2$S). One of hydrogen sulfide's most readily recognized properties is its odor, which is most frequently encountered as the offensive odor of rotten eggs. Hydrogen sulfide is toxic, but our noses can detect H$_2$S in extremely low, nontoxic concentrations. A sulfur-containing organic molecule, such as dimethyl sulfide, (CH$_3$)$_2$S, which is similarly odoriferous and can be detected by smell at a level of one part per trillion, is added to natural gas as a safety factor to give it a detectable odor.

Oxides, Oxyacids, and Oxyanions of Sulfur

Sulfur dioxide, formed when sulfur burns in air, has a choking odor and is poisonous. The gas is particularly toxic to lower organisms, such as fungi, so it is used to sterilize dried fruit and wine. At 1 atm and room temperature, SO$_2$ dissolves in water to produce a 1.6 M solution. The SO$_2$ solution is acidic, and we describe it as sulfurous acid (H$_2$SO$_3$).

Salts of SO$_3^{2-}$ (sulfites) and HSO$_3^-$ (hydrogen sulfites or bisulfites) are well known. Small quantities of Na$_2$SO$_3$ or NaHSO$_3$ are used as food additives to prevent bacterial spoilage. However, they are known to increase asthma symptoms in approximately 5% of asthmatics. Thus, all food products with sulfites must now carry a warning label disclosing their presence (**Figure 22.17**).

Although combustion of sulfur in air produces mainly SO$_2$, small amounts of SO$_3$ are also formed. The reaction produces chiefly SO$_2$ because the activation-energy barrier for oxidation to SO$_3$ is very high unless the reaction is catalyzed. Interestingly, the SO$_3$ by-product is used industrially to make H$_2$SO$_4$, which is the ultimate product of the reaction between SO$_3$ and water. In the manufacture of sulfuric acid, SO$_2$ is obtained by burning sulfur and then is oxidized to SO$_3$, using a catalyst such as V$_2$O$_5$ or platinum. The SO$_3$ is dissolved in H$_2$SO$_4$ because it does not dissolve quickly in water, and then the H$_2$S$_2$O$_7$ formed in this reaction, called pyrosulfuric acid, is added to water to form H$_2$SO$_4$:

$$SO_3(g) + H_2SO_4(l) \longrightarrow H_2S_2O_7(l) \qquad [22.30]$$

$$H_2S_2O_7(l) + H_2O(l) \longrightarrow 2\,H_2SO_4(l) \qquad [22.31]$$

Commercial sulfuric acid is 98% H$_2$SO$_4$. It is a dense, colorless, oily liquid. It is a strong acid, a good dehydrating agent (**Figure 22.18**), and a moderately good oxidizing agent.

$$C_{12}H_{22}O_{11}(s) \xrightarrow{\ H_2SO_4(aq)\ } 12\ C(s) + 11\ H_2O(l)$$

▲ Figure 22.18 Sulfuric acid dehydrates table sugar to produce elemental carbon.

Year after year, the production of sulfuric acid is the largest of any chemical produced in the United States. It is employed in some way in almost all manufacturing.

Sulfuric acid is a strong acid, but only the first hydrogen is completely ionized in aqueous solution:

$$H_2SO_4(aq) \longrightarrow H^+(aq) + HSO_4^-(aq) \tag{22.32}$$

$$HSO_4^-(aq) \rightleftharpoons H^+(aq) + SO_4^{2-}(aq) \quad K_a = 1.1 \times 10^{-2} \tag{22.33}$$

Consequently, sulfuric acid forms both sulfates (SO_4^{2-} salts) and bisulfates (or hydrogen sulfates, HSO_4^- salts). Bisulfate salts are common components of the "dry acids" used for adjusting the pH of swimming pools and hot tubs; they are also components of many toilet bowl cleaners.

Another important sulfur-containing ion is thiosulfate ion ($S_2O_3^{2-}$). The term *thio* indicates substitution of sulfur for oxygen. The structures of the sulfate and thiosulfate ions are compared in **Figure 22.19**.

Thiols, organic compounds with a —SH group bonded to carbon, play important roles in biology. When thiols are oxidized under mild conditions, they form disulfides, organic compounds with S—S bonds between carbon atoms that are analogous to peroxides. In the opposite fashion, disulfide bonds can be broken by reducing them to thiols. The shape and function of proteins can be modified by forming or breaking disulfide bonds between protein strands. Many proteins maintain their three-dimensional shape due to S–S crosslinks. Hair-straightening and curling technologies rely on the breaking and re-forming of disulfide bonds between the protein molecules of hair (**Figure 22.20**).

Go Figure What are the oxidation states of the sulfur atoms in the $S_2O_3^{2-}$ ion?

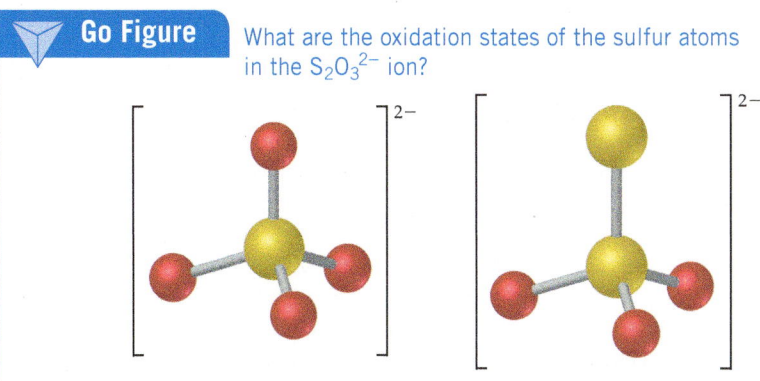

▲ Figure 22.19 Structures of the sulfate (left) and thiosulfate (right) ions.

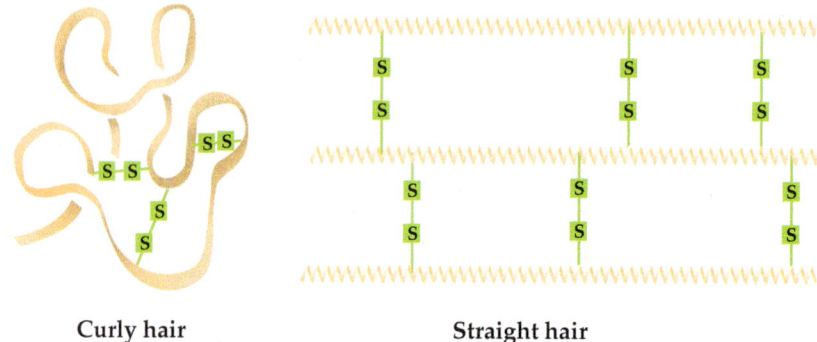

Curly hair   Straight hair

▲ **Figure 22.20 Hair straightening and curling.** People can alter the shape of their hair by first applying a reducing agent that breaks the disulfide bonds between protein strands. The hair is given its desired shape, and then an oxidizing agent is added to form new disulfide bonds to maintain the shape.

 Self-Assessment Exercises

SAE 22.18   What is the chemical formula for sulfurous acid? What is the oxidation state of sulfur in this compound? (**a**) H_2SO_3, +6 (**b**) HSO_3, +5 (**c**) H_2SO_3, +4 (**d**) H_2S, −2

SAE 22.19   Predict the molecular structure of sulfur tetrachloride, SCl_4. (**a**) square pyramidal (**b**) see-saw (**c**) linear (**d**) square-planar

SAE 22.20   Which of the following reactions best describes the reaction between hydrogen peroxide and selenium?

(**a**) $2 H_2O_2(aq) + Se(s) \longrightarrow SeO_2(s) + 2 H_2O(l)$
(**b**) $H_2O_2(aq) + Se(s) \longrightarrow SeO_2(s) + H_2O(l)$
(**c**) $H_2O_2(aq) + Se(s) \longrightarrow SeO_2(s) + 2 H_2O(l)$
(**d**) $H_2O_2(aq) + Se(s) \longrightarrow SeO_2(s) + H_2(g)$

SAE 22.21   For the reaction $RSH + HSR' \longrightarrow RSSR' + H_2$, where R and R′ are alkyl groups, what are the oxidation states of sulfur in the reactants and products, respectively? (**a**) 0, 0 (**b**) −1, −1 (**c**) −2, −2 (**d**) −2, −1 (**e**) −1, −2

22.7 | Nitrogen

Nitrogen constitutes 78% by volume of Earth's atmosphere, where it occurs as N_2 molecules. Although nitrogen is a key element in living organisms, compounds of nitrogen are not abundant in Earth's crust. The major natural deposits of nitrogen compounds are those of KNO_3 (saltpeter) in India and $NaNO_3$ (Chile saltpeter) in Chile and other desert regions of South America.

Properties of Nitrogen

Nitrogen is a colorless, odorless, tasteless gas composed of N_2 molecules. The N_2 molecule is very unreactive because of the strong triple bond between nitrogen atoms (the $N{\equiv}N$ bond enthalpy is 941 kJ/mol, nearly twice that for the bond in O_2.) (Table 8.3) When substances burn in air, they normally react with O_2 but not with N_2.

The electron configuration of the nitrogen atom is $[He]2s^2 2p^3$. The element exhibits all formal oxidation states from +5 to −3 (Table 22.6). The +5, 0, and −3 oxidation states are the most common and generally the most stable of these. Because nitrogen is more electronegative than all other elements except fluorine, oxygen, and chlorine, it exhibits positive oxidation states only in combination with these three elements.

Production and Uses of Nitrogen

Elemental nitrogen is obtained in commercial quantities by fractional distillation of liquid air. Because of its low reactivity, large quantities of N_2 are used as an inert gaseous blanket to exclude O_2 in food processing, manufacture of chemicals, metal fabrication, and production of electronic devices. Liquid N_2 is employed as a coolant to freeze foods rapidly.

The largest use of N_2 is in the manufacture of nitrogen-containing fertilizers. These fertilizers provide a source of *fixed* nitrogen—nitrogen that has been incorporated into compounds. We have previously discussed nitrogen fixation in the "Chemistry and Life" box on "Nitrogen Fixation and Nitrogenase" in Section 14.6 and in the "Chemistry and Sustainability" box on "The Haber Process: Feeding the World" in Section 15.2. Our

Learning Objectives

When you finish Section 22.7, you should be able to:

▶ Determine the oxidation state of nitrogen in compounds.

▶ Write balanced chemical equations for reactions of nitrogen-containing compounds.

TABLE 22.6   Oxidation States of Nitrogen

Oxidation State	Examples
+5	N_2O_5, HNO_3, NO_3^-
+4	NO_2, N_2O_4
+3	HNO_2, NO_2^-, NF_3
+2	NO
+1	N_2O, $H_2N_2O_2$, $N_2O_2^{2-}$, HNF_2
0	N_2
−1	NH_2OH, NH_2F
−2	N_2H_4
−3	NH_3, NH_4^+, NH_2^-

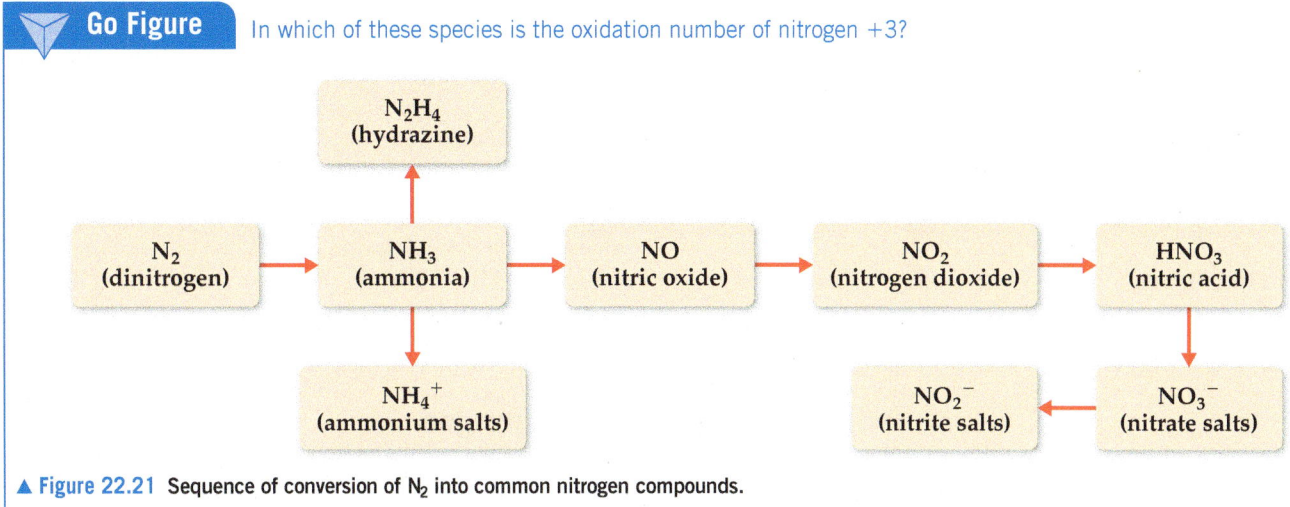

▲ Figure 22.21 Sequence of conversion of N_2 into common nitrogen compounds.

starting point in fixing nitrogen is the manufacture of ammonia via the Haber process. (Section 15.2) The ammonia can then be converted into a variety of useful, simple nitrogen-containing species (**Figure 22.21**).

Hydrogen Compounds of Nitrogen

Ammonia is one of the most important compounds of nitrogen. It is a colorless, toxic gas that has a characteristic irritating odor. As noted in previous discussions, the NH_3 molecule is basic ($K_b = 1.8 \times 10^{-5}$). (Section 16.7)

In the laboratory, NH_3 can be prepared by the action of NaOH on an ammonium salt. The NH_4^+ ion, which is the conjugate acid of NH_3, transfers a proton to OH^-. The resultant NH_3 is volatile and is driven from the solution by mild heating:

$$NH_4Cl(aq) + NaOH(aq) \longrightarrow NH_3(g) + H_2O(l) + NaCl(aq) \qquad [22.34]$$

Commercial production of NH_3 is achieved by the Haber process: (Section 15.2)

$$N_2(g) + 3\,H_2(g) \longrightarrow 2\,NH_3(g) \qquad [22.35]$$

About 75% is used for fertilizer.

Hydrazine (N_2H_4) is another important hydride of nitrogen. The hydrazine molecule contains an N—N single bond (**Figure 22.22**). It can be prepared by the reaction of ammonia with hypochlorite ion (OCl^-) in aqueous solution:

$$2\,NH_3(aq) + OCl^-(aq) \longrightarrow N_2H_4(aq) + Cl^-(aq) + H_2O(l) \qquad [22.36]$$

The reaction involves several intermediates, including chloramine (NH_2Cl). The poisonous NH_2Cl bubbles out of solution when household ammonia and chlorine bleach (which contains OCl^-) are mixed. This reaction is one reason for the frequently cited warning not to mix bleach and household ammonia.

Pure hydrazine is a strong and versatile reducing agent. The major use of hydrazine and compounds related to it, such as methylhydrazine (Figure 22.22), is as rocket fuel.

Go Figure

Is the N—N bond length in these molecules shorter or longer than the N—N bond length in N_2?

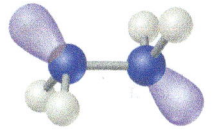

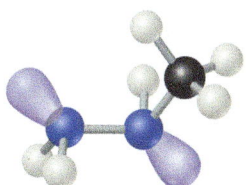

▲ Figure 22.22 Hydrazine (top, N_2H_4) and methylhydrazine (bottom, CH_3NHNH_2). The purple lobes represent the orbitals where the lone pairs of electrons reside.

Sample Exercise 22.5

Writing a Balanced Equation

Hydroxylamine (NH_2OH) reduces copper(II) to the free metal in acid solutions. Write a balanced equation for the reaction, assuming that N_2 is the oxidation product.

SOLUTION

Analyze We are asked to write a balanced oxidation–reduction equation in which NH_2OH is converted to N_2 and Cu^{2+} is converted to Cu.

Plan Because this is a redox reaction, the equation can be balanced by the method of half-reactions discussed in Section 20.2. Thus, we begin with two half-reactions, one involving the NH_2OH and N_2 and the other involving Cu^{2+} and Cu.

Continued

Solve

The unbalanced and incomplete half-reactions are

$$Cu^{2+}(aq) \longrightarrow Cu(s)$$

$$NH_2OH(aq) \longrightarrow N_2(g)$$

Balancing these equations as described in Section 20.2 gives

$$Cu^{2+}(aq) + 2\,e^- \longrightarrow Cu(s)$$

$$2\,NH_2OH(aq) \longrightarrow N_2(g) + 2\,H_2O(l) + 2\,H^+(aq) + 2\,e^-$$

Adding these half-reactions gives the balanced equation:

$$Cu^{2+}(aq) + 2\,NH_2OH(aq) \longrightarrow Cu(s) + N_2(g) + 2\,H_2O(l) + 2\,H^+(aq)$$

▶ **Practice Exercise**

Methylhydrazine, $N_2H_3CH_3(l)$, was used with the oxidizer dinitrogen tetroxide, $N_2O_4(l)$, to power the steering rockets of the Space Shuttle orbiter. The reaction of these two substances produces N_2, CO_2, and H_2O. Write a balanced equation for this reaction.

Oxides and Oxyacids of Nitrogen

Nitrogen forms three common oxides: N_2O (nitrous oxide), NO (nitric oxide), and NO_2 (nitrogen dioxide). It also forms two unstable oxides that we will not discuss: N_2O_3 (dinitrogen trioxide) and N_2O_5 (dinitrogen pentoxide).

Nitrous oxide (N_2O) is also known as laughing gas because a person becomes giddy after inhaling a small amount. This colorless gas was the first substance used as a general anesthetic. It is used as the compressed gas propellant in several aerosols and foams, such as in whipped cream.

Nitric oxide (NO) is also a colorless gas but, unlike N_2O, it is slightly toxic. It can be prepared in the laboratory by reduction of dilute nitric acid, using copper or iron as a reducing agent:

$$3\,Cu(s) + 2\,NO_3^-(aq) + 8\,H^+(aq) \longrightarrow 3\,Cu^{2+}(aq) + 2\,NO(g) + 4\,H_2O(l) \quad [22.37]$$

Nitric oxide is also produced by direct reaction of N_2 and O_2 at high temperatures. This reaction is a significant source of nitrogen oxide air pollutants. (Section 18.2) The direct combination of N_2 and O_2 is not used for commercial production of NO, however, because the yield is low, the equilibrium constant K_p at 2400 K being only 0.05. [See the "Chemistry and Sustainability" box on "Controlling Nitric Oxide Emissions."[(Section 15.7)]

The commercial route to NO (and hence to other oxygen-containing compounds of nitrogen) is via the catalytic oxidation of NH_3:

$$4\,NH_3(g) + 5\,O_2(g) \xrightarrow[850\,°C]{Pt\,catalyst} 4\,NO(g) + 6\,H_2O(g) \quad [22.38]$$

This reaction is the first step in the **Ostwald process**, by which NH_3 is converted commercially into nitric acid (HNO_3).

When exposed to air, nitric oxide reacts readily with O_2 (**Figure 22.23**):

$$2\,NO(g) + O_2(g) \longrightarrow 2\,NO_2(g) \quad [22.39]$$

▲ **Figure 22.23** Formation of $NO_2(g)$ as $NO(g)$ combines with $O_2(g)$ in the air.

When dissolved in water, NO_2 forms nitric acid:

$$3\,NO_2(g) + H_2O(l) \longrightarrow 2\,H^+(aq) + 2\,NO_3^-(aq) + NO(g) \qquad [22.40]$$

Nitrogen is both oxidized and reduced in this reaction, which means it disproportionates. The NO can be converted back into NO_2 by exposure to air (Equation 22.39) and thereafter dissolved in water to prepare more HNO_3.

NO is an important neurotransmitter in the human body. It causes the muscles that line blood vessels to relax, thus allowing an increased passage of blood (see the "Chemistry and Life" box on "Nitroglycerin, Nitric Oxide, and Heart Disease" at the end of this section).

Nitrogen dioxide (NO_2) is a yellow-brown gas (Figure 22.23). Like NO, it is a major constituent of smog. (Section 18.2) It is poisonous and has a choking odor. As discussed in Section 15.1, NO_2 and N_2O_4 exist in equilibrium:

$$2\,NO_2(g) \rightleftharpoons N_2O_4(g) \quad \Delta H° = -58\text{ kJ} \qquad [22.41]$$

The two common oxyacids of nitrogen are nitric acid (HNO_3) and nitrous acid (HNO_2) (Figure 22.24). *Nitric acid* is a strong acid. It is also a powerful oxidizing agent, as indicated by the standard reduction potential in the reaction

$$NO_3^-(aq) + 4\,H^+(aq) + 3\,e^- \longrightarrow NO(g) + 2\,H_2O(l) \quad E° = +0.96\text{ V} \quad [22.42]$$

Concentrated nitric acid attacks and oxidizes most metals except Au, Pt, Rh, and Ir. Its largest use is in the manufacture of NH_4NO_3 for fertilizers. It is also used in the production of plastics, drugs, and explosives. Among the explosives made from nitric acid are nitroglycerin, trinitrotoluene (TNT), and nitrocellulose. The following reaction occurs when nitroglycerin explodes:

$$4\,C_3H_5N_3O_9(l) \longrightarrow 6\,N_2(g) + 12\,CO_2(g) + 10\,H_2O(g) + O_2(g) \qquad [22.43]$$

All the products of this reaction contain very strong bonds and are gases. As a result, the reaction is very exothermic, and the volume of the products is far larger than the volume occupied by the reactant. Thus, the expansion resulting from the heat generated by the reaction produces the explosion.

Go Figure

Which is the shortest NO bond in these two molecules?

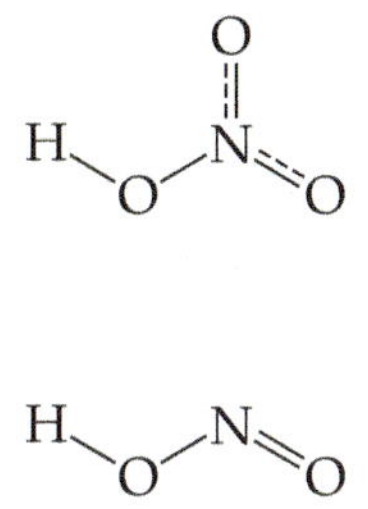

▲ Figure 22.24 Structures of nitric acid (top) and nitrous acid (bottom).

CHEMISTRY AND LIFE Nitroglycerin, Nitric Oxide, and Heart Disease

During the 1870s, an interesting observation was made in Alfred Nobel's dynamite factories. Workers who suffered from heart disease that caused chest pains when they exerted themselves found relief from the pains during the workweek. It quickly became apparent that nitroglycerin, present in the air of the factory, acted to enlarge blood vessels. Thus, this powerfully explosive chemical became a standard treatment for angina pectoris, the chest pains accompanying heart failure. It took more than 100 years to discover that nitroglycerin was converted in the vascular smooth muscle into NO, which is the chemical agent actually causing dilation of the blood vessels. In 1998, the Nobel Prize in Physiology or Medicine was awarded to Robert F. Furchgott, Louis J. Ignarro, and Ferid Murad for their discoveries of the detailed pathways by which NO acts in the cardiovascular system. It was a sensation that this simple, common air pollutant could exert important functions in mammals, including humans.

As useful as nitroglycerin is to this day in treating angina pectoris, it has a limitation in that prolonged administration results in development of tolerance, or desensitization, of the vascular muscle to further vasorelaxation by nitroglycerin. The bioactivation of nitroglycerin is the subject of active research in the hope that a means of circumventing desensitization can be found.

▲ Self-Assessment Exercises

SAE 22.22   Select the correct formula for sodium nitride and indicate the oxidation state of nitrogen in this compound. (**a**) Na_3N, $+1$ (**b**) Na_3NO_3, -3 (**c**) $NaNO_3$, $+5$ (**d**) Na_3N, -3

SAE 22.23   Select the balanced equation for the reaction of nitric oxide and oxygen.

(**a**) $N_2(g) + O_2(g) \longrightarrow 2\,NO_2(g)$
(**b**) $2\,NO(g) + O_2(g) \longrightarrow 2\,NO_2(g)$
(**c**) $2\,N_2(g) + 5\,O_2(g) \longrightarrow 2\,N_2O_5(g)$
(**d**) $2\,N_2O(g) \longrightarrow 2\,N_2(g) + O_2(g)$

22.8 | The Other Group 5A Elements: P, As, Sb, and Bi

⚠ Learning Objectives

When you are finished with Section 22.8, you should be able to:

▶ Compare and contrast the properties and reactivity of group 5A elements.

▶ Assign oxidation numbers to group 5A elements in their compounds.

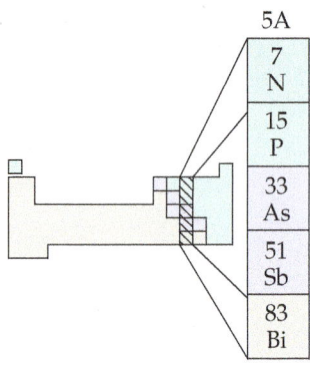

Of the other group 5A elements—phosphorus, arsenic, antimony, and bismuth—phosphorus has a central role in several aspects of biochemistry and environmental chemistry.

The group 5A elements have the outer-shell electron configuration ns^2np^3, with n ranging from 2 to 6. A noble-gas configuration is achieved by adding three electrons to form the -3 oxidation state. Ionic compounds containing X^{3-} ions are not common, however. More commonly, the group 5A element acquires an octet of electrons via covalent bonding and oxidation numbers ranging from -3 to $+5$.

Because of its lower electronegativity, phosphorus is found more frequently in positive oxidation states than is nitrogen. Furthermore, compounds in which phosphorus has the $+5$ oxidation state are not as strongly oxidizing as the corresponding compounds of nitrogen. Compounds in which phosphorus has a -3 oxidation state are much stronger reducing agents than are the corresponding nitrogen compounds.

Some properties of the group 5A elements are listed in Table 22.7. The variation in properties among group 5A elements is more striking than that seen in groups 6A and 7A. Nitrogen at the one extreme exists as a gaseous diatomic molecule, making it nonmetallic. At the other extreme, bismuth is a silvery-pink solid, frequently iridescent due to an oxide layer, that has most of the characteristics of a metal.

The values listed for $X-X$ single-bond enthalpies are not reliable because it is difficult to obtain such data from thermochemical experiments. However, there is no doubt about the general trend: a low value for the $N-N$ single bond, an increase at phosphorus, and then a gradual decline to arsenic and antimony. From observations of the elements in the gas phase, it is possible to estimate the $X\equiv X$ triple-bond enthalpies. Here, we see a trend that is different from that for the $X-X$ single bond. Nitrogen forms a much stronger triple bond than the other elements, and there is a steady decline in the triple-bond enthalpy down through the group. These data help us to appreciate why nitrogen alone of the group 5A elements exists as a diatomic molecule in its stable state at 25 °C. All the other elements exist in structural forms with single bonds between the atoms.

Occurrence, Isolation, and Properties of Phosphorus

Phosphorus occurs mainly in the form of phosphate minerals. The principal source of phosphorus is phosphate rock, which contains phosphate principally as $Ca_3(PO_4)_2$. The element is produced commercially by the reduction of calcium phosphate with carbon in the presence of SiO_2:

$$2\,Ca_3(PO_4)_2(s) + 6\,SiO_2(s) + 10\,C(s) \xrightarrow{1500\,°C} P_4(g) + 6\,CaSiO_3(l) + 10\,CO(g) \quad [22.44]$$

The phosphorus produced in this fashion is the allotrope known as white phosphorus. This form distills from the reaction mixture as the reaction proceeds.

Phosphorus exists in several allotropic forms. White phosphorus consists of P_4 tetrahedra (Figure 22.25). The bond angles in this molecule, 60°, are unusually small, so there is much strain in the bonding, which is consistent with the high reactivity of white phosphorus. This allotrope bursts spontaneously into flames if exposed to air. When heated in the absence of air to about 400 °C, white phosphorus is converted to a more stable allotrope

TABLE 22.7   Properties of the Group 5A Elements

Property	N	P	As	Sb	Bi
Atomic radius (Å)	0.71	1.07	1.19	1.39	1.48
First ionization energy (kJ/mol)	1402	1012	947	834	703
Electron affinity (kJ/mol)	> 0	−72	−78	−103	−91
Electronegativity	3.0	2.1	2.0	1.9	1.9
$X-X$ single-bond enthalpy (kJ/mol)*	163	200	150	120	—
$X\equiv X$ triple-bond enthalpy (kJ/mol)	941	490	380	295	192

*Approximate values only.

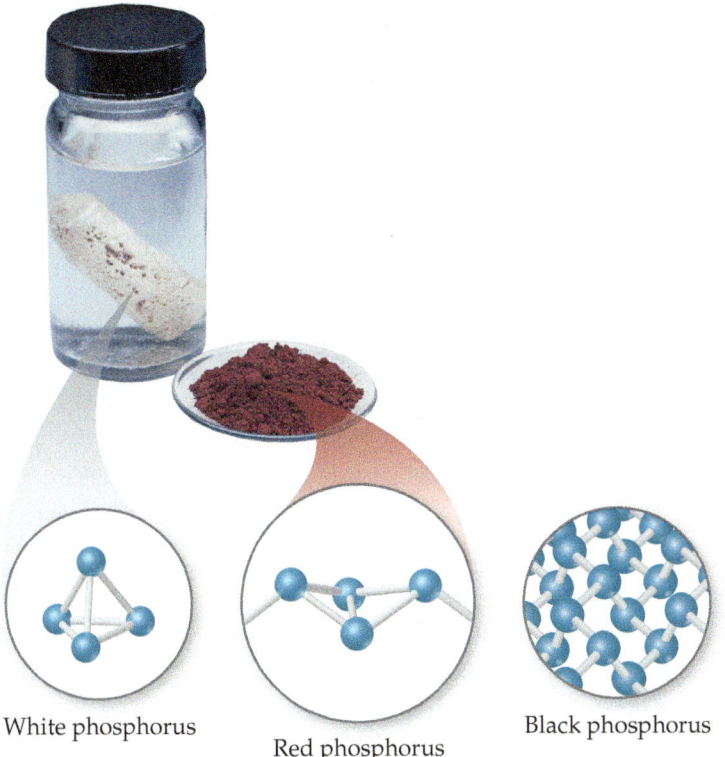

White phosphorus   Red phosphorus   Black phosphorus

▲ **Figure 22.25 White, red, and black phosphorus.** Despite the fact that they contain only phosphorus atoms, these forms of phosphorus differ greatly in reactivity. The white allotrope, which reacts violently with oxygen, must be stored under water so that it is not exposed to air. The much less reactive red and black forms do not need to be stored this way. Black phosphorus conducts electricity to some extent due to its structure.

known as red phosphorus, which does not ignite on contact with air. Red phosphorus is also considerably less poisonous than the white form. Another allotrope, black phosphorus, is an extended solid that is the thermodynamically stable form (by -39.3 kJ/mol) compared to white phosphorus. We denote elemental phosphorus as simply $P(s)$.

Phosphorus Halides

Phosphorus forms a wide range of compounds with the halogens, the most important of which are the trihalides and pentahalides. Phosphorus trichloride (PCl_3) is commercially the most significant of these compounds and is used to prepare a wide variety of products, including soaps, detergents, plastics, and insecticides.

Phosphorus chlorides, bromides, and iodides can be made by direct oxidation of elemental phosphorus with the elemental halogen. PCl_3, for example, which is a liquid at room temperature, is made by passing a stream of dry chlorine gas over white or red phosphorus:

$$2\,P(s) + 3\,Cl_2(g) \longrightarrow 2\,PCl_3(l) \qquad\qquad [22.45]$$

If excess chlorine gas is present, an equilibrium is established between PCl_3 and PCl_5.

$$PCl_3(l) + Cl_2(g) \rightleftharpoons PCl_5(s) \qquad\qquad [22.46]$$

The phosphorus halides react readily with water, and most fume in air because of reaction with water vapor. In the presence of excess water, the products are the corresponding phosphorus oxyacid and hydrogen halide:

$$PBr_3(l) + 3\,H_2O(l) \longrightarrow H_3PO_3(aq) + 3\,HBr(aq) \qquad\qquad [22.47]$$

$$PCl_5(l) + 4\,H_2O(l) \longrightarrow H_3PO_4(aq) + 5\,HCl(aq) \qquad\qquad [22.48]$$

Oxy Compounds of Phosphorus

Probably the most significant phosphorus compounds are those in which the element is combined with oxygen. Phosphorus(III) oxide (P_4O_6) is obtained by allowing white

 **Go Figure**

How do the electron domains about P in P_4O_6 differ from those about P in P_4O_{10}?

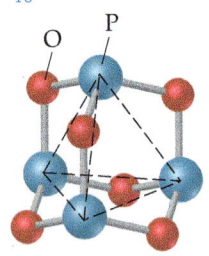

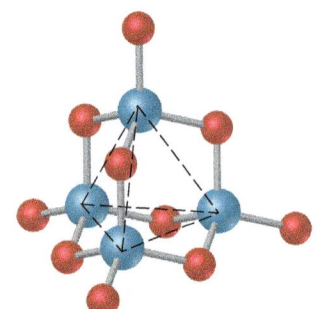

▲ **Figure 22.26 Structures of P_4O_6 (top) and P_4O_{10} (bottom).** The dashed lines outline the P_4 core in each molecule.

phosphorus to oxidize in a limited supply of oxygen. When oxidation takes place in the presence of excess oxygen, phosphorus(V) oxide (P_4O_{10}) forms. This compound is also readily formed by oxidation of P_4O_6. These two oxides represent the two most common oxidation states for phosphorus, +3 and +5. The structural relationship between P_4O_6 and P_4O_{10} is shown in **Figure 22.26**. Notice the resemblance these molecules have to the P_4 molecule (Figure 22.26); all three substances have a P_4 core.

Phosphorus(V) oxide is the anhydride of phosphoric acid (H_3PO_4), a weak triprotic acid. In fact, P_4O_{10} has a very high affinity for water and is consequently used as a drying agent. Phosphorus(III) oxide is the anhydride of phosphorous acid (H_3PO_3), a weak diprotic acid (**Figure 22.27**).*

One characteristic of phosphoric and phosphorous acids is their tendency to undergo *condensation reactions* when heated. (Section 12.6) For example, two H_3PO_4 molecules are joined by the elimination of one H_2O molecule to form $H_4P_2O_7$:

$$HO-\underset{\underset{OH}{|}}{\overset{\overset{O}{\|}}{P}}-OH \;+\; HO-\underset{\underset{OH}{|}}{\overset{\overset{O}{\|}}{P}}-OH \longrightarrow HO-\underset{\underset{OH}{|}}{\overset{\overset{O}{\|}}{P}}-O-\underset{\underset{OH}{|}}{\overset{\overset{O}{\|}}{P}}-OH \;+\; H_2O$$

These atoms are eliminated as H_2O [22.49]

Phosphoric acid, its salts, and "polyphosphates" such as the product in Equation 22.49 find their most important uses in detergents and fertilizers. The phosphates in detergents are often in the form of sodium tripolyphosphate ($Na_5P_3O_{10}$). The phosphate ions "soften" water by binding their oxygen groups to the metal ions that contribute to the hardness of water. This keeps the metal ions from interfering with the action of the detergent. The phosphates also keep the pH above 7 and thus prevent the detergent molecules from becoming protonated.

Most mined phosphate rock is converted to fertilizers. The $Ca_3(PO_4)_2$ in phosphate rock is insoluble ($K_{sp} = 2.0 \times 10^{-29}$). It is converted to a soluble form for use in fertilizers by treatment with sulfuric or phosphoric acid. The reaction with phosphoric acid yields $Ca(H_2PO_4)_2$:

$$Ca_3(PO_4)_2(s) + 4\,H_3PO_4(aq) \longrightarrow 3\,Ca^{2+}(aq) + 6\,H_2PO_4^-(aq) \qquad [22.50]$$

Although the solubility of $Ca(H_2PO_4)_2$ allows it to be assimilated by plants, it also allows it to be washed from the soil and into bodies of water, thereby contributing to water pollution. (Section 18.4)

Phosphorus compounds are important in biological systems. The element occurs in the phosphate groups in RNA and DNA, the molecules responsible for the control of protein biosynthesis and transmission of genetic information. It also occurs in adenosine triphosphate (ATP), which stores energy in biological cells and has the following structure:

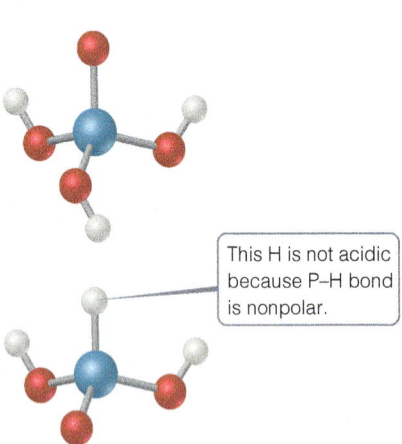

This H is not acidic because P–H bond is nonpolar.

▲ **Figure 22.27 Structures of H_3PO_4 (top) and H_3PO_3 (bottom).**

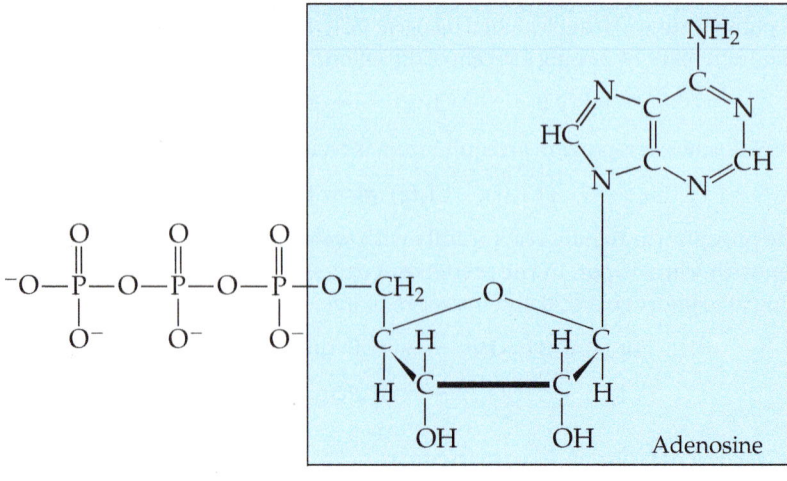

*Note that the element phosphorus (*FOS · for · us*) has a *-us* suffix, whereas the first word in the name phosphorous (*fos · FOR · us*) acid has an *-ous* suffix.

The P—O—P bond of the end phosphate group is broken by hydrolysis with water, forming adenosine diphosphate (ADP):

$$^-O-\overset{\overset{\displaystyle O}{\|}}{\underset{\underset{\displaystyle O^-}{|}}{P}}-O-\overset{\overset{\displaystyle O}{\|}}{\underset{\underset{\displaystyle O^-}{|}}{P}}-O-\overset{\overset{\displaystyle O}{\|}}{\underset{\underset{\displaystyle O^-}{|}}{P}}-O-\text{Adenosine} \ + \ H_2O \ \longrightarrow$$

ATP   [22.51]

$$HO-\overset{\overset{\displaystyle O}{\|}}{\underset{\underset{\displaystyle O^-}{|}}{P}}-O-\overset{\overset{\displaystyle O}{\|}}{\underset{\underset{\displaystyle O^-}{|}}{P}}-O-\text{Adenosine} \ + \ ^-O-\overset{\overset{\displaystyle O}{\|}}{\underset{\underset{\displaystyle O^-}{|}}{P}}-OH$$

ADP

This reaction releases 33 kJ of energy under standard conditions, but in the living cell, the Gibbs free energy change for the reaction is closer to -57 kJ/mol. The concentration of ATP inside a living cell is in the range of 1–10 mM, and a typical human metabolizes her or his body mass of ATP in one day! ATP is continually made from ADP and is continually converted back to ADP, releasing energy that can be harnessed by other cellular reactions.

Arsenic, Antimony, and Bismuth

The heavier members of group 5A are mostly known for their effects on human health. Arsenic in the form of As_2O_3, known as "white arsenic," was used for decades at a commercial rat poison and illegally as a way to poison people. The well-known Marsh test for arsenic, used from the 1840s through the 1970s in crime scene investigations, involves a redox reaction of samples containing white arsenic with sulfuric acid and zinc upon the application of heat:

$$As_2O_3(s) + 6\,Zn(s) + 6\,H_2SO_4(aq) \longrightarrow 2\,AsH_3(g) + 6\,ZnSO_4(aq) + 3\,H_2O(l) \quad [22.52]$$

In the Marsh test apparatus, a glass or ceramic bowl was placed near the reaction site; the hot gaseous arsine, AsH_3, would decompose to elemental arsenic there, creating a characteristic silvery-black "mirror" film on the bowl. Antimony reacts similar to arsenic in the Marsh test, but can be distinguished from As upon reaction of the mirror film with sodium hypochlorite: the arsenic mirror will dissolve upon reaction with sodium hypochlorite, but the antimony mirror will remain.

Arsenic, antimony, and bismuth react quite similarly to phosphorus to make oxyanions, halide compounds, and others. The most well-known compound of bismuth is bismuth subsalicylate, the pink colloidal solution known as Pepto-Bismol® that is used to treat minor gastrointestinal illnesses (see the "Putting Concepts Together" feature at the end of Section 7.7).

CHEMISTRY AND LIFE | Arsenic in Drinking Water

Arsenic, in the form of its oxides, has been known as a poison for centuries. The current Environmental Protection Agency (EPA) standard for arsenic in public water supplies is 10 ppb (equivalent to 10 μg/L). Most regions of the United States tend to have low to moderate (2–10 ppb) groundwater arsenic levels (Figure 22.28). The western region tends to have higher levels, coming mainly from natural geological sources in the area. Estimates, for example, indicate that 35% of water-supply wells in Arizona have arsenic concentrations above 10 ppb.

The problem of arsenic in drinking water in the United States is dwarfed by the problem in other parts of the world—especially in Bangladesh, where the problem is tragic. Historically, surface water

sources in that country have been contaminated with microorganisms, causing significant health problems. During the 1970s, international agencies, headed by the United Nations Children's Fund (UNICEF), began investing millions of dollars of aid money in Bangladesh for wells to provide "clean" drinking water. Unfortunately, no one tested the well water for the presence of arsenic; the problem was not discovered until the 1980s. The result has been the biggest outbreak of mass poisoning in history. Up to half of the country's estimated 10 million wells have arsenic concentrations above 50 ppb.

In water, the most common forms of arsenic are the arsenate ion and its protonated hydrogen anions (AsO_4^{3-}, $HAsO_4^{2-}$, and

Continued

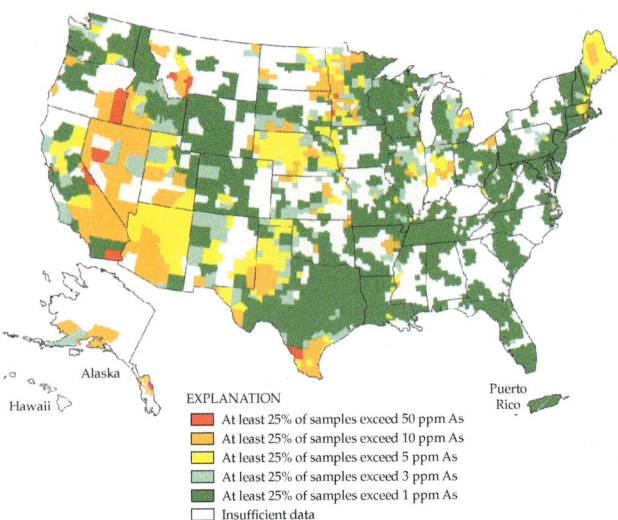

EXPLANATION

■ At least 25% of samples exceed 50 ppm As
■ At least 25% of samples exceed 10 ppm As
■ At least 25% of samples exceed 5 ppm As
■ At least 25% of samples exceed 3 ppm As
■ At least 25% of samples exceed 1 ppm As
□ Insufficient data

Alaska
Hawaii
Puerto Rico

▲ **Figure 22.28 Geographic distribution of arsenic in groundwater in the continental United States.**

$H_2AsO_4^-$) and the arsenite ion and its protonated forms (AsO_3^{3-}, $HAsO_3^{2-}$, $H_2AsO_3^-$, and H_3AsO_3). These species are collectively referred to by the oxidation number of the arsenic as arsenic(V) and arsenic(III), respectively. Arsenic(V) is more prevalent in oxygen-rich (aerobic) surface waters, whereas arsenic(III) is more likely to occur in oxygen-poor (anaerobic) groundwaters.

One of the challenges in determining the health effects of arsenic in drinking waters is the different chemistry of arsenic(V) and arsenic(III), as well as the different concentrations required for physiological responses in different individuals. In Bangladesh, skin lesions were the first sign of the arsenic problem. Statistical studies correlating arsenic levels with the occurrence of disease indicate a lung and bladder cancer risk arising from even low levels of arsenic.

The current technologies for removing arsenic perform most effectively when treating arsenic in the form of arsenic(V), so water treatment strategies require preoxidation of the drinking water. Once in the form of arsenic(V), there are a number of possible removal strategies. For example, Fe^{3+} can be added to precipitate $FeAsO_4$, which is then removed by filtration.

▲ Self-Assessment Exercises

SAE 22.24 PCl$_3$ is _____ stable than NCl$_3$ because phosphorus is _____ electronegative than nitrogen and therefore has a greater affinity for positive oxidation states. **(a)** more, more **(b)** less, less **(c)** less, more **(d)** more, less

SAE 22.25 Determine the oxidation state of phosphorus in $H_4P_2O_7$. **(a)** −3 **(b)** −5 **(c)** +5 **(d)** +3

SAE 22.26 Select the balanced equation that represents the reaction of phosphorus pentafluoride and water.

(a) $PF_5(g) + 4\,H_2O(l) \longrightarrow H_3PO_4(aq) + 5\,HF(aq)$
(b) $PF_3(g) + 3\,H_2O(l) \longrightarrow H_3PO_3(aq) + 3\,HF(aq)$
(c) $PF_5(g) + H_2O(l) \longrightarrow H_3PO_4(aq) + 5\,HF(aq)$
(d) $PF_3(g) + H_2O(l) \longrightarrow H_3PO_3(aq) + HF(aq)$

22.9 | Carbon

Carbon constitutes only 0.027% of Earth's crust. Although some carbon occurs in elemental form as graphite and diamond, most is found in combined form. Over half occurs in carbonate compounds. Carbon is also found in coal, petroleum, and natural gas. The importance of the element stems in large part from its occurrence in all living organisms: Life as we know it is based on carbon compounds.

Elemental Forms of Carbon

We have seen that carbon exists in several allotropic crystalline forms: graphite, diamond, fullerenes, carbon nanotubes, and graphene. Fullerenes, nanotubes, and graphene are discussed in Chapter 12; here we focus on graphite and diamond.

Graphite is a soft, black, slippery solid that has a metallic luster and conducts electricity. It consists of parallel sheets of sp^2-hybridized carbon atoms held together by dispersion forces [see Figure 12.28(b)]. (Section 12.5) *Diamond* is a clear, hard solid in which the carbon atoms form an sp^3-hybridized covalent network [see Figure 12.28(a)]. (Section 12.5) Diamond is denser than graphite ($d = 2.25\ \text{g/cm}^3$ for graphite; $d = 3.51\ \text{g/cm}^3$ for diamond). At approximately 100,000 atm at 3000 °C, graphite converts to diamond. In fact, almost any carbon-containing substance, if put under sufficiently high pressure, forms diamonds; scientists at General Electric in the 1950s used peanut butter to make diamonds. About 3×10^4 kg of industrial-grade diamonds are synthesized each year, mainly for use in cutting, grinding, and polishing tools.

Learning Objectives

When you have finished Section 22.9, you should be able to:

▶ Give the name of a carbon-containing compound, determine its chemical formula, and vice versa.

▶ Write balanced chemical equations for reactions involving carbon.

Graphite has a well-defined crystalline structure, but it also exists in two common amorphous forms: **carbon black** and **charcoal**. Carbon black is formed when hydrocarbons are heated in a very limited supply of oxygen, such as in this methane reaction:

$$CH_4(g) + O_2(g) \longrightarrow C(s) + 2 H_2O(g) \qquad [22.53]$$

Carbon black is used as a pigment in black inks; large amounts are also used in making automobile tires.

Charcoal is formed when wood is heated strongly in the absence of air. Charcoal has an open structure, giving it an enormous surface area per unit mass. "Activated charcoal," a pulverized form of charcoal whose surface is cleaned by heating with steam, is widely used to adsorb molecules. It is used in filters to remove offensive odors from air and colored or bad-tasting impurities from water.

Oxides of Carbon

Carbon forms two principal oxides: carbon monoxide (CO) and carbon dioxide (CO_2). *Carbon monoxide* is formed when carbon or hydrocarbons are burned in a limited supply of oxygen:

$$2 C(s) + O_2(g) \longrightarrow 2 CO(g) \qquad [22.54]$$

CO is a colorless, odorless, tasteless gas that is toxic because it binds to hemoglobin in the blood and thus interferes with oxygen transport. Low-level poisoning results in headaches and drowsiness; high-level poisoning can cause death.

Carbon monoxide is unusual in that it has a nonbonding pair of electrons on carbon: $:C{\equiv}O:$. It is isoelectronic with N_2, so you might expect CO to be equally unreactive. Moreover, both substances have high bond energies (1072 kJ/mol for $C{\equiv}O$ and 941 kJ/mol for $N{\equiv}N$). Because of the lower nuclear charge on carbon (compared with either N or O), however, the carbon nonbonding pair is not held as strongly as that on N or O. Consequently, CO is better able to function as a Lewis base than is N_2; for example, CO can coordinate its nonbonding pair to the iron of hemoglobin, displacing O_2, but N_2 cannot.

A CLOSER LOOK | **Carbon Fibers and Composites**

The properties of graphite are anisotropic; that is, they differ in different directions through the solid. Along the carbon planes, graphite possesses great strength because of the number and strength of the carbon–carbon bonds in this direction. The bonds between planes are relatively weak, however, making graphite weak in that direction.

Fibers of graphite can be prepared in which the carbon planes are aligned to varying extents parallel to the fiber axis. These fibers are lightweight (density of about 2 g/cm³) and chemically quite unreactive. The oriented fibers are made by first slowly pyrolyzing (decomposing by action of heat) organic fibers at about 150 to 300 °C. These fibers are then heated to about 2500 °C to graphitize them (convert amorphous carbon to graphite). Stretching the fiber during pyrolysis helps orient the graphite planes parallel to the fiber axis. More amorphous carbon fibers are formed by pyrolysis of organic fibers at lower temperatures (400 to 1200 °C). These amorphous materials, commonly called *carbon fibers*, are the type most often used in commercial materials.

Composite materials that take advantage of the strength, stability, and low density of carbon fibers are widely used. Composites are combinations of two or more materials. These materials are present as separate phases and are combined to form structures that take advantage of certain desirable properties of each component. In carbon composites, the graphite fibers are often woven into a fabric that is embedded in a matrix that binds them into a solid structure. The fibers transmit loads evenly throughout the matrix. The finished composite is thus stronger than any one of its components.

Carbon composite materials are used widely in a number of applications, including high-performance graphite sports equipment such as tennis racquets, golf clubs, and bicycle wheels (**Figure 22.29**). Heat-resistant composites are required for many aerospace applications, where carbon composites now find wide use.

▲ **Figure 22.29 Carbon composites in commercial products.**

Strong acid $CO_2(g)$

$CaCO_3$

▲ **Figure 22.30** CO_2 formation from the reaction between an acid and calcium carbonate in rock.

Carbon monoxide has several commercial uses. Because it burns readily, forming CO_2, it is employed as a fuel:

$$2\,CO(g) + O_2(g) \longrightarrow 2\,CO_2(g) \qquad \Delta H° = -566\,kJ \qquad [22.55]$$

Carbon monoxide is an important reducing agent, widely used in metallurgical operations to reduce metal oxides, such as the iron oxides:

$$Fe_3O_4(s) + 4\,CO(g) \longrightarrow 3\,Fe(s) + 4\,CO_2(g) \qquad [22.56]$$

Carbon dioxide is produced when carbon-containing substances are burned in excess oxygen, such as in this reaction involving ethanol:

$$C_2H_5OH(l) + 3\,O_2(g) \longrightarrow 2\,CO_2(g) + 3\,H_2O(g) \qquad [22.57]$$

It is also produced when many carbonates are heated:

$$CaCO_3(s) \xrightarrow{\Delta} CaO(s) + CO_2(g) \qquad [22.58]$$

In the laboratory, CO_2 can be produced by the action of acids on carbonates (**Figure 22.30**):

$$CO_3{}^{2-}(aq) + 2\,H^+(aq) \longrightarrow CO_2(g) + H_2O(l) \qquad [22.59]$$

Carbon dioxide is a colorless, odorless gas. It is a minor component of Earth's atmosphere but a major contributor to the greenhouse effect. (Section 18.2) Although it is not toxic, high concentrations of CO_2 increase respiration rate and can cause suffocation. It is readily liquefied by compression. When cooled at atmospheric pressure, however, CO_2 forms a solid rather than liquefying. The solid sublimes at atmospheric pressure at $-78\,°C$. This property makes solid CO_2, known as *dry ice*, valuable as a refrigerant. About half of the CO_2 consumed annually is used for refrigeration. The other major use of CO_2 is in the production of carbonated beverages. Large quantities are also used to manufacture *washing soda* ($Na_2CO_3 \cdot 10\,H_2O$), used to precipitate metal ions that interfere with the cleansing action of soap, and *baking soda* ($NaHCO_3$). Baking soda is so named because the following reaction occurs during baking:

$$NaHCO_3(s) + H^+(aq) \longrightarrow Na^+(aq) + CO_2(g) + H_2O(l) \qquad [22.60]$$

The $H^+(aq)$ is provided by vinegar, sour milk, or the hydrolysis of certain salts. The bubbles of CO_2 that form are trapped in the baking dough, causing it to rise.

Carbonic Acid and Carbonates

Carbon dioxide is moderately soluble in H_2O at atmospheric pressure. The resulting solution is moderately acidic because of the formation of carbonic acid (H_2CO_3):

$$CO_2(aq) + H_2O(l) \rightleftharpoons H_2CO_3(aq) \qquad [22.61]$$

Carbonic acid is a weak diprotic acid. Its acidic character causes carbonated beverages to have a sharp, slightly acidic taste.

Although carbonic acid is unstable and cannot be isolated, hydrogen carbonates (also called bicarbonates) and carbonates can be obtained by neutralizing carbonic acid solutions. Partial neutralization produces $HCO_3{}^-$, and complete neutralization gives $CO_3{}^{2-}$. The $HCO_3{}^-$ ion could in theory function either as an acid or as a base, but it is a stronger base than acid ($K_b = 2.3 \times 10^{-8}$; $K_a = 5.6 \times 10^{-11}$). The carbonate ion is much more strongly basic than bicarbonate ($K_b = 1.8 \times 10^{-4}$).

The principal carbonate minerals are calcite ($CaCO_3$), magnesite ($MgCO_3$), dolomite [$MgCa(CO_3)_2$], and siderite ($FeCO_3$). Calcite is the principal mineral in limestone and the main constituent of marble, chalk, pearls, coral reefs, and the shells of marine animals such as clams and oysters. Although $CaCO_3$ has low solubility in pure water, it dissolves readily in acidic solutions with evolution of CO_2:

$$CaCO_3(s) + 2\,H^+(aq) \rightleftharpoons Ca^{2+}(aq) + H_2O(l) + CO_2(g) \qquad [22.62]$$

Because water containing CO_2 is slightly acidic (Equation 22.61), $CaCO_3$ dissolves slowly in this medium:

$$CaCO_3(s) + H_2O(l) + CO_2(g) \longrightarrow Ca^{2+}(aq) + 2\,HCO_3^-(aq) \qquad [22.63]$$

This reaction occurs when surface waters move underground through limestone deposits. It is the principal way Ca^{2+} enters groundwater, producing *hard water*, (water with a high mineral content, containing especially Ca^{2+} and Mg^{2+} ions). If the limestone deposit is deep enough underground, dissolution of the limestone produces a cave.

One of the most important reactions of $CaCO_3$ is its decomposition into CaO and CO_2 at elevated temperatures (Equation 22.58). Because calcium oxide, known as lime or quicklime, reacts with water to form $Ca(OH)_2$, it is an important commercial base. It is also important in making mortar, the mixture of sand, water, and CaO used to bind bricks, blocks, or rocks together. Calcium oxide reacts with water and CO_2 to form $CaCO_3$, which binds the sand in the mortar:

$$CaO(s) + H_2O(l) \longrightarrow Ca^{2+}(aq) + 2\,OH^-(aq) \qquad [22.64]$$

$$Ca^{2+}(aq) + 2\,OH^-(aq) + CO_2(aq) \longrightarrow CaCO_3(s) + H_2O(l) \qquad [22.65]$$

Carbides

The binary compounds of carbon with metals, metalloids, and certain nonmetals are called **carbides**. The more active metals form *ionic carbides*, and the most common of these contain the *acetylide* ion (C_2^{2-}). This ion is isoelectronic with N_2, and its Lewis structure, $[:C \equiv C:]^{2-}$, has a carbon–carbon triple bond. The most important ionic carbide is calcium carbide (CaC_2), produced by the reduction of CaO with carbon at high temperature:

$$2\,CaO(s) + 5\,C(s) \longrightarrow 2\,CaC_2(s) + CO_2(g) \qquad [22.66]$$

The carbide ion is a very strong base that reacts with water to form acetylene ($H-C \equiv C-H$):

$$CaC_2(s) + 2\,H_2O(l) \longrightarrow Ca(OH)_2(aq) + C_2H_2(g) \qquad [22.67]$$

Calcium carbide is therefore a convenient solid source of acetylene, which is used in welding (Figure 22.11).

Interstitial carbides are formed by many transition metals. The carbon atoms occupy open spaces (interstices) between the metal atoms in a manner analogous to the interstitial hydrides. (Section 22.2) This process generally makes the metal harder. Tungsten carbide (WC), for example, is very hard and very heat-resistant and, thus, used to make cutting tools.

Covalent carbides are formed by boron and silicon. Silicon carbide (SiC), known as Carborundum®, is used as an abrasive and in cutting tools. Almost as hard as diamond, SiC has a diamond-like structure with alternating Si and C atoms.

 Self-Assessment Exercises

SAE 22.27 What is the chemical formula of magnesium hydrogen carbonate? (a) $MgHCO_3$ (b) $Mg(HCO_3)_2$ (c) $MgCO_3$ (d) MgH_2CO_3

SAE 22.28 Select the correct balanced equation for the complete combustion of propane, C_3H_8.

(a) $C_3H_8(g) + 5\,O_2(g) \longrightarrow O_2(g) + 4\,H_2O(l)$
(b) $C_3H_8(g) + 3\,O_2(g) \longrightarrow 3\,CO_2(g) + 2\,H_2O(l)$
(c) $C_3H_8(g) + 5\,O_2(g) \longrightarrow 3\,CO_2(g) + 4\,H_2O(l)$
(d) $C_3H_8(g) + H_2O(l) \longrightarrow 3\,CO_2(g)$

SAE 22.29 Which of the following statements is/are *true*?

(i) Bicarbonate ion can act as either an acid or a base.
(ii) Ionic carbide compounds are weak acids.
(iii) CaC_2 is an example of a covalent carbide.

(a) Only one statement of i, ii, and iii is true. (b) Statements i and ii are true. (c) Statements i and iii are true. (d) Statements ii and iii are true. (e) All three statements are true.

22.10 | The Other Group 4A Elements: Si, Ge, Sn, and Pb

The trend from nonmetallic to metallic character as we go down a family is strikingly evident in group 4A. Carbon is a nonmetal; silicon and germanium are metalloids; tin and lead are metals. In this section, we consider a few general characteristics of group 4A and then look more thoroughly at silicon.

General Characteristics of the Group 4A Elements

The group 4A elements possess the outer-shell electron configuration ns^2np^2, with n ranging from 2 to 6. The electronegativities of the elements are generally low (Table 22.8); carbides that formally contain C^{4-} ions are observed only in the case of a few compounds of carbon with very active metals. Formation of 4+ ions by electron loss is not observed for any of these elements; the ionization energies are too high. The +4 oxidation state is common, however, and is found in the vast majority of the compounds of the group 4A elements. The +2 oxidation state is found in the chemistry of germanium, tin, and lead, however, and it is the principal oxidation state for lead. Carbon usually forms a maximum of four bonds. The other members of the family are able to form more than four bonds. (Section 8.7)

Table 22.8 shows that the strength of a bond between two atoms of a given element decreases as we go down group 4A. Carbon–carbon bonds are quite strong. Carbon, therefore, has a striking ability to form compounds in which carbon atoms are bonded to one another in extended chains and rings, which accounts for the large number of organic compounds that exist. Other elements can form chains and rings, but these bonds are far less important in the chemistries of these other elements. The Si—Si bond strength (226 kJ/mol), for example, is much lower than the Si—O bond strength (386 kJ/mol). As a result, the chemistry of silicon is dominated by the formation of Si—O bonds, and Si—Si bonds play a minor role.

Occurrence and Preparation of Silicon

Silicon is the second most abundant element, after oxygen, in Earth's crust. It occurs in SiO_2 (the empirical formula for quartz and sand) and in an enormous variety of silicate minerals. The element is obtained by the reduction of molten silicon dioxide with carbon at high temperature:

$$SiO_2(l) + 2\,C(s) \longrightarrow Si(l) + 2\,CO(g) \qquad [22.68]$$

Elemental silicon has a diamond-like structure. Crystalline silicon is a gray metallic-looking solid that melts at 1410 °C. The element is a semiconductor, as we saw in Chapters 7 and 12, and is used to make solar cells and transistors for computer chips. To be used as a semiconductor, it must be extremely pure, possessing less than $10^{-7}\%$ (1 ppb) impurities. One method of purification is to treat the element with Cl_2 to form $SiCl_4$, a volatile liquid that is purified by fractional distillation and then converted back to elemental silicon by reduction with H_2:

$$SiCl_4(g) + 2\,H_2(g) \longrightarrow Si(s) + 4\,HCl(g) \qquad [22.69]$$

The process known as *zone refining* can further purify the element (Figure 22.31). As a heated coil is passed slowly along a silicon rod, a narrow band of the element is melted. As

Learning Objectives

When you finish Section 22.10, you should be able to:

▶ Determine the oxidation states of group 4A elements in compounds.

▶ Compare and contrast the properties and reactivity of compounds containing heavier group 4A elements to compounds containing carbon.

▶ Use the formula of a silicate mineral to determine its general structure.

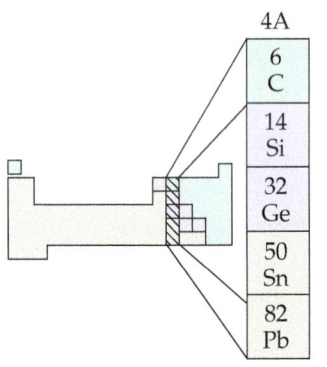

4A

| 6 C |
| 14 Si |
| 32 Ge |
| 50 Sn |
| 82 Pb |

Go Figure

What limits the range of temperatures you can use for zone refining of silicon?

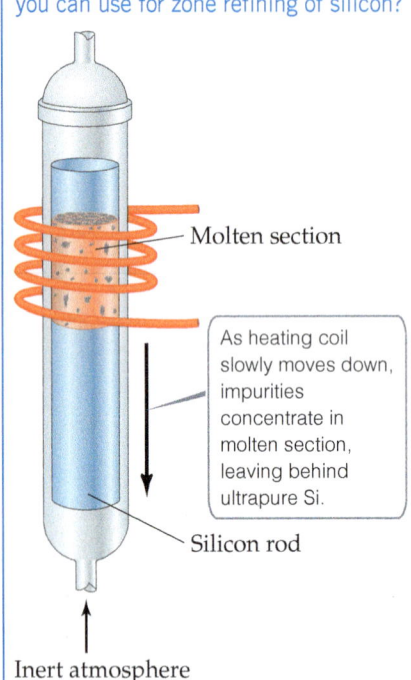

— Molten section

As heating coil slowly moves down, impurities concentrate in molten section, leaving behind ultrapure Si.

— Silicon rod

Inert atmosphere

▲ Figure 22.31 **Zone-refining apparatus for the production of ultrapure silicon.**

TABLE 22.8   Some Properties of the Group 4A Elements

Property	C	Si	Ge	Sn	Pb
Atomic radius (Å)	0.76	1.11	1.20	1.39	1.46
First ionization energy (kJ/mol)	1086	786	762	709	716
Electronegativity	2.5	1.8	1.8	1.8	1.9
X—X single-bond enthalpy (kJ/mol)	348	226	188	151	—

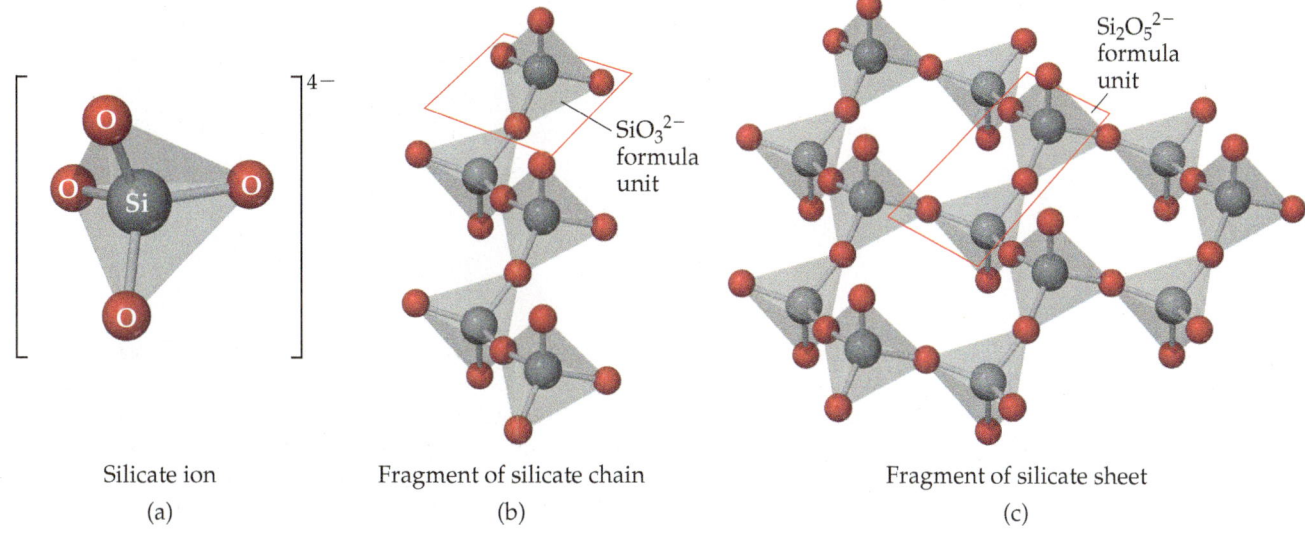

Silicate ion	Fragment of silicate chain	Fragment of silicate sheet
(a)	(b)	(c)

▲ **Figure 22.32 Silicate chains and sheets.**

the molten section is swept slowly along the length of the rod, the impurities concentrate in this section, following it to the end of the rod. The purified top portion of the rod crystallizes as 99.999999999% pure silicon.

Silicates

Silicon dioxide and other compounds that contain silicon and oxygen make up over 90% of Earth's crust. In **silicates**, a silicon atom is surrounded by four oxygens, and silicon is found in its most common oxidation state, $+4$. The orthosilicate ion, SiO_4^{4-}, is found in very few silicate minerals, but we can view it as the "building block" for many mineral structures. As **Figure 22.32** shows, adjacent tetrahedra can be linked by a common oxygen atom. Two tetrahedra joined in this way, called the *disilicate* ion, contain two Si atoms and seven O atoms. Silicon and oxygen are in the $+4$ and -2 oxidation states, respectively, in all silicates, so the overall charge of any silicate ion must be consistent with these oxidation states. For example, the charge on Si_2O_7 is $(2)(+4) + (7)(-2) = -6$; it is the $Si_2O_7^{6-}$ ion.

In most silicate minerals, silicate tetrahedra are linked together to form chains, sheets, or three-dimensional structures. We can connect two vertices of each tetrahedron to two other tetrahedra, for example, leading to an infinite chain with an $\cdots$ O—Si—O—Si $\cdots$ backbone as shown in Figure 22.32(b). Notice that each silicon in this structure has two unshared (terminal) oxygens and two shared (bridging) oxygens. The stoichiometry is then $2(1) + 2(1/2) = 3$ oxygens per silicon. Thus, the formula unit for this chain is SiO_3^{2-}. The mineral *enstatite* $(MgSiO_3)$ has this kind of structure, consisting of rows of single-strand silicate chains with Mg^{2+} ions between the strands to balance charge.

In Figure 22.32(c), each silicate tetrahedron is linked to three others, forming an infinite sheet structure. Each silicon in this structure has one unshared oxygen and three shared oxygens. The stoichiometry is then $1(1) + 3(\frac{1}{2}) = 2\frac{1}{2}$ oxygens per silicon. The simplest formula of this sheet is $Si_2O_5^{2-}$. The mineral *talc*, also known as talcum powder, has the formula $Mg_3(Si_2O_5)_2(OH)_2$ and is based on this sheet structure. The Mg^{2+} and OH^- ions lie between the silicate sheets. The slippery feel of talcum powder is due to the silicate sheets sliding relative to one another.

Asbestos is a general term applied to a group of fibrous silicate minerals. The structure of these minerals is either chains of silicate tetrahedra or sheets formed into rolls. The result is that the minerals have a fibrous character (**Figure 22.33**). Asbestos minerals were once widely used as thermal insulation, especially in high-temperature applications, because of the great chemical stability of the silicate structure. In addition, the fibers can be woven into asbestos cloth, which was once used for fireproof curtains and

▶ **Figure 22.33** Serpentine asbestos.

other applications. However, the fibrous structure of asbestos minerals poses a health risk because the fibers readily penetrate soft tissues, such as the lungs, where they can cause diseases, including cancer. The use of asbestos as a common building material has therefore been discontinued.

When all four vertices of each SiO_4 tetrahedron are linked to other tetrahedra, the structure extends in three dimensions. This linking of the tetrahedra forms quartz (SiO_2). Because the structure is locked together in a three-dimensional array much like diamond (Section 12.5), quartz is harder than strand- or sheet-type silicates.

Sample Exercise 22.6

Determining an Empirical Formula

The mineral *chrysotile* is a noncarcinogenic asbestos mineral that is based on the sheet structure shown in Figure 22.32(c). In addition to silicate tetrahedra, the mineral contains Mg^{2+} and OH^- ions. Analysis of the mineral shows that there are 1.5 Mg atoms per Si atom. What is the empirical formula for chrysotile?

SOLUTION

Analyze A mineral is described that has a sheet silicate structure with Mg^{2+} and OH^- ions to balance charge and 1.5 Mg for each 1 Si. We are asked to write the empirical formula for the mineral.

Plan As shown in Figure 22.32(c), the silicate sheet structure has the simplest formula $Si_2O_5^{2-}$. We first add Mg^{2+} to give the proper Mg:Si ratio. We then add OH^- ions to obtain a neutral compound.

Solve The observation that the Mg:Si ratio equals 1.5 is consistent with three Mg^{2+} ions per $Si_2O_5^{2-}$ unit. The addition of three Mg^{2+} ions would make $Mg_3(Si_2O_5)^{4+}$. In order to achieve charge balance in the mineral, there must be four OH^- per $Si_2O_5^{2-}$. Thus, the formula of chrysotile is $Mg_3(Si_2O_5)(OH)_4$. Because this is not reducible to a simpler formula, this is the empirical formula.

▶ **Practice Exercise**
The cyclosilicate ion consists of three silicate tetrahedra linked together in a ring. The ion contains three Si atoms and nine O atoms. What is the overall charge on the ion?

Glass

Quartz melts at approximately 1600 °C, forming a tacky liquid. In the course of melting, many silicon–oxygen bonds are broken. When the liquid cools rapidly, silicon–oxygen bonds are re-formed before the atoms are able to arrange themselves in a regular fashion. An amorphous solid, known as quartz glass or silica glass, results. Many substances can be added to SiO_2 to cause it to melt at a lower temperature. The common **glass** used in windows and bottles, known as soda-lime glass, contains CaO and Na_2O in addition to

SiO_2 from sand. The CaO and Na_2O are produced by heating two inexpensive chemicals, limestone ($CaCO_3$) and soda ash (Na_2CO_3), which decompose at high temperatures:

$$CaCO_3(s) \longrightarrow CaO(s) + CO_2(g) \qquad [22.70]$$

$$Na_2CO_3(s) \longrightarrow Na_2O(s) + CO_2(g) \qquad [22.71]$$

Other substances can be added to soda-lime glass to produce color or to change the properties of the glass in various ways. The addition of CoO, for example, produces the deep blue color of "cobalt glass." Replacing Na_2O with K_2O results in a harder glass that has a higher melting point. Replacing CaO with PbO results in a denser "lead crystal" glass with a higher refractive index. Lead crystal is used for decorative glassware; the higher refractive index gives this glass a particularly sparkling appearance. Addition of nonmetal oxides, such as B_2O_3 and P_4O_{10}, which form network structures related to the silicates, also changes the properties of the glass. Adding B_2O_3 creates a "borosilicate" glass with a higher melting point and a greater ability to withstand temperature changes. Such glasses, sold commercially under trade names such as Pyrex® and Kimax®, are used where resistance to thermal shock is important, such as in laboratory glassware or coffeemakers.

Silicones

Silicones consist of O—Si—O chains in which the two remaining bonding positions on each silicon are occupied by organic groups such as CH_3:

Depending on chain length and degree of cross-linking, silicones can be either oils or rubber-like materials. Silicones are nontoxic and have good stability toward heat, light, oxygen, and water. They are used commercially in a wide variety of products, including lubricants, car polishes, sealants, and gaskets. They are also used for water-proofing fabrics. When applied to a fabric, the oxygen atoms form hydrogen bonds with the molecules on the surface of the fabric. The hydrophobic (water-repelling) organic groups of the silicone are then left pointing away from the surface as a barrier.

Self-Assessment Exercises

SAE 22.30   What is the chemical formula for lead(II) chloride, and what is the oxidation state of lead in this compound? (**a**) $PbCl_2$, +2 (**b**) $PbCl_2$, −2 (**c**) PbCl, +1 (**d**) Pb_2Cl, +0.5

SAE 22.31   Which group 4A element has the highest first ionization energy? (**a**) germanium (**b**) tin (**c**) lead (**d**) silicon

SAE 22.32   In which of the following silicate minerals would you expect to find chains of corner-connected SiO_4 tetrahedra? (**a**) Na_2SiO_3 (**b**) Na_4SiO_4 (**c**) $Na_6Si_2O_7$ (**d**) $Mg_3Si_2O_5(OH)_4$

22.11 | Boron

Boron is the only group 3A element that can be considered nonmetallic; thus, it is our final element in this chapter. The element has an extended network structure with a melting point (2300 °C) that is intermediate between the melting points of carbon (3550 °C) and silicon (1410 °C). The electron configuration of boron is $[He]2s^22p^1$.

In the family of compounds called **boranes**, the molecules contain only boron and hydrogen. Because B is less electronegative than H, the B—H bond in these compounds is polarized, with H having the higher electron density. The simplest borane is BH_3. This molecule contains only six valence electrons and is therefore an exception to the octet rule. As a result, BH_3 reacts with itself to form *diborane* (B_2H_6). This reaction can be

Learning Objectives

When you finish Section 22.11, you should be able to:

▶ Predict bonding in boranes.

▶ Write balanced chemical equations for reactions involving boron-containing compounds.

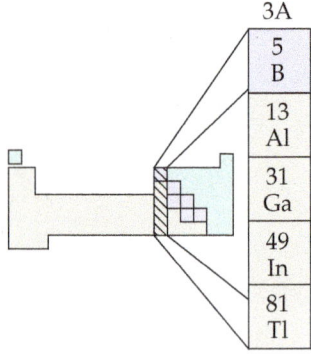

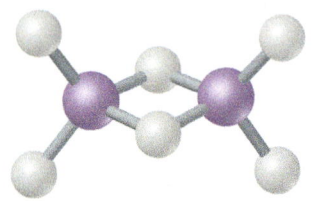

▲ **Figure 22.34** The structure of diborane (B_2H_6).

viewed as a Lewis acid–base reaction in which one B—H bonding pair of electrons in each BH_3 molecule is donated to the other. As a result, diborane is an unusual molecule in which hydrogen atoms form a bridge between two B atoms (**Figure 22.34**). Such hydrogens are called *bridging hydrogens.*

Sharing hydrogen atoms between the two boron atoms compensates somewhat for the deficiency in valence electrons around each boron. Nevertheless, diborane is an extremely reactive molecule, spontaneously flammable in air in a highly exothermic reaction:

$$B_2H_6(g) + 3\,O_2(g) \longrightarrow B_2O_3(s) + 3\,H_2O(g) \qquad \Delta H° = -2030\,kJ \qquad [22.72]$$

Boron and hydrogen form a series of anions called *borane anions.* Salts of the borohydride ion (BH_4^-) are widely used as reducing agents. For example, sodium borohydride $(NaBH_4)$ is a commonly used reducing agent for certain organic compounds.

The only important oxide of boron is boric oxide (B_2O_3). This substance is the anhydride of boric acid, which we may write as H_3BO_3 or $B(OH)_3$. Boric acid is so weak an acid $(K_a = 5.8 \times 10^{-10})$ that solutions of H_3BO_3 are used as an eyewash. Upon heating, boric acid loses water by a condensation reaction similar to that described for phosphorus in Section 22.8:

$$4\,H_3BO_3(s) \longrightarrow H_2B_4O_7(s) + 5\,H_2O(g) \qquad [22.73]$$

The diprotic acid $H_2B_4O_7$ is tetraboric acid. The hydrated sodium salt $Na_2B_4O_7 \cdot 10\,H_2O$, called borax, occurs in dry lake deposits in California and can also be prepared from other borate minerals. Solutions of borax are alkaline, and the substance is used in various laundry and cleaning products.

◢ Self-Assessment Exercises

SAE 22.33   In a molecule of diborane, B_2H_6, there are _____ valence electrons and _____ B—H bonds. (**a**) 16, 8 (**b**) 12, 8 (**c**) 12, 6 (**d**) 14, 7

SAE 22.34   Which of the following is the balanced equation that describes the reaction between diborane and oxygen?

(**a**) $B_2H_6 + 3\,O_2 \longrightarrow B_2O_3 + 3\,H_2O$
(**b**) $B_2H_6 + O_2 \longrightarrow B_2O_3 + 3\,H_2O$
(**c**) $2\,B_2H_6 + 6\,O_2 \longrightarrow 2\,B_2O_3 + 6\,H_2O$
(**d**) $B_2H_6 + 6\,O_2 \longrightarrow 2\,B_2O_3 + 6\,H_2O$

◢ Putting Concepts Together

The interhalogen compound BrF_3 is a volatile, straw-colored liquid. The compound exhibits appreciable electrical conductivity because of autoionization ("solv" refers to BrF_3 as the solvent):

$$2\,BrF_3(l) \rightleftharpoons BrF_2^+(solv) + BrF_4^-(solv)$$

(**a**) What are the molecular structures of the BrF_2^+ and BrF_4^- ions?
(**b**) The electrical conductivity of BrF_3 decreases with increasing temperature. Is the autoionization process exothermic or endothermic?
(**c**) One chemical characteristic of BrF_3 is that it acts as a Lewis acid toward fluoride ions. What do we expect will happen when KBr is dissolved in BrF_3?

SOLUTION

(**a**) The BrF_2^+ ion has $7 + 2(7) - 1 = 20$ valence-shell electrons. The Lewis structure for the ion is

$$\left[\ddot{\underset{\cdot\cdot}{F}} - \ddot{Br} - \ddot{\underset{\cdot\cdot}{F}}\right]^+$$

Because there are four electron domains around the central Br atom, the resulting electron domain geometry is tetrahedral (Section 9.2). Because bonding pairs of electrons occupy two of these domains, the molecular geometry is bent:

The BrF_4^- ion has $7 + 4(7) + 1 = 36$ electrons, leading to the Lewis structure:

Because there are six electron domains around the central Br atom in this ion, the electron-domain geometry is octahedral. The two nonbonding pairs of electrons are located opposite each other on the octahedron, leading to a square-planar molecular geometry:

(b) The observation that conductivity decreases as temperature increases indicates that fewer ions are present in the solution at the higher temperature. Thus, increasing the temperature causes the equilibrium to shift to the left. According to Le Châtelier's principle, this shift indicates that the reaction is exothermic as it proceeds from left to right. (Section 15.7)

(c) A Lewis acid is an electron-pair acceptor. (Section 16.1) The fluoride ion has four valence-shell electron pairs and can act as a Lewis base (an electron-pair donor). Thus, we can envision the following reaction occurring:

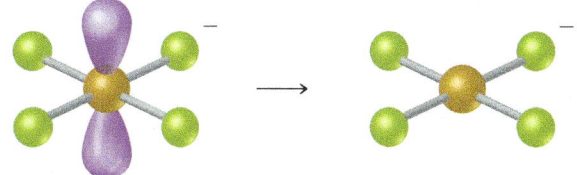

$$F^- \quad + \quad BrF_3 \quad \longrightarrow \quad BrF_4^-$$

Chapter Summary and Key Terms

THE PERIODIC TABLE AND CHEMICAL REACTIONS (SECTION 22.1) The periodic table is useful for organizing and remembering the descriptive chemistry of the elements. Among elements of a given group, size increases with increasing atomic number, and electronegativity and ionization energy decrease. The most nonmetallic elements are found in the upper right portion of the periodic table.

Among the nonmetallic elements, the first member of each group differs dramatically from the other members; it forms a maximum of four bonds to other atoms and exhibits a much greater tendency to form π bonds than the heavier elements in its group.

Because O_2 and H_2O are abundant in our world, we focus on two important and general reaction types as we discuss the nonmetals: oxidation by O_2 and proton-transfer reactions involving H_2O or aqueous solutions.

HYDROGEN (SECTION 22.2) Hydrogen has three isotopes: **protium** ($_1^1H$), **deuterium** ($_1^2H$), and **tritium** ($_1^3H$). Hydrogen is not a member of any particular periodic group, although it is usually placed above lithium. The hydrogen atom can either lose an electron, forming H^+, or gain one, forming H^- (the hydride ion). Because the $H-H$ bond is relatively strong, H_2 is fairly unreactive unless activated by heat or a catalyst. Hydrogen forms a very strong bond to oxygen, so the reactions of H_2 with oxygen-containing compounds usually lead to the formation of H_2O. The $H^+(aq)$ ion is able to oxidize many metals, forming $H_2(g)$. The electrolysis of water also forms $H_2(g)$.

The binary compounds of hydrogen are of three general types: **ionic hydrides** (formed by active metals), **metallic hydrides** (formed by transition metals), and **molecular hydrides** (formed by nonmetals). The ionic hydrides contain the H^- ion; because this ion is extremely basic, ionic hydrides react with H_2O to form H_2 and OH^-.

GROUP 8A: THE NOBLE GASES AND GROUP 7A: HALOGENS (SECTIONS 22.3 AND 22.4) The noble gases (group 8A) exhibit a very limited chemical reactivity because of the exceptional stability of their electron configurations. The xenon fluorides and oxides and KrF_2 are the best-established compounds of the noble gases.

The halogens (group 7A) occur as diatomic molecules. All except fluorine exhibit oxidation states varying from -1 to $+7$. Fluorine is the most electronegative element, so it is restricted to the oxidation states 0 and -1. The oxidizing power of the element (the tendency to form the -1 oxidation state) decreases as we proceed down the group.

The hydrogen halides are among the most useful compounds of these elements; these gases dissolve in water to form the hydrohalic acids, such as $HCl(aq)$. Hydrofluoric acid reacts with **silica**. The **interhalogens** are compounds formed between two different halogen elements. Chlorine, bromine, and iodine form a series of oxyacids, in which the halogen atom is in a positive oxidation state. These compounds and their associated oxyanions are strong oxidizing agents.

OXYGEN AND THE OTHER GROUP 6A ELEMENTS (SECTIONS 22.5 AND 22.6) Oxygen has two allotropes, O_2 and O_3 (ozone). Ozone is unstable compared to O_2, and it is a stronger oxidizing agent than O_2.

Most reactions of O_2 lead to oxides, compounds in which oxygen is in the -2 oxidation state. The soluble oxides of nonmetals generally produce acidic aqueous solutions; they are called **acidic anhydrides** or **acidic oxides.** In contrast, soluble metal oxides produce basic solutions and are called **basic anhydrides** or **basic oxides.** Many metal oxides that are insoluble in water dissolve in acid, accompanied by the formation of H_2O.

Peroxides contain $O-O$ bonds and oxygen in the -1 oxidation state. Peroxides are unstable, decomposing to O_2 and oxides. In such reactions, peroxides are simultaneously oxidized and reduced, a process called **disproportionation.** Superoxides contain the O_2^- ion in which oxygen is in the $-\frac{1}{2}$ oxidation state.

Sulfur is the most important of the other group 6A elements. It has several allotropic forms; the most stable one at room temperature consists of S_8 rings. Sulfur forms two oxides, SO_2 and SO_3, and both are important atmospheric pollutants. Sulfur trioxide is the anhydride of sulfuric acid, the most important sulfur compound and the most-produced industrial chemical. Sulfuric acid is a strong acid and a good dehydrating agent. Sulfur forms several oxyanions as well, including the SO_3^{2-} (sulfite), SO_4^{2-} (sulfate), and $S_2O_3^{2-}$ (thiosulfate) ions. Sulfur is found combined with many metals as a sulfide, in which sulfur is in the -2 oxidation state. These compounds often react with acids to form hydrogen sulfide (H_2S), which smells like rotten eggs.

NITROGEN AND THE OTHER GROUP 5A ELEMENTS (SECTIONS 22.7 AND 22.8) Nitrogen is found in the atmosphere as N_2 molecules. Molecular nitrogen is chemically very stable because of the strong $N\equiv N$ bond. Molecular nitrogen can be converted into ammonia via the Haber process. Once the ammonia is made, it can be converted into a variety of different compounds that exhibit nitrogen oxidation states ranging from -3 to $+5$. The most important industrial conversion of ammonia is the **Ostwald process,** in which ammonia is oxidized to nitric acid (HNO_3).

Nitrogen has three important oxides: nitrous oxide (N_2O), nitric oxide (NO), and nitrogen dioxide (NO_2). Nitrous acid (HNO_2) is a weak acid; its conjugate base is the nitrite ion (NO_2^-). Another important nitrogen compound is hydrazine (N_2H_4).

Phosphorus is the most important of the remaining group 5A elements. It occurs in nature as phosphate minerals. Phosphorus has several allotropes, including white phosphorus, which consists of P_4 tetrahedra. In reaction with the halogens, phosphorus forms trihalides PX_3 and pentahalides PX_5. These compounds undergo hydrolysis to produce an oxyacid of phosphorus and HX.

Phosphorus forms two oxides, P_4O_6 and P_4O_{10}. Their corresponding acids, phosphorous acid and phosphoric acid, undergo condensation reactions when heated. Phosphorus compounds are important in biochemistry and as fertilizers. Arsenic, antimony, and bismuth react similarly to phosphorus.

CARBON AND THE OTHER GROUP 4A ELEMENTS (SECTIONS 22.9 AND 22.10) The allotropes of carbon include diamond, graphite, fullerenes, carbon nanotubes, and graphene. Amorphous forms of graphite include **charcoal** and **carbon black.** Carbon forms two common oxides, CO and CO_2. Aqueous solutions of CO_2 produce the weak diprotic acid carbonic acid (H_2CO_3), which is the parent acid of hydrogen carbonate and carbonate salts. Binary compounds of carbon are called **carbides.** Carbides may be ionic, interstitial, or covalent. Calcium carbide (CaC_2) contains the strongly basic acetylide ion (C_2^{2-}), which reacts with water to form acetylene.

The other group 4A elements show great diversity in physical and chemical properties. Silicon, the second most abundant element in Earth's crust, is a semiconductor. It reacts with Cl_2 to form $SiCl_4$, a liquid at room temperature, a reaction that is used to help purify silicon from its native minerals. Silicon forms strong $Si-O$ bonds and therefore occurs in a variety of silicate minerals. Silica is SiO_2; **silicates** consist of SiO_4 tetrahedra, linked together at their vertices to form chains, sheets, or three-dimensional structures. The most common three-dimensional silicate is quartz (SiO_2). **Glass** is an amorphous (noncrystalline) form of SiO_2. Silicones contain $O-Si-O$ chains with organic groups bonded to the Si atoms. Like silicon, germanium is a metalloid; tin and lead are metallic.

BORON (SECTION 22.11) Boron is the only group 3A element that is a nonmetal. It forms a variety of compounds with hydrogen called boron hydrides, or **boranes.** Diborane (B_2H_6) has an unusual structure with two hydrogen atoms that bridge between the two boron atoms. Boranes react with oxygen to form boric oxide (B_2O_3), in which boron is in the $+3$ oxidation state. Boric oxide is the anhydride of boric acid (H_3BO_3). Boric acid readily undergoes condensation reactions.

Exam Prep

EP 22.1 Which description correctly describes a difference between the chemistry of oxygen and sulfur?

(a) Oxygen is a nonmetal and sulfur is a metalloid. **(b)** Oxygen can form more than four bonds, whereas sulfur cannot. **(c)** Sulfur has a higher electronegativity than oxygen. **(d)** Oxygen is better able to form π bonds than sulfur.

EP 22.2 When CaC_2 reacts with water, what carbon-containing compound forms? **(a)** CO **(b)** CO_2 **(c)** CH_4 **(d)** C_2H_2 **(e)** H_2CO_3

EP 22.3 What is the oxidation state of the H atom in methane? **(a)** -4 **(b)** -2 **(c)** 0 **(d)** $+1$ **(e)** $+4$

EP 22.4 Which of the elements Na, Rb, F, and Si has the smallest atomic radius? **(a)** F **(b)** Na **(c)** Rb **(d)** Si

EP 22.5 What is the maximum number of electrons hydrogen can have in its valence shell? **(a)** 2 **(b)** 6 **(c)** 8 **(d)** 10

EP 22.6 The steam reforming of methane produces hydrogen and what other product? **(a)** water **(b)** carbon monoxide **(c)** carbon dioxide **(d)** oxygen **(e)** propane

EP 22.7 Which element forms a metallic hydride with hydrogen? **(a)** Sr **(b)** As **(c)** P **(d)** Pd

EP 22.8 Which of the following choices is a metallic hydride? **(a)** BaH_2 **(b)** CuH **(c)** H_2Se **(d)** CsH **(e)** N_2H_2

EP 22.9 Predict the products for the following reaction:

$$Ni(s) + H_2SO_4(aq) \longrightarrow$$

(a) $Ni^{2+}(aq) + H_2(g) + SO_4^{2-}(aq)$

(b) $NiO(s) + H_2O(l) + SO_2(g)$

(c) $NiSO_4(s) + H_2(g)$

(d) $NiO_2(s) + H_2O(l) + SO(g)$

EP 22.10 Which of the following statements is or are *true*?

   (i) Hydrogen has three naturally occurring isotopes.

   (ii) All the isotopes of hydrogen have very similar properties.

   (iii) Protium is the most abundant isotope of hydrogen.

(a) Only statement i is true. **(b)** Statements ii and iii are true. **(c)** Statements i and iii are true. **(d)** Statements ii and iii are true. **(e)** All three statements are true.

EP 22.11 Compounds containing the XeF_3^+ ion have been characterized. Describe the electron-domain geometry and molecular geometry of this ion.

(a) trigonal planar, trigonal planar (b) tetrahedral, trigonal pyramidal (c) trigonal bipyramidal, T-shaped (d) tetrahedral, tetrahedral (e) octahedral, square planar

EP 22.12 Which compound has the largest (most positive) oxidation state on the halogen? (a) $KXeO_3F$ (b) $XeCl_2$ (c) $HBrO_3$ (d) IF_3

EP 22.13 Which of the following is (are) able to oxidize Cl^-? (a) F_2 (b) F^- (c) both Br_2 and I_2 (d) both Br^- and I^-

EP 22.14 Which acid is produced by the reaction of I_2O_5 with water? (a) HIO (b) HIO_2 (c) HIO_3 (d) HIO_4

EP 22.15 Which of these oxides is the most acidic? (a) N_2O (b) SiO_2 (c) P_4O_{10} (d) P_4O_6

EP 22.16 Predict the molecular geometry of $TeCl_2$. (a) tetrahedral (b) linear (c) trigonal pyramidal (d) bent

EP 22.17 The oxidation states of S in $H_2S_2O_7$ is _____ and the oxidation state of N in HNO_2 is _____. (a) +3, +3 (b) +6, +5 (c) +3, +5 (d) +6, +3 (e) +5, +2

EP 22.18 In power plants, hydrazine is used to prevent corrosion of the metal parts of steam boilers by the O_2 dissolved in the water. The hydrazine reacts with O_2 in water to give $N_2(g)$ and $H_2O(l)$. Write a balanced equation for this reaction.

(a) $N_2H_4(aq) + O_2(aq) \longrightarrow N_2(g) + H_2O(l)$

(b) $N_2H_4(aq) + O_2(aq) \longrightarrow N_2(g) + 2\,H_2O(l)$

(c) $N_2H_6(aq) + \frac{3}{2}\,O_2(aq) \longrightarrow N_2(g) + 3\,H_2O(l)$

(d) $2\,N_2H_6(aq) + 3\,O_2(aq) \longrightarrow 2\,N_2(g) + 6\,H_2O(l)$

EP 22.19 Which of the following statements is or are *true*?

(i) Trigonal bipyramidal molecular structures, with 5 atoms about the center atom, can form with either N or P as the center atom.

(ii) The properties of the group 5A elements are very similar to each other.

(iii) Compounds in which P is in the +5 oxidation state are much stronger oxidizing agents compared to analogous N-containing compounds.

(a) Only statement i is true. (b) Statements i and ii are true. (c) Statements i and iii are true. (d) Statements ii and iii are true. (e) None of the statements is true.

EP 22.20 Which oxyacid is produced when PF_3 reacts with water? (a) H_3FO_3 (b) H_3FO_4 (c) H_3PO_3 (d) H_3PO_4

EP 22.21 Which compound is a carbide? (a) SiC (b) $NaHCO_3$ (c) CO_2 (d) C_2H_2 (e) C_{60}

EP 22.22 What is the most common oxidation state of silicon in its compounds? (a) -4 (b) $+1$ (c) $+2$ (d) $+3$ (e) $+4$

EP 22.23 The mineral talc, also known as talcum powder, has the formula $Mg_3(Si_2O_5)_2(OH)_2$. What type of silicate structure does it form? (a) a 3D network like diamond (b) chains of linked SiO_4 tetrahedra (c) sheets of linked SiO_4 tetrahedra (d) chains of $-Si-O-Si-O-$ atoms with Mg(II) and hydroxide counterions

EP 22.24 In the mineral beryl, six silicate tetrahedra are connected to form a ring as shown here. The negative charge of this polyanion is balanced by Be^{2+} and Al^{3+} cations. If elemental analysis gives a Be:Si ratio of 1:2 and an Al:Si ratio of 1:3, then what is the empirical formula of beryl? (a) $Be_2Al_3Si_6O_{19}$ (b) $Be_3Al_2(SiO_4)_6$ (c) $Be_3Al_2Si_6O_{18}$ (d) $BeAl_2Si_6O_{15}$

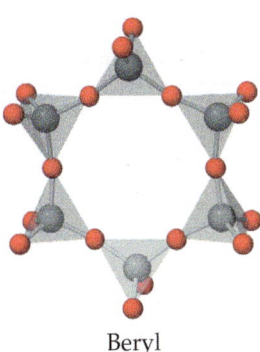

Beryl

EP 22.25 Polymers that have an $O-Si-O$ backbone and hydrocarbon groups on the Si are _____. The linking of SiO_4 tetrahedra in three dimensions forms _____. SiO_4 tetrahedra linked in sheets or chains form _____. (a) silicates, quartz, silicones (b) silicates, silicones, quartz (c) quartz, silicones, silicates (d) quartz, silicates, silicones (e) silicones, quartz, silicates

Exercises

Visualizing Concepts

22.1 Which statement identifies which of these two compounds is more stable and explains why? [Section 22.1]

$$H_2C=CH_2$$

$$H_2Si=SiH_2$$

(a) The carbon compound because C is less electronegative than Si

(b) The silicon compound because Si forms stronger sigma bonds than C

(c) The carbon compound because C forms stronger multiple bonds than Si

(d) The silicon compound because Si forms stronger pi bonds than C

22.2 (a) Identify the *type* of chemical reaction represented by the following diagram. (b) Place appropriate charges on the species on both sides of the equation. (c) Write the chemical equation for the reaction. [Section 22.1]

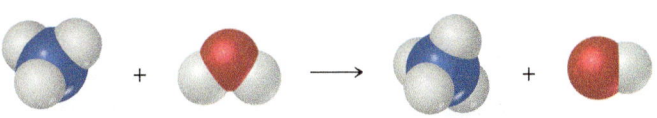

22.3 Which of the following species (there may be more than one) is/are likely to have the structure shown here?

(a) XeF_4 (b) BrF_4^+ (c) SiF_4 (d) $TeCl_4$ (e) $HClO_4$ (The colors do not reflect atom identities.) [Sections 22.3, 22.4, 22.6, and 22.10]

22.4 You have two glass bottles, one containing oxygen and one filled with nitrogen. How could you determine which one is which? [Sections 22.5 and 22.7]

22.5 Write the molecular formula and Lewis structure for each of the following oxides of nitrogen: [Section 22.7]

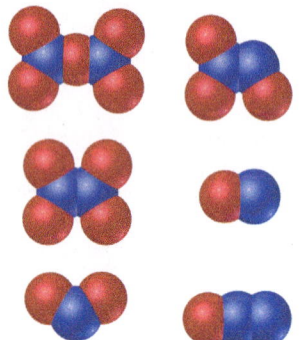

22.6 Which property of the group 6A elements might be the one depicted in the graph shown here? (a) electronegativity (b) first ionization energy (c) density (d) X—X single-bond enthalpy (e) electron affinity [Sections 22.5 and 22.6]

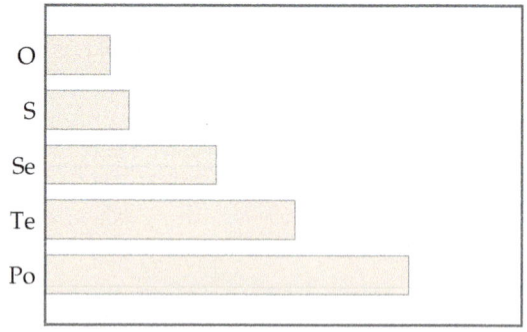

22.7 Identify the true statements concerning the atoms and ions of the group 6A elements. [Sections 22.5 and 22.6]

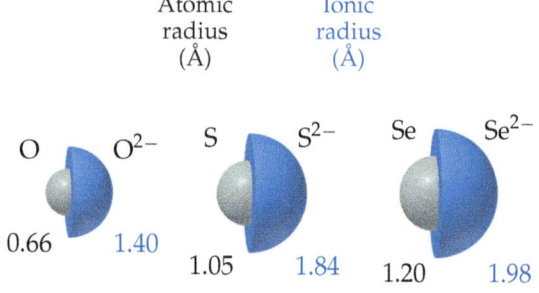

(a) The ionic radii are larger than the atomic radii because the ions have more electrons than their corresponding atoms.

(b) Atomic radii increase going down the group because of increasing nuclear charge.

(c) The ionic radii increase going down the group because of the increase in the principal quantum number of outermost electrons.

(d) Of these ions, Se^{2-} is the strongest base in water because it is the largest.

22.8 Which property of the third-row nonmetallic elements might be the one depicted in the graphic? (a) first ionization energy (b) atomic radius (c) electronegativity (d) melting point (e) X—X single-bond enthalpy [Sections 22.3, 22.4, 22.6, 22.8, and 22.10]

22.9 Which of the following compounds would you expect to be the most generally reactive, and why? (Each corner in these structures represents a CH_2 group.) [Section 22.8]

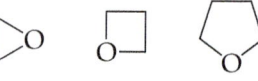

22.10 (a) Draw the Lewis structures for at least four species that have the general formula

$$\left[:X \equiv Y: \right]^n$$

where X and Y may be the same or different and n may have a value from +1 to −2. (b) Which of the compounds is likely to be the strongest Brønsted base? Explain. [Sections 22.1, 22.7, and 22.9]

Periodic Trends and Chemical Reactions (Section 22.1)

22.11 Identify each of the following elements as a metal, nonmetal, or metalloid: (a) phosphorus, (b) strontium, (c) manganese, (d) selenium, (e) sodium, (f) krypton.

22.12 Identify each of the following elements as a metal, nonmetal, or metalloid: (a) gallium, (b) molybdenum, (c) tellurium, (d) arsenic, (e) xenon, (f) ruthenium.

22.13 Consider the elements O, Ba, Co, Be, Br, and Se. From this list, select the element that (a) is most electronegative, (b) exhibits a maximum oxidation state of +7, (c) loses an electron most readily, (d) forms π bonds most readily, (e) is a transition metal, (f) is a liquid at room temperature and pressure.

22.14 Consider the elements Li, K, Cl, C, Ne, and Ar. From this list, select the element that (a) is most electronegative, (b) has the greatest metallic character, (c) most readily forms a positive ion, (d) has the smallest atomic radius, (e) forms π bonds most readily, (f) has multiple allotropes.

22.15 Which of the following statements are *true*?

(**a**) Both nitrogen and phosphorus can form a pentafluoride compound.

(**b**) Although CO is a well-known compound, SiO does not exist under ordinary conditions.

(**c**) Cl_2 is easier to oxidize than I_2.

(**d**) At room temperature, the stable form of oxygen is O_2, whereas that of sulfur is S_8.

22.16 Which of the following statements are *true*?

(**a**) Si can form an ion with six fluorine atoms, SiF_6^{2-}, whereas carbon cannot.

(**b**) Si can form three stable compounds containing two Si atoms each, Si_2H_2, Si_2H_4, and Si_2H_6.

(**c**) In HNO_3 and H_3PO_4 the central atoms, N and P, have different oxidation states.

(**d**) S is more electronegative than Se.

22.17 Complete and balance the following equations:

(**a**) $NaOCH_3(s) + H_2O(l) \longrightarrow$

(**b**) $CuO(s) + HNO_3(aq) \longrightarrow$

(**c**) $WO_3(s) + H_2(g) \xrightarrow{\Delta}$

(**d**) $NH_2OH(l) + O_2(g) \longrightarrow$

(**e**) $Al_4C_3(s) + H_2O(l) \longrightarrow$

22.18 Complete and balance the following equations:

(**a**) $Mg_3N_2(s) + H_2O(l) \longrightarrow$

(**b**) $C_3H_7OH(l) + O_2(g) \longrightarrow$

(**c**) $MnO_2(s) + C(s) \xrightarrow{\Delta}$

(**d**) $AlP(s) + H_2O(l) \longrightarrow$

(**e**) $Na_2S(s) + HCl(aq) \longrightarrow$

Hydrogen, the Noble Gases, and the Halogens (Sections 22.2, 22.3, and 22.4)

22.19 (**a**) Give the names and chemical symbols for the three isotopes of hydrogen. (**b**) List the isotopes in order of decreasing natural abundance. (**c**) Which hydrogen isotope is radioactive? (**d**) Write the nuclear equation for the radioactive decay of this isotope.

22.20 The physical properties of D_2O differ from those of H_2O because

(**a**) D has a different electron configuration than O.

(**b**) D is radioactive.

(**c**) D forms stronger bonds with O than H does.

(**d**) D is much more massive than H.

22.21 Give a reason why hydrogen might be placed with the group 1A elements of the periodic table.

22.22 Give a reason why hydrogen might be placed with the group 7A elements of the periodic table.

22.23 Complete and balance the following equations:

(**a**) $NaH(s) + H_2O(l) \longrightarrow$

(**b**) $Fe(s) + H_2SO_4(aq) \longrightarrow$

(**c**) $H_2(g) + Br_2(g) \longrightarrow$

(**d**) $Na(l) + H_2(g) \longrightarrow$

(**e**) $PbO(s) + H_2(g) \longrightarrow$

22.24 Write balanced equations for each of the following reactions (some of these are analogous to reactions shown in the chapter). (**a**) Aluminum metal reacts with acids to form hydrogen gas. (**b**) Steam reacts with magnesium metal to give magnesium oxide and hydrogen. (**c**) Manganese(IV) oxide is reduced to manganese(II) oxide by hydrogen gas. (**d**) Calcium hydride reacts with water to generate hydrogen gas.

22.25 Identify the following hydrides as ionic, metallic, or molecular: (**a**) BaH_2, (**b**) H_2Te, (**c**) $TiH_{1.7}$.

22.26 Identify the following hydrides as ionic, metallic, or molecular: (**a**) B_2H_6, (**b**) RbH, (**c**) $Th_4H_{1.5}$.

22.27 Describe two characteristics of hydrogen that are favorable for its use as a general energy source in vehicles.

22.28 The H_2/O_2 fuel cell converts elemental hydrogen and oxygen into water, producing, theoretically, 1.23 V. What is the most sustainable way to obtain hydrogen to run a large number of fuel cells? Explain.

22.29 Why does xenon form stable compounds with fluorine, whereas argon does not?

22.30 A friend tells you that the "neon" in neon signs is a compound of neon and aluminum. Can your friend be correct? Explain.

22.31 Write the chemical formula for each of the following, and indicate the oxidation state of the halogen or noble-gas atom in each: (**a**) calcium hypobromite, (**b**) bromic acid, (**c**) xenon trioxide, (**d**) perchlorate ion, (**e**) iodous acid, (**f**) iodine pentafluoride.

22.32 Write the chemical formula for each of the following compounds, and indicate the oxidation state of the halogen or noble-gas atom in each: (**a**) chlorate ion, (**b**) hydroiodic acid, (**c**) iodine trichloride, (**d**) sodium hypochlorite, (**e**) perchloric acid, (**f**) xenon tetrafluoride.

22.33 Name the following compounds and assign oxidation states to the halogens in them: (**a**) $Fe(ClO_3)_3$, (**b**) $HClO_2$, (**c**) XeF_6, (**d**) BrF_5, (**e**) $XeOF_4$, (**f**) HIO_3.

22.34 Name the following compounds and assign oxidation states to the halogens in them: (**a**) $KClO_3$, (**b**) $Ca(IO_3)_2$, (**c**) $AlCl_3$, (**d**) $HBrO_3$, (**e**) H_5IO_6, (**f**) XeF_4.

22.35 Explain each of the following observations: (**a**) At room temperature I_2 is a solid, Br_2 is a liquid, and Cl_2 and F_2 are both gases. (**b**) F_2 cannot be prepared by electrolytic oxidation of aqueous F^- solutions. (**c**) The boiling point of HF is much higher than those of the other hydrogen halides. (**d**) The halogens decrease in oxidizing power in the order $F_2 > Cl_2 > Br_2 > I_2$.

22.36 Explain the following observations: (**a**) For a given oxidation state, the acid strength of the oxyacid in aqueous solution decreases in the order chlorine > bromine > iodine. (**b**) Hydrofluoric acid cannot be stored in glass bottles. (**c**) HI cannot be prepared by treating NaI with sulfuric acid. (**d**) The interhalogen ICl_3 is known, but $BrCl_3$ is not.

Oxygen and the Other Group 6A Elements (Sections 22.5 and 22.6)

22.37 Write balanced equations for each of the following reactions. (**a**) When mercury(II) oxide is heated, it decomposes to form O_2 and mercury metal. (**b**) When copper(II) nitrate is heated strongly, it decomposes to form copper(II) oxide, nitrogen dioxide, and oxygen. (**c**) Lead(II) sulfide, $PbS(s)$, reacts with ozone to form $PbSO_4(s)$ and $O_2(g)$. (**d**) When heated in air, $ZnS(s)$ is converted to ZnO. (**e**) Potassium peroxide reacts with $CO_2(g)$ to give potassium carbonate and O_2. (**f**) Oxygen is converted to ozone in the upper atmosphere.

22.38 Complete and balance the following equations:

(**a**) $CaO(s) + H_2O(l) \longrightarrow$ (**b**) $Al_2O_3(s) + H^+(aq) \longrightarrow$

(**c**) $Na_2O_2(s) + H_2O(l) \longrightarrow$ (**d**) $N_2O_3(g) + H_2O(l) \longrightarrow$

(**e**) $KO_2(s) + H_2O(l) \longrightarrow$ (**f**) $NO(g) + O_3(g) \longrightarrow$

22.39 Predict whether each of the following oxides is acidic, basic, amphoteric, or neutral: (**a**) NO_2, (**b**) CO_2, (**c**) Al_2O_3, (**d**) CaO.

22.40 Select the more acidic member of each of the following pairs: (**a**) Mn_2O_7 and MnO_2, (**b**) SnO and SnO_2, (**c**) SO_2 and SO_3, (**d**) SiO_2 and SO_2, (**e**) Ga_2O_3 and In_2O_3, (**f**) SO_2 and SeO_2.

22.41 Write the chemical formula for each of the following compounds, and indicate the oxidation state of the group 6A element in each: (**a**) selenous acid, (**b**) potassium hydrogen sulfite, (**c**) hydrogen telluride, (**d**) carbon disulfide, (**e**) calcium sulfate, (**f**) cadmium sulfide, (**g**) zinc telluride.

22.42 Write the chemical formula for each of the following compounds, and indicate the oxidation state of the group 6A element in each: (**a**) sulfur tetrachloride, (**b**) selenium trioxide, (**c**) sodium thiosulfate, (**d**) hydrogen sulfide, (**e**) sulfuric acid, (**f**) sulfur dioxide, (**g**) mercury telluride.

22.43 In aqueous solution, hydrogen sulfide reduces (**a**) Fe^{3+} to Fe^{2+}, (**b**) Br_2 to Br^-, (**c**) MnO_4^- to Mn^{2+}, (**d**) HNO_3 to NO_2. In all cases, under appropriate conditions, the product is elemental sulfur. Write a balanced net ionic equation for each reaction.

22.44 An aqueous solution of SO_2 reduces (**a**) aqueous $KMnO_4$ to $MnSO_4(aq)$, (**b**) acidic aqueous $K_2Cr_2O_7$ to aqueous Cr^{3+}, (**c**) aqueous $Hg_2(NO_3)_2$ to mercury metal. Write balanced equations for these reactions.

22.45 Write the Lewis structure for each of the following species, and indicate the structure of each: (**a**) SeO_3^{2-}; (**b**) S_2Cl_2; (**c**) chlorosulfonic acid, HSO_3Cl (chlorine is bonded to sulfur).

22.46 The SF_5^- ion is formed when $SF_4(g)$ reacts with fluoride salts containing large cations, such as $CsF(s)$. Draw the Lewis structures for SF_4 and SF_5^-, and predict the molecular structure of each.

22.47 Write a balanced equation for each of the following reactions: (**a**) Sulfur dioxide reacts with water. (**b**) Solid zinc sulfide reacts with hydrochloric acid. (**c**) Elemental sulfur reacts with sulfite ion to form thiosulfate. (**d**) Sulfur trioxide is dissolved in sulfuric acid.

22.48 Write a balanced equation for each of the following reactions. (You may have to guess at one or more of the reaction products, but you should be able to make a reasonable guess, based on your study of this chapter.) (**a**) Hydrogen selenide can be prepared by reaction of an aqueous acid solution on aluminum selenide. (**b**) Sodium thiosulfate is used to remove excess Cl_2 from chlorine-bleached fabrics. The thiosulfate ion forms SO_4^{2-} and elemental sulfur, while Cl_2 is reduced to Cl^-.

Nitrogen and the Other Group 5A Elements (Sections 22.7 and 22.8)

22.49 Write the chemical formula for each of the following compounds, and indicate the oxidation state of nitrogen in each: (**a**) sodium nitrite, (**b**) ammonia, (**c**) nitrous oxide, (**d**) sodium cyanide, (**e**) nitric acid, (**f**) nitrogen dioxide, (**g**) nitrogen, (**h**) boron nitride.

22.50 Write the chemical formula for each of the following compounds, and indicate the oxidation state of nitrogen in each: (**a**) nitric oxide, (**b**) hydrazine, (**c**) potassium cyanide, (**d**) sodium nitrite, (**e**) ammonium chloride, (**f**) lithium nitride.

22.51 Write the Lewis structure for each of the following species, describe its geometry, and indicate the oxidation state of the nitrogen: (**a**) HNO_2, (**b**) N_3^-, (**c**) $N_2H_5^+$, (**d**) NO_3^-.

22.52 Write the Lewis structure for each of the following species, describe its geometry, and indicate the oxidation state of the nitrogen: (**a**) NH_4^+, (**b**) NO_2^-, (**c**) N_2O, (**d**) NO_2.

22.53 Complete and balance the following equations:
(**a**) $Mg_3N_2(s) + H_2O(l) \longrightarrow$
(**b**) $NO(g) + O_2(g) \longrightarrow$

(**c**) $N_2O_5(g) + H_2O(l) \longrightarrow$
(**d**) $NH_3(aq) + H^+(aq) \longrightarrow$
(**e**) $N_2H_4(l) + O_2(g) \longrightarrow$
Which ones of these are redox reactions?

22.54 Write a balanced net ionic equation for each of the following reactions: (**a**) Dilute nitric acid reacts with zinc metal with formation of nitrous oxide. (**b**) Concentrated nitric acid reacts with sulfur with formation of nitrogen dioxide. (**c**) Concentrated nitric acid oxidizes sulfur dioxide with formation of nitric oxide. (**d**) Hydrazine is burned in excess fluorine gas, forming NF_3. (**e**) Hydrazine reduces CrO_4^{2-} to $Cr(OH)_4^-$ in base (hydrazine is oxidized to N_2).

22.55 Write complete balanced half-reactions for (**a**) oxidation of nitrous acid to nitrate ion in acidic solution, (**b**) oxidation of N_2 to N_2O in acidic solution.

22.56 Write complete balanced half-reactions for (**a**) reduction of nitrate ion to NO in acidic solution, (**b**) oxidation of HNO_2 to NO_2 in acidic solution.

22.57 Write a molecular formula for each compound, and indicate the oxidation state of the group 5A element in each formula: (**a**) phosphorous acid, (**b**) pyrophosphoric acid, (**c**) antimony trichloride, (**d**) magnesium arsenide, (**e**) diphosphorus pentoxide, (**f**) sodium phosphate.

22.58 Write a chemical formula for each compound or ion, and indicate the oxidation state of the group 5A element in each formula: (**a**) phosphate ion, (**b**) arsenous acid, (**c**) antimony(III) sulfide, (**d**) calcium dihydrogen phosphate, (**e**) potassium phosphide, (**f**) gallium arsenide.

22.59 Account for the following observations: (**a**) Phosphorus forms a pentachloride, but nitrogen does not. (**b**) H_3PO_2 is a monoprotic acid. (**c**) Phosphonium salts, such as PH_4Cl, can be formed under anhydrous conditions, but they cannot be made in aqueous solution. (**d**) White phosphorus is more reactive than red phosphorus.

22.60 Account for the following observations: (**a**) H_3PO_3 is a diprotic acid. (**b**) Nitric acid is a strong acid, whereas phosphoric acid is weak. (**c**) Phosphate rock is ineffective as a phosphate fertilizer. (**d**) Phosphorus does not exist at room temperature as diatomic molecules, but nitrogen does. (**e**) Solutions of Na_3PO_4 are quite basic.

22.61 Write a balanced equation for each of the following reactions: (**a**) preparation of white phosphorus from calcium phosphate, (**b**) hydrolysis of PBr_3, (**c**) reduction of PBr_3 to P_4 in the gas phase, using H_2.

22.62 Write a balanced equation for each of the following reactions: (**a**) hydrolysis of PCl_5, (**b**) dehydration of phosphoric acid (also called orthophosphoric acid) to form pyrophosphoric acid, (**c**) reaction of P_4O_{10} with water.

Carbon, the Other Group 4A Elements, and Boron (Sections 22.9, 22.10, and 22.11)

22.63 Give the chemical formula for (**a**) hydrocyanic acid, (**b**) nickel tetracarbonyl, (**c**) barium bicarbonate, (**d**) calcium acetylide, (**e**) potassium carbonate.

22.64 Give the chemical formula for (**a**) carbonic acid, (**b**) sodium cyanide, (**c**) potassium hydrogen carbonate, (**d**) acetylene, (**e**) iron pentacarbonyl.

22.65 Complete and balance the following equations:
(**a**) $ZnCO_3(s) \xrightarrow{\Delta}$
(**b**) $BaC_2(s) + H_2O(l) \longrightarrow$
(**c**) $C_2H_2(g) + O_2(g) \longrightarrow$
(**d**) $CS_2(g) + O_2(g) \longrightarrow$
(**e**) $Ca(CN)_2(s) + HBr(aq) \longrightarrow$

22.66 Complete and balance the following equations:
(a) $CO_2(g) + OH^-(aq) \longrightarrow$
(b) $NaHCO_3(s) + H^+(aq) \longrightarrow$
(c) $CaO(s) + C(s) \xrightarrow{\Delta}$
(d) $C(s) + H_2O(g) \xrightarrow{\Delta}$
(e) $CuO(s) + CO(g) \longrightarrow$

22.67 Write a balanced equation for each of the following reactions: (a) Hydrogen cyanide is formed commercially by passing a mixture of methane, ammonia, and air over a catalyst at 800 °C. Water is a byproduct of the reaction. (b) Baking soda reacts with acids to produce carbon dioxide gas. (c) When barium carbonate reacts in air with sulfur dioxide, barium sulfate and carbon dioxide form.

22.68 Write a balanced equation for each of the following reactions: (a) Burning magnesium metal in a carbon dioxide atmosphere reduces the CO_2 to carbon. (b) In photosynthesis, solar energy is used to produce glucose ($C_6H_{12}O_6$) and O_2 from carbon dioxide and water. (c) When carbonate salts dissolve in water, they produce basic solutions.

22.69 Write the formulas for the following compounds, and indicate the oxidation state of the group 4A element or of boron in each: (a) boric acid, (b) silicon tetrabromide, (c) lead(II) chloride, (d) sodium tetraborate decahydrate (borax), (e) boric oxide, (f) germanium dioxide.

22.70 Write the formulas for the following compounds, and indicate the oxidation state of the group 4A element or of boron in each: (a) silicon dioxide, (b) germanium tetrachloride, (c) sodium borohydride, (d) stannous chloride, (e) diborane, (f) boron trichloride.

22.71 Select the member of group 4A that best fits each description: (a) has the lowest first ionization energy, (b) is found in oxidation states ranging from −4 to +4, (c) is most abundant in Earth's crust.

22.72 Select the member of group 4A that best fits each description: (a) forms chains to the greatest extent, (b) forms the most basic oxide, (c) is a metalloid that can form 2+ ions.

22.73 (a) What is the characteristic geometry about silicon in all silicate minerals? (b) Metasilicic acid has the empirical formula H_2SiO_3. Which of the structures shown in Figure 22.32 would you expect metasilicic acid to have?

22.74 Speculate as to why carbon forms carbonate rather than silicate analogs.

22.75 (a) Determine the number of calcium ions in the chemical formula of the mineral hardystonite, $Ca_xZn(Si_2O_7)$. (b) Determine the number of hydroxide ions in the chemical formula of the mineral pyrophyllite, $Al_2(Si_2O_5)_2(OH)_x$.

22.76 (a) Determine the number of sodium ions in the chemical formula of albite, $Na_xAlSi_3O_8$. (b) Determine the number of hydroxide ions in the chemical formula of tremolite, $Ca_2Mg_5(Si_4O_{11})_2(OH)_x$.

22.77 (a) How does the structure of diborane (B_2H_6) differ from that of ethane (C_2H_6)? (b) Explain why diborane adopts the geometry that it does. (c) What is the significance of the statement that the hydrogen atoms in diborane are described as "hydridic"?

22.78 Write a balanced equation for each of the following reactions: (a) Diborane reacts with water to form boric acid and molecular hydrogen. (b) Upon heating, boric acid undergoes a condensation reaction to form tetraboric acid. (c) Boron oxide dissolves in water to give a solution of boric acid.

Additional Exercises

22.79 Indicate whether each of the following statements is *true* or *false*. (a) $H_2(g)$ and $D_2(g)$ are allotropic forms of hydrogen. (b) ClF_3 is an interhalogen compound. (c) $MgO(s)$ is an acidic anhydride. (d) $SO_2(g)$ is an acidic anhydride. (e) $2\,H_3PO_4(l) \longrightarrow H_4P_2O_7(l) + H_2O(g)$ is an example of a condensation reaction. (f) Tritium is an isotope of the element hydrogen. (g) $2\,SO_2(g) + O_2(g) \longrightarrow 2\,SO_3(g)$ is an example of a disproportionation reaction.

22.80 Although the ClO_4^- and IO_4^- ions have been known for a long time, BrO_4^- was not synthesized until 1965. The ion was synthesized by oxidizing the bromate ion with xenon difluoride, producing xenon, hydrofluoric acid, and the perbromate ion. (a) Write the balanced equation for this reaction. (b) What are the oxidation states of Br in the Br-containing species in this reaction?

22.81 Write a balanced equation for the reaction of each of the following compounds with water: (a) $SO_2(g)$, (b) $Cl_2O_7(g)$, (c) $Na_2O_2(s)$, (d) $BaC_2(s)$, (e) $RbO_2(s)$ (f) $Mg_3N_2(s)$, (g) $NaH(s)$.

22.82 What is the anhydride for each of the following acids? (a) H_2SO_4 (b) $HClO_3$ (c) HNO_2 (d) H_2CO_3 (e) H_3PO_4

22.83 Hydrogen peroxide is capable of oxidizing (a) hydrazine to N_2 and H_2O, (b) SO_2 to SO_4^{2-}, (c) NO_2^- to NO_3^-, (d) $H_2S(g)$ to $S(s)$, (e) Fe^{2+} to Fe^{3+}. Write a balanced net ionic equation for each of these redox reactions.

22.84 A sulfuric acid plant produces a considerable amount of heat. This heat is used to generate electricity, which helps reduce operating costs. The synthesis of H_2SO_4 consists of three main chemical processes: (i) oxidation of S to SO_2, (ii) oxidation of SO_2 to SO_3, (iii) the dissolving of SO_3 in H_2SO_4 and the subsequent reaction with water to form H_2SO_4. (a) If the third process produces 130 kJ/mol, how much heat is produced in preparing a mole of H_2SO_4 from a mole of S? (b) How much heat is produced in preparing 5000 pounds of H_2SO_4?

22.85 (a) What is the oxidation state of P in PO_4^{3-} and of N in NO_3^-? (b) Why doesn't N form a stable NO_4^{3-} ion analogous to P?

22.86 (a) The P_4, P_4O_6, and P_4O_{10} molecules have a common structural feature of four P atoms arranged in a tetrahedron (Figures 22.25 and 22.26). Does this mean that the bonding between the P atoms is the same in all these cases? (b) Sodium trimetaphosphate ($Na_3P_3O_9$) and sodium tetrametaphosphate ($Na_4P_4O_{12}$) are used as water-softening agents. They contain cyclic $P_3O_9^{3-}$ and $P_4O_{12}^{4-}$ ions, respectively. Propose reasonable structures for these ions.

22.87 Write a balanced chemical reaction for the condensation reaction between H_3PO_4 molecules to form $H_5P_3O_{10}$.

22.88 Ultrapure germanium, like silicon, is used in semiconductors. Germanium of "ordinary" purity is prepared by the high-temperature reduction of GeO_2 with carbon. The Ge is converted to $GeCl_4$ by treatment with Cl_2 and then purified by distillation; $GeCl_4$ is then hydrolyzed in water to GeO_2 and reduced to the elemental form with H_2. The element is then zone refined. Write a balanced chemical equation for each of the chemical transformations in the course of forming ultrapure Ge from GeO_2.

22.89 When aluminum replaces up to half of the silicon atoms in SiO_2, a mineral class called feldspars results. The feldspars are the most abundant rock-forming minerals, comprising about 50% of the minerals in Earth's crust. Orthoclase is a feldspar in which Al replaces one-fourth of the Si atoms of SiO_2, and charge balance is completed by K^+ ions. Determine the chemical formula for orthoclase.

22.90 (a) Determine the charge of the aluminosilicate ion whose composition is $AlSi_3O_{10}$. (b) Using Figure 22.32, propose a reasonable description of the structure of this aluminosilicate.

22.91 (a) How many grams of H_2 can be stored in 100.0 kg of the alloy FeTi if the hydride $FeTiH_2$ is formed? (b) What volume does this quantity of H_2 occupy at STP? (c) If this quantity of hydrogen was combusted in air to produce liquid water, how much energy could be produced?

22.92 Using the thermochemical data in Table 22.1 and Appendix C, calculate the average Xe—F bond enthalpies in XeF_2, XeF_4, and XeF_6, respectively. What is the significance of the trend in these quantities?

22.93 Hydrogen gas has a higher fuel value than natural gas on a mass basis but not on a volume basis. Thus, hydrogen is not competitive with natural gas as a fuel transported long distances through pipelines. Calculate the heats of combustion of H_2 and CH_4 (the principal component of natural gas) (a) per mole of each, (b) per gram of each, (c) per cubic meter of each at STP. Assume $H_2O(l)$ as a product.

22.94 Using ΔG_f° for ozone from Appendix C, calculate the equilibrium constant for Equation 22.22 at 298.0 K, assuming no electrical input.

22.95 The solubility of Cl_2 in 100 g of water at STP is 310 cm^3. Assume that this quantity of Cl_2 is dissolved and equilibrated as follows:

$$Cl_2(aq) + H_2O \rightleftharpoons Cl^-(aq) + HClO(aq) + H^+(aq)$$

(a) If the equilibrium constant for this reaction is 4.7×10^{-4}, calculate the equilibrium concentration of HClO formed. (b) What is the pH of the final solution?

22.96 When ammonium perchlorate decomposes thermally, the products of the reaction are $N_2(g)$, $O_2(g)$, $H_2O(g)$, and $HCl(g)$. (a) Write a balanced equation for the reaction. (*Hint:* You might find it easier to use fractional coefficients for the products.) (b) Calculate the enthalpy change in the reaction per mole of NH_4ClO_4. The standard enthalpy of formation of $NH_4ClO_4(s)$ is −295.8 kJ. (c) When $NH_4ClO_4(s)$ is employed in solid-fuel booster rockets, it is packed with powdered aluminum. Given the high temperature needed for $NH_4ClO_4(s)$ decomposition and what the products of the reaction are, what role does the aluminum play? (d) Calculate the volume of all the gases that would be produced at STP, assuming complete reaction of one pound of ammonium perchlorate.

22.97 The dissolved oxygen present in any highly pressurized, high-temperature steam boiler can be extremely corrosive to its metal parts. Hydrazine, which is completely miscible with water, can be added to remove oxygen by reacting with it to form nitrogen and water. (a) Write the balanced equation for the reaction between gaseous hydrazine and oxygen. (b) Calculate the enthalpy change accompanying this reaction. (c) Oxygen in air dissolves in water to the extent of 9.1 ppm at 20 °C at sea level. How many grams of hydrazine are required to react with all the oxygen in 3.0×10^4 L (the volume of a small swimming pool) under these conditions?

22.98 One method proposed for removing SO_2 from the flue gases of power plants involves reaction with aqueous H_2S. Elemental sulfur is the product. (a) Write a balanced chemical equation for the reaction. (b) What volume of H_2S at 27 °C and 760 torr would be required to remove the SO_2 formed by burning 2.0 tons of coal containing 3.5% S by mass? (c) What mass of elemental sulfur is produced? Assume that all reactions are 100% efficient.

22.99 The maximum allowable concentration of $H_2S(g)$ in air is 20 mg per kilogram of air (20 ppm by mass). How many grams of FeS would be required to react with hydrochloric acid to produce this concentration at 1.00 atm and 25 °C in an average room measuring 12 ft $\times$ 20 ft $\times$ 8 ft? (Under these conditions, the average molar mass of air is 29.0 g/mol.)

22.100 The standard heats of formation of $H_2O(g)$, $H_2S(g)$, $H_2Se(g)$, and $H_2Te(g)$ are −241.8, −20.17, +29.7, and +99.6 kJ/mol, respectively. The enthalpies necessary to convert the elements in their standard states to one mole of gaseous atoms are 248, 277, 227, and 197 kJ/mol of atoms for O, S, Se, and Te, respectively. The enthalpy for dissociation of H_2 is 436 kJ/mol. Calculate the average H—O, H—S, H—Se, and H—Te bond enthalpies, and comment on their trend.

22.101 Manganese silicide has the empirical formula MnSi and melts at 1280 °C. It is insoluble in water but does dissolve in aqueous HF. (a) What type of compound do you expect MnSi to be: metallic, molecular, covalent-network, or ionic? (b) Write a likely balanced chemical equation for the reaction of MnSi with concentrated aqueous HF.

22.102 Hydrazine has been employed as a reducing agent for metals. Using standard reduction potentials, predict whether the following metals can be reduced to the metallic state by hydrazine under standard conditions in acidic solution: (a) Fe^{2+}, (b) Sn^{2+}, (c) Cu^{2+}, (d) Ag^+, (e) Cr^{3+}, (f) Co^{3+}.

22.103 Both dimethylhydrazine, $(CH_3)_2NNH_2$, and methylhydrazine, CH_3NHNH_2, have been used as rocket fuels. When dinitrogen tetroxide (N_2O_4) is used as the oxidizer, the products are H_2O, CO_2, and N_2. If the thrust of the rocket depends on the volume of the products produced, which of the substituted hydrazines produces a greater thrust per gram total mass of oxidizer plus fuel? Assume that both fuels generate the same temperature and that $H_2O(g)$ is formed.

22.104 Borazine, $(BH)_3(NH)_3$, is an analog of C_6H_6, benzene. It can be prepared from the reaction of diborane with ammonia, with hydrogen as another product; or from lithium borohydride and ammonium chloride, with lithium chloride and hydrogen as the other products. (a) Write balanced chemical equations for the production of borazine using both synthetic methods. (b) Draw the Lewis dot structure of borazine. (c) How many grams of borazine can be prepared from 2.00 L of ammonia at STP, assuming diborane is in excess?

Design an Experiment

You are given samples of five substances. At room temperature, three are colorless gases, one is a colorless liquid, and one is a white solid. You are told that the substances are NF_3, PF_3, PCl_3, PF_5, and PCl_5. Let's design experiments to determine which substance is which, using concepts from this and earlier chapters.

(a) Assuming that you don't have access to either the Internet or a handbook of chemistry (as is the case during your exams!), design experiments that would allow you to identify the substances. (b) Which of the substances could undergo reactions to add more atoms around the central atom? What types of reactions might you choose to test this hypothesis? (c) Based on what you know about intermolecular forces, which of the substances is likely the solid?

23

TRANSITION METALS AND COORDINATION CHEMISTRY

▲ **WATER LILIES AND THE JAPANESE BRIDGE** was painted by Claude Monet in 1899.

The colors of our world are beautiful, but to a chemist they are also informative—providing insights into the structure and bonding of matter. Compounds of the transition metals constitute an important group of colored substances, including the characteristic red color of oxygenated human blood. The red color is due to an iron-containing protein called hemoglobin that is found in red blood cells.

The colors generated by transition-metal compounds are among the most common we encounter in our world. Some transition-metal compounds are used as pigments; others produce the colors in glass and precious gems. The use of vivid green, yellow, and blue colors in the paintings of impressionists like Monet, Cezanne, and van Gogh was made possible by the development of synthetic pigments in the 1800s. Three such pigments used extensively by the impressionists were cobalt blue, $CoAl_2O_4$; chrome yellow, $PbCrO_4$; and emerald green, $Cu_4(CH_3COO)_2(AsO_2)_6$. In each case, the presence of a transition-metal ion is responsible for the color of the pigment—Co^{2+} in cobalt blue, Cr^{6+} in chrome yellow, and Cu^{2+} in emerald green. Unfortunately, emerald green also contains arsenic, which makes it toxic, and its use as a pigment was discontinued in the early 20th century.

The color of a given transition-metal compound depends on not only the transition-metal ion, but also the identity and geometry of the ions and/or molecules that surround it. To appreciate the importance of

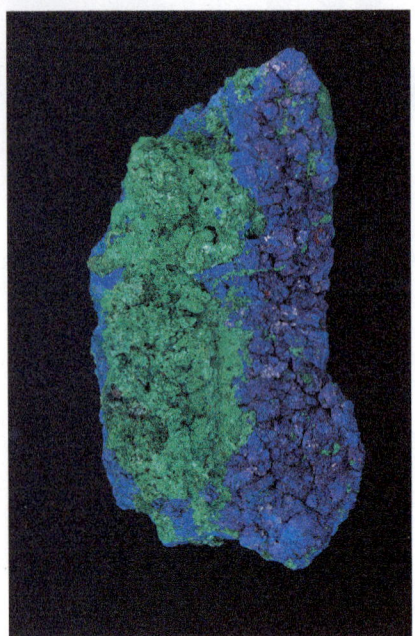

▲ **Figure 23.1 Crystals of blue azurite and green malachite.** These semiprecious stones were ground up and used as pigments in the Middle Ages and Renaissance, but they were eventually replaced with blue and green pigments that have superior chemical stability.

 Learning Objectives

When you finish **Section 23.1**, you should be able to:

▶ Predict the physical properties of transition metals based on their position in the periodic table.

▶ Determine the electron configuration of a transition-metal ion.

▶ Determine the oxidation state of a transition-metal ion in a compound.

▶ Describe the different magnetic properties that transition metals can exhibit.

the transition-metal ion surroundings, consider the colors of the minerals azurite, $Cu_3(CO_3)_2(OH)_2$, and malachite, $Cu_2(CO_3)(OH)_2$ (**Figure 23.1**). Both minerals contain Cu^{2+} ions coordinated by carbonate and hydroxide ions, but subtle differences in the local surroundings of the Cu^{2+} ion lead to the contrasting colors of these two minerals.

The importance of transition metals is not limited to biology and use as a source of color. Transition metals and their alloys are used as structural materials to make coins and jewelry, and as electronic conductors in wires and electronic devices. The presence of partially filled d-orbitals allows transition-metal compounds to act as catalysts and magnets. In earlier chapters, we saw that metal ions can function as Lewis acids, forming covalent bonds with molecules and ions that act as Lewis bases. (Section 16.1) We have previously encountered many ions and compounds that result from such interactions. We first mentioned hemoglobin in the "Chemistry and Life" box on "Sickle-Cell Anemia." (Section 13.6) In our discussion of solubility of ionic substances, we saw the formation of complex ions, such as $[Ag(NH_3)_2]^+$. (Section 17.5)

In this chapter, we take a closer look at many aspects of the transition-metal elements, including the rich and important chemistry associated with such complex assemblies of metal ions surrounded by molecules and ions. Metal compounds of this kind are called *coordination compounds*, and the branch of chemistry that focuses on them is called *coordination chemistry*.

23.1 | The Transition Metals

The part of the periodic table in which the d orbitals are being filled as we move left to right across a row is the home of the transition metals (**Figure 23.2**). (Section 6.8)

With some exceptions (such as platinum and gold), metallic elements are found in nature as solid inorganic compounds called **minerals**. Notice from **Table 23.1** that minerals are identified by common names rather than chemical names.

Most transition metals in minerals have oxidation states (Section 4.4) ranging from +1 to +4. To obtain a pure metal from its mineral, various chemical processes must be performed to reduce the metal to its elemental form (oxidation state = 0). **Metallurgy** is the science and technology of extracting metals from their natural sources and preparing them for practical use. It usually involves several steps: (1) mining, that is, removing the relevant *ore* (a mixture of minerals) from the ground, (2) concentrating the ore or otherwise preparing it for further treatment, (3) reducing the ore to obtain the free metal, (4) purifying the metal, and (5) mixing it with other elements to modify its properties. This last process produces an *alloy*, a metallic material composed of two or more elements. (Section 12.2)

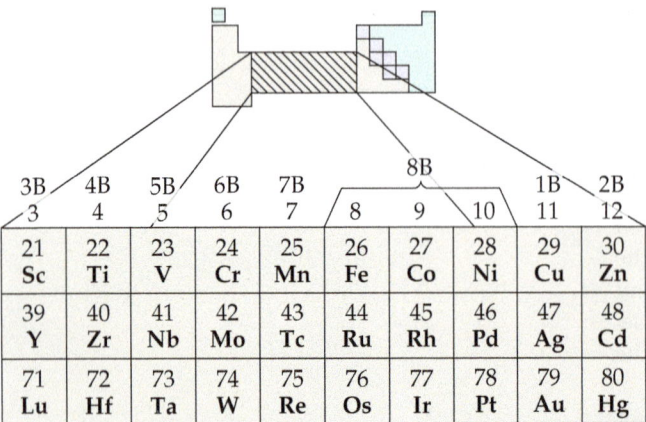

▲ **Figure 23.2 The position of the transition metals in the periodic table.** They are the B groups in periods 4, 5, and 6. The short-lived, radioactive transition metals from period 7 are not shown.

TABLE 23.1 Principal Mineral Sources of Some Transition Metals

Metal	Mineral	Mineral Composition
Chromium	Chromite	$FeCr_2O_4$
Cobalt	Cobaltite	$CoAsS$
Copper	Chalcocite	Cu_2S
	Chalcopyrite	$CuFeS_2$
	Malachite	$Cu_2(CO_3)(OH)_2$
Iron	Hematite	Fe_2O_3
	Magnetite	Fe_3O_4
Manganese	Pyrolusite	MnO_2
Mercury	Cinnabar	HgS
Molybdenum	Molybdenite	MoS_2
Titanium	Rutile	TiO_2
	Ilmenite	$FeTiO_3$
Zinc	Sphalerite	ZnS

TABLE 23.2   Properties of the Period 4 Transition Metals

Group	3B	4B	5B	6B	7B	8B			1B	2B
Element:	Sc	Ti	V	Cr	Mn	Fe	Co	Ni	Cu	Zn
Ground-state electron configuration	$3d^14s^2$	$3d^24s^2$	$3d^34s^2$	$3d^54s^1$	$3d^54s^2$	$3d^64s^2$	$3d^74s^2$	$3d^84s^2$	$3d^{10}4s^1$	$3d^{10}4s^2$
First ionization energy (kJ/mol)	631	658	650	653	717	759	758	737	745	906
Metallic radius (Å)	1.64	1.47	1.35	1.29	1.37	1.26	1.25	1.25	1.28	1.37
Density (g/cm³)	3.0	4.5	6.1	7.9	7.2	7.9	8.7	8.9	8.9	7.1
Melting point (°C)	1541	1660	1917	1857	1244	1537	1494	1455	1084	420
Crystal structure*	hcp	hcp	bcc	bcc	**	bcc	hcp	fcc	fcc	hcp

*Abbreviations for crystal structures are: hcp = hexagonal close packed, fcc = face-centered cubic, bcc = body-centered cubic. (Section 12.2)

**Manganese has a more complex crystal structure.

Physical Properties

Some physical properties of the period 4 (also known as "first-row") transition metals are listed in Table 23.2. The properties of the heavier transition metals vary similarly across periods 5 and 6.

Figure 23.3 shows the atomic radius observed in close-packed metallic structures as a function of group number.* The trends seen in the graph are a result of two competing forces. On the one hand, increasing effective nuclear charge favors a decrease in radius as we move left to right across each period. (Section 7.2) On the other hand, the metallic bonding strength increases until we reach the middle of each period and then decreases as we fill antibonding orbitals. (Section 12.3) As a general rule, a bond shortens as it becomes stronger. (Section 8.8) For groups 3B through 6B, these two effects work cooperatively, and the result is a marked decrease in radius. In elements to the right of group 6B, the two effects counteract each other, reducing the decrease and eventually leading to an increase in radius.

In general, atomic radii increase as we move down in a given group in the periodic table because of the increasing principal quantum number of the outer-shell electrons. (Section 7.3) However, Figure 23.3 shows that once we move beyond the group 3B elements, the period 5 and period 6 transition elements in a given group have virtually the same radii. In group 5B, for example, tantalum in period 6 has virtually the same radius as niobium in period 5. This interesting and important effect has its origin in the lanthanide series, elements 57 through 70. The filling of 4f orbitals through the lanthanide elements (Figure 6.28) causes a steady increase in the effective nuclear charge, producing a size decrease called the **lanthanide contraction**, which just offsets the increase we expect as we go from period 5 transition metals to period 6. Thus, the period 5 and period 6 transition metals in each group have very similar radii and chemical properties. For example, the group 4B metals zirconium (period 5) and hafnium (period 6) always occur together in nature and are very difficult to separate.

Electron Configurations and Oxidation States

As we saw in Figure 6.28, transition metals owe their location in the periodic table to the filling of the d subshells. Many of the chemical and physical properties of transition metals result from the unique characteristics of the d orbitals. For a given

Go Figure

Does the variation in radius of the transition metals follow the same trend as the effective nuclear charge on moving from left to right across the periodic table?

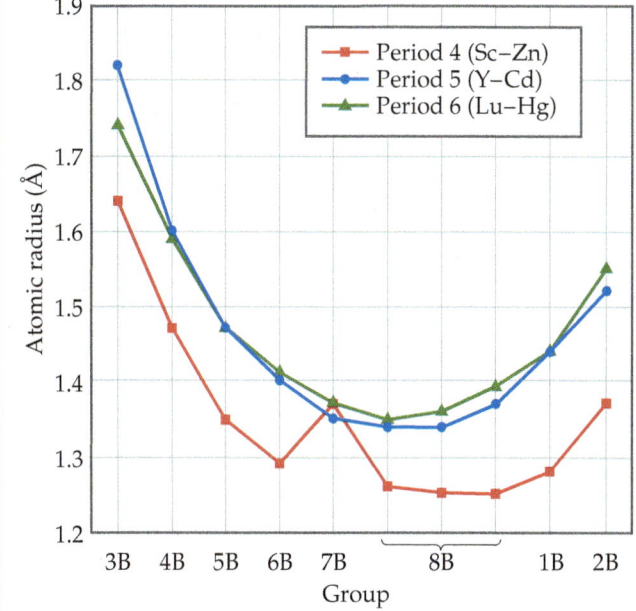

▲ **Figure 23.3** Radii of transition metals as a function of group number.

---

* Note that the radii defined in this way, often referred to as metallic radii, differ somewhat from the bonding atomic radii defined in Section 7.3.

transition-metal atom, the valence $(n - 1)d$ orbitals are smaller than the corresponding valence ns and np orbitals. In quantum-mechanical terms, the $(n - 1)d$ orbital wave functions drop off more rapidly as we move away from the nucleus than do the ns and np orbital wave functions. This characteristic feature of the d orbitals limits their interaction with orbitals on neighboring atoms, but not so much that they are insensitive to surrounding atoms. As a result, electrons in these orbitals behave sometimes like valence electrons and sometimes like core electrons. The details depend on location in the periodic table and the atom's environment.

When transition metals are oxidized, *they lose their outer s electrons before they lose electrons from the* d *subshell.* (Section 7.4) The electron configuration of Fe is $[Ar]3d^6 4s^2$, for example, whereas that of Fe^{2+} is $[Ar]3d^6$. Formation of Fe^{3+} requires loss of one $3d$ electron, giving $[Ar]3d^5$. Most transition-metal ions contain partially occupied d subshells, which are responsible in large part for three characteristics:

1. Transition metals often have more than one stable oxidation state.

2. Many transition-metal compounds are colored, as shown in **Figure 23.4**.

3. Transition metals and their compounds often exhibit magnetic properties.

Figure 23.5 shows the common nonzero oxidation states for the period 4 transition metals. The +2 oxidation state, which is common for most transition metals, is due to the loss of the two outer $4s$ electrons. This oxidation state is found for all these elements except Sc, where the 3+ ion with an [Ar] configuration is particularly stable.

Oxidation states above +2 are due to successive losses of $3d$ electrons. From Sc through Mn, the maximum oxidation state increases from +3 to +7, equaling in each case the total number of $4s$ plus $3d$ electrons in the atom. Thus, manganese, which has the electron configuration $[Ar]4s^2 3d^5$, has a maximum oxidation state of $2 + 5 = +7$. As we move to the right beyond Mn in Figure 23.5, the maximum oxidation state decreases. This decrease is due to the attraction of d orbital electrons to the nucleus, which increases faster than the attraction of the s orbital electrons to the nucleus as we move left to right across the periodic table. In other words, in each period the d electrons become more corelike as the atomic number increases. By the time we get to zinc, it is not possible to remove electrons from the $3d$ orbitals through chemical oxidation.

 Go Figure In which transition-metal ion of this group are the $3d$ orbitals completely filled?

▲ **Figure 23.4 Aqueous solutions of transition-metal ions.** Left to right: Co^{2+}, Ni^{2+}, Cu^{2+}, and Zn^{2+}. The counterion is nitrate in all cases.

In the transition metals of periods 5 and 6, the increased size of the 4d and 5d orbitals makes it possible to attain maximum oxidation states as high as +8, which is achieved in RuO_4 and OsO_4. In general, the maximum oxidation states are found only when the metals are combined with the most electronegative elements, especially O, F, and in some cases Cl.

Magnetism

The spin that an electron possesses gives the electron a *magnetic moment*, a property that causes the electron to behave like a tiny magnet. (Section 6.7) In a *diamagnetic* solid, defined as one in which all the electrons in the solid are paired, the spin-up and spin-down electrons cancel one another. (Section 9.8) Diamagnetic substances are generally described as being nonmagnetic, but when a diamagnetic substance is placed in a magnetic field, the motions of the electrons cause the substance to be very weakly repelled by the magnet. In other words, these supposedly nonmagnetic substances do show some very faint magnetic character in the presence of a magnetic field.

A substance in which the atoms or ions have one or more unpaired electrons is *paramagnetic*. (Section 9.8) In a paramagnetic solid, the electrons on one atom or ion do not influence the unpaired electrons on neighboring atoms or ions. As a result, the magnetic moments on the atoms or ions are randomly oriented and constantly changing direction, as shown in **Figure 23.6(a)**. When a paramagnetic substance is placed in a magnetic field, however, the magnetic moments tend to align parallel to one another, producing a net attractive interaction with the magnet. Thus, unlike a diamagnetic substance, which is weakly repelled by a magnetic field, a paramagnetic substance is attracted to a magnetic field.

When you think of a magnet, you probably picture a simple iron magnet. Iron exhibits **ferromagnetism**, a form of magnetism much stronger than paramagnetism. Ferromagnetism arises when the unpaired electrons of the atoms or ions in a solid are influenced by the orientations of the electrons in neighboring atoms or ions. The most stable (lowest-energy) arrangement is when the spins of electrons on adjacent atoms or ions are aligned in the same direction, as in Figure 23.6(b). When a ferromagnetic solid is placed in a magnetic field, the electrons align strongly in a direction parallel to the magnetic field. The attraction to the magnetic field that results is much stronger than that for a paramagnetic substance.

When a ferromagnet is removed from an external magnetic field, the interactions between the electrons cause the ferromagnetic substance to maintain a magnetic moment even in the absence of the external magnetic field. We then refer to it as a *permanent magnet* (**Figure 23.7**).

The only ferromagnetic transition metals are Fe, Co, and Ni, but many alloys also exhibit ferromagnetism, which is in some cases stronger than the ferromagnetism of the pure metals. Particularly powerful ferromagnetism is found in compounds containing both transition metals and lanthanide metals. Two of the most important examples are $SmCo_5$ and $Nd_2Fe_{14}B$.

Two additional types of magnetism involving ordered arrangements of unpaired electrons are depicted in Figure 23.6. In materials that exhibit **antiferromagnetism** [Figure 23.6(c)], the unpaired electrons on a given atom or ion align so that their spins are oriented in the direction opposite the spin direction on neighboring atoms. This means that the spin-up and spin-down

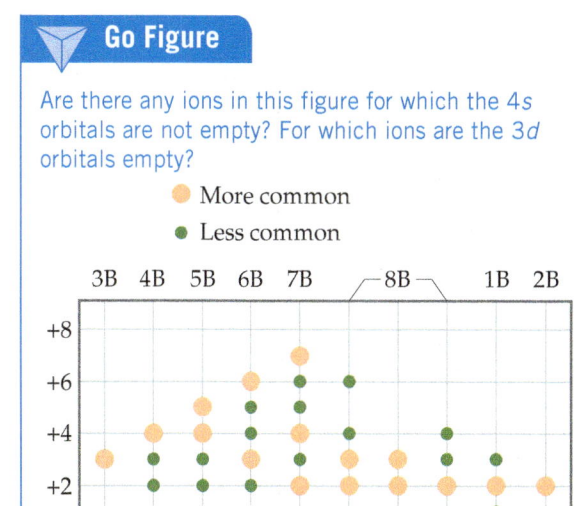

Go Figure

Are there any ions in this figure for which the 4s orbitals are not empty? For which ions are the 3d orbitals empty?

▲ **Figure 23.5** Nonzero oxidation states of the period 4 transition metals.

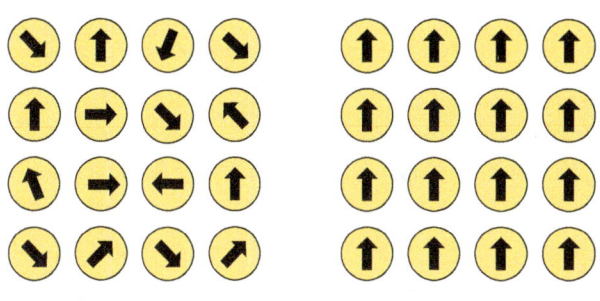

Go Figure

Describe how the representation shown for the paramagnetic material would change if the material were placed in a magnetic field.

Paramagnetic; spin direction is constantly changing and random in the absence of an external magnetic field

(a)

Ferromagnetic; spins align parallel to each other

(b)

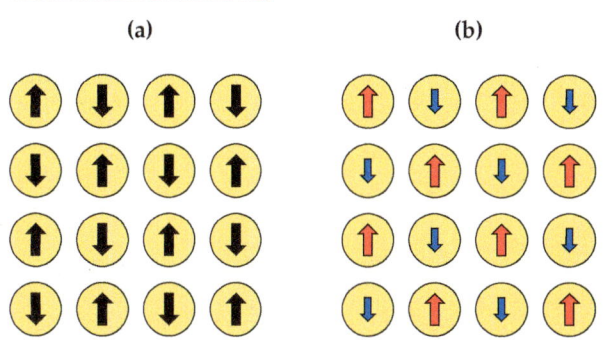

Antiferromagnetic; spins align in opposite directions and cancel each other

(c)

Ferrimagnetic; unequal spins align in opposite directions but do not completely cancel each other

(d)

▲ **Figure 23.6** The relative orientation of electron spins in various types of magnetic substances.

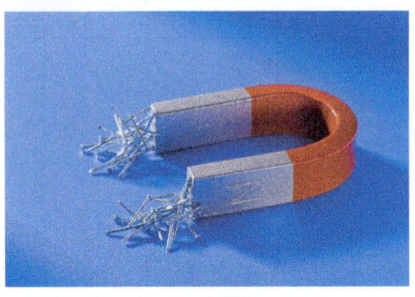

▲ **Figure 23.7 A permanent magnet.** Permanent magnets are made from ferromagnetic and ferrimagnetic materials.

electrons cancel each other. Examples of antiferromagnetic substances are chromium metal, FeMn alloys, and such transition-metal oxides as Fe_2O_3, $LaFeO_3$, and MnO.

A substance that exhibits **ferrimagnetism** [Figure 23.6(d)] has both ferromagnetic and antiferromagnetic characteristics. Like an antiferromagnet, the unpaired electrons align so that the spins in adjacent atoms or ions point in opposite directions. However, unlike an antiferromagnet, the net magnetic moments of the spin-up electrons are not fully canceled by the spin-down electrons. This can happen because the magnetic centers have different numbers of unpaired electrons (as is the case for $NiMnO_3$), because the number of magnetic sites aligned in one direction is larger than the number aligned in the other direction ($Y_3Fe_5O_{12}$), or because both of these conditions apply (Fe_3O_4). Because the magnetic moments do not cancel, the properties of ferrimagnetic materials are similar to the properties of ferromagnetic materials.

Ferromagnets, ferrimagnets, and antiferromagnets all become paramagnetic when heated above a critical temperature. This happens when the thermal energy is sufficient to overcome the forces determining the spin directions of the electrons—the special alignment of the electron spins is disrupted above the critical temperature, which varies from substance to substance. This temperature is called the *Curie temperature*, T_C, of the substance for ferromagnets and ferrimagnets and the *Néel temperature*, T_N, for antiferromagnets.

 Self-Assessment Exercises

SAE 23.1 Which of the following statements about the transition-metal elements is or are *true*?

(**i**) Most of the transition metals are found in nature as minerals.
(**ii**) The atomic radius of a period 6 transition metal is generally very similar to that of the period 5 transition metal immediately above it in the periodic table.
(**iii**) Transition-metal elements tend to form compounds in which they have negative oxidation states.

(**a**) Only statement i is true. (**b**) Statements i and ii are true. (**c**) Statements i and iii are true. (**d**) Statements ii and iii are true. (**e**) All three statements are true.

SAE 23.2 How many electrons are in the valence *d* orbitals of ground-state Zn^{2+}? (**a**) 2 (**b**) 6 (**c**) 8 (**d**) 10 (**e**) 12

SAE 23.3 What is the ground-state electron configuration of Mn^{2+}? (**a**) $[Ar]4s^23d^5$ (**b**) $[Ar]3d^5$ (**c**) $[Ar]4s^23d^3$ (**d**) $[Ar]4s^23d^7$

SAE 23.4 What is the oxidation state of chromium in Cr_2O_3? (**a**) −3 (**b**) 0 (**c**) +2 (**d**) +3 (**e**) +6

SAE 23.5 Which of the following transition-metal ions is expected to exhibit paramagnetism? (**a**) Sc^{3+} (**b**) Cu^+ (**c**) Fe^{3+} (**d**) Zn^{2+}

23.2 | Transition-Metal Complexes

The transition metals occur in many interesting and important molecular forms. Species that are assemblies of a central transition-metal ion bonded to a group of surrounding molecules or ions, such as $[Ag(NH_3)_2]^+$, $[CoCl_4]^{2-}$, and $[Fe(H_2O)_6]^{3+}$, are called **metal complexes**, or merely *complexes*.* If the complex carries a net charge, it is generally called a *complex ion*. (Section 17.5) Compounds that contain complexes are known as **coordination compounds**.

The molecules or ions that bond to the metal ion in a complex are known as **ligands** (from the Latin word *ligare*, "to bind"). There are two NH_3 ligands bonded to Ag^+ in the complex ion $[Ag(NH_3)_2]^+$, for instance, four Cl^- ligands bonded to Co^{2+} in $[CoCl_4]^{2-}$, and six H_2O ligands bonded to Fe^{3+} in $[Fe(H_2O)_6]^{3+}$. Each ligand functions as a Lewis base and so donates a pair of electrons to form the metal–ligand bond. (Section 16.1) Thus, every ligand has at least one lone pair of valence electrons. Four of the most frequently encountered ligands,

illustrate that most ligands are either polar molecules or anions. In forming a complex, the ligands are said to *coordinate* to the metal.

*Most of the coordination compounds we examine in this chapter contain transition-metal ions, although ions of other metals can also form complexes.

Learning Objectives

When you finish Section 23.2, you should be able to:

▶ Determine the coordination number and coordination sphere of a transition-metal complex.
▶ Describe how a metal–ligand bond is formed.

TABLE 23.3 Properties of Some Ammonia Complexes of Cobalt(III)

Original Formulation	Color	Ions per Formula Unit	"Free" Cl^- Ions per Formula Unit	Modern Formulation
$CoCl_3 \cdot 6 NH_3$	Orange	4	3	$[Co(NH_3)_6]Cl_3$
$CoCl_3 \cdot 5 NH_3$	Purple	3	2	$[Co(NH_3)_5Cl]Cl_2$
$CoCl_3 \cdot 4 NH_3$	Green	2	1	*trans*-$[Co(NH_3)_4Cl_2]Cl$
$CoCl_3 \cdot 4 NH_3$	Violet	2	1	*cis*-$[Co(NH_3)_4Cl_2]Cl$

Notice that, in writing the formula for complex ions, such as $[Fe(H_2O)_6]^{3+}$, we use square brackets to enclose the metal ion and all of the ligands. The overall charge of the complex ion is written outside the brackets. We expand more on this nomenclature shortly.

The Development of Coordination Chemistry: Werner's Theory

Because compounds of the transition metals are beautifully colored, the chemistry of these elements fascinated chemists even before the periodic table was introduced. During the late 1700s through the 1800s, the many coordination compounds that were isolated and studied had properties that were puzzling in light of the bonding theories prevailing at the time. Table 23.3, for example, lists a series of $CoCl_3$–NH_3 compounds that have strikingly different colors. Note that the third and fourth species have different colors, even though the originally assigned formula was the same for both, $CoCl_3 \cdot 4 NH_3$.

The modern formulations of the compounds, listed in Table 23.3, are based on various lines of experimental evidence. For example, all four compounds are strong electrolytes (Section 4.1) but yield different numbers of ions when dissolved in water. Dissolving $CoCl_3 \cdot 6 NH_3$ in water yields four ions per formula unit ($[Co(NH_3)_6]^{3+}$ plus three Cl^- ions), whereas $CoCl_3 \cdot 5 NH_3$ yields only three ions per formula unit ($[Co(NH_3)_5Cl]^{2+}$ and two Cl^- ions). Furthermore, the reaction of the compounds with excess aqueous silver nitrate leads to the precipitation of different amounts of $AgCl(s)$. When $CoCl_3 \cdot 6 NH_3$ is treated with excess $AgNO_3(aq)$, 3 mol of $AgCl(s)$ precipitate per mole of complex, which means all three Cl^- ions in the complex can react to form $AgCl(s)$. By contrast, when $CoCl_3 \cdot 5 NH_3$ is treated with excess $AgNO_3(aq)$, only 2 mol of $AgCl(s)$ precipitate per mole of complex, telling us that one of the Cl^- ions in the complex does not react with Ag^+. These results are summarized in Table 23.3.

In 1893, the Swiss chemist Alfred Werner (1866–1919) proposed a theory that successfully explained the observations shown in Table 23.3. In a theory that became the basis for understanding coordination chemistry, Werner proposed that any metal ion exhibits both a primary valence and a secondary valence. The *primary valence* is the oxidation state of the metal, which is +3 for the complexes in Table 23.3. The *secondary valence* is the number of atoms bonded to the metal ion, which is also called the **coordination number**. For these cobalt complexes, Werner deduced a coordination number of 6 with the ligands in an octahedral arrangement (Section 9.1) around the Co^{3+} ion.

Werner's theory provided a beautiful explanation for the results presented in Table 23.3. The NH_3 molecules are ligands bonded to the Co^{3+} ion (through the nitrogen atom as we will see later); if there are fewer than six NH_3 molecules, the remaining ligands are Cl^- ions. The central metal and the ligands bonded to it constitute the **coordination sphere** of the complex.

In writing the chemical formula for a coordination compound, Werner suggested using square brackets to signify the makeup of the coordination sphere in any given compound. He therefore proposed that $CoCl_3 \cdot 6 NH_3$ and $CoCl_3 \cdot 5 NH_3$ are better written as $[Co(NH_3)_6]Cl_3$ and $[Co(NH_3)_5Cl]Cl_2$, respectively. He further proposed that the chloride ions that are part of the coordination sphere are bonded so tightly that they do not dissociate when the complex is dissolved in water. Thus, dissolving $[Co(NH_3)_5Cl]Cl_2$ in water produces a $[Co(NH_3)_5Cl]^{2+}$ ion and two Cl^- ions.

Werner's ideas also explained why there are two forms of $CoCl_3 \cdot 4\,NH_3$. Using Werner's postulates, we write the formula as $[Co(NH_3)_4Cl_2]Cl$. As shown in **Figure 23.8**, there are two ways to arrange the ligands in the $[Co(NH_3)_4Cl_2]^+$ complex, called the *cis* and *trans* forms. In the cis form, the two chloride ligands occupy adjacent vertices of the octahedral arrangement. In *trans*-$[Co(NH_3)_4Cl_2]^+$, the two chlorides are opposite each other. It is this difference in positions of the Cl ligands that leads to two compounds, one violet and one green.

The insight Werner provided into the bonding in coordination compounds is even more remarkable when we realize that his theory predated Lewis's ideas of covalent bonding by more than 20 years! Because of his tremendous contributions to coordination chemistry, Werner was awarded the 1913 Nobel Prize in Chemistry.

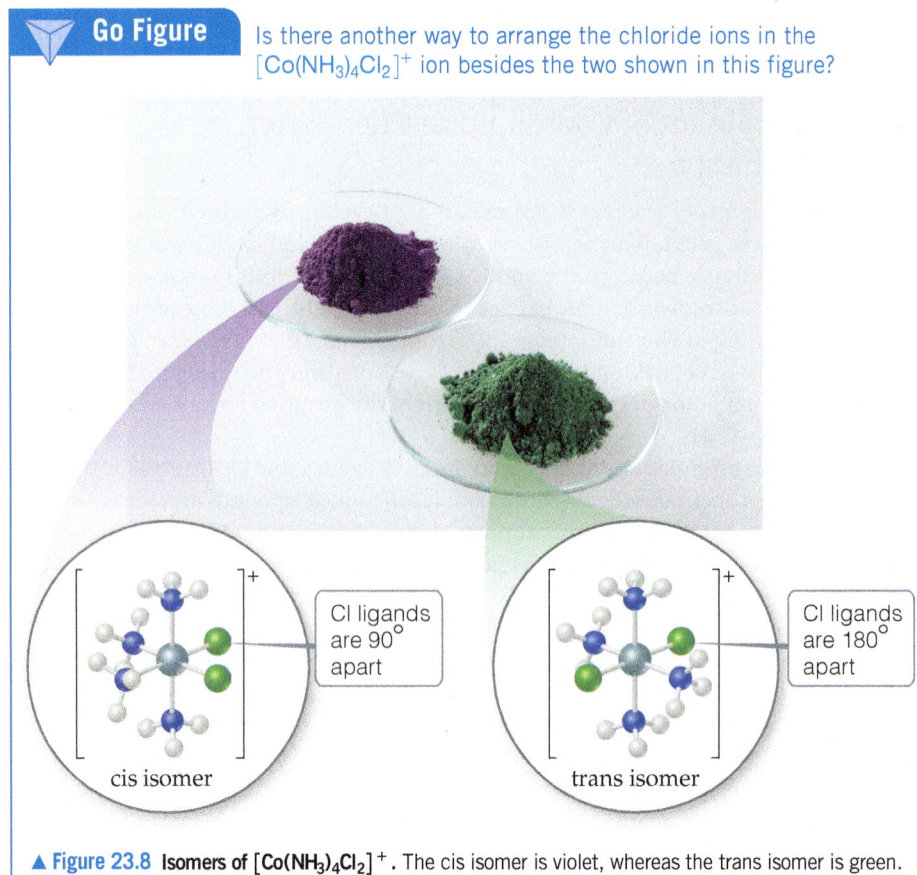

Go Figure Is there another way to arrange the chloride ions in the $[Co(NH_3)_4Cl_2]^+$ ion besides the two shown in this figure?

Cl ligands are 90° apart

cis isomer

Cl ligands are 180° apart

trans isomer

▲ **Figure 23.8 Isomers of $[Co(NH_3)_4Cl_2]^+$.** The cis isomer is violet, whereas the trans isomer is green.

Sample Exercise 23.1

Identifying the Coordination Sphere of a Complex

Palladium(II) tends to form complexes with coordination number 4. A compound has the composition $PdCl_2 \cdot 3\,NH_3$. **(a)** Write the formula for this compound that best shows the coordination structure. **(b)** When an aqueous solution of the compound is treated with excess $AgNO_3(aq)$, how many moles of $AgCl(s)$ are formed per mole of $PdCl_2 \cdot 3\,NH_3$?

SOLUTION

Analyze We are given the coordination number of Pd(II) and a chemical formula indicating that the complex contains NH_3 and Cl^-. We are asked to determine **(a)** which ligands are attached to Pd(II) in the compound and **(b)** how the compound behaves toward $AgNO_3$ in aqueous solution.

Plan **(a)** Because of their charge, the Cl^- ions can be either in the coordination sphere, where they are bonded directly to the metal, or outside the coordination sphere, where they are bonded ionically to the complex. The electrically neutral NH_3 ligands must be in the coordination sphere, and we are told that four ligands are

bonded to the Pd(II) ion. **(b)** Any chlorides in the coordination sphere do not precipitate as AgCl.

Solve

(a) By analogy to the ammonia complexes of cobalt(III) shown in Figure 23.8, we predict that the three NH_3 are ligands attached to the Pd(II) ion. The fourth ligand around Pd(II) is one chloride ion. The second chloride ion is not a ligand; it serves only as a *counterion* (a noncoordinating ion that balances charge) in the compound. We conclude that the formula showing the structure best is $[Pd(NH_3)_3Cl]Cl$.

(b) Because only the non-ligand Cl^- can react with Ag^+, we expect to produce 1 mol of $AgCl(s)$ per mole of complex. The balanced equation is

$$[Pd(NH_3)_3Cl]Cl(aq) + AgNO_3(aq) \longrightarrow$$
$$[Pd(NH_3)_3Cl]NO_3(aq) + AgCl(s)$$

This is a metathesis reaction (Section 4.2) in which one of the cations is the $[Pd(NH_3)_3Cl]^+$ complex ion.

▶ **Practice Exercise**
Predict the number of ions produced per formula unit when the compound $CoCl_2 \cdot 6\,H_2O$ dissolves in water to form an aqueous solution.

The Metal–Ligand Bond

The bond between a ligand and a metal ion forms as a result of a Lewis acid–base interaction. (Section 16.1) Because the ligands have available pairs of electrons, they can function as Lewis bases (electron-pair donors). Metal ions (particularly transition-metal ions) have empty valence orbitals, so they can act as Lewis acids (electron-pair acceptors). We can picture the bond between the metal ion and ligand as the result of their sharing a pair of electrons initially on the ligand:

$$Ag^+(aq) + 2\ :\!\!\underset{\underset{H}{|}}{\overset{\overset{H}{|}}{N}}\!\!-\!\!H(aq) \longrightarrow \left[H\!\!-\!\!\underset{\underset{H}{|}}{\overset{\overset{H}{|}}{N}}\!\!:\!Ag\!:\!\!\underset{\underset{H}{|}}{\overset{\overset{H}{|}}{N}}\!\!-\!\!H \right]^+ (aq)$$

The formation of metal–ligand bonds can markedly alter the properties we observe for the metal ion. A metal complex is a distinct chemical species that has physical and chemical properties different from those of the metal ion and ligands from which it is formed. As one example, **Figure 23.9** shows the color change that occurs when aqueous solutions of NCS^- (colorless) and Fe^{3+} (yellow) are mixed, forming $[Fe(H_2O)_5NCS]^{2+}$.

Complex formation can also significantly change other properties of metal ions, such as their ease of oxidation or reduction. Silver ion, for example, is readily reduced in water:

$$Ag^+(aq) + e^- \longrightarrow Ag(s) \qquad E°_{red} = +0.799\ V \qquad [23.2]$$

By comparison, the $[Ag(CN)_2]^-$ ion is much harder to reduce (more negative value of $E°_{red}$) because complexation by CN^- ions stabilizes silver in the +1 oxidation state:

$$[Ag(CN)_2]^-(aq) + e^- \longrightarrow Ag(s) + 2\ CN^-(aq) \quad E°_{red} = -0.31\ V \qquad [23.3]$$

Go Figure Does the coordination number of iron change during this reaction? Does the oxidation number of iron change?

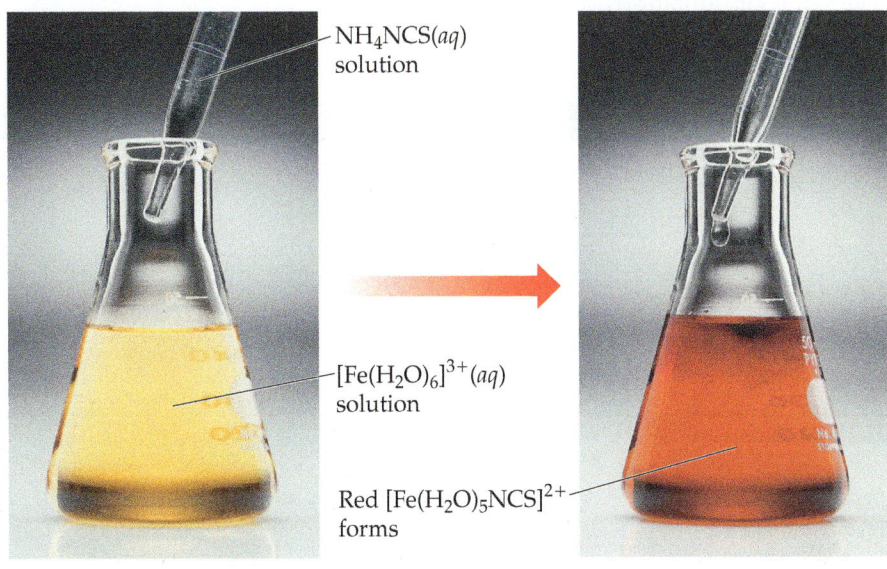

$NH_4NCS(aq)$ solution

$[Fe(H_2O)_6]^{3+}(aq)$ solution

Red $[Fe(H_2O)_5NCS]^{2+}$ forms

▲ **Figure 23.9** The reaction between $[Fe(H_2O)_6]^{3+}(aq)$ and $NCS^-(aq)$.

Hydrated metal ions are complexes in which the ligand is water. Thus, $Fe^{3+}(aq)$ consists largely of $[Fe(H_2O)_6]^{3+}$. (Section 16.9) It is important to realize that ligands can undergo reaction. For example, we saw in Figure 16.18 that a water molecule in $[Fe(H_2O)_6]^{3+}(aq)$ can be deprotonated to yield $[Fe(H_2O)_5OH]^{2+}(aq)$ and $H^+(aq)$. The iron ion retains its oxidation state; the coordinated hydroxide ligand, with a 1− charge, reduces the complex charge to 2+. Ligands can also be displaced from the coordination sphere by other ligands, if the incoming ligands bind more strongly to the metal ion than the original ones. For example, ligands such as NH_3, NCS^-, and CN^- can replace H_2O in the coordination sphere of metal ions, as we saw in Figure 23.9.

Charges, Coordination Numbers, and Geometries

The charge of a complex is the sum of the charges on the metal and on the ligands. In $[Cu(NH_3)_4]SO_4$ we can deduce the charge on the complex ion because we know that the sulfate ion has a 2− charge. Because the compound is electrically neutral, the complex ion must have a 2+ charge: $[Cu(NH_3)_4]^{2+}$. We can then use the charge of the complex ion to deduce the oxidation state of copper. Because the NH_3 ligands are uncharged molecules, the oxidation state of copper must be +2:

$$+2 + 4(0) = +2$$
$$[Cu(NH_3)_4]^{2+}$$

Recall that the number of atoms directly bonded to the metal atom in a complex is the *coordination number* of the complex. Thus, the copper ion in $[Cu(NH_3)_4]^{2+}$ has a coordination number of 4. Similarly, the silver ion in $[Ag(NH_3)_2]^+$ has a coordination number of 2, and the cobalt ion has a coordination number of 6 in all four complexes in Table 23.3.

Some metal ions exhibit only one observed coordination number. The coordination number of chromium(III) and cobalt(III) is invariably 6, for example, and that of platinum(II) is always 4. For most metals, however, the coordination number can be different for different ligands. In these complexes, the most common coordination numbers are 4 and 6.

The coordination number of a metal ion is often influenced by the relative sizes of the metal ion and the ligands. As the ligand gets larger, fewer of them can coordinate to the metal ion. Thus, iron(III) is able to coordinate to six fluorides in $[FeF_6]^{3-}$ but to only four chlorides in $[FeCl_4]^-$. Ligands that transfer substantial negative charge to the metal also produce reduced coordination numbers. For example, six ammonia molecules can coordinate to nickel(II), forming $[Ni(NH_3)_6]^{2+}$, but only four cyanide ions can coordinate to this ion, forming $[Ni(CN)_4]^{2-}$.

Sample Exercise 23.2

Determining the Oxidation State of a Metal in a Complex

What is the oxidation state of the metal in $[Rh(NH_3)_5Cl](NO_3)_2$?

SOLUTION

Analyze We are given the chemical formula of a coordination compound and asked to determine the oxidation state of its metal atom.

Plan To determine the oxidation state of Rh, we need to figure out what charges are contributed by the other groups. The overall charge is zero, so the oxidation state of the metal must balance the charge due to the rest of the compound.

Solve The NO_3 group is the nitrate anion, which has a 1− charge. The NH_3 ligands carry zero charge, and the Cl is a coordinated chloride ion, which has a 1− charge. The sum of all the charges must be zero:

$$x + 5(0) + (-1) + 2(-1) = 0$$
$$[Rh(NH_3)_5Cl](NO_3)_2$$

The oxidation state of rhodium, x, must therefore be +3.

▶ **Practice Exercise**

What is the charge of the complex formed by a platinum(II) metal ion to which two ammonia molecules and two bromide ions are coordinated?

The most common coordination geometries for coordination complexes are shown in **Figure 23.10**. Complexes in which the coordination number is 4 have two possible geometries—tetrahedral and square planar. The tetrahedral geometry is the more common of the two and is especially prevalent among non-transition metals [as we saw in our discussion of the VSEPR model (Section 9.2)]. The square-planar geometry is characteristic of most transition-metal ions with eight d electrons in the valence shell, such as platinum(II) and gold(III). Complexes with a coordination number of 6 almost always have an octahedral geometry. Even though the octahedron can be drawn as a square with one ligand above and another below the plane, all six vertices are equivalent.

Go Figure

In the drawings on the right-hand side, what does the solid wedge connecting atoms represent? What does the dashed wedge connecting atoms represent?

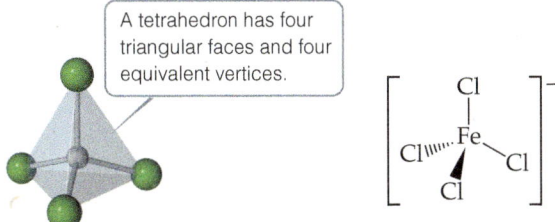

A tetrahedron has four triangular faces and four equivalent vertices.

Tetrahedral geometry

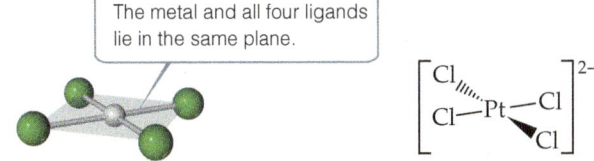

The metal and all four ligands lie in the same plane.

Square-planar geometry

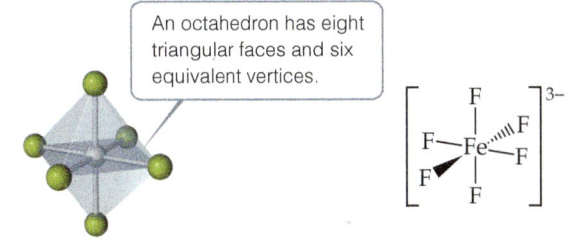

An octahedron has eight triangular faces and six equivalent vertices.

Octahedral geometry

▲ **Figure 23.10 Common geometries of coordination complexes.** In complexes having coordination number 4, the geometry is typically either tetrahedral or square planar. In complexes having coordination number 6, the geometry is nearly always octahedral.

 Self-Assessment Exercises

SAE 23.6   In the complex $[Co(NH_3)_5Br]Br_2$, the oxidation state of the cobalt atom is _____ and the coordination number is _____. **(a)** 0, 6 **(b)** 0, 8 **(c)** +2, 6 **(d)** +3, 5 **(e)** +3, 6

SAE 23.7   Which of the following statements about the complex $[Zn(NH_3)_4]Cl_2$ is *false*?

(a) The coordination number of the complex is 4.
(b) When an NH_3 ligand binds to the zinc atom, it is acting as a Lewis base.
(c) The geometry of the complex ion is octahedral.
(d) The oxidation state of the zinc atom is +2.
(e) Both of the Cl ions in the complex could be precipitated using $AgNO_3(aq)$.

23.3 | Common Ligands in Coordination Chemistry

The ligand atom that binds to the central metal ion in a coordination complex is called the **donor atom** of the ligand. Ligands having only one donor atom are called **monodentate ligands** (from the Latin, meaning "one-toothed"). These ligands are able to occupy only one site in a coordination sphere. Ligands having two donor atoms are **bidentate ligands** ("two-toothed"), and those having three or more donor atoms are **polydentate ligands** ("many-toothed"). In both bidentate and polydentate species, the multiple donor atoms can simultaneously bond to the metal ion, thereby occupying two or more sites in a coordination sphere. **Table 23.4** gives examples of all three types of ligands.

Because they appear to grasp the metal between two or more donor atoms, bidentate and polydentate ligands are also known as **chelating agents** (pronounced "KEE-lay-ting"; from the Greek *chele*, "claw").

One common chelating agent is the bidentate ligand *ethylenediamine*, $C_2N_2H_8$, which is denoted *en*:

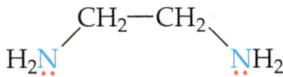

 Learning Objectives

When you finish Section 23.3, you should be able to:

▶ Distinguish between monodentate, bidentate, and polydentate ligands.
▶ For common ligands, identify the number and type of donor atoms that bind to the central metal atom in a coordination complex.

Each of the nitrogen atoms in an en ligand has one nonbonding electron pair. Thus, each of the N atoms can serve as a donor atom to a metal, and because they are sufficiently far apart, they can bind to the metal ion in adjacent positions. The complex ion $[Co(en)_3]^{3+}$, which contains three ethylenediamine ligands in the octahedral coordination sphere of

TABLE 23.4   Some Common Ligands

Ligand Type	Examples			
Monodentate	$H_2\ddot{O}$: Water	$:\ddot{F}:^-$ Fluoride ion	$[:C\equiv N:]^-$ Cyanide ion	$[:\ddot{O}-H]^-$ Hydroxide ion
	$:NH_3$ Ammonia	$:\ddot{C}l:^-$ Chloride ion	$[:\ddot{S}=C=\ddot{N}:]^-$ Thiocyanate ion	$[:\ddot{O}-N=\ddot{O}:]^-$ Nitrite ion

Bidentate

H₂C—CH₂ / H₂N NH₂
Ethylenediamine (en)

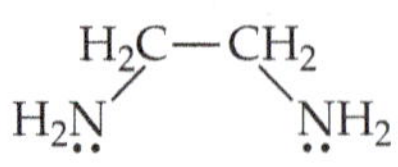

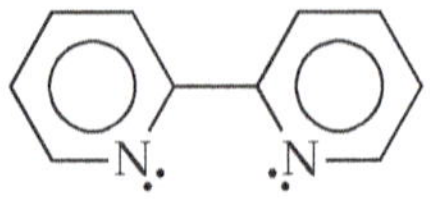

Bipyridine (bipy or bpy)

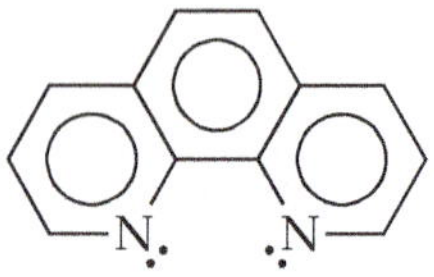

Ortho-phenanthroline (*o*-phen)

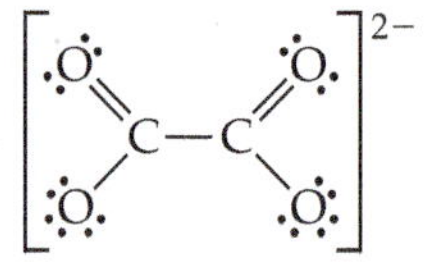

Oxalate ion

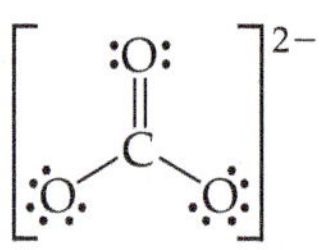

Carbonate ion

Polydentate

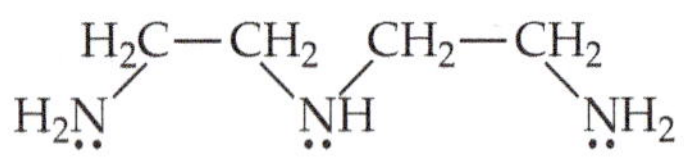

H₂C—CH₂ CH₂—CH₂ / H₂N NH NH₂
Diethylenetriamine

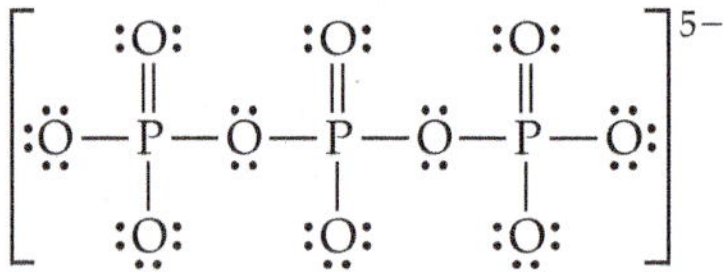

Triphosphate ion

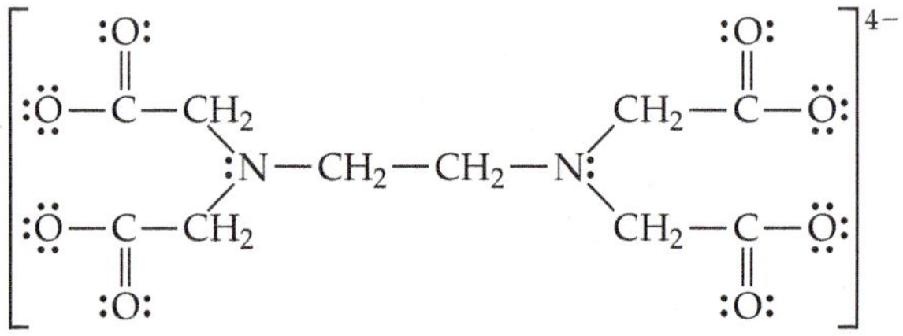

Ethylenediaminetetraacetate ion ($EDTA^{4-}$)

cobalt(III), is shown in **Figure 23.11**. Notice that in the image on the right the en is written in a shorthand notation as two nitrogen atoms connected by an arc.

The ethylenediaminetetraacetate ion, $[EDTA]^{4-}$, is an important polydentate ligand that has six donor atoms (two N atoms and four O atoms). It can wrap around a metal ion using all six donor atoms, as shown in **Figure 23.12**, although it sometimes binds to a metal using only five of its donor atoms.

In general, the complexes formed by chelating ligands (that is, bidentate and polydentate ligands) are more stable than the complexes formed by related monodentate ligands. The equilibrium formation constants for $[Ni(NH_3)_6]^{2+}$ and $[Ni(en)_3]^{2+}$ illustrate this observation:

$$[Ni(H_2O)_6]^{2+}(aq) + 6\,NH_3(aq) \rightleftharpoons [Ni(NH_3)_6]^{2+}(aq) + 6\,H_2O(l)$$

$$K_f = 1.2 \times 10^9 \qquad [23.4]$$

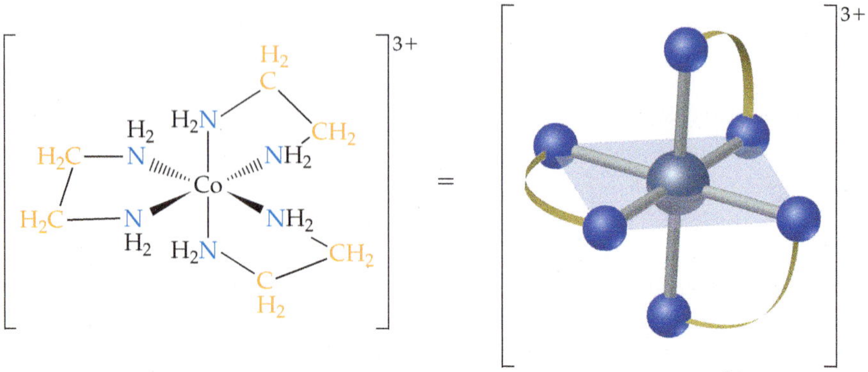

▶ **Figure 23.11 The [Co(en)₃]³⁺ ion.** The abbreviation en is used for the bidentate ethylenediamine ligand.

$[Co(en)_3]^{3+}$

$$[Ni(H_2O)_6]^{2+}(aq) + 3\,en(aq) \rightleftharpoons [Ni(en)_3]^{2+}(aq) + 6\,H_2O(l)$$

$$K_f = 6.8 \times 10^{17} \qquad [23.5]$$

Although the donor atom is nitrogen in both instances, $[Ni(en)_3]^{2+}$ has a formation constant that is more than 10^8 times larger than that of $[Ni(NH_3)_6]^{2+}$. This trend of generally larger formation constants for bidentate and polydentate ligands, known as the **chelate effect**, is examined in the "A Closer Look" box on "Entropy and the Chelate Effect" in this section.

Chelating agents are often used to prevent one or more of the customary reactions of a metal ion without removing the ion from solution. For example, a metal ion that interferes with a chemical analysis can often be complexed and its interference thereby removed. In a sense, the chelating agent hides the metal ion. For this reason, scientists sometimes refer to these ligands as *sequestering agents*. Phosphate ligands, such as sodium tripolyphosphate, $Na_5[OPO_2OPO_2OPO_3]$, are used to sequester Ca^{2+} and Mg^{2+} ions in hard water so that these ions cannot interfere with the action of soap or detergents.

Chelating agents are used in many prepared foods, such as salad dressings and frozen desserts, to complex trace metal ions that catalyze decomposition reactions. Chelating agents are used in medicine to remove toxic heavy metal ions that have been ingested, such as Hg^{2+}, Pb^{2+}, and Cd^{2+}. One method of treating lead poisoning, for example, is to administer $Na_2Ca(EDTA)$. The EDTA chelates the lead, allowing it to be removed from the body via urine.

Metals and Chelates in Living Systems

Ten of the 29 elements known to be essential for human life are transition metals [see the "Chemistry and Life" box on "Elements Required by Living Organisms" (Section 2.7)]. These 10 elements—V, Cr, Mn, Fe, Co, Ni, Cu, Zn, Mo, and Cd—form complexes with a variety of groups present in biological systems.

Although our bodies require only small quantities of metals, deficiencies can lead to serious illness. Iron deficiency, for example, can lead to *anemia*, in which the body lacks an adequate supply of healthy red blood cells, as is discussed in the "Chemistry and Life" box on "The Battle for Iron in Living Systems" in this section. A deficiency of manganese can lead to convulsive disorders. Some epilepsy patients have been helped by the addition of manganese to their diets.

Among the most important chelating agents in nature are those derived from the *porphine* molecule (**Figure 23.13**). This molecule can coordinate to a metal via its four nitrogen donor atoms. Once porphine bonds to a metal ion, the two H atoms on the nitrogens are displaced to form complexes called **porphyrins**. Two important porphyrins are *hemes*, in which the metal ion is Fe(II), and *chlorophylls*, with a Mg(II) central ion.

Figure 23.14 shows a schematic structure of myoglobin, a protein that contains one heme group. Myoglobin is a *globular protein*, one that folds into a compact, roughly spherical shape. Myoglobin is found in the cells of skeletal muscle, particularly in seals, whales, and porpoises. It stores oxygen in cells, one molecule of O_2 per myoglobin, until it is needed for metabolic activities. Hemoglobin, the protein that transports oxygen in human blood, is made up of four heme-containing subunits, each of which is very similar to myoglobin. One hemoglobin can bind up to four O_2 molecules.

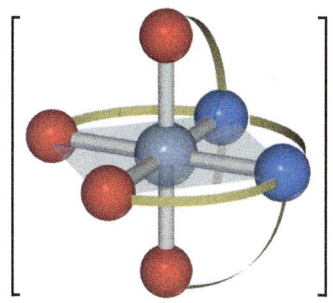

▲ Figure 23.12 **The complex ion** $[Co(EDTA)]^-$. The ligand is the polydentate ethylenediaminetetraacetate ion, whose unabbreviated structure is given in Table 23.2. This representation shows how the two N and four O donor atoms coordinate to cobalt.

Go Figure

How many carbon atoms are there in porphine? How many have sp^3 hybridization? How many have sp^2 hybridization?

Porphine

Heme b

Chlorophyll a

▲ Figure 23.13 **Porphine and two porphyrins, heme b and chlorophyll a.** Fe(II) and Mg(II) ions replace the two blue H atoms in porphine and bond with all four nitrogens in heme b and chlorophyll a, respectively.

▲ **Figure 23.14 Myoglobin.** This ribbon diagram does not show most of the atoms.

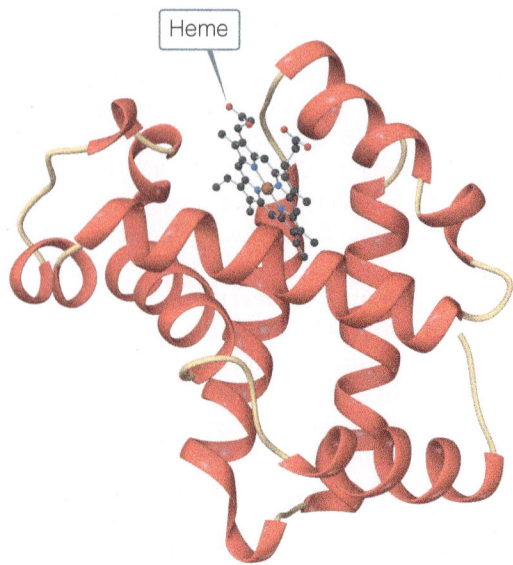

In both myoglobin and hemoglobin, the iron is coordinated to the four nitrogen atoms of a porphyrin and to a nitrogen atom from the protein chain (**Figure 23.15**). In hemoglobin, the sixth position around the iron is occupied either by O_2 (in oxyhemoglobin, the bright red form) or by water (in deoxyhemoglobin, the purplish red form). The oxy form is the one shown in Figure 23.15.

Carbon monoxide is poisonous because the equilibrium binding constant of human hemoglobin for CO is about 210 times greater than that for O_2. As a result, a relatively small quantity of CO can inactivate a substantial fraction of the hemoglobin in the blood by displacing the O_2 molecule from the heme-containing subunit. For example, a person breathing air that contains only 0.1% CO takes in enough CO after a few hours to convert up to 60% of the hemoglobin (Hb) into COHb, thereby reducing the blood's normal oxygen-carrying capacity by 60%.

Under normal conditions, a nonsmoker breathing unpolluted air has about 0.3 to 0.5% COHb in their blood. This amount arises mainly from the production of small quantities of CO in the course of normal body chemistry and from the small amount of CO present in clean air. Exposure to higher concentrations of CO causes the COHb level to increase, which in turn leaves fewer Hb sites to which O_2 can bind. If the level of COHb becomes too high, oxygen transport is effectively shut down and death occurs. Because CO is colorless and odorless, CO poisoning occurs with very little warning. Improperly ventilated combustion devices, such as kerosene lanterns and stoves, thus pose a potential health hazard.

Go Figure

What is the coordination number of iron in the heme unit shown here? What is the identity of the blue donor atoms in the heme?

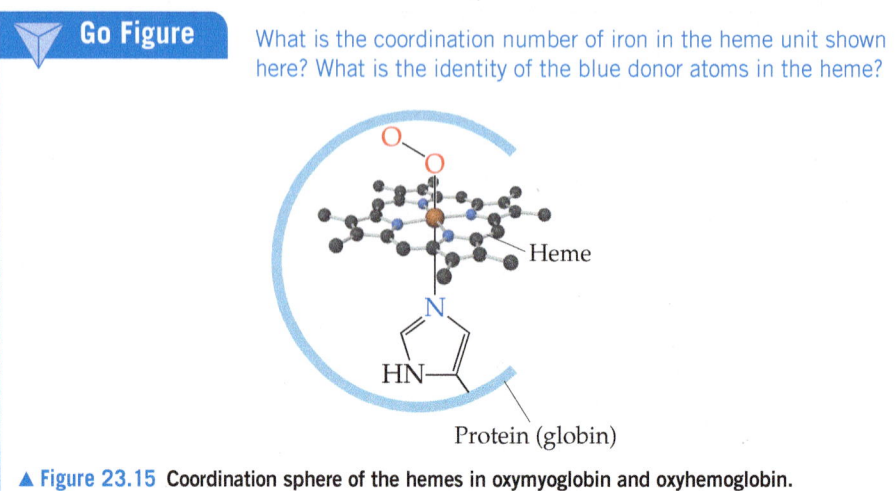

▲ **Figure 23.15** Coordination sphere of the hemes in oxymyoglobin and oxyhemoglobin.

A CLOSER LOOK | Entropy and the Chelate Effect

In our discussion of Gibbs free energy, we learned that chemical processes are favored by positive entropy changes and by negative enthalpy changes. (Section 19.5) The special stability associated with the formation of chelates, called the *chelate effect*, can be explained by comparing the entropy changes that occur with monodentate ligands with the entropy changes that occur with polydentate ligands.

We begin with the reaction in which two H_2O ligands of the square-planar Cu(II) complex $[Cu(H_2O)_4]^{2+}$ are replaced by monodentate NH_3 ligands at 27 °C:

$$[Cu(H_2O)_4]^{2+}(aq) + 2\,NH_3(aq) \rightleftharpoons$$
$$[Cu(H_2O)_2(NH_3)_2]^{2+}(aq) + 2\,H_2O(l)$$
$$\Delta H° = -46\,kJ;\quad \Delta S° = -8.4\,J/K;\quad \Delta G° = -43\,kJ$$

The thermodynamic data tell us about the relative abilities of H_2O and NH_3 to serve as ligands in this reaction. In general, NH_3 binds more tightly to metal ions than does H_2O, so this substitution reaction is exothermic ($\Delta H < 0$). The stronger bonding of the NH_3 ligands also causes the $[Cu(H_2O)_2(NH_3)_2]^{2+}$ ion to be more rigid, which is probably the reason $\Delta S°$ is slightly negative.

We can use Equation 19.20, $\Delta G° = -RT \ln K$, to calculate the equilibrium constant of the reaction at 27 °C. The result, $K = 3.1 \times 10^7$, tells us that the equilibrium lies far to the right, favoring replacement of H_2O by NH_3. For this equilibrium, therefore, the enthalpy change, $\Delta H° = -46\,kJ$, is large enough and negative enough to overcome the entropy change, $\Delta S° = -8.4\,J/K$.

Now let's use a single bidentate ethylenediamine (en) ligand in our substitution reaction:

$$[Cu(H_2O)_4]^{2+}(aq) + en(aq) \rightleftharpoons [Cu(H_2O)_2(en)]^{2+}(aq) + 2\,H_2O(l)$$
$$\Delta H° = -54\,kJ;\quad \Delta S° = +23\,J/K;\quad \Delta G° = -61\,kJ$$

The en ligand binds slightly more strongly to the Cu^{2+} ion than two NH_3 ligands, so the enthalpy change here ($-54\,kJ$) is slightly more negative than for $[Cu(H_2O)_2(NH_3)_2]^{2+}$ ($-46\,kJ$). There is a big difference in the entropy change, however: $\Delta S°$ is $-8.4\,J/K$ for the NH_3 reaction but $+23\,J/K$ for the en reaction. We can explain the positive $\Delta S°$ value using concepts discussed in Section 19.3. Because a single en ligand occupies two coordination sites, two molecules of H_2O are released when one en ligand bonds. Thus, there are three product molecules in the reaction but only two reactant molecules. The greater number of product molecules leads to the positive entropy change for the equilibrium.

The slightly more negative value of $\Delta H°$ for the en reaction ($-54\,kJ$ versus $-46\,kJ$) coupled with the positive entropy change leads to a much more negative value of $\Delta G°$ ($-61\,kJ$ for en, $-43\,kJ$ for NH_3) and thus a larger equilibrium constant: $K = 4.2 \times 10^{10}$.

We can combine our two equations using Hess's law (Section 5.6) to calculate the enthalpy, entropy, and free-energy changes that occur for en to replace ammonia as ligands on Cu(II):

$$[Cu(H_2O)_2(NH_3)_2]^{2+}(aq) + en(aq) \rightleftharpoons$$
$$[Cu(H_2O)_2(en)]^{2+}(aq) + 2\,NH_3(aq)$$
$$\Delta H° = (-54\,kJ) - (-46\,kJ) = -8\,kJ$$
$$\Delta S° = (+23\,J/K) - (-8.4\,J/K) = +31\,J/K$$
$$\Delta G° = (-61\,kJ) - (-43\,kJ) = -18\,kJ$$

Notice that at 27 °C, the entropic contribution ($-T\Delta S°$) to the free-energy change, $\Delta G° = \Delta H° - T\Delta S°$ (Equation 19.12), is negative and greater in magnitude than the enthalpic contribution ($\Delta H°$). The equilibrium constant for the substitution of two NH_3 ligands by one en ligand, 1.4×10^3, shows that the replacement of NH_3 by en is thermodynamically favorable.

The chelate effect is important in biochemistry and molecular biology. The additional thermodynamic stabilization provided by entropy effects helps stabilize biological metal–chelate complexes, such as porphyrins, and can allow changes in the oxidation state of the metal ion while retaining the structural integrity of the complex.
Related Exercise: 23.32

The **chlorophylls**, which are porphyrins that contain Mg(II) (Figure 23.13), are the key components in the conversion of solar energy into forms that can be used by living organisms. This process, called **photosynthesis**, occurs in the leaves of green plants:

$$6\,CO_2(g) + 6\,H_2O(l) \longrightarrow C_6H_{12}O_6(aq) + 6\,O_2(g) \qquad [23.6]$$

The formation of 1 mol of glucose, $C_6H_{12}O_6$, requires the absorption of 48 mol of photons from sunlight or other sources of light. Chlorophyll-containing pigments in the leaves of plants absorb the photons. Figure 23.13 shows that the chlorophyll molecule has a series of alternating, or *conjugated*, double bonds in the ring surrounding the metal ion. This system of conjugated double bonds makes it possible for chlorophyll to absorb light strongly in the visible region of the spectrum. As **Figure 23.16** shows, chlorophyll is green because it absorbs red light (maximum absorption at 655 nm) and blue light (maximum absorption at 430 nm) and transmits green light.

Photosynthesis is nature's solar energy-conversion machine, and thus the sustainability of all living systems on Earth depends on photosynthesis. The form of chlorophyll that facilitates photosynthesis, called *chlorophyll a*, is illustrated in Figure 23.13 and on the front cover of this book.

Go Figure

Which peak in this curve corresponds to the lowest-energy transition by an electron in a chlorophyll molecule?

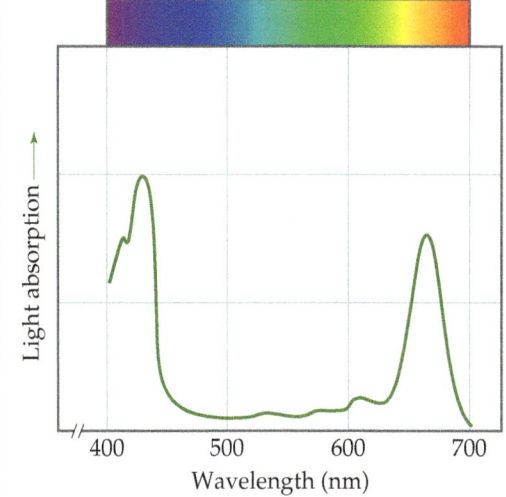

▲ **Figure 23.16** The absorption of sunlight by chlorophyll.

CHEMISTRY AND LIFE | The Battle for Iron in Living Systems

Because living systems have difficulty assimilating enough iron to satisfy their nutritional needs, iron-deficiency anemia is a common problem in humans. Chlorosis, an iron deficiency in plants that makes leaves turn yellow, is also commonplace.

Living systems have difficulty assimilating iron because most iron compounds found in nature are not very soluble in water. Microorganisms have adapted to this problem by secreting an iron-binding compound, called a *siderophore*, that forms an extremely stable water-soluble complex with iron(III). One such complex is *ferrichrome* (Figure 23.17). The iron-binding strength of a siderophore is so great that it can extract iron from iron oxides.

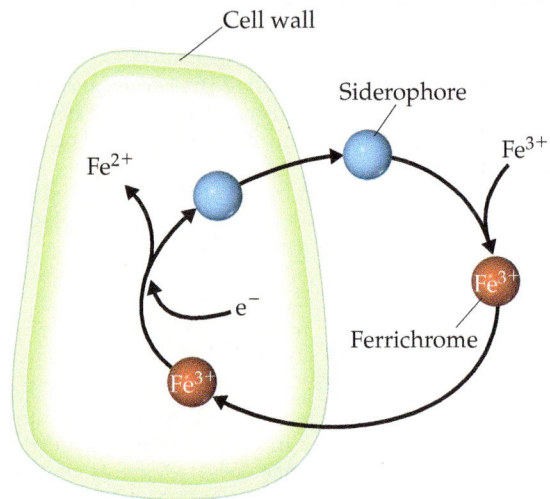

▲ **Figure 23.18 The iron-transport system of a bacterial cell.**

A bacterium that infects the blood requires a source of iron if it is to grow and reproduce. The bacterium excretes a siderophore into the blood to compete with transferrin for iron. The equilibrium constants for forming the iron complex are about the same for transferrin and siderophores. The more iron available to the bacterium, the more rapidly it can reproduce and thus the more harm it can do.

Several years ago, New Zealand clinics regularly gave iron supplements to infants soon after birth. However, the incidence of certain bacterial infections was eight times higher in treated than in untreated infants. Presumably, the presence of more iron in the blood than is absolutely necessary makes it easier for bacteria to obtain the iron needed for growth and reproduction.

In the United States, it is common medical practice to supplement infant formula with iron sometime during the first year of life. However, iron supplements are not necessary for infants who breast-feed because breast milk contains two specialized proteins, lactoferrin and transferrin, which provide sufficient iron while denying its availability to bacteria. Even for infants fed with infant formulas, supplementing with iron during the first several months of life may be ill-advised.

For bacteria to continue to multiply in the blood, they must synthesize new supplies of siderophores. Synthesis of siderophores in bacteria slows, however, as the temperature is increased above the normal body temperature of 37 °C and stops completely at 40 °C. This suggests that fever in the presence of an invading microbe is a mechanism used by the body to deprive bacteria of iron.

Related Exercise: 23.76

▲ **Figure 23.17 Ferrichrome.**

When ferrichrome enters a living cell, the iron it carries is removed through an enzyme-catalyzed reaction that reduces the strongly bonding iron(III) to iron(II), which is only weakly complexed by the siderophore (Figure 23.18). Microorganisms thus acquire iron by excreting a siderophore into their immediate environment and then taking the resulting iron complex into the cell.

In humans, iron is assimilated from food in the intestine. A protein called *transferrin* binds iron and transports it across the intestinal wall to distribute it to other tissues in the body. The normal adult body contains about 4 g of iron. At any one time, about 3 g of this iron is in the blood, mostly in the form of hemoglobin. Most of the remainder is carried by transferrin.

Self-Assessment Exercises

SAE 23.8 The oxalate ion, $C_2O_4^{2-}$, denoted *ox*, forms a complex ion with chromium: $[Cr(ox)_3]^{3-}$. Which of the following statements about this complex is *false*?

(a) The oxalate ion is a bidentate ligand.
(b) The coordination number for this complex is 3.
(c) The oxidation state of chromium in this complex is +3.
(d) The donor atoms in the ox ligand are oxygen atoms.
(e) The oxalate ion is an example of a chelating ligand.

SAE 23.9 *Chlorophyll a* is a porphyrin complex that is the principal substance involved in photosynthesis by plants. Which of the following statements about chlorophyll a is *true*?

(a) Chlorophyll a is an example of a transition-metal complex.
(b) The porphyrin is a bidentate ligand.
(c) Chlorophyll a can absorb visible light.
(d) The metal atom in chlorophyll a is in the +3 oxidation state.
(e) The metal atom at the center of chlorophyll a is manganese.

23.4 | Nomenclature and Isomerism in Coordination Chemistry

When complexes were first discovered, they were named after the chemist who originally prepared them. A few of these names persist; for example, the dark red substance $NH_4[Cr(NH_3)_2(NCS)_4]$ is still known as Reinecke's salt. Once the structures of complexes were more fully understood, it became possible to name them in a more systematic manner. Let's use two substances to illustrate how coordination compounds are named:

⚠ Learning Objectives

When you finish Section 23.4, you should be able to:

▶ Convert between the name of a coordination compound and its chemical formula.

▶ Describe the types of isomerization a coordination complex can exhibit.

▶ Describe how optical isomers of a complex differ from one another.

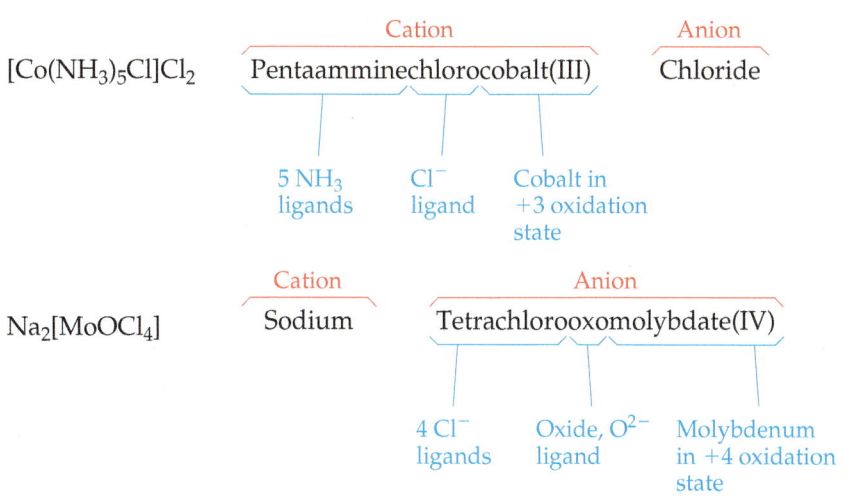

How to Name Coordination Compounds

1. *In naming complexes that are salts, the name of the cation is given before the name of the anion.* Thus, in $[Co(NH_3)_5Cl]Cl_2$ we name the $[Co(NH_3)_5Cl]^{2+}$ cation and then the Cl^-.

2. *In naming complex ions or molecules, the ligands are named before the metal. Ligands are listed in alphabetical order, regardless of their charges. Prefixes that give the number of ligands are not considered part of the ligand name in determining alphabetical order.* Thus, the $[Co(NH_3)_5Cl]^{2+}$ ion is pentaamminechlorocobalt(III). (Be careful to note, however, that the metal is written first in the chemical formula.)

3. *The names of anionic ligands end in the letter o, but electrically neutral ligands ordinarily bear the name of the molecules* (Table 23.5). Special names are used for H_2O (aqua), NH_3 (ammine), and CO (carbonyl). For example, $[Fe(CN)_2(NH_3)_2(H_2O)_2]^+$ is the diamminediaquadicyanoiron(III) ion.

4. *Greek prefixes* (di-, tri-, tetra-, penta-, hexa-) *are used to indicate the number of each kind of ligand when more than one is present. If the ligand contains a Greek prefix (for example, ethylenediamine) or is polydentate, the alternate prefixes* bis-, tris-, tetrakis-, pentakis-, *and* hexakis- *are used and the ligand name is placed in parentheses.* For example, the name for $[Co(en)_3]Br_3$ is tris(ethylenediamine)cobalt(III) bromide.

5. *If the complex is an anion, its name ends in* -ate. The compound $K_4[Fe(CN)_6]$ is potassium hexacyanoferrate(II), for example, and the ion $[CoCl_4]^{2-}$ is tetrachlorocobaltate(II) ion.

6. *The oxidation state of the metal is given in parentheses in Roman numerals following the name of the metal.* There is no space between the name of the metal and the parenthesis.

Three examples for applying these rules are

$[Ni[(NH_3)_6]Br_2$	Hexaamminenickel(II) bromide
$[Co(en)_2(H_2O)(CN)]Cl_2$	Aquacyanobis(ethylenediamine)cobalt(III) chloride
$Na_2[MoOCl_4]$	Sodium tetrachlorooxomolybdate(IV)

TABLE 23.5 Some Common Ligands and Their Names

Ligand	Name in Complexes	Ligand	Name in Complexes
Azide, N_3^-	Azido	Oxalate, $C_2O_4^{2-}$	Oxalato
Bromide, Br^-	Bromo	Oxide, O^{2-}	Oxo
Chloride, Cl^-	Chloro	Ammonia, NH_3	Ammine
Cyanide, CN^-	Cyano	Carbon monoxide, CO	Carbonyl
Fluoride, F^-	Fluoro	Ethylenediamine, en	Ethylenediamine
Hydroxide, OH^-	Hydroxo	Pyridine, C_5H_5N	Pyridine
Carbonate, CO_3^{2-}	Carbonato	Water, H_2O	Aqua

Sample Exercise 23.3
Naming Coordination Compounds

Name the compounds (**a**) $[Cr(H_2O)_4Cl_2]Cl$, (**b**) $K_4[Ni(CN)_4]$.

SOLUTION

Analyze We are given the chemical formulas for two coordination compounds and assigned the task of naming them.

Plan To name the complexes, we need to determine the ligands in the complexes, the names of the ligands, and the oxidation state of the metal ion. We then put the information together following the rules listed in the text.

Solve
(**a**) The ligands are four water molecules—tetraaqua—and two chloride ions—dichloro. By assigning all the oxidation states we know for this molecule, we see that the oxidation state of Cr is +3:

$$+3 + 4(0) + 2(-1) + (-1) = 0$$
$$[Cr(H_2O)_4Cl_2]Cl$$

Thus, we have chromium(III). Finally, the anion is chloride. The name of the compound is tetraaquadichlorochromium(III) chloride.

(**b**) The complex has four cyanide ion ligands, CN^-, which means tetracyano, and the oxidation state of the nickel is zero:

$$4(+1) + 0 + 4(-1) = 0$$
$$K_4[Ni(CN)_4]$$

Because the complex is an anion, the metal is indicated as nickelate(0). Putting these parts together and naming the cation first, we have potassium tetracyanonickelate(0).

▶ **Practice Exercise**
Write names for the compounds (**a**) $[Mo(NH_3)_3Br_3]NO_3$, (**b**) $(NH_4)_2[CuBr_4]$. (**c**) Write the formula for sodium diaquabis(oxalato)ruthenate(III).

Isomerism

When two or more compounds have the same composition but a different arrangement of atoms, we call them **isomers**. (Section 2.9) Here we consider two main kinds of isomers in coordination compounds: **structural isomers** (which have different bonds) and **stereoisomers** (which have the same bonds but different ways in which the ligands occupy the space around the metal center). Each of these classes also has subclasses, as shown in **Figure 23.19**.

Structural Isomerism

Many types of structural isomerism are known in coordination chemistry, including the two named in Figure 23.19: linkage isomerism and coordination-sphere isomerism. **Linkage isomerism** is a relatively rare but interesting type that arises when a particular ligand is capable of coordinating to a metal in two ways. The nitrite ion, NO_2^-, for example, can coordinate to a metal ion through either its nitrogen or one of its oxygens (**Figure 23.20**). When it coordinates through the nitrogen atom, the NO_2^- ligand is called *nitro*; when it coordinates through the oxygen atom, it is called *nitrito* and is generally written ONO^-. The isomers shown in Figure 23.20 have different properties. The nitro isomer is orange, for example, whereas the nitrito isomer is red.

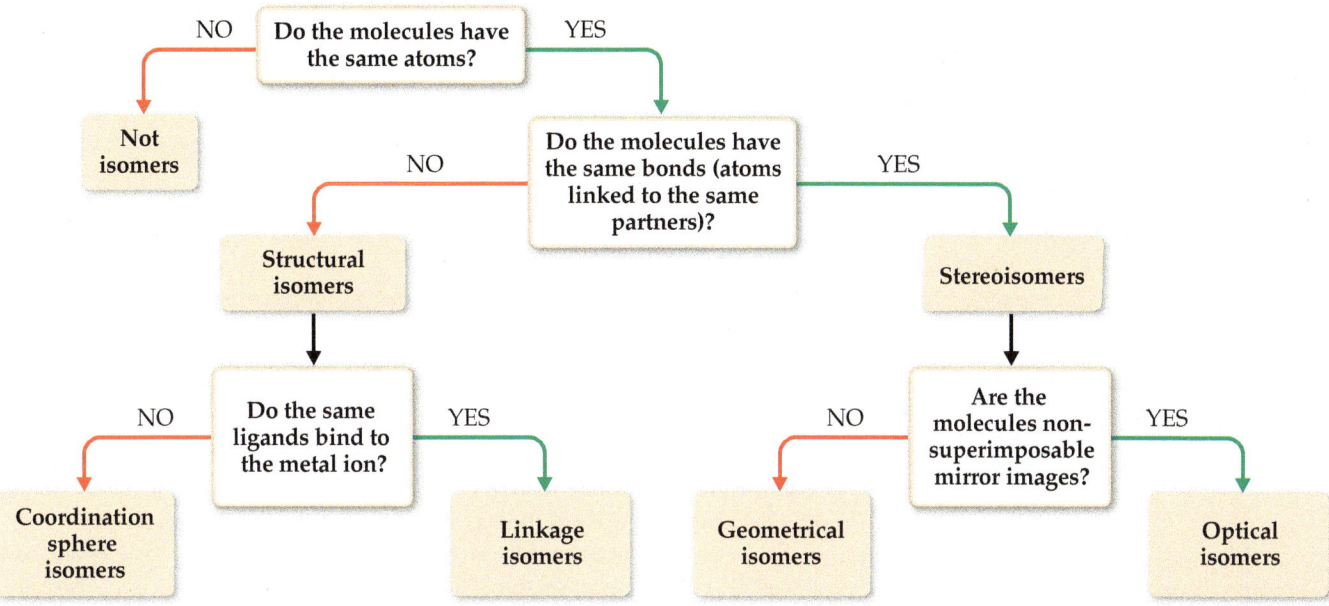

▲ **Figure 23.19 Forms of isomerism in coordination compounds.**

Go Figure What are the chemical formula and name for each of the complex ions in this figure?

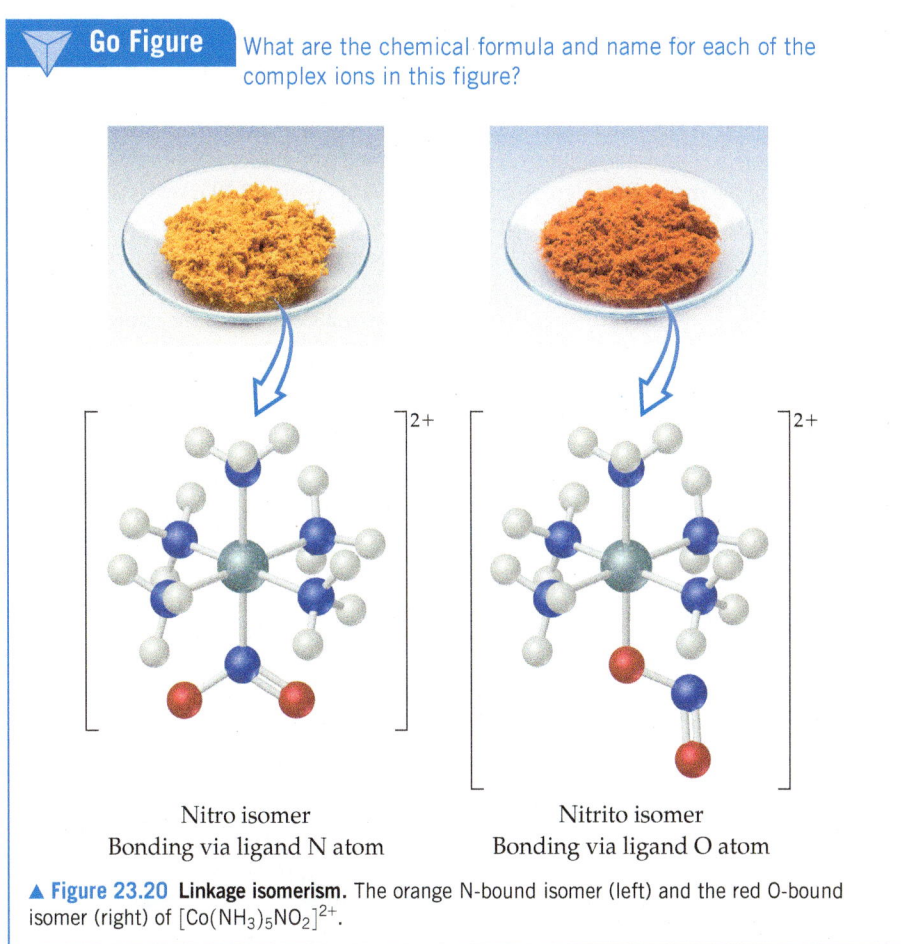

Nitro isomer
Bonding via ligand N atom

Nitrito isomer
Bonding via ligand O atom

▲ **Figure 23.20 Linkage isomerism.** The orange N-bound isomer (left) and the red O-bound isomer (right) of $[Co(NH_3)_5NO_2]^{2+}$.

Another ligand capable of coordinating through either of two donor atoms is thiocyanate, SCN^-, whose potential donor atoms are N and S.

Coordination-sphere isomers differ in which species in the complex act as ligands and which are outside the coordination sphere. For example, three isomers have the formula $CrCl_3(H_2O)_6$. When the ligands are six H_2O and the chloride ions are in the crystal lattice (as counterions), we have the violet compound $[Cr(H_2O)_6]Cl_3$. When the ligands

▼ **Go Figure**

Which of these isomers has a nonzero dipole moment?

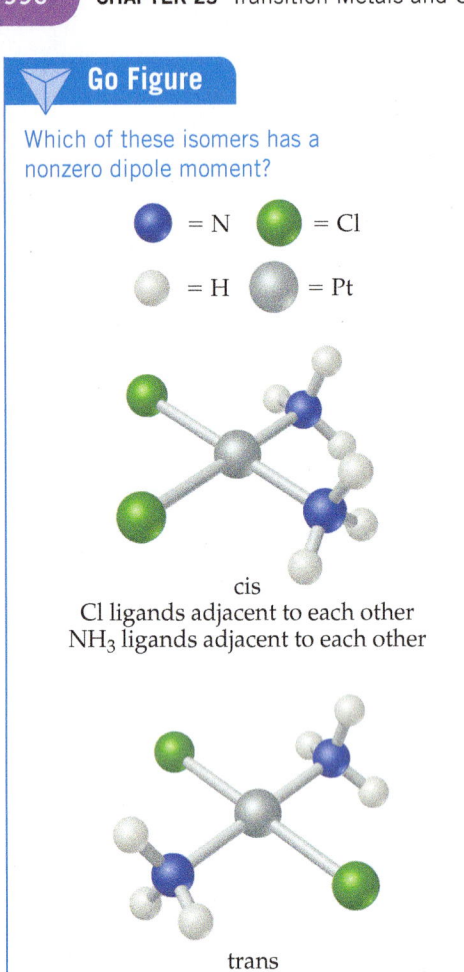

● = N   ● = Cl
○ = H   ● = Pt

cis
Cl ligands adjacent to each other
NH_3 ligands adjacent to each other

trans
Cl ligands on opposite sides of central atom
NH_3 ligands on opposite sides of central atom

▲ **Figure 23.21 Geometric isomerism.**

are five H_2O and one Cl^-, with the sixth H_2O and the two additional Cl^- ions out in the lattice, we have the green compound $[Cr(H_2O)_5Cl]Cl_2 \cdot H_2O$. The third isomer, $[Cr(H_2O)_4Cl_2]Cl \cdot 2 H_2O$, is also a green compound. In the two green compounds, either one or two water molecules have been displaced from the coordination sphere by chloride ions. The displaced H_2O molecules occupy a site in the crystal lattice.

Stereoisomerism

Stereoisomers have the same chemical bonds but different spatial arrangements. In the square-planar complex $[Pt(NH_3)_2Cl_2]$, for example, the chloro ligands can be either adjacent to or opposite each other (**Figure 23.21**). (We saw an earlier example of this type of isomerism in the cobalt complex of Figure 23.8, and we will return to that complex in a moment.) This form of stereoisomerism, in which the arrangement of the atoms is different but the same bonds are present, is called **geometric isomerism**. The isomer with like ligands in adjacent positions is the cis isomer, and the isomer with like ligands across from one another is the trans isomer.

Geometric isomers generally have different physical properties and may also have markedly different chemical reactivities. For example, *cis*-$[Pt(NH_3)_2Cl_2]$, also called *cisplatin*, is effective in the treatment of testicular, ovarian, and certain other cancers, whereas the trans isomer is ineffective. This is because cisplatin forms a chelate with two nitrogens of DNA, displacing the chloride ligands. The chloride ligands of the trans isomer are too far apart to form the N–Pt–N chelate with the nitrogen donors in DNA.

Geometric isomerism is also possible in octahedral complexes when two or more different ligands are present, as in the cis and trans tetraamminedichlorocobalt(III) ion in Figure 23.8. Because all four vertices of a tetrahedron are adjacent to one another, cis–trans isomerism is not observed in tetrahedral complexes.

The second type of stereoisomerism listed in Figure 23.19 is **optical isomerism**. Optical isomers, called **enantiomers**, are mirror images that cannot be superimposed on each other. They bear the same resemblance to each other that your left hand bears to your right hand. If you look at your left hand in a mirror, the image is identical to your right hand (**Figure 23.22**). No matter how hard you try, however, you cannot superimpose your two hands on each other. An example of a complex that exhibits this type of isomerism is the $[Co(en)_3]^{3+}$ ion. Figure 23.22 shows the two enantiomers of this complex and their mirror-image relationship. Just as there is no way that we can twist or turn our right hand to make it look identical to our left, so also there is no way to rotate one of these enantiomers to make it identical to the other. Molecules or ions that are not superimposable on their mirror image are said to be **chiral** (pronounced KY-rul).

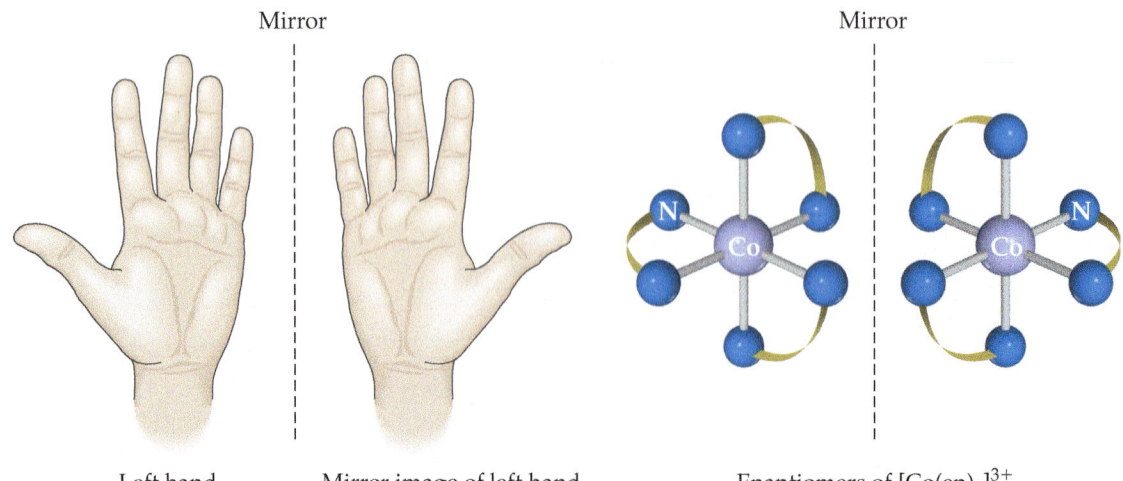

Left hand   Mirror image of left hand is identical to right hand   Enantiomers of $[Co(en)_3]^{3+}$

▲ **Figure 23.22 Optical isomerism.**

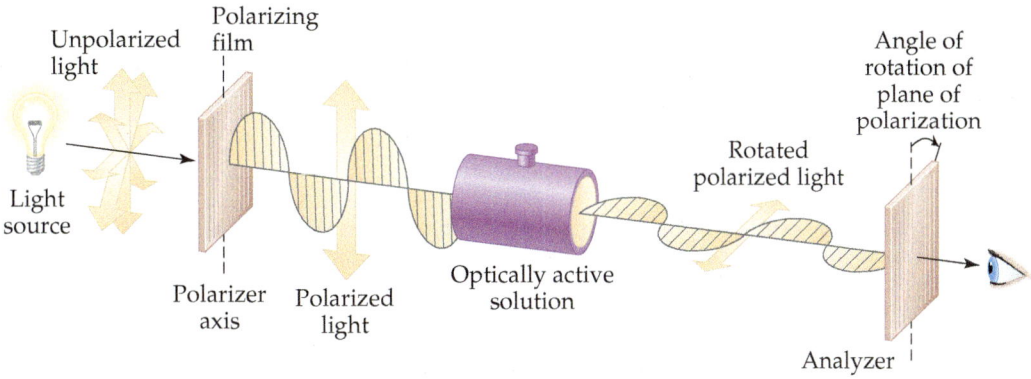

▲ **Figure 23.23** Using polarized light to detect optical activity.

The properties of two optical isomers differ only if the isomers are in a chiral environment—that is, an environment in which there is a sense of right- and left-handedness. A chiral enzyme, for example, might catalyze the reaction of one optical isomer but not the other. Consequently, one optical isomer may produce a specific physiological effect in the body, with its mirror image producing either a different effect or none at all. Chiral reactions are also extremely important in the synthesis of pharmaceuticals and other industrially important chemicals.

Optical isomers are usually distinguished from each other by their interaction with plane-polarized light. If light is polarized—for example, by being passed through a sheet of polarizing film—the electric-field vector of the light is confined to a single plane (**Figure 23.23**). If the polarized light is then passed through a solution containing one optical isomer, the plane of polarization is rotated either to the right or to the left. The isomer that rotates the plane of polarization to the right is **dextrorotatory**; it is the dextro, or *d*, isomer (Latin *dexter*, "right"). Its mirror image rotates the plane of polarization to the left; it is **levorotatory** and is the levo, or *l*, isomer (Latin *laevus*, "left"). The $[Co(en)_3]^{3+}$ isomer on the right in Figure 23.22 is found experimentally to be the *l* isomer of this ion. Its mirror image is the *d* isomer. Because of their effect on plane-polarized light, chiral molecules are said to be **optically active**.

Sample Exercise 23.4
Determining the Number of Geometric Isomers

The Lewis structure :C≡O: indicates that the CO molecule has two lone pairs of electrons. When CO binds to a transition-metal atom, it nearly always does so by using the C lone pair. How many geometric isomers are there for tetracarbonyldichloroiron(II)?

SOLUTION

Analyze We are given the name of a complex containing only monodentate ligands, and we must determine the number of isomers the complex can form.

Plan We can count the number of ligands to determine the coordination number of the Fe and then use the coordination number to predict the geometry. We can then either make a series of drawings with ligands in different positions to determine the number of isomers or deduce the number of isomers by analogy to cases we have discussed.

Solve The name indicates that the complex has four carbonyl (CO) ligands and two chloro (Cl⁻) ligands, so its formula is $Fe(CO)_4Cl_2$. The complex therefore has a coordination number of 6, and we can assume an octahedral geometry. Like $[Co(NH_3)_4Cl_2]^+$ (Figure 23.8), it has four ligands of one type and two of another. Consequently, there are two isomers possible: one with the Cl⁻ ligands across

the metal from each other (that is, separated by 180°), *trans*-$[Fe(CO)_4Cl_2]$, and one with the Cl⁻ ligands adjacent to each other (that is, separated by 90°), *cis*-$[Fe(CO)_4Cl_2]$.

Comment It is easy to overestimate the number of geometric isomers. Sometimes different orientations of a single isomer are incorrectly thought to be different isomers. If two structures can be rotated so that they are equivalent, they are not isomers of each other. The problem of identifying isomers is compounded by the difficulty we often have in visualizing three-dimensional molecules from their two-dimensional representations. It is sometimes easier to determine the number of isomers if we use three-dimensional models.

▶ **Practice Exercise**
How many isomers exist for the square-planar molecule $[Pt(NH_3)_2ClBr]$?

Sample Exercise 23.5

Predicting Whether a Complex Has Optical Isomers

Does either *cis*-[Co(en)$_2$Cl$_2$]$^+$ or *trans*-[Co(en)$_2$Cl$_2$]$^-$ have optical isomers?

SOLUTION

Analyze We are given the chemical formula for two geometric isomers and asked to determine whether either one has optical isomers. Because en is a bidentate ligand, we know that both complexes are octahedral and both have coordination number 6.

Plan We need to sketch the structures of the cis and trans isomers and their mirror images. We can draw the en ligand as two N atoms connected by an arc. If the mirror image cannot be superimposed on the original structure, the complex and its mirror image are optical isomers.

Solve The trans isomer of [Co(en)$_2$Cl$_2$]$^+$ and its mirror image are shown here. Notice that the mirror image of the isomer is identical to the original. Consequently *trans*-[Co(en)$_2$Cl$_2$]$^+$ does *not* exhibit optical isomerism.

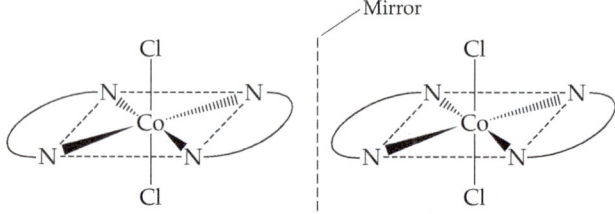

The cis isomer of [Co(en)$_2$Cl$_2$]$^+$ and its mirror image are shown here. In this case, the two cannot be superimposed on each other. Thus, the two cis structures are optical isomers (enantiomers). We say that *cis*-[Co(en)$_2$Cl$_2$]$^+$ is a chiral complex.

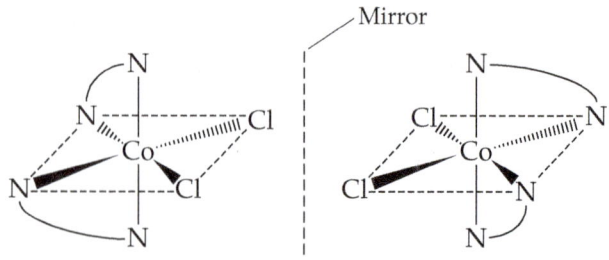

▶ **Practice Exercise**

Does the square-planar complex ion [Pt(NH$_3$)(N$_3$)ClBr]$^-$ have optical isomers? Explain your answer.

When a substance with optical isomers is prepared in the laboratory, the chemical environment during the synthesis is not usually chiral. Consequently, equal amounts of the two isomers are obtained, and the mixture is said to be **racemic**. A racemic mixture does not rotate polarized light because the rotatory effects of the two isomers cancel each other.

Self-Assessment Exercises

SAE 23.10   What is the correct formula for dichlorobis(ethylenediamine)cobalt(III) chloride? **(a)** [CoCl$_3$(en)]Cl **(b)** [Co(en)$_3$]Cl$_3$ **(c)** [CoCl$_3$(en)$_2$] **(d)** [CoCl$_2$(en)]Cl **(e)** [CoCl$_2$(en)$_2$]Cl

SAE 23.11   For which of the following are the name and formula correctly matched?

   (i)   Potassium tetrachloroplatinate(II): K$_2$[PtCl$_6$]
   (ii) Hexaaquamanganese(I) bromide: [Mn(H$_2$O)$_6$]Br
   (iii) Triamminetrichlorochromium(III): [Cr(NH$_3$)$_3$Cl$_3$]

(a) i only **(b)** ii only **(c)** i and ii **(d)** ii and iii **(e)** i, ii, and iii

SAE 23.12   Which of the following statements about isomers in coordination compounds is or are *true*?

   (i)   The octahedral complex [CoCl$_4$F$_2$]$^{3-}$ has two geometrical isomers.
   (ii) In order to form linkage isomers, a ligand must have two different donor atoms.
   (iii) An optical isomer rotates the plane of polarization of plane-polarized light.

(a) Only one statement is true. **(b)** Statements i and ii are true. **(c)** Statements i and iii are true. **(d)** Statements ii and iii are true. **(e)** All three statements are true.

SAE 23.13   Which of the following statements about the optical isomerism is *false*?

(a) Optical isomers are nonsuperimposable mirror images of one another.
(b) The two optical isomers of a chiral molecule will rotate plane-polarized light in opposite directions.
(c) The number and/or type of metal–ligand bonds must be different for two optical isomers of a coordination compound.
(d) A racemic mixture is one that has equal amounts of the two optical isomers of a compound.
(e) Some coordination compounds do not have optical isomers.

23.5 | Color and Magnetism in Coordination Chemistry

Studies of the colors and magnetic properties of transition-metal complexes have played an important role in the development of modern models for metal–ligand bonding. We discussed the various types of magnetic behavior of the transition metals in Section 23.1, and we discussed the interaction of radiant energy with matter in Section 6.3. Let's briefly examine the significance of these two properties for transition-metal complexes before we develop a model for metal–ligand bonding.

Color

In Figure 23.4, we saw the diverse range of colors seen in salts of transition-metal ions and their aqueous solutions. In general, the color of a complex depends on the identity of the metal ion, on its oxidation state, and on the ligands bonded to it. **Figure 23.24**, for instance, shows how the pale blue color characteristic of $[Cu(H_2O)_4]^{2+}$ changes to deep blue-violet as NH_3 ligands replace the H_2O ligands to form $[Cu(NH_3)_4]^{2+}$.

For a substance to have color we can see, it must absorb some portion of the spectrum of visible light. (Section 6.1) Absorption happens, however, only if the energy needed to move an electron in the substance from its ground state to an excited state corresponds to the energy of some portion of the visible spectrum. (Section 6.3) Thus, the particular energies of radiation a substance absorbs dictate the color we see for the substance.

When an object absorbs some portion of the visible spectrum, the color we perceive is the sum of the unabsorbed portions, which are either reflected or transmitted by the object and strike our eyes. (Opaque objects *reflect* light, and transparent ones *transmit* it.) If an object absorbs all wavelengths of visible light, none reaches our eyes and the object

> **Learning Objectives**
>
> **When you finish Section 23.5, you should be able to:**
>
> ► Predict the color of a coordination compound from its absorption spectrum.
>
> ► Recall that the magnetic behavior of coordination compounds will depend in part on the number of electrons on the metal atom in its particular oxidation state.

▼ **Go Figure**   Is the equilibrium binding constant of ammonia for Cu(II) likely to be larger or smaller than that of water for Cu(II)?

$[Cu(H_2O)_4]^{2+}(aq)$ $NH_3(aq)$ $[Cu(NH_3)_4]^{2+}(aq)$

▲ **Figure 23.24 The color of a coordination complex changes when the ligand changes.** When a small amount of concentrated $NH_3(aq)$ is added to a solution of $[Cu(H_2O)_4]^{2+}$, the color changes dramatically as NH_3 replaces H_2O in the coordination sphere.

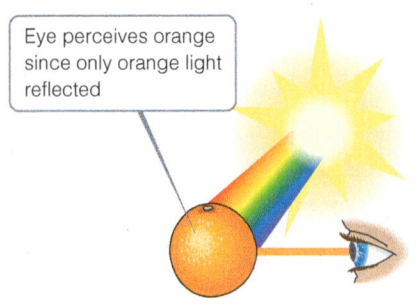

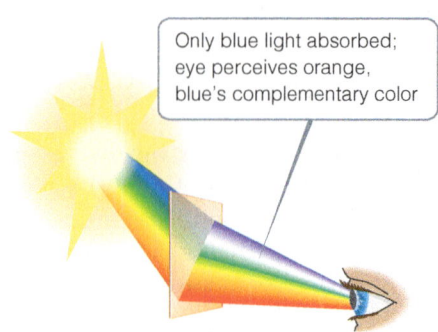

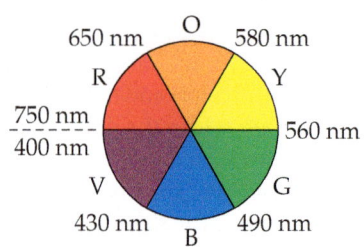

▲ **Figure 23.25 Two ways of perceiving the color orange.** An object appears orange either when it reflects orange light to the eye (left) or when it transmits to the eye all colors except blue, the complement of orange (middle). Complementary colors lie opposite to each other on an artist's color wheel (right).

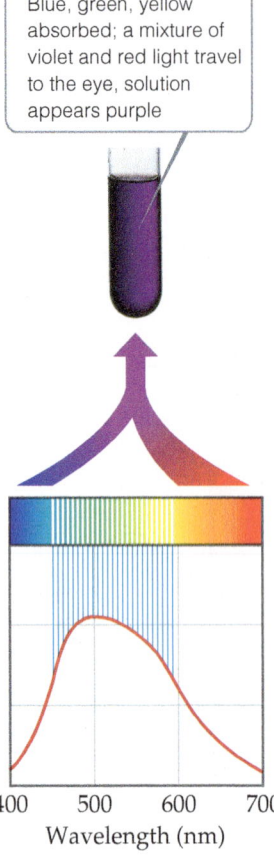

Go Figure

How would this absorbance spectrum change if you decreased the concentration of the $[Ti(H_2O)_6]^{3+}$ in solution?

Blue, green, yellow absorbed; a mixture of violet and red light travel to the eye, solution appears purple

▲ **Figure 23.26 The color of $[Ti(H_2O)_6]^{3+}$** A solution containing the $[Ti(H_2O)_6]^{3+}$ ion appears purple because, as its visible absorption spectrum shows, the solution does not absorb light from the violet and red ends of the spectrum. That unabsorbed light is what reaches our eyes.

appears black. If it absorbs no visible light, it is white if opaque or colorless if transparent. If it absorbs all but orange light, the orange light is what reaches our eye and therefore is the color we see.

An interesting phenomenon of vision is that we also perceive an orange color when an object absorbs only the blue portion of the visible spectrum and all the other colors strike our eyes. This is because orange and blue are **complementary colors**, which means that the removal of blue from white light makes the light look orange (and the removal of orange makes the light look blue).

Complementary colors can be determined with an artist's color wheel, which shows complementary colors on opposite sides (**Figure 23.25**). Orange and blue are complementary, as are red and green, and yellow and violet.

The amount of light absorbed by a sample as a function of wavelength is known as the sample's **absorption spectrum**. The visible absorption spectrum of a transparent sample can be determined using a spectrometer, as described in the "A Closer Look" box on "Using Spectroscopic Methods to Measure Reaction Rates: Beer's Law." (Section 14.2) The absorption spectrum of the ion $[Ti(H_2O)_6]^{3+}$ is shown in **Figure 23.26**. The absorption maximum is at 500 nm, but the graph shows that much of the yellow, green, and blue light is absorbed. Because the sample absorbs all of these colors, what we see is the unabsorbed red and violet light, which we perceive as purple (the color purple is classified as a tertiary color located between red and violet on an artist's color wheel).

Magnetism of Coordination Compounds

Many transition-metal complexes exhibit paramagnetism, as described in Sections 9.8 and 23.1. In such compounds, the metal ions possess some number of unpaired electrons. It is possible to determine experimentally the number of unpaired electrons per metal ion from the measured degree of paramagnetism, and experiments reveal some interesting comparisons.

Compounds of the complex ion $[Co(CN)_6]^{3-}$ have no unpaired electrons, for example, but compounds of the $[CoF_6]^{3-}$ ion have four unpaired electrons per metal ion. Both complexes contain Co(III) with a $3d^6$ electron configuration. (Section 7.4) As a result, there is a major difference in the ways in which the electrons are arranged in these two cases. Any successful bonding theory must explain this difference, and we present such a theory in the next section.

Sample Exercise 23.6
Relating Color Absorbed to Color Observed

The complex ion *trans*-$[Co(NH_3)_4Cl_2]^+$ absorbs light primarily in the red region of the visible spectrum (the most intense absorption is at 680 nm). What is the color of the complex ion?

SOLUTION

Analyze We need to relate the color absorbed by a complex (red) to the color observed for the complex.

Plan For an object that absorbs only one color from the visible spectrum, the color we see is complementary to the color absorbed. We can use the color wheel of Figure 23.25 to determine the complementary color.

Solve From Figure 23.25, we see that green is complementary to red, so the complex appears green.

Comment As noted in Section 23.2, this green complex helped Werner establish his theory of coordination (Table 23.3). The other geometric isomer of this complex, *cis*-$[Co(NH_3)_4Cl_2]^+$, absorbs yellow light and therefore appears violet.

▶ **Practice Exercise**
A certain transition-metal complex ion absorbs at 695 nm. Which color is this ion most likely to be—blue, yellow, green, or red?

Self-Assessment Exercises

SAE 23.14 The visible absorption spectrum of aqueous $[Ni(NH_3)_6]Cl_2$ has a strong absorption peak at 570 nm. What color is the solution? (**a**) blue-violet (**b**) yellow-orange (**c**) green (**d**) colorless

SAE 23.15 Which of the following statements about an aqueous solution of $[Zn(H_2O)_6]^{2+}$ is or are *true*?

   (**i**) In this complex, Zn is in the +2 oxidation state.
   (**ii**) In this complex, Zn has an $3d^{10}$ valence electron configuration.
   (**iii**) The complex will be paramagnetic.

(**a**) Only statement i is true. (**b**) Statements i and ii are true. (**c**) Statements i and iii are true. (**d**) Statements ii and iii are true. (**e**) All three statements are true.

23.6 | Crystal-Field Theory

Scientists have long recognized that many of the magnetic properties and colors of transition-metal complexes are related to the presence of *d* electrons in the metal cation. In this section, we consider a model for bonding in transition-metal complexes, called **crystal-field theory**, that accounts for many of the observed properties of these substances.* Because many predictions of crystal-field theory are qualitatively the same as those obtained with more advanced molecular-orbital theories, crystal-field theory is an excellent place to start in considering the electronic structure of coordination compounds.

As we have discussed, the interaction between a ligand and a metal ion is essentially a Lewis acid–base interaction in which the base—that is, the ligand—donates a pair of electrons to an empty orbital on the metal ion (**Figure 23.27**). Much of the attractive interaction between the metal ion and the ligands is due, however, to the electrostatic forces between the positive charge on the metal ion and negative charges on the ligands. An ionic ligand, such as Cl^- or SCN^-, experiences the usual cation–anion attraction. When the ligand is a neutral molecule, as in the case of H_2O or NH_3, the negative ends of these polar molecules, which contain a lone pair, are directed toward the metal ion. In this case, the attractive interaction is of the ion–dipole type. (Section 11.2) In either case, the ligands are attracted strongly toward the metal ion. Because of the metal–ligand electrostatic attraction, the energy of the complex is lower than the combined energy of the separated metal ion and ligands.

Learning Objectives

When you finish Section 23.6, you should be able to:

▶ Describe the splitting of the *d* orbitals of a metal ion in an octahedral crystal field.

▶ Describe how visible light can lead to a *d*–*d* transition in an octahedral complex.

▶ Relate the ability of a ligand to increase the crystal-field splitting to the energy gap, Δ, between the t_2 and *e* sets of *d* orbitals.

▶ Determine the electron configurations and magnetic properties of high-spin and low-spin octahedral complexes.

▶ Use crystal-field theory to determine the *d*-orbital splitting patterns and number of unpaired electrons in tetrahedral and square-planar complexes.

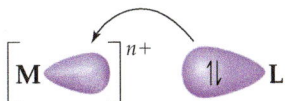

▲ **Figure 23.27 Metal–ligand bond formation.** The ligand acts as a Lewis base by donating its lone pair to an empty orbital on the metal ion. The bond that results is strongly polar with some covalent character.

---

*The name *crystal field* arose because the theory was first developed to explain the properties of solid crystalline materials. The theory applies equally well to complexes in solution, however.

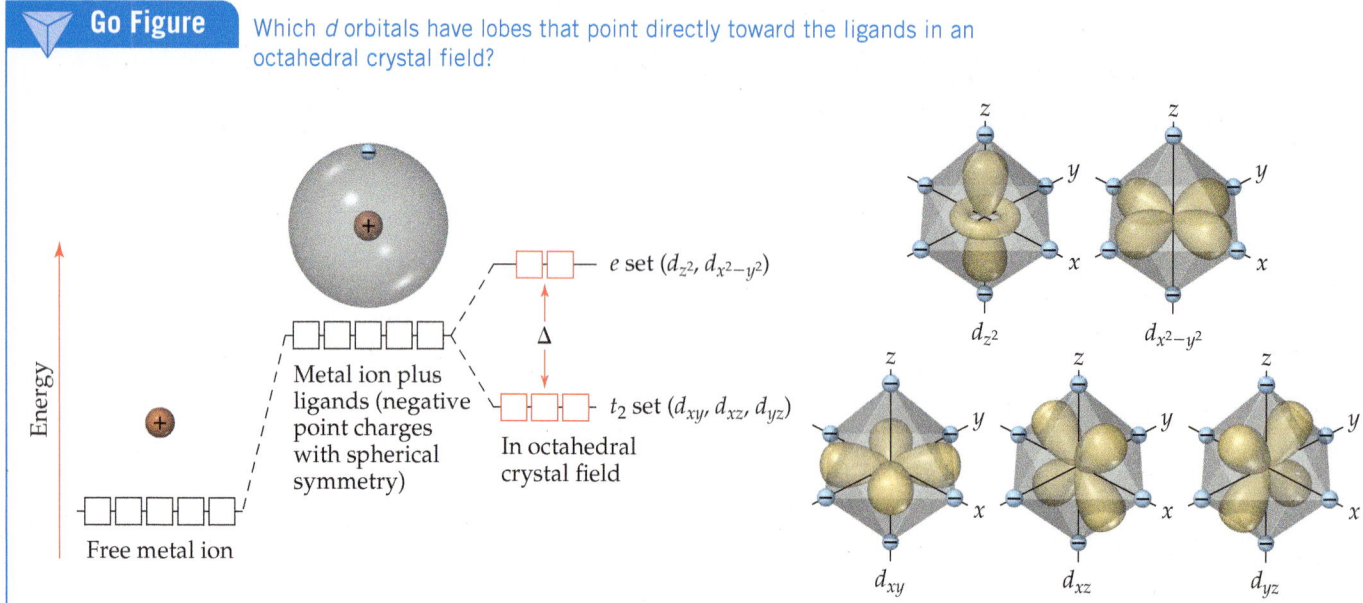

Go Figure   Which *d* orbitals have lobes that point directly toward the ligands in an octahedral crystal field?

▲ **Figure 23.28** **Energies of *d* orbitals in a free metal ion, a spherically symmetric crystal field, and an octahedral crystal field.** The degree of splitting of the *d* orbitals in the crystal fiend is the *crystal-field splitting energy*, denoted Δ.

Although the metal ion is attracted to the ligand electrons, the metal ion's *d* electrons are electrostatically repelled by the electrons on the ligands. Let's examine this effect more closely, specifically the case in which the ligands form an octahedral array around a metal ion that has coordination number 6.

In crystal-field theory, we consider the ligands to be negative points of charge that repel the negatively charged electrons in the *d* orbitals of the metal ion. The energy-level diagram in **Figure 23.28** shows how these ligand point charges affect the energies of the *d* orbitals. First, we imagine the complex as having all the ligand point charges uniformly distributed on the surface of a sphere centered on the metal ion. The *average* energy of the metal ion's *d* orbitals is raised by the presence of this uniformly charged sphere. Hence, the energies of all five *d* orbitals are raised by the same amount—the *d* orbitals remain *degenerate* under the influence of the negatively charged sphere. (Section 6.7)

This energy picture is only a first approximation, however, because the ligands are not distributed uniformly on a spherical surface and, therefore, do not approach the metal ion equally from every direction. Instead, we envision the six ligands approaching along *x*, *y*, and *z* axes, as shown on the right in Figure 23.28. This arrangement of ligands is called an *octahedral crystal field*. Because the *d* orbitals have different orientations and shapes, they do not all experience the same repulsion from the ligands and, therefore, do not all have the same energy under the influence of the octahedral crystal field. To see why, we must consider the shapes of the *d* orbitals and how their lobes are oriented relative to the ligands.

Figure 23.28 shows that the lobes of the d_{z^2} and $d_{x^2-y^2}$ orbitals are directed *along* the *x*, *y*, and *z* axes and so point directly toward the ligand point charges. The lobes of the d_{xy}, d_{xz}, and d_{yz} orbitals, however, are directed *between* the axes and so do not point directly toward the charges. The result of this difference in orientation—$d_{x^2-y^2}$ and d_{z^2} lobes point directly toward the ligand charges; d_{xy}, d_{xz}, and d_{yz} lobes do not—is that an electron residing in either the $d_{x^2-y^2}$ or d_{z^2} orbitals feels more repulsion from the negatively charged ligands than does an electron in the d_{xy}, d_{xz}, or d_{yz} orbitals. Hence, in an octahedral crystal field, the energy of the $d_{x^2-y^2}$ and d_{z^2} orbitals is higher than the energy of the d_{xy}, d_{xz}, and d_{yz} orbitals. This difference in energy is represented by the red boxes in the energy diagram of Figure 23.28.

It might seem like the energy of the $d_{x^2-y^2}$ orbital should be different from that of the d_{z^2} orbital because the $d_{x^2-y^2}$ has four lobes pointing at ligands and the d_{z^2} has only two lobes pointing at ligands. However, the d_{z^2} orbital does have electron

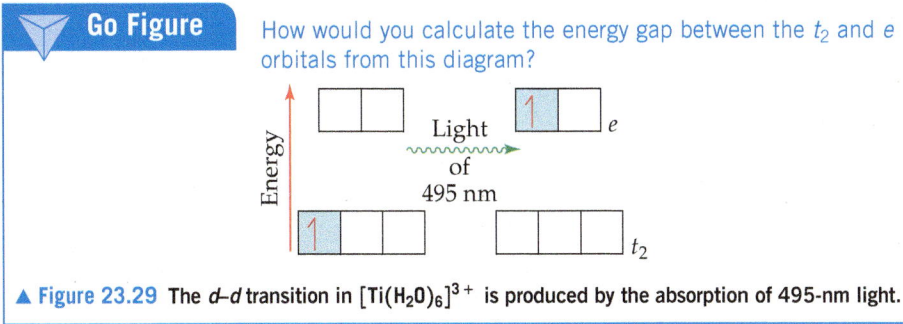

Go Figure

How would you calculate the energy gap between the t_2 and e orbitals from this diagram?

▲ **Figure 23.29** The *d–d* transition in $[Ti(H_2O)_6]^{3+}$ is produced by the absorption of 495-nm light.

density in the xy plane, represented by the ring encircling the point where the two lobes meet. More advanced theory proves that, in an octahedral crystal field, the $d_{x^2-y^2}$ and d_{z^2} orbitals both have the same energy, and the $d_{xy}, d_{xz},$ or d_{yz} orbitals all have the same energy.

The unequal repulsion of the d orbitals in an octahedral crystal field leads to the splitting pattern shown in Figure 23.28. The five d orbitals are split into two groups: a group of three orbitals at lower energy and a group of two orbitals at higher energy. The three lower-energy d orbitals are called the t_2 set of orbitals, and the two higher-energy ones are called the e set.* The energy gap Δ between the two sets is called the *crystal-field splitting energy*.

Crystal-field theory helps us account for the colors observed in transition-metal complexes. The energy gap Δ between the t_2 and e sets of d orbitals is of the same order of magnitude as the energy of a photon of visible light. It is therefore possible for a transition-metal complex to absorb visible light that excites an electron from a lower-energy (t_2) d orbital into a higher-energy (e) d orbital. In $[Ti(H_2O)_6]^{3+}$, for example, the Ti(III) ion has an $[Ar]3d^1$ electron configuration. (Recall from Section 7.4 that when determining the electron configurations of transition-metal ions, we remove the s electrons first.) Ti(III) is thus called a d^1 *ion*. In the ground state of $[Ti(H_2O)_6]^{3+}$, the single $3d$ electron resides in an orbital in the t_2 set (**Figure 23.29**). Absorption of 495-nm light excites this electron up to an orbital in the e set, generating the absorption spectrum shown in Figure 23.26. Because this transition involves exciting an electron from one set of d orbitals to the other, we call it a **d–d transition**. As noted previously, the absorption of visible radiation that produces this d–d transition causes the $[Ti(H_2O)_6]^{3+}$ ion to appear purple.

The magnitude of the crystal-field splitting energy and, consequently, the color of a complex depend on both the metal and the ligands. For example, we saw in Figure 23.4 that the color of $[M(H_2O)_6]^{2+}$ complexes changes from reddish-pink when the metal ion is Co^{2+}, to green for Ni^{2+}, to pale blue for Cu^{2+}. If we change the ligands in the $[Ni(H_2O)_6]^{2+}$ ion, the color also changes. $[Ni(NH_3)_6]^{2+}$ has a blue-violet color, while $[Ni(en)_3]^{2+}$ is purple (**Figure 23.30**). In a ranking called the **spectrochemical series**, ligands are arranged in order of their abilities to increase the crystal-field splitting energy, as in this abbreviated list:

—— Increasing Δ ⟶

$$Cl^- < F^- < H_2O < NH_3 < en < NO_2^-(\text{N-bonded}) < CN^-$$

The magnitude of Δ increases by roughly a factor of 2 from the far left to the far right of the spectrochemical series. Ligands at the low-Δ end of the spectrochemical series are called *weak-field ligands*; those at the high-Δ end are called *strong-field ligands*.

Let's take a closer look at the colors and crystal-field splitting as we vary the ligands for the series of Ni^{2+} complexes in Figure 23.30. Because the Ni atom has an $[Ar]3d^84s^2$ electron configuration, Ni^{2+} has the configuration $[Ar]3d^8$ and therefore is a d^8 ion. The t_2 set of orbitals holds six electrons, two in each orbital, while the last two electrons go

---

* The labels t_2 for the $d_{xy}, d_{xz},$ and d_{yz} orbitals and e for the d_{z^2} and $d_{x^2-y^2}$ orbitals come from application of a branch of mathematics called *group theory* to crystal-field theory. Group theory can be used to analyze the effects of symmetry on molecular properties.

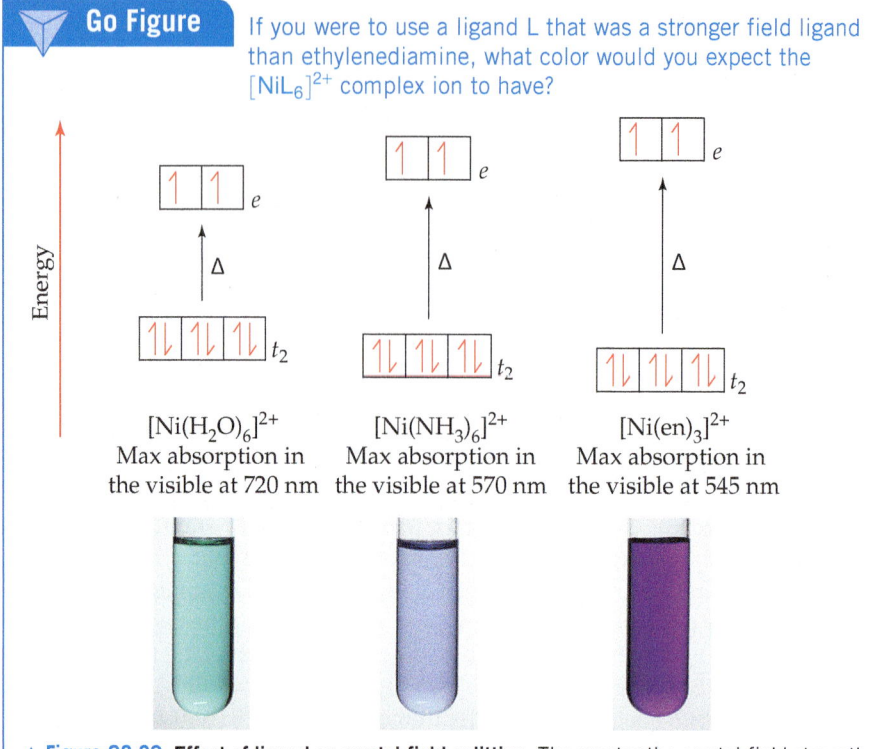

Go Figure

If you were to use a ligand L that was a stronger field ligand than ethylenediamine, what color would you expect the $[NiL_6]^{2+}$ complex ion to have?

$[Ni(H_2O)_6]^{2+}$
Max absorption in the visible at 720 nm

$[Ni(NH_3)_6]^{2+}$
Max absorption in the visible at 570 nm

$[Ni(en)_3]^{2+}$
Max absorption in the visible at 545 nm

▲ **Figure 23.30 Effect of ligand on crystal-field splitting.** The greater the crystal-field strength of the ligand, the greater the energy gap Δ it causes between the t_2 and e sets of the metal ion's orbitals. The increasing gap shifts the absorption maximum to shorter wavelengths.

into the e set of orbitals. Consistent with Hund's rule, each e orbital holds one electron and both electrons have the same spin. (Section 6.8)

As the ligand changes from H_2O to NH_3 to ethylenediamine, the spectrochemical series tells us that the crystal-field, Δ, exerted by the six ligands should increase. When there is more than one electron in the d orbitals, interactions between the electrons make the absorption spectra more complicated than the spectrum shown for $[Ti(H_2O)_6]^{3+}$ in Figure 23.26, which complicates the task of relating changes in Δ with color. With d^8 ions like Ni^{2+} three peaks are observed in the absorption spectra. Fortunately, for Ni^{2+} complexes we can simplify the analysis because only one of these three peaks falls in the visible region of the spectrum.* Because the energy separation Δ is increasing, the wavelength of the absorption peak should shift to a shorter wavelength. (Section 6.3) In the case of $[Ni(H_2O)_6]^{2+}$ the absorption peak in the visible part of the spectrum reaches a maximum near 720 nm, in the red region of the spectrum. So the complex ion takes the complementary color—green. For $[Ni(NH_3)_6]^{2+}$ the absorption peak reaches its maximum at 570 nm near the boundary between orange and yellow. The resulting color of the complex ion is a mixture of the complementary colors—blue and violet. Finally, for $[Ni(en)_3]^{2+}$ the peak shifts to an even shorter wavelength, 540 nm, which lies near the boundary between green and yellow. The resulting color purple is a mixture of the complementary colors red and violet.

Electron Configurations in Octahedral Complexes

Crystal-field theory helps us understand the magnetic properties and some important chemical properties of transition-metal ions. From Hund's rule, we expect electrons to always occupy the lowest-energy vacant orbitals first and to occupy a set of degenerate (same-energy) orbitals one at a time with their spins parallel. (Section 6.8) Thus, if we have a d^1, d^2, or d^3 octahedral complex, the electrons go into the lower-energy t_2 orbitals,

---

*The other two peaks fall in the infrared (IR) and ultraviolet (UV) regions of the spectrum. For $[Ni(H_2O)_6]^{2+}$ the IR peak is found at 1176 nm and the UV peak at 388 nm.

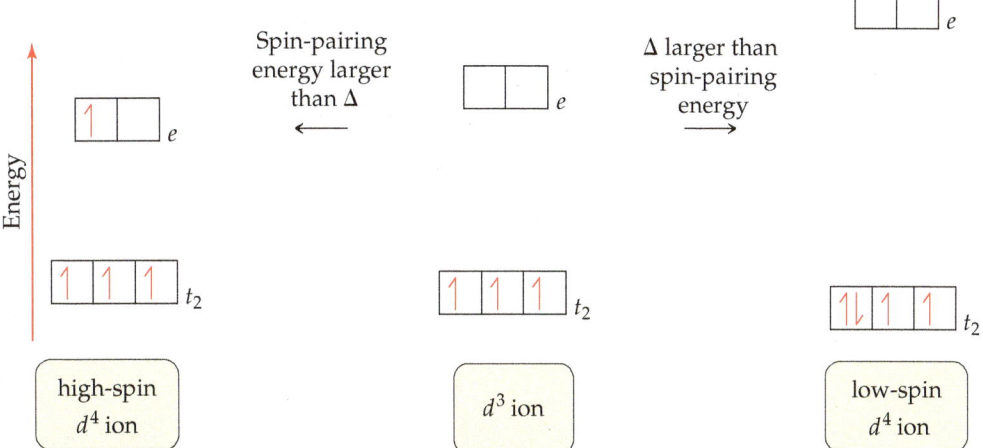

▲ **Figure 23.31 Two possibilities for adding a fourth electron to a d^3 octahedral complex.** Whether the fourth electron goes into a t_2 orbital or into an e orbital depends on the relative energies of the crystal-field splitting energy and the spin-pairing energy.

with their spins parallel. When a fourth electron must be added, we have the two choices shown in **Figure 23.31**: The electron can either go into an e orbital, where it will be the sole electron in the orbital or become the second electron in a t_2 orbital. Because the energy difference between the t_2 and e sets is the splitting energy Δ, the energy cost of going into an e orbital rather than a t_2 orbital is also Δ. Thus, the goal of filling lowest-energy available orbitals first is met by putting the electron in a t_2 orbital.

There is a penalty for doing this, however, because the electron must now be paired with the electron already occupying the orbital. The difference between the energy required to pair an electron in an occupied orbital and the energy required to place that electron in an empty orbital is called the **spin-pairing energy**. The spin-pairing energy arises from the fact that the electrostatic repulsion between two electrons that share an orbital (and so must have opposite spins) is greater than the repulsion between two electrons that are in different orbitals and have parallel spins.

In coordination complexes, the nature of the ligands and the charge on the metal ion often play major roles in determining which of the two electron arrangements shown in Figure 23.31 is used. In $[CoF_6]^{3-}$ and $[Co(CN)_6]^{3-}$, both ligands have a $1-$ charge. The F^- ion, however, is on the low end of the spectrochemical series, so it is a weak-field ligand. The CN^- ion is on the high end of the series and so is a strong-field ligand, which means it produces a larger energy gap Δ than the F^- ion. The splittings of the d-orbital energies in these two complexes are compared in **Figure 23.32**.

Cobalt(III) has an $[Ar]3d^6$ electron configuration, so both complexes in Figure 23.32 are d^6 complexes. Let's imagine that we add these six electrons one at a time to the d orbitals of the $[CoF_6]^{3-}$ ion. The first three go into the t_2 orbitals with their spins parallel. The fourth electron could pair up in one of the t_2 orbitals. The F^- ion is a weak-field ligand, however, and so the energy gap Δ between the t_2 set and the e set is small. In this case, the more stable arrangement is to place the fourth electron in one of the e orbitals. By the same energy argument, the fifth electron goes into the other e orbital. With all five d orbitals containing one electron, the sixth must pair up, and the energy needed to place the sixth electron in a t_2 orbital is less than that needed to place it in an e orbital. We end up with four t_2 electrons and two e electrons.

Figure 23.32 shows that the crystal-field splitting energy Δ is much larger in the $[Co(CN)_6]^{3-}$ complex. In this case, the spin-pairing energy is smaller than Δ, so the lowest-energy arrangement is to place all six electrons in the t_2 orbitals.

The $[CoF_6]^{3-}$ ion is called a **high-spin complex**; that is, the electrons are arranged so that they remain unpaired as much as possible. The $[Co(CN)_6]^{3-}$ ion is called a **low-spin complex**; that is, the electrons are arranged so that they remain paired as much as possible while still following Hund's rule. These two electronic arrangements can be

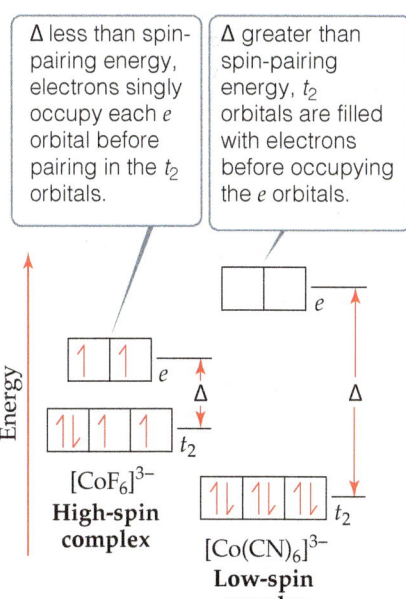

▲ **Figure 23.32 High-spin and low-spin complexes.** The high-spin $[CoF_6]^{3-}$ ion has a weak-field ligand and so a small Δ value. The low-spin $[Co(CN)_6]^{3-}$ ion has a strong-field ligand and so a large Δ value. Because $[CoF_6]^{3-}$ has unpaired electrons, it is paramagnetic, while $[Co(CN)_6]^{3-}$ is diamagnetic.

readily distinguished by measuring the magnetic properties of each complex. Experiments show that $[CoF_6]^{3-}$ has four unpaired electrons and is paramagnetic, whereas $[Co(CN)_6]^{3-}$ has no unpaired electrons and is diamagnetic. The absorption spectrum also shows peaks corresponding to the different values of Δ in these two complexes.

In the transition-metal ions of periods 5 and 6 (which have $4d$ and $5d$ valence electrons), the d orbitals are larger than in the period 4 ions (which have only $3d$ electrons). Thus, ions from periods 5 and 6 interact more strongly with ligands, resulting in a larger crystal-field splittings. *Consequently, metal ions in periods 5 and 6 are invariably low spin in an octahedral crystal field.*

Sample Exercise 23.7
The Spectrochemical Series, Crystal-Field Splitting, Color, and Magnetism

The compound hexaamminecobalt(III) chloride is diamagnetic and orange in color with a single absorption peak in its visible absorption spectrum. **(a)** What is the electron configuration of the cobalt(III) ion? **(b)** Is $[Co(NH_3)_6]^{3+}$ a high-spin complex or a low-spin complex? **(c)** Estimate the wavelength where you expect the absorption of light to reach a maximum? **(d)** What color and magnetic behavior would you predict for the complex ion $[Co(en)_3]^{3+}$?

SOLUTION

Analyze We are given the color and magnetic behavior of an octahedral complex containing Co with a +3 oxidation number. We need to use this information to determine its electron configuration, its spin state (low-spin or high-spin), and the color of light it absorbs. In part (d), we must use the spectrochemical series to predict how its properties will change if NH_3 is replaced by ethylenediamine (en).

Plan (a) From the oxidation number and the periodic table we can determine the number of valence electrons for Co(III), and from that we can determine the electron configuration. **(b)** The magnetic behavior can be used to determine whether this compound is a low-spin or high-spin complex. **(c)** Because there is a single peak in the visible absorption spectrum, the color of the compound should be complementary to the color of light that is absorbed most strongly. **(d)** Ethylenediamine is a stronger field ligand than NH_3, so we expect a larger Δ for $[Co(en)_3]^{3+}$ than for $[Co(NH_3)_6]^{3+}$.

Solve

(a) Co has an electron configuration of $[Ar]4s^23d^7$, and Co^{3+} has three fewer electrons than Co. Because transition-metal ions always lose their valence shell s electrons, the electron configuration of Co^{3+} is $[Ar]3d^6$.

(b) There are six valence electrons in the d orbitals. The filling of the t_2 and e orbitals for both high-spin and low-spin complexes is shown below. Because the compound is diamagnetic, we know all of the electrons must be paired up, which allows us to determine that $[Co(NH_3)_6]^{3+}$ is a low-spin complex.

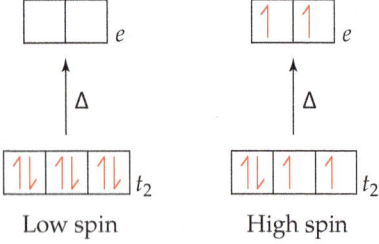

(c) We are told that the compound is orange and has a single absorption peak in the visible region of the spectrum. The compound must therefore absorb the complementary color of orange, which is blue. The blue region of the spectrum ranges from approximately 430 nm to 490 nm. As an estimate, we assume that the complex ion absorbs somewhere in the middle of the blue region, near 460 nm.

(d) Ethylenediamine is higher in the spectrochemical series than ammonia. Therefore, we expect a larger Δ for $[Co(en)_3]^{3+}$. Because Δ was already greater than the spin-pairing energy for $[Co(NH_3)_6]^{3+}$, we expect $[Co(en)_3]^{3+}$ to be a low-spin complex as well, with a d^6 configuration, so it will also be diamagnetic. The wavelength at which the complex absorbs light will shift to higher energy. If we assume a shift in the absorption maximum from blue to violet, the color of the complex will become yellow.

Comment The compound $[Co(en)_3]Cl_3$, which contains the $[Co(en)_3]^{3+}$ ion, was made and studied by Alfred Werner. This compound forms diamagnetic, golden-yellow crystals.

▶ **Practice Exercise**

Consider the colors of the ammonia complexes of Co^{3+} given in Table 23.3. Based on the change in color would you expect $[Co(NH_3)_5Cl]^{2+}$ to have a larger or smaller value of Δ than $[Co(NH_3)_6]^{3+}$? Is this prediction consistent with the spectrochemical series?

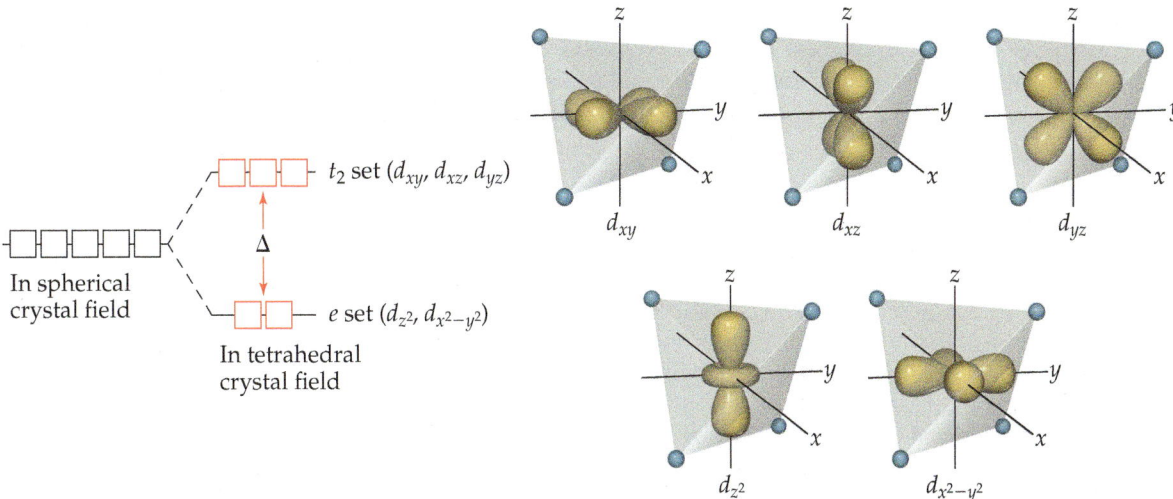

▲ **Figure 23.33 Energies of the *d* orbitals in a tetrahedral crystal field.** The splitting of the *e* and *t*$_2$ sets of orbitals is inverted with respect to the splitting associated with an octahedral crystal field. The crystal-field splitting energy Δ is smaller than it is in an octahedral crystal field.

Tetrahedral and Square-Planar Complexes

Thus far we have considered crystal-field theory only for complexes having an octahedral geometry. When there are only four ligands in a complex, the geometry is generally tetrahedral, except for the special case of d^8 metal ions, which we discuss in a moment.

The crystal-field splitting of *d* orbitals in tetrahedral complexes differs from that in octahedral complexes. Four equivalent ligands can interact with a central metal ion most effectively by approaching along the vertices of a tetrahedron. In this geometry, the lobes of the two *e* orbitals point toward the edges of the tetrahedron, exactly in between the ligands (**Figure 23.33**). This orientation keeps the $d_{x^2-y^2}$ and d_{z^2} as far from the ligand point charges as possible. Consequently, these two *d* orbitals experience less repulsion from the ligands and lie at lower energy than the other three *d* orbitals. The three t_2 orbitals do not point directly at the ligand point charges, but they do come closer to the ligands than the *e* set, and as a result they experience more repulsion and are higher in energy. As we see in Figure 23.33, the splitting of *d* orbitals in a tetrahedral geometry is the opposite of what we find for the octahedral geometry: the *e* orbitals are now *below* the t_2 orbitals. The crystal-field splitting energy Δ is much smaller for tetrahedral complexes than it is for comparable octahedral complexes, in part because there are fewer ligand point charges in the tetrahedral geometry, and in part because neither set of orbitals has lobes that point directly at the ligands. Calculations show that for the same metal ion and ligand set, Δ for the tetrahedral complex is only four-ninths as large as for the octahedral complex. For this reason, nearly all tetrahedral complexes are high spin; the crystal-field splitting energy is never large enough to overcome the spin-pairing energies.

In a square-planar complex, four ligands are arranged about the metal ion such that all five species are in the *xy* plane. The resulting energy levels of the *d* orbitals are illustrated in **Figure 23.34**. Note in particular that the d_{z^2} orbital is considerably lower in energy than the $d_{x^2-y^2}$ orbital. To understand why this is so, recall from Figure 23.28 that in an octahedral field the d_{z^2} orbital of the metal ion interacts with the ligands positioned above and below the *xy* plane. There are no ligands in these two positions in a square-planar complex, which means that the d_{z^2} orbital experiences less repulsion and so remains in a lower-energy, more stable state.

Square-planar complexes are characteristic of metal ions with a d^8 electron configuration. They are nearly always low spin, with the eight *d* electrons spin-paired to form a diamagnetic complex. This pairing leaves the $d_{x^2-y^2}$ orbital empty. Such an electronic arrangement is particularly common among the d^8 ions of periods 5 and 6, such as Pd^{2+}, Pt^{2+}, Ir^+, and Au^{3+}.

▼ **Go Figure**

For which *d* orbital(s) do the lobes point directly at the ligands in a square-planar crystal field?

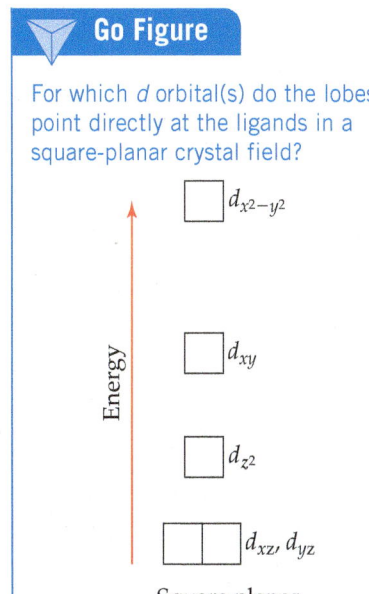

▲ **Figure 23.34 Energies of the *d* orbitals in a square-planar crystal field.**

Sample Exercise 23.8
Populating *d* Orbitals in Tetrahedral and Square-Planar Complexes

Nickel(II) complexes in which the metal coordination number is 4 can have either square-planar or tetrahedral geometry. $[NiCl_4]^{2-}$ is paramagnetic, and $[Ni(CN)_4]^{2-}$ is diamagnetic. One of these complexes is square planar, and the other is tetrahedral. Use the relevant crystal-field splitting diagrams in the text to determine which complex has which geometry.

SOLUTION

Analyze We are given two complexes containing Ni^{2+} and their magnetic properties. We are given two molecular geometry choices and asked to use crystal-field splitting diagrams from the text to determine which complex has which geometry.

Plan We need to determine the number of *d* electrons in Ni^{2+} and then use Figure 23.33 for the tetrahedral complex and Figure 23.34 for the square-planar complex.

Solve Nickel(II) has the electron configuration $[Ar]3d^8$. With very few exceptions, tetrahedral complexes are high spin and square-planar complexes are low spin. Therefore, the population of the *d* orbitals in the two geometries is

The tetrahedral complex has two unpaired electrons, and the square-planar complex has none. Hence, the tetrahedral complex must be paramagnetic and the square planar must be diamagnetic. Therefore, $[NiCl_4]^{2-}$ is tetrahedral, and $[Ni(CN)_4]^{2-}$ is square planar.

Comment Nickel(II) forms octahedral complexes more frequently than square-planar ones, whereas d^8 metals from periods 5 and 6 tend to favor square-planar coordination.

▶ **Practice Exercise**
Are there any diamagnetic tetrahedral complexes containing transition-metal ions with partially filled *d* orbitals? If so what electron count(s) leads to diamagnetism?

Tetrahedral:
$d_{x^2-y^2}$ (empty)
$t_2\ (d_{xy}, d_{yz}, d_{xz})$ — ↑↓ ↑ ↑
$e\ (d_{x^2-y^2}, d_{z^2})$ — ↑↓ ↑↓

Square planar:
d_{xy} — ↑↓
d_{z^2} — ↑↓
d_{xz}, d_{yz} — ↑↓ ↑↓

Crystal-field theory can be used to explain many observations in addition to those we have discussed. The theory is based on electrostatic interactions between ions and atoms, which essentially means ionic bonds. Many lines of evidence show, however, that the bonding in complexes must have some covalent character. Therefore, molecular-orbital theory (Sections 9.7 and 9.8) can also be used to describe the bonding in complexes, but the application of molecular-orbital theory to coordination compounds is beyond the scope of our discussion. Crystal-field theory, though not entirely accurate in all details, provides an adequate and useful first description of the electronic structure of complexes.

A CLOSER LOOK   Charge-Transfer Color

In the laboratory portion of your course, you have probably seen many colorful transition-metal compounds, including those shown in Figure 23.35. Many of these compounds are colored because of *d*–*d* transitions. Some colored complexes, however, including the violet permanganate ion, MnO_4^-, and the yellow chromate ion, CrO_4^{2-}, derive their color from a different type of excitation involving the *d* orbitals.

The permanganate ion strongly absorbs visible light, with a maximum absorption at 565 nm. Because violet is complementary to yellow, this strong absorption in the yellow portion of the

KMnO₄

K₂CrO₄

KClO₄

▲ **Figure 23.35 The colors of compounds can arise from charge-transfer transitions.** $KMnO_4$ and K_2CrO_4 are colored due to ligand-to-metal charge-transfer transitions in their anions. Higher-energy ultraviolet photons are needed to excite the charge-transfer transition in the perchlorate ion, so $KClO_4$ is white.

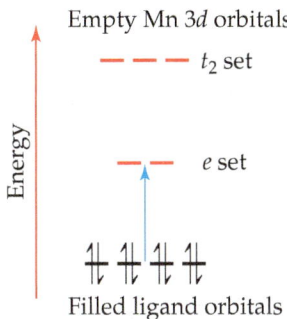

Empty Mn $3d$ orbitals

t_2 set

e set

Energy

Filled ligand orbitals

▲ Figure 23.36 **Ligand-to-metal charge-transfer transition in** MnO_4^-. As shown by the blue arrow, an electron is excited from a nonbonding orbital on O into one of the empty d orbitals on Mn.

visible spectrum is responsible for the violet color of salts and solutions of the ion. What is happening during this absorption of light? The MnO_4^- ion is a complex of Mn(VII). Mn(VII) has a $[Ar]3d^0$ electron configuration, so the absorption cannot be due to a d–d transition because there are no d electrons to excite! That does not mean, however, that the d orbitals are not involved in the transition. The excitation in the MnO_4^- ion is due to a *charge-transfer transition*, in which an electron on one oxygen ligand is excited into a vacant d orbital on the Mn ion (**Figure 23.36**).

In essence, an electron is transferred from a ligand to the metal, so this transition is called a *ligand-to-metal charge-transfer (LMCT) transition*.

An LMCT transition is also responsible for the color of the CrO_4^{2-}, which contains the Cr(VI) ion, also with an $[Ar]3d^0$ electron configuration.

Also shown in Figure 23.35 is a salt of the perchlorate ion, ClO_4^-. Like MnO_4^-, ClO_4^- is tetrahedral and has its central atom in the +7 oxidation state. However, because the Cl atom does not have low-lying d orbitals, exciting an electron from O to Cl requires a more energetic photon than it does in MnO_4^-. The first absorption for ClO_4^- is in the ultraviolet portion of the spectrum, so no visible light is absorbed and the salt appears white.

Other complexes exhibit charge-transfer excitations in which an electron from the metal atom is excited to an empty orbital on a ligand. Such an excitation is called a *metal-to-ligand charge-transfer (MLCT) transition*.

Charge-transfer transitions are generally more intense than d–d transitions. Many metal-containing pigments used for oil painting, such as cadmium yellow (CdS), chrome yellow ($PbCrO_4$), and red ochre (Fe_2O_3), have intense colors because of charge-transfer transitions.

Related Exercises: 23.84, 23.85

 Self-Assessment Exercises

SAE 23.16   Which of the following statements about crystal-field theory applied to octahedral complexes is *false*?

(a) In crystal-field theory, the ligands are modeled as point negative charges.

(b) In an octahedral crystal field, the ligands are assumed to lie on the $\pm x$, $\pm y$, and $\pm z$ axes.

(c) In an octahedral crystal field, the d orbitals split into a set of three orbitals at lower energy and a set of two orbitals at higher energy.

(d) In an octahedral crystal field, the d_{xy} orbital is higher in energy than the d_{xz} orbital.

SAE 23.17   As discussed in the text, a solution of $[Ti(H_2O)_6]^{3+}$ absorbs light of wavelength 495 nm. Which of the following statements about this observation is or are *true*?

(i)   Because the absorption is in the visible portion of the spectrum, the solution will have color.

(ii) $[Ti(H_2O)_6]^{3+}$ is an example of a d^2 transition-metal complex.

(iii)   The transition is an example of a d–d transition in which an electron is excited from the t_2 set of orbitals to the e set of orbitals.

(a) Only statement i is true. **(b)** Statements i and ii are true. **(c)** Statements i and iii are true. **(d)** Statements ii and iii are true. **(e)** All three statements are true.

SAE 23.18   Which of the following cobalt(III) complexes has the largest crystal-field splitting energy?

(a) $[Co(NH_3)_6]^{3+}$

(b) $[CoF_6]^{3-}$

(c) $[Co(en)_3]^{3+}$

(d) $[Co(H_2O)_6]^{3+}$

(e) $[Co(CN)_6]^{3-}$

SAE 23.19   Beakers A, B, and C contain solutions of octahedral complexes of the same transition metal in the same oxidation state, but with different ligands. Each solution undergoes the same d–d transition, and the colors of the solutions are yellow-orange for Beaker A, green for Beaker B, and blue-violet for Beaker C. What is the order of the solutions in Beakers A, B, and C from weakest- to strongest-field ligand?

(a) B < C < A

(b) A < C < B

(c) C < B < A

(d) A < B < C

SAE 23.20   Cobalt(II) forms both high-spin and low-spin octahedral complexes. Complete the following: The high-spin Co(II) complexes will have _____ unpaired electrons, whereas the low-spin complexes will have _____ unpaired electrons. **(a)** 1, 3 **(b)** 2, 1 **(c)** 3, 1 **(d)** 4, 0 **(e)** 5, 6

SAE 23.21   Which of the following statements about the application of crystal-field theory to tetrahedral and square-planar complexes is *false*?

(a) The splitting of the d orbitals in a tetrahedral complex has a set of two orbitals at lower energy and a set of three orbitals at higher energy.

(b) Tetrahedral complexes are almost always low-spin complexes.

(c) For a given ligand, the magnitude of the crystal-field splitting energy in a tetrahedral complex is less than that in an octahedral complex.

(d) In a square-planar complex, the energy of the d_{z^2} orbital is lower than that of the $d_{x^2-y^2}$ orbital.

(e) A d^8 square-planar complex has four filled d orbitals and one empty d orbital.

Putting Concepts Together

The oxalate ion has the Lewis structure shown in Table 23.4. (**a**) Show the geometry of the complex formed when this ion complexes with cobalt(II) to form $[Co(C_2O_4)(H_2O)_4]$. (**b**) Write the formula for the salt formed when three oxalate ions complex with Co(II), assuming that the charge-balancing cation is Na^+. (**c**) Sketch all the possible geometric isomers for the cobalt complex formed in part (b). Are any of these isomers chiral? Explain. (**d**) The equilibrium constant for the formation of the cobalt(II) complex produced by coordination of three oxalate anions, as in part (b), is 5.0×10^9, and the equilibrium constant for formation of the cobalt(II) complex with three molecules of *ortho*-phenanthroline (Table 23.4) is 9×10^{19}. From these results, what conclusions can you draw regarding the relative Lewis base properties of the two ligands toward cobalt(II)? (**e**) Using the approach described in Sample Exercise 17.16, calculate the concentration of free aqueous Co(II) ion in a solution initially containing 0.040 *M* oxalate (*aq*) and 0.0010 *M* Co^{2+}(*aq*).

SOLUTION

(**a**) The complex formed by coordination of one oxalate ion is octahedral:

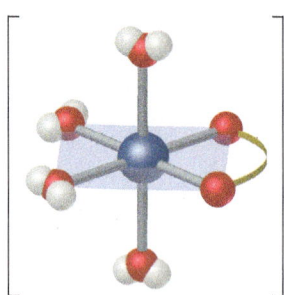

(**b**) Because the oxalate ion has a charge of 2−, the net charge of a complex with three oxalate anions and one Co^{2+} ion is 4−. Therefore, the coordination compound has the formula:

$Na_4[Co(C_2O_4)_3]$

(**c**) There is only one geometric isomer. The complex is chiral, however, in the same way the $[Co(en)_3]^{3+}$ complex is chiral (Figure 23.22). The two mirror images are not superimposable, so there are two enantiomers:

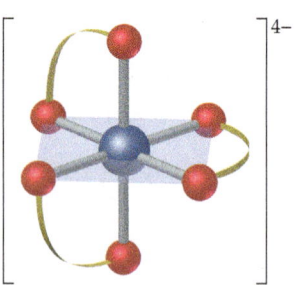 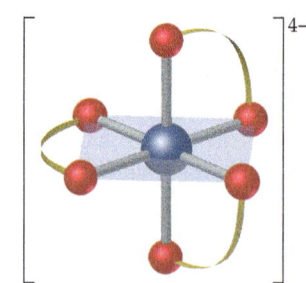

(**d**) The *ortho*-phenanthroline ligand is bidentate, like the oxalate ligand, so they both exhibit the chelate effect. Thus, we conclude that *ortho*-phenanthroline is a stronger Lewis base toward Co^{2+} than oxalate. This conclusion is consistent with what we learned about bases in Section 16.7, that nitrogen bases are generally stronger than oxygen bases. (Recall, for example, that NH_3 is a stronger base than H_2O.)

(**e**) The equilibrium we must consider involves 3 mol of oxalate ion (represented as ox^{2-}).

$Co^{2+}(aq) + 3\,ox^{2-}(aq) \rightleftharpoons [Co(ox)_3]^{4-}(aq)$

The formation-constant expression is:

$K_f = \dfrac{[[Co(ox)_3]^{4-}]}{[Co^{2+}][ox^{2-}]^3}$

Because K_f is so large, we can assume that essentially all the Co^{2+} is converted to the oxalato complex. Under that assumption, the final concentration of $[Co(ox)_3]^{4-}$ is 0.0010 *M* and that of oxalate ion is $[ox^{2-}] = (0.040) - 3(0.0010) = 0.037\,M$ (three ox^{2-} ions react with each Co^{2+} ion). We then have:

$[Co^{2+}] = x\,M,\ [ox^{2-}] \cong 0.037\,M,\ [[Co(ox)_3]^{4-}] \cong 0.0010\,M$

Inserting these values into the equilibrium-constant expression, and solving for x, we obtain $4 \times 10^{-9}\,M$. From this, we see that the oxalate has complexed all but a tiny fraction of the Co^{2+} in solution.

$K_f = \dfrac{(0.0010)}{x(0.037)^3} = 5 \times 10^9$

$x = 4 \times 10^{-9}\,M$

Chapter Summary and Key Terms

THE TRANSITION METALS (SECTION 23.1) Metallic elements are obtained from **minerals**, which are solid inorganic compounds found in nature. **Metallurgy** is the science and technology of extracting metals from the earth and processing them for further use. Transition metals are characterized by incomplete filling of the *d* orbitals. The presence of *d* electrons in transition elements leads to multiple oxidation states. As we proceed through the transition metals in a given row of the periodic table, the attraction between the nucleus and the valence electrons increases more markedly for *d* electrons than for *s* electrons. As a result, the later transition elements in a period tend to have lower oxidation states.

The atomic and ionic radii of period 5 transition metals are larger than those of period 4 metals. The transition metals of periods 5 and 6 have comparable atomic and ionic radii and are also similar in other properties. This similarity is due to the **lanthanide contraction**.

The presence of unpaired electrons in valence orbitals leads to magnetic behavior in transition metals and their compounds. In **ferromagnetic**, **ferrimagnetic**, and **antiferromagnetic** substances, the unpaired electron spins on atoms in a solid are affected by spins on neighboring atoms. In a ferromagnetic substance, the spins all point in the same direction. In an antiferromagnetic substance, the spins point in opposite directions and cancel one another. In a ferrimagnetic substance, the spins point in opposite directions but do not fully cancel. Ferromagnetic and ferrimagnetic substances are used to make permanent magnets.

TRANSITION-METAL COMPLEXES (SECTION 23.2) **Coordination compounds** are substances that contain **metal complexes**. Metal complexes contain metal ions bonded to several surrounding anions or molecules known as **ligands**. The metal ion and its ligands make up the **coordination sphere** of the complex. The number of atoms attached to the metal ion is the **coordination number** of the metal ion. The most common coordination numbers are 4 and 6; the most common coordination geometries are tetrahedral, square planar, and octahedral.

COMMON LIGANDS IN COORDINATION CHEMISTRY (SECTION 23.3) Ligands that occupy only one site in a coordination sphere are called **monodentate ligands**. The atom of the ligand that bonds to the metal ion is the **donor atom**. Ligands that have two donor atoms are **bidentate ligands**. **Polydentate ligands** have three or more donor atoms. Bidentate and polydentate ligands are also called **chelating agents**. In general, chelating agents form more stable complexes than do related monodentate ligands, an observation known as the **chelate effect**. Many biologically important molecules, such as the **porphyrins**, are complexes of chelating agents. A related group of plant pigments known as **chlorophylls** are important in **photosynthesis**, the process by which plants use solar energy to convert CO_2 and H_2O into carbohydrates.

NOMENCLATURE AND ISOMERISM IN COORDINATION CHEMISTRY (SECTION 23.4) In naming coordination compounds, the number and type of ligands attached to the metal ion are specified, as is the oxidation state of the metal ion. **Isomers** are compounds with the same composition but different arrangements of atoms and therefore different properties. **Structural isomers** differ in the bonding arrangements of the ligands. **Linkage isomerism** occurs when a ligand can coordinate to a metal ion through different donor atoms. **Coordination-sphere isomers** contain different ligands in the coordination sphere. **Stereoisomers** are isomers with the same chemical bonding arrangements but different spatial arrangements of ligands. The most common forms of stereoisomerism are **geometric isomerism** and **optical isomerism**. Geometric isomers differ from one another in the relative locations of donor atoms in the coordination sphere; the most common are cis and trans isomers. Geometric isomers differ from one another in their chemical and physical properties. Optical isomers are nonsuperimposable mirror images of each other. Optical isomers, or **enantiomers**, are **chiral**, meaning that they have a specific "handedness" and differ only in the presence of a chiral environment. Optical isomers can be distinguished from one another by their interactions with plane-polarized light; solutions of one isomer rotate the plane of polarization to the right (**dextrorotatory**), whereas solutions of its mirror image rotate the plane to the left (**levorotatory**). Chiral molecules, therefore, are **optically active**. A 50–50 mixture of two optical isomers does not rotate plane-polarized light and is said to be **racemic**.

COLOR AND MAGNETISM IN COORDINATION CHEMISTRY (SECTION 23.5) A substance has a particular color because it either reflects or transmits light of that color or absorbs light of the **complementary color**. The amount of light absorbed by a sample as a function of wavelength is known as its **absorption spectrum**. The light absorbed provides the energy to excite electrons to higher-energy states.

It is possible to determine the number of unpaired electrons in a complex from its degree of paramagnetism. Compounds with no unpaired electrons are diamagnetic.

CRYSTAL-FIELD THEORY (SECTION 23.6) **Crystal-field theory** successfully accounts for many properties of coordination compounds, including their color and magnetism. In crystal-field theory, the interaction between metal ion and ligand is viewed as electrostatic. Because some *d* orbitals point directly at the ligands whereas others point between them, the ligands split the energies of the metal *d* orbitals. For an octahedral complex, the *d* orbitals are split into a lower-energy set of three degenerate orbitals (the t_2 set) and a higher-energy set of two degenerate orbitals (the *e* set). Visible light can cause a ***d–d* transition**, in which an electron is excited from a lower-energy *d* orbital to a higher-energy *d* orbital. The **spectrochemical series** lists ligands in order of their ability to increase the split in *d*-orbital energies in octahedral complexes.

Strong-field ligands create a splitting of *d*-orbital energies that is large enough to overcome the **spin-pairing energy**. The *d* electrons then preferentially pair up in the lower-energy orbitals, producing a **low-spin complex**. When the ligands exert a weak crystal field, the splitting of the *d* orbitals is small. The electrons then occupy the higher-energy *d* orbitals in preference to pairing up in the lower-energy set, producing a **high-spin complex**. Transition-metal ions from periods 5 and 6 have large crystal-field splitting energies and adopt low-spin configurations in octahedral complexes.

Crystal-field theory also applies to tetrahedral and square-planar complexes, which leads to different *d*-orbital splitting patterns. In a tetrahedral crystal field, the splitting of the *d* orbitals results in a higher-energy t_2 set and a lower-energy *e* set, the opposite of the octahedral case. The splitting in a tetrahedral crystal field is much smaller than that in an octahedral crystal field, so tetrahedral complexes are nearly always high-spin complexes.

Exam Prep

EP 23.1 A certain transition-metal element exhibits the following properties: (i) Its maximum positive oxidation state is +5. (ii) Its metallic radius is greater than that of the element immediately to its right in the periodic table. (iii) Its metallic radius is nearly identical to that of the element immediately below it in the periodic table. What is the identity of the element? (**a**) vanadium (**b**) manganese (**c**) niobium (**d**) technetium (**e**) rhodium

EP 23.2 In which of the following compounds does the transition metal have the highest oxidation state?

(**a**) $[Co(NH_3)_4Cl_2]$

(**b**) $K_2[PtCl_6]$

(**c**) $Rb_3[MoO_3F_3]$

(**d**) $Na[Ag(CN)_2]$

(**e**) $K_4[Mn(CN)_6]$

EP 23.3 What is the ground-state valence electron configuration for Ru^{2+}? (**a**) $[Kr]5s^24d^6$ (**b**) $[Kr]4d^6$ (**c**) $[Kr]5s^24d^4$ (**d**) $[Kr]4d^8$

EP 23.4 Which type of magnetic behavior is described by the following statement? "In this phenomenon, the spins of the atoms and ions in the solid are all aligned parallel to one another." (**a**) paramagnetism (**b**) diamagnetism (**c**) ferrimagnetism (**d**) ferromagnetism (**e**) antiferromagnetism

EP 23.5 When the compound $RhCl_3 \cdot 4NH_3$ is dissolved in water and treated with excess $AgNO_3(aq)$, one mole of $AgCl(s)$ is formed for every mole of $RhCl_3 \cdot 4NH_3$. What is the correct way to write the formula of this compound?

(**a**) $[Rh(NH_3)_4Cl_3]$

(**b**) $[RhCl_3](NH_3)_4$

(**c**) $[Rh(NH_3)_4Cl]Cl_2$

(**d**) $[Rh(NH_3)_4]Cl_3$

(**e**) $[Rh(NH_3)_4Cl_2]Cl$

EP 23.6 Which of the following statements about metal–ligand bonding is or are *true*?

(**i**) The metal–ligand bond can be described as a Lewis acid–base interaction.

(**ii**) Ligands must have a negative charge.

(**iii**) To act as a ligand, a molecule or ion must have a lone pair of electrons.

(**a**) Only one statement is true. (**b**) Statements i and ii are true. (**c**) Statements i and iii are true. (**d**) Statements ii and iii are true. (**e**) All three statements are true.

EP 23.7 Which of the following statements about ligands is *false*?

(**a**) Bipyridine is a monodentate ligand.

(**b**) The chelate effect leads to a special stability of complexes formed with bidentate and polydentate ligands.

(**c**) The EDTA ligand has both nitrogen and oxygen donor atoms.

(**d**) The diethylenetriamine ligand has three donor atoms.

(**e**) A coordination complex with three bidentate ligands has a coordination number of six.

EP 23.8 Which of the following ligands does *not* have a nitrogen atom that can serve as a donor atom? (**a**) thiocyanate ion (**b**) ethylenediamine (**c**) bipyridine (**d**) nitrite ion (**e**) oxalate ion

EP 23.9 What is the name of the compound $[Rh(NH_3)_4Cl_2]Cl$?

(**a**) rhodium(III) tetraamminedichloro chloride (**b**) tetraammoniadichlororhodium(III) chloride (**c**) tetraamminedichlororhodium(III) chloride (**d**) tetraamminetrichloro-rhodium(III) (**e**) tetraamminedichlororhodium(II) chloride

EP 23.10 What is the chemical formula for the compound sodium tetracyanoplatinate(II)?

(**a**) $Na[Pt(CN)_4]$

(**b**) $Na_2[Pt(CN)_4]$

(**c**) $Na_4[Pt(CN)_4]$

(**d**) $Na_2[Pt(CN)_6]$

(**e**) $Na_4[Pt(CN)_6]$

EP 23.11 Which of the following molecules does *not* have a geometric isomer?

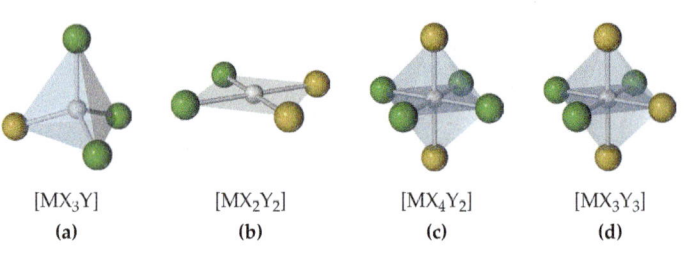

| $[MX_3Y]$ | $[MX_2Y_2]$ | $[MX_4Y_2]$ | $[MX_3Y_3]$ |
| (**a**) | (**b**) | (**c**) | (**d**) |

EP 23.12 Which of the following complexes has an optical isomer?

(**a**) tetrahedral $[CdBr_2Cl_2]^{2-}$

(**b**) octahedral $[CoCl_4(en)]^{2-}$

(**c**) octahedral $[Co(NH_3)_4Cl_2]^{2+}$

(**d**) tetrahedral $[Co(NH_3)BrClI]^-$

EP 23.13 A solution containing a certain transition-metal complex ion has the absorption spectrum shown here.

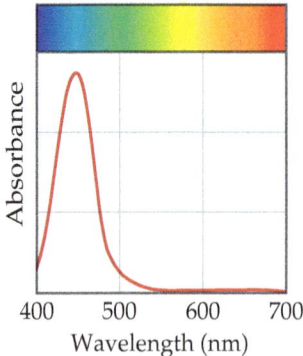

What color would you expect a solution containing this ion to be? (**a**) violet (**b**) blue (**c**) green (**d**) orange (**e**) red

EP 23.14 Which of the following statements about crystal-field theory is or are *true*?

(**i**) In an octahedral crystal field, the five d orbitals remain degenerate.

(**ii**) In an octahedral crystal field, the t_2 set of orbitals is lower in energy than the e set of orbitals.

(**iii**) The crystal-field splitting energy is the separation between the t_2 and e sets of orbitals.

(**a**) Only one statement is true. (**b**) Statements i and ii are true. (**c**) Statements i and iii are true. (**d**) Statements ii and iii are true. (**e**) All three statements are true.

EP 23.15 When a d^1 octahedral metal complex undergoes a d–d transition, the electron is excited from the _____ to the _____.

(**a**) t_2 set of orbitals, e set of orbitals (**b**) e set of orbitals, t_2 set of orbitals (**c**) t_2 set of orbitals, s orbital (**d**) s orbital, e set of orbitals (**e**) e set of orbitals, s orbital

EP 23.16 An aqueous solution of $[Co(H_2O)_6]^{3+}$ is blue. When the H_2O ligands are replaced with a certain ligand X, the solution turns green. When the H_2O ligands are replaced with a different ligand Y, the solution turns purple. What are the positions of ligands X and Y on the spectrochemical series relative to H_2O?

(**a**) $X < Y < H_2O$

(**b**) $H_2O < Y < X$

(**c**) $H_2O < X < Y$

(**d**) $Y < H_2O < X$

(**e**) $H_2O < X < Y$

EP 23.17 The energy-level diagram for the octahedral complex $[Fe(H_2O)_6]^{3+}$ is shown here. When the H_2O ligands are replaced with CN^- ligands, the low-spin complex $[Fe(CN)_6]^{3-}$ is formed. Which of the following will be different between $[Fe(H_2O)_6]^{3+}$ and $[Fe(CN)_6]^{3-}$?

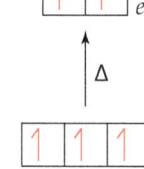

   (i) the color of the complex

  (ii) the number of unpaired electrons

 (iii) the crystal-field splitting energy (Δ)

(a) i only **(b)** i and ii **(c)** i and iii **(d)** ii and iii **(e)** i, ii, and iii

EP 23.18 Which of the following octahedral complex ions will have the fewest number of unpaired electrons?

(a) $[Cr(H_2O)_6]^{3+}$ **(d)** $[RhCl_6]^{3-}$

(b) $[V(H_2O)_6]^{3+}$ **(e)** $[Ni(NH_3)_6]^{2+}$

(c) $[FeF_6]^{3-}$

EP 23.19 How many unpaired electrons do you predict for the tetrahedral $[MnCl_4]^{2-}$ ion? **(a)** 1 **(b)** 2 **(c)** 3 **(d)** 4 **(e)** 5

EP 23.20 The crystal-field energy levels are given here for three different geometries. Which geometry corresponds to diagrams i, ii, and iii? **(a)** octahedral, tetrahedral, square planar **(b)** octahedral, square planar, tetrahedral **(c)** tetrahedral, octahedral, square planar **(d)** tetrahedral, square planar, octahedral **(e)** square planar, tetrahedral, octahedral

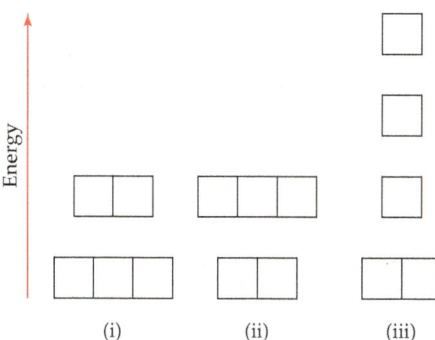

Exercises

Visualizing Concepts

23.1 The three graphs below show the variation in radius, effective nuclear charge, and maximum oxidation state for the transition metals of period 4. In each part below, identify which property is being plotted. [Section 23.1]

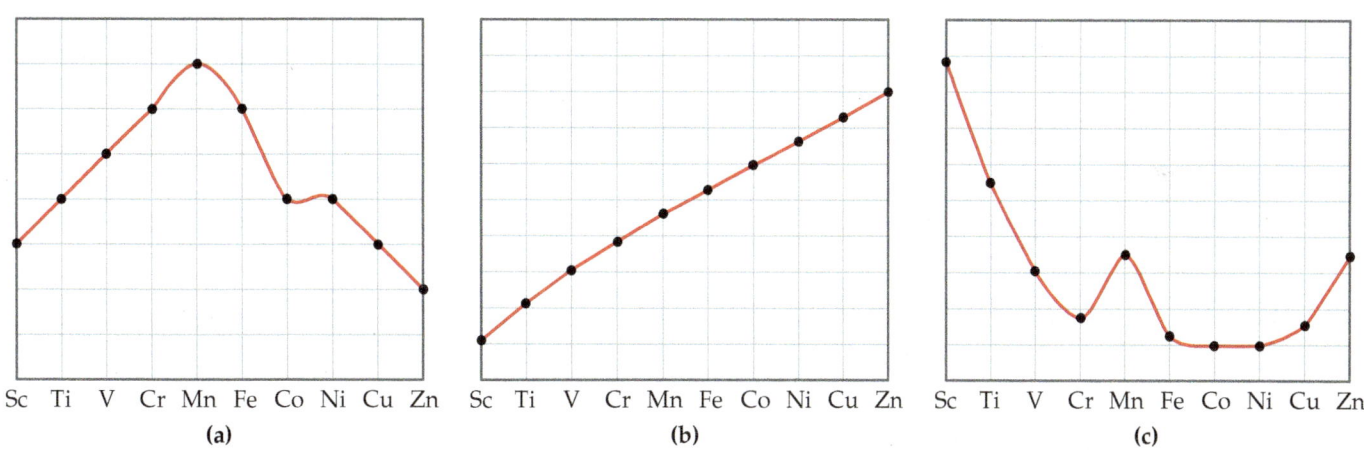

23.2 Draw the structure for $Pt(en)Cl_2$ and use it to answer the following questions: **(a)** What is the coordination number for platinum in this complex? **(b)** What is the coordination geometry? **(c)** What is the oxidation state of the platinum? **(d)** How many unpaired electrons are there? [Sections 23.2 and 23.6]

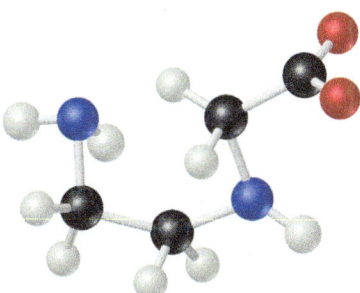

$NH_2CH_2CH_2NHCH_2CO_2^-$

23.3 Draw the Lewis structure for the ligand shown here. **(a)** Which atoms can serve as donor atoms? Classify this ligand as monodentate, bidentate, or polydentate. **(b)** How many of these ligands are needed to fill the coordination sphere in an octahedral complex? [Section 23.2]

23.4 Four-coordinate metals can have either a tetrahedral or a square-planar geometry; both possibilities are shown here for $[PtCl_2(NH_3)_2]$. **(a)** What is the name of this molecule? **(b)** Would the tetrahedral molecule have a geometric isomer? **(c)** Would the tetrahedral molecule be diamagnetic or paramagnetic? **(d)** Would the square-planar molecule have a geometric isomer? **(e)** Would the square-planar molecule be diamagnetic or paramagnetic? **(f)** Would determining the number of geometric isomers help you distinguish between the tetrahedral and square-planar geometries? **(g)** Would measuring the molecule's response to a magnetic field help you distinguish between the two geometries? [Sections 23.4–23.6]

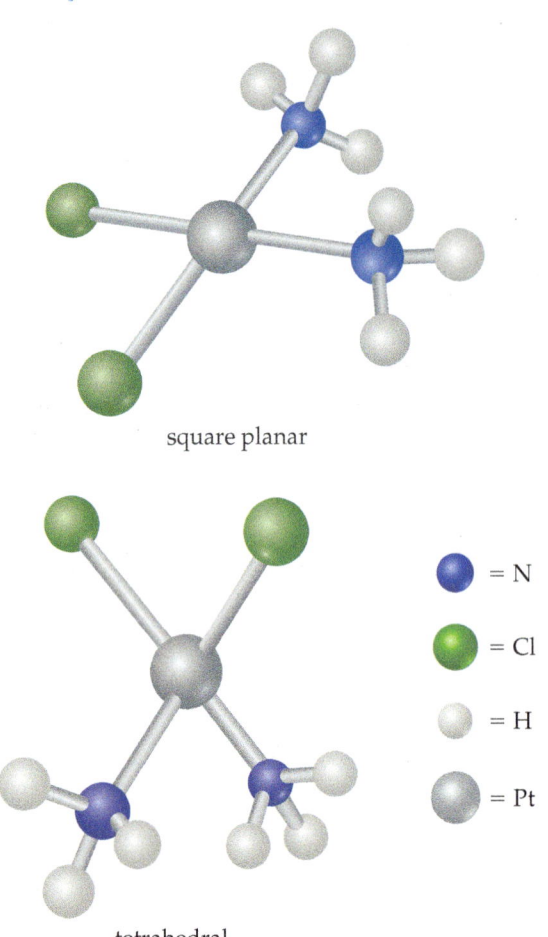

square planar

tetrahedral

= N

= Cl

= H

= Pt

23.5 There are two geometric isomers of octahedral complexes of the type MA_3X_3, where M is a metal and A and X are monodentate ligands. Of the complexes shown here, which are identical to (1) and which are the geometric isomers of (1)? [Section 23.4]

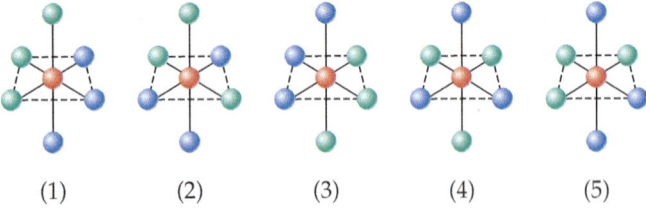

(1)   (2)   (3)   (4)   (5)

23.6 Which of the complexes shown here are chiral? [Section 23.4]

= Cr   = $NH_2CH_2CH_2NH_2$ = Cl   = NH_3

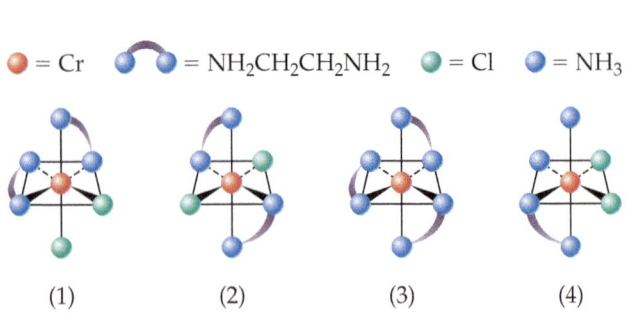

(1)   (2)   (3)   (4)

23.7 The solutions shown here each have an absorption spectrum with a single absorption peak like that shown in Figure 23.26. What color does each solution absorb most strongly? [Section 23.5]

23.8 Which of these crystal-field splitting diagrams represents: (**a**) a weak-field octahedral complex of Fe^{3+}, (**b**) a strong-field octahedral complex of Fe^{3+}, (**c**) a tetrahedral complex of Fe^{3+}, (**d**) a tetrahedral complex of Ni^{2+}? (The diagrams do not indicate the relative magnitudes of Δ.) [Section 23.6]

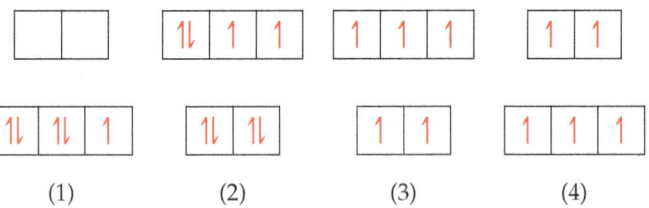

(1)   (2)   (3)   (4)

23.9 In the linear crystal-field shown here, the negative charges are on the z axis. Using Figure 23.28 as a guide, predict which of the following choices most accurately describes the splitting of the d orbitals in a linear crystal field? [Section 23.6]

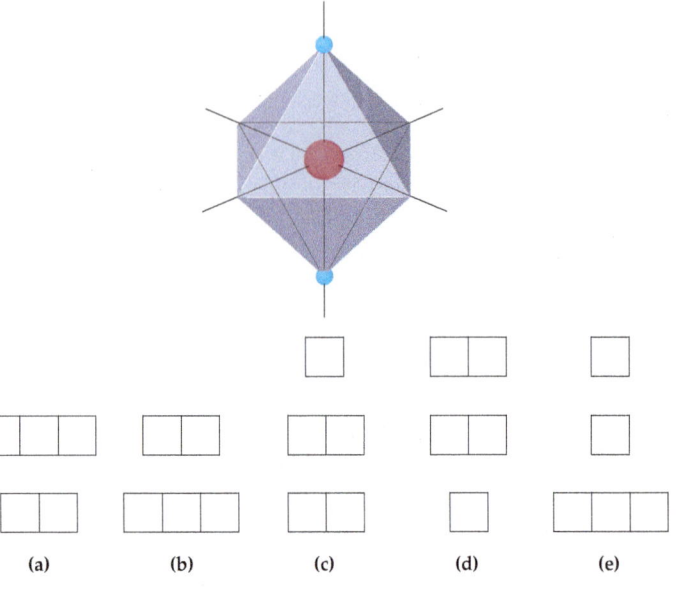

(a)   (b)   (c)   (d)   (e)

23.10 Two Fe(II) complexes are both low spin but have different ligands. A solution of one is green and a solution of the other is red. Which solution is likely to contain the complex that has the stronger-field ligand? [Section 23.6]

The Transition Metals (Section 23.1)

23.11 The lanthanide contraction explains which of the following periodic trends? (**a**) The atomic radii of the transition metals first decrease and then increase when moving horizontally across each period. (**b**) When forming ions, the period 4 transition metals lose their $4s$ electrons before their $3d$ electrons. (**c**) The radii of the period 5 transition metals (Y–Cd) are very similar to the radii of the period 6 transition metals (Lu–Hg).

23.12 Which periodic trend is partially responsible for the observation that the maximum oxidation state of the transition-metal elements peaks near groups 7B and 8B? (**a**) The number of valence electrons reaches a maximum at group 8B. (**b**) The effective nuclear charge increases on moving left across each period. (**c**) The radii of the transition-metal elements reach a minimum for group 8B, and as the size of the atoms decreases it becomes easier to remove electrons.

23.13 For each of the following compounds, determine the electron configuration of the transition-metal ion: (**a**) TiO, (**b**) TiO_2, (**c**) NiO, (**d**) ZnO.

23.14 Among the period 4 transition metals (Sc–Zn), which elements do *not* form ions where there are partially filled $3d$ orbitals?

23.15 Write out the ground-state electron configurations of (**a**) Ti^{3+}, (**b**) Ru^{2+}, (**c**) Au^{3+}, (**d**) Mn^{4+}.

23.16 How many electrons are in the valence d orbitals in these transition-metal ions? (**a**) Co^{3+} (**b**) Cu^+ (**c**) Cd^{2+} (**d**) Os^{3+}

23.17 Which type of substance is attracted by a magnetic field, a diamagnetic substance or a paramagnetic substance?

23.18 Which type of magnetic material cannot be used to make permanent magnets, a ferromagnetic substance, an antiferromagnetic substance, or a ferrimagnetic substance?

23.19 What kind of magnetism is exhibited by the following diagram?

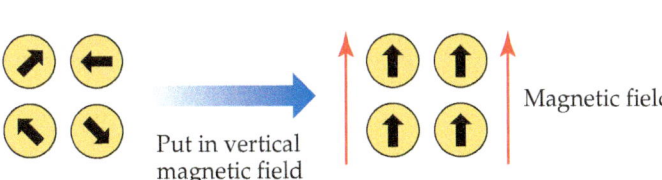

Put in vertical magnetic field

Magnetic field

23.20 The most important oxides of iron are magnetite, Fe_3O_4, and hematite, Fe_2O_3. (**a**) What are the oxidation states of iron in these compounds? (**b**) One of these iron oxides is ferrimagnetic, and the other is antiferromagnetic. Which iron oxide is more likely to be ferrimagnetic? Explain.

Transition-Metal Complexes (Section 23.2)

23.21 (**a**) Using Werner's definition of valence, which property is the same as oxidation number, *primary valence* or *secondary valence*? (**b**) What term do we normally use for the other type of valence? (**c**) Why can NH_3 serve as a ligand but BH_3 cannot?

23.22 Which species are more likely to act as ligands? (**a**) Positively charged ions or negatively charged ions (**b**) Neutral molecules that are polar or those that are nonpolar

23.23 A complex is written as $NiBr_2 \cdot 6\,NH_3$. (**a**) What is the oxidation state of the Ni atom in this complex? (**b**) What is the likely coordination number for the complex? (**c**) If the complex is treated with excess $AgNO_3(aq)$, how many moles of AgBr will precipitate per mole of complex?

23.24 Crystals of hydrated chromium(III) chloride are green, have an empirical formula of $CrCl_3 \cdot 6\,H_2O$, and are highly soluble, (**a**) Write the complex ion that exists in this compound. (**b**) If the complex is treated with excess $AgNO_3(aq)$, how many moles of AgCl will precipitate per mole of $CrCl_3 \cdot 6\,H_2O$ dissolved in solution? (**c**) Crystals of anhydrous chromium(III) chloride are violet and insoluble in aqueous solution. The coordination geometry of chromium in these crystals is octahedral, as is almost always the case for Cr^{3+}. How can this be the case if the ratio of Cr to Cl is not 1:6?

23.25 Indicate the coordination number and the oxidation number of the metal for each of the following complexes:

(**a**) $Na_2[CdCl_4]$

(**b**) $K_2[MoOCl_4]$

(**c**) $[Co(NH_3)_4Cl_2]Cl$

(**d**) $[Ni(CN)_5]^{3-}$

(**e**) $K_3[V(C_2O_4)_3]$

(**f**) $[Zn(en)_2]Br_2$

23.26 Indicate the coordination number and the oxidation number of the metal for each of the following complexes:

(**a**) $K_3[Co(CN)_6]$

(**b**) $Na_2[CdBr_4]$

(**c**) $[Pt(en)_3](ClO_4)_4$

(**d**) $[Co(en)_2(C_2O_4)]^+$

(**e**) $NH_4[Cr(NH_3)_2(NCS)_4]$

(**f**) $[Cu(bipy)_2I]I$

Common Ligands in Coordination Chemistry (Section 23.3)

23.27 For each of the following molecules or polyatomic ions, draw the Lewis structure and indicate if it can act as a monodentate ligand or a bidentate ligand, or is unlikely to act as a ligand at all: (**a**) ethylamine, $CH_3CH_2NH_2$, (**b**) trimethylphosphine, $P(CH_3)_3$, (**c**) carbonate, CO_3^{2-}, (**d**) ethane, C_2H_6.

23.28 For each of the following polydentate ligands, determine (i) the maximum number of coordination sites that the ligand can occupy on a single metal ion, and (ii) the number and type of donor atoms in the ligand: (**a**) ethylenediamine (en), (**b**) bipyridine (bipy), (**c**) the oxalate anion ($C_2O_4^{2-}$), (**d**) the 2– ion of the porphine molecule (Figure 23.13), (**e**) $[EDTA]^{4-}$.

23.29 Polydentate ligands can vary in the number of coordination positions they occupy. In each of the following, identify the polydentate ligand present and indicate the probable number of coordination positions it occupies:

(**a**) $[Co(NH_3)_4(o\text{-phen})]Cl_3$

(**b**) $[Cr(C_2O_4)(H_2O)_4]Br$

(**c**) $[Ca(EDTA)]^{2-}$

(**d**) $[Zn(en)_2](ClO_4)_2$

23.30 Indicate the likely coordination number of the metal in each of the following complexes:

(a) $[Rh(bipy)_3](NO_3)_3$

(b) $Na_3[Co(C_2O_4)_2Cl_2]$

(c) $[Cr(o\text{-phen})_3](CH_3COO)_3$

(d) $Na_2[Co(EDTA)Br]$

23.31 For each of the following pairs, identify the molecule or ion that is more likely to act as a ligand in a metal complex: (a) acetonitrile (CH_3CN) or ammonium (NH_4^+), (b) hydride (H^-) or hydronium (H_3O^+), (c) carbon monoxide (CO) or methane (CH_4).

23.32 Pyridine (C_5H_5N), abbreviated py, is the molecule

(a) Would you expect pyridine to act as a monodentate or bidentate ligand? (b) For the equilibrium reaction

$$[Ru(py)_4(bipy)]^{2+} + 2\,py \rightleftharpoons [Ru(py)_6]^{2+} + bipy$$

would you predict the equilibrium constant to be larger or smaller than one?

23.33 *True* or *false*? The following ligand can act as a bidentate ligand?

23.34 When silver nitrate is reacted with the molecular base *ortho*-phenanthroline, colorless crystals form that contain the transition-metal complex shown here. (a) What is the coordination geometry of silver in this complex? (b) Assuming no oxidation or reduction occurs during the reaction, what is charge of the complex shown here? (c) Do you expect that any nitrate ions will be present in the crystal? (d) Write a formula for the compound that forms in this reaction. (e) Use the accepted nomenclature to write the name of this compound.

Nomenclature and Isomerism in Coordination Chemistry (Section 23.4)

23.35 Write the formula for each of the following compounds, being sure to use brackets to indicate the coordination sphere:

(a) hexaamminechromium(III) nitrate

(b) tetraamminecarbonatocobalt(III) sulfate

(c) dichlorobis(ethylenediamine)platinum(IV) bromide

(d) potassium diaquatetrabromovanadate(III)

(e) bis(ethylenediamine)zinc(II) tetraiodomercurate(II)

23.36 Write the formula for each of the following compounds, being sure to use brackets to indicate the coordination sphere:

(a) tetraaquadibromomanganese(III) perchlorate

(b) bis(bipyridyl)cadmium(II) chloride

(c) potassium tetrabromo(*ortho*-phenanthroline)cobaltate(III)

(d) cesium diamminetetracyanochromate(III)

(e) tris(ethylenediamine)rhodium(III) tris(oxalato)-cobaltate(III)

23.37 Write the names of the following compounds, using the standard nomenclature rules for coordination complexes:

(a) $[Rh(NH_3)_4Cl_2]Cl$

(b) $K_2[TiCl_6]$

(c) $MoOCl_4$

(d) $[Pt(H_2O)_4(C_2O_4)]Br_2$

23.38 Write the names of the following coordination compounds:

(a) $[Cd(en)Cl_2]$

(b) $K_4[Mn(CN)_6]$

(c) $[Cr(NH_3)_5(CO_3)]Cl$

(d) $[Ir(NH_3)_4(H_2O)_2](NO_3)_3$

23.39 Consider the following three complexes:

(Complex 1) $[Co(NH_3)_4Br_2]Cl$

(Complex 2) $[Pd(NH_3)_2(ONO)_2]$

(Complex 3) $[V(en)_2Cl_2]^+$,

Which of the three complexes can have (a) geometric isomers, (b) linkage isomers, (c) optical isomers, (d) coordination-sphere isomers?

23.40 Consider the following three complexes:

(Complex 1) $[Co(NH_3)_5SCN]^{2+}$

(Complex 2) $[Co(NH_3)_3Cl_3]^{2+}$

(Complex 3) $CoClBr \cdot 5NH_3$

Which of the three complexes can have (a) geometric isomers, (b) linkage isomers, (c) optical isomers, (d) coordination-sphere isomers?

23.41 A four-coordinate complex MA_2B_2 is prepared and found to have two different isomers. Is it possible to determine from this information whether the complex is square planar or tetrahedral? If so, which is it?

23.42 Consider an octahedral complex MA_3B_3. How many geometric isomers are expected for this compound? Will any of the isomers be optically active? If so, which ones?

23.43 Determine if each of the following complexes exhibits geometric isomerism. If geometric isomers exist, determine how many there are. (a) tetrahedral $[Cd(H_2O)_2Cl_2]$, (b) square-planar $[IrCl_2(PH_3)_2]^-$, (c) octahedral $[Fe(o\text{-phen})_2Cl_2]^+$.

23.44 Determine if each of the following complexes exhibits geometric isomerism. If geometric isomers exist, determine how many there are. (a) $[Rh(bipy)(o\text{-phen})_2]^{3+}$, (b) $[Co(NH_3)_3(bipy)Br]^{2+}$, (c) square-planar $[Pd(en)(CN)_2]$.

23.45 Determine if each of the following metal complexes is chiral and therefore has an optical isomer: (a) tetrahedral $[Zn(H_2O)_2Cl_2]$, (b) octahedral *trans*-$[Ru(bipy)_2Cl_2]$, (c) octahedral *cis*-$[Ru(bipy)_2Cl_2]$.

23.46 Determine if each of the following metal complexes is chiral and therefore has an optical isomer: (a) square-planar $[Pd(en)(CN)_2]$, (b) octahedral $[Ni(en)(NH_3)_4]^{2+}$, (c) octahedral *cis*-$[V(en)_2ClBr]$.

Color and Magnetism in Coordination Chemistry; Crystal-Field Theory (Sections 23.5 and 23.6)

23.47 (a) If a complex absorbs light at 610 nm, what color would you expect the complex to be? (b) What is the energy in joules of a photon with a wavelength of 610 nm? (c) What is the energy of this absorption in kJ/mol?

23.48 (a) A complex absorbs photons with an energy of 4.51×10^{-19} J. What is the wavelength of these photons? (b) If this is the only place in the visible spectrum where the complex absorbs light, what color would you expect the complex to be?

23.49 Identify each of the following coordination complexes as either diamagnetic or paramagnetic:

(a) $[ZnCl_4]^{2-}$

(b) $[Pd(NH_3)_2Cl_2]$

(c) $[V(H_2O)_6]^{3+}$

(d) $[Ni(en)_3]^{2+}$

23.50 Identify each of the following coordination complexes as either diamagnetic or paramagnetic:

(a) $[Ag(NH_3)_2]^+$

(b) square-planar $[Cu(NH_3)_4]^{2+}$

(c) $[Ru(bipy)_3]^{2+}$

(d) $[CoCl_4]^{2-}$

23.51 If the lobes of a given d orbital point directly at the ligands, will an electron in that orbital have a higher or lower energy than an electron in a d orbital whose lobes *do not* point directly at the ligands?

23.52 The lobes of which d orbitals point directly between the ligands in (a) octahedral geometry, (b) tetrahedral geometry?

23.53 (a) Sketch a diagram that shows the definition of the *crystal-field splitting energy* (Δ) for an octahedral crystal field. (b) What is the relationship between the magnitude of Δ and the energy of the d–d transition for a d^1 complex? (c) Calculate Δ in kJ/mol if a d^1 complex has an absorption maximum at 545 nm.

23.54 As shown in Figure 23.26, the d–d transition of $[Ti(H_2O)_6]^{3+}$ produces an absorption maximum at a wavelength of about 500 nm. (a) What is the magnitude of Δ for $[Ti(H_2O)_6]^{3+}$ in kJ/mol? (b) How would the magnitude of Δ change if the H_2O ligands in $[Ti(H_2O)_6]^{3+}$ were replaced with NH_3 ligands?

23.55 The colors in the copper-containing minerals malachite, which is green and has an empirical formula of $Cu_2CO_3(OH)_2$, and azurite, which is blue and has an empirical formula of $Cu_3(CO_3)_2(OH)_2$, come from a single d–d transition in each compound. The compounds are sometimes found together in nature as shown in Figure 23.1. (a) What is the electron configuration of the copper ion in each mineral? (b) Based

on their colors, in which compound would you predict the crystal-field splitting Δ is larger?

23.56 The color and wavelength of the absorption maximum for $[Ni(H_2O)_6]^{2+}$, $[Ni(NH_3)_6]^{2+}$, and $[Ni(en)_3]^{2+}$ are given in Figure 23.30. The absorption maximum for the $[Ni(bipy)_3]^{2+}$ ion occurs at about 520 nm. (a) What color would you expect for the $[Ni(bipy)_3]^{2+}$ ion? (b) Based on these data, where would you put bipy in the spectrochemical series?

23.57 Give the number of (valence) d electrons associated with the central metal ion in each of the following complexes: (a) $K_3[TiCl_6]$, (b) $Na_3[Co(NO_2)_6]$, (c) $[Ru(en)_3]Br_3$, (d) $[Mo(EDTA)]ClO_4$, (e) $K_3[ReCl_6]$.

23.58 Give the number of (valence) d electrons associated with the central metal ion in each of the following complexes: (a) $K_3[Fe(CN)_6]$, (b) $[Mn(H_2O)_6](NO_3)_2$, (c) $Na[Ag(CN)_2]$, (d) $[Cr(NH_3)_4Br_2]ClO_4$, (e) $[Sr(EDTA)]^{2-}$.

23.59 A classmate says, "A weak-field ligand usually means the complex is high spin." Is your classmate correct? Explain.

23.60 For a given metal ion and set of ligands, is the crystal-field splitting energy larger for a tetrahedral or an octahedral geometry?

23.61 For each of the following metals, write the electronic configuration of the atom and its 2+ ion: (a) Mn, (b) Ru, (c) Rh. Draw the crystal-field energy-level diagram for the d orbitals of an octahedral complex, and show the placement of the d electrons for each 2+ ion, assuming a strong-field complex. How many unpaired electrons are there in each case?

23.62 For each of the following metals, write the electronic configuration of the atom and its 3+ ion: (a) Fe, (b) Mo, (c) Co. Draw the crystal-field energy-level diagram for the d orbitals of an octahedral complex, and show the placement of the d electrons for each 3+ ion, assuming a weak-field complex. How many unpaired electrons are there in each case?

23.63 Draw the crystal-field energy-level diagrams and show the placement of d electrons for each of the following: (a) $[Cr(H_2O)_6]^{2+}$ (four unpaired electrons), (b) $[Mn(H_2O)_6]^{2+}$ (a high-spin complex), (c) $[Ru(NH_3)_5(H_2O)]^{2+}$ (a low-spin complex), (d) $[IrCl_6]^{2-}$ (a low-spin complex), (e) $[Cr(en)_3]^{3+}$, (f) $[NiF_6]^{4-}$.

23.64 Draw the crystal-field energy-level diagrams and show the placement of electrons for the following complexes: (a) $[VCl_6]^{3-}$, (b) $[FeF_6]^{3-}$ (a high-spin complex), (c) $[Ru(bipy)_3]^{3+}$ (a low-spin complex), (d) $[NiCl_4]^{2-}$ (tetrahedral), (e) $[PtBr_6]^{2-}$, (f) $[Ti(en)_3]^{2+}$.

23.65 The complex $[Mn(NH_3)_6]^{2+}$ contains five unpaired electrons. Sketch the energy-level diagram for the d orbitals, and indicate the placement of electrons for this complex ion. Is the ion a high-spin or a low-spin complex?

23.66 The ion $[Fe(CN)_6]^{3-}$ has one unpaired electron, whereas $[Fe(NCS)_6]^{3-}$ has five unpaired electrons. From these results, what can you conclude about whether each complex is high spin or low spin? What can you say about the placement of NCS^- in the spectrochemical series?

Additional Exercises

23.67 The *Curie temperature* is the temperature at which a ferromagnetic solid switches from ferromagnetic to paramagnetic. For nickel, the Curie temperature is 354 °C. Knowing this, you tie a string to two paper clips made of nickel and hold the paper clips near a permanent magnet. The magnet attracts the paper clips, as shown in the photograph on the left. Now you heat one of the paper clips with a butane lighter, and the clip drops (right photograph). Explain what happened.

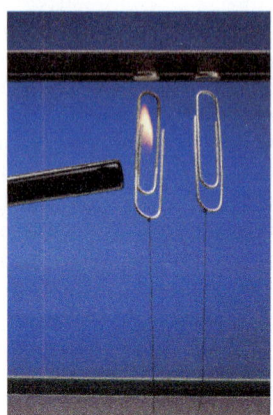

23.68 Explain why the transition metals in periods 5 and 6 have nearly identical radii in each group.

23.69 Based on the molar conductance values listed here for the series of platinum(IV) complexes, write the formula for each complex so as to show which ligands are in the coordination sphere of the metal. By way of example, the molar conductances of 0.050 M NaCl and BaCl$_2$ are 107 ohm^{-1} and 197 ohm^{-1}, respectively.

Complex	Molar Conductance (ohm^{-1})* of 0.050 M Solution
Pt(NH$_3$)$_6$Cl$_4$	523
Pt(NH$_3$)$_4$Cl$_4$	228
Pt(NH$_3$)$_3$Cl$_4$	97
Pt(NH$_3$)$_2$Cl$_4$	0
KPt(NH$_3$)Cl$_5$	108

*The ohm is a unit of resistance; conductance is the inverse of resistance.

23.70 **(a)** A compound with formula RuCl$_3 \cdot 5$ H$_2$O is dissolved in water, forming a solution that is approximately the same color as the solid. Immediately after forming the solution, the addition of excess AgNO$_3$(aq) forms 2 mol of solid AgCl per mole of complex. Write the formula for the compound, showing which ligands are likely to be present in the coordination sphere. **(b)** After a solution of RuCl$_3 \cdot 5$ H$_2$O has stood for about a year, addition of AgNO$_3$(aq) precipitates 3 mol of AgCl per mole of complex. What has happened in the ensuing time?

23.71 Sketch the structure of the complex in each of the following compounds and give the full compound name:

(a) *cis*-[Co(NH$_3$)$_4$(H$_2$O)$_2$](NO$_3$)$_2$

(b) Na$_2$[Ru(H$_2$O)Cl$_5$]

(c) *trans*-NH$_4$[Co(C$_2$O$_4$)$_2$(H$_2$O)$_2$]

(d) *cis*-[Ru(en)$_2$Cl$_2$]

23.72 Which complex ions in Exercise 23.71 have optical isomers?

23.73 The molecule *dimethylphosphinoethane* [(CH$_3$)$_2$PCH$_2$CH$_2$ P(CH$_3$)$_2$, which is abbreviated dmpe] is used as a ligand for some complexes that serve as catalysts. An octahedral complex that contains this ligand is Mo(CO)$_4$(dmpe). **(a)** Draw the Lewis structure for dmpe. **(b)** Is dmpe a monodentate, bidentate, or polydentate ligand? **(c)** What is the oxidation state of Mo in Mo(CO)$_4$(dmpe)? **(d)** Is Mo(CO)$_4$(dmpe) a chiral molecule?

23.74 The square-planar complex [Pt(en)Cl$_2$] only forms in one of two possible geometric isomers. Which isomer is not observed: cis or trans?

23.75 The acetylacetone ion forms very stable complexes with many metallic ions. It acts as a bidentate ligand, coordinating to the metal at two adjacent positions. Suppose that one of the CH$_3$ groups of the ligand is replaced by a CF$_3$ group, as shown here:

Trifluoromethyl acetylacetonate (tfac)

$$\left[CF_3-C\overset{\overset{\displaystyle H}{|}}{\underset{\underset{\displaystyle :\ddot{O}:}{\|}}{C}}C-CH_3 \right]^{-}$$

Sketch all possible isomers for the complex with three tfac ligands on cobalt(III). (You can use the symbol ◠◡ to represent the ligand.)

23.76 Which transition-metal atom is present in each of the following biologically important molecules: **(a)** hemoglobin, **(b)** chlorophylls, **(c)** siderophores, **(d)** hemocyanine.

23.77 Carbon monoxide, CO, is an important ligand in coordination chemistry. When CO is reacted with nickel metal, the product is [Ni(CO)$_4$], which is a toxic, pale yellow liquid. **(a)** What is the oxidation number for nickel in this compound? **(b)** Given that [Ni(CO)$_4$] is a diamagnetic molecule with a tetrahedral geometry, what is the electron configuration of nickel in this compound? **(c)** Write the name for [Ni(CO)$_4$] using the nomenclature rules for coordination compounds.

23.78 Some metal complexes have a coordination number of 5. One such complex is Fe(CO)$_5$, which adopts a *trigonal-bipyramidal* geometry (see Figure 9.8). **(a)** Write the name for Fe(CO)$_5$, using the nomenclature rules for coordination compounds. **(b)** What is the oxidation state of Fe in this compound? **(c)** Suppose one of the CO ligands is replaced with a CN$^-$ ligand, forming [Fe(CO)$_4$(CN)]$^-$. How many geometric isomers would you predict this complex could have?

23.79 Which of the following objects is chiral? **(a)** a left shoe **(b)** a slice of bread **(c)** a wood screw **(d)** a molecular model of Zn(en)Cl$_2$ **(e)** a typical golf club

23.80 The complexes [V(H$_2$O)$_6$]$^{3+}$ and [VF$_6$]$^{3-}$ are both known. **(a)** Draw the *d*-orbital energy-level diagram for V(III) octahedral complexes. **(b)** Which of the following gives rise to the colors of these complexes? **(i)** Excitation of an electron from the *e* orbitals to the t_2 orbitals. **(ii)** Excitation of an electron from the t_2 orbitals to the *e* orbitals. **(iii)** Excitation of a 4s electron to the t_2 orbitals. **(c)** Which of the two complexes would you expect to absorb light of higher energy?

23.81 One of the more famous species in coordination chemistry is the Creutz–Taube complex, discovered in 1969:

$$\left[(NH_3)_5RuN\bigcirc NRu(NH_3)_5 \right]^{5+}$$

It is named for Carol Creutz and 1983 Chemistry Nobel Laureate Henry Taube, the two scientists who discovered, it and initially studied its properties. The central ligand is pyrazine, a planar six-membered ring with nitrogens at opposite sides. (**a**) How can you account for the fact that the complex, which has only neutral ligands, has an odd overall charge? (**b**) The metal is in a low-spin configuration in both cases. Assuming octahedral coordination, draw the *d*-orbital energy-level diagram for each metal. (**c**) In many experiments, the two metal ions appear to be in exactly equivalent states. Can you think of a reason that this might appear to be so, recognizing that electrons move very rapidly compared to nuclei?

23.82 Solutions of $[Co(NH_3)_6]^{2+}$, $[Co(H_2O)_6]^{2+}$ (both octahedral), and $[CoCl_4]^{2-}$ (tetrahedral) are colored. One is pink, one is blue, and one is yellow. Based on the spectrochemical series and remembering that the energy splitting in tetrahedral complexes is normally much less than that in octahedral ones, assign a color to each complex.

23.83 Oxyhemoglobin, with an O_2 bound to iron, is a low-spin Fe(II) complex; deoxyhemoglobin, without the O_2 molecule, is a high-spin complex. (**a**) Assuming that the coordination environment about the metal is octahedral, how many unpaired electrons are centered on the metal ion in each case? (**b**) What ligand is coordinated to the iron in place of O_2 in deoxyhemoglobin? (**c**) Explain in a general way why the two forms of hemoglobin have different colors (hemoglobin is red, whereas deoxyhemoglobin has a bluish cast). (**d**) A 15-minute exposure to air containing 400 ppm of CO causes about 10% of the hemoglobin in the blood to be converted into the carbon monoxide complex called carboxyhemoglobin. What does this suggest about the relative equilibrium constants for binding of carbon monoxide and O_2 to hemoglobin? (**e**) CO is a strong-field ligand. What color might you expect carboxyhemoglobin to be?

23.84 Consider the tetrahedral anions VO_4^{3-} (orthovanadate ion), CrO_4^{2-} (chromate ion), and MnO_4^- (permanganate ion). (**a**) These anions are *isoelectronic*. What does this statement mean? (**b**) Would you expect these anions to exhibit *d–d* transitions? Explain. (**c**) As mentioned in the "A Closer Look" box on "Charge-Transfer Color" (Section 23.6), the violet color of MnO_4^- is due to a *ligand-to-metal charge transfer* (LMCT) transition. What is meant by this term? (**d**) The LMCT transition in MnO_4^- occurs at a wavelength of 565 nm. The CrO_4^{2-} ion is yellow. Is the wavelength of the LMCT transition for chromate larger or smaller than that for MnO_4^-? Explain. (**e**) The VO_4^{3-} ion is colorless. Do you expect the light absorbed by the LMCT to fall in the UV or the IR region of the electromagnetic spectrum? Explain your reasoning.

23.85 Given the colors observed for VO_4^{3-} (orthovanadate ion), CrO_4^{2-} (chromate ion), and MnO_4^- (permanganate ion) (see Exercise 23.84), what can you say about how the energy separation between the ligand orbitals and the empty *d* orbitals changes as a function of the oxidation state of the transition metal at the center of the tetrahedral anion?

23.86 The red color of ruby is due to the presence of Cr(III) ions at octahedral sites in the close-packed oxide lattice of Al_2O_3. Draw the crystal-field splitting diagram for Cr(III) in this environment. Suppose that the ruby crystal is subjected to high pressure. What do you predict for the variation in the wavelength of absorption of the ruby as a function of pressure? Explain.

23.87 In 2001, chemists at SUNY-Stony Brook succeeded in synthesizing the complex *trans*-$[Fe(CN)_4(CO)_2]^{2-}$, which could be a model of complexes that may have played a role in the

origin of life. (**a**) Sketch the structure of the complex. (**b**) The complex is isolated as a sodium salt. Write the complete name of this salt. (**c**) What is the oxidation state of Fe in this complex? How many *d* electrons are associated with the Fe in this complex? (**d**) Would you expect this complex to be high spin or low spin? Explain.

23.88 When Alfred Werner was developing the field of coordination chemistry, it was argued by some that the optical activity he observed in the chiral complexes he had prepared was due to the presence of carbon atoms in the molecule. To disprove this argument, Werner synthesized a chiral complex of cobalt that had no carbon atoms in it, and he was able to resolve it into its enantiomers. Design a cobalt(III) complex that would be chiral if it could be synthesized and that contains no carbon atoms. (It may not be possible to synthesize the complex you design, but we will not worry about that for now.)

23.89 Generally speaking, for a given metal and ligand, the stability of a coordination compound is greater for the metal in the +3 rather than the +2 oxidation state (for metals that form stable +3 ions in the first place). Suggest an explanation, keeping in mind the Lewis acid–base nature of the metal–ligand bond.

23.90 Many trace metal ions exist in the blood complexed with amino acids or small peptides. The anion of the amino acid glycine (gly),

$$H_2NCH_2\overset{\overset{\displaystyle O}{\|}}{C}-O^-$$

can act as a bidentate ligand, coordinating to the metal through nitrogen and oxygen atoms. How many isomers are possible for (**a**) $[Zn(gly)_2]$ (tetrahedral), (**b**) $[Pt(gly)_2]$ (square planar), (**c**) $[Co(gly)_3]$ (octahedral)? Sketch all possible isomers. Use the symbol ⌒ to represent the ligand.

23.91 The coordination complex $[Cr(CO)_6]$ forms colorless, diamagnetic crystals that melt at 90 °C. (**a**) What is the oxidation number of chromium in this compound? (**b**) Given that $[Cr(CO)_6]$ is diamagnetic, what is the electron configuration of chromium in this compound? (**c**) Given that $[Cr(CO)_6]$ is colorless, would you expect CO to be a weak-field or strong-field ligand? (**d**) Write the name for $[Cr(CO)_6]$ using the nomenclature rules for coordination compounds.

23.92 Metallic elements are essential components of many important enzymes operating within our bodies. *Carbonic anhydrase*, which contains Zn^{2+} in its active site, is responsible for rapidly interconverting dissolved CO_2 and bicarbonate ion, HCO_3^-. The zinc in carbonic anhydrase is tetrahedrally coordinated by three neutral nitrogen-containing groups and a water molecule. The coordinated water molecule has a pK_a of 7.5, which is crucial for the enzyme's activity. (**a**) Draw the active site geometry for the Zn(II) center in carbonic anhydrase, just writing "N" for the three neutral nitrogen ligands from the protein. (**b**) Compare the pK_a of carbonic anhydrase's active site with that of pure water; which species is more acidic? (**c**) When the coordinated water to the Zn(II) center in carbonic anhydrase is deprotonated, what ligands are bound to the Zn(II) center? Assume the three nitrogen ligands are unaffected. (**d**) The pK_a of $[Zn(H_2O)_6]^{2+}$ is 10. Suggest an explanation for the difference between this pK_a and that of carbonic anhydrase. (**e**) Would you expect carbonic anhydrase to have a deep color, like hemoglobin and other metal-ion-containing proteins do? Explain.

23.93 Two different compounds have the formulation $CoBr(SO_4) \cdot 5\,NH_3$. Compound A is dark violet, whereas compound B is red-violet. When compound A is treated with $AgNO_3(aq)$, no reaction occurs, whereas compound B reacts with $AgNO_3(aq)$ to form a white precipitate. When compound A is treated with $BaCl_2(aq)$, a white precipitate is formed, whereas compound B has no reaction with $BaCl_2(aq)$. (**a**) Is Co in the same oxidation state in these complexes? (**b**) What substance precipitates when compound B reacts with $AgNO_3(aq)$? (**c**) What substance precipitates when compound A reacts with $BaCl_2(aq)$? (**d**) Are compounds A and B isomers of one another? If so, which category from Figure 23.19 best describes the isomerism observed for these complexes? (**e**) Would compounds A and B be expected to be strong electrolytes, weak electrolytes, or nonelectrolytes?

23.94 The value of Δ for the $[CrF_6]^{3-}$ complex is 182 kJ/mol. Calculate the expected wavelength of the absorption corresponding to promotion of an electron from the lower-energy to the higher-energy d-orbital set in this complex. Should the complex absorb in the visible range?

Design an Experiment

Following a procedure found in a scientific paper, you go into the lab and attempt to prepare crystals of dichlorobis(ethylenediamine) cobalt(III) chloride. The paper states that this compound can be made by reacting $CoCl_2 \cdot 6\,H_2O$, an excess of ethylenediamine, O_2 from the air (which acts as an oxidizing agent), water, and concentrated hydrochloric acid. At the end of the reaction, you filter off the solution and are left with a green, crystalline product. (**a**) What experiment(s) could you perform to confirm that you have prepared $[CoCl_2(en)_2]Cl$ and not $[Co(en)_3]Cl_3$? (**b**) How could you verify that cobalt was present as Co^{3+} and determine the spin state of the cobalt complex in your product? (**c**) How many geometric isomers exist for $[CoCl_2(en)_2]Cl$? How could you determine if your product contains a single geometric isomer or a mixture of geometric isomers? (**d**) If the product does contain a single geometric isomer, how would you determine which one was present? (*Hint:* You may find the information in Table 23.3 to be helpful.)

24

THE CHEMISTRY OF LIFE: ORGANIC AND BIOLOGICAL CHEMISTRY

▲ **SPICE MARKET.** The many colors and tastes of these spices come from the organic compounds they contain.

Chemical substances can influence our health and behavior. Aspirin, also known as acetylsalicylic acid, relieves aches and pains. The caffeine that is in coffee and tea, and the ethanol that is in wine and beer, are well-known molecules that at the proper doses give us pleasure. Spices, as shown in the opening photograph, contain a host of molecules such as capsaicin (in chili peppers), cinnamaldehyde (in cinnamon), and vanillin (in the vanilla bean). Other plant-derived compounds include quinine, for treating malaria, and taxol, for treating many kinds of cancer.

Understanding how these molecules exert their effects and developing new molecules that can target disease and pain are important components of the modern chemical enterprise. This chapter is about the molecules, composed mainly of carbon, hydrogen, oxygen, and nitrogen, that bridge chemistry and biology.

More than 16 million carbon-containing compounds are known. The study of compounds whose molecules contain carbon constitutes the branch of chemistry known as **organic chemistry**. This term arose from the eighteenth-century belief that organic compounds could be formed only by living (that is, organic) systems. This idea was disproved by the German chemist Friedrich Wöhler (1800–1882). In 1828, he synthesized urea (H_2NCONH_2), an organic substance found in the urine of mammals, by heating ammonium cyanate (NH_4OCN), an inorganic (nonliving) substance. In chemistry, "organic compounds" are typically considered to be those that contain carbon-hydrogen bonds. Therefore, compounds such as ammonia or carbon dioxide are generally considered "inorganic."

Although not all organic molecules are formed by living systems, the chemistry of plants, animals, and other living organisms is based on organic substances. The study of the chemistry of living species is called *biological chemistry*, *chemical biology*, or **biochemistry**. The molecules that play important roles in biology share some common characteristics. Many are large because organisms build biomolecules from smaller, simpler substances readily available in the biosphere. The synthesis of large molecules requires energy because most of the reactions are endothermic. The ultimate source of this energy is the Sun. Animals have essentially no capacity for using solar energy directly and so depend on plant photosynthesis to supply the bulk of their energy needs. (Section 23.3) In addition to requiring large amounts of energy, living organisms are highly organized. In thermodynamic terms, this high degree of organization means that the entropy of living systems is much lower than that of the raw materials from which the systems formed. Thus, living systems must continuously work against the spontaneous tendency toward increased entropy. (Section 19.3)

In the "Chemistry and Life" essays that appear throughout this text, we have introduced some important biochemical applications of fundamental chemical ideas. In the second half of this chapter we delve a little deeper into the composition, structure, and properties of biomolecules.

24.1 | General Characteristics of Organic Molecules

Learning Objectives

When you finish Section 24.1, you should be able to:

▶ Predict the molecular geometry, orbital hybridization, and pattern of bonding in organic molecules from their structural formulas.

▶ Predict the dipole moments and solubilities of organic compounds from their molecular structure.

What is it about carbon that leads to the tremendous diversity in its compounds and allows it to play such crucial roles in biology and society? Let's start by considering some general features of organic molecules and, as we do, review principles we learned in previous chapters.

The Structures of Organic Molecules

Because carbon has four valence electrons ($[He]2s^2 2p^2$), it forms four bonds in virtually all its compounds. When all four bonds are single bonds, the electron pairs are disposed in a tetrahedral arrangement. (Section 9.2) In the hybridization model, the carbon $2s$ and $2p$ orbitals are then sp^3 hybridized. (Section 9.5) When there is one double bond, the geometry is trigonal planar and the orbitals at carbon are sp^2 hybridized. With a triple bond, the geometry is linear and the carbon is sp hybridized. Examples are shown in **Figure 24.1**.

Almost every organic molecule contains C—H bonds. Because the valence shell of H can hold only two electrons, hydrogen forms only one covalent bond. As a result, hydrogen atoms are always located on the *surface* of organic molecules, whereas the C—C bonds form the *backbone*, or *skeleton*, of the molecule, as in the propane molecule:

$$
\begin{array}{ccccccc}
 & H & & H & & H & \\
 & | & & | & & | & \\
H\!-\!\!\!&C&\!\!\!-\!\!\!&C&\!\!\!-\!\!\!&C&\!\!\!-\!H \\
 & | & & | & & | & \\
 & H & & H & & H &
\end{array}
$$

The Stability of Organic Compounds

Carbon forms strong bonds with a variety of elements, especially H, O, N, and the halogens. (Sections 5.8 and 8.8) Carbon also has an exceptional ability to bond to itself, forming a variety of molecules made up of chains or rings of carbon atoms. Most reactions with low or moderate activation energy (Section 14.4) begin when a region of high-electron density in

Go Figure What is the geometry around the bottom carbon atom in acetonitrile?

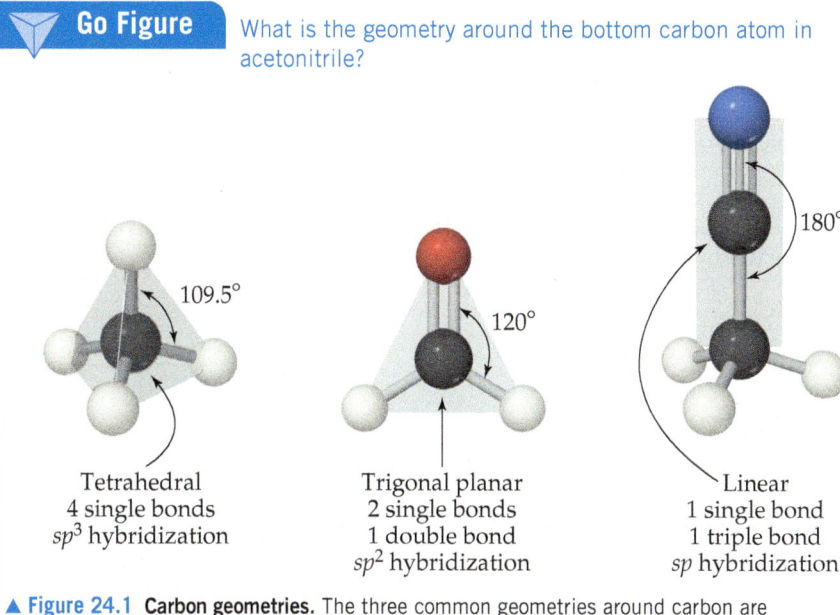

Tetrahedral
4 single bonds
sp^3 hybridization

Trigonal planar
2 single bonds
1 double bond
sp^2 hybridization

Linear
1 single bond
1 triple bond
sp hybridization

▲ **Figure 24.1 Carbon geometries.** The three common geometries around carbon are tetrahedral as in methane (CH_4), trigonal planar as in formaldehyde (CH_2O), and linear as in acetonitrile (CH_3CN). Notice that in all cases each carbon atom forms four bonds.

Go Figure

How would replacing OH groups on ascorbic acid with CH_3 groups affect the substance's solubility in (**a**) polar solvents and (**b**) nonpolar solvents?

one molecule encounters a region of low electron density in another molecule. The regions of high-electron density may be due to the presence of a multiple bond or to the more electronegative atom in a polar bond. Because of their strength (the C—C single bond enthalpy is 348 kJ/mol, and the C—H bond enthalpy is 413 kJ/mol (Tables 5.4 and 8.3) and lack of polarity, both C—C single bonds and C—H bonds are relatively unreactive. To better understand the implications of these facts, consider ethanol:

$$H-\underset{\underset{H}{|}}{\overset{\overset{H}{|}}{C}}-\underset{\underset{H}{|}}{\overset{\overset{H}{|}}{C}}-O-H$$

The differences in the electronegativity values of C (2.5), O (3.5) and H (2.1), indicate that the C—O and O—H bonds are quite polar. Thus, many reactions of ethanol involve these bonds while the hydrocarbon portion of the molecule remains intact. A group of atoms such as the C—O—H group, which determines how an organic molecule reacts (in other words, how the molecule *functions*), is called a **functional group**. The functional group is the center of reactivity in an organic molecule.

Solubility and Acid–Base Properties of Organic Compounds

In most organic substances, the nonpolar carbon–carbon and carbon–hydrogen bonds are most prevalent. For this reason, the overall polarity of organic molecules is often low, which makes them generally soluble in nonpolar solvents and not very soluble in water. (Section 13.3) Organic molecules that are soluble in polar solvents are those that have polar groups on the molecule surface, such as glucose and ascorbic acid (**Figure 24.2**). Organic molecules that have a long, nonpolar part bonded to a polar, ionic part, such as the stearate ion shown in Figure 24.2, function as *surfactants* and are used in soaps and detergents. (Section 13.6) The nonpolar part of the molecule extends into a nonpolar medium such as grease or oil, whereas the polar part extends into a polar medium such as water.

Many organic substances contain acidic or basic groups. The most important acidic organic substances are the carboxylic acids, which bear the functional group —COOH. (Sections 4.3 and 16.10) The most important basic organic substances are amines, which bear the —NH_2, —NHR, or —NR_2 groups, where R is an organic group made up of carbon and hydrogen atoms. (Section 16.7)

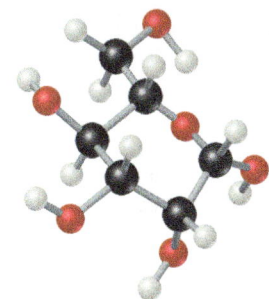

Glucose ($C_6H_{12}O_6$)

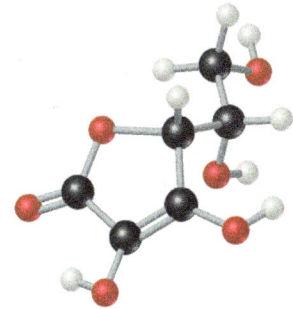

Ascorbic acid ($HC_6H_7O_6$)

Stearate ($C_{17}H_{35}COO^-$)

▲ **Figure 24.2** Organic molecules that are soluble in polar solvents.

Self-Assessment Exercises

SAE 24.1 In organic compounds we expect each carbon to form _____ bonds and each hydrogen to form _____ bond(s). (**a**) 4, 2 (**b**) 4, 1 (**c**) 6, 1 (**d**) 4, 4 (**e**) The number of bonds each element forms varies from one organic compound to another and cannot be generalized.

SAE 24.2 What is the geometry around each of the carbon atoms in ethylene, $H_2C=CH_2$? (**a**) tetrahedral (**b**) trigonal planar (**c**) linear (**d**) see-saw (**e**) trigonal pyramidal

SAE 24.3 In acetonitrile, CH_3CN, the $H-C-H$ bond angle is approximately _____ and the $C-C-N$ bond angle is approximately _____. (**a**) 109.5°, 180° (**b**) 180°, 120° (**c**) 109.5°, 120° (**d**) 120°, 180° (**2**) 120°, 120°

SAE 24.4 Which of the following molecules will be the least soluble in water? (**a**) CH_3NH_2 (**b**) $CH_3CH_2CH_2CH=CH_2$ (**c**) $HOCH_2CH_2OH$ (**d**) CH_3COOH (**e**) CH_2Cl_2

24.2 | Introduction to Hydrocarbons

Because organic compounds are so numerous, it is convenient to organize them into families that have structural similarities. The simplest class of organic compounds, *hydrocarbons*, are composed of only carbon and hydrogen. Because carbon is the only element capable of forming stable, extended chains of atoms bonded through single, double, or triple bonds, the number of different substances that can be made from these two elements is surprisingly large.

Hydrocarbons can be divided into four types, depending on the kinds of carbon–carbon bonds in their molecules.

- **Alkanes** contain only single $C-C$ bonds.
- **Alkenes**, also known as olefins, contain at least one $C=C$ double bond.
- **Alkynes** contain at least one $C\equiv C$ triple bond.
- **Aromatic hydrocarbons** have carbon atoms that are connected in a planar ring structure, joined by both σ and delocalized π bonds between carbon atoms. (Section 8.6)

An example of each type is shown in Table 24.1.

Learning Objectives

When you finish Section 24.2, you should be able to:

▶ Differentiate alkanes, alkenes, alkynes, and aromatic hydrocarbons from one another.

▶ Differentiate straight-chain alkanes, branched-chain alkanes, and cycloalkanes from one another.

▶ Interconvert between the structural formula of an alkane and its name.

▶ Identify alkanes that are structural isomers of each other.

▶ Write balanced chemical equations for the combustion of alkanes.

TABLE 24.1   The Four Hydrocarbon Types with Molecular Examples

Type		Example
Alkane	Ethane	CH_3CH_3
Alkene	Ethene (ethylene)	$CH_2=CH_2$
Alkyne	Ethyne (acetylene)	$CH\equiv CH$
Aromatic	Benzene	C_6H_6

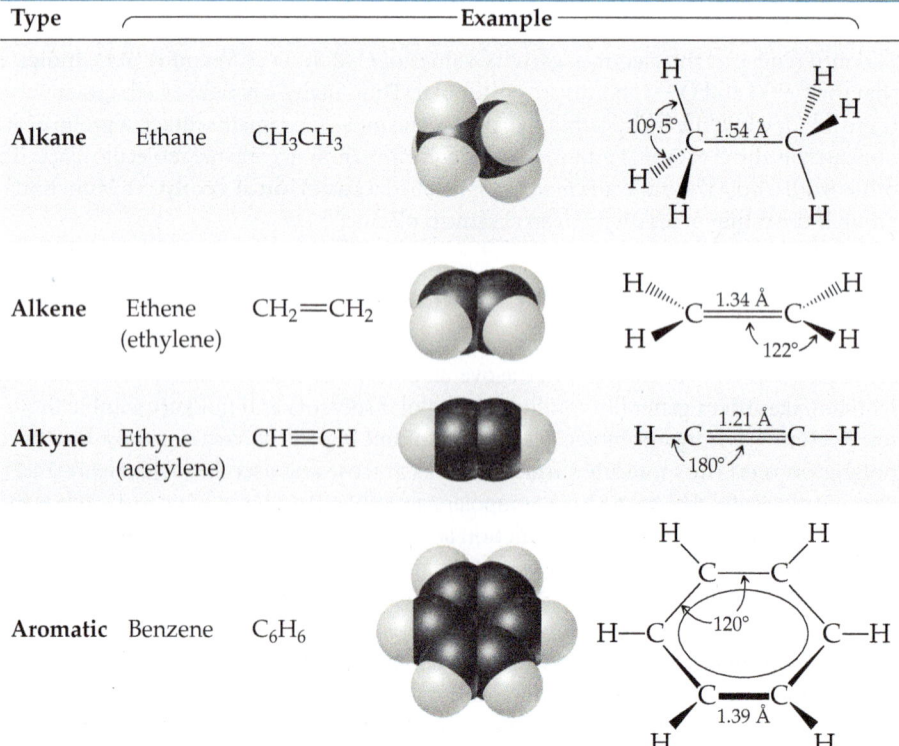

Each type of hydrocarbon exhibits different chemical behaviors, as we demonstrate shortly. The physical properties of all four types, however, are similar in many ways. Because hydrocarbon molecules are relatively nonpolar, they are almost completely insoluble in water but dissolve readily in nonpolar solvents. Their melting points and boiling points are determined by dispersion forces. (Section 11.2) As a result, hydrocarbons of very low molecular weight, such as C_2H_6 (bp = $-89\,°C$), are gases at room temperature; those of moderate molecular weight, such as C_6H_{14} (bp = $69\,°C$), are liquids; and those of high molecular weight, such as $C_{22}H_{46}$ (mp = $44\,°C$), are solids.

The molecular formulas, names, and boiling points of simple alkanes were discussed in Chapter 2, as was the convention for naming them. To aid in understanding the material that follows, you should review Section 2.9.

Several alkanes may be familiar to you because they are so widely used. Methane (CH_4) is a major component of natural gas. Ethane (C_2H_6) is also a component of natural gas and is a feedstock for making polyethylene. (Section 12.6) Propane (C_3H_8) is the major component of bottled gas used for home heating and cooking in areas where natural gas is unavailable. Butane (C_4H_{10}) is used in disposable lighters and in fuel canisters for gas camping stoves and lanterns. Alkanes with 5 to 12 carbon atoms per molecule are components of gasoline.

The formulas for the alkanes are written in a notation called *condensed structural formulas*. This notation reveals the way in which atoms are bonded to one another but does not require drawing in all the bonds. For example, the structural formula and the condensed structural formulas for butane (C_4H_{10}) are

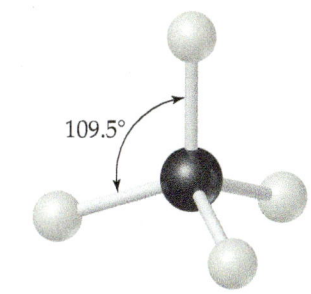

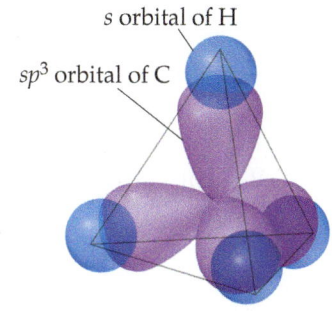

▲ **Figure 24.3** **Bonds about carbon in methane.** This tetrahedral molecular geometry is found around all carbon atoms in alkanes.

$$\begin{array}{cccc} H & H & H & H \\ | & | & | & | \\ H-C-C-C-C-H \\ | & | & | & | \\ H & H & H & H \end{array}$$

$$H_3C-CH_2-CH_2-CH_3$$

or

$$CH_3CH_2CH_2CH_3$$

Structures of Alkanes

According to the VSEPR model, the molecular geometry about each carbon atom in an alkane is tetrahedral. (Section 9.2) The bonding may be described as involving sp^3-hybridized orbitals on the carbon, as pictured in **Figure 24.3** for methane. (Section 9.5)

Rotation about a carbon–carbon single bond is relatively easy and occurs rapidly at room temperature. To visualize such rotation, imagine grasping either methyl group of the propane molecule in **Figure 24.4** and rotating the group relative to the rest of the molecule. Because motion of this sort occurs rapidly in alkanes, a long-chain alkane molecule is constantly undergoing motions that cause it to change its overall shape.

Alkanes with more than three carbon atoms exhibit structural isomerism. (Sections 2.9 and 23.4) As an example, the three structural isomers of pentane are shown in **Figure 24.5**. Those molecules where the carbon atoms connected in a continuous chain, such as *n*-pentane (the letter *n* indicates this is the "normal" structure), are referred to as *straight-chain* or *linear hydrocarbons*. All other alkanes contain one or more side chains that come off the linear carbon backbone of the molecule and are referred to as *branched-chain hydrocarbons*, 2-methylbutane (isopentane) and 2,2-dimethylpropane (neopentane) are examples of branched-chain hydrocarbons. Both straight- and branched-chain alkanes have the same molecular formula, C_nH_{2n+2}, where *n* is the number of carbon atoms in the molecule. The structural isomers of a given alkane differ from one another in physical properties, as can be seen from the melting and boiling points of the pentane isomers. The number of possible structural isomers increases rapidly with the number of carbon atoms in the alkane. There are 18 isomers with the molecular formula C_8H_{18}, for example, and 75 with the molecular formula $C_{10}H_{22}$.

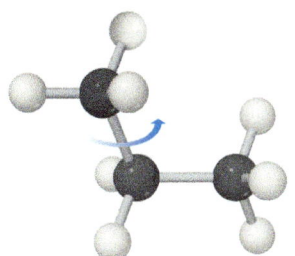

before rotation

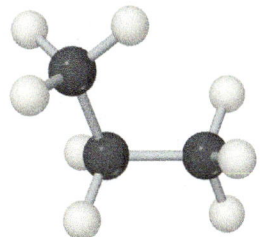

after rotation

▲ **Figure 24.4** **Rotation about a C—C bond occurs easily and rapidly in all alkanes.**

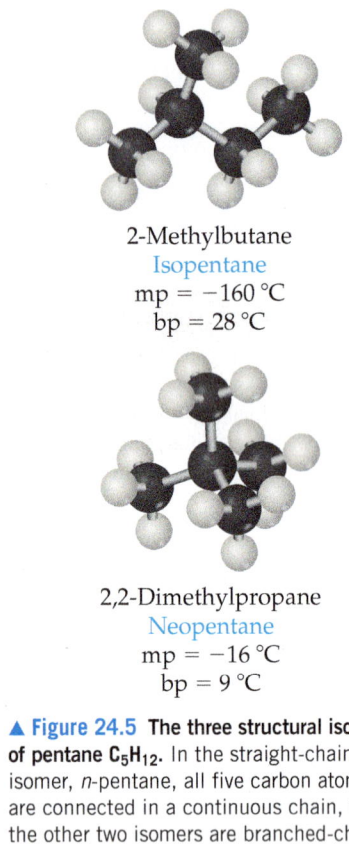

Pentane
n-Pentane
mp = −130 °C
bp = 36 °C

2-Methylbutane
Isopentane
mp = −160 °C
bp = 28 °C

2,2-Dimethylpropane
Neopentane
mp = −16 °C
bp = 9 °C

▲ **Figure 24.5** **The three structural isomers of pentane C₅H₁₂.** In the straight-chain isomer, *n*-pentane, all five carbon atoms are connected in a continuous chain, while the other two isomers are branched-chain alkanes. The common names are given in blue below the systematic names for each compound. Rules for naming alkanes are described in Section 2.9. The physical and chemical properties of isomers differ from each other, as exemplified here by the melting points (mp) and boiling points (bp).

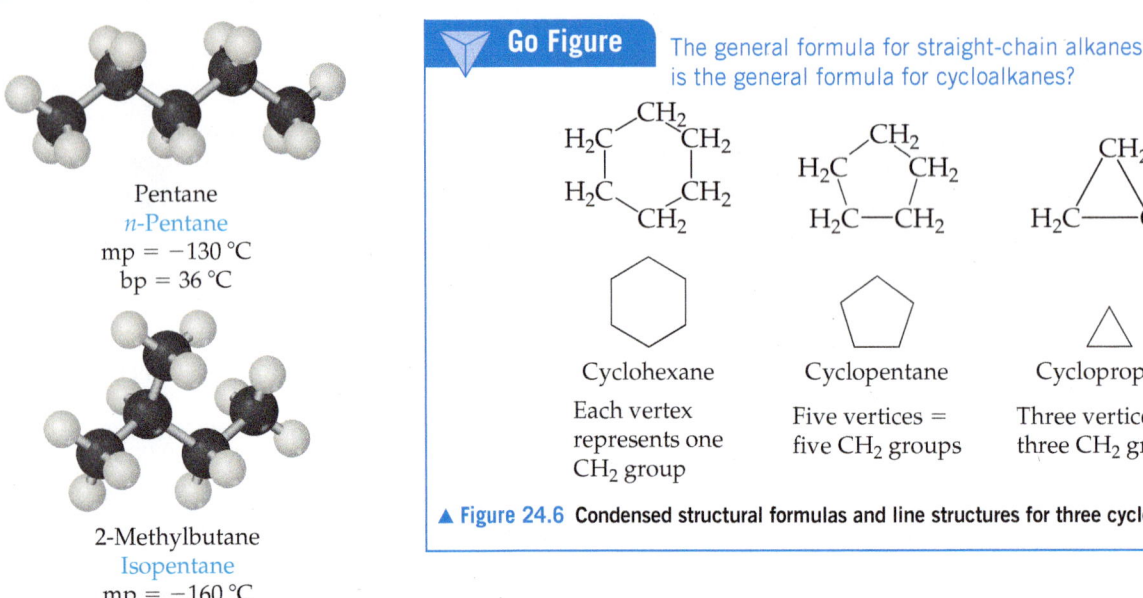

Go Figure The general formula for straight-chain alkanes is C_nH_{2n+2}. What is the general formula for cycloalkanes?

Cyclohexane
Each vertex represents one CH₂ group

Cyclopentane
Five vertices = five CH₂ groups

Cyclopropane
Three vertices = three CH₂ groups

▲ **Figure 24.6** Condensed structural formulas and line structures for three cycloalkanes.

Cycloalkanes

Alkanes that form rings, or cycles, are called **cycloalkanes**. As **Figure 24.6** illustrates, cycloalkane structures are sometimes drawn as *line structures*, which are polygons in which each corner represents a CH₂ group. This method of representation is similar to that used for benzene rings. (Section 8.6) (Remember from our discussion of benzene that each vertex in aromatic structures represents a CH group, not a CH₂ group.)

Carbon rings containing fewer than five carbon atoms are strained because the C—C—C bond angles must be less than the 109.5° tetrahedral angle. The amount of strain increases as the rings get smaller. In cyclopropane, which has the shape of an equilateral triangle, the angle is only 60°; this molecule is therefore much more reactive than propane, its straight-chain analog.

Reactions of Alkanes

Because they contain only C—C and C—H bonds, most alkanes are relatively unreactive. At room temperature, for example, they do not react with acids, bases, or strong oxidizing agents. Their low chemical reactivity, as noted in Section 24.1, is due primarily to the strength and lack of polarity of their C—C and C—H bonds.

Alkanes are not completely inert, however. One of their most commercially important reactions is *combustion* in the presence of oxygen leading to the formation of carbon dioxide and water, which is the basis of their use as fuels. (Section 3.2) For example, the complete combustion of ethane proceeds according to the following highly exothermic reaction:

$$2\,C_2H_6(g) + 7\,O_2(g) \longrightarrow 4\,CO_2(g) + 6\,H_2O(l) \quad \Delta H° = -2855 \text{ kJ}$$

Self-Assessment Exercises

SAE 24.5 How many hydrogen atoms does one molecule of 2,3-dimethylhexane have? **(a)** 8 **(b)** 12 **(c)** 14 **(d)** 18 **(e)** more than 18

SAE 24.6 Of the three compounds listed here, which are structural isomers?

(**i**) 3-ethylhexane
(**ii**) 4-methylhexane
(**iii**) 2,3,4-trimethylpentane

(**a**) i and ii (**b**) i and iii (**c**) ii and iii (**d**) i, ii and iii (**e**) All of these compounds have different molecular formulas, so no two are structural isomers of one another.

SAE 24.7 How would you classify a compound that has the molecular formula C_6H_{12} and contains only C—C and C—H single bonds? **(a)** straight-chain alkane **(b)** branched-chain alkane **(c)** cycloalkane **(d)** aromatic hydrocarbon

SAE 24.8 Which type or types of hydrocarbon do not contain π bonds? **(a)** alkanes **(b)** alkenes **(c)** alkynes **(d)** aromatic hydrocarbons **(e)** alkanes and aromatic hydrocarbons

SAE 24.9 The name of the compound with the structural formula shown here is _____. This compound has _____ structural isomers.

$$
\begin{array}{c}
H \\
| \\
H-C-H \\
\quad\quad | \quad\; H \quad H \\
\quad\quad | \quad\; | \quad\; | \\
H-C-C-C-H \\
\quad\; | \quad\; | \quad\; | \\
\quad\; H \quad H \quad H
\end{array}
$$

(a) 1-methylpropane, 1 (b) 1-methylpropane, 2 (c) n-butane, 0 (d) n-butane, 1 (e) n-butane, 2

SAE 24.10 Which reaction describes the combustion of n-pentane?
(a) $C_5H_{10}(l) \longrightarrow 5\,C(s) + 5\,H_2(g)$
(b) $2\,C_5H_{10}(l) + 15\,O_2(g) \longrightarrow 10\,CO_2(g) + 10\,H_2O(g)$
(c) $C_5H_{12}(l) + 8\,O_2(g) \longrightarrow 5\,CO_2(g) + 6\,H_2O(g)$
(d) $C_5H_{12}(l) + 5\,O_2(g) \longrightarrow 5\,CO_2(g) + 6\,H_2(g)$

24.3 | Alkenes, Alkynes, and Aromatic Hydrocarbons

Because alkanes have only single bonds, they contain the largest possible number of hydrogen atoms per carbon atom. As a result, they are called *saturated hydrocarbons*. Alkenes, alkynes, and aromatic hydrocarbons contain multiple carbon–carbon bonds (double bonds, triple bonds, or delocalized π bonds). As a result, they contain fewer hydrogen atoms than an alkane with the same number of carbon atoms. Collectively, they are called *unsaturated hydrocarbons*. On the whole, unsaturated molecules are more reactive than saturated ones.

Alkenes

Alkenes are unsaturated hydrocarbons that contain at least one C=C bond. The simplest alkene is $CH_2{=}CH_2$, called ethene (IUPAC) or ethylene (common name), which plays important roles as a plant hormone in seed germination and fruit ripening. The next member of the series is $CH_3{-}CH{=}CH_2$, called propene or propylene. Alkenes with four or more carbon atoms have several isomers. For example, the alkene C_4H_8 has the four structural isomers shown in **Figure 24.7**. Notice both their structures and their names.

The names of alkenes are based on the longest continuous chain of carbon atoms that contains the double bond. The chain is named by changing the ending of the name of the corresponding alkane from *-ane* to *-ene*. The compound on the left in Figure 24.7, for example, has a double bond as part of a three-carbon chain; thus, the parent alkene is propene.

> **Learning Objectives**
>
> **When you finish Section 24.3, you should be able to:**
> ▶ Interconvert between the structural formula and the name of an alkene or an alkyne.
> ▶ Identify structural isomers of alkenes and alkynes.
> ▶ Identify and name geometric isomers of alkenes and explain why such isomers are not usually encountered in alkanes or alkynes.
> ▶ Predict the products of addition reactions involving alkenes and alkynes.
> ▶ Describe the differences in the reactivities of alkenes and aromatic hydrocarbons.

Go Figure How many isomers are there for propene, C_3H_6?

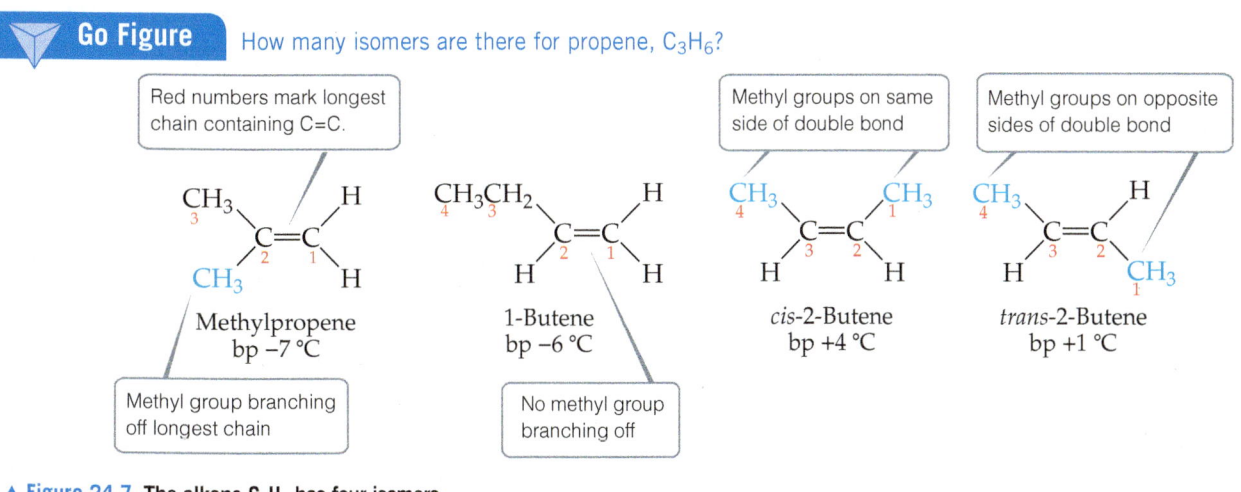

▲ **Figure 24.7** The alkene C_4H_8 has four isomers.

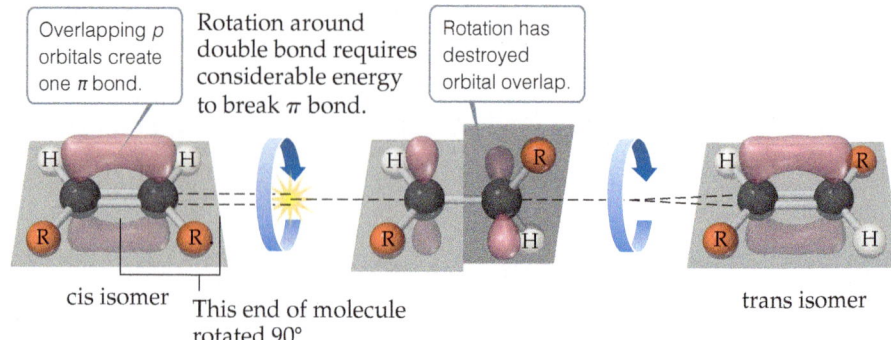

Overlapping *p* orbitals create one π bond.

Rotation around double bond requires considerable energy to break π bond.

Rotation has destroyed orbital overlap.

cis isomer

This end of molecule rotated 90°

trans isomer

▲ **Figure 24.8 Cis and trans isomers of alkenes.** Geometric isomers of alkenes exist because rotation about a carbon–carbon double bond requires too much energy to occur at ordinary temperatures.

The location of the double bond along an alkene chain is indicated by a prefix number that designates the double-bond carbon atom that is nearest an end of the chain. The chain is always numbered from the end that brings us to the double bond sooner and hence gives the smallest-numbered prefix. In propene, the only possible location for the double bond is between the first and second carbons; thus, a prefix indicating its location is unnecessary. For butene (Figure 24.7), there are two possible positions for the double bond, either after the first carbon (1-butene) or after the second carbon (2-butene).

If a substance contains two or more double bonds, the location of each is indicated by a numerical prefix, and the ending of the name is altered to identify the number of double bonds: diene (two), triene (three), and so forth. For example, $CH_2{=}CH{-}CH_2{-}CH{=}CH_2$ is 1,4-pentadiene.

The two isomers on the right in Figure 24.7 differ in the relative locations of their methyl groups. These two compounds are geometric isomers, compounds that have the same molecular formula and the same groups bonded to one another but differ in the spatial arrangement of these groups. (Section 23.4) In the cis isomer, the two methyl groups are on the same side of the double bond, whereas in the trans isomer they are on opposite sides. Geometric isomers possess distinct physical properties and can differ significantly from each other in their chemical behavior.

Geometric isomerism in alkenes arises because, unlike the C—C bond, the C=C bond resists twisting. Recall from Section 9.6 that the double bond between two carbon atoms consists of a σ and a π bond. **Figure 24.8** shows a cis alkene. The carbon–carbon bond axis and the bonds to the hydrogen atoms and to the alkyl groups (designated R) are all in a plane, and the *p* orbitals that form the π bond are perpendicular to that plane. As Figure 24.8 shows, rotation around the carbon–carbon double bond requires the π bond to be broken, a process that requires considerable energy (about 250 kJ/mol). Because rotation does not occur easily around the carbon–carbon bond, the cis and trans isomers of an alkene cannot readily interconvert; therefore, they are considered distinct compounds.

Sample Exercise 24.1
Drawing Isomers

Draw all the structural and geometric isomers of pentene, C_5H_{10}, that have an unbranched hydrocarbon chain.

SOLUTION

Analyze We are asked to draw all the isomers (both structural and geometric) for an alkene with a five-carbon chain.

Plan Because the compound is named pentene and not penta-diene or pentatriene, we know that the five-carbon chain contains only one carbon–carbon double bond. Thus, we begin by placing the double bond in various locations along the chain, remembering that the chain can be numbered from either end. After finding the different unique locations for the double bond, we consider whether the molecule can have cis and trans isomers.

Solve

There can be a double bond after either the first carbon (1-pentene) or second carbon (2-pentene). These are the only two possibilities because the chain can be numbered from either end. Thus, what we might erroneously call 3-pentene is actually 2-pentene, as seen by numbering the carbon chain from the other end:

$$\overset{1}{C}=\overset{2}{C}-\overset{3}{C}-\overset{4}{C}-\overset{5}{C}$$

$$\overset{1}{C}-\overset{2}{C}=\overset{3}{C}-\overset{4}{C}-\overset{5}{C}$$

$$\overset{1}{C}-\overset{2}{C}-\overset{3}{C}=\overset{4}{C}-\overset{5}{C} \quad \text{renumbered as} \quad \overset{5}{C}-\overset{4}{C}-\overset{3}{C}=\overset{2}{C}-\overset{1}{C}$$

$$\overset{1}{C}-\overset{2}{C}-\overset{3}{C}-\overset{4}{C}=\overset{5}{C} \quad \text{renumbered as} \quad \overset{5}{C}-\overset{4}{C}-\overset{3}{C}-\overset{2}{C}=\overset{1}{C}$$

Because the first C atom in 1-pentene is bonded to two H atoms, there are no cis–trans isomers. There are cis and trans isomers for 2-pentene, however. Thus, the three isomers for pentene are:

$$CH_2{=}CH{-}CH_2{-}CH_2{-}CH_3$$
1-Pentene

cis-2-Pentene

trans-2-Pentene

Comment Convince yourself that *cis*-3-pentene is identical to *cis*-2-pentene and that *trans*-3-pentene is identical to *trans*-2-pentene. The correct names, however, are *cis*-2-pentene and *trans*-2-pentene because they have smaller numbered prefixes.

▶ **Practice Exercise**
How many straight-chain isomers are there of hexene, C_6H_{12}?

Alkynes

Alkynes are unsaturated hydrocarbons containing one or more $C{\equiv}C$ bonds. The simplest alkyne is acetylene (C_2H_2, systematic name ethyne), a highly reactive molecule. When acetylene is burned in a stream of oxygen in an oxyacetylene torch, the flame reaches about 3200 K. Because alkynes in general are highly reactive, they are not as widely distributed in nature as alkenes. They are, however, important intermediates in many industrial processes.

Alkynes are named by identifying the longest continuous chain containing the triple bond and modifying the ending of the name of the corresponding alkane from *-ane* to *-yne*, as shown in Sample Exercise 24.2.

Sample Exercise 24.2
Naming Unsaturated Hydrocarbons

Name the following compounds:

(a) $CH_3CH_2CH_2{-}\overset{\overset{\displaystyle CH_3}{|}}{CH}$ ... CH_3
   ...C=C... H H

(b) $CH_3CH_2CH_2\overset{\overset{\displaystyle |}{CH}}{}{-}C{\equiv}CH$
   $CH_2CH_2CH_3$

SOLUTION

Analyze We are given the condensed structural formulas for an alkene and an alkyne and asked to name the compounds.

Plan In each case, the name is based on the number of carbon atoms in the longest continuous carbon chain that contains the multiple bond. In the alkene, care must be taken to indicate whether cis–trans isomerism is possible and, if so, which isomer is given.

Solve
(a) The longest continuous chain of carbons that contains the double bond is seven carbons long, so the parent hydrocarbon is heptene. Because the double bond begins at carbon 2 (numbering from the end closer to the double bond), we have 2-heptene. With a methyl group at carbon atom 4, we have 4-methyl-2-heptene. The geometrical configuration at the double bond is cis (that is, the alkyl groups are bonded

to the double bond on the same side). Thus, the full name is 4-methyl-*cis*-2-heptene.

(b) The longest continuous chain containing the triple bond has six carbons, so this compound is a derivative of hexyne. The triple bond comes after the first carbon (numbering from the right), making it 1-hexyne. The branch from the hexyne chain contains three carbon atoms, making it a propyl group. Because this substituent is located on C3 of the hexyne chain, the molecule is 3-propyl-1-hexyne.

▶ **Practice Exercise**
Draw the condensed structural formula for 4-methyl-2-pentyne.

Addition Reactions of Alkenes and Alkynes

The presence of carbon–carbon double or triple bonds in hydrocarbons markedly increases their chemical reactivity. The most characteristic reactions of alkenes and alkynes are **addition reactions**, in which a reactant is added to the two atoms that form the multiple bond. A simple example is the addition of elemental bromine to ethylene to produce 1,2-dibromoethane:

$$H_2C{=}CH_2 + Br_2 \longrightarrow \underset{\underset{\displaystyle Br \quad Br}{|\quad\;|}}{H_2C{-}CH_2} \qquad [24.1]$$

The π bond in ethylene is broken, and the electrons that formed the bond are used to form two σ bonds to the two bromine atoms. The σ bond between the carbon atoms is retained.

Addition of H_2 to an alkene converts it to an alkane:

$$CH_3CH{=}CHCH_3 + H_2 \xrightarrow{\text{Ni, 500 °C}} CH_3CH_2CH_2CH_3 \qquad [24.2]$$

The reaction between an alkene and H_2, referred to as *hydrogenation*, does not occur readily at ordinary temperatures and pressures. One reason for the lack of reactivity of H_2 toward alkenes is the stability of the H_2 bond. To promote the reaction, the reaction temperature must be raised (500 °C), and a catalyst (such as Ni) is used to assist in rupturing the H—H bond. We write such conditions over the reaction arrow to indicate they must be present in order for the reaction to occur. The most widely used catalysts are finely divided metals on which H_2 is adsorbed. (Section 14.6)

Hydrogen halides and water can also add to the double bond of alkenes, as in the following reactions of ethylene:

$$CH_2{=}CH_2 + HBr \longrightarrow CH_3CH_2Br \qquad [24.3]$$

$$CH_2{=}CH_2 + H_2 \xrightarrow{\text{H}_2\text{SO}_4} CH_3CH_2OH \qquad [24.4]$$

The addition of water is catalyzed by a strong acid, such as H_2SO_4.

The addition reactions of alkynes resemble those of alkenes, as shown in the following examples:

$$CH_3C{\equiv}CCH_3 + Cl_2 \longrightarrow \underset{\substack{\displaystyle CH_3 \qquad Cl}}{\overset{\substack{\displaystyle Cl \qquad CH_3}}{C{=}C}} \qquad [24.5]$$

2-Butyne   *trans*-2,3-Dichloro-2-butene

$$CH_3C{\equiv}CCH_3 + 2\,Cl_2 \longrightarrow \underset{\underset{\displaystyle Cl \quad\; Cl}{|\quad\;|}}{\overset{\overset{\displaystyle Cl \quad\; Cl}{|\quad\;|}}{CH_3{-}C{-}C{-}CH_3}} \qquad [24.6]$$

2-Butyne   2,2,3,3-Tetrachlorobutane

Sample Exercise 24.3
Predicting the Product of an Addition Reaction

Write the condensed structural formula for the product of the hydrogenation of 3-methyl-1-pentene.

SOLUTION

Analyze We are asked to predict the compound formed when a particular alkene undergoes hydrogenation (reaction with H_2) and to write the condensed structural formula of the product.

Plan To determine the condensed structural formula of the product, we must first write the condensed structural formula or Lewis structure of the reactant. In the hydrogenation of the alkene, H_2 adds to the double bond, producing an alkane.

Solve The name of the starting compound tells us that we have a chain of five C atoms with a double bond at one end (position 1) and a methyl group on C3:

$$\underset{\quad}{CH_2{=}CH{-}\underset{\overset{|}{CH_3}}{CH}{-}CH_2{-}CH_3}$$

Hydrogenation—the addition of two H atoms to the carbons of the double bond—leads to the following alkane:

$$CH_3-CH_2-\underset{\underset{CH_3}{|}}{CH}-CH_2-CH_3$$

Comment The longest chain in this alkane has five carbon atoms; the product is therefore 3-methylpentane.

▶ **Practice Exercise**
Addition of HCl to an alkene forms 2-chloropropane. What is the alkene?

A CLOSER LOOK | Mechanism of Addition Reactions

As the understanding of chemistry has grown, chemists have advanced from simply cataloging reactions that occur to explaining *how* they occur, by drawing the individual steps of a reaction based upon experimental and theoretical evidence. The sum of these steps is called a *reaction mechanism.* (Section 14.5)

The addition reaction between HBr and an alkene, for instance, is thought to proceed in two steps. In the first step, which is rate determining (Section 14.6), HBr attacks the electron-rich double bond, transferring a proton to one of the double-bond carbons. In the reaction of 2-butene with HBr, for example, the first step is

$$CH_3CH=CHCH_3 + HBr \longrightarrow \left[\underset{\underset{\underset{Br^{\delta-}}{|}}{\overset{\delta+}{CH_3CH}=CHCH_3}}{\overset{H}{\underset{|}{}}} \right]$$

$$\longrightarrow \overset{+}{CH_3CH}-CH_2CH_3 + \ Br^- \qquad [24.7]$$

The electron pair that formed the π bond is used to form the new C—H bond.

The second, faster step is addition of Br⁻ to the positively charged carbon. The bromide ion donates a pair of electrons to the carbon, forming the C—Br bond:

$$\overset{+}{CH_3CH}-CH_2CH_3 + \ Br^- \longrightarrow \left[\underset{\underset{Br^{\delta-}}{|}}{\overset{\delta+}{CH_3CH}}-CH_2CH_3 \right]$$

$$\longrightarrow \underset{\underset{Br}{|}}{CH_3CHCH_2CH_3} \qquad [24.8]$$

Because the rate-determining step involves both the alkene and the acid, the rate law for the reaction is second order, first order in the alkene and first order in HBr:

$$Rate = -\frac{\Delta[CH_3CH=CHCH_3]}{\Delta t} = k[CH_3CH=CHCH_3][HBr]$$

$$[24.9]$$

The energy profile for the reaction is shown in **Figure 24.9**. The first energy maximum represents the transition state in the first step, and the second maximum represents the transition state in the second step. The energy minimum represents the energies of the intermediate species, $CH_3\overset{+}{CH}-CH_2CH_3$ and Br^-.

To show electron movement in reactions like these, chemists often use curved arrows pointing in the direction of electron flow. For the addition of HBr to 2-butene, for example, the shifts in electron positions are shown as

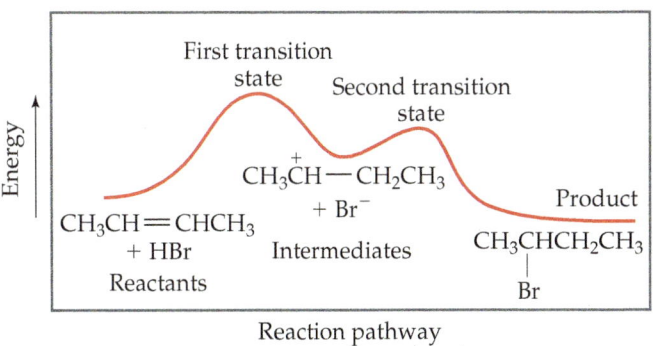

▲ **Figure 24.9** Energy profile for addition of HBr to 2-butene. The two maxima tell you that this is a two-step mechanism.

Aromatic Hydrocarbons

The simplest aromatic hydrocarbon, benzene (C_6H_6), is shown in **Figure 24.10** along with some other aromatic hydrocarbons. Benzene is the most important aromatic hydrocarbon, and most of our discussion focuses on it. When the C_6H_5-group is a substituent in a compound, it is known (somewhat confusingly) as the **phenyl** group.

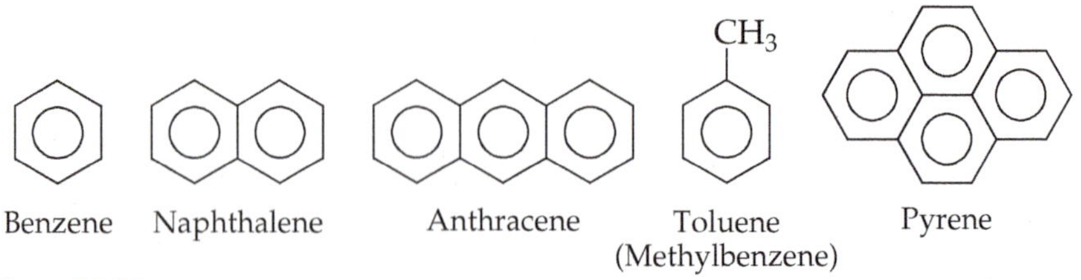

▲ **Figure 24.10 Line formulas and common names of several aromatic compounds.** The aromatic rings are represented by hexagons with a circle inscribed inside to denote delocalized π bonds. Each vertex represents a carbon atom. Each carbon is bound to three other atoms—either three carbons, or two carbons and a hydrogen—so that each carbon has the requisite four bonds.

If you draw one Lewis structure for benzene, you draw a ring that contains three $C=C$ double bonds and three single $C-C$ bonds. (Section 8.6) You might therefore expect benzene to resemble the alkenes and to be highly reactive. Benzene and the other aromatic hydrocarbons, however, are much more stable than alkenes because the π electrons are delocalized in the π orbitals. (Section 9.6)

We can estimate the stabilization of the π electrons in benzene by comparing the energy released upon forming cyclohexane by adding hydrogen to benzene, to cyclohexene (one double bond), and to 1,4-cyclohexadiene (two double bonds):

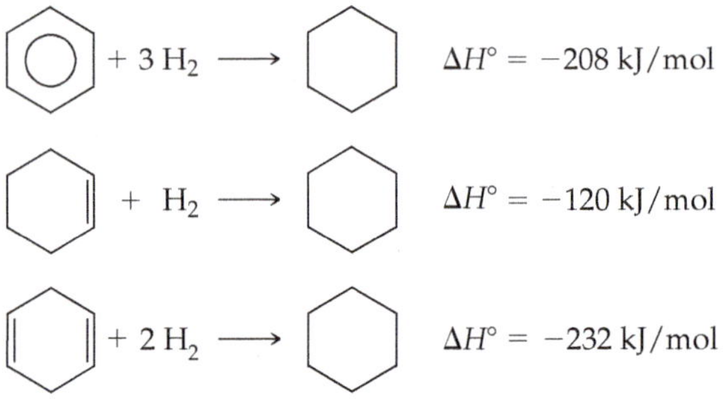

Notice that in the second and third reactions, we are hydrogenating one and two $C=C$ double bonds, respectively. It therefore appears that the energy released upon hydrogenation of each double bond is roughly 116–120 kJ/mol for each bond. Benzene contains the equivalent of three double bonds. We might expect, therefore, the energy released upon hydrogenation of the three double bonds in benzene to be roughly 348–360 kJ/mol, if benzene behaved as though it were "cyclohexatriene"—that is, if it behaved as though it had three isolated double bonds in a ring. Instead, the energy released is only 208 kJ, indicating that benzene is more stable than would be expected for three double bonds. The difference of 140–152 kJ/mol between the expected heat of hydrogenation and the observed heat of hydrogenation is due to stabilization of the π electrons through delocalization in the π orbitals that extend around the ring. Chemists call this stabilization energy the *resonance energy*.

Substitution Reactions of Aromatic Hydrocarbons

Although aromatic hydrocarbons are unsaturated, *they do not readily undergo addition reactions*. The delocalized π bonding causes aromatic compounds to behave quite differently from alkenes and alkynes. For example, benzene does not add Cl_2 or Br_2 to its double bonds under ordinary conditions. In contrast, aromatic hydrocarbons undergo **substitution reactions** relatively easily. In a substitution reaction, one hydrogen atom of the molecule is removed and replaced (substituted) by another atom or group of atoms. When benzene is warmed in a mixture of nitric and sulfuric acids, for example, one of the benzene hydrogens is replaced by the nitro group, NO_2:

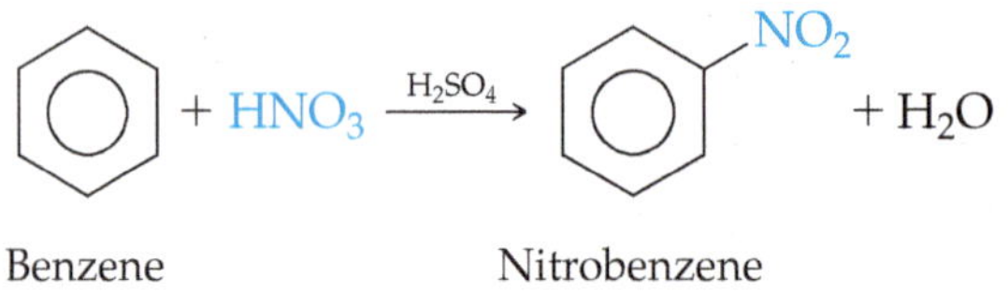

[24.10]

More vigorous treatment results in substitution of a second nitro group into the molecule:

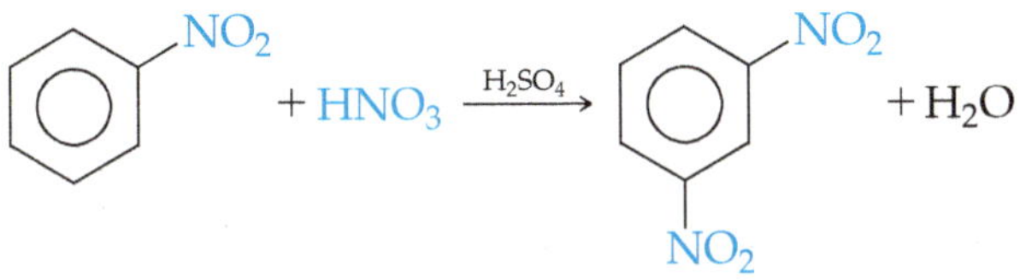

$$[24.11]$$

There are three isomers of benzene that contain two nitro groups—the 1,2- or *ortho*-isomer, the 1,3- or *meta*-isomer, and the 1,4- or *para*-isomer of dinitrobenzene:

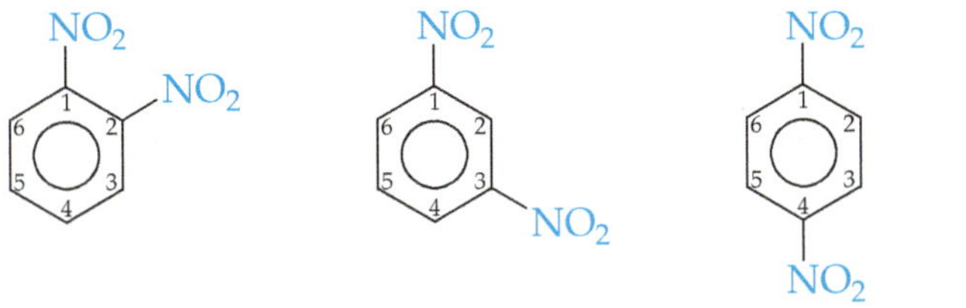

ortho-Dinitrobenzene	*meta*-Dinitrobenzene	*para*-Dinitrobenzene
1,2-Dinitrobenzene	1,3-Dinitrobenzene	1,4-Dinitrobenzene
mp 118 °C	mp 90 °C	mp 174 °C

In the reaction of Equation 24.11, the principal product is the *meta* isomer.

Bromination of benzene, carried out with $FeBr_3$ as a catalyst, is another substitution reaction:

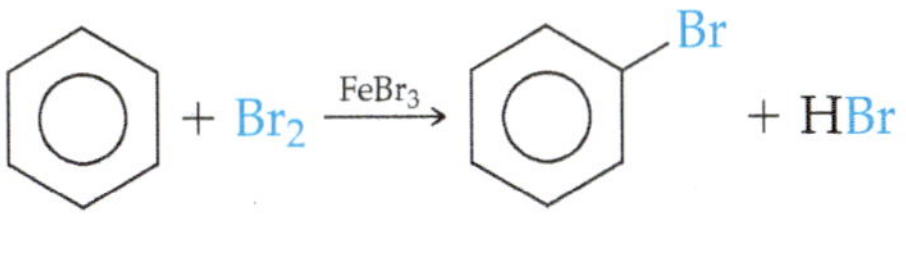

$$[24.12]$$

Benzene   Bromobenzene

In a similar substitution reaction, called the *Friedel–Crafts reaction*, alkyl groups can be substituted onto an aromatic ring by reacting an alkyl halide with an aromatic compound in the presence of $AlCl_3$ as a catalyst:

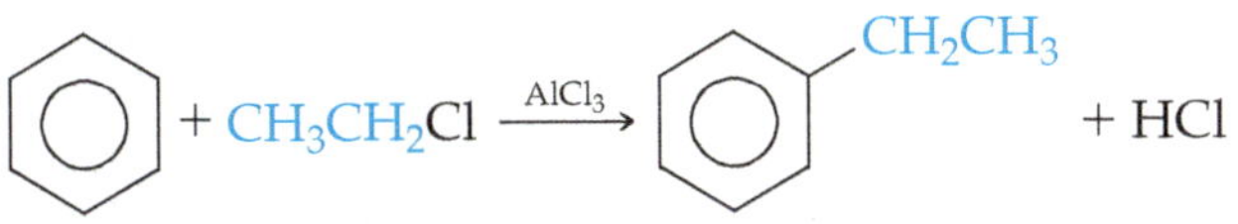

$$[24.13]$$

Benzene   Ethylbenzene

⚠ Self-Assessment Exercises

SAE 24.11 How many hydrogens does 2-bromo-3-hexene contain? (a) 6 (b) 9 (c) 11 (d) 14 (e) none of these

SAE 24.12 Which of the following compounds have geometrical isomers?

(i) 1,3-dibromo-2-butene
(ii) 1,1-dibromo-1-butene

(a) only i (b) only ii (c) both i and ii (d) Neither compound has geometric isomers.

SAE 24.13 Which of the following statements about addition reactions is *false*? (a) Alkenes, alkynes, and aromatic hydrocarbons undergo similar addition reactions. (b) Alkenes can undergo addition reactions with elemental halogens. (c) The addition of hydrogen

to an alkene produces an alkane. (d) The addition of water to an alkene produces an alcohol.

SAE 24.14 How many isomers, both structural and geometric, exist for an alkene with the molecular formula C_3H_5Cl? (a) two (b) three (c) four (d) more than four (e) There are no isomers, only one alkene has this molecular formula.

SAE 24.15 What is the product of the reaction between 3-methyl-2-pentene and elemental bromine? Neglect any possible geometrical isomerism. (a) 1,2-dibromo-3-methylpentane (b) 2,3-dibromo-3-methylpentane (c) 2-bromo-3-methylpentane (d) 3-bromo-3-methylpentane (e) 3-methyl-pentane

24.4 | Organic Functional Groups

Learning Objectives

When you finish Section 24.4, you should be able to:

▶ Recognize and name different organic functional groups from inspection of the structural formula of an organic compound.

▶ Identify the functional group(s) present in an organic compound from the name of the compound.

▶ Determine the products of simple chemical reactions involving organic compounds.

The $C\!=\!C$ double bonds of alkenes and the $C\!\equiv\!C$ triple bonds of alkynes are just two of many functional groups in organic molecules. As noted previously, these functional groups each undergo characteristic reactions, and the same is true of all other functional groups. Each kind of functional group often undergoes the same kinds of reactions in every molecule, regardless of the size and complexity of the molecule. Thus, the chemistry of an organic molecule is largely determined by the functional groups it contains. Table 24.2 lists the most common functional groups. Notice that, except for $C\!=\!C$ and $C\!\equiv\!C$, they all contain either O, N, or a halogen atom, X.

We can think of organic molecules as being composed of functional groups bonded to one or more alkyl groups. The alkyl groups, which are made of C—C and C—H single bonds, are the less reactive portions of the molecules. In describing general features of organic compounds, chemists often use the designation R to represent any alkyl group: methyl, ethyl, propyl, and so on. Alkanes, for example, which contain no functional group, are represented as R—H. Alcohols, which contain the functional group —OH, are represented as R—OH. If two or more different alkyl groups are present in a molecule, we designate them R, R′, R″, and so forth.

Alcohols

Alcohols are compounds in which one or more hydrogens of a parent hydrocarbon have been replaced by the functional group —OH, called either the *hydroxyl group* or the *alcohol group*. Note in Figure 24.11 that the name for an alcohol ends in *-ol*. The simple alcohols are named by changing the last letter in the name of the corresponding alkane to *-ol*. For example, ethan*e* becomes ethan*ol*. Where necessary, the location of the OH group is designated by a numeric prefix that indicates the number of the carbon atom bearing the OH group. Figure 24.12 shows several commercial products that consist entirely or in large part of an alcohol.

The O—H bond is polar, so alcohols are more soluble in polar solvents than are hydrocarbons. The —OH functional group can also participate in hydrogen bonding. As a result, the boiling points of alcohols are higher than those of their parent alkanes.

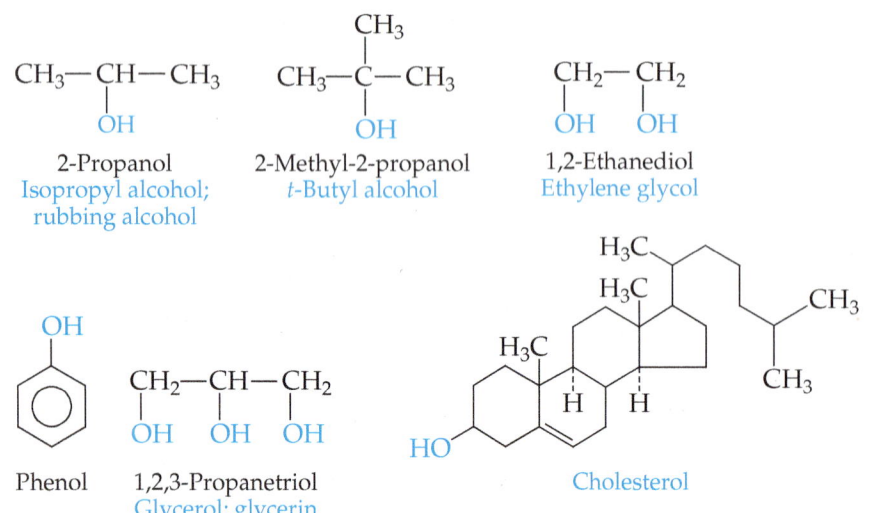

▲ **Figure 24.11 Condensed structural formulas of six important alcohols.** Common names are given in blue.

▲ **Figure 24.12 Everyday alcohols.** Many of the products we use every day—from rubbing alcohol to hair spray and antifreeze—are composed either entirely or mainly of alcohols.

TABLE 24.2   Common Functional Groups

Functional Group	Compound Type	Suffix	Example Structural Formula	Ball-and-stick Model	Systematic Name (Common Name)
C=C (Alkene group)	Alkene	-ene	H₂C=CH₂ structure		Ethene (Ethylene)
—C≡C—	Alkyne	-yne	H—C≡C—H		Ethyne (Acetylene)
—C—Ö—H	Alcohol	-ol	H—C—Ö—H		Methanol (Methyl alcohol)
—C—Ö—C—	Ether	ether	H—C—Ö—C—H		Methoxymethane (Dimethyl ether)
—C—Ẍ: (X = halogen)	Alkyl halide or haloalkane	-ide	H—C—Cl:		Chloromethane (Methyl chloride)
—C—N̈—	Amine	-amine	H—C—C—N̈—H		Ethylamine
—C—H with :O:	Aldehyde	-al	H—C—C—H		Ethanal (Acetaldehyde)
—C—C—C— with :O:	Ketone	-one	H—C—C—C—H		Propanone (Acetone)
—C—Ö—H with :O:	Carboxylic acid	-oic acid	H—C—C—Ö—H		Ethanoic acid (Acetic acid)
—C—Ö—C— with :O:	Ester	-oate	H—C—C—Ö—C—H		Methyl ethanoate (Methyl acetate)
—C—N̈— with :O:	Amide	-amide	H—C—C—N̈—H		Ethanamide (Acetamide)

The simplest alcohol, methanol (methyl alcohol), has many industrial uses and is produced on a large scale by heating carbon monoxide and hydrogen under pressure in the presence of a metal oxide catalyst:

$$CO(g) + 2\,H_2(g) \xrightarrow[400\,°C]{200-300\,atm} CH_3OH(g) \qquad [24.14]$$

Because methanol has a very high octane rating as an automobile fuel, it is used as a gasoline additive and as a fuel in its own right.

Ethanol (ethyl alcohol, C_2H_5OH) is a product of the fermentation of carbohydrates such as sugars and starches. In the absence of air, yeast cells convert these carbohydrates into ethanol and CO_2:

$$C_6H_{12}O_6(aq) \xrightarrow{yeast} 2\,C_2H_5OH(aq) + 2\,CO_2(g) \qquad [24.15]$$

In the process, the yeast cells derive energy necessary for growth. This reaction is carried out under carefully controlled conditions to produce beer, wine, and other beverages in which ethanol (called just "alcohol" in everyday language) is the active ingredient.

The simplest polyhydroxyl alcohol (an alcohol containing more than one OH group) is 1,2-ethanediol (ethylene glycol, $HOCH_2CH_2OH$), the major ingredient in automobile antifreeze. Another common polyhydroxyl alcohol is 1,2,3-propanetriol (glycerol, $HOCH_2CH(OH)CH_2OH$), a viscous liquid that dissolves readily in water and is used in cosmetics as a skin softener and in foods and candies to keep them moist.

Phenol is the simplest compound with an OH group attached to an aromatic ring. One of the most striking effects of the aromatic group is the greatly increased acidity of the OH group. Phenol is about 1 million times more acidic in water than a nonaromatic alcohol. Even so, it is not a very strong acid ($K_a = 1.3 \times 10^{-10}$). Phenol is used industrially to make plastics and dyes, and as a topical anesthetic in throat sprays.

Cholesterol, shown in Figure 24.11, is a biochemically important alcohol. The OH group forms only a small component of this molecule, so cholesterol is only slightly soluble in water (2.6 g/L of H_2O). Cholesterol is a normal and essential component of our bodies; when present in excessive amounts, however, it may precipitate from solution. It precipitates in the gallbladder to form crystalline lumps called *gallstones*. It may also precipitate against the walls of veins and arteries and thus contribute to high blood pressure and other cardiovascular problems.

Ethers

Compounds in which two hydrocarbon groups are bonded to one oxygen are called **ethers**. Ethers can be formed from two molecules of alcohol by eliminating a molecule of water. The reaction is catalyzed by sulfuric acid, which takes up water to remove it from the system:

$$CH_3CH_2-OH + H-OCH_2CH_3 \xrightarrow{H_2SO_4} CH_3CH_2-O-CH_2CH_3 + H_2O \quad [24.16]$$

A reaction in which water is eliminated from two substances is called a *condensation reaction*. (Sections 12.6 and 22.8)

Both diethyl ether and the cyclic ether tetrahydrofuran are common solvents for organic reactions:

$$CH_3CH_2-O-CH_2CH_3$$

$$\begin{array}{c} CH_2-CH_2 \\ | \quad\quad | \\ CH_2 \quad CH_2 \\ \diagdown O \diagup \end{array}$$

Diethyl ether   Tetrahydrofuran (THF)

Diethyl ether was formerly used as an anesthetic (known simply as "ether" in that context), but its use was discontinued because it had significant side effects.

Aldehydes and Ketones

Several of the functional groups listed in Table 24.2 contain the **carbonyl group**, $C=O$. This group, together with the atoms attached to its carbon, defines several important functional groups that we consider in this section.

In **aldehydes**, the carbonyl group has at least one hydrogen atom attached:

$$H-\overset{\overset{\displaystyle O}{\|}}{C}-H \qquad CH_3-\overset{\overset{\displaystyle O}{\|}}{C}-H$$

Methanal   Ethanal
Formaldehyde   Acetaldehyde

In **ketones**, the carbonyl group occurs at the interior of a carbon chain and is therefore flanked by carbon atoms:

$$CH_3-\overset{\overset{\displaystyle O}{\|}}{C}-CH_3 \qquad CH_3-\overset{\overset{\displaystyle O}{\|}}{C}-CH_2CH_3$$

Propanone   2-Butanone
Acetone   Methyl ethyl ketone

Testosterone

The systematic names of aldehydes contain -*al* and ketone names contain -*one*. Notice that testosterone has both alcohol and ketone groups; the ketone functional group dominates the molecular properties. Therefore, testosterone is considered a ketone first and an alcohol second, and its name reflects its ketone properties.

Many compounds found in nature contain an aldehyde or ketone functional group. Vanilla and cinnamon flavorings are naturally occurring aldehydes. Two isomers of the ketone carvone impart the characteristic flavors of spearmint leaves and caraway seeds.

Ketones are less reactive than aldehydes and are used extensively as solvents. Acetone (propanone), the most widely used ketone, is completely miscible with water, yet it dissolves a wide range of organic substances.

Carboxylic Acids and Esters

Carboxylic acids contain the *carboxyl* functional group, often written COOH. (Section 16.10) These weak acids are widely distributed in nature and are common in many fruits. They are also important in the manufacture of polymers used to make fibers, films, and paints. **Figure 24.13** shows the formulas of several carboxylic acids.

Go Figure Which of these substances have both a carboxylic acid functional group and an alcohol functional group?

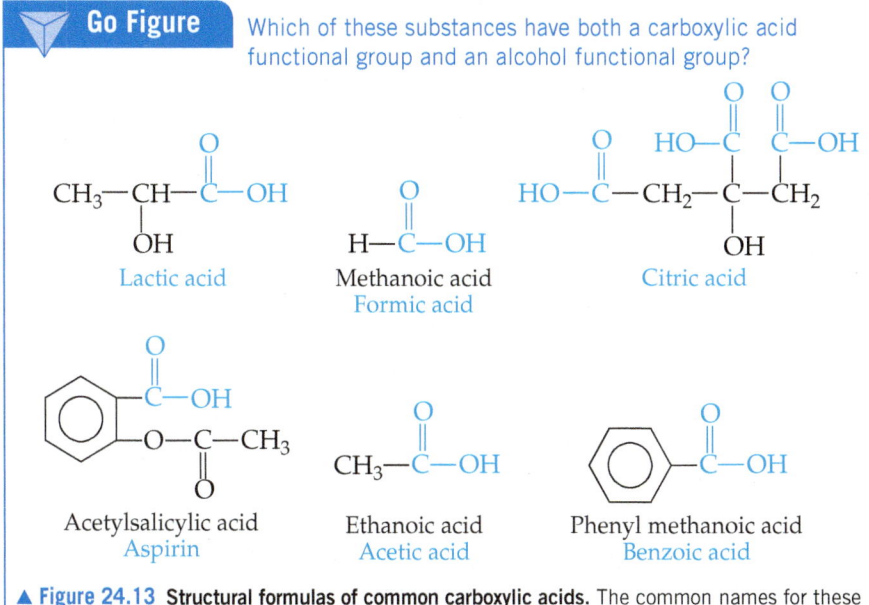

Lactic acid   Methanoic acid
Formic acid

Citric acid

Acetylsalicylic acid   Ethanoic acid   Phenyl methanoic acid
Aspirin   Acetic acid   Benzoic acid

▲ **Figure 24.13 Structural formulas of common carboxylic acids.** The common names for these acids are given in blue type.

The common names of many carboxylic acids are based on their historical origins. Formic acid, for example, was first prepared by extraction from ants; its name is derived from the Latin word for "ant," *formica*.

Carboxylic acids can be produced by oxidation of alcohols. Under appropriate conditions, the aldehyde may be isolated as the first product of oxidation, as in the sequence

$$CH_3CH_2OH + (O) \longrightarrow CH_3\overset{\overset{\displaystyle O}{\|}}{C}H + H_2O \qquad [24.17]$$

Ethanol   Acetaldehyde

$$CH_3\overset{\overset{\displaystyle O}{\|}}{C}H + (O) \longrightarrow CH_3\overset{\overset{\displaystyle O}{\|}}{C}OH \qquad [24.18]$$

Acetaldehyde   Acetic acid

where (O) represents any oxidant that can provide oxygen atoms. The air oxidation of ethanol to acetic acid is responsible for causing wines to turn sour, producing vinegar.

The oxidation processes in organic compounds are related to those oxidation reactions we studied in Chapter 20. Instead of counting electrons, the number of C—O bonds is usually considered to show the extent of oxidation of similar compounds. For example, methane can be oxidized to methanol; further oxidation leads to formaldehyde (methanal) and then formic acid (methanoic acid):

Methane   Methanol   Formaldehyde   Formic acid

From methane to formic acid, the number of C—O bonds increases from 0 to 3 (double bonds are counted as two). If you were to calculate the oxidation state of carbon in these compounds, it would range from −4 in methane (if H's are counted as +1) to +2 in formic acid, which is consistent with carbon being oxidized. The ultimate oxidation product of any organic compound, then, is CO_2, which is indeed the product of combustion reactions of carbon-containing compounds (CO_2 has 4 C—O bonds, and C has the oxidation state +4).

Aldehydes and ketones can be prepared by controlled oxidation of alcohols. Complete oxidation results in formation of CO_2 and H_2O, as in the burning of methanol:

$$2\,CH_3OH(g) + 3\,O_2(g) \longrightarrow 2\,CO_2(g) + 4\,H_2O(g)$$

Controlled partial oxidation to form other organic substances, such as aldehydes and ketones, is carried out by using various oxidizing agents, such as air, hydrogen peroxide (H_2O_2), ozone (O_3), and potassium dichromate ($K_2Cr_2O_7$).

Acetic acid can also be produced by the reaction of methanol with carbon monoxide in the presence of a rhodium catalyst:

$$CH_3OH + CO \xrightarrow{\text{catalyst}} CH_3-\overset{\overset{\displaystyle O}{\|}}{C}-OH \qquad [24.19]$$

This reaction is not an oxidation; it involves, in effect, the insertion of a carbon monoxide molecule between the CH_3 and OH groups. A reaction of this kind is called *carbonylation*.

Carboxylic acids can undergo condensation reactions with alcohols to form esters:

$$CH_3-\overset{\overset{\displaystyle O}{\|}}{C}-OH + HO-CH_2CH_3 \longrightarrow CH_3-\overset{\overset{\displaystyle O}{\|}}{C}-O-CH_2CH_3 + H_2O$$

Acetic acid   Ethanol   Ethyl acetate

[24.20]

Esters are compounds in which the H atom of a carboxylic acid is replaced by a carbon-containing group:

$$-\overset{\overset{\textstyle O}{\|}}{C}-O-\overset{|}{\underset{|}{C}}-$$

The name of any ester consists of the name of the group contributed by the alcohol followed by the name of the group contributed by the carboxylic acid, with the *-ic* replaced by *-ate*. For example, the ester formed from ethyl alcohol, CH_3CH_2OH, and butyric acid, $CH_3CH_2CH_2COOH$, is

$$CH_3CH_2CH_2\overset{\overset{\textstyle O}{\|}}{C}-OCH_2CH_3$$

Ethyl butyrate

Notice that the chemical formula generally has the group originating from the acid written first, which is opposite of the way the ester is named. Another example is isoamyl acetate, the ester formed from acetic acid and isoamyl alcohol. Isoamyl acetate smells like bananas or pears.

$$(CH_3)_2CHCH_2CH_2-O-\overset{\overset{\textstyle O}{\|}}{C}-CH_3$$

Isoamyl   Acetate

Many esters such as isoamyl acetate have pleasant odors and are largely responsible for the pleasing aromas of many fruits.

An ester treated with an acid or a base in aqueous solution is *hydrolyzed*; that is, the molecule is split into an alcohol and a carboxylic acid or its anion:

$$CH_3CH_2-\overset{\overset{\textstyle O}{\|}}{C}-O-CH_3 + OH^- \longrightarrow$$

Methyl propionate

$$CH_3CH_2-\overset{\overset{\textstyle O}{\|}}{C}-O^- + CH_3OH$$

Propionate   Methanol   [24.21]

The **hydrolysis** of an ester in the presence of a base is called **saponification**, a term that comes from the Latin word for soap, *sapon*. Naturally occurring esters include fats and oils, and in making soap an animal fat or a vegetable oil is boiled with a strong base. The resultant soap consists of a mixture of salts of long-chain carboxylic acids (called fatty acids), which form during the saponification reaction.

Sample Exercise 24.4

Naming Esters and Predicting Hydrolysis Products

In a basic aqueous solution, esters react with hydroxide ion to form the salt of the carboxylic acid and the alcohol from which the ester is constituted. Name each of the following esters, and indicate the products of their reaction with aqueous base.

(a) $C_6H_5-\overset{\overset{\textstyle O}{\|}}{C}-OCH_2CH_3$ (b) $CH_3CH_2CH_2-\overset{\overset{\textstyle O}{\|}}{C}-O-C_6H_5$

SOLUTION

Analyze We are given two esters and asked to name them and to predict the products formed when they undergo hydrolysis (split into an alcohol and carboxylate ion) in basic solution.

Plan Esters are formed by the condensation reaction between an alcohol and a carboxylic acid. To name an ester, we must analyze its structure and determine the identities of the alcohol and acid from which it is formed. We can identify the alcohol by adding an OH to the alkyl group attached to the O atom of the carboxyl (COO) group. We can identify the acid by adding an H to the O atom of the carboxyl group. We have learned that the first part of an ester name indicates the alcohol portion and the second indicates the acid portion. The name conforms to how the ester undergoes hydrolysis in base, reacting with base to form an alcohol and a carboxylate anion.

Solve

(a) This ester is derived from ethanol (CH_3CH_2OH) and benzoic acid (C_6H_5COOH). Its name is therefore ethyl benzoate. The net ionic equation for reaction of ethyl benzoate with hydroxide ion is

$$\text{C}_6\text{H}_5-\overset{\overset{\displaystyle O}{\|}}{C}-OCH_2CH_3(aq) \ + \ OH^-(aq) \ \longrightarrow$$

$$\text{C}_6\text{H}_5-\overset{\overset{\displaystyle O}{\|}}{C}-O^-(aq) \ + \ HOCH_2CH_3(aq)$$

The products are benzoate ion and ethanol.

(b) This ester is derived from phenol (C_6H_5) and butanoic acid (commonly called butyric acid) ($CH_3CH_2CH_2COOH$). The residue from the phenol is called the phenyl group. The ester is therefore named phenyl butyrate or phenyl butanoate. The net ionic equation for the reaction of phenyl butyrate with hydroxide ion is

$$CH_3CH_2CH_2\overset{\overset{\displaystyle O}{\|}}{C}-O-\text{C}_6\text{H}_5(aq) \ + \ OH^-(aq) \ \longrightarrow$$

$$CH_3CH_2CH_2\overset{\overset{\displaystyle O}{\|}}{C}-O^-(aq) \ + \ HO-\text{C}_6\text{H}_5(aq)$$

The products are butyrate ion and phenol.

▶ **Practice Exercise**

Write the condensed structural formula for the ester formed from propyl alcohol and propionic acid.

Amines and Amides

Amines are compounds in which one or more of the hydrogens of ammonia (NH_3) are replaced by an alkyl group:

$$CH_3CH_2NH_2 \qquad (CH_3)_3N \qquad \text{C}_6\text{H}_5-NH_2$$

Ethylamine   Trimethylamine   Phenylamine
                                                    Aniline

Amines are the most common organic bases. (Section 16.7) As we saw in the "Chemistry and Life" box on "Amines and Amine Hydrochlorides" in Section 16.8, many pharmaceutically active compounds are complex amines:

Cocaine   Morphine   Codeine

An amine with at least one H bonded to N can undergo a condensation reaction with a carboxylic acid to form an **amide**, which contains the carbonyl group ($C=O$) attached to N (Table 24.2):

$$CH_3\overset{O}{\overset{\|}{C}}-OH \ + \ H-N(CH_3)_2 \longrightarrow CH_3\overset{O}{\overset{\|}{C}}-N(CH_3)_2 \ + \ H_2O \quad [24.22]$$

We may consider the amide functional group to be derived from a carboxylic acid with an NRR', NH_2, or NHR' group replacing the OH of the acid, as in the following examples:

$$CH_3\overset{O}{\overset{\|}{C}}-NH_2$$

Ethanamide
Acetamide

Phenylmethanamide
Benzamide

N-(4-hydroxyphenyl)ethanamide
Acetaminophen

The amide linkage

$$R-\overset{O}{\overset{\|}{C}}-\underset{\underset{H}{|}}{N}-R'$$

where R and R' are organic groups, is the key functional group in proteins, as discussed in Section 24.7.

Self-Assessment Exercises

SAE 24.16 Consider the reaction ROH + R'OH $\longrightarrow$ ROR' + H_2O, where R and R' are generic alkyl groups. This is an example of _____ reaction and the product is _____. **(a)** a condensation, an ester **(b)** a condensation, an ether **(c)** an addition, an ester **(d)** an addition, an ether **(e)** a hydrolysis, a ketone

SAE 24.17 Name the compound formed in the condensation reaction between propionic acid and methanol. **(a)** methyl propionate **(b)** hexanoate **(c)** butyl methanoate **(d)** propyl methanoate

SAE 24.18 Which of the following substances is an ester? **(a)** 3-pentanone **(b)** 1-propylamine **(c)** ethyl acetate **(d)** propanal **(e)** formic acid

SAE 24.19 Which of the substances does *not* contain a carbon–oxygen double bond? **(a)** an ester **(b)** an amide **(c)** an amine **(d)** a ketone **(e)** an aldehyde

SAE 24.20 An amide is formed from the condensation reaction of _____ and _____. **(a)** a carboxylic acid, an alcohol **(b)** a carboxylic acid, an amine **(c)** an amine, an alcohol **(d)** an alcohol, an alcohol

24.5 | Chirality in Organic Chemistry

A molecule possessing a nonsuperimposable mirror image is said to be **chiral** (Greek *cheir*, "hand"). (Section 23.4) *Compounds containing carbon atoms with four different attached groups are inherently chiral.* A carbon atom with four different attached groups is called a *chiral center*. For example, consider 2-bromopentane:

Learning Objective

When you finish Section 24.5, you should be able to:

▶ Identify organic compounds that are chiral.

$$\underset{\underset{H}{|}}{CH_3-\overset{\overset{Br}{|}}{C}-CH_2CH_2CH_3}$$

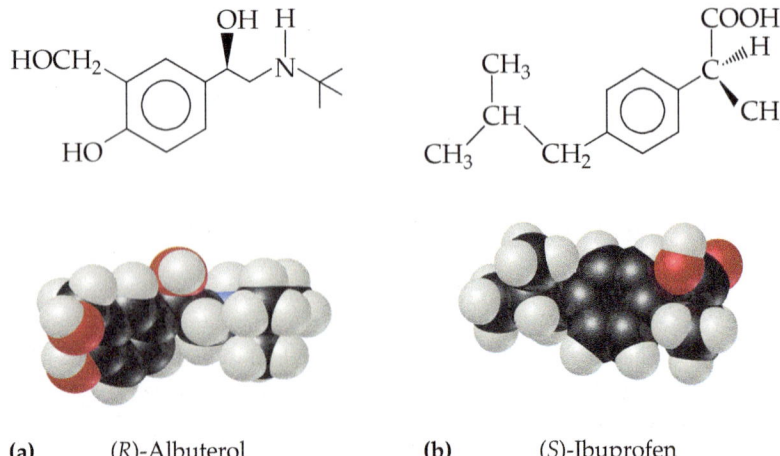

Go Figure If you replace Br with CH_3, will the compound be chiral?

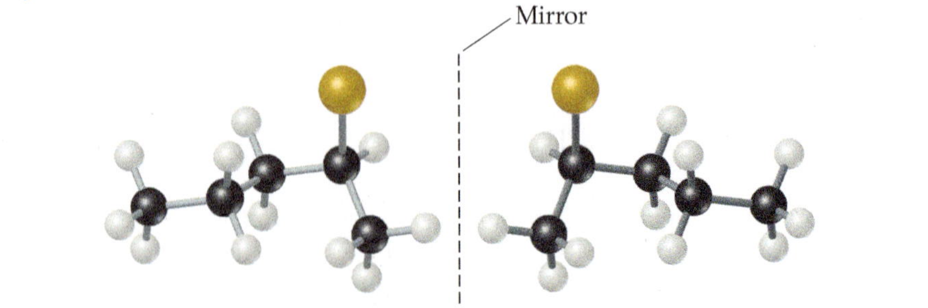

Mirror

▲ **Figure 24.14 The two enantiomeric forms of 2-bromopentane.** The mirror-image isomers are not superimposable on each other.

All four groups attached to the second carbon are different, making that carbon a chiral center. **Figure 24.14** illustrates the nonsuperimposable mirror images of this molecule. Imagine moving the molecule shown to the left of the mirror over to the right of the mirror. If you then turn it in every possible way, you should see that it cannot be superimposed on the molecule shown to the right of the mirror. Nonsuperimposable mirror images are called either *optical isomers* or *enantiomers*. (Section 23.4) Organic chemists use the labels *R* and *S* to distinguish the two forms. We will not go into the rules for deciding these labels.

The two members of an enantiomer pair have identical physical properties and identical chemical properties when they react with nonchiral reagents. Only in a chiral environment do they behave differently from each other. One interesting property of chiral substances is that their solutions may rotate the plane of polarized light, as explained in Section 23.4.

Chirality is common in organic compounds. It is often not observed, however, because when a chiral substance is synthesized in a typical reaction, the two enantiomers are formed in precisely the same quantity. The resulting mixture is called a *racemic mixture*, and it does not rotate the plane of polarized light because the two forms rotate the light to equal extents in opposite directions. (Section 23.4)

Many drugs are chiral compounds. When a drug is administered as a racemic mixture, often only one enantiomer has beneficial results. The other is often inert, or nearly so, or may even have a harmful effect. For example, the drug (*R*)-albuterol [**Figure 24.15(a)**] is a bronchodilator used to relieve the symptoms of asthma. The enantiomer (*S*)-albuterol is not only ineffective as a bronchodilator it actually counters the effects of (*R*)-albuterol. As another example, the nonsteroidal analgesic ibuprofen is a chiral molecule usually sold as the racemic mixture. However, a preparation consisting of just the more active enantiomer, (*S*)-ibuprofen [Figure 24.15(**b**)], relieves pain and reduces inflammation more rapidly than the racemic mixture. For this reason, the chiral version of the drug may in time come to replace the racemic one.

(a) (*R*)-Albuterol   **(b)** (*S*)-Ibuprofen

▲ **Figure 24.15** (*R*)-**Albuterol and (*S*)-ibuprofen. (a)** (*R*)-Albuterol, which acts as a bronchodilator in patients with asthma, is one member of an enantiomer pair. The other member, (*S*)-albuterol, has the OH group pointing down and does not have the same physiological effect. (**b**) For relieving pain and reducing inflammation, the ability of the (*S*) enantiomer of ibuprofen far exceeds that of the (*R*) isomer. In the (*R*) isomer, the positions of the H and CH_3 group on the far-right carbon are switched.

Self-Assessment Exercises

SAE 24.21 Which of the following statements is or are *true*?

   (i) To be chiral, an organic molecule must have a carbon atom with four different groups attached to it.

  (ii) To be chiral, an organic molecule must contain atoms other than C and H.

(iii) The two enantiomers of a chiral molecule are mirror images of one another.

(a) only i **(b)** both i and ii **(c)** both i and iii **(d)** both ii and iii **(e)** All three statements are true.

SAE 24.22 Draw out the structural formulas of the following organic compounds and use those drawings to determine which, if any, are chiral.

   (i) 2-chloro-3-methylpentane

  (ii) 2-bromo-2-chloropropane

(iii) 1-fluoro-2-chloropropane

(a) only i **(b)** both i and iii **(c)** both i and ii **(d)** both ii and iii **(e)** All three are chiral.

24.6 | Proteins

The functional groups discussed in Section 24.4 generate a vast array of molecules with very specific chemical reactivities. Nowhere is this specificity more apparent than in *biochemistry*—the chemistry of living organisms. The remainder of this chapter focuses on organic compounds that play a prominent role in plants, animals, fungi, and single-celled organisms like bacteria.

    As we move through the major classes of biomolecules, you will see some patterns emerging. Hydrogen bonding (Section 11.2), for example, is critical to the function of many biochemical systems, and the geometry of molecules (Section 9.1) can govern their biological importance and activity. Many of the large molecules in living systems are polymers (Section 12.6) of much smaller molecules. These **biopolymers** can be classified into three broad categories: *proteins*, *polysaccharides* (carbohydrates), and *nucleic acids*. *Lipids* are another common class of molecules in living systems, but they are usually large molecules, not biopolymers.

Learning Objectives

When you finish Section 24.6, you should be able to:

▶ Describe the features that are common to all amino acids as well as those that differentiate one amino acid from another.

▶ Interconvert between the structural formula and the name of a peptide or polypeptide.

▶ Analyze the structures of proteins, differentiating between primary, secondary, tertiary, and quaternary features of the structure.

Amino Acids

Proteins are macromolecules present in all living cells. About 50% of your body's dry mass is protein. Some proteins are structural components in animal tissues; they are a key part of skin, nails, cartilage, and muscles. Other proteins catalyze reactions, transport oxygen, serve as hormones to regulate specific body processes, and perform other tasks. To understand proteins, we must first become familiar with *amino acids*, the small molecules from which proteins are built.

    An **amino acid** is a molecule containing an amine group, $-NH_2$, and a carboxylic acid group, $-COOH$. The building blocks of all proteins are *α-amino acids*, where the α (alpha) indicates that the amino group is located on the carbon atom immediately adjacent to the carboxylic acid group. Thus, there is always one carbon atom between the amino group and the carboxylic acid group.

    The general formula for an α-amino acid is represented by

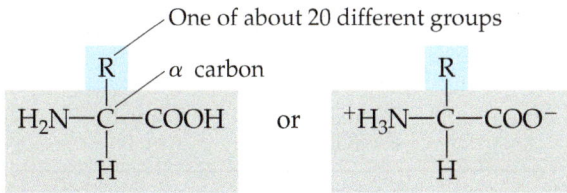

The doubly ionized form, called a *zwitterion*, usually predominates at near-neutral pH values. This form is a result of the transfer of a proton from the carboxylic acid group to the amine group. [See the "Chemistry and Life" box on "The Amphiprotic Behavior of Amino Acids" (Section 16.10)]

Amino acids differ from one another in the nature of their R groups. Twenty-two amino acids occur naturally in proteins, and **Figure 24.16** shows the 20 of these 22 found in humans. Our bodies can synthesize 11 of these 20 amino acids in sufficient amounts for our needs. The other 9 must be ingested and are called *essential amino acids* because they are necessary components of our diet.

The α-carbon atom of the amino acids, which is the carbon between the amino and carboxylate groups, has four different groups attached to it. The amino acids are thus chiral (except for glycine, which has two hydrogens attached to the central carbon). For historical reasons, the two enantiomeric forms of amino acids are often distinguished by

 Go Figure   Which group of amino acids has a net positive charge at pH 7?

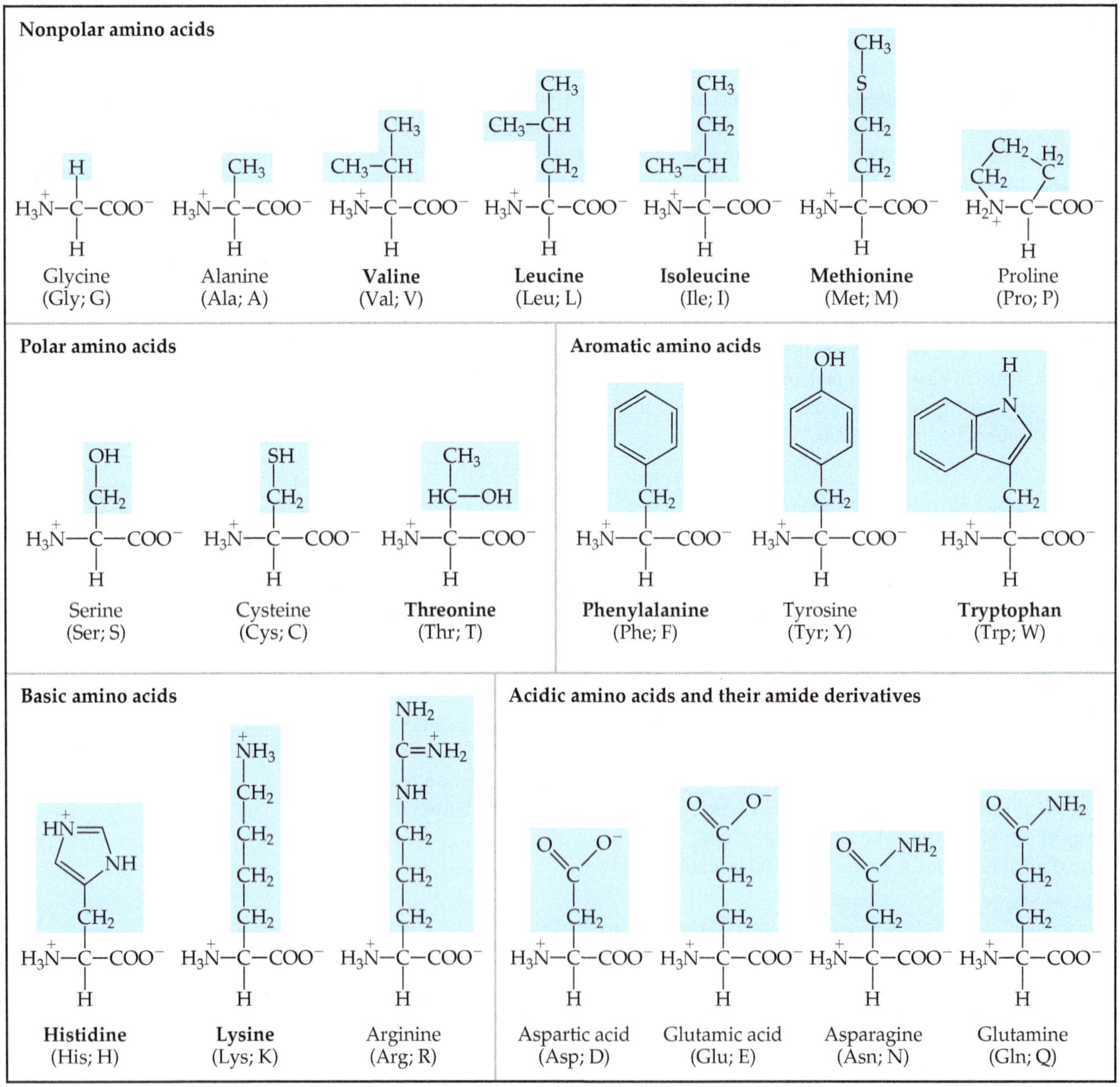

Nonpolar amino acids

Glycine (Gly; G) Alanine (Ala; A) **Valine (Val; V)** **Leucine (Leu; L)** **Isoleucine (Ile; I)** **Methionine (Met; M)** Proline (Pro; P)

Polar amino acids

Serine (Ser; S) Cysteine (Cys; C) **Threonine (Thr; T)**

Aromatic amino acids

Phenylalanine (Phe; F) Tyrosine (Tyr; Y) **Tryptophan (Trp; W)**

Basic amino acids

Histidine (His; H) **Lysine (Lys; K)** Arginine (Arg; R)

Acidic amino acids and their amide derivatives

Aspartic acid (Asp; D) Glutamic acid (Glu; E) Asparagine (Asn; N) Glutamine (Gln; Q)

▲ **Figure 24.16 The 20 amino acids found in the human body.** The blue screen identifies the different R groups for each amino acid. The acids are shown in the zwitterionic form in which they exist in water at near-neutral pH values. The names shown in bold are the nine essential amino acids that are necessary components of the human diet. Three-letter and one-letter abbreviations for each amino acid are shown in parentheses below its name.

the labels D (from the Latin *dexter*, "right") and L (from the Latin *laevus*, "left"). Nearly all the chiral amino acids found in living organisms have the L configuration at the chiral center. The principal exceptions are the proteins that make up the cell walls of bacteria, which can contain considerable quantities of the D isomers.

Polypeptides and Proteins

Amino acids are linked together into proteins by amide groups (Table 24.2):

$$R-\overset{\overset{\displaystyle O}{\|}}{C}-\underset{\underset{\displaystyle H}{|}}{N}-R$$

Each amide group is called a **peptide bond** when it is formed by amino acids. A peptide bond is formed by a condensation reaction between the carboxyl group of one amino acid and the amino group of another amino acid. Alanine and glycine, for example, form the dipeptide glycylalanine:

Glycine (Gly; G)   Alanine (Ala; A)

Glycylalanine (Gly–Ala; GA)   [24.23]

The amino acid that furnishes the carboxyl group for peptide-bond formation is named first, with a *-yl* ending; then the amino acid furnishing the amino group is named. Using the abbreviations shown in Figure 24.16, we can abbreviate glycylalanine as either Gly-Ala or GA. In this notation, it is understood that the unreacted amino group is on the left and the unreacted carboxyl group is on the right.

   The artificial sweetener *aspartame* (**Figure 24.17**) is the methyl ester of the dipeptide formed from the amino acids aspartic acid and phenylalanine.

Go Figure

How many chiral centers are there in one molecule of aspartame?

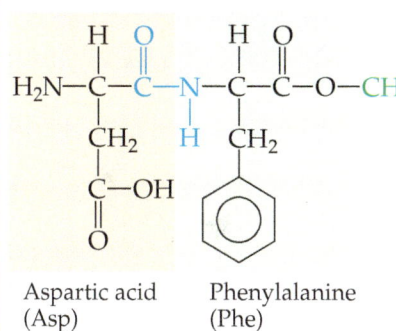

Aspartic acid (Asp)   Phenylalanine (Phe)

▲ **Figure 24.17 Sweet stuff.** The artificial sweetener aspartame is the methyl ester of a dipeptide.

Sample Exercise 24.5
Drawing the Structural Formula of a Tripeptide

Draw the structural formula for alanylglycylserine.

SOLUTION

Analyze We are given the name of a substance with peptide bonds and asked to write its structural formula.

Plan The name of this substance suggests that three amino acids—alanine, glycine, and serine—have been linked together, forming a *tripeptide*. Note that the ending *-yl* has been added to each amino acid except for the last one, serine. By convention, the sequence of amino acids in peptides and proteins is written from the nitrogen end to the carbon end: The first-named amino acid (alanine, in this case) has a free amino group, and the last-named one (serine) has a free carboxyl group.

Solve We first combine the carboxyl group of alanine with the amino group of glycine to form a peptide bond and then the

carboxyl group of glycine with the amino group of serine to form another peptide bond:

Amino group ────────→ Carboxyl group

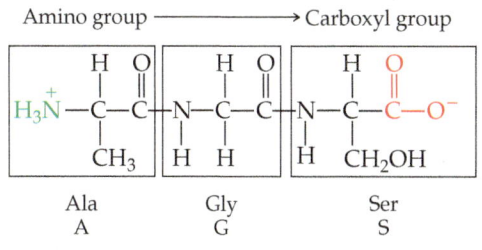

Ala   Gly   Ser
A   G   S

We can abbreviate this tripeptide as either Ala-Gly-Ser or AGS.

▶ **Practice Exercise**
Name the dipeptide and give the two ways of writing its
abbreviation.

$$\underset{\substack{|\\ HOCH_2}}{H_3\overset{+}{N}-\underset{}{C}}-\underset{}{\overset{\overset{O}{\|}}{C}}-\underset{\substack{|\\ H}}{N}-\underset{\substack{|\\ CH_2\\ |\\ COOH}}{\overset{H}{C}}-\overset{\overset{O}{\|}}{C}-O^-$$

Polypeptides are formed when a large number of amino acids (> 30) are linked together by peptide bonds. Proteins are linear (that is, unbranched) polypeptide molecules with molecular weights ranging from about 6000 to over 50 million amu. Because up to 22 different amino acids are linked together in proteins and because proteins consist of hundreds of amino acids, the number of possible arrangements of amino acids within proteins is virtually limitless.

Protein Structure

The sequence of amino acids from the "N terminus" (that is, the amino end) to the "C terminus" (the carboxylic acid end) along a protein chain is called its **primary structure** and gives the protein its unique identity. A change in even one amino acid can alter the biochemical characteristics of the protein. For example, sickle-cell anemia is a genetic disorder resulting from a single replacement in a protein chain in hemoglobin. The chain that is affected contains 146 amino acids. The substitution of an amino acid with a hydrocarbon side chain for one that has an acidic functional group in the side chain alters the solubility properties of the hemoglobin, and normal blood flow is impeded. [See the "Chemistry and Life" box on "Sickle-Cell Anemia" (Section 13.6)]

Proteins in living organisms are not simply long, flexible chains with random shapes. Rather, the chains self-assemble into structures based on the intermolecular forces we learned about in Chapter 11. This self-assembling leads to a protein's **secondary structure**, which refers to how segments of the protein chain are oriented in a regular pattern, as seen in Figure 24.18.

One of the most important and common secondary structure arrangements is the **α-helix** (alpha helix). The alpha helix is held in position by hydrogen bonds between amide H atoms and carbonyl O atoms in the main chain, not the side chains, of the protein. The pitch of the helix and its diameter must be such that (1) no bond angles are strained and (2) the N—H and C=O functional groups on adjacent turns are in proper position for hydrogen bonding. An arrangement of this kind is possible for some amino acids along the chain but not for others. Large protein molecules may contain segments of the chain that have the α-helical arrangement interspersed with sections in which the chain is in a random coil.

The other common secondary structure of proteins is the **β-sheet** (beta sheet). Beta sheets are made of two or more strands of peptides that hydrogen-bond from an amide H in one strand to a carbonyl O in the other strand. Like the alpha helix, the hydrogen bonding in beta sheets is between the peptide backbones, not the side chains.

Proteins are not active biologically unless they are in a particular shape in solution. The process by which the protein adopts its biologically active shape is called **folding**. The shape of a protein in its folded form—determined by all the bends, kinks, and sections of rodlike α-helical, β-sheet, or flexible coil components—is called the **tertiary structure**.

Globular proteins fold into a compact, roughly spherical shape. Globular proteins are generally soluble in water and are mobile within cells. They have nonstructural functions, such as combating the invasion of foreign objects, transporting and storing oxygen (hemoglobin and myoglobin), and acting as catalysts. The *fibrous proteins* form a second class of proteins. In these substances, the long coils align more or less in parallel to form long, water-insoluble fibers. Fibrous proteins provide structural integrity and strength to many kinds of tissue and are the main components of muscle, tendons, and hair. The largest known proteins, in excess of 27,000 amino acids long, are muscle proteins.

The tertiary structure of a protein is maintained by many different interactions. Certain foldings of the protein chain lead to lower-energy (more stable) arrangements than

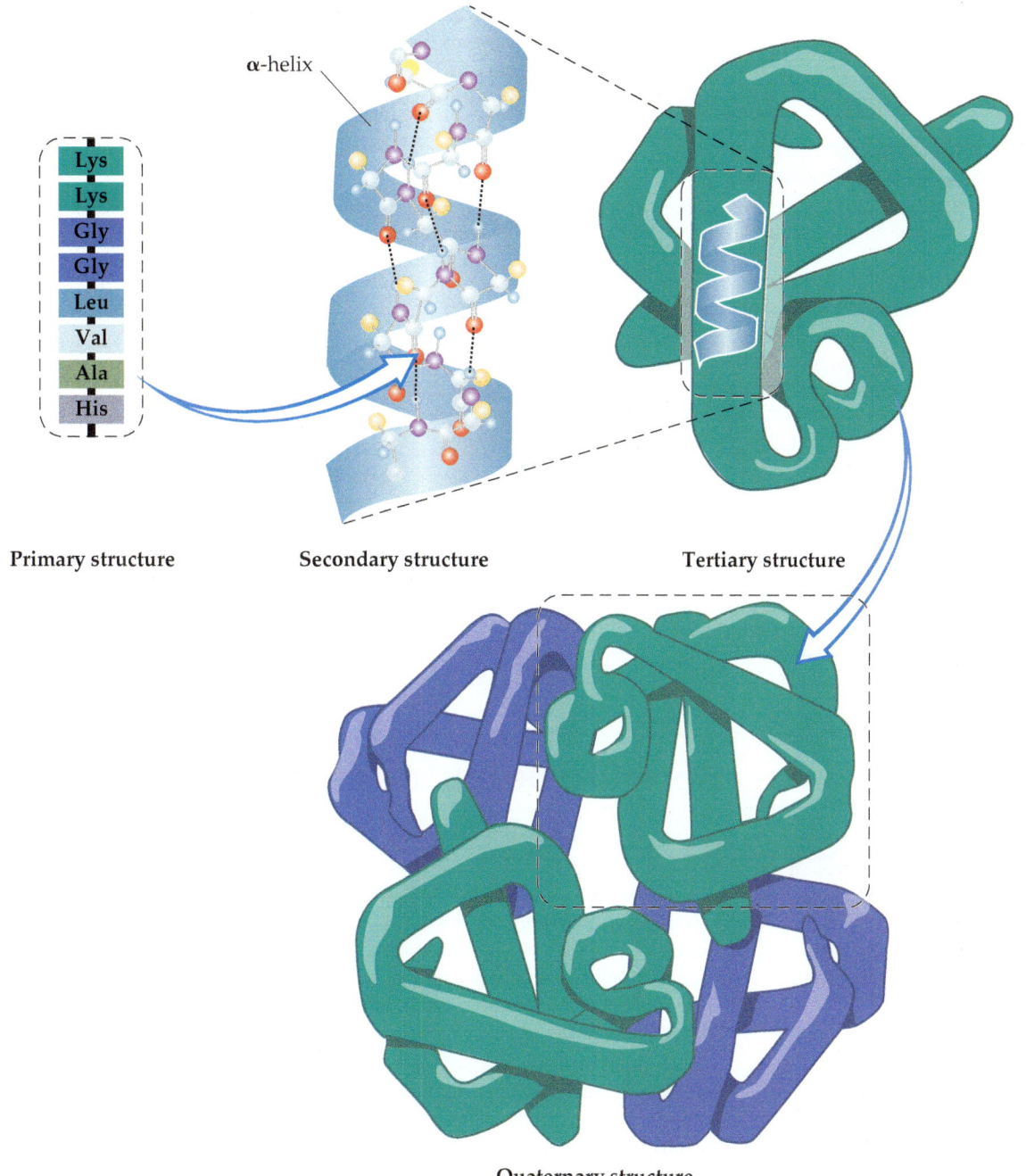

α-helix

| Lys |
| Lys |
| Gly |
| Gly |
| Leu |
| Val |
| Ala |
| His |

Primary structure   Secondary structure   Tertiary structure

Quaternary structure

▲ **Figure 24.18 The four levels of protein structure.** Amino acids, linked by amide bonds from the amine end to the acid end, can hydrogen-bond to form alpha helix or beta sheet secondary structures. These secondary structures fold into tertiary structures based on electrostatic and van der Waals interactions. Many proteins form quaternary structures in which multiple protein molecules associate to form dimers, trimers, or tetramers (as shown here).

do other folding patterns. For example, a globular protein dissolved in aqueous solution folds in such a way that the nonpolar hydrocarbon portions are tucked within the molecule, away from the polar water molecules. Most of the more polar acidic and basic side chains, however, project into the solution, where they can interact with water molecules through ion–dipole, dipole–dipole, or hydrogen-bonding interactions.

Some proteins are assemblies of more than one polypeptide chain. Each chain has its own tertiary structure, and two or more of these tertiary subunits may aggregate into a larger functional macromolecule. The way the tertiary subunits are arranged is called the **quaternary structure** of the protein (Figure 24.18). For example, hemoglobin, the oxygen-carrying protein of red blood cells, consists of four tertiary subunits. Each subunit contains a component called a heme with an iron atom that binds oxygen as depicted in Figure 23.15. The quaternary structure is maintained by the same types of interactions that maintain the tertiary structure.

One of the most fascinating current hypotheses in biochemistry is that misfolded proteins can cause infectious disease. These infectious misfolded proteins are called *prions*. The best example of a prion is the one thought to be responsible for mad cow disease, which can be transmitted to humans.

 Self-Assessment Exercises

SAE 24.23 Which chemical line structure represents an amino acid, where R is one of 20 side chains? (**a**) $H_2N-CHR-COOH$ (**b**) H_2N-CH_2-COOR (**c**) $HOOC-CH_2-CHR-NH_2$ (**d**) $HOOC-R-NH_2$

SAE 24.24 Which of the following statements about amino acids is or are *true*?

(**i**) All amino acids are chiral.
(**ii**) All amino acids have a net charge at pH 7.
(**iii**) Some amino acids are acidic and some are basic.

(**a**) only i (**b**) only ii (**c**) only iii (**d**) both i and iii (**e**) All three statements are true.

SAE 24.25 Which of the following statements about the peptide bond is or are *true*?

(**i**) Peptide bonds are formed from the hydrolysis reactions of amino acids.
(**ii**) The peptide bond contains a carbonyl group.
(**iii**) Any amino acid can make a peptide bond with either its amine group or its acid group.

(**a**) only i (**b**) only ii (**c**) only iii (**d**) both ii and iii (**e**) All three statements are true.

SAE 24.26 What is the name of the dipeptide shown here?

$$^-O-\overset{\overset{\displaystyle O}{\|}}{C}-\overset{\overset{\displaystyle H}{|}}{\underset{\underset{\displaystyle HS-CH_2}{|}}{C}}-\overset{\overset{\displaystyle H}{|}}{N}-\overset{\overset{\displaystyle O}{\|}}{C}-\overset{\overset{\displaystyle CH_3}{|}}{\underset{\underset{\displaystyle H}{|}}{C}}-NH_3^+$$

(**a**) glycylmethionine (G–A) (**b**) methionylalanine (M–A) (**c**) alanylcysteine (A–C) (**d**) cysteinylalanine (C–A) (**e**) alanylmethionine (A–M)

SAE 24.27 Hydrogen bonds play the dominant role in what aspect of a protein's structure? (**a**) primary structure (**b**) secondary structure (**c**) tertiary structure (**d**) quaternary structure

24.7 | Carbohydrates

Carbohydrates are an important class of naturally occurring substances found in both plant and animal matter. The name **carbohydrate** ("hydrate of carbon") comes from the empirical formulas for most substances in this class, which can be written as $C_x(H_2O)_y$. For example, **glucose**, the most abundant carbohydrate, has the molecular formula $C_6H_{12}O_6$, or $C_6(H_2O)_6$. Carbohydrates are not really hydrates of carbon; rather, they are polyhydroxy aldehydes and ketones. Glucose, for example, is a six-carbon aldehyde sugar, whereas *fructose*, the sugar that occurs widely in fruit, is a six-carbon ketone sugar (**Figure 24.19**).

The glucose molecule, having both alcohol and aldehyde functional groups and a reasonably long and flexible backbone, can form a six-member-ring structure, as shown in **Figure 24.20**. In fact, in an aqueous solution only a small percentage of the glucose molecules are in the open-chain form. Although the ring is often drawn as if it were planar, the molecules are actually nonplanar because of the tetrahedral bond angles around the C and O atoms of the ring.

Figure 24.20 shows that the ring structure of glucose can have two relative orientations. In the α form, the OH group on C1 and the CH_2OH group on C5 point in *opposite* directions, whereas in the β form, they point in the *same* direction. Although the difference between the α and β forms might seem small, it has enormous biological consequences, including the vast difference in properties between starch and cellulose.

Fructose can cyclize to form either five- or six-member rings. The five-member ring forms when the C5 hydroxyl group reacts with the C2 carbonyl group:

Learning Objectives

When you finish Section 24.7, you should be able to:

▶ Identify organic compounds that are carbohydrates from their molecular formula, structural formula, or condensed structural formula.

▶ Differentiate monosaccharides, disaccharides, and polysaccharides, and list common examples of each.

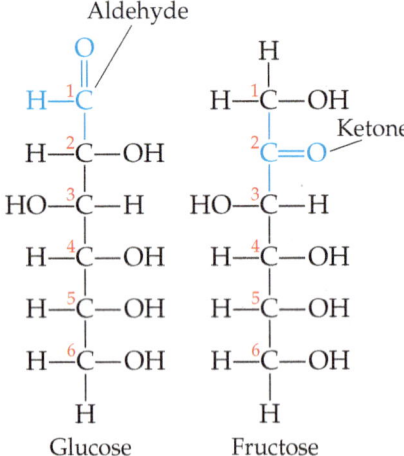

▲ **Figure 24.19** Linear structure of the carbohydrates glucose and fructose.

▲ **Figure 24.20** Cyclic glucose has an α form and a β form.

The six-member ring results from the reaction between the C6 hydroxyl group and the C2 carbonyl group.

Sample Exercise 24.6

Identifying Functional Groups and Chiral Centers in Carbohydrates

How many chiral carbon atoms are there in the open-chain form of glucose (Figure 24.19)?

SOLUTION

Analyze We are given the structure of glucose and asked to determine the number of chiral carbons in the molecule.

Plan A chiral carbon has four different groups attached (Section 24.5). We need to identify those carbon atoms in glucose.

Solve Carbons 2, 3, 4, and 5 each have four different groups attached to them:

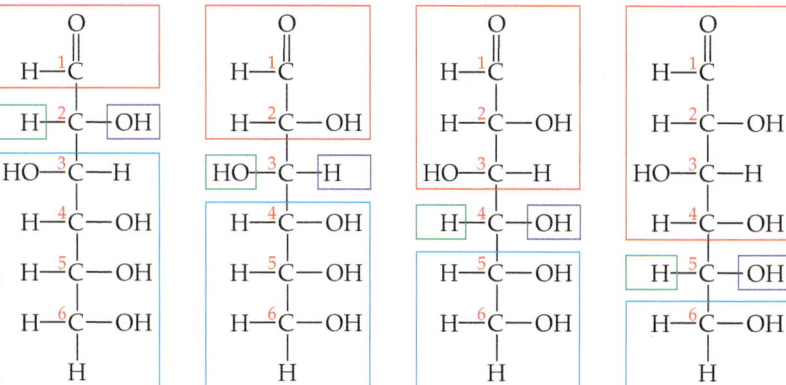

Thus, there are four chiral carbon atoms in the glucose molecule.

▶ **Practice Exercise**
   Name the functional groups present in the beta form of glucose.

Disaccharides

Both glucose and fructose are examples of **monosaccharides**, simple sugars that cannot be broken into smaller molecules by hydrolysis with aqueous acids. Two monosaccharide units can be linked together by a condensation reaction to form a **disaccharide**. The structures of two common disaccharides, *sucrose* (table sugar) and *lactose* (milk sugar), are shown in **Figure 24.21**.

The word *sugar* makes us think of sweetness. All sugars are sweet, but they differ in the degree of sweetness we perceive when we taste them. Sucrose is about six times sweeter than lactose, slightly sweeter than glucose, but only about half as sweet as fructose. Disaccharides

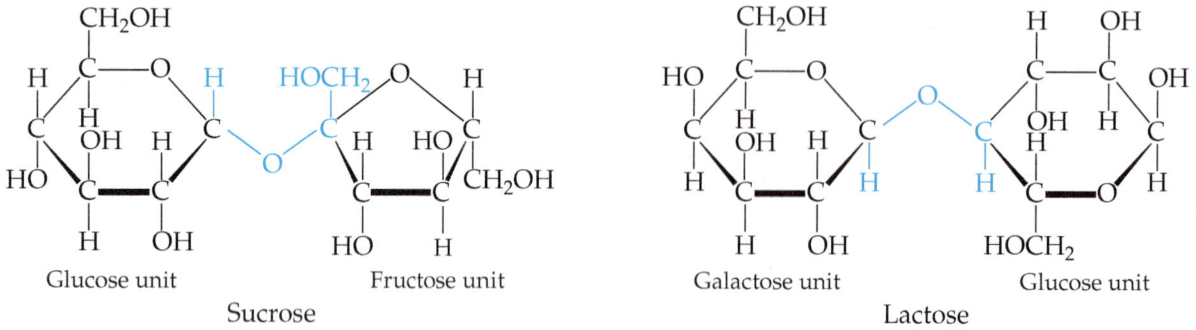

▲ **Figure 24.21 Two disaccharides.**

can be reacted with water (hydrolyzed) in the presence of an acid catalyst to form monosaccharides. When sucrose is hydrolyzed, the mixture of glucose and fructose that forms, called *invert sugar,** is sweeter to the taste than the original sucrose. The sweet syrup present in canned fruits and candies is largely invert sugar formed from hydrolysis of added sucrose.

Polysaccharides

Polysaccharides are made up of many monosaccharide units joined together. The most important polysaccharides are starch, glycogen, and cellulose, all three of which are formed from repeating glucose units.

Starch is not a pure substance. The term refers to a group of polysaccharides found in plants. Starches serve as a major method of food storage in plant seeds and tubers. Corn, potatoes, wheat, and rice all contain substantial amounts of starch. These plant products serve as major sources of needed food energy for humans. Enzymes in the digestive system catalyze the hydrolysis of starch to glucose.

Some starch molecules are unbranched chains, whereas others are branched. **Figure 24.22**(a) illustrates an unbranched starch structure. Notice, in particular, that the glucose units are in the α form, with the bridging oxygen atoms pointing in one direction and the CH$_2$OH groups pointing in the opposite direction.

Glycogen is a starch-like substance synthesized in the animal body. Glycogen molecules vary in molecular weight from about 5000 to more than 5 million amu. Glycogen acts as a kind of energy bank in the body. It is concentrated in the muscles and liver. In

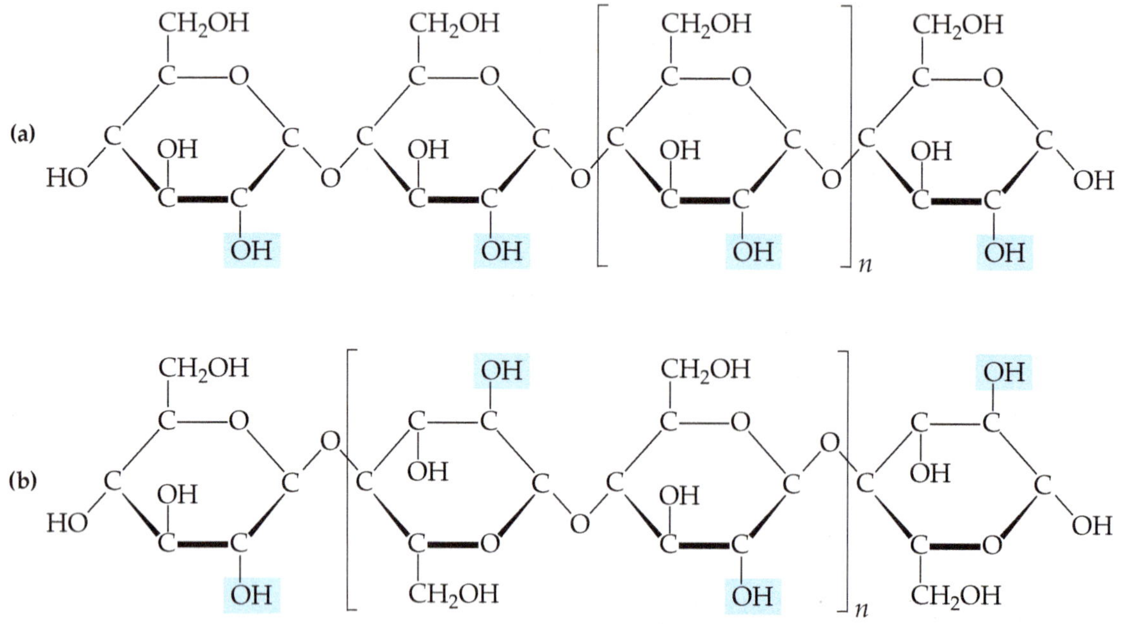

▲ **Figure 24.22 Structures of (a) starch and (b) cellulose.** Not all hydrogen atoms are shown.

---

*The term *invert sugar* comes from the fact that rotation of the plane of polarized light by the glucose-fructose mixture is in the opposite direction, or inverted, from that of the sucrose solution.

muscles, it serves as an immediate source of energy; in the liver, it serves as a storage place for glucose and helps to maintain a constant glucose level in the blood.

Cellulose [Figure 24.22(**b**)] forms the major structural unit of plants. Wood is about 50% cellulose; cotton fibers are almost entirely cellulose. Cellulose consists of an unbranched chain of glucose units, with molecular weights averaging more than 500,000 amu. At first glance, this structure looks very similar to that of starch. In cellulose, however, the glucose units are in the β form, with each bridging oxygen atom pointing in the same direction as the CH_2OH group in the ring to its left.

Because the individual glucose units have different relationships to one another in starch and cellulose, enzymes that readily hydrolyze starches do not hydrolyze cellulose. Thus, you might eat a kilogram of cellulose and receive no caloric value from it, even though the heat of combustion per unit mass is essentially the same for both cellulose and starch. A kilogram of starch, in contrast, would represent a substantial caloric intake. The difference is that the starch is hydrolyzed to glucose, which is eventually oxidized with the release of energy. However, enzymes in the body do not readily hydrolyze cellulose, so it passes through the digestive system relatively unchanged. Many bacteria contain enzymes, called cellulases, that hydrolyze cellulose. These bacteria are present in the digestive systems of grazing animals, such as cattle, that use cellulose for food.

 ## Self-Assessment Exercises

SAE 24.28 Which of the compounds shown here are carbohydrates?

(i)

(ii)

(iii)

(**a**) only i (**b**) only ii (**c**) only iii (**d**) both ii and iii (**e**) All three molecules are carbohydrates.

SAE 24.29 The open form and one possible ring form of fructose are shown here:

Both forms have hydroxy groups, just like alcohols. We can further specify the functional groups in each molecule by noting that the open form is _____ and the ring form is _____.
(**a**) an aldehyde, an aldehyde (**b**) an aldehyde, an ether (**c**) a ketone, a ketone (**d**) a ketone, an ester (**e**) a ketone, an ether

SAE 24.30 Which of the following compounds is a disaccharide? (**a**) glucose (**b**) sucrose (**c**) fructose (**d**) cellulose (**e**) starch

SAE 24.31 What is the relationship between the monosaccharides glucose and fructose (see Figure 24.19 for drawings of the open forms of these two molecules)?

(**a**) The two molecules are structural isomers.
(**b**) The two molecules are geometric isomers.
(**c**) The two molecules are optical isomers.
(**d**) The two molecules are not isomers.

24.8 | Lipids

Lipids are a diverse class of nonpolar biological molecules used by organisms for long-term energy storage (fats, oils) and as elements of biological structures (phospholipids, cell membranes, waxes).

⚠ **Learning Objectives**

When you finish Section 24.8, you should be able to:

▶ Describe the characteristic features of lipids.

▶ Distinguish various types of lipids from one another, including saturated fats, unsaturated fats, fatty acids, and phospholipids.

Fats

Fats are lipids derived from glycerol and fatty acids. Glycerol is an alcohol with three OH groups. Fatty acids are carboxylic acids (RCOOH) in which R is a hydrocarbon chain, usually 15 to 19 carbon atoms in length. Glycerol and fatty acids undergo condensation reactions to form ester linkages as shown in **Figure 24.23**. Three fatty acid molecules join to a glycerol. Although the three fatty acids in a fat can be the same, as they are in Figure 24.23, it is also possible that a fat contains three different fatty acids.

Lipids with saturated fatty acids are called saturated fats and are commonly solids at room temperature (such as butter and shortening). Unsaturated fats contain one or more double bonds in their carbon–carbon chains. The cis and trans nomenclature we learned for alkenes applies: Trans fats have H atoms on the opposite sides of the $C=C$ double bond, whereas cis fats have H atoms on the same sides of the $C=C$ double bond. Unsaturated fats (such as olive oil and peanut oil) are usually liquid at room temperature and are more often found in plants. For example, the major component (approximately 60 to 80%) of olive oil is oleic acid, cis-$CH_3(CH_2)_7CH=CH(CH_2)_7COOH$. Oleic acid is an example of a *monounsaturated* fatty acid, meaning it has only one carbon–carbon double bond in the chain. In contrast, *polyunsaturated* fatty acids have more than one carbon–carbon double bond in the chain.

For humans, trans fats are not nutritionally required, which is why some governments are moving to ban them in foods. How, then, do trans fats end up in our food? The process that converts unsaturated fats (such as oils) into saturated fats (such as shortening) is hydrogenation. (Section 24.3) The by-products of this hydrogenation process include trans fats.

Some of the fatty acids essential for human health must be available in our diets because our metabolism cannot synthesize them. These essential fatty acids are ones that have the carbon–carbon double bonds either three carbons or six carbons away from the —CH_3 end of the chain. These are called omega-3 and omega-6 fatty acids, where *omega* refers to the last carbon in the chain (the carboxylic acid carbon is considered the first, or alpha, one).

▼ **Go Figure** What structural features of a fat molecule cause it to be insoluble in water?

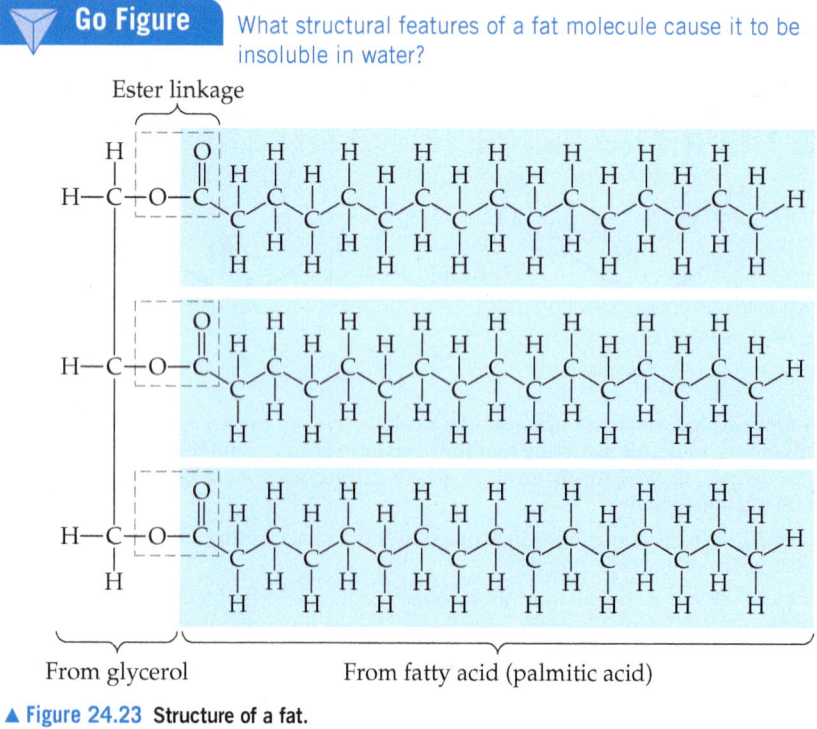

▲ Figure 24.23 **Structure of a fat.**

Go Figure Why do phospholipids form bilayers but not monolayers in water?

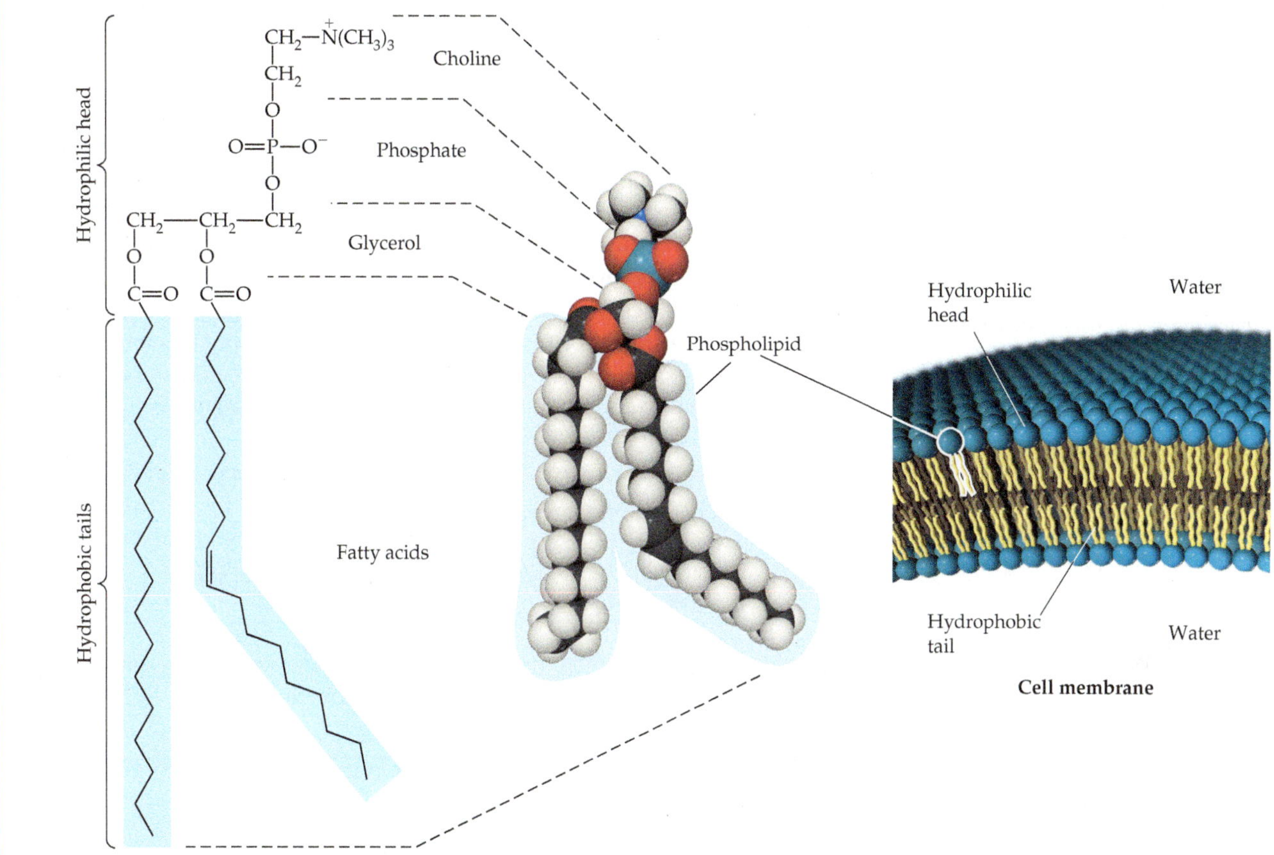

Choline

Phosphate

Glycerol

Phospholipid

Fatty acids

Hydrophilic head

Hydrophobic tails

Hydrophilic head

Water

Hydrophobic tail

Water

Cell membrane

▲ **Figure 24.24 Structure of a phospholipid and a cell membrane.** Living cells are encased in membranes typically made of phospholipid bilayers. The bilayer structure is stabilized by the favorable interactions of the hydrophobic tails of the phospholipids, which point away from both the water inside the cell and the water outside the cell, while the charged head groups face the two water environments.

Phospholipids

Phospholipids are similar in chemical structure to fats but have only two fatty acids attached to a glycerol. The third alcohol group of glycerol is joined to a phosphate group (Figure 24.24). The phosphate group can also be attached to a small charged or polar group, such as choline, as shown in the figure. The diversity in phospholipids is based on differences in their fatty acids and in the groups attached to the phosphate group.

In water, phospholipids cluster together, with their charged polar heads facing the water and their nonpolar tails facing inward. The phospholipids thus form a bilayer that is a key component of cell membranes (Figure 24.24).

Self-Assessment Exercises

SAE 24.32 When glycerol and a fatty acid undergo a _____ reaction, the products are a _____ and water? **(a)** condensation, fat **(b)** condensation, phospholipid **(c)** hydrogenation, fat **(d)** hydrogenation, phospholipid **(e)** hydrolysis, fat

SAE 24.33 Which of the following statements about lipids is or are *true*?

  (i) Unsaturated fats contain both C=C and C=O double bonds.
  (ii) Cell membranes are made from fatty acids.
 (iii) Saturated fats can be further subdivided into trans and cis fats.

(a) only i **(b)** only ii **(c)** only iii **(d)** both i and ii **(e)** both ii and iii

24.9 | Nucleic Acids

Learning Objectives

When you finish Section 24.9, you should be able to:

▶ Describe the composition and structures of nucleic acids, in both the single-strand and double-helix forms.

▶ Explain the differences between deoxyribonucleic acids (DNAs) and ribonucleic acids (RNAs).

▶ Determine the sequence of bases that are complementary on two strands of DNA or RNA, and describe the process of DNA replication.

Nucleic acids are a class of biopolymers that are the chemical carriers of an organism's genetic information. **Deoxyribonucleic acids (DNAs)** are huge molecules whose molecular weights may range from 6 to 16 million amu. **Ribonucleic acids (RNAs)** are smaller molecules, with molecular weights in the range of 20,000 to 40,000 amu. Whereas DNA is found primarily in the nucleus of the cell, RNA is found mostly outside the nucleus in the *cytoplasm*, the nonnuclear material enclosed by the cell membrane. DNA stores the genetic information of the cell and specifies which proteins the cell can synthesize. RNA carries the information stored by DNA out of the cell nucleus into the cytoplasm, where the information is used in protein synthesis.

The monomers of nucleic acids, called **nucleotides**, are formed from a five-carbon sugar, a nitrogen-containing organic base, and a phosphate group. An example is shown in Figure 24.25.

The five-carbon sugar in RNA is *ribose*, and that in DNA is *deoxyribose*:

Ribose   Deoxyribose

Deoxyribose differs from ribose only in having one fewer oxygen atom at carbon 2.

There are five nitrogen-containing bases in nucleic acids:

Adenine (A)   Guanine (G)   Cytosine (C)   Thymine (T)   Uracil (U)
DNA   DNA   DNA   DNA   RNA
RNA   RNA   RNA

The first three bases shown here are found in both DNA and RNA. Thymine occurs only in DNA, whereas uracil occurs only in RNA. In either nucleic acid, each base is attached to a five-carbon sugar through a bond to the nitrogen atom shown in color.

NH₂
N-containing base unit
Phosphate unit
Five-carbon sugar unit

▶ **Figure 24.25 A nucleotide.** Structure of deoxyadenylic acid, the nucleotide formed from phosphoric acid, the sugar deoxyribose, and the organic base adenine.

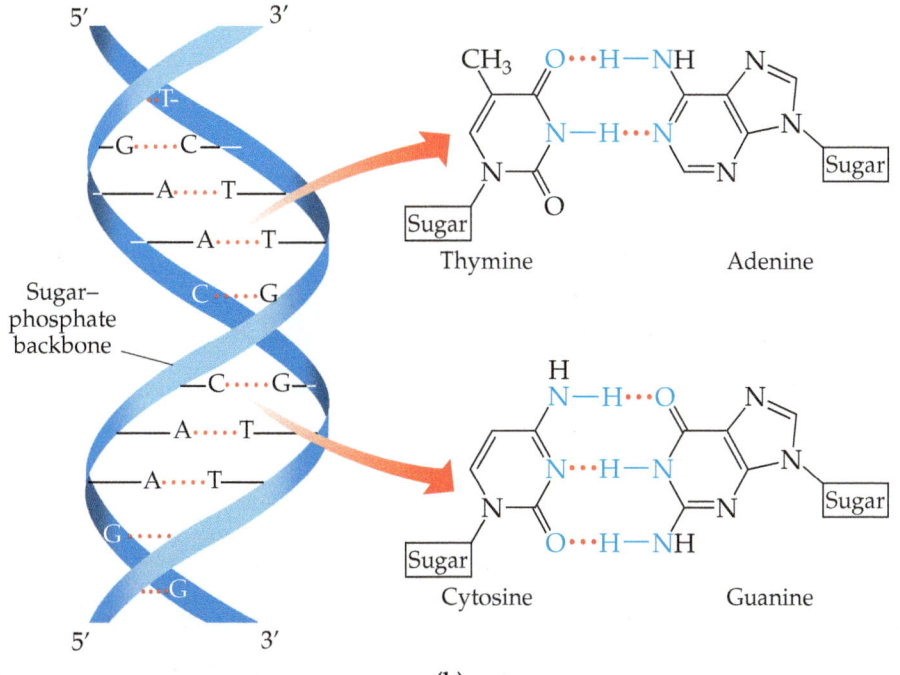

The nucleic acids RNA and DNA are *polynucleotides* formed by condensation reactions between a phosphoric acid OH group on one nucleotide and a sugar OH group on another nucleotide. Thus, the polynucleotide strand has a backbone consisting of alternating sugar and phosphate groups with the bases extending off the chain as side groups (**Figure 24.26**). The carbons in the sugars are numbered 1′, 2′, and so forth, as shown in Figure 24.26. Just as proteins have a sequence of amino acids from the N terminus to the C terminus, nucleic acids have a sequence of bases that start from the 5′ end of the sugar–phosphate backbone and go to the 3′ end.

The DNA strands wind together in an antiparallel **double helix** [**Figure 24.27(a)**]. The two strands are held together by attractions between bases (represented by T, A, C, and G). These attractions involve dispersion forces, dipole–dipole forces, and hydrogen bonds. (Section 11.2) As shown in Figure 24.27(**b**), the structures of thymine and adenine

Go Figure Which pair of complementary bases, AT or GC, would you expect to be more difficult to dissociate from each other?

(a) (b)

▲ **Figure 24.27** **DNA and bonding between complementary bases.** (a) The DNA double helix, showing the sugar–phosphate backbone as a pair of ribbons and dotted lines to indicate hydrogen bonding between the complementary bases. (b) Structures of the complementary base pairs in DNA.

CHEMISTRY AND LIFE   COVID-19 Vaccines

In 2019, a virus known as SARS-CoV-2 was first reported in humans. This virus led to the disease known as COVID-19, which to date has killed millions of people on the planet. This virus consists of a single strand of RNA, approximately 30,000 nucleotides long, enveloped in a protein coat (Figure 24.28). There are three types of protein on the surface of the SARS-CoV-2 virus, which are referred to as the spike (S), membrane (M), and envelope (E) proteins. The S protein is the one that recognizes a common protein on the cell membrane of the host, docks the virus there, and then fuses the virus with the cell membrane. Once inside the cell, the virus uses the host's cellular machinery to replicate its RNA over and over.

Understanding how this virus works has been key to developing vaccines to combat it. The current Pfizer-BioNTech and Moderna vaccines are based on a type of RNA called messenger RNA or mRNA for short. mRNA is a single-stranded RNA molecule that is complementary to a strand of DNA that codes for a specific gene. Normally, the mRNA molecules are made in the cell nucleus and then migrate into the cytoplasm, where the translation machinery of the cell binds to them, reads the code on the mRNA, and makes a specific protein.

Both the Pfizer-BioNTech and Moderna vaccines contain mRNA molecules that carry the instructions to produce a modified spike protein. In the vaccine, the mRNA is mixed with lipids and salts to stabilize it. Once injected, the body's cells take up the mRNA and produce the modified spike protein, which is subsequently displayed on the surface of the cell. The modified protein contains proline residues in certain spots that make the protein look like it has just bound to the outer surface of the cell membrane but has not yet fused with the membrane. The immune system detects these modified (and harmless) versions of the spike protein and generates antibodies against it. These antibodies, large proteins themselves, circulate in the body ready to combat real SARS-CoV-2 viruses that may enter the body, thereby providing immunity.

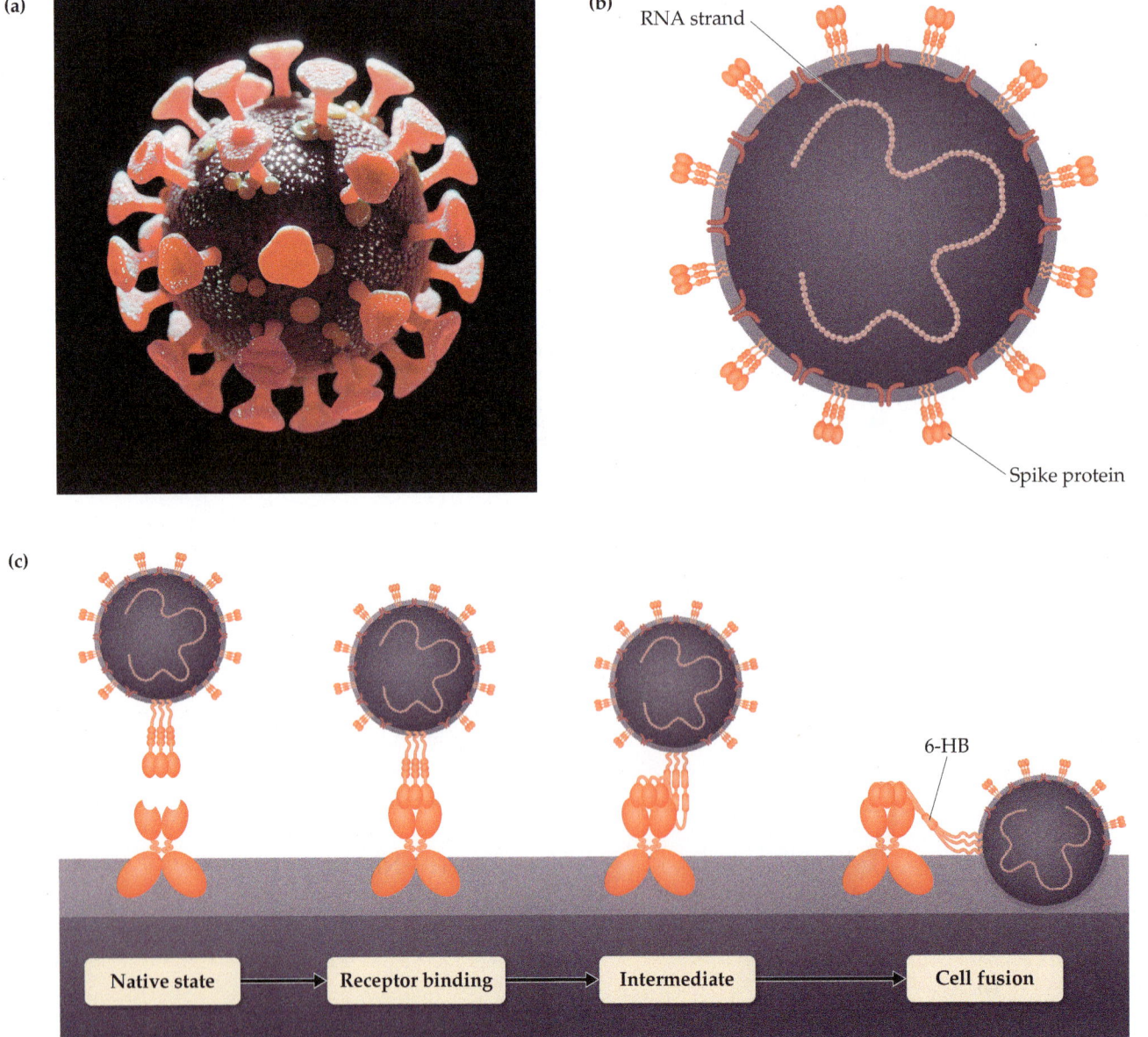

(a)

(b)

RNA strand

Spike protein

(c)

6-HB

Native state → Receptor binding → Intermediate → Cell fusion

▲ **Figure 24.28 The SARS-CoV-2 virus. (a)** A microscopic image of the virus where the spike proteins on the surface are clearly visible. **(b)** A schematic view of the virus. **(c)** The mechanism by which the virus docks and then penetrates the cell membrane to release its RNA into the cell.

The traditional approach to vaccines is to introduce either dead/inactivated virus cells or weakened virus cells to trigger an immune response. The Sinovac and Sinopharm COVID-19 vaccines, developed in China, take this approach by using inactivated SARS-CoV-2 virus cells. The Johnson & Johnson (Jannsen) vaccine uses yet another approach, called a virus vector, to provoke an immune response. In this technology, a piece of DNA encoded with instructions to create modified spike protein is embedded inside an unrelated virus (a virus similar to the one that causes the common cold). The virus delivers this piece of DNA to your cells upon which the cell machinery builds modified spike proteins. From there the process unfolds very much as it does with the mRNA vaccines.

The COVID-19 vaccines were developed at an unprecedentedly rapid pace. Most vaccines take 10–15 years to develop, test, and gain all the necessary government approvals. The fastest vaccination development prior to the outbreak of COVID-19 was the mumps vaccine, which was developed in the 1960s in a mere 4 years. The rapid development and effectiveness of mRNA and viral vector vaccines to protect against this deadly disease are a triumph of modern science, built upon decades of fundamental biochemical research that has saved countless lives.

make them perfect partners for hydrogen bonding. Likewise, cytosine and guanine form ideal hydrogen-bonding partners. We say that thymine and adenine are *complementary* to each other and cytosine and guanine are *complementary* to each other. In the double-helix structure, therefore, each thymine on one strand is opposite an adenine on the other strand, and each cytosine is opposite a guanine. The double-helix structure with complementary bases on the two strands is the key to understanding how DNA functions.

The two strands of DNA unwind during cell division, and new complementary strands are constructed on the unraveling strands (**Figure 24.29**). This process results in two identical double-helix DNA structures, each containing one strand from the original structure and one new strand. This replication allows genetic information to be transmitted when cells divide.

The structure of DNA is also the key to understanding protein synthesis, the means by which viruses infect cells, and many other problems of central importance to modern biology. These themes are beyond the scope of this book. If you take courses in the life sciences, however, you will learn a good deal about such matters.

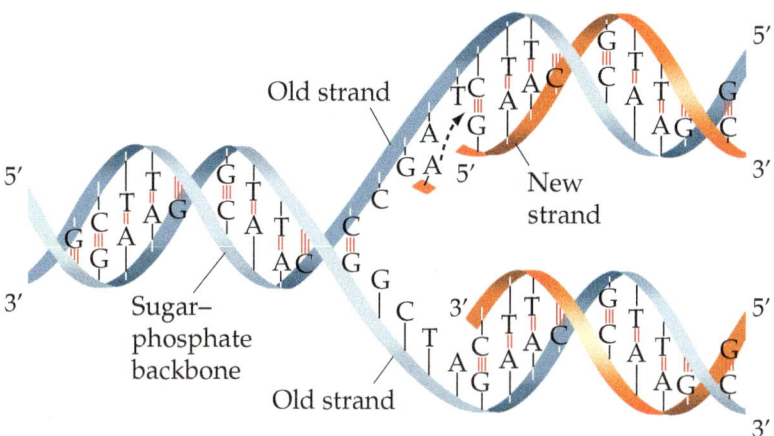

▲ **Figure 24.29 DNA replication.** The original DNA double helix partially unwinds, and new nucleotides line up on each strand in complementary fashion. Hydrogen bonds help align the new nucleotides with the original DNA chain. When the new nucleotides are joined by condensation reactions, two identical double-helix DNA molecules result.

 Self-Assessment Exercises

SAE 24.34 Which of the following molecular fragments is *not* part of the structure of a nucleic acid? **(a)** nitrogen-containing base **(b)** amino acid **(c)** phosphate group **(d)** a sugar ring containing five carbon atoms

SAE 24.35 Which base is found in RNA but not in DNA? **(a)** adenine **(b)** guanine **(c)** cytosine **(d)** thymine **(e)** uracil

SAE 24.36 Which of the following statements describing the structure and/or function of RNA is *false*?

(a) RNA is found primarily inside the cell nucleus.
(b) RNA carries information stored by DNA.
(c) RNA is used in protein synthesis.
(d) RNA is generally a smaller molecule than DNA.
(e) The sugar rings in RNA have an additional hydroxyl group compared to the sugar rings in DNA.

SAE 24.37 A short strand of DNA is found to have the sequence 5′–TTGCA–3′. What is the sequence of the complementary DNA strand? **(a)** 3′–TTGCA–5′ **(b)** 3′–TTGGG–5′ **(c)** 3′–GGATC–5′ **(d)** 3′–AACGT–5′

Putting Concepts Together

Pyruvic acid,

$$CH_3-\overset{\overset{\displaystyle O}{\|}}{C}-\overset{\overset{\displaystyle O}{\|}}{C}-OH$$

is formed in the body from carbohydrate metabolism. In muscles, it is reduced to lactic acid in the course of exertion. The acid-dissociation constant for pyruvic acid is 3.2×10^{-3}. **(a)** Why does pyruvic acid have a larger acid-dissociation constant than acetic acid? **(b)** Would you expect pyruvic acid to exist primarily as the neutral acid or as dissociated ions in muscle tissue, assuming a pH of 7.4 and an initial acid concentration of 2×10^{-4} M? **(c)** What would you predict for the solubility properties of pyruvic acid? Explain. **(d)** What is the hybridization of each carbon atom in pyruvic acid? **(e)** Assuming H atoms as the reducing agent, write a balanced chemical equation for the reduction of pyruvic acid to lactic acid (Figure 24.13). (Although H atoms do not exist as such in biochemical systems, biochemical reducing agents deliver hydrogen for such reductions.)

SOLUTION

(a) The acid-dissociation constant for pyruvic acid should be somewhat greater than that of acetic acid because the carbonyl function on the α-carbon atom of pyruvic acid exerts an electron-withdrawing effect on the carboxylic acid group. In the C—O—H bond system, the electrons are shifted from H, facilitating loss of the H as a proton. (Section 16.10)

(b) To determine the extent of ionization, we first set up the ionization equilibrium and equilibrium-constant expression. Using HPv as the symbol for the acid, we have

$$HPv \rightleftharpoons H^+ + Pv^-$$

$$K_a = \frac{[H^+][Pv^-]}{[HPv]} = 3.2 \times 10^{-3}$$

Let $[Pv^-] = x$. Then the concentration of undissociated acid is $2 \times 10^{-4} - x$. The concentration of $[H^+]$ is fixed at 4.0×10^{-8} M (the antilog of the pH value). Substituting, we obtain

$$3.2 \times 10^{-3} = \frac{(4.0 \times 10^{-8})(x)}{(2 \times 10^{-4} - x)}$$

Solving for x, we obtain

$$x = [Pv^-] = 2 \times 10^{-4}\, M$$

This is the initial concentration of acid, which means that essentially all the acid has dissociated. We might have expected this result because the acid is quite dilute and the acid-dissociation constant is fairly high.

(c) Pyruvic acid should be quite soluble in water because it has polar functional groups and a small hydrocarbon component. We would predict it would be soluble in polar organic solvents, especially ones that contain oxygen. In fact, pyruvic acid dissolves in water, ethanol, and diethyl ether.

(d) The methyl group carbon has sp^3 hybridization. The carbon of the carbonyl group has sp^2 hybridization because of the double bond to oxygen. Similarly, the carboxylic acid carbon is sp^2 hybridized.

(e) The balanced chemical equation for this reaction is

$$\overset{\overset{\displaystyle O}{\|}}{CH_3CCOOH} + 2\,(H) \longrightarrow CH_3\overset{\overset{\displaystyle OH}{|}}{\underset{\underset{\displaystyle H}{|}}{C}}COOH$$

Essentially, the ketonic functional group has been reduced to an alcohol.

STRATEGIES FOR SUCCESS | What Now?

If you are reading this box, you have made it to the end of our text. We congratulate you on the tenacity and dedication that you have exhibited to make it this far!

As an epilogue, we offer the ultimate study strategy in the form of a question: What do you plan to do with the knowledge of chemistry that you have gained thus far in your studies? Many of you will enroll in additional courses in chemistry as part of your required curriculum. For others, this will be the last formal course in chemistry that you will take. Regardless of the career path you plan to take—whether it is chemistry, one of the biomedical fields, engineering, the liberal arts, or another field—we hope that this text has increased your appreciation of the chemistry in the world around you. As you will observe, chemistry is everywhere, from food and pharmaceuticals to solar cells and sports equipment.

We have also tried to give you a sense that chemistry is a dynamic, continuously changing science. Research chemists synthesize new compounds, develop new reactions, uncover chemical properties that were previously unknown, find new applications for known compounds, and refine theories. The understanding of biological systems in terms of the underlying chemistry has become increasingly important as new levels of complexity are uncovered. Solving the global challenges of sustainable energy and clean water requires the work of many chemists. We encourage you to participate in the fascinating world of chemical research by enrolling in an undergraduate research program. Given all the answers that chemists seem to have, you may be surprised at the large number of questions that they still find to ask.

Finally, we hope you have enjoyed using this textbook. We certainly enjoyed sharing our experiences and perspectives on the fascinating world of chemistry with you. We truly believe it to be the central science, one that benefits all who learn about it and from it.

Chapter Summary and Key Terms

GENERAL CHARACTERISTICS OF ORGANIC COMPOUNDS (INTRODUCTION AND SECTION 24.1) This chapter introduces **organic chemistry**, which is the study of compounds containing carbon–hydrogen bonds, and **biochemistry**, which is the study of the chemistry of living organisms. C—C single bonds and C—H bonds tend to have low reactivity. Those bonds that have a high-electron density (such as multiple bonds or bonds with an atom of high electronegativity) tend to be the sites of reactivity in an organic compound. These sites of reactivity are called **functional groups**.

INTRODUCTION TO HYDROCARBONS (SECTION 24.2) The simplest types of organic compounds are hydrocarbons, those composed of only carbon and hydrogen. There are four major kinds of hydrocarbons: alkanes, alkenes, alkynes, and aromatic hydrocarbons. **Alkanes** are composed of only C—H and C—C single bonds. **Alkenes** contain one or more carbon–carbon double bonds. **Alkynes** contain one or more carbon–carbon triple bonds. **Aromatic hydrocarbons** contain cyclic arrangements of carbon atoms bonded through both σ and delocalized π bonds. Alkanes are saturated hydrocarbons; the others are unsaturated.

Alkanes may form straight-chain, branched-chain, and cyclic arrangements. Isomers are substances that possess the same molecular formula but differ in the arrangements of atoms. In **structural isomers**, the bonding arrangements of the atoms differ. Different isomers are given different systematic names. The naming of hydrocarbons is based on the longest continuous chain of carbon atoms in the structure. The locations of **alkyl groups**, which branch off the chain, are specified by numbering along the carbon chain.

Alkanes with ring structures are called **cycloalkanes**. Alkanes are relatively unreactive. They do, however, undergo combustion in air, and their chief use is as a source of heat produced by combustion reactions.

ALKENES, ALKYNES, AND AROMATIC HYDROCARBONS (SECTION 24.3) The names of alkenes and alkynes are based on the longest continuous chain of carbon atoms that contains the multiple bond, and the location of the multiple bond is specified by a numerical prefix. Alkenes exhibit not only structural isomerism but geometric (cis–trans) isomerism as well. In **geometric isomers**, the bonds are the same, but the molecules have different geometries. Geometric isomerism is possible in alkenes because rotation about the C=C double bond is restricted.

Alkenes and alkynes readily undergo **addition reactions** to the carbon–carbon multiple bonds. Addition reactions are difficult to carry out with aromatic hydrocarbons, but **substitution reactions** are easily accomplished in the presence of catalysts.

ORGANIC FUNCTIONAL GROUPS (SECTION 24.4) The chemistry of organic compounds is dominated by the nature of their functional groups. The functional groups we have considered are

R, R' and R" represent hydrocarbon groups—for example, methyl (CH_3) or phenyl (C_6H_5).

Alcohols are hydrocarbon derivatives containing one or more OH groups. **Ethers** are formed by a condensation reaction of two molecules of alcohol. Several functional groups contain the **carbonyl (C=O) group**, including **aldehydes, ketones, carboxylic acids, esters**, and **amides**. Aldehydes and ketones can be produced by the oxidation of certain alcohols. Further oxidation of the aldehydes produces carboxylic acids. Carboxylic acids can form esters by a condensation reaction with alcohols, or they can form amides by a condensation reaction with amines. Esters undergo **hydrolysis** (**saponification**) in the presence of strong bases.

CHIRALITY IN ORGANIC CHEMISTRY (SECTION 24.5) Molecules that possess nonsuperimposable mirror images are said to be **chiral**. The two nonsuperimposable forms of a chiral molecule are called *enantiomers*. In carbon compounds, a chiral center is created when all four groups bonded to a central carbon atom are different. Many of the molecules occurring in living systems, such as the amino acids, are chiral and exist in nature in only one enantiomeric form. Many drugs of importance in human medicine are chiral, and the enantiomers may produce very different biochemical effects.

PROTEINS (SECTION 24.6) Many of the molecules that are essential for life are large natural polymers that are constructed from smaller molecules called monomers. Three of these **biopolymers** are considered in this chapter: proteins, polysaccharides (carbohydrates), and nucleic acids.

Proteins are polymers of **amino acids**. They are the major structural materials in animal systems. All naturally occurring proteins are formed from 22 amino acids, although only 20 are common. The amino acids are linked by **peptide bonds**. A **polypeptide** is a polymer formed by linking many amino acids by peptide bonds.

Amino acids are chiral substances. Usually, only one of the enantiomers is found to be biologically active. Protein structure is determined by the sequence of amino acids in the chain (its **primary structure**), the intramolecular interactions within the chain (its **secondary structure**), and the overall shape of the complete molecule (its **tertiary structure**). Two important secondary structures are the **α-helix** and the **β-sheet**. The process by which a protein assumes its biologically active tertiary structure is called **folding**. Sometimes several proteins aggregate together to form a **quaternary structure**.

CARBOHYDRATES AND LIPIDS (SECTIONS 24.7 AND 24.8) **Carbohydrates**, which are polyhydroxy aldehydes and ketones, are the major structural constituents of plants and are a source of energy in both plants and animals. **Glucose** is the most common **monosaccharide**, or simple sugar. Two monosaccharides can be linked together by means of a condensation reaction to form a **disaccharide**. **Polysaccharides** are complex carbohydrates made up of many monosaccharide units joined together. The three most important polysaccharides are **starch** and **cellulose**, both of which are found in plants, and **glycogen**, which is found in mammals.

Lipids are compounds formed from condensation reactions between glycerol and fatty acids. They include fats and **phospholipids**. Fatty acids can be saturated, unsaturated, cis, or trans, depending on their chemical formulas and structures.

NUCLEIC ACIDS (SECTION 24.9) **Nucleic acids** are biopolymers that carry the genetic information necessary for cell reproduction; they also control cell development through control of protein synthesis. The building blocks of these biopolymers are **nucleotides**. There are two types of nucleic acids, **ribonucleic acid (RNA)** and

deoxyribonucleic acid (DNA). These substances consist of a polymeric backbone of alternating phosphate and ribose or deoxyribose sugar groups with organic bases attached to the sugar molecules. The DNA polymer is a double-stranded helix (**double helix**) held together by hydrogen bonding between matching organic bases situated across from one another on the two strands. The hydrogen bonding between specific base pairs is the key to gene replication and protein synthesis.

Exam Prep

EP 24.1 Predict the bond angle of the $O-C-O$ bond in the following molecule:

$$CH_3CH_2\overset{\overset{\displaystyle O}{\|}}{C}-OH$$

(**a**) exactly 90°

(**b**) slightly greater than 90°

(**c**) exactly 120°

(**d**) slightly less than 120°

(**e**) slightly greater than 120°

EP 24.2 How many distinct locations are there for a double bond in a six-carbon linear chain? (**a**) 1 (**b**) 2 (**c**) 3 (**d**) 4 (**e**) 5

EP 24.3 Which of the following compounds does *not* exist? (**a**) 1,3,5,7-octatetraene (**b**) *cis*-2-butane (**c**) *trans*-3-hexene (**d**) 1-propene (**e**) *cis*-4-decene

EP 24.4 How many hydrogen atoms are in 3-ethyl-2,3-dimethylpentane? (**a**) 5 (**b**) 9 (**c**) 12 (**d**) 18 (**e**) 20

EP 24.5 What is the relationship between 2,3-dimethylpentane and 3-methylhexane?

(**a**) They are geometric isomers.

(**b**) They are structural isomers.

(**c**) They are optical isomers (enantiomers).

(**d**) They are *not* isomers.

EP 24.6 The following compound is a(n) _____ with a molecular formula of _____.

(**a**) aromatic hydrocarbon, C_6ClOH

(**b**) aromatic hydrocarbon, C_6H_4ClOH

(**c**) aromatic hydrocarbon, C_6H_8ClOH

(**d**) alkene, C_6H_4ClOH

(**e**) alkene, C_6H_8ClOH

EP 24.7 Which of the following molecules is capable of geometric isomerism?

 (**i**)   1-butene

 (**ii**)   2-butene

(**a**) only i (**b**) only ii (**c**) both i and ii (**d**) neither i nor ii

EP 24.8 What product is formed from the hydrogenation of 2-methylpropene? (**a**) propane (**b**) butane (**c**) 2-methylbutane (**d**) 2-methylpropane (**e**) 2-methylpropyne

EP 24.9 Which types of organic compounds readily undergo addition reactions?

 (**i**)   alkenes

 (**ii**)   alkynes

 (**iii**)   aromatic hydrocarbons

(**a**) only i (**b**) only iii (**c**) both i and ii (**d**) both i and iii (**e**) All three readily undergo addition reactions.

EP 24.10 The amino acid glycine contains which functional groups?

$$H-\overset{\overset{\displaystyle H}{|}}{N}-\overset{\overset{\displaystyle H}{|}}{\underset{\underset{\displaystyle H}{|}}{C}}-\overset{\overset{\displaystyle O}{\|}}{C}-O-H$$

(**a**) amine, alcohol

(**b**) amide, alcohol

(**c**) amine, carboxylic acid

(**d**) amide, carboxylic acid

(**e**) amine, alkene, alcohol

EP 24.11 A condensation reaction between _____ and _____ results in the formation of an ester.

(**a**) an aldehyde, alcohol

(**b**) an alcohol, alcohol

(**c**) a carboxylic acid, alcohol

(**d**) a carboxylic acid, amine

(**e**) a carboxylic acid, carboxylic acid

EP 24.12 How many chiral carbons are in the following molecule?

$$\underset{H}{\overset{CH_3CH_2}{\diagup}}C=C\underset{H}{\overset{CH_2\overset{\overset{\displaystyle CH_3}{|}}{CH}CH_2CH_3}{\diagup}}$$

(**a**) 1 (**b**) 2 (**c**) 3 (**d**) 5 (**e**) 8

EP 24.13 How many nitrogen atoms are in the tripeptide Arg-Asp-Gly? (**a**) 3 (**b**) 4 (**c**) 5 (**d**) 6 (**e**) 7

EP 24.14 Which statement describing the structure and composition of different amino acids is *false*?

(**a**) The carboxylic acid and amine groups are always bonded to the α-carbon.

(**b**) All amino acids can form zwitterions.

(**c**) The hybridization at the α-carbon is always sp^3.

(**d**) The α-carbon is usually, but not always, a chiral center.

(**e**) The $-R$ group contains only carbon and hydrogen, but the number of atoms in this group differs from one amino acid to another.

EP 24.15 If you heat a protein to a sufficiently high temperature to break intramolecular hydrogen bonds, but not covalent bonds, will the α-helical and/or β-sheet structure be retained?

(**a**) The α-helical structure will remain, but not the β-sheet structure.

(**b**) The β-sheet structure will remain, but not the α-helical structure.

(**c**) Both the α-helical structure and the β-sheet structure will be retained.

(**d**) Neither the α-helical structure nor the β-sheet structure will be retained.

EP 24.16 Which of the following molecules is a carbohydrate?

OH OH	OH	O
| |	|	||
CH_3CHCH_2	$HOCH_2CHCCH_2OH$	$HOCH_2CH_2CCH_2OH$
|	||	
OH	O	
(i)	**(ii)**	**(iii)**

(a) only i **(b)** only ii **(c)** only iii **(d)** both i and ii **(e)** both ii and iii

EP 24.17 How many chiral carbon atoms are there in the open-chain form of fructose?

Fructose

(a) 0 **(b)** 1 **(c)** 2 **(d)** 3 **(e)** 4

EP 24.18 What is the relationship between the monosaccharides glucose and galactose (see Figure 24.21 for drawings of the ring forms of these two molecules)?

(a) The two molecules are structural isomers.

(b) The two molecules are geometric isomers.

(c) The two molecules are optical isomers.

(d) The two molecules are *not* isomers.

EP 24.19 Disaccharides, peptides, fats, and polynucleotides are all formed by what type of reaction? **(a)** addition **(b)** oxidation **(c)** condensation **(d)** hydrogenation **(e)** carbonylation

EP 24.20 Which of the following statements is or are *true*?

 (i) Fats are lipids derived from glycerol and fatty acids.

 (ii) Trans fats are saturated fats.

 (iii) Phospholipids are a key component of cell membranes because they form a bilayer containing a polar and nonpolar end.

(a) only i **(b)** both i and ii **(c)** both i and iii **(d)** both ii and iii **(e)** All three statements are true.

EP 24.21 Which class of biomolecules will be the least soluble in water? **(a)** proteins **(b)** amino acids **(c)** carbohydrates **(d)** lipids **(e)** nucleic acids

EP 24.22 In the double-helix structure of DNA, which base forms hydrogen bonds with thymine? **(a)** adenine **(b)** guanine **(c)** cytosine **(d)** uracil

EP 24.23 Imagine a single DNA strand with the following base sequence: 5′–GACCTTA–3′. What is the base sequence of the complementary strand?

(a) 3′–GACCTTA–5′ **(b)** 3′–AGTTCCA–5′ **(c)** 3′–TCAAGGC–5′ **(d)** 3′–CTGGAAT–5′

EP 24.24 Which of the following comparisons of DNAs and RNAs is *false*?

(a) DNAs and RNAs contain different sugar molecules in their backbone.

(b) DNAs and RNAs are built from the same nitrogen-containing bases.

(c) DNA molecules are normally much larger than RNA molecules.

(d) DNAs are typically found in the cell nucleus, while RNAs are found outside the cell nucleus.

(e) DNAs and RNAs are both biopolymers.

Exercises

Visualizing Concepts

24.1 Give the systematic name for each of these hydrocarbons. [Sections 24.2, 24.3]

24.2 Which of these molecules is unsaturated? [Section 24.3]

24.3 **(a)** Which of these molecules most readily undergoes an addition reaction? **(b)** Which of these molecules is aromatic? **(c)** Which of these molecules most readily undergoes a substitution reaction? [Section 24.3]

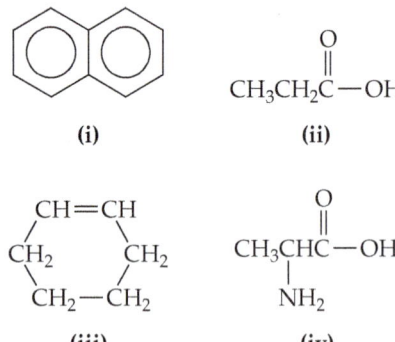

(i)

$$CH_3CH_2\overset{\displaystyle O}{\overset{\displaystyle \|}{C}}-OH$$

(ii)

(iii)

$$CH_3\overset{\displaystyle O}{\underset{\displaystyle NH_2}{\overset{\displaystyle \|}{\underset{\displaystyle |}{CHC}}}}-OH$$

(iv)

24.4 **(a)** Which of these compounds would you expect to have the highest boiling point? **(b)** Which of these compounds is the most oxidized? **(c)** Which of these compounds, if any, is an ether? **(d)** Which of these compounds, if any, is an ester? **(e)** Which of these compounds, if any, is a ketone? [Section 24.4]

$$\overset{\displaystyle O}{\overset{\displaystyle \|}{CH_3CH}}$$ $$CH_3CH_2OH$$ $$CH_3C\equiv CH$$ $$\overset{\displaystyle O}{\overset{\displaystyle \|}{HCOCH_3}}$$

(i)   (ii)   (iii)   (iv)

24.5 For each of the following compounds, state its systematic name, its functional group(s), and whether or not it is chiral. [Sections 24.2, 24.4]

$$CH_3\underset{\displaystyle NH_3^+}{\overset{\displaystyle CH_3}{\underset{\displaystyle |}{\overset{\displaystyle |}{CHCHC}}}}\overset{\displaystyle O}{\overset{\displaystyle \|}{C}}-O^-$$

(a)

$$\overset{\displaystyle O}{\overset{\displaystyle \|}{C}}-OH$$ with Cl (b)

$$CH_3CH_2CH=CHCH_3$$ $$CH_3CH_2CH_3$$

(c)   (d)

24.6 From examination of the molecular models i–v, choose the substance that **(a)** can be hydrolyzed to form a solution containing glucose, **(b)** is capable of forming a zwitterion, **(c)** is one of the four bases present in DNA, **(d)** reacts with an acid to form an ester, **(e)** is a lipid. [Sections 24.6–24.9]

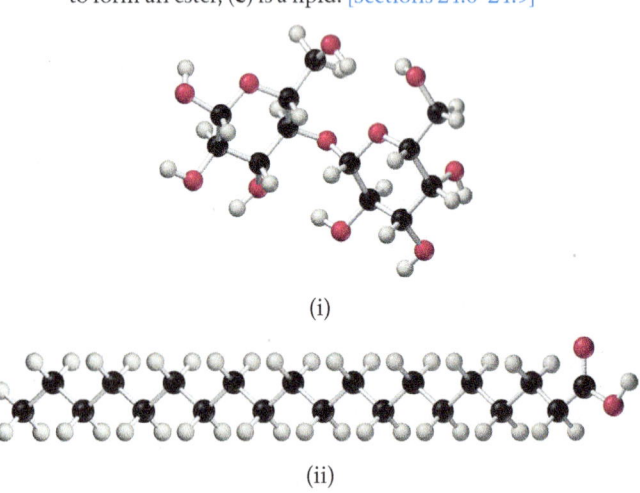

(i)

(ii)

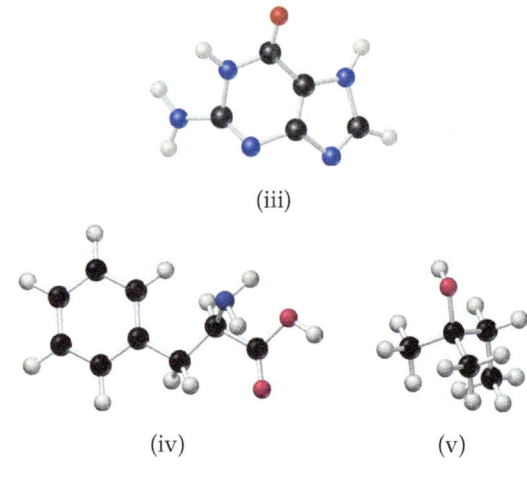

(iii)

(iv)   (v)

Introduction to Organic Compounds; Hydrocarbons (Sections 24.1 and 24.2)

24.7 Indicate whether each statement is *true* or *false*. **(a)** Butane contains carbons that are sp^2 hybridized. **(b)** Cyclohexane is another name for benzene. **(c)** The isopropyl group contains three sp^3-hybridized carbons. **(d)** Olefin is another name for alkyne.

24.8 Indicate whether each statement is *true* or *false*. **(a)** Pentane has a higher molar mass than hexane. **(b)** The longer the linear alkyl chain for straight-chain hydrocarbons, the higher the boiling point. **(c)** The local geometry around the alkyne group is linear. **(d)** Propane has two structural isomers.

24.9 Predict the ideal values for the bond angles about each carbon atom in the following molecule. Indicate the hybridization of orbitals for each carbon.

$$CH_3CCCH_2COOH$$

24.10 Identify the carbon atom(s) in the structure shown that has (have) each of the following hybridizations: **(a)** sp^3, **(b)** sp, **(c)** sp^2.

$$N\equiv C-CH_2-CH_2-CH=CH-\underset{\displaystyle \underset{\displaystyle H}{\overset{\displaystyle |}{\underset{\displaystyle |}{C}}=O}}{\overset{\displaystyle |}{CHOH}}$$

24.11 For each of the following hydrocarbons, state how many carbon atoms are in each molecule:

(a) methane   **(d)** neopentane

(b) decane   **(e)** acetylene

(c) 2-methylhexane

24.12 *True* or *false*: The weaker a single bond in a molecule, the greater the chance it will be the site of a reaction (compared to stronger single bonds in the molecule).

24.13 Indicate whether each statement is *true* or *false*. **(a)** Alkanes do not contain any carbon–carbon multiple bonds. **(b)** Cyclobutane contains a four-membered ring. **(c)** Alkenes contain carbon–carbon triple bonds. **(d)** Alkynes contain carbon–carbon double bonds. **(e)** Pentane is a saturated hydrocarbon, but 1-pentene is an unsaturated hydrocarbon. **(f)** Cyclohexane is an aromatic hydrocarbon. **(g)** The methyl group contains one less hydrogen atom than methane.

24.14 What structural features help us identify a compound as **(a)** an alkane, **(b)** a cycloalkane, **(c)** an alkene, **(d)** an alkyne, **(e)** a saturated hydrocarbon, **(f)** an aromatic hydrocarbon?

24.15 Give the name or condensed structural formula, as appropriate:

(a)
$$CH_3CHCH_3$$
$$CHCH_2CH_2CH_2CH_3$$
$$CH_3$$

(b) 2,2-dimethylpentane

(c) 4-ethyl-1,1-dimethylcyclohexane

(d) $(CH_3)_2CHCH_2CH_2C(CH_3)_3$

(e) $CH_3CH_2CH(C_2H_5)CH_2CH_2CH_2CH_3$

24.16 Give the name or condensed structural formula, as appropriate:

(a) 3-phenylpentane

(b) 2,3-dimethylhexane

(c) 3,3-dimethyloctane

(d) $CH_3CH_2CH(CH_3)CH_2CH(CH_3)_2$

(e) ⬦—CH_3

Alkenes, Alkynes, and Aromatic Hydrocarbons (Section 24.3)

24.17 (a) Is C_4H_6 a saturated or unsaturated hydrocarbon?

(b) Are all alkynes unsaturated?

24.18 (a) Is the compound $CH_3CH{=}CH_2$ saturated or unsaturated? Explain. (b) What is wrong with the formula $CH_3CH_2CH{=}CH_3$?

24.19 Give the molecular formula of a hydrocarbon containing five carbon atoms that is (a) an alkane, (b) a cycloalkane, (c) an alkene, (d) an alkyne.

24.20 Give the molecular formula of a hydrocarbon containing six carbon atoms that is (a) a cyclic alkane, (b) a cyclic alkene, (c) a linear alkyne, (d) an aromatic hydrocarbon.

24.21 Enediynes are a class of compounds that include some antibiotic drugs. Draw the structure of an "enediyne" fragment that contains six carbons in a row.

24.22 Give the general formula for any cyclic alkene—that is, a cyclic hydrocarbon with one double bond.

24.23 Write the condensed structural formulas for two alkenes and one alkyne that all have the molecular formula C_6H_{10}.

24.24 Draw all the possible noncyclic structural isomers of C_5H_{10}. Name each compound.

24.25 Name or write the condensed structural formula for the following compounds:

(a) *trans*-2-pentene

(b) 2,5-dimethyl-4-octene

(c)
$$CH_3$$
$$CH_3CH_2 \quad\quad CH_2CHCH_2CH_3$$
$$C{=}C$$
$$H \quad\quad H$$

(d)
Br
(benzene ring)
Br

(e) $HC{\equiv}CCH_2CCH_3$ with CH_2CH_3 and CH_3 groups

24.26 Name or write the condensed structural formula for the following compounds:

(a) 4-methyl-2-pentene

(b) *cis*-2,5-dimethyl-3-hexene

(c) *ortho*-dimethylbenzene

(d) $HC{\equiv}CCH_2CH_3$

(e) $trans - CH_3CH{=}CHCH_2CH_2CH_2CH_3$

24.27 Indicate whether each statement is *true* or *false*. (a) Two geometric isomers of pentane are *n*-pentane and neopentane. (b) Alkenes can have cis and trans isomers around the carbon–carbon double bond. (c) Alkynes can have cis and trans isomers around the carbon–carbon triple bond.

24.28 Draw all structural and geometric isomers of butene and name them.

24.29 Indicate whether each of the following molecules is capable of geometrical isomerism. For those that are, draw the structures: (a) 1,1-dichloro-1-butene, (b) 2,4-dichloro-2-butene, (c) 1,4-dichlorobenzene, (d) 4,4-dimethyl-2-pentyne.

24.30 Draw the three distinct geometric isomers of 2,4-hexadiene.

24.31 (a) *True* or *false*: Alkenes undergo addition reactions, and aromatic hydrocarbons undergo substitution reactions. (b) Using condensed structural formulas, write the balanced equation for the reaction of 2-pentene with Br_2 and name the resulting compound. Is this an addition or a substitution reaction? (c) Write a balanced chemical equation for the reaction of Cl_2 with benzene to make *para*-dichlorobenzene in the presence of $FeCl_3$ as a catalyst. Is this an addition or a substitution reaction?

24.32 Using condensed structural formulas, write a balanced chemical equation for each of the following reactions: (a) hydrogenation of cyclohexene, (b) addition of H_2O to *trans*-2-pentene using H_2SO_4 as a catalyst (two products), (c) reaction of 2-chloropropane with benzene in the presence of $AlCl_3$.

24.33 (a) When cyclopropane is treated with HI, 1-iodopropane is formed. A similar type of reaction does not occur with cyclopentane or cyclohexane. Suggest an explanation for cyclopropane's reactivity. (b) Suggest a method of preparing ethylbenzene, starting with benzene and ethylene as the only organic reagents.

24.34 (a) One test for the presence of an alkene is to add a small amount of bromine, which is a red-brown liquid, and look for the disappearance of the red-brown color. This test does not work for detecting the presence of an aromatic hydrocarbon. Explain. (b) Write a series of reactions leading to *para*-bromoethylbenzene, beginning with benzene and using other reagents as needed. What isomeric side products might also be formed?

24.35 The rate law for addition of Br_2 to an alkene is first order in Br_2 and first order in the alkene. Does this information suggest that the mechanism of addition of Br_2 to an alkene proceeds in the same manner as for addition of HBr? Explain.

24.36 Describe the intermediate that is thought to form in the addition of a hydrogen halide to an alkene, using cyclohexene as the alkene in your description.

24.37 The molar heat of combustion of gaseous cyclopropane is -2089 kJ/mol; that for gaseous cyclopentane is -3317 kJ/mol. Calculate the heat of combustion per CH_2 group in the two cases, and account for the difference.

24.38 The heat of combustion of decahydronaphthalene ($C_{10}H_{18}$) is -6286 kJ/mol. The heat of combustion of naphthalene ($C_{10}H_8$) is -5157 kJ/mol. [In both cases, $CO_2(g)$ and $H_2O(l)$ are the products.] Using these data and data in Appendix C, calculate the heat of hydrogenation and the resonance energy of naphthalene.

Functional Groups and Chirality (Sections 24.4 and 24.5)

24.39 (**a**) Which of the following compounds, if any, is an ether? (**b**) Which compound, if any, is an alcohol? (**c**) Which compound, if any, would produce a basic solution if dissolved in water? (Assume solubility is not a problem). (**d**) Which compound, if any, is a ketone? (**e**) Which compound, if any, is an aldehyde?

(i) H_3C-CH_2-OH

(ii) $H_3C-\overset{H}{\underset{}{N}}-CH_2CH=CH_2$

(ii)

(iv)

(v) $CH_3CH_2CH_2CH_2CHO$

(vi) $CH_3C \equiv CCH_2COOH$

24.40 Identify the functional groups in each of the following compounds:

(a)

(b)

(c)

(d)

(e)

(f) $CH_3CH_2CH_2CH_2$ $CH_2CH_2CH_2CH_3$

24.41 Draw the molecular structure for (**a**) an aldehyde that is an isomer of acetone, (**b**) an ether that is an isomer of 1-propanol.

24.42 (**a**) Give the empirical formula and structural formula for a cyclic ether containing four carbon atoms in the ring. (**b**) Write the structural formula for a straight-chain compound that is a structural isomer of your answer to part (a).

24.43 The IUPAC name for a carboxylic acid is based on the name of the hydrocarbon with the same number of carbon atoms. The ending *-oic* is appended, as in ethanoic acid, which is the IUPAC name for acetic acid. Draw the structure of the following acids: (**a**) methanoic acid, (**b**) pentanoic acid, (**c**) 2-chloro-3-methyldecanoic acid.

24.44 Aldehydes and ketones can be named in a systematic way by counting the number of carbon atoms (including the carbonyl carbon) that they contain. The name of the aldehyde or ketone is based on the hydrocarbon with the same number of carbon atoms. The ending *-al* for aldehyde or *-one* for ketone is added as appropriate. Draw the structural formulas for the following aldehydes or ketones: (**a**) propanal, (**b**) 2-pentanone, (**c**) 3-methyl-2-butanone, (**d**) 2-methylbutanal.

24.45 Draw the condensed structure of the compounds formed by condensation reactions between (**a**) benzoic acid and ethanol, (**b**) ethanoic acid and methylamine, (**c**) acetic acid and phenol. Name the compound in each case.

24.46 Draw the condensed structures of the compounds formed from (**a**) butanoic acid and methanol, (**b**) benzoic acid and 2-propanol, (**c**) propanoic acid and dimethylamine. Name the compound in each case.

24.47 Write a balanced chemical equation using condensed structural formulas for the saponification (base hydrolysis) of (**a**) methyl propionate, (**b**) phenyl acetate.

24.48 Write a balanced chemical equation using condensed structural formulas for (**a**) the formation of butyl propionate from the appropriate acid and alcohol, (**b**) the saponification (base hydrolysis) of methyl benzoate.

24.49 Pure acetic acid is a viscous liquid, with high melting and boiling points (16.7 °C and 118 °C) compared to compounds of similar molecular weight. Suggest an explanation.

24.50 *Acetic anhydride* is formed from two acetic acid molecules, in a condensation reaction that involves the removal of a molecule of water. Write the chemical equation for this process and show the structure of acetic anhydride.

24.51 Write the condensed structural formula for each of the following compounds: (**a**) 2-pentanol, (**b**) 1,2-propanediol, (**c**) ethyl acetate, (**d**) diphenyl ketone, (**e**) methyl ethyl ether.

24.52 Write the condensed structural formula for each of the following compounds: (**a**) 2-ethyl-1-hexanol, (**b**) methyl phenyl ketone, (**c**) *para*-bromobenzoic acid, (**d**) butyl ethyl ether, (**e**) *N,N*-dimethylbenzamide.

24.53 How many chiral carbons are in 2-bromo-2-chloro-3-methylpentane? (**a**) 0 (**b**) 1 (**c**) 2 (**d**) 3 (**e**) 4 or more

24.54 Is 3-chloro-3-methylhexane chiral?

Proteins (Section 24.6)

24.55 (**a**) Draw the chemical structure of a generic amino acid, using R for the side chain. (**b**) When amino acids react to form proteins, do they do so via substitution, addition, or condensation reactions? (**c**) Draw the bond that links amino acids together in proteins. What is this called?

24.56 Indicate whether each statement is *true* or *false*. (**a**) Tryptophan is an aromatic amino acid. (**b**) Lysine is positively charged at pH 7. (**c**) Asparagine has two amide bonds. (**d**) Isoleucine and leucine are enantiomers. (**e**) Valine is probably more water-soluble than arginine.

24.57 Draw the two possible dipeptides formed by condensation reactions between histidine and aspartic acid.

24.58 Write a chemical equation for the formation of methionylglycine from the constituent amino acids.

24.59 (**a**) Draw the condensed structure of the tripeptide Gly-Gly-His. (**b**) How many different tripeptides can be made from the amino acids glycine and histidine? Give the abbreviations for each of these tripeptides, using the three-letter and one-letter codes for the amino acids.

24.60 (a) What amino acids would be obtained by hydrolysis of the following tripeptide?

$$H_2NCHCNHCHCNHCHCOH$$

$$(CH_3)_2CH \quad H_2COH \quad H_2CCH_2COH$$

(b) How many different tripeptides can be made from glycine, serine, and glutamic acid? Give the abbreviation for each of these tripeptides, using the three-letter codes and one-letter codes for the amino acids.

24.61 Indicate whether each statement is *true* or *false*. **(a)** The sequence of amino acids in a protein, from the amine end to the acid end, is called the primary structure of the protein. **(b)** Alpha helix and beta sheet structures are examples of quaternary protein structure. **(c)** It is impossible for more than one protein to bind to another and make a higher-order structure.

24.62 Indicate whether each statement is *true* or *false*: **(a)** In the alpha helical structure of proteins, hydrogen bonding occurs between the side chains (R groups). **(b)** Dispersion forces, not hydrogen bonding, holds beta sheet structures together.

Carbohydrates and Lipids (Sections 24.7 and 24.8)

24.63 Indicate whether each statement is *true* or *false*: **(a)** Disaccharides are a type of carbohydrate. **(b)** Sucrose is a monosaccharide. **(c)** All carbohydrates have the formula $C_nH_{2m}O_m$.

24.64 (a) Are α-glucose and β-glucose enantiomers? **(b)** Show the condensation of two glucose molecules to form a disaccharide with an α linkage. **(c)** Repeat part (b) but with a β linkage.

24.65 (a) What is the empirical formula of cellulose? **(b)** What is the monomer that forms the basis of the cellulose polymer? **(c)** What bond connects the monomer units in cellulose: amide, acid, ether, ester, or alcohol?

24.66 (a) What is the empirical formula of starch? **(b)** What is the monomer that forms the basis of the starch polymer? **(c)** What bond connects the monomer units in starch: amide, acid, ether, ester, or alcohol?

24.67 The structural formula for the open-chain form of D-mannose is

$$
\begin{array}{c}
O \\
\| \\
CH \\
HO-C-H \\
HO-C-H \\
H-C-OH \\
H-C-OH \\
CH_2OH
\end{array}
$$

(a) Is this molecule a sugar? **(b)** How many chiral carbons are present in the molecule? **(c)** Draw the structure of the six-member-ring form of this molecule.

24.68 The structural formula for the open-chain form of galactose is

$$
\begin{array}{c}
O \\
\| \\
CH \\
H-C-OH \\
HO-C-H \\
HO-C-H \\
H-C-OH \\
CH_2OH
\end{array}
$$

(a) Is this molecule a sugar? **(b)** How many chiral carbons are present in the molecule? **(c)** Draw the structure of the six-member-ring form of this molecule.

24.69 Indicate whether each statement is *true* or *false*: **(a)** Fat molecules contain amide bonds. **(b)** Phospholipids can be zwitterions. **(c)** Phospholipids form bilayers in water in order to have their long hydrophobic tails interact favorably with each other, leaving their polar heads to the aqueous environment.

24.70 Indicate whether each statement is *true* or *false*: **(a)** If you use data from Table 8.3 on bond enthalpies, you can show that the more C—H bonds a molecule has compared to C—O and O—H bonds, the more energy it can store. **(b)** Trans fats are saturated. **(c)** Fatty acids are long-chain carboxylic acids. **(d)** Monounsaturated fatty acids have one C—C single bond in the chain, while the rest are double or triple bonds.

Nucleic Acids (Section 24.9)

24.71 Adenine and guanine are members of a class of molecules known as *purines*; they have two rings in their structure. Thymine and cytosine, on the other hand, are *pyrimidines* and have only one ring in their structure. Predict which have larger dispersion forces in aqueous solution, the purines or the pyrimidines.

24.72 A nucleoside consists of an organic base of the kind shown in Section 24.9, bound to ribose or deoxyribose. Draw the structure for deoxyguanosine, formed from guanine and deoxyribose.

24.73 What is the DNA sequence for the molecule shown here?

24.74 You are working in a biotechnology lab and are analyzing DNA. You obtain a sample of a short dodecamer of DNA that contains 12 base pairs. (a) What must the ratio of adenine to thymine be in your sample? (b) What must the ratio of cytosine to guanine be in your sample? (c) Assume the counterions present in your DNA solution are sodium ions. How many sodium ions must there be per dodecamer? Assume the 5' end phosphates each bear a −1 charge.

24.75 Imagine a single DNA strand containing a section with the following base sequence: 5'-GCATTGGC-3'. What is the base sequence of the complementary strand?

24.76 Which statement best explains the chemical differences between DNA and RNA? (a) DNA has two different sugars in its sugar–phosphate backbone, but RNA only has one. (b) Thymine is one of the DNA bases, whereas RNA's corresponding base is thymine minus a methyl group. (c) The RNA sugar–phosphate backbone contains fewer oxygen atoms than DNA's backbone. (d) DNA forms double helices but RNA cannot.

Additional Exercises

24.77 Draw the condensed structural formulas for two different molecules with the formula C_3H_4O.

24.78 How many structural isomers are there for a five-member straight carbon chain with one double bond? For a six-member straight carbon chain with two double bonds?

24.79 (a) Draw the condensed structural formulas for the cis and trans isomers of 2-pentene. (b) Can cyclopentene exhibit cis–trans isomerism? Explain. (c) Does 1-pentyne have enantiomers? Explain.

24.80 If a molecule is an "ene-one," what functional groups must it have?

24.81 Identify each of the functional groups in these molecules:

(a)

(Responsible for the odor of cucumbers)

(b)

(Quinine — an antimalarial drug)

(c)

(Indigo — a blue dye)

(d)

(Acetaminophen — aka Tylenol)

24.82 For the molecules shown in 24.81, (a) Which one(s) of them, if any, would produce a basic solution if dissolved in water? (b) Which one(s) of them, if any, would produce an acidic solution if dissolved in water? (c) Which of them is the most water-soluble?

24.83 Write a condensed structural formula for each of the following: (a) an acid with the formula $C_4H_8O_2$, (b) a cyclic ketone with the formula C_5H_8O, (c) a dihydroxy compound with the formula $C_3H_8O_2$, (d) a cyclic ester with the formula $C_5H_8O_2$.

24.84 Draw each molecule given its name and the following information. (a) Nitroglycerin, also known as 1,2,3-trinitroxypropane, the active ingredient in dynamite and a medication administered to people having a heart attack, (Hint: The nitroxy group is the conjugate base of nitric acid.) (b) Putrescine, also known as 1,4-diaminobutane, the compound responsible for the odor of putrefying fish, (c) Cyclohexanone, the precursor to Nylon, (d) 1,1,2,2-tetrafluoroethene, the precursor to Teflon, (e) Oleic acid, also known as cis-9-octanedecenoic acid, a monounsaturated fatty acid found in many fats and oils. Draw the correct isomer.

24.85 Indole smells terrible in high concentrations but has a pleasant floral-like odor when highly diluted. Its structure is

The molecule is planar, and the nitrogen is a very weak base, with $K_b = 2 \times 10^{-12}$. Explain how this information indicates that the indole molecule is aromatic.

24.86 For each of the following molecules, identify its functional groups and how many chiral centers it contains.

(a) $HOCH_2CH_2\overset{\displaystyle O}{\overset{\displaystyle \|}{C}}CH_2OH$

(b) $HOCH_2\overset{\displaystyle OH}{\overset{\displaystyle |}{C}H}CCH_2OH$ with $\overset{\displaystyle }{\underset{\displaystyle O}{\|}}$

(c) $HO\overset{\displaystyle O}{\overset{\displaystyle \|}{C}}\overset{\displaystyle CH_3}{\overset{\displaystyle |}{CH}}CHCH_2H_5$ with NH_2

24.87 Which of the following peptides have a net positive charge at pH 7? (a) Gly-Ser-Lys (b) Pro-Leu-Ile (c) Phe-Tyr-Asp

24.88 Glutathione is a tripeptide found in most living cells. Partial hydrolysis yields Cys-Gly and Glu-Cys. What structures are possible for glutathione?

24.89 Monosaccharides can be categorized in terms of the number of carbon atoms (pentoses have five carbons and hexoses have six carbons) and according to whether they contain an aldehyde (*aldo-* prefix, as in aldopentose) or ketone group (*keto-* prefix, as in ketopentose). Classify glucose and fructose in this way.

24.90 Can a DNA strand bind to a complementary RNA strand? Explain.

24.91 An organic compound is analyzed and found to contain 66.7% carbon, 11.2% hydrogen, and 22.1% oxygen by mass. The compound boils at 79.6 °C. At 100 °C and 0.970 atm, the vapor has a density of 2.28 g/L. The compound has a carbonyl group and cannot be oxidized to a carboxylic acid. Suggest a structure for the compound.

24.92 An unknown substance is found to contain only carbon and hydrogen. It is a liquid that boils at 49 °C at 1 atm pressure. Upon analysis, it is found to contain 85.7% carbon and 14.3% hydrogen by mass. At 100 °C and 735 torr, the vapor of this unknown has a density of 2.21 g/L. When it is dissolved in hexane solution and bromine water is added, no reaction occurs. What is the identity of the unknown compound?

Design an Experiment

Quaternary structures of proteins arise if two or more smaller polypeptides or proteins associate with each other to make a much larger protein structure. The association is due to the same hydrogen bonding, electrostatic, and dispersion forces we have seen before. Hemoglobin, the protein used to transport oxygen molecules in our blood, is an example of a protein that has quaternary structure. Hemoglobin is a tetramer; it is made of four smaller polypeptides, two "alphas" and two "betas." (These names do not imply anything about the number of alpha helices or beta sheets in the individual polypeptides.) Design a set of experiments that would provide sound evidence that hemoglobin exists as a tetramer and not as one enormous polypeptide chain.

MATHEMATICAL OPERATIONS

A.1 | Exponential Notation

The numbers used in chemistry are often either extremely large or extremely small. Such numbers are conveniently expressed in the form

$$N \times 10^n$$

where N is a number between 1 and 10, and n is the exponent. Some examples of this *exponential notation*, which is also called *scientific notation*, follow.

1,200,000 is 1.2×10^6 (read "one point two multi ten to the sixth power")

0.000604 is 6.04×10^{-4} (read "six point zero four times ten to the negative fourth power")

A positive exponent, as in the first example, tells us how many times a number must be multiplied by 10 to give the long form of the number:

$$1.2 \times 10^6 = 1.2 \times 10 \times 10 \times 10 \times 10 \times 10 \times 10 \quad \text{(six tens)}$$

$$= 1,200,000$$

It is also convenient to think of the *positive exponent* as the number of places the decimal point must be moved to the *left* to obtain a number greater than 1 and less than 10. For example, if we begin with 3450 and move the decimal point three places to the left, we end up with 3.45×10^3.

In a related fashion, a negative exponent tells us how many times we must divide a number by 10 to give the long form of the number.

$$6.04 \times 10^{-4} = \frac{6.04}{10 \times 10 \times 10 \times 10} = 0.000604$$

It is convenient to think of the *negative exponent* as the number of places the decimal point must be moved to the *right* to obtain a number greater than 1 but less than 10. For example, if we begin with 0.0048 and move the decimal point three places to the right, we end up with 4.8×10^{-3}.

In the system of exponential notation, with each shift of the decimal point one place to the right, the exponent *decreases* by 1:

$$4.8 \times 10^{-3} = 48 \times 10^{-4}$$

Similarly, with each shift of the decimal point one place to the left, the exponent *increases* by 1:

$$4.8 \times 10^{-3} = 0.48 \times 10^{-2}$$

Many scientific calculators have a key labeled EXP or EE, which is used to enter numbers in exponential notation. To enter the number 5.8×10^3 on such a calculator, the key sequence is

$$\boxed{5}\boxed{\cdot}\boxed{8}\boxed{\text{EXP}}\,(\text{or}\,\boxed{\text{EE}}\,)\boxed{3}$$

On some calculators the display will show 5.8, then a space, followed by 03, the exponent. On other calculators, a small 10 is shown with an exponent 3.

To enter a negative exponent, use the key labeled $+/-$. For example, to enter the number 8.6×10^{-5}, the key sequence is

$$\boxed{8}\boxed{\cdot}\boxed{6}\boxed{\text{EXP}}\boxed{+/-}\boxed{5}$$

When entering a number in exponential notation, do not key in the 10 if you use the EXP or EE button.

In working with exponents, it is important to recall that $10^0 = 1$. The following rules are useful for carrying exponents through calculations.

1. **Addition and Subtraction** In order to add or subtract numbers expressed in exponential notation, the powers of 10 must be the same.

$$(5.22 \times 10^4) + (3.21 \times 10^2) = (522 \times 10^2) + (3.21 \times 10^2)$$
$$= 525 \times 10^2 \quad \text{(3 significant figures)}$$
$$= 5.25 \times 10^4$$
$$(6.25 \times 10^{-2}) - (5.77 \times 10^{-3}) = (6.25 \times 10^{-2}) - (0.577 \times 10^{-2})$$
$$= 5.67 \times 10^{-2} \quad \text{(3 significant figures)}$$

When you use a calculator to add or subtract, you need not be concerned with having numbers with the same exponents because the calculator automatically takes care of this matter.

2. **Multiplication and Division** When numbers expressed in exponential notation are multiplied, the exponents are added; when numbers expressed in exponential notation are divided, the exponent of the denominator is subtracted from the exponent of the numerator.

$$(5.4 \times 10^2)(2.1 \times 10^3) = (5.4)(2.1) \times 10^{2+3}$$
$$= 11 \times 10^5$$
$$= 1.1 \times 10^6$$
$$(1.2 \times 10^5)(3.22 \times 10^{-3}) = (1.2)(3.22) \times 10^{5+(-3)} = 3.9 \times 10^2$$

$$\frac{3.2 \times 10^5}{6.5 \times 10^2} = \frac{3.2}{6.5} \times 10^{5-2} = 0.49 \times 10^3 = 4.9 \times 10^2$$

$$\frac{5.7 \times 10^7}{8.5 \times 10^{-2}} = \frac{5.7}{8.5} \times 10^{7-(-2)} = 0.67 \times 10^9 = 6.7 \times 10^8$$

3. **Powers and Roots** When numbers expressed in exponential notation are raised to a power, the exponents are multiplied by the power. When the roots of numbers expressed in exponential notation are taken, the exponents are divided by the root.

$$(1.2 \times 10^5)^3 = (1.2)^3 \times 10^{5 \times 3}$$
$$= 1.7 \times 10^{15}$$
$$\sqrt[3]{2.5 \times 10^6} = \sqrt[3]{2.5} \times 10^{6/3}$$
$$= 1.3 \times 10^2$$

Scientific calculators usually have keys labeled x^2 and $\sqrt{x}$ for squaring and taking the square root of a number, respectively. To take higher powers or roots, many calculators have y^x and $\sqrt[x]{y}$ (or INV y^x) keys. For example, to perform the operation $\sqrt[3]{7.5 \times 10^{-4}}$ on such a calculator, you would key in 7.5×10^{-4}, press the $\sqrt[x]{y}$ key (or the INV and then the y^x keys), enter the root, 3, and finally press $=$. The result is 9.1×10^{-2}.

Sample Exercise 1
Using Exponential Notation

Perform each of the following operations, using your calculator where possible:

(a) Write the number 0.0054 in standard exponential notation. (b) $(5.0 \times 10^{-2}) + (4.7 \times 10^{-3})$ (c) $(5.98 \times 10^{12})(2.77 \times 10^{-5})$
(d) $\sqrt[4]{1.75 \times 10^{-12}}$

SOLUTION

(a) Because we move the decimal point three places to the right to convert 0.0054 to 5.4, the exponent is -3:

$$5.4 \times 10^{-3}$$

Scientific calculators are generally able to convert numbers to exponential notation using one or two keystrokes; frequently "SCI" for "scientific notation" will convert a number into exponential notation. Consult your instruction manual to see how this operation is accomplished on your calculator.

(b) To add these numbers longhand, we must convert them to the same exponent.

$$(5.0 \times 10^{-2}) + (0.47 \times 10^{-2}) = (5.0 + 0.47) \times 10^{-2} = 5.5 \times 10^{-2}$$

(c) Performing this operation longhand, we have

$$(5.98 \times 2.77) \times 10^{12-5} = 16.6 \times 10^{7} = 1.66 \times 10^{8}$$

(d) To perform this operation on a calculator, we enter the number, press the $\sqrt[x]{y}$ key (or the INV and y^x keys), enter 4, and press the $=$ key. The result is 1.15×10^{-3}.

▶ **Practice Exercise**

Perform the following operations:
(a) Write 67,000 in exponential notation, showing two significant figures.
(b) $(3.378 \times 10^{-3}) - (4.97 \times 10^{-5})$
(c) $(1.84 \times 10^{15})(7.45 \times 10^{-2})$
(d) $(6.67 \times 10^{-8})^3$

A.2 | Logarithms

Common Logarithms

The common, or base-10, logarithm (abbreviated log) of any number is the power to which 10 must be raised to equal the number. For example, the common logarithm of 1000 (written log 1000) is 3 because raising 10 to the third power gives 1000.

$$10^3 = 1000, \text{ therefore, } \log 1000 = 3$$

Further examples are

$$\log 10^5 = 5$$
$$\log 1 = 0 \quad \text{Remember that } 10^0 = 1$$
$$\log 10^{-2} = -2$$

In these examples the common logarithm can be obtained by inspection. However, it is not possible to obtain the logarithm of a number such as 31.25 by inspection. The logarithm of 31.25 is the number x that satisfies the following relationship:

$$10^x = 31.25$$

Most electronic calculators have a key labeled LOG that can be used to obtain logarithms. For example, on many calculators we obtain the value of log 31.25 by entering 31.25 and pressing the LOG key. We obtain the following result:

$$\log 31.25 = 1.4949$$

Notice that 31.25 is greater than 10 (10^1) and less than 100 (10^2). The value for log 31.25 is accordingly between log 10 and log 100, that is, between 1 and 2.

Significant Figures and Common Logarithms

For the common logarithm of a measured quantity, the number of digits after the decimal point equals the number of significant figures in the original number. For example, if 23.5 is a measured quantity (three significant figures), then log 23.5 = 1.371 (three significant figures after the decimal point).

Antilogarithms

The process of determining the number that corresponds to a certain logarithm is known as obtaining an *antilogarithm*. It is the reverse of taking a logarithm. For example, we saw previously that log 23.5 = 1.371. This means that the antilogarithm of 1.371 equals 23.5.

$$\log 23.5 = 1.371$$

$$\text{antilog } 1.371 = 23.5$$

The process of taking the antilog of a number is the same as raising 10 to a power equal to that number.

$$\text{antilog } 1.371 = 10^{1.371} = 23.5$$

Many calculators have a key labeled 10^x that allows you to obtain antilogs directly. On others, it will be necessary to press a key labeled INV (for *inverse*), followed by the LOG key.

Natural Logarithms

Logarithms based on the number e are called natural, or base e, logarithms (abbreviated ln). The natural log of a number is the power to which e (which has the value $2.71828\ldots$) must be raised to equal the number. For example, the natural log of 10 equals 2.303.

$$e^{2.303} = 10, \text{ therefore } \ln 10 = 2.303$$

Your calculator probably has a key labeled LN that allows you to obtain natural logarithms. For example, to obtain the natural log of 46.8, you enter 46.8 and press the LN key.

$$\ln 46.8 = 3.846$$

The natural antilog of a number is e raised to a power equal to that number. If your calculator can calculate natural logs, it will also be able to calculate natural antilogs. On some calculators there is a key labeled e^x that allows you to calculate natural antilogs directly; on others, it will be necessary to first press the INV key followed by the LN key. For example, the natural antilog of 1.679 is given by

$$\text{Natural antilog } 1.679 = e^{1.679} = 5.36$$

The relation between common and natural logarithms is as follows:

$$\ln a = 2.303 \log a$$

Notice that the factor relating the two, 2.303, is the natural log of 10, which we calculated earlier.

Mathematical Operations Using Logarithms

Because logarithms are exponents, mathematical operations involving logarithms follow the rules for the use of exponents. For example, the product of z^a and z^b (where z is any number) is given by

$$z^a \cdot z^b = z^{(a+b)}$$

Similarly, the logarithm (either common or natural) of a product equals the *sum* of the logs of the individual numbers.

$$\log ab = \log a + \log b \qquad \ln ab = \ln a + \ln b$$

For the log of a quotient,

$$\log(a/b) = \log a - \log b \qquad \ln(a/b) = \ln a - \ln b$$

Using the properties of exponents, we can also derive the rules for the logarithm of a number raised to a certain power.

$$\log a^n = n \log a \qquad \ln a^n = n \ln a$$

$$\log a^{1/n} = (1/n)\log a \qquad \ln a^{1/n} = (1/n)\ln a$$

pH Problems

One of the most frequent uses for common logarithms in general chemistry is in working pH problems. The pH is defined as $-\log[H^+]$, where $[H^+]$ is the hydrogen ion concentration of a solution. (Section 16.4) The following sample exercise illustrates this application.

Sample Exercise 2
Using Logarithms

(a) What is the pH of a solution whose hydrogen ion concentration is 0.015 M?
(b) If the pH of a solution is 3.80, what is its hydrogen ion concentration?

SOLUTION

(1) We are given the value of $[H^+]$. We use the LOG key of our calculator to calculate the value of $\log[H^+]$. The pH is obtained by changing the sign of the value obtained. (Be sure to change the sign *after* taking the logarithm.)

$$[H^+] = 0.015$$

$$\log[H^+] = -1.82 \quad \text{(2 significant figures)}$$

$$pH = -(-1.82) = 1.82$$

(2) To obtain the hydrogen ion concentration when given the pH, we must take the antilog of $-pH$.

$$pH = -\log[H^+] = 3.80$$

$$\log[H^+] = -3.80$$

$$[H^+] = \text{antilog}(-3.80) = 10^{-3.80} = 1.6 \times 10^{-4}\, M$$

▶ **Practice Exercise**
Perform the following operations:
(a) $\log(2.5 \times 10^{-5})$
(b) $\ln 32.7$
(c) antilog -3.47
(d) $e^{-1.89}$

A.3 | Quadratic Equations

An algebraic equation of the form $ax^2 + bx + c = 0$ is called a *quadratic equation*. The two solutions to such an equation are given by the quadratic formula:

$$x = \frac{-b \pm \sqrt{b^2 - 4ac}}{2a}$$

Many calculators today can calculate the solutions to a quadratic equation with one or two keystrokes. Most of the time, x corresponds to the concentration of a chemical species in solution. Only one of the solutions will be a positive number, and that is the one you should use; a "negative concentration" has no physical meaning.

Sample Exercise 3
Using the Quadratic Formula

Find the values of x that satisfy the equation $2x^2 + 4x = 1$.

SOLUTION

To solve the given equation for x, we must first put it in the form

$$ax^2 + bx + c = 0$$

and then use the quadratic formula. If

$$2x^2 + 4x = 1$$

then

$$2x^2 + 4x - 1 = 0$$

Using the quadratic formula, where $a = 2$, $b = 4$, and $c = -1$, we have

$$x = \frac{-4 \pm \sqrt{4^2 - 4(2)(-1)}}{2(2)}$$

$$= \frac{-4 \pm \sqrt{16 + 8}}{4} = \frac{-4 \pm \sqrt{24}}{4} = \frac{-4 \pm 4.899}{4}$$

The two solutions are

$$x = \frac{0.899}{4} = 0.225 \quad \text{and} \quad x = \frac{-8.899}{4} = -2.225$$

If this was a problem in which x represented a concentration, we would say $x = 0.225$ (in the appropriate units), since a negative number for concentration has no physical meaning.

A.4 | Graphs

Often the clearest way to represent the interrelationship between two variables is to graph them. Usually, the variable that is being experimentally varied, called the *independent variable*, is shown along the horizontal axis (x-axis). The variable that responds

TABLE A.1 **Interrelation between Pressure and Temperature**

Temperature (°C)	Pressure (atm)
20.0	0.120
30.0	0.124
40.0	0.128
50.0	0.132

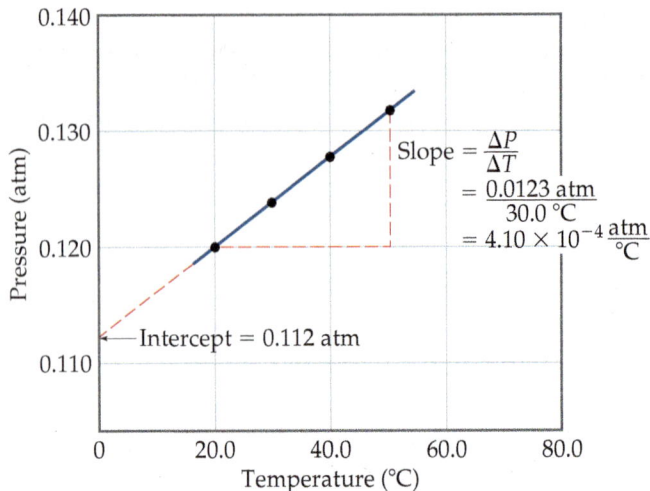

▲ **Figure A.1** A graph of pressure versus temperature yields a straight line for the data.

to the change in the independent variable, called the *dependent variable*, is then shown along the vertical axis (*y*-axis). For example, consider an experiment in which we vary the temperature of an enclosed gas and measure its pressure. The independent variable is temperature and the dependent variable is pressure. The data shown in **Table A.1** can be obtained by means of this experiment. These data are shown graphically in **Figure A.1**. The relationship between temperature and pressure is linear. The equation for any straight-line graph has the form

$$y = mx + b$$

where *m* is the slope of the line and *b* is the intercept with the *y*-axis. In the case of Figure A.1, we could say that the relationship between temperature and pressure takes the form

$$P = mT + b$$

where *P* is pressure in atm and *T* is temperature in °C. As shown in Figure A.1, the slope is 4.10×10^{-4} atm/°C, and the intercept—the point where the line crosses the *y*-axis—is 0.112 atm. Therefore, the equation for the line is

$$P = \left(4.10 \times 10^{-4} \frac{\text{atm}}{°C}\right)T + 0.112 \text{ atm}$$

A.5 | Standard Deviation

The standard deviation from the mean, *s*, is a common method for describing precision in experimentally determined data. We define the standard deviation as

$$s = \sqrt{\frac{\sum_{i=1}^{N}(x_i - \bar{x})^2}{N - 1}}$$

where *N* is the number of measurements, $\bar{x}$ is the average (also called the mean) of the measurements, and x_i represent the individual measurements. Electronic calculators with built-in statistical functions can calculate *s* directly by inputting the individual measurements.

A smaller value of *s* indicates a higher precision, meaning that the data are more closely clustered around the average. The standard deviation has statistical significance. If a large number of measurements is made, 68% of the measured values is expected to be within one standard deviation of the average, assuming only random errors are associated with the measurements.

Sample Exercise 4

Calculating an Average and Standard Deviation

The percent carbon in a sugar is measured four times: 42.01%, 42.28%, 41.79%, and 42.25%. Calculate (**a**) the average and (**b**) the standard deviation for these measurements.

SOLUTION

(**a**) The average is found by adding the quantities and dividing by the number of measurements:

$$\bar{x} = \frac{42.01 + 42.28 + 41.79 + 42.25}{4} = \frac{168.33}{4} = 42.08$$

(**b**) The standard deviation is found using the preceding equation:

$$s = \sqrt{\frac{\sum\limits_{i=1}^{N}(x_i - \bar{x})^2}{N - 1}}$$

Let's tabulate the data so the calculation of $\sum\limits_{i=1}^{N}(x_i - \bar{x})^2$ can be seen clearly.

Percent C	Difference between Measurement and Average, $(x_i - x)$	Square of Difference, $(x_i - x)^2$
42.01	$42.01 - 42.08 = -0.07$	$(-0.07)^2 = 0.005$
42.28	$42.28 - 42.08 = 0.20$	$(0.20)^2 = 0.040$
41.79	$41.79 - 42.08 = -0.29$	$(-0.29)^2 = 0.084$
42.25	$42.25 - 42.08 = 0.17$	$(0.17)^2 = 0.029$

The sum of the quantities in the last column is

$$\sum_{i=1}^{N}(x_i - \bar{x})^2 = 0.005 + 0.040 + 0.084 + 0.029 = 0.16$$

Thus, the standard deviation is

$$s = \sqrt{\frac{\sum\limits_{i=1}^{N}(x_i - \bar{x})^2}{N - 1}} = \sqrt{\frac{0.16}{4 - 1}} = \sqrt{\frac{0.16}{3}} = \sqrt{0.053} = 0.23$$

Based on these measurements, it would be appropriate to represent the measured percent carbon as 42.08 ± 0.23.

PROPERTIES OF WATER

Density:	0.99987 g/mL at 0 °C
	1.00000 g/mL at 4 °C
	0.99707 g/mL at 25 °C
	0.95838 g/mL at 100 °C
Heat (enthalpy) of fusion:	6.008 kJ/mol at 0 °C
Heat (enthalpy) of vaporization:	44.94 kJ/mol at 0 °C
	44.02 kJ/mol at 25 °C
	40.67 kJ/mol at 100 °C
Ion-product constant, K_w:	1.14×10^{-15} at 0 °C
	1.01×10^{-14} at 25 °C
	5.47×10^{-14} at 50 °C
Specific heat:	2.092 J/g-K = 2.092 J/g·°C for ice at −3 °C
	4.184 J/g-K = 4.184 J/g·°C for water at 25 °C
	1.841 J/g-K = 1.841 J/g·°C for steam at 100 °C

Vapor Pressure (torr) at Different Temperatures

T(°C)	P	T(°C)	P	T(°C)	P	T(°C)	P
0	4.58	21	18.65	35	42.2	92	567.0
5	6.54	22	19.83	40	55.3	94	610.9
10	9.21	23	21.07	45	71.9	96	657.6
12	10.52	24	22.38	50	92.5	98	707.3
14	11.99	25	23.76	55	118.0	100	760.0
16	13.63	26	25.21	60	149.4	102	815.9
17	14.53	27	26.74	65	187.5	104	875.1
18	15.48	28	28.35	70	233.7	106	937.9
19	16.48	29	30.04	80	355.1	108	1004.4
20	17.54	30	31.82	90	525.8	110	1074.6

THERMODYNAMIC QUANTITIES FOR SELECTED SUBSTANCES AT 298.15 K (25 °C)

Substance	ΔH_f° (kJ/mol)	ΔG_f° (kJ/mol)	S° (J/mol-K)	Substance	ΔH_f° (kJ/mol)	ΔG_f° (kJ/mol)	S° (J/mol-K)
Aluminum				$C_2H_2(g)$	227.4	210.0	200.9
Al(s)	0	0	28.32	$C_2H_4(g)$	52.4	68.3	219.4
$AlCl_3(s)$	−705.6	−630.0	109.3	$C_2H_6(g)$	−84.68	−32.78	229.5
$Al_2O_3(s)$	−1669.8	−1576.5	51.00	$C_3H_8(g)$	−103.85	−24.16	269.9
				$C_4H_{10}(g)$	−124.73	−15.71	310.0
Barium				$C_4H_{10}(l)$	−147.6	−15.0	231.0
Ba(s)	0	0	63.2	$C_6H_6(g)$	82.9	129.7	269.2
$BaCO_3(s)$	−1216.3	−1137.6	112.1	$C_6H_6(l)$	49.0	124.5	173.3
BaO(s)	−553.5	−525.1	70.42	$CH_3OH(g)$	−201.2	−161.9	237.6
Beryllium				$CH_3OH(l)$	−238.4	−166.1	127.2
Be(s)	0	0	9.44	$C_2H_5OH(g)$	−235.1	−168.5	282.7
BeO(s)	−608.4	−579.1	13.77	$C_2H_5OH(l)$	−277.0	−174.0	160.7
$Be(OH)_2(s)$	−905.8	−817.9	50.21	$C_6H_{12}O_6(s)$	−1273.02	−910.4	212.1
Bromine				CO(g)	−110.5	−137.2	197.9
Br(g)	111.8	82.38	174.9	$CO_2(g)$	−393.5	−394.4	213.6
$Br^-(aq)$	−120.9	−102.8	80.71	$CH_3COOH(l)$	−487.0	−392.4	159.8
$Br_2(g)$	30.71	3.14	245.3	**Cesium**			
$Br_2(l)$	0	0	152.3	Cs(g)	76.50	49.53	175.6
HBr(g)	−36.29	−53.34	198.7	Cs(l)	2.09	0.03	92.07
Calcium				Cs(s)	0	0	85.15
Ca(g)	179.3	145.5	154.8	CsCl(s)	−442.8	−414.4	101.2
Ca(s)	0	0	41.4	**Chlorine**			
$CaCO_3(s, \text{calcite})$	−1207.1	−1128.76	92.88	Cl(g)	121.7	105.7	165.2
$CaCl_2(s)$	−795.8	−748.1	104.6	$Cl^-(aq)$	−167.2	−131.2	56.5
$CaF_2(s)$	−1219.6	−1167.3	68.87	$Cl_2(g)$	0	0	222.96
CaO(s)	−635.1	−604.0	39.7	HCl(aq)	−167.2	−131.2	56.5
$Ca(OH)_2(s)$	−986.2	−898.5	83.4	HCl(g)	−92.31	−95.27	186.69
$CaSO_4(s)$	−1434.0	−1321.8	106.7	**Chromium**			
Carbon				Cr(g)	397.5	352.6	174.2
C(g)	718.4	672.9	158.0	Cr(s)	0	0	23.6
C(s, diamond)	1.88	2.87	2.43	$Cr_2O_3(s)$	−1139.7	−1058.1	81.2
C(s, graphite)	0	0	5.74	**Cobalt**			
$CCl_4(g)$	−106.7	−64.0	309.4	Co(g)	439	393	179
$CCl_4(l)$	−139.3	−68.6	214.4	Co(s)	0	0	28.4
$CF_4(g)$	−679.9	−635.1	262.3				
$CH_4(g)$	−74.6	−50.5	186.3				

Substance	ΔH_f° (kJ/mol)	ΔG_f° (kJ/mol)	S° (J/mol-K)
Copper			
Cu(g)	338.4	298.6	166.3
Cu(s)	0	0	33.30
$CuCl_2(s)$	−205.9	−161.7	108.1
CuO(s)	−156.1	−128.3	42.59
$Cu_2O(s)$	−170.7	−147.9	92.36
Fluorine			
F(g)	80.0	61.9	158.7
$F^-(aq)$	−332.6	−278.8	−13.8
$F_2(g)$	0	0	202.8
HF(g)	−273.30	−275.40	173.78
Hydrogen			
H(g)	217.94	203.26	114.60
$H^+(aq)$	0	0	0
$H^+(g)$	1536.2	1517.0	108.9
$H_2(g)$	0	0	130.68
Iodine			
I(g)	106.60	70.16	180.66
$I^-(g)$	−55.19	−51.57	111.3
$I_2(g)$	62.25	19.37	260.57
$I_2(s)$	0	0	116.73
HI(g)	26.50	1.79	206.6
Iron			
Fe(g)	415.5	369.8	180.5
Fe(s)	0	0	27.31
$Fe^{2+}(aq)$	−87.86	−84.93	113.4
$Fe^{3+}(aq)$	−47.69	−10.54	293.3
$FeCl_2(s)$	−341.8	−302.3	117.9
$FeCl_3(s)$	−400	−334	142.2
FeO(s)	−271.9	−255.2	60.75
$Fe_2O_3(s)$	−822.16	−740.98	89.96
$Fe_3O_4(s)$	−1117.1	−1014.2	146.4
$FeS_2(s)$	−171.5	−160.1	52.92
Lead			
Pb(s)	0	0	68.85
$PbBr_2(s)$	−277.4	−260.7	161
$PbCO_3(s)$	−699.1	−625.5	131.0
$Pb(NO_3)_2(aq)$	−421.3	−246.9	303.3
$Pb(NO_3)_2(s)$	−451.9	—	—
PbO(s)	−217.3	−187.9	68.70
Lithium			
Li(g)	159.3	126.6	138.8
Li(s)	0	0	29.09

Substance	ΔH_f° (kJ/mol)	ΔG_f° (kJ/mol)	S° (J/mol-K)
$Li^+(aq)$	−278.5	−273.4	12.2
$Li^+(g)$	685.7	648.5	133.0
LiCl(s)	−408.3	−384.0	59.30
Magnesium			
Mg(g)	147.1	112.5	148.6
Mg(s)	0	0	32.51
$MgCl_2(s)$	−641.6	−592.1	89.6
MgO(s)	−601.8	−569.6	26.8
$Mg(OH)_2(s)$	−924.7	−833.7	63.24
Manganese			
Mn(g)	280.7	238.5	173.6
Mn(s)	0	0	32.0
MnO(s)	−385.2	−362.9	59.7
$MnO_2(s)$	−519.6	−464.8	53.14
$MnO_4^-(aq)$	−541.4	−447.2	191.2
Mercury			
Hg(g)	60.83	31.76	174.89
Hg(l)	0	0	77.40
$HgCl_2(s)$	−230.1	−184.0	144.5
$Hg_2Cl_2(s)$	−264.9	−210.5	192.5
Nickel			
Ni(g)	429.7	384.5	182.1
Ni(s)	0	0	29.9
$NiCl_2(s)$	−305.3	−259.0	97.65
NiO(s)	−239.7	−211.7	37.99
Nitrogen			
N(g)	472.7	455.5	153.3
$N_2(g)$	0	0	191.61
$NH_3(aq)$	−80.29	−26.50	111.3
$NH_3(g)$	−45.94	−16.41	192.8
$NH_4^+(aq)$	−132.5	−79.31	113.4
$N_2H_4(g)$	95.40	159.4	238.5
$NH_4CN(s)$	0.4	—	—
$NH_4Cl(s)$	−314.4	−203.0	94.6
$NH_4NO_3(s)$	−365.6	−184.0	151
NO(g)	90.37	86.71	210.62
$NO_2(g)$	33.84	51.84	240.45
$N_2O(g)$	81.6	103.59	220.0
$N_2O_4(g)$	9.66	98.28	304.3
NOCl(g)	52.6	66.3	264
$HNO_3(aq)$	−206.6	−110.5	146
$HNO_3(g)$	−134.3	−73.94	266.4

Substance	ΔH_f° (kJ/mol)	ΔG_f° (kJ/mol)	S° (J/mol-K)	Substance	ΔH_f° (kJ/mol)	ΔG_f° (kJ/mol)	S° (J/mol-K)
Oxygen				**Scandium**			
$O(g)$	247.5	230.1	161.0	$Sc(g)$	377.8	336.1	174.7
$O_2(g)$	0	0	205.15	$Sc(s)$	0	0	34.6
$O_3(g)$	142.3	163.4	237.6	**Selenium**			
$OH^-(aq)$	−230.0	−157.3	−10.7	$H_2Se(g)$	29.7	15.9	219.0
$H_2O(g)$	−241.82	−228.57	188.83	**Silicon**			
$H_2O(l)$	−285.83	−237.13	69.91	$Si(g)$	368.2	323.9	167.8
$H_2O_2(g)$	−136.10	−105.48	232.9	$Si(s)$	0	0	18.7
$H_2O_2(l)$	−187.8	−120.4	109.6	$SiC(s)$	−73.22	−70.85	16.61
Phosphorus				$SiCl_4(l)$	−640.1	−572.8	239.3
$P(g)$	316.4	280.0	163.2	$SiO_2(s, \text{quartz})$	−910.9	−856.5	41.84
$P_2(g)$	144.3	103.7	218.1	**Silver**			
$P_4(g)$	58.9	24.4	280	$Ag(s)$	0	0	42.55
$P_4(s, \text{red})$	−17.46	−12.03	22.85	$Ag^+(aq)$	105.90	77.11	73.93
$P_4(s, \text{white})$	0	0	41.08	$AgCl(s)$	−127.0	−109.70	96.11
$PCl_3(g)$	−288.07	−269.6	311.7	$Ag_2O(s)$	−31.05	−11.20	121.3
$PCl_3(l)$	−319.6	−272.4	217	$AgNO_3(s)$	−124.4	−33.41	140.9
$PF_5(g)$	−1594.4	−1520.7	300.8	**Sodium**			
$PH_3(g)$	5.4	13.4	210.2	$Na(g)$	107.7	77.3	153.7
$P_4O_6(s)$	−1640.1	—	—	$Na(s)$	0	0	51.30
$P_4O_{10}(s)$	−2940.1	−2675.2	228.9	$Na^+(aq)$	−240.1	−261.9	59.0
$POCl_3(g)$	−542.2	−502.5	325	$Na^+(g)$	609.3	574.3	148.0
$POCl_3(l)$	−597.0	−520.9	222	$NaBr(aq)$	−360.6	−364.7	141.00
$H_3PO_4(aq)$	−1288.3	−1142.6	158.2	$NaBr(s)$	−361.4	−349.3	86.82
Potassium				$Na_2CO_3(s)$	−1130.8	−1047.7	136.0
$K(g)$	89.99	61.17	160.2	$NaCl(aq)$	−407.1	−393.0	115.5
$K(s)$	0	0	64.67	$NaCl(g)$	−181.4	−201.3	229.8
$K^+(aq)$	−252.4	−283.3	102.5	$NaCl(s)$	−411.1	−384.1	72.11
$K^+(g)$	514.2	481.2	154.5	$NaHCO_3(s)$	−947.7	−851.8	102.1
$KCl(s)$	−435.9	−408.3	82.7	$NaNO_3(aq)$	−446.2	−372.4	207
$KClO_3(s)$	−391.2	−289.9	143.0	$NaNO_3(s)$	−467.9	−367.0	116.5
$KClO_3(aq)$	−349.5	−284.9	265.7	$NaOH(aq)$	−469.6	−419.2	49.8
$K_2CO_3(s)$	−1150.18	−1064.58	155.44	$NaOH(s)$	−425.6	−379.5	64.46
$KI(s)$	−327.9	−322.9	106.4	$Na_2SO_4(s)$	−1387.1	−1270.2	149.6
$KNO_3(s)$	−492.70	−393.13	132.9	**Strontium**			
$K_2O(s)$	−363.2	−322.1	94.14	$SrO(s)$	−592.0	−561.9	54.9
$KO_2(s)$	−284.5	−240.6	122.5	$Sr(g)$	164.4	110.0	164.6
$K_2O_2(s)$	−495.8	−429.8	113.0	**Sulfur**			
$KOH(s)$	−424.7	−378.9	78.91	$S(s, \text{rhombic})$	0	0	31.88
$KOH(aq)$	−482.4	−440.5	91.6	$S_8(g)$	102.3	49.7	430.9
Rubidium				$SO_2(g)$	−296.9	−300.4	248.5
$Rb(g)$	85.8	55.8	170.0	$SO_3(g)$	−395.2	−370.4	256.2
$Rb(s)$	0	0	76.78	$SO_4^{2-}(aq)$	−909.3	−744.5	20.1
$RbCl(s)$	−430.5	−412.0	92				
$RbClO_3(s)$	−392.4	−292.0	152				

Substance	ΔH_f° (kJ/mol)	ΔG_f° (kJ/mol)	S° (J/mol-K)	Substance	ΔH_f° (kJ/mol)	ΔG_f° (kJ/mol)	S° (J/mol-K)
$SOCl_2(l)$	−245.6	—	—	Vanadium			
$H_2S(g)$	−20.17	−33.01	205.6	$V(g)$	514.2	453.1	182.2
$H_2SO_4(aq)$	−909.3	−744.5	20.1	$V(s)$	0	0	28.9
$H_2SO_4(l)$	−814.0	−689.9	156.1	Zinc			
Titanium				$Zn(g)$	130.7	95.2	160.9
$Ti(g)$	468	422	180.3	$Zn(s)$	0	0	41.63
$Ti(s)$	0	0	30.76	$ZnCl_2(s)$	−415.1	−369.4	111.5
$TiCl_4(g)$	−763.2	−726.8	354.9	$ZnO(s)$	−348.0	−318.2	43.9
$TiCl_4(l)$	−804.2	−728.1	221.9				
$TiO_2(s)$	−944.7	−889.4	50.29				

AQUEOUS EQUILIBRIUM CONSTANTS

TABLE D.1 Dissociation Constants for Acids at 25 °C

Name	Formula	K_{a1}	K_{a2}	K_{a3}
Acetic acid	CH_3COOH (or $HC_2H_3O_2$)	1.8×10^{-5}		
Arsenic acid	H_3AsO_4	5.6×10^{-3}	1.0×10^{-7}	3.0×10^{-12}
Arsenous acid	H_3AsO_3	5.1×10^{-10}		
Ascorbic acid	$H_2C_6H_6O_6$	8.0×10^{-5}	1.6×10^{-12}	
Benzoic acid	C_6H_5COOH (or $HC_7H_5O_2$)	6.3×10^{-5}		
Boric acid	H_3BO_3	5.8×10^{-10}		
Butanoic acid	C_3H_7COOH (or $HC_4H_7O_2$)	1.5×10^{-5}		
Carbonic acid	H_2CO_3	4.3×10^{-7}	5.6×10^{-11}	
Chloroacetic acid	$CH_2ClCOOH$ (or $HC_2H_2O_2Cl$)	1.4×10^{-3}		
Chlorous acid	$HClO_2$	1.0×10^{-2}		
Citric acid	$HOOCC(OH)(CH_2COOH)_2$ (or $H_3C_6H_5O_7$)	7.4×10^{-4}	1.7×10^{-5}	4.0×10^{-7}
Cyanic acid	$HCNO$	3.5×10^{-4}		
Formic acid	$HCOOH$ (or $HCHO_2$)	1.8×10^{-4}		
Hydroazoic acid	HN_3	1.9×10^{-5}		
Hydrocyanic acid	HCN	4.9×10^{-10}		
Hydrofluoric acid	HF	6.8×10^{-4}		
Hydrogen chromate ion	$HCrO_4^-$	3.0×10^{-7}		
Hydrogen peroxide	H_2O_2	2.4×10^{-12}		
Hydrogen selenate ion	$HSeO_4^-$	2.2×10^{-2}		
Hydrogen sulfide	H_2S	9.5×10^{-8}	1×10^{-19}	
Hypobromous acid	$HBrO$	2.5×10^{-9}		
Hypochlorous acid	$HClO$	3.0×10^{-8}		
Hypoiodous acid	HIO	2.3×10^{-11}		
Iodic acid	HIO_3	1.7×10^{-1}		
Lactic acid	$CH_3CH(OH)COOH$ (or $HC_3H_5O_3$)	1.4×10^{-4}		
Malonic acid	$CH_2(COOH)_2$ (or $H_2C_3H_2O_4$)	1.5×10^{-3}	2.0×10^{-6}	
Nitrous acid	HNO_2	4.5×10^{-4}		
Oxalic acid	$(COOH)_2$ (or $H_2C_2O_4$)	5.9×10^{-2}	6.4×10^{-5}	
Paraperiodic acid	H_5IO_6	2.8×10^{-2}	5.3×10^{-9}	
Phenol	C_6H_5OH (or HC_6H_5O)	1.3×10^{-10}		
Phosphoric acid	H_3PO_4	7.5×10^{-3}	6.2×10^{-8}	4.2×10^{-13}
Propionic acid	C_2H_5COOH (or $HC_3H_5O_2$)	1.3×10^{-5}		
Pyrophosphoric acid	$H_4P_2O_7$	3.0×10^{-2}	4.4×10^{-3}	2.1×10^{-7}
Selenous acid	H_2SeO_3	2.3×10^{-3}	5.3×10^{-9}	
Sulfuric acid	H_2SO_4	Strong acid	1.2×10^{-2}	
Sulfurous acid	H_2SO_3	1.7×10^{-2}	6.4×10^{-8}	
Tartaric acid	$HOOC(CHOH)_2COOH$ (or $H_2C_4H_4O_6$)	1.0×10^{-3}	1.4×10^{-5}	

TABLE D.2   Dissociation Constants for Bases at 25 °C

Name	Formula	K_b
Ammonia	NH_3	1.8×10^{-5}
Aniline	$C_6H_5NH_2$	4.3×10^{-10}
Dimethylamine	$(CH_3)_2NH$	5.4×10^{-4}
Ethylamine	$C_2H_5NH_2$	6.4×10^{-4}
Hydrazine	H_2NNH_2	1.3×10^{-6}
Hydroxylamine	$HONH_2$	1.1×10^{-8}
Methylamine	CH_3NH_2	4.4×10^{-4}
Pyridine	C_5H_5N	1.7×10^{-9}
Trimethylamine	$(CH_3)_3N$	6.4×10^{-5}

TABLE D.3   Solubility-Product Constants for Compounds at 25 °C

Name	Formula	K_{sp}	Name	Formula	K_{sp}
Barium carbonate	$BaCO_3$	5.0×10^{-9}	Lead(II) fluoride	PbF_2	3.6×10^{-8}
Barium chromate	$BaCrO_4$	2.1×10^{-10}	Lead(II) sulfate	$PbSO_4$	6.3×10^{-7}
Barium fluoride	BaF_2	1.7×10^{-6}	Lead(II) sulfide*	PbS	3×10^{-28}
Barium oxalate	BaC_2O_4	1.6×10^{-6}	Magnesium hydroxide	$Mg(OH)_2$	1.8×10^{-11}
Barium sulfate	$BaSO_4$	1.1×10^{-10}	Magnesium carbonate	$MgCO_3$	3.5×10^{-8}
Cadmium carbonate	$CdCO_3$	1.8×10^{-14}	Magnesium oxalate	MgC_2O_4	8.6×10^{-5}
Cadmium hydroxide	$Cd(OH)_2$	2.5×10^{-14}	Manganese(II) carbonate	$MnCO_3$	5.0×10^{-10}
Cadmium sulfide*	CdS	8×10^{-28}	Manganese(II) hydroxide	$Mn(OH)_2$	1.6×10^{-13}
Calcium carbonate (calcite)	$CaCO_3$	4.5×10^{-9}	Manganese(II) sulfide*	MnS	2×10^{-53}
Calcium chromate	$CaCrO_4$	7.1×10^{-4}	Mercury(I) chloride	Hg_2Cl_2	1.2×10^{-18}
Calcium fluoride	CaF_2	3.9×10^{-11}	Mercury(I) iodide	Hg_2I_2	$1.1 \times 10^{-1.1}$
Calcium hydroxide	$Ca(OH)_2$	6.5×10^{-6}	Mercury(II) sulfide*	HgS	2×10^{-53}
Calcium phosphate	$Ca_3(PO_4)_2$	2.0×10^{-29}	Nickel(II) carbonate	$NiCO_3$	1.3×10^{-7}
Calcium sulfate	$CaSO_4$	2.4×10^{-5}	Nickel(II) hydroxide	$Ni(OH)_2$	6.0×10^{-16}
Chromium(III) hydroxide	$Cr(OH)_3$	6.7×10^{-31}	Nickel(II) sulfide*	NiS	3×10^{-20}
Cobalt(II) carbonate	$CoCO_3$	1.0×10^{-10}	Silver bromate	$AgBrO_3$	5.5×10^{-13}
Cobalt(II) hydroxide	$Co(OH)_2$	1.3×10^{-15}	Silver bromide	$AgBr$	5.0×10^{-13}
Cobalt(II) sulfide*	CoS	5×10^{-22}	Silver carbonate	Ag_2CO_3	8.1×10^{-12}
Copper(I) bromide	$CuBr$	5.3×10^{-9}	Silver chloride	$AgCl$	1.8×10^{-10}
Copper(II) carbonate	$CuCO_3$	2.3×10^{-10}	Silver chromate	Ag_2CrO_4	1.2×10^{-12}
Copper(II) hydroxide	$Cu(OH)_2$	4.8×10^{-20}	Silver iodide	AgI	8.3×10^{-17}
Copper(II) sulfide*	CuS	6×10^{-37}	Silver sulfate	Ag_2SO_4	1.5×10^{-5}
Iron(II) carbonate	$FeCO_3$	2.1×10^{-11}	Silver sulfide*	Ag_2S	6×10^{-51}
Iron(II) hydroxide	$Fe(OH)_2$	7.9×10^{-16}	Strontium carbonate	$SrCO_3$	9.3×10^{-10}
Lanthanum fluoride	LaF_3	2×10^{-19}	Tin(II) sulfide*	SnS	1×10^{-26}
Lanthanum iodate	$La(IO_3)_3$	7.4×10^{-14}	Zinc carbonate	$ZnCO_3$	1.0×10^{-10}
Lead(II) carbonate	$PbCO_3$	7.4×10^{-14}	Zinc hydroxide	$Zn(OH)_2$	3.0×10^{-16}
Lead(II) chloride	$PbCl_2$	1.7×10^{-5}	Zinc oxalate	ZnC_2O_4	2.7×10^{-8}
Lead(II) chromate	$PbCrO_4$	2.8×10^{-13}	Zinc sulfide*	ZnS	2×10^{-25}

*For a solubility equilibrium of the type $MS(s) + H_2O(l) \rightleftharpoons M^{2+}(aq) + HS^-(aq) + OH^-(aq)$

STANDARD REDUCTION POTENTIALS AT 25 °C

Half-Reaction	$E°(V)$	Half-Reaction	$E°(V)$
$Ag^+(aq) + e^- \longrightarrow Ag(s)$	+0.80	$2 H_2O(l) + 2 e^- \longrightarrow H_2(g) + 2 OH^-(aq)$	−0.83
$AgBr(s) + e^- \longrightarrow Ag(s) + Br^-(aq)$	+0.10	$HO_2^-(aq) + H_2O(l) + 2 e^- \longrightarrow 3 OH^-(aq)$	+0.88
$AgCl(s) + e^- \longrightarrow Ag(s) + Cl^-(aq)$	+0.22	$H_2O_2(aq) + 2 H^+(aq) + 2 e^- \longrightarrow 2 H_2O(l)$	+1.78
$Ag(CN)_2^-(aq) + e^- \longrightarrow Ag(s) + 2 CN^-(aq)$	−0.31	$Hg_2^{2+}(aq) + 2 e^- \longrightarrow 2 Hg(l)$	+0.79
$Ag_2CrO_4(s) + 2 e^- \longrightarrow 2 Ag(s) + CrO_4^{2-}(aq)$	+0.45	$2 Hg^{2+}(aq) + 2 e^- \longrightarrow Hg_2^{2+}(aq)$	+0.92
$AgI(s) + e^- \longrightarrow Ag(s) + I^-(aq)$	−0.15	$Hg^{2+}(aq) + 2 e^- \longrightarrow Hg(l)$	+0.85
$Ag(S_2O_3)_2^{3-}(aq) + e^- \longrightarrow Ag(s) + 2 S_2O_3^{2-}(aq)$	+0.01	$I_2(s) + 2 e^- \longrightarrow 2 I^-(aq)$	+0.54
$Al^{3+}(aq) + 3 e^- \longrightarrow Al(s)$	−1.66	$2 IO_3^-(aq) + 12 H^+(aq) + 10 e^- \longrightarrow I_2(s) + 6 H_2O(l)$	+1.20
$H_3AsO_4(aq) + 2 H^+(aq) + 2 e^- \longrightarrow H_3AsO_3(aq) + H_2O(l)$	+0.56	$K^+(aq) + e^- \longrightarrow K(s)$	−2.92
$Ba^{2+}(aq) + 2 e^- \longrightarrow Ba(s)$	−2.90	$Li^+(aq) + e^- \longrightarrow Li(s)$	−3.05
$BiO^+(aq) + 2 H^+(aq) + 3 e^- \longrightarrow Bi(s) + H_2O(l)$	+0.32	$Mg^{2+}(aq) + 2 e^- \longrightarrow Mg(s)$	−2.37
$Br_2(l) + 2 e^- \longrightarrow 2 Br^-(aq)$	+1.06	$Mn^{2+}(aq) + 2 e^- \longrightarrow Mn(s)$	−1.18
$2 BrO_3^-(aq) + 12 H^+(aq) + 10 e^- \longrightarrow Br_2(l) + 6 H_2O(l)$	+1.52	$MnO_2(s) + 4 H^+(aq) + 2 e^- \longrightarrow Mn^{2+}(aq) + 2 H_2O(l)$	+1.23
$2 CO_2(g) + 2 H^+(aq) + 2 e^- \longrightarrow H_2C_2O_4(aq)$	−0.49	$MnO_4^-(aq) + 8 H^+(aq) + 5 e^- \longrightarrow Mn^{2+}(aq) + 4 H_2O(l)$	+1.51
$Ca^{2+}(aq) + 2 e^- \longrightarrow Ca(s)$	−2.87	$MnO_4^-(aq) + 2 H_2O(l) + 3 e^- \longrightarrow MnO_2(s) + 4 OH^-(aq)$	+0.59
$Cd^{2+}(aq) + 2 e^- \longrightarrow Cd(s)$	−0.40	$HNO_2(aq) + H^+(aq) + e^- \longrightarrow NO(g) + H_2O(l)$	+1.00
$Ce^{4+}(aq) + e^- \longrightarrow Ce^{3+}(aq)$	+1.61	$N_2(g) + 4 H_2O(l) + 4 e^- \longrightarrow 4 OH^-(aq) + N_2H_4(aq)$	−1.16
$Cl_2(g) + 2 e^- \longrightarrow 2 Cl^-(aq)$	+1.36	$N_2(g) + 5 H^+(aq) + 4 e^- \longrightarrow N_2H_5^+(aq)$	−0.23
$2 HClO(aq) + 2 H^+(aq) + 2 e^- \longrightarrow Cl_2(g) + 2 H_2O(l)$	+1.63	$NO_3^-(aq) + 4 H^+(aq) + 3 e^- \longrightarrow NO(g) + 2 H_2O(l)$	+0.96
$ClO^-(aq) + H_2O(l) + 2 e^- \longrightarrow Cl^-(aq) + 2 OH^-(aq)$	+0.89	$Na^+(aq) + e^- \longrightarrow Na(s)$	−2.71
$2 ClO_3^-(aq) + 12 H^+(aq) + 10 e^- \longrightarrow Cl_2(g) + 6 H_2O(l)$	+1.47	$Ni^{2+}(aq) + 2 e^- \longrightarrow Ni(s)$	−0.28
$Co^{2+}(aq) + 2 e^- \longrightarrow Co(s)$	−0.28	$O_2(g) + 4 H^+(aq) + 4 e^- \longrightarrow 2 H_2O(l)$	+1.23
$Co^{3+}(aq) + e^- \longrightarrow Co^{2+}(aq)$	+1.84	$O_2(g) + 2 H_2O(l) + 4 e^- \longrightarrow 4 OH^-(aq)$	+0.40
$Cr^{3+}(aq) + 3 e^- \longrightarrow Cr(s)$	−0.74	$O_2(g) + 2 H^+(aq) + 2 e^- \longrightarrow H_2O_2(aq)$	+0.68
$Cr^{3+}(aq) + e^- \longrightarrow Cr^{2+}(aq)$	−0.41	$O_3(g) + 2 H^+(aq) + 2 e^- \longrightarrow O_2(g) + H_2O(l)$	+2.07
$Cr_2O_7^{2-}(aq) + 14 H^+(aq) + 6 e^- \longrightarrow 2 Cr^{3+}(aq) + 7 H_2O(l)$	+1.33	$Pb^{2+}(aq) + 2 e^- \longrightarrow Pb(s)$	−0.13
$CrO_4^{2-}(aq) + 4 H_2O(l) + 3 e^- \longrightarrow$ $Cr(OH)_3(s) + 5 OH^-(aq)$	−0.13	$PbO_2(s) + HSO_4^-(aq) + 3 H^+(aq) + 2 e^- \longrightarrow$ $PbSO_4(s) + 2 H_2O(l)$	+1.69
$Cu^{2+}(aq) + 2 e^- \longrightarrow Cu(s)$	+0.34	$PbSO_4(s) + H^+(aq) + 2 e^- \longrightarrow Pb(s) + HSO_4^-(aq)$	−0.36
$Cu^{2+}(aq) + e^- \longrightarrow Cu^+(aq)$	+0.15	$PtCl_4^{2-}(aq) + 2 e^- \longrightarrow Pt(s) + 4 Cl^-(aq)$	+0.73
$Cu^+(aq) + e^- \longrightarrow Cu(s)$	+0.52	$S(s) + 2 H^+(aq) + 2 e^- \longrightarrow H_2S(g)$	+0.14
$CuI(s) + e^- \longrightarrow Cu(s) + I^-(aq)$	−0.19	$H_2SO_3(aq) + 4 H^+(aq) + 4 e^- \longrightarrow S(s) + 3 H_2O(l)$	+0.45
$F_2(g) + 2 e^- \longrightarrow 2 F^-(aq)$	+2.87	$HSO_4^-(aq) + 3 H^+(aq) + 2 e^- \longrightarrow H_2SO_3(aq) + H_2O(l)$	+0.17
$Fe^{2+}(aq) + 2 e^- \longrightarrow Fe(s)$	−0.44	$Sn^{2+}(aq) + 2 e^- \longrightarrow Sn(s)$	−0.14
$Fe^{3+}(aq) + e^- \longrightarrow Fe^{2+}(aq)$	+0.77	$Sn^{4+}(aq) + 2 e^- \longrightarrow Sn^{2+}(aq)$	+0.15
$Fe(CN)_6^{3-}(aq) + e^- \longrightarrow Fe(CN)_6^{4-}(aq)$	+0.36	$VO_2^+(aq) + 2 H^+(aq) + e^- \longrightarrow VO^{2+}(aq) + H_2O(l)$	+1.00
$2 H^+(aq) + 2 e^- \longrightarrow H_2(g)$	0.00	$Zn^{2+}(aq) + 2 e^- \longrightarrow Zn(s)$	−0.76

STANDARD REDUCTION
POTENTIALS AT 25 °C

ANSWERS TO RED EXERCISES

Answers to Go Figures, Practice Exercises, and Self Assessment Exercises are available as PDF files within Mastering Chemistry. We invite instructors to share these resources as needed.

Chapter 1 **1.1 (a)** Four **(b)** six **(c)** hydrogen **(d)** 22 **1.2 (a)** Pure element: i **(b)** mixture of elements: v, vi **(c)** pure compound: iv **(d)** mixture of an element and a compound: ii, iii **1.4 (a)** Homogeneous mixture **(b)** Yes, brass is a solution. **1.5 (iv)** A gas is turning into a solid. **1.7** Filtration **1.8 (a)** The aluminum sphere is lightest, then nickel, then silver. **(b)** The platinum cube is smallest, then gold, then lead. **1.10 (a)** 7.5 cm; two significant figures (sig figs) **(b)** 72 mi/hr (inner scale, two sig figs) or 115 km/hr (outer scale, three sig figs) **1.12** 464 jelly beans. The mass of an average bean has 2 decimal places and 3 sig figs. The number of beans then has 3 sig figs, by the rules for multiplication and division. **1.13** Statement (c) is false. **1.15 (a)** Heterogeneous mixture **(b)** homogeneous mixture (heterogeneous if there are undissolved particles) **(c)** pure substance **(d)** pure substance. **1.17 (a)** S **(b)** Au **(c)** K **(d)** Cl **(e)** Cu **(f)** uranium **(g)** nickel **(h)** sodium **(i)** aluminum **(j)** silicon **1.19** C is a compound; it contains both carbon and oxygen. A is a compound; it contains at least carbon and oxygen. B is not defined by the data given; it is probably also a compound because few elements exist as white solids. **1.21** Physical properties: silvery white; lustrous; melting point $= 649\,°C$, boiling point $= 1105\,°C$; density at $20\,°C = 1.738\,g/cm^3$; pounded into sheets; drawn into wires; good conductor. Chemical properties: burns in air; reacts with Cl_2. **1.23 (a)** Chemical **(b)** physical **(c)** physical **(d)** chemical **(e)** chemical **1.25 (a)** Filtration **(b)** chromatography **(c)** distillation **1.27** Statements (i) and (iii) are true. **1.29 (a)** $1.9 \times 10^5\,J$ **(b)** $4.6 \times 10^4\,cal$ **(c)** 14 m/s **1.31 (a)** Kinetic energy **(b)** Potential energy decreases. **(c)** increase **1.33 (a)** 1×10^{-1} **(b)** 1×10^{-2} **(c)** 1×10^{-15} **(d)** 1×10^{-6} **(e)** 1×10^{6} **(f)** 1×10^{3} **(g)** 1×10^{-9} **(h)** 1×10^{-3} **(i)** 1×10^{-12} **1.35 (a)** $22\,°C$ **(b)** $422.1\,°F$ **(c)** 506 K **(d)** $107\,°F$ **(e)** 1600 K **(f)** $-459.67\,°F$ **1.37 (a)** 1.62 g/mL. Tetrachloroethylene, 1.62 g/mL, is more dense than water, 1.00 g/mL; tetrachloroethylene will sink rather than float on water. **(b)** 11.7 g **1.39 (a)** Calculated density = 0.86 g/mL. The substance is probably toluene, density = 0.866 g/mL. **(b)** 89.2 mL benzene **(c)** 1.11×10^3 g nickel **1.41 (a)** 940 Tg **(b)** 1.1×10^{11} gal gasoline **(c)** 520 teraliters (TL) of CO_2 **1.43** Exact: (b), (d), and (f) **1.45 (a)** 2 **(b)** 4 **(c)** 4 **(d)** 3 **(e)** 3 **(f)** 1 **1.47 (a)** 1.025×10^2 **(b)** 6.570×10^2 **(c)** 8.543×10^{-3} **(d)** 2.579×10^{-4} **(e)** -3.572×10^{-2} **1.49 (a)** 17.00 **(b)** 812.0 **(c)** 8.23×10^3 **(d)** 8.69×10^{-2} **1.51 (a)** 3.80×10^5 **(b)** 7.91×10^6 **(c)** 1.68×10^2 **(d)** -9.45 **1.53** 5 significant figures

1.55 (a) $\dfrac{1 \times 10^{-3}\,m}{1\,mm} \times \dfrac{1\,nm}{1 \times 10^{-9}\,m}$ **(b)** $\dfrac{1 \times 10^{-3}\,g}{1\,mg} \times \dfrac{1\,kg}{1000\,g}$

(c) $\dfrac{1000\,m}{1\,km} \times \dfrac{1\,cm}{1 \times 10^{-2}\,m} \times \dfrac{1\,in.}{2.54\,cm} \times \dfrac{1\,ft}{12\,in.}$ **(d)** $\dfrac{(2.54)^3\,cm^3}{1^3\,in.^3}$

1.57 (a) 54.7 km/hr **(b)** 1.3×10^3 gal **(c)** 46.0 m **(d)** 0.984 in/hr **1.59 (a)** 4.32×10^5 s **(b)** 88.5 m **(c)** \$0.499/L **(d)** 46.6 km/hr **(e)** 1.420 L/s **(f)** $707.9\,cm^3$ **1.61 (a)** 1.2×10^2 L **(b)** 5×10^2 mg **(c)** 19.9 mi/gal (2×10^1 mi/gal for 1 significant figure) **(d)** 1.81 kg **1.63** 0.18 g CO **1.65 (a)** $1208\,lb/ft^3$ **(b)** 232.2 g **1.67** \$1 × 10^5 **1.71** 8.47 g O **1.73 (a)** Set 1, 34.44; set 2, 34.52. Based on the average, set 1 is more accurate. **(b)** The average deviation for set 1 is 0.02 and for set 2 is 0.03. Set 1 is more precise. **1.74 (a)** Volume **(b)** area **(c)** volume **(d)** density **(e)** time **(f)** length **(g)** temperature **1.78** Substances (c), (d), (e), (g) and (h) are pure or nearly pure.

1.80 (a) 1.13×10^5 quarters **(b)** 6.41×10^5 g **(c)** $\$2.83 \times 10^4$ **(d)** 9.28×10^8 stacks **1.84** The most dense liquid, Hg, will sink; the least dense, cyclohexane, will float; H_2O will be in the middle. **1.88** density of solid $= 1.63\,g/mL$ **1.94** The inner diameter of the tube is 1.71 cm. **1.96** Statements (i) and (iii) are true. **1.99 (a)** Volume $= 0.050\,mL$ **(b)** surface area $= 12.4\,m^2$ **(c)** 99.99% of the mercury was removed. **(d)** The spongy material weighs 17.7 mg after exposure to mercury.

Chapter 2 **2.1 (a)** $(-)$ **(b)** increase **(c)** decrease **2.4** The particle is an ion. $^{32}_{16}S^{2-}$ **2.6** Formula: IF_5; name: iodine pentafluoride; the compound is molecular. **2.8** Only $Ca(NO_3)_2$, calcium nitrate, is consistent with the diagram. **2.10 (a)** In the presence of an electric field, there is electrostatic attraction between the negatively charged oil drops and the positively charged plate as well as electrostatic repulsion between the negatively charged oil drops and the negatively charged plate. These electrostatic forces oppose the force of gravity and change the rate of fall of the drops. **(b)** Each individual drop has a different number of electrons associated with it. If the combined electrostatic forces are greater than the force of gravity, the drop moves up. **2.11 (a)** mass O/mass C = 2.66 **(b)** mass O/mass C = 1.33 **(c)** CO **2.13 (a)** 0.5711 g O/1 g N; 1.142 g O/1 g N; 2.284 g O/1 g N; 2.855 g O/1 g N **(b)** The numbers in part **(a)** obey the *law of multiple proportions*. Multiple proportions arise because atoms are the indivisible entities combining, as stated in Dalton's atomic theory. **2.15** Neutrons were discovered last. **2.17** (a) is correct – only statement (i) is true. **2.19 (a)** 0.135 nm; 1.35×10^2 or 135 pm **(b)** 3.70×10^6 Au atoms **(c)** $1.03 \times 10^{-23}\,cm^3$ **2.21 (a)** Proton, neutron, electron **(b)** proton $= 1+$, neutron $= 0$, electron $= 1-$ **(c)** The neutron is most massive. (The neutron and proton have very similar masses.) **(d)** The electron is least massive. **2.23 (a)** 5 protons, 5 neutrons, 5 electrons **(b)** $^{11}_{6}C$ **(c)** $^{11}_{5}B$ **(d)** The atom in part **(c)** is an isotope of ^{10}B. **2.25 (a)** Atomic number is the number of protons in the nucleus of an atom. Mass number is the total number of nuclear particles, protons plus neutrons, in an atom. **(b)** atomic number **(c)** mass number **2.27 (a)** ^{40}Ar: 18 p, 22 n, 18 e **(b)** ^{65}Zn: 30 p, 35 n, 30 e **(c)** ^{70}Ga: 31 p, 39 n, 31 e **(d)** ^{80}Br: 35 p, 45 n, 35 e **(e)** ^{184}W: 74 p, 110 n, 74 e **(f)** ^{243}Am: 95 p, 148 n, 95e

2.29

Symbol	^{79}Br	^{55}Mn	^{112}Cd	^{222}Rn	^{207}Pb
Protons	35	25	48	86	82
Neutrons	44	30	64	136	125
Electrons	35	25	48	86	82
Mass no.	79	55	112	222	207

2.31 (a) $^{196}_{78}Pt$ **(b)** $^{84}_{36}Kr$ **(c)** $^{75}_{33}As$ **(d)** $^{24}_{12}Mg$ **2.33** Answer (b) is the best choice. Each B atom will have the mass of one of the naturally occurring isotopes, while the "atomic weight" is an average value. **2.35** 63.55 amu **2.37 (a)** In Thomson's cathode ray tube, the charged particles are electrons. In a mass spectrometer, the charged particles are positively charged ions (cations). **(b)** The *x*-axis label is *m/z* (the ratio of mass to charge of the particles) and the *y*-axis label is signal intensity. **(c)** The Cl^{2+} ion will be deflected more. **2.39 (a)** average atomic mass $= 24.31$ amu

(b)

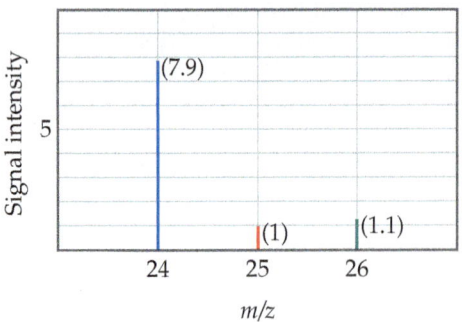

2.41 (a) Cr, 24 (metal) **(b)** He, 2 (nonmetal) **(c)** P, 15 (nonmetal) **(d)** Zn, 30 (metal) **(e)** Mg, 12 (metal) **(f)** Br, 35 (nonmetal) **(g)** As, 33 (metalloid) **2.43 (a)** K, alkali metals (metal) **(b)** I, halogens (nonmetal) **(c)** Mg, alkaline earth metals (metal) **(d)** Ar, noble gases (nonmetal) **(e)** S, chalcogens (nonmetal) **2.45 (a)** C_4H_{10} is the molecular formula for both compounds. **(b)** C_2H_5 is the empirical formula for both compounds. **(c)** structural **2.47** From left to right: molecular, N_2H_4, empirical, NH_2; molecular, N_2H_2, empirical, NH, molecular and empirical, NH_3 **2.49 (a)** $AlBr_3$ **(b)** C_4H_5 **(c)** C_2H_4O **(d)** P_2O_5 **(e)** C_3H_2Cl **(f)** BNH_2 **2.51 (a)** 6 **(b)** 10 **(c)** 12

2.53

(a) C_2H_6O,

(b) C_2H_6O,

(c) CH_4O,

(d) PF_3,

2.55

Symbol	$^{59}Co^{3+}$	$^{80}Se^{2-}$	$^{192}Os^{2+}$	$^{200}Hg^{2+}$
Protons	27	34	76	80
Neutrons	32	46	116	120
Electrons	24	36	74	78
Net Charge	3+	2−	2+	2+

2.57 (a) Mg^{2+} **(b)** Al^{3+} **(c)** K^+ **(d)** S^{2-} **(e)** F^- **2.59 (a)** GaF_3, gallium(III) fluoride **(b)** LiH, lithium hydride **(c)** AlI_3, aluminum iodide **(d)** K_2S, potassium sulfide **2.61 (a)** $CaBr_2$ **(b)** K_2CO_3 **(c)** $Al(CH_3COO)_3$ **(d)** $(NH_4)_2SO_4$ **(e)** $Mg_3(PO_4)_2$

2.63

Ion	K^+	NH_4^+	Mg^{2+}	Fe^{3+}
Cl^-	KCl	NH_4Cl	$MgCl_2$	$FeCl_3$
OH^-	KOH	NH_4OH	$Mg(OH)_2$	$Fe(OH)_3$
CO_3^{2-}	K_2CO_3	$(NH_4)_2CO_3$	$MgCO_3$	$Fe_2(CO_3)_3$
PO_4^{3-}	K_3PO_4	$(NH_4)_3PO_4$	$Mg_3(PO_4)_2$	$FePO_4$

2.65 Molecular: **(a)** B_2H_6 **(b)** CH_3OH **(f)** NOCl **(g)** NF_3. Ionic: **(c)** $LiNO_3$ **(d)** Sc_2O_3 **(e)** CsBr **(f)** Ag_2SO_4 **2.67 (a)** ClO_2^- **(b)** Cl^- **(c)** ClO_3^- **(d)** ClO_4^- **(e)** ClO^- **2.69 (a)** calcium, 2+; oxide, 2− **(b)** sodium, 1+; sulfate, 2− **(c)** potassium, 1+; perchlorate, 1− **(d)** iron, 2+, nitrate, 1− **(e)** chromium, 3+; hydroxide, 1− **2.71 (a)** lithium oxide **(b)** iron(III) chloride (ferric chloride) **(c)** sodium hypochlorite **(d)** calcium sulfite **(e)** copper(II) hydroxide (cupric hydroxide) **(f)** iron(II) nitrate (ferrous nitrate) **(g)** calcium acetate **(h)** chromium(III) carbonate (chromic carbonate) **(i)** potassium chromate **(j)** ammonium sulfate **2.73 (a)** $Al(OH)_3$ **(b)** K_2SO_4 **(c)** Cu_2O **(d)** $Zn(NO_3)_2$ **(e)** $HgBr_2$ **(f)** $Fe_2(CO_3)_3$ **(g)** NaBrO **2.75 (a)** Bromic acid **(b)** hydrobromic acid **(c)** phosphoric acid **(d)** HClO **(e)** HIO_3 **(f)** H_2SO_3 **2.77 (a)** Sulfur hexafluoride **(b)** selenium dichloride **(c)** xenon trioxide **(d)** N_2O_4 **(e)** HCN **(f)** P_4S_6 **2.79 (a)** $ZnCO_3$, ZnO, CO_2 **(b)** HF, SiO_2, SiF_4, H_2O **(c)** SO_2, H_2O, H_2SO_3 **(d)** PH_3 **(e)** $HClO_4$, Cd, $Cd(ClO_4)_2$ **(f)** VBr_3 **2.81 (a)** A hydrocarbon is a compound composed of the elements hydrogen and carbon only.

(b)

molecular formula C_8H_{18}, empirical formula C_4H_9

2.83 (a) —OH

(b)

2.85 The correct answer is (c). The incorrect formula for methanol, CH_4OH, should be corrected to CH_3OH. **2.87** Structures c and d are the same molecule. **2.89 (a)** 2-methylhexane **(b)** 4-ethyl-2, 4-dimethyldecane **(c)** $CH_3CH_2CH_2CH_2CH_2CH(CH_3)_2$ **(d)** $CH_3CH_2CH_2CH_2CH(CH_2CH_3)CH(CH_3)CH(CH_3)_2$ **2.91** Statement (d) is correct. **2.94 (a)** 2 protons, 1 neutron, 2 electrons **(b)** tritium, ^3H, is more massive. **(c)** A precision of 1×10^{-27} g would be required to differentiate between $^3H^+$ and $^3He^+$. **2.97 (a)** $^{16}_8O$, $^{17}_8O$, $^{18}_8O$ **(b)** None of them contain 10 electrons; they all have 8. **(c)** 9 **2.99 (a)** $^{69}_{31}Ga$, 31 protons, 38 neutrons; $^{71}_{31}Ga$, 31 protons, 40 neutrons **(b)** $^{69}_{31}Ga$, 60.3%, $^{71}_{31}Ga$, 39.7%. **2.101 (a)** 5 significant figures **(b)** An electron is 0.05444% of the mass of an ^1H atom. **2.106 (a)** nickel(II) oxide, 2+ **(b)** manganese(IV) oxide, 4+ **(c)** chromium(III) oxide, 3+ **(d)** molybdenum(VI) oxide, 6+ **2.109 (a)** Perbromate ion **(b)** selenite ion **(c)** AsO_4^{3-} **(d)** $HTeO_4^-$ **2.112 (a)** Potassium nitrate **(b)** sodium carbonate **(c)** calcium oxide **(d)** hydrochloric acid **(e)** magnesium sulfate **(f)** magnesium hydroxide

Chapter 3   **3.1** Equation (a) best fits the diagram. **3.3 (a)** NO_2 **(b)** No, because we have no way of knowing whether the empirical and molecular formulas are the same. NO_2 represents the simplest ratio of atoms in a molecule but not the only possible molecular formula. **3.5 (a)** $C_2H_5NO_2$ **(b)** 75.0 g/mol **(c)** 1.332 mol glycine **(d)** Mass %N in glycine is 18.7%. **3.7 (a)** $N_2 + 3H_2 \longrightarrow 2NH_3$. **(b)** H_2 is the limiting reactant. **(c)** 6 NH_3 molecules can be made. **(d)** One N_2 molecule is left over. **3.9 (a)** False **(b)** true **(c)** false **3.11 (a)** $2CO(g) + O_2(g) \longrightarrow 2CO_2(g)$ **(b)** $N_2O_5(g) + H_2O(l) \longrightarrow 2HNO_3(aq)$ **(c)** $CH_4(g) + 4Cl_2(g) \longrightarrow CCl_4(l) + 4HCl(g)$ **(d)** $Zn(OH)_2(s) + 2HNO_3(aq) \longrightarrow Zn(NO_3)_2(aq) + 2H_2O(l)$ **3.13 (a)** $Al_4C_3(s) + 12H_2O(l) \longrightarrow 4Al(OH)_3(s) + 3CH_4(g)$ **(b)** $2C_5H_{10}O_2(l) + 13O_2(g) \longrightarrow 10CO_2(g) + 10H_2O(g)$ **(c)** $2Fe(OH)_3(s) + 3H_2SO_4(aq) \longrightarrow Fe_2(SO_4)_3(aq) + 6H_2O(l)$ **(d)** $Mg_3N_2(s) + 4H_2SO_4(aq) \longrightarrow 3MgSO_4(aq) + (NH_4)_2SO_4(aq)$

3.15 (a) $CaC_2(s) + 2 H_2O(l) \longrightarrow Ca(OH)_2(aq) + C_2H_2(g)$
(b) $2 KClO_3(s) \overset{\Delta}{\longrightarrow} 2 KCl(s) + 3 O_2(g)$
(c) $Zn(s) + H_2SO_4(aq) \longrightarrow ZnSO_4(aq) + H_2(g)$
(d) $PCl_3(l) + 3 H_2O(l) \longrightarrow H_3PO_3(aq) + 3 HCl(aq)$
(e) $3 H_2S(g) + 2 Fe(OH)_3(s) \longrightarrow Fe_2S_3(s) + 6 H_2O(g)$
3.17 (a) NaBr **(b)** solid **(c)** 2 **3.19 (a)** $Mg(s) + Cl_2(g) \longrightarrow MgCl_2(s)$
(b) $BaCO_3(s) \overset{\Delta}{\longrightarrow} BaO(s) + CO_2(g)$
(c) $C_8H_8(l) + 10 O_2(g) \longrightarrow 8 CO_2(g) + 4 H_2O(l)$
(d) $C_2H_6O(g) + 3 O_2(g) \longrightarrow 2 CO_2(g) + 3 H_2O(l)$
3.21 (a) $2 C_3H_6(g) + 9 O_2(g) \longrightarrow 6 CO_2(g) + 6 H_2O(g)$ combustion
(b) $NH_4NO_3(s) \longrightarrow N_2O(g) + 2 H_2O(g)$ decomposition
(c) $C_5H_6O(l) + 6 O_2(g) \longrightarrow 5 CO_2(g) + 3 H_2O(g)$ combustion
(d) $N_2(g) + 3 H_2(g) \longrightarrow 2 NH_3(g)$ combination
(e) $K_2O(s) + H_2O(l) \longrightarrow 2KOH(aq)$ combination **3.23 (a)** 63.0 amu
(b) 158.0 amu **(c)** 310.3 amu **(d)** 60.1 amu **(e)** 235.7 amu **(f)** 392.3 amu
(g) 137.5 amu **3.25 (a)** 16.8% **(b)** 16.1% **(c)** 21.1% **(d)** 28.8% **(e)** 27.2%
(f) 26.5% **3.27 (a)** 79.2% **(b)** 63.2% **(c)** 64.6% **3.29 (a)** False **(b)** true
(c) false **(d)** true **3.31** 23 g Na contains 1 mol of atoms; 0.5 mol H_2O
contains 1.5 mol atoms; 6.0×10^{23} N_2 molecules contain 2 mol of
atoms. **3.33** 4.4×10^{25} kg. One mole of people weighs 7.4 times as
much as Earth. **3.35 (a)** 35.9 g $C_{12}H_{22}O_{11}$ **(b)** 0.75766 mol $Zn(NO_3)_2$
(c) 6.0×10^{17} CH_3CH_2OH molecules **(d)** 2.47×10^{23} N atoms
3.37 (a) 0.373 g $(NH_4)_3PO_4$ **(b)** 5.737×10^{-3} mol Cl^- **(c)** 0.248 g
$C_8H_{10}N_4O_2$ **(d)** 387 g cholesterol/mol **3.39 (a)** Molar mass = 162.3 g
(b) 3.08×10^{-5} mol allicin **(c)** 1.86×10^{19} allicin molecules
(d) 3.71×10^{19} S atoms **3.41 (a)** 2.500×10^{21} H atoms **(b)** 2.083×10^{20}
$C_6H_{12}O_6$ molecules **(c)** 3.460×10^{-4} mol $C_6H_{12}O_6$ **(d)** 0.06227 g
$C_6H_{12}O_6$ **3.43** 3.2×10^{-8} mol C_2H_3Cl/L; 1.9×10^{16} molecules/L
3.45 (a) C_2H_6O **(b)** Fe_2O_3 **(c)** CH_2O **3.47 (a)** $CSCl_2$ **(b)** C_3OF_6 **(c)** Na_3AlF_6
3.49 31 g/mol **3.51 (a)** C_6H_{12} **(b)** NH_2Cl **3.53 (a)** Empirical formula,
CH; molecular formula, C_8H_8 **(b)** empirical formula, $C_4H_5N_2O$;
molecular formula, $C_8H_{10}N_4O_2$ **(c)** empirical formula and molecular
formula, $NaC_5H_8O_4N$ **3.55 (a)** C_7H_8 **(b)** The empirical and molecular
formulas are $C_{10}H_{20}O$. **3.57** Empirical formula, C_4H_8O; molecular
formula, $C_8H_{16}O_2$ **3.59** $x = 10$; $Na_2CO_3 \cdot 10H_2O$
3.61 (a) 2.40 mol HF **(b)** 5.25 g NaF **(c)** 0.610 g Na_2SiO_3
3.63 (a) $Al(OH)_3(s) + 3HCl(aq) \longrightarrow AlCl_3(aq) + 3H_2O(l)$
(b) 0.701 g HCl **(c)** 0.855 g $AlCl_3$; 0.347 g H_2O
(d) Mass of reactants = 0.500 g + 0.701 g = 1.201 g;
mass of products = 0.855 g + 0.347 g = 1.202 g.
Mass is conserved, within the precision of the data.
3.65 (a) $Al_2S_3(s) + 6H_2O(l) \longrightarrow 2 Al(OH)_3(s) + 3 H_2S(g)$
(b) 14.7 g $Al(OH)_3$ **3.67 (a)** 2.25 mol N_2 **(b)** 15.5 g NaN_3 **(c)** 548 g NaN_3
3.69 (a) 5.50×10^{-3} mol Al **(b)** $2 Al(s) + 3 Br_2(l) \longrightarrow 2 AlBr_3(s)$
(c) 1.47 g $AlBr_3$ **3.71** 1.25×10^5 kJ **3.73 (a)** The *limiting reactant*
determines the maximum number of product moles resulting from
a chemical reaction; any other reactant is an *excess reactant*. **(b)** The
limiting reactant regulates the amount of products because it is
completely used up during the reaction; no more product can be
made when one of the reactants is unavailable. **(c)** Combining ratios
are molecule and mole ratios. Since different molecules have different
masses, comparing initial masses of reactants will not provide a
comparison of numbers of molecules or moles.
3.75 (a) $2 C_2H_5OH + 6 O_2 \longrightarrow 4 CO_2 + 6 H_2O$ [This equation
corresponds to the mixture of reactants in the box, but it does not have
the simplest ratio of coefficients. Divide all coefficients by 2 to obtain
$C_2H_5OH + 3 O_2 \longrightarrow 2 CO_2 + 3 H_2O$] **(b)** C_2H_5OH limits **(c)** If the
reaction goes to completion, there will be four molecules of CO_2, six
molecules of H_2O, zero molecules of C_2H_5OH and one molecule of O_2.
3.77 NaOH is the limiting reactant; 0.925 mol Na_2CO_3 can be produced;
0.075 mol CO_2 remains. **3.79 (a)** $NaHCO_3$ is the limiting reactant.
(b) 0.524 g CO_2 **(c)** 0.238 g citric acid remains **3.81** 0.00 g $AgNO_3$ (limiting
reactant), 1.94 g Na_2CO_3, 4.06 g Ag_2CO_3, 2.50 g $NaNO_3$ **3.83 (a)** The

theoretical yield is 60.3 g C_6H_5Br. **(b)** 70.1% yield **3.85** 28 g S_8 actual
yield **3.87 (a)** $C_2H_4O_2(l) + 2 O_2(g) \longrightarrow 2 CO_2(g) + 2 H_2O(l)$
(b) $Ca(OH)_2(s) \longrightarrow CaO(s) + H_2O(g)$ **(c)** $Ni(s) + Cl_2(g) \longrightarrow NiCl_2(s)$
3.91 (a) 4.8×10^{-20} g CdSe **(b)** 150 Cd atoms **(c)** 8.4×10^{-19} g CdSe
(d) 2.6×10^3 Cd atom **3.95** $C_8H_8O_3$ **3.99 (a)** 1.19×10^{-5} mol NaI
(b) 8.1×10^{-3} g NaI **3.103** 7.5 mol H_2 and 4.5 mol N_2 present initially
3.108 6.46×10^{24} O atoms **3.112 (a)** $S(s) + O_2(g) \longrightarrow SO_2(g)$;
$SO_2(g) + CaO(s) \longrightarrow CaSO_3(s)$ **(b)** 7.9×10^7 g CaO
(c) 1.7×10^8 g $CaSO_3$

Chapter 4   **4.1** Diagram **(c)** represents Li_2SO_4 **4.5** $BaCl_2$
4.7 (c) is correct **4.9** This reaction is **(c)**, a redox reaction.
4.13 (a) False. Electrolyte solutions conduct electricity because *ions*
are moving through the solution. **(b)** True. Because ions are mobile
in solution, the added presence of uncharged molecules does not
inhibit conductivity. **4.15** Statement **(b)** is most correct.
4.17 (a) $FeCl_2(aq) \longrightarrow Fe^{2+}(aq) + 2 Cl^-(aq)$
(b) $HNO_3(aq) \longrightarrow H^+(aq) + NO_3^-(aq)$
(c) $(NH_4)_2SO_4(aq) \longrightarrow 2 NH_4^+(aq) + SO_4^{2-}(aq)$
(d) $Ca(OH)_2(aq) \longrightarrow Ca^{2+}(aq) + 2 OH^-(aq)$
4.19 HCOOH molecules, H^+ ions, and $HCOO^-$ ions;
$HCOOH(aq) \rightleftharpoons H^+(aq) + HCOO^-(aq)$ **4.21 (a)** Soluble
(b) insoluble **(c)** soluble **(d)** soluble **(e)** soluble **4.23 (a)** $Na_2CO_3(aq) +$
$2 AgNO_3(aq) \longrightarrow Ag_2CO_3(s) + 2 NaNO_3(aq)$ **(b)** No precipitate
(c) $FeSO_4(aq) + Pb(NO_3)_2(aq) \longrightarrow PbSO_4(s) + Fe(NO_3)_2(aq)$
4.25 (a) K^+, SO_4^{2-} **(b)** Li^+, NO_3^- **(c)** NH_4^+, Cl^- **4.27** Only Pb^{2+} could
be present. **4.29** Answer **(c)** is correct. **4.31** The 0.20 M HI(aq) is
most acidic **(b)**. **4.33 (a)** False. H_2SO_4 is a diprotic acid; it has two
ionizable hydrogen atoms. **(b)** False. HCl is a strong acid. **(c)** False.
CH_3OH is a molecular nonelectrolyte. **4.35 (a)** Acid, mixture of ions
and molecules (weak electrolyte) **(b)** none of the above, entirely
molecules (nonelectrolyte) **(c)** salt, entirely ions (strong electrolyte)
(d) base, entirely ions (strong electrolyte) **4.37 (a)** H_2SO_3, weak
electrolyte **(b)** CH_3CH_2OH, nonelectrolyte **(c)** NH_3, weak electrolyte
(d) $KClO_3$, strong electrolyte **(e)** $Cu(NO_3)_2$, strong electrolyte
4.39 (a) $2 HBr(aq) + Ca(OH)_2(aq) \longrightarrow CaBr(aq) + 2 H_2O(l)$;
$H^+(aq) + OH^-(aq) \longrightarrow H_2O(l)$;
(b) $Cu(OH)_2(s) + 2 HClO_4(aq) \longrightarrow Cu(ClO_4)_2(aq) + 2 H_2O(l)$;
$Cu(OH)_2(s) + 2 H^+(aq) \longrightarrow 2 H_2O(l) + Cu^{2+}(aq)$
(c) $Al(OH)_3(s) + 3 HNO_3(aq) \longrightarrow Al(NO_3)_3(aq) + 3 H_2O(l)$;
$Al(OH)_3(s) + 3 H^+(aq) \longrightarrow 3 H_2O(l) + Al^{3+}(aq)$
4.41 (a) $CdS(s) + H_2SO_4(aq) \longrightarrow CdSO_4(aq) + H_2S(g)$;
$CdS(s) + 2H^+(aq) \longrightarrow H_2S(g) + Cd^{2+}(aq)$
(b) $MgCO_3(s) + 2 HClO_4(aq) \longrightarrow Mg(ClO_4)_2(aq) + H_2O(l) + CO_2(g)$;
$MgCO_3(s) + 2 H^+(aq) \longrightarrow H_2O(l) + CO_2(g) + Mg^{2+}(aq)$
4.43 $MgCO_3(s) + 2 HCl(aq) \longrightarrow MgCl_2(aq) + H_2O(l) + CO_2(g)$;
$MgCO_3(s) + 2 H^+(aq) \longrightarrow Mg^{2+}(aq) + H_2O(l) + CO_2(g)$;
$MgO(s) + 2 HCl(aq) \longrightarrow MgCl_2(aq) + H_2O(l)$;
$MgO(s) + 2 H^+(aq) \longrightarrow Mg^{2+}(aq) + H_2O(l)$;
$Mg(OH)_2(s) + 2 H^+(aq) \longrightarrow Mg^{2+}(aq) + 2 H_2O(l)$
4.45 (a) False **(b)** true **4.47** Metals in region A are most easily
oxidized. Nonmetals in region D are least easily oxidized.
4.49 (a) +4 **(b)** +4 **(c)** +7 **(d)** +1 **(e)** +3 **(f)** −1 **4.51 (a)** H is
oxidized; N is reduced **(b)** Fe is reduced; Al is oxidized **(c)** Cl is
reduced; I is oxidized **(d)** S is oxidized; O is reduced
4.53 (a) $Sn(s) + 2 HCl(aq) \longrightarrow SnCl_2(aq) + H_2(g)$;
$Sn(s) + 2 H^+(aq) \longrightarrow Sn^{2+}(aq) + H_2(g)$
(b) $2 Al(s) + 6 HCOOH(aq) \longrightarrow 2 Al(HCOO)_3(aq) + 3 H_2(g)$;
$2 Al(s) + 6 HCOOH(aq) \longrightarrow 2 Al^{3+}(aq) + 6 HCOO^-(aq) + 3 H_2(g)$
4.55 (a) $Fe(s) + Cu(NO_3)_2(aq) \longrightarrow Fe(NO_3)_2(aq) + Cu(s)$
(b) NR **(c)** $Sn(s) + 2 HBr(aq) \longrightarrow SnBr_2(aq) + H_2(g)$
(d) NR **(e)** $2 Al(s) + 3 CoSO_4(aq) \longrightarrow Al_2(SO_4)_3(aq) + 3 Co(s)$

4.57 (a) i. $Zn(s) + Cd^{2+}(aq) \longrightarrow Cd(s) + Zn^{2+}(aq)$;
ii. $Cd(s) + Ni^{2+}(aq) \longrightarrow Ni(s) + Cd^{2+}(aq)$ **(b)** Cd must be below
Zn in the activity series. **(c)** Cd must be above Ni in the activity
series. **4.59 (a)** Intensive; the ratio of amount of solute to total
amount of solution is the same, regardless of how much solution is
present. **(b)** The term 0.50 mol HCl defines an amount (~18 g) of the
pure substance HCl. The term $0.50\ M$ HCl is a ratio; it indicates that
there is 0.50 mol of HCl solute in 1.0 liter of solution. **4.61 (a)** $1.17\ M$
$ZnCl_2$ **(b)** 0.158 mol H^+ **(c)** 58.3 mL of $6.00\ M$ NaOH **4.63** 16 g $Na^+(aq)$
4.65 BAC of $0.08 = 0.02\ M\ CH_3CH_2OH$ (alcohol) **4.67 (a)** 316 g
ethanol **(b)** 401 mL ethanol **4.69 (a)** $0.15\ M\ K_2CrO_4$ has the highest
K^+ concentration. **(b)** 30.0 mL of $0.15\ M\ K_2CrO_4$ has more K^+ ions.
4.71 (a) $0.25\ M\ Na^+$, $0.25\ M\ NO_3^-$ **(b)** $1.3 \times 10^{-2}\ M\ Mg^{2+}$, 1.3×10^{-2}
$M\ SO_4^{2-}$ **(c)** $0.0150\ M\ C_6H_{12}O_6$ **(d)** $0.111\ M\ Na^+$, $0.111\ M\ Cl^-$,
$0.0292\ M\ NH_4^+$, $0.0146\ M\ CO_3^{2-}$ **4.73 (a)** 16.9 mL $14.8\ M\ NH_3$
(b) $0.296\ M\ NH_3$ **4.75 (a)** The ratio of drug molecules to cancer cells
is 4.5×10^6 **4.77** $1.398\ M\ CH_3COOH$ **4.79 (a)** 20.0 mL of $0.15\ M$
HCl **(b)** 0.224 g KCl **(c)** The KCl reagent is virtually free relative to
the HCl solution. The KCl analysis is more cost-effective.
4.81 (a) 38.0 mL of $0.115\ M\ HClO_4$ **(b)** 769 mL of $0.128\ M$ HCl
(c) $0.408\ M\ AgNO_3$ **(d)** 0.275 g KOH **4.83** 27 g $NaHCO_3$ **4.85 (a)** Molar
mass of metal hydroxide is 103 g/mol. **(b)** Rb^+ **4.87 (a)** $NiSO_4(aq) +$
$2\ KOH(aq) \longrightarrow Ni(OH)_2(s) + K_2SO_4(aq)$ **(b)** $Ni(OH)_2$ **(c)** KOH is
the limiting reactant. **(d)** 0.927 g $Ni(OH)_2$ **(e)** $0.0667\ M\ Ni^{2+}(aq)$,
$0.0667\ M\ K^+(aq)$, $0.100\ M\ SO_4^{2-}(aq)$ **4.89** 91.39% $Mg(OH)_2$
4.91 (a) $U(s) + 2\ ClF_3(g) \longrightarrow UF_6(g) + Cl_2(g)$ **(b)** This is not a
metathesis reaction. **(c)** It is a redox reaction.
4.95 (a) $Al(OH)_3(s) + 3\ H^+(aq) \longrightarrow Al^{3+}(aq) + 3\ H_2O(l)$
(b) $Mg(OH)_2(s) + 2\ H^+(aq) \longrightarrow Mg^{2+}(aq) + 2\ H_2O(l)$
(c) $MgCO_3(s) + 2\ H^+(aq) \longrightarrow Mg^{2+}(aq) + H_2O(l) + CO_2(g)$
(d) $NaAl(CO_3)(OH)_2(s) + 4\ H^+(aq) \longrightarrow Na^+(aq) + Al^{3+}(aq)$
$+ 3\ H_2O(l) + CO_2(g)$ **(e)** $CaCO_3(s) + 2\ H^+(aq) \longrightarrow$
$Ca^{2+}(aq) + H_2O(l) + CO_2(g)$ [In **(c)**, **(d)** and **(e)**, one could
also write the equation for formation of bicarbonate, e.g.,
$MgCO_3(s) + H^+(aq) \longrightarrow Mg^{2+} + HCO_3^-(aq)$.]
4.100 (a) $2.055\ M\ Sr(OH)_2$ **(b)** $2\ HNO_3(aq) + Sr(OH)_2(aq) \longrightarrow$
$Sr(NO_3)_2(aq) + 2\ H_2O(l)$ **(c)** $2.62\ M\ HNO_3$
4.106 (a) $Mg(OH)_2(s) + 2\ HNO_3(aq) \longrightarrow Mg(NO_3)_2(aq) + 2\ H_2O(l)$
(b) HNO_3 is the limiting reactant. **(c)** 0.130 mol $Mg(OH)_2$, 0 mol
HNO_3, and 0.00250 mol $Mg(NO_3)_2$ are present.

Chapter 5   5.1 (a) The food that the caterpillar eats. (b) As
the caterpillar climbs, part of the food energy it uses is released
as heat, so $q < 0$. **(c)** Yes, the caterpillar does work as it moves its
mass up against the force of gravity. **(d)** Yes, the amount of work is
determined by the specific way (path) that the caterpillar moves.
(e) No, the change in potential energy depends only on the initial
and final positions of the caterpillar. **5.4 (a)** (iii) **(b)** none of them
(c) all of them **5.6 (a)** The sign of w is (+). **(b)** The sign of q is (−).
(c) The sign of w is positive and the sign of q is negative, so we
cannot absolutely determine the sign of ΔE. It is likely that the heat
lost is much smaller than the work done on the system, so the sign
of ΔE is probably positive. **5.9 (a)** $N_2(g) + O_2(g) \longrightarrow 2\ NO(g)$.
Since $\Delta V = 0$, $w = 0$. **(b)** $\Delta H = \Delta H_f = 90.37$ kJ. The definition of
a formation reaction is one where elements combine to form one
mole of a single product. The enthalpy change for such a reaction is
the enthalpy of formation. **5.12** Statements I and III are true.
5.13 (a) $E_{el} = -4.3 \times 10^{-18}$ J **(b)** $\Delta E_{el} = 4.1 \times 10^{-18}$ J **(c)** The
electrostatic potential energy of the system increases (becomes less
negative) as the separation between the oppositely charged particles
increases. **5.15 (a)** $F_{el} = -2.3 \times 10^{-8}$ N **(b)** $F_g = 1.0 \times 10^{-47}$ N
(c) The electrostatic force of attraction is 2.3×10^{39} times larger.
5.17 $w = 2.6 \times 10^{-18}$ J. **5.19 (a)** Matter cannot leave or enter a
closed system. **(b)** Neither matter nor energy can leave or enter an

isolated system. **(c)** Any part of the universe not part of the system
is called the surroundings. **5.21 (a)** According to the first law of
thermodynamics, energy is conserved. **(b)** The *internal energy*
(E) of a system is the sum of all the kinetic and potential energies
of the system components. **(c)** The internal energy of a closed
system increases when work is done on the system and when heat is
transferred to the system. **5.23 (a)** $\Delta E = -0.077$ kJ, endothermic
(b) $\Delta E = -22.1$ kJ, exothermic **5.25 (a)** Since no work is done by
the system in case (2), the gas will absorb most of the energy as heat;
the case (2) gas will have the higher temperature. **(b)** In case (1)
energy will be used to do work on the surroundings $(-w)$, but some
will be absorbed as heat $(+q)$. In case (2) $w = 0$ and q is (+). **(c)** ΔE
is greater for case (2) because the entire 100 J increases the internal
energy of the system rather than a part of the energy doing work on
the surroundings. **5.27** Your change in elevation and your change
in gravitational potential energy are state functions. The other two
quantities depend on the specific path taken. **5.29 (a)** For Path 1,
$q = +470$ J. **(b)** For Path 2, $w = +110$ J. **5.31** $w = -51$ J
5.33 Statements I and II are true. **5.35 (a)** We must know either
the temperature, T, or the values of P and ΔV in order to calculate
ΔE from ΔH. **(b)** ΔE is larger than ΔH. **(c)** Because the value of Δn
is negative, the quantity $(-P\Delta V)$ is positive. We add a positive
quantity to ΔH to calculate ΔE, so ΔE must be larger.
5.37 $\Delta E = 1.47$ kJ; $\Delta H = 0.824$ kJ
5.39 (a) $C_2H_5OH(l) + 3O_2(g) \longrightarrow 3\ H_2O + 2\ CO_2(g)$, $\Delta H = -1235$ kJ

(b) $C_2H_5OH(l) + 3\ O_2(g)$

$\Delta H = -1235$ kJ

$3\ H_2O(g) + 2\ CO_2(g)$

5.41 (a) $\Delta H = -142.3$ kJ/mol $O_3(g)$ **(b)** $2\ O_3(g)$ has the higher
enthalpy. **5.43 (a)** Exothermic **(b)** -87.9 kJ heat transferred **(c)** 15.7 g
MgO produced **(d)** 602 kJ heat absorbed **5.45 (a)** -29.5 kJ **(b)** -4.11 kJ
(c) 60.6 J **5.47 (a)** $\Delta H = +2248$ kJ **(b)** $\Delta H = -4496$ kJ **(c)** The exothermic
forward reaction is more likely to be thermodynamically favored.
(d) Vaporization is endothermic. If the product were $H_2O(g)$,
the reaction would have a less negative ΔH. **5.49 (a)** J/mol-°C or
J/mol − K **(b)** J/g-°C or J/g-K **(c)** To calculate heat capacity from
specific heat, the mass of the particular piece of copper pipe must be
known. **5.51 (a)** 4.184 J/g-K **(b)** 75.40 J/mol-°C **(c)** 774 J/°C **(d)** 904 kJ
5.53 (a) 2.93×10^3 J **(b)** It will require more heat to increase the
temperature of one mole of ethanol, $C_2H_5OH(l)$, by a certain amount
than to increase the temperature of one mole of water, $H_2O(l)$, by
the same amount. **5.55 (a)** 5.31 kJ **(b)** -41.6 kJ/mol NaOH
5.57 (a) $\Delta H_{rxn} = -25.5$ kJ/g $C_6H_4O_2$ **(b)** -2.75×10^3 kJ/mol $C_6H_4O_2$
5.59 (a) Total heat capacity of the calorimeter = 8.74 kJ/°C
(b) 30.0 kJ/g **(c)** If water is lost, the total heat capacity of the calorimeter
will decrease. **5.61 (a)** +90 kJ
(b)

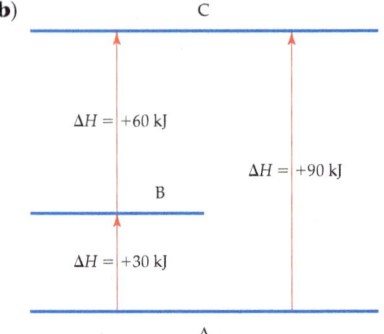

5.63 $\Delta H = -1300.0$ kJ **5.65** $\Delta H = -2.49 \times 10^3$ kJ **5.67** $\Delta H° = +201.9$ kJ
5.69 (a) *Standard conditions* for enthalpy changes are $P = 1$ atm and some common temperature, usually 298 K. **(b)** *Enthalpy of formation* is the enthalpy change that occurs when a compound is formed from its component elements. **(c)** *Standard enthalpy of formation*, $\Delta H_f°$, is the enthalpy change that accompanies formation of one mole of a substance from elements in their standard states.
5.71 (a) $N_2(g) + \frac{1}{2}O_2(g) \longrightarrow N_2O(g)$, $\Delta H_f° = 81.6$ kJ
(b) $Fe(s) + \frac{3}{2}Cl_2(g) \longrightarrow FeCl_3(s)$, $\Delta H_f° = -400$ kJ
(c) $P_4(s, \text{white}) + 5\,O_2(g) \longrightarrow P_4O_{10}(s)$, $\Delta H_f° = -2940.1$ kJ
(d) $Ca(s) + H_2(g) + O_2(g) \longrightarrow Ca(OH)_2(s)$, $\Delta H_f° = -986.2$ kJ
5.73 $\Delta H_{rxn}° = -847.6$ kJ **5.75 (a)** $\Delta H_{rxn}° = -508.3$ kJ
(b) $\Delta H_{rxn}° = +88.3$ kJ **(c)** $\Delta H_{rxn}° = +246$ kJ **(d)** $\Delta H_{rxn}° = -196.1$ kJ
5.77 $\Delta H_f° = -248$ kJ
5.79 (a) $C_8H_{18}(l) + \frac{25}{2}O_2(g) \longrightarrow 8\,CO_2(g) + 9\,H_2O(g)$
(b) $\Delta H_f° = -259.5$ kJ
5.81 (a) $C_2H_5OH(l) + 3\,O_2(g) \longrightarrow 2\,CO_2(g) + 3\,H_2O(g)$
(b) $\Delta H_{rxn}° = -1235$ kJ **(c)** 2.11×10^4 kJ/L heat produced **(d)** 0.071284 g CO_2/kJ heat emitted **5.83** Statements I and III are true. **5.85 (a)** ΔH for the reaction is $+1312$ kJ. $D(C-Cl) = 1312/4 = +328$ kJ **(b)** The difference between the value calculated in part **(a)** and the value from Table **5.4** is zero, to three significant figures.
5.87 (a) $\Delta H = -103$ kJ **(b)** $\Delta H = -1295$ kJ **5.89 (a)** ΔH for reaction calculated using bond enthalpies is -485 kJ. **(b)** The estimate from part **(a)** is less negative or larger than the true reaction enthalpy. **(c)** ΔH for the reaction calculated using enthalpies of formation is -572 kJ. **5.91 (a)** *Fuel value* is the amount of energy produced when 1 g of a substance (fuel) is combusted. **(b)** 5 g of fat **(c)** These products of metabolism are expelled as waste via the alimentary tract, $H_2O(l)$ primarily in urine and feces, and $CO_2(g)$ as gas when breathing. **5.93 (a)** 108 or 1×10^2 Cal/serving **(b)** Sodium does not contribute to the calorie content of the food because it is not metabolized by the body. **5.95** 59.7 Cal
5.97 (a) $\Delta H_{comb} = -1850$ kJ/mol C_3H_4, -1926 kJ/mol C_3H_6, -2044 kJ/mol C_3H_8 **(b)** $\Delta H_{comb} = -4.616 \times 10^4$ kJ/kg C_3H_4, -4.578×10^4 kJ/kg C_3H_6, -4.635×10^4 kJ/kg C_3H_8 **(c)** These three substances yield nearly identical quantities of heat per unit mass, but propane is marginally higher than the other two. **5.99** 1.0×10^{12} kg $C_6H_{12}O_6$/yr **5.101 (a)** $E_{el} = 1.8 \times 10^2$ J **(b)** If the spheres are released, they will move away from each other. **(c)** v = 19 m/s **5.103** The spontaneous air bag reaction is probably exothermic, with $-\Delta H$ and thus $-q$. When the bag inflates, work is done by the system, so the sign of w is also negative. **5.107** $\Delta H = 38.95$ kJ; $\Delta E = 36.48$ kJ **5.108** 1.8×10^4 bricks **5.112 (a)** $\Delta H_{rxn}° = -353.0$ kJ **(b)** 1.2 g Mg needed **5.115** $\Delta H = -445$ kJ **5.117 (a)** $\Delta H° = -633.2$ kJ **(b)** 3 mol of acetylene gas has greater enthalpy. **(c)** Fuel values are 50 kJ/g $C_2H_2(g)$, 42 kJ/g $C_6H_6(l)$. **5.119 (a)** Enthalpies of combustion: For $C_4H_6(g)$, $\Delta H = -2543$ kJ/mol; for $C_4H_8(g)$, $\Delta H = -2719$ kJ/mol; for $C_4H_{10}(g)$, $\Delta H = -2878$ kJ/mol. **(b)** Fuel values: For $C_4H_6(g)$, fuel value = 47.0 kJ/g; for $C_4H_8(g)$, fuel value = 48.5 kJ/g; for $C_4H_{10}(g)$, fuel value = 49.5 kJ/g. **(c)** Percentage H by mass: For $C_4H_6(g)$, 11.2 % H; for $C_4H_8(g)$, 14.4% H; for $C_4H_{10}(g)$, 17.3% H. **(d)** The fuel value increases with increasing percentage of hydrogen by mass.

Chapter 6

6.2 (a) 0.1 m or 10 cm **(b)** No. Visible radiation has wavelengths much shorter than 0.1 m. **(c)** Energy and wavelength are inversely proportional. Photons of the longer 0.1-m radiation have less energy than visible photons. **(d)** Radiation with $\lambda = 0.1$ m is in the low-energy portion of the microwave region. The appliance is probably a microwave oven. **6.5 (a)** Increase **(b)** decrease **6.9 (a)** $l = 1$ **(b)** $3p_y$ **(c)** (iii) **6.13 (a)** Meters **(b)** 1/second **(c)** meters/second **6.15 (a)** True **(b)** False. Ultraviolet light has shorter wavelengths than visible light. **(c)** False. X-rays travel at the same speed as microwaves. **(d)** False. Electromagnetic radiation and

sound waves travel at different speeds. **6.17** Wavelength of X-rays < ultraviolet < green light < red light < infrared < radio waves **6.19 (a)** 3.0×10^{13} s^{-1} **(b)** 5.45×10^{-7} m = 545 nm. **(c)** The radiation in **(b)** is visible; the radiation in **(a)** is not. **(d)** 1.50×10^4 m **6.21** 4.6×10^{14} s^{-1}; red. **6.23** (iii) **6.25 (a)** 1.95×10^{-19} J **(b)** 4.81×10^{-19} J **(c)** 328 nm **6.27 (a)** $\lambda = 3.3$ μm, $E = 6.0 \times 10^{-20}$ J; $\lambda = 0.154$ nm, $E = 1.29 \times 10^{-15}$ J **(b)** The 3.3-μm photon is in the infrared region and the 0.154-nm photon is in the X-ray region; the X-ray photon has the greater energy. **6.29 (a)** 6.11×10^{-19} J/photon **(b)** 368 kJ/mol **(c)** 1.64×10^{15} photons **(d)** 368 kJ/mol **6.31 (a)** The $\sim 1 \times 10^{-6}$ m radiation is in the infrared portion of the spectrum. **(b)** 8.1×10^{16} photons/s **6.33 (a)** $E_{min} = 7.22 \times 10^{-19}$ J **(b)** $\lambda = 275$ nm **(c)** $E_{120} = 1.66 \times 10^{-18}$ J. The excess energy of the 120-nm photon is converted into the kinetic energy of the emitted electron. $E_k = 9.3 \times 10^{-19}$ J/electron. **6.35** When an electron in a hydrogen atom transitions from $n = 1$ to $n = 3$, the atom "expands". **6.37 (a)** Emitted **(b)** absorbed **(c)** emitted **6.39 (a)** $E_2 = -5.45 \times 10^{-19}$ J; $E_6 = -0.606 \times 10^{-19}$ J; $\Delta E = 4.84 \times 10^{-19}$ J; $\lambda = 410$ nm **(b)** visible, violet **6.41 (a)** (ii) **(b)** $n_i = 3$, $n_f = 2$; $\lambda = 6.56 \times 10^{-7}$ m; this is the red line at 656 nm. $n_i = 4$, $n_f = 2$; $\lambda = 4.86 \times 10^{-7}$ m; this is the blue-green line at 486 nm. $n_i = 5$, $n_f = 2$; $\lambda = 4.34 \times 10^{-7}$ m; this is the blue-violet line at 434 nm. **6.43 (a)** Ultraviolet region **(b)** $n_i = 7$, $n_f = 1$ **6.45** The order of increasing frequency of light absorbed is: $n = 4$ to $n = 9$; $n = 3$ to $n = 6$; $n = 2$ to $n = 3$; $n = 1$ to $n = 2$ **6.47 (a)** $\lambda = 5.6 \times 10^{-37}$ m **(b)** $\lambda = 2.65 \times 10^{-34}$ m **(c)** $\lambda = 2.3 \times 10^{-13}$ m **(d)** $\lambda = 1.51 \times 10^{-11}$ m **6.49** 3.16×10^3 m/s **6.51 (a)** $\Delta x \geq 4 \times 10^{-27}$ m **(b)** $\Delta x \geq 3 \times 10^{-10}$ m **6.53 (a)** False **(b)** false **6.55 (a)** $n = 4$, $l = 3, 2, 1, 0$ **(b)** $l = 2$, $m_l = -2, -1, 0, 1, 2$ **(c)** $m_l = 2$, $l \geq 2$ or $l = 2, 3$ or 4 **6.57 (a)** $3p$: $n = 3$, $l = 1$ **(b)** $2s$: $n = 2$, $l = 0$ **(c)** $4f$: $n = 4$, $l = 3$ **(d)** $5d$: $n = 5$, $l = 2$ **6.59 (a)** $2, 1, 0, -1, -2$ **(b)** $\frac{1}{2}, -\frac{1}{2}$ **6.61 (a)** Impossible, $1p$ **(b)** possible **(c)** possible **(d)** impossible, $2d$ **6.63**

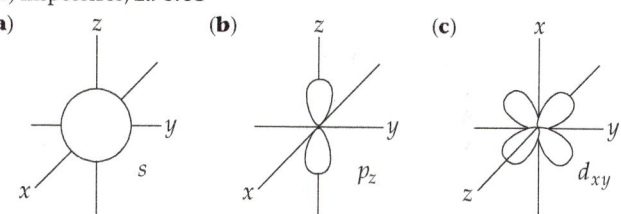

(a) s **(b)** p_z **(c)** d_{xy}

6.65 (a) The $4s$ orbital has three radial nodes. **(b)** Like all p orbitals, the $2p_x$ orbital has a single nodal plane. **(c)** The most probable distance of an electron from the nucleus in a $2s$ orbital is smaller than for an electron in a $3s$ orbital. **(d)** $1s < 2p < 3d < 4f < 6s$ **6.67 (a)** For the He$^+$ ion, the $2s$ and $2p$ orbitals have the same energy. **(b)** Yes. In a helium atom, the $2s$ orbital is lower in energy than the $2p$ orbital. **6.69 (a)** No. Both configurations follow the Pauli exclusion principle. **(b)** No. Both configurations follow Hund's rule. **(c)** No. In the absence of a magnetic field, we cannot say which configuration has the lower energy. **6.71 (a)** 6 **(b)** 10 **(c)** 2 **(d)** 14 **6.73 (a)** "Valence electrons" are those involved in chemical bonding. They are part or all of the outer-shell electrons listed after the core. **(b)** "Core electrons" are inner-shell electrons that have the electron configuration of the nearest noble-gas element. **(c)** Each box represents an orbital. **(d)** Each half-arrow in an orbital diagram represents an electron. The direction of the half-arrow represents electron spin. **6.75 (a)** Cs, [Xe]$6s^1$ **(b)** Ni, [Ar]$4s^2 3d^8$ **(c)** Se, [Ar]$4s^2 3d^{10} 4p^4$ **(d)** Cd, [Kr]$5s^2 4d^{10}$ **(e)** U, [Rn]$5f^3 6d^1 7s^2$ **(f)** Pb, [Xe]$6s^2 4f^{14} 5d^{10} 6p^2$ **6.77 (a)** Be, 0 unpaired electrons **(b)** O, 2 unpaired electrons **(c)** Cr, 6 unpaired electrons **(d)** Te, 2 unpaired electrons **6.79 (a)** The fifth electron would fill the $2p$ subshell before the $3s$. **(b)** Either the core is [He], or the outer electron configuration should be $3s^2 3p^3$. **(c)** The $3p$ subshell would fill before the $3d$. **6.81 (a)** $\lambda_A = 3.6 \times 10^{-8}$ m, $\lambda_B = 8.0 \times 10^{-8}$ m **(b)** $\nu_A = 8.4 \times 10^{15}$ s^{-1}, $\nu_B = 3.7 \times 10^{15}$ s^{-1} **(c)** A, ultraviolet; B, ultraviolet **6.84** 35.0 min

6.86 1.6×10^{18} photons **6.91 (a)** The Paschen series lies in the infrared. **(b)** $n_i = 4$, $\lambda = 1.87 \times 10^{-6}$ m; $n_i = 5$, $\lambda = 1.28 \times 10^{-6}$ m; $n_i = 6$, $\lambda = 1.09 \times 10^{-6}$ m **6.95** $\lambda = 10.6$ pm **6.97 (a)** The nodal plane of the p_z orbital is the xy-plane. **(b)** The two nodal planes of the d_{xy} orbital are the ones where $x = 0$ and $y = 0$. These are the yz- and xz-planes. **(c)** The two nodal planes of the $d_{x^2-y^2}$ orbital are the ones that bisect the x- and y-axes and contain the z-axis. **6.100 (a)** Br: $[\text{Ar}]4s^2 3d^{10} 4p^5$, 1 unpaired electron **(b)** Ga: $[\text{Ar}]$ $4s^2 3d^{10} 4p^1$, 1 unpaired electron **(c)** Hf: $[\text{Xe}]6s^2 4f^{14} 5d^2$, 2 unpaired electrons **(d)** Sb: $[\text{Kr}]5s^2 4d^{10} 5p^3$, 3 unpaired electrons **(e)** Bi: $[\text{Xe}]6s^2 4f^{14} 5d^{10} 6p^3$, 3 unpaired electrons **(f)** Sg: $[\text{Rn}]7s^2 5f^{14} 6d^4$, 4 unpaired electrons **6.103 (a)** 1.7×10^{28} photons **(b)** 34 s

Chapter 7 **7.2** The largest brown sphere is Br^-, the intermediate blue one is Br, and the smallest red one is F. **7.5 (a)** The bonding atomic radius of A, r_A, is $d_1/2$; $r_x = d_2 - (d_1/2)$. **(b)** The length of the X—X bond is $2r_x$ or $2d_2 - d_1$. **7.8 (a)** X $+ 2F_2 \longrightarrow XF_4$ **(b)** If X were a metal, the compound would be ionic and the charge on X would be $4+$. **(c)** If X is a nonmetal, it could be C, which is similar in size to F, and indeed carbon tetrafluoride is a known compound. **7.9** The p-block. **7.11 (a)** Of the elements listed, only Fe was known before 1700. **(b)** The seven metals known in ancient times, Fe, Cu, Ag, Sn, Au, Hg, and Pb, are mostly near the bottom of the activity series, Table **4.5**. **7.13** For elements 1–18, H, Li, and Na have minimum values of Z_{eff}; Ne and Ar have maximum values. **7.15 (a)** For both Na and K, $Z_{\text{eff}} = 1$. **(b)** For both Na and K, $Z_{\text{eff}} = 2.2$. **(c)** Slater's rules give values closer to the detailed calculations: Na, 2.51; K, 3.49. **(d)** Both approximations give the same value of Z_{eff} for Na and K; neither accounts for the gradual increase in Z_{eff} moving down a group. **(e)** Following the trend from detailed calculations, we predict a Z_{eff} value of approximately 4.5. **7.17** The $n = 3$ electrons in Kr experience a greater effective nuclear charge and thus have a greater probability of being closer to the nucleus. **7.19** The best answer is **(c)**. **7.21 (a)** 1.37 Å **(b)** The distance between W atoms will decrease. **7.23** From the sum of the atomic radii, As—I = 2.58 Å. This is very close to the experimental value of 2.55 Å. **7.25 (a)** Cs $>$ K $>$ Li **(b)** Pb $>$ Sn $>$ Si **(c)** N $>$ O $>$ F **7.27 (a)** False **(b)** true **(c)** false **7.29** Ga^{3+}: none; Zr^{4+}: Kr; Mn^{7+}: Ar; I^-: Xe; Pb^{2+}: Hg **7.31 (a)** Na^+ **(b)** F^-, $Z_{\text{eff}} = 7$; Na^+, $Z_{\text{eff}} = 9$ **(c)** S = 4.15; F^-, $Z_{\text{eff}} = 4.85$; Na^+, $Z_{\text{eff}} = 6.85$ **(d)** For isoelectronic ions, as nuclear charge (Z) increases, effective nuclear charge (Z_{eff}) increases and ionic radius decreases. **7.33 (a)** Cl $<$ S $<$ K **(b)** $\text{K}^+ < \text{Cl}^- < \text{S}^{2-}$ **(c)** The neutral K atom has the largest radius because the n-value of its outer electron is larger than the n-value of valence electrons in S and Cl. The K^+ ion is smallest because, in an isoelectronic series, the ion with the largest Z has the smallest ionic radius. **7.35 (a)** True. O^{2-} is larger than O because the increase in electron repulsions that accompanies addition of an electron causes the electron cloud to expand. **(b)** False. S^{2-} is larger than O^{2-} because for particles with like charges, size increases going down a family. **(c)** True. S^{2-} is larger than K^+ because the two ions are isoelectronic and K^+ has the larger Z and Z_{eff}. **(d)** True. K^+ is larger than Ca^{2+} because the two ions are isoelectronic and Ca^{2+} has the larger Z and Z_{eff}. **7.37 (a)** $1s^2 2s^2 2p^6 3s^2 3p^6 3d^7$, or $[\text{Ar}]3d^7$ **(b)** $[\text{Kr}]5s^2 4d^{10}$ **(c)** $1s^2 2s^2 2p^6 3s^2 3p^6 4s^2 3d^{10} 4p^6$, which is $[\text{Kr}]$, which is a noble-gas configuration **(d)** $[\text{Kr}]4d^{10}$ **(e)** $1s^2 2s^2 2p^6 3s^2 3p^6$, which is $[\text{Ar}]$, a noble-gas configuration. **7.39** The correct answer is (e), more than one of the choices (both Ni^{2+} and Pt^{2+} are d^8 systems). **7.41 (a)** Second electron affinity of Cl: $\text{Cl}^- + 1\text{e}^- \longrightarrow \text{Cl}^{2-}(g)$. **(b)** We predict that the second electron affinity of chlorine will be positive, since adding an electron to the chloride ion (a noble gas configuration) is not favorable. It is probably not possible to directly measure this quantity, because a positive value indicates that Cl^{2-} ion is unstable and will not form. **7.43** $\text{Al}(g) \longrightarrow \text{Al}^+(g) + 1\text{e}^-$; $\text{Al}^+(g) \longrightarrow \text{Al}^{2+}(g) + 1\text{e}^-$; $\text{Al}^{2+}(g) \longrightarrow \text{Al}^{3+}(g) + 1\text{e}^-$. The process

for the first ionization energy requires the least amount of energy. **7.45** Of these three elements, Li has the highest second ionization energy. Both Li^+ and K^+ have the stable electron configurations of noble gases, but the unshielded $1s$ electron of Li^+ is much closer to the nucleus and requires more energy to remove. **7.47 (a)** The larger the atom, the smaller its first ionization energy. **(b)** Of the nonradioactive elements, He has the largest first ionization energy. **(c)** Of the nonradioactive elements, Cs has the smallest first ionization energy. **7.49 (a)** Cl **(b)** Ca **(c)** K **(d)** Ge **(e)** Sn **7.51** The electron affinity of K^+ is more negative. The electron-electron repulsions created by adding an electron to a neutral K atom causes the electron affinity of K to be greater (less negative) than that of K^+. **7.53 (a)** Ionization energy (I_1) of Ne: $\text{Ne}(g) \longrightarrow \text{Ne}^+(g) + 1\text{e}^-$; $[\text{He}]2s^2 2p^6 \longrightarrow [\text{He}]2s^2 2p^5$; electron affinity ($E_1$) of F: $\text{F}(g) + 1\text{e}^- \longrightarrow \text{F}^-(g)$; $[\text{He}]2s^2 2p^5 \longrightarrow [\text{He}]2s^2 2p^6$. **(b)** I_1 of Ne is positive; E_1 of F is negative. **(c)** One process is apparently the reverse of the other, with one important difference. Ne has a greater Z and Z_{eff}, so we expect I_1 for Ne to be somewhat greater in magnitude and opposite in sign to E_1 for F. **7.55 (a)** Decrease **(b)** Increase **(c)** The smaller the first ionization energy of an element, the greater the metallic character of that element. The trends in **(a)** and **(b)** are the opposite of the trends in ionization energy. **7.57** True. When forming ions, all metals form cations. The only nonmetallic element that forms cations is the metalloid Sb, which is likely to have significant metallic character. **7.59** Ionic: SnO_2, Al_2O_3, Li_2O, Fe_2O_3; molecular: CO_2, H_2O. Ionic compounds are formed by combining a metal and a nonmetal; molecular compounds are formed by two or more nonmetals. **7.61** MnO will react more readily with HCl. **7.63 (a)** Dichlorine heptoxide **(b)** $2\,\text{Cl}_2(g) + 7\,\text{O}_2(g) \longrightarrow 2\,\text{Cl}_2\text{O}_7(l)$ **(c)** Cl_2O_7 is an acidic oxide, so it will be more reactive to base, OH^-. **(d)** The oxidation state of Cl in Cl_2O_7 is $+7$; the corresponding electron configuration for Cl is $[\text{He}]2s^2 2p^6$ or $[\text{Ne}]$. **7.65 (a)** $\text{BaO}(s) + \text{H}_2\text{O}(l) \longrightarrow \text{Ba(OH)}_2(aq)$ **(b)** $\text{FeO}(s) + 2\,\text{HClO}_4(aq) \longrightarrow \text{Fe(ClO}_4)_2(aq) + \text{H}_2\text{O}(l)$ **(c)** $\text{SO}_3(g) + \text{H}_2\text{O}(l) \longrightarrow \text{H}_2\text{SO}_4(aq)$ **(d)** $\text{CO}_2(g) + 2\,\text{NaOH}(aq) \longrightarrow \text{Na}_2\text{CO}_3(aq) + \text{H}_2\text{O}(l)$ **7.67** K $>$ Ca $>$ Mg, based on the order of ionization energies. **7.69 (a)** $2\,\text{K}(s) + \text{Cl}_2(g) \longrightarrow 2\,\text{KCl}(s)$ **(b)** $\text{SrO}(s) + \text{H}_2\text{O}(l) \longrightarrow \text{Sr(OH)}_2(aq)$ **(c)** $4\,\text{Li}(s) + \text{O}_2(g) \longrightarrow 2\,\text{Li}_2\text{O}(s)$ **(d)** $2\,\text{Na}(s) + \text{S}(l) \longrightarrow \text{Na}_2\text{S}(s)$ **7.71 (a)** The reactions of the alkali metals with hydrogen and with a halogen are redox reactions. Both hydrogen and the halogen gain electrons and are reduced. (The alkali metal loses electrons and is oxidized.) $\text{Ca}(s) + \text{F}_2(g) \longrightarrow \text{CaF}_2(s)$; $\text{Ca}(s) + \text{H}_2(g) \longrightarrow \text{CaH}_2(s)$. **(b)** The oxidation number of Ca in both products is $+2$. The electron configuration is that of Ar, $[\text{Ne}]3s^2 3p^6$. **7.73 (a)** Br, $[\text{Ar}]4s^2 4p^5$; Cl, $[\text{Ne}]3s^2 3p^5$ **(b)** Br and Cl are in the same group, and both adopt a $1-$ ionic charge. **(c)** The question should have asked you to predict the relative first ionization energy. The ionization energy of Br is smaller than that of Cl, because the $4p$ valence electrons in Br are farther from to the nucleus and less tightly held than the $3p$ electrons of Cl. **(d)** Both react slowly with water to form HX + HOX. **(e)** The question should have asked you to predict the relative electron affinity. The electron affinity of Br is less negative than that of Cl, because the electron added to the $4p$ orbital in Br is farther from the nucleus and less tightly held than the electron added to the $3p$ orbital of Cl. **(f)** The question should have asked you to predict the relative atomic radius. The atomic radius of Br is larger than that of Cl, because the $4p$ valence electrons in Br are farther from the nucleus and less tightly held than the $3p$ electrons of Cl. **7.75 (a)** The term *inert* was dropped because it no longer described all the Group 8A elements. **(b)** In the 1960s, scientists discovered that Xe would react with substances having a strong tendency to remove electrons, such as F_2. Thus, Xe could not be categorized as an "inert" gas. **(c)** The group is now called the noble gases. **7.77 (a)** $2 O_3(g) \longrightarrow 3 O_2(g)$ **(b)** $Xe(g) + F_2(g) \longrightarrow XeF_2(g)$; $Xe(g) + 2 F_2(g) \longrightarrow XeF_4(s)$; $Xe(g) + 3 F_2(g) \longrightarrow XeF_6(s)$ **(c)** $S(s) + H_2(g) \longrightarrow H_2S(g)$ **(d)** $2 F_2(g) + 2 H_2O(l) \longrightarrow 4 HF(aq) + O_2(g)$ **7.79** Up to $Z = 82$, there are three instances where atomic weights are reversed relative to atomic numbers: Ar and K; Co and Ni; Te and I. **7.81 (a)** 5+ **(b)** 4.8+ **(c)** Shielding is greater for $3p$ electrons, owing to penetration by $3s$ electrons, so Z_{eff} for $3p$ electrons is less than that for $3s$ electrons. **(d)** The first electron lost is a $3p$ electron because it has a smaller Z_{eff} and experiences less attraction for the nucleus than a $3s$ electron does. **7.84 (a)** The As—Cl distance is 2.24 Å. **(b)** The predicted As—Cl bond length is 2.21 Å **7.86 (a)** Chalcogens, -2; halogens, -1. **(b)** The family with the larger value is: atomic radii, chalcogens; ionic radii of the most common oxidation state, chalcogens; first ionization energy, halogens; second ionization energy, halogens **7.89** C: $1s^2 2s^2 2p^2$. I_1 through I_4 represent loss of the $2p$ and $2s$ electrons in the outer shell of the atom. The values of $I_1 - I_4$ increase as expected. I_5 and I_6 represent loss of the $1s$ core electrons. These $1s$ electrons are much closer to the nucleus and experience the full nuclear charge, so the values of I_5 and I_6 are significantly greater than $I_1 - I_4$. **7.94 (a)** Cl^-, K^+ **(b)** Mn^{2+}, Fe^{3+} **(c)** Sn^{2+}, Sb^{3+} **7.96 (a)** For both H and the alkali metals, the added electron will complete an ns subshell, so shielding and repulsion effects will be similar. For the halogens, the electron is added to an np subshell, so the energy change is likely to be quite different. **(b)** True. The electron configuration of H is $1s^1$. The single $1s$ electron experiences no repulsion from other electrons and feels the full unshielded nuclear charge. The outer electrons of all other elements that form compounds are shielded by a spherical inner core of electrons and are less strongly attracted to the nucleus, resulting in larger bonding atomic radii. **(c)** Both H and the halogens have large ionization energies. The relatively large effective nuclear charge experienced by np electrons of the halogens is similar to the unshielded nuclear charge experienced by the H $1s$ electron. For the alkali metals, the ns electron being removed is effectively shielded by the core electrons, so ionization energies are low. **(d)** ionization energy of hydride, $H^-(g) \longrightarrow H(g) + 1e^-$ **(e)** electron affinity of hydrogen, $H(g) + 1 e^- \longrightarrow H^-(g)$. The value for the ionization energy of hydride is equal in magnitude but opposite in sign to the electron affinity of hydrogen. **7.99** The most likely product is **(a)**. **7.103** Electron configuration, $[Rn]7s^2 5f^{14} 6d^{10} 7p^5$; first ionization energy, 805 kJ/mol; electron affinity, -235 kJ/mol; atomic size, 1.65 Å; common oxidation state, -1. **7.106 (a)** Li, $[He]2s^1$; $Z_{eff} \approx 1+$ **(b)** $I_1 \approx 5.45 \times 10^{-19}$ J/atom ≈ 328 kJ/mol **(c)** The estimated value of 328 kJ/mol is less than the Table **7.4** value of 520 kJ/mol. Our estimate for Z_{eff} was a lower limit; the [He] core electrons do not perfectly shield the $2s$ electron from the nuclear charge. **(d)** Based on the experimental ionization energy, $Z_{eff} = 1.26$. This value is greater than the estimate from part **(a)** but agrees well with the "Slater" value of 1.3 and is consistent with the explanation in part **(c)**. **7.108 (a)** Mg_3N_2 **(b)** $Mg_3N_2(s) + 3 H_2O(l) \longrightarrow 3 MgO(s) + 2 NH_3(g)$; the driving force is the production of $NH_3(g)$. **(c)** 17% Mg_3N_2 **(d)** $3 Mg(s) + 2 NH_3(g) \longrightarrow Mg_3N_2(s) + 3 H_2(g)$. NH_3 is the limiting reactant and 0.46 g H_2 is formed. **(e)** $\Delta H^\circ_{rxn} = -368.70$ kJ

Chapter 8   **8.1 (a)** Group 4A or 14 **(b)** Group 2A or 2
(c) Group 5A or 15 **8.4 (a)** Co **(b)** $[Ar]4s^2 3d^7$. **8.9 (a)** Bond 3 **(b)** bond 2 **(c)** bond 1 **8.11 (a)** False **(b)** three **(c)** four **8.13 (a)** Si, $1s^2 2s^2 2p^6 3s^2 3p^2$ **(b)** four **(c)** The $3s$ and $3p$ electrons are valence electrons. **8.15 (a)** $\cdot \dot{Al} \cdot$ **(b)** $:\ddot{Br} \cdot$ **(c)** $:\ddot{Ar}:$ **(d)** $\cdot \dot{Sr}$

8.17 $\overset{\frown}{\underset{}{Mg}} + :\ddot{O}: \longrightarrow Mg^{2+} + \left[:\ddot{O}:\right]^{2-}$

(b) two **(c)** Mg loses electrons. **8.19 (a)** AlF_3 **(b)** K_2S **(c)** Y_2O_3 **(d)** Mg_3N_2 **8.21 (a)** Rb^+, $[Ar]4s^2 3d^{10} 4p^6 = [Kr]$, noble-gas configuration **(b)** Rh^{3+}, $[Kr]4d^6$ **(c)** P^{3-}, $[Ne]3s^2 3p^6 = [Ar]$, noble-gas configuration

(d) Sc^{3+}, $[Ne]3s^2 3p^6 = [Ar]$, noble-gas configuration **(e)** S^{2-}, $[Ne]3s^2 3p^6 = [Ar]$, noble-gas configuration **(f)** V^{2+}, $[Ar]3d^3$ **8.23 (a)** Endothermic **(b)** $NaCl(s) \longrightarrow Na^+(g) + Cl^-(g)$ **(c)** Salts like NaCl, that have singly charged ions, will have smaller lattice energies compared with salts like CaO, that have doubly charged ions. **8.25 (a)** Na^+, 1+; Ca^{2+}, 2+ **(b)** F^-, 1−; O^{2-}, 2− **(c)** CaO will have the larger lattice energy. **(d)** We expect the lattice energy of ScN to be slightly less than 8.10×10^3 kJ **8.27 (a)** K—F, 2.71 Å; Na—Cl, 2.83 Å; Na—Br, 2.98 Å; Li—Cl, 2.57 Å **(b)** LiCl > KF > NaCl > NaBr **(c)** From Table **8.2**: LiCl, 1030 kJ; KF, 808 kJ; NaCl, 788 kJ; NaBr, 732 kJ. The predictions from ionic radii are correct. **8.29** Statement **(a)** is the best explanation. **8.31** The lattice energy of KI(s) is +649 kJ/mol. **8.33 (a)** The bonding in (iii) and (iv) is likely to be covalent. **(b)** Covalent, because it is a gas at room temperature and below.

8.35

(a) 4 **(b)** 7 **(c)** 8 **(d)** 8 **(e)** 4 **8.37 (a)** $:\ddot{O}=\ddot{O}:$ **(b)** four bonding electrons (two bonding electron pairs) **(c)** An O=O double bond is shorter than an O—O single bond. The greater the number of shared electron pairs between two atoms, the shorter the distance between the atoms. **8.39** Statements (ii) and (iii) are true. **8.41 (a)** Mg **(b)** S **(c)** C **(d)** As **8.43** The bonds in **(a)**, **(c)**, and **(d)** are polar. The more electronegative element in each polar bond is **(a)** F **(c)** O **(d)** I. **8.45 (a)** The calculated charge on H and Br is 0.12e. **(b)** Decrease **8.47 (a)** SiF_4, molecular, silicon tetrachloride; LaF_3, ionic, lanthanum(III) fluoride **(b)** $FeCl_2$, ionic, iron(II) chloride; $ReCl_6$, molecular (metal in high oxidation state), rhenium hexachloride. **(c)** $PbCl_4$, molecular (by contrast to the distinctly ionic RbCl), lead tetrachloride; RbCl, ionic, rubidium chloride **8.49**

8.51 Statement **(b)** is most true. Keep in mind that when it is necessary to place more than an octet of electrons around an atom in order to minimize formal charge, there may not be a "best" Lewis structure. **8.53** Formal charges are shown on the Lewis structures; oxidation numbers are listed below each structure.

8.55 (a) $:\ddot{O}—S=\ddot{O}$
(b) O_3 is isoelectronic with SO_2; both have 18 valence electrons. **(c)** Yes, there are multiple equivalent resonance structures for the

molecule. (**d**) Because each S—O bond has partial double-bond character, the S—O bond length in SO_2 should be shorter than an S—O single bond but longer than an S—O double bond. **8.57** The more electron pairs shared by two atoms, the shorter the bond. Thus, the C—O bond lengths vary in the order $CO < CO_2 < CO_3^{2-}$. **8.59** Statements (i) and (iii) are true. **8.61** (**a**) AsF_5 is an exception to the octet rule. (**b**) BCl_3 is an exception to the octet rule. **8.63** Assume that the dominant structure is the one that minimizes formal charge. Following this guideline, only ClO^- obeys the octet rule. ClO, ClO_2^-, ClO_3^-, and ClO_4^- do not obey the octet rule.

ClO, ·C̈l=Ö   ClO⁻, [:C̈l—Ö:]⁻

ClO_2^-, [Ö=C̈l—Ö:]⁻ ClO_3^-, [Ö=C̈l—Ö / ‖ / :O:]⁻

ClO_4^-, [:Ö: / ‖ / Ö=Cl=Ö / ‖ / :O:]⁻

8.65 (**a**) H—P̈—H (with H below)   (**b**) H—Al—H (with H below)

(**b**) Does not obey the octet rule. Central Al has only 6 electrons

(**c**) [:N≡N—N̈:]⁻ ⟷ [:N̈—N≡N:]⁻ ⟷ [:N̈=N=N̈:]⁻

(**d**) :C̈l: | :C̈l—C—H | H   (**e**) [:F: / :F̈—Sn—F̈: (with F's around)]²⁻

(**e**) Does not obey octet rule. Central Sn has 12 electrons.

8.67 (**a**) :C̈l—Be—C̈l: (formal charges 0 0 0)

This structure violates the octet rule.

(**b**) C̈l=Be=C̈l ⟷ :C̈l—Be≡Cl ⟷ Cl≡Be—C̈l: (formal charges 1 2 1   0 2 2   2 2 0)

(**c**) Formal charges are minimized on the structure that violates the octet rule; this form is probably dominant.

8.69 (**a**) :Ö: | :Ö—S—Ö—H | :Ö: | H   (**b**) :O: ‖ :O=S—Ö—H | :O: | H

8.71 (**a**) $\Delta H = -304$ kJ (**b**) $\Delta H = -82$ kJ (**c**) $\Delta H = -467$ kJ **8.73** Only statement (e) is true. **8.75** The Ca—O bond will be stronger than the Na—Cl bond, because the ion charges are greater. **8.77** A triple C—C bond **8.79** (**a**) No (**b**) bond 2 (**c**) bond 1 **8.87** (**a**) B—O. The most polar bond will be formed by the two elements with the greatest difference in electronegativity. (**b**) Te—I. These elements have the two largest covalent radii among this group. (**c**) TeI_2. The octet rule is satisfied for all three atoms. (**d**) P_2O_3. Each P atom needs to share 3 e⁻ and each O atom 2 e⁻ to achieve an octet. And B_2O_3. Although this is not a purely ionic compound, it can be understood in terms of gaining and losing electrons to achieve a noble-gas

configuration. If each B atom were to lose 3 e⁻ and each O atom were to gain 2 e⁻, charge balance and the octet rule would be satisfied. **8.88** (**a**) 0.162 e (**b**) the O atom

(**c**) +1 ·C̈l—Ö: −1   0 ·C̈l=Ö 0   (**d**) 0

8.91

	Isomer A	Isomer B	Isomer C
Number of single bonds	5	6	5
Number of double bonds	2	0	2
Number of triple bonds	0	1	0
Number of nonbonding pairs	2	2	2

8.93 (**a**) +1 (**b**) −1 (**c**) +1 (assuming the odd electron is on N) (**d**) 0 (**e**) +3 **8.95** (**a**) The leftmost structure, with a N—N triple bond, leads to the most favorable formal charges. (**b**) The rightmost structure, with two double bonds, is most consistent with the observed bond lengths. **8.99** (**a**) $\Delta H = +40$ kJ/mol, ethanol has the lower enthalpy. (**b**) $\Delta H = -83$ kJ/mol, acetaldehyde has the lower enthalpy. (**c**) $\Delta H = +82$ kJ/mol, cyclopentene has the lower enthalpy. (**d**) $\Delta H = -55$ kJ/mol, acetonitrile has the lower enthalpy. **8.103** (**a**) $C_2H_3Cl_3O_2$ (**b**) $C_2H_3Cl_3O_2$

(**c**)   :C̈l::Ö—H | :C̈l—C—C—Ö—H | :C̈l: H

Chapter 9   **9.1** Removing an atom from the equatorial plane of the trigonal bipyramid in Figure **9.3** creates a seesaw shape. **9.3** (**a**) Two electron-domain geometries, linear and trigonal bipyramidal (**b**) one electron-domain geometry, trigonal bipyramidal (**c**) one electron-domain geometry, octahedral (**d**) one electron domain geometry, octahedral (**e**) one electron domain geometry, octahedral (**f**) one electron-domain geometry, trigonal bipyramidal (This triangular pyramid is an unusual molecular geometry not listed in Table **9.3**. It could occur if the equatorial substituents on the trigonal bipyramid were extremely bulky, causing the nonbonding electron pair to occupy an axial position.) **9.5** (**a**) Zero. Moving from left to right along the x-axis of the plot, the distance between the Cl atoms increases. At very large separation, the potential energy of interaction approaches zero. (**b**) The Cl—Cl bond distance is approximately 2.0 Å. The Cl—Cl bond energy is approximately 240 kJ/mol. (**c**) Weaker. Under extreme pressure, the Cl—Cl bond gets shorter. The potential energy of the atom pair increases and the bond gets weaker. **9.11** (**a**) i, Two s atomic orbitals; ii, two p atomic orbitals overlapping end to end; iii, two p atomic orbitals overlapping side to side (**b**) i, σ-type MO; ii, σ-type MO; iii, π-type MO (**c**) i, antibonding; ii, bonding; iii, antibonding (**d**) i, the nodal plane is between the atom centers, perpendicular to the interatomic axis and equidistant from each atom. ii, there are two nodal planes; both are perpendicular to the interatomic axis. One is left of the left atom and the second is right of the right atom. iii, there are two nodal planes; one is between the atom centers, perpendicular to the interatomic axis and equidistant from each atom. The second contains the interatomic axis and is perpendicular to the first. **9.13** (**a**) The $2p_x$ orbital (**b**) Statements (i) and (iii) are true. **9.15** (**a**) Tetrahedral (**b**) bent (**c**) No, an AB_2 molecule with A=B double bonds would have a linear molecular geometry. **9.17** (**a**) Trigonal pyramidal (**b**) tetrahedral (**c**) one **9.19** (**a**) Octahedral (**b**) octahedral (**c**) square planar **9.21** (**a**) No effect on molecular shape (**b**) 1 nonbonding pair on P influences molecular shape (**c**) no effect (**d**) no effect (**e**) 1 nonbonding pair on S influences molecular shape **9.21** (**a**) 2 (**b**) 1 (**c**) none (**d**) 3 **9.23** (**a**) Tetrahedral, tetrahedral (**b**) trigonal bipyramidal, T-shaped (**c**) octahedral, square pyramidal (**d**) octahedral, square planar **9.25** (**a**) Linear, linear (**b**) tetrahedral, trigonal pyramidal (**c**) trigonal bipyramidal, seesaw (**d**) octahedral, octahedral (**e**) tetrahedral, tetrahedral (**f**) linear, linear **9.27** (**a**) i, trigonal planar; ii, tetrahedral; iii, trigonal bipyramidal (**b**) i, 0; ii, 1; iii, 2 (**c**) N and P (**d**) Cl (or Br or I). This T-shaped molecular geometry arises from a trigonal-bipyramidal electron-domain geometry with 2 nonbonding domains. Assuming each F atom has 3 nonbonding domains and forms only single bonds with A, A must have 7 valence electrons and be in or below the third row of the periodic table to produce these electron-domain and molecular geometries. **9.29** (**a**) 1, less than $109.5°$; 2, less than $109.5°$ (**b**) 3, different than $109.5°$; 4, less than $109.5°$ (**c**) 5, $180°$ (**d**) 6, slightly more than $120°$; 7, less than $109.5°$; 8, different than $109.5°$ **9.31** (**a**) NH_4^+ has zero nonbonding electron pairs on N and the largest bond angles. (**b**) NH_2^- has two nonbonding electron pairs on N and the smallest bond angles. **9.33** (**a**) PF_4^-, BrF_4^- and ClF_4^- (**b**) AlF_4^- (**c**) BrF_4^- (**d**) PF_4^- and ClF_4^+ **9.35** Statements (i) and (iii) are true. **9.37** (**a**) Yes. (**b**) Statement (iii) correctly describes the direction of the dipole moment. **9.39** (**a**) Nonpolar. The polar B—F bonds are arranged in a symmetrical trigonal planar geometry. (**b**) No. The added nonbonding electron pair requires that the electron domain geometry is tetrahedral and the shape is a trigonal pyramid. (**c**) Yes. In BF_2Cl, the bond dipoles do not cancel. **9.41** (**a**) IF (**d**) PCl_3 and (**f**) IF_5 are polar. **9.43** (**a**) Lewis structures

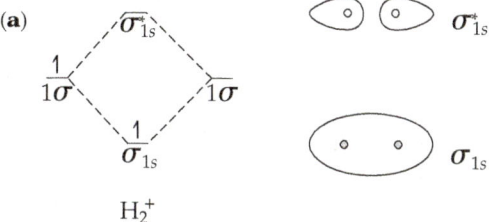

Molecular geometries

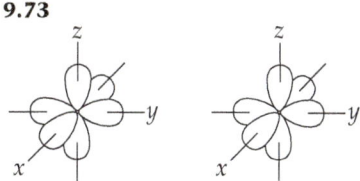

Polar   Nonpolar   Polar

(**b**) The middle isomer has a zero net dipole moment. (**c**) C_2H_3Cl has only one isomer, and it has a dipole moment. **9.45** (**a**) True (**b**) false (**c**) false (**d**) true (**e**) true **9.47** (**a**) False (**b**) true (**c**) false (**d**) false **9.49** (**a**) B, $[He]2s^22p^1$ (**b**) F, $[He]2s^22p^5$ (**c**) sp^2 (**d**) A single $2p$ orbital is unhybridized. It lies perpendicular to the trigonal plane of the sp^2 hybrid orbitals. **9.51** (**a**) sp^2 (**b**) sp^3 (**c**) sp (**d**) sp^3 **9.53** Left, no hybrid orbitals discussed in this chapter form angles of $90°$ with each other; p atomic orbitals are perpendicular to each other; center, $109.5°$, sp^3; right, $120°$, sp^2 **9.55** (**a**) True (**b**) false (**c**) true (**d**) true

9.57

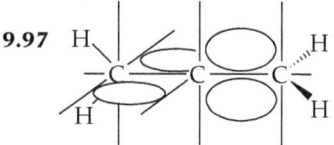

(**b**) sp^3, sp^2, sp (**c**) nonplanar, planar, planar (**d**) 7 σ, 0 π; 5 σ, 1 π; 3 σ, 2 π **9.59** (**a**) 18 valence electrons (**b**) 16 valence electrons form σ bonds. (**c**) 2 valence electrons form π bonds. (**d**) No valence electrons are nonbonding. (**e**) The left and central C atoms are sp^2 hybridized; the right C atom is sp^3 hybridized. **9.61** (**a**) ~$109.5°$ about the leftmost C, sp^3; ~$120°$ about the right-hand C, sp^2 (**b**) The doubly bonded O can be viewed as sp^2, and the other as sp^3; the nitrogen is sp^3 with bond angles less than $109.5°$. (**c**) nine σ bonds, one π bond **9.63** (**a**) True (**b**) true (**c**) true (**d**) false

9.65

(**a**) $\left[\begin{array}{c} :\ddot{O}: \\ | \\ H - C = \ddot{O}: \end{array}\right]^{1-}$

(**b**) sp^2
(**c**) Yes, there is one other resonance structure.

$\left[\begin{array}{c} :\ddot{O}: \\ || \\ H - C - \ddot{O}: \end{array}\right]^{1-}$

(**d**) There are four electrons in the π system of the ion. **9.67** (**a**) Linear (**b**) The two central C atoms each have trigonal planar geometry with ~ $120°$ bond angles about them. The C and O atoms lie in a plane with the H atoms free to rotate in and out of this plane. (**c**) The molecule is planar with ~ $120°$ bond angles about the two N atoms. **9.69** (**a**) True (**b**) true (**c**) true (**d**) false (**e**) true

9.71

(**a**)
σ_{1s}^*

σ_{1s}

H_2^+

(**b**) There is one electron in H_2^+. (**c**) σ_{1s}^1 (**d**) BO $= \frac{1}{2}$ (**e**) Fall apart. If the single electron in H_2^+ is excited to the σ_{1s}^* orbital, its energy is higher than the energy of an H $1s$ atomic orbital and H_2^+ will decompose into a hydrogen atom and a hydrogen ion. (**f**) Statement (**i**) is correct.

9.73

(**a**) 1 σ bond (**b**) 2 π bonds (**c**) 1 σ^* and 2 π^* **9.75** (**a**) Two (**b**) one (**c**) $\frac{1}{2}$ (**d**) Na_2^+ (**e**) Na_2 **9.77** (**a**) True (**b**) false (**c**) true (**d**) true **9.79** (**a**) B_2^+, $\sigma_{2s}^2\sigma_{2s}^{*2}\pi_{2p}^1$, increase (**b**) Li_2^+, $\sigma_{1s}^2\sigma_{1s}^{*2}\sigma_{2s}^1$, increase (**c**) N_2^+, $\sigma_{2s}^2\sigma_{2s}^{*2}\pi_{2p}^4\sigma_{2p}^1$, increase (**d**) Ne_2^{2+}, $\sigma_{2s}^2\sigma_{2s}^{*2}\sigma_{2p}^2\pi_{2p}^4\pi_{2p}^{*4}$, decrease **9.81** CN, $\sigma_{2s}^2\sigma_{2s}^{*2}\sigma_{2p}^2\pi_{2p}^3$, bond order $= 2.5$; CN^+, $\sigma_{2s}^2\sigma_{2s}^{*2}\sigma_{2p}^2\pi_{2p}^2$, bond order $= 2.0$; CN^-, $\sigma_{2s}^2\sigma_{2s}^{*2}\sigma_{2p}^2\pi_{2p}^4$, bond order $= 3.0$. (**a**) CN^- (**b**) CN, CN^+ **9.83** (**a**) $3s$, $3p_x$, $3p_y$, $3p_z$ (**b**) π_{3p} (**c**) 2 (**d**) If the MO diagram for P_2 is similar to that of N_2, P_2 will have no unpaired electrons and be diamagnetic. **9.86** (**a**) Two (**b**) answer (iii), T-shaped **9.90** (**a**) 2 σ bonds, 2 π bonds (**b**) 3 σ bonds, 4 π bonds (**c**) 3 σ bonds, 1 π bond (**d**) 4 σ bonds, 1 π bond **9.92** (**a**) Square pyramidal (**b**) octahedral (**c**) answer (iii), Group 7A

9.97

(**a**) sp hybridization (**b**) The molecule is nonplanar. (**c**) Allene has no dipole moment. (**d**) The bonding in allene would not be described as delocalized. The π electron clouds of the two adjacent C=C are mutually perpendicular, so there is no overlap and no delocalization of π electrons. **9.100** (**a**) All O atoms have sp^2 hybridization. (**b**) The two σ bonds are formed by overlap of sp^2 hybrid orbitals, the π bond is formed by overlap of atomic p orbitals, one nonbonded pair is in a p atomic orbital and the other five nonbonded pairs are in sp^2 hybrid orbitals. (**c**) unhybridized p atomic orbitals (**d**) four, two from the π bond and two from the nonbonded pair in the p atomic orbital **9.104** (**a**) σ antibonding (**b**) one (**c**) $\frac{1}{2}$ (**d**) answer (iv), longer and weaker **9.109** (**a**) The π_{2p} molecular orbital (**b**) the π_{2p}^* molecular orbital

(c) The C—C bond is weaker in the excited state because an electron has been excited from a bonding MO to an antibonding MO. (d) Yes, the molecule would be easier to twist in the excited state because the π component of the C—C bond has been destroyed.
9.111 (a) $2 SF_4(g) + O_2(g) \longrightarrow 2 OSF_4(g)$

(b)

(c) $\Delta H = -551$ kJ, exothermic (d) The electron-domain geometry is trigonal bipyramidal. The O atom can be either equatorial or axial.

(e) In the structure on the left, there are 3 equatorial and 1 axial fluorine atoms. In the structure on the right, there are 2 equatorial and 2 axial fluorine atoms.

9.113

(a)

(The structure on the right does not minimize formal charges and will make a minor contribution to the true structure.)

(b)

Both resonance structures predict the same bond angles. (c) The two extreme Lewis structures predict different bond lengths. These bond length estimates assume that the structure minimizing formal charge makes a larger contribution to the true structure. C—O, 1.28 Å; C=N, 1.33 Å; C—N, 1.43 Å; C—H, 1.07 Å (d) The molecule will have a dipole moment. The C=N and C=O bond dipoles are opposite each other, but they are not equal. In addition, there are nonbonding electron pairs that are not directly opposite each other and will not cancel.

Chapter 10   **10.1** It would be much easier to drink from
a straw on Mars. When a straw is placed in a glass of liquid, the atmosphere exerts equal pressure inside and outside the straw. When we drink through a straw, we withdraw air, thereby reducing the pressure on the liquid inside. If only 0.007 atm is exerted on the liquid in the glass, a very small reduction in pressure inside the straw will cause the liquid to rise. **10.3** At the same temperature, volume and the lower pressure, the container would have half as many particles as at the higher pressure, so answer (b) is correct. **10.5** For a fixed amount of gas at constant pressure, the volume and temperature are proportional to each other. This is Charles' Law. Therefore, graph (a) is correct. **10.7** (a) $P_{red} < P_{yellow} < P_{blue}$ (b) $P_{red} = 0.28$ atm; $P_{yellow} = 0.42$ atm; $P_{blue} = 0.70$ atm **10.9** (a) Curve B is helium. (b) Curve B corresponds to the higher temperature. (c) The root mean square speed is highest. **10.11** The $NH_4Cl(s)$ ring will form at location a. **10.13** Statement (c) is false. Gaseous molecules are so far apart that there is no barrier to mixing, regardless of the identity of the molecules. **10.15** (a) 2.6×10^2 lb/in.^{2} (b) 1.8×10^3 kPa (c) 18 atm **10.17** (a) 8.20 m (b) 1.4 atm **10.19** (a) 0.349 atm (b) 265 mm Hg

(c) 3.53×10^4 Pa (d) 0.353 bar (e) 5.13 psi **10.21** (a) $P = 773.4$ torr (b) $P = 1.018$ atm **10.23** (i) 0.31 atm (ii) 1.88 atm (iii) 0.136 atm **10.25** The action in (c) would double the pressure **10.27** (a) Boyle's Law, $PV =$ constant or $P_1V_1 = P_2V_2$, at constant V, $P_1/P_2 = 1$; Charles' Law, $V/T =$ constant or $V_1/T_1 = V_2/T_2$, at constant V, $T_1/T_2 = 1$; then $P_1/T_1 = P_2/T_2$ or $P/T =$ constant. Amonton's law is that pressure and Kelvin temperature are directly proportional at constant volume. (b) 34.7 psi **10.29** (a) STP stands for standard temperature, 0 °C (or 273 K), and standard pressure, 1 atm. (b) 22.4 L (c) 24.5 L (d) 0.08315 L-bar/mol-K **10.31** Flask A contains the gas with a molar mass of 30 g/mol and flask B contains the gas with a molar mass of 60 g/mol.

10.33

P	V	n	T
200 atm	1.00 L	0.500 mol	48.7 K
0.300 atm	0.250 L	3.05×10^{-3} mol	27 °C
650 torr	11.2 L	0.333 mol	350 K
10.3 atm	585 mL	0.250 mol	295 K

10.35 8.2×10^2 kg He **10.37** (a) 5.5×10^{22} molecules (b) 6.5 kg air **10.39** (a) 91 atm (b) 2.3×10^2 L **10.41** $p = 4.9$ atm **10.43** (a) 29.8 g Cl_2 (b) 9.42 L (c) 501 K (d) 2.28 atm **10.45** (a) $n = 2 \times 10^{-4}$ mol O_2 (b) The roach needs 8×10^{-3} mol O_2 in 48 h, approximately 100% of the O_2 in the jar. **10.47** The density of a gas increases with increasing molar mass. The order of increasing density is : HF (20 g/mol) < CO (28 g/mol) < N_2O (44 g/mol) < Cl_2 (71 g/mol) **10.49** (c) Because the helium atoms are of lower mass than the average air molecule, the helium gas is less dense than air. The balloon thus weighs less than the air displaced by its volume. **10.51** (a) $d = 1.77$ g/L (b) molar mass $= 80.1$ g/mol **10.53** molar mass $= 89.4$ g/mol **10.55** 4.1×10^{-9} g Mg **10.57** (a) 21.4 L CO_2 (b) 40.7 L O_2 **10.59** 0.402 g Zn **10.61** (a) When the stopcock is opened, the volume occupied by $N_2(g)$ increases from 2.0 L to 5.0 L. $P_{N_2} = 0.40$ atm (b) When the gases mix, the volume of $O_2(g)$ increases from 3.0 L to 5.0 L. $P_{O_2} = 1.2$ atms (c) $P_t = 1.6$ atm **10.63** (a) $P_{He} = 1.87$ atm $P_{Ne} = 0.807$ atm, $P_{Ar} = 0.269$ atm, (b) $P_t = 2.95$ atm, **10.65** $X_{CO_2} = 0.000407$ **10.67** $P_{CO_2} = 0.305$ atm, $P_t = 1.232$ atm **10.69** $P_{CO_2} = 0.9$ atm **10.71** 2.5 mole % O_2 **10.73** $P_t = 2.47$ atm **10.75** (a) Decrease (b) increase (c) decrease **10.77** The root-mean-square speed of WF_6 is approximately 9 times slower than that of He. **10.79** (a) Average kinetic energy of the molecules increases. (b) Root mean square speed of the molecules increases. (c) Strength of an average impact with the container walls increases. (d) Total collisions of molecules with walls per second increases. **10.81** (a) In order of increasing speed and decreasing molar mass: HBr < NF_3 < SO_2 < CO < Ne (b) $u_{NF_3} = 324$ m/s (c) The most probable speed of an ozone molecule in the stratosphere is 306 m/s. **10.83** Statements (a) and (d) are true. **10.85** The order of increasing rate of effusion is $^2H^{37}CI < {}^1H^{37}CI < {}^2H^{35}CI^1 < H^{35}CI$. **10.87** As_4S_6 **10.89** (a) Non-ideal-gas behavior is observed at very high pressures and low temperatures. (b) The real volumes of gas molecules and attractive intermolecular forces between molecules cause gases to behave nonideally. **10.91** Statement (b) is true. **10.93** From the value of b for Xe, the nonbonding radius is 2.72 Å. From Figure **7.7**, the bonding atomic radius of Xe is 1.40 Å. We expect the bonding radius of an atom to be smaller than its nonbonding radius, but our calculated value is nearly twice as large. **10.97** $P = 0.43$ mm Hg **10.100** (a) Molar mass of the unknown gas is 100.4 g/mol (b) We assume that the gases behave ideally, and that P, V and T are constant. **10.102** (a) 0.00378 mol O_2 (b) 0.0345 g C_8H_{18} **10.104** 42.2 g O_2 **10.106** $T_2 = 687$ °C **10.112** (a) 44.58% C, 6.596% H, 16.44% Cl, 32.38% N (b) $C_8H_{14}N_5Cl$ (c) Molar mass of the compound is required in order to determine molecular formula when the empirical formula is known. **10.114 (a)** $NH_3(g)$ remains after reaction. **(b)** $P = 0.957$ atm **(c)** 7.33 g NH_4Cl **10.117 (a)** $P_{IF_3} = 0.515$ atm **(b)** $X_{IF_5} = 0.544$

(c)

(d) Total mass in the flask is 20.00 g; mass is conserved.

Chapter 11

11.1 (a) The diagram best describes a liquid. **(b)** In the diagram, particles are close together, mostly touching, but there is no regular arrangement or order. This rules out a gaseous sample, where the particles are far apart, and a crystalline solid, which has a regular repeating structure in all three directions. **11.5** In its final state, methane is a gas at 185 °C. **11.8 (a)** Propanol can engage in hydrogen bonding. **(b)** While both molecules are somewhat polar, we expect propanol to have a larger dipole moment because of its O—H bond. **(c)** Propanol boils at 97.2 °C, while ethyl methyl ether boils at 10.8 °C. The stronger intermolecular forces in propanol cause it to have the higher boiling point. **11.11 (a)** Solid < liquid < gas **(b)** gas < liquid < solid **(c)** Matter in the gaseous state is most easily compressed because particles are far apart and there is much empty space. **11.13 (a)** It increases. Kinetic energy is the energy of motion. As melting occurs, the motion of atoms relative to each other increases. **(b)** It increases somewhat. The density of liquid lead is less than the density of solid lead. The smaller density means a greater sample volume and greater average distance between atoms in three dimensions. **11.15 (a)** The molar volumes of Cl_2 and NH_3 are nearly the same because they are both gases. **(b)** On cooling to 160 K, both compounds condense from the gas phase to the solid-state, so we expect a significant decrease in the molar volume. **(c)** The molar volumes are 0.0351 L/mol Cl_2 and 0.0203 L/mol NH_3 **(d)** Solid-state molar volumes are not as similar as those in the gaseous state, because most of the empty space is gone and molecular characteristics determine properties. $Cl_2(s)$ is heavier, has a longer bond distance and weaker intermolecular forces, so it has a significantly larger molar volume than $NH_3(s)$. **(e)** There is little empty space between molecules in the liquid state, so we expect their molar volumes to be closer to those in the solid state than those in the gaseous state. **11.17 (a)** London dispersion forces **(b)** dipole–dipole forces **(c)** hydrogen bonding **11.19 (a)** SO_2, dipole–dipole and London dispersion forces **(b)** CH_3COOH, London dispersion, dipole–dipole, and hydrogen bonding **(c)** H_2S, dipole–dipole and London dispersion forces (but not hydrogen bonding) **11.21 (a)** In order of increasing polarizability: $CH_4 < SiH_4 < SiCl_4 < GeCl_4 < GeBr_4$ **(b)** The magnitudes of London dispersion forces and thus the boiling points of molecules increase as polarizability increases. The order of increasing boiling points is the order of increasing polarizability given in **(a)**. **11.23 (a)** H_2S **(b)** CO_2 **(c)** GeH_4 **11.25** Both rodlike butane molecules and spherical 2-methylpropane molecules experience dispersion forces. The larger contact surface between butane molecules facilitates stronger forces and produces a higher boiling point. **11.27 (a)** A molecule must contain H atoms, bound to either N, O, or F atoms, in order to participate in hydrogen bonding with like molecules. **(b)** CH_3NH_2 and CH_3OH **11.29 (a)** Replacing a hydroxyl hydrogen with a CH_3 group eliminates hydrogen bonding in that part of the molecule. This reduces the strength of intermolecular forces and leads to a lower boiling point. **(b)** $CH_3OCH_2CH_2OCH_3$ is a larger, more polarizable molecule with stronger London dispersion forces and thus a higher boiling point.

11.3

Physical Property	H₂O	H₂S
Normal boiling point, °C	100.00	−60.7
Normal melting point, °C	0.00	−85.5

(a) Based on its much higher normal melting point and boiling point, H_2O has much stronger intermolecular forces. **(b)** H_2O has hydrogen bonding, while H_2S has dipole–dipole forces. Both molecules have London dispersion forces. **11.33** SO_4^{2-} has a greater negative charge than BF_4^-, so ion–ion electrostatic attractions are greater in sulfate salts and they are less likely to form liquids. **11.35 (a)** As temperature increases, surface tension decreases; they are inversely related. **(b)** As temperature increases, viscosity decreases; they are inversely related. **(c)** The same attractive forces that cause surface molecules to be difficult to separate (high surface tension) cause molecules elsewhere in the sample to resist movement relative to each other (high viscosity). **11.37 (a)** Diagram (ii) shows stronger adhesive forces between the surface and the liquid. **(b)** Diagram (i) represents water on a nonpolar surface. **(c)** Diagram (ii) represents water on a polar surface. **11.39 (a)** The three molecules have similar structures and experience the same types of intermolecular forces. As molar mass increases, the strength of dispersion forces increases and the boiling points, surface tension, and viscosities all increase. **(b)** Ethylene glycol has an —OH group at both ends of the molecule. This greatly increases the possibilities for hydrogen bonding; the overall intermolecular attractive forces are greater and the viscosity of ethylene glycol is much greater. **(c)** Water has the highest surface tension but lowest viscosity because it is the smallest molecule in the series. There is no hydrocarbon chain to inhibit their strong attraction to molecules in the interior of the drop, resulting in high surface tension. The absence of an alkyl chain also means the molecules can move around each other easily, resulting in the low viscosity. **11.41 (a)** Melting, endothermic **(b)** evaporation, endothermic **(c)** deposition, exothermic **(d)** condensation, exothermic **11.43 (a)** Melting, (s) ⟶ (l) **(b)** endothermic **(c)** Heat of vaporization is usually larger than heat of fusion. **11.45** 2.3×10^3 g H_2O **11.47 (a)** 39.3 kJ **(b)** 60 kJ **11.49 (a)** False **(b)** true **(c)** false **(d)** true **11.51** Properties **(c)** intermolecular attractive forces, **(d)** temperature and **(e)** density of the liquid affect vapor pressure of a liquid. **11.53 (a)** $CBr_4 < CHBr_3 < CH_2Br_2 < CH_2Cl_2 < CH_3Cl < CH_4$. **(b)** $CH_4 < CH_3Cl < CH_2Cl_2 < CH_2Br_2 < CHBr_3 < CBr_4$ **(c)** By analogy to attractive forces in HCl, the trend will be dominated by dispersion forces, even though four of the molecules are polar. The order of increasing boiling point is the order of increasing molar mass and increasing strength of dispersion forces. **11.55 (a)** The temperature of the water in the two pans is the same. **(b)** Vapor pressure does not depend on either volume or surface area of the liquid. At the same temperature, the vapor pressures of water in the two containers are the same. **11.57 (a)** Approximately 48 °C **(b)** approximately 340 torr **(c)** approximately 17 °C **(d)** approximately 1000 torr **11.59 (a)** Yes, provided the temperature is less than the critical temperature. **(b)** No, at temperatures above the critical temperature a substance can only be a gas or a supercritical fluid. **(c)** No, at pressures below the triple point a substance can only be a solid or a gas. **(d)** It depends on the slope of the melting curve. For most substances the melting curve slopes to the right slightly, and in such cases the liquid cannot exist at temperatures below the triple point **11.61 (a)** $H_2O(g)$ will condense to $H_2O(s)$ at approximately 4 torr; at a higher pressure, perhaps 5 atm or so, will melt to form $H_2O(l)$. **(b)** At 100 °C and

0.50 atm, water is in the vapor phase. As it cools, water vapor condenses to the liquid at approximately 82 °C, the temperature where the vapor pressure of liquid water is 0.50 atm. Further cooling results in freezing at approximately 0 °C. The freezing point of water increases with decreasing pressure, so at 0.50 atm the freezing temperature is very slightly above 0 °C. **11.63 (a)** 24 K **(b)** Neon sublimes at pressures less than the triple point pressure, approximately 0.5 atm. **(c)** No **11.65 (a)** Methane on the surface of Titan is likely to exist in both solid and liquid forms. **(b)** As pressure decreases upon moving away from the surface of Titan, $CH_4(l)$ (at −178 °C) will vaporize to $CH_4(g)$, and $CH_4(s)$ (at temperatures below −180 °C) will sublime to $CH_4(g)$. **11.67** In a nematic liquid crystalline phase, molecules are aligned along their long axes, but the molecular ends are not aligned. Molecules are free to translate in all dimensions, but they cannot tumble or rotate out of the molecular plane, or the order of the nematic phase is lost and the sample becomes an ordinary liquid. In an ordinary liquid, molecules are randomly oriented and free to move in any direction. **11.69 (a)** True **(b)** false **(c)** true **(d)** false **(e)** false **(f)** true **11.71 (a)** endothermic **(b)** decrease **11.73** nematic **11.75 (a)** Decrease **(b)** increase **(c)** increase **(d)** increase **(e)** increase **(f)** increase **(g)** increase **11.78 (a)** The *cis* isomer has stronger dipole-dipole forces; the *trans* isomer is nonpolar. **(b)** The *cis* isomer boils at 60.3 °C and the *trans* isomer boils at 47.5 °C. **11.80 (a)** Four, all of them **(b)** three, benzene is nonpolar **(c)** one, phenol **(d)** Bromine is larger and more polarizable than chlorine, so the dispersion forces in bromobenzene are stronger than those in chlorobenzene and bromobenzene has the higher boiling point. **(e)** Phenol exhibits hydrogen bonding, which is the strongest intermolecular interaction among covalent molecules. **11.83** A plot of number of carbon atoms versus boiling point indicates that the boiling point of C_8H_{18} is approximately 130 °C. The more carbon atoms in the hydrocarbon, the longer the chain, the more polarizable the electron cloud, the higher the boiling point. **11.85 (a)** Evaporation is an endothermic process. The heat required to vaporize sweat is absorbed from your body, helping to keep it cool. **(b)** The vacuum pump reduces the pressure of the atmosphere above the water until atmospheric pressure equals the vapor pressure of water and the water boils. Boiling is an endothermic process, and the temperature drops if the system is not able to absorb heat from the surroundings fast enough. As the temperature of the water decreases, the water freezes. **11.89** At low Antarctic temperatures, molecules in the liquid crystalline phase have less kinetic energy due to temperature, and the applied voltage may not be sufficient to overcome orienting forces among the ends of molecules. If some or all of the molecules do not rotate when the voltage is applied, the display will not function properly.

11.93

```
        CH2                 CH2    CH2
CH3         CH3   CH3          CH2    CH3
  (i) M = 44          (ii) M = 72
```

```
                                    Br
                                    |
                                    CH
                              CH3      CH3
                         (iii) M = 123
```

```
    O
    ||
    C                 CH2    Br         CH2    OH
CH3    CH3      CH3      CH2      CH3      CH2
 (iv) M = 58      (v) M = 123       (vi) M = 60
```

(a) Molar mass: Compounds (**i**) and (ii) have similar rodlike structures. The longer chain in (ii) leads to greater molar mass, stronger London dispersion forces, and higher heat of vaporization.

(b) Molecular shape: Compounds (iii) and (v) have the same chemical formula and molar mass but different molecular shapes. The more rodlike shape of (v) leads to more contact between molecules, stronger dispersion forces, and higher heat of vaporization. **(c)** Molecular polarity: Compound (iv) has a smaller molar mass than (ii) but a larger heat of vaporization, which must be due to the presence of dipole–dipole forces. **(d)** Hydrogen bonding interactions: Molecules (v) and (vi) have similar structures. Even though (v) has larger molar mass and dispersion forces, hydrogen bonding causes (vi) to have the higher heat of vaporization.

Chapter 12 **12.1** The red-orange compound is more likely to be a semiconductor and the white one an insulator. The red-orange compound absorbs light in the visible spectrum (red-orange is reflected, so blue-green is absorbed), while the white compound does not. This indicates that the red-orange compound has a lower energy electron transition than the white one. Semiconductors have lower energy electron transitions than insulators. **12.5 (a)** The structure is hexagonal close-packed. **(b)** The coordination number, CN, is twelve. **(c)** CN(1) = 9, CN(2) = 6. **12.7** Fragment (**b**) is more likely to give rise to electrical conductivity. Fragment (**b**) has a delocalized π system, in which electrons are free to move. Mobile electrons are required for electrical conductivity. **12.9** We expect linear polymer (**a**), with ordered regions, to be more crystalline and to have a higher melting point than branched polymer (**b**). **12.11** Statement (**b**) is the best explanation. **12.13 (a)** Hydrogen bonding, dipole-dipole forces, London dispersion forces **(b)** covalent chemical bonds **(c)** ionic bonds **(d)** metallic bonds **12.15 (a)** Ionic **(b)** metallic **(c)** covalent-network (It could also be characterized as ionic with some covalent character to the bonds.) **(d)** molecular **(e)** molecular **(f)** molecular **12.17** Metallic, because of its melting point, conductivity, and insolubility in water

12.19 (a) 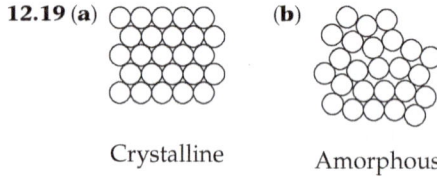 **(b)**

Crystalline   Amorphous

12.21

Two-dimensional structure	(i)	(ii)
(a) unit cell	A B	A B
(b) γ, a, b	$\lambda = 90°$, $a = b$	$\lambda = 120°$, $a = b$
(c) lattice type	square	hexagonal

12.23 Tetragonal **12.25 (e)** both triclinic and rhombohedral **12.27 (b)** 2 **12.29 (a)** Primitive hexagonal unit cell **(b)** NiAs **12.31** Potassium. A body-centered cubic structure has more empty space than a face-centered cubic one. The more empty space, the less dense the solid. We expect the element with the lowest density, potassium, to adopt the body-centered cubic structure. **12.33 (a)** Structure types A and C have equally dense packing and are more densely packed than structure type B. **(b)** Structure type B is least densely packed. **12.35 (a)** The radius of an Ir atom is 1.355 Å.

(b) The density of Ir is 22.67 g/cm^3 **12.37 (a)** The radius of a Ca atom is 1.976 Å. **(b)** The density of Ca is 1.526 g/cm^3 **12.39 (a)** 4 Al atoms per unit cell **(b)** coordination number = 12 **(c)** a = 4.04 Å or 4.04 × 10^{-8} cm **(d)** density = 2.71 g/cm^3 **12.41** Statement **(b)** is false. **12.43 (a)** Interstitial alloy **(b)** substitutional alloy **(c)** intermetallic compound **12.45 (a)** True **(b)** false **(c)** false **12.47** Au$_{2.8}$Ag$_{1.0}$Cu$_{1.1}$ **12.49 (a)** True **(b)** false **(c)** false **(d)** false

12.51

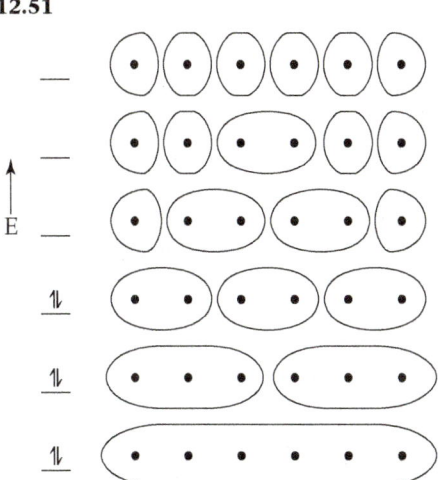

(a) Six AOs require six MOs **(b)** zero nodes in the lowest energy orbital **(c)** five nodes in highest energy orbital **(d)** two nodes in the HOMO **(e)** three nodes in the LUMO **(f)** The HOMO-LUMO energy gap for the six-atom diagram is smaller than the one for the four-atom diagram. In general, the more atoms in the chain, the smaller the HOMO-LUMO energy gap. **12.53 (a)** Ag is more ductile. Mo has stronger metallic bonding, a stiffer lattice, and is less susceptible to distortion. **(b)** Zn is more ductile. Si is a covalent-network solid with a stiffer lattice than metallic Zn. **12.55** The order of increasing melting points is Y < Zr < Nb < Mo. Moving from Y to Mo, the number of valence electrons, occupancy of the bonding band, and strength of metallic bonding increase. Stronger metallic bonding requires more energy to break bonds and mobilize atoms, resulting in higher melting points. **12.57 (a)** SrTiO$_3$ **(b)** Six **(c)** Each Sr atom is coordinated to a total of twelve O atoms in the eight unit cells that contain the Sr atom. **12.59** The density of MnS is 4.056 g/cm^3 **12.61 (a)** 7.711 g/cm^3 **(b)** We expect Se^{2-} to have a larger ionic radius than S^{2-}, so HgSe will occupy a larger volume and the unit cell edge will be longer. **(c)** The density of HgSe is 8.241 g/cm^3. The greater mass of Se accounts for the greater density of HgSe. **12.63 (a)** Cs$^+$ and I$^-$ have the most similar radii and will adopt the CsCl-type structure. The radii of Na$^+$ and I$^-$ are somewhat different; NaI will adopt the NaCl-type structure. The radii of Cu$^+$ and are very different; CuI has the ZnS-type structure. **(b)** CsI, 8; NaI, 6; CuI, 4 **12.65 (a)** 6 **(b)** 3 **(c)** 6 **12.67 (a)** False **(b)** true **12.69 (a)** Ionic solids are much more likely to dissolve in water. **(b)** Covalent-network solids can become considerably better conductors of electricity via chemical substitution. **12.71 (a)** CdS **(b)** GaN **(c)** GaAs **12.73** Si, **(d)** is the best choice; we want an element that has more valence electrons than Ga. **12.75 (a)** A 1.1 eV photon corresponds to a wavelength of 1.1 × 10^{-6} m, or 1100 nm **(b)** According to the figure, Si can absorb all wavelengths in the visible portion of the solar spectrum. **(c)** Si absorbs wavelengths less than 1,100 nm. This corresponds to approximately 80–90% of the total area under the curve. **12.77** The emitted light has a wavelength of 713 nm. This is red light in the visible region of the electromagnetic spectrum. **12.79** The band gap is approximately 1.85 eV, which corresponds to

a wavelength of 672 nm. **12.81 (a)** A monomer is a small molecule with low molecular mass that can be joined with other monomers to form a polymer. They are the repeating units of a polymer. **(b)** ethene (also known as ethylene) **12.83** Reasonable values for a polymer's molecular weight are 10,000 amu, 100,000 amu, and 1,000,000 amu. **12.85 (d)** shows the polyester repeat unit.

12.87

$$\begin{array}{c}\text{Cl}\\ \\ \text{Cl}\end{array}\!\!C=C\!\!\begin{array}{c}\text{H}\\ \\ \text{H}\end{array}$$

12.89

HOOC—⬡—COOH   NH$_2$—⬡—NH$_2$

12.91 (a) Flexibility of the molecular chains causes flexibility of the bulk polymer. Flexibility is enhanced by molecular features that inhibit order, such as branching, and diminished by features that encourage order, such as cross-linking or delocalized π electron density. **(b)** Less flexible **12.93** Low degree of crystallinity **12.95** 1–100 nm. **12.97 (a)** False. As particle size decreases, the band gap increases. **(b)** False. As particle size decreases, the band gap increases and hence wavelength of emitted light decreases. **12.99** 2.47 × 10^5 Au atoms **12.101** Statement **(b)** is correct. **12.109** The wavelength that corresponds to a photon with this energy is 564 nm. **12.111 (a)** Zinc sulfide, ZnS **(b)** covalent **(c)** In the solid, each Si is bound to four C atoms in a tetrahedral arrangement, and each C is bound to four Si atoms in a tetrahedral arrangement, producing an extended three-dimensional network, much like diamond. SiC is high-melting because melting requires breaking covalent Si—C bonds, which takes a huge amount of thermal energy. It is hard because the three-dimensional lattice resists any deformation that would weaken the Si—C bonding network. **12.122 (a)** 109° **(b)** 120° **(c)** atomic p orbitals **12.126 (a)** 2.50 × 10^{22} Si atoms (To 1 sig fig, the result is 2 × 10^{22} Si atoms.) **(b)** 1.29 × 10^{-3} mg P (1.29 μg P) **12.128** 32 Si atoms

Chapter 13   13.1 (a) < **(b)** < **(c) 13.3 (a)** No **(b)** The ionic solid with the smaller lattice energy will be more soluble in water. **13.7** Vitamin B$_6$ is more water soluble. Vitamin E is more fat soluble. **13.9 (a)** Yes, the *molarity* changes with a change in temperature. **(b)** No, *molality* does not change with change in temperature. **13.11** The volume inside the balloon will be 0.5 L, assuming perfect osmosis across the semipermeable membrane. **13.13 (a)** False **(b)** false **(c)** true **13.15 (a)** Dispersion **(b)** hydrogen bonding **(c)** ion–dipole **(d)** dipole–dipole **13.17** Very soluble **(b)** ΔH_{mix} will be the largest negative number. In order for ΔH_{soln} to be negative, the magnitude of ΔH_{mix} must be greater than the magnitude of ($\Delta H_{solute} + \Delta H_{solvent}$). **13.19 (a)** ΔH_{solute} **(b)** ΔH_{mix} **13.21 (a)** ΔH_{soln} is nearly zero. Since the solute and solvent experience very similar London dispersion forces, the energy required to separate them individually and the energy released when they are mixed are approximately equal. $\Delta H_{solute} + \Delta H_{solvent} \approx -\Delta H_{mix}$. **(b)** The entropy of the system increases when heptane and hexane form a solution. From part **(a)**, the enthalpy of mixing is nearly zero, so the increase in entropy is the driving force for mixing in all proportions. **13.23 (a)** Supersaturated **(b)** The bits of glass scraped from the vessel act as a seed crystal, a place where solute molecules can align to form a crystal. The excess chromium nitrate is crystallizing out. **(c)** 116 g crystals form **13.25 (a)** Unsaturated **(b)** saturated **(c)** saturated **(d)** unsaturated **13.27 (a)** We expect the liquids water and glycerol to be miscible in all proportions. **(b)** Hydrogen bonding, dipole-dipole forces, London dispersion forces **13.29** Toluene, $C_6H_5CH_3$, is the best solvent for nonpolar solutes. Without polar groups or nonbonding electron pairs, it forms only dispersion interactions with itself and other molecules. **13.31 (a)** Carbon tetrachloride **(b)** water **13.33 (a)** CCl_4 is more soluble because dispersion forces among nonpolar CCl_4 molecules are similar to dispersion forces in hexane. **(b)** C_6H_6 is a nonpolar hydrocarbon and will be more soluble in the similarly nonpolar hexane. **(c)** The long, rodlike hydrocarbon chain of octanoic acid forms strong dispersion interactions and causes it to be more soluble in hexane. **13.35 (a)** False **(b)** true **(c)** false **(d)** true **13.37** $S_{He} = 5.6 \times 10^{-4}\,M$, $S_{N_2} = 9.0 \times 10^{-4}\,M$ **13.39 (a)** 2.15% Na_2SO_4 by mass **(b)** 3.15 ppm Ag **13.41 (a)** $X_{CH_3OH} = 0.0427$ **(b)** 7.35% CH_3OH by mass **(c)** 2.48 m CH_3OH **13.43 (a)** $1.46 \times 10^{-2}\,M$ $Mg(NO_3)_2$ **(b)** 1.12 M $LiClO_4 \cdot 3H_2O$ **(c)** 0.350 M HNO_3 **13.45 (a)** 4.70 m C_6H_6 **(b)** 0.235 m NaCl **13.47 (a)** 43.01% H_2SO_4 by mass **(b)** $X_{H_2SO_4} = 0.122$ **(c)** 7.69 m H_2SO_4 **(d)** 5.827 M H_2SO_4 **13.49 (a)** $X_{CH_3OH} = 0.227$ **(b)** 7.16 m CH_3OH **(c)** 4.58 M CH_3OH **13.51 (a)** 0.150 mol $SrBr_2$ **(b)** 1.56×10^{-2} mol KCl **(c)** 4.44×10^{-2} mol $C_6H_{12}O_6$ **13.53 (a)** Weigh out 1.3 g KBr, dissolve in water, dilute with stirring to 0.75 L. **(b)** Weigh out 2.62 g KBr, dissolve it in 122.38 g H_2O to make exactly 125 g of 0.180 m solution. **13.55** 71% HNO_3 by mass **13.57 (a)** 3.82 m Zn **(b)** 26.8 M Zn **13.59 (a)** Dissolve 244 g KBr in water, dilute with stirring to 1.85 L. **(b)** Dissolve 214 g $Pb(NO_3)_2$ in water, dilute with stirring to 1.20 L. **13.61** The minimum pressure required to initiate reverse osmosis is greater than 5.1 atm. **13.63 (a)** False **(b)** true **(c)** true **(d)** false **13.65** The vapor pressure of both solutions is 17.5 torr. Because these two solutions are so dilute, they have essentially the same vapor pressure. Generally, the less concentrated solution, the one with fewer moles of solute per kilogram of solvent, will have the higher vapor pressure. **13.67 (a)** $P_{H_2O} = 186.4$ torr **(b)** 78.9 g $C_3H_8O_2$ **13.69 (a)** $X_{Eth} = 0.2812$ **(b)** $P_{soln} = 238$ torr **(c)** X_{Eth} in vapor $= 0.472$ **13.71 (a)** 106.3 °C **(b)** $i = 1.6$ **13.73** 0.050 m LiBr $< 0.120\,m$ glucose $< 0.050\,m$ $Zn(NO_3)_2$ **13.75 (a)** $T_f = -115.0\,°C$, $T_b = 78.7\,°C$ **(b)** $T_f = -67.3\,°C$, $T_b = 64.2\,°C$ **(c)** $T_f = -0.4\,°C$, $T_b = 100.1\,°C$ **(d)** $T_f = -0.6\,°C$, $T_b = 100.2\,°C$ **13.77** 167 g $C_2H_6O_2$ **13.79** $\Pi = 0.0168$ atm $= 12.7$ torr **13.81** The approximate molar mass of adrenaline is 1.8×10^2 g. **13.83** Molar mass of lysozyme $= 1.39 \times 10^4$ g **13.85** $i = 2.8$ **13.87 (a)** emulsion **(b)** sol **(c)** solid foam **13.89** Choice **(d)**, $CH_3(CH_2)_{11}COONa$, is the best emulsifying agent. The long hydrocarbon chain will interact with the hydrophobic component, while the ionic end will interact with the hydrophilic component, as well as stabilize the colloid. **13.91 (a)** No. The hydrophobic or hydrophilic nature of the protein will determine which electrolyte at which concentration will be the most effective precipitating agent. **(b)** Stronger. If a protein has been "salted out", protein-protein interactions are stronger than protein-solvent interactions and solid protein forms. **(c)** The first hypothesis seems plausible, since ion-dipole interactions among electrolytes and water molecules are stronger than dipole-dipole and hydrogen bonding interactions between water and protein molecules. But, we also know that ions adsorb on the surface of a hydrophobic colloid; the second hypothesis also seems plausible. If we could measure the charge and adsorbed water content of protein molecules as a function of salt concentration, then we could distinguish between these two hypotheses. **13.93 (a)** Hydrochloride **(b)** free base **(c)** 0.492 M free base **(d)** 7.36 M hydrochloride **(e)** 275 mL of 12.0 M HCl **13.96 (a)** $k_{Rn} = 7.27 \times 10^{-3}$ mol/L-atm **(b)** $P_{Rn} = 1.1 \times 10^{-4}$ atm; $S_{Rn} = 8.1 \times 10^{-7}\,M$ **13.101 (a)** 2.69 m LiBr **(b)** $X_{LiBr} = 0.0994$ **(c)** 81.1% LiBr by mass **13.102** $X_{H_2O} = 0.808$; 0.0273 mol ions; 0.0137 mol NaCl; 0.798 g NaCl **13.105 (a)** $-0.6\,°C$ **(b)** $-0.4\,°C$ **13.108 (a)**, CF_4, $1.7 \times 10^{-4}\,m$; $CClF_3$, $9 \times 10^{-4}\,m$; CCl_2F_2, $2.3 \times 10^{-2}\,m$; $CHClF_2$, $3.5 \times 10^{-2}\,m$ **(b)** dipole moment **(c)** 3.9×10^{-4} mol O_2

Chapter 14

14.1 The rate of the combustion reaction in the cylinder depends on the surface area of the droplets in the spray. The smaller the droplets, the greater the surface area exposed to oxygen, the faster the combustion reaction. In the case of a clogged injector, larger droplets lead to slower combustion. Uneven combustion in the various cylinders can cause the engine to run roughly and decrease fuel economy. **14.3** Equation (iv) **(b)** rate $= -\Delta[B]/\Delta t = \frac{1}{2}\Delta[A]/\Delta t$ **14.9** (1) Total potential energy of the reactants (2) E_a, activation energy of the reaction (3) ΔE, net energy change for the reaction (4) total potential energy of the products **14.12 (a)** $NO_2 + F_2 \longrightarrow NO_2F + F$; $NO_2 + F \longrightarrow NO_2F$ **(b)** $2NO_2 + F_2 \longrightarrow 2NO_2F$ **(c)** F (atomic fluorine) is the intermediate **(d)** rate $= k[NO_2][F_2]$ **14.15 (a)** Net reaction: $AB + AC \longrightarrow BA_2 + C$ **(b)** A is the intermediate. **(c)** A_2 is the catalyst. **14.17 (a)** *Reaction rate* is the change in the amount of products or reactants in a given amount of time. **(b)** Rates depend on concentration of reactants, surface area of reactants, temperature, and activation energy/presence of catalyst. **(c)** No. The stoichiometry of the reaction (mole ratios of reactants and products) must be known to relate rate of disappearance of reactants to rate of appearance of products.

14.19

Time (min)	Mol A	(a) Mol B	[A] (mol/L)	Δ[A] (mol/L)	(b) Rate (M/s)
0	0.065	0.000	0.65		
10	0.051	0.014	0.51	-0.14	2.3×10^{-4}
20	0.042	0.023	0.42	-0.09	1.5×10^{-4}
30	0.036	0.029	0.36	-0.06	1.0×10^{-4}
40	0.031	0.034	0.31	-0.05	0.8×10^{-4}

(c) $\Delta[B]_{avg}/\Delta t = 1.3 \times 10^{-4}\,M/s$

14.21

(a)

Time (s)	Time Interval (s)	Concentration (M)	ΔM	Rate (M/s)
0		0.0165		
2,000	2,000	0.0110	-0.0055	28×10^{-7}
5,000	3,000	0.00591	-0.0051	17×10^{-7}
8,000	3,000	0.00314	-0.00277	9.3×10^{-7}
12,000	4,000	0.00137	-0.00177	4.43×10^{-7}
15,000	3,000	0.00074	-0.00063	2.1×10^{-7}

(b) The average rate of reaction is $1.05 \times 10^{-6}\,M/s$ **(c)** The average rate between $t = 2000$ and $t = 12{,}000$ s ($9.63 \times 10^{-7}\,M/s$) is greater than the average rate between $t = 8{,}000$ and $t = 15{,}000$ s ($3.43 \times 10^{-7}\,M/s$). **(d)** From the slopes of the tangents to the graph, the rates are $12 \times 10^{-7}\,M/s$ at 5000 s, $5.8 \times 10^{-7}\,M/s$ at 8000 s. **14.23 (a)** $-\Delta[H_2O_2]/\Delta t = \Delta[H_2]/\Delta t = \Delta[O_2]/\Delta t$

(b) $-\frac{1}{2}\Delta[N_2O]/\Delta t = \frac{1}{2}\Delta N_2/\Delta t = \Delta[O_2]/\Delta t$

(c) $-\Delta[N_2]/\Delta t = -1/3[H_2]/\Delta t = -1/2\Delta[H_2][NH_3]/\Delta t$ **(d)** $-\Delta[C_2H_5NH_2]/\Delta t = \Delta[C_2H_4]//\Delta t = \Delta[NH_3]/\Delta t$ **14.25 (a)** $-\Delta[O_2]/\Delta t = 0.24$ mol/s; $\Delta[H_2O]/\Delta t = 0.48$ mol/s **(b)** P_{total} decreases by 28 torr/min. **14.27 (a)** If [A] doubles, there is no change in the rate or the rate constant. **(b)** The reaction is zero order in A, second order in B, and second order overall. **(c)** units of $k = M^{-1}s^{-1}$ **14.29 (a)** Rate $= k[N_2O_5]$ **(b)** Rate $= 1.16 \times 10^{-4}\,M/s$ **(c)** When the concentration of N_2O_5 doubles, the rate doubles. **(d)** When the concentration of N_2O_5 is halved, the rate is halved.

14.31 (a, b) $k = 1.7 \times 10^2\,M^{-1}s^{-1}$ (c) If $[OH^-]$ is tripled, the rate triples. (d) If $[OH^-]$ and $[CH_3Br]$ both triple, the rate increases by a factor of 9. **14.33** (a) Rate $= k[OCl^-][I^-]$ (b) $k = 60\,M^{-1}s^{-1}$. (c) Rate $= 6.0 \times 10^{-5}\,M/s$ **14.35** (a) Rate $= k[BF_3][NH_3]$ (b) The reaction is second order overall. (c) $k_{avg} = 3.41\,M^{-1}s^{-1}$ (d) $0.170\,M/s$ **14.37** (a) Rate $= k[NO]^2[Br_2]$ (b) $k_{avg} = 1.2 \times 10^4\,M^{-2}s^{-1}$ (c) $\frac{1}{2}\Delta[NOBr]/\Delta t = -\Delta[Br_2]/\Delta t$ (d) $-\Delta[Br_2]/\Delta t = 8.4\,M/s$ **14.39** (a) A graph of $\ln[A]$ versus time yields a straight line for a first-order reaction. (b) On a graph of $\ln[A]$ versus time, the rate constant is the (–slope) of the straight line. **14.41**(a) $k = 3.0 \times 10^{-6}\,s^{-1}$ (b) $t_{1/2} = 3.2 \times 10^4\,s$ **14.43** (a) $p = 30$ torr (b) $t = 51\,s$ **14.45** Plot $(\ln P_{SO_2Cl_2})$ versus time, $k = -slope = 2.19 \times 10^{-5}\,s^{-1}$ **14.47** (a) The plot of $1/[A]$ versus time is linear, so the reaction is second order in $[A]$. (b) $k = 0.040\,M^{-1}\,min^{-1}$ (c) $t_{1/2} = 38$ min **14.49** (a) The plot of $1/[NO_2]$ versus time is linear, so the reaction is second order in NO_2. (b) $k = slope = 10\,M^{-1}s^{-1}$ (c) rate at $0.200\,M = 0.400\,M/s$, rate at $0.100\,M = 0.100\,M/s$ rate at $0.050\,M = 0.025\,M/s$ **14.51** (a) The energy of the collision and the orientation of the molecules when they collide determine whether a reaction will occur. (b) The rate constant usually increases with an increase in reaction temperature. (c) The fraction of molecules with energy greater than the activation energy changes most dramatically with temperature. Frequency of collision and the orientation factor are lumped into the frequency factor, A, which is considered to be constant with temperature. **14.53** $f = 4.94 \times 10^{-2}$ At 400 K, approximately 1 out of 20 molecules has this kinetic energy.

14.55 (a)

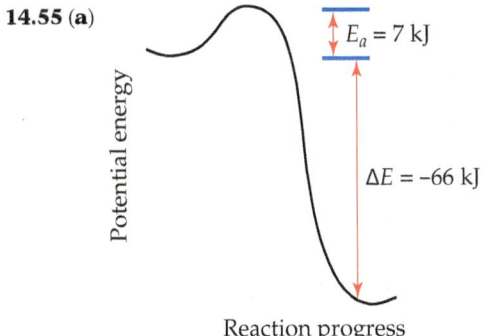

$E_a = 7$ kJ

$\Delta E = -66$ kJ

Reaction progress

(b) $E_a(reverse) = 73$ kJ **14.57** (a) False (b) false (c) true **14.59** The order of slowest reaction to fastest reaction is: rate (c) < rate (a) < rate (b). **14.61** (a) $k = 1.1\,s^{-1}$ (b) $k = 13\,s^{-1}$ (c) The method in parts (a) and (b) assumes that the collision model and thus the Arrhenius equation describe the kinetics of the reactions. That is, activation energy is constant over the temperature range under consideration. **14.63** A plot of $\ln k$ versus $1/T$ has a slope of -5.64×10^3; $E_a = -R(slope) = 47.5$ kJ/mol **14.65** (a) An *elementary reaction* is a process that occurs as a single event; the order is given by the coefficients in the balanced equation for the reaction. (b) A *unimolecular* elementary reaction involves only one reactant molecule; a *bimolecular* elementary reaction involves two reactant molecules. (c) A *reaction mechanism* is a series of elementary reactions that describes how an overall reaction occurs and explains the experimentally determined rate law. (d) A *rate-determining step* is the slow step in a reaction mechanism. It limits the overall reaction rate. **14.67** (a) Unimolecular, rate $= k[Cl_2]$ (b) bimolecular, rate $= k[OCl^{-1}][H_2O]$ (c) bimolecular, rate $= k[NO][Cl_2]$ **14.69** (a) Two intermediates, B and C. (b) three transition states (c) C $\longrightarrow$ D is fastest. (d) ΔE is positive. **14.71** (a) $H_2(g) + 2\,ICl(g) \longrightarrow I_2(g) + 2\,HCl(g)$ (b) HI is the intermediate. (c) If the first step is slow, the observed rate law is rate $= k[H_2][ICl]$. **14.73** (a) The two-step mechanism is consistent with the data, assuming that the second step is rate determining. (b) No. The linear

plot guarantees that the overall rate law will include $[NO]^2$. Since the data were obtained at constant $[Cl_2]$, we have no information about reaction order with respect to $[Cl_2]$. **14.75** (a) A catalyst is a substance that changes (usually increases) the speed of a chemical reaction without undergoing a permanent chemical change itself. (b) A homogeneous catalyst is in the same phase as the reactants, while a hetereogeneous catalyst is in a different phase. (c) A catalyst has no effect on the overall enthalpy change for a reaction, but it does affect activation energy. It can also affect the frequency factor.

14.77

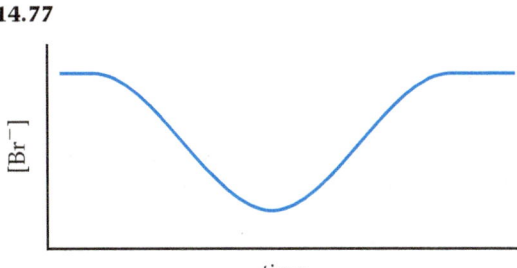

[Br⁻] vs time

14.79 (a) Multiply the coefficients in the first reaction by 2 and sum. (b) $NO_2(g)$ is a catalyst. (c) $NO(g)$ is an intermediate. (d) This is an example of homogeneous catalysis. **14.81** (a) Use of chemically stable supports makes it possible to obtain very large surface areas per unit mass of the precious metal catalyst because the metal can be deposited in a very thin, even monomolecular, layer on the surface of the support. (b) The greater the surface area of the catalyst, the more reaction sites, the greater the rate of the catalyzed reaction. **14.83** To put two D atoms on a single carbon, it is necessary that one of the already existing C—H bonds in ethylene be broken while the molecule is adsorbed, so that the H atom moves off as an adsorbed atom and is replaced by a D atom. This requires a larger activation energy than simply adsorbing C_2H_4 and adding one D atom to each carbon. **14.85** Carbonic anhydrase lowers the activation energy of the reaction by 42 kJ. **14.87** (a) The catalyzed reaction is approximately 10,000,000 times faster at 25 °C (b) The catalyzed reaction is 180,000 times faster at 125 °C. **14.91** (a) Rate $= 4.7 \times 10^{-5}\,M/s$ (b, c) $k = 0.84\,M^{-2}s^{-1}$ (d) If the $[NO]$ is increased by a factor of 1.8, the rate would increase by a factor of 3.2. **14.95** (a) The reaction is second order in NO_2. (b) If $[NO_2]_0 = 0.100\,M$ and $[NO_2]_t = 0.025\,M$, use the integrated form of the second-order rate equation to solve for t. $t = 48\,s$ **14.99** (a) The half-life of ^{241}Am is 4.3×10^2 yr, that of ^{125}I is 63 days (b) ^{125}I decays at a much faster rate. (c) 0.13 mg of each isotope remains after 3 half-lives. (d) The amount of ^{241}Am remaining after 4 days is 1.00 mg. The amount of ^{125}I remaining after 4 days is 0.96 mg. **14.103** (a) The plot of $1/[C_5H_6]$ versus time is linear and the reaction is second order. (b) $k = 0.167\,M^{-1}\,s^{-1}$ **14.107** (a) When the two elementary reactions are added, $N_2O_2(g)$ appears on both sides and cancels, resulting in the overall reaction. $2\,NO(g) + H_2(g) \longrightarrow N_2O(g) + H_2O(g)$ (b) First reaction, $-[NO]\Delta t = k[NO]^2$, second reaction, $-[H_2]/\Delta t = k[H_2]N_2O_2$ (c) N_2O_2 is the intermediate. (d) Because $[H_2]$ appears in the rate law, the second step must be slow relative to the first. **14.110** (a) $Cl_2(g) + CHCl_3(g) \longrightarrow HCl(g) + CCl_4(g)$ (b) $Cl(g)$, $CCl_3(g)$ (c) reaction 1, unimolecular; reaction 2, bimolecular; reaction 3, bimolecular (d) Reaction 2 is rate determining. (e) Rate $= k[CHCl_3][Cl_2]^{1/2}$. **14.115** The enzyme must lower the activation energy by 22 kJ in order to make it useful. **14.120** (a) $k = 8 \times 10^7\,M^{-1}s^{-1}$

(b) :N=O:

:Ö=N—F̈: ⟷ (:Ö—N=F̈:)

(c) NOF is bent with a bond angle of approximately 120°

(d)
$$\left[\begin{array}{c} O{=}N \\ F{-}F \end{array} \right]$$

(e) The electron-deficient NO molecule is attracted to electron-rich F_2 so the driving force for formation of the transition state is greater than simple random collisions.

14.123 (a)

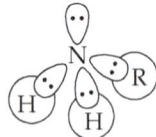

(b) A reactant that is attracted to the lone pair of electrons on nitrogen will produce a tetrahedral intermediate. This can be a moiety with a full, partial, or even transient positive charge.

Chapter 15
15.1 (a) $k_f > k_r$ **(b)** The equilibrium constant is greater than one. **15.3** K is greater than one. **15.5** $A(g) + B(g) \rightleftharpoons AB(g)$ has the larger equilibrium constant. **15.6** From the smallest to the largest equilibrium constant, (c) < (b) < (a). **15.7** Statements (a) and (c) are true. **15.8** Statement (b) is true. **15.11 (a)** The value of K_p at 500 K is less than the value of K_p at 300 K. **(b)** K_c decreases as T increases, so the reaction is exothermic. **15.13 (a)** $K = 8.1 \times 10^{-3}$. **(b)** At equilibrium, the partial pressure of A is greater than the partial pressure of B. **15.15 (a)** $K_c = [N_2O][NO_2]/[NO]^3$; homogeneous **(b)** $K_c = [CS_2][H_2]^4/[CH_4][H_2S]^2$; homogeneous **(c)** $K_c = [CO]^4/[Ni(CO)_4]$; heterogeneous **(d)** $K_c = [H^+][F^-]/[HF]$; homogeneous **(e)** $K_c = [Ag^+]^2/[Zn^{2+}]$; heterogeneous **(f)** $K_c = [H^+][OH^-]$; homogeneous **(g)** $K_c = [H_2O]^2[O_2]/[H_2O_2]^2$; heterogeneous **15.17 (a)** Mostly reactants **(b)** mostly products **15.19 (a)** True **(b)** true **(c)** false **15.21** $K_p = 1.0 \times 10^{-3}$ **15.23 (a)** The equilibrium favors NO and Br_2 at this temperature. **(b)** $K_c = 77$ **(c)** $K_c = 8.8$ **15.25 (a)** $K_p = 0.541$ **(b)** $K_p = 3.42$ **(c)** $K_c = 281$ **15.27** $K_c = 0.14$ **15.29 (a)** $K_p = P_{O_2}$ **(b)** $K_c = [Hg(solv)^4][O_2(solv)]$ **15.31** $K_c = 10.5$ **15.33 (a)** $K_p = 51$ **(b)** $K_c = 2.1 \times 10^3$ **15.35 (a)** $[H_2] = 0.012\,M$, $[N_2] = 0.019\,M$, $[H_2O] = 0.138\,M$ **(b)** $K_c = 653.7 = 7 \times 10^2$ **15.37 (a)** $P_{CO_2} = 4.10$ atm, $P_{H_2} = 2.05$ atm, $P_{H_2O} = 3.28$ atm, **(b)** $P_{CO_2} = 3.87$ atm, $P_{H_2} = 1.82$ atm, $P_{CO} = 0.23$ atm **(c)** $K_p = 0.11$ **(d)** $K_c = 0.11$ **15.39** $K_c = 2.0 \times 10^4$ **15.41 (a)** True **(b)** true **(c)** false **15.43 (a)** $Q = 1.1 \times 10^{-8}$, the reaction will proceed to the left. **(b)** $Q = 5.5 \times 10^{-12}$, the reaction will proceed to the right. **(c)** $Q = 2.19 \times 10^{-10}$, the mixture is at equilibrium. **15.45** $P_{Cl_2} = 5.0$ atm **15.47** $[Br_2] = 0.00767M$, Br $= 0.00282\,M$, 0.0451 g Br(g) **15.49** $[I] = 2.10 \times 10^{-5}\,M$, $[I_2] = 1.43 \times 10^{-5}\,M$, 0.0362 g I_2 **15.51 (a)** $[NH_3] = [H_2S] = 0.011\,M$ **(b)** 0.56 g **15.53** $[Ca^{2+}] = [CrO_4^{2-}] = 0.027\,M$ **15.55** $[NO] = = 0.002\,M$, $[N_2] = [O_2] = 0.087\,M$ **15.57** $[Br_2] = 0.020\,M$, $[Cl_2] = 0.12\,M$, $[BrCl] = 0.13\,M$ **15.59 (a)** $P_{CH_3I} = P_{HI} = 0.422$ torr, $P_{CH_4} = 104.7$ torr, $P_{I_2} = 7.54$ torr **15.61 (a)** Shift equilibrium to the right **(b)** decrease the value of K **(c)** shift equilibrium to the left **(d)** no effect **(e)** no effect **(f)** shift equilibrium to the right **15.63 (a)** No effect **(b)** no effect **(c)** no effect **(d)** increase equilibrium constant **(e)** no effect **15.65 (a)** Δn does not equal zero, so a change in volume at constant temperature will affect the fraction of products in the equilibrium mixture. **(b)** $\Delta H° = -155.7$ kJ **(c)** The reaction is exothermic, so the equilibrium constant will decrease with increasing temperature. **15.67** An increase in pressure by decreasing the volume favors formation of ozone. **15.69 (a)** Endothermic **(b)** The equilibrium constant increases. **(c)** The forward rate constant increases by a larger amount

than the reverse rate constant. **15.73** $K_p = 24.7$; $K_c = 3.67 \times 10^{-3}$ **15.76 (a)** $P_{Br_2} = 0.161$ atm $P_{NO} = 0.628$ atm, $P_{NOBr} = 0.179$ atm; $K_c = 0.0643$ **(b)** $P_t = 0.968$ atm, **(c)** 10.49 g NOBr **15.79** At equilibrium, $P_{IBr} = 0.021$ atm, $P_{I_2} = P_{Br_2} = 1.9 \times 10^{-3}$ atm **15.82** $K_p = 4.33$, $K_c = 0.0480$ **15.86** $[CO_2] = [H_2] = 0.264\,M$, $[CO] = [H_2O] = 0.236\,M$ **15.89 (a)** The fraction of CCl_4 that is converted to C and Cl_2 is 0.26 (26%). **(b)** At equilibrium, $P_{CCl_4} = 1.47$ atm and $P_{Cl_2} = 1.06$ atm. **15.91** Following the addition of 0.200 mol of HI, the equilibrium partial pressures are $P_{H_2} = P_{I_2} = 1.61$ atm and $P_{HI} = 11.2$ atm. **15.94** At equilibrium, $[H_4IO_6^-] = 0.0015\,M$

Chapter 16
16.1 (a) HCl, the H^+ donor, is the Brønsted–Lowry acid. NH_3, the H^+ acceptor, is the Brønsted–Lowry base. **(b)** HCl, the electron pair acceptor, is the Lewis acid. NH_3, the electron pair donor, is the Lewis base. **16.6 (a)** HY is a strong acid. There are no neutral HY molecules in solution, only H^+ cations and Y^- anions. **(b)** HX has the smallest K_a value. It has most neutral acid molecules and fewest ions. **(c)** HX has fewest H^+ and highest pH. **16.12 (a)** Molecule **(b)** is more acidic because its conjugate base is resonance-stabilized and the ionization equilibrium favors the more stable products. **(b)** Increasing the electronegativity of X increases the strength of both acids. As X becomes more electronegative and attracts more electron density, the O—H bond becomes weaker, more polar, and more likely to be ionized. An electronegative X group also stabilizes the anionic conjugate bases by delocalizing the negative charge. The equilibria favor products and the values of K_a increase. **16.13** HCl is the Brønsted–Lowry acid; NH_3 is the Brønsted–Lowry base. **16.15 (a)** CH_3COOH **(b)** H_2O **(c)** $CH_3NH_3^+$ **16.17 (a)** Acid, $Fe(ClO_4)_3$ or Fe^{3+}, base, H_2O **(b)** Acid, H_2O; base, CN^- **(c)** Acid, BF_3; base, $(CH_3)_3\,N$ **(d)** Acid, HIO; base, NH_2^- **16.19 (a) (i)** IO_3^- **(ii)** NH_3 **(b) (i)** OH^- **(ii)** H_3PO_4

16.21

Acid	+	Base	$\rightleftharpoons$	Conjugate Acid	+	Conjugate Base
(a) $NH_4^+(aq)$		$CN^-(aq)$		HCN(aq)		$NH_3(aq)$
(b) $H_2O(l)$		$(CH_3)_3N(aq)$		$(CH_3)_3NH^+(aq)$		$OH^-(aq)$
(c) HCOOH(aq)		$PO_4^{3-}(aq)$		$HPO_4^{2-}(aq)$		$HCOO^-(aq)$

16.23 (a) Acid: $HSO_3^-(aq) + H_2O(l) \rightleftharpoons SO_3^{2-}(aq) + H_3O^+(aq)$ Base: $HSO_3^-(aq) + H_2O(l) \rightleftharpoons H_2SO_3(aq) + OH^-(aq)$ **(b)** H_2SO_3 is the conjugate acid of $H_2SO_3^{2-}$. SO_3^{2-} is the conjugate base of HSO_3^- **16.25 (a)** CH_3COOH, weak base; CH_3COOH, weak acid **(b)** HCO_3^-, weak base; H_2CO_3 weak acid **(c)** O_2^-, strong base; OH^-, strong base **(d)** $C1^-$, negligible base; HCl, strong acid **(e)** NH_3, weak base; NH_4^+, weak acid **16.27 (a)** HBr **(b)** F^- **16.25 (a)** $OH^-(aq) + OH^-(aq)$, the equilibrium lies to the right. **(b)** $H_2S^-(aq) + CH_3COO^-(aq)$, the equilibrium lies to the right. **(c)** $HNO_3(aq) + OH^-(aq)$, the equilibrium lies to the left. **16.31** Statement (ii) is correct. **16.33 (a)** $[H^+] = 2.2 \times 10^{-11}\,M$, basic **(b)** $[H^+] = 1.1 \times 10^{-6}\,M$, acidic **(c)** $[H^+] = 1.1 \times 10^{-8}\,M$, basic **16.35** $[H^+] = [OH^-] = 3.5 \times 10^{-8}\,M$ **16.37 (a)** $[H^+]$ changes by a factor of 100. **(b)** $[H^+]$ changes by a factor of 3.2

16.39

$[H^+]$	$[OH^-]$	pH	pOH	Acidic or Basic
$7.5 \times 10^{-3}\,M$	$1.3 \times 10^{-12}\,M$	2.12	11.88	acidic
$2.8 \times 10^{-5}\,M$	$3.6 \times 10^{-10}\,M$	4.56	9.44	acidic
$5.6 \times 10^{-9}\,M$	$1.8 \times 10^{-6}\,M$	8.25	5.75	basic
$5.0 \times 10^{-9}\,M$	$2.0 \times 10^{-6}\,M$	8.30	5.70	basic

16.41 $[H^+] = 4.0 \times 10^{-8}\,M$, $[OH^-] = 6.0 \times 10^{-7}\,M$, pOH $= 6.22$
16.43 (a) Acidic **(b)** The range of possible integer pH values for the solution is 4-6. **(c)** Methyl red would help determine the pH of the solution more precisely. **16.45 (a)** True **(b)** true **(c)** false
16.47 (a) $[H^+] = 8.5 \times 10^{-3}\,M$, pH $= 2.07$
(b) $[H^+] = 0.0419\,M$, pH $= 1.377$
(c) $[H^+] = 0.0250\,M$, pH $= 1.602$
(d) $[H^+] = 0.167\,M$, pH $= 0.778$
16.49 (a) $[OH^-] = 3.0 \times 10^{-3}\,M$, pH $= 11.48$
(b) $[OH^-] = 0.3758\,M$, pH $= 13.5750$
(c) $[OH^-] = 8.75 \times 10^{-5}\,M$, pH $= 9.942$
(d) $[OH^-] = 0.17\,M$, pH $= 13.23$
16.47 $3.2 \times 10^{-3}\,M$ NaOH
16.53 (a) $HBrO_2(aq) \rightleftharpoons H^+(aq) + BrO_2^-(aq)$,
$K_a = [H^+][BrO_2^-]/[HBrO_2]$;
$HBrO_2(aq) + H_2O(l) \rightleftharpoons H_3O^+(aq) + BrO_2^-(aq)$
$K_a = [H_3O^+][BrO_2^-]/[HBrO_2]$
(b) $C_2H_5COOH(aq) \rightleftharpoons H^+(aq) + C_2H_5COO^-(aq)$
$K_a = [H^+][C_2H_5COO^-]/[C_2H_5COOH]$;
$C_2H_5COOH(aq) + H_2O(l) \rightleftharpoons H_3O^+(aq) + C_2H_5COO^-(aq)$
$K_a = [H_3O^+][C_2H_5COO^-]/[C_2H_5COOH]$
16.55 $K_a = 1.4 \times 10^{-4}$ **16.57** $[H^+] = [ClCH_2COO^-] = 0.0110\,M$, $[ClCH_2COOH] = 0.089\,M$, $K_a = 1.4 \times 10^{-3}$
16.59 $0.089\,M$ CH_3COOH
16.61 $[H^+] = [C_6H_5COO^-] = 1.8 \times 10^{-3}\,M$,
$[C_6H_5COOH] = 0.048\,M$
16.63 (a) $[H^+] = 1.1 \times 10^{-3}\,M$, pH $= 2.95$
(b) $[H^+] = 1.7 \times 10^{-4}\,M$, pH $= 3.76$
(c) $[OH^-] = 1.4 \times 10^{-5}\,M$, pH $= 9.15$
16.65 $[H^+] = 2.0 \times 10^{-2}\,M$, pH $= 1.71$
16.67 (a) $[H^+] = 2.8 \times 10^{-3}\,M$, 0.69% ionization
(b) $[H^+] = 1.4 \times 10^{-3}\,M$, 1.4% ionization
(c) $[H^+] = 8.7 \times 10^{-4}\,M$, 2.2% ionization
16.69 (a) $[H^+] = 5.1 \times 10^{-3}\,M$, pH $= 2.30$. **(b)** Yes. We started the calculation by assuming that only the first step made a significant contribution to $[H^+]$ and pH. Calculation proved this assumption to be true. Next we assumed $[H^+]$ from the first ionization was small relative to $0.040\,M$ citric acid; this assumption was not valid. Finally we assumed that additional ionization of $[H_2C_6H_5O_7^{2-}]$ was small, which was true. **(c)** $[C_6H_5O_7^{3-}]$ is much less than $[H^+]$.
16.71 (a) $HONH_3^+$ **(b)** When hydroxylamine acts as a base, the nitrogen atom accepts a proton. **(c)** In hydroxylamine, O and N are the atoms with nonbonding electron pairs; in the neutral molecule both have zero formal charges. Nitrogen is less electronegative than oxygen and more likely to share a lone pair of electrons with an incoming (and electron-deficient) H^+. The resulting cation with the $+1$ formal charge on N is more stable than the one with the $+1$ formal charge on O.
16.73 (a) $(CH_3)_2NH(aq) + H_2O(l) \rightleftharpoons (CH_3)_2NH_2^+(aq) + OH^-(aq)$;
$K_b = [(CH_3)_2NH_2^+][OH^-]/[(CH_3)_2NH]$
(b) $CO_3^{2-}(aq) + H_2O(l) \rightleftharpoons HCO_3^-(aq) + OH^-(aq)$;
$K_b = [HCO_3^-][OH^-]/[(CH_3)_2NH]$
(c) $HCOO^-(aq) + H_2O(l) \rightleftharpoons HCOOH(aq) + OH^-(aq)$;
$K_b = [HCOOH][OH^-]/[HCOO^-]$
16.75 From the quadratic formula, $[OH^-] = 6.6 \times 10^{-3}\,M$, pH $= 11.82$. **16.77 (a)** $[C_{10}H_{15}ON] = 0.033\,M$, $[C_{10}H_{15}ONH^+] = [OH^+] = 2.1 \times 10^{-3}\,M$ **(b)** $K_b = 1.4 \times 10^{-4}$
16.79 (a) $C_6H_5OH(aq) + H_2O(l) \rightleftharpoons H_3O^+(aq) + C_6H_5O^-(aq)$
(b) $K_b = 7.7 \times 10^{-5}$ **(c)** Phenol is a stronger acid than water.
16.81 (a) Acetic acid is stronger. (b) Hypochlorite ion is the stronger base. **(c)** For CH_3COO^-, $K_b = 5.6 \times 10^{-10}$; for ClO^-, $K_b = 3.3 \times 10^{-7}$.
16.83 (a) $[OH^-] = 6.3 \times 10^{-4}\,M$, pH $= 10.80$
(b) $[OH^-] = 9.2 \times 10^{-5}\,M$, pH $= 9.96$
(c) $[OH^-] = 3.3 \times 10^{-6}\,M$, pH $= 8.52$

16.85 $4.5\,M$ $NaCH_3COO$ **16.87 (a)** Acidic **(b)** acidic **(c)** basic **(d)** neutral **(e)** acidic **16.89 (a)** Cu^{2+}, higher cation charge **(b)** Fe^{3+}, higher cation charge **(c)** Al^{3+}, smaller cation radius, same charge
16.91 K_b for the anion of the unknown salt is 1.4×10^{-11}; K_b for F^- is 1.5×10^{-11}; The unknown salt is NaF. **16.93 (a)** HNO_3 is a stronger acid than HNO_2 **(b)** H_2S is a stronger acid than H_2O. **(c)** H_2SO_4 is a stronger acid than H_2SeO_4. **(d)** CCl_3COOH is stronger than CH_3COOH **16.95 (a)** BrO^- **(b)** BrO^- **(c)** HPO_4^{2-}
16.97 (a) True **(b)** False. In a series of acids that have the same central atom, acid strength increases with the number of nonprotonated oxygen atoms bonded to the central atom. **(c)** False. H_2Te is a stronger acid than H_2S because the H—Te bond is longer, weaker, and more easily ionized than the H—S bond.
16.93 $NH_3(aq) + H_2O(l) \rightleftharpoons NH_4^+(aq) + OH^-(aq)$. Ammonia, NH_3, acts as an Arrhenius base because it increases the concentration of hydroxide ion, OH^-, in aqueous solution. It acts like a Brønsted-Lowry base because it is a proton, H^+, acceptor. It acts like a Lewis base because it is an electron pair donor. **16.101** $K = 3.3 \times 10^7$
16.105 pH $- 7.01$ (not 5.40, from the typical calculation, which does not make sense.) Usually we assume that $[H^+]$ and $[OH^-]$ from the autoionization of water do not contribute to the overall $[H^+]$ and $[OH^-]$. However, for acid or base solute concentrations less than $1 \times 10^{-6}\,M$, the autoionization of water produces significant $[H^+]$ and $[OH^-]$ and we must consider it when calculating pH.
16.108 (a) $pK_b = 9.16$ **(b)** pH $= 3.07$ **(c)** pH $= 8.77$
16.112 Nicotine, protonated; caffeine, neutral base; strychnine, protonated; quinine, protonated **16.115** 6.0×10^{13} H^+ ions
16.117 (a) To the precision of the reported data, the pH of rainwater 40 years ago was 5.4, no different from the pH today. With extra significant figures, $[H^+] = 3.61 \times 10^{-6}\,M$, pH $= 5.443$ **(b)** A 20.0-L bucket of today's rainwater contains 0.02 L (with extra significant figures, 0.0200 L) of dissolved CO_2.

Chapter 17 **17.1** The middle box has the highest pH. For equal amounts of acid HX, the greater the amount of conjugate base X^-, the smaller the amount of H^+ and the higher the pH. **17.7 (a)** The red curve corresponds to the more concentrated acid solution. **(b)** On the titration curve of a weak acid, pH $= pK_a$ at the volume halfway to the equivalence point. Reading the pK_a values from the two curves, the red curve has the smaller pK_a and the larger K_a. **17.9 (a)** The right-most diagram represents the solubility of $BaCO_3$ as HNO_3 is added. **(b)** The left-most diagram represents the solubility of $BaCO_3$ as Na_2CO_3 is added. **(c)** The center diagram represents the solubility of $BaCO_3$ as $NaNO_3$ is added. **17.13** Statement **(a)** is most correct.
17.15 (a) $[H^+] = 1.8 \times 10^{-5}\,M$, pH $= 4.73$ **(b)** $[OH^-] = 4.8 \times 10^{-5}\,M$, pH $= 9.68$ **(c)** $[H^+] = 1.4 \times 10^{-5}\,M$, pH $= 4.87$ **17.17 (a)** 4.5% ionization **(b)** 0.018% ionization **17.19** Only solution **(a)** is a buffer.
17.21 (a) pH $= 3.82$ **(b)** pH $= 3.96$ **17.23 (a)** pH $= 5.26$
(b) $Na^+(aq) + CH_3COO^-(aq) + H^+(aq) + Cl^-(aq) \longrightarrow CH_3COOH(aq) + Na^+(aq) + Cl^-(aq)$
(c) $CH_3COOH(aq) + Na^+(aq) + OH^-(aq) \longrightarrow CH_3COO^-(aq) + H_2O(l) + Na^+(aq)$
17.25 (a) pH $= 1.58$ **(b)** 36 g NaF **17.27 (a)** pH $= 4.86$
(b) pH $= 5.0$ **(c)** pH $= 4.71$ **17.29 (a)** $[HCO_3^-]/[H_2CO_3] = 11$
(b) $[HCO_3^-]/[H_2CO_3] = 5.4$ **17.31** 360 mL of $0.10\,M$ HCOONa, 640 mL of $0.10\,M$ HCOOH **17.33 (a)** Curve B **(b)** pH at the approximate equivalence point of curve A $= 8.0$, pH at the approximate equivalence point of curve B $= 7.0$ **(c)** For equal volumes of A and B, the concentration of acid B is greater, since it requires a larger volume of base to reach the equivalence point. **(d)** The pK_a value of the weak acid is approximately 4.5 **17.35 (a)** False **(b)** true **(c)** true
17.37 (a) Above pH 7 **(b)** below pH 7 **(c)** at pH 7 **17.39** The second color change of thymol blue is in the correct pH range to show the

17.41 (a) 42.4 mL NaOH soln **(b)** 35.0 mL NaOH soln **(c)** 29.8 mL NaOH soln **17.43 (a)** pH = 1.54 **(b)** pH = 3.30 **(c)** pH = 7.00 **(d)** pH = 10.69 **(e)** pH = 12.74 **17.45 (a)** pH = 2.78 **(b)** pH = 4.74 **(c)** pH = 6.58 **(d)** pH = 8.81 **(e)** pH = 11.03 **(f)** pH = 12.42 **17.47 (a)** pH = 7.00 **(b)** $[HONH_3^+]$ = 0.100 M, pH = 3.52 **(c)** $[C_6H_5NH_3^+]$ = 0.100 M, pH = 2.82 **17.49 (a)** True **(b)** false **(c)** false **(d)** true **17.51** $K_{sp} = [Ag^+][I^-]$; $K_{sp} = [Sr^{2+}][SO_4^{2-}]$; $K_{sp} = [Fe^{2+}][OH^-]^2$; $K_{sp} = [Hg_2^{2+}][Br^-]^2$ **17.53 (a)** $K_{sp} = 7.63 \times 10^{-9}$ **(b)** $K_{sp} = 2.7 \times 10^{-9}$ **(c)** 5.3×10^{-4} mol Ba(IO_3)_2/L **17.55** $K = 2.3 \times 10^{-9}$ **17.57 (a)** 7.1×10^{-7} mol AgBr/L **(b)** 1.7×10^{-11} mol AgBr/L **(c)** 5.0×10^{-12} mol AgBr/L **17.59 (a)** The amount of $CaF_2(s)$ on the bottom of the beaker increases. **(b)** The $[Ca^{2+}]$ in solution increases. **(c)** The $[F^-]$ in solution decreases. **17.61 (a)** 1.4×10^3 g Mn(OH)_2/L **(b)** 0.014 g/L **(c)** 3.6×10^{-7} g/L **17.63** More soluble in acid: **(a)** $ZnCO_3$ **(b)** ZnS **(d)** AgCN **(e)** Ba_3(PO_4)_2 **17.65** $[Ni^{2+}] = 1.3 \times 10^{-6} M$, $[Ni(NH_3)_6^{2+}] = 0.0964 M$ **17.67 (a)** 9.1×10^{-9} mol AgI per L pure water **(b)** $K = K_{sp} \times K_f = 8 \times 10^4$ **(c)** 0.0500 mol AgI per L 0.100 M NaCN **17.69 (a)** $Q < K_{sp}$; no Ca(OH)_2 precipitates **(b)** $Q < K_{sp}$; no Ag_2SO_4 precipitates **17.71** pH = 11.5 **17.73** AgI will precipitate first, at $[I^-] = 4.2 \times 10^{-13} M$. **17.75** AgCl will precipitate first. **17.77** Ag^+ (Group 1) and Mg^{2+} (Group 4) are definitely absent. Al^{3+} (Group 3) is definitely present and Na^+ (Group 5) is possibly present. **17.79 (a)** Make the solution acidic with 0.2 M HCl; saturate with H_2S. CdS will precipitate; ZnS will not. **(b)** Add excess base; Fe(OH)_3(s) precipitates, but Cr^{3+} forms the soluble complex $Cr(OH)_4^-$. **(c)** Add $(NH_4)_2HPO_4$; Mg^{2+} precipitates as $MgNH_4PO_4$; K^+ remains soluble. **(d)** Add 6 M HCl; precipitate Ag^+ as AgCl(s); Mn^{2+} remains soluble. **17.81 (a)** Base is required to increase $[PO_4^{3-}]$ so that the solubility product of the metal phosphates of interest is exceeded and the phosphate salts precipitate. **(b)** K_{sp} for the cations in group 3 is much larger; to exceed K_{sp}, a higher $[S^{2-}]$ is required. **(c)** They should all redissolve in strongly acidic solution. **17.83** Choice (b) is correct. **17.89 (a)** 6.5 mL glacial acetic acid, 5.25 g CH_3COONa **17.91 (a)** The molar mass of the acid is 94.6 g/mol. **(b)** $K_a = 1.4 \times 10^{-3}$ **17.97 (a)** $[Pb^{2+}] = 2.7 \times 10^{-7} M$ **(b)** 56 ppb Pb^{2+} **(c)** The solubility of $PbCO_3$ increases as pH is lowered. **(d)** A saturated solution of lead carbonate, with a lead concentration of 56 ppb, exceeds the EPA acceptable lead level of 15 ppb. **17.102** The solubility of Mg(OH)_2 in 0.50 M NH_4Cl is 0.11 mol/L **17.108** $[OH^-] \geq 1.0 \times 10^{-2} M$ or pH $\geq$ 12.02 **17.115 (a)** $[Ag^+]$ from AgCl is 1.4×10^3 ppb or 1.4 ppm. **(b)** $[Ag^+]$ from AgBr is 76 ppb. **(c)** $[Ag^+]$ from AgI is 0.98 ppb. AgBr would maintain $[Ag^+]$ in the correct range.

Chapter 18

18.1 (a) A volume greater than 22.4 L **(b)** No. The relative volumes of one mole of an ideal gas at 50 km and 85 km depend on the temperature and pressure at the two altitudes. From Figure **18.1**, the gas will occupy a much larger volume at 85 km than at 50 km. **(c)** We expect gases to behave most ideally in the thermosphere, around the stratopause and in the troposphere at low altitude. **18.3 (a)** A = troposphere, 0–10 km; B = stratosphere, 12–50 km; C = mesosphere, 50–85 km **(b)** Ozone is a pollutant in the troposphere and filters UV radiation in the stratosphere. **(c)** The troposphere **(d)** Only region C in the diagram is involved in an aurora borealis, assuming a narrow "boundary" between the stratosphere and mesosphere at 50 km. **(e)** The concentration of water vapor is greatest near Earth's surface in region A and decreases with altitude. Water's single bonds are susceptible to photodissociation in regions B and C, so its concentration is likely to be very low in these regions. The relative concentration of CO_2, with strong double bonds, increases in regions B and C, because it is less susceptible to photodissociation. **18.4** The Sun **18.6** $CO_2(g)$

dissolves in seawater to form $H_2CO_3(aq)$. The basic pH of the ocean encourages ionization of $H_2CO_3(aq)$ to form $HCO_3^-(aq)$ and $CO_3^{2-}(aq)$. Under the correct conditions, carbon is removed from the ocean as $CaCO_3(s)$ (sea shells, coral, chalk cliffs). As carbon is removed, more $CO_2(g)$ dissolves to maintain the balance of complex and interacting acid-base and precipitation equilibria. **18.7** Above ground, evaporation of petroleum gases and hydrogen sulfide at the well head, evaporation of volatile petroleum products and organic compounds from waste ponds, as well as leakage and overflow from the ponds are potential sources of contamination. Below ground, petroleum gases and fracking liquid can migrate into groundwater, both deep and shallow aquifers. **18.9 (a)** Its temperature profile **(b)** troposphere, 0 to 12 km; stratosphere, 12 to 50 km; mesosphere, 50 to 85 km; thermosphere, 85 to 110 km **18.11 (a)** The partial pressure of O_3 is 3.0×10^{-7} atm (2.2×10^{-4} torr). **(b)** 7.3×10^{15} O_3 molecules/1.0 L air **18.13** 8.7×10^{16} CO molecules/1.0 L air **18.15 (a)** 570 nm **(b)** visible electromagnetic radiation **18.17 (a)** *Photodissociation* is cleavage of a bond such that two neutral species are produced. *Photoionization* is absorption of a photon with sufficient energy to eject an electron, producing an ion and the ejected electron. **(b)** Photoionization of O_2 requires 1205 kJ/mol. Photodissociation requires only 495 kJ/mol. At lower elevations, high-energy short-wavelength solar radiation has already been absorbed. Below 90 km, the increased concentration of O_2 and the availability of longer-wavelength radiation cause the photodissociation process to dominate. **18.19 (a)** A wavelength of 145 nm is in the ultraviolet portion of the electromagnetic spectrum. **(b)** The energy of one mole of 145-nm photons is 826 kJ. This is more than enough energy to photodissociate O_2, but not enough energy to photoionize O_2. **18.21** Ozone depletion reactions, which involve only O_3, O_2, or O (oxidation state = 0), do not involve a change in oxidation state for oxygen atoms. Reactions involving ClO and one of the oxygen species with a zero oxidation state do involve a change in the oxidation state of oxygen atoms. **18.23 (a)** A chlorofluorocarbon is a compound that contains chlorine, fluorine, and carbon, while a hydrofluorocarbon is a compound that contains hydrogen, fluorine, and carbon. An HFC contains hydrogen in place of the chlorine present in a CFC. **(b)** HFCs are potentially less harmful to the ozone layer than CFCs because their photodissociation does not produce Cl atoms, which catalyze the destruction of ozone. **18.25 (a)** The maximum wavelength of a single photon that will break a C—F bond is 247 nm. The maximum wavelength of a single photon that will break a C—Cl bond is 365 nm. **(b)** We don't expect the photodissociation of C—F bonds to be significant in the lower atmosphere. (The photodissociation of C—Cl bonds will be significant.) **18.27** $2 NO_2(g) + H_2O(l) \rightleftharpoons HNO_2 + HNO_3(aq)$; $2 NO(g) + O_2 + H_2O(l) \rightleftharpoons HNO_2(aq) + HNO_3$ **18.29 (a)** $H_2SO_4(aq) + CaCO_3(s) \longrightarrow CaSO_4(s) + H_2O(l) + CO_2(g)$ **(b)** The $CaSO_4(s)$ would be much less reactive with acidic solution, since it would require a strongly acidic solution to shift the relevant equilibrium to the right: $CaSO_4(s) + 2H^+(aq) \rightleftharpoons Ca^{2+}(aq) + 2HSO_4^-(aq)$. $CaSO_4$ would protect $CaCO_3$ from attack by acid rain, but it would not provide the structural strength of limestone. **18.31 (a)** Ultraviolet **(b)** 357 kJ/mol **(c)** The average C—H bond energy from Table **8.3** is 413 kJ/mol. The C—H bond energy in CH_2O, 357 kJ/mol, is less than the "average" C—H bond energy.

(d)

$$H\!-\!\overset{\displaystyle :\!\overset{..}{O}\!:}{\underset{\|}{C}}\!-\!H + h\nu \longrightarrow H\!-\!\overset{\displaystyle :\!\overset{..}{O}\!:}{\underset{\|}{C}}\!\cdot + H\!\cdot$$

18.33 (a) 235 W/m². **(b)** Four sources, surface radiation, evapotranspiration, incoming solar radiation and convective heating, transfer energy to the atmosphere. Surface radiation makes the largest contribution and convective heating the smallest. (**c**) Of the 519 W/m² absorbed by the atmosphere, 324 W/m² are radiated back to the surface. The percentage is 62.4%. **18.35** 0.099 M Na⁺ **18.37** (**a**) 3.22 × 10³ gH₂O (**b**) The final temperature is 43.4 °C. **18.39** 4.361 × 10⁵ g CaO **18.41** (**a**) The total cation charge is 5.866 = 5.9 C; the total anion charge is 5.847 = 5.8 C. The two numbers vary in the third significant figure. This is not surprising, because the molarities of the various ions are given to two significant figures. **18.43** (**a**) $CO_2(g)$, HCO_3^-, $H_2O(l)$, SO_4^{2-}, NO_3^-, HPO_4^{2-}, $H_2PO_4^-$ (**b**) $CH_4(g)$, $H_2S(g)$, $NH_3(g)$, $PH_3(g)$, **18.45** 25.1 g O₂ **18.47** 0.42 mol Ca(OH)₂, 0.18 mol Na₂CO₃ **18.49** (**a**) *Trihalomethanes* are the by-products of water chlorination; they contain one central carbon atom bound to one hydrogen and three halogen atoms.

(**b**)

H
|
Cl—C—Cl
|
Cl

H
|
Cl—C—Br
|
Cl

18.51 The fewer steps in a process, the less waste is generated. Processes with fewer steps require less energy at the site of the process and for subsequent cleanup or disposal of waste. **18.53** (**a**) H₂O (**b**) It is better to prevent waste than to treat it. Atom economy. Less hazardous chemical synthesis and inherently safer for accident prevention. Catalysis and design for energy efficiency. Raw materials should be renewable. **18.55** (**a**) Water as a solvent, by criteria 5, 7, and 12. (**b**) Reaction temperature of 500 K, by criteria 6, 12, and 1. (**c**) Sodium chloride as a by-product, according to criteria 1, 3, and 12. **18.59** Multiply equations 18.7 and 18.9 by a factor of 2. Add these two equations to a third equation, $O(g) + O(g) \longrightarrow O_2(g)$. 2 Cl (g) and 2 ClO (g) cancel from each side of the resulting equation to produce equation 18.10. **18.62** (**a**) A CFC has C—Cl bonds and C—F bonds. In an HFC, the C—Cl bonds are replaced by C—H bonds. (**b**) Stratospheric lifetime is significant because, the longer a halogen-containing molecule exists in the stratosphere, the greater the likelihood that it will encounter light with energy sufficient to dissociate a carbon-halogen bond. Free halogen atoms are the bad actors in ozone destruction. (**c**) HFCs have replaced CFCs because it is infrequent that light with energy sufficient to dissociate a C—F bond will reach an HFC molecule. F atoms are much less likely than Cl atoms to be produced by photodissociation in the stratosphere. (**d**) The main disadvantage of HFCs as replacements for CFCs is that they are potent greenhouse gases. **18.64** (**a**) $CH_4(g) + 2 O_2(g) \longrightarrow CO_2(g) + 2 H_2O(g)$ (**b**) $2 CH_4(g) + 3 O_2(g) \longrightarrow 2 CO_2(g) + 4 H_2O(g)$ (**c**) 9.5 L of dry air **18.68** 7.1 × 10⁸ m² **18.70** (**a**) CO_3^{2-} is a relatively strong Brønsted–Lowry base and produces OH⁻ in aqueous solution. If $[OH^-(aq)]$ is sufficient for the reaction quotient to exceed K_{sp} for Mg(OH)₂, the solid will precipitate. (**b**) At these ion concentrations, $Q > K_{sp}$ and Mg(OH)₂ will precipitate. **18.72** (**a**) 2.5 × 10⁷ ton CO₂, 4.2 × 10⁵ ton SO₂ (**b**) 4.3 × 10⁵ ton CaSO₃ **18.74** (**a**) H—Ö—H $\longrightarrow$ H· + ·Ö—H (**b**) 258 nm (**c**) The overall reaction is $O_3(g) + O(g) \longrightarrow 2 O_2(g)$. OH$(g)$ is the catalyst in the overall reaction because it is consumed and then reproduced. **18.75** The enthalpy change for the first step is −141 kJ, for the second step, −249 kJ, for the overall reaction, −390 kJ. **18.77** (**a**) Rate = $k[O_3][H]$ (**b**) k_{avg} = 1.13 × 10⁴⁴ M^{-1} s⁻¹ **18.80** (**a**) Process (**i**) is greener because it involves neither the toxic reactant phosgene nor the by-product HCl. (**b**) Reaction (**i**): C in CO₂ is linear with sp hybridization; C in R—N=C=O is linear with sp hybridization; C in the urethane monomer is trigonal planar with sp^2 hybridization. Reaction (ii): C in COCl₂ is trigonal planar with sp^2 hybridization; C in R—N=C=O is linear with sp hybridization; C in the urethane monomer is trigonal planar with

sp^2 hybridization. (**c**) The greenest way to promote formation of the isocyanate is to remove by-product, either water or HCl, from the reaction mixture.

Chapter 19

19.1 (**a**)

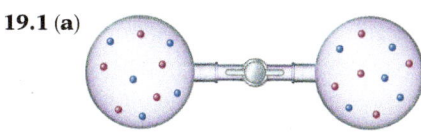

(**b**) $\Delta H = 0$ for mixing ideal gases. ΔS is positive because the disorder of the system increases. (**c**) The process is spontaneous and therefore irreversible. (**d**) Since $\Delta H = 0$, the process does not affect the entropy of the surroundings. **19.4** Both ΔH and ΔS are positive. The net change in the chemical reaction is the breaking of five blue-blue bonds. Enthalpies for bond breaking are always positive. There are twice as many molecules of gas in the products, so ΔS is positive for this reaction. **19.7** Statements (**b**) and (**c**) are true. **19.10** (**a**) True (**b**) false (**c**) false (**d**) false (**e**) true **19.11** Spontaneous: a, b, c, d; nonspontaneous: e **19.13** (**a**) True, assuming that the forward and reverse processes occur at the same conditions. (**b**) false (**c**) false (**d**) true (**e**) false **19.15** (**a**) Endothermic (**b**) above 100 °C (**c**) below 100 °C (**d**) at 100 °C **19.17** Statement (a) is true because ΔE is a state function. Statements (b) and (c) are true because it is specified that the reversible path has been taken in both directions. **19.19** (**a**) The melting of water is spontaneous at temperatures above 0 °C. (**b**) The process is irreversible. (**c**) The process is endothermic—the ice absorbs heat when it melts, so $\Delta H > 0$. (**d**) The entropy of liquid water is greater than that of ice, so $\Delta S > 0$ for the process. **19.21** (**a**) True (**b**) false (**c**) true (**d**) false **19.23** (**a**) Entropy increases. (**b**) 89.2 J/K **19.25** (**a**) False (**b**) true (**c**) false **19.27** (**a**) Positive ΔS (**b**) ΔS = 1.02 J/K (**c**) Statements (i) and (iii) are true. **19.29** (**a**) Yes, the expansion is spontaneous. (**b**) As the ideal gas expands into the vacuum, there is nothing for it to "push back," so no work is done. Mathematically, $w = -P_{ext}\Delta V$. Since the gas expands into a vacuum, $P_{ext} = 0$ and $w = 0$. (**c**) The "driving force" for the expansion of the gas is the increase in entropy. **19.31** (**a**) An increase in temperature produces more available microstates for a system. (**b**) A decrease in volume produces fewer available microstates for a system. (**c**) Going from liquid to gas, the number of available microstates increases. **19.33** (**a**) ΔS is positive. (**b**) S of the system clearly increases in 19.11 (**a**), (**b**), and (**e**); it clearly decreases in 19.11 (**c**). The definition of the system in (**d**) is problematic. **19.35** S increases in (**a**) and (**c**); S decreases in (**b**). **19.37** (**a**) False (**b**) true (**c**) false (**d**) true **19.39** (**a**) Ar(g) (**b**) He(g) at 1.5 atm (**c**) 1 mol of Ne(g) in 15.0 L (**d**) CO₂(g) **19.41** (**a**) $\Delta S < 0$ (**b**) $\Delta S > 0$ (**c**) $\Delta S < 0$ (**d**) $\Delta S \approx 0$ **19.43** Statements (**b**) and (**c**) are true. **19.45** (**a**) Br₂(g) (**b**) C₂H₅OH(l) (**c**) SiCl₄(l) (**d**) Fe(s) at 500 °C **19.47** (**a**) Sc(g) will have the higher standard enthalpy at 25 °C. Sc(s), 34.6 J/mol − K; Sc(g), 174.7 J/mol − K. (**b**) NH₃(g) will have the higher standard enthalpy at 25 °C. NH₃(g), 192.5 J/mol − K; NH₃(aq), 111.3 J/mol − K (**c**) O₃(g) will have the higher standard enthalpy at 25 °C. O₂(g), 205.0 J/K; O₃(g), 237.6 J/K. (**d**) C (graphite) will have the higher standard enthalpy at 25 °C. C (diamond), 2.43 J/mol-K; C (graphite), 5.69 J/mol-K. **19.49** $\Delta S > 0$ for the dissolution of NH₄NO₃(s) in water. **19.51** (**a**) $\Delta S°$ = −120.6 J/K. $\Delta S°$ is negative because there are fewer moles of gas in the products. (**b**) $\Delta S°$ = +176.6 J/K. $\Delta S°$ is positive because there are more moles of gas in the products. (**c**) $\Delta S°$ = +152.39 J/K. $\Delta S°$ is positive because the product contains more total particles and more moles of gas. (**d**) $\Delta S°$ = +91.9 J/K. $\Delta S°$ is positive because there are more moles of gas in the products. **19.53** (**a**) Yes. $\Delta G = \Delta H - T\Delta S$ (**b**) No. If ΔG is positive, the process is nonspontaneous. (**c**) No. There is no relationship between ΔG and rate of reaction. **19.55** (**a**) Exothermic

(b) $\Delta S°$ is negative; the reaction leads to a decrease in disorder.
(c) $\Delta G° = -9.9$ kJ (d) If all reactants and products are present in their standard states, the reaction is spontaneous in the forward direction at this temperature. **19.57** (a) $\Delta H° = -546.60$ kJ, $\Delta S° = 14.1$ J/K, $\Delta G° = -550.80$ kJ, $\Delta G° = \Delta H° - T\Delta S° = -550.80$ kJ
(b) $\Delta H° = -106.7$ kJ, $\Delta S° = -142.3$ J/K $\Delta G° = -64.0$ kJ, $\Delta G° = \Delta H° - T\Delta S° = -64.3$ kJ (c) $\Delta H° = -508.3$ kJ, $\Delta S° = -179$ J/K, $\Delta G° = -465.8$ kJ, $\Delta G° = \Delta H° - T\Delta S° = -455.0$ kJ. The discrepancy in $\Delta G°$ values is due to experimental uncertainties in the tabulated thermodynamic data. (d) $\Delta H° = -165.9$ kJ, $\Delta S° = 1.3$ J/K, $\Delta G° = -166.1$ kJ, $\Delta G° = \Delta H° - T\Delta S° = -166.3$ kJ
19.59 (a) $\Delta G° = -140.0$ kJ, spontaneous (b) $\Delta G° = +104.70$ kJ, nonspontaneous (c) $\Delta G° = +146$ kJ, nonspontaneous
(d) $\Delta G° = -156.7$ kJ, spontaneous
19.61 (a) $2\,C_8H_{18}(l) + 25\,O_2(g) \longrightarrow 16\,CO_2(g) + 18\,H_2O(l)$
(b) Because $\Delta S°$ is positive, $\Delta G°$ is more negative than $\Delta H°$.
19.63 (a) The forward reaction is spontaneous at low temperatures but becomes nonspontaneous at higher temperatures. (b) The reaction is nonspontaneous in the forward direction at all temperatures.
(c) The forward reaction is nonspontaneous at low temperatures but becomes spontaneous at higher temperatures.**19.65** $\Delta S > 60.8$ J/K
19.67 (a) $T = 330$ K (b) nonspontaneous**19.69** (a) $\Delta H° = 155.7$ kJ, $\Delta S° = 171.4$ kJ. Since $\Delta S°$ is positive, $\Delta G°$ becomes more negative with increasing temperature. (b) $\Delta G° = 19$ kJ. The reaction is not spontaneous under standard conditions at 800 K (c) $\Delta G° = -15.7$ kJ. The reaction is spontaneous under standard conditions at 1000 K. **19.71** (a) $T_b = 353.6$ K $= 80.4\,°C$ (b) From the *NIST Chemistry Webbook* (https://webbook.nist.gov/chemistry/), $T_b = 353.3$ K. The values are remarkably close; the small difference is due to deviation from ideal behavior by $C_6H_6(g)$ and experimental uncertainty in the boiling point measurement and the thermodynamic data.
19.73 (a) $C_2H_2(g) + \frac{5}{2}O_2(g) \longrightarrow 2\,CO_2(g) + H_2O(l)$ (b) -1300.2 kJ of heat produced/mol C_2H_2 burned (c) $w_{max} = -1235.9$ kJ/mol C_2H_2
19.75 (a) ΔG decreases; it becomes more negative. (b) ΔG increases; it becomes more positive. (c) ΔG increases; it becomes more positive.
19.77 (a) $\Delta G° = -5.40$ kJ (b) $\Delta G = 0.30$ kJ **19.79** (a) $\Delta G° = -15.79$ kJ, $K = 590$ (b) $\Delta G° = 8.2$ kJ, $K = 0.036$ (c) $\Delta G° = -500.3$ kJ, $K = 5 \times 10^{87}$ **19.81** $\Delta H° = 269.3$ kJ, $\Delta S° = 0.1719$ kJ/K
(a) $P_{CO_2} = 6.0 \times 10^{-39}$ atm (b) $P_{CO_2} = 1.6 \times 10^{-4}$ atm
19.83 (a) $HNO_2(aq) \rightleftharpoons H^+(aq) + NO_2^-(aq)$ (b) $\Delta G° = 19.1$ kJ
(c) $\Delta G = 0$ at equilibrium (d) $\Delta G = -2.7$ kJ **19.85** (a) The thermodynamic quantities T, E, and S are state functions. (b) The quantities q and w depend on the path taken. (c) There is only one *reversible* path between states. (d) $\Delta E = q_{rev} + w_{max}$, $\Delta S = q_{rev}/T$.
19.88 Statements (a) and (b) are true.
19.93 (a) $\frac{1}{2}N_2(g) + \frac{3}{2}H_2(g) \longrightarrow NH_3(g)$; $C(s) + \frac{1}{2}O_2(g) \longrightarrow CO(g)$
(b) In the first reaction, $\Delta G_f°$ will be more positive (less negative) than $\Delta H_f°$. (c) Condition (iii), when $\Delta S_f°$ is negative. **19.98** (a) For the oxidation of glucose, $\Delta H° = -2803.0$ kJ and $\Delta G° = -2878.8$ kJ.
(b) For the anaerobic decomposition of glucose, $\Delta H° = -68.0$ kJ and $\Delta G° = -226.4$ kJ. (c) $K = 5 \times 10^{39}$. **19.103** (a) For any given total pressure, the condition of equal moles of the two gases can be achieved at some temperature. For individual gas pressures of 1 atm and a total pressure of 2 atm, the mixture is at equilibrium at 328.5 K or 55.5 °C. (b) 333.0 K or 60 °C (c) 374.2 K or 101.2 °C

Chapter 20   20.1 In this analogy, the electron is analogous to the proton (H^+). Redox reactions can be viewed as electron-transfer reactions, just as acid-base reactions can be viewed as proton-transfer reactions. Oxidizing agents are themselves reduced; they gain electrons. A strong oxidizing agent would be analogous

to a strong base. **20.3** (a) The process represents oxidation. (b) The electrode is the anode. (c) When a neutral atom loses a valence electron, the radius of the resulting cation is smaller than the radius of the neutral atom. **20.7** (a) The sign of $\Delta G°$ is positive. (b) The equilibrium constant is less than one. (c) No. An electrochemical cell based on this reaction cannot accomplish work on its surroundings.
20.10 (a) Zinc is the anode. (b) The energy density of the silver oxide battery is most similar to the nickel-cadmium battery. The molar masses of the electrode materials and the cell potentials for these two batteries are most similar. **20.13** (a) *Oxidation* is the loss of electrons. (b) Electrons appear on the products' side (right side).
(c) The *oxidant* is the reactant that is reduced. (d) An *oxidizing agent* is the substance that promotes oxidation; it is the oxidant.
20.15 (a) True (b) false (c) true **20.17** (a) (i) Reactants: I, +5; O, -2; C +2; O, -2. Products: I, 0; C, +4; O, -2 (ii) The total number of electrons transferred is 10. (b) (i) Reactants: Hg, +2; N, -2; H, +1. Products: Hg, 0; N, 0; H, +1. (ii) The total number of electrons transferred is 4. (c) (i) Reactants: H, +1; S, -2; N, +5; O, -2. Products: S, 0; N, +2; O, -2; H, +1; O, -2. (ii) The total electrons transferred is 6. **20.19** (a) No oxidation-reduction (b) Iodine is oxidized from -1 to +5; chlorine is reduced from +1 to -1. (c) Sulfur is oxidized from +4 to +6; nitrogen is reduced from +5 to +2.
20.21 (a) $TiCl_4(g) + 2\,Mg(l) \longrightarrow Ti(s) + 2\,MgCl_2(l)$ (b) $Mg(l)$ is oxidized; $TiCl_4(g)$ is reduced. (c) $Mg(l)$ is the reductant; $TiCl_4(g)$ is the oxidant. **20.23** (a) $Sn^{2+}(aq) \longrightarrow Sn^{4+}(aq) + 2\,e^-$, oxidation
(b) $TiO_2(s) + 4\,H^+(aq) + 2\,e^- \longrightarrow Ti^{2+}(aq) + 2\,H_2O(l)$, reduction
(c) $ClO_3^-(aq) + 6\,H^+(aq) + 6\,e^- \longrightarrow Cl_2(aq) + 3\,H_2O(l)$, reduction
(d) $N_2(g) + 8\,H^+(aq) + 6\,e^- \longrightarrow 2\,NH_4^+(aq)$, reduction
20.25 (a) $O_2(g) + 2\,H_2O(l) + 4\,e^- \longrightarrow 4\,OH^-(aq)$, reduction
(b) $Mn^{2+}(aq) + 4\,OH^-(aq) \longrightarrow MnO_2(s) + 2\,H_2O(l) + 2\,e^-$, oxidation
(c) $Cr(OH)_3(s) + 5\,OH^-(aq) \longrightarrow CrO_4^{2-}(aq) + 4\,H_2O(l) + 3\,e^-$, oxidation (d) $N_2H_4(aq) + 4\,OH^-(aq) \longrightarrow N_2(g) + 4\,H_2O(l) + 4\,e^-$, oxidation **20.27** (a) $Cr_2O_7^{2-}(aq) + I^-(aq) + 8\,H^+(aq) \longrightarrow 2\,Cr^{3+}(aq) + IO_3^-(aq) + 4\,H_2O(l)$; oxidizing agent, $Cr_2O_7^{2-}$; reducing agent, I^- (b) $I_2(s) + 5\,OCl^-(aq) + H_2O(l) \longrightarrow 2\,IO_3^-(aq) + 5\,Cl^-(aq) + 2\,H^+(aq)$; oxidizing agent, OCl^-; reducing agent, I_2 (c) $2\,MnO_4^-(aq) + Br^-(aq) + H_2O(l) \longrightarrow 2\,MnO_2(s) + BrO_3^-(aq) + 2\,OH^-(aq)$; oxidizing agent, MnO_4^-; reducing agent, Br^-

20.27 (a) $3\,NO_2^-(aq) + Cr_2O_7^{2-}(aq) + 8\,H^+(aq) \longrightarrow 2\,Cr^{3+}(aq) + 3\,NO_3^-(aq) + 4\,H_2O(l)$; oxidizing agent, $Cr_2O_7^{2-}$; reducing agent, NO_2^- (b) $2\,Cr_2O_7^{2-}(aq) + 3\,CH_3OH(aq) + 16\,H^+(aq) \longrightarrow 3\,HCOOH(aq) + 4\,Cr^{3+}(aq) + 11\,H_2O(l)$; oxidizing agent, $Cr_2O_7^{2-}$; reducing agent, CH_3OH (c) $NO_2^-(aq) + 2\,Al(s) + 2\,H_2O(l) \longrightarrow NH_4^+(aq) + 2\,AlO_2^-(aq)$; oxidizing agent, NO_2^-; reducing agent, Al **20.31** (a) False (b) true (c) true **20.33** (a) Fe(s) is oxidized, $Ag^+(aq)$ is reduced.
(b) $Ag^+(aq) + e^- \longrightarrow Ag(s)$; $Fe(s) \longrightarrow Fe^{2+}(aq) + 2e^-$. (c) Fe(s) is the anode, Ag(s) is the cathode. (d) Fe(s) is negative; Ag(s) is positive.
(e) Electrons flow from the Fe electrode ($-$) toward the Ag electrode ($+$).
(f) Cations migrate toward the Ag(s) cathode; anions migrate toward the Fe(s) anode. **20.35** (a) One *volt* is the potential energy difference required to impart 1 J of energy to a charge of 1 coulomb. (b) Yes. All voltaic cells involve spontaneous redox reactions that produce positive cell potentials or emfs. **20.37** (a) $2\,H^+(aq) + 2\,e^- \longrightarrow H_2(g)$
(b) $H_2(g) \longrightarrow 2\,H^+(aq) + 2\,e^-$ (c) A *standard* hydrogen electrode, SHE, is a hydrogen electrode where the components are at standard conditions, $1\,M\,H^+(aq)$ and $H_2(g)$ at 1 atm, such that $E° = 0$ V.
(c) The platinum foil in a SHE serves as an inert electron carrier and a solid reaction surface. **20.39** (a) $Cr^{2+}(aq) \longrightarrow Cr^{3+}(aq) + e^-$; $Tl^{3+}(aq) + 2\,e^- \longrightarrow Tl^+(aq)$ (b) $E_{red}° = 0.78$ V

23.53 (a)

$\overline{}\ \overline{}$ $d_{x^2y^2}, d_{z^2}$

Δ

$\overline{}\ \overline{}\ \overline{}$ d_{xy}, d_{xz}, d_{yz}

(b) The magnitude of Δ and the energy of the d-d transition for a d^1 complex are equal. **(c)** $\Delta = 220\,\text{kJ/mol}$ **23.55 (a)** Both minerals contain Cu^{2+}, $[Ar]3d^9$. **(b)** Azurite will probably have the larger Δ. It absorbs orange visible light, which has shorter wavelengths than the red light absorbed by malachite. **23.57 (a)** Ti^{3+}, d^1 **(b)** Co^{3+}, d^6 **(c)** Ru^{3+}, d^5 **(d)** Mo^{5+}, d^1 **(e)** Re^{3+}, d^4 **23.59** Yes. A weak-field ligand leads to a small Δ value and a small d-orbital splitting energy. If the splitting energy of a complex is smaller than the energy required to pair electrons in an orbital, the complex is high spin. **23.61 (a)** Mn, $[Ar]4s^23d^5$; Mn^{2+}, $[Ar]3d^5$; 1 unpaired electron

$\boxed{}\ \boxed{}$

$\boxed{\uparrow\downarrow}\ \boxed{\uparrow\downarrow}\ \boxed{\uparrow}$

(b) Ru, $[Kr]5s^14d^7$; Ru^{2+}, $[Kr]4d^6$; 0 unpaired electrons

$\boxed{}\ \boxed{}$

$\boxed{\uparrow\downarrow}\ \boxed{\uparrow\downarrow}\ \boxed{\uparrow\downarrow}$

(c) Rh, $[Kr]5s^14d^8$; Rh^{2+}, $[Kr]4d^7$; 1 unpaired electron

$\boxed{\uparrow}\ \boxed{}$

$\boxed{\uparrow\downarrow}\ \boxed{\uparrow\downarrow}\ \boxed{\uparrow\downarrow}$

23.63 All complexes in this exercise are six-coordinate octahedral.

(a) $\boxed{\uparrow}\ \boxed{}$
$\boxed{\uparrow}\ \boxed{\uparrow}\ \boxed{\uparrow}$
d^4, high spin

(b) $\boxed{\uparrow}\ \boxed{\uparrow}$
$\boxed{\uparrow}\ \boxed{\uparrow}\ \boxed{\uparrow}$
d^5, high spin

(c) $\boxed{}\ \boxed{}$
$\boxed{\uparrow\downarrow}\ \boxed{\uparrow\downarrow}\ \boxed{\uparrow\downarrow}$
d^6, low spin

(d) $\boxed{}\ \boxed{}$
$\boxed{\uparrow\downarrow}\ \boxed{\uparrow\downarrow}\ \boxed{\uparrow}$
d^5, low spin

(e) $\boxed{}\ \boxed{}$
$\boxed{\uparrow}\ \boxed{\uparrow}\ \boxed{\uparrow}$
d^3

(f) $\boxed{\uparrow}\ \boxed{\uparrow}$
$\boxed{\uparrow\downarrow}\ \boxed{\uparrow\downarrow}\ \boxed{\uparrow\downarrow}$
d^8

23.65

$\boxed{\uparrow}\ \boxed{\uparrow}$

$\boxed{\uparrow}\ \boxed{\uparrow}\ \boxed{\uparrow}$

high spin

23.69 $[Pt(NH_3)_6]Cl_4$; $[Pt(NH_3)_4Cl_2]Cl_2$; $[Pt(NH_3)_3Cl_3]Cl$; $[Pt(NH_3)_2Cl_4]$; $K[Pt(NH_3)Cl_5]$

23.73 (a)

```
      H       H  H        H
      |       |  |        |
  H — C — P̈ — C — C — P̈ — C — H
      |   |   |  |   |    |
      H   |   H  H   |    H
        H—C—H    H—C—H
          |        |
          H        H
```

(b) The dmpe ligand is bidentate, binding through the two P atoms. **(c)** The oxidation state of Mo is zero. **(d)** $Mo(CO)_4(dmpe)$ is *not* chiral as it can be superimposed on its mirror image. **23.76 (a)** Iron **(b)** magnesium **(c)** iron **(d)** copper **23.78 (a)** Pentacarbonyliron(0) **(b)** The oxidation state of iron must be zero. **(c)** Two. One isomer has CN in an axial position and the other has it in an equatorial position.

23.80

(a) $\boxed{}\ \boxed{}$

Δ

$\boxed{\uparrow}\ \boxed{\uparrow}\ \boxed{}$
d^2

(b) Statement (ii) explains the colors of these complexes: an electron is excited from the t_2 orbitals to the e orbitals. **(c)** $[V(H_2O)_6]^{3+}$ will absorb light with higher energy because it has a larger Δ than $[VF_6]^{3-}$. H_2O is in the middle of the spectrochemical series and causes a larger Δ than F^-, a weak-field ligand. **23.82** $[Co(NH_3)_6]^{3+}$, yellow; $[Co(H_2O)_6]^{2+}$, pink; $[CoCl_4]^{2-}$, blue

23.83 (a)

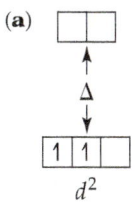

(b) sodium dicarbonyltetracyanoferrate(II) **(c)** $+2$, 6 d-electrons **(d)** We expect the complex to be low spin. Cyanide (and carbonyl) are high on the spectrochemical series, which means the complex will have a large Δ splitting, characteristic of low-spin complexes. **23.93 (a)** Yes, the oxidation state of Co is $+3$ in both complexes. **(b)** When compound B reacts with $AgNO_3(aq)$, white $AgBr(s)$ precipitates. **(c)** When compound A reacts with $BaCl_2(aq)$, white $BaSO_4(s)$ precipitates. **(d)** Compounds A and B are coordination sphere isomers. **(e)** Both compounds are strong electrolytes. **23.94** $\Delta E = 3.02 \times 10^{-19}\,\text{J/photon}$, $\lambda = 657\,\text{nm}$. The complex will absorb in the visible around 660 nm and appear blue-green.

Chapter 24

24.1 Structures **(c)** and **(d)** are the same molecule. **24.7 (a)** False **(b)** false **(c)** true **(d)** false **24.9** Numbering from the right on the condensed structural formula, C1 has trigonal-planar electron-domain geometry, 120° bond angles, and sp^2 hybridization; C2 and C5 have tetrahedral electron-domain geometry, 109° bond angles, and sp^3 hybridization; C3 and C4 have linear electron-domain geometry, 180° bond angles, and sp hybridization. **24.11 (a)** One **(b)** ten **(c)** seven **(d)** five **(e)** two **24.13 (a)** True **(b)** true **(c)** false **(d)** false **(e)** true **(f)** false **(g)** true **24.15 (a)** 2,3-dimethylheptane **(b)** $CH_3CH_2CH_2C(CH_3)_3$

(c)

```
  H3C        CH3
     \   C  /
      \ /  \
   H2C       CH2
    |          |
   H2C       CH2
      \       /
        CH
        |
      CH2CH3
```

(d) 2,2,5-trimethylhexane **(e)** methylcyclobutane **24.17 (a)** Unsaturated **(b)** Yes, all alkynes are unsaturated. **24.19 (a)** $CH_3CH_2CH_2CH_2CH_3$, C_5H_{12}

(b)

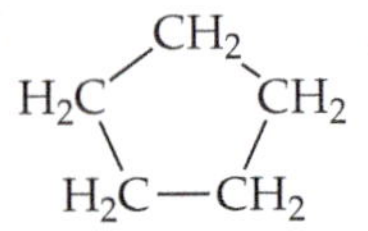

, C_5H_{10}

(c) $CH_2 = = CHCH_2\ CH_2\ CH_3$, C_5H_{10}
(d) $HC\equiv CCH_2CH_2CH_3$, C_5H_8
24.21 One possible structure is $CH\equiv C-CH=CH-C\equiv CH$
24.23 There are at least 46 structural isomers with the formula C_6H_{10}. A few of them are

$CH_3CH_2CH_2CH_2C\equiv CH$ $CH_3CH_2CH_2C\equiv CCH_3$

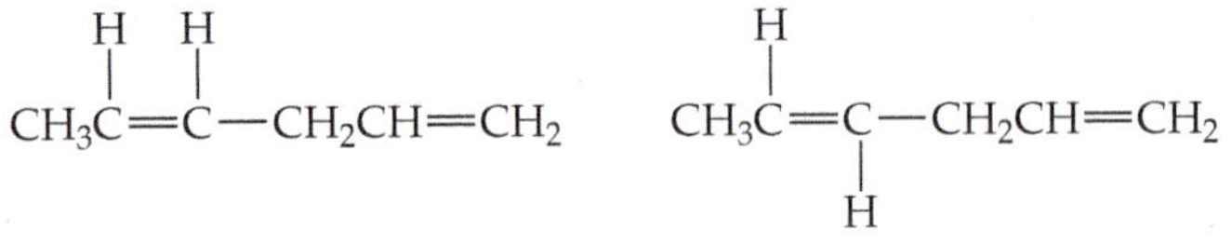

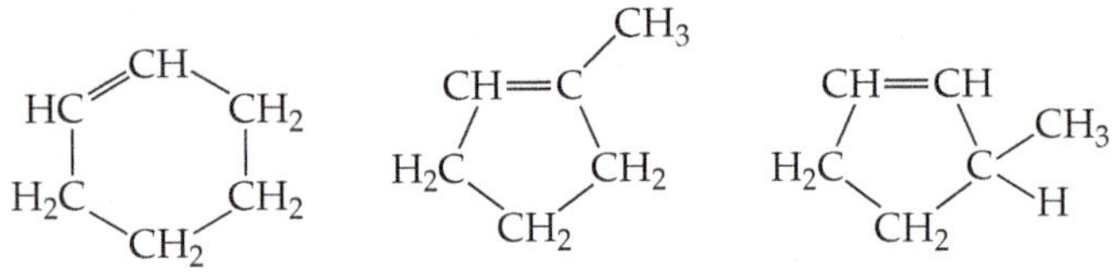

24.25

(a)

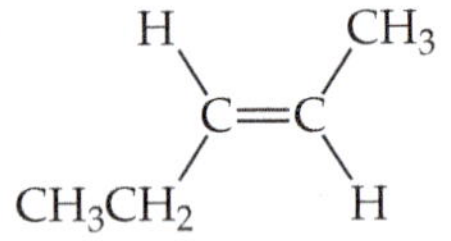

(b)

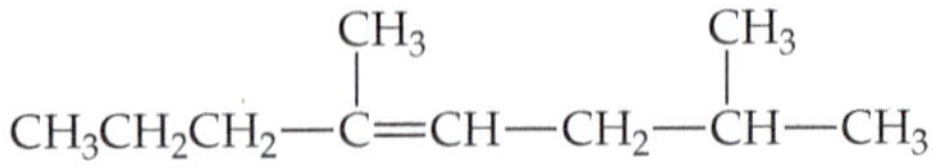

$CH_3CH_2CH_2-\overset{\overset{\displaystyle CH_3}{|}}{C}=CH-CH_2-\overset{\overset{\displaystyle CH_3}{|}}{CH}-CH_3$

(c) *cis*-6-methyl-3-octene **(d)** *para*-dibromobenzene **(e)** 4,4-dimethyl-1-hexyne **24.27 (a)** True **(b)** true **(c)** false **24.29 (a)** No

(b)

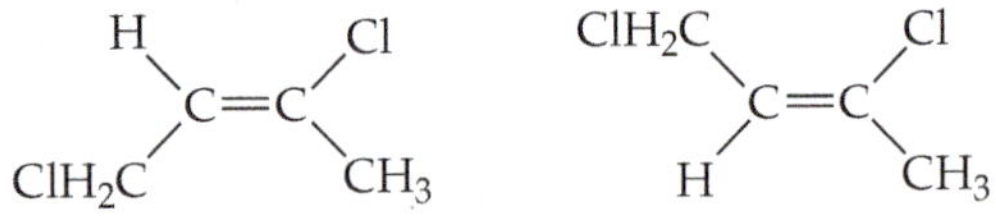

(c) no **(d)** no **24.31 (a)** True

(b) $CH_3CH_2CH=CH-CH_3 + Br_2 \longrightarrow$
             2-pentene

                         $CH_3CH_2CH(Br)CH(Br)CH_3$
                         2, 3-dibromopentane

This is an addition reaction.

(c)

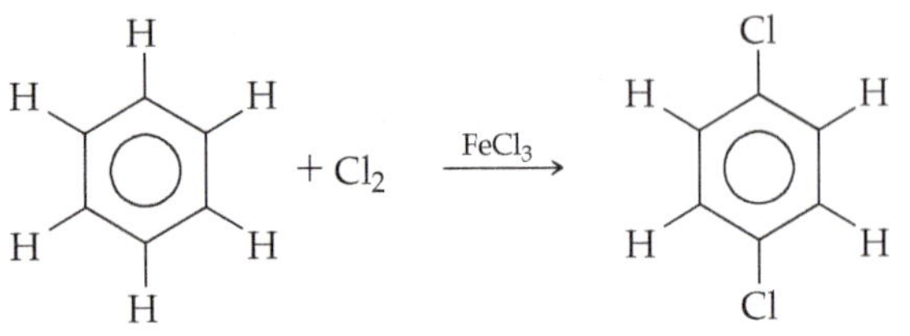

$C_6H_6 + Cl_2 \xrightarrow{FeCl_3} C_6H_4Cl_2$

This is an substitution reaction.

24.33 The 60° C—C—C angles in the cyclopropane ring cause strain that provides a driving force for reactions that result in ring opening. There is no comparable strain in the five-membered or six-membered rings. **(b)** $C_2H_4(g) + HBr(g) \longrightarrow CH_3CH_2Br(l)$;

$C_6H_6(l) + CH_3CH_2Br(l) \xrightarrow{AlCl_3} C_6H_5CH_2CH_3(l) + HBr(g)$

24.35 Yes, this information suggests (but does not prove) that the reactions proceed in the same manner. That the two rate laws are first order in both reactants and second order overall indicates that the activated complex in the rate-determining step in each mechanism is bimolecular and contains one molecule of each reactant. This is usually an indication that the mechanisms are the same, but it does not rule out the possibility of different fast steps or a different order of elementary steps. **24.37** ΔH_{comb}/mol CH_2 for cyclopropane = 696.3 kJ, for cyclopentane = 663.4 kJ. $\Delta H_{comb}/CH_2$ group for cyclopropane is greater because C_3H_6 contains a strained ring. When combustion occurs, the strain is relieved and the stored energy is released. **24.39 (a)** (iii) **(b)** (i) **(c)** (ii) **(d)** (iv) **(e)** (v) **24.41 (a)** Propionaldehyde (or propanal):

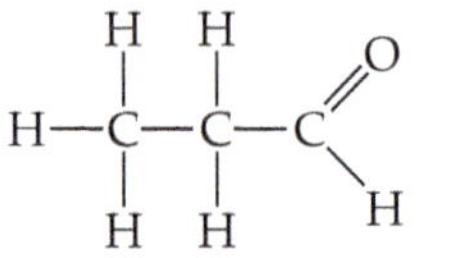

(b) ethylmethyl ether:

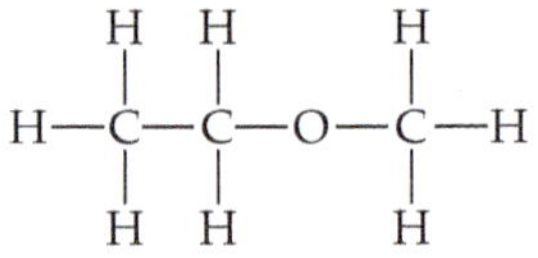

24.43

(a)

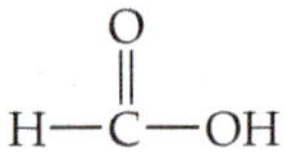

(b)

or

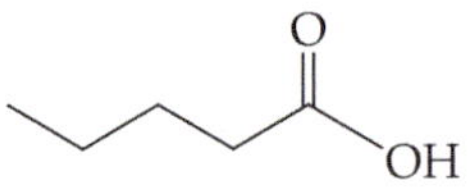

(c)

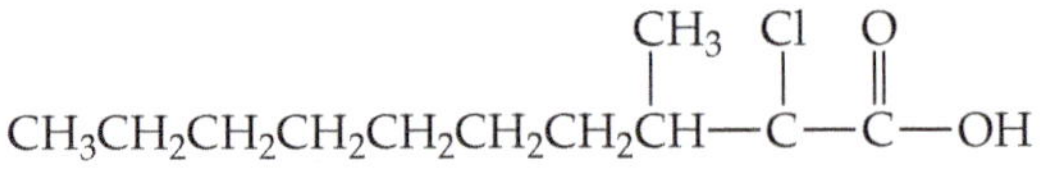

or

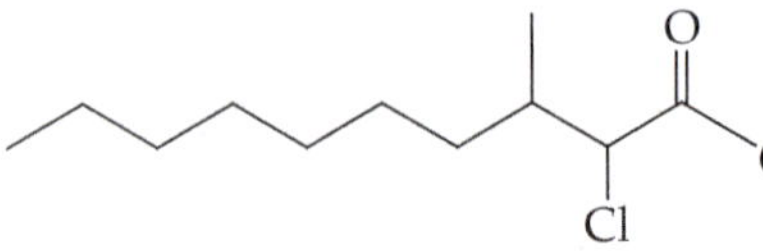

24.45

(a)

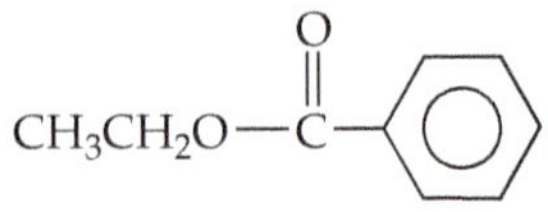

Ethylbenzoate

(b)

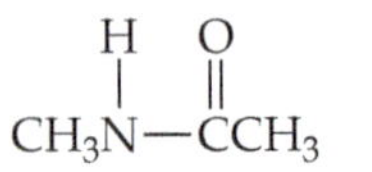

N-methylethanamide or
N-methylacetamide

(c)

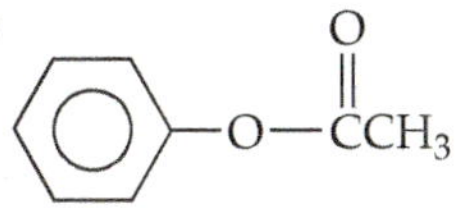

Phenylacetate

24.47

(a)

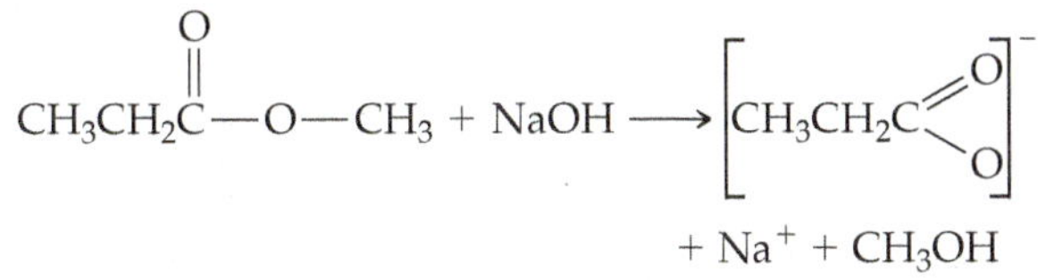

(b)

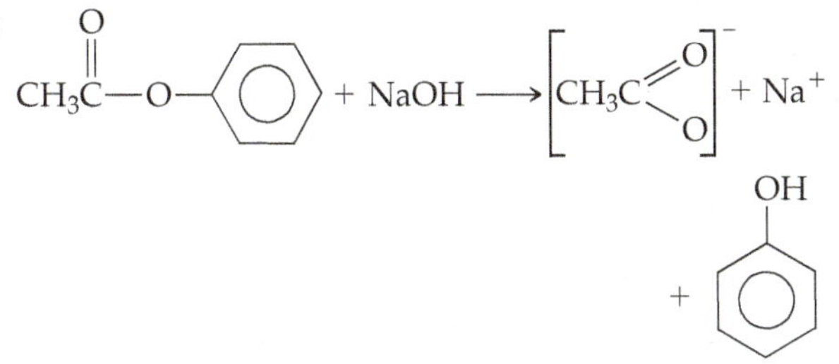

24.49 High melting and boiling points are indicators of strong intermolecular forces in the bulk substance. The presence of both —OH and —C=O groups in pure acetic acid leads us to conclude that it will be a strongly hydrogen-bonded substance. That the melting and boiling points of pure acetic acid are both higher than those of water, a substance we know to be strongly hydrogen-bonded, supports this conclusion. **24.51 (a)** $CH_3CH_2CH_2CH(OH)CH_3$ **(b)** $CH_3CH(OH)CH_2OH$

(c)

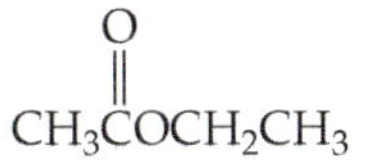

(d)

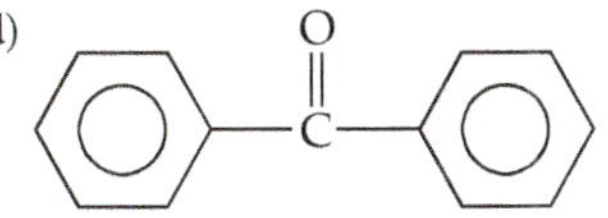

(e) $CH_3OCH_2CH_3$

24.53 (c) There are two chiral carbon atoms in the molecule.

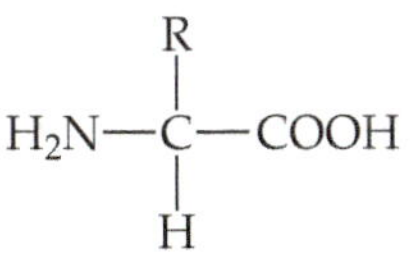

24.55 (a)

(b) In protein formation, amino acids undergo a condensation reaction between the amino group of one molecule and the carboxylic acid group of another to form the amide linkage.

(c) The bond that links amino acids in proteins is called the peptide bond. It is shown in bold in the figure below.

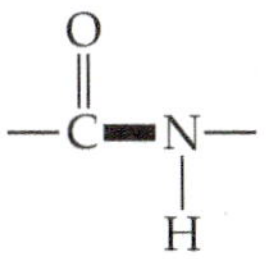

24.57

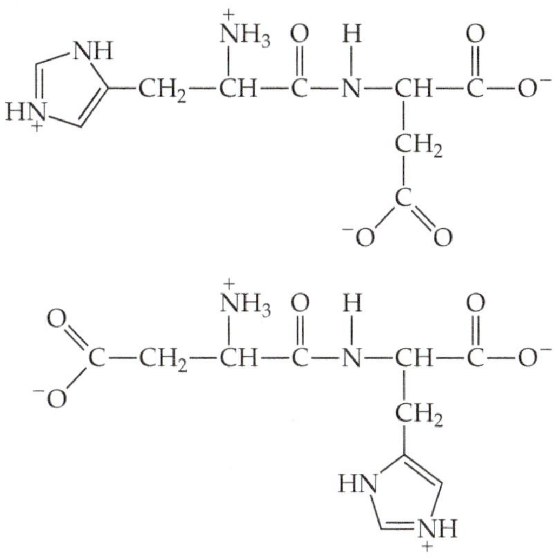

24.59 (a)

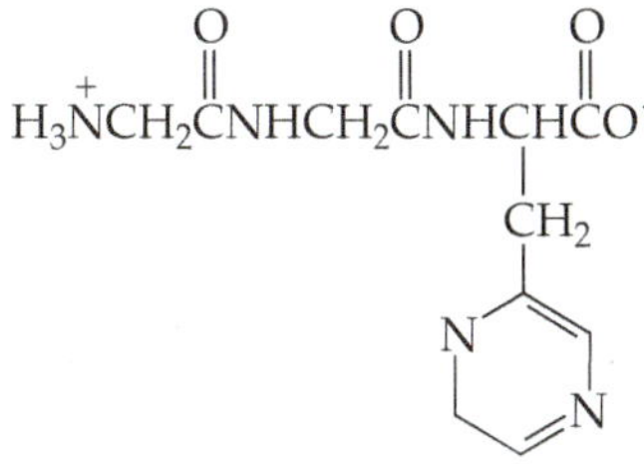

(b) Three tripeptides are possible: Gly-Gly-His, GGH; Gly-His-Gly, GHG; His-Gly-Gly, HGG **24.61 (a)** True **(b)** false **(c)** false **24.63 (a)** True **(b)** false **(c)** true **24.65 (a)** The empirical formula of cellulose is $C_6H_{10}O_5$. **(b)** The six-membered ring form of glucose is the monomer unit that is the basis of the polymer cellulose. **(c)** Ether linkages connect the glucose monomer units in cellulose. **24.67 (a)** Yes. **(b)** Four. In the linear form of mannose, the aldehydic carbon is C1. Carbon atoms 2, 3, 4, and 5 are chiral because they each carry four different groups. **(c)** Both the α (left) and β (right) forms are possible.

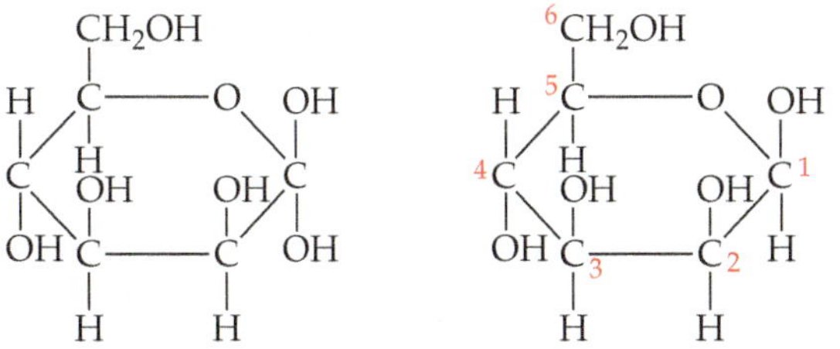

24.69 (a) False **(b)** true **(c)** true **24.71** *Purines*, with the larger electron cloud and molar mass, will have larger dispersion forces than *pyrimidines* in aqueous solution. **24.73** 5′–TACG–3′ **24.75** The complimentary strand for 5′–GCATTGGC–3′ is 3′–CGTAACCG–5′.

24.77

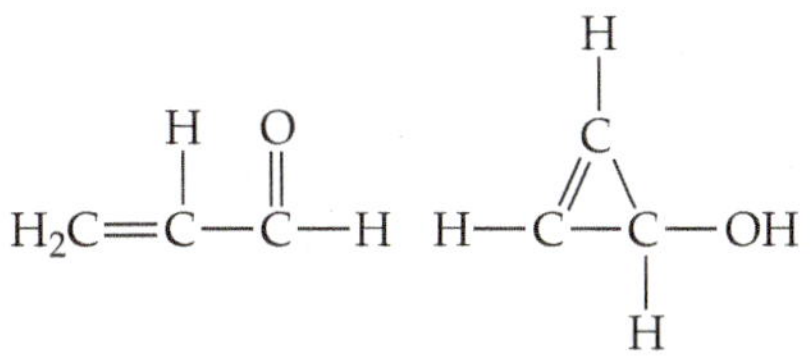

24.86 (a) One ketone and two alcohol functional groups; no chiral centers **(b)** One ketone and three alcohol functional groups; one chiral center **(c)** One carboxylic acid and one amine functional group; two chiral centers.

ANSWERS TO EXAM PREP QUESTIONS

Chapter 1   **EP 1.1(b)** molecules of a mixture of two different elements **EP 1.2 (c)** A solid can be readily compressed. **EP 1.3 (a)** the air that we breathe. **EP 1.4 (d)** carbon **EP 1.5 (b)** physical, extensive **EP 1.6 (a)** Only one of the three is a physical change.
EP 1.7 (d) Distillation is a technique that depends on the differences in the tendencies of substances to form gases. **EP 1.8 (e)** All three statements are correct. **EP 1.9 (c)** Its kinetic energy increases and its gravitational potential energy decreases. **EP 1.10 (d)** 4.0×10^6 cg
EP 1.11 (b) C < A < B **EP 1.12 (a)** 19 bars **EP 1.13 (c)** 15.5 kg
EP 1.14 (d) 1000 J **EP 1.15 (b)** 2 **EP 1.16 (c)** Statements i and iii are correct. **EP 1.17 (c)** 4 **EP 1.18 (d)** 4 **EP 1.19 (e)** 2.9×10^2 g
EP 1.20 (c) 3 **EP 1.21 (b)** 236 m/s **EP 1.22 (b)** 0.961 kg/L
EP 1.23 (b) 9.02 mg

Chapter 2   **EP 2.1 (b)** the volume of space occupied by the electrons of the atom **EP 2.2 (d)** ^{162}Ho **EP 2.3 (c)** $^{63}_{30}$Cu
EP 2.4 (b) ^{63}Cu must be more abundant than ^{65}Cu **EP 2.5 (c)** Se
EP 2.6 (d) C_4O_2, C_2O **EP 2.7 (e)** Ce^{4+} **EP 2.8** (a) Ti^{4+} **EP 2.9 (a)** CBr$_4$
EP 2.10 (d) N **EP 2.11 (a)** ClO$_2^-$, chlorate **EP 2.12 (c)** Fe$_2$O$_3$, diiron trioxide **EP 2.13 (b)** nitrous acid, HNO$_3$ **EP 2.14 (a)** CS_2
EP 2.15 (d) 2-methylbutane **EP 2.16 (c)** C_3H_7 **EP 2.17 (b)** 2

Chapter 3   **EP 3.1 (d)** 6 **EP 3.2 (d)** 4
EP 3.3 (b) $HCl(g) + NH_3(g) \longrightarrow NH_4Cl(s)$
EP 3.4 (d) $2\,Ag_2O(s) \longrightarrow 4\,Ag(s) + O_2(g)$
EP 3.5 (b) $2\,C_2H_4(OH)_2(l) + 5\,O_2(g) \longrightarrow 4\,CO_2(g) + 6\,H_2O(g)$
EP 3.6 (a) 310.2 amu **EP 3.7 (b)** 17.1% **EP 3.8 (c)** 50 g sodium chloride **EP 3.9 (b)** 2+ **EP 3.10 (c)** 1.91×10^{23} **EP 3.11 (d)** C_6H_{12}
EP 3.12 (d) $C_4H_8O_2$ **EP 3.13 (a)** 3.18 g **EP 3.14 (d)** 3.63 g
EP 3.15 (d) 14 mol $CH_3OH(l)$ **EP 3.16 (a)** 1.20 g **EP 3.17 (b)** 96%
EP 3.18 (b) O_2, 106 g **EP 3.19 (b)** 72.6%

Chapter 4   **EP 4.1 (e)** 3.0. **EP 4.2 (b)** CaCO$_3$ **EPQ 4.3** (e) No precipitate will form. **EP 4.4 (a)** There is no reaction; all possible products are soluble. **EP 4.5 (b)** hydrobromic acid **EP 4.6 (d)** sodium nitrate
EP 4.7 (d) $NH_3(aq) + H^+(aq) \longrightarrow NH_4^+(aq)$ **EP 4.8 (d)** H_2O_2
EP 4.9 (a) Zinc is oxidized, and copper ion is reduced.
EP 4.10 (b) lithium **EP 4.11 (c)** 3.91×10^{-2} M **EP 4.12 (e)** 2:1
EP 4.13 (d) 1.84 M **EP 4.14 (b)** 8.75 mL **EP 4.15 (b)** 19.5 mg
EP 4.16 (c) 0.163 M **EP 4.17 (e)** 81.0%

Chapter 5   **EP 5.1 (c)** The electrostatic potential energy between potassium and bromide ions is positive. **EP 5.2 (c)** $q < 0$, $w > 0$, the sign of ΔE cannot be determined from the information given
EP 5.3 (c) i and iii are true. **EP 5.4 (d)** Because ΔH relates to heat, the value of ΔH depends on the path taken between two states.
EP 5.5 (b) -1.37 L-atm **EP 5.6 (d)** exothermic, negative **EP 5.7 (c)** i and iii are true. **EP 5.8 (b)** -181 kJ **EP 5.9 (c)** decreases, increases
EP 5.10 (d) -464 kJ > mol **EP 5.11 (b)** 6.42 kJ > °C
EP 5.12 (d) -171.0 kJ **EP 5.13 (c)** 131.3 kJ
EP 5.14 (c) $H_2(g) + \frac{1}{2}O_2(g) \longrightarrow H_2O(l) \qquad \Delta H = -286$ kJ
EP 5.15 (b) -196.0 kJ **EP 5.16 (c)** $2\Delta H_f^\circ[SO_3] = \Delta H_{rxn}^\circ + 2\Delta H_f^\circ[SO_2]$
EP 5.17 (e) All three statements are true. **EP 5.18 (a)** 242 kJ
EP 5.19 (a) 2 g carbohydrate and 0.1 g fat **EP 5.20 (a)** A < B < C

Chapter 6   **EP 6.1(d)** 34.5 μm **EP 6.2 (c)** frequency, speed
EP 6.3 (b) 0.16 m **EP 6.4 (b)** Electrons will be emitted, and the maximum kinetic energies of those electrons will be greater than those emitted when irradiated with the red laser pointer.

EP 6.5 (e) $E = N_A \dfrac{hc}{\lambda}$ **EP 6.6 (b)** They all have $n_f = 2$. **EP 6.7 (b)** the uncertainty principle **EP 6.8 (a)** i < iii = ii **EP 6.9 (b)** 2.3×10^{-10} m **EP 6.10 (d)** both (i) and (iii) **EP 6.11 (c)** The electron moves in a circular orbit around the nucleus. **EP 6.12 (d)** $n = 4, l = 3$
EP 6.13 (b) both (i) and (ii) **EP 6.14 (c)** 1, 2, 3 **EP 6.15 (c)** both (i) and (iii) **EP 6.16 (b)** p **EP 6.17 (d)** electrons 2 and 3, which are degenerate and lower in energy than electron 1 **EP 6.18 (d)** 10
EP 6.19 (b) only Ca **EP 6.20 (d)** $[Xe]6s^2 4f^{14} 6d^6$

Chapter 7   **EP 7.1 (a)** 1+ **EP 7.2 (c)** 1.86 Å
EP 7.3 (d) O < N < P < Ge **EP 7.4 (b)** F < Cl < S^{2-} < Se^{2-}
EP 7.5 (a) Sr^{2+} < Rb$^+$ < Br$^-$ < Se^{2-} < Te^{2-}
EP 7.6 (e) $Br^{2+}(g) \longrightarrow Br^{3+}(g) + e^-$ **EP 7.7 (d)** Statements (ii) and (iii) are true **EP 7.8 (a)** $[Kr]4d^4$ **EP 7.9 (c)** it is likely to be a halogen
EP 7.10 (d) decreases, increases **EP 7.11 (d)** If the nuclear charge of atom A is greater than that of atom B, then the effective nuclear charge of A is greater than that of B **EP 7.12 (b)** increases, decreases
EP 7.13 (d) Si **EP 7.14 (d)** Ag **EP 7.15 (e)** M(OH)$_3$(aq) **EP 7.16 (c)** 3
EP 7.17 (e) All three statements are true **EP 7.18 (b)** S **EP 7.19 (b)** The first ionization energy of cesium is less than that of sodium
EP 7.20 (d) $Sr^{2+}(aq) + H_2(g) + O_2(g)$; (e) $Sr^{2+}(aq) + OH^-(aq) + H_2(g)$

Chapter 8   **EP 8.1 (c)** Si **EP 8.2 (e)** The attraction between cations and anions increases as the radii of the ions increase
EP 8.3 (b) ScN > MgO > NaCl > CsI **EP 8.4 (b)** Ca **EP 8.5 (c)** In forming the Cl—Cl bond, an electron is completely transferred from one Cl atom to the other **EP 8.6 (b)** H_2S **EP 8.7 (e)** All three statements are true **EP 8.8 (b)** Fluorine has a large first ionization energy and a large electron affinity **EP 8.9 (a)** H—F **EP 8.10 (d)** 4.39 D
EP 8.11 (d) W is in a higher oxidation state in WF$_6$, which increases the degree of covalent bonding **EP 8.12 (e)** more than one of a, b, c, d
EP 8.13 (a) 0 **EP 8.14 (b)** 6 **EP 8.15 (c)** 2 **EP 8.16 (b)** Because of multiple resonance structures, the six carbon-carbon bonds in benzene, C_6H_6, all have the same length **EP 8.17 (b)** NO$_2^-$ and CO$_3^{2-}$
EP 8.18 (c) $\ddot{\text{F}}$—$\ddot{\text{C}}$—$\ddot{\text{F}}$ **EP 8.19 (a)** SF$_4$ **EP 8.20 (e)** 1, 2

Chapter 9   **EP 9.1 (c)** The shape of PF$_3$ can be obtained by removing one atom from a tetrahedral arrangement of atoms
EP 9.2 (c) 3 **EP 9.3 (a)** H_2S has a trigonal planar electron-domain geometry **EP 9.4 (d)** Statements ii and iii are true **EP 9.5 (a)** 109.5° and 109.5° **EP 9.6 (c)** Statements i and iii are true. **EP 9.7 (b)** trigonal planar **EP 9.8 (a)** The bond is formed from the overlap of a $3p$ orbital on one Cl atom with a $3s$ orbital on the other Cl atom **EP 9.9 (b)** Hybrid orbitals can allow us to make multiple equivalent bonds when atomic orbitals can't do so **EP 9.10 (c)** sp, 180 **EP 9.11 (c)** O_3
EP 9.12 (c) i and iii **EP 9.13 (e)** 4, 4 **EP 9.14 (d)** 8 **EP 9.15 (d)** σ^*_{1s}, antibonding **EP 9.16 (b)** H_2^+ and H_2^- **EP 9.17 (a)** The number of molecular orbitals for a molecule is less than the number of atomic orbitals that are used to make them **EP 9.18 (e)** 1, ½
EP 9.19 (b) Statements i and ii are true **EP 9.20 (e)** O_2, paramagnetic
EP 9.21 (e) $F_2^- < C_2^{2+} < O_2^- < N_2^-$

Chapter 10   **EP 10.1 (b)** Gases are incompressible
EP 10.2 (d) 10.3 m **EP 10.3 (b)** 95.6 mm **EP 10.4 (d)** 1000 atm
EP 10.5 (c) (i) and (iii) **EP 10.6 (c)** If you plot V versus $1/P$ for a fixed quantity of a gas at constant temperature, you will get a straight line with a positive slope **EP 10.7 (a)** It doubles **EP 10.8 (b)** 1.9 L
EP 10.9 (c) 1.4 L **EP 10.10 (e)** 9.39×10^5 g **EP 10.11 (b)** 27 psi

EP 10.12 (**c**) 0.76 L **EP 10.13** (**a**) 0.92 g/L **EP 10.14** (**c**) 44.1 g/mol
EP 10.15 (**d**) 0.47 atm **EP 10.16** (**d**) 4.9 atm **EP 10.17** (**b**) 0.33
EP 10.18 (**b**) The C_3H_8 and CH_4 molecules have the same average
kinetic energy **EP 10.19** (**b**) Gas molecules move in circles
EP 10.20 (**a**) four, will not change **EP 10.21** (**d**) 335 m/s
EP 10.22 (**d**) both (ii) and (iii) **EP 10.23** (**b**) 37.2%
EP 10.24 (**a**) Underestimate by 12.38 atm **EP 10.25** (**e**) (ii), (iii), and (iv)

Chapter 11
EP 11.1 (**d**) fills the entire volume of the
container it occupies **EP 11.2** (**b**) only (ii) **EP 11.3** (**a**) only (i)
EP 11.4 (**a**) only (i) **EP 11.5** (**a**) dimethyl sulfoxide CH_3
EP 11.6 (**d**) hydrogen peroxide, H_2O_2 **EP 11.7** (**d**) ionic bonding,
ion–dipole interactions **EP 11.8** (**c**) $H_2 < O_2 < H_2CO < H_2O < CaO$
EP 11.9 (**c**) $CH_4 < Ar < Cl_2 < CH_3COOH$ **EP 11.10** (**b**) The ethylene
glycol molecule can form more hydrogen bonds per molecule than
1-propanol **EP 11.11** (**c**) vapor pressure **EP 11.12** (**d**) U-shaped, stronger
EP 11.13 (**d**) endothermic, larger **EP 1.14** (**d**) heat of vaporization
and specific heat of $H_2O(l)$ **EP 111.15** (**c**) its ability to assume the
shape and volume of its container **EP 11.16** (**b**) its vapor pressure equals
the pressure of the surrounding atmosphere **EP 11.17** (**d**) both (i) and (ii)
EP 11.18 (d) The piston is at a higher position than it was before the
heat was added **EP 11.19** (**c**) The normal boiling point of substance
A is higher than that of substance B **EP 11.20** (**c**) phenol C_6H_5OH
EP 11.21 (**a**) high temperature, high pressure **EP 1.22** (**a**) It sublimes
at about $-200\,°C$ **EP 1.23** (**b**) solid < smectic A liquid crystal <
nematic liquid crystal < liquid **EP 1.24** (**e**) rigid rod-shaped
molecules

Chapter 12
EP 12.1 (**d**) covalent-network **EP 12.2** (**a**) only i
EP 12.3 (b) lattice = square, atoms per unit cell = 1 A + 2 B
EP 12.4 (**a**) primitive cubic **EP 12.5** (**d**) rhombohedral **EP 12.6** (b) Ir
EP 12.7 (**c**) 6 **EP 12.8** (c) 1.45 Å **EP 12.9** (**e**) 78.5% **EP 12.10**
(**b**) Nonmetallic elements are typically present in both interstitial and
substitutional alloys **EP 12.11** (**e**) the low melting points of metals at
the very end of the transition series, like cadmium (Cd) and mercury (Hg)
EP 12.12 (**a**) Cs **EP 12.13** (**c**) Ni_3Sn **EP 12.14** (**d**) 2.39 g/cm³
EP 12.15 (**d**) neither i nor ii **EP 12.16** (**e**) The atoms in the crystal
have a large number (8–12) of nearest neighbors. **EP 12.17** (**e**) In
general, the more polar the bonds are in compound semiconductors,
the smaller the band gap. **EP 12.18** (**c**) Si:Al

Chapter 13
EP 13.1 (**a**) increases, increases **EP 13.2** (**b**) acetone
(CH_3COCH_3) dissolved in H_2O **EP 13.3** (**a**) only (i) **EP 13.4**
(**b**) $|\Delta H_{solute}| < |\Delta H_{solvent}| < |\Delta H_{mix}|$ **EP 13.5** (**d**) equal to, saturated
EP 13.6 (**c**) settle to the bottom without dissolving, leaving the
concentration of the solution unchanged **EP 13.7** (**a**) benzene (C_6H_6)
EP 13.8 (**d**) both (i) and (ii) **EP 13.9** (**a**) argon, 0.26 **EP 13.10**
(**c**) 0.057 mol/L **EP 13.11** (**b**) 230 ppb **EP 13.12** (**d**) 2.91%
EP 13.13 (**b**) 1.82×10^{-5} **EP 13.14** (**b**) 0.022 M **EP 13.15** (**e**) 34 m
EP 13.16 (b) The vapor pressure will be higher in the container with
the glucose solution **EP 13.17** (**c**) 0.125 mol **EP 13.18** (**b**) 0.15 m NaCl
EP 13.19 (**d**) the solute–solute, solvent–solvent and solute–solvent
interactions are all assumed to be the same strength
EP 13.20 (**b**) hypotonic, hemolysis **EP 13.21** (**b**) only (ii)
EP 13.22 (**d**) norfenefrine ($C_8H_{11}NO_2$) **EP 13.23** (**e**) 8
EP 13.24 (**d**) 199 kg **EP 13.25** (**e**) both (i) and (iii) **EP 13.26** (**d**) solid
emulsion **EP 13.27** (**c**) both (i) and (ii)

Chapter 14
EP 14.1 (**d**) size of reaction vessel
EP 14.2 (**b**) Statements (i) and (ii) are true **EP 14.3** (**d**) 2.7×10^{-5} M/s
EP 14.4 (**c**) $2A \longrightarrow B + 3C$ **EP 14.5** (**e**) None of statements i, ii,
and iii is true **EP 14.6** (**a**) $M^{-\frac{1}{2}}\,s^{-1}$ **EP 14.7** (**d**) does not change;
increases by a factor of four **EP 14.8** (**c**) 4600 **EP 14.9** (**b**) Statements

(i) and (ii) are true **EP 14.10** (**c**) 11 h **EP 14.11** (**a**) Only (ii)
EP 14.12 (**d**) 2.3×10^{-1} s^{-1} **EP 14.13** (**c**) Statements (i) and (iii) are true
EP 14.14 (**e**) $A + B \longrightarrow X + Z$ **EP 14.15** (**a**) $C + C \longrightarrow K + Z$
EP 14.16 (**c**) Rate $= k[P]^2[Q]$ **EP 14.17** (**c**) C **EP 14.18** (**d**) A catalyst
cannot affect the mechanism of a reaction **EP 14.19** (**d**) The number
of reactions per second done by her enzyme was 15 times that of the
standard catalyst.

Chapter 15
EP 15.1 (**d**) i and ii **EP 15.2** (**c**) $K_c = \dfrac{[SO_3]^2}{[SO_2]^2[O_2]}$

EP 15.3 (**d**) 1.88×10^3

EP 15.4 (**b**)

EP 15.5 (**d**) $Ni(CO)_4(g) \rightleftharpoons Ni(s) + 4\,CO(g)$ **EP 15.6** (**d**) Statements
ii and iii are true **EP 15.7** (**d**) 1.8×10^2 **EP 15.8** (e) Neither $[Ag^+]$ nor
$[Cl^-]$ will change **EP 15.9** (**d**) increasing the volume of the vessel
EP 15.10 (a) 0.0410 **EP 15.11** (**d**) 1.4 **EP 15.12** (**b**) 4.7×10^{-2}
EP 15.13 (**c**) Statements i and ii are true **EP 15.14** (**b**) 0.36 atm
EP 15.15 (**a**) 0.57 atm **EP 15.16** (**d**) increase, decrease, leave unchanged
EP 15.17 (**a**) increasing the volume of the reaction vessel
EP 15.18 (**b**) The equilibrium will not shift **EP 15.19** (**c**) At a fixed
temperature, a catalyst can shift the equilibrium toward the creation
of more products

Chapter 16
EP 16.1 (**b**) acceptor, donor **EP 16.2** (**b**) HSO_4^-
and H_2O **EP 16.3** (**a**) Arrhenius **EP 16.4** (**d**) All three statements are
true. **EP 16.5** (**c**) i and iii **EP 16.6** (**b**) O^{2-} **EP 16.7** (**e**) iii < ii < i
EP 16.8 (**c**) 2.5×10^{-7} M **EP 16.9** (**a**) 1.0×10^{-8} M **EP 16.10** (**d**) less
than 7, equal to **EP 16.11** (**d**) 11.83 **EP 16.12** (**d**) less than, 3.47
EP 16.13 (**c**) iii < i < ii **EP 16.14** (**c**) 12.48 **EP 16.15** (**e**) iii < ii < i
EP 16.16 (**b**) B < C < A **EP 16.17** (**c**) 6.7×10^{-5} **EP 16.18** (**e**) 9.0%
EP 16.19 (**a**) 2.30 **EP 16.20** (**e**) 2.64 **EP 16.21** (**c**) **EP 16.22** (b) 2.32
EP 16.23 (**a**) 0.012 **EP 16.24** (**c**) 9.52 **EP 16.25** (**d**) aniline and sulfite
ion **EP 16.26** (**c**) 6.3×10^{-5} **EP 16.27** (**a**) 0.38 **EP 16.28** (**b**) weakest
acid $HClO < C_5H_5NH^+ < CH_2ClCOOH$ strongest acid
EP 16.29 (**a**) weakest base $HCOO^- < BrO^- < (CH_3)_3N$ strongest base
EP 16.30 (**e**) iii < ii < i **EP 16.31** (b) $NaHSO_4$ and NaH_2PO_4
EP 16.32 (**b**) $HOI < HIO_2 < HBrO_2 < HClO_2 < HClO_3$

Chapter 17
EP 17.1 (**d**) If you add the soluble salt KA to a
solution of HA that is at equilibrium, the pH would increase
EP 17.2 (**e**) 1.73×10^{-4} M **EP 17.3** (**c**) The acid and base concentrations
must be equal **EP 17.4** (**a**) 0.174 **EP 17.5** (**a**) 60.7 g **EP 17.6** (**c**) C
EP 17.7 (**c**) The pH at the equivalence point is 7.00 **EP 17.8** (**d**) less
than 7, NH_4^+ **EP 17.9** (**d**) The conjugate base that is formed at the
equivalence point reacts with water **EP 17.10** (**c**) 1.60
EP 17.11 (**c**) $[Ag^+]^3[PO_4^{3-}]$ **EP 17.12** (**e**) 1.40×10^{-37}
EP 17.13 (**b**) cobalt(II) hydroxide, $K_{sp} = 1.3 \times 10^{-15}$
EP 17.14 (**d**) Raising the temperature of the solution
EP 17.15 (**e**) none of these actions **EP 17.16** (**c**) The precipitate was
chromium hydroxide, which then reacted with more hydroxide to
produce a soluble complex ion, $Cr(OH)_4^-(aq)$ **EP 17.17** (**a**) larger, will
EP 17.18 (d) aluminum **EP 17.19** (d) 10.52

Chapter 18
EP 18.1 (**a**) troposphere **EP 18.2** (**d**) thermosphere
EP 18.3 (**b**) $O_2 > CO_2 > CH_4$ **EP 18.4** (**d**) 1×10^{13} molecules/cm³
EP 18.5 (**e**) 6.24 mm Hg **EP 18.6** (**a**) 620 nm **EP 18.7** (**b**) -146 kJ/mol
EP 18.8 (**b**) catalyst, intermediate, product **EP 18.9** (**e**) 5.6,
$CO_2(g) + H_2O(l) \longrightarrow H_2CO_3(aq)$ **EP 18.10** (**c**) O_3

EP 18.11 (b) 5.7 kg **EP 18.12 (b)** only ii **EP 18.13 (c)** They contain more than two atoms. **EP 18.14 (c)** endothermic, exothermic **EP 18.15 (a)** calcium oxide **EP 18.16 (c)** both N and P **EP 18.17 (b)** When $Al_2(SO_4)_3$ is added, it reacts with a variety of cations to form sulfate precipitates. These precipitates settle slowly, carrying suspended particles down with them. **EP 18.18 (c)** All waste should be carefully disposed of after it has been created. **EP 18.19 (d)** both ii and iii

Chapter 19
EP 19.1 (d) Equilibrium is achieved in a closed system when the rate of iron oxidation is equal to the rate of iron(III) oxide reduction. **EP 19.2 (c)** i and iii **EP 19.3 (d)** The value of ΔS for a system depends on the path taken between two states of the system. **EP 19.4 (a)** Yes, because the heat transferred from the system has a negative sign. **EP 19.5 (e)** All three statements are true. **EP 19.6 (e)** one, two, less than **EP 19.7 (c)** i and iii **EP 19.8 (b)** 1 mol of $H_2(g)$ at 100 °C and 0.5 atm **EP 19.9 (c)** At $T = 0$ K, a pure crystalline solid has zero possible microstates. **EP 19.10 (d)** Standard molar entropies can have positive and negative values. **EP 19.11 (a)** 326.4 J/K **EP 19.12 (c)** All spontaneous reactions have a negative free-energy change. **EP 19.13 (b)** -192.7 kJ **EP 19.14 (a)** The reaction is spontaneous at all temperatures **EP 19.15 (d)** 193 °C **EP 19.16 (c)** $+34,000$ J **EP 19.17 (e)** $+2.2$ kJ **EP 19.18 (e)** $+265$ kJ/mol **EP 19.19 (d)** $K = 23.2$

Chapter 20
EP 20.1 (c) HNO_3 **EP 20.2 (e)** both ii and iii **EP 20.3 (a)** $Br^-(aq)$ **EP 20.4 (c)** only iii. **EP 20.5 (d)** seven on the product side **EP 20.6 (a)** one on the reactant side **EP 20.7 (b)** The electrode is gaining mass and cations from the salt bridge are flowing into the half-cell. **EP 20.8 (b)** oxidation, anode **EP 20.9 (a)** -0.35 V **EP 20.10 (c)** $+1.08$ V **EP 20.11 (d)** cells 1 and 2 **EP 20.12 (b)** $Cl_2(g)$ **EP 20.13 (d)** Cl_2 and Br_2 **EP 20.14 (d)** both ii and iii **EP 20.15 (d)** -0.13 V **EP 20.16 (d)** 2.5×10^4 **EP 20.17 (b)** 0.55 V **EP 20.18 (c)** 4.1×10^{-2} M **EP 20.19 (c)** 0.46 V, cathode **EP 20.20 (c)** As the battery discharges, K^+ migrate toward the anode and OH^- migrate toward the cathode to maintain charge balance. **EP 20.21 (c)** It allows H^+ ions to migrate from the anode to the cathode. **EP 20.22 (b)** An increase in the pH. **EP 20.23 (c)** Water is reduced to H_2 gas at the cathode more easily than Na^+ ions are reduced to sodium metal. **EP 20.24 (c)** 62 min

Chapter 21
EP 21.1 (b) uranium-234 **EP 21.2 (b)** Statements i and ii are true. **EP 21.3 (a)** alpha decay followed by beta emission **EP 21.4 (d)** manganese-55 **EP 21.5 (b)** $^{134}_{55}$Cs **EP 21.6 (d)** iodine-131 **EP 21.7 (e)** $^{239}_{93}$Np **EP 21.8 (b)** 6.0 h **EP 21.9 (d)** 45.5 yr **EP 21.10 (c)** 2340 yr **EP 21.11 (a)** 2.27×10^6 kJ **EP 21.12 (c)** $3.19553534 \times 10^{-11}$ J **EP 21.13 (a)** $^{144}_{54}$Xe **EP 21.14 (e)** All three statements are true.

Chapter 22
EP 22.1 (d) Oxygen is better able to form π bonds than sulfur **EP 22.2 (d)** C_2H_2 **EP 22.3 (d)** $+1$ **EP 22.4 (a)** F **EP 22.5 (a)** 2 **EP 22.6 (b)** carbon monoxide **EP 22.7 (d)** Pd **EP 22.8 (b)** CuH **EP 22.9 (a)** $Ni^{2+}(aq) + H_2(g) + SO_4{}^{2-}(aq)$ **EP 22.10 (c)** Statements i and iii are true **EP 22.10 (c)** trigonal bipyramidal, T-shaped **EP 22.12 (c)** $HBrO_3$ **EP 22.13 (a)** F_2 **EP 22.14 (c)** HIO_3 **EP 22.15 (c)** P_4O_{10} **EP 22.16 (d)** bent **EP 22.17 (d)** $+6, +3$ **EP 22.18 (b)** $N_2H_4(aq) + O_2(aq) \longrightarrow N_2(g) + 2\,H_2O(l)$ **EP 22.19 (e)** None of the statements is true **EP 22.20 (c)** H_3PO_3 **EP 22.21 (a)** SiC **EP 22.22 (e)** $+4$ **EP 22.23 (c)** sheets of linked SiO_4 tetrahedra **EP 22.24 (c)** $Be_3Al_2Si_6O_{18}$ **EP. 22.25 (e)** silicones, quartz, silicates

Chapter 23
EP 23.1 (c) niobium **EP 23.2 (c)** $Rb_3[MoO_3F_3]$ **EP 23.3 (b)** $[Kr]4d^6$ **EP 23.4 (d)** ferromagnetism **EP 23.5 (e)** $[Rh(NH_3)_4Cl_2]Cl$ **EP 23.6 (c)** Statements i and iii are true. **EP 23.7 (a)** Bipyridine is a monodentate ligand. **EP 23.8 (e)** oxalate ion **EP 23.9 (c)** tetraamminedichlororhodium(III) chloride **EP 23.10 (b)** $Na_2[Pt(CN)_4]$

EP 23.11 (a)

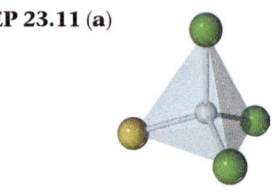

$[MX_3Y]$

EP 23.12 (d) tetrahedral $[Co(NH_3)BrClI]^-$ **EP 23.13 (d)** orange **EP 23.14 (d)** Statements ii and iii are true. **EP 23.15 (a)** t_2 set of orbitals, e set of orbitals **EP 23.16 (c)** $X < H_2O < Y$ **EP 23.17 (e)** i, ii, and iii **EP 23.18 (d)** $[RhCl_6]^{3-}$ **EP 23.19 (e)** 5 **EP 23.20 (a)** octahedral, tetrahedral, square planar

Chapter 24
EP 24.1 (e) slightly greater than 120° **EP 24.2 (c)** 3 **EP 24.3 (b)** *cis*-2-butane **EP 24.4 (e)** 20 **EP 24.5 (b)** They are structural isomers **EP 24.6 (b)** aromatic hydrocarbon, C_6H_4ClOH **EP 24.7 (b)** only ii **EP 24.8 (d)** 2-methylpropane **EP 24.9 (c)** both i and ii **EP 24.10 (c)** amine, carboxylic acid **EP 24.11 (c)** a carboxylic acid, alcohol **EP 24.12 (a)** 1 **EP 24.13 (d)** 6 **EP 24.14 (e)** The —R group contains only carbon and hydrogen, but the number of atoms in this group differs from one amino acid to another **EP 24.15 (d)** Neither the α-helical structure nor the β-sheet structure will be retained **EP 24.16 (b)** only ii **EP 24.17 (d)** 3 **EP 24.18 (b)** The two molecules are geometric isomers **EP 24.19 (c)** condensation **EP 24.20 (c)** both i and iii **EP 24.21 (d)** lipids **EP 24.22 (a)** adenine **EP 24.23 (d)** 3′–CTGGAAT–5′ **EP 24.24 (b)** DNAs and RNAs are built from the same nitrogen containing bases

GLOSSARY

absolute zero The temperature at which all thermal motion ceases: 0 K on the Kelvin scale and $-273.15\ °C$ on the Celsius scale.

absorption spectrum A pattern of variation in the amount of light absorbed by a sample as a function of wavelength.

accuracy A measure of how closely individual measurements agree with the correct, or "true," value.

acid A substance that is able to donate a H^+ ion (a proton) and, hence, increases the concentration of $H^+(aq)$ when it dissolves in water.

acid-dissociation constant (K_a) An equilibrium constant that expresses the extent to which an acid transfers a proton to solvent water.

acidic anhydride (acidic oxide) An oxide that forms an acid when added to water; soluble nonmetal oxides are acidic anhydrides.

acidic oxide (acidic anhydride) An oxide that reacts either with a base to form a salt or with water to form an acid.

acid rain Rainwater that has become excessively acidic because of absorption of pollutant oxides, notably SO_3, produced by human activities.

actinide element Element in which the $5f$ orbitals are only partially occupied.

activated complex (transition state) The particular arrangement of atoms found at the top of the potential-energy barrier as a reaction proceeds from reactants to products.

activation energy (E_a) The minimum energy needed for reaction; the height of the energy barrier to formation of products.

active site Specific site on a heterogeneous catalyst or an enzyme where catalysis occurs.

activity The decay rate of a radioactive material, generally expressed as the number of disintegrations per unit time.

activity series A list of metals in order of decreasing ease of oxidation.

addition polymerization Polymerization in which monomers are coupled through their multiple bonds.

addition reaction A reaction in which a reagent adds to the two carbon atoms of a carbon–carbon multiple bond.

adsorption The binding of molecules to a surface.

alcohol An organic compound obtained by substituting a hydroxyl group ($-OH$) for a hydrogen on a hydrocarbon.

aldehyde An organic compound that contains a carbonyl group ($C=O$) to which at least one hydrogen atom is attached.

alkali metals Members of group 1A in the periodic table.

alkaline earth metals Members of group 2A in the periodic table.

alkanes Compounds of carbon and hydrogen containing only carbon–carbon single bonds.

alkenes Hydrocarbons containing one or more carbon–carbon double bonds.

alkyl group A group that is formed by removing a hydrogen atom from an alkane.

alkynes Hydrocarbons containing one or more carbon–carbon triple bonds.

alloy A substance that has the characteristic properties of a metal and contains more than one element. Often there is one principal metallic component, with other elements present in smaller amounts. Alloys may be homogeneous or heterogeneous.

alpha emission A type of radioactive decay in which an atomic nucleus emits an alpha particle and thereby transforms (or "decays") into an atom with a mass number 4 less and atomic number 2 less.

alpha (α) helix A protein structure in which the protein is coiled in the form of a helix with hydrogen bonds between $C=O$ and $N-H$ groups on adjacent turns.

alpha particles Particles that are identical to helium-4 nuclei, consisting of two protons and two neutrons, symbol ^4_2He or $^4_2\alpha$.

amide An organic compound that has an NR_2 group attached to a carbonyl, where R may be H or a hydrocarbon group.

amine A compound that has the general formula R_3N, where R may be H or a hydrocarbon group.

amino acid A carboxylic acid that contains an amino ($-NH_2$) group attached to the carbon atom adjacent to the carboxylic acid ($-COOH$) functional group.

amorphous solid A solid whose molecular arrangement lacks the order (regularly repeating long-range pattern) of a crystal.

amphiprotic Refers to the capacity of a substance to either add or lose a proton (H^+).

amphoteric oxides and hydroxides Oxides and hydroxides that are only slightly soluble in water but that dissolve in either acidic or basic solutions.

angstrom A common non-SI unit of length, denoted Å, that is used to measure atomic dimensions: $1Å = 10^{-10}$ m.

anion A negatively charged ion.

anode An electrode at which oxidation occurs.

antibonding molecular orbital A molecular orbital in which electron density is concentrated outside the region between the two nuclei of bonded atoms. Such orbitals, designated as σ^* or π^*, are less stable (of higher energy) than bonding molecular orbitals.

antiferromagnetism A form of magnetism in which unpaired electron spins on adjacent sites point in opposite directions and cancel each other's effects.

aqueous solution A solution in which water is the solvent.

aromatic hydrocarbons Hydrocarbon compounds that contain a planar, cyclic arrangement of carbon atoms linked by both σ and delocalized π bonds.

Arrhenius equation An equation that relates the rate constant for a reaction to the frequency factor, A, the activation energy, E_a, and the temperature, T: $k = Ae^{-E_a/RT}$. In its logarithmic form it is written $\ln k = -E_a/RT + \ln A$.

atmosphere (atm) A unit of pressure equal to 760 torr; 1 atm = 101.325 kPa.

atom The smallest representative particle of an element. The almost infinitesimally small building blocks of matter.

atomic mass unit (amu) A unit based on the value of exactly 12 amu for the mass of the isotope of carbon that has six protons and six neutrons in the nucleus; 1 amu = 1.66054×10^{-24} g.

atomic number The number of protons in the nucleus of an atom of an element.

atomic radius An estimate of the size of an atom. See bonding atomic radius.

atomic weight The average mass of the atoms of an element in atomic mass units (amu); it is numerically equal to the mass in grams of one mole of the element.

autoionization The process whereby water spontaneously forms low concentrations of $H^+(aq)$ and $OH^-(aq)$ ions by proton transfer from one water molecule to another.

Avogadro's hypothesis A statement that equal volumes of gases at the same temperature and pressure contain equal numbers of molecules.

Avogadro's law A statement that the volume of a gas maintained at constant temperature and pressure is directly proportional to the number of moles of the gas.

Avogadro's number (N_A) The number of ^{12}C atoms in exactly 12 g of ^{12}C; it equals 6.022×10^{23} mol^{-1}.

ball-and-stick model Model that depicts atoms as spheres and bonds as sticks.

band An array of closely spaced molecular orbitals occupying a continuous range of energy.

band gap The energy gap between a fully occupied valence band and an empty conduction band.

band structure The electronic structure of a bulk solid.

bar A unit of pressure equal to 10^5 Pa.

base A substance that is an H^+ acceptor; a base produces an excess of $OH^-(aq)$ ions when it dissolves in water.

base-dissociation constant (K_b) An equilibrium constant that expresses the extent to which a base reacts with solvent water, accepting a proton and forming $OH^-(aq)$.

basic anhydride (basic oxide) An oxide that forms a base when added to water; soluble metal oxides are basic anhydrides.

basic oxide (basic anhydride) An oxide that either reacts with water to form a base or reacts with an acid to form a salt and water.

battery A portable, self-contained electrochemical power source that contains one or more voltaic cells.

becquerel The SI unit of radioactivity. It corresponds to one nuclear disintegration per second.

Beer's law The light absorbed by a substance (A) equals the product of its extinction coefficient (ε), the path length through which the light passes (b), and the molar concentration of the substance (c): $A = \varepsilon bc$.

beta emission A nuclear decay process where a beta particle is emitted from the nucleus; also called beta decay.

beta particles Energetic electrons emitted from the nucleus, symbol $_{-1}^{0}e$ or β^-.

beta sheet A structural form of protein in which sheets are made of two or more strands of peptides that hydrogen-bond from an amide H in one strand to a carbonyl O in the other strand.

bidentate ligand A ligand that contains two donor atoms, each with a nonbonding electron pair, that can coordinate to a metal.

bimolecular reaction An elementary reaction that involves two molecules.

biochemistry The study of the chemistry of living systems.

biodegradable Organic material that bacteria are able to oxidize.

biopolymers Three broad categories of polymers found in living organisms: proteins, polysaccharides (carbohydrates), and nucleic acids.

body-centered lattice A crystal lattice in which the lattice points are located at the center and corners of each unit cell.

bomb calorimeter A device for measuring the heat evolved in the combustion of a substance under constant-volume conditions.

bond angles The angles made by the lines joining the nuclei of the atoms in a molecule.

bond dipole The dipole moment that is due only to unequal electron sharing between two atoms in a covalent bond.

bond enthalpy The enthalpy change, ΔH, required to break a particular bond when the substance is in the gas phase.

bonding atomic radius The radius of an atom as defined by the distances separating it from other atoms to which it is chemically bonded.

bonding molecular orbital A molecular orbital in which the electron density is concentrated in the internuclear region. The energy of a bonding molecular orbital is lower than the energy of the separate atomic orbitals from which it forms.

bonding pair In a Lewis structure, a pair of electrons that is shared by two atoms.

bond length The distance between the centers of two bonded atoms.

bond order The number of bonding electron pairs shared between two atoms, minus the number of antibonding electron pairs: bond order = (number of bonding electrons − number of antibonding electrons)/2.

bond polarity A measure of the degree to which the electrons are shared unequally between two atoms in a chemical bond.

boranes Molecules containing only boron and hydrogen; covalent hydrides of boron.

Born–Haber cycle A thermodynamic cycle based on Hess's law that relates the lattice energy of an ionic substance to its enthalpy of formation and to other measurable quantities.

Boyle's law A law stating that at constant temperature, the product of the volume and pressure of a given amount of gas is a constant.

Brønsted–Lowry acid A substance (molecule or ion) that acts as a proton donor.

Brønsted–Lowry base A substance (molecule or ion) that acts as a proton acceptor.

Brownian motion The random motion of colloidal particles in solution due to collisions with solvent molecules.

buffer capacity The amount of acid or base a buffer can neutralize before the pH begins to change appreciably.

buffered solution (buffer) A solution that undergoes a limited change in pH upon addition of a small amount of acid or base.

calorie A unit of energy; it is the amount of energy needed to raise the temperature of 1 g of water by 1 °C from 14.5 °C to 15.5 °C. A related unit is the joule: 1 cal = 4.184 J.

calorimeter An apparatus that measures the heat released or absorbed in a chemical or physical process.

calorimetry The experimental measurement of heat produced in chemical and physical processes.

capillary action The process by which a liquid rises in a tube because of a combination of adhesion to the walls of the tube and cohesion between liquid particles.

carbide A binary compound of carbon with a metal or metalloid.

carbohydrates A class of substances formed from polyhydroxy aldehydes or ketones.

carbon black An amorphous form of carbon.

carbonyl group The C=O double bond, a characteristic feature of several organic functional groups, such as ketones and aldehydes.

carboxylic acid An acid that contains the —COOH functional group.

catalyst A substance that changes the speed of a chemical reaction without itself undergoing a permanent chemical change in the process.

cathode An electrode at which reduction occurs.

cathode rays Streams of electrons that are produced when a high voltage is applied to electrodes in an evacuated tube.

cathodic protection A means of protecting a metal against corrosion by making it the cathode in a voltaic cell. This can be achieved by attaching a more easily oxidized metal, which serves as an anode, to the metal to be protected.

cation A positively charged ion.

cell potential The potential difference between the cathode and anode in an electrochemical cell; it is measured in volts: 1 V = 1 J/C. Also called electromotive force.

cellulose A polysaccharide of glucose; it is the major structural element in plant matter.

Celsius scale A temperature scale on which water freezes at 0° and boils at 100° at sea level.

chain reaction A series of reactions in which one reaction initiates the next.

changes of state Physical changes of matter from one state to a different one, for example, from a gas to a liquid.

charcoal An amorphous form of carbon produced when wood is heated strongly in a deficiency of air.

Charles's law A law stating that at constant pressure, the volume of a given quantity of gas is proportional to absolute temperature.

chelate effect The generally larger formation constants for bidentate and polydentate ligands as compared with the corresponding *monodentate* ligands.

chelating agent Bidentate and polydentate ligands capable of occupying two or more sites in the coordination sphere.

chemical bond A strong attractive force that exists between atoms in a molecule.

chemical changes Processes in which a substance is transformed into a chemically different substance; also called chemical reactions.

chemical equation A representation of a chemical reaction using the chemical formulas of the reactants and products; a balanced chemical equation contains equal numbers of atoms of each element on both sides of the equation.

chemical equilibrium A state of dynamic balance in which the rate of formation of the products of a reaction from the reactants equals the rate of formation of the reactants from the products; at equilibrium the concentrations of the reactants and products remain constant.

chemical formula A notation that uses chemical symbols with numerical subscripts to convey the relative proportions of atoms of the different elements in a substance.

chemical kinetics The area of chemistry concerned with the speeds, or rates, at which chemical reactions occur.

chemical nomenclature The rules used in naming substances.

chemical properties Properties that describe the way a substance may change, or *react*, to form other substances.

chemical reactions Processes in which a substance is transformed into a chemically different substance; also called chemical changes.

chemistry The scientific discipline that studies matter, its properties, and the changes that matter undergoes.

chiral A term describing a molecule or an ion that cannot be superimposed on its mirror image.

chlorofluorocarbons Ozone-damaging substances, principally $CFCl_3$ and CF_2Cl_2, that were once widely used as propellants in spray cans. They do not occur in nature.

chlorophyll A plant pigment that plays a major role in conversion of solar energy to chemical energy in photosynthesis.

cholesteric liquid crystal A liquid crystal having molecules arranged in layers, with their long axes parallel to the other molecules within the same layer. Moving from one layer to the next, the orientation of the molecules rotates by a fixed angle, resulting in a spiral pattern.

coal A naturally occurring solid containing hydrocarbons of high molecular weight, as well as compounds containing sulfur, oxygen, and nitrogen.

colligative property A property of a solvent (vapor-pressure lowering, freezing-point lowering, boiling-point elevation, osmotic pressure) that depends on the total concentration of solute particles present.

collision model A model of reaction rates based on the idea that molecules must collide to react; it explains the factors influencing reaction rates in terms of the frequency of collisions, the number of collisions with energies exceeding the activation energy, and the probability that the collisions occur with suitable orientations.

colloids (colloidal dispersions) Mixtures containing particles larger than normal solutes but small enough to remain suspended in the dispersing medium.

combination reaction A chemical reaction in which two or more substances combine to form a single product.

combustion reaction A chemical reaction that proceeds with evolution of heat and usually also a flame; most combustion involves reaction with oxygen, as in the burning of a match.

common-ion effect A shift of an equilibrium induced by an ion common to the equilibrium. Whenever a weak electrolyte and a strong electrolyte containing a common ion are together in solution, the weak electrolyte ionizes less than it would if it were alone in solution.

complementary colors Colors that, when mixed in proper proportions, appear white or colorless. For example, orange and blue are complementary colors that form white light when mixed; when blue is removed, the light looks orange.

complete ionic equation A chemical equation in which dissolved strong electrolytes (such as dissolved ionic compounds) are written as separate ions.

complex ion (complex) An assembly of a metal ion and the Lewis bases bonded to it.

compound A substance composed of two or more elements united chemically in definite proportions.

concentration The quantity of solute present in a given quantity of solvent or solution.

concentration cell A voltaic cell containing the same electrolyte and the same electrode materials in both the anode and cathode compartments. The emf of the cell is derived from a difference in the concentrations of the same electrolyte solutions in the compartments.

condensation polymerization Polymerization in which two molecules are joined to form a larger molecule by elimination of a small molecule, such as H_2O.

condensation reaction A reaction in which two molecules are joined to form a larger molecule by elimination of a small molecule, such as H_2O.

conduction band A band of unoccupied antibonding molecular orbitals lying higher in energy than the occupied valence band and distinctly separated from it.

conjugate acid A substance formed by addition of a proton to a Brønsted–Lowry base.

conjugate acid–base pair An acid and a base, such as H_2O and OH^-, that differ only in the presence or absence of a proton.

conjugate base A substance formed by the loss of a proton from a Brønsted–Lowry acid.

continuous spectrum A spectrum that contains radiation distributed over all wavelengths.

conversion factor A fraction whose numerator and denominator are the same quantity expressed in different units.

coordination compound A compound containing a coordination complex.

coordination number The number of adjacent atoms in a crystal structure to which an atom is directly bonded. In a complex, the coordination number of the metal ion is the number of atoms bonded to the metal ion.

coordination sphere The central metal ion and its surrounding ligands.

coordination-sphere isomers Structural isomers that differ in which species in the complex are ligands and which are outside the coordination sphere in the solid.

copolymer A complex polymer resulting from the polymerization of two or more chemically different monomers.

core electrons The electrons that are not in the outermost shell of an atom.

corrosion Spontaneous redox reactions in which a metal is oxidized by some substance in its environment and converted to an unwanted compound.

covalent bond A bond formed between two or more atoms by a sharing of electrons.

covalent-network solids Solids that are held together by an extended network of covalent bonds.

critical mass The amount of fissionable material necessary to maintain a nuclear chain reaction.

critical point The temperature and pressure beyond which the liquid and gas phases are indistinguishable from each other.

critical pressure The pressure at which a gas at its critical temperature is converted to a liquid state.

critical temperature The highest temperature at which a distinct liquid phase can form. The critical temperature increases with an increase in the magnitude of intermolecular forces.

cross-linking A method of stiffening polymers by introducing chemical bonds between chains.

crystal-field theory A theory that accounts for the colors and the magnetic and other properties of transition-metal complexes in terms of the splitting of the energies of metal ion d orbitals by the electrostatic interaction with the ligands.

crystal lattice The geometrical pattern of points on which the unit cells are arranged; in effect, an abstract (that is, not real) scaffolding for the crystal structure.

crystalline solid (crystal) A solid whose internal arrangement of atoms, molecules, or ions possesses a regularly repeating pattern throughout the solid.

crystallinity A measure of the extent to which a solid maintains the regular repeating pattern of atoms seen in a crystalline solid.

crystallization The process in which molecules, ions, or atoms come together to form a crystalline solid. The opposite of a solid dissolving in a solvent to form a solution.

cubic close packing A crystal structure where the atoms are packed together as close as possible, and the close-packed layers of atoms adopt a three-layer repeating pattern that leads to a face-centered cubic unit cell.

curie A measure of radioactivity: 1 curie $= 3.7 \times 10^{10}$ nuclear disintegrations per second.

cycloalkanes Saturated hydrocarbons of general formula C_nH_{2n} in which the carbon atoms form a closed ring.

Dalton's law of partial pressures A law stating that the total pressure of a mixture of gases is the sum of the pressures that each gas would exert if it were present alone.

d–d transition The transition of an electron in a transition-metal compound from a lower-energy d orbital to a higher-energy d orbital.

decomposition reaction A chemical reaction in which a single compound reacts to give two or more products.

degenerate A situation in which two or more orbitals have the same energy.

delocalized electrons Electrons that are spread over a number of atoms in a molecule or a crystal rather than localized on a single atom or a pair of atoms.

density The amount of mass in a unit volume of a substance.

deoxyribonucleic acid (DNA) A polynucleotide in which the sugar component is deoxyribose.

derived unit An SI unit obtained by multiplication or division of one or more of the SI base units.

desalination The removal of salts from seawater, brine, or brackish water to make it fit for human consumption.

deuterium The isotope of hydrogen whose nucleus contains a proton and a neutron: $_1^2$H.

dextrorotatory, or merely dextro or d A term used to label a chiral molecule that rotates the plane of polarization of plane-polarized light to the right (clockwise).

diamagnetism A type of magnetism that causes a substance with no unpaired electrons to be weakly repelled from a magnetic field.

diatomic molecule A molecule composed of only two atoms.

diffusion The spreading of one substance throughout a space or throughout a second substance.

dilution The process of preparing a less concentrated solution from a more concentrated one by adding solvent.

dimensional analysis A method of problem solving in which units are multiplied together or divided into each other along with the numerical values. Dimensional analysis ensures that the final answer of a calculation has the desired units.

dipole A molecule with one end having a partial negative charge and the other end having a partial positive charge.

dipole–dipole interactions A force that becomes significant when polar molecules come in close contact with one another. The force is attractive when the positive end of one polar molecule approaches the negative end of another.

dipole moment A measure of the separation and magnitude of the positive and negative charges in polar molecules.

disaccharide Two monosaccharide units linked together by a condensation reaction.

dispersion forces Intermolecular forces resulting from attractions between induced dipoles. Also called London dispersion forces.

displacement reaction A reaction in which an ion in solution is displaced (replaced) through oxidation of an element.

disproportionation reaction A reaction in which an element is simultaneously oxidized and reduced.

distillation A separation process that depends on the different boiling points of substances.

donor atom The atom of a ligand that bonds to the central metal ion in a coordination complex.

doping The process of adding controlled amounts of impurity atoms to a material. For example, incorporation of P into Si.

double bond A covalent bond involving two electron pairs.

double helix The structure for DNA that involves the winding of two DNA polynucleotide chains together in a helical arrangement. The two strands of the double helix are complementary in that the organic bases on the two strands are paired for optimal hydrogen bond interaction.

dynamic equilibrium A state of balance in which opposing processes occur at the same rate.

effective nuclear charge The net positive charge experienced by an electron in a many-electron atom; this charge is not the full nuclear charge because there is some shielding of the nucleus by the other electrons in the atom.

effusion The escape of a gas through an orifice or hole.

elastomer A material that can undergo a substantial change in shape via stretching, bending, or compression and return to its original shape upon release of the distorting force.

electrochemistry The branch of chemistry that deals with the relationships between electricity and chemical reactions.

electrolysis reaction A reaction in which a nonspontaneous redox reaction is brought about by the passage of current under a sufficient external electrical potential. The devices in which electrolysis reactions occur are called electrolytic cells.

electrolyte A solute that produces ions in solution; an electrolytic solution conducts an electric current.

electrolytic cell A device in which a nonspontaneous oxidation–reduction reaction is caused to occur by passage of current under a sufficient external electrical potential.

electromagnetic radiation (radiant energy) A form of energy that has wave characteristics and that propagates through a vacuum at the characteristic speed of $3.00 \times 10^8 \, \text{m/s}$.

electrometallurgy The use of electrolysis to reduce or refine metals.

electromotive force (emf) A measure of the driving force, or *electrical pressure*, for the completion of an electrochemical reaction. Electromotive force is measured in volts: $1 \, \text{V} = 1 \, \text{J/C}$. Also called the cell potential.

electron A negatively charged subatomic particle found outside the atomic nucleus; it is a part of all atoms.

electron affinity The energy change that occurs when an electron is added to a gaseous atom or ion.

electron capture A mode of radioactive decay in which an inner-shell orbital electron is captured by the nucleus.

electron configuration The arrangement of electrons in the orbitals of an atom or molecule

electron density The probability of finding an electron at any particular point in an atom; this probability is equal to ψ^2, the square of the wave function. Also called the probability density.

electron domain In the VSEPR model, a region about a central atom in which an electron pair is concentrated.

electron-domain geometry The three-dimensional arrangement of the electron domains around an atom according to the VSEPR model.

electronegativity A measure of the ability of an atom that is bonded to another atom to attract electrons to itself.

electronic charge The negative charge carried by an electron; it has a magnitude of $1.602 \times 10^{-19} \, \text{C}$.

electronic structure The arrangement of electrons in an atom or molecule.

electron-sea model A model for the behavior of electrons in metals.

electron shell A collection of orbitals that have the same value of n. For example, the orbitals with $n = 3$ (the $3s$, $3p$, and $3d$ orbitals) comprise the third shell.

electron spin A property of the electron that makes it behave as though it were a tiny magnet. The electron behaves as if it were spinning on its axis; electron spin is quantized.

element A substance that cannot be decomposed into simpler substances.

elemental composition The percentage composition of an element in a substance.

elementary reaction A process in a chemical reaction that occurs in a single step. An overall chemical reaction consists of one or more elementary reactions or steps.

empirical formula A chemical formula that shows the kinds of atoms and their relative numbers in a substance in the smallest possible whole-number ratios.

enantiomers Molecules of a chiral substance that are nonsuperimpoable mirror images of each other.

endothermic process A process in which a system absorbs heat from its surroundings.

energy The capacity to do work or to transfer heat.

energy-level diagram A diagram that shows the energies of molecular orbitals relative to the atomic orbitals from which they are derived. Also called a **molecular-orbital diagram**.

enthalpy A quantity defined by the relationship $H = E + PV$; the enthalpy change, ΔH, for a reaction that occurs at constant pressure is the heat evolved or absorbed in the reaction: $\Delta H = q_p$.

enthalpy of formation The enthalpy change that accompanies the formation of a substance from the most stable forms of its component elements.

enthalpy of reaction The enthalpy change associated with a chemical reaction.

entropy A measure of the degree of *randomness* or *disorder* associated with a system.

enzyme A protein molecule that acts to catalyze specific biochemical reactions.

equilibrium constant The numerical value of the equilibrium-constant expression for a system at equilibrium. The equilibrium constant is usually denoted by K_p for gas-phase systems or K_c for solution-phase systems.

equilibrium-constant expression The expression that describes the relationship between the activities (usually expressed as concentrations or partial pressures) of the substances present in a chemical reaction at equilibrium.

equivalence point The point in a titration at which the added solute reacts completely with the solute present in the solution.

ester An organic compound that has an –OR group attached to a carbonyl; it is the product of a reaction between a carboxylic acid and an alcohol.

ether A compound in which two hydrocarbon groups are bonded to one oxygen.

exchange (metathesis) reaction A reaction between compounds that when written as a molecular equation appears to involve the exchange of ions between the two reactants.

excited state A higher energy state than the ground state.

exothermic process A process in which a system releases heat to its surroundings.

extensive property A property that depends on the amount of sample considered; for example, mass or volume.

face-centered lattice A crystal lattice in which the lattice points are located at the faces and corners of each unit cell.

Faraday constant (F) The magnitude of charge of one mole of electrons: 96,485 C/mol.

f-block metals Lanthanide and actinide elements in which the $4f$ or $5f$ orbitals are partially occupied.

ferrimagnetism A form of magnetism in which unpaired electron spins on different-type ions point in opposite directions but do not fully cancel out.

ferromagnetism A form of magnetism in which unpaired electron spins align parallel to one another.

first law of thermodynamics A statement that energy is conserved in any process. One way to express the law is that the change in internal energy, ΔE, of a system in any process is equal to the heat, q, added to the system, plus the work, w, done on the system by its surroundings: $\Delta E = q + w$.

first-order reaction A reaction in which the reaction rate is proportional to the concentration of a single reactant, raised to the first power.

fission The splitting of a large nucleus into two smaller ones.

folding The process by which a protein adopts its biologically active shape.

force A push or a pull.

formal charge The charge an atom (in a molecule) would have if each bonding electron pair in the molecule were shared equally between its two atoms.

formation constant The equilibrium constant for formation of a complex ion from the metal ion and Lewis bases (ligands) present in solution.

formula weight The sum of the atomic weights (AW) of the atoms in the chemical formula of the substance. For example, the formula weight of NO_2 (46.0 amu) is the sum of the masses of one nitrogen atom and two oxygen atoms.

fossil fuels Coal, oil, and natural gas, which are presently our major sources of energy.

fracking The practice in which water laden with sand and other materials is pumped at high pressure into rock formations to release natural gas and other petroleum materials.

free energy (Gibbs free energy, G) A thermodynamic state function that gives a criterion for spontaneous change in terms of enthalpy and entropy: $G = H - TS$.

free radical A substance with one or more unpaired electrons.

frequency The number of times per second that one complete wavelength passes a given point.

frequency factor (A) A term in the Arrhenius equation that is related to the frequency of collision and the probability that the collisions are favorably oriented for reaction.

fuel cell A voltaic cell that utilizes the oxidation of a conventional fuel, such as H_2 or CH_4, in the cell reaction.

fuel value The energy released when 1 g of a substance is combusted.

functional group An atom or group of atoms that imparts characteristic chemical properties to an organic compound.

fusion The joining of two light nuclei to form a more massive one.

galvanic cell See **voltaic (galvanic) cell**.

gamma radiation Energetic electromagnetic radiation emanating from the nucleus of a radioactive atom.

gas Matter that has no fixed volume or shape; it uniformly fills its container.

gas constant (R) The constant of proportionality in the ideal-gas equation.

geometric isomerism A form of isomerism in which compounds with the same type and number of atoms and the same chemical bonds have different spatial arrangements of these atoms and bonds.

Gibbs free energy A thermodynamic state function that combines enthalpy and entropy, in the form $G = H - TS$. For a change occurring at constant temperature and pressure, the change in free energy is $\Delta G = \Delta H - T\Delta S$.

glass An amorphous solid formed by fusion of SiO_2, CaO, and Na_2O. Other oxides may also be used to form glasses with differing characteristics.

glucose A polyhydroxy aldehyde whose formula is $CH_2OH(CHOH)_4CHO$; it is the most important of the monosaccharides.

glycogen The general name given to a group of polysaccharides of glucose that are synthesized in mammals and used to store energy from carbohydrates.

Graham's law A law stating that the rate of effusion of a gas is inversely proportional to the square root of its molecular weight.

gray (Gy) The SI unit for radiation dose corresponding to the absorption of 1 J of energy per kilogram of biological material; $1\,Gy = 100\,rads$.

green chemistry Chemistry that promotes the design and application of chemical products and processes that are compatible with human health and that preserve the environment.

greenhouse gases Gases in an atmosphere that absorb and emit infrared radiation (radiant heat), "trapping" heat in the atmosphere.

ground state The lowest-energy, or most stable, state.

group Elements that are in the same column of the periodic table; elements within the same group or family exhibit similarities in their chemical behavior.

Haber process The catalyst system and conditions of temperature and pressure developed by Fritz Haber and coworkers for the formation of NH_3 from H_2 and N_2.

half-life The time required for the concentration of a reactant substance to decrease to half its initial value; the time required for half of a sample of a particular radioisotope to decay.

half-reaction An equation for either an oxidation or a reduction that explicitly shows the electrons involved, for example, $Zn^{2+}(aq) + 2\,e^- \longrightarrow Zn(s)$.

halogens Members of group 7A in the periodic table.

heat The energy that causes the temperature of an object to increase. The flow of energy from a body at higher temperature to one at lower temperature when they are placed in thermal contact.

heat capacity The quantity of heat required to raise the temperature of a sample of matter by 1 °C (or 1 K).

heat of fusion The enthalpy change, ΔH, for melting a solid.

heat of sublimation The enthalpy change, ΔH, for vaporization of a solid.

heat of vaporization The enthalpy change, ΔH, for vaporization of a liquid.

Henderson–Hasselbalch equation The relationship among the pH, pK_a, and the concentrations of acid and conjugate base in an aqueous solution: $\mathrm{pH} = \mathrm{p}K_a + \log\dfrac{[\text{base}]}{[\text{acid}]}$.

Henry's law A law stating that the concentration of a gas in a solution, S_g, is proportional to the pressure of gas over the solution: $S_g = kP_g$.

Hess's law If a reaction is carried out in a series of steps, ΔH for the overall reaction equals the sum of the enthalpy changes for the individual steps.

heterogeneous alloy An alloy in which the components are not distributed uniformly; instead, two or more distinct phases with characteristic compositions are present.

heterogeneous catalyst A catalyst that is in a different phase from that of the reactant substances.

heterogeneous equilibrium The equilibrium established between substances in two or more different phases, for example, between a gas and a solid or between a solid and a liquid.

hexagonal close packing A crystal structure where the atoms are packed together as closely as possible. The close-packed layers adopt a two-layer repeating pattern, which leads to a primitive hexagonal unit cell.

high-spin complex A complex whose electrons are arranged to give the maximum number of unpaired electrons.

hole A vacancy in the valence band of a semiconductor, created by doping.

homogeneous catalyst A catalyst that is in the same phase as the reactant substances.

homogeneous equilibrium The equilibrium established between reactant and product substances that are all in the same phase.

Hund's rule A rule stating that electrons occupy degenerate orbitals in such a way as to maximize the number of electrons with the same spin. In other words, each orbital has one electron placed in it before pairing of electrons in orbitals occurs.

hybridization The mathematical mixing of different types of atomic orbitals to produce a set of equivalent hybrid orbitals.

hybrid orbital An orbital that results from the mixing of different kinds of atomic orbitals on the same atom. For example, an sp^3 hybrid results from the mixing, or hybridizing, of one s orbital and three p orbitals.

hydration Solvation when the solvent is water.

hydride ion An ion formed by the addition of an electron to a hydrogen atom: H^-.

hydrocarbons Compounds composed of only carbon and hydrogen.

hydrogen bonding Bonding that results from intermolecular attractions between molecules containing hydrogen bonded to an electronegative element. The most important examples involve OH, NH, and HF bonds.

hydrolysis A reaction with water. When a cation or anion reacts with water, it changes the pH.

hydronium ion (H_3O^+) The predominant form of the proton in aqueous solution.

hydrophilic Water attracting. The term is often used to describe a type of colloid.

hydrophobic Water repelling. The term is often used to describe a type of colloid.

hypervalent A compound with more than an octet of electrons around the central atom.

hypothesis A model or tentative explanation of a series of observations.

ideal gas A hypothetical gas whose pressure, volume, and temperature behavior is completely described by the ideal-gas equation.

ideal-gas equation An equation of state for gases that embodies Boyle's law, Charles's law, and Avogadro's hypothesis in the form $PV = nRT$.

ideal solution A solution where the solvent-solute, solute-solute, and solvent-solvent interactions are assumed to all be the same strength. An ideal solution perfectly obeys Raoult's Law.

immiscible liquids Liquids that do not dissolve in one another to a significant extent.

indicator A substance added to a solution that changes color when the added solute reacts with the solute present in solution. The most common type of indicator is an acid–base indicator whose color changes as a function of pH.

instantaneous rate The reaction rate at a particular time as opposed to the average rate over an interval of time.

insulators Materials that do not conduct electricity.

intensive property A property that does not depend on the amount of sample being examined, for example, density.

interhalogens Compounds formed between two different halogen elements. Examples include IBr and BrF_3.

intermediate A substance formed in one elementary step of a multistep mechanism and consumed in another; it is neither a reactant nor an ultimate product of the overall reaction.

intermetallic compound A homogeneous alloy with definite properties and a fixed composition.

intermolecular forces Forces that exist between molecules.

internal energy The sum of all the kinetic and potential energies of the components of the system. When a system undergoes a change, the change in internal energy, ΔE, is defined as the heat, q, added to the system, plus the work, w, done on the system by its surroundings: $\Delta E = q + w$.

interstitial alloy An alloy formed when solute atoms occupy interstitial positions in the "holes" between solvent atoms.

ion Electrically charged atom or group of atoms (polyatomic ion); ions can be positively or negatively charged, depending on whether electrons are lost (positive) or gained (negative) by the atoms.

ion–dipole force The force that exists between an ion and a polar molecule that possesses a permanent dipole moment.

ionic bond A bond formed by the electrostatic attractions between oppositely charged ions. The ions are formed from atoms by transfer of one or more electrons.

ionic compound A compound composed of cations and anions.

ionic hydrides Compounds formed when hydrogen reacts with alkali metals and also the heavier alkaline earths (Ca, Sr, and Ba); these compounds contain the hydride ion, H^-.

ionic solids Solids that are held together by the mutual electrostatic attraction between cations and anions.

ionization energy The minimum energy required to remove an electron from the ground state of an isolated gaseous atom or ion.

ionizing radiation Radiation that has sufficient energy to remove an electron from a molecule, thereby ionizing it.

ion-product constant For water, K_w is the product of the hydrogen ion and hydroxide ion concentrations: $[H^+][OH^-] = K_w = 1.0 \times 10^{-14}$ at 25 °C.

irreversible process A process that leaves the surroundings somehow changed when the system is restored to its original state.

isoelectronic series A series of atoms, ions, or molecules having the same number of electrons.

isomers Compounds whose molecules have the same overall composition but different structures.

isothermal process A process that occurs at constant temperature.

isotopes Atoms of the same element containing different numbers of neutrons and therefore having different masses.

joule (J) The SI unit of energy, 1 kg-m^2/s^2. A related unit is the calorie: 4.184 J = 1 cal.

Kelvin scale The SI temperature scale; the SI unit for temperature is the kelvin. Zero on the Kelvin scale corresponds to −273.15 °C.

ketone A compound in which the carbonyl group ($C=O$) occurs at the interior of a carbon chain and is therefore flanked by carbon atoms.

kinetic energy The energy that an object possesses by virtue of its motion.

kinetic-molecular theory of gases A set of assumptions about the nature of gases. These assumptions, when translated into mathematical form, yield the ideal-gas equation.

lanthanide contraction The gradual decrease in atomic and ionic radii with increasing atomic number among the lanthanide elements, atomic numbers 57 through 70. The decrease arises because of a gradual increase in effective nuclear charge through the lanthanide series.

lanthanide (rare earth) element Element in which the $4f$ subshell is only partially occupied.

lattice energy The energy required to separate completely the ions in an ionic solid into gaseous ions.

lattice points Points in a crystal, all of which have identical environments.

lattice vectors The vectors $\boldsymbol{a}$, $\boldsymbol{b}$, and $\boldsymbol{c}$ that define each lattice point in a crystal lattice. Beginning from any lattice point, it is possible to move to any other lattice point by adding together whole-number multiples of the two lattice vectors.

law of conservation of mass The total mass of materials present after a chemical reaction is the same as the total mass present before the reaction.

law of constant composition A law that states that the elemental composition of a compound is always the same; also called the **law of definite proportions**.

law of definite proportions A law that states that the elemental composition of a substance is always the same; also called the **law of constant composition**.

law of mass action The rules by which the equilibrium constant is expressed in terms of the concentrations of reactants and products, in accordance with the balanced chemical equation for the reaction.

law of multiple proportions If two elements A and B combine to form more than one compound, the masses of B that can combine with a given mass of A are in the ratio of small whole numbers.

Le Châtelier's principle A principle stating that when we disturb a system at chemical equilibrium, the relative concentrations of reactants and products shift so as to partially undo the effects of the disturbance.

levorotatory, or merely levo or *l* A term used to label a chiral molecule that rotates the plane of polarization of plane-polarized light to the left (counterclockwise).

Lewis acid An electron-pair acceptor.

Lewis base An electron-pair donor.

Lewis structure A representation of covalent bonding in a molecule that is drawn using Lewis symbols. Shared electron pairs are shown as lines, and unshared electron pairs are shown as pairs of dots. Only the valence-shell electrons are shown.

Lewis symbol (electron-dot symbol) The chemical symbol for an element, with a dot for each valence electron.

ligand An ion or molecule that coordinates to a metal atom or to a metal ion to form a complex.

limiting reactant (limiting reagent) The reactant that is completely consumed in a reaction; the amount of product that can form is limited by the complete consumption of the limiting reactant.

line spectrum A spectrum that contains radiation at only certain specific wavelengths.

linkage isomers Structural isomers of coordination compounds in which a ligand is capable of coordinating to a metal in two ways.

lipid A nonpolar biological molecule used by organisms for long-term energy storage or as a structural component.

liquid Matter that has a distinct volume independent of its container.

liquid crystal A substance that exhibits a viscous, milky state between the liquid and solid states.

lock-and-key model A model of enzyme action in which the substrate molecule is pictured as fitting rather specifically into the active site on the enzyme. It is assumed that in being bound to the active site, the substrate is somehow activated for reaction.

low-spin complex A metal complex in which the electrons are paired in lower-energy orbitals.

macroporous Solids having pores that can be seen with an optical microscope.

magic numbers Numbers of protons and neutrons that result in very stable nuclei.

main-group elements Elements in the s and p blocks of the periodic table.

mass A measure of the amount of material in an object. In SI units, mass is measured in kilograms.

mass defect The difference between the mass of a nucleus and the total masses of the individual nucleons that it contains.

mass number The sum of the number of protons and neutrons in the nucleus of a particular atom.

mass percentage The number of grams of solute in each 100 g of solution.

mass spectrometer An instrument used to measure the precise masses and relative amounts of atomic and molecular ions.

matter Anything that occupies space and has mass; the physical material of the universe.

matter waves The term used to describe the wave characteristics of a moving particle.

mean free path The average distance traveled by a gas molecule between collisions.

mesoporous Solids having pore sizes in the range 2 to 50 nm.

metal complex Species that are assemblies of a central metal ion bonded to a group of surrounding molecules or ions, such as $[Ag(NH_3)_2]^+$ and $[Fe(H_2O)_6]^{3+}$.

metallic bond Bonding, usually in solid metals, in which the bonding electrons are delocalized and therefore, relatively free to move throughout the three-dimensional structure.

metallic character The extent to which an element exhibits the physical and chemical properties characteristic of metals, for example, luster, malleability, ductility, and good thermal and electrical conductivity.

metallic elements (metals) Elements that are usually solids at room temperature, exhibit high electrical and heat conductivity, and appear lustrous. Most of the elements in the periodic table are metals.

metallic hydrides Compounds formed when hydrogen reacts with metals; these compounds contain the hydride ion, H^-.

metallic solids Solids that are composed of metal atoms held together by a delocalized "sea" of collectively shared valence electrons.

metalloids Elements that lie along the diagonal line separating the metals from the nonmetals in the periodic table; the properties of metalloids are intermediate between those of metals and nonmetals.

metallurgy The science of extracting metals from their natural sources by a combination of chemical and physical processes. It is also concerned with the properties and structures of metals and alloys.

metathesis (exchange) reaction A reaction in which two substances react through an exchange of their component ions: $AX + BY \longrightarrow AY + BX$. Precipitation and acid–base neutralization reactions are examples of metathesis reactions.

metric system A system of measurement used in science and in most countries. The meter and the gram are examples of metric units.

microporous Refers to solids that have pores up to 2 nm in size.

microstate A single possible arrangement of the positions and kinetic energies of the molecules when the molecules are in a specific thermodynamic state.

mineral A solid, inorganic substance occurring in nature, such as calcium carbonate, which occurs as calcite.

miscible liquids Liquids that mix in all proportions.

mixture A combination of two or more substances in which each substance retains its own chemical identity.

molal boiling-point-elevation constant (K_b) A constant characteristic of a particular solvent that gives the increase in boiling point as a function of solution molality: $\Delta T_b = iK_b m$.

molal freezing-point-depression constant (K_f) A constant characteristic of a particular solvent that gives the decrease in freezing point as a function of solution molality: $\Delta T_f = -iK_f m$.

molality The concentration of a solution expressed as moles of solute per kilogram of solvent; abbreviated m.

molar heat capacity The heat required to raise the temperature of one mole of a substance by 1 °C.

molarity The concentration of a solution expressed as moles of solute per liter of solution; abbreviated M.

molar mass The mass of one mole of a substance in grams; it is numerically equal to the formula weight in atomic mass units.

mole A collection of Avogadro's number (6.022×10^{23}) of objects; for example, a mole of H_2O is 6.022×10^{23} H_2O molecules.

molecular compound A compound that consists of molecules.

molecular equation A chemical equation in which the formula for each substance is written without regard for whether it is an electrolyte or a nonelectrolyte.

molecular formula A chemical formula that indicates the actual number of atoms of each element in one molecule of a substance.

molecular geometry The arrangement in space of the atoms of a molecule.

molecular hydrides Compounds formed when hydrogen reacts with nonmetals and metalloids.

molecularity The number of molecules that participate as reactants in an elementary reaction.

molecular orbital (MO) An allowed state for an electron in a molecule. According to molecular-orbital theory, a molecular orbital is entirely analogous to an atomic orbital, which is an allowed state for an electron in an atom. Most bonding molecular orbitals can be classified as σ or π, depending on the disposition of electron density with respect to the internuclear axis.

molecular-orbital diagram A diagram that shows the energies of molecular orbitals relative to the atomic orbitals from which they are derived; also called an **energy-level diagram**.

molecular-orbital theory A theory that accounts for the allowed states for electrons in molecules by using specific wave functions.

molecular solids Solids that are composed of molecules held together by their intermolecular forces.

molecular weight The sum of the atomic weights (AW) of the atoms represented by their chemical formula for a molecule.

molecule A chemical combination of two or more atoms.

mole fraction The ratio of the number of moles of one component of a mixture to the total moles of all components; abbreviated X, with a subscript to identify the component.

momentum The product of the mass, m, and velocity, v, of an object.

monodentate ligand A ligand that binds to the metal ion via a single donor atom. It occupies one position in the coordination sphere.

monomers Molecules with low molecular weights, which can be joined together (polymerized) to form a polymer.

monosaccharide A simple sugar, most commonly containing six carbon atoms. The joining together of monosaccharide units by condensation reactions results in formation of polysaccharides.

motif In a crystal, the group of atoms associated with each lattice point.

nanomaterial A solid whose dimensions range from 1 to 100 nm and whose properties differ from those of a bulk material with the same composition.

natural gas A naturally occurring mixture of gaseous hydrocarbon compounds composed of hydrogen and carbon.

nematic liquid crystalline phase A liquid crystal in which the molecules are aligned in the same general direction, along their long axes, but in which the ends of the molecules are not aligned.

Nernst equation An equation that relates the cell emf, E, to the standard emf, $E°$, and the reaction quotient, Q: $E = E° - (RT/nF) \ln Q$.

net ionic equation A chemical equation for a solution reaction in which soluble strong electrolytes are written as ions and spectator ions are omitted.

neutralization reaction A reaction in which an acid and a base react in stoichiometrically equivalent amounts; the neutralization reaction between an acid and a metal hydroxide produces water and a salt.

neutron An electrically neutral particle found in the nucleus of an atom; it has approximately the same mass as a proton.

noble gases Members of group 8A in the periodic table.

nodal plane A plane where the electron density in an atom or a molecule is zero. Both atomic orbitals and molecular orbitals can have nodal planes.

node Points in an atom at which the electron density is zero. Radial nodes are spherical surfaces, while angular nodes can be planes or cones.

nonbonding pair In a Lewis structure a pair of electrons assigned completely to one atom; also called a lone pair.

nonelectrolyte A substance that does not ionize in water and consequently dissolves to form a nonconducting solution.

nonionizing radiation Radiation that does not have sufficient energy to remove an electron from a molecule.

nonmetallic elements (nonmetals) Elements in the upper right corner of the periodic table; nonmetals differ from metals in their physical and chemical properties.

nonpolar covalent bond A covalent bond in which the electrons are shared equally.

normal boiling point The boiling point at 1 atm pressure.

normal melting point The melting point at 1 atm pressure.

nuclear binding energy The energy required to decompose an atomic nucleus into its component protons and neutrons.

nuclear disintegration series A series of nuclear reactions that begins with an unstable nucleus and terminates with a stable one; also called a radioactive series.

nuclear model Model of the atom with a very small and extremely dense nucleus containing most of the mass (protons and neutrons) and with electrons in the space outside the nucleus.

nuclear reaction Reaction in which transformations of atomic nuclei occur.

nuclear transmutation A conversion of one kind of nucleus to another.

nucleic acids Polymers of high molecular weight that carry genetic information and control protein synthesis.

nucleon Particles found in the nucleus of an atom; protons and neutrons.

nucleotide Monomers of nucleic acids formed from a five-carbon sugar, a nitrogen-containing organic base, and a phosphate group. Nucleotides form linear polymers called DNA and RNA, which are involved in protein synthesis and cell reproduction.

nucleus The very small, very dense, positively charged portion of an atom; it is composed of protons and neutrons.

octet rule A rule stating that atoms tend to gain, lose, or share electrons until they are surrounded by eight valence electrons.

optical isomerism A form of isomerism in which the two forms of a compound (stereoisomers) are nonsuperimposable mirror images. Also called **enantiomers**.

optically active A substance that possesses the ability to rotate the plane of polarized light.

orbital An allowed energy state of an electron in the quantum-mechanical model of the atom; the term *orbital* is also used to describe the spatial distribution of electron density. An orbital is defined by the values of three quantum numbers: n, l, and m_l

orbital diagram A representation of an atomic orbital drawn as a box with one or two half-arrows inside representing electrons.

organic chemistry The study of carbon-containing compounds, typically containing carbon–hydrogen bonds.

osmosis The net movement of solvent through a semipermeable membrane toward the solution with greater solute concentration.

osmotic pressure The pressure that must be applied to a solution to stop osmosis from pure solvent into the solution.

Ostwald process An industrial process used to make nitric acid from ammonia. The NH_3 is catalytically oxidized by O_2 to form NO; NO in air is oxidized to NO_2; HNO_3 is formed in a disproportionation reaction when NO_2 dissolves in water.

overall reaction order The sum of the reaction orders of all the reactants appearing in the rate expression when the rate can be expressed as rate $= k[A]^a[B]^b$ ....

overlap The extent to which atomic orbitals on different atoms share the same region of space. When the overlap between two orbitals is large, a strong bond may be formed.

oxidation A process in which a substance loses one or more electrons.

oxidation number (oxidation state) A positive or negative whole number assigned to an element in a molecule or ion on the basis of a set of formal rules; to some degree it reflects the positive or negative character of that atom.

oxidation–reduction (redox) reaction A reaction in which electrons are transferred from one reactant to another and the oxidation states of these atoms change.

oxidizing agent, or oxidant The substance that is reduced and thereby causes the oxidation of some other substance in an oxidation–reduction reaction.

oxyacid A compound in which one or more OH groups, and possibly additional oxygen atoms, are bonded to a central atom.

oxyanion A polyatomic anion that contains one or more oxygen atoms.

ozone The name given to O_3, an allotrope of oxygen.

paramagnetism A property that a substance possesses if it contains one or more unpaired electrons. A paramagnetic substance is attracted to a magnetic field.

partial pressure The pressure exerted by a particular gas in a mixture.

particle accelerator A device that uses strong magnetic and electrostatic fields to accelerate charged particles.

parts per billion (ppb) The concentration of a solution in grams of solute per 10^9 (billion) grams of solution; equals micrograms of solute per liter of solution for aqueous solutions.

parts per million (ppm) The concentration of a solution in grams of solute per 10^6 (million) grams of solution; equals milligrams of solute per liter of solution for aqueous solutions.

pascal (Pa) The SI unit of pressure: $1\ Pa = 1\ N/m^2$.

Pauli exclusion principle A rule stating that no two electrons in an atom may have the same four quantum numbers (n, l, m_l, and m_s). As a reflection of this principle, there can be no more than two electrons in any one atomic orbital.

peptide bond An amide group formed by amino acids. A bond formed between two amino acids.

percent ionization The percent of a substance that undergoes ionization on dissolution in water. The term applies to solutions of weak acids and bases.

percent yield The ratio of the actual (experimental) yield of a product to its theoretical (calculated) yield, multiplied by 100.

period The row of elements that lie in a horizontal row in the periodic table.

periodic table The arrangement of elements in order of increasing atomic number, with elements having similar properties placed in vertical columns.

perspective drawing A model that uses wedges and dashed lines to depict bonds that are not in the plane of the paper.

petroleum A naturally occurring combustible liquid composed of hundreds of hydrocarbons and other organic compounds.

pH The negative log in base 10 of the hydrogen ion concentration: $pH = -\log[H^+]$.

pH titration curve A graph of pH as a function of added titrant.

phase change The conversion of a substance from one state of matter to another. The phase changes we consider are melting and freezing (solid $\rightleftharpoons$ liquid), sublimation and deposition (solid $\rightleftharpoons$ gas), and vaporization and condensation (liquid $\rightleftharpoons$ gas).

phase diagram A graphic representation of the equilibria among the solid, liquid, and gaseous phases of a substance as a function of temperature and pressure.

phospholipid A form of lipid molecule that contains charged phosphate groups.

photochemical smog A complex mixture of undesirable substances produced by the action of sunlight on an urban atmosphere polluted with automobile emissions. The major starting ingredients are nitrogen oxides and organic substances.

photodissociation The rupture of a chemical bond resulting from absorption of a photon by a molecule.

photoelectric effect The emission of electrons from a metal surface induced by light.

photoionization The removal of an electron from an atom or molecule by absorption of light.

photon The smallest increment (a quantum) of radiant energy; a photon of light with frequency ν has an energy equal to $h\nu$.

photosynthesis The process that occurs in plant leaves by which light energy is used to convert carbon dioxide and water to carbohydrates and oxygen.

physical changes Changes in physical appearance (such as a phase change) that occur with no change in chemical composition.

physical properties Properties that can be measured without changing the composition of a substance, for example, color and freezing point.

pi (π) bond A covalent bond in which electron density is concentrated above and below the internuclear axis; produced by the sideways overlap of p orbitals.

pi (π) molecular orbital A molecular orbital that concentrates the electron density on opposite sides of an imaginary line that passes through the nuclei.

Planck constant (h) The constant that relates the energy and frequency of a photon, $E = h\nu$. Its value is 6.626×10^{-34} J-s.

plastic A polymeric solid that can be formed into particular shapes by application of heat and pressure.

polar covalent bond A covalent bond in which the electrons are not shared equally.

polarizability The ease with which the electron cloud of an atom or a molecule is distorted by an outside influence, thereby inducing a dipole moment.

polar molecule A molecule in which the centers of positive and negative charge do not coincide; a molecule that possesses a nonzero dipole moment.

polyatomic ion An electrically charged group of two or more atoms.

polydentate ligand A ligand in which three or more donor atoms can coordinate to the same metal ion.

polymer Large molecules with long chains of atoms (usually carbon), where the atoms within a given chain are connected by covalent bonds and adjacent chains are held to one another largely by weaker intermolecular forces.

polypeptide A polymer of amino acids formed when a large number of amino acids (>30) are linked together by peptide bonds.

polyprotic acid A substance capable of dissociating more than one proton in water; H_2SO_4 is an example.

polysaccharide A substance made up of many monosaccharide units joined together.

porphyrin A complex derived from the porphine molecule.

positron A particle with the same mass as an electron but with a positive charge, $_{+1}^{0}e$, or β^+.

positron emission A nuclear decay process where a positron, a particle with the same mass as an electron but with a positive charge, symbol $_{+1}^{0}e$, or β^+, is emitted from the nucleus.

potential energy The energy that an object possesses as a result of its position with respect to another object.

precipitate An insoluble substance that forms in, and separates from, a solution.

precipitation reaction A reaction that occurs between substances in solution in which one of the products is insoluble.

precision The closeness of agreement among several measurements of the same quantity.

pressure A measure of the force exerted on a unit area. In chemistry, pressure is often expressed in units of atmospheres (atm) or torr: 760 torr = 1 atm; in SI units pressure is expressed in pascals (Pa).

pressure–volume (PV) work Work performed by expansion of a gas against a resisting pressure.

primary cell A voltaic cell that cannot be recharged.

primary structure The sequence of amino acids along a protein chain.

primitive lattice A crystal lattice in which the lattice points are located only at the corners of each unit cell.

principal quantum number The integer, n, associated with orbits and energy levels in the Bohr atom.

probability density (ψ^2) A value that represents the probability that an electron will be found at a given point in space. Also called **electron density**.

product A substance produced in a chemical reaction; it appears to the right of the arrow in a chemical equation.

property Any characteristic that allows us to recognize a particular type of matter and to distinguish it from other types.

protein Macromolecules present in all living cells; a biopolymer formed from amino acids.

protium The most common isotope of hydrogen.

proton A positively charged subatomic particle found in the nucleus of an atom.

pure substance Matter that has a fixed composition and distinct properties.

qualitative analysis The determination of the presence or absence of a particular substance in a mixture.

quantitative analysis The determination of the amount of a given substance that is present in a sample.

quantum The smallest increment of radiant energy that may be absorbed or emitted; the magnitude of radiant energy is $h\nu$.

quaternary structure The structure of a protein resulting from the clustering of several individual protein chains into a final specific shape.

racemic mixture A mixture of equal amounts of the dextrorotatory and levorotatory forms of a chiral molecule. A racemic mixture will not rotate the plane of polarized light.

rad A measure of the energy absorbed from radiation by tissue or other biological material; 1 rad = transfer of 1×10^{-2} J of energy per kilogram of material.

radial probability function The probability that the electron will be found at a certain distance from the nucleus.

radioactive Possessing radioactivity, the spontaneous disintegration of an unstable atomic nucleus with accompanying emission of radiation.

radioactive decay chain A series of nuclear reactions that begins with an unstable nucleus and terminates with a stable one. Also called **nuclear disintegration series**.

radioisotope An isotope that is radioactive; that is, it is undergoing nuclear changes with emission of radiation.

radionuclide A radioactive nuclide.

radiotracer A radioisotope that can be used to trace the path of an element in a chemical system.

Raoult's law A law stating that the partial pressure exerted by solvent vapor above the solution, $P_{solution}$, equals the product of the mole fraction of the solvent, $X_{solvent}$, and the vapor pressure of the pure solvent, $P^{\circ}_{solvent}$. $P_{solution} = X_{solvent}P^{\circ}_{solvent}$.

rare earth element See **lanthanide (rare earth) element**.

rate constant A constant of proportionality between the reaction rate and the concentrations of reactants that appear in the rate law.

rate-determining step The slowest elementary step in a reaction mechanism.

rate law An equation that relates the reaction rate to the concentrations of reactants (and sometimes of products also).

reactant A starting substance in a chemical reaction; it appears to the left of the arrow in a chemical equation.

reaction mechanism A detailed picture, or model, of how the reaction occurs; that is, the order in which bonds are broken and formed and the changes in relative positions of the atoms as the reaction proceeds.

reaction order The power to which the concentration of a reactant is raised in a rate law.

reaction quotient (Q) The value that is obtained when concentrations of reactants and products are inserted into the equilibrium expression. If the concentrations are equilibrium concentrations, $Q = K$; otherwise, $Q \neq K$.

reaction rate The speed at which a chemical reaction occurs; a measure of the decrease in concentration of a reactant or the increase in concentration of a product with time.

redox (oxidation–reduction) reaction A reaction in which certain atoms undergo changes in oxidation states. The substance increasing in oxidation state is oxidized; the substance decreasing in oxidation state is reduced.

reducing agent, or reductant The substance that is oxidized and thereby causes the reduction of some other substance in an oxidation–reduction reaction.

reduction A process in which a substance gains one or more electrons.

rem A measure of the biological damage caused by radiation; rems = rads × RBE.

renewable energy sources Energy such as solar energy, wind energy, and hydroelectric energy derived from essentially inexhaustible sources.

representative (main-group) element An element from within the s and p blocks of the periodic table.

resonance structures (resonance forms) Individual Lewis structures in cases where two or more Lewis structures are equally good descriptions of a single molecule. The resonance structures in such an instance are "averaged" to give a more accurate description of the real molecule.

reverse osmosis The process by which water molecules move under high pressure through a semipermeable membrane from the more concentrated to the less concentrated solution.

reversible process A process in which the system can be restored to its original condition with no change to the surroundings.

ribonucleic acid (RNA) A polynucleotide in which ribose is the sugar component.

root-mean-square (rms) speed (μ_{rms}) The square root of the average of the squared speeds of the gas molecules in a gas sample.

rotational motion Movement of a molecule as it spins about an axis.

salinity A measure of the salt content of seawater, brine, or brackish water. It is equal to the mass in grams of dissolved salts present in 1 kg of seawater.

salt Any ionic compound whose cation comes from a base (for example, Na^+ from NaOH) and whose anion comes from an acid (for example, Cl^- from HCl).

saponification Hydrolysis of an ester in the presence of a base.

saturated solution A solution in which undissolved solute and dissolved solute are in equilibrium.

scientific law A statement of what is always observed to happen, to the best of our knowledge.

scientific method The general process of advancing scientific knowledge by making experimental observations and by formulating hypotheses, theories, and laws.

secondary cell A voltaic cell that can be recharged.

secondary structure Refers to how segments of the protein chain are oriented in a regular pattern.

second law of thermodynamics A statement that relates the change in the entropy of the universe (ΔS_{univ}) to the reversibility of the process. For a reversible process, $\Delta S_{univ} = 0$. For an irreversible process, $\Delta S_{univ} > 0$. Because spontaneous processes are irreversible, the second law means that in any spontaneous process the entropy of the universe increases.

second-order reaction A reaction in which the rate depends either on a reactant concentration raised to the second power or on the concentrations of two reactants each raised to the first power.

semiconductor A material that has electrical conductivity between that of a metal and that of an insulator.

sigma (σ) bond A covalent bond in which electron density is concentrated along the internuclear axis.

sigma (σ) molecular orbital A molecular orbital that centers the electron density about an imaginary line passing through two nuclei.

significant figures The digits that indicate the precision with which a measurement is made; all digits of a measured quantity are significant, including the last digit, which is uncertain.

silica Common name for silicon dioxide.

silicates Compounds containing silicon and oxygen, structurally based on SiO_4 tetrahedra.

single bond A covalent bond involving one electron pair.

SI units The preferred metric units for use in science.

smectic A and smectic C liquid crystals Liquid crystals in which the molecules maintain the long-axis alignment seen in nematic crystals, but in addition pack into layers.

solid Matter that has both a definite shape and a definite volume.

solubility The amount of a substance that dissolves in a given quantity of solvent at a given temperature to form a saturated solution.

solubility-product constant (solubility product, K_{sp}) An equilibrium constant related to the equilibrium between a solid salt and its ions in solution. It provides a quantitative measure of the solubility of a slightly soluble salt.

solute A substance dissolved in a solvent to form a solution; it is normally the component of a solution present in the smaller amount.

solution A mixture of substances that has a uniform composition; a homogeneous mixture.

solvation The clustering of solvent molecules around a solute particle.

solvent The dissolving medium of a solution; it is normally the component of a solution present in the greater amount.

space-filling model Model that shows relative sizes of the atoms.

specific heat (C_s) The heat capacity of 1 g of a substance; the heat required to raise the temperature of 1 g of a substance by 1 °C.

spectator ions Ions that go through a reaction unchanged and that appear on both sides of the complete ionic equation.

spectrochemical series A list of ligands arranged in order of their abilities to increase the crystal-field splitting energy.

spectrum The distribution among various wavelengths of the radiant energy emitted or absorbed by an object.

spin magnetic quantum number (m_s) A quantum number associated with the electron spin; it may have values of $+\frac{1}{2}$ or $-\frac{1}{2}$.

spin-pairing energy The difference between the energy required to pair an electron in an occupied orbital and the energy required to place that electron in an empty orbital.

spontaneous process A process that is capable of proceeding in a given direction, as written or described, without needing to be driven by an outside source of energy. A process may be spontaneous even though it is very slow.

standard atmospheric pressure Defined as 760 torr or, in SI units, 101.325 kPa.

standard emf, also called the standard cell potential ($E°$) The emf of a cell when all reagents are at standard conditions.

standard enthalpy change ($\Delta H°$) The change in enthalpy in a process when all reactants and products are in their stable forms at 1 atm pressure and a specified temperature, commonly 25 °C.

standard enthalpy of formation ($\Delta H_f°$) The change in enthalpy that accompanies the formation of one mole of a substance from its elements, with all substances in their standard states.

standard free energy of formation ($\Delta G_f°$) The change in free energy associated with the formation of a substance from its elements under standard conditions.

standard hydrogen electrode (SHE) An electrode based on the half-reaction $2 H^+(1 M) + 2 e^- \longrightarrow H_2(1 atm)$. The standard electrode potential of the standard hydrogen electrode is defined as 0 V.

standard molar entropy ($S°$) The entropy value for a mole of a substance in its standard state.

standard reduction potential ($E°_{red}$) The potential of a reduction half-reaction under standard conditions, measured relative to the standard hydrogen electrode. A standard reduction potential is also called a standard electrode potential.

standard solution A solution of known concentration.

standard temperature and pressure (STP) Defined as 0 °C and 1 atm pressure; frequently used as reference conditions for a gas.

starch The general name given to a group of polysaccharides that acts as energy-storage substances in plants.

state function A property of a system that is determined by its state or condition and not by how it got to that state; its value is fixed when temperature, pressure, composition, and physical form are specified; P, V, T, E, and H are state functions.

states of matter The three forms that matter can assume: solid, liquid, and gas.

stereoisomers Compounds possessing the same formula and bonding arrangement but differing in the spatial arrangements of the atoms.

stoichiometry The relationships among the quantities of reactants and products involved in chemical reactions.

stratosphere The region of Earth's atmosphere from 10 km to 50 km above the surface.

strong acid An acid that ionizes completely in water.

strong base A base that ionizes completely in water.

strong electrolyte A substance (strong acids, strong bases, and most salts) that is completely ionized in solution.

structural formula A formula that shows not only the number and kinds of atoms in the molecule but also the arrangement (connections) of the atoms.

structural isomers Compounds possessing the same formula but differing in the bonding arrangements of the atoms.

subatomic particles Particles such as protons, neutrons, and electrons that are smaller than an atom.

subshell One or more orbitals with the same set of quantum numbers n and l. For example, we speak of the $2p$ subshell ($n = 2, l = 1$), which is composed of three orbitals ($2p_x$, $2p_y$, and $2p_z$).

substitutional alloy An alloy formed when atoms of the solute in a solid solution occupy positions normally occupied by a solvent atom.

substitution reactions Reactions in which one atom (or group of atoms) replaces another atom (or group) within a molecule; substitution reactions are typical for alkanes and aromatic hydrocarbons.

substrate A substance that undergoes a reaction at the active site in an enzyme.

supercritical fluid The state that exists when liquid and gas phases are indistinguishable from each other as the temperature exceeds the critical temperature and the pressure exceeds the critical pressure.

supercritical mass An amount of fissionable material larger than the critical mass.

supersaturated solution A solution containing more solute than an equivalent saturated solution.

surface tension The energy required to increase the surface area of a liquid by a unit amount.

surroundings In thermodynamics, everything that lies outside the system that we study.

system In thermodynamics, the portion of the universe that we single out for study. We must be careful to state exactly what the system contains and what transfers of energy it may have with its surroundings.

temperature A measure of the hotness or coldness of an object; a physical property that determines the direction of heat flow.

termolecular reaction An elementary reaction that involves three molecules. Termolecular reactions are rare.

tertiary structure The overall shape of a large protein, specifically, the manner in which sections of the protein fold back upon themselves or intertwine.

theoretical yield The quantity of product that is calculated to form when all of the limiting reagent reacts.

theory A tested model or explanation that has predictive powers and that accounts for all available observations.

thermochemistry The relationship between chemical reactions and energy changes.

thermodynamics The study of energy and its transformation.

thermonuclear reaction Another name for fusion reactions; reactions in which two light nuclei are joined to form a more massive one.

thermoplastic A polymeric material that can be readily reshaped by application of heat and pressure.

thermosetting plastic A plastic that is formed by irreversible chemical processes and therefore is not readily reshaped by application of heat and pressure.

third law of thermodynamics A law stating that the entropy of a pure, crystalline solid at absolute zero temperature is zero: $S(0\,\text{K}) = 0$.

titration The process of reacting a solution of unknown concentration with one of known concentration (a standard solution).

torr A unit of pressure (1 torr $= 1$ mm Hg).

transition elements (transition metals) Elements in which the d orbitals are partially occupied.

transition state (activated complex) The particular arrangement of reactant and product molecules at the point of maximum energy in the rate-determining step of a reaction.

translational motion Movement in which an entire molecule moves in a definite direction.

transuranium elements Elements that follow uranium in the periodic table.

triple bond A covalent bond involving three electron pairs.

triple point The temperature at which solid, liquid, and gas phases coexist in equilibrium.

tritium The isotope of hydrogen whose nucleus contains a proton and two neutrons.

troposphere The region of Earth's atmosphere extending from the surface to about 10 km altitude.

Tyndall effect The scattering of a beam of visible light by the particles in a colloidal dispersion.

uncertainty principle A principle stating there is an inherent uncertainty in the precision with which we can simultaneously specify the position and momentum of a particle. This uncertainty is significant only for particles of extremely small mass, such as electrons.

unimolecular reaction An elementary reaction that involves a single molecule.

unit cell The smallest portion of a crystal that reproduces the structure of the entire crystal when repeated in different directions in space. It is the repeating unit or building block of the crystal lattice.

unsaturated solution A solution containing less solute than a saturated solution.

valence band A band of closely spaced bonding molecular orbitals that is essentially fully occupied by electrons.

valence-bond theory A model of chemical bonding in which an electron-pair bond is formed between two atoms by the overlap of orbitals on the two atoms.

valence electrons The outermost electrons of an atom; those that occupy orbitals not occupied in the nearest noble-gas element of lower atomic number. The valence electrons are the ones the atom uses in bonding.

valence orbitals Orbitals that contain the outer-shell electrons of an atom.

valence-shell electron-pair repulsion (VSEPR) model A model that accounts for the geometric arrangements of shared and unshared electron pairs around a central atom in terms of the repulsions between electron pairs.

van der Waals equation An equation of state for nonideal gases that is based on adding corrections to the ideal-gas equation. The correction terms account for intermolecular forces of attraction and for the volumes occupied by the gas molecules themselves.

van't Hoff factor The number of fragments that a solute breaks into for a particular solvent.

vapor Gaseous state of any substance that normally exists as a liquid or solid.

vapor pressure The pressure exerted by a vapor in equilibrium with its liquid or solid phase.

vibrational motion Movement of the atoms within a molecule in which they move periodically toward and away from one another.

viscosity A measure of the resistance of fluids to flow.

volatile Refers to liquids that evaporate readily.

voltaic (galvanic) cell A device in which a spontaneous oxidation–reduction reaction occurs with the passage of electrons through an external circuit.

vulcanization The process of cross-linking polymer chains in rubber.

watt A unit of power; $1\,W = 1\,J/s$.

wave function A mathematical description of an allowed energy state (an orbital) for an electron in the quantum-mechanical model of the atom; it is usually symbolized by the Greek letter ψ.

wavelength The distance between identical points (such as two adjacent peaks or two adjacent troughs) on successive waves.

weak acid An acid that only partly ionizes in water.

weak base A base that only partly ionizes in water.

weak electrolyte A substance that only partly ionizes in solution.

work The energy transferred when a force exerted on an object causes a displacement of that object.

zeolite A class of aluminosilicates that occur naturally and can also be synthesized.

zero-order reaction A reaction in which the rate of disappearance of A is independent of [A].

PHOTO AND ART CREDITS

FRONTMATTER p. ii SJ Travel Photo and Video/Shutterstock; **p. ix (left)** GL Archive/Alamy Stock Photo; **p. ix (right)** Neirfy/Shutterstock; **p. x (top left)** Food and Drink Photos/AlamyStock Photo; **p. x (bottom left)** EpicStockMedia/Shutterstock; **p. x (right)** Richard Megna/Fundamental Photographs; **p. xi (left)** Simon's passion 4 Travel/Shutterstock; **p. xi (right)** Cla78/Shutterstock; **p. xii (top left)** Alexander van Driessche; **p. xii (bottom left)** nrqemi/Shutterstock **p. xii (right)** Jim Lozouski/Shutterstock; **p. xiii (left)** Ljupco Smokovski/Shutterstock; **p. xiii (right)** demarcomedia/Shutterstock; **p. xiv (top left)** Ido Meirovich/Shutterstock; **p. xiv (bottom left)** Xvision/Moment/Getty Images; **p. xiv (right)** GALA Images/Alamy Stock Photo; **p. xv (left)** Kolpakova Svetlana/Shutterstock; **p. xv (right)** Deb22/Shutterstock; **p. xvi (top left)** argus/Shutterstock; **p. xvi (bottom left)** Molekuul/123RF GB Ltd.; **p. xvi (right)** National Renewable Energy Laboratory; **p. xvii (left)** Stocktrek Images/Getty Images; **p. xvii (right)** Anton Gvozdikov/Shutterstock; **p. xviii (left)** Art Collection 3/Alamy Stock Photo; **p. xviii (right)** Pikoso.kz/Shutterstock. **CHAPTER 1 Opener** GL Archive/Alamy Stock Photo; **1.2** Colleen Michaels/Shutterstock; **1.6** Charles D. Winters/Science Source; **1.7a** spe/Shutterstock; **1.7b** Richard Megna/Fundamental Photographs; **1.10** Richard Megna/Fundamental Photographs; **1.11** Richard Megna/Fundamental Photographs; **1.14a** Wirestock Creators/Shutterstock; **1.14b** Zoom-Zoom/Getty Images; **1.16** United Nations Department of Global Communications https://www.un.org/sustainabledevelopment/ (The content of this publication has not been approved by the United Nations and does not reflect the views of the United Nations or its officials or Member States) Pearson Education Inc. supports the Sustainable Development Goals (SDGs); **Table 1.3** FOXTROT ©2008 Bill Amend. Reprinted with permission of ANDREWS MCMEEL SYNDICATION. All rights reserved; **1.17** Duplass/Shutterstock; **1.23** Mettler-Toledo; **EX 1.4** Jose Gil/Shutterstock; **EX 1.5** rodimov/Shutterstock; **EX 1.12** Christine Glade/Shutterstock; **EX 1.22** Dinodia Photos/Alamy Stock Photo; **EX 1.53** Pencil case/Shutterstock; **EX 1.54** Josef Bosak/Shutterstock; **EX 1.72** Richard Megna/Fundamental Photographs. **CHAPTER 2 Opener** Neirfy/Shutterstock; **2.1** 1814 painting by Joseph Allen; **2.2** Drs. Ali Yazdani & Daniel J. Hornbaker/Science Source; **2.3** Richard Megna/Fundamental Photographs; **2.6** Pictorial Press Ltd/Alamy Stock Photo; **2.15** Richard Megna/Fundamental Photographs; **2.19** EllieB Photography/Getty Images; **2.21** Martyn F. Chillmaid/Science Source. **CHAPTER 3 Opener** Food and Drink Photos/AlamyStock Photo; **3.5** Richard Megna/Fundamental Photographs; **3.6** Caspar Benson/fStop Images/Getty Images; **3.7** Richard Megna/Fundamental Photographs; **3.8** Sara Sadler/Alamy Stock Photo; **3.10** Richard Megna/Fundamental Photographs; **EX 3.62** Damian Dovarganes/AP Images; **EX 3.69** Richard Megna/Fundamental Photographs; **EX 3.79** W1zzard/Getty Images. **CHAPTER 4 Opener** EpicStockMedia/Shutterstock; **4.3** Richard Megna/Fundamental Photographs; **4.8** Richard Megna/Fundamental Photographs; **4.10a** Markthai/Getty Images; **4.10b** Imseco/Getty Images; **4.10c** Triffitt/Getty Images; **4.11** Richard Megna/Fundamental Photographs; **4.13** Peticolas/Richard Megna/Fundamental Photographs; **4.15** Richard Megna/Fundamental Photographs; **4.17** Richard Megna/

Fundamental Photographs; **EX 4.7** Turtle Rock Scientific/Science Source; **EX 4.92** Richard Megna/Fundamental Photographs. **CHAPTER 5 Opener** Richard Megna/Fundamental Photographs; **5.1a** YlinPhoto/Shutterstock; **5.1b** Lightwork/Shutterstock; **5.8** Richard Megna/Fundamental Photographs; **5.14** Charles D. Winters/Science Source; **5.20** Rocketclips, Inc./Shutterstock; **5.27** Bloomberg/Getty Images; **EX 5.1** Rick & Nora Bower/Alamy Stock Photo. **CHAPTER 6 Opener** Simon's passion 4 Travel/Shutterstock; **6.1** pakete/123rf.com; **6.5** Iacopo Giangrandi; **6.8** Hipix/Alamy Stock Photo; **6.10** Richard Megna/Fundamental Photographs; **6.14** Lawrence Berkeley NATL LAB/MCT/Newscom; **6.17** Sonsedska Yuliia/Shutterstock; **6.26** Medical Body Scans/Science Source; **EX 6.4** Bierchen/Shutterstock, Stocktrek Images/Getty Images, AP Images; **EX 6.5** Yakobchuk/Getty Images. **CHAPTER 7 Opener** Cla78/Shutterstock; **7.15** Achim Prill/Getty Images; **7.17** Richard Megna/Fundamental Photographs; **7.18** Richard Treptow/Science Source; **7.19** Richard Megna/Fundamental Photographs; **7.20** Jeff J Daly/AlamyStock Photo; **7.21** Charles D. Winters/Science Source; **7.22** Richard Megna/Fundamental Photographs; **7.23** David Taylor/Science Source, Andrew Lambert/Science Source; **7.24** Dolce Vita/Shutterstock; **7.25** Bruce Bursten; **7.26** Richard Megna/Fundamental Photographs; **7.28** Richard Megna/Fundamental Photographs. **CHAPTER 8 Opener** Alexander van Driessche; **8.1** Tobik/Shutterstock. **CHAPTER 9 Opener** nrqemi/Shutterstock **9.5** Kristen Brochmann/Fundamental Photographs; **9.28** Science Photo Library/Alamy Stock Photo; **9.45** Richard Megna/Fundamental Photographs **CHAPTER 10 Opener** Jim Lozouski/Shutterstock; **10.10** EyeEm/Alamy Stock Photo; **10.15** Richard Megna/Fundamental Photographs; **10.16** Danita Delimont/Alamy Stock Photo. **CHAPTER 11 Opener** Ljupco Smokovski/Shutterstock; **11.1 (left to right)** Charles D. Winters/Science Source, sciencephotos/Alamy Stock Photo, Richard Megna/Fundamental Photographs; **11.10** Ted Kinsman/Science Source; **11.15** Fundamental Photographs; **11.16** Hermann Eisenbeiss/Science Source; **11.17** Fundamental Photographs; **11.25** AmaPhoto/Shutterstock; **11.31** Fundamental Photographs. **CHAPTER 12 Opener** demarcomedia/Shutterstock; **12.2 (top to bottom)** Photo Fun/Shutterstock, mahirart/Shutterstock; **12.16** (DoITPoMS) Dissemination of IT for the Promotion of Materials Science; **12.32** Francesco Zerilli/Alamy Stock Photo; **12.41** SPL/Science Source; **12.42** SPL/Science Source; **12.43** funkyfood London - Paul Williams/Alamy Stock Photo; **12.44a** Maen CG/Shutterstock; **12.44b** Zoonar Gmbh/Alamy Stock Photo; **12.44c** Vinaches, P.; Schwanke, A.J.; Lopes, C.W.; Souza, I.M.S.; Villarroel-Rocha, J.; Sapag, K.; Pergher, S.B.C. Incorporation of Brazilian Diatomite in the Synthesis of An MFI Zeolite. Molecules 2019, 24, 1980. https://doi.org/10.3390/molecules24101980; **EX 12.1** Richard Megna/Fundamental Photographs; **EX 12.5** lillisphotography/Getty Images; **EX 12.10** Jinghong Li. **CHAPTER 13 Opener** Ido Meirovich/Shutterstock; **13.5** Richard Megna/Fundamental Photographs; **13.6** Richard Megna/Fundamental Photographs; **13.7** Richard Megna/Fundamental Photographs; **13.8** Richard Megna/Fundamental Photographs; **13.14** Charles D. Winters/Science Source;

13.19 Michael Utech/Getty Images; **13.26** DuPont External Affairs; **13.27** Richard Megna/Fundamental Photographs; **13.31** Eye of Science/Science Source; **EP 13.26** Ildi Papp/Shutterstock. **CHAPTER 14 Opener** Xvision/Moment/Getty Images; **14.1** (top to bottom) Michael Dalton/Fundamental Photographs, Richard Megna/Fundamental Photographs; **14.12** Richard Megna/Fundamental Photographs; **14.21** Richard Megna/Fundamental Photographs; **14.25** Richard Megna/Fundamental Photographs; **14.29** Nigel Cattlin/Alamy Stock Photo; **14.30** Heikki Saukkoma/Shutterstock. **CHAPTER 15 Opener** GALA Images/Alamy Stock Photo; **15.1** Richard Megna/Fundamental Photographs; **15.4** Joseph Kreiss/Shutterstock; **15.14** Richard Megna/Fundamental Photographs. **CHAPTER 16 Opener** Kolpakova Svetlana/Shutterstock; **16.3** Chip Clark/Fundamental Photographs; **16.7** Charles D. Winters/Science Source; **16.9** Richard Megna/Fundamental Photographs; **16.17** Richard Megna/Fundamental Photographs. **CHAPTER 17 Opener** Deb22/Shutterstock; **17.1** Thermo Fisher Scientific; **17.5** Pietro M. Motta/Silvia Correr/Science Source; **17.20** Richard Megna/Fundamental Photographs; **17.22** Richard Megna/Fundamental Photographs; **EX 17.2** Richard Megna/Fundamental Photographs. **CHAPTER 18 Opener** argus/Shutterstock; **18.2** V. Belov/Shutterstock; **18.6** U.S. Geological Survey Library; **18.7** NASA; **18.10** Beijingstory/Getty Images; **18.15** Grigorii Pisotsckii/Shutterstock; **18.18** U.S. Geological Survey Library; **18.19** Nagel Photography/Shutterstock; **18.22** Damsea/Shutterstock. **CHAPTER 19 Opener** Molekuul/123RF GB Ltd.; **19.1** Fundamental Photographs; **19.3** Fundamental Photographs; **19.7** Drescher/ullstein bild/Getty Images. **CHAPTER 20 Opener** National Renewable Energy Laboratory; **20.1** Fundamental Photographs; **20.2** Fundamental Photographs; **20.3** Fundamental Photographs; **20.4** Richard Megna/Fundamental Photographs; **20.7** Ryuivst/iStock/Getty Images; **20.12** Photos.com/Getty Images; **20.15** Azoor Wildlife Photo/Alamy Stock Photo; **20.22** Eye35 stock/Alamy Stock Photo. **CHAPTER 21 Opener** Stocktrek Images/Getty Images; **21.4** David Parker/Science Source; **21.5** Doe Photo/Alamy Stock Photo; **21.8** (left to right) Don Murray/Getty Images, Dario Lo Presti/Shutterstock; **21.11** Susan Landau; **21.18** Los Alamos National Laboratory; **21.25** Science History Images/Alamy Stock Photo; **p. 932** Ted Kinsman/Science Source. **CHAPTER 22 Opener** Anton Gvozdikov/Shutterstock; **22.5** Richard Megna/Fundamental Photographs; **22.10** NASA; **22.11** Lisa F. Young/Shutterstock; **22.12** Richard Megna/Fundamental Photographs; **22.13** Maksym Gorpenyuk/Shutterstock; **22.15** Penny Tweedie/Alamy Stock Photo; **22.16** Fundamental Photographs; **22.17** Richard Megna/Fundamental Photographs; **22.18** Kristen Brochmann/Fundamental Photographs; **22.25** Richard Megna/Fundamental Photographs; **22.29** Marekuliasz/Shutterstock, (inset top left) Chris H. Galbraith/Shutterstock, (inset top right) Mikhail Bakunovich/Shutterstock, (inset bottom left) Mezzotint/Shutterstock; **22.33** Arena Creative/Shutterstock. **CHAPTER 23 Opener** Art Collection 3/Alamy Stock Photo; **23.1** John Cancalosi/Alamy Stock Photo; **23.4** Fundamental Photographs; **23.7** Pat_Hastings/Shutterstock; **23.8** Fundamental Photographs; **23.9** Fundamental Photographs; **23.20** Fundamental Photographs; **23.24** Fundamental Photographs; **23.30** Fundamental Photographs; **EX 23.7** Gino Santa Maria/Shutterstock. **CHAPTER 24 Opener** Pikoso.kz/Shutterstock; **24.12** Fundamental Photographs; **24.28a** Dotted Yeti/Shutterstock.

INDEX

Note: Page references with *f* indicate a figure on that page; *t* indicates table; *n* indicates note.

Common Ions

Positive Ions (Cations)

1 +
ammonium (NH_4^+)
cesium (Cs^+)
copper(I) or cuprous (Cu^+)
hydrogen (H^+)
lithium (Li^+)
potassium (K^+)
silver (Ag^+)
sodium (Na^+)

2 +
barium (Ba^{2+})
cadmium (Cd^{2+})
calcium (Ca^{2+})
chromium(II) or chromous (Cr^{2+})
cobalt(II) or cobaltous (Co^{2+})
copper(II) or cupric (Cu^{2+})
iron(II) or ferrous (Fe^{2+})
lead(II) or plumbous (Pb^{2+})
magnesium (Mg^{2+})
manganese(II) or manganous (Mn^{2+})
mercury(I) or mercurous (Hg_2^{2+})

mercury(II) or mercuric (Hg^{2+})
strontium (Sr^{2+})
nickel(II) (Ni^{2+})
tin(II) or stannous (Sn^{2+})
zinc (Zn^{2+})

3 +
aluminum (Al^{3+})
chromium(III) or chromic (Cr^{3+})
iron(III) or ferric (Fe^{3+})

Negative Ions (Anions)

1 −
acetate (CH_3COO^- or $C_2H_3O_2^-$)
bromide (Br^-)
chlorate (ClO_3^-)
chloride (Cl^-)
cyanide (CN^-)
dihydrogen phosphate ($H_2PO_4^-$)
fluoride (F^-)
hydride (H^-)
hydrogen carbonate or
   bicarbonate (HCO_3^-)

hydrogen sulfite or bisulfite (HCO_3^-)
hydroxide (OH^-)
iodide (I^-)
nitrate (NO_3^-)
nitrite (NO_2^-)
perchlorate (ClO_4^-)
permanganate (MnO_4^-)
thiocyanate (SCN^-)

2 −
carbonate (CO_3^{2-})
chromate (CrO_4^{2-})
dichromate ($Cr_2O_7^{2-}$)
hydrogen phosphate (HPO_4^{2-})
oxide (O^{2-})
peroxide (O_2^{2-})
sulfate (SO_4^{2-})
sulfide (S^{2-})
sulfite (SO_3^{2-})

3 −
arsenate (AsO_4^{3-})
phosphate (PO_4^{3-})

Fundamental Constants*

Atomic mass constant	1 amu	$= 1.660539067 \times 10^{-27}$ kg
	1 g	$= 6.02214076 \times 10^{23}$ amu (exact)
Avogadro's number†	N_A	$= 6.02214076 \times 10^{23}$/mol (exact)
Boltzmann constant	k	$= 1.380649 \times 10^{-23}$ J/K (exact)
Electron charge	e	$= 1.602176634 \times 10^{-19}$ C (exact)
Faraday constant	F	$= 9.648533212 \times 10^4$ C/mol
Gas constant	R	$= 0.082057366$ L-atm/mol-K
		$= 8.3144626$ J/mol-K
Mass of electron	m_e	$= 5.485799091 \times 10^{-4}$ amu
		$= 9.109383702 \times 10^{-31}$ kg
Mass of neutron	m_n	$= 1.008664916$ amu
		$= 1.674927498 \times 10^{-27}$ kg
Mass of proton	m_p	$= 1.007276467$ amu
		$= 1.672621923 \times 10^{-27}$ kg
Pi	π	$= 3.1415926536$
Planck constant	h	$= 6.62607015 \times 10^{-34}$ J-s (exact)
Speed of light in vacuum	c	$= 2.99792458 \times 10^8$ m/s (exact)

*Many of the fundamental constants were redefined as exact quantities in 2019 by the International Bureau of Weights and Measures (BIPM). These and other fundamental constants are listed at the National Institute of Standards and Technology (NIST) website: http://physics.nist.gov/cuu/Constants/index.html

†Avogadro's number is also referred to as the Avogadro constant. The latter term is the name adopted by agencies such as the International Union of Pure and Applied Chemistry (IUPAC) and the National Institute of Standards and Technology (NIST), but "Avogadro's number" remains in widespread usage and is used in most places in this book.